NEW! *Interactive Physiology*® 10-SYSTEM SUITE
WITH GRADABLE QUIZZES
The Best Way to Learn the Hardest Part of A&P

Interactive Physiology 10-System Suite (IP-10) significantly enriches teaching and learning environments by providing an audio/visual presentation of complex topics. For use as both a teaching tool in the classroom and a study tool for students, IP-10 features full-color animations and video, both with sound, that thoroughly demonstrate difficult physiology concepts, many of which occur at the cellular and molecular level. Extensive interactive games and Gradable Quizzes reinforce the material.

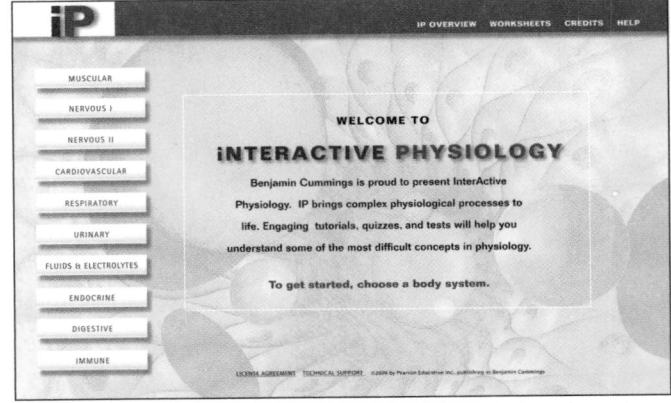

Modules:

- **Muscular System**
- **Nervous System I**
- **Nervous System II**
- **Cardiovascular System**
- **Respiratory System**
- **Urinary System**
- **Fluids & Electrolytes**
- **Endocrine System**
- **Digestive System**
- **NEW! Immune System**

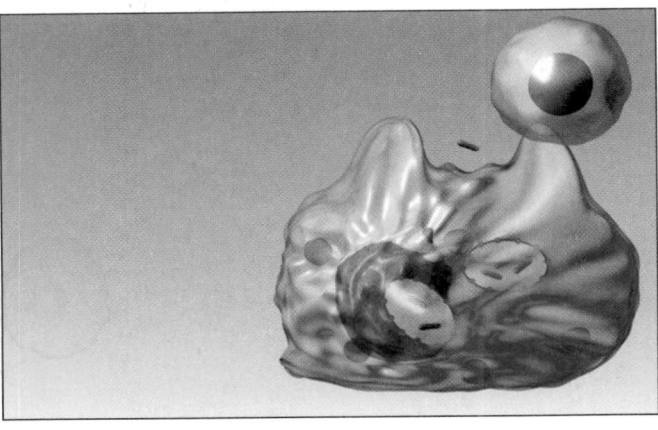

Images from the new 10th module on the Immune System

NEW! PHYSIOLOGY ANIMATIONS WITH GRADABLE QUIZZES

The Physiology Animations help students visualize difficult physiology concepts. They are available on both the instructor's Media Manager for classroom presentation and the myA&P™ Companion Website for student learning and practice. Gradable Quizzes reinforce the material.

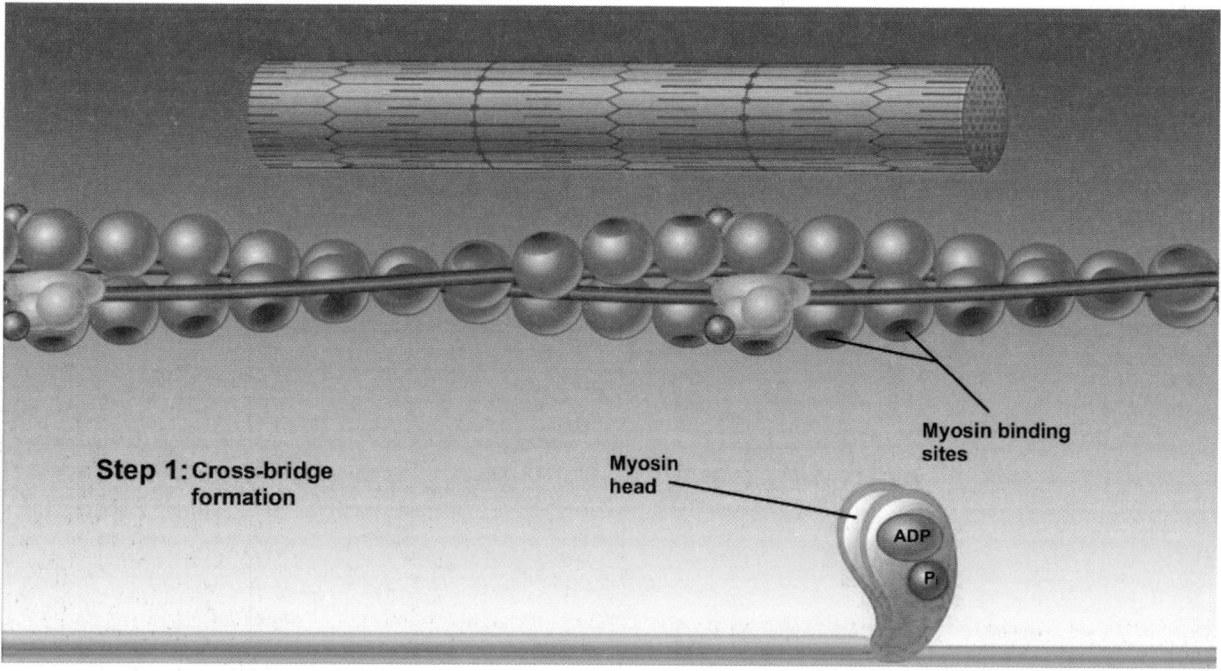

Image from the new Physiology Animations

myA&P
www.myaandp.com

A wealth of resources and tools, all in one convenient location

myA&P™

This companion website for *Fundamentals of Anatomy & Physiology,* **Eighth Edition** includes chapter guides, learning activities, self-study quizzes, *Interactive Physiology* (**IP**), brand-new anatomy and physiology animations with quizzes, interactive flashcards, bone review, muscle review, histology review, case studies, web links, and a glossary with pronunciations. This greatly expanded website provides access to the *Get Ready for A&P* **Media Update**. It also includes a password-protected Instructor's Resource Section.

myA&P with CourseCompass™

myA&P with CourseCompass combines the strength of the content of the myA&P website with state-of-the-art eLearning tools. CourseCompass is a nationally hosted, dynamic, interactive online course management system powered by Blackboard, the leading platform for Internet-based learning tools. This easy-to-use and customizable program enables professors to tailor content and functionality to meet individual course needs. The course includes all of the content found on the myA&P website together with course management functionality.

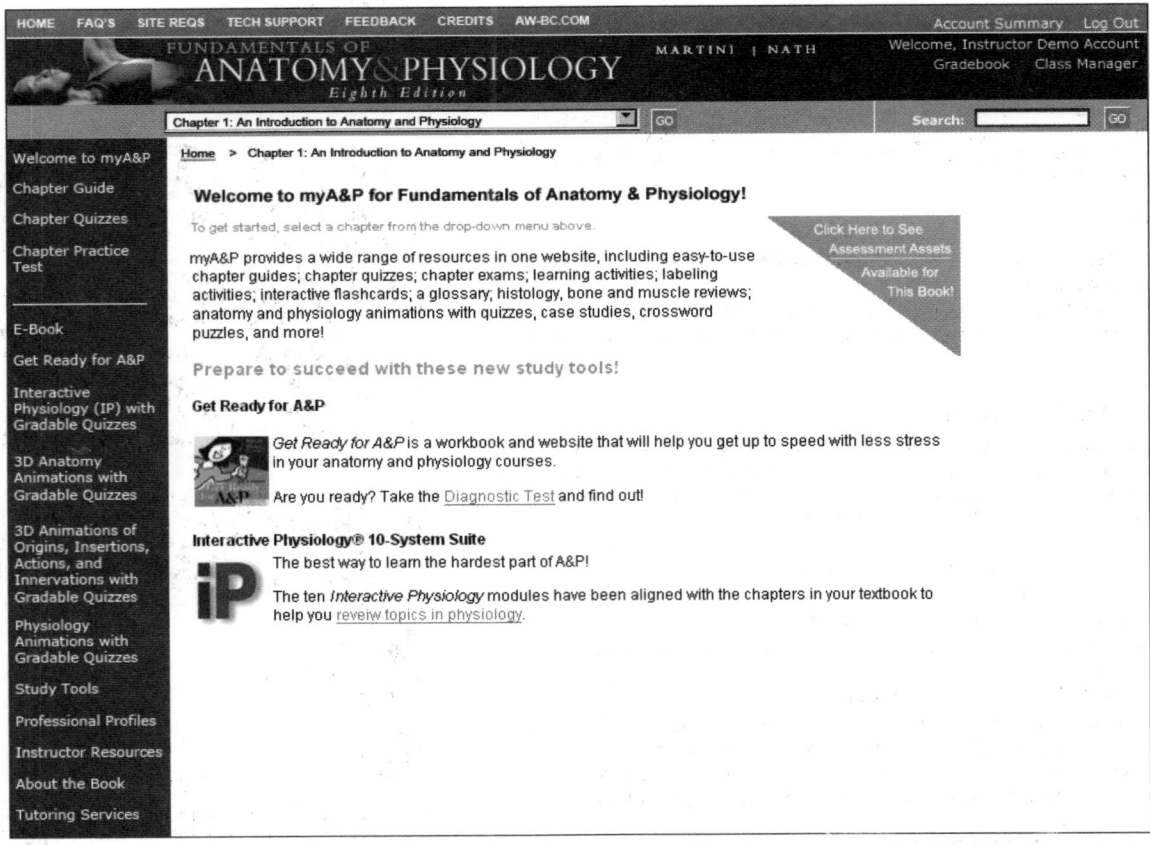

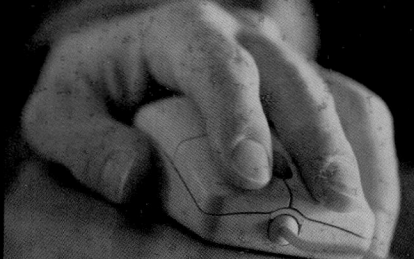

www.myaandp.com

Your steps to success.

STEP 1: Register

All you need to get started is a valid email address and the access code below. To register, simply:

1. Go to www.myaandp.com.
2. Click **"Students"** under **"First-time users."**
3. Under **"myA&P,"** click **"I already have an access code."**
4. Click the appropriate book cover. Cover must match the textbook edition being used for your class.
5. Click **"Register"** under **"First Time User?"**
6. Read the **License Agreement** and **Private Policy**. If you accept, click **"I Accept."**
7. Leave **"No, I Am a New User"** selected.
8. Using a coin, scratch off the silver coating below to reveal your access code. *Do not use a knife or other sharp object, which can damage the code.*
9. Enter your access code in lowercase or uppercase, without the dashes.
10. Follow the on-screen instructions to complete registration.

    During registration, you will establish a personal login name and password to use for logging into the website. You will also be sent a registration confirmation email that contains your login name and password.

Your Access Code is:

Note: If there is no silver foil covering the access code, it may already have been redeemed, and therefore may no longer be valid. In that case, you can purchase access online using a major credit card. To do so, go to www.myaandp.com, click "students" under "First Time Users," click "I need to buy access" under "myA&P," click the cover of your textbook, then click **"Buy Now,"** and follow the on-screen instructions.

STEP 2: Log in

1. Go to www.myaandp.com. and click **"myA&P"** under **"Returning Users."**
2. Click the appropriate book cover.
3. Under **"Established User?"** Enter the login name and password that you created during registration. *If unsure of this information, refer to your registration confirmation email.*
4. Click **"Log In."**

STEP 3: (Optional) Join a class

Instructors have the option of creating an online class for you to use with this website. If your instructor decides to do this, you'll need to complete the following steps using the Class ID your instructor provides you. By "joining a class," you enable your instructor to view the scored results of your work on the website in his or her online gradebook.

To join a class:

1. Log into the website. For instructions, see **"STEP 2: Log in."**
2. Click **"Join a Class."**
3. Enter your instructor's **"Class ID"** and then click **"Next."**
4. At the Confirm Class page you will see your instructor's name and class information. If this information is correct, click **"Next."**
5. Click **"Enter Class Now"** from the Class Confirmation page.

- *To confirm your enrollment in the class, check for your instructor and class name at the top right of the page. You will be sent a class enrollment confirmation email.*
- *As you complete activities on the website from now through the class end date, your results will post to your instructor's gradebook, in addition to appearing in your personal view of the Results Reporter.*

To log into the class later, follow the instructions under **"STEP 2: Log in."**

Got technical questions?

Customer Technical Support: To obtain support, please visit us online anytime at http://247.aw.com where you can search our knowledgebase for common solutions, view product alerts, and review all options for additional assistance.

SITE REQUIREMENTS

For the latest updates on Site Requirements, go to www.myaandp.com, click "myA&P" under "Returning Users." Pick your book and click the lower right link to "Site Requirements." Or you can click the "Books Available" tab at the top of the page, click your book cover, then click the lower right link "Site Requirements."

WINDOWS

OS: Windows 2000, XP
Resolution: 1024 x 768
Plugins: Latest Version of Flash/QuickTime/Shockwave (as needed)
Browsers: Internet Explorer 6.x; Firefox 1.x
Internet connection: 56K minimum

MACINTOSH
OS: 10.2.4, 10.3.2
Resolution: 1024 x 768
Plugins: Latest Version of Flash/QuickTime/Shockwave (as needed)
Browsers: Firefox 1.x; Safari 1.3
Internet connection: 56K minimum

Register and log in

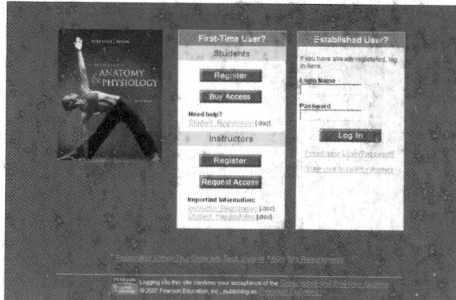

Join a class

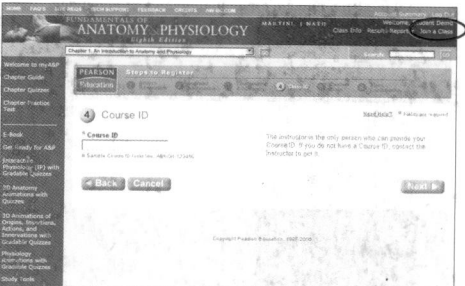

Important: Please read the Subscription and End-User License agreement, accessible from the book website's login page, before using the *myA&P* website. By using the website, you indicate that you have read, understood, and accepted the terms of this agreement.

FUNDAMENTALS OF
ANATOMY
& PHYSIOLOGY

Eighth Edition

with

Edwin F. Bartholomew, M.S.
Contributor

William C. Ober, M.D.
Art Coordinator and Illustrator

Claire W. Garrison, R.N.
Illustrator

Kathleen Welch, M.D.
Clinical Consultant

Ralph T. Hutchings
Biomedical Photographer

Frederic H. Martini, Ph.D.
University of Hawaii at Manoa

Judi L. Nath, Ph.D.
Lourdes College

PEARSON

Benjamin
Cummings

San Francisco   Boston   New York
Cape Town   Hong Kong   London   Madrid   Mexico City
Montreal   Munich   Paris   Singapore   Sydney   Tokyo   Toronto

Executive Editor: *Leslie Berriman*
Project Editor: *Katy German*
Development Manager: *Claire Alexander*
Development Editor: *Alan Titche*
Art Development Manager: *Mark Ong*
Art Development Editor: *Blakely Kim*
Associate Editor: *Robin Pille*
Editorial Assistant: *Kelly Reed*
Director, Media Development: *Lauren Fogel*
Media Producers: *Sarah Young-Dualan, Aimee Pavy, Erik Fortier*
Managing Editor: *Deborah Cogan*
Production Supervisor: *Caroline Ayres*
Copyeditor: *Michael Rossa*

Production Management and Compositor: *S4Carlisle Publishing Services*
Art Coordinator: *Jean Lake*
Design Manager: *Mark Ong*
Interior Designer: *Gary Hespenheide*
Cover Designer: *Yvo Reizebos*
Director, Image Resource Center: *Melinda Patelli*
Image Rights and Permissions Manager: *Zina Arabia*
Senior Manufacturing Buyer: *Stacey Weinberger*
Marketing Manager: *Gordon Lee*
Text printer: *RR Donnelly*
Cover printer: *Phoenix Color*
Cover Photo Credit: *David Sacks/Getty*

http://www.aw-bc.com

ISBN 10: 0-321-53910-9 (PIE Edition)
ISBN 13: 978-0321-53910-6 (PIE Edition)

Text and Illustration Team

Authors

Frederic (Ric) Martini, Ph.D.

Dr. Martini received his Ph.D. from Cornell University in comparative and functional anatomy for work on the pathophysiology of stress. His publications include journal articles and contributed chapters, technical reports, and magazine articles. He is the lead author of six other undergraduate texts on anatomy and physiology or anatomy, and has a feature column, "Back to Basics," in Physician License and Practice Today, a quarterly publication for medical students and residents. He is currently affiliated faculty of the University of Hawaii at Manoa and remains connected with the Shoals Marine Laboratory, a joint venture between Cornell University and the University of New Hampshire. Dr. Martini has been active in the Human Anatomy and Physiology Society (HAPS) for 15 years, and was a member of the committee that established the course curriculum guidelines for A&P. He is now a President Emeritus of HAPS after serving as President-Elect, President, and Past-President over 2004–2007. He is also a member of the American Physiological Society, the American Association of Anatomists, the Society for Integrative and Comparative Biology, the Australia/New Zealand Association of Clinical Anatomists, the Hawaii Academy of Science, the American Association for the Advancement of Science, and the International Society of Vertebrate Morphologists.

Judi L. Nath, Ph.D.

Dr. Nath is a professor of Biology and Health Sciences at Lourdes College in Sylvania, Ohio, where she teaches anatomy and physiology, medical terminology, pharmacology, and pathophysiology. After receiving her bachelor's degrees in biology and German from Bowling Green State University, she continued on at Bowling Green to earn her master's degree and her Ph.D. from the University of Toledo.

Dr. Nath is devoted to her students and strives to convey the intricacies of science in a captivating way that students find meaningful, interactive, and exciting. She won Lourdes' "Faculty Excellence" award, granted by the college to recognize her effective teaching, scholarship, and community service. Dr. Nath has also served as biology department chair.

Being a member of the Human Anatomy and Physiology Society (HAPS) is one of her passions, and she has been elected to office twice and has served on several committees, including the Curriculum and Instruction Committee, which was charged with developing outcomes that match the prescribed curriculum standards established by HAPS.

She also holds professional memberships in the American Association of Anatomists (AAA), the Society for College Science Teaching (SCST), and the National Science Teachers Association (NSTA).

She is the sole author of *Using Medical Terminology: A Practical Approach* (published by Lippincott Williams & Wilkins), the first book to use a "foreign language/total immersion" approach to teaching medical terminology within the context of applied anatomy and physiology. Dr. Nath has also written ancillary materials, including lecture outlines and test items, for several Martini A&P textbooks.

In her spare time Judi takes pleasure in bicycling, playing piano, and enjoying life with her husband, Mike, and their dogs.

Contributor

Edwin F. Bartholomew, M.S.

Edwin F. Bartholomew received his undergraduate degree from Bowling Green State University in Ohio and his M.S. from the University of Hawaii. His interests range widely, from human anatomy and physiology to paleontology, the marine environment, and "backyard" aquaculture. Mr. Bartholomew has taught human anatomy and physiology at both the secondary and undergraduate levels and a wide variety of other science courses (from botany to zoology) at Maui Community College. He is presently teaching at historic Lahainaluna High School, the oldest high school west of the Rockies. He has written journal articles, a weekly newspaper column, and many magazine articles. Working with Dr. Martini, he coauthored *Essentials of Anatomy & Physiology Structure and Function of the Human Body* and *The Human Body in Health and Disease* (all published by Benjamin Cummings). Mr. Bartholomew is a member of the Human Anatomy and Physiology Society, the National Association of Biology Teachers, the National Science Teachers Association, the Hawaii Science Teachers Association, and the American Association for the Advancement of Science.

Illustrators

William C. Ober, M.D.

Dr. William C. Ober received his undergraduate degree from Washington and Lee University and his M.D. from the University of Virginia. While in medical school, he also studied in the Department of Art as Applied to Medicine at Johns Hopkins University. After graduation, Dr. Ober completed a residency in Family Practice and later was on the faculty at the University of Virginia in the Department of Family Medicine. He is currently an Affiliate Professor of Biology at Washington and Lee University and is part of the Core Faculty at Shoals Marine Laboratory, where he teaches Biological Illustration every summer. The textbooks illustrated by Medical & Scientific Illustration have won numerous design and illustration awards.

Claire W. Garrison, R.N.

Claire W. Garrison, R.N., B.A., practiced pediatric and obstetric nursing before turning to medical illustration as a full-time career. She returned to school at Mary Baldwin College where she received her degree with distinction in studio art. Following a five-year apprenticeship, she has worked as Dr. Ober's partner in Medical & Scientific Illustration since 1986. She is on the Core Faculty at Shoals Marine Laboratory and co-teaches the Biological Illustration course.

Clinical Consultant

Kathleen Welch, M.D.

Dr. Welch received her M.D. from the University of Washington in Seattle and did her residency in Family Practice at the University of North Carolina in Chapel Hill. For two years, she served as Director of Maternal and Child Health at the LBJ Tropical Medical Center in American Samoa and subsequently was a member of the Department of Family Practice at the Kaiser Permanente Clinic in Lahaina, Hawaii. She has been in private practice since 1987. Dr. Welch is a Fellow of the American Academy of Family Practice and a member of the Hawaii Medical Association and the Human Anatomy and Physiology Society. With Dr. Martini, she has coauthored both a textbook on anatomy and physiology and the *A&P Applications Manual* available as a supplement to the eighth edition of *Fundamentals of Anatomy & Physiology*.

Photographer

Ralph T. Hutchings

Mr. Hutchings was associated with Royal College of Surgeons for 20 years. An engineer by training, he has focused for years on photographing the structure of the human body. The result has been a series of color atlases, including the *Color Atlas of Human Anatomy*, the *Color Atlas of Surface Anatomy,* and *The Human Skeleton* (all published by Mosby-Yearbook Publishing). For his anatomical portrayal of the human body, the International Photographers Association has chosen Mr. Hutchings as the best photographer of humans in the twentieth century. He lives in North London, where he tries to balance the demands of his photographic assignments with his hobbies of early motor cars and airplanes.

Preface

In this Eighth Edition of *Fundamentals of Anatomy & Physiology,* I am pleased to introduce a new coauthor, Dr. Judi Nath. I have known Judi for over a decade as a colleague and friend, and over the years we have corresponded extensively, discussing ways to improve the text. She is a dedicated and experienced teacher and has contributed greatly to this edition.

Together, we have built on the strengths of this text with the goal of making the Eighth Edition even more student-focused than previous editions. Our classroom experience, combined with in-depth student and instructor comments, led us to rework the overall presentation of the material; as a result, it is now easier to follow the main threads and navigate through the chapters. New features include chapter-opening Learning Outcomes coordinated with the new sentence-style chapter headings that not only introduce new topics but also teach. A new and simple Tips & Tricks feature gives students easy analogies and mnemonic devices to help them remember terms and information. Every figure in the book has been reviewed with an eye toward improving it as a teaching and learning tool. Many figures were reformatted, revised, or entirely redrawn to make the structures clearer, the distinguishing colors more contrasting, and the perspectives more 3-dimensional. In addition, labels were repositioned to make them easier to follow when studying.

The broad changes to this edition are presented in the **Overall Changes in the Eighth Edition** section below, which is followed by a more detailed section, **Chapter-by-Chapter Changes in the Eighth Edition.** A visual tour of the book follows in the remaining pages of the Preface.

The Student Package

Each new student copy of the text is automatically packaged with:

- *Martini's Atlas of the Human Body,* which has been expanded to include new cadaver images and scans and made easier to use with new headings that indicate the regions of the body;
- the *Interactive Physiology*® 10-System Suite (IP-10) CD, which includes the brand-new and eagerly awaited Tenth Module on the Immune System; and

- **access to the dramatically upgraded myA&P™ companion website,** which features the new IP-10, the new 3D Anatomy Animations with Gradable Quizzes, the new 3D Animations of Origins, Insertions, Actions, and Innervations with Gradable Quizzes, the new Physiology Animations with Gradable Quizzes, the new *Get Ready for A&P Media Update,* Study Tools, Chapter Quizzes, Chapter Practice Tests, and a new Gradebook.

Convenient references in the chapters of the book to the Atlas, IP, and myA&P help students seamlessly integrate each valuable component into their study routine.

In response to requests for simplifying the book package, the revised *A&P Applications Manual* is no longer automatically packaged with the book. The *A&P Applications Manual* is available separately (see page xx).

Overall Changes in the Eighth Edition

The following is a list of the new features, revised features, and other key revisions in this Eighth Edition.

- **NEW Learning Outcomes** are chapter-opening numbered lists that indicate to students what they should be able to do after studying the chapter. A&P educators have begun to shift from general "objectives" (that focus on what students "need to know") to these more concrete "learning outcomes" (that focus on what students "need to be able to do"). This feature puts this edition in sync with current trends in the A&P teaching curriculum that address the specific needs of A&P students. The Learning Outcomes can be coordinated with departmental plans for student assessment.

- **NEW Sentence-style chapter headings** do more than introduce new topics: They state the core fact or concept that will be presented in the section. These numbered chapter headings are directly correlated to the numbered Learning Outcomes on the chapter-opening page. The two elements together help students readily see and learn the core chapter content.

- **NEW and REVISED Checkpoints,** which had been called "Concept Checks" in previous editions, are located predictably throughout the chapter to close out each topic. With both the name and purpose of this feature having been honed, each Checkpoint now checks not only student understanding of concepts but also of facts—according to what students most need at a precise point in a chapter. The Checkpoints appear consistently throughout each chapter and reinforce the Learning Outcomes presented on the chapter-opening page, resulting in a systematic integration of the Learning Outcomes over the course of the chapter. The Checkpoint questions are numbered sequentially throughout each chapter so that students can easily check their understanding at each "pause" against the corresponding answers located in the blue Answers tab at the end of the book.

- **NEW** **Tips & Tricks** boxes are brief and concrete learning tools that give students simple analogies and easy mnemonic devices to help them remember facts and concepts.

- **REVISED** **The A&P Top 100** boxes, strategically placed throughout the narrative, include the core facts and concepts necessary to understanding the basics of anatomy and physiology—the nuggets that students should remember many years after taking their A&P course, regardless of their chosen profession. Titled "The 100 Keys" in the Seventh Edition, this feature has been revised and numbered to make it more useful to students.

- **NEW** **Did You Know . . .?** To engage students at the very beginning of each chapter, the new Did You Know . . .? "factoid" relates to the new chapter-opening photo and provides intriguing and fun information about the human body.

- **NEW** and **REVISED** **Clinical Notes** have been updated with the latest available clinical information and statistics, and one new Clinical Note has been added to every chapter. Definitions and descriptions of pathological processes were revised as needed.

- **REVISED** The **Study Outline**, at the end of each chapter, has been modified to be part of the new three-pronged organization of the material across each chapter:

  (1) The numbered Learning Outcomes on the chapter-opening page correlate to

  (2) the numbered Chapter Headings within the chapters that correspond to

  (3) the numbered headings in the Study Outline at the end of the chapter.

  References to the *Interactive Physiology* (IP) media program now appear at the beginning of the Study Outline (when relevant) and remind students which IP modules and topics can help them review key chapter material. IP has been used by over one million students, and many of them have told us that IP really helped them succeed in their A&P course. Students can access IP via the IP-10 CD packaged with their new book or via the myA&P™ companion website.

- **NEW** and **REVISED** The **Review Questions**, also at the end of the chapter, include new questions based on chapter figures that have been modified to ensure student visualization of key concepts. Many other Review Questions are new or revised. A reference to the myA&P companion website is placed at the beginning of the Review Questions to remind students that additional study and assessment materials can be found there. The 3-level learning system (Level 1: factual questions; Level 2: conceptual problems; Level 3: analytical exercises) is carried through to the chapter quizzes on the myA&P companion website (and to the chapters of the substantially revised printed Study Guide).

- **REVISED** **Eponymous terms** have, in most cases, been replaced with the preferred terms in *Terminologia Anatomica: International Anatomical Terminology (TA)*, the standard reference book of anatomical terminology created jointly by the Federative Committee on Anatomical Terminology (FCAT) and the International Federation of Associations of Anatomists (IFAA). Because terms other than those in *TA* are still in common usage, we present those other terms parenthetically, where appropriate. Furthermore, following the recommendations in the American Medical Association's *Manual of Style,* we use the non-possessive forms of eponyms in most instances; when the term appeared too foreign to us without the apostrophe "s", however, we let past convention rule.

- **REVISED** and **IMPROVED** **art program** builds on illustrations that have won multiple awards and are generally recognized as the most "teachable" and "learnable" of any illustrations in any anatomy and physiology textbook. Many of the **Side-by-Side Figures** have been reformatted to present, even more clearly, illustrations paired with cadaver photographs (and others paired with micrographs). All physiology figures, including the **Step-By-Step Figures**, have been revised in this edition, either largely or in more subtle ways, to improve the flow of process and the ease of information access. The orientation icons and arrows in the **Macro-to-Micro Figures** have been revised to lead students more clearly and directly from larger to smaller structures. In response to enthusiastic feedback from instructors and students, the **Navigator Figures** have been restored in this edition to clarify complex physiological processes. The 75 new **Illustration-over-Photo Figures** help students see structures in the context and proportions of the actual human body. The labels and leaders of virtually every anatomical figure in the book were reoriented and placed to make them visually cleaner and easier to follow. The figure locator dots have been replaced by green-colored figure callouts to make returning to the printed text easier after looking at the art. Finally, increased color contrast and increased dimensionality help students see, understand, and appreciate structures and functions.

- NEW **Clear and bright tabs** indicate the unit (by location along the edge of the page) and the related body system (by color). Chapter numbers are printed on the tabs. These tabs make it easier to navigate around the textbook. The tabbing system continues at the back of the book, where the Appendix, Answers, Glossary, and Index are each tagged with easy-to-find colors.

- **Concept Links**, signaled with blue chain link icons, have been retained from previous editions and alert students to material that is related to, or builds upon, previous discussions. Each link refers students to a page number for quickly reviewing the relevant material from an earlier chapter.

Chapter-by-Chapter Changes in the Eighth Edition

This annotated Table of Contents provides select examples of the revision highlights in each chapter of the Eighth Edition.

Chapter 1: An Introduction to Anatomy and Physiology
- New Clinical Note: Homeostasis and Disease
- New Figure 1–11: Ventral Body Cavity and Subdivisions

Chapter 2: The Chemical Level of Organization
- New Clinical Note: Fatty Acids and Health
- Figure 2–8: The Activities of Water Molecules in Aqueous Solutions: Revised to enhance consistency with other figures
- Figure 2–23: The Structure of Nucleic Acids: Revised to show the proper helical orientation of DNA

Chapter 3: The Cellular Level of Organization
- New Clinical Note: Telomerase, Aging, and Cancer
- Clinical Note: Parkinson Disease: Relocated from Chapter 14
- Table 3–1: Organelles of a Representative Cell: Reorganized for clarity
- Figure 3–1: The Anatomy of a Model Cell: Revised to include proteasomes

Chapter 4: The Tissue Level of Organization
- New Clinical Note: Problems with Serous Membranes
- Figure 4–2: Intercellular Connections: Terminology related to intercellular connections revised

Chapter 5: The Integumentary System
- New Clinical Note: Skin Abnormalities
- New Figure 5–7: Rickets
- New Figure 5–15: A Keloid

Chapter 6: Osseous Tissue and Bone Structure
- New Clinical Note: Heterotopic Bone Formation
- New Figure 6–12: Heterotopic Bone Formation
- New Figure 6–14: Examples of Abnormal Bone Development: Added to indicate pituitary growth failure and Marfan syndrome

Chapter 7: The Axial Skeleton
- New Clinical Note: Kyphosis, Lordosis, and Scoliosis
- New Figure 7–17: Abnormal Curvatures of the Spine

Chapter 8: The Appendicular Skeleton
- New Clinical Note: Carpal Tunnel Syndrome

Chapter 9: Articulations
- New Clinical Note: Knee Injuries
- Figure 9–12: The Knee Joint: Layout altered for clarity

Chapter 10: Muscle Tissue
- New Clinical Note: Rigor Mortis
- New Clinical Note: Delayed-Onset Muscle Soreness
- Figure 10–5: Sarcomere Structure: Terminology revised
- Figure 10–18: Isotonic and Isometic Contractions: Terminology revised

Chapter 11: The Muscular System
- New Clinical Note: Intramuscular Injections

Chapter 12: Neural Tissue
- New Clinical Note: Rabies
- New Figure 12–4: An Introduction to Neuroglia
- Figures 12–11: Gated Channels and 12–12: Graded Potentials: Terminology revised

Chapter 13: The Spinal Cord, Spinal Nerves, and Spinal Reflexes
- New Clinical Note: Shingles
- New Figure 13–9: Shingles
- New Figure 13–16: The Classification of Reflexes

Chapter 14: The Brain and Cranial Nerves
- New Clinical Note: Epidural and Subdural Hemorrhages
- New Clinical Note: Disconnection Syndrome
- Clinical Note: Parkinson Disease: relocated to Chapter 3

- Figures 14–l2: Ventricles of the Brain, 14–4: The Formation and Circulation of Cerebrospinal Fluid, 14–8: The Mesencephalon, and 14–10: The Hypothalamus in Sagittal Section: Terminology revised (mesencephalic aqueduct to aqueduct of midbrain)

Chapter 15: Neural Integration I: Sensory Pathways and the Somatic Nervous System

- New Clinical Note: Assessment of Tactile Sensitivities

Chapter 16: Neural Integration II: The Autonomic Nervous System and Higher-Order Functions

- New Clinical Note: Categorizing Nervous System Disorders
- Figure 16–11:Visceral Reflexes: Caption revised to aid the reader

Chapter 17: The Special Senses

- New Clinical Note: Motion Sickness
- Figure 17–20: Terminology revised
- Figure 17–26b: The Cochlea: Additional structures labeled
- Figure 17–31: Pathways for Auditory Sensations: Revised figure to show bilateral auditory pathways

Chapter 18: The Endocrine System

- Text and art: Revised to reflect the designation of suprarenal gland as the primary term and adrenal gland as a secondary term

Chapter 19: Blood

- New Clinical Note: Abnormal Hemoglobin
- Figure 19–2: The Anatomy of Red Blood Cells: Reorganized to enhance its depiction of RBCs.
- New Figure 19–4: Sickling in Red Blood Cells

Chapter 20: The Heart

- New Clinical Note: Abnormal Conditions Affecting Cardiac Output
- Figure 20–5: Cardiac Muscle Cells: Terminology revised

Chapter 21: Blood Vessels and Circulation

- New Clinical Note: Edema

Chapter 22: The Lymphoid System and Immunity

- Chapter has been re-titled using Lymphoid instead of Lymphatic to reflect internationally accepted terminology for the system
- New Clinical Note: Graft Rejection and Immunosuppression
- Figure 22–2: Lymphatic Capillaries: Art re-colored; Part b redrawn for clarity

- Figure 22–4: The Relationship between the Lymphatic Ducts and the Venous System: Photo added to give students a true anatomical perspective
- Text and art: Revised to reflect the designation of thymic corpuscle as the primary term and Hassal corpuscle as a secondary term.
- Figure 22-20: Step 2: The Sensitization and Activation of B Cells, Activation: Labels corrected

Chapter 23: The Respiratory System

- New Clinical Note: Pneumothorax
- New Clinical Note: Carbon Monoxide Poisoning
- Text and art: Revised to reflect the designations of pneumocyte type I as the primary term and alveolar epithelial cell as a secondary term, and pneumocyte type II as the primary term and septal cell as a secondary term

Chapter 24: The Digestive System

- New Clinical Note: Colorectal Cancer
- New Clinical Note: Inflammatory and Infectious Disorders of the Digestive System
- Figure 24–9: Primary and Secondary Dentitions: Redrawn to add mandible and maxillae with unerupted teeth
- Text and art: Changed the designation of plica to the formal term plica circulares

Chapter 25: Metabolism and Energetics

- New Clinical Note: Induced Hypothermia
- Table 25–2: Basic Food Groups of the 2005 Dietary Guidelines and Their General Effects on Health, and Figure 25–13: The MyPyramid Plan: Revised to include the new USDA Food Pyramid information
- Tables 25–3: Minerals and Mineral Reserves, and 25–5: The Water-Soluble Vitamins: Revised to use the latest terms for dietary intakes, minimums, and recommended amounts

Chapter 26: The Urinary System

- New Clinical Note: Diuretics
- Text and art: Revised to reflect the designation of nephron loop as the primary term and loop of Henle as a secondary term

Chapter 27: Fluid, Electrolyte, and Acid–Base Balance

- New Clinical Note: Water and Weight Loss
- Figure 27–1: Revised for clarity
- Table 27–2: Electrolyte Balance for Average Adult: Updated ranges of Ion and Normal ECF concentrations

Chapter 28: The Reproductive System

- New Clinical Note: Ovarian Cancer

- Clinical Notes: Prostatic Hypertrophy and Prostate Cancer: Rewritten with current statistics

- Clinical Note: Breast Cancer: Rewritten with current statistics

- Text and art: Revised to reflect the designations of seminal gland(s) as the primary term and seminal vesicle(s) as a secondary term, and nurse cell as the primary term and sustentacular cell as a secondary term

- Figures 28–7: Spermatogenesis, and 28–15: Oogenesis: Revised to reflect that DNA replication occurs before meiosis and that synapsis and tetrad formation occur within meiosis

Chapter 29: Development and Inheritance

- New Clinical Note: Chromosomal Abnormalities and Genetic Analysis

- The Human Genome Project and Beyond section: Updated with information regarding new study directions

---

Turn the page for a visual tour of the media and art programs. And, if you have any comments or suggestions, we'd like to hear from you. Please get in touch with us through the e-mail addresses below.

Frederic H. Martini
martini@maui.net

Judi L. Nath
JudiNath@bex.net

NEW! *Interactive Physiology*® 10-SYSTEM SUITE
WITH GRADABLE QUIZZES

The Best Way to Learn the Hardest Part of A&P

Interactive Physiology **10-System Suite (IP-10)** significantly enriches teaching and learning environments by providing an audio/visual presentation of complex topics. For use as both a teaching tool in the classroom and a study tool for students, IP-10 features full-color animations and video, both with sound, that thoroughly demonstrate difficult physiology concepts, many of which occur at the cellular and molecular level. Extensive interactive games and Gradable Quizzes reinforce the material.

Modules:

- Muscular System
- Nervous System I
- Nervous System II
- Cardiovascular System
- Respiratory System
- Urinary System
- Fluids & Electrolytes
- Endocrine System
- Digestive System
- NEW! Immune System

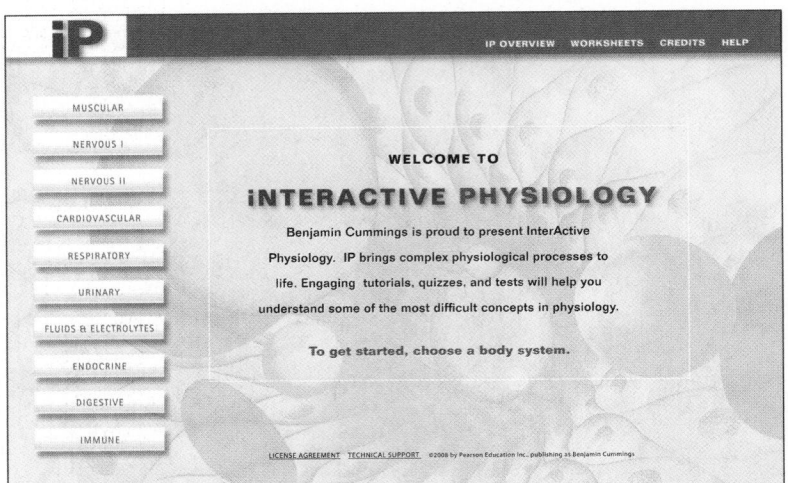

Gradebook

IP-10 now has **Gradable Quizzes** via the Gradebook feature on the website!

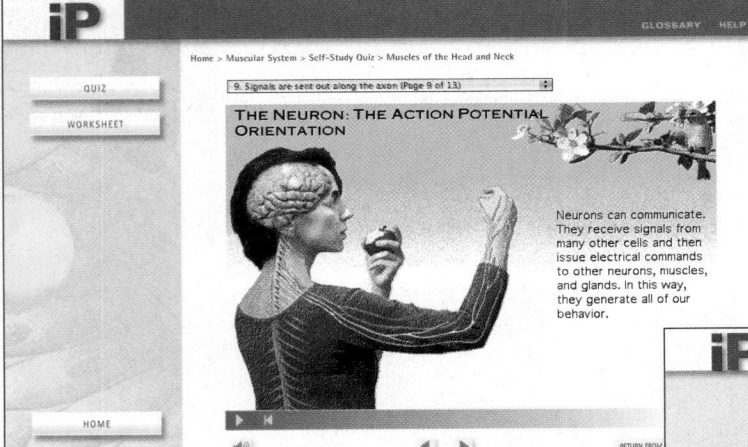

Improved Navigation

The new *Interactive Physiology* interface has been re-designed for more intuitive navigation.

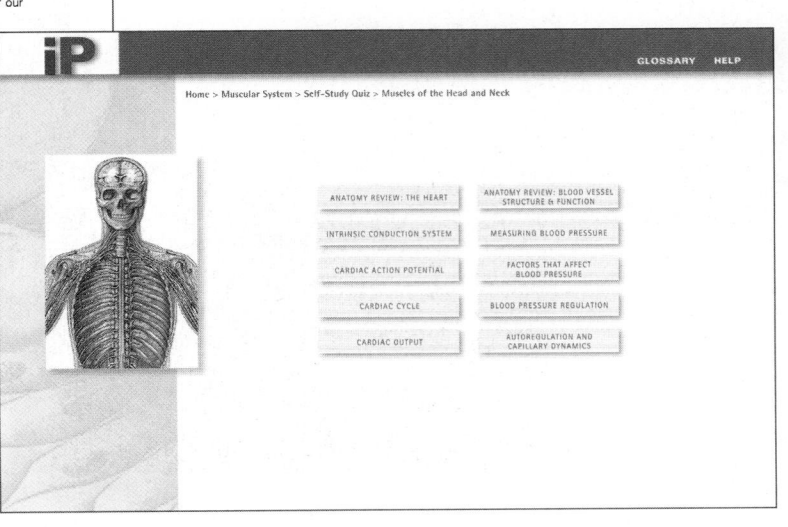

NEW! Immune System Module

Helpful Analogy

To aid students in understanding the functions of the Immune System, the new Immune System module is built around the analogy of a medieval castle and its defense systems.

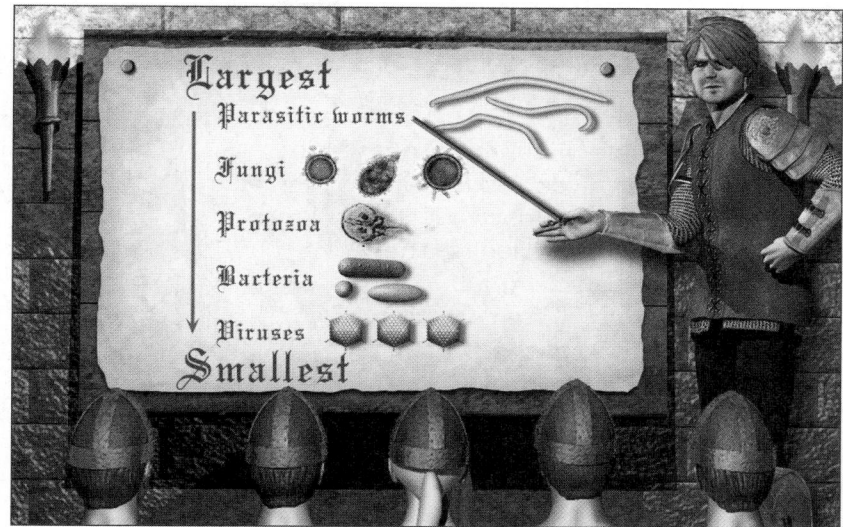

NEW! PAL™ 2.0

The Ultimate Anatomy Practice Tool

Practice Anatomy Lab (PAL) 2.0 is an indispensable virtual anatomy practice tool that gives students 24/7 access to the most widely used lab specimens, including the **human cadaver, anatomical models, histology slides, cat, and fetal pig**. Each of the five specimen modules includes hundreds of images as well as interactive tools for reviewing the specimens, hearing the names of anatomical structures, and taking multiple choice quizzes and fill-in-the-blank lab practical exams. Specimen images are also linked to animations.

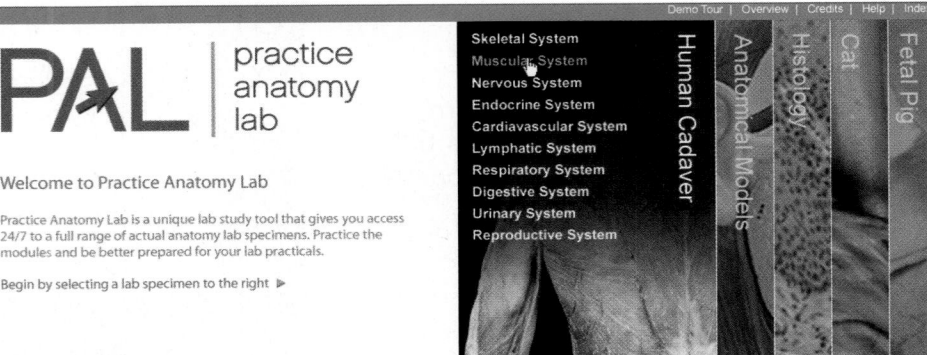

NEW! PAL 2.0 features:

- The all-new Human Cadaver module includes hundreds of specially-commissioned cadaver photos that can be found in all three activity sections in the module. This module also provides a **fully rotatable human skull** and 17 other rotatable skeletal structures.

- **3D Animations of Origins, Insertions, Actions, and Innervations** depict more than 65 muscles in the Human Cadaver module. These 3D Animations are viewable as students learn and hear the name of a muscle, thereby giving students **a one-stop learning experience**. Quizzes for the Animations are included.

- A greatly expanded Histology module includes more images of different types of tissues.

- Two new body systems, Endocrine and Lymphatic, have been added to the Human Cadaver and Anatomical Models modules.

Specially-commissioned Cadaver Photos

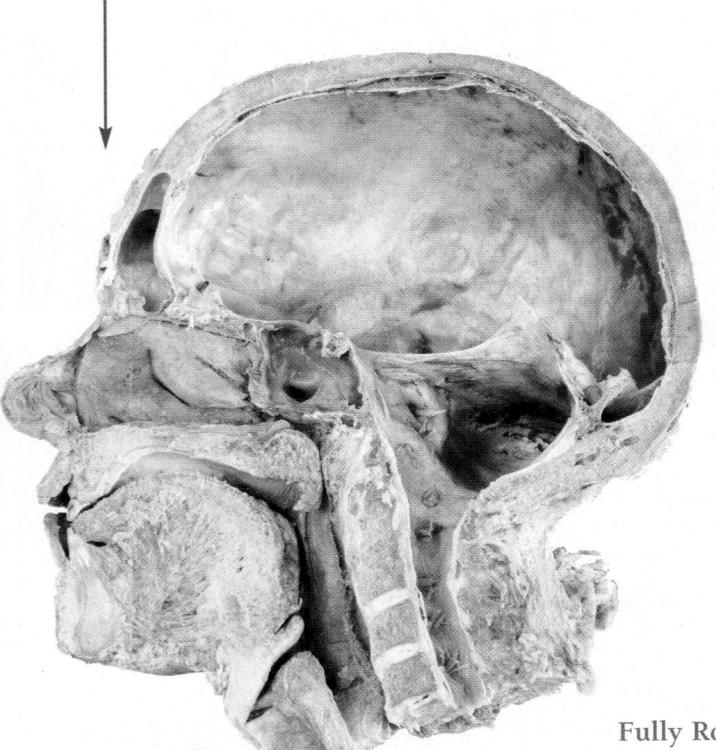

Fully Rotatable Human Skull

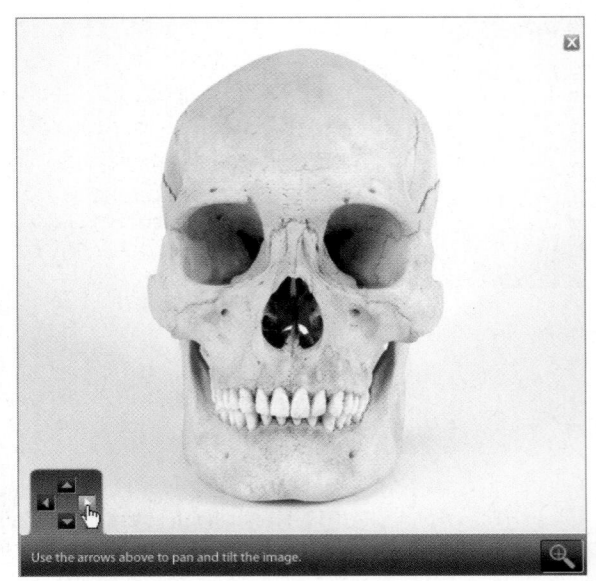

The PAL 2.0 CD-ROM is available for purchase.

Everything Students Need to Prepare for Their Lab Practical

Three Activity Sections

Self-Review allows students to review the names of structures, see structures highlighted, hear pronunciations of anatomical terms, and turn labels on and off.

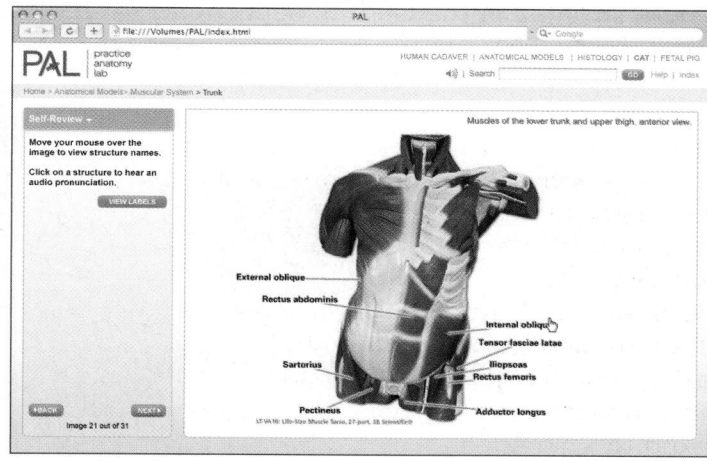

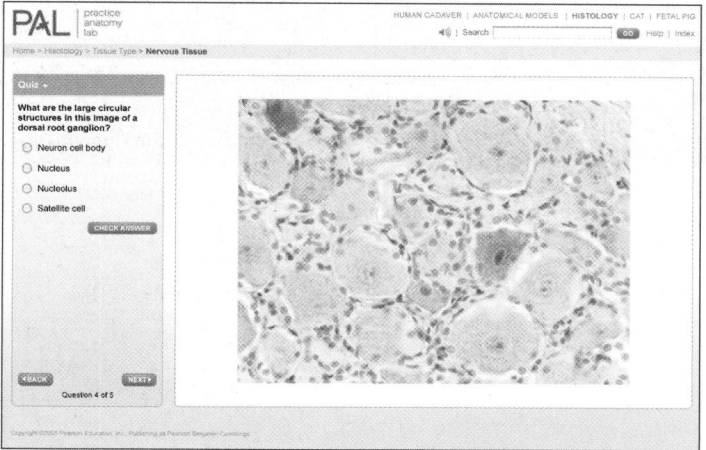

Quizzes let students quiz themselves on their knowledge of anatomical structures and functional anatomy. Students can receive immediate feedback to their answers, or they can wait to see a summary of results with the correct answers next to the images.

Lab Practicals simulate a real lab practical exam. Fill-in-the-blank exam questions ask students to identify and spell the names of a series of structures.

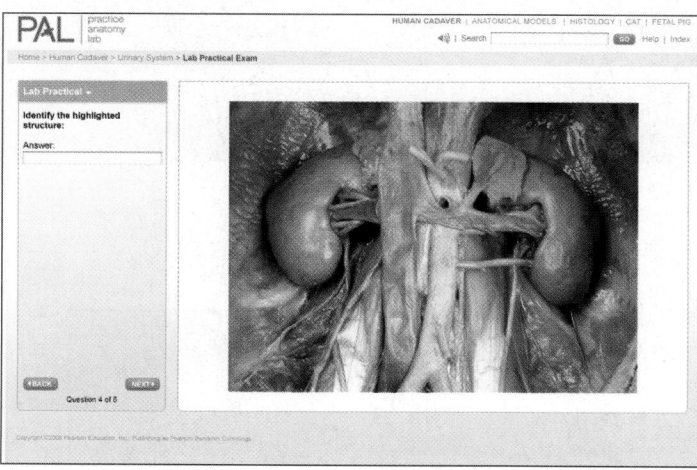

The **Instructor's Resource DVD for PAL 2.0** includes labeled and unlabeled images from PAL 2.0 in both JPEG and PowerPoint® formats. The PowerPoint Label Edit feature, with editable labels and leader lines, is built into every image. PowerPoint image slides include embedded links to the relevant 3D Animations of Origins, Insertions, Actions, and Innervations (see next page).

NEW! 3D ANIMATIONS OF ORIGINS, INSERTIONS, ACTIONS, AND INNERVATIONS WITH GRADABLE QUIZZES

Link from more than 65 muscle images in the PAL™ 2.0
Human Cadaver module to these 3D Animations of Origins,
Insertions, Actions, and Innervations. Quizzes for the Animations
are included in the Quizzes section of PAL 2.0 (see previous page).

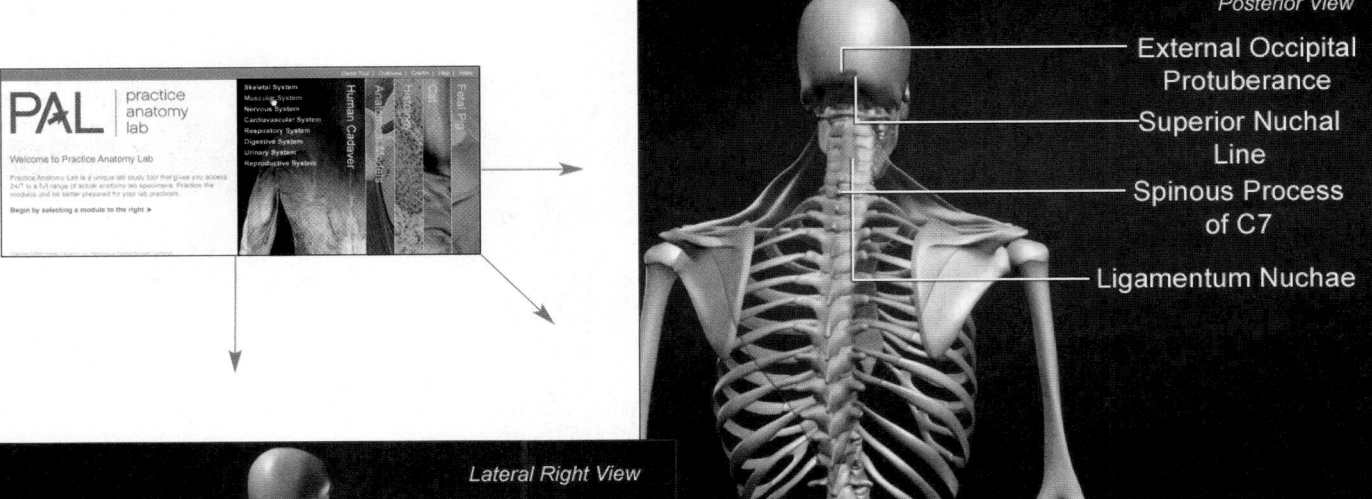

Posterior View

External Occipital
Protuberance

Superior Nuchal
Line

Spinous Process
of C7

Ligamentum Nuchae

Lateral Right View

**Available on PAL 2.0, myA&P,
and the instructor's Media
Manager. The Gradable Quizzes
are on myA&P.**

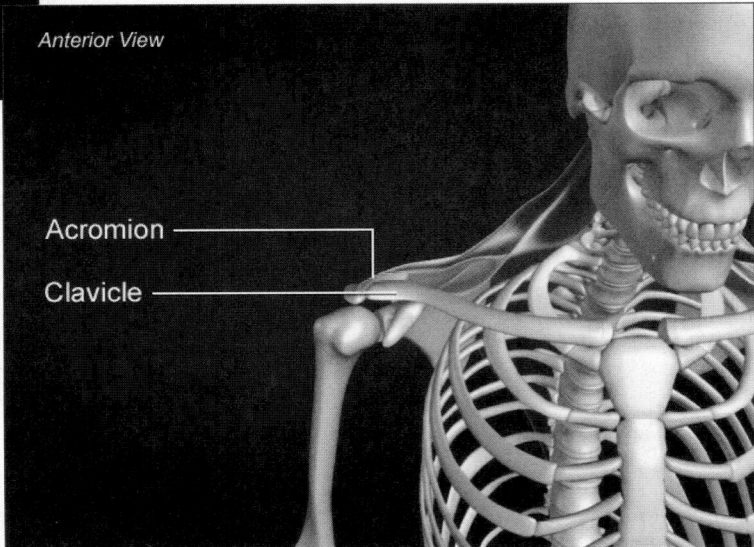

Anterior View

Acromion

Clavicle

NEW! 3D ANATOMY ANIMATIONS WITH GRADABLE QUIZZES
NEW! PHYSIOLOGY ANIMATIONS WITH GRADABLE QUIZZES

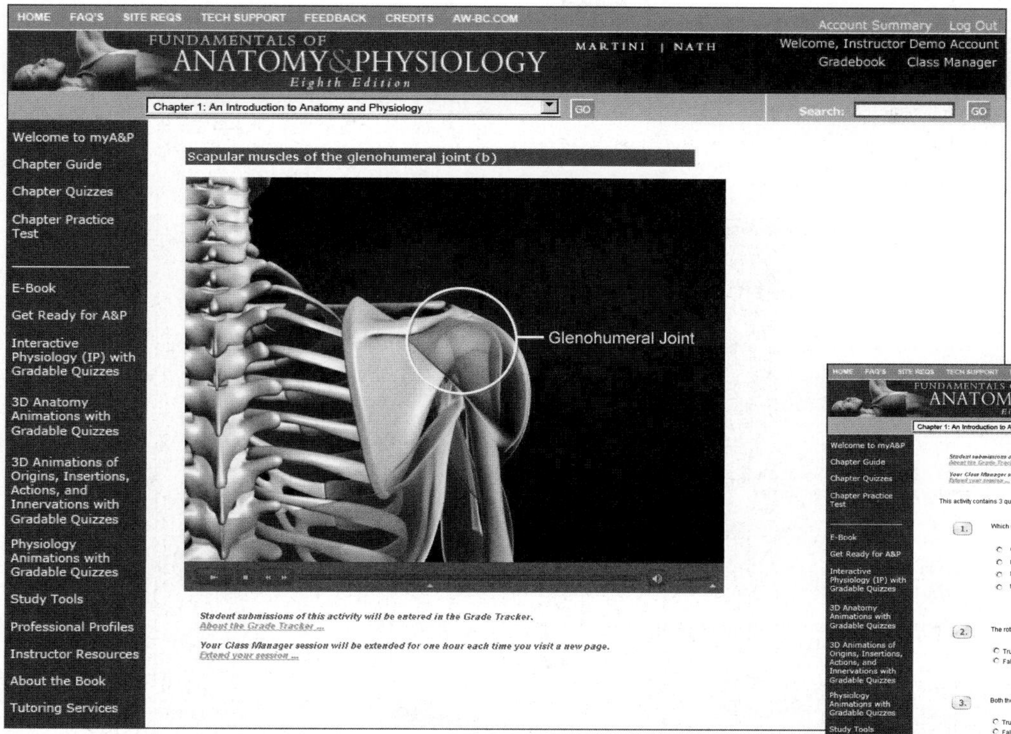

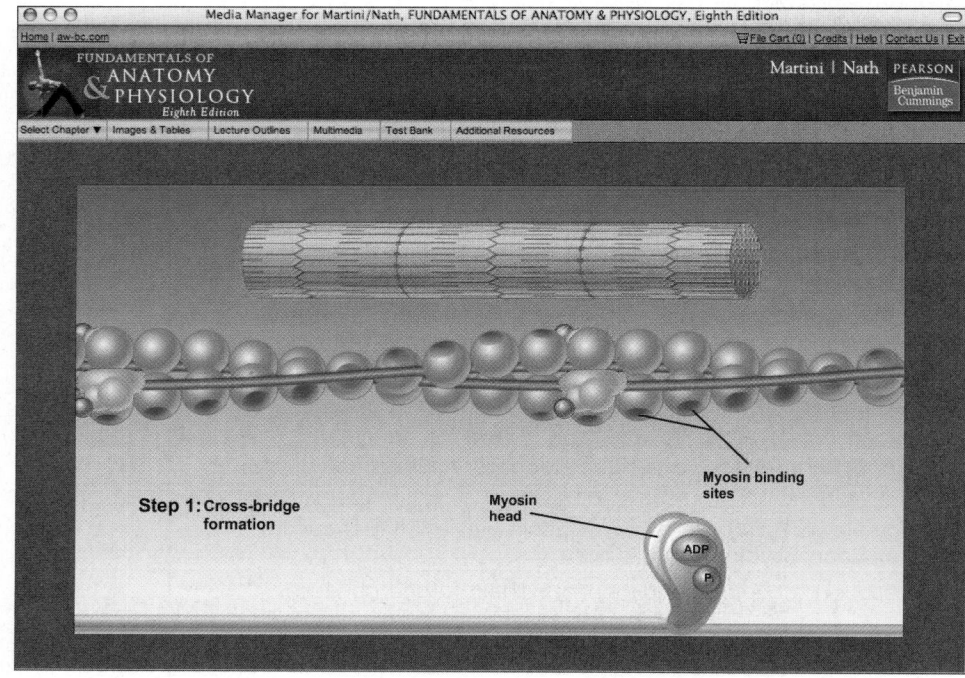

The 50 3D Anatomy Animations with Gradable Quizzes help students understand muscle and joint actions and appear on **both** the myA&P™ Companion Website and on the instructor's Media Manager (see next spread).

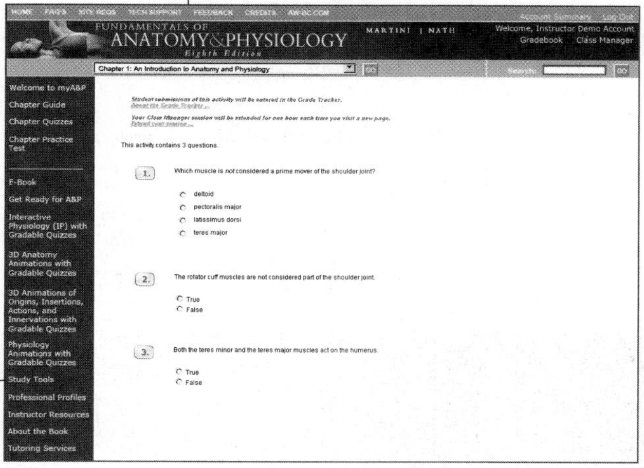

The Physiology Animations with Gradable Quizzes help students visualize difficult physiology concepts and appear on **both** the myA&P Companion Website and on the instructor's Media Manager.

myA&P™ Companion Website

Learning through Practice

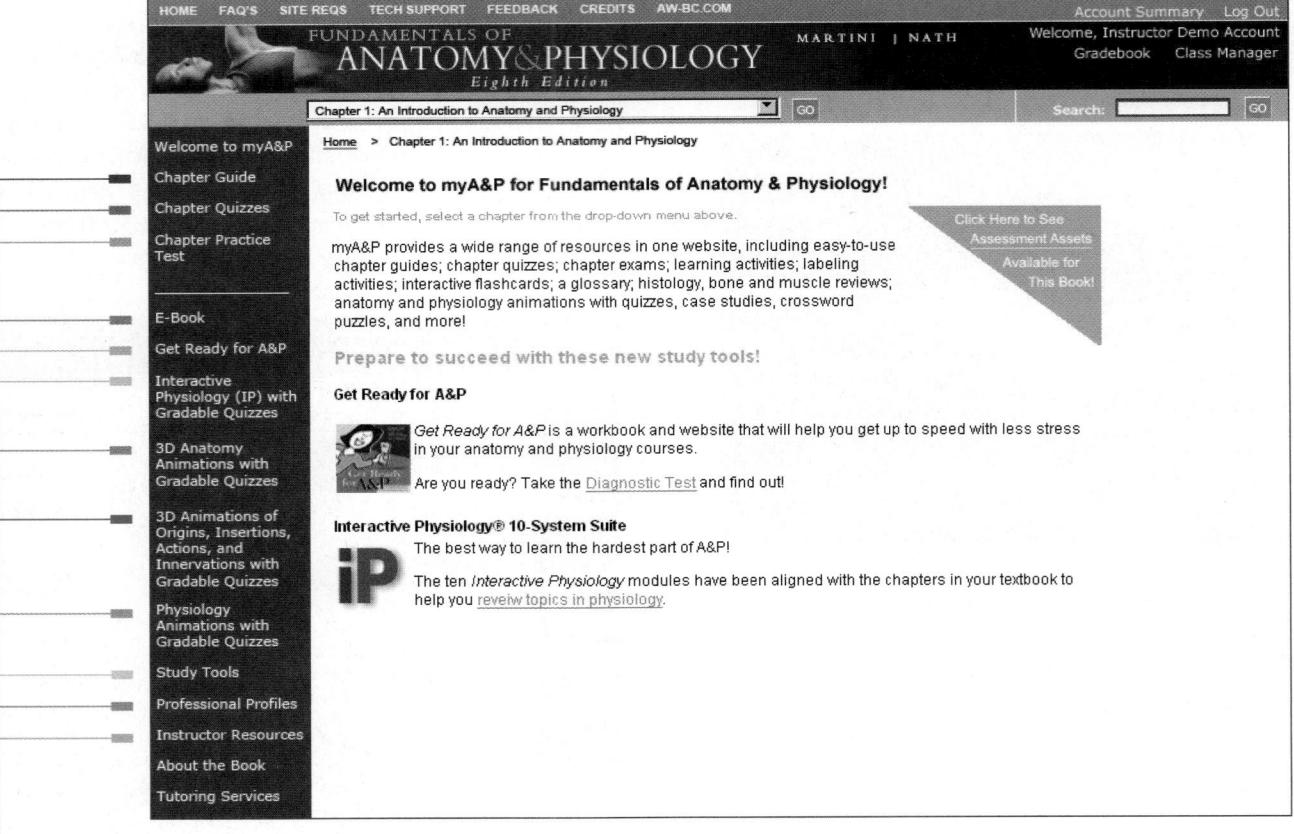

Features of myA&P Companion Website:

- **Chapter Guides** organize all chapter-specific activities and assessments by learning unit on one page.
- **Chapter Quizzes** now include more visuals for labeling practice.
- **Chapter Practice Tests** assess students' overall understanding of the chapter and include a timer feature.
- **The E-book** is a full electronic version of *Fundamentals of Anatomy & Physiology*, **Eighth Edition**.
- *Get Ready for A&P* **Media Update** includes diagnostic quizzes and access to the *Get Ready for A&P* website so students can get up to speed with the basics.
- *Interactive Physiology®* **10-System Suite (IP-10) with Gradable Quizzes** provides audio/visual understanding of complex topics (see previous spread).
- **3D Anatomy Animations with Gradable Quizzes** help students understand muscle and joint actions and include 50 animations (see previous page).
- **3D Animations of Origins, Insertions, Actions, and Innervations with Gradable Quizzes** illustrate and test students on these topics (see previous spread).
- **Physiology Animations with Gradable Quizzes** help students visualize difficult physiology concepts (see previous page).
- **Study Tools** include Histology Review, Bone Review, Muscle Review, Flashcards, Glossary with Pronunciations, Case Studies, and more.
- **Professional Profiles** help students explore and visualize their future careers.
- **A Gradebook** tracks students' progress and grades.

NEW! MEDIA MANAGER 2.1

All Instructor Media Resources in One Convenient Place

The **Media Manager** organizes all instructor media resources by chapter into **one convenient and easy-to-use package**.

New to Media Manager 2.1:

- **A "shopping cart" functionality** that allows instructors to quickly search, select, and download any item

- **"Quiz Game" chapter reviews** (in PRS-enabled Clicker Question format) that encourage student interaction and facilitate discussion

- **More than 50 new 3D Anatomy Animations with Quizzes** (in PRS-enabled Clicker Question format) that focus on hard-to-visualize anatomical concepts, especially muscle actions and joint movements (see previous spread)

- **3D Animations of Origins, Insertions, Actions, and Innervations with Quizzes** (in PRS-enabled Clicker Question format) that help students understand these topics (see previous spread)

- **Physiology Animations with Quizzes** (in PRS-enabled Clicker Question format) that help students visualize difficult concepts (see previous spread)

Customizable Art

Label Edit feature

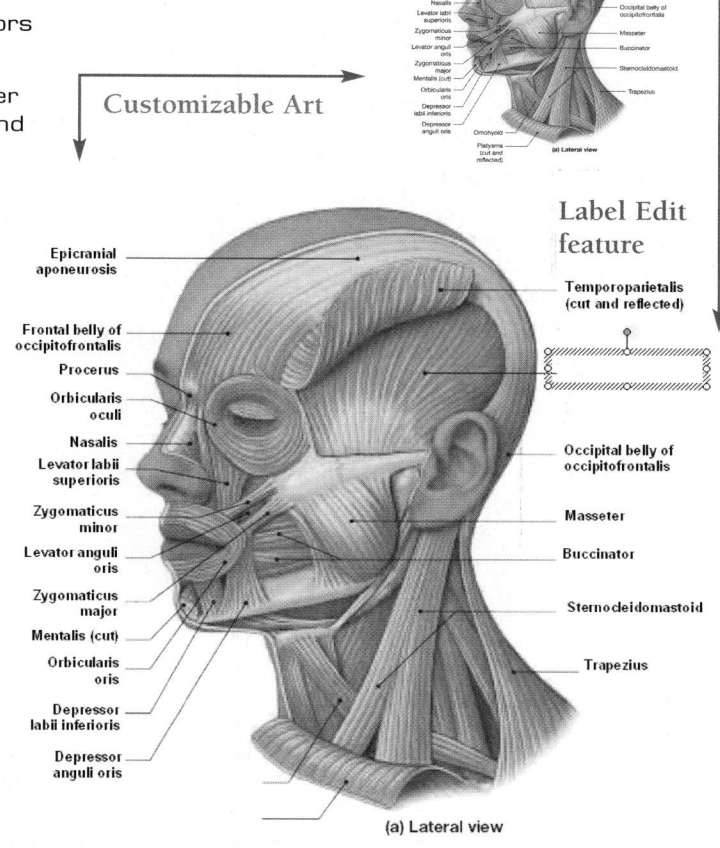

(a) Lateral view

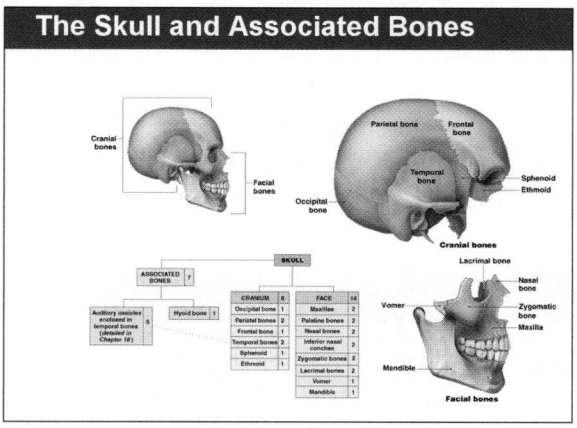

↑ Customizable PowerPoint® Lecture Slides

Quiz Game Chapter Reviews

The Media Manager also includes:

- **Customizable Art** including all illustrations and photos from the textbook. The optional **Label Edit** feature allows instructors to choose which structures to present.

- **All tables** from the textbook.

- **A bank of additional images** (not in the textbook) including MRI/CT scans, histology slides, and muscle origins and insertions.

- **All images from *Martini's Atlas of the Human Body*.**

- Select *Interactive Physiology*® **(IP) slides.**

- **Active Lecture Questions** that stimulate effective classroom discussions and check comprehension, in PRS-enabled Clicker Question format.

- **Customizable PowerPoint®** lecture slides, available for every chapter, combining lecture notes, illustrations, photos, tables, and animations.

- **The Computerized Test Bank.**

ART THAT TEACHES

"Side-by-Side" Figures

Multiple views of the same structure or tissue allow students to compare an illustrator's rendering with a photo of the actual structure or tissue as it may be seen in a laboratory or operating room.

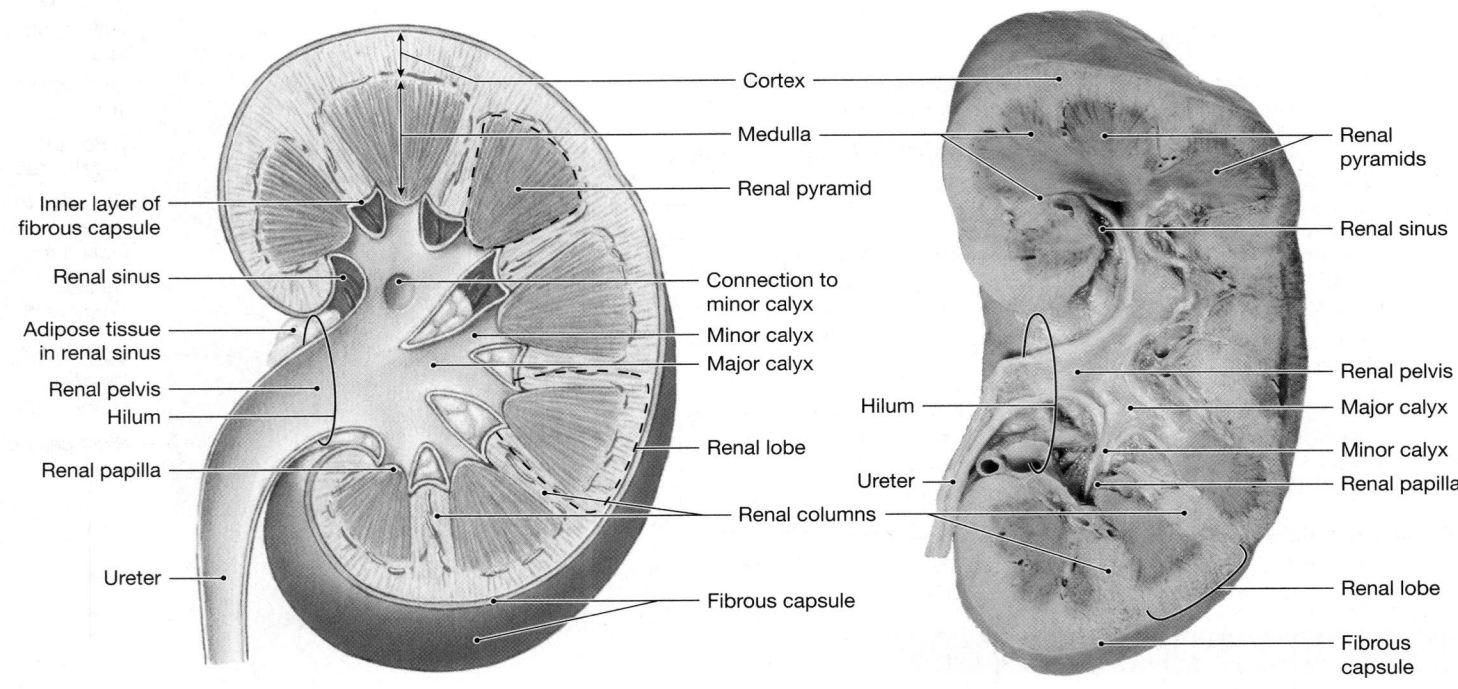

"Step-by-Step" Figures

These figures break down multifaceted processes into numbered step-by-step illustrations that coordinate with the authors' narrative descriptions.

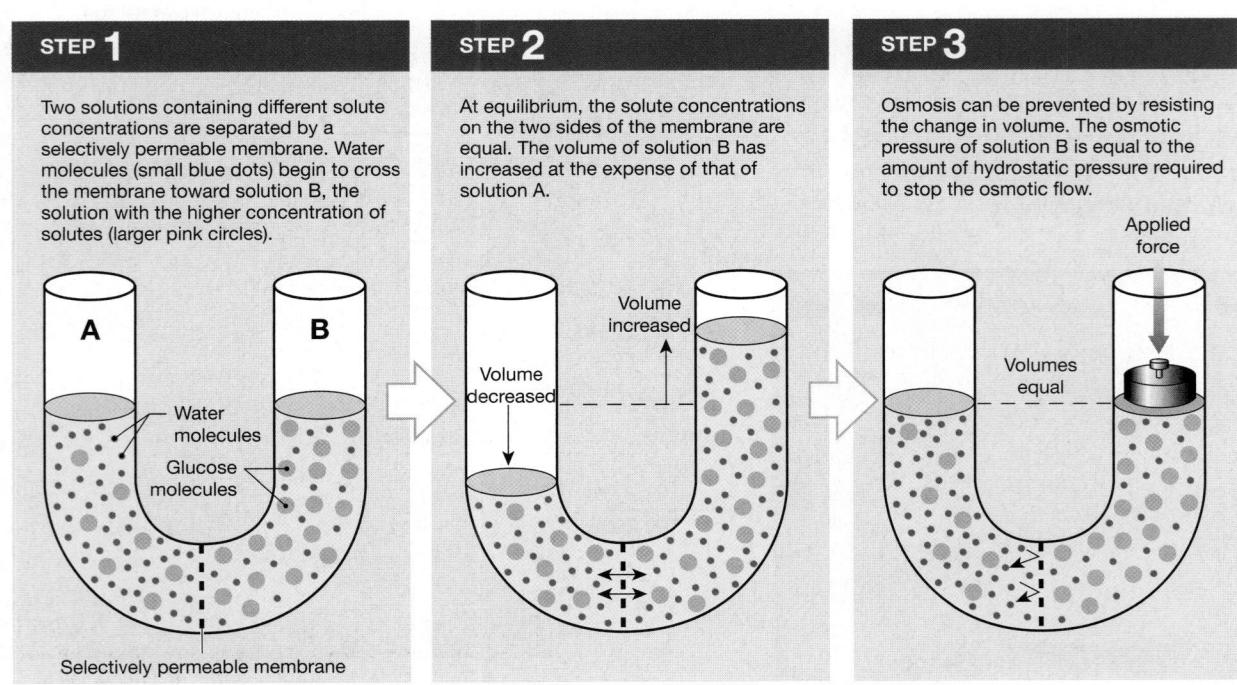

STEP 1

Two solutions containing different solute concentrations are separated by a selectively permeable membrane. Water molecules (small blue dots) begin to cross the membrane toward solution B, the solution with the higher concentration of solutes (larger pink circles).

STEP 2

At equilibrium, the solute concentrations on the two sides of the membrane are equal. The volume of solution B has increased at the expense of that of solution A.

STEP 3

Osmosis can be prevented by resisting the change in volume. The osmotic pressure of solution B is equal to the amount of hydrostatic pressure required to stop the osmotic flow.

"Macro-to-Micro" Figures

These figures help students to bridge the gap between familiar and unfamiliar structures by sequencing anatomical views from whole organs or other structures to their smaller parts.

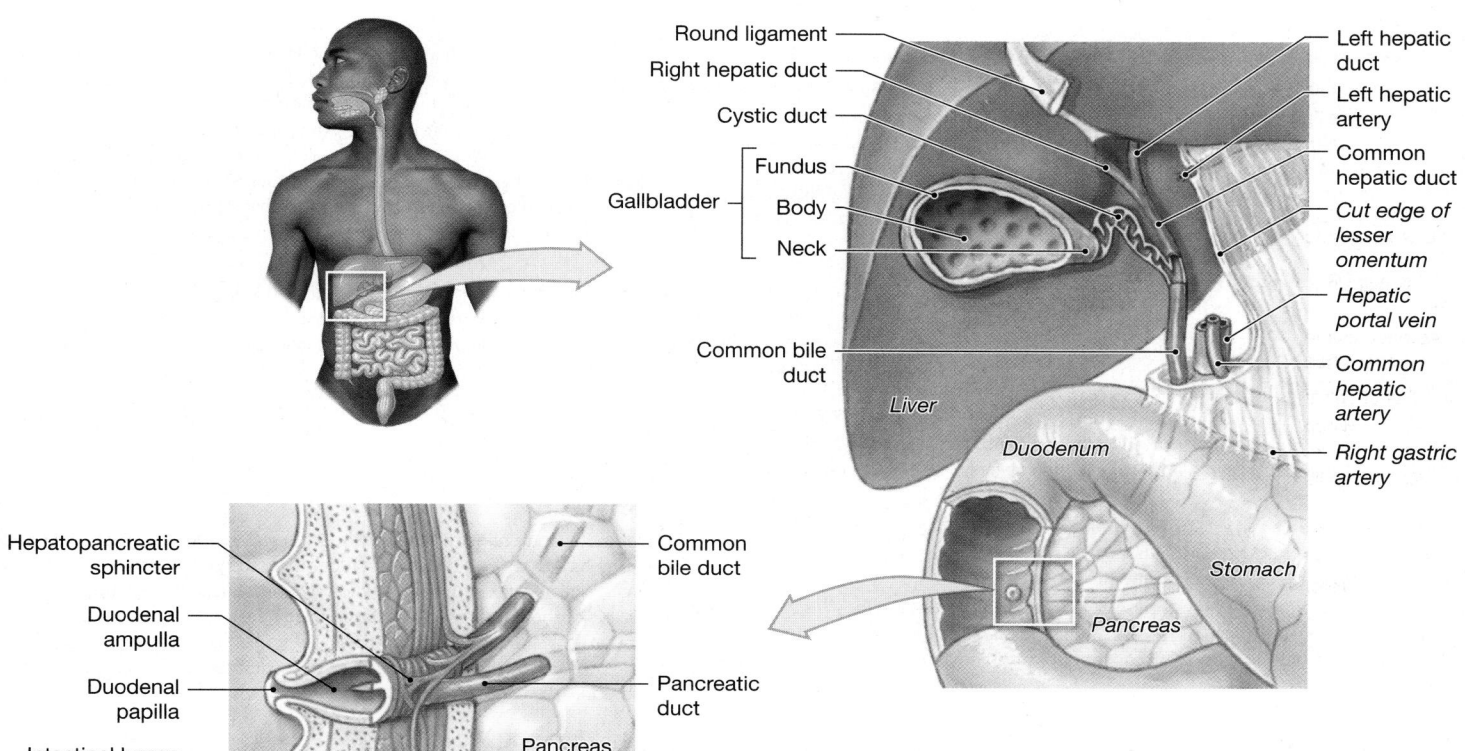

"Navigator" Figures

These figures are "big picture" overviews that appear at the start of discussions of complex physiological processes and re-appear as needed to help students follow the process.

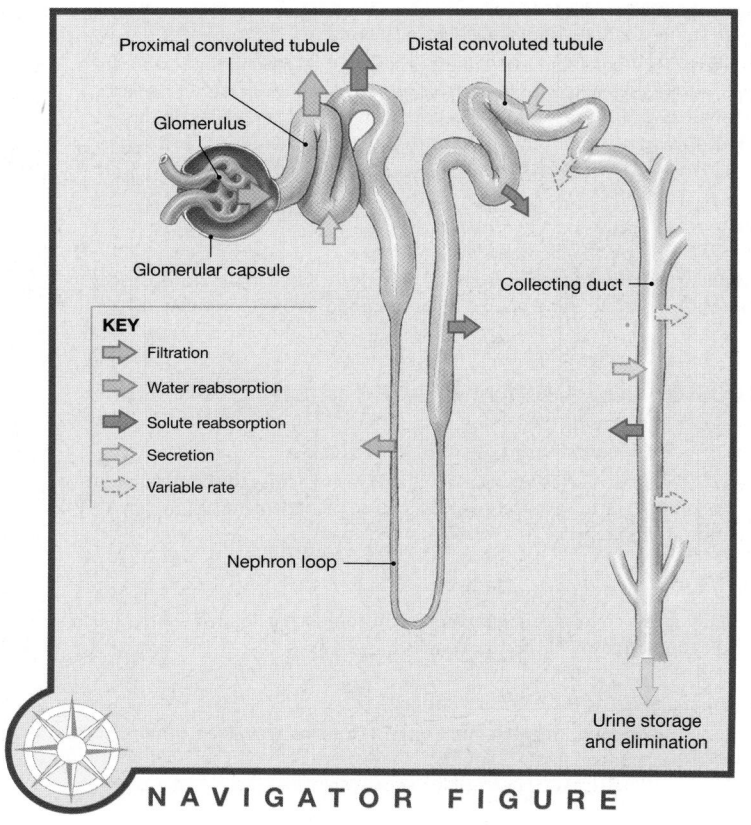

NAVIGATOR FIGURE

SUPPLEMENTS PACKAGE

Instructor Supplements

Media Manager 2.1 *(includes Computerized Test Bank)*
978-0-321-51230-7 • 0-321-51230-8

This media tool organizes all instructor media resources by chapter into one convenient and easy-to-use package and includes:

- **A DVD of all chapter-specific material**.
- **A DVD of Additional Resources,** such as the *Interactive Physiology*® 10-System Suite (IP-10) PowerPoint® slides, the 3D Anatomy Animations with Quizzes, the 3D Animations of Origins, Insertions, Actions, and Innervations with Quizzes, the Physiology Animations with Quizzes, a bank of additional images comprised of MRI/CT scans, histology slides, and muscle origins and insertions, and the images from *Martini's Atlas of the Human Body*.
- **A CD of the Computerized Test Bank**.
- **A fold-out Quick Reference Guide** that displays all assets by chapter.

Instructor Visual Guide
978-0-321-51229-1 • 0-321-51229-4

The Instructor Visual Guide is a printed and bound collection of thumbnails of the images and media on the Media Manager. With this take-anywhere supplement, instructors can plan lectures when away from their computers.

Instructor Manual
978-0-321-51097-6 • 0-321-51097-6
by Alan D. Magid, Charles Kazilek, and Kim Cooper

This useful resource includes a wealth of materials to help instructors organize their lectures, such as lecture and lab ideas, analogies, vocabulary aids, applications, common student misconceptions/problems, and classroom demonstrations.

Printed Test Bank
978-0-321-51098-3 • 0-321-51098-4
by Alan D. Magid, Judi L. Nath and Michael G. Wood

The test bank of more than 3,000 questions, organized around the textbook's 3-level learning system, helps instructors design a variety of tests and quizzes. More than 25% of the test questions have been revised or replaced in this edition. This supplement is the print version of the Computerized Test Bank that is in the Media Manager package.

Transparency Acetates
978-0-321-49410-8 • 0-321-49410-5

Illustrations, photos, and tables from the text are included in the Transparency Acetates. Complex figures are broken out for proper projected display.

Website and Course Management Programs

myA&P™ www.myaandp.com
The myA&P website serves as a one-stop gateway to access the companion website (myA&P) or the CourseCompass™ course management system (myA&P with CourseCompass).

WebCT www.aw-bc.com/webct
This open-access course management system contains preloaded content, such as testing and assessment question banks.

Blackboard www.aw-bc.com/blackboard
This open-access course management system contains preloaded content, such as testing and assessment question banks.

Student Supplements

Included with every new copy of the textbook:

Martini's Atlas of the Human Body
by Frederic H. Martini

The Atlas supplements the textbook with an abundant collection of anatomy photographs (194), radiology scans (53), and embryology summaries (21). The anatomy photographs, taken by Ralph Hutchings, the renowned biomedical photographer, extend the anatomy photographs that appear in the book. The radiology scans help students visualize deep anatomical structures and encourage them to become familiar with the types of images seen in a clinical setting. The embryology summaries depict the developmental origins of major tissues, organs, and body systems.

Interactive Physiology 10-System Suite (IP-10) CD-ROM
IP-10 helps students understand the hardest part of A&P (i.e., physiology) and now includes the eagerly-awaited 10th module on the Immune System.

Available separately or within a special-priced value package

Get Ready for A&P Media Update
978-0-321-49326-2 • 0-321-49326-5
by Lori K. Garrett

This workbook and online tool helps students prepare for the two-semester A&P course. The hands-on workbook gets students up to speed with basic study skills, math skills, basic chemistry, cell biology, anatomical terminology, and other basics of the human body. Each topic area includes a pre-test, guided explanation, interactive quizzes and exercises, and end-of-chapter cumulative tests. The online component includes self-study quizzes, tutorials, animations, and additional resources for review.

A&P Applications Manual
978-0-321-51310-6 • 0-321-51310-X
by Frederic H. Martini and Kathleen Welch

This manual contains extensive discussions on clinical topics and disorders and helps students apply the concepts of anatomy and physiology to daily life and is especially valuable for students who are planning careers in the health professions.

Study Card: Body System Overview
978-0-321-53601-3 • 0-321-53601-0

The updated Study Card is a six-panel laminated card showing all body systems and their organs and functions.

Study Guide
978-0-321-51231-4 • 0-321-51231-6
by Charles M. Seiger

The Study Guide includes a variety of review activities, including multiple choice questions, labeling exercises, and concept maps–all organized by the three-level learning system used in the book.

Practice Anatomy Lab (PAL™) 2.0 CD-ROM
978-0-321-54725-5 • 0-321-54725-X

PAL 2.0 is an indispensable virtual anatomy practice tool that gives students 24/7 access to the most widely used lab specimens, including the human cadaver, anatomical models, histology slides, cat, and fetal pig. Each of the five specimen modules includes hundreds of images as well as interactive tools for reviewing the specimens, hearing the names of anatomical structures, and taking multiple choice quizzes and fill-in-the-blank lab practical exams. Specimen images are also linked to animations.

Acknowledgments

This textbook is not the product of any single individual. It represents a group effort, and the members of the group deserve to be acknowledged.

Foremost on our thank you list are the instructors who reviewed the published text of the Seventh Edition and the manuscript of the Eighth Edition to offer the suggestions that helped guide us through the revision process. To them, we express our sincere thanks and best wishes:

R. Michael Anson, The Community College of Baltimore County
Nishi Bryska, University of North Carolina at Charlotte
Brad Caldwell, Greenville Technical College
Jorge D. Cortese, Durham Technical Community College
Brian D. Feige, Mott Community College
Lorraine Findlay, Nassau Community College
Lauren Goodwyn, Borough of Manhattan Community College, CUNY
Chaya Gopalan, St. Louis Community College
Jean Helgeson, Collin County Community College
Stephen T. Kabrhel, The Community College of Baltimore County
Marian Leal, Sacred Heart University
John Martin, Clark College
Maire McShane, Montgomery County Community College
Judith M. Megaw, Indian River Community College
Zvi Ostrin, Hostos Community College, CUNY
Andrew J. Petto, University of Wisconsin—Milwaukee
Mary L. Puglia, Central Arizona College
Julie Rosenheimer, Texas Tech University
Dee Ann S. Sato, Cypress College
Claudia Stanescu, University of Arizona
Robert J. Sullivan, Marist College
Janis Thompson, Lorain County Community College
Vernon L. Wiersema, Houston Community College
Elizabeth T. Wise, Lourdes College

The accuracy and currency of the clinical material in this edition and in the *Applications Manual* in large part reflect the work of my wife, Kathleen Welch, M.D. Her professionalism and concern for practicality and common sense make the clinical information especially relevant for today's students.

Virtually without exception, reviewers stressed the importance of accurate, integrated, and visually attractive illustrations in aiding the students to understand essential material. The revision of the art program was directed by Bill Ober, M.D., and Claire Garrison, R.N. Their suggestions about topics of clinical importance, presentation sequence, and revisions to the proposed art were of incalculable value to us and to the project. The illustration program for this edition was further enhanced by the efforts of two other talented individuals. Jim Gibson designed most of the new features in the art program and consulted on the design and layout of the individual figures. His talents have helped produce an illustration program that is attractive, cohesive, and easy to understand. Anita Hylton helped to create the new photo/art combinations that enable better orientation and a greater sense of realism in important anatomical figures.

Many of the text's illustrations include color photographs or micrographs collected from a variety of sources. The striking anatomical photos in the text, in *Martini's Atlas of the Human Body*, in the *Laboratory Manual for Anatomy & Physiology* by Michael G. Wood, and in *Laboratory Investigations in Anatomy & Physiology* by Stephen N. Sarikas are the work of Ralph Hutchings, whose efforts on this project are deeply appreciated.

A special thanks goes to Alan D. Magid of the Duke University School of Medicine who was our Technical Editor and whose expertise was invaluable throughout this edition.

We also express our appreciation to the editors and support staff at Benjamin Cummings. First on the list is Katy German, Project Editor, who somehow managed to handle every crisis and kept things moving in the right direction. Her support, hard work, and patience are deeply appreciated.

We owe special thanks to our editor, Leslie Berriman, for her creativity and dedication. Her vision helped shape this book in countless ways.

Alan Titche, our Development Editor, played a vital role in fashioning the Eighth Edition. We could not have survived this process without him, and his unfailing attention to detail and quality made a tremendous difference in the final package.

Thanks are also due to the rest of the editorial team, including Robin Pille and Kelly Reed, for their assistance with the integrated supplements package.

This book would not exist without the extraordinary dedication of the Production team, especially Debbie Cogan and Caroline Ayres, who solved so many problems under pressure with unfailing good cheer. The design process was headed by Design Manager Mark Ong, and we want to thank him along with Gary Hespenheide, the interior designer, and Yvo Riezebos, the cover designer, for the strong look and navigability of the book.

We also want to thank the media producers for their creative efforts on the media supplements and for producing the new media components incorporated into the myA&P companion website, the course management programs, and the instructor's Media Manager: Sarah Young-Dualan, Aimee Pavy, and Erik Fortier.

We would also like to express our gratitude to Linda Davis, President, Frank Ruggirello, Editorial Director, and Lauren Fogel, Director of Media Development for their continuing support of this project.

Finally, thanks to Gordon Lee, Marketing Manager, and the Pearson Science sales reps who work tirelessly to present the book, associated media, and other materials to anatomy and physiology instructors.

To help improve future editions, we encourage you to send any pertinent information, suggestions, or comments about the organization or content of this textbook to us directly, using the e-mail addresses below. We will deeply appreciate any and all comments and suggestions and will carefully consider them in the preparation of the Ninth Edition.

Frederic H. Martini
martini@maui.net

Judi L. Nath
JudiNath@bex.net

Brief Contents

UNIT 1: LEVELS OF ORGANIZATION

Chapter 1   An Introduction to Anatomy and Physiology   1

Chapter 2   The Chemical Level of Organization   28

Chapter 3   The Cellular Level of Organization   66

Chapter 4   The Tissue Level of Organization   112

UNIT 2: SUPPORT AND MOVEMENT

Chapter 5   The Integumentary System   157

Chapter 6   Osseous Tissue and Bone Structure   184

Chapter 7   The Axial Skeleton   211

Chapter 8   The Appendicular Skeleton   246

Chapter 9   Articulations   267

Chapter 10   Muscle Tissue   293

Chapter 11   The Muscular System   335

UNIT 3: CONTROL AND REGULATION

Chapter 12   Neural Tissue   386

Chapter 13   The Spinal Cord, Spinal Nerves, and Spinal Reflexes   429

Chapter 14   The Brain and Cranial Nerves   460

Chapter 15   Neural Integration I: Sensory Pathways and the Somatic Nervous System   506

Chapter 16   Neural Integration II: The Autonomic Nervous System and Higher-Order Functions   528

Chapter 17   The Special Senses   561

Chapter 18   The Endocrine System   603

UNIT 4: FLUIDS AND TRANSPORT

Chapter 19   Blood   650

Chapter 20   The Heart   681

Chapter 21   Blood Vessels and Circulation   719

Chapter 22   The Lymphoid System and Immunity   776

UNIT 5: ENVIRONMENTAL EXCHANGE

Chapter 23   The Respiratory System   825

Chapter 24   The Digestive System   874

Chapter 25   Metabolism and Energetics   929

Chapter 26   The Urinary System   965

Chapter 27   Fluid, Electrolyte, and Acid–Base Balance   1009

UNIT 6: CONTINUITY OF LIFE

Chapter 28   The Reproductive System   1041

Chapter 29   Development and Inheritance   1086

Contents

Preface   v

UNIT 1: LEVELS OF ORGANIZATION

Chapter 1
An Introduction to Anatomy and Physiology   1

An Introduction to Studying the Human Body   2

1-1 Anatomy and physiology directly affect your life   2

1-2 Good study strategies are crucial for success   2

1-3 Anatomy is structure, and physiology is function   4

1-4 Anatomy and physiology are closely integrated   5
  Anatomy   6
  Physiology   7

1-5 Levels of organization progress from molecules to a complete organism   8

1-6 Homeostasis is the tendency toward internal balance   12

1-7 Negative feedback opposes variations from normal, whereas positive feedback exaggerates them   14
  The Role of Negative Feedback in Homeostasis   14
  The Role of Positive Feedback in Homeostasis   14
  Systems Integration, Equilibrium, and Homeostasis   16

1-8 Anatomical terms describe body regions, anatomical positions and directions, and body sections   17
  Superficial Anatomy   17
  Sectional Anatomy   18

1-9 Body cavities protect internal organs and allow them to change shape   22
  The Thoracic Cavity   23
  The Abdominopelvic Cavity   24

Chapter Review   25
Clinical Notes
  The Visible Human Project   7
  Homeostasis and Disease   13

Chapter 2
The Chemical Level of Organization   28

An Introduction to the Chemical Level of Organization   29

2-1 Atoms are the basic particles of matter   29
  Atomic Structure   29
  Elements and Isotopes   30
  Atomic Weights   31
  Electrons and Energy Levels   31

2-2 Chemical bonds are forces formed by atom interactions   32
  Ionic Bonds   32
  Covalent Bonds   34
  Hydrogen Bonds   35
  States of Matter   35
  Molecular Weights   36

2-3 Decomposition, synthesis, and exchange reactions are important chemical reactions in physiology   36
  Basic Energy Concepts   36

Focus Chemical Notation   37
  Types of Chemical Reactions   38

2-4 Enzymes catalyze specific biochemical reactions by lowering a reaction's activation energy   39

2-5 Inorganic compounds usually lack carbon, and organic compounds always contain carbon   40

2-6 Physiological systems depend on water   41
  The Properties of Aqueous Solutions   41
  Colloids and Suspensions   42

2-7 Body fluid pH is vital for homeostasis   43

2-8 Acids, bases, and salts are inorganic compounds with important physiological roles   44
  Salts   45
  Buffers and pH Control   45

2-9 Carbohydrates contain carbon, hydrogen, and oxygen in a 1:2:1 ratio   45
  Monosaccharides   45
  Disaccharides and Polysaccharides   46

2-10 Lipids contain a carbon-to-hydrogen ratio of 1:2 48

Fatty Acids 48

Eicosanoids 49

Glycerides 50

Steroids 50

Phospholipids and Glycolipids 51

2-11 Proteins are formed from amino acids and contain carbon, hydrogen, oxygen, and nitrogen 51

Protein Structure 53

Protein Shape 54

Enzyme Function 56

Glycoproteins and Proteoglycans 57

2-12 DNA and RNA are nucleic acids 57

Structure of Nucleic Acids 58

RNA and DNA 58

2-13 ATP is a high-energy compound used by cells 59

2-14 Chemicals form functional units called cells 60

Chapter Review 62
Clinical Notes

Solute Concentrations 43

Fatty Acids and Health 49

Chapter 3

The Cellular Level of Organization

66

An Introduction to Cells 67

3-1 The plasma membrane separates the cell from its surrounding environment and performs various functions 67

Membrane Lipids 70

Membrane Proteins 70

Membrane Carbohydrates 71

3-2 Organelles within the cytoplasm perform particular functions 72

The Cytosol 72

The Organelles 72

3-3 The nucleus contains DNA and enzymes essential for controlling cellular activities 82

Contents of the Nucleus 83

Information Storage in the Nucleus 83

3-4 DNA controls protein synthesis, cell structure, and cell function 84

The Role of Gene Activation in Protein Synthesis 84

The Transcription of mRNA 84

Translation 86

How the Nucleus Controls Cell Structure and Function 88

3-5 Diffusion is a passive transport mechanism facilitating membrane passage 89

Diffusion 89

Diffusion across Plasma Membranes 90

3-6 Carrier-mediated and vesicular transport mechanisms facilitate membrane passage 94

Carrier-Mediated Transport 94

Vesicular Transport 96

3-7 The transmembrane potential results from the unequal distribution of ions across the plasma membrane 99

3-8 Stages of a cell's life cycle include interphase, mitosis, and cytokinesis 100

Interphase 100

Mitosis 102

Cytokinesis 103

The Mitotic Rate and Energy Use 103

3-9 Several growth factors affect the cell life cycle 104

3-10 Tumors and cancers are characterized by abnormal cell growth and division 105

3-11 Differentiation is cellular specialization as a result of gene activation or repression 106

Chapter Review 107
Clinical Notes

Inheritable Mitochondrial Disorders 81

DNA Fingerprinting 84

Mutations 88

Drugs and the Plasma Membrane 91

Telomerase, Aging, and Cancer 106

Parkinson Disease 107

Chapter 4

The Tissue Level of Organization

111

An Introduction to the Tissue Level of Organization 112

4-1 The four tissue types are epithelial, connective, muscle, and neural 112

4-2 Epithelial tissue covers body surfaces, lines cavities and tubular structures, and serves essential functions 112

Functions of Epithelial Tissue 113

Specializations of Epithelial Cells 113

Maintaining the Integrity of Epithelia 114

4-3 Cell shape and number of layers determine the classification of epithelia 116

Classification of Epithelia 116

Glandular Epithelia 121

4-4 Connective tissue provides a protective structural framework for other tissue types 124

Classification of Connective Tissues 124

Connective Tissue Proper 124

4-5 Cartilage and bone provide a strong supporting framework 131

Cartilage 131

Bone 132

4-6 Membranes are physical barriers of four types: mucous, serous, cutaneous, and synovial   135

Mucous Membranes   135

Serous Membranes   136

The Cutaneous Membrane   136

Synovial Membranes   137

4-7 Connective tissues create the internal framework of the body   137

4-8 The three types of muscle tissue are skeletal, cardiac, and smooth   138

Skeletal Muscle Tissue   138

Cardiac Muscle Tissue   140

Smooth Muscle Tissue   140

4-9 Neural tissue responds to stimuli and conducts electrical impulses throughout the body   140

4-10 The response to tissue injury involves inflammation and regeneration   141

Inflammation   142

Regeneration   143

4-11 With advancing age, tissue repair declines and cancer rates increase   143

Aging and Tissue Structure   143

Aging and Cancer Incidence   143

Chapter Review   144

Clinical Notes

Exfoliative Cytology   118

Marfan Syndrome   126

Problems with Serous Membranes   136

SYSTEMS OVERVIEW 148

UNIT 2: SUPPORT AND MOVEMENT

Chapter 5

The Integumentary System 157

An Introduction to the Integumentary System   158

5-1 The epidermis is composed of strata (layers) with various functions   159

Stratum Germinativum   160

Stratum Spinosum   161

Stratum Granulosum   161

Stratum Lucidum   161

Stratum Corneum   161

5-2 Factors influencing skin color are epidermal pigmentation and dermal circulation   162

The Role of Epidermal Pigmentation   162

The Role of Dermal Circulation   163

5-3 Sunlight causes epidermal cells to convert a steroid into vitamin D$_3$ 165

5-4 Epidermal growth factor has several effects on the epidermis and epithelia   165

5-5 The dermis is the tissue layer that supports the epidermis   166

Dermal Strength and Elasticity   166

Lines of Cleavage   166

The Dermal Blood Supply   167

Innervation of the Skin   168

5-6 The hypodermis is tissue beneath the dermis that connects it to underlying tissues   168

5-7 Hair is composed of keratinized dead cells that have been pushed to the surface   169

Hair Production   169

The Hair Growth Cycle   171

Types of Hairs   171

Hair Color   171

5-8 Sebaceous glands and sweat glands are exocrine glands found in the skin   172

Sebaceous (Oil) Glands   172

Sweat Glands   172

Other Integumentary Glands   173

Control of Glandular Secretions and the Homeostatic Role of the Integument   174

5-9 Nails are keratinized epidermal cells that protect the tips of fingers and toes   174

5-10 Several steps are involved in repairing the integument following an injury   175

5-11 Effects of aging include dermal thinning, wrinkling, and reduced melanocyte activity   177

5-12 The integumentary system provides protection for all other body systems   178

Chapter Review   181

Clinical Notes

Skin Cancers, Melanomas, and Sunblocks   164

Ulcers   167

Liposuction   168

Skin Abnormalities   178

Burns and Grafts   180

Chapter 6

Osseous Tissue and Bone Structure 184

An Introduction to the Skeletal System   185

6-1 The skeletal system has five primary functions   185

6-2 Bones are classified according to shape and structure, and feature surface markings   185

Bone Shapes   185
Bone Markings (Surface Features)   186
Bone Structure   188

6-3 Bone is composed of matrix and several types of cells: osteocytes, osteoblasts, osteoprogenitor cells, and osteoclasts   189
The Matrix of Bone   189
The Cells of Bone   189

6-4 Compact bone contains parallel osteons, and spongy bone contains trabeculae   191
The Structure of Compact Bone   191
The Structure of Spongy Bone   192
The Periosteum and Endosteum   193

6-5 Ossification and appositional growth are mechanisms of bone formation and enlargement   194
Endochondral Ossification   195
Intramembranous Ossification   198
The Blood and Nerve Supplies to Bone   198

6-6 Bone growth and development depend on a balance between bone formation and bone resorption   200

6-7 Exercise, hormones, and nutrition affect bone development and the skeletal system   200
The Effects of Exercise on Bone   201
Hormonal and Nutritional Effects on Bone   201

6-8 Calcium plays a critical role in bone physiology   203
The Skeleton as a Calcium Reserve   203
Hormones and Calcium Balance   203

6-9 A fracture is a crack or break in a bone   205

Focus Types of Fractures   206

6-10 Osteopenia has a widespread effect on aging skeletal tissue   207

Chapter Review   208
Clinical Notes
Heterotopic Bone Formation   195
Abnormal Bone Growth and Development   202

Chapter 7

The Axial Skeleton   211

An Introduction to the Axial Skeleton   212

7-1 The 80 bones of the longitudinal axis make up the axial skeleton   212

7-2 The skull is composed of 8 cranial bones and 14 facial bones   213

Focus The Individual Bones of the Skull   218

7-3 Foramina and fissures of the skull serve as passageways for nerves and vessels   226

7-4 An orbital complex contains each eye, and the nasal complex encloses the nasal cavities   228

The Orbital Complexes   228
The Nasal Complex   228

7-5 Fontanelles are non-ossified areas between cranial bones that allow for brain growth   229

7-6 The vertebral column has four spinal curves   231
Spinal Curvature   231
Vertebral Anatomy   232

7-7 The five vertebral regions are the cervical, thoracic, lumbar, sacral, and coccygeal   234
Cervical Vertebrae   234
Thoracic Vertebrae   236
Lumbar Vertebrae   237
The Sacrum   237
The Coccyx   239

7-8 The thoracic cage protects organs in the chest and provides sites for muscle attachment   239
The Ribs   239
The Sternum   242

Chapter Review   243
Clinical Notes
Temporomandibular Joint Syndrome   226
Craniostenosis   230
Kyphosis, Lordosis, and Scoliosis   232

Chapter 8

The Appendicular Skeleton   246

An Introduction to the Appendicular Skeleton   247

8-1 The pectoral girdle attaches to the upper limbs and consists of the clavicles and scapulae   247
The Clavicles   247
The Scapulae   247

8-2 The upper limbs are adapted for freedom of movement   250
The Humerus   250
The Ulna   251
The Radius   252
The Carpal Bones   253
The Metacarpal Bones and Phalanges   254

8-3 The pelvic girdle attaches to the lower limbs and consists of two coxal bones   254
The Pelvic Girdle   254
The Pelvis   256

8-4 The lower limbs are adapted for locomotion and support   258
The Femur   258
The Patella   259
The Tibia   259
The Fibula   260

The Tarsal Bones 260

The Metatarsal Bones and Phalanges 261

8-5 Sex differences and age account for individual skeletal variation 262

Chapter Review 264

Clinical Notes

Carpal Tunnel Syndrome 254

Congenital Talipes Equinovarus 262

Chapter 9

Articulations *267*

An Introduction to Articulations 268

9-1 Joints are categorized according to their range of motion or anatomical organization 268

Synarthroses (Immovable Joints) 269

Amphiarthroses (Slightly Movable Joints) 269

Diarthroses (Freely Movable Joints) 269

9-2 Synovial joints are freely movable articulations (diarthroses) containing synovial fluid 270

Articular Cartilages 270

Synovial Fluid 270

Accessory Structures 271

Factors That Stabilize Synovial Joints 271

9-3 Anatomical and functional properties of synovial joints enable various skeletal movements 272

Describing Dynamic Motion 272

Types of Movements at Synovial Joints 273

Types of Synovial Joints 277

9-4 Intervertebral discs and ligaments are structural components of intervertebral articulations 277

Intervertebral Discs 277

Intervertebral Ligaments 277

Vertebral Movements 279

9-5 The shoulder is a ball-and-socket joint, and the elbow is a hinge joint 281

The Shoulder Joint 281

The Elbow Joint 282

9-6 The hip is a ball-and-socket joint, and the knee is a hinge joint 283

The Hip Joint 283

The Knee Joint 283

9-7 With advancing age, arthritis and other degenerative changes impair joint mobility 286

9-8 The skeletal system supports and stores energy and minerals for other body systems 288

Chapter Review 290

Clinical Notes

Bursitis and Bunions 271

Knee Injuries 286

Chapter 10

Muscle Tissue *293*

An Introduction to Muscle Tissue 294

10-1 Skeletal muscle performs six major functions 294

10-2 A skeletal muscle contains muscle tissue, connective tissues, blood vessels, and nerves 294

Organization of Connective Tissues 294

Blood Vessels and Nerves 295

10-3 Skeletal muscle fibers have distinctive features 296

The Sarcolemma and Transverse Tubules 297

Myofibrils 297

The Sarcoplasmic Reticulum 298

Sarcomeres 298

Sliding Filaments and Muscle Contraction 301

10-4 Communication between the nervous system and skeletal muscles occurs at the neuromuscular junction 304

The Control of Skeletal Muscle Activity 304

Excitation-Contraction Coupling 306

Relaxation 310

10-5 Sarcomere shortening and muscle fiber stimulation produce tension 311

Tension Production by Muscle Fibers 311

Tension Production by Skeletal Muscles 314

Motor Units and Tension Production 315

10-6 ATP is the energy source for muscle contraction 318

ATP and CP Reserves 318

ATP Generation 319

Energy Use and the Level of Muscular Activity 320

Muscle Fatigue 320

The Recovery Period 322

Hormones and Muscle Metabolism 322

10-7 Muscle fiber type and physical conditioning determine muscle performance capabilities 323

Types of Skeletal Muscle Fibers 323

Muscle Performance and the Distribution of Muscle Fibers 324

Muscle Hypertrophy and Atrophy 324

Physical Conditioning 325

10-8 Cardiac muscle tissue differs structurally and functionally from skeletal muscle tissue 326

Structural Characteristics of Cardiac Muscle Tissue 326

Functional Characteristics of Cardiac Muscle Tissue 327

10-9 Smooth muscle tissue differs structurally and functionally from skeletal muscle tissue 328

Structural Characteristics of Smooth Muscle Tissue 328

Functional Characteristics of Smooth Muscle Tissue 330

Chapter Review 331

Clinical Notes
Tetanus   304
Rigor Mortis   311
Delayed-Onset Muscle Soreness   326

Chapter 11 _____

The Muscular System   *335*

An Introduction to the Muscular System   336

11-1 Fascicle arrangement is correlated with muscle power and range of motion   336
Parallel Muscles   337
Convergent Muscles   337
Pennate Muscles   337
Circular Muscles   337

11-2 The three classes of levers increase muscle efficiency   337

11-3 Muscle origins are at the fixed end of muscles, whereas insertions are at the movable end of muscles   339
Origins and Insertions   339
Actions   340

11-4 Descriptive terms are used to name skeletal muscles   341
Location in the Body   341
Origin and Insertion   341
Fascicle Organization   341
Relative Position   341
Structural Characteristics   341
Action   341
Axial and Appendicular Muscles   342

11-5 Axial muscles are muscles of the head and neck, vertebral column, trunk, and pelvic floor   345
Muscles of the Head and Neck   345
Muscles of the Vertebral Column   351
Oblique and Rectus Muscles   355
Muscles of the Pelvic Floor   356

11-6 Appendicular muscles are muscles of the shoulders, upper limbs, pelvic girdle, and lower limbs   359
Muscles of the Shoulders and Upper Limbs   359
Muscles of the Pelvis and Lower Limbs   371

11-7 With advancing age, the size and power of muscle tissue decrease   378

11-8 Exercise produces responses in multiple body systems   380

Chapter Review   383
Clinical Notes
Hernia   359
Intramuscular Injections   371
Compartment Syndrome   381

UNIT 3: CONTROL AND REGULATION

Chapter 12 _____

Neural Tissue   *386*

An Introduction to Neural Tissue   387

12-1 The nervous system has anatomical and functional divisions   387
The Anatomical Divisions of the Nervous System   387
The Functional Divisions of the Nervous System   387

12-2 Neurons are nerve cells specialized for intercellular communication   388
The Structure of Neurons   388
The Classification of Neurons   390

12-3 CNS and PNS neuroglia support and protect neurons   392
Neuroglia of the Central Nervous System   392
Neuroglia of the Peripheral Nervous System   395
Neural Responses to Injuries   397

12-4 The transmembrane potential is the electrical potential of the cell's interior relative to its surroundings   398
The Transmembrane Potential   399
Changes in the Transmembrane Potential   403
Graded Potentials   404

12-5 An action potential is a nerve impulse   407
The All-or-None Principle   407
Generation of Action Potentials   407
Propagation of Action Potentials   410

12-6 Axon diameter, in addition to myelin, affects propagation speed   413

12-7 At synapses, communication occurs among neurons or between neurons and other cells   413
Synaptic Activity   413
General Properties of Synapses   414
Cholinergic Synapses   414

12-8 Neurotransmitters and neuromodulators have various functions   417
The Activities of Other Neurotransmitters   417
Neuromodulators   418
How Neurotransmitters and Neuromodulators Work   418

12-9 Information processing by individual neurons involves integrating excitatory and inhibitory stimuli   419
Postsynaptic Potentials   419
Presynaptic Inhibition and Presynaptic Facilitation   424
The Rate of Generation of Action Potentials   424

Chapter Review   426
Clinical Notes
Rabies   390
Tumors   395
Demyelination   397

Chapter 13
The Spinal Cord, Spinal Nerves, and Spinal Reflexes

429

An Introduction to the Spinal Cord, Spinal Nerves, and Spinal Reflexes   430

13-1 The brain and spinal cord make up the central nervous system, and the cranial nerves and spinal nerves constitute the peripheral nervous system   430

13-2 The spinal cord is surrounded by three meninges and conveys sensory and motor information   431
Gross Anatomy of the Spinal Cord   431
Spinal Meninges   433

13-3 Gray matter is the region of integration and command initiation, and white matter carries information from place to place   435
Organization of Gray Matter   435
Organization of White Matter   437

13-4 Spinal nerves form plexuses that are named according to their level of emergence from the vertebral canal   437
Anatomy of Spinal Nerves   437
Peripheral Distribution of Spinal Nerves   438
Nerve Plexuses   440

13-5 Neuronal pools are functional groups of interconnected neurons   446

13-6 Reflexes are rapid, automatic responses to stimuli   448
The Reflex Arc   448
Classification of Reflexes   449

13-7 Spinal reflexes vary in complexity   450
Monosynaptic Reflexes   450
Polysynaptic Reflexes   452

13-8 The brain can affect spinal cord-based reflexes   454
Voluntary Movements and Reflex Motor Patterns   455
Reinforcement and Inhibition   455

Chapter Review   456
Clinical Notes
Spinal Anesthesia   434
Shingles   440

Chapter 14
The Brain and Cranial Nerves

460

An Introduction to the Brain and Cranial Nerves   461

14-1 The brain has several principal structures, each with specific functions   461
Major Brain Regions and Landmarks   461
Embryology of the Brain   462
Ventricles of the Brain   463

14-2 The brain is protected and supported by the cranial meninges, cerebrospinal fluid, and the blood-brain barrier   464
The Cranial Meninges   464
Cerebrospinal Fluid   465
The Blood Supply to the Brain   466

14-3 The medulla oblongata, which is continuous with the spinal cord, contains vital centers   469

14-4 The pons contains nuclei and tracts that carry or relay sensory and motor information   472

14-5 The cerebellum coordinates learned and reflexive patterns of muscular activity at the subconscious level   472

14-6 The mesencephalon regulates auditory and visual reflexes and controls alertness   474

14-7 The diencephalon integrates sensory information with motor output at the subconscious level   475
The Thalamus   475
The Hypothalamus   476

14-8 The limbic system is a group of tracts and nuclei with various functions   478

14-9 The cerebrum, the largest region of the brain, contains motor, sensory, and association areas   480
The Cerebral Cortex   480
The White Matter of the Cerebrum   481
The Basal Nuclei   482
Motor and Sensory Areas of the Cortex   484

Focus Cranial Nerves   490

14-10 Cranial reflexes involve sensory and motor fibers of cranial nerves   501

Chapter Review   502
Clinical Notes
Epidural and Subdural Hemorrhages   466
Aphasia and Dyslexia   488
Disconnection Syndrome   488

Chapter 15
Neural Integration I: Sensory Pathways and the Somatic Nervous System

506

An Introduction to Sensory Pathways and the Somatic Nervous System   507

15-1 Sensory information from all parts of the body is routed to the somatosensory cortex   507

15-2 Sensory receptors connect our internal and external environments with the nervous system   507
The Detection of Stimuli   508
The Interpretation of Sensory Information   509
Adaptation   509

15-3 General sensory receptors can be classified by the type of stimulus that excites them 510

Nociceptors 510
Thermoreceptors 511
Mechanoreceptors 511
Chemoreceptors 514

15-4 Separate pathways carry somatic sensory and visceral sensory information 514

Somatic Sensory Pathways 515
Visceral Sensory Pathways 519

15-5 The somatic nervous system is an efferent division that controls skeletal muscles 519

The Corticospinal Pathway 520
The Medial and Lateral Pathways 522
The Basal Nuclei and Cerebellum 523
Levels of Processing and Motor Control 524

Chapter Review 525
Clinical Notes
Assessment of Tactile Sensitivities 513
Cerebral Palsy 522
Amyotrophic Lateral Sclerosis 524
Anencephaly 525

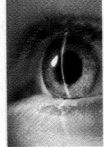

Chapter 16

Neural Integration II: The Autonomic Nervous System and Higher-Order Functions

528

An Introduction to the Autonomic Nervous System and Higher-Order Functions 529

16-1 The autonomic nervous system, composed of the sympathetic and parasympathetic divisions, is involved in the unconscious regulation of visceral functions 529

Organization of the ANS 529
Divisions of the ANS 531

16-2 The sympathetic division consists of preganglionic neurons and ganglionic neurons involved in using energy and increasing metabolic rate 532

Organization and Anatomy of the Sympathetic Division 533
Sympathetic Activation 536

16-3 Stimulation of sympathetic neurons leads to the release of various neurotransmitters 536

Sympathetic Stimulation and the Release of NE and E 537
Sympathetic Stimulation and the Release of ACh and NO 538
Summary: The Sympathetic Division 538

16-4 The parasympathetic division consists of preganglionic neurons and ganglionic neurons involved in conserving energy and lowering metabolic rate 538

Organization and Anatomy of the Parasympathetic Division 539
Parasympathetic Activation 541

16-5 Stimulation of parasympathetic neurons leads to the release of the neurotransmitter ACh 541

Neurotransmitter Release 541
Membrane Receptors and Responses 541
Summary: The Parasympathetic Division 542

16-6 The sympathetic and parasympathetic divisions interact, creating dual innervation 543

Anatomy of Dual Innervation 543
Autonomic Tone 545

16-7 Visceral reflexes play a role in the integration and control of autonomic functions 547

Visceral Reflexes 547
Higher Levels of Autonomic Control 548
The Integration of SNS and ANS Activities 549

16-8 Higher-order functions include memory and states of consciousness 550

Memory 550
States of Consciousness 552

16-9 Neurotransmitters influence brain chemistry and behavior 554

16-10 Aging produces various structural and functional changes in the nervous system 554

16-11 The nervous system is closely integrated with other body systems 556

Chapter Review 558
Clinical Notes
Amnesia 551
Alzheimer Disease 555
Categorizing Nervous System Disorders 556

Chapter 17

The Special Senses

561

An Introduction to the Special Senses 562

17-1 Olfaction, the sense of smell, involves olfactory receptors responding to chemical stimuli 562

Olfactory Receptors 563
Olfactory Pathways 563
Olfactory Discrimination 563

17-2 Gustation, the sense of taste, involves taste receptors responding to chemical stimuli 564

Taste Receptors 564
Gustatory Pathways 564
Gustatory Discrimination 564

17-3 Internal eye structures contribute to vision, while accessory eye structures provide protection 566

Accessory Structures of the Eye 566
The Eye 569

17-4 Photoreceptors respond to light and change it into electrical signals essential to visual physiology 578

    Visual Physiology 578

    The Visual Pathways 583

17-5 Equilibrium sensations originate within the inner ear, while hearing involves the detection and interpretation of sound waves 585

    Anatomy of the Ear 585

    Equilibrium 588

    Hearing 592

Chapter Review 599

Clinical Notes

    Diabetic Retinopathy 571

    Detached Retina 573

    Glaucoma 574

    Accommodation Problems 577

    Motion Sickness 591

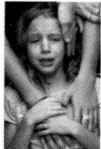

Chapter 18

The Endocrine System 603

An Introduction to the Endocrine System 604

18-1 Homeostasis is preserved through intercellular communication 604

18-2 The endocrine system regulates physiological processes through the binding of hormones to receptors 606

    Classes of Hormones 606

    Secretion and Distribution of Hormones 609

    Mechanisms of Hormone Action 609

    Control of Endocrine Activity by Endocrine Reflexes 611

18-3 The bilobed pituitary gland is an endocrine organ that releases nine peptide hormones 614

    The Adenohypophysis 614

    The Neurohypophysis 618

    Summary: The Hormones of the Pituitary Gland 619

18-4 The thyroid gland lies inferior to the larynx and requires iodine for hormone synthesis 620

    Thyroid Follicles and Thyroid Hormones 622

    Functions of Thyroid Hormones 623

    The C Cells of the Thyroid Gland and Calcitonin 624

18-5 The four parathyroid glands, embedded in the posterior surface of the thyroid gland, secrete parathyroid hormone to elevate plasma Ca^{2+} 625

18-6 The suprarenal glands, consisting of a cortex and medulla, cap the superior borders of the kidneys and secrete several hormones 627

    The Suprarenal Cortex 627

    The Suprarenal Medulla 629

18-7 The pineal gland, attached to the third ventricle, secretes melatonin 630

18-8 The pancreas, located within the abdominopelvic cavity, is both an exocrine organ and endocrine gland 631

    The Pancreatic Islets 631

    Insulin 633

    Glucagon 633

18-9 Many organs have secondary endocrine functions 634

    The Intestines 635

    The Kidneys 635

    The Heart 637

    The Thymus 637

    The Gonads 637

    Adipose Tissue 638

18-10 Hormones interact to produce coordinated physiological responses 638

    Role of Hormones in Growth 640

    The Hormonal Responses to Stress 640

    The Effects of Hormones on Behavior 642

    Aging and Hormone Production 642

18-11 Extensive integration occurs between the endocrine system and other body systems 644

Chapter Review 646

Clinical Notes

    Diabetes Insipidus 618

    Diabetes Mellitus 634

    Endocrine Disorders 639

    Hormones and Athletic Performance 643

UNIT 4: FLUIDS AND TRANSPORT

Chapter 19

Blood 650

An Introduction to the Cardiovascular System 651

19-1 Blood has several important functions and unique physical characteristics 651

19-2 Plasma, the fluid portion of blood, contains significant quantities of plasma proteins 653

    The Composition of Plasma 653

    Plasma Proteins 653

19-3 Red blood cells, formed by erythropoiesis, contain hemoglobin that can be recycled 655

    Abundance of RBCs 655

    Structure of RBCs 655

    Hemoglobin 656

    RBC Formation and Turnover 657

    RBC Production 660

19-4 The ABO blood types and Rh system are based on antigen–antibody responses 662

    Cross-Reactions in Transfusions 662

    Testing for Transfusion Compatibility 664

19-5 The various types of white blood cells contribute to the body's defenses   666
WBC Circulation and Movement   666
Types of WBCs   667
The Differential Count and Changes in WBC Profiles   669
WBC Production   669

19-6 Platelets, disc-shaped structures formed from megakaryocytes, function in the clotting process   672
Platelet Functions   672
Platelet Production   673

19-7 Hemostasis involves vascular spasm, platelet plug formation, and blood coagulation   673
The Vascular Phase   673
The Platelet Phase   674
The Coagulation Phase   674
Fibrinolysis   677

Chapter Review   678
Clinical Notes
Collecting Blood for Analysis   653
Plasma Expanders   654
Abnormal Hemoglobin   658
Hemolytic Disease of the Newborn   664

Chapter 20

The Heart   681

An Introduction to the Cardiovascular System   682

20-1 The heart is a four-chambered organ, supplied by the coronary circulation, that pumps oxygen-poor blood to the lungs and oxygen-rich blood to the rest of the body   682
The Pericardium   684
Superficial Anatomy of the Heart   684
The Heart Wall   684
Internal Anatomy and Organization   685
Connective Tissues and the Cardiac Skeleton   692
The Blood Supply to the Heart   692

20-2 The conducting system distributes electrical impulses through the heart, and an electrocardiogram records the associated electrical events   695
Cardiac Physiology   695
The Conducting System   696
The Electrocardiogram   698
Contractile Cells   700

20-3 Events during a complete heartbeat constitute a cardiac cycle   703
Phases of the Cardiac Cycle   703
Pressure and Volume Changes in the Cardiac Cycle   704
Heart Sounds   706

20-4 Cardiodynamics examines the factors that affect cardiac output   707
Overview: Factors Affecting Cardiac Output   708
Factors Affecting the Heart Rate   709
Factors Affecting the Stroke Volume   711
Summary: The Control of Cardiac Output   713
The Heart and the Cardiovascular System   714

Chapter Review   716
Clinical Notes
Coronary Artery Disease   694
Myocardial Infarction   702
Abnormal Conditions Affecting Cardiac Output   709

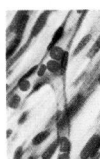

Chapter 21

Blood Vessels and Circulation   719

An Introduction to Blood Vessels and Circulation   720

21-1 Arteries, arterioles, capillaries, venules, and veins differ in size, structure, and functional properties   720
The Structure of Vessel Walls   720
Differences between Arteries and Veins   721
Arteries   722
Capillaries   724
Veins   728
The Distribution of Blood   729

21-2 Pressure and resistance determine blood flow and affect rates of capillary exchange   729
Pressure   730
Total Peripheral Resistance   730
An Overview of Cardiovascular Pressures   731
Capillary Pressures and Capillary Exchange   734

21-3 Cardiovascular regulatory mechanisms involve autoregulation, neural mechanisms, and endocrine responses   737
Autoregulation of Blood Flow within Tissues   738
Neural Mechanisms   739
Hormones and Cardiovascular Regulation   742

21-4 The cardiovascular system adapts to physiological stress and maintains a special vascular supply to the brain, heart, and lungs   744
The Cardiovascular Response to Exercise   744
The Cardiovascular Response to Hemorrhaging   745
Vascular Supply to Special Regions   747

21-5 The pulmonary and systemic circuits of the cardiovascular system exhibit three general functional patterns   748

21-6 In the pulmonary circuit, deoxygenated blood enters the lungs in arteries, and oxygenated blood leaves the lungs via veins   749

21-7 The systemic circuit carries oxygenated blood from the left ventricle to tissues and organs other than the pulmonary exchange surfaces, and returns deoxygenated blood to the right atrium   750

Systemic Arteries   750
Systemic Veins   758

21-8 Modifications of fetal and maternal cardiovascular systems promote the exchange of materials, and independence is achieved at birth   767

Placental Blood Supply   767
Fetal Circulation in the Heart and Great Vessels   767
Cardiovascular Changes at Birth   767

21-9 Aging affects the blood, heart, and blood vessels   770

21-10 The cardiovascular system is both anatomically and functionally linked to all other systems   770

Chapter Review   772
Clinical Notes

Arteriosclerosis   724
Edema   737
Congenital Cardiovascular Problems   768

Chapter 22

The Lymphoid System and Immunity   776

An Introduction to the Lymphoid System and Immunity   777

22-1 Anatomical barriers and defense mechanisms constitute nonspecific defense, and lymphocytes provide specific defense   777

22-2 Lymphatic vessels, lymphocytes, lymphoid tissues, and lymphoid organs function in body defenses   777

Functions of the Lymphoid System   777
Lymphatic Vessels   778
Lymphocytes   780
Lymphoid Tissues   783
Lymphoid Organs   783
The Lymphoid System and Body Defenses   788

22-3 Nonspecific defenses do not discriminate between potential threats and respond the same regardless of the invader   789

Physical Barriers   789
Phagocytes   789
Immunological Surveillance   791
Interferons   793
Complement   793
Inflammation   793
Fever   796

22-4 Specific defenses (immunity) respond to individual threats and are either cell-mediated or antibody-mediated   796

Forms of Immunity   796
Properties of Immunity   797
An Introduction to the Immune Response   798

22-5 T cells play a role in the initiation, maintenance, and control of the immune response   798

Antigen Presentation   799
Antigen Recognition   800
Activation of CD8 T Cells   801
Activation of CD4 T Cells   802

22-6 B cells respond to antigens by producing specific antibodies   804

B Cell Sensitization and Activation   804
Antibody Structure   805
Primary and Secondary Responses to Antigen Exposure   808
Summary of the Immune Response   809

22-7 Immunological competence enables a normal immune response; abnormal responses result in immune disorders   811

The Development of Immunological Competence   812

Focus   Hormones of the Immune System   813

Immune Disorders   815
Stress and the Immune Response   816

22-8 The immune response diminishes with advancing age   817

22-9 For all body systems, the lymphoid system provides defenses against infection and returns tissue fluid to the circulation   817

Chapter Review   820
Clinical Notes

Cancer and the Lymphoid System   786
Graft Rejection and Immunosuppression   803
AIDS   819

UNIT 5: ENVIRONMENTAL EXCHANGE

Chapter 23

The Respiratory System   825

An Introduction to the Respiratory System   826

23-1 The respiratory system, organized into an upper respiratory system and a lower respiratory system, has several basic functions   826

Functions of the Respiratory System   826
Organization of the Respiratory System   826

23-2 Located outside the thoracic cavity, the upper respiratory system consists of the nose, nasal cavity, paranasal sinuses, and pharynx   829

The Nose, Nasal Cavity, and Paranasal Sinuses   829
The Pharynx   831

23-3 Composed of cartilages, ligaments, and muscles, the larynx produces sound 831

Cartilages and Ligaments of the Larynx 832

Sound Production 833

The Laryngeal Musculature 834

23-4 The trachea and primary bronchi convey air to and from the lungs 834

The Trachea 834

The Primary Bronchi 834

23-5 Enclosed by a pleural membrane, the lungs are paired organs containing alveoli, which permit gaseous exchange 835

Lobes and Surfaces of the Lungs 835

The Bronchi 837

The Bronchioles 837

Alveolar Ducts and Alveoli 839

The Blood Supply to the Lungs 841

The Pleural Cavities and Pleural Membranes 841

23-6 External respiration and internal respiration allow gaseous exchange within the body 842

23-7 Pulmonary ventilation—the exchange of air between the atmosphere and the lungs—involves pressure changes, muscle movement, and respiratory rates and volumes 843

The Movement of Air 843

Pressure Changes during Inhalation and Exhalation 844

The Mechanics of Breathing 847

Respiratory Rates and Volumes 849

23-8 Gas exchange depends on the partial pressures of gases and the diffusion of molecules 851

The Gas Laws 851

Diffusion and Respiratory Function 852

23-9 Most oxygen is transported bound to hemoglobin; and carbon dioxide is transported in three ways: as carbonic acid, bound to hemoglobin, or dissolved in plasma 855

Oxygen Transport 855

Carbon Dioxide Transport 858

Summary: Gas Transport 859

23-10 Neurons in the medulla oblongata and pons, along with respiratory reflexes, control respiration 860

Local Regulation of Gas Transport and Alveolar Function 860

The Respiratory Centers of the Brain 861

Respiratory Reflexes 862

Voluntary Control of Respiration 865

Changes in the Respiratory System at Birth 865

23-11 Respiratory performance declines with age 866

23-12 The respiratory system provides oxygen to, and eliminates carbon dioxide from, other organ systems 866

Chapter Review 869

Clinical Notes

Breakdown of the Respiratory Defense System 829

Pneumothorax 847

Decompression Sickness 853

Blood Gas Analysis 854

Carbon Monoxide Poisoning 856

Emphysema and Lung Cancer 867

Chapter 24

The Digestive System 874

An Introduction to the Digestive System 875

24-1 The digestive system, consisting of the digestive tract and accessory organs, has overlapping food utilization functions 875

Functions of the Digestive System 875

The Digestive Organs and the Peritoneum 876

Histological Organization of the Digestive Tract 877

The Movement of Digestive Materials 880

Control of Digestive Function 881

24-2 The oral cavity contains the tongue, salivary glands, and teeth, each with specific functions 882

The Tongue 883

Salivary Glands 884

The Teeth 885

24-3 The pharynx is a passageway between the oral cavity and esophagus 888

24-4 The esophagus is a muscular tube that transports solids and liquids from the pharynx to the stomach 888

Histology of the Esophagus 889

Swallowing 889

24-5 The stomach is a J-shaped organ that receives the bolus from the esophagus and aids in chemical and mechanical digestion 891

Anatomy of the Stomach 891

Regulation of Gastric Activity 894

Digestion and Absorption in the Stomach 897

24-6 The small intestine digests and absorbs nutrients, and associated glandular organs assist with the digestive process 898

The Small Intestine 898

Histology of the Small Intestine 899

Intestinal Secretions 901

Intestinal Movements 901

The Pancreas 901

The Liver 903

The Gallbladder 908

The Coordination of Secretion and Absorption 909

24-7 The large intestine is divided into three parts with regional specialization 910

The Cecum 910

The Colon 911

The Rectum 913

Histology of the Large Intestine 913

Physiology of the Large Intestine 913

24-8 Digestion is the mechanical and chemical alteration of food that allows the absorption and use of nutrients 916

The Processing and Absorption of Nutrients 916
Carbohydrate Digestion and Absorption 916
Lipid Digestion and Absorption 919
Protein Digestion and Absorption 919
Water Absorption 920
Ion Absorption 920

24-9 Many age-related changes affect digestion and absorption 921

24-10 The digestive system is extensively integrated with other body systems 922

Chapter Review 924
Clinical Notes

Peritonitis 877
Epithelial Renewal and Repair 879
Mumps 885
Gastritis and Peptic Ulcers 894
Pancreatitis 903
Cirrhosis 907
Colorectal Cancer 913
Inflammatory and Infectious Disorders of the Digestive System 922

Chapter 25
Metabolism and Energetics
929

An Introduction to Metabolism and Energetics 930

25-1 Metabolism refers to all the chemical reactions that occur in the body 930

25-2 Carbohydrate metabolism involves glycolysis, ATP production, and gluconeogenesis 932

Glycolysis 932
Mitochondrial ATP Production 934
Energy Yield of Glycolysis and Cellular Respiration 939
Gluconeogenesis 940

25-3 Lipid metabolism involves lipolysis, beta-oxidation, and the transport and distribution of lipids as free fatty acids and lipoproteins 941

Lipid Catabolism 941
Lipid Synthesis 944
Lipid Transport and Distribution 944

25-4 Protein catabolism involves transamination and deamination, whereas protein synthesis involves amination and transamination 946

Amino Acid Catabolism 946
Protein Synthesis 947

25-5 The body experiences two patterns of metabolic activity: the absorptive and postabsorptive states 949

25-6 Adequate nutrition is necessary to prevent deficiency disorders and ensure physiological functioning 950

Food Groups and the MyPyramid Plan 951
Nitrogen Balance 952
Minerals 952
Vitamins 953
Diet and Disease 955

25-7 Metabolic rate is the average caloric expenditure, and thermoregulation involves balancing heat-producing and heat-losing mechanisms 955

Energy Gains and Losses 955
Thermoregulation 957

Chapter Review 962
Clinical Notes

Carbohydrate Loading 940
Dietary Fats and Cholesterol 943
Vitamins 954
Alcohol—A Risky Diversion 956
Induced Hypothermia 960
Thermoregulatory Disorders 961

Chapter 26
The Urinary System
965

An Introduction to the Urinary System 966

26-1 Consisting of the kidneys, ureters, urinary bladder, and urethra, the urinary system has three primary functions 966

26-2 Kidneys are highly vascular structures containing functional units called nephrons, which perform filtration, reabsorption, and secretion 967

Sectional Anatomy of the Kidneys 969
Blood Supply and Innervation of the Kidneys 969
The Nephron 969

26-3 Different segments of the nephron form urine by filtration, reabsorption, and secretion 976

Basic Processes of Urine Formation 976
An Overview of Renal Function 978

26-4 Hydrostatic and colloid osmotic pressures influence glomerular filtration pressure, which in turn affects the glomerular filtration rate 980

Filtration Pressures 980
The Glomerular Filtration Rate 981
Control of the GFR 982

26-5 Countercurrent multiplication and the influence of antidiuretic hormone and aldosterone affect reabsorption and secretion 984

Reabsorption and Secretion at the PCT 984
The Nephron Loop and Countercurrent Multiplication 986
Reabsorption and Secretion at the DCT 988
Reabsorption and Secretion along the Collecting System 989
The Control of Urine Volume and Osmotic Concentration 990
The Function of the Vasa Recta 993

The Composition of Normal Urine   993

Summary: Renal Function   993

26-6 Urine is transported via the ureters, stored in the bladder, and eliminated through the urethra, aided by the micturition reflex   997

The Ureters   997

The Urethra   999

The Micturition Reflex and Urination   1001

26-7 Age-related changes affect kidney function and the micturition reflex   1002

26-8 The urinary system is one of several body systems involved in waste excretion   1003

Chapter Review   1005

Clinical Notes

Analysis of Renal Blood Flow   969

Glomerulonephritis   975

Diuretics   991

Renal Failure and Kidney Transplantation   998

Urinary Obstruction   1002

Chapter 27
Fluid, Electrolyte, and Acid–Base Balance   1009

An Introduction to Fluid, Electrolyte, and Acid–Base Balance   1010

27-1 Fluid balance, electrolyte balance, and acid–base balance are interrelated and essential to homeostasis   1010

27-2 The ECF and ICF make up the fluid compartments, which also contain cations and anions   1011

The ECF and the ICF   1012

Basic Concepts in the Regulation of Fluids and Electrolytes   1012

An Overview of the Primary Regulatory Hormones   1013

The Interplay between Fluid Balance and Electrolyte Balance   1014

27-3 Hydrostatic and osmotic pressures regulate the movement of water and electrolytes to maintain fluid balance   1014

Fluid Movement within the ECF   1015

Fluid Gains and Losses   1015

Fluid Shifts   1016

27-4 Balance of the electrolytes sodium, potassium, calcium, and chloride is essential for maintaining homeostasis   1017

Sodium Balance   1018

Potassium Balance   1020

Balance of Other Electrolytes   1021

27-5 In acid–base balance, regulation of hydrogen ions in body fluids involves buffer systems and renal and respiratory compensatory mechanisms   1023

The Importance of pH Control   1023

Types of Acids in the Body   1023

Mechanisms of pH Control   1024

Maintenance of Acid–Base Balance   1027

27-6 Respiratory acidosis/alkalosis and metabolic acidosis/alkalosis are classes of acid–base balance disturbances   1030

Respiratory Acidosis   1030

Respiratory Alkalosis   1032

Metabolic Acidosis   1033

Metabolic Alkalosis   1034

The Detection of Acidosis and Alkalosis   1034

27-7 Aging affects several aspects of fluid, electrolyte, and acid–base balance   1036

Chapter Review   1037

Clinical Notes

Water and Weight Loss   1017

Athletes and Salt Loss   1020

UNIT 6: CONTINUITY OF LIFE

Chapter 28
The Reproductive System   1041

An Introduction to the Reproductive System   1042

28-1 Basic reproductive system structures are gonads, ducts, accessory glands and organs, and external genitalia   1042

28-2 Spermatogenesis occurs in the testes, and hormones from the hypothalamus, adenohypophysis, and testes control male reproductive functions   1042

The Testes   1043

Spermatogenesis   1047

The Anatomy of a Spermatozoon   1050

The Male Reproductive Tract   1051

The Accessory Glands   1054

Semen   1055

The External Genitalia   1056

Hormones and Male Reproductive Function   1056

28-3 Oogenesis occurs in the ovaries, and hormones from the pituitary gland and gonads control female reproductive functions   1060

The Ovaries   1060

The Uterine Tubes   1065

The Uterus   1065

The Vagina   1070

The External Genitalia   1071

The Mammary Glands   1071

Hormones and the Female Reproductive Cycle   1074

Summary: Hormonal Regulation of the Female Reproductive Cycle   1075

28-4 The autonomic nervous system influences male and female sexual function   1077

   Male Sexual Function   1078

   Female Sexual Function   1078

28-5 With age, decreasing levels of reproductive hormones cause functional changes   1079

   Menopause   1079

   The Male Climacteric   1079

28-6 The reproductive system secretes hormones affecting growth and metabolism of all body systems   1080

Chapter Review   1081
Clinical Notes

   Dehydroepiandrosterone (DHEA)   1058

   Prostatic Hypertrophy and Prostate Cancer   1059

   Ovarian Cancer   1065

   Breast Cancer   1073

Chapter 29
Development and Inheritance   1086

An Introduction to Development and Inheritance   1087

29-1 Development, marked by various stages, is a continuous process that occurs from fertilization to maturity   1087

29-2 Fertilization—the fusion of a secondary oocyte and a spermatozoon—forms a zygote   1087

   The Oocyte at Ovulation   1088

   Oocyte Activation   1088

29-3 Gestation consists of three stages of prenatal development: the first, second, and third trimesters   1090

29-4 Cleavage, implantation, placentation, and embryogenesis are critical events of the first trimester   1091

   Cleavage and Blastocyst Formation   1091

   Implantation   1091

   Placentation   1095

   Embryogenesis   1098

29-5 During the second and third trimesters, maternal organ systems support the developing fetus, and the uterus undergoes structural and functional changes   1102

   Pregnancy and Maternal Systems   1104

   Structural and Functional Changes in the Uterus   1104

29-6 Labor consists of the dilation, expulsion, and placental stages   1105

   Stages of Labor   1105

   Premature Labor   1107

   Difficult Deliveries   1107

   Multiple Births   1107

29-7 Postnatal stages are the neonatal period, infancy, childhood, adolescence, maturity, and senescence   1107

   The Neonatal Period, Infancy, and Childhood   1108

   Adolescence and Maturity   1109

   Senescence   1111

29-8 Genes and chromosomes determine patterns of inheritance   1112

   Patterns of Inheritance   1112

   Sources of Individual Variation   1115

   Sex-Linked Inheritance   1116

   The Human Genome Project and Beyond   1117

Chapter Review   1120
Clinical Notes

   Gestational Trophoblastic Neoplasia   1093

   Abortion   1104

   Chromosomal Abnormalities and Genetic Analysis   1119

Appendix

   Weights and Measures

   Periodic Table

   Normal Physiological Values

Answers to Checkpoint and Review Questions

Glossary

   Eponyms in Common Use

   Key Terms

Credits

Index

An Introduction to Anatomy and Physiology

Did you know...?
The resiliency of the human body is demonstrated by its ability to tolerate a broad range of environmental conditions.

Learning Outcomes

After completing this chapter, you should be able to do the following:

1-1 Explain the importance of studying anatomy and physiology.

1-2 Identify basic study skill strategies to use in this course.

1-3 Define anatomy and physiology, describe the origins of anatomical and physiological terms, and explain the significance of *Terminologia Anatomica (International Anatomical Terminology)*.

1-4 Explain the relationship between anatomy and physiology, and describe various specialties of each discipline.

1-5 Identify the major levels of organization in organisms, from the simplest to the most complex, and identify major components of each organ system.

1-6 Explain the concept of homeostasis.

1-7 Describe how negative feedback and positive feedback are involved in homeostatic regulation, and explain the significance of homeostasis.

1-8 Use anatomical terms to describe body sections, body regions, and relative positions.

1-9 Identify the major body cavities and their subdivisions, and describe the functions of each.

Clinical Notes

The Visible Human Project p. 7
Homeostasis and Disease p. 13

An Introduction to Studying the Human Body

This textbook will serve as an introduction to the inner workings of your body, providing information about both its structure and its function. Many of the students who use this book are preparing for careers in health-related fields—but regardless of your career choice, you will find the information within these pages relevant to your future. You do, after all, live in a human body! Being human, you most likely have a seemingly insatiable curiosity—and few subjects arouse so much curiosity as our own bodies. The study of anatomy and physiology will provide answers to many questions regarding the functioning of your body in both health and disease.

Although we will be focusing on the human body, the principles we will learn apply to other living things as well. Our world contains an enormous diversity of living organisms that vary widely in appearance and lifestyle. One aim of biology—the science of life—is to discover the unity and the patterns that underlie this diversity, and thereby shed light on what we have in common with other living things.

Animals can be classified according to their shared characteristics, and birds, fish, and humans are members of a group called the *vertebrates*, characterized by a segmented vertebral column. The shared characteristics and organizational patterns provide useful clues about how these animals have evolved over time. Many of the complex structures and functions of the human body discussed in this text have distant evolutionary origins. When we compare the particular adaptations of human beings with those of other creatures, we find two important principles: there are obvious structural and functional similarities among vertebrates, and form follows function.

This chapter explores the structural and functional characteristics of living things. It includes the levels of organization that anatomical structures and physiological processes display, and discusses *homeostasis*, the goal of physiological regulation and the key to survival in a changing environment.

1-1 Anatomy and physiology directly affect your life

Welcome to the field of anatomy and physiology, the study of body structures and functions. In this course, you will discover how your body works under normal and abnormal conditions. This course will also be important because it serves as the foundation for understanding all other basic life sciences, and for making common sense decisions about your own life. Basic knowledge of normal physiological function, for example, will prove useful whenever you or a friend or relative becomes ill. Our study of anatomy and physiology will

devote considerable time to explaining how the body responds to normal and abnormal conditions and maintains homeostasis, a relatively constant internal environment. As we proceed, you will see how your body copes with injury, disease, or anything that threatens homeostasis.

Anatomy is considered the oldest medical science. Egyptian drawings from 1600 BCE illustrating basic knowledge of blood vessels demonstrate that we have always been fascinated with the human body. Since that time, techniques for studying the human body have evolved, enabling us to describe the locations and functions of body parts. Over the last two decades, the most rapid progress has been made in the field of molecular biology, which investigates processes at the level of individual genes and incorporates principles of biology, chemistry, genetics, and biochemistry. By enhancing our understanding of molecular biology, we are learning how the body works at the most fundamental level and revealing the underlying basis for many disorders and diseases.

Medical science expands continuously and affects our everyday lives. We are flooded with health information from the popular press, news media, and advertisements. As a result, medical terms are a part of our common language, and we owe it to ourselves to understand them. You will find that this course significantly expands your vocabulary and your understanding of the origins and meanings of medical terms.

> **CHECKPOINT**
>
> 1. Identify the oldest medical science.
> 2. Why is studying human anatomy and physiology important?
>
> See the blue Answers tab at the end of the book.

1-2 Good study strategies are crucial for success

Completing this course successfully will require you to work hard and often. Although you are going to spend lots of time reading this textbook, reading alone won't be enough—you need to develop good study skills and strategies.

Before going to lectures and labs, read the material so that it and the terms used by the instructor don't come as a complete surprise. During class, take notes on notebook paper *and* in your textbook, especially near the relevant illustrations. You might also find it useful to record the lectures. After class, reread the textbook while referring to your class notes. If something doesn't make sense, don't be shy about asking for help. Write yourself a note so that you will be sure to ask your instructor about it.

Memorization alone is not enough, and waiting until the night before an exam to begin studying is courting disaster. Success in an A&P course is like constructing a house—it is an ongoing process that starts with building a solid foundation and then advances in small steps, each of which builds on the previous one. Using your time in lectures and labs to the fullest will go a long way in making this course a positive experience. Lab exercises will greatly enhance your understanding of the topics covered in lecture, so do not consider them separate activities but, instead, educational tools that go hand-in-hand with what you learn in lecture. Here are some practical tips for success in this course:

- **Attend all lectures, labs, and study sessions.** Ask questions and participate in discussions.

- **Read your lecture and laboratory assignments before going to class or lab.** You'll understand class/lab better if you do.

- **Devote a block of time each day to your A&P course.** There are no shortcuts. (Sorry.) You won't get the grade you want and the knowledge you need if you don't put in the time and do the work. This requires preparation throughout the term.

- **Set up a study schedule and stick to it.** Create your study schedule during the first week of class.

- **Do not procrastinate!** Do not do all your studying the night before the exam! Actually STUDY the material several times throughout the week. Marathon study sessions are often counterproductive. Expect to push yourself and stretch your limits.

- **Approach the information in different ways.** For example, you might visualize the information, talk it over with or "teach" a fellow student, or spend additional time in lab asking questions of your lab instructor.

- **Develop the skill of memorization, and practice it regularly.** Memorization is an important skill, and an integral part of the course. You are going to have to memorize all sorts of things—among them muscle names, directional terms, and the names of bones and brain parts. Realize that this is an important study skill, and that the more you practice, the better you will be at remembering terms and definitions. We will give you tips and tricks along the way to help you keep the information in mind.

- **As soon as you experience difficulty with the course, seek assistance.** Do not wait until the end of the term when it is too late to salvage your grade.

This textbook is an investment in your future; you owe it to yourself to use it wisely. It contains significant information that is useful beyond the bounds of this course.

Approach your textbook differently from a novel. That is, you might have to read it more slowly, with a critical eye to detail, while taking notes along the way. This book was designed with student-friendly resources, not only to enhance your study, but also to ensure your success in the course. However, you will succeed only by being an active learner and *using* these features as you work through the textbook:

- **Learning Outcomes.** The first page of every chapter includes a list of Learning Outcomes that shows what you should be able to do by the end of the chapter. Each Learning Outcome corresponds to a main section in the chapter, and the Learning Outcome and the heading of that main section share a number (such as 1-1, 1-2, 1-3) for easy reference. That numbering system repeats in the Study Outline at the end of the chapter to make studying and reviewing as straightforward as possible.

- **Illustrations and Photos.** The art program is designed to complement the text and provide visual aids for understanding complex topics. Figure references in the narrative are bright green to function as placeholders to help you return to reading after viewing a figure. Every time you see a colored figure reference, pause and refer to the figure. The text and art work together—you need to integrate them as you study.

- **Pronunciation Guides.** We have provided a pronunciation guide for each new term. When you read the term, stop and say it aloud until you are familiar with the sound. Afterward, use the term at every opportunity. If you don't use it, you are likely to forget it!

- **Checkpoint Questions.** Whenever you encounter a Checkpoint, stop and answer the questions before going on to the next section. If you cannot answer the questions quickly and correctly, reread the section. Do not go on until you understand the answers, because each topic builds upon the previous one.

- **The A&P Top 100.** Carefully read The A&P Top 100 because these brief boxes present the core concepts, principles, and facts that you should remember years after your A&P course. In other words, regardless of your career choice, the information in this feature pertains to you!

- **Tips & Tricks.** Consider the Tips & Tricks as tools to help your memorization. They present easy analogies and mnemonic devices to better enable you to retain information.

- **Clinical Notes.** Think of the Clinical Notes sections as applications to the real world. They show how what you are reading applies to life beyond your textbook and why learning this information is important.

- **Chain Link Icons.** Watch for the chain link icon ∞. This icon appears when material relates to topics presented earlier and offers specific page numbers to facilitate review.

- **End-of-Chapter Study and Review Materials.** After completing each chapter, read the Study Outline and work through the end-of-chapter Review Questions to make sure that you have a solid understanding of the chapter material. Answers to the Checkpoint Questions (see above) and Review Questions are found in the blue Answers tab at the end of the book.

- **Systems Overview Section.** Between Chapters 4 and 5 (the end of the four introductory chapters and the beginning of the body system chapters), the Systems Overview section presents each of the body systems, including summaries of body organs and functions. This overview of the body systems prepares you for the body system chapters that will follow, and it is also a useful reference as you move forward in the textbook. The section can be easily found via the deep yellow band at the edges of its pages.

- **System in Perspective Summaries.** The body functions as an integrated unit rather than as a set of isolated, independent systems. When you learn about a new body system, such as the digestive system, try to relate it to what you learned about another body system earlier, such as the nervous system. The System in Perspective summaries, which appear at the end of each set of chapters on a body system, provide excellent, illustrated reviews of the basic functions of each body system. You can use these to "see the big picture" and to quiz yourself.

- **Colored Tabs.** Colored tabs are located on the edges of the pages to make it easy for you to find your place in the textbook. The color of the tab indicates one of the eleven body systems, such as light brown for the integumentary system and red for the cardiovascular system. The location of the tab along the edge of the page indicates one of the six units, such as the top of the page for Unit 1 and one notch below that location for Unit 2. Chapter numbers are printed on the tabs. Using the tabbing system, you will be able to navigate easily through the textbook as you read and study.

- **End-of-Book Reference Sections.** Also indicated with easy-to-find colored tabs are four important sections at the back of the book: **Appendix**, **Answers**, **Glossary**, and **Index**. You'll refer regularly to each of these sections as you move through the chapters of the book, and the brightly colored tabs will help you get to where you want to go in no time.

Your new textbook package comes with several supplementary learning tools that are referenced in the chapters of your textbook: the Interactive Physiology® (IP) CD, an access

code for myA&P™, and *Martini's Atlas of the Human Body*. These references are called out with distinctive designs and colors for easy integration with your studies. Here is how these supplements can help you succeed in your course:

- **The Interactive Physiology® (IP) CD** contains detailed animations and engaging quizzes to help you learn the most difficult concepts in your A&P course. This resource will help you advance beyond memorization to a genuine understanding of complex topics. Interactive Physiology icons that refer to this CD are found in the end-of-chapter sections throughout the textbook.

- **myA&P™** is your one-stop online location for accessing great media that will help you increase your understanding of anatomy and physiology and raise your A&P test scores. This website includes IP (described above) as well as chapter quizzes, practice tests, anatomy animations, physiology animations, histology review, bone review, muscle review, flashcards with pronunciations, and many other useful features.

- *Martini's Atlas of the Human Body* is a full-color, spiral-bound book of vivid anatomy photographs, radiology images, and embryology summaries. The anatomy images are directly related to the illustrations in your textbook, and figure captions in the book will direct you to the *Atlas* so that you can compare artistic presentations to cadaver dissections or medical scans. You can also use the *Atlas* as a reference in the lab.

Some additional resources are also available to you: the *Get Ready for A&P!* paperback book and website, the *A&P Applications Manual* with more extensive clinical material, and the *Study Guide*, which provides abundant exercises for additional practice. See descriptions of these resources in the Preface to this textbook and get in touch with the publisher if you would like to learn more.

CHECKPOINT

3. **Identify several strategies for success in this course.**

4. **List the resources packaged with your textbook to assist you in learning anatomy and physiology.**

See the blue Answers tab at the end of the book.

1-3 Anatomy is structure, and physiology is function

People have always been interested in the inner workings of the human body. The word *anatomy* has Greek roots, as do many other anatomical terms and phrases that originated more than 1500 years ago. **Anatomy**, which means "a cutting open," is the study of internal and external structures of the body and the physical relationships among body parts. In contrast, **physiology**, another Greek term, is the study of how living organisms perform their vital functions. Thus, someone studying anatomy might, for example, examine how a particular muscle attaches to the skeleton, whereas someone studying physiology might consider how a muscle contracts or what forces a contracting muscle exerts on the skeleton. Because you will be studying anatomy and physiology throughout this book, it is appropriate that we spend some time at the outset taking a closer look at the relationships between these sciences.

Early anatomists faced serious problems in communication. Stating that a bump is "on the back," for example, does not give very precise information about its location. So anatomists created maps of the human body. Prominent anatomical structures serve as landmarks, distances are measured in centimeters or inches, and specialized directional terms are used. In effect, anatomy uses a special language that must be learned almost at the start of your study.

That special language, called **medical terminology**, involves the use of word roots, prefixes, suffixes, and combining forms to construct terms related to the body in health and disease. Many of the anatomical and physiological terms you will encounter in this textbook are derived from Greek or Latin. Learning the word parts used in medical terminology will greatly assist in the study of anatomy and physiology, and in preparation for any health-related career.

There are four basic building blocks of medical terms. *Word roots* are the basic, meaningful parts of a term that cannot be broken down into another term with another definition. *Prefixes* are word elements that are attached to the beginning of words to modify their meaning but cannot stand alone. *Suffixes* are word elements or letters added to the end of a word or word part to form another term. *Combining forms* are independent words or word roots that occur in combination with words, prefixes, suffixes, or other combining forms to build a new term. The table inside the back cover of your textbook lists commonly used word roots, prefixes, suffixes, and combining forms.

To illustrate the concept of building medical terms, consider the word *anatomy*, derived from the Greek root *anatome*, meaning dissection. The prefix *ana-* means up, while the suffix *-tomy* means to cut. Hence, *anatomy* means to "cut up" or dissect. Another commonly used term is *pathology*. Breaking this word into its fundamental elements reveals its meaning. The prefix *path-* refers to disease (the Greek term for disease is *pathos*), while the suffix *-ology* means "study of." Therefore, pathology is the study of disease.

A familiarity with Latin and Greek word roots and patterns makes anatomical terms more understandable. As the text introduces new terms, it will provide notes on pronunciation and relevant word roots.

Latin and Greek terms are not the only ones that have been imported into the anatomical vocabulary over the centuries, and this vocabulary continues to expand. Many anatomical structures and clinical conditions were initially named after either the discoverer or, in the case of diseases, the most famous victim. Over the last 100 years, most of these commemorative names, or *eponyms*, have been replaced by more precise terms. The Glossary includes a table listing the most important eponyms and related historical details.

It is important for scientists throughout the world to use the same name for each body structure, so in 1998, two scientific organizations—the Federative Committee on Anatomical Terminology and the 56 member associations of the International Associations of Anatomists—published *International Anatomical Terminology (Terminologia Anatomica, or TA)*. Although Latin continues to be the language of anatomy, this reference provides an English equivalent term for each anatomical structure. The *TA* serves as a worldwide official standard of anatomical vocabulary, and we have used it as our standard in preparing this text.

> **CHECKPOINT**
> 5. Define anatomy.
> 6. Define physiology.
> 7. Describe medical terminology.
> 8. Define eponym.
> 9. Name the book that serves as the international standard for anatomical vocabulary.
>
> See the blue Answers tab at the end of the book.

1-4 Anatomy and physiology are closely integrated

Anatomy and physiology are closely integrated, both theoretically and practically. Anatomical information provides clues about functions, and physiological mechanisms can be explained only in terms of the underlying anatomy. This is a very important concept: *All specific functions are performed by specific structures.* The link between structure and function is always present, but not always understood. For example, although the anatomy of the heart was clearly described in the 15th century, almost 200 years passed before the heart's pumping action was demonstrated.

Anatomists and physiologists approach the relationship between structure and function from different perspectives. To understand the difference, consider a simple nonbiological analogy. Suppose that an anatomist and a physiologist were asked to examine a pickup truck and report their findings. The anatomist might begin by measuring and photographing the various parts of the truck and, if possible, taking it apart and putting it back together. The anatomist could then explain its key *structural* relationships—for example, how the pistons are seated in the engine cylinders, how the drive shaft is connected to the pistons, how the transmission links the drive shaft to the axles, and thus to the wheels. The physiologist also would note the relationships among the truck's components, but his or her primary focus would be on *functional* characteristics, such as how the combustion of gasoline in the cylinders moves the pistons up and down and causes the drive shaft to rotate, and how the transmission conveys this motion to the axles and wheels so that the car moves. Additionally, he or she might also study the amount of power that the engine could generate, the amount of force transmitted to the wheels in different gears, and so on.

This text will introduce anatomical structures and the physiological processes that make human life possible. The basic approach will be to start with the descriptive anatomy (appearance, size, shape, location, weight, and color) before considering the related functions. Sometimes the organs within an organ system perform very diverse functions, and in those cases the functions of each individual organ will be considered separately. A good example is the discussion of the digestive system, where you will learn about the functions of the salivary glands in one section, and the functions of the tongue in another. In other systems, the organs work together so extensively that the physiological discussion is presented in a block, after the system's anatomy has been described. The lymphoid system and the cardiovascular system are examples of this approach.

Knowledge of the anatomy and physiology of the healthy human body will enable you to understand important mechanisms of disease and will help you make intelligent decisions about personal health.

The A&P Top 100

#1 All physiological functions are performed by specific anatomical structures. These functions follow the same physical and mechanical principles that can be seen in the world at large.

Anatomy

How you look at things often determines what you see; you get a very different view of your neighborhood from a satellite photo than when standing in your front yard. Your method of observation has an equally dramatic effect on your understanding of the structure of the human body. Based on the degree of structural detail under consideration, anatomy can be divided into *gross (macroscopic) anatomy* and *microscopic anatomy*. Other anatomical specialties focus on specific processes, such as respiration, or medical applica-

tions, such as *surgical anatomy*, which deals with landmarks on the body that are useful during surgical procedures.

Anatomy is a dynamic field. Despite centuries of observation and dissection, new information and interpretations occur frequently. As recently as 1996, researchers working on the Visible Human database described a facial muscle that had previously been overlooked. The Clinical Note on the Visible Human Project describes the origins and uses of one of the most powerful tools in modern anatomy.

Gross Anatomy

Gross anatomy, or *macroscopic anatomy*, involves the examination of relatively large structures and features usually visible with the unaided eye. There are many different forms of gross anatomy:

- *Surface anatomy* is the study of general form and superficial markings.

- *Regional anatomy* focuses on the anatomical organization of specific areas of the body, such as the head, neck, or trunk. Many advanced courses in anatomy stress a regional approach, because it emphasizes the spatial relationships among structures already familiar to students.

- *Systemic anatomy* is the study of the structure of **organ systems**, which are groups of organs that function together in a coordinated manner. Examples include the *skeletal system*, composed primarily of bones; the *muscular system*, made up of skeletal muscles; and the *cardiovascular system*, consisting of the heart, blood, and vessels, which distribute oxygen and nutrients throughout the body. Introductory texts such as this book take a systemic anatomy approach because that approach clarifies functional relationships among the component organs. The text will introduce the 11 organ systems in the human body later in the chapter.

- *Developmental anatomy* describes the changes in form that occur between conception and physical maturity. Because developmental anatomy considers anatomical structures over such a broad range of sizes (from a single cell to an adult human), the techniques of developmental anatomists are similar to those used in gross anatomy and in microscopic anatomy. The most extensive structural changes occur during the first two months of development. The study of these early developmental processes is called **embryology** (em-brē-OL-ō-jē).

- *Clinical anatomy* includes a number of subspecialties important in clinical practice. Examples include *pathological anatomy* (anatomical features that change during illness), *radiographic anatomy* (anatomical structures seen using specialized imaging techniques), and *surgical anatomy* (anatomical landmarks important in surgery).

CLINICAL NOTE

The Visible Human Project

When Joseph Paul Jernigan, a condemned criminal, was executed in 1993 for a particularly brutal murder in Texas, neither he nor the state of Texas could have predicted the impact his death would have on medical research and education. Before his execution, Jernigan had donated his body to the State Anatomical Board of Texas. Because of his age (39), size (1.76 m [5'10"] and 90.4 kg [199 lb]), and good physical health prior to his death, his body was selected for the Visible Human Project. This project, funded by the U.S. National Library of Medicine (NLM), was designed to create accurate, computerized three-dimensional versions of the human body that can be visualized and explored from multiple perspectives. The data sets, or "visible humans," could be studied and manipulated in ways that are impossible using real bodies.

Using Jernigan as the basis for a "virtual male," Dr. Victor Spitzer and colleagues at the University of Colorado generated the data set. After they took serial CT and MRI scans of the body, it was frozen and then sectioned at 1-mm intervals. As each slice was generated, the researchers took high-resolution digital images and 70-mm color photographs. Shortly after image generation and processing were completed in 1995, the researchers generated a data set of a female using a 59-year-old woman who had died of natural causes and donated her body to the Anatomy Board of Maryland. The processing methods were similar, except the sections were taken at 0.33 mm, providing roughly three times the resolution of the male data set.

By combining x-rays, MRI, and CT scans with digitized images of cross sections through the body, the Visible Human data sets include an impressive volume of information. The digitized images, in computerized format, can be accessed through the Internet at the NLM/NIH Website. (www.nlm.nih.gov)

Because the three-dimensional relationships among anatomical structures can be manipulated and seen from any perspective, CT scans and MRI images of patients can often be directly compared to the sectional images of the "visible humans." The data have subsequently been used to generate a variety of enhanced images and interactive educational projects. Images based on the Visible Human Project are scattered throughout this book.

Microscopic Anatomy

Microscopic anatomy deals with structures that cannot be seen without magnification, and thus the boundaries of microscopic anatomy are established by the limits of the equipment used. With a dissecting microscope you can see tissue structure; with a light microscope, you can see basic details of cell structure; with an electron microscope, you can see individual molecules that are only a few nanometers (billionths of a meter) across.

Microscopic anatomy includes two major subdivisions: cytology and histology. **Cytology** (sī-TOL-o-jē) is the analysis of the internal structure of individual **cells**, the simplest units of life. Cells are composed of chemical substances in various combinations, and our lives depend on the chemical processes occurring in the trillions of cells in the body. For this reason, we consider basic chemistry (Chapter 2) before we examine cell structure (Chapter 3). **Histology** (his-TOL-ō-jē) is the examination of **tissues**—groups of specialized cells and cell products that work together to perform specific functions (Chapter 4). Tissues combine to form **organs**, such as the heart, kidney, liver, or brain. Many organs are easily examined without a microscope, so at the organ level we cross the boundary from microscopic anatomy to gross anatomy. As we proceed through the text, we will consider details at all levels, from macroscopic to microscopic. (Readers unfamiliar with the terms used to describe measurements and weights should consult the reference tables in the Appendix.)

Physiology

As noted earlier, physiology is the study of the function of anatomical structures; **human physiology** is the study of the functions of the human body. These functions are complex and much more difficult to examine than most anatomical structures. As a result, there are even more specialties in physiology than in anatomy, including the following:

- *Cell physiology*, the study of the functions of cells, is the cornerstone of human physiology. Cell physiology considers events at the chemical and molecular levels—both chemical processes within cells and chemical interactions between cells.

- *Organ physiology* is the study of the physiology of specific organs. An example is *cardiac physiology*, the study of heart function.

- *Systemic physiology* includes all aspects of the functioning of specific organ systems. Cardiovascular physiology, respiratory physiology, and reproductive physiology are examples of systemic physiology.

- *Pathological physiology* is the study of the effects of diseases on organ or system functions. Modern medicine depends on an understanding of both normal physiology and pathological physiology. You will find extensive information on clinically important topics in subsequent chapters.

Physicians normally use a combination of anatomical, physiological, chemical, and psychological information when they evaluate patients. When a patient presents symptoms to a physician, the physician will look at the structures affected (gross anatomy), perhaps collect a fluid or tissue sample (microscopic anatomy) for analysis, and ask questions to determine what alterations from normal functioning the patient is experiencing. Think back to your last trip to a doctor's office. Not only did the attending physician examine your body, noting any anatomical abnormalities, but he or she also evaluated your physiological processes by asking questions, observing your movements, listening to your body sounds, taking your temperature, and perhaps requesting chemical analyses of fluids such as blood or urine. In evaluating all these observations to reach a diagnosis, physicians rely on a logical framework based on the *scientific method*. The scientific method is at the core of all scientific thought, including medical diagnosis.

CHECKPOINT

10. Describe how anatomy and physiology are closely related.

11. What is the difference between gross anatomy and microscopic anatomy?

12. Identify several specialties of physiology.

13. Why is it difficult to separate anatomy from physiology?

See the blue Answers tab at the end of the book.

1-5 Levels of organization progress from molecules to a complete organism

Over the next three chapters, we will consider events and structures at several interdependent levels of organization. These levels of organization, which are illustrated in **Figure 1–1**, are the following:

- *The Chemical (or Molecular) Level.* Atoms, the smallest stable units of matter, can combine to form molecules with complex shapes. Even at this simplest level, form determines function: The functional properties of a particular molecule are determined by its unique three-dimensional shape and atomic components. We explore this level of organization in Chapter 2.

- *The Cellular Level.* Molecules can interact to form various types of organelles, each type of which has specific

functions. Organelles are structural and functional components of cells, the smallest living units in the body. Interactions among protein filaments, for example, produce the contractions of muscle cells in the heart. We examine the cellular level of organization in Chapter 3.

- *The Tissue Level.* A *tissue* is a group of cells working together to perform one or more specific functions. Heart muscle cells, or cardiac muscle cells (*cardium*, heart), interact with other types of cells and with extracellular materials to form *cardiac muscle tissue*. We consider the tissue level of organization in Chapter 4.

- *The Organ Level.* Organs consist of two or more tissues working in combination to perform several functions. Layers of cardiac muscle tissue, in combination with *connective tissue*, another type of tissue, form the bulk of the wall of the *heart*, a hollow, three-dimensional organ.

- *The Organ System Level.* Organs interact in **organ systems**. Each time it contracts, the heart pushes blood into a network of blood vessels. Together, the heart, blood, and blood vessels form the cardiovascular system, one of 11 organ systems in the body.

- *The Organism Level.* An organism—in this case, a human—is the highest level of organization. All organ systems of the body must work together to maintain the life and health of the organism.

The organization at each level determines not only the structural characteristics, but also the functions, of higher levels. For example, the arrangement of atoms and molecules at the chemical level creates the protein filaments that, at the cellular level, give cardiac muscle cells the ability to contract powerfully. At the tissue level, these cells are linked, forming cardiac muscle tissue. The structure of the tissue ensures that the contractions are coordinated, producing a heartbeat. When that beat occurs, the internal anatomy of the heart, an organ, enables it to function as a pump. The heart is filled with blood and connected to the blood vessels, and the pumping action circulates blood through the vessels of the cardiovascular system. Through interactions with the respiratory, digestive, urinary, and other systems, the cardiovascular system performs a variety of functions essential to the survival of the organism.

Something that affects a system will ultimately affect each of the system's components. For example, the heart cannot pump blood effectively after massive blood loss. If the heart cannot pump and blood cannot flow, oxygen and nutrients cannot be distributed. Very soon, the cardiac muscle tissue begins to break down as individual muscle cells die from oxygen and nutrient starvation. These changes will not be restricted to the cardiovascular system; all cells, tissues, and organs in the body will be damaged. **Figure 1–2** introduces the 11 interdependent, interconnected organ systems in the human body.

Figure 1–1 **Levels of Organization.** Interacting atoms form molecules that combine in the protein filaments of a heart muscle cell. Such cells interlock, creating heart muscle tissue, which constitutes most of the walls of the heart, a three-dimensional organ. The heart is but one component of the cardiovascular system, which also includes the blood and blood vessels. The various organ systems must work together to maintain life at the organism level.

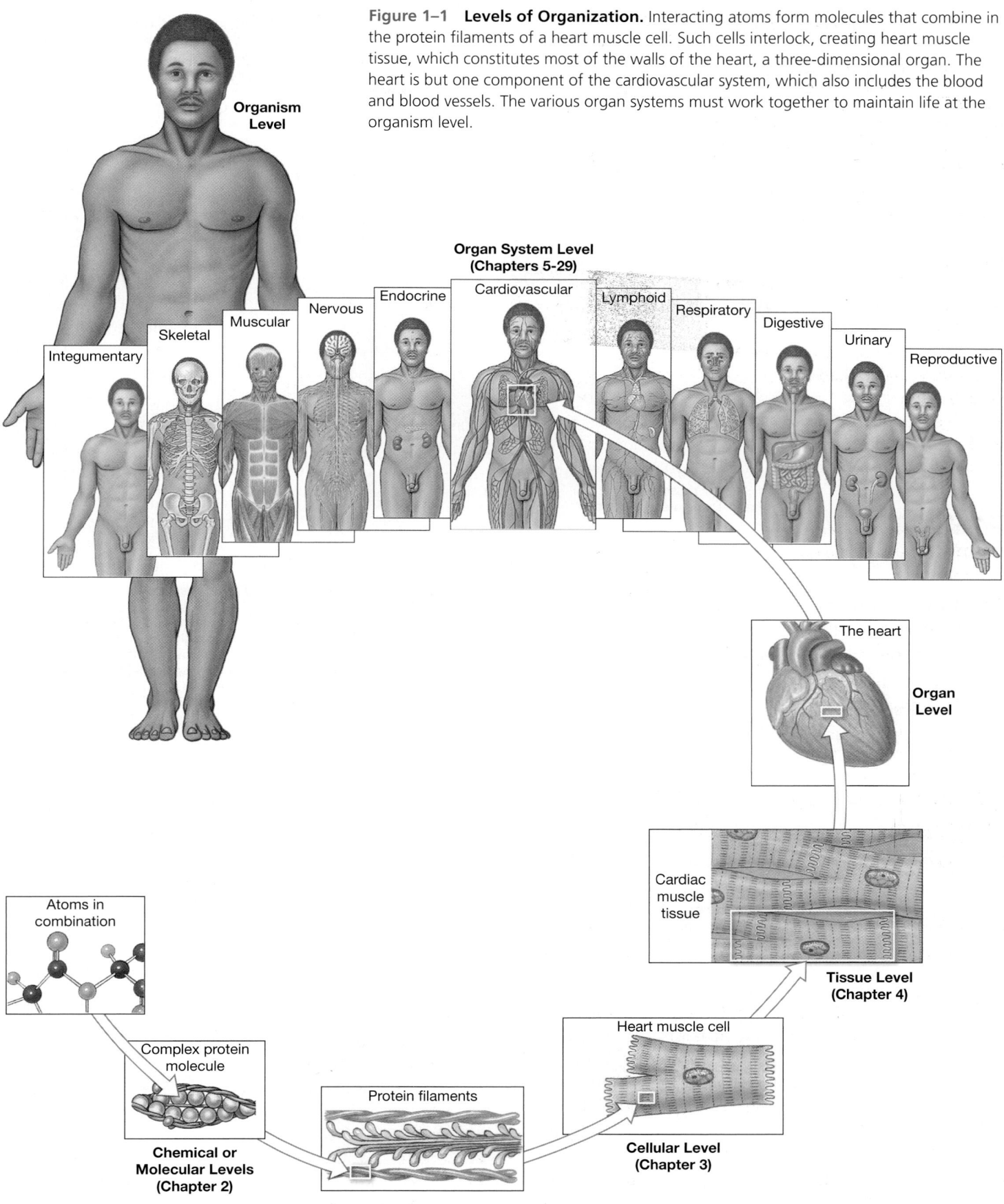

Organism Level

Organ System Level (Chapters 5-29)

Integumentary · Skeletal · Muscular · Nervous · Endocrine · Cardiovascular · Lymphoid · Respiratory · Digestive · Urinary · Reproductive

The heart

Organ Level

Cardiac muscle tissue

Tissue Level (Chapter 4)

Atoms in combination

Complex protein molecule

Chemical or Molecular Levels (Chapter 2)

Protein filaments

Heart muscle cell

Cellular Level (Chapter 3)

THE INTEGUMENTARY SYSTEM

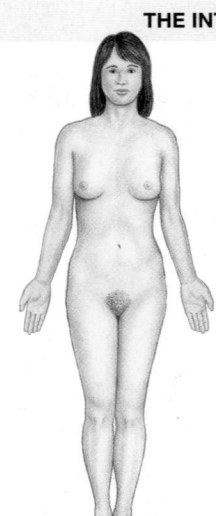

Major Organs:
- Skin
- Hair
- Sweat glands
- Nails

Functions:
- Protects against environmental hazards
- Helps regulate body temperature
- Provides sensory information

THE SKELETAL SYSTEM

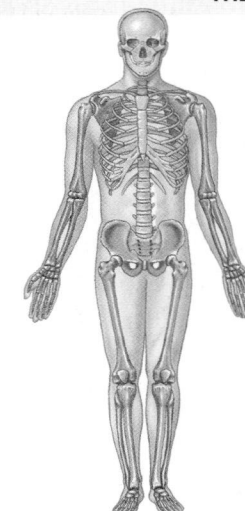

Major Organs:
- Bones
- Cartilages
- Associated ligaments
- Bone marrow

Functions:
- Provides support and protection for other tissues
- Stores calcium and other minerals
- Forms blood cells

THE MUSCULAR SYSTEM

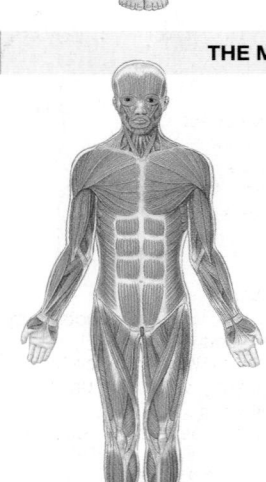

Major Organs:
- Skeletal muscles and associated tendons

Functions:
- Provides movement
- Provides protection and support for other tissues
- Generates heat that maintains body temperature

THE NERVOUS SYSTEM

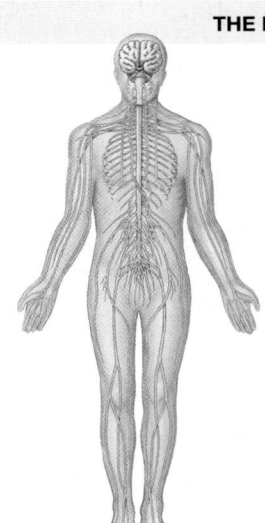

Major Organs:
- Brain
- Spinal cord
- Peripheral nerves
- Sense organs

Functions:
- Directs immediate responses to stimuli
- Coordinates or moderates activities of other organ systems
- Provides and interprets sensory information about external conditions

THE ENDOCRINE SYSTEM

Major Organs:
- Pituitary gland
- Thyroid gland
- Pancreas
- Suprarenal glands
- Gonads (testes and ovaries)
- Endocrine tissues in other systems

Functions:
- Directs long-term changes in the activities of other organ systems
- Adjusts metabolic activity and energy use by the body
- Controls many structural and functional changes during development

THE CARDIOVASCULAR SYSTEM

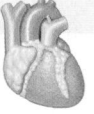

Major Organs:
- Heart
- Blood
- Blood vessels

Functions:
- Distributes blood cells, water, and dissolved materials, including nutrients, waste products, oxygen, and carbon dioxide
- Distributes heat and assists in control of body temperature

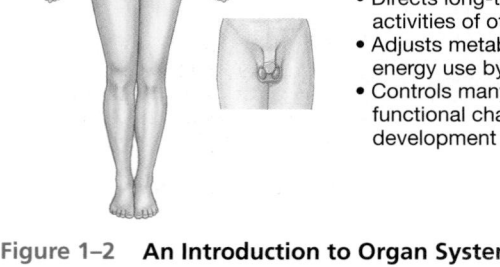

Figure 1–2   **An Introduction to Organ Systems.**

1 INTRODUCTION

THE LYMPHOID SYSTEM

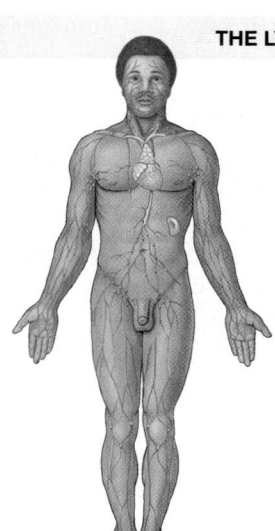

Major Organs:
- Spleen
- Thymus
- Lymphatic vessels
- Lymph nodes
- Tonsils

Functions:
- Defends against infection and disease
- Returns tissue fluids to the bloodstream

THE RESPIRATORY SYSTEM

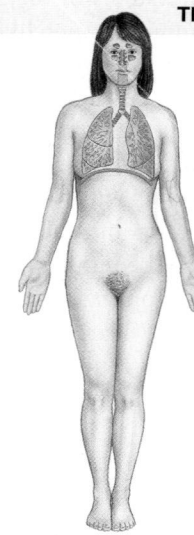

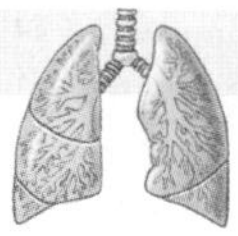

Major Organs:
- Nasal cavities
- Sinuses
- Larynx
- Trachea
- Bronchi
- Lungs
- Alveoli

Functions:
- Delivers air to alveoli (sites in lungs where gas exchange occurs)
- Provides oxygen to bloodstream
- Removes carbon dioxide from bloodstream
- Produces sounds for communication

THE DIGESTIVE SYSTEM

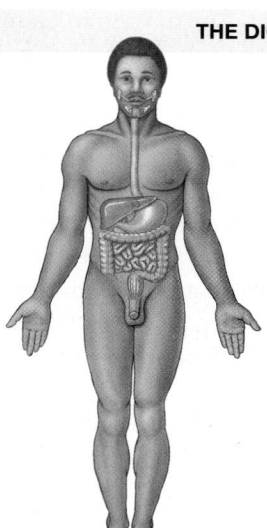

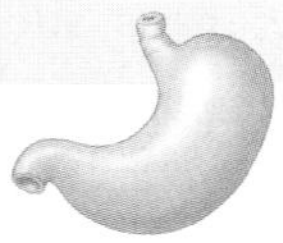

Major Organs:
- Teeth
- Tongue
- Pharynx
- Esophagus
- Stomach
- Small intestine
- Large intestine
- Liver
- Gallbladder
- Pancreas

Functions:
- Processes and digests food
- Absorbs and conserves water
- Absorbs nutrients (ions, water, and the breakdown products of dietary sugars, proteins, and fats)
- Stores energy reserves

THE URINARY SYSTEM

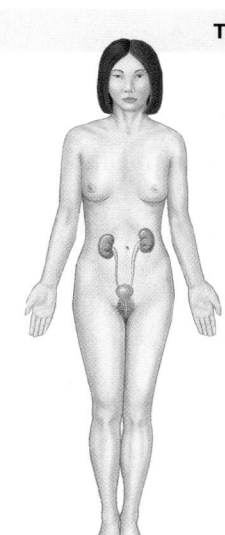

Major Organs:
- Kidneys
- Ureters
- Urinary bladder
- Urethra

Functions:
- Excretes waste products from the blood
- Controls water balance by regulating volume of urine produced
- Stores urine prior to voluntary elimination
- Regulates blood ion concentrations and pH

THE MALE REPRODUCTIVE SYSTEM

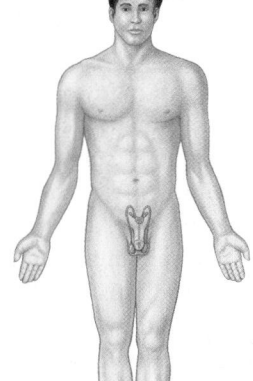
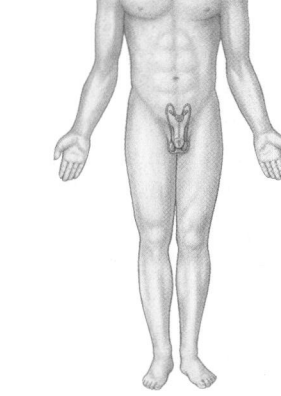

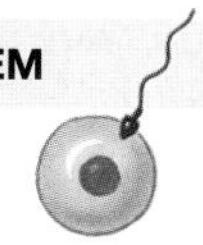

Major Organs:
- Testes
- Epididymides
- Ductus deferens
- Seminal vesicles
- Prostate gland
- Penis
- Scrotum

Functions:
- Produces male sex cells (sperm), suspending fluids, and hormones
- Sexual intercourse

THE FEMALE REPRODUCTIVE SYSTEM

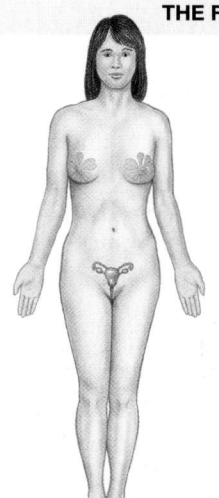

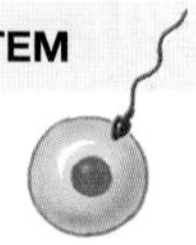

Major Organs:
- Ovaries
- Uterine tubes
- Uterus
- Vagina
- Labia
- Clitoris
- Mammary glands

Functions:
- Produces female sex cells (oocytes) and hormones
- Supports developing embryo from conception to delivery
- Provides milk to nourish newborn infant
- Sexual intercourse

Figure 1–2   An Introduction to Organ Systems (continued)

The cells, tissues, organs, and organ systems of the body coexist in a relatively small, shared environment, much like the inhabitants of a large city. Just as city dwellers breathe the same air and drink the water provided by the local water company, cells in the human body absorb oxygen and nutrients from the fluids that surround them. If a city is blanketed in smog or its water supply is contaminated, the inhabitants will become ill. Similarly, if the body fluid composition becomes abnormal, cells will be injured or destroyed. Suppose the temperature or salt content of the blood changes. The effect on the heart could range from the need for a minor adjustment (heart muscle tissue contracts more often, raising the heart rate) to a total disaster (the heart stops beating, so the individual dies).

The A&P Top 100

#2 The body can be divided into 11 organ systems, but all work together and the boundaries between them aren't absolute.

CHECKPOINT

14. Identify major levels of organization from the simplest to the most complex.

15. Identify the organ systems of the body and cite some major structures of each.

16. At which level of biological organization does a histologist investigate structures?

See the blue Answers tab at the end of the book.

1-6 Homeostasis is the tendency toward internal balance

Various physiological mechanisms act to prevent harmful changes in the composition of body fluids and the environment inside our cells. **Homeostasis** (*homeo*, unchanging + *stasis*, standing) refers to the existence of a stable internal environment. Maintaining homeostasis is absolutely vital to an organism's survival; failure to maintain homeostasis soon leads to illness or even death.

The principle of homeostasis is the central theme of this text and the foundation of all modern physiology. *Homeostatic regulation* is the adjustment of physiological systems to preserve homeostasis. Physiological systems have evolved to maintain homeostasis in an environment that is often inconsistent, unpredictable, and potentially dangerous. An understanding of homeostatic regulation is crucial to making accurate predictions about the body's responses to both normal and abnormal conditions.

Two general mechanisms are involved in homeostatic regulation: autoregulation and extrinsic regulation.

1. **Autoregulation**, or *intrinsic regulation*, occurs when a cell, a tissue, an organ, or an organ system adjusts its activities automatically in response to some environmental change. For example, when oxygen levels decline in a tissue, the cells release chemicals that dilate local blood vessels. This dilation increases the rate of blood flow and provides more oxygen to the region.

2. **Extrinsic regulation** results from the activities of the nervous system or endocrine system, two organ systems that control or adjust the activities of many other systems simultaneously. For example, when you exercise, your nervous system issues commands that increase your heart rate so that blood will circulate faster. Your nervous system also reduces blood flow to less active organs, such as the digestive tract. The oxygen in circulating blood is thus available to the active muscles, where it is needed most.

In general, the nervous system directs rapid, short-term, and very specific responses. When you accidentally set your hand on a hot stove, the heat produces a painful, localized disturbance of homeostasis. Your nervous system responds by ordering the immediate contraction of specific muscles that will pull your hand away from the stove. These contractions last only as long as the neural activity continues, usually a matter of seconds.

In contrast, the endocrine system releases chemical messengers called *hormones*, which affect tissues and organs throughout the body. Even though the responses may not be immediately apparent, they may persist for days or weeks. Examples of homeostatic regulation dependent on endocrine function include the long-term regulation of blood volume and composition, and the adjustment of organ system function during starvation. The endocrine system also plays a major role in growth and development: It is responsible for the changes that take place in your body as you mature and age.

Regardless of the system involved, the function of homeostatic regulation is always to keep the characteristics of the internal environment within certain limits. A homeostatic regulatory mechanism consists of three parts: (1) a **receptor**, a sensor that is sensitive to a particular environmental change, or *stimulus*; (2) a **control center**, or *integration center*, which receives and processes the information supplied by the receptor, and which sends out commands; and (3) an **effector**, a cell or organ that responds to the commands of the control center and whose activity either opposes or enhances the stimulus. You are probably already familiar with comparable regulatory mechanisms, such as the thermostat in your house or apartment (**Figure 1–3**).

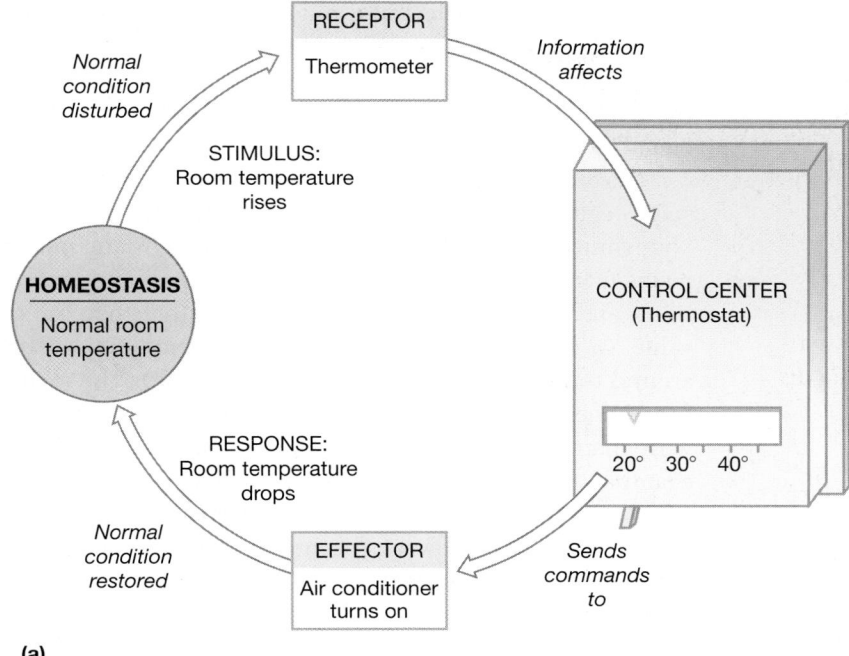

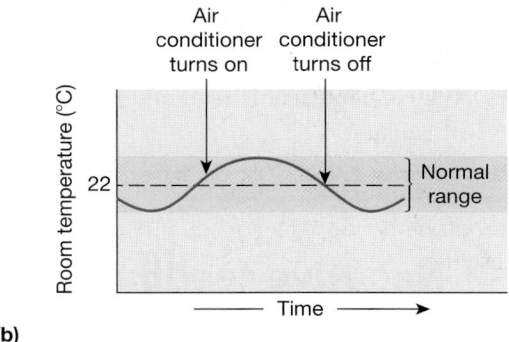

Figure 1–3 **The Control of Room Temperature. (a)** In response to input from a receptor (a thermometer), a thermostat (the control center) triggers an effector response (either an air conditioner or a heater) that restores normal temperature. In this case, when room temperature rises above the set point, the thermostat turns on the air conditioner, and the temperature returns to normal. **(b)** With this regulatory system, room temperature oscillates around the set point.

The thermostat is the control center; it receives information about room temperature from an internal or remote thermometer (a receptor). The dial on the thermostat establishes the *set point*, or desired value, which in this case is the temperature you select. (In our example, the set point is 22°C, or about 72°F.) The function of the thermostat is to keep room temperature within acceptable limits, usually within a degree or so of the set point. In summer, the thermostat accomplishes this function by controlling an air conditioner (an effector). When the temperature at the thermometer rises above the acceptable range, the thermostat turns on the air conditioner, which then cools the room (**Figure 1–3b**); when the temperature at the thermometer returns to the set point, the thermostat turns off the air conditioner. The control is not precise; the room is large, and the thermostat is located on just one wall. Over time, the temperature in the center of the room oscillates around the set point. The essential feature of temperature control by thermostat can be summarized very simply: A variation outside the desired range triggers an automatic response that corrects the situation. This method of homeostatic regulation is called *negative feedback*, because an effector activated by the control center opposes, or *negates*, the original stimulus. Negative feedback thus tends to minimize change, keeping variation in key body systems within limits that are compatible with our long-term survival.

CLINICAL NOTE

Homeostasis and Disease   The human body is amazingly effective in maintaining homeostasis. Nevertheless, an infection, an injury, or a genetic abnormality can sometimes have effects so severe that homeostatic mechanisms can't fully compensate for them. One or more characteristics of the internal environment may then be pushed outside of normal limits. When this happens, organ systems begin to malfunction, producing a state known as illness or disease.

An understanding of normal homeostatic mechanisms usually enables one to draw conclusions about what might be responsible for the signs and symptoms that are characteristic of many diseases. *Symptoms* are subjective— things that a person experiences and describes but that aren't otherwise detectable or measurable, such as pain, nausea, and anxiety. A *sign*, by contrast, is an objectively observable or measurable physical indication of a disease, such as a rash, a swelling, a fever, or sounds of abnormal breathing. Nowadays, technological aids can reveal many additional signs that would not be evident to a physician's unaided senses: an unusual shape on an x-ray or MRI scan, or an elevated concentration of a particular chemical in a blood test. Many aspects of human health, disease, and treatment are described in this textbook.

CHECKPOINT

17. Define homeostasis.

18. Which general mechanism of homeostatic regulation always involves the nervous or endocrine system?

19. Why is homeostatic regulation important to an organism?

See the blue Answers tab at the end of the book.

1-7 Negative feedback opposes variations from normal, whereas positive feedback exaggerates them

In this section we examine the roles of positive and negative feedback in homeostasis before considering the roles of organ systems in regulating homeostasis.

The Role of Negative Feedback in Homeostasis

Most homeostatic regulatory mechanisms involve **negative feedback**. An important example is the control of body temperature, a process called *thermoregulation*. In thermoregulation, the relationship between heat loss, which occurs primarily at the body surface, and heat production, which occurs in all active tissues, is altered.

In the homeostatic control of body temperature (**Figure 1–4**), the control center is in the *hypothalamus*, a region of the brain. This control center receives information from two sets of temperature receptors, one in the skin and the other within the hypothalamus. At the normal set point, body temperature (as measured with an oral thermometer) will be approximately 37°C (98.6°F). If body temperature rises above 37.2°C, activity in the control center targets two effectors: (1) muscle tissue in the walls of blood vessels supplying the skin and (2) sweat glands. The muscle tissue relaxes and the blood vessels dilate, increasing blood flow through vessels near the body surface; the sweat glands accelerate their secretion. The skin then acts like a radiator by losing heat to the environment, and the evaporation of sweat speeds the process. As body temperature returns to normal, temperature

at the hypothalamus declines, and the thermoregulatory control center becomes less active. Superficial blood flow and sweat gland activity then decrease to previous levels, although body temperature declines past the set point as the secreted sweat evaporates.

Negative feedback is the primary mechanism of homeostatic regulation, and it provides long-term control over the body's internal conditions and systems. Homeostatic mechanisms using negative feedback normally ignore minor variations, and they maintain a normal *range* rather than a fixed value. In the previous example, body temperature oscillated around the set-point temperature (**Figure 1–4b**). The regulatory process itself is dynamic, because the set point may vary with changing environments or differing activity levels. For example, when you are asleep, your thermoregulatory set point is lower, whereas when you work outside on a hot day (or when you have a fever), it is higher. Thus, body temperature can vary from moment to moment or from day to day for any individual, due to either (1) small oscillations around the set point or (2) changes in the set point. Comparable variations occur in all other aspects of physiology.

The variability among individuals is even greater than that within an individual. Each of us has homeostatic set points determined by genetic factors, age, gender, general health, and environmental conditions. It is therefore impractical to define "normal" homeostatic conditions very precisely. By convention, physiological values are reported either as average values obtained by sampling a large number of individuals, or as a range that includes 95 percent or more of the sample population. For example, for 95 percent of healthy adults, body temperature ranges between 36.7–37.2°C (98.1–98.9°F). However, 5 percent of healthy adults have resting body temperatures that are below 36.7°C or above 37.2°C. These temperatures are perfectly normal for them, and the variations have no clinical significance. Physicians must keep this variability in mind when they review lab reports or clinical discussions, because unusual values—even those outside the "normal" range—may represent individual variation rather than disease.

The Role of Positive Feedback in Homeostasis

In **positive feedback**, an initial stimulus produces a response that exaggerates or enhances the change in the original conditions, rather than opposing it. You seldom encounter positive feedback in your daily life, simply because it tends to produce extreme responses. For example, suppose that the thermostat in **Figure 1–3a** was accidentally connected to a heater rather than to an air conditioner. Now, when room temperature exceeds the set point, the thermostat turns on the heater, causing a further rise in room temperature. Room temperature will

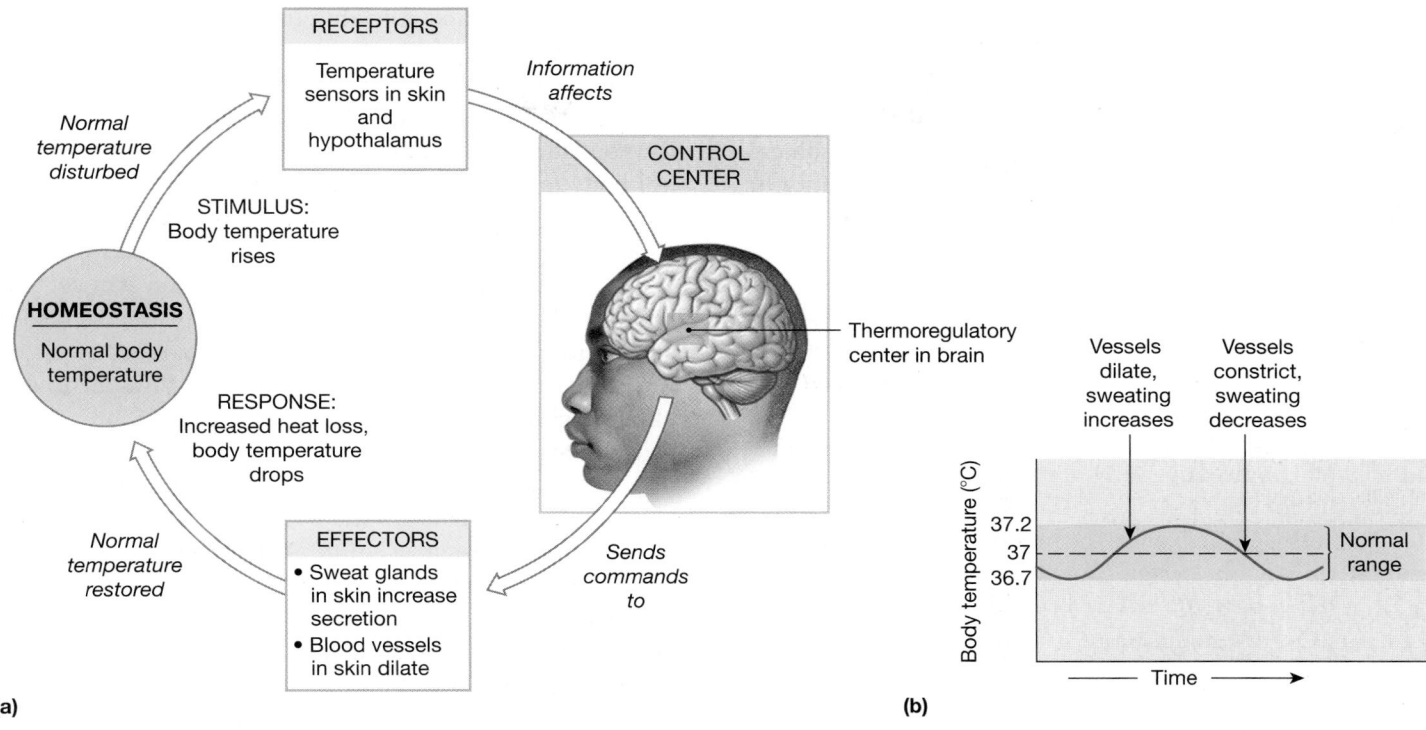

Figure 1–4 **Negative Feedback in the Control of Body Temperature.** In negative feedback, a stimulus produces a response that opposes or negates the original stimulus. **(a)** Events in the regulation of body temperature, which are comparable to those shown in *Figure 1–3*. A control center in the brain (the hypothalamus) functions as a thermostat with a set point of 37°C. If body temperature exceeds 37.2°C, heat loss is increased through enhanced blood flow to the skin and increased sweating. **(b)** The thermoregulatory center keeps body temperature oscillating within an acceptable range, usually between 36.7° and 37.2°C.

continue to increase until someone switches off the thermostat, turns off the heater, or intervenes in some other way. This kind of escalating cycle is often called a *positive feedback loop*.

In the body, positive feedback loops are typically found when a potentially dangerous or stressful process must be completed quickly before homeostasis can be restored. For example, the immediate danger from a severe cut is loss of blood, which can lower blood pressure and reduce the efficiency of the heart. The body's response to blood loss is diagrammed in **Figure 1–5**. Damage to cells in the blood vessel

wall releases chemicals that begin the process of blood clotting. As clotting gets under way, each step releases chemicals that accelerate the process. This escalating process is a positive feedback loop that ends with the formation of a blood clot, which patches the vessel wall and stops the bleeding. Blood clotting will be examined more closely in Chapter 19. Labor and delivery, another example of positive feedback in action, will be discussed in Chapter 29.

The human body is amazingly effective in maintaining homeostasis. Nevertheless, an infection, an injury, or a genetic

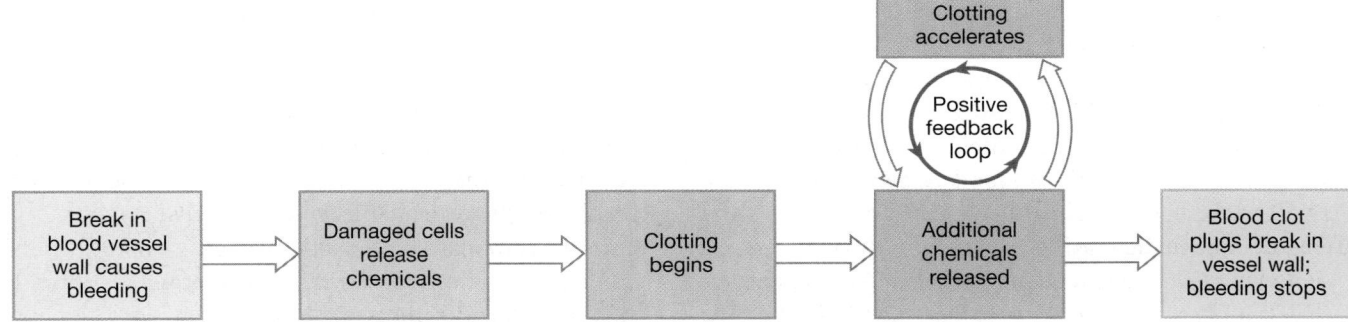

Figure 1–5 **Positive Feedback: Blood Clotting.** Positive feedback loops are important in accelerating processes that must proceed to completion rapidly. In this example, positive feedback accelerates the clotting process until a blood clot forms and stops the bleeding.

abnormality can sometimes have effects so severe that homeostatic mechanisms can't fully compensate for them. One or more characteristics of the internal environment may then be pushed outside normal limits. When this happens, organ systems begin to malfunction, producing a state known as illness, or **disease**. Chapters 5–29 devote considerable attention to the mechanisms responsible for a variety of human diseases.

Systems Integration, Equilibrium, and Homeostasis

Homeostatic regulation controls aspects of the internal environment that affect every cell in the body. No single organ system has total control over any of these aspects; such control requires the coordinated efforts of multiple organ systems. In later chapters we will explore the functions of each organ system and see how the systems interact to preserve homeostasis. Table 1–1 lists the roles of various organ systems in regulating several important physiological characteristics that are subject to homeostatic control. Note that in each case such regulation involves several organ systems.

A **state of equilibrium** exists when opposing processes or forces are in balance. In the case of body temperature, a state of equilibrium exists when the rate of heat loss equals the rate of heat production. Each physiological system functions to maintain a state of equilibrium that keeps vital conditions within normal limits. This is often called a state of *dynamic equilibrium* because physiological systems are continually adapting and adjusting to changing conditions. For example, when muscles become more active, more heat is produced. More heat must then be lost at the skin surface to reestablish a state of equilibrium before body temperature rises outside normal limits. Yet the adjustments made to control body temperature have other consequences: The sweating that increases heat loss at the skin surface increases losses of both water and salts. Other systems must then compensate for these losses and reestablish an equilibrium state for water and salts. This is a general pattern: Any adjustments made by one physiological system have direct and indirect effects on a variety of other systems. The maintenance of homeostasis is like a juggling act that keeps lots of balls in the air.

Although each organ system interacts with and is, in turn, dependent on other organ systems, it is much easier for introductory students to learn the basics of anatomy and physiology one system at a time. Although Chapters 5–29 are organized around individual systems, remember that these systems all work together. The 11 *Systems in Perspective* figures in later chapters will help reinforce this message; each provides an overview of one system's functions and summarizes its functional relationships with other systems.

The A&P Top 100

#4 A state of equilibrium exists when opposing processes or forces are in balance. The body always tries to maintain equilibrium because the failure to do so leads to disease and/or death.

TABLE 1–1   The Roles of Organ Systems in Homeostatic Regulation

Internal Characteristic	Primary Organ Systems Involved	Functions of the Organ Systems
Body temperature	Integumentary system	Heat loss
	Muscular system	Heat production
	Cardiovascular system	Heat distribution
	Nervous system	Coordination of blood flow, heat production, and heat loss
Body fluid composition		
Nutrient concentration	Digestive system	Nutrient absorption, storage, and release
	Cardiovascular system	Nutrient distribution
	Urinary system	Control of nutrient loss in the urine
Oxygen, carbon dioxide levels	Respiratory system	Absorption of oxygen, elimination of carbon dioxide
	Cardiovascular system	Internal transport of oxygen and carbon dioxide
Body fluid volume	Urinary system	Elimination or conservation of water from the blood
	Digestive system	Absorption of water; loss of water in feces
	Integumentary system	Loss of water through perspiration
	Cardiovascular system	Distribution of water
Waste product concentration	Urinary system	Elimination of waste products from the blood
	Digestive system	Elimination of waste products by the liver in feces
	Cardiovascular system	Transport of waste products to sites of excretion
Blood pressure	Cardiovascular system	Pressure generated by the heart moves blood through blood vessels
	Nervous system and endocrine system	Adjustments in heart rate and blood vessel diameter can raise or lower blood pressure

20. **Explain the function of negative feedback systems.**

21. **What happens to the body when homeostasis breaks down?**

22. **Explain how a positive feedback system works.**

23. **Why is positive feedback helpful in blood clotting but unsuitable for the regulation of body temperature?**

24. **Define equilibrium.**

25. **When the body continuously adapts by utilizing homeostatic systems, it is said to be in a state of _____ equilibrium.**

See the blue Answers tab at the end of the book.

1-8 Anatomical terms describe body regions, anatomical positions and directions, and body sections

Anatomists use anatomical terms to describe body regions, relative positions and directions, and body sections, as well as major body cavities and their subdivisions. In the following sections we will introduce the terms used in superficial anatomy and sectional anatomy.

Superficial Anatomy

Superficial anatomy involves locating structures on or near the body surface. A familiarity with anatomical landmarks (palpable structures), anatomical regions (specific areas used for reference purposes), and terms for anatomical directions will make the material in subsequent chapters more understandable. As you encounter new terms, create your own mental maps from the information provided in the accompanying anatomical illustrations.

Anatomical Landmarks

Important anatomical landmarks are presented in **Figure 1–6**. Understanding the terms and their etymology (origins) will help you remember both the location of a particular structure and its name. For example, *brachium* refers to the arm; later we will consider the *brachialis muscle* and the *brachial artery*, which are (as their names suggest) in the arm.

The A&P Top 100

#5 Anatomical descriptions refer to an individual in the anatomical position: standing, with the hands at the sides, palms facing forward, and feet together.

Standard anatomical illustrations show the human form in the **anatomical position**. When the body is in this position, the hands are at the sides with the palms facing forward, and the feet are together. **Figure 1–6a** shows an individual in the anatomical position as seen from the front (an *anterior view*), **Figure 1–6b** from the back (a *posterior view*). Unless otherwise noted, all descriptions in this text refer to the body in the anatomical position. A person lying down in the anatomical position is said to be **supine** (soo-PĪN) when face up, and **prone** when face down.

Tips&Tricks

Supine means up. In order to carry a bowl of *soup*, your hand must be in the *supine* position.

Anatomical Regions

Major anatomical regions of the body are listed in Table 1–2. But to describe a general area of interest or injury, anatomists and clinicians often need broader terms in addition to specific landmarks. Two methods are used to map the surface of the abdomen and pelvis.

Clinicians refer to four **abdominopelvic quadrants** (**Figure 1–7a**) formed by a pair of imaginary perpendicular lines that intersect at the umbilicus (navel). This simple method provides useful references for the description of aches, pains, and injuries. The location can help the physician determine the possible cause; for example, tenderness in the right lower quadrant (RLQ) is a symptom of appendicitis, whereas tenderness in the right upper quadrant (RUQ) may indicate gallbladder or liver problems.

Anatomists prefer more precise terms to describe the location and orientation of internal organs. They recognize nine **abdominopelvic regions** (**Figure 1–7b**). **Figure 1–7c** shows the relationships among quadrants, regions, and internal organs.

Tips&Tricks

The imaginary lines dividing the abdominopelvic regions resemble a tic-tac-toe game.

Anatomical Directions

Figure 1–8 and Table 1–3 introduce the principal directional terms and some examples of their use. There are many different terms, and some can be used interchangeably. For example, *anterior* refers to the front of the body when viewed in the anatomical position; in humans, this term is equivalent to *ventral*, which refers to the belly. Before you read further, analyze the table in detail, and practice using these terms. If you are familiar with the basic vocabulary, the descriptions in subsequent chapters will be easier to follow. When reading anatomical descriptions, you will find it useful to remember that the terms *left* and *right* always refer to the *left* and *right* sides of the *subject*, not of the observer.

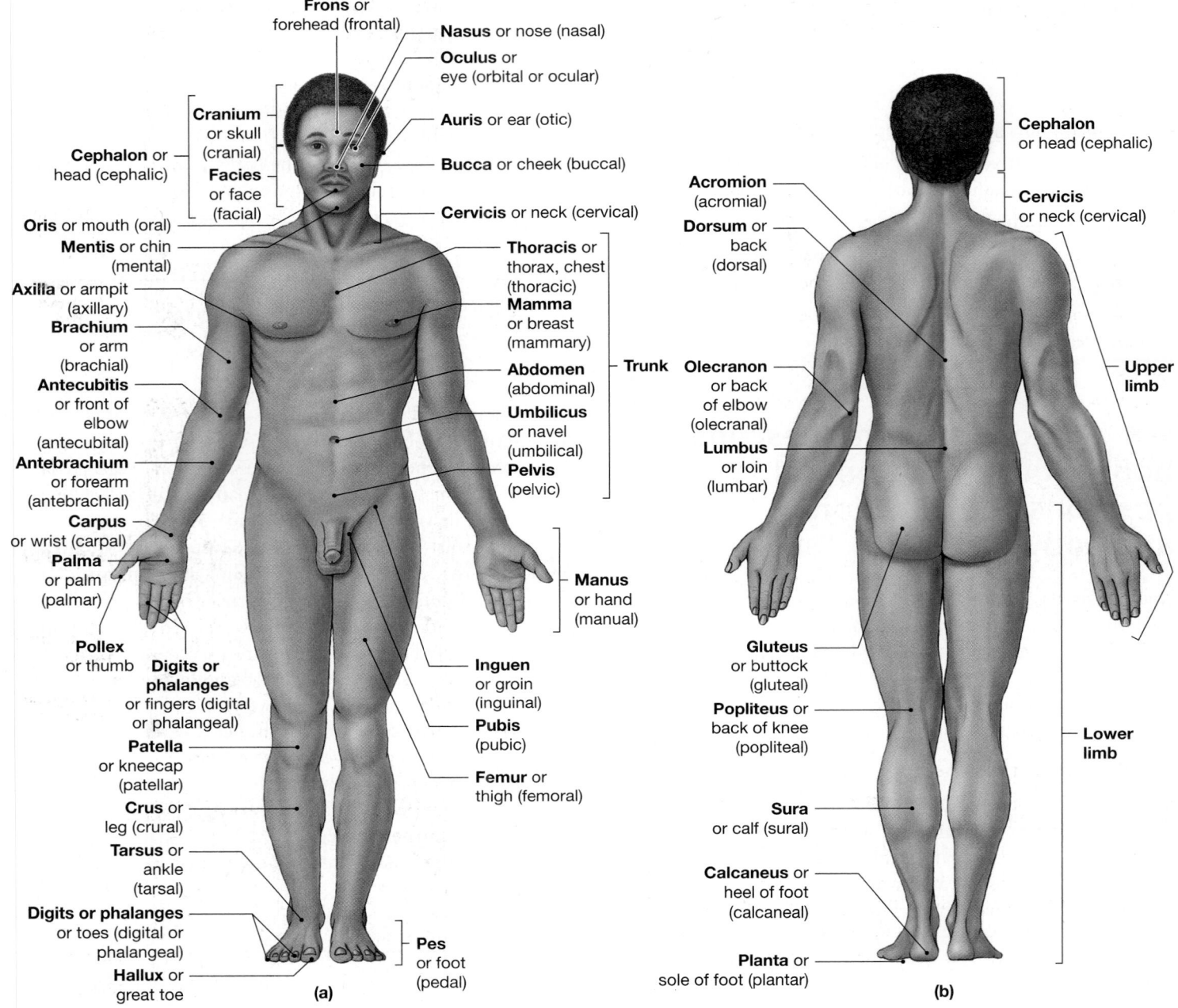

Frons or forehead (frontal)

Nasus or nose (nasal)

Oculus or eye (orbital or ocular)

Cranium or skull (cranial)

Auris or ear (otic)

Cephalon or head (cephalic)

Facies or face (facial)

Bucca or cheek (buccal)

Oris or mouth (oral)

Cervicis or neck (cervical)

Mentis or chin (mental)

Axilla or armpit (axillary)

Brachium or arm (brachial)

Antecubitis or front of elbow (antecubital)

Antebrachium or forearm (antebrachial)

Carpus or wrist (carpal)

Palma or palm (palmar)

Pollex or thumb

Digits or phalanges or fingers (digital or phalangeal)

Patella or kneecap (patellar)

Crus or leg (crural)

Tarsus or ankle (tarsal)

Digits or phalanges or toes (digital or phalangeal)

Hallux or great toe

Thoracis or thorax, chest (thoracic)

Mamma or breast (mammary)

Abdomen (abdominal)

Umbilicus or navel (umbilical)

Pelvis (pelvic)

Trunk

Manus or hand (manual)

Inguen or groin (inguinal)

Pubis (pubic)

Femur or thigh (femoral)

Pes or foot (pedal)

(a)

Acromion (acromial)

Dorsum or back (dorsal)

Olecranon or back of elbow (olecranal)

Lumbus or loin (lumbar)

Gluteus or buttock (gluteal)

Popliteus or back of knee (popliteal)

Sura or calf (sural)

Calcaneus or heel of foot (calcaneal)

Planta or sole of foot (plantar)

Cephalon or head (cephalic)

Cervicis or neck (cervical)

Upper limb

Lower limb

(b)

Figure 1–6 **Anatomical Landmarks.** Anatomical terms are shown in boldface type, common names in plain type, and anatomical adjectives in parentheses.

Sectional Anatomy

Sometimes the only way to understand the relationships among the parts of a three-dimensional object is to slice through it and look at the internal organization.

An understanding of sectional views is particularly important now that electronic imaging techniques enable us to see inside the living body. Although these views are sometimes difficult to interpret, it is worth spending the time required to understand what they show. Once you are able to interpret sectional views, you will have a good mental model for studying the anatomy and physiology of a particular region or system. Radiologists and other medical professionals responsible for interpreting medical scans spend much of their time analyzing sectional views of the body.

TABLE 1–2 Regions of the Human Body (see Figure 1–6)	
Structure	**Region**
Cephalon (head)	Cephalic region
Cervicis (neck)	Cervical region
Thoracis (thorax or chest)	Thoracic region
Brachium (arm)	Brachial region
Antebrachium (forearm)	Antebrachial region
Carpus (wrist)	Carpal region
Manus (hand)	Manual region
Abdomen	Abdominal region
Lumbus (loin)	Lumbar region
Gluteus (buttock)	Gluteal region
Pelvis	Pelvic region
Pubis (anterior pelvis)	Pubic region
Inguen (groin)	Inguinal region
Femur (thigh)	Femoral region
Crus (anterior leg)	Crural region
Sura (calf)	Sural region
Tarsus (ankle)	Tarsal region
Pes (foot)	Pedal region
Planta (sole)	Plantar region

Planes and Sections

Any slice through a three-dimensional object can be described in reference to three **sectional planes**, as indicated in Figure 1–9 and Table 1–4. A *plane* is an axis; three planes are needed to describe any three-dimensional object. A *section* is a single view or slice along one of these planes. The **transverse plane** lies at right angles to the long axis of the body, dividing it into *superior* and *inferior* portions. A cut in this plane is called a **transverse section**, or *cross section*. The **frontal plane** (or *coronal plane*) and the **sagittal plane** are parallel to the long axis of the body. The frontal plane extends from side to side, dividing the body into *anterior* and *posterior* portions. The sagittal plane extends from the front to back, dividing the body into left and right portions. A cut that passes along the midline and divides the body into left and right halves is a *midsagittal section*, or *median section*; a cut parallel to the midsagittal line is a *parasagittal section*. The atlas that accompanies this text contains images of sections taken through the body in various planes. You will be referred to these images later in the text for comparison with specific figure illustrations. Unless otherwise noted, all anatomical diagrams that present cross sectional views of the body are oriented as though the subject were supine with the observer standing at the feet and looking toward the head.

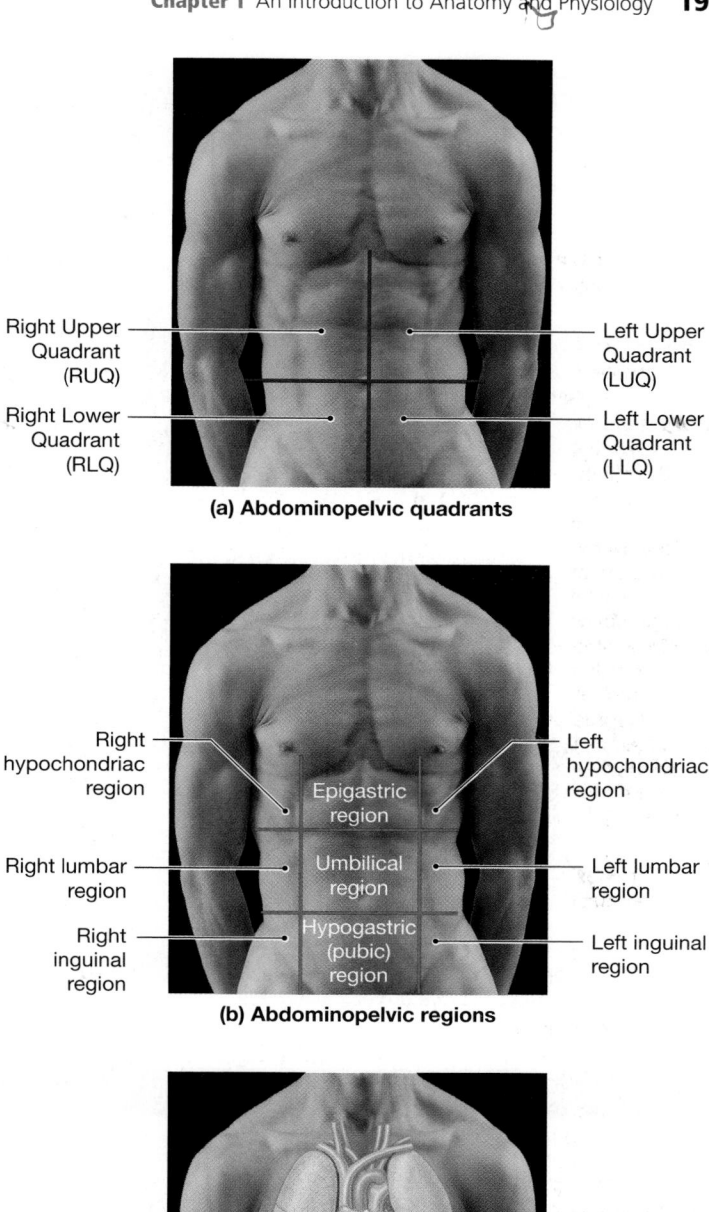

Right Upper Quadrant (RUQ) — Left Upper Quadrant (LUQ)
Right Lower Quadrant (RLQ) — Left Lower Quadrant (LLQ)

(a) Abdominopelvic quadrants

Right hypochondriac region — Left hypochondriac region
Epigastric region
Right lumbar region — Umbilical region — Left lumbar region
Right inguinal region — Hypogastric (pubic) region — Left inguinal region

(b) Abdominopelvic regions

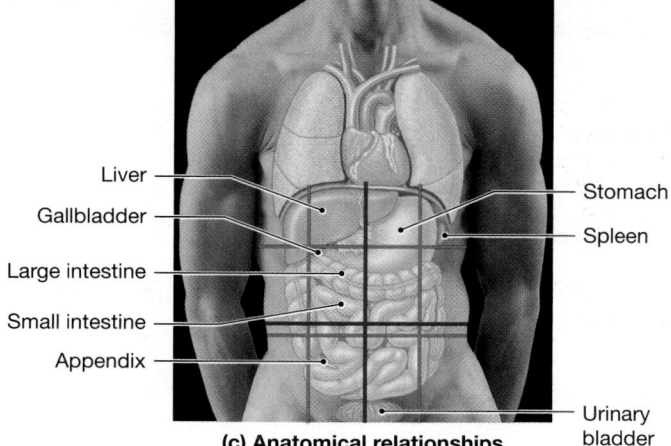

Liver — Stomach
Gallbladder — Spleen
Large intestine
Small intestine
Appendix
Urinary bladder

(c) Anatomical relationships

Figure 1–7 Abdominopelvic Quadrants and Regions. (a) The four abdominopelvic quadrants, formed by two perpendicular lines that intersect at the umbilicus. The terms for these quadrants, or their abbreviations, are most often used in clinical discussions. **(b)** The nine abdominopelvic regions, which provide more precise regional descriptions. **(c)** The relationship between the abdominopelvic quadrants and regions and the locations of the internal organs.

Figure 1–8 **Directional References. (a)** A lateral view. **(b)** An anterior view. Important directional terms used in this text are indicated by arrows; definitions and descriptions are given in Table 1–3.

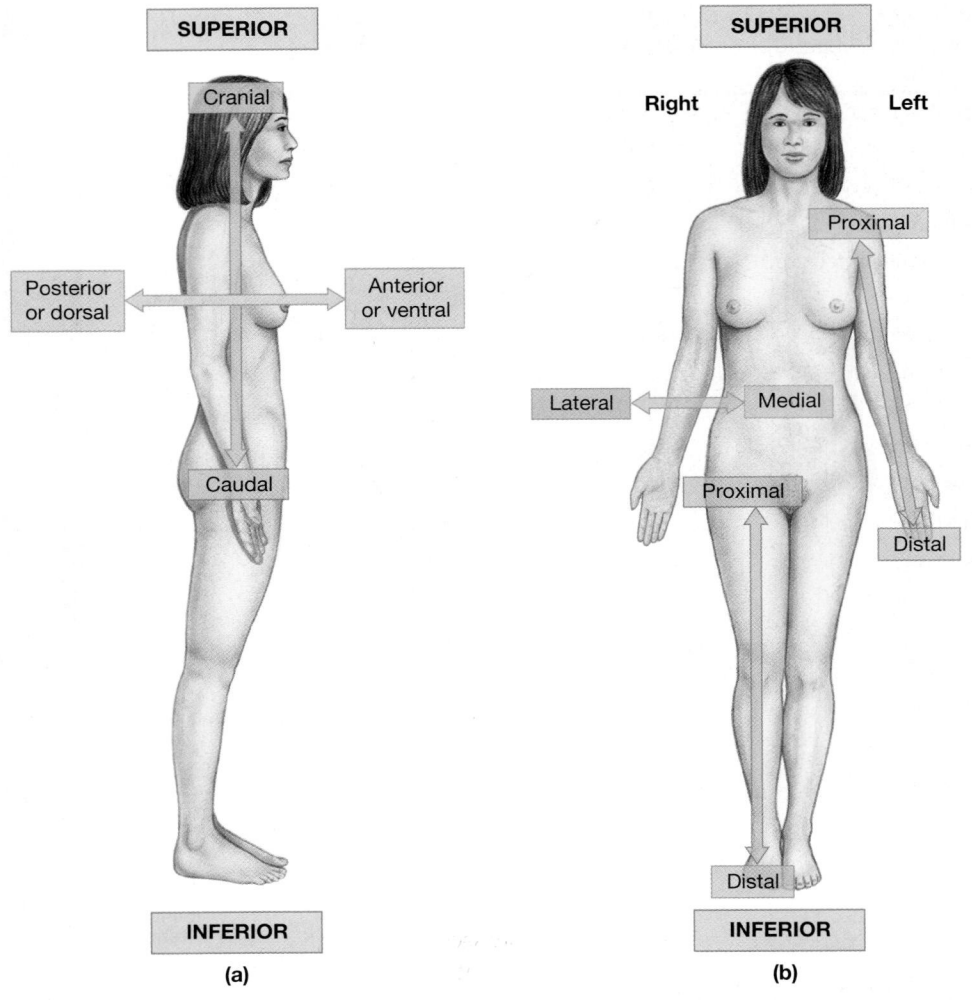

TABLE 1–3	Directional Terms (see Figure 1–8)	
Term	**Region or Reference**	**Example**
Anterior	The front; before	The navel is on the *anterior* surface of the trunk.
Ventral	The belly side (equivalent to anterior when referring to human body)	The navel is on the *ventral* surface of the trunk.
Posterior or **dorsal**	The back; behind	The shoulder blade is located *posterior* to the rib cage.
Cranial or **cephalic**	The head	The *cranial*, or *cephalic*, border of the pelvis is on the side toward the head rather than toward the thigh.
Superior	Above; at a higher level (in human body, toward the head)	In humans, the cranial border of the pelvis is *superior* to the thigh.
Caudal	The tail (coccyx in humans)	The hips are *caudal* to the waist.
Inferior	Below; at a lower level	The knees are *inferior* to the hips.
Medial	Toward the body's longitudinal axis; toward the midsagittal plane	The *medial* surfaces of the thighs may be in contact; moving medially from the arm across the chest surface brings you to the sternum.
Lateral	Away from the body's longitudinal axis; away from the midsagittal plane	The thigh articulates with the *lateral* surface of the pelvis; moving laterally from the nose brings you to the cheeks.
Proximal	Toward an attached base	The thigh is *proximal* to the foot; moving proximally from the wrist brings you to the elbow.
Distal	Away from an attached base	The fingers are *distal* to the wrist; moving distally from the elbow brings you to the wrist.
Superficial	At, near, or relatively close to the body surface	The skin is *superficial* to underlying structures.
Deep	Farther from the body surface	The bone of the thigh is *deep* to the surrounding skeletal muscles.

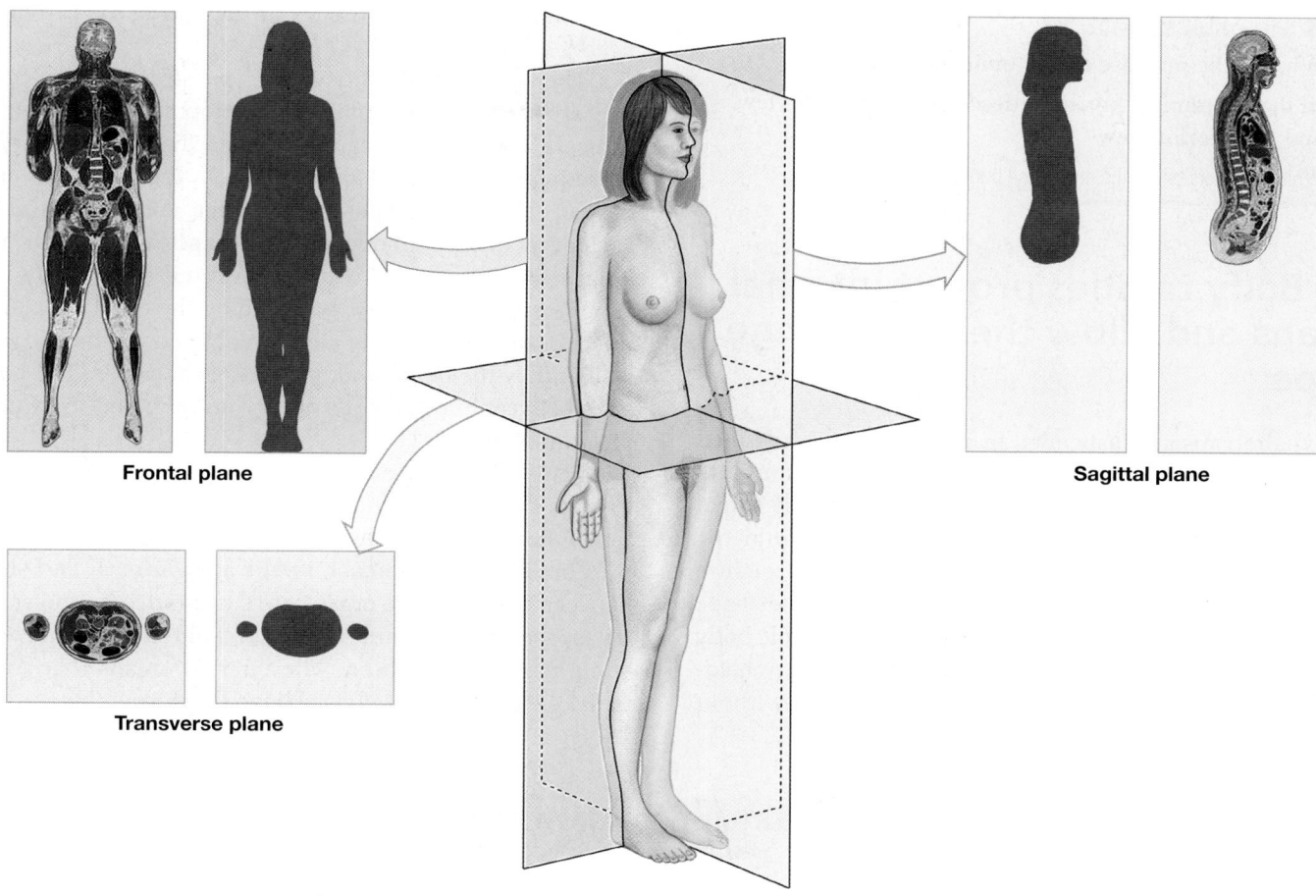

Figure 1–9   Sectional Planes. The three primary sectional planes, which are defined and described in Table 1–4. The photos of sectional images were derived from the Visible Human data set.

TABLE 1–4 Terms That Indicate Sectional Planes (see Figure 1–9)			
Orientation of Plane	**Plane**	**Directional Reference**	**Description**
Perpendicular to long axis	Transverse or horizontal	Transversely or horizontally	A *transverse*, or *horizontal*, *section* separates superior and inferior portions of the body. A cut in this plane is called a *cross section*.
Parallel to long axis	Sagittal	Sagittally	A *sagittal section* separates right and left portions. You examine a sagittal section, but you section sagittally.
	Midsagittal		In a *midsagittal section* or *median section*, the plane passes through the midline, dividing the body into right and left sides.
	Parasagittal		A *parasagittal section,* which is a cut parallel to the midsagittal plane, separates the body into right and left portions of unequal size.
	Frontal or coronal	Frontally or coronally	A *frontal*, or *coronal*, *section* separates anterior and posterior portions of the body; *coronal* usually refers to sections passing through the skull.

CHECKPOINT

CHECKPOINT

26. What is the purpose of anatomical terms?

27. In the anatomical position, describe an anterior view and a posterior view.

See the blue Answers tab at the end of the book.

1-9 Body cavities protect internal organs and allow them to change shape

Many vital organs are suspended in internal chambers called *body cavities*. These cavities have two essential functions: (1) They protect delicate organs, such as the brain and spinal cord, from accidental shocks, and cushion them from the thumps and bumps that occur when we walk, jump, or run; and (2) they permit significant changes in the size and shape of internal organs. For example, because they are inside body cavities, the lungs, heart, stomach, intestines, urinary bladder, and many other organs can expand and contract without

distorting surrounding tissues or disrupting the activities of nearby organs.

The *ventral body cavity*, or *coelom* (SĒ-lōm; *koila*, cavity), appears early in embryological development. It contains organs of the respiratory, cardiovascular, digestive, urinary, and reproductive systems (**Figure 1–10**). As these internal organs develop, their relative positions change, and the ventral body cavity is gradually subdivided. The **diaphragm** (DĪ-uh-fram), a flat muscular sheet, divides the ventral body cavity into a superior **thoracic cavity**, bounded by the chest wall, and an inferior **abdominopelvic cavity**, enclosed by the abdominal wall and by the bones and muscles of the pelvis. The boundaries between the divisions of the ventral body cavity are depicted in **Figure 1–11**.

Many of the organs in the thoracic and abdominopelvic cavities change size and shape as they perform their functions. For example, the lungs inflate and deflate as you breathe, and your stomach swells at each meal and shrinks between meals. These organs are surrounded by moist internal spaces that permit expansion and limited movement while preventing friction. The internal organs that are partially or completely enclosed by these cavities are called

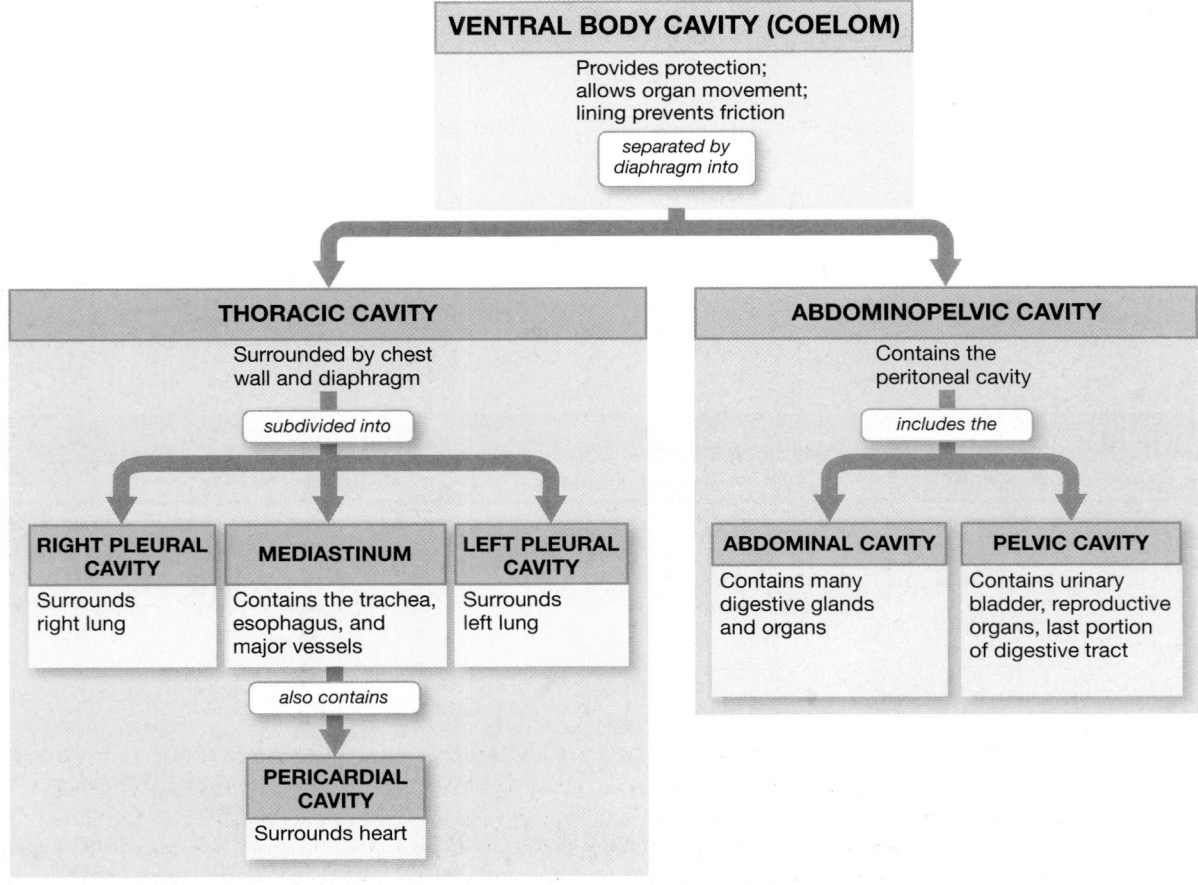

Figure 1–10   **Relationships Among the Subdivisions of the Ventral Body Cavity.**

viscera (VIS-e-ruh). A delicate layer called a *serous membrane* lines the walls of these internal cavities and covers the surfaces of the enclosed viscera. A watery fluid that coats the opposing surfaces and reduces friction moistens serous membranes. The portion of a serous membrane that covers a visceral organ is called the *visceral* layer; the opposing layer that lines the inner surface of the body wall or chamber is called the *parietal* layer.

We will now take a closer look at the anatomy of the thoracic and abdominopelvic cavities.

The Thoracic Cavity

The thoracic cavity contains the lungs and heart; associated organs of the respiratory, cardiovascular, and lymphoid systems; the inferior portions of the esophagus; and the thymus. The boundaries of the thoracic cavity are established by the muscles and bones of the chest wall and the diaphragm (**Figure 1–11a**). The thoracic cavity is subdivided into the left and right **pleural cavities**, separated by the **mediastinum** (mē-dē-as-TĪ-num or mē-dē-AS-ti-num) (**Figure 1–11c**). Each pleural cavity, which

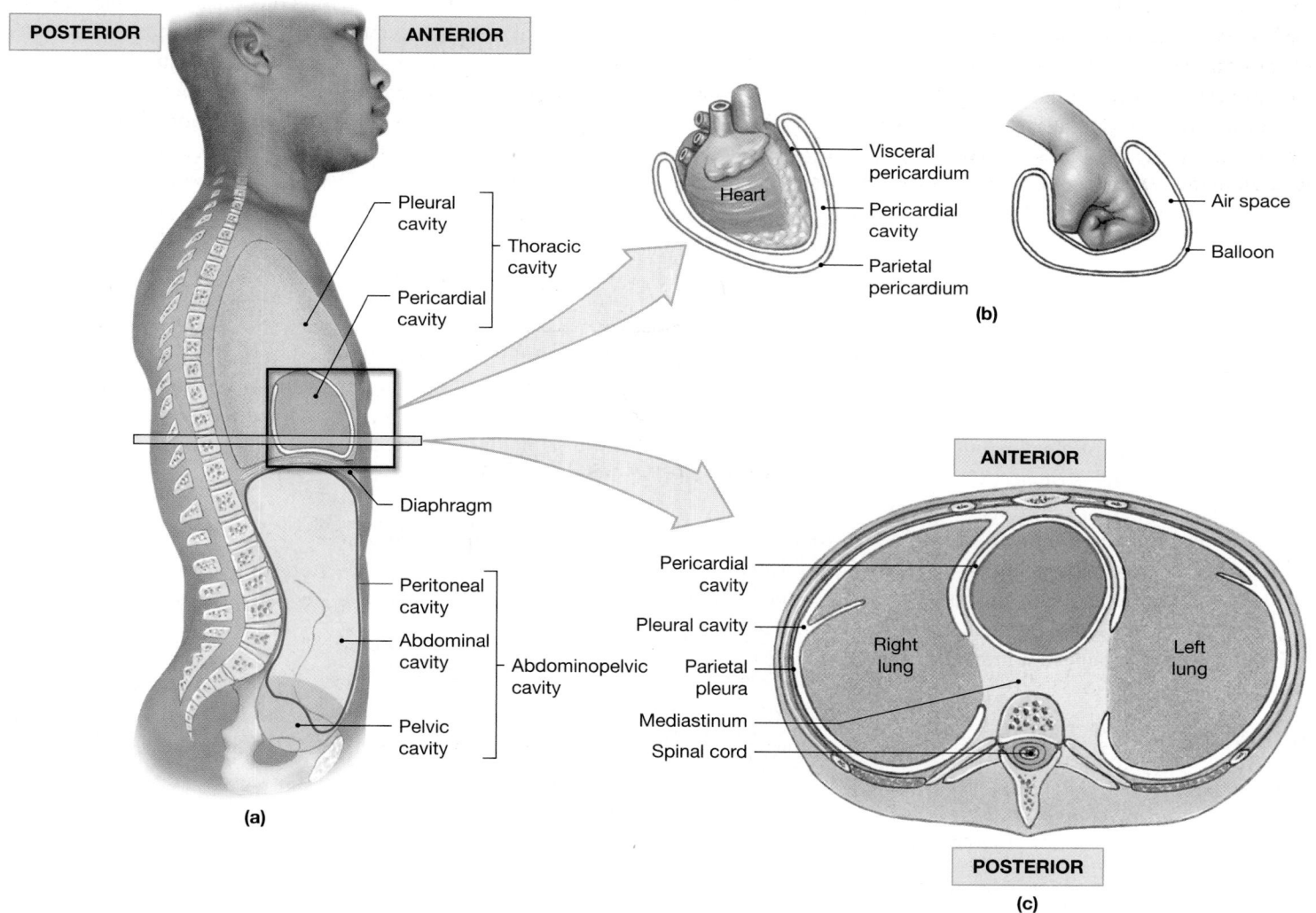

Figure 1–11   The Ventral Body Cavity and Its Subdivisions. (a) A lateral view showing the ventral body cavity, which is divided by the muscular diaphragm into a superior thoracic (chest) cavity and an inferior abdominopelvic cavity. **(b)** The heart is suspended within the pericardial cavity like a fist pushed into a balloon. The attachment site, corresponding to the wrist of the hand, lies at the connection between the heart and major blood vessels. **(c)** A transverse section through the ventral body cavity, showing the central location of the pericardial cavity within the thoracic cavity. Notice how the mediastinum divides the thoracic cavity into two pleural cavities. Note that this transverse or cross sectional view is oriented as though the observer were standing at the subject's feet and looking toward the subject's head.

contains a lung, is lined by a shiny, slippery serous membrane that reduces friction as the lung expands and recoils during respiration. The serous membrane lining a pleural cavity is called a *pleura* (PLOOR-ah). The *visceral pleura* covers the outer surfaces of a lung, whereas the *parietal pleura* covers the mediastinal surface and the inner body wall.

The mediastinum consists of a mass of connective tissue that surrounds, stabilizes, and supports the esophagus, trachea, and thymus, as well as the major blood vessels that originate or end at the heart. The mediastinum also contains the **pericardial cavity**, a small chamber that surrounds the heart. The relationship between the heart and the pericardial cavity resembles that of a fist pushing into a balloon (**Figure 1–11b**). The wrist corresponds to the *base* (attached portion) of the heart, and the balloon corresponds to the serous membrane that lines the pericardial cavity. The serous membrane associated with the heart is called the *pericardium* (peri-, around + cardium, heart). The layer covering the heart is the *visceral pericardium*, and the opposing surface is the *parietal pericardium*. During each beat, the heart changes in size and shape. The pericardial cavity permits these changes, and the slippery pericardial lining prevents friction between the heart and adjacent structures in the thoracic cavity.

The Abdominopelvic Cavity

The abdominopelvic cavity extends from the diaphragm to the pelvis. It is subdivided into a superior **abdominal cavity** and an inferior **pelvic cavity** (**Figure 1–11a**). The abdominopelvic cavity contains the *peritoneal* (per-i-tō-NĒ-al) *cavity*, a chamber lined by a serous membrane known as the *peritoneum* (per-i-tō-NĒ-um). The *parietal peritoneum* lines the inner surface of the body wall. A narrow space containing a small amount of fluid separates the parietal peritoneum from the *visceral peritoneum*, which covers the enclosed organs. You are probably already aware of the movements of the organs in this cavity. Who has not had at least one embarrassing moment when the contraction of a digestive organ produced a movement of liquid or gas and a

gurgling or rumbling sound? The peritoneum allows the organs of the digestive system to slide across one another without damage to themselves or the walls of the cavity.

The abdominal cavity extends from the inferior surface of the diaphragm to the level of the superior margins of the pelvis. This cavity contains the liver, stomach, spleen, small intestine, and most of the large intestine. (The positions of most of these organs are shown in **Figure 1–7c**.) The organs are partially or completely enclosed by the peritoneal cavity, much as the heart and lungs are enclosed by the pericardial and pleural cavities, respectively. A few organs, such as the kidneys and pancreas, lie between the peritoneal lining and the muscular wall of the abdominal cavity. Those organs are said to be *retroperitoneal* (retro, behind).

The pelvic cavity is the portion of the ventral body cavity inferior to the abdominal cavity. The bones of the pelvis form the walls of the pelvic cavity, and a layer of muscle forms its floor. The pelvic cavity contains the distal portion of the large intestine, the urinary bladder, and various reproductive organs. The pelvic cavity of females, for example, contains the ovaries, uterine tubes, and uterus; in males, it contains the prostate gland and seminal glands. The pelvic cavity also contains the inferior portion of the peritoneal cavity. The peritoneum covers the ovaries and the uterus in females, as well as the superior portion of the urinary bladder in both sexes.

This chapter provided an overview of the locations and functions of the major components of each organ system. It also introduced the vocabulary you need to follow more detailed anatomical descriptions in later chapters. Many of the figures in those chapters contain images produced by modern clinical imaging procedures. You will find numerous examples in the *Atlas*.

CHECKPOINT

28. **Name two essential functions of body cavities.**

29. **Identify the subdivisions of the ventral body cavity.**

See the blue Answers tab at the end of the book.

Related Clinical Terms

abdominopelvic quadrant: One of four divisions of the anterior abdominal surface.

abdominopelvic region: One of nine divisions of the anterior abdominal surface.

CT, CAT (computerized [axial] tomography): An imaging technique that uses x-rays to reconstruct the body's three-dimensional structure.

disease: A malfunction of organs or organ systems resulting from a failure of homeostatic regulation.

DSA (digital subtraction angiography): A technique used to monitor blood flow through specific organs, such as the brain, heart, lungs, or kidneys. X-rays are taken before and after a radiopaque dye is

administered, and a computer "subtracts" details common to both images. The result is a high-contrast image showing the distribution of the dye.

embryology: The study of structural changes during the first two months of development.

histology: The study of tissues.

MRI (magnetic resonance imaging): An imaging technique that employs a magnetic field and radio waves to portray subtle structural differences.

PET (positron emission tomography) scan: An imaging technique that shows the chemical functioning, as well as the structure, of an organ.

radiologist: A physician who specializes in performing and analyzing radiological procedures.

sign: A physical manifestation of a disease that can be measured and observed through sight, hearing, or touch.

spiral-CT: A method of processing computerized tomography data to provide rapid, three-dimensional images of internal organs.

symptom: A patient's perception of a change in normal body function.

ultrasound: An imaging technique that uses brief bursts of high-frequency sound waves reflected by internal structures.

x-rays: High-energy radiation that can penetrate living tissues.

Chapter Review
Study Outline

An Introduction to Studying the Human Body p. 2

1. Biology is the study of life. One of its goals is to discover the unity and the patterns that underlie the diversity of organisms.

1-1 Anatomy and physiology directly affect your life p. 2

2. This course will help you discover how your body works under normal and abnormal conditions, by serving as a base for understanding other life sciences and expanding your vocabulary.

1-2 Good study strategies are crucial to success p. 2

3. Your success in your A&P course requires developing good study skills.

4. Your textbook and textbook package contain a diversity of features and resources to support your efforts to be an active learner.

1-3 Anatomy is structure, and physiology is function p. 4

5. **Anatomy** is the study of internal and external structures of the body and the physical relationships among body parts. **Physiology** is the study of how living organisms perform their vital functions. All physiological functions are performed by specific structures.

6. **Medical terminology** is the use of prefixes, suffixes, word roots, and combining forms to construct anatomical, physiological, or medical terms.

7. *Terminologia Anatomica (International Anatomical Terminology)* was used as the standard in preparing your textbook.

1-4 Anatomy and physiology are closely integrated p. 5

8. All specific functions are performed by specific structures.

9. In **gross** (*macroscopic*) **anatomy**, we consider features that are visible without a microscope. This field includes *surface anatomy* (general form and superficial markings); *regional anatomy* (anatomical organization of specific areas of the body); and *systemic anatomy* (structure of organ systems). In *developmental anatomy*, we examine the changes in form that occur between conception and physical maturity. In *embryology*, we study developmental processes that occur during the first two months of development. *Clinical anatomy* includes anatomical subspecialties important to the practice of medicine.

10. The limits of *microscopic anatomy* are established by the equipment used. In **cytology**, we analyze the internal structure of individual cells. In **histology**, we examine **tissues**, groups of cells that perform specific functions. Tissues combine to form **organs**, anatomical units with multiple functions.

11. Human physiology is the study of the functions of the human body. It is based on *cell physiology*, the study of the functions of cells. In *organ physiology*, we study the physiology of specific organs. In *systemic physiology*, we consider all aspects of the functioning of specific organ systems. In *pathological physiology*, we study the effects of diseases on organ or system functions.

1-5 Levels of organization progress from molecules to a complete organism p. 8

12. Anatomical structures and physiological mechanisms occur in a series of interacting levels of organization. (*Figure 1–1*)

13. The 11 **organ systems** of the body are the integumentary, skeletal, muscular, nervous, endocrine, cardiovascular, lymphoid, respiratory, digestive, urinary, and reproductive systems. (*Figure 1–2*)

1-6 Homeostasis is the tendency toward internal balance p. 12

14. **Homeostasis** is the existence of a stable environment within the body.

15. Physiological systems preserve homeostasis through *homeostatic regulation*.

16. **Autoregulation** occurs when a cell, tissue, organ, or organ system adjusts its activities automatically in response to some environmental change. **Extrinsic regulation** results from the activities of the nervous system or endocrine system.

17. Homeostatic regulation mechanisms usually involve a **receptor** that is sensitive to a particular stimulus; a **control center**, which receives and processes the information supplied by the receptor, and then sends out commands; and an **effector** that responds to the commands of the control center and whose activity either opposes or enhances the stimulus. (*Figure 1–3*)

1-7 Negative feedback opposes variations from normal, whereas positive feedback exaggerates them p. 14

18. **Negative feedback** is a corrective mechanism involving an action that directly opposes a variation from normal limits. (*Figure 1–4*)

19. In **positive feedback**, an initial stimulus produces a response that exaggerates or enhances the change in the original conditions, creating a *positive feedback loop*. (*Figure 1–5*)

20. No one organ system has total control over the body's internal environment; all organ systems work in concert. (*Table 1–1*)

1-8 Anatomical terms describe body regions, anatomical positions and directions, and body sections p. 17

21. Standard anatomical illustrations show the human form in the **anatomical position**. If the figure is shown lying down, it can be either **supine** (face up) or **prone** (face down). (*Figure 1–6*)
22. **Abdominopelvic quadrants** and **abdominopelvic regions** represent two approaches to describing anatomical regions of that portion of the body. (*Figure 1–7; Table 1–2*)
23. The use of special directional terms provides clarity for the description of anatomical structures. (*Figure 1–8; Table 1–3*)
24. The three **sectional planes** (**transverse plane**; **frontal**, or *coronal*, **plane**; and **sagittal plane**) describe relationships among the parts of the three-dimensional human body. (*Figure 1–9; Table 1–4*)

1-9 Body cavities protect internal organs and allow them to change shape p. 22

25. *Body cavities* protect delicate organs and permit significant changes in the size and shape of internal organs. The **ventral body cavity**, or *coelom*, surrounds developing respiratory, cardiovascular, digestive, urinary, and reproductive organs. (*Figure 1–10*)
26. The **diaphragm** divides the ventral body cavity into the (superior) **thoracic** and (inferior) **abdominopelvic cavities**. The thoracic cavity consists of two **pleural cavities** (each containing a lung) separated by the **mediastinum**. Within the mediastinum is the **pericardial cavity**, which contains the heart. The abdominopelvic cavity consists of the **abdominal cavity** and the **pelvic cavity** and contains the *peritoneal cavity*, a chamber lined by the *peritoneum*, a *serous membrane*. (*Figure 1–11*)

Review Questions

See the blue Answers tab at the end of the book.

myA&P Access more review material online at **myA&P™** (www.myaandp.com). There, you'll find chapter guides, chapter quizzes, practice tests, animations, flashcards, a glossary with pronunciations, *Interactive Physiology*® (IP) exercises and quizzes, and more to help you succeed in the course.

LEVEL 1   Reviewing Facts and Terms

1. Label the directional terms in the figures below.

(a) _____
(b) _____
(c) _____
(d) _____
(e) _____
(f) _____
(g) _____
(h) _____
(i) _____
(j) _____

Match each numbered item with the most closely related lettered item. Use letters for answers in the spaces provided.

____ 2. cytology
____ 3. physiology
____ 4. histology
____ 5. metabolism
____ 6. homeostasis
____ 7. muscle
____ 8. heart
____ 9. endocrine
____ 10. temperature regulation
____ 11. labor and delivery
____ 12. supine
____ 13. prone
____ 14. divides ventral body cavity
____ 15. abdominopelvic cavity
____ 16. pericardium

(a) study of tissues
(b) constant internal environment
(c) face up
(d) study of functions
(e) positive feedback
(f) organ system
(g) study of cells
(h) negative feedback
(i) serous membrane
(j) all chemical activity in body
(k) diaphragm
(l) tissue
(m) peritoneal cavity
(n) organ
(o) face down

17. The following is a list of six levels of organization that make up the human body:
    1. tissue   2. cell
    3. organ   4. molecule
    5. organism   6. organ system
    The correct order, from the smallest to the largest level, is
    (a) 2, 4, 1, 3, 6, 5.   (b) 4, 2, 1, 3, 6, 5.
    (c) 4, 2, 1, 6, 3, 5.   (d) 4, 2, 3, 1, 6, 5.
    (e) 2, 1, 4, 3, 5, 6.
18. The study of the structure of tissue is called
    (a) gross anatomy.   (b) cytology.
    (c) histology.   (d) organology.
19. The increasingly forceful labor contractions during childbirth are an example of
    (a) receptor activation.   (b) effector shutdown.
    (c) negative feedback.   (d) positive feedback.
20. Failure of homeostatic regulation in the body results in
    (a) autoregulation.   (b) extrinsic regulation.
    (c) disease.   (d) positive feedback.
21. A plane through the body that passes perpendicular to the long axis of the body and divides the body into a superior and an inferior section is a
    (a) sagittal section.   (b) transverse section.
    (c) coronal section.   (d) frontal section.
22. In which body cavity would you find each of the following organs?
    (a) heart
    (b) small intestine, large intestine
    (c) lung
    (d) kidneys
23. The mediastinum is the region between the
    (a) lungs and heart.   (b) two pleural cavities.
    (c) chest and abdomen.   (d) heart and pericardium.

LEVEL 2   Reviewing Concepts

24. (a) Define *anatomy*.
    (b) Define *physiology*.
25. The two primary subdivisions of the ventral body cavity are the
    (a) pleural cavity and pericardial cavity.
    (b) coelom and peritoneal cavity.
    (c) pleural cavity and peritoneal cavity.
    (d) thoracic cavity and abdominopelvic cavity.

26. What distinguishes autoregulation from extrinsic regulation?
27. Describe the position of the body when it is in the anatomical position.
28. Which sectional plane could divide the body so that the face remains intact?
    (a) sagittal plane   (b) coronal plane
    (c) equatorial plane   (d) midsagittal plane
    (e) parasagittal plane
29. Which the following is *not* an example of negative feedback?
    (a) Increased pressure in the aorta triggers mechanisms to lower blood pressure.
    (b) A rise in blood calcium levels triggers the release of a hormone that lowers blood calcium levels.
    (c) A rise in estrogen during the menstrual cycle increases the number of progesterone receptors in the uterus.
    (d) Increased blood sugar stimulates the release of a hormone from the pancreas that stimulates the liver to store blood sugar.

LEVEL 3   Critical Thinking and Clinical Applications

30. The hormone *calcitonin* is released from the thyroid gland in response to increased levels of calcium ions in the blood. If this hormone is controlled by negative feedback, what effect would calcitonin have on blood calcium levels?
31. It is a warm day and you feel a little chilled. On checking your temperature, you find that your body temperature is 1.5 degrees C below normal. Suggest some possible reasons for this situation.

2

The Chemical Level of Organization

Learning Outcomes

After completing this chapter, you should be able to do the following:

2-1 **Describe an atom and how atomic structure affects interactions between atoms.**

2-2 **Compare the ways in which atoms combine to form molecules and compounds.**

2-3 **Use chemical notation to symbolize chemical reactions and distinguish among the major types of chemical reactions that are important for studying physiology.**

2-4 **Describe the crucial role of enzymes in metabolism.**

2-5 **Distinguish between organic and inorganic compounds.**

2-6 **Explain how the chemical properties of water make life possible.**

2-7 **Discuss the importance of pH and the role of buffers in body fluids.**

2-8 **Describe the physiological roles of inorganic compounds.**

2-9 **Discuss the structures and functions of carbohydrates.**

2-10 **Discuss the structures and functions of lipids.**

2-11 **Discuss the structures and functions of proteins.**

2-12 **Discuss the structures and functions of nucleic acids.**

2-13 **Discuss the structures and functions of high-energy compounds.**

2-14 **Explain the relationship between chemicals and cells.**

Clinical Notes

Solute Concentrations p. 43
Fatty Acids and Health p. 49

An Introduction to the Chemical Level of Organization

This chapter considers the structure of atoms, the basic chemical building blocks. It also shows how atoms can be combined to form increasingly complex structures.

2-1 Atoms are the basic particles of matter

Our study of the human body begins at the chemical level of organization. *Chemistry* is the science that deals with the structure of *matter*, defined as anything that takes up space and has mass. *Mass*, the amount of material in matter, is a physical property that determines the weight of an object in Earth's gravitational field. For our purposes, the mass of an object is the same as its weight. However, the two are not always equivalent: In orbit you would be weightless, but your mass would remain unchanged.

The smallest stable units of matter are called **atoms**. Air, elephants, oranges, oceans, rocks, and people are all composed of atoms in varying combinations. The unique characteristics of each object, living or nonliving, result from the types of atoms involved and the ways those atoms combine and interact.

Atoms are composed of **subatomic particles**. Although many different subatomic particles exist, only three are important for understanding the chemical properties of matter: *protons, neutrons,* and *electrons.* Protons and neutrons are similar in size and mass, but **protons** (p^+) have a positive electrical charge, whereas **neutrons** (n or n^0) are electrically *neutral,* or uncharged. **Electrons** (e^-) are much lighter than protons—only 1/1836 as massive—and bear a negative electrical charge. The mass of an atom is therefore determined primarily by the number of protons and neutrons in the nucleus, the central region of an atom. The mass of a large object, such as your body, is the sum of the masses of all the component atoms.

The A&P Top 100

#6 All matter is composed of atoms in various combinations. Cellular physiology is based on rules governing the interactions of these atoms.

Atomic Structure

Atoms normally contain equal numbers of protons and electrons. The number of protons in an atom is known as its **atomic number**. *Hydrogen* (H) is the simplest atom, with an atomic number of 1. Thus, an atom of hydrogen contains one proton, and one electron as well. Hydrogen's proton is located in the center of the atom and forms the nucleus. Hydrogen atoms seldom contain neutrons, but when neutrons are present, they are also located in the nucleus. All atoms other than hydrogen have both neutrons and protons in their nuclei.

Electrons travel around the nucleus at high speed, forming a spherical **electron cloud**. We often illustrate atomic structure in the simplified form shown in **Figure 2–1a**. In this two-dimensional representation, the electrons occupy a circular **electron shell**. One reason an electron tends to remain in its electron shell is that the negatively charged electron is attracted to the positively charged proton. The attraction between opposite electrical charges is an example of an *electrical force*. As you will see in later chapters, electrical forces are involved in many physiological processes.

The dimensions of the electron cloud determine the overall size of the atom. To get an idea of the scale involved, consider that if the nucleus were the size of a tennis ball, the electron cloud of a hydrogen atom would have a radius of 10 m

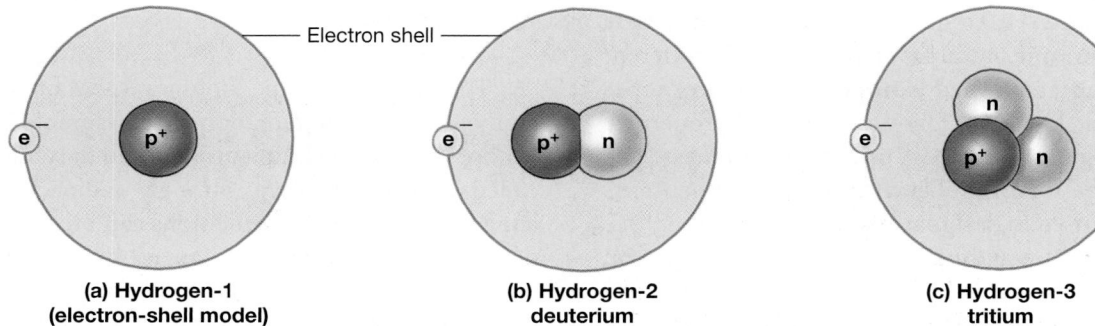

| (a) Hydrogen-1 (electron-shell model) | (b) Hydrogen-2 deuterium | (c) Hydrogen-3 tritium |

Figure 2–1 The Structure of Hydrogen Atoms. Three forms of hydrogen atoms are depicted using the two-dimensional electron-shell model, which indicates the location of the electron cloud surrounding the nucleus. **(a)** A typical hydrogen nucleus contains a proton and no neutrons. **(b)** A deuterium (^{2}H) nucleus contains a proton and a neutron. **(c)** A tritium (^{3}H) nucleus contains a pair of neutrons in addition to the proton.

(roughly 6 miles!). In reality, atoms are so small that atomic measurements are most conveniently reported in *nanometers* (nan-OM-e-ter; nan-ō-MĒ-ter) (*nm*). One nanometer is 10^{-9} meter (0.000000001 m). The very largest atoms approach 0.5 nm in diameter (0.0000000005 m, or 0.00000002 in.).

Elements and Isotopes

An **element** is a pure substance composed of atoms of only one kind; because atoms are the smallest particles of an element that still retain the characteristics of that element, each element has uniform composition and properties. Each element includes all the atoms that have the same number of protons, and thus the same atomic number. Only 92 elements exist in nature, although about two dozen additional elements have been created through nuclear reactions in research laboratories. Every element has a chemical symbol, an abbreviation recognized by scientists everywhere. Most of the symbols are easily connected with the English names of the elements (O for oxygen, N for nitrogen, C for carbon, and so on), but a few are abbreviations of their Latin names. For example, the symbol for sodium, Na, comes from the Latin word *natrium*. (The *periodic table* in the Appendix gives the chemical symbol and atomic number of each element.)

Because atomic nuclei are unaltered by ordinary chemical processes, elements cannot be changed or broken down into simpler substances in chemical reactions. Thus, an atom of carbon always remains an atom of carbon, regardless of the chemical events in which it may take part.

The relative contributions of the 13 most abundant elements in the human body to total body weight are shown in Table 2–1. The human body also contains atoms of another 14 elements—called *trace elements*—that are present in very small amounts.

The atoms of a single element can differ in the number of neutrons in the nucleus. For example, although most hydrogen nuclei consist of a single proton, 0.015 percent also contain one neutron, and a very small percentage contain two neutrons (**Figure 2–1**). Atoms of the same element whose nuclei contain the same number of protons, but different numbers of neutrons, are called **isotopes**. Different isotopes of an element have essentially identical chemical properties, and so are indistinguishable except on the basis of mass. The **mass number**—the total number of protons plus neutrons in the nucleus—is used to designate isotopes. Thus, the three isotopes of hydrogen are hydrogen-1, or ^1H, with one proton and one electron (**Figure 2–1a**); hydrogen-2, or ^2H, also known as *deuterium*, with one proton, one electron, and one neutron (**Figure 2–1b**); and hydrogen-3, or ^3H, also known as *tritium*, with one proton, one electron, and two neutrons (**Figure 2–1c**).

TABLE 2–1 Principal Elements in the Human Body	
Element (% of total body weight)	**Significance**
Oxygen, O (65)	A component of water and other compounds; gaseous form is essential for respiration
Carbon, C (18.6)	Found in all organic molecules
Hydrogen, H (9.7)	A component of water and most other compounds in the body
Nitrogen, N (3.2)	Found in proteins, nucleic acids, and other organic compounds
Calcium, Ca (1.8)	Found in bones and teeth; important for membrane function, nerve impulses, muscle contraction, and blood clotting
Phosphorus, P (1.0)	Found in bones and teeth, nucleic acids, and high-energy compounds
Potassium, K (0.4)	Important for proper membrane function, nerve impulses, and muscle contraction
Sodium, Na (0.2)	Important for blood volume, membrane function, nerve impulses, and muscle contraction
Chlorine, Cl (0.2)	Important for blood volume, membrane function, and water absorption
Magnesium, Mg (0.06)	A cofactor for many enzymes
Sulfur, S (0.04)	Found in many proteins
Iron, Fe (0.007)	Essential for oxygen transport and energy capture
Iodine, I (0.0002)	A component of thyroid hormones
Trace elements: silicon (Si), fluorine (F), copper (Cu), manganese (Mn), zinc (Zn), selenium (Se), cobalt (Co), molybdenum (Mo), cadmium (Cd), chromium (Cr), tin (Sn), aluminum (Al), boron (B), and vanadium (V)	Some function as cofactors; the functions of many trace elements are poorly understood

The nuclei of some isotopes spontaneously emit subatomic particles or radiation in measurable amounts. Such isotopes are called **radioisotopes**, and the emission process is called *radioactive decay*. Strongly radioactive isotopes are dangerous, because the emissions can break molecules apart, destroy cells, and otherwise damage living tissues. Weakly radioactive isotopes are sometimes used in diagnostic procedures to monitor the structural or functional characteristics of internal organs.

Radioisotopes differ in how rapidly they decay. The decay rate of a radioisotope is commonly expressed in terms of its

half-life: the time required for half of a given amount of the isotope to decay. The half-lives of radioisotopes range from fractions of a second to billions of years.

Atomic Weights

A typical *oxygen* atom, which has an atomic number of 8, contains eight protons and eight neutrons. The mass number of this isotope is therefore 16. The mass numbers of other isotopes of oxygen depend on the number of neutrons present. Mass numbers are useful because they tell us the number of subatomic particles in the nuclei of different atoms. However, they do not tell us the *actual* mass of the atoms. For example, they do not take into account the masses of the electrons or the slight difference between the mass of a proton and that of a neutron. The actual mass of an atom is known as its **atomic weight**.

The unit used to express atomic weight is the **dalton** (also known as the *atomic mass unit*, or *amu*). One dalton is very close to the weight of a single proton or neutron. Thus, the atomic weight of the most common isotope of hydrogen is very close to 1, and that of the most common isotope of oxygen is very close to 16.

The atomic weight of an element is an average mass number that reflects the proportions of different isotopes. That is, the atomic weight of an element is usually very close to the mass number of the most common isotope of that element. For example, the atomic number of hydrogen is 1, but the atomic weight of hydrogen is 1.0079, primarily because some hydrogen atoms (0.015 percent) have a mass number of 2, and even fewer have a mass number of 3. The atomic weights of the elements are included in the Appendix.

Atoms participate in chemical reactions in fixed numerical ratios. To form water, for example, exactly two atoms of hydrogen combine with one atom of oxygen. But individual atoms are far too small and too numerous to be counted, so chemists use a unit called the *mole*. For any element, a **mole** (abbreviated *mol*) is a quantity with a weight in grams equal to that element's atomic weight. The mole is useful because one mole of a given element always contains the same number of atoms as one mole of any other element. That number (called *Avogadro's number*) is 6.023×10^{23}, or about 600 billion trillion. Expressing relationships in moles rather than in grams makes it much easier to keep track of the relative numbers of atoms in chemical samples and processes. For example, if a report stated that a sample contains 0.5 mol of hydrogen atoms and 0.5 mol of oxygen atoms, you would know immediately that they were present in equal numbers. That would not be so evident if the report stated that there were 0.505 g of hydrogen atoms and 8.00 g of oxygen atoms. Most chemical analyses and clinical laboratory tests report data in moles (mol), millimoles (mmol—1/1000 mol, or 10^{-3}

mol), or micromoles (μmol—1/1,000,000 mol, or 10^{-6} mol). (Additional information on metric weights and measures can be found in the Appendix.)

Electrons and Energy Levels

Atoms are electrically neutral; every positively charged proton is balanced by a negatively charged electron. Thus, each increase in the atomic number is accompanied by a comparable increase in the number of electrons traveling around the nucleus. Within the electron cloud, electrons occupy an orderly series of energy levels. Although the electrons in an energy level may travel in complex patterns around the nucleus, for our purposes the patterns can be diagrammed as a series of concentric electron shells. The first electron shell (the one closest to the nucleus) corresponds to the lowest energy level.

Each energy level is limited in the number of electrons it can hold. The first energy level can hold at most two electrons, and for our purposes, the next two levels can each hold up to eight electrons. Note that the maximum number of electrons that may occupy shells 1–3 corresponds to the number of elements in rows 1–3, respectively, of the periodic table. The electrons in an atom occupy successive shells in an orderly manner: The first energy level is filled before any electrons enter the second, and the second energy level is filled before any electrons enter the third.

The outermost energy level forms the "surface" of the atom and is called the valence shell. The number of electrons in this level determines the chemical properties of the element. Atoms with unfilled energy levels are unstable—that is, they will react with other atoms, usually in ways that give them full outer electron shells. In contrast, atoms with a filled outermost energy level are stable and therefore do not readily react with other atoms.

As indicated in **Figure 2–2a**, a hydrogen atom has one electron in the first energy level, and that level is thus unfilled. A hydrogen atom readily reacts with other atoms. A helium atom has two electrons in its first energy level (**Figure 2–2b**). Because its outer energy level is filled, a helium atom is very stable; it will not ordinarily react with other atoms. A lithium atom has three electrons (**Figure 2–2c**). Its first energy level can hold only two of them, so lithium has a single electron in a second, unfilled energy level. As you would expect, lithium is extremely reactive. The second energy level is filled in a neon atom, which has an atomic number of 10 (**Figure 2–2d**). Neon atoms, like helium atoms and other elements in the last column of the periodic table, are thus very stable. The atoms that are most important to biological systems are unstable, because those atoms interact to form larger structures (Table 2–1). (Further information on these and other elements is given in the Appendix.)

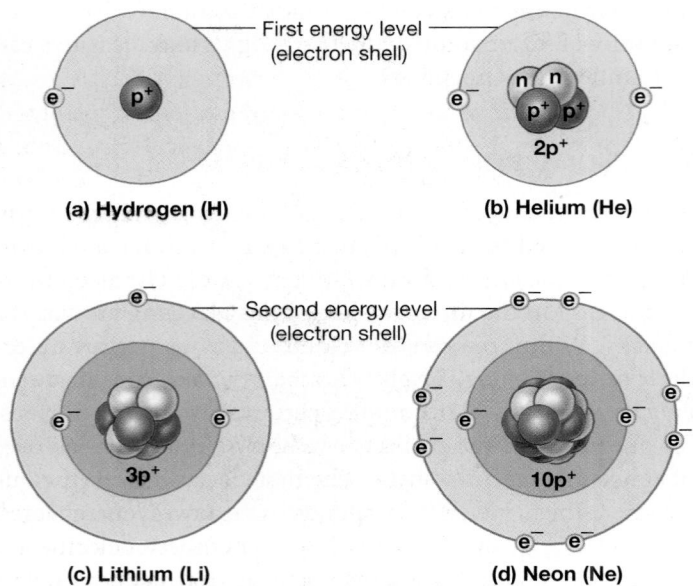

Figure 2–2 **The Arrangement of Electrons into Energy Levels.** (a) A typical hydrogen atom has one proton and one electron. The electron orbiting the nucleus occupies the first energy level, diagrammed as an electron shell. (b) An atom of helium has two protons, two neutrons, and two electrons. The two electrons orbit in the same energy level. (c) A lithium atom has three protons, four neutrons, and three electrons. The first energy level can hold only two electrons, so the third electron occupies a second energy level. (d) A neon atom has 10 protons, 10 neutrons, and 10 electrons. The second level can hold up to eight electrons; thus, both the first and second energy levels are filled.

CHECKPOINT

1. Define atom.

2. Atoms of the same element that have different numbers of neutrons are called _____.

3. How is it possible for two samples of hydrogen to contain the same number of atoms, yet have different weights?

See the blue Answers tab at the end of the book.

2-2 Chemical bonds are forces formed by atom interactions

Elements that do not readily participate in chemical processes are said to be *inert*. Helium, neon, and argon have filled outermost energy levels. These elements are called *inert gases*, because their atoms neither react with one another nor combine with atoms of other elements. Elements with unfilled outermost energy levels, such as hydrogen and lithium, are called *reactive*, because they readily interact or combine with other atoms. In doing so, these atoms achieve stability by gaining, losing, or sharing electrons to fill their outermost energy level. The interactions often involve the formation of **chemical bonds**, which hold the participating atoms together once the reaction has ended. In the sections that follow, we will consider three basic types of chemical bonds: *ionic bonds*, *covalent bonds*, and *hydrogen bonds*.

When chemical bonding occurs, the result is the creation of new chemical entities called *molecules* and *compounds*. The term **molecule** refers to any chemical structure consisting of atoms held together by covalent bonds. A **compound** is a pure chemical substance made up of atoms of two or more different elements, regardless of the type of bond joining them. The two categories overlap, but they aren't the same. Not all molecules are compounds, because some molecules consist of atoms of only one element. (Two oxygen atoms, for example, can be joined by a covalent bond to form a molecule of oxygen.) And not all compounds consist of molecules, because some compounds, such as ordinary salt (sodium chloride) are held together by ionic rather than covalent bonds. Many substances, however, belong to both categories. Water is a compound because it contains two different elements—hydrogen and oxygen—and it consists of molecules, because the hydrogen and oxygen atoms are held together by covalent bonds. As we will see in subsequent sections, most biologically important compounds, from water to DNA, are molecular.

Regardless of the type of bonding involved, a chemical compound has properties that can be quite different from those of its components. For example, a mixture of hydrogen gas and oxygen gas can explode, but the explosion is a chemical reaction that produces liquid water, a compound used to put out fires.

The human body consists of countless molecules and compounds, so it is a challenge to describe these substances and their varied interactions. Fortunately, chemists rely on a standardized system of *chemical notation*. The very useful rules of this system are listed in "FOCUS: Chemical Notation," p. 37.

Ionic Bonds

As the name implies, ionic bonds form between ions. **Ions** are atoms or molecules that carry an electric charge, either positive or negative. Ions with a positive charge (+) are called **cations** (KAT-ī-onz); ions with a negative charge (−) are called **anions** (AN-ī-onz). **Ionic bonds** are chemical bonds created by the electrical attraction between anions and cations.

Tips&Tricks

Think of the *t* in ca*t*ion as a plus sign (+) to remember that a ca*t*ion has a positive charge, and think of the *n* in a*n*ion as standing for *n*egative to remember that a*n*ions have a *n*egative charge.

Ions have an unequal number of protons and electrons. Atoms become ions by losing or gaining electrons. We assign a value of $+1$ to the charge on a proton; the charge on an electron is -1. When the number of protons is equal to the number of electrons, an atom is electrically neutral. An atom that loses an electron becomes a cation with a charge of $+1$, because it then has one proton that lacks a corresponding electron. Losing a second electron would give the cation a charge of $+2$. Adding an extra electron to a neutral atom produces an anion with a charge of -1; adding a second electron gives the anion a charge of -2.

In the formation of an ionic bond,

• one atom—the *electron donor*—loses one or more electrons and becomes a cation, with a positive $(+)$ charge.

• another atom—the *electron acceptor*—gains those same electrons and becomes an anion, with a negative $(-)$ charge.

• attraction between the opposite charges then draws the two ions together.

The formation of an ionic bond is illustrated in **Figure 2–3a.** The sodium atom diagrammed in STEP 1 has an atomic number of 11, so this atom normally contains 11 protons and 11 electrons. (Because neutrons are electrically neutral, their presence has no effect on the formation of ions or ionic bonds.) Electrons fill the first and second energy levels, and a single electron occupies the outermost level. Losing that one electron would give the sodium atom a full outermost energy level—the second level—and would produce a **sodium ion**, with a charge of $+1$. (The chemical shorthand for a sodium ion is Na^+.) But a sodium atom cannot simply throw away the electron: The electron must be donated to an electron acceptor. A chlorine atom has seven electrons in its outermost energy level, so it needs only one electron to achieve stability. A sodium atom can provide the extra electron. In the process (STEP 1), the chlorine atom becomes a **chloride ion** (Cl^-) with a charge of -1.

Both atoms have now become stable ions with filled outermost energy levels. But the two ions do not move apart after the electron transfer, because the positively charged sodium ion is attracted to the negatively charged chloride ion (STEP 2). The combination of oppositely charged ions forms an *ionic compound*—in this case, **sodium chloride**, otherwise known as table salt (STEP 3). Large numbers of sodium and chloride ions interact to form highly structured crystals, held together by the strong electrical attraction of oppositely charged ions (**Figure 2–3b**). Note that ionic compounds, because they consist of an aggregation of ions rather than covalently bonded atoms, are not called molecules. Although

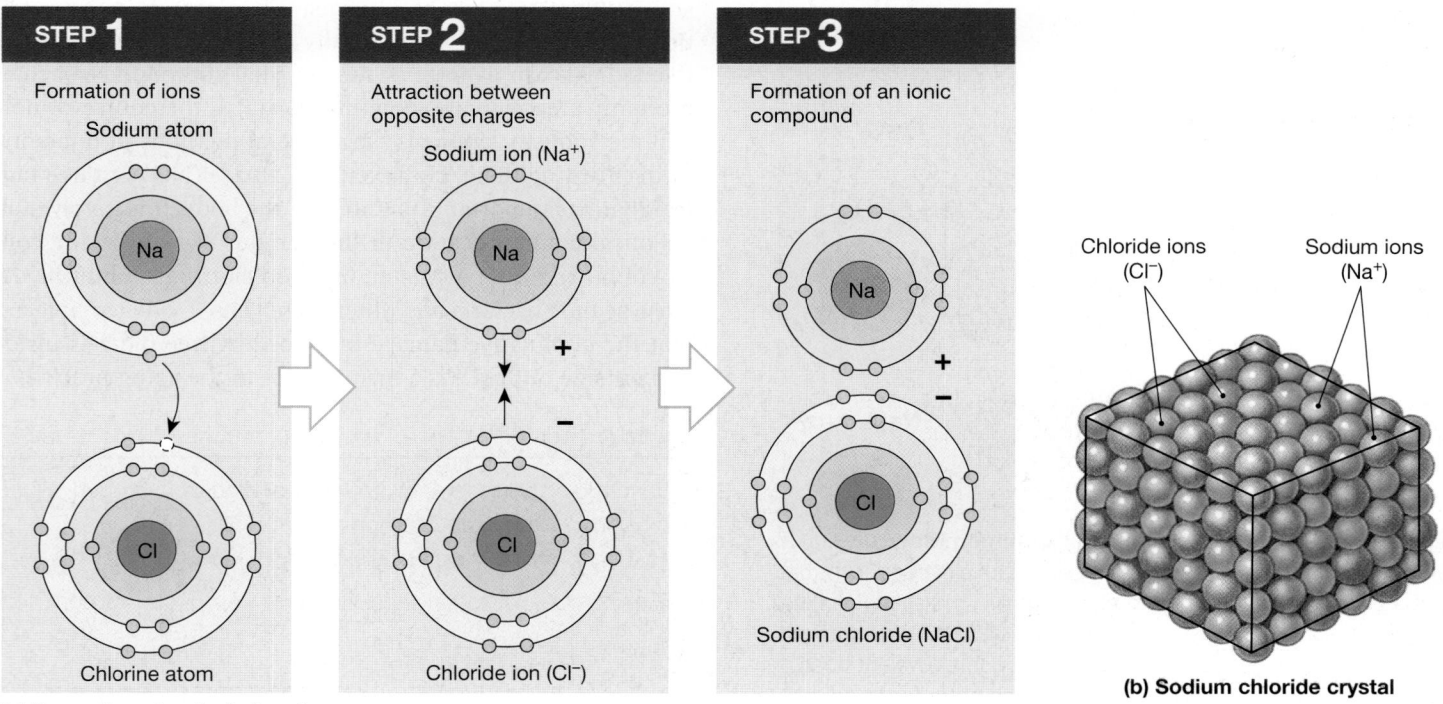

(a) Formation of an ionic bond

(b) Sodium chloride crystal

Figure 2–3   The Formation of Ionic Bonds. (a) STEP 1: A sodium (Na) atom loses an electron, which is accepted by a chlorine (Cl) atom. **STEP 2:** Because the sodium (Na^+) and chloride (Cl^-) ions have opposite charges, they are attracted to one another. **STEP 3:** The association of sodium and chloride ions forms the ionic compound sodium chloride. **(b)** Large numbers of sodium and chloride ions form a crystal of sodium chloride (table salt).

sodium chloride and other ionic compounds are common in body fluids, they are not present as intact crystals. When placed in water, many ionic compounds dissolve, and some of the component anions and cations separate.

Covalent Bonds

Some atoms can complete their outer electron shells not by gaining or losing electrons, but by sharing electrons with other atoms. Such sharing creates **covalent** (kō-VĀ-lent) **bonds** between the atoms involved.

Individual hydrogen atoms, as diagrammed in **Figure 2–2a**, do not exist in nature. Instead, we find hydrogen molecules. Molecular hydrogen consists of a pair of hydrogen atoms (**Figure 2–4**). In chemical shorthand, molecular hydrogen is indicated by H_2, where H is the chemical symbol for hydrogen, and the subscript 2 indicates the number of atoms. Molecular hydrogen is a gas that is present in the atmosphere in very small quantities. The two hydrogen atoms share their electrons,

MOLECULE	ELECTRON-SHELL MODEL AND STRUCTURAL FORMULA
Hydrogen (H_2)	H–H
Oxygen (O_2)	O=O
Carbon dioxide (CO_2)	O=C=O
Nitric oxide (NO)	N=O

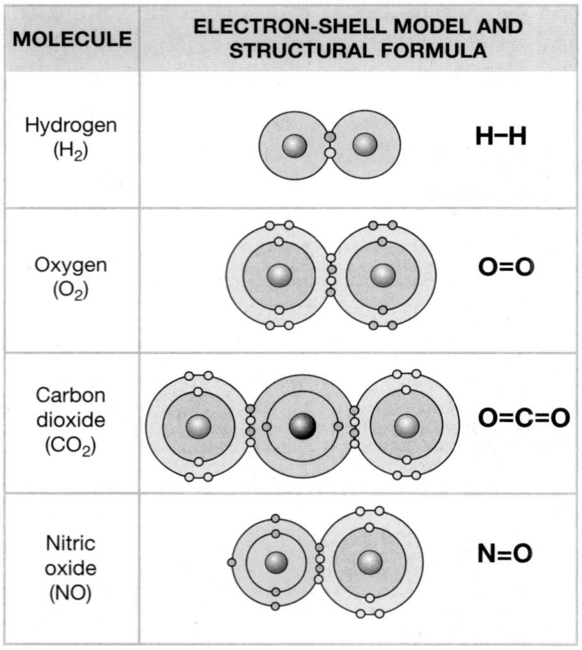

Figure 2–4   Covalent Bonds in Four Common Molecules. In a hydrogen molecule, two hydrogen atoms share electrons such that each atom has a filled outermost electron shell. This sharing creates a single covalent bond. In an oxygen molecule, two oxygen atoms share two pairs of electrons. The result is a double covalent bond. In a carbon dioxide molecule, a central carbon atom forms double covalent bonds with two oxygen atoms. A nitric oxide molecule is held together by a double covalent bond, but the outer electron shell of the nitrogen atom requires an additional electron to be complete. Thus, nitric oxide is a free radical, which reacts readily with another atom or molecule.

and each electron whirls around both nuclei. The sharing of one pair of electrons creates a **single covalent bond** (—).

Oxygen, with an atomic number of 8, has two electrons in its first energy level and six in its second. The oxygen atoms diagrammed in **Figure 2–4** attain a stable electron configuration by sharing two pairs of electrons, thereby forming a **double covalent bond**. In a structural formula, a double covalent bond is represented by two lines (=). Molecular oxygen (O_2) is an atmospheric gas that is very important to most organisms. Our cells would die without a relatively constant supply of oxygen.

In our bodies, chemical processes that consume oxygen generally also produce **carbon dioxide** (CO_2) as a waste product. Each of the oxygen atoms in a carbon dioxide molecule forms double covalent bonds with the carbon atom.

A triple covalent bond, such as the one joining two nitrogen molecules (N_2), is indicated by three lines (≡). Molecular nitrogen accounts for roughly 79 percent of our planet's atmosphere, but our cells ignore it completely. In fact, deep-sea divers live for long periods while breathing artificial air that does not contain nitrogen. (We will discuss the reasons for eliminating nitrogen under these conditions in Chapter 23.) Covalent bonds usually form molecules in which the outer energy levels of the atoms involved are complete. An ion or molecule that contains unpaired electrons in its outermost energy level is called a *free radical*. Free radicals are highly reactive. Almost as fast as it forms, a free radical enters additional reactions that are typically destructive. For example, free radicals can damage or destroy vital compounds, such as proteins. Free radicals sometimes form in the course of normal metabolism, but cells have several methods of removing or inactivating them. However, *nitric oxide* (NO) is a free radical that has important functions in the body. It is involved in chemical communication in the nervous system, in the control of blood vessel diameter, in blood clotting, and in the defense against bacteria and other pathogens. Evidence suggests that the cumulative damage produced by free radicals inside and outside our cells is a major factor in the aging process.

Tips&Tricks

Remember this mnemonic for the bonding of hydrogen, oxygen, nitrogen, and carbon atoms: HONC 1234. **H**ydrogen shares *1* pair of electrons (H–), **o**xygen shares *2* pairs (–O–), **n**itrogen shares *3* pairs ($-\overset{|}{N}-$), and **c**arbon shares *4* pairs ($-\overset{|}{\underset{|}{C}}-$).

Nonpolar Covalent Bonds

Covalent bonds are very strong, because the shared electrons hold the atoms together. In typical covalent bonds the atoms remain electrically neutral, because each shared electron spends just as much time "at home" as away. (If you and a friend were tossing a pair of baseballs back and forth as fast as you could, on average, each of you would have just one baseball.) Many covalent bonds involve an equal sharing of electrons. Such bonds—which occur, for instance, between two atoms of the same type—are called **nonpolar covalent bonds**. Nonpolar covalent bonds are very common; those involving carbon atoms form most of the structural components of the human body.

Polar Covalent Bonds

Covalent bonds involving different types of atoms may instead involve an unequal sharing of electrons, because the elements differ in how strongly they attract electrons. An unequal sharing of electrons creates a **polar covalent bond**. For example, in a molecule of water (**Figure 2–5**), an oxygen atom forms covalent bonds with two hydrogen atoms. The oxygen nucleus has a much stronger attraction for the shared electrons than the hydrogen atoms do. As a result, the electrons spend more time orbiting the oxygen nucleus than orbiting the hydrogen nuclei. Because the oxygen atom has two extra electrons most of the time, it develops a slight (partial) negative charge, indicated by δ^-. At the same time, each hydrogen atom develops a slight (partial) positive charge, δ^+, because its electron is away much of the time. (Suppose you and a friend were tossing a pair of baseballs back and forth, but one of you returned them back as fast as

possible while the other held onto them for a while before throwing them back. One of you would now, on average, have more than one baseball, and the other would have less than one.) The unequal sharing of electrons makes polar covalent bonds somewhat weaker than nonpolar covalent bonds. Polar covalent bonds often create *polar molecules*— molecules that have positive and negative ends. Polar molecules have very interesting properties; we will consider the characteristics of the most important polar molecule in the body, water, in a later section.

Hydrogen Bonds

Covalent and ionic bonds tie atoms together to form molecules and/or compounds. Other, comparatively weak forces also act between adjacent molecules, and even between atoms within a large molecule. The most important of these weak attractive forces is the **hydrogen bond**. A hydrogen bond is the attraction between a δ^+ on the hydrogen atom of a polar covalent bond and a δ^- on an oxygen, nitrogen, or fluorine atom of another polar covalent bond. The polar covalent bond containing the oxygen, nitrogen, or fluorine atom can be in a different molecule from, or in the same molecule as, the hydrogen atom. Hydrogen bonds are too weak to create molecules, but they can change molecular shapes or pull molecules together. For example, hydrogen bonding occurs between water molecules (**Figure 2–6**). At the water surface, the attraction between molecules slows the rate of evaporation and creates the phenomenon known as surface tension. **Surface tension** acts as a barrier that keeps small objects from entering the water. For example, it allows insects to walk across the surface of a pond or puddle. Similarly, small objects such as dust particles are prevented from touching the surface of the eye by the surface tension in a layer of tears. At the cellular level, hydrogen bonds affect the shapes and properties of complex molecules, such as proteins and nucleic acids (including DNA), and they may also determine the three-dimensional relationships between molecules.

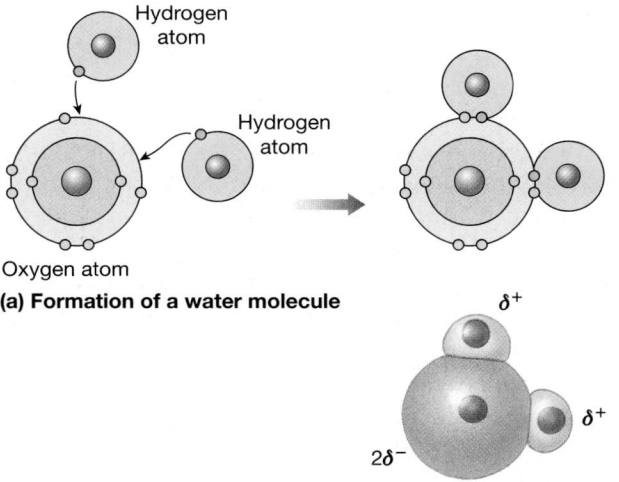

(a) Formation of a water molecule

Hydrogen atom

Hydrogen atom

Oxygen atom

δ^+

δ^+

$2\delta^-$

(b) Charges on a water molecule

Figure 2–5   Polar Covalent Bonds and the Structure of Water.
(a) In forming a water molecule, an oxygen atom completes its outermost energy level by sharing electrons with a pair of hydrogen atoms. The sharing is unequal, because the oxygen atom holds the electrons more tightly than do the hydrogen atoms. **(b)** Because the oxygen atom has two extra electrons much of the time, it develops a slight negative charge, and the hydrogen atoms become weakly positive. The bonds in a water molecule are polar covalent bonds.

States of Matter

Most matter in our environment exists in one of three states: solid, liquid, or gas. *Solids* maintain their volume and their shape at ordinary temperatures and pressures. A lump of granite, a brick, and a textbook are solid objects. *Liquids* have a constant volume, but no fixed shape. The shape of a liquid is determined by the shape of its container. Water, coffee, and soda are liquids. A *gas* has neither a constant volume nor a fixed shape. Gases can be compressed or expanded; unlike liquids they will fill a container of any size. The air of our atmosphere is the gas with which we are most familiar.

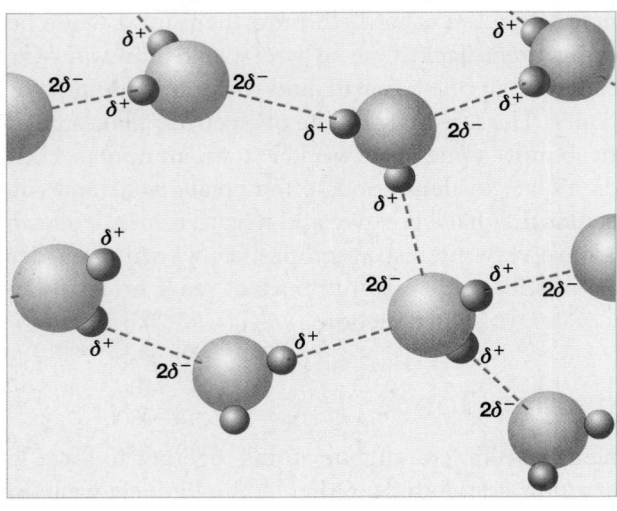

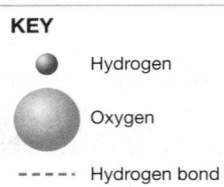

KEY

● Hydrogen

● Oxygen

----- Hydrogen bond

Figure 2–6 **Hydrogen Bonds between Water Molecules.**
The hydrogen atoms of a water molecule have a slight positive
charge, and the oxygen atom has a slight negative charge. *(See
Figure 2–5b.)* The distances between these molecules have been
exaggerated for clarity.

Whether a particular substance is a solid, a liquid, or a gas
depends on the degree of interaction among its atoms or molecules. The particles of a solid are placed tightly together, while
those of a gas are very far apart. Water is the only substance that
occurs as a solid (ice), a liquid (water), and a gas (water vapor)
at temperatures compatible with life. Water exists in the liquid
state over a broad range of temperatures primarily because of
hydrogen bonding among the water molecules. We will talk
more about the unusual properties of water in a later section.

Molecular Weights

The **molecular weight** of a molecule is the sum of the atomic
weights of its component atoms. It follows from the definition
of the mole given previously that the molecular weight of a molecule in grams is equal to the weight of one mole of molecules.
Molecular weights are important because you can neither handle individual molecules nor easily count the billions of molecules involved in chemical reactions in the body. Using
molecular weights, you can calculate the quantities of reactants
needed to perform a specific reaction and determine the amount
of product generated. For example, suppose you want to create
water from hydrogen and oxygen according to the equation

$$2H_2 + O_2 \longrightarrow 2H_2O$$

The first step would be to calculate the molecular weights involved. The atomic weight of hydrogen is close to 1.0, so one
hydrogen molecule (H_2) has a molecular weight near 2.0.
Oxygen has an atomic weight of about 16, so the molecular
weight of an oxygen molecule (O_2) is roughly 32. Thus you
would combine 4 g of hydrogen with 32 g of oxygen to produce 36 g of water. You could also work with ounces, pounds,
or tons, as long as the proportions remained the same.

CHECKPOINT

4. Define chemical bond and identify several types of
chemical bonds.

5. Which kind of bond holds atoms in a water molecule
together? What attracts water molecules to one
another?

6. Both oxygen and neon are gases at room temperature.
Oxygen combines readily with other elements, but
neon does not. Why?

See the blue Answers tab at the end of the book.

2-3 Decomposition, synthesis, and exchange reactions are important chemical reactions in physiology

Cells remain alive and functional by controlling chemical reactions. In a **chemical reaction**, new chemical bonds form
between atoms, or existing bonds between atoms are broken.
These changes occur as atoms in the reacting substances, or
reactants, are rearranged to form different substances, or
products. (See "FOCUS: Chemical Notation," p. 37.)

In effect, each cell is a chemical factory. Growth, maintenance and repair, secretion, and contraction all involve complex chemical reactions. Cells use chemical reactions to
provide the energy needed to maintain homeostasis and to
perform essential functions. All of the reactions under way in
the cells and tissues of the body at any given moment constitute its **metabolism** (me-TAB-ō-lizm).

Basic Energy Concepts

An understanding of some basic relationships between matter and energy is essential for any discussion of chemical reactions. **Work** is the movement of an object or a change in the
physical structure of matter. In your body, work includes
movements like walking or running, and also the synthesis of

Focus

Chemical Notation

Before we can consider the specific compounds that occur in the human body, we must be able to describe chemical compounds and reactions effectively. The use of sentences to describe chemical structures and events often leads to confusion. A simple form of "chemical shorthand" makes communication much more efficient. The chemical shorthand we will use is known as **chemical notation**. Chemical notation enables us to describe complex events briefly and precisely; its rules are summarized in Table 2–2.

TABLE 2–2   Rules of Chemical Notation

1. The symbol of an element indicates one atom of that element:

   H = one atom of hydrogen

   O = one atom of oxygen

2. A number preceding the symbol of an element indicates more than one atom of that element:

   2H = two atoms of hydrogen

   2O = two atoms of oxygen

3. A subscript following the symbol of an element indicates a molecule with that number of atoms of that element:

   H_2 = hydrogen molecule, composed of two hydrogen atoms

   O_2 = oxygen molecule, composed of two oxygen atoms

   H_2O = water molecule, composed of two hydrogen atoms and one oxygen atom

4. In a description of a chemical reaction, the participants at the start of the reaction are called reactants, and the reaction generates one or more products. An arrow indicates the direction of the reaction, from reactants (usually on the left) to products (usually on the right). In the following reaction, two atoms of hydrogen combine with one atom of oxygen to produce a single molecule of water:

   $2H + O \longrightarrow H_2O$

5. A superscript plus or minus sign following the symbol of an element indicates an ion. A single plus sign indicates a cation with a charge of +1. (The original atom has lost one electron.) A single minus sign indicates an anion with a charge of −1. (The original atom has gained one electron.) If more than one electron has been lost or gained, the charge on the ion is indicated by a number preceding the plus or minus sign:

   Na^+ = sodium ion [the sodium atom has lost one electron]

   Cl^- = chloride ion [the chlorine atom has gained one electron]

   Ca^{2+} = calcium ion [the calcium atom has lost two electrons]

6. Chemical reactions neither create nor destroy atoms; they merely rearrange atoms into new combinations. Therefore, the numbers of atoms of each element must always be the same on both sides of the equation for a chemical reaction. When this is the case, the equation is balanced:

   Unbalanced : $H_2 + O_2 \longrightarrow H_2O$

   Balanced : $2H_2 + O_2 \longrightarrow 2H_2O$

organic (carbon containing) molecules and the conversion of liquid water to water vapor (evaporation). **Energy** is the capacity to perform work; movement or physical change cannot occur unless energy is provided. The two major types of energy are kinetic energy and potential energy:

1. **Kinetic energy** is the energy of motion—energy that can be transferred to another object and perform work. When you fall off a ladder, it is kinetic energy that does the damage.

2. **Potential energy** is stored energy—energy that has the potential to do work. It may derive from an object's position (you standing on a ladder) or from its physical or chemical structure (a stretched spring or a charged battery).

Kinetic energy must be used in climbing the ladder, in stretching the spring, or in charging the battery. The resulting potential energy is converted back into kinetic energy when you descend, the spring recoils, or the battery discharges. The kinetic energy can then be used to perform work. For example, in an MP3 player, the chemical potential energy stored in small batteries is converted to kinetic energy that vibrates the sound-producing membranes in headphones or external speakers.

Energy cannot be destroyed; it can only be converted from one form to another. A conversion between potential energy and kinetic energy is never 100 percent efficient. Each time an energy exchange occurs, some of the energy is released in the form of heat. *Heat* is an increase in random molecular motion; the temperature of an object is proportional to the average kinetic energy of its molecules. Heat can never be completely converted to work or any other form of energy, and cells cannot capture it or use it to perform work.

Cells perform work as they synthesize complex molecules and move materials into, out of, and within the cell. The cells of a skeletal muscle at rest, for example, contain potential energy in the form of the positions of protein filaments and the covalent bonds between molecules inside the cells. When a muscle contracts, it performs work; potential energy is converted into kinetic energy, and heat is released. The amount of heat is proportional to the amount of work done. As a result, when you exercise, your body temperature rises.

The A&P Top 100

#7 When energy is exchanged, heat is produced. Heat raises local temperatures, but cells cannot capture it or use it to perform work.

Types of Chemical Reactions

Three types of chemical reactions are important to the study of physiology: decomposition reactions, synthesis reactions, and exchange reactions.

Decomposition Reactions

Decomposition is a reaction that breaks a molecule into smaller fragments. You could diagram a simple *decomposition reaction* as:

$$AB \longrightarrow A + B$$

Decomposition reactions occur outside cells as well as inside them. For example, a typical meal contains molecules of fats, sugars, and proteins that are too large and too complex to be absorbed and used by your body. Decomposition reactions in the digestive tract break these molecules down into smaller fragments before absorption begins.

Decomposition reactions involving water are important in the breakdown of complex molecules in the body. In **hydrolysis** (hī-DROL-i-sis; *hydro-*, water + *lysis*, dissolution), one of the bonds in a complex molecule is broken, and the components of a water molecule (H and OH) are added to the resulting fragments:

$$A–B + H_2O \longrightarrow A–H + HO–B$$

Collectively, the decomposition reactions of complex molecules within the body's cells and tissues are referred to as **catabolism** (ka-TAB-ō-lizm; *katabole*, a throwing down). When a covalent bond—a form of potential energy—is broken, it releases kinetic energy that can perform work. By harnessing the energy released in this way, cells perform vital functions such as growth, movement, and reproduction.

Synthesis Reactions

Synthesis (SIN-the-sis) is the opposite of decomposition. A *synthesis reaction* assembles smaller molecules into larger molecules. A simple synthetic reaction could be diagrammed as:

$$A + B \longrightarrow AB$$

Synthesis reactions may involve individual atoms or the combination of molecules to form even larger products. The formation of water from hydrogen and oxygen molecules is a synthesis reaction. Synthesis always involves the formation of new chemical bonds, whether the reactants are atoms or molecules.

Dehydration synthesis, or *condensation*, is the formation of a complex molecule by the removal of a water molecule:

$$A–H + HO–B \longrightarrow A–B + H_2O$$

Dehydration synthesis is therefore the opposite of hydrolysis. We will encounter examples of both reactions in later sections.

Collectively, the synthesis of new molecules within the body's cells and tissues is known as **anabolism** (a-NAB-ō-lizm; *anabole*, a throwing upward). Because it takes energy to create a chemical bond, anabolism is usually considered an "uphill" process. Cells must balance their energy budgets, with catabolism providing the energy to support anabolism and other vital functions.

Tips&Tricks

To remember the difference between *anabolism* (synthesis) and *catabolism* (decomposition), relate the terms to words you already know: *Anabolic* steroids are used to build up muscle tissue, while both *catastrophe* and *catabolism* involve destruction (breakdown).

Exchange Reactions

In an **exchange reaction**, parts of the reacting molecules are shuffled around to produce new products:

$$AB + CD \longrightarrow AD + CB$$

Although the reactants and products contain the same components (A, B, C, and D), those components are present in different combinations. In an exchange reaction, the reactant molecules AB and CD must break apart (a decomposition) before they can interact with each other to form AD and CB (a synthesis).

Reversible Reactions

Chemical reactions are (at least theoretically) reversible, so if $A + B \longrightarrow AB$, then $AB \longrightarrow A + B$. Many important biological reactions are freely reversible. Such reactions can be represented as an equation:

$$A + B \rightleftharpoons AB$$

This equation indicates that, in a sense, two reactions are occurring simultaneously, one a synthesis $(A + B \longrightarrow AB)$ and the other a decomposition $(AB \longrightarrow A + B)$. At equilibrium, the rates at which the two reactions proceed are in balance. As fast as one molecule of AB forms, another degrades into $A + B$.

Tips&Tricks

Jell-O provides an observable example of a reversible reaction. Once Jell-O has been refrigerated, the gelatin sets up and forms a solid; if it sits without refrigeration for too long, it reverts to a liquid again.

The result of a disturbance in the equilibrium condition can be predicted. In our example, the rate at which the synthesis reaction proceeds is directly proportional to the frequency of encounters between A and B. In turn, the frequency of encounters depends on the degree of crowding: You are much more likely to run into another person in a crowded room than in a room that is almost empty. The addition of more AB molecules will increase the rate of conversion of AB to A and B. The amounts of A and B will then increase, leading to an increase in the rate of the reverse reaction—the formation of AB from A and B. Eventually, an equilibrium is again established.

#8 Things tend to even out, unless something prevents this from happening. Reversible reactions quickly reach equilibrium, in which opposing reaction rates are balanced. If reactants are added or removed, reaction rates change until a new equilibrium is established.

CHECKPOINT

7. The chemical shorthand used to describe chemical compounds and reactions effectively is known as _____.

8. Using the rules for chemical notation, write the molecular formula for glucose, a compound composed of six carbon atoms, 12 hydrogen atoms, and six oxygen atoms.

9. Identify and describe three types of chemical reactions important to human physiology.

10. In cells, glucose, a six-carbon molecule, is converted into two three-carbon molecules by a reaction that releases energy. How would you classify this reaction?

See the blue Answers tab at the end of the book.

2-4 Enzymes catalyze specific biochemical reactions by lowering a reaction's activation energy

Most chemical reactions do not occur spontaneously, or they occur so slowly that they would be of little value to cells. Before a reaction can proceed, enough energy must be provided to activate the reactants. The amount of energy required to start a reaction is called the **activation energy**. Although many reactions can be activated by changes in temperature or acidity, such changes are deadly to cells. For example, every day your cells break down complex sugars as part of your normal metabolism. Yet to break down a complex sugar in a laboratory, you must boil it in an acidic solution. Your cells don't have that option; temperatures that high and solutions that corrosive would immediately destroy living tissues. Instead, your cells use special proteins called *enzymes* to perform most of the complex synthesis and decomposition reactions in your body.

Enzymes promote chemical reactions by lowering the activation energy requirements (**Figure 2–7**). In doing so, they make it possible for chemical reactions, such as the breakdown of sugars, to proceed under conditions compatible with life. Enzymes belong to a class of substances called **catalysts** (KAT-uh-lists; *katalysis*, dissolution), compounds that accelerate chemical reactions without themselves being permanently changed or consumed. A cell makes an enzyme

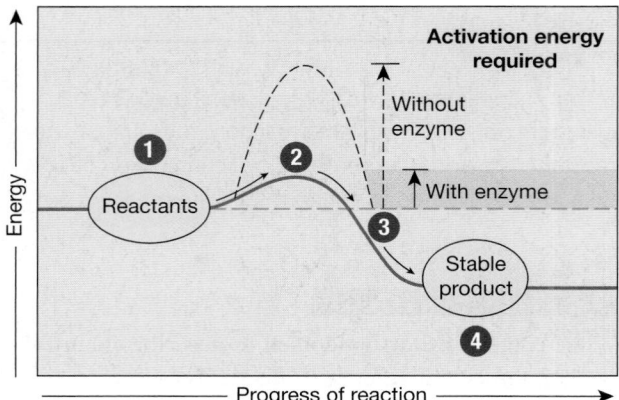

Figure 2–7 **The Effect of Enzymes on Activation Energy.**
Enzymes lower the activation energy required for a reaction to proceed readily (in order, from 1–4) under conditions in the body.

molecule to promote a specific reaction. Enzymatic reactions, which are reversible, can be written as:

$$A + B \xrightleftharpoons{\text{enzyme}} AB$$

Although the presence of an appropriate enzyme can accelerate a reaction, an enzyme affects only the rate of the reaction, not its direction or the products that are formed. An enzyme cannot bring about a reaction that would otherwise be impossible. Enzymatic reactions are generally reversible, and they proceed until an equilibrium becomes established.

The complex reactions that support life proceed in a series of interlocking steps, each controlled by a specific enzyme. Such a reaction sequence is called a *pathway*. A synthetic pathway can be diagrammed as:

$$A \xrightleftharpoons[\text{Step 1}]{\text{enzyme 1}} B \xrightleftharpoons[\text{Step 2}]{\text{enzyme 2}} C \xrightleftharpoons[\text{Step 3}]{\text{enzyme 3}} \text{ and so on.}$$

In many cases, the steps in the synthetic pathway differ from those in the decomposition pathway, and separate enzymes are often involved.

It takes activation energy to start a chemical reaction, but once it has begun, the reaction as a whole may absorb or release energy, generally in the form of heat, as it proceeds to completion. If the amount of energy released is greater than the activation energy needed to start the reaction, there will be a net release of energy. Reactions that release energy are said to be **exergonic** or **exothermic** (*exo-*, outside). If more energy is required to begin the reaction than is released as it proceeds, the reaction as a whole will absorb energy. Such reactions are called **endergonic** or **endothermic** (*endo-*, inside). Exergonic reactions are relatively common in the body; they are responsible for generating the heat that maintains your body temperature.

The A&P Top 100

#9 Enzymes must be present for most life-sustaining chemical reactions to occur.

2-5 Inorganic compounds usually lack carbon, and organic compounds always contain carbon

Although the human body is very complex, it contains relatively few elements (Table 2–1). But knowing the identity and quantity of each element in the body will not help you understand the body any more than studying the alphabet will help you understand this textbook. Just as only 26 letters can be combined to form thousands of different words in this book, only about 26 elements combine to form thousands of different chemical compounds in our bodies. As we saw in Chapter 1, these compounds make up the living cells that constitute the framework of the body and carry on all its life processes. So it is impossible to understand the structure and functioning of the human body without learning about the major classes of chemical compounds.

We will next turn our attention to nutrients and metabolites. **Nutrients** are the essential elements and molecules normally obtained from the diet. **Metabolites** (me-TAB-ō-līts; *metabole*, change), a much larger group, include all the molecules (nutrients included) that can be synthesized or broken down by chemical reactions inside our bodies. Nutrients and metabolites can be broadly categorized as either inorganic or organic. **Inorganic compounds** generally do not contain carbon and hydrogen atoms as their primary structural ingredients, whereas carbon and hydrogen always form the basis for **organic compounds**.

The most important inorganic compounds in the body are (1) carbon dioxide, a by-product of cell metabolism; (2) oxygen, an atmospheric gas required in important metabolic reactions; (3) water, which accounts for most of our body weight; and (4) inorganic acids, bases, and salts—compounds held together partially or completely by ionic bonds. In this section, we will be concerned primarily with water, its properties, and how those properties establish the conditions necessary for life. Most of the other inorganic molecules and compounds in the body exist in association with water, the primary component of our body fluids. Both carbon dioxide and oxygen, for example, are gas molecules that are transported in body fluids, and all the inorganic acids, bases, and salts we will discuss are dissolved in body fluids.

2-6 Physiological systems depend on water

Water, H_2O, is the most important constituent of the body, accounting for up to two-thirds of total body weight. A change in the body's water content can have fatal consequences because virtually all physiological systems will be affected.

Although water is familiar to everyone, it has some highly unusual properties. These properties are a direct result of the hydrogen bonding that occurs between adjacent water molecules:

1. *Solubility.* A remarkable number of inorganic and organic molecules are soluble, meaning they will dissolve or break up in water. The individual particles become distributed within the water, and the result is a solution—a uniform mixture of two or more substances. The medium in which other atoms, ions, or molecules are dispersed is called the **solvent**; the dispersed substances are the **solutes**. In *aqueous solutions*, water is the solvent. The solvent properties of water are so important that we will consider them further in the next section.

2. *Reactivity.* In our bodies, chemical reactions occur in water, and water molecules are also participants in some reactions. Hydrolysis and dehydration synthesis are two examples noted earlier in the chapter.

3. *High Heat Capacity. Heat capacity* is the ability to absorb and retain heat. Water has an unusually high heat capacity, because water molecules in the liquid state are attracted to one another through hydrogen bonding. Important consequences of this attraction include the following:

   • The temperature of water must be quite high before individual molecules have enough energy to break free and become water vapor, a gas. Consequently, water stays in the liquid state over a broad range of environmental temperatures, and the freezing and boiling points of water are far apart.

   • Water carries a great deal of heat away with it when it finally does change from a liquid to a gas. This feature accounts for the cooling effect of perspiration on the skin.

   • An unusually large amount of heat energy is required to change the temperature of 1 g of water by 1°C. As a result, a large mass of water changes temperature only slowly. This property is called *thermal inertia*. Because water accounts for up to two-thirds of the weight of the human body, thermal inertia helps stabilize body temperature.

4. *Lubrication.* Water is an effective lubricant because there is little friction between water molecules. Thus, if even a thin layer of water separates two opposing surfaces, friction between those surfaces will be greatly re-

duced. (That is why driving on wet roads can be tricky; your tires may start sliding on a layer of water rather than maintaining contact with the road.) Within joints such as the knee, an aqueous solution prevents friction between the opposing surfaces. Similarly, a small amount of fluid in the ventral body cavities prevents friction between internal organs, such as the heart or lungs, and the body wall. ∞ p. 23

The A&P Top 100

#10 Water accounts for most of your body weight; proteins, the key structural and functional components of cells, and nucleic acids, which control cell structure and function, work only in aqueous solutions.

The Properties of Aqueous Solutions

Water's chemical structure makes it an unusually effective solvent (**Figure 2–8**). The bonds in a water molecule are oriented such that the hydrogen atoms are relatively close together. As a result, the water molecule has positive and negative poles (**Figure 2–8a**). A water molecule is therefore called a *polar molecule*, or a *dipole*.

Many inorganic compounds are held together partially or completely by ionic bonds. In water, these compounds undergo *ionization* (ī-on-i-ZĀ-shun), or *dissociation* (di-sō-sē-Ā-shun). In this process, ionic bonds are broken as the individual ions interact with the positive or negative ends of polar water molecules (**Figure 2–8b**). The result is a mixture of cations and anions surrounded by water molecules. The water molecules around each ion form a *hydration sphere*.

An aqueous solution containing anions and cations will conduct an electrical current. Cations $(+)$ move toward the negative side, or negative *terminal*, and anions $(-)$ move toward the positive terminal. Electrical forces across plasma membranes affect the functioning of all cells, and small electrical currents carried by ions are essential to muscle contraction and nerve function. Chapters 10 and 12 will discuss these processes in more detail.

Electrolytes and Body Fluids

Soluble inorganic molecules whose ions will conduct an electrical current in solution are called **electrolytes** (e-LEK-trō-līts). Sodium chloride is an electrolyte. The dissociation of electrolytes in blood and other body fluids releases a variety of ions. Table 2–3 lists important electrolytes and the ions released by their dissociation.

Changes in the concentrations of electrolytes in body fluids will disturb almost every vital function. For example, declining potassium levels will lead to a general muscular paralysis, and rising concentrations will cause weak and irregular heartbeats. The concentrations of ions in body fluids are

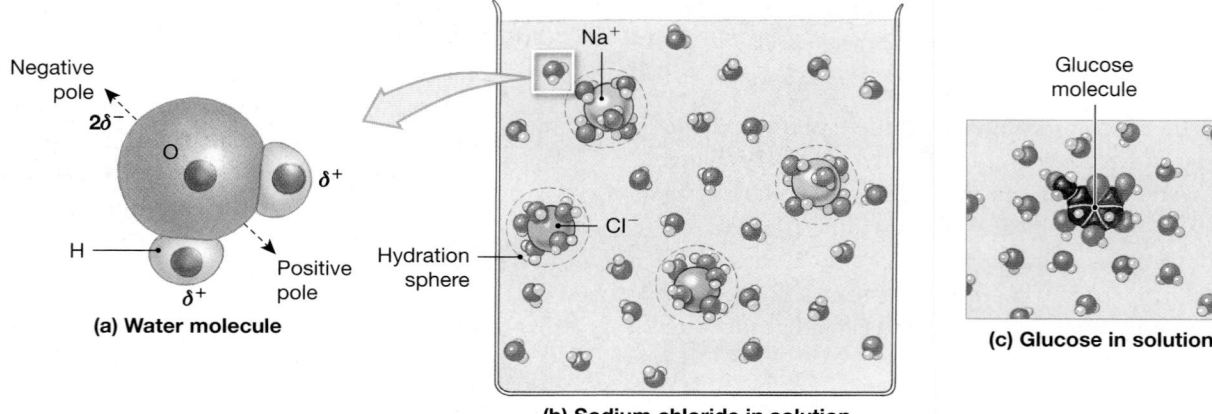

(a) Water molecule

(b) Sodium chloride in solution

(c) Glucose in solution

Figure 2–8 The Activities of Water Molecules in Aqueous Solutions. **(a)** In a water molecule, oxygen forms polar covalent bonds with two hydrogen atoms. Because both hydrogen atoms are at one end of the molecule, it has an uneven distribution of charges, creating positive and negative poles. **(b)** Ionic compounds, such as sodium chloride, dissociate in water as the polar water molecules break the ionic bonds in the large crystal structure. Each ion in solution is surrounded by water molecules, creating hydration spheres. **(c)** Hydration spheres also form around an organic molecule containing polar covalent bonds. If the molecule binds water strongly, as does glucose, it will be carried into solution—in other words, it will dissolve. Note that the molecule does not dissociate, as occurs for ionic compounds.

TABLE 2–3	Important Electrolytes That Dissociate in Body Fluids	
Electrolyte		**Ions Released**
NaCl (sodium chloride)	$\longrightarrow$	$Na^+ + Cl^-$
KCl (potassium chloride)	$\longrightarrow$	$K^+ + Cl^-$
CaPO₄ (calcium phosphate)	$\longrightarrow$	$Ca^{2+} + PO_4^{2-}$
NaHCO₃ (sodium bicarbonate)	$\longrightarrow$	$Na^+ + HCO_3^-$
MgCl₂ (magnesium chloride)	$\longrightarrow$	$Mg^{2+} + 2Cl^-$
Na₂HPO₄ (sodium hydrogen phosphate)	$\longrightarrow$	$2Na^+ + HPO_4^{2-}$
Na₂SO₄ (sodium sulfate)	$\longrightarrow$	$2Na^+ + SO_4^{2-}$

carefully regulated, primarily by the coordination of activities at the kidneys (ion excretion), the digestive tract (ion absorption), and the skeletal system (ion storage or release).

Hydrophilic and Hydrophobic Compounds

Some organic molecules contain polar covalent bonds, which also attract water molecules. The hydration spheres that form may then carry these molecules into solution (**Figure 2–8c**). Molecules such as *glucose*, an important soluble sugar, that interact readily with water molecules in this way are called hydrophilic (hī-drō-FIL-ik; *hydro-*, water + *philos*, loving).

Many other organic molecules either lack polar covalent bonds or have very few. Such molecules do not have positive and negative terminals, and are said to be *nonpolar*. When nonpolar molecules are exposed to water, hydration spheres do not form and the molecules do not dissolve. Molecules that do not readily interact with water are called **hydrophobic** (hī-drō-FŌ-bik; *hydro-*, water + *phobos*, fear). Among the most familiar hydrophobic molecules are fats and oils of all kinds. Body fat deposits, for example, consist of large, hydrophobic droplets trapped in the watery interior of cells. Gasoline and heating oil are hydrophobic molecules not found in the body; when accidentally discharged into lakes or oceans, they form tenacious oil slicks instead of dissolving.

Tips&Tricks

To distinguish between hydrophobic and hydrophilic, remember that a phobia is a fear of something, and that -philic ends with "lic," which resembles "like."

Colloids and Suspensions

Body fluids may contain large and complex organic molecules, such as proteins and protein complexes, that are held in solution by their association with water molecules (**Figure 2–8c**). A solution containing dispersed proteins or other large molecules is called a **colloid**. Liquid Jell-O is a familiar viscous (thick) colloid.

The particles or molecules in a colloid will remain in solution indefinitely. A **suspension** contains even larger particles that will, if undisturbed, settle out of solution due to the force of gravity. Stirring beach sand into a bucket of water creates a temporary suspension that will last only until the sand settles to the bottom. Whole blood is another temporary suspension, because the blood cells are suspended in the blood plasma. If clotting is prevented, the cells in a blood sample will gradually settle to the bottom of the container. Measuring that settling rate, or "sedimentation rate," is a common laboratory test.

CLINICAL NOTE

Solute Concentrations

The **concentration** of a substance is the amount of that substance in a specified volume of solvent. Physiologists and clinicians often monitor inorganic and organic solute concentrations in body fluids such as blood or urine. Each solute has a normal range of values (see the Appendix), and variations outside this range may indicate disease. Many solutes are participants in biochemical reactions, and as noted earlier, the concentrations of reactants and products in a chemical reaction directly affect reaction rates.

Solute concentrations can be expressed in several ways. In one method, we express the number of solute atoms, molecules, or ions in a specific volume of solution. Values are reported in moles per liter (mol/L, or M) or millimoles per liter

(mmol/L, or mM). A concentration expressed in these units is referred to as the *molarity* of the solution. (Recall that a mole is a quantity of any substance having a weight in grams equal to the atomic or molecular weight of that substance.) Physiological concentrations in clinical lab reports are most often reported in millimoles per liter.

You can report concentrations in terms of molarity only when you know the molecular weight of the ion or molecule in question. When the chemical structure is unknown or when you are dealing with a complex mixture of materials, concentration is expressed in terms of the weight of material dissolved in a unit volume of solution. Values are then reported in milligrams (mg) or grams (g) per deciliter (dL, or 100 mL). This is the method used, for example, in reporting the concentration of plasma proteins in a blood sample.

CHECKPOINT

14. **Explain how the chemical properties of water make life possible.**

See the blue Answers tab at the end of the book.

2-7 Body fluid pH is vital for homeostasis

A hydrogen atom involved in a chemical bond or participating in a chemical reaction can easily lose its electron, to become a hydrogen ion, H^+. Hydrogen ions are extremely reactive in solution. In excessive numbers, they will break chemical bonds, change the shapes of complex molecules, and generally disrupt cell and tissue functions. As a result, the concentration of hydrogen ions in body fluids must be regulated precisely.

A few hydrogen ions are normally present even in a sample of pure water, because some of the water molecules dissociate spontaneously, releasing cations and anions. The dissociation of water is a reversible reaction that can be represented as:

$$H_2O \rightleftharpoons H^+ + OH^-$$

The dissociation of one water molecule yields a hydrogen ion and a *hydroxide* (hī-DROK-sīd) *ion*, OH^-.

Very few water molecules ionize in pure water, and the number of hydrogen and hydroxide ions is small. The quantities are usually reported in moles, making it easy to keep track of the relative numbers of hydrogen and hydroxide ions. One liter of pure water contains about 0.0000001 mol of hydrogen

ions and an equal number of hydroxide ions. In other words, the concentration of hydrogen ions in a solution of pure water is 0.0000001 mol per liter. This can be written as:

$$[H^+] = 1 \times 10^{-7} \, mol/L$$

The brackets around the H^+ signify "the concentration of," another example of chemical notation.

The hydrogen ion concentration in body fluids is so important to physiological processes that a special shorthand is used to express it. The **pH** of a solution is defined as the negative logarithm of the hydrogen ion concentration in moles per liter. Thus, instead of using the equation $[H^+] = 1 \times 10^{-7}$ mol/L, we say that the pH of pure water is $-(-7)$, or 7. Using pH values saves space, but always remember that the pH number is an *exponent* and that the pH scale is logarithmic. For instance, a pH of 6 ($[H^+] = 1 \times 10^{-6}$, or 0.000001) means that the concentration of hydrogen ions is *10 times as great* as it is at a pH of 7 ($[H^+] = 1 \times 10^{-7}$, or 0.0000001). For common liquids, the pH scale, included in **Figure 2–9**, ranges from 0 to 14.

Although pure water has a pH of 7, solutions display a wide range of pH values, depending on the nature of the solutes involved:

- A solution with a pH of 7 is said to be **neutral**, because it contains equal numbers of hydrogen and hydroxide ions.

- A solution with a pH below 7 is **acidic** (a-SI-dik), meaning that it contains more hydrogen ions than hydroxide ions.

- A pH above 7 is **basic**, or *alkaline* (AL-kuh-lin), meaning that it has more hydroxide ions than hydrogen ions.

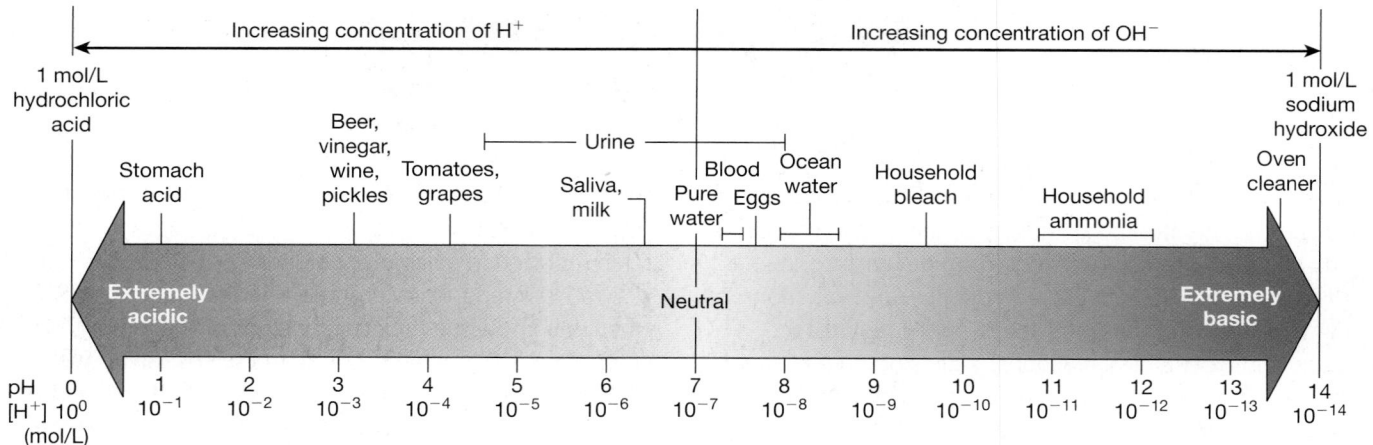

Figure 2–9 **pH and Hydrogen Ion Concentration.** The pH scale is logarithmic; an increase or decrease of one unit corresponds to a tenfold change in H^+ concentration.

The pH of blood normally ranges from 7.35 to 7.45. Abnormal fluctuations in pH can damage cells and tissues by breaking chemical bonds, changing the shapes of proteins, and altering cellular functions. *Acidosis* is an abnormal physiological state caused by low blood pH (below 7.35); a pH below 7 can produce coma. *Alkalosis* results from an abnormally high pH (above 7.45); a blood pH above 7.8 generally causes uncontrollable and sustained skeletal muscle contractions.

The A&P Top 100

#11 The pH of body fluids is a measure of how many free hydrogen ions are in solution. Excessive hydrogen ions (a low pH) can damage cells and tissues, cause changes in the shapes and functions of proteins, and disrupt normal physiological systems. While a high pH has adverse effects, problems due to a low pH are more common.

CHECKPOINT

15. **Define pH, and explain how the pH scale relates to acidity and alkalinity.**

16. **What is the significance of pH in physiological systems?**

See the blue Answers tab at the end of the book.

2-8 Acids, bases, and salts are inorganic compounds with important physiological roles

The body contains both inorganic and organic *acids* and *bases* that may cause acidosis or alkalosis, respectively. An **acid** is any solute that dissociates in solution and releases

hydrogen ions, thereby lowering the pH. Because a hydrogen atom that loses its electron consists solely of a proton, hydrogen ions are often referred to simply as protons, and acids as *proton donors*.

A *strong acid* dissociates completely in solution, and the reaction occurs essentially in one direction only. *Hydrochloric acid* (HCl) is a representative strong acid; in water, it ionizes as follows:

$$HCl \longrightarrow H^+ + Cl^-$$

The stomach produces this powerful acid to assist in the breakdown of food. Hardware stores sell HCl under the name muriatic acid, for cleaning concrete and swimming pools.

A **base** is a solute that removes hydrogen ions from a solution and thereby acts as a *proton acceptor*. In solution, many bases release a hydroxide ion (OH^-). Hydroxide ions have a strong affinity for hydrogen ions and react quickly with them to form water molecules. A *strong base* dissociates completely in solution. *Sodium hydroxide*, NaOH, is a strong base; in solution, it releases sodium ions and hydroxide ions:

$$NaOH \longrightarrow Na^+ + OH^-$$

Strong bases have a variety of industrial and household uses. Drain openers (Drano) and lye are two familiar examples.

Weak acids and *weak bases* fail to dissociate completely. At equilibrium, a significant number of molecules remain intact in the solution. For the same number of molecules in solution, weak acids and weak bases therefore have less of an impact on pH than do strong acids and strong bases. *Carbonic acid* (H_2CO_3) is a weak acid found in body fluids. In solution, carbonic acid reversibly dissociates into a hydrogen ion and a *bicarbonate ion*, HCO_3^-:

$$H_2CO_3 \rightleftharpoons H^+ + HCO_3^-$$

Salts

A **salt** is an ionic compound consisting of any cation except a hydrogen ion, and any anion except a hydroxide ion. Because they are held together by ionic bonds, many salts dissociate completely in water, releasing cations and anions. For example, sodium chloride (table salt) dissociates immediately in water, releasing Na^+ and Cl^-. Sodium and chloride are the most abundant ions in body fluids. However, many other ions are present in lesser amounts as a result of the dissociation of other inorganic compounds. Ionic concentrations in the body are regulated by mechanisms we will describe in Chapters 26 and 27.

The ionization of sodium chloride does not affect the local concentrations of hydrogen ions or hydroxide ions, so NaCl, like many salts, is a "neutral" solute. Through their interactions with water molecules, however, other salts may indirectly affect the concentrations of H^+ and OH^- ions. Thus, the dissociation of some salts makes a solution slightly acidic or slightly basic.

Buffers and pH Control

Buffers are compounds that stabilize the pH of a solution by removing or replacing hydrogen ions. *Buffer systems* typically involve a weak acid and its related salt, which functions as a weak base. For example, the carbonic acid–bicarbonate buffer system (which will be detailed in Chapter 27) consists of carbonic acid and sodium bicarbonate, $NaHCO_3$, otherwise known as baking soda. Buffers and buffer systems in body fluids help maintain the pH within normal limits. The pH of several body fluids is included in **Figure 2–9**.

The use of antacids such as Alka-Seltzer provides one illustration of the type of reaction that occurs in buffer systems. Alka-Seltzer uses sodium bicarbonate to neutralize excess hydrochloric acid in the stomach. Note that the effects of neutralization are most evident when you add a strong acid to a strong base. For example, by adding hydrochloric acid to sodium hydroxide, you neutralize both the strong acid and the strong base:

$$HCl + NaOH \longrightarrow H_2O + NaCl$$

This reaction produces water and a salt—in this case, the neutral salt sodium chloride.

CHECKPOINT

17. Define the following terms: acid, base, and salt.

18. How does an antacid help decrease stomach discomfort?

See the blue Answers tab at the end of the book.

2-9 Carbohydrates contain carbon, hydrogen, and oxygen in a 1:2:1 ratio

Carbohydrates are molecules classified as a kind of organic compound. Organic compounds always contain the elements carbon and hydrogen, and generally oxygen as well. Many organic molecules are made up of long chains of carbon atoms linked by covalent bonds. The carbon atoms typically form additional covalent bonds with hydrogen or oxygen atoms and, less commonly, with nitrogen, phosphorus, sulfur, iron, or other elements.

Many organic molecules are soluble in water. Although the previous discussion focused on inorganic acids and bases, there are also important organic acids and bases. For example, *lactic acid* is an organic acid, generated by active muscle tissues, that must be neutralized by the carbonic acid–bicarbonate buffer system to prevent a potentially dangerous pH decline in body fluids.

Although organic compounds are diverse, certain groupings of atoms occur again and again, even in very different types of molecules. These *functional groups* greatly influence the properties of any molecule they are part of. Table 2–4 details the functional groups you will encounter in this chapter.

A **carbohydrate** is an organic molecule that contains carbon, hydrogen, and oxygen in a ratio near 1:2:1. Familiar carbohydrates include the *sugars* and *starches* that make up roughly half of the typical U.S. diet. Carbohydrates typically account for approximately 1 percent of total body weight. Although they may have other functions, carbohydrates are most important as energy sources that are catabolized rather than stored. We will focus on *monosaccharides*, *disaccharides*, and *polysaccharides*.

Monosaccharides

A *simple sugar*, or **monosaccharide** (mon-ō-SAK-uh-rīd; *mono-*, single + *sakcharon*, sugar), is a carbohydrate containing from three to seven carbon atoms. A monosaccharide can be called a *triose* (three-carbon), *tetrose* (four-carbon), *pentose* (five-carbon), *hexose* (six-carbon), or *heptose* (seven-carbon). The hexose **glucose** (GLOO-kōs), $C_6H_{12}O_6$, is the most important metabolic "fuel" in the body. The atoms in a glucose molecule may form either a straight chain (**Figure 2–10a**) or a ring (**Figure 2–10b**). In the body, the ring form is more common. A three-dimensional model shows the arrangement of atoms in the ring most clearly (**Figure 2–10c**).

The three-dimensional structure of an organic molecule is an important characteristic, because it usually determines the molecule's fate or function. Some molecules have the

TABLE 2–4 Important Functional Groups of Organic Compounds

Functional Group	Structural Formula*	Importance	Examples
Carboxylic acid, —COOH	R . . . —C=O with OH	Acts as an acid, releasing H^+ to become R—COO^-	Fatty acids, amino acids
Amino group, —NH_2	R—N with H and H	Can accept or release H^+, depending on pH; can form bonds with other molecules	Amino acids
Hydroxyl group, —OH	R—O—H	Strong bases dissociate to release hydroxide ions [OH^-]; may link molecules through dehydration synthesis (condensation)	Carbohydrates, fatty acids, amino acids
Phosphate group, —PO_4	R—O—P(=O)—O^- with O^-	May link other molecules to form larger structures; may store energy in high-energy bonds	Phospholipids, nucleic acids, high-energy compounds

*The term *R group* is used to denote the rest of the molecule, whatever that might be. The R group is also known as a *side chain*.

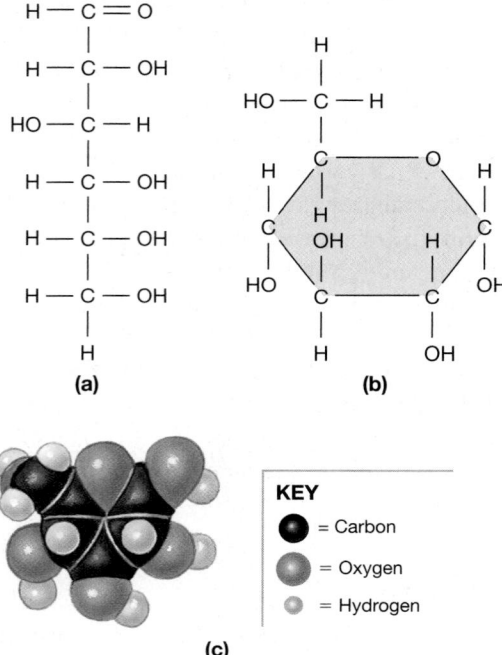

Figure 2–10 The Structure of Glucose. (a) The structural formula of the straight-chain form. **(b)** The structure of the ring form, the most common form in nature. **(c)** A three-dimensional model that shows the organization of the atoms in the ring.

KEY
● = Carbon
● = Oxygen
○ = Hydrogen

same molecular formula—in other words, the same types and numbers of atoms—but different structures. Such molecules are called **isomers**. The body usually treats different isomers as distinct molecules. For example, the monosaccharides glucose and fructose are isomers. *Fructose* is a hexose found in many fruits and in secretions of the male reproductive tract. Although its chemical formula, $C_6H_{12}O_6$,

is the same as that of glucose, the arrangement of its atoms differs from that of glucose. As a result, separate enzymes and reaction sequences control its breakdown and synthesis. Monosaccharides such as glucose and fructose dissolve readily in water and are rapidly distributed throughout the body by blood and other body fluids.

Disaccharides and Polysaccharides

Carbohydrates other than simple sugars are complex molecules composed of monosaccharide building blocks. Two monosaccharides joined together form a **disaccharide** (dī-SAK-uh-rīd; *di-*, two). Disaccharides such as *sucrose* (table sugar) have a sweet taste and, like monosaccharides, are quite soluble in water. The formation of sucrose (**Figure 2–11a**) involves dehydration synthesis, a process introduced earlier in the chapter. Dehydration synthesis, or condensation, links molecules together by the removal of a water molecule. The breakdown of sucrose into simple sugars is an example of hydrolysis, the functional opposite of dehydration synthesis (**Figure 2–11b**).

Many foods contain disaccharides, but all carbohydrates except monosaccharides must be disassembled through hydrolysis before they can provide useful energy. Most popular junk foods, such as candies and sodas, abound in monosaccharides (commonly fructose) and disaccharides (generally sucrose). Some people cannot tolerate sugar for medical reasons; others avoid it in an effort to control their weight. (Excess sugars are converted to fat for long-term storage.) Many such people use *artificial sweeteners* in their foods and beverages. These compounds have a very sweet taste, but they either cannot be broken down in the body or are used in insignificant amounts.

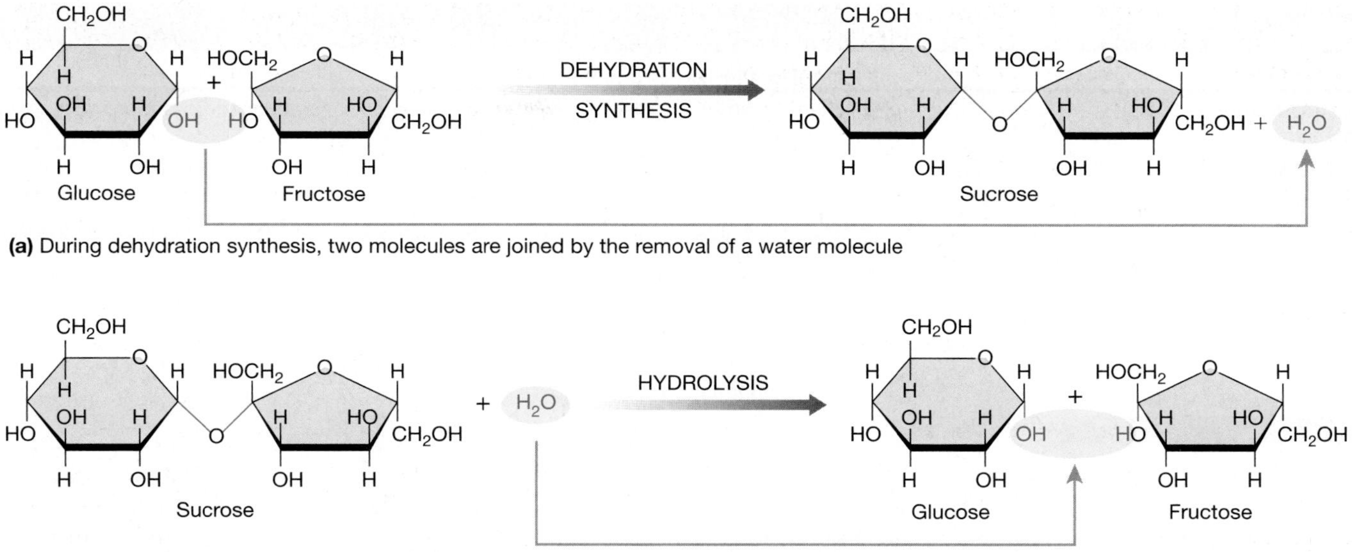

CH₂OH ... Glucose + Fructose → DEHYDRATION SYNTHESIS → Sucrose + H₂O

(a) During dehydration synthesis, two molecules are joined by the removal of a water molecule

Sucrose + H₂O → HYDROLYSIS → Glucose + Fructose

(b) Hydrolysis reverses the steps of dehydration synthesis; a complex molecule is broken down by the addition of a water molecule

Figure 2–11 **The Formation and Breakdown of Complex Sugars. (a)** Formation of the disaccharide sucrose through dehydration synthesis. During this reaction, two monosaccharides are joined by the removal of a water molecule. **(b)** Breakdown of sucrose into simple sugars by hydrolysis. During this reaction, which reverses the steps of dehydration synthesis, a complex sugar is broken down by the addition of a water molecule. Enzymes perform both these reactions.

More-complex carbohydrates result when repeated dehydration synthesis reactions add additional monosaccharides or disaccharides. These large molecules are called **polysaccharides** (pol-ē-SAK-uh-rīdz; *poly-*, many). Polysaccharide chains can be straight or highly branched. *Cellulose*, a structural component of many plants, is a polysaccharide that our bodies cannot digest because the particular linkages between the glucose molecules cannot be cleaved by enzymes in the body. Foods such as celery, which contains cellulose, water, and little else, contribute bulk to digestive wastes but are useless as a source of energy. (In fact, you expend more energy when you chew a stalk of celery than you obtain by digesting it.)

Starches are large polysaccharides formed from glucose molecules. Most starches are manufactured by plants. Your digestive tract can break these molecules into monosaccharides. Starches such as those in potatoes and grains are a major dietary energy source.

The polysaccharide **glycogen** (GLĪ-kō-jen), or *animal starch*, has many side branches consisting of chains of glucose molecules (**Figure 2–12**). Like most other starches, glycogen does not dissolve in water or other body fluids. Muscle cells make and store glycogen. When muscle cells have a high demand for glucose, glycogen molecules are broken down; when the demand is low, they absorb glucose from the bloodstream and rebuild glycogen reserves. Table 2–5 summarizes information about the carbohydrates.

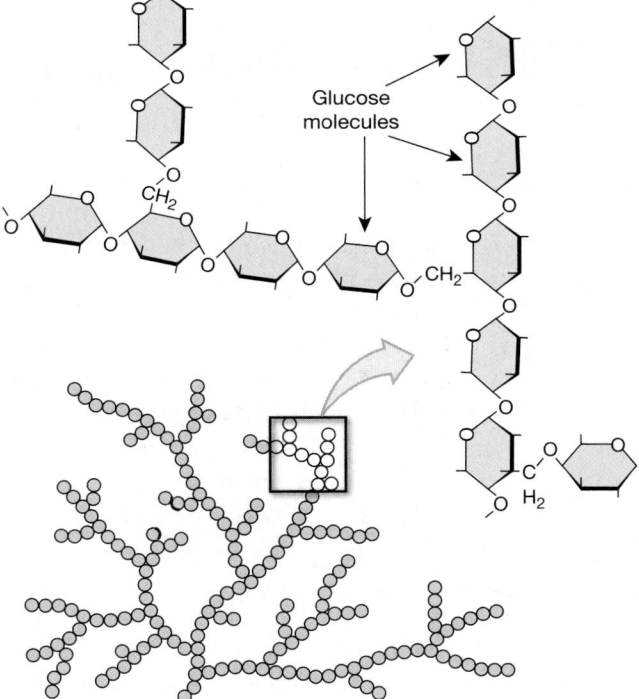

Figure 2–12 **The Structure of the Polysaccharide Glycogen.** Liver and muscle cells store glucose as the polysaccharide glycogen, a long, branching chain of glucose molecules. This figure uses a different method of representing a carbon ring structure: At each corner of the hexagon is a carbon atom. The position of an oxygen atom in each glucose ring is shown.

TABLE 2–5 Carbohydrates in the Body

Structural Class	Examples	Primary Function	Remarks
Monosaccharides (simple sugars)	Glucose, fructose	Energy source	Manufactured in the body and obtained from food; distributed in body fluids
Disaccharides	Sucrose, lactose, maltose	Energy source	Sucrose is table sugar, lactose is in milk, and maltose is malt sugar; all must be broken down to monosaccharides before absorption
Polysaccharides	Glycogen	Storage of glucose	Glycogen is in animal cells; other starches and cellulose are within or surround plant cells

CHECKPOINT

19. A food contains organic molecules with the elements C, H, and O in a ratio of 1:2:1. What class of compounds do these molecules belong to, and what are their major functions in the body?

See the blue Answers tab at the end of the book.

The A&P Top 100

#12 Carbohydrates are important as a quick source of energy and as integral components of membranes. Lipids form membranes within and around cells that restrict the diffusion of many solutes. Fats are large lipids important as energy reserves.

2-10 Lipids contain a carbon-to-hydrogen ratio of 1:2

Like carbohydrates, **lipids** (*lipos*, fat) contain carbon, hydrogen, and oxygen, and the carbon-to-hydrogen ratio is typically near 1:2. However, lipids contain much less oxygen than do carbohydrates with the same number of carbon atoms. The hydrogen-to-oxygen ratio is therefore very large; a representative lipid, such as lauric acid, has a formula of $C_{12}H_{24}O_2$. Lipids may also contain small quantities of phosphorus, nitrogen, or sulfur. Familiar lipids include *fats*, *oils*, and *waxes*. Most lipids are insoluble in water, but special transport mechanisms carry them in the circulating blood.

Lipids form essential structural components of all cells. In addition, lipid deposits are important as energy reserves. On average, lipids provide roughly twice as much energy as carbohydrates do, gram for gram, when broken down in the body. When the supply of lipids exceeds the demand for energy, the excess is stored in fat deposits. For this reason, there has been great interest in developing *fat substitutes* that provide less energy, but have the same taste and texture as lipids.

Lipids normally account for 12–18 percent of the total body weight of adult men, and 18–24 percent of that of adult women. Many kinds of lipids exist in the body. We will consider five classes of lipids: *fatty acids*, *eicosanoids*, *glycerides*, *steroids*, and *phospholipids and glycolipids*.

Fatty Acids

Fatty acids are long carbon chains with hydrogen atoms attached. One end of the carbon chain always bears a *carboxylic* (kar-bok-SIL-ik) *acid group*, —COOH (Table 2–4). The name *carboxyl* should help you remember that a carbon and a hydroxyl (—OH) group are the important structural features of fatty acids. The carbon chain attached to the carboxylic acid group is known as the hydrocarbon *tail* of the fatty acid. **Figure 2–13a** shows a representative fatty acid, *lauric acid*, found in coconut oil and oils of the laurel evergreen.

When a fatty acid is in solution, only the carboxyl end associates with water molecules, because that is the only hydrophilic portion of the molecule. The hydrocarbon tail is hydrophobic, so fatty acids have a very limited solubility in water. In general, the longer the hydrocarbon tail, the lower the solubility of the molecule.

Fatty acids may be either saturated or unsaturated (**Figure 2–13b**). These terms refer to the number of hydrogen atoms bound to the carbon atoms in the hydrocarbon tail. In a *saturated* fatty acid, each carbon atom in the tail has four single covalent bonds (**Figure 2–13a**). Within the tail, two of those bonds are adjacent to carbon atoms, and the other two bind hydrogen atoms; the carbon atom at the distal end of the tail binds three hydrogen atoms. In an *unsaturated* fatty acid, one or more of the single covalent bonds between the carbon atoms has been replaced by a double covalent bond. As a result, the carbon atoms involved will each bind only one hydrogen atom rather than two. This changes both the shape of the hydrocarbon tail and the way the fatty acid is metabolized. A *monounsaturated* fatty acid has a single double bond in the hydrocarbon tail. A *polyunsaturated* fatty acid contains multiple double bonds.

CLINICAL NOTE

Fatty Acids and Health

Humans love fatty foods. The smooth, creamy texture of fatty substances, together with their appealing taste, make fats a welcome part of our diet. Unfortunately, a diet containing large amounts of saturated fatty acids has been shown to increase the risk of heart disease and other cardiovascular problems. Saturated fats are found in such popular foods as fatty meat and dairy products (including such favorites as butter, cheese, and ice cream).

Some unsaturated fats, by contrast, are thought to decrease the risk of heart disease. Vegetable oils contain a mixture of monounsaturated and polyunsaturated fatty acids. Recent studies indicate that monounsaturated fats may be more effective than polyunsaturated fats in lowering the risk of heart disease. According to current research, perhaps the healthiest choices are olive and canola oils, which contain particularly abundant quantities of oleic acid,

an 18-carbon monounsaturated fatty acid. Surprisingly, compounds called *trans* fatty acids (TFA), produced during the manufacturing of some margarines and vegetable shortenings, appear to increase the risk of heart disease. U.S. Food and Drug Administration (FDA) guidelines now require that *trans* fatty acids be declared in the nutrition label of foods and dietary supplements.

Inuit have lower rates of heart disease than do other populations, even though the typical Inuit diet is high in fats and cholesterol. Interestingly, the main fatty acids in the Inuit diet are of the omega-3 type, which means they have an unsaturated bond three carbons before the last (or omega) carbon, a position known as "omega minus 3." Fish flesh and fish oils, a substantial portion of the Inuit diet, contain an abundance of omega-3 fatty acids. Why the presence of omega-3 fatty acids in the diet reduces the risks of heart disease, rheumatoid arthritis, and other inflammatory diseases is not yet apparent, but it is a research topic of great interest.

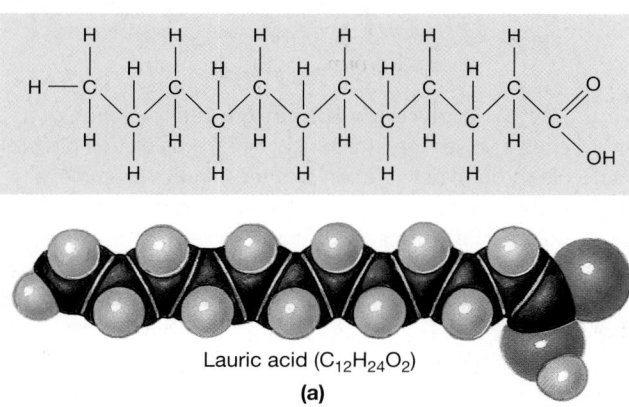

Lauric acid ($C_{12}H_{24}O_2$)
(a)

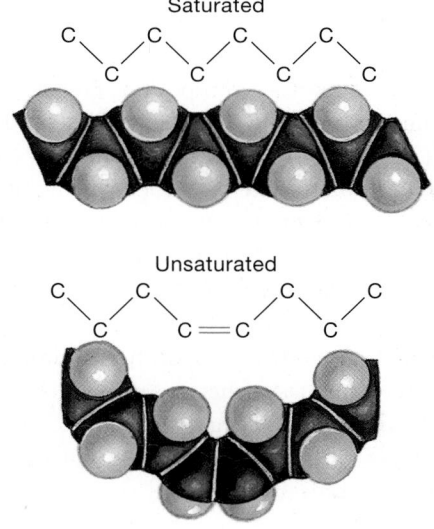

Saturated

Unsaturated

(b)

Figure 2–13 Fatty Acids. (a) Lauric acid demonstrates two structural characteristics common to all fatty acids: a long chain of carbon atoms and a carboxylic acid group (—COOH) at one end. **(b)** A fatty acid is either saturated (has single covalent bonds only) or unsaturated (has one or more double covalent bonds). The presence of a double bond causes a sharp bend in the molecule.

Eicosanoids

Eicosanoids (ī-KŌ-sa-noydz) are lipids derived from *arachidonic* (ah-rak-i-DON-ik) *acid*, a fatty acid that must be absorbed in the diet because it cannot be synthesized by the body. The two major classes of eicosanoids are *leukotrienes* and *prostaglandins*. Leukotrienes are produced primarily by cells involved with coordinating the responses to injury or disease. We will consider leukotrienes in Chapters 18 and 22. We consider only prostaglandins here, because virtually all tissues synthesize and respond to them.

Prostaglandins (pros-tuh-GLAN-dinz) are short-chain fatty acids in which five of the carbon atoms are joined in a ring (**Figure 2–14**). These compounds are released by cells to coordinate or direct local cellular activities, and they are extremely powerful even in minute quantities. The effects of prostaglandins vary with their structure and the site of their release. Prostaglandins released by damaged tissues, for example, stimulate nerve endings and produce the sensation of pain (Chapter 15). Those released in the uterus help trigger the start of labor contractions (Chapter 29).

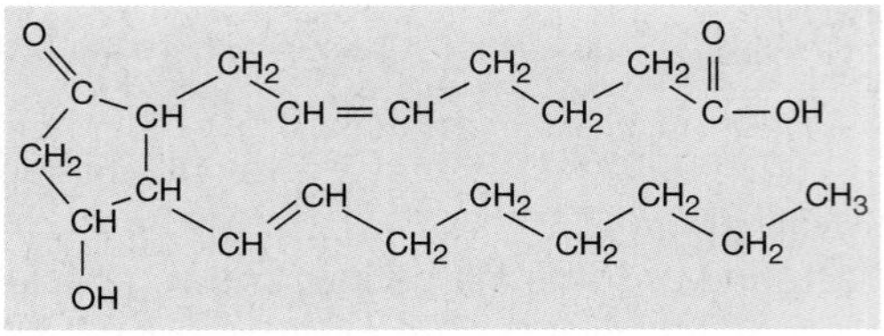

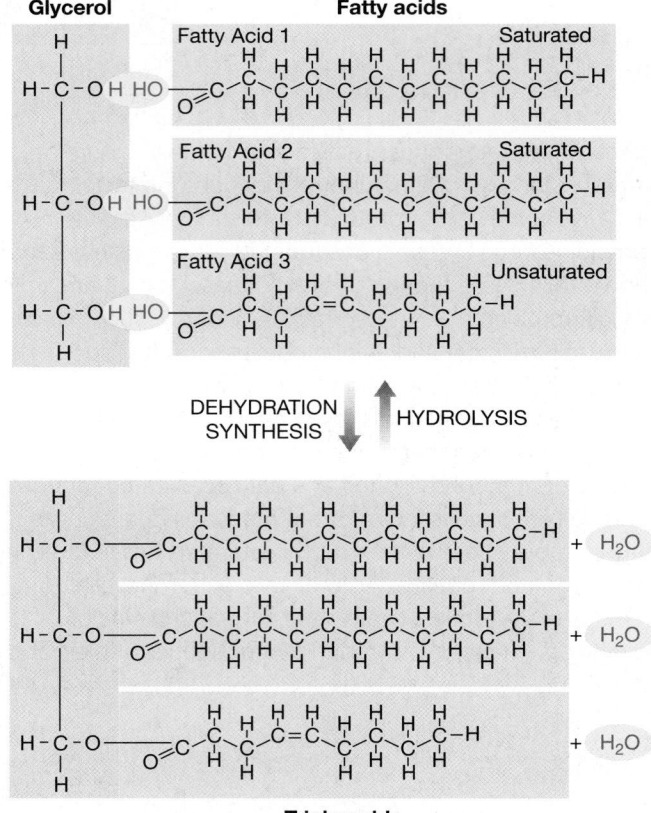

Figure 2–14 **Prostaglandins.** Prostaglandins are unusual short-chain fatty acids.

The body uses several types of chemical messengers. Those that are produced in one part of the body but have effects on distant parts are called *hormones*. Hormones are distributed throughout the body in the bloodstream, whereas most prostaglandins affect only the area in which they are produced. As a result, prostaglandins are often called *local hormones*. The distinction is not a rigid one, however, as some prostaglandins also enter the bloodstream and affect other areas to some degree. We will discuss hormones and prostaglandins in Chapter 18.

Glycerides

Unlike monosaccharides, individual fatty acids cannot be strung together in a chain by dehydration synthesis. But they can be attached to another compound, **glycerol** (GLIS-er-ol), through a similar reaction. The result is a lipid known as a **glyceride** (GLIS-er-īd). Dehydration synthesis can produce a **monoglyceride** (mon-ō-GLI-ser-īd), consisting of glycerol plus one fatty acid. Subsequent reactions can yield a **diglyceride** (glycerol + two fatty acids) and then a **triglyceride** (glycerol + three fatty acids), as in **Figure 2–15**. Hydrolysis breaks the glycerides into fatty acids and glycerol. Comparing **Figure 2–15** with **Figure 2–11** shows that dehydration synthesis and hydrolysis operate the same way, whether the molecules involved are carbohydrates or lipids.

Triglycerides, also known as *triacylglycerols* or *neutral fats*, have three important functions:

1. *Energy Source.* Fatty deposits in specialized sites of the body represent a significant energy reserve. In times of need, the triglycerides are disassembled by hydrolysis, yielding fatty acids that can be broken down to provide energy.

2. *Insulation.* Fat deposits under the skin serve as insulation, slowing heat loss to the environment. Heat loss across a layer of lipids is only about one-third that through other tissues.

3. *Protection.* A fat deposit around a delicate organ such as a kidney provides a cushion that protects against shocks or jolts.

Figure 2–15 **Triglyceride Formation.** The formation of a triglyceride involves the attachment of fatty acids to a glycerol molecule through dehydration synthesis. In this example, a triglyceride is formed by the attachment of one unsaturated and two saturated fatty acids to a glycerol molecule.

Triglycerides are stored in the body as lipid droplets within cells. The droplets absorb and accumulate lipid-soluble vitamins, drugs, or toxins that appear in body fluids. This accumulation has both positive and negative effects. For example, the body's lipid reserves retain both valuable lipid-soluble vitamins (A, D, E, K) and potentially dangerous lipid-soluble pesticides, such as DDT.

Steroids

Steroids are large lipid molecules that share a distinctive carbon framework (**Figure 2–16**). They differ in the functional groups that are attached to the basic structure. The steroid **cholesterol** (koh-LES-ter-ol; *chole-*, bile (+) *stereos*, solid) and related steroids are important for the following reasons:

- All animal plasma membranes contain cholesterol (**Figure 2–16a**). Cells need cholesterol to maintain their plasma membranes, as well as for cell growth and division.

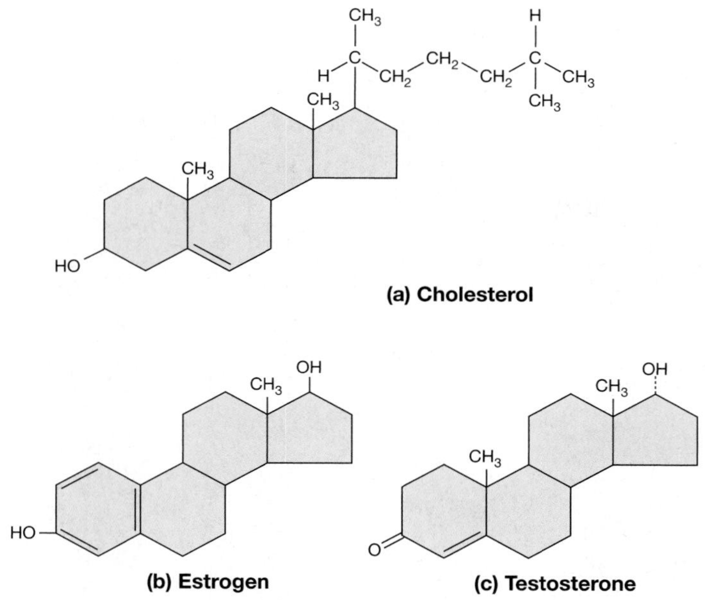

(a) Cholesterol

(b) Estrogen

(c) Testosterone

Figure 2–16 Steroids. All steroids share a complex four-ring structure. Individual steroids differ in the side chains attached to the carbon rings.

- Steroid hormones are involved in the regulation of sexual function. Examples include the sex hormones, such as the *estrogens* and *testosterone* (**Figure 2–16b,c**).

- Steroid hormones are important in the regulation of tissue metabolism and mineral balance. Examples include the hormones of the adrenal cortex, called *corticosteroids*, and *calcitriol*, a hormone important in the regulation of calcium ion concentrations in the body.

- Steroid derivatives called *bile salts* are required for the normal processing of dietary fats. Bile salts are produced in the liver and secreted in bile. They interact with lipids in the intestinal tract and facilitate the digestion and absorption of lipids.

Cholesterol is obtained in two ways: (1) by absorption from animal products in the diet and (2) by synthesis within the body. Liver, fatty meat, cream, and egg yolks are especially rich dietary sources of cholesterol. A diet high in cholesterol can be harmful, because a strong link exists between high blood cholesterol levels and heart disease. Current nutritional advice suggests limiting cholesterol intake to less than 300 mg per day. This amount represents a 40 percent reduction for the average adult in the United States. Unfortunately, because the body can synthesize cholesterol as well, blood cholesterol levels can be difficult to control by dietary restriction alone. We will examine the connection between blood cholesterol levels and heart disease more closely in later chapters.

Phospholipids and Glycolipids

Phospholipids (FOS-fō-lip-idz) and **glycolipids** (GLĪ-kō-lip-idz) are structurally related, and our cells can synthesize both types of lipids, primarily from fatty acids. In a *phospho*lipid, a *phosphate group* (PO_4^{3-}) links a diglyceride to a nonlipid group (**Figure 2–17a**). In a *glyco*lipid, a carbohydrate is attached to a diglyceride (**Figure 2–17b**). Note that placing -*lipid* last in these names indicates that the molecule consists primarily of lipid.

The long hydrocarbon tails of phospholipids and glycolipids are hydrophobic, but the opposite ends, the nonlipid *heads*, are hydrophilic. In water, large numbers of these molecules tend to form droplets, or *micelles* (mī-SELZ), with the hydrophilic portions on the outside (**Figure 2–17c**). Most meals contain a mixture of lipids and other organic molecules, and micelles form as the food breaks down in your digestive tract. In addition to phospholipids and glycolipids, micelles may contain other insoluble lipids, such as steroids, glycerides, and long-chain fatty acids.

Cholesterol, phospholipids, and glycolipids are called *structural lipids*, because they help form and maintain intracellular structures called membranes. At the cellular level, *membranes* are sheets or layers composed primarily of hydrophobic lipids. A membrane surrounds each cell and separates the aqueous solution inside the cell from the aqueous solution in the extracellular environment. A variety of internal membranes subdivide the interior of the cell into specialized compartments, each with a distinctive chemical nature and, as a result, a different function.

The types of lipids and their characteristics are summarized in Table 2–6.

CHECKPOINT

20. Describe lipids.

21. Which lipids would you find in human plasma membranes?

See the blue Answers tab at the end of the book.

2-11 Proteins are formed from amino acids and contain carbon, hydrogen, oxygen, and nitrogen

Proteins are the most abundant organic components of the human body and in many ways the most important. The human body contains many different proteins, and they account for about 20 percent of total body weight. All proteins contain carbon, hydrogen, oxygen, and nitrogen; smaller quantities of sulfur and phosphorus may also be present.

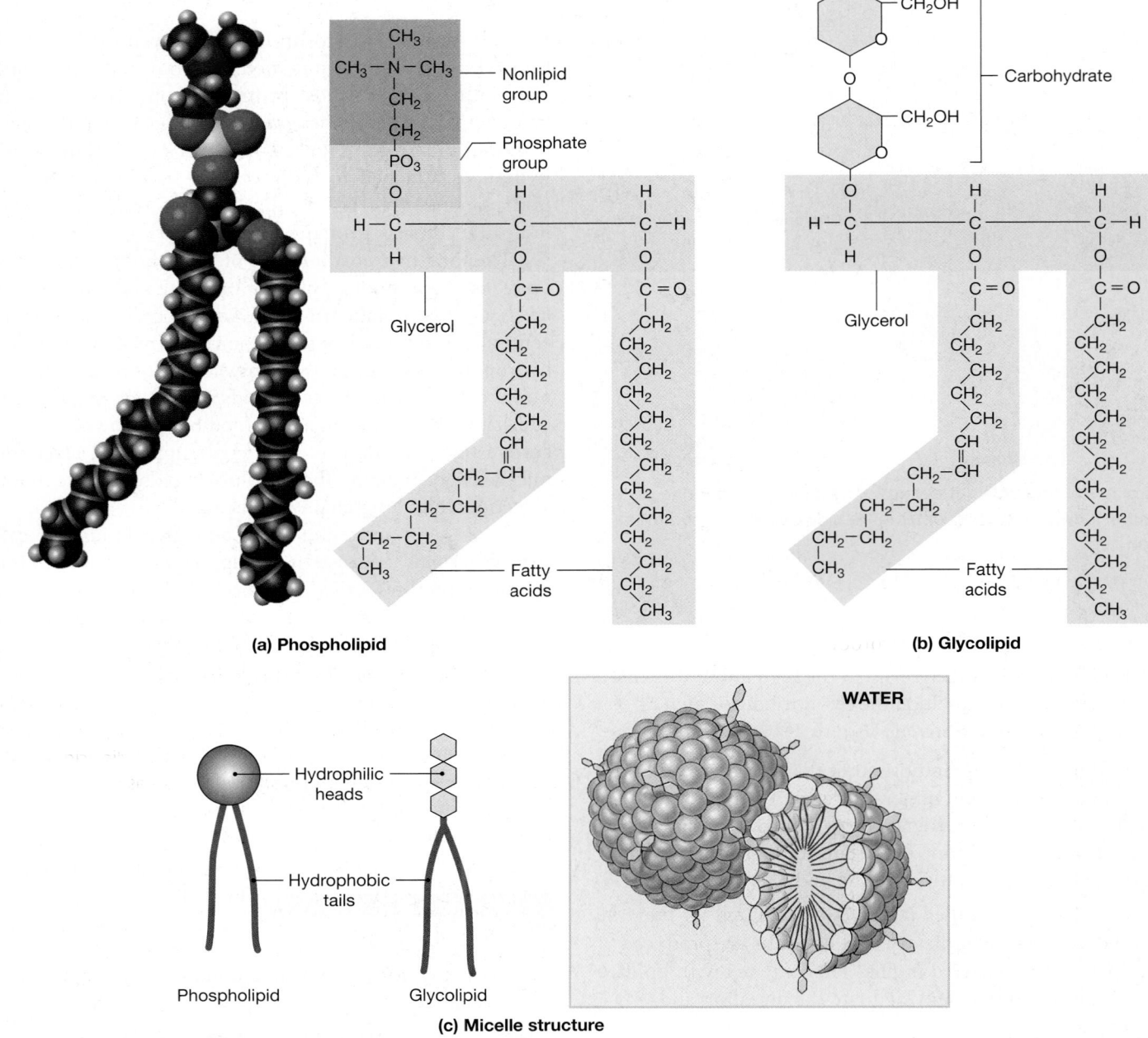

(a) Phospholipid

(b) Glycolipid

(c) Micelle structure

Figure 2–17 **Phospholipids and Glycolipids. (a)** The phospholipid *lecithin*. In a phospholipid, a phosphate group links a nonlipid molecule to a diglyceride. **(b)** In a glycolipid, a carbohydrate is attached to a diglyceride. **(c)** In large numbers, phospholipids and glycolipids form micelles, with the hydrophilic heads facing the water molecules, and the hydrophobic tails on the inside of each droplet.

Proteins perform a variety of essential functions, which can be classified into seven major categories:

1. *Support. Structural proteins* create a three-dimensional framework for the body, providing strength, organization, and support for cells, tissues, and organs.

2. *Movement. Contractile proteins* are responsible for muscular contraction; related proteins are responsible for the movement of individual cells.

3. *Transport.* Insoluble lipids, respiratory gases, special minerals such as iron, and several hormones cannot be transported in the blood, unless they are first bound to special *transport proteins*. Other specialized proteins transport materials from one part of a cell to another.

4. *Buffering.* Proteins provide a considerable *buffering* action and thereby help prevent dangerous changes in pH in cells and tissues.

TABLE 2–6 **Representative Lipids and Their Functions in the Body**

Lipid Type	Example(s)	Primary Function(s)	Remarks
Fatty acids	Lauric acid	Energy source	Absorbed from food or synthesized in cells; transported in the blood
Eicosanoids	Prostaglandins, leukotrienes	Chemical messengers coordinating local cellular activities	Prostaglandins are produced in most body tissues
Glycerides	Monoglycerides, diglycerides, triglycerides	Energy source, energy storage, insulation, and physical protection	Stored in fat deposits; must be broken down to fatty acids and glycerol before they can be used as an energy source
Steroids	Cholesterol, estrogen, testosterone	Structural component of plasma membranes, hormones, digestive secretions in bile	All have the same carbon ring framework
Phospholipids, glycolipids	Lecithin (a phospholipid)	Structural components of plasma membranes	Derived from fatty acids and nonlipid components

5. *Metabolic Regulation. Enzymes* are proteins that accelerate chemical reactions in cells. The sensitivity of enzymes to environmental factors is extremely important in controlling the pace and direction of metabolic operations.

6. *Coordination and Control.* Protein *hormones* can influence the metabolic activities of every cell in the body or affect the function of specific organs or organ systems.

7. *Defense.* The tough, waterproof proteins of the skin, hair, and nails protect the body from environmental hazards. Proteins called *antibodies*, components of the *immune response*, help protect us from disease. After an injury, special *clotting proteins* restrict bleeding to the cardiovascular system.

The A&P Top 100

#13 Proteins are the most abundant organic components of the body. Proteins determine cell shapes and the properties of tissues. Almost all cell functions are performed by proteins and by interactions between proteins and their local environment.

Protein Structure

Proteins consist of long chains of organic molecules called **amino acids**. Twenty different amino acids occur in significant quantities in the body. A typical protein contains 1000 amino acids; the largest protein complexes have 100,000 or more. Each amino acid consists of five components (**Figure 2–18**):

- a central carbon atom
- a hydrogen atom
- an *amino group* (—NH₂)
- a *carboxylic acid group* (—COOH)
- a variable group, known as an *R group* or *side chain*

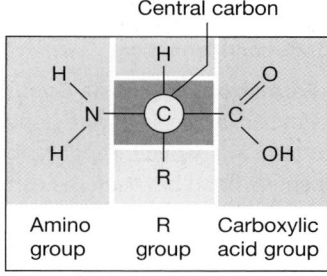

Central carbon

Amino group R group Carboxylic acid group

Structure of an amino acid

Figure 2–18 **Amino Acids.** Each amino acid consists of a central carbon atom to which four different groups are attached: a hydrogen atom, an amino group (—NH₂), a carboxylic acid group (—COOH), and a variable side group designated as R.

The name *amino acid* refers to the presence of the *amino* group and the carboxylic *acid* group, which all amino acids have in common. The different R groups distinguish one amino acid from another, giving each its own chemical properties.

All amino acids are relatively small water-soluble molecules. In the normal pH range of body fluids, the carboxylic acid groups on many amino acids release hydrogen ions. When this occurs, the carboxylic acid group changes from —COOH to —COO⁻, and the amino acid becomes negatively charged.

Figure 2–19 shows how dehydration synthesis can link two representative amino acids: *glycine* and *alanine*. This reaction creates a covalent bond between the carboxylic acid group of one amino acid and the amino group of another. Such a bond is known as a **peptide bond**. Molecules consisting of amino acids held together by peptide bonds are called **peptides**. The molecule created in this example is called a *dipeptide*, because it contains two amino acids.

The chain can be lengthened by the addition of more amino acids. Attaching a third amino acid produces a *tripeptide*.

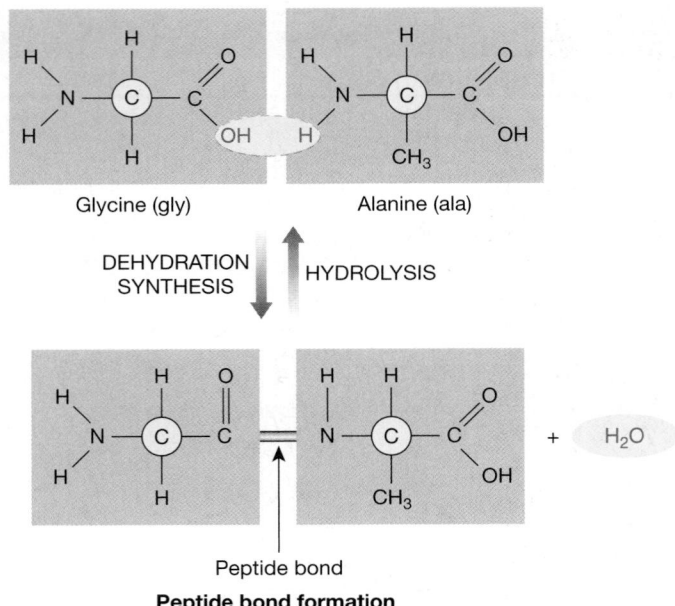

Glycine (gly)   Alanine (ala)

DEHYDRATION SYNTHESIS   HYDROLYSIS

+ H_2O

Peptide bond

Peptide bond formation

Figure 2–19 **The Formation of Peptide Bonds.** In this example, a peptide bond links the amino acids glycine (for which R = H) and alanine (R = CH_3) to form a dipeptide. Peptides form as dehydration synthesis creates a peptide bond between the carboxylic acid group of one amino acid and the amino group of another.

Tripeptides and larger peptide chains are called **polypeptides**. Polypeptides containing more than 100 amino acids are usually called proteins. Familiar proteins include *hemoglobin* in red blood cells and *keratin* in fingernails and hair. Because most proteins contain side groups that are negatively charged, the entire protein acts as an anion. For that reason, they are often indicated by the abbreviation Pr^-.

Protein Shape

The characteristics of a particular protein are determined in part by the R groups on its component amino acids. But the properties of a protein are more than just the sum of the properties of its parts, for polypeptides can have highly complex shapes. Proteins can have four levels of structural complexity (**Figure 2–20**):

1. **Primary structure** is the sequence of amino acids along the length of a single polypeptide (**Figure 2–20a**).

2. **Secondary structure** results from bonds between atoms at different parts of the polypeptide chain. Hydrogen bonding, for example, may create either a simple spiral, known as an *alpha-helix*, or a flat *pleated sheet* (**Figure 2–20b**). Which forms depends on the sequence of amino acids in the peptide chain and where hydrogen bonding occurs along the peptide. The alpha-helix is the

most common form, but a given polypeptide chain may have both helical and pleated sections.

3. **Tertiary structure** is the complex coiling and folding that gives a protein its final three-dimensional shape (**Figure 2–20c**). Tertiary structure results primarily from interactions between the polypeptide chain and the surrounding water molecules, and to a lesser extent from interactions between the R groups of amino acids in different parts of the molecule. Most such interactions are relatively weak. One, however, is very strong: the *disulfide bond*, a covalent bond that may form between two molecules of the amino acid *cysteine* located at different sites along the chain. Disulfide bonds create permanent loops or coils in a polypeptide chain.

4. **Quaternary structure** is the interaction between individual polypeptide chains to form a protein complex (**Figure 2–20d**). Each of the polypeptide subunits has its own secondary and tertiary structures. The protein *hemoglobin* contains four globular subunits. Hemoglobin is found within red blood cells, where it binds and transports oxygen. In *keratin* and *collagen*, three alpha-helical polypeptides are wound together like the strands of a rope. Keratin is the tough, water-resistant protein at the surface of the skin and in nails and hair. Collagen is the most abundant structural protein and is found in skin, bones, cartilages, and tendons; collagen fibers form the framework that supports cells in most tissues.

Fibrous and Globular Proteins

Proteins fall into two general structural classes on the basis of their overall shape and properties:

- **Fibrous proteins** form extended sheets or strands. These shapes are usually the product of secondary structure (for proteins that exhibit the pleated-sheet configuration) or quaternary structure (for keratin and collagen). Fibrous proteins are tough, durable, and generally insoluble in water; in the body, they usually play structural roles.

- **Globular proteins** are compact, generally rounded, and readily enter an aqueous solution. The unique shape of each globular protein is the product of its tertiary structure. *Myoglobin*, a protein in muscle cells, is a globular protein, as is hemoglobin, the oxygen-carrying pigment in red blood cells. Many enzymes, hormones, and other molecules that circulate in the bloodstream are globular proteins, as are the enzymes that control chemical reactions inside cells. These proteins can function only so long as they remain in solution.

Protein Shape and Function

Proteins are extremely versatile and have a variety of functions. The shape of a protein determines its functional prop-

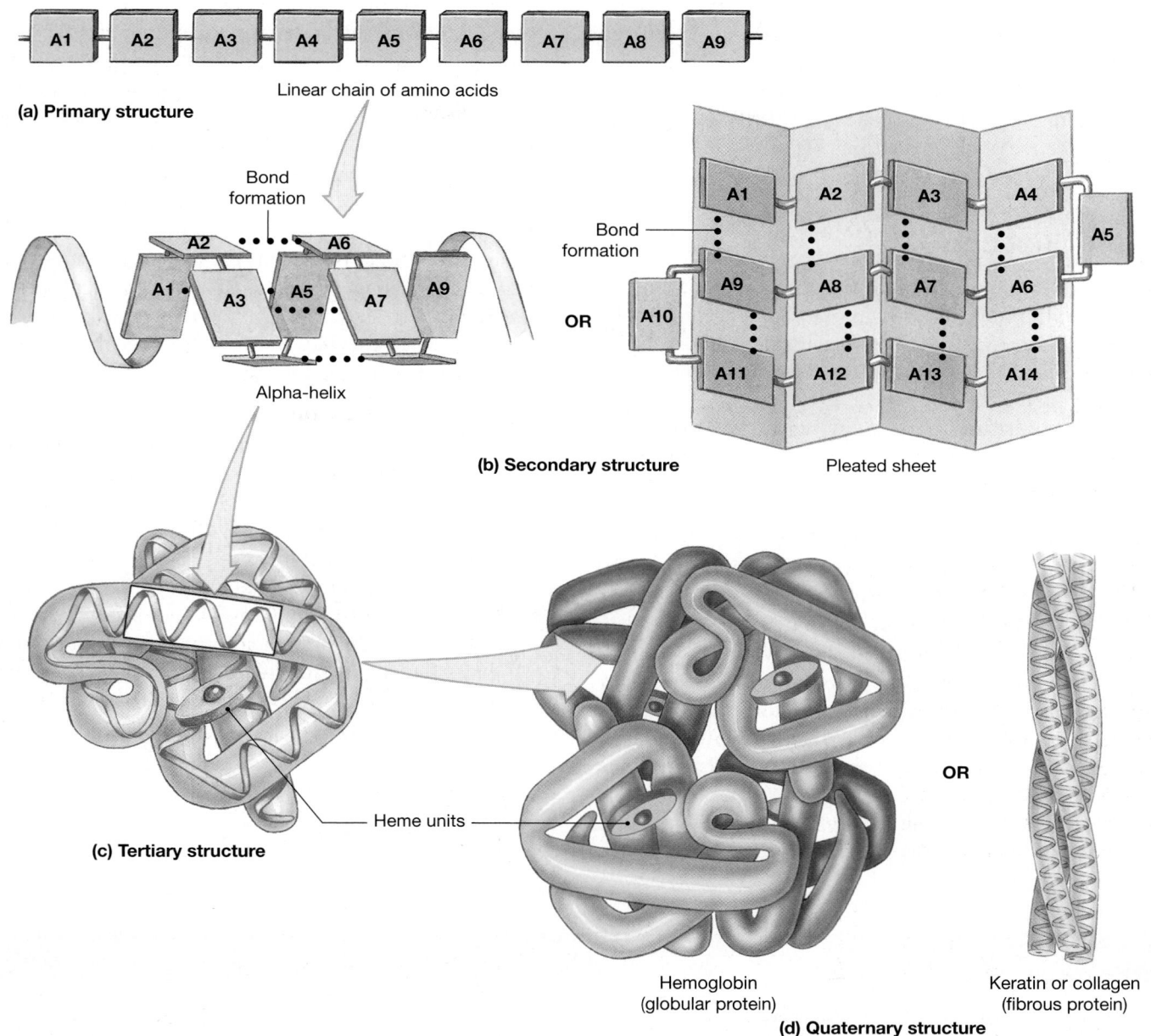

(a) Primary structure

Linear chain of amino acids

Bond formation

Alpha-helix

(b) Secondary structure

Bond formation

OR

Pleated sheet

(c) Tertiary structure

Heme units

Hemoglobin
(globular protein)

OR

Keratin or collagen
(fibrous protein)

(d) Quaternary structure

Figure 2–20 **Protein Structure. (a)** The primary structure of a polypeptide is the sequence of amino acids (A1, A2, A3, and so on) along its length. **(b)** Secondary structure is primarily the result of hydrogen bonding along the length of the polypeptide chain. Such bonding often produces a simple spiral (an alpha-helix) or a flattened arrangement known as a pleated sheet. **(c)** Tertiary structure is the coiling and folding of a polypeptide. Within the cylindrical segments of this globular protein, the polypeptide chain is arranged in an alpha-helix. **(d)** Quaternary structure develops when separate polypeptide subunits interact to form a larger molecule. A single hemoglobin molecule contains four globular subunits. Hemoglobin transports oxygen in the blood; the oxygen binds reversibly to the heme units. In keratin and collagen, three fibrous subunits intertwine. Keratin is a tough, water-resistant protein in skin, hair, and nails. Collagen is the principal extracellular protein in most organs.

erties, and the ultimate determinant of shape is the sequence of amino acids. The 20 common amino acids can be linked in an astonishing number of combinations, creating proteins of enormously varied shape and function. Changing the identity of only one of the 10,000 or more amino acids in a protein can significantly alter the protein's functional properties. For ex-

ample, several cancers and *sickle cell anemia*, a blood disorder, result from single changes in the amino acid sequences of complex proteins.

The tertiary and quaternary shapes of complex proteins depend not only on their amino acid sequence, but also on the local environmental conditions. Small changes in the ionic composition, temperature, or pH of their surroundings can thus affect the function of proteins. Protein shape can also be affected by hydrogen bonding to other molecules in solution. The significance of these factors is most striking when we consider the function of enzymes, for these proteins are essential to the metabolic operations occurring in every one of our cells.

Enzyme Function

Among the most important of all the body's proteins are the enzymes, first introduced earlier in this chapter. These molecules catalyze the reactions that sustain life: Almost everything that happens inside the human body does so because a specific enzyme makes it possible.

The reactants in enzymatic reactions are called **substrates**. As in other types of chemical reactions, the interactions among substrates yield specific products. Before an enzyme can function as a catalyst—to accelerate a chemical reaction without itself being permanently changed or consumed—the substrates must bind to a special region of the enzyme. This region, called the **active site**, is typically a groove or pocket into which one or more substrates nestle, like a key fitting into a lock. Weak electrical attractive forces, such as hydrogen bonding, reinforce the physical fit. The tertiary or quaternary structure of the enzyme molecule determines the shape of the active site. Although enzymes are proteins, any organic or inorganic compound that will bind to the active site can be a substrate.

Figure 2–21 presents one example of enzyme structure and function. Substrates bind to the enzyme at its active site

(STEP 1). Substrate binding typically results in a temporary, reversible change in the shape of the protein; this change may further the reaction by placing physical stresses on the substrate molecules. The enzyme then promotes product formation (STEP 2). The completed product then detaches from the active site (STEP 3), and the enzyme is free to repeat the process. Enzymes work quickly, cycling rapidly between substrates and products. For example, an enzyme providing energy during a muscular contraction performs its reaction sequence 100 times per second; hydrolytic enzymes can work even faster, breaking down almost 20,000 molecules a second!

Figure 2–21 shows an enzyme that catalyzes a synthesis reaction. Other enzymes may catalyze decomposition reactions or exchange reactions. Regardless of the reaction they catalyze, all enzymes share three basic characteristics:

1. *Specificity.* Each enzyme catalyzes only one type of reaction, a characteristic called **specificity**. An enzyme's specificity is determined by the ability of its active sites to bind only to substrates with particular shapes and charges. Thus, differences in enzyme structure that neither affect the active site nor change the response of the enzyme to substrate binding do not affect enzyme function. In fact, different tissues typically contain enzymes that differ slightly in structure, but catalyze the same reaction. Such enzyme variants are called **isozymes**.

2. *Saturation Limits.* The rate of an enzymatic reaction is directly related to the concentrations of substrate molecules and enzymes. An enzyme molecule must encounter appropriate substrates before it can catalyze a reaction; the higher the substrate concentration, the more frequent en-

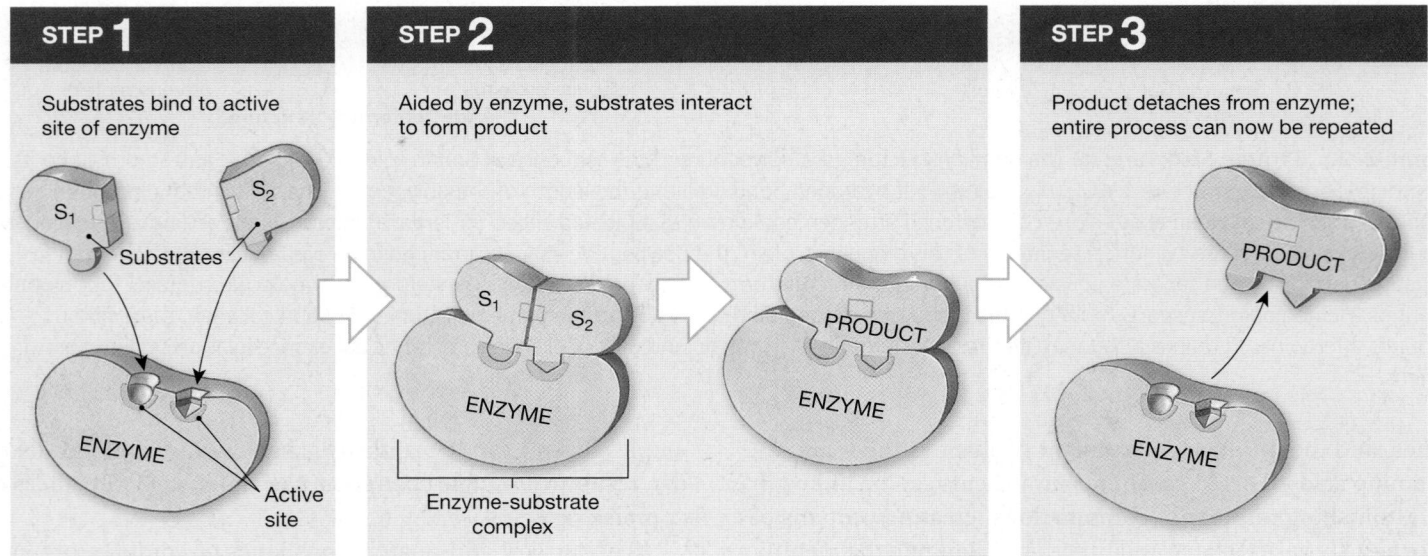

STEP **1**	STEP **2**	STEP **3**
Substrates bind to active site of enzyme	Aided by enzyme, substrates interact to form product	Product detaches from enzyme; entire process can now be repeated

S_1 S_2 Substrates · ENZYME · Active site

S_1 S_2 · ENZYME · Enzyme-substrate complex

PRODUCT · ENZYME

PRODUCT · ENZYME

Figure 2–21 **A Simplified View of Enzyme Structure and Function.** Each enzyme contains a specific active site somewhere on its exposed surface.

counters will be. When substrate concentrations are high enough that every enzyme molecule is cycling through its reaction sequence at top speed, further increases in substrate concentration will not affect the rate of reaction unless additional enzyme molecules are provided. The substrate concentration required to have the maximum rate of reaction is called the *saturation limit*. An enzyme that has reached its saturation limit is said to be **saturated**. To increase the reaction rate further, the cell must increase the number of enzyme molecules available. This is one important way that cells promote specific reactions.

3. *Regulation.* Each cell contains an assortment of enzymes, and any particular enzyme may be active under one set of conditions and inactive under another. Virtually anything that changes the tertiary or quaternary shape of an enzyme can turn it "on" or "off" and thereby control reaction rates inside the cell. Because the change is immediate, enzyme activation or inactivation is an important method of short-term control over reaction rates and pathways. Here we will consider only one example of enzyme regulation: the presence or absence of *cofactors*.

Cofactors and Enzyme Function

A **cofactor** is an ion or a molecule that must bind to the enzyme before substrates can also bind. Without a cofactor, the enzyme is intact but nonfunctional; with the cofactor, the enzyme can catalyze a specific reaction. Examples of cofactors include ions such as calcium (Ca^{2+}) and magnesium (Mg^{2+}), which bind at the enzyme's active site. Cofactors may also bind at other sites, as long as they produce a change in the shape of the active site that makes substrate binding possible.

Coenzymes are nonprotein organic molecules that function as cofactors. Our bodies convert many vitamins into essential coenzymes. *Vitamins*, detailed in Chapter 25, are structurally related to lipids or carbohydrates, but have unique functional roles. Because the human body cannot synthesize most of the vitamins it needs, you must obtain them from your diet.

Effects of Temperature and pH on Enzyme Function

Each enzyme works best at specific temperatures and pH values. As temperatures rise, protein shape changes and enzyme function deteriorates. Eventually the protein undergoes **denaturation**, a change in tertiary or quaternary structure that makes it nonfunctional. You see permanent denaturation when you fry an egg. As the temperature rises, the proteins in the clear white portion surrounding the yolk denature. Eventually, the proteins become completely and irreversibly denatured, forming an insoluble white mass. Death occurs at very high body temperatures (above 43°C, or 110°F) because the denaturation of structural proteins and

enzymes soon causes irreparable damage to organs and organ systems. However, this denaturation can be reversed if the temperature is reduced before the individual dies.

Enzymes are equally sensitive to changes in pH. *Pepsin*, an enzyme that breaks down proteins in stomach contents, works best at a pH of 2.0 (strongly acidic). Your small intestine contains *trypsin*, another enzyme that attacks proteins. Trypsin works only in an alkaline environment, with an optimum pH of 7.7 (weakly basic).

Glycoproteins and Proteoglycans

Glycoproteins (GLĪ-kō-prō-tēnz) and **proteoglycans** (prō-tē-ō-GLĪ-kanz) are combinations of protein and carbohydrate molecules. Glyco*proteins* are large proteins with small carbohydrate groups attached. These molecules may function as enzymes, antibodies, hormones, or protein components of plasma membranes. Glycoproteins in plasma membranes play a major role in the identification of normal versus abnormal cells, as well as in the initiation and coordination of the immune response (Chapter 22). Glycoprotein secretions called *mucins* absorb water to form **mucus**. Mucus coats the surfaces of the reproductive and digestive tracts, providing lubrication. Proteo*glycans* are large polysaccharide molecules linked by polypeptide chains. The proteoglycans in tissue fluids give them a syrupy consistency.

CHECKPOINT

22. Describe a protein.

23. How does boiling a protein affect its structural and functional properties?

See the blue Answers tab at the end of the book.

2-12 DNA and RNA are nucleic acids

Nucleic (noo-KLĀ-ik) **acids** are large organic molecules composed of carbon, hydrogen, oxygen, nitrogen, and phosphorus. Nucleic acids store and process information at the molecular level, inside cells. The two classes of nucleic acid molecules are **deoxyribonucleic** (dē-oks-ē-rī-bō-noo-KLĀ-ik) **acid**, or **DNA**, and **ribonucleic** (rī-bō-noo-KLĀ-ik) **acid**, or **RNA**. As we will see, these two classes of nucleic acids differ in composition, structure, and function.

The DNA in our cells determines our inherited characteristics, including eye color, hair color, and blood type. DNA affects all aspects of body structure and function, because DNA molecules encode the information needed to build proteins. By directing the synthesis of structural proteins, DNA controls the shape and physical characteristics of our bodies. By controlling the manufacture of enzymes, DNA regulates not only protein synthesis, but all aspects of cellular metabolism, including the creation and destruction of lipids, carbohydrates, and other vital molecules.

Several forms of RNA cooperate to manufacture specific proteins by using the information provided by DNA. We will detail the functional relationships between DNA and RNA in Chapter 3.

The A&P Top 100

#14 The DNA in the cell contains the information needed to construct all of the proteins in the body.

Structure of Nucleic Acids

A nucleic acid consists of one or two long chains that are formed by dehydration synthesis. The individual subunits are called **nucleotides** (**Figure 2–22**). Each nucleotide has three components: (1) a *pentose* (five-carbon sugar) attached to both (2) a phosphate group and (3) a **nitrogenous** (nitrogen-containing) **base**. The pentose is either *ribose* (in RNA) or *deoxyribose* (in DNA). Five nitrogenous bases occur in nucleic acids: **adenine (A)**, **guanine (G)**, **cytosine (C)**, **thymine (T)**, and **uracil (U)** (**Figure 2–22b,c**). Adenine and guanine are double-ringed molecules called *purines*; the other three bases are single-ringed molecules called *pyrimidines*. Both RNA and DNA contain adenine, guanine, and cytosine. Uracil occurs only in RNA and thymine only in DNA.

A nucleotide forms when a phosphate group binds to a pentose already attached to a nitrogenous base. In the formation of a nucleic acid, dehydration synthesis then attaches the phosphate group of one nucleotide to the sugar of another. The "backbone" of a nucleic acid molecule is thus a linear sugar-to-phosphate-to-sugar sequence, with the nitrogenous bases projecting to one side (**Figure 2–23**). The primary role of nucleic acids is the storage and transfer of information—specifically, information essential to the synthesis of proteins within our cells. Regardless of whether we are speaking of DNA or RNA, it is the sequence of nitrogenous bases that carries the information.

RNA and DNA

Important structural differences distinguish RNA from DNA. A molecule of RNA consists of a single chain of nucleotides (**Figure 2–23a**). Its shape depends on the order of the nucleotides and the interactions among them. Our cells have three types of RNA: (1) *messenger RNA (mRNA)*, (2) *transfer RNA (tRNA)*, and (3) *ribosomal RNA (rRNA)*. These types have different shapes and functions, but all three are required for the synthesis of proteins, as you will see in Chapter 3.

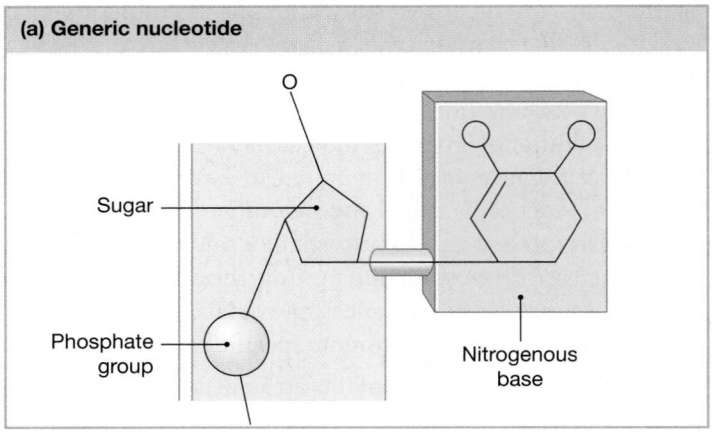

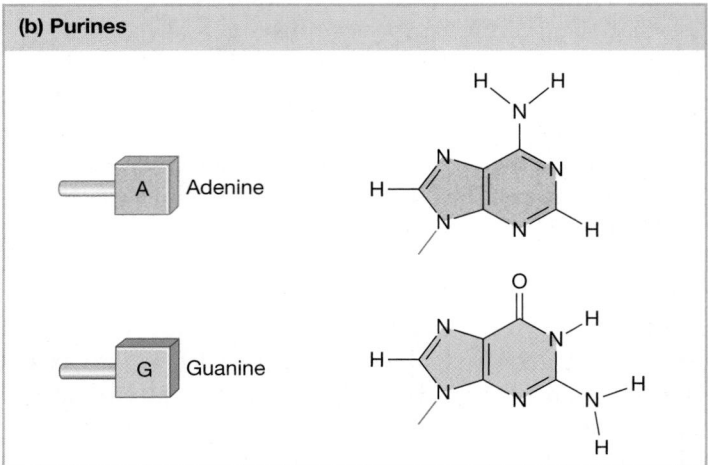

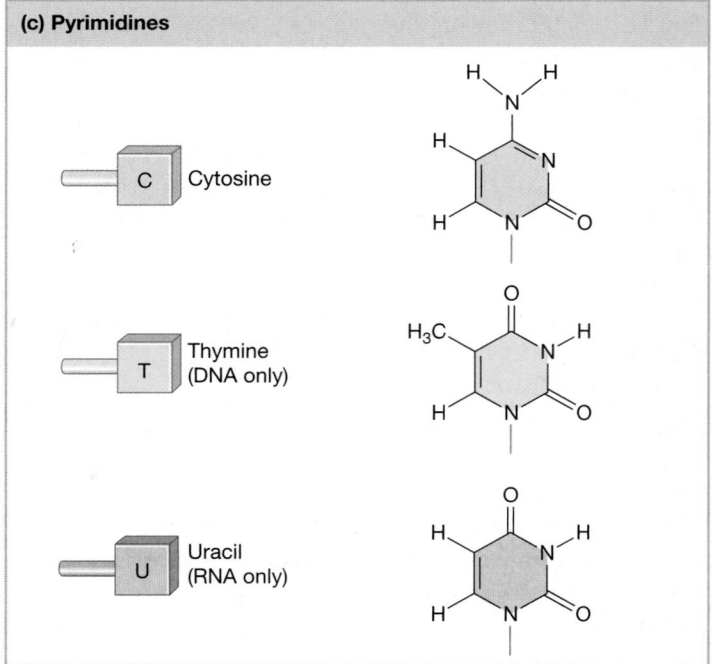

Figure 2–22   Nucleotides and Nitrogenous Bases. (a) General structure of a nucleotide. The nitrogenous base involved may be a purine or a pyrimidine. **(b)** Purines. **(c)** Pyrimidines.

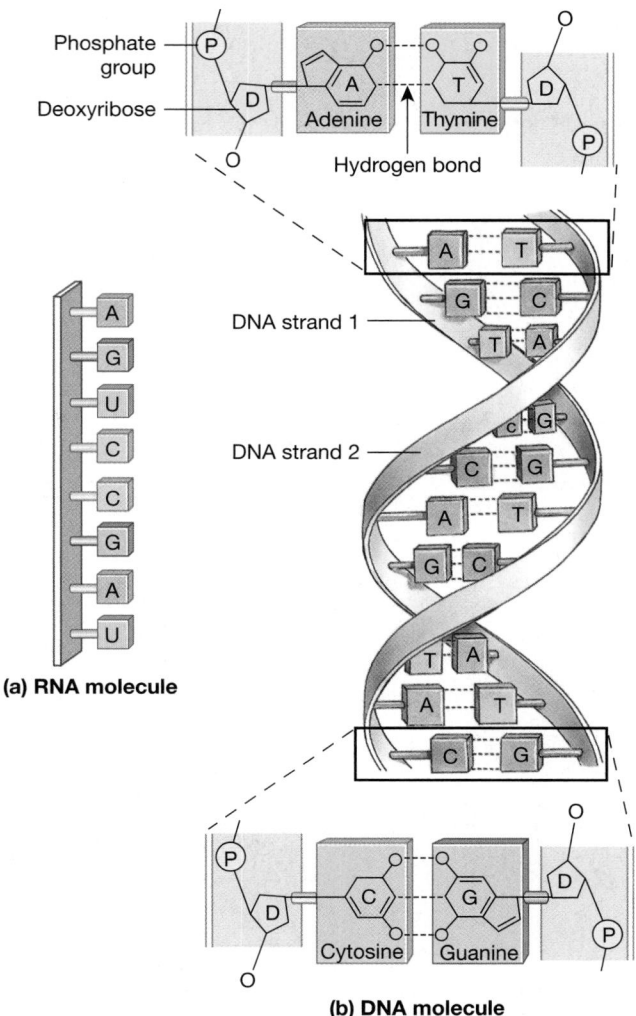

Phosphate group
Deoxyribose
Adenine
Thymine
Hydrogen bond

DNA strand 1

DNA strand 2

(a) RNA molecule

Cytosine   Guanine

(b) DNA molecule

Figure 2–23   The Structure of Nucleic Acids. Nucleic acids are long chains of nucleotides. Each molecule starts at the sugar of the first nucleotide and ends at the phosphate group of the last member of the chain. **(a)** An RNA molecule has a single nucleotide chain. Its shape is determined by the sequence of nucleotides and by the interactions among them. **(b)** A DNA molecule has a pair of nucleotide chains linked by hydrogen bonding between complementary base pairs.

A DNA molecule consists of a *pair* of nucleotide chains (**Figure 2–23b**). Hydrogen bonding between opposing nitrogenous bases holds the two strands together. The shapes of the nitrogenous bases allow adenine to bond only to thymine, and cytosine to bond only to guanine. As a result, the combinations adenine–thymine (A–T) and cytosine–guanine (C–G) are known as **complementary base pairs**, and the two nucleotide chains of the DNA molecule are known as **complementary strands**. Through a sequence of events described in Chapter 3, the cell uses one of the two complementary DNA strands to provide the information needed to synthesize a specific protein. The two strands of DNA twist around one another in a double helix that resembles a spiral staircase. Each step of the staircase corresponds to one complementary base pair (**Figure 2–23b**). Table 2–7 compares RNA with DNA.

CHECKPOINT

24. Describe a nucleic acid.

25. A large organic molecule composed of the sugar ribose, nitrogenous bases, and phosphate groups is which kind of nucleic acid?

See the blue Answers tab at the end of the book.

2-13 ATP is a high-energy compound used by cells

To perform their vital functions, cells must use energy obtained by breaking down organic substrates (catabolism). To be useful, that energy must be transferred from molecule to molecule or from one part of the cell to another.

The usual method of energy transfer involves the creation of *high-energy bonds* by enzymes within cells. A high-energy

TABLE 2–7	Comparison of RNA with DNA	
Characteristic	**RNA**	**DNA**
Sugar	Ribose	Deoxyribose
Nitrogenous bases	Adenine (A)	Adenine
	Guanine (G)	Guanine
	Cytosine (C)	Cytosine
	Uracil (U)	Thymine (T)
Number of nucleotides in typical molecule	Varies from fewer than 100 nucleotides to about 50,000	Always more than 45 million
Shape of molecule	Varies with hydrogen bonding along the length of the strand; three main types (mRNA, rRNA, tRNA)	Paired strands coiled in a double helix
Function	Performs protein synthesis as directed by DNA	Stores genetic information that controls protein synthesis

bond is a covalent bond whose breakdown releases energy the cell can use directly. In your cells, a high-energy bond generally connects a phosphate group (PO_4^{3-}) to an organic molecule. The resulting product is called a **high-energy compound**. Most high-energy compounds are derived from nucleotides, the building blocks of nucleic acids.

The attachment of a phosphate group to another molecule is called **phosphorylation** (fos-for-i-LĀ-shun). This process does not necessarily produce high-energy bonds. The creation of a high-energy compound requires (1) a phosphate group, (2) enzymes capable of catalyzing the reactions involved, and (3) suitable organic substrates to which the phosphate can be added.

The most important such substrate is the nucleotide *adenosine monophosphate (AMP)*. Attaching a second phosphate group produces **adenosine diphosphate (ADP)**. A significant energy input is required to convert AMP to ADP, and the second phosphate is attached by a high-energy bond. Even more energy is required to add a third phosphate and thereby create the high-energy compound **adenosine triphosphate**, or **ATP** (**Figure 2–24**).

The conversion of ADP to ATP is the most important method of energy storage in our cells; the reversion of ATP to ADP is the most important method of energy release. The relationships involved can be diagrammed as:

$$\text{ADP} + \text{phosphate group} + \text{energy} \rightleftharpoons \text{ATP} + H_2O$$

The conversion of ATP to ADP requires an enzyme known as **adenosine triphosphatase**, or **ATPase**. Throughout life, cells continuously generate ATP from ADP and use the energy provided by ATP to perform vital functions, such as the synthesis of proteins or the contraction of muscles.

Although ATP is the most abundant high-energy compound, there are others—typically, other nucleotides that have undergone phosphorylation. For example, *guanosine triphosphate (GTP)* and *uridine triphosphate (UTP)* are nucleotide-based high-energy compounds that transfer energy in specific enzymatic reactions.

Table 2–8 summarizes the inorganic and organic compounds covered in this chapter.

CHECKPOINT

26. Describe ATP.

27. What molecule is produced by the phosphorylation of ADP?

See the blue Answers tab at the end of the of book.

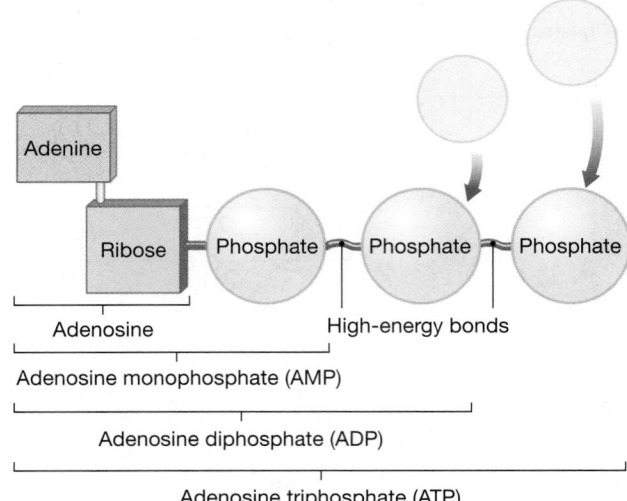

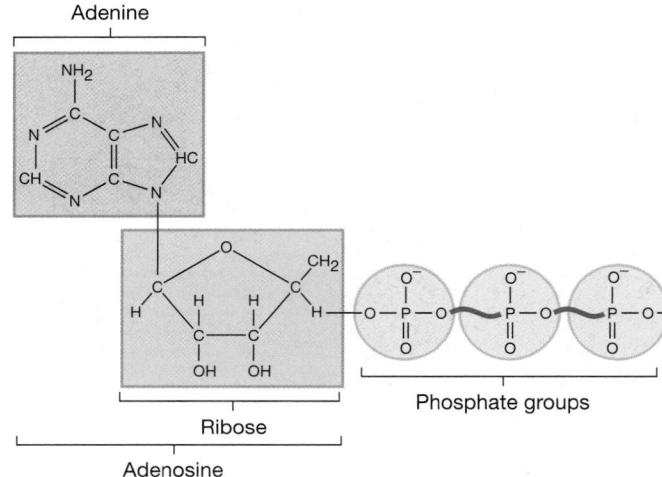

Figure 2–24   The Structure of ATP. A molecule of ATP is formed by attaching two phosphate groups to the nucleotide adenosine monophosphate. These two phosphate groups are connected by high-energy bonds incorporating energy released by catabolism. Cells most often obtain quick energy to power cellular operations by removing one phosphate group from ATP, forming ADP (adenosine diphosphate). ADP can later be reconverted to ATP, and the cycle repeated.

2-14 Chemicals form functional units called cells

The human body is more than a collection of chemicals. The biochemical building blocks discussed in this chapter form functional units called *cells*. ∞ p. 7 Each cell behaves like a miniature organism, responding to internal and external stimuli. This is possible only because cells are dynamic structures

SUMMARY TABLE 2–8   Classes of Inorganic and Organic Compounds

Class	Building Blocks	Sources	Functions
INORGANIC			
Water	Hydrogen and oxygen atoms	Absorbed from the diet or generated by metabolism	Solvent; transport medium for dissolved materials and heat; cooling through evaporation; medium for chemical reactions; reactant in hydrolysis
Acids, bases, salts	H^+, OH^-, various anions and cations	Obtained from the diet or generated by metabolism	Structural components; buffers; sources of ions
Dissolved gases	O, C, N, and other atoms	Atmosphere, metabolism	O_2: required for cellular metabolism CO_2: generated by cells as a waste product NO: chemical messenger in cardiovascular, nervous, and lymphoid systems
ORGANIC			
Carbohydrates	C, H, O, in some cases N; CHO in a 1:2:1 ratio	Obtained from the diet or manufactured in the body	Energy source; some structural role when attached to lipids or proteins; energy storage
Lipids	C, H, O, in some cases N or P; CHO not in 1:2:1 ratio	Obtained from the diet or manufactured in the body	Energy source; energy storage; insulation; structural components; chemical messengers; protection
Proteins	C, H, O, N, commonly S	20 common amino acids; roughly half can be manufactured in the body, others must be obtained from the diet	Catalysts for metabolic reactions; structural components; movement; transport; buffers; defense; control and coordination of activities
Nucleic acids	C, H, O, N, and P; nucleotides composed of phosphates, sugars, and nitrogenous bases	Obtained from the diet or manufactured in the body	Storage and processing of genetic information
High-energy compounds	Nucleotides joined to phosphates by high-energy bonds	Synthesized by all cells	Storage or transfer of energy

that adapt to changes in their environment. Such adaptation may involve changes in the chemical organization of the cell—changes that are easily made because organic molecules other than DNA are temporary rather than permanent components of the cell. Their continuous removal and replacement are part of the process of **metabolic turnover**.

Most of the organic molecules in the cell are replaced at intervals ranging from hours to months. The average time between synthesis and breakdown is known as the *turnover time*. Table 2–9 lists the turnover times of the organic components of representative cells.

In the next chapter we will learn more about the functions of these organic components as we explore the cellular level of organization.

TABLE 2–9   Turnover Times

Cell Type	Component	Average Recycling Time*
Liver	Total protein Enzymes	5–6 days 1 hour to several days, depending on the enzyme
	Glycogen Cholesterol	1–2 days 5–7 days
Muscle cell	Total protein Glycogen	30 days 12–24 hours
Neuron	Phospholipids Cholesterol	200 days 100+ days
Fat cell	Triglycerides	15–20 days

*Most values were obtained from studies on mammals other than humans.

The A&P Top 100

#15 Everything falls apart eventually, some things faster than others. Your survival depends on preventative maintenance; your body must recycle and renew most of its chemical components at intervals ranging from minutes to years. Although it has a relatively high energy cost, metabolic turnover lets your body grow and change, adapting to new conditions and activities.

CHECKPOINT

28. Identify biochemical building blocks discussed in this chapter that are the components of cells.

29. Define metabolic turnover.

See the blue Answers tab at the end of the book.

Related Clinical Terms

cholesterol: A steroid, important in the structure of cellular membranes, that, in high concentrations, increases the risk of heart disease.

mole (mol), millimole (mmol): A quantity of an element or compound that has a weight in grams equal to its atomic or molecular weight, respectively; 1 mmol = 0.001 mol.

omega-3 fatty acids: Fatty acids, abundant in fish flesh and fish oils, that have a double bond three carbons away from the end of the hydrocarbon chain. Their presence in the diet has been linked to reduced risks of heart disease and other conditions.

radioisotopes: Isotopes with unstable nuclei that spontaneously emit subatomic particles or radiation in measurable amounts.

Chapter Review
Study Outline

An Introduction to the Chemical Level of Organization p. 29

1. Chemicals combine to form complex structures.

2-1 Atoms are the basic particles of matter p. 29

2. Atoms are the smallest units of matter. They consist of **protons**, **neutrons**, and **electrons**. (*Figure 2–1*)

3. The number of protons in an atom is its **atomic number**. Each **element** includes all the atoms that have the same number of protons and thus the same atomic number.

4. Within an atom, an **electron cloud** surrounds the nucleus. (*Figure 2–1; Table 2–1*)

5. The **mass number** of an atom is the total number of protons and neutrons in its nucleus. **Isotopes** are atoms of the same element whose nuclei contain different numbers of neutrons.

6. Electrons occupy an orderly series of **energy levels**, commonly illustrated as **electron shells**. The electrons in the outermost energy level determine an element's chemical properties. (*Figure 2–2*)

2-2 Chemical bonds are forces formed by atom interactions p. 32

7. Atoms can combine through chemical reactions that create **chemical bonds**. A **molecule** is any chemical structure consisting of atoms held together by covalent bonds. A **compound** is a chemical substance made up of atoms of two or more elements.

8. An **ionic bond** results from the attraction between **ions**, atoms that have gained or lost electrons. **Cations** are positively charged; **anions** are negatively charged. (*Figure 2–3*)

9. Atoms that share electrons to form a molecule are held together by **covalent bonds**. A sharing of one pair of electrons is a **single covalent bond**; a sharing of two pairs is a **double covalent bond**. A bond with equal sharing of electrons is a **nonpolar covalent bond**; a bond with unequal sharing of electrons is a **polar covalent bond**. (*Figures 2–4, 2–5*)

10. A **hydrogen bond** is a weak, but important, force that can affect the shapes and properties of molecules. (*Figure 2–6*)

11. Matter can exist as a *solid*, a *liquid*, or a *gas*, depending on the nature of the interactions among the component atoms or molecules.

12. The molecular weight of a molecule is the sum of the atomic weights of its component atoms.

13. The rules of **chemical notation** are used to describe chemical compounds and reactions. (*Table 2–2*)

2-3 Decomposition, synthesis, and exchange reactions are important chemical reactions in physiology p. 36

14. A chemical reaction occurs when **reactants** are rearranged to form one or more **products**. Collectively, all the **chemical reactions** in the body constitute its **metabolism**. Through metabolism, cells capture, store, and use energy to maintain homeostasis and to perform essential functions.

15. **Work** is the movement of an object or a change in the physical structure of matter. **Energy** is the capacity to perform work.

16. **Kinetic energy** is the energy of motion. **Potential energy** is stored energy that results from the position or structure of an object. Conversions from potential to kinetic energy (or vice versa) are not 100 percent efficient; every such energy conversion releases *heat*.

17. A chemical reaction is classified as a **decomposition**, a **synthesis**, or an **exchange reaction**.

18. Cells gain energy to power their functions by **catabolism**, the breakdown of complex molecules. Much of this energy supports **anabolism**, the synthesis of new molecules.

19. All chemical reactions are theoretically reversible. At **equilibrium**, the rates of two opposing reactions are in balance.

2-4 Enzymes catalyze specific biochemical reactions by lowering a reaction's activation energy p. 39

20. **Activation energy** is the amount of energy required to start a reaction. **Enzymes** are **catalysts**—compounds that accelerate chemical reactions without themselves being permanently changed or consumed. Enzymes promote chemical reactions by lowering the activation energy requirements. (*Figure 2–7*)

21. **Exergonic** (exothermic) reactions release energy; **endergonic** (endothermic) reactions absorb energy.

2-5 Inorganic compounds usually lack carbon, and organic compounds always contain carbon p. 40

22. **Nutrients** are the essential elements and molecules normally obtained from the diet; **metabolites** are molecules that can be synthesized or broken down by chemical reactions inside our bodies. Nutrients and metabolites can be broadly categorized as either **inorganic** or **organic compounds**.

2-6 Physiological systems depend on water p. 41

23. Water is the most important constituent of the body.
24. A **solution** is a uniform mixture of two or more substances. It consists of a medium, or **solvent**, in which atoms, ions, or molecules of another substance, or **solute**, are individually dispersed. In *aqueous solutions*, water is the solvent. (*Figure 2–8*)
25. Many inorganic compounds, called **electrolytes**, undergo *ionization*, or *dissociation*, in water to form ions. (*Figure 2–8; Table 2–3*) Molecules that interact readily with water molecules are called **hydrophilic**; those that do not are called **hydrophobic**.

2-7 Body fluid pH is vital for homeostasis p. 43

26. The **pH** of a solution indicates the concentration of hydrogen ions it contains. Solutions are classified as **neutral**, **acidic**, or **basic** (*alkaline*) on the basis of pH. (*Figure 2–9*)

2-8 Acids, bases, and salts are inorganic compounds with important physiological roles p. 44

27. An **acid** releases hydrogen ions; a **base** removes hydrogen ions from a solution. *Strong acids* and *strong bases* ionize completely, whereas *weak acids* and *weak bases* do not.
28. A **salt** is an electrolyte whose cation is not hydrogen (H^+) and whose anion is not hydroxide (OH^-).
29. **Buffers** remove or replace hydrogen ions in solution. Buffers and *buffer systems* in body fluids maintain the pH within normal limits.

2-9 Carbohydrates contain carbon, hydrogen, and oxygen in a 1:2:1 ratio p. 45

30. Carbon and hydrogen are the main constituents of **organic compounds**, which generally contain oxygen as well. The properties of the different classes of organic compounds are due to the presence of *functional groups* of atoms. (*Table 2–4*)
31. **Carbohydrates** are most important as an energy source for metabolic processes. The three major types of carbohydrates are **monosaccharides** (*simple sugars*), **disaccharides**, and **polysaccharides**. Disaccharides and polysaccharides form from monosaccharides by **dehydration synthesis**. (*Figures 2–10 to 2–12; Table 2–5*)

2-10 Lipids contain a carbon-to-hydrogen ratio of 1:2 p. 48

32. **Lipids** include *fats*, *oils*, and *waxes*; most are water-insoluble molecules. The five important classes of lipids are **fatty acids**, **eicosanoids**, **glycerides**, **steroids**, and **phospholipids and glycolipids**. (*Figures 2–13 to 2–17; Table 2–6*)
33. **Triglycerides** (*neutral fats*) consist of three fatty acid molecules attached by dehydration synthesis to a molecule of **glycerol**. **Diglycerides** consist of two fatty acids and glycerol. **Monoglycerides** consist of one fatty acid plus glycerol. (*Figure 2–15*)
34. Steroids (1) are components of plasma membranes, (2) include sex hormones and hormones regulating metabolic activities, and (3) are important in lipid digestion. (*Figure 2–16*)

2-11 Proteins are formed from amino acids and contain carbon, hydrogen, oxygen, and nitrogen p. 51

35. **Proteins** perform a variety of essential functions in the body. Seven important types of proteins are *structural proteins*, *contractile proteins, transport proteins, buffering proteins, enzymes, hormones,* and *antibodies*.
36. Proteins are chains of **amino acids**. Each amino acid consists of an *amino group*, a *carboxylic acid group*, a hydrogen atom, and an *R group* (*side chain*) attached to a central carbon atom. A **polypeptide** is a linear sequence of amino acids held together by **peptide bonds**; **proteins** are polypeptides containing over 100 amino acids. (*Figures 2–18, 2–19*)
37. The four levels of protein structure are **primary structure** (amino acid sequence), **secondary structure** (amino acid interactions, such as hydrogen bonds), **tertiary structure** (complex folding, *disulfide bonds*, and interaction with water molecules), and **quaternary structure** (formation of protein complexes from individual subunits). **Fibrous proteins**, such as *keratin* and *collagen*, are elongated, tough, durable, and generally insoluble. **Globular proteins**, such as *myoglobin*, are generally rounded and water-soluble. (*Figure 2–20*)
38. The reactants in an enzymatic reaction, called **substrates**, interact to yield a product by binding to the enzyme's **active site**. **Cofactors** are ions or molecules that must bind to the enzyme before substrate binding can occur. **Coenzymes** are organic cofactors commonly derived from *vitamins*. (*Figure 2–21*)
39. The shape of a protein determines its functional characteristics. Each protein works best at an optimal combination of temperature and pH and will undergo temporary or permanent **denaturation** at temperatures or pH values outside the normal range.

2-12 DNA and RNA are nucleic acids p. 57

40. **Nucleic acids** store and process information at the molecular level. The two kinds of nucleic acids are **deoxyribonucleic acid (DNA)** and **ribonucleic acid (RNA)**. (*Figures 2–22, 2–23; Table 2–7*)
41. Nucleic acids are chains of **nucleotides**. Each nucleotide contains a sugar, a phosphate group, and a **nitrogenous base**. The sugar is *ribose* in RNA and *deoxyribose* in DNA. DNA is a two-stranded double helix containing the nitrogenous bases **adenine**, **guanine**, **cytosine**, and **thymine**. RNA consists of a single strand; it contains **uracil** instead of thymine.

2-13 ATP is a high-energy compound used by cells p. 59

42. Cells store energy in the *high-energy bonds* of **high-energy compounds**. The most important high-energy compound is **ATP (adenosine triphosphate)**. Cells make ATP by adding a phosphate group to **ADP (adenosine diphosphate)** through **phosphorylation**. When ATP is broken down to ADP and phosphate, energy is released. The cell can use this energy to power essential activities. (*Figure 2–24; Summary Table 2–8*)

2-14 Chemicals form functional units called cells p. 60

43. Biochemical building blocks form functional units called *cells*.
44. The continuous removal and replacement of cellular organic molecules (other than DNA), a process called **metabolic turnover**, allows cells to change and to adapt to changes in their environment. (*Table 2–9*)

Review Questions

See the blue Answers tab at the end of the book.

 Access more review material online at myA&P™ (www.myaandp.com). There, you'll find chapter guides, chapter quizzes, practice tests, animations, flashcards, a glossary with pronunciations, *Interactive Physiology*® (IP) exercises and quizzes, and more to help you succeed in the course.

LEVEL 1 Reviewing Facts and Terms

1. An oxygen atom has eight protons. (**a**) Sketch in the arrangement of electrons around the nucleus of the oxygen atom in the following diagram. (**b**) How many more electrons will it take to fill the outermost energy level? _____.

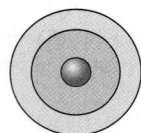

Oxygen atom

2. What is the following type of decomposition reaction called?

$$A-B-C-D + H_2O \rightarrow A-B-C-H + HO-D$$

3. The lightest of an atom's main constituents
   (**a**) carries a negative charge.
   (**b**) carries a positive charge.
   (**c**) plays no part in the atom's chemical reactions.
   (**d**) is found only in the nucleus.

4. Isotopes of an element differ from each other in the number of
   (**a**) protons in the nucleus.
   (**b**) neutrons in the nucleus.
   (**c**) electrons in the outer shells.
   (**d**) a, b, and c are all correct.

5. The number and arrangement of electrons in an atom's outer energy level determines the atom's
   (**a**) atomic weight.
   (**b**) atomic number.
   (**c**) molecular weight.
   (**d**) chemical properties.

6. All organic compounds in the human body contain all of the following elements *except*
   (**a**) hydrogen.   (**b**) oxygen.
   (**c**) carbon.   (**d**) calcium.
   (**e**) both a and d.

7. A substance containing atoms of different elements that are bonded together is called a(n)
   (**a**) molecule.   (**b**) compound.
   (**c**) mixture.   (**d**) isotope.
   (**e**) solution.

8. All the chemical reactions that occur in the human body are collectively referred to as
   (**a**) anabolism.   (**b**) catabolism.
   (**c**) metabolism.   (**d**) homeostasis.

9. Which of the following equations illustrates a typical decomposition reaction?
   (**a**) $A + B \rightarrow AB$
   (**b**) $AB + CD \rightarrow AD + CB$
   (**c**) $2A_2 + B_2 \rightarrow 2A_2B$
   (**d**) $AB \rightarrow A + B$

10. The speed, or rate, of a chemical reaction is influenced by
    (**a**) the presence of catalysts.
    (**b**) the temperature.
    (**c**) the concentration of the reactants.
    (**d**) a, b, and c are all correct.

11. A pH of 7.8 in the human body typifies a condition referred to as
    (**a**) acidosis.   (**b**) alkalosis.
    (**c**) dehydration.   (**d**) homeostasis.

12. A(n) _____ is a solute that dissociates to release hydrogen ions, and a(n) _____ is a solute that removes hydrogen ions from solution.
    (**a**) base, acid   (**b**) salt, base
    (**c**) acid, salt   (**d**) acid, base

13. Chemical reactions in the human body are controlled by special catalytic molecules called
    (**a**) enzymes.   (**b**) cytozymes.
    (**c**) cofactors.   (**d**) activators.
    (**e**) cytochromes.

14. Which the following is *not* a function of a protein?
    (**a**) support
    (**b**) transport
    (**c**) metabolic regulation
    (**d**) storage of genetic information
    (**e**) movement

15. Complementary base pairing in DNA includes the pairs
    (**a**) adenine–uracil and cytosine–guanine.
    (**b**) adenine–thymine and cytosine–guanine.
    (**c**) adenine–guanine and cytosine–thymine.
    (**d**) guanine–uracil and cytosine–thymine.

16. What are the three stable fundamental particles in atoms?

17. What four major classes of organic compounds are found in the body?

18. List three important functions of triglycerides (neutral fats) in the body.

19. List seven major functions performed by proteins.

20. (**a**) What three basic components make up a nucleotide of DNA?
    (**b**) What three basic components make up a nucleotide of RNA?

21. What three components are required to create the high-energy compound ATP?

LEVEL 2 Reviewing Concepts

22. If a polypeptide contains 10 peptide bonds, how many amino acids does it contain?
    (**a**) 0   (**b**) 5
    (**c**) 10   (**d**) 11
    (**e**) 12

23. A dehydration synthesis reaction between glycerol and a single fatty acid would yield a(n)
    (**a**) micelle.   (**b**) omega-3 fatty acid.
    (**c**) monoglyceride.   (**d**) diglyceride.
    (**e**) triglyceride.

24. Explain how enzymes function in chemical reactions.

25. What is a salt? How does a salt differ from an acid or a base?

26. Explain the differences among nonpolar covalent bonds, polar covalent bonds, and ionic bonds.

27. In an exergonic reaction,
    (a) large molecules are broken down into smaller ones.
    (b) small molecules are assembled into larger ones.
    (c) molecules are rearranged to form new molecules.
    (d) molecules move from reactants to products and back.
    (e) energy is released during the reaction.
28. The hydrogen bonding that occurs in water is responsible for all of the following, *except*
    (a) the high boiling point of water.
    (b) the low freezing point of water.
    (c) the ability of water to dissolve nonpolar substances.
    (d) the ability of water to dissolve inorganic salts.
    (e) the surface tension of water.
29. A sample that contains an organic molecule has the following constituents: carbon, hydrogen, oxygen, nitrogen, and phosphorus. Is the molecule more likely to be a carbohydrate, a lipid, a protein, or a nucleic acid?

Level 3   Critical Thinking and Clinical Applications

30. An atom of the element calcium has 20 protons and 20 neutrons. Determine the following information about calcium:
    (a) number of electrons
    (b) atomic number
    (c) atomic weight
    (d) number of electrons in each energy level
31. A certain reaction pathway consists of four steps. How would decreasing the amount of enzyme that catalyzes the second step affect the amount of product produced at the end of the pathway?
32. An important buffer system in the human body involves carbon dioxide (CO_2) and bicarbonate ion (HCO_3^-) in the reversible reactions

$$CO_2 + H_2O \rightleftharpoons H_2CO_3 \rightleftharpoons H^+ + HCO_3^-$$

If a person becomes excited and exhales large amounts of CO_2, how will the pH of the person's body be affected?

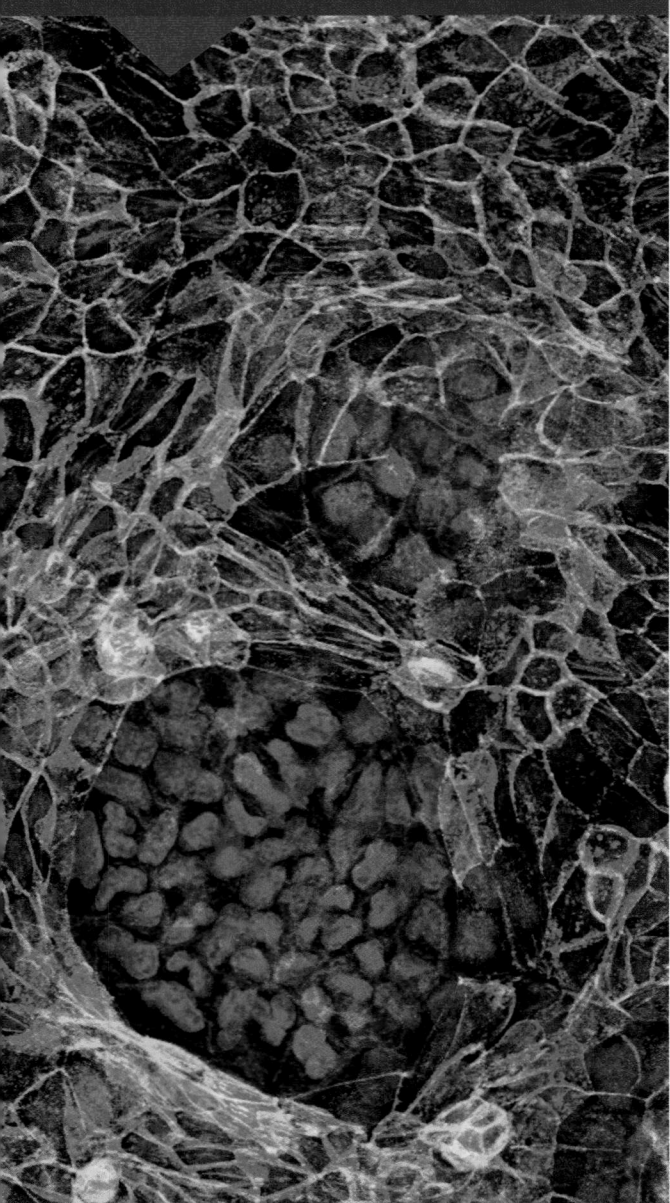

3

The Cellular Level of Organization

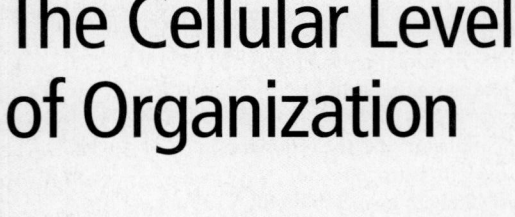

Learning Outcomes

After completing this chapter, you should be able to do the following:

3-1 List the functions of the plasma membrane and the structural features that enable it to perform those functions.

3-2 Describe the organelles of a typical cell, and indicate the specific functions of each.

3-3 Explain the functions of the cell nucleus and discuss the nature and importance of the genetic code.

3-4 Summarize the role of DNA in protein synthesis, cell structure, and cell function.

3-5 Describe the processes of cellular diffusion and osmosis, and explain their role in physiological systems.

3-6 Describe carrier-mediated transport and vesicular transport mechanisms, which cells use to facilitate the absorption or removal of specific substances.

3-7 Explain the origin and significance of the transmembrane potential.

3-8 Describe the stages of the cell life cycle, including mitosis, interphase, and cytokinesis, and explain their significance.

3-9 Discuss the regulation of the cell life cycle.

3-10 Discuss the relationship between cell division and cancer.

3-11 Define differentiation, and explain its importance.

Clinical Notes

Inheritable Mitochondrial Disorders p. 81
DNA Fingerprinting p. 84
Mutations p. 88
Drugs and the Plasma Membrane p. 91
Telomerase, Aging, and Cancer p. 106
Parkinson Disease p. 107

An Introduction to Cells

This chapter relates how combinations of chemicals form *cells*, the smallest living units in the human body. It also describes the chemical events that sustain life, most of which occur inside cells.

Cells are very small indeed—a typical cell is only about 0.1 mm in diameter. As a result, no one could actually examine the structure of a cell until relatively effective microscopes were invented in the 17th century. In 1665, Robert Hooke inspected thin slices of cork and found that they consisted of millions of small, irregular units. In describing his observations, Hooke used the term *cell* because the many small, bare spaces he saw reminded him of the rooms, or cells, in a monastery or prison. Although Hooke saw only the outlines of the cells, and not the cells themselves, he stimulated considerable interest in the microscopic world and in the nature of cellular life. The research that he began more than 342 years ago has, over time, produced the *cell theory* in its current form. The basic concepts of this theory can be summarized as follows:

- Cells are the building blocks of all plants and animals.
- All cells come from the division of preexisting cells.
- Cells are the smallest units that perform all vital physiological functions.
- Each cell maintains homeostasis at the cellular level.

Homeostasis at the level of the tissue, organ, organ system, and organism reflects the combined and coordinated actions of many cells.

The human body contains trillions of cells, and all our activities—from running to thinking—result from the combined and coordinated responses of millions or even billions of cells. Many insights into human physiology arose from studies of the functioning of individual cells. What we have learned over the last 50 years has given us a new understanding of cellular physiology and the mechanisms of homeostatic control. Today, the study of cellular structure and function, or **cytology**, is part of the broader discipline of **cell biology**, which incorporates aspects of biology, chemistry, and physics.

The human body contains two general classes of cells: sex cells and somatic cells. **Sex cells** (also called *germ cells* or *reproductive cells*) are either the *sperm* of males or the *oocytes* of females. The fusion of a sperm and an oocyte at fertilization is the first step in the creation of a new individual. **Somatic cells** (*soma*, body) include all the other cells in the human body. In this chapter, we focus on somatic cells; we will discuss sex cells in Chapters 28 and 29, which describe the reproductive system and development, respectively.

In the rest of this chapter, we describe the structure of a typical somatic cell, consider some of the ways in which cells interact with their environment, and discuss how somatic cells reproduce. It is important to keep in mind that the "typical" somatic cell is like the "average" person: Any description masks enormous individual variations. Our model cell, shown in **Figure 3–1,** shares features with most cells of the body, without being identical to any one. Table 3–1 summarizes the structures and functions of the cell's parts.

Our model cell is surrounded by a watery medium known as the **extracellular fluid**. The extracellular fluid in most tissues is called **interstitial** (in-ter-STISH-ul) **fluid** (*interstitium*, something standing between). A *plasma membrane (cell membrane)* separates the cell contents, or *cytoplasm*, from the extracellular fluid. The cytoplasm can itself be subdivided into (1) the *cytosol*, a liquid, and (2) intracellular structures collectively known as *organelles* (or-ga-NELZ; "little organs").

3-1 The plasma membrane separates the cell from its surrounding environment and performs various functions

We begin our look at the anatomy of cells by discussing the first structure you encounter when viewing cells through a microscope. The outer boundary of the cell is the **plasma membrane**, also called the **cell membrane** or **plasmalemma** (*lemma*, husk). Its general functions include the following:

- *Physical Isolation.* The plasma membrane is a physical barrier that separates the inside of the cell from the surrounding extracellular fluid. Conditions inside and outside the cell are very different, and those differences must be maintained to preserve homeostasis. For example, the plasma membrane keeps enzymes and structural proteins inside the cell.

- *Regulation of Exchange with the Environment.* The plasma membrane controls the entry of ions and nutrients, such as glucose; the elimination of wastes; and the release of secretions.

- *Sensitivity to the Environment.* The plasma membrane is the first part of the cell affected by changes in the composition, concentration, or pH of the extracellular fluid. It also contains a variety of receptors that allow the cell to recognize and respond to specific molecules in its environment. For instance, the plasma membrane may receive chemical signals from other cells. The binding of just one molecule may trigger the activation or deactivation of enzymes that affect many cellular activities.

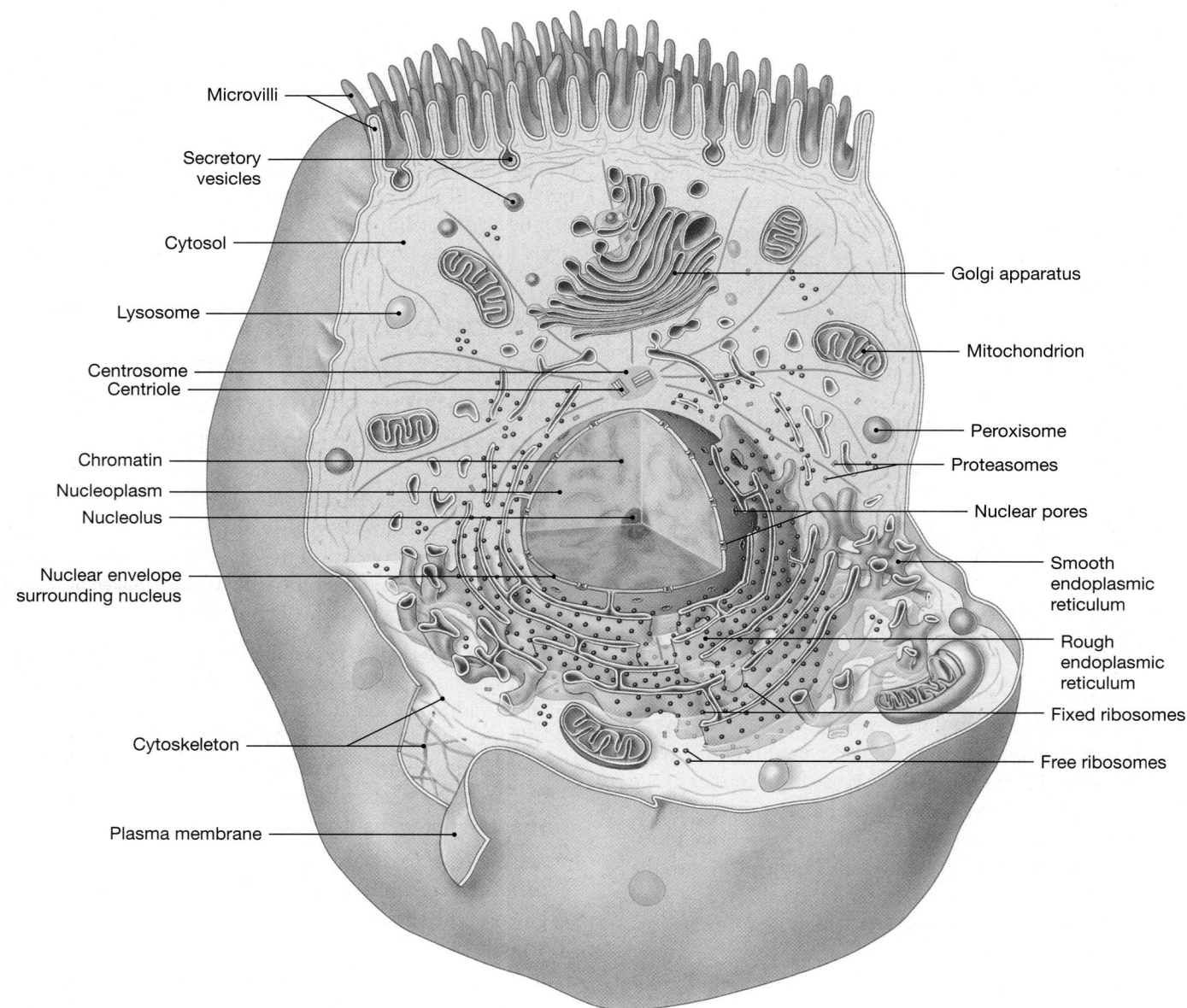

Figure 3–1 **The Anatomy of a Model Cell.** See Table 3–1 for a summary of the functions associated with the various cell structures.

TABLE 3–1 Organelles of a Representative Cell

Appearance	Structure	Composition	Function(s)
	PLASMA MEMBRANE	Lipid bilayer containing phospholipids, steroids, proteins, and carbohydrates	Isolation; protection; sensitivity; support; controls entry and exit of materials
	CYTOSOL	Fluid component of cytoplasm	Distributes materials by diffusion
NONMEMBRANOUS ORGANELLES			
	Cytoskeleton Microtubule Microfilament	Proteins organized in fine filaments or slender tubes	Strength and support; movement of cellular structures and materials
	Microvilli	Membrane extensions containing microfilaments	Increase surface area to facilitate absorption of extracellular materials
	Centrosome Centrioles	Cytoplasm containing two centrioles at right angles; each centriole is composed of 9 microtubule triplets in a 9 + 0 array	Essential for movement of chromosomes during cell division; organization of microtubules in cytoskeleton
	Cilia	Membrane extensions containing microtubule doublets in a 9 + 2 array	Movement of materials over cell surface
	Ribosomes	RNA + proteins; fixed ribosomes bound to rough endoplasmic reticulum, free ribosomes scattered in cytoplasm	Protein synthesis
	Proteasomes	Hollow cylinders of proteolytic enzymes with regulatory proteins at ends	Breakdown and recycling of damaged or abnormal intracellular proteins
MEMBRANOUS ORGANELLES			
	Mitochondria	Double membrane, with inner membrane folds (cristae) enclosing important metabolic enzymes	Produce 95% of the ATP required by the cell
	Endoplasmic reticulum (ER)	Network of membranous channels extending throughout the cytoplasm	Synthesis of secretory products; intracellular storage and transport
	Rough ER	Has ribosomes bound to membranes	Modification and packaging of newly synthesized proteins
	Smooth ER	Lacks attached ribosomes	Lipid and carbohydrate synthesis
	Golgi apparatus	Stacks of flattened membranes (cisternae) containing chambers	Storage, alteration, and packaging of secretory products and lysosomal enzymes
	Lysosome	Vesicles containing digestive enzymes	Intracellular removal of damaged organelles or pathogens
	Peroxisome	Vesicles containing degradative enzymes	Catabolism of fats and other organic compounds; neutralization of toxic compounds generated in the process
	NUCLEUS Nuclear envelope Nuclear pore	Nucleoplasm containing nucleotides, enzymes, nucleoproteins, and chromatin; surrounded by double membrane (nuclear envelope)	Control of metabolism; storage and processing of genetic information; control of protein synthesis
	Nucleolus	Dense region in nucleoplasm containing DNA and RNA	Site of rRNA synthesis and assembly of ribosomal subunits

- *Structural Support.* Specialized connections between plasma membranes, or between membranes and extracellular materials, give tissues stability. For example, the cells at the surface of the skin are bound together, while those in the deepest layers are attached to extracellular protein fibers in underlying tissues.

The plasma membrane is extremely thin and delicate, ranging from 6 to 10 nm in thickness (**Figure 3–2**). This membrane contains lipids, proteins, and carbohydrates.

Membrane Lipids

Although lipids form most of the surface area of the plasma membrane, they account for only about 42 percent of its weight. The plasma membrane is called a **phospholipid bilayer**, because the phospholipid molecules in it form two layers. Recall from Chapter 2 that a phospholipid has both a hydrophilic end (the phosphate portion) and a hydrophobic end (the lipid portion). ∞ p. 51 In each half of the bilayer, the phospholipids lie with their hydrophilic heads at the membrane surface and their hydrophobic tails on the inside. Thus, the hydrophilic heads of the two layers are in contact with the aqueous environments on either side of the membrane—the interstitial fluid on the outside and the cytosol on the inside—and the hydrophobic tails form the interior of the membrane. The lipid bilayer also contains cholesterol and small quantities of other lipids, but these have relatively little effect on the general properties of the plasma membrane.

Notice the similarities in lipid organization between the plasma membrane and a micelle (see **Figure 2–17c**, p. 52). Ions and water-soluble compounds cannot enter the interior of a micelle, because the lipid tails of the phospholipid molecules are hydrophobic and will not associate with water molecules. For the same reason, water and solutes cannot cross the lipid portion of the plasma membrane. Thus, the hydrophobic compounds in the center of the membrane isolate the cytoplasm from the surrounding fluid environment. Such isolation is important because the composition of cytoplasm is very different from that of extracellular fluid, and the cell cannot survive if the differences are eliminated.

Membrane Proteins

Proteins, which are much denser than lipids, account for roughly 55 percent of the weight of a plasma membrane. There are two general structural classes of membrane pro-

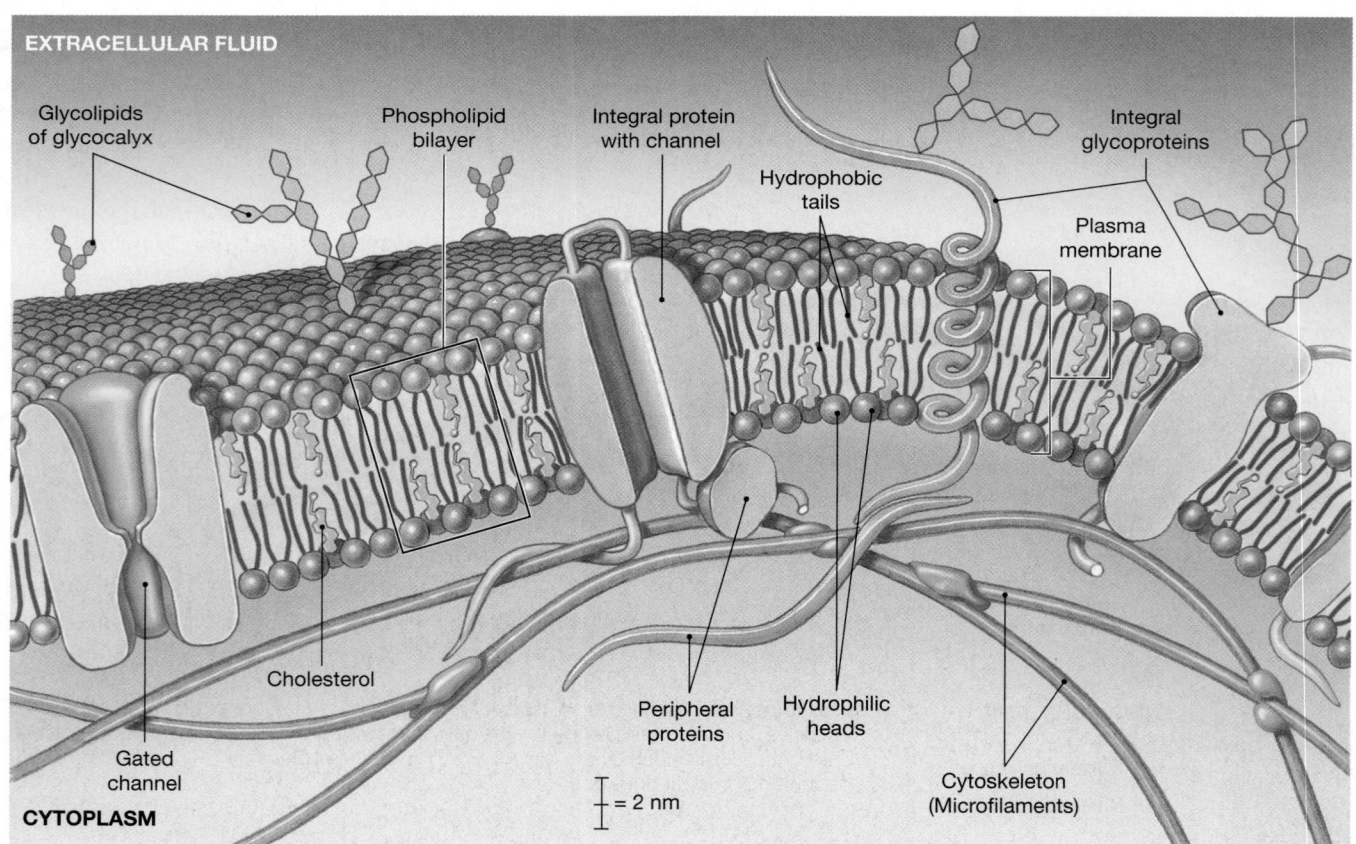

Figure 3–2   The Plasma Membrane.

teins (**Figure 3–2**). **Integral proteins** are part of the membrane structure and cannot be removed without damaging or destroying the membrane. Most integral proteins span the width of the membrane one or more times, and are therefore known as *transmembrane proteins*. **Peripheral proteins** are bound to the inner or outer surface of the membrane and (like Post-it notes) are easily separated from it. Integral proteins greatly outnumber peripheral proteins.

Membrane proteins may have a variety of specialized functions. Examples of important types of functional proteins include the following:

1. *Anchoring Proteins.* **Anchoring proteins** attach the plasma membrane to other structures and stabilize its position. Inside the cell, membrane proteins are bound to the *cytoskeleton*, a network of supporting filaments in the cytoplasm. Outside the cell, other membrane proteins may attach the cell to extracellular protein fibers or to another cell.

2. *Recognition Proteins (Identifiers).* The cells of the immune system recognize other cells as normal or abnormal based on the presence or absence of characteristic **recognition proteins**. Many important recognition proteins are glycoproteins. ∞ p. 57 (We will discuss one group, the *MHC proteins* involved in the immune response, in Chapter 22.)

3. *Enzymes.* Enzymes in plasma membranes may be integral or peripheral proteins. They catalyze reactions in the extracellular fluid or in the cytosol, depending on the location of the protein and its active site. For example, dipeptides are broken down into amino acids by enzymes on the exposed membranes of cells that line the intestinal tract.

4. *Receptor Proteins.* **Receptor proteins** in the plasma membrane are sensitive to the presence of specific extracellular molecules called **ligands** (LĪ-gandz). A ligand can be anything from a small ion, like calcium (Ca^{2+}), to a relatively large and complex hormone. A receptor protein exposed to an appropriate ligand will bind to it, and that binding may trigger changes in the activity of the cell. For example, the binding of the hormone *insulin* to a specific membrane receptor protein is the key step that leads to an increase in the rate of glucose absorption by the cell. Plasma membranes differ in the type and number of receptor proteins they contain, and these differences account for their differing sensitivities to hormones and other potential ligands.

5. *Carrier Proteins.* **Carrier proteins** bind solutes and transport them across the plasma membrane. Carrier proteins may require ATP as an energy source. ∞ p. 59 For example, virtually all cells have carrier proteins that can bring glucose into the cytoplasm without expending ATP, but these cells must expend ATP to transport ions such as sodium and calcium across the plasma membrane and out of the cytoplasm.

6. *Channels.* Some integral proteins contain a central pore, or **channel**, that forms a passageway completely across the plasma membrane. The channel permits the movement of water and small solutes across the plasma membrane. Ions do not dissolve in lipids, so they cannot cross the phospholipid bilayer. Thus, ions and other small water-soluble materials can cross the membrane only by passing through channels. Many channels are highly specific—that is, they permit the passage of only one particular ion. The movement of ions through channels is involved in a variety of physiological mechanisms. Although channels account for about 0.2 percent of the total surface area of the membrane, they are extremely important in such physiological processes as nerve impulse transmission and muscle contraction, described in Chapters 10 and 12.

Membranes are neither rigid nor uniform. At each location, the inner and outer surfaces of the plasma membrane may differ in important respects. For example, some cytoplasmic enzymes are found only on the inner surface of the membrane, and some receptors are found exclusively on its outer surface. Some embedded proteins are always confined to specific areas of the plasma membrane. These areas, called *rafts*, mark the location of anchoring proteins and some kinds of receptor proteins. Yet because membrane phospholipids are fluid at body temperature, many other integral proteins drift across the surface of the membrane like ice cubes in a bowl of punch. In addition, the composition of the entire plasma membrane can change over time, because large areas of the membrane surface are continually being removed and recycled in the process of metabolic turnover. ∞ p. 61

Membrane Carbohydrates

Carbohydrates account for roughly 3 percent of the weight of a plasma membrane. The carbohydrates in the plasma membrane are components of complex molecules such as *proteoglycans, glycoproteins*, and *glycolipids*. ∞ pp. 51, 57 The carbohydrate portions of these large molecules extend beyond the outer surface of the membrane, forming a layer known as the **glycocalyx** (glī-kō-KĀ-liks; *calyx*, cup). The glycocalyx has a variety of important functions, including the following:

- *Lubrication and Protection.* The glycoproteins and glycolipids form a viscous layer that lubricates and protects the plasma membrane.

- *Anchoring and Locomotion.* Because the components are sticky, the glycocalyx can help anchor the cell in place. It also participates in the locomotion of specialized cells.

- *Specificity in Binding.* Glycoproteins and glycolipids can function as receptors, binding specific extracellular compounds. Such binding can alter the properties of the cell surface and indirectly affect the cell's behavior.

- *Recognition.* Glycoproteins and glycolipids are recognized as normal or abnormal by cells involved with the immune response. The characteristics of the glycocalyx are genetically determined. The body's immune system recognizes its own membrane glycoproteins and glycolipids as "self" rather than as "foreign." This recognition system keeps your immune system from attacking your cells, while still enabling it to recognize and destroy invading pathogens.

The plasma membrane serves as a barrier between the cytosol and the extracellular fluid. If the cell is to survive, dissolved substances and larger compounds must be permitted to move across this barrier. Metabolic wastes must be able to leave the cytosol, and nutrients must be able to enter the cell. The structure of the plasma membrane is ideally suited to this need for selective transport. We will discuss selective transport and other membrane functions further, after we have completed our overview of cellular anatomy.

CHECKPOINT

1. List the general functions of the plasma membrane.
2. Identify the components of the plasma membrane that enable it to perform its characteristic functions.
3. Which component of the plasma membrane is primarily responsible for the membrane's ability to form a physical barrier between the cell's internal and external environments?
4. Which type of integral protein allows water and small ions to pass through the plasma membrane?

See the blue Answers tab at the end of the book.

3-2 Organelles within the cytoplasm perform particular functions

Cytoplasm is a general term for the material located between the plasma membrane and the membrane surrounding the nucleus. A colloid with a consistency that varies between that of thin maple syrup and almost-set gelatin, cytoplasm contains many more proteins than does extracellular fluid. ∞ p. 42 As an indication of the importance of proteins to the cell, roughly 30 percent of a typical cell's weight can be attributed to proteins. The cytoplasm contains cytosol and organelles. **Cytosol**, or *intracellular fluid*, contains dissolved nutrients, ions, soluble and insoluble proteins, and waste products. **Organelles** are structures suspended within the cytosol that perform specific functions within the cell.

The Cytosol

The most important differences between cytosol and extracellular fluid are as follows:

1. The concentration of potassium ions is much higher in the cytosol than in the extracellular fluid. Conversely, the concentration of sodium ions is much lower in the cytosol than in the extracellular fluid.

2. The cytosol contains a much higher concentration of suspended proteins than does extracellular fluid. Many of the proteins are enzymes that regulate metabolic operations; others are associated with the various organelles. The consistency of the cytosol is determined in large part by the enzymes and cytoskeletal proteins.

3. The cytosol usually contains small quantities of carbohydrates, and small reserves of amino acids and lipids. The extracellular fluid is a transport medium only, and no reserves are stored there. The carbohydrates in the cytosol are broken down to provide energy, and the amino acids are used to manufacture proteins. Lipids, in particular triglycerides, are used primarily as a source of energy when carbohydrates are unavailable.

Both the cytosol and the extracellular fluid within tissues (*interstitial fluid*) may contain masses of insoluble materials. In the cytosol, these masses are known as **inclusions**. Among the most common inclusions are stored nutrients, such as glycogen granules in liver or in skeletal muscle cells, and lipid droplets in fat cells. Other common inclusions are pigment granules, such as the brown pigment *melanin* and the orange pigment *carotene*. Examples of insoluble materials in interstitial fluids include melanin in the skin and mineral deposits in bone.

The Organelles

Organelles are the internal structures that perform most of the tasks that keep a cell alive and functioning normally. Each organelle has specific functions related to cell structure, growth, maintenance, and metabolism. Cellular organelles can be divided into two broad categories, nonmembranous and membranous. **Nonmembranous organelles** are not completely enclosed by membranes, and all of their components are in direct contact with the cytosol. **Membranous organelles** are isolated from the cytosol by phospholipid membranes, just as the plasma membrane isolates the cytosol from the extracellular fluid.

The cell's nonmembranous organelles include the *cytoskeleton*, *microvilli*, *centrioles*, *cilia*, *ribosomes*, and *proteasomes*. Membranous organelles include the *endoplasmic reticulum*, the *Golgi apparatus*, *lysosomes*, *peroxisomes*, and *mitochondria*. The *nucleus*, also surrounded by a membranous envelope—and therefore, strictly speaking, a membranous organelle—has so many vital functions that we will consider it in a separate section.

The Cytoskeleton

The **cytoskeleton** functions as the cell's skeleton. It provides an internal protein framework that gives the cytoplasm strength and flexibility. The cytoskeleton of all cells includes *microfilaments*, *intermediate filaments*, and *microtubules*. Muscle cells contain these cytoskeletal elements plus *thick filaments*. The filaments of the cytoskeleton form a dynamic network. The organizational details remain poorly understood, because the network is extremely delicate and thus hard to study intact. **Figure 3–3a** is based on our current knowledge of cytoskeletal structure.

We will consider only a few of the many functions of the cytoskeleton in this section. In addition to the functions described here, the cytoskeleton plays a role in the metabolic organization of the cell by determining where in the cytoplasm key enzymatic reactions occur and where specific proteins are synthesized. For example, many intracellular enzymes—especially those involved with metabolism and energy production, and the ribosomes and RNA molecules responsible for the synthesis of proteins—are attached to the microfilaments and microtubules of the cytoskeleton. The varied metabolic functions of the cytoskeleton are now a subject of intensive research.

Microfilaments   The smallest of the cytoskeletal elements are the **microfilaments**. These protein strands are generally less than 6 nm in diameter. Typical microfilaments are composed of the protein **actin**. In most cells, actin filaments are common in the periphery of the cell, but relatively rare in the region immediately surrounding the nucleus. In cells that form a layer or lining, such as the lining of the intestinal tract, actin filaments also form a layer, the *terminal web*, just inside the membrane at the exposed surface of the cell (**Figure 3–3a**).

Microfilaments have three major functions:

1. Microfilaments anchor the cytoskeleton to integral proteins of the plasma membrane. They provide the cell additional mechanical strength and attach the plasma membrane to the enclosed cytoplasm.

2. Microfilaments, interacting with other proteins, determine the consistency of the cytoplasm. Where microfilaments form a dense, flexible network, the cytoplasm has a gelatinous consistency; where they are widely dispersed, the cytoplasm is more fluid.

3. Actin can interact with the protein *myosin* to produce active movement of a portion of a cell or to change the shape of the entire cell.

Intermediate Filaments   The protein composition of **intermediate filaments** varies among cell types. These filaments, which range from 7 to 11 nm in diameter, are intermediate in size between microfilaments and thick filaments. Intermediate filaments (1) strengthen the cell and help maintain its shape, (2) stabilize the positions of organelles, and (3) stabilize the position of the cell with respect to surrounding cells through specialized attachment to the plasma membrane. Intermediate filaments, which are insoluble, are the most durable of the cytoskeletal elements. Many cells contain specialized intermediate filaments with unique functions. For example, the keratin fibers in superficial layers of the skin are intermediate filaments that make these layers strong and able to resist stretching. ∞ p. 54

Microtubules   Most cells contain **microtubules**, hollow tubes built from the globular protein **tubulin**. Microtubules are the largest components of the cytoskeleton, with diameters of about 25 nm. Microtubules extend outward into the periphery of the cell from a region near the nucleus called the *centrosome* (**Figure 3–1**). The number and distribution of microtubules in the cell can change over time. Each microtubule forms by the aggregation of tubulin molecules, growing out from its origin at the centrosome. The entire structure persists for a time and then disassembles into individual tubulin molecules again. At any given moment, roughly half of the tubulin molecules in the cell are tied up in microtubules, and the rest are awaiting recycling.

Microtubules have the following functions:

1. Microtubules form the primary components of the cytoskeleton, giving the cell strength and rigidity and anchoring the position of major organelles.

2. The disassembly of microtubules provides a mechanism for changing the shape of the cell, perhaps assisting in cell movement.

3. Microtubules can serve as a kind of monorail system to move vesicles or other organelles within the cell. The movement is effected by proteins called *molecular motors*. These proteins, which bind to the structure being moved, also bind to a microtubule and move along its length. The direction of movement depends on which of several known motor proteins is involved. For example, the molecular motors *kinesin* and *dynein* carry materials in opposite directions: Kinesin moves toward one end of a microtubule, dynein toward the other. Regardless of the

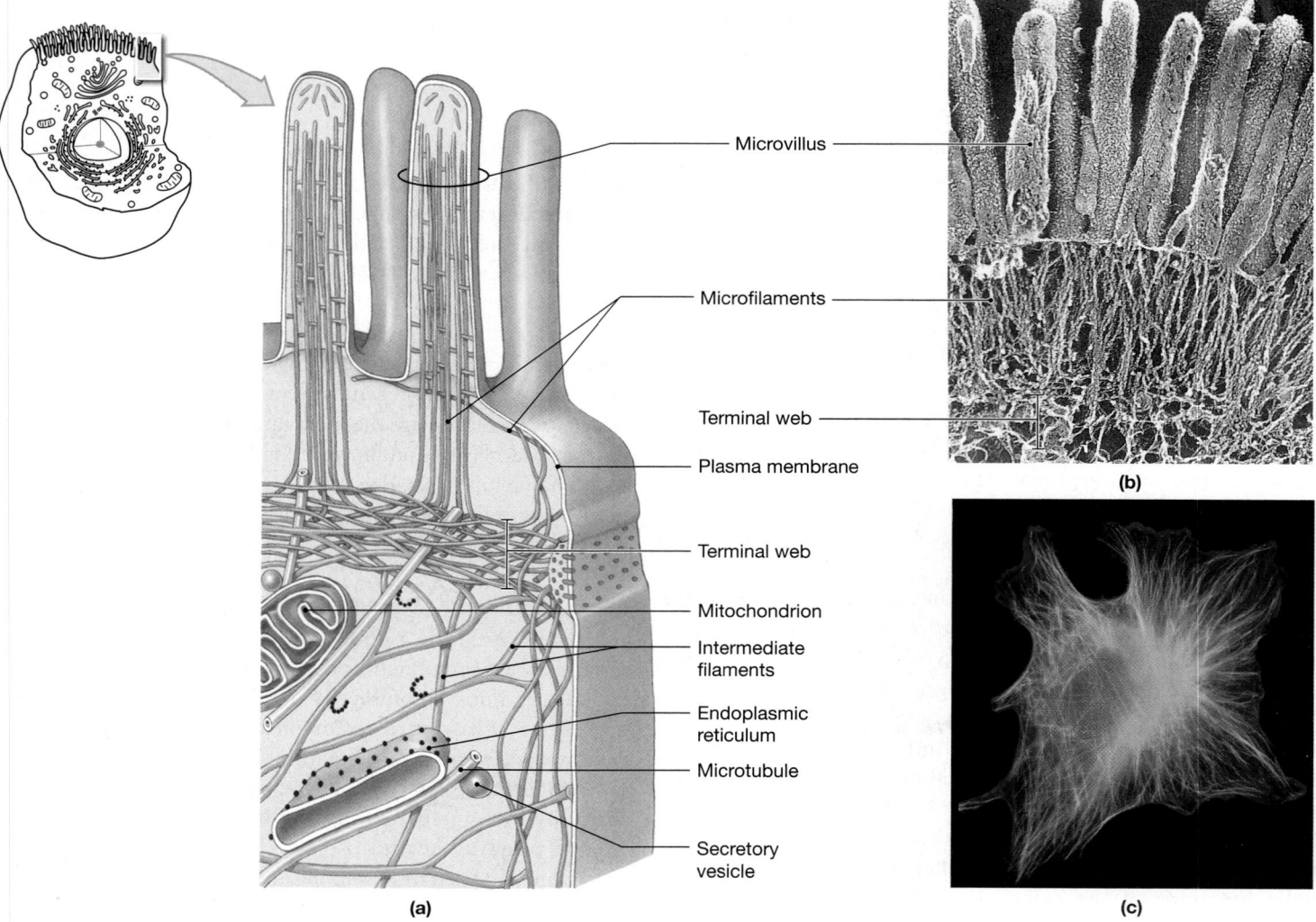

Figure 3–3 **The Cytoskeleton.** **(a)** The cytoskeleton provides strength and structural support for the cell and its organelles. Interactions between cytoskeletal components are also important in moving organelles and in changing the shape of the cell. **(b)** The microfilaments and microvilli of an intestinal cell. Such an image, produced by a scanning electron microscope, is called an SEM. **(c)** Microtubules in a living cell, as seen after special fluorescent labeling (LM × 3200).

direction of transport or the nature of the motor, the process requires ATP and is essential to normal cellular function.

4. During cell division, microtubules form the *spindle apparatus*, which distributes duplicated chromosomes to opposite ends of the dividing cell. We will consider this process in more detail in a later section.

5. Microtubules form structural components of organelles, such as *centrioles* and *cilia*.

Thick Filaments   **Thick filaments** are relatively massive bundles of subunits composed of the protein **myosin**. Thick filaments, which may reach 15 nm in diameter, appear only in muscle cells, where they interact with actin filaments to produce powerful contractions.

Microvilli

Many cells have small, finger-shaped projections of the plasma membrane on their exposed surfaces (**Figure 3–3b**). These projections, called **microvilli**, greatly increase the surface area of the cell exposed to the extracellular environment. Accordingly, they cover the surfaces of cells that are actively absorbing materials from the extracellular fluid, such as the cells lining the digestive tract. Microvilli have extensive connections with the cytoskeleton: A core of microfilaments stiffens each microvillus and anchors it to the cytoskeleton at the terminal web.

Centrioles

All animal cells capable of undergoing cell division contain a pair of **centrioles**, cylindrical structures composed of short microtubules (**Figure 3–4a**). The microtubules form nine

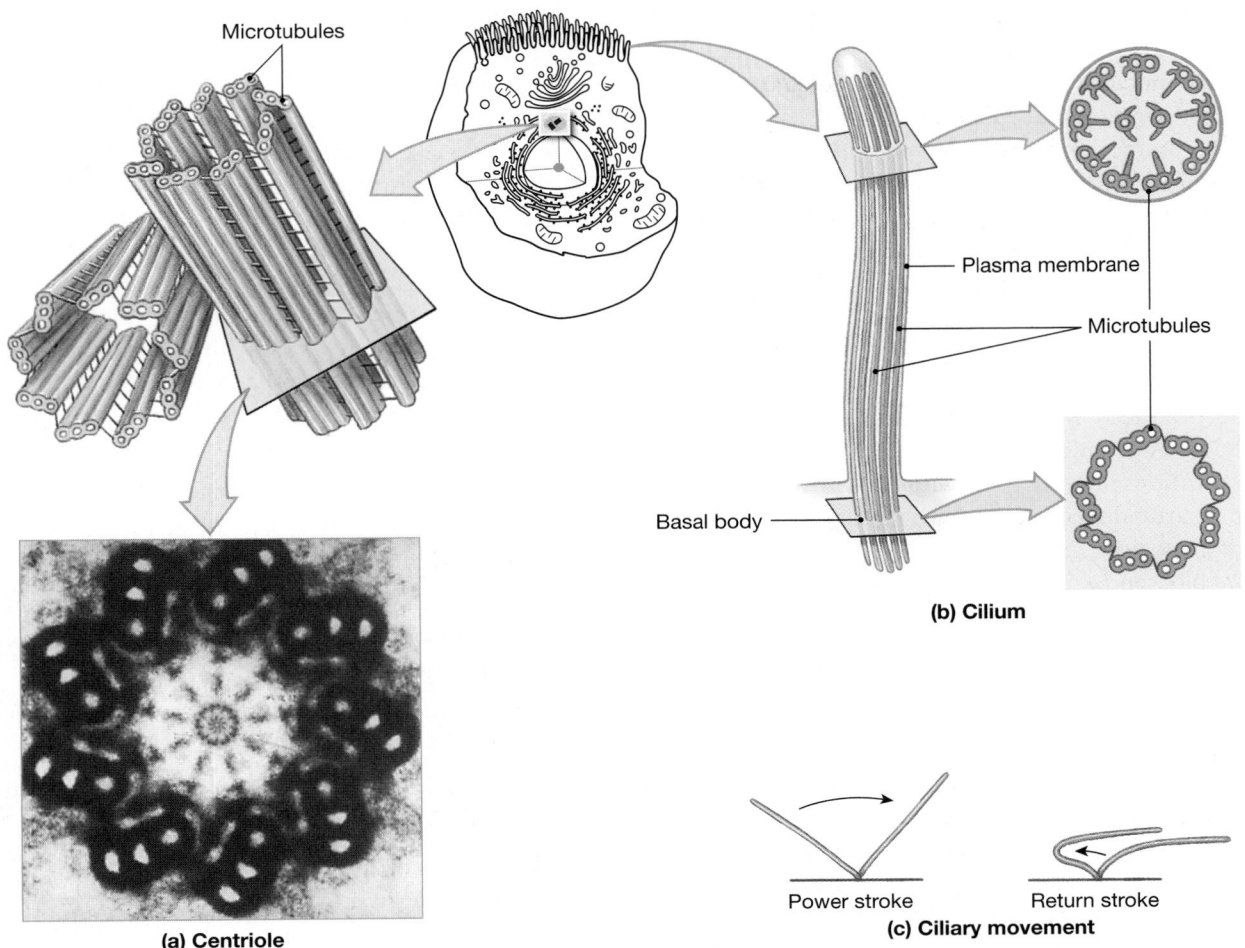

Figure 3–4   Centrioles and Cilia. (a) A centriole consists of nine microtubule triplets (known as a 9 + 0 array). A pair of centrioles oriented at right angles to one another occupies the centrosome. The photograph, produced by a transmission electron microscope, is called a *TEM*. **(b)** A cilium contains nine pairs of microtubules surrounding a central pair (9 + 2 array). The basal body to which the cilium is anchored has a structure similar to that of a centriole. **(c)** Action of a single cilium. During the power stroke, the cilium is relatively stiff; during the return stroke, it bends and returns to its original position.

groups, three in each group. Each of these nine "triplets" is connected to its nearest neighbors on either side. Because there are no central microtubules, this organization is called a *9 + 0 array*. (An axial structure with radial spokes leading toward the microtubular groups has also been observed, but its function is not known.)

During cell division, the centrioles form the spindle apparatus associated with the movement of DNA strands. Mature red blood cells, skeletal muscle cells, cardiac muscle cells, and typical neurons have no centrioles; as a result, these cells are incapable of dividing.

Centrioles are intimately associated with the cytoskeleton. The **centrosome**, the cytoplasm surrounding the centrioles, is the heart of the cytoskeletal system. Microtubules of the cytoskeleton generally begin at the centrosome and radiate through the cytoplasm.

Cilia

Cilia (singular, *cilium*) are relatively long, slender extensions of the plasma membrane. They are found on cells lining the respiratory tract, on cells lining the reproductive tract, and at various other locations in the body. Cilia have an internal arrangement similar to that of centrioles. However, in cilia, nine *pairs* of microtubules (rather than triplets) surround a central pair (**Figure 3–4b**)—an organization known as a *9 + 2 array*. The microtubules are anchored to a compact **basal body** situated just beneath the cell surface. The organization of microtubules in the basal body resembles the 9 + 0 array of a centriole: nine triplets with no central pair.

Cilia are important because they can "beat" rhythmically to move fluids or secretions across the cell surface (**Figure 3–4c**). The cilium is relatively stiff during the effective *power stroke* and flexible during the *return stroke*. The ciliated cells along your trachea beat their cilia in synchronized waves to move sticky mucus and trapped dust particles toward the throat and away from delicate respiratory surfaces. If the cilia are damaged or immobilized by heavy smoking or a metabolic problem, the cleansing action is lost and the irritants will no longer be removed. As a result, a chronic cough and respiratory infections develop. Ciliated cells also move oocytes along the uterine tubes, and waft sperm from the testes into the male reproductive tract.

Ribosomes

Proteins are produced within cells, using information provided by the DNA of the nucleus. The organelles responsible for protein synthesis are called **ribosomes**. The number of ribosomes in a particular cell varies with the type of cell and its demand for new proteins. For example, liver cells, which manufacture blood proteins, contain far more ribosomes than do fat cells (adipocytes), which primarily synthesize lipids.

Individual ribosomes are not visible with the light microscope. In an electron micrograph, they appear as dense granules approximately 25 nm in diameter. Each ribosome is roughly 60 percent RNA and 40 percent protein.

A functional ribosome consists of two subunits that are normally separate and distinct. One is called a **small ribosomal subunit** and the other a **large ribosomal subunit**. These subunits contain special proteins and **ribosomal RNA (rRNA)**, one of the RNA types introduced in Chapter 2. ∞ p. 58 Before protein synthesis can begin, a small and a large ribosomal subunit must join together with a strand of *messenger RNA* (*mRNA*, another type of RNA).

Two major types of functional ribosomes are found in cells: free ribosomes and fixed ribosomes (**Figure 3–1**). **Free ribosomes** are scattered throughout the cytoplasm. The proteins they manufacture enter the cytosol. **Fixed ribosomes** are attached to the *endoplasmic reticulum (ER)*, a membranous organelle. Proteins manufactured by fixed ribosomes enter the ER, where they are modified and packaged for secretion. We will examine ribosomal structure and functions in later sections, when we discuss the endoplasmic reticulum and protein synthesis.

Proteasomes

Free ribosomes produce proteins within the cytoplasm; the smaller proteasomes remove them. **Proteasomes** are organelles that contain an assortment of protein-digesting enzymes, or *proteases*. Cytoplasmic enzymes attach chains of *ubiquitin*, a molecular "tag," to proteins destined for recycling. Tagged proteins are quickly transported into the proteasome. Once inside, they are rapidly disassembled into amino acids and small peptides, which can be released into the cytoplasm.

Proteasomes are responsible for removing and recycling damaged or denatured proteins, and for breaking down abnormal proteins, such as those produced within cells infected by viruses. They also play a key role in the immune response, as we will see in Chapter 22.

Table 3–1 provides a review of the characteristics of nonmembranous organelles.

The Endoplasmic Reticulum

The **endoplasmic reticulum** (en-dō-PLAZ-mik re-TIK-ū-lum), or **ER**, is a network of intracellular membranes connected to the *nuclear envelope*, which surrounds the nucleus. The name *endoplasmic reticulum* is very descriptive. *Endo-* means "within," *plasm* refers to the cytoplasm, and a *reticulum* is a network. The ER has four major functions:

1. *Synthesis.* Specialized regions of the ER synthesize proteins, carbohydrates, and lipids.

2. *Storage.* The ER can store synthesized molecules or materials absorbed from the cytosol without affecting other cellular operations.

3. *Transport.* Materials can travel from place to place in the ER.

4. *Detoxification.* Drugs or toxins can be absorbed by the ER and neutralized by enzymes within it.

The ER forms hollow tubes, flattened sheets, and chambers called **cisternae** (sis-TUR-nē; singular, *cisterna*, a reservoir for water). Two types of ER exist: *smooth endoplasmic reticulum* and *rough endoplasmic reticulum* (**Figure 3–5**).

Smooth Endoplasmic Reticulum The term "smooth" refers to the fact that no ribosomes are associated with the smooth endoplasmic reticulum (SER). The SER has the following functions, all associated with the synthesis of lipids and carbohydrates:

- Synthesis of the phospholipids and cholesterol needed for maintenance and growth of the plasma membrane, ER, nuclear membrane, and Golgi apparatus in all cells

- Synthesis of steroid hormones, such as *androgens* and *estrogens* (the dominant sex hormones in males and in females, respectively) in the reproductive organs

- Synthesis and storage of glycerides, especially triacylglycerides, in liver cells and adipocytes

- Synthesis and storage of glycogen in skeletal muscle and liver cells

In muscle cells, neurons, and many other types of cells, the SER also adjusts the composition of the cytosol by absorbing and storing ions, such as Ca^{2+}, or larger molecules. In addition, the SER in liver and kidney cells is responsible for the detoxification or inactivation of drugs.

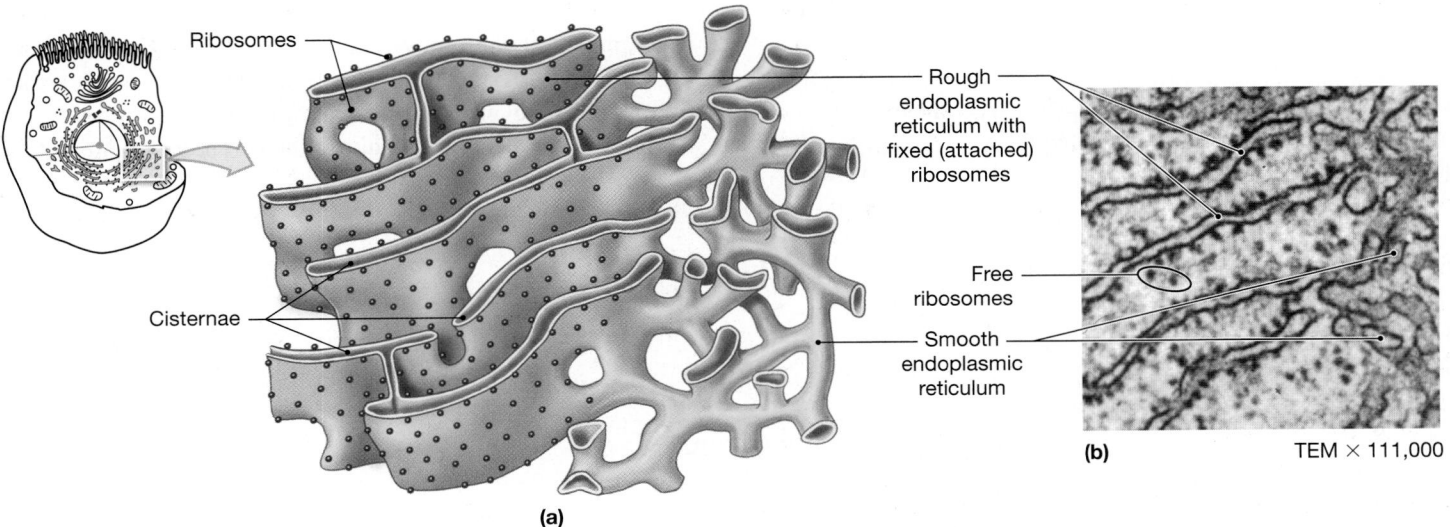

Figure 3–5 **The Endoplasmic Reticulum. (a)** The three-dimensional relationships between the rough and smooth endoplasmic reticula.
(b) Rough endoplasmic reticulum and free ribosomes in the cytoplasm of a cell.

Rough Endoplasmic Reticulum

The rough endoplasmic reticulum (RER) functions as a combination workshop and shipping depot. It is where many newly synthesized proteins are chemically modified and packaged for export to their next destination, the *Golgi apparatus.*

The ribosomes on the outer surface of the rough endoplasmic reticulum are fixed ribosomes (**Figure 3–5**). Their presence gives the RER a beaded, grainy, or rough appearance. Both free and fixed ribosomes synthesize proteins using instructions provided by messenger RNA. The new polypeptide chains produced at fixed ribosomes are released into the cisternae of the RER. Inside the RER, each protein assumes its secondary and tertiary structures. ∞ p. 54 Some of the proteins are enzymes that will function inside the endoplasmic reticulum. Other proteins are chemically modified by the attachment of carbohydrates, creating glycoproteins. Most of the proteins and glycoproteins produced by the RER are packaged into small membranous sacs that pinch off from the tips of the cisternae. These **transport vesicles** subsequently deliver their contents to the Golgi apparatus.

The amount of endoplasmic reticulum and the proportion of RER to SER vary with the type of cell and its ongoing activities. For example, pancreatic cells that manufacture digestive enzymes contain an extensive RER, but the SER is relatively small. The situation is just the reverse in reproductive system cells that synthesize steroid hormones.

The Golgi Apparatus

When a transport vesicle carries a newly synthesized protein or glycoprotein that is destined for export from the cell, it travels from the ER to an organelle that looks a bit like a stack of dinner plates. This organelle, the **Golgi** (GŌL-jē) **apparatus** (**Figure 3–6**), typically consists of five or six flattened membranous discs called *cisternae.* A single cell may contain several of these organelles, most often near the nucleus.

The Golgi apparatus has three major functions: It (1) modifies and packages secretions, such as hormones or enzymes, for release through exocytosis; (2) renews or modifies the plasma membrane; and (3) packages special enzymes within vesicles for use in the cytosol.

Figure 3–7a diagrams the role of the Golgi apparatus in packaging secretions. Some proteins and glycoproteins synthesized in the RER are delivered to the Golgi apparatus by transport vesicles. The vesicles generally arrive at a cisterna known as the *forming* (or *cis*) *face.* The transport vesicles then fuse with the Golgi membrane, emptying their contents into the cisternae. Inside the Golgi apparatus, enzymes modify the arriving proteins and glycoproteins. For example, the enzymes may change the carbohydrate structure of a glycoprotein, or they may attach a phosphate group, sugar, or fatty acid to a protein.

One of the unique things about this organelle is that compounds that enter the cisternae are constantly in motion, traveling up the stack from the ER toward the plasma membrane. Small vesicles move material from one cisterna to the next. Ultimately, the product arrives at the *maturing* (or *trans*) *face*, which usually faces the cell surface. Three types of vesicles that carry materials away from the Golgi apparatus form at the maturing face:

1. *Secretory Vesicles.* **Secretory vesicles** contain secretions that will be discharged from the cell. These vesicles fuse with the plasma membrane and empty their contents into the extracellular environment (**Figure 3–7b**). This

process, known as *exocytosis*, is discussed in greater detail later in this chapter.

2. *Membrane Renewal Vesicles.* When vesicles produced at the Golgi apparatus fuse with the surface of the cell, they are adding new lipids and proteins to the plasma membrane. At the same time, other areas of the plasma membrane are being removed and recycled. The Golgi apparatus can thus change the properties of the plasma membrane over time. For example, new glycoprotein receptors can be added, making the cell more sensitive to a

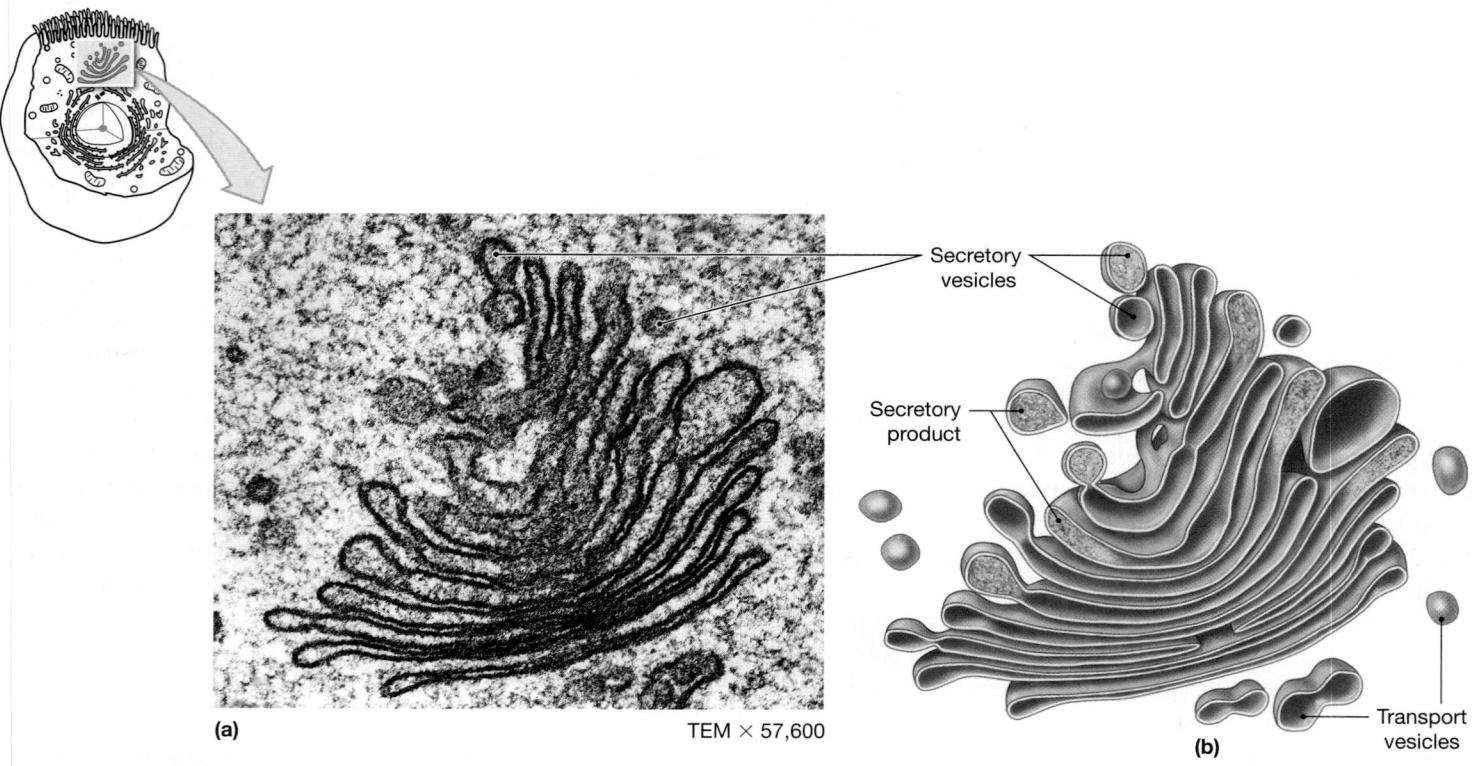

(a)   TEM × 57,600

(b)

Figure 3–6   The Golgi Apparatus. (a) A sectional view of the Golgi apparatus of an active secretory cell, produced by a transmission electron microscope (TEM) corresponding to **(b)**, a three-dimensional view of the Golgi apparatus with a cut edge.

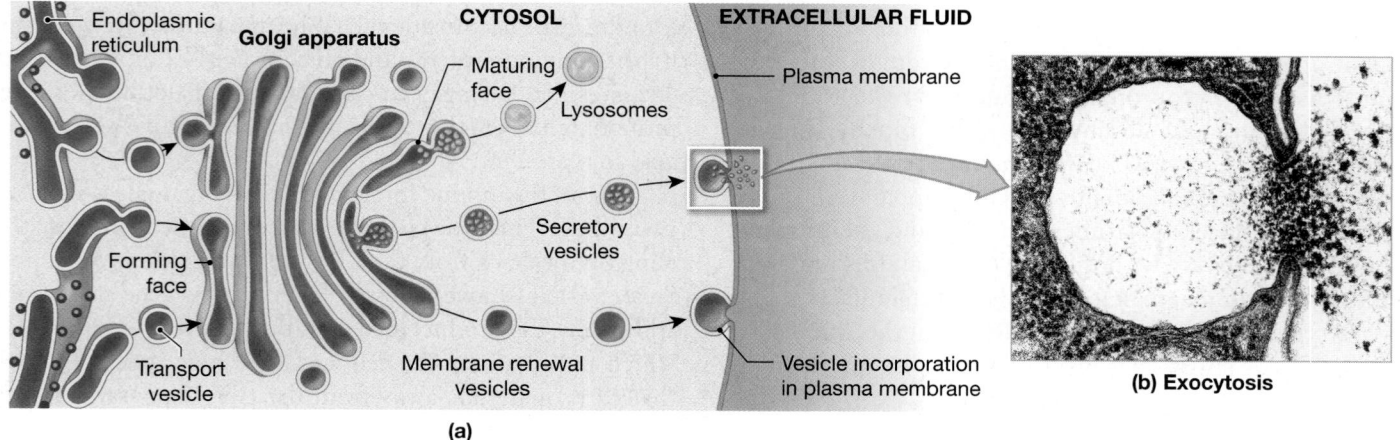

(a)

(b) Exocytosis

Figure 3–7   Functions of the Golgi Apparatus. (a) Transport vesicles carry the secretory product from the endoplasmic reticulum to the Golgi apparatus (simplified to clarify the relationships between the membranes). Small vesicles move membrane and materials between the Golgi cisternae. At the maturing face, three functional categories of vesicles develop. Lysosomes, which remain in the cytoplasm, are vesicles filled with digestive enzymes. Secretory vesicles carry the secretion from the Golgi to the cell surface, where exocytosis releases the contents into the extracellular fluid. Other vesicles add surface area and integral proteins to the plasma membrane. **(b)** Exocytosis at the surface of a cell.

particular stimulus. Alternatively, receptors can be removed and not replaced, making the cell less sensitive to specific ligands. Such changes can profoundly alter the sensitivity and functions of the cell.

3. *Lysosomes.* Vesicles called *lysosomes* that remain in the cytoplasm contain digestive enzymes. Their varied functions will be detailed next.

Lysosomes

Cells often need to break down and recycle large organic molecules and even complex structures like organelles. The breakdown process requires the use of powerful enzymes, and it often generates toxic chemicals that could damage or kill the cell. **Lysosomes** (LĪ-sō-sōmz; *lyso-*, dissolution + *soma*, body) are special vesicles that provide an isolated environment for potentially dangerous chemical reactions. These vesicles, produced at the Golgi apparatus, contain digestive enzymes. Lysosomes are small, often spherical bodies with contents that look dense and dark in electron micrographs (**Figure 3–7b**).

Lysosomes have several functions (**Figure 3–8**). *Primary lysosomes* contain inactive enzymes. When these lysosomes fuse with the membranes of damaged organelles (such as mitochondria or fragments of the ER), the enzymes are activated and *secondary lysosomes* are formed. These enzymes then break down the lysosomal contents. The cytosol reabsorbs re-

leased nutrients, and the remaining material is eliminated from the cell by exocytosis.

Lysosomes also function in the destruction of bacteria (as well as liquids and organic debris) that enter the cell from the extracellular fluid. The cell encloses these substances in a small portion of the plasma membrane, which is then pinched off to form a transport vesicle in the cytoplasm. (This method of transporting substances into the cell, called *endocytosis*, will be discussed later in this chapter.) When a primary lysosome fuses with the vesicle, activated enzymes in the secondary lysosome break down the contents and release usable substances, such as sugars or amino acids. In this way, the cell both protects itself against harmful substances and obtains valuable nutrients.

Lysosomes also perform essential cleanup and recycling functions inside the cell. For example, when muscle cells are inactive, lysosomes gradually break down their contractile proteins. (This mechanism accounts for the reduction in muscle mass seen among retired athletes.) The process is usually precisely controlled, but in a damaged or dead cell, the regulatory mechanism fails. Lysosomes then disintegrate, releasing enzymes that become activated within the cytosol. These enzymes rapidly destroy the cell's proteins and organelles in a process called **autolysis** (aw-TOL-i-sis; *auto-*, self). We do not know how to control lysosomal activities or why the enclosed enzymes do not digest the lysosomal walls unless the cell, is damaged.

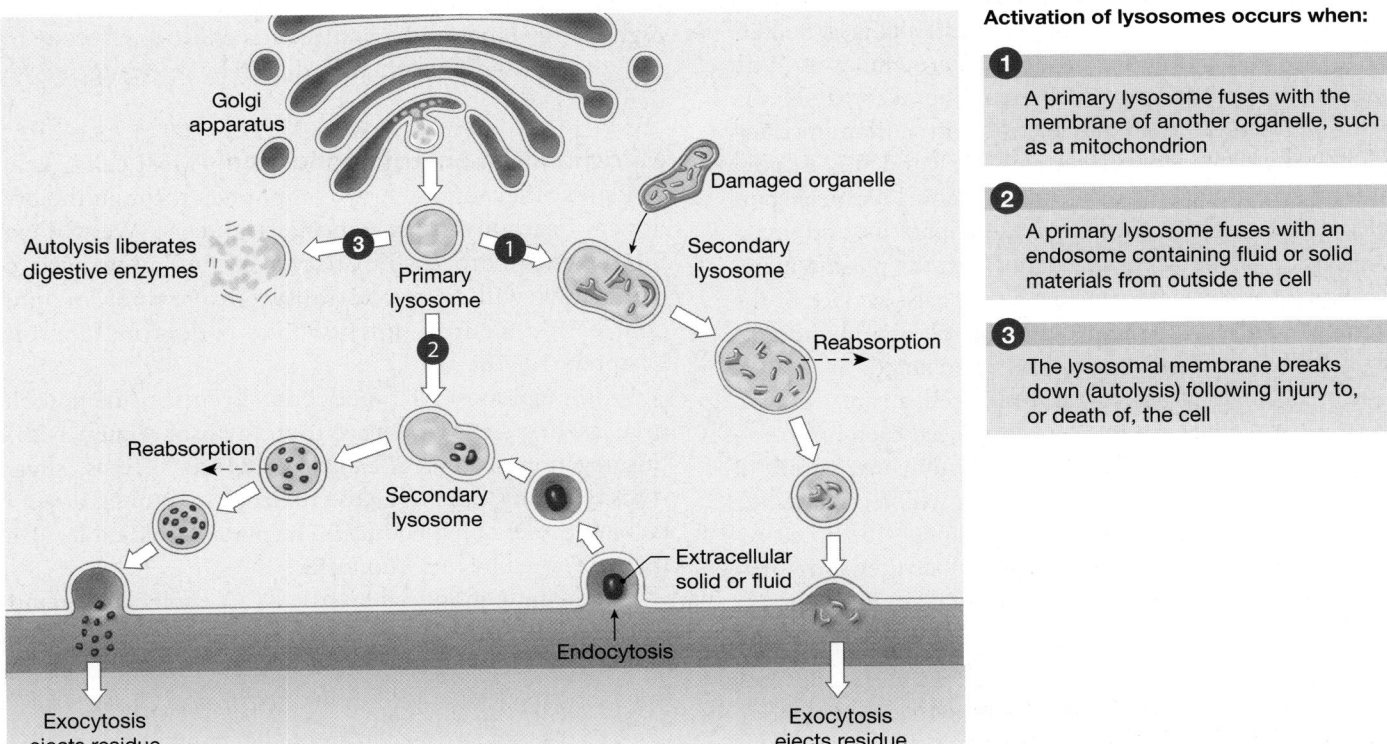

Activation of lysosomes occurs when:

1 A primary lysosome fuses with the membrane of another organelle, such as a mitochondrion

2 A primary lysosome fuses with an endosome containing fluid or solid materials from outside the cell

3 The lysosomal membrane breaks down (autolysis) following injury to, or death of, the cell

Figure 3–8 Lysosome Functions. Primary lysosomes, formed at the Golgi apparatus, contain inactive enzymes. Activation may occur under any of three basic conditions indicated here.

Problems with lysosomal enzyme production cause more than 30 serious diseases affecting children. In these conditions, called *lysosomal storage diseases*, the lack of a specific lysosomal enzyme results in the buildup of waste products and debris normally removed and recycled by lysosomes. Affected individuals may die when vital cells, such as those of the heart, can no longer function.

Peroxisomes

Peroxisomes are smaller than lysosomes and carry a different group of enzymes. In contrast to lysosomes, which are produced at the Golgi apparatus, new peroxisomes are produced by the growth and subdivision of existing peroxisomes. Their enzymes are produced at free ribosomes and transported from the cytosol into the peroxisomes by carrier proteins.

Peroxisomes absorb and break down fatty acids and other organic compounds. As they do so, peroxisomes generate hydrogen peroxide (H_2O_2), a potentially dangerous free radical. ∞ p. 34 Other enzymes within the peroxisome then break down the hydrogen peroxide to oxygen and water. Peroxisomes thus protect the cell from the potentially damaging effects of free radicals produced during catabolism. While these organelles are present in all cells, their numbers are highest in metabolically active cells, such as liver cells.

Membrane Flow

When the temperature changes markedly, you change your clothes. Similarly, when a cell's environment changes, it alters the structure and properties of its plasma membrane. With the exception of mitochondria, all membranous organelles in the cell are either interconnected or in communication through the movement of vesicles. The RER and SER are continuous and are connected to the nuclear envelope. Transport vesicles connect the ER with the Golgi apparatus, and secretory vesicles link the Golgi apparatus with the plasma membrane. Finally, vesicles forming at the exposed surface of the cell remove and recycle segments of the plasma membrane. This continuous movement and exchange is called **membrane flow**. In an actively secreting cell, an area equal to the entire membrane surface may be replaced each hour.

Membrane flow is an example of the dynamic nature of cells. It provides a mechanism by means of which cells change the characteristics of their plasma membranes—the lipids, receptors, channels, anchors, and enzymes—as they grow, mature, or respond to a specific environmental stimulus.

The A&P Top 100

#16 Cells are the basic structural and functional units of life that respond directly to the environment to help maintain homeostasis. Over time, they are capable of changing their internal structure and physiological functions.

Mitochondria

Cells, like other living things, require energy to carry out the functions of life. The organelles responsible for energy production are the **mitochondria** (mī-tō-KON-drē-uh; singular, *mitochondrion*; *mitos*, thread + *chondrion*, granule). These small structures vary widely in shape, from long and slender to short and fat. The number of mitochondria in a particular cell varies with the cell's energy demands. Red blood cells lack mitochondria altogether, whereas these organelles may account for 30 percent of the volume of a heart muscle cell.

Mitochondria have an unusual double membrane (**Figure 3–9a**). The outer membrane surrounds the organelle. The inner membrane contains numerous folds called **cristae**. Cristae increase the surface area exposed to the fluid contents, or **matrix**, of the mitochondrion. Metabolic enzymes in the matrix catalyze the reactions that provide energy for cellular functions.

Most of the chemical reactions that release energy occur in the mitochondria, but most of the cellular activities that require energy occur in the surrounding cytoplasm. Cells must therefore store energy in a form that can be moved from place to place. Recall from Chapter 2 that cellular energy is stored and transferred in the form of *high-energy bonds*, such as those that attach a phosphate group (PO_4^{3-}) to adenosine diphosphate (ADP), creating the high-energy compound *adenosine triphosphate* (ATP). Cells can break the high-energy bond under controlled conditions, reconverting ATP to ADP and phosphate and thereby releasing energy for the cell's use.

Mitochondrial Energy Production Most cells generate ATP and other high-energy compounds through the breakdown of carbohydrates, especially glucose. We will examine the entire process in Chapter 25, but a few basic concepts now will help you follow discussions of muscle contraction, neuron function, and endocrine function in Chapters 10–18.

Although most ATP production occurs inside mitochondria, the first steps take place in the cytosol (**Figure 3–9b**). In this reaction sequence, called **glycolysis** (*glycos*, sugar + -*lysis*, splitting), each glucose molecule is broken down into two molecules of *pyruvic acid*. The pyruvic acid molecules are then absorbed by mitochondria.

In the mitochondrial matrix, a CO_2 molecule is removed from each absorbed pyruvic acid molecule; the remainder enters the **tricarboxylic acid cycle**, or **TCA cycle** (also known as the *Krebs cycle* and the *citric acid cycle*). The TCA cycle is an enzymatic pathway that systematically breaks down the absorbed pyruvic acid in the presence of oxygen. The remnants of pyruvic acid molecules contain carbon, oxygen, and hydrogen atoms. The carbon and oxygen atoms

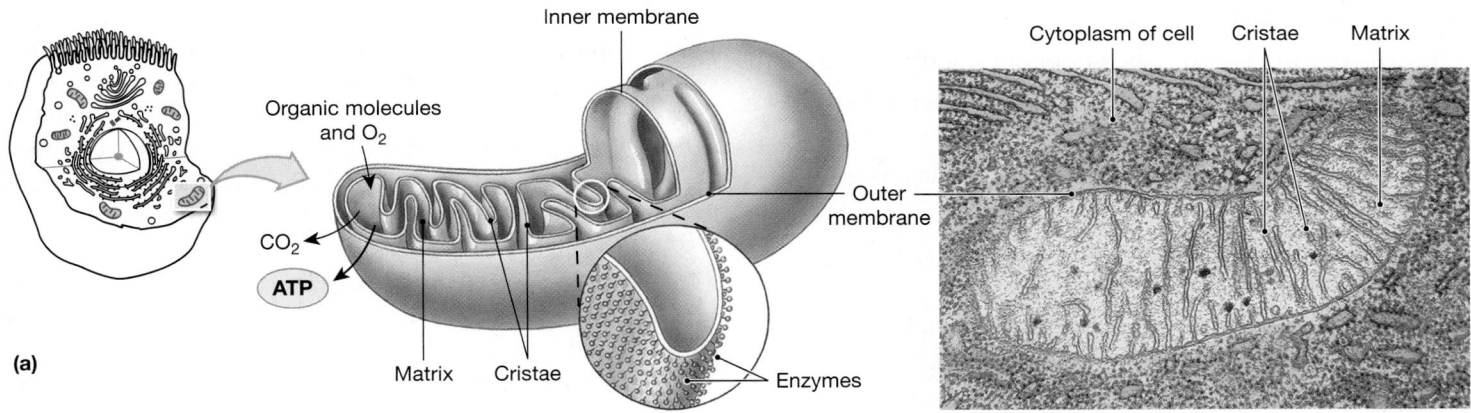

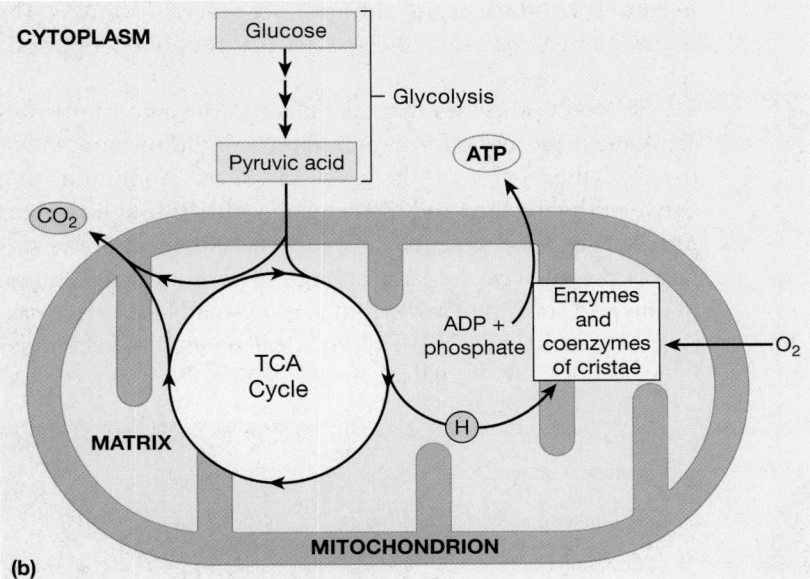

TEM × 46,332

Figure 3–9   Mitochondria. (a) The three-dimensional organization and a color-enhanced TEM of a typical mitochondrion in section. **(b)** An overview of the role of mitochondria in energy production. Mitochondria absorb short carbon chains (such as pyruvic acid) and oxygen and generate carbon dioxide and ATP.

are released as carbon dioxide, which diffuses out of the cell. The hydrogen atoms are delivered to carrier proteins in the cristae. The electrons from the hydrogen atoms are then removed and passed along a chain of coenzymes. The energy released during these steps performs the enzymatic conversion of ADP to ATP. ∞ p. 60

Because mitochondrial activity requires oxygen, this method of ATP production is known as **aerobic metabolism** (*aer*, air + *bios*, life), or *cellular respiration*. Aerobic metabolism in mitochondria produces about 95 percent of the ATP needed to keep a cell alive. (Enzymatic reactions in the cytosol produce the rest.)

The A&P Top 100

#17 Mitochondria provide most of the energy needed to keep your cells (and you) alive. They require oxygen and organic substrates, and they generate carbon dioxide and ATP.

CLINICAL NOTE

Inheritable Mitochondrial Disorders   Several inheritable disorders result from abnormal mitochondrial activity. While not totally self-sufficient, mitochondria do carry their own DNA and manufacture many of their own proteins under the direction of the genes on this DNA. The mitochondria involved in congenital diseases contain abnormal DNA, and the enzymes they produce reduce the efficiency of ATP production. Cells throughout the body may be affected, but symptoms involving muscle cells, neurons, and the receptor cells in the eye are most common, because these cells have especially high energy demands.

5. Differentiate between the cytoplasm and the cytosol.

6. What are the major differences between cytosol and extracellular fluid?

7. Identify the nonmembranous organelles, and cite a function of each.

8. Identify the membranous organelles, and cite their functions.

9. What does the presence of many mitochondria imply about a cell's energy requirements?

10. Certain cells in the ovaries and testes contain large amounts of smooth endoplasmic reticulum (SER). Why?

See the blue Answers tab at the end of the book.

3-3 The nucleus contains DNA and enzymes essential for controlling cellular activities

The **nucleus** is usually the largest and most conspicuous structure in a cell; under a light microscope, it is often the only organelle visible. The nucleus serves as the control center for cellular operations. A single nucleus stores all the information needed to direct the synthesis of the more than 100,000 different proteins in the human body. The nucleus determines the structure of the cell and what functions it can perform by controlling which proteins are synthesized, under what circumstances, and in what amounts. A cell without a nucleus cannot repair itself, and it will disintegrate within three or four months.

Most cells contain a single nucleus, but exceptions exist. For example, skeletal muscle cells have many nuclei, whereas mature red blood cells have none. **Figure 3–10** details the structure of a typical nucleus. Surrounding the nucleus and separating it from the cytosol is a **nuclear envelope**, a double membrane with its two layers separated by a narrow **perinuclear space** (*peri-*, around). At several locations, the nuclear envelope is connected to the rough endoplasmic reticulum (**Figure 3–1**).

To direct processes that take place in the cytosol, the nucleus must receive information about conditions and activities in other parts of the cell. Chemical communication between the nucleus and the cytosol occurs through **nuclear pores**. These pores, which cover about 10 percent of the surface of the nucleus, are large enough to permit the movement of ions and small molecules, but are too small for the free passage of proteins or DNA. Each nuclear pore contains regula-

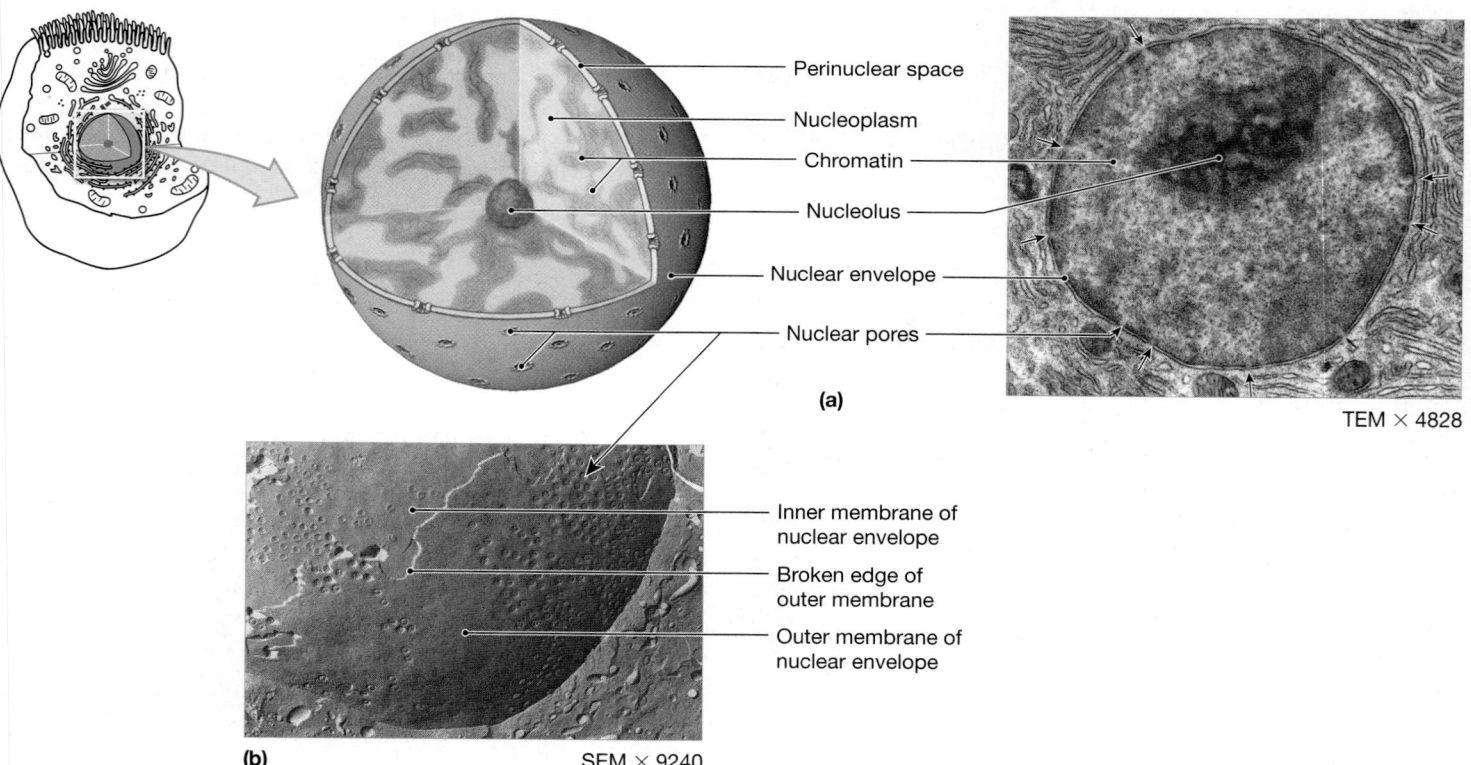

Perinuclear space
Nucleoplasm
Chromatin
Nucleolus
Nuclear envelope
Nuclear pores

(a)

TEM × 4828

Inner membrane of nuclear envelope
Broken edge of outer membrane
Outer membrane of nuclear envelope

(b)   SEM × 9240

Figure 3–10   The Nucleus. (a) Important nuclear structures. **(b)** This cell was frozen and then broken apart to make its internal structures visible. The technique, called *freeze fracture* or *freeze-etching*, provides a unique perspective on the internal organization of cells. The nuclear envelope and nuclear pores are visible. The fracturing process broke away part of the outer membrane of the nuclear envelope, and the cut edge of the nucleus can be seen.

tory proteins that govern the transport of specific proteins and RNA into or out of the nucleus.

Contents of the Nucleus

The fluid contents of the nucleus are called the *nucleoplasm*. The nucleoplasm contains the **nuclear matrix**, a network of fine filaments that provides structural support and may be involved in the regulation of genetic activity. The nucleoplasm also contains ions, enzymes, RNA and DNA nucleotides, small amounts of RNA, and DNA.

Most nuclei contain several dark-staining areas called **nucleoli** (noo-KLĒ-ō-lī; singular, *nucleolus*). Nucleoli are transient nuclear organelles that synthesize ribosomal RNA. They also assemble the ribosomal subunits, which reach the cytoplasm by carrier-mediated transport at the nuclear pores. Nucleoli are composed of RNA, enzymes, and proteins called **histones**. The nucleoli form around portions of DNA that contain the instructions for producing ribosomal proteins and RNA when those instructions are being carried out. Nucleoli are most prominent in cells that manufacture large amounts of proteins, such as liver, nerve, and muscle cells, because those cells need large numbers of ribosomes.

It is the DNA in the nucleus that stores the instructions for protein synthesis. Interactions between the DNA and the histones help determine the information available to the cell at any moment. The organization of DNA within the nucleus is shown in **Figure 3–11**. At intervals, the DNA strands wind around the histones, forming a complex known as a **nucleosome**. Such winding allows a great deal of DNA to be packaged in a small space. The entire chain of nucleosomes may coil around other proteins. The degree of coiling varies depending on whether cell division is under way. In cells that are not dividing, the nucleosomes are loosely coiled within the nucleus, forming a tangle of fine filaments known as **chromatin**. Chromatin gives the nucleus a clumped, grainy appearance. Just before cell division begins, the coiling becomes tighter, forming distinct structures called **chromosomes** (*chroma*, color). In humans, the nuclei of somatic cells contain 23 pairs of chromosomes. One member of each pair is derived from the mother, and one from the father.

The A&P Top 100

#18 The nucleus contains the genetic instructions needed to synthesize the proteins that determine cell structure and function. This information is stored in chromosomes, which consist of DNA and various proteins involved in controlling and accessing the genetic information.

Information Storage in the Nucleus

As we saw in Chapter 2, each protein molecule consists of a unique sequence of amino acids. ∞ p. 53 Any "recipe" for a

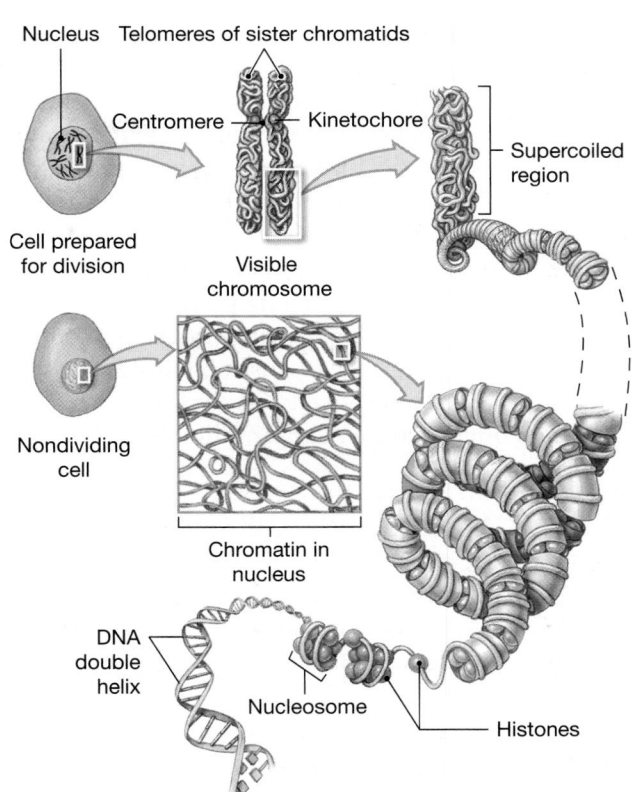

Figure 3–11 The Organization of DNA within the Nucleus. DNA strands are coiled around histones to form nucleosomes. Nucleosomes form coils that may be very tight or rather loose. In cells that are not dividing, the DNA is loosely coiled, forming a tangled network known as chromatin. When the coiling becomes tighter, as it does in preparation for cell division, the DNA becomes visible as distinct structures called chromosomes.

protein, therefore, must specify the order of amino acids in the polypeptide chain. This information is stored in the chemical structure of the DNA strands in the nucleus. The chemical "language" the cell uses is known as the **genetic code**. An understanding of the genetic code has enabled us to determine how cells build proteins and how various structural and functional characteristics are inherited from generation to generation.

To understand how the genetic code works, recall the basic structure of nucleic acids described in Chapter 2. ∞ p. 58 A single DNA molecule consists of a pair of DNA strands held together by hydrogen bonding between complementary nitrogenous bases. Information is stored in the sequence of nitrogenous bases along the length of the DNA strands. Those nitrogenous bases are adenine (A), thymine (T), cytosine (C), and guanine (G). The genetic code is called a *triplet code*, because a sequence of three nitrogenous bases specifies the identity of a single amino acid. Thus, the information encoded in the sequence of nitrogenous bases must be read in groups of three. For example, the triplet thymine–guanine–thymine (TGT) on one DNA strand (the *coding strand*) codes for the amino acid cysteine. More than one triplet may represent the same amino acid, however. For example, the DNA triplet thymine–guanine–cytosine (TGC) also codes for cysteine.

A **gene** is the functional unit of heredity; it contains all the DNA triplets needed to produce specific proteins. The number of triplets in a gene depends on the size of the polypeptide represented. A relatively short polypeptide chain might require fewer than 100 triplets, whereas the instructions for building a large protein might involve 1000 or more triplets. Not all of the DNA molecule carries instructions for proteins; some segments contain instructions for the synthesis of transfer RNA or ribosomal RNA, some have a regulatory function, and others have no apparent function.

CLINICAL NOTE

DNA Fingerprinting Every nucleated somatic cell in the body carries a set of 46 chromosomes that are copies of the set formed at fertilization. Not all the DNA of these chromosomes codes for proteins, however; a significant percentage of DNA segments have no known function. Some of the "useless" segments contain the same nucleotide sequence repeated over and over. The number of segments and the number of repetitions vary among individuals. The chance that any two individuals, other than identical twins, will have the same pattern of repeating DNA segments is less than one in 9 billion. Individuals can therefore be identified on the basis of their DNA pattern, just as they can on the basis of a fingerprint. Skin scrapings, blood, semen, hair, or other tissues can be used as the DNA source. Information from *DNA fingerprinting* has been used to convict (and to acquit) people accused of violent crimes, such as rape or murder. The science of molecular biology has thus become a useful addition to the crime-fighting arsenal.

CHECKPOINT

11. Describe the contents and structure of the nucleus.

12. What is a gene?

See the blue Answers tab at the end of the book.

3-4 DNA controls protein synthesis, cell structure, and cell function

We begin this section by examining the major events of protein synthesis: gene activation, transcription, and translation. We then consider how the nucleus controls cell structure and function.

The Role of Gene Activation in Protein Synthesis

Each DNA molecule contains thousands of genes and therefore holds the information needed to synthesize thousands of proteins. Normally, the genes are tightly coiled, and bound histones keep the genes inactive. Before a gene can affect a cell, the portion of the DNA molecule containing that gene must be uncoiled and the histones temporarily removed.

The factors controlling this process, called **gene activation**, are only partially understood. We know, however, that every gene contains segments responsible for regulating its own activity. In effect, these are triplets that say "do (or do not) read this message," "message starts here," or "message ends here." The "read me," "don't read me," and "start" signals form a special region of DNA called the *promoter*, or control segment, at the start of each gene. Each gene ends with a "stop" signal. Gene activation begins with the temporary disruption of the weak bonds between the nitrogenous bases of the two DNA strands and the removal of the histone that guards the promoter.

After the complementary strands have separated and the histone has been removed, the enzyme **RNA polymerase** binds to the promoter of the gene. This binding is the first step in the process of **transcription**, the production of RNA from a DNA template. The term *transcription* is appropriate, as it means "to copy" or "rewrite." All three types of RNA are formed through the transcription of DNA, but we will focus here on the transcription of mRNA, which carries the information needed to synthesize proteins. **Messenger RNA (mRNA)** is absolutely vital, because the DNA cannot leave the nucleus. Instead, its information is copied to messenger RNA, which *can* leave the nucleus and carry the information to the cytoplasm, where protein synthesis occurs.

The Transcription of mRNA

The two DNA strands in a gene are complementary. The strand containing the triplets that specify the sequence of amino acids in the polypeptide is the **coding strand**. The other strand, called the **template strand**, contains complementary triplets that will be used as a template for mRNA production. The resulting mRNA will have a nucleotide sequence identical to that of the coding strand, but with uracil substituted for thymine. **Figure 3–12** illustrates the steps in transcription:

Step 1 Once the DNA strands have separated and the promoter has been exposed, transcription can begin. The key event is the attachment of RNA polymerase to the template strand.

Step 2 RNA polymerase promotes hydrogen bonding between the nitrogenous bases of the template strand and

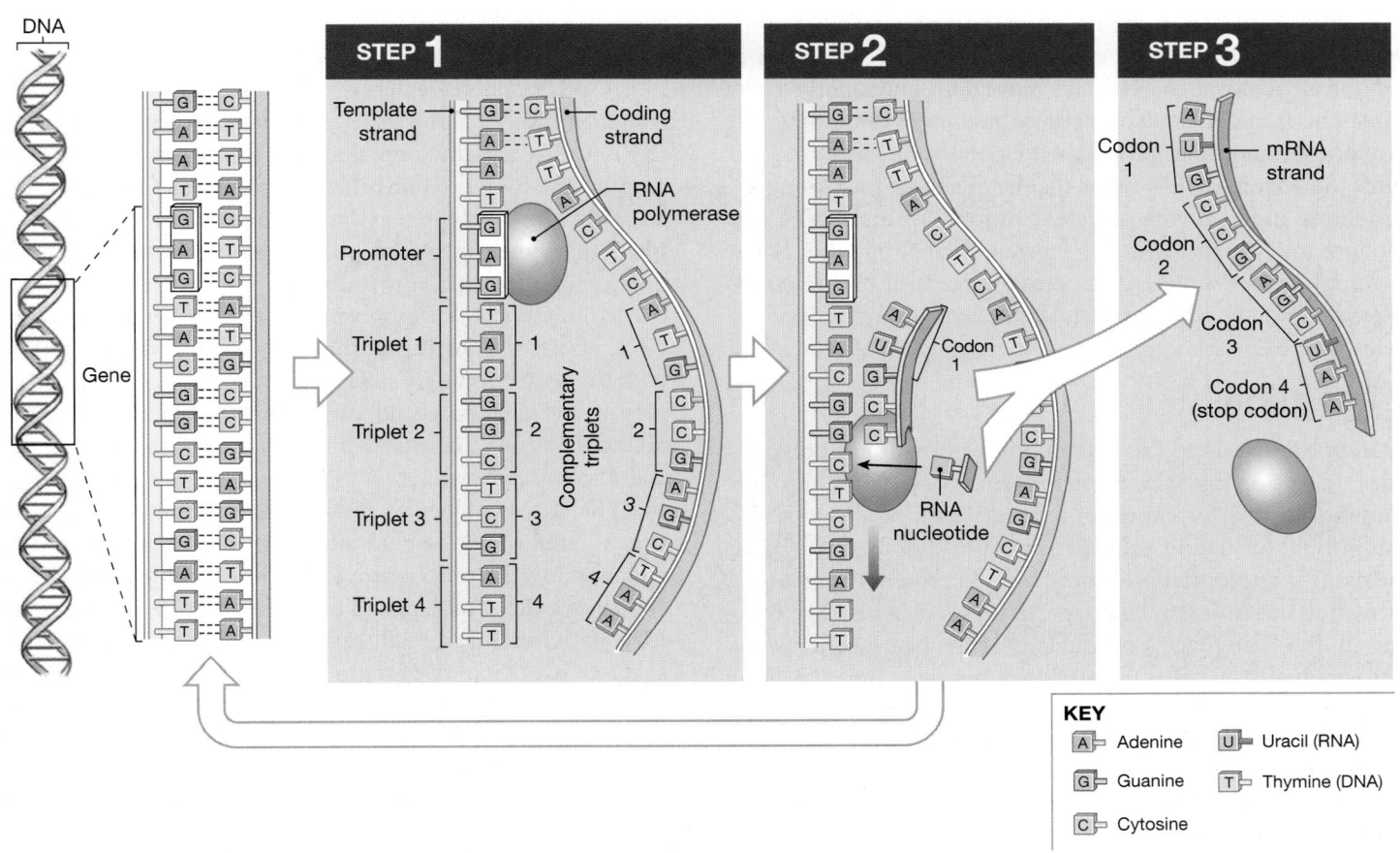

Figure 3–12 mRNA Transcription. A small portion of a single DNA molecule, containing a single gene available for transcription. STEP 1: The two DNA strands separate, and RNA polymerase binds to the promoter of the gene. STEP 2: The RNA polymerase moves from one nucleotide to another along the length of the template strand. At each site, complementary RNA nucleotides form hydrogen bonds with the DNA nucleotides of the template strand. The RNA polymerase then strings the arriving nucleotides together into a strand of mRNA. STEP 3: On reaching the stop signal at the end of the gene, the RNA polymerase and the mRNA strand detach, and the two DNA strands reassociate.

complementary nucleotides in the nucleoplasm. This enzyme begins at a "start" signal in the promoter region. It then strings nucleotides together by covalent bonding. The RNA polymerase interacts with only a small portion of the template strand at any one time as it travels along the DNA strand. The complementary strands separate in front of the enzyme as it moves one nucleotide at a time, and they reassociate behind it. The enzyme collects additional nucleotides and attaches them to the growing chain. The nucleotides involved are those characteristic of RNA, not of DNA; RNA polymerase can attach adenine, guanine, cytosine, or uracil, but never thymine. Thus, wherever an A occurs in the DNA strand, the polymerase will attach a U rather than a T to the growing mRNA strand. In this way, RNA polymerase assembles a complete strand of mRNA. The nucleotide sequence of the template strand determines the nucleotide sequence of the mRNA strand. Thus, each DNA triplet corresponds to a sequence of three nucleotide bases in the mRNA strand. Such a three-base mRNA se-

quence is called a **codon** (KŌ-don). Codons contain nitrogenous bases that are complementary to those of the triplets in the template strand. For example, if the DNA triplet is TCG, the corresponding mRNA codon will be AGC. This method of copying ensures that the mRNA exactly matches the coding strand of the gene.

Step 3 At the "stop" signal, the enzyme and the mRNA strand detach from the DNA strand, and transcription ends. The complementary DNA strands now complete their reassociation as hydrogen bonding occurs between complementary base pairs.

Each gene includes a number of triplets that are not needed to build a functional protein. As a result, the mRNA strand assembled during transcription, sometimes called immature mRNA or *pre-mRNA*, must be "edited" before it leaves the nucleus to direct protein synthesis. In this **RNA processing**, nonsense regions, called **introns**, are snipped out, and the remaining, coding segments, or **exons**, are spliced together.

The process creates a much shorter, functional strand of mRNA that then enters the cytoplasm through a nuclear pore.

Intron removal is extremely important and tightly regulated. This is understandable because an error in the editing will produce an abnormal protein with potentially disastrous results. Moreover, we now know that by changing the editing instructions and removing different introns, a single gene can produce mRNAs that code for several different proteins. Some introns, however, act as enzymes to catalyze their own removal. How this variable editing is regulated remains a mystery.

Translation

Protein synthesis is the assembling of functional polypeptides in the cytoplasm. Protein synthesis occurs through **translation**, the formation of a linear chain of amino acids, using the information provided by an mRNA strand. Again, the name is appropriate: To *translate* is to present the same information in a different language; in this case, a message written in the "language" of nucleic acids (the sequence of nitrogenous bases) is translated by ribosomes into the "language" of proteins (the sequence of amino acids in a polypeptide chain). Each mRNA codon designates a particular amino acid to be incorporated into the polypeptide chain.

The amino acids are provided by **transfer RNA (tRNA)**, a relatively small and mobile type of RNA. Each tRNA molecule binds and delivers an amino acid of a specific type. More than 20 kinds of transfer RNA exist—at least one for each of the amino acids used in protein synthesis.

A tRNA molecule has a tail that binds an amino acid. Roughly midway along its length, the nucleotide chain of the tRNA forms a tight loop that can interact with an mRNA strand. The loop contains three nitrogenous bases that form an **anticodon**. During translation, the anticodon bonds complementarily with an appropriate mRNA codon. The base sequence of the anticodon indicates the type of amino acid carried by the tRNA. For example, a tRNA with the anticodon GGC always carries the amino acid *proline*, whereas a tRNA with the anticodon CGG carries *alanine*. Table 3–2 lists examples of several codons and anticodons that specify individual amino acids and summarizes the relationships among DNA, codons, and anticodons.

The tRNA molecules thus provide the physical link between codons and amino acids. During translation, each codon along the mRNA strand binds a complementary anticodon on a tRNA molecule. Thus, if the mRNA has the codons AUG–CCG–AGC, it will bind to tRNAs with anticodons UAC–GGC–UCG. The amino acid sequence of the polypeptide chain created is determined by the sequence of delivery by tRNAs, and that sequence depends on the arrangement of codons along the mRNA strand. In this case, the amino acid sequence in the resulting polypeptide would be methionine–proline–serine.

The translation process is illustrated in **Figure 3–13**:

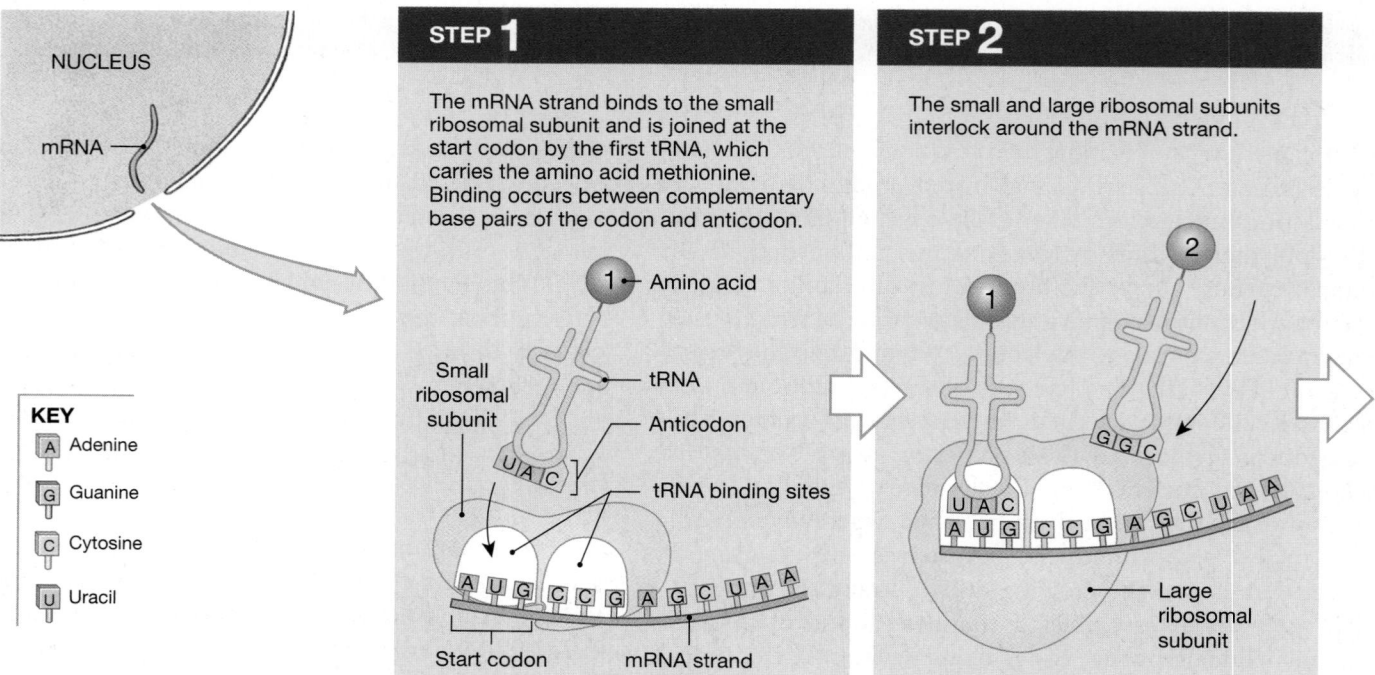

Figure 3–13 **The Process of Translation.** For clarity, the components are not drawn to scale and their three-dimensional relationships have been simplified.

TABLE 3–2 Examples of the Triplet Code

DNA Triplets

Coding Strand	Template Strand	mRNA Codon	tRNA Anticodon	Amino Acid
TTT	AAA	UUU	AAA	Phenylalanine
TTA	AAT	UUA	AAU	Leucine
TGT	ACA	UGU	ACA	Cysteine
GTT	CAA	GUU	CAA	Valine
ATG	TAC	AUG	UAC	Methionine
AGC	TCG	AGC	UCG	Serine
CCG	GGC	CCG	GGC	Proline
GCC	CGG	GCC	CGG	Alanine

Step 1 Translation begins as the mRNA strand binds to a small ribosomal subunit. The first codon, or *start codon*, of the mRNA strand always has the base sequence AUG. It binds a tRNA with the complementary anticodon sequence UAC. This tRNA, which carries the amino acid *methionine*, attaches to the first of two tRNA binding sites on the small ribosomal subunit. (The initial methionine will be removed from the finished protein.)

Step 2 When this tRNA binding occurs, a large ribosomal subunit joins the complex to create a complete ribosome. The mRNA strand nestles in the gap between the small and the large ribosomal subunits.

Step 3 A second tRNA now arrives at the second tRNA binding site of the ribosome, and its anticodon binds to the next codon of the mRNA strand.

Step 4 Enzymes of the large ribosomal subunit then break the linkage between the tRNA and its amino acid. At the same time, the enzymes attach the amino acid to its neighbor by means of a peptide bond. The ribosome then moves one codon down the mRNA strand. The cycle is then repeated with the arrival of another molecule of tRNA. The tRNA stripped of its amino acid drifts away. It will soon bind to another amino acid and be available to participate in protein synthesis again.

Step 5 The polypeptide chain continues to grow by the addition of amino acids until the ribosome reaches a "stop" signal, or *stop codon*, at the end of the mRNA strand. The ribosomal subunits now detach, leaving an intact strand of mRNA and a completed polypeptide.

Translation proceeds swiftly, producing a typical protein in about 20 seconds. The mRNA strand remains intact, and it can interact with other ribosomes to create additional copies of the same polypeptide chain. The process does not continue indefinitely, however, because after a few minutes to a few hours, mRNA strands are broken down and the nucleotides are recycled. However, large numbers of protein chains can be produced during that time. Although only two mRNA codons are "read" by a ribosome at any one time, the entire strand may

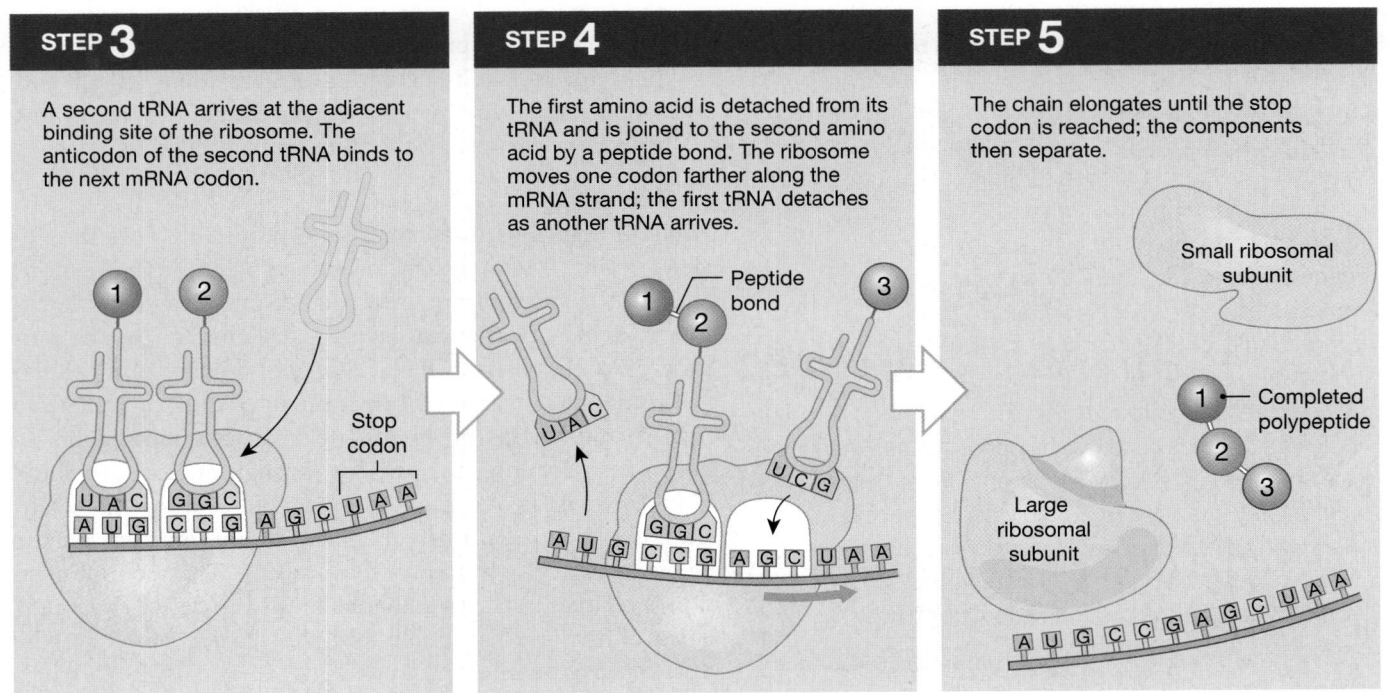

STEP 3 A second tRNA arrives at the adjacent binding site of the ribosome. The anticodon of the second tRNA binds to the next mRNA codon.

STEP 4 The first amino acid is detached from its tRNA and is joined to the second amino acid by a peptide bond. The ribosome moves one codon farther along the mRNA strand; the first tRNA detaches as another tRNA arrives.

STEP 5 The chain elongates until the stop codon is reached; the components then separate.

contain thousands of codons. As a result, many ribosomes can bind to a single mRNA strand. At any moment, each ribosome will be reading a different part of the same message, but each will end up constructing a copy of the same protein as the others. The arrangement is similar to a line of people who make identical choices at a buffet lunch; all the people will assemble the same meal, but each person is always a step behind the person ahead. A series of ribosomes attached to the same mRNA strand is called a *polyribosome*, or *polysome*.

The A&P Top 100

#19 Genes are the functional units of DNA that contain the instructions for making one or more proteins. The creation of specific proteins involves multiple enzymes and three types of RNA.

CLINICAL NOTE

Mutations **Mutations** are permanent changes in a cell's DNA that affect the nucleotide sequence of one or more genes. The simplest is a *point mutation*, a change in a single nucleotide that affects one codon. The triplet code has some flexibility, because several different codons can specify the same amino acid. But a point mutation that produces a codon that specifies a different amino acid will usually change the structure of the completed protein. A single change in the amino acid sequence of a structural protein or enzyme can prove fatal. Certain cancers and two potentially lethal blood disorders discussed in Chapter 19, *thalassemia* and *sickle cell anemia*, result from variations in a single nucleotide.

Several hundred inherited disorders have been traced to abnormalities in enzyme or protein structure that reflect single changes in nucleotide sequence. More elaborate mutations, such as additions or deletions of nucleotides, can affect multiple codons in one gene or in several adjacent genes, or they can affect the structure of one or more chromosomes.

Most mutations occur during DNA replication, when cells are duplicating their DNA in preparation for cell division. A single cell, a group of cells, or an entire individual may be affected. This last prospect occurs when the changes are made early in development. For example, a mutation affecting the DNA of an individual's sex cells will be inherited by that individual's children. Our understanding of genetic structure is opening the possibility of diagnosing and correcting some of these problems.

The A&P Top 100

#20 A mutation is a change in the nucleotide sequence of a gene. Mutations can occur at any time, due to chemical or radiation exposure, but they also occur during DNA replication. Mistakes in copying are usually detected and corrected, but any that persist may alter or disrupt gene function.

How the Nucleus Controls Cell Structure and Function

As noted previously, the DNA of the nucleus controls the cell by directing the synthesis of specific proteins. Through the control of protein synthesis, virtually every aspect of cell structure and function can be regulated. Two levels of control are involved:

1. The DNA of the nucleus has *direct* control over the synthesis of structural proteins, such as cytoskeletal components, membrane proteins (including receptors), and secretory products. By issuing appropriate instructions, in the form of mRNA strands, the nucleus can alter the internal structure of the cell, its sensitivity to substances in its environment, or its secretory functions to meet changing needs.

2. The DNA of the nucleus has *indirect* control over all other aspects of cellular metabolism, because it regulates the synthesis of enzymes. By ordering or stopping the production of appropriate enzymes, the nucleus can regulate all metabolic activities and functions of the cell. For example, the nucleus can accelerate the rate of glycolysis by increasing the number of needed enzymes in the cytoplasm.

This brings us to a central question: How does the nucleus "know" what genes to activate? Although we don't have all the answers, we know that in many cases gene activation or deactivation is triggered by changes in the surrounding cytoplasm. Such changes in the intracellular environment can, in turn, affect the nucleoplasm enough to turn specific genes on or off. Alternatively, messengers or hormones may enter the nucleus through nuclear pores and bind to specific receptors or promoters along the DNA strands. Thus, continual chemical communication occurs between the cytoplasm and the nucleus. That communication is relatively selective, thanks to the restrictive characteristics of the nuclear pores and the barrier posed by the nuclear envelope.

Of course, continual communication also occurs between the cytoplasm and the extracellular fluid across the plasma membrane, and what crosses the plasma membrane today may alter gene activity tomorrow. In the next section, we will examine how the plasma membrane selectively regulates the passage of materials in and out of the cell.

3-5 Diffusion is a passive transport mechanism facilitating membrane passage

The plasma membrane is a barrier that isolates the cytoplasm from the extracellular fluid. Because the plasma membrane is an effective barrier, conditions inside the cell can be considerably different from conditions outside the cell. However, the barrier cannot be perfect, because cells are not self-sufficient. Each day they require nutrients to provide the energy they need to stay alive and function normally. They also generate waste products that must be eliminated. Whereas your body has passageways and openings for nutrients, gases, and wastes, the cell is surrounded by a continuous, relatively uniform membrane. So how do materials—whether nutrients or waste products—get across the plasma membrane without damaging it or reducing its effectiveness as a barrier? To answer this question, we must take a closer look at the structure and function of the plasma membrane.

Permeability is the property of the plasma membrane that determines precisely which substances can enter or leave the cytoplasm. A membrane through which nothing can pass is described as **impermeable**. A membrane through which any substance can pass without difficulty is **freely permeable**. The permeability of plasma membranes lies somewhere between those extremes, so plasma membranes are said to be **selectively permeable**.

A selectively permeable membrane permits the free passage of some materials and restricts the passage of others. The distinction may be based on size, electrical charge, molecular shape, lipid solubility, or other factors. Cells differ in their permeabilities, depending on what lipids and proteins are present in the plasma membrane and how these components are arranged.

Passage across the membrane is either passive or active. *Passive processes* move ions or molecules across the plasma membrane with no expenditure of energy by the cell. *Active processes* require that the cell expend energy, generally in the form of ATP.

Transport processes are also categorized by the mechanism involved. The three major categories are diffusion, carrier-mediated transport, and vesicular transport. *Diffusion*, which results from the random motion and collisions of ions and molecules, is a passive process and will be considered first.

Diffusion

Ions and molecules are constantly in motion, colliding and bouncing off one another and off obstacles in their paths. The movement is random: A molecule can bounce in any direction. One result of this continuous random motion is that, over time, the molecules in any given space will tend to become evenly distributed. This distribution process is called **diffusion**. As the molecules move around, there will be a net movement of material from areas of higher concentration to areas of lower concentration. The difference between the high and low concentrations is a **concentration gradient** (and thus a potential energy gradient). Diffusion tends to eliminate that gradient.

After the gradient has been eliminated, the molecular motion continues, but net movement no longer occurs in any particular direction. (For convenience, we restrict use of the term *diffusion* to the directional movement that eliminates concentration gradients—a process sometimes called *net diffusion*.) Because diffusion tends to spread materials from a region of higher concentration to one of lower concentration, it is often described as proceeding "down a concentration gradient" or "downhill."

Diffusion in air and water is slow, and it is most important over very small distances. However, an everyday example can give you a mental image of how diffusion works. Consider a colored sugar cube dropped in water (**Figure 3–14**). Placing the cube in a large volume of clear water establishes a steep concentration gradient for the ingredients as they dissolve: The sugar and dye concentration is high near the cube and negligible elsewhere. As time passes, the sugar and dye molecules spread through the solution until they are distributed evenly.

Diffusion is important in body fluids, because it tends to eliminate local concentration gradients. For example, every cell in the body generates carbon dioxide, and the intracellular concentration is relatively high. Carbon dioxide concentrations are

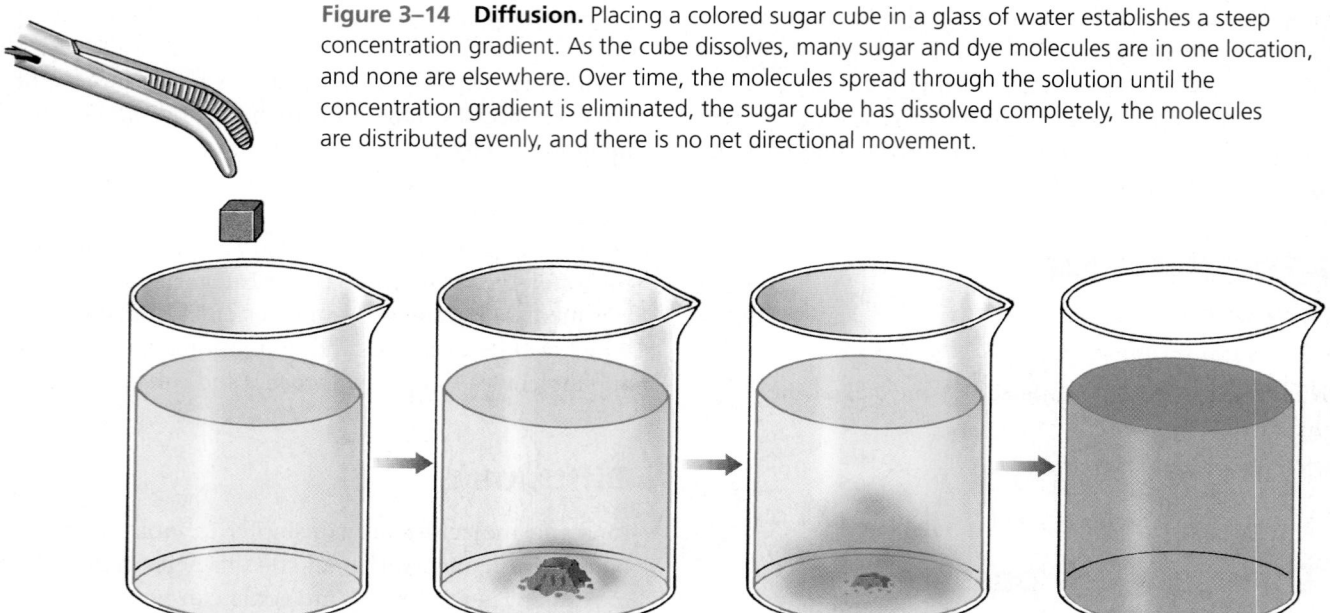

Figure 3–14 Diffusion. Placing a colored sugar cube in a glass of water establishes a steep concentration gradient. As the cube dissolves, many sugar and dye molecules are in one location, and none are elsewhere. Over time, the molecules spread through the solution until the concentration gradient is eliminated, the sugar cube has dissolved completely, the molecules are distributed evenly, and there is no net directional movement.

lower in the surrounding interstitial fluid, and lower still in the circulating blood. Because plasma membranes are freely permeable to carbon dioxide, it can diffuse down its concentration gradient—traveling from the cell's interior into the interstitial fluid and then into the bloodstream, for eventual delivery to the lungs.

To be effective, the diffusion of nutrients, waste products, and dissolved gases must keep pace with the demands of active cells. Important factors that influence diffusion rates include the following:

- *Distance.* The shorter the distance, the more quickly concentration gradients are eliminated. In the human body, diffusion distances are generally small. For example, few cells are farther than 25 μm from a blood vessel.

- *Molecule Size.* Ions and small organic molecules, such as glucose, diffuse more rapidly than do large proteins.

- *Temperature.* The higher the temperature, the faster the diffusion rate. Diffusion proceeds somewhat more rapidly at human body temperature (about 37°C, or 98.6°F) than at cooler environmental temperatures.

- *Gradient Size.* The larger the concentration gradient, the faster diffusion proceeds. When cells become more active, the intracellular concentration of oxygen declines. This change increases the concentration gradient for oxygen between the inside of the cell (relatively low) and the interstitial fluid outside (relatively high). The rate of oxygen diffusion into the cell then increases.

- *Electrical Forces.* Opposite electrical charges (+ and −) attract each other; like charges (+ and + *or* − and −) repel each other. The interior of the plasma membrane has a net negative charge relative to the exterior surface, due in part to the high concentration of proteins in the cell. This negative charge tends to pull positive ions from the extracellular fluid into the cell, while opposing the entry of negative ions. For example, interstitial fluid contains higher concentrations of sodium ions (Na^+) and chloride ions (Cl^-) than does cytosol. Diffusion of the positively charged sodium ions into the cell is therefore favored by both the concentration gradient, or *chemical gradient*, and the electrical gradient. In contrast, diffusion of the negatively charged chloride ions into the cell is favored by the chemical gradient, but opposed by the electrical gradient. For any ion, the net result of the chemical and electrical forces acting on it is called the *electrochemical gradient.*

Diffusion across Plasma Membranes

In extracellular fluids, water and dissolved solutes diffuse freely. A plasma membrane, however, acts as a barrier that selectively restricts diffusion: Some substances pass through easily, while others cannot penetrate the membrane. An ion or a molecule can diffuse across a plasma membrane only by (1) crossing the lipid portion of the membrane or (2) passing through a membrane channel (**Figure 3–15**).

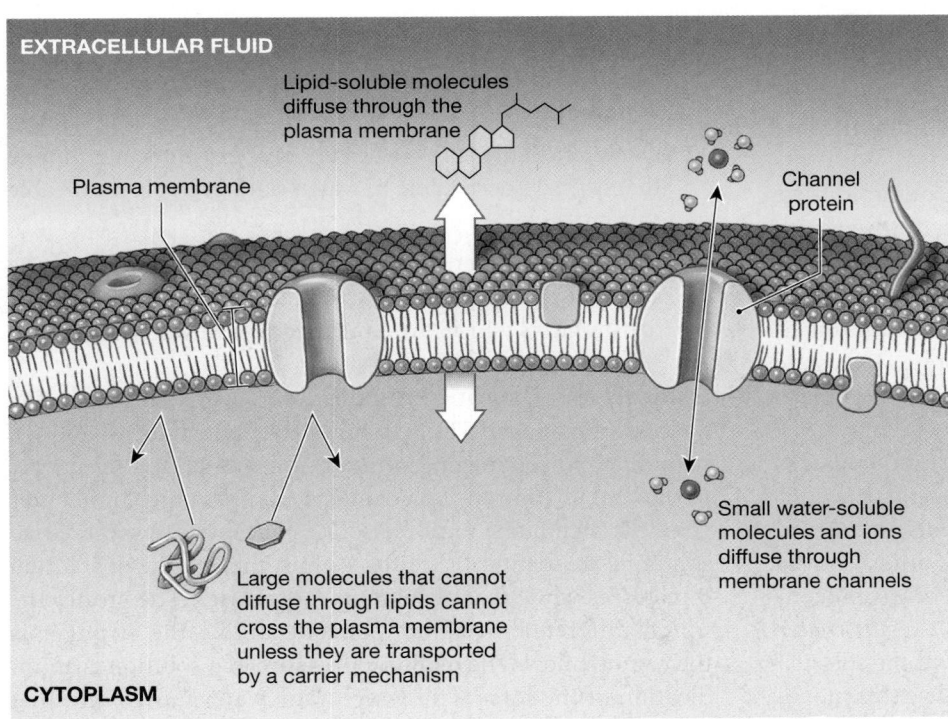

Simple Diffusion Alcohol, fatty acids, and steroids can enter cells easily, because they can diffuse through the lipid portions of the membrane. Dissolved gases, such as oxygen and carbon dioxide, and lipid-soluble drugs also enter and leave our cells by diffusing through the phospholipid bilayer. The situation is more complicated for ions and water-soluble compounds, which are not lipid-soluble. To enter or leave the cytoplasm, these substances must pass through a membrane channel.

CLINICAL NOTE

Drugs and the Plasma Membrane Many clinically important drugs affect the plasma membrane. For some anesthetics, such as chloroform, ether, halothane, and nitrous oxide, potency is directly correlated with its lipid solubility. Presumably, high lipid solubility accelerates the drug's entry into cells and enhances its ability to block ion channels or change other properties of plasma membranes and thereby reduce the sensitivity of neurons and muscle cells. However, some common anesthetics have relatively low lipid solubility. For example, the local anesthetics, *procaine* and *lidocaine*, affect nerve cells by blocking sodium channels in their plasma membranes; this blockage reduces or eliminates the responsiveness of these cells to painful (or any other) stimuli. Although procaine and lidocaine are both effective local anesthetics, procaine has very low lipid solubility.

Channel-Mediated Diffusion Membrane channels are very small passageways created by transmembrane proteins. On average, the channel is about 0.8 nm in diameter. Water molecules can enter or exit freely, but even a small organic molecule, such as glucose, is too big to fit through the channels. Whether an ion can transit a particular membrane channel depends on many factors, including the size and charge of the ion, the size of the hydration sphere, and interactions between the ion and the channel walls. The mechanics of diffusion through membrane channels is therefore more complex than simple diffusion. For example, the rate at which a particular ion diffuses across the membrane can be limited by the availability of suitable channels. However, for many ions, including sodium, potassium, and chloride, movement across the plasma membrane occurs at rates comparable to those one would predict if relying on simple diffusion.

Osmosis: A Special Case of Diffusion

The net diffusion of water across a membrane is so important that it is given a special name: **osmosis** (oz-MŌ-sis; *osmos*, thrust). For convenience, we will always use the term *osmosis* for the movement of water, and the term *diffusion* for the movement of solutes.

Intracellular and extracellular fluids are solutions that contain a variety of dissolved materials. Each solute diffuses as though it were the only material in solution. The diffusion of sodium ions, for example, occurs only in response to the existence of a concentration gradient for sodium. A concentration gradient for another ion will have no effect on the rate or direction of sodium ion diffusion.

Some solutes diffuse into the cytoplasm, others diffuse out, and a few (such as proteins) are unable to diffuse across the plasma membrane at all. Yet if we ignore the individual identities and simply count ions and molecules, we find that the *total* concentration of dissolved ions and molecules on either side of the plasma membrane stays the same. This state of equilibrium persists because a typical plasma membrane is freely permeable to water.

To understand the basis for such equilibrium, consider that whenever a solute concentration gradient exists, a concentration gradient for *water* exists also. Because dissolved solute molecules occupy space that would otherwise be taken up by water molecules, the higher the solute concentration, the lower the water concentration. As a result, *water molecules tend to flow across a membrane toward the solution containing the higher solute concentration*, because this movement is down the concentration gradient for water. Water movement will continue until water concentrations—and thus solute concentrations—are the same on either side of the membrane.

Remember these three characteristics of osmosis:

1. Osmosis is the movement of water molecules across a membrane.

2. Osmosis occurs across a selectively permeable membrane that is freely permeable to water, but not freely permeable to solutes.

3. In osmosis, water flows across a membrane toward the solution that has the higher concentration of solutes, because that is where the concentration of water is lower.

Osmosis and Osmotic Pressure   **Figure 3–16** diagrams the process of osmosis. STEP 1 shows two solutions (A and B), with different solute concentrations, separated by a selectively permeable membrane. As osmosis occurs, water molecules cross the membrane until the solute concentrations in the two solutions are identical (STEP 2). Thus, the volume of solution B increases while that of solution A decreases. The greater the initial difference in solute concentrations, the stronger is the osmotic flow. The **osmotic pressure** of a solution is an indication of the force with which pure water moves into that solution as a result of its solute concentration. We can measure a solution's osmotic pressure in several ways. For example, an opposing pressure can prevent the osmotic flow of water into the solution. Pushing against a fluid generates **hydrostatic**

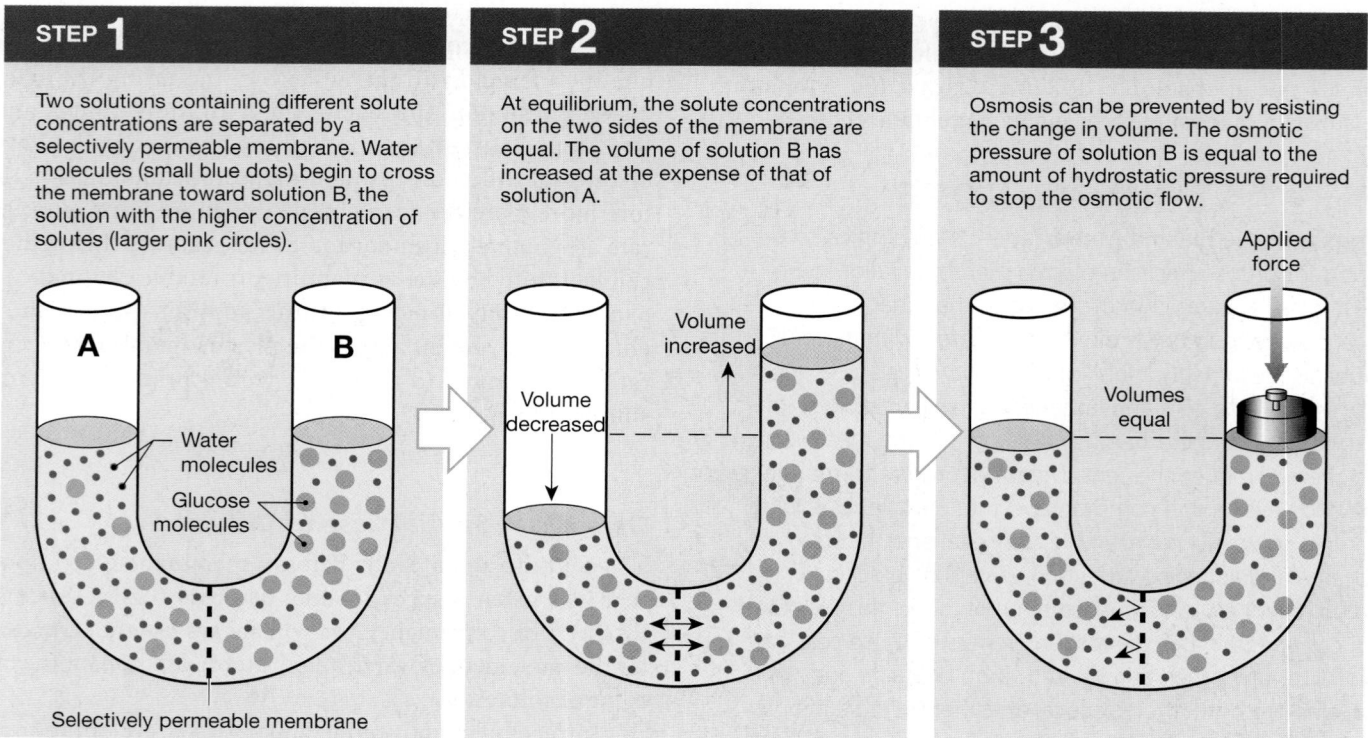

FIGURE 3–16   **Osmosis.** The osmotic pressure of solution B is equal to the amount of hydrostatic pressure required to stop the osmotic flow.

pressure. In STEP 3, hydrostatic pressure opposes the osmotic pressure of solution B, so no net osmotic flow occurs.

Osmosis eliminates solute concentration differences much more quickly than solute diffusion. In large part this is because water molecules cross a membrane through abundant water channels called *aquaporins*, which exceed the number of solute channels, through which water can also pass. This difference results in a higher membrane permeability for water compared to solutes.

Osmolarity and Tonicity

The total solute concentration in an aqueous solution is the solution's **osmolarity**, or **osmotic concentration**. The nature of the solutes, however, is often as important as the total osmolarity. Therefore, when we describe the effects of various osmotic solutions on cells, we usually use the term **tonicity** instead of osmolarity. A solution that does not cause an osmotic flow of water into or out of a cell is called **isotonic** (*iso-*, same + *tonos*, tension).

Although often used interchangeably, the terms *osmolarity* and *tonicity* do not always mean the same thing. Osmolarity refers to the solute concentration of the solution, while tonicity is a description of how the solution affects a cell. Consider a solution that has the same osmolarity as the intracellular fluid, but a higher concentration of one or more individual ions. If any of those ions can cross the plasma membrane and diffuse into the cell, the osmolarity of the intracellular fluid will increase, and that of the extracellular solution will decrease. Os-

mosis will then occur, moving water into the cell. If the process continues, the cell will gradually inflate like a water balloon. In this case, the extracellular solution and the intracellular fluid were initially equal in osmolarity, but they were not isotonic.

Figure 3–17a shows a red blood cell in an isotonic solution. If a red blood cell is in a **hypotonic** solution, water will flow into the cell, causing it to swell up like a balloon (**Figure 3–17b**). The cell may eventually burst, releasing its contents. This event is **hemolysis** (*haima*, blood + *lysis*, dissolution). A cell in a **hypertonic** solution will lose water by osmosis. As it does, the cell shrivels and dehydrates. The shrinking of red blood cells is called **crenation** (**Figure 3–17c**).

It is often necessary to give patients large volumes of fluid to combat severe blood loss or dehydration. One fluid frequently administered is a 0.9 percent (0.9 g/dL) solution of sodium chloride (NaCl). This solution, which approximates the normal osmotic concentration of extracellular fluids, is called *normal saline*. It is used because sodium and chloride are the most abundant ions in the extracellular fluid. Little net movement of either ion across plasma membranes occurs; thus, normal saline is essentially isotonic with respect to body cells. An alternative treatment involves the use of an isotonic saline solution containing *dextran*, a carbohydrate that cannot cross plasma membranes. The dextran molecules elevate the osmolarity of the blood, and as osmosis draws water into the blood vessels from the extracellular fluid, blood volume increases further.

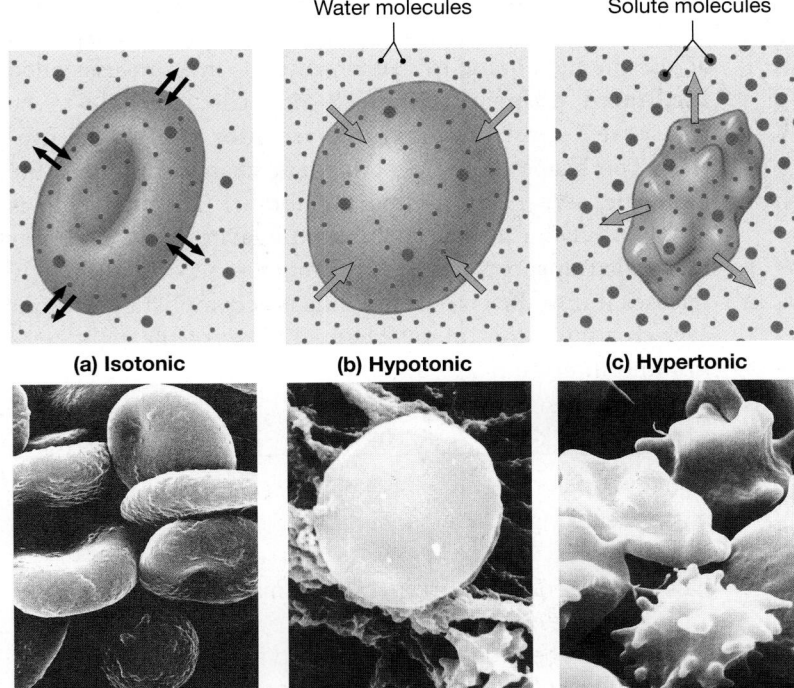

Water molecules

Solute molecules

(a) Isotonic   **(b) Hypotonic**   **(c) Hypertonic**

SEMs × 833

Figure 3–17   Osmotic Flow across a Plasma Membrane. Black arrows indicate an equilibrium with no net water movement. Blue arrows indicate the direction of osmotic water movement. **(a)** In an isotonic saline solution, no osmotic flow occurs, and these red blood cells appear normal. **(b)** Immersion in a hypotonic saline solution results in the osmotic flow of water into the cells. The swelling may continue until the plasma membrane ruptures, or lyses. **(c)** Exposure to a hypertonic solution results in the movement of water out of the cells. The red blood cells shrivel and become crenated.

The A&P Top 100

#21 *Things tend to even out*, unless something—like a plasma membrane—prevents this from happening. In the absence of a plasma membrane, or across a freely permeable membrane, diffusion will quickly eliminate concentration gradients. Osmosis acts to eliminate concentration gradients across membranes that are permeable only to water.

CHECKPOINT

16. Name the three major categories of cellular transport processes.

17. What is meant by "selectively permeable" when referring to a plasma membrane?

18. Define diffusion.

19. How would a decrease in the concentration of oxygen in the lungs affect the diffusion of oxygen into the blood?

20. Define osmosis.

21. Some pediatricians recommend using a 10 percent salt solution as a nasal spray to relieve congestion in infants with stuffy noses. What effect would such a solution have on the cells lining the nasal cavity, and why?

See the blue Answers tab at the end of the book.

3-6 Carrier-mediated and vesicular transport mechanisms facilitate membrane passage

In this section we consider two additional major categories of ways by which substances are taken into or removed from cells: carrier-mediated transport and vesicular transport. *Carrier-mediated transport* requires the presence of specialized integral membrane proteins. It can be passive or active, depending on the substance transported and the nature of the transport mechanism. *Vesicular transport* involves the movement of materials within small membranous sacs, or *vesicles*. Vesicular transport is always an active process.

Carrier-Mediated Transport

In **carrier-mediated transport**, integral proteins bind specific ions or organic substrates and carry them across the plasma membrane. All forms of carrier-mediated transport have the following characteristics, which they share with enzymes:

1. *Specificity.* Each carrier protein in the plasma membrane will bind and transport only certain substances. For example, the carrier protein that transports glucose will not transport other simple sugars.

2. *Saturation Limits.* The availability of substrate molecules and carrier proteins limits the rate of transport into or out of the cell, just as enzymatic reaction rates are limited by the availability of substrates and enzymes. When all the available carrier proteins are operating at maximum speed, the carriers are said to be *saturated*. The rate of transport cannot increase further, regardless of the size of the concentration gradient.

3. *Regulation.* Just as enzyme activity often depends on the presence of cofactors, the binding of other molecules, such as hormones, can affect the activity of carrier proteins. Hormones thus provide an important means of coordinating carrier protein activity throughout the body. The interplay between hormones and plasma membranes will be examined when we consider the endocrine system (Chapter 18) and metabolism (Chapter 25).

Many examples of carrier-mediated transport involve the movement of a single substrate molecule across the plasma membrane. A few carrier mechanisms transport more than one substrate at a time. In **cotransport**, or *symport*, the carrier transports two substances in the same direction simultaneously, either into or out of the cell. In **countertransport**, or *antiport*, one substance moves into the cell and the other moves out.

We will consider two major examples of carrier-mediated transport here: *facilitated diffusion* and *active transport*.

Facilitated Diffusion

Many essential nutrients, such as glucose and amino acids, are insoluble in lipids and too large to fit through membrane channels. These substances can be passively transported across the membrane by carrier proteins in a process called **facilitated diffusion** (Figure 3–18). The molecule to be transported must first bind to a **receptor site** on the protein. The shape of the protein then changes, moving the molecule across the plasma membrane. The molecule is now within the plasma membrane bilayer. Next, a carrier molecule transports it across the membrane, and ultimately releases it into the cytoplasm. This is accomplished without ever creating a continuous channel between the cell's exterior and interior.

Tips&Tricks

To understand the mechanism of cellular channels, think about entering a store that has a double set of automated sliding doors. When you near the first set of automated doors, it opens. As you step onto the mat between the sets of doors, the doors behind you close, trapping you in the vestibule. When you take another step forward, the second set of doors opens, enabling you to enter the store.

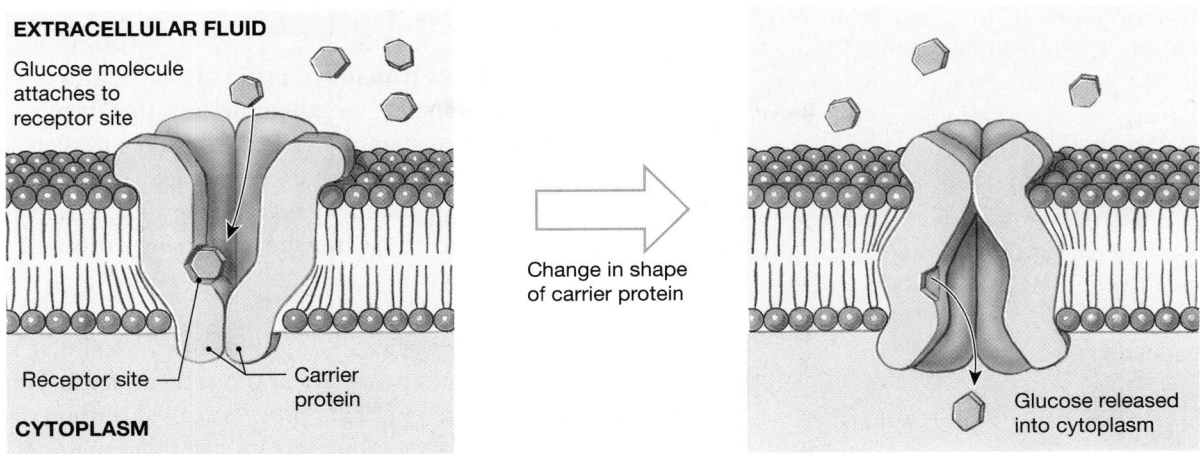

EXTRACELLULAR FLUID

Glucose molecule attaches to receptor site

Change in shape of carrier protein

Receptor site — Carrier protein

CYTOPLASM

Glucose released into cytoplasm

FIGURE 3–18   Facilitated Diffusion. In facilitated diffusion, an extracellular molecule, such as glucose, binds to a specific receptor site on a carrier protein. This binding permits the molecule to diffuse across the membrane.

As in the case of simple or channel-mediated diffusion, no ATP is expended in facilitated diffusion: The molecules simply move from an area of higher concentration to one of lower concentration. However, once the carrier proteins are saturated, the rate of transport cannot increase, regardless of further increases in the concentration gradient.

All cells move glucose across their membranes through facilitated diffusion. However, several different carrier proteins are involved. In muscle cells, fat cells, and many other types of cell, the glucose transporter functions only when stimulated by the hormone *insulin*. Inadequate production of this hormone is one cause of *diabetes mellitus*, a metabolic disorder that we will discuss in Chapter 18.

Active Transport

In **active transport**, a high-energy bond (in ATP or another high-energy compound) provides the energy needed to move ions or molecules across the membrane. Despite the energy cost, active transport offers one great advantage: It is not dependent on a concentration gradient. As a result, the cell can import or export specific substrates, *regardless of their intracellular or extracellular concentrations*.

All cells contain carrier proteins called **ion pumps**, which actively transport the cations sodium (Na^+), potassium (K^+), calcium (Ca^{2+}), and magnesium (Mg^{2+}) across their plasma membranes. Specialized cells can transport additional ions, such as iodide (I^-), chloride (Cl^-), and iron (Fe^{2+}). Many of these carrier proteins move a specific cation or anion in one direction only, either into or out of the cell. In a few instances, one carrier protein will move more than one kind of ion at the same time. If countertransport occurs, the carrier protein is called an **exchange pump**.

The Sodium–Potassium Exchange Pump   Sodium and potassium ions are the principal cations in body fluids.

Sodium ion concentrations are high in the extracellular fluids, but low in the cytoplasm. The distribution of potassium in the body is just the opposite: low in the extracellular fluids and high in the cytoplasm. As a result, sodium ions slowly diffuse into the cell, and potassium ions diffuse out through leak channels. Homeostasis within the cell depends on the ejection of sodium ions and the recapture of lost potassium ions. This exchange is accomplished through the activity of a **sodium–potassium exchange pump**. The carrier protein involved in the process is called *sodium–potassium ATPase*.

The sodium–potassium exchange pump exchanges intracellular sodium for extracellular potassium (**Figure 3–19**). On average, for each ATP molecule consumed, three sodium ions are ejected and two potassium ions are reclaimed by the cell. If ATP is readily available, the rate of transport depends on the concentration of sodium ions in the cytoplasm. When the concentration rises, the pump becomes more active. The energy demands are impressive: Sodium–potassium ATPase may use up to 40 percent of the ATP produced by a resting cell!

Secondary Active Transport   In **secondary active transport**, the transport mechanism itself does not require energy from ATP, but the cell often needs to expend ATP at a later time to preserve homeostasis. As does facilitated transport, a secondary active transport mechanism moves a specific substrate down its concentration gradient. Unlike the proteins in facilitated transport, however, these carrier proteins can also move another substrate at the same time, without regard to its concentration gradient. In effect, the concentration gradient for one substance provides the driving force needed by the carrier protein, and the second substance gets a "free ride."

The concentration gradient for sodium ions most often provides the driving force for cotransport mechanisms that move materials into the cell. For example, sodium-linked cotransport is important in the absorption of glucose and amino acids along the intestinal tract. Although the initial transport activity proceeds without direct energy expenditure, the cell must expend ATP to pump the arriving sodium ions out of the cell by using the sodium–potassium exchange pump (**Figure 3–20**). Sodium ions are also involved with many countertransport mechanisms. Sodium–calcium countertransport is responsible for keeping intracellular calcium ion concentrations very low.

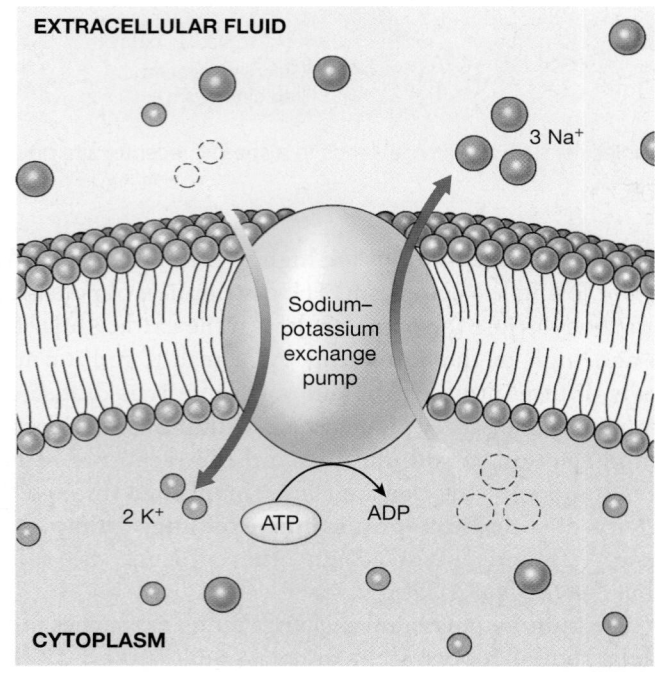

Figure 3–19 **The Sodium–Potassium Exchange Pump.** The operation of the sodium–potassium exchange pump is an example of active transport. For each ATP converted to ADP, the protein called sodium–potassium ATPase carries three Na$^+$ out of the cell and two K$^+$ into the cell.

Vesicular Transport

In **vesicular transport**, materials move into or out of the cell in **vesicles**, small membranous sacs that form at, or fuse with, the plasma membrane. Because tiny droplets of fluid and solutes are transported rather than single molecules, this process is also known as *bulk transport*. The two major categories of vesicular transport are *endocytosis* and *exocytosis*.

Endocytosis

As we saw earlier in this chapter, extracellular materials can be packaged in vesicles at the cell surface and imported into the cell (p. 79). This process, called **endocytosis**, involves relatively large volumes of extracellular material and requires energy in the form of ATP. The three major types of endocytosis are (1) *receptor-mediated endocytosis*, (2) *pinocytosis*, and (3) *phagocytosis*. All three are active processes that require energy in the form of ATP.

Vesicles produced by receptor-mediated endocytosis or by pinocytosis are called *endosomes*; those produced by phagocytosis are called *phagosomes*. The contents of endosomes and phagosomes remain isolated from the cytoplasm, trapped within the vesicle. The movement of materials into the surrounding cytoplasm may involve active transport, simple or facilitated diffusion, or the destruction of the vesicle membrane.

Receptor-Mediated Endocytosis A selective process, **receptor-mediated endocytosis** involves the formation of small vesicles at the surface of the membrane. This process produces vesicles that contain a specific target molecule in high concentrations. Receptor-mediated endocytosis begins when materials in the extracellular fluid bind to receptors on the membrane surface (**Figure 3–21**). Most receptor molecules are glycoproteins, and each binds to a specific ligand, or tar-

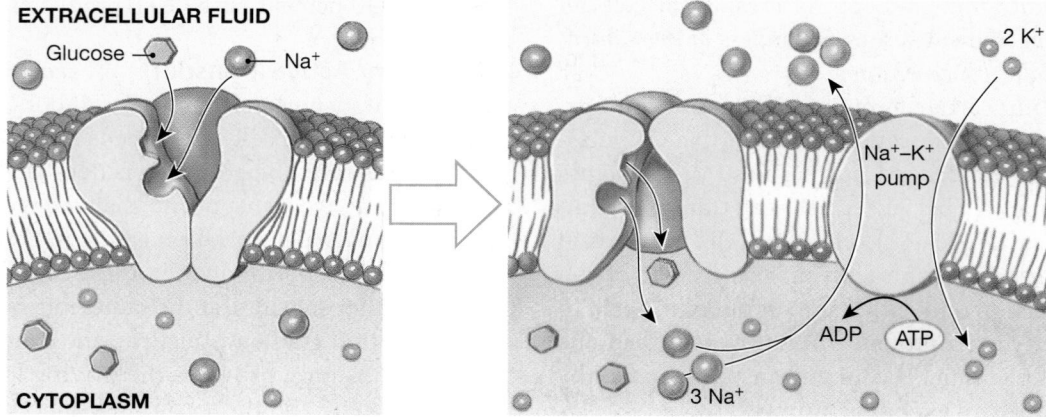

Figure 3–20 **Secondary Active Transport.** In secondary active transport, glucose transport by a carrier protein will occur only after the carrier has bound two sodium ions. In three cycles, three glucose molecules and six sodium ions are transported into the cytoplasm. The cell then pumps the sodium ions across the plasma membrane via the sodium–potassium exchange pump, at a cost of two ATP molecules.

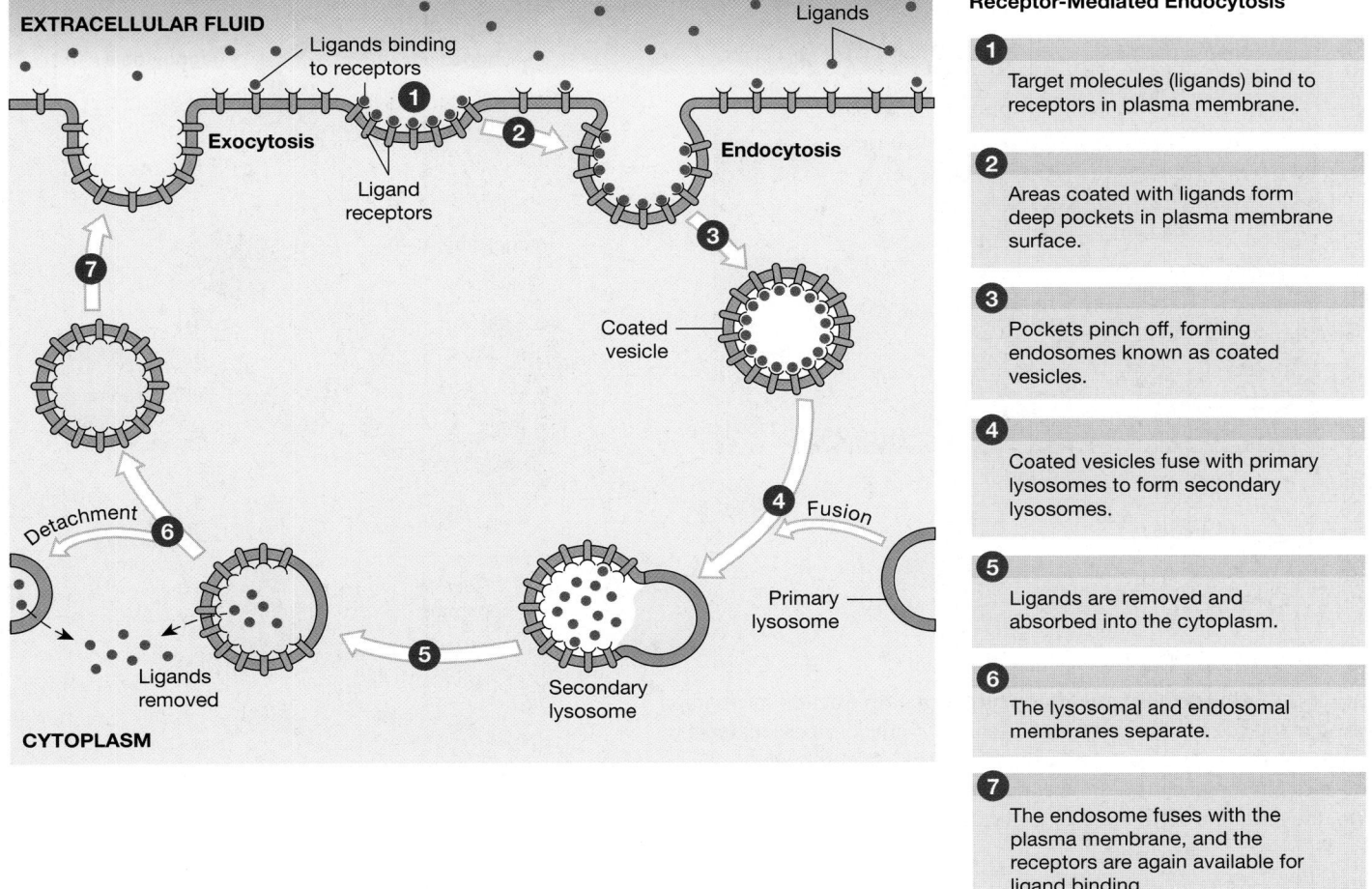

Receptor-Mediated Endocytosis

1 Target molecules (ligands) bind to receptors in plasma membrane.

2 Areas coated with ligands form deep pockets in plasma membrane surface.

3 Pockets pinch off, forming endosomes known as coated vesicles.

4 Coated vesicles fuse with primary lysosomes to form secondary lysosomes.

5 Ligands are removed and absorbed into the cytoplasm.

6 The lysosomal and endosomal membranes separate.

7 The endosome fuses with the plasma membrane, and the receptors are again available for ligand binding.

Figure 3–21   **Receptor-Mediated Endocytosis.**

get, such as a transport protein or a hormone. Some receptors are distributed widely over the surface of the plasma membrane; others are restricted to specific regions or in depressions on the cell surface.

Receptors bound to ligands cluster together. Once an area of the plasma membrane has become covered with ligands, it forms grooves or pockets that move to one area of the cell and then pinch off to form an endosome. The endosomes produced in this way are called **coated vesicles**, because they are surrounded by a protein–fiber network that originally carpeted the inner membrane surface beneath the receptor–ligand clusters. This coating is essential to endosome formation and movement. Inside the cell, the coated vesicles fuse with primary lysosomes filled with digestive enzymes, creating secondary lysosomes (p. 79). The lysosomal enzymes then free the ligands from their receptors, and the ligands enter the cytosol by diffusion or active transport. The vesicle membrane detaches from the secondary lysosome and returns to the cell surface, where its receptors are available to bind more ligands.

Many important substances, including cholesterol and iron ions (Fe^{2+}), are distributed through the body attached to spe-

cial transport proteins. These proteins are too large to pass through membrane pores, but they can and do enter cells by receptor-mediated endocytosis.

Pinocytosis   "Cell drinking," or **pinocytosis** (pi-nō-sī-TŌ-sis), is the formation of endosomes filled with extracellular fluid. This process is not as selective as receptor-mediated endocytosis, because no receptor proteins are involved. The target appears to be the fluid contents in general, rather than specific bound ligands. In pinocytosis, a deep groove or pocket forms in the plasma membrane and then pinches off (**Figure 3–22a**). The steps involved in the formation and fate of an endosome created by pinocytosis are similar to the steps in receptor-mediated endocytosis, except that ligand binding is not the trigger.

Phagocytosis   "Cell eating," or **phagocytosis** (fag-ō-sī-TŌ-sis), produces phagosomes containing solid objects that may be as large as the cell itself. In this process, cytoplasmic extensions called **pseudopodia** (soo-dō-PŌ-dē-ah; *pseudo-*, false + *podon*, foot; singular *pseudopodium*) surround the object, and their

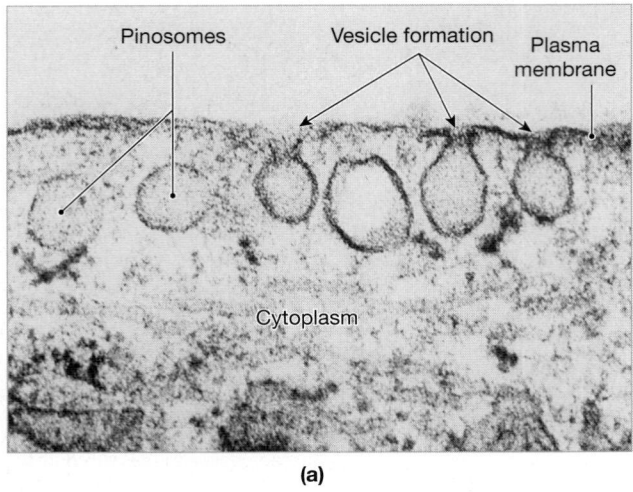

(a)

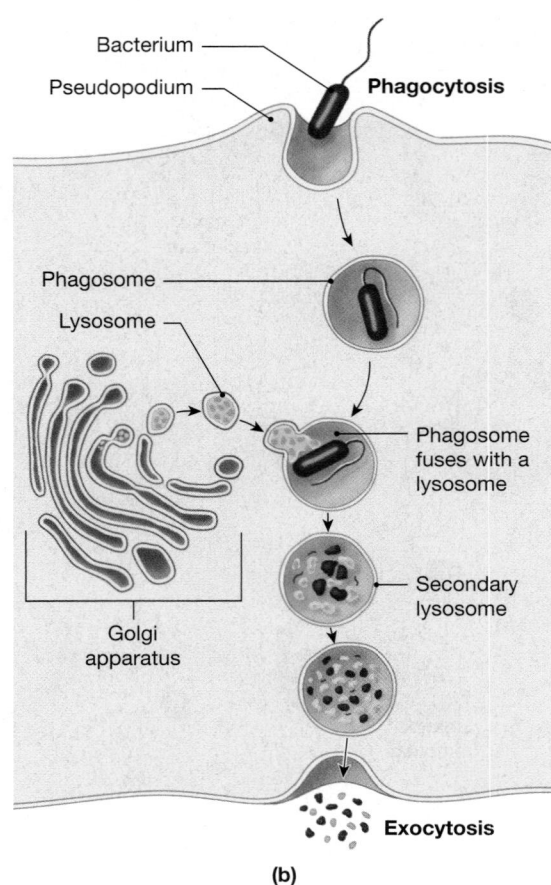

(b)

Figure 3–22 Pinocytosis and Phagocytosis. (a) An electron micrograph showing pinocytosis at the surface of a cell. **(b)** In phagocytosis, material is brought into the cell enclosed in a phagosome that is subsequently exposed to lysosomal enzymes. After nutrients are absorbed from the vesicle, the residue is discharged by exocytosis.

membranes fuse to form a phagosome (**Figure 3–22b**). This vesicle then fuses with many lysosomes, whereupon its contents are digested by lysosomal enzymes. Although most cells display pinocytosis, phagocytosis is performed only by specialized cells, such as the *macrophages* that protect tissues by engulfing bacteria, cell debris, and other abnormal materials.

Exocytosis

Exocytosis (ek-sō-sī-TŌ-sis), introduced in our discussion of the Golgi apparatus (p. 77), is the functional reverse of endocytosis. In exocytosis, a vesicle created inside the cell fuses with the plasma membrane and discharges its contents into the extracellular environment (**Figure 3–22b**). The ejected material may be secretory products, such as mucins or hormones, or waste products, such as those accumulating in endocytic vesicles. In a few specialized cells, endocytosis produces vesicles on one side of the cell that are discharged through exocytosis on the opposite side. This method of bulk transport is common in cells lining capillaries, which use a combination of pinocytosis and exocytosis

to transfer fluid and solutes from the bloodstream into the surrounding tissues.

Many different mechanisms are moving materials into and out of the cell at any moment. Before proceeding further, review and compare the mechanisms summarized in Table 3–3.

CHECKPOINT

22. Describe the process of carrier-mediated transport.

23. During digestion in the stomach, the concentration of hydrogen ions (H^+) within the stomach rises to many times that in cells lining the stomach. Which transport process must be operating?

24. Describe endocytosis.

25. Describe exocytosis.

26. When they encounter bacteria, certain types of white blood cells engulf the bacteria and bring them into the cell. What is this process called?

See the blue Answers tab at the end of the book.

SUMMARY TABLE 3–3 Mechanisms Involved in Movement across Plasma Membranes

Mechanism	Process	Factors Affecting Rate	Substances Involved (Sites)
Diffusion (includes simple diffusion and channel-mediated diffusion)	Molecular movement of solutes; direction determined by relative concentrations	Size of gradient; size of molecules; charge; lipid solubility; temperature; additional factors apply to channel-mediated diffusion	Small inorganic ions; lipid-soluble materials (all cells)
Osmosis	Movement of water molecules toward solution containing relatively higher solute concentration; requires selectively permeable membrane	Concentration gradient; opposing osmotic or hydrostatic pressure; number of aquaporins (water channels)	Water only (all cells)
Carrier-Mediated Transport:			
Facilitated diffusion	Carrier proteins passively transport solutes across a membrane down a concentration gradient	Size of gradient, temperature, and availability of carrier protein	Glucose and amino acids (all cells, but several different regulatory mechanisms exist)
Active transport	Carrier proteins actively transport solutes across a membrane, often against a concentration gradient	Availability of carrier, substrates, and ATP	Na^+, K^+, Ca^{2+}, Mg^{2+} (all cells); other solutes by specialized cells
Secondary active transport	Carrier proteins passively transport two solutes, with one (normally Na^+) moving down its concentration gradient; the cell must later expend ATP to eject the Na^+	Availability of carrier, substrates, and ATP	Glucose and amino acids (specialized cells); iodide
Vesicular Transport:			
Endocytosis	Creation of membranous vesicles containing fluid or solid material	Stimulus and mechanics incompletely understood; requires ATP	Fluids, nutrients (all cells); debris, pathogens (specialized cells)
Exocytosis	Fusion of vesicles containing fluids or solids (or both) with the plasma membrane	Stimulus and mechanics incompletely understood; requires ATP	Fluids, debris (all cells)

3-7 The transmembrane potential results from the unequal distribution of ions across the plasma membrane

As noted previously, the inside of the plasma membrane has a slight negative charge with respect to the outside. The cause is a slight excess of positively charged ions outside the plasma membrane, and a slight excess of negatively charged ions (especially proteins) inside the plasma membrane. This unequal charge distribution is created by differences in the permeability of the membrane to various ions, as well as by active transport mechanisms.

Although the positive and negative charges are attracted to each other and would normally rush together, they are kept apart by the phospholipid membrane. When positive and negative charges are held apart, a **potential difference** is said to exist between them. We refer to the potential difference across a plasma membrane as the **transmembrane potential**.

The unit of measurement of potential difference is the *volt* (V). Most cars, for example, have 12-V batteries. The transmembrane potentials of cells are much smaller, typically in the vicinity of 0.07 V. Such a value is usually expressed as 70 mV, or 70 *millivolts* (thousandths of a volt). The transmembrane potential in an undisturbed cell is called the **resting potential**. Each type of cell has a characteristic resting potential between −10 mV (−0.01 V) and −100 mV (−0.1 V), with the minus sign signifying that the inside of the plasma membrane contains an excess of negative charges compared with the outside. Examples include fat cells (−40 mV), thyroid cells (−50 mV), neurons (−70 mV), skeletal muscle cells (−85 mV), and cardiac muscle cells (−90 mV).

If the lipid barrier were removed, the positive and negative charges would rush together and the potential difference would be eliminated. The plasma membrane thus acts like a dam across a stream. Just as a dam resists the water pressure that builds up on the upstream side, a plasma membrane resists electrochemical forces that would otherwise drive ions into or out of the cell. The water retained behind a dam and the ions held on either side of the plasma membrane have *potential energy*—stored energy that can be released to do work. People have designed many ways to use the potential energy stored behind a dam—for example, turning a mill wheel or a turbine. Similarly, cells have ways of utilizing the potential energy stored in the transmembrane potential. For example, it is the transmembrane potential that makes possible the transmission of information in the nervous system, and thus our perceptions and thoughts. As we will see in later chapters, changes in the transmembrane potential also trigger the contraction of muscles and the secretion of glands.

CHECKPOINT

27. What is the transmembrane potential, and in what units is it expressed?

28. If the plasma membrane were freely permeable to sodium ions (Na$^+$), how would the transmembrane potential be affected?

See the blue Answers tab at the end of the book.

3-8 Stages of a cell's life cycle include interphase, mitosis, and cytokinesis

The period between fertilization and physical maturity involves tremendous changes in organization and complexity. At fertilization, a single cell is all there is; at maturity, your body has roughly 75 trillion cells. This amazing transformation involves a form of cellular reproduction called **cell division**. The division of a single cell produces a pair of **daughter cells**, each half the size of the original. Before dividing, each of the daughter cells will grow to the size of the original cell.

Even when development is complete, cell division continues to be essential to survival. Cells are highly adaptable, but they can be damaged by physical wear and tear, toxic chemicals, temperature changes, and other environmental stresses. And, like individuals, cells age. The life span of a cell varies from hours to decades, depending on the type of cell and the stresses involved. Many cells apparently self-destruct after a certain period of time as a result of the activation of specific "suicide genes" in the nucleus. The genetically controlled death of cells is called **apoptosis** (ap-op-TŌ-sis or ap-ō-TŌ-sis; *apo-,* separated from + *ptosis,* a falling). Several genes involved in the regulation of this process have been identified. For example, a gene called *bcl-2* appears to prevent apoptosis and to keep a cell alive and functional. If something interferes with the function of this gene, the cell self-destructs.

Because a typical cell does not live nearly as long as a typical person, cell populations must be maintained over time by cell division. For cell division to be successful, the genetic material in the nucleus must be duplicated accurately, and one

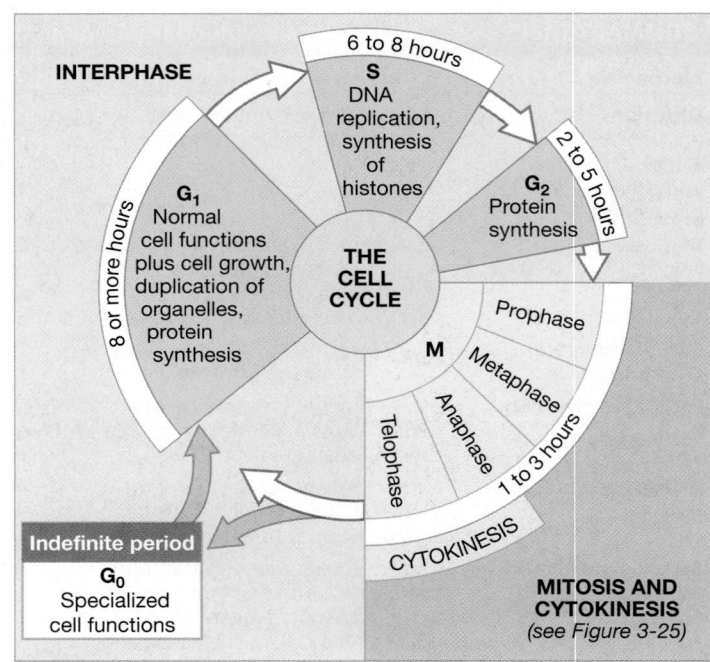

Figure 3–23 **The Cell Life Cycle.**

copy must be distributed to each daughter cell. The duplication of the cell's genetic material is called **DNA replication**, and nuclear division is called **mitosis** (mī-TŌ-sis). Mitosis occurs during the division of somatic cells. The production of sex cells involves a different process, **meiosis** (mī-Ō-sis), described in Chapter 28.

Figure 3–23 depicts the life cycle of a typical cell. That life cycle includes a fairly brief period of mitosis alternating with an *interphase* of variable duration.

Interphase

Most cells spend only a small part of their time actively engaged in cell division. Somatic cells spend the majority of their functional lives in a state known as **interphase**. During interphase, a cell performs all its normal functions and, if necessary, prepares for cell division. In a cell preparing to divide, interphase can be divided into the G_1, S, and G_2 *phases*. An interphase cell in the G_0 *phase* is not preparing for division, but is performing all of the other functions appropriate for that particular cell type. Some mature cells, such as skeletal muscle cells and most neurons, remain in G_0 indefinitely and never divide. In contrast, *stem cells*, which divide repeatedly with very brief interphase periods, never enter G_0.

The G_1 Phase

A cell that is ready to divide first enters the **G_1 phase**. In this phase, the cell makes enough mitochondria, cytoskeletal elements, endoplasmic reticula, ribosomes, Golgi membranes, and cytosol for two functional cells. Centriole replication begins in G_1 and commonly continues until G_2. In cells dividing at top speed, G_1 may last just 8–12 hours. Such cells pour all their energy into mitosis, and all other activities cease. If G_1 lasts for days, weeks, or months, preparation for mitosis occurs as the cells perform their normal functions.

The S Phase

When the activities of G_1 have been completed, the cell enters the **S phase**. Over the next 6–8 hours, the cell duplicates its chromosomes. This involves DNA replication and the synthesis of histones and other proteins in the nucleus. The goal of DNA replication, which occurs in cells preparing to undergo either mitosis or meiosis, is to copy the genetic information in the nucleus. The cell ends up with two identical sets of chromosomes. In mitosis, one set is given to each of the two daughter cells.

DNA Replication   Each DNA molecule consists of a pair of DNA strands joined by hydrogen bonding between complementary nitrogenous bases. ∞ p. 59 **Figure 3–24** diagrams DNA replication. The process begins when enzymes called *helicases* unwind the strands and disrupt the weak bonds between the bases. As the strands unwind, molecules of **DNA polymerase** bind to the exposed nitrogenous bases. This enzyme (1) promotes bonding between the nitrogenous bases of the DNA strand and complementary DNA nucleotides dissolved in the nucleoplasm and (2) links the nucleotides by covalent bonds.

Many molecules of DNA polymerase work simultaneously along the DNA strands (**Figure 3–24**). DNA polymerase can work in only one direction along a strand of DNA, but the two strands in a DNA molecule are oriented in oppo-

site directions. As a result, the DNA polymerase on one strand works toward the site where the strands are unzipping, but those on the other strand work away from it. As the two original strands gradually separate, the DNA polymerase bound to one strand (the upper strand in the figure) adds nucleotides to make a single, continuous complementary copy of that strand. This copy grows toward the "zipper" from right to left, adding nucleotides 1 through 9 in sequence; the 1 is added first, then 2 to the left of 1, and so on.

DNA polymerase on the other original strand, however, can work only away from the unzipping site. In the lower strand in **Figure 3–24**, the first DNA polymerase to bind to it must work from left to right, adding nucleotides in the sequence $1 \longrightarrow 2 \longrightarrow 3 \longrightarrow 4 \longrightarrow 5$. But as the original strands continue to unzip, additional nucleotides are continuously exposed. This molecule of DNA polymerase cannot go into reverse; it can only continue working from left to right. Thus, a second molecule of DNA polymerase must bind closer to the point of unzipping and assemble a complementary copy that grows in the sequence $6 \longrightarrow 7 \longrightarrow 8 \longrightarrow 9$, until it bumps into the segment created by the first DNA polymerase. The two segments are then spliced together by enzymes called **ligases** (LĪ-gās-ez; *liga*, to tie). Eventually, the unzipping completely separates the original strands. The copying ends, the last splicing is done, and two identical DNA molecules have formed.

The G₂ Phase

Once DNA replication has ended, there is a brief (2–5 hours) **G₂ phase** devoted to last-minute protein synthesis and to the completion of centriole replication. The cell then enters the **M phase**, and mitosis begins.

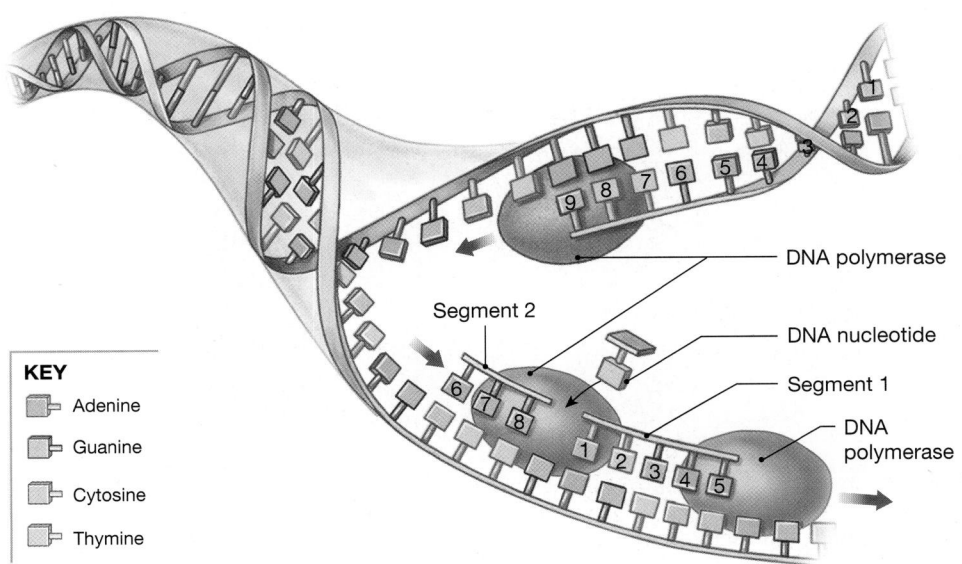

KEY

◻ Adenine
◻ Guanine
◻ Cytosine
◻ Thymine

— DNA polymerase

Segment 2

— DNA nucleotide

— Segment 1

— DNA polymerase

Figure 3–24   DNA Replication. In DNA replication, the DNA strands unwind, and DNA polymerase begins attaching complementary DNA nucleotides along each strand. On one original strand, the complementary copy is produced as a continuous strand. Along the other original strand, the copy begins as a series of short segments spliced together by ligases. This process ultimately produces two identical copies of the original DNA molecule.

Mitosis

Mitosis separates the duplicated chromosomes of a cell into two identical nuclei. The term *mitosis* specifically refers to the division and duplication of the cell's nucleus; division of the cytoplasm to form two distinct new cells involves a separate, but related, process known as **cytokinesis** (sī-tō-ki-NĒ-sis; *cyto-*, cell + *kinesis*, motion). **Figure 3–25** depicts interphase and summarizes the four stages of mitosis: *prophase* (early and late), *metaphase*, *anaphase*, and *telophase*. Bear in mind that, although we describe mitosis in stages, it is really one smooth, continuous process.

Stage 1: Prophase

Prophase (PRŌ-fāz; *pro*, before) begins when the chromosomes coil so tightly that they become visible as individual structures under a light microscope. As a result of DNA replication during the S phase, two copies of each chromosome now exist. Each copy, called a **chromatid** (KRŌ-ma-tid), is physically connected to its duplicate copy at a single point, the **centromere** (SEN-trō-mēr). The centromere is surrounded by a protein complex known as the **kinetochore** (ki-NE-tō-kor). (These structures can be seen in **Figure 3–11.**)

As the chromosomes appear, the nucleoli disappear. The disappearance occurs in late prophase, often called *prometaphase*. Around this time, the two pairs of centrioles replicated during the $G_1 - G_2$ period move toward opposite poles of the nucleus. An array of microtubules called **spindle fibers** extends between the centriole pairs. Smaller microtubules called *astral rays* radiate into the surrounding cytoplasm. Late in prophase, the nuclear envelope disappears. The spindle fibers now form among the chromosomes, and the kinetochore of each chromatid becomes attached to a spindle fiber. Once that attachment occurs, the spindle fiber is called a *chromosomal microtubule*.

Stage 2: Metaphase

Metaphase (MET-a-fāz; *meta*, after) begins as the chromatids move to a narrow central zone called the **metaphase plate**. Metaphase ends when all the chromatids are aligned in the plane of the metaphase plate.

Stage 3: Anaphase

Anaphase (AN-a-fāz; *ana-*, apart) begins when the centromere of each chromatid pair splits and the chromatids separate. The two **daughter chromosomes** are now pulled

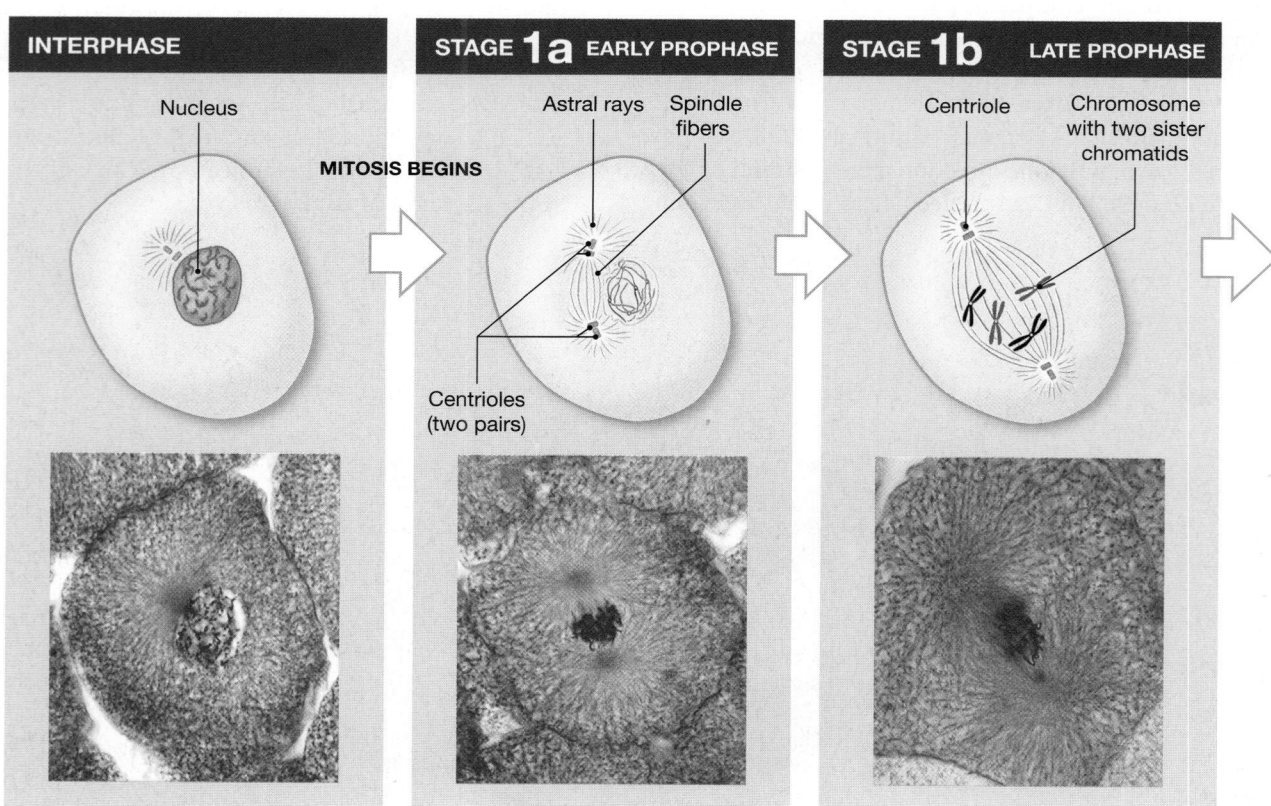

Figure 3–25 **Interphase, Mitosis, and Cytokinesis.** Diagrammatic and microscopic views of representative cells undergoing cell division. All are light micrographs (LM × 450).

toward opposite ends of the cell along the chromosomal microtubules. This movement involves an interaction between the kinetochore and the microtubule. Anaphase ends when the daughter chromosomes arrive near the centrioles at opposite ends of the cell.

Stage 4: Telophase

During **telophase** (TĒL-ō-fāz; *telo-*, end), each new cell prepares to return to the interphase state. The nuclear membranes re-form, the nuclei enlarge, and the chromosomes gradually uncoil. Once the chromosomes have relaxed and the fine filaments of chromatin become visible again, the nucleoli reappear and the nuclei resemble those of interphase cells. This stage marks the end of mitosis.

Tips&Tricks

In order to remember the correct sequence of events during mitosis, imagine the contour rug in front of your toilet as the P-MAT, for **p**rophase, **m**etaphase, **a**naphase, and **t**elophase.

The A&P Top 100

#22 Mitosis is the duplication of the chromosomes in the nucleus and their separation into two identical sets in the process of somatic cell division.

Cytokinesis

Cytokinesis is the cytoplasmic division of the daughter cells. This process usually begins in late anaphase. As the daughter chromosomes approach the ends of the spindle apparatus, the cytoplasm constricts along the plane of the metaphase plate, forming a *cleavage furrow*. Cytokinesis continues throughout telophase and is usually completed sometime after a nuclear membrane has re-formed around each daughter nucleus. The completion of cytokinesis marks the end of cell division, creating two separate and complete cells, each surrounded by its own plasma membrane.

The Mitotic Rate and Energy Use

The preparations for cell division that occur between G_1 and the M phase are difficult to recognize in a light micrograph.

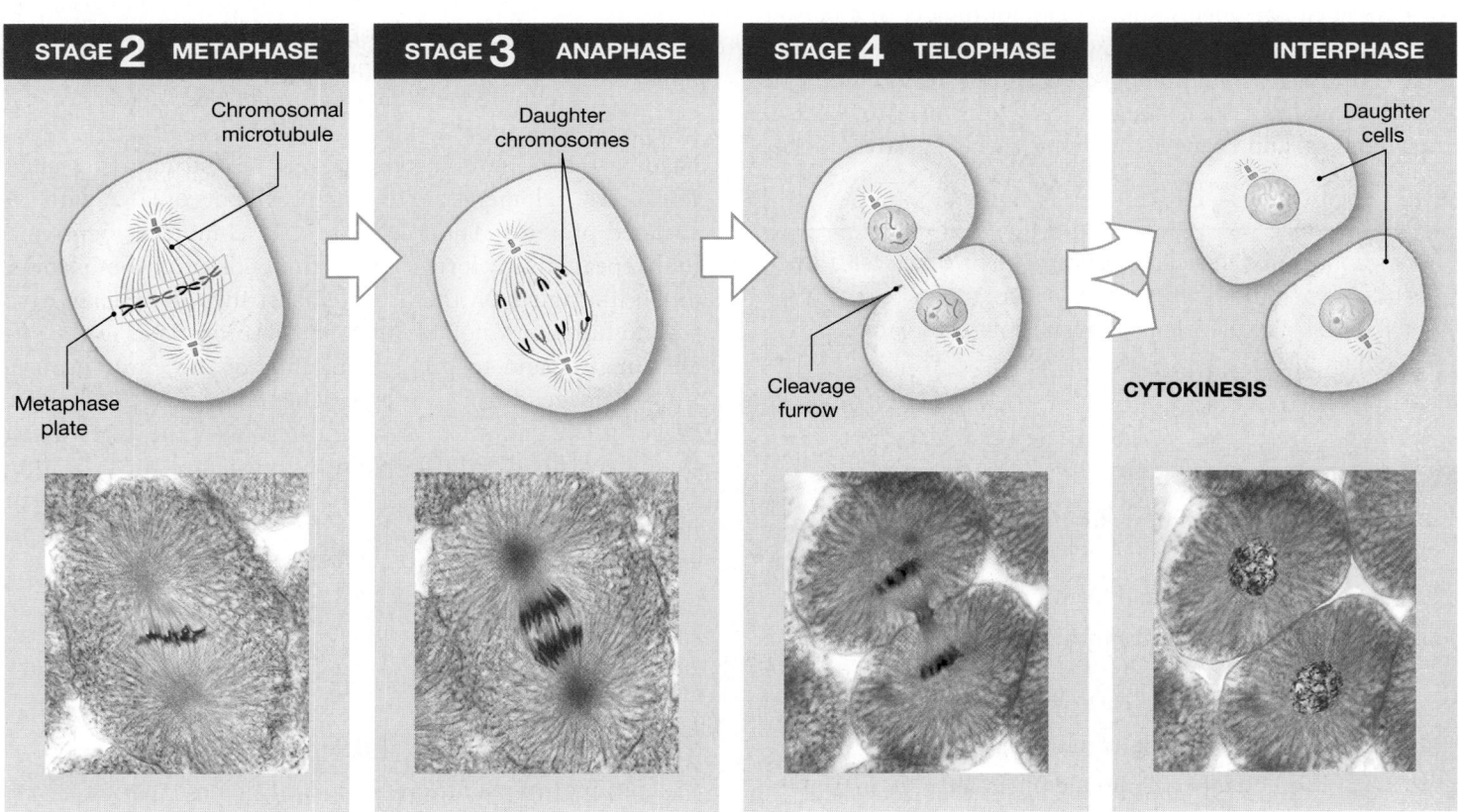

STAGE **2** METAPHASE — Chromosomal microtubule — Metaphase plate

STAGE **3** ANAPHASE — Daughter chromosomes

STAGE **4** TELOPHASE — Cleavage furrow

INTERPHASE — Daughter cells — **CYTOKINESIS**

However, the start of mitosis is easy to recognize, because the chromosomes become condensed and highly visible. The frequency of cell division can thus be estimated by the number of cells in mitosis at any time. As a result, we often use the term **mitotic rate** when we discuss rates of cell division. In general, the longer the life expectancy of a cell type, the slower the mitotic rate. Long-lived cells, such as muscle cells and neurons, either never divide or do so only under special circumstances. Other cells, such as those covering the surface of the skin or the lining of the digestive tract, are subject to attack by chemicals, pathogens, and abrasion. They survive for only days or even hours. Special cells called **stem cells** maintain these cell populations through repeated cycles of cell division.

Stem cells are relatively unspecialized; their only function is the production of daughter cells. Each time a stem cell divides, one of its daughter cells develops functional specializations while the other prepares for further stem cell divisions. The rate of stem cell division can vary with the type of tissue and the demand for new cells. In heavily abraded skin, stem cells may divide more than once a day, but stem cells in adult connective tissues may remain inactive for years.

Dividing cells use an unusually large amount of energy. For example, they must synthesize new organic materials and move organelles and chromosomes within the cell. All these processes require ATP in substantial amounts. Cells that do not have adequate energy sources cannot divide. In a person who is starving, normal cell growth and maintenance grind to a halt. For this reason, prolonged starvation stunts childhood growth, slows wound healing, lowers resistance to disease, thins the skin, and changes the lining of the digestive tract.

Tips&Tricks

When considering the relative length of time that a cell spends in interphase compared to mitosis, think about taking a test. You prepare a long time (interphase) for something that happens quickly (mitosis).

CHECKPOINT

29. Give the biological terms for (a) cellular reproduction and (b) cellular death.

30. Describe interphase, and identify its stages.

31. A cell is actively manufacturing enough organelles to serve two functional cells. This cell is probably in what phase of its life cycle?

32. Define mitosis, and list its four stages.

33. What would happen if spindle fibers failed to form in a cell during mitosis?

See the blue Answers tab at the end of the book.

3-9 Several growth factors affect the cell life cycle

In normal tissues, the rate of cell division balances the rate of cell loss or destruction. Mitotic rates are genetically controlled, and many different stimuli may be responsible for activating genes that promote cell division. Some of the stimuli are internal, and many cells set their own pace of mitosis and cell division. An important internal trigger is the level of **M-phase promoting factor (MPF)**, also known as *maturation-promoting factor*. MPF is assembled from two parts: a cell division cycle protein called *Cdc2* and a second protein called *cyclin*. Cyclin levels climb as the cell life cycle proceeds. When levels are high enough, MPF appears in the cytoplasm and mitosis gets under way.

Various extracellular compounds—generally, peptides—can stimulate the division of specific types of cells. These compounds include several hormones and a variety of **growth factors**. Table 3–4 lists some of the stimulatory compounds and their target tissues; we will discuss these hormones and factors in later chapters.

Genes that inhibit cell division have recently been identified. Such genes are known as *repressor genes*. One gene, called *p53*, controls a protein that resides in the nucleus and activates genes that direct the production of growth-inhibiting factors inside the cell. Roughly half of all cancers are associated with abnormal forms of the *p53* gene.

There are indications that in humans, the *number* of cell divisions performed by a cell and its descendants is regulated at the chromosome level by structures called **telomeres**. Telomeres are terminal segments of DNA with associated proteins. These DNA-protein complexes bend and fold repeatedly to form caps at the ends of chromosomes, much like aglets on the tips of shoestrings. Telomeres have several functions, notably to attach chromosomes to the nuclear matrix and to protect the ends of the chromosomes from damage during mitosis. The telomeres themselves, however, are subject to wear and tear over the years. Each time a cell divides during adult life, some of the repeating segments break off, and the telomeres get shorter. When they get too short, repressor gene activity tells the cell to stop dividing.

CHECKPOINT

34. Define growth factor, and identify several growth factors that affect cell division.

See the blue Answers tab at the end of the book.

TABLE 3–4 Representative Chemical Factors Affecting Cell Division

Factor	Source(s)	Effect(s)	Target(s)
M-phase promoting factor (maturation-promoting factor)	Forms within cytoplasm from Cdc2 and cyclin	Triggers start of mitosis	Regulatory mechanism active in all dividing cells
Growth hormone	Adenohypophysis	Stimulation of growth, cell division, differentiation	All cells, especially in epithelia and connective tissues
Prolactin	Adenohypophysis	Stimulation of cell growth, division, development	Gland and duct cells of mammary glands
Nerve growth factor (NGF)	Salivary glands; other sources suspected	Stimulation of nerve cell repair and development	Neurons and neuroglia
Epidermal growth factor (EGF)	Duodenal glands; other sources suspected	Stimulation of stem cell divisions and epithelial repairs	Epidermis
Fibroblast growth factor (FGF)	Unknown	Division and differentiation of fibroblasts and related cells	Connective tissues
Erythropoietin	Kidneys (primary source)	Stimulation of stem cell divisions and maturation of red blood cells	Bone marrow
Thymosins and related compounds	Thymus	Stimulation of division and differentiation of lymphocytes (especially T cells)	Thymus and other lymphoid tissues and organs
Chalones	Many tissues	Inhibition of cell division	Cells in the immediate area

3-10 Tumors and cancers are characterized by abnormal cell growth and division

When the rates of cell division and growth exceed the rate of cell death, a tissue begins to enlarge. A **tumor,** or *neoplasm,* is a mass or swelling produced by abnormal cell growth and division. In a *benign tumor,* the cells usually remain within the epithelium (one of the four primary tissue types) or a connective-tissue capsule. Such a tumor seldom threatens an individual's life and can usually be surgically removed if its size or position disturbs tissue function.

Cells in a *malignant tumor* no longer respond to normal control mechanisms. These cells do not remain confined within the epithelium or a connective tissue capsule, but spread into surrounding tissues. The tumor of origin is called the *primary tumor* (or *primary neoplasm*), and the spreading process is called **invasion**. Malignant cells may also travel to distant tissues and organs and establish *secondary tumors.* This dispersion, called **metastasis** (me-TAS-ta-sis; *meta-,* after + *stasis,* standing still), is very difficult to control.

Cancer is an illness characterized by mutations that disrupt normal control mechanisms and produce potentially malignant cells. Cancer develops in the series of steps diagrammed in **Figure 3–26**. Initially, the cancer cells are restricted to the primary tumor. In most cases, all the cells in the tumor are the daughter cells of a single malignant cell. Normal cells often become malignant when a mutation occurs in a gene involved with cell growth, differentiation, or division. The modified genes are called **oncogenes** (ON-kō-gēnz; *oncos,* tumor).

Cancer cells gradually lose their resemblance to normal cells. They change shape and typically become abnormally large or small. At first, the growth of the primary tumor distorts the tissue, but the basic tissue organization remains intact. Metastasis begins with invasion as tumor cells "break out" of the primary tumor and invade the surrounding tissue. They may then enter the lymphoid system and accumulate in nearby lymph nodes. When metastasis involves the penetration of blood vessels, the cancer cells circulate throughout the body.

Responding to cues that are as yet unknown, cancer cells in the bloodstream ultimately escape out of blood vessels to establish secondary tumors at other sites. These tumors are extremely active metabolically, and their presence stimulates the growth of blood vessels into the area. The increased circulatory supply provides additional nutrients to the cancer cells and further accelerates tumor growth and metastasis.

As malignant tumors grow, organ function begins to deteriorate. The malignant cells may no longer perform their original functions, or they may perform normal functions in an abnormal way. For example, endocrine cancer cells may produce normal hormones, but in excessively large amounts. Cancer cells do not use energy very efficiently. They grow and multiply at the expense of healthy tissues, competing for space and nutrients with normal cells. This competition contributes to the starved appearance of many patients in the late stages of cancer. Death may occur as a result of the compression of vital organs when nonfunctional cancer cells have

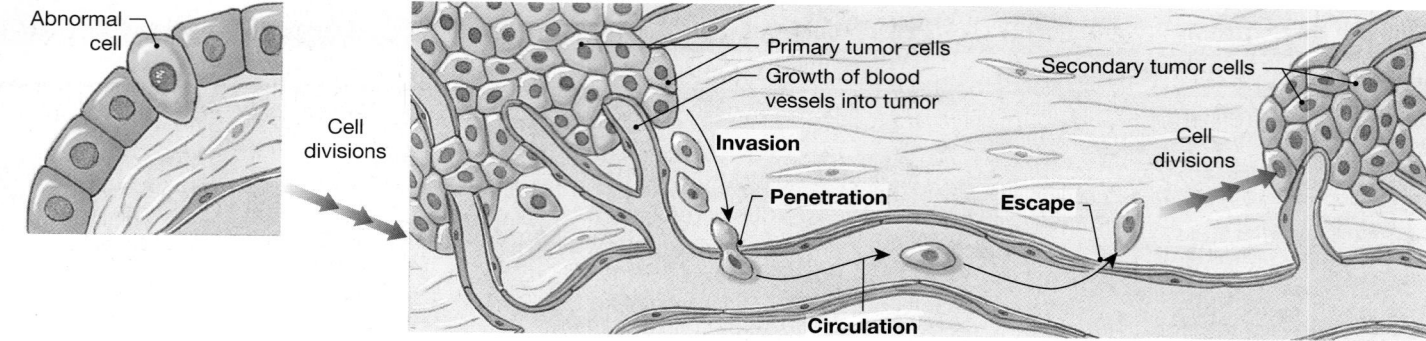

Figure 3–26 **The Development of Cancer.**

killed or replaced the healthy cells in those organs, or when the cancer cells have starved normal tissues of essential nutrients. We will return to the subject of cancer in later chapters that deal with specific systems.

CLINICAL NOTE

Telomerase, Aging, and Cancer   Each telomere contains a sequence of roughly 8000 nitrogenous bases, but they are multiple copies of the same base sequence, TTAGGG, repeated over and over again. Telomeres are not formed by DNA polymerase; instead, they are created by an enzyme called *telomerase*. Telomerase is functional early in life, but by adulthood it has become inactive. As a result, the telomere segments lost during each mitotic division are not replaced. Eventually, shortening of the telomere reaches a point at which the cell ceases to divide.

This mechanism is clearly a major factor in the aging process, since many of the signs of age result from the gradual loss of functional stem cell populations. Experiments are in progress to determine whether activating telomerase (or a suspected alternative repair enzyme) can forestall or reverse the effects of aging. This would seem to be a very promising area of research. Activate telomerase, and halt aging—sounds good, doesn't it? Unfortunately, there's a catch: In adults, telomerase activation is a key step in the development of cancer.

If for some reason a cell with short telomeres does *not* respond normally to repressor genes, it will continue to divide. The result is mechanical damage to the DNA strands, chromosomal abnormalities, and mutations. Interestingly, one of the first consequences of such damage is the abnormal activation of telomerase. Once this occurs, the abnormal cells can continue dividing indefinitely. Telomerase is active in at least 90 percent of all cancer cells. Research is therefore under way to find out how to turn off telomerase that has been improperly activated.

CHECKPOINT

35. An illness characterized by mutations that disrupt normal control mechanisms and produce potentially malignant cells is termed _____.

36. Define metastasis.

See the blue Answers tab at the end of the book.

3-11 Differentiation is cellular specialization as a result of gene activation or repression

An individual's liver cells, fat cells, and neurons all contain the same set of chromosomes and genes, but in each case a different set of genes has been turned *off*. In other words, liver cells and fat cells differ because liver cells have one set of genes accessible for transcription, and fat cells another.

When a gene is functionally eliminated, the cell loses the ability to produce a particular protein—and thus to perform any functions involving that protein. Each time another gene switches off, the cell's functional abilities become more restricted. This specialization process is called **differentiation**.

Fertilization produces a single cell with all its genetic potential intact. Repeated cell divisions follow, and differentiation begins as the number of cells increases. Differentiation produces specialized cells with limited capabilities. These cells form organized collections known as *tissues*, each with discrete functional roles. In Chapter 4, we will examine the structure and function of tissues and will consider the role of tissue interactions in the maintenance of homeostasis.

CLINICAL NOTE

Parkinson Disease In most cases, differentiation is irreversible: Once genes are turned off, they won't be turned back on. However, some cells, such as stem cells, are relatively undifferentiated. These cells can differentiate into any of several different types of cell, depending on local conditions. For example, when more nutrients are consumed than the body can use, stem cells in many parts of the body can differentiate into fat cells. Researchers are gradually discovering what chemical cues are responsible for controlling the differentiation of specific cell types. The ability to take a person's stem cells and create new heart muscle cells or neurons on demand may one day revolutionize the treatment of heart attacks and strokes.

Parkinson disease, a neurodegenerative disease characterized by progressive degeneration and loss of dopamine (DA)-producing neurons, may be the first disorder suited to stem cell implantation treatment. (Dopamine is one kind of neurotransmitter, a substance that one neuron releases to communicate with other neurons.) Several laboratories have demonstrated success in inducing embryonic stem cells to differentiate into cells that function as dopamine-producing neurons. Studies on animal models show that stem cell–derived DA neurons reinnervated the brains of rats with Parkinson disease, released dopamine, and improved motor function.

The A&P Top 100

#24 All cells in your body (except sex cells, which form sperm or oocytes) contain the same 46 chromosomes. What makes one cell type different from another type is which genes are active, and which are inactive.

CHECKPOINT

37. **Define differentiation.**

See the blue Answers tab at the end of the book.

Related Clinical Terms

benign tumor: A mass or swelling in which the cells usually remain within a connective-tissue capsule; rarely life threatening.

cancer: An illness caused by mutations leading to the uncontrolled growth and replication of affected cells.

dextran: A carbohydrate that cannot cross plasma membranes; commonly administered in solution to patients after blood loss or dehydration.

DNA fingerprinting: Identifying an individual on the basis of repeating nucleotide sequences in his or her DNA.

invasion: The spread of cancer cells from a primary tumor into surrounding tissues.

malignant tumor: A mass or swelling in which the cells no longer respond to normal control mechanisms, but divide rapidly.

metastasis: The spread of malignant cells into distant tissues and organs, where secondary tumors subsequently develop.

normal saline: A sodium chloride (NaCl) solution that approximates the normal osmotic concentration of extracellular fluids.

oncogene: A cancer-causing gene created by a somatic mutation in a normal gene involved with growth, differentiation, or cell division.

primary tumor (*primary neoplasm*)**:** The mass of cells in which a cancer cell initially developed.

secondary tumor: A colony of cancerous cells formed by metastasis.

tumor (*neoplasm*)**:** A mass or swelling produced by abnormal cell growth and division.

Chapter Review
Study Outline

An Introduction to Cells p. 67

1. Contemporary *cell theory* incorporates several basic concepts: (1) Cells are the building blocks of all plants and animals; (2) cells are produced by the division of preexisting cells; (3) cells are the smallest units that perform all vital physiological functions; and (4) each cell maintains homeostasis at the cellular level (*Figure 3–1*)

2. **Cytology**, the study of cellular structure and function, is part of **cell biology**.

3. The human body contains two types of cells: **sex cells** (*sperm* and *oocytes*) and **somatic cells** (all other cells). (*Figure 3–1; Table 3–1*)

3-1 The plasma membrane separates the cell from its surrounding environment and performs various functions p. 67

4. A typical cell is surrounded by **extracellular fluid**— specifically, the **interstitial fluid** of the tissue. The cell's outer

boundary is the **plasma membrane** (*cell membrane*, or **plasmalemma**).

5. The plasma membrane's functions include physical isolation, regulation of exchange with the environment, sensitivity to the environment, and structural support. (*Figure 3–2*)

6. The plasma membrane, which is a **phospholipid bilayer**, contains other lipids, proteins, and carbohydrates.

7. **Integral proteins** are part of the membrane itself; **peripheral proteins** are attached to, but can separate from, the membrane.

8. Membrane proteins can act as anchors (**anchoring proteins**), identifiers (**recognition proteins**), enzymes, receptors (**receptor proteins**), carriers (**carrier proteins**), or **channels**.

9. The **glycocalyx** on the outer cell surface is formed by the carbohydrate portions of *proteoglycans*, *glycoproteins*, and *glycolipids*. Functions include lubrication and protection, anchoring and locomotion, specificity in binding, and recognition.

3-2 Organelles within the cytoplasm perform particular functions p. 72

10. The **cytoplasm** contains the fluid **cytosol** and the **organelles** suspended in the cytosol.

11. Cytosol differs from extracellular fluid in composition and in the presence of **inclusions**.

12. **Nonmembranous organelles** are not completely enclosed by membranes, and all of their components are in direct contact with the cytosol. They include the *cytoskeleton*, *microvilli*, *centrioles*, *cilia*, *ribosomes*, and *proteasomes*. (*Table 3–1*)

13. **Membranous organelles** are surrounded by phospholipid membranes that isolate them from the cytosol. They include the *endoplasmic reticulum*, the *Golgi apparatus*, *lysosomes*, *peroxisomes*, *mitochondria*, and *nucleus*. (*Table 3–1*)

14. The **cytoskeleton** gives the cytoplasm strength and flexibility. It has four components: **microfilaments** (typically made of **actin**), **intermediate filaments**, **microtubules** (made of **tubulin**), and **thick filaments** (made of **myosin**). (*Figure 3–3*)

15. **Microvilli** are small projections of the plasma membrane that increase the surface area exposed to the extracellular environment. (*Figure 3–3*)

16. **Centrioles** direct the movement of chromosomes during cell division and organize the cytoskeleton. The **centrosome** is the cytoplasm surrounding the centrioles. (*Figure 3–4*)

17. **Cilia**, anchored by a **basal body**, beat rhythmically to move fluids or secretions across the cell surface. (*Figure 3–4*)

18. **Ribosomes**, responsible for manufacturing proteins, are composed of a **small** and a **large ribosomal subunit**, both of which contain **ribosomal RNA (rRNA)**. **Free ribosomes** are in the cytoplasm, and **fixed ribosomes** are attached to the endoplasmic reticulum. (*Figure 3–1*)

19. **Proteasomes** remove and break down damaged or abnormal proteins that have been tagged with *ubiquitin*.

20. The **endoplasmic reticulum (ER)** is a network of intracellular membranes that function in synthesis, storage, transport, and detoxification. The ER forms hollow tubes, flattened sheets, and chambers called **cisternae**. **Smooth endoplasmic reticulum (SER)** is involved in lipid synthesis; **rough endoplasmic reticulum (RER)** contains ribosomes on its outer surface and forms **transport vesicles**. (*Figure 3–5*)

21. The **Golgi apparatus** forms **secretory vesicles** and new membrane components, and packages *lysosomes*. Secretions are discharged from the cell by exocytosis. (*Figures 3–6, 3–7*)

22. **Lysosomes**, vesicles filled with digestive enzymes, are responsible for the **autolysis** of injured cells. (*Figures 3–6, 3–8*)

23. **Peroxisomes** carry enzymes that neutralize potentially dangerous free radicals.

24. **Membrane flow** refers to the continuous movement and recycling of the membrane among the ER, vesicles, the Golgi apparatus, and the plasma membrane.

25. **Mitochondria** are responsible for ATP production through aerobic metabolism. The **matrix**, or fluid contents of a mitochondrion, lies inside the **cristae**, or folds of an inner membrane. (*Figure 3–9*)

3-3 The nucleus contains DNA and enzymes essential for controlling cellular activities p. 82

26. The **nucleus** is the control center of cellular operations. It is surrounded by a **nuclear envelope** (a double membrane with a **perinuclear space**), through which it communicates with the cytosol by way of **nuclear pores**. (*Figures 3–1, 3–10*)

27. The nucleus contains a supportive **nuclear matrix**; one or more **nucleoli** typically are present.

28. The nucleus controls the cell by directing the synthesis of specific proteins, using information stored in **chromosomes**, which consist of DNA bound to **histones**. In nondividing cells, DNA and associated proteins form a tangle of filaments called **chromatin**. (*Figure 3–11*)

29. The cell's information storage system, the **genetic code**, is called a *triplet code* because a sequence of three nitrogenous bases specifies the identity of a single amino acid. Each **gene** contains all the DNA triplets needed to produce a specific polypeptide chain.

3-4 DNA controls protein synthesis, cell structure, and cell function p. 84

30. As **gene activation** begins, **RNA polymerase** must bind to the gene.

31. **Transcription** is the production of RNA from a DNA template. After transcription, a strand of **messenger RNA (mRNA)** carries instructions from the nucleus to the cytoplasm. (*Figure 3–12*)

32. During **translation**, a functional polypeptide is constructed using the information contained in the sequence of **codons** along an mRNA strand. The sequence of codons determines the sequence of amino acids in the polypeptide.

33. By complementary base pairing of **anticodons** to mRNA codons, **transfer RNA (tRNA)** molecules bring amino acids to the ribosomal complex. (*Figure 3–13; Table 3–2*)

34. The DNA of the nucleus has both direct and indirect control over protein synthesis.

3-5 Diffusion is a passive transport mechanism facilitating membrane passage p. 89

35. The **permeability** of a barrier such as the plasma membrane is an indication of the barrier's effectiveness. Nothing can pass through an **impermeable** barrier; anything can pass through a **freely permeable** barrier. Plasma membranes are **selectively permeable**.

36. **Diffusion** is the net movement of material from an area of higher concentration to an area of lower concentration. Diffusion occurs until the **concentration gradient** is eliminated. (*Figures 3–14, 3–15*)

37. Most lipid-soluble materials and gases freely diffuse across the phospholipid bilayer of the plasma membrane. Water and small ions rely on channel-mediated diffusion through a passageway within a transmembrane protein.

38. **Osmosis** is the net flow of water across a membrane in response to differences in osmotic pressure. **Osmotic pressure** is the force of water movement into a solution resulting from solute concentration. **Hydrostatic pressure** can oppose osmotic pressure. (*Figure 3–16*)

39. **Tonicity** describes the effects of osmotic solutions on cells. A solution that does not cause an osmotic flow is **isotonic**. A solution that causes water to flow into a cell is **hypotonic** and can lead to **hemolysis** of red blood cells. A solution that causes water to flow out of a cell is **hypertonic** and can lead to **crenation**. (*Figure 3–17*)

3-6 Carrier-mediated and vesicular transport mechanisms facilitate membrane passage p. 94

40. Carrier-mediated transport involves the binding and transporting of specific ions by integral proteins. **Cotransport** moves two substances in the same direction; **countertransport** moves them in opposite directions.

41. In **facilitated diffusion**, compounds are transported across a membrane after binding to a **receptor site** within the channel of a carrier protein. (*Figure 3–18*)

42. **Active transport** mechanisms consume ATP and are not dependent on concentration gradients. Some **ion pumps** are **exchange pumps**. **Secondary active transport** may involve cotransport or countertransport. (*Figures 3–19, 3–20*)

43. In **vesicular transport**, materials move into or out of the cell in membranous **vesicles**. Movement into the cell is accomplished through **endocytosis**, an active process that can take three forms: **receptor-mediated endocytosis** (by means of **coated vesicles**), **pinocytosis**, or **phagocytosis** (using **pseudopodia**). The ejection of materials from the cytoplasm is accomplished by **exocytosis**. (*Figures 3–21, 3–22; Summary Table 3–3*)

3-7 The transmembrane potential results from the unequal distribution of ions across the plasma membrane p. 99

44. The **potential difference**, measured in volts, between the two sides of a plasma membrane is a **transmembrane potential**. The transmembrane potential in an undisturbed cell is the cell's **resting potential**.

3-8 Stages of a cell's life cycle include interphase, mitosis, and cytokinesis p. 100

45. **Cell division** is the reproduction of cells. **Apoptosis** is the genetically controlled death of cells. **Mitosis** is the nuclear division of somatic cells. Sex cells are produced by **meiosis**. (*Figure 3–23*)

46. Most somatic cells spend the majority of their time in **interphase**, which includes the G_1, **S** (**DNA replication**), and G_2 **phases**. (*Figures 3–23, 3–24*)

47. Mitosis proceeds in four stages: **prophase**, **metaphase**, **anaphase**, and **telophase**. (*Figure 3–25*)

48. During **cytokinesis**, the cytoplasm is divided and cell division ends. (*Figure 3–25*)

49. In general, the longer the life expectancy of a cell type, the slower is the **mitotic rate**. **Stem cells** undergo frequent mitosis to replace other, more specialized cells.

3-9 Several growth factors affect the cell life cycle p. 104

50. A variety of **growth factors** can stimulate cell division and growth. (*Table 3–4*)

3-10 Tumors and cancers are characterized by abnormal cell growth and division p. 105

51. Produced by abnormal cell growth and division, a **tumor**, or **neoplasm**, can be **benign** or **malignant**. Malignant cells may spread locally (by **invasion**) or to distant tissues and organs (through **metastasis**). The resultant illness is called **cancer**. Malignancy is often caused by modified genes called **oncogenes**. (*Figure 3–26*)

3-11 Differentiation is cellular specialization as a result of gene activation or repression p. 106

52. **Differentiation**, a process of specialization, results from the inactivation of particular genes in different cells, producing populations of cells with limited capabilities. Specialized cells form organized collections called *tissues*, each of which has certain functional roles.

Review Questions

See the blue Answers tab at the end of the book.

myA&P Access more review material online at **myA&P™** (www.myaandp.com). There, you'll find chapter guides, chapter quizzes, practice tests, animations, flashcards, a glossary with pronunciations, *Interactive Physiology*® (IP) exercises and quizzes, and more to help you succeed in the course.

LEVEL 1   Reviewing Facts and Terms

1. In the following diagram, identify the type of solution (hypertonic, hypotonic, or isotonic) in which the red blood cells are immersed.

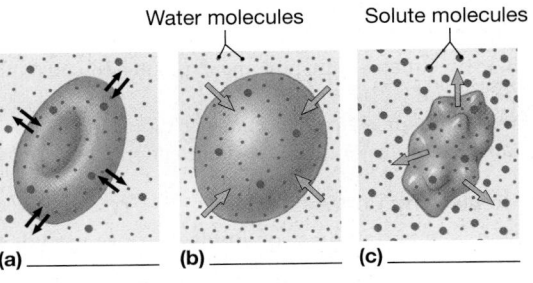

Water molecules   Solute molecules

(a) _____   (b) _____   (c) _____

2. The process that transports solid objects such as bacteria into the cell is called
   (a) pinocytosis.
   (b) phagocytosis.
   (c) exocytosis.
   (d) receptor-mediated endocytosis.
   (e) channel-mediated transport.

3. Plasma membranes are said to be
   (a) impermeable.   (b) freely permeable.
   (c) selectively permeable. (d) actively permeable.
   (e) slightly permeable.

4. _____ ion concentrations are high in the extracellular fluids, and _____ ion concentrations are high in the cytoplasm.
   (a) Calcium, magnesium   (b) Chloride, sodium
   (c) Potassium, sodium   (d) Sodium, potassium

5. In a resting transmembrane potential, the inside of the cell is _____, and the cell exterior is _____.
   (a) slightly negative, slightly positive
   (b) slightly positive, slightly negative
   (c) slightly positive, neutral
   (d) slightly negative, neutral

6. The organelle responsible for a variety of functions centering around the synthesis of lipids and carbohydrates is
   (a) the Golgi apparatus.
   (b) the rough endoplasmic reticulum.
   (c) the smooth endoplasmic reticulum.
   (d) mitochondria.

7. The construction of a functional polypeptide by using the information in an mRNA strand is
   (a) translation.   (b) transcription.
   (c) replication.   (d) gene activation.

8. Our somatic cell nuclei contain _____ pairs of chromosomes.
   (a) 8   (b) 16
   (c) 23   (d) 46

9. The movement of water across a membrane from an area of low solute concentration to an area of higher solute concentration is known as
   (a) osmosis.   (b) active transport.
   (c) diffusion.   (d) facilitated transport.
   (e) filtration.

10. The interphase of the cell life cycle is divided into
    (a) prophase, metaphase, anaphase, and telophase.
    (b) G_0, G_1, S, and G_2.
    (c) mitosis and cytokinesis.
    (d) all of these.

11. List the four basic concepts that make up modern-day cell theory.

12. What are four general functions of the plasma membrane?

13. What are the primary functions of membrane proteins?

14. By what three major transport mechanisms do substances get into and out of cells?

15. List four important factors that influence diffusion rates.

16. What are the four major functions of the endoplasmic reticulum?

LEVEL 2 Reviewing Concepts

17. Diffusion is important in body fluids, because it tends to
    (a) increase local concentration gradients.
    (b) eliminate local concentration gradients.
    (c) move substances against concentration gradients.
    (d) create concentration gradients.

18. Microvilli are found
    (a) mostly in muscle cells.
    (b) on the inside of plasma membranes.
    (c) in large numbers on cells that secrete hormones.
    (d) in cells that are actively engaged in absorption.
    (e) only on cells lining the reproductive tract.

19. When a cell is placed in a(n) _____ solution, the cell will lose water through osmosis. This process results in the _____ of red blood cells.
    (a) hypotonic, crenation   (b) hypertonic, crenation
    (c) isotonic, hemolysis   (d) hypotonic, hemolysis

20. Suppose that a DNA segment has the following nucleotide sequence: CTC–ATA–CGA–TTC–AAG–TTA. Which nucleotide sequences would a complementary mRNA strand have?
    (a) GAG–UAU–GAU–AAC–UUG–AAU
    (b) GAG–TAT–GCT–AAG–TTC–AAT
    (c) GAG–UAU–GCU–AAG–UUC–AAU
    (d) GUG–UAU–GGA–UUG–AAC–GGU

21. How many amino acids are coded in the DNA segment in Review Question 21?
    (a) 18   (b) 9
    (c) 6   (d) 3

22. The sodium–potassium exchange pump
    (a) is an example of facilitated diffusion.
    (b) does not require the input of cellular energy in the form of ATP.
    (c) moves the sodium and potassium ions along their concentration gradients.
    (d) is composed of a carrier protein located in the plasma membrane.
    (e) is not necessary for the maintenance of homeostasis.

23. If a cell lacked ribosomes, it would not be able to
    (a) move.   (b) synthesize proteins.
    (c) produce DNA.   (d) metabolize sugar.
    (e) divide.

24. List, in sequence, the phases of the interphase stage of the cell life cycle, and briefly describe what happens in each.

25. List the stages of mitosis, and briefly describe the events that occur in each.

26. (a) What is cytokinesis?
    (b) What is the role of cytokinesis in the cell cycle?

LEVEL 3 Critical Thinking and Clinical Applications

27. The transport of a certain molecule exhibits the following characteristics: (1) The molecule moves down its concentration gradient; (2) at concentrations above a given level, the rate of transport does not increase; and (3) cellular energy is not required for transport to occur. Which transport process is at work?

28. Solutions A and B are separated by a selectively permeable barrier. Over time, the level of fluid on side A increases. Which solution initially had the higher concentration of solute?

29. A molecule that blocks the ion channels in integral proteins in the plasma membrane would interfere with
    (a) cell recogniton.
    (b) the movement of lipid soluble molecules.
    (c) the ability of the plasma membrane to depolarize.
    (d) the ability of protein hormones to stimulate the cell.
    (e) the cell's ability to divide.

30. What is the benefit of having some of the cellular organelles enclosed by a membrane similar to the plasma membrane?

The Tissue Level of Organization

Did you know...?
Pathologists view thin slices of stained tissue under the microscope to detect structural changes caused by trauma or disease.

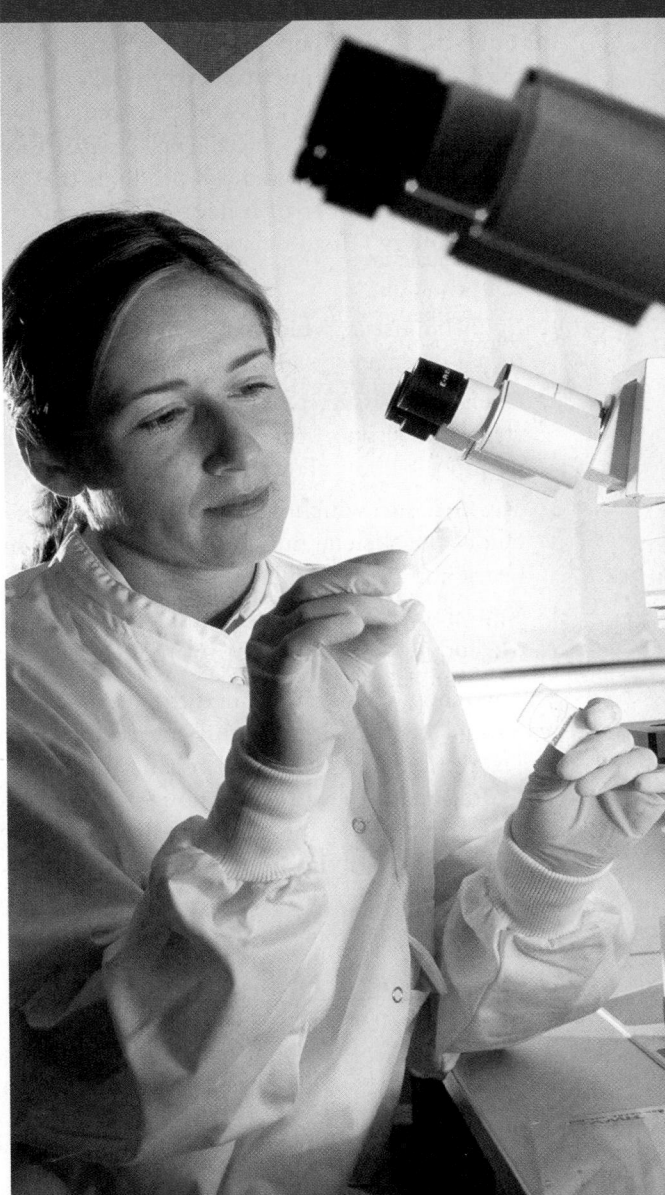

Learning Outcomes

After completing this chapter, you should be able to do the following:

4-1 Identify the four major types of tissues in the body and describe their roles.

4-2 Discuss the types and functions of epithelial tissue.

4-3 Describe the relationship between form and function for each type of epithelium.

4-4 Compare the structures and functions of the various types of connective tissues.

4-5 Describe how cartilage and bone function as a supporting connective tissue.

4-6 Explain how epithelial and connective tissues combine to form four types of membranes, and specify the functions of each.

4-7 Describe how connective tissue establishes the framework of the body.

4-8 Describe the three types of muscle tissue and the special structural features of each type.

4-9 Discuss the basic structure and role of neural tissue.

4-10 Describe how injuries affect the tissues of the body.

4-11 Describe how aging affects the tissues of the body.

Clinical Notes

Exfoliative Cytology p. 118

Marfan Syndrome p. 126

Problems with Serous Membranes p. 136

An Introduction to the Tissue Level of Organization

This chapter discusses how a variety of cell types arranged in various combinations form *tissues*, structures with discrete structural and functional properties. Tissues in combination form *organs*, such as the heart or liver, and in turn organs can be grouped into 11 *organ systems*. The *Systems Overview* at the end of this chapter provides a broad summary of these organ systems and the major organs associated with them.

4-1 The four tissue types are epithelial, connective, muscle, and neural

Although the human body contains trillions of cells, differentiation produces only about 200 types of cells. To work efficiently, several different types of cells must coordinate their efforts. Cells organized into layers create **tissues**—collections of specialized cells and cell products that perform a relatively limited number of functions. The study of tissues is called **histology**. Histologists recognize four basic types of tissue:

1. *Epithelial tissue*, which covers exposed surfaces, lines internal passageways and chambers, and forms glands.

2. *Connective tissue*, which fills internal spaces, provides structural support for other tissues, transports materials within the body, and stores energy reserves.

3. *Muscle tissue*, which is specialized for contraction and includes the skeletal muscles of the body, the muscle of the heart, and the muscular walls of hollow organs.

4. Neural tissue, which carries information from one part of the body to another in the form of electrical impulses.

This chapter will introduce the basic characteristics of these tissues. You will need this information to understand the descriptions of organs and organ systems in later chapters. Additionally, a working knowledge of basic histology will help you make the connections between anatomical structures and their physiological functions. ATLAS: Embryology Summary 1: The Formation of Tissues

The A&P Top 100

#25 Tissues are collections of cells organized into groups or layers to perform specific functions. Four types of tissue—epithelial, connective, muscle, and neural—form all body structures.

4-2 Epithelial tissue covers body surfaces, lines cavities and tubular structures, and serves essential functions

It is convenient to begin our discussion with **epithelial tissue**, because it includes the surface of your skin, a relatively familiar feature. Epithelial tissue includes *epithelia* and *glands*. **Epithelia** (ep-i-THĒ-lē-a; singular, *epithelium*) are layers of cells that cover internal or external surfaces. **Glands** are structures that produce fluid secretions; they are either attached to or derived from epithelia.

Epithelia cover every exposed surface of the body. Epithelia form the surface of the skin and line the digestive, respiratory, reproductive, and urinary tracts—in fact, they line all passageways that communicate with the outside world. The more delicate epithelia line internal cavities and passageways, such as the chest cavity, fluid-filled spaces in the brain, the inner surfaces of blood vessels, and the chambers of the heart.

Epithelia have several important characteristics:

- *Cellularity.* Epithelia are composed almost entirely of cells bound closely together by interconnections known as *cell junctions*. In other tissue types, the cells are often widely separated by extracellular materials.

- *Polarity.* An epithelium has an exposed surface, which faces the exterior of the body or some internal space, and a base, which is attached to adjacent tissues. The term **polarity** refers to the presence of structural and functional differences between the exposed and attached surfaces. In an epithelium consisting of a single layer of cells, exposed (*apical*) and attached (*basal*) surfaces differ in terms of membrane structure and function, and between the two surfaces the organelles and other cytoplasmic components are unevenly distributed.

- *Attachment.* The base of an epithelium is bound to a thin **basal lamina**, or *basement membrane*. The basal lamina is a complex structure produced by the basal surface of the epithelium and the underlying connective tissue.

- *Avascularity.* Epithelia are **avascular** (ā-VAS-kū-lar; *a-*, without + *vas*, vessel); that is, they lack blood vessels. Epithelial cells must therefore obtain nutrients by diffusion or absorption across either the exposed or the attached epithelial surface.

- *Regeneration.* Epithelial cells that are damaged or lost at the exposed surface are continuously replaced through the divisions of stem cells in the epithelium. Although regeneration is a characteristic of other tissues as well, the rates of cell division and replacement are typically much higher in epithelia than in other tissues.

Functions of Epithelial Tissue

Epithelia perform four essential functions:

1. *Provide Physical Protection.* Epithelia protect exposed and internal surfaces from abrasion, dehydration, and destruction by chemical or biological agents.

2. *Control Permeability.* Any substance that enters or leaves your body must cross an epithelium. Some epithelia are relatively impermeable; others are easily crossed by compounds as large as proteins. Many epithelia contain the molecular "machinery" needed for selective absorption or secretion. The epithelial barrier can be regulated and modified in response to stimuli. For example, hormones can affect the transport of ions and nutrients through epithelial cells. Even physical stress can alter the structure and properties of epithelia; for example, calluses form on your hands when you do manual labor for a while.

3. *Provide Sensation.* Most epithelia are extremely sensitive to stimulation, because they have a large sensory nerve supply. These sensory nerves continually provide information about the external and internal environments. For example, the lightest touch of a mosquito will stimulate sensory neurons that tell you where to swat. A *neuroepithelium* is an epithelium that is specialized to perform a particular sensory function; neuroepithelia contain sensory cells that provide the sensations of smell, taste, sight, equilibrium, and hearing.

4. *Produce Specialized Secretions.* Epithelial cells that produce secretions are called *gland cells.* Individual gland cells are typically scattered among other cell types in an epithelium. In a **glandular epithelium**, most or all of the epithelial cells produce secretions, which are either discharged onto the surface of the epithelium (to provide physical protection or temperature regulation) or released into the surrounding interstitial fluid and blood (to act as chemical messengers).

Specializations of Epithelial Cells

Epithelial cells have several structural specializations that distinguish them from other body cells. For the epithelium as a whole to perform the functions just listed, individual epithelial cells may be specialized for (1) the movement of fluids over the

epithelial surface, providing protection and lubrication; (2) the movement of fluids through the epithelium, to control permeability; or (3) the production of secretions that provide physical protection or act as chemical messengers. Specialized epithelial cells generally possess a strong polarity; one common type of epithelial polarity is shown in **Figure 4–1**.

The cell is often divided into two functional regions: (1) the *apical surface,* where the cell is exposed to an internal or external environment; and (2) the *basolateral surfaces,* which include both the base, where the cell attaches to underlying epithelial cells or deeper tissues, and the sides, where the cell contacts its neighbors.

Many epithelial cells that line internal passageways have *microvilli* on their exposed surfaces. ∞ p. 74 Just a few may be present, or microvilli may carpet the entire surface. Microvilli are especially abundant on epithelial surfaces where absorption and secretion take place, such as along portions of the digestive and urinary tracts. The epithelial cells in these locations are transport specialists; each cell has at least 20 times more surface area than it would have if it lacked microvilli.

Cilia are characteristic of surfaces covered by a **ciliated epithelium**. A typical ciliated cell contains about 250 cilia that beat in a coordinated fashion. As though on an escalator, substances are moved over the epithelial surface by the synchronized beating of the cilia. The ciliated epithelium that lines the respiratory tract, for example, moves mucus up from

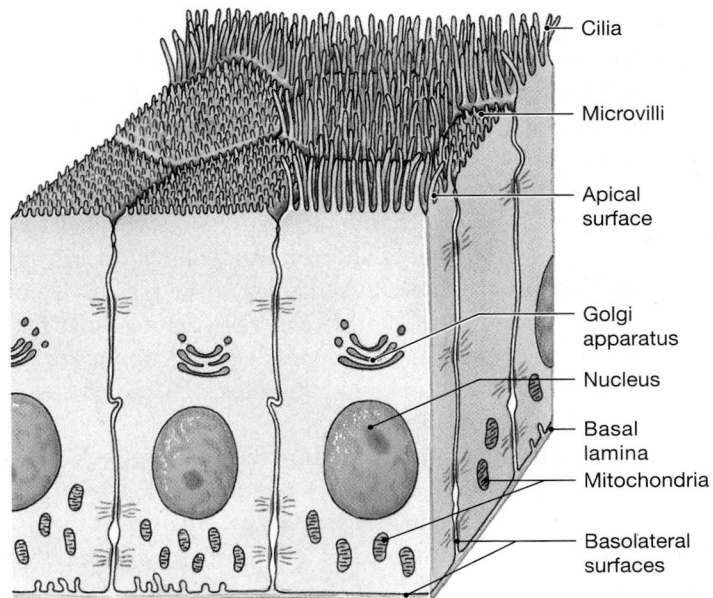

Figure 4–1 The Polarity of Epithelial Cells. Many epithelial cells have an uneven distribution of organelles between the free surface (here, the top) and the basal lamina. Often, the free surface bears microvilli; sometimes it has cilia. In some epithelia, such as the lining of the kidney tubules, mitochondria are concentrated near the base of the cell, probably to provide energy for the cell's transport activities.

the lungs and toward the throat. The sticky mucus traps inhaled particles, including dust, pollen, and pathogens; the ciliated epithelium carries the mucus and the trapped debris to the throat, where they can be swallowed or expelled by coughing. Injury to the cilia or to the epithelial cells, most commonly by abrasion or exposure to toxic compounds such as nicotine and carbon monoxide in cigarette smoke, can stop ciliary movement and block the protective flow of mucus, possibly leading to infection or disease.

Maintaining the Integrity of Epithelia

To be effective as a barrier, an epithelium must form a complete cover or lining. Three factors help maintain the physical integrity of an epithelium: (1) intercellular connections, (2) attachment to the basal lamina, and (3) epithelial maintenance and repair.

Intercellular Connections

Cells in an epithelium are firmly attached to one another, and the epithelium as a unit is attached to extracellular fibers of the clear layer of the basal lamina. Many cells in your body form permanent or temporary bonds with other cells or extracellular material. Epithelial cells, however, are specialists in intercellular connection (**Figure 4–2**).

Intercellular connections involve either extensive areas of opposing plasma membranes or specialized attachment sites, discussed shortly. Large areas of opposing plasma membranes are interconnected by transmembrane proteins called **cell adhesion molecules (CAMs)**, which bind to each other and to extracellular materials. For example, CAMs on the basolateral surface of an epithelium help bind the cell to the underlying basal lamina. The membranes of adjacent cells may also be bonded by **intercellular cement**, a thin layer of proteoglycans that contain polysaccharide derivatives known as *glycosaminoglycans*, most notably **hyaluronan** (*hyaluronic acid*).

Cell junctions are specialized areas of the plasma membrane that attach a cell to another cell or to extracellular materials. The three most common types of cell junctions are (1) occluding (tight) junctions, (2) gap junctions, and (3) macula adherens (desmosomes).

At an **occluding** or **tight junction**, the lipid portions of the two plasma membranes are tightly bound together by interlocking membrane proteins (**Figure 4–2b**). Inferior to the occluding junctions, a continuous *adhesion belt* forms a band that encircles cells and binds them to their neighbors. The bands are attached to the microfilaments of the terminal web. ∞ p. 73 This kind of attachment is so tight that these junctions largely prevent the passage of water and solutes between the cells. When the epithelium lines a tube, such as the intestinal tract, the apical surfaces of the epithelial cells are exposed to the space inside the tube, a passageway called the **lumen** (LOO-men). Occluding junctions effectively isolate the contents of

the lumen from the basolateral surfaces of the cell. For example, occluding junctions near the apical surfaces of cells that line the digestive tract help keep enzymes, acids, and wastes in the lumen from reaching the basolateral surfaces and digesting or otherwise damaging the underlying tissues and organs.

Some epithelial functions require rapid intercellular communication. At a **gap junction** (**Figure 4–2c**), two cells are held together by interlocking junctional proteins called *connexons*. Because these are channel proteins, they form a narrow passageway that lets small molecules and ions pass from cell to cell. Gap junctions are common among epithelial cells, where the movement of ions helps coordinate functions such as the beating of cilia. Gap junctions are also common in other tissues. For example, gap junctions in cardiac muscle tissue and smooth muscle tissue are essential to the coordination of muscle cell contractions.

Most epithelial cells are subject to mechanical stresses—stretching, bending, twisting, or compression—so they must have durable interconnections. At a **macula adherens** or **desmosome** (DEZ-mō-sōm; *desmos*, ligament + *soma*, body), CAMs and proteoglycans link the opposing plasma membranes. Macula adherens are very strong and can resist stretching and twisting.

A typical macula adherens is formed by two cells. Within each cell is a complex known as a *dense area*, which is connected to the cytoskeleton (**Figure 4–2d**). It is this connection to the cytoskeleton that gives the macula adherens—and the epithelium—its strength. For example, macula adherens are abundant between cells in the superficial layers of the skin. As a result, damaged skin cells are usually lost in sheets rather than as individual cells. (That is why your skin peels rather than comes off as a powder after a sunburn.)

There are two types of macula adherens:

- *Spot desmosomes* are small discs connected to bands of intermediate filaments. The intermediate filaments function to stabilize the shape of the cell.

- *Hemidesmosomes* resemble half of a spot desmosome. Rather than attaching one cell to another, a hemidesmosome attaches a cell to extracellular filaments in the basal lamina (**Figure 4–2e**). This attachment helps stabilize the position of the epithelial cell and anchors it to underlying tissues.

Attachment to the Basal Lamina

Not only do epithelial cells hold onto one another, but they also remain firmly connected to the rest of the body. The inner surface of each epithelium is attached to a two-part basal lamina. The layer closer to the epithelium, the *clear layer* or *lamina lucida* (LAM-i-nah LOO-si-dah; *lamina*, thin layer + *lucida*, clear), contains glycoproteins and a network of fine protein filaments (**Figure 4–2e**). Secreted by the adjacent layer of epithelial cells, the clear layer acts as a barrier that restricts

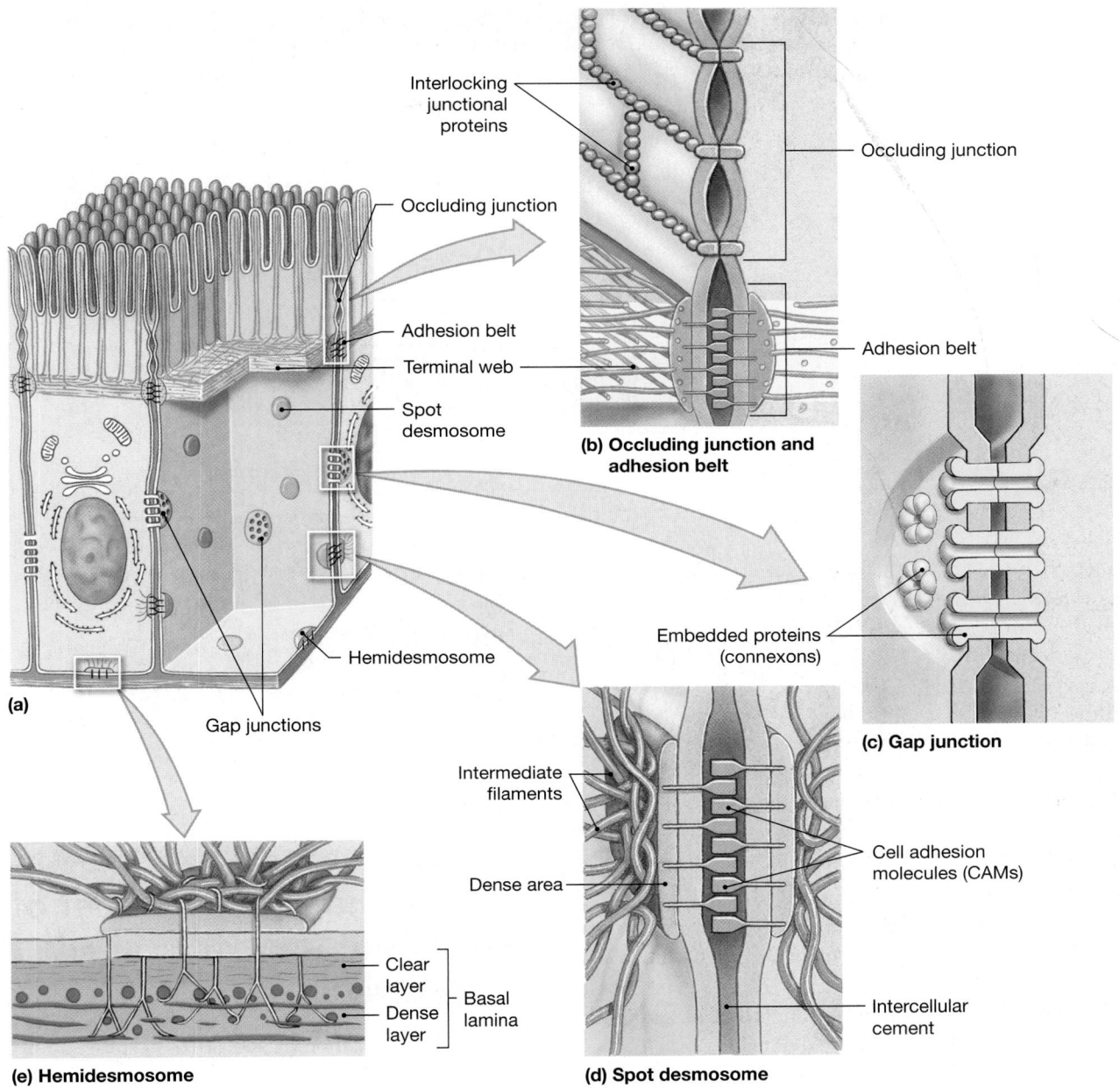

Figure 4–2 Intercellular Connections. (a) A diagrammatic view of an epithelial cell, showing the major types of intercellular connections. **(b)** An occluding junction is formed by the fusion of the outer layers of two plasma membranes. Occluding junctions prevent the diffusion of fluids and solutes between the cells. A continuous adhesion belt lies deep to the occluding junction. This belt is tied to the microfilaments of the terminal web. **(c)** Gap junctions permit the free diffusion of ions and small molecules between two cells. **(d)** A spot desmosome ties adjacent cells together. **(e)** Hemidesmosomes attach a cell to extracellular structures, such as the protein fibers in the basal lamina.

the movement of proteins and other large molecules from the underlying connective tissue into the epithelium.

The deeper portion of the basal lamina, the *dense layer* (formerly called the *lamina densa*), contains bundles of coarse protein fibers produced by connective tissue cells. The dense layer gives the basement membrane its strength. Attachments between the fibers of the clear layer and those of the dense layer hold the two layers together, and hemidesmo-

somes attach the epithelial cells to the composite basal lamina. The dense layer also acts as a filter that determines what substances can diffuse between the adjacent tissues and the epithelium.

Epithelial Maintenance and Repair

Epithelial cells lead hard lives, for they are exposed to disruptive enzymes, toxic chemicals, pathogenic bacteria, and

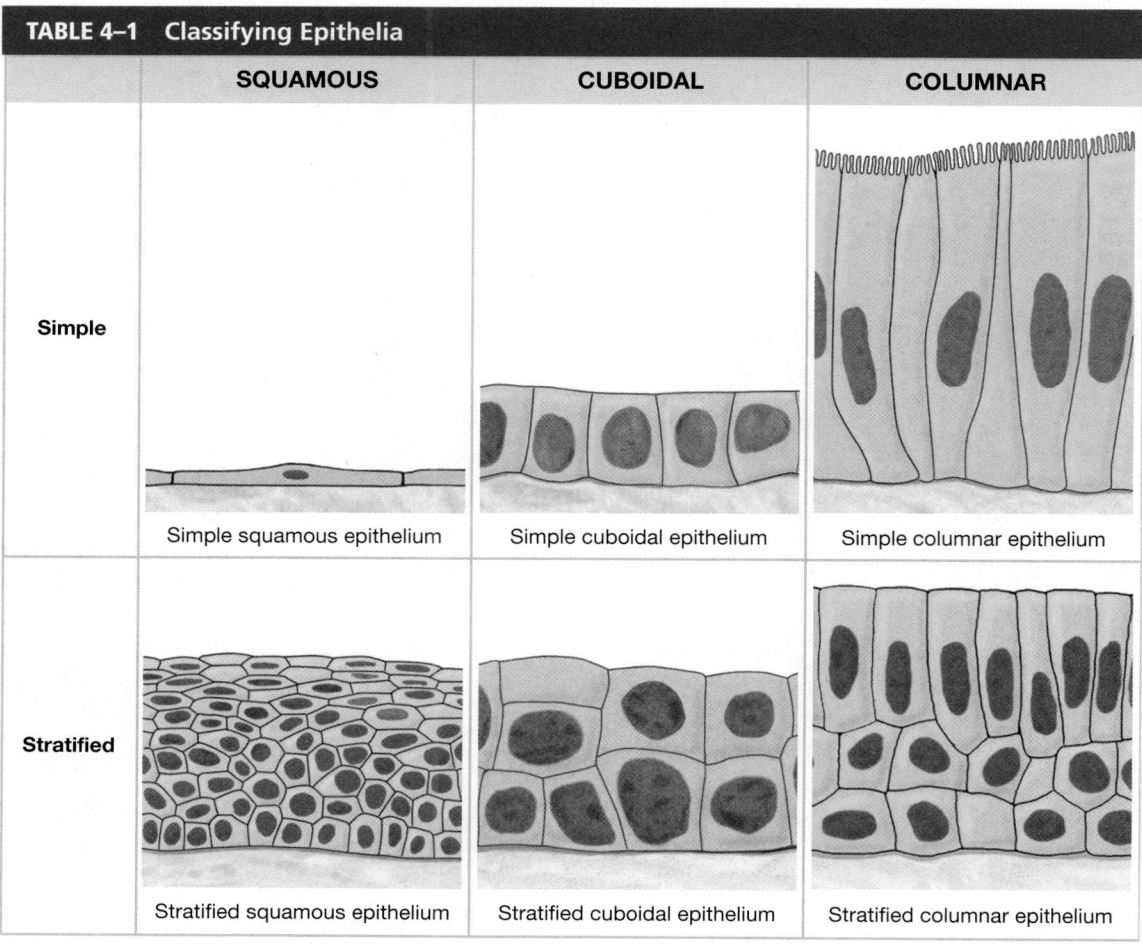

TABLE 4–1 Classifying Epithelia

	SQUAMOUS	CUBOIDAL	COLUMNAR
Simple	Simple squamous epithelium	Simple cuboidal epithelium	Simple columnar epithelium
Stratified	Stratified squamous epithelium	Stratified cuboidal epithelium	Stratified columnar epithelium

mechanical abrasion. Consider the lining of the small intestine, where epithelial cells are exposed to a variety of enzymes and abraded by partially digested food. In this extreme environment, an epithelial cell may last just a day or two before it is shed or destroyed. The only way the epithelium can maintain its structure over time is by the continual division of *stem cells.* ∞ p. 104 Most stem cells, also called **germinative cells,** are located near the basal lamina, in a relatively protected location.

ATLAS: Embryology Summary 2: The Development of Epithelia

4-3 Cell shape and number of layers determine the classification of epithelia

There are many different specialized types of epithelia. You can easily sort these into categories based on (1) the cell shape, and (2) the number of cell layers between the base and the exposed surface of the epithelium. Using these two criteria—cell shape and number of cell layers—we can describe almost every epithelium in the body (Table 4–1).

Classification of Epithelia

Three epithelial cell shapes are identified: *squamous, cuboidal,* and *columnar.* For classification purposes, one looks at the superficial cells in a section perpendicular to both the exposed surface and the basal lamina. In sectional view, squamous cells appear thin and flat, cuboidal cells look like little boxes, and columnar cells are tall and relatively slender rectangles.

Once you have determined whether the superficial cells are squamous, cuboidal, or columnar, you then look at the number of cell layers. There are only two options: *simple* or *stratified*.

If only one layer of cells covers the basal lamina, that layer is a **simple epithelium**. Simple epithelia are necessarily thin. All the cells have the same polarity, so the distance from the nucleus to the basal lamina does not change from one cell to the next. Because they are so thin, simple epithelia are also fragile. A single layer of cells cannot provide much mechanical protection, so simple epithelia are located only in protected areas inside the body. They line internal compartments and passageways, including the ventral body cavities, the heart chambers, and blood vessels.

Simple epithelia are also characteristic of regions in which secretion or absorption occurs, such as the lining of the intestines and the gas-exchange surfaces of the lungs. In these places, thinness is an advantage, for it reduces the time required for materials to cross the epithelial barrier.

In a **stratified epithelium**, several layers of cells cover the basal lamina. Stratified epithelia are generally located in areas that need protection from mechanical or chemical stresses, such as the surface of the skin and the lining of the mouth.

Squamous Epithelia

The cells in a **squamous epithelium** (SKWĀ-mus; *squama*, plate or scale) are thin, flat, and somewhat irregular in shape, like pieces of a jigsaw puzzle (**Figure 4–3**). From the surface, the cells resemble fried eggs laid side by side. In sectional view, the disc-shaped nucleus occupies the thickest portion of each cell.

A **simple squamous epithelium** is the body's most delicate type of epithelium. This type of epithelium is located in protected regions where absorption or diffusion takes place, or where a slick, slippery surface reduces friction. Examples are the respiratory exchange surfaces (*alveoli*) of the lungs, the lining of the ventral body cavities (**Figure 4–3a**), and the

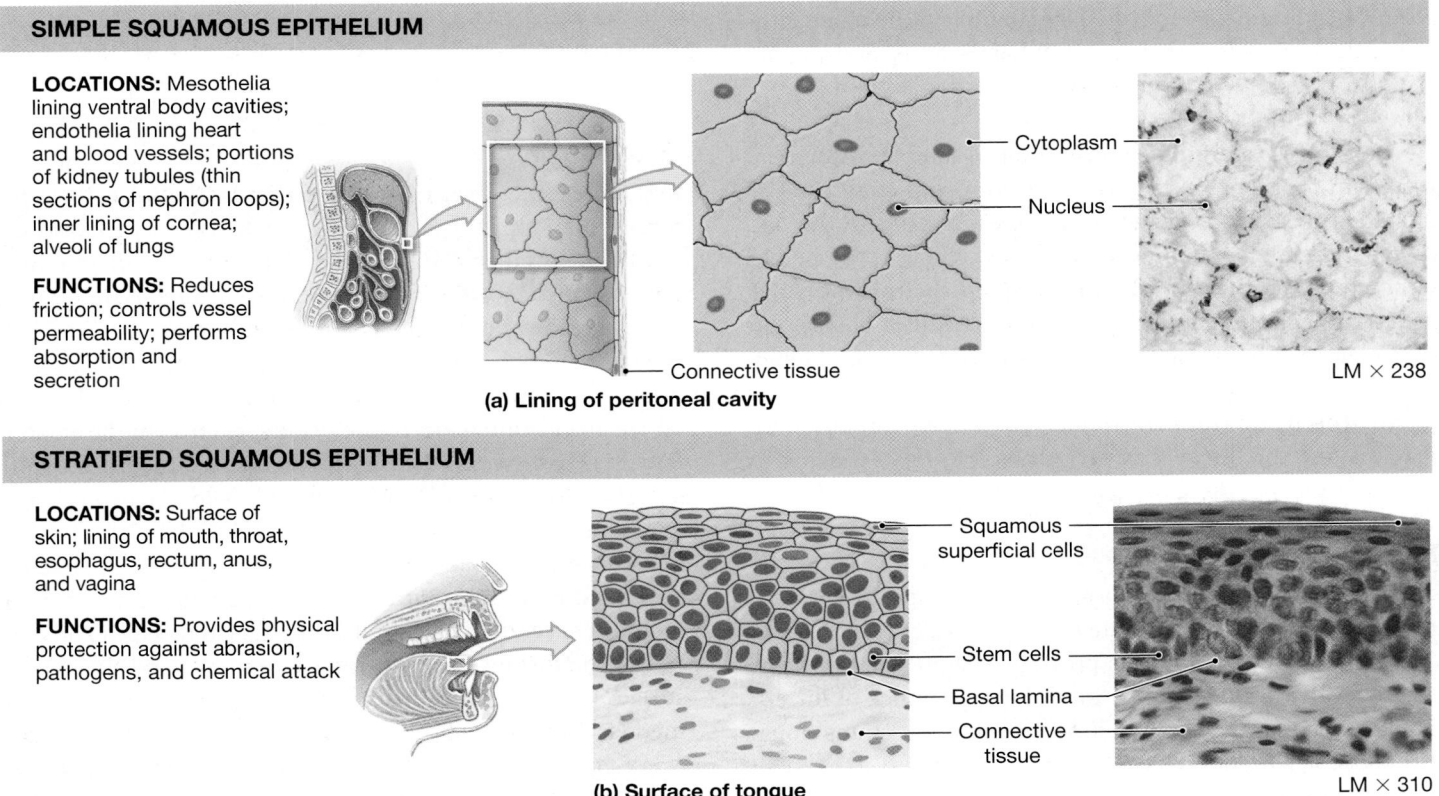

SIMPLE SQUAMOUS EPITHELIUM

LOCATIONS: Mesothelia lining ventral body cavities; endothelia lining heart and blood vessels; portions of kidney tubules (thin sections of nephron loops); inner lining of cornea; alveoli of lungs

FUNCTIONS: Reduces friction; controls vessel permeability; performs absorption and secretion

— Cytoplasm
— Nucleus
— Connective tissue

LM × 238

(a) Lining of peritoneal cavity

STRATIFIED SQUAMOUS EPITHELIUM

LOCATIONS: Surface of skin; lining of mouth, throat, esophagus, rectum, anus, and vagina

FUNCTIONS: Provides physical protection against abrasion, pathogens, and chemical attack

— Squamous superficial cells
— Stem cells
— Basal lamina
— Connective tissue

LM × 310

(b) Surface of tongue

Figure 4–3   Squamous Epithelia. (a) A superficial view of the simple squamous epithelium (mesothelium) that lines the peritoneal cavity. The three-dimensional drawing shows the epithelium in superficial and sectional views. **(b)** Sectional and diagrammatic views of the stratified squamous epithelium that covers the tongue.

lining of the heart and blood vessels. Smooth linings are extremely important; for example, any irregularity in the lining of a blood vessel may result in the formation of a potentially dangerous blood clot.

Special names have been given to the simple squamous epithelia that line chambers and passageways that do not communicate with the outside world. The simple squamous epithelium that lines the ventral body cavities is a **mesothelium** (mez-ō-THĒ-lē-um; *mesos*, middle). The pleura, peritoneum, and pericardium each contain a superficial layer of mesothelium. The simple squamous epithelium lining the inner surface of the heart and all blood vessels is an **endothelium** (en-dō-THĒ-lē-um; *endo-*, inside).

A **stratified squamous epithelium** (**Figure 4–3b**) is generally located where mechanical stresses are severe. The cells form a series of layers, like the layers in a sheet of plywood. The surface of the skin and the lining of the mouth, esophagus, and anus are areas where this type of epithelium protects against physical and chemical attacks. On exposed body surfaces, where mechanical stress and dehydration are potential problems, apical layers of epithelial cells are packed with filaments of the protein *keratin*. As a result, superficial layers are both tough and water resistant; the epithelium is said to be *keratinized*. A *nonkeratinized* stratified squamous epithelium resists abrasion, but will dry out and deteriorate unless kept moist. Nonkeratinized stratified squamous epithelia are situated in the oral cavity, pharynx, esophagus, anus, and vagina.

Cuboidal Epithelia

The cells of a **cuboidal epithelium** resemble hexagonal boxes. (In typical sectional views they appear square.) The spherical nuclei are near the center of each cell, and the distance between adjacent nuclei is roughly equal to the height of the epithelium. A **simple cuboidal epithelium** provides limited protection and occurs where secretion or absorption takes place. Such an epithelium lines portions of the kidney tubules (**Figure 4–4a**).

Stratified cuboidal epithelia are relatively rare; they are located along the ducts of sweat glands (**Figure 4–4b**) and in the larger ducts of the mammary glands.

Transitional Epithelia

A **transitional epithelium** (**Figure 4–4c**) is unusual because, unlike most epithelia, it tolerates repeated cycles of stretching and recoiling (returning to its previous shape) without damage. It is called transitional because the appearance of the epithelium changes as stretching occurs. A transitional epithelium is situated in regions of the urinary system, such as the urinary bladder, where large changes in volume occur. In an empty urinary bladder, the epithelium seems to have many layers, and the superficial cells are typically plump and cuboidal. The multilayered appearance results from over-

crowding. In a full urinary bladder, when the volume of urine has stretched the lining to its limits, the epithelium appears flattened, and more like a simple epithelium.

CLINICAL NOTE

Exfoliative Cytology *Exfoliative cytology* (eks-FŌ-lē-a-tiv; *ex-* from + *folium*, leaf) is the study of cells shed or removed from epithelial surfaces. The cells are examined for a variety of reasons—for example, to check for cellular changes that indicate cancer, or for genetic screening of a fetus. Cells are collected by sampling the fluids that cover the epithelia lining the respiratory, digestive, urinary, or reproductive tract; by removing fluid from one of the ventral body cavities; or by removing cells from an epithelial surface.

One common sampling procedure is called a *Pap test*, named after Dr. George Papanicolaou, who pioneered its use. The most familiar Pap test is that for cervical cancer; it involves the scraping of cells from the tip of the *cervix*, the portion of the uterus that projects into the vagina.

Amniocentesis is another important test based on exfoliative cytology. In this procedure, shed epithelial cells are collected from a sample of *amniotic fluid*, which surrounds and protects a developing fetus. Examination of these cells can determine whether the fetus has a genetic abnormality, such as *Down syndrome*, that affects the number or structure of chromosomes.

Columnar Epithelia

In a typical sectional view, **columnar epithelial cells** appear rectangular. In reality, the densely packed cells are hexagonal, but they are taller and more slender than cells in a cuboidal epithelium (**Figure 4–5**). The elongated nuclei are crowded into a narrow band close to the basal lamina. The height of the epithelium is several times the distance between adjacent nuclei. A **simple columnar epithelium** is typically found where absorption or secretion occurs, such as in the small intestine (**Figure 4–5a**). In the stomach and large intestine, the secretions of simple columnar epithelia protect against chemical stresses.

Portions of the respiratory tract contain a **pseudostratified columnar epithelium**, a columnar epithelium that includes several types of cells with varying shapes and functions. The distances between the cell nuclei and the exposed surface vary, so the epithelium appears to be layered, or stratified (**Figure 4–5b**). It is not truly stratified, though, because every epithelial cell contacts the basal lamina. Pseudostratified columnar epithelial cells typically possess cilia. Epithelia of this type line most of the nasal cavity, the trachea (windpipe), the bronchi (branches of the trachea leading to the lungs), and portions of the male reproductive tract.

SIMPLE CUBOIDAL EPITHELIUM

LOCATIONS: Glands; ducts; portions of kidney tubules; thyroid gland

FUNCTIONS: Limited protection, secretion, absorption

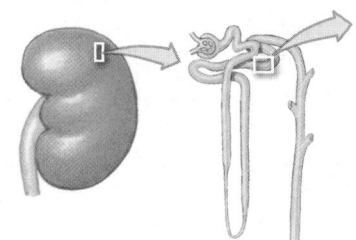

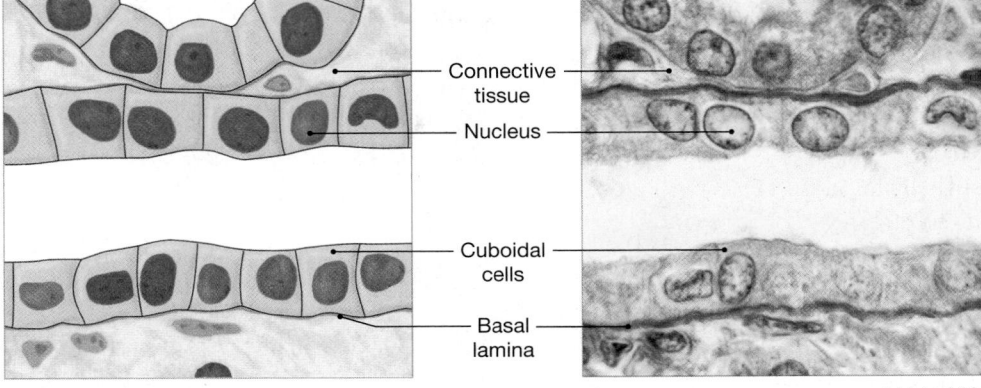

Connective tissue

Nucleus

Cuboidal cells

Basal lamina

(a) Kidney tubule

LM × 1426

STRATIFIED CUBOIDAL EPITHELIUM

LOCATIONS: Lining of some ducts (rare)

FUNCTIONS: Protection, secretion, absorption

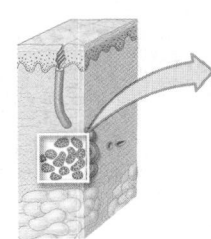

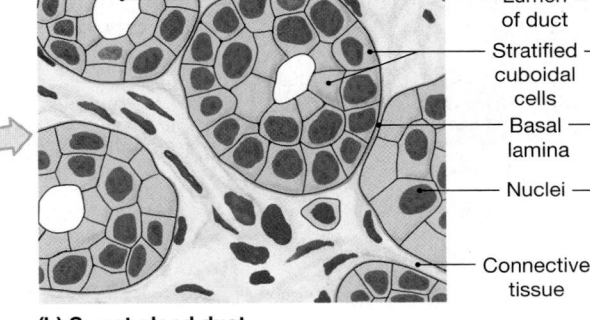

Lumen of duct

Stratified cuboidal cells

Basal lamina

Nuclei

Connective tissue

(b) Sweat gland duct

LM × 1413

TRANSITIONAL EPITHELIUM

LOCATIONS: Urinary bladder; renal pelvis; ureters

FUNCTIONS: Permits expansion and recoil after stretching

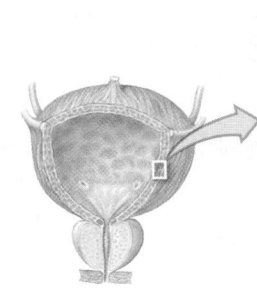

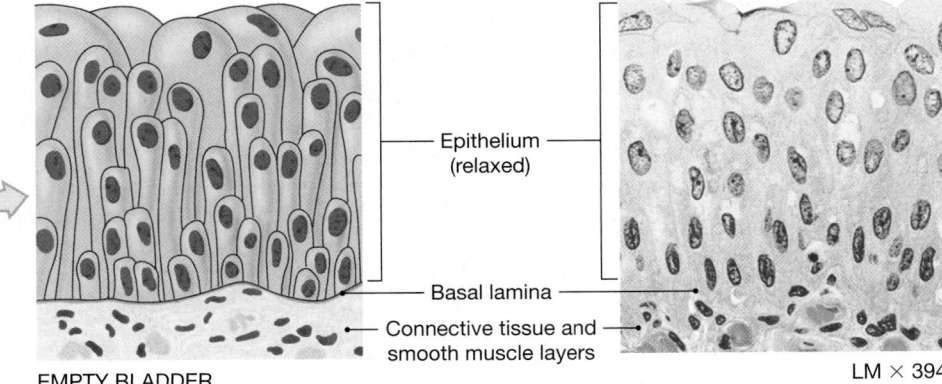

Epithelium (relaxed)

Basal lamina

Connective tissue and smooth muscle layers

EMPTY BLADDER

LM × 394

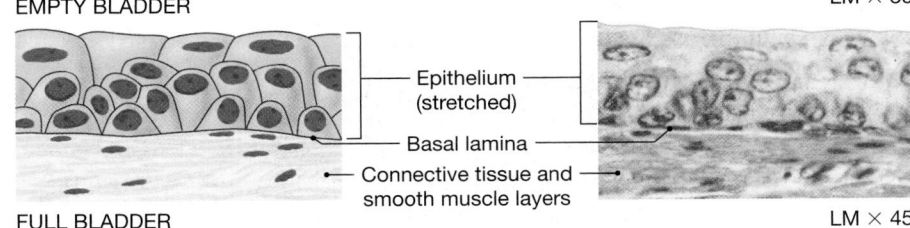

Epithelium (stretched)

Basal lamina

Connective tissue and smooth muscle layers

FULL BLADDER

(c) Urinary bladder

LM × 454

Figure 4–4 Cuboidal and Transitional Epithelia. (a) A section through the simple cuboidal epithelial cells of a kidney tubule. **(b)** A sectional view of the stratified cuboidal epithelium that lines a sweat gland duct in the skin. **(c)** Top: The lining of the empty urinary bladder, showing a transitional epithelium in the relaxed state. Bottom: The lining of the full bladder, showing the effects of stretching on the appearance of cells in the epithelium.

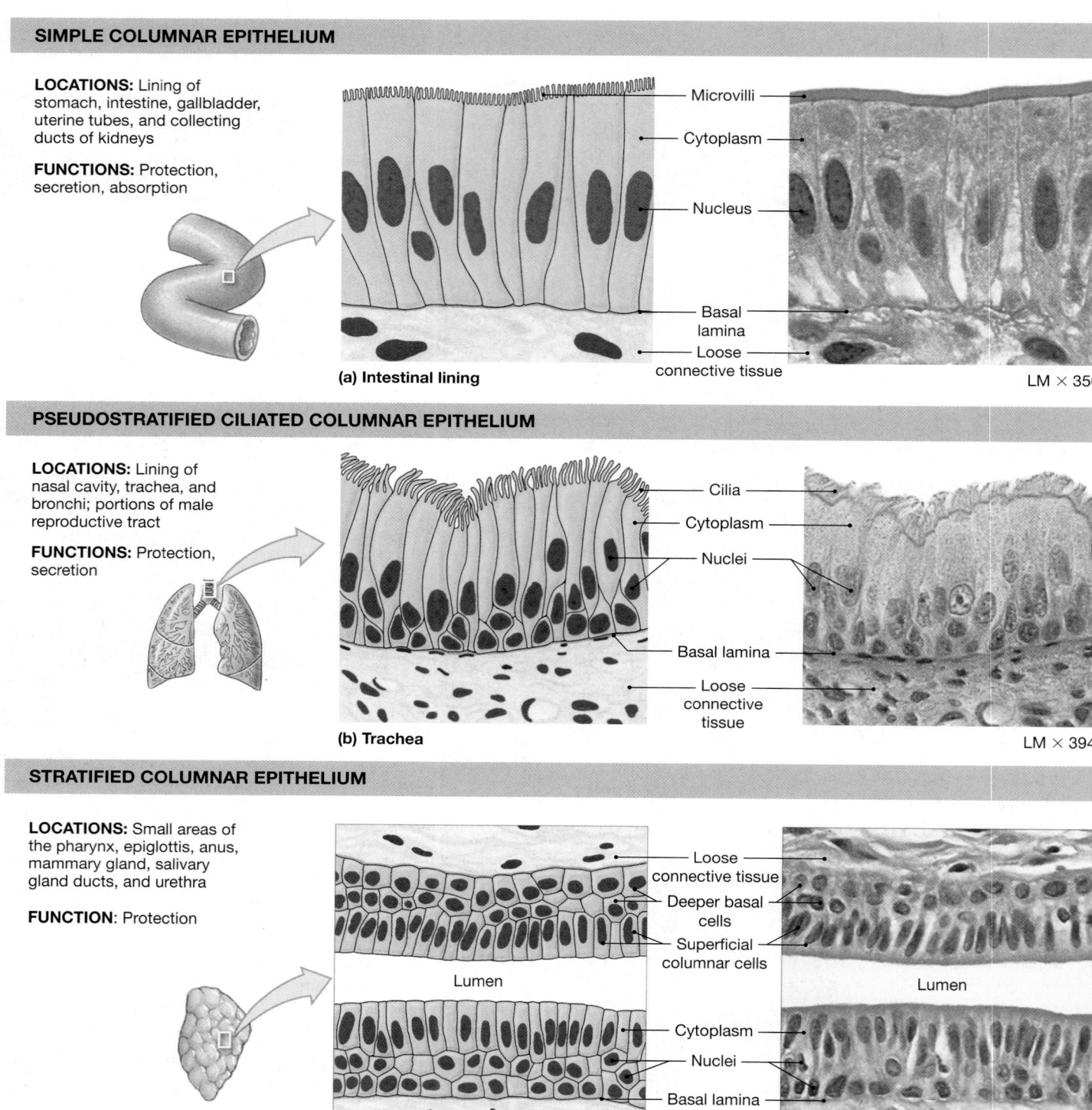

SIMPLE COLUMNAR EPITHELIUM

LOCATIONS: Lining of stomach, intestine, gallbladder, uterine tubes, and collecting ducts of kidneys

FUNCTIONS: Protection, secretion, absorption

Microvilli

Cytoplasm

Nucleus

Basal lamina

Loose connective tissue

(a) Intestinal lining

LM × 350

PSEUDOSTRATIFIED CILIATED COLUMNAR EPITHELIUM

LOCATIONS: Lining of nasal cavity, trachea, and bronchi; portions of male reproductive tract

FUNCTIONS: Protection, secretion

Cilia

Cytoplasm

Nuclei

Basal lamina

Loose connective tissue

(b) Trachea

LM × 394

STRATIFIED COLUMNAR EPITHELIUM

LOCATIONS: Small areas of the pharynx, epiglottis, anus, mammary gland, salivary gland ducts, and urethra

FUNCTION: Protection

Loose connective tissue

Deeper basal cells

Superficial columnar cells

Lumen

Lumen

Cytoplasm

Nuclei

Basal lamina

(c) Salivary gland duct

LM × 175

Figure 4–5   Columnar Epithelia. Note the thickness of the epithelium and the location and orientation of the nuclei. **(a)** The simple columnar epithelium lining the small intestine. **(b)** The pseudostratified ciliated columnar epithelium of the respiratory tract. Note that despite the uneven layering of the nuclei, all of the cells contact the basal lamina. **(c)** A stratified columnar epithelium occurs along some large ducts, such as this salivary gland duct.

Stratified columnar epithelia are relatively rare, providing protection along portions of the pharynx, epiglottis, anus, and urethra, as well as along a few large excretory ducts. The epithelium has either two layers (**Figure 4–5c**) or multiple layers. In the latter case, only the superficial cells are columnar.

Now that we have considered the classes of epithelia, we turn to a specialized type of epithelium: glandular epithelia.

Glandular Epithelia

Many epithelia contain gland cells that are specialized for secretion. Collections of epithelial cells (or structures derived from epithelial cells) that produce secretions are called *glands*. They range from scattered cells to complex glandular organs. Some of these glands, called **endocrine glands**, release their secretions into the interstitial fluid. Others, known as **exocrine glands**, release their secretions into passageways called **ducts** that open onto an epithelial surface.

Endocrine Glands

An endocrine gland produces *endocrine* (*endo-*, inside + *krinein*, to separate) *secretions*, which are released directly into the surrounding interstitial fluid. These secretions, also called *hormones*, enter the bloodstream for distribution throughout the body. Hormones regulate or coordinate the activities of various tissues, organs, and organ systems. Examples of endocrine glands include the thyroid gland and the pituitary gland. Because their secretions are not released into ducts, endocrine glands are often called *ductless glands*.

Endocrine cells may be part of an epithelial surface, such as the lining of the digestive tract, or they may be found in separate organs, such as the pancreas, thyroid gland, thymus, and pituitary gland. We will consider endocrine cells, organs, and hormones further in Chapter 18.

Exocrine Glands

Exocrine glands produce *exocrine* (*exo-*, outside) *secretions*, which are discharged onto an epithelial surface. Most exocrine secretions reach the surface through tubular ducts, which empty onto the skin surface or onto an epithelium lining an internal passageway that communicates with the exterior. Examples of exocrine secretions delivered to epithelial surfaces by ducts are enzymes entering the digestive tract, perspiration on the skin, tears in the eyes, and milk produced by mammary glands.

Exocrine glands exhibit several different methods of secretion; therefore, they are classified by their mode and type of secretion, and by the structure of the gland cells and associated ducts.

Modes of Secretion   A glandular epithelial cell releases its secretions by (1) merocrine secretion, (2) apocrine secretion, or (3) holocrine secretion.

In **merocrine secretion** (MER-u-krin; *meros*, part), the product is released from secretory vesicles by exocytosis (**Figure 4–6a**). This is the most common mode of secretion. One type of merocrine secretion, *mucin*, mixes with water to form **mucus**, an effective lubricant, a protective barrier, and a sticky trap for foreign particles and microorganisms. The mucous secretions of the salivary glands coat food and reduce friction during swallowing. In the skin, merocrine sweat glands produce the watery perspiration that helps cool you on a hot day.

Apocrine secretion (AP-ō-krin; *apo-*, off) involves the loss of cytoplasm as well as the secretory product (**Figure 4–6b**). The apical portion of the cytoplasm becomes packed with secretory vesicles and is then shed. Milk production in the mammary glands involves a combination of merocrine and apocrine secretions.

Merocrine and apocrine secretions leave a cell relatively intact and able to continue secreting. **Holocrine secretion** (HOL-ō-krin; *holos*, entire), by contrast, destroys the gland cell. During holocrine secretion, the entire cell becomes packed with secretory products and then bursts (**Figure 4–6c**), releasing the secretion, but killing the cell. Further secretion depends on the replacement of destroyed gland cells by the division of stem cells. Sebaceous glands, associated with hair follicles, produce an oily hair coating by means of holocrine secretion.

Types of Secretions   Exocrine glands are also categorized by the types of secretion produced:

1. *Serous glands* secrete a watery solution that contains enzymes. The parotid salivary glands are serous glands.

2. *Mucous glands* secrete mucins that hydrate to form mucus. The sublingual salivary glands and the submucosal glands of the small intestine are mucous glands.

3. *Mixed exocrine glands* contain more than one type of gland cell and may produce two different exocrine secretions, one serous and the other mucous. The submandibular salivary glands are mixed exocrine glands.

Gland Structure   The final method of classifying exocrine glands is by structure. In epithelia that have independent, scattered gland cells, the individual secretory cells are called

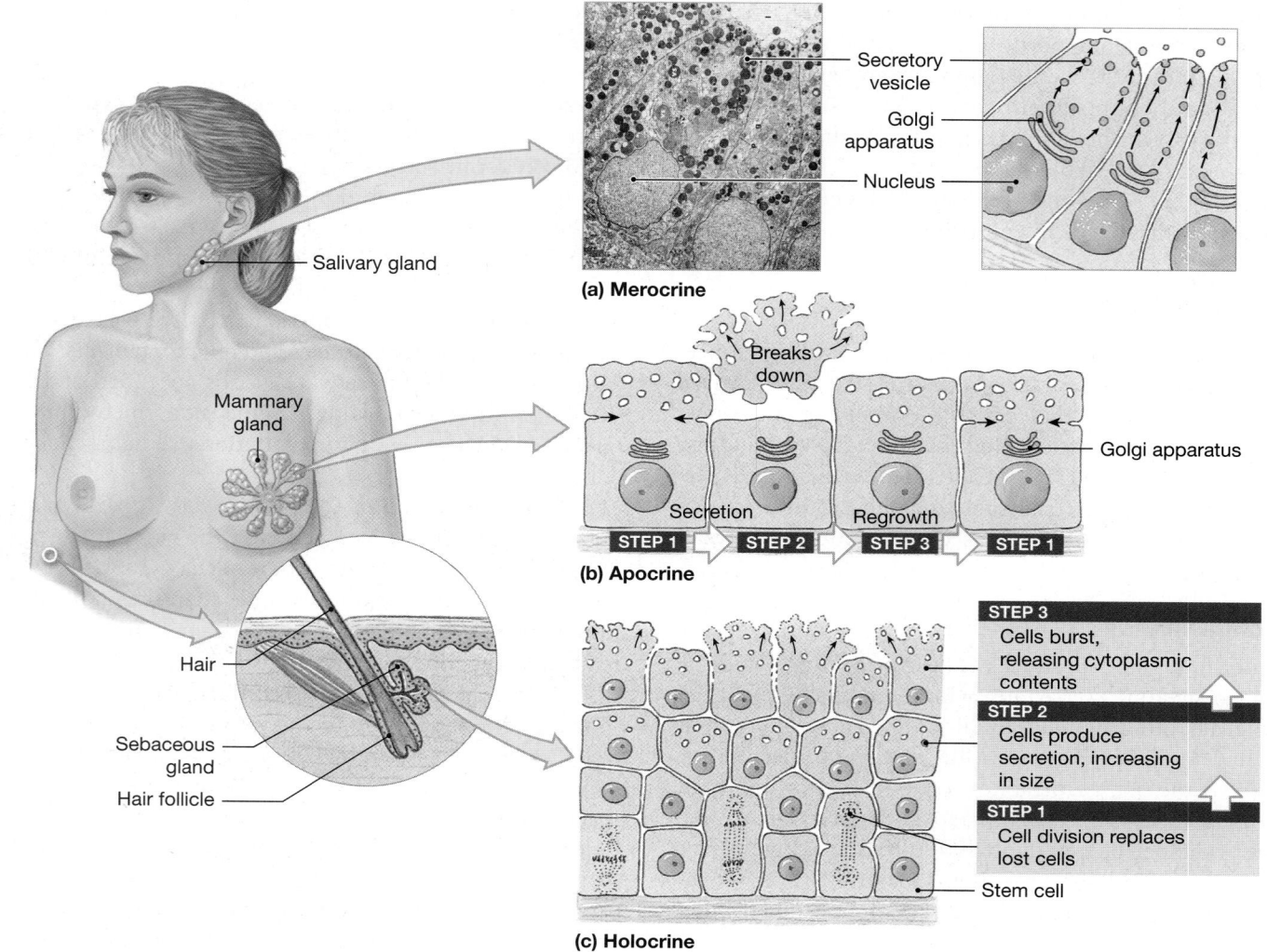

Figure 4–6   Modes of Glandular Secretion. (a) In merocrine secretion, secretory vesicles are discharged at the apical surface of the gland cell by exocytosis. **(b)** Apocrine secretion involves the loss of apical cytoplasm. Inclusions, secretory vesicles, and other cytoplasmic components are shed in the process. The gland cell then undergoes growth and repair before it releases additional secretions. **(c)** Holocrine secretion occurs as superficial gland cells burst. Continued secretion involves the replacement of these cells through the mitotic division of underlying stem cells.

unicellular glands. **Multicellular glands** include glandular epithelia and aggregations of gland cells that produce exocrine or endocrine secretions.

The only **unicellular exocrine glands** in the body are **mucous (goblet) cells**, which secrete mucins. Mucous cells are scattered among other epithelial cells. Both the pseudostratified ciliated columnar epithelium that lines the trachea and the columnar epithelium of the small and large intestines have an abundance of mucous cells.

The simplest **multicellular exocrine gland** is a *secretory sheet*, in which gland cells form an epithelium that releases secretions into an inner compartment. The continuous secretion of mucin-secreting cells that line the stomach, for instance, protects that organ from its own acids and enzymes. Most other multicellular exocrine glands are in pockets set back from the epithelial surface; their secretions travel through one or more ducts to the surface. Examples include the salivary glands, which produce mucins and digestive enzymes.

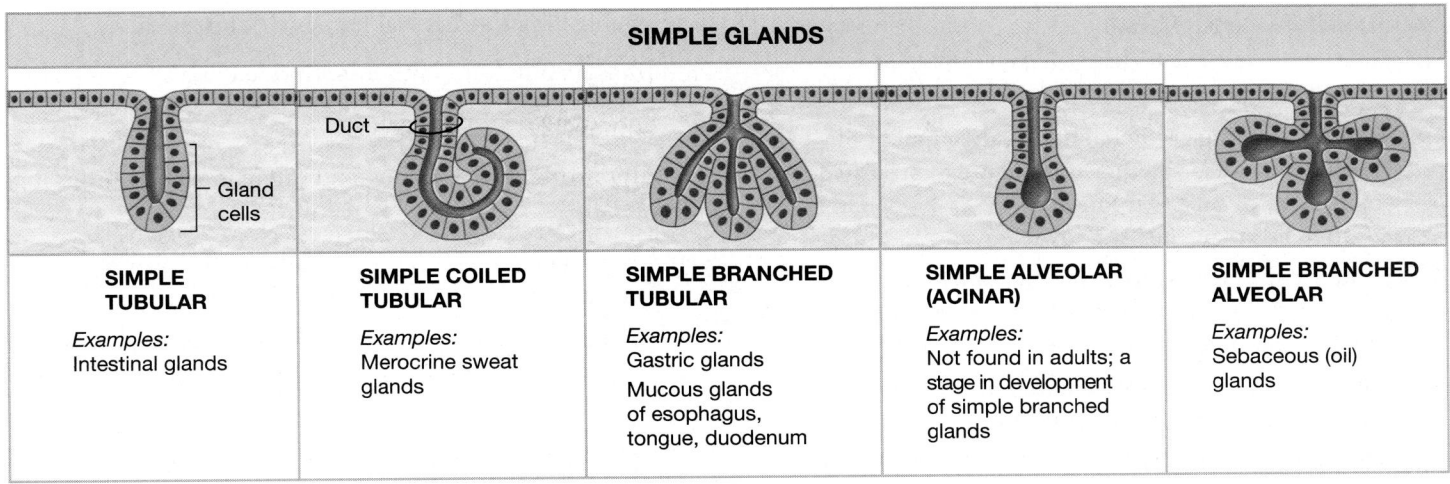

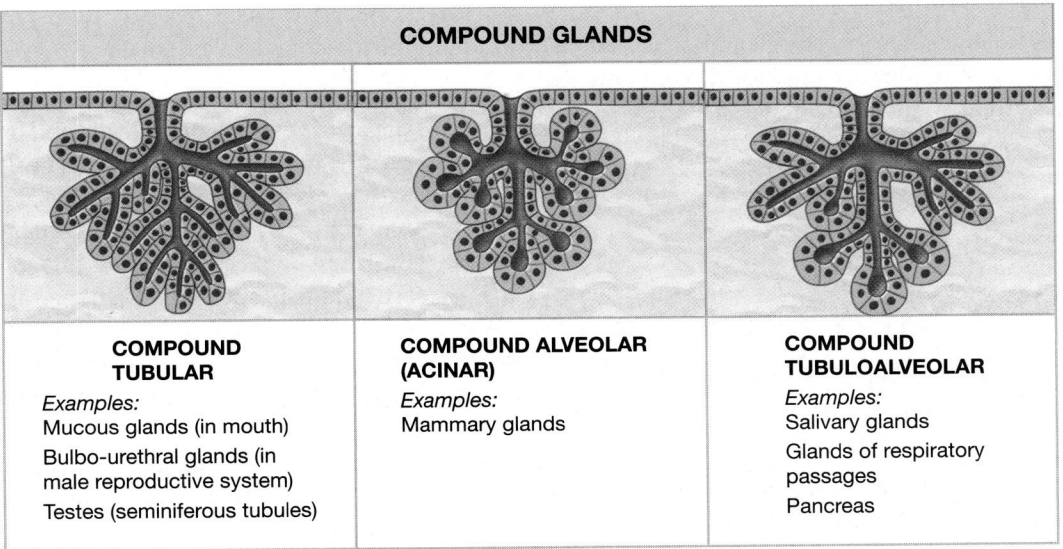

Figure 4–7   A Structural Classification of Exocrine Glands.

Three characteristics are used to describe the structure of multicellular exocrine glands (**Figure 4–7**):

1. *The Structure of the Duct.*   A gland is *simple* if it has a single duct that does not divide on its way to the gland cells. The gland is *compound* if the duct divides one or more times on its way to the gland cells.

2. *The Shape of the Secretory Portion of the Gland.*   Glands whose glandular cells form tubes are *tubular*; the tubes may be straight or coiled. Those that form blind pockets are *alveolar* (al-VĒ-ō-lar; *alveolus*, sac) or *acinar* (AS-i-nar; *acinus*, chamber). Glands whose secretory cells form both tubes and pockets are called *tubuloalveolar* or *tubuloacinar*.

3. *The Relationship between the Ducts and the Glandular Areas.*   A gland is *branched* if several secretory areas (tubular or acinar) share a duct. ("Branched" refers to the glandular areas and not to the duct.)

The vast majority of glands in the body produce either exocrine or endocrine secretions. However, a few complex organs, including the digestive tract and the pancreas, produce both kinds of secretions. We will consider the organization of these glands in Chapters 18 and 24.

8. Identify the three cell shapes characteristic of epithelial cells.

9. When classifying epithelial tissue, the number of layers determines whether it is simple or stratified. A single layer of cells is termed _____, whereas multiple layers of cells is known as _____.

10. Using a light microscope, you examine a tissue and see a simple squamous epithelium on the outer surface. Can this be a sample of the skin surface?

11. Why do the pharynx, esophagus, anus, and vagina have a similar epithelial organization?

12. Name the two primary types of glandular epithelia.

13. The secretory cells of sebaceous glands fill with secretions and then rupture, releasing their contents. Which mode of secretion is this?

14. A gland has no ducts to carry the glandular secretions, and the gland's secretions are released directly into the extracellular fluid. Which type of gland is this?

See the blue Answers tab at the end of the book.

4-4 Connective tissue provides a protective structural framework for other tissue types

It is impossible to discuss epithelial tissue without mentioning an associated type of tissue: **connective tissue**. Recall that the dense layer of the basal lamina of all epithelial tissues is created by connective tissue; in essence, connective tissue connects the epithelium to the rest of the body. Other connective tissues include bone, fat, and blood, which provide structure, store energy reserves, and transport materials throughout the body. Connective tissues vary widely in appearance and function, but they all share three basic components: (1) specialized cells, (2) extracellular protein fibers, and (3) a fluid known as **ground substance**. The extracellular fibers and ground substance together constitute the **matrix**, which surrounds the cells. Whereas cells make up the bulk of epithelial tissue, the matrix typically accounts for most of the volume of connective tissues. ATLAS: Embryology Summary 3: The Origins of Connective Tissues

Connective tissues are situated throughout the body, but are never exposed to the outside environment. Many connective tissues are highly vascular (that is, they have many blood vessels) and contain sensory receptors that detect pain, pressure, temperature, and other stimuli. Among the specific functions of connective tissues are the following:

- Establishing a structural framework for the body.
- Transporting fluids and dissolved materials.
- Protecting delicate organs.
- Supporting, surrounding, and interconnecting other types of tissue.
- Storing energy reserves, especially in the form of triacylglycerols.
- Defending the body from invading microorganisms.

Classification of Connective Tissues

Connective tissues are classified on the basis of their physical properties. The three general categories of connective tissue are connective tissue proper, fluid connective tissues, and supporting connective tissues.

1. **Connective tissue proper** includes those connective tissues with many types of cells and extracellular fibers in a syrupy ground substance. This broad category contains a variety of connective tissues that are divided into (a) *loose connective tissues* and (b) *dense connective tissues* based on the number of cell types present, and on the relative properties and proportions of fibers and ground substance. Both *adipose tissue* or fat (a loose connective tissue) and *tendons* (a dense connective tissue) are connective tissue proper, but they have very different structural and functional characteristics.

2. **Fluid connective tissues** have distinctive populations of cells suspended in a watery matrix that contains dissolved proteins. Two types exist: *blood* and *lymph*.

3. **Supporting connective tissues** differ from connective tissue proper in having a less diverse cell population and a matrix containing much more densely packed fibers. Supporting connective tissues protect soft tissues and support the weight of part or all of the body. The two types of supporting connective tissues are *cartilage* and *bone*. The matrix of cartilage is a gel whose characteristics vary with the predominant type of fiber. The matrix of bone is said to be **calcified**, because it contains mineral deposits (primarily calcium salts) that provide rigidity.

Connective Tissue Proper

Connective tissue proper contains extracellular fibers, a viscous (syrupy) ground substance, and a varied cell population (**Figure 4–8**). Some of the cells, including *fibroblasts, fibrocytes,*

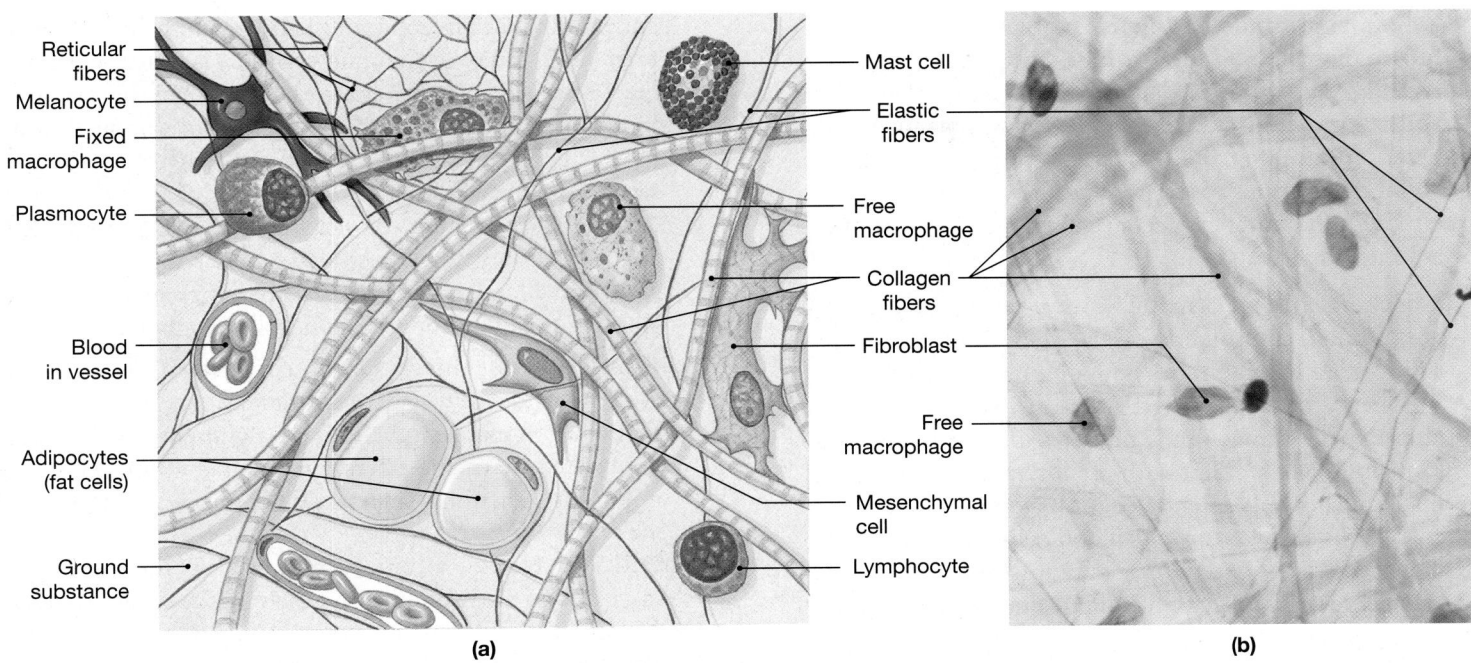

Figure 4–8 **The Cells and Fibers of Connective Tissue Proper.** Diagrammatic and histological views of the cell types and fibers of connective tissue proper. (Microphages, not shown, are common only in damaged or abnormal tissues.)

adipocytes, and *mesenchymal cells*, function in local mainte-nance, repair, and energy storage. These cells are permanent residents of the connective tissue. Other cells, including *macrophages*, *mast cells*, *lymphocytes*, *plasmocytes (plasma cells)*, and *microphages*, defend and repair damaged tissues. These cells are not permanent residents; they migrate through healthy connective tissues and aggregate at sites of tissue in-jury. The number of cells and cell types in a tissue at any mo-ment varies with local conditions.

Components of Connective Tissue Proper

The Cell Populations Fibroblasts (FĪ-brō-blasts) are one of the two most abundant permanent residents of connective tis-sue proper, and the only cells that are *always* present in it. Fi-broblasts secrete hyaluronan (a polysaccharide derivative) and proteins. (Recall that hyaluronan is one of the ingredients in the intercellular cement that helps lock epithelial cells together.) In connective tissue proper, extracellular fluid, hyaluronan, and proteins interact to form the proteoglycans that make ground substance viscous. Each fibroblast also secretes protein subunits that assemble to form large extracellular fibers. ∞ p. 54

In addition to fibroblasts, connective tissue proper may contain several other cell types:

- **Fibrocytes** (FĪ-brō-sīts) are the second most abundant fixed cell in connective tissue proper and differentiate from fibroblasts. These stellate cells maintain the connective tissue fibers of connective tissue proper.

- **Macrophages** (MAK-rō-fā-jez; *phagein*, to eat) are large amoeboid cells scattered throughout the matrix. These scavengers engulf pathogens or damaged cells that enter the tissue. (The name literally means "big eater.") Although not abundant, macrophages are important in mobilizing the body's defenses. When stimulated, they release chemicals that activate the immune system and attract large numbers of additional macrophages and other cells involved in tissue defense. The two classes of macrophage are *fixed macrophages*, which spend long periods in a tissue, and *free macrophages*, which migrate rapidly through tissues. In effect, fixed macrophages provide a "frontline" defense that can be reinforced by the arrival of free macrophages and other specialized cells.

- **Adipocytes** (AD-i-pō-sīts) are also known as *adipose cells*, or fat cells. A typical adipocyte contains a single, enormous lipid droplet. The nucleus, other organelles, and cytoplasm are squeezed to one side, making a sectional view of the cell resemble a class ring. The number of fat cells varies from one type of connective tissue to another, from one region of the body to another, and among individuals.

- **Mesenchymal cells** are stem cells that are present in many connective tissues. These cells respond to local injury or infection by dividing to produce daughter cells that differentiate into fibroblasts, macrophages, or other connective tissue cells.

- **Melanocytes** (me-LAN-ō-sīts) synthesize and store the brown pigment **melanin** (MEL-a-nin), which gives tissues a dark color. Melanocytes are common in the epithelium of the skin, where they play a major role in determining skin color. Melanocytes are also abundant in connective tissues of the eye and the dermis of the skin, although the number present differs by body region and among individuals.

- **Mast cells** are small, mobile connective tissue cells that are common near blood vessels. The cytoplasm of a mast cell is filled with granules containing **histamine** (HIS-tuh-mēn) and **heparin** (HEP-uh-rin). Histamine, released after injury or infection, stimulates local inflammation. (You are likely familiar with the inflammatory effects of histamine; people often take antihistamines to reduce cold symptoms.) *Basophils*, blood cells that enter damaged tissues and enhance the inflammation process, also contain granules of histamine and heparin.

- **Lymphocytes** (LIM-fō-sīts) migrate throughout the body, traveling through connective tissues and other tissues. Their numbers increase markedly wherever tissue damage occurs. Some lymphocytes may develop into **plasmocytes**, formerly called **plasma cells**, which produce *antibodies*—proteins involved in defending the body against disease.

- **Microphages** (*neutrophils* and *eosinophils*) are phagocytic blood cells that normally move through connective tissues in small numbers. When an infection or injury occurs, chemicals released by macrophages and mast cells attract numerous microphages to the site.

Connective Tissue Fibers   Three types of fibers occur in connective tissue: *collagen*, *reticular*, and *elastic*. Fibroblasts form all three by secreting protein subunits that interact in the matrix. Fibrocytes are responsible for maintaining these connecture tissue fibers.

1. **Collagen fibers** are long, straight, and unbranched. They are the most common fibers in connective tissue proper. Each collagen fiber consists of a bundle of fibrous protein subunits wound together like the strands of a rope. Like a rope, a collagen fiber is flexible, but it is stronger than steel when pulled from either end. *Tendons*, which connect skeletal muscles to bones, consist almost entirely of collagen fibers. Typical *ligaments* are similar to tendons, but they connect one bone to another. Tendons and ligaments can withstand tremendous forces. Uncontrolled muscle contractions or skeletal movements are more likely to break a bone than to snap a tendon or a ligament.

2. **Reticular fibers** (*reticulum*, network) contain the same protein subunits as do collagen fibers, but arranged differently. Thinner than collagen fibers, reticular fibers form a branching, interwoven framework that is tough, yet flexible. Because they form a network rather than share a common alignment, reticular fibers resist forces applied from many directions. This interwoven network, called a *stroma*, stabilizes the relative positions of the functional cells, or **parenchyma** (pa-RENG-ki-ma), of organs such as the liver. Reticular fibers also stabilize the positions of an organ's blood vessels, nerves, and other structures, despite changing positions and the pull of gravity.

3. **Elastic fibers** contain the protein *elastin*. Elastic fibers are branched and wavy. After stretching, they will return to their original length. **Elastic ligaments**, which are dominated by elastic fibers, are relatively rare but have important functions, such as interconnecting vertebrae.

Ground Substance   Ground substance fills the spaces between cells and surrounds connective tissue fibers (**Figure 4–8**). In connective tissue proper, ground substance is clear, colorless, and viscous (due to the presence of proteoglycans and glycoproteins). ∞ p. 57 Ground substance is dense enough that bacteria have trouble moving through it—imagine swimming in molasses. This density slows the spread of pathogens and makes them easier for phagocytes to catch.

CLINICAL NOTE

Marfan Syndrome   *Marfan syndrome* is an inherited condition caused by the production of an abnormally weak form of *fibrillin*, a glycoprotein that imparts strength and elasticity to connective tissues. Because most organs contain connective tissues, the effects of this defect are widespread. The most visible sign of Marfan syndrome involves the skeleton; most individuals with the condition are tall and have abnormally long limbs and fingers. The most serious consequences involve the cardiovascular system; roughly 90 percent of people with Marfan syndrome have structural abnormalities in their cardiovascular system. The most dangerous possibility is that the weakened elastic connective tissues in the walls of major arteries, such as the aorta, may burst, causing a sudden, fatal loss of blood.

Embryonic Connective Tissues

Mesenchyme, or *embryonic connective tissue*, is the first connective tissue to appear in a developing embryo. Mesenchyme contains an abundance of star-shaped stem cells (mesenchymal cells) separated by a matrix with very fine protein filaments (**Figure 4–9a**). Mesenchyme gives rise to all other connective tissues. **Mucous connective tissue** (**Figure 4–9b**), or Wharton's jelly, is a loose connective tissue found in many parts of the embryo, including the umbilical cord.

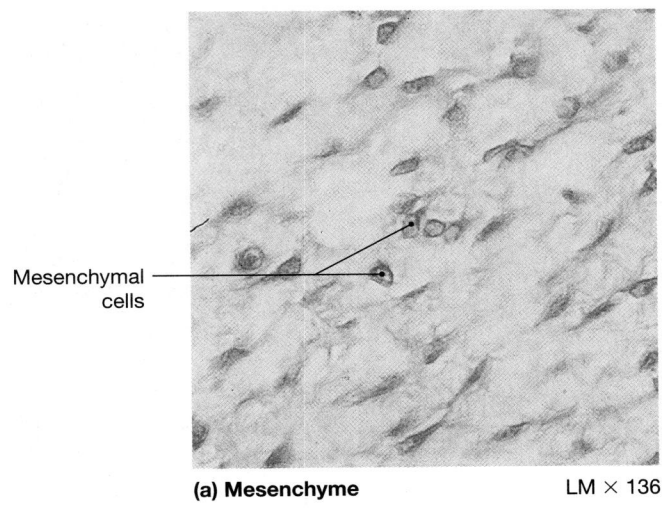

(a) Mesenchyme LM × 136

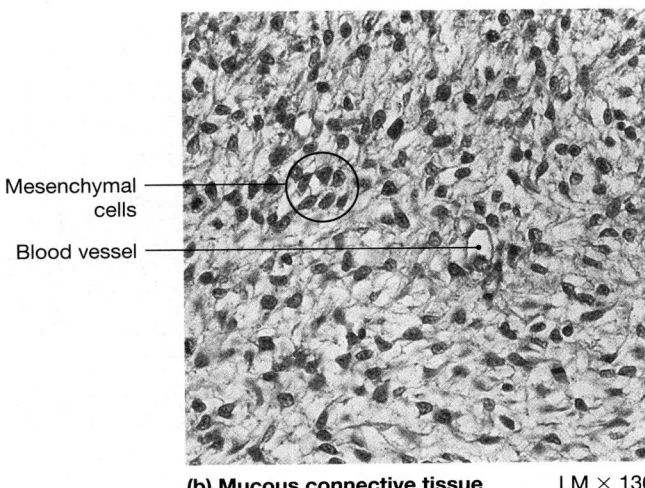

Mesenchymal cells

Blood vessel

(b) Mucous connective tissue LM × 136

Figure 4–9   Connective Tissues in Embryos. (a) Mesenchyme, the first connective tissue to appear in an embryo. **(b)** Mucous connective tissue, which is derived from mesenchyme. Shown here is mucous connective tissue in the umbilical cord of a fetus, where it is also known as *Wharton's jelly.*

Adults have neither form of embryonic connective tissue. However, many adult connective tissues contain scattered mesenchymal stem cells that can assist in tissue repair after an injury.

Loose Connective Tissues

Loose connective tissues are the "packing materials" of the body. They fill spaces between organs, cushion and stabilize specialized cells in many organs, and support epithelia. These tissues surround and support blood vessels and nerves, store lipids, and provide a route for the diffusion of materials. Loose connective tissues include mucous connective tissue in embryos and *areolar tissue, adipose tissue,* and *reticular tissue* in adults.

Areolar Tissue   Areolar tissue (*areola*, little space) is the least specialized connective tissue in adults. It may contain

all the cells and fibers of any connective tissue proper in a very loosely organized array **(Figure 4–8)**. Areolar tissue has an open framework. A viscous ground substance accounts for most of its volume and absorbs shocks. Because its fibers are loosely organized, areolar tissue can distort without damage. The presence of elastic fibers makes it resilient, so areolar tissue returns to its original shape after external pressure is relieved.

Areolar tissue forms a layer that separates the skin from deeper structures. In addition to providing padding, the elastic properties of this layer allow a considerable amount of independent movement. Thus, if you pinch the skin of your arm, you will not affect the underlying muscle. Conversely, contractions of the underlying muscle do not pull against your skin; as the muscle bulges, the areolar tissue stretches. Because this tissue has an extensive blood supply, the areolar tissue layer under the skin is a common injection site for drugs.

The capillaries (thin-walled blood vessels) in areolar tissue deliver oxygen and nutrients and remove carbon dioxide and waste products. They also carry wandering cells to and from the tissue. Epithelia commonly cover areolar tissue; fibrocytes maintain the dense layer of the basal lamina that separates the two kinds of tissue. The epithelial cells rely on the diffusion of oxygen and nutrients across that membrane from capillaries in the underlying connective tissue.

Adipose Tissue   The distinction between areolar tissue and fat, or **adipose tissue**, is somewhat arbitrary. Adipocytes account for most of the volume of adipose tissue **(Figure 4–10a)**, but only a fraction of the volume of areolar tissue. Adipose tissue provides padding, absorbs shocks, acts as an insulator to slow heat loss through the skin, and serves as packing or filler around structures. Adipose tissue is common under the skin of the flanks (between the last rib and the hips), buttocks, and breasts. It fills the bony sockets behind the eyes, surrounds the kidneys, and is common beneath the mesothelial lining of the pericardial and abdominal cavities.

Most of the adipose tissue in the body is called **white fat**, because it has a pale, yellow-white color. In infants and young children, however, the adipose tissue between the shoulder blades, around the neck, and possibly elsewhere in the upper body is highly vascularized, and the individual adipocytes contain numerous mitochondria. Together, these characteristics give the tissue a deep, rich color from which the name **brown fat** is derived. When these cells are stimulated by the nervous system, lipid breakdown accelerates. The cells do not capture the energy that is released. Instead, it is absorbed by the surrounding tissues as heat. The heat warms the circulating blood, which distributes the heat throughout the body. In this way, an infant can increase metabolic heat generation by 100 percent very quickly. (In adults, who have little if any brown fat, body temperature is elevated primarily by shivering.)

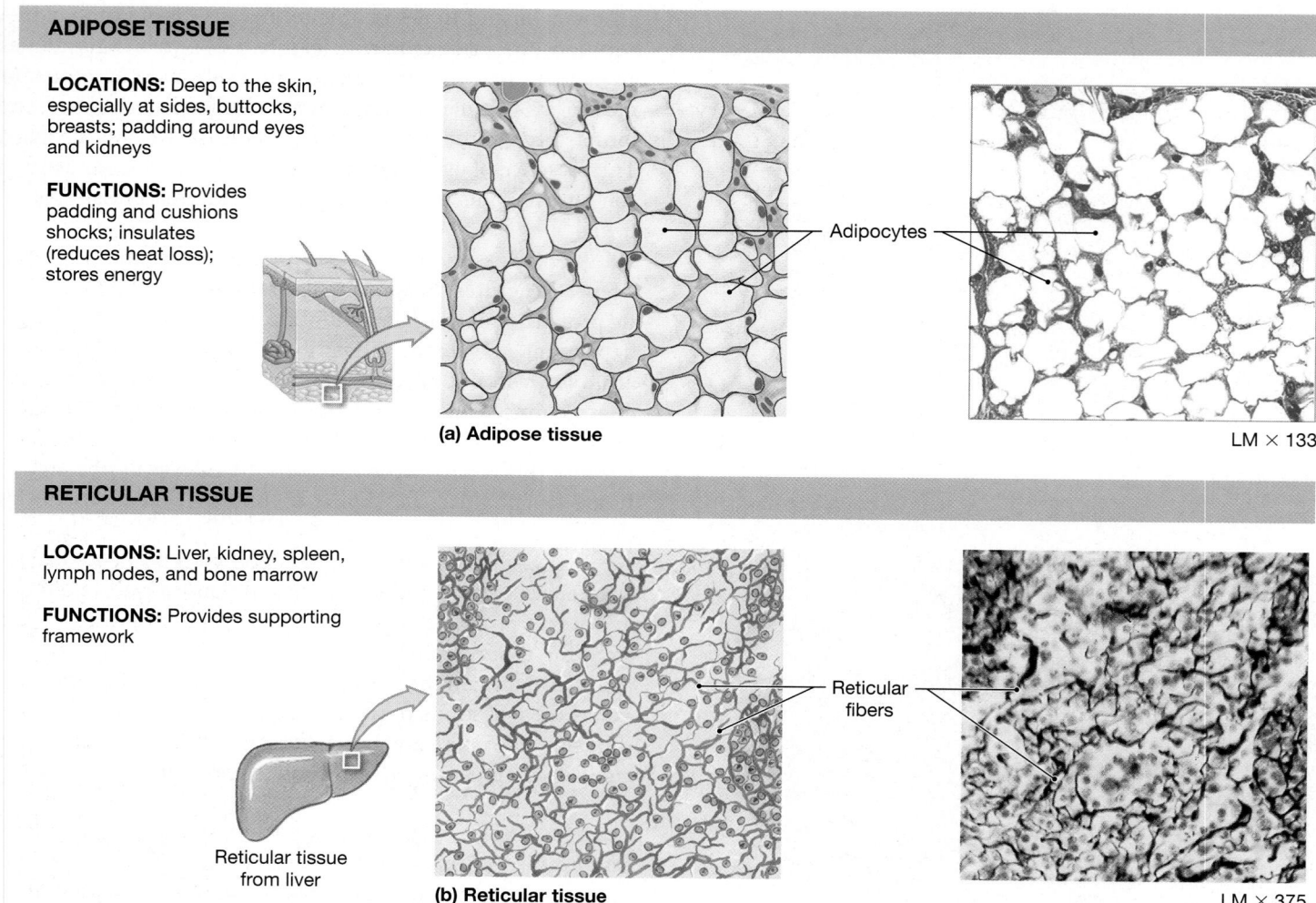

ADIPOSE TISSUE

LOCATIONS: Deep to the skin, especially at sides, buttocks, breasts; padding around eyes and kidneys

FUNCTIONS: Provides padding and cushions shocks; insulates (reduces heat loss); stores energy

Adipocytes

(a) Adipose tissue

LM × 133

RETICULAR TISSUE

LOCATIONS: Liver, kidney, spleen, lymph nodes, and bone marrow

FUNCTIONS: Provides supporting framework

Reticular tissue from liver

Reticular fibers

(b) Reticular tissue

LM × 375

Figure 4–10   Adipose and Reticular Tissues. (a) Adipose tissue is a loose connective tissue dominated by adipocytes. In standard histological preparations, the tissue looks empty because the lipids in the fat cells dissolve in the alcohol used in tissue processing. **(b)** Reticular tissue has an open framework of reticular fibers, which are usually very difficult to see because of the large numbers of cells around them.

Adipocytes are metabolically active cells; their lipids are constantly being broken down and replaced. When nutrients are scarce, adipocytes deflate like collapsing balloons as their lipids are broken down and the fatty acids released to support metabolism. Because the cells are not killed but merely reduced in size, the lost weight can easily be regained in the same areas of the body. In adults, adipocytes are incapable of dividing. The number of fat cells in peripheral tissues is established in the first few weeks of a newborn's life, perhaps in response to the amount of fats in the diet. However, that is not the end of the story, because loose connective tissues also contain mesenchymal cells. If circulating-lipid levels are chronically elevated, the mesenchymal cells will divide, giving rise to cells that differentiate into fat cells. As a result, areas of areolar tissue can become adipose tissue in times of nutritional plenty, even in adults.

In the procedure known as **liposuction**, unwanted adipose tissue is surgically removed. Because adipose tissue can regenerate through the differentiation of mesenchymal cells, liposuction provides only a temporary and potentially risky solution to the problem of excess weight.

Reticular Tissue   As mentioned earlier, organs such as the spleen and liver contain **reticular tissue**, in which reticular fibers create a complex three-dimensional stroma (**Figure 4–10b**). The stroma supports the parenchyma (functional cells) of these organs. This fibrous framework is also found in the lymph nodes and bone marrow. Fixed macrophages, fibroblasts, and fibrocytes are associated with the reticular fibers, but these cells are seldom visible, because the organs are dominated by specialized cells with other functions.

Dense Connective Tissues

Most of the volume of **dense connective tissues** is occupied by fibers. Dense connective tissues are often called **collagenous** (ko-LAJ-e-nus) **tissues**, because collagen fibers are the dominant type of fiber in them. The body has two types of dense connective tissues: dense regular connective tissue and dense irregular connective tissue.

In **dense regular connective tissue**, the collagen fibers are parallel to each other, packed tightly, and aligned with the forces applied to the tissue. **Tendons** are cords of dense regu-

lar connective tissue that attach skeletal muscles to bones (**Figure 4–11a**). The collagen fibers run along the longitudinal axis of the tendon and transfer the pull of the contracting muscle to the bone. **Ligaments** resemble tendons, but connect one bone to another or stabilize the positions of internal organs. An **aponeurosis** (AP-ō-noo-RŌ-sis; plural,

DENSE REGULAR CONNECTIVE TISSUE

LOCATIONS: Between skeletal muscles and skeleton (tendons and aponeuroses); between bones or stabilizing positions of internal organs (ligaments); covering skeletal muscles; deep fasciae

FUNCTIONS: Provides firm attachment; conducts pull of muscles; reduces friction between muscles; stabilizes relative positions of bones

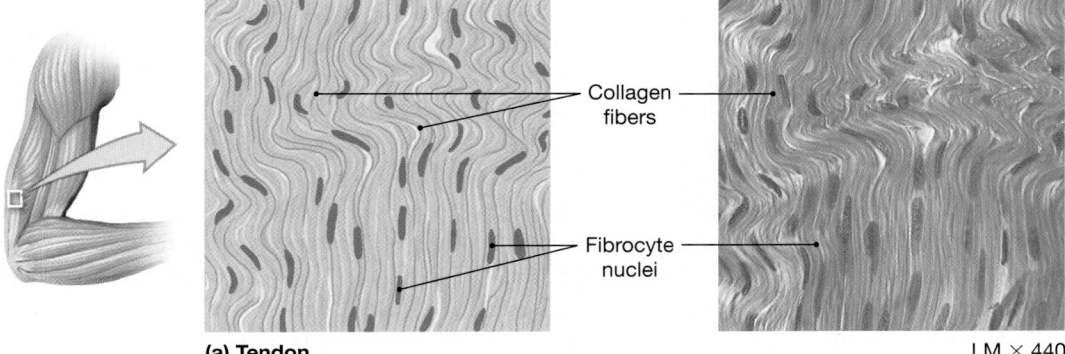

(a) Tendon

LM × 440

DENSE IRREGULAR CONNECTIVE TISSUE

LOCATIONS: Capsules of visceral organs; periostea and perichondria; nerve and muscle sheaths; dermis

FUNCTIONS: Provides strength to resist forces applied from many directions; helps prevent overexpansion of organs such as the urinary bladder

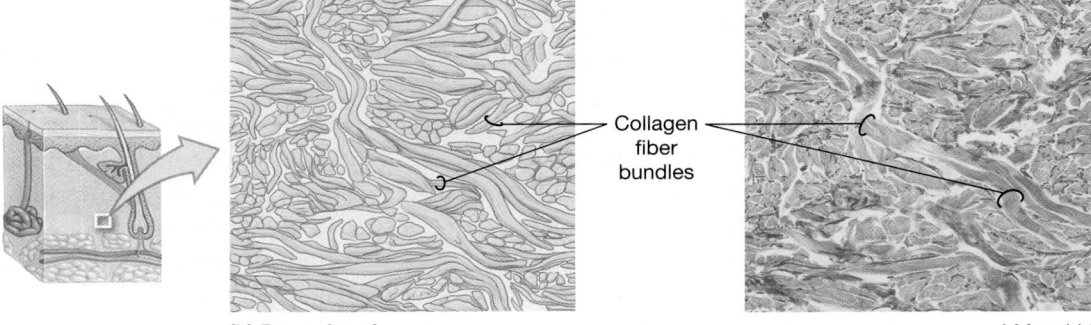

(b) Deep dermis

LM × 111

ELASTIC TISSUE

LOCATIONS: Between vertebrae of the spinal column (ligamentum flavum and ligamentum nuchae); ligaments supporting penis; ligaments supporting transitional epithelia; in blood vessel walls

FUNCTIONS: Stabilizes positions of vertebrae and penis; cushions shocks; permits expansion and contraction of organs

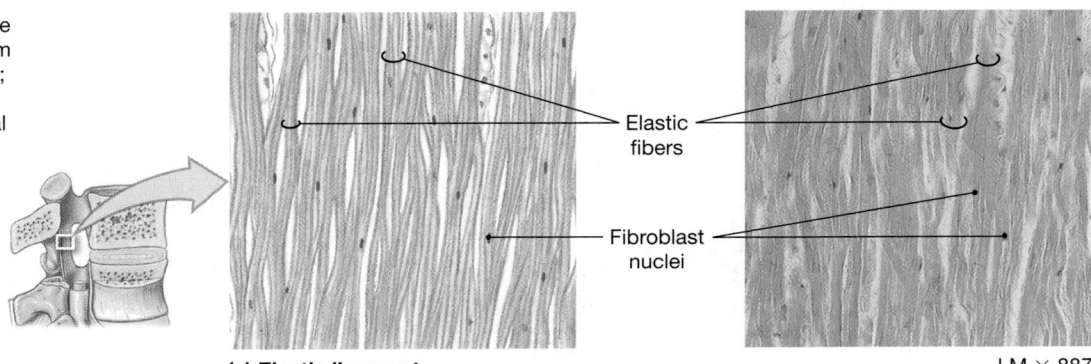

(c) Elastic ligament

LM × 887

Figure 4–11 **Dense Connective Tissues.** (a) The dense regular connective tissue in a tendon. Notice the densely packed, parallel bundles of collagen fibers. The fibrocyte nuclei are flattened between the bundles. (b) The deep dermis of the skin contains a thick layer of dense irregular connective tissue. (c) An elastic ligament is an example of elastic tissue. Elastic ligaments extend between the vertebrae of the vertebral column. The bundles are fatter than those of a tendon or a ligament composed of collagen.

aponeuroses) is a tendinous sheet that attaches a broad, flat muscle to another muscle or to several bones of the skeleton. It can also stabilize the positions of tendons and ligaments. Aponeuroses are associated with large muscles of the skull, lower back, and abdomen, and with the tendons and ligaments of the palms of the hands and the soles of the feet. Large numbers of fibroblasts are scattered among the collagen fibers of tendons, ligaments, and aponeuroses.

In contrast, the fibers in **dense irregular connective tissue** form an interwoven meshwork in no consistent pattern (**Figure 4–11b**). These tissues strengthen and support areas subjected to stresses from many directions. A layer of dense irregular connective tissue gives skin its strength. Cured leather (animal skin) is an excellent illustration of the interwoven nature of this tissue. Except at joints, dense irregular connective tissue forms a sheath around cartilages (the *perichondrium*) and bones (the *periosteum*). Dense irregular connective tissue also forms a thick fibrous layer called a **capsule**, which surrounds internal organs such as the liver, kidneys, and spleen and encloses the cavities of joints.

Dense regular and dense irregular connective tissues contain variable amounts of elastic fibers. When elastic fibers outnumber collagen fibers, the tissue has a springy, resilient nature that allows it to tolerate cycles of extension and recoil. Abundant elastic fibers are present in the connective tissue that supports transitional epithelia, in the walls of large blood vessels such as the aorta, and around the respiratory passageways.

Elastic tissue is a dense regular connective tissue dominated by elastic fibers. Elastic ligaments, which are almost completely dominated by elastic fibers, help stabilize the positions of the vertebrae of the spinal column (**Figure 4–11c**).

Fluid Connective Tissues

Blood and *lymph* are connective tissues with distinctive collections of cells. The fluid matrix that surrounds the cells also includes many types of suspended proteins that do not form insoluble fibers under normal conditions.

In **blood**, the watery matrix is called **plasma**. Plasma contains blood cells and fragments of cells, collectively known as *formed elements* (**Figure 4–12**). There are three types of formed elements: red blood cells, white blood cells, and platelets.

A single cell type—the **red blood cell**, or **erythrocyte** (e-RITH-rō-sīt; *erythros*, red + *cyte*, cell)—accounts for almost half the volume of blood and is the reason we associate the color red with blood. Red blood cells are responsible for the transport of oxygen (and, to a lesser degree, of carbon dioxide) in the blood.

Blood also contains small numbers of **white blood cells**, or **leukocytes** (LOO-kō-sīts; *leuko-*, white). White blood cells include the phagocytic microphages (*neutrophils* and *eosinophils*), *basophils*, *lymphocytes*, and *monocytes*, cells related to the macrophages found in other connective tissues. White blood cells are important components of the immune system, which protects the body from infection and disease.

The third type of formed element in blood consists not of whole cells, but of tiny membrane-enclosed packets of cytoplasm called **platelets**. These cell fragments, which contain enzymes and special proteins, function in the clotting response that seals breaks in the endothelial lining.

Recall from Chapter 3 that the human body contains a large volume of extracellular fluid. This fluid includes three major subdivisions: *plasma*, *interstitial fluid*, and *lymph*. Plasma is normally confined to the vessels of the cardiovascular system, and contractions of the heart keep it in motion. **Arteries** carry blood away from the heart and into the tissues of the body. In those tissues, blood pressure forces water and small solutes out of the bloodstream across the walls of **capillaries**, the smallest blood vessels. This is the origin of the interstitial fluid that bathes the body's cells. The remaining blood flows from the capillaries into **veins** that return it to the heart.

Lymph forms as interstitial fluid enters **lymphatic vessels**. As fluid passes along the lymphatic vessels, cells of the immune system monitor the composition of the lymph and

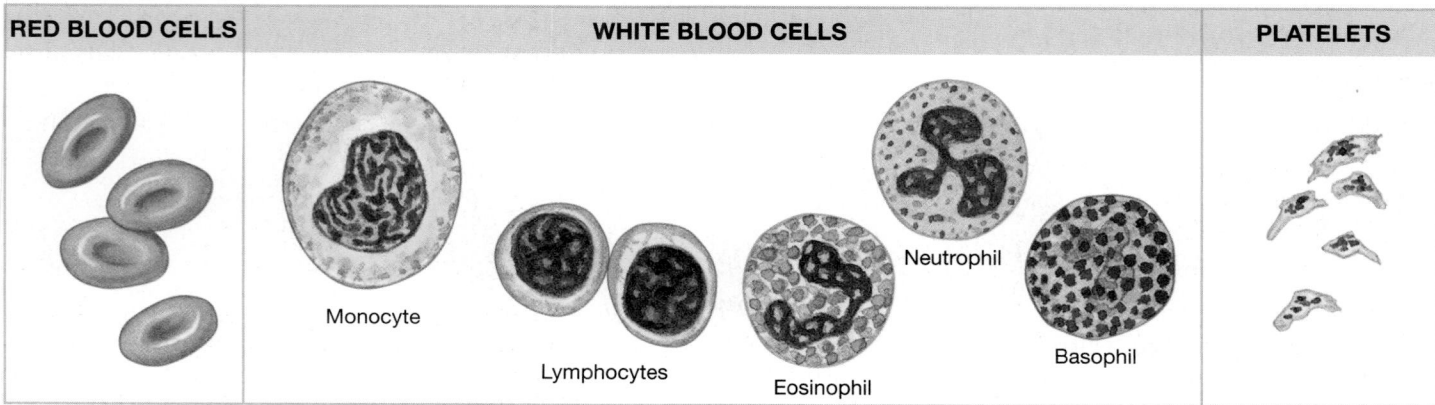

RED BLOOD CELLS	WHITE BLOOD CELLS	PLATELETS

Monocyte

Lymphocytes

Eosinophil

Neutrophil

Basophil

Figure 4–12 Formed Elements of the Blood.

respond to signs of injury or infection. The lymphatic vessels ultimately return the lymph to large veins near the heart. This recirculation of fluid—from the cardiovascular system, through the interstitial fluid, to the lymph, and then back to the cardiovascular system—is a continuous process that is essential to homeostasis. It helps eliminate local differences in the levels of nutrients, wastes, or toxins; maintains blood volume; and alerts the immune system to infections that may be under way in peripheral tissues.

15. Identify several functions of connective tissues.

16. List the three categories of connective tissues.

17. Identify the populations of cells found in connective tissue proper.

18. Lack of vitamin C in the diet interferes with the ability of fibroblasts to produce collagen. What effect might this interference have on connective tissue?

19. Many allergy sufferers take antihistamines to relieve their allergy symptoms. Which types of cell produce the molecule that this medication blocks?

20. Which type of connective tissue contains primarily triacylglycerols?

21. Which two types of connective tissue have a fluid matrix?

See the blue Answers tab at the end of the book.

4-5 Cartilage and bone provide a strong supporting framework

Cartilage and *bone* are called supporting connective tissues because they provide a strong framework that supports the rest of the body. In these connective tissues, the matrix contains numerous fibers and, in bone, deposits of insoluble calcium salts.

Cartilage

The matrix of **cartilage** is a firm gel that contains polysaccharide derivatives called **chondroitin sulfates** (kon-DROY-tin; *chondros*, cartilage). Chondroitin sulfates form complexes with proteins in the ground substance, producing proteoglycans. Cartilage cells, or **chondrocytes** (KON-drō-sīts), are the only cells in the cartilage matrix. They occupy small chambers known as **lacunae** (la-KOO-nē; *lacus*, lake). The physical properties of cartilage depend on the proteoglycans of the matrix, and on the type and abundance of extracellular fibers.

Unlike other connective tissues, cartilage is avascular, so all exchange of nutrients and waste products must occur by diffusion through the matrix. Blood vessels do not grow into cartilage because chondrocytes produce a chemical that discourages their formation. This chemical, named **antiangiogenesis factor** (*anti-*, against + *angeion*, vessel + *genno*, to produce), is now being tested as a potential anticancer agent.

A cartilage is generally set apart from surrounding tissues by a fibrous **perichondrium** (per-i-KON-drē-um; *peri-*, around). The perichondrium contains two distinct layers: an outer, fibrous region of dense irregular connective tissue, and an inner, cellular layer. The fibrous layer provides mechanical support and protection and attaches the cartilage to other structures. The cellular layer is important to the growth and maintenance of the cartilage.

Cartilage Growth

Cartilage grows by two mechanisms: *interstitial growth* and *appositional growth* (**Figure 4–13**).

In **interstitial growth**, chondrocytes in the cartilage matrix undergo cell division, and the daughter cells produce additional matrix (**Figure 4–13a**). This process enlarges the cartilage from within. Interstitial growth is most important during development. The process begins early in embryonic development and continues through adolescence.

In **appositional growth**, new layers of cartilage are added to the surface (**Figure 4–13b**). In this process, cells of the inner layer of the perichondrium undergo repeated cycles of division. The innermost cells then differentiate into immature chondrocytes, which begin producing cartilage matrix. As they become surrounded by and embedded in new matrix, they differentiate into mature chondrocytes. Appositional growth gradually increases the size of the cartilage by adding to its outer surface.

Both interstitial and appositional growth occur during development, although interstitial growth contributes more to the mass of the adult cartilage. Neither interstitial nor appositional growth occurs in the cartilages of normal adults. However, appositional growth may occur in unusual circumstances, such as after cartilage has been damaged or excessively stimulated by *growth hormone* from the pituitary gland. Minor damage to cartilage can be repaired by appositional growth at the damaged surface. After more severe damage, the injured portion of the cartilage will be replaced by a dense fibrous patch.

Types of Cartilage

The body contains three major types of cartilage: hyaline cartilage, elastic cartilage, and fibrous cartilage.

1. **Hyaline cartilage** (HĪ-uh-lin; *hyalos*, glass) is the most common type of cartilage. Except inside joint cavities, a

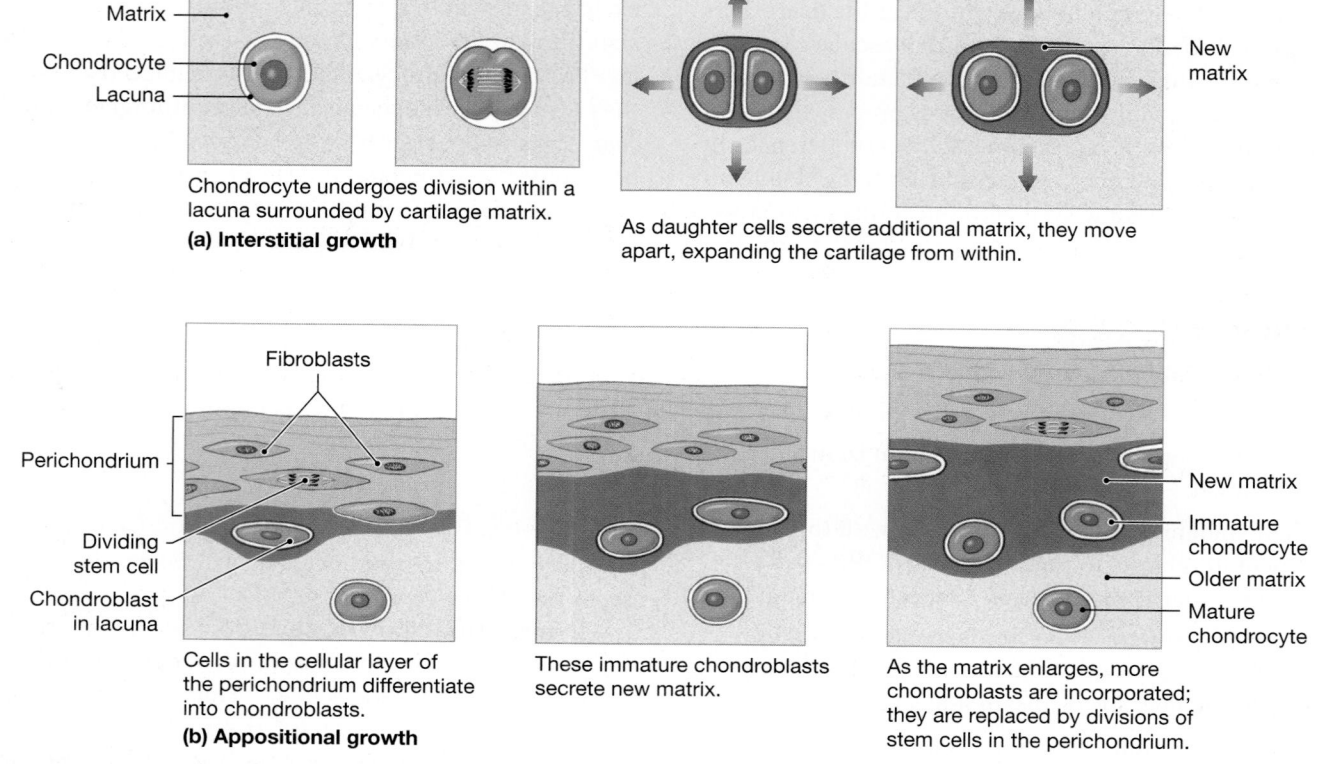

Matrix

Chondrocyte

Lacuna

Chondrocyte undergoes division within a lacuna surrounded by cartilage matrix.
(a) Interstitial growth

As daughter cells secrete additional matrix, they move apart, expanding the cartilage from within.

New matrix

Fibroblasts

Perichondrium

Dividing stem cell

Chondroblast in lacuna

Cells in the cellular layer of the perichondrium differentiate into chondroblasts.
(b) Appositional growth

These immature chondroblasts secrete new matrix.

As the matrix enlarges, more chondroblasts are incorporated; they are replaced by divisions of stem cells in the perichondrium.

New matrix

Immature chondrocyte

Older matrix

Mature chondrocyte

Figure 4–13 **The Growth of Cartilage. (a)** In interstitial growth, the cartilage expands from within as chondrocytes in the matrix divide, grow, and produce new matrix. **(b)** In appositional growth, the cartilage grows at its external surface as fibroblasts in the cellular layer of the perichondrium differentiate into chondrocytes.

dense perichondrium surrounds hyaline cartilages. The matrix of hyaline cartilage contains closely packed collagen fibers, making it tough but somewhat flexible. Because the fibers are not in large bundles and do not stain darkly, they are not always apparent in light microscopy (**Figure 4–14a**). Examples in adults include the connections between the ribs and the sternum; the nasal cartilages and the supporting cartilages along the conducting passageways of the respiratory tract; and the *articular cartilages*, which cover opposing bone surfaces within many joints, such as the elbow and knee.

2. **Elastic cartilage** (**Figure 4–14b**) contains numerous elastic fibers that make it extremely resilient and flexible. These cartilages usually have a yellowish color on gross dissection. Elastic cartilage forms the external flap (the *auricle*, or *pinna*) of the outer ear; the epiglottis; a passageway to the middle ear cavity (the *auditory tube*); and small cartilages in the larynx (the *cuneiform cartilages*).

3. **Fibrous cartilage**, or **fibrocartilage**, has little ground substance, and its matrix is dominated by densely interwoven collagen fibers (**Figure 4–14c**), making this tissue extremely durable and tough. Pads of fibrous cartilage lie between the spinal vertebrae, between the pubic bones of

the pelvis, and around tendons and within or around joints. In these positions, fibrous cartilage resists compression, absorbs shocks, and prevents damaging bone-to-bone contact. Cartilage heals poorly, and damaged fibrous cartilage in joints such as the knee can interfere with normal movements.

Several complex joints, including the knee, contain both hyaline cartilage and fibrous cartilage. The hyaline cartilage covers bony surfaces, and fibrous cartilage pads in the joint prevent contact between bones during movement. Injuries to these pads do not heal, and after repeated or severe damage, joint mobility is severely reduced. Surgery generally produces only a temporary or incomplete repair.

Bone

Because we will examine the detailed histology of **bone**, or **osseous** (OS-ē-us; *os*, bone) **tissue**, in Chapter 6, here we focus only on significant differences between cartilage and bone. The volume of ground substance in bone is very small. Roughly two-thirds of the matrix of bone consists of a mixture of calcium salts—primarily calcium phosphate, with lesser amounts of calcium carbonate. The rest of the matrix is dominated by

HYALINE CARTILAGE

LOCATIONS: Between tips of ribs and bones of sternum; covering bone surfaces at synovial joints; supporting larynx (voice box), trachea, and bronchi; forming part of nasal septum

FUNCTIONS: Provides stiff but somewhat flexible support; reduces friction between bony surfaces

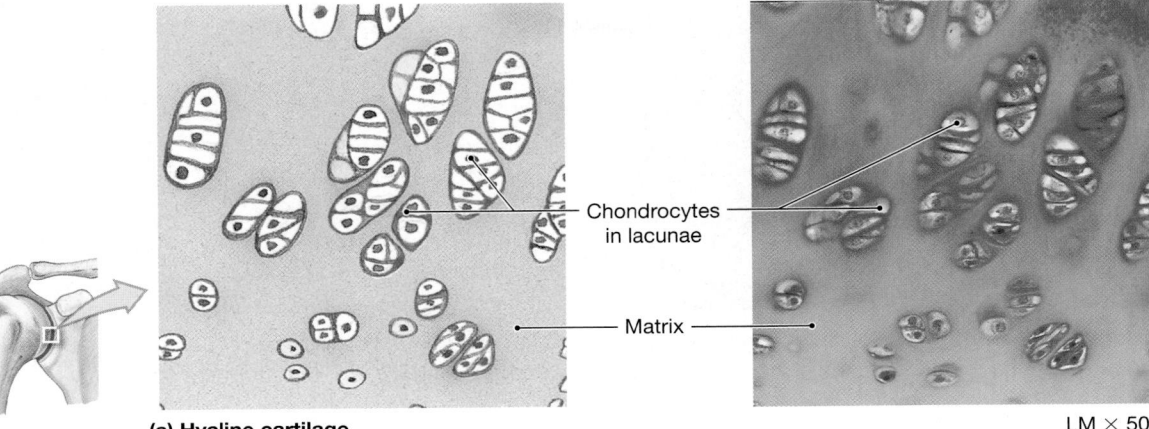

Chondrocytes in lacunae

Matrix

(a) Hyaline cartilage

LM × 500

ELASTIC CARTILAGE

LOCATIONS: Auricle of external ear; epiglottis; auditory tube; cuneiform cartilages of larynx

FUNCTIONS: Provides support, but tolerates distortion without damage and returns to original shape

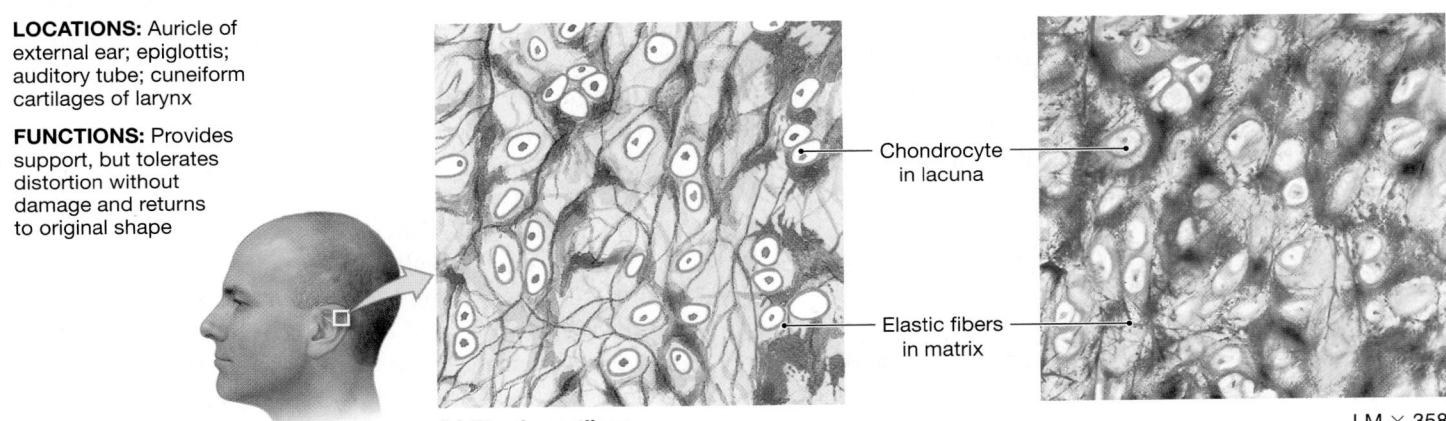

Chondrocyte in lacuna

Elastic fibers in matrix

(b) Elastic cartilage

LM × 358

FIBROUS CARTILAGE

LOCATIONS: Pads within knee joint; between pubic bones of pelvis; intervertebral discs

FUNCTIONS: Resists compression; prevents bone-to-bone contact; limits relative movement

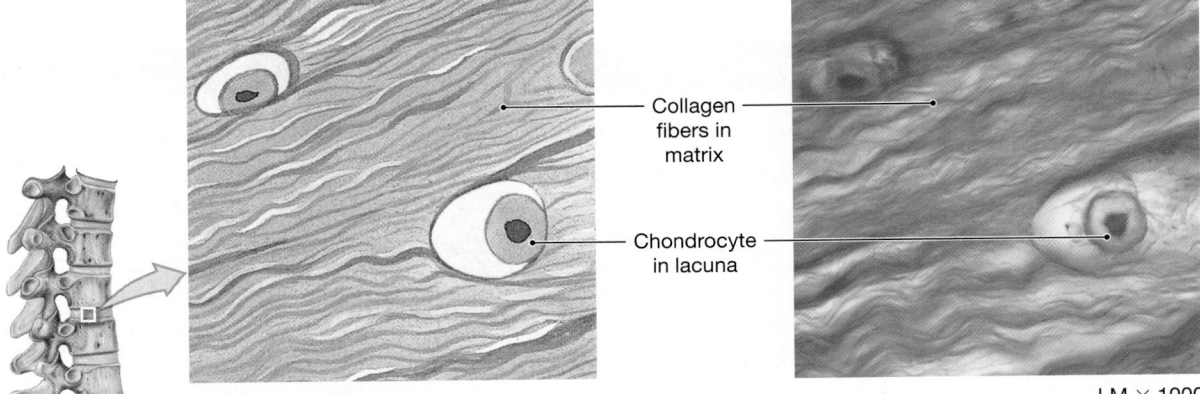

Collagen fibers in matrix

Chondrocyte in lacuna

(c) Fibrous cartilage

LM × 1000

Figure 4–14   The Types of Cartilage.   **(a)** Hyaline cartilage. Note the translucent matrix and the absence of prominent fibers. **(b)** Elastic cartilage. The closely packed elastic fibers are visible between the chondrocytes. **(c)** Fibrous cartilage. The collagen fibers are extremely dense, and the chondrocytes are relatively far apart.

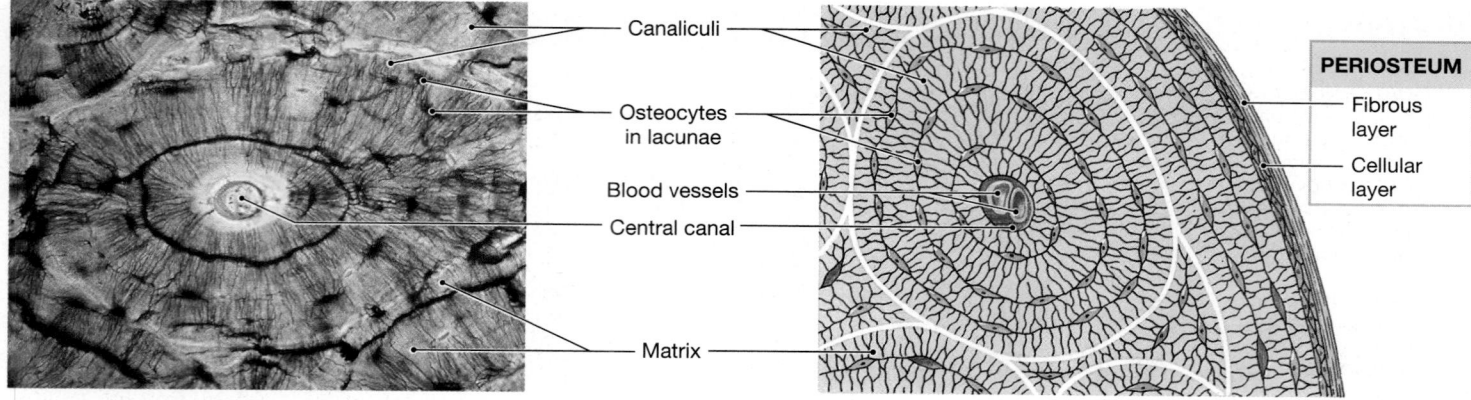

Figure 4–15   Bone. The osteocytes in bone are generally organized in groups around a central space that contains blood vessels. Bone dust produced during preparation of the section fills the lacunae and the central canal, making them appear dark in the micrograph.

collagen fibers. This combination gives bone truly remarkable properties. By themselves, calcium salts are hard but rather brittle, whereas collagen fibers are stronger but relatively flexible. In bone, the presence of the minerals surrounding the collagen fibers produces a strong, somewhat flexible combination that is highly resistant to shattering. In its overall properties, bone can compete with the best steel-reinforced concrete. In essence, the collagen fibers in bone act like the steel reinforcing rods, and the mineralized matrix acts like the concrete.

Figure 4–15 shows the general organization of osseous tissue. Lacunae in the matrix contain **osteocytes** (OS-tē-ō-sīts), or bone cells. The lacunae are typically organized around blood vessels that branch through the bony matrix. Although diffusion cannot occur through the hard matrix, osteocytes communicate with the blood vessels and with one another by means of slender cytoplasmic extensions. These extensions run through long, slender passageways in the matrix called

canaliculi (kan-a-LIK-ū-lē; little canals). These passageways form a branching network for the exchange of materials between blood vessels and osteocytes.

Except in joint cavities, where they are covered by a layer of hyaline cartilage, bone surfaces are sheathed by a **periosteum** (per-ē-OS-tē-um), a layer composed of fibrous (outer) and cellular (inner) layers. The periosteum assists in the attachment of a bone to surrounding tissues and to associated tendons and ligaments. The cellular layer functions in appositional bone growth and participates in repairs after an injury. Unlike cartilage, bone undergoes extensive remodeling throughout life, and complete repairs can be made even after severe damage has occurred. Bones also respond to the stresses placed on them, growing thicker and stronger with exercise and becoming thin and brittle with inactivity.

Table 4–2 summarizes the similarities and differences between cartilage and bone.

TABLE 4–2 A Comparison of Cartilage and Bone		
Characteristic	**Cartilage**	**Bone**
Structural Features		
Cells	Chondrocytes in lacunae	Osteocytes in lacunae
Ground substance	Chondroitin sulfate (in proteoglycan) and water	A small volume of liquid surrounding insoluble crystals of calcium salts (calcium phosphate and calcium carbonate)
Fibers	Collagen, elastic, and reticular fibers (proportions vary)	Collagen fibers predominate
Vascularity	None	Extensive
Covering	Perichondrium (two layers)	Periosteum (two layers)
Strength	Limited: bends easily, but hard to break	Strong: resists distortion until breaking point
Metabolic Features		
Oxygen demands	Low	High
Nutrient delivery	By diffusion through matrix	By diffusion through cytoplasm and fluid in canaliculi
Growth	Interstitial and appositional	Appositional only
Repair capabilities	Limited	Extensive

22. Identify the two types of supporting connective tissue.

23. Why does bone heal faster than cartilage?

24. If a person has a herniated intervertebral disc, which type of cartilage has been damaged?

See the blue Answers tab at the end of the book.

4-6 Membranes are physical barriers of four types: mucous, serous, cutaneous, and synovial

A membrane is a physical barrier. There are many different types of anatomical membranes—you encountered plasma membranes in Chapter 3, and you will find many other kinds of membranes in later chapters. The membranes we are concerned with here line or cover body surfaces. Each consists of an epithelium supported by connective tissue. Four such membranes occur in the body: (1) *mucous membranes*, (2) *serous membranes*, (3) the *cutaneous membrane*, and (4) *synovial membranes* (**Figure 4–16**).

Mucous Membranes

Mucous membranes, or **mucosae** (mū-KŌ-sē), line passageways and chambers that communicate with the exterior, including those in the digestive, respiratory, reproductive, and urinary tracts (**Figure 4–16a**). The epithelial surfaces of these passageways must be kept moist to reduce friction and, in many cases, facilitate absorption or secretion. The epithelial surfaces are lubricated either by mucus, produced by mucous

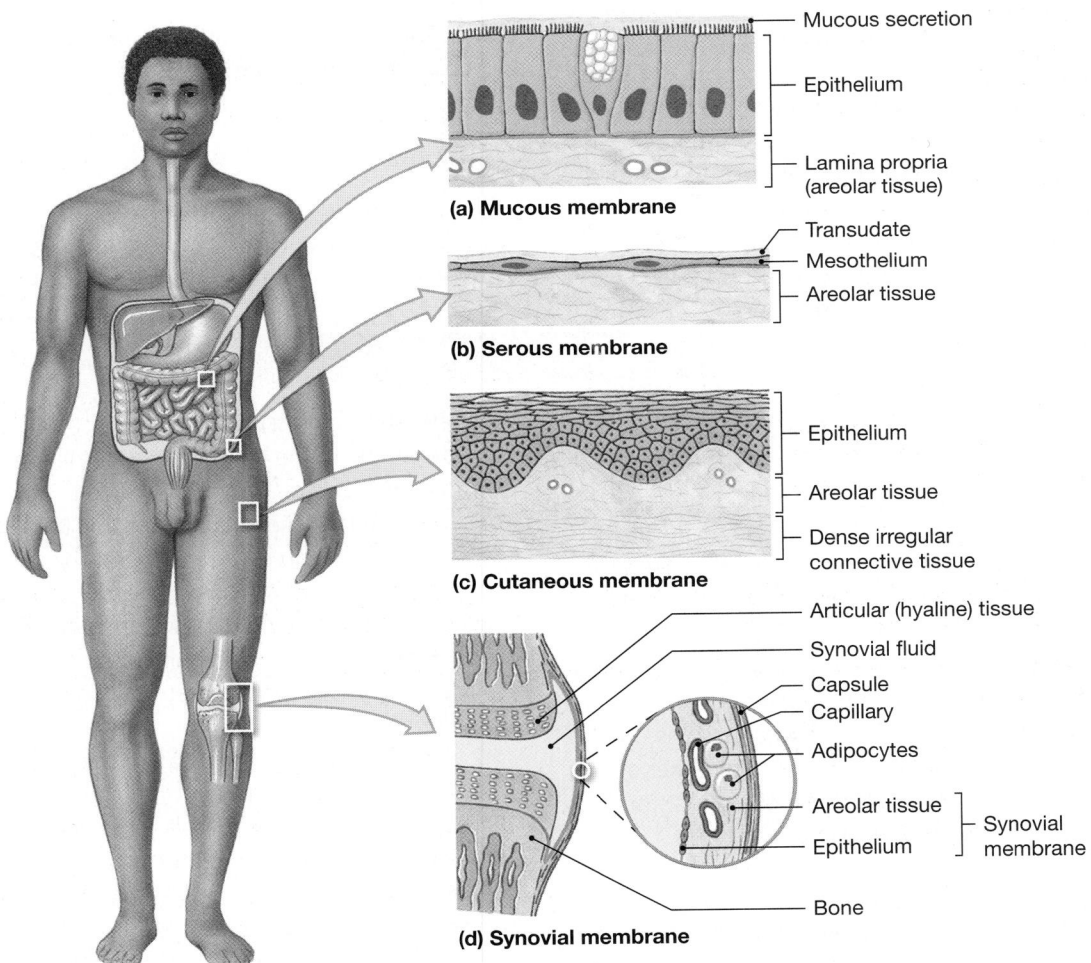

(a) Mucous membrane
Mucous secretion
Epithelium
Lamina propria (areolar tissue)

(b) Serous membrane
Transudate
Mesothelium
Areolar tissue

(c) Cutaneous membrane
Epithelium
Areolar tissue
Dense irregular connective tissue

(d) Synovial membrane
Articular (hyaline) tissue
Synovial fluid
Capsule
Capillary
Adipocytes
Areolar tissue
Epithelium
} Synovial membrane
Bone

Figure 4–16 **Membranes. (a)** Mucous membranes are coated with the secretions of mucous glands. These membranes line the digestive, respiratory, urinary, and reproductive tracts. **(b)** Serous membranes line the ventral body cavities (the peritoneal, pleural, and pericardial cavities). **(c)** The cutaneous membrane, or skin, covers the outer surface of the body. **(d)** Synovial membranes line joint cavities and produce the fluid within the joint.

cells or multicellular glands, or by exposure to fluids such as urine or semen. The areolar tissue component of a mucous membrane is called the **lamina propria** (PRŌ-prē-uh). We will consider the organization of specific mucous membranes in greater detail in later chapters.

Many mucous membranes are lined by simple epithelia that perform absorptive or secretory functions, such as the simple columnar epithelium of the digestive tract. However, other types of epithelia may be involved. For example, a stratified squamous epithelium is part of the mucous membrane of the mouth, and the mucous membrane along most of the urinary tract has a transitional epithelium.

Serous Membranes

Serous membranes line the sealed, internal subdivisions of the ventral body cavity—cavities that are not open to the exterior. These membranes consist of a mesothelium supported by areolar tissue (**Figure 4–16b**). As you may recall from Chapter 1, the three types of serous membranes are (1) the *pleura*, which lines the pleural cavities and covers the lungs; (2) the *peritoneum*, which lines the peritoneal cavity and covers the surfaces of the enclosed organs; and (3) the *pericardium*, which lines the pericardial cavity and covers the heart. ∞ p. 24 Serous membranes are very thin, but they are firmly attached to the body wall and to the organs they cover. When looking at an organ such as the heart or stomach, you

are really seeing the tissues of the organ through a transparent serous membrane.

Each serous membrane can be divided into a *parietal portion*, which lines the inner surface of the cavity, and an opposing *visceral portion*, or **serosa**, which covers the outer surfaces of visceral organs. These organs often move or change shape as they perform their various functions, and the parietal and visceral surfaces of a serous membrane are in close contact at all times. Thus, the primary function of any serous membrane is to minimize friction between the opposing parietal and visceral surfaces. Friction is kept to a minimum because mesothelia are very thin and permeable; tissue fluids continuously diffuse onto the exposed surface, keeping it moist and slippery.

The fluid formed on the surfaces of a serous membrane is called a *transudate* (TRAN-sū-dāt; *trans-*, across). In healthy individuals, the total volume of transudate is extremely small—just enough to prevent friction between the walls of the cavities and the surfaces of internal organs. However, after an injury or in certain disease states, the volume of transudate may increase dramatically, complicating existing medical problems or producing new ones.

The Cutaneous Membrane

The **cutaneous membrane**, or skin, covers the surface of the body. It consists of a stratified squamous epithelium and a layer of areolar tissue reinforced by underlying dense

---

CLINICAL NOTE

Problems with Serous Membranes

Several clinical conditions, including infection and chronic irritation, can cause the abnormal buildup of fluid in a ventral body cavity.

Pleuritis, or *pleurisy*, is an inflammation of the pleural cavities. At first, the membranes become rough, and the opposing membranes may scratch against one another, producing pain and a sound known as a *pleural rub*. In general, friction between opposing layers of serous membranes may promote the formation of adhesions—fibrous connections that lock the membranes together and eliminate the friction. Adhesions also severely restrict the movement of the affected organ or organs and may compress blood vessels or nerves. However, adhesions seldom form between the serous membranes of the pleural cavities. More commonly, continued inflammation and rubbing lead to a gradual increase in fluid production to levels well above normal. Fluid then accumulates in the pleural cavities, producing a condition known as *pleural effusion*. Pleural effusion is also caused by heart conditions

that elevate the pressure in blood vessels of the lungs. As fluids build up in the pleural cavities, the lungs are compressed, making breathing difficult. The combination of severe pleural effusion and heart disease can be lethal.

Pericarditis is an inflammation of the pericardium. This condition may lead to pericardial effusion, an abnormal accumulation of the fluid in the pericardial cavity. When sudden or severe, the fluid buildup can seriously reduce the efficiency of the heart and restrict blood flow through major vessels.

Peritonitis, an inflammation of the peritoneum, can follow infection of, or injury to, the peritoneal lining. Peritonitis is a potential complication of any surgical procedure in which the peritoneal cavity is opened. Liver disease, kidney disease, or heart failure can cause an increase in the rate of fluid movement through the peritoneal lining. *Ascites* (a-SĪ-tēz), the accumulation of peritoneal fluid, creates a characteristic abdominal swelling. The distortion of internal organs by the contained fluid can lead to symptoms such as heartburn, indigestion, and low-back pain.

irregular connective tissue (**Figure 4–16c**). In contrast to serous and mucous membranes, the cutaneous membrane is thick, relatively waterproof, and usually dry. We will examine the cutaneous membrane further in Chapter 5.

Synovial Membranes

Adjacent bones often interact at joints, or *articulations*. At an articulation, the two articulating bones are very close together if not in contact. Joints that permit significant amounts of movement are complex structures. Such a joint is surrounded by a fibrous capsule, and the ends of the articulating bones lie within a *joint cavity* filled with **synovial** (si-NŌ-vē-ul) **fluid**. (**Figure 4–16d**). The synovial fluid is produced by a **synovial membrane**, which lines the joint cavity. A synovial membrane consists of an extensive area of areolar tissue containing a matrix of interwoven collagen fibers, proteoglycans, and glycoproteins. An incomplete layer of macrophages and specialized fibroblasts separates the areolar tissue from the joint cavity. These cells regulate the composition of the synovial fluid. Although this lining is often called an epithelium, it differs from true epithelia in four respects: (1) It develops within a connective tissue, (2) no basal lamina is present, (3) gaps of up to 1 mm may separate adjacent cells, and (4) fluid and solutes are continuously exchanged between the synovial fluid and capillaries in the underlying connective tissue.

Even though the adjacent ends of the bones are covered by a smooth layer of articular cartilage, the surfaces must be lubricated to keep friction from damaging the opposing surfaces. The necessary lubrication is provided by the synovial fluid, which is similar in composition to the ground substance in loose connective tissues. Synovial fluid circulates from the areolar tissue into the joint cavity and percolates through the articular cartilages, providing oxygen and nutrients to the chondrocytes. Joint movement is important in stimulating the formation and circulation of synovial fluid: If a synovial joint is immobilized for long periods, the articular cartilages and the synovial membrane undergo degenerative changes.

> **CHECKPOINT**
>
> 25. Identify the four types of membranes found in the body.
>
> 26. Which cavities in the body are lined by serous membranes?
>
> 27. The lining of the nasal cavity is normally moist, contains numerous mucous cells, and rests on a layer of connective tissue called the lamina propria. Which type of membrane is this?
>
> See the blue Answers tab at the end of the book.

4-7 Connective tissues create the internal framework of the body

Connective tissues provide the internal structure of the body. Layers of connective tissue surround and support the organs within the subdivisions of the ventral body cavity and connect them to the rest of the body. These layers (1) provide strength and stability, (2) maintain the relative positions of internal organs, and (3) provide a route for the distribution of blood vessels, lymphatic vessels, and nerves. **Fasciae** (FASH-ē-ē; singular, *fascia*) are connective tissue layers and wrappings that support and surround organs. We can divide the fasciae into three types of layers: the superficial fascia, the deep fascia, and the subserous fascia (**Figure 4–17**).

1. The **superficial fascia** (FASH-ē-uh), or **subcutaneous layer** (*sub-*, below + *cutis*, skin), is also termed the *hypodermis* (*hypo*, below + *dermis*, skin). This layer of areolar tissue and fat separates the skin from underlying tissues and organs, provides insulation and padding, and lets the skin and underlying structures move independently.

2. The **deep fascia** consists of dense irregular connective tissue. The organization of the fibers resembles that of plywood: In each layer all the fibers run in the same direction, but the orientation of the fibers changes from layer to layer. This arrangement helps the tissue resist forces applied from many directions. The tough capsules that surround most organs, including the kidneys and the organs in the thoracic and peritoneal cavities, are bound to the deep fascia. The perichondrium around cartilages, the periosteum around bones and the ligaments that interconnect them, and the connective tissues of muscle (including tendons) are also connected to the deep fascia. The dense connective tissue components are interwoven. For example, the deep fascia around a muscle blends into the tendon, whose fibers intermingle with those of the periosteum. This arrangement creates a strong, fibrous network and ties structural elements together.

3. The **subserous fascia** is a layer of areolar tissue that lies between the deep fascia and the serous membranes that line body cavities. Because this layer separates the serous membranes from the deep fascia, movements of muscles or muscular organs do not severely distort the delicate lining.

> **CHECKPOINT**
>
> 28. A sheet of tissue has many layers of collagen fibers that run in different directions in successive layers. Which type of tissue is this?
>
> See the blue Answers tab at the end of the book.

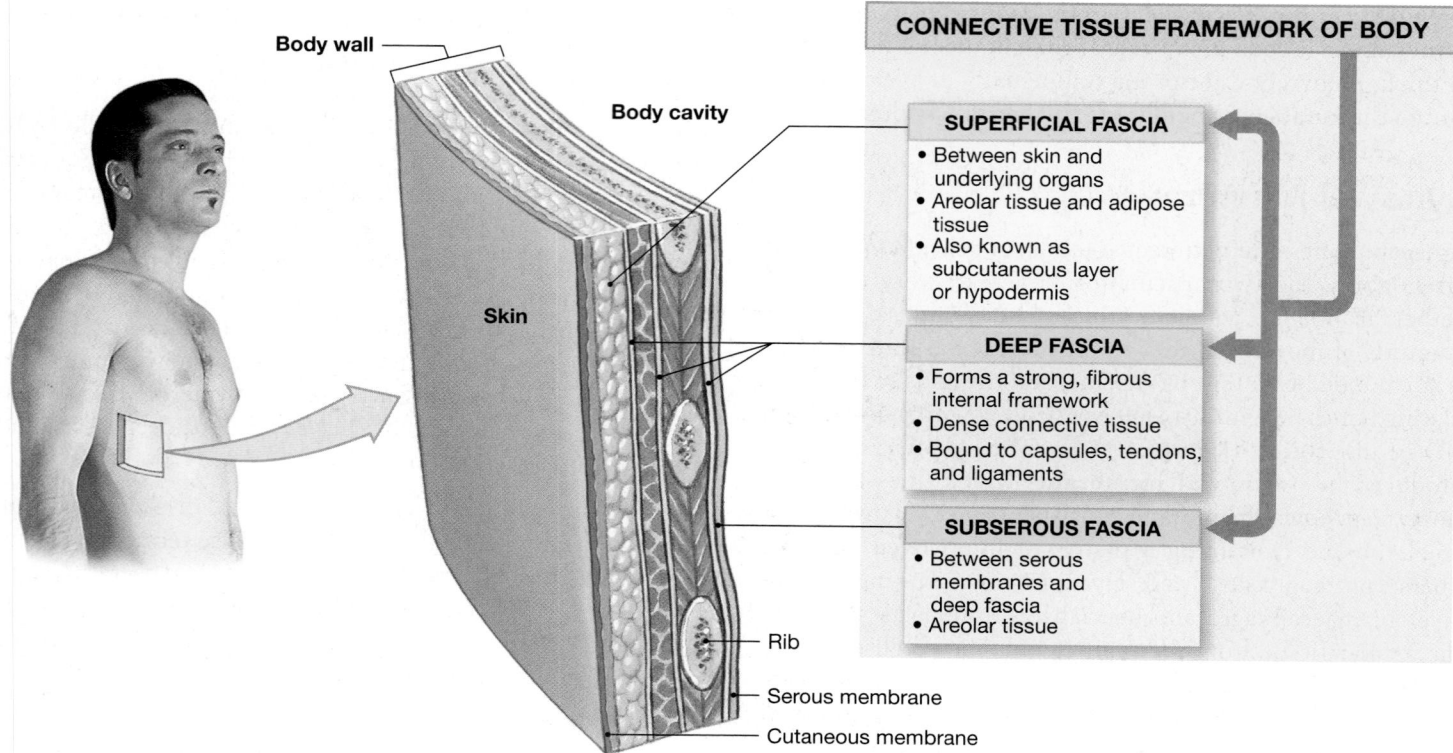

Figure 4–17 **The Fasciae.** The relationships among the connective tissue elements in the body.

4-8 The three types of muscle tissue are skeletal, cardiac, and smooth

Epithelia cover surfaces and line passageways; connective tissues support weight and interconnect parts of the body. Together, these tissues provide a strong, interwoven framework within which the organs of the body can function. Several vital functions involve movement of one kind or another—movement of materials along the digestive tract, movement of blood around the cardiovascular system, or movement of the body from one place to another. Movement is produced by **muscle tissue**, which is specialized for contraction. Muscle cells possess organelles and properties distinct from those of other cells.

There are three types of muscle tissue: (1) *skeletal muscle*, which forms the large muscles responsible for gross body movements and locomotion; (2) *cardiac muscle*, found only in the heart and responsible for the circulation of blood; and (3) *smooth muscle*, found in the walls of visceral organs and a variety of other locations, where it provides elasticity, contractility, and support. The contraction mechanism is similar in all three types of muscle tissue, but the muscle cells differ in internal organization. We will examine only general characteristics at this point, because each type of muscle is described more fully in Chapter 10.

Skeletal Muscle Tissue

Skeletal muscle tissue contains very large muscle cells—up to 0.3 m (1 ft) or more in length. Because the individual muscle cells are relatively long and slender, they are usually called **muscle fibers**. Each muscle fiber is described as *multinucleate*, because it has several hundred nuclei distributed just inside the plasma membrane (**Figure 4–18a**). Skeletal muscle fibers are incapable of dividing, but new muscle fibers are produced through the divisions of **myosatellite cells** (*satellite cells*), stem cells that persist in adult skeletal muscle tissue. As a result, skeletal muscle tissue can at least partially repair itself after an injury.

As noted in Chapter 3, the cytoskeleton contains actin and myosin filaments. ∞ p. 73 In skeletal muscle fibers, however, these filaments are organized into repeating groups that give the cells a *striated*, or banded, appearance. The *striations*, or bands, are readily apparent in light micrographs. Skeletal muscle fibers do not usually contract unless stimulated by nerves, and the nervous system provides voluntary control over their activities. Thus, skeletal muscle is called **striated voluntary muscle.**

Tips & Tricks

Associate the sound of the word **stri**ated with the sound of the word **stri**ped.

SKELETAL MUSCLE TISSUE

Cells are long, cylindrical, striated, and multinucleate.

LOCATIONS: Combined with connective tissues and neural tissue in skeletal muscles

FUNCTIONS: Moves or stabilizes the position of the skeleton; guards entrances and exits to the digestive, respiratory, and urinary tracts; generates heat; protects internal organs

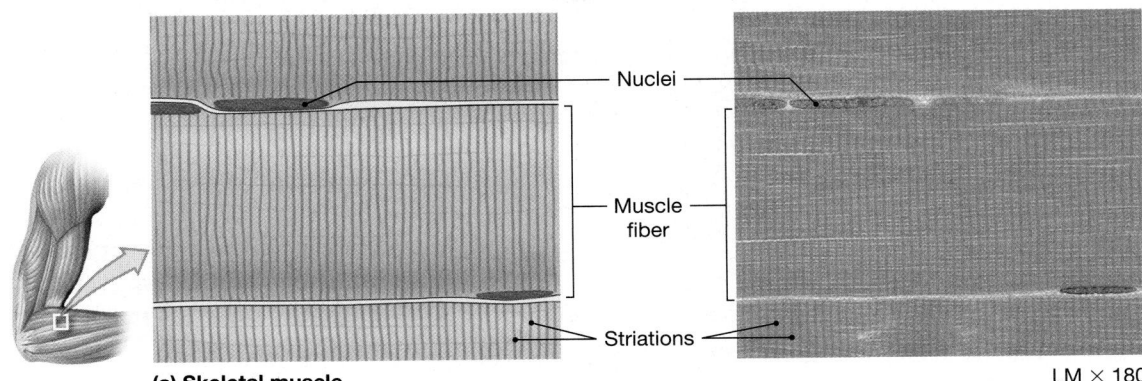

(a) Skeletal muscle

Nuclei

Muscle fiber

Striations

LM × 180

CARDIAC MUSCLE TISSUE

Cells are short, branched, and striated, usually with a single nucleus; cells are interconnected by intercalated discs.

LOCATION: Heart

FUNCTIONS: Circulates blood; maintains blood (hydrostatic) pressure

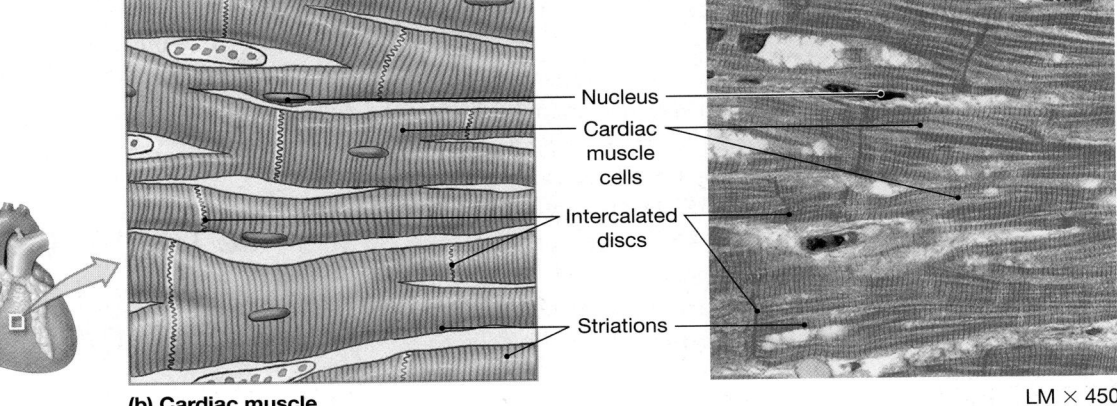

(b) Cardiac muscle

Nucleus

Cardiac muscle cells

Intercalated discs

Striations

LM × 450

SMOOTH MUSCLE TISSUE

Cells are short, spindle-shaped, and nonstriated, with a single, central nucleus

LOCATIONS: Found in the walls of blood vessels and in digestive, respiratory, urinary, and reproductive organs

FUNCTIONS: Moves food, urine, and reproductive tract secretions; controls diameter of respiratory passageways; regulates diameter of blood vessels

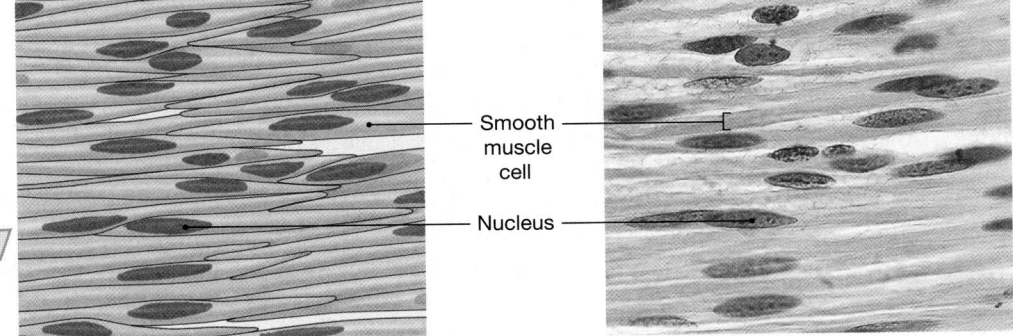

(c) Smooth muscle

Smooth muscle cell

Nucleus

LM × 235

Figure 4–18 **Muscle Tissue. (a)** Skeletal muscle fibers are large unbranched cells with multiple, peripherally located nuclei and prominent striations (banding). **(b)** Cardiac muscle cells differ from skeletal muscle fibers in three major ways: They are smaller, they branch, and they have one centrally placed nucleus. The striations of both skeletal and cardiac muscle cells result from an organized array of actin and myosin filaments. **(c)** Smooth muscle cells are small and spindle shaped, with a central nucleus. They lack branches and striations.

A *skeletal muscle* is an organ of the muscular system, and although muscle tissue predominates, it contains all four types of body tissue. Within a skeletal muscle, adjacent skeletal muscle fibers are tied together by collagen and elastic fibers that blend into the attached tendon or aponeurosis. The tendon or aponeurosis conducts the force of contraction, often to a bone of the skeleton. Thus, when the muscles contract, they pull on the attached bone, producing movement.

Cardiac Muscle Tissue

Cardiac muscle tissue is located only in the heart. A typical cardiac muscle cell, also known as a **cardiocyte**, is smaller than a skeletal muscle cell (**Figure 4–18b**). A typical cardiac muscle cell has one centrally positioned nucleus, but some cardiocytes have as many as five. Prominent striations resemble those of skeletal muscle; the actin and myosin filaments are arranged the same way in both cell types.

Cardiac muscle cells form extensive connections with one another. As a result, cardiac muscle tissue consists of a branching network of interconnected muscle cells. The connections occur at specialized regions known as **intercalated discs**. At an intercalated disc, the membranes are locked together by macula adherens, intercellular cement, and gap junctions. Ion movement through gap junctions helps synchronize the contractions of the cardiac muscle cells, and the macula adherens and intercellular cement lock the cells together during a contraction. Cardiac muscle tissue has a very limited ability to repair itself. Although some cardiac muscle cells do divide after an injury to the heart, the repairs are incomplete and some heart function is usually lost.

Cardiac muscle cells do not rely on nerve activity to start a contraction. Instead, specialized cardiac muscle cells called *pacemaker cells* establish a regular rate of contraction. Although the nervous system can alter the rate of pacemaker cell activity, it does not provide voluntary control over individual cardiac muscle cells. Therefore, cardiac muscle is called **striated involuntary muscle**.

Smooth Muscle Tissue

Smooth muscle tissue is located in the walls of blood vessels, around hollow organs such as the urinary bladder, and in layers around the respiratory, circulatory, digestive, and reproductive tracts. A smooth muscle cell is a small, spindle-shaped cell with tapering ends and a single, oval nucleus (**Figure 4–18c**). Smooth muscle cells can divide; hence, smooth muscle tissue can regenerate after an injury.

The actin and myosin filaments in smooth muscle cells are organized differently from those of skeletal and cardiac muscles. One result of this difference is that smooth muscle tissue has no striations. Smooth muscle cells may contract on their own, with gap junctions between adjacent cells coordinating the contractions of individual cells. The contraction of some smooth muscle tissue can be controlled by the nervous system, but contractile activity is not under voluntary control. (Imagine the degree of effort that would be required to exert conscious control over the smooth muscles along the 8 m [26 ft.] of digestive tract, not to mention the miles of blood vessels!) Because the nervous system usually does not provide voluntary control over smooth muscle contractions, smooth muscle is known as **nonstriated involuntary muscle**.

29. Identify the three types of muscle tissue in the body.
30. Which type of muscle tissue has small, tapering cells with single nuclei and no obvious striations?
31. If skeletal muscle cells in adults are incapable of dividing, how is skeletal muscle repaired?

See the blue Answers tab at the end of the book.

4-9 Neural tissue responds to stimuli and conducts electrical impulses throughout the body

Neural tissue, which is also known as *nervous tissue* or *nerve tissue*, is specialized for the conduction of electrical impulses from one region of the body to another. Ninety-eight percent of the neural tissue in the body is concentrated in the brain and spinal cord, which are the control centers of the nervous system.

Neural tissue contains two basic types of cells: (1) **neurons** (NOOR-onz; *neuro*, nerve) and (2) several kinds of supporting cells, collectively called **neuroglia** (noo-ROG-lē-uh or noo-rō-GLĒ-uh), or *glial cells* (*glia*, glue). Our conscious and unconscious thought processes reflect the communication among neurons in the brain. Such communication involves the propagation of electrical impulses, in the form of changes in the transmembrane potential. Information is conveyed both by the frequency and by the pattern of the impulses. Neuroglia support and repair neural tissue and supply nutrients to neurons.

The longest cells in your body are neurons, many of which are as much as a meter (39 in.) long. Most neurons cannot divide under normal circumstances, so they have a very limited ability to repair themselves after injury. A typical neuron has a large **cell body** with a large nucleus and a prominent nucleolus (**Figure 4–19**). Extending from the cell

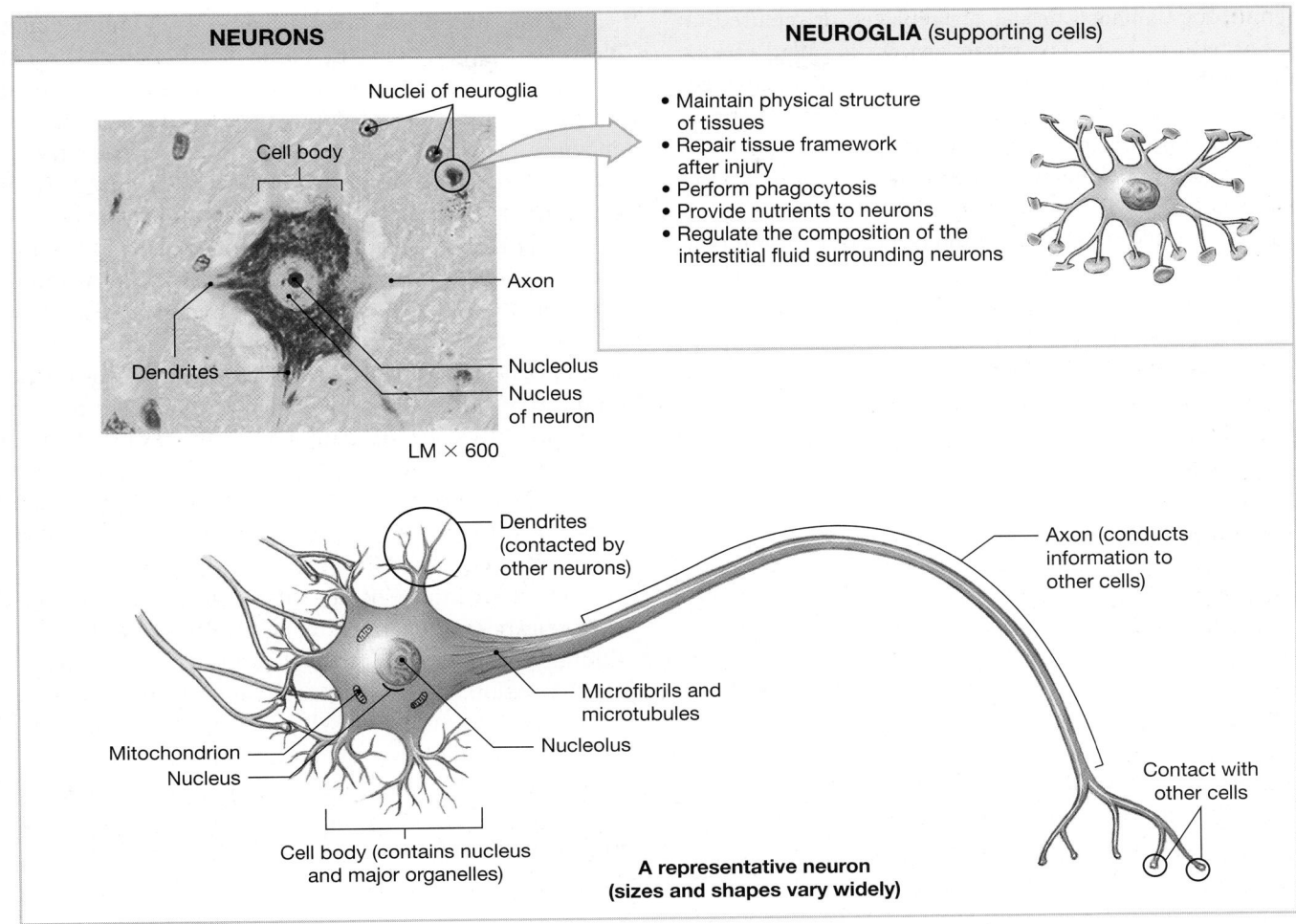

NEURONS	NEUROGLIA (supporting cells)

Nuclei of neuroglia

Cell body

Axon

Dendrites

Nucleolus

Nucleus of neuron

LM × 600

- Maintain physical structure of tissues
- Repair tissue framework after injury
- Perform phagocytosis
- Provide nutrients to neurons
- Regulate the composition of the interstitial fluid surrounding neurons

Dendrites (contacted by other neurons)

Axon (conducts information to other cells)

Microfibrils and microtubules

Nucleolus

Mitochondrion

Nucleus

Cell body (contains nucleus and major organelles)

Contact with other cells

**A representative neuron
(sizes and shapes vary widely)**

Figure 4–19   Neural Tissue.

body are many branching processes (projections or outgrowths) termed **dendrites** (DEN-drīts; *dendron*, a tree), and one **axon**. The dendrites receive information, typically from other neurons, and the axon conducts that information to other cells. Because axons tend to be very long and slender, they are also called **nerve fibers**. In Chapter 12 we will further examine the properties of neural tissue.

Tips&Tricks

To remember the direction of electrical impulses in a neuron, associate the "t" in dendri**t**e with "**t**o" and the "a" in **a**xon with "**a**way."

CHECKPOINT

32. **A tissue contains irregularly-shaped cells with many fibrous projections, some several centimeters long. These are probably which type of cell?**

See the blue Answers tab at the end of the book.

4-10 The response to tissue injury involves inflammation and regeneration

Tissues are not isolated; they combine to form organs with diverse functions. Therefore, any injury affects several types of tissue simultaneously. These tissues must respond in a coordinated way to preserve homeostasis. The restoration of homeostasis after a tissue has been injured involves two related processes: inflammation and regeneration.

Immediately after an injury, the area is isolated while damaged cells, tissue components, and any dangerous microorganisms are cleaned up. This process, which coordinates the activities of several types of tissue, is called **inflammation**, or the **inflammatory response**. It produces several familiar signs and symptoms of injury, including swelling, redness, warmth, and pain. Inflammation may also result from the presence of pathogens, such as harmful bacteria, within the tissues; the presence of these pathogens constitutes an **infection**.

Second, the damaged tissues are replaced or repaired to restore normal function. The repair process is called **regeneration**. Inflammation and regeneration are controlled at the tissue level. The two processes overlap; isolation establishes a framework that guides the cells responsible for reconstruction, and repairs are under way well before cleanup operations have ended.

Next we consider the basics of the repair process after an injury. At this time, we will focus on the interaction among different tissues. Our example includes two connective tissues (areolar tissue and blood), an epithelium (the endothelia of blood vessels), a muscle tissue (smooth muscle in the vessel walls), and neural tissue (sensory nerve endings). In later chapters, especially Chapters 5 and 22, we will examine inflammation and regeneration in more detail.

Inflammation

Many stimuli—including impact, abrasion, distortion, chemical irritation, infection by pathogenic organisms (such as bacteria or viruses), and extreme temperatures (hot or cold)—can produce inflammation. Each of these stimuli kills cells, damages fibers, or injures the tissue in some other way. Such changes alter the chemical composition of the interstitial fluid: Damaged cells release prostaglandins, proteins, and potassium ions, and the injury itself may have introduced foreign proteins or pathogens into the body.

Tissue conditions soon become even more abnormal. **Necrosis** (ne-KRŌ-sis), the tissue destruction that occurs after cells have been damaged or killed, begins several hours after the original injury. The damage is caused by lysosomal enzymes. Through widespread autolysis, lysosomes release enzymes that first destroy the injured cells and then attack surrounding tissues. ∞ p. 79 The result may be an accumulation of debris, fluid, dead and dying cells, and necrotic tissue components collectively known as **pus**. An accumulation of pus in an enclosed tissue space is an **abscess**.

These tissue changes trigger the inflammatory response by stimulating mast cells—connective tissue cells introduced on p. 126. **Figure 4–20** depicts the events set in motion by the activation of mast cells. Although in this example the injury has occurred in areolar tissue, the process would be basically the same after an injury to any connective tissue proper. Because all organs have connective tissues, inflammation can occur anywhere in the body.

When an injury occurs that damages fibers and cells, mast cells release a variety of chemicals. These chemicals, including histamine and prostaglandins, trigger changes in local circulation. In response, the smooth muscle tissue that surrounds local blood vessels relaxes, and the vessels *dilate*, or enlarge in diameter. This dilation increases blood flow through the tissue, turning the region red and making it warm to the touch. The combination of abnormal tissue conditions and chemicals released by mast cells stimulates sensory nerve

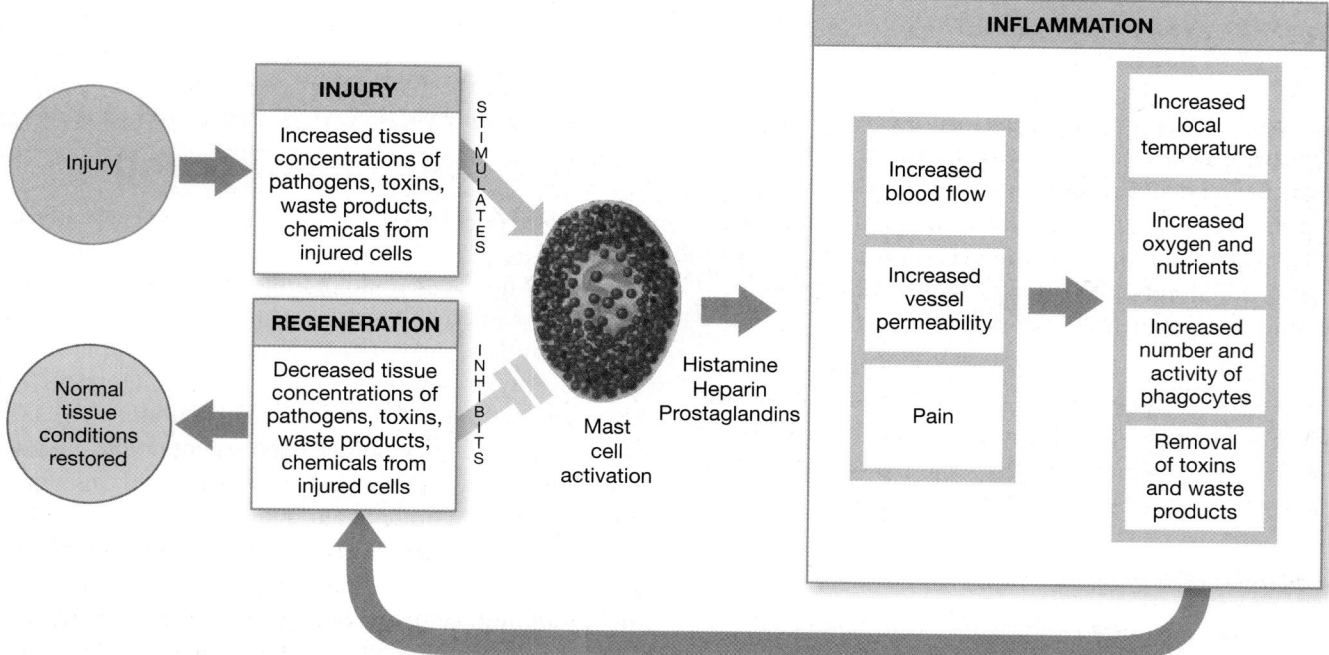

Figure 4–20   An Introduction to Inflammation. Inflammation is triggered by mast cell activation in response to tissue injury. It creates the conditions under which the regeneration of tissues can occur.

endings that produce sensations of pain. At the same time, the chemicals released by mast cells make the endothelial cells of local capillaries more permeable. Plasma, including blood proteins, now diffuses into the injured tissue, so the area becomes swollen.

The increased blood flow accelerates the delivery of nutrients and oxygen and the removal of dissolved waste products and toxic chemicals. It also brings white blood cells to the region. These phagocytic cells migrate to the site of the injury and assist in defense and cleanup operations. Macrophages and microphages protect the tissue from infection and perform cleanup by engulfing both debris and bacteria. Over a period of hours to days, the cleanup process generally succeeds in eliminating the inflammatory stimulus.

Regeneration

By the time the inflammation phase is over, the situation is under control and no further damage will occur. However, many cells in the area either have already died or will soon die as a consequence of the original injury. As tissue conditions return to normal, fibrocytes move into the necrotic area, laying down a network of collagen fibers that stabilizes the injury site. This process produces a dense, collagenous framework known as *scar tissue* or *fibrous tissue*. Over time, scar tissue is usually "remodeled" and gradually assumes a more normal appearance. The cell population in the area gradually increases; some cells migrate to the site, and others are produced by the division of mesenchymal stem cells.

Each organ has a different ability to regenerate after injury—an ability that can be directly linked to the pattern of tissue organization in the injured organ. Epithelia, connective tissues (except cartilage), and smooth muscle tissue usually regenerate well, whereas other muscle tissues and neural tissue regenerate relatively poorly if at all. The skin, which is dominated by epithelia and connective tissues, regenerates rapidly and completely after injury. (We will consider the process in Chapter 5.) In contrast, damage to the heart is much more serious. Although the connective tissues of the heart can be repaired, the majority of damaged cardiac muscle cells are replaced only by fibrous tissue. The permanent replacement of normal tissue by fibrous tissue is called *fibrosis* (fī-BRŌ-sis). Fibrosis in muscle and other tissues may occur in response to injury, disease, or aging.

CHECKPOINT

33. **Identify the two phases in the response to tissue injury.**

See the blue Answers tab at the end of the book.

4-11 With advancing age, tissue repair declines and cancer rates increase

In this section we briefly consider two important effects of aging on tissues: the body's decreased ability to repair damage to tissues, and an increase in the occurrence of cancer.

Aging and Tissue Structure

Tissues change with age, and the speed and effectiveness of tissue repairs decrease. Repair and maintenance activities throughout the body slow down; the rate of energy consumption in general declines. All these changes reflect various hormonal alterations occurring with age, often coupled with a reduction in physical activity and the adoption of a more sedentary lifestyle. These factors combine to alter the structure and chemical composition of many tissues. Epithelia get thinner and connective tissues more fragile. Individuals bruise easily and bones become brittle; joint pain and broken bones are common in the elderly. Because cardiac muscle cells and neurons are not normally replaced, cumulative damage can eventually cause major health problems, such as cardiovascular disease or a deterioration in mental functioning.

In later chapters, we will consider the effects of aging on specific organs and systems. Some of these effects are genetically programmed. For example, the chondrocytes of older individuals produce a slightly different form of proteoglycan than do those of younger people. This difference probably accounts for the thinner and less resilient cartilage of older people. In some cases, the tissue degeneration can be temporarily slowed or even reversed. Age-related reduction in bone strength, a condition called *osteoporosis*, typically results from a combination of inactivity, low dietary calcium levels, and a reduction in circulating sex hormones. A program of exercise that includes weight-bearing activity, calcium supplements, and medication can generally maintain healthy bone structure for many years.

Aging and Cancer Incidence

Cancer rates increase with age, and roughly 25 percent of all people in the United States develop cancer at some point in their lives. It has been estimated that 70–80 percent of cancer cases result from chemical exposure, environmental factors, or some combination of the two, and 40 percent of those cancers are caused by cigarette smoke. Each year in the United States, more than 500,000 individuals die of cancer, making it second only to heart disease as a cause of death. We discussed the development and growth of cancer in Chapter 3. ∞ p. 105

This chapter concludes the introductory portion of this text. In combination, the four basic tissue types described here form all of the organs and systems discussed in subsequent chapters. The *Systems Overview* that follows this chapter will help you make the transition from atoms, molecules, cells, and tissues to organ systems. One of the most important themes in this text is that organ systems interact continuously—they do not function in isolation. Thus, to understand specifics about one system, you need to know something about all of the others. The *Systems Overview* section provides a general orientation in more detail than was possible in Chapter 1. You will find this section useful as a reference throughout the remainder of the text.

CHECKPOINT

34. **Identify some age-related factors that affect tissue repair and structure.**

See the blue Answers tab at the end of the book.

Related Clinical Terms

abscess: The accumulation of pus within an enclosed tissue space.

adhesions: Restrictive fibrous connections that can result from surgery, infection, or other injuries to serous membranes.

anaplasia: An irreversible change in the size and shape of tissue cells.

antiangiogenesis factor: A secretion, produced by chondrocytes, that inhibits the growth of blood vessels.

ascites: The accumulation of fluid in the peritoneal cavity, usually caused by liver or kidney disease, malnutrition, or heart failure.

dysplasia: A reversible change in the normal shape, size, and organization of tissue cells.

exfoliative cytology: The study of cells shed or collected from epithelial surfaces.

liposuction: A surgical procedure to remove unwanted adipose tissue by sucking it out through a tube.

metaplasia: A reversible structural change that alters the character of a tissue.

necrosis: Tissue destruction that occurs after cells have been injured or destroyed; a result of the release of lysosomal enzymes through autolysis.

pathologists: Physicians who specialize in the study of disease processes in tissues and body fluids.

pericarditis: An inflammation of the pericardial lining that may lead to the accumulation of pericardial fluid (a *pericardial effusion*).

peritonitis: An inflammation of the peritoneum after infection or injury.

pleural effusion: The accumulation of fluid within the pleural cavities as a result of chronic infection or inflammation of the pleura.

pleuritis (*pleurisy*): An inflammation of the pleural cavities. This condition may cause the production of a sound known as a *pleural rub*.

regeneration: The repairing of injured tissues that follows injury and inflammation.

Chapter Review
Study Outline

An Introduction to the Tissue Level of Organization p. 112

1. Tissues are structures with discrete structural and functional properties that combine to form organs.

4-1 The four tissue types are epithelial, connective, muscle, and neural p. 112

2. **Tissues** are collections of specialized cells and cell products that perform a relatively limited number of functions. The four *tissue types* are *epithelial tissue, connective tissue, muscle tissue,* and *neural tissue.*

3. **Histology** is the study of tissues.

4-2 Epithelial tissue covers body surfaces, lines cavities and tubular structures, and serves essential functions p. 112

4. **Epithelial tissue** includes epithelia and glands. An **epithelium** is an **avascular** layer of cells that forms a barrier that provides protection and regulates permeability. **Glands** are secretory structures derived from epithelia. Epithelial cells may show **polarity**, an uneven distribution of cytoplasmic components.

5. **A basal lamina** attaches epithelia to underlying connective tissues.

6. Epithelia provide physical protection, control permeability, provide sensation, and produce specialized secretions. Gland cells are epithelial cells that produce secretions. In **glandular epithelia**, most cells produce secretions.

7. Epithelial cells are specialized to perform secretory or transport functions and to maintain the physical integrity of the epithelium. (*Figure 4–1*)

8. Many epithelial cells have microvilli.

9. The coordinated beating of the cilia on a **ciliated epithelium** moves materials across the epithelial surface.

10. Cells can attach to other cells or to extracellular protein fibers by means of **cell adhesion molecules (CAMs)** or at specialized attachment sites called **cell junctions**. The three major types of cell junctions are **occluding junctions (tight junctions)**, **gap junctions**, and **macula adherens (desmosomes)**. (*Figure 4–2*)

11. The inner surface of each epithelium is connected to a two-part basal lamina consisting of a *clear layer* (*lamina lucida*) and a *dense layer* (*lamina densa*). Divisions by **germinative cells** continually replace the short-lived epithelial cells.

4-3 Cell shape and number of layers determine the classification of epithelia p. 116

12. Epithelia are classified on the basis of the number of cell layers and the shape of the cells at the apical surface.

13. A **simple epithelium** has a single layer of cells covering the basal lamina; a **stratified epithelium** has several layers. The cells in a **squamous epithelium** are thin and flat. Cells in a **cuboidal epithelium** resemble hexagonal boxes; those in a **columnar epithelium** are taller and more slender. (*Table 4–1; Figures 4–3 to 4–5*)

14. Epithelial cells (or structures derived from epithelial cells) that produce secretions are called *glands*. **Exocrine glands** discharge secretions onto the body surface or into **ducts**, which communicate with the exterior. *Hormones*, the secretions of **endocrine glands**, are released by gland cells into the surrounding interstitial fluid.

15. A glandular epithelial cell may release its secretions by merocrine, apocrine, or holocrine modes. In **merocrine secretion**, the most common mode, the product is released through exocytosis. **Apocrine secretion** involves the loss of both the secretory product and cytoplasm. Unlike the other two methods, **holocrine secretion** destroys the gland cell, which becomes packed with secretions and then bursts. (*Figure 4–6*)

16. In epithelia that contain scattered gland cells, individual secretory cells are called **unicellular glands**. **Multicellular glands** are organs that contain glandular epithelia that produce exocrine or endocrine secretions.

17. Exocrine glands can be classified on the basis of structure as **unicellular exocrine glands (mucous cells)** or as **multicellular exocrine glands**. Multicellular exocrine glands can be further classified according to structure. (*Figure 4–7*)

4-4 Connective tissue provides a protective structural framework for other tissue types p. 124

18. **Connective tissues** are internal tissues with many important functions: establishing a structural framework; transporting fluids and dissolved materials; protecting delicate organs; supporting, surrounding, and interconnecting tissues; storing energy reserves; and defending the body from microorganisms.

19. All connective tissues contain specialized cells and a **matrix**, composed of extracellular protein fibers and a **ground substance**.

20. **Connective tissue proper** is connective tissue that contains varied cell populations and fiber types surrounded by a syrupy ground substance. (*Figure 4–8*)

21. **Fluid connective tissues** have distinctive populations of cells suspended in a watery matrix that contains dissolved proteins. The two types of fluid connective tissues are *blood* and *lymph*.

22. **Supporting connective tissues** have a less diverse cell population than connective tissue proper and a dense matrix with closely packed fibers. The two types of supporting connective tissues are *cartilage* and *bone*.

23. Connective tissue proper contains fibers, a viscous ground substance, and a varied population of cells, including **fibroblasts, fibrocytes, macrophages, adipocytes, mesenchymal cells, melanocytes, mast cells, lymphocytes,** and **microphages**.

24. The three types of fibers in connective tissue are **collagen fibers, reticular fibers**, and **elastic fibers**.

25. The first connective tissue to appear in an embryo is **mesenchyme**, or *embryonic connective tissue*.

26. Connective tissue proper is classified as either **loose connective tissue** or **dense connective tissue**. Loose connective tissues are mesenchyme and **mucous connective tissues** in the embryo; *areolar tissue*; *adipose tissue*, including **white fat** and **brown fat**; and **reticular tissue**. Most of the volume in dense connective tissue consists of fibers. The two types of dense connective tissue are **dense regular connective tissue** and **dense irregular connective tissue** in the adult. (*Figures 4–9 to 4–11*)

27. **Blood** and **lymph** are connective tissues that contain distinctive collections of cells in a fluid matrix.

28. Blood contains *formed elements*: **red blood cells** (*erythrocytes*), **white blood cells** (*leukocytes*), and **platelets**. The watery matrix of blood is called **plasma**. (*Figure 4–12*)

29. **Arteries** carry blood away from the heart and toward **capillaries**, where water and small solutes move into the interstitial fluid of surrounding tissues. **Veins** return blood to the heart.

30. Lymph forms as interstitial fluid enters the **lymphatic vessels**, which return lymph to the cardiovascular system.

4-5 Cartilage and bone provide a strong supporting framework p. 131

31. Cartilage and bone are called supporting connective tissues because they support the rest of the body.

32. The matrix of cartilage is a firm gel that contains **chondroitin sulfates** (used to form proteoglycans) and cells called **chondrocytes**. Chondrocytes occupy chambers called **lacunae**. A fibrous **perichondrium** separates cartilage from surrounding tissues. The three types of cartilage are **hyaline cartilage**, **elastic cartilage**, and **fibrous cartilage**. (*Figure 4–14*)

33. Chondrocytes rely on diffusion through the avascular matrix to obtain nutrients.

34. **Cartilage** grows by two mechanisms: **interstitial growth** and **appositional growth**. (*Figure 4–13*)

35. **Bone**, or **osseous tissue**, consists of **osteocytes**, little ground substance, and a dense, mineralized matrix. Osteocytes are situated in lacunae. The matrix consists of calcium salts and collagen fibers, giving it unique properties. (*Figure 4–15; Table 4–2*)

36. Osteocytes depend on diffusion through **canaliculi** for nutrient intake.

37. Each bone is surrounded by a **periosteum** with fibrous and cellular layers.

4-6 Membranes are physical barriers of four types: mucous, serous, cutaneous, and synovial p. 135

38. Membranes form a barrier or interface. Epithelia and connective tissues combine to form membranes that cover and protect other structures and tissues. (*Figure 4–16*)

39. **Mucous membranes** line cavities that communicate with the exterior. They contain areolar tissue called the **lamina propria**.

40. **Serous membranes** line the body's sealed internal cavities. They form a fluid called a *transudate*.
41. The **cutaneous membrane**, or skin, covers the body surface.
42. **Synovial membranes** form an incomplete lining within the cavities of synovial joints.

4-7 Connective tissues create the internal framework of the body p. 137

43. Internal organs and systems are tied together by a network of connective tissue proper. This network consists of the **superficial fascia** (the **subcutaneous layer**, or *hypodermis*, separating the skin from underlying tissues and organs), the **deep fascia** (dense connective tissue), and the **subserous fascia** (the layer between the deep fascia and the serous membranes that line body cavities). (*Figure 4–17*)

4-8 The three types of muscle tissue are skeletal, cardiac, and smooth p. 138

44. **Muscle tissue** is specialized for contraction. (*Figure 4–18*)
45. The cells of **skeletal muscle tissue** are *multinucleate*. Skeletal muscle, or **striated voluntary muscle**, produces new fibers by the division of **myosatellite cells**.
46. **Cardiocytes**, the cells of **cardiac muscle tissue**, occur only in the heart. Cardiac muscle, or **striated involuntary muscle**, relies on *pacemaker cells* for regular contraction.
47. **Smooth muscle tissue**, or **nonstriated involuntary muscle**, is not striated. Smooth muscle cells can divide and therefore regenerate after injury has occurred.

4-9 Neural tissue responds to stimuli and conducts electrical impulses throughout the body p. 140

48. **Neural tissue** conducts electrical impulses, which convey information from one area of the body to another.

49. Cells in neural tissue are either neurons or neuroglia. **Neurons** transmit information as electrical impulses. Several kinds of **neuroglia** exist, and their basic functions include supporting neural tissue and helping supply nutrients to neurons. (*Figure 4–19*)
50. A typical neuron has a **cell body**, **dendrites**, and an **axon**, or **nerve fiber**. The axon carries information to other cells.

4-10 The response to tissue injury involves inflammation and regeneration p. 141

51. Any injury affects several types of tissue simultaneously, and they respond in a coordinated manner. After an injury, homeostasis is restored by two processes: inflammation and regeneration.
52. **Inflammation**, or the **inflammatory response**, isolates the injured area while damaged cells, tissue components, and any dangerous microorganisms (which could cause **infection**) are cleaned up. **Regeneration** is the repair process that restores normal function. (*Figure 4–20*)

4-11 With advancing age, tissue repair declines and cancer rates increase p. 143

53. Tissues change with age. Repair and maintenance become less efficient, and the structure and chemical composition of many tissues are altered.
54. The incidence of cancer increases with age, with roughly three-quarters of all cases caused by exposure to chemicals or by other environmental factors, such as cigarette smoke.

Review Questions

See the blue Answers tab at the end of the book.

myA&P Access more review material online at **myA&P™** (www.myaandp.com). There, you'll find chapter guides, chapter quizzes, practice tests, animations, flashcards, a glossary with pronunciations, *Interactive Physiology*® (IP) exercises and quizzes, and more to help you succeed in the course.

LEVEL 1   Reviewing Facts and Terms

1. Identify the six categories of epithelial tissue shown in the drawing to the right.

   (a) _____   (b) _____
   (c) _____   (d) _____
   (e) _____   (f) _____

2. Collections of specialized cells and cell products that perform a relatively limited number of functions are called
   (a) cellular aggregates.   (b) tissues.
   (c) organs.   (d) organ systems.
   (e) organisms.

3. Tissue that is specialized for contraction is
   (a) epithelial tissue.   (b) muscle tissue.
   (c) connective tissue.   (d) neural tissue.

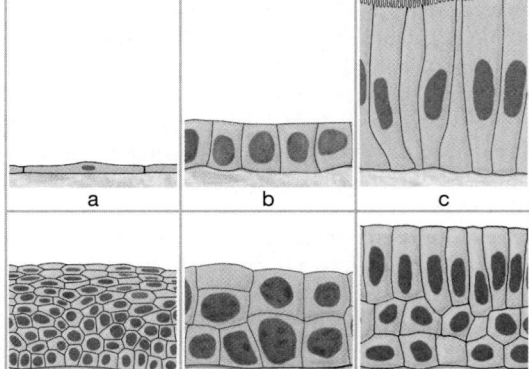

4. A type of junction common in cardiac and smooth muscle tissues is the
   (a) hemidesmosome.   (b) basal junction.
   (c) occluding junction.   (d) gap junction.

5. The most abundant connections between cells in the superficial layers of the skin are
   (a) connexons.   (b) gap junctions.
   (c) macula adherens.   (d) occluding junctions.

6. _____ membranes have an epithelium that is stratified and supported by dense connective tissue.
   (a) Synovial   (b) Serous
   (c) Cutaneous   (d) Mucous

7. Mucous secretions that coat the passageways of the digestive and respiratory tracts result from _____ secretion.
   (a) apocrine   (b) merocrine
   (c) holocrine   (d) endocrine

8. Matrix is a characteristic of which type of tissue?
   (a) epithelial   (b) neural
   (c) muscle   (d) connective

9. Functions of connective tissue include
   (a) establishing a structural framework for the body.
   (b) storing energy reserves.
   (c) providing protection for delicate organs.
   (d) all of these.
   (e) a and c only.

10. Which of the following epithelia most easily permits diffusion?
    (a) stratified squamous
    (b) simple squamous
    (c) transitional
    (d) simple columnar

11. The three major types of cartilage in the body are
    (a) collagen, reticular, and elastic.
    (b) areolar, adipose, and reticular.
    (c) hyaline, elastic, and fibrous.
    (d) tendons, reticular, and elastic.

12. The primary function of serous membranes in the body is to
    (a) minimize friction between opposing surfaces.
    (b) line cavities that communicate with the exterior.
    (c) perform absorptive and secretory functions.
    (d) cover the surface of the body.

13. The type of cartilage growth characterized by adding new layers of cartilage to the surface is
    (a) interstitial growth.
    (b) appositional growth.
    (c) intramembranous growth.
    (d) longitudinal growth.

14. Tissue changes with age can result from
    (a) hormonal changes.
    (b) increased need for sleep.
    (c) improper nutrition.
    (d) all of these.
    (e) a and c only.

15. Axons, dendrites, and a cell body are characteristic of cells located in
    (a) neural tissue.
    (b) muscle tissue.
    (c) connective tissue.
    (d) epithelial tissue.

16. The repair process necessary to restore normal function in damaged tissues is
    (a) isolation.
    (b) regeneration.
    (c) reconstruction.
    (d) all of these.

17. What are the four essential functions of epithelial tissue?

18. Differentiate between endocrine and exocrine glands.

19. By what three methods do various glandular epithelial cells release their secretions?

20. List three basic components of connective tissues.

21. What are the four kinds of membranes composed of epithelial and connective tissue that cover and protect other structures and tissues in the body?

22. What two cell populations make up neural tissue? What is the function of each?

LEVEL 2 Reviewing Concepts

23. What is the difference between an exocrine secretion and an endocrine secretion?

24. A significant structural feature in the digestive system is the presence of occluding junctions near the exposed surfaces of cells lining the digestive tract. Why are these junctions so important?

25. Describe the fluid connective tissues in the human body. What are the main differences between fluid connective tissues and supporting connective tissues?

26. Why are infections always a serious threat after a severe burn or abrasion?

27. A layer of glycoproteins and a network of fine protein filaments that prevents the movement of proteins and other large molecules from the connective tissue to the epithelium describes
    (a) interfacial canals.
    (b) the basal lamina.
    (c) the reticular lamina.
    (d) areolar tissue.
    (e) squamous epithelium.

28. Why does damaged cartilage heal slowly?
    (a) Chondrocytes cannot be replaced if killed, and other cell types must take their place.
    (b) Cartilage is avascular, so nutrients and other molecules must diffuse to the site of injury.
    (c) Damaged cartilage becomes calcified, thus blocking the movement of materials required for healing.
    (d) Chondrocytes divide more slowly than other cell types, delaying the healing process.
    (e) Damaged collagen cannot be quickly replaced, thereby slowing the healing process.

29. List the similarities and differences among the three types of muscle tissue.

LEVEL 3 Critical Thinking and Clinical Applications

30. Assuming that you had the necessary materials to perform a detailed chemical analysis of body secretions, how could you determine whether a secretion was merocrine or apocrine?

31. During a lab practical, a student examines a tissue that is composed of densely packed protein fibers that are running parallel and form a cord. There are no striations, but small nuclei are visible. The student identifies the tissue as skeletal muscle. Why is the student's choice wrong, and what tissue is he probably observing?

32. While in a chemistry lab, Jim accidentally spills a small amount of a caustic chemical on his arm. What changes in the characteristics of the skin would you expect to observe, and what would cause these changes?

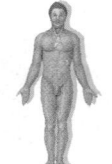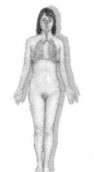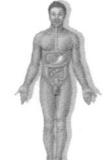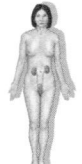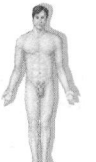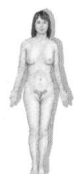

Systems Overview

Our perspective has gradually changed over the preceding chapters. Atoms can only be imagined or indirectly examined through experimental procedures. Cellular details often escape detection unless an electron microscope is used. Tissue structure, however, can be examined with a light microscope, and based on your experiences in the laboratory you may already be able to identify some tissues with the unaided eye. For example, once you have handled adipose tissue, with its lumpy, greasy texture, it would be difficult to mistake it for any other type of tissue.

Organs are combinations of tissues that perform complex functions. A great deal of information concerning organs and organ structure can be obtained by dissection and direct examination. In organ systems, several organs work together in a coordinated fashion. We can easily observe the functions of intact organ systems as they perform, direct, or moderate the activities of individual human beings. As a result, at the start of this course you probably knew much more about the major organs and systems than you did about cells and tissue structure.

Figure 1 presents four views of the composition of the human body, reflecting the changes in our perspective over the last four chapters. In Chapter 2 the body was treated as a collection of chemical elements (**Figure 1a**) that combine to form molecules. Cells, described in Chapter 3, are composed of organic molecules, inorganic molecules, ions, and water. **Figure 1b** indicates the proportions of water and organic molecules in the body as a whole.

Chapter 4 described the association of roughly 200 types of cells in four types of body tissues (**Figure 1c**). These four tissue types combine to form thousands of different organs. Some are quite large and distinctive; the liver, an organ of the

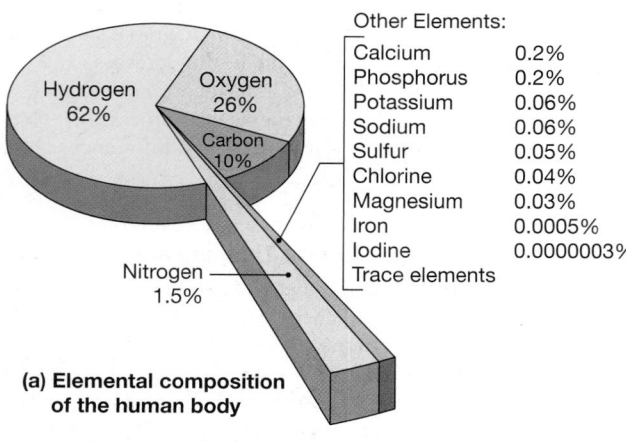

Other Elements:

Calcium	0.2%
Phosphorus	0.2%
Potassium	0.06%
Sodium	0.06%
Sulfur	0.05%
Chlorine	0.04%
Magnesium	0.03%
Iron	0.0005%
Iodine	0.0000003%
Trace elements	

Hydrogen 62%
Oxygen 26%
Carbon 10%
Nitrogen 1.5%

(a) Elemental composition of the human body

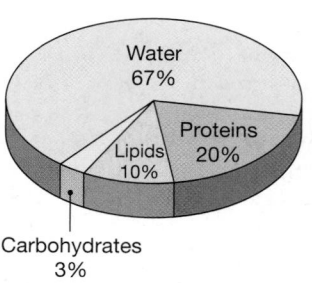

Water 67%
Proteins 20%
Lipids 10%
Carbohydrates 3%

(b) Molecular composition of the human body

FIGURE 1   **Composition of the Human Body.**

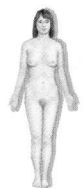
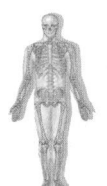

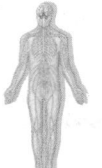
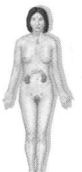
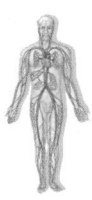

digestive tract, weighs about 1.6 kg (3.5 lb), and some skeletal muscles can be even larger. Other organs are tiny and far more numerous; the skin contains roughly 3 million sweat glands that are barely large enough to see without a magnifying glass. Regardless of their size, all of these organs contain all four tissue types, although the proportions vary from organ to organ. For example, all four tissue types contribute extensively to the structure of the stomach, but most of the heart is composed of cardiac muscle tissue. **Figure 1d** characterizes the human body at the organ system level, the focus of the rest of this textbook.

Despite their structural and functional differences, all organ systems share certain characteristics:

1. *Specialization for performing a limited number of functions.* In other words, there is a division of labor among organ systems.

2. *Functional independence* in responding to local environmental stimuli.

3. *Dependence on other organ systems* for nutrient supply, oxygen, and waste removal.

4. *Integration of activity* through neural and hormonal mechanisms.

Figure 2, on the pages that follow, summarizes the components, organization, and functions of the 11 organ systems in the human body. This information will provide a framework for later chapters dealing with specific systems. Refer to them while you work through those chapters, whenever you need a reminder of the general functions or locations of specific organs.

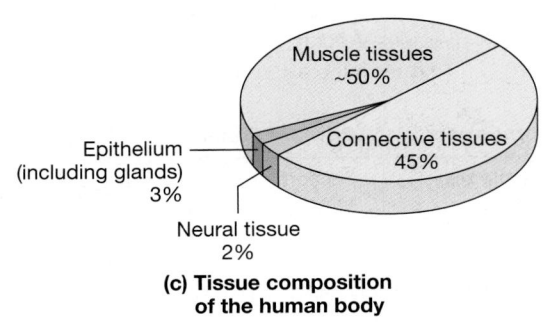

(c) Tissue composition of the human body

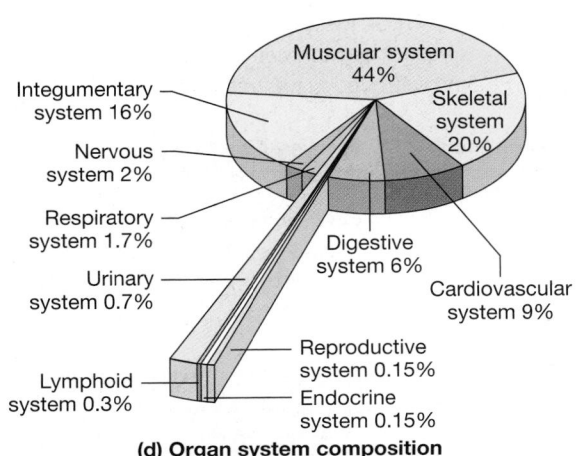

(d) Organ system composition of the human body by weight

FIGURE 1 (*continued*)

Figure 2 Organ System Components and Functions.

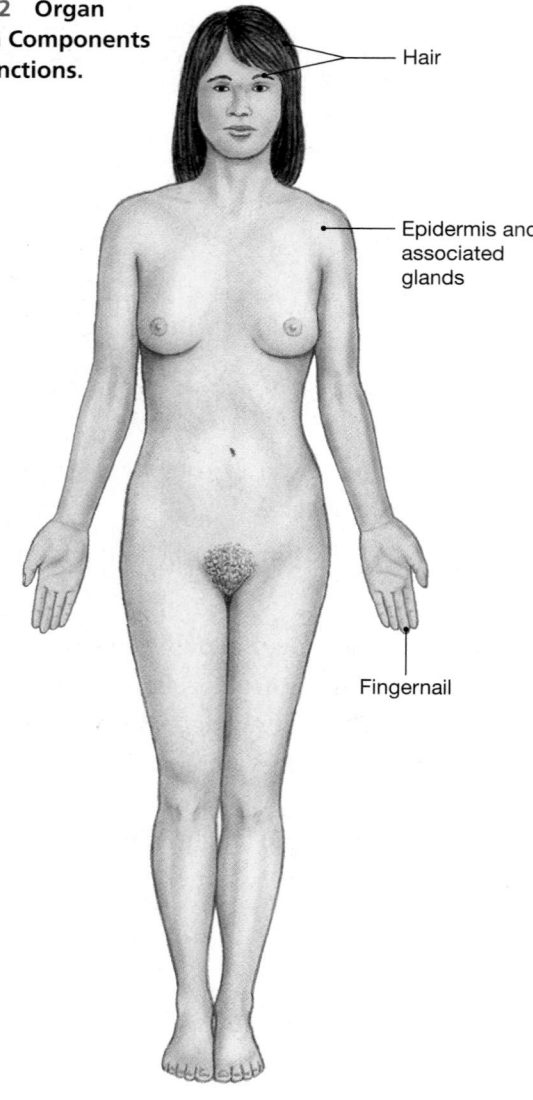

Hair

Epidermis and associated glands

Fingernail

(a) THE INTEGUMENTARY SYSTEM
Protects against environmental hazards; helps control body temperature

Organ/Component	Primary Functions
Cutaneous Membrane	
Epidermis	Covers surface; protects deeper tissues
Dermis	Nourishes epidermis; provides strength; contains exocrine glands
Hair Follicles	Produce hair; innervation provides sensation
Hairs	Provide some protection for head
Sebaceous Glands	Secrete lipid coating that lubricates hair shaft and epidermis
Sweat Glands	Produce perspiration for evaporative cooling
Nails	Protect and stiffen distal tips of digits
Sensory Receptors	Provide sensations of touch, pressure, temperature, pain
Subcutaneous Layer	Stores lipids; attaches skin to deeper structures; cushions underlying tissues

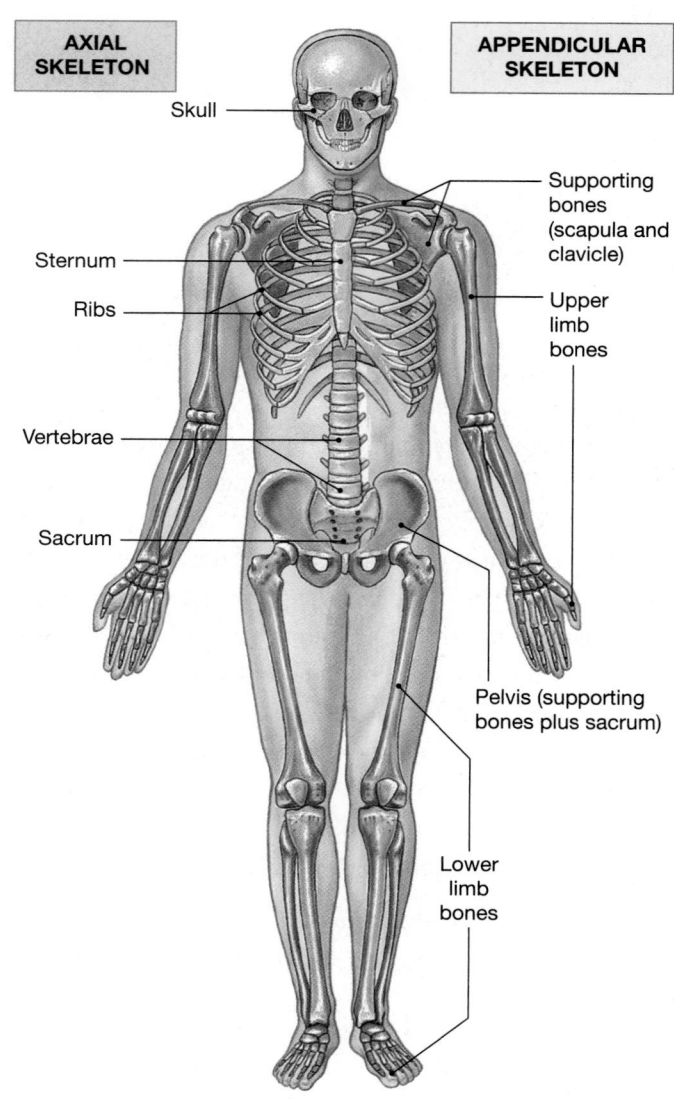

AXIAL SKELETON

APPENDICULAR SKELETON

Skull

Sternum

Ribs

Vertebrae

Sacrum

Supporting bones (scapula and clavicle)

Upper limb bones

Pelvis (supporting bones plus sacrum)

Lower limb bones

(b) THE SKELETAL SYSTEM
Provides support; protects tissues; stores minerals; forms blood

Organ/Component	Primary Functions
Bones, Cartilages, and Joints	Support, protect soft tissues; bones store minerals
Axial Skeleton (skull, vertebrae, ribs, sternum, sacrum, cartilages, and ligaments)	Protects brain, spinal cord, sense organs, and soft tissues of thoracic cavity; supports the body weight over the lower limbs
Appendicular Skeleton (supporting bones, cartilages, and ligaments of the limbs)	Provides internal support and positioning of the limbs; supports and moves axial skeleton
Bone Marrow	Acts as primary site of blood cell production (red blood cells, white blood cells); stores lipid reserves

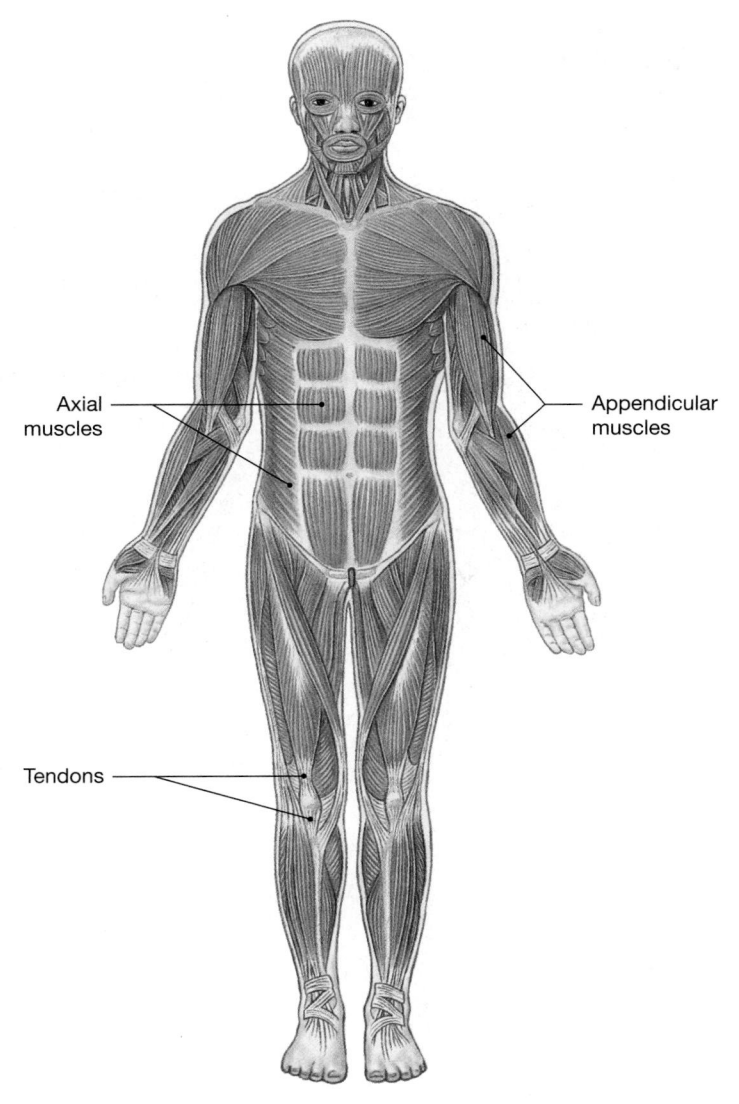

Axial muscles

Appendicular muscles

Tendons

(c) THE MUSCULAR SYSTEM
Produces movement and locomotion;
provides support; generates heat

Organ/Component	Primary Functions
Skeletal Muscles (700)	Provide skeletal movement; control entrances and exits of digestive tract; produce heat; support skeletal position; protect soft tissues
Axial Muscles	Support and position axial skeleton
Appendicular Muscles	Support, move, and brace limbs
Tendons, Aponeuroses	Harness forces of contraction to perform specific tasks

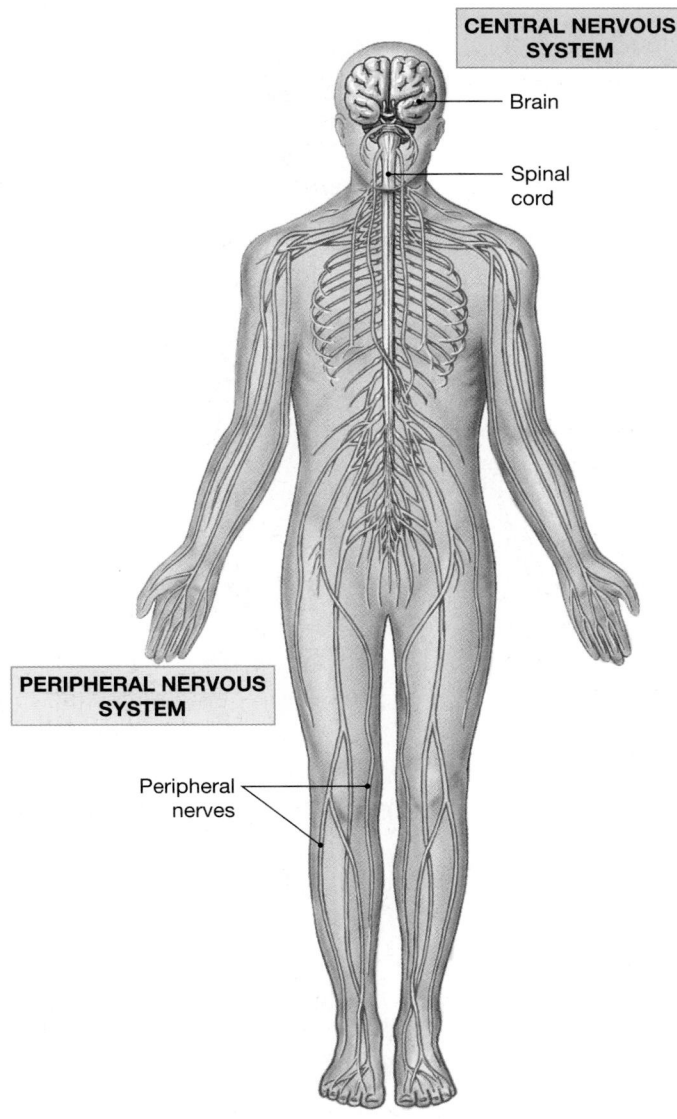

CENTRAL NERVOUS SYSTEM

Brain

Spinal cord

PERIPHERAL NERVOUS SYSTEM

Peripheral nerves

(d) THE NERVOUS SYSTEM
Directs immediate response to stimuli, usually by coordinating
the activities of other organ systems

Organ/Component	Primary Functions
Central Nervous System (CNS)	Acts as control center for nervous system; processes information; provides short-term control over activities of other systems
Brain	Performs complex integrative functions; controls both voluntary and autonomic activities
Spinal Cord	Relays information to and from brain; performs less-complex integrative functions; directs many simple involuntary activities
Peripheral Nervous System (PNS)	Links CNS with other systems and with sense organs

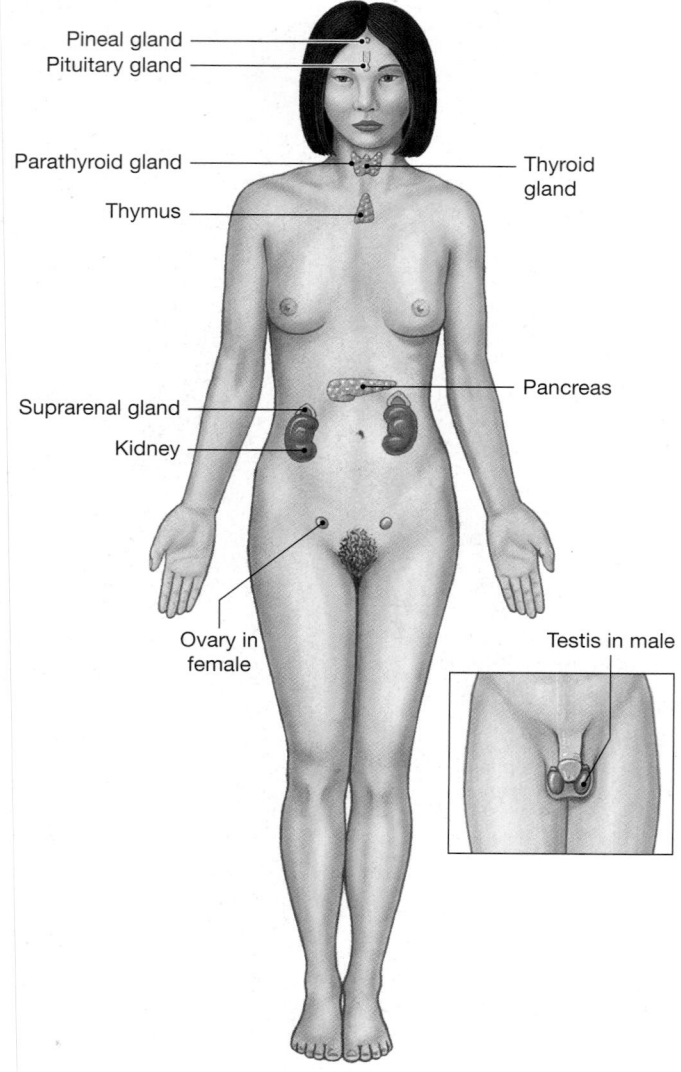

(e) THE ENDOCRINE SYSTEM
Directs long-term changes in other organ systems

Organ/Component	Primary Functions
Pineal Gland	May control timing of reproduction and set day-night rhythms
Pituitary Gland	Controls other endocrine glands; regulates growth and fluid balance
Thyroid Gland	Controls tissue metabolic rate; regulates calcium levels
Parathyroid Glands	Regulate calcium levels (with thyroid)
Thymus	Controls maturation of lymphocytes
Suprarenal Glands	Adjust water balance, tissue metabolism, cardiovascular and respiratory activity
Kidneys	Control red blood cell production and assist in calcium regulation
Pancreas	Regulates blood glucose levels
Gonads	
Testes	Support male sexual characteristics and reproductive functions [see part k]
Ovaries	Support female sexual characteristics and reproductive functions [see part l]

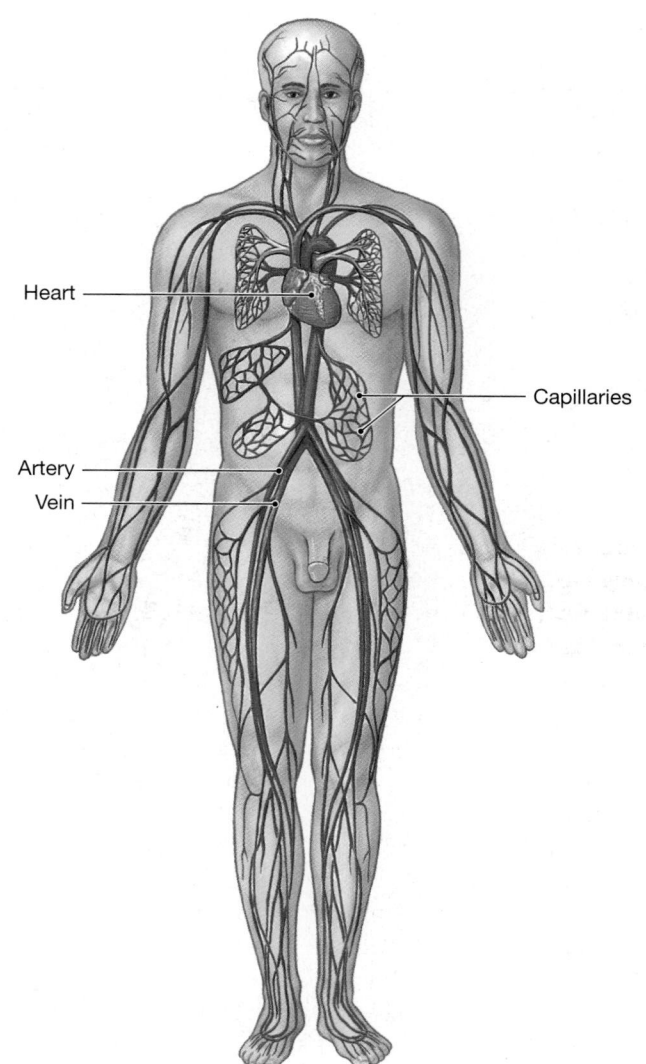

(f) THE CARDIOVASCULAR SYSTEM
Transports cells and dissolved materials, including nutrients, wastes, and gases

Organ/Component	Primary Functions
Heart	Propels blood; maintains blood pressure
Blood Vessels	Distribute blood around the body
Arteries	Carry blood from heart to capillaries
Capillaries	Permit diffusion between blood and interstitial fluids
Veins	Return blood from capillaries to the heart
Blood	Transports oxygen, carbon dioxide, and blood cells; delivers nutrients and hormones; removes waste products; assists in temperature regulation and defense against disease

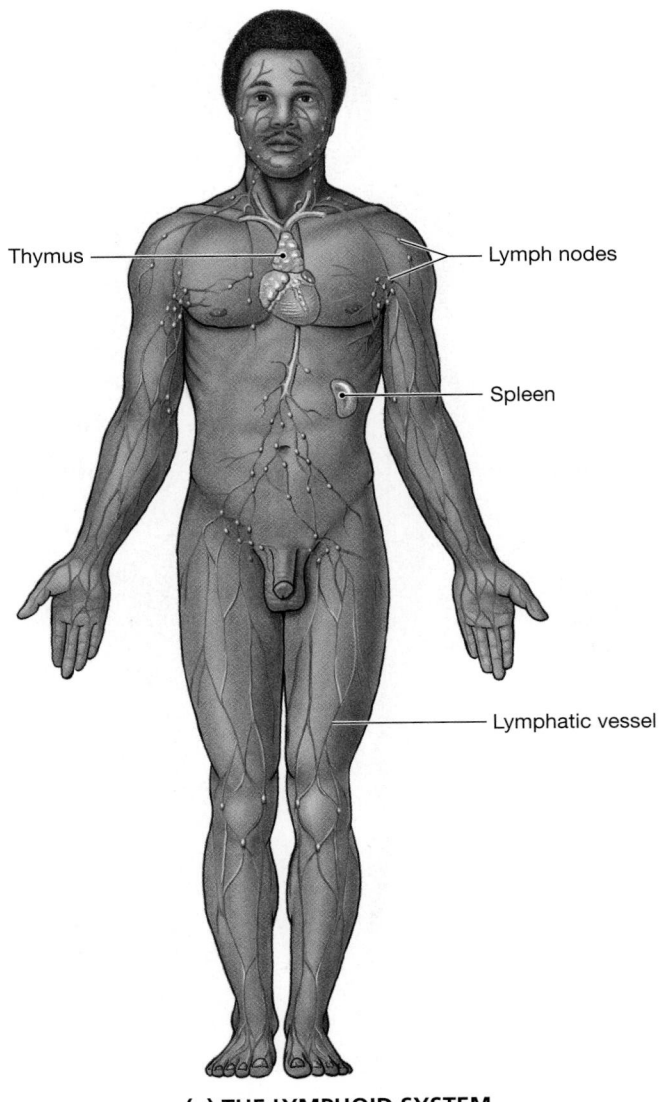

(g) THE LYMPHOID SYSTEM
Defends against infection and disease;
returns tissue fluid to the bloodstream

Organ/Component	Primary Functions
Lymphatic Vessels	Carry lymph (water and proteins) and lymphocytes from peripheral tissues to veins of the cardiovascular system
Lymph Nodes	Monitor the composition of lymph; macrophages engulf pathogens; stimulate immune response
Spleen	Monitors circulating blood; macrophages engulf pathogens; stimulates immune response
Thymus	Controls development and maintenance of one class of lymphocytes (T cells)

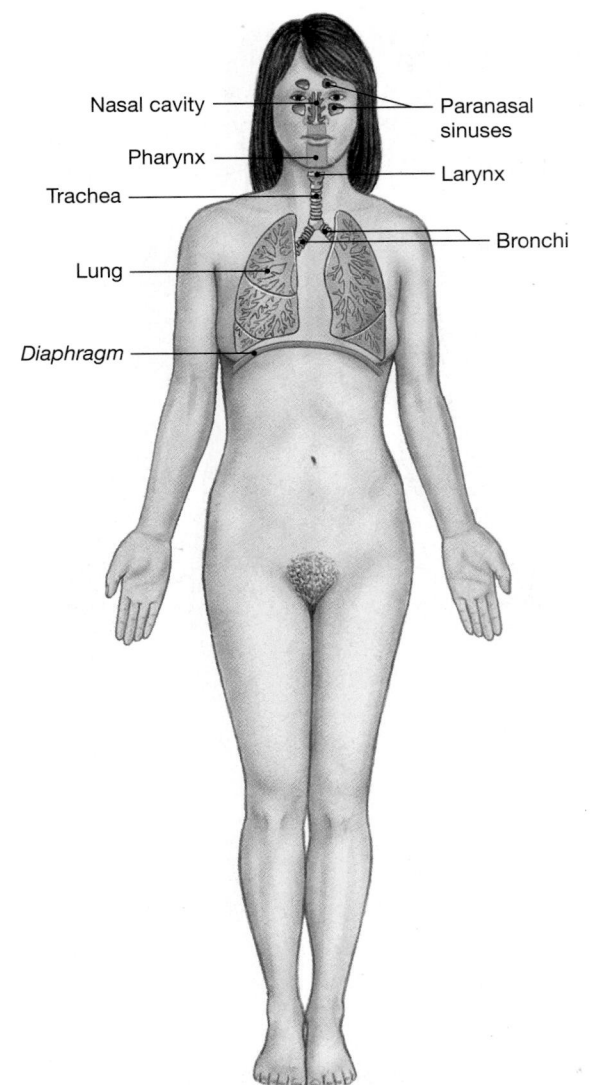

(h) THE RESPIRATORY SYSTEM
Delivers air to sites where gas exchange can
occur between the air and circulating blood

Organ/Component	Primary Functions
Nasal Cavities, Paranasal Sinuses	Filter, warm, humidify air; detect smells; lessen weight of skull
Pharynx	Conducts air to larynx; is a chamber shared with the digestive tract *[see part i]*
Larynx	Protects opening to trachea and contains vocal cords
Trachea	Filters air, traps particles in mucus; cartilages keep airway open
Bronchi	(Same functions as trachea)
Lungs	Responsible for air movement through volume changes during movements of ribs and diaphragm; include airways and alveoli
Alveoli	Act as sites of gas exchange between air and blood

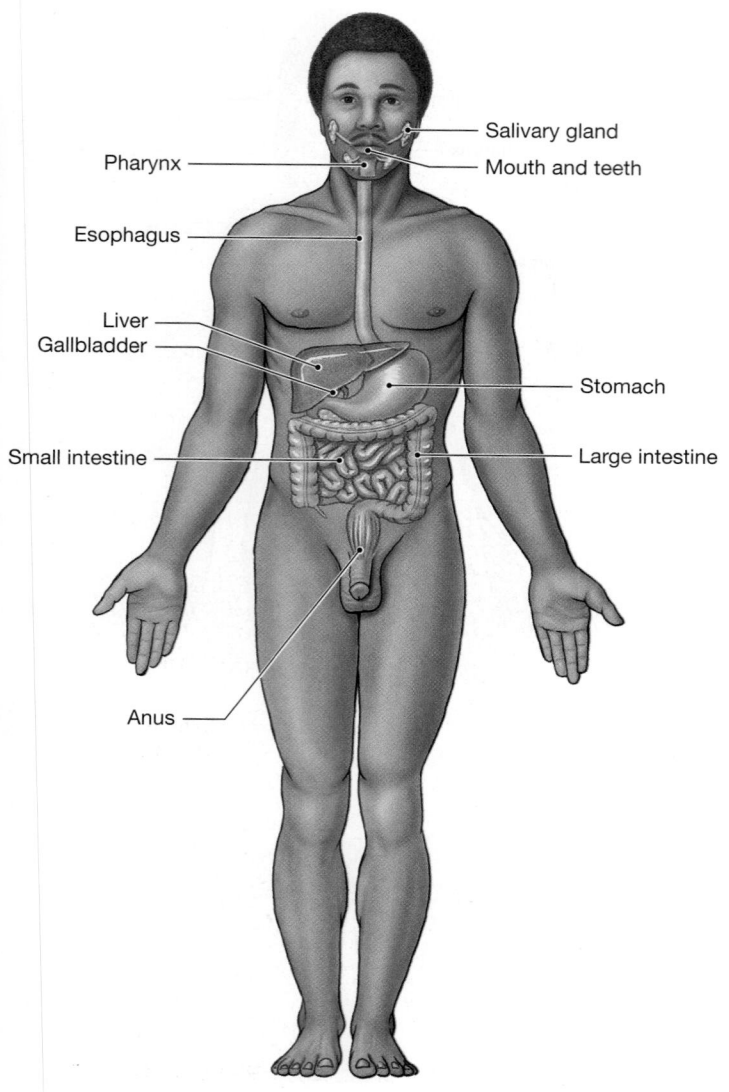

(i) THE DIGESTIVE SYSTEM
Processes food; absorbs nutrients; eliminates waste products

Organ/Component	Primary Functions
Salivary Glands	Provide buffers and lubrication; produce enzymes that begin digestion
Mouth and Pharynx	Conduct solid foods and liquids to esophagus; mouth contains teeth; pharynx is a chamber shared with respiratory tract *[see part h]*
Esophagus	Delivers food to stomach
Stomach	Secretes acids and enzymes
Small Intestine	Secretes digestive enzymes, buffers, and hormones; absorbs nutrients
Liver	Secretes bile; regulates nutrient composition of blood
Gallbladder	Stores bile for release into small intestine
Pancreas	Secretes digestive enzymes and buffers; contains endocrine cells *[see part e]*
Large Intestine	Removes water from fecal material; stores wastes

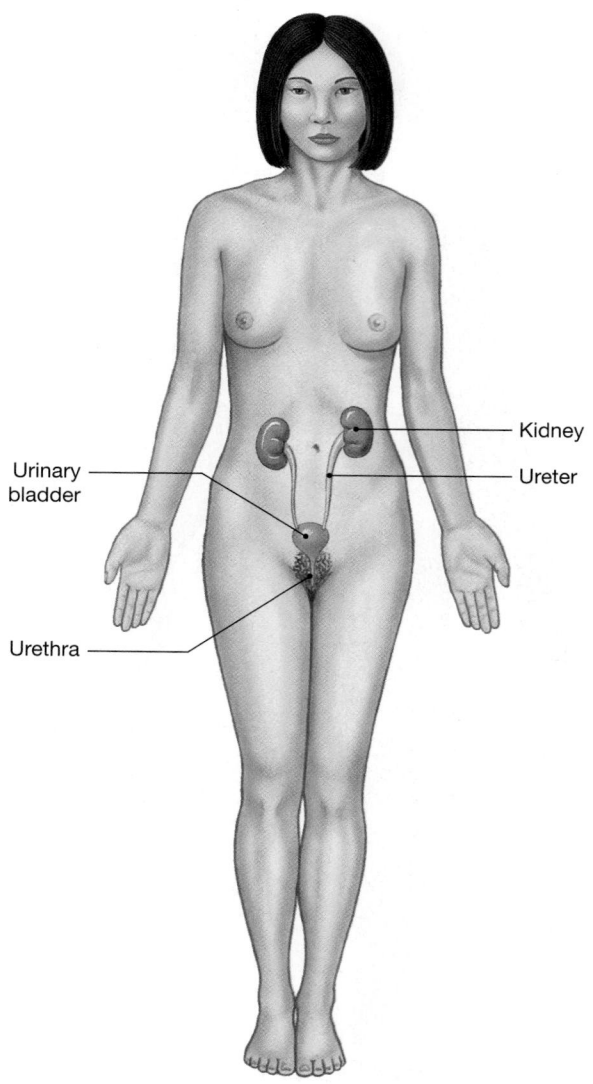

(j) THE URINARY SYSTEM
Eliminates excess water, salts, and waste products

Organ/Component	Primary Functions
Kidneys	Form and concentrate urine; regulate blood pH and ion concentrations; perform endocrine functions *[see part e]*
Ureters	Conduct urine from kidneys to urinary bladder
Urinary Bladder	Stores urine for eventual elimination
Urethra	Conducts urine to exterior

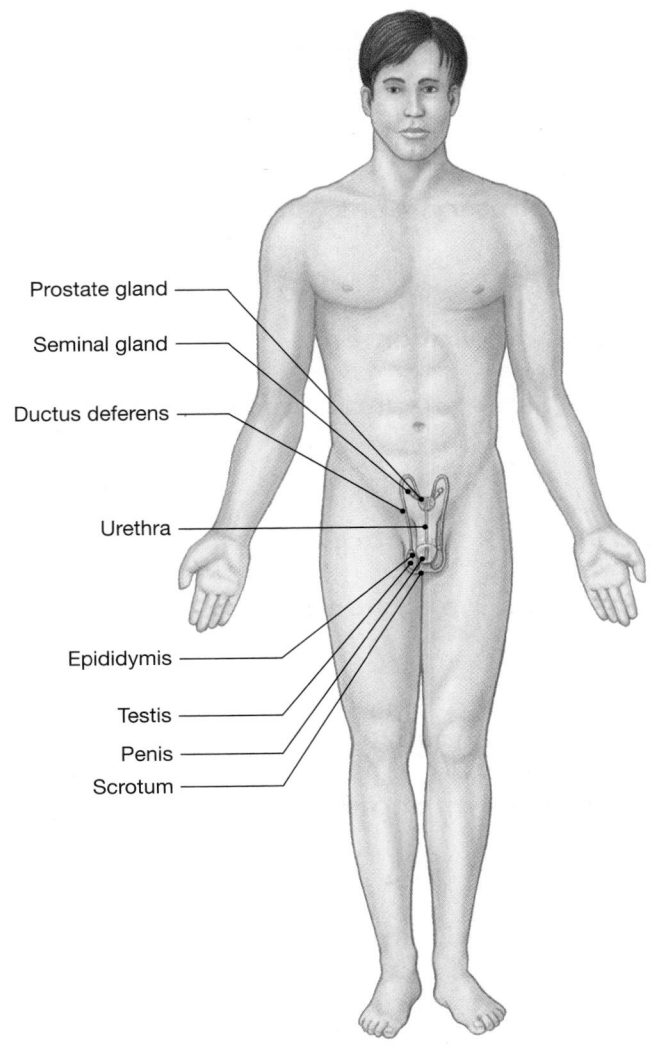

Prostate gland

Seminal gland

Ductus deferens

Urethra

Epididymis

Testis

Penis

Scrotum

(k) THE MALE REPRODUCTIVE SYSTEM
Produces sex cells and hormones

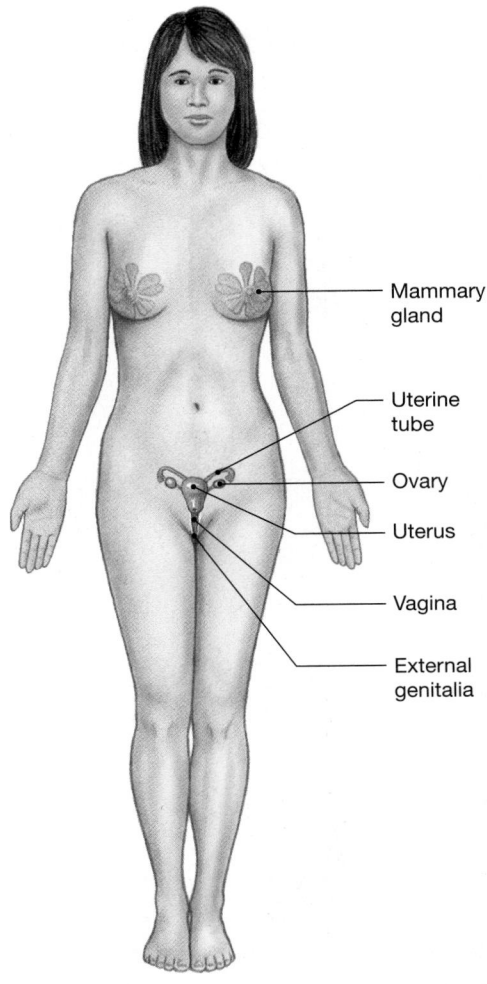

Mammary gland

Uterine tube

Ovary

Uterus

Vagina

External genitalia

(l) THE FEMALE REPRODUCTIVE SYSTEM
Produces sex cells and hormones

Organ/Component	Primary Functions
Testes	Produce sperm and hormones *(see part e)*
Accessory Organs **Epididymides**	Act as sites of sperm maturation in each testis
Ductus Deferens (Sperm Duct)	Conducts sperm between epididymis and prostate gland
Seminal Glands	Secrete fluid that makes up much of the volume of semen
Prostate Gland	Secretes fluid and enzymes
Urethra	Conducts semen to exterior
External Genitalia **Penis**	Contains erectile tissue; deposits sperm in vagina of female; produces pleasurable sensations during sexual activities
Scrotum	Surrounds the testes and controls their temperature

Organ/Component	Primary Functions
Ovaries	Produce oocytes and hormones *(see part e)*
Uterine Tubes	Deliver oocyte or embryo to uterus; normal site of fertilization
Uterus	Site of embryonic development and exchange between maternal and embryonic bloodstreams
Vagina	Site of sperm deposition; acts as birth canal during delivery; provides passageway for fluids during menstruation
External Genitalia **Clitoris**	Contains erectile tissue; produces pleasurable sensations during sexual activities
Labia	Contain glands that lubricate entrance to vagina
Mammary Glands	Produce milk that nourishes newborn infant

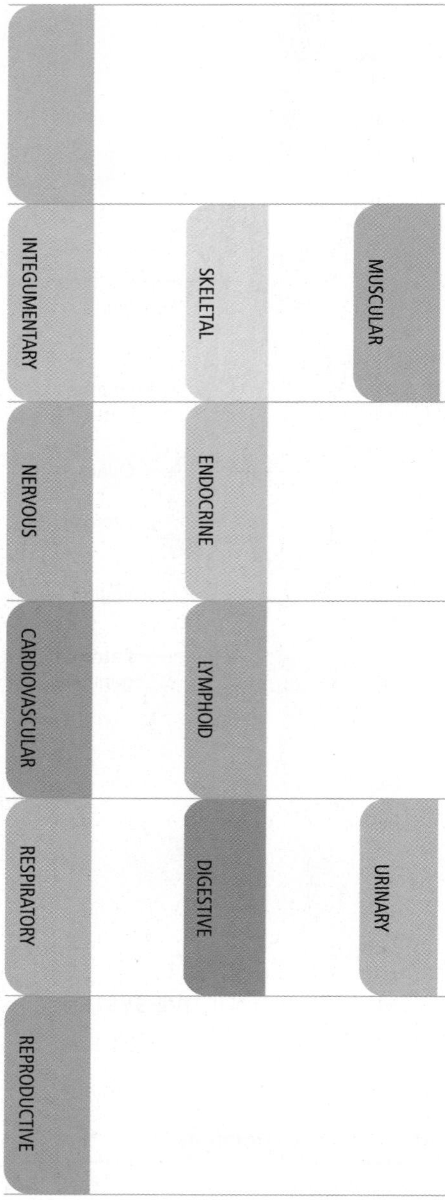

INTEGUMENTARY			**Unit 1: Levels of Organization** **Chapters 1–4** **Introduction, Chemistry, Cells, and Tissues**

INTEGUMENTARY | SKELETAL | MUSCULAR

Unit 2: Support and Movement
Chapters 5–11
Integumentary, Skeletal, and Muscular Systems

NERVOUS | ENDOCRINE

Unit 3: Control and Regulation
Chapters 12–18
Nervous and Endocrine Systems

CARDIOVASCULAR | LYMPHOID

Unit 4: Fluids and Transport
Chapters 19–22
Cardiovascular and Lymphoid Systems

RESPIRATORY | DIGESTIVE | URINARY

Unit 5: Environmental Exchange
Chapters 23–27
Respiratory, Digestive, and Urinary Systems

REPRODUCTIVE

Unit 6: Continuity of Life
Chapters 28 and 29
Reproductive System

Welcome to the rest of the textbook

Chapter 4 concluded the introductory portion of this textbook, and the Systems Overview reviewed the transition from atoms, molecules, cells, and tissues to organ systems. One of the most important themes in this book is that organ systems interact continuously—they do not function in isolation.

To help you navigate more easily, the chapters dealing with body systems have been collected into Units that share functional characteristics. As indicated here, the specific Unit is indicated by the placement of the thumb tab on the margin of the page. The body systems within each Unit are distinguished by the color of the thumb tab.

You will find this overview page useful as a reference throughout the remainder of the book.

5

The Integumentary System

Learning Outcomes

After completing this chapter, you should be able to do the following:

5-1 Describe the main structural features of the epidermis, and explain the functional significance of each.

5-2 Explain what accounts for individual differences in skin color, and discuss the response of melanocytes to sunlight exposure.

5-3 Describe the interaction between sunlight and vitamin D_3 production.

5-4 Describe the roles of epidermal growth factor.

5-5 Describe the structure and functions of the dermis.

5-6 Describe the structure and functions of the hypodermis.

5-7 Describe the mechanisms that produce hair, and explain the structural basis for hair texture and color.

5-8 Discuss the various kinds of glands in the skin, and list the secretions of those glands.

5-9 Describe the anatomical structure of nails, and explain how they are formed.

5-10 Explain how the skin responds to injury and repairs itself.

5-11 Summarize the effects of aging on the skin.

5-12 Give examples of interactions between the integumentary system and each of the other organ systems.

Clinical Notes

Skin Cancers, Melanomas, and Sunblocks p. 164

Ulcers p. 167

Liposuction p. 168

Skin Abnormalities p. 178

Burns and Grafts p. 180

An Introduction to the Integumentary System

This chapter considers the many and varied functions of the skin, the organ system with which you are probably most familiar. No other organ system is as accessible, large, and underappreciated as the integumentary system. Often referred to simply as the **integument** (in-TEG-ū-ment), this system accounts for about 16 percent of your total body weight. Its surface, 1.5–2 m² in area, is continually abraded, attacked by microorganisms, irradiated by sunlight, and exposed to environmental chemicals. The integumentary system is your body's first line of defense against an often hostile environment—the place where you and the outside world meet.

The integumentary system has two major components: the **cutaneous membrane** or skin, and the **accessory structures** (Figure 5–1).

1. The cutaneous membrane has two components: the **epidermis** (*epi-*, above) or superficial epithelium, and the **dermis**, an underlying area of connective tissues.

2. The accessory structures include hair, nails, and multicellular exocrine glands. These structures are located primarily in the dermis and protrude through the epidermis to the skin surface.

The integument does not function in isolation. An extensive network of blood vessels branches through the dermis, and sensory receptors that monitor touch, pressure, temperature, and pain provide valuable information to the central nervous system about the state of the body. Deep to the dermis, the loose connective tissue of the **hypodermis**, also known as the superficial fascia or *subcutaneous layer*, separates the integument from the deep fascia around other organs, such as muscles and bones. ∞ p. 137 Although the hypodermis is often considered separate from the integument, we will consider it in this chapter because its connective tissue fibers are interwoven with those of the dermis.

The general functions of the skin and hypodermis include the following:

- *Protection* of underlying tissues and organs against impact, abrasion, fluid loss, and chemical attack.

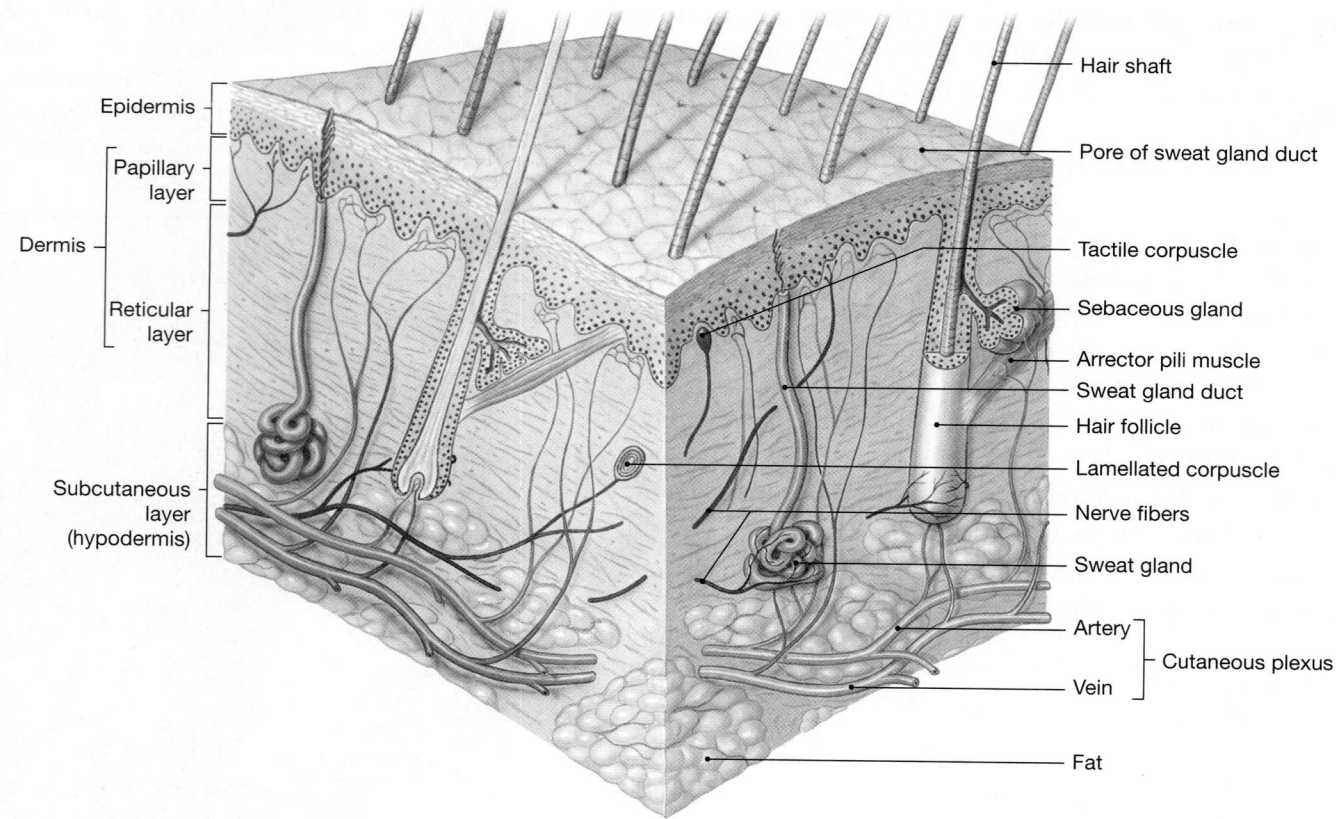

Figure 5–1 **The Components of the Integumentary System.** This diagrammatic section of skin illustrates the relationships among the two components of the cutaneous membrane (epidermis and dermis) and the accessory structures of the integumentary system (with the exception of nails, shown in *Figure 5–13*).

- *Excretion* of salts, water, and organic wastes by integumentary glands.
- *Maintenance* of normal body temperature through either insulation or evaporative cooling, as needed.
- *Production* of melanin, which protects underlying tissue from ultraviolet radiation.
- *Production* of keratin, which protects against abrasion and serves as a water repellent.
- *Synthesis of vitamin D₃,* a steroid that is subsequently converted to calcitriol, a hormone important to normal calcium metabolism.
- *Storage* of lipids in adipocytes in the dermis and in adipose tissue in the subcutaneous layer.
- *Detection* of touch, pressure, pain, and temperature stimuli, and the relaying of that information to the nervous system. (These *general senses*, which provide information about the external environment, will be considered further in Chapter 15.)

In the next several sections we will consider the various components of the integument.

5-1 The epidermis is composed of strata (layers) with various functions

The epidermis consists of a stratified squamous epithelium. Recall from Chapter 4 that such an epithelium provides considerable mechanical protection and also helps keep microorganisms outside the body. ∞ p. 118 Like all other epithelia,

the epidermis is avascular. Because there are no local blood vessels, epidermal cells rely on the diffusion of nutrients and oxygen from capillaries within the dermis. As a result, the epidermal cells with the highest metabolic demands are found close to the basal lamina, where the diffusion distance is short. The superficial cells, far removed from the source of nutrients, are either inert or dead.

The epidermis is dominated by **keratinocytes** (ke-RAT-i-nō-sīts), the body's most abundant epithelial cells. These cells, which form several layers, contain large amounts of the protein *keratin* (discussed shortly). **Thin skin** (**Figure 5–2a,b**), which covers most of the body surface, contains four layers of keratinocytes, and is about as thick as the wall of a plastic sandwich bag (roughly 0.08 mm). **Thick skin** (**Figure 5–2c**), which occurs on the palms of the hands and the soles of the feet, contains a fifth layer, the *stratum lucidum*, and because it has a much thicker superficial layer (the *stratum corneum*) it is about as thick as a standard paper towel (roughly 0.5 mm). Note that the terms *thick* and *thin* refer to the relative thickness of the epidermis, not to the integument as a whole.

Figure 5–3 shows the layers of keratinocytes in a section of the epidermis in an area of thick skin. The boundaries between the layers are often difficult to see in a standard light micrograph. You will notice that the various layers have Latin names. The word *stratum* (plural, *strata*) means "layer"; the rest of the name refers to the function or appearance of the layer. The strata, in order from the basal lamina toward the free surface, are the *stratum germinativum*, the *stratum spinosum*, the *stratum granulosum*, the *stratum lucidum*, and the *stratum corneum*.

Tips&Tricks

*Strat*um means layer, as in **strat**ified.

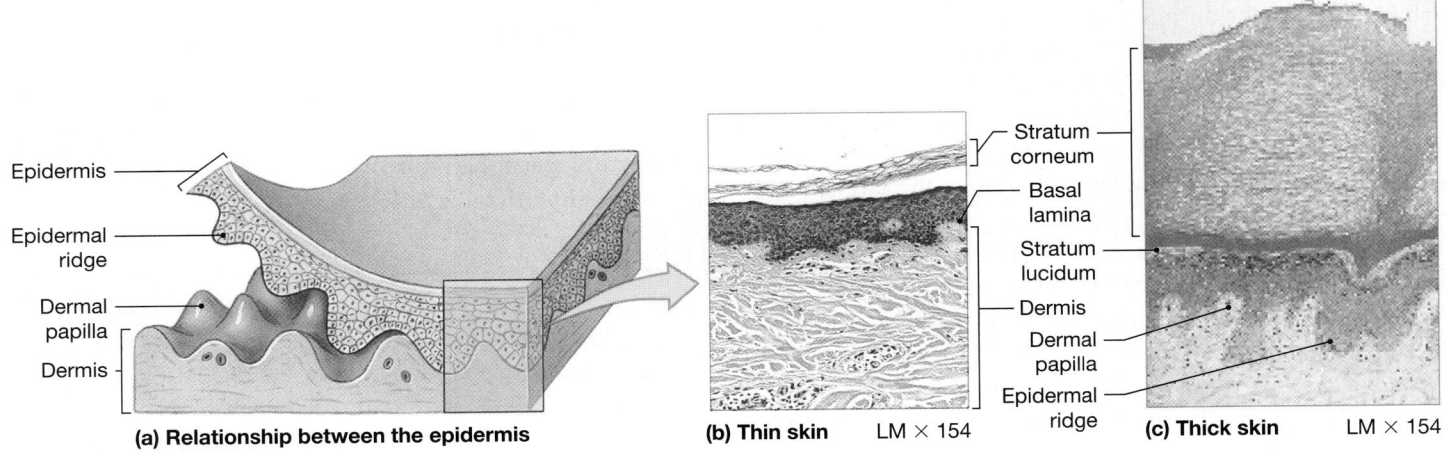

(a) Relationship between the epidermis and dermis

(b) Thin skin   LM × 154

Stratum corneum
Basal lamina
Stratum lucidum
Dermis
Dermal papilla
Epidermal ridge

(c) Thick skin   LM × 154

Figure 5–2   The Basic Organization of the Epidermis. (a) A drawing showing the structural relationship and interface between the epidermis and underlying dermis. The proportions of the various layers differ with the location sampled. **(b)** A micrograph of thin skin, which covers most of the exposed body surface. **(c)** A micrograph of thick skin, which covers the surfaces of the palms and soles.

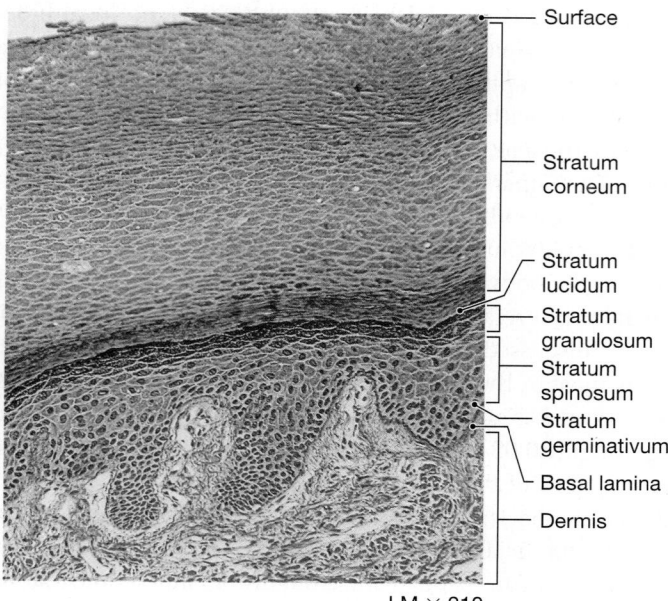

LM × 210

Figure 5–3 **The Structure of the Epidermis.** A portion of the epidermis in thick skin, showing the major layers of stratified epidermal cells. Note that Langerhans cells cannot be distinguished in standard histological preparations.

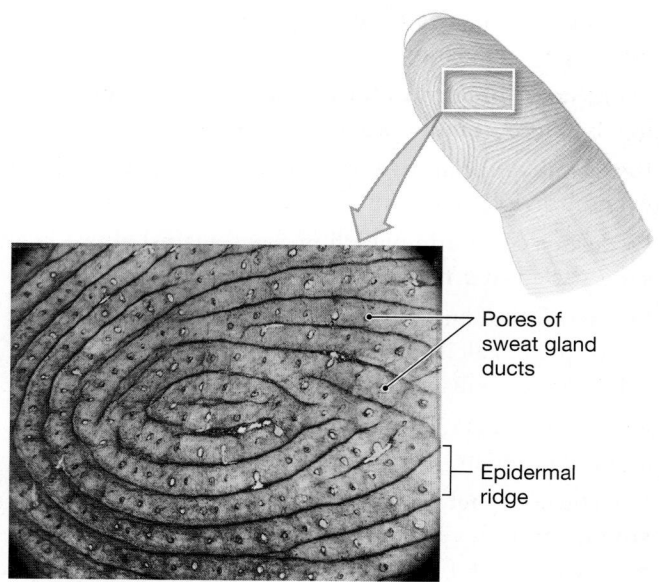

Figure 5–4 **The Epidermal Ridges of Thick Skin.** Fingerprints reveal the pattern of epidermal ridges. This scanning electron micrograph shows the ridges on a fingertip. The pits are the openings into the ducts of merocrine sweat glands. [©R. G. Kessel and R. H. Kardon, *Tissues and Organs: A Text-Atlas of Scanning Electron Microscopy*, W. H. Freeman & Co., 1979. All Rights Reserved.]

Stratum Germinativum

The innermost epidermal layer is the **stratum germinativum** (STRA-tum jer-mi-na-TĒ-vum), or *stratum basale* (**Figure 5–3**). Hemidesmosomes attach the cells of this layer to the basal lamina that separates the epidermis from the areolar tissue of the adjacent dermis. ∞ p. 114 The stratum germinativum and the underlying dermis interlock, increasing the strength of the bond between the epidermis and dermis. The stratum germinativum forms **epidermal ridges**, which extend into the dermis and are adjacent to dermal projections called **dermal papillae** (singular, *papilla*; a nipple-shaped mound) that project into the epidermis (**Figure 5–2a**). These ridges and papillae are significant because the strength of the attachment is proportional to the surface area of the basal lamina: The more and deeper the folds, the larger the surface area becomes.

The contours of the skin surface follow the ridge patterns, which vary from small conical pegs (in thin skin) to the complex whorls seen on the thick skin of the palms and soles. Ridges on the palms and soles increase the surface area of the skin and increase friction, ensuring a secure grip. Ridge shapes are genetically determined. The pattern of your epidermal ridges is unique and does not change during your lifetime. The ridge patterns on the tips of the fingers are the basis

of fingerprints (**Figure 5–4**), which have been used to identify individuals in criminal investigations for more than a century.

Basal cells, or *germinative cells*, dominate the stratum germinativum. Basal cells are stem cells whose divisions replace the more superficial keratinocytes that are lost or shed at the epithelial surface. Skin surfaces that lack hair also contain specialized epithelial cells known as *Merkel cells* scattered among the cells of the stratum germinativum. Merkel cells are sensitive to touch; when compressed, they release chemicals that stimulate sensory nerve endings. (The skin contains many other kinds of sensory receptors, as we will see in later sections.) The brown tones of skin result from the synthetic activities of pigment cells called *melanocytes,* ∞ p. 126 which are distributed throughout the stratum germinativum, with cell processes extending into more superficial layers.

Tips&Tricks

Associate the word *germinativum* with *germinate*, which means to sprout or grow. Just as blades of grass sprout upward and extend beyond the soil, the daughter cells of stem cells dividing in the stratum germinativum are pushed toward the skin surface; on the way, they elongate, acquire organelles, and mature.

Stratum Spinosum

Each time a stem cell divides, one of the daughter cells is pushed superficial to the stratum germinativum into the **stratum spinosum** (**Figure 5–3**), which consists of 8 to 10 layers of keratinocytes bound together by desmosomes. ∞ p. 114 The name *stratum spinosum*, which means "spiny layer," refers to the fact that the cells look like miniature pincushions in standard histological sections. They look that way because the keratinocytes were processed with chemicals that shrank the cytoplasm but left the cytoskeletal elements and desmosomes intact. Some of the cells entering this layer from the stratum germinativum continue to divide, further increasing the thickness of the epithelium. The stratum spinosum also contains *dendritic (Langerhans) cells*, which participate in the immune response by stimulating a defense against (1) microorganisms that manage to penetrate the superficial layers of the epidermis and (2) superficial skin cancers. Dendritic cells and other cells of the immune response will be considered in Chapter 22.

Stratum Granulosum

The region superficial to the stratum spinosum is the **stratum granulosum**, or "grainy layer" (**Figure 5–3**). The stratum granulosum consists of three to five layers of keratinocytes derived from the stratum spinosum. By the time cells reach this layer, most have stopped dividing and have started making large amounts of the proteins **keratin** (KER-a-tin; *keros*, horn) and **keratohyalin** (ker-a-tō-HĪ-a-lin). Keratin, a tough, fibrous protein, is the basic structural component of hair and nails in humans. ∞ p. 118 As keratin fibers develop, the cells grow thinner and flatter, and their membranes thicken and become less permeable. Keratohyalin forms dense cytoplasmic granules that promote dehydration of the cell as well as aggregation and cross-linking of the keratin fibers. The nuclei and other organelles then disintegrate, and the cells die. Further dehydration creates a tightly interlocked layer of cells that consist of keratin fibers surrounded by keratohyalin.

Stratum Lucidum

In the thick skin of the palms and soles, a glassy **stratum lucidum** ("clear layer") covers the stratum granulosum (**Figure 5–3**). The cells in the stratum lucidum are flattened, densely packed, largely devoid of organelles, and filled with keratin.

Stratum Corneum

At the exposed surface of both thick skin and thin skin is the **stratum corneum** (KOR-nē-um; *cornu*, horn) (**Figure 5–3**). It normally contains 15 to 30 layers of keratinized cells. **Keratinization**, or *cornification*, is the formation of protective, superficial layers of cells filled with keratin. This process occurs on all exposed skin surfaces except the anterior surfaces of the eyes. The dead cells in each layer of the stratum corneum remain tightly interconnected by desmosomes. The connections are so secure that keratinized cells are generally shed in large groups or sheets rather than individually.

It takes 15 to 30 days for a cell to move from the stratum germinativum to the stratum corneum. The dead cells generally remain in the exposed stratum corneum for an additional two weeks before they are shed or washed away. This arrangement places the deeper portions of the epithelium and underlying tissues beneath a protective barrier of dead, durable, and expendable cells. Normally, the surface of the stratum corneum is relatively dry, so it is unsuitable for the growth of many microorganisms. Maintenance of this barrier involves coating the surface with lipid secretions from sebaceous glands.

The stratum corneum is water resistant, but not waterproof. Water from interstitial fluids slowly penetrates the surface, to be evaporated into the surrounding air. You lose roughly 500 mL (about 1 pt) of water in this way each day. The process is called **insensible perspiration**, because you are unable to see or feel the water loss. In contrast, you are usually very aware of the **sensible perspiration** produced by active sweat glands. Damage to the epidermis can increase the rate of insensible perspiration. If the damage breaks connections between superficial and deeper layers of the epidermis, fluid will accumulate in pockets, or *blisters*, within the epidermis. (Blisters also form between the epidermis and dermis if the basal lamina is damaged.) If damage to the stratum corneum reduces its effectiveness as a water barrier, the rate of insensible perspiration skyrockets, and a potentially dangerous fluid loss occurs. This is a serious consequence of severe burns and a complication in the condition known as *xerosis* (excessively dry skin).

When the skin is immersed in water, osmotic forces may move water into or out of the epithelium. ∞ p. 91 Sitting in a freshwater bath causes water to move into the epidermis, because fresh water is hypotonic (has fewer dissolved materials) compared with body fluids. The epithelial cells may swell to four times their normal volumes, a phenomenon particularly noticeable in the thickly keratinized areas of the palms and soles. Swimming in the ocean reverses the direction of osmotic flow; because the ocean is a hypertonic solution, water leaves the body, crossing the epidermis from the underlying tissues. The process is slow, but long-term exposure to seawater endangers survivors of a shipwreck by accelerating dehydration.

1. Identify the layers of the epidermis.

2. Dandruff is caused by excessive shedding of cells from the outer layer of skin in the scalp. Thus, dandruff is composed of cells from which epidermal layer?

3. A splinter that penetrates to the third layer of the epidermis of the palm is lodged in which layer?

4. Why does swimming in fresh water for an extended period cause epidermal swelling?

5. Some criminals sand the tips of their fingers so as not to leave recognizable fingerprints. Would this practice permanently remove fingerprints? Why or why not?

See the blue Answers tab at the end of the book.

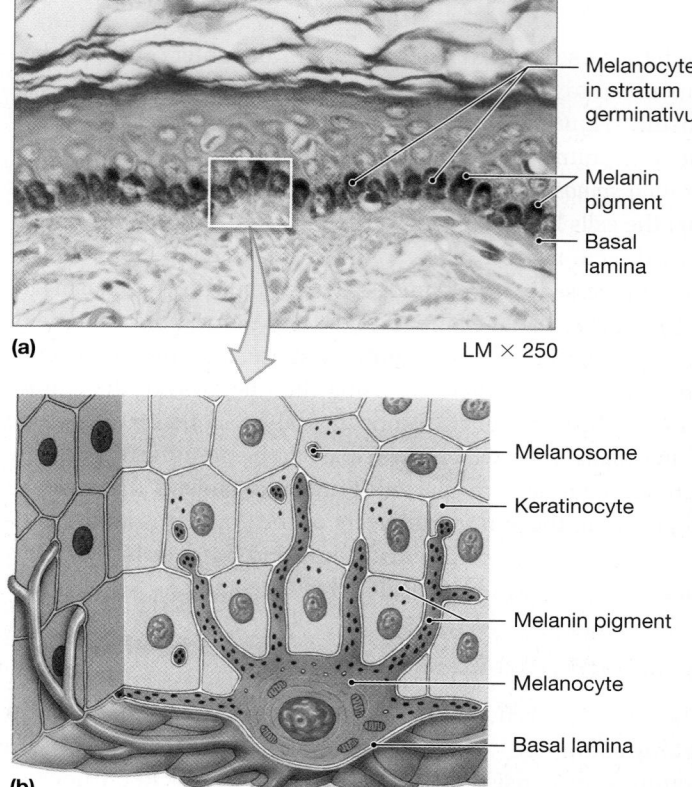

(a)   LM × 250

- Melanocytes in stratum germinativum
- Melanin pigment
- Basal lamina

(b)

- Melanosome
- Keratinocyte
- Melanin pigment
- Melanocyte
- Basal lamina

Figure 5–5 **Melanocytes.** The **(a)** micrograph and **(b)** accompanying drawing indicate the location and orientation of melanocytes in the stratum germinativum of a dark-skinned person.

5-2 Factors influencing skin color are epidermal pigmentation and dermal circulation

In this section we examine how pigments in the epidermis and blood flow in the dermis influence skin color.

The Role of Epidermal Pigmentation

The epidermis contains variable quantities of two pigments: carotene and melanin.

Carotene (KAR-uh-tēn) is an orange-yellow pigment that normally accumulates in epidermal cells. It is most apparent in cells of the stratum corneum of light-skinned individuals, but it also accumulates in fatty tissues in the deep dermis and subcutaneous layer. Carotene is found in a variety of orange vegetables, such as carrots and squashes, and thus the skin of individuals with a special fondness for carrots can actually turn orange from an overabundance of carotene. The color change is very striking in pale-skinned individuals, but less obvious in people with darker skin pigmentation. Carotene can be converted to vitamin A, which is required for both the normal maintenance of epithelia and the synthesis of photoreceptor pigments in the eye.

Melanin is a brown, yellow-brown, or black pigment produced by melanocytes, pigment cells introduced in Chapter 4. The **melanocytes** involved are located in the stratum germinativum, squeezed between or deep to the epithelial cells (**Figure 5–5**). Melanocytes manufacture melanin from the amino acid *tyrosine*, and package it in intracellular vesicles called *melanosomes*. These vesicles travel within the processes of melanocytes and are transferred intact to kerat-

inocytes. The transfer of pigmentation colors the keratinocyte temporarily, until the melanosomes are destroyed by fusion with lysosomes. In individuals with pale skin, this transfer occurs in the stratum germinativum and stratum spinosum, and the cells of more superfical layers lose their pigmentation. In dark-skinned people, the melanosomes are larger and the transfer may occur in the stratum granulosum as well; skin pigmentation is thus darker and more persistent.

The ratio of melanocytes to germinative cells ranges between 1:4 and 1:20, depending on the region of the body. The skin covering most areas of the body has about 1000 melanocytes per square millimeter. The cheeks and forehead, the nipples, and the genital region (the scrotum of males and the labia majora of females) have higher concentrations (about 2000 per square millimeter). The differences in skin pigmentation among individuals do not reflect different numbers of melanocytes, but merely different levels of synthetic activity. Even the melanocytes of *albino* individuals are distributed normally, although the cells are incapable of producing melanin. There can also be localized differences in the rates of melanin production by your melanocytes. *Freckles* are small pigmented areas on relatively pale skin. These spots, which typically have an irregular border, represent the areas serviced by melanocytes that are producing larger-than-average amounts of melanin. Freckles tend to be most abundant on surfaces such as the face, probably owing to its greater exposure to the sun. *Lentigos* are similar to freckles, but have regular borders and contain abnormal melanocytes. *Senile lentigos*, or *liver spots*, are variably pigmented areas that develop on sun-exposed skin in older individuals with pale skin.

The melanin in keratinocytes protects your epidermis and dermis from the harmful effects of sunlight, which contains significant amounts of **ultraviolet (UV) radiation**. A small amount of UV radiation is beneficial, because it stimulates the epidermal production of a compound required for calcium ion homeostasis (a process discussed in a later section). However, UV radiation can also damage DNA, causing mutations and promoting the development of cancer. Within keratinocytes, melanosomes become concentrated in the region around the nucleus, where the melanin pigments act like a sunshade to provide some UV protection for the DNA in those cells.

UV radiation can also produce some immediate effects—burns, which if severe can damage both the epidermis and the dermis. Thus, the presence of pigment layers in the epidermis helps protect both epidermal and dermal tissues. However, although melanocytes respond to UV exposure by increasing their activity, the response is not rapid enough to prevent sunburn the first day you spend at the beach. Melanin synthesis accelerates slowly, peaking about 10 days after the initial exposure. Individuals of any skin color can suffer sun damage to the integument, but dark-skinned individuals have greater initial protection against the effects of UV radiation.

Over time, cumulative damage to the integument by UV exposure can harm fibroblasts, causing impaired maintenance of the dermis. The resulting structural alterations lead to premature wrinkling. In addition, skin cancers can develop from chromosomal damage in germinative cells or melanocytes. One of the major consequences of the global depletion of the ozone layer in Earth's upper atmosphere is likely the sharp increase in the rates of skin cancers (such as *malignant melanoma*) that has been seen in recent years. Such increased cancer rates have been reported in Australia, which has already experienced a significant loss of ozone, as well as in the United States, Canada, and parts of Europe, which have experienced a more moderate ozone loss. For this reason, limiting UV exposure through a combination of protective clothing and sunscreens (or, better yet, sunblocks) is recommended during outdoor activities.

The Role of Dermal Circulation

Blood contains red blood cells filled with the pigment *hemoglobin*, which binds and transports oxygen in the bloodstream. When bound to oxygen, hemoglobin is bright red, giving capillaries in the dermis a reddish tint that is most apparent in lightly pigmented individuals. If those vessels are dilated, the red tones become much more pronounced. For example, your skin becomes flushed and red when your body temperature rises because the superficial blood vessels dilate so that the skin can act like a radiator and lose heat. ∞ p. 14

When its blood supply is temporarily reduced, the skin becomes relatively pale; a light-skinned individual who is frightened may "turn white" as a result of a sudden drop in blood supply to the skin. During a sustained reduction in circulatory supply, the oxygen levels in the tissues decline, and under these conditions, hemoglobin releases oxygen and turns a much darker red. Seen from the surface, the skin then takes on a bluish coloration called **cyanosis** (sī-uh-NŌ-sis; *kyanos*, blue). In individuals of any skin color, cyanosis is most apparent in areas of very thin skin, such as the lips or beneath the nails. It can occur in response to extreme cold or as a result of cardiovascular or respiratory disorders, such as heart failure or severe asthma. Parenting guides often advise parents to watch for "blue lips," because it is a sign that their young children are cold.

Because the skin is easily observed, changes in skin appearance can be useful in diagnosing diseases that primarily affect other body systems. Several diseases can produce secondary effects on skin color and pigmentation:

- In *jaundice* (JAWN-dis), the liver is unable to excrete bile, so a yellowish pigment accumulates in body fluids. In advanced stages, the skin and whites of the eyes turn yellow.

- Some tumors affecting the pituitary gland result in the secretion of large amounts of *melanocyte-stimulating hormone (MSH)*. This hormone causes a darkening of the skin, as if the individual has an extremely deep bronze tan.

- In *Addison disease*, the pituitary gland secretes large quantities of *adrenocorticotropic hormone (ACTH)*, which is structurally similar to MSH. The effect of ACTH on skin color is similar to that of MSH.

- In *vitiligo* (vit-i-LĪ-gō), individuals lose their melanocytes. The condition develops in about 1 percent of the population, and its incidence increases among individuals with thyroid gland disorders, Addison disease, or several other disorders. It is suspected that vitiligo develops when the immune defenses malfunction and antibodies attack normal melanocytes. The primary problem with vitiligo is cosmetic, especially for individuals with darkly pigmented skin.

CLINICAL NOTE

Skin Cancers, Melanomas, and Sunblocks

Almost everyone has several benign tumors of the skin; moles and warts are common examples. However, **skin cancers**, which are more dangerous, are the most common form of cancer.

An *actinic keratosis* is a scaly area on sun-damaged skin. It is an indication that sun damage has occurred, but it is not a sign of skin cancer. In contrast, *basal cell carcinoma* (**Figure 5–6a**), a cancer that originates in the stratum germinativum, is the most common skin cancer. Roughly two-thirds of these cancers appear in body areas subjected to chronic UV exposure. Researchers have identified genetic factors that predispose people to this condition. *Squamous cell carcinomas* are less common, but almost totally restricted to areas of sun-exposed skin. Metastasis seldom occurs in squamous cell carcinomas and virtually never in basal cell carcinomas, and most people survive these cancers. The usual treatment involves the surgical removal of the tumor, and 95 percent of patients survive for five years or longer after treatment. (This statistic, the 5-year survival rate, is a common method of reporting long-term outcomes.)

Unlike these common and seldom life-threatening cancers, *malignant melanomas* (mel-a-NŌ-muz) (**Figure 5–6b**) are extremely dangerous. In this condition, cancerous melanocytes grow rapidly and metastasize through the lymphoid system. The outlook for long-term survival is in many cases determined by how early the condition is diagnosed. If the cancer is detected early, while it is still localized, the 5-year survival rate is 99 percent; if it is not detected until extensive metastasis has occurred, the survival rate drops to 14 percent.

To detect melanoma at an early stage, you must examine your skin, and you must know what to look for. The mnemonic ABCD makes it easy to remember this cancer's key characteristics:

- **A** is for *asymmetry:* Melanomas tend to be irregular in shape. Typically, they are raised; they may also ooze or bleed.

- **B** is for *border:* The border of a melanoma is generally indistinct, irregular, and in some cases notched.

- **C** is for *color:* A melanoma is generally mottled, with any combination of tan, brown, black, red, pink, white, and blue tones.

- **D** is for *diameter:* Any skin growth more than about 5 mm (0.2 in.) in diameter, or roughly the area covered by the eraser on a pencil, is dangerous.

Fair-skinned individuals who live in the tropics are most susceptible to all forms of skin cancer, because their melanocytes are unable to shield them from UV radiation. Sun damage can be prevented by avoiding exposure to the sun during the middle of the day and by using a sunblock (not a tanning oil) before any sun exposure. This practice also delays the cosmetic problems of aging and wrinkling. *Everyone* who spends any time out in the sun should choose a broad-spectrum sunblock with a sun protection factor (SPF) of at least 15; blondes, redheads, and people with very pale skin are better off with an SPF of 20 to 30. (The risks are the same for those who spend time in a tanning salon or tanning bed.) The protection offered by these "sunscreens" is afforded by both organic molecules that absorb UV radiation and inorganic pigments that absorb, scatter, and reflect UV rays. The higher the SPF factor, the more of these chemicals the product contains, and the fewer UV rays are able to penetrate to the skin's surface. Wearing a hat with a brim and panels to shield the neck and face, long pants, and long-sleeved shirts provide added protection.

The use of sunblocks will be even more important as the ozone gas in the upper atmosphere is further destroyed by our industrial emissions. Ozone absorbs UV radiation before it reaches Earth's surface; in doing so, ozone assists the melanocytes in preventing skin cancer. Australia, the continent that is most affected by the depletion of ozone above the South Pole (the "ozone hole"), is already reporting an increased incidence of skin cancers.

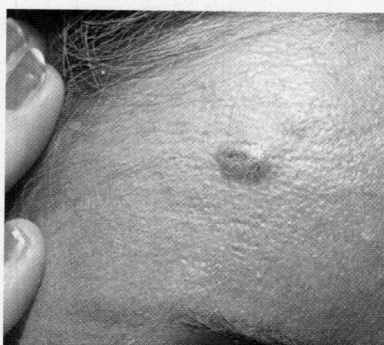

(a) Basal cell carcinoma

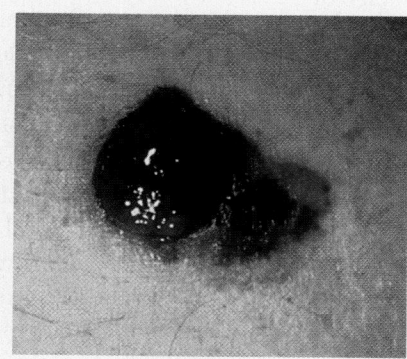

(b) Melanoma

Figure 5–6 **Skin Cancers. (a)** Basal cell carcinoma. **(b)** Melanoma.

CHECKPOINT

6. Name the two pigments contained in the epidermis.

7. Why does exposure to sunlight or sunlamps darken skin?

8. Why does the skin of a fair-skinned person appear red during exercise in hot weather?

See the blue Answers tab at the end of the book.

5-3 Sunlight causes epidermal cells to convert a steroid into vitamin D₃

Although too much sunlight can damage epithelial cells and deeper tissues, limited exposure to sunlight is beneficial. When exposed to ultraviolet radiation, epidermal cells in the stratum spinosum and stratum germinativum convert a cholesterol-related steroid into **cholecalciferol** (kō-le-kal-SIF-er-ol), or **vitamin D₃**. The liver then converts cholecalciferol into an intermediary product used by the kidneys to synthesize the hormone **calcitriol** (kal-si-TRĪ-ol). Calcitriol is essential for the normal absorption of calcium and phosphorus by the small intestine; an inadequate supply leads to impaired bone maintenance and growth.

The term *vitamin* is usually reserved for essential organic nutrients that must be obtained from the diet because the body either cannot make them or makes them in insufficient amounts. If present in the diet, cholecalciferol can be absorbed by the digestive tract, and if the skin cannot make enough cholecalciferol, a dietary supply will maintain normal bone development. Under these circumstances, dietary cholecalciferol acts like a vitamin, and this accounts for the alternative name for cholecalciferol: *vitamin D₃*. If cholecalciferol cannot be produced by the skin and is not included in the diet, bone development is abnormal and bone maintenance is inadequate. For example, children who live in areas with overcast skies and whose diet lacks cholecalciferol can have abnormal bone development. This condition, called *rickets* (**Figure 5–7**), has largely been eliminated in the United States because dairy companies are required to add cholecalciferol, usually identified as "vitamin D," to the milk sold in grocery stores. In Chapter 6, we will consider the hormonal control of bone growth in greater detail.

CHECKPOINT

9. Explain the relationship between sunlight exposure and vitamin D₃ synthesis.

10. In some cultures, women must be covered completely, except for their eyes, when they go outside. Explain why these women may develop bone problems later in life.

See the blue Answers tab at the end of the book.

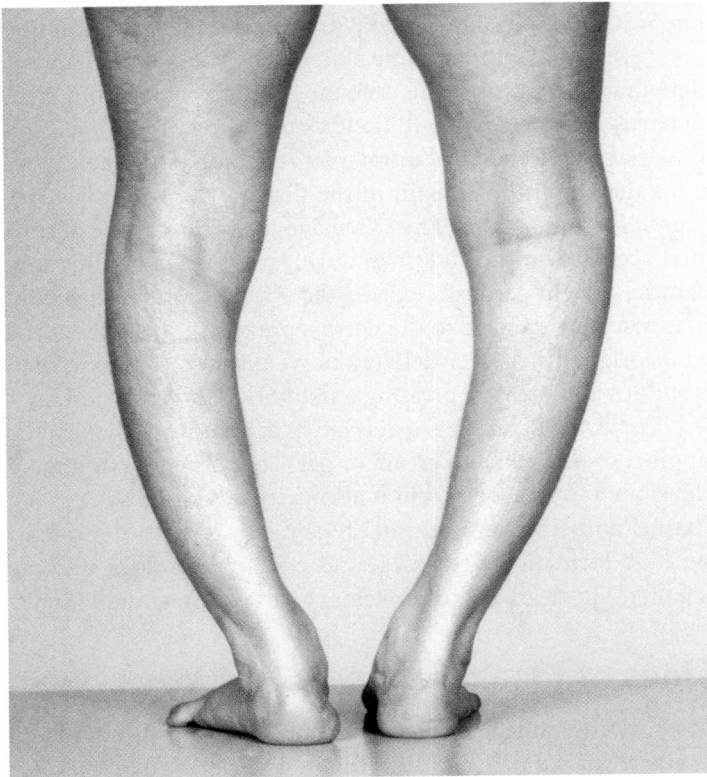

Figure 5–7 Rickets. Rickets, a disease caused by vitamin D₃ deficiency, results in the bending of abnormally weak and flexible bones under the weight of the body, plus other structural changes.

5-4 Epidermal growth factor has several effects on the epidermis and epithelia

Epidermal growth factor (EGF) is one of the peptide growth factors introduced in Chapter 3. ∞ p. 104 Although named for its effects on the epidermis, we now know that EGF has widespread effects on epithelia throughout the body, and that it is produced by the salivary glands and glands of the duodenum. Among the roles of EGF are the following:

- Promoting the divisions of germinative cells in the stratum germinativum and stratum spinosum

- Accelerating the production of keratin in differentiating keratinocytes

- Stimulating epidermal development and epidermal repair after injury

- Stimulating synthetic activity and secretion by epithelial glands

In the procedure known as tissue culture, cells are grown under laboratory conditions for experimental or therapeutic use. Epidermal growth factor has such a pronounced effect that it can be used in tissue culture to stimulate the growth and division of epidermal cells (or other epithelial cells). It is now possible to grow sheets of epidermal cells for use in the treatment of severe or extensive burns. The burned areas can be covered by epidermal sheets "grown" from a small sample of intact skin from another part of the burn victim's body. (We will consider this treatment later in the chapter when we discuss burns.)

The A&P Top 100

#26 The epidermis is a multilayered, flexible, self-repairing barrier that prevents fluid loss, provides protection from UV radiation, produces vitamin D_3, and resists damage from abrasion, chemicals, and pathogens.

CHECKPOINT

11. **Name the sources of epidermal growth factor in the body.**

12. **Identify some roles of epidermal growth factor pertaining to the epidermis.**

See the blue Answers tab at the end of the book.

5-5 The dermis is the tissue layer that supports the epidermis

The dermis lies between the epidermis and the subcutaneous layer (**Figure 5–1**). In this section we will discuss the organization and properties of the dermis, and dermal circulation and innervation. The dermis has two major components: (1) a superficial *papillary layer* and (2) a deeper *reticular layer* (**Figure 5–1**).

The **papillary layer**, which consists of areolar tissue, contains the capillaries, lymphatics, and sensory neurons that supply the surface of the skin. The papillary layer derives its name from the dermal papillae that project between the epidermal ridges (**Figure 5–2a**).

The **reticular layer**, deep to the papillary layer, consists of an interwoven meshwork of dense irregular connective tissue containing both collagen and elastic fibers. ∞ p. 130 Bundles of collagen fibers extend superficially beyond the reticular layer to blend into those of the papillary layer, so the boundary between the two layers is indistinct. Collagen fibers of the reticular layer also extend into the deeper subcutaneous layer. In addition to extracellular protein fibers, the dermis contains all the cells of connective tissue proper. ∞ p. 125 Accessory organs of epidermal origin, such as hair follicles and sweat glands, extend into the dermis. In addition, the reticular and papillary layers of the dermis contain networks of blood vessels, lymphatic vessels, and nerve fibers (**Figure 5–1**).

Because of the abundance of sensory receptors in the skin, regional infection or inflammation can be very painful. **Dermatitis** (der-muh-TĪ-tis) is an inflammation of the skin that primarily involves the papillary layer. The inflammation typically begins in a part of the skin exposed to infection or irritated by chemicals, radiation, or mechanical stimuli. Dermatitis may cause no discomfort, or it may produce an annoying itch, as in poison ivy. Other forms of the condition can be quite painful, and the inflammation can spread rapidly across the entire integument.

Dermal Strength and Elasticity

The presence of two types of fibers—*collagen fibers*, which are very strong and resist stretching but are easily bent or twisted, and *elastic fibers*, which permit stretching and then recoil to their original length—enables the dermis to tolerate limited stretching. The elastic fibers provide flexibility, and the collagen fibers limit that flexibility to prevent damage to the tissue.

The water content of the skin also helps maintain its flexibility and resilience, properties collectively known as *skin turgor*. One of the signs of dehydration is the loss of skin turgor, revealed by pinching the skin on the back of the hand. A dehydrated dermis will remain peaked when pinched, whereas hydrated skin will flatten out. Aging, hormones, and the destructive effects of ultraviolet radiation permanently reduce the amount of elastin in the dermis; the result is wrinkles and sagging skin. The extensive distortion of the dermis that occurs over the abdomen during pregnancy or after substantial weight gain can exceed the elastic limits of the skin. The resulting damage to the dermis prevents it from recoiling to its original size after delivery or weight loss. The skin then wrinkles and creases, creating a network of **stretch marks**.

Tretinoin (Retin-A) is a derivative of vitamin A that can be applied to the skin as a cream or gel. This drug was originally developed to treat acne, but it also increases blood flow to the dermis and stimulates dermal repair. As a result, the rate of wrinkle formation decreases, and existing wrinkles become smaller. The degree of improvement varies among individuals.

Lines of Cleavage

Most of the collagen and elastic fibers at any location are arranged in parallel bundles oriented to resist the forces applied to the skin during normal movement. The resulting pattern of fiber bundles establishes **lines of cleavage** of the skin (**Figure 5–8**). Lines of cleavage are clinically significant: A cut

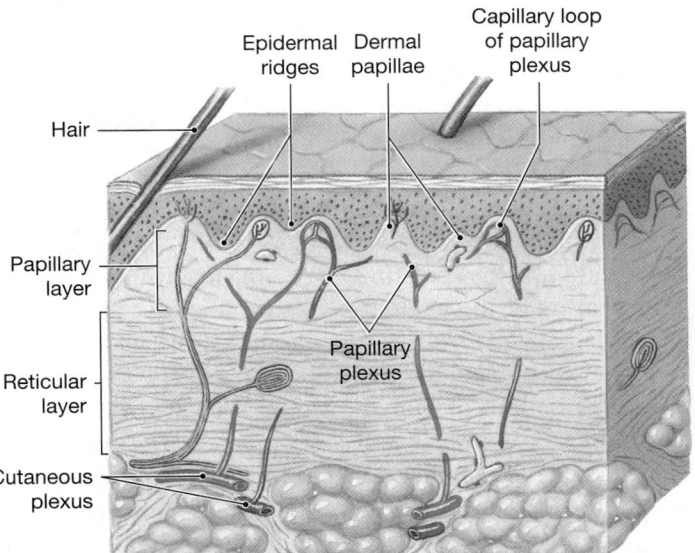

Figure 5–9 **Dermal Circulation.** Shown are the cutaneous and papillary plexuses.

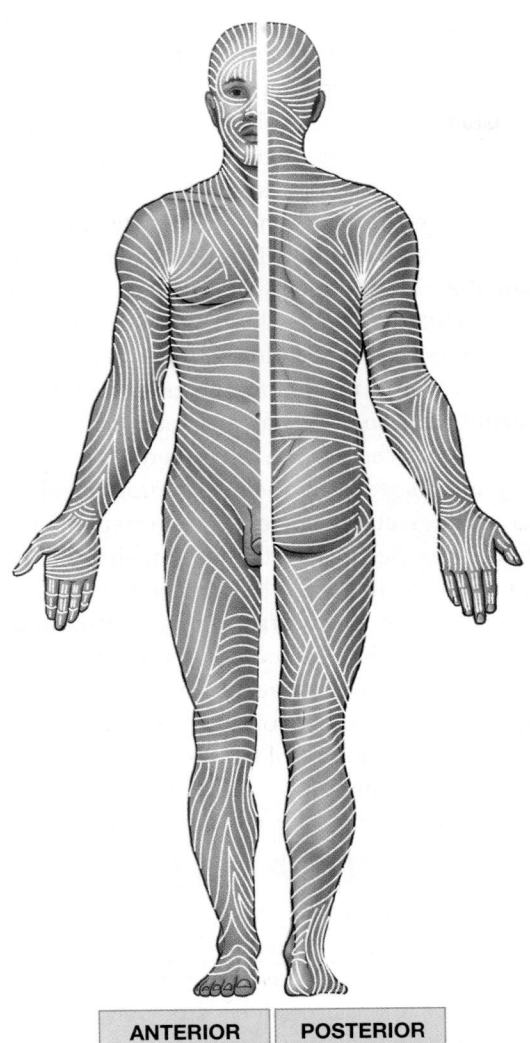

ANTERIOR | POSTERIOR

Figure 5–8 **Lines of Cleavage of the Skin.** Lines of cleavage follow the pattern of fiber bundles in the skin. They reflect the orientation of collagen fiber bundles in the dermis.

parallel to a cleavage line will usually remain closed and heal with little scarring, whereas a cut at right angles to a cleavage line will be pulled open as severed elastic fibers recoil and will result in greater scarring. For these reasons, surgeons choose to make neat incisions parallel to the lines of cleavage.

The Dermal Blood Supply

Arteries supplying the skin form networks in the subcutaneous layer along its border with the reticular layer of the dermis. This network is called the *cutaneous plexus* (**Figure 5–9**). Tributaries of these arteries supply both the adipose tissues of the subcutaneous layer and the tissues of the integument. As small arteries travel toward the epidermis, branches supply the hair follicles, sweat glands, and other structures in the

dermis. On reaching the papillary layer, the small arteries form another branching network, the *papillary plexus*, which provides arterial blood to capillary loops that follow the contours of the epidermis–dermis boundary (**Figure 5–9**). These capillaries empty into a network of small veins that form a venous plexus deep to the papillary plexus. This network is in turn connected to a larger venous plexus in the subcutaneous layer. Trauma to the skin often results in a *contusion*, or bruise. As a result of the rupture of dermal blood vessels, blood leaks into the dermis, and the area develops the familiar "black and blue" color.

CLINICAL NOTE

Ulcers Problems with dermal circulation affect both the epidermis and the dermis. An *ulcer* is a localized shedding of an epithelium. *Decubitis ulcers*, or *bedsores*, affect patients whose circulation is restricted, especially when a splint, a cast, or lying in bed continuously compresses superficial blood vessels. Such sores most commonly affect the skin covering joints or bony prominences, where dermal blood vessels are pressed against deeper structures. The chronic lack of circulation kills epidermal cells, removing a barrier to bacterial infection; eventually, dermal tissues deteriorate as well. (Cell death and tissue destruction, or *necrosis*, can occur in any tissue deprived of adequate blood flow.) Bedsores can be prevented or treated by frequently changing the position of the body or by placing patients in specially designed beds containing deflating and inflating air coils; both approaches vary the pressures applied to local blood vessels.

Innervation of the Skin

The integument is filled with sensory receptors, and anything that comes in contact with the skin—from the lightest touch of a mosquito to the weight of a loaded backpack—initiates a nerve impulse that can reach our conscious awareness. Nerve fibers in the skin control blood flow, adjust gland secretion rates, and monitor sensory receptors in the dermis and the deeper layers of the epidermis. We have already noted that the deeper layers of the epidermis contain Merkel cells; sensory terminals known as *tactile discs* monitor these cells. The epidermis also contains the extensions of sensory neurons that provide sensations of pain and temperature. The dermis contains similar receptors, as well as other, more specialized receptors. Examples shown in **Figure 5–1** include receptors sensitive to light touch—*tactile corpuscles*, located in dermal papillae—and receptors sensitive to deep pressure and vibration—*lamellated corpuscles*, in the reticular layer.

Even this partial list of the receptors found in the skin is enough to highlight the importance of the integument as a sensory structure. We will return to this topic in Chapter 15, where we consider not only what receptors are present, but how they function.

The A&P Top 100

#27 The dermis provides mechanical strength, flexibility, and protection for underlying tissues. It is highly vascular and contains a variety of sensory receptors that provide information about the external environment.

CHECKPOINT

13. Describe the location of the dermis.

14. Where are the capillaries and sensory neurons that supply the epidermis located?

15. What accounts for the ability of the dermis to undergo repeated stretching?

See the blue Answers tab at the end of the book.

5-6 The hypodermis is tissue beneath the dermis that connects it to underlying tissues

The connective tissue fibers of the reticular layer are extensively interwoven with those of the subcutaneous layer, or **hypodermis**. The boundary between the two is generally indistinct (**Figure 5–1**). Although the hypodermis is not a part of the integument, it is important in stabilizing the position of the skin in relation to underlying tissues, such as skeletal muscles or other organs, while permitting independent movement.

The hypodermis consists of areolar and adipose tissues and is quite elastic. Only its superficial region contains large arteries and veins. The venous circulation of this region contains a substantial amount of blood, and much of this volume will shift to the general circulation if these veins constrict. For that reason, the skin is often described as a *blood reservoir*. The rest of the hypodermis contains a limited number of capillaries and no vital organs. This last characteristic makes **subcutaneous injection**—by means of a **hypodermic needle**—a useful method of administering drugs.

Most infants and small children have extensive "baby fat," which provides extra insulation and helps reduce heat loss. Subcutaneous fat also serves as a substantial energy reserve and as a shock absorber for the rough-and-tumble activities of our early years. As we grow, the distribution of subcutaneous fat changes. The greatest changes occur in response to circulating sex hormones. Beginning at puberty, men accumulate subcutaneous fat at the neck, on the arms, along the lower back, and over the buttocks. In contrast, women accumulate subcutaneous fat at the breasts, buttocks, hips, and thighs. In adults of either gender, the subcutaneous layer of the backs of the hands and the upper surfaces of the feet contain few fat cells, whereas distressing amounts of adipose tissue can accumulate in the abdominal region, producing a prominent "potbelly."

CLINICAL NOTE

Liposuction   The accumulation of excessive amounts of adipose tissue increases the risks of diabetes, stroke, and other serious conditions. Dietary restrictions and increased activity levels are often successful in promoting weight loss and reducing these risks. However, a "quick fix" is often promised by a surgical procedure called **liposuction** or **lipoplasty**. In this procedure, subcutaneous adipose tissue is removed through a tube inserted deep to the skin. Adipose tissue tears relatively easily, and suction applied to the tube rips chunks of adipose tissue from the body. After liposuction, the skin is loose fitting, and until it recoils, a tight-fitting garment is usually worn. Liposuction is relatively common and is increasing in popularity. In 2006, liposuction was the leading cosmetic surgical procedure, and 403,684 procedures were performed in the United States.

Although it might sound like an easy way to remove unwanted fat, in practice, liposuction can be dangerous. There are risks from anesthesia, bleeding (adipose tissue is quite vascular), sensory loss, infection, and fluid loss. The death rate from liposuction procedures is 1 in 5000, very high for what is basically cosmetic surgery that provides only a temporary solution to a chronic problem. As noted in Chapter 4, unless there are changes in diet and lifestyle, the damaged adipose tissue will repair itself, and areas of areolar tissue will convert to adipose tissue. Over time, the surgery will have to be repeated.

5-7 Hair is composed of keratinized dead cells that have been pushed to the surface

Hair and several other structures—hair follicles, sebaceous and sweat glands, and nails—are considered accessory structures of the integument. During embryological development, these structures originate from the epidermis, so they are also known as *epidermal derivatives*. Although located in the dermis, they project through the epidermis to the integumentary surface.

ATLAS: Embryology Summary 5: The Development of the Integumentary System

Hairs project above the surface of the skin almost everywhere, except over the sides and soles of the feet, the palms of the hands, the sides of the fingers and toes, the lips, and portions of the external genitalia. The human body has about 2.5 million hairs, and 75 percent of them are on the general body surface, not on the head. Hairs are nonliving structures produced in organs called **hair follicles**.

The hairs and hair follicles on your body have important functions. The roughly 500,000 hairs on your head protect your scalp from ultraviolet radiation, help cushion light impacts to the head, and insulate the skull. The hairs guarding the entrances to your nostrils and external ear canals help prevent the entry of foreign particles and insects, and your eyelashes perform a similar function for the surface of the eye. Eyebrows are also important because they help keep sweat out of your eyes. However, hairs are also extremely important as sensory receptors.

Figure 5–10 illustrates important details about the structure of hairs and hair follicles. Each hair follicle opens onto the surfaces of the epidermis but extends deep into the dermis and usually into the hypodermis. Deep to the epidermis, each follicle is wrapped in a dense connective tissue sheath. A **root hair plexus** of sensory nerves surrounds the base of each hair follicle (**Figure 5–10a**). As a result, you can feel the movement of the shaft of even a single hair. This sensitivity provides an early-warning system that may help prevent injury; for example, you may be able to swat a mosquito before it reaches your skin.

A bundle of smooth muscle cells forms the **arrector pili** (a-REK-tor PI-lē; plural, *arrectores pilorum*) muscle, which extends from the papillary layer of the dermis to the connective tissue sheath surrounding the hair follicle. When stimu-lated, the arrector pili muscle contracts, pulling on the follicle and forcing the hair to stand erect. Contraction may be the result of emotional states, such as fear or rage, or a response to cold, producing "goose bumps." In a furry mammal, this action increases the thickness of its insulating coat. Although humans do not receive any comparable insulating benefits, the response persists.

Each hair is a long, cylindrical structure that extends outward, past the epidermal surface (**Figure 5–10a,b**). The **hair root**—the portion that anchors the hair into the skin—begins at the base of the hair, at the *hair bulb*, and extends distally to the point at which the internal organization of the hair is complete, about halfway to the skin surface. The **hair shaft**, part of which we see on the surface, extends from this halfway point to the exposed tip of the hair.

Hair Production

Hair production begins at the base of a hair follicle (**Figure 5–10b,c**). Here a mass of epithelial cells forms a cap, called the **hair bulb**, that surrounds a small **hair papilla**, a peg of connective tissue containing capillaries and nerves. The superficial cells of the hair bulb are responsible for producing the hair; they form a layer called the **hair matrix**. Basal cells near the center of the hair matrix divide, producing daughter cells that are gradually pushed toward the surface. Daughter cells closest to the center of the matrix form the **medulla**, or core, of the hair. Daughter cells farther from the center of the hair matrix form the **cortex**, an intermediate layer. Those at the edges of the hair matrix form the **cuticle**, which will be the surface of the hair.

As cell divisions continue at the hair matrix, the daughter cells are pushed toward the surface of the skin, and the hair gets longer. Keratinization is completed by the time these cells approach the surface. At this level, which corresponds to the start of the hair shaft, the cells of the medulla, cortex, and cuticle are dead, and the keratinization process is at an end. The medulla contains a flexible **soft keratin**; the cortex and cuticle contain thick layers of **hard keratin**, which give the hair its stiffness. The cuticle consists of overlapping layers of dead, flattened, heavily keratinized cells.

The epithelial cells of the follicle walls are organized into several concentric layers (**Figure 5–10c,d**). Moving outward from the hair cuticle, these layers include:

- The **internal root sheath**, which surrounds the hair root and the deeper portion of the shaft. The internal root sheath is produced by the cells at the periphery of the hair matrix. The cells of this sheath disintegrate quickly. This layer does not extend the entire length of the follicle.

- The **external root sheath**, which in longitudinal section extends from the skin surface to the hair matrix. Over most of that distance, it has all the cell layers found in

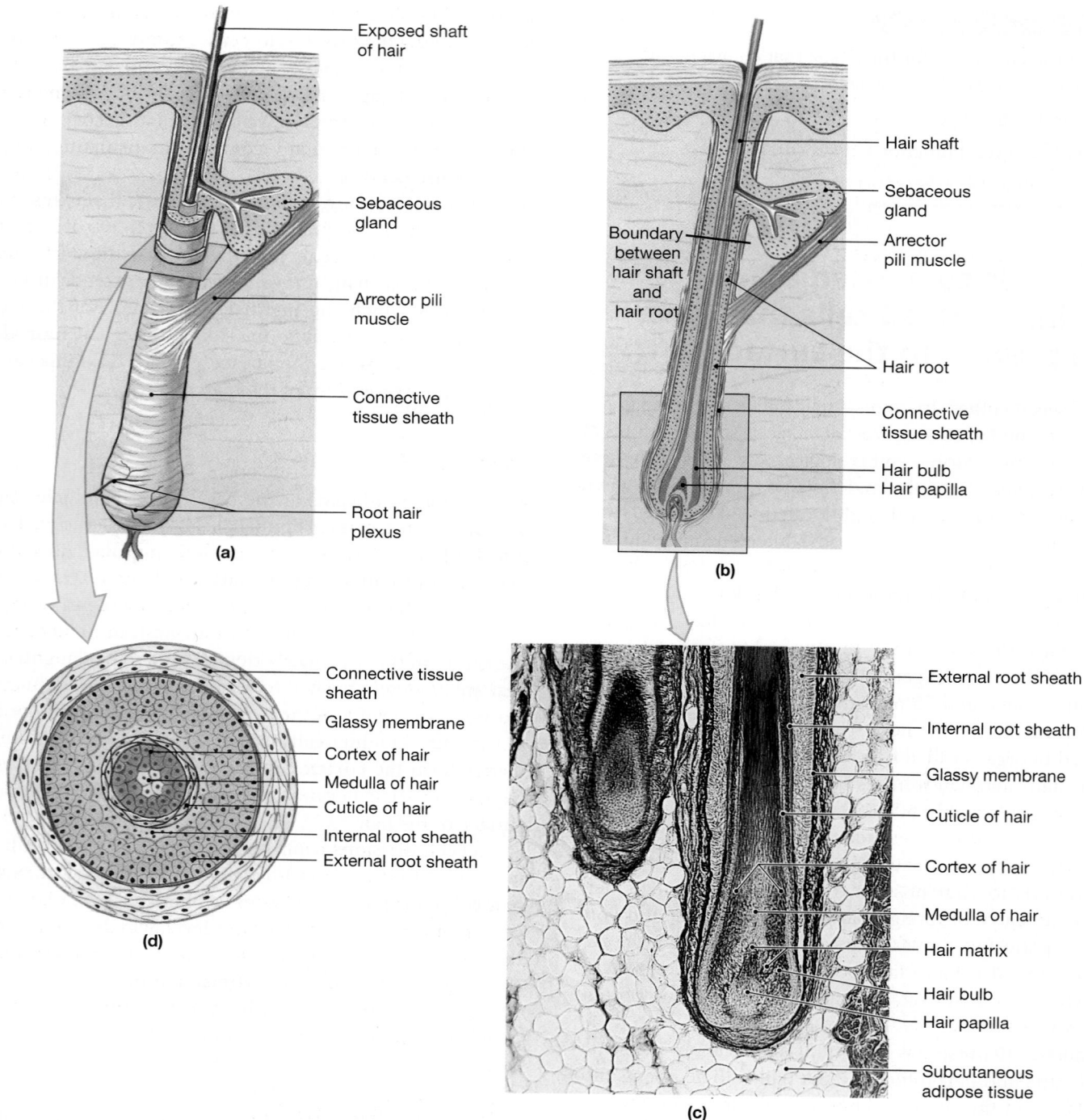

Figure 5–10 **Hair Follicles and Hairs.** (a) A single hair follicle, showing the associated accessory structures; a superficial view of the deeper portions of the follicle illustrates the connective tissue sheath and the root hair plexus. (b) A diagrammatic sectional view along the long axis of a hair follicle. (c) A longitudinal section through two hair follicles, showing the base of the follicle and the matrix and papilla at the root of the hair. (d) A cross section through a hair follicle and a hair, near the junction between the hair root and hair shaft.

the superficial epidermis. However, where the external root sheath joins the hair matrix, all the cells resemble those of the stratum germinativum.

- The **glassy membrane**, a thickened clear layer wrapped in a dense connective tissue sheath. This membrane is in contact with the surrounding connective tissues of the dermis.

The Hair Growth Cycle

Hairs grow and are shed according to a **hair growth cycle**. A hair in the scalp grows for two to five years, at a rate of about 0.33 mm per day. Variations in the growth rate and in the duration of the hair growth cycle account for individual differences in the length of uncut hair.

While hair is growing, the cells of the hair root absorb nutrients and incorporate them into the hair structure. As a result, clipping or collecting hair for analysis can be helpful in diagnosing several disorders. For example, hairs of individuals with lead poisoning or other heavy-metal poisoning contain high levels of those metal ions. Hair samples containing nucleated cells can also be used for identification purposes through DNA fingerprinting. ∞ p. 84

As it grows, the root is firmly attached to the matrix of the follicle. At the end of the growth cycle, the follicle becomes inactive. The hair is now termed a **club hair**. The follicle gets smaller, and over time the connections between the hair matrix and the club hair root break down. When another cycle begins, the follicle produces a new hair; the old club hair is pushed to the surface and is shed.

If you are a healthy adult, you lose about 50 hairs from your head each day. Sustained losses of more than 100 hairs per day generally indicate that a net loss of hairs is under way, and noticeable hair loss will eventually result. Temporary increases in hair loss can result from drugs, dietary factors, radiation, an excess of vitamin A, high fever, stress, or hormonal factors related to pregnancy. In males, changes in the level of the sex hormones circulating in the blood can affect the scalp, causing a shift in the type of hair produced (discussed shortly), beginning at the temples and the crown of the head. This alteration is called *male pattern baldness*. Some cases of male pattern baldness respond to drug therapies, such as the topical application of *minoxidil (Rogaine)*.

Types of Hairs

Hairs first appear after roughly three months of embryonic development. These hairs, collectively known as *lanugo* (la-NOO-gō), are extremely fine and unpigmented. Most lanugo hairs are shed before birth. They are replaced by one of two types of hairs in the adult integument: vellus hairs or terminal hairs. **Vellus hairs** are the fine "peach fuzz" hairs located over much of the body surface. **Terminal hairs** are heavy, more deeply pigmented, and sometimes curly. The hairs on your head, including your eyebrows and eyelashes, are terminal hairs that are present throughout life. Hair follicles may alter the structure of the hairs in response to circulating hormones. For example, vellus hairs are present at the armpits, pubic area, and limbs until puberty; thereafter, the follicles produce terminal hairs, in response to circulating sex hormones.

Tips&Tricks

Associate the word *vellus* ("peach fuzz") with "velvet."

Hair Color

Variations in hair color reflect differences in structure and variations in the pigment produced by melanocytes at the hair papilla. Different forms of melanin give a dark brown, yellow-brown, or red color to the hair. These structural and biochemical characteristics are genetically determined, but hormonal and environmental factors also influence the condition of your hair. As pigment production decreases with age, hair color lightens. White hair results from the combination of a lack of pigment and the presence of air bubbles in the medulla of the hair shaft. As the proportion of white hairs increases, the individual's overall hair color is described as gray. Because hair itself is dead and inert, any changes in its coloration are gradual. We are able to change our hair color by using harsh chemicals that disrupt the cuticle and permit dyes to enter and stain the cortex and medulla. These color treatments damage the hair by disrupting the protective cuticle layer and dehydrating and weakening the hair shaft. As a result, the hair becomes thin and brittle. Cream rinses, conditioners, and oil treatments attempt to reduce the effects of this structural damage by rehydrating and recoating the shaft.

CHECKPOINT

19. Describe a typical strand of hair.

20. What happens when the arrector pili muscle contracts?

21. Once a burn on the forearm that destroys the epidermis and extensive areas of the deep dermis heals, will hair grow again in the affected area?

See the blue Answers tab at the end of the book.

5-8 Sebaceous glands and sweat glands are exocrine glands found in the skin

In this section we examine the structure and function of various integumentary system accessory structures that produce exocrine secretions, with a focus on sebaceous glands and sweat glands.

Sebaceous (Oil) Glands

Sebaceous (se-BĀ-shus) **glands**, or *oil glands*, are holocrine glands that discharge an oily lipid secretion into hair follicles (**Figure 5–11**). Sebaceous glands that communicate with a single follicle share a duct and thus are classified as simple branched alveolar glands. ∞ p. 123 The gland cells produce large quantities of lipids as they mature. The lipid product is released through holocrine secretion, a process that involves the rupture of the secretory cells. ∞ p. 121

The lipids released from gland cells enter the lumen (open passageway) of the gland. The arrector pili muscles that erect the hair then contract, squeezing the sebaceous gland and forcing the lipids into the hair follicle and onto the surface of the skin. The secreted lipid product, called **sebum** (SĒ-bum), is a mixture of triacylglycerides, cholesterol, proteins, and electrolytes. Sebum inhibits the growth of bacteria, lubricates and protects the keratin of the hair shaft, and conditions the surrounding skin. Keratin is a tough protein, but dead, keratinized cells become dry and brittle once exposed to the environment. It is interesting to reflect on our custom of washing and shampooing to remove the oily secretions of sebaceous glands, only to add other lipids to the hair in the form of conditioners, and to the skin in the form of creams and lotions.

Sebaceous follicles are large sebaceous glands that are not associated with hair follicles; their ducts discharge sebum directly onto the epidermis (**Figure 5–11**). Sebaceous follicles are located on the face, back, chest, nipples, and external genitalia.

Surprisingly, sebaceous glands are very active during the last few months of fetal development. Their secretions, mixed with shed epidermal cells, form a protective superficial layer—the *vernix caseosa*—that coats the skin surface. Sebaceous gland activity all but ceases after birth, but it increases again at puberty in response to rising levels of sex hormones.

Seborrheic dermatitis is an inflammation around abnormally active sebaceous glands, most often those of the scalp. The affected area becomes red and oily, and increased epidermal scaling occurs. In infants, mild cases are called *cradle cap*. Seborrheic dermatitis is a common cause of dandruff in adults. Anxiety, stress, and fungal or bacterial infections can aggravate the problem.

Sweat Glands

The skin contains two types of sweat glands, or **sudoriferous glands** (*sudor*, sweat): *apocrine sweat glands* and *merocrine sweat glands* (**Figure 5–12**).

Apocrine Sweat Glands

In the armpits (axillae), around the nipples, and in the pubic region, **apocrine sweat glands** secrete their products into hair follicles (**Figure 5–12a**). These coiled, tubular glands produce a sticky, cloudy, and potentially odorous secretion. The name *apocrine* was originally chosen because it was thought the gland cells use an apocrine method of secretion. ∞ p. 121 Although we now know that they rely on merocrine secretion, the name has not changed.

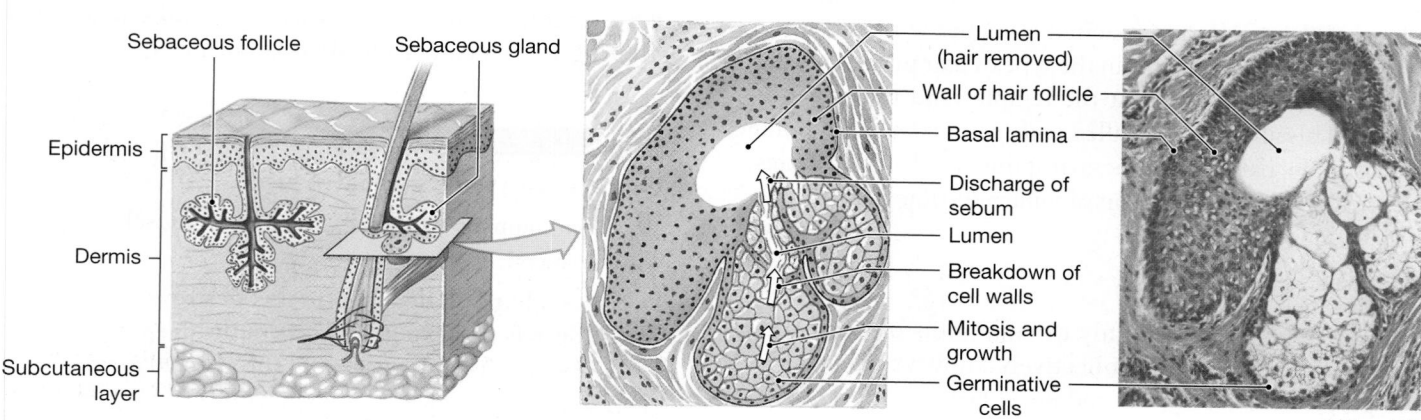

Sebaceous follicle   Sebaceous gland

Epidermis

Dermis

Subcutaneous layer

Lumen (hair removed)

Wall of hair follicle

Basal lamina

Discharge of sebum

Lumen

Breakdown of cell walls

Mitosis and growth

Germinative cells

LM × 150

Figure 5–11 **The Structure of Sebaceous Glands and Sebaceous Follicles.**

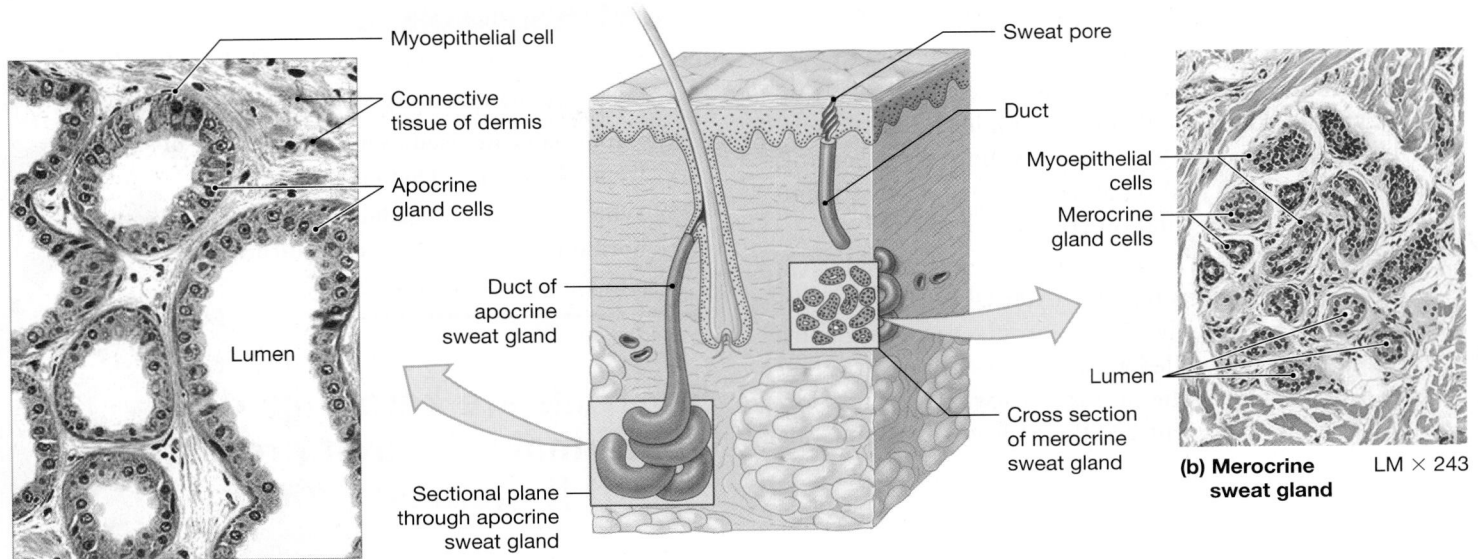

(a) Apocrine sweat gland LM × 459

(b) Merocrine sweat gland LM × 243

Figure 5–12   **Sweat Glands. (a)** Apocrine sweat glands, which secrete a thick, odorous fluid into hair follicles. **(b)** Merocrine sweat glands, which discharge a watery fluid onto the surface of the skin.

Apocrine sweat glands begin secreting at puberty. The sweat produced is a nutrient source for bacteria, which intensify its odor. Surrounding the secretory cells in these (and other exocrine) glands are special **myoepithelial cells** (*myo-*, muscle), which by contracting squeeze the gland and discharge the accumulated secretion into the hair follicles. The secretory activities of the gland cells and the contractions of myoepithelial cells are controlled by the nervous system and by circulating hormones.

Merocrine (Eccrine) Sweat Glands

Merocrine sweat glands are also known as **eccrine** (EK-rin) **sweat glands**. These are coiled, tubular glands that discharge their secretions directly onto the surface of the skin (**Figure 5–12b**). Eccrine sweat glands are far more numerous and widely distributed than apocrine sweat glands. The adult integument contains 2–5 million merocrine sweat glands, which are smaller than apocrine sweat glands and do not extend as deeply into the dermis. The palms and soles have the highest numbers, with the palm possessing an estimated 500 merocrine sweat glands per square centimeter (3000 per square inch).

As noted earlier, the sweat produced by merocrine sweat glands is called sensible perspiration. Sweat is 99 percent water, but it also contains some electrolytes (chiefly sodium chloride), a number of organic nutrients, a peptide with antibiotic properties, and various waste products. It has a pH of 4.0–6.8, and the presence of sodium chloride gives sweat a salty taste. (See the Appendix for a complete analysis of the composition of normal sweat.)

The functions of merocrine sweat gland activity include the following:

- *Cooling the Surface of the Skin to Reduce Body Temperature.* This is the primary function of sensible perspiration. The degree of secretory activity is regulated by neural and hormonal mechanisms; when all the merocrine sweat glands are working at their maximum, the rate of perspiration can exceed a gallon per hour, and dangerous fluid and electrolyte losses can occur. For this reason, athletes participating in endurance sports must drink fluids at regular intervals.

- *Excreting Water and Electrolytes.* A number of metabolized drugs are excreted as well.

- *Providing Protection from Environmental Hazards.* Sweat dilutes harmful chemicals in contact with the skin and discourages the growth of microorganisms in two ways: (1) by either flushing them from the surface or making it difficult for them to adhere to the epidermal surface, and (2) through the action of *dermicidin*, a small peptide that has powerful antibiotic properties.

Other Integumentary Glands

As we have seen, merocrine sweat glands are widely distributed across the body surface, sebaceous glands are located wherever there are hair follicles, and apocrine sweat glands are located in relatively restricted areas. The skin also contains a variety of specialized glands that are restricted to specific locations. Two examples of particular importance are the following:

1. The **mammary glands** of the breasts are anatomically related to apocrine sweat glands. A complex interaction between sex hormones and pituitary hormones controls their development and secretion. We will discuss mammary gland structure and function in Chapter 28.

2. Ceruminous (se-ROO-mi-nus) glands are modified sweat glands in the passageway of the external ear. Their secretions combine with those of nearby sebaceous glands, forming a mixture called cerumen, or earwax. Together with tiny hairs along the ear canal, earwax helps trap foreign particles, preventing them from reaching the eardrum.

Control of Glandular Secretions and the Homeostatic Role of the Integument

The autonomic nervous system (ANS) controls the activation and deactivation of sebaceous glands and apocrine sweat glands at the subconscious level. Regional control is not possible; the commands issued by the ANS affect all the glands of that type, everywhere on the body surface. Merocrine sweat glands are much more precisely controlled, and the amount of secretion and the area of the body involved can vary independently. For example, when you are nervously awaiting an anatomy and physiology exam, only your palms may begin to sweat.

As we noted earlier, the primary function of sensible perspiration is to cool the surface of the skin and to reduce body temperature. When the environmental temperature is high, this is a key component of *thermoregulation*, the process of maintaining temperature homeostasis. When you sweat in the hot sun, all your merocrine glands are working together. The blood vessels beneath your epidermis are dilated and filled with blood, your skin reddens, and the surface of your skin is warm and wet. As the moisture evaporates, your skin cools. If your body temperature subsequently falls below normal, sensible perspiration ceases, blood flow to the skin is reduced, and the skin surface cools and dries, releasing little heat into the environment. Chapter 1 introduced the negative feedback mechanisms of thermoregulation (pp. 14–15); additional details will be found in Chapter 25.

The A&P Top 100

#28 The skin plays a major role in controlling body temperature by acting as a radiator. The heat delivered by the dermal circulation is removed primarily by the evaporation of sensible perspiration.

CHECKPOINT

22. Identify two types of exocrine glands found in the skin.

23. What are the functions of sebaceous secretions?

24. Deodorants are used to mask the effects of secretions from which type of skin gland?

25. Which type of skin gland is most affected by the hormonal changes that occur during puberty?

See the blue Answers tab at the end of the book.

5-9 Nails are keratinized epidermal cells that protect the tips of fingers and toes

Nails protect the exposed dorsal surfaces of the tips of the fingers and toes (**Figure 5–13a**). They also help limit distortion of the digits when they are subjected to mechanical stress—

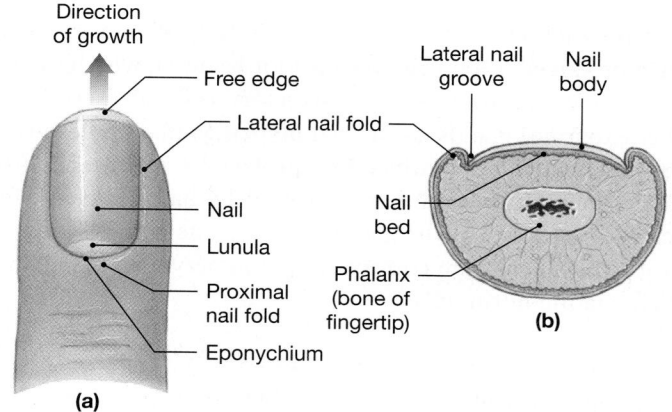

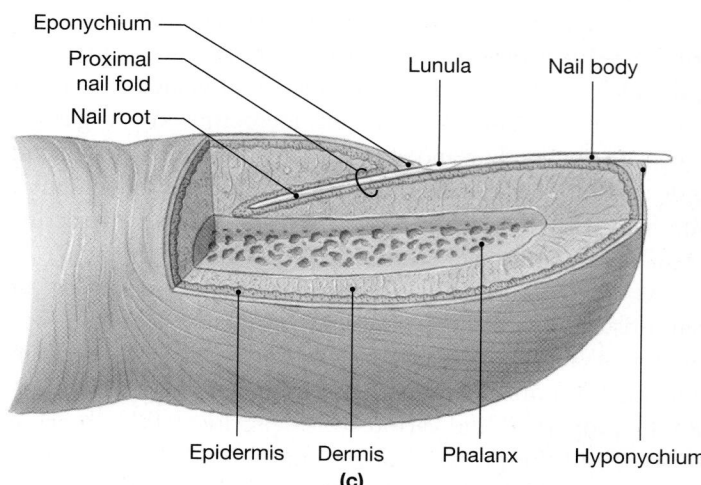

Figure 5–13 **The Structure of a Nail. (a)** A superficial view. **(b)** A cross-sectional view. **(c)** A longitudinal section.

for example, when you run or grasp objects. The **nail body**, the visible portion of the nail, covers an area of epidermis called the **nail bed** (**Figure 5–13b**). The nail body is recessed deep to the level of the surrounding epithelium and is bounded on either side by **lateral nail grooves** (depressions) and **lateral nail folds**. The **free edge** of the nail—the distal portion that continues past the nail bed—extends over the **hyponychium** (hī-pō-NIK-ē-um), an area of thickened stratum corneum (**Figure 5–13c**).

Nail production occurs at the **nail root**, an epidermal fold not visible from the surface. The deepest portion of the nail root lies very close to the bone of the fingertip. A portion of the stratum corneum of the nail root extends over the exposed nail, forming the **eponychium** (ep-ō-NIK-ē-um; *epi-*, over + *onyx*, nail), or **cuticle**. Underlying blood vessels give the nail its characteristic pink color. Near the root, these vessels may be obscured, leaving a pale crescent known as the **lunula** (LOO-nū-la; *luna*, moon) (**Figure 5–13a**).

The body of the nail consists of dead, tightly compressed cells packed with keratin. The cells producing the nails can be affected by conditions that alter body metabolism, so changes in the shape, structure, or appearance of the nails can provide useful diagnostic information. For example, the nails may turn yellow in individuals who have chronic respiratory disorders, thyroid gland disorders, or AIDS. Nails may become pitted and distorted as a result of *psoriasis* (a condition marked by rapid stem cell division in the stratum germinativum), and concave as a result of some blood disorders.

CHECKPOINT

26. Describe a typical fingernail.

27. What term is used to describe the thickened stratum corneum underlying the free edge of a nail?

28. Where does nail growth occur?

See the blue Answers tab at the end of the book.

5-10 Several steps are involved in repairing the integument following an injury

The integumentary system displays a significant degree of functional independence—it often responds directly and automatically to local influences without the involvement of the nervous or endocrine systems. For example, when the skin is continually subjected to mechanical stresses, stem cells in the stratum germinativum divide more rapidly, and the depth of the epithelium increases. That is why calluses form on your palms when you perform manual labor.

A more dramatic display of local regulation can be seen after an injury to the skin. The skin can regenerate effectively, even after considerable damage has occurred, because stem cells persist in both the epithelial and connective tissue components. Germinative cell divisions replace lost epidermal cells, and mesenchymal cell divisions replace lost dermal cells. The process can be slow. When large surface areas are involved, problems of infection and fluid loss complicate the situation. The relative speed and effectiveness of skin repair vary with the type of wound involved. A slender, straight cut, or *incision*, may heal relatively quickly compared with a deep scrape, or *abrasion*, which involves a much greater surface area to be repaired.

Figure 5–14 illustrates the four stages in the regeneration of the skin after an injury. When damage extends through the epidermis and into the dermis, bleeding generally occurs (STEP 1). The blood clot, or **scab**, that forms at the surface temporarily restores the integrity of the epidermis and restricts the entry of additional microorganisms into the area (STEP 2). The bulk of the clot consists of an insoluble network of *fibrin*, a fibrous protein that forms from blood proteins during the clotting response. The clot's color reflects the presence of trapped red blood cells. Cells of the stratum germinativum undergo rapid divisions and begin to migrate along the edges of the wound in an attempt to replace the missing epidermal cells. Meanwhile, macrophages patrol the damaged area of the dermis, phagocytizing any debris and pathogens.

If the wound occupies an extensive area or involves a region covered by thin skin, dermal repairs must be under way before epithelial cells can cover the surface. Divisions by fibroblasts and mesenchymal cells produce mobile cells that invade the deeper areas of injury. Endothelial cells of damaged blood vessels also begin to divide, and capillaries follow the fibroblasts, enhancing circulation. The combination of blood clot, fibroblasts, and an extensive capillary network is called **granulation tissue**.

Over time, deeper portions of the clot dissolve, and the number of capillaries declines. Fibroblast activity leads to the appearance of collagen fibers and typical ground substance (STEP 3). The repairs do not restore the integument to its original condition, however, because the dermis will contain an abnormally large number of collagen fibers and relatively few blood vessels. Severely damaged hair follicles, sebaceous or sweat glands, muscle cells, and nerves are seldom repaired, and they too are replaced by fibrous tissue. The formation of this rather inflexible, fibrous, noncellular **scar tissue** completes the repair process but fails to restore the tissue to its original condition (STEP 4).

We do not know what regulates the extent of scar tissue formation, and the process is highly variable. For example, surgical procedures performed on a fetus do not leave scars, perhaps because damaged fetal tissues do not produce the same types of growth factors that adult tissues do. In some adults,

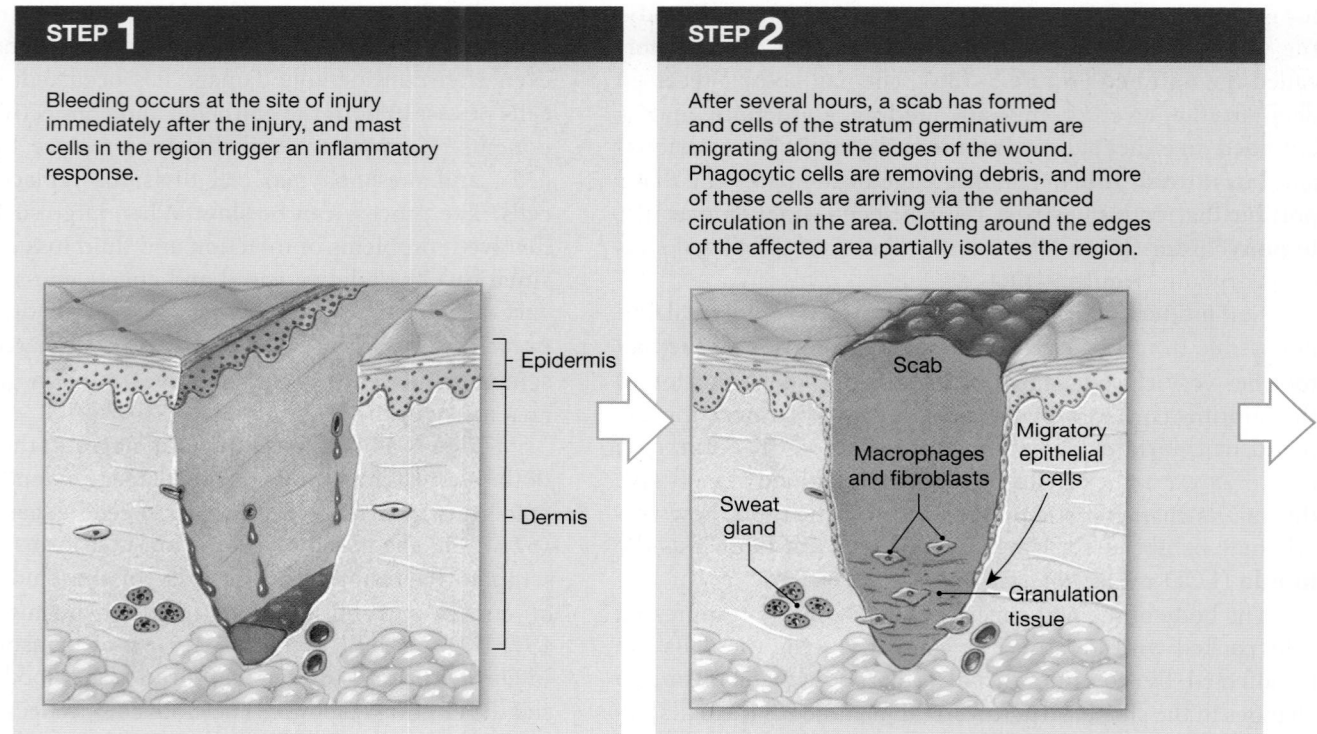

Figure 5–14 **Repair of Injury to the Integument.** STEP 1 is the inflammatory phase; STEP 2 is the migratory phase; STEP 3 is the proliferation phase; STEP 4 is the maturation phase.

most often those with dark skin, scar tissue formation may continue beyond the requirements of tissue repair. The result is a thickened mass of scar tissue that begins at the site of injury and grows into the surrounding dermis. This thick, raised area of scar tissue, called a **keloid** (KĒ-loyd), is covered by a shiny, smooth epidermal surface (**Figure 5–15**). Keloids most commonly develop on the upper back, shoulders, anterior chest, or earlobes. They are harmless; indeed some aboriginal cultures intentionally produce keloids as a form of body decoration.

Furthermore, people in societies around the world adorn the skin with culturally significant markings of one kind or another. Tattoos, piercings, keloids and other scar patterns, and even high-fashion makeup are all used to "enhance" the appearance of the integument. Scarification is performed in several African cultures, resulting in a series of complex, raised scars on the skin. Polynesian cultures have long preferred tattoos as a sign of status and beauty. A dark pigment is inserted deep within the dermis of the skin by tapping on a needle, shark tooth, or bit of bone. Because the pigment is inert, if infection does not occur (a potentially serious complication), the markings remain for the life of the individual, clearly visible through the overlying epidermis. American popular culture has recently rediscovered tattoos as a fashionable form of body adornment. The colored inks that are commonly used are less durable than those used by the

Polynesians, and today's tattoos can gradually fade or lose their definition.

Tattoos can now be partially or completely removed. The removal process takes time (10 or more sessions may be required to remove a large tattoo), and scars often remain. To remove the tattoo, an intense, narrow beam of light from a laser breaks down the ink molecules in the dermis. Each blast

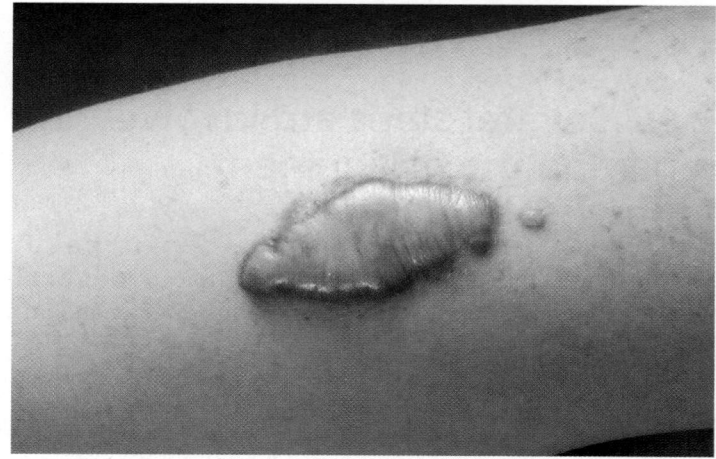

Figure 5–15 **A Keloid.** Keloids are areas of raised fibrous scar tissue.

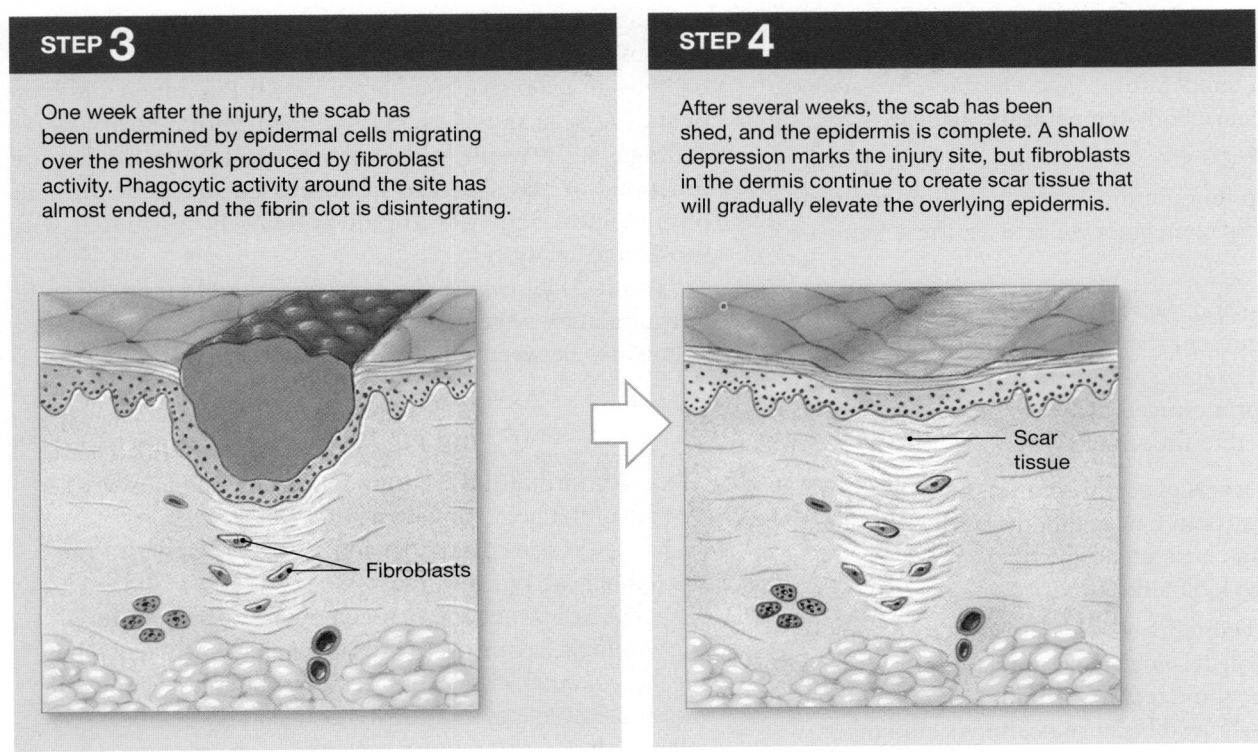

STEP 3

One week after the injury, the scab has been undermined by epidermal cells migrating over the meshwork produced by fibroblast activity. Phagocytic activity around the site has almost ended, and the fibrin clot is disintegrating.

Fibroblasts

STEP 4

After several weeks, the scab has been shed, and the epidermis is complete. A shallow depression marks the injury site, but fibroblasts in the dermis continue to create scar tissue that will gradually elevate the overlying epidermis.

Scar tissue

of the laser that destroys the ink also burns the surrounding dermal tissue. Although the burns are minor, they accumulate and result in the formation of localized scar tissue.

CHECKPOINT

29. **What term describes the combination of fibrin clots, fibroblasts, and the extensive network of capillaries in healing tissue?**

30. **Why can skin regenerate effectively even after considerable damage?**

See the blue Answers tab at the end of the book.

5-11 Effects of aging include dermal thinning, wrinkling, and reduced melanocyte activity

Aging affects all the components of the integumentary system:

- The epidermis thins as germinative cell activity declines, and the connections between the epidermis and dermis weaken, making older people more prone to injury, skin tears, and skin infections.

- The number of dendritic (Langerhans) cells decreases to about 50 percent of levels seen at maturity (roughly age 21). This decrease may reduce the sensitivity of the immune system and further encourage skin damage and infection.

- Vitamin D_3 production declines by about 75 percent. The result can be reduced calcium and phosphate absorption, eventually leading to muscle weakness and a reduction in bone strength and density.

- Melanocyte activity declines, and in light-skinned individuals the skin becomes very pale. With less melanin in the skin, people become more sensitive to exposure to the sun and more likely to experience sunburn.

- Glandular activity declines. The skin becomes dry and often scaly, because sebum production is reduced. Merocrine sweat glands are also less active, and with impaired perspiration, older people cannot lose heat as fast as younger people can. Thus, the elderly are at greater risk of overheating in warm environments.

- The blood supply to the dermis is reduced. Reduction in blood flow makes the skin become cool, which in turn can stimulate thermoreceptors, making a person feel cold even in a warm room. However, because reduced circulation and sweat gland function in the elderly lessens their ability to lose body heat, overexertion or exposure to high temperatures (such as those in a sauna or hot tub) can cause body temperatures to soar dangerously high.

- Hair follicles stop functioning or produce thinner, finer hairs. With decreased melanocyte activity, these hairs are gray or white.

- The dermis thins, and the elastic fiber network decreases in size. The integument therefore becomes weaker and less resilient, and sagging and wrinkling occur. These effects are most pronounced in areas of the body that have been exposed to the sun.

- With changes in levels of sex hormones, secondary sexual characteristics in hair and body-fat distribution begin to fade. As a consequence, people age 90–100 of both sexes tend to look alike.

- Skin repairs proceed more slowly. Thus, whereas repairs to an uninfected blister might take three to four weeks in a young adult, the same repairs could take six to eight weeks at age 65–75. And because healing occurs more slowly, recurring infections may result.

CHECKPOINT

31. Older individuals do not tolerate the summer heat as well as they did when they were young, and they are more prone to heat-related illness. What accounts for these changes?

32. Why does hair turn white or gray with age?

See the blue Answers tab at the end of the book.

5-12 The integumentary system provides protection for all other body systems

The integumentary system forms the external surface of the body and provides protection from dehydration, environmental chemicals, and impacts. The integument is sepa-

rated and insulated from the rest of the body by the subcutaneous layer, but is interconnected with the rest of the body by an extensive circulatory network of blood and lymphatic vessels and is richly supplied with sensory nerve endings. As a result, although the protective mechanical functions of the skin can be discussed independently, its physiological activities are always closely integrated with those of other systems.

Figure 5–16 reviews the components and functions of the integumentary system and diagrams the major functional relationships between that system and other systems.

CLINICAL NOTE

Skin Abnormalities Because the skin is the most visible organ of the body, abnormalities are easily recognized. Changes in skin color, tone, and overall condition commonly accompany disease and can assist in the diagnosis of conditions involving other systems. A bruise for example, is a swollen, discolored area where blood has leaked through vessel walls. Extensive bruising without any obvious cause may indicate a blood-clotting disorder; and yellowish skin and mucous membranes may signify *jaundice*, which generally indicates some type of liver disorder. The general condition of the skin can also be significant. In addition to color changes, changes in the flexibility, elasticity, dryness, or sensitivity of the skin commonly follow malfunctions in other organ systems.

CHECKPOINT

33. Identify the common purpose that the integumentary system serves for all body systems.

34. Why is the skin important to muscular system activity?

See the blue Answers tab at the end of the book.

THE INTEGUMENTARY SYSTEM IN PERSPECTIVE

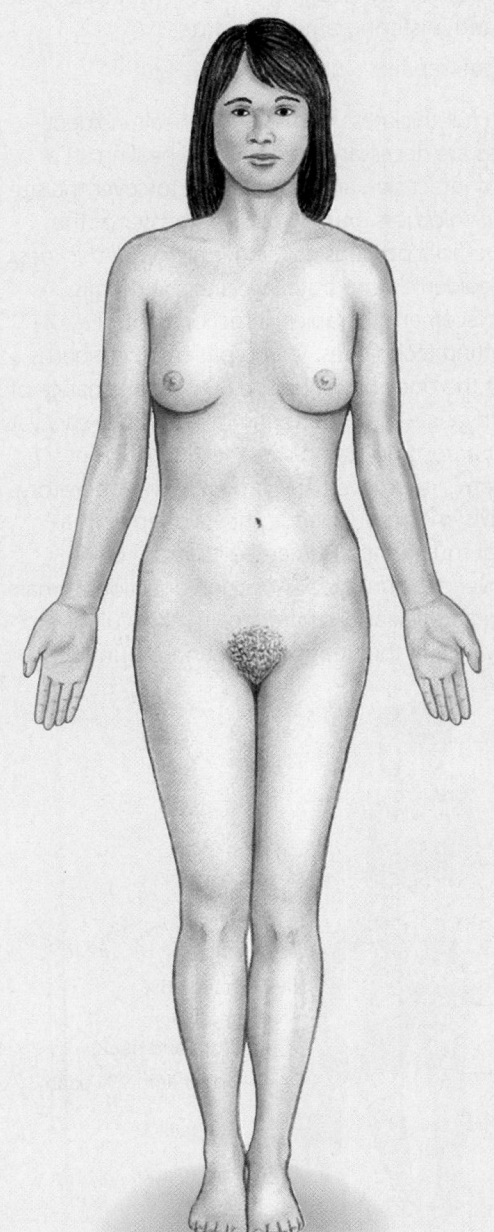

Skeletal System

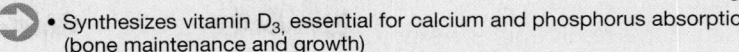

- Provides structural support
- Synthesizes vitamin D_3, essential for calcium and phosphorus absorption (bone maintenance and growth)

Muscular System

- Contractions of skeletal muscles pull against skin of face, producing facial expressions important in communication
- Synthesizes vitamin D_3, essential for normal calcium absorption (calcium ions play an essential role in muscle contraction)

Nervous System

- Controls blood flow and sweat gland activity for thermoregulation; stimulates contraction of arrector pili muscles to elevate hairs
- Receptors in dermis and deep epidermis provide sensations of touch, pressure, vibration, temperature, and pain

Endocrine System

- Sex hormones stimulate sebaceous gland activity; male and female sex hormones influence growth, distribution of subcutaneous fat, and apocrine sweat gland activity; suprarenal hormones alter dermal blood flow and help mobilize lipids from adipocytes
- Synthesizes vitamin D_3, precursor of calcitriol

Cardiovascular System

- Provides oxygen and nutrients; delivers hormones and cells of immune system; carrries away carbon dioxide, waste products, and toxins; provides heat to maintain normal skin temperature
- Stimulation of mast cells produces localized changes in blood flow and capillary permeability

Lymphoid System

- Assists in defending the integument by providing additional macrophages and mobilizing lymphocytes
- Provides physical barriers that prevent entry of pathogens; dendritic (Langerhans) cells and macrophages resist infection; mast cells trigger inflammation and initiate the immune response

Respiratory System

- Provides oxygen and eliminates carbon dioxide
- Hairs guard entrance to nasal cavity

Digestive System

- Provides nutrients for all cells and lipids for storage by adipocytes
- Synthesizes vitamin D_3, needed for absorption of calcium and phosphorus

Urinary System

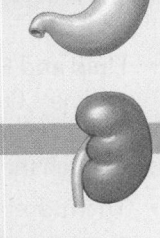

- Excretes waste products; maintains normal pH and ion composition of body fluids
- Assists in excretion of water and solutes; keratinized epidermis limits fluid loss through skin

Reproductive System

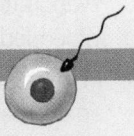

- Sex hormones affect hair distribution, adipose tissue distribution in subcutaneous layer, and mammary gland development
- Covers external genitalia; provides sensations that stimulate sexual behaviors; mammary gland secretions provide nourishment for newborn infant

FOR ALL SYSTEMS

Provides mechanical protection against environmental hazards

Figure 5–16   The Integumentary System in Perspective.

Burns and Grafts

Burns are significant injuries in that they can damage the integrity of large areas of the skin. The integumentary functions of protection, excretion, maintenance, sensory detection, and storage and synthesis of vitamin D are all compromised. Burns result from the exposure of skin to heat, friction, radiation, electrical shock, or strong chemical agents. The severity of the burn depends on the depth of penetration and the total area affected.

First- and second-degree burns are also called *partial-thickness burns*, because damage is restricted to the superficial layers of the skin. Only the surface of the epidermis is affected by a *first-degree burn*. In this type of burn, which includes most sunburns, the skin reddens and can be painful. The redness, a sign called **erythema** (er-i-THĒ-muh), results from inflammation of the sun-damaged tissues. In a *second-degree burn*, the entire epidermis and perhaps some of the dermis are damaged. Accessory structures such as hair follicles and glands are generally not affected, but blistering, pain, and swelling occur. If the blisters rupture at the surface, infection can easily develop. Healing typically takes one to two weeks, and some scar tissue may form.

Full-thickness burns, or *third-degree burns*, destroy the epidermis and dermis, extending into subcutaneous tissues. Despite swelling, these burns are less painful than second-degree burns, because sensory nerves are destroyed along with accessory structures, blood vessels, and other dermal components. Extensive third-degree burns cannot repair themselves, because granulation tissue cannot form and epithelial cells are unable to cover the injury. As a result, the affected area remains open to infection. Extensive third-degree burns often require skin grafts, discussed shortly.

Each year in the United States, roughly 10,000 people die from the effects of burns. The larger the area burned, the more significant are the effects on integumentary function. **Figure 5–17** presents a standard reference for calculating the percentage of total surface area damaged. Burns that cover more than 20 percent of the skin surface represent serious threats to life, because they affect the following functions:

- **Fluid and Electrolyte Balance.** Even areas with partial-thickness burns lose their effectiveness as barriers to fluid and electrolyte losses. In full-thickness burns, the rate of fluid loss through the skin may reach five times the normal level.

- **Thermoregulation.** Increased fluid loss means increased evaporative cooling. As a result, more energy must be expended to keep body temperature within acceptable limits.

- **Protection from Infection.** The dampness of the epidermal surface, resulting from uncontrolled fluid loss, encourages bacterial growth. If the skin is broken at a blister or at the site of a third-degree burn, infection is likely. Widespread bacterial infection, or **sepsis** (*septikos*, rotting), is the leading cause of death in burn victims. Effective treatment of full-thickness burns focuses on the following four procedures:

1. Replacing lost fluids and electrolytes.
2. Providing sufficient nutrients to meet increased metabolic demands for thermoregulation and healing.
3. Preventing infection by cleaning and covering the burn while administering antibiotic drugs.
4. Assisting tissue repair.

Because large full-thickness burns cannot heal unaided, surgical procedures are necessary to encourage healing. In a **skin graft**, areas of intact skin are transplanted to cover the site of the burn. A *split-thickness graft* involves a transfer of the epidermis and superficial portions of the dermis; a *full-thickness graft* involves the epidermis and both layers of the dermis.

With fluid-replacement therapies, infection control methods, and grafting techniques, young patients with burns over 80 percent of the body have about a 50 percent chance of recovery. Recent advances in cell culturing may improve survival rates further. After a postage-stamp-sized section of undamaged epidermis is removed and grown in the laboratory, germinative cell divisions produce large sheets of epidermal cells—up to several square meters in area—that can be transplanted to cover the burn area. Although questions remain about the strength and flexibility of the repairs, skin cultivation is a substantial advance in the treatment of serious burns.

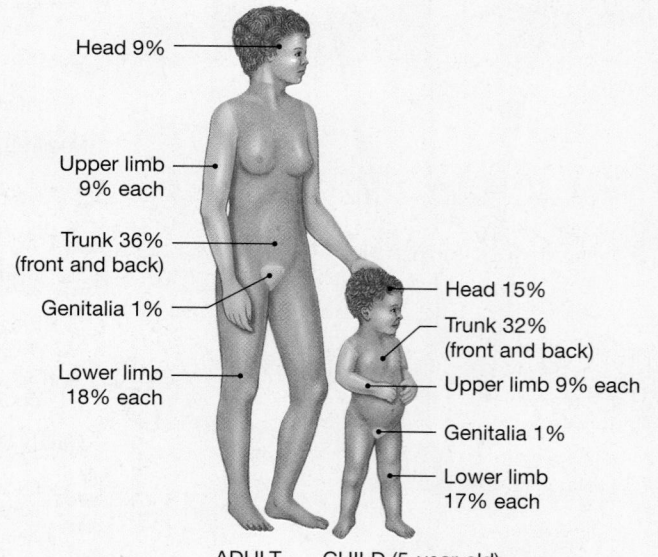

Head 9%
Upper limb 9% each
Trunk 36% (front and back)
Genitalia 1%
Lower limb 18% each

Head 15%
Trunk 32% (front and back)
Upper limb 9% each
Genitalia 1%
Lower limb 17% each

ADULT   CHILD (5-year-old)

Figure 5–17   A Quick Method of Estimating the Percentage of Surface Area Affected by Burns. This method is called the *rule of nines*, because the surface area in adults is divided into multiples of 9. The rule must be modified for children, because their proportions are quite different.

Related Clinical Terms

basal cell carcinoma: A skin cancer caused by malignant stem cells within the stratum germinativum.

cutaneous anthrax: An infection resulting when spores of the anthrax bacterium enter a break in the skin.

cyanosis: Bluish skin color as a result of reduced oxygenation of the blood in superficial vessels.

decubitis ulcers (*bedsores*)**:** Ulcers that occur in areas subject to restricted circulation; are especially common in bedridden people.

dermatitis: An inflammation of the skin that primarily involves the papillary region of the dermis.

granulation tissue: A combination of fibrin, fibroblasts, and capillaries that forms during tissue repair after inflammation.

hypodermic needle: A needle used to administer drugs via subcutaneous injection.

keloid: A thickened area of scar tissue covered by a shiny, smooth epidermal surface.

male pattern baldness: Hair loss in an adult male due to changes in levels of circulating testosterone.

malignant melanoma: A skin cancer originating in melanocytes.

scab: A blood clot that forms at the surface of a wound to the skin.

seborrheic dermatitis: An inflammation around abnormally active sebaceous glands.

sepsis: A dangerous, widespread bacterial infection; the leading cause of death in burn patients.

skin graft: The transplantation of a section of skin (either partial thickness or full thickness) to cover an extensive injury, such as a third-degree burn.

squamous cell carcinoma: A skin cancer resulting from chronic exposure to excessive amounts of UV radiation in sunlight.

ulcer: A localized area where the epithelium has been eroded or lost.

Chapter Review
Study Outline

An Introduction to the Integumentary System p. 158

1. The **integument**, or **integumentary system**, consists of the **cutaneous membrane** or **skin** (which includes the **epidermis** and the **dermis**) and the **accessory structures**. Beneath the dermis lies the **hypodermis** or subcutaneous layer. (*Figure 5–1*)

2. Functions of the integument include *protection, excretion, temperature maintenance, vitamin D₃ synthesis, nutrient storage,* and *sensory detection.*

5-1 The epidermis is composed of strata (layers) with various functions p. 159

3. **Thin skin**, consisting of four layers of **keratinocytes**, covers most of the body. Heavily abraded body surfaces may be covered by **thick skin** containing five layers of keratinocytes. (*Figure 5–2*)

4. The epidermis provides mechanical protection, prevents fluid loss, and helps keep microorganisms out of the body.

5. Cell divisions in the **stratum germinativum**, the innermost epidermal layer, replace more superficial cells. (*Figure 5–3*)

6. As epidermal cells age, they pass through the **stratum spinosum**, the **stratum granulosum**, the **stratum lucidum** (in thick skin), and the **stratum corneum**. In the process, they accumulate large amounts of **keratin**. Ultimately, the cells are shed. (*Figure 5–3*)

7. **Epidermal ridges**, interlocked with **dermal papillae** of the underlying dermis, improve the gripping ability of the palms and soles and increase the skin's sensitivity. (*Figure 5–4*)

8. *Dendritic cells* in the stratum spinosum are part of the immune system. *Merkel cells* in the stratum germinativum provide sensory information about objects that touch the skin.

5-2 Factors influencing skin color are epidermal pigmentation and dermal circulation p. 162

9. The color of the epidermis depends on two factors: dermal blood supply and epidermal pigmentation.

10. The epidermis contains the pigments **carotene** and **melanin**. **Melanocytes**, which produce melanin, protect us from **ultraviolet (UV) radiation**. (*Figure 5–5*)

11. Interruptions of the dermal blood supply or poor oxygenation in the lungs can lead to **cyanosis**.

5-3 Sunlight causes epidermal cells to convert a steroid into vitamin D₃ p. 165

12. Epidermal cells synthesize **vitamin D₃** or **cholecalciferol**, when exposed to the UV radiation in sunlight.

5-4 Epidermal growth factor has several effects on the epidermis and epithelia p. 165

13. **Epidermal growth factor (EGF)** promotes growth, division, and repair of the epidermis and the secretion of epithelial glands.

5-5 The dermis is the tissue layer that supports the epidermis p. 166

14. The dermis consists of the superficial **papillary layer** and the deeper **reticular layer**. (*Figures 5–1, 5–2*)

15. The papillary layer of the dermis contains blood vessels, lymphatics, and sensory nerves that supply the epidermis. The reticular layer consists of a meshwork of collagen and elastic fibers oriented to resist tension in the skin.

16. Extensive distension of the dermis can cause **stretch marks**.

17. The pattern of collagen and elastic fiber bundles forms **lines of cleavage**. (*Figure 5–8*)

18. Arteries to the skin form the **cutaneous plexus** and the **papillary plexus** in the hypodermis and the papillary dermis, respectively. (*Figure 5–9*)

19. Integumentary sensory receptors detect both light touch and pressure.

5-6 The hypodermis is tissue beneath the dermis that connects it to underlying tissues p. 168

20. The **hypodermis**, or subcutaneous layer, stabilizes the skin's position against underlying organs and tissues. (*Figure 5–1*)

5-7 Hair is composed of keratinized dead cells that have been pushed to the surface p. 169

21. **Hairs** originate in complex organs called **hair follicles**. Each hair has a **root** and a **shaft**. At the base of the root are a **hair papilla**, surrounded by a **hair bulb**, and a **root hair plexus** of sensory nerves. Hairs have a **medulla**, or core of soft keratin, surrounded by a **cortex** of hard keratin. The **cuticle** is a superficial layer of dead cells that protects the hair. (*Figure 5–10*)

22. Our bodies have both **vellus hairs** ("peach fuzz") and heavy **terminal hairs**. A hair that has stopped growing is called a **club hair**.

23. Each **arrector pili** muscle can erect a single hair. (*Figure 5–10*)

24. Our hairs grow and are shed according to the **hair growth cycle**. A typical hair on the head grows for two to five years and is subsequently shed.

5-8 Sebaceous glands and sweat glands are exocrine glands found in the skin p. 172

25. A typical **sebaceous gland** discharges waxy **sebum** into a lumen and, ultimately, into a hair follicle. **Sebaceous follicles** are large sebaceous glands that discharge sebum directly onto the epidermis. (*Figure 5–11*)

26. The two types of sweat glands, or **sudoriferous glands**, are apocrine and merocrine sweat glands. **Apocrine sweat glands** produce an odorous secretion. The more numerous **merocrine**, or **eccrine**, **sweat glands** produce a watery secretion known as sensible perspiration. (*Figure 5–12*)

27. **Mammary glands** of the breasts are structurally similar to apocrine sweat glands. **Ceruminous glands** in the ear produce a waxy substance called **cerumen**.

5-9 Nails are keratinized epidermal cells that protect the tips of fingers and toes p. 174

28. The **nail body** of a nail covers the **nail bed**. Nail production occurs at the **nail root**, which is overlain by the **cuticle**, or **eponychium**. The **free edge** of the nail extends over the **hyponychium**. (*Figure 5–13*)

5-10 Several steps are involved in repairing the integument following an injury p. 175

29. Based on the division of stem cells, the skin can regenerate effectively even after considerable damage. The process begins with bleeding and includes the formation of a **scab**, **granulation tissue**, and **scar tissue**. (*Figure 5–14*)

5-11 Effects of aging include dermal thinning, wrinkling, and reduced melanocyte activity p. 177

30. With aging, the integument thins, blood flow decreases, cellular activity decreases, and repairs occur more slowly.

5-12 The integumentary system provides protection for all other body systems p. 178

31. The integumentary system has extensive anatomical and physiological connections with other body systems. (*Figure 5–16*)

Review Questions

See the blue Answers tab at the end of the book.

myA&P™ Access more review material online at **myA&P**™ (www.myaandp.com). There, you'll find chapter guides, chapter quizzes, practice tests, animations, flashcards, a glossary with pronunciations, *Interactive Physiology*® (IP) exercises and quizzes, and more to help you succeed in the course.

LEVEL 1 Reviewing Facts and Terms

1. Identify the different portions (a–d) of the cutaneous membrane and the underlying layer of loose connective tissue (e) in the diagram to the right.
   (a) _____ (b) _____ (c) _____
   (d) _____ (e) _____

2. The two major components of the integumentary system are
   (a) the cutaneous membrane and the accessory structures.
   (b) the epidermis and the hypodermis.
   (c) the hair and the nails.
   (d) the dermis and the subcutaneous layer.

3. Beginning at the basal lamina and traveling toward the free surface, the epidermis includes the following layers:
   (a) corneum, lucidum, granulosum, spinosum, germinativum.
   (b) granulosum, lucidum, spinosum, germinativum, corneum.
   (c) germinativum, spinosum, granulosum, lucidum, corneum.
   (d) lucidum, granulosum, spinosum, germinativum, corneum.

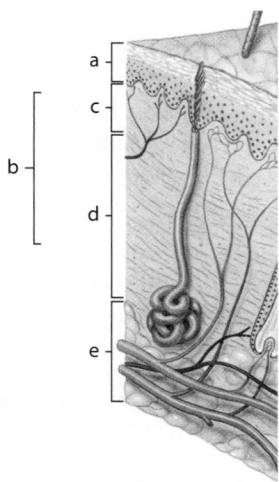

4. Each of the following is a function of the integumentary system, *except*
   (a) protection of underlying tissue.
   (b) excretion of salts and wastes.
   (c) maintenance of body temperature.
   (d) synthesis of vitamin C.
   (e) storage of nutrients.

5. Exposure of the skin to ultraviolet light
   (a) can result in increased numbers of melanocytes forming in the skin.
   (b) can result in decreased melanin production in melanocytes.
   (c) can cause destruction of vitamin D$_3$.
   (d) can result in damage to the DNA of cells in the stratum germinativum.
   (e) has no effect on the skin cells.

6. The two major components of the dermis are the
   (a) superficial fascia and cutaneous membrane.
   (b) epidermis and hypodermis.
   (c) papillary layer and reticular layer.
   (d) stratum germinativum and stratum corneum.

7. The cutaneous plexus and papillary plexus consist of
   (a) a network of arteries providing the dermal blood supply.
   (b) a network of nerves providing dermal sensations.
   (c) specialized cells for cutaneous sensations.
   (d) gland cells that release cutaneous secretions.

8. The accessory structures of the integument include the
   (a) blood vessels, glands, muscles, and nerves.
   (b) Merkel cells, lamellated corpuscles, and tactile corpuscles.
   (c) hair, skin, and nails.
   (d) hair follicles, nails, sebaceous glands, and sweat glands.

9. The portion of the hair follicle where cell divisions occur is the
   (a) shaft.   (b) matrix.
   (c) root hair plexus.   (d) cuticle.

10. The two types of exocrine glands in the skin are
    (a) merocrine and sweat glands.
    (b) sebaceous and sweat glands.
    (c) apocrine and sweat glands.
    (d) eccrine and sweat glands.

11. Apocrine sweat glands can be controlled by
    (a) the autonomic nervous system.
    (b) regional control mechanisms.
    (c) the endocrine system.
    (d) both a and c.

12. The primary function of sensible perspiration is to
    (a) get rid of wastes.
    (b) protect the skin from dryness.
    (c) maintain electrolyte balance.
    (d) reduce body temperature.

13. The stratum corneum of the nail root, which extends over the exposed nail, is called the
    (a) hyponychium.   (b) eponychium.
    (c) lunula.   (d) cerumen.

14. Muscle weakness and a reduction in bone strength in the elderly result from decreased
    (a) vitamin D$_3$ production.   (b) melanin production.
    (c) sebum production.   (d) dermal blood supply.

15. In which layer(s) of the epidermis does cell division occur?

16. What is the function of the arrector pili muscles?

17. What widespread effects does epidermal growth factor (EGF) have on the integument?

18. What two major layers constitute the dermis, and what components are in each layer?

19. List the four stages in the regeneration of the skin after an injury.

LEVEL 2 Reviewing Concepts

20. How do insensible perspiration and sensible perspiration differ?

21. In clinical practice, drugs can be delivered by diffusion across the skin; this delivery method is called transdermal administration. Why are fat-soluble drugs more suitable for transdermal administration than drugs that are water soluble?

22. In our society, a tan body is associated with good health. However, medical research constantly warns about the dangers of excessive exposure to the sun. What are the benefits of a tan?

23. Why is it important for a surgeon to choose an incision pattern according to the lines of cleavage of the skin?

24. The fibrous protein that is responsible for the strength and water resistance of the skin surface is
    (a) collagen.   (b) eleidin.
    (c) keratin.   (d) elastin.
    (e) keratohyalin.

25. The darker an individual's skin color,
    (a) the more melanocytes she has in her skin.
    (b) the more layers she has in her epidermis.
    (c) the more melanin her melanocytes produce.
    (d) the more superficial her blood vessels are.

26. In order for bacteria on the skin to cause an infection in the skin, they must accomplish all of the following, *except*
    (a) survive the bactericidal components of sebum.
    (b) avoid being flushed from the surface of the skin by sweat.
    (c) penetrate the stratum corneum.
    (d) penetrate to the level of the capillaries.
    (e) escape the dendritic (Langerhans) cells.

LEVEL 3 Critical Thinking and Clinical Applications

27. In the elderly, blood supply to the dermis is reduced and sweat glands are less active. This combination of factors would most affect
    (a) the ability to thermoregulate.
    (b) the ability to heal injured skin.
    (c) the ease with which the skin is injured.
    (d) the physical characteristics of the skin.
    (e) the ability to grow hair.

28. Two patients are brought to the emergency room. One has cut his finger with a knife; the other has stepped on a nail. Which wound has a greater chance of becoming infected? Why?

29. Exposure to optimum amounts of sunlight is necessary for proper bone maintenance and growth in children.
    (a) What does sunlight do to promote bone maintenance and growth?
    (b) If a child lives in an area where exposure to sunlight is rare because of pollution or overcast skies, what can be done to minimize impaired maintenance and growth of bone?

30. One of the factors to which lie detectors respond is an increase in skin conductivity due to the presence of moisture. Explain the physiological basis for the use of this indicator.

31. Many people change the natural appearance of their hair, either by coloring it or by altering the degree of curl in it. Which layers of the hair do you suppose are affected by the chemicals added during these procedures? Why are the effects of the procedures not permanent?

6

Osseous Tissue and Bone Structure

Learning Outcomes

After completing this chapter, you should be able to do the following:

6-1 Describe the primary functions of the skeletal system.

6-2 Classify bones according to shape and internal organization, giving examples of each type, and explain the functional significance of each of the major types of bone markings.

6-3 Identify the cell types in bone, and list their major functions.

6-4 Compare the structures and functions of compact bone and spongy bone.

6-5 Compare the mechanisms of endochondral ossification and intramembranous ossification.

6-6 Describe the remodeling and homeostatic mechanisms of the skeletal system.

6-7 Discuss the effects of exercise, hormones, and nutrition on bone development and on the skeletal system.

6-8 Explain the role of calcium as it relates to the skeletal system.

6-9 Describe the types of fractures, and explain how fractures heal.

6-10 Summarize the effects of the aging process on the skeletal system.

Clinical Notes

Heterotopic Bone Formation 195
Abnormal Bone Growth and Development 202

An Introduction to the Skeletal System

This chapter expands upon the introduction to bone, presented in Chapter 4, by examining the mechanisms involved with the growth, remodeling, and repair of the skeleton. Skeletal elements are more than just props, or racks from which muscles hang; they have a great variety of vital functions. In addition to supporting the weight of the body, bones work together with muscles to maintain body position and to produce controlled, precise movements. Without the skeleton to pull against, contracting muscle fibers could not make us sit, stand, walk, or run.

6-1 The skeletal system has five primary functions

The skeletal system includes the bones of the skeleton and the cartilages, ligaments, and other connective tissues that stabilize or interconnect the bones. This system has five primary functions:

1. *Support.* The skeletal system provides structural support for the entire body. Individual bones or groups of bones provide a framework for the attachment of soft tissues and organs.

2. *Storage of Minerals and Lipids.* As we will learn in Chapter 25, minerals are inorganic ions that contribute to the osmotic concentration of body fluids. Minerals also participate in various physiological processes, and several are important as enzyme cofactors. Calcium is the most abundant mineral in the human body. The calcium salts of bone are a valuable mineral reserve that maintains normal concentrations of calcium and phosphate ions in body fluids. In addition to acting as a mineral reserve, the bones of the skeleton store energy reserves as lipids in areas filled with *yellow marrow.*

3. *Blood Cell Production.* Red blood cells, white blood cells, and other blood elements are produced in *red marrow,* which fills the internal cavities of many bones. We will describe the role of bone marrow in blood cell formation when we examine the cardiovascular and lymphoid systems (Chapters 19 and 22).

4. *Protection.* Many soft tissues and organs are surrounded by skeletal elements. The ribs protect the heart and lungs, the skull encloses the brain, the vertebrae shield the spinal cord, and the pelvis cradles delicate digestive and reproductive organs.

5. *Leverage.* Many bones function as levers that can change the magnitude and direction of the forces generated by skeletal muscles. The movements produced range from the precise motion of a fingertip to changes in the position of the entire body.

Chapters 6–9 describe the structure and function of the skeletal system. We begin by describing bone, or osseous tissue, a supporting connective tissue introduced in Chapter 4. ∞ p. 132 All of the features and properties of the skeletal system ultimately depend on the unique and dynamic properties of bone. The bone specimens that you study in lab or that you are familiar with from skeletons of dead animals are only the dry remains of this living tissue. They bear the same relationship to the bone in a living organism as a kiln-dried 2-by-4 does to a living oak.

CHECKPOINT

1. Name the five primary functions of the skeletal system.

See the blue Answers tab at the end of the book.

6-2 Bones are classified according to shape and structure, and feature surface markings

A bone may be classified by its general shape or by its internal tissue organization. Before considering specific bones of the skeleton, you must be familiar with both classification schemes.

Bone Shapes

The typical adult skeleton contains 206 major bones, which we can divide into six broad categories according to their individual shapes (**Figure 6–1**):

1. **Long bones** are relatively long and slender (**Figure 6–1a**). Long bones are located in the arm and forearm, thigh and leg, palms, soles, fingers, and toes. The femur, the long bone of the thigh, is the largest and heaviest bone in the body.

2. **Flat bones** have thin roughly parallel surfaces. Flat bones form the roof of the skull (**Figure 6–1b**), the sternum, the ribs, and the scapulae. They provide protection for underlying soft tissues and offer an extensive surface area for the attachment of skeletal muscles.

3. **Sutural bones,** or *Wormian bones,* are small, flat, irregularly shaped bones between the flat bones of the skull (**Figure 6–1c**). There are individual variations in the number, shape, and position of the sutural bones. Their borders are like pieces of a jigsaw puzzle, and they range in size from a grain of sand to a quarter.

4. **Irregular bones** have complex shapes with short, flat, notched, or ridged surfaces (**Figure 6–1d**). The spinal

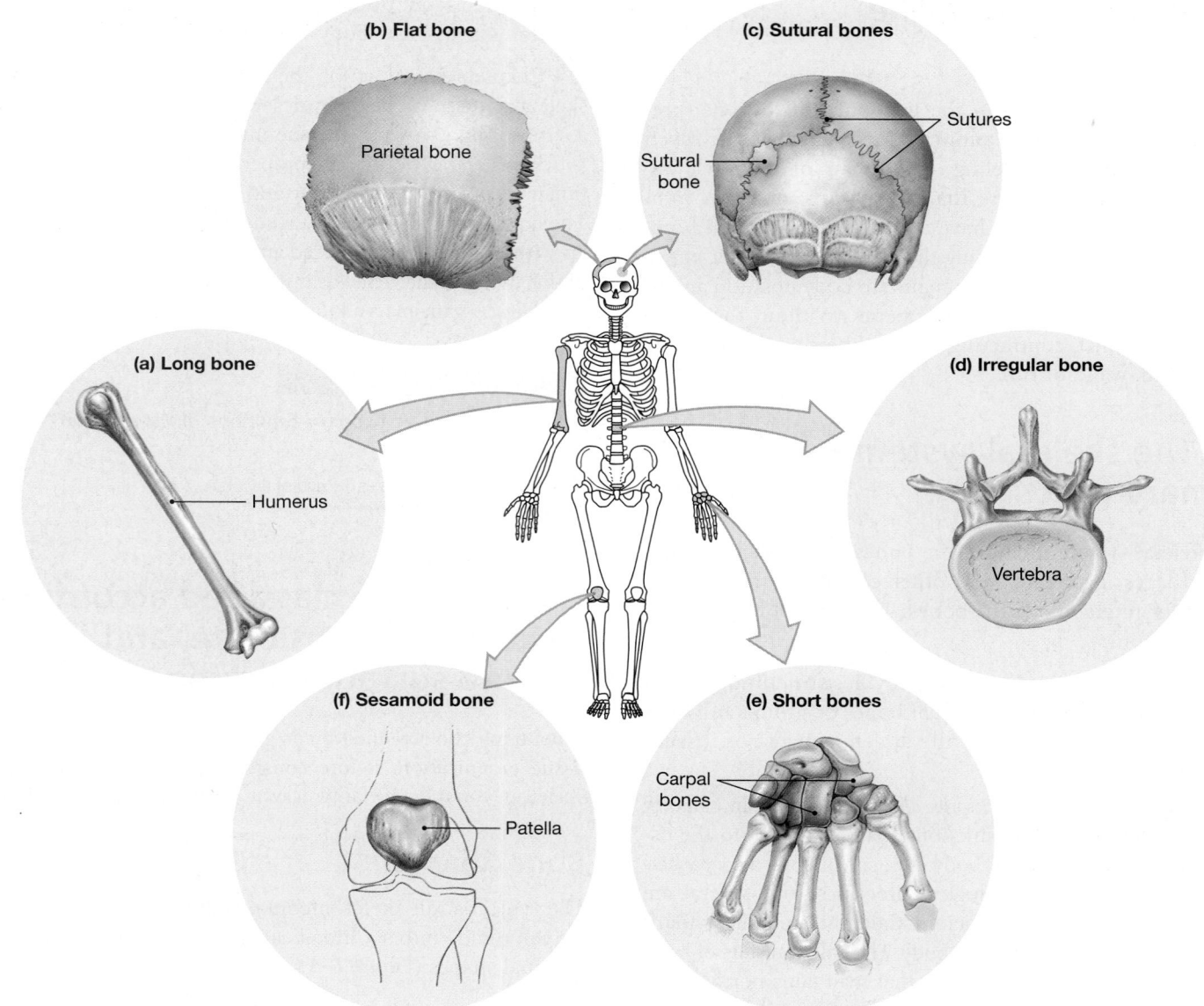

Figure 6–1   A Classification of Bones by Shape.

vertebrae, the bones of the pelvis, and several skull bones are irregular bones.

5. **Short bones** are small and boxy (**Figure 6–1e**). Examples of short bones include the carpal bones (wrists) and tarsal bones (ankles).

6. **Sesamoid bones** are generally small, flat, and shaped somewhat like a sesame seed (**Figure 6–1f**). They develop inside tendons and are most commonly located near joints at the knees, the hands, and the feet. Everyone has sesamoid *patellae* (pa-TEL-ē; singular, *patella*, a small shallow dish), or kneecaps, but individuals vary in the location and abundance of other sesamoid bones. This variation, among others, accounts for disparities in the total number of bones in the skeleton. (Sesamoid bones may form in at least 26 locations.)

Bone Markings (Surface Features)

Each bone in the body has characteristic external and internal features. Elevations or projections form where tendons and ligaments attach, and where adjacent bones articulate (that is, at joints). Depressions, grooves, and tunnels in bone indicate sites where blood vessels or nerves lie alongside or penetrate the bone. Detailed examination of these **bone markings**, or *surface features*, can yield an abundance of anatomical information. For example, anthropologists, criminologists, and pathologists can often determine the size, age, sex, and general appearance of an individual on the basis of incomplete skeletal remains.

Table 6–1 presents an introduction to the prominent surface features of bones, using specific anatomical terms to describe the

6 SKELETAL

TABLE 6–1   An Introduction to Bone Surface Features

General Description	Anatomical Term	Definition
Elevations and projections (general)	**Process**	Any projection or bump
	Ramus	An extension of a bone making an angle with the rest of the structure
Processes formed where tendons or ligaments attach	**Trochanter**	A large, rough projection
	Tuberosity	A smaller, rough projection
	Tubercle	A small, rounded projection
	Crest	A prominent ridge
	Line	A low ridge
	Spine	A pointed or narrow process
Processes formed for articulation with adjacent bones	**Head**	The expanded articular end of an epiphysis, separated from the shaft by a neck
	Neck	A narrow connection between the epiphysis and the diaphysis
	Condyle	A smooth, rounded articular process
	Trochlea	A smooth, grooved articular process shaped like a pulley
	Facet	A small, flat articular surface
Depressions	**Fossa**	A shallow depression
	Sulcus	A narrow groove
Openings	**Foramen**	A rounded passageway for blood vessels or nerves
	Canal or Meatus	A passageway through the substance of a bone
	Fissure	An elongated cleft
	Sinus or Antrum	A chamber within a bone, normally filled with air

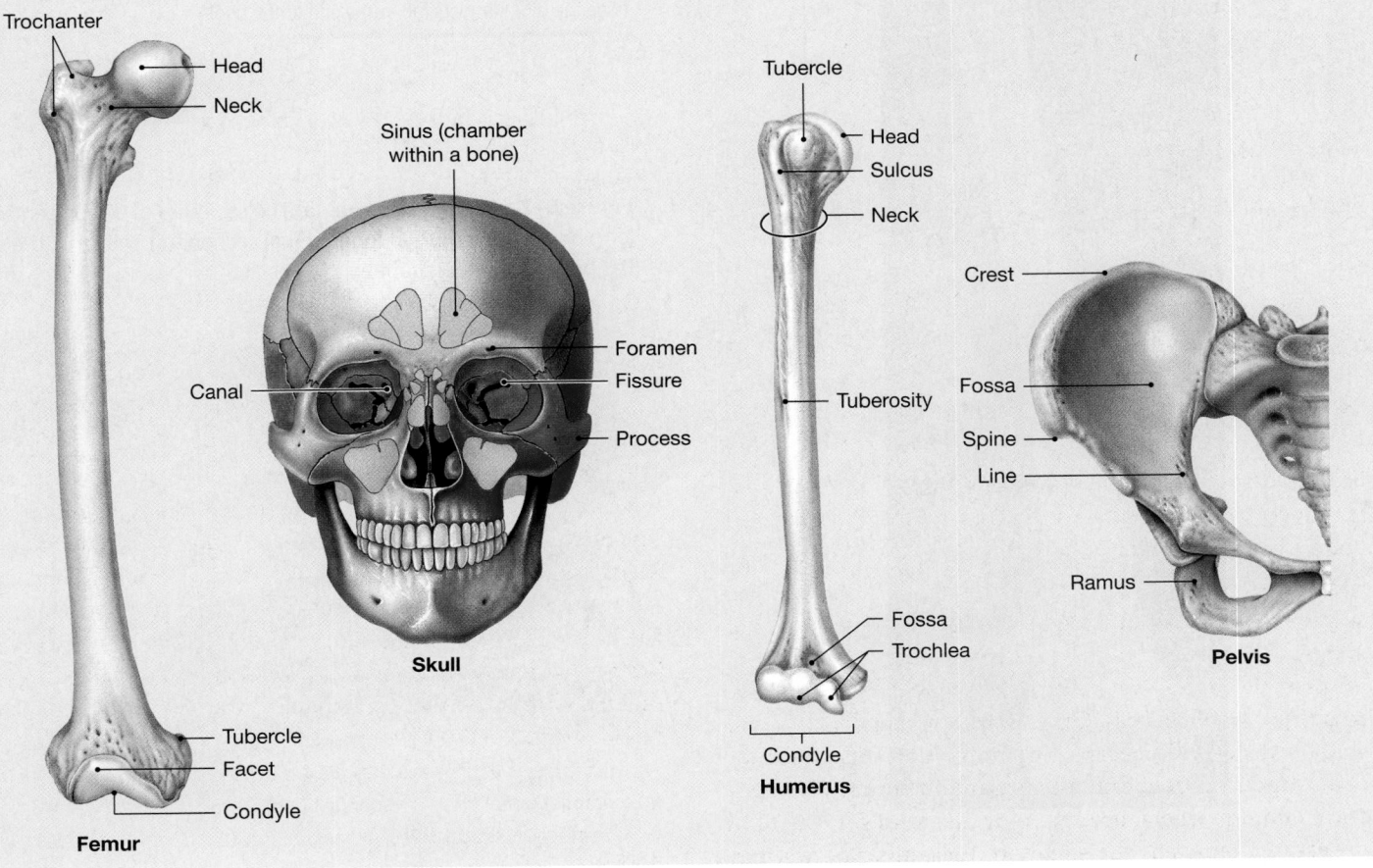

various projections, depressions, and openings. These markings provide fixed landmarks that can help us determine the position of the soft-tissue components of other organ systems.

Bone Structure

Figure 6–2a introduces the anatomy of the femur, a representative long bone with an extended tubular shaft, or **diaphysis** (dī-AF-i-sis). At each end is an expanded area known as the **epiphysis** (ē-PIF-i-sis). The diaphysis is connected to each epiphysis at a narrow zone known as the **metaphysis** (me-TAF-i-sis; *meta*, between). The wall of the diaphysis consists of a layer of compact bone, or *dense bone*. Compact bone, which is relatively solid, forms a sturdy protective layer that surrounds a central space called the **medullary cavity** (*medulla*, innermost part), or *marrow cavity*. The epiphyses

consist largely of spongy bone, also called *cancellous* (KAN-se-lus) or *trabecular bone*. Spongy bone consists of an open network of struts and plates that resembles latticework with a thin covering, or **cortex**, of compact bone. This superficial layer covering spongy bone is also known as *cortical bone*.

Figure 6–2b details the structure of a flat bone from the skull, such as one of the *parietal bones*. A flat bone resembles a spongy bone sandwich, with layers of compact bone covering a core of spongy bone. Within the cranium, the layer of spongy bone between the layers of compact bone is called the *diploë* (DIP-lō-ē; *diplous*, twofold). Although red bone marrow is present within the spongy bone, there is no large medullary cavity as in the diaphysis of a long bone.

Many people think of the skeleton as a rather dull collection of bony props. This is far from the truth. Our bones are complex, dynamic organs that constantly change to adapt to the demands we place on them. We will now consider the histological organization of a typical bone.

CHECKPOINT

2. **Identify the six broad categories for classifying a bone according to shape.**

3. **Define bone marking (surface feature).**

See the blue Answers tab at the end of the book.

Figure 6–2 **Bone Structure. (a)** The structure of a representative long bone (the femur) in longitudinal section. **(b)** The structure of a flat bone (the parietal bone).

Spongy bone

Epiphysis

Metaphysis

Compact bone

Diaphysis (shaft)

Medullary cavity

Metaphysis

Epiphysis

(a)

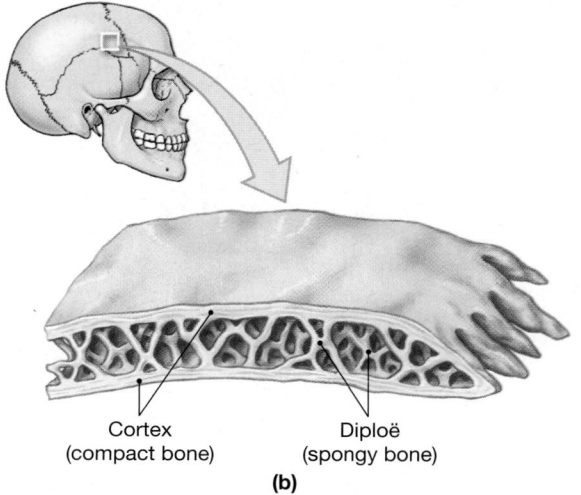

Cortex (compact bone)

Diploë (spongy bone)

(b)

6-3 Bone is composed of matrix and several types of cells: osteocytes, osteoblasts, osteoprogenitor cells, and osteoclasts

Osseous tissue is a supporting connective tissue. (You may wish to review the sections on dense connective tissues, cartilage, and bone in Chapter 4.) ∞ pp. 128–134 Like other connective tissues, osseous tissue contains specialized cells and a matrix consisting of extracellular protein fibers and a ground substance. The matrix of bone tissue is solid and sturdy, owing to the deposition of calcium salts around the protein fibers.

In Chapter 4, which introduced the organization of bone tissue, we discussed the following four characteristics of bone:

1. The matrix of bone is very dense and contains deposits of calcium salts.

2. The matrix contains bone cells, or *osteocytes*, within pockets called *lacunae*. (The spaces that chondrocytes occupy in cartilage are also called lacunae. ∞ p. 131) The lacunae of bone are typically organized around blood vessels that branch through the bony matrix.

3. *Canaliculi*, narrow passageways through the matrix, extend between the lacunae and nearby blood vessels, forming a branching network for the exchange of nutrients, waste products, and gases.

4. Except at joints, the outer surfaces of bones are covered by a *periosteum*, which consists of outer fibrous and inner cellular layers.

We now take a closer look at the organization of the matrix and cells of bone.

The Matrix of Bone

Calcium phosphate, $Ca_3(PO_4)_2$, accounts for almost two-thirds of the weight of bone. Calcium phosphate interacts with calcium hydroxide, $Ca(OH)_2$, to form crystals of **hydroxyapatite**, $Ca_{10}(PO_4)_6(OH)_2$. As they form, these crystals incorporate other calcium salts, such as calcium carbonate ($CaCO_3$), and ions such as sodium, magnesium, and fluoride. Roughly one-third of the weight of bone is contributed by collagen fibers. Cells account for only 2 percent of the mass of a typical bone.

Calcium phosphate crystals are very hard, but relatively inflexible and quite brittle. They can withstand compression, but are likely to shatter when exposed to bending, twisting, or sudden impacts. Collagen fibers, by contrast, are remarkably strong; when subjected to tension (pull), they are stronger than steel. Flexible as well as tough, they can easily tolerate twisting and bending, but offer little resistance to compression. When compressed, they simply bend out of the way.

The composition of the matrix in compact bone is the same as that in spongy bone. The collagen fibers provide an organic framework on which hydroxyapatite crystals can form. These crystals form small plates and rods that are locked into the collagen fibers at regular angles. The result is a protein–crystal combination that possesses the flexibility of collagen and the compressive strength of hydroxyapatite crystals. The protein–crystal interactions allow bone to be strong, somewhat flexible, and highly resistant to shattering. In its overall properties, bone is on a par with the best steel-reinforced concrete. In fact, bone is far superior to concrete, because it can undergo remodeling (cycles of bone formation and resorption) as needed and can repair itself after injury.

The Cells of Bone

Although osteocytes are most abundant, bone contains four types of cells: osteocytes, osteoblasts, osteoprogenitor cells, and osteoclasts (**Figure 6–3**).

Osteocytes (OS-tē-ō-sīts) (*osteo-*, bone + *-cyte*, cell) are mature bone cells that account for most of the cell population. Each osteocyte occupies a lacuna, a pocket sandwiched between layers of matrix. The layers are called **lamellae** (lah-MEL-lē; singular, *lamella*, a thin plate). Osteocytes cannot divide, and a lacuna never contains more than one osteocyte. Narrow passageways called **canaliculi** penetrate the lamellae, radiating through the matrix and connecting lacunae with one another and with sources of nutrients, such as the central canal.

Canaliculi contain cytoplasmic extensions of osteocytes. Neighboring osteocytes are linked by gap junctions, which permit the exchange of ions and small molecules, including nutrients and hormones, between the cells. The interstitial fluid that surrounds the osteocytes and their extensions provides an additional route for the diffusion of nutrients and waste products.

Osteocytes have two major functions:

1. *Osteocytes maintain the protein and mineral content of the surrounding matrix.* This is not a static process, as there is continual turnover of matrix components. Osteocytes secrete chemicals that dissolve the adjacent matrix, and the minerals released enter the circulation. Osteocytes then rebuild the matrix, stimulating the deposition of new hydroxyapatite crystals. The turnover rate varies from bone to bone; we will consider this process further in a later section.

2. *Osteocytes participate in the repair of damaged bone.* If released from their lacunae, osteocytes can convert to a less specialized type of cell, such as an osteoblast or an osteoprogenitor cell.

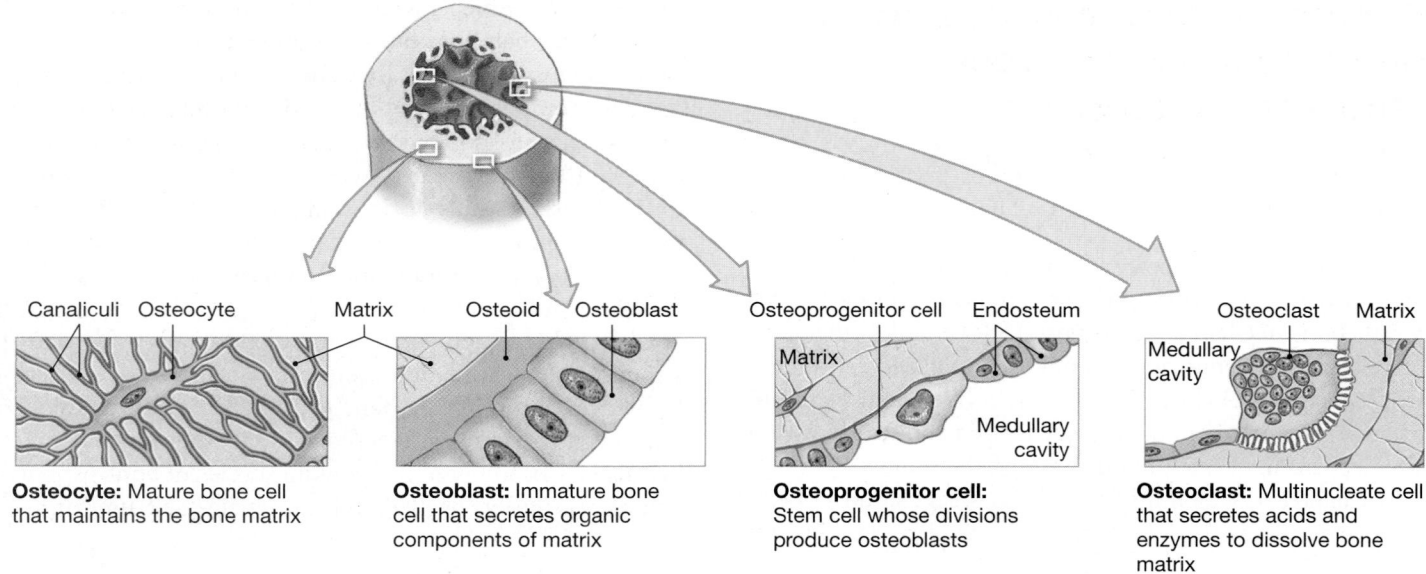

Canaliculi Osteocyte   Matrix   Osteoid Osteoblast   Osteoprogenitor cell Endosteum   Osteoclast   Matrix

Osteocyte: Mature bone cell that maintains the bone matrix

Osteoblast: Immature bone cell that secretes organic components of matrix

Osteoprogenitor cell: Stem cell whose divisions produce osteoblasts

Osteoclast: Multinucleate cell that secretes acids and enzymes to dissolve bone matrix

Figure 6–3   Types of Bone Cells.

Osteoblasts (OS-tē-ō-blasts; *blast*, precursor) produce new bone matrix in a process called **osteogenesis** (os-tē-ō-JEN-e-sis; *gennan*, to produce). Osteoblasts make and release the proteins and other organic components of the matrix. Before calcium salts are deposited, this organic matrix is called **osteoid** (OS-tē-oyd). Osteoblasts also assist in elevating local concentrations of calcium phosphate above its solubility limit, thereby triggering the deposition of calcium salts in the organic matrix. This process converts osteoid to bone. Osteocytes develop from osteoblasts that have become completely surrounded by bone matrix.

Bone contains small numbers of mesenchymal cells called **osteoprogenitor** (os-tē-ō-prō-JEN-i-tor) **cells** (*progenitor*, ancestor). These squamous stem cells divide to produce daughter cells that differentiate into osteoblasts. Osteoprogenitor cells maintain populations of osteoblasts and are important in the repair of a *fracture* (a break or a crack in a bone). Osteoprogenitor cells are located in the inner, cellular layer of the periosteum; in an inner layer, or *endosteum*, that lines medullary cavities; and in the lining of passageways, containing blood vessels, that penetrate the matrix of compact bone.

Osteoclasts (OS-tē-ō-clasts; *clast*, to break) are cells that remove and recycle bone matrix. These are giant cells with 50 or more nuclei. Osteoclasts are not related to osteoprogenitor cells or their descendants. Instead, they are derived from the same stem cells that produce monocytes and macrophages. Acids and proteolytic (protein-digesting) enzymes secreted by osteoclasts dissolve the matrix and release the stored minerals. This erosion process, called **osteolysis** (os-tē-OL-i-sis; *osteo-*, bone + *lysis*, dissolution) or *resorption*, is important in

the regulation of calcium and phosphate concentrations in body fluids.

In living bone, osteoclasts are constantly removing matrix, and osteoblasts are always adding to it. The balance between the opposing activities of osteoblasts and osteoclasts is very important. When osteoclasts remove calcium salts faster than osteoblasts deposit them, bones weaken. When osteoblast activity predominates, bones become stronger and more massive. This opposition causes some interesting differences in skeletal components among individuals. Those who subject their bones to muscular stress through weight training or strenuous exercise develop not only stronger muscles, but also stronger bones. Alternatively, declining muscular activity due to immobility leads to a reduction in bone mass at sites of muscle attachment. We will investigate this phenomenon further in a later section of the chapter.

CHECKPOINT

4. Mature bone cells are known as _____, bone-building cells are called _____, and _____ are bone-resorbing cells.

5. How would the compressive strength of a bone be affected if the ratio of collagen to hydroxyapatite increased?

6. If the activity of osteoclasts exceeds the activity of osteoblasts in a bone, how will the mass of the bone be affected?

See the blue Answers tab at the end of the book.

6-4 Compact bone contains parallel osteons, and spongy bone contains trabeculae

In this section, we examine the structures of compact and spongy bone in detail. Additionally, two bone layers, the periosteum and endosteum, will be discussed.

The Structure of Compact Bone

The basic functional unit of mature compact bone is the **osteon** (OS-tē-on), or *Haversian system* (**Figures 6–4** and **6–5a**). In an osteon, the osteocytes are arranged in concentric layers around a **central canal**, or *Haversian canal*. This canal contains one or more blood vessels (normally a capillary and a *venule*, a very small vein) that carry blood to and from the osteon. Central canals generally run parallel to the surface of the bone. Other passageways, known as **perforating canals** or *canals of Volkmann*, extend roughly perpendicular to the surface. Blood vessels in these canals supply blood to osteons deeper in the bone and to tissues of the medullary cavity.

The lamellae of each osteon form a series of nested cylinders around the central canal. In transverse section, these *concentric lamellae* create a targetlike pattern, with the central canal as the bull's-eye. Collagen fibers within each lamella form a spiral that adds strength and resiliency. Canaliculi radiating through the lamellae interconnect the lacunae of the osteons with one another and with the central canal. *Interstitial lamellae* fill in the spaces between the osteons in compact bone. These lamellae are remnants of osteons whose matrix components have been almost completely recycled by osteoclasts. *Circumferential lamellae* (*circum-*, around + *ferre*, to bear) are found at the outer and inner surfaces of the bone, where they are covered by the periosteum and endosteum, respectively (**Figure 6–5a,b**). These lamellae are produced during the growth of the bone, and this process will be described in a later section.

Tips&Tricks

To understand the relationship of collagen fibers to bone matrix, envision the placement of reinforcing steel rods (rebar) in concrete. Like the rebar in concrete, collagen adds strength to bone.

Compact bone is thickest where stresses arrive from a limited range of directions. All osteons in compact bone are aligned the same, making such bones very strong when stressed along the axis of alignment. You might think of a single osteon as a drinking straw with very thick walls: When you attempt to push the ends of the straw together or to pull them apart, the straw is quite strong. But if you hold the ends and push from the side, the straw will bend sharply with relative ease.

The osteons in the diaphysis of a long bone are parallel to the long axis of the shaft. Thus, the shaft does not bend, even when extreme forces are applied to either end. (The femur can withstand 10–15 times the body's weight without breaking.) Yet a much smaller force applied to the side of the shaft can break the femur. The majority of breaks in this bone are caused by a sudden sideways force, such as occurs in a fall or auto accident.

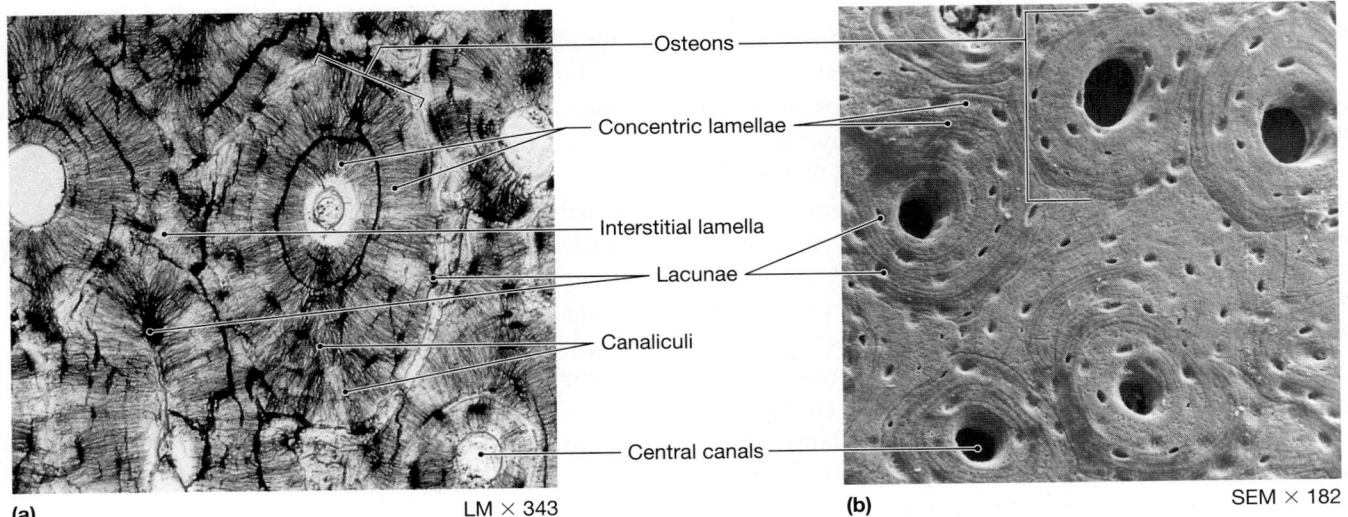

Osteons

Concentric lamellae

Interstitial lamella

Lacunae

Canaliculi

Central canals

(a)   LM × 343

(b)   SEM × 182

Figure 6–4   The Histology of Compact Bone. (a) A thin section through compact bone. By this procedure, the intact matrix and central canals appear white, and the lacunae and canaliculi are shown in black. **(b)** Several osteons in compact bone. [© R. G. Kessel and R. H. Kardon, *Tissues and Organs: A Text-Atlas of Scanning Electron Microscopy*, W. H. Freeman & Co., 1979. All Rights Reserved.]

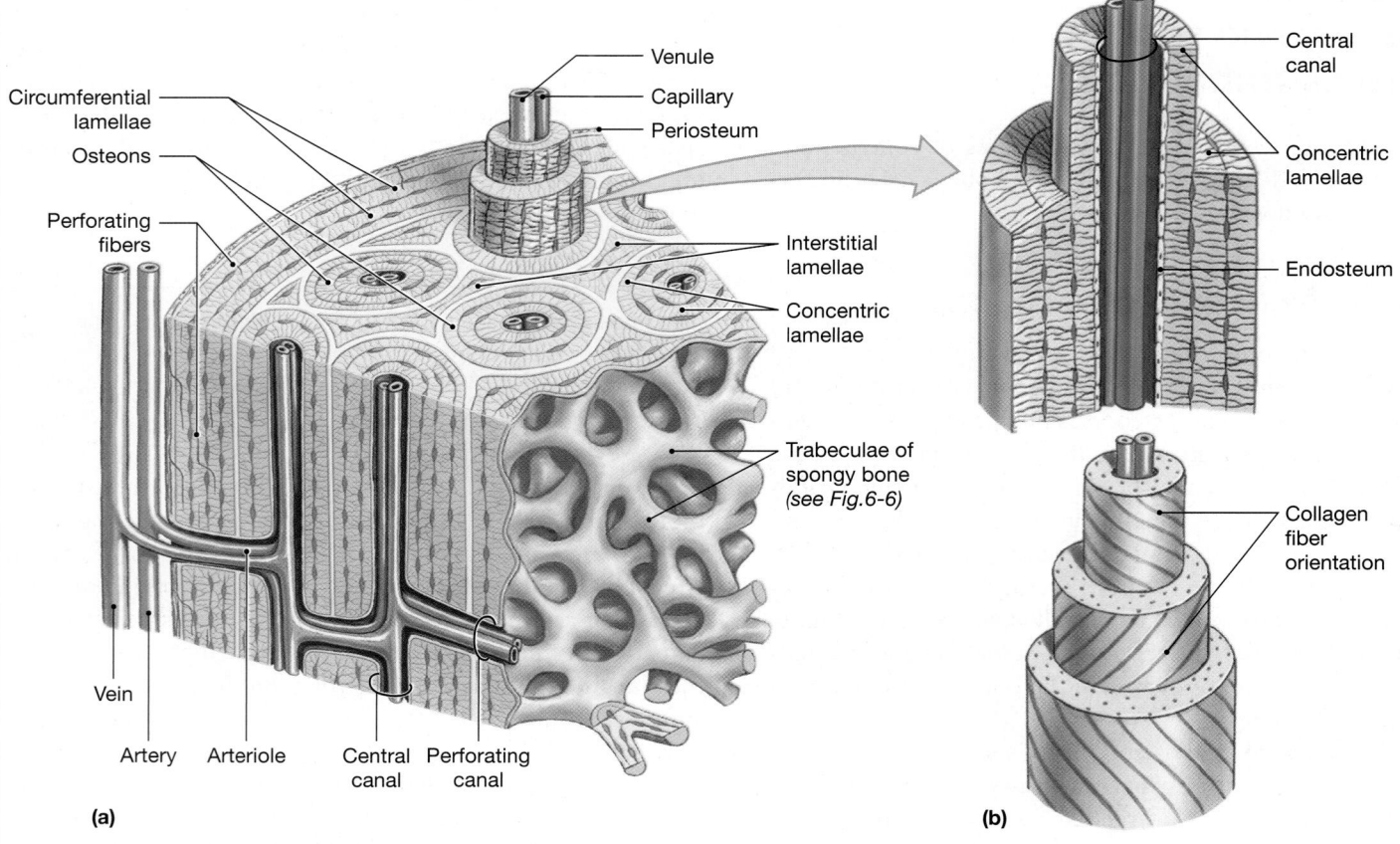

Figure 6–5 **The Structure of Compact Bone.** (a) The organization of osteons and lamellae in compact bone. (b) The orientation of collagen fibers in adjacent lamellae.

The Structure of Spongy Bone

In spongy bone, lamellae are not arranged in osteons. The matrix in spongy bone forms struts and plates called **trabeculae** (tra-BEK-ū-lē) (**Figure 6–6**). The thin trabeculae branch, creating an open network. There are no capillaries or venules in the matrix of spongy bone. Nutrients reach the osteocytes by diffusion along canaliculi that open onto the surfaces of trabeculae. Red marrow is found between the trabeculae of spongy bone, and blood vessels within this tissue deliver nutrients to the trabeculae and remove wastes generated by the osteocytes.

Spongy bone is located where bones are not heavily stressed or where stresses arrive from many directions. The trabeculae are oriented along stress lines and are cross-braced extensively. In addition to being able to withstand stresses applied from many directions, spongy bone is much lighter than compact bone. Spongy bone thus reduces the weight of the skeleton and thereby makes it easier for muscles to move the bones. Finally, the framework of trabeculae supports and protects the cells of the bone marrow. Spongy bone within the epiphyses of long bones, such as the femur, and the interior

of other large bones such as the sternum and ilium, contains **red bone marrow** responsible for blood cell formation. At other sites, spongy bone may contain **yellow bone marrow—** adipose tissue important as an energy reserve.

Figure 6–7 shows the distribution of forces applied to the femur, and illustrates the functional relationship between compact bone and spongy bone. The head of the femur articulates with a corresponding socket on the lateral surface of the pelvis. At the proximal epiphysis of the femur, trabeculae transfer forces from the pelvis to the compact bone of the femoral shaft, across the hip joint; at the distal epiphysis, trabeculae transfer weight from the shaft to the leg, across the knee joint. The femoral head projects medially, and the body weight compresses the medial side of the shaft. However, because the force is applied off center, the bone must also resist the tendency to bend into a lateral bow. So while the medial portion of the shaft is under compression, the lateral portion of the shaft, which resists this bending, is placed under a stretching load, or *tension*. Because the center of the bone is subjected to neither compression nor tension, the presence of the medullary cavity does not reduce the bone's strength.

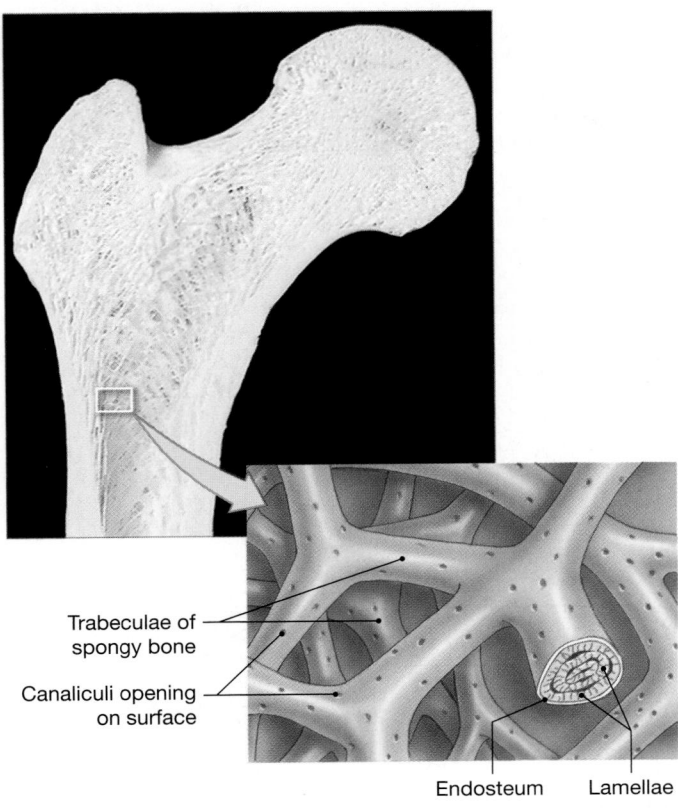

Figure 6–6   **The Structure of Spongy Bone.**

Trabeculae of spongy bone
Canaliculi opening on surface
Endosteum   Lamellae

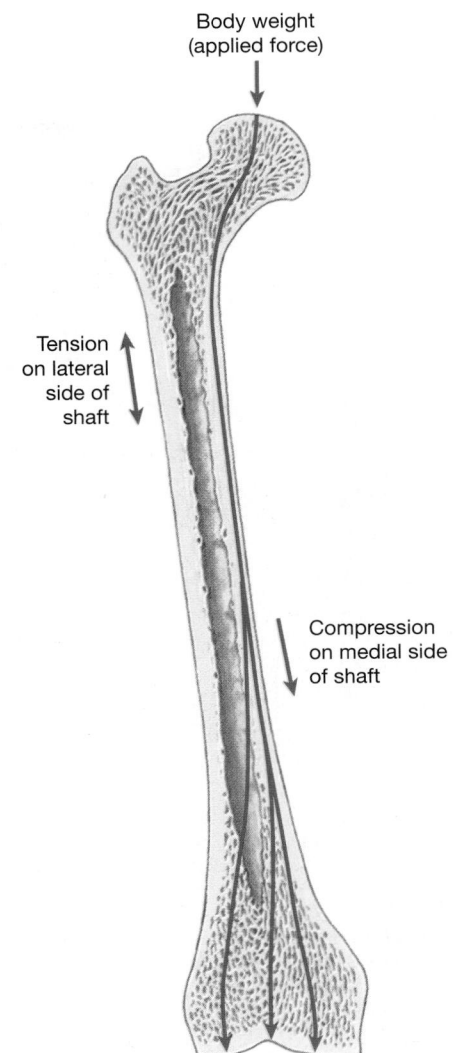

Body weight (applied force)

Tension on lateral side of shaft

Compression on medial side of shaft

Figure 6–7   The Distribution of Forces on a Long Bone. The femur, or thigh bone, has a diaphysis (shaft) with walls of compact bone and epiphyses filled with spongy bone. The body weight is transferred to the femur at the hip joint. Because the hip joint is off center relative to the axis of the shaft, the body weight is distributed along the bone such that the medial (inner) portion of the shaft is compressed and the lateral (outer) portion is stretched.

The Periosteum and Endosteum

Except within joint cavities, the superficial layer of compact bone that covers all bones is wrapped by a **periosteum**, a membrane with a fibrous outer layer and a cellular inner layer (**Figure 6–8a**). The periosteum (1) isolates the bone from surrounding tissues, (2) provides a route for the circulatory and nervous supply, and (3) actively participates in bone growth and repair.

Near joints, the periosteum becomes continuous with the connective tissues that lock the bones together. At a synovial joint, the periosteum is continuous with the joint capsule. The fibers of the periosteum are also interwoven with those of the tendons attached to the bone. As the bone grows, these tendon fibers are cemented into the circumferential lamellae by osteoblasts from the cellular layer of the periosteum. Collagen fibers incorporated into bone tissue from tendons and ligaments, as well as from the superficial periosteum, are called *perforating fibers (Sharpey fibers)*. This method of attachment bonds the tendons and ligaments into the general structure of the bone, providing a much stronger attachment than would otherwise be possible. An extremely powerful pull on a tendon or ligament will usually break a bone rather than snap the collagen fibers at the bone surface.

The **endosteum**, an incomplete cellular layer, lines the medullary cavity (**Figure 6–8b**). This layer, which is active during bone growth, repair, and remodeling, covers the trabeculae of spongy bone and lines the inner surfaces of the central canals. The endosteum consists of a simple flattened layer of osteoprogenitor cells that covers the bone matrix, generally without any intervening connective tissue fibers. Where the cellular layer is not complete, the matrix is exposed. At these exposed sites, osteoclasts and osteoblasts can remove or deposit matrix components. The osteoclasts generally occur in

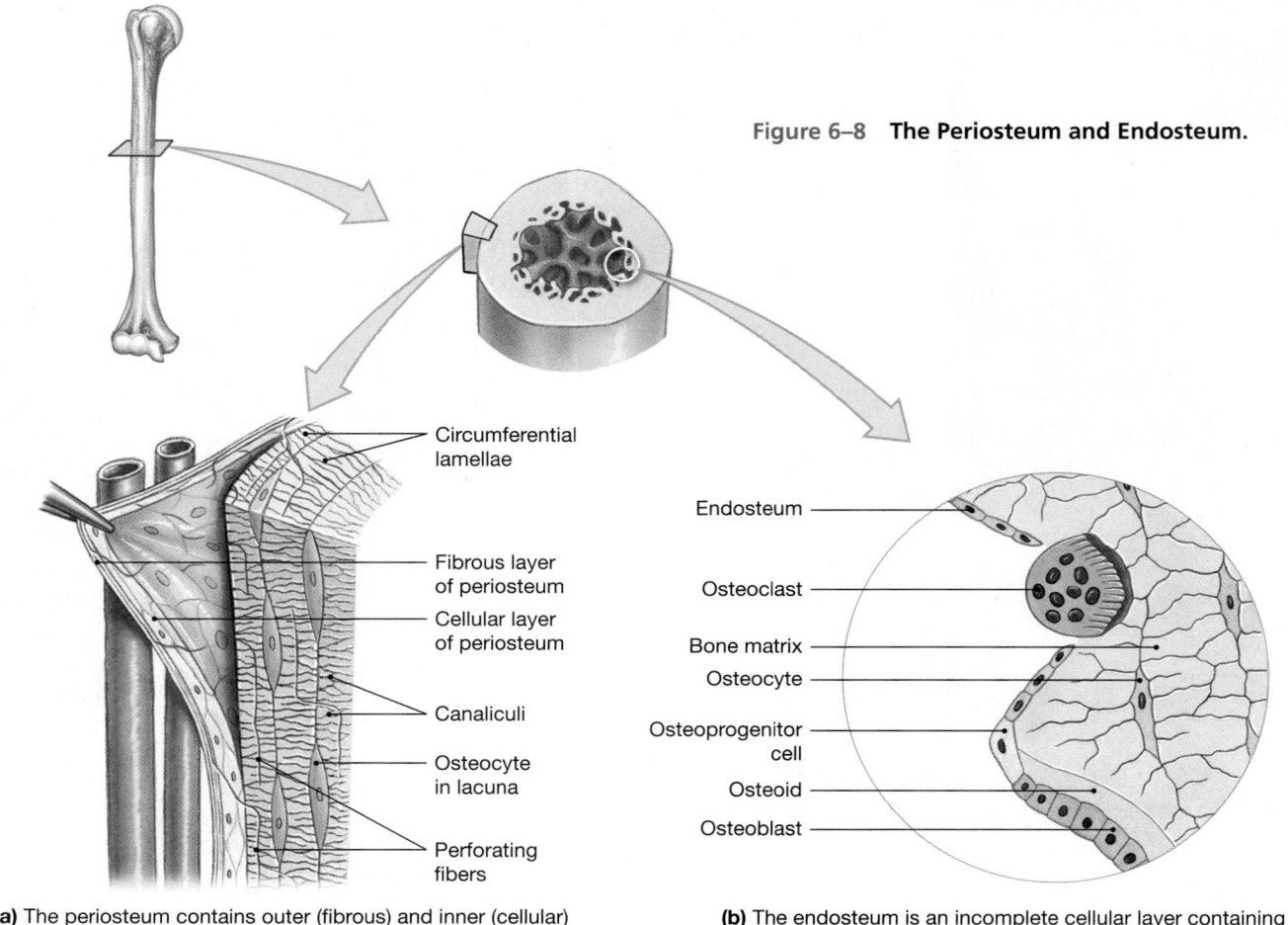

Figure 6–8 **The Periosteum and Endosteum.**

Circumferential lamellae

Fibrous layer of periosteum

Cellular layer of periosteum

Canaliculi

Osteocyte in lacuna

Perforating fibers

Endosteum

Osteoclast

Bone matrix

Osteocyte

Osteoprogenitor cell

Osteoid

Osteoblast

(a) The periosteum contains outer (fibrous) and inner (cellular) layers. Collagen fibers of the periosteum are continuous with those of the bone, adjacent joint capsules, and attached tendons and ligaments.

(b) The endosteum is an incomplete cellular layer containing osteoblasts, osteoprogenitor cells, and osteoclasts.

shallow depressions (*osteoclastic crypts*) that they have eroded into the matrix.

CHECKPOINT

7. **Compare the structures and functions of compact bone and spongy bone.**

8. **A sample of bone has lamellae, which are not arranged in osteons. Is the sample most likely taken from the epiphysis or diaphysis?**

See the blue Answers tab at the end of the book.

6-5 Ossification and appositional growth are mechanisms of bone formation and enlargement

The growth of the skeleton determines the size and proportions of your body. The bony skeleton begins to form about six weeks after fertilization, when the embryo is approximately 12 mm (0.5 in.) long. (At this stage, the existing skeletal elements are cartilaginous.) During subsequent development, the bones undergo a tremendous increase in size. Bone growth continues through adolescence, and portions of the skeleton generally do not stop growing until roughly age 25. In this section, we consider the physical process of osteogenesis (bone formation) and bone growth. The process of replacing other tissues with bone is called **ossification**. The term refers specifically to the formation of bone. The process of **calcification**—the deposition of calcium salts—occurs during ossification, but it can also occur in other tissues. When calcification occurs in tissues other than bone, the result is a calcified tissue (such as calcified cartilage) that does not resemble bone.

Two major forms of ossification exist: endochondral and intramembranous. In *endochondral ossification*, bone replaces existing cartilage. In *intramembranous ossification*, bone develops directly from mesenchyme or fibrous connective tissue.

CLINICAL NOTE

Heterotopic Bone Formation

In response to abnormal stresses, bone may form anywhere in the dermis or within tendons, around joints, in the kidneys, or in skeletal muscles. Dermal bones forming in abnormal locations are called *heterotopic bones* (*hetero-*, different + *topos*, place), or *ectopic bones* (*ektos*, outside). These bones can form in very odd places, such as the testes or the whites of the eyes. Physical or chemical events can stimulate the abnormal development of osteoblasts in normal connective tissues, such as sesamoid bones developing within tendons or near points of friction and pressure. Bone can also form within a large blood clot or the site of an injury or within portions of the dermis subjected to chronic abuse.

Persons with the rare genetic disease *Fibrodysplasia Ossificans Progressiva* (FOP) form normal bone in the wrong places after minor injury and provide the most dramatic demonstrations of heterotopic bone formation. The muscles of the back, neck, and upper limbs are gradually replaced by bone. The extent of the conversion can be seen in **Figure 6–9**. **Figure 6-9a** shows the skeleton of a healthy adult male; **Figure 6-9b** shows the skeleton of an adult male with advanced FOP. Several of the vertebrae have fused into a solid mass, and major muscles of the back, shoulders, and hips have undergone extensive ossification. Treatment can be problematic, because any surgical excision may trigger more ossification.

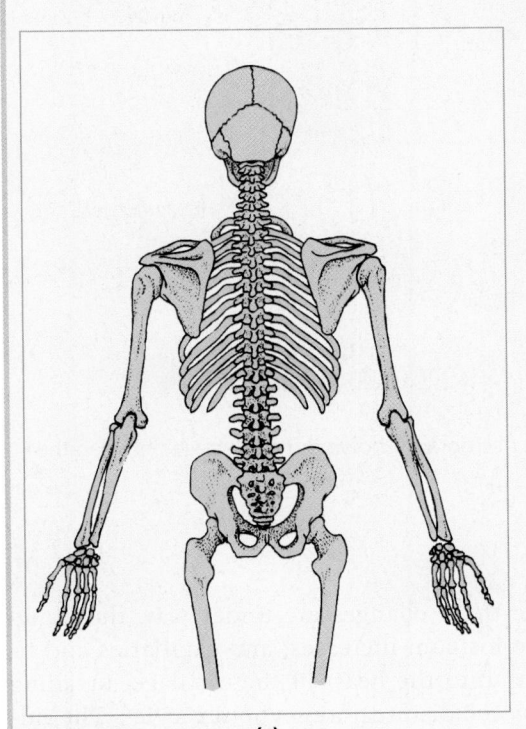

(a)

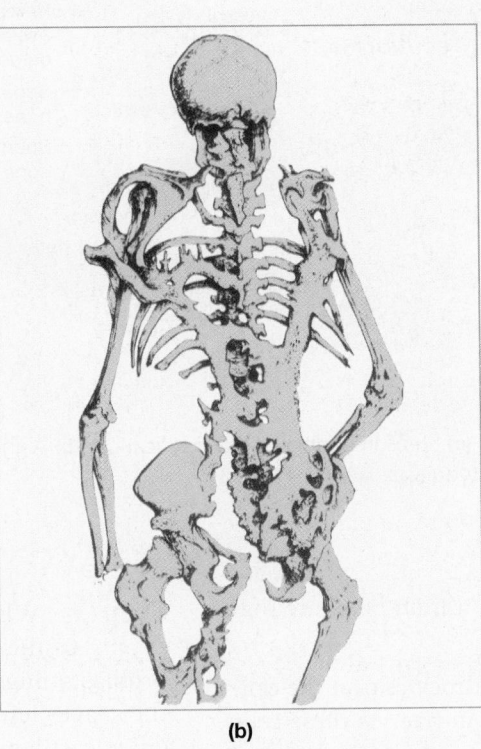

(b)

Figure 6–9 Heterotopic Bone Formation. (a) The skeleton of a healthy adult male, posterior view. **(b)** The skeleton of a 39-year-old man with advanced Fibrodysplasia Ossificans Progressiva (FOP).

Endochondral Ossification

During development, most bones originate as hyaline cartilages that are miniature models of the corresponding bones of the adult skeleton. These cartilage models are gradually converted to bone through the process of **endochondral** (en-dō-KON-drul) **ossification** (*endo-*, inside + *chondros*, cartilage).

As an example, consider the steps in limb bone development. By the time an embryo is six weeks old, the proximal bone of the limb—either the humerus (arm) or femur (thigh)—is present but composed entirely of hyaline cartilage. This cartilage model continues to grow by expansion of the cartilage matrix (*interstitial growth*) and the production of new cartilage at the outer surface (*appositional growth*). ∞ p. 131

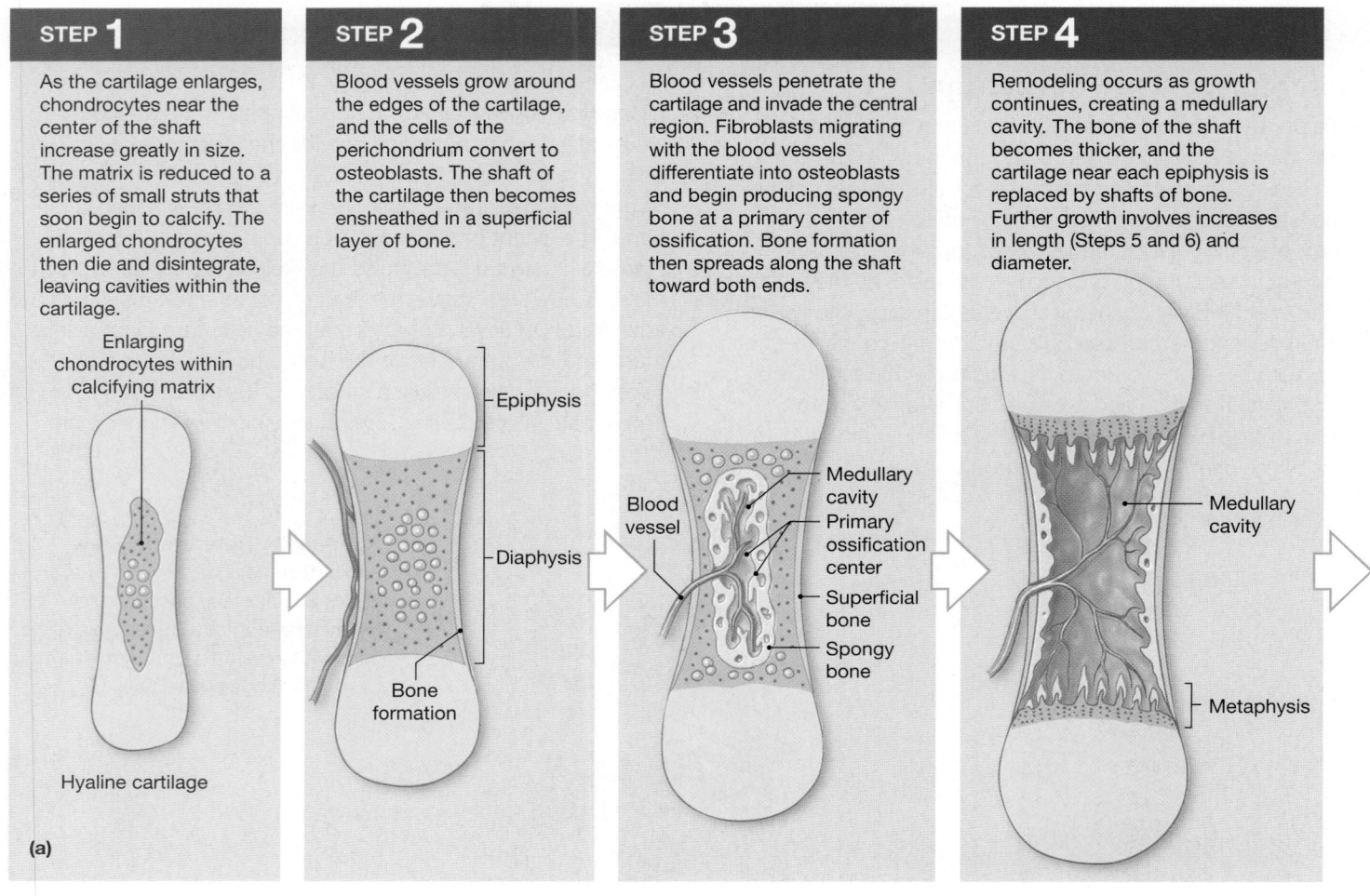

STEP 1

As the cartilage enlarges, chondrocytes near the center of the shaft increase greatly in size. The matrix is reduced to a series of small struts that soon begin to calcify. The enlarged chondrocytes then die and disintegrate, leaving cavities within the cartilage.

Enlarging chondrocytes within calcifying matrix

Hyaline cartilage

(a)

STEP 2

Blood vessels grow around the edges of the cartilage, and the cells of the perichondrium convert to osteoblasts. The shaft of the cartilage then becomes ensheathed in a superficial layer of bone.

Epiphysis

Diaphysis

Bone formation

STEP 3

Blood vessels penetrate the cartilage and invade the central region. Fibroblasts migrating with the blood vessels differentiate into osteoblasts and begin producing spongy bone at a primary center of ossification. Bone formation then spreads along the shaft toward both ends.

Blood vessel

Medullary cavity

Primary ossification center

Superficial bone

Spongy bone

STEP 4

Remodeling occurs as growth continues, creating a medullary cavity. The bone of the shaft becomes thicker, and the cartilage near each epiphysis is replaced by shafts of bone. Further growth involves increases in length (Steps 5 and 6) and diameter.

Medullary cavity

Metaphysis

Figure 6–10 **Endochondral Ossification. (a)** Steps in endochondral ossification. **(b)** A light micrograph showing the interface between the degenerating cartilage and the advancing osteoblasts. ATLAS: Plate 90

Steps in the growth and ossification of a limb bone are diagrammed in **Figure 6–10**:

Step 1   As the cartilage enlarges, chondrocytes near the center of the shaft begin to increase greatly in size. As these cells enlarge, their lacunae expand and the matrix is reduced to a series of thin struts that soon begin to calcify. The enlarged chondrocytes are now deprived of nutrients, because diffusion cannot occur through calcified cartilage. These chondrocytes become surrounded by calcified cartilage, die, and disintegrate.

Step 2   Blood vessels grow into the perichondrium surrounding the shaft of the cartilage. (We introduced the structure of the perichondrium and its role in cartilage formation in Chapter 4. ∞ pp. 131–132) The cells of the inner layer of the perichondrium in this region then differentiate into osteoblasts and begin producing a thin layer of bone around the shaft of the cartilage. The perichondrium is now technically a periosteum, because it covers bone rather than cartilage.

Step 3   While these changes are under way, the blood supply to the periosteum increases, and capillaries and fibroblasts migrate into the heart of the cartilage, invading the spaces left by the disintegrating chondrocytes. The calcified cartilaginous matrix breaks down; the fibroblasts differentiate into osteoblasts that replace it with spongy bone. Bone development begins at this site, called the **primary ossification center**, and spreads toward both ends of the cartilaginous model. While the diameter of the diaphysis is small, it is filled with spongy bone and there is no medullary cavity.

Step 4   As the bone enlarges, osteoclasts appear and begin eroding the trabeculae in the center of the diaphysis, creating a medullary cavity. Further growth involves two distinct processes: an increase in length, and an enlargement in diameter by appositional growth. (We will consider appositional growth in the next subsection.)

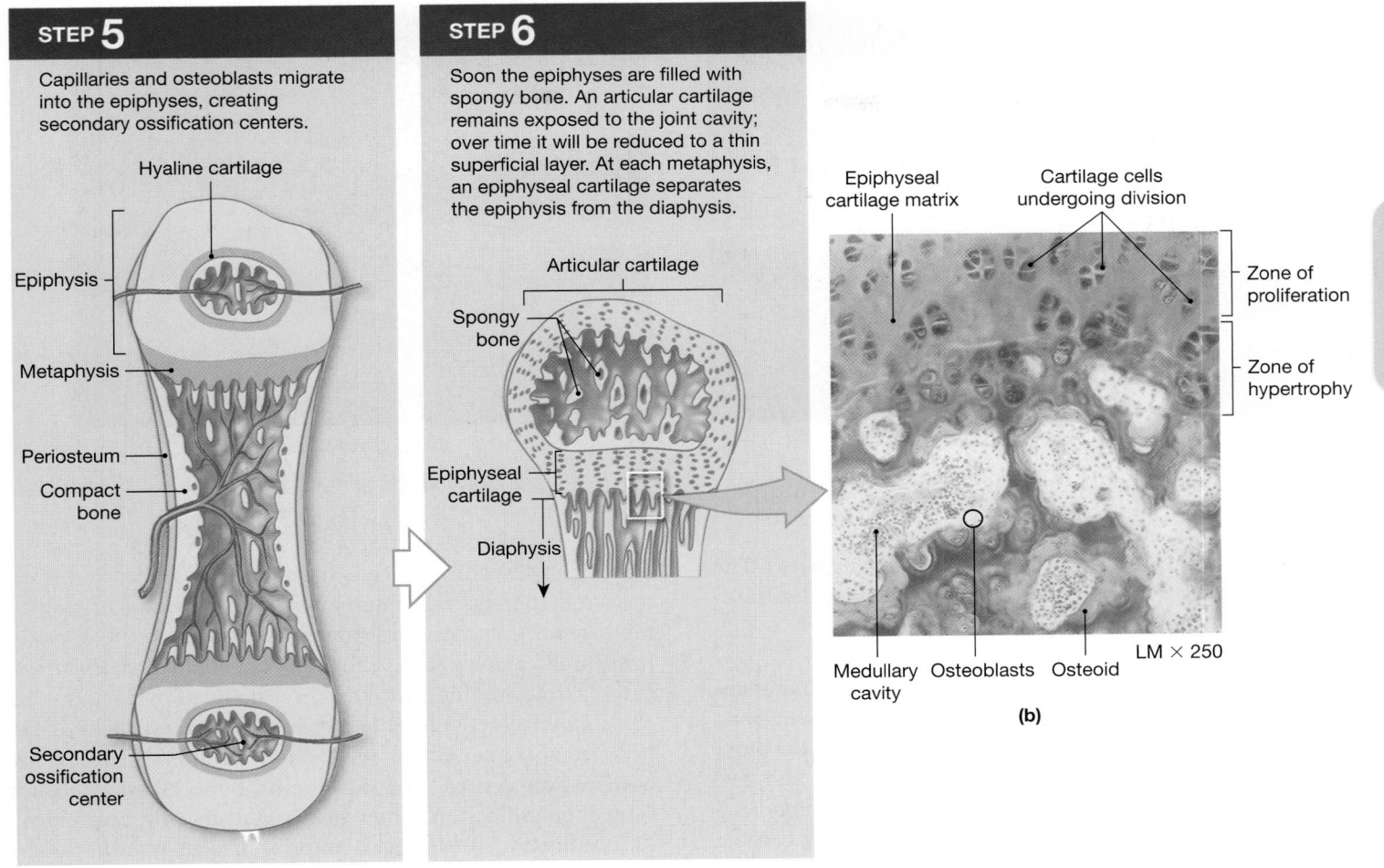

STEP 5

Capillaries and osteoblasts migrate into the epiphyses, creating secondary ossification centers.

Hyaline cartilage

Epiphysis

Metaphysis

Periosteum

Compact bone

Secondary ossification center

STEP 6

Soon the epiphyses are filled with spongy bone. An articular cartilage remains exposed to the joint cavity; over time it will be reduced to a thin superficial layer. At each metaphysis, an epiphyseal cartilage separates the epiphysis from the diaphysis.

Articular cartilage

Spongy bone

Epiphyseal cartilage

Diaphysis

Epiphyseal cartilage matrix

Cartilage cells undergoing division

Zone of proliferation

Zone of hypertrophy

Medullary cavity   Osteoblasts   Osteoid

LM × 250

(b)

Step 5 The next major change occurs when the centers of the epiphyses begin to calcify. Capillaries and osteoblasts migrate into these areas, creating **secondary ossification centers**. The time of appearance of secondary ossification centers varies from one bone to another and from individual to individual. Secondary ossification centers may occur at birth in both ends of the humerus (arm), femur (thigh), and tibia (leg), but the ends of some other bones, such as those of the fingers and toes, remain cartilaginous until early adulthood.

Step 6 The epiphyses eventually become filled with spongy bone. A thin cap of the original cartilage model remains exposed to the joint cavity as the **articular cartilage**. This cartilage prevents damaging bone-to-bone contact within the joint. At the metaphysis, a relatively narrow cartilaginous region called the **epiphyseal cartilage**, or *epiphyseal plate*, now separates the epiphysis from the diaphysis. **Figure 6–10b** shows the interface between the degenerating cartilage and the advancing osteoblasts.

As long as the epiphyseal cartilage continues to grow at its epiphyseal surface, the bone will continue to increase in length.

On the shaft side, osteoblasts continuously invade the cartilage and replace it with bone. On the epiphyseal side, new cartilage is continuously added. The osteoblasts are therefore moving toward the epiphysis, which is being pushed away by the expansion of the epiphyseal cartilage. The situation is like a pair of joggers, one in front of the other. As long as they are running at the same speed, the one in back will never catch the one in front, no matter how far they travel. The osteoblasts don't catch up to the epiphysis, as long as both the osteoblasts and the epiphysis "run away" from the primary ossification center at the same rate. Meanwhile, the bone grows longer and longer.

At puberty, the combination of rising levels of sex hormones, growth hormone, and thyroid hormones stimulates bone growth dramatically. Osteoblasts now begin producing bone faster than chondrocytes are producing new epiphyseal cartilage. As a result, the osteoblasts "catch up" and the epiphyseal cartilage gets narrower and narrower until it ultimately disappears. The timing of this event can be monitored by comparing the width of the epiphyseal cartilages in successive x-rays. In adults, the former location of this cartilage is often detectable in x-rays as a distinct **epiphyseal line**, which

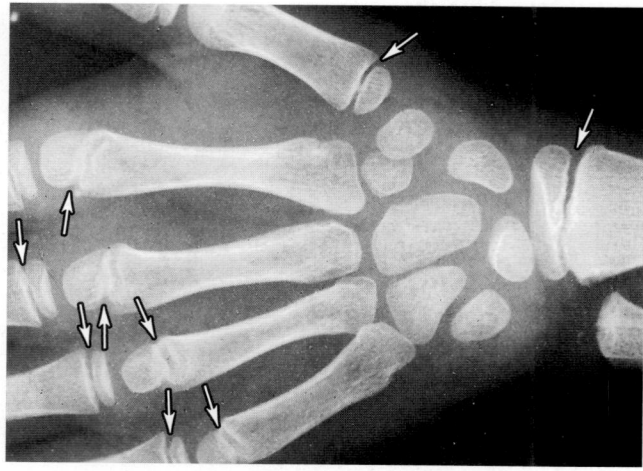

(a) Epiphyseal cartilages

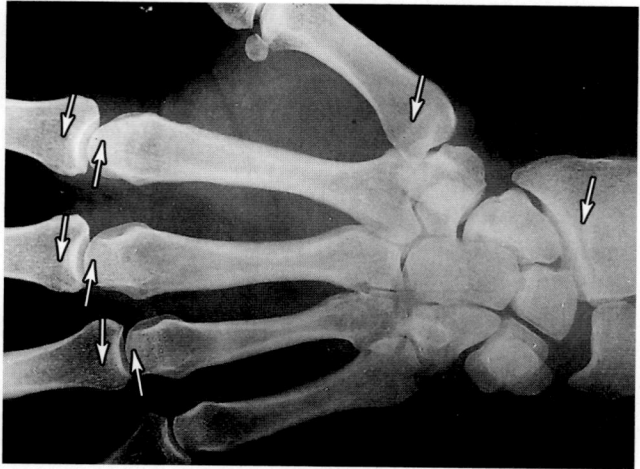

(b) Epiphyseal lines

Figure 6–11 **Bone Growth at an Epiphyseal Cartilage. (a)** An x-ray of growing epiphyseal cartilages (arrows). **(b)** Epiphyseal lines in an adult (arrows).

remains after epiphyseal growth has ended (**Figure 6–11**). The completion of epiphyseal growth is called *epiphyseal closure*.

Appositional Growth

A superficial layer of bone forms early in endochondral ossification (**Figure 6–10a,** STEP 2). Thereafter, the developing bone increases in diameter through appositional growth at the outer surface. In this process, cells of the inner layer of the periosteum differentiate into osteoblasts and deposit superficial layers of bone matrix. Eventually, these osteoblasts become surrounded by matrix and differentiate into osteocytes. Over much of the surface, appositional growth adds a series of layers that form circumferential lamellae. In time, the deepest circumferential lamellae are recycled and replaced by osteons typical of compact bone. However, blood vessels and collagen fibers of the periosteum can sometimes become enclosed within the matrix produced by osteoblasts. Osteons may then form around the smaller vessels. While bone matrix is being added to the outer surface of the growing bone, osteoclasts are removing bone matrix at the inner surface, albeit at a slower rate. As a result, the medullary cavity gradually enlarges as the bone gets larger in diameter.

Intramembranous Ossification

Intramembranous (in-tra-MEM-bra-nus) **ossification** begins when osteoblasts differentiate within a mesenchymal or fibrous connective tissue. This type of ossification is also called *dermal ossification* because it normally occurs in the deeper layers of the dermis. The bones that result are called **dermal bones**. Examples of dermal bones are the flat bones of the skull, the mandible (lower jaw), and the clavicle (collarbone).

The steps in the process of intramembranous ossification (**Figure 6–12**) can be summarized as follows:

Step 1 Mesenchymal cells first cluster together and start to secrete the organic components of the matrix. The resulting osteoid then becomes mineralized through the crystallization of calcium salts. (The enzyme *alkaline phosphatase* plays a role in this process.) As calcification occurs, the mesenchymal cells differentiate into osteoblasts. The location in a tissue where ossification begins is called an **ossification center**. The developing bone grows outward from the ossification center in small struts called **spicules**. As ossification proceeds, it traps some osteoblasts inside bony pockets; these cells differentiate into osteocytes. Meanwhile, mesenchymal cell divisions continue to produce additional osteoblasts.

Step 2 Bone growth is an active process, and osteoblasts require oxygen and a reliable supply of nutrients. Blood vessels begin to grow into the area. As spicules meet and fuse together, some of these blood vessels become trapped within the developing bone.

Step 3 Initially, the intramembranous bone consists only of spongy bone. Subsequent remodeling around trapped blood vessels can produce osteons typical of compact bone. As the rate of growth slows, the connective tissue around the bone becomes organized into the fibrous layer of the periosteum. The osteoblasts closest to the bone surface become less active, but remain as the inner, cellular layer of the periosteum.

The Blood and Nerve Supplies to Bone

In order for bones to grow and be maintained, they require an extensive blood supply. Therefore, osseous tissue is highly vascular. In a typical bone such as the humerus, three major sets of blood vessels develop (**Figure 6–13**):

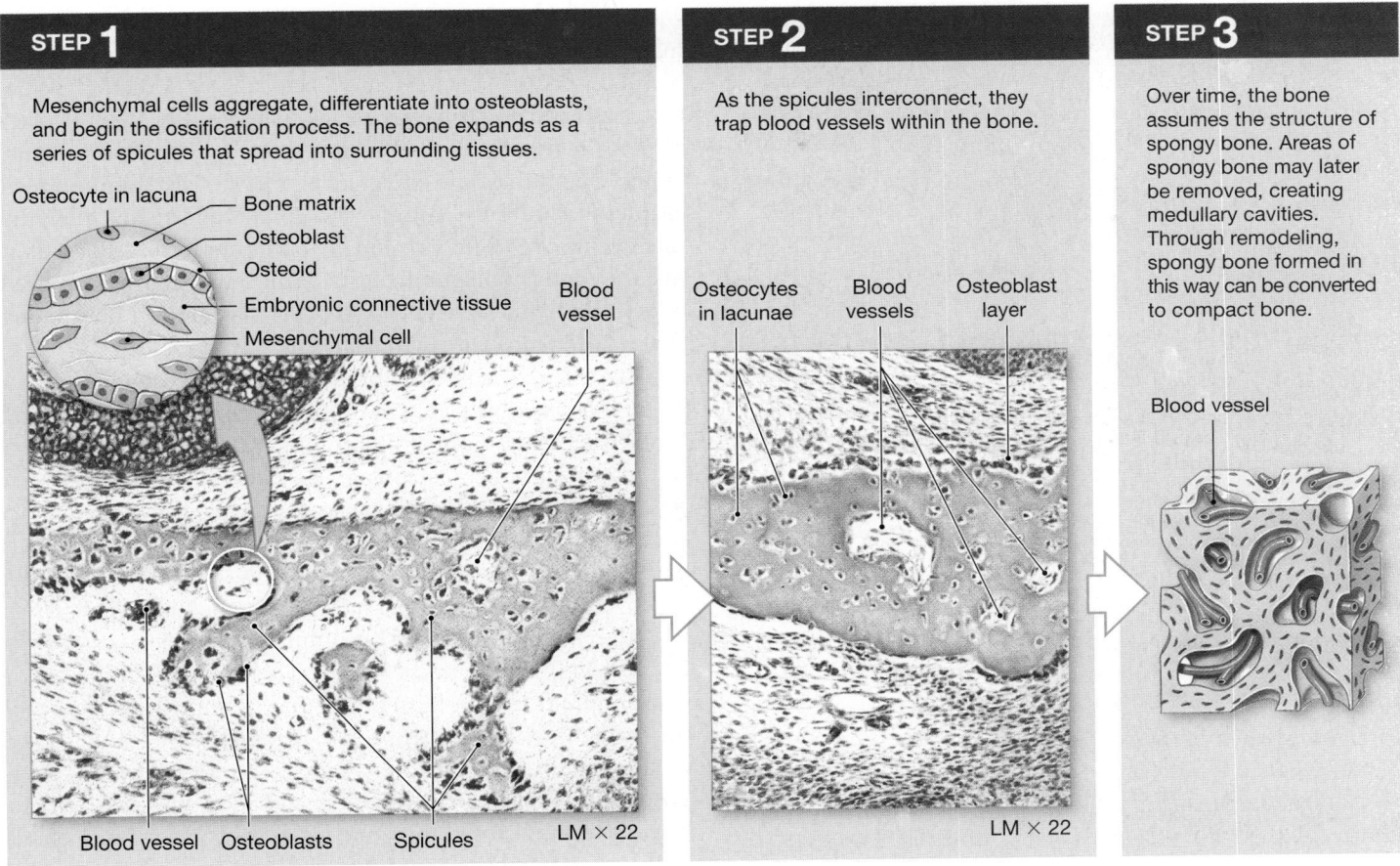

STEP 1

Mesenchymal cells aggregate, differentiate into osteoblasts, and begin the ossification process. The bone expands as a series of spicules that spread into surrounding tissues.

Osteocyte in lacuna — Bone matrix
— Osteoblast
— Osteoid
— Embryonic connective tissue
— Mesenchymal cell

Blood vessel

Blood vessel   Osteoblasts   Spicules   LM × 22

STEP 2

As the spicules interconnect, they trap blood vessels within the bone.

Osteocytes in lacunae   Blood vessels   Osteoblast layer

LM × 22

STEP 3

Over time, the bone assumes the structure of spongy bone. Areas of spongy bone may later be removed, creating medullary cavities. Through remodeling, spongy bone formed in this way can be converted to compact bone.

Blood vessel

6 SKELETAL

Figure 6–12   Intramembranous Ossification.

1. *The Nutrient Artery and Vein.* The blood vessels that supply the diaphysis form by invading the cartilage model as endochondral ossification begins. Most bones have only one *nutrient artery* and one *nutrient vein*, but a few bones, including the femur, have more than one of each. The vessels enter the bone through one or more round passageways called *nutrient foramina* in the diaphysis. Branches of these large vessels form smaller perforating canals and extend along the length of the shaft into the osteons of the surrounding cortex.

2. *Metaphyseal Vessels.* The *metaphyseal vessels* supply blood to the inner (diaphyseal) surface of each epiphyseal cartilage, where that cartilage is being replaced by bone.

3. *Periosteal Vessels.* Blood vessels from the periosteum provide blood to the superficial osteons of the shaft. During endochondral bone formation, branches of periosteal vessels also enter the epiphyses, providing blood to the secondary ossification centers.

Following the closure of the epiphyses, all three sets of vessels become extensively interconnected.

The periosteum also contains an extensive network of lymphatic vessels and sensory nerves. The lymphatics collect lymph from branches that enter the bone and reach individual osteons via the perforating canals. The sensory nerves penetrate the cortex with the nutrient artery to innervate the endosteum, medullary cavity, and epiphyses. Because of the rich sensory innervation, injuries to bones are usually very painful.

In the next section, we examine the maintenance and replacement of mineral reserves in the adult skeleton.

CHECKPOINT

9. During intramembranous ossification, which type(s) of tissue is (are) replaced by bone?

10. In endochondral ossification, what is the original source of osteoblasts?

11. How could x-rays of the femur be used to determine whether a person has reached full height?

See the blue Answers tab at the end of the book.

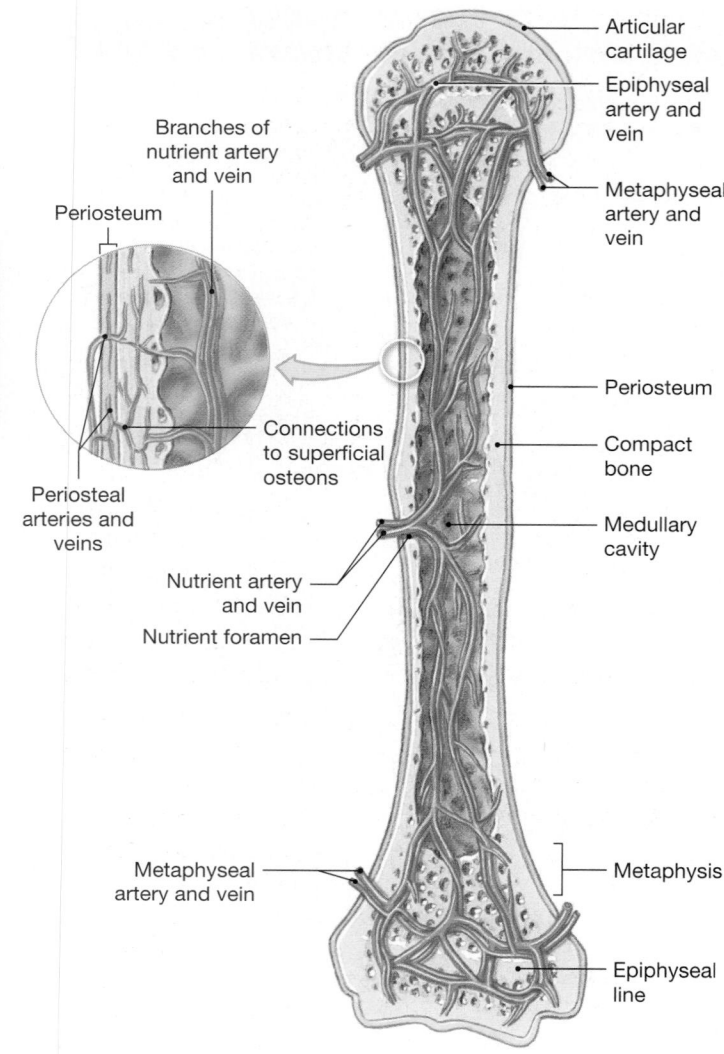

Figure 6–13   **The Blood Supply to a Mature Bone.**

surrounding calcium salts. Osteoclasts and osteoblasts also remain active, even after the epiphyseal cartilages have closed. Normally, their activities are balanced: As quickly as osteoblasts form one osteon, osteoclasts remove another by osteolysis. The turnover rate of bone is quite high. In young adults, almost one-fifth of the adult skeleton is recycled and replaced each year. Not every part of every bone is affected equally; the rate of turnover differs regionally and even locally. For example, the spongy bone in the head of the femur may be replaced two or three times each year, whereas the compact bone along the shaft remains largely unchanged.

Because of their biochemical similarity to calcium, heavy-metal ions such as lead, strontium, or cobalt, or more exotic forms such as uranium or plutonium, can be incorporated into the matrix of bone. Osteoblasts do not differentiate between these heavy-metal ions and calcium, and any heavy-metal ions present in the bloodstream will be deposited into the bone matrix. Some of these ions are potentially dangerous, and the turnover of bone matrix can have detrimental health effects as ions that are absorbed and accumulated are released into the circulation over a period of years. This was one of the major complications in the aftermath of the Chernobyl nuclear reactor incident in 1986. Radioactive compounds released in the meltdown of the reactor were deposited into the bones of exposed individuals. Over time, the radiation released by their own bones has caused thyroid cancers and may result in cases of leukemia and other potentially fatal cancers.

The A&P Top 100

#29 Bone is continually remodeled, recycled, and replaced. The rate of turnover varies from bone to bone and from moment to moment. When deposition exceeds removal, bones get stronger; when removal exceeds deposition, bones get weaker.

6-6 Bone growth and development depend on a balance between bone formation and bone resorption

The organic and mineral components of the bone matrix are continuously being recycled and renewed through the process of **remodeling.** Bone remodeling goes on throughout life, as part of normal bone maintenance. Remodeling can replace the matrix but leave the bone as a whole unchanged, or it may change the shape, internal architecture, or mineral content of the bone. Through this remodeling process, older mineral deposits are removed from bone and released into the circulation at the same time that circulating minerals are being absorbed and deposited.

Bone remodeling involves an interplay among the activities of osteocytes, osteoblasts, and osteoclasts. In adults, osteocytes are continuously removing and replacing the

CHECKPOINT

12. **Describe bone remodeling.**

13. **Explain how heavy-metal ions could be incorporated into bone matrix.**

See the blue Answers tab at the end of the book.

6-7 Exercise, hormones, and nutrition affect bone development and the skeletal system

In this section we direct our attention to the factors that have the most important effects on the processes of bone remodeling.

The Effects of Exercise on Bone

The turnover and recycling of minerals give each bone the ability to adapt to new stresses. The sensitivity of osteoblasts to electrical events has been theorized as the mechanism that controls the internal organization and structure of bone. Whenever a bone is stressed, the mineral crystals generate minute electrical fields. Osteoblasts are apparently attracted to these electrical fields and, once in the area, begin to produce bone. This finding has led to the successful use of small electrical fields in stimulating the repair of severe fractures.

Because bones are adaptable, their shapes reflect the forces applied to them. For example, bumps and ridges on the surface of a bone mark the sites where tendons are attached. If muscles become more powerful, the corresponding bumps and ridges enlarge to withstand the increased forces. Heavily stressed bones become thicker and stronger, whereas bones that are not subjected to ordinary stresses become thin and brittle. Regular exercise is therefore an important stimulus for maintaining normal bone structure. Champion weight lifters have massive bones with thick, prominent ridges where muscles attach. In nonathletes (especially "couch potatoes"), moderate amounts of physical activity and weight-bearing activities are essential for stimulating normal bone maintenance and maintaining adequate bone strength.

Degenerative changes in the skeleton occur after relatively brief periods of inactivity. For example, you may use a crutch to take weight off an injured leg while you wear a cast. After a few weeks, your unstressed bones will lose up to a third of their mass. The bones rebuild just as quickly when you resume normal weight loading. However, the removal of calcium salts can be a serious health hazard both for astronauts remaining in a weightless environment and for bedridden or paralyzed patients who spend months or years without stressing their skeleton.

The A&P Top 100

#30 What you don't use, you lose. The stresses applied to bones during physical activity are essential to maintaining bone strength and bone mass.

Hormonal and Nutritional Effects on Bone

Normal bone growth and maintenance depend on a combination of nutritional and hormonal factors:

- Normal bone growth and maintenance cannot occur without a constant dietary source of calcium and phosphate salts. Lesser amounts of other minerals, such as magnesium, fluoride, iron, and manganese, are also required.

- The hormone *calcitriol*, synthesized in the kidneys, is essential for normal calcium and phosphate ion absorption in the digestive tract. Calcitriol is synthesized from a related steroid, *cholecalciferol* (vitamin D$_3$), which may be produced in the skin or absorbed from the diet. ∞ p. 165

- Adequate levels of vitamin C must be present in the diet. This vitamin, which is required for certain key enzymatic reactions in collagen synthesis, also stimulates osteoblast differentiation. One of the signs of vitamin C deficiency—a condition called *scurvy*—is a loss of bone mass and strength.

- Three other vitamins have significant effects on bone structure. Vitamin A, which stimulates osteoblast activity, is particularly important for normal bone growth in children. Vitamins K and B$_{12}$ are required for the synthesis of proteins in normal bone.

- *Growth hormone*, produced by the pituitary gland, and *thyroxine*, from the thyroid gland, stimulate bone growth. Growth hormone stimulates protein synthesis and cell growth throughout the body. Thyroxine stimulates cell metabolism and increases the rate of osteoblast activity. In proper balance, these hormones maintain normal activity at the epiphyseal cartilages until roughly the time of puberty.

- At puberty, rising levels of sex hormones (*estrogens* in females and *androgens* in males) stimulate osteoblasts to produce bone faster than the rate at which epiphyseal cartilage expands. Over time, the epiphyseal cartilages narrow and eventually close. The timing of epiphyseal closure differs from bone to bone and from individual to individual. The toes may complete ossification by age 11, but parts of the pelvis or the wrist may continue to enlarge until roughly age 25. Differences in male and female sex hormones account for significant variations in body size and proportions. Because estrogens cause faster epiphyseal closure than do androgens, women are generally shorter than men at maturity.

Two other hormones—*calcitonin* (kal-si-TŌ-nin), from the thyroid gland, and *parathyroid hormone*, from the parathyroid gland—are important in the homeostatic control of calcium and phosphate levels in body fluids. We consider the interactions of these hormones in the next section. The major hormones affecting the growth and maintenance of the skeletal system are summarized in Table 6–2.

The skeletal system is unique in that it persists after life, providing clues to the sex, lifestyle, and environmental conditions experienced by the individual. Not only do the bones reflect the physical stresses placed on the body, but they also provide clues concerning the person's health and diet. By using the appearance, strength, and composition of bone, forensic scientists and physical anthropologists can detect features characteristic of hormonal deficiencies. (For more on the effects of hormones on bone growth, see the Clinical Note "Abnormal Bone Growth and Development.") Combining the physical clues provided by the skeleton with modern molecular techniques, such as DNA fingerprinting, can provide a wealth of information.

TABLE 6–2 Hormones Involved in the Regulation of Bone Growth and Maintenance

Hormone	Primary Source	Effects on Skeletal System
Calcitriol	Kidneys	Promotes calcium and phosphate ion absorption along the digestive tract
Growth hormone	Pituitary gland	Stimulates osteoblast activity and the synthesis of bone matrix
Thyroxine	Thyroid gland (follicle cells)	With growth hormone, stimulates osteoblast activity and the synthesis of bone matrix
Sex hormones	Ovaries (estrogens) Testes (androgens)	Stimulate osteoblast activity and the synthesis of bone matrix
Parathyroid hormone	Parathyroid glands	Stimulates osteoclast (and osteoblast) activity; elevates calcium ion concentrations in body fluids
Calcitonin	Thyroid gland (C cells)	Inhibits osteoclast activity; promotes calcium loss at kidneys; reduces calcium ion concentrations in body fluids

CLINICAL NOTE

Abnormal Bone Growth and Development

A variety of endocrine or metabolic problems can result in characteristic skeletal changes. In pituitary growth failure, or pituitary dwarfism (**Figure 6-14a**), inadequate production of growth hormone leads to reduced epiphyseal cartilage activity and abnormally short bones. This condition is becoming increasingly rare in the United States, because children can be treated with synthetic human growth hormone.

Gigantism results from an overproduction of growth hormone before puberty. (The world record for height is 272 cm, or 8 ft, 11 in., reached by Robert Wadlow, of Alton, Illinois, who died at age 22 in 1940. Wadlow weighed 216 kg, or 475 lb.) If growth hormone levels rise abnormally after epiphyseal cartilages close, the skeleton does not grow

longer, but bones get thicker, especially those in the face, jaw, and hands. Cartilage growth and alterations in soft-tissue structure lead to changes in physical features, such as the contours of the face. These physical changes occur in the disorder called **acromegaly**.

Several inherited metabolic conditions that affect many systems influence the growth and development of the skeletal system. These conditions produce characteristic variations in body proportions. For example, many individuals with **Marfan syndrome** are very tall and have long, slender limbs (**Figure 6–14b**), due to excessive cartilage formation at the epiphyseal cartilages. Although this is an obvious physical distinction, the characteristic body proportions are not in themselves dangerous. However, the underlying mutation, which affects the structure of connective tissue throughout the body, commonly causes life-threatening cardiovascular problems.

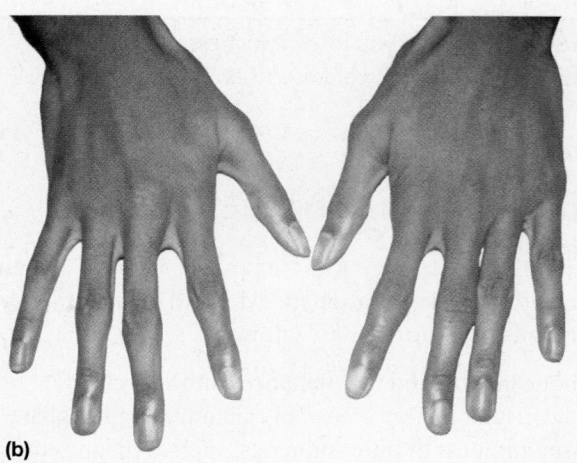

(a)

(b)

Figure 6–14 **Examples of Abnormal Bone Development.** (a) Pituitary growth failure (pituitary dwarfism). (b) Marfan syndrome.

14. Why would you expect the arm bones of a weight lifter to be thicker and heavier than those of a jogger?

15. A child who enters puberty several years later than the average age is generally taller than average as an adult. Why?

16. A 7-year-old child has a pituitary gland tumor involving the cells that secrete growth hormone (GH), resulting in increased levels of GH. How will this condition affect the child's growth?

See the blue Answers tab at the end of the book.

6-8 Calcium plays a critical role in bone physiology

Next we discuss the dynamic relationship between calcium and the skeletal system, and the role of hormones in calcium balance in the body.

The Skeleton as a Calcium Reserve

The chemical analysis shown in **Figure 6–15** reveals the importance of bones as mineral reservoirs. For the moment, we will focus on the homeostatic regulation of calcium ion concentration in body fluids; we will consider other minerals in later chapters. Calcium is the most abundant mineral in the human body. A typical human body contains 1–2 kg (2.2–4.4 lb) of calcium, with roughly 99 percent of it deposited in the skeleton.

Calcium ions play a role in a variety of physiological processes, so the body must tightly control calcium ion concentrations in order to prevent damage to essential physiological systems. Even small variations from the normal concentration affect cellular operations; larger changes can cause a clinical crisis. Calcium ions are particularly important to both the membranes and the intracellular activities of neurons and muscle cells, especially cardiac muscle cells. If the calcium concentration of body fluids increases by 30 percent, neurons and muscle cells become relatively unresponsive. If calcium levels decrease by 35 percent, neurons become so excitable that convulsions can occur. A 50 percent reduction in calcium concentration generally causes death. Calcium ion concentration is so closely regulated, however, that daily fluctuations of more than 10 percent are highly unusual.

Hormones and Calcium Balance

Calcium ion homeostasis is maintained by a pair of hormones with opposing effects. These hormones, parathyroid hormone and calcitonin, coordinate the storage, absorption, and excretion of calcium ions. Three target sites and functions are involved: (1) the bones (storage), (2) the digestive tract (absorption), and (3) the kidneys (excretion). **Figure 6–16a** indicates factors that elevate calcium levels in the blood; **Figure 6–16b** indicates factors that depress blood calcium levels.

When calcium ion concentrations in the blood fall below normal, cells of the **parathyroid glands**, embedded in the thyroid gland in the neck, release **parathyroid hormone (PTH)** into the bloodstream. Parathyroid hormone has three major effects, all of which increase blood calcium levels:

1. *Stimulating osteoclast activity and enhancing the recycling of minerals by osteocytes.* (PTH also stimulates osteoblast activity, but to a lesser degree.)

2. *Increasing the rate of intestinal absorption of calcium ions* by enhancing the action of calcitriol. Under normal circumstances, calcitriol is always present, and parathyroid hormone controls its effect on the intestinal epithelium.

3. *Decreasing the rate of excretion of calcium ions at the kidneys.*

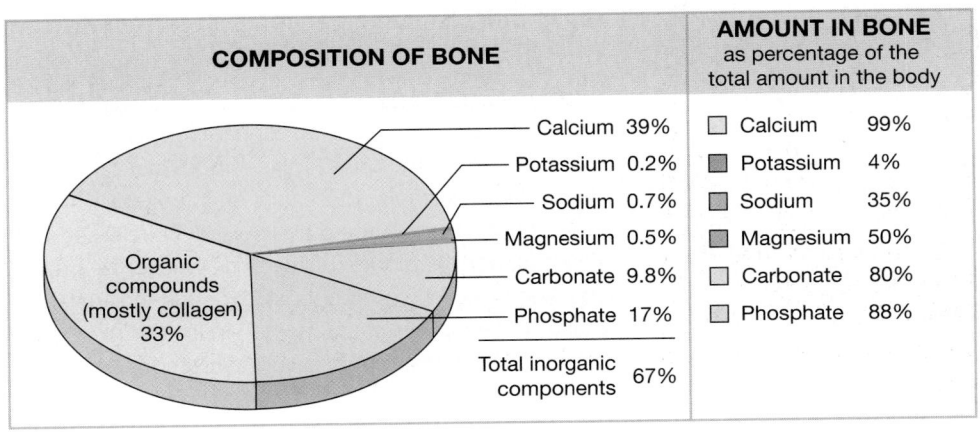

COMPOSITION OF BONE		AMOUNT IN BONE as percentage of the total amount in the body	
Calcium	39%	Calcium	99%
Potassium	0.2%	Potassium	4%
Sodium	0.7%	Sodium	35%
Magnesium	0.5%	Magnesium	50%
Organic compounds (mostly collagen) 33%	Carbonate	9.8%	Carbonate	80%
	Phosphate	17%	Phosphate	88%
	Total inorganic components	67%		

Figure 6–15 **A Chemical Analysis of Bone.**

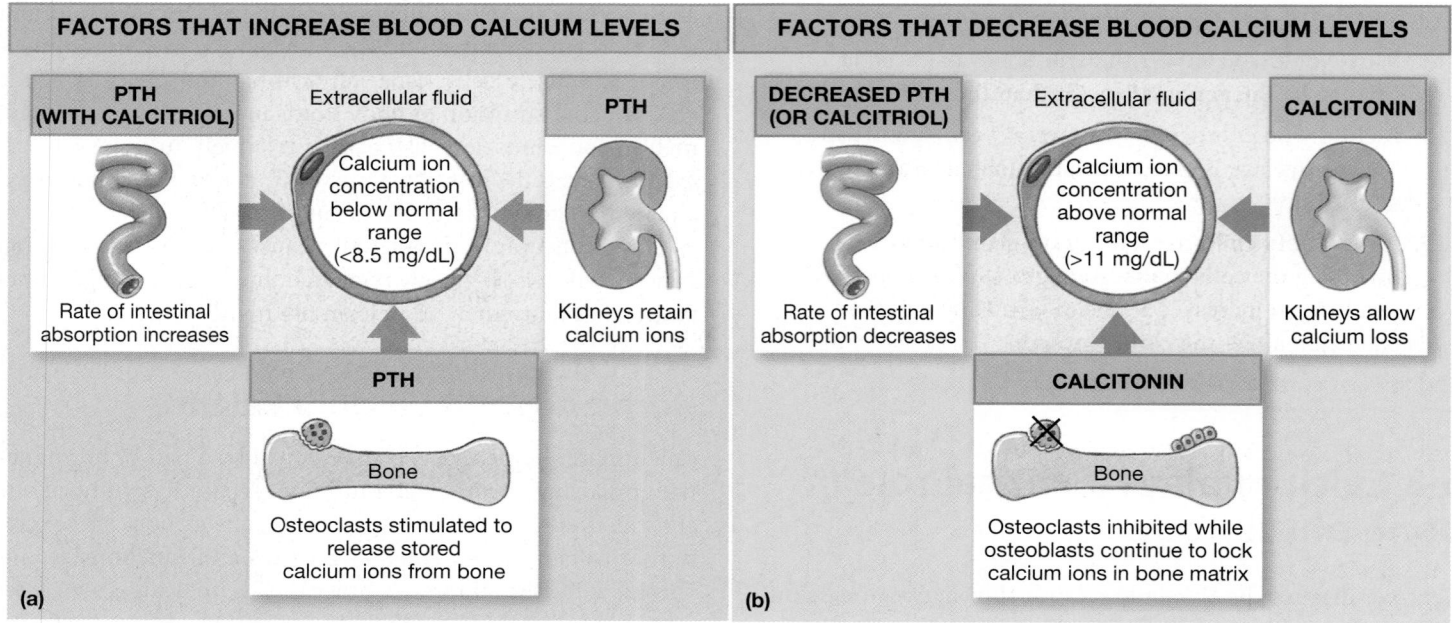

Figure 6–16   Factors That Alter the Concentration of Calcium Ions in Body Fluids.

Under these conditions, more calcium ions enter body fluids, and losses are restricted. The calcium ion concentration increases to normal levels, and homeostasis is restored.

If the calcium ion concentration of the blood instead rises above normal, special cells (*parafollicular cells*, or *C cells*) in the thyroid gland secrete **calcitonin**. This hormone has two major functions, which together act to decrease calcium ion concentrations in body fluids:

1. *Inhibiting osteoclast activity.*

2. *Increasing the rate of excretion of calcium ions at the kidneys.*

Under these conditions, less calcium *enters* body fluids because osteoclasts leave the mineral matrix alone. More calcium *leaves* body fluids because osteoblasts continue to produce new bone matrix while calcium ion excretion at the kidneys accelerates. The net result is a decline in the calcium ion concentration of body fluids, restoring homeostasis.

By providing a calcium reserve, the skeleton plays the primary role in the homeostatic maintenance of normal calcium ion concentrations of body fluids. This function can have a direct effect on the shape and strength of the bones in the skeleton. When large numbers of calcium ions are mobilized in body fluids, the bones become weaker; when calcium salts are deposited, the bones become denser and stronger.

Because the bone matrix contains protein fibers as well as mineral deposits, changes in mineral content do not necessarily affect the shape of the bone. In *osteomalacia* (os-tē-ō-ma-LĀ-shē-uh; *malakia*, softness), the bones appear normal, although they are weak and flexible owing to poor mineralization. *Rickets*, a form of osteomalacia affecting children, generally results from a vitamin D₃ deficiency caused by inadequate skin exposure to sunlight and an inadequate dietary supply of the vitamin. ∞ p. 165 The bones of children with rickets are so poorly mineralized that they become very flexible. Because the walls of each femur can no longer resist the tension and compression forces applied by the body weight (**Figure 6–7**), the bones bend laterally and affected individuals develop a bow-legged appearance. In the United States, homogenized milk is fortified with vitamin D specifically to prevent rickets.

The A&P Top 100

#31 Each day calcium and phosphate ions circulating in the blood are lost in the urine. To keep body fluid concentrations stable, those ions must be replaced; if they aren't obtained from the diet, they will be released from the skeleton, and the bones will become weaker as a result. If you want to keep your bones strong, you must exercise and make sure your diet contains vitamin D₃ and plenty of calcium—at least enough to compensate for daily excretion.

17. Identify the hormones involved in stimulating and inhibiting the release of calcium ions from bone matrix.

18. Why does a child who has rickets have difficulty walking?

19. What effect would increased PTH secretion have on blood calcium levels?

20. How does calcitonin help lower the calcium ion concentration of blood?

See the blue Answers tab at the end of the book.

6-9 A fracture is a crack or break in a bone

Despite its mineral strength, bone can crack or even break if subjected to extreme loads, sudden impacts, or stresses from unusual directions. The damage produced constitutes a fracture. (See "FOCUS: Types of Fractures," p. 206.) Most fractures heal even after severe damage, provided that the blood supply and the cellular components of the endosteum and periosteum survive. Steps in the repair process are illustrated in **Figure 6–17**:

Step 1 In even a small fracture, many blood vessels are broken and extensive bleeding occurs. A large blood clot, or **fracture hematoma**, soon closes off the injured vessels and leaves a fibrous meshwork in the damaged area. The disruption of circulation kills osteocytes around the fracture, broadening the area affected. Dead bone soon extends along the shaft in either direction from the break.

Step 2 In adults, the cells of the periosteum and endosteum are normally relatively inactive. When a fracture occurs, the cells of the intact endosteum and periosteum undergo rapid cycles of cell division, and the daughter cells migrate into the fracture zone. An **external callus** (*callum*, hard skin), or enlarged collar of cartilage and bone, forms and encircles the bone at the level of the fracture. An extensive **internal callus** organizes within the medullary cavity and between the broken ends of the

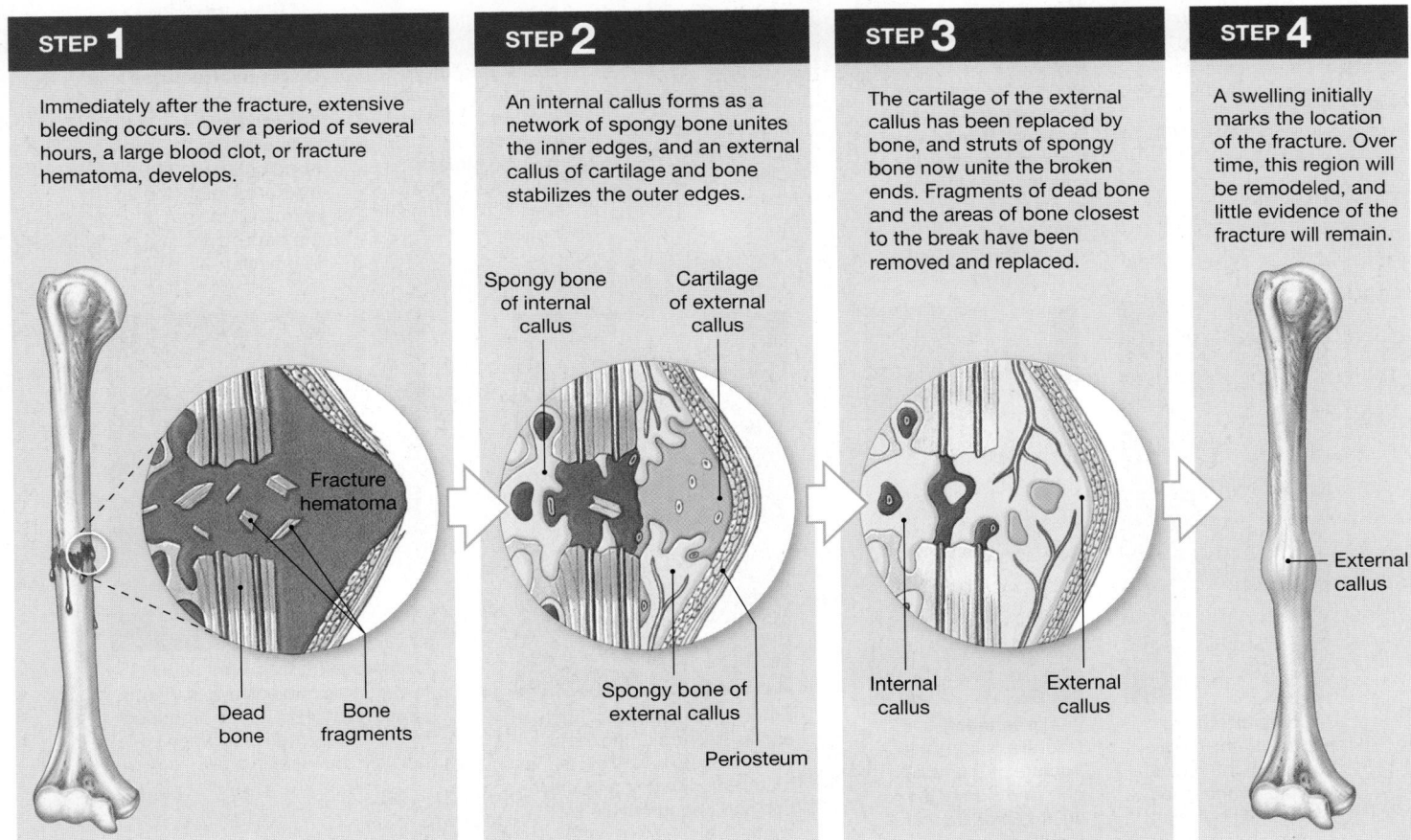

STEP 1

Immediately after the fracture, extensive bleeding occurs. Over a period of several hours, a large blood clot, or fracture hematoma, develops.

Fracture hematoma

Dead bone

Bone fragments

STEP 2

An internal callus forms as a network of spongy bone unites the inner edges, and an external callus of cartilage and bone stabilizes the outer edges.

Spongy bone of internal callus

Cartilage of external callus

Spongy bone of external callus

Periosteum

STEP 3

The cartilage of the external callus has been replaced by bone, and struts of spongy bone now unite the broken ends. Fragments of dead bone and the areas of bone closest to the break have been removed and replaced.

Internal callus

External callus

STEP 4

A swelling initially marks the location of the fracture. Over time, this region will be remodeled, and little evidence of the fracture will remain.

External callus

Figure 6–17 Steps in the Repair of a Fracture.

Types of Fractures

Fractures are named according to their external appearance, their location, and the nature of the crack or break in the bone. Important types of fractures are illustrated here by representative x-rays. The broadest general categories are closed fractures and open fractures. **Closed**, or *simple*, fractures are completely internal. They can be seen only on x-rays, because they do not involve a break in the skin. Closed fractures are usually relatively simple to treat, as the surrounding tissues keep the broken ends of the bone aligned. **Open**, or *compound*, fractures project through the skin. These fractures, which are obvious on inspection, are more dangerous than closed fractures, due to the possibility of infection or uncontrolled bleeding. A *transverse fracture* involves a break at right angles to the long axis of a bone, and a *comminuted fracture* involves a shattering of the affected bone(s). Many fractures fall into more than one category, because the terms overlap. Representative examples of the common types of fractures are shown in **Figure 6–18**.

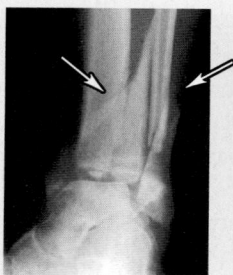

A **Pott fracture** occurs at the ankle and affects both bones of the leg.

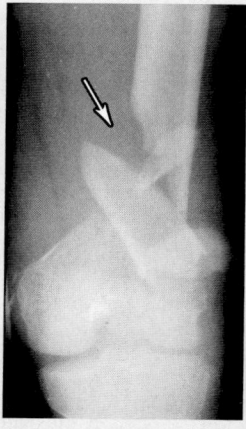

Comminuted fractures, such as this fracture of the femur, shatter the affected area into a multitude of bony fragments.

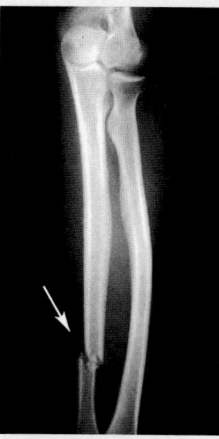

Transverse fractures, such as this fracture of the ulna, break a bone shaft across its long axis.

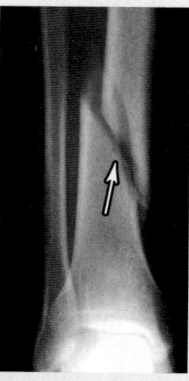

Spiral fractures, such as this fracture of the tibia, are produced by twisting stresses that spread along the length of the bone.

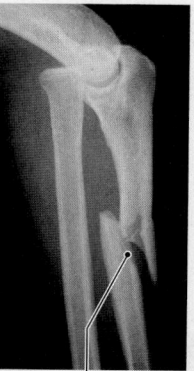

Displaced fractures produce new and abnormal bone arrangements; **nondisplaced fractures** retain the normal alignment of the bones or fragments.

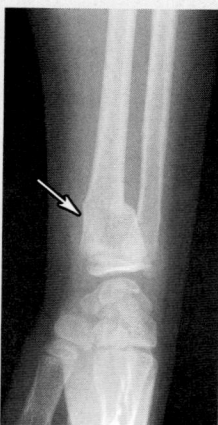

A **Colles fracture**, a break in the distal portion of the radius, is typically the result of reaching out to cushion a fall.

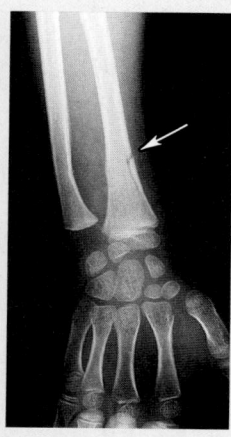

In a **greenstick fracture**, such as this fracture of the radius, only one side of the shaft is broken, and the other is bent. This type of fracture generally occurs in children, whose long bones have yet to ossify fully.

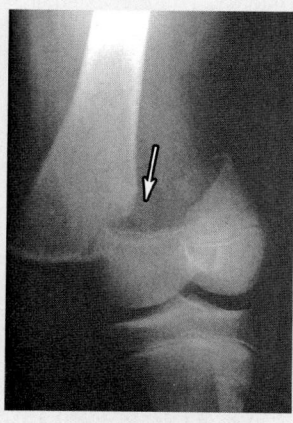

Epiphyseal fractures, such as this fracture of the femur, tend to occur where the bone matrix is undergoing calcification and chondrocytes are dying. A clean transverse fracture along this line generally heals well. Unless carefully treated, fractures between the epiphysis and the epiphyseal cartilage can permanently stop growth at this site.

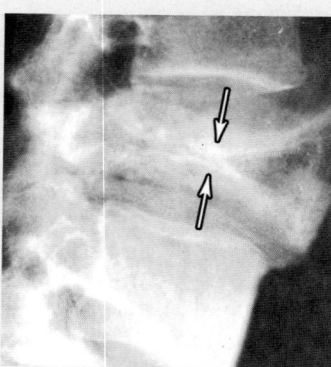

Compression fractures occur in vertebrae subjected to extreme stresses, such as those produced by the forces that arise when you land on your seat in a fall.

Figure 6–18 Major Types of Fractures.

shaft. At the center of the external callus, cells differentiate into chondrocytes and produce blocks of hyaline cartilage. At the edges of each callus, the cells differentiate into osteoblasts and begin creating a bridge between the bone fragments on either side of the fracture. At this point, the broken ends have been temporarily stabilized.

Step 3   As the repair continues, osteoblasts replace the central cartilage of the external callus with spongy bone. When this conversion is complete, the external and internal calluses form an extensive and continuous brace at the fracture site. Struts of spongy bone now unite the broken ends. The surrounding area is gradually reshaped as fragments of dead bone are removed and replaced. The ends of the fracture are now held firmly in place and can withstand normal stresses from muscle contractions. If the fracture required external support in the form of a cast, that support can be removed at this stage.

Step 4   Osteoclasts and osteoblasts continue to remodel the region of the fracture for a period ranging from four months to well over a year. When the remodeling is complete, the bone of the calluses is gone and only living compact bone remains. The repair may be "good as new" and leave no indications that a fracture ever occurred, or the bone may be slightly thicker and stronger than normal at the fracture site. Under comparable stresses, a second fracture will generally occur at a different site.

CHECKPOINT

21. **List the steps involved in fracture repair, beginning at the onset of the bone break.**

22. **At which point in fracture repair would you find an external callus?**

See the blue Answers tab at the end of the book.

6-10 Osteopenia has a widespread effect on aging skeletal tissue

The bones of the skeleton become thinner and weaker as a normal part of the aging process. Inadequate ossification is called **osteopenia** (os-tē-ō-PĒ-nē-uh; *penia*, lacking), and all of us become slightly osteopenic as we age. This reduction in bone mass begins between ages 30 and 40. Over that period, osteoblast activity begins to decline, while osteoclast activity continues at previous levels. Once the reduction begins, women lose roughly 8 percent of their skeletal mass every decade, whereas the skeletons of men deteriorate at about 3 percent per decade. Not all parts of the skeleton are equally affected. Epiphyses, vertebrae, and the jaws lose more mass than other sites, resulting in fragile limbs, reduction in height, and loss of teeth.

When the reduction in bone mass is sufficient to compromise normal function, the condition is known as **osteoporosis** (os-tē-ō-po-RŌ-sis; *porosus*, porous). The fragile bones that result are likely to break when exposed to stresses that younger individuals could easily tolerate. For example, a hip fracture can occur when a ninety-year-old simply tries to stand. Any fractures that occur in aged individuals lead to a loss of independence and an immobility that further weakens the skeleton. The extent of the loss of spongy bone mass due to osteoporosis is shown in **Figure 6–19**; the reduction in compact and cortical bone mass is equally severe.

Sex hormones are important in maintaining normal rates of bone deposition. Over age 45, an estimated 29 percent of

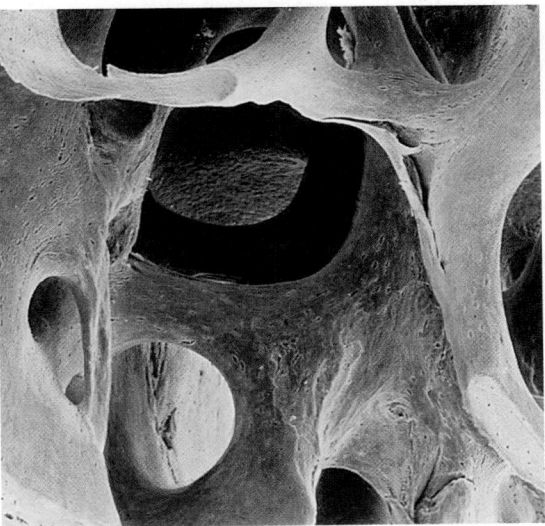

(a) Normal spongy bone   SEM × 25

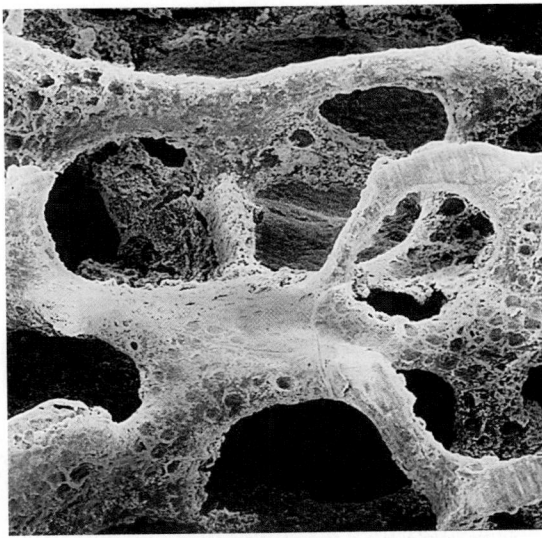

(b) Spongy bone in osteoporosis   SEM × 21

Figure 6–19   The Effects of Osteoporosis on Spongy Bone.
(a) Normal spongy bone from the epiphysis of a young adult.
(b) Spongy bone from a person with osteoporosis.

women and 18 percent of men have osteoporosis. In women, the condition accelerates after menopause, owing to a decline in circulating estrogens. Because men continue to produce androgens until relatively late in life, severe osteoporosis is less common in men below age 60 than in women of that same age group.

Osteoporosis can also develop as a secondary effect of many cancers. Cancers of the bone marrow, breast, or other tissues release a chemical known as **osteoclast-activating**

factor. This compound increases both the number and activity of osteoclasts and produces severe osteoporosis.

CHECKPOINT

23. Define osteopenia.

24. Why is osteoporosis more common in older women than in older men?

See the blue Answers tab at the end of the book.

Related Clinical Terms

acromegaly: A condition caused by excess secretion of growth hormone after puberty. Skeletal abnormalities develop, affecting the cartilages and various small bones.

external callus: A toughened layer of connective tissue that encircles and stabilizes a bone at a fracture site.

fracture: A crack or break in a bone.

fracture hematoma: A large blood clot that closes off the injured vessels around a fracture and leaves a fibrous meshwork in the damaged area of bone; the first step in fracture repair.

gigantism: A condition resulting from an overproduction of growth hormone before puberty.

internal callus: A bridgework of bone trabeculae that unites the broken ends of a bone on the marrow side of a fracture.

Marfan syndrome: An inherited condition linked to defective production of *fibrillin*, a connective tissue glycoprotein. Extreme height and long, slender limbs are the most obvious physical indications of Marfan syndrome; cardiovascular problems are the most dangerous aspects of the condition.

osteoclast-activating factor: A compound, released by cancers of the bone marrow, breast, or other tissues, that produces severe osteoporosis.

osteomalacia: A softening of bone due to a decrease in its mineral content.

osteopenia: Inadequate ossification, leading to thinner, weaker bones.

osteoporosis: A reduction in bone mass to a degree that compromises normal function.

pituitary growth failure (*pituitary dwarfism*): A disorder caused by inadequate production of growth hormone prior to puberty.

rickets: A childhood disorder that reduces the amount of calcium salts in the skeleton; typically characterized by a bow-legged appearance, because the leg bones bend under the body's weight.

scurvy: A condition involving weak, brittle bones as a result of a vitamin C deficiency.

Chapter Review

Study Outline

An Introduction to the Skeletal System p. 185

1. Skeletal elements have a variety of purposes, such as providing a framework for body posture and allowing for precise movements.

6-1 The skeletal system has five primary functions p. 185

2. The skeletal system includes the bones of the skeleton and the cartilages, ligaments, and other connective tissues that stabilize or connect the bones. The functions of the skeletal system include support, storage of minerals and lipids, blood cell production, protection, and leverage.

6-2 Bones are classified according to shape and structure, and feature surface markings p. 185

3. Bones may be categorized as **long bones**, **flat bones**, **sutural bones** (*Wormian bones*), **irregular bones**, **short bones**, and **sesamoid bones**. (*Figure 6–1*)

4. Each bone has characteristic **bone markings**, including elevations, projections, depressions, grooves, and tunnels. (*Table 6–1*)

5. The two types of bone tissue are compact (*dense*) bone and spongy (*cancellous*) bone.

6. A representative long bone has a **diaphysis**, **epiphyses**, **metaphyses**, **articular cartilages**, and a **medullary cavity**. (*Figure 6–2*)

7. The medullary cavity and spaces within spongy bone contain either **red bone marrow** (for blood cell formation) or **yellow bone marrow** (for lipid storage).

6-3 Bone is composed of matrix and several types of cells: osteocytes, osteoblasts, osteoprogenitor cells, and osteoclasts p. 189

8. Osseous tissue is a supporting connective tissue with a solid matrix and ensheathed by a *periosteum*.

9. Bone matrix consists largely of crystals of **hydroxyapatite**; the minerals are deposited in **lamellae**.

10. **Osteocytes**, located in *lacunae*, are mature bone cells. Adjacent osteocytes are interconnected by **canaliculi**. **Osteoblasts** synthesize the bony matrix by **osteogenesis**. **Osteoclasts** dissolve the bony matrix through **osteolysis**. **Osteoprogenitor cells** differentiate into osteoblasts. (*Figure 6–3*)

6-4 Compact bone contains parallel osteons, and spongy bone contains trabeculae p. 191

11. The basic functional unit of compact bone is the **osteon**, containing osteocytes arranged around a **central canal**. **Perforating canals** extend perpendicularly to the bone surface. (*Figures 6–4, 6–5*)

12. Compact bone is located where stresses come from a limited range of directions, such as along the diaphysis of long bones.

13. Spongy bone contains **trabeculae**, typically in an open network. (*Figure 6–6*)

14. Spongy bone is located where stresses are few or come from many directions, such as at the epiphyses of long bones. (*Figure 6–7*)

15. A bone is covered by a **periosteum** and lined with an **endosteum**. (*Figure 6–8*)

6-5 Ossification and appositional growth are mechanisms of bone formation and enlargement p. 194

16. **Ossification** is the process of converting other tissues to bone. **Calcification** is the process of depositing calcium salts within a tissue.

17. **Endochondral ossification** begins with a cartilage model that is gradually replaced by bone at the metaphysis. In this way, bone length increases. (*Figure 6–9*)

18. The timing of *closure* of the **epiphyseal cartilage** differs among bones and among individuals. (*Figure 6–10*)

19. Bone diameter increases through *appositional growth*.

20. **Intramembranous ossification** begins when osteoblasts differentiate within connective tissue. The process produces dermal bones. Such ossification begins at an **ossification center**. (*Figure 6–11*)

21. Three major sets of blood vessels provide an extensive supply of blood to bone. (*Figure 6–13*)

6-6 Bone growth and development depend on a balance between bone formation and bone resorption p.200

22. The organic and mineral components of bone are continuously recycled and renewed through **remodeling**.

6-7 Exercise, hormones, and nutrition affect bone development and the skeletal system p. 200

23. The shapes and thicknesses of bones reflect the stresses applied to them.

24. Normal osteogenesis requires a reliable source of minerals, vitamins, and hormones.

25. *Growth hormone* and *thyroxine* stimulate bone growth. Calcitonin and parathyroid hormone control blood calcium levels. (*Table 6–2*)

6-8 Calcium plays a critical role in bone physiology p. 203

26. Calcium is the most abundant mineral in the human body; roughly 99 percent of it is located in the skeleton. (*Figure 6–15*)

27. Interactions among the bones, digestive tract, and kidneys affect the calcium ion concentration. (*Figure 6–16*)

28. Two hormones, **calcitonin** and **parathyroid hormone (PTH)**, regulate calcium ion homeostasis. Calcitonin leads to a decline in the calcium concentration in body fluids, whereas parathyroid hormone increases the calcium concentration in body fluids. (*Figure 6–16*)

6-9 A fracture is a crack or break in a bone p. 205

29. A break or crack in a bone is a **fracture**. The repair of a fracture involves the formation of a **fracture hematoma**, an **external callus**, and an **internal callus**. (*Figure 6–17*)

6-10 Osteopenia has a widespread effect on aging skeletal tissue p. 207

30. The effects of aging on the skeleton include **osteopenia** and **osteoporosis**. (*Figure 6–19*)

Review Questions

See the blue Answers tab at the end of the book.

 Access more review material online at **myA&P**™ (www.myaandp.com). There, you'll find chapter guides, chapter quizzes, practice tests, animations, flashcards, a glossary with pronunciations, *Interactive Physiology*® (IP) exercises and quizzes, and more to help you succeed in the course.

LEVEL 1   Reviewing Facts and Terms

1. Blood cell formation occurs in
   (a) yellow bone marrow.
   (b) red bone marrow.
   (c) the matrix of bone tissue.
   (d) the ground substance of bones.

2. Two-thirds of the weight of bone is accounted for by
   (a) crystals of calcium phosphate.
   (b) collagen fibers.
   (c) osteocytes.
   (d) calcium carbonate.

3. The membrane found wrapping the bones, except at the joint cavity, is the
   (a) periosteum.   (b) endosteum.
   (c) perforating fibers. (d) a, b, and c are correct.

4. The basic functional unit of compact bone is the Haversian system or
   (a) osteocyte.   (b) osteoclast.
   (c) osteon.   (d) osseous matrix.
   (e) osseous lamellae.

5. The vitamins essential for normal adult bone maintenance and repair are
   (a) A and E.   (b) C and D.
   (c) B and E.   (d) B complex and K.

6. The hormones that coordinate the storage, absorption, and excretion of calcium ions are
   (a) growth hormone and thyroxine.
   (b) calcitonin and parathyroid hormone.
   (c) calcitriol and cholecalciferol.
   (d) estrogens and androgens.

7. Classify the bones in the following diagram according to their shape.

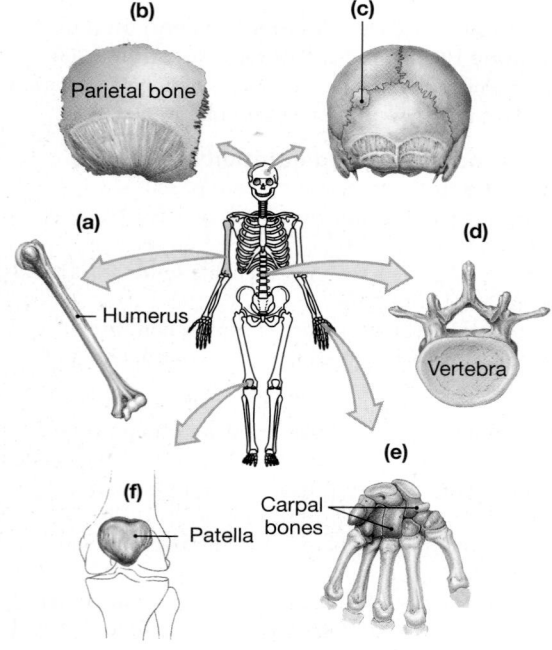

(b) Parietal bone
(c)
(a) Humerus
(d) Vertebra
(f) Patella Carpal bones
(e)

(a) _____   (b) _____
(c) _____   (d) _____
(e) _____   (f) _____

8. The presence of an epiphyseal line indicates
   (a) epiphyseal growth has ended.
   (b) epiphyseal growth is just beginning.
   (c) growth of bone diameter is just beginning.
   (d) the bone is fractured at the location.
   (e) no particular event.

9. The *primary* reason that osteoporosis accelerates after menopause in women is
   (a) reduced levels of circulating estrogens.
   (b) reduced levels of vitamin C.
   (c) diminished osteoclast activity.
   (d) increased osteoblast activity.

10. The nonpathologic loss of bone that occurs with aging is called
    (a) osteomyelitis.   (b) osteoporosis.
    (c) osteopenia.   (d) osteoitis.
    (e) osteomalacia.

11. What are the five primary functions of the skeletal system?
12. List the four distinctive cell populations of osseous tissue.
13. What are the primary parts of a typical long bone?
14. What is the primary difference between endochondral ossification and intramembranous ossification?
15. List the organic and inorganic components of bone matrix.
16. (a) What nutritional factors are essential for normal bone growth and maintenance?
    (b) What hormonal factors are necessary for normal bone growth and maintenance?
17. Which three organs or tissues interact to assist in the regulation of calcium ion concentration in body fluids?

18. What major effects of parathyroid hormone oppose those of calcitonin?

LEVEL 2 Reviewing Concepts

19. If spongy bone has no osteons, how do nutrients reach the osteocytes?
20. Why are stresses or impacts to the side of the shaft in a long bone more dangerous than stress applied to the long axis of the shaft?
21. Why do extended periods of inactivity cause degenerative changes in the skeleton?
22. What are the functional relationships between the skeleton, on the one hand, and the digestive and urinary systems, on the other?
23. Why would a physician concerned about the growth patterns of a young child request an x-ray of the hand?
24. Why does a second fracture in the same bone tend to occur at a site different from that of the first fracture?
25. The process of bone growth at the epiphyseal cartilage is similar to
    (a) intramembranous ossification.
    (b) endochondral ossification.
    (c) the process of osteopenia.
    (d) the process of healing a fracture.
    (e) the process of calcification.
26. How might bone markings be useful in identifying the remains of an individual who was shot and killed years ago?

LEVEL 3 Critical Thinking and Clinical Applications

27. While playing on her swing set, 10-year-old Sally falls and breaks her right leg. At the emergency room, the doctor tells her parents that the proximal end of the tibia where the epiphysis meets the diaphysis is fractured. The fracture is properly set and eventually heals. During a routine physical when she is 18, Sally learns that her right leg is 2 cm shorter than her left, probably because of her accident. What might account for this difference?
28. Which of the following conditions would you possibly observe in a child who is suffering from rickets?
    (a) abnormally short limbs
    (b) abnormally long limbs
    (c) oversized facial bones
    (d) bowed legs
    (e) weak, brittle bones
29. Frank does not begin puberty until he is 16. What effect would you predict this will have on his stature?
    (a) Frank will probably be taller than if he had started puberty earlier.
    (b) Frank will probably be shorter than if he had started puberty earlier.
    (c) Frank will probably be a dwarf.
    (d) Frank will have bones that are heavier than normal.
    (e) The late onset of puberty will have no effect on Frank's stature.
30. In physical anthropology, cultural conclusions can be drawn from a thorough examination of the skeletons of ancient peoples. What sorts of clues might bones provide as to the lifestyles of those individuals?

The Axial Skeleton

7

Learning Outcomes

After completing this chapter, you should be able to do the following:

7-1 Identify the bones of the axial skeleton, and specify their functions.

7-2 Identify the bones of the cranium and face, and explain the significance of the markings on the individual bones.

7-3 Identify the foramina and fissures of the skull, and cite the major structures using the passageways.

7-4 Describe the structure and functions of the orbital complex, nasal complex, and paranasal sinuses.

7-5 Describe the key structural differences among the skulls of infants, children, and adults.

7-6 Identify and describe the curvatures of the spinal column, and indicate the function of each.

7-7 Identify the vertebral regions, and describe the distinctive structural and functional characteristics of vertebrae in each region.

7-8 Explain the significance of the articulations between the thoracic vertebrae and the ribs, and between the ribs and sternum.

Clinical Notes

Temporomandibular Joint (TMJ) Syndrome p. 226
Craniostenosis p. 230
Kyphosis, Lordosis, and Scoliosis p. 232

An Introduction to the Axial Skeleton

In this chapter, we turn our attention to the functional anatomy of the bones that form the longitudinal axis of the body. These bones include the skull and associated bones, the thoracic cage, and the vertebral column.

7-1 The 80 bones of the longitudinal axis make up the axial skeleton

The **axial skeleton** forms the longitudinal axis of the body (**Figure 7–1**). The axial skeleton has 80 bones, roughly 40 percent of the bones in the human body. The axial components are as follows:

- The *skull* (8 *cranial bones* and 14 *facial bones*).

- Bones associated with the skull (6 *auditory ossicles* and the *hyoid bone*).

- The *vertebral column* (24 *vertebrae*, the *sacrum*, and the *coccyx*).

- The *thoracic cage* (the *sternum* and 24 *ribs*).

The axial skeleton provides a framework that supports and protects the brain, the spinal cord, and the organs in the ventral body cavities. It also provides an extensive surface area for the attachment of muscles that (1) adjust the positions of the head, neck, and trunk; (2) perform respiratory movements; and (3) stabilize or position parts of the **appendicular skeleton**, which supports the limbs. The joints of the axial skeleton permit limited movement, but they are very strong and heavily reinforced with ligaments.

We will now consider each of the components of the axial skeleton, beginning with the skull.

CHECKPOINT

1. Identify the bones of the axial skeleton.
2. List the primary functions of the axial skeleton.

See the blue Answers tab at the end of the book.

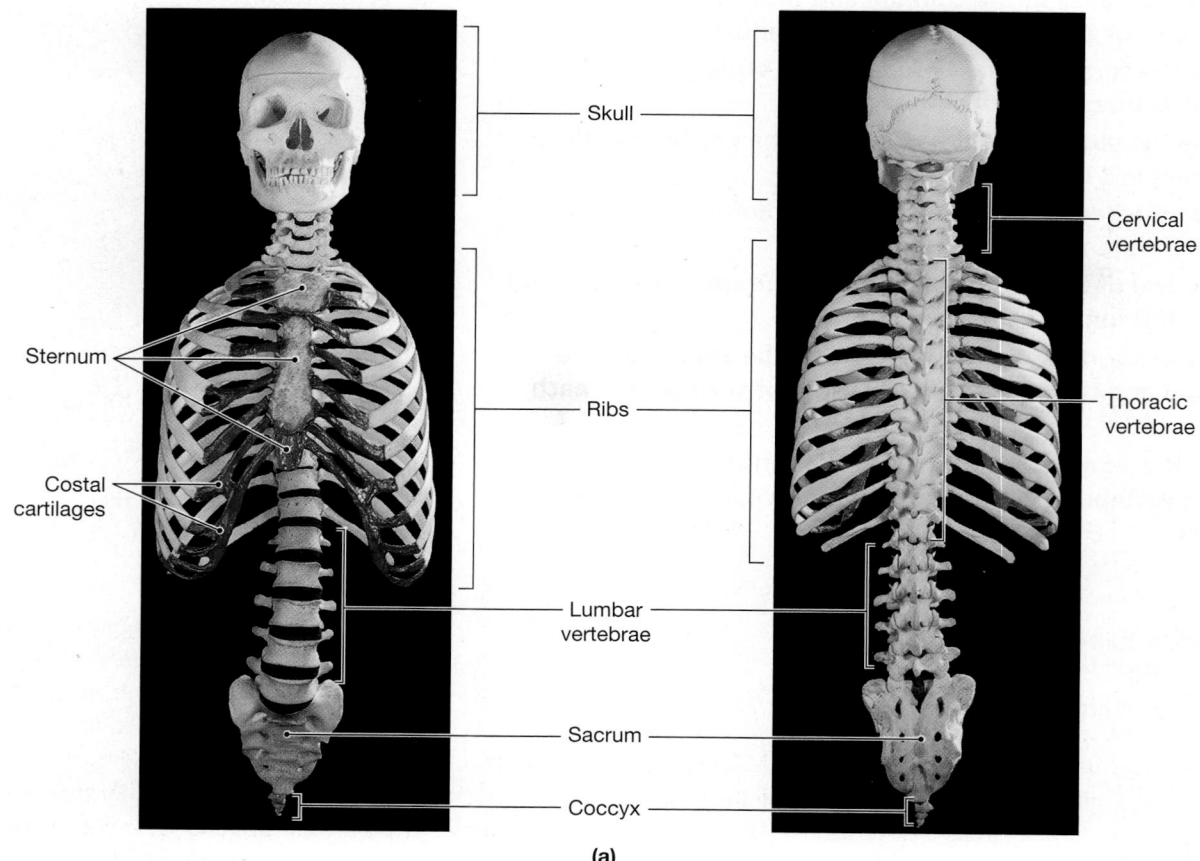

(a)

Figure 7–1 **The Axial Skeleton. (a)** Anterior and posterior views. The bones associated with the skull are not visible.

Figure 7–1 The Axial Skeleton *(continued)*. **(b)** An anterior view of the entire skeleton, with the axial components highlighted. The numbers in the boxes indicate the number of bones in the adult skeleton. ATLAS: Plates 1a,b

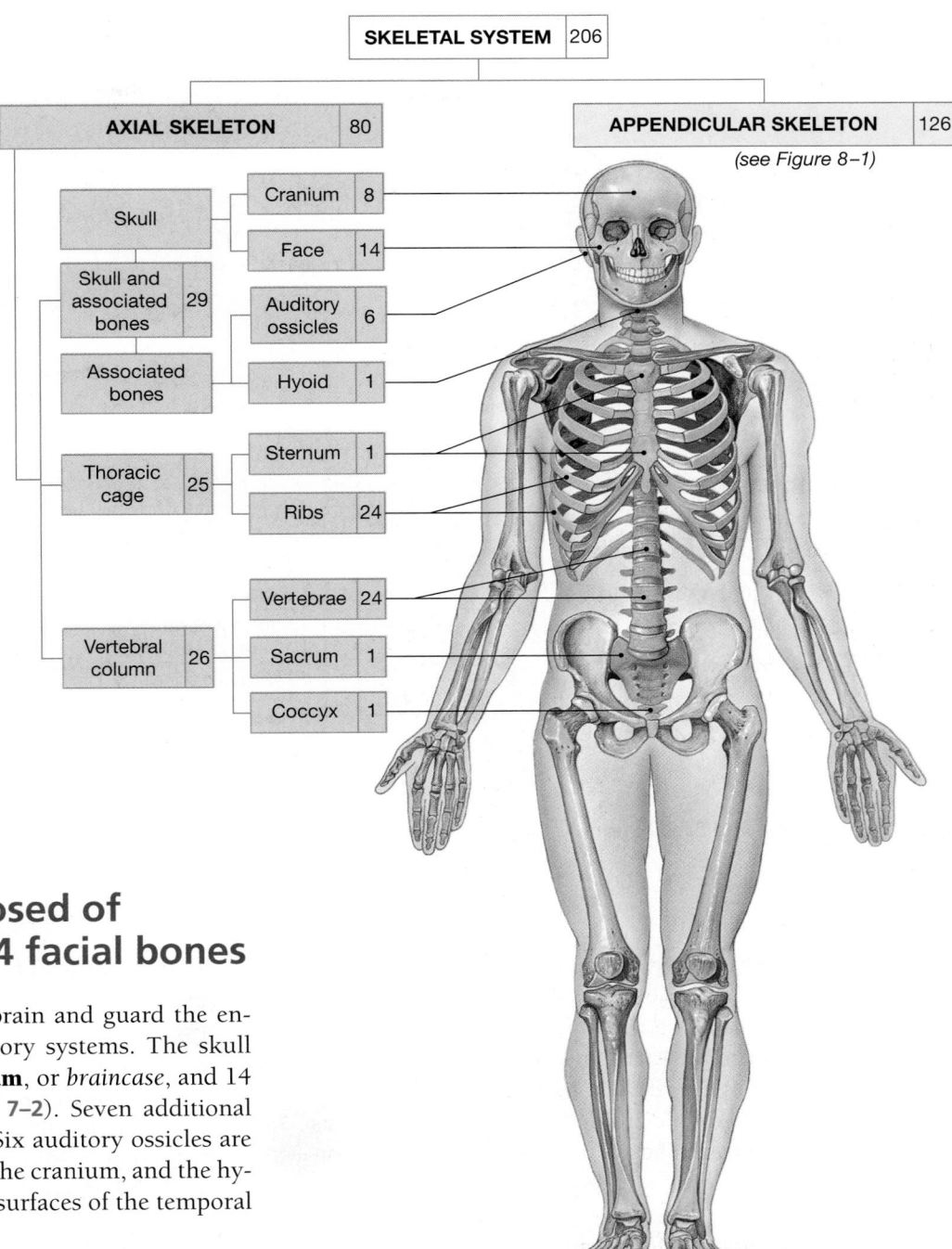

(b)

7-2 The skull is composed of 8 cranial bones and 14 facial bones

The bones of the **skull** protect the brain and guard the entrances to the digestive and respiratory systems. The skull contains 22 bones: 8 form the **cranium**, or *braincase*, and 14 are associated with the face (**Figure 7–2**). Seven additional bones are associated with the skull: Six auditory ossicles are situated within the *temporal bones* of the cranium, and the hyoid bone is connected to the inferior surfaces of the temporal bones by a pair of ligaments.

The cranium consists of 8 **cranial bones**: the *occipital bone*, *frontal bone*, *sphenoid*, *ethmoid*, and the paired *parietal* and *temporal bones*. Together, the cranial bones enclose the **cranial cavity**, a fluid-filled chamber that cushions and supports the brain. Blood vessels, nerves, and membranes that stabilize the position of the brain are attached to the inner surface of the cranium. Its outer surface provides an extensive area for the attachment of muscles that move the eyes, jaws, and head. A joint between the occipital bone and the first vertebra of the neck stabilizes the positions of the brain and spinal cord, while the joints between the vertebrae of the neck permit a wide range of head movements.

If the cranium is the house where the brain resides, the *facial complex* is the front porch. **Facial bones** protect and support the entrances to the digestive and respiratory tracts. The superficial facial bones (the paired *maxillae*, *lacrimal*, *nasal*, and *zygomatic bones*, and the *mandible*) (**Figure 7–2**) provide areas for the attachment of muscles that control facial expressions and assist in manipulating food. The deeper facial bones (the paired *palatine bones* and *inferior nasal conchae*, and the single median *vomer*) help separate the oral and nasal cavities, increase

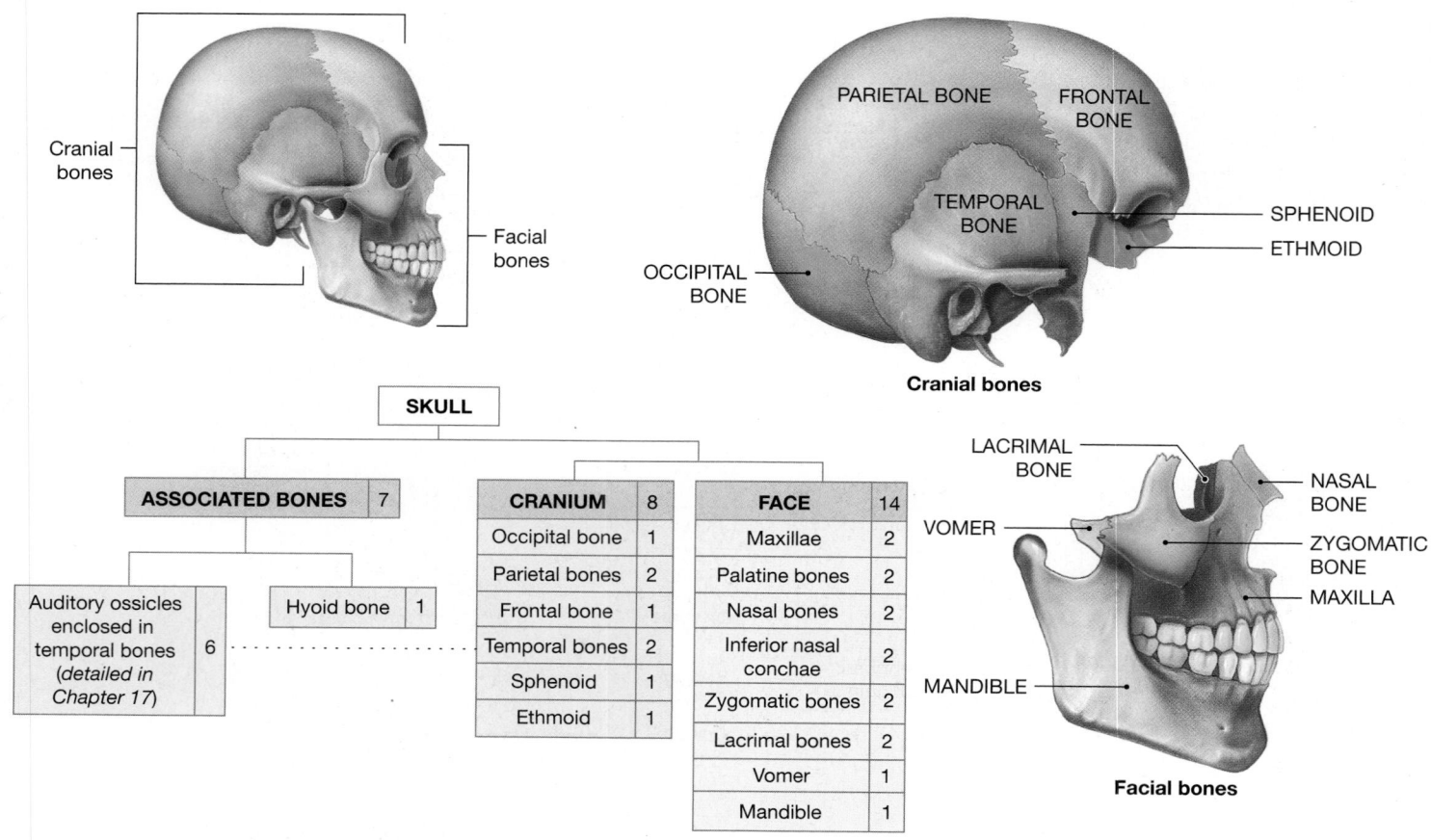

Figure 7–2 **Cranial and Facial Subdivisions of the Skull.** The seven associated bones are not illustrated.

the surface area of the nasal cavities, or help form the **nasal septum** (*septum*, wall), which subdivides the nasal cavity.

Several bones of the skull contain air-filled chambers called **sinuses**. Sinuses have two major functions: (1) They make a bone much lighter than it would otherwise be, and (2) the mucous membrane lining them produces mucus that moistens and cleans the air in and adjacent to the sinus. We will consider the sinuses as we discuss specific bones.

Joints, or *articulations*, form where two bones interconnect. Except where the mandible contacts the cranium, the connections between the skull bones of adults are immovable joints called **sutures**. At a suture, bones are tied firmly together with dense fibrous connective tissue. Each suture of the skull has a name, but at this point you need to know only four major sutures:

1. *Lambdoid Suture.* The **lambdoid** (LAM-doyd) **suture** (Greek *lambda*, L + *eidos*, shape) arches across the posterior surface of the skull (**Figure 7–3a**). This suture separates the occipital bone from the two parietal bones. One

or more **sutural bones** (*Wormian bones*) may be present along the lambdoid suture. ∞ p. 185

2. *Coronal Suture.* The **coronal suture** attaches the frontal bone to the parietal bones of either side (**Figure 7–3b**). The occipital, parietal, and frontal bones form the **calvaria** (kal-VA-rē-uh), or skullcap. A cut through the body that parallels the coronal suture produces a *coronal section*, or *frontal section* (see **Figure 1–9**, p. 21).

3. *Sagittal Suture.* The **sagittal suture** extends from the lambdoid suture to the coronal suture, between the parietal bones (**Figure 7–3b**). A cut along the midline of the suture produces a midsagittal section; a slice that parallels the sagittal suture produces a parasagittal section. ∞ p. 19

4. *Squamous Sutures.* A **squamous** (SKWĀ-mus) **suture** on each side of the skull forms the boundary between the temporal bone and the parietal bone of that side. **Figure 7–3a** shows the intersection between the squamous sutures and the lambdoid suture. **Figure 7–3c** shows the path of the squamous suture on the right side of the skull.

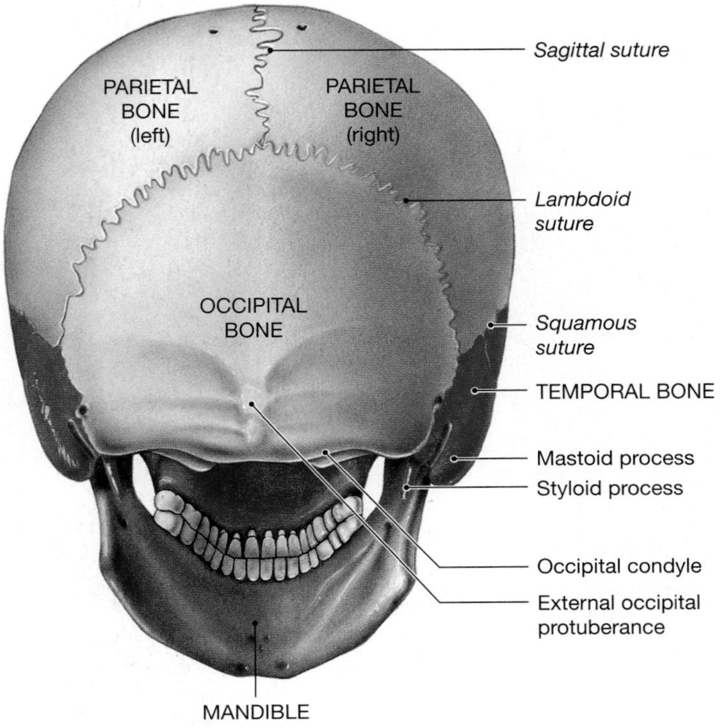

7 SKELETAL

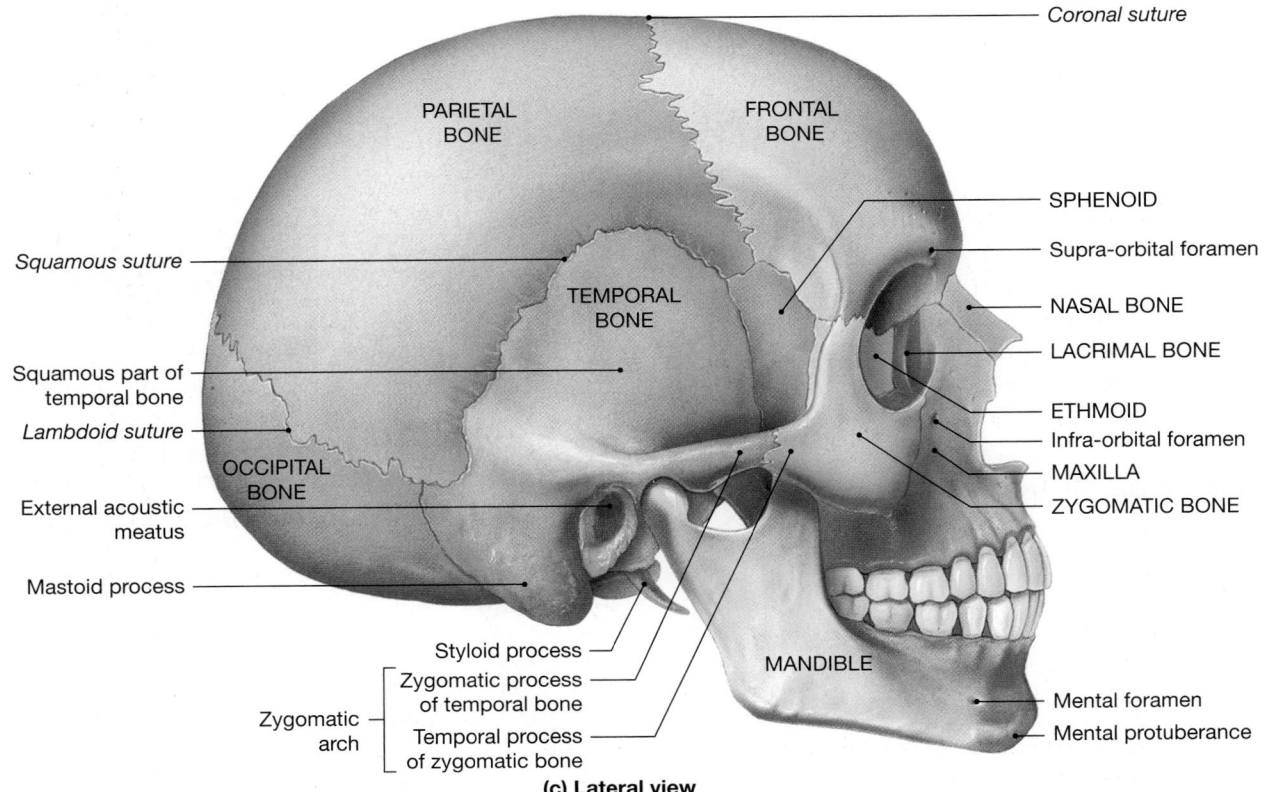

(a) Posterior view

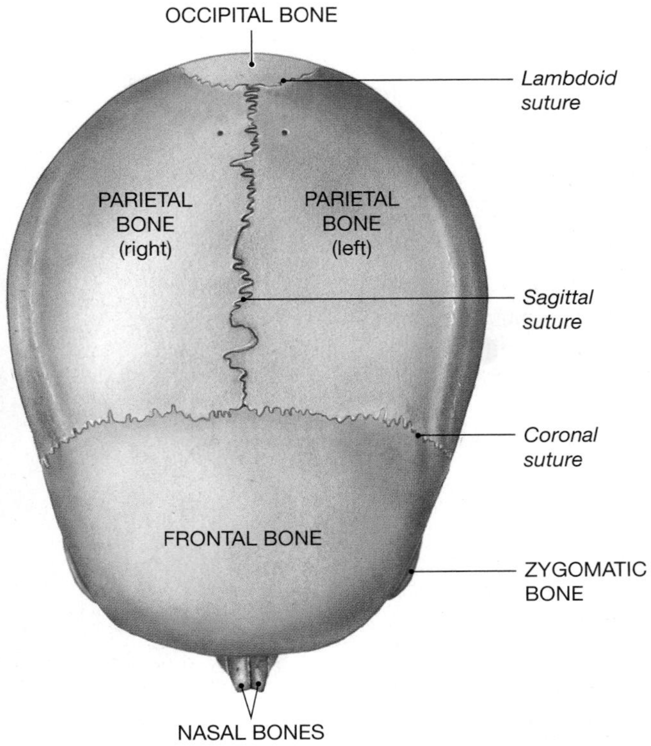

- Sagittal suture
- PARIETAL BONE (left)
- PARIETAL BONE (right)
- Lambdoid suture
- OCCIPITAL BONE
- Squamous suture
- TEMPORAL BONE
- Mastoid process
- Styloid process
- Occipital condyle
- External occipital protuberance
- MANDIBLE

(b) Superior view

- OCCIPITAL BONE
- Lambdoid suture
- PARIETAL BONE (right)
- PARIETAL BONE (left)
- Sagittal suture
- Coronal suture
- FRONTAL BONE
- ZYGOMATIC BONE
- NASAL BONES

(c) Lateral view

- Coronal suture
- PARIETAL BONE
- FRONTAL BONE
- Squamous suture
- Squamous part of temporal bone
- Lambdoid suture
- OCCIPITAL BONE
- External acoustic meatus
- Mastoid process
- TEMPORAL BONE
- SPHENOID
- Supra-orbital foramen
- NASAL BONE
- LACRIMAL BONE
- ETHMOID
- Infra-orbital foramen
- MAXILLA
- ZYGOMATIC BONE
- MANDIBLE
- Mental foramen
- Mental protuberance
- Styloid process
- Zygomatic arch
  - Zygomatic process of temporal bone
  - Temporal process of zygomatic bone

Figure 7–3   The Adult Skull.   ATLAS: Plates 4a,b; 5a–e

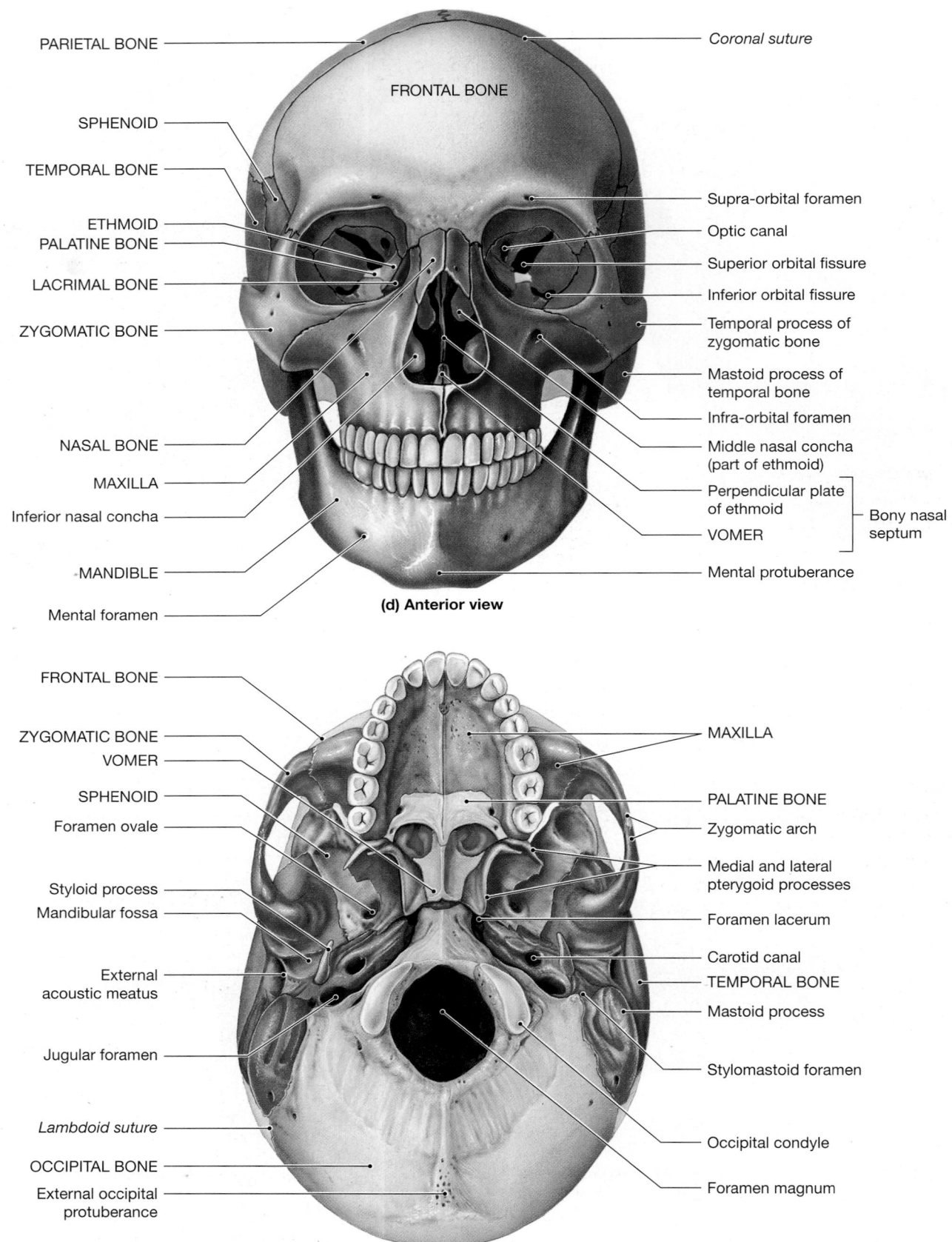

PARIETAL BONE

FRONTAL BONE

Coronal suture

SPHENOID

TEMPORAL BONE

ETHMOID
PALATINE BONE

LACRIMAL BONE

ZYGOMATIC BONE

Supra-orbital foramen

Optic canal

Superior orbital fissure

Inferior orbital fissure

Temporal process of
zygomatic bone

Mastoid process of
temporal bone

Infra-orbital foramen

Middle nasal concha
(part of ethmoid)

Perpendicular plate
of ethmoid

VOMER

Bony nasal
septum

NASAL BONE

MAXILLA

Inferior nasal concha

MANDIBLE

Mental foramen

Mental protuberance

(d) Anterior view

FRONTAL BONE

ZYGOMATIC BONE

VOMER

SPHENOID

Foramen ovale

Styloid process

Mandibular fossa

External
acoustic meatus

Jugular foramen

Lambdoid suture

OCCIPITAL BONE

External occipital
protuberance

MAXILLA

PALATINE BONE

Zygomatic arch

Medial and lateral
pterygoid processes

Foramen lacerum

Carotid canal

TEMPORAL BONE

Mastoid process

Stylomastoid foramen

Occipital condyle

Foramen magnum

(e) Inferior view

Figure 7–3 **The Adult Skull** *(continued).*

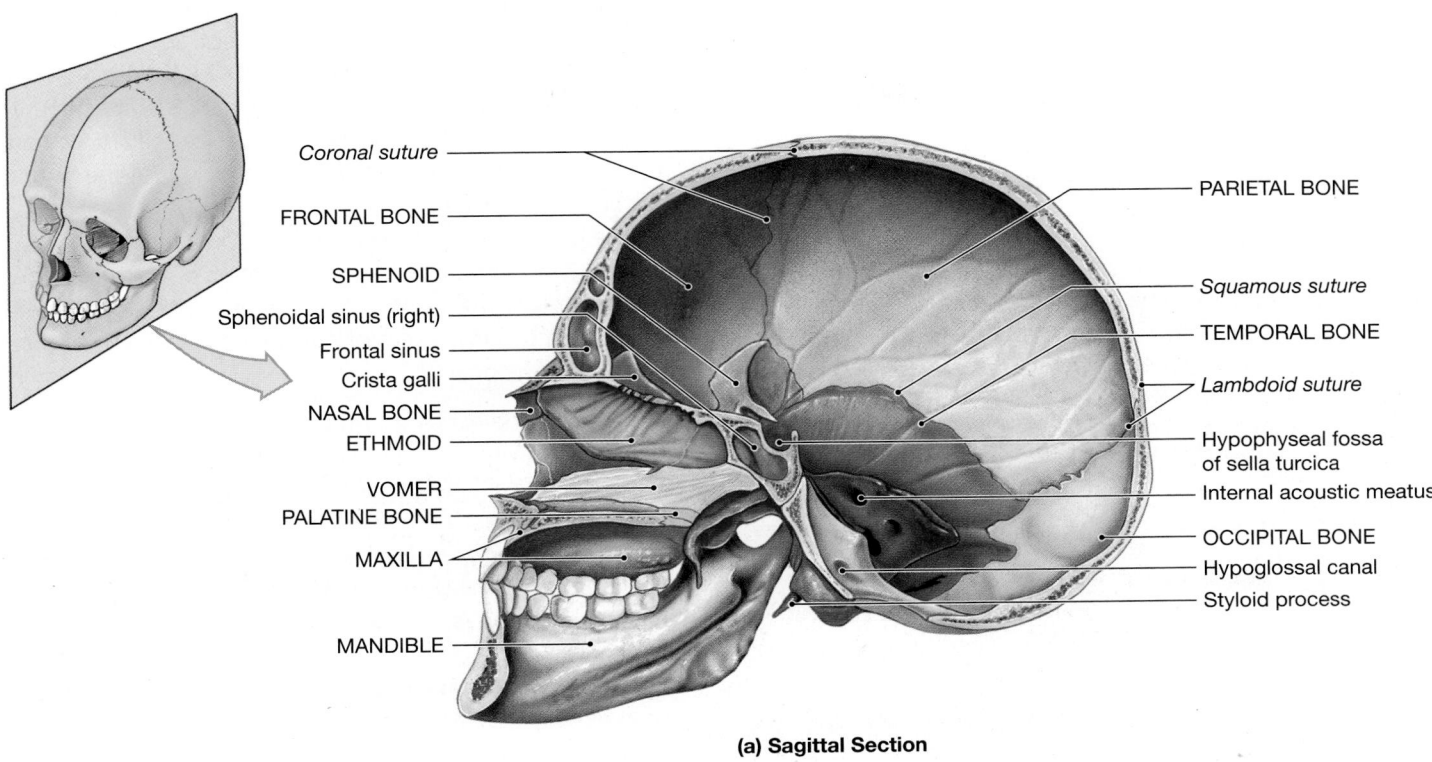

Coronal suture

FRONTAL BONE

SPHENOID

Sphenoidal sinus (right)

Frontal sinus

Crista galli

NASAL BONE

ETHMOID

VOMER

PALATINE BONE

MAXILLA

MANDIBLE

PARIETAL BONE

Squamous suture

TEMPORAL BONE

Lambdoid suture

Hypophyseal fossa of sella turcica

Internal acoustic meatus

OCCIPITAL BONE

Hypoglossal canal

Styloid process

(a) Sagittal Section

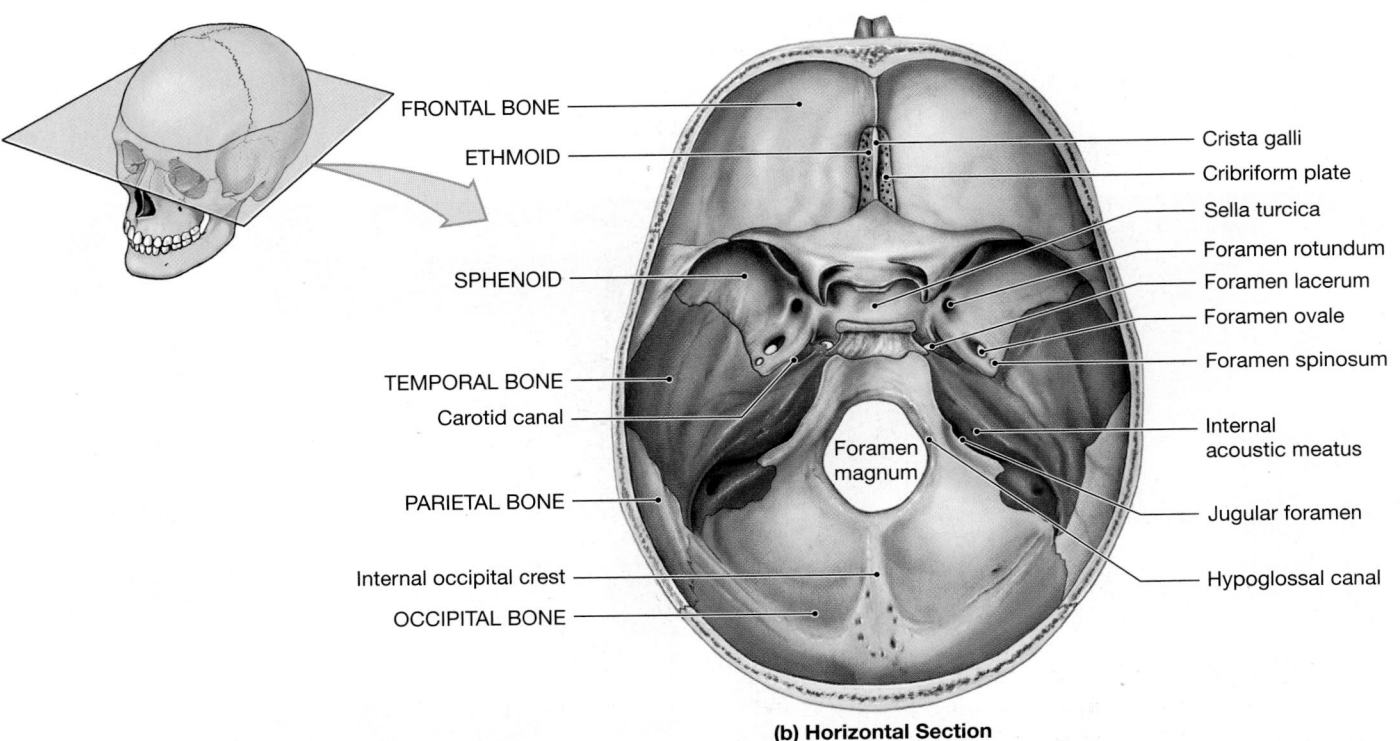

FRONTAL BONE

ETHMOID

SPHENOID

TEMPORAL BONE

Carotid canal

PARIETAL BONE

Internal occipital crest

OCCIPITAL BONE

Crista galli

Cribriform plate

Sella turcica

Foramen rotundum

Foramen lacerum

Foramen ovale

Foramen spinosum

Internal acoustic meatus

Jugular foramen

Hypoglossal canal

Foramen magnum

(b) Horizontal Section

Figure 7–4   The Sectional Anatomy of the Skull.   **(a)** Medial view of a sagittal section through the skull. **(b)** Superior view of a horizontal section through the skull, showing the floor of the cranial cavity. Compare with part **(a)** and with *Figure 7–3e*. ATLAS: Plates 4c; 6; 7a,b

The Individual Bones of the Skull

Each of these bones can be explored further, using the related images in the *Atlas*. Foramina and fissures are present for the passage of vessels and nerves. The vessels are detailed in Chapter 21; the nerves are shown in the Focus box on cranial nerves in Chapter 14.

Cranial Bones

The Occipital Bone (Figure 7–5a)

General Functions: The **occipital bone** forms much of the posterior and inferior surfaces of the cranium.

Articulations: The occipital bone articulates with the parietal bones, the temporal bones, the sphenoid, and the first cervical vertebra (the atlas) (**Figures 7–3a–c,e** and **7–4**).

Regions/Landmarks: The **external occipital protuberance** is a small bump at the midline on the inferior surface.

The **external occipital crest**, which begins at the external occipital protuberance, marks the attachment of a ligament that helps stabilize the vertebrae of the neck.

The **occipital condyles** are the site of articulation between the skull and the first vertebra of the neck.

The *inferior* and *superior nuchal* (NOO-kul) *lines* are ridges that intersect the occipital crest. They mark the attachment sites of muscles and ligaments that stabilize the articulation at the occipital condyles and balance the weight of the head over the vertebrae of the neck.

The concave internal surface of the occipital bone (**Figure 7–4a**) closely follows the contours of the brain. The grooves follow the paths of major blood vessels, and the ridges mark the attachment sites of membranes that stabilize the position of the brain.

Foramina: The **foramen magnum** (**Figure 7–4b**) connects the cranial cavity with the spinal cavity, which is enclosed by the vertebral column. This foramen surrounds the connection between the brain and spinal cord.

The **jugular foramen** lies between the occipital bone and the temporal bone (**Figure 7–3e**). The *internal jugular vein* passes through this foramen, carrying venous blood from the brain.

The **hypoglossal canals** (**Figure 7–4b**) begin at the lateral base of each occipital condyle and end on the inner surface of the occipital bone near the foramen magnum. The *hypoglossal nerves*, cranial nerves that control the tongue muscles, pass through these canals.

The Parietal Bones (Figure 7–5b)

General Functions: The **parietal bones** form part of the superior and lateral surfaces of the cranium.

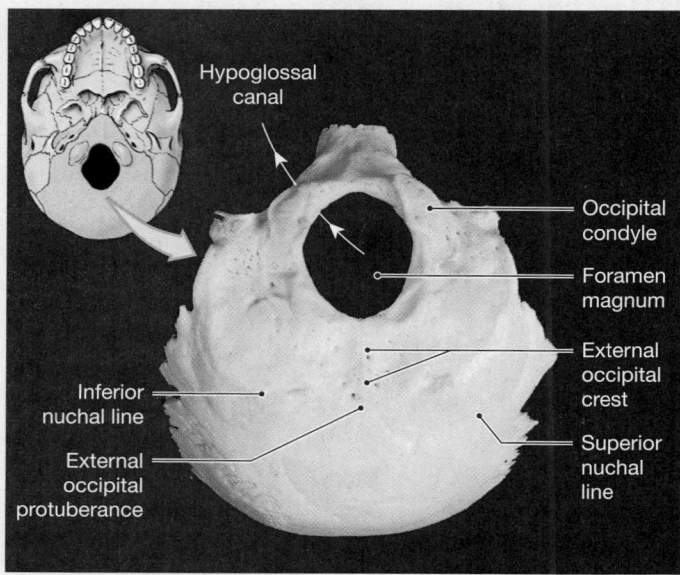

(a) Occipital bone, inferior view

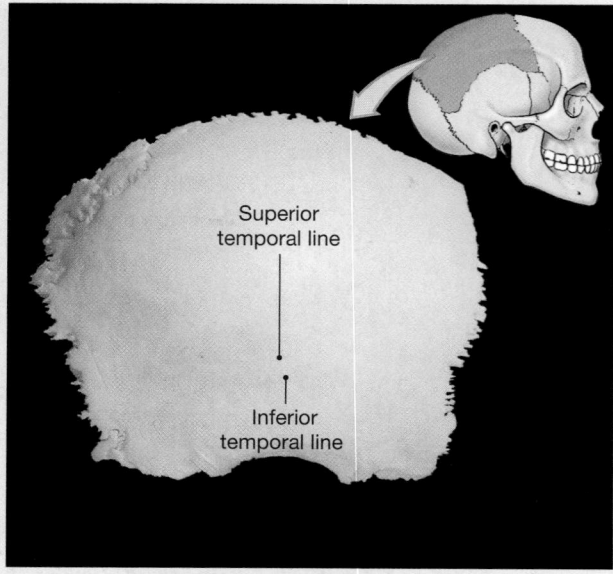

(b) Right parietal bone, lateral view

Figure 7–5   The Occipital and Parietal Bones.

Articulations: The parietal bones articulate with one another and with the occipital, temporal, frontal, and sphenoid bones (**Figures 7–3a–d** and **7–4**).

Regions/Landmarks: The *superior* and *inferior temporal lines* are low ridges that mark the attachment sites of the *temporalis muscle*, a large muscle that closes the mouth.

Grooves on the inner surface of the parietal bones mark the paths of cranial blood vessels (**Figure 7–4a**).

The Frontal Bone (Figure 7–6a,b)

General Functions: The **frontal bone** forms the anterior portion of the cranium and the roof of the *orbits* (eye sockets). Mucous secretions of the *frontal sinuses* within this bone help flush the surfaces of the nasal cavities.

Articulations: The frontal bone articulates with the parietal, sphenoid, ethmoid, nasal, lacrimal, maxillary, and zygomatic bones (**Figures 7–3b–e** and **7–4**).

Regions/Landmarks: The **frontal squama**, or forehead, forms the anterior, superior portion of the cranium and provides surface area for the attachment of facial muscles.

The *superior temporal line* is continuous with the superior temporal line of the parietal bone.

The **supra-orbital margin** is a thickening of the frontal bone that helps protect the eye.

The **lacrimal fossa** on the superior and lateral surface of the orbit is a shallow depression that marks the location of the *lacrimal* (tear) *gland*, which lubricates the surface of the eye.

The **frontal sinuses** are extremely variable in size and time of appearance. They generally appear after age 6, but some people never develop them. We will describe the frontal sinuses and other sinuses of the cranium and face in a later section.

Foramina: The **supra-orbital foramen** provides passage for blood vessels that supply the eyebrow, eyelids, and frontal sinuses. In some cases, this foramen is incomplete; the vessels then cross the orbital rim within a **supra-orbital notch**.

Remarks: During development, the bones of the cranium form by the fusion of separate centers of ossification. At birth, the fusions are not yet complete: Two frontal bones articulate along the *frontal (metopic) suture*. Although the suture generally disappears by age 8 as the bones fuse, the adult skull commonly retains traces of the suture line. This suture, or what remains of it, runs down the center of the frontal squama.

The Temporal Bones (Figure 7–7a,b)

General Functions: The **temporal bones** (1) form part of both the lateral walls of the cranium and the *zygomatic*

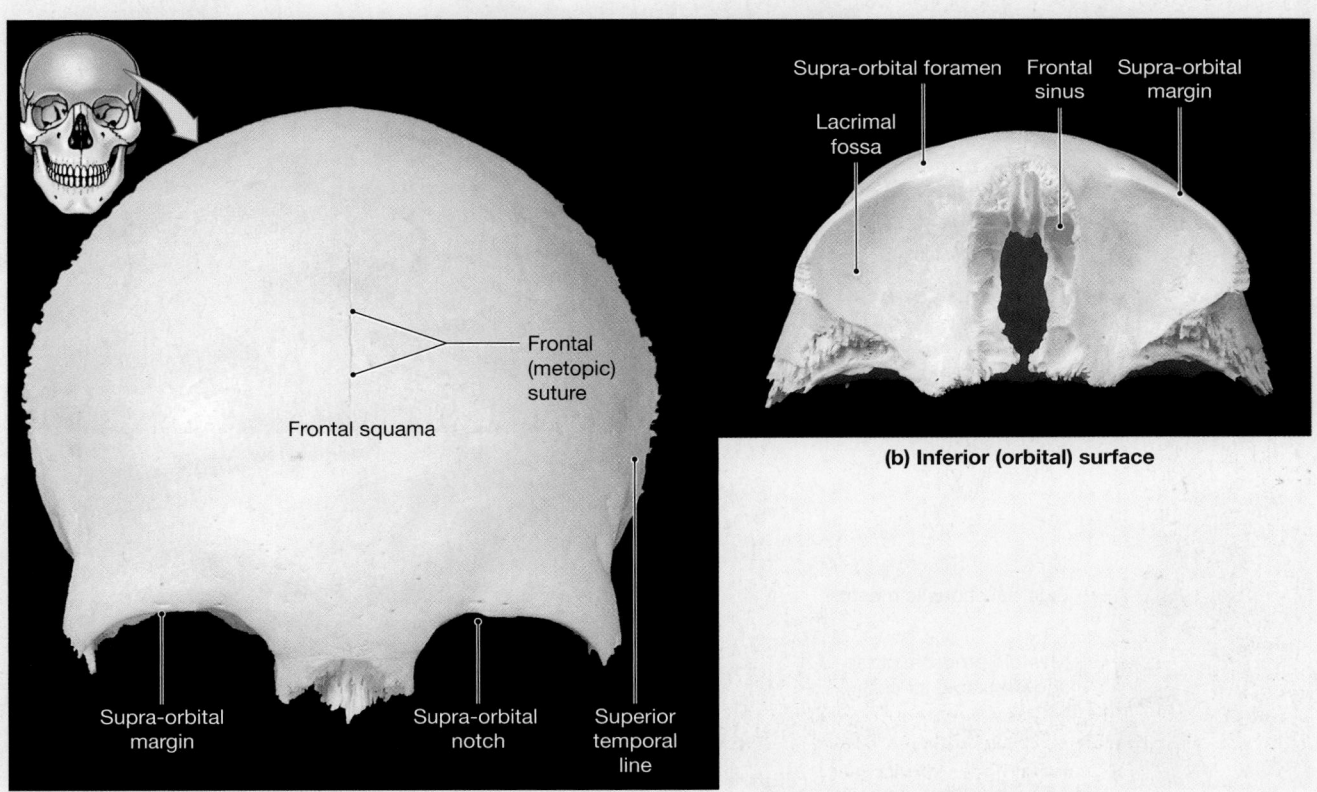

(a) Anterior surface

(b) Inferior (orbital) surface

Figure 7–6 The Frontal Bone.

arches, (2) form the only articulations with the mandible, (3) surround and protect the sense organs of the inner ear, and (4) are attachment sites for muscles that close the jaws and move the head.

Articulations: The temporal bones articulate with the zygomatic, sphenoid, parietal, and occipital bones of the cranium and with the mandible (**Figures 7–3** and **7–4**).

Regions/Landmarks: The **squamous part**, or *squama*, of the temporal bone is the convex, irregular surface that borders the squamous suture.

The **zygomatic process**, inferior to the squamous portion, articulates with the *temporal process* of the zygomatic bone. Together, these processes form the **zygomatic arch**, or cheekbone (**Figure 7–3c,e**).

The **mandibular fossa** on the inferior surface marks the site of articulation with the mandible.

The **mastoid process** is an attachment site for muscles that rotate or extend the head. *Mastoid air cells,* small interconnected cavities within the mastoid process (**Figure 7–7c**), are connected to the middle ear cavity. If pathogens invade the mastoid air cells, *mastoiditis* develops. Symptoms include severe earaches, fever, and swelling behind the ear.

The **styloid** (STĪ-loyd; *stylos*, pillar) **process**, near the base of the mastoid process, is attached to ligaments that support the hyoid bone and to the tendons of several muscles associated with the hyoid bone, the tongue, and the pharynx.

The **petrous part** of the temporal bone, located on its internal surface, encloses the structures of the *inner ear*—sense organs that provide information about hearing and balance. The **auditory ossicles** are located in the *tympanic cavity*, or *middle ear*, a cavity within the petrous part. These tiny bones—three on each side—transfer sound vibrations from the delicate *tympanic membrane*, or eardrum, to the inner ear. (We will discuss these bones and their functions in Chapter 17.)

Foramina (**Figure 7–3e**): The *jugular foramen*, between the temporal and occipital bones, provides passage for the internal jugular vein.

The **carotid canal** provides passage for the internal carotid artery, a major artery to the brain. As it leaves the carotid canal, the internal carotid artery passes through the anterior portion of the foramen lacerum.

The **foramen lacerum** (LA-se-rum; *lacerare*, to tear) is a jagged slit extending between the sphenoid and the petrous portion of the temporal bone and containing hyaline cartilage and small arteries that supply the inner surface of the cranium.

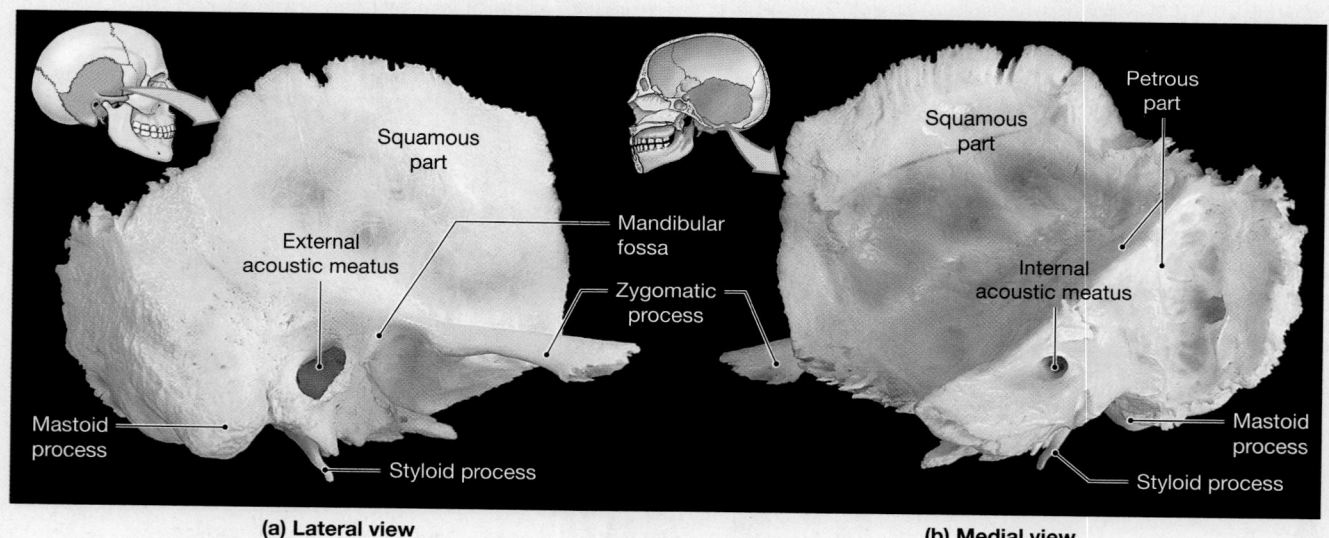

(a) Lateral view

(b) Medial view

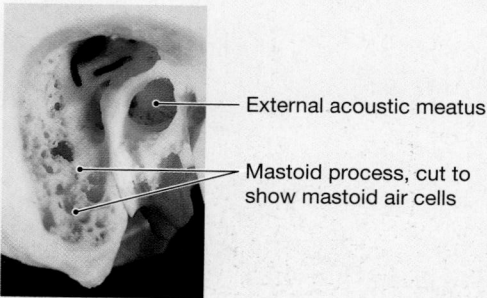

(c) Mastoid air cells

Figure 7–7   **The Temporal Bones. (a,b)** The right temporal bone. **(c)** A cutaway view of the mastoid air cells.

The *auditory tube*, an air-filled passageway that connects the pharynx to the tympanic cavity, passes through the posterior portion of the foramen lacerum.

The **external acoustic meatus**, or *external acoustic canal*,[1] on the lateral surface ends at the tympanic membrane (which disintegrates during the preparation of a dried skull).

The **stylomastoid foramen** lies posterior to the base of the styloid process. The *facial nerve* passes through this foramen to control the facial muscles.

The **internal acoustic meatus**, or *internal acoustic canal*,[1] begins on the medial surface of the petrous part of the temporal bone. It carries blood vessels and nerves to the inner ear and conveys the facial nerve to the stylomastoid foramen.

[1] The names for these passageways vary widely; the terms *acoustic* and *auditory* are used interchangeably, as are *meatus* and *canal*.

The Sphenoid (Figure 7–8a,b)

General Functions: The **sphenoid**, or *sphenoidal bone*, forms part of the floor of the cranium, unites the cranial and facial bones, and acts as a cross brace that strengthens the sides of the skull. Mucous secretions of the *sphenoidal sinuses* within this bone help clean the surfaces of the nasal cavities.

Articulations: The sphenoid articulates with the ethmoid and the frontal, occipital, parietal, and temporal bones of the cranium and the palatine bones, zygomatic bones, maxillae, and vomer of the face (**Figures 7–3c–e** and **7–4**).

Regions/Landmarks: The shape of the sphenoid has been compared to a bat with its wings extended. Although this bone is relatively large, much of it is hidden by more superficial bones.

The **body** forms the central axis of the sphenoid.

The **sella turcica** (TUR-si-kuh), or Turkish saddle, is a bony, saddle-shaped enclosure on the superior surface of the body. The **hypophyseal** (hī-pō-FIZ-ē-ul) **fossa** is the depression within the sella turcica. The *pituitary gland* occupies this fossa.

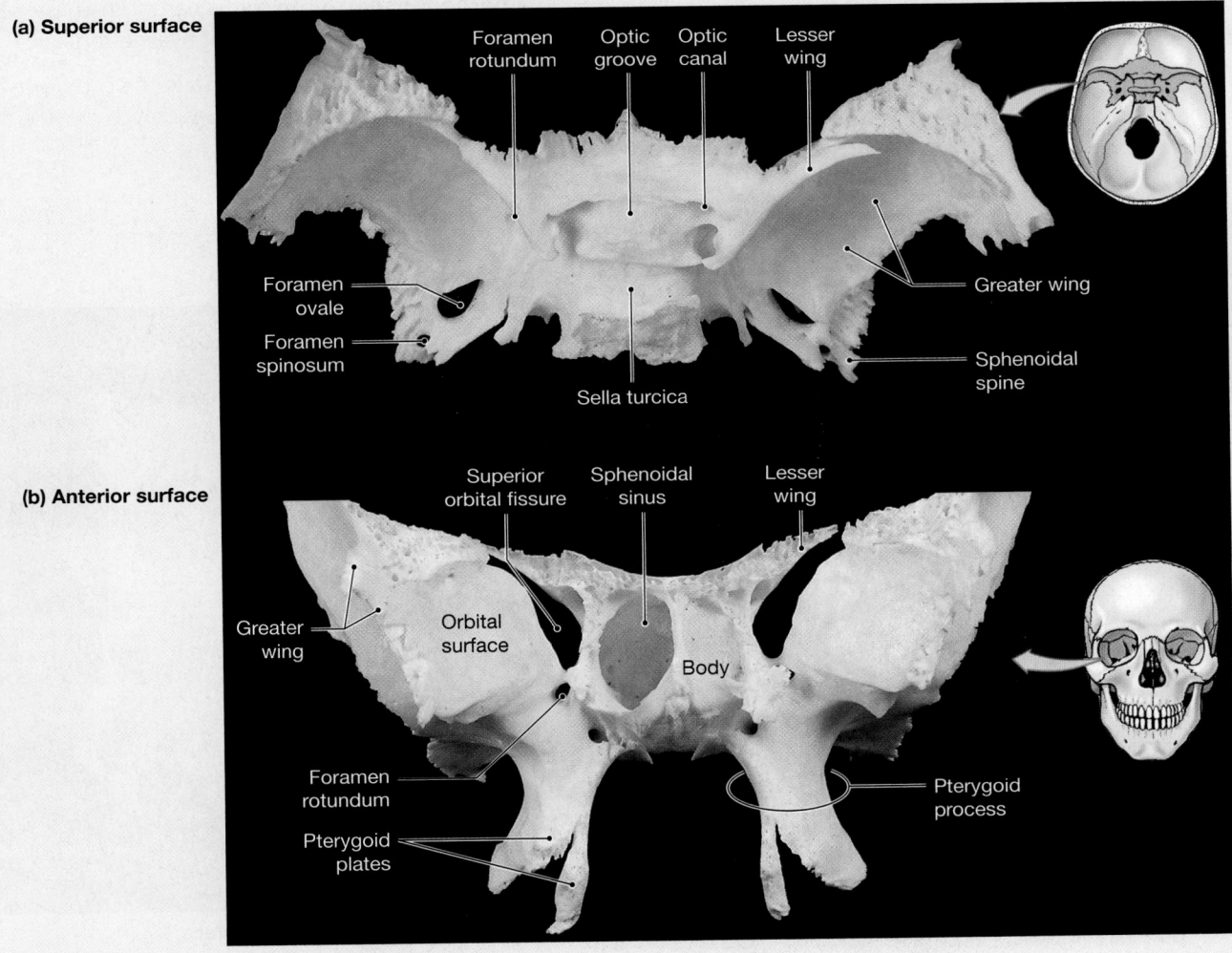

(a) Superior surface

Foramen rotundum — Optic groove — Optic canal — Lesser wing

Foramen ovale

Foramen spinosum

Sella turcica

Greater wing

Sphenoidal spine

(b) Anterior surface

Superior orbital fissure — Sphenoidal sinus — Lesser wing

Greater wing

Orbital surface

Body

Foramen rotundum

Pterygoid plates

Pterygoid process

Figure 7–8   The Sphenoid.

The **sphenoidal sinuses** are on either side of the body, inferior to the sella turcica.

The **lesser wings** extend horizontally anterior to the sella turcica.

The **greater wings** extend laterally from the body and form part of the cranial floor. A sharp *sphenoidal spine* lies at the posterior, lateral corner of each greater wing. Anteriorly, each greater wing contributes to the posterior wall of the orbit.

The **pterygoid** (TER-i-goyd; *pterygion*, wing) **processes** are vertical projections that originate on either side of the body. Each pterygoid process forms a pair of *pterygoid plates*, which are attachment sites for muscles that move the mandible and soft palate.

Foramina: The **optic canals** permit passage of the optic nerves from the eyes to the brain.

A **superior orbital fissure**, **foramen rotundum**, **foramen ovale** (ō-VAH-lē), and **foramen spinosum** penetrate each greater wing. These passages carry blood vessels and nerves to the orbit, face, jaws, and membranes of the cranial cavity, respectively.

The Ethmoid (Figure 7–9a,b)

General Functions: The **ethmoid**, or *ethmoidal bone*, forms the anteromedial floor of the cranium, the roof of the nasal cavity, and part of the nasal septum and medial orbital wall.

Mucous secretions from a network of sinuses, or *ethmoidal air cells*, within this bone flush the surfaces of the nasal cavities.

Articulations: The ethmoid articulates with the frontal bone and sphenoid of the cranium and with the nasal, lacrimal, palatine, and maxillary bones and the inferior nasal conchae and vomer of the face (**Figures 7–3c,d** and **7–4**).

Regions/Landmarks: The ethmoid has three parts: (1) the cribriform plate, (2) the paired lateral masses, and (3) the perpendicular plate.

The **cribriform plate** (*cribrum*, sieve) forms the anteromedial floor of the cranium and the roof of the nasal cavity. The **crista galli** (*crista*, crest + *gallus*, chicken; cock's comb) is a bony ridge that projects superior to the cribriform plate. The *falx cerebri*, a membrane that stabilizes the position of the brain, attaches to this ridge.

The **lateral masses** contain the **ethmoidal labyrinth**, which consists of the interconnected **ethmoidal air cells** that open into the nasal cavity on each side. The **superior nasal conchae** (KONG-kē; singular, *concha*, a snail shell) and the **middle nasal conchae** are delicate projections of the lateral masses.

The **perpendicular plate** forms part of the nasal septum, along with the vomer and a piece of hyaline cartilage.

Foramina: The **olfactory foramina** in the cribriform plate permit passage of the olfactory nerves, which provide the sense of smell.

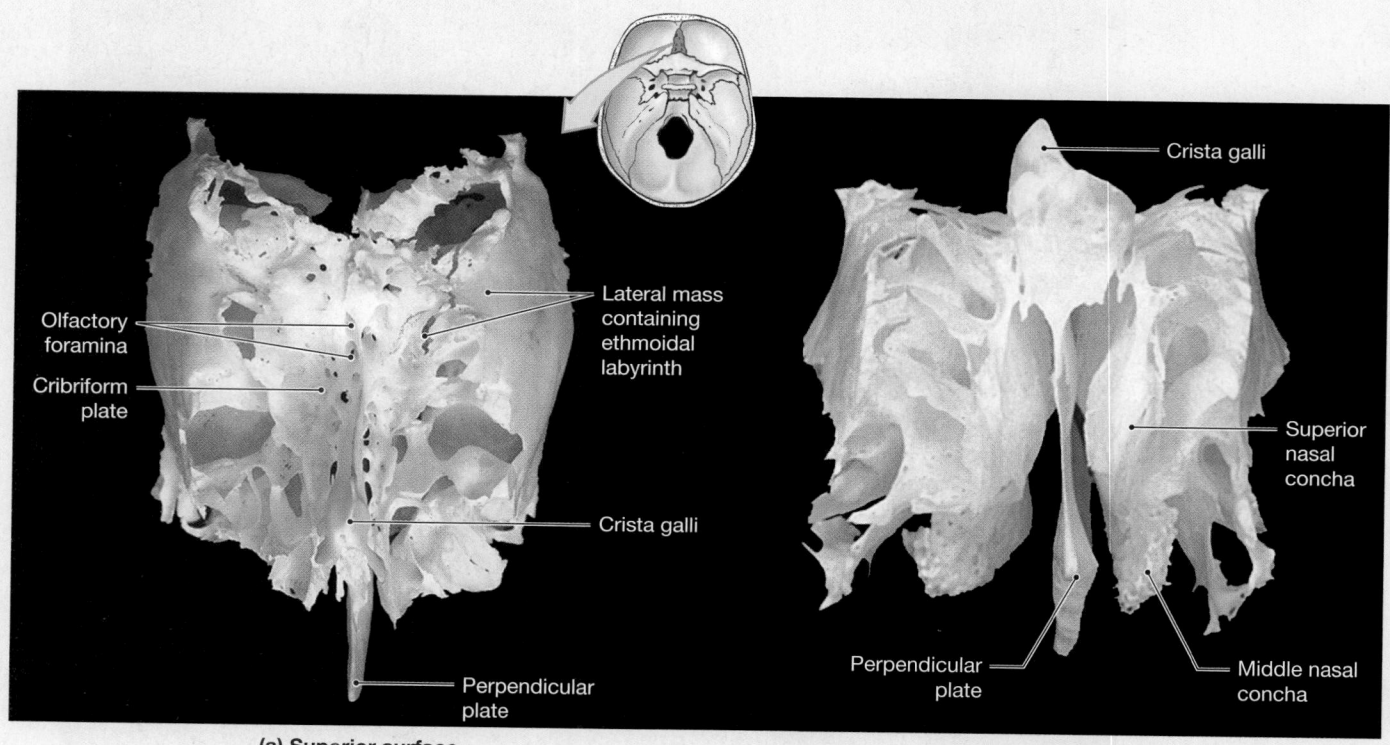

Olfactory foramina

Cribriform plate

Crista galli

Lateral mass containing ethmoidal labyrinth

Crista galli

Perpendicular plate

(a) Superior surface

Superior nasal concha

Perpendicular plate

Middle nasal concha

(b) Posterior surface

Figure 7–9   The Ethmoid.

Remarks: *Olfactory* (smell) *receptors* are located in the epithelium that covers the inferior surfaces of the cribriform plate, the medial surfaces of the superior nasal conchae, and the superior portion of the perpendicular plate.

The nasal conchae break up the airflow in the nasal cavity, creating swirls, turbulence, and eddies that have three major functions: (1) The swirling throws any particles in the air against the sticky mucus that covers the walls of the nasal cavity; (2) the turbulence slows air movement, providing time for warming, humidification, and dust removal before the air reaches more delicate portions of the respiratory tract; and (3) the eddies direct air toward the superior portion of the nasal cavity, adjacent to the cribriform plate, where the olfactory receptors are located.

Facial Bones

The Maxillae (Figure 7–10a,b)

General Functions: The **maxillae**, or *maxillary bones*, support the upper teeth and form the inferior orbital rim, the lateral margins of the external nares, the upper jaw, and most of the hard palate. The *maxillary sinuses* in these bones produce mucous secretions that flush the inferior surfaces of the nasal cavities. The maxillae are the largest facial bones, and the maxillary sinuses are the largest sinuses.

Articulations: The maxillae articulate with the frontal bones and ethmoid, with one another, and with all the other facial bones except the mandible (**Figures 7–3c–e** and **7–4a**).

Regions/Landmarks: The **orbital rim** protects the eye and other structures in the orbit. The *anterior nasal spine* is found at the anterior portion of the maxilla, at its articulation with the maxilla of the other side. It is an attachment point for the cartilaginous anterior portion of the nasal septum.

The **alveolar process** that borders the mouth supports the upper teeth.

The **palatine processes** form most of the **hard palate**, or bony roof of the mouth. One type of *cleft palate*, a developmental disorder, results when the maxillae fail to meet along the midline of the hard palate. ATLAS: Embryology Summary 6: The Development of the Skull

The **maxillary sinuses** lighten the portion of the maxillae superior to the teeth.

The **nasolacrimal canal**, formed by a maxilla and lacrimal bone, protects the *lacrimal sac* and the *nasolacrimal duct*, which carries tears from the orbit to the nasal cavity.

Foramina: The **infra-orbital foramen** marks the path of a major sensory nerve that reaches the brain via the foramen rotundum of the sphenoid.

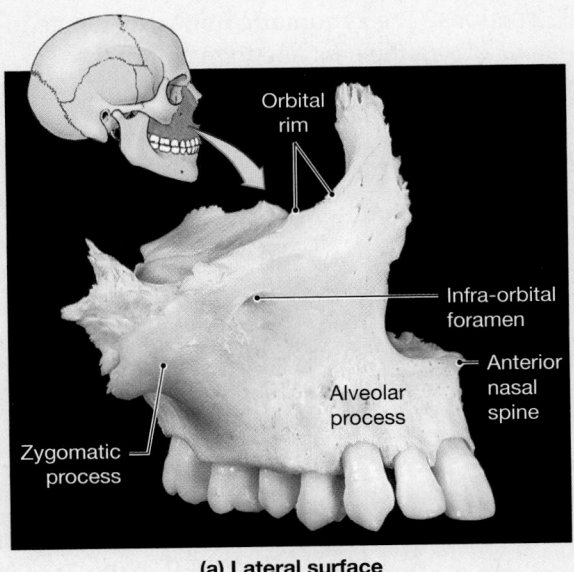

(a) Lateral surface

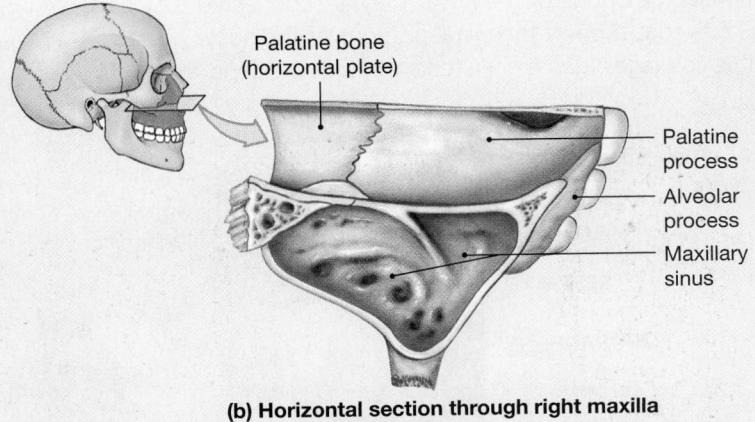

(b) Horizontal section through right maxilla and palatine bone

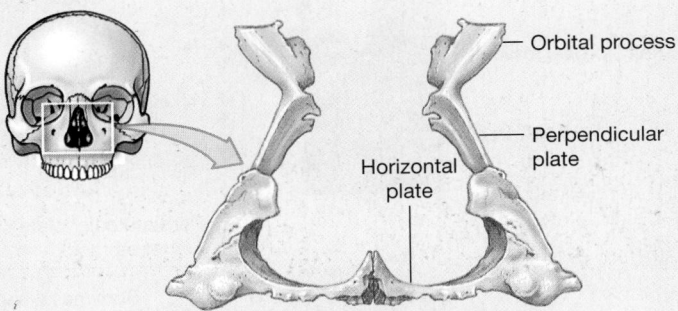

(c) Anterior view of palatine bones

Figure 7–10 The Maxillae and Palatine Bones. (a) An anterolateral view of the right maxilla. **(b)** Superior view of a horizontal section; note the size and orientation of the maxillary sinus. **(c)** An anterior view of the two palatine bones. ATLAS: Plates 8a–d; 12d

The **inferior orbital fissure** (Figure 7–3d), which lies between the maxilla and the sphenoid, permits passage of cranial nerves and blood vessels.

The Palatine Bones (Figure 7–10b,c)

General Functions: The **palatine bones** form the posterior portion of the hard palate and contribute to the floor of each orbit.

Articulations: The palatine bones articulate with one another, with the maxillae, with the sphenoid and ethmoid, with the inferior nasal conchae, and with the vomer (Figures 7–3e and 7–4a).

Regions/Landmarks: The palatine bones are roughly L-shaped. The **horizontal plate** forms the posterior part of the hard palate; the **perpendicular plate** extends from the horizontal plate to the **orbital process**, which forms part of the floor of the orbit. This process contains a small sinus that normally opens into the sphenoidal sinus.

Foramina: Small blood vessels and nerves supplying the roof of the mouth penetrate the lateral portion of the horizontal plate.

The Nasal Bones (Figure 7–11)

General Functions: The **nasal bones** support the superior portion of the bridge of the nose. They are connected to cartilages that support the distal portions of the nose. These flexible cartilages, and associated soft tissues, extend to the superior border of the **external nares** (NA-rēz; singular, *naris*), the entrances to the nasal cavity.

Articulations: The paired nasal bones articulate with one another, with the ethmoid, and with the frontal bone and maxillae (Figures 7–3b–d and 7–4a).

The Vomer (Figure 7–11)

General Functions: The **vomer** forms the inferior portion of the bony nasal septum.

Articulations: The vomer articulates with the maxillae, sphenoid, ethmoid, and palatine bones, and with the cartilaginous part of the nasal septum, which extends into the fleshy part of the nose (Figures 7–3d,e and 7–4a).

The Inferior Nasal Conchae (Figure 7–11)

General Functions: The **inferior nasal conchae** create turbulence in air passing through the nasal cavity, and increase the epithelial surface area to promote warming and humidification of inhaled air.

Articulations: The inferior nasal conchae articulate with the maxillae, ethmoid, palatine, and lacrimal bones (Figure 7–3d).

The Zygomatic Bones (Figure 7–11)

General Functions: The **zygomatic bones** contribute to the rim and lateral wall of the orbit and form part of the zygomatic arch.

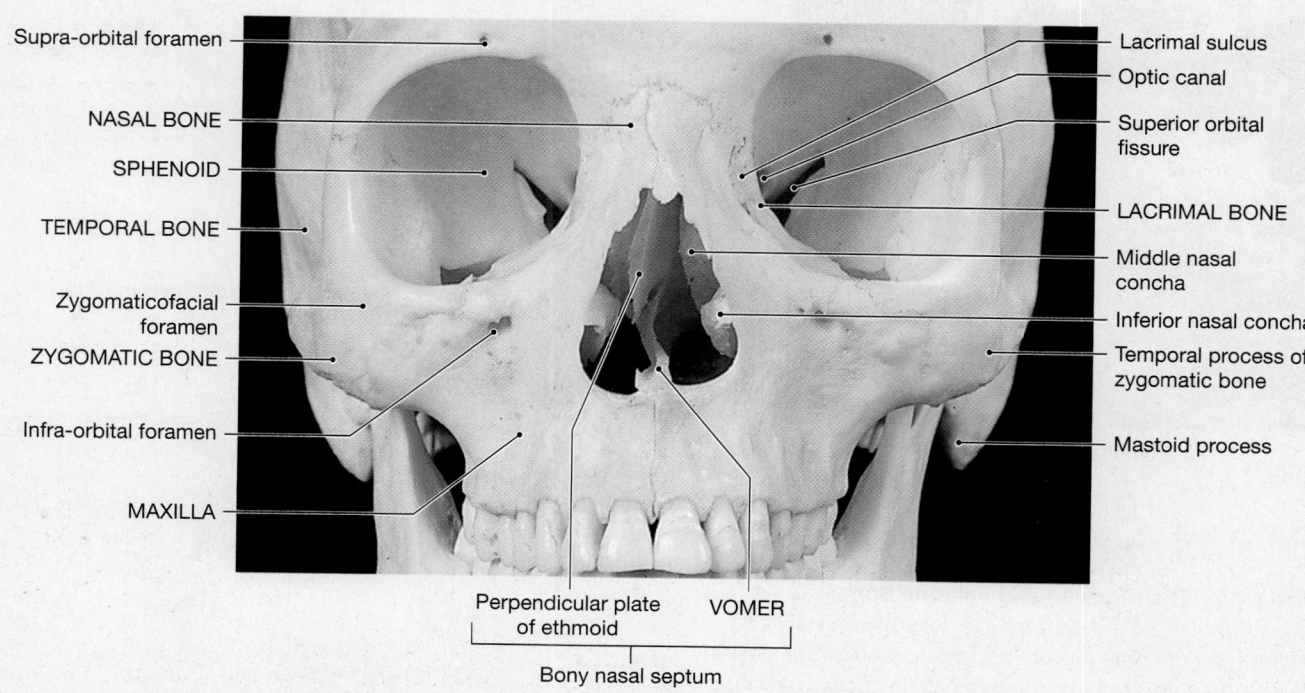

Figure 7–11   The Smaller Bones of the Face.

Articulations: The zygomatic bones articulate with the maxillae, and the sphenoid, frontal, and temporal bones (**Figure 7–3b–e**).

Regions/Landmarks: The **temporal process** curves posteriorly to meet the zygomatic process of the temporal bone.

Foramina: The **zygomaticofacial foramen** on the anterior surface of each zygomatic bone carries a sensory nerve that innervates the cheek.

The Lacrimal Bones (Figure 7–11)

General Functions: The **lacrimal bones** form part of the medial wall of the orbit.

Articulations: The lacrimal bones—the smallest facial bones—articulate with the frontal bones and maxillae, and with the ethmoid (**Figure 7–3c,d**).

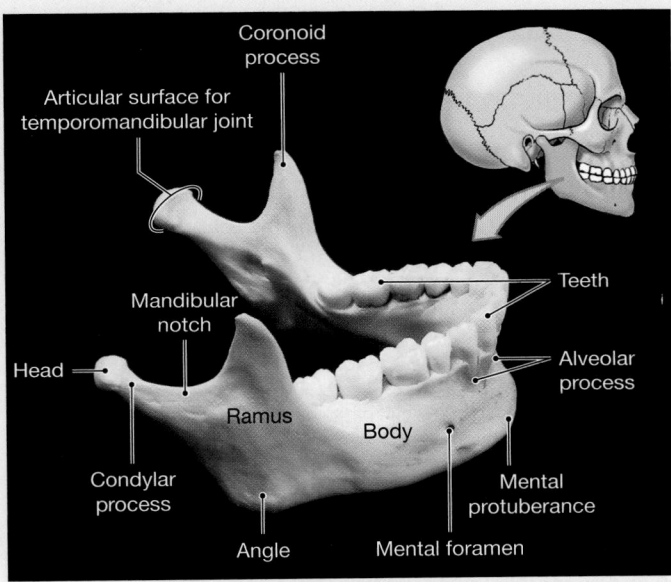

(a) Lateral view

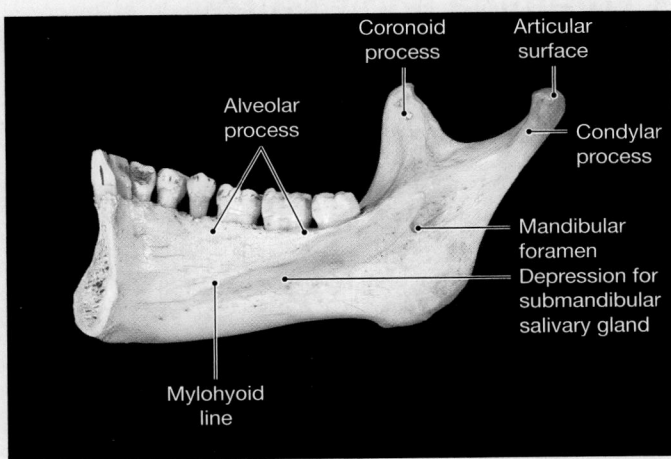

(b) Medial view

Regions/Landmarks: The **lacrimal sulcus**, a groove along the anterior, lateral surface of the lacrimal bone, marks the location of the lacrimal sac. The lacrimal sulcus leads to the nasolacrimal canal, which begins at the orbit and opens into the nasal cavity. As noted earlier, this canal is formed by the lacrimal bone and the maxilla.

The Mandible (Figure 7–12a,b)

General Functions: The **mandible** forms the lower jaw.

Articulations: The mandible articulates with the mandibular fossae of the temporal bones (**Figures 7–3c,e** and **7–7a**).

Regions/Landmarks: The **body** of the mandible is the horizontal portion of that bone.

The **alveolar process** supports the lower teeth.

The *mental protuberance* (*mentalis*, chin) is the attachment site for several facial muscles.

A prominent depression on the medial surface marks the position of the *submandibular salivary gland*.

The *mylohyoid line* marks the insertion of the *mylohyoid muscle*, which supports the floor of the mouth.

Figure 7–12   The Mandible and Hyoid Bone. (a) A lateral and slightly superior view of the mandible. **(b)** A medial view of the right mandible. **(c)** An anterior view of the hyoid bone.

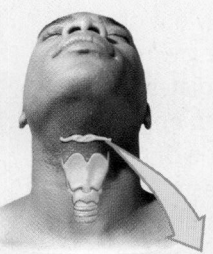

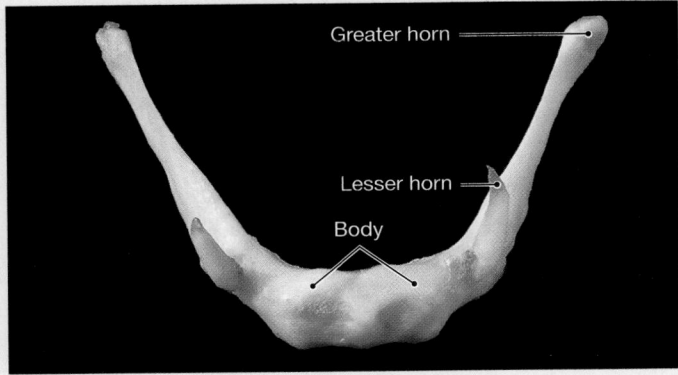

(c) Anterior-superior view

The **ramus** of the mandible is the ascending part that begins at the *mandibular angle* on either side. On each ramus:

1. The **condylar process** articulates with the temporal bone at the *temporomandibular joint*.
2. The **coronoid** (KOR-ō-noyd) **process** is the insertion point for the *temporalis muscle*, a powerful muscle that closes the jaws.
3. The **mandibular notch** is the depression that separates the condylar and coronoid processes.

Foramina: The **mental foramina** are openings for nerves that carry sensory information from the lips and chin to the brain.

The **mandibular foramen** is the entrance to the *mandibular canal*, a passageway for blood vessels and nerves that service the lower teeth. Dentists typically anesthetize the sensory nerve that enters this canal before they work on the lower teeth.

The Hyoid Bone (Figure 7–12c)

General Functions: The **hyoid bone** supports the larynx and is the attachment site for muscles of the larynx, pharynx, and tongue.

Articulations: *Stylohyoid ligaments* connect the *lesser horns* to the styloid processes of the temporal bones.

Regions/Processes: The **body** of the hyoid is an attachment site for muscles of the larynx, tongue, and pharynx.

The **greater horns**, or *greater cornua*, help support the larynx and are attached to muscles that move the tongue.

The **lesser horns**, or *lesser cornua*, are attached to the stylohyoid ligaments; from these ligaments, the hyoid and larynx hang beneath the skull like a child's swing from the limb of a tree.

CLINICAL NOTE

Temporomandibular Joint Syndrome   The *temporomandibular joint (TMJ)*, between each temporal bone and the mandible, is quite mobile, allowing your jaw to move while you chew or talk. The disadvantage of such mobility is that your jaw can easily be dislocated by forceful forward or lateral displacement. The connective tissue sheath, or *capsule*, that surrounds the joint is relatively loose, and the opposing bone surfaces are separated by a pad of fibrous cartilage. In **TMJ syndrome**, or *myofacial pain syndrome*, the mandible is pulled slightly out of alignment, generally by spasms in one of the jaw muscles. The individual experiences facial pain that radiates around the ear on the affected side and an inability to open the mouth fully. TMJ syndrome is a repeating cycle of muscle spasm → misalignment → pain → muscle spasm. It has been linked to involuntary behaviors, such as grinding of the teeth during sleep *(bruxism)*, and to emotional stress. Treatment focuses on breaking the cycle of muscle spasm and pain and, when necessary, providing emotional support. The application of heat to the affected joint, coupled with the use of anti-inflammatory drugs, local anesthetics, or both, may help. If teeth grinding is suspected, special mouth guards may be worn during sleep.

CHECKPOINT

3. In which bone is the foramen magnum located?
4. Tomás suffers a blow to the skull that fractures the right superior lateral surface of his cranium. Which bone is fractured?
5. Which bone contains the depression called the sella turcica? What is located in this depression?
6. Identify the facial bones.

See the blue Answers tab at the end of the book.

7-3 Foramina and fissures of the skull serve as passageways for nerves and vessels

Table 7–1 summarizes information about the foramina and fissures introduced thus far. This reference will be especially important to you in later chapters when you study the nervous and cardiovascular systems.

7 SKELETAL

TABLE 7–1   A Key to the Foramina and Fissures of the Skull

Bone	Foramen/Fissure	Major Structures Using Passageway	
		Neural Tissue*	Vessels and Other Structures
OCCIPITAL BONE	Foramen magnum	Medulla oblongata (most caudal portion of brain) and accessory nerve (N XI), which provides motor control over several neck and back muscles	Vertebral arteries to brain; supporting membranes around central nervous system
	Hypoglossal canal	Hypoglossal nerve (N XII) provides motor control to muscles of the tongue	
With temporal bone	Jugular foramen	Glossopharyngeal nerve (N IX), vagus nerve (N X), accessory nerve (N XI). N IX provides taste sensation; N X is important for visceral functions; N XI innervates important muscles of the back and neck	Internal jugular vein; important vein returning blood from brain to heart
FRONTAL BONE	Supra-orbital foramen (or notch)	Supra-orbital nerve, sensory branch of ophthalmic nerve, innervating the eyebrow, eyelid, and frontal sinus	Supra-orbital artery delivers blood to same region
LACRIMAL BONE	Lacrimal sulcus, nasolacrimal canal (with maxilla)		Lacrimal sac and tear duct; drains into nasal cavity
TEMPORAL BONE	Stylomastoid foramen	Facial nerve (N VII) provides motor control of facial muscles	
	Carotid canal		Internal carotid artery supplies blood to brain
	External acoustic meatus		Air in meatus conducts sound to eardrum
	Internal acoustic meatus	Vestibulocochlear nerve (N VIII) from sense organs for hearing and balance. Facial nerve (N VII) enters here, exits at stylomastoid foramen	Internal acoustic artery supplies blood to inner ear
SPHENOID	Optic canal	Optic nerve (N II) brings information from the eye to the brain	Ophthalmic artery brings blood into orbit
	Superior orbital fissure	Oculomotor nerve (N III), trochlear nerve (N IV), ophthalmic branch of trigeminal nerve (N V), abducens nerve (N VI). Ophthalmic nerve provides sensory information about eye and orbit; other nerves control muscles that move the eye	Ophthalmic vein returns blood from orbit
	Foramen rotundum	Maxillary branch of trigeminal nerve (N V) provides sensation from the face	
	Foramen ovale	Mandibular branch of trigeminal nerve (N V) controls the muscles that move the lower jaw and provides sensory information from that area	
	Foramen spinosum		Vessels to membranes around central nervous system
With temporal and occipital bones	Foramen lacerum		Internal carotid artery after leaving carotid canal; auditory tube; small vessels; hyaline cartilage
With maxilla	Inferior orbital fissure	Maxillary branch of trigeminal nerve (N V); See Foramen rotundum	
ETHMOID	Olfactory foramina	Olfactory nerve (N I) provides sense of smell	
MAXILLA	Infra-orbital foramen	Infra-orbital nerve, maxillary branch of trigeminal nerve (N V) from the inferior orbital fissure to face	Infra-orbital artery with same distribution
MANDIBLE	Mental foramen	Mental nerve, sensory branch of the mandibular nerve, provides sensation from the chin and lips	Mental vessels to chin and lips
	Mandibular foramen	Inferior alveolar nerve, sensory branch of mandibular nerve, provides sensation from the gums, teeth	Inferior alveolar vessels supply same region
ZYGOMATIC BONE	Zygomaticofacial foramen	Zygomaticofacial nerve, sensory branch of maxillary nerve to cheek	

*Twelve pairs of cranial nerves, numbered N I–XII, exist. Their functions and distribution are detailed in Chapter 14.

7. Identify the bone containing the mental foramen, and list the structures using this passageway.

8. Identify the bone containing the optic canal, and cite the structures using this passageway.

9. Name the foramina found in the ethmoid bone.

See the blue Answers tab at the end of the book.

7-4 An orbital complex contains each eye, and the nasal complex encloses the nasal cavities

The facial bones not only protect and support the openings of the digestive and respiratory systems, but also protect the delicate sense organs responsible for vision and smell. Together, certain cranial bones and facial bones form an *orbital complex*, which surrounds each eye, and the *nasal complex*, which surrounds the nasal cavities.

The Orbital Complexes

The **orbits** are the bony recesses that contain the eyes. Each orbit is formed by the seven bones of the **orbital complex** (**Figure 7–13**). The frontal bone forms the roof, and the maxilla provides most of the orbital floor. The orbital rim and the first portion of the medial wall are formed by the maxilla, the lacrimal bone, and the lateral mass of the ethmoid. The lateral mass articulates with the sphenoid and a small process of the palatine bone. Several prominent foramina and fissures penetrate the sphenoid or lie between it and the maxilla. Laterally, the sphenoid and maxilla articulate with the zygomatic bone, which forms the lateral wall and rim of the orbit.

The Nasal Complex

The **nasal complex** (**Figure 7–14**) includes the bones that enclose the nasal cavities and the *paranasal sinuses*, air-filled chambers connected to the nasal cavities. The frontal bone, sphenoid, and ethmoid form the superior wall of the nasal cavities. The lateral walls are formed by the maxillae and the lacrimal bones (not shown), the ethmoid (the superior and middle nasal conchae), and the inferior nasal conchae. Much of the anterior margin of the nasal cavity is formed by the soft tissues of the nose, but the bridge of the nose is supported by the maxillae and nasal bones.

The sphenoid, ethmoid, frontal bone, palatine bone, and maxillae contain the **paranasal sinuses**. **Figure 7–14a** shows the location of the frontal and sphenoidal sinuses. Ethmoidal air cells and maxillary sinuses are shown in **Figure 7–14b**. (The tiny palatine sinuses, not shown, generally open into the sphenoidal sinuses.) The paranasal sinuses lighten the skull bones and provide an extensive area of mucous epithelium.

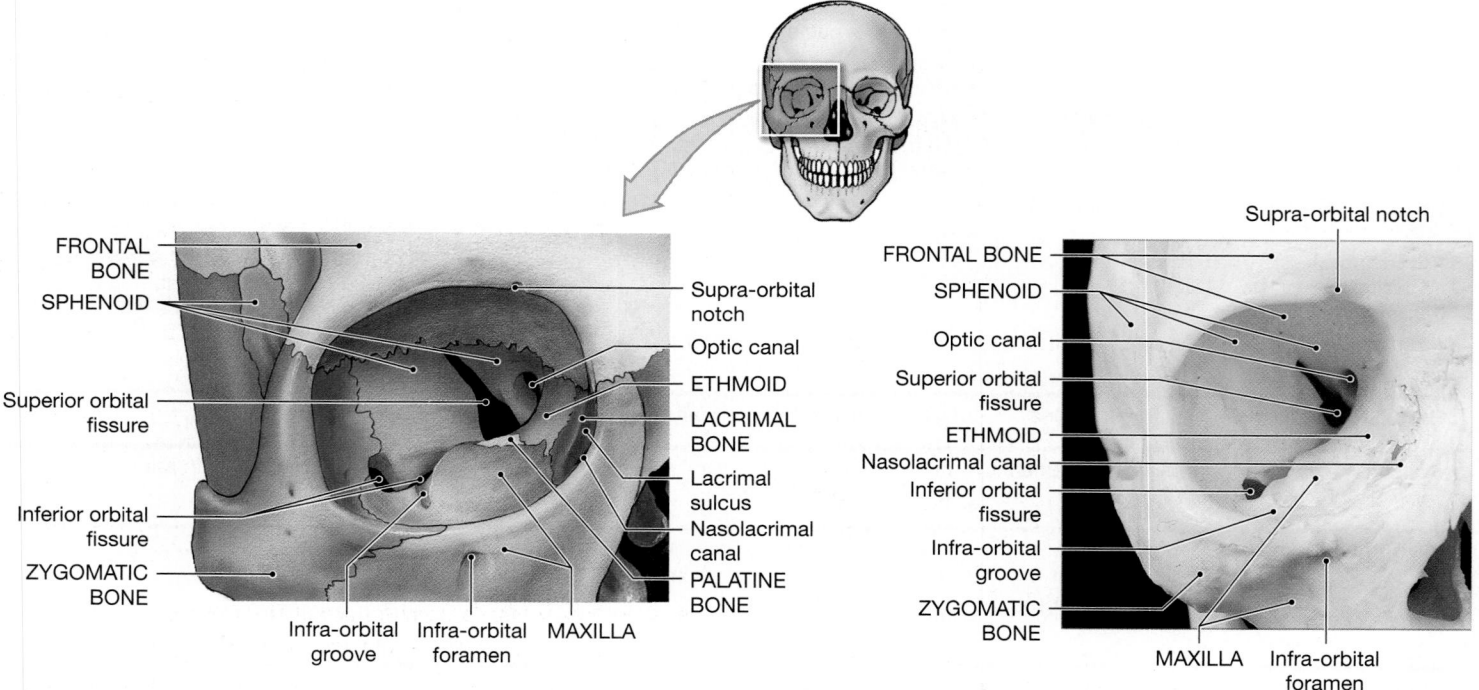

Figure 7–13 **The Orbital Complex.** The right orbital region. ATLAS: Plate 5f

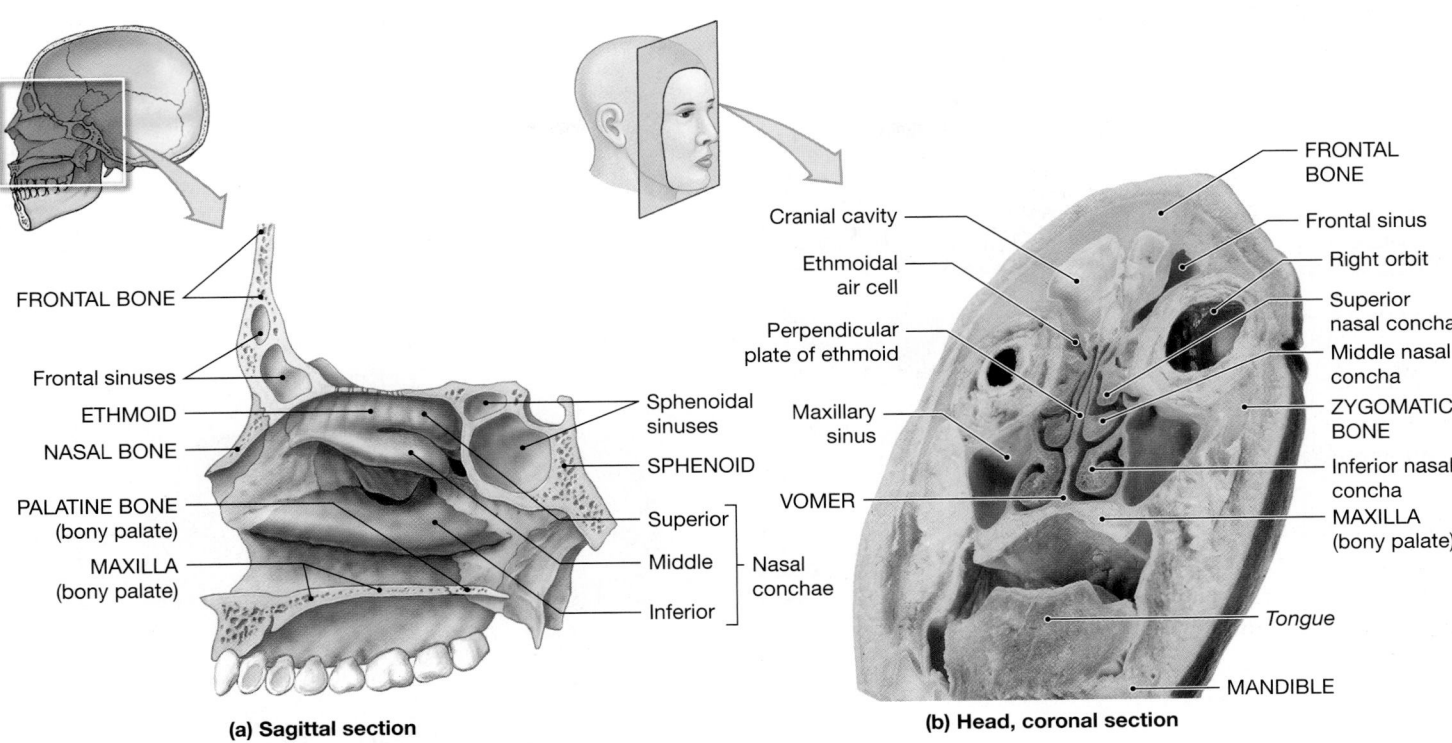

(a) Sagittal section

(b) Head, coronal section

Figure 7–14   The Nasal Complex. (a) A sagittal section through the skull, with the nasal septum removed to show major features of the wall of the right nasal cavity. The sphenoidal sinuses are visible. **(b)** A frontal section through the ethmoidal air cells and maxillary sinuses, part of the paranasal sinuses. ATLAS: Plates 11b; 12d; 13b,g

The mucous secretions are released into the nasal cavities. The ciliated epithelium passes the mucus back toward the throat, where it is eventually swallowed or expelled by coughing. Incoming air is humidified and warmed as it flows across this thick carpet of mucus. Foreign particulate matter, such as dust or microorganisms, becomes trapped in the sticky mucus and is then swallowed or expelled. This mechanism helps protect the more delicate portions of the respiratory tract.

CHECKPOINT

10. **Identify the bones of the orbital complex.**

11. **Identify the bones of the nasal complex.**

12. **Identify the bones containing the paranasal sinuses.**

See the blue Answers tab at the end of the book.

7-5 Fontanelles are non-ossified areas between cranial bones that allow for brain growth

Many different centers of ossification are involved in the formation of the skull. As development proceeds, the centers fuse, producing a smaller number of composite bones. For example, the sphenoid begins as 14 separate ossification centers. At birth, fusion has not been completed: There are two frontal bones, four occipital bones, and several sphenoid and temporal elements.

The skull organizes around the developing brain. As the time of birth approaches, the brain enlarges rapidly. Although the bones of the skull are also growing, they fail to keep pace. At birth, the cranial bones are connected by areas of fibrous connective tissue (**Figure 7–15**). The connections are quite flexible, so the skull can be distorted without damage. Such distortion normally occurs during delivery, and the changes in head shape ease the passage of the infant through the birth canal. The largest fibrous areas between the cranial bones are known as **fontanelles** (fon-tuh-NELZ; sometimes spelled *fontanels*):

• The *anterior fontanelle* is the largest fontanelle. It lies at the intersection of the frontal, sagittal, and coronal sutures in the anterior portion of the skull.

• The *occipital fontanelle* is at the junction between the lambdoid and sagittal sutures.

• The *sphenoidal fontanelles* are at the junctions between the squamous sutures and the coronal suture.

• The *mastoid fontanelles* are at the junctions between the squamous sutures and the lambdoid suture.

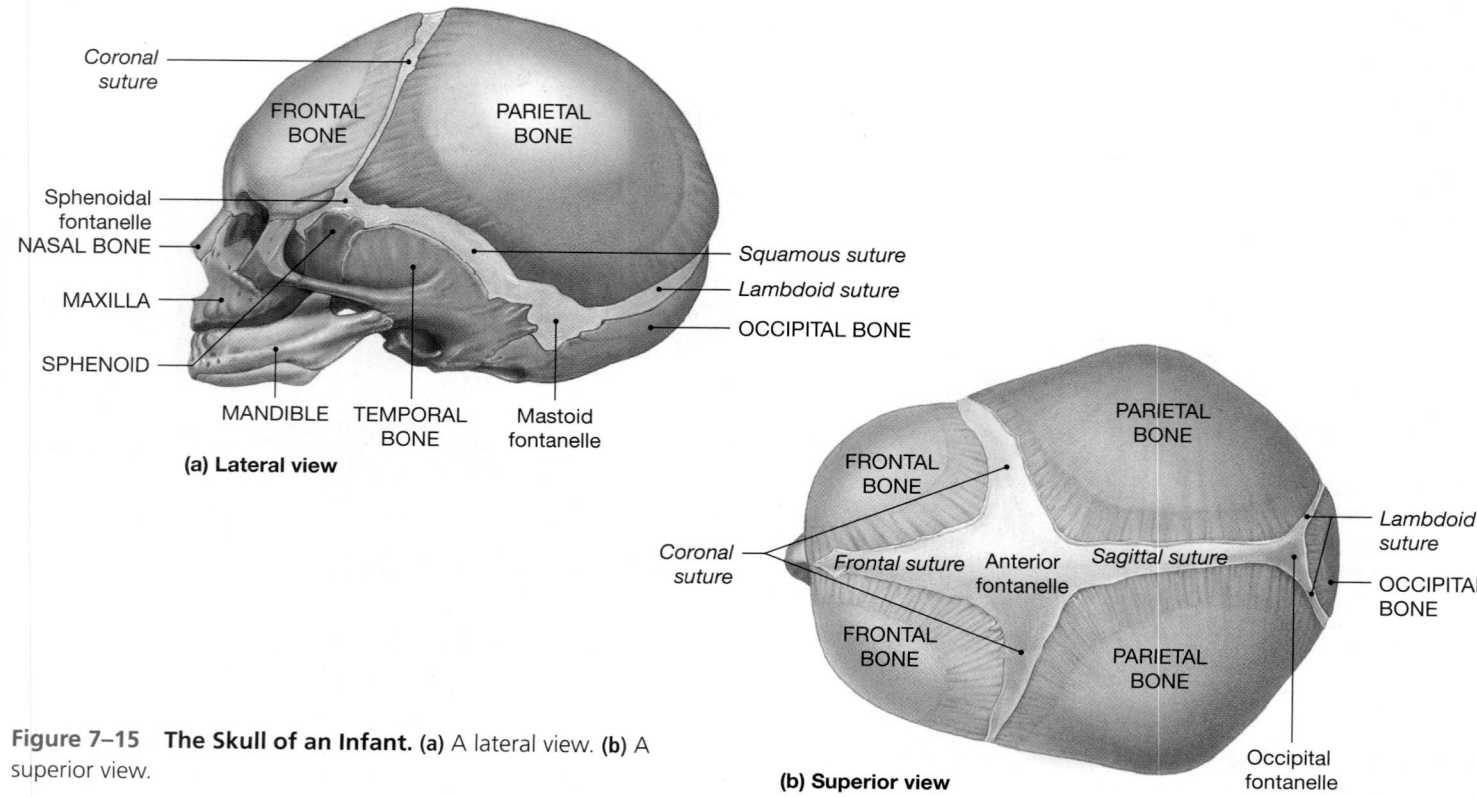

Figure 7–15 **The Skull of an Infant.** **(a)** A lateral view. **(b)** A superior view.

The anterior fontanelle is often referred to as the "soft spot" on newborns, and is often the only fontanelle easily seen by new parents. Because it is composed of fibrous connective tissue and covers a major blood vessel, the anterior fontanelle pulses as the heart beats. This fontanelle is sometimes used to determine whether an infant is dehydrated, as the surface becomes indented when blood volume is low.

The occipital, sphenoidal, and mastoid fontanelles disappear within a month or two after birth. The anterior fontanelle generally persists until the child is nearly 2 years old. Even after the fontanelles disappear, the bones of the skull remain separated by fibrous connections.

The skulls of infants and adults differ in terms of the shape and structure of cranial elements. This difference accounts for variations in proportions as well as in size. The most significant growth in the skull occurs before age 5, because at that time the brain stops growing and the cranial sutures develop. As a result, the cranium of a young child, compared with the skull as a whole, is relatively larger than that of an adult. The growth of the cranium is generally coordinated with the expansion of the brain. If one or more sutures form before the brain stops growing, the skull will be abnormal in shape, size, or both.

CLINICAL NOTE

Craniostenosis Unusual distortions of the skull result from *craniostenosis* (krā-nē-ō-sten-Ō-sis; *stenosis*, narrowing), the premature closure of one or more fontanelles. As the brain continues to enlarge, the rest of the skull distorts to accommodate it. A long and narrow head is produced by early closure of the sagittal suture, whereas a very broad skull results if the coronal suture forms prematurely. Early closure of all cranial sutures restricts the development of the brain, and surgery must be performed to prevent brain damage. If brain enlargement stops due to genetic or developmental abnormalities, however, skull growth ceases as well. This condition, which results in an undersized head, is called *microcephaly* (mī-krō-SEF-uh-lē; *micro-*, small + *cephalon*, head).

CHECKPOINT

13. **Define fontanelle, and identify the major fontanelles.**

14. **What purpose does a fontanelle serve?**

See the blue Answers tab at the end of the book.

7-6 The vertebral column has four spinal curves

The rest of the axial skeleton consists of the vertebral column, ribs, and sternum. The adult **vertebral column**, or *spine*, consists of 26 bones: the **vertebrae** (24), the **sacrum**, and the **coccyx** (KOK-siks), or tailbone. The vertebrae provide a column of support, bearing the weight of the head, neck, and trunk and ultimately transferring the weight to the appendicular skeleton of the lower limbs. The vertebrae also protect the spinal cord and help maintain an upright body position, as in sitting or standing. The total length of the vertebral column of an adult averages 71 cm (28 in.). We begin this section by examining the curvature of the vertebral column; then we consider the basics of vertebral anatomy.

Spinal Curvature

The vertebral column is not straight and rigid. A lateral view shows four **spinal curves** (**Figure 7–16**): the (1) **cervical curve**, (2) **thoracic curve**, (3) **lumbar curve**, and (4) **sacral curve**.

You may have noticed that an infant's body axis forms a loose comma or a C, with the back curving posteriorly. The C shape results from the thoracic and sacral curves. These are called **primary curves**, because they appear late in fetal development, or **accommodation curves**, because they accommodate the thoracic and abdominopelvic viscera. The primary curves are present in the vertebral column at birth. The lumbar and cervical curves, known as **secondary curves**, do not appear until several months after birth. These curves are also called **compensation curves**, because they help shift the weight to permit an upright posture. The cervical curve develops as the infant learns to balance the weight of the head on the vertebrae of the neck. The lumbar curve balances the weight of the trunk over the lower limbs; it develops with the ability to stand. Both compensations become accentuated as the toddler learns to walk and run. All four curves are fully developed by age 10.

When you stand, the weight of your body must be transmitted through the vertebral column to the hips and ultimately to the lower limbs. Yet most of your body weight lies anterior to the vertebral column. The various curves bring that weight in line with the body axis. Consider what you do automatically when standing with a heavy object hugged to

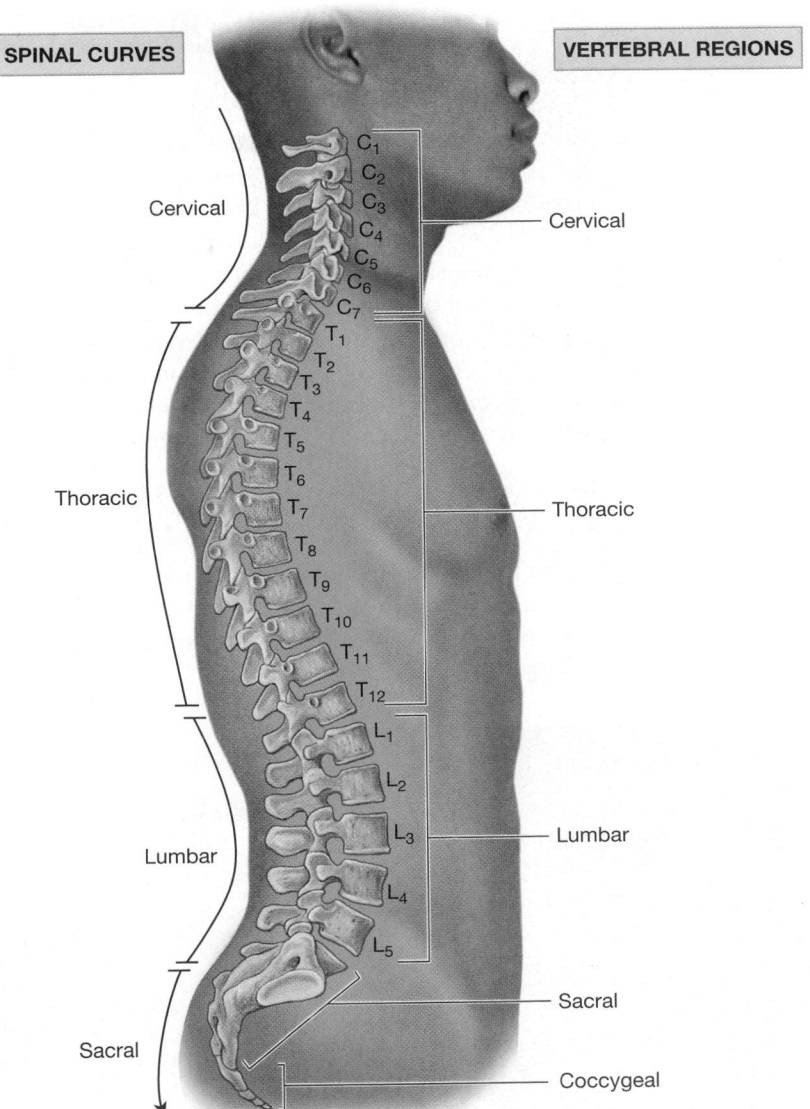

SPINAL CURVES   **VERTEBRAL REGIONS**

Cervical — Cervical

C_1
C_2
C_3
C_4
C_5
C_6
C_7
T_1
T_2
T_3
T_4
T_5
T_6
T_7
T_8
T_9
T_{10}
T_{11}
T_{12}

Thoracic — Thoracic

L_1
L_2
L_3
L_4
L_5

Lumbar — Lumbar

Sacral — Sacral

— Coccygeal

Figure 7–16   The Vertebral Column. The major regions of the adult vertebral column; notice the four spinal curves. ATLAS: Plate 2b

your chest. You avoid toppling forward by exaggerating the lumbar curve and by keeping the weight back toward the body axis. This posture can lead to discomfort at the base of the spinal column. For example, many women in the last three months of pregnancy develop chronic back pain from the changes in lumbar curvature that must adjust for the increasing weight of the fetus. In many parts of the world, people often balance heavy objects on their head. This practice increases the load on the vertebral column, but the spinal curves are not affected because the weight is aligned with the axis of the spine.

CLINICAL NOTE

Kyphosis, Lordosis, and Scoliosis

The vertebral column, composed of multiple bones and joints, must move, balance, and support the trunk and head. Conditions or events that damage the bones, muscles, and/or nerves can result in distorted shapes and impaired function. In **kyphosis** (kī-FŌ-sis; *kyphos*, hump-backed, bent), the normal thoracic curvature becomes exaggerated posteriorly, producing a "round-back" appearance (**Figure 7–17a**). This condition can be caused by (1) osteoporosis with compression fractures affecting the anterior portions of vertebral bodies, (2) chronic contractions in muscles that insert on the vertebrae, or (3) abnormal vertebral growth.

In **lordosis** (lor-DŌ-sis; *lordosis*, a bending backward), or "swayback," both the abdomen and buttocks protrude abnormally (**Figure 7–17b**). The cause is an anterior exaggeration of the lumbar curvature. This may occur during pregnancy or result from abdominal obesity or weakness in the muscles of the abdominal wall.

Scoliosis (skō-lē-Ō-sis; *scoliosis*, crookedness) is an abnormal lateral curvature of the spine (**Figure 7–17c**). This lateral deviation can occur in one or more of the movable vertebrae. Scoliosis is the most common distortion of the spinal curvature. This condition may result from developmental problems (such as incomplete vertebral formation), from damage to vertebral bodies, or from muscular paralysis affecting one side of the back (as in some cases of polio). In four out of five cases, the structural or functional cause of the abnormal spinal curvature is impossible to determine. This idiopathic scoliosis generally appears in girls during adolescence, when periods of growth are most rapid. Small curves may later stabilize once growth is complete. For larger curves, bracing may prevent progression. Severe cases can be treated through surgical straightening with implanted metal rods or cables.

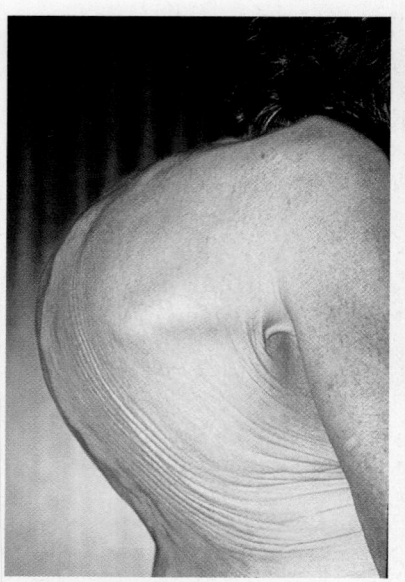

(a) Kyphosis

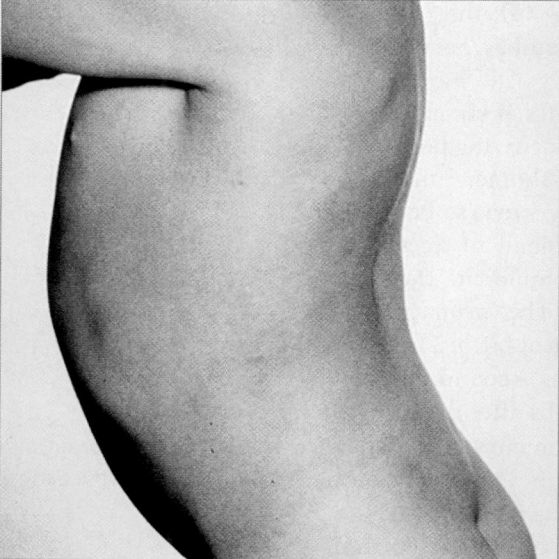

(b) Lordosis

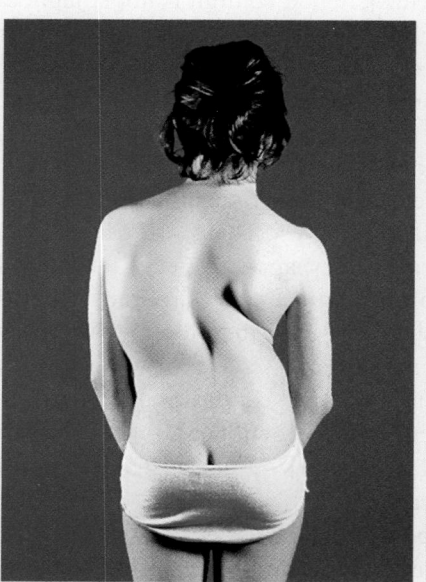

(c) Scoliosis

Figure 7–17   **Abnormal Curvatures of the Spine.**

Vertebral Anatomy

Each vertebra consists of three basic parts: (1) a *vertebral body*, (2) a *vertebral arch*, and (3) *articular processes* (**Figure 7–18a**).

The **vertebral body**, or *centrum* (plural, *centra*), is the part of a vertebra that transfers weight along the axis of the vertebral column (**Figure 7–18a,b,e**). The bodies of adjacent vertebrae are interconnected by ligaments, but are separated by pads of fibrous cartilage, the **intervertebral discs**.

The **vertebral arch** forms the posterior margin of each **vertebral foramen** (**Figure 7–18a,c**). The vertebral arch has walls, called **pedicles** (PED-i-kulz), and a roof, formed by flat layers called **laminae** (LAM-i-nē; singular, *lamina*, a thin plate). The pedicles arise along the posterior and lateral margins of the body. The laminae on either side extend dorsally and medially to complete the roof. Together, the vertebral foramina of successive vertebrae form the **vertebral canal**, which encloses the spinal cord (**Figure 7–18e**).

A **spinous process** projects posteriorly from the point where the vertebral laminae fuse to complete the vertebral arch. You can see—and feel—the spinous processes through the skin of the back when the spine is flexed. **Transverse processes** project laterally or dorsolaterally on both sides from the point where the laminae join the pedicles. These processes are sites of muscle attachment, and they may also articulate with the ribs.

Like the transverse processes, the **articular processes** arise at the junction between the pedicles and the laminae. A **superior** and an **inferior articular process** lie on each side of the vertebra. The superior articular processes articulate with the inferior articular processes of a more superior verte-

bra (or the occipital condyles, in the case of the first cervical vertebra). The inferior articular processes articulate with the superior articular processes of a more inferior vertebra (or the sacrum, in the case of the last lumbar vertebra). Each articular process has a smooth concave surface called an **articular facet**. The superior processes have articular facets on their dorsal surfaces, whereas the inferior processes articulate along their ventral surfaces.

Adjacent vertebral bodies are separated by intervertebral discs, and gaps separate the pedicles of successive vertebrae. These gaps, called **intervertebral foramina**, permit the passage of nerves running to or from the enclosed spinal cord.

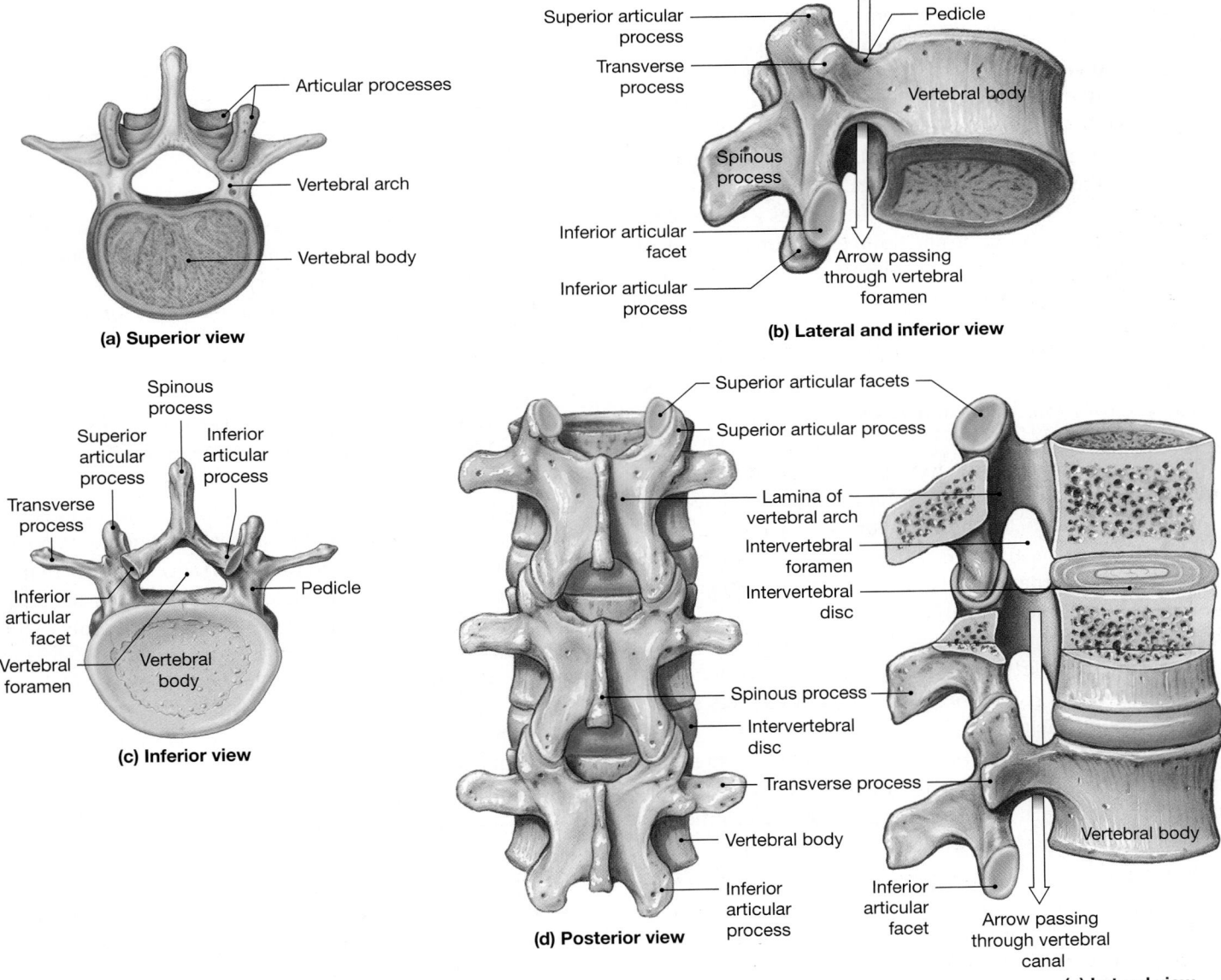

Figure 7–18   Vertebral Anatomy. (a) The major components of a typical vertebra. **(b)** A lateral and slightly inferior view of a vertebra. **(c)** An inferior view of a vertebra. **(d)** A posterior view of three articulated vertebrae. **(e)** A lateral and sectional view of three articulated vertebrae.

15. What is the importance of the secondary curves of the spine?

16. When you run your finger along a person's spine, what part of the vertebrae are you feeling just beneath the skin?

See the blue Answers tab at the end of the book.

7-7 The five vertebral regions are the cervical, thoracic, lumbar, sacral, and coccygeal

As we saw in **Figure 7–16**, the vertebral column is divided into cervical, thoracic, lumbar, sacral, and coccygeal regions. Seven **cervical vertebrae** (C_1–C_7) constitute the neck and extend inferiorly to the trunk. Twelve **thoracic vertebrae** (T_1–T_{12}) form the superior portion of the back; each articulates with one or more pairs of ribs. Five **lumbar vertebrae** (L_1–L_5) form the inferior portion of the back; the fifth articulates with the sacrum, which in turn articulates with the coccyx. The cervical, thoracic, and lumbar regions consist of individual vertebrae. During development, the sacrum originates as a group of five vertebrae, and the coccyx begins as three to five very small vertebrae. In general, the vertebrae of the sacrum are completely fused by age 25–30. Ossification of the distal coccygeal vertebrae is not complete before puberty, and thereafter fusion occurs at a variable pace. ATLAS: Embryology Summary 7: The Development of the Vertebral Column

Note that when referring to a specific vertebra, we use a capital letter to indicate the vertebral region: C, T, L, S, and Co indicate the cervical, thoracic, lumbar, sacral, and coccygeal regions, respectively. In addition, we use a subscript number to indicate the relative position of the vertebra within that region, with 1 indicating the vertebra closest to the skull. For example, C_3 is the third cervical vertebra; C_1 is in contact with the skull. Similarly, L_4 is the fourth lumbar vertebra; L_1 is in contact with T_{12} (**Figure 7–16**). We will use this shorthand throughout the text.

Although each vertebra bears characteristic markings and articulations, we will focus on the general characteristics of each region, and on how regional variations determine the vertebral group's function.

Cervical Vertebrae

Most mammals—whether giraffes, whales, mice, or humans—have seven cervical vertebrae (**Figure 7–19**). The cervical vertebrae are the smallest in the vertebral column and extend from the occipital bone of the skull to the thorax. The body of a cervical vertebra is small compared with the size of the vertebral foramen (**Figure 7–19b**). At this level, the spinal cord still contains most of the axons that connect the brain to the rest of the body. The diameter of the spinal cord decreases as you proceed caudally along the vertebral canal, and so does the diameter of the vertebral arch. However, cervical vertebrae support only the weight of the head, so the vertebral body can be relatively small and light. As you continue toward the sacrum, the loading increases and the vertebral bodies gradually enlarge.

In a typical cervical vertebra, the superior surface of the body is concave from side to side, and it slopes, with the anterior edge inferior to the posterior edge (**Figure 7–19c**). Vertebra C_1 has no spinous process. The spinous processes of the other cervical vertebrae are relatively stumpy, generally shorter than the diameter of the vertebral foramen. In the case of vertebrae C_2–C_6, the tip of each spinous process bears a prominent notch (**Figure 7–19b**). A notched spinous process is said to be **bifid** (BĪ-fid).

Laterally, the transverse processes are fused to the **costal processes**, which originate near the ventrolateral portion of the vertebral body. The costal and transverse processes encircle prominent, round **transverse foramina**. These passageways protect the *vertebral arteries* and *vertebral veins*, important blood vessels that service the brain.

The preceding description is adequate for identifying the cervical vertebrae C_3–C_6. The first two cervical vertebrae are unique, and the seventh is modified; these vertebrae are described shortly. The interlocking bodies of articulated C_3–C_7 permit more flexibility than do those of other regions. Table 7–2 includes a summary of the features of these cervical vertebrae.

Compared with the cervical vertebrae, your head is relatively massive. It sits atop the cervical vertebrae like a soup bowl on the tip of a finger. With this arrangement, small muscles can produce significant effects by tipping the balance one way or another. But if you change position suddenly, as in a fall or during rapid acceleration (a jet takeoff) or deceleration (a car crash), the balancing muscles are not strong enough to stabilize the head. A dangerous partial or complete dislocation of the cervical vertebrae can result, with injury to muscles and ligaments and potential injury to the spinal cord. The term **whiplash** is used to describe such an injury, because the movement of the head resembles the cracking of a whip.

The Atlas (C_1)

The **atlas**, cervical vertebra C_1 (**Figure 7–19d**), holds up the head, articulating with the occipital condyles of the skull. This vertebra is named after Atlas, who, according to Greek myth, holds the world on his shoulders. The articulation between the occipital condyles and the atlas is a joint that permits you to nod (such as when you indicate "yes"). The atlas can easily be dis-

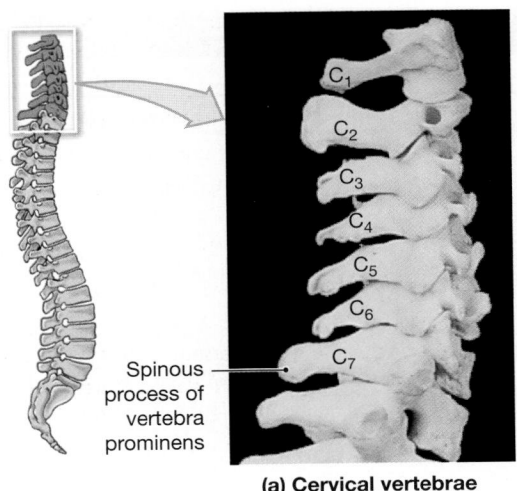

(a) Cervical vertebrae

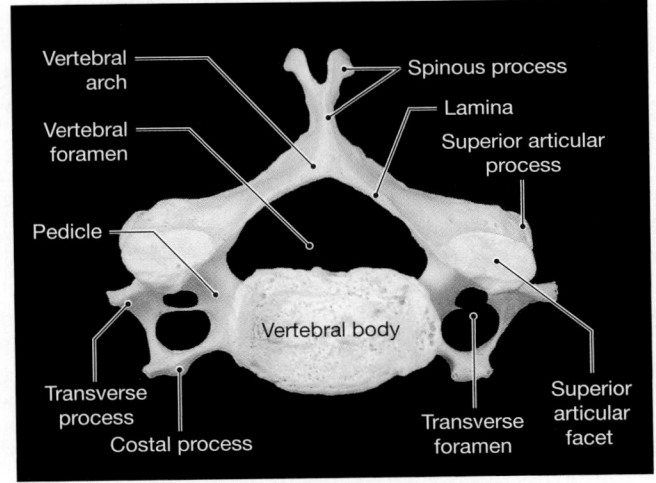

(b) Typical cervical vertebra (superior view)

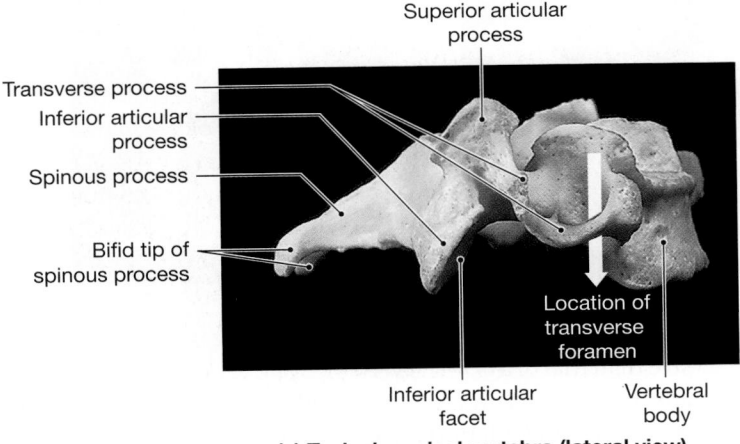

(c) Typical cervical vertebra (lateral view)

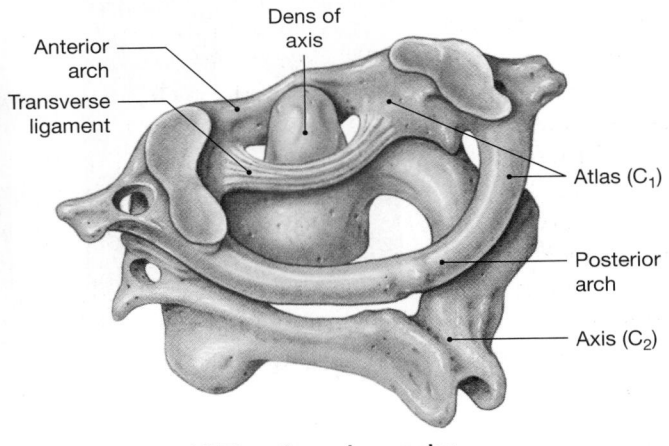

(d) The atlas–axis complex

Figure 7–19 The Cervical Vertebrae. (a) A lateral view of the cervical vertebrae, C_1–C_7. **(b)** A superior view of a representative cervical vertebra showing characteristics of C_3–C_6. Notice the typical features listed in *Table 7–2*. **(c)** A lateral view of the same vertebra. **(d)** The atlas (C_1) and axis (C_2). ATLAS: Plates 20b; 21a–e

tinguished from other vertebrae by (1) the lack of a body and spinous process and (2) the presence of a large, round vertebral foramen bounded by **anterior** and **posterior arches**.

The atlas articulates with the second cervical vertebra, the *axis*. This articulation permits rotation (as when you shake your head to indicate "no").

The Axis (C_2)

During development, the body of the atlas fuses to the body of the second cervical vertebra, called the **axis** (C_2) (**Figure 7–19d**). This fusion creates the prominent **dens** (DENZ; *dens*, tooth), or *odontoid* (ō-DON-toyd; *odontos*, tooth) *process*, of the axis. A transverse ligament binds the dens to the inner surface of the atlas, forming a pivot for rotation of the atlas and skull. Important muscles controlling the position of the head and neck attach to the especially robust spinous process of the axis.

In children, the fusion between the dens and axis is incomplete. Impacts or even severe shaking can cause dislocation of the dens and severe damage to the spinal cord. In adults, a blow to the base of the skull can be equally dangerous, because a dislocation of the axis–atlas joint can force the dens into the base of the brain, with fatal results.

Tips&Tricks

To remember the difference between the atlas and the axis and their respective movements, consider Greek mythology and the Earth. In this case, the head (Earth) is held up by Atlas and is capable of nodding "yes" movements; the Earth rotates on its axis, and the axis allows for us to shake our heads in a "no" movement.

TABLE 7–2 Regional Differences in Vertebral Structure and Function

Feature	Type (Number)		
	Cervical Vertebrae (7)	Thoracic Vertebrae (12)	Lumbar Vertebrae (5)
Location	Neck	Chest	Inferior portion of back
Body	Small, oval, curved faces	Medium, heart-shaped, flat faces; facets for rib articulations	Massive, oval, flat faces
Vertebral foramen	Large	Smaller	Smallest
Spinous process	Long; split tip; points inferiorly	Long, slender; not split; points inferiorly	Blunt, broad; points posteriorly
Transverse processes	Have transverse foramina	All but two [T_{11}, T_{12}] have facets for rib articulations	Short; no articular facets or transverse foramina
Functions	Support skull, stabilize relative positions of brain and spinal cord, and allow controlled head movement	Support weight of head, neck, upper limbs, and chest; articulate with ribs to allow changes in volume of thoracic cage	Support weight of head, neck, upper limbs, and trunk

Typical appearance (superior view)

The Vertebra Prominens (C_7)

The transition from one vertebral region to another is not abrupt, and the last vertebra of one region generally resembles the first vertebra of the next. The **vertebra prominens**, or seventh cervical vertebra (C_7), has a long, slender spinous process (**Figure 7–19a**) that ends in a broad tubercle that you can feel through the skin at the base of the neck. This vertebra is the interface between the cervical curve, which arches anteriorly, and the thoracic curve, which arches posteriorly (**Figure 7–16**). The transverse processes of C_7 are large, providing additional surface area for muscle attachment. The **ligamentum nuchae** (lig-uh-MEN-tum NOO-kē; *nucha*, nape), a stout elastic ligament, begins at the vertebra prominens and extends to an insertion along the occipital crest of the skull. Along the way, it attaches to the spinous processes of the other cervical vertebrae. When your head is upright, this ligament acts like the string on a bow, maintaining the cervical curvature without muscular effort. If you have bent your neck forward, the elasticity in the ligamentum nuchae helps return your head to an upright position.

Thoracic Vertebrae

There are 12 thoracic vertebrae (**Figure 7–20**). A typical thoracic vertebra has a distinctive heart-shaped body that is more mas-

sive than that of a cervical vertebra. The vertebral foramen is relatively smaller, and the long, slender spinous process projects posteriorly and inferiorly. The spinous processes of T_{10}, T_{11}, and T_{12} increasingly resemble those of the lumbar region as the transition between the thoracic and lumbar curves approaches. Because the inferior thoracic and lumbar vertebrae carry so much weight, the transition between the thoracic and lumbar curves is difficult to stabilize. As a result, compression fractures or compression–dislocation fractures incurred after a hard fall tend to involve the last thoracic and first two lumbar vertebrae.

Each thoracic vertebra articulates with ribs along the dorsolateral surfaces of the body. The **costal facets** on the vertebral bodies articulate with the heads of the ribs. The location and structure of the articulations vary somewhat among thoracic vertebrae (**Figure 7–20a**). Vertebrae T_1–T_8 each articulate with two pairs of ribs, so their vertebral bodies have two costal facets (*superior* and *inferior*) on each side. Vertebrae T_9–T_{11} have a single costal facet on each side, and each vertebra articulates with a single pair of ribs.

The transverse processes of vertebrae T_1–T_{10} are relatively thick and contain **transverse costal facets** for rib articulation (**Figure 7–20b,c**). Thus, rib pairs 1 through 10 contact their vertebrae at two points: a costal facet and a transverse costal facet. Table 7–2 summarizes the features of thoracic vertebrae.

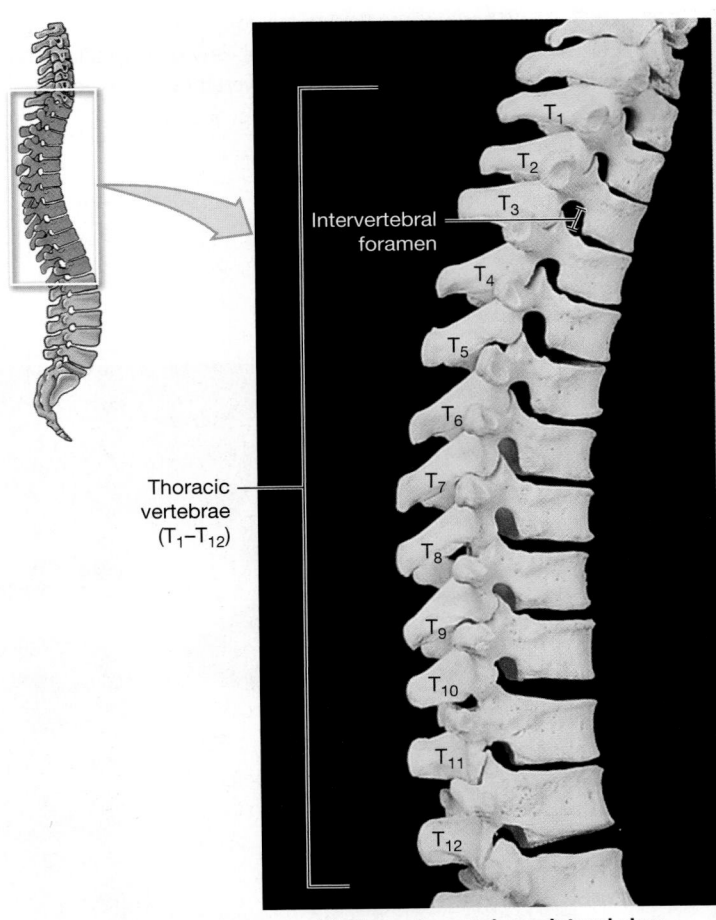

(a) Thoracic vertebrae, lateral view

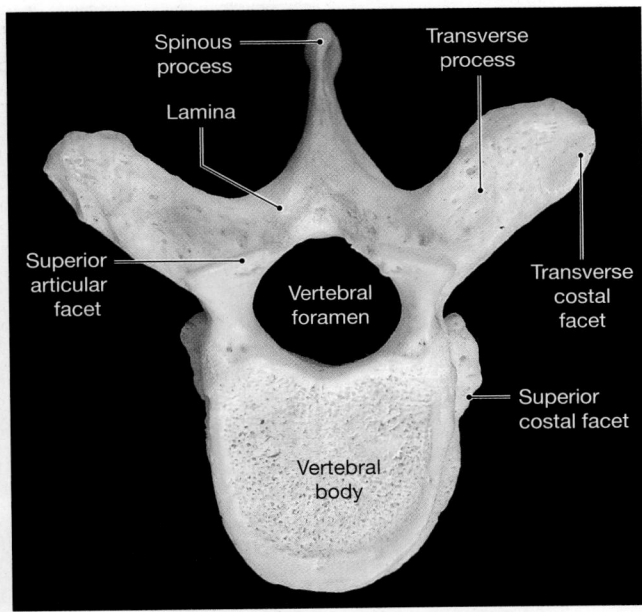

(b) Thoracic vertebra, superior view

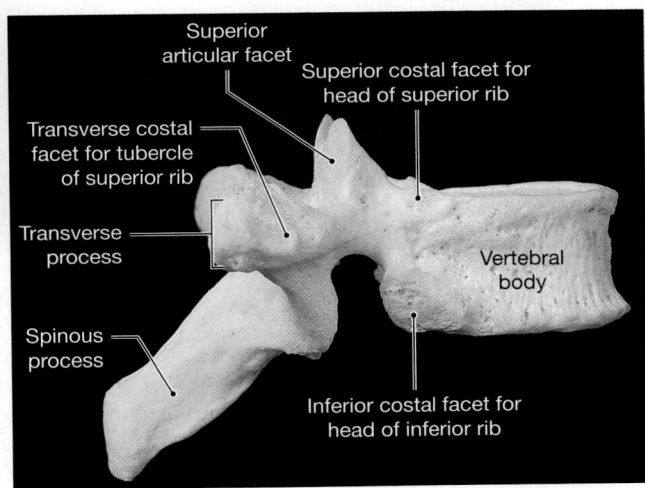

(c) Thoracic vertebra, lateral view

Figure 7–20   The Thoracic Vertebrae. (a) A lateral view of the thoracic region of the vertebral column. The vertebra prominens (C_7) resembles T_1, but lacks facets for rib articulation. Vertebra T_{12} resembles the first lumbar vertebra (L_1) but has a facet for rib articulation. **(b)** Thoracic vertebra, superior view. **(c)** Thoracic vertebra, lateral view. Notice the characteristic features listed in *Table 7–2*. ATLAS: Plates 22a–c

Lumbar Vertebrae

The five lumbar vertebrae are the largest vertebrae. The body of a typical lumbar vertebra (**Figure 7–21**) is thicker than that of a thoracic vertebra, and the superior and inferior surfaces are oval rather than heart shaped. Other noteworthy features are that (1) lumbar vertebrae do not have costal facets; (2) the slender transverse processes, which lack transverse costal facets, project dorsolaterally; (3) the vertebral foramen is triangular; (4) the stumpy spinous processes project dorsally; (5) the superior articular processes face medially ("up and in"); and (6) the inferior articular processes face laterally ("down and out").

The lumbar vertebrae bear the most weight. Their massive spinous processes provide surface area for the attachment of lower back muscles that reinforce or adjust the lumbar curve. Table 7–2 summarizes the characteristics of lumbar vertebrae.

Tips&Tricks

To remember the number of bones in the first three spinal curves, think about mealtimes. You eat breakfast at 7 a.m. (7 cervical vertebrae), lunch at 12 p.m. (12 thoracic vertebrae), and dinner at 5 p.m. (5 lumbar vertebrae).

The Sacrum

The sacrum consists of the fused components of five sacral vertebrae. These vertebrae begin fusing shortly after puberty and,

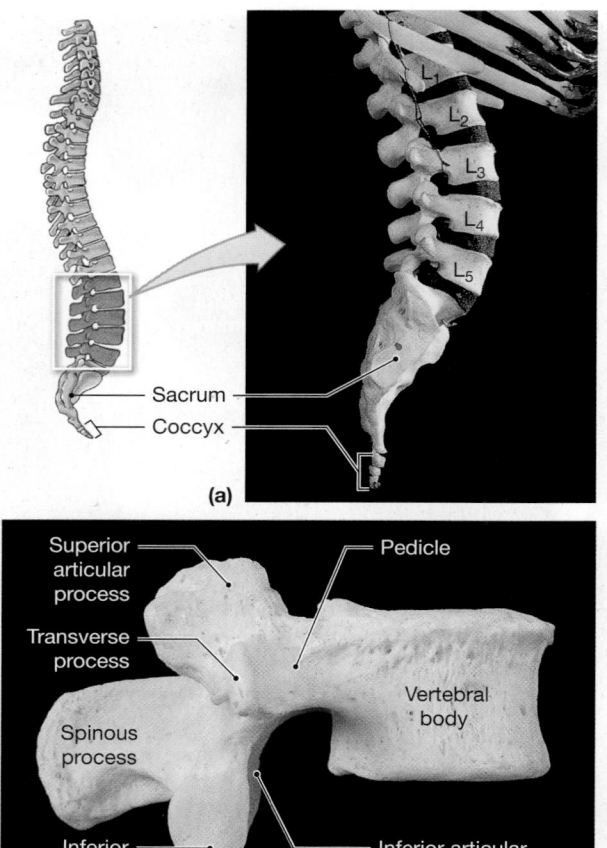

(a)

Sacrum
Coccyx

Figure 7–21 **The Lumbar Vertebrae. (a)** A lateral view of the lumbar vertebrae and sacrum. **(b)** A lateral view of a typical lumbar vertebra. **(c)** A superior view of the same vertebra. ATLAS: Plates 23a–c

Superior articular process
Pedicle
Transverse process
Vertebral body
Spinous process
Inferior articular facet
Inferior articular process

(b) Lateral view

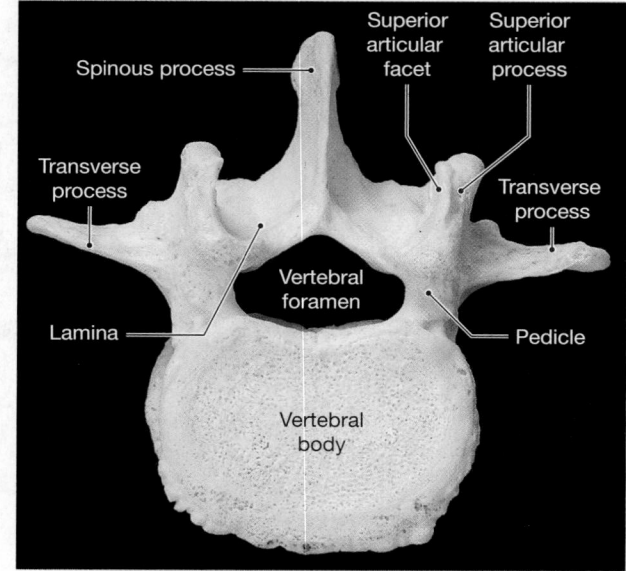

Spinous process
Superior articular facet
Superior articular process
Transverse process
Transverse process
Vertebral foramen
Lamina
Pedicle
Vertebral body

(c) Superior view

in general, are completely fused at age 25–30. The sacrum protects the reproductive, digestive, and urinary organs and, via paired articulations, attaches the axial skeleton to the pelvic girdle of the appendicular skeleton (**Figure 7–1b**).

The broad posterior surface of the sacrum (**Figure 7–22a**) provides an extensive area for the attachment of muscles, especially those that move the thigh. The superior articular processes of the first sacral vertebra articulate with the last lumbar vertebra. The **sacral canal** is a passageway that begins between these articular processes and extends the length of the sacrum. Nerves and membranes that line the vertebral canal in the spinal cord continue into the sacral canal.

The **median sacral crest** is a ridge formed by the fused spinous processes of the sacral vertebrae. The laminae of the fifth sacral vertebra fail to contact one another at the midline; they form the **sacral cornua** (KOR-nū-uh; singular, *cornu*; *cornua*, horns). These ridges form the margins of the **sacral hiatus** (hī-Ā-tus), the opening at the inferior end of the sacral canal. This opening is covered by connective tissues. Four pairs of **sacral foramina** open on either side of the median sacral crest. The intervertebral foramina of the fused sacral vertebrae open into these passageways. The **lateral sacral**

crest on each side is a ridge that represents the fused transverse processes of the sacral vertebrae. The sacral crests provide surface area for the attachment of muscles.

The sacrum is curved, with a convex posterior surface (**Figure 7–22b**). The degree of curvature is more pronounced in males than in females. The **auricular surface** is a thickened, flattened area lateral and anterior to the superior portion of the lateral sacral crest. The auricular surface is the site of articulation with the pelvic girdle (the *sacroiliac joint*). The **sacral tuberosity** is a roughened area between the lateral sacral crest and the auricular surface. It marks the attachment site of ligaments that stabilize the sacroiliac joint.

The subdivisions of the sacrum are most clearly seen in anterior view (**Figure 7–22c**). The narrow, inferior portion is the sacral **apex**, whereas the broad superior surface forms the **base**. The **sacral promontory**, a prominent bulge at the anterior tip of the base, is an important landmark in females during pelvic examinations and during labor and delivery. Prominent *transverse lines* mark the former boundaries of individual vertebrae that fuse during the formation of the sacrum. At the base of the sacrum, a broad sacral **ala**, or *wing*, extends on either side. The anterior and superior surfaces of each ala provide an exten-

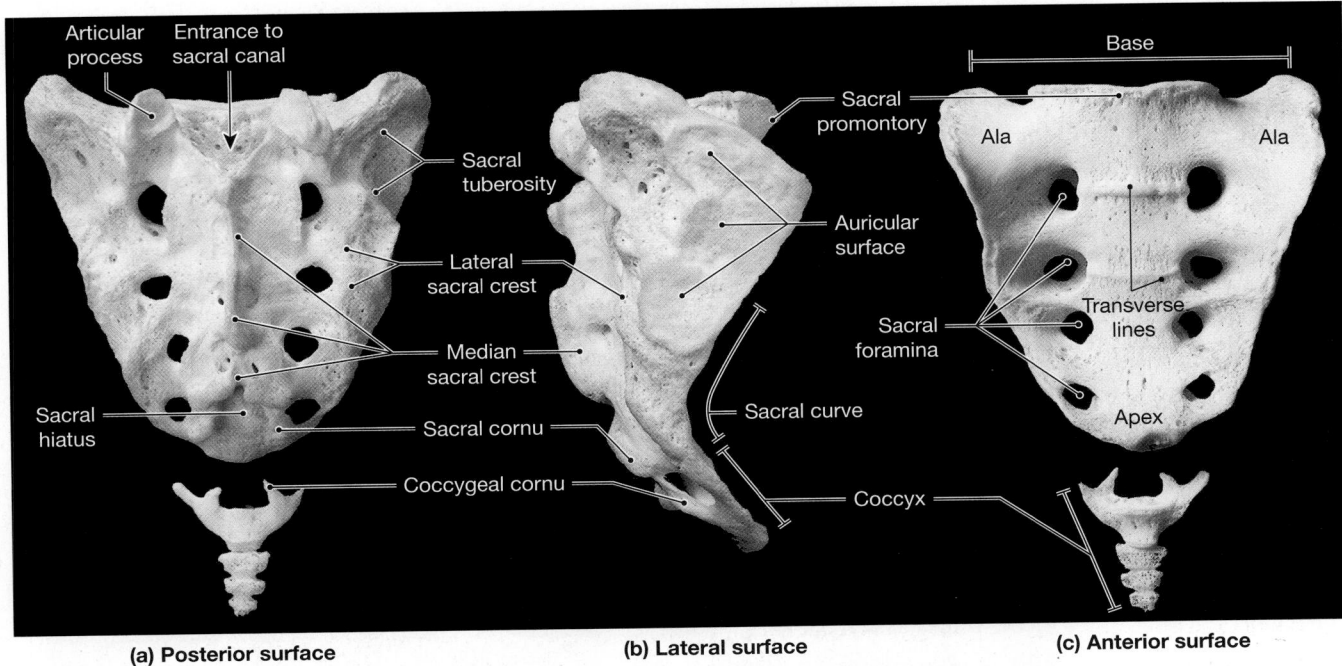

(a) Posterior surface   (b) Lateral surface   (c) Anterior surface

Figure 7–22   The Sacrum and Coccyx. (a) A posterior view. **(b)** A lateral view from the right side. **(c)** An anterior view.

sive area for muscle attachment. At the apex, a flattened area marks the site of articulation with the coccyx.

The Coccyx

The small coccyx consists of three to five (typically, four) coccygeal vertebrae that have generally begun fusing by age 26 (**Figure 7–22**). The coccyx provides an attachment site for a number of ligaments and for a muscle that constricts the anal opening. The first two coccygeal vertebrae have transverse processes and unfused vertebral arches. The prominent laminae of the first coccygeal vertebrae are known as the **coccygeal cornua**. These laminae curve to meet the sacral cornua. The coccygeal vertebrae do not fuse completely until late in adulthood. In very old people, the coccyx may fuse with the sacrum.

17. Why does the vertebral column of an adult have fewer vertebrae than that of a newborn?

18. Joe suffered a hairline fracture at the base of the dens. Which bone is fractured, and where is it located?

19. Examining a human vertebra, you notice that, in addition to the large foramen for the spinal cord, two smaller foramina are on either side of the bone in the region of the transverse processes. From which region of the vertebral column is this vertebra?

20. Why are the bodies of the lumbar vertebrae so large?

See the blue Answers tab at the end of the book.

7-8 The thoracic cage protects organs in the chest and provides sites for muscle attachment

The skeleton of the chest, or **thoracic cage** (**Figure 7–23**), provides bony support to the walls of the thoracic cavity. It consists of the thoracic vertebrae, the ribs, and the sternum (breastbone). The ribs and the sternum form the *rib cage*, whose movements are important in respiration. The thoracic cage as a whole serves two functions:

1. It protects the heart, lungs, thymus, and other structures in the thoracic cavity.

2. It serves as an attachment point for muscles involved in (1) respiration, (2) maintenance of the position of the vertebral column, and (3) movements of the pectoral girdle and upper limbs.

The Ribs

Ribs, or *costae*, are elongate, curved, flattened bones that originate on or between the thoracic vertebrae and end in the wall of the thoracic cavity. Each of us, regardless of sex, has 12 pairs of ribs (**Figure 7–23**). The first seven pairs are called **true ribs**, or *vertebrosternal ribs*. They reach the anterior body wall and are connected to the sternum by separate cartilaginous extensions, the **costal cartilages**. Beginning with the first rib, the vertebrosternal ribs gradually increase in length and in radius of curvature.

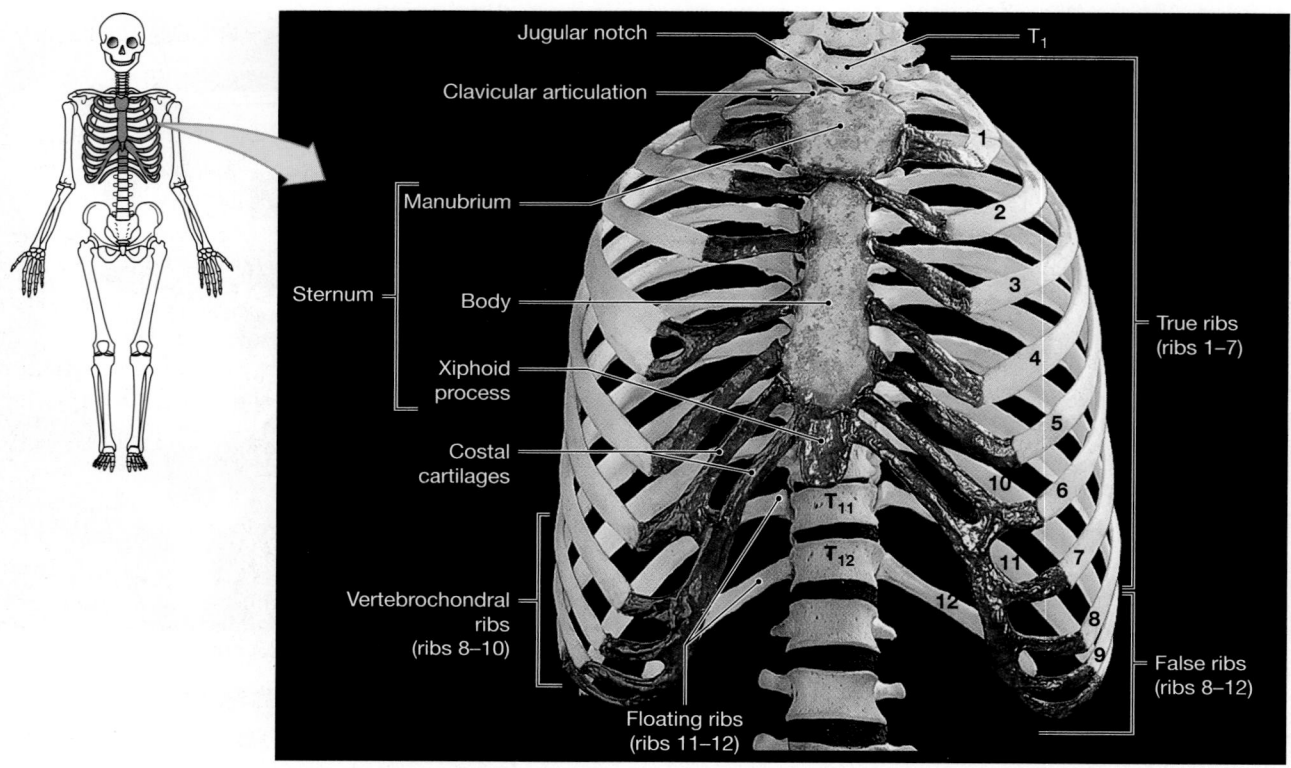

(a) Anterior view

(b) Posterior view

Figure 7–23 **The Thoracic Cage. (a)** An anterior view, showing the costal cartilages and the sternum. **(b)** A posterior view, showing the articulations of the ribs and vertebrae. ATLAS: Plate 22b

Ribs 8–12 are called **false ribs**, because they do not attach directly to the sternum. The costal cartilages of ribs 8–10, the *vertebrochondral ribs*, fuse together and merge with the cartilages of rib pair 7 before they reach the sternum (**Figure 7–23a**). The last two pairs of ribs (11 and 12) are called *floating ribs*, because they have no connection with the sternum, or *vertebral ribs*, because they are attached only to the vertebrae (**Figure 7–23b**) and muscles of the body wall.

Figure 7–24a shows the superior surface of a typical rib. The *vertebral end* of the rib articulates with the vertebral column at the **head**, or *capitulum* (ka-PIT-ū-lum). A ridge divides the articular surface of the head into superior and inferior articular facets (**Figure 7–24b**). From the head, a short **neck** leads to the **tubercle**, or *tuberculum* (too-BER-kū-lum), a small elevation that projects dorsally. The inferior portion of the tubercle contains an articular facet that contacts the transverse process of the thoracic vertebra. Ribs 1 and 10 originate at costal facets on vertebrae T_1 and T_{10}, respectively, and their tubercular facets articulate with the transverse costal facets on those vertebrae. The heads of ribs 2–9 articulate with costal facets on two adjacent vertebrae; their tubercular facets articulate with the transverse costal facets of the inferior member of the vertebral pair. Ribs 11 and 12, which originate at T_{11} and T_{12}, do not have tubercular facets and do not contact the transverse processes of

T_{11} or T_{12}. The difference in rib orientation can be seen by comparing **Figure 7–20a** with **Figure 7–23b**.

The bend, or *angle*, of the rib is the site where the tubular **body**, or *shaft*, begins curving toward the sternum. The internal rib surface is concave, and a prominent *costal groove* along its inferior border marks the path of nerves and blood vessels. The superficial surface is convex and provides an attachment site for muscles of the pectoral girdle and trunk. The *intercostal muscles*, which move the ribs, are attached to the superior and inferior surfaces.

With their complex musculature, dual articulations at the vertebrae, and flexible connection to the sternum, the ribs are quite mobile. Note how the ribs curve away from the vertebral column to angle inferiorly (**Figure 7–23**). A typical rib acts as if it were the handle on a bucket, lying just below the horizontal plane. Pushing the handle down forces it inward; pulling it up swings it outward (**Figure 7–24c**). In addition, because of the curvature of the ribs, the same movements change the position of the sternum. Depression of the ribs pulls the sternum inward, whereas elevation moves it outward. As a result, movements of the ribs affect both the width and the depth of the thoracic cage, increasing or decreasing its volume accordingly. The role of these movements in breathing will be discussed further in Chapter 23.

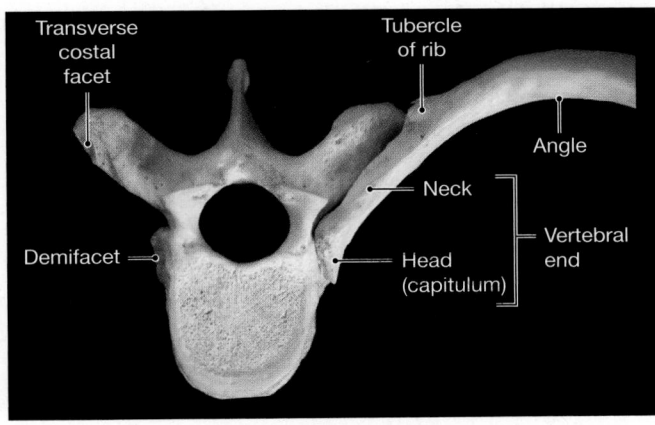

(a) Superior view

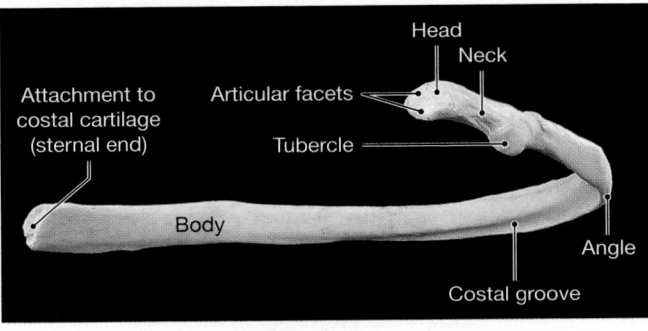

(b) Posterior view

Figure 7–24 The Ribs. (a) Details of rib structure and the articulations between the ribs and thoracic vertebrae. **(b)** A posterior view of the head of a representative rib from the right side (ribs 2–9). **(c)** The effect of rib movement on the thoracic cavity, similar to the movement of a bucket handle. ATLAS: Plates 22a,b

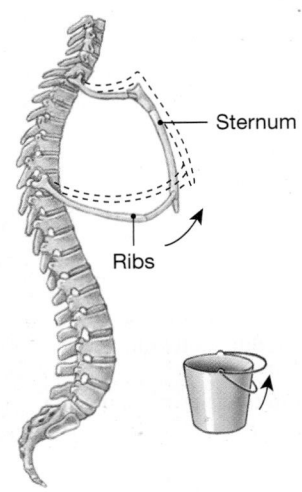

(c) Movement of curved ribs affects the volume of the thoracic cavity

The ribs can bend and move to cushion shocks and absorb blows, but severe or sudden impacts can cause painful rib fractures. Because the ribs are tightly bound in connective tissues, a cracked rib can heal without a cast or splint. But compound fractures of the ribs can send bone splinters or fragments into the thoracic cavity, with potential damage to internal organs.

Surgery on the heart, lungs, or other organs in the thorax typically involves entering the thoracic cavity. The mobility of the ribs and the cartilaginous connections with the sternum allow the ribs to be temporarily moved out of the way. "Rib spreaders" are used to push the ribs apart in much the same way that a jack lifts a car off the ground for a tire change. If more extensive access is required, the cartilages of the sternum can be cut and the entire sternum folded out of the way. Once the sternum is replaced, scar tissue reunites the cartilages, and the ribs heal fairly rapidly.

The Sternum

The adult **sternum**, or breastbone, is a flat bone that forms in the anterior midline of the thoracic wall (**Figure 7–23a**). The sternum has three components:

1. The broad, triangular **manubrium** (ma-NOO-brē-um) articulates with the *clavicles* (collarbones) and the cartilages of the first pair of ribs. The manubrium is the widest and most superior portion of the sternum. Only the first pair of ribs is attached by cartilage to this portion of the sternum. The **jugular notch**, located between the clavicular articulations, is a shallow indentation on the superior surface of the manubrium.

2. The tongue-shaped **body** attaches to the inferior surface of the manubrium and extends inferiorly along the midline. Individual costal cartilages from rib pairs 2–7 are attached to this portion of the sternum.

3. The xiphoid (ZI-foyd) process, the smallest part of the sternum, is attached to the inferior surface of the body. The muscular *diaphragm* and *rectus abdominis muscles* attach to the xiphoid process.

Ossification of the sternum begins at 6 to 10 ossification centers, and fusion is not complete until at least age 25. Before that age, the sternal body consists of four separate bones. In adults, their boundaries appear as a series of transverse lines crossing the sternum. The xiphoid process is generally the last sternal component to ossify and fuse. Its connection to the sternal body can be broken by impact or by strong pressure, creating a spear of bone that can severely damage the liver. The xiphoid process is used as a palpable landmark during the administration of cardiopulmonary resuscitation (CPR), and CPR training strongly emphasizes proper hand positioning to reduce the chances of breaking ribs or the xiphoid process.

The A&P Top 100

#32 The axial skeleton protects the brain, spinal cord, and visceral organs of the chest. The vertebrae conduct the body weight to the lower limbs; the inferior vertebrae are larger and stronger because they bear the most weight.

CHECKPOINT

21. **How are true ribs distinguished from false ribs?**

22. **Improper administration of cardiopulmonary resuscitation (CPR) can result in a fracture of which bone(s)?**

23. **What are the main differences between vertebrosternal ribs and vertebrochondral ribs?**

See the blue Answers tab at the end of the book.

Related Clinical Terms

craniostenosis: The premature closure of one or more fontanelles, which can lead to unusual distortions of the skull and brain damage.

deviated (nasal) septum: A bent nasal septum that slows or prevents sinus drainage.

microcephaly: An undersized head resulting from genetic or developmental abnormalities.

sinusitis: Inflammation and congestion of the sinuses.

spina bifida: A condition resulting from the failure of the vertebral laminae to unite during development; commonly associated with developmental abnormalities of the brain and spinal cord.

TMJ syndrome: A painful condition resulting from a misalignment of the mandible at the temporomandibular joint.

whiplash: An injury caused by displacement of the cervical vertebrae during a sudden change in body position.

Chapter Review

Study Outline

7-1   The 80 bones of the longitudinal axis make up the axial skeleton p. 212

1. The skeletal system consists of the axial skeleton and the appendicular skeleton. The **axial skeleton** can be divided into the **skull**, the **auditory ossicles** and **hyoid bone**, the **vertebral column**, and the **thoracic cage**. (*Figure 7–1*)

2. The **appendicular skeleton** includes the pectoral and pelvic girdles, which support the upper and lower limbs.

7-2   The skull is composed of 8 cranial bones and 14 facial bones p. 213

3. The **skull** consists of the **cranium** and the bones of the face. The cranium, composed of **cranial bones**, encloses the **cranial cavity**. The **facial bones** protect and support the entrances to the digestive and respiratory tracts. (*Figure 7–2*)

4. Prominent superficial landmarks on the skull include the **lambdoid, coronal, sagittal,** and **squamous sutures**. (*Figure 7–3*)

5. The bones of the cranium are the **occipital bone**, the two **parietal bones**, the **frontal bone**, the two **temporal bones**, the **sphenoid**, and the **ethmoid**. (*Figures 7–2 to 7–9*)

6. The occipital bone surrounds the **foramen magnum**. (*Figures 7–3 to 7–5*)

7. The frontal bone contains the **frontal sinuses**. (*Figures 7–4, 7–6*)

8. The **auditory ossicles** are located in a cavity within the temporal bone. (*Figure 7–7*)

9. The bones of the face are the **maxillae**, the **palatine bones**, the **nasal bones**, the **vomer**, the **inferior nasal conchae**, the **zygomatic bones**, the **lacrimal bones**, and the **mandible**. (*Figures 7–2 to 7–4, 7–10 to 7–12*)

10. The left and right maxillae, or *maxillary bones*, are the largest facial bones; they form the upper jaw and most of the **hard palate**. (*Figures 7–3, 7–4, 7–10*)

11. The palatine bones are small L-shaped bones that form the posterior portions of the hard palate and contribute to the floor of the orbital cavities. (*Figures 7–3, 7–4, 7–10*)

12. The paired nasal bones extend to the superior border of the **external nares**. (*Figures 7–3, 7–4, 7–11*)

13. The vomer forms the inferior portion of the **nasal septum**. (*Figures 7–3, 7–4, 7–11*)

14. The **temporal process** of the zygomatic bone articulates with the **zygomatic process** of the temporal bone to form the **zygomatic arch**. (*Figures 7–3, 7–7, 7–11*)

15. The paired lacrimal bones, the smallest bones of the face, are situated medially in each **orbit**. (*Figures 7–3, 7–11*)

16. The mandible is the bone of the lower jaw. (*Figures 7–3, 7–4, 7–12*)

17. The **hyoid bone**, suspended by *stylohyoid ligaments*, supports the larynx. (*Figure 7–12*)

7-3   Foramina and fissures of the skull serve as passageways for nerves and vessels p. 226

18. The foramina and fissures of the adult skull are summarized in *Table 7–1*.

7-4   An orbital complex contains each eye, and the nasal complex encloses the nasal cavities p. 228

19. Seven bones form each **orbital complex**. (*Figure 7–13*)

20. The **nasal complex** includes the bones that enclose the nasal cavities and the **paranasal sinuses**, hollow airways that connect with the nasal passages. (*Figure 7–14*)

7-5   Fontanelles are non-ossified fibrous areas between cranial bones that allow for brain growth p. 229

21. Fibrous connective tissue **fontanelles** permit the skulls of infants and children to continue growing after birth. (*Figure 7–15*)

7-6   The vertebral column has four spinal curves p. 231

22. The **vertebral column** consists of the vertebrae, sacrum, and coccyx. We have 7 **cervical vertebrae** (the first articulates with the skull), 12 **thoracic vertebrae** (which articulate with the ribs), and 5 **lumbar vertebrae** (the last articulates with the sacrum). The **sacrum** and **coccyx** consist of fused vertebrae. (*Figure 7–16*)

23. The spinal column has four **spinal curves**. The **thoracic** and **sacral curves** are called **primary**, or **accommodation**, **curves**; the **lumbar** and **cervical curves** are known as **secondary**, or **compensation**, **curves**. (*Figure 7–16*)

24. A typical vertebra has a **vertebral body** and a **vertebral arch**, and articulates with adjacent vertebrae at the **superior** and **inferior articular processes**. (*Figure 7–18*)

25. Adjacent vertebrae are separated by **intervertebral discs**. Spaces between successive **pedicles** form the **intervertebral foramina**. (*Figure 7–18*)

7-7   The five vertebral regions are the cervical, thoracic, lumbar, sacral, and coccygeal p. 234

26. Cervical vertebrae are distinguished by the shape of the body, the relative size of the vertebral foramen, the presence of **costal processes** with **transverse foramina**, and notched **spinous processes**. These vertebrae include the **atlas**, **axis**, and **vertebra prominens**. (*Figure 7–19; Table 7–2*)

27. Thoracic vertebrae have a distinctive heart-shaped body; long, slender spinous processes; and articulations for the ribs. (*Figures 7–20, 7–23; Table 7–2*)

28. The lumbar vertebrae are the most massive and least mobile of the vertebrae; they are subjected to the greatest strains. (*Figure 7–21; Table 7–2*)

29. The sacrum protects reproductive, digestive, and urinary organs and articulates with the pelvic girdle and with the fused elements of the coccyx. (*Figure 7–22*)

7-8   The thoracic cage protects organs in the chest and provides sites for muscle attachment p. 239

30. The skeleton of the **thoracic cage** consists of the thoracic vertebrae, the ribs, and the sternum. The **ribs** and **sternum** form the *rib cage*. (*Figure 7–23*)

31. Ribs 1–7 are **true**, or *vertebrosternal*, **ribs**. Ribs 8–12 are called **false ribs**; they include the *vertebrochondral ribs* (ribs 8–10) and two pairs of *floating (vertebral) ribs* (ribs 11–12). A typical rib has a **head**, or *capitulum*; a **neck**; a **tubercle**, or *tuberculum*; an *angle*; and a **body**, or *shaft*. A *costal groove* marks the path of nerves and blood vessels. (*Figures 7–23, 7–24*)

32. The sternum consists of the **manubrium**, **body**, and **xiphoid process**. (*Figure 7–23*)

Review Questions

See the blue Answers tab at the end of the book.

myA&P Access more review material online at **myA&P™** (www.myaandp.com). There, you'll find chapter guides, chapter quizzes, practice tests, animations, flashcards, a glossary with pronunciations, *Interactive Physiology*® (IP) exercises and quizzes, and more to help you succeed in the course.

LEVEL 1   Reviewing Facts and Terms

1. Identify the cranial and facial bones in the diagram below.

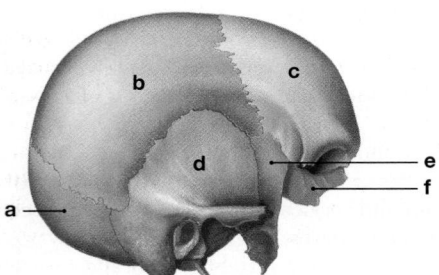

Cranial bones

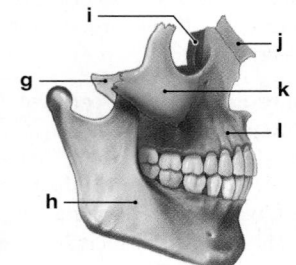

Facial bones

(a) _____   (b) _____
(c) _____   (d) _____
(e) _____   (f) _____
(g) _____   (h) _____
(i) _____   (j) _____
(k) _____   (l) _____

2. Which of the following lists contains *only* facial bones?
   (a) mandible, maxilla, nasal, zygomatic
   (b) frontal, occipital, zygomatic, parietal
   (c) occipital, sphenoid, temporal, lacrimal
   (d) frontal, parietal, occipital, sphenoid

3. The unpaired facial bones include the
   (a) lacrimal and nasal.
   (b) vomer and mandible.
   (c) maxilla and mandible.
   (d) zygomatic and palatine.

4. The boundaries between skull bones are immovable joints called
   (a) foramina.   (b) fontanelles.
   (c) lacunae.   (d) sutures.

5. The joint between the frontal and parietal bones is correctly called the _____ suture.
   (a) parietal   (b) lambdoid
   (c) squamous   (d) coronal

6. Blood vessels that drain blood from the head pass through the
   (a) jugular foramina.
   (b) hypoglossal canals.
   (c) stylomastoid foramina.
   (d) mental foramina.
   (e) lateral canals.

7. For each of the following vertebrae, indicate its vertebral region.

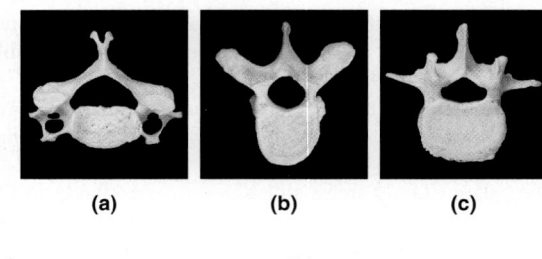

(a)   (b)   (c)

(a) _____   (b) _____
(c) _____

8. Cervical vertebrae can usually be distinguished from other vertebrae by the presence of
   (a) transverse processes.
   (b) transverse foramina.
   (c) demifacets on the centrum.
   (d) the vertebra prominens.
   (e) large spinous processes.

9. The side walls of the vertebral foramen are formed by the
   (a) centrum of the vertebra.
   (b) spinous process.
   (c) pedicles.
   (d) laminae.
   (e) transverse processes.

10. The part(s) of the vertebra that transfer(s) weight along the axis of the vertebral column is (are) the
    (a) vertebral arch.   (b) lamina.
    (c) pedicles.   (d) body.
11. Which eight bones make up the cranium?
12. What seven bones constitute the orbital complex?
13. What is the primary function of the vomer?
14. Which bones contain the paranasal sinuses?

LEVEL 2   Reviewing Concepts

15. What is the relationship between the temporal bone and the ear?
16. What is the relationship between the ethmoid and the nasal cavity?
17. Describe how ribs function in breathing.
18. Why is it important to keep your back straight when you lift a heavy object?
19. The atlas (C_1) can be distinguished from the other vertebrae by
    (a) the presence of anterior and posterior vertebral arches.
    (b) the lack of a body.
    (c) the presence of superior facets and inferior articular facets.
    (d) all of these.
20. What purpose do the fontanelles serve during birth?
21. The secondary spinal curves
    (a) help position the body weight over the legs.
    (b) accommodate the thoracic and abdominopelvic viscera.
    (c) include the thoracic curvature.
    (d) do all of these.
    (e) do only a and c.

22. When you rotate your head to look to one side,
    (a) the atlas rotates on the occipital condyles.
    (b) C_1 and C_2 rotate on the other cervical vertebrae.
    (c) the atlas rotates on the dens of the axis.
    (d) the skull rotates the atlas.
    (e) all cervical vertebrae rotate.
23. Improper administration of CPR (cardiopulmonary resuscitation) can force the _____ into the liver.
    (a) floating ribs
    (b) lumbar vertebrae
    (c) manubrium of the sternum
    (d) costal cartilage
    (e) xiphoid process

LEVEL 3   Critical Thinking and Clinical Applications

24. Jane has an upper respiratory infection and begins to feel pain in her teeth. This is a good indication that the infection is located in the
    (a) frontal sinuses.   (b) sphenoid bone.
    (c) temporal bone.   (d) maxillary sinuses.
    (e) zygomatic bones.
25. While working at an excavation, an archaeologist finds several small skull bones. She examines the frontal, parietal, and occipital bones and concludes that the skulls are those of children not yet 1 year old. How can she tell their ages from an examination of these bones?
26. Mary is in her last month of pregnancy and is suffering from lower back pains. Since she is carrying excess weight in front of her, she wonders why her back hurts. What would you tell her?

8

The Appendicular Skeleton

Did you know...?
Humans are the only mammals with both opposable thumbs and little fingers.

Learning Outcomes

After completing this chapter, you should be able to do the following:

8-1 Identify the bones that form the pectoral girdle, their functions, and their superficial features.

8-2 Identify the bones of the upper limbs, their functions, and their superficial features.

8-3 Identify the bones that form the pelvic girdle, their functions, and their superficial features.

8-4 Identify the bones of the lower limbs, their functions, and their superficial features.

8-5 Summarize sex differences and age-related changes in the human skeleton.

Clinical Notes

Carpal Tunnel Syndrome p. 254
Congenital Talipes Equinovarus p. 262

An Introduction to the Appendicular Skeleton

This chapter focuses on the functional anatomy of the bones of the limbs and of the supporting elements (or girdles) that connect the limbs to the body. In the previous chapter, we discussed the 80 bones of the axial skeleton. Appended to these bones are the remaining 60 percent of the bones that make up the skeletal system. The **appendicular skeleton** includes the bones of the limbs and the supporting elements, or *girdles,* that connect them to the trunk (**Figure 8–1**). To appreciate the role that the appendicular skeleton plays in your life, make a mental list of all the things you have done with your arms or legs in the past day. Standing, walking, writing, turning pages, eating, dressing, shaking hands, waving—the list quickly becomes unwieldy. Your axial skeleton protects and supports internal organs and participates in vital functions, such as respiration. But it is your appendicular skeleton that lets you manipulate objects and move from place to place.

The appendicular skeleton is dominated by the long bones that support the limbs. ∞ p. 185 Each long bone shares common features with other long bones. For example, one epiphysis is usually called the *head,* the diaphysis is called the *shaft,* and the head and shaft are normally separated by a *neck.* For simplicity, an illustration of a single bone will have labels that do not include the name of the bone. Thus, a photo of the humerus will have the label *head* rather than *head of the humerus* or *humeral head.* When more than one bone is shown, the label will use the complete name to avoid confusion. The descriptions in this chapter emphasize surface features that either have functional importance (such as the attachment sites for skeletal muscles and the paths of major nerves and blood vessels) or provide landmarks that define areas and locate structures of the body.

8-1 The pectoral girdle attaches to the upper limbs and consists of the clavicles and scapulae

Each arm articulates (that is, forms a joint) with the trunk at the **pectoral girdle**, or *shoulder girdle* (**Figure 8–1**). The pectoral girdle consists of two S-shaped *clavicles* (KLAV-i-kulz; collarbones) and two broad, flat *scapulae* (SKAP-ū-lē; singular, *scapula,* SKAP-ū-luh; shoulder blades). The medial, anterior end of each clavicle articulates with the manubrium of the sternum. ∞ p. 242 These articulations are the *only* direct connections between the pectoral girdle and the axial skeleton. Skeletal muscles support and position the scapulae, which have no direct bony or ligamentous connections to the thoracic cage. As a result, the shoulders are extremely mobile, but not very strong.

Movements of the clavicles and scapulae position the shoulder joints and provide a base for arm movement. The shoulder joints are positioned and stabilized by skeletal muscles that extend between the axial skeleton and the pectoral girdle. Once the joints are in position, other skeletal muscles, including several that originate on the pectoral girdle, move the upper limbs.

The surfaces of the scapulae and clavicles are extremely important as sites for muscle attachment. Bony ridges and flanges mark the attachment sites of major muscles. Other bone markings, such as sulci or foramina, indicate the positions of nerves that control the muscles, or the passage of blood vessels that nourish the muscles and bones. Next we examine the bones of the pectoral girdle in detail.

The Clavicles

The **clavicles** are S-shaped bones that originate at the superior, lateral border of the manubrium of the sternum, lateral to the jugular notch (**Figure 8–2a**). From the roughly pyramidal **sternal end**, each clavicle curves laterally and posteriorly for roughly half its length. It then forms a smooth posterior curve to articulate with a process of the scapula, the *acromion* (a-KRŌ-mē-on). The flat, **acromial end** of the clavicle is broader than the sternal end (**Figure 8–2b,c**).

The smooth, superior surface of the clavicle lies just beneath the skin. The acromial end has a rough inferior surface that bears prominent lines and tubercles (**Figure 8–2c**). These surface features are attachment sites for muscles and ligaments of the shoulder. The combination of the direction of curvature and the differences between superior and inferior surfaces make it relatively easy to tell a left clavicle from a right clavicle.

You can explore the connection between the clavicles and sternum. With your fingers in your jugular notch (the depression just superior to the manubrium), locate the clavicle on either side and find the *sternoclavicular joints* where the sternum articulates with the clavicles. These are the only articulations between the pectoral girdle and the axial skeleton. When you move your shoulders, you can feel the sternal ends of the clavicles change their positions.

The clavicles are relatively small and fragile, and therefore fractures of the clavicle are fairly common. For example, a simple fall can fracture a clavicle if you land on your hand with your arm outstretched. Fortunately, in view of the clavicle's vulnerability, most clavicular fractures heal rapidly without a cast.

The Scapulae

The anterior surface of the **body** of each **scapula** forms a broad triangle (**Figure 8–3a**). The three sides of the triangle are the **superior border**; the **medial border**, or *vertebral border*; and the **lateral border**, or *axillary border* (*axilla,* armpit).

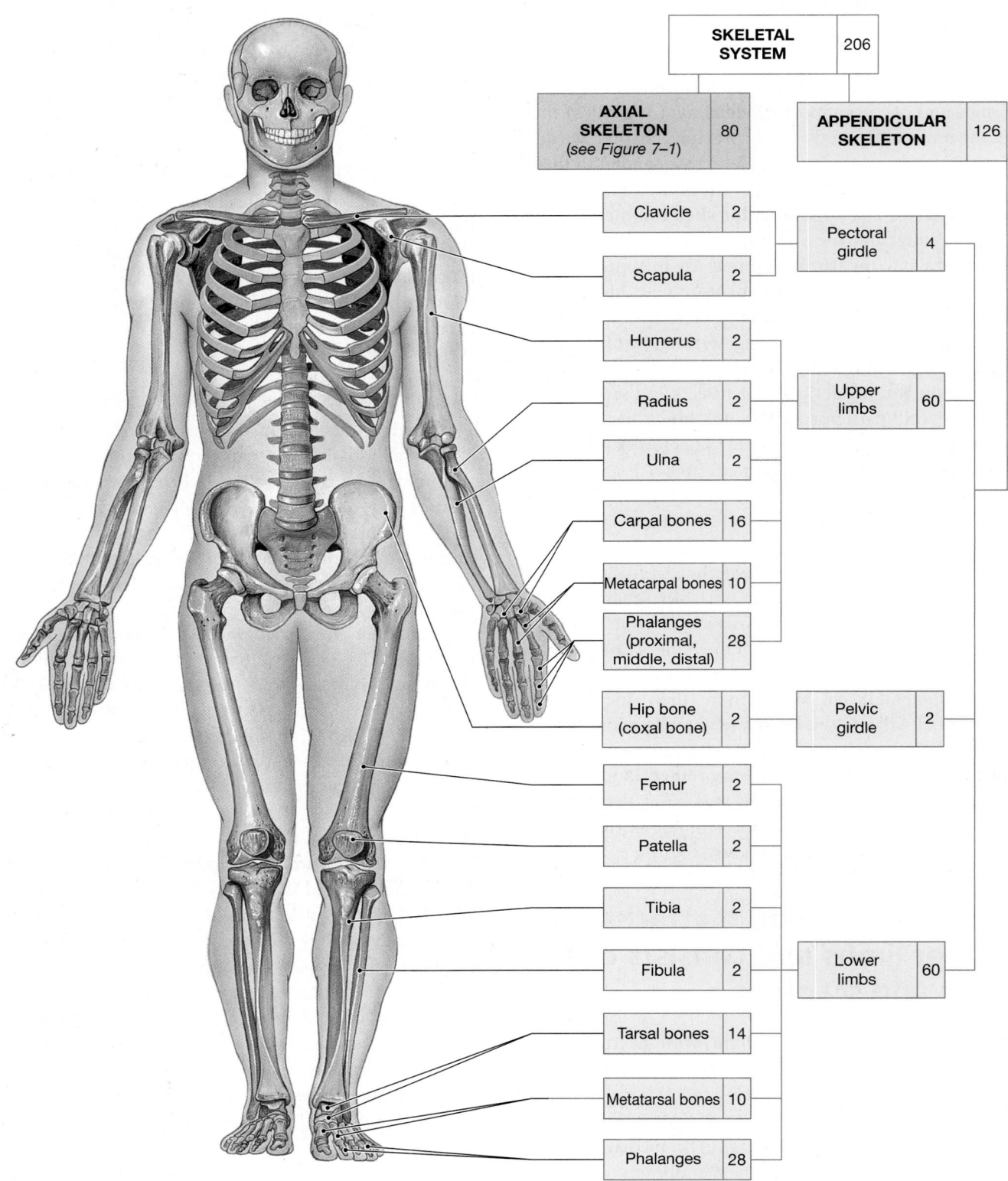

Figure 8–1   The Appendicular Skeleton. An anterior view of the skeleton, detailing the appendicular components. The numbers in the boxes indicate the total number of bones of each type or within each category. ATLAS: Plate 1a,b

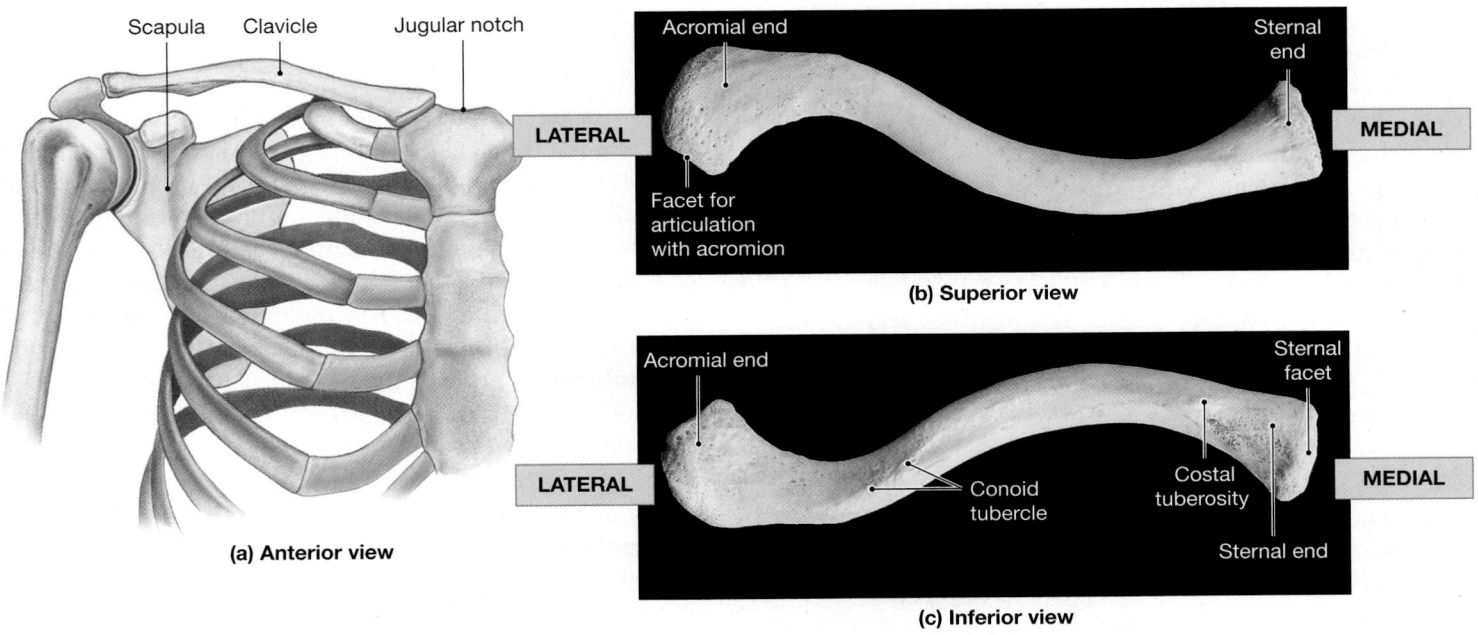

Figure 8–2   The Clavicle. (a) The position of the clavicle within the pectoral girdle, anterior view. **(b)** Superior and **(c)** inferior views of the right clavicle. Stabilizing ligaments attach to the conoid tubercle and the costal tuberosity. ATLAS: Plate 26a,b

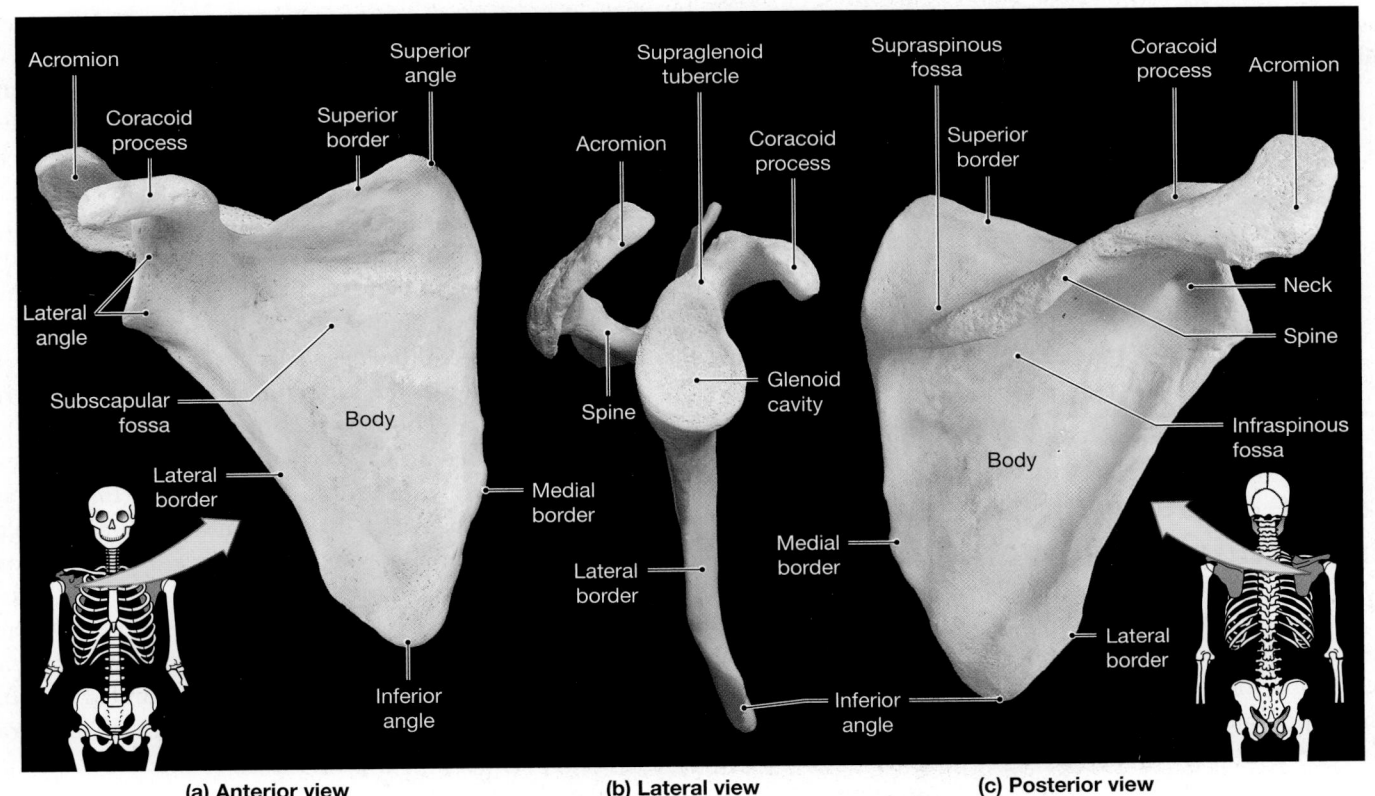

Figure 8–3   The Scapula. (a) Anterior, **(b)** lateral, and **(c)** posterior views of the right scapula. ATLAS: Plate 26a,b

Muscles that position the scapula attach along these edges. The corners of the triangle are called the *superior angle*, the *inferior angle*, and the *lateral angle*. The lateral angle, or *head* of the scapula, forms a broad process that supports the cup-shaped **glenoid cavity** (**Figure 8–3b**). At the glenoid cavity, the scapula articulates with the *humerus*, the proximal bone of the upper limb. This articulation is the shoulder joint, also known as the *glenohumeral joint*. The anterior surface of the body of the scapula is relatively smooth and concave. The depression in the anterior surface is called the **subscapular fossa**.

Two large scapular processes extend beyond the margin of the glenoid cavity (**Figure 8–3b**) superior to the head of the humerus. The smaller, anterior projection is the **coracoid** (KOR-uh-koyd) **process**. The **acromion** is the larger, posterior process. If you run your fingers along the superior surface of the shoulder joint, you will feel this process. The acromion articulates with the clavicle at the *acromioclavicular joint*. Both the coracoid process and the acromion are attached to ligaments and tendons associated with the shoulder joint.

The acromion is continuous with the **scapular spine** (**Figure 8–3c**), a ridge that crosses the posterior surface of the scapular body before ending at the medial border. The scapular spine divides the convex posterior surface of the body into two regions. The area superior to this spine constitutes the **supraspinous fossa** (*supra*, above); the region inferior to the spine is the **infraspinous fossa** (*infra*, beneath). The entire posterior surface is marked by small ridges and lines where smaller muscles attach to the scapula.

CHECKPOINT

1. Name the bones of the pectoral girdle.
2. Why would a broken clavicle affect the mobility of the scapula?
3. Which bone articulates with the scapula at the glenoid cavity?

See the blue Answers tab at the end of the book.

8-2 The upper limbs are adapted for freedom of movement

The skeleton of the upper limbs consists of the bones of the arms, forearms, wrists, and hands. Notice that in anatomical descriptions, the term *arm* refers only to the proximal portion of the upper limb (from shoulder to elbow), not to the entire limb. We will examine the bones of the right upper limb. The arm, or *brachium*, contains one bone, the **humerus**, which extends from the scapula to the elbow.

The Humerus

At the proximal end of the humerus, the round **head** articulates with the scapula (**Figure 8–4**). The prominent **greater tubercle** is a rounded projection on the lateral surface of the epiphysis, near the margin of the humeral head. The greater tubercle establishes the lateral contour of the shoulder. You can verify its position by feeling for a bump situated a few centimeters from the tip of the acromion. The **lesser tubercle** is a smaller projection that lies on the anterior, medial surface of the epiphysis, separated from the greater tubercle by the **intertubercular groove**, or *intertubercular sulcus*. Both tubercles are important sites for muscle attachment; a large tendon runs along the groove. Lying between the tubercles and the articular surface of the head, the **anatomical neck** marks the extent of the joint capsule. The narrower **surgical neck** corresponds to the metaphysis of the growing bone. The name reflects the fact that fractures typically occur at this site.

The proximal shaft of the humerus is round in section. The **deltoid tuberosity** is a large, rough elevation on the lateral surface of the shaft, approximately halfway along its length. It is named after the *deltoid muscle*, which attaches to it.

On the posterior surface, the deltoid tuberosity ends at the **radial groove** (**Figure 8–4b**). This depression marks the path of the *radial nerve*, a large nerve that provides both sensory information from the posterior surface of the limb and motor control over the large muscles that straighten the elbow. Distal to the radial groove, the posterior surface of the humerus is relatively flat. Near the distal articulation with the bones of the forearm, the shaft expands to either side at the **medial** and **lateral epicondyles**. *Epicondyles* are processes that develop proximal to an articulation and provide additional surface area for muscle attachment. The *ulnar nerve* crosses the posterior surface of the medial epicondyle. A blow at the posteromedial surface of the elbow joint can strike this nerve and produce a temporary numbness and paralysis of muscles on the anterior surface of the forearm. Because of the odd sensation, this area is sometimes called the *funny bone*.

At the **condyle**, the humerus articulates with the *radius* and the *ulna*, the bones of the forearm (*antebrachium*). The condyle is divided into two articular regions: the trochlea and the capitulum (**Figure 8–4a**). The **trochlea** (*trochlea*, a pulley), the spool-shaped medial portion of the condyle, extends from the base of the **coronoid** (*corona*, crown) **fossa** on the anterior surface to the **olecranon** (ō-LEK-ruh-non) **fossa** on the posterior surface (**Figure 8–4b**). These depressions accept projections from the ulnar surface as the elbow approaches the limits of its range of motion. The rounded **capitulum** forms the lateral surface of the condyle. A shallow **radial fossa** superior to the capitulum accommodates a portion of the radial head as the forearm approaches the humerus. The prominent lateral head and the differences between the lateral

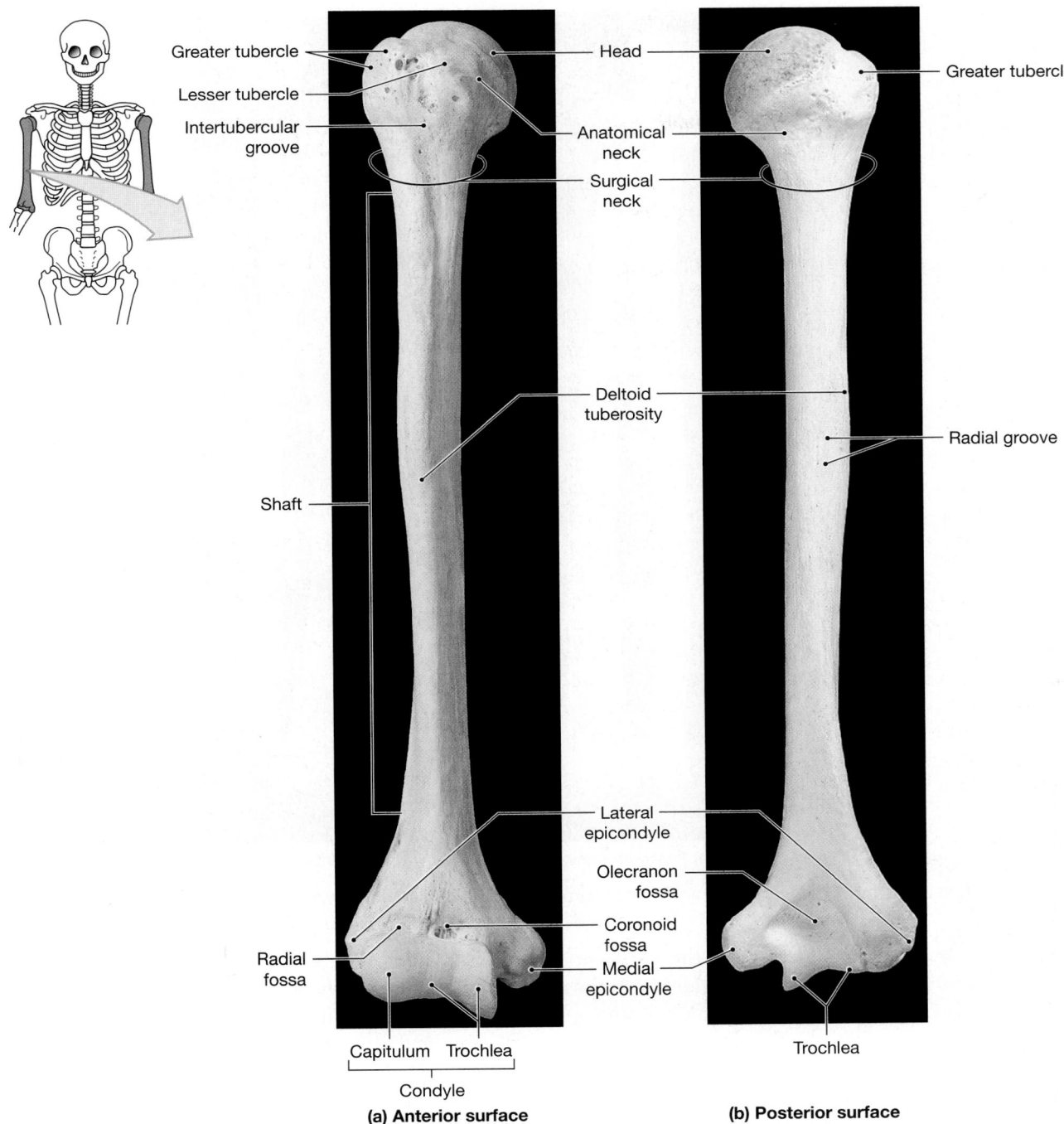

Greater tubercle
Lesser tubercle
Intertubercular groove
Shaft
Radial fossa
Capitulum   Trochlea
Condyle
(a) Anterior surface

Head
Anatomical neck
Surgical neck
Deltoid tuberosity
Lateral epicondyle
Olecranon fossa
Coronoid fossa
Medial epicondyle

Greater tubercle
Radial groove
Trochlea
(b) Posterior surface

Figure 8–4   The Humerus. (a) The anterior and **(b)** posterior surfaces of the right humerus. ATLAS: Plates 31; 34a–d

and medial condyles make it relatively easy to tell a left humerus from a right humerus.

The Ulna

The *ulna* and *radius* are parallel bones that support the forearm. In the anatomical position, the **ulna** lies medial to the radius. The **olecranon**, the superior end of the ulna, is the point of the elbow (**Figure 8–5a**). On the anterior surface of the proximal

epiphysis (**Figure 8–5b**), the **trochlear notch** of the ulna articulates with the trochlea of the humerus at the elbow joint.

Tips&Tricks

The memory tool for olecranon involves the letter "*l*": The knob you feel at the elbow is the olecranon of the ulna. You can remember that the trochlear notch is a feature of the ulna because this notch forms a U in lateral view.

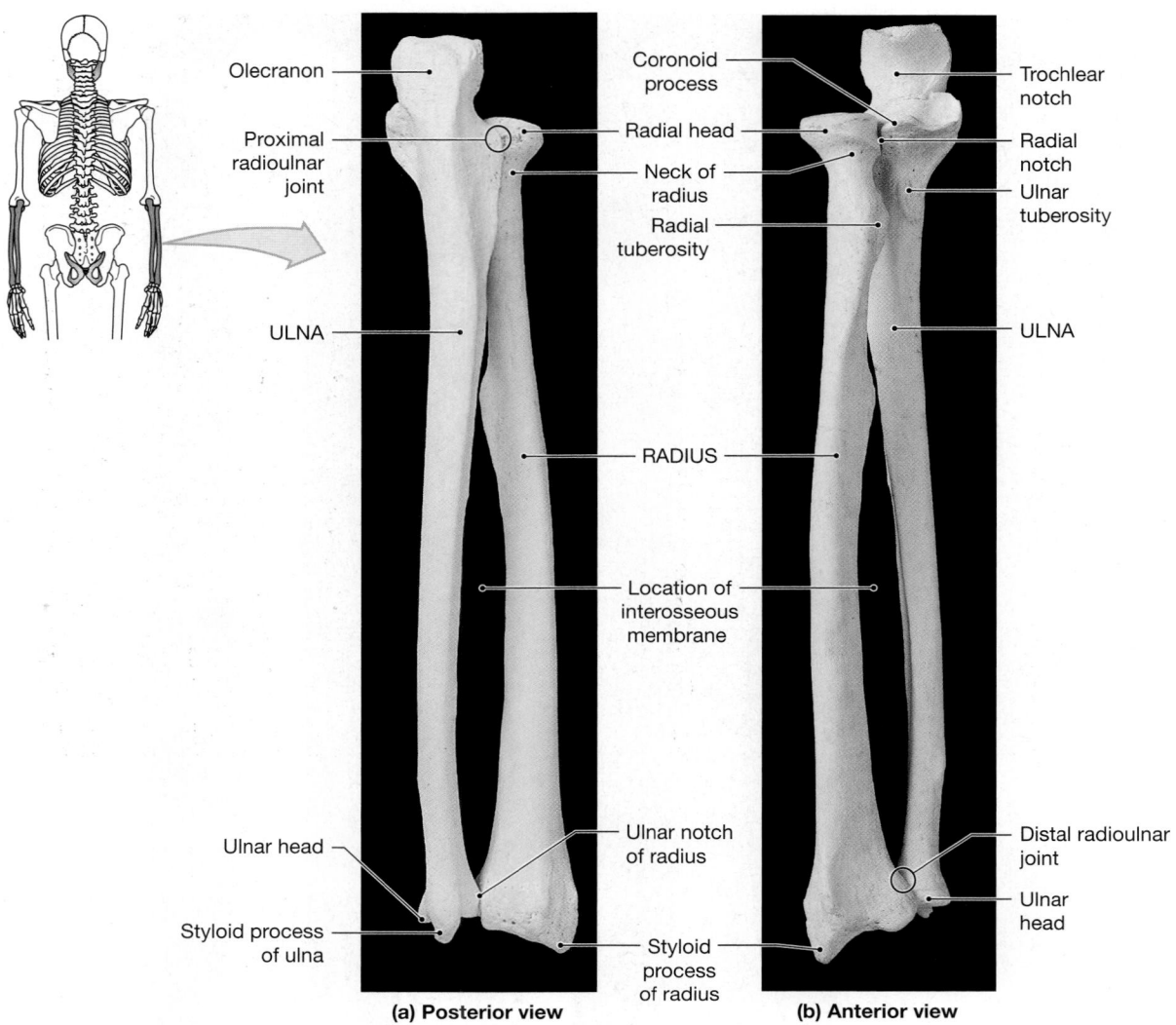

Figure 8–5 **The Radius and Ulna.** The right radius and ulna in **(a)** posterior and **(b)** anterior views. ATLAS: Plates 31; 35f; 36a,b

The olecranon forms the superior lip of the trochlear notch, and the **coronoid process** forms its inferior lip. At the limit of **extension**, with the forearm and arm forming a straight line, the olecranon swings into the olecranon fossa on the posterior surface of the humerus. At the limit of *flexion,* a movement that decreases the angle between the articulating bones, the arm and forearm form a tight V and the coronoid process projects into the coronoid fossa on the anterior humeral surface. Lateral to the coronoid process, a smooth **radial notch** accommodates the head of the radius at the *proximal radioulnar joint.*

Viewed in cross section, the shaft of the ulna is roughly triangular. The **interosseous membrane**, a fibrous sheet, connects the lateral margin of the ulna to the radius. Near the wrist, the shaft of the ulna narrows before ending at a disc-shaped **ulnar head**, or *head of the ulna.* The posterior, lateral surface of the ulnar head bears a short **styloid process** (*styloid,* long and pointed). A triangular *articular disc* attaches

to the styloid process; this cartilage separates the ulnar head from the bones of the wrist. The lateral surface of the ulnar head articulates with the distal end of the radius to form the *distal radioulnar joint.*

The Radius

The **radius** is the lateral bone of the forearm (**Figure 8–5**). The disc-shaped **radial head**, or *head of the radius,* articulates with the capitulum of the humerus. During flexion, the radial head swings into the radial fossa of the humerus. A narrow neck extends from the radial head to the **radial tuberosity**, which marks the attachment site of the *biceps brachii muscle,* a large muscle on the anterior surface of the arm. The shaft of the radius curves along its length. It also enlarges, and the distal portion of the radius is considerably larger than the distal portion of the ulna. The **ulnar notch** on the medial surface of the distal end of the radius marks the site of

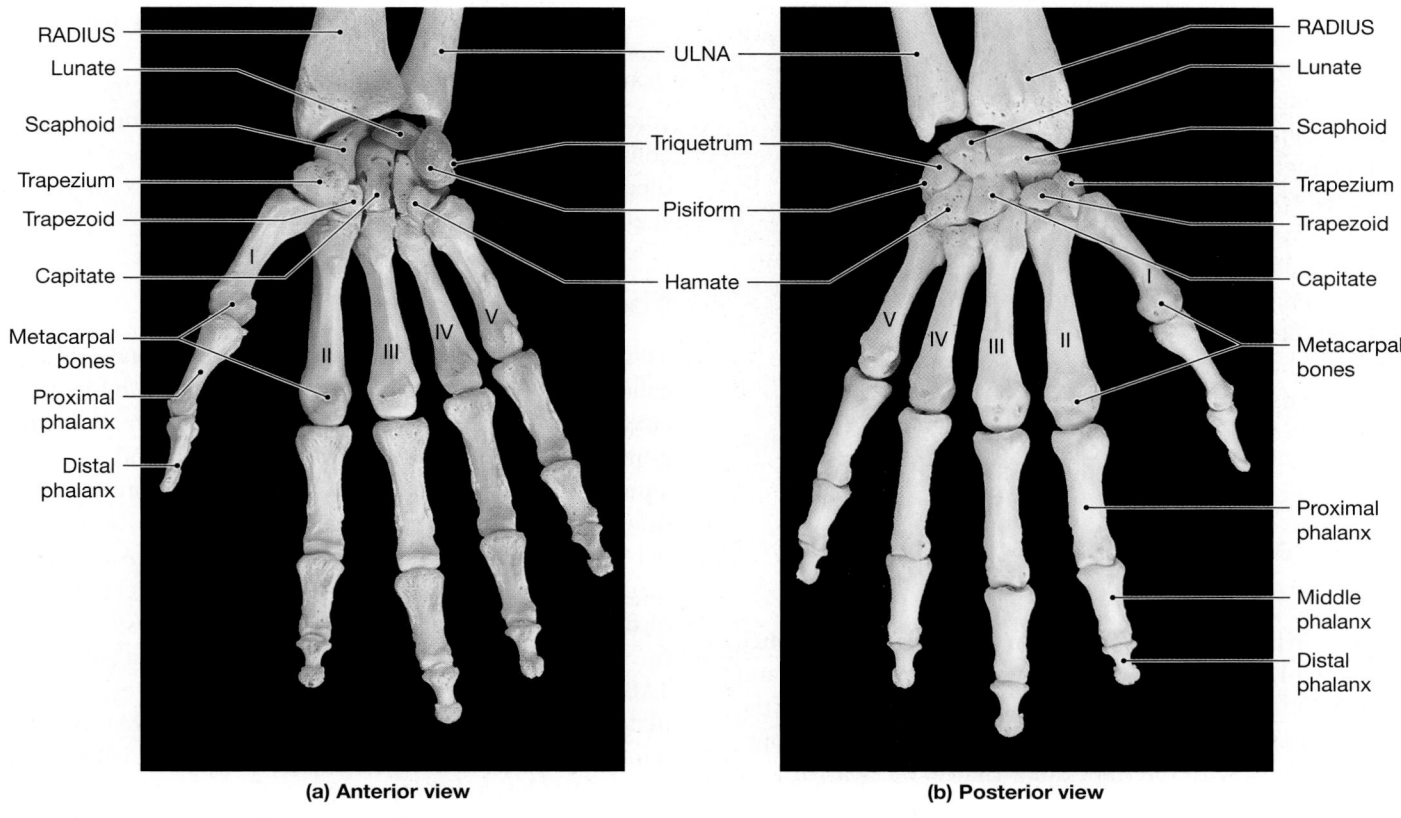

(a) Anterior view   **(b) Posterior view**

Figure 8–6   Bones of the Wrist and Hand. (a) Anterior and **(b)** posterior views of the right hand. ATLAS: Plate 38a,b

articulation with the head of the ulna. The distal end of the radius articulates with the bones of the wrist. The **styloid process** on the lateral surface of the radius helps stabilize this joint. If you are looking at an isolated radius or ulna, you can quickly identify whether it is left or right by finding the radial notch (ulna) or ulnar notch (radius) and remembering that the radius lies lateral to the ulna.

The Carpal Bones

The *carpus*, or wrist, contains eight **carpal bones**. These bones form two rows, one with four **proximal carpal bones** and the other with four **distal carpal bones**.

The proximal carpal bones are the scaphoid, lunate, triquetrum, and pisiform (**Figure 8–6a**).

- The **scaphoid** (*skaphe*, boat) is the proximal carpal bone on the lateral border of the wrist; it is the carpal bone closest to the styloid process of the radius.

- The comma-shaped **lunate** (*luna*, moon) lies medial to the scaphoid and, like the scaphoid, articulates with the radius.

- The **triquetrum** (*triquetrus*, three-cornered) is a small pyramid-shaped bone medial to the lunate. The triquetrum articulates with the articular disc that separates the ulnar head from the wrist.

- The small, pea-shaped **pisiform** (PIS-i-form; *pisum*, pea) sits anterior to the triquetrum. The distal carpal bones are the trapezium, trapezoid, capitate, and hamate (**Figure 8–6b**).

- The **trapezium** (*trapezion*, four-sided with no parallel sides) is the lateral bone of the distal row; its proximal surface articulates with the scaphoid.

- The wedge-shaped **trapezoid** lies medial to the trapezium. Like the trapezium, it has a proximal articulation with the scaphoid.

- The **capitate** (*caput*, head), the largest carpal bone, sits between the trapezoid and the hamate.

- The **hamate** (*hamatum*, hooked) is the medial distal carpal bone.

Tips&Tricks

It may help you to identify the eight carpal bones if you remember the sentence "Sam Likes To Push The Toy Car Hard." In lateral-to-medial order, the first four words stand for the proximal carpal bones (scaphoid, lunate, triquetrum, pisiform), and the last four stand for the distal carpal bones (trapezium, trapezoid, capitate, hamate).

CLINICAL NOTE

Carpal Tunnel Syndrome The carpal bones articulate with one another at joints that permit limited sliding and twisting. Ligaments interconnect the carpal bones and help stabilize the wrist joint. The tendons of muscles that flex the fingers pass across the anterior surface of the wrist, sandwiched between the intercarpal ligaments and a broad, superficial transverse ligament called the *flexor retinaculum*. Inflammation of the connective tissues between the flexor retinaculum and the carpal bones can compress the tendons and the adjacent *median nerve,* producing pain, weakness, and reduced wrist mobility. This condition is called *carpal tunnel syndrome.*

The Metacarpal Bones and Phalanges

Five **metacarpal** (met-uh-KAR-pul; *metacarpus,* hand) **bones** articulate with the distal carpal bones and support the hand (**Figure 8–6**). Roman numerals I–V are used to identify the metacarpal bones, beginning with the lateral metacarpal bone, which articulates with the trapezium. Hence, metacarpal I articulates with the proximal bone of the thumb.

Distally, the metacarpal bones articulate with the proximal finger bones. Each hand has 14 finger bones, or **phalanges** (fa-LAN-jēz; singular, *phalanx*). The first finger, known as the **pollex** (POL-eks), or thumb, has two phalanges (*proximal* and *distal*). Each of the other fingers has three phalanges (*proximal, middle,* and *distal*).

CHECKPOINT

4. List all the bones of the upper limb.

5. The rounded projections on either side of the elbow are parts of which bone?

6. Which bone of the forearm is lateral in the anatomical position?

7. Bill accidentally fractures his first distal phalanx with a hammer. Which finger is broken?

See the blue Answers tab at the end of the book.

8-3 The pelvic girdle attaches to the lower limbs and consists of two coxal bones

Because they must withstand the stresses involved in weight bearing and locomotion, the bones of the **pelvic girdle** are more massive than those of the pectoral girdle. For similar reasons, the bones of the lower limbs are more massive than those of the upper limbs.

In this section we first consider the pelvic girdle, which consists of the two hip bones. Then we examine the *pelvis,* a composite structure that includes the hip bones of the appendicular skeleton plus the sacrum and coccyx of the axial skeleton. ∞ pp. 237–239

The Pelvic Girdle

The pelvic girdle consists of the paired hip bones, which are called the **coxal bones**, or *innominate bones*. Each hip bone, or coxal bone, forms by the fusion of three bones: an **ilium** (IL-ē-um; plural, *ilia*), an **ischium** (IS-kē-um; plural, *ischia*), and a **pubis** (PŪ-bis) (**Figure 8–7**). The ilia have a sturdy articulation with the auricular surfaces of the sacrum, attaching the pelvic girdle to the axial skeleton. ∞ p. 232 Anteriorly, the medial surfaces of the hip bones are interconnected by a pad of fibrous cartilage at a joint called the *pubic symphysis*. On the lateral surface of each hip bone, the **acetabulum** (as-e-TAB-ū-lum; *acetabulum,* vinegar cup), a concave socket, articulates with the head of the femur (**Figure 8–7a**). A ridge of bone forms the lateral and superior margins of the acetabulum, which has a diameter of about 5 cm (2 in.). The anterior and inferior portion of the ridge is incomplete; the gap is called the **acetabular notch**. The smooth, cup-shaped articular surface of the acetabulum is the **lunate surface**.

The ilium, ischium, and pubis meet inside the acetabulum, as though it were a pie sliced into three pieces. Superior to the acetabulum, the ilium forms a broad, curved surface that provides an extensive area for the attachment of muscles, tendons, and ligaments (**Figure 8–7a**). Landmarks along the margin of the ilium include the *iliac spines,* which mark the attachment sites of important muscles and ligaments; the *gluteal lines,* which mark the attachment of large hip muscles; and the **greater sciatic** (sī-AT-ik) **notch**, through which a major nerve (the *sciatic nerve*) reaches the lower limb.

The ischium forms the posterior, inferior portion of the acetabulum. Posterior to the acetabulum, the prominent **ischial spine** projects superior to the *lesser sciatic notch,* through which blood vessels, nerves, and a small muscle pass. The **ischial tuberosity**, a roughened projection, is located at the posterior and lateral edge of the ischium. When you are seated, the ischial tuberosities bear your body's weight.

The narrow **ischial ramus** (branch) continues until it meets the **inferior ramus** of the pubis. The inferior pubic ramus extends between the ischial ramus and the *pubic tubercle,* a small, elevated area anterior and lateral to the pubic symphysis. There the inferior pubic ramus meets the **superior ramus** of the pubis, which originates near the acetabulum. The

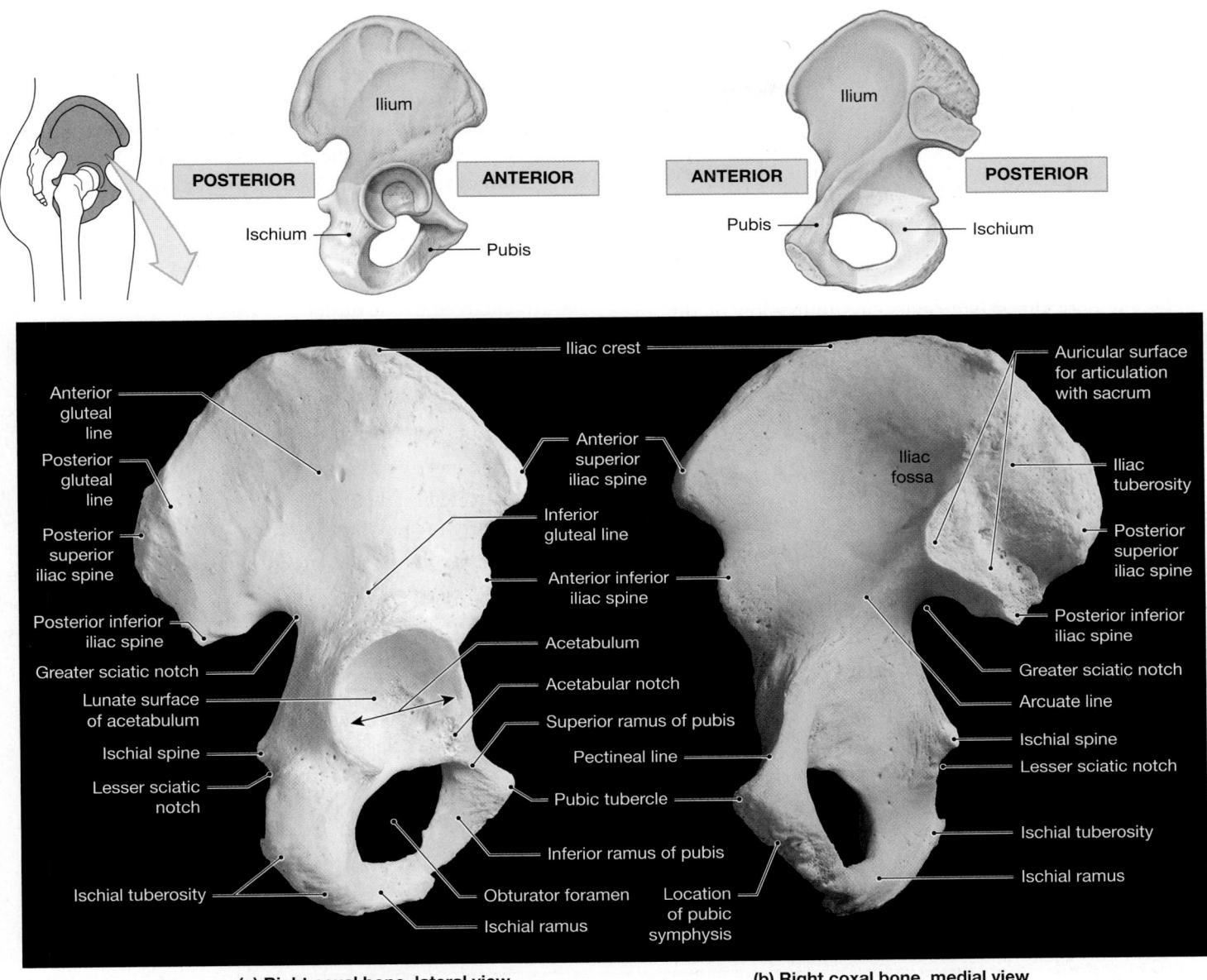

(a) Right coxal bone, lateral view

(b) Right coxal bone, medial view

Figure 8–7 **The Right Coxal Bone.** The left and right coxal bones constitute the pelvic girdle.

anterior, superior surface of the superior ramus bears the *pectineal line*, a ridge that ends at the pubic tubercle. The pubic and ischial rami encircle the **obturator** (OB-tū-rā-tor) **foramen**, a space that is closed by a sheet of collagen fibers whose inner and outer surfaces provide a firm base for the attachment of muscles of the hip.

The broadest part of the ilium extends between the **arcuate line**, which is continuous with the pectineal line, and the **iliac crest** (**Figure 8–7b**). These prominent ridges mark the attachments of ligaments and muscles. The area between the arcuate line and the iliac crest forms a shallow depression known as the **iliac fossa**. The concave surface of the iliac

fossa helps support the abdominal organs and provides additional area for muscle attachment.

In medial view, the anterior and medial surface of the pubis contains a roughened area that marks the site of articulation with the pubis of the opposite side (**Figure 8–7b**). At this articulation—the **pubic symphysis**—the two pubic bones are attached to a median fibrous cartilage pad. Posteriorly, the **auricular surface** of the ilium articulates with the auricular surface of the sacrum at the *sacroiliac joint.* ∞ p. 232 Ligaments arising at the **iliac tuberosity**, a roughened area superior to the auricular surface, stabilize this joint.

The Pelvis

Figure 8–8 shows anterior and posterior views of the **pelvis**, which consists of the two coxal bones, the sacrum, and the coccyx. An extensive network of ligaments connects the lateral borders of the sacrum with the iliac crest, the ischial tuberosity, the ischial spine, and the arcuate line. Other ligaments tie the ilia to the posterior lumbar vertebrae. These interconnections increase the stability of the pelvis.

The pelvis may be divided into the **true** (*lesser*) **pelvis** and the **false** (*greater*) **pelvis** (**Figure 8–9a,b**). The true pelvis encloses the *pelvic cavity*, a subdivision of the abdominopelvic cavity. ∞ p. 22 The superior limit of the true pelvis is a line that extends from either side of the base of the sacrum, along the arcuate line and pectineal line to the pubic symphysis. The bony edge of the true pelvis is called the **pelvic brim**, or *linea terminalis*, and the enclosed space is the **pelvic inlet**. The false pelvis consists of the expanded, bladelike portions of each ilium superior to the pelvic brim.

The **pelvic outlet** is the opening bounded by the coccyx, the ischial tuberosities, and the inferior border of the pubic symphysis (**Figure 8–9b,c**). The region bounded by the inferior edges of the pelvis is called the *perineum* (per-i-NĒ-um). Perineal muscles form the floor of the pelvic cavity and support the organs in the true pelvis.

The shape of the pelvis of a female is somewhat different from that of a male (**Figure 8–10**). Some of the differences are the result of variations in body size and muscle mass. For example, in females, the pelvis is generally smoother and lighter and has less-prominent markings. Females have other skeletal adaptations for childbearing, including:

- An enlarged pelvic outlet.
- A broader pubic angle (the inferior angle between the pubic bones), greater than 100°.
- Less curvature on the sacrum and coccyx, which, in males, arcs into the pelvic outlet.
- A wider, more circular pelvic inlet.
- A relatively broad pelvis that does not extend as far superiorly (a "low pelvis").
- Ilia that project farther laterally, but do not extend as far superior to the sacrum.

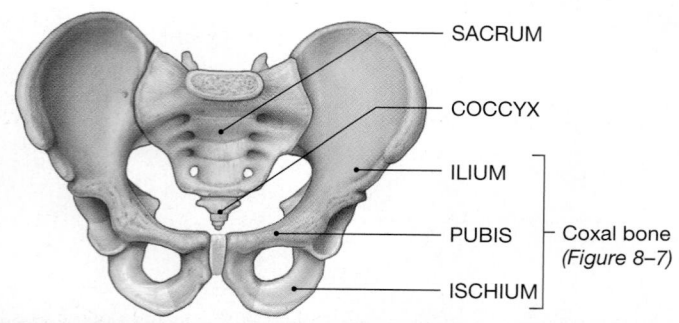

Figure 8–8 **The Pelvis.** The pelvis of an adult male. (See *Figure 7–22, p. 239,* for a detailed view of the sacrum and coccyx.)

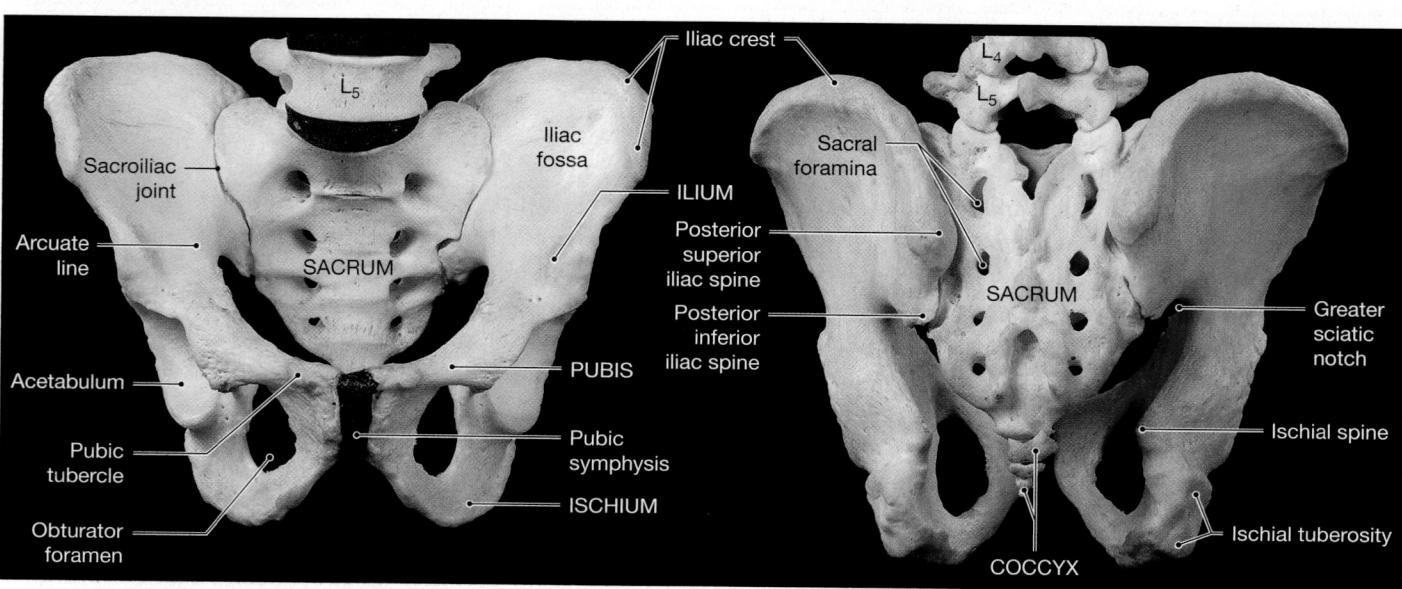

(a) Anterior view

(b) Posterior view

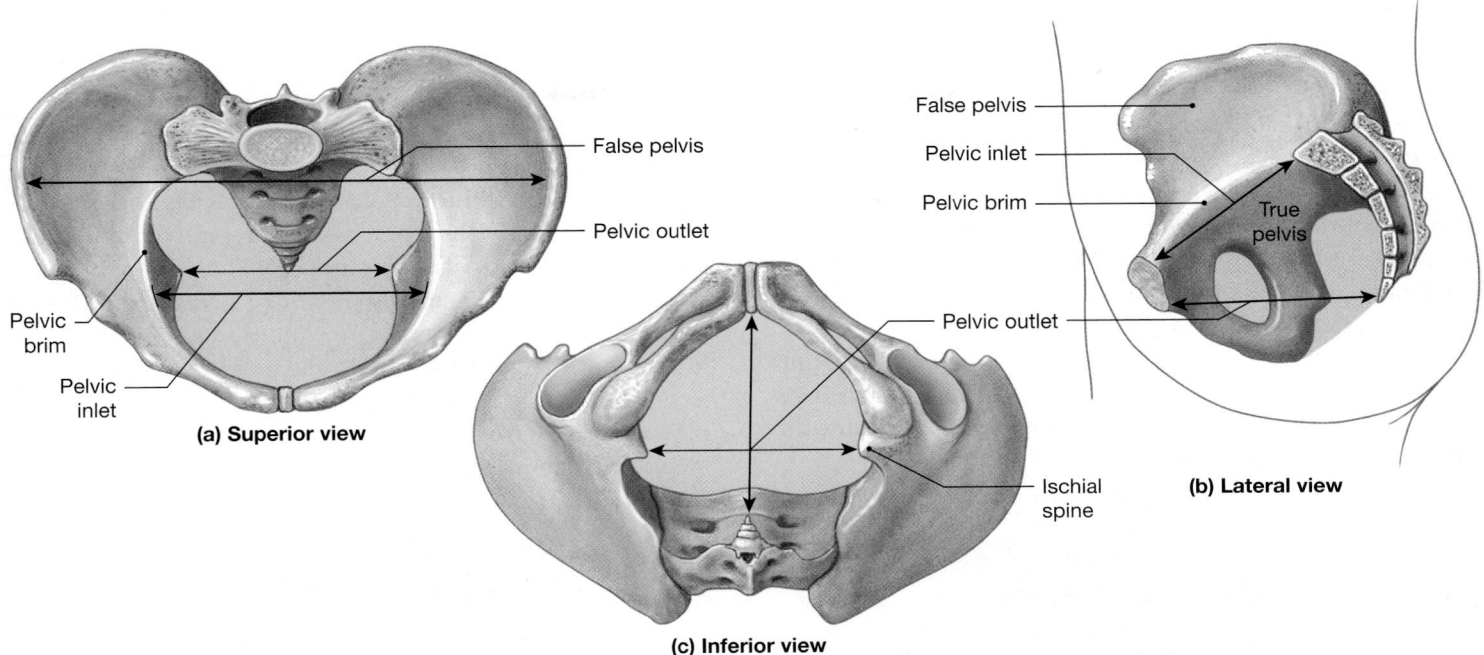

Figure 8–9   Divisions of the Pelvis. (a) The pelvic brim, pelvic inlet, and pelvic outlet. **(b)** The boundaries of the true (lesser) pelvis and the false (greater) pelvis. **(c)** The limits of the pelvic outlet.

These adaptations are related to the support of the weight of the developing fetus and uterus, and the passage of the newborn through the pelvic outlet during delivery. In addition, the hormone *relaxin*, produced during pregnancy, loosens the pubic symphysis and sacroiliac ligaments, allowing relative movement between the hip bones that can further increase the size of the pelvic inlet and outlet.

C H E C K P O I N T

8. Name the bones of the pelvic girdle.

9. Which three bones make up a coxal bone?

10. How is the pelvis of females adapted for childbearing?

11. When you are seated, which part of the pelvis bears your body's weight?

See the blue Answers tab at the end of the book.

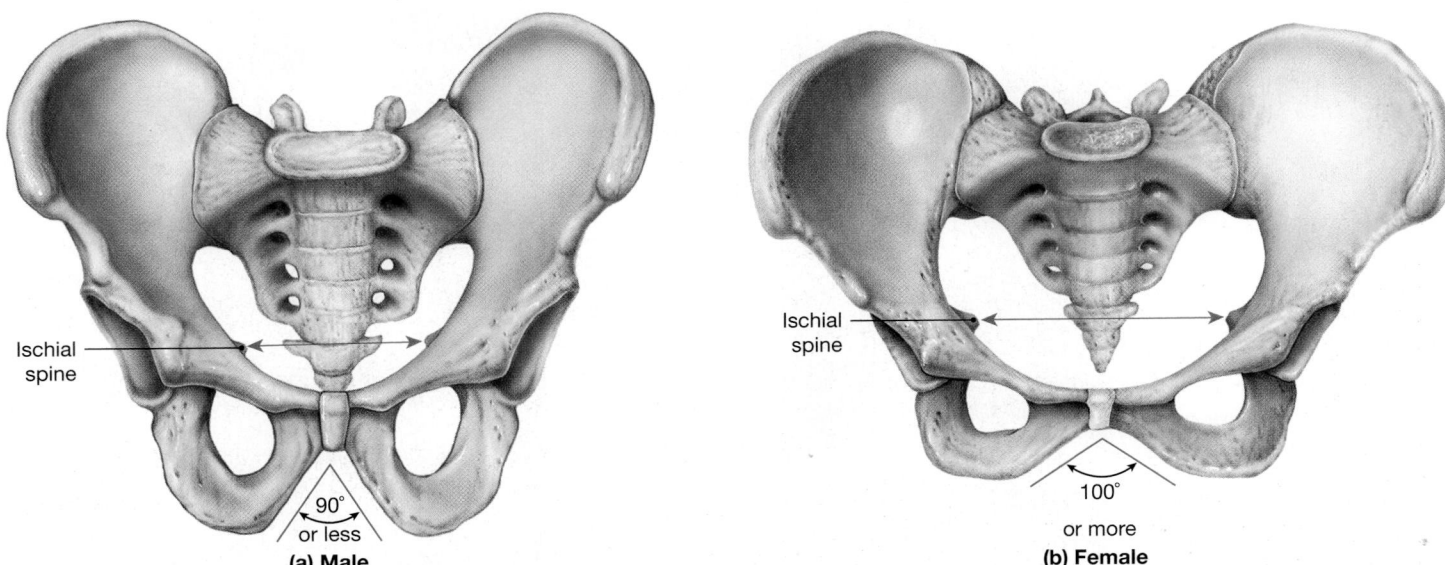

Figure 8–10   Anatomical Differences in the Pelvis of a Male and a Female. Representative pelvises of a male **(a)** and a female **(b)** in anterior view. Notice the much sharper pubic angle (indicated by the black arrows) and the smaller pelvic outlet (red arrows) in the pelvis of a male as compared with that of a female.

8-4 The lower limbs are adapted for locomotion and support

The skeleton of each lower limb consists of a *femur* (thigh), a *patella* (kneecap), a *tibia* and a *fibula* (leg), and the tarsal bones, metatarsal bones, and phalanges of the foot. Once again, anatomical terminology differs from common usage. In anatomical terms, *leg* refers only to the distal portion of the limb, not to the entire lower limb. Thus, we will use *thigh* and *leg*, rather than *upper leg* and *lower leg*.

The functional anatomy of the lower limbs differs from that of the upper limbs, primarily because the lower limbs transfer the body weight to the ground. We now examine the bones of the right lower limb.

The Femur

The **femur** is the longest and heaviest bone in the body (**Figure 8–11**). It articulates with the coxal bone at the hip joint and with the tibia of the leg at the knee joint. The rounded epiphysis, or **femoral head**, articulates with the pelvis at the acetabulum. A ligament attaches the acetabulum to the femur at the **fovea capitis**, a small pit in the center of the femoral head. The **neck** of the femur joins the **shaft** at an angle of about 125°. The **greater** and **lesser trochanters** are large, rough projections that originate at the junction of the neck and shaft. The greater trochanter projects laterally; the lesser trochanter projects posteriorly and medially. These trochanters develop where large tendons attach to the femur. On the anterior surface of the femur, the

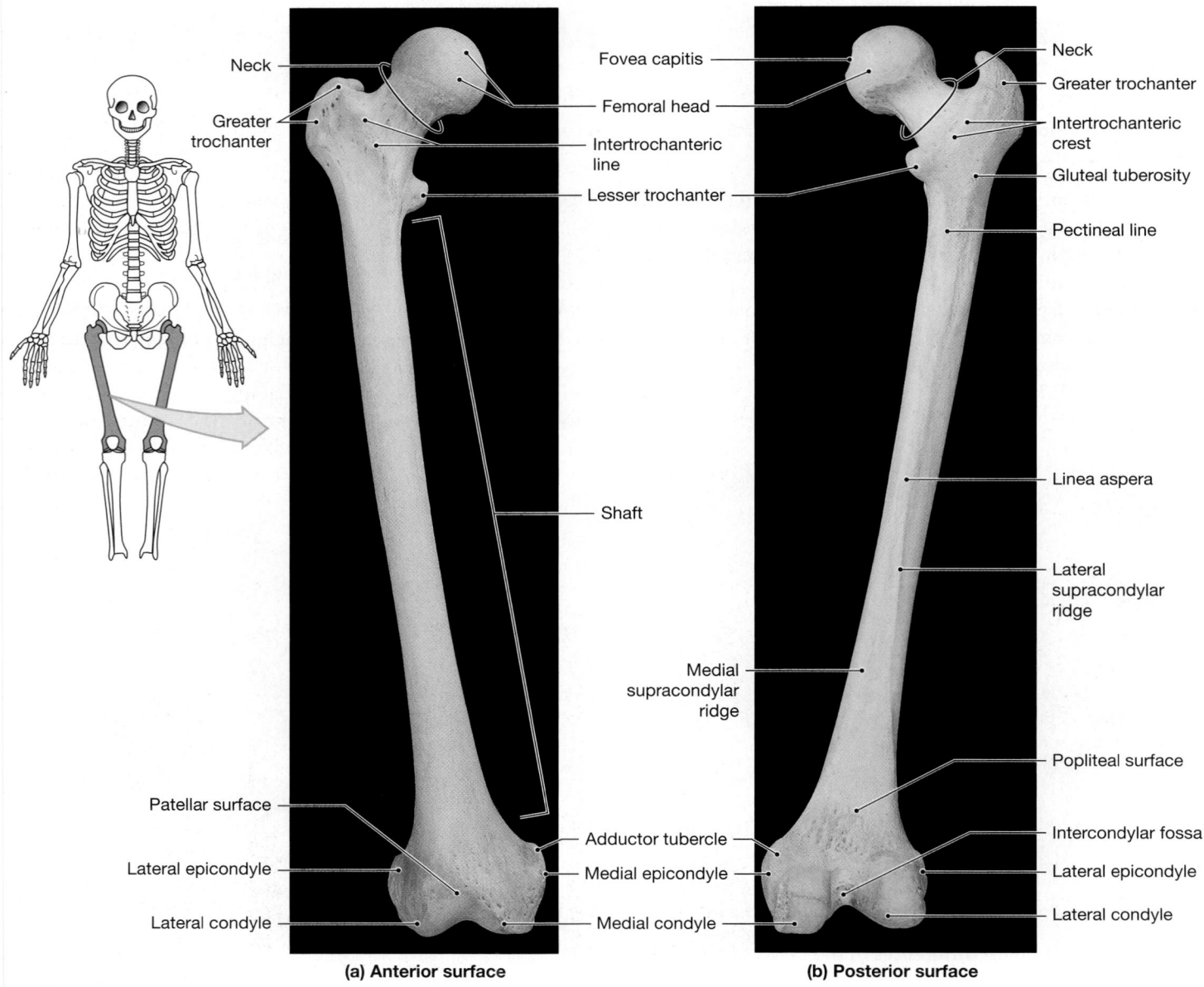

(a) Anterior surface

(b) Posterior surface

Figure 8–11 **The Femur.** Bone markings on the right femur as seen from the **(a)** anterior and **(b)** posterior surfaces. ATLAS: Plates 32; 75a–d; 77

raised **intertrochanteric** (in-ter-trō-kan-TER-ik) **line** marks the edge of the articular capsule. This line continues around to the posterior surface as the **intertrochanteric crest**.

The **linea aspera** (*aspera*, rough), a prominent elevation, runs along the center of the posterior surface of the femur, marking the attachment site of powerful hip muscles (**Figure 8–11b**). As it approaches the knee joint, the linea aspera divides into a pair of ridges that continue to the **medial** and **lateral epicondyles**. These smoothly rounded projections form superior to the **medial** and **lateral condyles**, which participate in the knee joint. The two condyles are separated by a deep **intercondylar fossa**.

The medial and lateral condyles extend across the inferior surface of the femur, but the intercondylar fossa does not reach the anterior surface (**Figure 8–11a**). The anterior and inferior surfaces of the two condyles are separated by the **patellar surface**, a smooth articular surface over which the patella glides.

The Patella

The **patella** is a large sesamoid bone that forms within the tendon of the *quadriceps femoris*, a group of muscles that extend (straighten) the knee. The patella has a rough, convex anterior surface and a broad **base** (**Figure 8–12a**). The roughened surface reflects the attachment of the quadriceps tendon (anterior and superior surfaces) and the *patellar ligament* (anterior and inferior surfaces). The patellar ligament connects the **apex** of the patella to the tibia. The posterior patellar surface (**Figure 8–12b**) presents two concave facets for articulation with the medial and lateral condyles of the femur. The patellae are cartilaginous at birth, but start to ossify after the individual begins walking, as thigh and leg movements become more powerful. Ossification usually begins at age 2 or 3 and ends at roughly the time of puberty.

Normally, the patella glides across the patellar surface of the femur. Its direction of movement is superior–inferior (up and down), not medial–lateral (side to side). *Runner's knee*, or *patellofemoral stress syndrome*, develops from improper tracking of the patella across the patellar surface. In this syndrome, the patella is forced outside its normal track, so that it shifts laterally; the movement is often associated with increased compression forces or with lateral muscles in the quadriceps group overpowering the medial muscles. Running on hard or slanted surfaces (such as the intertidal area of a beach or the shoulder of a road) and inadequate arch support are often responsible. The misalignment puts lateral pressure on the knee, resulting in swelling and tenderness after exercise.

The Tibia

The **tibia** (TIB-ē-uh), or shinbone, is the large medial bone of the leg (**Figure 8–13a**). The medial and lateral condyles of the femur articulate with the **medial** and **lateral tibial condyles** at the proximal end of the tibia. The **intercondylar eminence** is a ridge that separates the condyles (**Figure 8–13b**). The anterior surface of the tibia near the condyles bears a prominent, rough **tibial tuberosity**, which you can feel through the skin. This tuberosity marks the attachment of the patellar ligament.

The **anterior margin** is a ridge that begins at the tibial tuberosity and extends distally along the anterior tibial surface. You can also easily feel the anterior margin of the tibia through the skin. As it approaches the ankle joint, the tibia broadens, and the medial border ends in the **medial malleolus** (ma-LĒ-o-lus; *malleolus*, hammer), a large process familiar to you as the medial projection at the ankle. The inferior surface of the tibia articulates with the proximal bone of the ankle; the medial malleolus provides medial support for this joint.

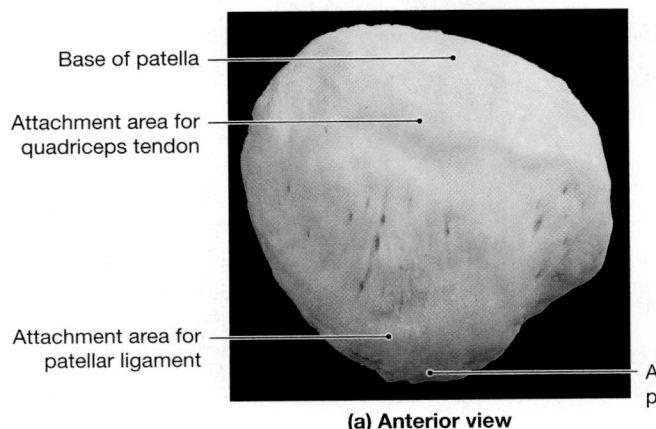

Base of patella

Attachment area for quadriceps tendon

Attachment area for patellar ligament

Apex of patella

(a) Anterior view

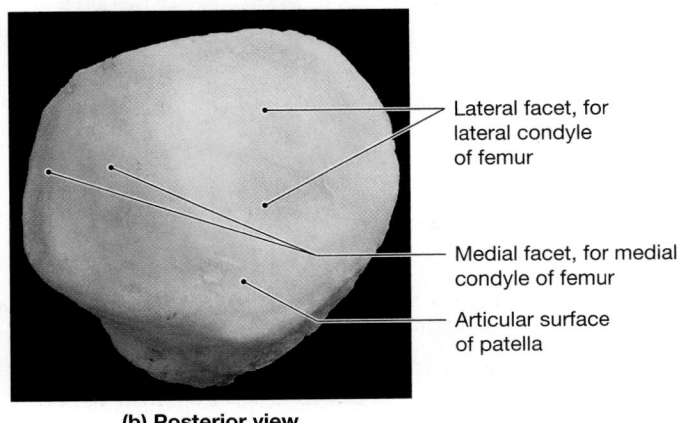

Lateral facet, for lateral condyle of femur

Medial facet, for medial condyle of femur

Articular surface of patella

(b) Posterior view

Figure 8–12   The Right Patella.

The Fibula

The slender **fibula** (FIB-ū-luh) parallels the lateral border of the tibia (**Figure 8–13a,b**). The head of the fibula articulates with the tibia. The articular facet is located on the anterior, inferior surface of the lateral tibial condyle. The medial border of the thin shaft is bound to the tibia by the **interosseous membrane**, which extends to the lateral margin of the tibia. This membrane helps stabilize the positions of the two bones and provides additional surface area for muscle attachment.

As its relatively small diameter suggests, the fibula does not help transfer weight to the ankle and foot. In fact, it does not even articulate with the femur. However, the fibula is important as a site for the attachment of muscles that move the foot and toes. In addition, the distal tip of the fibula extends lateral to the ankle joint. This fibular process, the **lateral malleolus**, provides lateral stability to the ankle. However, forceful movement of the foot outward and backward can dis-

locate the ankle, breaking both the lateral malleolus of the fibula and the medial malleolus of the tibia. This injury is called a *Pott fracture*. ∞ p. 206

Tips&Tricks

To remember how to distinguish the fibula from the tibia, consider the "fib" and "l" parts of the word *fibula*. To tell a small **l**ie is to **fib**. The **fib**ula is smaller than the tibia, and is also **l**ateral to it.

The Tarsal Bones

The ankle, or *tarsus*, consists of seven **tarsal bones** (**Figure 8–14**). The large **talus** transmits the weight of the body from the tibia toward the toes. The articulation between the talus and the tibia occurs across the superior and medial surfaces of the **trochlea**, a pulley-shaped articular process. The lateral

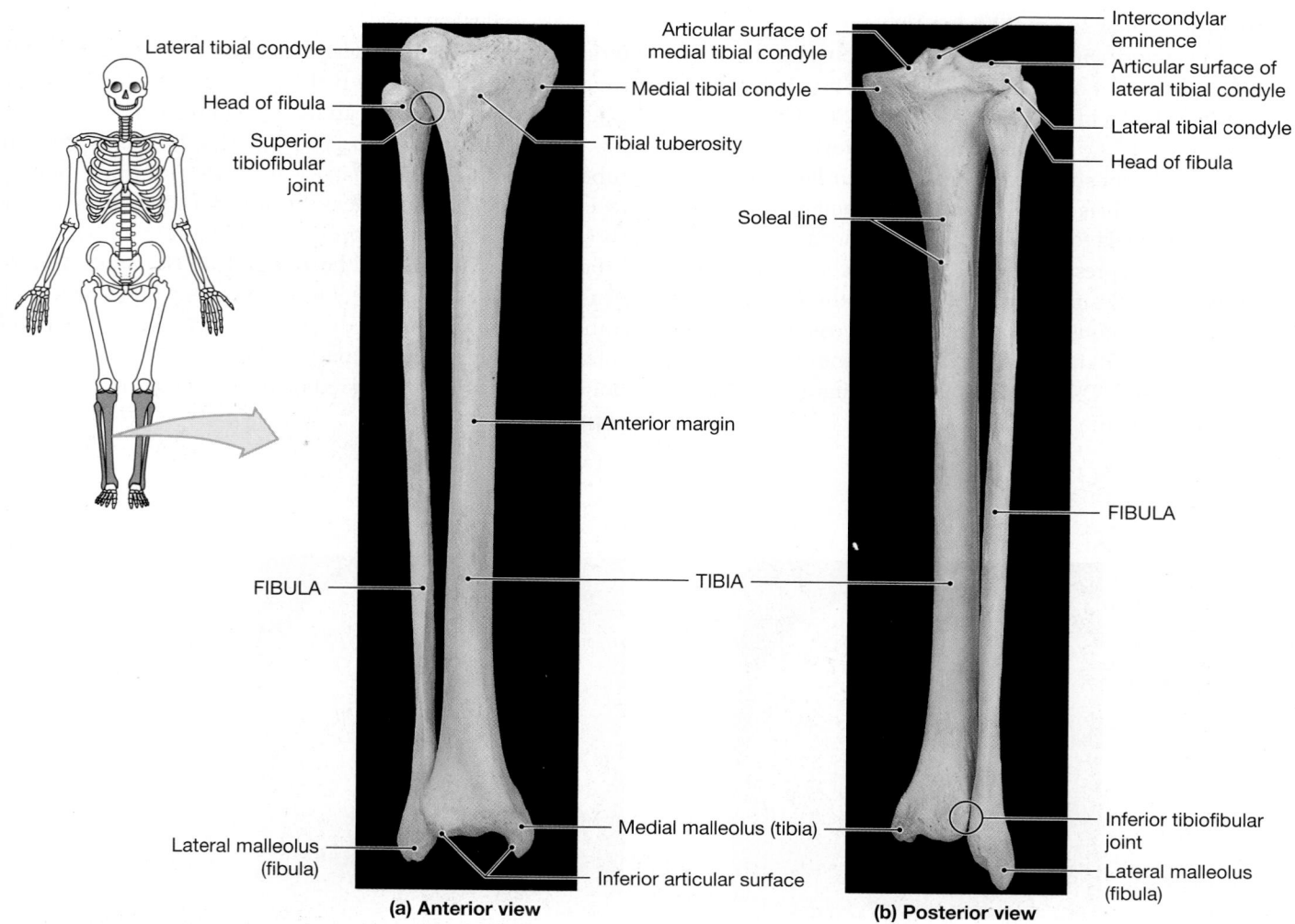

(a) Anterior view

(b) Posterior view

Figure 8–13 **The Tibia and Fibula. (a)** Anterior and **(b)** posterior views of the right tibia and fibula. ATLAS: Plates 32; 80a,b; 83a,b

surface of the trochlea articulates with the lateral malleolus of the fibula.

Tips & Tricks

To remember where the talus is located, think that the **t**alus is on **t**op of the foot and articulates with the **t**ibia.

The **calcaneus** (kal-KĀ-nē-us), or heel bone, is the largest of the tarsal bones. When you stand normally, most of your weight is transmitted from the tibia, to the talus, to the calcaneus, and then to the ground. The posterior portion of the calcaneus is a rough, knob-shaped projection. This is the attachment site for the *calcaneal tendon* (*Achilles tendon* or *calcanean tendon*), which arises at the calf muscles. If you are standing, these strong muscles can lift the heel off the ground so that you stand on tiptoes. The superior and anterior surfaces of the calcaneus bear smooth facets for articulation with other tarsal bones.

The **cuboid** articulates with the anterior surface of the calcaneus. The **navicular** is anterior to the talus, on the medial side of the ankle. It articulates with the talus and with the three *cuneiform* (kū-NĒ-i-form) *bones*. These are wedge-shaped bones arranged in a row, with articulations between them. They are named according to their position: **medial cuneiform**, **intermediate cuneiform**, and **lateral cuneiform**. Proximally, the cuneiform bones articulate with the anterior surface of the navicular. The lateral cuneiform also articulates with the medial surface of the cuboid. The distal surfaces of the cuboid and the cuneiform bones articulate with the metatarsal bones of the foot.

Tips & Tricks

To remember the names of the tarsal bones in the order presented in this text, try the memory aid "Tom Can Control Not Much In Life."

The Metatarsal Bones and Phalanges

The **metatarsal bones** are five long bones that form the distal portion of the foot, or *metatarsus* (**Figure 8–14**). The metatarsal bones are identified by Roman numerals I–V, proceeding from medial to lateral across the sole. Proximally, metatarsal bones I–III articulate with the three cuneiform bones, and metatarsal

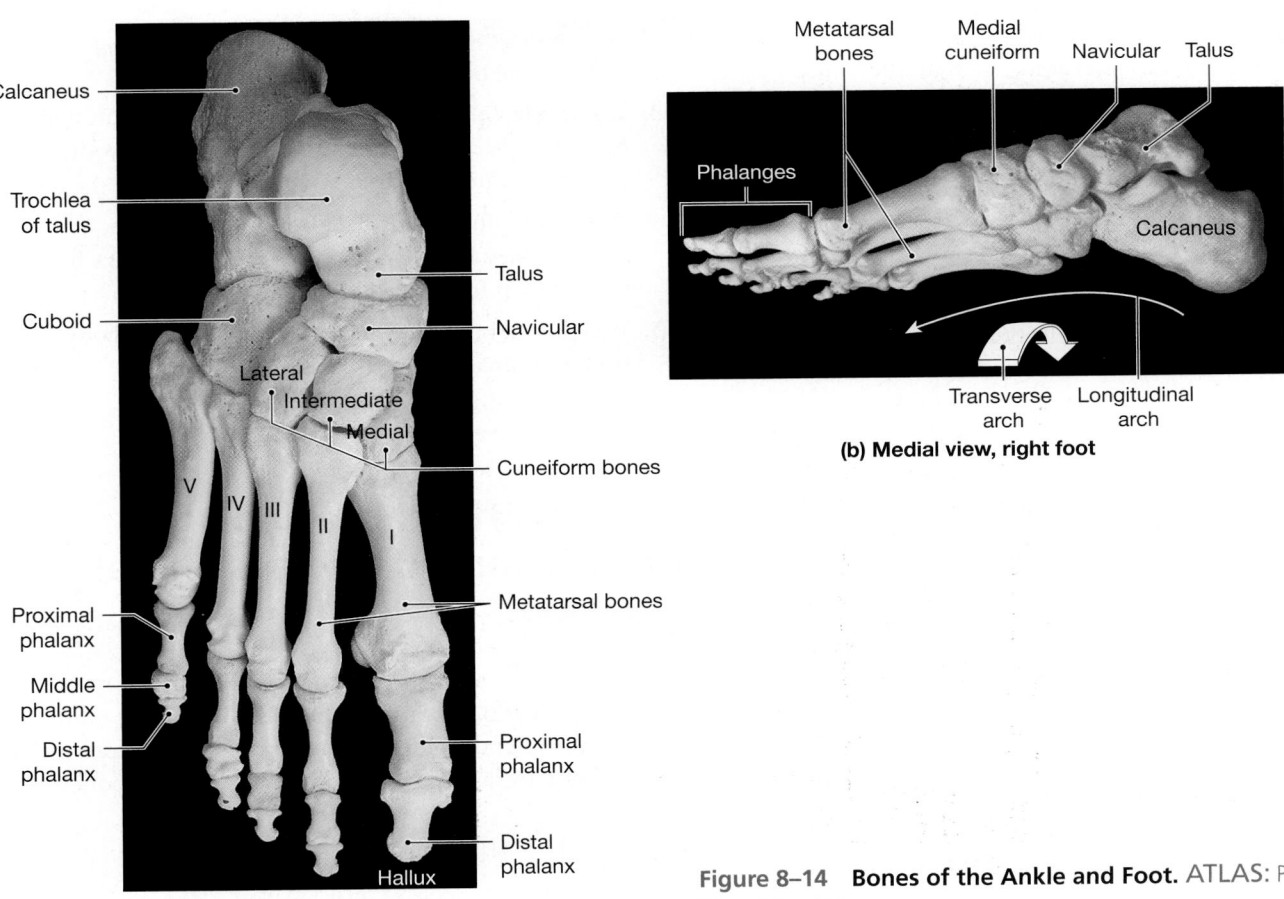

(a) Superior view, right foot

(b) Medial view, right foot

Figure 8–14   Bones of the Ankle and Foot. ATLAS: Plates 32; 85a; 86a,c; 87a–c; 88

bones IV and V articulate with the cuboid. Distally, each metatarsal bone articulates with a different proximal phalanx. The **phalanges**, or toe bones (**Figure 8–14**), have the same anatomical organization as the fingers. The toes contain 14 phalanges. The **hallux**, or great toe, has two phalanges (*proximal and distal*), and the other four toes have three phalanges apiece (*proximal*, *middle*, and *distal*).

Running, while beneficial to overall health, places the foot bones under more stress than does walking. *Stress fractures* are hairline fractures that develop in bones subjected to repeated shocks or impacts. Stress fractures of the foot usually involve one of the metatarsal bones. These fractures are caused either by improper placement of the foot while running or by poor arch support. In a fitness regime that includes street running, it is essential to provide proper support for the bones of the foot. An entire running-shoe market has arisen around the amateur runner's need for good arch support.

Arches of the Foot

Weight transfer occurs along the **longitudinal arch** of the foot (**Figure 8–14b**). Ligaments and tendons maintain this arch by tying the calcaneus to the distal portions of the metatarsal bones. However, the lateral, or *calcaneal*, portion of the longitudinal arch has much less curvature than the medial, *talar* portion, in part because the talar portion has considerably more elasticity. As a result, the medial plantar surface of the foot remains elevated, so that the muscles, nerves, and blood vessels that supply the inferior surface are not squeezed between the metatarsal bones and the ground. In the condition known as *flatfeet*, normal arches are lost ("fall") or never form.

The elasticity of the talar portion of the longitudinal arch absorbs the shocks from sudden changes in weight loading. For example, the stresses that running or ballet dancing places on the toes are cushioned by the elasticity of this portion of the arch. The degree of curvature changes from the medial to the lateral borders of the foot, so a **transverse arch** also exists.

When you stand normally, your body weight is distributed evenly between the calcaneus and the distal ends of the metatarsal bones. The amount of weight transferred forward depends on the position of the foot and the placement of one's body weight. During *flexion* at the ankle, a movement also called *dorsiflexion*, all your body weight rests on the calcaneus—as when you "dig in your heels." During *extension* at the ankle, also known as *plantar flexion*, the talus and calcaneus transfer your weight to the metatarsal bones and phalanges through the more anterior tarsal bones; this occurs when you stand on tiptoe.

The A&P Top 100

#33 The pectoral girdle is highly mobile and stabilized primarily by muscles; the pelvic girdle is more massive, stronger, and far less mobile.

CHECKPOINT

12. Identify the bones of the lower limb.

13. The fibula neither participates in the knee joint nor bears weight. When it is fractured, however, walking becomes difficult. Why?

14. While jumping off the back steps at his house, 10-year-old Joey lands on his right heel and breaks his foot. Which foot bone is most likely broken?

15. Which foot bone transmits the weight of the body from the tibia toward the toes?

See the blue Answers tab at the of the book.

8-5 Sex differences and age account for individual skeletal variation

A comprehensive study of a human skeleton can reveal important information about the individual. We can estimate a person's muscular development and muscle mass from the appearance of various ridges and from the general bone mass.

Details such as the condition of the teeth or the presence of healed fractures give an indication of the individual's medical history. Two important details, sex and age, can be determined or closely estimated on the basis of measurements indicated in Tables 8–1 and 8–2. In some cases, the skeleton may provide clues about the individual's nutritional state, handedness, and even occupation. ATLAS: Embryology Summary 8: The Development of the Appendicular Skeleton

Table 8–1 identifies characteristic differences between the skeletons of males and females, but not every skeleton shows every feature in classic detail. Many differences, including markings on the skull, cranial capacity, and general skeletal features, reflect differences in average body size, muscle mass, and muscular strength. The general changes in the skeletal system that take place with age are summarized in Table 8–2. Note that these changes begin at age 3 months and continue throughout life. The epiphyseal cartilages, for example, begin to fuse at about age 3, and degenerative changes in

the normal skeletal system, such as a reduction in mineral content in the bony matrix, typically do not begin until age 30–45. The timing of epiphyseal closure is a key factor determining adult body size. Young people whose long bones are still growing should avoid very heavy weight training, because they risk crushing the epiphyseal cartilages and thus shortening their stature.

CHECKPOINT

16. Compare and contrast the bones of males and females with respect to weight and surface markings.

17. An anthropologist discovered several bones in a deep grave. After close visual inspection and careful measurements, what sort of information could the bones reveal?

See the blue Answers tab at the end of the book.

TABLE 8–1 Sex Differences in the Human Skeleton

Region and Feature	Male (compared with female)	Female (compared with male)
SKULL		
General appearance	Heavier, rougher	Lighter, smoother
Forehead	More sloping	More vertical
Sinuses	Larger	Smaller
Cranium	About 10% larger (average)	About 10% smaller
Mandible	Larger, more robust	Smaller, lighter
Teeth	Larger	Smaller
PELVIS		
General appearance	Narrower, more robust, rougher	Broader, lighter, smoother
Pelvic inlet	Heart shaped	Oval to round shaped
Iliac fossa	Deeper	Shallower
Ilium	More vertical; extends farther superior to sacroiliac joint	Less vertical; less extension superior to sacral articulation
Angle inferior to pubic symphysis	Under 90°	100° or more (Figure 8–10)
Acetabulum	Directed laterally	Faces slightly anteriorly as well as laterally
Obturator foramen	Oval	Triangular
Ischial spine	Points medially	Points posteriorly
Sacrum	Long, narrow triangle with pronounced sacral curvature	Broad, short triangle with less curvature
Coccyx	Points anteriorly	Points inferiorly
OTHER SKELETAL ELEMENTS		
Bone weight	Heavier	Lighter
Bone markings	More prominent	Less prominent

TABLE 8–2 Age-Related Changes in the Skeleton

Region and Feature	Event(s)	Age (Years)
GENERAL SKELETON		
Bony matrix	Reduction in mineral content; increased risk of osteoporosis	Begins at age 30–45; values differ for males versus females between ages 45 and 65; similar reductions occur in both sexes after age 65
Markings	Reduction in size, roughness	Gradual reduction with increasing age and decreasing muscular strength and mass
SKULL		
Fontanelles	Closure	Completed by age 2
Frontal suture	Fusion	2–8
Occipital bone	Fusion of ossification centers	1–4
Styloid process	Fusion with temporal bone	12–16
Hyoid bone	Complete ossification and fusion	25–30
Teeth	Loss of "baby teeth"; appearance of secondary dentition; eruption of permanent molars	Detailed in Chapter 24 (digestive system)
Mandible	Loss of teeth; reduction in bone mass; change in angle at mandibular notch	Accelerates in later years (60+)
VERTEBRAE		
Curvature	Development of major curves	3 months–10 years
Intervertebral discs	Reduction in size, percentage contribution to height	Accelerates in later years (60+)
LONG BONES		
Epiphyseal cartilages	Fusion	Begins about age 3; ranges vary, but general analysis permits determination of approximate age
PECTORAL AND PELVIC GIRDLES		
Epiphyses	Fusion	Relatively narrow ranges of ages (e.g., 14–16, 16–18, 22–25) increase accuracy of age estimates

Related Clinical Terms

carpal tunnel syndrome: An inflammation of the tissues at the anterior wrist, causing compression of adjacent tendons and the median nerve. Symptoms are pain, numbness, weakness, and a reduction in wrist mobility.

congenital talipes equinovarus (*clubfoot*)**:** A congenital deformity affecting one or both feet. It develops secondary to abnormalities in muscular development.

flatfeet: The loss or absence of a longitudinal arch.

Chapter Review
Study Outline

An Introduction to the Appendicular Skeleton p. 247

1. The **appendicular skeleton** includes the bones of the upper and lower limbs and the pectoral and pelvic girdles, which connect the limbs to the trunk. (*Figure 8–1*)

8-1 The pectoral girdle attaches to the upper limbs and consists of the clavicles and scapulae p. 247

2. Each upper limb articulates with the trunk via the **pectoral girdle**, or *shoulder girdle*, which consists of two **scapulae** and two **clavicles**.

3. On each side, a clavicle and scapula position the shoulder joint, help move the upper limb, and provide a base for muscle attachment. (*Figures 8–2, 8–3*)

4. Both the **coracoid process** and the **acromion** of the scapula are attached to ligaments and tendons associated with the shoulder joint. (*Figure 8–3*)

8-2 The upper limbs are adapted for freedom of movement p. 250

5. The scapula articulates with the **humerus** at the shoulder (*glenohumeral*) joint. The **greater** and **lesser tubercles** of

the humerus are important sites of muscle attachment. (*Figure 8–4*)

6. The humerus articulates with the **radius** and **ulna**, the bones of the forearm, at the elbow joint. (*Figure 8–5*)

7. The **carpal bones** of the wrist, or *carpus*, form two rows. The distal row articulates with the five **metacarpal bones**. Four of the fingers contain three **phalanges**; the **pollex** (thumb) has only two phalanges. (*Figure 8–6*)

8-3 The pelvic girdle attaches to the lower limbs and consists of two coxal bones p. 254

8. The bones of the **pelvic girdle** are more massive than those of the pectoral girdle.

9. The pelvic girdle consists of two **coxal bones**. Each coxal bone forms through the fusion of an **ilium**, an **ischium**, and a **pubis**. (*Figure 8–7*)

10. The ilium is the largest hip bone. Inside the **acetabulum**, the ilium is fused to the ischium (posteriorly) and the pubis (anteriorly). The *pubic symphysis* limits movement between the pubic bones of the left and right hip bones. (*Figures 8–7, 8–8*)

11. The **pelvis** consists of the hip bones, the sacrum, and the coccyx. It is subdivided into the **false** (*greater*) **pelvis** and the **true** (*lesser*) **pelvis**. (*Figures 8–8 to 8–10*)

8-4 The lower limbs are adapted for locomotion and support p. 258

12. The **femur** is the longest and heaviest bone in the body. It articulates with the **tibia** at the knee joint. (*Figures 8–11, 8–13*)

13. The **patella** is a large sesamoid bone. (*Figure 8–12*)

14. The **fibula** parallels the tibia laterally. (*Figure 8–13*)

15. The *tarsus*, or ankle, has seven **tarsal bones**. (*Figure 8–14*)

16. The basic organizational pattern of the **metatarsal bones** and **phalanges** of the foot resembles that of the hand. All the toes have three phalanges, except for the **hallux**, which has two. (*Figure 8–14*)

17. When a person stands normally, most of the body weight is transferred to the **calcaneus**, and the rest is passed on to the five metatarsal bones. Weight transfer occurs along the **longitudinal arch**; there is also a **transverse arch**. (*Figure 8–14*)

8-5 Sex differences and age account for individual skeletal variation p. 262

18. Studying a human skeleton can reveal important information, such as the person's weight, sex, body size, muscle mass, and age. (*Tables 8–1, 8–2*)

19. Age-related changes take place in the skeletal system. These changes begin at about age 1 and continue throughout life. (*Table 8–2*)

Review Questions

See the blue Answers tab at the end of the book.

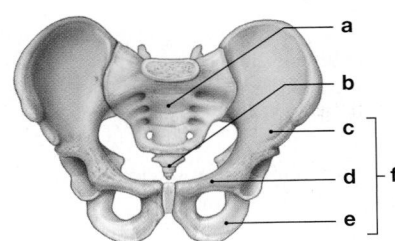

Access more review material online at **myA&P™** (**www.myaandp.com**). There, you'll find chapter guides, chapter quizzes, practice tests, animations, flashcards, a glossary with pronunciations, *Interactive Physiology*® (IP) exercises and quizzes, and more to help you succeed in the course.

LEVEL 1 Reviewing Facts and Terms

1. In the following photographs of the scapula, identify the three views (a–c) and the indicated bone markings (d–g).

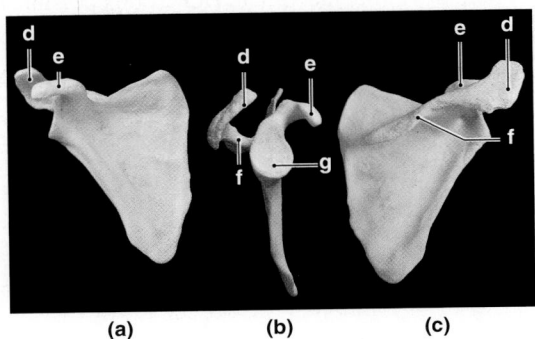

(a) _____ (b) _____
(c) _____ (d) _____
(e) _____ (f) _____
(g) _____

2. In the following drawing of the pelvis, label the structures indicated.

(a) _____ (b) _____
(c) _____ (d) _____
(e) _____ (f) _____

3. Which of the following is primarily responsible for stabilizing, positioning, and bracing the pectoral girdle?
   (a) tendons   (b) ligaments
   (c) the joint shape (d) muscles
   (e) the shape of the bones within the joint

4. In anatomical position, the ulna lies
   (a) medial to the radius.
   (b) lateral to the radius.
   (c) inferior to the radius.
   (d) superior to the radius.

5. The point of the elbow is actually the _____ of the ulna.
   (a) styloid process   (b) olecranon
   (c) coronoid process (d) trochlear notch

6. The bones of the hand articulate distally with the
   - (a) carpal bones.
   - (b) ulna and radius.
   - (c) metacarpal bones.
   - (d) phalanges.
7. The epiphysis of the femur articulates with the pelvis at the
   - (a) pubic symphysis.
   - (b) acetabulum.
   - (c) sciatic notch.
   - (d) obturator foramen.
8. What is the name of the flexible sheet that interconnects the radius and ulna (and the tibia and fibula)?
9. Name the components of each coxal bone.
10. Which seven bones make up the ankle (tarsus)?

LEVEL 2   Reviewing Concepts

11. The presence of tubercles on bones indicates the positions of
    - (a) tendons and ligaments.
    - (b) muscle attachment.
    - (c) ridges and flanges.
    - (d) a and b.
12. At the glenoid cavity, the scapula articulates with the proximal end of the
    - (a) humerus.
    - (b) radius.
    - (c) ulna.
    - (d) femur.
13. All of the following structural characteristics of the pelvic girdle adapt it to the role of bearing the weight of the body, *except*
    - (a) heavy bones.
    - (b) strong and stable articulating surfaces.
    - (c) the arrangement of bursae around the joints.
    - (d) limited range of movement at some of the joints within the pelvic girdle.
    - (e) the arrangement of ligaments surrounding the joints.
14. The large foramen between the pubic and ischial rami is the
    - (a) foramen magnum.
    - (b) suborbital foramen.
    - (c) acetabulum.
    - (d) obturator foramen.
15. Which of the following is an adaption for childbearing?
    - (a) inferior angle of 100° or more between the pubic bones
    - (b) a relatively broad, low pelvis
    - (c) less curvature of the sacrum and coccyx
    - (d) All of these are correct.
16. The fibula
    - (a) forms an important part of the knee joint.
    - (b) articulates with the femur.
    - (c) helps to bear the weight of the body.
    - (d) provides lateral stability to the ankle.
    - (e) does (a) and (b).

17. The tarsal bone that accepts weight and distributes it to the heel or toes is the
    - (a) cuneiform.
    - (b) calcaneus.
    - (c) talus.
    - (d) navicular.
18. What is the difference in skeletal structure between the pelvic girdle and the pelvis?
19. Jack injures himself playing hockey, and the physician who examines him informs him that he has dislocated his pollex. What part of Jack's body did he injure?
    - (a) his arm
    - (b) his leg
    - (c) his hip
    - (d) his thumb
    - (e) his shoulder
20. Why would a self-defense instructor advise a student to strike an assailant's clavicle?
21. The pelvis
    - (a) protects the upper abdominal organs.
    - (b) contains bones from both the axial and appendicular skeletons.
    - (c) is composed of the coxal bone, sacrum, and coccyx.
    - (d) does all of these.
    - (e) does (b) and (c).
22. Why is the tibia, but not the fibula, involved in the transfer of weight to the ankle and foot?
23. In determining the age of a skeleton, all of the following pieces of information would be helpful *except*
    - (a) the number of cranial sutures.
    - (b) the size and roughness of the markings of the bones.
    - (c) the presence or absence of fontanelles.
    - (d) the presence or absence of epiphyseal cartilages.
    - (e) the types of minerals deposited in the bones.

LEVEL 3   Critical Thinking and Clinical Applications

24. Why would a person suffering from osteoporosis be more likely to suffer a broken hip than a broken shoulder?
25. While Fred, a fireman, is fighting a fire in a building, part of the ceiling collapses, and a beam strikes him on his left shoulder. He is rescued, but has a great deal of pain in his shoulder. He cannot move his arm properly, especially in the anterior direction. His clavicle is not broken, and his humerus is intact. What is the probable nature of Fred's injury?
26. Archaeologists find the pelvis of a primitive human and are able to identify the sex, the relative age, and some physical characteristics of the individual. How is this possible from only the pelvis?

Articulations

9

Did you know...?
There are about 230 joints in the human body; some people have joints that are unusually mobile.

Learning Outcomes

After completing this chapter, you should be able to do the following:

9-1 Contrast the major categories of joints, and explain the relationship between structure and function for each category.

9-2 Describe the basic structure of a synovial joint, and describe common synovial joint accessory structures and their functions.

9-3 Describe how the anatomical and functional properties of synovial joints permit dynamic movements of the skeleton.

9-4 Describe the articulations between the vertebrae of the vertebral column.

9-5 Describe the structure and function of the shoulder joint and the elbow joint.

9-6 Describe the structure and function of the hip joint and the knee joint.

9-7 Describe the effects of aging on articulations, and discuss the most common age-related clinical problems for articulations.

9-8 Explain the functional relationships between the skeletal system and other body systems.

Clinical Notes

Bursitis and Bunions p. 271
Knee Injuries p. 286

An Introduction to Articulations

This chapter considers the ways bones interact wherever they interconnect. In the last two chapters, you have become familiar with the individual bones of the skeleton. These bones provide strength, support, and protection for softer tissues of the body. However, your daily life demands more of the skeleton—it must also facilitate and adapt to body movements. Think of your activities in a typical day: You breathe, talk, walk, sit, stand, and change positions innumerable times. In each case, your skeleton is directly involved. Because the bones of the skeleton are relatively inflexible, movements can occur only at **articulations**, or joints, where two bones interconnect. The characteristic structure of a joint determines the type and amount of movement that may occur. Each joint reflects a compromise between the need for strength and the need for mobility.

This chapter compares the relationships between articular form and function. We will consider several examples that range from the relatively immobile but very strong (the intervertebral articulations) to the highly mobile but relatively weak (the shoulder).

9-1 Joints are categorized according to their range of motion or anatomical organization

Two classification methods are used to categorize joints. The first—the one we will use in this chapter—is a functional scheme because it is based on the amount of movement possible, a property known as the *range of motion* (ROM). Each functional group is further subdivided primarily on the basis of the anatomical structure of the joint (Table 9–1):

1. An *immovable joint* is a **synarthrosis** (sin-ar-THRŌ-sis; *syn*, together + *arthros*, joint). A synarthrosis can be *fibrous* or *cartilaginous*, depending on the nature of the connection. Over time, the two bones may fuse.

2. A *slightly movable joint* is an **amphiarthrosis** (am-fē-ar-THRŌ-sis; *amphi*, on both sides). An amphiarthrosis is either *fibrous* or *cartilaginous*, depending on the nature of the connection between the opposing bones.

3. A *freely movable joint* is a **diarthrosis** (dī-ar-THRŌ-sis; *dia*, through), or *synovial joint*. Diarthroses are subdivided according to the nature of the movement permitted.

TABLE 9–1 Functional and Structural Classifications of Articulations

Functional Category	Structural Category and Type	Description	Example(s)
Synarthrosis (no movement)	*Fibrous*		
	Suture	Fibrous connections plus interlocking projections	Between the bones of the skull
	Gomphosis	Fibrous connections plus insertion in alveolar process	Between the teeth and jaws
	Cartilaginous		
	Synchondrosis	Interposition of cartilage plate	Epiphyseal cartilages
	Bony fusion		
	Synostosis	Conversion of other articular form to a solid mass of bone	Portions of the skull, epiphyseal lines
Amphiarthrosis (little movement)	*Fibrous*		
	Syndesmosis	Ligamentous connection	Between the tibia and fibula
	Cartilaginous		
	Symphysis	Connection by a fibrous cartilage pad	Between right and left pubic bones of pelvis; between adjacent vertebral bodies along vertebral column
Diarthrosis (free movement)	*Synovial*	Complex joint bounded by joint capsule and containing synovial fluid	Numerous; subdivided by range of movement (*Figure 9–6*)
Planes of Movement	**Monaxial**	Permits movement in one plane	Elbow, ankle
	Biaxial	Permits movement in two planes	Ribs, wrist
	Triaxial	Permits movement in all three planes	Shoulder, hip

The second classification scheme relies solely on the anatomical organization of the joint, without regard to the degree of movement permitted. In this framework, joints are classified as *bony*, *fibrous*, *cartilaginous*, or *synovial* (Table 9–2).

The two classification schemes are loosely correlated. Many anatomical patterns are seen among immovable or slightly movable joints, but there is only one type of freely movable joint—synovial joints—and all synovial joints are diarthroses. We will use the functional classification rather than the anatomical one because our primary interest is how joints work.

TABLE 9–2   A Structural Classification of Articulations

Structural Category	Structural Type	Functional Category
Bony fusion	Synostosis	Synarthrosis
Fibrous joint	Suture	Synarthrosis
	Gomphosis	Synarthrosis
	Syndesmosis	Amphiarthrosis
Cartilaginous joint	Synchondrosis	Synarthrosis
	Symphysis	Amphiarthrosis
Synovial joint	Monaxial Biaxial Triaxial }	Diarthroses

Synarthroses (Immovable Joints)

At a synarthrosis, the bony edges are quite close together and may even interlock. These extremely strong joints are located where movement between the bones must be prevented. There are four major types of synarthrotic joints:

1. *Sutures.* A **suture** (*sutura*, a sewing together) is a synarthrotic joint located only between the bones of the skull. The edges of the bones are interlocked and bound together at the suture by dense fibrous connective tissue.

2. *Gomphoses.* A **gomphosis** (gom-FŌ-sis; *gomphosis*, a bolting together) is a synarthrosis that binds the teeth to bony sockets in the maxillae and mandible. The fibrous connection between a tooth and its socket is a *periodontal* (per-ē-ō-DON-tal) *ligament* (*peri*, around + *odontos*, tooth).

3. *Synchondroses.* A **synchondrosis** (sin-kon-DRŌ-sis; *syn*, together + *chondros*, cartilage) is a rigid, cartilaginous bridge between two articulating bones. The cartilaginous connection between the ends of the first pair of vertebrosternal ribs and the sternum is a synchondrosis. Another example is the epiphyseal cartilage, which in a growing long bone connects the diaphysis to the epiphysis. ∞ p. 197

4. *Synostoses.* A **synostosis** (sin-os-TŌ-sis) is a totally rigid, immovable joint created when two bones fuse and the boundary between them disappears. The *frontal*, or *metopic, suture* of the frontal bone and the epiphyseal lines of mature long bones are synostoses. ∞ pp. 197, 219

Amphiarthroses (Slightly Movable Joints)

An amphiarthrosis permits more movement than a synarthrosis, but is much stronger than a freely movable joint. The articulating bones are connected by collagen fibers or cartilage. There are two major types of amphiarthrotic joints:

1. At a **syndesmosis** (sin-dez-MŌ-sis; *desmos*, a band or ligament), bones are connected by a ligament. One example is the distal articulation between the tibia and fibula (see **Figure 8–13**, p. 260).

2. At a **symphysis**, or *symphyseal joint*, the articulating bones are separated by a wedge or pad of fibrous cartilage. The articulation between the bodies of vertebrae (at the *intervertebral disc*) and the connection between the two pubic bones (the *pubic symphysis*) are examples of symphyses.

Diarthroses (Freely Movable Joints)

Diarthroses, or **synovial** (si-NŌ-vē-ul) **joints**, permit a wider range of motion than do other types of joints. They are typically located at the ends of long bones, such as those of the upper and lower limbs. A synovial joint (**Figure 9–1**) is surrounded by a fibrous **articular capsule**, and a *synovial membrane* lines the walls of the articular cavity. This membrane does not cover the articulating surfaces within the joint. Recall that a synovial membrane consists of areolar tissue covered by an incomplete epithelial layer. The synovial fluid that fills the joint cavity originates in the areolar tissue of the synovial membrane. ∞ p. 137 We will now consider the major features of synovial joints.

> ### CHECKPOINT
>
> 1. Name and describe the three types of joints as classified by the amount of movement possible.
>
> 2. What characteristics do typical synarthrotic and amphiarthrotic joints share?
>
> 3. In a newborn, the large bones of the skull are joined by fibrous connective tissue. Which type of joints are these? The bones later grow, interlock, and form immovable joints. Which type of joints are these?
>
> See the blue Answers tab at the end of the book.

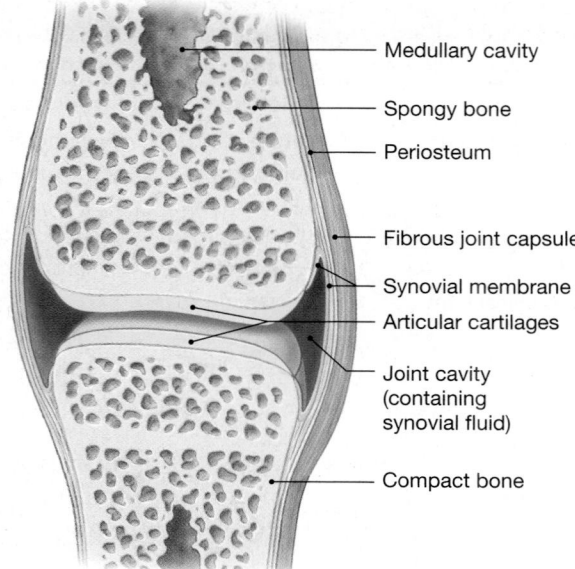

(a) Synovial joint, sagittal section

Medullary cavity
Spongy bone
Periosteum
Fibrous joint capsule
Synovial membrane
Articular cartilages
Joint cavity (containing synovial fluid)
Compact bone

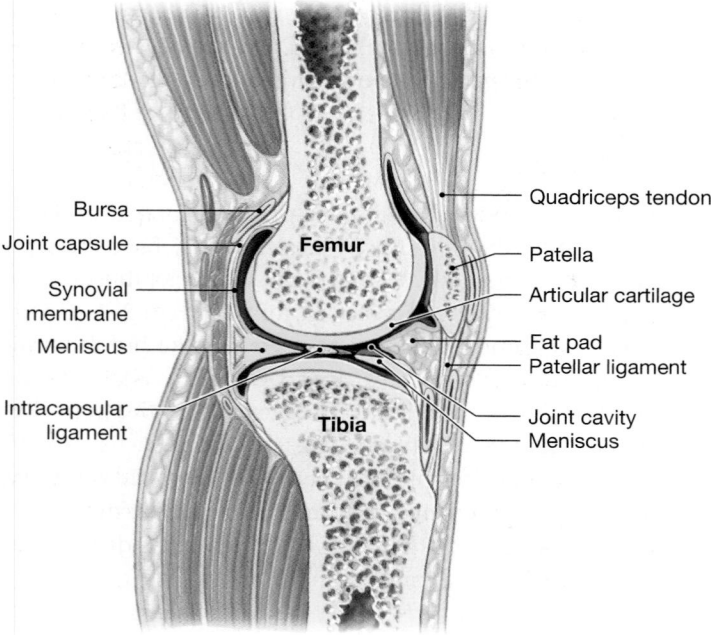

(b) Knee joint, sagittal section

Bursa
Joint capsule
Synovial membrane
Meniscus
Intracapsular ligament
Femur
Tibia
Quadriceps tendon
Patella
Articular cartilage
Fat pad
Patellar ligament
Joint cavity
Meniscus

Figure 9–1   The Structure of a Synovial Joint. (a) A diagrammatic view of a simple articulation. **(b)** A simplified sectional view of the knee joint.

9-2 Synovial joints are freely movable articulations (diarthroses) containing synovial fluid

In this section we examine several components of synovial joints: the articular cartilages, synovial fluid, and various accessory structures.

Articular Cartilages

Under normal conditions, the bony surfaces at a synovial joint cannot contact one another, because the articulating surfaces are covered by special **articular cartilages**. Articular cartilages resemble hyaline cartilages elsewhere in the body. ∞ p. 131 However, articular cartilages have no perichondrium (the fibrous sheath described in Chapter 4), and the matrix contains more water than that of other cartilages.

The surfaces of the articular cartilages are slick and smooth. This feature alone can reduce friction during movement at the joint. However, even when pressure is applied across a joint, the smooth articular cartilages do not touch one another, because they are separated by a thin film of synovial fluid within the joint cavity (**Figure 9–1a**). This fluid acts as a lubricant, minimizing friction.

Normal synovial joint function cannot continue if the articular cartilages are damaged. When such damage occurs, the matrix may begin to break down. The exposed surface will then change from a slick, smooth-gliding surface to a rough feltwork of bristly collagen fibers. This feltwork drastically increases friction at the joint.

Synovial Fluid

Synovial fluid resembles interstitial fluid, but contains a high concentration of proteoglycans secreted by fibroblasts of the synovial membrane. Even in a large joint such as the knee, the total quantity of synovial fluid in a joint is normally less than 3 mL. A clear, viscous solution with the consistency of heavy molasses, the synovial fluid within a joint has three primary functions:

1. *Lubrication.* The articular cartilages behave like sponges filled with synovial fluid. When part of an articular cartilage is compressed, some of the synovial fluid is squeezed out of the cartilage and into the space between the opposing surfaces. This thin layer of fluid markedly reduces friction between moving surfaces, just as a thin film of water reduces friction between a car's tires and a highway. When the compression stops, synovial fluid is sucked back into the articular cartilages.

2. *Nutrient Distribution.* The synovial fluid in a joint must circulate continuously to provide nutrients and a waste-disposal route for the chondrocytes of the articular cartilages. It circulates whenever the joint moves, and the compression and reexpansion of the articular cartilages pump synovial fluid into and out of the cartilage matrix. As the synovial fluid flows through the areolar tissue of the synovial membrane, waste products are absorbed and additional nutrients are obtained by diffusion across capillary walls.

3. *Shock Absorption.* Synovial fluid cushions shocks in joints that are subjected to compression. For example, your hip, knee, and ankle joints are compressed as you walk and are more severely compressed when you jog or run. When the pressure across a joint suddenly increases, the synovial fluid lessens the shock by distributing it evenly across the articular surfaces and outward to the articular capsule.

Accessory Structures

Synovial joints may have a variety of accessory structures, including pads of cartilage or fat, ligaments, tendons, and bursae (**Figure 9–1b**).

Cartilages and Fat Pads

In several joints, including the knee (**Figure 9–1b**), menisci and fat pads may lie between the opposing articular surfaces. A **meniscus** (me-NIS-kus; a crescent; plural, *menisci*) is a pad of fibrous cartilage situated between opposing bones within a synovial joint. Menisci, or *articular discs*, may subdivide a synovial cavity, channel the flow of synovial fluid, or allow for variations in the shapes of the articular surfaces.

Fat pads are localized masses of adipose tissue covered by a layer of synovial membrane. They are commonly superficial to the joint capsule (**Figure 9–1b**). Fat pads protect the articular cartilages and act as packing material for the joint. When the bones move, the fat pads fill in the spaces created as the joint cavity changes shape.

Ligaments

The capsule that surrounds the entire joint is continuous with the periostea of the articulating bones. **Accessory ligaments** support, strengthen, and reinforce synovial joints. *Intrinsic ligaments,* or *capsular ligaments,* are localized thickenings of the joint capsule. *Extrinsic ligaments* are separate from the joint capsule. These ligaments may be located either inside or outside the joint capsule, and are called *intracapsular* or *extracapsular* ligaments, respectively.

Ligaments are very strong. In a **sprain**, a ligament is stretched to the point at which some of the collagen fibers are torn, but the ligament as a whole survives and the joint is not damaged. With excessive force, one of the attached bones usually breaks before the ligament tears. In general, a broken bone heals much more quickly and effectively than does a torn ligament, because ligaments have no direct blood supply and thus must derive essential substances by diffusion.

Tendons

Although not part of the articulation itself, tendons passing across or around a joint may limit the joint's range of motion and provide mechanical support for it. For example, tendons associated with the muscles of the arm provide much of the bracing for the shoulder joint.

Bursae

Bursae (BUR-sē; singular, *bursa,* a pouch) are small, fluid-filled pockets in connective tissue. They contain synovial fluid and are lined by a synovial membrane. Bursae may be connected to the joint cavity or separate from it. They form where a tendon or ligament rubs against other tissues. Located around most synovial joints, including the shoulder joint, bursae reduce friction and act as shock absorbers. *Synovial tendon sheaths* are tubular bursae that surround tendons where they cross bony surfaces. Bursae may also appear deep to the skin, covering a bone or lying within other connective tissues exposed to friction or pressure. Bursae that develop in abnormal locations, or because of abnormal stresses, are called *adventitious bursae.*

CLINICAL NOTE

Bursitis and Bunions When bursae become inflamed, causing pain in the affected area whenever the tendon or ligament moves, the condition is called **bursitis**. Inflammation can result from the friction due to repetitive motion, pressure over the joint, irritation by chemical stimuli, infection, or trauma. Bursitis associated with repetitive motion typically occurs at the shoulder; musicians, golfers, baseball pitchers, and tennis players may develop bursitis there. The most common pressure-related bursitis is a **bunion**. Bunions may form over the base of the great toe as a result of friction and distortion of the first metatarsophalangeal joint by tight shoes, especially narrow shoes with pointed toes.

We have special names for bursitis at other locations, indicating the occupations most often associated with them. In "housemaid's knee," which accompanies prolonged kneeling, the affected bursa lies between the patella and the skin. The condition of "student's elbow" is a form of bursitis that can result from propping your head up with your arm on a desk while you read your anatomy and physiology textbook.

Factors That Stabilize Synovial Joints

A joint cannot be both highly mobile and very strong. The greater the range of motion at a joint, the weaker it becomes. A synarthrosis, the strongest type of joint, permits no movement, whereas a diarthrosis, such as the shoulder, is far weaker but permits a broad range of movement. Any mobile diarthrosis will be damaged by movement beyond its normal range of motion. Several factors are responsible for limiting the range of motion, stabilizing the joint, and reducing the chance of injury:

- The collagen fibers of the joint capsule and any accessory, extracapsular, or intracapsular ligaments.
- The shapes of the articulating surfaces and menisci, which may prevent movement in specific directions.
- The presence of other bones, skeletal muscles, or fat pads around the joint.
- Tension in tendons attached to the articulating bones. When a skeletal muscle contracts and pulls on a tendon, movement in a specific direction may be either encouraged or opposed.

The pattern of stabilizing structures varies among joints. For example, the hip joint is stabilized by the shapes of the bones (the head of the femur projects into the acetabulum), a heavy capsule, intracapsular and extracapsular ligaments, tendons, and massive muscles. It is therefore very strong and stable. In contrast, the elbow, another stable joint, gains its stability primarily from the interlocking of the articulating bones; the capsule and associated ligaments provide additional support. In general, the more stable the joint, the more restricted is its range of motion. The shoulder joint, the most mobile synovial joint, relies only on the surrounding ligaments, muscles, and tendons for stability. It is thus fairly weak.

When reinforcing structures cannot protect a joint from extreme stresses, a **dislocation**, or **luxation** (luk-SĀ-shun), results. In a dislocation, the articulating surfaces are forced out of position. The displacement can damage the articular cartilages, tear ligaments, or distort the joint capsule. Although the *inside* of a joint has no pain receptors, nerves that monitor the capsule, ligaments, and tendons are quite sensitive, so dislocations are very painful. The damage accompanying a partial dislocation, or **subluxation** (sub-luk-SĀ-shun), is less severe. People who are "double jointed" have joints that are weakly stabilized. Although their joints permit a greater range of motion than do those of other individuals, they are more likely to suffer partial or complete dislocations.

CHECKPOINT

4. Describe the components of a synovial joint, and identify the functions of each.

5. Define subluxation.

6. Why would improper circulation of synovial fluid lead to the degeneration of articular cartilages in the affected joint?

See the blue Answers tab at the end of the book.

9-3 Anatomical and functional properties of synovial joints enable various skeletal movements

To understand human movement, you must be aware of the relationship between structure and function at each articulation. To *describe* human movement, you need a frame of reference that enables accurate and precise communication. We can classify the synovial joints according to their anatomical and functional properties. To demonstrate the basis for that classification, we will use a simple model to describe the movements that occur at a typical synovial joint.

Describing Dynamic Motion

Take a pencil (or a pen) as your model, and stand it upright on the surface of a desk or table (**Figure 9–2a**). The pencil represents a bone, and the desktop represents an articular surface. A little imagination and a lot of twisting, pushing, and pulling will demonstrate that there are only three ways to move the model. Considering them one at a time will provide a frame of reference for us to analyze complex movements:

Possible Movement 1: The pencil point can move. If you hold the pencil upright, without securing the point, you can push the pencil point across the surface. This kind of motion, *gliding* (**Figure 9–2b**), is an example of **linear motion**. You could slide the point forward or backward, from side to side, or diagonally. However you move the pencil, the motion can be described by using two lines of reference (axes). One line represents forward–backward motion, the other left–right movement. For example, a simple movement along one axis could be described as "forward 1 cm" or "left 2 cm." A diagonal movement could be described with both axes, as in "backward 1 cm and to the right 2.5 cm."

Possible Movement 2: The pencil shaft can change its angle with the surface. With the tip held in position, you can move the free (eraser) end of the pencil forward and backward, from side to side, or at some intermediate angle. These movements, which change the angle between the shaft and the desktop, are examples of **angular motion** (**Figure 9–2c**). We can describe such motion by the angle the pencil shaft makes with the surface.

Any angular movement can be described with reference to the same two axes (forward–backward, left–right) and the angular change (in degrees). In one instance, however, a special term is used to describe a complex angular movement. Grasp the pencil eraser and

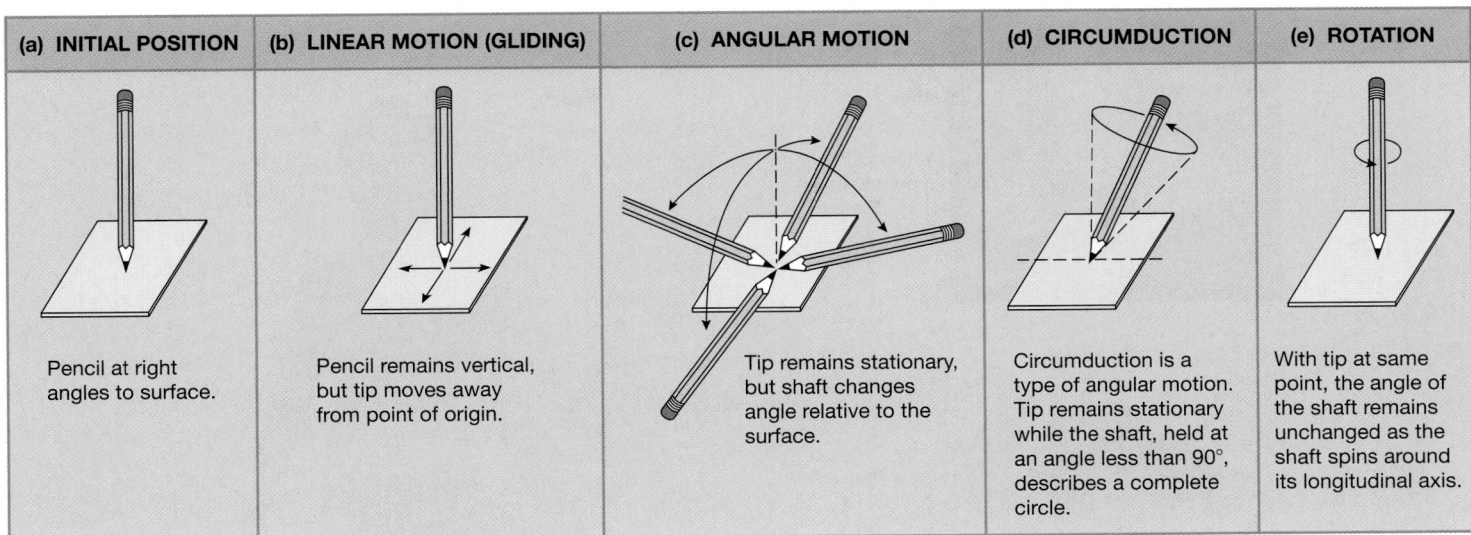

(a) INITIAL POSITION	(b) LINEAR MOTION (GLIDING)	(c) ANGULAR MOTION	(d) CIRCUMDUCTION	(e) ROTATION
Pencil at right angles to surface.	Pencil remains vertical, but tip moves away from point of origin.	Tip remains stationary, but shaft changes angle relative to the surface.	Circumduction is a type of angular motion. Tip remains stationary while the shaft, held at an angle less than 90°, describes a complete circle.	With tip at same point, the angle of the shaft remains unchanged as the shaft spins around its longitudinal axis.

Figure 9–2   A Simple Model of Articular Motion.

move the pencil in any direction until it is no longer vertical. Now swing the eraser through a complete circle (**Figure 9–2d**). This movement, which corresponds to the path of your arm when you draw a large circle on a chalkboard, is very difficult to describe. Anatomists avoid the problem by using a special term, **circumduction** (sir-kum-DUK-shun; *circum*, around), for this type of angular motion.

Possible Movement 3: The pencil shaft can rotate. If you keep the shaft vertical and the point at one location, you can still spin the pencil around its longitudinal axis. This movement is called **rotation** (**Figure 9–2e**). Several articulations permit partial rotation, but none can rotate freely. Such a movement would hopelessly tangle the blood vessels, nerves, and muscles that cross the joint.

An articulation that permits movement along only one axis is called **monaxial** (mon-AKS-ē-ul). In the pencil model, if an articulation permits only angular movement in the forward–backward plane or prevents any movement other than rotation around its longitudinal axis, it is monaxial. If movement can occur along two axes, the articulation is **biaxial** (bī-AKS-ē-ul). If the pencil could undergo angular motion in the forward–backward *and* left–right planes, but not rotation, it would be biaxial. The most mobile joints permit a combination of angular movement and rotation. These joints are said to be **triaxial** (trī-AKS-ē-ul).

Joints that permit gliding allow only small amounts of movement. These joints may be called *nonaxial*, because they permit only small sliding movements, or *multiaxial*, because sliding may occur in any direction.

Types of Movements at Synovial Joints

In descriptions of motion at synovial joints, phrases such as "bend the leg" or "raise the arm" are not sufficiently precise. Anatomists use descriptive terms that have specific meanings. We will consider these motions with reference to the basic categories of movement discussed previously: linear motion (gliding), angular motion, and rotation.

Linear Motion (Gliding)

In **gliding**, two opposing surfaces slide past one another, as in possible movement 1. Gliding occurs between the surfaces of articulating carpal bones, between tarsal bones, and between the clavicles and the sternum. The movement can occur in almost any direction, but the amount of movement is slight, and rotation is generally prevented by the capsule and associated ligaments.

Angular Motion

Examples of angular motion include *flexion, extension, abduction, adduction,* and *circumduction* (**Figure 9–3**). Descriptions of these movements are based on reference to an individual in the anatomical position. ∞ p. 17

Flexion and Extension. **Flexion** (FLEK-shun) is movement in the anterior–posterior plane that reduces the angle between the articulating elements. **Extension** occurs in the same plane, but it increases the angle between articulating elements (**Figure 9–3a**). These terms are usually applied to the movements of the long bones of the limbs, but they are also used to describe movements of the axial skeleton. For example, when you bring your head toward your chest, you flex the intervertebral

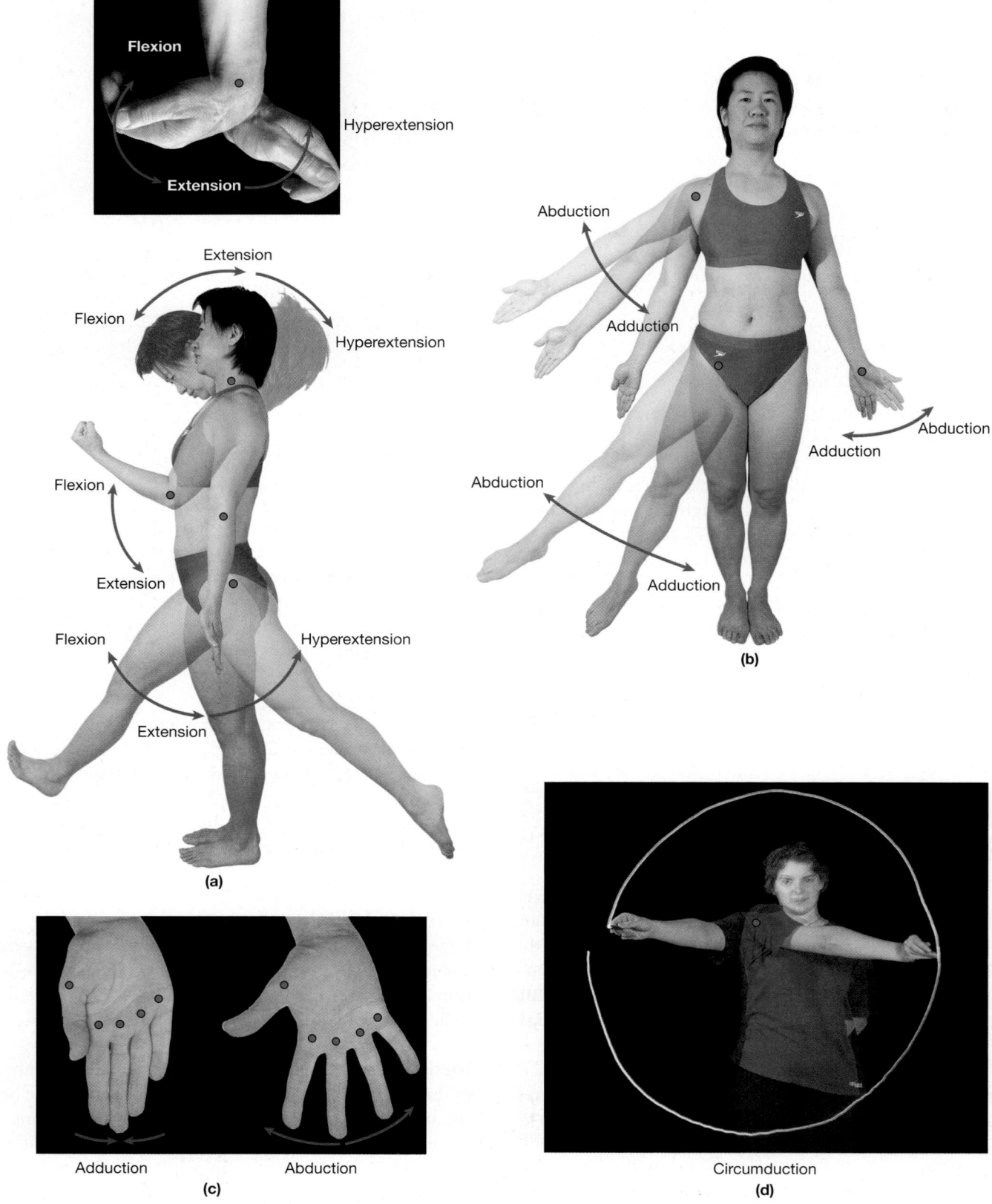

Figure 9–3 Angular Movements. The red dots indicate the locations of the joints involved in the movements illustrated.

joints of the neck. When you bend down to touch your toes, you flex the intervertebral joints of the spine. Extension reverses these movements, returning you to the anatomical position. When a person is in the anatomical position, all of the major joints of the axial and appendicular skeletons (except the ankle) are at full extension. (Special terms used to describe movements of the ankle joint are introduced shortly.)

Flexion of the shoulder joint or hip joint moves the limbs anteriorly, whereas extension moves them posteriorly. Flexion of the wrist joint moves the hand anteriorly, and extension moves it posteriorly. In each of these examples, extension can be continued past the anatomical position. Extension past the anatomical position is called **hyperextension** (**Figure 9–3a**). When you hyperextend your neck, you can gaze at the ceiling. Hyperextension of many joints, such as the elbow or the knee, is prevented by ligaments, bony processes, or soft tissues.

Abduction and Adduction. **Abduction** (*ab*, from) is movement *away from the longitudinal axis of the body* in the frontal plane (**Figure 9–3b**). For example, swinging the upper limb to the side is abduction of the limb. Moving it back to the anatomical position constitutes **adduction** (*ad*, to). Adduction of the wrist moves the heel of the hand and fingers *toward* the body, whereas abduction moves them farther away. Spreading the fingers or toes apart abducts them, because they move *away from* a central digit (**Figure 9–3c**). Bringing them together constitutes adduction. (Fingers move toward or away from the middle finger; toes move toward or away from the second toe.) Abduction and adduction always refer to movements of the appendicular skeleton, not to those of the axial skeleton.

Tips&Tricks

When someone is **abduct**ed, they are taken away, just as **abduct**ion takes the limb away from the body. During **add**uction, the limb is **add**ed to the body.

Circumduction. We introduced a special type of angular motion, circumduction, in our model. Moving your arm in a loop is circumduction (**Figure 9–3d**), as when you draw a large circle on a chalkboard. Your hand moves in a circle, but your arm does not rotate.

Rotation

Rotational movements are also described with reference to a figure in the anatomical position. Rotation of the head may involve **left rotation** or **right rotation** (**Figure 9–4a**). Limb rotation may be described by reference to the longitudinal axis of the trunk. During **medial rotation**, also known as *internal rotation* or *inward rotation*, the anterior surface of a limb turns toward the long axis of the trunk (**Figure 9–4a**). The reverse movement is called **lateral rotation**, *external rotation*, or *outward rotation*.

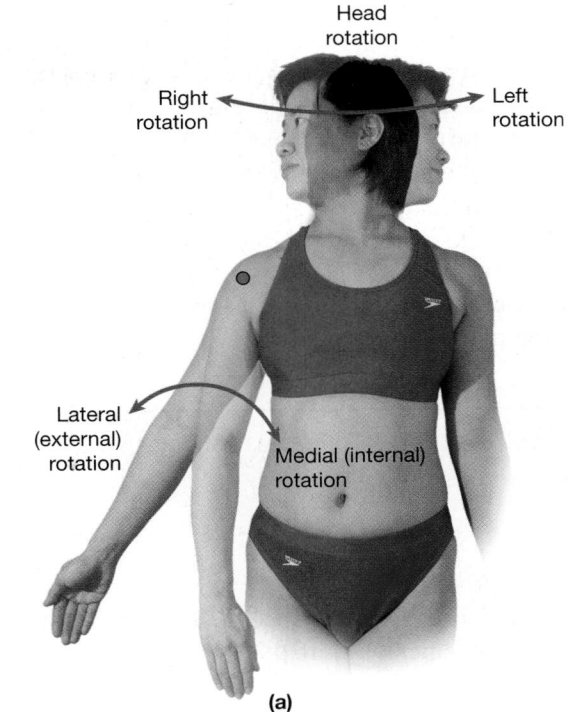

(a)

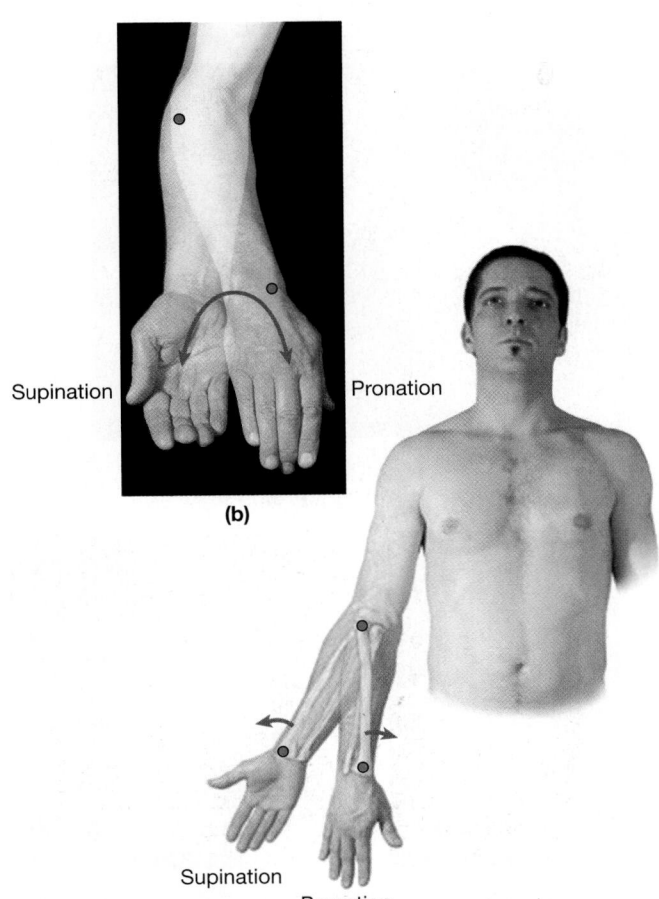

(b)

Figure 9–4   Rotational Movements.

The proximal articulation between the radius and the ulna permits rotation of the radial head. As the shaft of the radius rotates, the distal epiphysis of the radius rolls across the anterior surface of the ulna. This movement, called **pronation** (prō-NĀ-shun), turns the wrist and hand from palm facing front to palm facing back (**Figure 9–4b**). The opposing movement, in which the palm is turned anteriorly, is **supination** (soo-pi-NĀ-shun). The forearm is supinated in the anatomical position. This view makes it easier to follow the path of the blood vessels, nerves, and tendons, which rotate with the radius during pronation.

Tips&Tricks

In order to carry a bowl of *soup*, the hand must be *sup*inated.

Special Movements

Several special terms apply to specific articulations or unusual types of movement (**Figure 9–5**):

- **Inversion** (*in*, into + *vertere*, to turn) is a twisting motion of the foot that turns the sole inward, elevating

the medial edge of the sole. The opposite movement is called **eversion** (ē-VER-zhun; *e*, out).

- **Dorsiflexion** is flexion at the ankle joint and elevation of the sole, as when you dig in your heel. **Plantar flexion** (*planta*, sole), the opposite movement, extends the ankle joint and elevates the heel, as when you stand on tiptoe. However, it is also acceptable (and simpler) to use "flexion and extension at the ankle," rather than "dorsiflexion and plantar flexion."

- **Opposition** is movement of the thumb toward the surface of the palm or the pads of other fingers. Opposition enables you to grasp and hold objects between your thumb and palm. It involves movement at the first carpometacarpal and metacarpophalangeal joints. Flexion at the fifth metacarpophalangeal joint can assist this movement.

- **Protraction** entails moving a part of the body anteriorly in the horizontal plane. **Retraction** is the reverse movement. You protract your jaw when you grasp your upper lip with your lower teeth, and you protract your clavicles when you cross your arms.

- **Elevation** and **depression** occur when a structure moves in a superior or an inferior direction, respectively. You

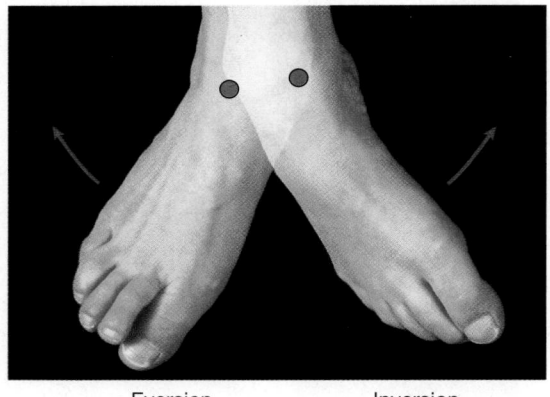

Eversion   Inversion

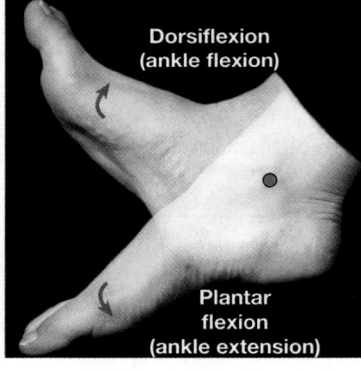

Dorsiflexion (ankle flexion)

Plantar flexion (ankle extension)

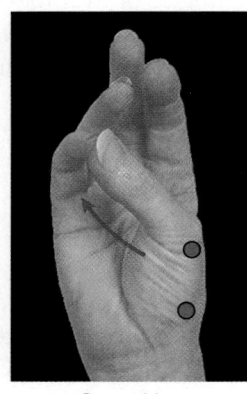

Opposition

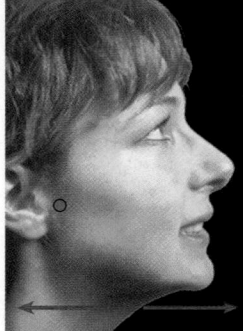

Retraction   Protraction

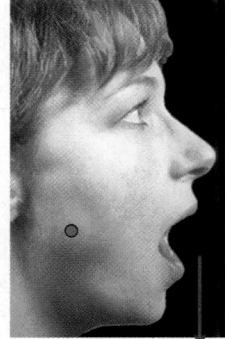

Depression

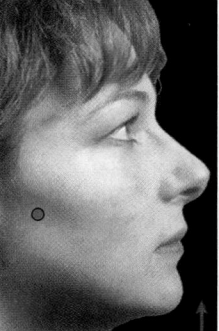

Elevation

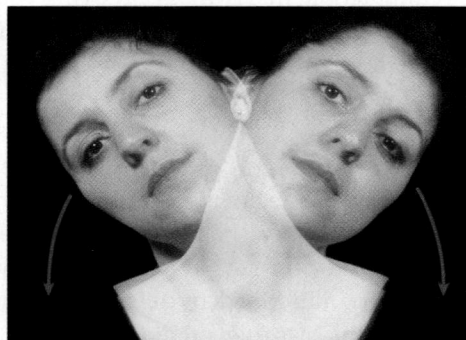

Lateral flexion

Figure 9–5 **Special Movements.**

depress your mandible when you open your mouth; you elevate your mandible as you close your mouth. Another familiar elevation occurs when you shrug your shoulders.

- **Lateral flexion** occurs when your vertebral column bends to the side. This movement is most pronounced in the cervical and thoracic regions.

Types of Synovial Joints

Synovial joints are described as *gliding, hinge, pivot, ellipsoid, saddle,* or *ball-and-socket joints* on the basis of the shapes of the articulating surfaces. Each type of joint permits a different type and range of motion. **Figure 9–6** lists the structural categories and the types of movement each permits.

- **Gliding joints**, also called *planar joints*, have flattened or slightly curved faces. The relatively flat articular surfaces slide across one another, but the amount of movement is very slight. Although rotation is theoretically possible at such a joint, ligaments usually prevent or restrict such movement.

- **Hinge joints** permit angular motion in a single plane, like the opening and closing of a door.

- **Pivot joints** also are monaxial, but they permit only rotation.

- In an **ellipsoid joint**, or *condylar joint*, an oval articular face nestles within a depression in the opposing surface. With such an arrangement, angular motion occurs in two planes: along or across the length of the oval.

- **Saddle joints**, or *sellaris joints*, fit together like a rider in a saddle. Each articular face is concave along one axis and convex along the other. This arrangement permits angular motion, including circumduction, but prevents rotation.

- In a **ball-and-socket joint**, the round head of one bone rests within a cup-shaped depression in another. All combinations of angular and rotational movements, including circumduction, can be performed at ball-and-socket joints.

The A&P Top 100

#34 A joint cannot be both highly mobile and very strong. The greater the mobility, the weaker the joint, because mobile joints rely on muscular and ligamentous support rather than solid bone-to-bone connections.

CHECKPOINT

7. Identify the types of synovial joints based on the shapes of the articulating surfaces.

8. When you do jumping jacks, which lower limb movements are necessary?

9. Which movements are associated with hinge joints?

See the blue Answers tab at the end of the book.

9-4 Intervertebral discs and ligaments are structural components of intervertebral articulations

The articulations between the superior and inferior articular processes of adjacent vertebrae—called *intervertebral articulations*—are gliding joints that permit small movements associated with flexion and rotation of the vertebral column (**Figure 9–7**). Little gliding occurs between adjacent vertebral bodies. From axis to sacrum, the vertebrae are separated and cushioned by pads of fibrous cartilage called **intervertebral discs**. Thus, the bodies of vertebrae form symphyseal joints. Intervertebral discs and symphyseal joints are found neither in the sacrum or coccyx, where vertebrae have fused, nor between the first and second cervical vertebrae. The first cervical vertebra has no vertebral body and no intervertebral disc; the only articulation between the first two cervical vertebrae is a pivot joint that permits much more rotation than do the symphyseal joints between other cervical vertebrae.

Intervertebral Discs

Each intervertebral disc has a tough outer layer of fibrous cartilage, the **anulus fibrosus** (AN-ū-lus fi-BRŌ-sus). The collagen fibers of this layer attach the disc to the bodies of adjacent vertebrae. The anulus fibrosus surrounds the **nucleus pulposus** (pul-PŌ-sus), a soft, elastic, gelatinous core (**Figure 9–7**). The nucleus pulposus gives the disc resiliency and enables it to absorb shocks.

Movement of the vertebral column compresses the nucleus pulposus and displaces it in the opposite direction. This displacement permits smooth gliding movements between vertebrae while maintaining their alignment. The discs make a significant contribution to an individual's height: They account for roughly one-quarter the length of the vertebral column superior to the sacrum. As we grow older, the water content of the nucleus pulposus in each disc decreases. The discs gradually become less effective as cushions, and the chances of vertebral injury increase. Water loss from the discs also causes shortening of the vertebral column, accounting for the characteristic decrease in height with advancing age.

Intervertebral Ligaments

Numerous ligaments are attached to the bodies and processes of all vertebrae, binding them together and stabilizing the vertebral column (**Figure 9–7**). Ligaments interconnecting adjacent vertebrae include the following:

- The *anterior longitudinal ligament*, which connects the anterior surfaces of adjacent vertebral bodies.

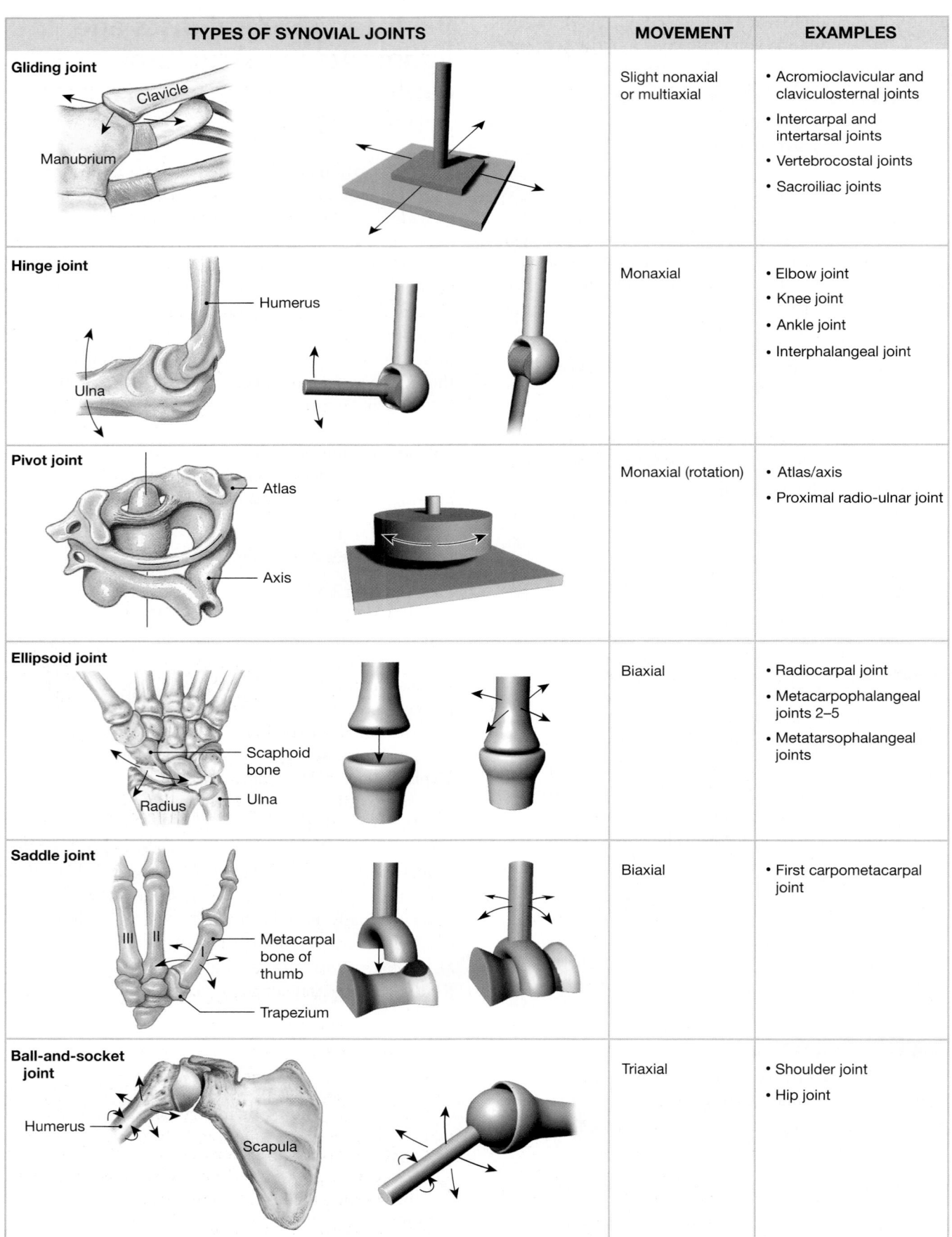

TYPES OF SYNOVIAL JOINTS	MOVEMENT	EXAMPLES
Gliding joint — Clavicle, Manubrium	Slight nonaxial or multiaxial	• Acromioclavicular and claviculosternal joints • Intercarpal and intertarsal joints • Vertebrocostal joints • Sacroiliac joints
Hinge joint — Humerus, Ulna	Monaxial	• Elbow joint • Knee joint • Ankle joint • Interphalangeal joint
Pivot joint — Atlas, Axis	Monaxial (rotation)	• Atlas/axis • Proximal radio-ulnar joint
Ellipsoid joint — Scaphoid bone, Radius, Ulna	Biaxial	• Radiocarpal joint • Metacarpophalangeal joints 2–5 • Metatarsophalangeal joints
Saddle joint — III, II, Metacarpal bone of thumb, Trapezium	Biaxial	• First carpometacarpal joint
Ball-and-socket joint — Humerus, Scapula	Triaxial	• Shoulder joint • Hip joint

Figure 9–6 Movements at Synovial Joints.

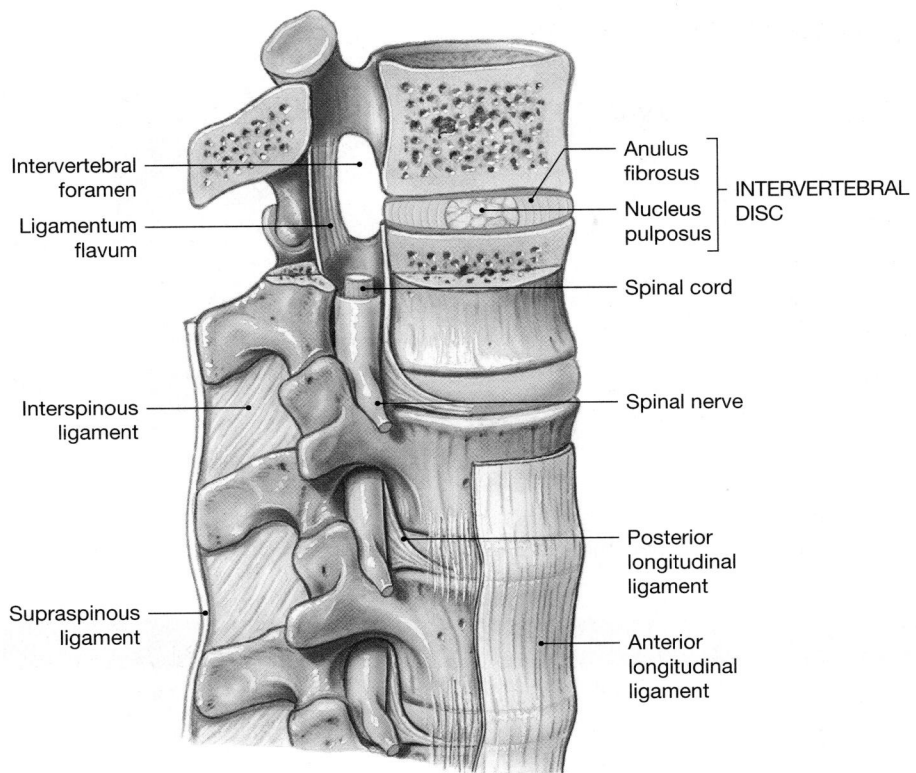

Intervertebral foramen

Ligamentum flavum

Interspinous ligament

Supraspinous ligament

Anulus fibrosus

Nucleus pulposus

INTERVERTEBRAL DISC

Spinal cord

Spinal nerve

Posterior longitudinal ligament

Anterior longitudinal ligament

Figure 9–7 **Intervertebral Articulations.** ATLAS: Plates 20b; 23c

- The *posterior longitudinal ligament*, which parallels the anterior longitudinal ligament and connects the posterior surfaces of adjacent vertebral bodies.
- The *ligamentum flavum* (plural, *ligamenta flava*), which connects the laminae of adjacent vertebrae.
- The *interspinous ligament*, which connects the spinous processes of adjacent vertebrae.
- The *supraspinous ligament*, which interconnects the tips of the spinous processes from C$_7$ to the sacrum. The *ligamentum nuchae*, which extends from vertebra C$_7$ to the base of the skull, is continuous with the supraspinous ligament. ∞ p. 236

Tips&Tricks

The inter**spinous** ligament and supra**spinous** ligament get their names because they are attached to the **spinous** processes of the vertebrae.

If the posterior longitudinal ligaments are weakened, as often occurs with advancing age, the compressed nucleus pulposus may distort the anulus fibrosus, forcing it partway into the vertebral canal. This condition is called a **slipped disc** (**Figure 9–8a**), although the disc does not actually slip. If the nucleus pulposus breaks through the anulus fibrosus, it too may protrude into the vertebral canal. This condition is

called a **herniated disc** (**Figure 9–8b**). When a disc herniates, sensory nerves are distorted, and the protruding mass can also compress the nerves passing through the adjacent intervertebral foramen.

Vertebral Movements

The following movements can occur across the intervertebral joints of the vertebral column: (1) flexion, or bending anteriorly; (2) extension, or bending posteriorly; (3) lateral flexion, or bending laterally; and (4) rotation. Table 9–3 summarizes information about intervertebral and other articulations of the axial skeleton.

CHECKPOINT

10. What types of movements can occur across intervertebral joints?

11. Which regions of the vertebral column lack intervertebral discs? Explain why the absence of discs is significant.

12. Which vertebral movements are involved in (a) bending forward, (b) bending to the side, and (c) moving the head to signify "no"?

See the blue Answers tab at the end of the book.

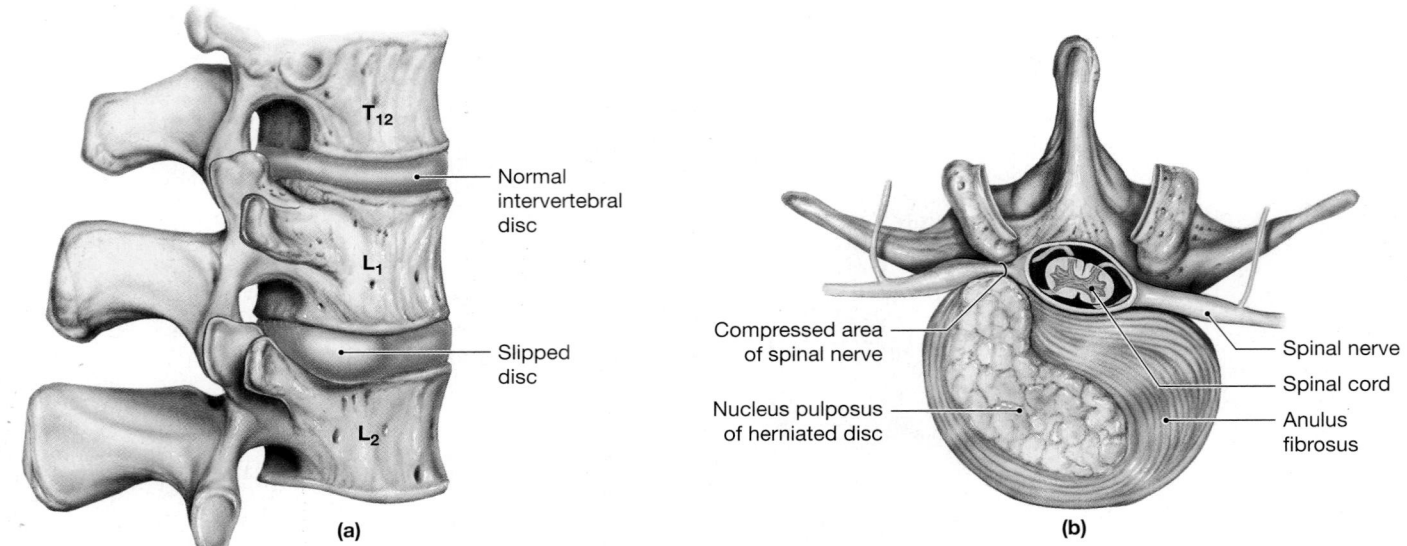

Figure 9–8   Damage to the Intervertebral Discs. (a) A lateral view of the lumbar region of the spinal column, showing a distorted intervertebral disc (a "slipped" disc). **(b)** A sectional view through a herniated disc, showing the release of the nucleus pulposus and its effect on the spinal cord and adjacent spinal nerves.

SUMMARY TABLE 9–3   Articulations of the Axial Skeleton

	Element	Joint	Type of Articulation	Movement(s)
	SKULL			
	Cranial and facial bones of skull	Various	Synarthroses (suture or synostosis)	None
	Maxilla/teeth and mandible/teeth	Alveolar	Synarthrosis (gomphosis)	None
	Temporal bone/mandible	Temporomandibular	Combined gliding joint and hinge diarthrosis	Elevation, depression, and lateral gliding
	VERTEBRAL COLUMN			
	Occipital bone/atlas	Atlanto-occipital	Ellipsoid diarthrosis	Flexion/extension
	Atlas/axis	Atlanto-axial	Pivot diarthrosis	Rotation
	Other vertebral elements	Intervertebral (between vertebral bodies)	Amphiarthrosis (symphysis)	Slight movement
		Intervertebral (between articular processes)	Gliding diarthrosis	Slight rotation and flexion/extension
	L5/sacrum	Between L5 body and sacral body	Amphiarthrosis (symphysis)	Slight movement
		Between inferior articular processes of L5 and articular processes of sacrum	Gliding diarthrosis	Slight flexion/extension
	Sacrum/coxal bone	Sacroiliac	Gliding diarthrosis	Slight movement
	Sacrum/coccyx	Sacrococcygeal	Gliding diarthrosis (may become fused)	Slight movement
	Coccygeal bones		Synarthrosis (synostosis)	No movement
	THORACIC CAGE			
	Bodies of T1–T12 and heads of ribs	Costovertebral	Gliding diarthrosis	Slight movement
	Transverse processes of T1–T10	Costovertebral	Gliding diarthrosis	Slight movement
	Ribs and costal cartilages		Synarthrosis (synchondrosis)	No movement
	Sternum and first costal cartilage	Sternocostal (1st)	Synarthrosis (synchondrosis)	No movement
	Sternum and costal cartilages 2–7	Sternocostal (2nd–7th)	Gliding diarthrosis*	Slight movement

*Commonly converts to synchondrosis in elderly individuals.

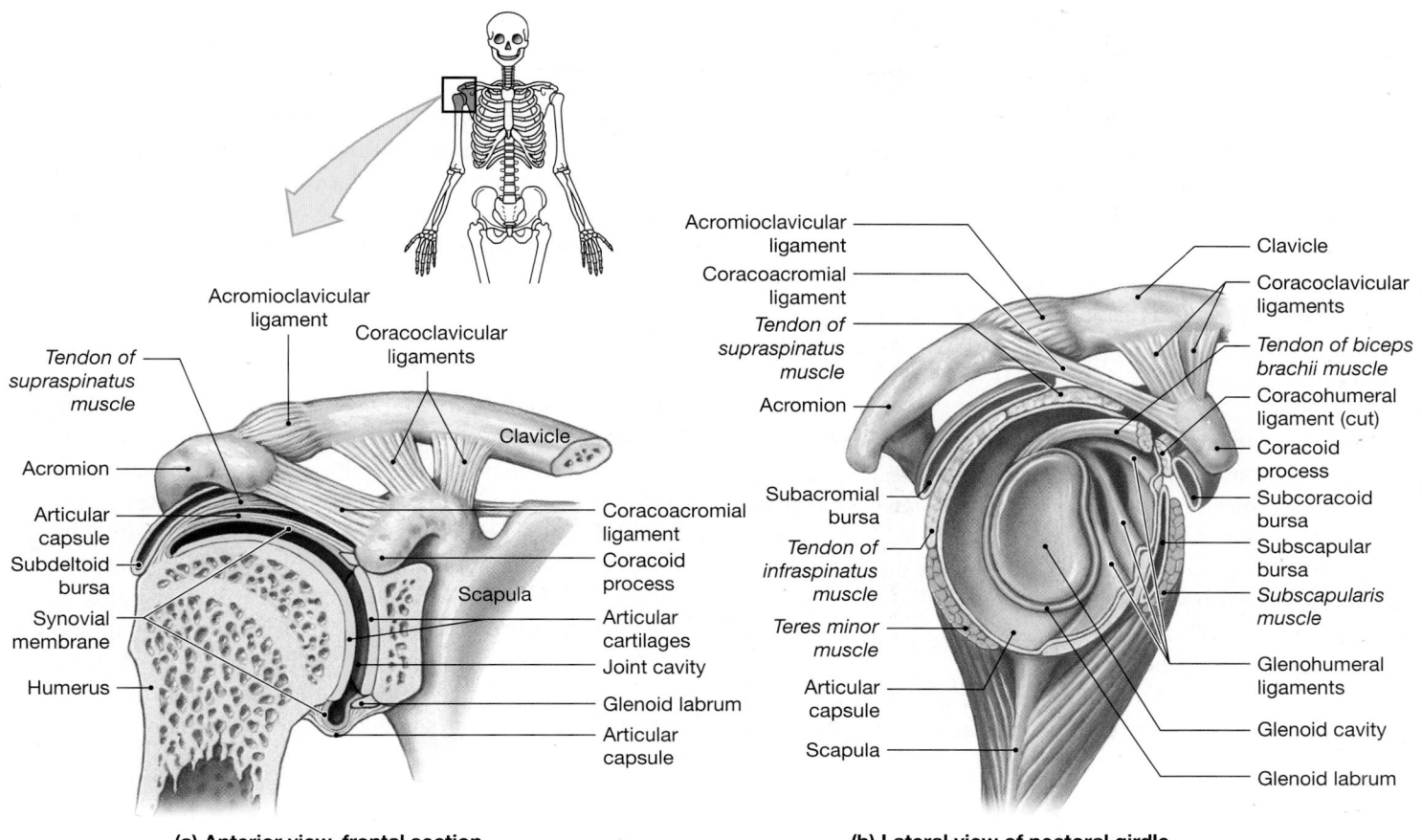

Acromioclavicular ligament
Coracoclavicular ligaments

Tendon of supraspinatus muscle

Acromion

Articular capsule

Subdeltoid bursa

Synovial membrane

Humerus

Clavicle

Coracoacromial ligament

Coracoid process

Scapula

Articular cartilages

Joint cavity

Glenoid labrum

Articular capsule

(a) Anterior view, frontal section

Acromioclavicular ligament
Coracoacromial ligament
Tendon of supraspinatus muscle
Acromion

Subacromial bursa

Tendon of infraspinatus muscle

Teres minor muscle

Articular capsule

Scapula

Clavicle

Coracoclavicular ligaments

Tendon of biceps brachii muscle

Coracohumeral ligament (cut)

Coracoid process

Subcoracoid bursa

Subscapular bursa

Subscapularis muscle

Glenohumeral ligaments

Glenoid cavity

Glenoid labrum

(b) Lateral view of pectoral girdle

Figure 9–9   The Shoulder Joint. (a) A sectional view showing major structural features. **(b)** A lateral view of the shoulder joint with the humerus removed. ATLAS: Plate 27d

9-5 The shoulder is a ball-and-socket joint, and the elbow is a hinge joint

In this section we consider the structure and function of two major joints in the upper limb: the shoulder joint and elbow joint.

The Shoulder Joint

The shoulder joint, or *glenohumeral joint*, permits the greatest range of motion of any joint. Because it is also the most frequently dislocated joint, it provides an excellent demonstration of the principle that stability must be sacrificed to obtain mobility.

This joint is a ball-and-socket diarthrosis formed by the articulation of the head of the humerus with the glenoid cavity of the scapula (**Figure 9–9a**). The extent of the glenoid cavity is increased by a fibrouscartilaginous **glenoid labrum** (*labrum*, lip or edge), which continues beyond the bony rim and deepens the socket (**Figure 9–9b**). The relatively loose articular capsule extends from the scapula, proximal to the glenoid labrum, to the anatomical neck of the humerus. Somewhat oversized, the articular capsule permits an extensive range of motion. The bones of the pectoral girdle provide some stabil-

ity to the superior surface, because the acromion and coracoid process project laterally superior to the head of the humerus. However, most of the stability at this joint is provided by the surrounding skeletal muscles, with help from their associated tendons and various ligaments. Bursae reduce friction between the tendons and other tissues at the joint.

The major ligaments that help stabilize the shoulder joint are the *glenohumeral, coracohumeral, coracoacromial, coracoclavicular,* and *acromioclavicular ligaments.* The acromioclavicular ligament reinforces the capsule of the acromioclavicular joint and supports the superior surface of the shoulder. A **shoulder separation** is a relatively common injury involving partial or complete dislocation of the acromioclavicular joint. This injury can result from a blow to the superior surface of the shoulder. The acromion is forcibly depressed while the clavicle is held back by strong muscles.

The muscles that move the humerus do more to stabilize the shoulder joint than do all the ligaments and capsular fibers combined. Muscles originating on the trunk, pectoral girdle, and humerus cover the anterior, superior, and posterior surfaces of the capsule. The tendons of the *supraspinatus, infraspinatus, teres minor,* and *subscapularis muscles* reinforce the joint capsule and limit range of movement. These muscles, known as the muscles of the *rotator cuff*, are the primary mechanism for supporting the shoulder joint and limiting its range of movement. Damage to the rotator cuff typically occurs when individuals are engaged in sports that place severe strains on the shoulder. White-water kayakers, baseball pitchers, and quarterbacks are all at high risk for rotator cuff injuries.

Tips&Tricks

The rotator cuff muscles can be remembered by using the acronym **SITS**, for **s**upraspinatus, **i**nfraspinatus, **t**eres minor, and **s**ubscapularis.

The anterior, superior, and posterior surfaces of the shoulder joint are reinforced by ligaments, muscles, and tendons, but the inferior capsule is poorly reinforced. As a result, a dislocation caused by an impact or a violent muscle contraction is most likely to occur at this site. Such a dislocation can tear the inferior capsular wall and the glenoid labrum. The healing process typically leaves a weakness that increases the chances for future dislocations.

As at other joints, bursae at the shoulder reduce friction where large muscles and tendons pass across the joint capsule. The shoulder has a relatively large number of important bursae, such as the *subacromial bursa*, the *subdeltoid bursa*, the *subcoracoid bursa*, and the *subscapular bursa* (**Figure 9–9a,b**). A tendon of the biceps brachii muscle runs through the shoulder joint. As it passes through the articular capsule, it is surrounded by a tubular bursa that is continuous with the joint cavity. Inflammation of any of these extracapsular bursae can restrict motion and produce the painful symptoms of bursitis (p. 271).

The Elbow Joint

The elbow joint is a complex hinge joint that involves the humerus, radius, and ulna (**Figure 9–10**). The largest and strongest articulation at the elbow is the *humero-ulnar joint*, where the trochlea of the humerus articulates with the trochlear notch of the ulna. This joint works like a door hinge, with physical limitations imposed on the range of motion. In the case of the elbow, the shape of the trochlear notch of the ulna determines the plane of movement, and the combination of the notch and the olecranon limits the degree of extension permitted. At the smaller *humeroradial joint*, the capitulum of the humerus articulates with the head of the radius.

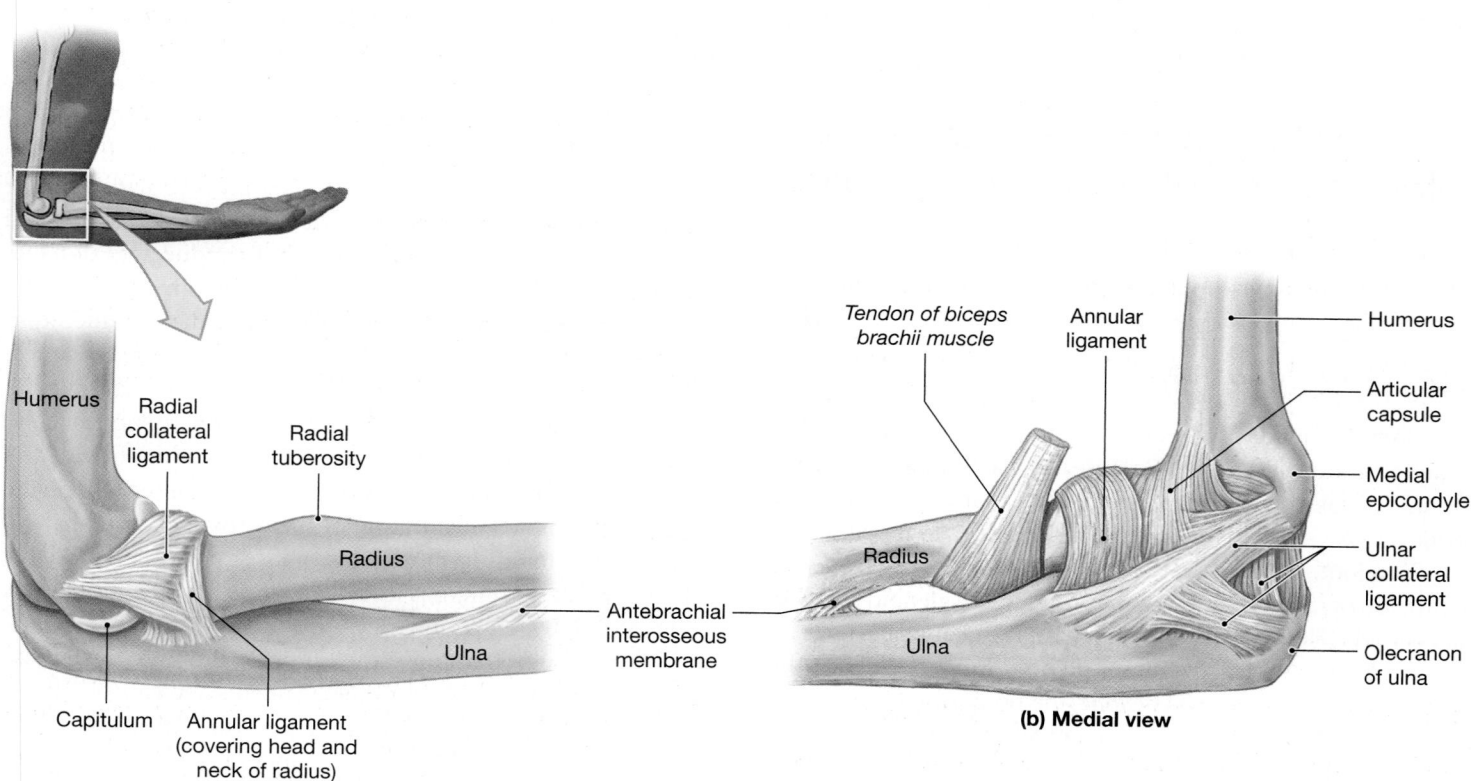

Figure 9–10 **The Elbow Joint. The right elbow joint.** (a) A lateral view. (b) A medial view, showing ligaments that stabilize the joint.
ATLAS: Plates 35a–g

Muscles that extend the elbow attach to the rough surface of the olecranon. These muscles are primarily under the control of the *radial nerve*, which passes along the radial groove of the humerus. ∞ p. 250 The large *biceps brachii muscle* covers the anterior surface of the arm. Its tendon is attached to the radius at the radial tuberosity. Contraction of this muscle produces supination of the forearm and flexion at the elbow.

The elbow joint is extremely stable because (1) the bony surfaces of the humerus and ulna interlock, (2) a single, thick articular capsule surrounds both the humero-ulnar and proximal radio-ulnar joints, and (3) the articular capsule is reinforced by strong ligaments. The *radial collateral ligament* stabilizes the lateral surface of the elbow joint (**Figure 9–10a**). It extends between the lateral epicondyle of the humerus and the *annular ligament*, which binds the head of the radius to the ulna. The medial surface of the elbow joint is stabilized by the *ulnar collateral ligament*, which extends from the medial epicondyle of the humerus anteriorly to the coronoid process of the ulna and posteriorly to the olecranon (**Figure 9–10b**).

Despite the strength of the capsule and ligaments, the elbow can be damaged by severe impacts or unusual stresses. For example, if you fall on your hand with a partially flexed elbow, contractions of muscles that extend the elbow may break the ulna at the center of the trochlear notch. Less violent stresses can produce dislocations or other injuries to the elbow, especially if epiphyseal growth has not been completed. For example, parents in a rush may drag a toddler along behind them exerting an upward, twisting pull on the elbow joint that can result in a partial dislocation known as *nursemaid's elbow*.

CHECKPOINT

13. Which tissues or structures provide most of the stability for the shoulder joint?

14. Would a tennis player or a jogger be more likely to develop inflammation of the subscapular bursa? Why?

15. A football player received a blow to the upper surface of his shoulder, causing a shoulder separation. What does this mean?

16. Terry suffers an injury to his forearm and elbow. After the injury, he notices an unusually large degree of motion between the radius and the ulna at the elbow. Which ligament did Terry most likely damage?

See the blue Answers tab at the end of the book.

9-6 The hip is a ball-and-socket joint, and the knee is a hinge joint

Next we examine two representative articulations in the lower limb: the hip joint and the knee joint.

The Hip Joint

The hip joint is a sturdy ball-and-socket diarthrosis that permits flexion and extension, adduction and abduction, circumduction, and rotation. **Figure 9–11** introduces the structure of the hip joint. The acetabulum, a deep fossa, accommodates the head of the femur. ∞ p. 258 Within the acetabulum, a fibrous cartilage pad extends like a horseshoe to either side of the acetabular notch (**Figure 9–11a**). The *acetabular labrum*, a projecting rim of fibrous cartilage, increases the depth of the joint cavity.

The articular capsule of the hip joint is extremely dense and strong. It extends from the lateral and inferior surfaces of the pelvic girdle to the intertrochanteric line and intertrochanteric crest of the femur, enclosing both the head and neck of the femur. ∞ p. 259 This arrangement helps keep the femoral head from moving too far from the acetabulum.

Four broad ligaments reinforce the articular capsule (**Figure 9–11**). Three of them—the *iliofemoral*, *pubofemoral*, and *ischiofemoral ligaments*—are regional thickenings of the capsule. The *transverse acetabular ligament* crosses the acetabular notch, filling in the gap in the inferior border of the acetabulum. A fifth ligament, the *ligament of the femoral head*, or *ligamentum teres* (*teres*, long and round), originates along the transverse acetabular ligament (**Figure 9–11a**) and attaches to the fovea capitis, a small pit at the center of the femoral head. ∞ p. 258 This ligament tenses only when the hip is flexed and the thigh is undergoing lateral rotation. Much more important stabilization is provided by the bulk of the surrounding muscles, aided by ligaments and capsular fibers.

The combination of an almost complete bony socket, a strong articular capsule, supporting ligaments, and muscular padding makes the hip joint an extremely stable joint. The head of the femur is well supported, but the ball-and-socket joint is not directly aligned with the weight distribution along the shaft. Stress must be transferred at an angle from the joint, along the relatively thin femoral neck to the length of the femur. ∞ p. 192 Fractures of the femoral neck or between the greater and lesser trochanters of the femur are more common than hip dislocations. As we noted in Chapter 6, hip fractures are relatively common in elderly individuals with severe osteoporosis. ∞ p. 207

The Knee Joint

The hip joint passes weight to the femur, and the knee joint transfers the weight from the femur to the tibia. The shoulder is mobile; the hip, stable; and the knee . . . ? If you had to choose one word, it would probably be "complicated." Although the knee joint functions as a hinge, the articulation is far more complex than the elbow or even the ankle. The rounded condyles of the femur roll across the superior surface of the tibia, so the points of contact are constantly changing. The joint permits flexion, extension, and very limited rotation.

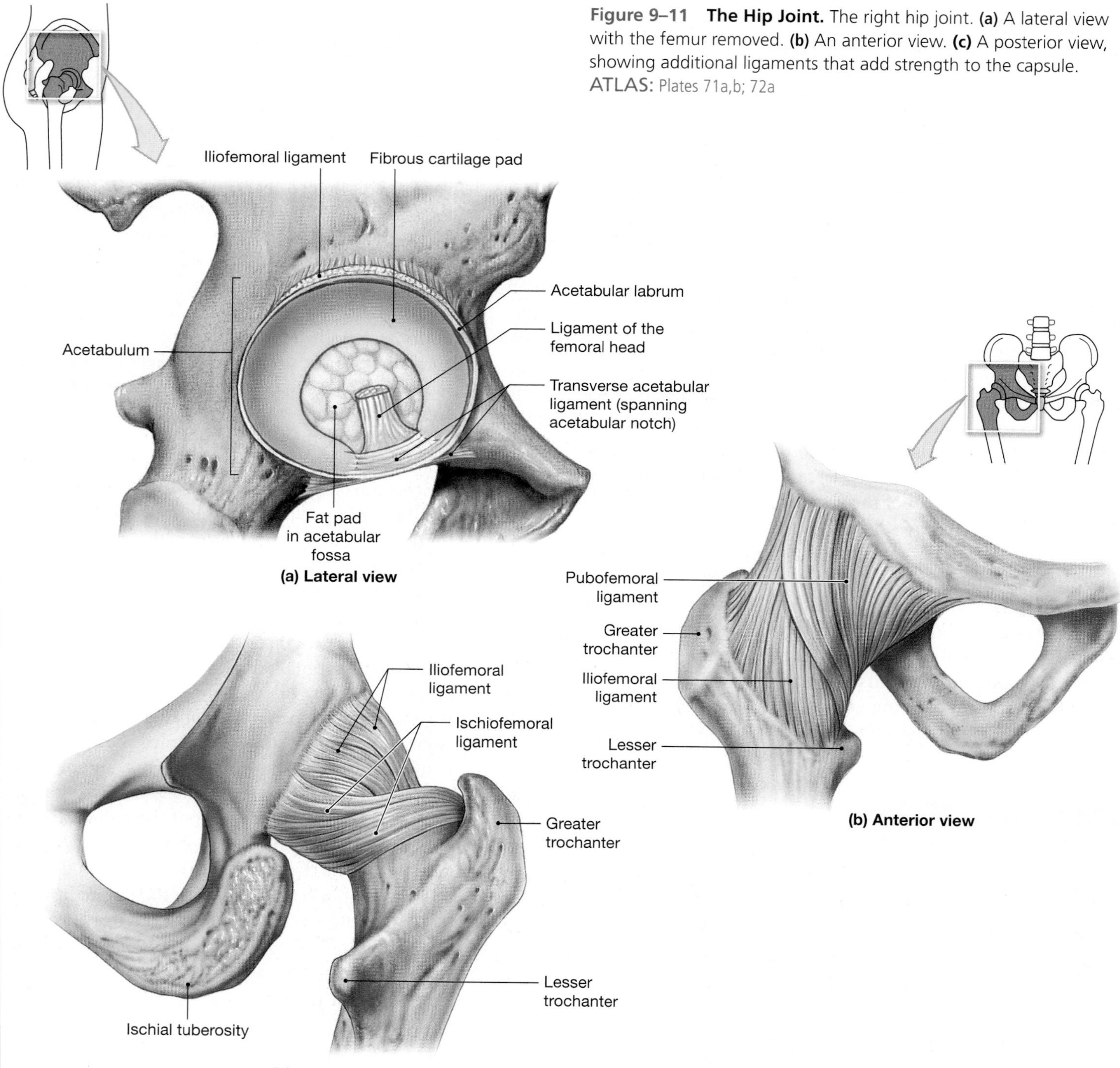

Figure 9–11 **The Hip Joint.** The right hip joint. **(a)** A lateral view with the femur removed. **(b)** An anterior view. **(c)** A posterior view, showing additional ligaments that add strength to the capsule. ATLAS: Plates 71a,b; 72a

Iliofemoral ligament   Fibrous cartilage pad

Acetabular labrum

Ligament of the femoral head

Transverse acetabular ligament (spanning acetabular notch)

Acetabulum

Fat pad in acetabular fossa

(a) Lateral view

Iliofemoral ligament

Ischiofemoral ligament

Greater trochanter

Lesser trochanter

Ischial tuberosity

(c) Posterior view

Pubofemoral ligament

Greater trochanter

Iliofemoral ligament

Lesser trochanter

(b) Anterior view

The knee joint contains three separate articulations: two between the femur and tibia (medial condyle to medial condyle, and lateral condyle to lateral condyle) and one between the patella and the patellar surface of the femur (**Figure 9–12**).

The Articular Capsule and Joint Cavity

The articular capsule at the knee joint is thin and in some areas incomplete, but it is strengthened by various ligaments and the tendons of associated muscles. A pair of fibrous cartilage pads, the **medial** and **lateral menisci**, lie between the femoral

and tibial surfaces (**Figures 9–1b** and **9–12c,d**). The menisci (1) act as cushions, (2) conform to the shape of the articulating surfaces as the femur changes position, and (3) provide lateral stability to the joint. Prominent fat pads cushion the margins of the joint and assist the many bursae in reducing friction between the patella and other tissues.

Supporting Ligaments

A complete dislocation of the knee is very rare, largely because seven major ligaments stabilize the knee joint:

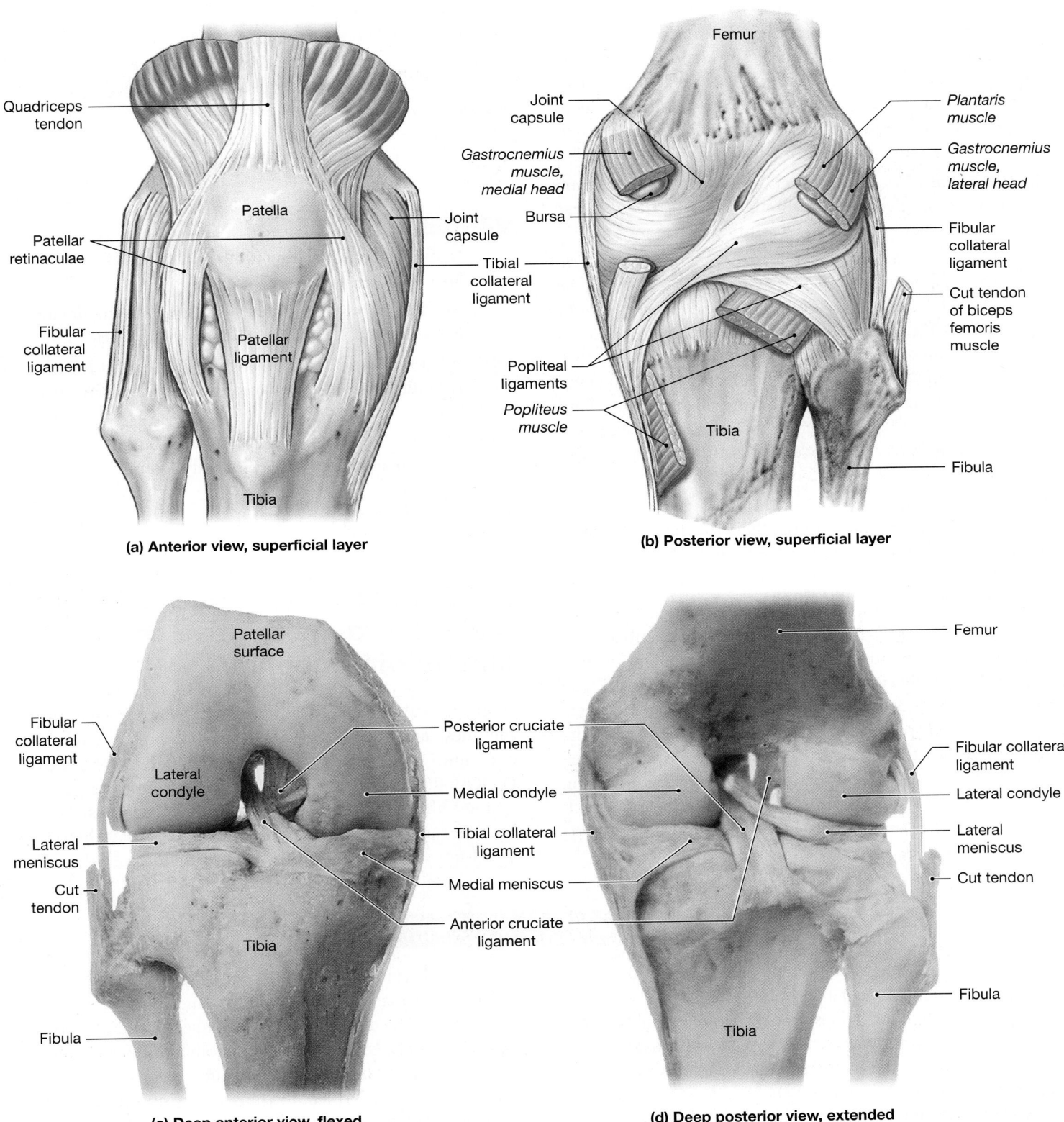

(a) **Anterior view, superficial layer**

(b) **Posterior view, superficial layer**

(c) **Deep anterior view, flexed**

(d) **Deep posterior view, extended**

Figure 9–12 **The Knee Joint.** The right knee. Superficial anterior **(a)** and posterior **(b)** views of the extended knee joint. **(c)** An anterior view, at full flexion. **(d)** A deep posterior view, at full extension. ATLAS: Plates 78a–i; 79a,b; 80a,b

9 SKELETAL

1. The tendon from the muscles responsible for extending the knee passes over the anterior surface of the joint (**Figure 9–12a**). The patella is embedded in this tendon, and the *patellar ligament* continues to its attachment on the anterior surface of the tibia. The patellar ligament and two ligamentous bands known as the *patellar retinaculae* support the anterior surface of the knee joint.

2,3. Two *popliteal ligaments* extend between the femur and the heads of the tibia and fibula (**Figure 9–12b**). These ligaments reinforce the knee joint's posterior surface.

4,5. Inside the joint capsule, the *anterior cruciate ligament* (ACL) and *posterior cruciate ligament* (PCL) attach the intercondylar area of the tibia to the condyles of the femur (**Figure 9–12c,d**). *Anterior* and *posterior* refer to the sites of origin of these ligaments on the tibia. They cross one another as they proceed to their destinations on the femur. (The term *cruciate* is derived from the Latin word *crucialis*, meaning a cross.) The ACL and the PCL limit the anterior and posterior movement of the femur and maintain the alignment of the femoral and tibial condyles.

6,7. The *tibial collateral ligament* reinforces the medial surface of the knee joint, and the *fibular collateral ligament* reinforces the lateral surface (**Figure 9–12**). These ligaments tighten only at full extension, the position in which they stabilize the joint.

At full extension, a slight lateral rotation of the tibia tightens the anterior cruciate ligament and jams the lateral meniscus between the tibia and femur. The knee joint is essentially locked in the extended position. With the joint locked, a person can stand for prolonged periods without using (and tiring) the muscles that extend the knee. Unlocking the knee joint requires muscular contractions that medially rotate the tibia or laterally rotate the femur. If the locked knee is struck from the side, the lateral meniscus can tear and the supporting ligaments can be seriously damaged.

The knee joint is structurally complex and is subjected to severe stresses in the course of normal activities. Painful knee injuries are all too familiar to both amateur and professional athletes. Treatment is often costly and prolonged, and repairs seldom make the joint "good as new."

Table 9–4 summarizes information about the articulations of the appendicular skeleton.

CHECKPOINT

17. Name the bones making up the shoulder joint and knee joint, respectively.

18. At what site are the iliofemoral ligament, pubofemoral ligament, and ischiofemoral ligament located?

19. What signs and symptoms would you expect in an individual who has damaged the menisci of the knee joint?

20. Why is "clergyman's knee" (a type of bursitis) common among carpet layers and roofers?

See the blue Answers tab at the end of the book.

9-7 With advancing age, arthritis and other degenerative changes impair joint mobility

Joints are subjected to heavy wear and tear throughout our lifetimes, and problems with joint function are relatively common, especially in older individuals. **Rheumatism** (ROO-muh-tiz-um) is a general term that indicates pain and stiffness affecting the skeletal system, the muscular system, or both. Several major forms of rheumatism exist. **Arthritis** (ar-THRĪ-tis) encompasses all the rheumatic diseases that affect

CLINICAL NOTE

Knee Injuries Athletes place tremendous stresses on their knees. Ordinarily, the medial and lateral menisci move as the position of the femur changes. Placing a lot of weight on the knee while it is partially flexed can trap a meniscus between the tibia and femur, resulting in a break or tear in the cartilage. In the most common injury, the lateral surface of the leg is driven medially, tearing the medial meniscus. In addition to being quite painful, the torn cartilage may restrict movement at the joint. It can also lead to chronic problems and the development of a "trick knee"—a knee that feels unstable. Sometimes the meniscus can be heard and felt popping in and out of position when the knee is extended.

Less common knee injuries involve tearing one or more stabilizing ligaments or damaging the patella. Torn ligaments can be difficult to correct surgically, and healing is slow. The patella can be injured in a number of ways. If the leg is immobilized (as it might be in a football pileup) while you try to extend the knee, the muscles are powerful enough to pull the patella apart. Impacts to the anterior surface of the knee can also shatter the patella. Knee injuries may lead to chronic painful arthritis that impairs walking. Total knee replacement surgery is rarely performed on young people, but they are becoming increasingly common among elderly patients with severe arthritis.

SUMMARY TABLE 9–4   Articulations of the Appendicular Skeleton

	Element	Joint	Type of Articulation	Movements
	ARTICULATIONS OF THE PECTORAL GIRDLE AND UPPER LIMB			
	Sternum/clavicle	Sternoclavicular	Gliding diarthrosis*	Protraction/retraction, elevation/depression, slight rotation
	Scapula/clavicle	Acromioclavicular	Gliding diarthrosis	Slight movement
	Scapula/humerus	Shoulder, or glenohumeral	Ball-and-socket diarthrosis	Flexion/extension, adduction/abduction, circumduction, rotation
	Humerus/ulna and humerus/radius	Elbow (humero-ulnar and humeroradial)	Hinge diarthrosis	Flexion/extension
	Radius/ulna	Proximal radio-ulnar	Pivot diarthrosis	Rotation
		Distal radio-ulnar	Pivot diarthrosis	Pronation/supination
	Radius/carpal bones	Radiocarpal	Ellipsoid diarthrosis	Flexion/extension, adduction/abduction, circumduction
	Carpal bone to carpal bone	Intercarpal	Gliding diarthrosis	Slight movement
	Carpal bone to metacarpal bone (I)	Carpometacarpal of thumb	Saddle diarthrosis	Flexion/extension, adduction/abduction, circumduction, opposition
	Carpal bone to metacarpal bone (II–V)	Carpometacarpal	Gliding diarthrosis	Slight flexion/extension, adduction/abduction
	Metacarpal bone to phalanx	Metacarpophalangeal	Ellipsoid diarthrosis	Flexion/extension, adduction/abduction, circumduction
	Phalanx/phalanx	Interphalangeal	Hinge diarthrosis	Flexion/extension
	ARTICULATIONS OF THE PELVIC GIRDLE AND LOWER LIMB			
	Sacrum/ilium of coxal bone	Sacroiliac	Gliding diarthrosis	Slight movement
	Coxal bone/coxal bone	Pubic symphysis (symphysis)	Amphiarthrosis	None†
	Coxal bone/femur	Hip	Ball-and-socket diarthrosis	Flexion/extension, adduction/abduction, circumduction, rotation
	Femur/tibia	Knee	Complex, functions as hinge	Flexion/extension, limited rotation
	Tibia/fibula	Tibiofibular (proximal)	Gliding diarthrosis	Slight movement
		Tibiofibular (distal)	Gliding diarthrosis and amphiarthrotic syndesmosis	Slight movement
	Tibia and fibula with talus	Ankle, or talocrural	Hinge diarthrosis	Flexion/extension (dorsiflexion/plantar flexion)
	Tarsal bone to tarsal bone	Intertarsal	Gliding diarthrosis	Slight movement
	Tarsal bone to metatarsal bone	Tarsometatarsal	Gliding diarthrosis	Slight movement
	Metatarsal bone to phalanx	Metatarsophalangeal	Ellipsoid diarthrosis	Flexion/extension, adduction/abduction
	Phalanx/phalanx	Interphalangeal	Hinge diarthrosis	Flexion/extension

*A "double gliding joint," with two joint cavities separated by an articular cartilage.

†During pregnancy, hormones weaken the symphysis and permit movement important to childbirth; see Chapter 29.

synovial joints. Arthritis always involves damage to the articular cartilages, but the specific cause can vary. For example, arthritis can result from bacterial or viral infection, injury to the joint, metabolic problems, or severe physical stresses.

Osteoarthritis (os-tē-ō-ar-THRĪ-tis), also known as *degenerative arthritis* or *degenerative joint disease (DJD)*, generally affects individuals age 60 or older. Osteoarthritis can result from cumulative wear and tear at the joint surfaces or from genetic factors affecting collagen formation. In the U.S. population, 25 percent of women and 15 percent of men over age 60 show signs of this disease.

Rheumatoid arthritis is an inflammatory condition that affects roughly 0.5–1.0 percent of the adult population. At least some cases occur when the immune response mistakenly attacks the joint tissues. Such a condition, in which the body attacks its own tissues, is called an *autoimmune disease*. Allergies, bacteria, viruses, and genetic factors have all been proposed as contributing to or triggering the destructive inflammation.

In *gouty arthritis*, or *crystal arthritis*, crystals of uric acid form within the synovial fluid of joints. The accumulation of crystals of uric acid over time eventually interferes with normal movement. This form of arthritis is named after the metabolic disorder known as *gout*, discussed further in Chapter 25. In gout, the crystals are derived from uric acid (a metabolic waste product), and the joint most often affected is the metatarsal–phalangeal joint of the great toe. Gout is relatively rare, but other forms of gouty arthritis are much more common—some degree of calcium salt deposition occurs in the joints in 30–60 percent of those over age 85. The cause is unknown, but the condition appears to be linked to age-related changes in the articular cartilages.

Regular exercise, physical therapy, and drugs that reduce inflammation (such as aspirin) can often slow the progress of osteoarthritis. Surgical procedures can realign or redesign the affected joint. In extreme cases involving the hip, knee, elbow, or shoulder, the defective joint can be replaced by an artificial one.

Degenerative changes comparable to those seen in arthritis may result from joint immobilization. When motion ceases, so does the circulation of synovial fluid, and the cartilages begin to degenerate. **Continuous passive motion (CPM)** of any injured joint appears to encourage the repair process by improving the circulation of synovial fluid. The movement is often performed by a physical therapist or a machine during the recovery process.

With age, bone mass decreases and bones become weaker, so the risk of fractures increases. ∞ p. 207 If osteoporosis develops, the bones may weaken to the point at which fractures occur in response to stresses that could easily be tolerated by normal bones. Hip fractures are among the most dangerous fractures seen in elderly people, with or without osteoporosis. These fractures, most often involving individuals over age 60, may be accompanied by hip dislocation or by pelvic fractures.

Although severe hip fractures are most common among those over age 60, in recent years the frequency of hip fractures has increased dramatically among young, healthy professional athletes.

> ### CHECKPOINT
> 21. Define rheumatism.
> 22. Define arthritis.
> See the blue Answers tab at the end of the book.

9-8 The skeletal system supports and stores energy and minerals for other body systems

Although the bones you study in the lab may seem to be rigid and unchanging structures, the living skeleton is dynamic and undergoes continuous remodeling. The balance between osteoblast and osteoclast activity is delicate and subject to change at a moment's notice. When osteoblast activity predominates, bones thicken and strengthen; when osteoclast activity predominates, bones get thinner and weaker. The balance between bone formation and bone recycling varies with (1) the age of the individual, (2) the physical stresses applied to the bone, (3) circulating hormone levels, (4) rates of calcium and phosphorus absorption and excretion, and (5) genetic or environmental factors. Most of these variables involve some interaction between the skeletal system and other body systems.

In fact, the skeletal system is intimately associated with other systems. For instance, the bones of the skeleton are attached to the muscular system, extensively connected to the cardiovascular and lymphoid systems, and largely under the physiological control of the endocrine system. The digestive and urinary systems also play important roles in providing the calcium and phosphate minerals needed for bone growth. In return, the skeleton represents a reserve of calcium, phosphate, and other minerals that can compensate for reductions in the dietary supply of those ions. **Figure 9–13** reviews the components and functions of the skeletal system, and diagrams the major functional relationships between that system and other systems.

> ### CHECKPOINT
> 23. Describe the functional relationship between the skeletal system and the digestive system.
> 24. Describe the functional relationship between the skeletal system and all other body systems.
> See the blue Answers tab at the end of the book.

Related Clinical Terms

arthritis: A group of rheumatic diseases that affect synovial joints. Arthritis always involves damage to the articular cartilages, but the specific cause can vary. The diseases of arthritis are usually classified as either *degenerative* or *inflammatory*.

bunion: The most common pressure-related bursitis, involving a tender nodule formed around bursae over the base of the great toe.

bursitis: An inflammation of a bursa, causing pain and swelling whenever the associated tendon or ligament moves.

continuous passive motion (CPM): A therapeutic procedure involving the passive movement of an injured joint to stimulate the circulation of synovial fluid. The goal is to prevent degeneration of the articular cartilages.

herniated disc: A condition caused by intervertebral compression severe enough to rupture the anulus fibrosus and release the nucleus pulposus, which may protrude beyond the intervertebral space.

luxation: A dislocation; a condition in which the articulating surfaces are forced out of position.

THE SKELETAL SYSTEM IN PERSPECTIVE

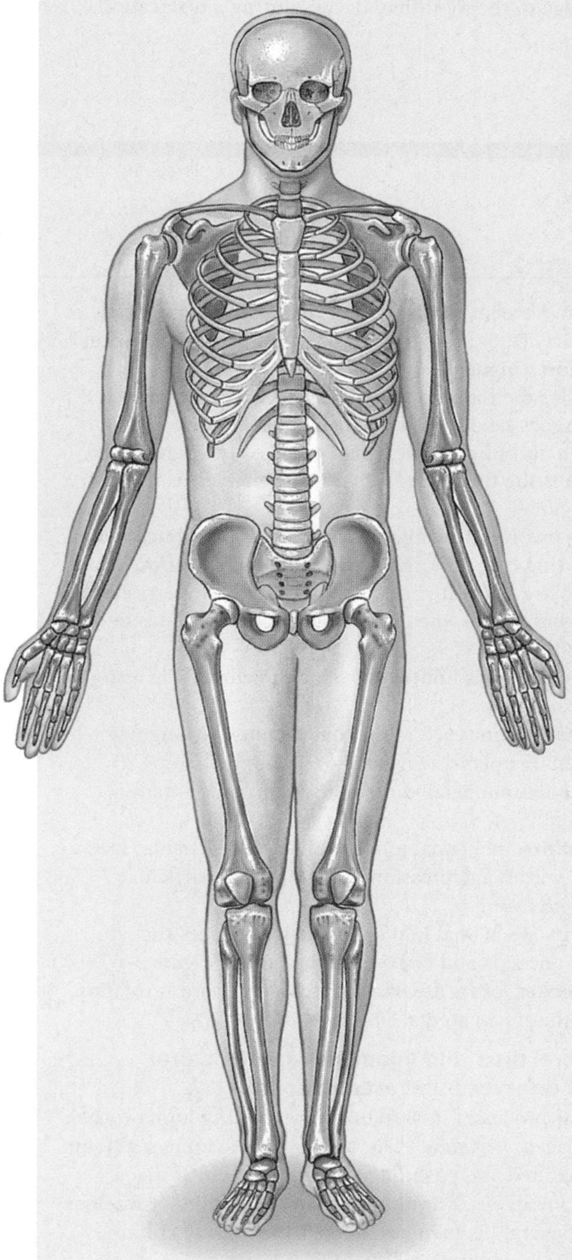

Integumentary System

- Synthesizes vitamin D₃, essential for calcium and phosphorus absorption (bone maintenance and growth)
- Provides structural support

Muscular System

- Stabilizes bone positions; tension in tendons stimulates bone growth and maintenance
- Provides calcium needed for normal muscle contraction; bones act as levers to produce body movements

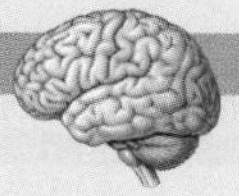

Nervous System

- Regulates bone position by controlling muscle contractions
- Provides calcium for neural function; protects brain, spinal cord; receptors at joints provide information about body position

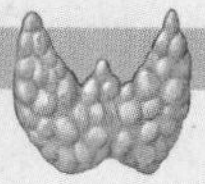

Endocrine System

- Skeletal growth regulated by growth hormone, thyroid hormones, and sex hormones; calcium mobilization regulated by parathyroid hormone and calcitonin
- Protects endocrine organs, especially in brain, chest, and pelvic cavity

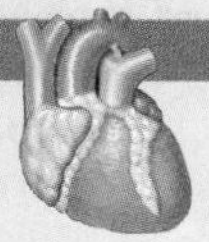

Cardiovascular System

- Provides oxygen, nutrients, hormones, blood cells; removes waste products and carbon dioxide
- Provides calcium needed for cardiac muscle contraction, blood cells produced in bone marrow; axial skeleton protects heart and great vessels

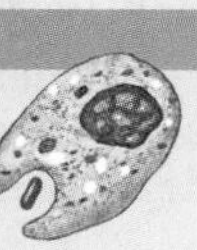

Lymphoid System

- Lymphocytes assist in the defense and repair of bone following injuries
- Lymphocytes and other cells of the immune response are produced and stored in bone marrow

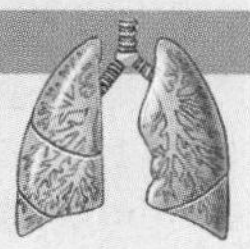

Respiratory System

- Provides oxygen and eliminates carbon dioxide
- Movements of ribs important in breathing; axial skeleton surrounds and protects lungs

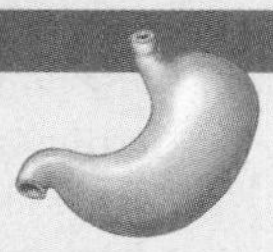

Digestive System

- Provides nutrients, calcium, and phosphate
- Ribs protect portions of liver, stomach, and intestines

Urinary System

- Conserves calcium and phosphate needed for bone growth; disposes of waste products
- Axial skeleton provides some protection for kidneys and ureters; pelvis protects urinary bladder and proximal urethra

Reproductive System

- Sex hormones stimulate growth and maintenance of bones; surge of sex hormones at puberty causes acceleration of growth and closure of epiphyseal cartilages
- Pelvis protects reproductive organs of female, protects portion of ductus deferens and accessory glands in males

FOR ALL SYSTEMS

Provides mechanical support; stores energy reserves; stores calcium and phosphate reserves

Figure 9–13 **Functional Relationships between the Skeletal System and Other Systems.**

osteoarthritis (*degenerative arthritis* or *degenerative joint disease, DJD*)**:** An arthritic condition resulting from either cumulative wear and tear on joint surfaces or a genetic predisposition. In the U.S. population, 25 percent of women and 15 percent of men over age 60 show signs and symptoms of this disease.

rheumatism: A general term that indicates pain and stiffness affecting the skeletal system, the muscular system, or both.

rheumatoid arthritis: An inflammatory arthritis that affects roughly 0.5–1.0 percent of the adult U.S. population. The cause is uncertain, although allergies, bacteria, viruses, and genetic factors have all been proposed. The primary sign is *synovitis*—swelling and inflammation of the synovial membrane.

shoulder separation: The partial or complete dislocation of the acromioclavicular joint.

slipped disc: A common name for a condition caused by the distortion of an intervertebral disc. The distortion applies pressure to spinal nerves, causing pain and limiting range of motion.

sprain: A condition in which a ligament is stretched to the point at which some of the collagen fibers are torn. The ligament remains functional, and the structure of the joint is not affected.

subluxation: A partial dislocation; the displacement of articulating surfaces sufficient to cause discomfort, but resulting in less physical damage to the joint than occurs during a dislocation.

Chapter Review
Study Outline

An Introduction to Articulations p. 268

1. **Articulations** (joints) exist wherever two bones interconnect.

9-1 Joints are categorized according to their range of motion or anatomical organization p. 268

2. *Immovable joints* are **synarthroses**; *slightly movable joints* are **amphiarthroses**; and joints that are *freely movable* are called **diarthroses** or **synovial joints**. (*Table 9–1*)

3. Alternatively, joints are classified structurally, as *bony, fibrous, cartilaginous,* or *synovial*. (*Table 9–2*)

4. The four major types of synarthroses are a **suture** (skull bones bound together by dense connective tissue), a **gomphosis** (teeth bound to bony sockets by *periodontal ligaments*), a **synchondrosis** (two bones joined by a rigid cartilaginous bridge), and a **synostosis** (two bones completely fused).

5. The two major types of amphiarthroses are a **syndesmosis** (bones connected by a ligament) and a **symphysis** (bones separated by fibrous cartilage).

6. The bony surfaces at diarthroses are enclosed within an **articular capsule** lined by a synovial membrane.

9-2 Synovial joints are freely movable articulations (diarthroses) containing synovial fluid p. 270

7. The bony surfaces within a synovial joint are covered by **articular cartilages**, and lubricated by **synovial fluid**.

8. Accessory synovial structures include **menisci**, or *articular discs*; **fat pads**; **accessory ligaments**; **tendons**; and **bursae**. (*Figure 9–1*)

9. A **dislocation** occurs when articulating surfaces are forced out of position.

9-3 Anatomical and functional properties of synovial joints enable various skeletal movements p. 272

10. The possible types of articular movements are **linear motion (gliding)**, **angular motion**, and **rotation**. (*Figure 9–2*)

11. Joints are called **monaxial**, **biaxial**, or **triaxial**, depending on the planes of movement they allow.

12. In **gliding**, two opposing surfaces slide past one another.

13. Important terms that describe angular motion are **flexion**, **extension**, **hyperextension**, **abduction**, **adduction**, and **circumduction**. (*Figure 9–3*)

14. Rotational movement can be **left** or **right**, **medial** (*internal*) or **lateral** (*external*), or, in the bones of the forearm, **pronation** or **supination**. (*Figure 9–4*)

15. Movements of the foot include **inversion** and **eversion**. The ankle undergoes flexion and extension, also known as **dorsiflexion** and **plantar flexion**, respectively. (*Figure 9–5*)

16. **Opposition** is the thumb movement that enables us to grasp objects. (*Figure 9–5*)

17. **Protraction** involves moving something anteriorly; **retraction** involves moving it posteriorly. **Depression** and **elevation** occur when we move a structure inferiorly and superiorly, respectively. **Lateral flexion** occurs when the vertebral column bends to one side. (*Figure 9–5*)

18. **Gliding joints** permit limited movement, generally in a single plane. (*Figure 9–6*)

19. **Hinge joints** are monaxial joints that permit only angular movement in one plane. (*Figure 9–6*)

20. **Pivot joints** are monaxial joints that permit only rotation. (*Figure 9–6*)

21. **Ellipsoid joints** are biaxial joints with an oval articular face that nestles within a depression in the opposing articular surface. (*Figure 9–6*)

22. **Saddle joints** are biaxial joints with articular faces that are concave on one axis and convex on the other. (*Figure 9–6*)

23. **Ball-and-socket joints** are triaxial joints that permit rotation as well as other movements. (*Figure 9–6*)

9-4 Intervertebral discs and ligaments are structural components of intervertebral articulations p. 277

24. The articular processes of vertebrae form gliding joints with those of adjacent vertebrae. The bodies form symphyseal joints that are separated and cushioned by **intervertebral discs**, which contain an outer **anulus fibrosus** and an inner **nucleus pulposus**. Several ligaments stabilize the vertebral column. (*Figures 9–7, 9–8; Summary Table 9–3*)

9-5 The shoulder is a ball-and-socket joint, and the elbow is a hinge joint p. 280

25. The shoulder joint, or *glenohumeral joint*, is formed by the glenoid cavity and the head of the humerus. This articulation

permits the greatest range of motion of any joint. It is a ball-and-socket diarthrosis with various stabilizing ligaments. Strength and stability are sacrificed in favor of mobility. (*Figure 9–9; Summary Table 9–4*)

26. The elbow joint permits only flexion–extension. It is a hinge diarthrosis whose capsule is reinforced by strong ligaments. (*Figure 9–10; Summary Table 9–4*)

9-6 The hip is a ball-and-socket joint, and the knee is a hinge joint p. 283

27. The hip joint is a ball-and-socket diarthrosis formed by the union of the acetabulum with the head of the femur. The joint permits flexion–extension, adduction–abduction, circumduction, and rotation; it is stabilized by numerous ligaments. (*Figure 9–11; Summary Table 9–4*)

28. The knee joint is a hinge joint formed by the union of the condyles of the femur with the superior condylar surfaces of the tibia. The joint permits flexion–extension and limited rotation,

and it has various supporting ligaments. (*Figure 9–12; Summary Table 9–4*)

9-7 With advancing age, arthritis and other degenerative changes impair joint mobility p. 287

29. Problems with joint function are relatively common, especially in older individuals. **Rheumatism** is a general term for pain and stiffness affecting the skeletal system, the muscular system, or both; several major forms exist. **Arthritis** encompasses all the rheumatic diseases that affect synovial joints. Both conditions become increasingly common with age.

9-8 The skeletal system supports and stores energy and minerals for other body systems p. 288

30. The skeletal system interacts extensively with the muscular, cardiovascular, lymphoid, digestive, urinary, and endocrine systems. (*Figure 9–13*)

Review Questions

See the blue Answers tab at the end of the book.

 Access more review material online at **myA&P™** (www.myaandp.com). There, you'll find chapter guides, chapter quizzes, practice tests, animations, flashcards, a glossary with pronunciations, *Interactive Physiology*® (IP) exercises and quizzes, and more to help you succeed in the course.

LEVEL 1   Reviewing Facts and Terms

1. Label the structures in the following illustration of a synovial joint.

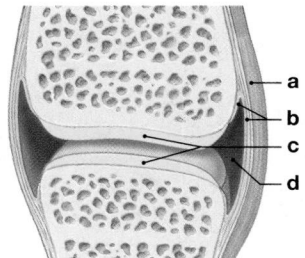

   (a) _____   (b) _____
   (c) _____   (d) _____

2. A synarthrosis located between the bones of the skull is a
   (a) symphysis.   (b) syndesmosis.
   (c) synchondrosis.   (d) suture.

3. The articulation between adjacent vertebral bodies is a
   (a) syndesmosis.   (b) symphysis.
   (c) synchondrosis.   (d) synostosis.

4. The anterior articulation between the two pubic bones is a
   (a) synchondrosis.   (b) synostosis.
   (c) symphysis.   (d) synarthrosis.

5. Joints typically located between the ends of adjacent long bones are
   (a) synarthroses.   (b) amphiarthroses.
   (c) diarthroses.   (d) symphyses.

6. The function of the articular cartilage is
   (a) to reduce friction.
   (b) to prevent bony surfaces from contacting one another.
   (c) to provide lubrication.
   (d) both a and b.

7. Which of the following is *not* a function of synovial fluid?
   (a) shock absorption
   (b) nutrient distribution
   (c) maintenance of ionic balance
   (d) lubrication of the articular surfaces
   (e) waste disposal

8. The structures that limit the range of motion of a joint and provide mechanical support across or around the joint are
   (a) bursae.   (b) tendons.
   (c) menisci.   (d) a, b, and c.

9. A partial dislocation of an articulating surface is a
   (a) circumduction.   (b) hyperextension.
   (c) subluxation.   (d) supination.

10. Abduction and adduction always refer to movements of the
   (a) axial skeleton.   (b) appendicular skeleton.
   (c) skull.   (d) vertebral column.

11. Rotation of the forearm that makes the palm face posteriorly is
   (a) supination.   (b) pronation.
   (c) proliferation.   (d) projection.

12. A saddle joint permits _____ movement but prevents _____ movement.
   (a) rotational, gliding   (b) angular, linear
   (c) linear, rotational   (d) angular, rotational

13. Standing on tiptoe is an example of _____ at the ankle.
   (a) elevation   (b) flexion
   (c) extension   (d) retraction

14. Examples of monaxial joints, which permit angular movement in a single plane, are
   (a) the intercarpal and intertarsal joints.
   (b) the shoulder and hip joints.
   (c) the elbow and knee joints.
   (d) all of these.

15. Decreasing the angle between bones is termed
    (a) flexion.   (b) extension.
    (c) abduction.   (d) adduction.
    (e) hyperextension.
16. Movements that occur at the shoulder and the hip represent
    the actions that occur at a _____ joint.
    (a) hinge   (b) ball-and-socket
    (c) pivot   (d) gliding
17. The anulus fibrosus and nucleus pulposus are structures
    associated with the
    (a) intervertebral discs.   (b) knee and elbow.
    (c) shoulder and hip.   (d) carpal and tarsal bones.
18. Subacromial, subcoracoid, and subscapular bursae reduce
    friction in the _____ joint.
    (a) hip   (b) knee
    (c) elbow   (d) shoulder
19. Although the knee joint is only one joint, it resembles
    _____ separate joints.
    (a) 2   (b) 3
    (c) 4   (d) 5
    (e) 6

LEVEL 2   Reviewing Concepts

20. The hip is an extremely stable joint because it has
    (a) a complete bony socket.
    (b) a strong articular capsule.
    (c) supporting ligaments.
    (d) all of these.
21. Dislocations involving synovial joints are usually prevented
    by all of the following *except*
    (a) structures such as ligaments that stabilize and support the
        joint.
    (b) the position of bursae that limits the degree of movement.
    (c) the presence of other bones that prevent certain movements.
    (d) the position of muscles and fat pads that limits the degree of
        movement.
    (e) the shape of the articular surface.

22. How does a meniscus (articular disc) function in a joint?
23. Partial or complete dislocation of the acromioclavicular joint
    is called a(n) _____.
24. How do articular cartilages differ from other cartilages in the
    body?
25. Differentiate between a slipped disc and a herniated disc.
26. How would you explain to your grandmother the
    characteristic decrease in height with advancing age?
27. List the six different types of diarthroses, and give an example
    of each.

LEVEL 3   Critical Thinking and Clinical Applications

28. While playing tennis, Dave "turns his ankle." He experiences
    swelling and pain. After being examined, he is told that he has
    no ruptured ligaments and that the structure of the ankle is
    not affected. On the basis of the signs and symptoms and the
    examination results, what happened to Dave's ankle?
29. Joe injures his knee during a football practice such that the
    synovial fluid in the knee joint no longer circulates normally. The
    physician who examines him tells him that they have to
    reestablish circulation of the synovial fluid before the articular
    cartilages become damaged. Why?
30. When playing a contact sport, which injury would you expect
    to occur more frequently, a dislocated shoulder or a dislocated
    hip? Why?

Muscle Tissue 10

Did you know...?
If all the body muscles were pulling in one direction, they would develop 25 tons of force.

Learning Outcomes

After completing this chapter, you should be able to do the following:

10-1 Specify the functions of skeletal muscle tissue.

10-2 Describe the organization of muscle at the tissue level.

10-3 Explain the characteristics of skeletal muscle fibers, and identify the structural components of a sarcomere.

10-4 Identify the components of the neuromuscular junction, and summarize the events involved in the neural control of skeletal muscles.

10-5 Describe the mechanism responsible for tension production in a muscle fiber, and compare the different types of muscle contraction.

10-6 Describe the mechanisms by which muscle fibers obtain the energy to power contractions.

10-7 Relate the types of muscle fibers to muscle performance, and distinguish between aerobic and anaerobic endurance.

10-8 Identify the structural and functional differences between skeletal muscle fibers and cardiac muscle cells.

10-9 Identify the structural and functional differences between skeletal muscle fibers and smooth muscle cells, and discuss the roles of smooth muscle tissue in systems throughout the body.

Clinical Notes

Tetanus p. 304
Rigor Mortis p. 311
Delayed-Onset Muscle Soreness p. 326

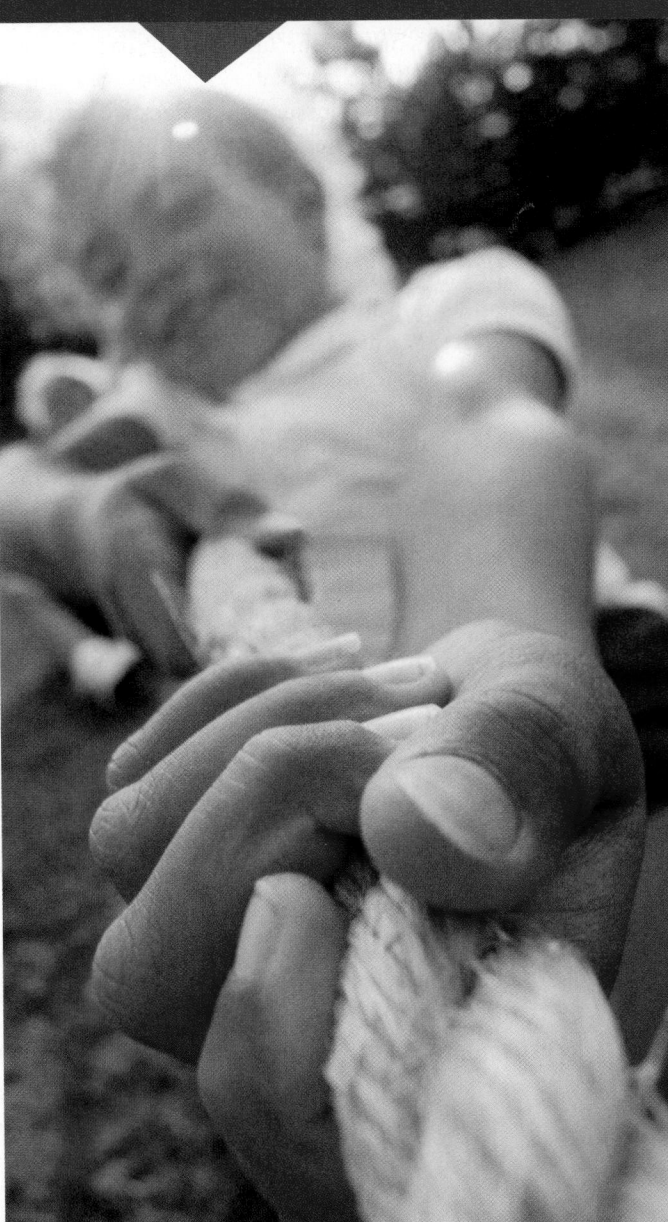

An Introduction to Muscle Tissue

This chapter discusses one of the four primary tissue types, with particular attention to skeletal muscle tissue. It also examines the histological and physiological characteristics of skeletal muscle cells, and relates those features to the functional properties of the entire tissue.

10-1 Skeletal muscle performs six major functions

Muscle tissue, one of the four primary types of tissue, consists chiefly of muscle cells that are highly specialized for contraction. Three types of muscle tissue exist: (1) *skeletal muscle*, (2) *cardiac muscle*, and (3) *smooth muscle*. ∞ p. 138 Without these muscle tissues, nothing in the body would move, and no body movement could occur. Skeletal muscle tissue moves the body by pulling on bones of the skeleton, making it possible for us to walk, dance, bite an apple, or play the ukulele. Cardiac muscle tissue pushes blood through the circulatory system. Smooth muscle tissue pushes fluids and solids along the digestive tract and regulates the diameters of small arteries, among other functions.

This chapter primarily describes the structure and function of skeletal muscle tissue, in preparation for our discussion of the muscular system (Chapter 11). This chapter also provides an overview of the differences among skeletal, cardiac, and smooth muscle tissues.

Skeletal muscles are organs composed primarily of skeletal muscle tissue, but they also contain connective tissues, nerves, and blood vessels. Each cell in skeletal muscle tissue is a single muscle *fiber*. Skeletal muscles are directly or indirectly attached to the bones of the skeleton. Our skeletal muscles perform the following six functions:

1. *Produce Skeletal Movement.* Skeletal muscle contractions pull on tendons and move the bones of the skeleton. The effects range from simple motions such as extending the arm or breathing, to the highly coordinated movements of swimming, skiing, or typing.

2. *Maintain Posture and Body Position.* Tension in our skeletal muscles maintains body posture—for example, holding your head still when you read a book or balancing your body weight above your feet when you walk. Without constant muscular activity, we could neither sit upright nor stand.

3. *Support Soft Tissues.* The abdominal wall and the floor of the pelvic cavity consist of layers of skeletal muscle. These muscles support the weight of visceral organs and shield internal tissues from injury.

4. *Guard Entrances and Exits.* The openings of the digestive and urinary tracts are encircled by skeletal muscles. These muscles provide voluntary control over swallowing, defecation, and urination.

5. *Maintain Body Temperature.* Muscle contractions require energy; whenever energy is used in the body, some of it is converted to heat. The heat released by working muscles keeps body temperature in the range required for normal functioning.

6. *Store Nutrient Reserves.* When the diet contains inadequate proteins or calories, the contractile proteins in skeletal muscles are broken down, and the amino acids released into the circulation. Some of these amino acids can be used by the liver to synthesize glucose; others can be broken down to provide energy.

We will begin our discussion with the functional anatomy of a typical skeletal muscle, with particular emphasis on the microscopic structural features that make contractions possible.

CHECKPOINT

1. **Identify the three types of muscle tissue.**

2. **Identify the six major functions of skeletal muscle.**

See the blue Answers tab at the end of the book.

10-2 A skeletal muscle contains muscle tissue, connective tissues, blood vessels, and nerves

Figure 10–1 illustrates the organization of a representative skeletal muscle. In the next section we will examine skeletal muscle tissue in detail; here we consider how connective tissues are organized in skeletal muscle, and how skeletal muscles are supplied with blood vessels and nerves.

Organization of Connective Tissues

Three layers of connective tissue are part of each muscle: (1) an epimysium, (2) a perimysium, and (3) an endomysium. These layers and the relationships among them are diagrammed in **Figure 10–1**.

The entire muscle is surrounded by the **epimysium** (ep-i-MIZ-ē-um; *epi-*, on + *mys*, muscle), a dense layer of collagen fibers. The epimysium separates the muscle from surrounding tissues and organs. It is connected to the deep fascia, a dense connective tissue layer.

The connective tissue fibers of the **perimysium** (per-i-MIZ-ē-um; *peri-*, around) divide the skeletal muscle into a series of compartments, each containing a bundle of muscle fibers called a **fascicle** (FAS-i-kl; *fasciculus*, a bundle). In addition to possessing collagen and elastic fibers, the perimysium contains blood vessels and nerves that maintain blood flow and innervate the muscle fibers within the fascicles. Each fascicle receives branches of these blood vessels and nerves.

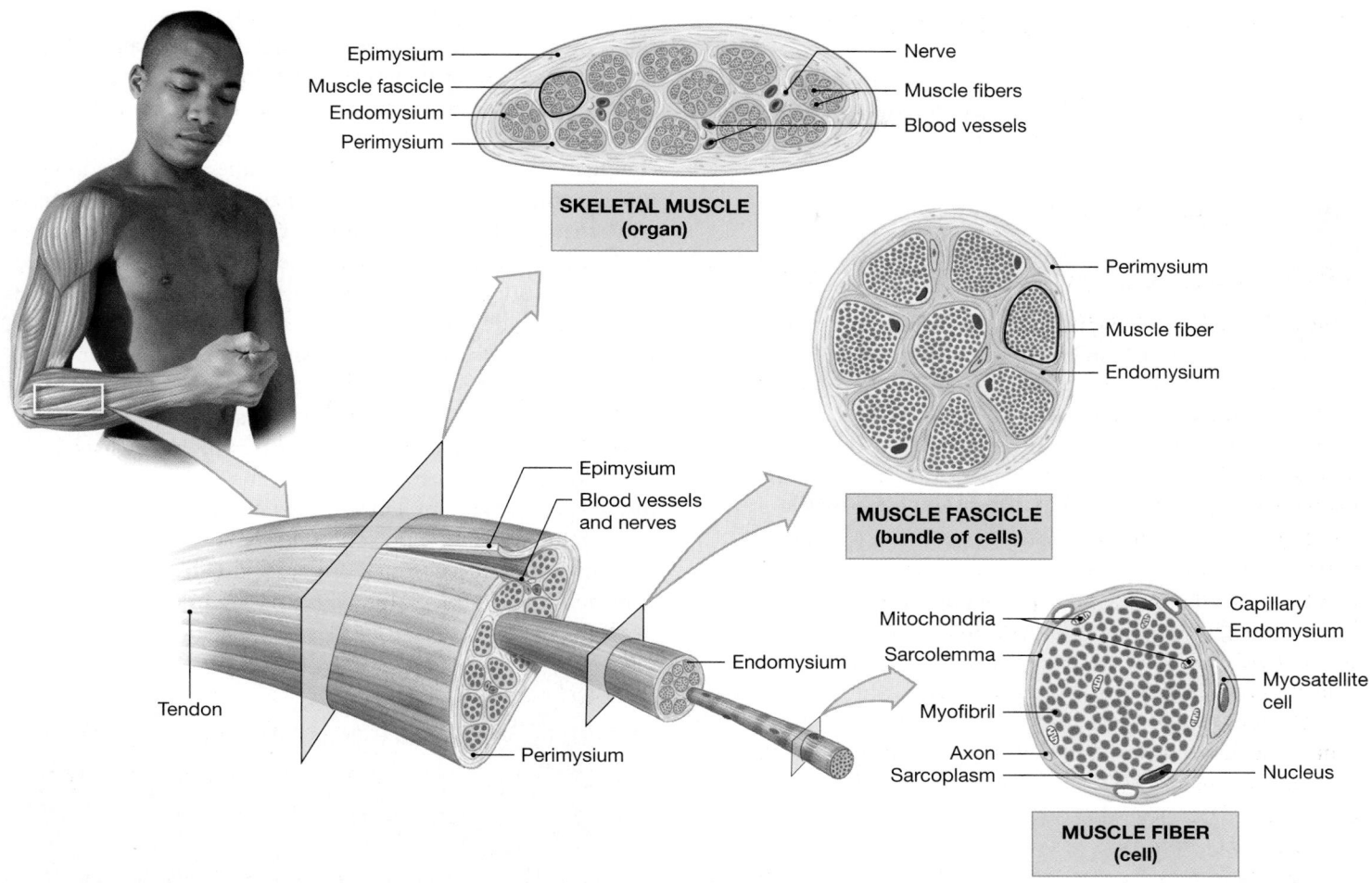

Figure 10–1 **The Organization of Skeletal Muscles.** A skeletal muscle consists of fascicles (bundles of muscle fibers) enclosed by the epimysium. The bundles are separated by connective tissue fibers of the perimysium, and within each bundle the muscle fibers are surrounded by the endomysium. Each muscle fiber has many superficial nuclei, as well as mitochondria and other organelles (*See Figure 10–3*).

Within a fascicle, the delicate connective tissue of the **endomysium** (en-dō-MIZ-ē-um; *endo-*, inside) surrounds the individual skeletal muscle cells, or *muscle fibers*, and loosely interconnects adjacent muscle fibers. This flexible, elastic connective tissue layer contains (1) capillary networks that supply blood to the muscle fibers; (2) **myosatellite cells**, embryonic stem cells that function in the repair of damaged muscle tissue; and (3) nerve fibers that control the muscle. All these structures are in direct contact with the individual muscle fibers. ∞ p. 138

The collagen fibers of the perimysium and endomysium are interwoven and blend into one another. At each end of the muscle, the collagen fibers of the epimysium, perimysium, and endomysium come together to form either a bundle known as a **tendon**, or a broad sheet called an **aponeurosis** (ap-ō-noo-RŌ-sis). Tendons and aponeuroses usually attach skeletal muscles to bones. Where they contact the bone, the collagen fibers extend into the bone matrix, providing a firm attachment. As a result, any contraction of the muscle will exert a pull on the attached bone (or bones).

Blood Vessels and Nerves

The connective tissues of the endomysium and perimysium contain the blood vessels and nerves that supply the muscle fibers. Muscle contraction requires tremendous quantities of energy. An extensive vascular network delivers the necessary oxygen and nutrients and carries away the metabolic wastes generated by active skeletal muscles. The blood vessels and the nerve supply generally enter the muscle together and follow the same branching course through the perimysium. Within the endomysium, arterioles supply blood to a capillary network that services the individual muscle fiber.

Skeletal muscles contract only under stimulation from the central nervous system. Axons, or *nerve fibers*, penetrate the epimysium, branch through the perimysium, and enter the endomysium to innervate individual muscle fibers. Skeletal muscles are often called voluntary muscles, because we have voluntary control over their contractions. Many skeletal muscles may also be controlled at a subconscious level. For example, skeletal muscles involved with breathing, such as the *diaphragm*, usually work outside our conscious awareness.

Next, we will examine the microscopic structure of a typical skeletal muscle fiber and relate that microstructure to the physiology of the contraction process.

10-3 Skeletal muscle fibers have distinctive features

Skeletal muscle fibers are quite different from the "typical" cells we described in Chapter 3. One obvious difference is size:

Skeletal muscle fibers are enormous. A muscle fiber from a thigh muscle could have a diameter of 100 μm and a length equal to the distance between the tendons at either end (up to 30 cm, or 12 in.). A second obvious difference is that skeletal muscle fibers are *multinucleate*: Each contains hundreds of nuclei just internal to the plasma membrane. The genes in these nuclei control the production of enzymes and structural proteins required for normal muscle contraction, and the more copies of these genes, the faster these proteins can be produced.

The distinctive features of size and multiple nuclei are related. During development, groups of embryonic cells called **myoblasts** (*myo-*, muscle + *blastos*, formative cell or germ) fuse, forming individual multinucleate skeletal muscle fibers (**Figure 10–2**). Each nucleus in a skeletal muscle fiber reflects the contribution of a single myoblast. Some myoblasts, however, do not fuse with developing muscle fibers. These unfused cells remain in adult skeletal muscle tissue as the myosatellite cells seen in **Figures 10–1** and **10–2a**. After an injury, myosatellite cells may enlarge, divide, and fuse with damaged muscle fibers, thereby assisting in the repair of the tissue.

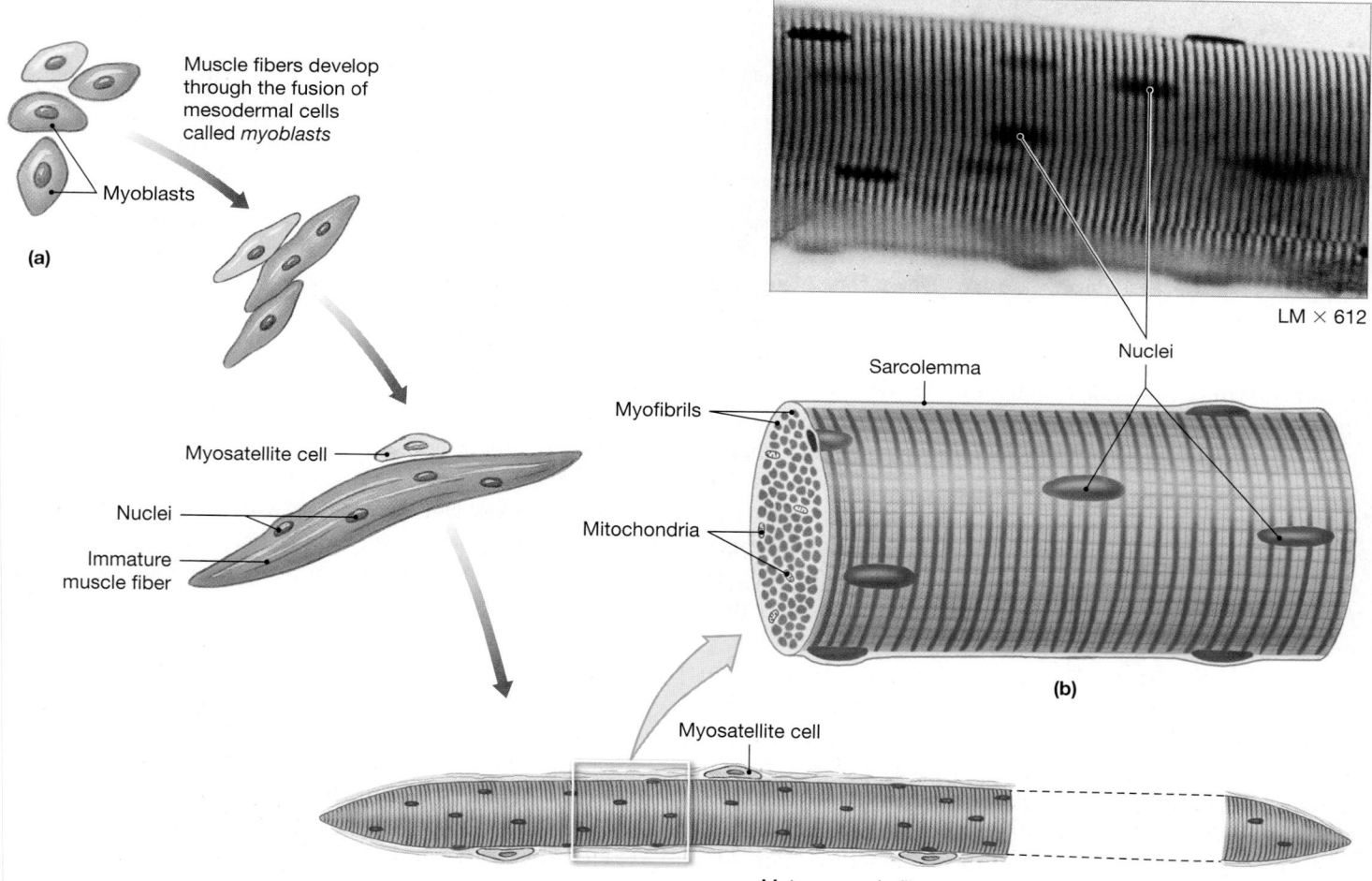

Figure 10–2   The Formation of a Multinucleate Skeletal Muscle Fiber. (a) The formation of a muscle fiber by the fusion of myoblasts. **(b)** A diagrammatic view and a micrograph of one muscle fiber.

The Sarcolemma and Transverse Tubules

The **sarcolemma** (sar-kō-LEM-uh; *sarkos*, flesh + *lemma*, husk), or plasma membrane of a muscle fiber, surrounds the **sarcoplasm** (SAR-kō-plazm), or cytoplasm of the muscle fiber (**Figure 10–3**). Like other plasma membranes, the sarcolemma has a characteristic transmembrane potential due to the unequal distribution of positive and negative charges across the membrane. ∞ p. 99 In a skeletal muscle fiber, a sudden change in the transmembrane potential is the first step that leads to a contraction.

Even though a skeletal muscle fiber is very large, all regions of the cell must contract simultaneously. Thus, the signal to contract must be distributed quickly throughout the interior of the cell. This signal is conducted through the transverse tubules. **Transverse tubules**, or **T tubules**, are narrow tubes that are continuous with the sarcolemma and extend into the sarcoplasm at right angles to the cell surface (**Figure 10–3**). Filled with extracellular fluid, T tubules form passageways through the muscle fiber, like a network of tunnels through a mountain. The T tubules have the same gen-

eral properties as the sarcolemma, so electrical impulses conducted by the sarcolemma travel along the T tubules into the cell interior. These impulses, or *action potentials*, are the triggers for muscle fiber contraction.

Myofibrils

Inside the muscle fiber, branches of the transverse tubules encircle cylindrical structures called **myofibrils** (**Figure 10–3**). A myofibril is 1–2 μm in diameter and as long as the entire cell. Each skeletal muscle fiber contains hundreds to thousands of myofibrils.

Myofibrils consist of bundles of protein filaments called **myofilaments**. Two types of myofilaments were introduced in Chapter 3): **Thin filaments** are composed primarily of actin, whereas **thick filaments** are composed primarily of myosin. ∞ pp. 73, 74 In addition, myofibrils contain *titin*, elastic myofilaments associated with the thick filaments. (We will consider the role of titin later in the chapter.)

Myofibrils, which can actively shorten, are responsible for skeletal muscle fiber contraction. At each end of the skeletal

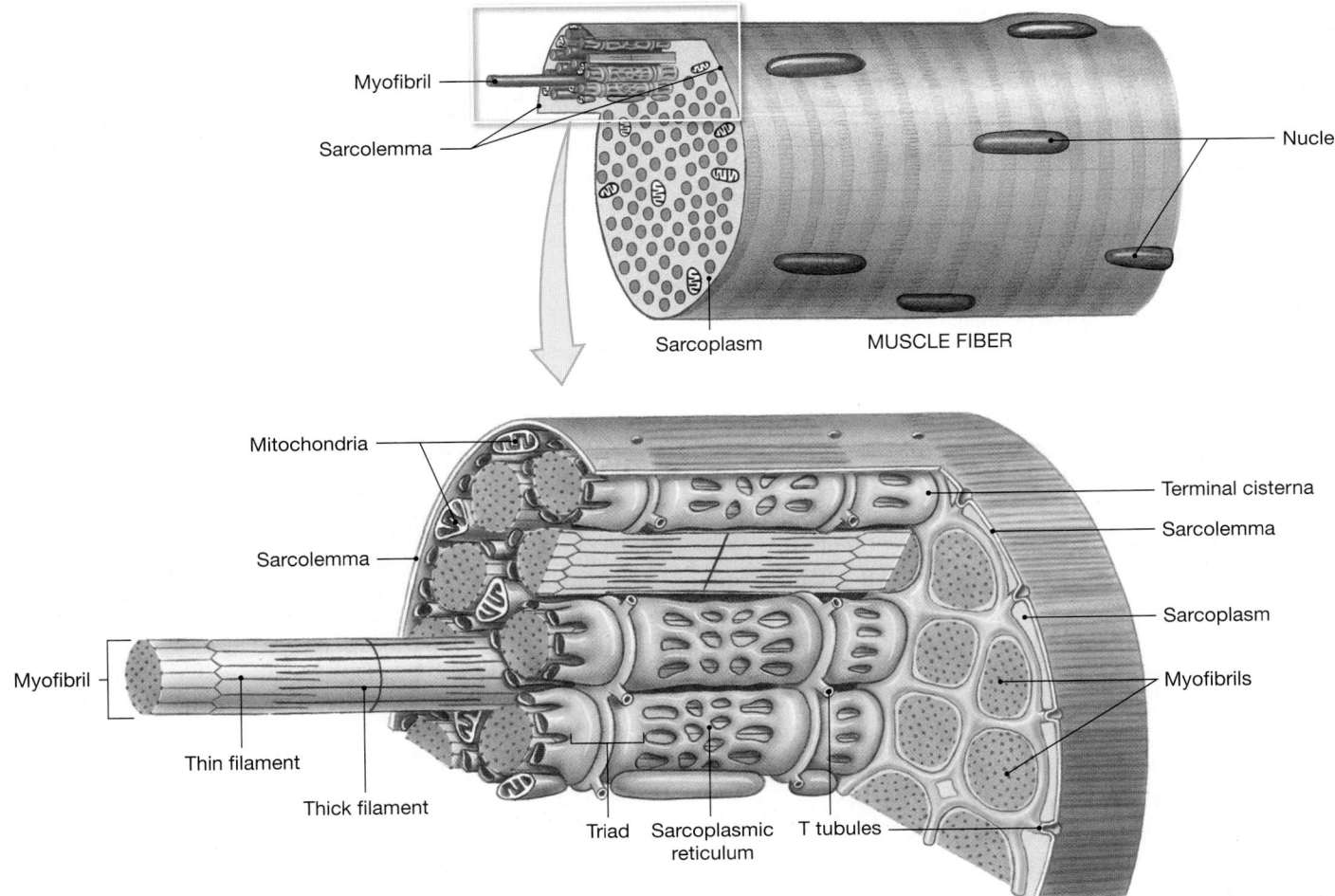

Figure 10–3   The Structure of a Skeletal Muscle Fiber. The internal organization of a muscle fiber.

muscle fiber, the myofibrils are anchored to the inner surface of the sarcolemma. In turn, the outer surface of the sarcolemma is attached to collagen fibers of the tendon or aponeurosis of the skeletal muscle. As a result, when the myofibrils contract, the entire cell shortens and pulls on the tendon. Scattered among and around the myofibrils are mitochondria and granules of glycogen, the storage form of glucose. Glucose breakdown through glycolysis and mitochondrial activity provides the ATP required by short-duration, maximum intensity muscular contractions. ∞ p. 80

The Sarcoplasmic Reticulum

Wherever a transverse tubule encircles a myofibril, the tubule is tightly bound to the membranes of the sarcoplasmic reticulum. The **sarcoplasmic reticulum (SR)** is a membrane complex related to the smooth endoplasmic reticulum of other cells. In skeletal muscle fibers, the SR forms a tubular network around each individual myofibril, fitting over it like a lacy shirt sleeve (**Figure 10–3**). On either side of a T tubule, the tubules of the SR enlarge, fuse, and form expanded chambers called **terminal cisternae** (sis-TUR-nē). The combination of a pair of terminal cisternae plus a transverse tubule is known as a **triad**. Although the membranes of the triad are tightly bound together, their fluid contents are separate and distinct.

In Chapter 3, we noted the existence of special ion pumps that keep the intracellular concentration of calcium ions (Ca^{2+}) very low. ∞ p. 95 Most cells pump the calcium ions across their plasma membranes and into the extracellular fluid. Although skeletal muscle fibers do pump Ca^{2+} out of the cell in this way, they also remove calcium ions from the sarcoplasm by actively transporting them into the terminal cisternae of the sarcoplasmic reticulum. The sarcoplasm of a resting skeletal muscle fiber contains very low concentrations of Ca^{2+} around 10^{-7} mmol/L. The free Ca^{2+} concentration levels inside the terminal cisternae may be as much as 1000 times higher. In addition, cisternae contain the protein *calsequestrin*, which reversibly binds Ca^{2+}. Including both the free calcium and the bound calcium, the total concentration of Ca^{2+} inside cisternae can be 40,000 times that of the surrounding sarcoplasm.

A muscle contraction begins when stored calcium ions are released into the sarcoplasm. These ions then diffuse into individual contractile units called sarcomeres.

Sarcomeres

As we have seen, myofibrils are bundles of thin and thick filaments. These myofilaments are organized into repeating functional units called **sarcomeres** (SAR-kō-mērz; *sarkos*, flesh + *meros*, part).

A myofibril consists of approximately 10,000 sarcomeres, end to end. Each sarcomere has a resting length of about 2 μm. Sarcomeres are the smallest functional units of the muscle fiber. Interactions between the thick and thin filaments of sarcomeres are responsible for muscle contraction. A sarcomere contains (1) thick filaments, (2) thin filaments, (3) proteins that stabilize the positions of the thick and thin filaments, and (4) proteins that regulate the interactions between thick and thin filaments.

Differences in the size, density, and distribution of thick filaments and thin filaments account for the banded appearance of each myofibril. Each sarcomere has dark bands (**A bands**) and light bands (**I bands**) (**Figure 10–4**). The names of these bands are derived from *anisotropic* and *isotropic*, which refer to their appearance when viewed under polarized light.

Tips&Tricks

You can remember that actin forms the backbone of a thin filament by associating the "tin" in ac*tin* with the word *thin*. The th*I*n filaments look l*I*ght and form the *I* band; the **A** bands are d**A**rk.

The A Band

The thick filaments are located at the center of a sarcomere, in the A band. The length of the A band is equal to the length of a typical thick filament. The A band, which also includes portions of thin filaments, contains the following three subdivisions (**Figure 10–4**):

1. *The M Line.* The central portion of each thick filament is connected to its neighbors by proteins of the **M line**. These dark-staining proteins help stabilize the positions of the thick filaments.

2. *The H band.* In a resting sarcomere, the **H band**, or *H zone*, is a lighter region on either side of the M line. The H band contains thick filaments, but no thin filaments.

3. *The Zone of Overlap.* In the **zone of overlap**, thin filaments are situated between the thick filaments. In this region, each thin filament is surrounded by three thick filaments, and each thick filament is surrounded by six thin filaments.

The cross-sectional views in **Figure 10–5** should help you visualize these features of the three-dimensional structure of the sarcomere.

The I Band

Each I band, which contains thin filaments but not thick filaments, extends from the A band of one sarcomere to the A

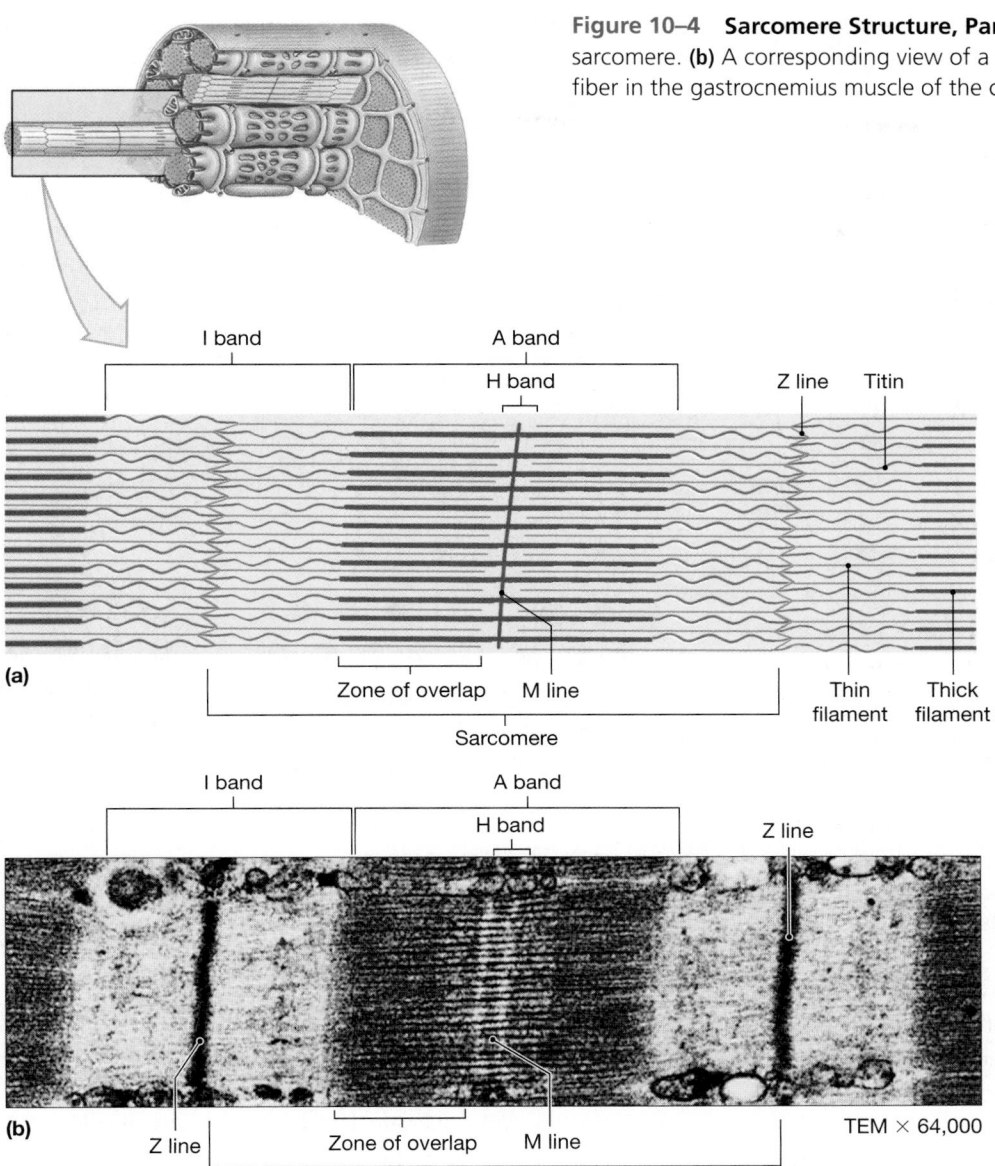

Figure 10–4 Sarcomere Structure, Part I. (a) A longitudinal section of a sarcomere. **(b)** A corresponding view of a sarcomere in a myofibril from a muscle fiber in the gastrocnemius muscle of the calf.

band of the next sarcomere (**Figure 10–4**). **Z lines** mark the boundary between adjacent sarcomeres. The Z lines consist of proteins called *actinins*, which interconnect thin filaments of adjacent sarcomeres. From the Z lines at either end of the sarcomere, thin filaments extend toward the M line and into the zone of overlap. Strands of the elastic protein **titin** extend from the tips of the thick filaments to attachment sites at the Z line (**Figures 10–4a** and **10–5**). Titin helps keep the thick and thin filaments in proper alignment and aids in restoring resting sarcomere length; it also helps the muscle fiber resist extreme stretching that would otherwise disrupt the contraction mechanism.

Two transverse tubules encircle each sarcomere, and triads are located in the zones of overlap, at the edges of the A band (**Figure 10–3**). As a result, calcium ions released by the SR enter the regions where thick and thin filaments can interact.

Each Z line is surrounded by a meshwork of intermediate filaments that interconnect adjacent myofibrils. The myofibrils closest to the sarcolemma are bound to attachment sites on the inside of the membrane. Because the Z lines of all the myofibrils are aligned in this way, the muscle fiber as a whole has a banded appearance (**Figure 10–2b**). These bands, or *striations*, are visible with the light micro-

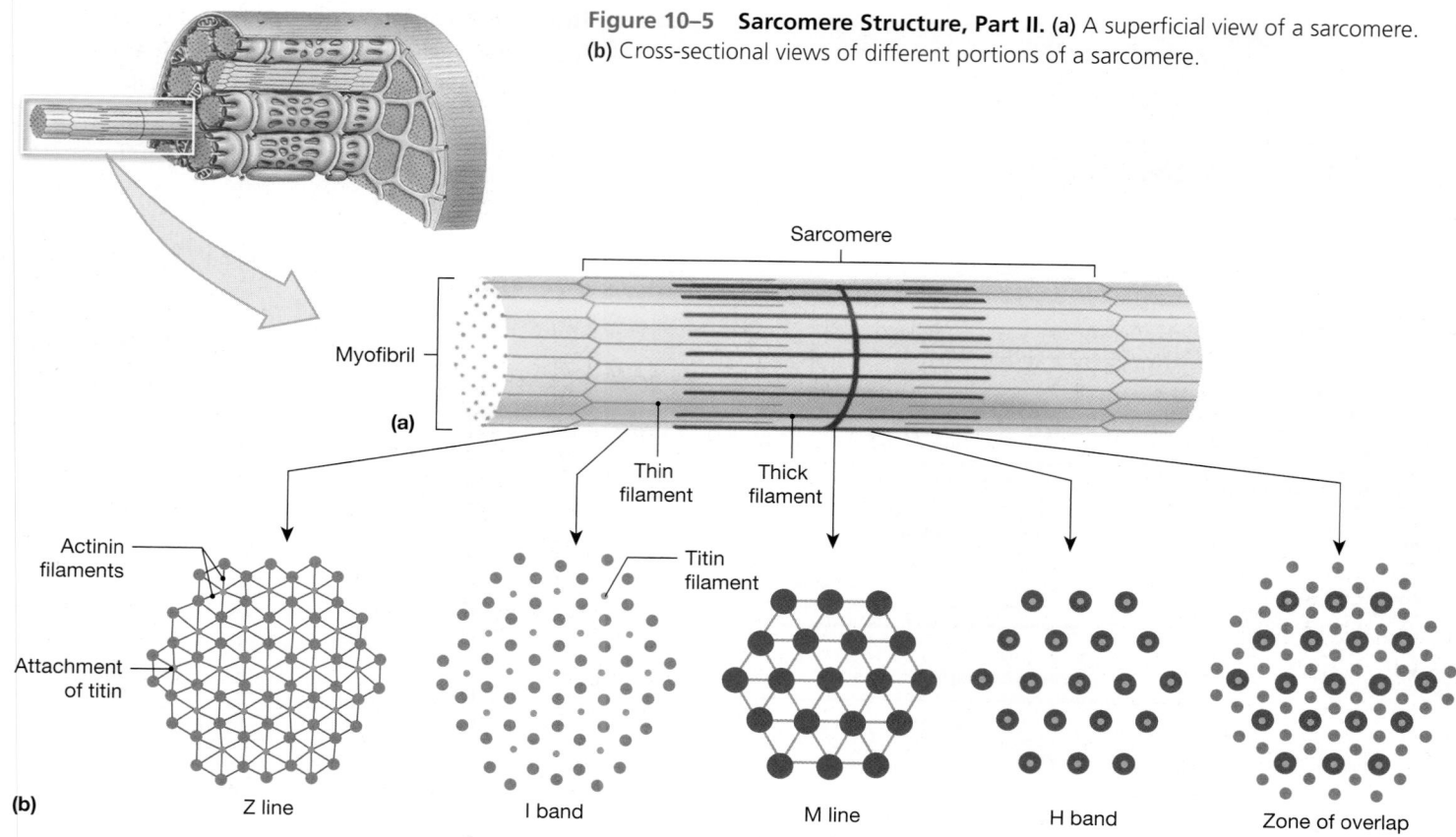

Figure 10–5 **Sarcomere Structure, Part II. (a)** A superficial view of a sarcomere. **(b)** Cross-sectional views of different portions of a sarcomere.

Sarcomere

Myofibril

(a)

Thin filament

Thick filament

Actinin filaments

Attachment of titin

Titin filament

(b)

Z line

I band

M line

H band

Zone of overlap

scope, so skeletal muscle tissue is also known as striated muscle. ∞ p. 138

Figure 10–6 reviews the levels of organization we have discussed so far. We now consider the molecular structure of the myofilaments responsible for muscle contraction.

Thin Filaments

A typical thin filament is 5–6 nm in diameter and 1 μm in length (**Figure 10–7a**). A single thin filament contains four proteins: F-actin, nebulin, tropomyosin, and troponin (**Figure 10–7b**).

Filamentous actin, or **F-actin**, is a twisted strand composed of two rows of 300–400 individual globular molecules of **G-actin** (**Figure 10–7b**). A long strand of **nebulin** extends along the F-actin strand in the cleft between the rows of G-actin molecules. Nebulin holds the F-actin strand together; as thin filaments develop, the length of the nebulin molecule probably determines the length of the F-actin strand. Each G-actin molecule contains an **active site** that can bind to myosin much as a substrate molecule binds to the active site of an enzyme. ∞ p. 56 Under resting conditions, myosin binding is prevented by the *troponin–tropomyosin complex*.

Strands of **tropomyosin** (trō-pō-MĪ-ō-sin; *trope*, turning) cover the active sites on G-actin and prevent actin–myosin interaction. A tropomyosin molecule is a double-stranded pro-

tein that covers seven active sites. It is bound to one molecule of troponin midway along its length. A **troponin** (TRŌ-pō-nin) molecule consists of three globular subunits. One subunit binds to tropomyosin, locking them together as a troponin–tropomyosin complex; a second subunit binds to one G-actin, holding the troponin–tropomyosin complex in position; the third subunit has a receptor that binds two calcium ions. In a resting muscle, intracellular Ca^{2+} concentrations are very low, and that binding site is empty.

A contraction cannot occur unless the position of the troponin–tropomyosin complex changes, exposing the active sites on F-actin. The necessary change in position occurs when calcium ions bind to receptors on the troponin molecules.

At either end of the sarcomere, the thin filaments are attached to the Z line (**Figure 10–7a**). Although it is called a "line" because it looks like a dark line on the surface of the myofibril, the Z line in sectional view is more like a disc with an open meshwork (**Figure 10–5b**). For this reason, the Z line is often called the *Z disc*.

Thick Filaments

Thick filaments are 10–12 nm in diameter and 1.6 μm long (**Figure 10–7c**). A thick filament contains roughly 300 myosin molecules, each made up of a pair of myosin subunits twisted around one another (**Figure 10–7d**). The long **tail** is bound to

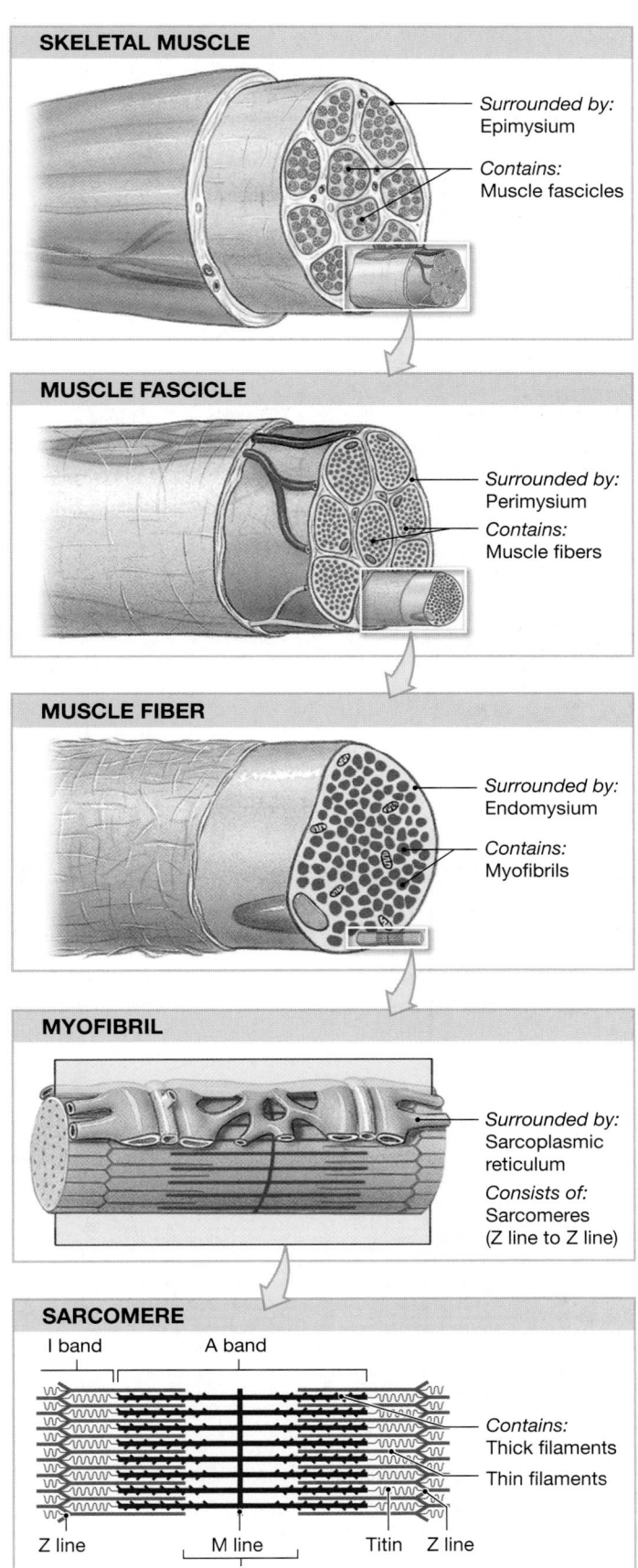

Figure 10–6 Levels of Functional Organization in a Skeletal Muscle.

other myosin molecules in the thick filament. The free **head**, which projects outward toward the nearest thin filament, has two globular protein subunits. When the myosin heads interact with thin filaments during a contraction, they are known as **cross-bridges**. The connection between the head and the tail functions as a hinge that lets the head pivot at its base. When pivoting occurs, the head swings toward or away from the M line. As we will see in a later section, this pivoting is the key step in muscle contraction.

All the myosin molecules are arranged with their tails pointing toward the M line (**Figure 10–7c**). The H band includes a central region where there are no myosin heads. Elsewhere, the myosin heads are arranged in a spiral, each facing one of the surrounding thin filaments.

Each thick filament has a core of titin. On either side of the M line, a strand of titin extends the length of the thick filament and then continues across the I band to the Z line on that side. The portion of the titin strand exposed within the I band is elastic and will recoil after stretching. In the normal resting sarcomere, the titin strands are completely relaxed; they become tense only when some external force stretches the sarcomere.

Sliding Filaments and Muscle Contraction

When a skeletal muscle fiber contracts, (1) the H bands and I bands get smaller, (2) the zones of overlap get larger, (3) the Z lines move closer together, and (4) the width of the A band remains constant (**Figure 10–8a**). These observations make sense only if the thin filaments are sliding toward the center of each sarcomere, alongside the thick filaments; this explanation is known as the **sliding filament theory**. The contraction weakens with the disappearance of the I bands, at which point the Z lines are in contact with the ends of the thick filaments.

During a contraction, sliding occurs in every sarcomere along the myofibril. As a result, the myofibril gets shorter. Because myofibrils are attached to the sarcolemma at each Z line and at either end of the muscle fiber, when myofibrils get shorter, so does the muscle fiber.

Tips&Tricks

The sliding of filaments during a muscle contraction is similar to the way an accordion operates. To better understand the actions of sliding filaments during a muscle contraction, hold your hands in front of you, palms toward your body and thumbs sticking straight up. Now move your hands together, so that the fingers of one hand move in between the fingers of the other hand. Your fingers represent thin and thick filaments, and your thumbs the Z lines. Notice that finger length stays the same, but your thumbs move closer together.

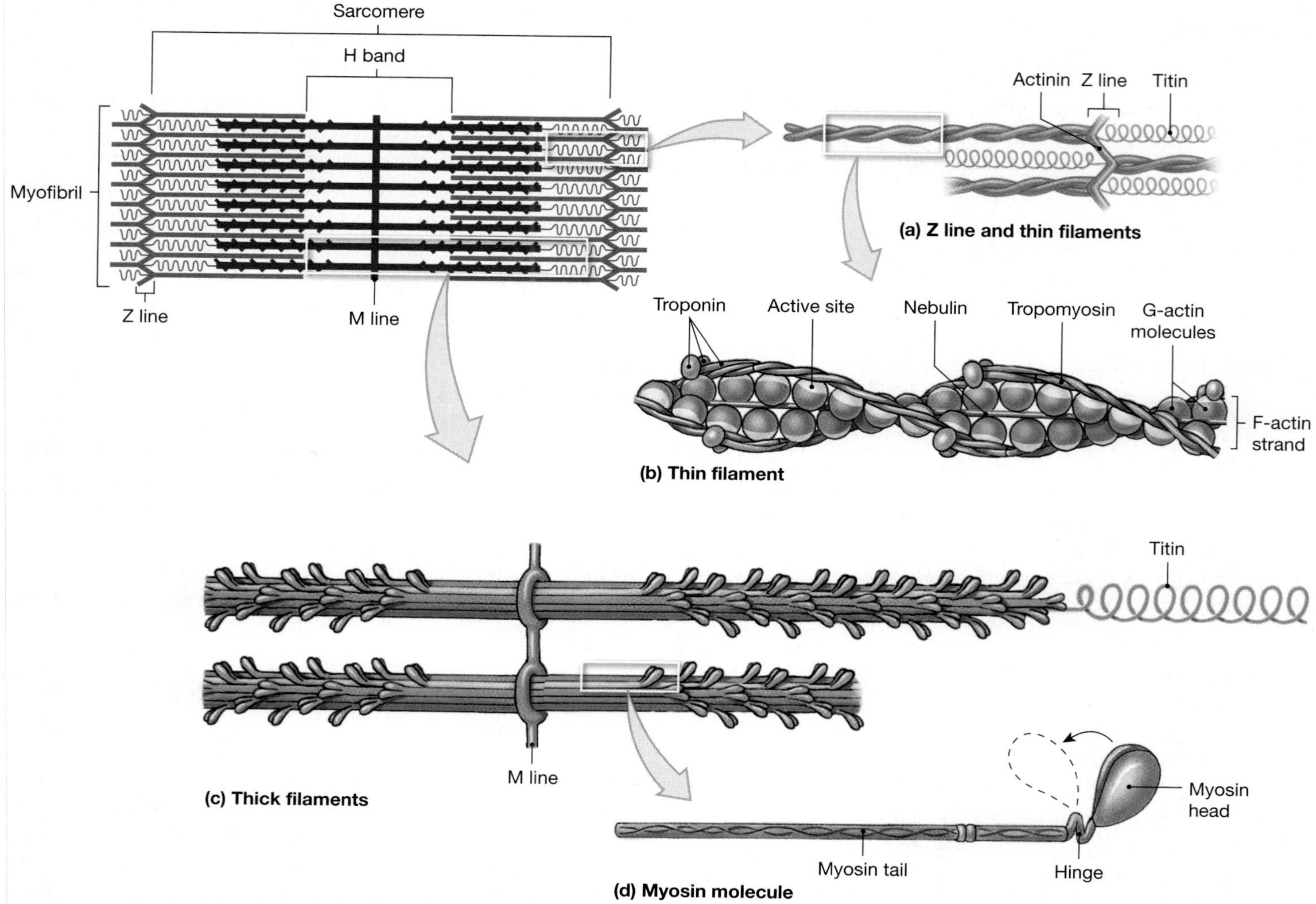

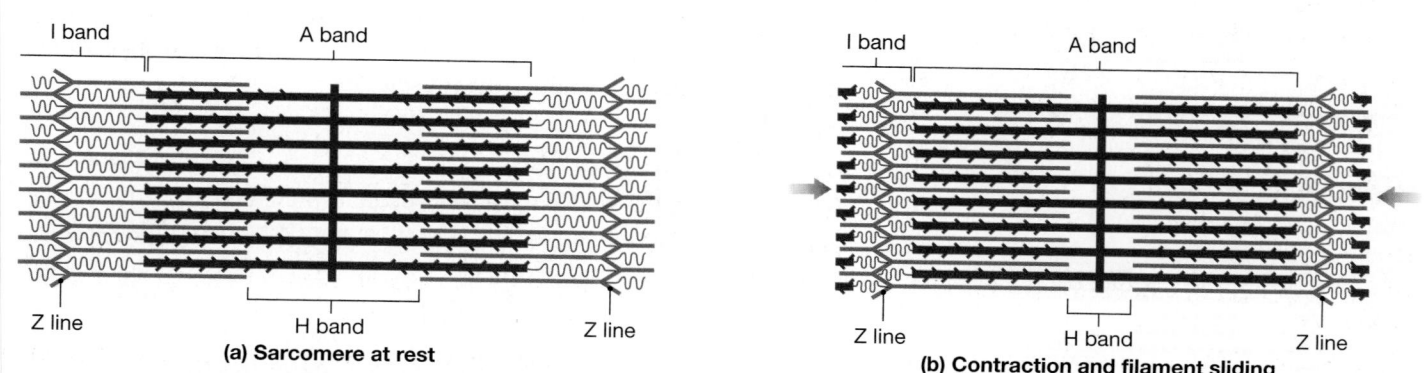

Figure 10–7   Thick and Thin Filaments. (a) The gross structure of a thin filament, showing the attachment at the Z line. **(b)** The organization of G-actin subunits in an F-actin strand, and the position of the troponin–tropomyosin complex. **(c)** The structure of thick filaments, showing the orientation of the myosin molecules. **(d)** The structure of a myosin molecule.

Figure 10–8   Changes in the Appearance of a Sarcomere during the Contraction of a Skeletal Muscle Fiber. (a) During a contraction, the A band stays the same width, but the Z lines move closer together and the I band gets smaller. **(b)** When the ends of a myofibril are free to move, the sarcomeres shorten simultaneously and the ends of the myofibril are pulled toward its center.

You now know *how* the myofilaments in a sarcomere change position during a contraction, but not *why* these changes occur. To understand this process in detail, we must take a closer look at the contraction process and its regulation. But before we can understand how muscles contract and how those contractions are harnessed to do what you want them to do, we have to understand some basic physical principles that apply to muscle cells.

When muscle cells contract, they pull on the attached tendon fibers the way a line of people might pull on a rope. The pull, called *tension*, is an active force: Energy must be expended to produce it. Tension is applied *to* some object, whether a rope, a rubber band, or a book on a tabletop.

Tension applied to an object tends to pull the object toward the source of the tension. However, before movement can occur, the applied tension must overcome the object's *resistance* (or *load*), a passive force that opposes movement. The amount of resistance can depend on the weight of the object, its shape, friction, and other factors. When the applied tension exceeds the resistance, the object moves. In contrast, *compression*, or a push applied to an object, tends to force the object away from the source of the compression. Again, no movement can occur until the applied compression exceeds the resistance of the object. Muscle cells can use energy to shorten and generate tension, through interactions between thick and thin filaments, but not to lengthen and generate compression. In other words, muscle cells can pull, but they cannot push.

With that background, we can now investigate the mechanics of muscle contraction in some detail. **Figure 10–9** provides an overview of the "big picture" we will be examining.

- Normal skeletal muscle is under neural control: Contraction occurs only when skeletal muscle fibers are activated by neurons whose cell bodies are in the central nervous system (brain and spinal cord). A neuron can activate a muscle fiber through stimulation of its sarcolemma. What follows is called *excitation–contraction coupling*.
- The first step in excitation–contraction coupling is the release of calcium ions from the cisternae of the sarcoplasmic reticulum.
- The calcium ions then trigger interactions between thick filaments and thin filaments, resulting in muscle fiber contraction and the consumption of energy in the form of ATP.
- These filament interactions produce active tension.

We turn next to the first aspect of "the big picture": how skeletal muscle is under neural control.

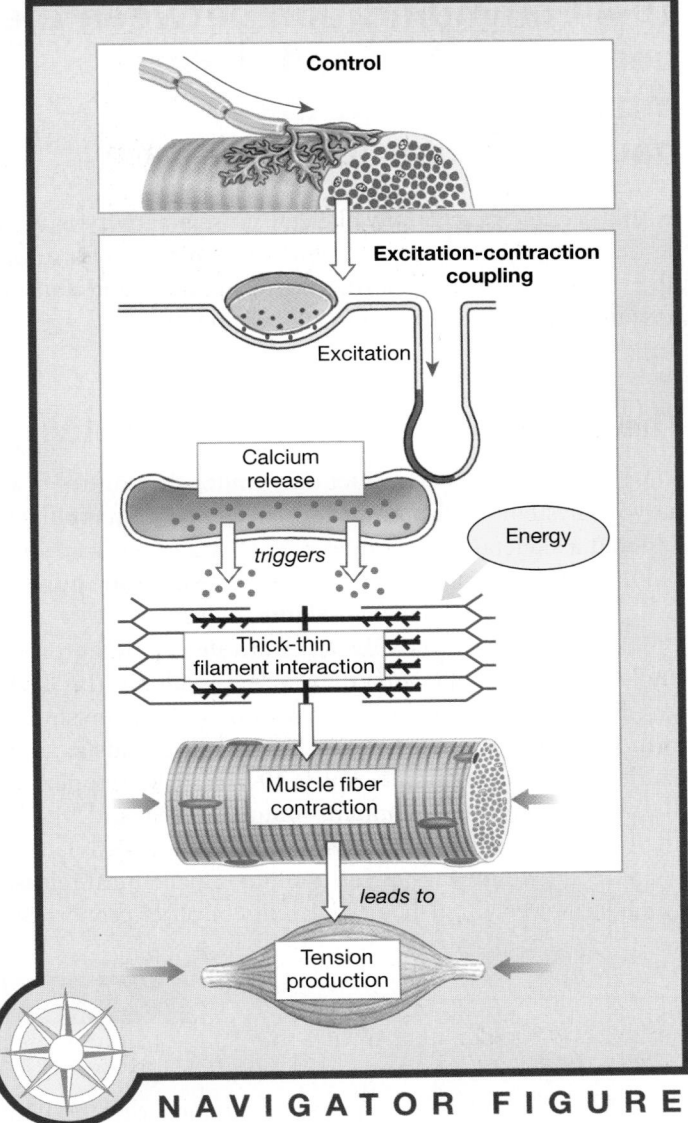

NAVIGATOR FIGURE

Figure 10–9 **An Overview of Skeletal Muscle Contraction.** The major factors are indicated here as a series of interrelated steps and processes. Each factor will be described further in a related section of the text. A simplified version of this figure will appear in later figures as a Navigator icon; its presence indicates that we are taking another step in the discussion.

CHECKPOINT

5. Describe the structural components of a sarcomere.
6. Why do skeletal muscle fibers appear striated when viewed through a light microscope?
7. Where would you expect the greatest concentration of Ca^{2+} in resting skeletal muscle to be?

See the blue Answers tab at the end of the book.

10-4 Communication between the nervous system and skeletal muscles occurs at the neuromuscular junction

In this section we see how skeletal muscle activity is under neural control, examine how neural stimulation of a muscle fiber is coupled to the contraction of the fiber, and consider how muscle fibers relax following contraction.

The Control of Skeletal Muscle Activity

Skeletal muscle fibers contract only under the control of the nervous system. Communication between the nervous system and a skeletal muscle fiber occurs at a specialized intercellular connection known as a **neuromuscular junction (NMJ)**, or *myoneural junction* (**Figure 10–10a**).

Each skeletal muscle fiber is controlled by a neuron at a single neuromuscular junction midway along the fiber's length. A single axon branches within the perimysium to form a number of fine branches. Each branch ends at an expanded **synaptic terminal** (**Figure 10–10b**). The cytoplasm of the synaptic terminal contains mitochondria and vesicles filled with molecules of **acetylcholine** (as-ē-til-KŌ-lēn), or **ACh**. Acetylcholine is a *neurotransmitter*, a chemical released by a neuron to change the permeability or other properties of another plasma membrane. In this case, the release of ACh from the synaptic terminal can alter the permeability of the sarcolemma and trigger the contraction of the muscle fiber.

The **synaptic cleft**, a narrow space, separates the synaptic terminal of the neuron from the opposing sarcolemmal surface. This surface, which contains membrane receptors that bind ACh, is known as the **motor end plate**. The motor end plate has deep creases called *junctional folds*, which increase its surface area and thus the number of available ACh receptors. The synaptic cleft and the sarcolemma also contain molecules of the enzyme **acetylcholinesterase** (**AChE**, or *cholinesterase*), which breaks down ACh.

A neuron stimulates a muscle fiber through a series of steps (**Figure 10–10c**):

Step 1 *The Arrival of an Action Potential.* The stimulus for ACh release is the arrival of an electrical impulse, or **action potential**, at the synaptic terminal. An action potential is a sudden change in the transmembrane potential that travels along the length of the axon.

Step 2 *The Release of ACh.* When the action potential reaches the synaptic terminal, permeability changes in the membrane trigger the exocytosis of ACh into the synaptic cleft. This is accomplished when vesicles in the synaptic terminal fuse with the plasma membrane of the neuron.

Step 3 *ACh Binding at the Motor End Plate.* ACh molecules diffuse across the synaptic cleft and bind to ACh receptors on the surface of the sarcolemma at the motor end

---

CLINICAL NOTE

Tetanus

Children are often told to be careful around rusty nails. But it's not the rust or the nail, but instead an infection by the very common bacterium *Clostridium tetani* that causes the disease called **tetanus**. Although this disease and the normal muscle response to rapid neural stimulation share the same name, the mechanisms involved are very different.

Clostridium bacteria occur in soil and virtually everywhere else in the environment, but they can thrive only in tissues with low oxygen levels. For this reason, a deep puncture wound, such as that from a nail, is much more likely to result in tetanus than a shallow, open cut that bleeds freely. When active in body tissues, these bacteria release a powerful toxin that affects the central nervous system. Motor neurons, which control skeletal muscles throughout the body, are particularly sensitive to it. The toxin suppresses the mechanism that inhibits motor neuron activity. The result is a sustained, powerful contraction of skeletal muscles throughout the body.

The incubation period (the time between exposure and the development of symptoms) is generally less than 2 weeks. The most common early complaints are headache, muscle stiffness, and difficulty in swallowing. Because it soon becomes difficult to open the mouth, the disease is also called *lockjaw*. Widespread muscle spasms typically develop within 2 or 3 days of the initial symptoms and continue for a week before subsiding. After 2–4 weeks, patients who survive recover with no aftereffects.

Severe tetanus has a 40–60 percent mortality rate; that is, for every 100 people who develop severe tetanus, 40 to 60 die. Approximately 500,000 cases of tetanus occur worldwide each year, but only about 100 of them occur in the United States, thanks to an effective immunization program. (Recommended immunization involves a "tetanus shot" followed by a booster shot every 10 years.) In unimmunized patients, severe symptoms can be prevented by early administration of an antitoxin—in most cases, *human tetanus immune globulin*. However, this treatment does not reduce symptoms that have already appeared.

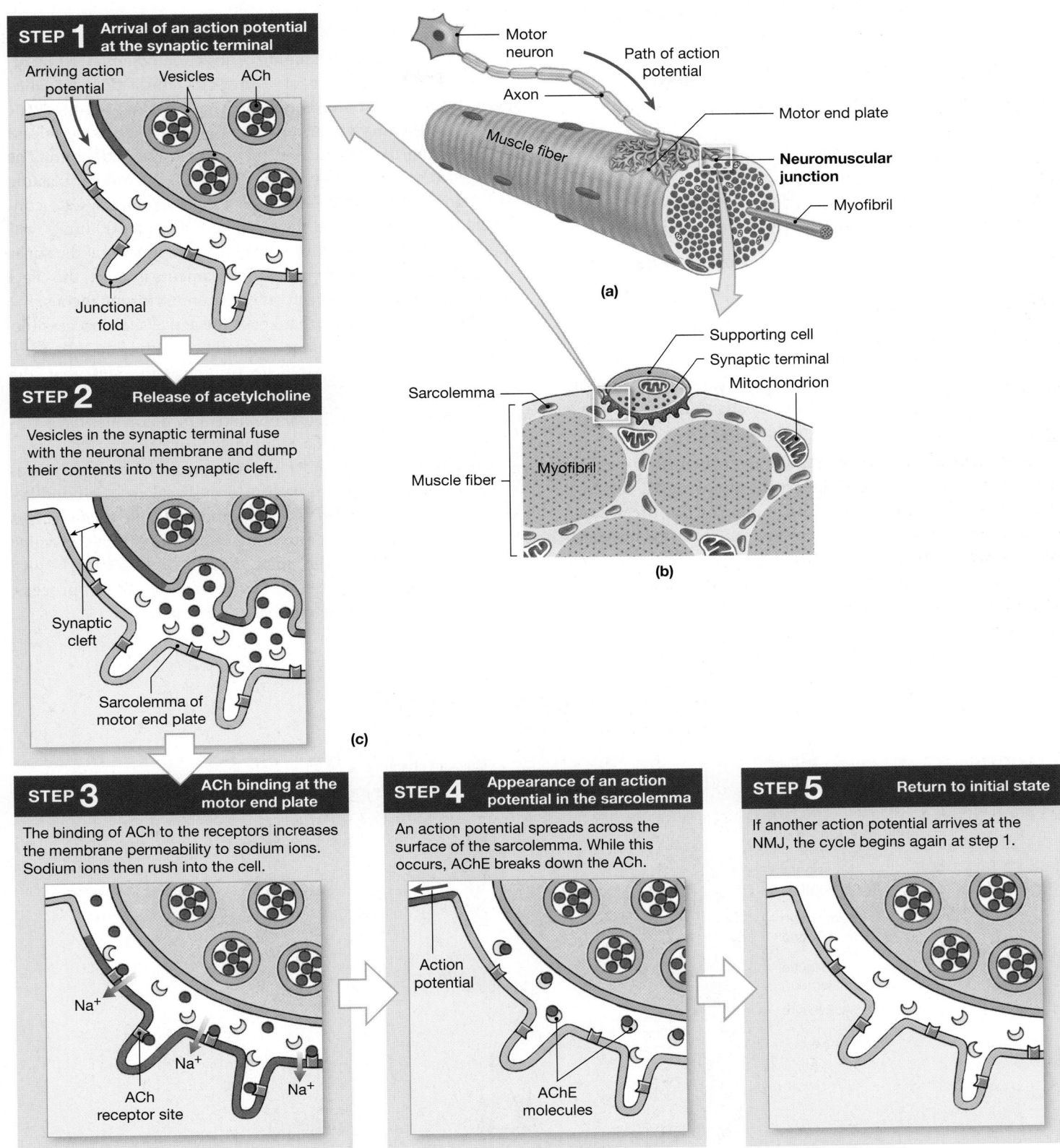

STEP 1 Arrival of an action potential at the synaptic terminal

Arriving action potential   Vesicles   ACh

Junctional fold

STEP 2 Release of acetylcholine

Vesicles in the synaptic terminal fuse with the neuronal membrane and dump their contents into the synaptic cleft.

Synaptic cleft

Sarcolemma of motor end plate

(a)

Motor neuron

Path of action potential

Axon

Motor end plate

Muscle fiber

Neuromuscular junction

Myofibril

Supporting cell

Synaptic terminal

Mitochondrion

Sarcolemma

Muscle fiber — Myofibril

(b)

(c)

STEP 3 ACh binding at the motor end plate

The binding of ACh to the receptors increases the membrane permeability to sodium ions. Sodium ions then rush into the cell.

Na⁺

Na⁺

ACh receptor site

Na⁺

STEP 4 Appearance of an action potential in the sarcolemma

An action potential spreads across the surface of the sarcolemma. While this occurs, AChE breaks down the ACh.

Action potential

AChE molecules

STEP 5 Return to initial state

If another action potential arrives at the NMJ, the cycle begins again at step 1.

Figure 10–10   Skeletal Muscle Innervation. (a) A diagrammatic view of a neuromuscular junction. **(b)** Details of the neuromuscular junction. **(c)** Changes at the motor end plate that trigger an action potential in the sarcolemma.

plate. ACh binding changes the permeability of the motor end plate to sodium ions. Recall from Chapter 3 that the extracellular fluid contains a high concentration of sodium ions, whereas sodium ion concentrations inside the cell are very low. ∞ p. 72 When the membrane permeability to sodium increases, sodium ions rush into the sarcoplasm. This influx continues until AChE removes the ACh from the cleft. As the concentration of ACh in the cleft falls, bound ACh diffuses off the receptors and into the cleft.

Step 4 *Appearance of an Action Potential in the Sarcolemma.* The sudden inrush of sodium ions results in the generation of an action potential in the sarcolemma. This electrical impulse originates at the edges of the motor end plate, sweeps across the entire membrane surface, and travels inward along each T tubule. The arrival of an action potential at the synaptic terminal thus leads to the appearance of an action potential in the sarcolemma.

Step 5 *Return to Initial State.* Even before the action potential has spread across the entire sarcolemma, the ACh has been broken down by AChE. Some of the breakdown products will be absorbed by the synaptic terminal and used to resynthesize ACh for subsequent release. This sequence of events can now be repeated should another action potential arrive at the synaptic terminal.

Excitation–Contraction Coupling

The link between the generation of an action potential in the sarcolemma and the start of a muscle contraction is called **excitation–contraction coupling**. This coupling occurs at the triads. On reaching a triad, an action potential triggers the release of Ca^{2+} from the cisternae of the sarcoplasmic reticulum. The change in the permeability of the SR to Ca^{2+} is temporary, lasting only about 0.03 second. Yet within a millisecond, the Ca^{2+} concentration in and around the sarcomere reaches 100 times resting levels. Because the terminal cisternae are situated at the zones of overlap, where the thick and thin filaments interact, the effect of calcium release on the sarcomere is almost instantaneous.

Troponin is the lock that keeps the active sites inaccessible. Calcium is the key to that lock. Recall from **Figure 10–7** that troponin binds to actin and to tropomyosin, and that the tropomyosin molecules cover the active sites and prevent interactions between thick filaments and thin filaments. Each troponin molecule also has a binding site for calcium; this site is empty when the muscle fiber is at rest. Calcium binding changes the shape of the troponin molecule and weakens the bond between tropinin and actin. The troponin molecule then changes position, rolling the tropomyosin strand away from the active sites (**Figure 10–11**). At this point, the **contraction cycle** begins.

Table 10–1 provides a review of the contraction process, from ACh release to the end of the contraction.

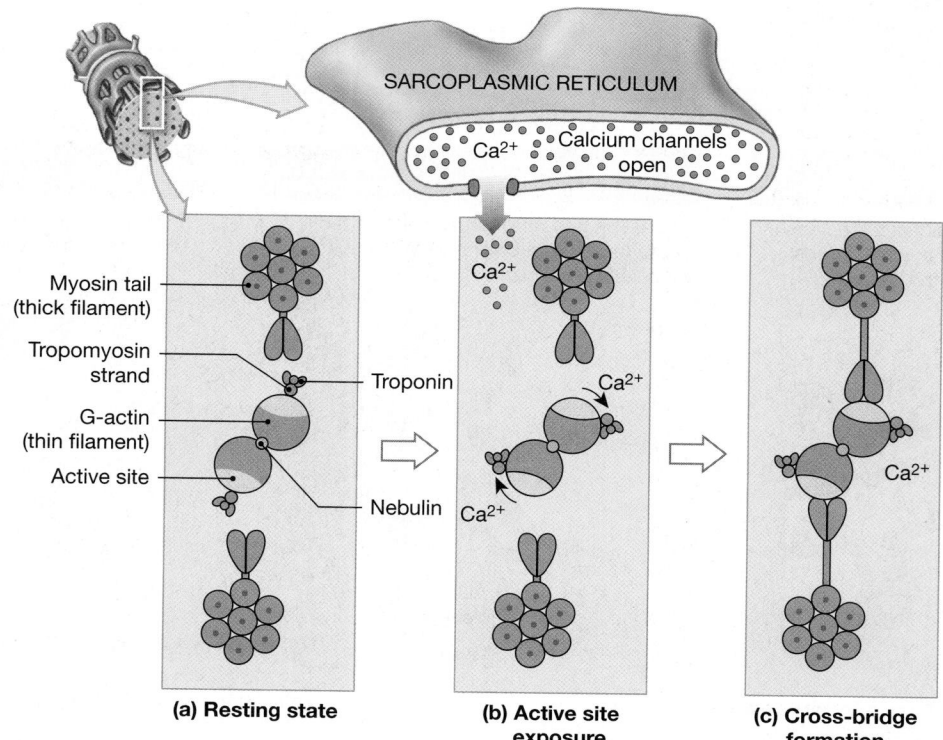

Figure 10–11 **The Exposure of Active Sites. (a)** In a resting sarcomere, the tropomyosin strands cover the active sites on the thin filaments, preventing cross-bridge formation. **(b)** When calcium ions enter the sarcomere, they bind to troponin, which rotates and swings the tropomyosin away from the active sites. **(c)** Cross-bridge formation then occurs, and the contraction cycle begins.

SUMMARY TABLE 10–1   Steps Involved In Skeletal Muscle Contraction

STEPS IN INITIATING MUSCLE CONTRACTION

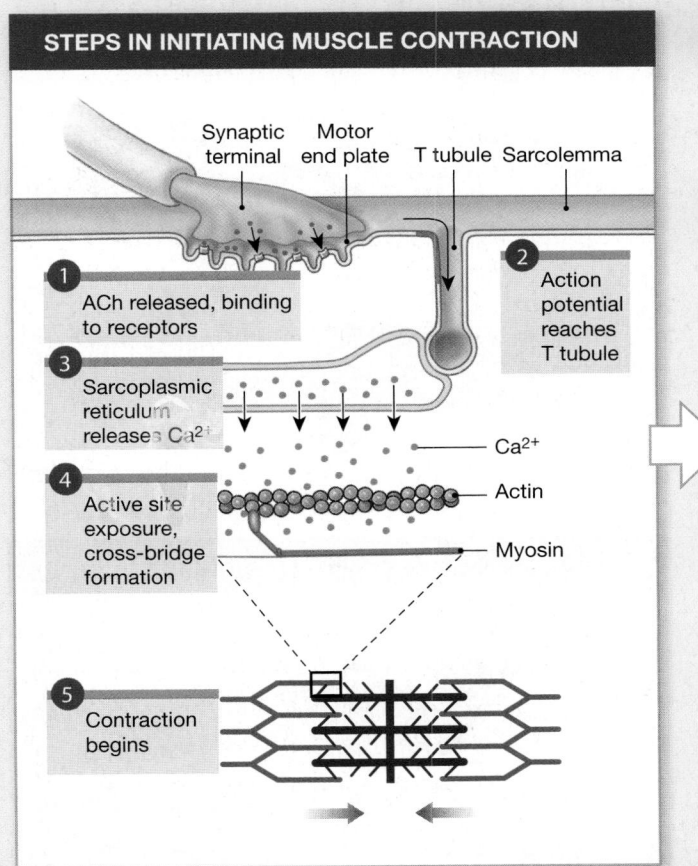

STEPS IN MUSCLE RELAXATION

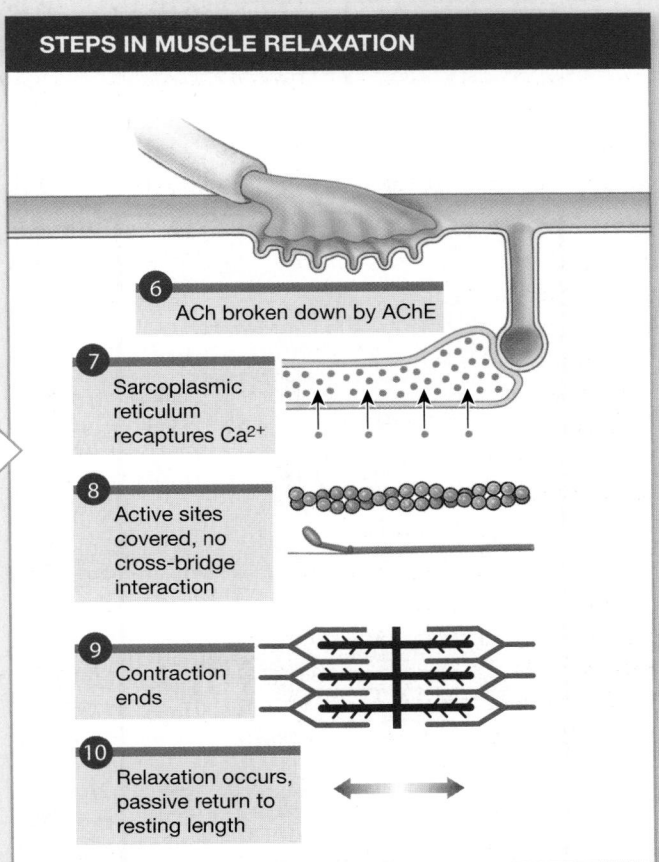

STEPS THAT INITIATE A CONTRACTION:

1. At the neuromuscular junction (NMJ), ACh released by the synaptic terminal binds to receptors on the sarcolemma.

2. The resulting change in the transmembrane potential of the muscle fiber leads to the production of an action potential that spreads across the entire surface of the muscle fiber and along the T tubules.

3. The sarcoplasmic reticulum (SR) releases stored calcium ions, increasing the calcium concentration of the sarcoplasm in and around the sarcomeres.

4. Calcium ions bind to troponin, producing a change in the orientation of the troponin–tropomyosin complex that exposes active sites on the thin (actin) filaments. Cross-bridges form when myosin heads bind to active sites on F-actin.

5. The contraction begins as repeated cycles of cross-bridge binding, pivoting, and detachment occur, powered by the hydrolysis of ATP. These events produce filament sliding, and the muscle fiber shortens.

STEPS THAT END A CONTRACTION:

6. Action potential generation ceases as ACh is broken down by acetylcholinesterase (AChE).

7. The SR reabsorbs calcium ions, and the concentration of calcium ions in the sarcoplasm declines.

8. When calcium ion concentrations approach normal resting levels, the troponin–tropomyosin complex returns to its normal position. This change re-covers the active sites and prevents further cross-bridge interaction.

9. Without cross-bridge interactions, further sliding cannot take place, and the contraction ends.

10. Muscle relaxation occurs, and the muscle returns passively to its resting length.

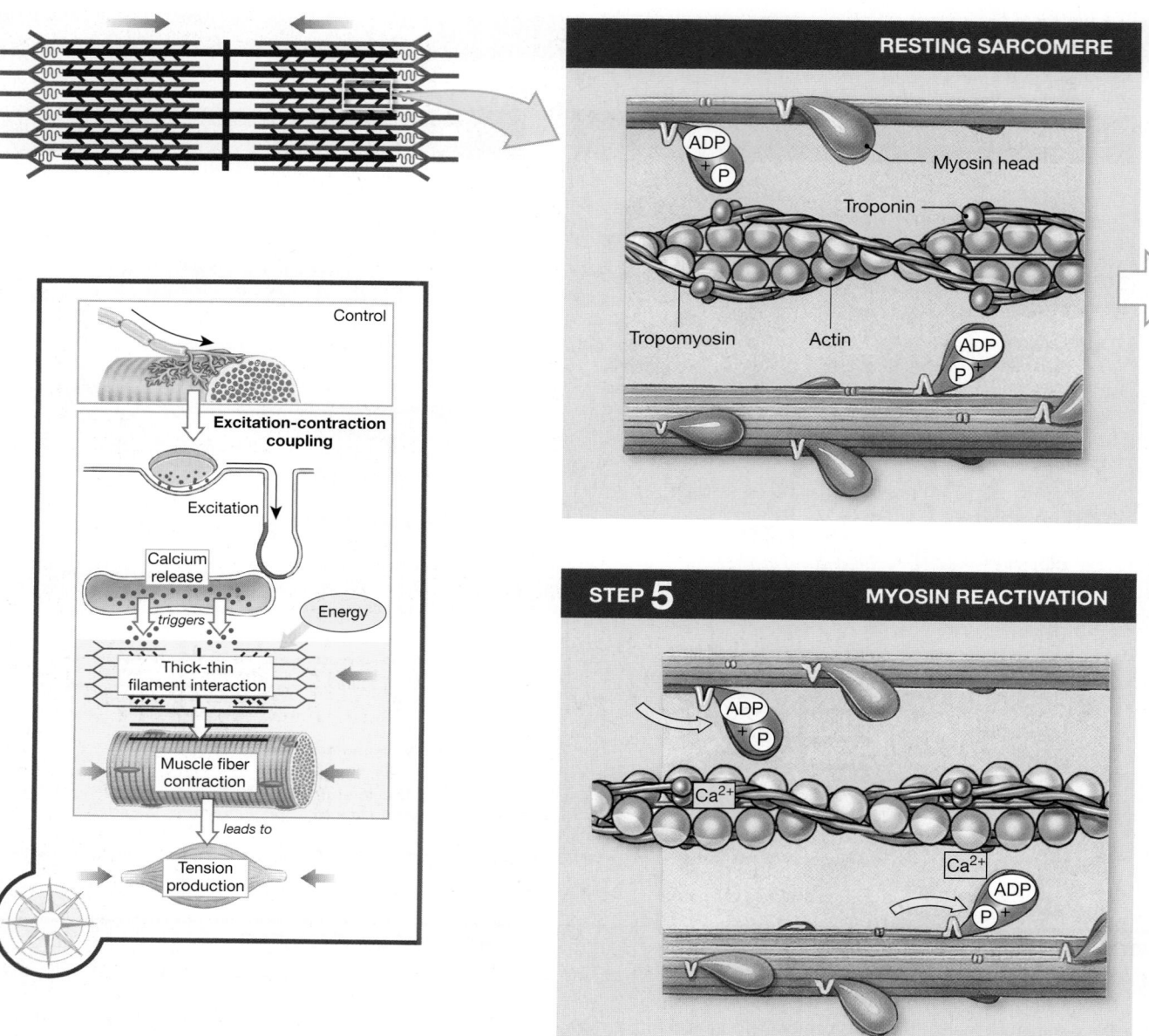

Figure 10–12 **The Contraction Cycle.**

The Contraction Cycle

Figure 10–12 details the molecular events that occur during the contraction cycle. In the resting sarcomere, each myosin head is already "energized"—charged with the energy that will be used to power a contraction. The myosin head functions as ATPase, an enzyme that can break down ATP. ⚬ p. 60 At the start of the contraction cycle, each myosin head has already split a molecule of ATP and stored the energy released in the process. The breakdown products, ADP and phosphate (often represented as P), remain bound to the myosin head.

The contraction cycle involves five interlocking steps (**Figure 10–12**):

Step 1 *Exposure of Active Sites.* The calcium ions entering the sarcoplasm bind to troponin. This binding weakens the bond between the troponin–tropomyosin complex and actin. The troponin molecule then changes position, rolling the tropomyosin molecule away from the active sites on actin and allowing interaction with the energized myosin heads.

10 MUSCULAR

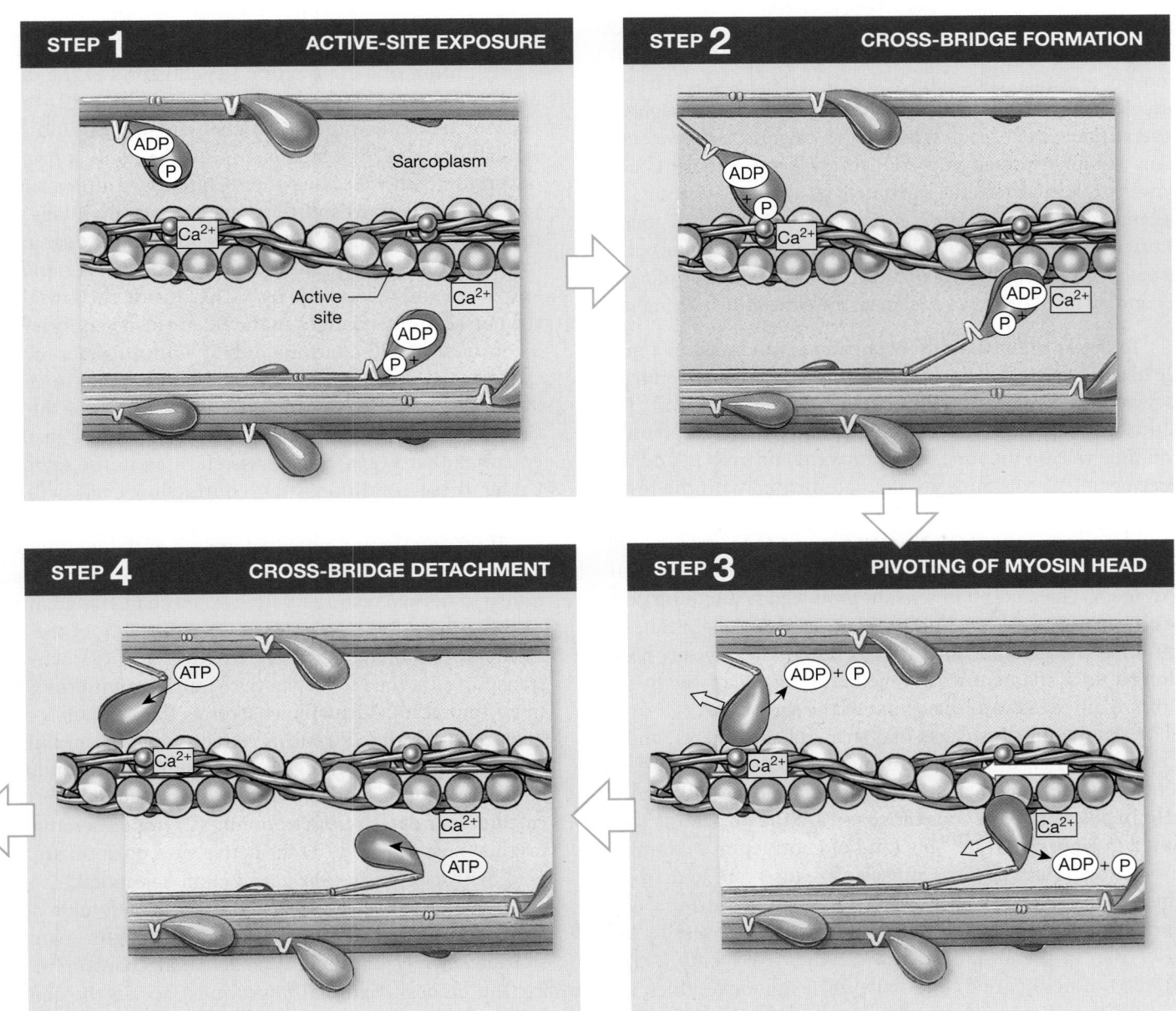

STEP **1** — ACTIVE-SITE EXPOSURE

STEP **2** — CROSS-BRIDGE FORMATION

STEP **4** — CROSS-BRIDGE DETACHMENT

STEP **3** — PIVOTING OF MYOSIN HEAD

Step 2 *Formation of Cross-Bridges.* When the active sites are exposed, the energized myosin heads bind to them, forming cross-bridges.

Step 3 *Pivoting of Myosin Heads.* In the resting sarcomere, each myosin head points away from the M line. In this position, the myosin head is "cocked" like the spring in a mousetrap. Cocking the myosin head requires energy, which is obtained by breaking down ATP into ADP and a phosphate group. In the cocked position, both the ADP and

the phosphate are still bound to the myosin head. After cross-bridge formation, the stored energy is released as the myosin head pivots toward the M line. This action is called the *power stroke*; when it occurs, the ADP and phosphate group are released.

Step 4 *Detachment of Cross-Bridges.* When another ATP binds to the myosin head, the link between the active site on the actin molecule and the myosin head is broken. The active site is now exposed and able to form another cross-bridge.

Step 5 *Reactivation of Myosin.* Myosin reactivation occurs when the free myosin head splits the ATP into ADP and a phosphate group. The energy released in this process is used to recock the myosin head. The entire cycle will now be repeated, several times each second, as long as calcium ion concentrations remain elevated and ATP reserves are sufficient. Each power stroke shortens the sarcomere by about 0.5 percent, and because all the sarcomeres contract together, the entire muscle shortens at the same rate. The speed at which shortening occurs depends on the cycling rate (the number of power strokes per second): The higher the resistance, the slower the cycling rate.

To better understand how tension is produced in a muscle fiber, imagine that you are on a tug-of-war team. You reach forward, grab the rope with both hands, and pull it in. This action corresponds to cross-bridge attachment and pivoting. You then release the rope, reach forward and grab it, and pull once again. Your actions are not coordinated with the rest of your team; if everyone let go at the same time, your opponents would pull the rope away. So at any given time, some people are reaching and grabbing, some are pulling, and others are letting go. The amount of tension produced is thus a function of how many people are pulling at any given moment. The situation is comparable in a muscle fiber; the myosin heads along a thick filament work together in a similar way to pull a thin filament toward the center of the sarcomere.

Each myofibril consists of a string of sarcomeres, and in a contraction all of the thin filaments are pulled toward the centers of the sarcomeres. If neither end of the myofibril is held in position, both ends move toward the middle, as illustrated in **Figure 10–13a**. This kind of contraction seldom occurs in an intact skeletal muscle, because one end of the muscle (the *origin*) is usually fixed in position during a contraction, while the other end (the *insertion*) moves. In that case, the free end moves toward the fixed end (**Figure 10–13b**). If neither end of the myofibril can move, thick and thin filament interactions consume energy and generate tension, but sliding cannot occur. This kind of contraction, called *isometric*, will be the topic of a later section.

Relaxation

The duration of a contraction depends on (1) the duration of stimulation at the neuromuscular junction, (2) the presence of free calcium ions in the sarcoplasm, and (3) the availability of ATP. A single stimulus has only a brief effect on a muscle fiber, because the ACh released after a single action potential arrives at the synaptic terminal does not remain intact for long. Whether it is bound to the sarcolemma or free in the synaptic cleft, the released ACh is rapidly broken down and inactivated by AChE. Inside the muscle fiber, the permeability changes in the SR are also very brief. Thus, a contraction will continue only if additional action potentials arrive at the synaptic terminal in rapid succession. When they do, the continual release of ACh into the synaptic cleft produces a series of action potentials in the sarcolemma that keeps Ca^{2+} levels elevated in the sarcoplasm. Under these conditions, the contraction cycle will be repeated over and over.

If just one action potential arrives at the neuromuscular junction, Ca^{2+} concentrations in the sarcoplasm will quickly return to normal resting levels. Two mechanisms are involved in this process: (1) active Ca^{2+} transport across the plasma membrane into the extracellular fluid and (2) active Ca^{2+} transport into the SR. Of the two, transport into the SR is far more important. Virtually as soon as the calcium ions have been released, the SR returns to its normal permeability and begins actively absorbing Ca^{2+} from the surrounding sarcoplasm. As Ca^{2+} concentrations in the sarcoplasm fall, (1) calcium ions detach from troponin, (2) troponin returns to its original position, and (3) the active sites on actin are re-covered by tropomyosin. The contraction has ended.

Once the contraction has ended, the sarcomere does not automatically return to its original length. Sarcomeres shorten actively, but there is no active mechanism for reversing the process. External forces must act on the contracted muscle fiber to stretch the myofibrils and sarcomeres to their original dimensions. We will describe those forces in a later section.

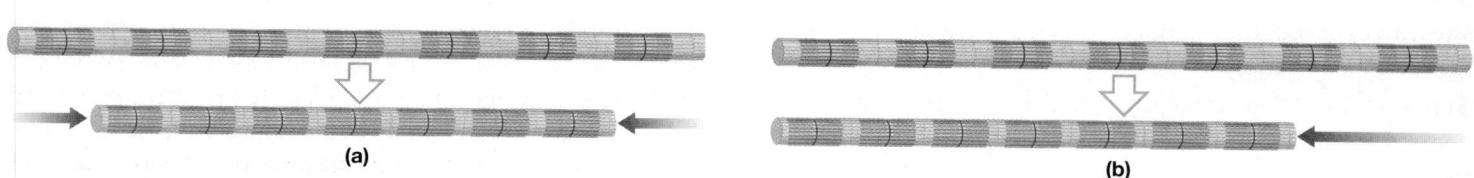

(a)   (b)

Figure 10–13 **Shortening during a Contraction. (a)** When both ends are free to move, the ends of a contracting muscle fiber move toward the center of the muscle fiber. **(b)** When one end of a myofibril is fixed in position, and the other end free to move, the free end is pulled toward the fixed end.

CLINICAL NOTE

Rigor Mortis   When death occurs, circulation ceases and the skeletal muscles are deprived of nutrients and oxygen. Within a few hours, the skeletal muscle fibers have run out of ATP and the sarcoplasmic reticulum becomes unable to pump Ca^{2+} out of the sarcoplasm. Calcium ions diffusing into the sarcoplasm from the extracellular fluid or leaking out of the sarcoplasmic reticulum then trigger a sustained contraction. Without ATP, the cross-bridges cannot detach from the active sites, so skeletal muscles throughout the body become locked in the contracted position. Because all the skeletal muscles are involved, the individual becomes "stiff as a board." This physical state—**rigor mortis**—lasts until the lysosomal enzymes released by autolysis break down the Z lines and titin filaments starting at 15–25 hours after death and ending 72 or more hours later. The timing is dependent on environmental factors, such as temperature. Forensic pathologists can estimate the time of death on the basis of the degree of rigor mortis and environmental conditions.

The A&P Top 100

#35 Skeletal muscle fibers shorten as thin filaments interact with thick filaments and sliding occurs. The trigger for contraction is calcium ions released by the SR when the muscle fiber is stimulated by a motor neuron. Contraction is an active process; relaxation and the return to resting length is entirely passive.

CHECKPOINT

8. Describe the neuromuscular junction.

9. How would a drug that blocks acetylcholine release affect muscle contraction?

10. What would happen to a resting skeletal muscle if the sarcolemma suddenly became very permeable to Ca^{2+}

11. Predict what would happen to a muscle if the motor end plate failed to produce acetylcholinesterase.

See the blue Answers tab at the end of the book.

10-5 Sarcomere shortening and muscle fiber stimulation produce tension

When sarcomeres shorten in a contraction, they shorten the muscle fiber. This shortening exerts tension on the connective tissue fibers attached to the muscle fiber. The tension produced by an individual muscle fiber can vary, and in the next section we will consider the specific factors involved. In a subsequent section, we will see that the tension produced by an entire skeletal *muscle* can vary even more widely, because not only can tension production vary among the individual muscle fibers, but the number of stimulated muscle fibers can change from moment to moment.

Tension Production by Muscle Fibers

The amount of tension produced by an individual muscle fiber ultimately depends on the number of pivoting cross-bridges. There is no mechanism to regulate the amount of tension produced in that contraction by changing the number of contracting sarcomeres. When calcium ions are released, they are released from all triads in the muscle fiber. Thus, a muscle fiber is either "on" (producing tension) or "off" (relaxed). Tension production at the level of the individual muscle fiber *does* vary, depending on (1) the fiber's resting length at the time of stimulation, which determines the degree of overlap between thick and thin filaments; and (2) the frequency of stimulation, which affects the internal concentration of calcium ions and thus the amount bound to troponins.

Length–Tension Relationships

When many people pull on a rope, the amount of tension produced is proportional to the number of people pulling. Similarly, in a skeletal muscle fiber, the amount of tension generated during a contraction depends on the number of pivoting cross-bridges in each of the myofibrils. The number of cross-bridges that can form, in turn, depends on the degree of overlap between thick filaments and thin filaments within these sarcomeres. When the muscle fiber is stimulated to contract, only myosin heads in the zones of overlap can bind to active sites and produce tension. The tension produced by the entire muscle fiber can thus be related to the structure of individual sarcomeres.

A sarcomere works most efficiently within an optimal range of lengths (**Figure 10–14a**). When the resting sarcomere length is within this range, the maximum number of cross-bridges can form, and the tension produced is highest. If the resting sarcomere length falls outside the range—if the sarcomere is compressed and shortened, or stretched and lengthened—it cannot produce as much tension when stimulated. This is because the amount of tension produced is largely determined by the number of cross-bridges that form. An increase in sarcomere length reduces the tension produced by reducing the size of the zone of overlap and the number of potential cross-bridge interactions (**Figure 10–14b**). When the zone of overlap is reduced to zero, thin and thick filaments cannot interact at all. Under these conditions, the muscle fiber cannot produce any active tension, and a contraction cannot occur. Such extreme stretch-

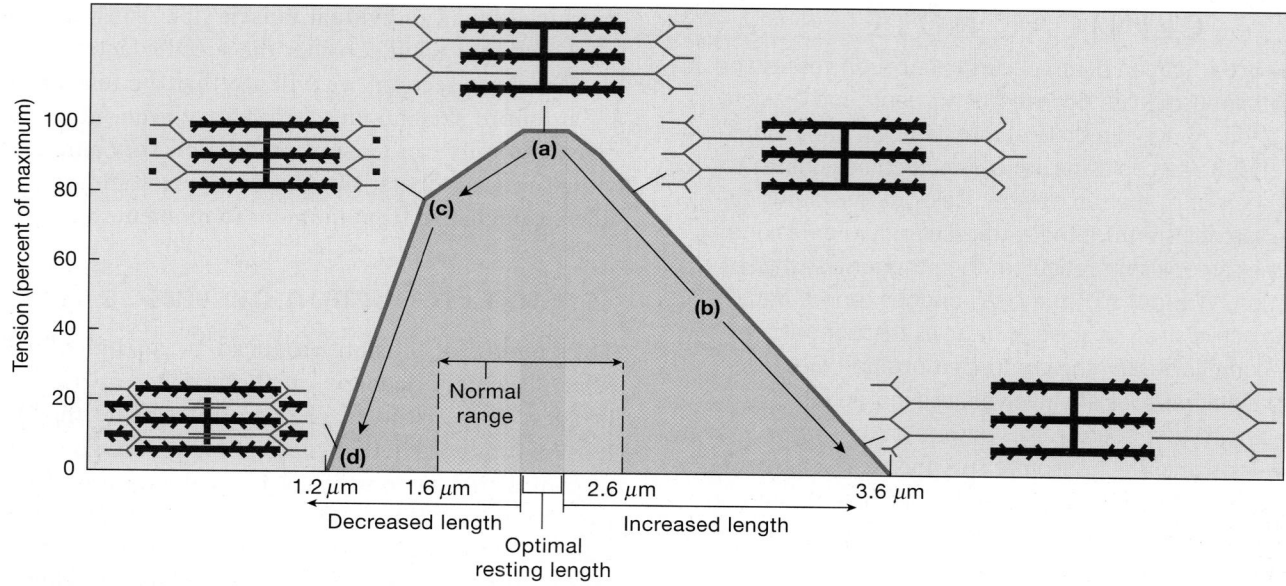

Figure 10–14   The Effect of Sarcomere Length on Active Tension. (a) Maximum tension is produced when the zone of overlap is large but the thin filaments do not extend across the sarcomere's center. **(b)** If the sarcomeres are stretched too far, the zone of overlap is reduced or disappears, and cross-bridge interactions are reduced or cannot occur. **(c)** At short resting lengths, thin filaments extending across the center of the sarcomere interfere with the normal orientation of thick and thin filaments, reducing tension production. **(d)** When the thick filaments contact the Z lines, the sarcomere cannot shorten—the myosin heads cannot pivot and tension cannot be produced. The width of the light purple area represents the normal operating range of sarcomere lengths.

ing of a muscle fiber is normally opposed by the titin filaments in the muscle fiber, which tie the thick filaments to the Z lines, and by the surrounding connective tissues, which limit the degree of muscle stretch.

A decrease in the resting sarcomere length reduces efficiency because the stimulated sarcomere cannot shorten very much before the thin filaments extend across the center of the sarcomere and collide with or overlap the thin filaments of the opposite side (**Figure 10–14c**). This disrupts the precise three-dimensional relationship between thick and thin filaments (**Figure 10–5**) and interferes with the binding of myosin heads to active sites and the propagation of the action potential along the transverse tubules. Because the number of cross-bridges is reduced, tension declines in the stimulated muscle fiber. Tension production falls to zero when the resting sarcomere is as short as it can be (**Figure 10–14d**). At this point, the thick filaments are jammed against the Z lines and the sarcomere cannot shorten further. Although cross-bridge binding can still occur, the myosin heads cannot pivot and generate tension, because the thin filaments cannot move.

In summary, skeletal muscle fibers contract most forcefully when stimulated over a narrow range of lengths. The normal range of sarcomere lengths in the body is 75 to 130 percent of the optimal length (**Figure 10–14**). The arrangement of skeletal muscles, connective tissues, and bones nor-

mally prevents extreme compression or excessive stretching. For example, straightening your elbow stretches your *biceps brachii muscle*, but the bones and ligaments of the elbow stop this movement before the muscle fibers stretch too far. During an activity such as walking, in which muscles contract and relax cyclically, muscle fibers are stretched to a length very close to "ideal" before they are stimulated to contract. When muscles must contract over a larger range of resting lengths, they often "team up" to improve efficiency. (We will discuss the mechanical principles involved in Chapter 11.)

The Frequency of Stimulation

A single stimulation produces a single contraction, or *twitch*, that may last 7–100 milliseconds, depending on the muscle stimulated. Although muscle twitches can be produced by electrical stimulation in a laboratory, they are too brief to be part of any normal activity. The duration of a contraction can be extended by repeated stimulation, and a muscle fiber undergoing such a sustained contraction produces more tension than it does in a single twitch. To understand why, we need to take a closer look at tension production during a twitch and then follow the changes that occur as the rate of stimulation increases. This is a subject with real importance, as all consciously and subconsciously directed muscular activities—

standing, walking, running, reaching, and so forth—involve sustained muscular contractions rather than twitches.

Twitches. A **twitch** is a single stimulus–contraction–relaxation sequence in a muscle fiber. Twitches vary in duration, depending on the type of muscle, its location, internal and external environmental conditions, and other factors. Twitches in an eye muscle fiber can be as brief as 7.5 msec, but a twitch in a muscle fiber from the *soleus*, a small calf muscle, lasts about 100 msec. **Figure 10–15a** is a **myogram**, or graph, of twitch tension development in muscle fibers from various skeletal muscles.

Figure 10–15b details the phases of a 40-msec twitch in a muscle fiber from the *gastrocnemius muscle*, a prominent calf

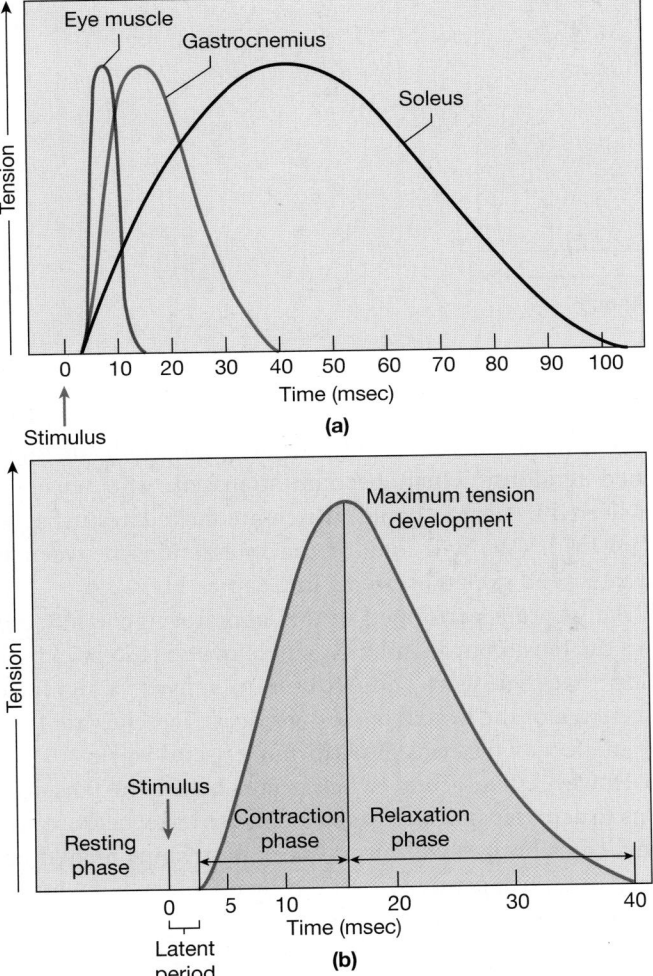

Figure 10–15 The Development of Tension in a Twitch. (a) A myogram showing differences in tension over time for a twitch in different skeletal muscles. **(b)** The details of tension over time for a single twitch in the gastrocnemius muscle. Notice the presence of a latent period, which corresponds to the time needed for the conduction of an action potential and the subsequent release of calcium ions by the sarcoplasmic reticulum.

muscle. A single twitch can be divided into a *latent period*, a *contraction phase*, and a *relaxation phase*:

1. The **latent period** begins at stimulation and typically lasts about 2 msec. During this period, the action potential sweeps across the sarcolemma, and the sarcoplasmic reticulum releases calcium ions. The muscle fiber does not produce tension during the latent period, because the contraction cycle has yet to begin.

2. In the **contraction phase**, tension rises to a peak. As the tension rises, calcium ions are binding to troponin, active sites on thin filaments are being exposed, and cross-bridge interactions are occurring. For this muscle fiber, the contraction phase ends roughly 15 msec after stimulation.

3. The **relaxation phase** lasts about 25 msec. During this period, calcium levels are falling, active sites are being covered by tropomyosin, and the number of active cross-bridges is declining as they detach. As a result, tension falls to resting levels.

Treppe. If a skeletal muscle is stimulated a second time immediately after the relaxation phase has ended, the resulting contraction will develop a slightly higher maximum tension than did the contraction after the first stimulus. The increase in peak tension indicated in **Figure 10–16a** will continue over the first 30–50 stimulations. Thereafter, the amount of tension produced will remain constant. Because the tension rises in stages, like the steps in a staircase, this phenomenon is called **treppe** (TREP-eh, German for staircase). Most skeletal muscles don't demonstrate treppe. The rise is thought to result from a gradual increase in the concentration of calcium ions in the sarcoplasm, in part because the ion pumps in the sarcoplasmic reticulum have too little time to recapture the ions between stimulations.

Wave Summation and Incomplete Tetanus. If a second stimulus arrives before the relaxation phase has ended, a second, more powerful contraction occurs. The addition of one twitch to another in this way constitutes the **summation of twitches**, or **wave summation** (**Figure 10–16b**). The duration of a single twitch determines the maximum time available to produce wave summation. For example, if a twitch lasts 20 msec (1/50 sec), subsequent stimuli must be separated by less than 20 msec—a stimulation rate of more than 50 stimuli per second. This rate is usually expressed in terms of *stimulus frequency*, which is the number of stimuli per unit time. In this instance, a stimulus frequency of greater than 50 per second produces wave summation, whereas a stimulus frequency of less than 50 per second will produce individual twitches and treppe.

If the stimulation continues and the muscle is never allowed to relax completely, tension will rise until it reaches a peak value roughly four times the maximum produced by

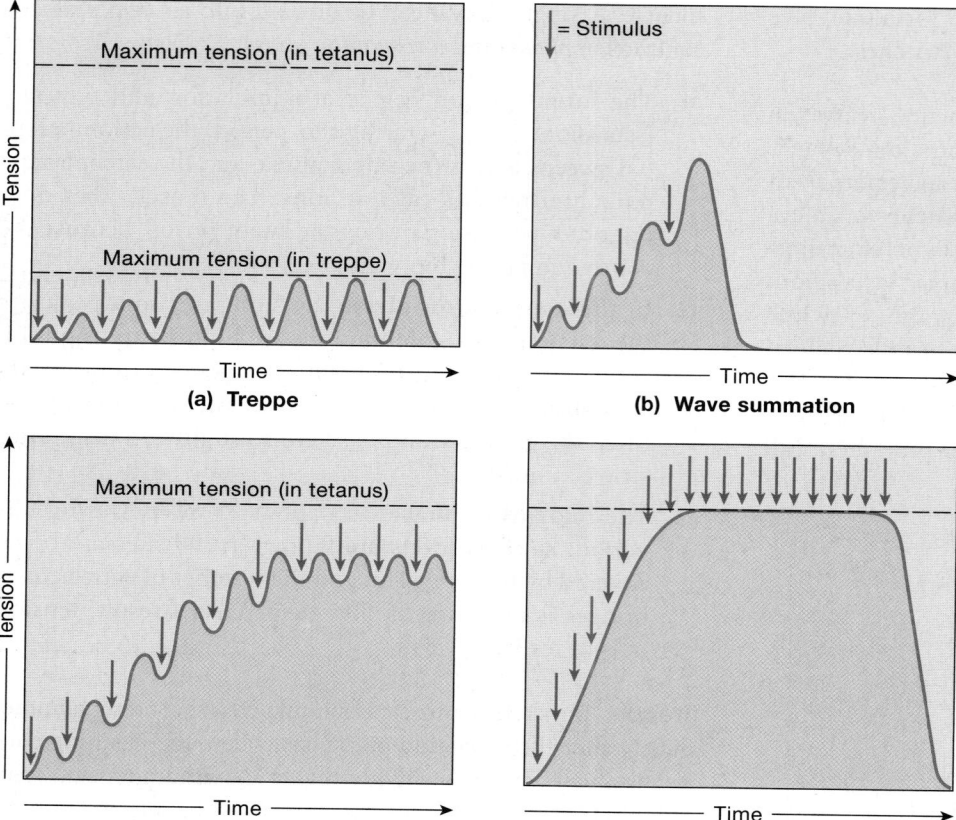

(a) Treppe

(b) Wave summation

(c) Incomplete tetanus

(d) Complete tetanus

Figure 10–16 **Effects of Repeated Stimulations. (a)** Treppe is an increase in peak tension with each successive stimulus delivered shortly after the completion of the relaxation phase of the preceding twitch. **(b)** Wave summation occurs when successive stimuli arrive before the relaxation phase has been completed. **(c)** Incomplete tetanus occurs if the stimulus frequency increases further. Tension production rises to a peak, and the periods of relaxation are very brief. **(d)** During complete tetanus, the stimulus frequency is so high that the relaxation phase is eliminated; tension plateaus at maximal levels.

treppe (**Figure 10–16c**). A muscle producing almost peak tension during rapid cycles of contraction and relaxation is in **incomplete tetanus** (*tetanos*, convulsive tension).

Complete Tetanus. Complete tetanus occurs when a higher stimulation frequency eliminates the relaxation phase (**Figure 10–16d**). During complete tetanus, action potentials arrive so rapidly that the sarcoplasmic reticulum does not have time to reclaim the calcium ions. The high Ca^{2+} concentration in the cytoplasm prolongs the contraction, making it continuous.

Tension Production by Skeletal Muscles

Now that you are familiar with the basic mechanisms of muscle contraction at the level of the individual muscle fiber, we can begin to examine the performance of skeletal muscles—the organs of the muscular system. In this section, we will consider the coordinated contractions of an entire population of skeletal muscle fibers. The amount of tension produced in the skeletal muscle *as a whole* is determined by (1) the tension produced by the stimulated muscle fibers and (2) the total number of muscle fibers stimulated.

As muscle fibers actively shorten, they pull on the attached tendons, which become stretched. The tension is transferred in turn to bones, which are moved against an external load. (We will look at the interactions between the muscular and skeletal systems in Chapter 11.)

A myogram performed in the laboratory generally measures the tension in a tendon. A single twitch is so brief in duration that there isn't enough time to activate a significant percentage of the available cross-bridges. Twitches are therefore ineffective in terms of performing useful work.

However, if a second twitch occurs before the tension returns to zero, tension will peak at a higher level, because additional cross-bridges will form. Think of pushing a child on a swing: You push gently to start the swing moving; if you push harder the second time, the child swings higher because the energy of the second push is added to the energy remaining from the first. Each successive contraction begins before the tension has fallen to resting levels, so the tension continues to rise until it peaks. During a tetanic contraction, there is enough time for essentially all of the potential cross-bridges to form, and tension peaks. Muscles are rarely used this way in the body, but they can be made to contract tetanically in the laboratory.

Motor Units and Tension Production

The amount of tension produced by the muscle as a whole is the sum of the tensions generated by the individual muscle fibers, since they are all pulling together. Thus, you can control the amount of tension produced by a skeletal muscle by controlling the number of stimulated muscle fibers.

A typical skeletal muscle contains thousands of muscle fibers. Although some motor neurons control a few muscle fibers, most control hundreds of them. All the muscle fibers controlled by a single motor neuron constitute a **motor unit**. The size of a motor unit is an indication of how fine the control of movement can be. In the muscles of the eye, where precise control is extremely important, a motor neuron may control 4–6 muscle fibers. We have much less precise control over our leg muscles, where a single motor neuron may control 1000–2000 muscle fibers. The muscle fibers of each motor unit are intermingled with those of other motor units (**Figure 10–17a**). Because of this intermingling, the direction of pull exerted on the tendon does not change when the number of activated motor units changes.

When you decide to perform a specific arm movement, specific groups of motor neurons in the spinal cord are stimulated. The contraction begins with the activation of the smallest motor units in the stimulated muscle. These motor units generally contain muscle fibers that contract relatively slowly. As the movement continues, larger motor units containing faster and more powerful muscle fibers are activated, and tension production rises steeply. The smooth, but steady, increase in muscular tension produced by increasing the number of active motor units is called **recruitment**.

Peak tension production occurs when all motor units in the muscle contract in a state of complete tetanus. Such powerful contractions do not last long, however, because the individual muscle fibers soon use up their available energy reserves. During a sustained contraction, motor units are activated on a rotating basis, so some of them are resting and recovering while others are actively contracting. In this "relay team" approach, called *asynchronous motor unit summation*, each motor unit can recover somewhat before it is stimulated again (**Figure 10–17b**). As a result, when your muscles contract for sustained periods, they produce slightly less than maximal tension.

The A&P Top 100

#36 All voluntary muscle contractions and intentional movements involve the sustained contractions of skeletal muscle fibers. The force generated can be increased by increasing the frequency of motor neuron action potentials or by recruiting additional motor units.

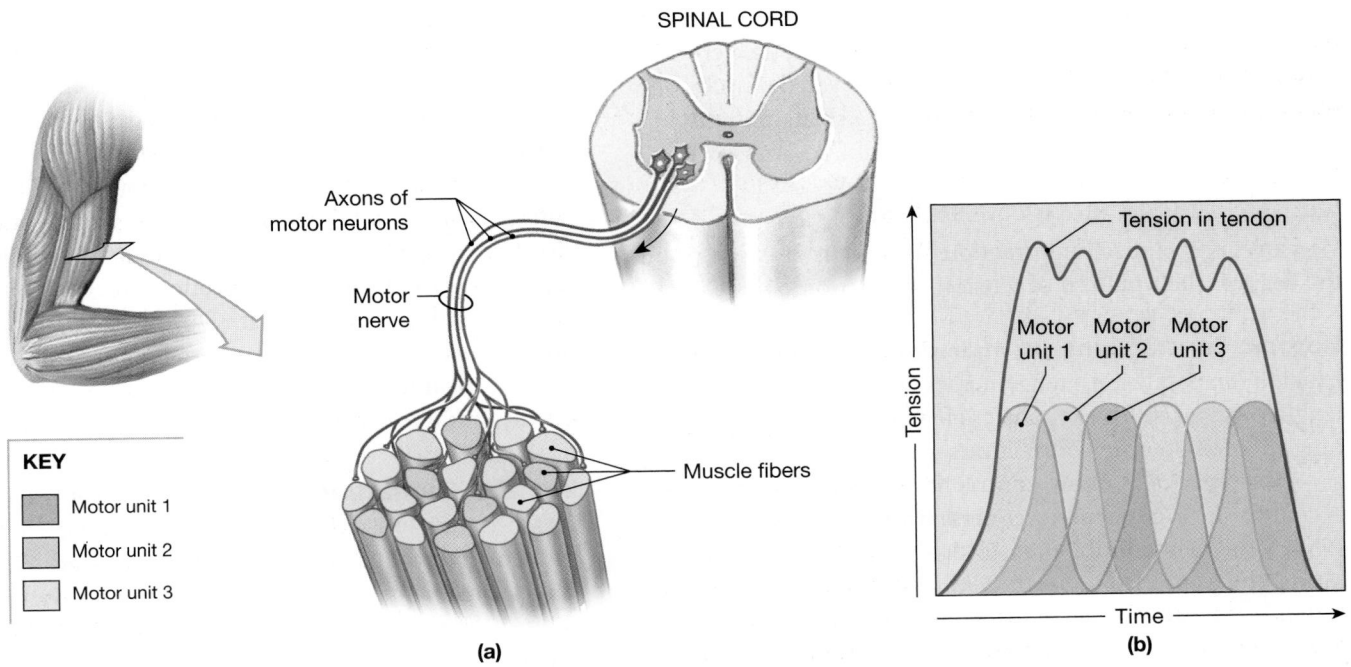

Figure 10–17 The Arrangement and Activity of Motor Units in a Skeletal Muscle. (a) Muscle fibers of different motor units are intermingled, so the forces applied to the tendon remain roughly balanced regardless of which motor units are stimulated. **(b)** The tension applied to the tendon remains relatively constant, even though individual motor units cycle between contraction and relaxation.

Muscle Tone

In any skeletal muscle, some motor units are always active, even when the entire muscle is not contracting. Their contractions do not produce enough tension to cause movement, but they do tense and firm the muscle. This resting tension in a skeletal muscle is called **muscle tone**. A muscle with little muscle tone appears limp and flaccid, whereas one with moderate muscle tone is firm and solid. The identity of the stimulated motor units changes constantly, so individual muscle fibers can relax while a constant tension is maintained in the attached tendon.

Resting muscle tone stabilizes the positions of bones and joints. For example, in muscles involved with balance and posture, enough motor units are stimulated to produce the tension needed to maintain body position. Muscle tone also helps prevent sudden, uncontrolled changes in the positions of bones and joints. In addition to bracing the skeleton, the elastic nature of muscles and tendons lets skeletal muscles act as shock absorbers that cushion the impact of a sudden bump or shock. Heightened muscle tone accelerates the recruitment process during a voluntary contraction, because some of the motor units are already stimulated. Strong muscle tone also makes skeletal muscles appear firm and well defined, even at rest.

Activated muscle fibers use energy, so the greater the muscle tone, the higher the "resting" rate of metabolism. Increasing this rate is one of the significant effects of exercise in a weight-loss program. In such a program, you lose weight when your daily energy use exceeds your daily energy intake in food. Although exercise consumes energy very quickly, the period of activity is usually quite brief. In contrast, elevated muscle tone increases resting energy consumption by a small amount, but the effects are cumulative, and they continue 24 hours per day.

Isotonic and Isometric Contractions

We can classify muscle contractions as *isotonic* or *isometric* on the basis of their pattern of tension production.

Isotonic Contractions. In an **isotonic contraction** (*iso-*, equal + *tonos*, tension), tension rises and the skeletal muscle's length changes. Lifting an object off a desk, walking, and running involve isotonic contractions.

Two types of isotonic contractions exist: concentric and eccentric. In a **concentric contraction**, the muscle tension *exceeds* the load and the muscle *shortens*. Consider a skeletal muscle that is 1 cm^2 in cross-sectional area and can produce roughly 4 kg (8.8 lb.)of tension in complete tetanus. If we hang a load of 2 kg (4.4 lb.) from that muscle and stimulate it, the muscle will shorten (**Figure 10–18a**). Before the muscle can shorten, the cross-bridges must produce enough tension to overcome the load—in this case, the 2-kg weight. During this initial period, tension in the muscle fibers rises until the

tension in the tendon exceeds the load. As the muscle shortens, the tension in the skeletal muscle remains constant at a value that just exceeds the load (**Figure 10–18b**). The term *isotonic* originated from this type of experiment.

In the body, however, the situation is more complicated. Muscles are not always positioned directly above the load, and they are attached to bones rather than to static weights. Changes in the relative positions of the muscle and the articulating bones, the effects of gravity, and other mechanical and physical factors interact to increase or decrease the load the muscle must overcome as a movement proceeds. Nevertheless, during a concentric contraction, the peak tension produced exceeds that load.

The speed of shortening varies with the difference between the amount of tension produced and the size of the load. If all the motor units are stimulated and the load is relatively small, the muscle will shorten very quickly. In contrast, if the muscle barely produces enough tension to overcome the load, it will shorten very slowly.

In an **eccentric contraction**, the peak tension developed is *less than* the load, and the muscle *elongates* owing to the contraction of another muscle or the pull of gravity. Think of a tug-of-war team trying to stop a moving car. Although everyone pulls as hard as they can, the rope slips through their fingers. The speed of elongation depends on the difference between the amount of tension developed by the active muscle fibers and the size of the load. In our analogy, the team might slow down a small car, but would have little effect on a large truck.

Eccentric contractions are very common, and they are an important part of a variety of movements. In these movements, you exert precise control over the amount of tension produced. By varying the tension in an eccentric contraction, you can control the rate of elongation, just as you can vary the tension in a concentric contraction. During physical training, people commonly perform cycles of concentric and eccentric contractions, as when you hold a weight in your hand and slowly perform flexion and extension at the elbow. The flexion involves concentric contractions that exceed the resistance posed by the weight. During extension, the same muscles are active, but the contractions are eccentric. The tension produced isn't sufficient to overcome the force of gravity, but it is enough to control the speed of movement.

Isometric Contractions. In an **isometric contraction** (*metric*, measure), the muscle as a whole does not change length, and the tension produced never exceeds the load. **Figure 10–18c** shows what happens if we attach a weight heavier than 4 kg to the experimental muscle and then stimulate the muscle. Although cross-bridges form and tension rises to peak values, the muscle cannot overcome the resistance of the weight and so cannot shorten (**Figure 10–18d**). Examples of isometric contractions include carrying a bag of groceries and holding our heads up. Many of the reflexive muscle contractions that keep your

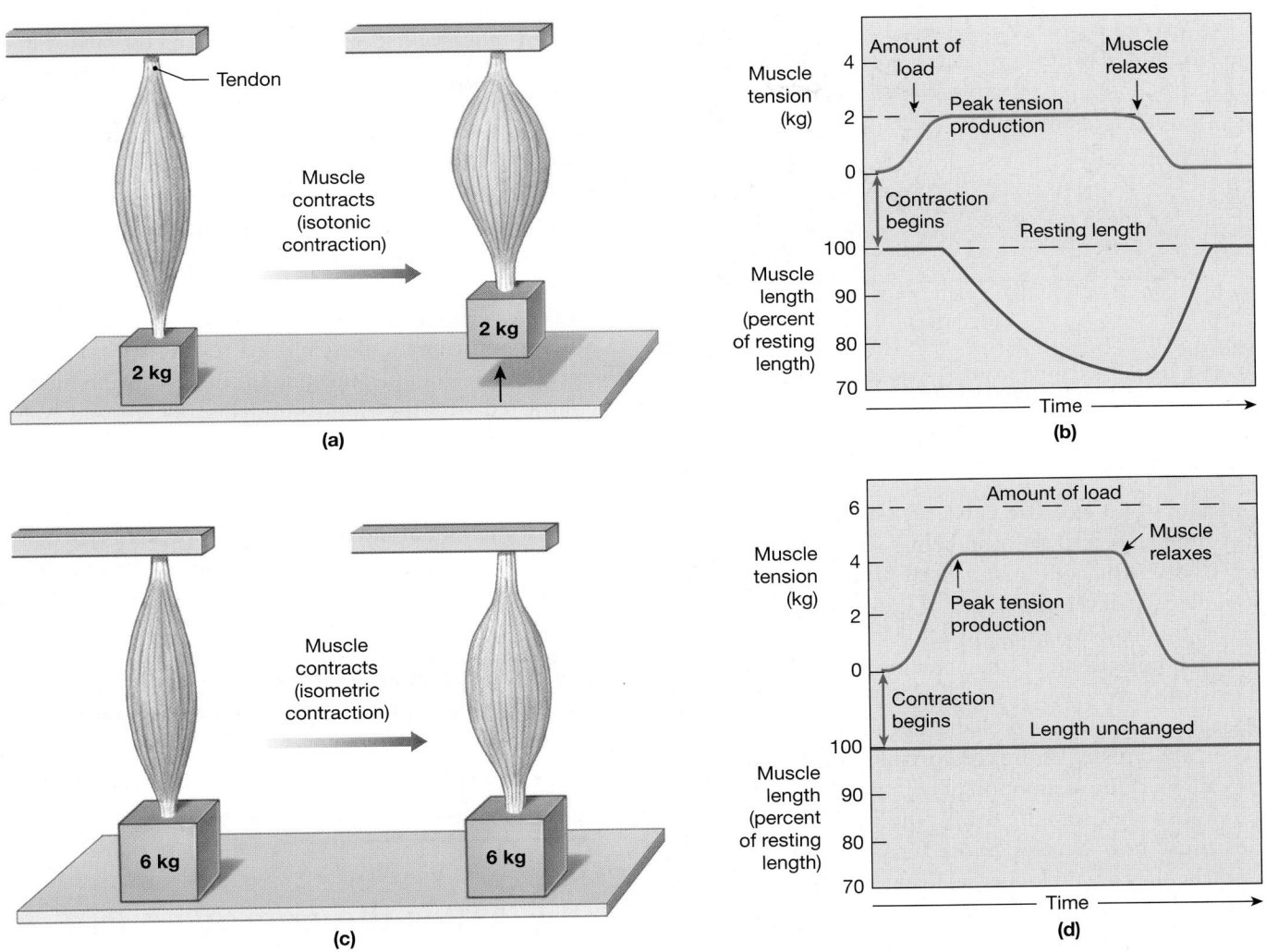

Figure 10–18 Isotonic and Isometric Contractions. (a,b) This muscle is attached to a weight less than its peak tension capabilities. On stimulation, it develops enough tension to lift the weight. Tension remains constant for the duration of the contraction, although the length of the muscle changes. This is an example of isotonic contraction. **(c,d)** The same muscle is attached to a weight that exceeds its peak tension capabilities. On stimulation, tension will rise to a peak, but the muscle as a whole cannot shorten. This is an isometric contraction.

body upright when you stand or sit involve isometric contractions of muscles that oppose the force of gravity.

Notice that when you perform an isometric contraction, the contracting muscle bulges, but not as much as it does during an isotonic contraction. In an isometric contraction, although the muscle *as a whole* does not shorten, the individual muscle fibers shorten as connective tissues stretch. The muscle fibers cannot shorten further, because the tension does not exceed the load.

Normal daily activities therefore involve a combination of isotonic and isometric muscular contractions. As you sit and read this text, isometric contractions of postural muscles stabilize your vertebrae and maintain your upright position. When you turn a page, the movements of your arm, forearm, hand, and fingers are produced by a combination of concentric and eccentric isotonic contractions.

Tips&Tricks

During an *isotonic* contraction such as lifting a baby, muscle tension remains constant while muscle length changes; during an *isometric* contraction such as holding a baby at arm's length, muscle length remains constant but muscle tension changes.

Load and Speed of Contraction

You can lift a light object more rapidly than you can lift a heavy one because load and the speed of contraction are inversely related. If the load is less than the tension produced, a concentric isotonic contraction will occur; the muscle will shorten. The heavier the load, the longer it takes for the

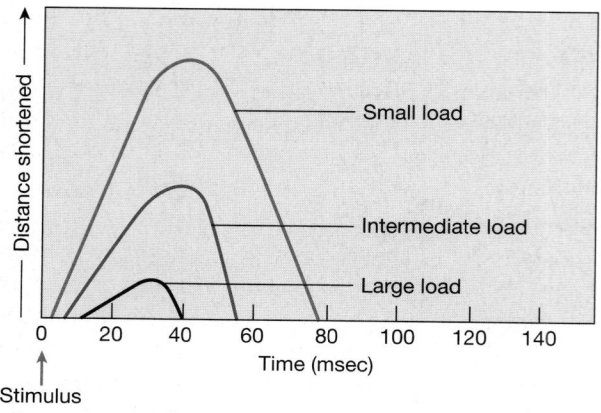

Figure 10–19 **Load and Speed of Contraction.** The heavier the load on a muscle, the longer it will take for the muscle to begin to shorten and the less the muscle will shorten.

movement to begin, because muscle tension (which increases gradually) must exceed the load before shortening can occur (**Figure 10–19**). The contraction itself proceeds more slowly. At the molecular level, the speed of cross-bridge pivoting is reduced as the load increases.

For each muscle, an optimal combination of tension and speed exists for any given resistance. If you have ever ridden a 21-speed bicycle, you are probably already aware of this fact. When you are cruising along comfortably, your thigh and leg muscles are working at an optimal combination of speed and tension. When you start up a hill, the load increases. Your muscles must now develop more tension, and they move more slowly; they are no longer working at optimal efficiency. If you then shift to a lower gear, the load on your muscles decreases and their speed increases, and the muscles are once again working efficiently.

Muscle Relaxation and the Return to Resting Length

As we noted earlier, there is no active mechanism for muscle fiber elongation. The sarcomeres in a muscle fiber can shorten and develop tension, but the power stroke cannot be reversed to push the Z lines farther apart. After a contraction, a muscle fiber returns to its original length through a combination of elastic forces, opposing muscle contractions, and gravity.

Elastic Forces. When the contraction ends, some of the energy initially "spent" in stretching the tendons and distorting intracellular organelles is recovered as they recoil or rebound to their original dimensions. This elasticity gradually helps return the muscle fiber to its original resting length.

Opposing Muscle Contractions. The contraction of opposing muscles can return a muscle to its resting length more quickly than elastic forces can. Consider the muscles of the

arm that flex or extend the elbow. Contraction of the *biceps brachii muscle* on the anterior part of the arm flexes the elbow; contraction of the *triceps brachii muscle* on the posterior part of the arm extends the elbow. When the biceps brachii muscle contracts, the triceps brachii muscle is stretched. When the biceps brachii muscle relaxes, contraction of the triceps brachii muscle extends the elbow and stretches the muscle fibers of the biceps brachii muscle to their original length.

Gravity. Gravity may assist opposing muscle groups in quickly returning a muscle to its resting length after a contraction. For example, imagine the biceps brachii muscle fully contracted with the elbow pointed at the ground. When the muscle relaxes, gravity will pull the forearm down and stretch the muscle. Although gravity can provide assistance in stretching muscles, some active muscle tension is needed to control the rate of movement and to prevent damage to the joint. In the previous example, eccentric contraction of the biceps brachii muscle can control the movement.

CHECKPOINT

12. **Why does a muscle that has been overstretched produce less tension?**

13. **Can a skeletal muscle contract without shortening?**

See the blue Answers tab at the end of the book.

10-6 ATP is the energy source for muscle contraction

A single muscle fiber may contain 15 billion thick filaments. When that muscle fiber is actively contracting, each thick filament breaks down roughly 2500 ATP molecules per second. Because even a small skeletal muscle contains thousands of muscle fibers, the ATP demands of a contracting skeletal muscle are enormous. In practical terms, the demand for ATP in a contracting muscle fiber is so high that it would be impossible to have all the necessary energy available as ATP before the contraction begins. Instead, a resting muscle fiber contains only enough ATP and other high-energy compounds to sustain a contraction until additional ATP can be generated. Throughout the rest of the contraction, the muscle fiber will generate ATP at roughly the same rate as it is used.

ATP and CP Reserves

The primary function of ATP is the transfer of energy from one location to another rather than the long-term storage of energy. At rest, a skeletal muscle fiber produces more ATP than it needs. Under these conditions, ATP transfers energy to creatine. *Creatine* (KRĒ-uh-tēn) is a small molecule that muscle cells assemble from fragments of amino acids. The energy transfer creates another high-energy compound, **creatine phosphate (CP)**, or *phosphorylcreatine*:

$$\text{ATP + creatine} \rightarrow \text{ADP + creatine phosphate}$$

During a contraction, each myosin head breaks down ATP, producing ADP and a phosphate group. The energy stored in creatine phosphate is then used to "recharge" ADP, converting it back to ATP through the reverse reaction:

$$\text{ADP + creatine phosphate} \rightarrow \text{ATP + creatine}$$

The enzyme that facilitates this reaction is **creatine phosphokinase (CPK or CK)**. When muscle cells are damaged, CPK leaks across the plasma membranes and into the bloodstream. Thus, a high blood concentration of CPK usually indicates serious muscle damage.

The energy reserves of a representative muscle fiber are indicated in Table 10–2. A resting skeletal muscle fiber contains about six times as much creatine phosphate as ATP, but when a muscle fiber is undergoing a sustained contraction, these energy reserves are exhausted in only about 15 seconds. The muscle fiber must then rely on other mechanisms to generate ATP from ADP.

ATP Generation

As we saw in Chapter 3, most cells in the body generate ATP through (1) aerobic metabolism in mitochondria and (2) glycolysis in the cytoplasm. ∞ p. 80

Aerobic Metabolism

Aerobic metabolism normally provides 95 percent of the ATP demands of a resting cell. In this process, mitochondria absorb oxygen, ADP, phosphate ions, and organic substrates (such as pyruvic acid) from the surrounding cytoplasm. The substrates then enter the *TCA (tricarboxylic acid) cycle* (also known as the *citric acid cycle* or the *Krebs cycle*), an enzymatic pathway that breaks down organic molecules. The carbon atoms are released as carbon dioxide. The hydrogen atoms are

shuttled to respiratory enzymes in the inner mitochondrial membrane, where their electrons are removed. After a series of intermediate steps, the protons and electrons are combined with oxygen to form water. Along the way, large amounts of energy are released and used to make ATP. The entire process is very efficient: For each molecule of pyruvic acid "fed" into the TCA cycle, the cell gains 17 ATP molecules.

Resting skeletal muscle fibers rely almost exclusively on the aerobic metabolism of fatty acids to generate ATP. These fatty acids are absorbed from the circulation. When the muscle starts contracting, the mitochondria begin breaking down molecules of pyruvic acid instead of fatty acids. The pyruvic acid is provided by the enzymatic pathway of glycolysis, which breaks down glucose in the cytoplasm. The glucose can come either from the surrounding interstitial fluid or through the breakdown of glycogen reserves within the sarcoplasm. Because a typical skeletal muscle fiber contains large amounts of glycogen, the shift from fatty acid metabolism to glucose metabolism makes it possible for the cell to continue contracting for an extended period, even without an external source of nutrients.

Glycolysis

Glycolysis is the breakdown of glucose to pyruvic acid in the cytoplasm of a cell. It is an **anaerobic process**, because it does not require oxygen. Glycolysis provides a net gain of 2 ATP molecules and generates 2 pyruvic acid molecules from each glucose molecule. The ATP produced by glycolysis is therefore only a small fraction of that produced by aerobic metabolism, in which the breakdown of the 2 pyruvic acid molecules in mitochondria would generate 34 ATP molecules. Thus, when energy demands are relatively low and oxygen is readily available, glycolysis is important only because it provides the substrates for aerobic metabolism. Yet, because it can proceed in the absence of oxygen, glycolysis becomes an important source of energy when energy demands are at a maximum and the availability of oxygen limits the rate of mitochondrial ATP production.

The glucose broken down under these conditions is obtained primarily from the reserves of glycogen in the sarcoplasm. Glycogen is a polysaccharide chain of glucose molecules. ∞ p. 47 Typical skeletal muscle fibers contain

TABLE 10–2	Sources of Energy Stored in a Typical Muscle Fiber			
Energy Stored as	**Utilized through**	**Initial Quantity**	**Number of Twitches Supported by Each Energy Source Alone**	**Duration of Isometric Tetanic Contraction Supported by Each Energy Source Alone**
ATP	ATP ⟶ ADP + P	3 mmol	10	2 sec
CP	ADP ⟶ ATP + C	20 mmol	70	15 sec
Glycogen	Glycolysis (anaerobic)	100 mmol	670	130 sec
	Aerobic metabolism		12,000	2400 sec (40 min)

large glycogen reserves, which may account for 1.5 percent of the total muscle weight. When the muscle fiber begins to run short of ATP and CP, enzymes split the glycogen molecules, releasing glucose, which can be used to generate more ATP. When energy demands are low and oxygen is abundant, glycolysis provides substrates for anaerobic metabolism, and aerobic metabolism provides the ATP needed for contraction. However, during peak periods of muscular activity, energy demands are extremely high and oxygen supplies are very limited. Under these conditions, glycolysis provides most of the ATP needed to sustain muscular contraction.

Energy Use and the Level of Muscular Activity

As the level of muscular activity increases, the pattern of energy production and use changes:

- In a resting skeletal muscle (**Figure 10–20a**), the demand for ATP is low. More than enough oxygen is available for the mitochondria to meet that demand, and they produce a surplus of ATP. The extra ATP is used to build up reserves of CP and glycogen. Resting muscle fibers absorb fatty acids and glucose delivered by the bloodstream. The fatty acids are broken down in the mitochondria, and the ATP that is generated is used to convert creatine to creatine phosphate and glucose to glycogen.

- At moderate levels of activity (**Figure 10–20b**), the demand for ATP increases. This demand is met by the mitochondria. As the rate of mitochondrial ATP production rises, so does the rate of oxygen consumption. Oxygen availability is not a limiting factor, because oxygen can diffuse into the muscle fiber fast enough to meet mitochondrial needs. But all the ATP produced is needed by the muscle fiber, and no surplus is available. The skeletal muscle now relies primarily on the aerobic metabolism of pyruvic acid to generate ATP. The pyruvic acid is provided by glycolysis, which breaks down glucose molecules obtained from glycogen in the muscle fiber. If glycogen reserves are low, the muscle fiber can also break down other substrates, such as lipids or amino acids. As long as the demand for ATP can be met by mitochondrial activity, the ATP provided by glycolysis makes a relatively minor contribution to the total energy budget of the muscle fiber.

- At peak levels of activity (**Figure 10–20c**), ATP demands are enormous and mitochondrial ATP production rises to a maximum. This maximum rate is determined by the availability of oxygen, and oxygen cannot diffuse into the muscle fiber fast enough to enable the mitochondria to produce the required ATP. At peak levels of exertion, mitochondrial activity can provide only about one-third of the ATP needed. The remainder is produced through glycolysis. When glycolysis produces pyruvic acid faster than it can be utilized by the mitochondria, pyruvic acid levels rise in the sarcoplasm. Under these conditions, pyruvic acid is converted to **lactic acid**, a related three-carbon molecule.

The anaerobic process of glycolysis enables the cell to generate additional ATP when the mitochondria are unable to meet the current energy demands. However, anaerobic energy production has drawbacks. First, the lactic acid produced is an organic acid that dissociates in body fluids into a hydrogen ion and a negatively charged *lactate ion*. Thus, production of lactic acid can lower the intracellular pH. Buffers in the sarcoplasm can resist pH shifts, but these mechanisms are limited. Eventually, changes in pH will alter the functional characteristics of key enzymes so that the muscle fiber cannot continue to contract. Moreover, glycolysis is a relatively inefficient way to generate ATP. Under anaerobic conditions, each glucose molecule generates 2 pyruvic acid molecules, which are converted to lactic acid. In return, the cell gains 2 ATP molecules through glycolysis. Had those 2 pyruvic acid molecules been catabolized aerobically in a mitochondrion, the cell would have produced 34 additional ATP.

Muscle Fatigue

An active skeletal muscle is said to be **fatigued** when it can no longer continue to perform at the required level of activity. Many factors are involved in promoting muscle fatigue. For example, muscle fatigue has been correlated with (1) depletion of metabolic reserves within the muscle fibers; (2) damage to the sarcolemma and sarcoplasmic reticulum; (3) a decline in pH within the muscle fibers and the muscle as a whole, decreasing calcium ion binding to troponin and altering enzyme activities; and (4) a sense of weariness and a reduction in the desire to continue the activity, due to the effects of low blood pH and pain on the brain. Muscle fatigue is cumulative—the effects become more pronounced as more neurons and muscle fibers are affected. The result is a gradual reduction in the capabilities and performance of the entire skeletal muscle.

If a muscle fiber is contracting at moderate levels and ATP demands can be met through aerobic metabolism, fatigue will not occur until glycogen, lipid, and amino acid reserves are depleted. This type of fatigue affects the muscles of endurance athletes, such as marathon runners, after hours of exertion.

When a muscle produces a sudden, intense burst of activity at peak levels, most of the ATP is provided by glycolysis. After just seconds to minutes, the rising lactic acid levels

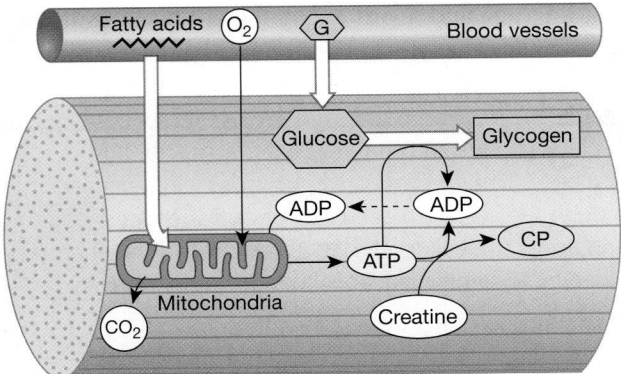

(a) Resting muscle: Fatty acids are catabolized; the ATP produced is used to build energy reserves of ATP, CP, and glycogen.

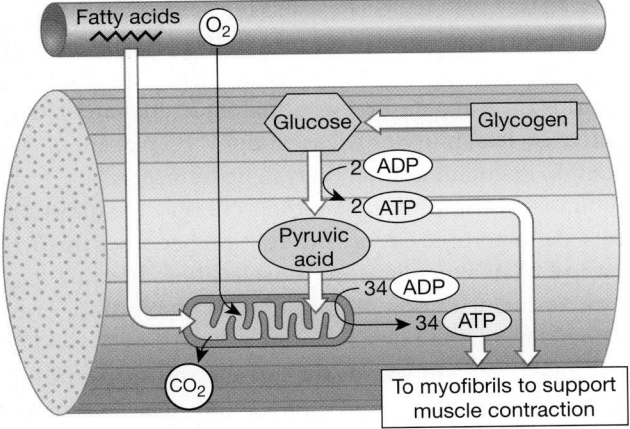

(b) Moderate activity: Glucose and fatty acids are catabolized; the ATP produced is used to power contraction.

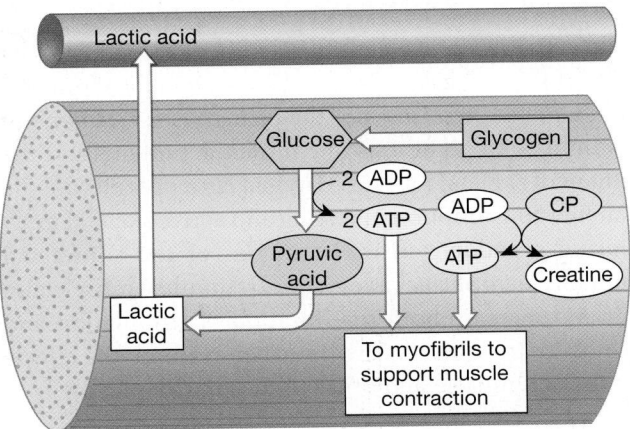

(c) Peak activity: Most ATP is produced through glycolysis, with lactic acid as a by-product. Mitochondrial activity (not shown) now provides only about one-third of the ATP consumed.

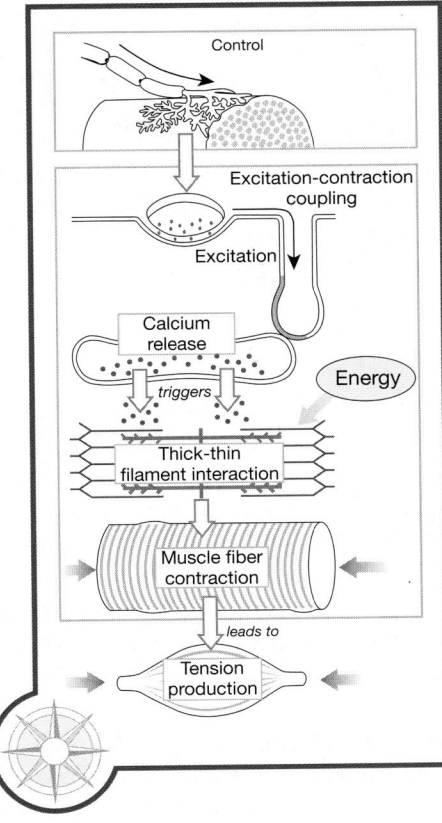

Figure 10–20 **Muscle Metabolism.** **(a)** A resting muscle breaks down fatty acids by aerobic metabolism to make ATP. Surplus ATP is used to build reserves of creatine phosphate (CP) and glycogen. **(b)** At moderate activity levels, mitochondria can meet ATP demands through the aerobic metabolism of fatty acids and glucose. **(c)** At peak activity levels, mitochondria cannot get enough oxygen to meet ATP demands. Most of the ATP is provided by glycolysis, leading to the production of lactic acid.

lower the tissue pH, and the muscle can no longer function normally. Sprinters get this type of muscle fatigue. We will return to the topics of fatigue, athletic training, and metabolic activity later in the chapter.

Normal muscle function requires (1) substantial intracellular energy reserves, (2) a normal circulatory supply, (3) normal blood oxygen levels, and (4) blood pH within normal limits. Anything that interferes with any of these factors will promote premature muscle fatigue. For example, reduced blood flow from tight clothing, heart problems, or blood loss slows the delivery of oxygen and nutrients, accelerates the buildup of lactic acid, and promotes muscle fatigue.

10 MUSCULAR

The Recovery Period

When a muscle fiber contracts, conditions in the sarcoplasm change. Energy reserves are consumed, heat is released, and, if the contraction was at peak levels, lactic acid is generated. In the **recovery period**, the conditions in muscle fibers are returned to normal, preexertion levels. After a period of moderate activity, it may take several hours for muscle fibers to recover. After sustained activity at higher levels, complete recovery can take a week.

Lactic Acid Removal and Recycling

Glycolysis enables a skeletal muscle to continue contracting even when mitochondrial activity is limited by the availability of oxygen. As we have seen, however, lactic acid production is not an ideal way to generate ATP. It squanders the glucose reserves of the muscle fibers, and it is potentially dangerous because lactic acid can lower the pH of the blood and tissues.

During the recovery period, when oxygen is available in abundance, lactic acid can be recycled by conversion back to pyruvic acid. The pyruvic acid can then be used either by mitochondria to generate ATP or as a substrate for enzyme pathways that synthesize glucose and rebuild glycogen reserves.

During exertion, lactic acid diffuses out of muscle fibers and into the bloodstream. The process continues after the exertion has ended, because intracellular lactic acid concentrations are still relatively high. The liver absorbs the lactic acid and converts it to pyruvic acid. Roughly 30 percent of these pyruvic acid molecules are broken down in the TCA cycle, providing the ATP needed to convert the other pyruvic acid molecules to glucose. (We will cover these processes more fully in Chapter 25.) The glucose molecules are then released into the circulation, where they are absorbed by skeletal muscle fibers and used to rebuild their glycogen reserves. This shuffling of lactic acid to the liver and glucose back to muscle cells is called the **Cori cycle**.

The Oxygen Debt

During the recovery period, the body's oxygen demand remains elevated above normal resting levels. The more ATP required, the more oxygen will be needed. The amount of oxygen required to restore normal, preexertion conditions is called the **oxygen debt**, or *excess postexercise oxygen consumption (EPOC)*.

Most of the additional oxygen consumption occurs in skeletal muscle fibers, which must restore ATP, creatine phosphate, and glycogen concentrations to their former levels, and in liver cells, which generate the ATP needed to convert excess lactic acid to glucose. However, several other tissues also increase their rate of oxygen consumption and ATP generation during the recovery period. For example, sweat glands increase their secretory activity until normal body temperature is restored. While the oxygen debt is being re-

paid, breathing rate and depth are increased. As a result, you continue to breathe heavily long after you stop exercising.

The A&P Top 100

#37 Skeletal muscles at rest metabolize fatty acids and store glycogen. During light activity, muscles can generate ATP through the aerobic breakdown of carbohydrates, lipids, or amino acids. At peak levels of activity, most of the energy is provided by anaerobic reactions that generate lactic acid as a by-product.

Heat Production and Loss

Muscular activity generates substantial amounts of heat. During a catabolic process, such as the breakdown of glycogen or the reactions of glycolysis, a muscle fiber captures only a portion of the released energy. ∞ p. 38 The rest is released as heat. A resting muscle fiber relying on aerobic metabolism captures about 42 percent of the energy released in catabolism. The other 58 percent warms the sarcoplasm, interstitial fluid, and circulating blood. Active skeletal muscles release roughly 85 percent of the heat needed to maintain normal body temperature.

When muscles become active, their energy consumption skyrockets. As anaerobic energy production becomes the primary method of ATP generation, muscle fibers become less efficient at capturing energy. At peak levels of exertion, only about 30 percent of the released energy is captured as ATP; the remaining 70 percent warms the muscle and surrounding tissues. Body temperature soon climbs, and heat loss at the skin accelerates through mechanisms introduced in Chapters 1 and 5. ∞ pp. 14, 174

Hormones and Muscle Metabolism

Metabolic activities in skeletal muscle fibers are adjusted by hormones of the endocrine system. *Growth hormone* from the pituitary gland and *testosterone* (the primary sex hormone in males) stimulate the synthesis of contractile proteins and the enlargement of skeletal muscles. *Thyroid hormones* elevate the rate of energy consumption in resting and active skeletal muscles. During a sudden crisis, hormones of the suprarenal gland, notably *epinephrine* (adrenaline), stimulate muscle metabolism and increase both the duration of stimulation and the force of contraction. (We will further examine the effects of hormones on muscle and other tissues in Chapter 18.)

CHECKPOINT

14. How do muscle cells continuously synthesize ATP?

15. What is muscle fatigue?

16. Define oxygen debt.

See the blue Answers tab at the end of the book.

10-7 Muscle fiber type and physical conditioning determine muscle performance capabilities

Muscle performance can be considered in terms of **force**, the maximum amount of tension produced by a particular muscle or muscle group, and **endurance**, the amount of time during which the individual can perform a particular activity. Here we consider the factors that determine the performance capabilities of any skeletal muscle: the type, distribution, and size of muscle fibers in the muscle, and physical conditioning or training.

Types of Skeletal Muscle Fibers

The human body has three major types of skeletal muscle fibers: *fast fibers*, *slow fibers*, and *intermediate fibers* (Table 10–3).

Fast Fibers

Most of the skeletal muscle fibers in the body are called **fast fibers**, because they can reach peak twitch tension in 0.01 sec or less after stimulation. Fast fibers are large in diameter and contain densely packed myofibrils, large glycogen reserves, and relatively few mitochondria. The tension produced by a muscle fiber is directly proportional to the number of myofibrils, so muscles dominated by fast fibers produce powerful contractions. However, fast fibers fatigue rapidly because their contractions use ATP in massive amounts, and there are relatively few mitochondria to generate ATP. As a result, prolonged activity is supported primarily by anaerobic metabolism. Other names used to refer to these muscle fibers include *white muscle fibers*, *fast-twitch glycolytic fibers*, *fast fatigue (FF) fibers*, and *Type II-B fibers*.

Slow Fibers

Slow fibers have only about half the diameter of fast fibers and take three times as long to reach peak tension after stimulation. These fibers are specialized to enable them to continue contracting for extended periods, long after a fast fiber would have become fatigued. The most important specializations improve mitochondrial performance.

One of the main characteristics of slow muscle fibers is that they are surrounded by a more extensive network of capillaries than is typical of fast muscle tissue; thus, they have a dramatically higher oxygen supply to support mitochondrial activity. Slow fibers also contain the red pigment **myoglobin** (MĪ-ō-glō-bin). This globular protein is structurally related to hemoglobin, the red oxygen-carrying pigment in blood. Both myoglobin and hemoglobin reversibly bind oxygen molecules. Although other muscle fiber types contain small amounts of myoglobin, it is most abundant in slow fibers. As a result, resting slow fibers contain substantial oxygen reserves that can be mobilized during a contraction. Because slow fibers have both an extensive capillary supply and a high concentration of myoglobin, skeletal muscles dominated by slow fibers are dark red. Slow fibers are also known as *red muscle fibers*, *slow-twitch oxidative fibers*, *slow oxidative (SO) fibers*, and *Type I fibers*.

With oxygen reserves and a more efficient blood supply, the mitochondria of slow fibers can contribute more ATP during contraction. In addition, the cross-bridges in slow fibers cycle more slowly than those of fast fibers, and this reduces demand for ATP. Thus, slow fibers are less dependent on anaerobic metabolism than are fast fibers. Some of the mitochondrial energy production involves the breakdown of stored lipids rather than glycogen, so glycogen reserves of slow fibers are smaller than those of fast fibers. Slow fibers

TABLE 10–3 Properties of Skeletal Muscle Fiber Types			
Property	**Slow**	**Intermediate**	**Fast**
Cross-sectional diameter	Small	Intermediate	Large
Tension	Low	Intermediate	High
Contraction speed	Slow	Fast	Fast
Fatigue resistance	High	Intermediate	Low
Color	Red	Pink	White
Myoglobin content	High	Low	Low
Capillary supply	Dense	Intermediate	Scarce
Mitochondria	Many	Intermediate	Few
Glycolytic enzyme concentration in sarcoplasm	Low	High	High
Substrates used for ATP generation during contraction	Lipids, carbohydrates, amino acids (aerobic)	Primarily carbohydrates (anaerobic)	Carbohydrates (anaerobic)
Alternative names	Type I, S (slow), red, SO (slow oxidative), slow-twitch oxidative	Type II-A, FR (fast resistant), fast-twitch oxidative	Type II-B, FF (fast fatigue), white, fast-twitch glycolytic

also contain more mitochondria than do fast fibers. **Figure 10–21** compares the appearance of fast and slow fibers.

Intermediate Fibers

Most properties of **intermediate fibers** are intermediate between those of fast fibers and slow fibers. In appearance, intermediate fibers most closely resemble fast fibers, for they contain little myoglobin and are relatively pale. They have a more extensive capillary network around them, however, and are more resistant to fatigue than are fast fibers. Intermediate fibers are also known as *fast-twitch oxidative fibers, fast resistant (FR) fibers,* and *Type II-A fibers.*

Muscle Performance and the Distribution of Muscle Fibers

The percentages of fast, intermediate, and slow fibers in a skeletal muscle can be quite variable. In muscles that contain a mixture of fast and intermediate fibers, the proportion can change with physical conditioning. For example, if a muscle is used repeatedly for endurance events, some of the fast fibers will develop the appearance and functional capabilities of intermediate fibers. The muscle as a whole will thus become more resistant to fatigue.

Muscles dominated by fast fibers appear pale and are often called **white muscles**. Chicken breasts contain "white meat" because chickens use their wings only for brief intervals, as when fleeing from a predator, and the power for flight comes from the anaerobic process of glycolysis in the fast fibers of their breast muscles. As we saw earlier, the extensive blood vessels and myoglobin in slow fibers give these fibers a reddish color; muscles dominated by slow fibers are therefore known as **red muscles**. Chickens walk around all day, and these movements are powered by aerobic metabolism in the slow fibers of the "dark meat" of their legs.

Most human muscles contain a mixture of fiber types and so appear pink. However, there are no slow fibers in muscles of the eye or hand, where swift, but brief, contractions are required. Many back and calf muscles are dominated by slow fibers; these muscles contract almost continuously to maintain an upright posture. The percentage of fast versus slow fibers in each muscle is genetically determined. As noted earlier, the ratio of intermediate fibers to fast fibers can increase as a result of athletic training.

Muscle Hypertrophy and Atrophy

As a result of repeated, exhaustive stimulation, muscle fibers develop more mitochondria, a higher concentration of glycolytic enzymes, and larger glycogen reserves. Such muscle fibers have more myofibrils than do fibers that are less used, and each myofibril contains more thick and thin filaments. The net effect is **hypertrophy**, or an enlargement of the stimulated muscle. The number of muscle fibers does not change significantly, but the muscle as a whole enlarges because each muscle fiber increases in diameter.

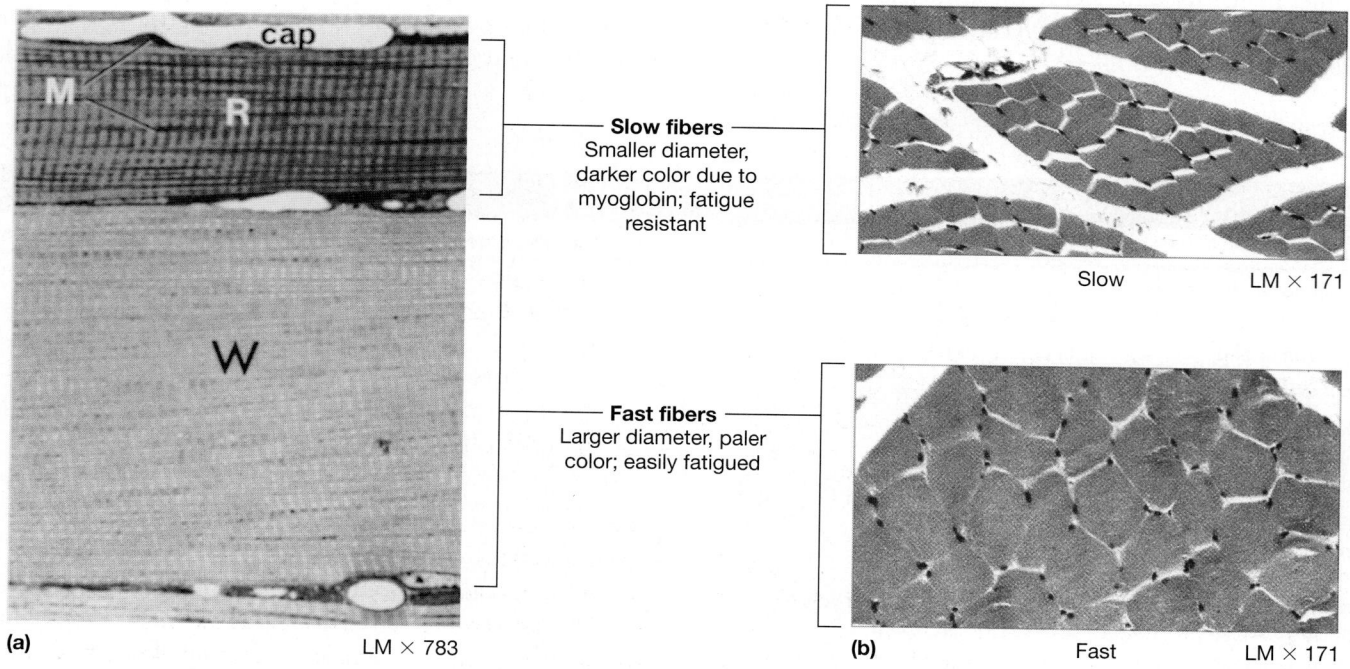

Figure 10–21 **Fast versus Slow Fibers.** **(a)** A longitudinal section, showing more mitochondria (M) and a more extensive capillary supply (cap) in a slow fiber (R, for red) than in a fast fiber (W, for white). **(b)** Cross sections of both types of fibers. Note the larger diameter of fast fibers.

Hypertrophy occurs in muscles that have been repeatedly stimulated to produce near-maximal tension. The intracellular changes that occur increase the amount of tension produced when these muscles contract. The muscles of a bodybuilder are excellent examples of muscular hypertrophy.

A skeletal muscle that is not regularly stimulated by a motor neuron loses muscle tone and mass. The muscle becomes flaccid, and the muscle fibers become smaller and weaker. This reduction in muscle size, tone, and power is called **atrophy**. Individuals paralyzed by spinal injuries or other damage to the nervous system will gradually lose muscle tone and size in the areas affected. Even a temporary reduction in muscle use can lead to muscular atrophy; you can easily observe this effect by comparing "before and after" limb muscles in someone who has worn a cast. Muscle atrophy is initially reversible, but dying muscle fibers are not replaced. In extreme atrophy, the functional losses are permanent. That is why physical therapy is crucial for people who are temporarily unable to move normally. Direct electrical stimulation by an external device can substitute for nerve stimulation and prevent or reduce muscle atrophy.

Because skeletal muscles depend on motor neurons for stimulation, disorders that affect the nervous system can indirectly affect the muscular system. In *polio*, a virus attacks motor neurons in the spinal cord and brain, causing muscular paralysis and atrophy.

Physical Conditioning

Physical conditioning and training schedules enable athletes to improve both power and endurance. In practice, the training schedule varies, depending on whether the activity is supported primarily by aerobic or anaerobic energy production.

Anaerobic endurance is the length of time muscular contraction can continue to be supported by glycolysis and by the existing energy reserves of ATP and CP. Anaerobic endurance is limited by (1) the amount of ATP and CP available, (2) the amount of glycogen available for breakdown, and (3) the ability of the muscle to tolerate the lactic acid generated during the anaerobic period. Typically, the onset of muscle fatigue occurs within 2 minutes of the start of maximal activity.

Activities that require above-average levels of anaerobic endurance include a 50-meter dash or swim, pole vaulting, and competitive weight lifting. These activities involve the contractions of fast fibers. The energy for the first 10–20 seconds of activity comes from the ATP and CP reserves of the cytoplasm. As these reserves dwindle, glycogen breakdown and glycolysis provide additional energy. Athletes training to improve anaerobic endurance perform frequent, brief, intensive workouts that stimulate muscle hypertrophy.

Aerobic endurance is the length of time a muscle can continue to contract while supported by mitochondrial activities. Aerobic endurance is determined primarily by the availability of substrates for aerobic respiration, which muscle fibers can obtain by breaking down carbohydrates, lipids, or amino acids. Initially, many of the nutrients catabolized by muscle fibers are obtained from reserves in the sarcoplasm. Prolonged aerobic activity, however, must be supported by nutrients provided by the circulating blood.

During exercise, blood vessels in the skeletal muscles dilate, increasing blood flow and thus bringing more oxygen and nutrients to the active muscle tissue. Warm-up periods are therefore important because they both stimulate circulation in the muscles before the serious workout begins and activate the enzyme that breaks down glycogen to release glucose. Warm-ups also increase muscle temperature and thus accelerate the contraction process substantially. Because glucose is a preferred energy source, endurance athletes such as marathon runners typically "load" or "bulk up" on carbohydrates for the three days before an event. They may also consume glucose-rich "sports drinks" during a competition. (We will consider the risks and benefits of these practices in Chapter 25.)

Training to improve aerobic endurance generally involves sustained low levels of muscular activity. Examples include jogging, distance swimming, and other exercises that do not require peak tension production. Improvements in aerobic endurance result from two factors:

1. *Alterations in the Characteristics of Muscle Fibers.* The composition of fast and slow fibers in each muscle is genetically determined, and individual differences are significant. These variations affect aerobic endurance, because a person with more slow fibers in a particular muscle will be better able to perform under aerobic conditions than will a person with fewer. However, skeletal muscle cells respond to changes in the pattern of neural stimulation. Fast (Type II-B) fibers trained for aerobic competition develop the characteristics of intermediate (Type II-A) fibers, and this change improves aerobic endurance.

2. *Improvements in Cardiovascular Performance.* Cardiovascular activity affects muscular performance by delivering oxygen and nutrients to active muscles. Physical training alters cardiovascular function by accelerating blood flow, thus improving oxygen and nutrient availability. Another important benefit of endurance training is increased capillarity, providing better blood flow at the cellular level. (We will examine factors involved in improving cardiovascular performance in Chapter 21.)

Aerobic activities do not promote muscle hypertrophy. Many athletes train using a combination of aerobic and anaerobic exercises so that their muscles will enlarge and both anaerobic and aerobic endurance will improve. These athletes alternate an aerobic activity, such as swimming, with sprinting or weight lifting. The combination, known as *interval training* or *cross-training*, is particularly useful for sports, such as tennis, which involve brief periods of anaerobic effort.

CLINICAL NOTE

Delayed-Onset Muscle Soreness

You have probably experienced muscle soreness the day after a period of physical exertion. Considerable controversy exists over the source and significance of this pain, which is known as *delayed-onset muscle soreness* (DOMS) and has several interesting characteristics:

- DOMS is distinct from the soreness you experience immediately after you stop exercising. The initial short-term soreness is probably related to the biochemical events associated with muscle fatigue.

- DOMS generally begins several hours after the exercise period and may last 3 or 4 days.

- The amount of DOMS is highest when the activity involves eccentric contractions (in which a muscle elongates despite producing tension). Activities dominated by concentric or isometric contractions produce less soreness.

- Levels of CPK and myoglobin are elevated in the blood, indicating damage to muscle plasma membranes. The nature of the activity (eccentric, concentric, or isometric) has no effect on these levels, nor can the levels be used to predict the degree of soreness experienced.

Three mechanisms have been proposed to explain DOMS:

1. Small tears may exist in the muscle tissue, leaving muscle fibers with damaged membranes. The sarcolemma of each damaged muscle fiber permits the loss of enzymes, myoglobin, and other chemicals that may stimulate nearby pain receptors.

2. The pain may result from muscle spasms in the affected skeletal muscles. In some studies, stretching the muscle involved after exercise can reduce the degree of soreness.

3. The pain may result from tears in the connective tissue framework and tendons of the skeletal muscle.

Some evidence supports each of these mechanisms, but it is unlikely that any one tells the entire story. For example, muscle fiber damage is certainly supported by biochemical findings, but if that were the only factor, the type of activity and level of circulating enzymes would be correlated with the level of pain experienced, and such is not the case.

The A&P Top 100

#38 What you don't use, you lose. Muscle tone is an indication of the chronic background level of activity in the motor units in skeletal muscles. When inactive for days or weeks, muscles become flaccid, and the muscle fibers break down their contractile proteins and grow smaller and weaker. If inactive for long periods, muscle fibers may atrophy and be replaced by fibrous tissue.

CHECKPOINT

17. Identify the three types of skeletal muscle fibers.

18. Why would a sprinter experience muscle fatigue before a marathon runner would?

19. Which activity would be more likely to create an oxygen debt: swimming laps or lifting weights?

20. Which type of muscle fibers would you expect to predominate in the large leg muscles of someone who excels at endurance activities, such as cycling or long-distance running?

See the blue Answers tab at the end of the book.

10-8 Cardiac muscle tissue differs structurally and functionally from skeletal muscle tissue

We introduced **cardiac muscle tissue** in Chapter 4 and briefly compared its properties with those of other types of muscle. **Cardiac muscle cells**, also called *cardiocytes* or *cardiac myocytes*, are found only in the heart. We begin a more detailed examination of cardiac muscle tissue by considering its structural characteristics.

Structural Characteristics of Cardiac Muscle Tissue

Like skeletal muscle fibers, cardiac muscle cells contain organized myofibrils, and the presence of many aligned sarcomeres gives the cells a striated appearance. But there are important structural differences between skeletal muscle fibers and cardiac muscle cells, including the following:

- Cardiac muscle cells are relatively small, averaging 10–20 μm in diameter and 50–100 μm in length.

- A typical cardiac muscle cell (**Figure 10–22a,b**) has a single, centrally placed nucleus, although a few may have two or more nuclei. Unlike skeletal muscle, cardiac myocytes are typically branched.

- The T tubules in a cardiac muscle cell are short and broad, and there are no triads (**Figure 10–22c**). The T tubules encircle the sarcomeres at the Z lines rather than at the zones of overlap.

- The SR of a cardiac muscle cell lacks terminal cisternae, and its tubules contact the plasma membrane as well as the T tubules (**Figure 10–22c**). As in skeletal muscle fibers, the appearance of an action potential triggers the release of calcium from the SR and the contraction of sarcomeres; it also increases the permeability of the sarcolemma to extracellular calcium ions.

- Cardiac muscle cells are almost totally dependent on aerobic metabolism to obtain the energy they need to continue contracting. Energy reserves are maintained in the form of glycogen and lipid inclusions. The sarcoplasm of a cardiac muscle cell contains large numbers of mitochondria and abundant reserves of myoglobin that store the oxygen needed to break down those energy reserves during times of peak activity.

- Each cardiac muscle cell contacts several others at specialized sites known as **intercalated** (in-TER-ka-lā-ted) **discs**. ∞ p. 140 Intercalated discs play a vital role in the function of cardiac muscle, as we will see next.

Intercalated Discs

At an intercalated disc (**Figure 10–22a,b**), the plasma membranes of two adjacent cardiac muscle cells are extensively intertwined and bound together by gap junctions and desmosomes. ∞ p. 114 These connections help stabilize the relative positions of adjacent cells and maintain the three-dimensional structure of the tissue. The gap junctions allow ions and small molecules to move from one cell to another. This arrangement creates a direct electrical connection between the two muscle cells. An action potential can travel across an intercalated disc, moving quickly from one cardiac muscle cell to another.

Myofibrils in the two interlocking muscle cells are firmly anchored to the membrane at the intercalated disc. Because their myofibrils are essentially locked together, the two muscle cells can "pull together" with maximum efficiency. Because the cardiac muscle cells are mechanically, chemically, and electrically connected to one another, the entire tissue resembles a single, enormous muscle cell. For this reason, cardiac muscle has been called a *functional syncytium* (sin-SISH-ē-um; a fused mass of cells).

Functional Characteristics of Cardiac Muscle Tissue

In Chapter 20, we will examine cardiac muscle physiology in detail; here, we will briefly summarize four major functional specialties of cardiac muscle:

1. Cardiac muscle tissue contracts without neural stimulation. This property is called **automaticity**. The timing of

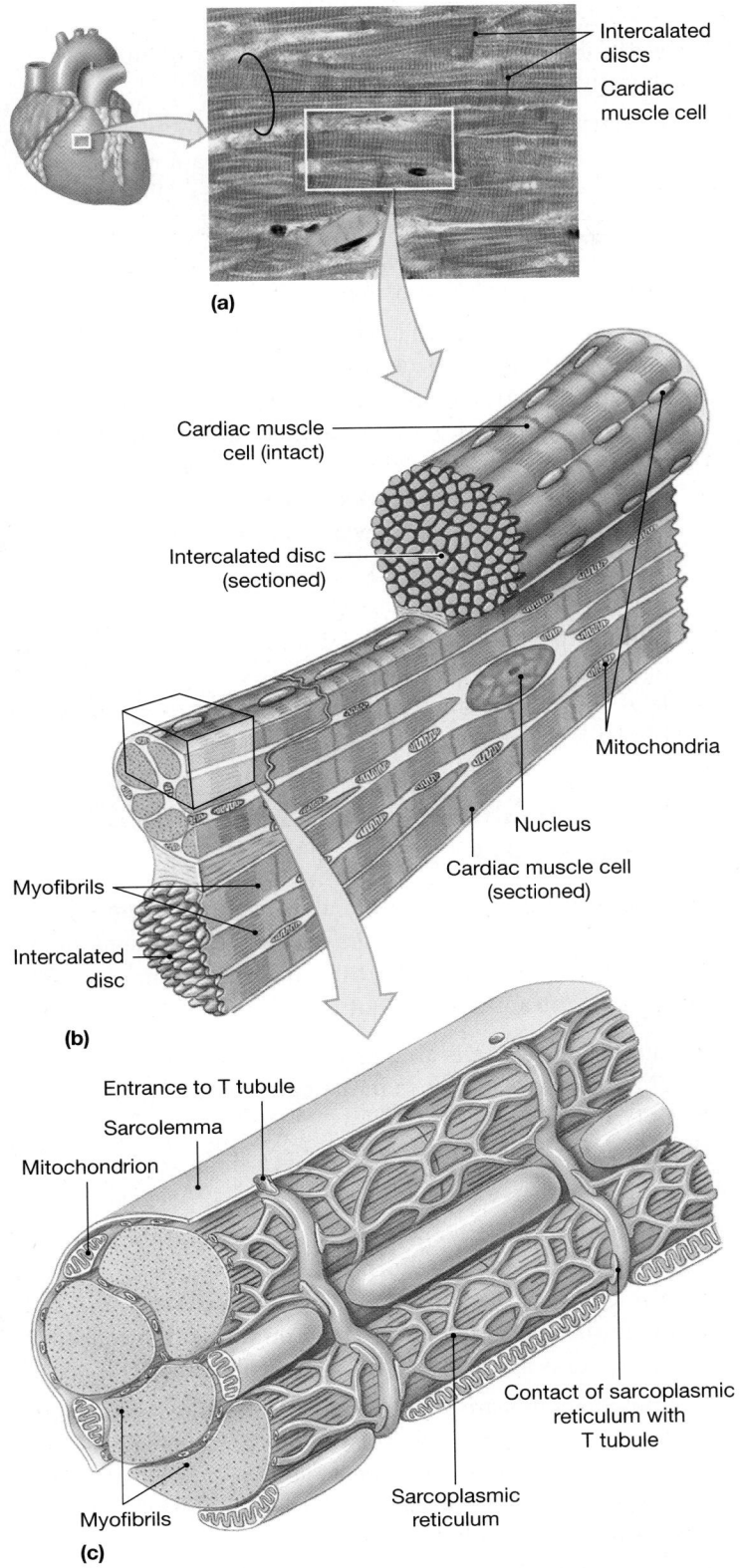

(a)

Cardiac muscle cell (intact)

Intercalated disc (sectioned)

Mitochondria

Nucleus

Cardiac muscle cell (sectioned)

Myofibrils

Intercalated disc

(b)

Entrance to T tubule

Sarcolemma

Mitochondrion

Contact of sarcoplasmic reticulum with T tubule

Sarcoplasmic reticulum

Myofibrils

(c)

Intercalated discs

Cardiac muscle cell

Figure 10–22 Cardiac Muscle Tissue. (a) A light micrograph of cardiac muscle tissue. Notice the striations and the intercalated discs. **(b,c)** The structure of a cardiac muscle cell; compare with *Figure 10–3*.

contractions is normally determined by specialized cardiac muscle cells called **pacemaker cells**.

2. Innervation by the nervous system can alter the pace established by the pacemaker cells and adjust the amount of tension produced during a contraction.

3. Cardiac muscle cell contractions last roughly 10 times as long as do those of skeletal muscle fibers. They also have longer refractory periods and do not readily fatigue.

4. The properties of cardiac muscle plasma membranes differ from those of skeletal muscle fiber membranes. As a result, individual twitches cannot undergo wave summation, and cardiac muscle tissue cannot produce tetanic contractions. This difference is important, because a heart in a sustained tetanic contraction could not pump blood.

CHECKPOINT

21. **Compare and contrast skeletal muscle tissue and cardiac muscle tissue.**

22. **What feature of cardiac muscle tissue allows the heart to act as a functional syncytium?**

See the blue Answers tab at the end of the book.

10-9 Smooth muscle tissue differs structurally and functionally from skeletal muscle tissue

Smooth muscle tissue forms sheets, bundles, or sheaths around other tissues in almost every organ. Smooth muscles around blood vessels regulate blood flow through vital organs. In the digestive and urinary systems, rings of smooth muscle, called *sphincters*, regulate the movement of materials along internal passageways. Smooth muscles play a variety of other roles in various body systems.

- **Integumentary System:** Smooth muscles around blood vessels regulate the flow of blood to the superficial dermis; smooth muscles of the arrector pili elevate hairs. ∞ p. 169

- **Cardiovascular System:** Smooth muscles encircling blood vessels control the distribution of blood and help regulate blood pressure.

- **Respiratory System:** Smooth muscle contraction or relaxation alters the diameters of the respiratory passageways and changes the resistance to airflow.

- **Digestive System:** Extensive layers of smooth muscle in the walls of the digestive tract play an essential role in

moving materials along the tract. Smooth muscle in the walls of the gallbladder contract to eject bile into the digestive tract.

- **Urinary System:** Smooth muscle tissue in the walls of small blood vessels alters the rate of filtration in the kidneys. Layers of smooth muscle in the walls of the ureters transport urine to the urinary bladder; the contraction of the smooth muscle in the wall of the urinary bladder forces urine out of the body.

- **Reproductive System:** Layers of smooth muscle help move sperm along the reproductive tract in males and cause the ejection of glandular secretions from the accessory glands into the reproductive tract. In females, layers of smooth muscle help move oocytes (and perhaps sperm) along the reproductive tract, and contraction of the smooth muscle in the walls of the uterus expels the fetus at delivery.

Figure 10–23a shows typical smooth muscle tissue as seen by light microscopy. Smooth muscle tissue differs from both skeletal and cardiac muscle tissues in structure and function (Table 10–4).

Structural Characteristics of Smooth Muscle Tissue

Actin and myosin are present in all three types of muscle tissue. In skeletal and cardiac muscle cells, these proteins are organized in sarcomeres, with thin and thick filaments. The internal organization of a smooth muscle cell is very different:

- Smooth muscle cells are relatively long and slender, ranging from 5 to 10 μm in diameter and from 30 to 200 μm in length.

- Each cell is spindle shaped and has a single, centrally located nucleus.

- A smooth muscle fiber has no T tubules, and the sarcoplasmic reticulum forms a loose network throughout the sarcoplasm. Smooth muscle cells also lack myofibrils and sarcomeres. As a result, this tissue also has no striations and is called **nonstriated** muscle.

- Thick filaments are scattered throughout the sarcoplasm of a smooth muscle cell. The myosin proteins are organized differently than in skeletal or cardiac muscle cells, and smooth muscle cells have more myosin heads per thick filament.

- The thin filaments in a smooth muscle cell are attached to **dense bodies**, structures distributed throughout the sarcoplasm in a network of intermediate filaments composed of the protein *desmin* (**Figure 10–23b**).

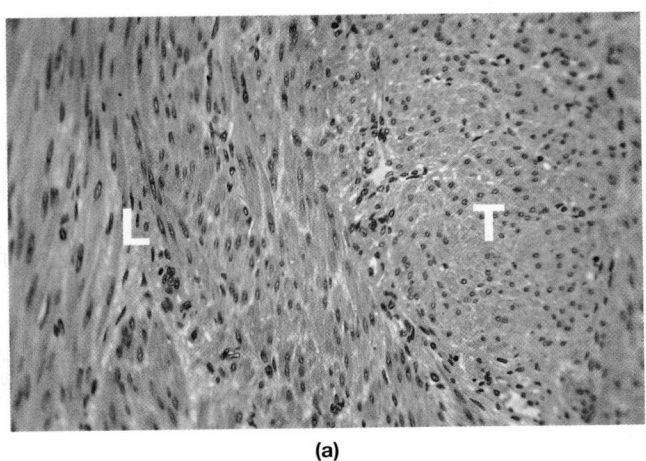

(a)

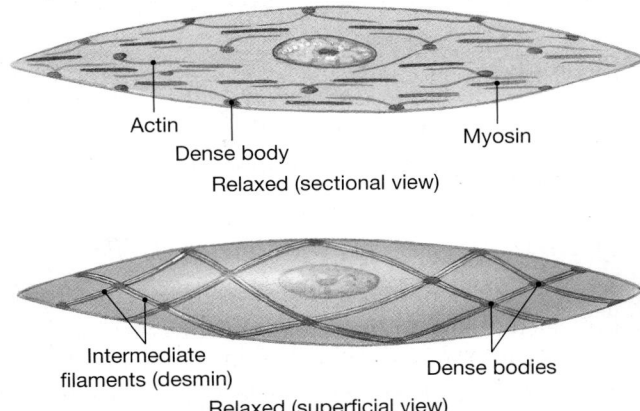

Actin
Dense body
Myosin
Relaxed (sectional view)

Intermediate
filaments (desmin)
Dense bodies
Relaxed (superficial view)

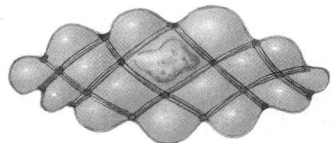

Contracted (superficial view)
(b)

Figure 10–23 **Smooth Muscle Tissue.** **(a)** Many visceral organs contain several layers of smooth muscle tissue oriented in different directions. Here, a single sectional view shows smooth muscle cells in both longitudinal (L) and transverse (T) sections. **(b)** A single relaxed smooth muscle cell is spindle shaped and has no striations. Note the changes in cell shape as contraction occurs.

Some of the dense bodies are firmly attached to the sarcolemma. The dense bodies and intermediate filaments anchor the thin filaments such that, when sliding occurs between thin and thick filaments, the cell shortens. Dense bodies are not arranged in straight lines, so when a contraction occurs, the muscle cell twists like a corkscrew.

- Adjacent smooth muscle cells are bound together at dense bodies, transmitting the contractile forces from cell to cell throughout the tissue.
- Although smooth muscle cells are surrounded by connective tissue, the collagen fibers never unite to form tendons or aponeuroses, as they do in skeletal muscles.

TABLE 10–4 **A Comparison of Skeletal, Cardiac, and Smooth Muscle Tissues**

Property	Skeletal Muscle	Cardiac Muscle	Smooth Muscle
Fiber dimensions (diameter × length)	$100 \ \mu m$ × up to 30 cm	$10–20 \ \mu m$ × $50–100 \ \mu m$	$5–10 \ \mu m$ × $30–200 \ \mu m$
Nuclei	Multiple, near sarcolemma	Generally single, centrally located	Single, centrally located
Filament organization	In sarcomeres along myofibrils	In sarcomeres along myofibrils	Scattered throughout sarcoplasm
SR	Terminal cisternae in triads at zones of overlap	SR tubules contact T tubules at Z lines	Dispersed throughout sarcoplasm, no T tubules
Control mechanism	Neural, at single neuromuscular junction	Automaticity (pacemaker cells)	Automaticity (pacesetter cells), neural or hormonal control
Ca^{2+} source	Release from SR	Extracellular fluid and release from SR	Extracellular fluid and release from SR
Ca^{2+} regulation	Troponin on thin filaments	Troponin on thin filaments	Calmodulin on myosin heads
Contraction	Rapid onset; may be tetanized; rapid fatigue	Slower onset; cannot be tetanized; resistant to fatigue	Slow onset; may be tetanized; resistant to fatigue
Energy source	Aerobic metabolism at moderate levels of activity; glycolysis (anaerobic during peak activity)	Aerobic metabolism, usually lipid or carbohydrate substrates	Primarily aerobic metabolism

Functional Characteristics of Smooth Muscle Tissue

Smooth muscle tissue differs from other muscle tissue in (1) excitation–contraction coupling, (2) length–tension relationships, (3) control of contractions, and (4) smooth muscle tone.

Excitation–Contraction Coupling

The trigger for smooth muscle contraction is the appearance of free calcium ions in the cytoplasm. On stimulation, a blast of calcium ions enters the cell from the extracellular fluid, and additional calcium ions are released by the sarcoplasmic reticulum. The net result is a rise in calcium ion concentrations throughout the cell. Once in the sarcoplasm, the calcium ions interact with **calmodulin**, a calcium-binding protein. Calmodulin then activates the enzyme **myosin light chain kinase**, which in turn enables the attachment of myosin heads to actin. This mechanism is quite different from that in skeletal and cardiac muscles, in which the trigger for contraction is the binding of calcium ions to troponin.

Length–Tension Relationships

Because the thick and thin filaments are scattered and are not organized into sarcomeres in smooth muscle, tension development and resting length are not directly related. A stretched smooth muscle soon adapts to its new length and retains the ability to contract on demand. This ability to function over a wide range of lengths is called **plasticity**. Smooth muscle can contract over a range of lengths four times greater than that of skeletal muscle. Plasticity is especially important in digestive organs, such as the stomach, that undergo great changes in volume. Despite the lack of sarcomere organization, smooth muscle contractions can be just as powerful as those of skeletal muscles. Like skeletal muscle fibers, smooth muscle cells can undergo sustained contractions.

Control of Contractions

Many smooth muscle cells are not innervated by motor neurons, and the neurons that do innervate smooth muscles are not under voluntary control. Smooth muscle cells are categorized as either multiunit or visceral. **Multiunit smooth muscle cells** are innervated in motor units comparable to those of skeletal muscles, but each smooth muscle cell may be connected to more than a single motor neuron. In contrast, many **visceral smooth muscle cells** lack a direct contact with any motor neuron.

Multiunit smooth muscle cells resemble skeletal muscle fibers and cardiac muscle cells in that neural activity produces an action potential that is propagated over the sarcolemma. However, the contractions of these smooth muscle cells occur more slowly than those of skeletal or cardiac muscle cells.

Multiunit smooth muscle cells are located in the iris of the eye, where they regulate the diameter of the pupil; along portions of the male reproductive tract; within the walls of large arteries; and in the arrector pili muscles of the skin. Multiunit smooth muscle cells do not typically occur in the digestive tract.

Visceral smooth muscle cells are arranged in sheets or layers. Within each layer, adjacent muscle cells are connected by gap junctions. As a result, whenever one muscle cell contracts, the electrical impulse that triggered the contraction can travel to adjacent smooth muscle cells. The contraction therefore spreads in a wave that soon involves every smooth muscle cell in the layer. The initial stimulus may be the activation of a motor neuron that contacts one of the muscle cells in the region. But smooth muscle cells also contract or relax in response to chemicals, hormones, local concentrations of oxygen or carbon dioxide, or physical factors such as extreme stretching or irritation.

Many visceral smooth muscle networks show rhythmic cycles of activity in the absence of neural stimulation. These cycles are characteristic of the smooth muscle cells in the wall of the digestive tract, where **pacesetter cells** undergo spontaneous depolarization and trigger the contraction of entire muscular sheets. Visceral smooth muscle cells are located in the walls of the digestive tract, the gallbladder, the urinary bladder, and many other internal organs.

Smooth Muscle Tone

Both multiunit and visceral smooth muscle tissues have a normal background level of activity, or smooth muscle tone. The regulatory mechanisms just detailed stimulate contraction and increase muscle tone. Neural, hormonal, or chemical factors can also stimulate smooth muscle relaxation, producing a decrease in muscle tone. For example, smooth muscle cells at the entrances to capillaries regulate the amount of blood flow into each vessel. If the tissue becomes starved for oxygen, the smooth muscle cells relax, whereupon blood flow increases, delivering additional oxygen. As conditions return to normal, the smooth muscle regains its normal muscle tone.

CHECKPOINT

23. Identify the structural characteristics of smooth muscle tissue.

24. Why are cardiac and smooth muscle contractions more affected by changes in extracellular Ca^{2+} than are skeletal muscle contractions?

25. Why can smooth muscle contract over a wider range of resting lengths than skeletal muscle can?

See the blue Answers tab at the end of the book.

Related Clinical Terms

botulism: A severe, potentially fatal paralysis of skeletal muscles, resulting from the consumption of a bacterial toxin.

Duchenne muscular dystrophy (DMD): One of the most common and best understood of the muscular dystrophies.

muscular dystrophies: A varied collection of inherited diseases that produce progressive muscle weakness and deterioration.

myasthenia gravis: A general muscular weakness resulting from a reduction in the number of ACh receptors on the motor end plate.

polio: A disease resulting from the destruction of motor neurons by a certain virus and characterized by the paralysis and atrophy of motor units.

rigor mortis: A state following death during which muscles are locked in the contracted position, making the body extremely stiff.

tetanus: A disease in which sustained, powerful contractions of skeletal muscles are stimulated by the action of a bacterial toxin.

Chapter Review

Study Outline

iP Use *Interactive Physiology*® (IP) —available on the IP CD packaged with your new textbook and online at **myA&P**™ (www.myaandp.com)—to help you understand difficult physiological concepts. Follow these navigation paths on IP for concepts in this chapter:

- Muscular System: Sliding Filament Theory
- Muscular System: The Neuromuscular Junction
- Muscular System: Contraction of Motor Units
- Muscular System: Contraction of Whole Muscles
- Muscular System: Muscle Metabolism
- Muscular System: Anatomy Review: Skeletal Muscle Tissue

10-1 Skeletal muscle performs six major functions p. 294

1. The three types of muscle tissue are *skeletal muscle, cardiac muscle,* and *smooth muscle.*
2. **Skeletal muscles** attach to bones directly or indirectly. Their functions are to (1) produce skeletal movement, (2) maintain posture and body position, (3) support soft tissues, (4) guard entrances and exits, (5) maintain body temperature, and (6) store nutrient reserves.

10-2 A skeletal muscle contains muscle tissue, connective tissues, blood vessels, and nerves p. 294

3. Each muscle cell or fiber is surrounded by an **endomysium.** Bundles of muscle fibers are sheathed by a **perimysium,** and the entire muscle is covered by an **epimysium.** At the ends of the muscle are **tendons** or **aponeuroses** that attach the muscle to bones. (*Figure 10–1*)
4. The perimysium and endomysium contain the blood vessels and nerves that supply the muscle fibers.

10-3 Skeletal muscle fibers have distinctive features p. 296

5. A skeletal muscle fiber has a **sarcolemma,** or plasma membrane; **sarcoplasm** (cytoplasm); and **sarcoplasmic reticulum (SR),** similar to the smooth endoplasmic reticulum of other cells. **Transverse (T) tubules** and **myofibrils** aid in contraction. Filaments in a myofibril are organized into repeating functional units called **sarcomeres.** (*Figures 10–2 to 10–6*)
6. **Myofilaments** called **thin filaments** and **thick filaments** form myofibrils. (*Figures 10–2 to 10–6*)

7. Thin filaments consist of **F-actin, nebulin, tropomyosin,** and **troponin.** Tropomyosin molecules cover **active sites** on the **G-actin** subunits that form the F-actin strand. Troponin binds to G-actin and tropomyosin and holds the tropomyosin in position. (*Figure 10–7*)
8. Thick filaments consist of a bundle of myosin molecules around a **titin** core. Each **myosin** molecule has an elongated **tail** and a globular **head,** which forms **cross-bridges** during contraction. In a resting muscle cell, the attachment of myosin heads to active sites on G-actin is prevented by tropomyosin. (*Figure 10–7*)
9. The relationship between thick and thin filaments changes as a muscle fiber contracts. (*Figure 10–8*)

10-4 Communication between the nervous system and skeletal muscles occurs at the neuromuscular junction p. 304

10. When muscle cells contract, they create *tension* and pull on the attached tendons. (*Figure 10–9*)
11. The activity of a muscle fiber is controlled by a neuron at a **neuromuscular** (*myoneural*) **junction (NMJ).** (*Figure 10–10*)
12. When an **action potential** arrives at the **synaptic terminal, acetylcholine (ACh)** is released into the **synaptic cleft.** The binding of ACh to receptors on the opposing *junctional folds* leads to the generation of an action potential in the sarcolemma. (*Figure 10–10*)
13. **Excitation–contraction coupling** occurs as the passage of an action potential along a T tubule triggers the release of Ca²⁺ from the cisternae of the SR at triads. (*Figure 10–11*)
14. Release of Ca²⁺ initiates a **contraction cycle** of attachment, pivoting, detachment, and return. The calcium ions bind to troponin, which changes position and moves tropomyosin away from the active sites of actin. Cross-bridges of myosin heads then bind to actin. Next, each myosin head pivots at its base, pulling the actin filament toward the center of the sarcomere. (*Figures 10–12/10–13*)
15. Acetylcholinesterase (AChE) breaks down ACh and limits the duration of muscle stimulation. (*Summary Table 10–1*)

10-5 Sarcomere shortening and muscle fiber stimulation produce tension p. 311

16. The amount of tension produced by a muscle fiber depends on the number of cross-bridges formed.

17. Skeletal muscle fibers can contract most forcefully when stimulated over a narrow range of resting lengths. *(Figure 10–14)*
18. A **twitch** is a cycle of contraction and relaxation produced by a single stimulus. *(Figure 10–15)*
19. Repeated stimulation at a slow rate produces **treppe**, a progressive increase in twitch tension. *(Figure 10–16)*
20. Repeated stimulation before the relaxation phase ends may produce **summation of twitches (wave summation)**, in which one twitch is added to another; **incomplete tetanus**, in which tension peaks and falls at intermediate stimulus rates; or **complete tetanus**, in which the relaxation phase is eliminated by very rapid stimuli. *(Figure 10–16)*
21. The number and size of a muscle's **motor units** determine how precisely controlled its movements are. *(Figure 10–17)*
22. Resting **muscle tone** stabilizes bones and joints.
23. Normal activities generally include both **isotonic contractions** (in which the tension in a muscle rises and the length of the muscle changes) and **isometric contractions** (in which tension rises, but the length of the muscle remains constant). *(Figure 10–18)*
24. Load and speed of contraction are inversely related. *(Figure 10–19)*
25. The return to resting length after a contraction may involve elastic forces, the contraction of opposing muscle groups, and gravity.

10-6 ATP is the energy source for muscle contraction p. 318

26. Muscle contractions require large amounts of energy. *(Table 10–2)*
27. **Creatine phosphate (CP)** can release stored energy to convert ADP to ATP. *(Table 10–2)*
28. At rest or at moderate levels of activity, **aerobic metabolism** can provide most of the ATP needed to support muscle contractions.
29. At peak levels of activity, the cell relies heavily on the **anaerobic process** of **glycolysis** to generate ATP, because the mitochondria cannot obtain enough oxygen to meet the existing ATP demands.
30. As muscular activity changes, the pattern of energy production and use changes. *(Figure 10–20)*
31. A fatigued muscle can no longer contract, because of a drop in pH due to the buildup of **lactic acid**, the exhaustion of energy resources, or other factors.
32. The **recovery period** begins immediately after a period of muscle activity and continues until conditions inside the muscle have returned to preexertion levels. The **oxygen debt**, or *excess postexercise oxygen consumption (EPOC)*, created during exercise is the amount of oxygen required during the recovery period to restore the muscle to its normal condition.
33. Circulating hormones may alter metabolic activities in skeletal muscle fibers.

10-7 Muscle fiber type and physical conditioning determine muscle performance capabilities p. 323

34. The three types of skeletal muscle fibers are **fast fibers**, **slow fibers**, and **intermediate fibers**. *(Table 10–3; Figure 10–21)*

35. Fast fibers, which are large in diameter, contain densely packed myofibrils, large glycogen reserves, and relatively few mitochondria. They produce rapid and powerful contractions of relatively brief duration. *(Figure 10–21)*
36. Slow fibers are about half the diameter of fast fibers and take three times as long to contract after stimulation. Specializations such as abundant mitochondria, an extensive capillary supply, and high concentrations of **myoglobin** enable slow fibers to continue contracting for extended periods. *(Figure 10–21)*
37. Intermediate fibers are very similar to fast fibers, but have a greater resistance to fatigue.
38. Muscles dominated by fast fibers appear pale and are called **white muscles**.
39. Muscles dominated by slow fibers are rich in myoglobin and appear as **red muscles**.
40. Training to develop anaerobic endurance can lead to **hypertrophy** (enlargement) of the stimulated muscles.
41. **Anaerobic endurance** is the time over which muscular contractions can be sustained by glycolysis and reserves of ATP and CP.
42. **Aerobic endurance** is the time over which a muscle can continue to contract while supported by mitochondrial activities.

10-8 Cardiac muscle tissue differs structurally and functionally from skeletal muscle tissue p. 326

43. **Cardiac muscle tissue** is located only in the heart. **Cardiac muscle cells** are small; have one centrally located nucleus; have short, broad T tubules; and are dependent on aerobic metabolism. **Intercalated discs** are found where neighboring plasma membranes connect. *(Figure 10–22; Table 10–4)*
44. Cardiac muscle cells contract without neural stimulation (**automaticity**), and their contractions last longer than those of skeletal muscle.
45. Because cardiac muscle twitches do not exhibit wave summation, cardiac muscle tissue cannot produce tetanic contractions.

10-9 Smooth muscle tissue differs structurally and functionally from skeletal muscle tissue p. 328

46. **Smooth muscle tissue** is nonstriated, involuntary muscle tissue.
47. Smooth muscle cells lack sarcomeres and the resulting striations. The thin filaments are anchored to **dense bodies**. *(Figure 10–23; Table 10–4)*
48. Smooth muscle contracts when calcium ions interact with **calmodulin**, which activates **myosin light chain kinase**.
49. Smooth muscle functions over a wide range of lengths (**plasticity**).
50. In **multiunit smooth muscle cells**, each smooth muscle cell acts relatively independently of other smooth muscle cells in the organ. **Visceral smooth muscle cells** are not always innervated by motor neurons. Neurons that innervate smooth muscle cells are not under voluntary control.

Review Questions

See the blue Answers tab at the end of the book.

 Access more review material online at **myA&P™** (www.myaandp.com). There, you'll find chapter guides, chapter quizzes, practice tests, animations, flashcards, a glossary with pronunciations, *Interactive Physiology®* (IP) exercises and quizzes, and more to help you succeed in the course.

LEVEL 1 Reviewing Facts and Terms

1. Identify the structures in the following figure.

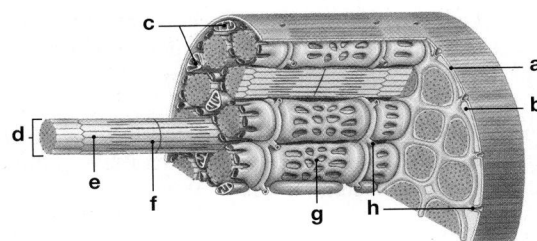

  (a) _____   (b) _____
  (c) _____   (d) _____
  (e) _____   (f) _____
  (g) _____   (h) _____

2. The connective tissue coverings of a skeletal muscle, listed from superficial to deep, are
   (a) endomysium, perimysium, and epimysium.
   (b) endomysium, epimysium, and perimysium.
   (c) epimysium, endomysium, and perimysium.
   (d) epimysium, perimysium, and endomysium.

3. The command to contract is distributed deep into a muscle fiber by the
   (a) sarcolemma.   (b) sarcomere.
   (c) transverse tubules.   (d) myotubules.
   (e) myofibrils.

4. The detachment of the myosin cross-bridges is directly triggered by
   (a) the repolarization of T tubules.
   (b) the attachment of ATP to myosin heads.
   (c) the hydrolysis of ATP.
   (d) calcium ions.

5. A muscle producing near-peak tension during rapid cycles of contraction and relaxation is said to be in
   (a) incomplete tetanus.   (b) treppe.
   (c) complete tetanus.   (d) a twitch.

6. The type of contraction in which the tension rises, but the load does not move, is
   (a) a wave summation.   (b) a twitch.
   (c) an isotonic contraction. (d) an isometric contraction.

7. Which of the following statements about myofibrils is *not* correct?
   (a) Each skeletal muscle fiber contains hundreds to thousands of myofibrils.
   (b) Myofibrils contain repeating units called sarcomeres.
   (c) Myofibrils extend the length of a skeletal muscle fiber.
   (d) Filaments consist of bundles of myofibrils.
   (e) Myofibrils are attached to the plasma membrane at both ends of a muscle fiber.

8. An action potential can travel quickly from one cardiac muscle cell to another because of the presence of
   (a) gap junctions.
   (b) tight junctions.
   (c) intercalated discs.
   (d) both a and c.

9. List the three types of muscle tissue in the body.

10. What three layers of connective tissue are part of each muscle? What functional role does each layer play?

11. The _____ contains vesicles filled with acetylcholine.
    (a) synaptic terminal
    (b) motor end plate
    (c) neuromuscular junction
    (d) synaptic cleft
    (e) transverse tubule

12. What structural feature of a skeletal muscle fiber is responsible for conducting action potentials into the interior of the cell?

13. What five interlocking steps are involved in the contraction process?

14. What two factors affect the amount of tension produced when a skeletal muscle contracts?

15. What forms of energy reserves do resting skeletal muscle fibers contain?

16. What two mechanisms are used to generate ATP from glucose in muscle cells?

17. What is the calcium-binding protein in smooth muscle tissue?

LEVEL 2 Reviewing Concepts

18. An activity that would require anaerobic endurance is
    (a) a 50-meter dash.
    (b) a pole vault.
    (c) a weight-lifting competition.
    (d) all of these.

19. Areas of the body where you would *not* expect to find slow fibers include the
    (a) back and calf muscles.
    (b) eye and hand.
    (c) chest and abdomen.
    (d) a, b, and c.

20. During relaxation, muscles return to their original length because of all of the following *except*
    (a) actin and myosin actively pushing away from one another.
    (b) the contraction of opposing muscles.
    (c) the pull of gravity.
    (d) the elastic nature of the sarcolemma.
    (e) elastic forces.

21. According to the length–tension relationship,
    (a) longer muscles can generate more tension than shorter muscles.
    (b) the greater the zone of overlap in the sarcomere, the greater the tension the muscle can develop.
    (c) the greatest tension is achieved in sarcomeres where actin and myosin initially do not overlap.
    (d) there is an optimum range of actin and myosin overlap that will produce the greatest amount of tension.
    (e) both b and d are correct.

22. For each portion of a myogram tracing a twitch in a stimulated calf muscle fiber, describe the events that occur within the muscle.
23. What three processes are involved in repaying the oxygen debt during a muscle's recovery period?
24. How does cardiac muscle tissue contract without neural stimulation?
25. Atracurium is a drug that blocks the binding of ACh to receptors. Give an example of a site where such binding normally occurs, and predict the physiological effect of this drug.
26. Explain why a murder victim's time of death can be estimated according to the flexibility or rigidity of the body.
27. Which of the following activities would employ isometric contractions?
    (a) flexing the elbow
    (b) chewing food
    (c) maintaining an upright posture
    (d) running
    (e) writing

LEVEL 3   Critical Thinking and Clinical Applications

28. Many potent insecticides contain toxins, called organophosphates, that interfere with the action of the enzyme acetylcholinesterase. Ivan is using an insecticide containing organophosphates and is very careless. Because he does not use gloves or a dust mask, he absorbs some of the chemical through his skin and inhales a large amount as well. What signs would you expect to observe in Ivan as a result of organophosphate poisoning?
29. Linda's father suffers an apparent heart attack and is rushed to the emergency room of the local hospital. The doctor on call tells her that he has ordered some blood work and that he will be able to tell if her father actually had an attack by looking at the blood levels of CPK, LDH, and cardiac troponin. Why would the level of these enzymes help to indicate if a person suffered a heart attack?
30. Bill broke his leg in a football game, and after 6 weeks the cast is finally removed. As he steps down from the examination table, he loses his balance and falls. Why?

The Muscular System

11

Did you know...?
The human body consists of almost 700 muscles, which constitute about 40% of our body weight.

Learning Outcomes

After completing this chapter, you should be able to do the following:

11-1 Describe the arrangement of fascicles in the various types of muscles, and explain the resulting functional differences.

11-2 Describe the classes of levers, and explain how they make muscles more efficient.

11-3 Predict the actions of a muscle on the basis of the relative positions of its origin and insertion, and explain how muscles interact to produce or oppose movements.

11-4 Explain how the name of a muscle can help identify its location, appearance, or function.

11-5 Identify the principal axial muscles of the body, plus their origins, insertions, actions, and innervation.

11-6 Identify the principal appendicular muscles of the body, plus their origins, insertions, actions, and innervation, and compare the major functional differences between the upper and lower limbs.

11-7 Identify age-related changes of the muscular system.

11-8 Explain the functional relationship between the muscular system and other body systems, and explain the role of exercise in producing various responses in other body systems.

Clinical Notes

Hernia p. 359
Intramuscular Injections p. 371
Compartment Syndrome p. 381

An Introduction to the Muscular System

This chapter describes the gross anatomy of the muscular system and considers functional relationships between muscles and bones of the body. Although most skeletal muscle fibers contract at similar rates and shorten to the same degree, variations in microscopic and macroscopic organization can dramatically affect the power, range, and speed of movement produced when muscles contract.

11-1 Fascicle arrangement is correlated with muscle power and range of motion

Muscle fibers in a skeletal muscle form bundles called *fascicles*. ∞ p. 294 The muscle fibers in a single fascicle are parallel, but the organization of fascicles in skeletal muscles can vary, as can the relationship between the fascicles and the associated tendon. Based on the patterns of fascicle organization, skeletal muscles can be classified as *parallel muscles, convergent muscles, pennate muscles*, and *circular muscles* (**Figure 11–1**).

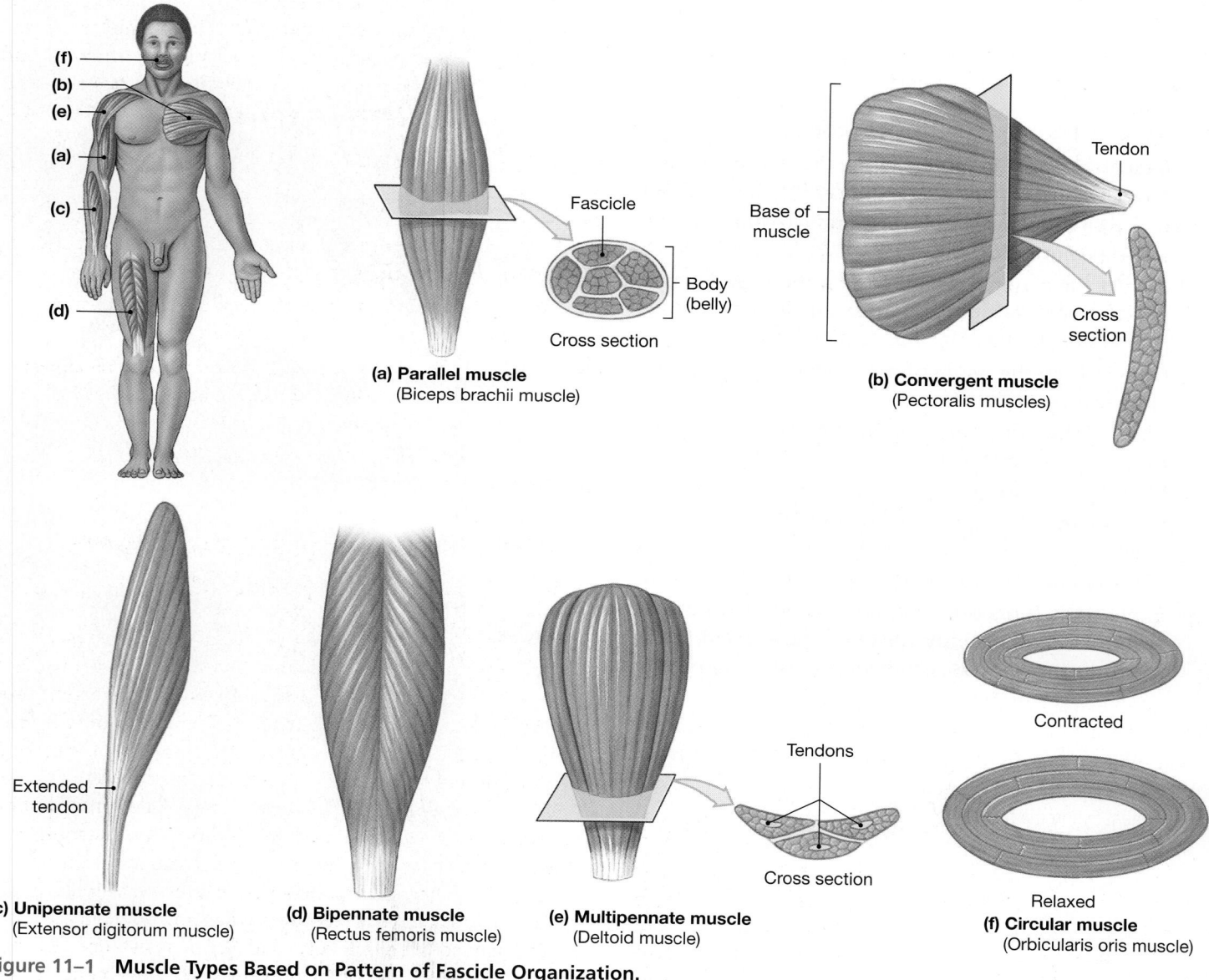

(a) **Parallel muscle**
(Biceps brachii muscle)

Fascicle
Body (belly)
Cross section

(b) **Convergent muscle**
(Pectoralis muscles)

Base of muscle
Tendon
Cross section

(c) **Unipennate muscle**
(Extensor digitorum muscle)

Extended tendon

(d) **Bipennate muscle**
(Rectus femoris muscle)

(e) **Multipennate muscle**
(Deltoid muscle)

Tendons
Cross section

(f) **Circular muscle**
(Orbicularis oris muscle)

Contracted
Relaxed

Figure 11–1   Muscle Types Based on Pattern of Fascicle Organization.

Parallel Muscles

In a **parallel muscle**, the fascicles are parallel to the long axis of the muscle. Most of the skeletal muscles in the body are parallel muscles. Some are flat bands with broad attachments (*aponeuroses*) at each end; others are plump and cylindrical, with tendons at one or both ends. In the latter case, the muscle is spindle shaped (**Figure 11–1a**), with a central **body**, also known as the *belly*, or *gaster* (GAS-ter; stomach). The *biceps brachii muscle* of the arm is a parallel muscle with a central body. When a parallel muscle contracts, it shortens and gets larger in diameter. You can see the bulge of the contracting biceps brachii muscle on the anterior surface of your arm when you flex your elbow.

A skeletal muscle fiber can contract until it has shortened by roughly 30 percent. Because the muscle fibers in a parallel muscle are parallel to the long axis of the muscle, when those fibers contract together the entire muscle shortens by about 30 percent. Thus, if the muscle is 10 cm (3.94 in.) long and one end is held in place, the other end will move 3 cm when the muscle contracts. The tension developed during this contraction depends on the total number of myofibrils the muscle contains. ∞ p. 315 Because the myofibrils are distributed evenly through the sarcoplasm of each cell, we can use the cross-sectional area of the resting muscle to estimate the tension. For each 6.45 cm^2 (1 in.^{2}) in cross-sectional area, a parallel muscle can develop approximately 23 kg (50 lb) of isometric tension.

Convergent Muscles

In a **convergent muscle**, muscle fascicles extending over a broad area converge on a common attachment site (**Figure 11–1b**). The muscle may pull on a tendon, an aponeurosis, or a slender band of collagen fibers known as a **raphe** (RĀ-fē; seam). The muscle fibers typically spread out, like a fan or a broad triangle, with a tendon at the apex. Examples include the prominent *pectoralis muscles* of the chest. A convergent muscle is versatile, because the stimulation of different portions of the muscle can change the direction of pull. However, when the entire muscle contracts, the muscle fibers do not pull as hard on the attachment site as would a parallel muscle of the same size. This is because convergent muscle fibers pull in different directions, rather than all pulling in the same direction.

Pennate Muscles

In a **pennate muscle** (*penna*, feather), the fascicles form a common angle with the tendon. Because the muscle fibers pull at an angle, contracting pennate muscles do not move their tendons as far as parallel muscles do. But a pennate muscle contains more muscle fibers—and thus more myofibrils—than does a parallel muscle of the same size, and so produces more tension.

If all the muscle fibers are on the same side of the tendon, the pennate muscle is *unipennate*. The *extensor digitorum muscle*, a forearm muscle that extends the finger joints, is unipennate (**Figure 11–1c**). More commonly, a pennate muscle has fibers on both sides of the tendon. Such a muscle is called *bipennate*. The *rectus femoris muscle*, a prominent muscle that extends the knee, is bipennate (**Figure 11–1d**). If the tendon branches within a pennate muscle, the muscle is said to be *multipennate*. The triangular *deltoid muscle* of the shoulder is multipennate (**Figure 11–1e**).

Circular Muscles

In a **circular muscle**, or **sphincter** (SFINK-ter), the fascicles are concentrically arranged around an opening or a recess. When the muscle contracts, the diameter of the opening decreases. Circular muscles guard entrances and exits of internal passageways such as the digestive and urinary tracts. An example is the *orbicularis oris muscle* of the mouth (**Figure 11–1f**).

CHECKPOINT

1. Based on patterns of fascicle organization, name the four classifications of skeletal muscle tissue.

2. Why does a pennate muscle generate more tension than does a parallel muscle of the same size?

3. Which type of fascicle arrangement would you expect in a muscle guarding the opening between the stomach and the small intestine?

See the blue Answers tab at the end of the book.

11-2 The three classes of levers increase muscle efficiency

Skeletal muscles do not work in isolation. For muscles attached to the skeleton, the nature and site of the connection determine the force, speed, and range of the movement produced. These characteristics are interdependent, and the relationships can explain a great deal about the general organization of the muscular and skeletal systems.

The force, speed, or direction of movement produced by contraction of a muscle can be modified by attaching the muscle to a lever. A lever is a rigid structure—such as a board, a crowbar, or a bone—that moves on a fixed point called the fulcrum. A lever moves when an applied force (AF) is sufficient to overcome any load (L) or resistance (R) that would otherwise oppose or prevent such movement. In the body, each bone is a lever and each joint is a fulcrum, and muscles provide the applied force. The load can vary from the weight of an object held in the hand to the weight of a limb or the weight of the entire body, depending on the situation. The important thing about levers is that they can change (1) the direction of an applied force, (2) the distance and speed of movement produced by an applied force, and (3) the effective strength of an applied force.

There are three classes of levers, and examples of each are found in the human body (**Figure 11–2**). A seesaw or teeter-totter is an example of a **first-class lever**. In such a lever, the fulcrum (F) lies between the applied force (AF) and the load (L). The body has few first-class levers. One, involved with extension of the neck, is shown in **Figure 11–2a**.

In a **second-class lever** (**Figure 11–2b**), the load is located between the applied force and the fulcrum. A familiar example

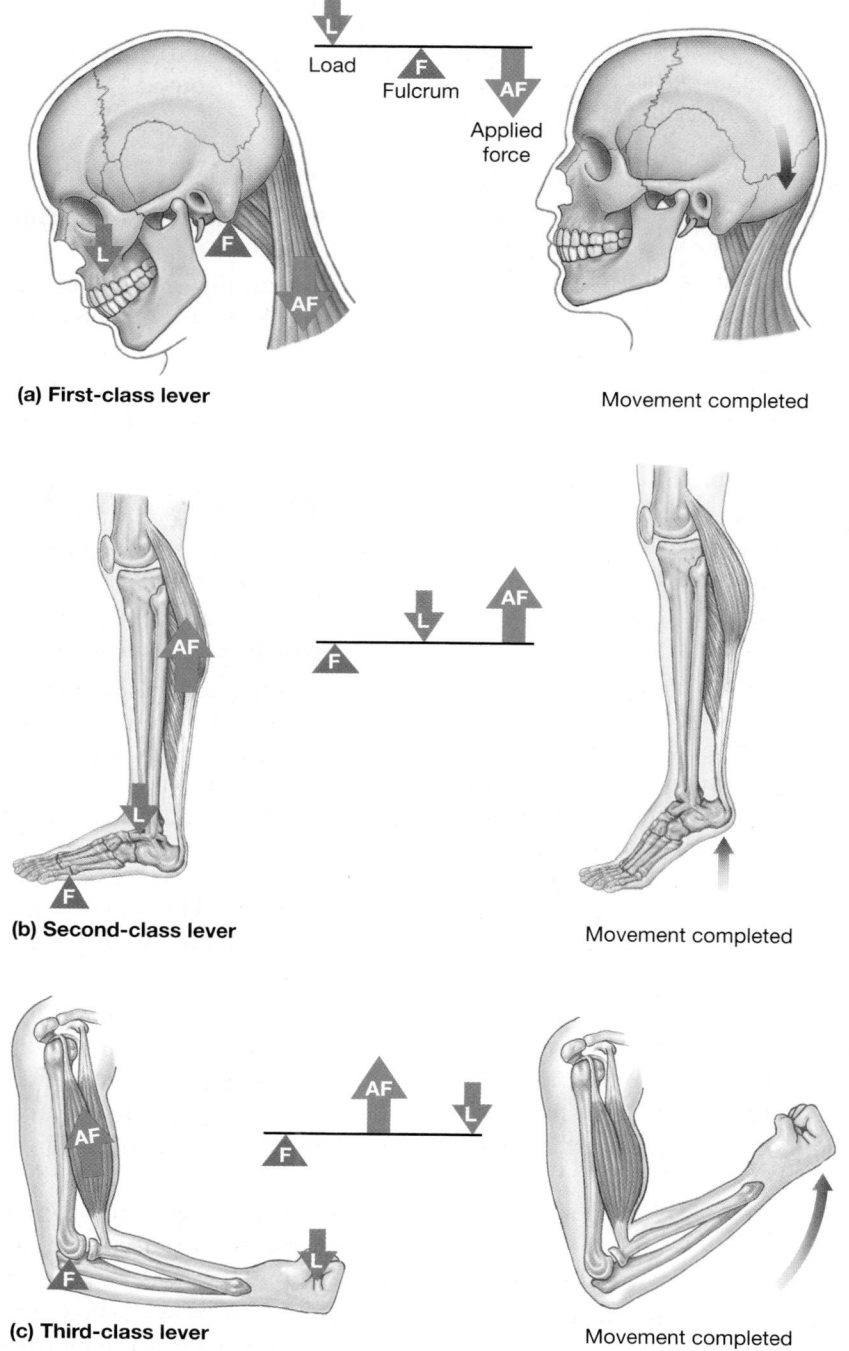

(a) First-class lever

Movement completed

(b) Second-class lever

Movement completed

(c) Third-class lever

Movement completed

Figure 11–2   The Three Classes of Levers. (a) In a first-class lever, the applied force and the load are on opposite sides of the fulcrum. **(b)** In a second-class lever, the load lies between the applied force and the fulcrum. **(c)** In a third-class lever, the force is applied between the load and the fulcrum.

is a loaded wheelbarrow. The weight is the load, and the upward lift on the handle is the applied force. Because in this arrangement the force is always farther from the fulcrum than the load is, a small force can move a larger weight. That is, the effective force is increased. Notice, however, that when a force moves the handle, the load moves more slowly and covers a shorter distance. Thus the effective force is increased at the expense of speed and distance. The body has few second-class levers. Ankle extension (plantar flexion) by the calf muscles involves a second-class lever (**Figure 11–2b**).

Third-class levers are the most common levers in the body. In this lever system, a force is applied between the load and the fulcrum (**Figure 11–2c**). The effect is the reverse of that for a second-class lever: Speed and distance traveled are increased at the expense of effective force. In the example shown (the biceps brachii muscle, which flexes the elbow), the load is six times farther from the fulcrum than is the applied force. The effective force is reduced to the same degree. The muscle must generate 180 kg (396 lb) of tension at its attachment to the forearm to support 30 kg (66 lb) held in the hand. However, the distance traveled and the speed of movement are increased by that same 6:1 ratio: The load will travel 45 cm (18 in.) when the point of attachment moves 7.5 cm (2.9 in.).

Although not every muscle operates as part of a lever system, the presence of levers provides speed and versatility far in excess of what we would predict on the basis of muscle physiology alone. Skeletal muscle fibers resemble one another closely, and their abilities to contract and generate tension are quite similar. Consider a skeletal muscle that can shorten 1 cm (0.39 in.) while it exerts a 10-kg (22-lb) pull. Without using a lever, this muscle would be performing efficiently only when moving a 10-kg (22-lb) weight a distance of 1 cm (0.39 in.). By using a lever, however, the same muscle operating at the same efficiency could move 20 kg (44 lb) a distance of 0.5 cm (0.2 in.), 5 kg (11 lb) a distance of 2 cm (0.79 in.), or 1 kg (2.2 lb) a distance of 10 cm (3.9 in.).

The A&P Top 100

#39 Most skeletal muscles can shorten to roughly 70 percent of their "ideal" resting length. The power, speed, and range of body movements results from differences in fascicle arrangement and in the positions of muscle attachments at the joints involved.

CHECKPOINT

4. Define a lever, and describe the three classes of lever.

5. The joint between the occipital bone of the skull and the first cervical vertebra (atlas) is part of which class of lever system?

See the blue Answers tab at the end of the book.

11-3 Muscle origins are at the fixed end of muscles, whereas insertions are at the movable end of muscles

This chapter focuses on the functional anatomy of skeletal muscles and muscle groups. You must learn a number of new terms, and this section attempts to help you understand them. It may also help you to create a vocabulary list from the terms introduced. Once you are familiar with the basic terminology, the names and actions of skeletal muscles are easily understood.

Origins and Insertions

In Chapter 10 we noted that when both ends of a myofibril are free to move, the ends move toward the center during a contraction. ∞ p. 310 In the body, however, the ends of a skeletal muscle are always attached to other structures that limit their movement. In most cases one end is fixed in position, and during a contraction the other end moves toward the fixed end. The place where the fixed end attaches to a bone, cartilage, or connective tissue is called the **origin** of the muscle. The site where the movable end attaches to another structure is called the **insertion** of the muscle. The origin is typically proximal to the insertion. When a muscle contracts, it produces a specific **action**, or movement. Actions are described using the terms introduced in Chapter 9 (flexion, extension, adduction, and so forth).

As an example, consider the *gastrocnemius muscle*, a calf muscle that extends from the distal portion of the femur to the calcaneus. As **Figure 11–2b** shows, when the gastrocnemius muscle contracts, it pulls the calcaneus toward the knee. As a result, we say that the gastrocnemius muscle has its origin at the femur and its insertion at the calcaneus; its action can be described as "extension at the ankle" or "plantar flexion."

The decision as to which end is the origin and which is the insertion is usually based on movement from the anatomical position. Part of the fun of studying the muscular system is that you can actually do the movements and think about the muscles involved. As a result, laboratory discussions of the muscular system tend to resemble disorganized aerobics classes.

When the origins and insertions cannot be determined easily on the basis of movement from the anatomical position, other rules are used. If a muscle extends between a broad aponeurosis and a narrow tendon, the aponeurosis is the origin and the tendon is the insertion. If several tendons are at one end and just one is at the other, the muscle has multiple origins and a single insertion. These simple rules cannot cover every situation. Knowing which end is the origin and which is the insertion is ultimately less important than knowing where the two ends attach and what the muscle accomplishes when it contracts.

Most muscles originate at a bone, but some originate at a connective tissue sheath or band. Examples of these sheaths or bands include *intermuscular septa* (components of the deep fascia that may separate adjacent skeletal muscles), *tendinous inscriptions* that join muscle fibers to form long muscles such as the *rectus abdominis*, the interosseous membranes of the forearm or leg, and the fibrous sheet that spans the obturator foramen of the pelvis.

Actions

Almost all skeletal muscles either originate or insert on the skeleton. When a muscle moves a portion of the skeleton, that movement may involve flexion, extension, adduction, abduction, protraction, retraction, elevation, depression, rotation, circumduction, pronation, supination, inversion, eversion, lateral flexion, or opposition. (Before proceeding, you may want to review the discussions of planes of motion and **Figures 9–2** to **9–5**.) ∞ pp. 273–276

Actions can be described in one of two ways. The first, used by most undergraduate textbooks and references such as *Gray's Anatomy*, describes actions in terms of the bone or region affected. Thus, a muscle such as the biceps brachii muscle is said to perform "flexion of the forearm." The second way, of increasing use among specialists such as kinesiologists and physical therapists, identifies the joint(s) involved. In this approach, the action of the biceps brachii muscle would be "flexion at (or of) the elbow." Both approaches are valid, and each has its advantages. In general, we will use the latter approach.

When complex movements occur, muscles commonly work in groups rather than individually. Their cooperation improves the efficiency of a particular movement. For example, large muscles of the limbs produce flexion or extension over an extended range of motion. Although these muscles cannot produce powerful movements at full extension due to the relative positions of the articulating bones, they are generally paired with one or more smaller muscles that provide assistance until the larger muscle can perform at maximum efficiency. At the start of the movement, the smaller muscle is producing maximum tension, while the larger muscle is producing minimum tension. The importance of this smaller "assistant" decreases as the movement proceeds and the effectiveness of the primary muscle increases.

Based on their functions, muscles are described as follows:

- An **agonist**, or **prime mover**, is a muscle whose contraction is chiefly responsible for producing a particular movement. The biceps brachii muscle is an agonist that produces flexion at the elbow.

- An **antagonist** is a muscle whose action opposes that of a particular agonist. The *triceps brachii muscle* is an agonist that extends the elbow. It is therefore an antagonist of the biceps brachii muscle, and the biceps brachii is an antagonist of the triceps brachii. Agonists and antagonists are functional opposites; if one produces flexion, the other will produce extension. When an agonist contracts to produce a particular movement, the corresponding antagonist will be stretched, but it will usually not relax completely. Instead, it will contract eccentrically, with just enough tension to control the speed of the movement and ensure its smoothness. ∞ p. 316 You may find it easiest to learn about muscles in agonist–antagonist pairs (flexors–extensors, abductors–adductors) that act at a specific joint. This method highlights the functions of the muscles involved, and it can help organize the information into a logical framework. The tables in this chapter are arranged to facilitate such an approach.

- When a **synergist** (*syn-*, together + *ergon*, work) contracts, it helps a larger agonist work efficiently. Synergists may provide additional pull near the insertion or may stabilize the point of origin. Their importance in assisting a particular movement may change as the movement progresses. In many cases, they are most useful at the start, when the agonist is stretched and unable to develop maximum tension. For example, the *latissimus dorsi muscle* is a large trunk muscle that extends, adducts, and medially rotates the arm at the shoulder joint. A much smaller muscle, the *teres* (TER-ēz) *major muscle*, assists in starting such movements when the shoulder joint is at full flexion. Synergists may also assist an agonist by preventing movement at another joint and thereby stabilizing the origin of the agonist. Such synergists are called **fixators**.

CHECKPOINT

6. Define the term *synergist* as it relates to muscle action.

7. The *gracilis muscle* is attached to the anterior surface of the tibia at one end, and to the pubis and ischium of the pelvis at the other. When the muscle contracts, flexion occurs at the hip. Which attachment point is the muscle's origin?

8. Muscle A abducts the humerus, and muscle B adducts the humerus. What is the relationship between these two muscles?

See the blue Answers tab at the end of the book.

11-4 Descriptive terms are used to name skeletal muscles

Except for the *platysma* and the *diaphragm*, the complete names of all skeletal muscles include the term *muscle*. Although the full name, such as the biceps brachii muscle, will usually appear in the text, for simplicity only the descriptive name (biceps brachii) will be used in figures and tables.

You need not learn every one of the approximately 700 muscles in the human body, but you will have to become familiar with the most important ones. Fortunately, the names anatomists assigned to the muscles include descriptive terms that can help you remember the names and identify the muscles. When faced with a new muscle name, it is helpful to first identify the descriptive portions of the name. The name of a muscle may include descriptive information about its location in the body, origin and insertion, fascicle organization, relative position, structural characteristics, and action.

Location in the Body

Table 11–1 includes a useful summary of muscle terminology, including terms that designate specific regions of the body. Regional terms are most common as modifiers that help identify individual muscles. In a few cases, a muscle is such a prominent feature of a body region that a name referring to the region alone will identify it. Examples include the *temporalis muscle* of the head and the *brachialis* (brā-kē-A-lis) *muscle* of the arm.

Origin and Insertion

Many muscle names include terms for body locales that tell you the specific origin and insertion of each muscle. In such cases, the first part of the name indicates the origin, the second part the insertion. The *genioglossus muscle*, for example, originates at the chin (*geneion*) and inserts in the tongue (*glossus*). The names may be long and difficult to pronounce, but Table 11–1 and the anatomical terms introduced in Chapter 1 can help you identify and remember them. ∞ pp. 18–24

Fascicle Organization

A muscle name may refer to the orientation of the muscle fascicles within a particular skeletal muscle. **Rectus** means "straight," and rectus muscles are parallel muscles whose fibers generally run along the long axis of the body. Because we have several rectus muscles, the name typically includes a second term that refers to a precise region of the body. For ex-

ample, the *rectus abdominis muscle* is located on the abdomen, and the *rectus femoris muscle* on the thigh. Other common directional indicators include **transversus** and **oblique**, for muscles whose fibers run across or at an oblique angle to the longitudinal axis of the body, respectively.

Relative Position

Muscles visible at the body surface are often called **externus** or **superficialis**, whereas deeper muscles are termed **internus** or **profundus**. Superficial muscles that position or stabilize an organ are called **extrinsic**; muscles located entirely within an organ are **intrinsic**.

Structural Characteristics

Some muscles are named after distinctive structural features. The biceps brachii muscle, for example, has two tendons of origin (*bi-*, two + *caput*, head); the triceps brachii muscle has three; and the *quadriceps group*, four. Shape is sometimes an important clue to the name of a muscle. For example, the *trapezius* (tra-PĒ-zē-us), *deltoid, rhomboid* (ROM-boyd), and *orbicularis* (or-bik-ū-LĀ-ris) muscles look like a trapezoid, a triangle, a rhomboid, and a circle, respectively. Many terms refer to muscle size. Long muscles are called **longus** (long) or **longissimus** (longest), and **teres** muscles are both long and round. Short muscles are called **brevis**. Large ones are called **magnus** (big), **major** (bigger), or **maximus** (biggest); small ones are called **minor** (smaller) or **minimus** (smallest).

Action

Many muscles are named *flexor, extensor, pronator, abductor,* and so on. These are such common actions that the names almost always include other clues as to the appearance or location of the muscle. For example, the *extensor carpi radialis longus muscle* is a long muscle along the radial (lateral) border of the forearm. When it contracts, its primary function is extension at the carpus (wrist).

A few muscles are named after the specific movements associated with special occupations or habits. The *sartorius* (sar-TOR-ē-us) *muscle*, the longest in the body, is active when you cross your legs. Before sewing machines were invented, a tailor would sit on the floor cross-legged, and the name of this muscle was derived from *sartor*, the Latin word for "tailor." The *buccinator* (BUK-si-nā-tor) *muscle* on the face compresses the cheeks—when, for example, you purse your lips and blow forcefully. *Buccinator* translates as "trumpet player." Another facial muscle, the *risorius* (ri-SOR-ē-us) *muscle*, was supposedly named after the mood expressed. However, the Latin word *risor* means "laughter"; a more appropriate description for the effect would be "grimace."

TABLE 11–1 Muscle Terminology

Terms Indicating Specific Regions of the Body*	Terms Indicating Position, Direction, or Fascicle Organization	Terms Indicating Structural Characteristics of the Muscle	Terms Indicating Actions
Abdominis (abdomen)	Anterior (front)	**Nature of Origin**	**General**
Anconeus (elbow)	Externus (superficial)	Biceps (two heads)	Abductor
Auricularis (auricle of ear)	Extrinsic (outside)	Triceps (three heads)	Adductor
Brachialis (brachium)	Inferioris (inferior)	Quadriceps (four heads)	Depressor
Capitis (head)	Internus (deep, internal)		Extensor
Carpi (wrist)	Intrinsic (inside)		Flexor
Cervicis (neck)	Lateralis (lateral)	**Shape**	Levator
Cleido-/-clavius (clavicle)	Medialis/medius (medial, middle)	Deltoid (triangle)	Pronator
Coccygeus (coccyx)	Oblique	Orbicularis (circle)	Rotator
Costalis (ribs)	Posterior (back)	Pectinate (comblike)	Supinator
Cutaneous (skin)	Profundus (deep)	Piriformis (pear-shaped)	Tensor
Femoris (femur)	Rectus (straight, parallel)	Platy- (flat)	
Genio- (chin)	Superficialis (superficial)	Pyramidal (pyramid)	
Glosso-/-glossal (tongue)	Superioris (superior)	Rhomboid	**Specific**
Hallucis (great toe)	Transversus (transverse)	Serratus (serrated)	Buccinator (trumpeter)
Ilio- (ilium)		Splenius (bandage)	Risorius (laughter)
Inguinal (groin)		Teres (long and round)	Sartorius (like a tailor)
Lumborum (lumbar region)		Trapezius (trapezoid)	
Nasalis (nose)			
Nuchal (back of neck)			
Oculo- (eye)		**Other Striking Features**	
Oris (mouth)		Alba (white)	
Palpebrae (eyelid)		Brevis (short)	
Pollicis (thumb)		Gracilis (slender)	
Popliteus (posterior to knee)		Lata (wide)	
Psoas (loin)		Latissimus (widest)	
Radialis (radius)		Longissimus (longest)	
Scapularis (scapula)		Longus (long)	
Temporalis (temples)		Magnus (large)	
Thoracis (thoracic region)		Major (larger)	
Tibialis (tibia)		Maximus (largest)	
Ulnaris (ulna)		Minimus (smallest)	
Uro- (urinary)		Minor (smaller)	
		-tendinosus (tendinous)	
		Vastus (great)	

*For other regional terms, refer to Figure 1–6, p. 18, which deals with anatomical landmarks.

Axial and Appendicular Muscles

The separation of the skeletal system into axial and appendicular divisions provides a useful guideline for subdividing the muscular system as well:

1. The **axial muscles** arise on the axial skeleton. They position the head and spinal column and also move the rib cage, assisting in the movements that make breathing possible. They do not play a role in movement or support of either the pectoral or pelvic girdle or the limbs. This category encompasses roughly 60 percent of the skeletal muscles in the body.

2. The **appendicular muscles** stabilize or move components of the appendicular skeleton and include the remaining 40 percent of all skeletal muscles.

Figure 11–3 provides an overview of the major axial and appendicular muscles of the human body. These are superficial muscles, which tend to be relatively large. The superficial muscles cover deeper, smaller muscles that cannot be seen unless the overlying muscles are either removed or *reflected*—that is, cut and pulled out of the way. Later figures that show deep muscles in specific regions will indicate for the sake of clarity whether superficial muscles have been removed or reflected.

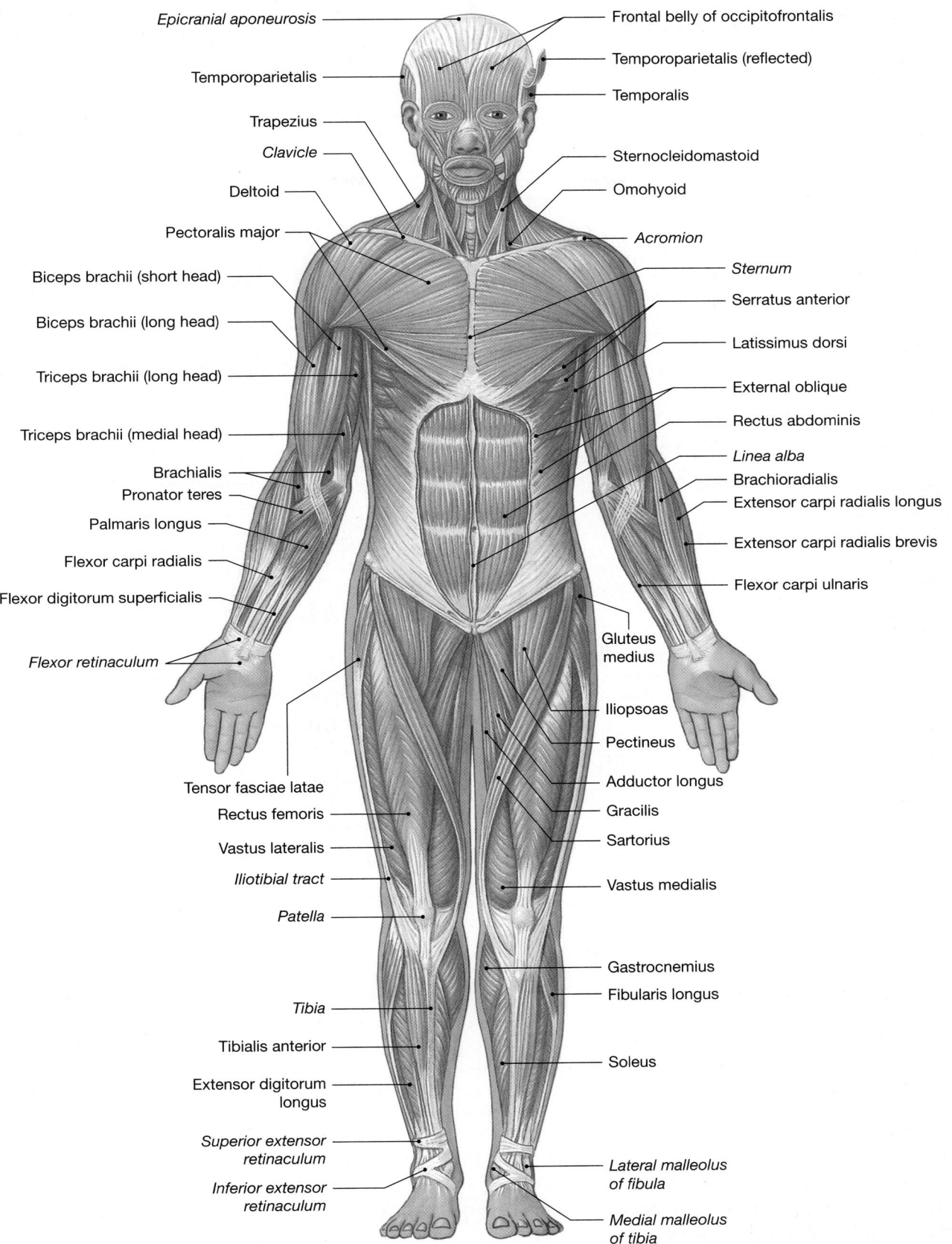

Epicranial aponeurosis — Frontal belly of occipitofrontalis

Temporoparietalis — Temporoparietalis (reflected)

Temporalis

Trapezius

Clavicle — Sternocleidomastoid

Deltoid — Omohyoid

Pectoralis major — Acromion

Biceps brachii (short head) — Sternum

Biceps brachii (long head) — Serratus anterior

Triceps brachii (long head) — Latissimus dorsi

Triceps brachii (medial head) — External oblique

— Rectus abdominis

Brachialis — Linea alba

Pronator teres — Brachioradialis

Palmaris longus — Extensor carpi radialis longus

Flexor carpi radialis — Extensor carpi radialis brevis

Flexor digitorum superficialis — Flexor carpi ulnaris

Flexor retinaculum — Gluteus medius

Iliopsoas

Pectineus

Tensor fasciae latae — Adductor longus

Rectus femoris — Gracilis

Vastus lateralis — Sartorius

Iliotibial tract — Vastus medialis

Patella

Gastrocnemius

Fibularis longus

Tibia

Tibialis anterior — Soleus

Extensor digitorum longus

Superior extensor retinaculum — Lateral malleolus of fibula

Inferior extensor retinaculum — Medial malleolus of tibia

11 MUSCULAR

Figure 11–3 An Overview of the Major Skeletal Muscles. (a) An anterior view. ATLAS: Plates 1a; 39a–d

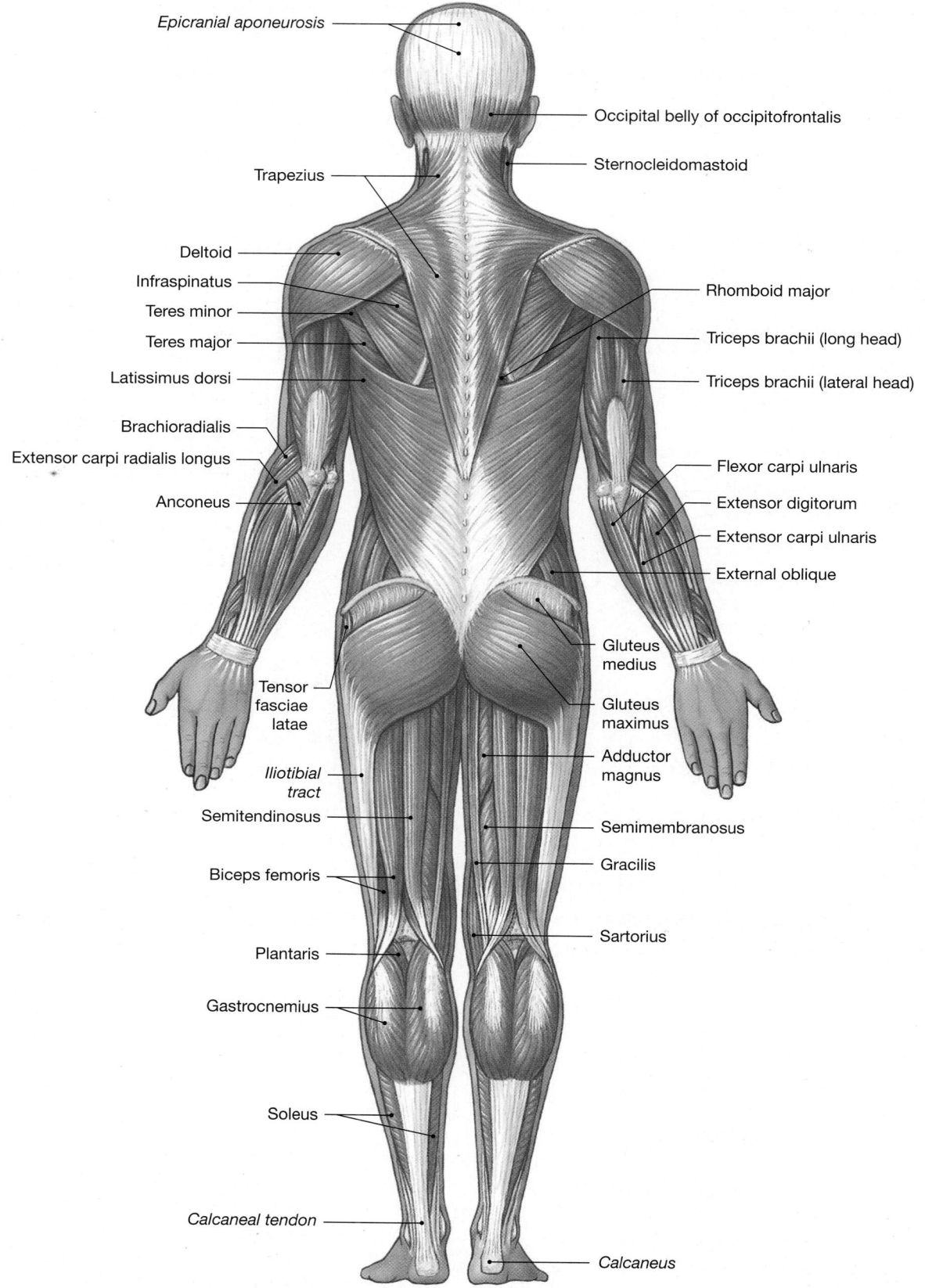

Epicranial aponeurosis

Occipital belly of occipitofrontalis

Sternocleidomastoid

Trapezius

Deltoid

Infraspinatus

Teres minor

Teres major

Latissimus dorsi

Rhomboid major

Triceps brachii (long head)

Triceps brachii (lateral head)

Brachioradialis

Extensor carpi radialis longus

Anconeus

Flexor carpi ulnaris

Extensor digitorum

Extensor carpi ulnaris

External oblique

Tensor fasciae latae

Gluteus medius

Gluteus maximus

Iliotibial tract

Semitendinosus

Adductor magnus

Biceps femoris

Semimembranosus

Gracilis

Plantaris

Sartorius

Gastrocnemius

Soleus

Calcaneal tendon

Calcaneus

Figure 11–3 **An Overview of the Major Skeletal Muscles. (continued) (b)** A posterior view. ATLAS: Plates 1b; 40a,b

Paying attention to patterns of origin, insertion, and action, we will now examine representatives of both muscular divisions. The discussion assumes that you already understand skeletal anatomy. As you examine the figures in this chapter, you will find that some bony and cartilaginous landmarks are labeled to provide orientation. These labels are shown in italics, to differentiate these landmarks from the muscles and tendons that are the primary focus of each figure. Should you need further review of skeletal anatomy, figure captions in this chapter indicate the relevant figures in Chapters 7, 8, and 9.

The tables that follow also contain information about the innervation of the individual muscles. **Innervation** is the distribution of nerves to a region or organ; the tables indicate the nerves that control each muscle. Many of the muscles of the head and neck are innervated by cranial nerves, which originate at the brain and pass through the foramina of the skull. Alternatively, spinal nerves are connected to the spinal cord and pass through the intervertebral foramina. For example, spinal nerve L_1 passes between vertebrae L_1 and L_2. Spinal nerves may form a complex network (a plexus) after exiting the spinal cord; one branch of this network may contain axons from several spinal nerves. Thus, many tables identify the spinal nerves involved as well as the names of the peripheral nerves.

> **CHECKPOINT**
>
> 9. Identify the kinds of descriptive information used to name skeletal muscles.
>
> 10. What does the name *flexor carpi radialis longus* tell you about this muscle?
>
> See the blue Answers tab at the end of the book.

11-5 Axial muscles are muscles of the head and neck, vertebral column, trunk, and pelvic floor

The axial muscles fall into logical groups on the basis of location, function, or both. The groups do not always have distinct anatomical boundaries. For example, a function such as extension of the vertebral column involves muscles along its entire length and movement at each of the intervertebral joints. We will discuss the axial muscles in four groups:

1. *The Muscles of the Head and Neck.* This group includes muscles that move the face, tongue, and larynx. They are therefore responsible for verbal and nonverbal communication—laughing, talking, frowning, smiling, whistling, and so on. You also use these muscles while eating—

especially in sucking and chewing—and even while looking for food, as some of them control your eye movements. The group does not include muscles of the neck that are involved with movements of the vertebral column.

2. *The Muscles of the Vertebral Column.* This group includes numerous flexors, extensors, and rotators of the vertebral column.

3. *The Oblique and Rectus Muscles.* This group forms the muscular walls of the thoracic and abdominopelvic cavities between the first thoracic vertebra and the pelvis. In the thoracic area these muscles are partitioned by the ribs, but over the abdominal surface they form broad muscular sheets. The neck also has oblique and rectus muscles. Although they do not form a complete muscular wall, they share a common developmental origin with the oblique and rectus muscles of the trunk.

4. *The Muscles of the Pelvic Floor.* These muscles extend between the sacrum and pelvic girdle. The group forms the *perineum,* a muscular sheet that closes the pelvic outlet.

Muscles of the Head and Neck

We can divide the muscles of the head and neck into several functional groups. The *muscles of facial expression,* the *muscles of mastication* (chewing), the *muscles of the tongue,* and the *muscles of the pharynx* originate on the skull or hyoid bone. Muscles involved with sight and hearing also are based on the skull. Here, we will consider the *extrinsic eye muscles*—those associated with movements of the eye. We will discuss the intrinsic eye muscles, which control the diameter of the pupil and the shape of the lens, and the tiny skeletal muscles associated with the auditory ossicles, in Chapter 17. In the neck, the *extrinsic muscles of the larynx* adjust the position of the hyoid bone and larynx. We will examine the intrinsic laryngeal muscles, including those of the vocal cords, in Chapter 23.

Muscles of Facial Expression

The muscles of facial expression originate on the surface of the skull (**Figure 11–4** and Table 11–2). At their insertions, the fibers of the epimysium are woven into those of the superficial fascia and the dermis of the skin: Thus, when they contract, the skin moves.

The largest group of facial muscles is associated with the mouth. The **orbicularis oris** muscle constricts the opening, and other muscles move the lips or the corners of the mouth. The **buccinator** muscle has two functions related to eating (in addition to its importance to musicians). During chewing, it cooperates with the masticatory muscles by moving food back across the teeth from the *vestibule,* the space inside the cheeks. In infants, the buccinator provides suction for suckling at the breast.

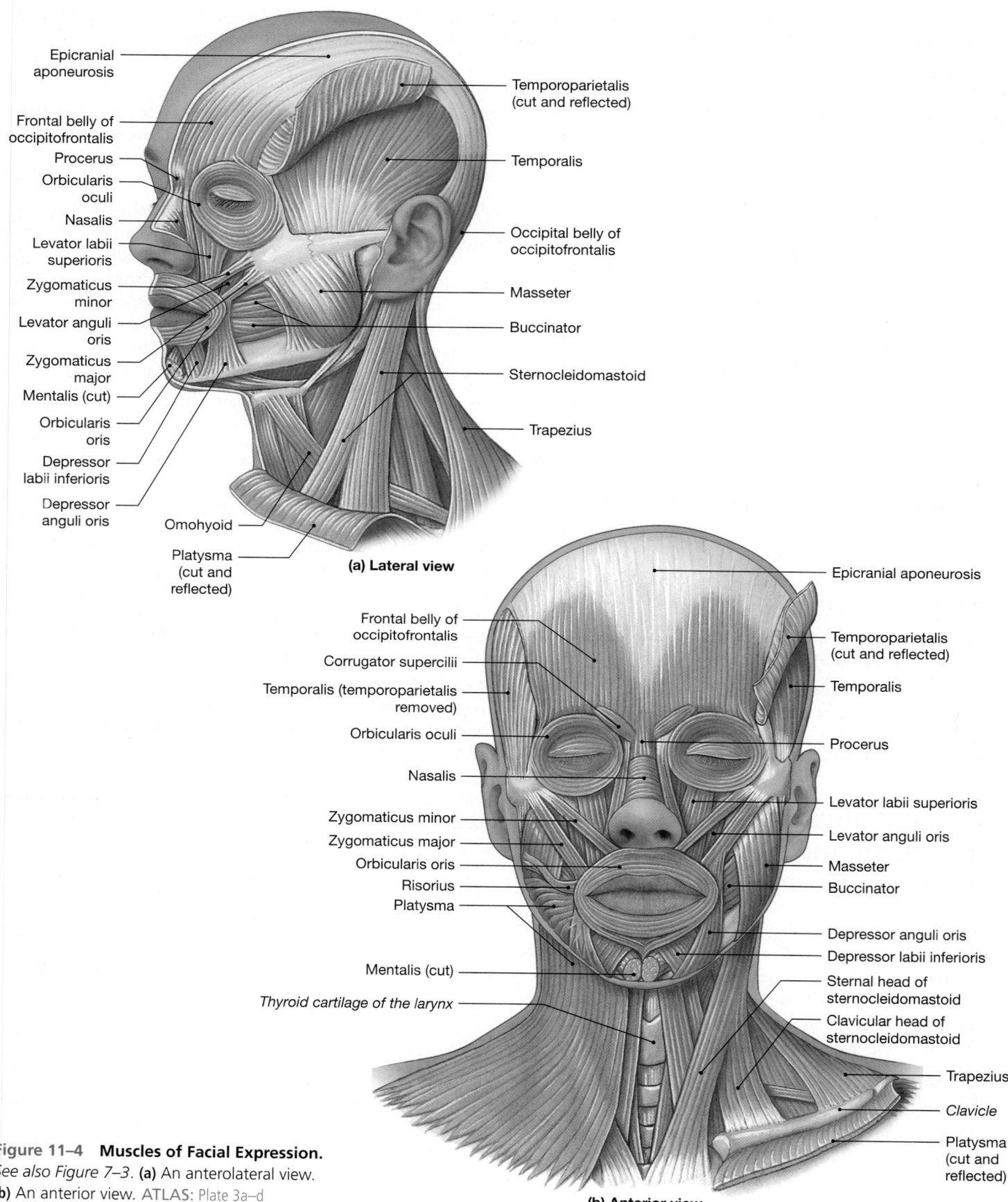

Epicranial aponeurosis

Frontal belly of occipitofrontalis

Procerus

Orbicularis oculi

Nasalis

Levator labii superioris

Zygomaticus minor

Levator anguli oris

Zygomaticus major

Mentalis (cut)

Orbicularis oris

Depressor labii inferioris

Depressor anguli oris

Omohyoid

Platysma (cut and reflected)

Temporoparietalis (cut and reflected)

Temporalis

Occipital belly of occipitofrontalis

Masseter

Buccinator

Sternocleidomastoid

Trapezius

(a) Lateral view

Frontal belly of occipitofrontalis

Corrugator supercilii

Temporalis (temporoparietalis removed)

Orbicularis oculi

Nasalis

Zygomaticus minor

Zygomaticus major

Orbicularis oris

Risorius

Platysma

Mentalis (cut)

Thyroid cartilage of the larynx

Epicranial aponeurosis

Temporoparietalis (cut and reflected)

Temporalis

Procerus

Levator labii superioris

Levator anguli oris

Masseter

Buccinator

Depressor anguli oris

Depressor labii inferioris

Sternal head of sternocleidomastoid

Clavicular head of sternocleidomastoid

Trapezius

Clavicle

Platysma (cut and reflected)

(b) Anterior view

Figure 11–4 Muscles of Facial Expression.
See also Figure 7–3. **(a)** An anterolateral view.
(b) An anterior view. ATLAS: Plate 3a–d

TABLE 11–2 Muscles of Facial Expression (Figure 11–4)

Region/Muscle	Origin	Insertion	Action	Innervation
MOUTH				
Buccinator	Alveolar processes of maxilla and mandible	Blends into fibers of orbicularis oris	Compresses cheeks	Facial nerve (N VII)*
Depressor labii inferioris	Mandible between the anterior midline and the mental foramen	Skin of lower lip	Depresses lower lip	As above
Levator labii superioris	Inferior margin of orbit, superior to the infra-orbital foramen	Orbicularis oris	Elevates upper lip	As above
Levator anguli oris	Maxilla below the infra-orbital foramen	Corner of mouth	Elevates mouth corner	As above
Mentalis	Incisive fossa of mandible	Skin of chin	Elevates and protrudes lower lip	As above
Orbicularis oris	Maxilla and mandible	Lips	Compresses, purses lips	As above
Risorius	Fascia surrounding parotid salivary gland	Angle of mouth	Draws corner of mouth to the side	As above
Depressor anguli oris	Anterolateral surface of mandibular body	Skin at angle of mouth	Depresses corner of mouth	As above
Zygomaticus major	Zygomatic bone near zygomaticomaxillary suture	Angle of mouth	Retracts and elevates corner of mouth	As above
Zygomaticus minor	Zygomatic bone posterior to zygomaticotemporal suture	Upper lip	Retracts and elevates upper lip	As above
EYE				
Corrugator supercilii	Orbital rim of frontal bone near nasal suture	Eyebrow	Pulls skin inferiorly and anteriorly; wrinkles brow	As above
Levator palpebrae superioris (Figure 11–5)	Tendinous band around optic foramen	Upper eyelid	Elevates upper eyelid	Oculomotor nerve (N III)**
Orbicularis oculi	Medial margin of orbit	Skin around eyelids	Closes eye	Facial nerve (N VII)
NOSE				
Procerus	Nasal bones and lateral nasal cartilages	Aponeurosis at bridge of nose and skin of forehead	Moves nose, changes position and shape of nostrils	As above
Nasalis	Maxilla and alar cartilage of nose	Bridge of nose	Compresses bridge, depresses tip of nose; elevates corners of nostrils	As above
EAR				
Temporoparietalis	Fascia around external ear	Epicranial aponeurosis	Tenses scalp, moves auricle of ear	As above
SCALP (EPICRANIUM)				
Occipitofrontalis		Skin of eyebrow and bridge of nose	Raises eyebrows, wrinkles forehead	As above
Frontal belly	Epicranial aponeurosis			
Occipital belly	Epicranial aponeurosis	Epicranial aponeurosis	Tenses and retracts scalp	As above
NECK				
Platysma	Superior thorax between cartilage of 2nd rib and acromion of scapula	Mandible and skin of cheek	Tenses skin of neck; depresses mandible	As above

*An uppercase N and Roman numerals refer to a cranial nerve.

**This muscle originates in association with the extrinsic eye muscles, so its innervation is unusual.

Smaller groups of muscles control movements of the eyebrows and eyelids, the scalp, the nose, and the external ear. The **epicranium** (ep-i-KRĀ-nē-um; *epi-*, on + *kranion*, skull), or scalp, contains the **temporoparietalis** muscle and the **occipitofrontalis** muscle, which has a *frontal belly* and an *occipital belly*. The two bellies are separated by the **epicranial aponeurosis**, a thick, collagenous sheet. The **platysma** (pla-TIZ-muh; *platy*, flat) covers the anterior surface of the neck, extending from the base of the neck to the periosteum of the mandible and the fascia at the corner of the mouth. One of the effects of aging is the loss of muscle tone in the platysma, resulting in a looseness of the skin of the anterior throat.

Extrinsic Eye Muscles

Six extrinsic eye muscles, also known as the **extra-ocular muscles** or *oculomotor muscles*, originate on the surface of the orbit and control the position of each eye. These muscles, shown in **Figure 11–5** and detailed in Table 11–3, are the **inferior rectus**, **medial rectus**, **superior rectus**, **lateral rectus**, **inferior oblique**, and **superior oblique** muscles.

Muscles of Mastication

The muscles of mastication (**Figure 11–6** and Table 11–4) move the mandible at the temporomandibular joint. The large **masseter** muscle is the strongest jaw muscle. The **temporalis**

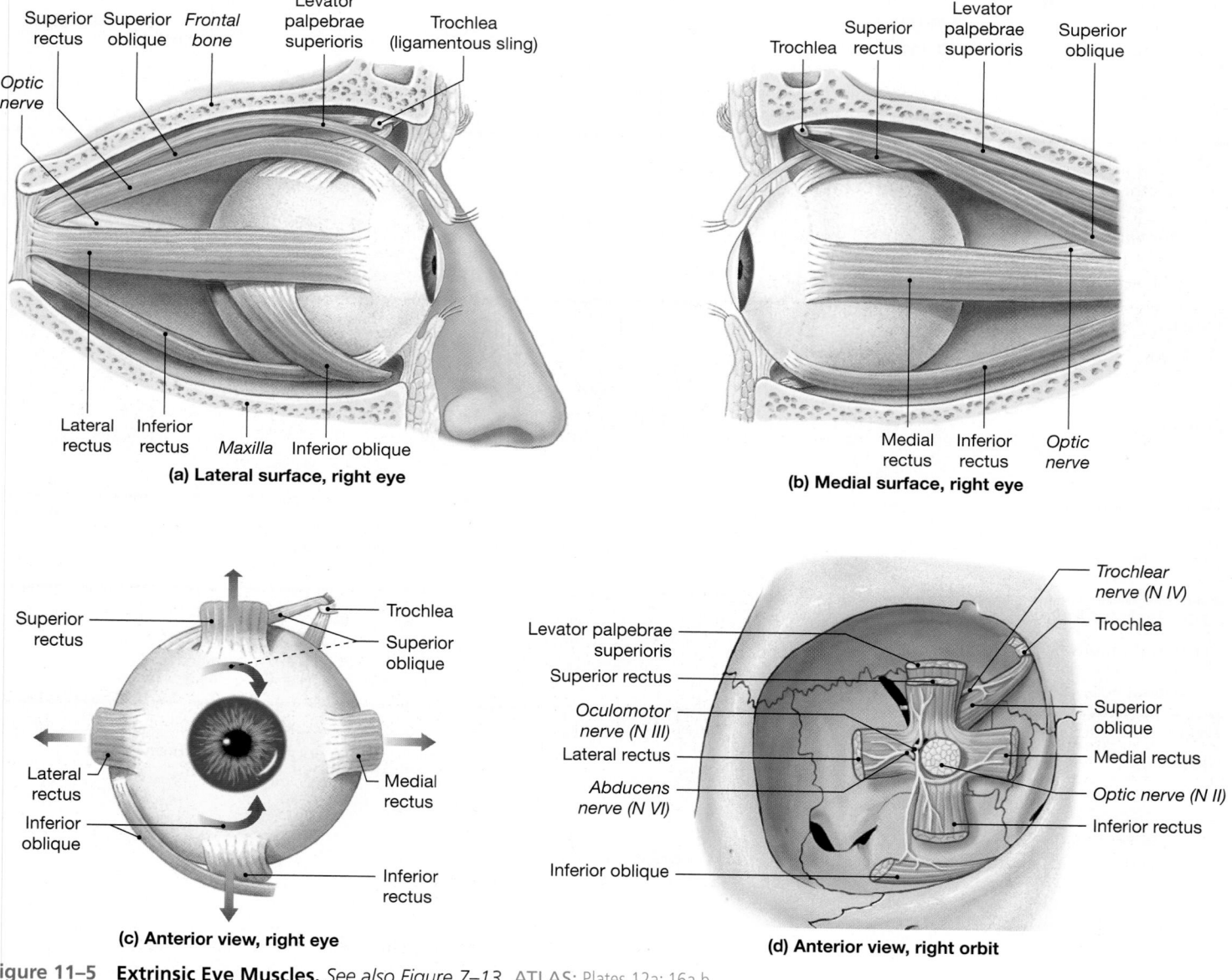

(a) Lateral surface, right eye

(b) Medial surface, right eye

(c) Anterior view, right eye

(d) Anterior view, right orbit

Figure 11–5 **Extrinsic Eye Muscles.** *See also Figure 7–13.* ATLAS: Plates 12a; 16a,b

11 MUSCULAR

TABLE 11–3 Extrinsic Eye Muscles (Figure 11–5)

Muscle	Origin	Insertion	Action	Innervation
Inferior rectus	Sphenoid around optic canal	Inferior, medial surface of eyeball	Eye looks down	Oculomotor nerve (N III)
Medial rectus	As above	Medial surface of eyeball	Eye looks medially	As above
Superior rectus	As above	Superior surface of eyeball	Eye looks up	As above
Lateral rectus	As above	Lateral surface of eyeball	Eye looks laterally	Abducens nerve (N VI)
Inferior oblique	Maxilla at anterior portion of orbit	Inferior, lateral surface of eyeball	Eye rolls, looks up and laterally	Oculomotor nerve (N III)
Superior oblique	Sphenoid around optic canal	Superior, lateral surface of eyeball	Eye rolls, looks down and laterally	Trochlear nerve (N IV)

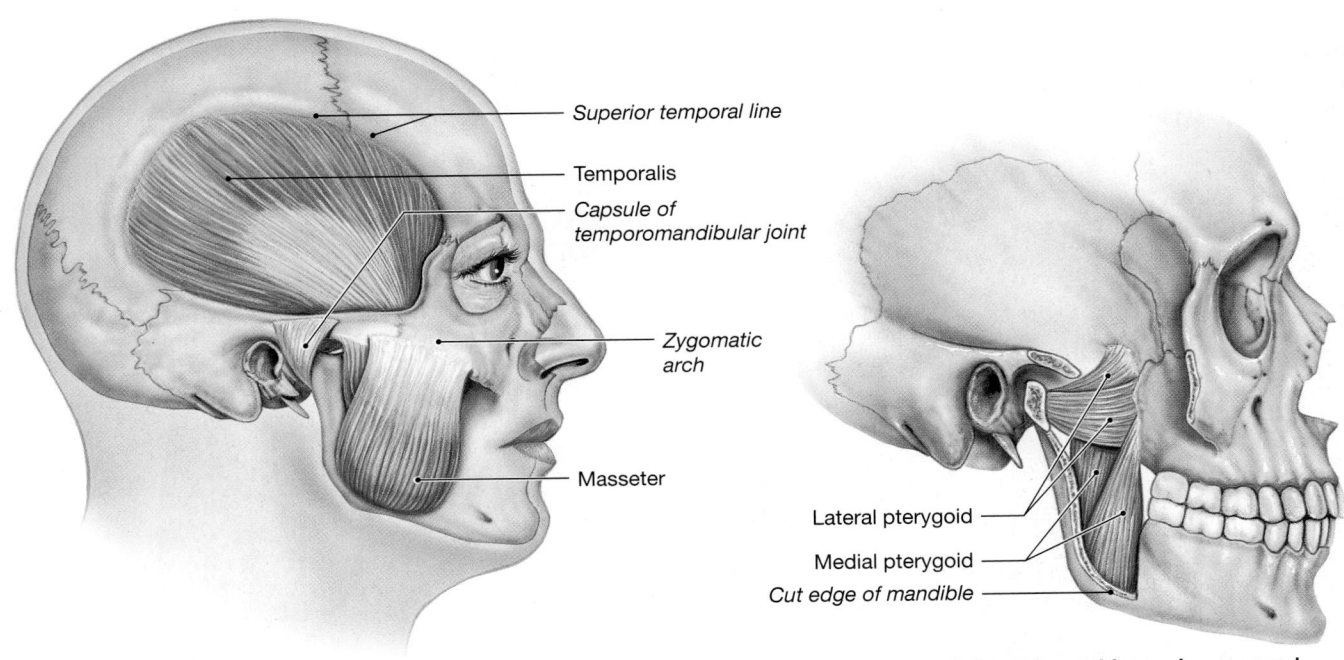

(a) Lateral view

(b) Lateral view, pterygoid muscles exposed

Figure 11–6 Muscles of Mastication. (a) The temporalis muscle passes medial to the zygomatic arch to insert on the coronoid process of the mandible. The masseter inserts on the angle and lateral surface of the mandible. **(b)** The location and orientation of the pterygoid muscles can be seen after the overlying muscles, along with a portion of the mandible, are removed. *See also Figures 7–3 and 7–12.* ATLAS: Plate 3c,d

TABLE 11–4 Muscles of Mastication (Figure 11–6)

Muscle	Origin	Insertion	Action	Innervation
Masseter	Zygomatic arch	Lateral surface of mandibular ramus	Elevates mandible and closes the jaws	Trigeminal nerve (N V), mandibular branch
Temporalis	Along temporal lines of skull	Coronoid process of mandible	Elevates mandible	As above
Pterygoids (medial and lateral)	Lateral pterygoid plate	Medial surface of mandibular ramus	*Medial:* Elevates the mandible and closes the jaws, or performs lateral excursion	As above
			Lateral: Opens jaws, protrudes mandible, or performs lateral excursion	As above

muscle assists in elevation of the mandible. You can feel these muscles in action by gritting your teeth while resting your hand on the side of your face below and then above the zygomatic arch. The **pterygoid** muscles, used in various combinations, can elevate, depress, or protract the mandible or slide it from side to side, a movement called *lateral excursion*. These movements are important in making efficient use of your teeth while you chew foods of various consistencies.

Tips&Tricks

The medial and lateral pterygoid muscles are named for their origin on the pterygoid ("winged") portion of the sphenoid bone. The *ptero*dactyl was a prehistoric winged reptile.

Muscles of the Tongue

The muscles of the tongue have names ending in *glossus*, the Greek word for "tongue." The **palatoglossus** muscle originates at the palate, the **styloglossus** muscle at the styloid process of the temporal bone, the **genioglossus** muscle at the chin, and the **hyoglossus** muscle at the hyoid bone (**Figure 11–7**). These muscles, used in various combinations, move the tongue in the delicate and complex patterns necessary for speech, and manipulate food within the mouth in preparation for swallowing (Table 11–5).

Muscles of the Pharynx

The muscles of the pharynx (**Figure 11–8** and Table 11–6) are responsible for initiating the swallowing process. The **pharyngeal constrictor** muscles (*superior*, *middle*, and *inferior*) move materials into the esophagus by constricting

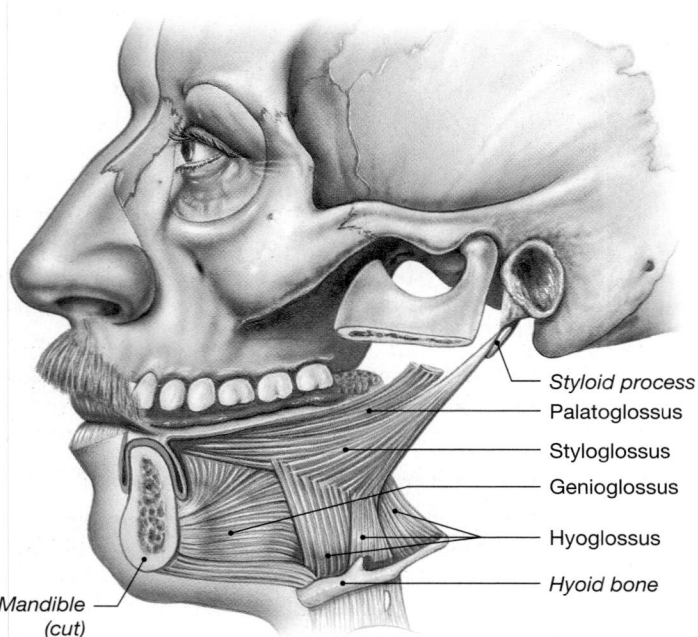

Figure 11–7 **Muscles of the Tongue.** *See also Figure 7–3.*

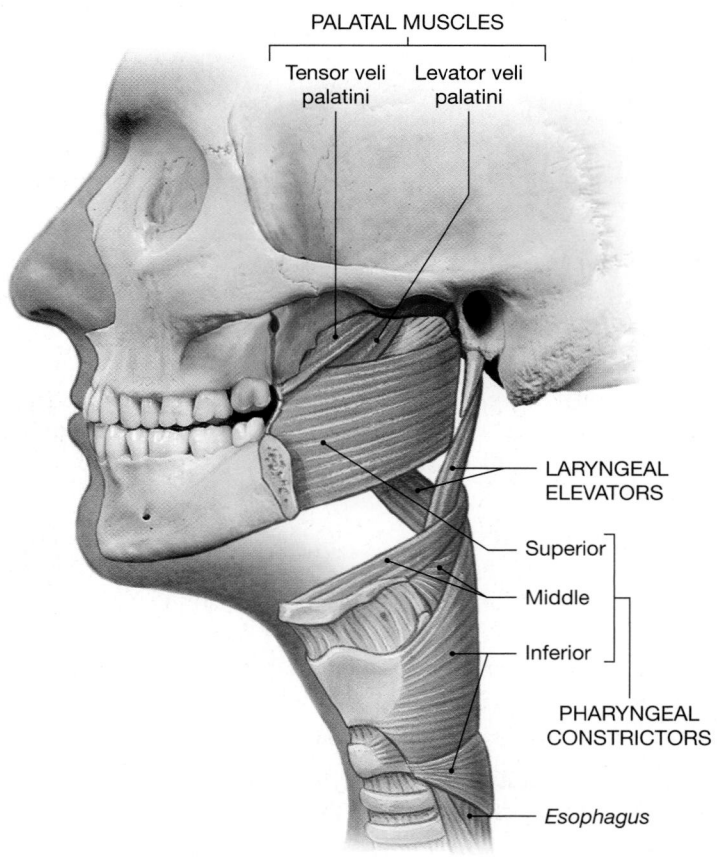

Figure 11–8 **Muscles of the Pharynx.** A lateral view. *See also Figure 7–4.*

TABLE 11–5	Muscles of the Tongue (Figure 11–7)			
Muscle	**Origin**	**Insertion**	**Action**	**Innervation**
Genioglossus	Medial surface of mandible around chin	Body of tongue, hyoid bone	Depresses and protracts tongue	Hypoglossal nerve (N XII)
Hyoglossus	Body and greater horn of hyoid bone	Side of tongue	Depresses and retracts tongue	As above
Palatoglossus	Anterior surface of soft palate	As above	Elevates tongue, depresses soft palate	Internal branch of accessory nerve (N XI)
Styloglossus	Styloid process of temporal bone	Along the side to tip and base of tongue	Retracts tongue, elevates side	Hypoglossal nerve (N XII)

TABLE 11–6 Muscles of the Pharynx (Figure 11–8)

Muscle	Origin	Insertion	Action	Innervation
PHARYNGEAL CONSTRICTORS				
Superior constrictor	Pterygoid process of sphenoid, medial surfaces of mandible	Median raphe attached to occipital bone	Constricts pharynx to propel bolus into esophagus	Branches of pharyngeal plexus (N X)
Middle constrictor	Horns of hyoid bone	Median raphe	As above	As above
Inferior constrictor	Cricoid and thyroid cartilages of larynx	As above	As above	As above
LARYNGEAL ELEVATORS*	Ranges from soft palate, to cartilage around inferior portion of auditory tube, to styloid process of temporal bone	Thyroid cartilage	Elevate larynx	Branches of pharyngeal plexus (N IX and N X)
PALATAL MUSCLES				
Levator veli palatini	Petrous part of temporal bone; tissues around the auditory tube	Soft palate	Elevates soft palate	Branches of pharyngeal plexus (N X)
Tensor veli palatini	Sphenoidal spine; tissues around the auditory tube	As above	As above	Trigeminal nerve (N V)

*Refers to the palatopharyngeus, salpingopharyngeus, and stylopharyngeus, assisted by the thyrohyoid, geniohyoid, stylohyoid, and hyoglossus muscles, discussed in Tables 11–5 and 11–7.

the pharyngeal walls. The **laryngeal elevator** muscles elevate the larynx. The two **palatal muscles**—the *tensor veli palatini* and the *levator veli palatini*—elevate the soft palate and adjacent portions of the pharyngeal wall and also pull open the entrance to the auditory tube. As a result, swallowing repeatedly can open the entrance to the auditory tube and help you adjust to pressure changes when you fly or dive.

Anterior Muscles of the Neck

The anterior muscles of the neck include (1) five muscles that control the position of the larynx, (2) muscles that depress the mandible and tense the floor of the mouth, and (3) muscles that provide a stable foundation for muscles of the tongue and pharynx (**Figure 11–9** and Table 11–7). The **digastric** (dī-GAS-trik) muscle has two bellies, as the name implies (*di-*, two + *gaster*, stomach). One belly extends from the chin to the hyoid bone; the other continues from the hyoid bone to the mastoid portion of the temporal bone. Depending on which belly contracts and whether fixator muscles are stabilizing the position of the hyoid bone, the digastric muscle can open the mouth by depressing the mandible, or it can elevate the larynx by raising the hyoid bone. The digastric muscle overlies the broad, flat **mylohyoid** muscle, which provides a muscular floor to the mouth, aided by the deeper **geniohyoid** muscles that extend between the hyoid bone and the chin. The **stylohyoid** muscle forms a muscular connection between the hyoid bone and the styloid process of the skull. The **sternocleidomastoid** (ster-nō-klī-dō-MAS-toyd) muscle extends from the clavicle and the sternum to the mastoid region of the skull (**Figures 11–4** and **11–9**). The **omohyoid** (ō-mō-HĪ-oyd) muscle attaches to the scapula, the clavicle and first rib, and the hyoid bone. The other members of this group are straplike muscles that extend between the sternum and larynx (*sternothyroid*) or hyoid bone (*sternohyoid*), and between the larynx and hyoid bone (*thyrohyoid*).

Muscles of the Vertebral Column

The muscles of the vertebral column are covered by more superficial back muscles, such as the trapezius and latissimus dorsi muscles (**Figure 11–3b**). The **erector spinae** muscles, or *spinal extensors*, include superficial and deep layers. The superficial layer can be divided into **spinalis**, **longissimus**, and **iliocostalis** groups (**Figure 11–10** and Table 11–8). In the inferior lumbar and sacral regions, the boundary between the longissimus and iliocostalis muscles is indistinct. When contracting together, the erector spinae extend the vertebral column. When the muscles on only one side contract, the result is lateral flexion of the vertebral column.

Deep to the spinalis muscles, smaller muscles interconnect and stabilize the vertebrae. These muscles include the **semispinalis** group; the **multifidus** muscle; and the **interspinales**, **intertransversarii**, and **rotatores** muscles (**Figure 11–10c**). In various combinations, they produce slight extension or rotation of the vertebral column. They are also important in making delicate adjustments in the positions of individual vertebrae, and they stabilize adjacent vertebrae. If injured, these muscles can start a cycle of pain → muscle stimulation → contraction → pain. Resultant pressure on adjacent spinal nerves can lead to sensory losses and mobility limitions. Many of the warm-up and stretching exercises recommended before athletic activity are intended to prepare these small but very important muscles for their supporting role.

The muscles of the vertebral column include many dorsal extensors, but few ventral flexors. The vertebral column does

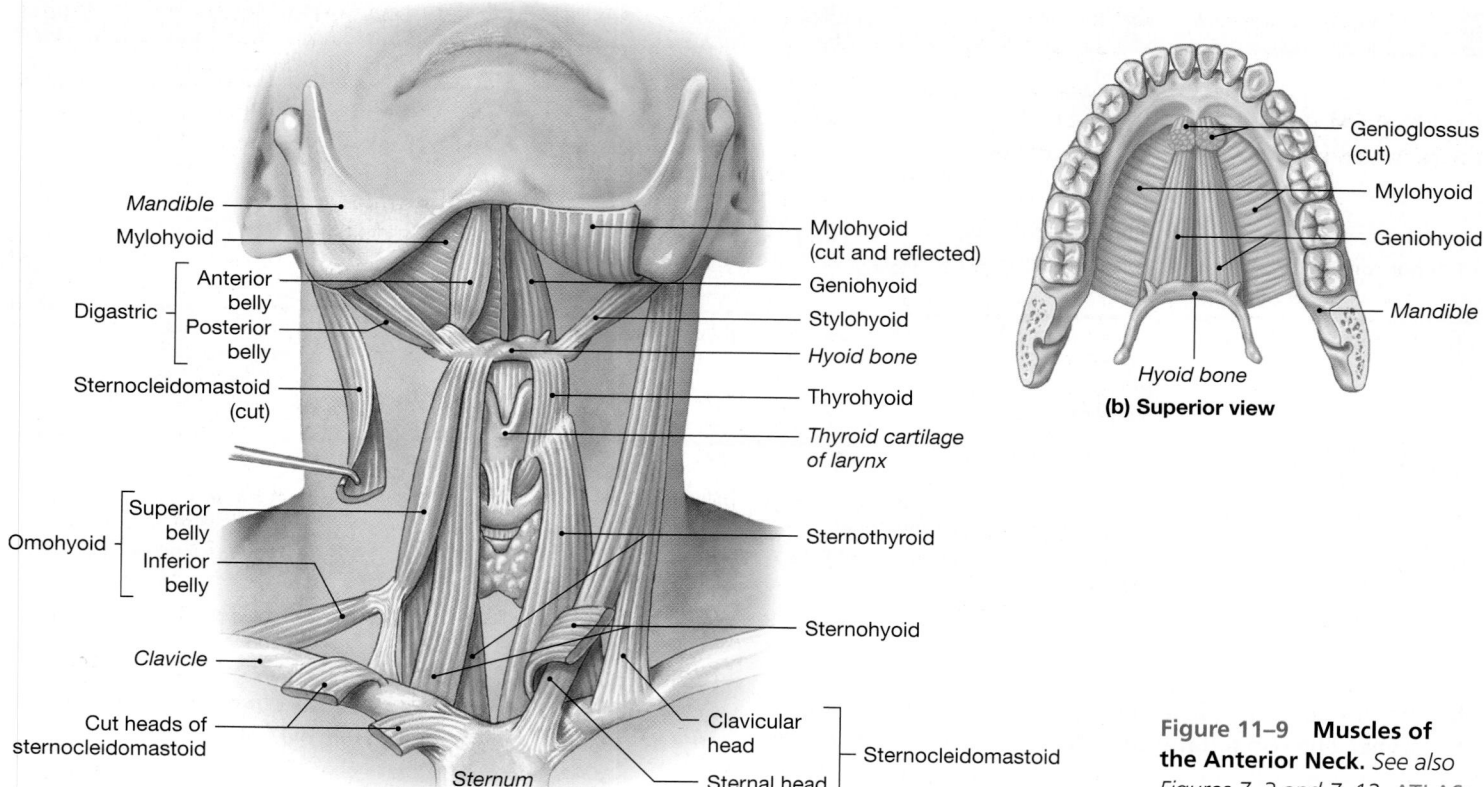

Mandible
Mylohyoid

Digastric
  Anterior belly
  Posterior belly

Sternocleidomastoid (cut)

Omohyoid
  Superior belly
  Inferior belly

Clavicle

Cut heads of sternocleidomastoid

Sternum

(a) Anterior view

Mylohyoid (cut and reflected)
Geniohyoid
Stylohyoid
Hyoid bone
Thyrohyoid
Thyroid cartilage of larynx

Sternothyroid

Sternohyoid

Clavicular head
Sternal head
Sternocleidomastoid

Genioglossus (cut)
Mylohyoid
Geniohyoid
Mandible

Hyoid bone
(b) Superior view

Figure 11–9 **Muscles of the Anterior Neck.** *See also Figures 7–3 and 7–12.* ATLAS: Plates 3a–d; 17; 18a–c; 25

TABLE 11–7 Anterior Muscles of the Neck (Figure 11–9)

Muscle	Origin	Insertion	Action	Innervation
Digastric	Two bellies: *anterior* from inferior surface of mandible at chin; *posterior* from mastoid region of temporal bone	Hyoid bone	Depresses mandible or elevates larynx	*Anterior belly*: Trigeminal nerve (N V), mandibular branch *Posterior belly*: Facial nerve (N VII)
Geniohyoid	Medial surface of mandible at chin	Hyoid bone	As above and pulls hyoid bone anteriorly	Cervical nerve C_1 via hypoglossal nerve (N XII)
Mylohyoid	Mylohyoid line of mandible	Median connective tissue band (raphe) that runs to hyoid bone	Elevates floor of mouth and hyoid bone or depresses mandible	Trigeminal nerve (N V), mandibular branch
Omohyoid (superior and inferior bellies united at central tendon anchored to clavicle and first rib)	Superior border of scapula near scapular notch	Hyoid bone	Depresses hyoid bone and larynx	Cervical spinal nerves C_2–C_3
Sternohyoid	Clavicle and manubrium	Hyoid bone	As above	Cervical spinal nerves C_1–C_3
Sternothyroid	Dorsal surface of manubrium and first costal cartilage	Thyroid cartilage of larynx	As above	As above
Stylohyoid	Styloid process of temporal bone	Hyoid bone	Elevates larynx	Facial nerve (N VII)
Thyrohyoid	Thyroid cartilage of larynx	Hyoid bone	Elevates thyroid, depresses hyoid bone	Cervical spinal nerves C_1–C_2 via hypoglossal nerve (N XII)
Sternocleido-mastoid	Two bellies: *clavicular head* attaches to sternal end of clavicle; *sternal head* attaches to manubrium	Mastoid region of skull and lateral portion of superior nuchal line	Together, they flex the neck; alone, one side bends head toward shoulder and turns face to opposite side	Accessory nerve (N XI) and cervical spinal nerves (C_2–C_3) of cervical plexus

not need a massive series of flexor muscles, because (1) many of the large trunk muscles flex the vertebral column when they contract, and (2) most of the body weight lies anterior to the vertebral column, so gravity tends to flex the spine. However, a few spinal flexors are associated with the anterior surface of the vertebral column. In the neck, the **longus capitis** and the **longus colli** muscles rotate or flex the neck, depending on whether the muscles of one or both sides are contracting (**Figure 11–10b**). In the lumbar region, the large **quadratus lumborum** muscles flex the vertebral column and depress the ribs (**Figure 11–10a**).

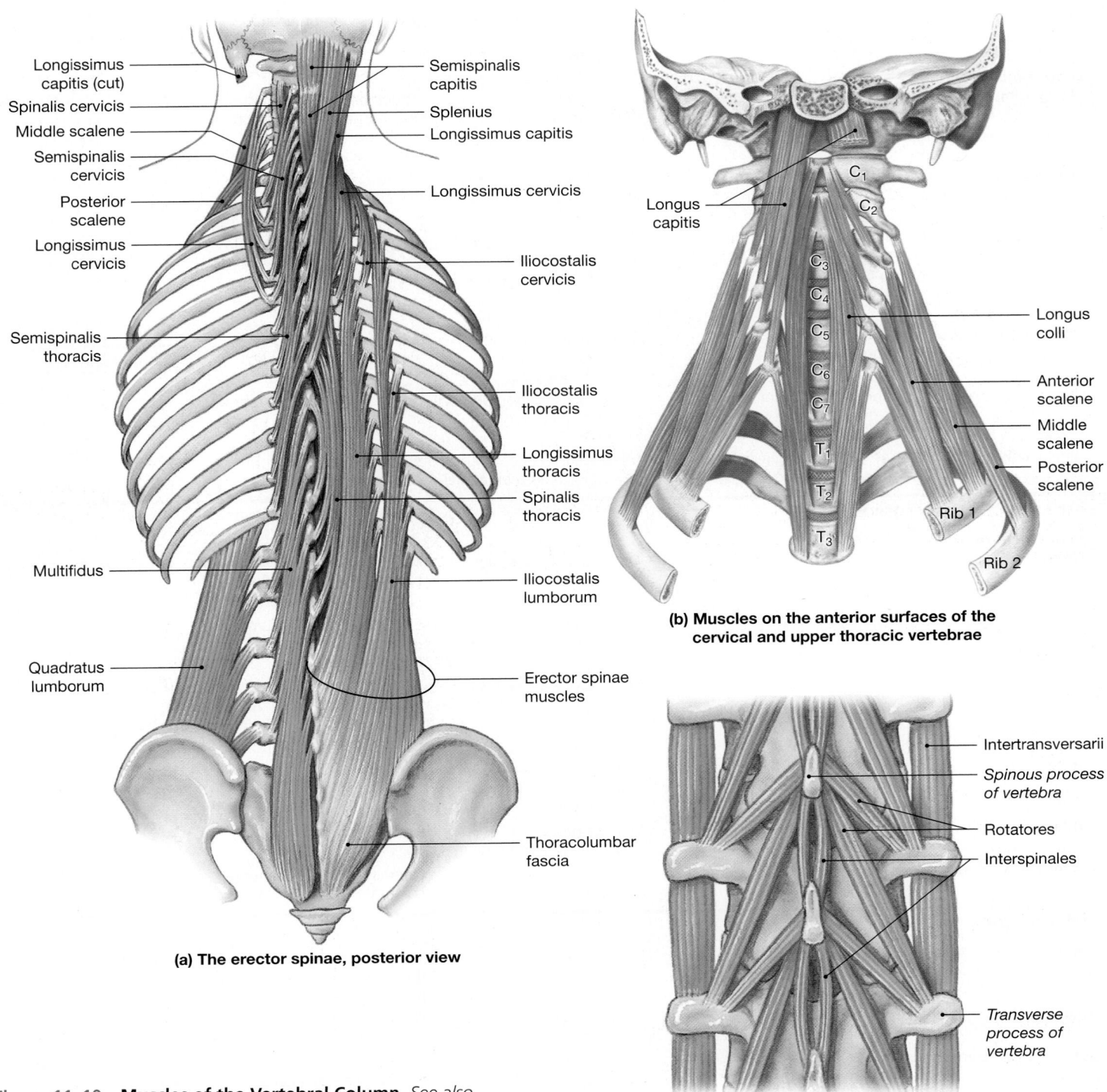

(a) The erector spinae, posterior view

(b) Muscles on the anterior surfaces of the cervical and upper thoracic vertebrae

(c) Intervertebral muscles, posterior view

Figure 11–10 **Muscles of the Vertebral Column.** *See also Figures 7–1a and 7–22.*

TABLE 11–8 Muscles of the Vertebral Column (Figure 11–10)

Group and Muscle(s)	Origin	Insertion	Action	Innervation
SUPERFICIAL LAYER				
Splenius (Splenius capitis, splenius cervicis)	Spinous processes and ligaments connecting inferior cervical and superior thoracic vertebrae	Mastoid process, occipital bone of skull, and superior cervical vertebrae	Together, the two sides extend neck; alone, each rotates and laterally flexes neck to that side	Cervical spinal nerves
Erector spinae *Spinalis group*				
Spinalis cervicis	Inferior portion of ligamentum nuchae and spinous process of C_7	Spinous process of axis	Extends neck	As above
Spinalis thoracis	Spinous processes of inferior thoracic and superior lumbar vertebrae	Spinous processes of superior thoracic vertebrae	Extends vertebral column	Thoracic and lumbar spinal nerves
Longissimus group				
Longissimus capitis	Transverse processes of inferior cervical and superior thoracic vertebrae	Mastoid process of temporal bone	Together, the two sides extend head; alone, each rotates and laterally flexes neck to that side	Cervical and thoracic spinal nerves
Longissimus cervicis	Transverse processes of superior thoracic vertebrae	Transverse processes of middle and superior cervical vertebrae	As above	As above
Longissimus thoracis	Broad aponeurosis and transverse processes of inferior thoracic and superior lumbar vertebrae; joins iliocostalis	Transverse processes of superior vertebrae and inferior surfaces of ribs	Extends vertebral column; alone, each produces lateral flexion to that side	Thoracic and lumbar spinal nerves
Iliocostalis group				
Iliocostalis cervicis	Superior borders of vertebrosternal ribs near the angles	Transverse processes of middle and inferior cervical vertebrae	Extends or laterally flexes neck, elevates ribs	Cervical and superior thoracic spinal nerves
Iliocostalis thoracis	Superior borders of inferior seven ribs medial to the angles	Upper ribs and transverse process of last cervical vertebra	Stabilizes thoracic vertebrae in extension	Thoracic spinal nerves
Iliocostalis lumborum	Iliac crest, sacral crests, and spinous processes	Inferior surfaces of inferior seven ribs near their angles	Extends vertebral column, depresses ribs	Inferior thoracic and lumbar spinal nerves
DEEP LAYER (TRANSVERSOSPINALIS)				
Semispinalis group				
Semispinalis capitis	Articular processes of inferior cervical and transverse processes of superior thoracic vertebrae	Occipital bone, between nuchal lines	Together, the two sides extend head; alone, each extends and laterally flexes neck	Cervical spinal nerves
Semispinalis cervicis	Transverse processes of T_1–T_5 or T_6	Spinous processes of C_2–C_5	Extends vertebral column and rotates toward opposite side	As above
Semispinalis thoracis	Transverse processes of T_6–T_{10}	Spinous processes of C_5–T_4	As above	Thoracic spinal nerves
Multifidus	Sacrum and transverse processes of each vertebra	Spinous processes of the third or fourth more superior vertebrae	As above	Cervical, thoracic, and lumbar spinal nerves
Rotatores	Transverse processes of each vertebra	Spinous processes of adjacent, more superior vertebra	As above	As above
Interspinales	Spinous processes of each vertebra	Spinous processes of more superior vertebra	Extends vertebral column	As above
Intertransversarii	Transverse processes of each vertebra	Transverse process of more superior vertebra	Laterally flexes the vertebral column	As above
SPINAL FLEXORS				
Longus capitis	Transverse processes of cervical vertebrae	Base of the occipital bone	Together, the two sides flex the neck; alone, each rotates head to that side	Cervical spinal nerves
Longus colli	Anterior surfaces of cervical and superior thoracic vertebrae	Transverse processes of superior cervical vertebrae	Flexes or rotates neck; limits hyperextension	As above
Quadratus lumborum	Iliac crest and iliolumbar ligament	Last rib and transverse processes of lumbar vertebrae	Together, they depress ribs; alone, each side laterally flexes vertebral column	Thoracic and lumbar spinal nerves

Oblique and Rectus Muscles

The oblique and rectus muscles lie within the body wall, between the spinous processes of vertebrae and the ventral midline (**Figures 11–3** and **11–11** and Table 11–9). The oblique muscles compress underlying structures or rotate the vertebral column, depending on whether one or both sides contract. The rectus muscles are important flexors of the vertebral column, acting in opposition to the erector spinae. The oblique and rectus muscles share embryological origins; we can divide these groups into cervical, thoracic, and abdominal regions.

The oblique group includes the **scalene** muscles of the neck (**Figure 11–10b**) and the **intercostal** and **transversus** muscles of the thorax (**Figure 11–11a,b**). The scalene muscles (*anterior*, *middle*, and *posterior*) elevate the first two ribs and assist in flexion of the neck. In the thorax, the oblique muscles extend between the ribs, with the **external intercostal** muscles covering the **internal intercostal** muscles. Both groups of intercostal muscles aid in respiratory movements of the ribs. A small **transversus thoracis** muscle crosses the inner surface of the rib cage and is separated from the pleural cavity by the parietal pleura, a *serous membrane.* p. 136 The sternum occupies the

11 MUSCULAR

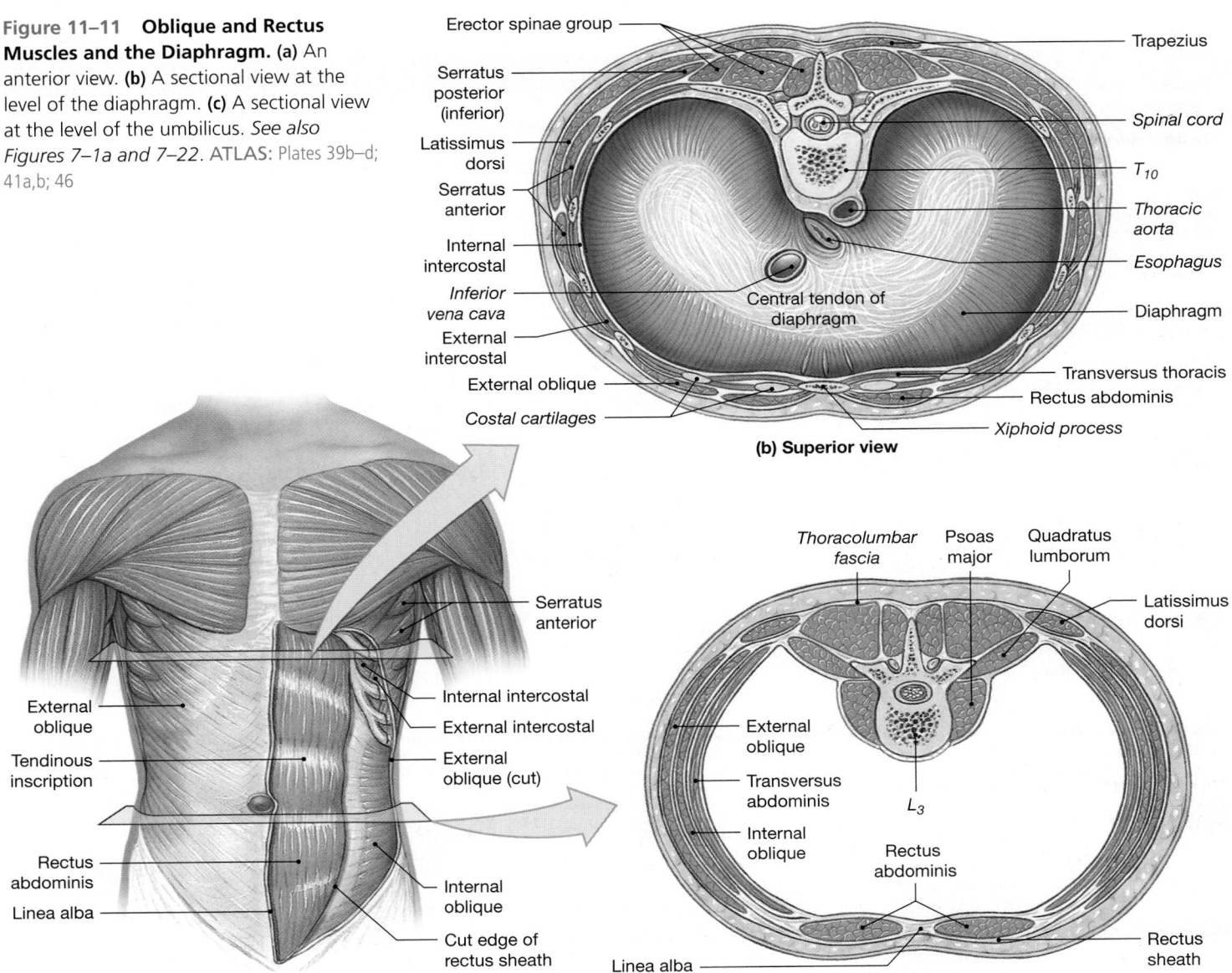

Figure 11–11 Oblique and Rectus Muscles and the Diaphragm. (a) An anterior view. **(b)** A sectional view at the level of the diaphragm. **(c)** A sectional view at the level of the umbilicus. *See also Figures 7–1a and 7–22.* ATLAS: Plates 39b–d; 41a,b; 46

(b) Superior view

Erector spinae group
Serratus posterior (inferior)
Latissimus dorsi
Serratus anterior
Internal intercostal
Inferior vena cava
External intercostal
External oblique
Costal cartilages
Trapezius
Spinal cord
T₁₀
Thoracic aorta
Esophagus
Diaphragm
Transversus thoracis
Rectus abdominis
Xiphoid process
Central tendon of diaphragm

(a) Anterior view

Serratus anterior
Internal intercostal
External intercostal
External oblique (cut)
Internal oblique
Cut edge of rectus sheath
External oblique
Tendinous inscription
Rectus abdominis
Linea alba

(c) Horizontal section view

Thoracolumbar fascia   Psoas major   Quadratus lumborum
Latissimus dorsi
External oblique
Transversus abdominis
Internal oblique
L₃
Rectus abdominis
Rectus sheath
Linea alba

place where we might otherwise expect thoracic rectus muscles to be.

The same basic pattern of musculature extends unbroken across the abdominopelvic surface (**Figure 11–11a,c**). Here, the muscles are called the **external oblique**, **internal oblique**, **transversus abdominis**, and **rectus abdominis** muscles (the "abs"). The rectus abdominis muscle inserts at the xiphoid process and originates near the pubic symphysis. This muscle is longitudinally divided by the **linea alba** (white line), a median collagenous partition (**Figure 11–3a**). The rectus abdominis muscle is separated into segments by transverse bands of collagen fibers called **tendinous inscriptions**. Each segment contains muscle fibers that extend longitudinally, originating and inserting on the tendinous inscriptions. Due to the bulging of enlarged muscle fibers between the tendinous inscriptions, bodybuilders often refer to the rectus abdominis as the "six-pack."

The Diaphragm

The term *diaphragm* refers to any muscular sheet that forms a wall. When used without a modifier, however, **diaphragm**, or *diaphragmatic muscle*, specifies the muscular partition that separates the abdominopelvic and thoracic cavities (**Figure 11–11b**). We include this muscle here because it develops in association with the other muscles of the chest wall. The diaphragm is a major respiratory muscle.

Muscles of the Pelvic Floor

The muscles of the pelvic floor (**Figure 11–12** and Table 11–10) extend from the sacrum and coccyx to the ischium and pubis. These muscles (1) support the organs of the pelvic cavity, (2) flex the sacrum and coccyx, and (3) control the movement of materials through the urethra and anus.

TABLE 11–9 Oblique and Rectus Muscles (Figure 11–11)

Group and Muscle(s)	Origin	Insertion	Action	Innervation*
OBLIQUE GROUP				
Cervical region **Scalenes (anterior, middle, and posterior)**	Transverse and costal processes of cervical vertebrae	Superior surfaces of first two ribs	Elevate ribs or flex neck	Cervical spinal nerves
Thoracic region **External intercostals**	Inferior border of each rib	Superior border of more inferior rib	Elevate ribs	Intercostal nerves (branches of thoracic spinal nerves)
Internal intercostals	Superior border of each rib	Inferior border of the preceding rib	Depress ribs	As above
Transversus thoracis	Posterior surface of sternum	Cartilages of ribs	As above	As above
Serratus posterior superior (Figure 11–13a)	Spinous processes of C_7–T_3 and ligamentum nuchae	Superior borders of ribs 2–5 near angles	Elevates ribs, enlarges thoracic cavity	Thoracic nerves (T_1–T_4)
inferior	Aponeurosis from spinous processes of T_{10}–L_3	Inferior borders of ribs 8–12	Pulls ribs inferiorly; also pulls outward, opposing diaphragm	Thoracic nerves (T_9–T_{12})
Abdominal region **External oblique**	External and inferior borders of ribs 5–12	Linea alba and iliac crest	Compresses abdomen, depresses ribs, flexes or bends spine	Intercostal, iliohypogastric, and ilioinguinal nerves
Internal oblique	Thoracolumbar fascia and iliac crest	Inferior ribs, xiphoid process, and linea alba	As above	As above
Transversus abdominis	Cartilages of ribs 6–12, iliac crest, and thoracolumbar fascia	Linea alba and pubis	Compresses abdomen	As above
RECTUS GROUP				
Cervical region	See muscles in Table 11–6			
Thoracic region **Diaphragm**	Xiphoid process, cartilages of ribs 4–10, and anterior surfaces of lumbar vertebrae	Central tendinous sheet	Contraction expands thoracic cavity, compresses abdominopelvic cavity	Phrenic nerves (C_3–C_5)
Abdominal region **Rectus abdominis**	Superior surface of pubis around symphysis	Inferior surfaces of costal cartilages (ribs 5–7) and xiphoid process	Depresses ribs, flexes vertebral column, compresses abdomen	Intercostal nerves (T_7–T_{12})

*Where appropriate, spinal nerves involved are given in parentheses.

11 MUSCULAR

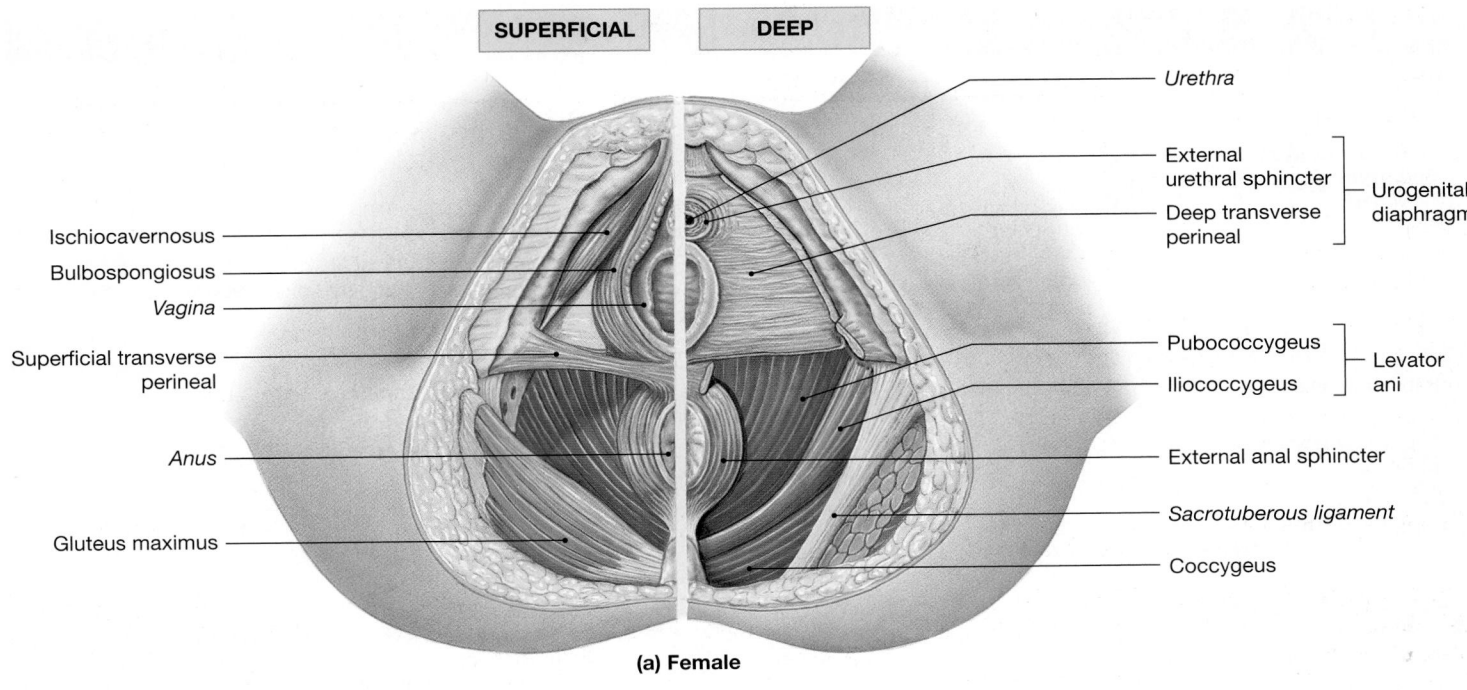

(a) Female

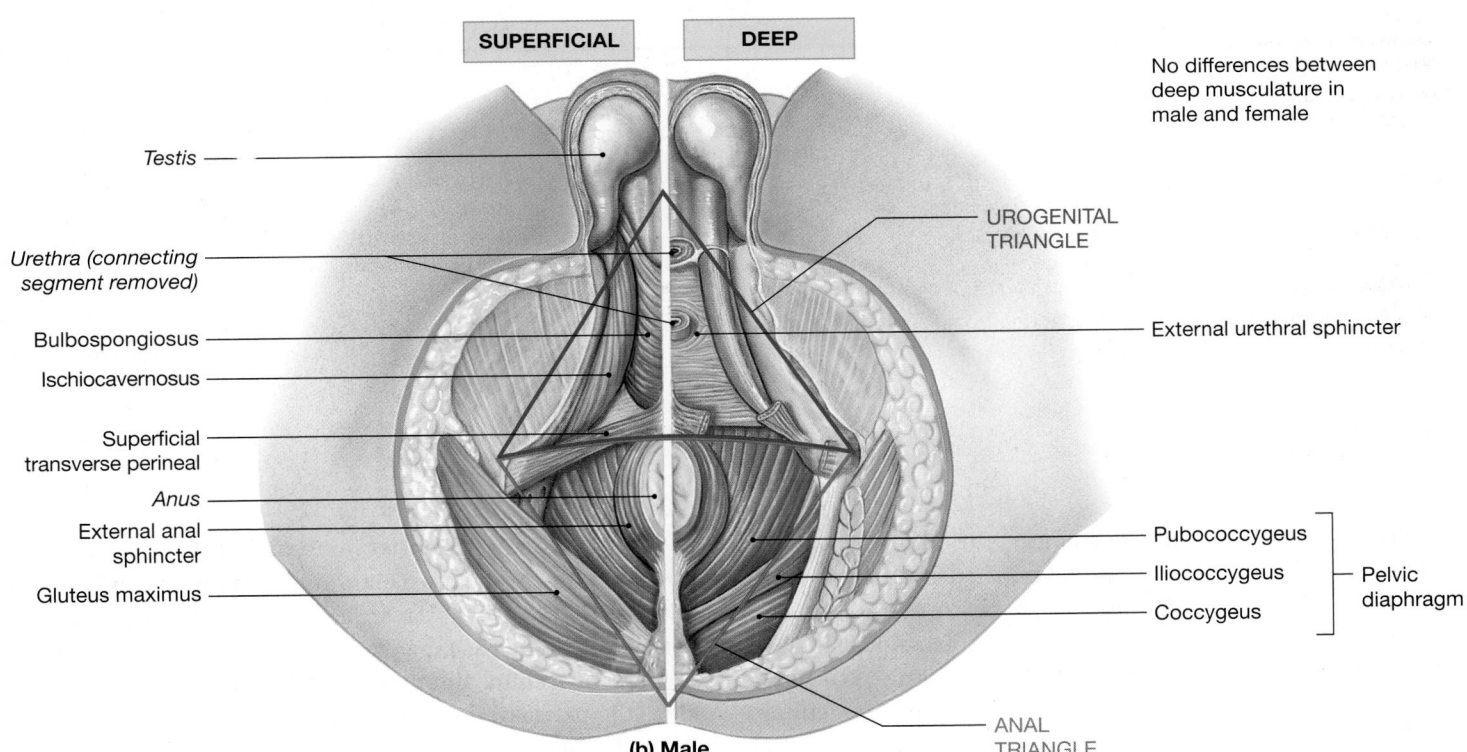

(b) Male

Figure 11–12 **Muscles of the Pelvic Floor.** *See also Figures 7–1a, 8–8, and 8–9.*

TABLE 11–10 Muscles of the Pelvic Floor (Figure 11–12)

Group and Muscle(s)	Origin	Insertion	Action	Innervation*
UROGENITAL TRIANGLE				
Superficial muscles				
Bulbospongiosus				
Males	Collagen sheath at base of penis; fibers cross over urethra	Median raphe and central tendon of perineum	Compresses base and stiffens penis; ejects urine or semen	Pudendal nerve, perineal branch (S_2–S_4)
Females	Collagen sheath at base of clitoris; fibers run on either side of urethral and vaginal opening	Central tendon of perineum	Compresses and stiffens clitoris; narrows vaginal opening	As above
Ischiocavernosus	Ischial ramus and tuberosity	Pubic symphysis anterior to base of penis or clitoris	Compresses and stiffens penis or clitoris	As above
Superficial transverse perineal	Ischial ramus	Central tendon of perineum	Stabilizes central tendon of perineum	As above
Deep muscles				
Urogenital diaphragm				
Deep transverse perineal	Ischial ramus	Median raphe of urogenital diaphragm	As above	As above
External urethral sphincter:				
Males	Ischial and pubic rami	To median raphe at base of penis; inner fibers encircle urethra	Closes urethra; compresses prostate and bulbo-urethral glands	As above
Females	Ischial and pubic rami	To median raphe; inner fibers encircle urethra	Closes urethra; compresses vagina and greater vestibular glands	As above
ANAL TRIANGLE				
Pelvic diaphragm				
Coccygeus	Ischial spine	Lateral, inferior borders of sacrum and coccyx	Flexes coccygeal joints; tenses and supports pelvic floor	Inferior sacral nerves (S_4–S_5)
Levator ani				
Iliococcygeus	Ischial spine, pubis	Coccyx and median raphe	Tenses floor of pelvis; flexes coccygeal joints; elevates and retracts anus	Pudendal nerve (S_2–S_4)
Pubococcygeus	Inner margins of pubis	As above	As above	As above
External anal sphincter	Via tendon from coccyx	Encircles anal opening	Closes anal opening	Pudendal nerve, hemorrhoidal branch (S_2–S_4)

*Where appropriate, spinal nerves involved are given in parentheses.

The boundaries of the **perineum**, the muscular sheet that forms the pelvic floor, are established by the inferior margins of the pelvis. A line drawn between the ischial tuberosities divides the perineum into two triangles: an anterior **urogenital triangle** and a posterior **anal triangle** (**Figure 11–12b**). The superficial muscles of the urogenital triangle are the muscles of the external genitalia. They cover deeper muscles that strengthen the pelvic floor and encircle the urethra. These muscles constitute the **urogenital diaphragm** (**Figure 11–12a**), a deep muscular layer that extends between the pubic bones.

An even more extensive muscular sheet, the **pelvic diaphragm**, forms the muscular foundation of the anal triangle (**Figure 11–12b**). This layer, covered by the urogenital diaphragm, extends as far as the pubic symphysis.

The urogenital and pelvic diaphragms do not completely close the pelvic outlet, for the urethra, vagina, and anus pass through them to open on the external surface. Muscular sphincters surround the passageways, and the external sphincters permit voluntary control of urination and defecation. Muscles, nerves, and blood vessels also pass through the pelvic outlet as they travel to or from the lower limbs.

CLINICAL NOTE

Hernia

When your abdominal muscles contract forcefully, the pressure in your abdominopelvic cavity can skyrocket, and that pressure is applied to internal organs. If you exhale at the same time, the pressure is relieved, because your diaphragm can move upward as your lungs deflate. But during vigorous isometric exercises or when you lift a heavy weight while holding your breath, pressure in the abdominopelvic cavity can rise high enough to cause a variety of problems, among them a hernia.

A **hernia** develops whenever an organ protrudes through an abnormal opening. The most common hernias are inguinal and diaphragmatic hernias. **Inguinal hernias** typically occur in males, at the *inguinal canal*, the site where blood vessels, nerves, and reproductive ducts pass through the abdominal wall to reach the testes. Elevated abdominal pressure can force open the inguinal canal and push a portion of the intestine into the pocket created. **Diaphragmatic hernias** develop when visceral organs, such as a portion of the stomach, are forced into the left pleural cavity. If herniated structures become trapped or twisted, surgery may be required to prevent serious complications, such as intestinal blockage or tissue necrosis due to the interruption of blood flow.

CHECKPOINT

11. Describe the location of axial muscles.

12. If you were contracting and relaxing your masseter muscle, what would you probably be doing?

13. Which facial muscle would you expect to be well developed in a trumpet player?

14. Why can swallowing help alleviate the pressure sensations at the eardrum when you are in an airplane that is changing altitude?

15. Damage to the external intercostal muscles would interfere with what important process?

16. If someone hit you in your rectus abdominis muscle, how would your body position change?

17. After spending an afternoon carrying heavy boxes from his basement to his attic, Joe complains that the muscles in his back hurt. Which muscle(s) is (are) most likely sore?

See the blue Answers tab at the end of the book.

11-6 Appendicular muscles are muscles of the shoulders, upper limbs, pelvic girdle, and lower limbs

The appendicular musculature positions and stabilizes the pectoral and pelvic girdles and moves the upper and lower limbs. There are two major groups of appendicular muscles: (1) *the muscles of the shoulders and upper limbs* and (2) *the*

muscles of the pelvis and lower limbs. The functions and required ranges of motion are very different between these groups. In addition to increasing the mobility of the arms, the muscular connections between the pectoral girdle and the axial skeleton must act as shock absorbers. For example, while you jog, you can still perform delicate hand movements, because the muscular connections between the axial and appendicular components of the skeleton smooth out the bounces in your stride. In contrast, the pelvic girdle has evolved to transfer weight from the axial to the appendicular skeleton. Rigid, bony articulations are essential, because the emphasis is on strength rather than versatility, and a muscular connection would reduce the efficiency of the transfer. **Figure 11–13** provides an introduction to the organization of the appendicular muscles of the trunk. The larger appendicular muscles are often used as sites for drug injection and vaccination delivery.

Muscles of the Shoulders and Upper Limbs

Muscles associated with the shoulders and upper limbs can be divided into four groups: (1) *muscles that position the pectoral girdle*, (2) *muscles that move the arm*, (3) *muscles that move the forearm and hand*, and (4) *muscles that move the hand and fingers*.

Muscles That Position the Pectoral Girdle

The large, superficial **trapezius** muscles, commonly called the "traps," cover the back and portions of the neck, reaching to the base of the skull. These muscles originate along the midline of the neck and back and insert on the clavicles and the scapular spines (**Figures 11–13** and **11–14a**). The trapezius muscles are innervated by more than one nerve (Table 11–11), and specific regions can be made to contract independently. As a result, their actions are quite varied.

Removing the trapezius muscle reveals the **rhomboid** and **levator scapulae** muscles (**Figure 11–14a**). These muscles are attached to the dorsal surfaces of the cervical and thoracic vertebrae. They insert along the vertebral border of each scapula, between the superior and inferior angles. Contraction of a rhom-

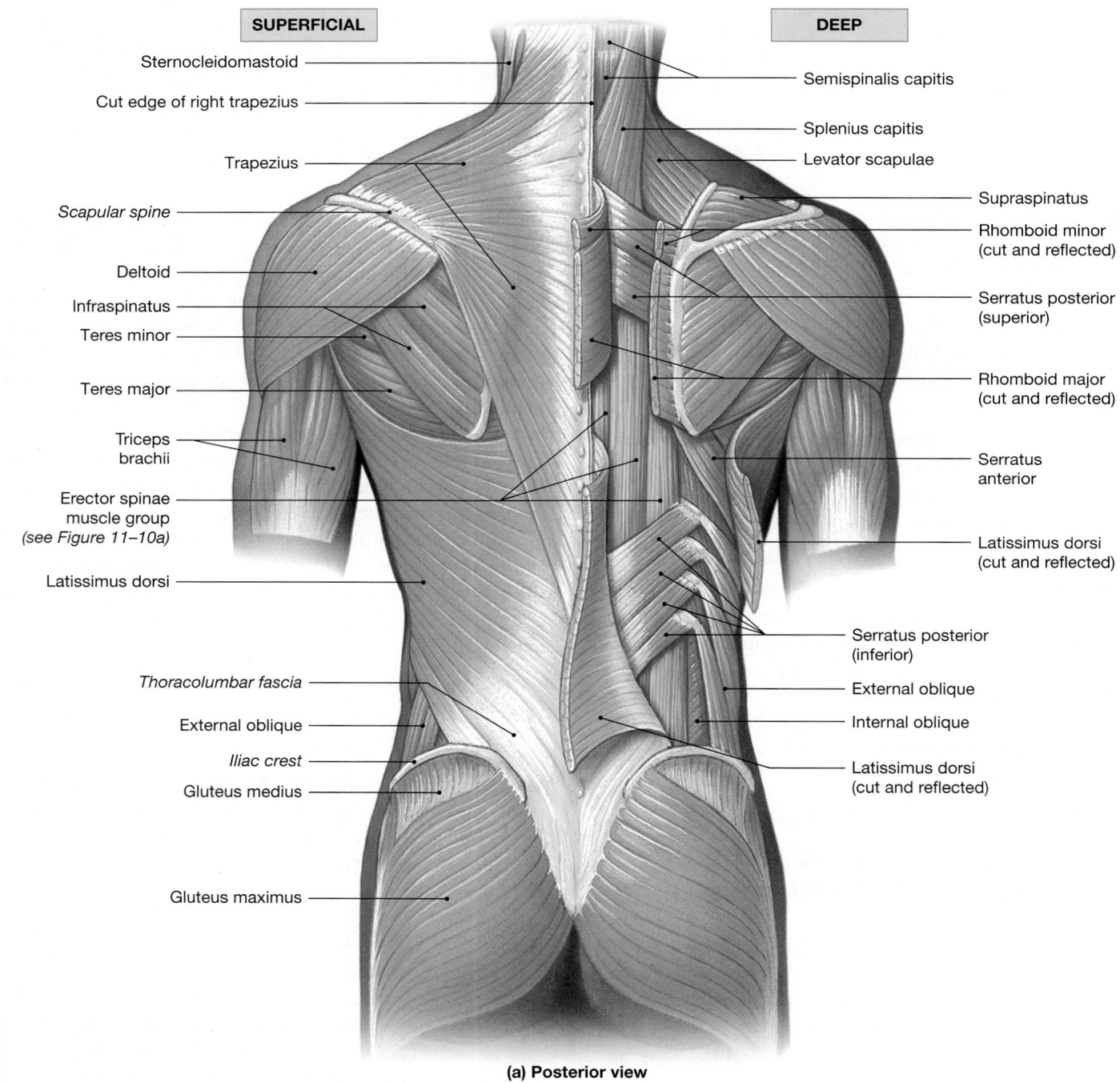

SUPERFICIAL

DEEP

Sternocleidomastoid

Cut edge of right trapezius

Trapezius

Scapular spine

Deltoid

Infraspinatus

Teres minor

Teres major

Triceps brachii

Erector spinae muscle group *(see Figure 11–10a)*

Latissimus dorsi

Thoracolumbar fascia

External oblique

Iliac crest

Gluteus medius

Gluteus maximus

Semispinalis capitis

Splenius capitis

Levator scapulae

Supraspinatus

Rhomboid minor (cut and reflected)

Serratus posterior (superior)

Rhomboid major (cut and reflected)

Serratus anterior

Latissimus dorsi (cut and reflected)

Serratus posterior (inferior)

External oblique

Internal oblique

Latissimus dorsi (cut and reflected)

(a) Posterior view

Figure 11–13   An Overview of the Appendicular Muscles of the Trunk. ATLAS: Plate 40a,b

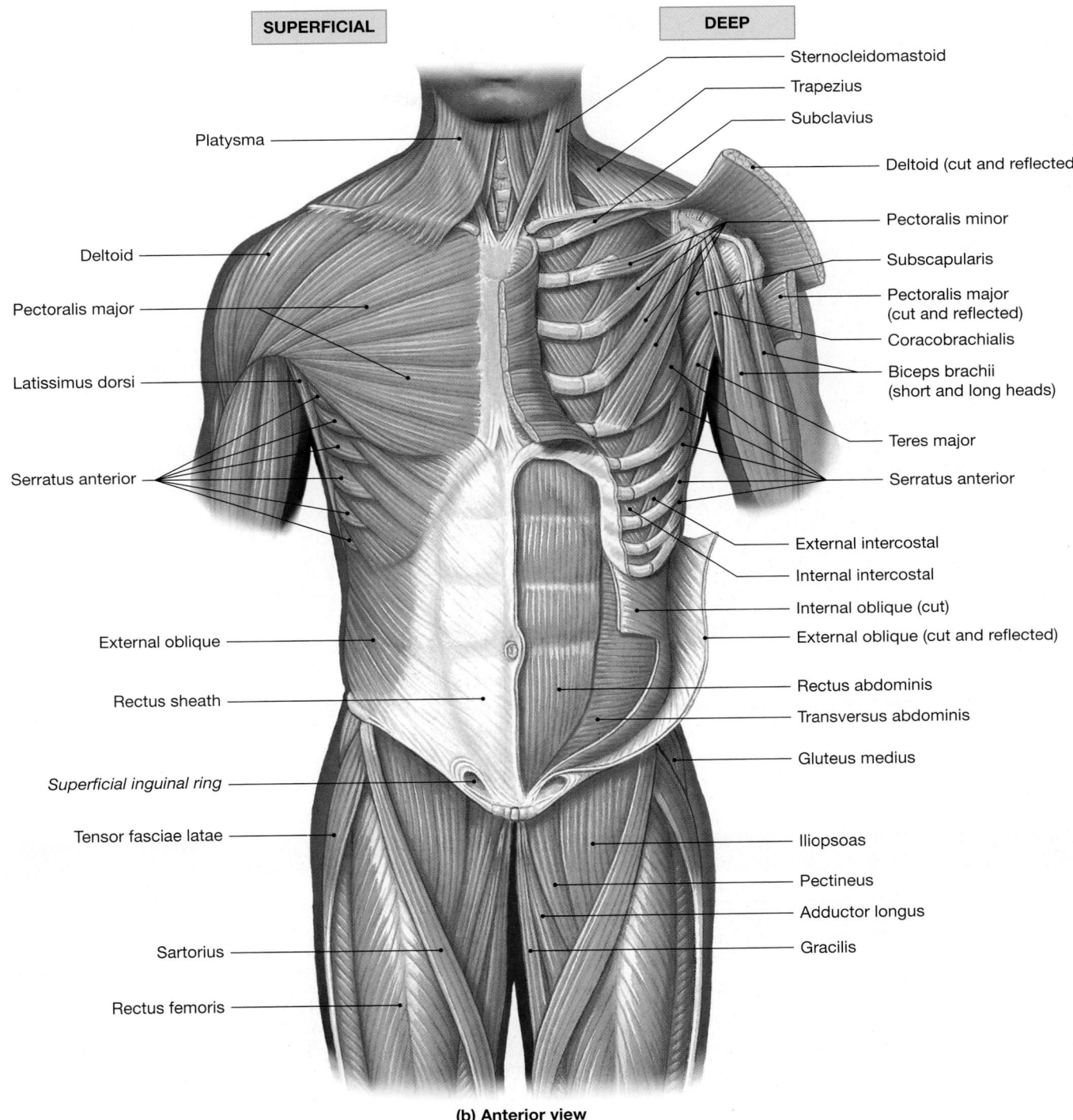

SUPERFICIAL   DEEP

Platysma

Deltoid

Pectoralis major

Latissimus dorsi

Serratus anterior

External oblique

Rectus sheath

Superficial inguinal ring

Tensor fasciae latae

Sartorius

Rectus femoris

Sternocleidomastoid
Trapezius
Subclavius
Deltoid (cut and reflected)
Pectoralis minor
Subscapularis
Pectoralis major
(cut and reflected)
Coracobrachialis
Biceps brachii
(short and long heads)
Teres major
Serratus anterior
External intercostal
Internal intercostal
Internal oblique (cut)
External oblique (cut and reflected)
Rectus abdominis
Transversus abdominis
Gluteus medius
Iliopsoas
Pectineus
Adductor longus
Gracilis

11 MUSCULAR

(b) Anterior view

Figure 11–13   An Overview of the Appendicular Muscles of the Trunk. *(continued)* ATLAS: Plates 25; 39b

boid muscle adducts (retracts) the scapula on that side. The levator scapulae muscle, as its name implies, elevates the scapula.

On the chest, the **serratus anterior** muscle originates along the anterior surfaces of several ribs (**Figures 11–3** and

11–14a,b). This fan-shaped muscle inserts along the anterior margin of the vertebral border of the scapula. When the serratus anterior muscle contracts, it abducts (protracts) the scapula and swings the shoulder anteriorly.

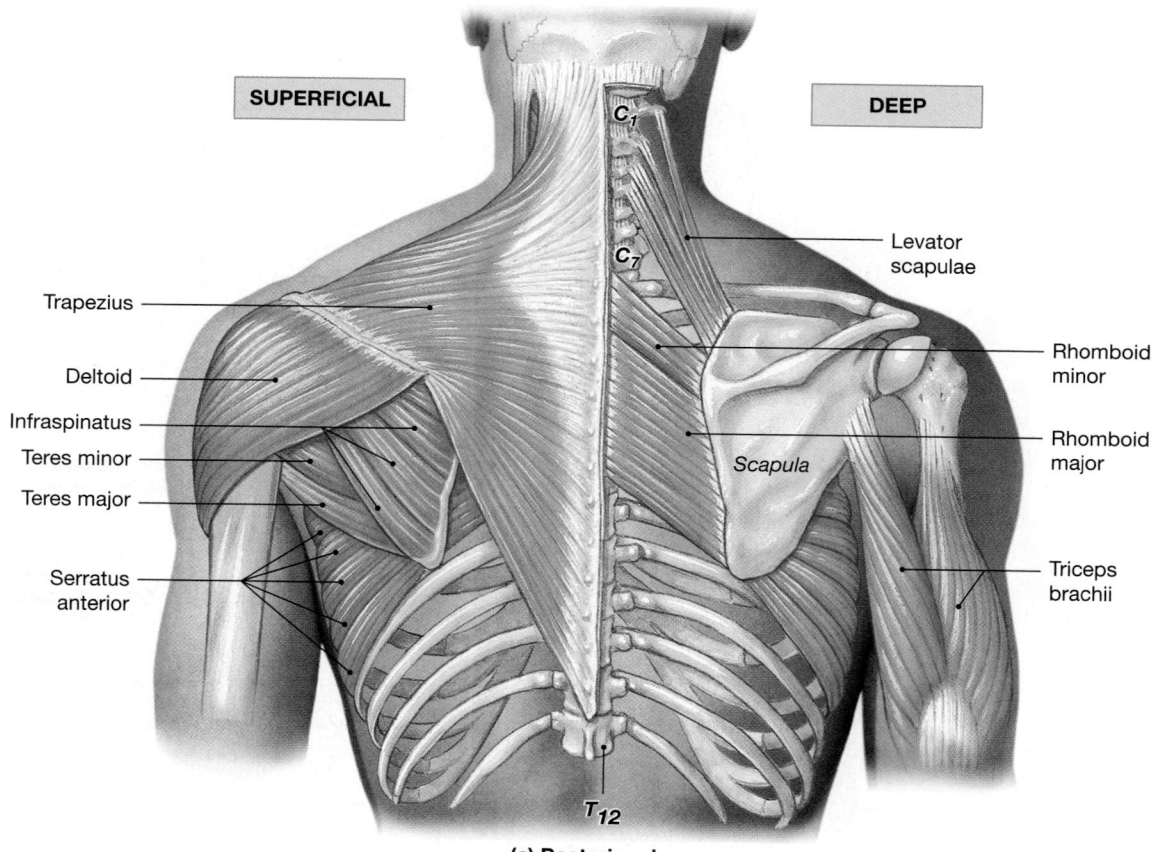

SUPERFICIAL

DEEP

C_1

C_7

T_{12}

Trapezius

Deltoid

Infraspinatus

Teres minor

Teres major

Serratus anterior

Levator scapulae

Rhomboid minor

Rhomboid major

Scapula

Triceps brachii

(a) Posterior view

Figure 11–14 **Muscles That Position the Pectoral Girdle.** *See also Figures 8–2, 8–3, 8–4, and 9–9.* ATLAS: Plates 27b; 40a–b

	TABLE 11–11 Muscles That Position the Pectoral Girdle (Figures 11–13, 11–14)			
Muscle	**Origin**	**Insertion**	**Action**	**Innervation***
Levator scapulae	Transverse processes of first four cervical vertebrae	Vertebral border of scapula near superior angle	Elevates scapula	Cervical nerves C_3–C_4 and dorsal scapular nerve (C_5)
Pectoralis minor	Anterior-superior surfaces of ribs 3–5	Coracoid process of scapula	Depresses and protracts shoulder; rotates scapula so glenoid cavity moves inferiorly (downward rotation); elevates ribs if scapula is stationary	Medial pectoral nerve(C_8, T_1)
Rhomboid major	Spinous processes of superior thoracic vertebrae	Vertebral border of scapula from spine to inferior angle	Adducts scapula and performs downward rotation	Dorsal scapular nerve (C_5)
Rhomboid minor	Spinous processes of vertebrae C_7–T_1	Vertebral border of scapula near spine	As above	As above
Serratus anterior	Anterior and superior margins of ribs 1–8 or 1–9	Anterior surface of vertebral border of scapula	Protracts shoulder; rotates scapula so glenoid cavity moves superiorly (upward rotation)	Long thoracic nerve (C_5–C_7)
Subclavius	First rib	Clavicle (inferior border)	Depresses and protracts shoulder	Nerve to subclavius (C_5–C_6)
Trapezius	Occipital bone, ligamentum nuchae, and spinous processes of thoracic vertebrae	Clavicle and scapula (acromion and scapular spine)	Depends on active region and state of other muscles; may (1) elevate, retract, depress, or rotate scapula upward, (2) elevate clavicle, or (3) extend neck	Accessory nerve (N XI) and cervical spinal nerves (C_3–C_4)

*Where appropriate, spinal nerves involved are given in parentheses.

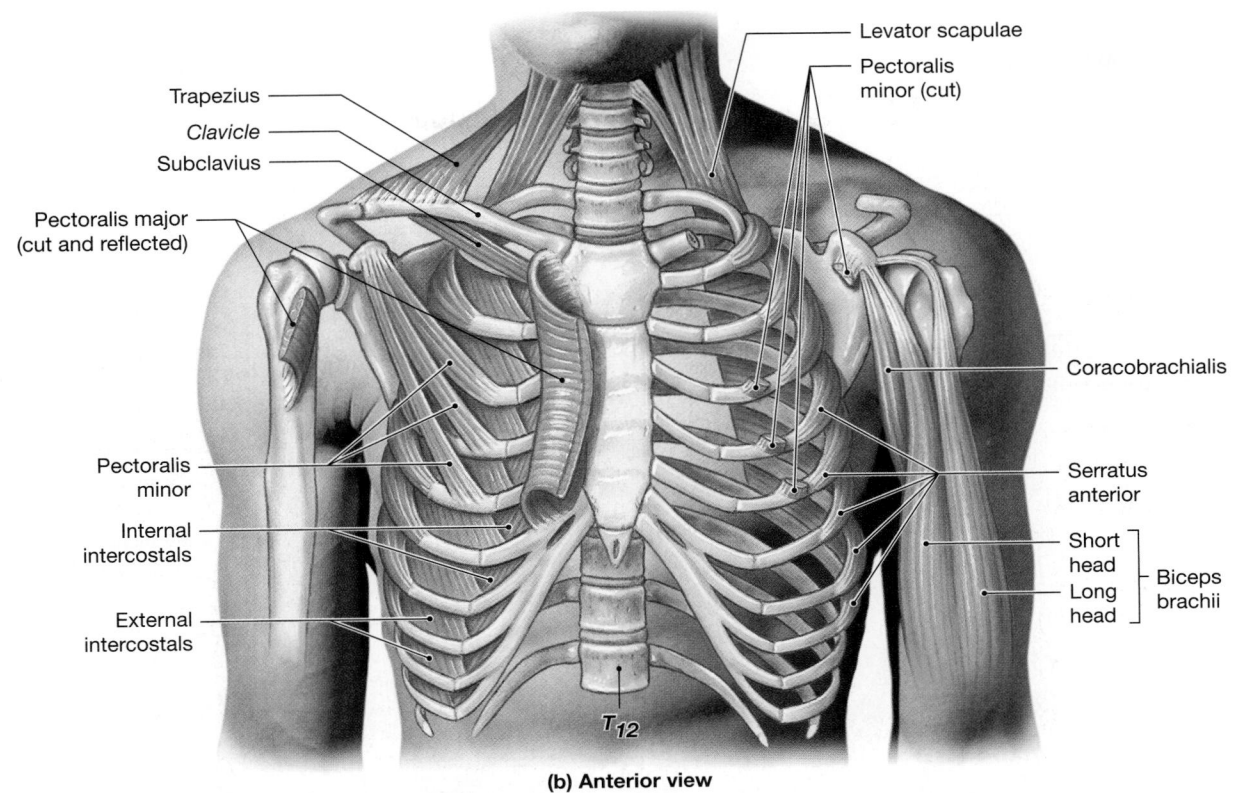

Trapezius
Clavicle
Subclavius
Pectoralis major (cut and reflected)
Pectoralis minor
Internal intercostals
External intercostals
Levator scapulae
Pectoralis minor (cut)
Coracobrachialis
Serratus anterior
Short head
Long head
Biceps brachii
T_{12}

(b) Anterior view

Figure 11–14   Muscles That Position the Pectoral Girdle. (continued) *See also Figures 8–2, 8–3, 8–4, and 9–9.* ATLAS: Plates 39a–d; 40a–b

Two other deep chest muscles arise along the ventral surfaces of the ribs on either side. The **subclavius** (sub-KLĀ-vē-us; *sub-*, below + *clavius*, clavicle) muscle inserts on the inferior border of the clavicle (**Figure 11–14b**). When it contracts, it depresses and protracts the scapular end of the clavicle. Because ligaments connect this end to the shoulder joint and scapula, those structures move as well. The **pectoralis** (pek-tō-RA-lis) **minor** muscle attaches to the coracoid process of the scapula. The contraction of this muscle generally complements that of the subclavius muscle.

Muscles That Move the Arm

The muscles that move the arm (**Figures 11–13, 11–14** and **11–15**) are easiest to remember when they are grouped by their actions at the shoulder joint (Table 11–12). The **deltoid** muscle is the major abductor, but the **supraspinatus** (soo-pra-spī-NĀ-tus) muscle assists at the start of this movement. The **subscapularis** and **teres major** muscles produce medial rotation at the shoulder, whereas the **infraspinatus** and the **teres minor** muscles produce lateral rotation. All these mus-

cles originate on the scapula. The small **coracobrachialis** (KOR-uh-kō-brā-kē-A-lis) muscle is the only muscle attached to the scapula that produces flexion and adduction at the shoulder (**Figure 11–15a**).

Tips&Tricks

The supraspinatus and infraspinatus muscles are named for their locations above and below the spine of the scapula, respectively, not because they are located on the backbone.

The **pectoralis major** muscle extends between the anterior portion of the chest and the crest of the greater tubercle of the humerus. The **latissimus dorsi** (la-TIS-i-mus DOR-sē) muscle extends between the thoracic vertebrae at the posterior midline and the intertubercular groove of the humerus (**Figure 11–15b**). The pectoralis major muscle produces flexion at the shoulder joint, and the latissimus dorsi muscle produces extension. These muscles, commonly known as the "pecs" and the "lats," can also work together to produce adduction and medial rotation of the humerus at the shoulder.

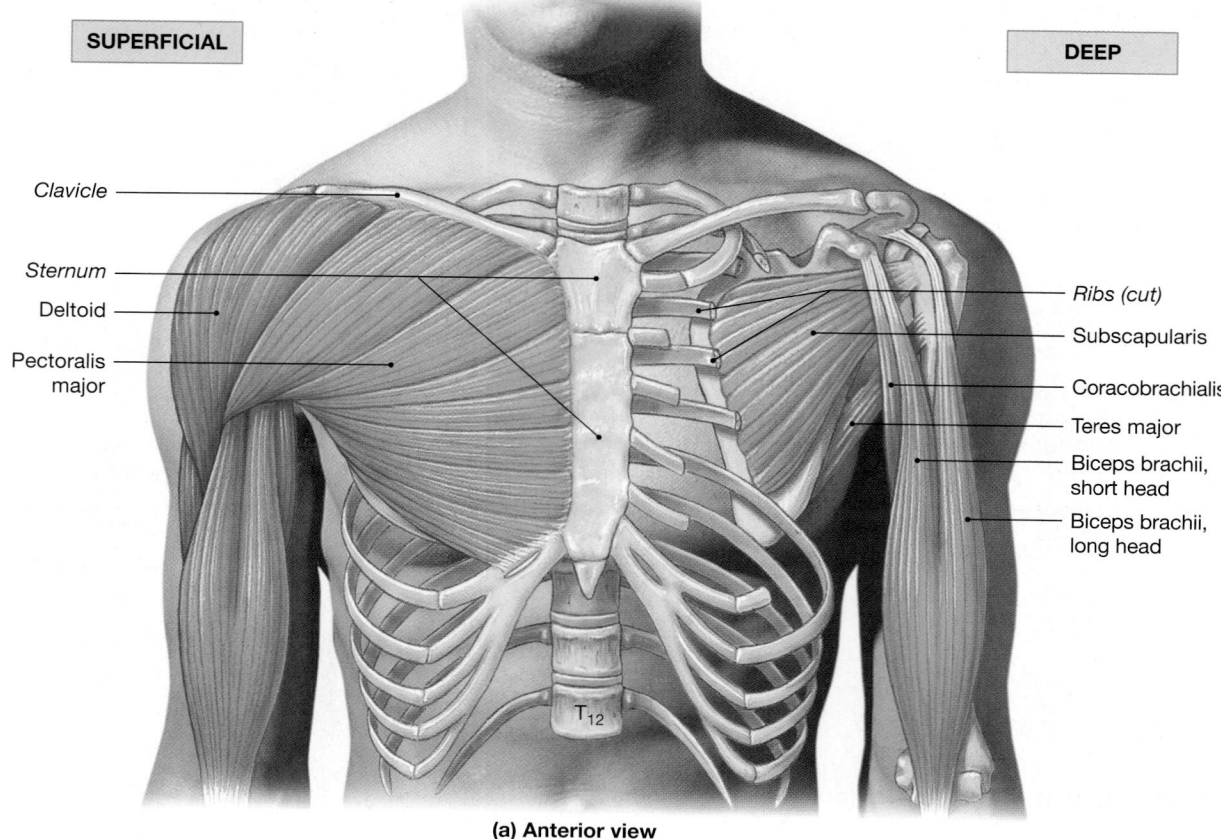

(a) Anterior view

SUPERFICIAL

DEEP

Clavicle

Sternum

Deltoid

Pectoralis
major

Ribs (cut)

Subscapularis

Coracobrachialis

Teres major

Biceps brachii,
short head

Biceps brachii,
long head

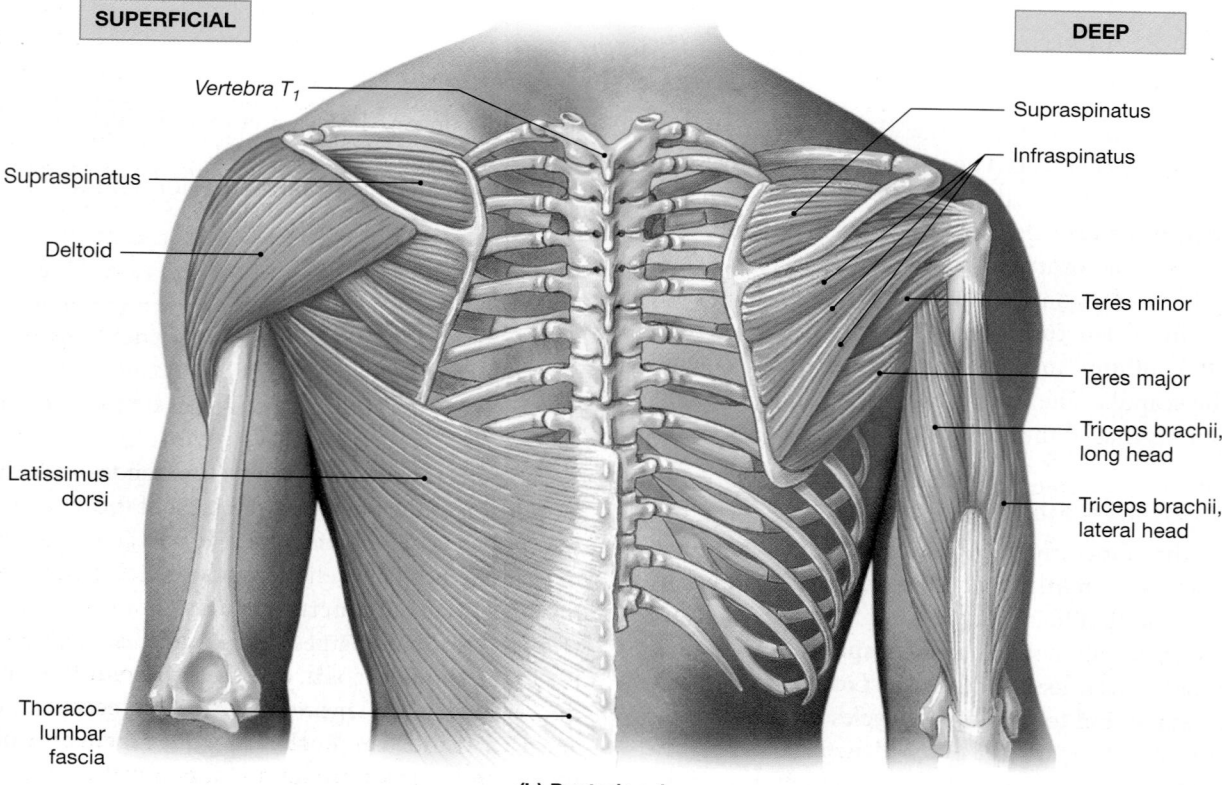

(b) Posterior view

SUPERFICIAL

DEEP

Vertebra T₁

Supraspinatus

Deltoid

Latissimus
dorsi

Thoraco-
lumbar
fascia

Supraspinatus

Infraspinatus

Teres minor

Teres major

Triceps brachii,
long head

Triceps brachii,
lateral head

Figure 11–15 **Muscles That Move the Arm.** *See also Figures 7–22, 8–3, and 9–9.* ATLAS: Plates 39a–d; 40a–b

TABLE 11–12 Muscles That Move the Arm (Figures 11–13, 11–14 to 11–15)

Muscle	Origin	Insertion	Action	Innervation*
Deltoid	Clavicle and scapula (acromion and adjacent scapular spine)	Deltoid tuberosity of humerus	*Whole muscle*: abduction at shoulder; *anterior part*: flexion and medial rotation; *posterior part*: extension and lateral rotation	Axillary nerve (C_5–C_6)
Supraspinatus	Supraspinous fossa of scapula	Greater tubercle of humerus	Abduction at the shoulder	Suprascapular nerve (C_5)
Subscapularis	Subscapular fossa of scapula	Lesser tubercle of humerus	Medial rotation at shoulder	Subscapular nerves (C_5–C_6)
Teres major	Inferior angle of scapula	Passes medially to reach the medial lip of intertubercular groove of humerus	Extension, adduction, and medial rotation at shoulder	Lower subscapular nerve (C_5–C_6)
Infraspinatus	Infraspinous fossa of scapula	Greater tubercle of humerus	Lateral rotation at shoulder	Suprascapular nerve (C_5–C_6)
Teres minor	Lateral border of scapula	Passes laterally to reach the greater tubercle of humerus	Lateral rotation at shoulder	Axillary nerve (C_5)
Coracobrachialis	Coracoid process	Medial margin of shaft of humerus	Adduction and flexion at shoulder	Musculocutaneous nerve (C_5–C_7)
Pectoralis major	Cartilages of ribs 2–6, body of sternum, and inferior, medial portion of clavicle	Crest of greater tubercle and lateral lip of intertubercular groove of humerus	Flexion, adduction, and medial rotation at shoulder	Pectoral nerves (C_5–T_1)
Latissimus dorsi	Spinous processes of inferior thoracic and all lumbar vertebrae, ribs 8–12, and thoracolumbar fascia	Floor of intertubercular groove of the humerus	Extension, adduction, and medial rotation at shoulder	Thoracodorsal nerve (C_6–C_8)
Triceps brachii (long head)	*See Table 11–13*			

*Where appropriate, spinal nerves involved are given in parentheses.

Collectively, the supraspinatus, infraspinatus, subscapularis, and teres minor muscles and their associated tendons form the **rotator cuff**. Sports that involve throwing a ball, such as baseball or football, place considerable strain on the rotator cuff, and rotator cuff injuries are relatively common.

Tips&Tricks

The acronym SITS is useful in remembering the four muscles of the rotator cuff.

Muscles That Move the Forearm and Hand

Although most of the muscles that insert on the forearm and hand originate on the humerus, the biceps brachii and triceps brachii muscles are noteworthy exceptions. The **biceps brachii** muscle and the *long head* of the **triceps brachii** muscle originate on the scapula and insert on the bones of the forearm (**Figure 11–16**). The triceps brachii muscle inserts on the olecranon. Contraction of the triceps brachii muscle extends the elbow, as when you do push-ups. The biceps brachii muscle inserts on the radial tuberosity, a roughened bump on the anterior surface of the radius. ∞ p. 252 Contraction of the biceps brachii muscle flexes the elbow and supinates the forearm. With the forearm pronated (palm facing back), the biceps

brachii muscle cannot function effectively. As a result, you are strongest when you flex your elbow with a supinated forearm; the biceps brachii muscle then makes a prominent bulge.

The biceps brachii muscle plays an important role in the stabilization of the shoulder joint. The short head originates on the coracoid process and provides support to the posterior surface of the capsule. The long head originates at the supraglenoid tubercle, inside the shoulder joint. ∞ p. 249 After crossing the head of the humerus, it passes along the intertubercular groove. In this position, the tendon helps to hold the head of the humerus within the glenoid cavity while arm movements are under way.

More muscles are shown in **Figure 11–16** and listed in Table 11–13. As you study these muscles, notice that, in general, the extensor muscles lie along the posterior and lateral surfaces of the arm, whereas the flexors are on the anterior and medial surfaces. Connective tissue partitions separate major muscle groups, dividing the muscles into *compartments* that are discussed further on p. 381.

The **brachialis** and **brachioradialis** (BRĀ-kē-ō-rā-dē-A-lis) muscles flex the elbow and are opposed by the **anconeus** muscle and the triceps brachii muscle, respectively.

The **flexor carpi ulnaris**, **flexor carpi radialis**, and **palmaris longus** muscles are superficial muscles that work together to produce flexion of the wrist. The flexor carpi ra-

Figure 11–16 **Muscles That Move the Forearm and Hand.** Superficial muscles are shown in **(a)** posterior and **(b)** anterior views. Deeper muscles are shown in the sectional views and in *Figure 11–18. See also Figures 8–4, 8–5, and 9–10.* ATLAS: Plates 27a–c; 29a; 30; 33a–d; 37a,b

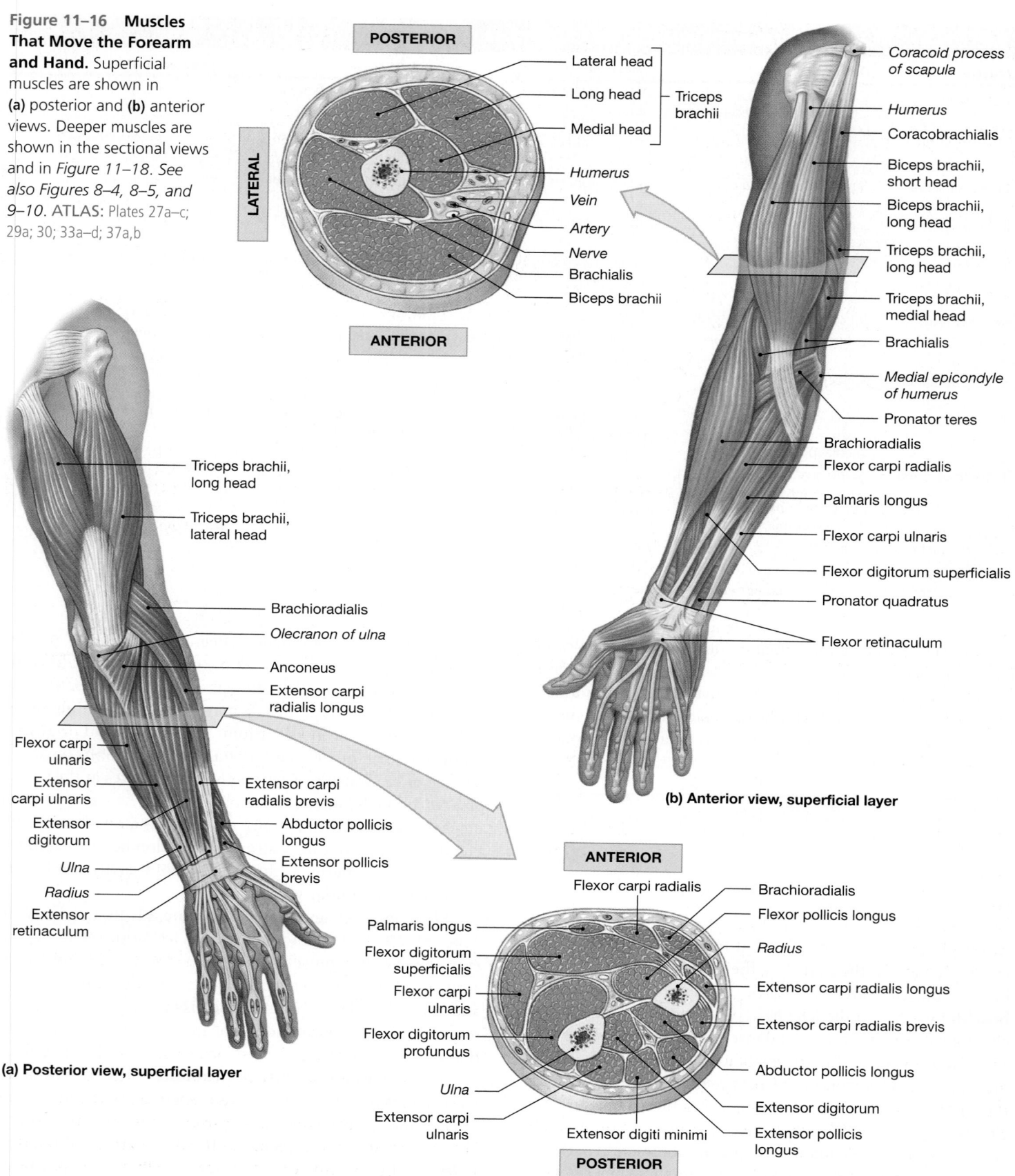

POSTERIOR

Lateral head
Long head — Triceps brachii
Medial head

Humerus
Vein
Artery
Nerve
Brachialis
Biceps brachii

LATERAL

ANTERIOR

Coracoid process of scapula
Humerus
Coracobrachialis
Biceps brachii, short head
Biceps brachii, long head
Triceps brachii, long head
Triceps brachii, medial head
Brachialis
Medial epicondyle of humerus
Pronator teres
Brachioradialis
Flexor carpi radialis
Palmaris longus
Flexor carpi ulnaris
Flexor digitorum superficialis
Pronator quadratus
Flexor retinaculum

(b) Anterior view, superficial layer

Triceps brachii, long head
Triceps brachii, lateral head
Brachioradialis
Olecranon of ulna
Anconeus
Extensor carpi radialis longus
Flexor carpi ulnaris
Extensor carpi ulnaris
Extensor digitorum
Ulna
Radius
Extensor retinaculum
Extensor carpi radialis brevis
Abductor pollicis longus
Extensor pollicis brevis

(a) Posterior view, superficial layer

ANTERIOR

Flexor carpi radialis
Palmaris longus
Flexor digitorum superficialis
Flexor carpi ulnaris
Flexor digitorum profundus
Ulna
Extensor carpi ulnaris
Extensor digiti minimi

Brachioradialis
Flexor pollicis longus
Radius
Extensor carpi radialis longus
Extensor carpi radialis brevis
Abductor pollicis longus
Extensor digitorum
Extensor pollicis longus

POSTERIOR

TABLE 11–13 Muscles That Move the Forearm and Hand (Figure 11–16)

Muscle	Origin	Insertion	Action	Innervation
ACTION AT THE ELBOW				
Flexors				
Biceps brachii	*Short head* from the coracoid process; *long head* from the supraglenoid tubercle (both on the scapula)	Tuberosity of radius	Flexion at elbow and shoulder; supination	Musculocutaneous nerve (C$_5$–C$_6$)
Brachialis	Anterior, distal surface of humerus	Tuberosity of ulna	Flexion at elbow	As above and radial nerve (C$_7$–C$_8$)
Brachioradialis	Ridge superior to the lateral epicondyle of humerus	Lateral aspect of styloid process of radius	As above	Radial nerve (C$_5$–C$_6$)
Extensors				
Anconeus	Posterior, inferior surface of lateral epicondyle of humerus	Lateral margin of olecranon on ulna	Extension at elbow	Radial nerve (C$_7$–C$_8$)
Triceps brachii				
lateral head	Superior, lateral margin of humerus	Olecranon of ulna	As above	Radial nerve (C$_6$–C$_8$)
long head	Infraglenoid tubercle of scapula	As above	As above, plus extension and adduction at the shoulder	As above
medial head	Posterior surface of humerus inferior to radial groove	As above	Extension at elbow	As above
PRONATORS/SUPINATORS				
Pronator quadratus	Anterior and medial surfaces of distal portion of ulna	Anterolateral surface of distal portion of radius	Pronation	Median nerve (C$_8$–T$_1$)
Pronator teres	Medial epicondyle of humerus and coronoid process of ulna	Midlateral surface of radius	As above	Median nerve (C$_6$–C$_7$)
Supinator	Lateral epicondyle of humerus, annular ligament, and ridge near radial notch of ulna	Anterolateral surface of radius distal to the radial tuberosity	Supination	Deep radial nerve (C$_6$–C$_8$)
ACTION AT THE HAND				
Flexors				
Flexor carpi radialis	Medial epicondyle of humerus	Bases of second and third metacarpal bones	Flexion and abduction at wrist	Median nerve (C$_6$–C$_7$)
Flexor carpi ulnaris	Medial epicondyle of humerus; adjacent medial surface of olecranon and anteromedial portion of ulna	Pisiform, hamate, and base of fifth metacarpal bone	Flexion and adduction at wrist	Ulnar nerve (C$_8$–T$_1$)
Palmaris longus	Medial epicondyle of humerus	Palmar aponeurosis and flexor retinaculum	Flexion at wrist	Median nerve (C$_6$–C$_7$)
Extensors				
Extensor carpi radialis longus	Lateral supracondylar ridge of humerus	Base of second metacarpal bone	Extension and abduction at wrist	Radial nerve (C$_6$–C$_7$)
Extensor carpi radialis brevis	Lateral epicondyle of humerus	Base of third metacarpal bone	As above	As above
Extensor carpi ulnaris	Lateral epicondyle of humerus; adjacent dorsal surface of ulna	Base of fifth metacarpal bone	Extension and adduction at wrist	Deep radial nerve (C$_6$–C$_8$)

dialis muscle flexes and *ab*ducts, and the flexor carpi ulnaris muscle flexes and *ad*ducts. *Pitcher's arm* is an inflammation at the origins of the flexor carpi muscles at the medial epicondyle of the humerus. This condition results from forcibly flexing the wrist just before releasing a baseball.

The **extensor carpi radialis** muscles and the **extensor carpi ulnaris** muscle have a similar relationship to that between

the flexor carpi muscles. That is, the extensor carpi radialis muscles produce extension and *ab*duction, whereas the extensor carpi ulnaris muscle produces extension and *ad*duction.

The **pronator teres** and **supinator** muscles originate on both the humerus and ulna. These muscles rotate the radius without either flexing or extending the elbow. The **pronator quadratus** muscle originates on the ulna and assists the

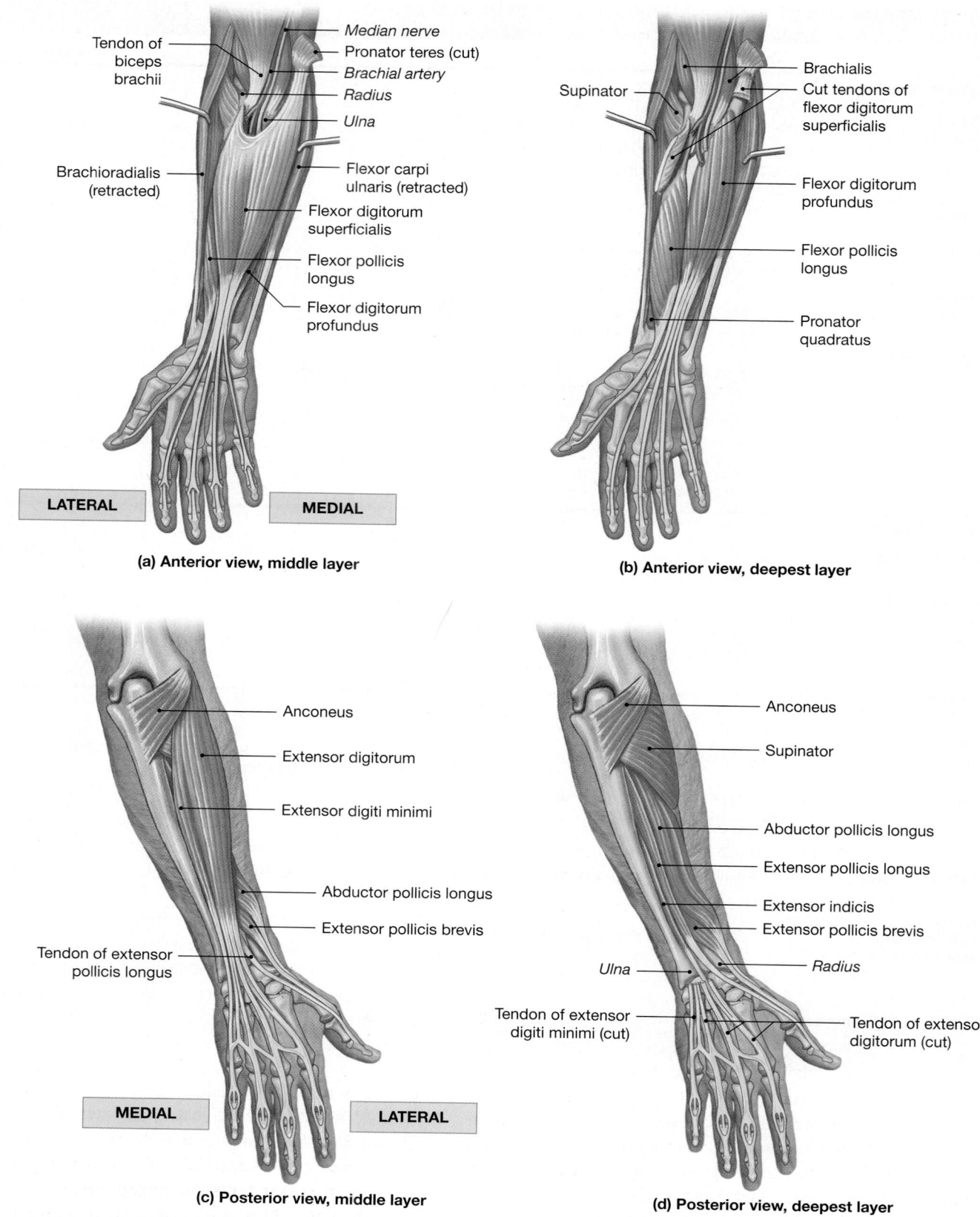

Tendon of biceps brachii

Median nerve
Pronator teres (cut)
Brachial artery
Radius
Ulna

Brachioradialis (retracted)

Flexor carpi ulnaris (retracted)
Flexor digitorum superficialis
Flexor pollicis longus
Flexor digitorum profundus

LATERAL

MEDIAL

(a) Anterior view, middle layer

Supinator

Brachialis
Cut tendons of flexor digitorum superficialis
Flexor digitorum profundus
Flexor pollicis longus
Pronator quadratus

(b) Anterior view, deepest layer

Anconeus
Extensor digitorum
Extensor digiti minimi
Abductor pollicis longus
Extensor pollicis brevis

Tendon of extensor pollicis longus

MEDIAL

LATERAL

(c) Posterior view, middle layer

Anconeus
Supinator
Abductor pollicis longus
Extensor pollicis longus
Extensor indicis
Extensor pollicis brevis

Ulna

Radius

Tendon of extensor digiti minimi (cut)

Tendon of extensor digitorum (cut)

(d) Posterior view, deepest layer

Figure 11–17 **Muscles That Move the Hand and Fingers.** Middle and deep muscle layers of the right forearm; for superficial muscles, *see Figure 11–15. See also Figure 8–5.*

TABLE 11–14 Muscles That Move the Hand and Fingers (Figure 11–17)

Muscle	Origin	Insertion	Action	Innervation
Abductor pollicis longus	Proximal dorsal surfaces of ulna and radius	Lateral margin of first metacarpal bone	Abduction at joints of thumb and wrist	Deep radial nerve (C_6–C_7)
Extensor digitorum	Lateral epicondyle of humerus	Posterior surfaces of the phalanges, fingers 2–5	Extension at finger joints and wrist	Deep radial nerve (C_6–C_8)
Extensor pollicis brevis	Shaft of radius distal to origin of adductor pollicis longus	Base of proximal phalanx of thumb	Extension at joints of thumb; abduction at wrist	Deep radial nerve (C_6–C_7)
Extensor pollicis longus	Posterior and lateral surfaces of ulna and interosseous membrane	Base of distal phalanx of thumb	As above	Deep radial nerve (C_6–C_8)
Extensor indicis	Posterior surface of ulna and interosseous membrane	Posterior surface of phalanges of index finger (2), with tendon of extensor digitorum	Extension and adduction at joints of index finger	As above
Extensor digiti minimi	Via extensor tendon to lateral epicondyle of humerus and from intermuscular septa	Posterior surface of proximal phalanx of little finger (5)	Extension at joints of little finger	As above
Flexor digitorum superficialis	Medial epicondyle of humerus; adjacent anterior surfaces of ulna and radius	Midlateral surfaces of middle phalanges of fingers 2–5	Flexion at proximal interphalangeal, metacarpophalangeal, and wrist joints	Median nerve (C_7–T_1)
Flexor digitorum profundus	Medial and posterior surfaces of ulna, medial surface of coronoid process, and interosseus membrane	Bases of distal phalanges of fingers 2–5	Flexion at distal interphalangeal joints and, to a lesser degree, proximal interphalangeal joints and wrist	Palmar interosseous nerve, from median nerve, and ulnar nerve (C_8–T_1)
Flexor pollicis longus	Anterior shaft of radius, interosseous membrane	Base of distal phalanx of thumb	Flexion at joints of thumb	Median nerve (C_8–T_1)

pronator teres muscle in opposing the actions of the supinator or biceps brachii muscles. The muscles involved in pronation and supination are shown in **Figure 11–17**. During pronation, the tendon of the biceps brachii muscle rotates with the radius. As a result, this muscle cannot assist in flexion of the elbow when the forearm is pronated.

Muscles That Move the Hand and Fingers

Several superficial and deep muscles of the forearm flex and extend the finger joints (**Figure 11–17** and Table 11–14). These relatively large muscles end before reaching the wrist, and only their tendons cross the articulation, ensuring maximum mobility at both the wrist and hand. The tendons that cross the dorsal and ventral surfaces of the wrist pass through **synovial tendon sheaths**, elongated bursae that reduce friction. ∞ p. 271

The muscles of the forearm provide strength and crude control of the hand and fingers. These muscles are known as the *extrinsic muscles of the hand*. Fine control of the hand involves small *intrinsic muscles*, which originate on the carpal and metacarpal bones. No muscles originate on the phalanges, and only tendons extend across the distal joints of the fingers. The intrinsic muscles of the hand are detailed in **Figure 11–18** and Table 11–15.

The fascia of the forearm thickens on the posterior surface of the wrist, forming the **extensor retinaculum** (ret-i-NAK-ū-lum), a wide band of connective tissue. The extensor retinaculum holds the tendons of the extensor muscles in place. On the anterior surface, the fascia also thickens to form another wide band of connective tissue, the **flexor retinaculum**, which stabilizes the tendons of the flexor muscles. Inflammation of the retinacula and synovial tendon sheaths can restrict movement and irritate the distal portions of the *median nerve*, a mixed (sensory and motor) nerve that innervates the hand. This condition, known as *carpal tunnel syndrome*, causes chronic pain.

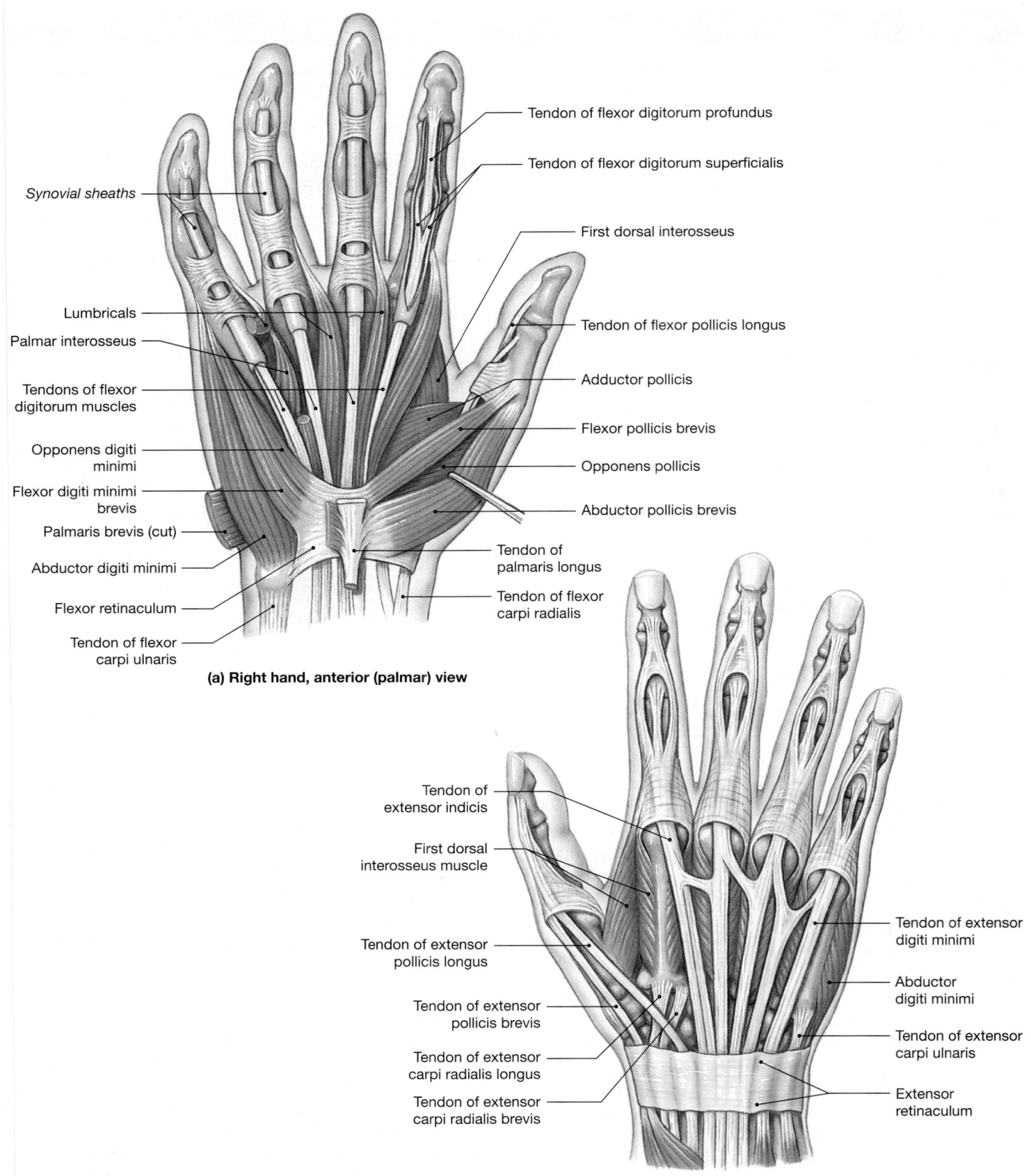

Synovial sheaths

Lumbricals

Palmar interosseus

Tendons of flexor digitorum muscles

Opponens digiti minimi

Flexor digiti minimi brevis

Palmaris brevis (cut)

Abductor digiti minimi

Flexor retinaculum

Tendon of flexor carpi ulnaris

Tendon of flexor digitorum profundus

Tendon of flexor digitorum superficialis

First dorsal interosseus

Tendon of flexor pollicis longus

Adductor pollicis

Flexor pollicis brevis

Opponens pollicis

Abductor pollicis brevis

Tendon of palmaris longus

Tendon of flexor carpi radialis

(a) Right hand, anterior (palmar) view

Tendon of extensor indicis

First dorsal interosseus muscle

Tendon of extensor pollicis longus

Tendon of extensor pollicis brevis

Tendon of extensor carpi radialis longus

Tendon of extensor carpi radialis brevis

Tendon of extensor digiti minimi

Abductor digiti minimi

Tendon of extensor carpi ulnaris

Extensor retinaculum

(b) Right hand, posterior (dorsal) view

Figure 11–18   **Intrinsic Muscles of the Hand.** *See also Figure 8–6.* ATLAS: Plates 37b; 38c–f

TABLE 11–15 Intrinsic Muscles of the Hand (Figure 11–18)

Muscle	Origin	Insertion	Action	Innervation
Adductor pollicis	Metacarpal and carpal bones	Proximal phalanx of thumb	Adduction of thumb	Ulnar nerve, deep branch (C_8–T_1)
Opponens pollicis	Trapezium and flexor retinaculum	First metacarpal bone	Opposition of thumb	Median nerve (C_6–C_7)
Palmaris brevis	Palmar aponeurosis	Skin of medial border of hand	Moves skin on medial border toward midline of palm	Ulnar nerve, superficial branch (C_8)
Abductor digiti minimi	Pisiform	Proximal phalanx of little finger	Abduction of little finger and flexion at its metacarpophalangeal joint	Ulnar nerve, deep branch (C_8–T_1)
Abductor pollicis brevis	Transverse carpal ligament, scaphoid, and trapezium	Radial side of base of proximal phalanx of thumb	Abduction of thumb	Median nerve (C_6–C_7)
Flexor pollicis brevis	Flexor retinaculum, trapezium, capitate, and ulnar side of first metacarpal bone	Radial and ulnar sides of proximal phalanx of thumb	Flexion and adduction of thumb	Branches of median and ulnar nerves
Flexor digiti minimi brevis	Hamate	Proximal phalanx of little finger	Flexion at joints of little finger	Ulnar nerve, deep branch (C_8–T_1)
Opponens digiti minimi	As above	Fifth metacarpal bone	Opposition of fifth metacarpal bone	As above
Lumbrical (4)	Tendons of flexor digitorum profundus	Tendons of extensor digitorum to digits 2–5	Flexion at metacarpophalangeal joints 2–5; extension at proximal and distal interphalangeal joints, digits 2–5	No. 1 and no. 2 by median nerve; no. 3 and no. 4 by ulnar nerve, deep branch
Dorsal interosseus (4)	Each originates from opposing faces of two metacarpal bones (I and II, II and III, III and IV, IV and V)	Bases of proximal phalanges of fingers 2–4	Abduction at metacarpophalangeal joints of fingers 2 and 4; flexion at metacarpophalangeal joints; extension at interphalangeal joints	Ulnar nerve, deep branch (C_8–T_1)
Palmar interosseus* (3–4)	Sides of metacarpal bones II, IV, and V	Bases of proximal phalanges of fingers 2, 4, and 5	Adduction at metacarpophalangeal joints of fingers 2, 4, and 5; flexion at metacarpophalangeal joints; extension at interphalangeal joints	As above

*The deep, medial portion of the flexor pollicis brevis originating on the first metacarpal bone is sometimes called the *first palmar interosseus muscle*; it inserts on the ulnar side of the phalanx and is innervated by the ulnar nerve.

CLINICAL NOTE

Intramuscular Injections Drugs are commonly injected into muscle or adipose tissues rather than directly into the bloodstream. An **intramuscular (IM) injection** introduces a fairly large amount of a drug, which will then enter the circulation gradually. The drug is introduced into the mass of a large skeletal muscle. Uptake is generally faster and accompanied by less tissue irritation than when drugs are administered *intradermally* or *subcutaneously* (injected into the dermis or subcutaneous layer, respectively). Depending on the size of the muscle, up to 5 mL of fluid may be injected at one time, and multiple injections are possible.

Bulky muscles that contain few large vessels or nerves are ideal sites. The gluteus medius muscle or the posterior, lateral, superior part of the gluteus maximus muscle is commonly selected. The deltoid muscle of the arm, about 2.5 cm (1 in.) distal to the acromion, is another effective site. Probably most satisfactory from a technical point of view is the vastus lateralis muscle of the thigh; an injection into this thick muscle will not encounter vessels or nerves, but may cause pain later when the muscle is used in walking.

Muscles of the Pelvis and Lower Limbs

The pelvic girdle is tightly bound to the axial skeleton, permitting little relative movement. In our discussion of the axial musculature, we therefore encountered few muscles that can influence the position of the pelvis. The muscles that position the lower limbs can be divided into three functional groups: (1) *muscles that move the thigh*, (2) *muscles that move the leg*, and (3) *muscles that move the foot and toes*.

Muscles That Move the Thigh

Table 11–16 lists the muscles that move the thigh. **Gluteal muscles** cover the lateral surfaces of the ilia (**Figure 11–13a** and **Figure 11–19a,b,c**). The **gluteus maximus** muscle is the largest and most posterior of the gluteal muscles. Its origin includes parts of the ilium; the sacrum, coccyx, and associated ligaments; and the thoracolumbar fascia (**Figure 11–13**). Acting alone, this massive muscle produces extension and lateral rotation at the hip joint. The gluteus maximus shares an insertion with the **tensor fasciae latae** (FAH-shē-āy LAH-tāy) muscle, which originates on the iliac crest and the anterior superior iliac spine. Together, these muscles pull on the **iliotibial** (il-ē-ō-TIB-ē-ul) **tract**, a band of collagen fibers that extends along the lateral surface of the thigh and inserts on the tibia. This tract provides a lateral brace for the knee that becomes particularly important when you balance on one foot.

The **gluteus medius** and **gluteus minimus** muscles (**Figure 11–19b,c**) originate anterior to the origin of the gluteus maximus muscle and insert on the greater trochanter of the femur. The anterior gluteal line on the lateral surface of the ilium marks the boundary between these muscles.

The **lateral rotators** originate at or inferior to the horizontal axis of the acetabulum. There are six lateral rotator muscles in all, of which the **piriformis** (pir-i-FOR-mis) muscle and the **obturator** muscles are dominant (**Figure 11–19c,d**).

The **adductors** (**Figure 11–19c,d**) originate inferior to the horizontal axis of the acetabulum. This muscle group includes the **adductor magnus**, **adductor brevis**, **adductor longus**, **pectineus** (pek-ti-NĒ-us), and **gracilis** (GRAS-i-lis) muscles. All but the adductor magnus originate both anterior and inferior to the joint, so they perform hip flexion as well as adduction. The adductor magnus muscle can produce either adduction and flexion or adduction and extension, depending on the region stimulated. The adductor magnus muscle can also produce medial or lateral rotation at the hip. The other muscles, which insert on low ridges along the posterior surface of the femur, produce medial rotation. When an athlete suffers a *pulled groin*, the problem is a *strain*—a muscle tear or break—in one of these adductor muscles.

The internal surface of the pelvis is dominated by a pair of muscles. The large **psoas** (SŌ-us) **major** muscle originates alongside the inferior thoracic and lumbar vertebrae, and its insertion lies on the lesser trochanter of the femur. Before reaching this insertion, its tendon merges with that of the **iliacus** (il-Ī-ah-kus) muscle, which nestles within the iliac fossa. These two powerful hip flexors are often referred to collectively as the **iliopsoas** (il-ē-ō-SŌ-us) muscle.

Muscles That Move the Leg

As in the upper limb, muscle distribution in the lower limb exhibits a pattern: Extensor muscles are located along the anterior and lateral surfaces of the leg, and flexors lie along the posterior and medial surfaces (**Figure 11–20** and Table 11–17). As in the upper limb, sturdy connective tissue partitions divide the lower limb into separate muscular compartments. Although the flexors and adductors originate on the pelvic girdle, most extensors originate on the femoral surface.

The *flexors of the knee* include the **biceps femoris**, **semimembranosus** (sem-ē-mem-bra-NŌ-sus), **semitendinosus** (sem-ē-ten-di-NŌ-sus), and **sartorius** muscles (**Figure 11–20**). These muscles originate along the edges of the pelvis and insert on the tibia and fibula. The sartorius muscle is the only knee

flexor that originates superior to the acetabulum, and its insertion lies along the medial surface of the tibia. When the sartorius contracts, it produces flexion at the knee and lateral rotation at the hip—for example, when you cross your legs.

Because the biceps femoris, semimembranosus, and semitendinosus muscles originate on the pelvic surface inferior and posterior to the acetabulum, their contractions produce not only flexion at the knee, but also extension at the hip. These three muscles are often called the **hamstrings**. A *pulled hamstring* is a relatively common sports injury caused by a strain affecting one of the hamstring muscles.

Tips & Tricks

To remember that three muscles make up the *ham*strings, think "the three little pigs." These three muscles are portions of the cut of meat sold as ham.

The knee joint can be locked at full extension by a slight lateral rotation of the tibia. ⊃ p. 286 The small **popliteus** (pop-LI-tē-us) muscle originates on the femur near the lateral condyle and inserts on the posterior tibial shaft (**Figure 11–21a**). When flexion is initiated, this muscle contracts to produce a slight medial rotation of the tibia that unlocks the knee joint.

Collectively, four *knee extensors*—the three **vastus muscles**, which originate along the shaft of the femur, and the **rectus femoris muscle**—make up the **quadriceps femoris** (the "quads"). Together, the vastus muscles cradle the rectus femoris muscle the way a bun surrounds a hot dog (**Figure 11–20c**). All four muscles insert on the patella via the quadriceps tendon. The force of their contraction is relayed to the tibial tuberosity by way of the patellar ligament. The rectus femoris muscle originates on the anterior inferior iliac spine and the superior acetabular rim—so in addition to extending the knee, it assists in flexion of the hip.

Tips & Tricks

Think of the quadriceps muscles as "the four at the fore."

Muscles That Move the Foot and Toes

The extrinsic muscles that move the foot and toes are shown in **Figure 11–21** and listed in Table 11–18. Most of the muscles that move the ankle produce the plantar flexion involved with walking and running movements. The **gastrocnemius** (gas-trok-NĒ-mē-us; *gaster*, stomach + *kneme*, knee) muscle of the calf is an important plantar flexor, but the slow muscle fibers of the underlying **soleus** (SŌ-lē-us) muscle are better suited for making continuous postural adjustments against the pull of gravity. These muscles are best seen in posterior and lateral views (**Figure 11–21a,b**). The gastrocnemius muscle arises from two heads located on the medial and lateral epicondyles of the femur just proximal to the knee. The *fabella*, a sesamoid

11 MUSCULAR

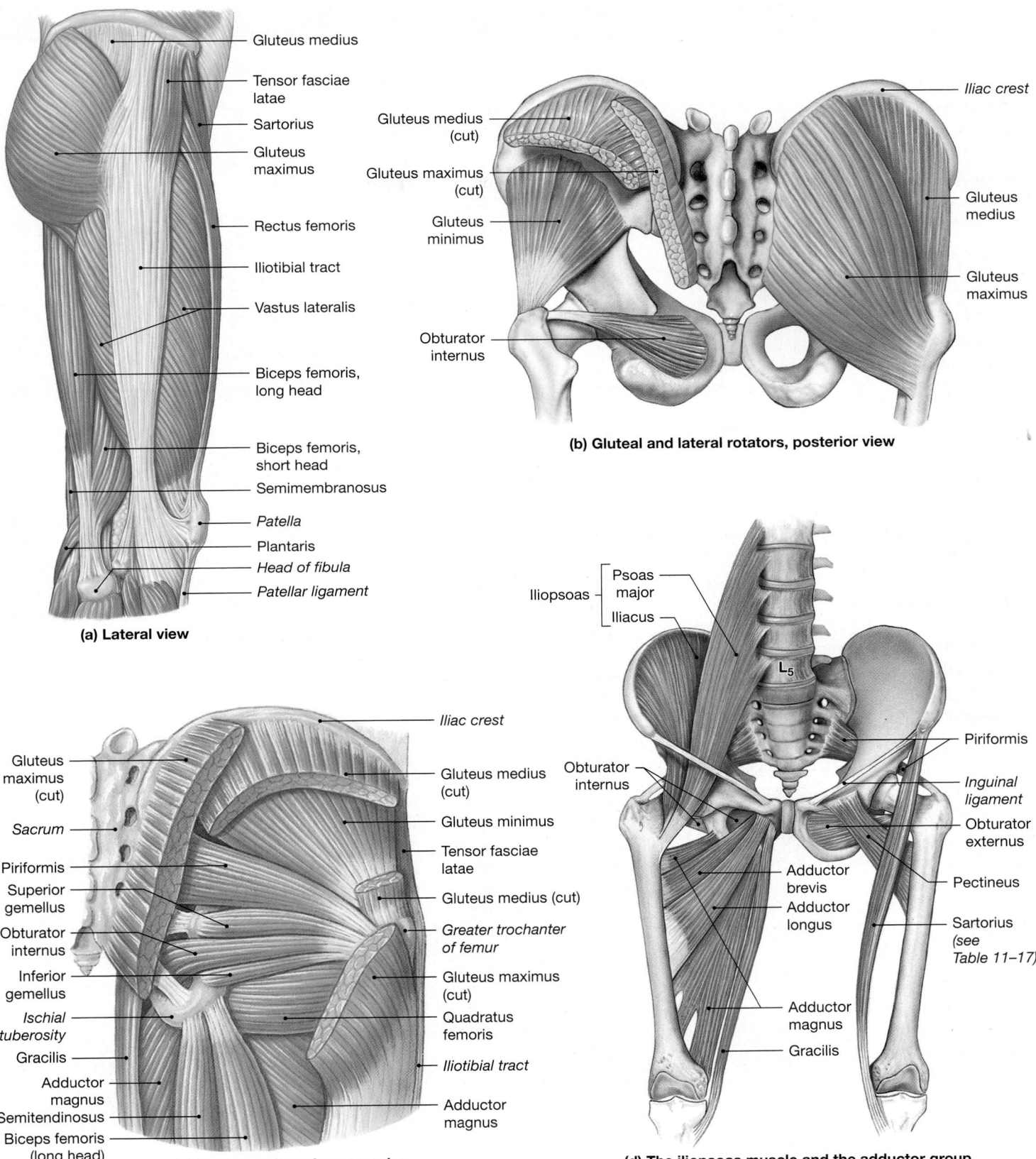

(a) Lateral view

Gluteus medius
Tensor fasciae latae
Sartorius
Gluteus maximus
Rectus femoris
Iliotibial tract
Vastus lateralis
Biceps femoris, long head
Biceps femoris, short head
Semimembranosus
Patella
Plantaris
Head of fibula
Patellar ligament

(b) Gluteal and lateral rotators, posterior view

Gluteus medius (cut)
Gluteus maximus (cut)
Gluteus minimus
Obturator internus
Iliac crest
Gluteus medius
Gluteus maximus

(c) Posterior view, deep muscles

Gluteus maximus (cut)
Sacrum
Piriformis
Superior gemellus
Obturator internus
Inferior gemellus
Ischial tuberosity
Gracilis
Adductor magnus
Semitendinosus
Biceps femoris (long head)
Iliac crest
Gluteus medius (cut)
Gluteus minimus
Tensor fasciae latae
Gluteus medius (cut)
Greater trochanter of femur
Gluteus maximus (cut)
Quadratus femoris
Iliotibial tract
Adductor magnus

(d) The iliopsoas muscle and the adductor group

Iliopsoas
Psoas major
Iliacus
L₅
Obturator internus
Piriformis
Inguinal ligament
Obturator externus
Pectineus
Sartorius (*see Table 11–17*)
Adductor brevis
Adductor longus
Adductor magnus
Gracilis

Figure 11–19   Muscles That Move the Thigh. *See also Figures 8–7, 8–8, 8–11, and 9–11.* ATLAS: Plates 68a–c; 72a,b; 73a,b

TABLE 11–16 Muscles That Move the Thigh (Figure 11–19)

Group and Muscle(s)	Origin	Insertion	Action	Innervation*
GLUTEAL GROUP				
Gluteus maximus	Iliac crest, posterior gluteal line, and lateral surface of ilium; sacrum, coccyx, and thoracolumbar fascia	Iliotibial tract and gluteal tuberosity of femur	Extension and lateral rotation at hip	Inferior gluteal nerve (L_5–S_2)
Gluteus medius	Anterior iliac crest of ilium, lateral surface between posterior and anterior gluteal lines	Greater trochanter of femur	Abduction and medial rotation at hip	Superior gluteal nerve (L_4–S_1)
Gluteus minimus	Lateral surface of ilium between inferior and anterior gluteal lines	As above	As above	As above
Tensor fasciae latae	Iliac crest and lateral surface of anterior superior iliac spine	Iliotibial tract	Flexion and medial rotation at hip; tenses fascia lata, which laterally supports the knee	As above
LATERAL ROTATOR GROUP				
Obturators (externus and internus)	Lateral and medial margins of obturator foramen	Trochanteric fossa of femur (externus); medial surface of greater trochanter (internus)	Lateral rotation at hip	Obturator nerve (externus: L_3–L_4) and special nerve from sacral plexus (internus: L_5–S_2)
Piriformis	Anterolateral surface of sacrum	Greater trochanter of femur	Lateral rotation and abduction at hip	Branches of sacral nerves (S_1–S_2)
Gemelli (superior and inferior)	Ischial spine and tuberosity	Medial surface of greater trochanter with tendon of obturator internus	Lateral rotation at hip	Nerves to obturator internus and quadratus femoris
Quadratus femoris	Lateral border of ischial tuberosity	Intertrochanteric crest of femur	As above	Special nerve from sacral plexus (L_4–S_1)
ADDUCTOR GROUP				
Adductor brevis	Inferior ramus of pubis	Linea aspera of femur	Adduction, flexion, and medial rotation at hip	Obturator nerve (L_3–L_4)
Adductor longus	Inferior ramus of pubis anterior to adductor brevis	As above	As above	As above
Adductor magnus	Inferior ramus of pubis posterior to adductor brevis and ischial tuberosity	Linea aspera and adductor tubercle of femur	Adduction at hip; superior part produces flexion and medial rotation; inferior part produces extension and lateral rotation	Obturator and sciatic nerves
Pectineus	Superior ramus of pubis	Pectineal line inferior to lesser trochanter of femur	Flexion, medial rotation, and adduction at hip	Femoral nerve (L_2–L_4)
Gracilis	Inferior ramus of pubis	Medial surface of tibia inferior to medial condyle	Flexion at knee; adduction and medial rotation at hip	Obturator nerve (L_3–L_4)
ILIOPSOAS GROUP				
Iliacus	Iliac fossa of ilium	Femur distal to lesser trochanter; tendon fused with that of psoas major	Flexion at hip	Femoral nerve (L_2–L_3)
Psoas major	Anterior surfaces and transverse processes of vertebrae (T_{12}–L_5)	Lesser trochanter in company with iliacus	Flexion at hip or lumbar intervertebral joints	Branches of the lumbar plexus (L_2–L_3)

*Where appropriate, spinal nerves involved are given in parentheses.

11 MUSCULAR

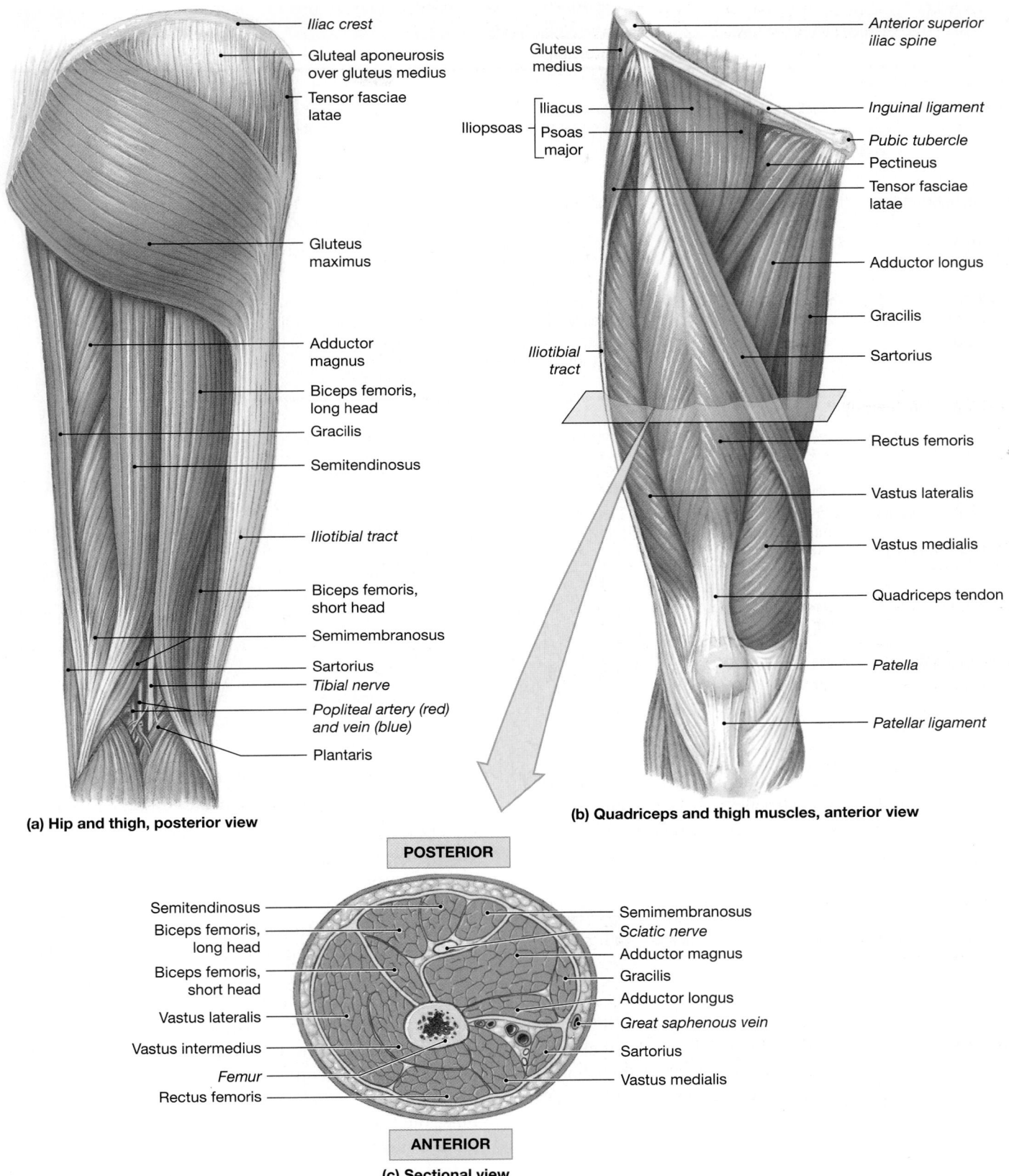

(a) Hip and thigh, posterior view

- Iliac crest
- Gluteal aponeurosis over gluteus medius
- Tensor fasciae latae
- Gluteus maximus
- Adductor magnus
- Biceps femoris, long head
- Gracilis
- Semitendinosus
- *Iliotibial tract*
- Biceps femoris, short head
- Semimembranosus
- Sartorius
- *Tibial nerve*
- *Popliteal artery (red) and vein (blue)*
- Plantaris

(b) Quadriceps and thigh muscles, anterior view

- Gluteus medius
- Iliopsoas — Iliacus / Psoas major
- *Iliotibial tract*
- *Anterior superior iliac spine*
- *Inguinal ligament*
- *Pubic tubercle*
- Pectineus
- Tensor fasciae latae
- Adductor longus
- Gracilis
- Sartorius
- Rectus femoris
- Vastus lateralis
- Vastus medialis
- Quadriceps tendon
- *Patella*
- *Patellar ligament*

POSTERIOR

- Semitendinosus
- Biceps femoris, long head
- Biceps femoris, short head
- Vastus lateralis
- Vastus intermedius
- *Femur*
- Rectus femoris
- Semimembranosus
- *Sciatic nerve*
- Adductor magnus
- Gracilis
- Adductor longus
- *Great saphenous vein*
- Sartorius
- Vastus medialis

ANTERIOR

(c) Sectional view

Figure 11–20 **Muscles That Move the Leg.** *See also Figures 8–11 to 8–13, and 9–12.* ATLAS: Plates 69a,b; 70b; 72a,b; 74; 76a,b; 78b–g

TABLE 11–17	Muscles That Move the Leg (Figure 11–20)			
Muscle	**Origin**	**Insertion**	**Action**	**Innervation***
FLEXORS OF THE KNEE				
Biceps femoris	Ischial tuberosity and linea aspera of femur	Head of fibula, lateral condyle of tibia	Flexion at knee; extension and lateral rotation at hip	Sciatic nerve; tibial portion (S_1–S_3; to long head) and common fibular branch (L_5–S_2; to short head)
Semimembranosus	Ischial tuberosity	Posterior surface of medial condyle of tibia	Flexion at knee; extension and medial rotation at hip	Sciatic nerve (tibial portion; L_5–S_2)
Semitendinosus	As above	Proximal, medial surface of tibia near insertion of gracilis	As above	As above
Sartorius	Anterior superior iliac spine	Medial surface of tibia near tibial tuberosity	Flexion at knee; flexion and lateral rotation at hip	Femoral nerve (L_2–L_3)
Popliteus	Lateral condyle of femur	Posterior surface of proximal tibial shaft	Medial rotation of tibia (or lateral rotation of femur); flexion at knee	Tibial nerve (L_4–S_1)
EXTENSORS OF THE KNEE				
Rectus femoris	Anterior inferior iliac spine and superior acetabular rim of ilium	Tibial tuberosity via patellar ligament	Extension at knee; flexion at hip	Femoral nerve (L_2–L_4)
Vastus intermedius	Anterolateral surface of femur and linea aspera (distal half)	As above	Extension at knee	As above
Vastus lateralis	Anterior and inferior to greater trochanter of femur and along linea aspera (proximal half)	As above	As above	As above
Vastus medialis	Entire length of linea aspera of femur	As above	As above	As above

*Where appropriate, spinal nerves involved are given in parentheses.

bone, is occasionally present within the lateral head of the gastrocnemius muscle. The gastrocnemius and soleus muscles share a common tendon, the **calcaneal tendon**, commonly known as the *Achilles tendon* or *calcanean tendon*.

Tips&Tricks

The soleus is so named because it resembles the flatfish we call sole.

The term "Achilles tendon" comes from Greek mythology. Achilles was a warrior who was invincible but for one vulnerable spot: the calcaneal tendon. His mother had dipped him in the River Styx as an infant to make him invulnerable, but she held him by the ankle and forgot to dip the heel of his foot, which became his point of weakness. This oversight proved fatal for Achilles, who was killed in battle by an arrow through the tendon that now bears his name. Outside mythology, damage to the calcaneal tendon isn't a fatal problem. Although it is among the largest, strongest tendons in the body, its rupture is

relatively common. The applied forces increase markedly during rapid acceleration or deceleration; sprinters can rupture the calcaneal tendon pushing off from the starting blocks, and the elderly often snap this tendon during a stumble or fall. Surgery may be necessary to reposition and reconnect the broken ends of the tendon to promote healing.

Deep to the gastrocnemius and soleus muscles lie a pair of **fibularis** muscles, or *peroneus* muscles (**Figure 11–21b,c**). The fibularis muscles produce eversion and extension (plantar flexion) at the ankle. Inversion is caused by the contraction of the **tibialis** (tib-ē-A-lis) muscles. The large **tibialis anterior** muscle (**Figure 11–21b,d**) flexes the ankle and opposes the gastrocnemius muscle.

Important digital muscles originate on the surface of the tibia, the fibula, or both (**Figure 11–21b,c,d**). Large synovial tendon sheaths surround the tendons of the tibialis anterior, **extensor digitorum longus**, and **extensor hallucis longus** muscles, where they cross the ankle joint. The positions of these sheaths are stabilized by superior and inferior **extensor retinacula** (**Figure 11–21b,d**).

11 MUSCULAR

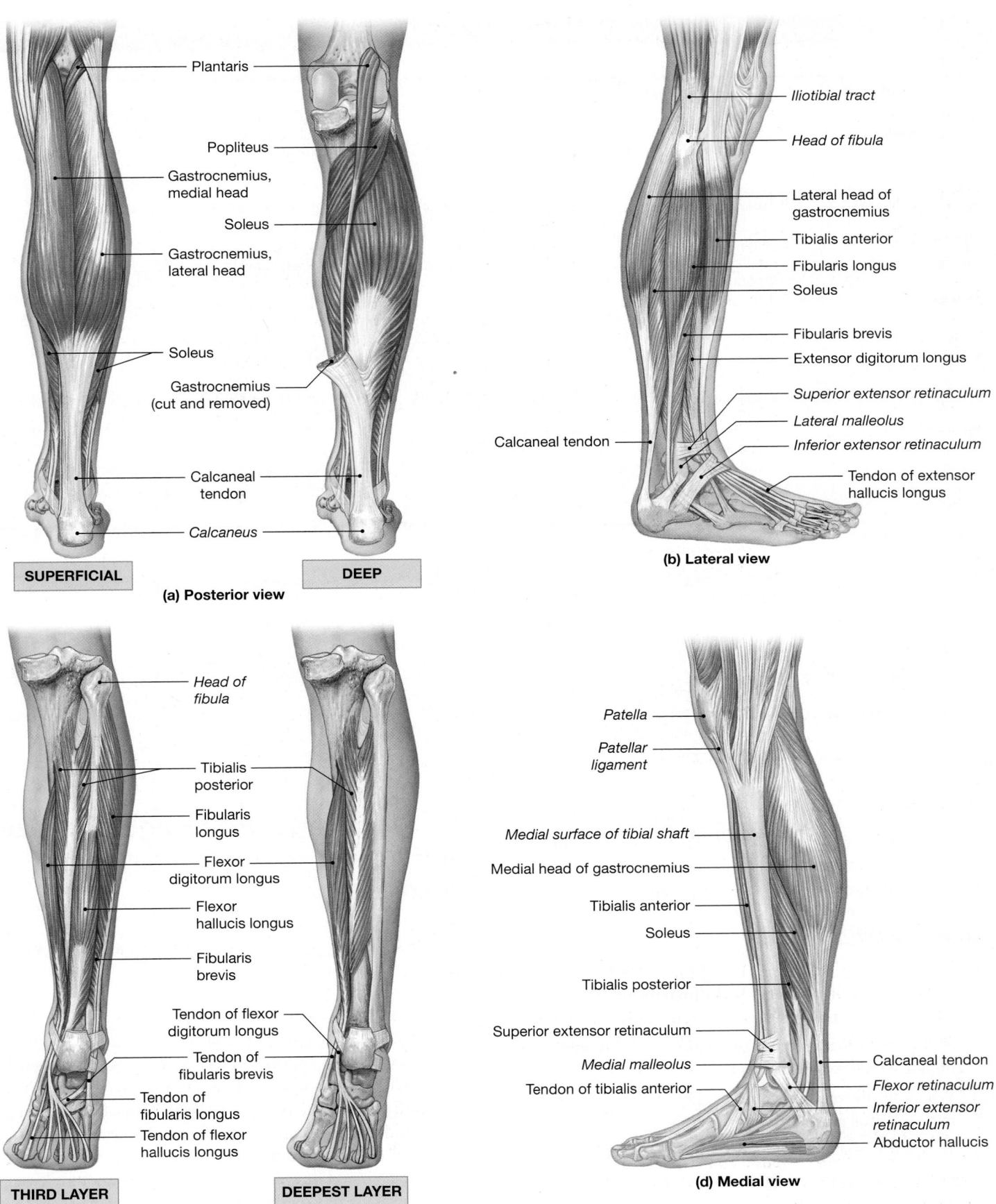

Plantaris

Popliteus

Gastrocnemius, medial head

Soleus

Gastrocnemius, lateral head

Soleus

Gastrocnemius (cut and removed)

Calcaneal tendon

Calcaneus

SUPERFICIAL

DEEP

(a) Posterior view

Iliotibial tract

Head of fibula

Lateral head of gastrocnemius

Tibialis anterior

Fibularis longus

Soleus

Fibularis brevis

Extensor digitorum longus

Superior extensor retinaculum

Lateral malleolus

Inferior extensor retinaculum

Tendon of extensor hallucis longus

Calcaneal tendon

(b) Lateral view

Head of fibula

Tibialis posterior

Fibularis longus

Flexor digitorum longus

Flexor hallucis longus

Fibularis brevis

Tendon of flexor digitorum longus

Tendon of fibularis brevis

Tendon of fibularis longus

Tendon of flexor hallucis longus

THIRD LAYER

DEEPEST LAYER

(c) Posterior view

Patella

Patellar ligament

Medial surface of tibial shaft

Medial head of gastrocnemius

Tibialis anterior

Soleus

Tibialis posterior

Superior extensor retinaculum

Medial malleolus

Tendon of tibialis anterior

Calcaneal tendon

Flexor retinaculum

Inferior extensor retinaculum

Abductor hallucis

(d) Medial view

Figure 11–21 **Extrinsic Muscles That Move the Foot and Toes.** *See also Figures 8–13 and 8–14.* ATLAS: Plates 81a,b; 82a,b; 84a,b

TABLE 11–18 Extrinsic Muscles That Move the Foot and Toes (Figure 11–21)

Muscle	Origin	Insertion	Action	Innervation
ACTION AT THE ANKLE				
Flexors (Dorsiflexors)				
Tibialis anterior	Lateral condyle and proximal shaft of tibia	Base of first metatarsal bone and medial cuneiform bone	Flexion (dorsiflexion) at ankle; inversion of foot	Deep fibular nerve (L_4–S_1)
Extensors (Plantar flexors)				
Gastrocnemius	Femoral condyles	Calcaneus via calcaneal tendon	Extension (plantar flexion) at ankle; inversion of foot; flexion at knee	Tibial nerve (S_1–S_2)
Fibularis brevis	Midlateral margin of fibula	Base of fifth metatarsal bone	Eversion of foot and extension (plantar flexion) at ankle	Superficial fibular nerve (L_4–S_1)
Fibularis longus	Lateral condyle of tibia, head and proximal shaft of fibula	Base of first metatarsal bone and medial cuneiform bone	Eversion of foot and extension (plantar flexion) at ankle; supports longitudinal arch	As above
Plantaris	Lateral supracondylar ridge	Posterior portion of calcaneus	Extension (plantar flexion) at ankle; flexion at knee	Tibial nerve (L_4–S_1)
Soleus	Head and proximal shaft of fibula and adjacent posteromedial shaft of tibia	Calcaneus via calcaneal tendon (with gastrocnemius)	Extension (plantar flexion) at ankle	Sciatic nerve, tibial branch (S_1–S_2)
Tibialis posterior	Interosseous membrane and adjacent shafts of tibia and fibula	Tarsal and metatarsal bones	Adduction and inversion of foot; extension (plantar flexion) at ankle	As above
ACTION AT THE TOES				
Digital flexors				
Flexor digitorum longus	Posteromedial surface of tibia	Inferior surfaces of distal phalanges, toes 2–5	Flexion at joints of toes 2–5	Sciatic nerve, tibial branch (L_5–S_1)
Flexor hallucis longus	Posterior surface of fibula	Inferior surface, distal phalanx of great toe	Flexion at joints of great toe	As above
Digital extensors				
Extensor digitorum longus	Lateral condyle of tibia, anterior surface of fibula	Superior surfaces of phalanges, toes 2–5	Extension at joints of toes 2–5	Deep fibular nerve (L_4–S_1)
Extensor hallucis longus	Anterior surface of fibula	Superior surface, distal phalanx of great toe	Extension at joints of great toe	As above

Intrinsic muscles of the foot originate on the tarsal and metatarsal bones (**Figure 11–22** and Table 11–19). Their contractions move the toes and contribute to the maintenance of the longitudinal arch of the foot. ∞ p. 262

CHECKPOINT

18. Shrugging your shoulders uses which muscles?

19. Baseball pitchers sometimes suffer from rotator cuff injuries. Which muscles are involved in this type of injury?

20. An injury to the flexor carpi ulnaris muscle would impair which two movements?

21. Which leg movement would be impaired by injury to the obturator muscle?

22. To what does a "pulled hamstring" refer?

23. How would a torn calcaneal tendon affect movement of the foot?

See the blue Answers tab at the end of the book.

11-7 With advancing age, the size and power of muscle tissue decrease

The effects of aging on the muscular system can be summarized as follows:

- *Skeletal Muscle Fibers Become Smaller in Diameter.* This reduction in size reflects primarily a decrease in the number of myofibrils. In addition, the muscle fibers contain smaller ATP, CP, and glycogen reserves and less myoglobin. The overall effect is a reduction in skeletal muscle size, strength, and endurance, combined with a tendency to fatigue rapidly. Because cardiovascular performance also decreases with age, blood flow to active muscles does not increase with exercise as rapidly as it does in younger people. These factors interact to produce decreases of 30–50 percent in anaerobic and aerobic performance by age 65.

11 MUSCULAR

Figure 11–22 **Intrinsic Muscles of the Foot.** *See also Figure 8–14.* ATLAS: Plates 84a; 85a,b; 86c; 87a–c; 89

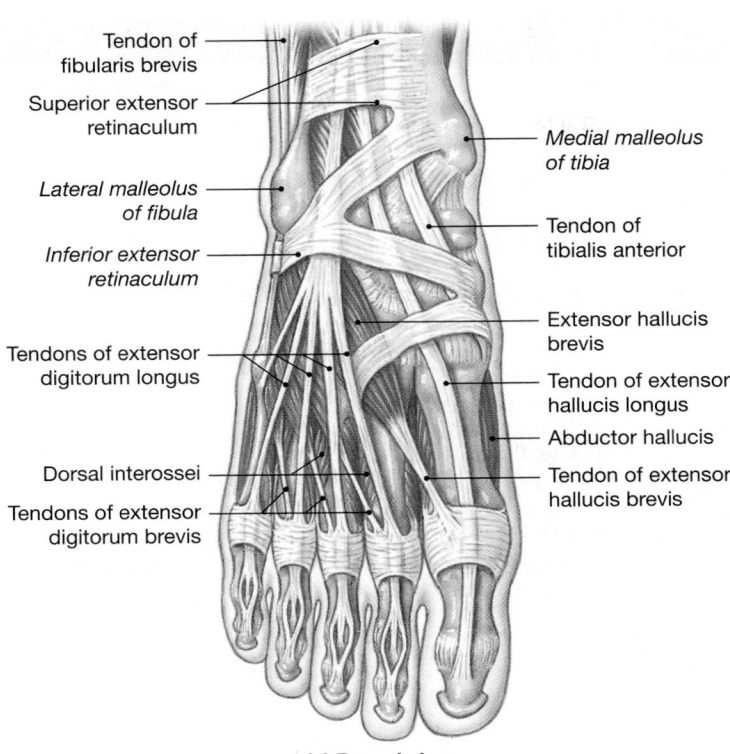

Tendon of fibularis brevis
Superior extensor retinaculum
Lateral malleolus of fibula
Inferior extensor retinaculum
Tendons of extensor digitorum longus
Dorsal interossei
Tendons of extensor digitorum brevis

Medial malleolus of tibia
Tendon of tibialis anterior
Extensor hallucis brevis
Tendon of extensor hallucis longus
Abductor hallucis
Tendon of extensor hallucis brevis

(a) Dorsal view

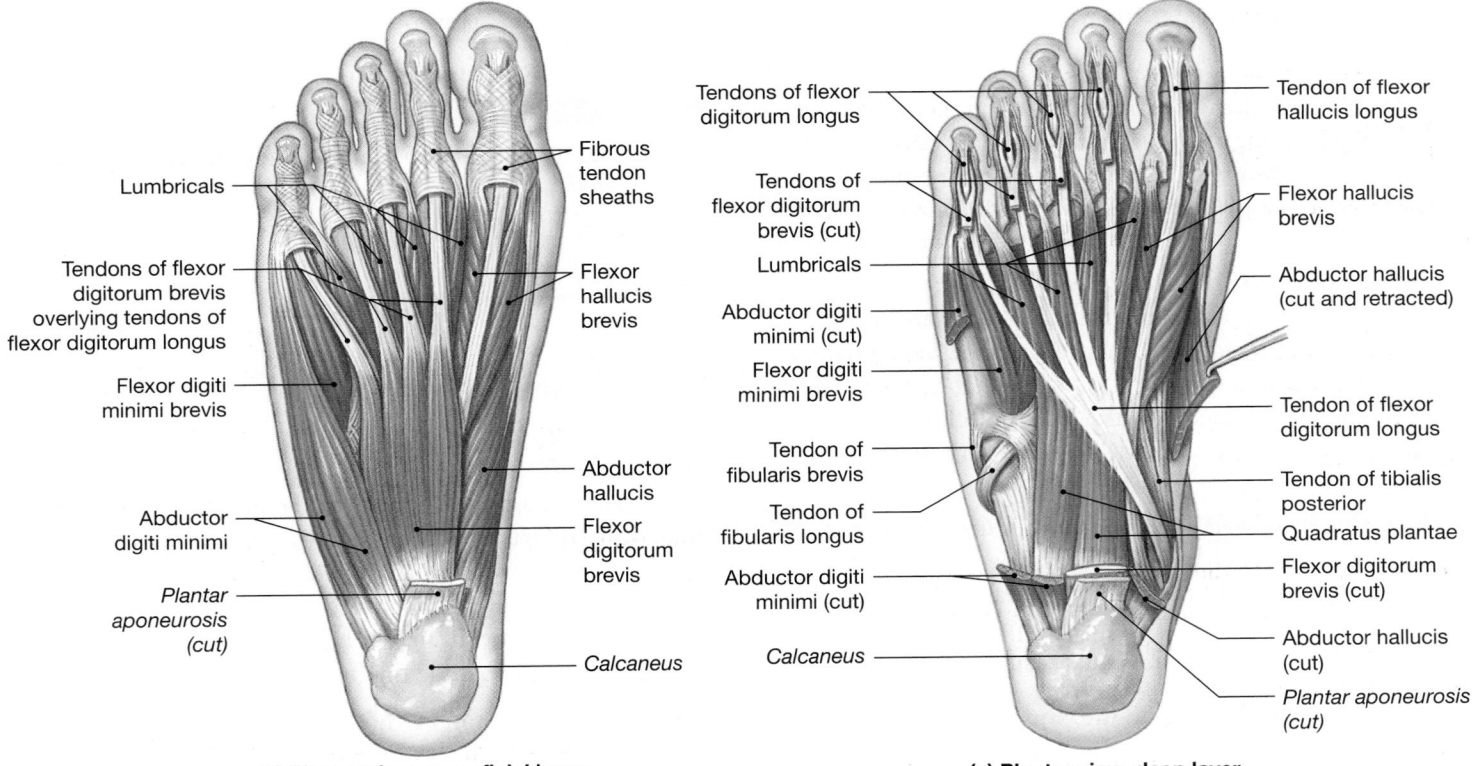

Lumbricals
Tendons of flexor digitorum brevis overlying tendons of flexor digitorum longus
Flexor digiti minimi brevis
Abductor digiti minimi
Plantar aponeurosis (cut)

Fibrous tendon sheaths
Flexor hallucis brevis
Abductor hallucis
Flexor digitorum brevis
Calcaneus

(b) Plantar view, superficial layer

Tendons of flexor digitorum longus
Tendons of flexor digitorum brevis (cut)
Lumbricals
Abductor digiti minimi (cut)
Flexor digiti minimi brevis
Tendon of fibularis brevis
Tendon of fibularis longus
Abductor digiti minimi (cut)
Calcaneus

Tendon of flexor hallucis longus
Flexor hallucis brevis
Abductor hallucis (cut and retracted)
Tendon of flexor digitorum longus
Tendon of tibialis posterior
Quadratus plantae
Flexor digitorum brevis (cut)
Abductor hallucis (cut)
Plantar aponeurosis (cut)

(c) Plantar view, deep layer

TABLE 11–19	Intrinsic Muscles of the Foot (Figure 11–22)			
Muscle	**Origin**	**Insertion**	**Action**	**Innervation**
Extensor digitorum brevis	Calcaneus (superior and lateral surfaces)	Dorsal surfaces of toes 1–4	Extension at metatarsophalangeal joints of toes 1–4	Deep fibular nerve (L_5–S_1)
Abductor hallucis	Calcaneus (tuberosity on inferior surface)	Medial side of proximal phalanx of great toe	Abduction at metatarsophalangeal joint of great toe	Medial plantar nerve (L_4–L_5)
Flexor digitorum brevis	As above	Sides of middle phalanges, toes 2–5	Flexion at proximal interphalangeal joints of toes 2–5	As above
Abductor digiti minimi	As above	Lateral side of proximal phalanx, toe 5	Abduction at metatarsophalangeal joint of toe 5	Lateral plantar nerve (L_4–L_5)
Quadratus plantae	Calcaneus (medial, inferior surfaces)	Tendon of flexor digitorum longus	Flexion at joints of toes 2–5	As above
Lumbrical (4)	Tendons of flexor digitorum longus	Insertions of extensor digitorum longus	Flexion at metatarsophalangeal joints; extension at proximal interphalangeal joints of toes 2–5	Medial plantar nerve (1), lateral plantar nerve (2–4)
Flexor hallucis brevis	Cuboid and lateral cuneiform bones	Proximal phalanx of great toe	Flexion at metatarsophalangeal joint of great toe	Medial plantar nerve (L_4–L_5)
Adductor hallucis	Bases of metatarsal bones II–IV and plantar ligaments	As above	Adduction at metatarsophalangeal joint of great toe	Lateral plantar nerve (S_1–S_2)
Flexor digiti minimi brevis	Base of metatarsal bone V	Lateral side of proximal phalanx of toe 5	Flexion at metatarsophalangeal joint of toe 5	As above
Dorsal interosseus (4)	Sides of metatarsal bones	Medial and lateral sides of toe 2; lateral sides of toes 3 and 4	Abduction at metatarsophalangeal joints of toes 3 and 4	As above
Plantar interosseus (3)	Bases and medial sides of metatarsal bones	Medial sides of toes 3–5	Adduction at metatarsophalangeal joints of toes 3–5	As above

- *Skeletal Muscles Become Less Elastic.* Aging skeletal muscles develop increasing amounts of fibrous connective tissue, a process called **fibrosis**. Fibrosis makes the muscle less flexible, and the collagen fibers can restrict movement and circulation.

- *Tolerance for Exercise Decreases.* A lower tolerance for exercise results in part from the tendency toward rapid fatigue and in part from the reduction in thermoregulatory ability described in Chapter 5. ∞ p. 174 Individuals over age 65 cannot eliminate the heat their muscles generate during contraction as effectively as younger people can and thus are subject to overheating.

- *The Ability to Recover from Muscular Injuries Decreases.* The number of satellite cells steadily decreases with age, and the amount of fibrous tissue increases. As a result, when an injury occurs, repair capabilities are limited. Scar tissue formation is the usual result.

To be in good shape late in life, you must be in *very* good shape early in life. Regular exercise helps control body weight, strengthens bones, and generally improves the quality of life at all ages. Extremely demanding exercise is not as important as regular exercise. In fact, extreme exercise in the elderly can damage tendons, bones, and joints.

11-8 Exercise produces responses in multiple body systems

To operate at maximum efficiency, the muscular system must be supported by many other systems. The changes that occur during exercise provide a good example of such interaction. As noted earlier, active muscles consume oxygen and generate carbon dioxide and heat. The effects of exercise on various body systems include the following:

- *Cardiovascular System:* Blood vessels in active muscles and the skin dilate, and heart rate increases. These adjustments accelerate oxygen and nutrient delivery to and carbon dioxide removal from the muscle, and bring heat to the skin for radiation into the environment.

CLINICAL NOTE

Compartment Syndrome

In the limbs, the interconnections among the superficial fascia, the deep fascia of the muscles, and the periostea of the appendicular skeleton are quite substantial. The muscles within a limb are in effect isolated in **compartments** formed by dense collagenous sheets (**Figure 11–23**). Blood vessels and nerves traveling to specific muscles within the limb enter and branch within the appropriate compartments.

When a crushing injury, severe contusion, or muscle strain occurs, the blood vessels in one or more compartments may be damaged. The compartments become swollen with blood and fluid leaked from damaged vessels. Additionally, injured muscle cells swell. The connective-tissue partitions are very strong; the accumulated fluid cannot escape, so pressure rises within the affected compartments. Eventually, the pressure can become so high that it compresses regional blood vessels, eliminating the circulatory supply to the muscles and nerves of the compartment. This compression produces **ischemia** (is-KĒ-mē-uh), or "blood starvation," known in this case as **compartment syndrome**.

Emergency measures for relieving the pressure include slicing into the compartment along its longitudinal axis and inserting a drain. If such steps are not taken, the contents of the compartment will suffer severe damage. Nerves in the affected compartment will be destroyed after 2–4 hours of ischemia, although they can regenerate to some degree if the circulation is restored. After 6 hours or more, the muscle tissue will also be destroyed, and no regeneration can occur. The muscles will be replaced by scar tissue, and shortening of the connective tissue fibers may result in *contracture*, a

permanent contraction of an entire muscle following the atrophy of individual muscle fibers.

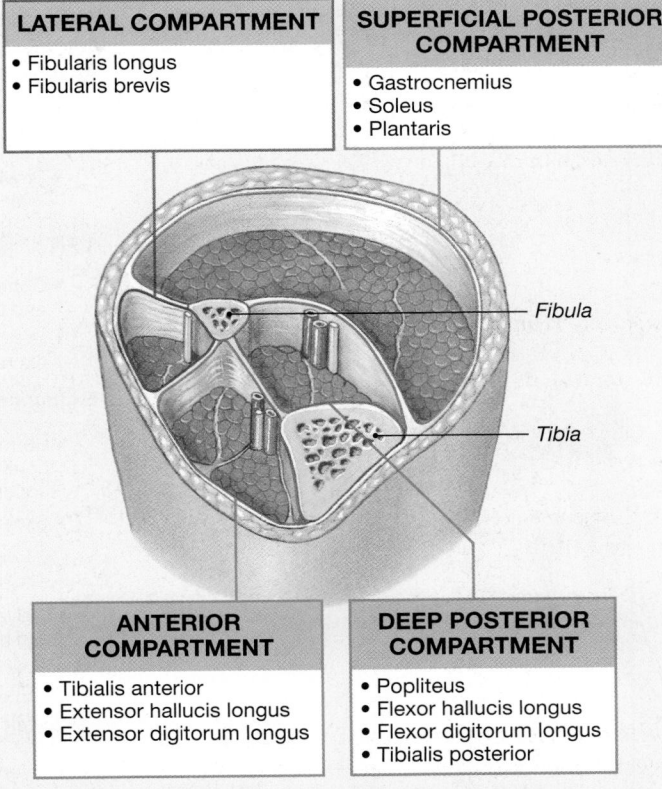

LATERAL COMPARTMENT	**SUPERFICIAL POSTERIOR COMPARTMENT**
• Fibularis longus • Fibularis brevis	• Gastrocnemius • Soleus • Plantaris

Fibula

Tibia

ANTERIOR COMPARTMENT	**DEEP POSTERIOR COMPARTMENT**
• Tibialis anterior • Extensor hallucis longus • Extensor digitorum longus	• Popliteus • Flexor hallucis longus • Flexor digitorum longus • Tibialis posterior

Figure 11–23 **Musculoskeletal Compartments.** A section through the leg, with the muscles partially removed. A section through the thigh or arm would show a comparable arrangement of dense connective-tissue partitions.

- *Respiratory System:* Respiratory rate and depth of respiration increase. Air moves into and out of the lungs more quickly, keeping pace with the increased rate of blood flow through the lungs.

- *Integumentary System:* Blood vessels dilate, and sweat gland secretion increases. This combination promotes evaporation at the skin surface and removes the excess heat generated by muscular activity.

- *Nervous and Endocrine Systems:* The above responses of other systems are directed and coordinated through neural and hormonal adjustments in heart rate, respiratory rate, sweat gland activity, and mobilization of stored nutrient reserves.

Even when the body is at rest, the muscular system has extensive interactions with other systems. **Figure 11–24** sum-

marizes the range of interactions between the muscular system and other vital systems.

CHECKPOINT

26. **What major function does the muscular system perform for the body as a whole?**

27. **Identify the physiological effects of exercise on the cardiovascular, respiratory, and integumentary systems, and indicate the relationship between those physiological effects and the nervous and endocrine systems.**

See the blue Answers tab at the end of the book.

THE MUSCULAR SYSTEM IN PERSPECTIVE

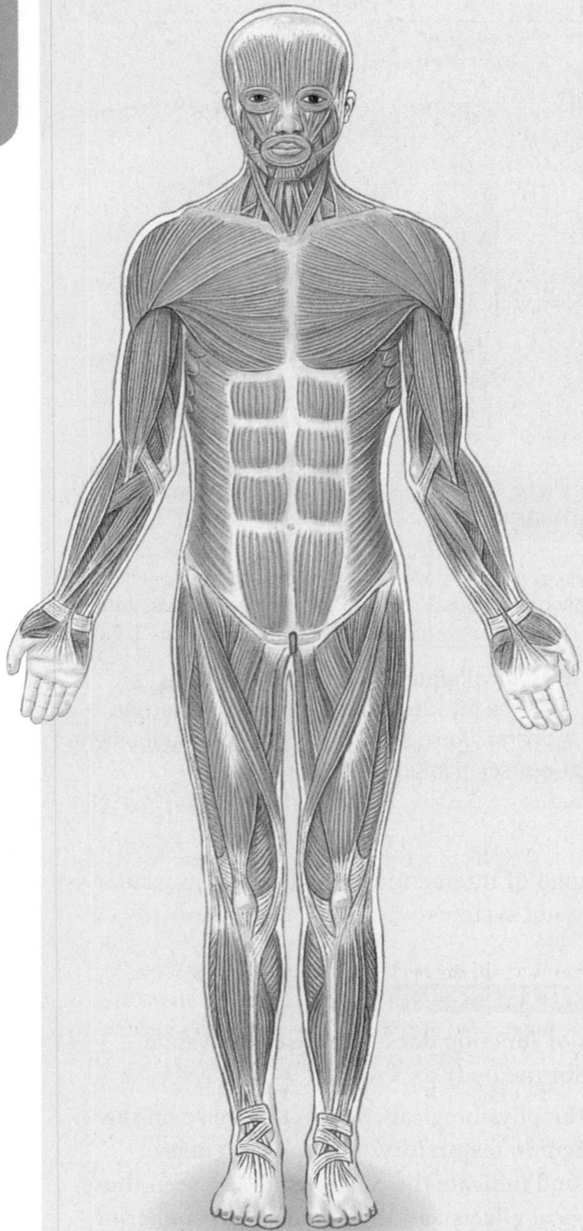

Integumentary System

- Removes excess body heat; synthesizes vitamin D_3 for calcium and phosphate absorption; protects underlying muscles
- Skeletal muscles pulling on skin of face produce facial expressions

Skeletal System

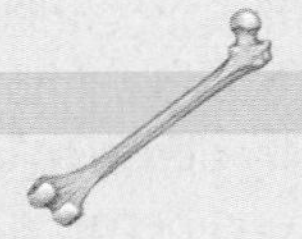

- Maintains normal calcium and phosphate levels in body fluids; supports skeletal muscles; provides sites of attachment
- Provides movement and support; stresses exerted by tendons maintain bone mass; stabilizes bones and joints

Nervous System

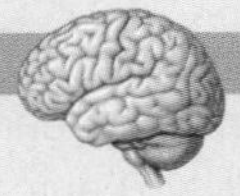

- Controls skeletal muscle contractions; adjusts activities of respiratory and cardiovascular systems during periods of muscular activity
- Muscle spindles monitor body position; facial muscles express emotion; intrinsic laryngeal muscles make speech possible

Endocrine System

- Hormones adjust muscle metabolism and growth; parathyroid hormone and calcitonin regulate calcium and phosphate ion concentrations; hormones release glucose and lipids from storage sites
- Skeletal muscles provide protection for some endocrine organs

Cardiovascular System

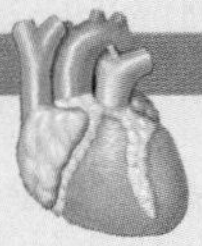

- Delivers oxygen and nutrients; removes carbon dioxide, lactic acid, and heat
- Skeletal muscle contractions assist in moving blood through veins; protects deep blood vessels

Lymphoid System

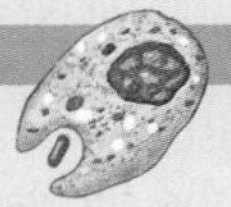

- Defends skeletal muscles against infection and assists in tissue repairs after injury
- Protects superficial lymph nodes and the lymphatic vessels in the abdominopelvic cavity

Respiratory System

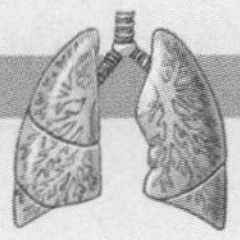

- Provides oxygen and eliminates carbon dioxide
- Muscles generate carbon dioxide; muscles control entrances to respiratory tract, fill and empty lungs, control airflow through larynx, and produce sounds

Digestive System

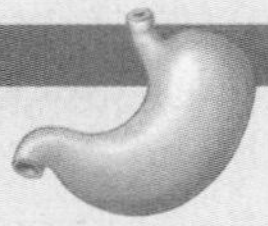

- Provides nutrients; liver regulates blood glucose and fatty acid levels and removes lactic acid from circulation
- Protects and supports soft tissues in abdominal cavity; controls entrances to and exits from digestive tract

Urinary System

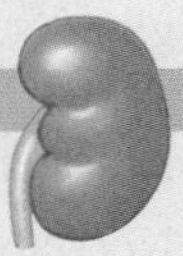

- Removes waste products of protein metabolism; assists in regulation of calcium and phosphate concentrations
- External sphincter controls urination by constricting urethra

Reproductive System

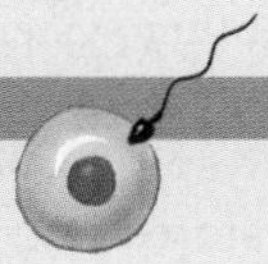

- Reproductive hormones accelerate skeletal muscle growth
- Contractions of skeletal muscles eject semen from male reproductive tract; muscle contractions during sex produce pleasurable sensations

FOR ALL SYSTEMS
Generates heat that maintains normal body temperature

Figure 11–24   Functional Relationships between the Muscular System and Other Systems.

Related Clinical Terms

carpal tunnel syndrome: An inflammation of the sheath surrounding the flexor tendons of the palm; leads to nerve compression, pain, and weakness.

compartment syndrome: Ischemia resulting from accumulated blood and fluid trapped within a musculoskeletal compartment.

diaphragmatic hernia (*hiatal hernia*): A hernia that occurs when abdominal organs are forced into the thoracic cavity.

fibrosis: The formation of fibrous connective tissue; in muscles, the replacement of muscle tissue by fibrous connective tissue makes muscles weaker and less flexible.

hernia: A condition wherein an organ or a body part protrudes through an abnormal opening.

inguinal hernia: A condition in which the inguinal canal enlarges and abdominal contents are forced into it.

intramuscular (IM) injection: The administration of a drug by injecting it into the mass of a large skeletal muscle.

ischemia: Insufficient blood supply ("blood starvation") resulting from the compression of regional blood vessels.

rotator cuff: The muscles that surround the shoulder joint; a common site of sports injuries.

Chapter Review

Study Outline

11-1 Fascicle arrangement is correlated with muscle power and range of motion p. 336

1. Structural variations among skeletal muscles affect their power, range, and speed of movement.
2. A muscle can be classified as a **parallel muscle**, **convergent muscle**, **pennate muscle**, or **circular muscle (sphincter)** according to the arrangement of fibers and fascicles in it. A pennate muscle may be *unipennate*, *bipennate*, or *multipennate*. (*Figure 11–1*)

11-2 The three classes of levers increase muscle efficiency p. 337

3. A **lever** is a rigid structure that moves around a fixed point called the **fulcrum**. Levers can change the direction and effective strength of an applied force, and the distance and speed of the movement such a force produces.
4. Levers are classified as **first-class**, **second-class**, or **third-class levers**. Third-class levers are the most common levers in the body. (*Figure 11–2*)

11-3 Muscle origins are at the fixed end of muscles, whereas insertions are at the movable end of muscles p. 339

5. Each muscle can be identified by its *origin*, *insertion*, and *action*.
6. The site of attachment of the fixed end of a muscle is called the **origin**; the site where the movable end of the muscle attaches to another structure is called the **insertion**.
7. The movement produced when a muscle contracts is its **action**.
8. According to the function of its action, a muscle can be classified as an **agonist**, or **prime mover**; an **antagonist**; a **synergist**; or a **fixator**.

11-4 Descriptive terms are used to name skeletal muscles p. 341

9. The names of muscles commonly provide clues to their body region, origin and insertion, fascicle organization, relative position, structural characteristics, and action. (*Table 11–1*)
10. The **axial musculature** arises on the axial skeleton; it positions the head and spinal column and moves the rib cage. The **appendicular musculature** stabilizes or moves components of the appendicular skeleton. (*Figure 11–3*)

11. Innervation refers to the distribution of nerves that control a region or organ, including a muscle.

11-5 Axial muscles are muscles of the head and neck, vertebral column, trunk, and pelvic floor p. 345

12. The axial muscles fall into logical groups on the basis of location, function, or both.
13. The principal muscles of facial expression are the **orbicularis oris**, **buccinator**, and **occipitofrontalis** muscles and the **platysma**. (*Figure 11–4; Table 11–2*)
14. Six extrinsic eye muscles (**extra-ocular muscles** or *oculomotor muscles*) control eye movements: the **inferior** and **superior rectus** muscles, the **lateral** and **medial rectus** muscles, and the **inferior** and **superior oblique** muscles. (*Figure 11–5; Table 11–3*)
15. The muscles of mastication (chewing) are the **masseter**, **temporalis**, and **pterygoid** muscles. (*Figure 11–6; Table 11–4*)
16. The muscles of the tongue are necessary for speech and swallowing and assist in mastication. They are the **palatoglossus**, **styloglossus**, **genioglossus**, and **hyoglossus** muscles. (*Figure 11–7; Table 11–5*)
17. The muscles of the pharynx constrict the pharyngeal walls (**pharyngeal constrictors**), elevate the larynx (**laryngeal elevators**), or raise the soft palate (**palatal muscles**). (*Figure 11–8; Table 11–6*)
18. The anterior muscles of the neck control the position of the larynx, depress the mandible, and provide a foundation for the muscles of the tongue and pharynx. The neck muscles include the **digastric** and **sternocleidomastoid** muscles and seven muscles that originate or insert on the hyoid bone. (*Figure 11–9; Table 11–7*)
19. The superficial muscles of the spine can be classified into the **spinalis**, **longissimus**, and **iliocostalis** groups. (*Figure 11–10; Table 11–8*)
20. Other muscles of the spine include the **longus capitis** and **longus colli** muscles of the neck, the small intervertebral muscles of the deep layer, and the **quadratus lumborum** muscle of the lumbar region. (*Figure 11–10; Table 11–8*)
21. The oblique muscles include the **scalene** muscles and the **intercostal** and **transversus** muscles. The **external** and **internal intercostal** muscles are important in respiratory

movements of the ribs. Also important to respiration is the **diaphragm**. (*Figures 11–10, 11–11; Table 11–9*)

22. The **perineum** can be divided into an anterior **urogenital triangle** and a posterior **anal triangle**. The pelvic floor consists of the **urogenital diaphragm** and the **pelvic diaphragm**. (*Figure 11–12; Table 11–10*)

11-6 Appendicular muscles are muscles of the shoulders, upper limbs, pelvic girdle, and lower limbs p. 359

23. The **trapezius** muscle affects the positions of the shoulder girdle, head, and neck. Other muscles inserting on the scapula include the **rhomboid**, **levator scapulae**, **serratus anterior**, **subclavius**, and **pectoralis minor** muscles. (*Figures 11–13, 11–15; Table 11–11*)

24. The **deltoid** and the **supraspinatus** muscles are important abductors. The **subscapularis** and **teres major** muscles produce medial rotation at the shoulder; the **infraspinatus** and **teres minor** muscles produce lateral rotation; and the **coracobrachialis** muscle produces flexion and adduction at the shoulder. (*Figures 11–13, 11–14, 11–15; Table 11–12*)

25. The **pectoralis major** muscle flexes the shoulder joint, and the **latissimus dorsi** muscle extends it. (*Figures 11–13, 11–14, 11–15; Table 11–12*)

26. The actions of the **biceps brachii** muscle and the **triceps brachii** muscle (long head) affect the elbow joint. The **brachialis** and **brachioradialis** muscles flex the elbow, opposed by the **anconeus** muscle. The **flexor carpi ulnaris**, **flexor carpi radialis**, and **palmaris longus** muscles cooperate to flex the wrist. They are opposed by the **extensor carpi radialis** muscles and the **extensor carpi ulnaris** muscle. The **pronator teres** and **pronator quadratus** muscles pronate the forearm and are opposed by the **supinator** muscle. (*Figures 11–15 to 11–18; Tables 11–13 to 11–15*)

27. **Gluteal muscles** cover the lateral surfaces of the ilia. The largest is the **gluteus maximus** muscle, which shares an insertion with the **tensor fasciae latae**. Together, these muscles pull on the **iliotibial tract**. (*Figures 11–13, 11–19; Table 11–16*)

28. The **piriformis** muscle and the **obturator** muscles are the most important **lateral rotators**. The **adductors** can produce a variety of movements. (*Figure 11–19; Table 11–16*)

29. The **psoas major** and **iliacus** muscles merge to form the **iliopsoas** muscle, a powerful flexor of the hip. (*Figures 11–19, 11–20; Table 11–16*)

30. The flexors of the knee include the **biceps femoris**, **semimembranosus**, and **semitendinosus** muscles (the three **hamstrings**) and the **sartorius** muscle. The **popliteus** muscle unlocks the knee joint. (*Figures 11–20, 11–21; Table 11–17*)

31. Collectively, the knee extensors are known as the **quadriceps femoris**. This group consists of the three **vastus** muscles and the **rectus femoris** muscle. (*Figure 11–20; Table 11–17*)

32. The **gastrocnemius** and **soleus** muscles produce plantar flexion (ankle extension). A pair of **fibularis** muscles produces eversion as well as extension (plantar flexion) at the ankle. (*Figure 11–21; Table 11–18*)

33. Smaller muscles of the calf and shin position the foot and move the toes. Precise control of the phalanges is provided by muscles originating at the tarsal and metatarsal bones. (*Figure 11–22; Table 11–19*)

11-7 With advancing age, the size and power of muscle tissue decrease p. 378

34. With aging, the size and power of all muscle tissues decrease. Skeletal muscles undergo **fibrosis**, the tolerance for exercise decreases, and repair of injuries slows.

11-8 Exercise produces responses in multiple body systems p. 380

35. Exercise illustrates the integration of the muscular system with the cardiovascular, respiratory, integumentary, nervous, and endocrine systems. (*Figure 11–24*)

Review Questions

See the blue Answers tab at the end of the book.

myA&P Access more review material online at **myA&P**™ (www.myaandp.com). There, you'll find chapter guides, chapter quizzes, practice tests, animations, flashcards, a glossary with pronunciations, *Interactive Physiology*® (IP) exercises and quizzes, and more to help you succeed in the course.

LEVEL 1   Reviewing Facts and Terms

1. Name the three pennate muscles in the following figure, and for each muscle indicate the type of pennate muscle based on the relationship of muscle fibers to the tendon.
   (a) _____
   (b) _____
   (c) _____

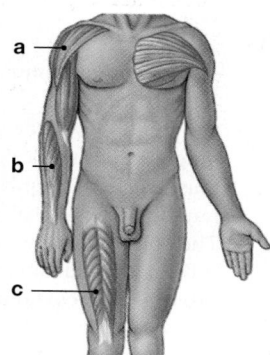

2. Label the three visible muscles of the rotator cuff in the following posterior view of the deep muscles that move the arm.
   (a) _____
   (b) _____
   (c) _____

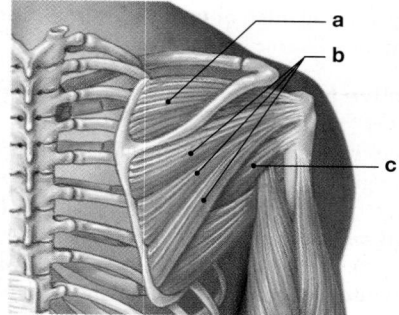

3. The bundles of muscle fibers within a skeletal muscle are called
   - (a) muscles.
   - (b) fascicles.
   - (c) fibers.
   - (d) myofilaments.
   - (e) groups.

4. Levers make muscle action more versatile by all of the following, *except*
   - (a) changing the location of the muscle's insertion.
   - (b) changing the speed of movement produced by an applied force.
   - (c) changing the distance of movement produced by an applied force.
   - (d) changing the strength of an applied force.
   - (e) changing the direction of an applied force.

5. The more movable end of a muscle is the
   - (a) insertion.
   - (b) belly.
   - (c) origin.
   - (d) proximal end.
   - (e) distal end.

6. The muscles of facial expression are innervated by cranial nerve
   - (a) VII.
   - (b) V.
   - (c) IV.
   - (d) VI.

7. The strongest masticatory muscle is the _____ muscle.
   - (a) pterygoid
   - (b) masseter
   - (c) temporalis
   - (d) mandible

8. The muscle that rotates the eye medially is the _____ muscle.
   - (a) superior oblique
   - (b) inferior rectus
   - (c) medial rectus
   - (d) lateral rectus

9. Important flexors of the vertebral column that act in opposition to the erector spinae are the _____ muscles.
   - (a) rectus abdominis
   - (b) longus capitis
   - (c) longus colli
   - (d) scalene

10. The major extensor of the elbow is the _____ muscle.
    - (a) triceps brachii
    - (b) biceps brachii
    - (c) deltoid
    - (d) subscapularis

11. The muscles that rotate the radius without producing either flexion or extension of the elbow are the _____ muscles.
    - (a) brachialis and brachioradialis
    - (b) pronator teres and supinator
    - (c) biceps brachii and triceps brachii
    - (d) a, b, and c

12. The powerful flexors of the hip are the _____ muscles.
    - (a) piriformis
    - (b) obturator
    - (c) pectineus
    - (d) iliopsoas

13. Knee extensors known as the quadriceps consist of the
    - (a) three vastus muscles and the rectus femoris muscle.
    - (b) biceps femoris, gracilis, and sartorius muscles.
    - (c) popliteus, iliopsoas, and gracilis muscles.
    - (d) gastrocnemius, tibialis, and peroneus muscles.

14. List the four fascicle organizations that produce the different patterns of skeletal muscles.

15. What is an aponeurosis? Give two examples.

16. Which four muscle groups make up the axial musculature?

17. What three functions are accomplished by the muscles of the pelvic floor?

18. On which bones do the four rotator cuff muscles originate and insert?

19. What three functional groups make up the muscles of the lower limbs?

LEVEL 2   Reviewing Concepts

20. Of the following actions, the one that illustrates that of a second-class lever is
    - (a) knee extension.
    - (b) ankle extension (plantar flexion).
    - (c) flexion at the elbow.
    - (d) a, b, and c.

21. Compartment syndrome can result from all of the following *except*
    - (a) compressing a nerve in the wrist.
    - (b) compartments swelling with blood due to an injury involving blood vessels.
    - (c) torn ligaments in a given compartment.
    - (d) pulled tendons in the muscles of a given compartment.
    - (e) torn muscles in a particular compartment.

22. A(n) _____ develops when an organ protrudes through an abnormal opening.

23. Elongated bursae that reduce friction and surround the tendons that cross the dorsal and ventral surfaces of the wrist form _____.

24. The muscles of the vertebral column include many dorsal extensors but few ventral flexors. Why?

25. Why does a convergent muscle exhibit more versatility when contracting than does a parallel muscle?

26. Why can a pennate muscle generate more tension than can a parallel muscle of the same size?

27. Why is it difficult to lift a heavy object when the elbow is at full extension?

28. Which types of movements are affected when the hamstrings are injured?

LEVEL 3   Critical Thinking and Clinical Applications

29. Mary sees Jill coming toward her and immediately contracts her frontalis and procerus muscles. She also contracts her right levator labii muscles. Is Mary glad to see Jill? How can you tell?

30. Mary's newborn is having trouble suckling. The doctor suggests that it may be a problem with a particular muscle. What muscle is the doctor probably referring to?
    - (a) orbicularis oris
    - (b) buccinator
    - (c) masseter
    - (d) risorius
    - (e) zygomaticus

31. While unloading her car trunk, Amy pulls a muscle and as a result has difficulty moving her arm. The doctor in the emergency room tells her that she pulled her pectoralis major. Amy tells you that she thought the pectoralis major was a chest muscle and doesn't understand what that has to do with her arm. What would you tell her?

12

Neural Tissue

Did you know...?
Sixty percent of the sensory neurons innervate the forehead and hands.

Learning Outcomes

After completing this chapter, you should be able to do the following:

12-1 Describe the anatomical and functional divisions of the nervous system.

12-2 Sketch and label the structure of a typical neuron, describe the functions of each component, and classify neurons on the basis of their structure and function.

12-3 Describe the locations and functions of the various types of neuroglia.

12-4 Explain how the resting potential is created and maintained.

12-5 Describe the events involved in the generation and propagation of an action potential.

12-6 Discuss the factors that affect the speed with which action potentials are propagated.

12-7 Describe the structure of a synapse, and explain the mechanism involved in synaptic activity.

12-8 Describe the major types of neurotransmitters and neuromodulators, and discuss their effects on postsynaptic membranes.

12-9 Discuss the interactions that enable information processing to occur in neural tissue.

Clinical Notes

Rabies p. 390
Tumors p. 395
Demyelination p. 397

An Introduction to Neural Tissue

This chapter considers the structure of neural tissue and introduces the basic principles of neurophysiology. The nervous system includes all the neural tissue in the body. ∞ p. 141 The basic functional units of the nervous system are individual cells called **neurons**. Supporting cells, or **neuroglia** (noo-RŌG-lē-uh or noo-rō-GLĒ-uh; *glia*, glue), separate and protect the neurons, provide a supportive framework for neural tissue, act as phagocytes, and help regulate the composition of the interstitial fluid. Neuroglia, also called *glial cells*, far outnumber neurons.

Neural tissue, with supporting blood vessels and connective tissues, forms the organs of the nervous system: the brain; the spinal cord; the receptors in complex sense organs, such as the eye and ear; and the *nerves* that link the nervous system with other systems. In the *Systems Overview* section (pp. 148–156), we introduced the two major anatomical divisions of the nervous system: the central nervous system and the peripheral nervous system.

12-1 The nervous system has anatomical and functional divisions

In this section we provide an overview of the nervous system. We begin with anatomical and functional perspectives.

The Anatomical Divisions of the Nervous System

Viewed anatomically, the nervous system has two divisions: the central nervous system and the peripheral nervous system. The **central nervous system (CNS)** consists of the spinal cord and brain. These are complex organs that include not only neural tissue, but also blood vessels and the various connective tissues that provide physical protection and support. The CNS is responsible for integrating, processing, and coordinating sensory data and motor commands. Sensory data convey information about conditions inside or outside the body. Motor commands control or adjust the activities of peripheral organs, such as skeletal muscles. When you stumble, for example, the CNS integrates information regarding your balance and the position of your limbs and then coordinates your recovery by sending motor commands to appropriate skeletal muscles—all in a split second and without conscious effort. The CNS—specifically, the brain—is also the seat of higher functions, such as intelligence, memory, learning, and emotion.

The **peripheral nervous system (PNS)** includes all the neural tissue outside the CNS. The PNS delivers sensory information to the CNS and carries motor commands to peripheral tissues and systems. Bundles of axons, or *nerve fibers*, carry sensory information and motor commands in the PNS. Such bundles, with associated blood vessels and connective tissues, are called *peripheral nerves*, or simply **nerves**. Nerves connected to the brain are called **cranial nerves**; those attached to the spinal cord are called **spinal nerves**.

The Functional Divisions of the Nervous System

The PNS is divided into afferent and efferent divisions. The **afferent division** (*ad*, to + *ferre*, to carry) of the PNS brings sensory information to the CNS from receptors in peripheral tissues and organs. **Receptors** are sensory structures that either detect changes in the internal environment or respond to the presence of specific stimuli. There are complex receptor organs, such as the eye or ear; at the cellular level, receptors range from the dendrites (slender cytoplasmic extensions) of single cells to complex organs. Receptors may be neurons or specialized cells of other tissues. ∞ p. 168

The **efferent division** (*effero*, to bring out) of the PNS carries motor commands from the CNS to muscles, glands, and adipose tissue. These target organs, which respond by *doing* something, are called **effectors**. The efferent division has both somatic and autonomic components.

Tips&Tricks

To distinguish between afferent and efferent, associate the "a" in **a**fferent with the "a" in **a**ccessing, and the "e" in **e**fferent with the "e" in **e**xiting.

- The **somatic nervous system (SNS)** controls skeletal muscle contractions. *Voluntary* contractions are under conscious control. For example, you exert conscious control over your arm as you raise a full glass of water to your lips. *Involuntary* contractions may be simple, automatic responses or complex movements, but they are controlled at the subconscious level, outside your awareness. For instance, if you accidentally place your hand on a hot stove, you will withdraw it immediately, usually before you even notice any pain. This type of automatic response is called a **reflex**.

- The **autonomic nervous system (ANS)**, or *visceral motor system*, provides automatic regulation of smooth muscle, cardiac muscle, and glandular secretions at the subconscious level. The ANS includes a *sympathetic division* and a *parasympathetic division*, which commonly have antagonistic effects. For example, activity of the sympathetic division accelerates the heart rate, whereas parasympathetic activity slows the heart rate.

Now that we have completed a brief orientation on the nervous system as a whole, we can examine the structure of neural tissue and the functional principles that govern neural activities. We begin by considering neurons, the basic functional units of the nervous system.

CHECKPOINT

1. **Identify the two anatomical divisions of the nervous system.** CNS, PNS

2. **Identify the two functional divisions of the peripheral nervous system, and cite their primary functions.** Afferent: to CNS Efferent: from CNS

3. **Identify the two components of the efferent division of the PNS.** Somatic / Autonomic

4. **What would be the effect of damage to the afferent division of the PNS?** not experience sensory stimuli

See the blue Answers tab at the end of the book.

12-2 Neurons are nerve cells specialized for intercellular communication

In this section, we will examine the structure of a representative neuron before considering the structural and functional classifications of neurons.

The Structure of Neurons

Figure 12–1 shows the structure of a representative neuron. Neurons have a variety of shapes. The one shown is a *multipolar neuron*, the most common type of neuron in the central nervous system. Each multipolar neuron has a large *cell body* that is connected to a single, elongate *axon* and several short, branched *dendrites*.

The Cell Body

The **cell body**, or *soma* (plural, *somata*), contains a relatively large, round nucleus with a prominent nucleolus (**Figure 12–1**). The cytoplasm surrounding the nucleus constitutes the **perikaryon** (per-i-KAR-ē-on; *peri*, around +

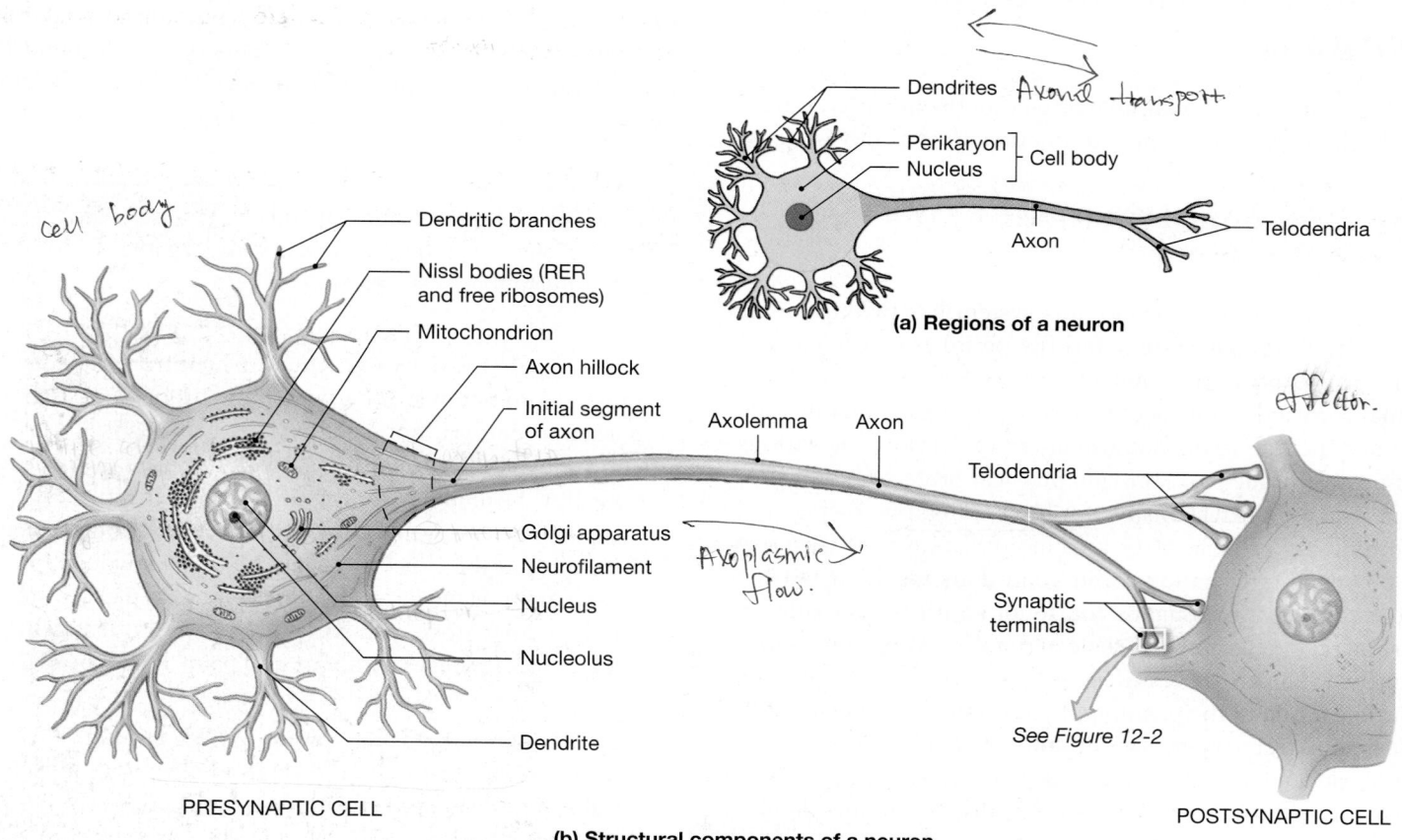

(a) Regions of a neuron

PRESYNAPTIC CELL

(b) Structural components of a neuron

POSTSYNAPTIC CELL

Figure 12–1   The Anatomy of a Multipolar Neuron. (a) The general structure of a neuron and its primary components. **(b)** A more detailed view of a neuron, showing major organelles.

karyon, nucleus). The cytoskeleton of the perikaryon contains **neurofilaments** and **neurotubules**, which are similar to the intermediate filaments and microtubules of other types of cells. Bundles of neurofilaments, called **neurofibrils**, extend into the dendrites and axon, providing internal support for these slender processes.

The perikaryon contains organelles that provide energy and synthesize organic materials, especially the chemical neurotransmitters that are important in cell-to-cell communication. ∞ p. 304 The numerous mitochondria, free and fixed ribosomes, and membranes of rough endoplasmic reticulum (RER) give the perikaryon a coarse, grainy appearance. Mitochondria generate ATP to meet the high energy demands of an active neuron. The ribosomes and RER synthesize proteins. Some areas of the perikaryon contain clusters of RER and free ribosomes. These regions, which stain darkly, are called *Nissl bodies,* because they were first described by the German microscopist Franz Nissl. Nissl bodies account for the gray color of areas containing neuron cell bodies—the *gray matter* seen in gross dissection.

Most neurons lack centrioles, important organelles involved in the organization of the cytoskeleton and the movement of chromosomes during mitosis. ∞ p. 102 As a result, typical CNS neurons cannot divide; thus, they cannot be replaced if lost to injury or disease. Although neural stem cells persist in the adult nervous system, they are typically inactive except in the nose, where the regeneration of olfactory (smell) receptors maintains our sense of smell, and in the *hippocampus,* a portion of the brain involved with memory storage. The control mechanisms that trigger neural stem cell activity are now being investigated, with the goal of preventing or reversing neuron loss due to trauma, disease, or aging.

Dendrites and Axons

A variable number of slender, sensitive processes known as **dendrites** extend out from the cell body (**Figure 12–1**). Typical dendrites are highly branched, and each branch bears fine 0.5- to 1-μm-long studded processes called *dendritic spines.* In the CNS, a neuron receives information from other neurons primarily at the dendritic spines, which represent 80–90 percent of the neuron's total surface area.

An **axon** is a long cytoplasmic process capable of propagating an electrical impulse known as an *action potential.* ∞ p. 304 The **axoplasm** (AK-sō-plazm), or cytoplasm of the axon, contains neurofibrils, neurotubules, small vesicles, lysosomes, mitochondria, and various enzymes. The axoplasm is surrounded by the **axolemma** (*lemma,* husk), a specialized portion of the plasma membrane. In the CNS, the axolemma may be exposed to the interstitial fluid or covered by the processes of neuroglia. The base, or **initial segment**, of the axon in a multipolar neuron is attached to the cell body at a thickened region known as the **axon hillock** (**Figure 12–1**).

An axon may branch along its length, producing side branches collectively known as **collaterals**. Collaterals enable a single neuron to communicate with several other cells. The main axon trunk and any collaterals end in a series of fine extensions, or **telodendria** (tel-ō-DEN-drē-uh; *telo-,* end + *dendron,* tree) (**Figure 12–1**). The telodendria of an axon end at **synaptic terminals**.

The Synapse

Each synaptic terminal is part of a **synapse**, a specialized site where the neuron communicates with another cell (**Figure 12–2**). Every synapse involves two cells: (1) the *presynaptic cell,* which

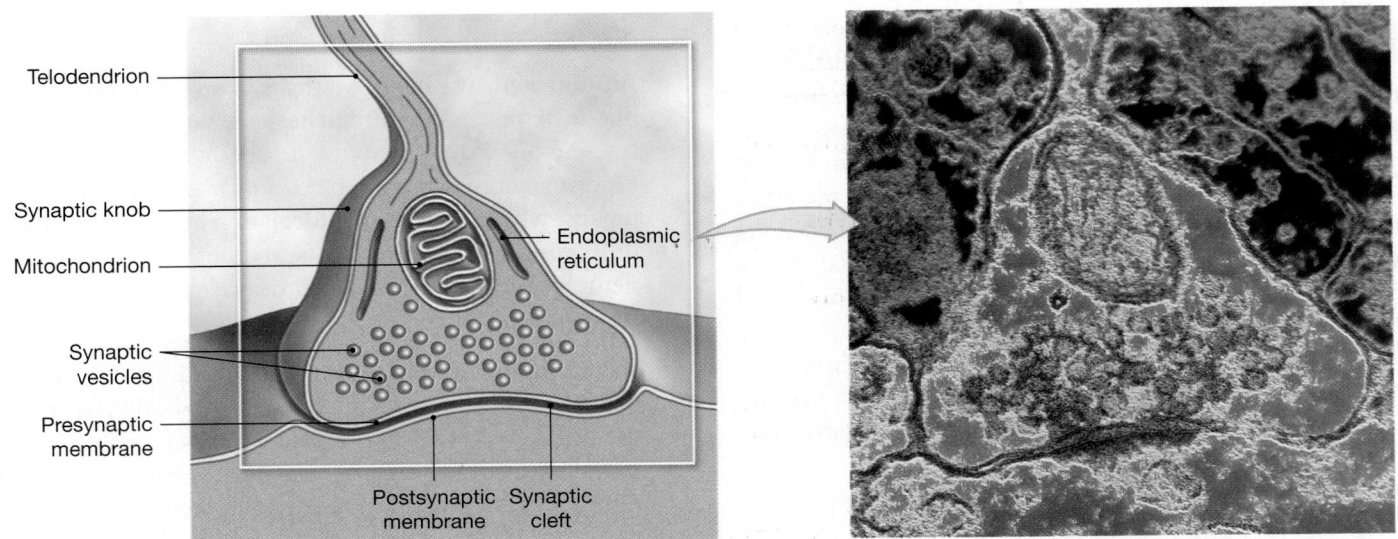

Figure 12–2 **The Structure of a Typical Synapse.** A diagrammatic view (at left) and a micrograph (at right) of a typical synapse between two neurons. (TEM, color enhanced, ×222,000).

includes the synaptic terminal and sends a message; and (2) the *postsynaptic cell*, which receives the message. The communication between cells at a synapse most commonly involves the release of chemicals called **neurotransmitters** by the synaptic terminal. These chemicals, released by the presynaptic cell, affect the activity of the postsynaptic cell. As we saw in Chapter 10, this release is triggered by electrical events, such as the arrival of an action potential. Neurotransmitters are typically packaged in *synaptic vesicles.*

The presynaptic cell is usually a neuron. (Specialized receptor cells may form synaptic connections with the dendrites of neurons, a process that will be described in Chapter 15.) The postsynaptic cell can be either a neuron or another type of cell. When one neuron communicates with another, the synapse may occur on a dendrite, on the cell body, or along the length of the axon of the receiving cell. A synapse between a neuron and a muscle cell is called a **neuromuscular junction**. ∞ p. 304 At a **neuroglandular junction**, a neuron controls or regulates the activity of a secretory (gland) cell. Neurons also innervate a variety of other cell types, such as adipocytes (fat cells). We will consider the nature of that innervation in later chapters.

The structure of the synaptic terminal varies with the type of postsynaptic cell. A relatively simple, round **synaptic knob** occurs where the postsynaptic cell is another neuron.[1] At a synapse, a narrow *synaptic cleft* separates the **presynaptic membrane**, where neurotransmitters are released, from the **postsynaptic membrane**, which bears receptors for neurotransmitters (**Figure 12–2**). The synaptic terminal at a neuromuscular junction is much more complex. We will primarily consider the structure of synaptic knobs in this chapter, leaving the details of other types of synaptic terminals to later chapters.

Each synaptic knob contains mitochondria, portions of the endoplasmic reticulum, and thousands of vesicles filled with neurotransmitter molecules. Breakdown products of neurotransmitters released at the synapse are reabsorbed and reassembled at the synaptic knob, which also receives a continuous supply of neurotransmitters synthesized in the cell body, along with enzymes and lysosomes. These materials travel the length of the axon along neurotubules, pulled along by "molecular motors," called *kinesin* and *dynein*, that run on ATP. The movement of materials between the cell body and synaptic knobs is called **axoplasmic transport**. Some materials travel slowly, at rates of a few millimeters per day. This transport mechanism is known as the "slow stream." Vesicles containing neurotransmitters move much more rapidly, traveling in the "fast stream" at 5–10 mm per hour.

Axoplasmic transport occurs in both directions. The flow of materials from the cell body to the synaptic knob is *anterograde* (AN-ter-ō-grād; *antero-*, forward) *flow,* carried by kinesin. At the same time, other substances are being transported toward the cell body in *retrograde* (RET-rō-grād) *flow* (*retro*, backward), carried by dynein. If debris or unusual chemicals appear in the synaptic knob, retrograde flow soon delivers them to the cell body. The arriving materials may then alter the activity of the cell by turning appropriate genes on or off.

CLINICAL NOTE

Rabies   *Rabies* is perhaps the most dramatic example of a clinical condition directly related to retrograde flow. A bite from a rabid animal injects the rabies virus into peripheral tissues, where virus particles quickly enter synaptic knobs and peripheral axons. Retrograde flow then carries the virus into the CNS, with fatal results. Many toxins (including heavy metals), some pathogenic bacteria, and other viruses also bypass CNS defenses by exploiting axoplasmic transport.

The Classification of Neurons

Neurons can be grouped by structure or by function.

Structural Classification of Neurons

Neurons are classified as anaxonic, bipolar, unipolar, or multipolar on the basis of the relationship of the dendrites to the cell body and the axon (**Figure 12–3**):

- **Anaxonic** (an-ak-SON-ik) **neurons** are small and have no anatomical features that distinguish dendrites from axons; all the cell processes look alike. Anaxonic neurons are located in the brain and in special sense organs. Their functions are poorly understood.

- **Bipolar neurons** have two distinct processes—one dendritic process that branches extensively at its distal tip, and one axon—with the cell body between the two. Bipolar neurons are rare, but occur in special sense organs, where they relay information about sight, smell, or hearing from receptor cells to other neurons. Bipolar neurons are small; the largest measure less than 30 μm from end to end.

- In a **unipolar neuron**, or *pseudounipolar neuron*, the dendrites and axon are continuous—basically, fused—and the cell body lies off to one side. In such a neuron, the initial segment lies where the dendrites converge. The rest of the process, which carries action potentials, is usually considered to be an axon. Most sensory neurons of the peripheral nervous system are unipolar. Their axons may extend a meter or more, ending at synapses in the central

---

[1] The term *synaptic knob* is widely recognized and will be used throughout this text. However, the same structures are also called terminal buttons, terminal boutons, end bulbs, or neuropods.

ANAXONIC NEURON	BIPOLAR NEURON	UNIPOLAR NEURON	MULTIPOLAR NEURON

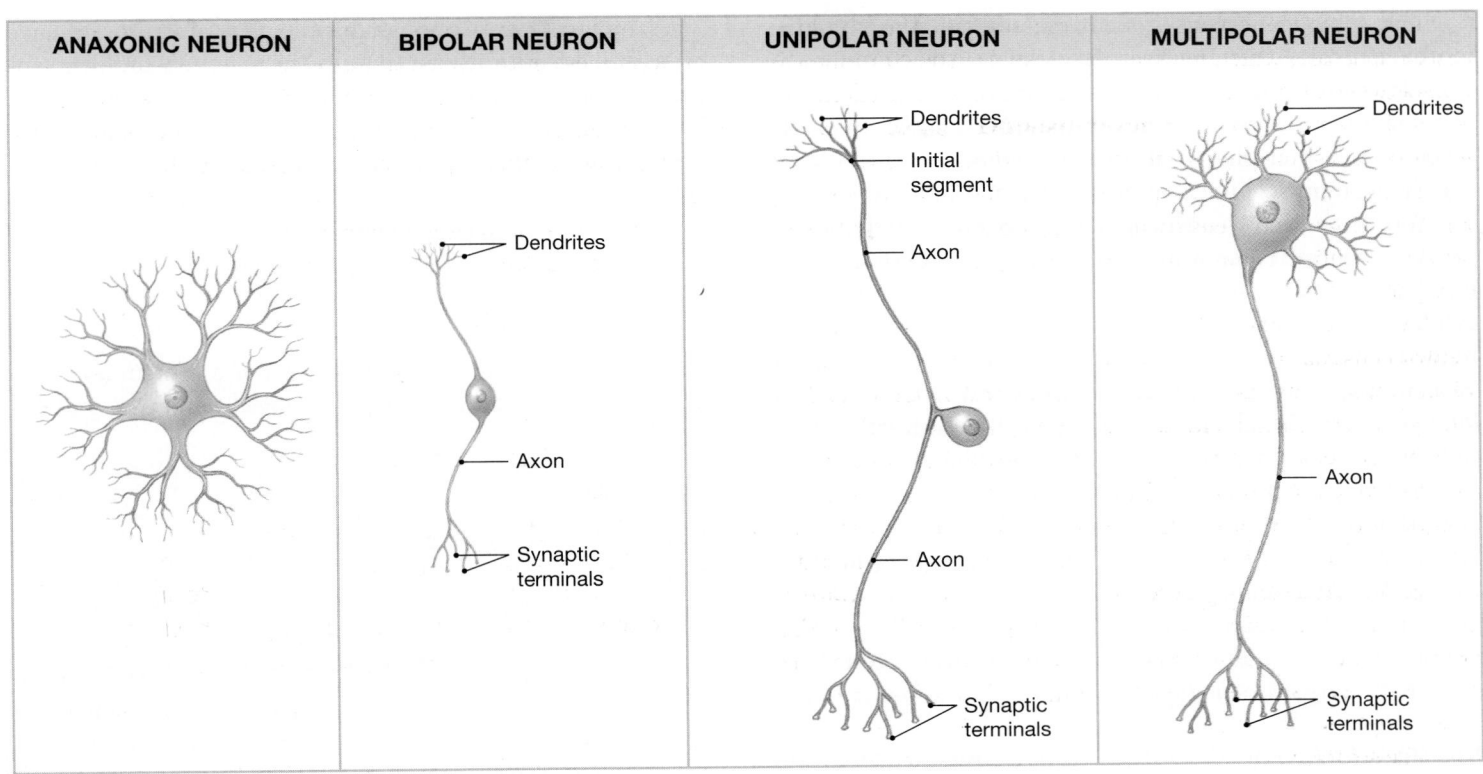

Figure 12–3   A Structural Classification of Neurons. The neurons are not drawn to scale; typical anaxonic neurons and bipolar neurons are much smaller than typical unipolar or multipolar neurons.

nervous system. The longest are those carrying sensations from the tips of the toes to the spinal cord.

- **Multipolar neurons** have two or more dendrites and a single axon. These are the most common neurons in the CNS. All the motor neurons that control skeletal muscles, for example, are multipolar neurons. The axons of multipolar neurons can be as long as those of unipolar neurons; the longest carry motor commands from the spinal cord to small muscles that move the toes.

Functional Classification of Neurons

Alternatively, neurons are categorized by function as (1) sensory neurons, (2) motor neurons, or (3) interneurons.

Sensory Neurons. Sensory neurons, or *afferent neurons*, form the afferent division of the PNS. They deliver information from sensory receptors to the CNS. The cell bodies of sensory neurons are located in peripheral *sensory ganglia*. (A *ganglion* is a collection of neuron cell bodies in the PNS.) Sensory neurons are unipolar neurons with processes, known as **afferent fibers**, that extend between a sensory receptor and the central nervous system (spinal cord or brain). The human body's 10 million or so sensory neurons collect information concerning the external or internal environment. **Somatic sensory neurons** monitor the outside world and our position

within it; **visceral sensory neurons** monitor internal conditions and the status of other organ systems.

Sensory receptors are either the processes of specialized sensory neurons or cells monitored by sensory neurons. Receptors are broadly categorized as follows:

- **Interoceptors** (*intero-*, inside) monitor the digestive, respiratory, cardiovascular, urinary, and reproductive systems and provide sensations of taste, deep pressure, and pain.

- **Exteroceptors** (*extero-*, outside) provide information about the external environment in the form of touch, temperature, or pressure sensations and the more complex senses of sight, smell, and hearing.

- **Proprioceptors** (prō-prē-ō-SEP-torz) monitor the position and movement of skeletal muscles and joints.

Motor Neurons. Motor neurons, or *efferent neurons*, form the efferent division of the PNS. These neurons carry instructions from the CNS to peripheral effectors in a peripheral tissue, organ, or organ system. The human body has about half a million motor neurons. Axons traveling away from the CNS are called **efferent fibers**. As noted earlier, the two major efferent systems are the somatic nervous system (SNS) and the autonomic (visceral) nervous system (ANS).

The somatic nervous system includes all the **somatic motor neurons** that innervate skeletal muscles. You have conscious control over the activity of the SNS. The cell body of a somatic motor neuron lies in the CNS, and its axon extends into the periphery within a peripheral nerve to innervate skeletal muscle fibers at neuromuscular junctions.

You do not have conscious control over the activities of the ANS. **Visceral motor neurons** innervate all peripheral effectors other than skeletal muscles—that is, smooth muscle, cardiac muscle, glands, and adipose tissue throughout the body. The axons of visceral motor neurons in the CNS innervate a second set of visceral motor neurons in peripheral *autonomic ganglia*. The neurons whose cell bodies are located in those ganglia innervate and control peripheral effectors.

To get from the CNS to a visceral effector such as a smooth muscle cell, the signal must travel along one axon, be relayed across a synapse, and then travel along a second axon to its final destination. The axons extending from the CNS to an autonomic ganglion are called *preganglionic fibers*; axons connecting the ganglion cells with the peripheral effectors are known as *postganglionic fibers*.

Interneurons. The 20 billion or so **interneurons**, or *association neurons*, outnumber all other types of neurons combined. Although most are located within the brain and spinal cord, some are in autonomic ganglia. Interneurons are responsible for both the distribution of sensory information and the coordination of motor activity. One or more interneurons are situated between sensory neurons and motor neurons; the more complex the response to a given stimulus, the more interneurons are involved. Interneurons are also involved with all higher functions, such as memory, planning, and learning.

We next turn our attention to the neuroglia, cells that support and protect the neurons.

CHECKPOINT

5. Name structural components of a typical neuron.
6. Classify neurons according to their structure.
7. Classify neurons according to their function.
8. Are unipolar neurons in a tissue sample more likely to function as sensory neurons or motor neurons?

See the blue Answers tab at the end of the book.

12-3 CNS and PNS neuroglia support and protect neurons

Neuroglia are abundant and diverse, and they account for roughly half of the volume of the nervous system. The organization of neural tissue in the CNS differs from that in the PNS, primarily because of the greater variety of neuroglial cell types in the CNS. Although histological descriptions have been available for the past century, the technical problems involved in isolating and manipulating individual glial cells have limited our understanding of their functions. **Figure 12–4** summarizes what we know about the major neuroglia populations in the CNS and PNS.

Neuroglia of the Central Nervous System

The central nervous system has four types of neuroglia: (1) *ependymal cells*, (2) *astrocytes*, (3) *oligodendrocytes*, and (4) *microglia* (**Figure 12–5**).

Ependymal Cells

A fluid-filled central passageway extends along the longitudinal axis of the spinal cord and brain. This passageway is filled with **cerebrospinal fluid (CSF)**, which also surrounds the brain and spinal cord. This fluid, which circulates continuously, provides a protective cushion and transports dissolved gases, nutrients, wastes, and other materials. The diameter of the internal passageway varies from one region to another. The narrow passageway in the spinal cord is called the *central canal* (**Figure 12–5a,b**). In several regions of the brain, the passageway forms enlarged chambers called *ventricles*. The central canal and ventricles are lined by **ependymal cells**, which form an epithelium known as the **ependyma** (ep-EN-di-muh).

During embryonic development and early childhood, the free surfaces of ependymal cells are covered with cilia. The cilia persist in adults only within the ventricles of the brain, where they assist in the circulation of the CSF. In other areas, the ependymal cells typically have scattered microvilli. In a few parts of the brain, specialized ependymal cells participate in the secretion of the CSF. Other regions of the ependyma may have sensory functions, such as monitoring the composition of the CSF. It appears that the ependyma in adults contains stem cells that can divide to produce additional neurons. The specific regulatory mechanisms involved are now being investigated.

Unlike typical epithelial cells, ependymal cells have slender processes that branch extensively and make direct contact with neuroglia in the surrounding neural tissue. The functions of these connections are not known. During early embryonic development, stem cells line the central canal and ventricles; the divisions of these stem cells give rise to neurons and all CNS neuroglia other than microglia.

Astrocytes

Astrocytes (AS-trō-sīts; *astro-*, star + *cyte*, cell), are the largest and most numerous neuroglia in the CNS (**Figure 12–5b**). These cells have a variety of functions, many of them poorly understood:

- *Maintaining the Blood–Brain Barrier.* Compounds dissolved in the circulating blood do not have free access

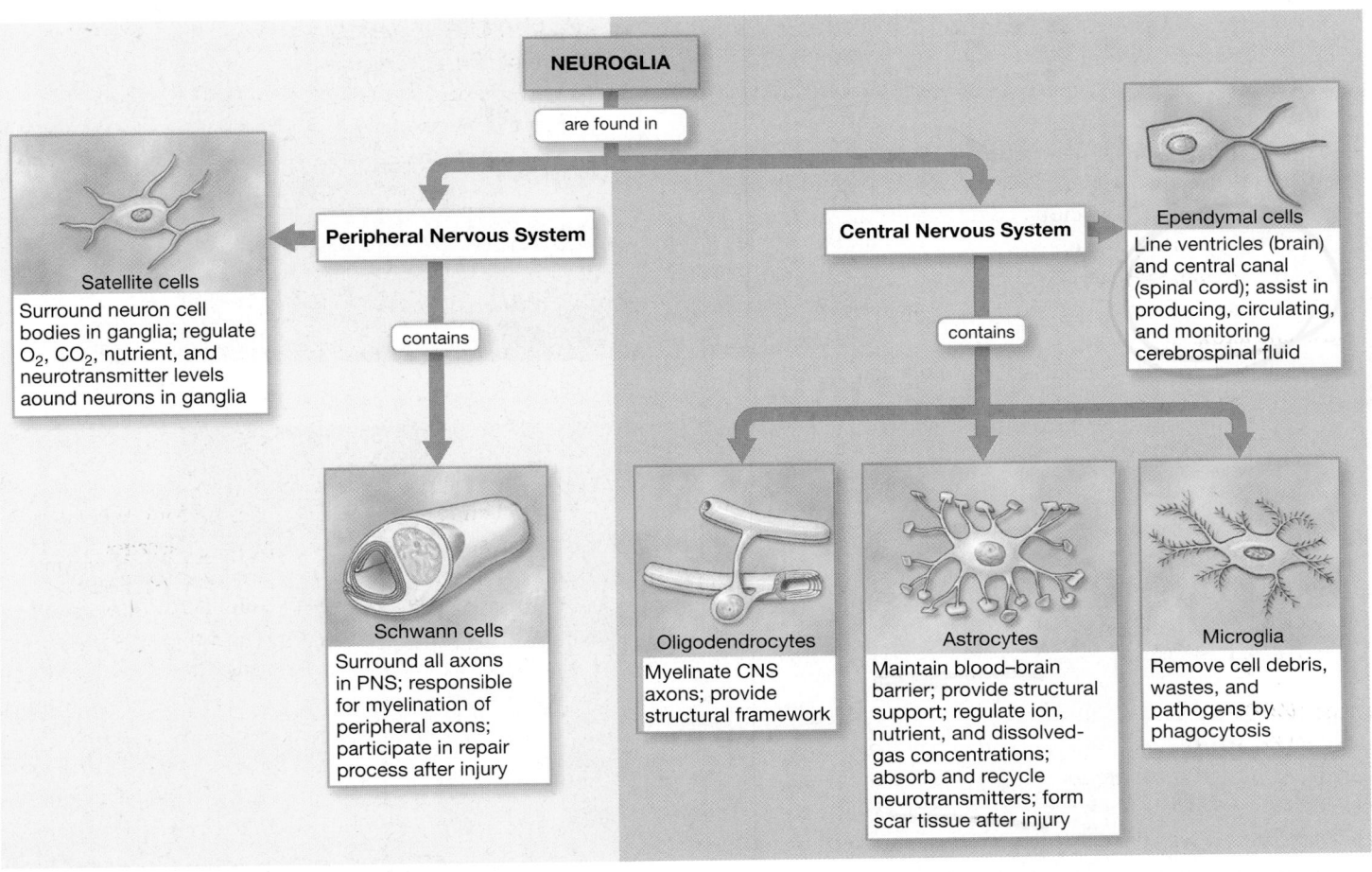

Figure 12–4 An Introduction to Neuroglia.

to the interstitial fluid of the CNS. Neural tissue must be physically and biochemically isolated from the general circulation, because hormones, amino acids, or other chemicals in the blood can alter neuron function. The endothelial cells lining CNS capillaries control the chemical exchange between the blood and interstitial fluid. These cells create a **blood–brain barrier** (BBB) that isolates the CNS from the general circulation.

The slender cytoplasmic extensions of astrocytes end in expanded "feet," processes that wrap around capillaries. Astrocytic processes form a complete blanket around the capillaries, interrupted only where other neuroglia come in contact with the capillary walls. Chemicals secreted by astrocytes are somehow responsible for maintaining the special permeability characteristics of endothelial cells. (We will discuss the blood–brain barrier further in Chapter 14.)

- *Creating a Three-Dimensional Framework for the CNS.* Astrocytes are packed with microfilaments that extend across the breadth of the cell and its processes. This extensive cytoskeletal reinforcement assists astrocytes in

providing a structural framework for the neurons of the brain and spinal cord.

- *Repairing Damaged Neural Tissue.* In the CNS, damaged neural tissue seldom regains normal function. However, astrocytes moving into an injury site can make structural repairs that stabilize the tissue and prevent further injury. We will consider neural damage and subsequent repair in a later section.

- *Guiding Neuron Development.* Astrocytes in the embryonic brain appear to be involved in directing both the growth and interconnection of developing neurons.

- *Controlling the Interstitial Environment.* Astrocytes appear to adjust the composition of interstitial fluid by several means: (1) regulating the concentration of sodium ions, potassium ions, and carbon dioxide; (2) providing a "rapid-transit system" for the transport of nutrients, ions, and dissolved gases between capillaries and neurons; (3) controlling the volume of blood flow through the capillaries; (4) absorbing and recycling some neurotransmitters; and (5) releasing chemicals that enhance or suppress communication across synaptic terminals.

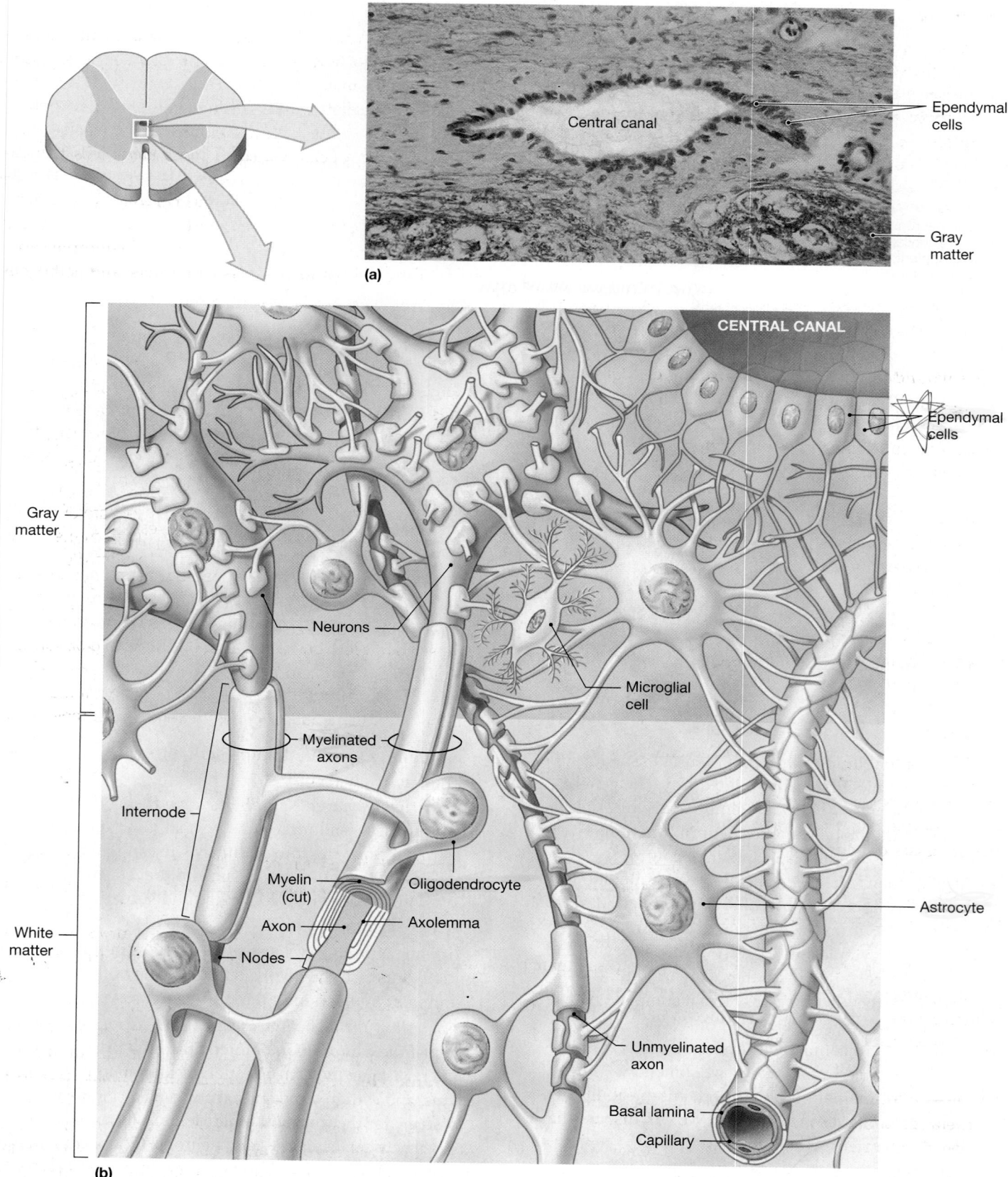

Figure 12–5 **Neuroglia in the CNS. (a)** Light micrograph showing the ependymal lining of the central canal of the spinal cord. (LM × 236) **(b)** A diagrammatic view of neural tissue in the CNS, showing relationships between neuroglia and neurons.

Oligodendrocytes

Like astrocytes, **oligodendrocytes** (ol-i-gō-DEN-drō-sīts; *oligo-*, few) possess slender cytoplasmic extensions, but the cell bodies of oligodendrocytes are smaller, and have fewer processes, than astrocytes (**Figure 12–5b**). The processes of oligodendrocytes generally are in contact with the exposed surfaces of neurons; the functions of processes ending at the neuron cell body have yet to be determined. Much more is known about the processes that end on the surfaces of axons. Many axons in the CNS are completely sheathed in these processes, which insulate them from contact with the extra-cellular fluid.

Near the tip of each process, the axolemma expands to *plasma membrane around axon* form an enormous membranous pad, and the cytoplasm there becomes very thin. This flattened "pancake" some-how gets wound around the axon, forming concentric lay-ers of plasma membrane (**Figure 12–5b**). The membranous wrapping of electrical insulation, called **myelin** (MĪ-e-lin), increases the speed at which an action potential travels along the axon (the mechanism will be described in a later section).

Many oligodendrocytes cooperate in the formation of a **myelin sheath** along the length of an axon. Such an axon is said to be **myelinated**. Each oligodendrocyte myelinates seg-ments of several axons. The relatively large areas of the axon that are thus wrapped in myelin are called **internodes** (*inter*, between). Internodes are typically 1–2 mm in length. The small gaps of a few micrometers that separate adjacent inter-nodes are called **nodes**, or *nodes of Ranvier* (rahn-vē-Ā). When an axon branches, the branches originate at nodes.

In dissection, myelinated axons appear glossy white, pri-marily because of the lipids within the myelin. As a result, re-gions dominated by myelinated axons constitute the **white matter** of the CNS. Not all axons in the CNS are myelinated, however. **Unmyelinated** axons may not be completely covered by the processes of neuroglia. Such axons are common where relatively short axons and collaterals form synapses with densely packed neuron cell bodies. Areas containing neuron cell bodies, dendrites, and unmyelinated axons have a dusky gray color, and they constitute the **gray matter** of the CNS.

In sum, oligodendrocytes play a role in structural organi-zation by tying clusters of axons together, and these neuroglia also improve the functional performance of neurons by wrap-ping axons within a myelin sheath.

Tips & Tricks

The overall color of CNS tissue is related to its structure and function. **Gr**ay matter has a **gr**eat concentration of neuron cell bodies and is a region of inte**gr**ation. **Wh**ite matter has a **wh**ole lot of myelinated axons and **wh**isks nerve impulses.

Microglia

The least numerous and smallest neuroglia in the CNS are **microglia** (mī-KRŌG-lē-uh). Their slender cytoplasmic processes have many fine branches (**Figure 12–5b**). These cells are capable of migrating through neural tissue. Microglia appear early in embryonic development, originating from mesodermal stem cells related to those stem cells that pro-duce monocytes and macrophages. ∞ pp. 126, 131 Mi-croglia migrate into the CNS as the nervous system forms. Thereafter, they remain isolated in neural tissue, where in ef-fect they act as a wandering police force and janitorial service by engulfing cellular debris, waste products, and pathogens.

CLINICAL NOTE

Tumors Tumors of the brain, spinal cord, and associated membranes result in approximately 90,000 deaths in the United States each year. Tumors that originate in the central nervous system are called *primary CNS tumors*, to distinguish them from *secondary CNS tumors*, which arise from the metastasis (spread) of cancer cells that originate elsewhere. Roughly 75 percent of CNS tumors are primary tumors. In adults, primary CNS tumors result from the divisions of abnormal neuroglia rather than from the divisions of abnormal neurons, because typical neurons in adults cannot divide. However, through the divisions of stem cells, neurons increase in number until children reach age 4. As a result, primary CNS tumors involving abnormal neurons can occur in young children. Symptoms of CNS tumors vary with the location affected. Treatment may involve surgery, radiation, chemotherapy, or a combination of these procedures.

Neuroglia of the Peripheral Nervous System

As previously noted, the cell bodies of neurons in the PNS are clustered in masses called **ganglia** (singular, *ganglion*). Neu-ronal cell bodies and most axons in the PNS are completely insulated from their surroundings by the processes of neu-roglia. The two types of neuroglia in the PNS are called satel-lite cells and Schwann cells.

Satellite cells, or *amphicytes* (AM-fi-sīts), surround neu-ron cell bodies in ganglia; they regulate the environment around the neurons, much as astrocytes do in the CNS. **Schwann cells**, or *neurilemmal cells* (*neurilemmocytes*), form a sheath around peripheral axons (**Figure 12–6**). Wherever a Schwann cell covers an axon, the outer surface of the Schwann cell is called the **neurilemma** (noor-i-LEM-uh). Most axons in the PNS, whether myelinated or unmyelinated, are shielded from contact with interstitial fluids by Schwann cells.

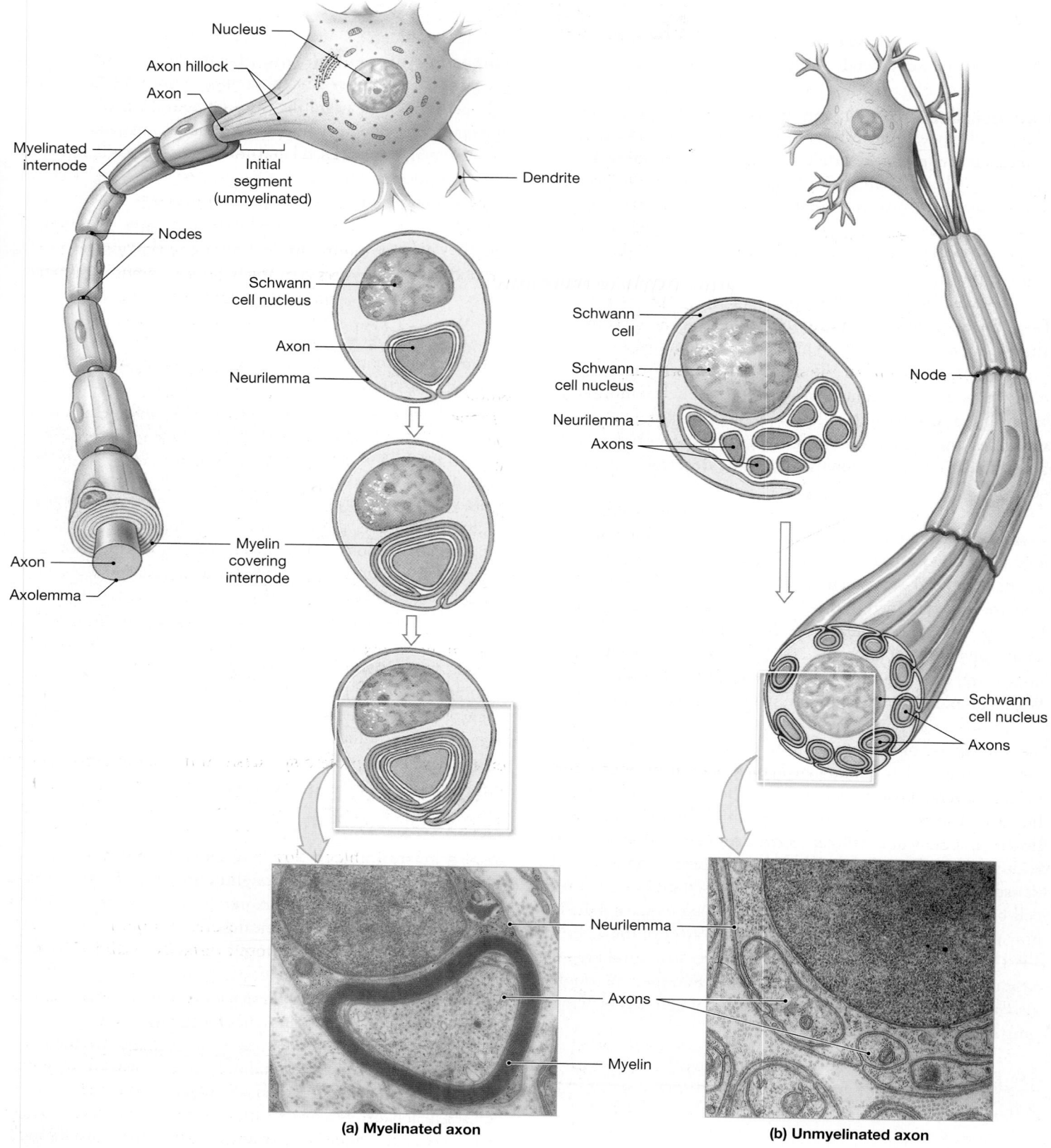

(a) **Myelinated axon**

(b) **Unmyelinated axon**

Figure 12–6   Schwann Cells and Peripheral Axons. (a) A myelinated axon, showing the organization of Schwann cells along the length of the axon. Also shown are stages in the formation of a myelin sheath by a single Schwann cell along a portion of a single axon (compare with myelin in the CNS, shown in *Figure 12–5b*). **(b)** The enclosing of a group of unmyelinated axons by a single Schwann cell. A series of Schwann cells is required to cover the axons along their entire length.

CLINICAL NOTE

Demyelination

Demyelination is the progressive destruction of myelin sheaths, both in the CNS and in the PNS. The result is a loss of sensation and motor control that leaves affected regions numb and paralyzed. Many unrelated conditions that result in the destruction of myelin can cause symptoms of demyelination. Several important demyelinating disorders are *heavy-metal poisoning, diphtheria, multiple sclerosis (MS)*, and *Guillain–Barré syndrome.*

Chronic exposure to heavy-metal ions, such as arsenic, lead, or mercury, can cause **heavy-metal poisoning**, leading to damage of neuroglia and to demyelination. As demyelination occurs, the affected axons deteriorate, and the condition becomes irreversible.

Diphtheria (dif-THĒ-rē-uh; *diphthera*, leather + *-ia*, disease) is a disease that results from a bacterial infection of the respiratory tract or skin. In the nervous system, diphtheria toxin damages Schwann cells and destroys myelin sheaths in the PNS. The resulting demyelination leads to sensory and motor problems that can ultimately produce a fatal paralysis. Because an effective vaccine (which is frequently combined with the tetanus vaccine) exists, cases are relatively rare in countries with adequate health care.

Multiple sclerosis (skler-Ō-sis; *sklerosis*; hardness), or **MS**, is a disease characterized by recurrent incidents of demyelination that affects axons in the optic nerve, brain, and spinal cord. Common signs and symptoms include partial loss of vision and problems with speech, balance, and general motor coordination, including bowel and urinary bladder control. The time between incidents and degree of recovery vary from case to case. In about one-third of all cases, the disorder is progressive, and each incident leaves a greater degree of functional impairment. The first attack typically occurs in individuals 30–40 years old; the incidence among women is 1.5 times that among men. In some patients, corticosteroid or interferon injections have slowed the progression of the disease.

Guillain–Barré (ghee-yan bah-ray) **syndrome** is an autoimmune disorder characterized by demyelination of peripheral nerves. The severity of the signs and symptoms, which include weakness or tingling of the legs that spreads to the arms, typically increases to paralysis; when breathing is affected, the patient is placed on a ventilator. The disorder affects both sexes equally and each year afflicts 1 of every 100,000 Americans. It appears that a virus triggers the syndrome, because the onset is usually within a few days or weeks of a respiratory or gastrointestinal infection, or occasionally after surgery or immunization. Most patients fully recover, but some continue to have residual weakness.

Whereas an oligodendrocyte in the CNS may myelinate portions of several adjacent axons (**Figure 12–5b**), a Schwann cell can myelinate only one segment of a single axon (**Figure 12–6a**). However, a Schwann cell can *enclose* segments of several unmyelinated axons (**Figure 12–6b**). A series of Schwann cells is required to enclose an axon along its entire length.

The A&P Top 100

#40 Neurons perform all of the communication, information processing, and control functions of the nervous system. Neuroglia outnumber neurons and have functions essential to the survival and functionality of neurons and to preserving the physical and biochemical structure of neural tissue.

Neural Responses to Injuries

A neuron responds to injury in a very limited, stereotyped fashion. In the cell body, the Nissl bodies disperse and the nucleus moves away from its centralized location as the cell increases its rate of protein synthesis. If the neuron recovers its functional abilities, it will regain its normal appearance. The key to recovery appears to be events in the axon. If, for example, the pressure applied during a crushing injury produces a local decrease in blood flow and oxygen availability, the affected axonal membrane becomes unexcitable. If the pressure is alleviated after an hour or two, the neuron will recover within a few weeks, but more severe or prolonged pressure produces effects similar to those caused by cutting the axon.

In the PNS, Schwann cells participate in the repair of damaged nerves. In the process known as **Wallerian degeneration** (**Figure 12–7**), the axon distal to the injury site degenerates, and macrophages migrate into the area to phagocytize the debris. The Schwann cells do not degenerate; instead, they proliferate and form a solid cellular cord that follows the path of the original axon. As the neuron recovers, its axon grows into the site of injury, and the Schwann cells wrap around the axon.

If the axon continues to grow into the periphery alongside the appropriate cord of Schwann cells, it may eventually reestablish its normal synaptic contacts. However, if it stops growing or wanders off in some new direction, normal function

Site of injury

STEP 1
Fragmentation of axon and myelin occurs in distal stump.
Axon
Myelin
Proximal stump
Distal stump
STEP 2
Schwann cells form cord, grow into cut, and unite stumps.
Macrophages engulf degenerating axon and myelin.
Schwann cell
Macrophage
STEP 3
Axon sends buds into network of Schwann cells
and then starts growing along cord of Schwann cells.
STEP 4
Axon continues to grow into distal stump and is enclosed
by Schwann cells.

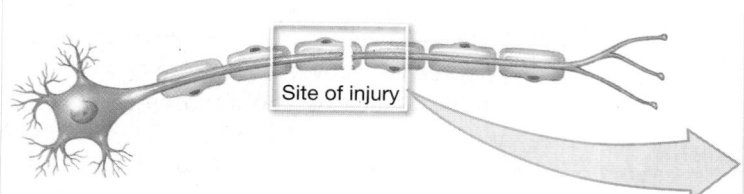

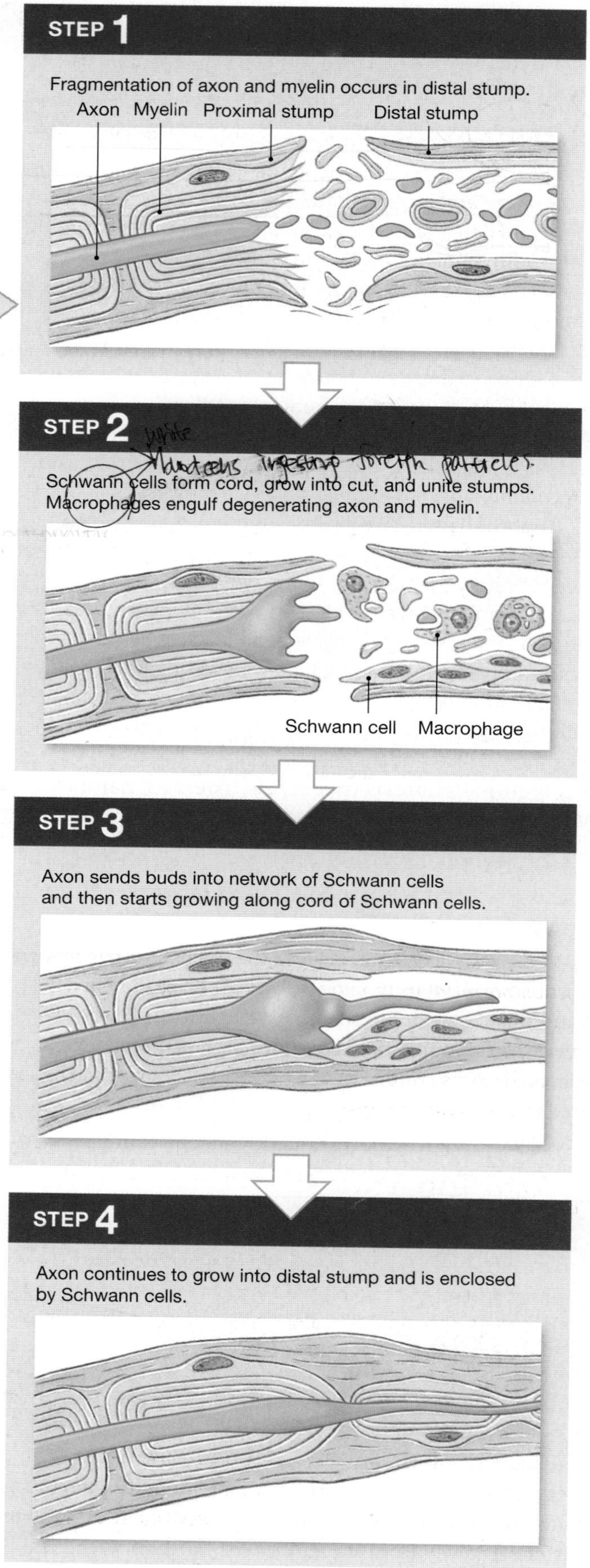

Figure 12–7 Peripheral Nerve Regeneration after Injury.

will not return. The growing axon is most likely to arrive at its appropriate destination if the cut edges of the original nerve bundle remain in contact.

Limited regeneration can occur in the CNS, but the situation is more complicated because (1) many more axons are likely to be involved, (2) astrocytes produce scar tissue that can prevent axon growth across the damaged area, and (3) astrocytes release chemicals that block the regrowth of axons.

CHECKPOINT

9. **Identify the neuroglia of the central nervous system.** ependyma, astrocytes, oligodendrocytes & microglia.
10. **Identify the neuroglia of the peripheral nervous system.** satellite & schwann.
11. **Which type of neuroglia would occur in larger-than-normal numbers in the brain tissue of a person with a CNS infection?** microglia - pathogens by phagocytosis.

See the blue Answers tab at the end of the book.

12-4 The transmembrane potential is the electrical potential of the cell's interior relative to its surroundings

In Chapter 3, we introduced the concepts of the *transmembrane potential* and the *resting potential*, two characteristic features of all cells. ∞ p. 99 In this discussion, we will focus on the membrane properties of neurons; many of the principles discussed also apply to other types of cells.

Figure 12–8 introduces the important membrane processes we will be examining.

- All living cells have a transmembrane potential that varies from moment to moment depending on the activities of the cell. The *resting potential* is the

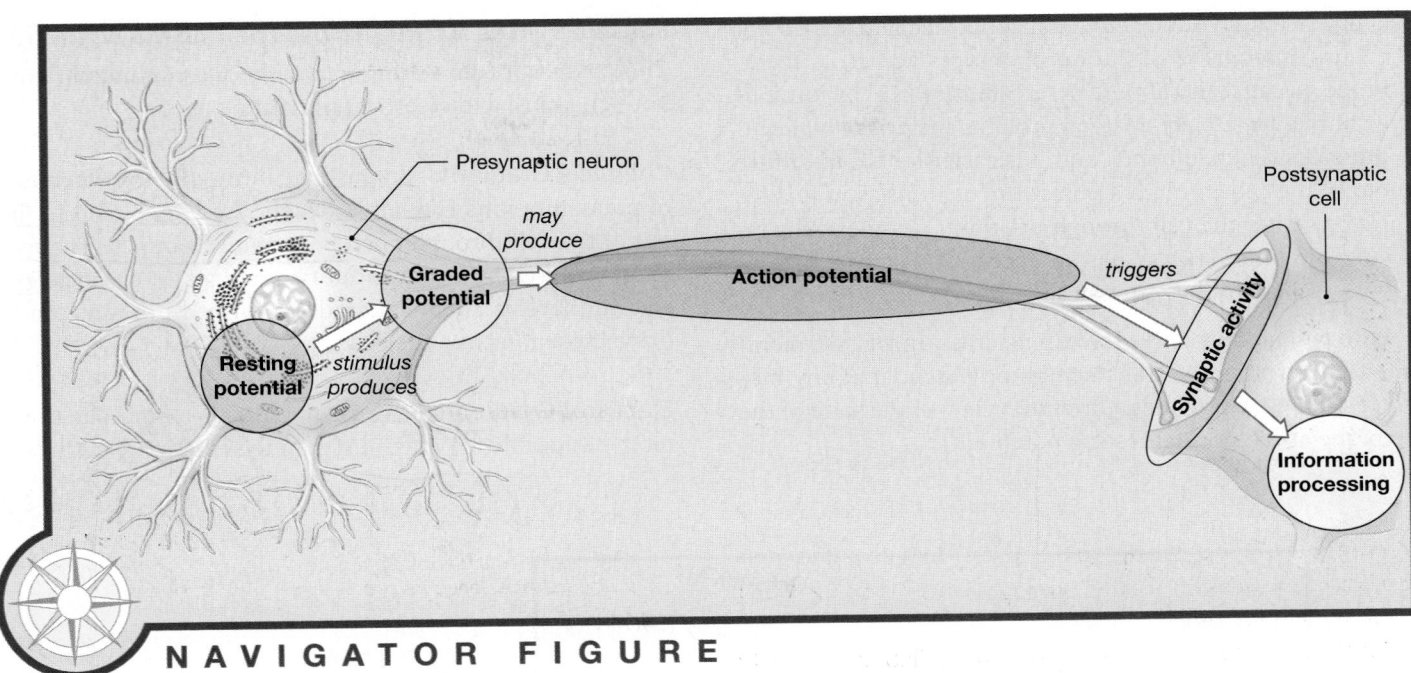

NAVIGATOR FIGURE

Figure 12–8 **An Overview of Neural Activities.** The important membrane processes are shown in order of their presentation in the text. This figure will be repeated, in simplified form, as a Navigator icon in other figures whenever we are changing topics.

transmembrane potential of a resting cell. All neural activities begin with a change in the resting potential of a neuron.

- A typical stimulus produces a temporary, localized change in the resting potential. The effect, which decreases with distance from the stimulus, is called a *graded potential*.

- If the graded potential is sufficiently large, it triggers an *action potential* in the membrane of the axon. An action potential is an electrical impulse that is propagated along the surface of an axon and does not diminish as it moves away from its source. This impulse travels along the axon to one or more synapses.

- *Synaptic activity* then produces graded potentials in the plasma membrane of the postsynaptic cell. The process typically involves the release of neurotransmitters, such as ACh, by the presynaptic cell. These compounds bind to receptors on the postsynaptic plasma membrane, changing its permeability. The mechanism is comparable to that of the neuromuscular junction, described in Chapter 10. ∞ p. 304

- The response of the postsynaptic cell ultimately depends on what the stimulated receptors do and what other stimuli are influencing the cell at the same time. The integration of stimuli at the level of the individual cell is the simplest form of *information processing* in the nervous system.

When you understand each of the foregoing processes, you will know how neurons process information and communicate with one another and with peripheral effectors.

The Transmembrane Potential

Chapter 3 introduced three important concepts regarding the transmembrane potential:

1. The intracellular fluid (cytosol) and extracellular fluid differ markedly in ionic composition. The extracellular fluid (ECF) contains high concentrations of sodium ions (Na^+) and chloride ions (Cl^-), whereas the cytosol contains high concentrations of potassium ions (K^+) and negatively charged proteins.

2. If the plasma membrane were freely permeable, diffusion would continue until all the ions were evenly distributed across the membrane and a state of equilibrium existed. But an even distribution does not occur, because cells have selectively permeable membranes. ∞ p. 89 Ions cannot freely cross the lipid portions of the plasma membrane; they can enter or leave the cell only through membrane channels. Many kinds of membrane channels exist, each with its own properties. At the resting potential, or transmembrane potential of an undisturbed cell, ion movement occurs through *leak channels*—membrane channels that are always open. ∞ pp. 70–71 Active transport mechanisms also move specific ions into or out of the cell.

3. The cell's passive and active transport mechanisms do not ensure an equal distribution of charges across its membrane, because membrane permeability varies by ion. For example, negatively charged proteins inside the cell cannot cross the membrane, and it is easier for K^+ to diffuse out of the cell through a potassium channel than it is for Na^+ to enter the cell through a sodium channel. As a result, the inner surface has an excess of negative charges with respect to the outer surface.

Both passive and active forces act across the plasma membrane to determine the transmembrane potential at any moment. **Figure 12–9** provides a brief overview of the state of the membrane at the normal resting potential.

Passive Forces Acting across the Plasma Membrane

The passive forces acting across the plasma membrane are both chemical and electrical in nature.

Chemical Gradients. Because the intracellular concentration of potassium ions is relatively high, these ions tend to move out of the cell through open potassium channels. The movement is driven by a concentration gradient, or *chemical gradient*. Similarly, a chemical gradient for sodium ions tends to drive those ions *into* the cell.   more → less

Electrical Gradients. Because the plasma membrane is much more permeable to potassium than to sodium, potassium ions

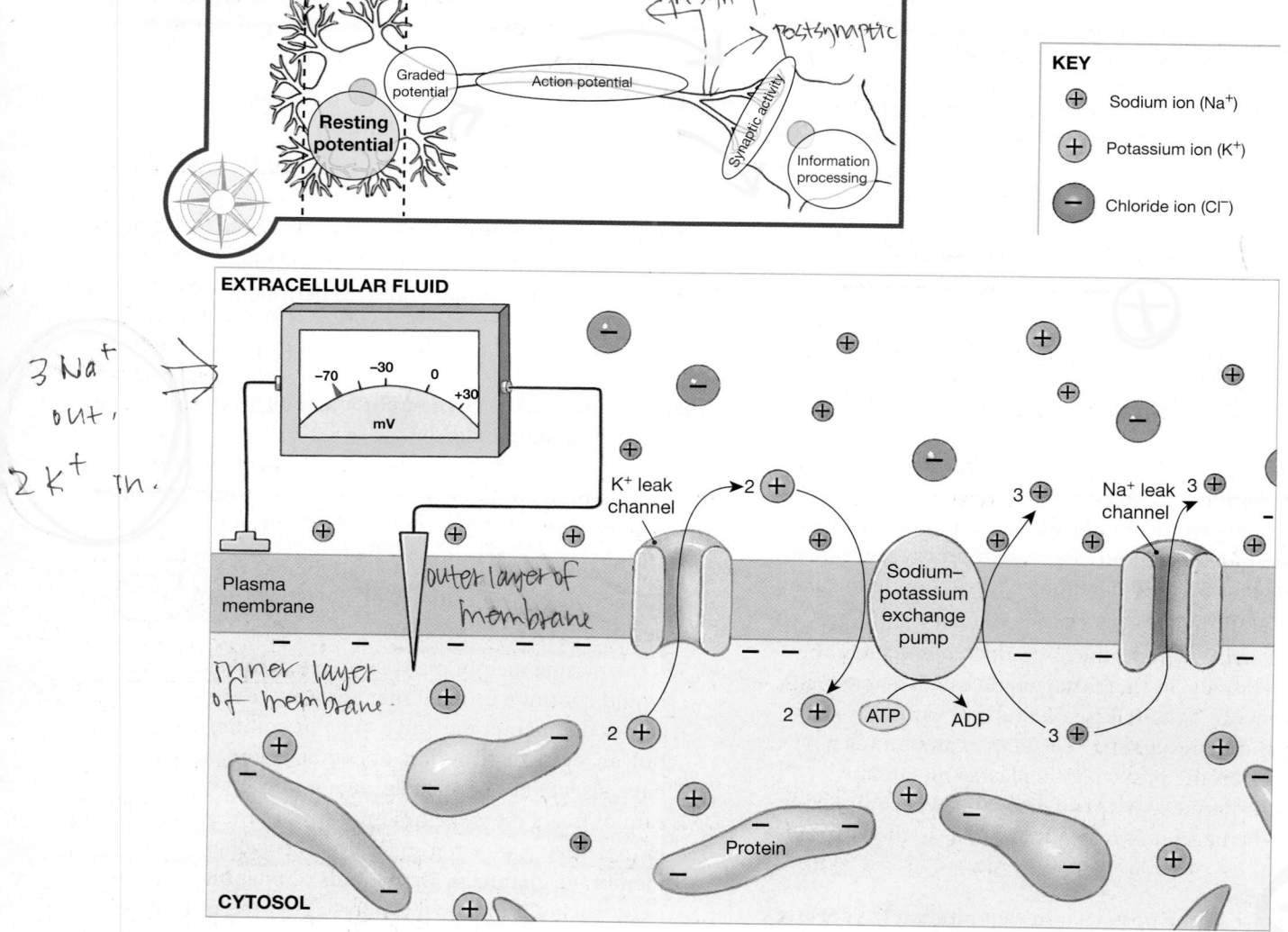

Figure 12–9   An Introduction to the Resting Potential. The resting potential is the transmembrane potential of an undisturbed cell. The phospholipid bilayer of the plasma membrane is represented by a simple blue band. The Navigator icon highlights the resting potential to indicate "You are here!"

leave the cytoplasm more rapidly than sodium ions enter. As a result, the cytosol along the interior of the membrane exhibits a net loss of positive charges, leaving an excess of negatively charged proteins. At the same time, the extracellular fluid near the outer surface of the plasma membrane displays a net gain of positive charges. The positive and negative charges are separated by the plasma membrane, which restricts the free movement of ions. Whenever positive and negative ions are held apart, a *potential difference* arises.

The size of a potential difference is measured in volts (V) or millivolts (mV; thousandths of a volt). The resting potential varies widely, depending on the type of cell, but averages about 0.07 V for many cells, including most neurons. We will use this value in our discussion, usually expressing it as −70 mV (**Figure 12–9**). The minus sign signifies that the inner surface of the plasma membrane is negatively charged with respect to the exterior.

Positive and negative charges attract one another. If nothing separates them, oppositely charged ions will move together and eliminate the potential difference between them. A movement of charges to eliminate a potential difference is called a **current**. If a barrier (such as a plasma membrane) separates the oppositely charged ions, the amount of current depends on how easily the ions can cross the membrane. The **resistance** of the membrane is a measure of how much the membrane restricts ion movement. If the resistance is high, the current is very small, because few ions can cross the membrane. If the resistance is low, the current is very large, because ions flood across the membrane. The resistance of a plasma membrane can be changed by the opening or closing of ion channels. The ensuing changes result in currents that bring ions into or out of the cytoplasm.

The Electrochemical Gradient.

Electrical gradients can either reinforce or oppose the chemical gradient for each ion. The **electrochemical gradient** for a specific ion is the sum of the chemical and electrical forces acting on that ion across the plasma membrane. The electrochemical gradients for K^+ and Na^+ are the primary factors affecting the resting potential of most cells, including neurons. We will consider the forces acting on each ion independently.

The intracellular concentration of potassium ions is relatively high, whereas the extracellular concentration is very low. Therefore, the chemical gradient for potassium ions tends to drive them out of the cell, as indicated by the black arrow in **Figure 12–10a**. However, the electrical gradient opposes this movement, because K^+ inside and outside of the cell are attracted to the negative charges on the inside of the plasma membrane, and repelled by the positive charges on the outside of the plasma membrane. The size and direction of the electrical gradient is indicated by the white arrow in **Figure 12–10a**. The chemical gradient is strong enough to overpower the elec-

trical gradient, but this weakens the force driving K^+ out of the cell; the net driving force is represented by the gray arrow.

If the plasma membrane were freely permeable to K^+ but impermeable to other positively charged ions, potassium ions would continue to leave the cell until the electrical gradient (opposing the exit of K^+ from the cell) was as strong as the chemical gradient (driving K^+ out of the cell). The transmembrane potential at which there is no net movement of a particular ion across the plasma membrane is called the *equilibrium potential* for that ion. For potassium ions, this equilibrium occurs at a transmembrane potential of about −90 mV, as illustrated in **Figure 12–10b**. The resting membrane potential is typically −70 mV, a value close to the equilibrium potential for K^+; the difference is due primarily to the continuous leakage of Na^+ into the cell.

Tips&Tricks

To remember the sign of the resting membrane potential, associate **N**egative with i**N**side and p**O**sitive with the **O**utside.

The sodium ion concentration in the extracellular fluid is relatively high, whereas that inside the cell is extremely low. As a result, there is a strong chemical gradient driving Na^+ into the cell (the black arrow in **Figure 12–10c**). In addition, the extracellular sodium ions are attracted by the excess of negative charges on the inner surface of the plasma membrane, and the relative size and direction of this electrical gradient is indicated by the white arrow in **Figure 12–10c**. This means that electrical forces and chemical forces drive Na^+ into the cell, and the net driving force is represented by the gray arrow.

If the plasma membrane were freely permeable to Na^+, these ions would continue to enter the plasma until the interior of the plasma membrane contained enough excess positive charges to reverse the electrical gradient. In other words, ion movement would continue until the interior developed such a strongly positive charge that repulsion between the positive charges would prevent any further net movement of Na^+ into the cell. The equilibrium potential for Na^+ is approximately +66 mV, as illustrated in **Figure 12–10d**. The resting potential is nowhere near that value, because the resting membrane permeability to Na^+ is very low, and because ion pumps in the plasma membrane are able to eject sodium ions as fast as they cross the membrane.

An electrochemical gradient is a form of *potential energy*. ∞ p. 99 Potential energy is stored energy—the energy of position, as exists in a stretched spring, a charged battery, or water behind a dam. Without a plasma membrane, diffusion would eliminate all electrochemical gradients. In effect, the plasma membrane acts like a dam across a river. Without the dam, water would simply respond to gravity and flow downstream,

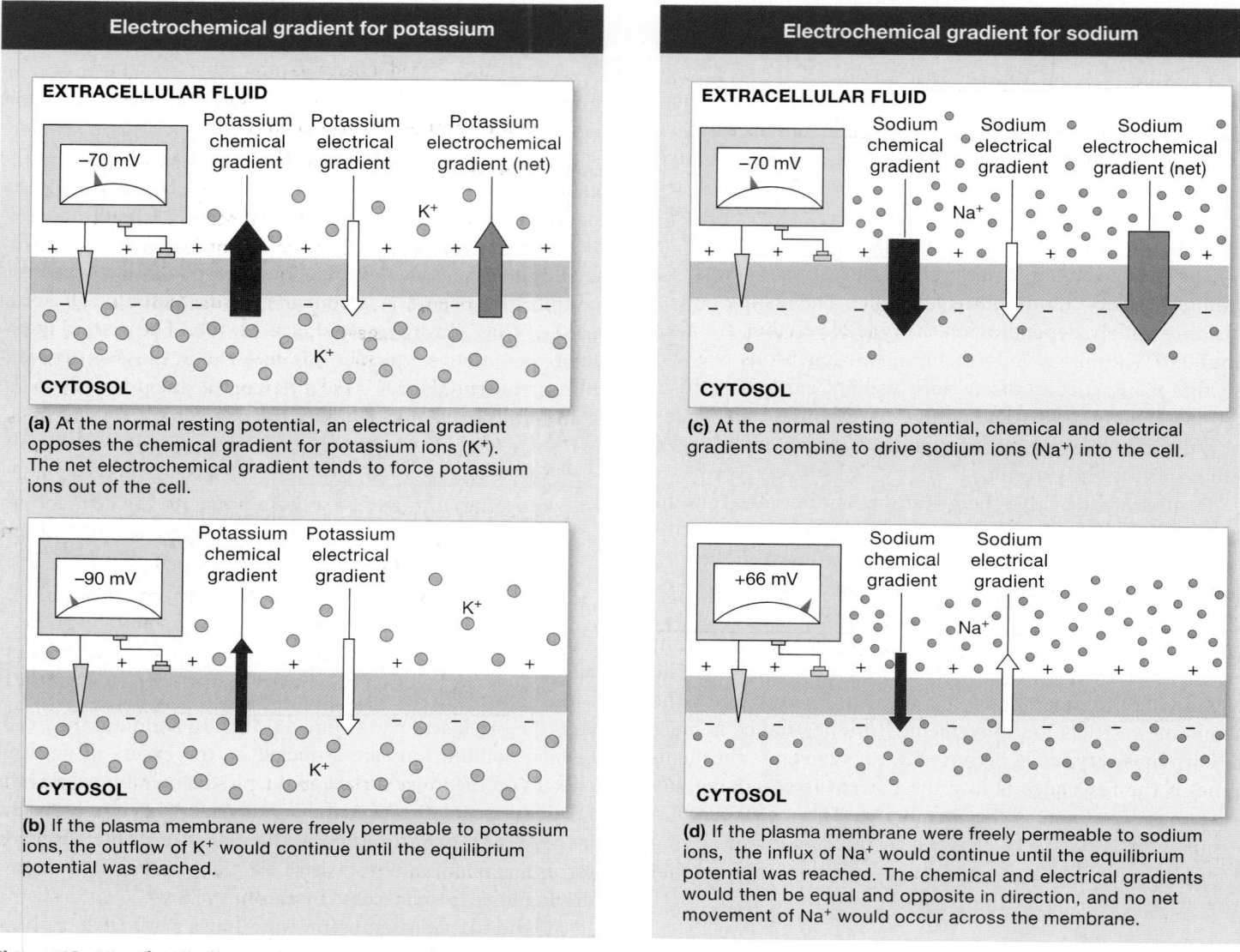

Figure 12–10   Electrochemical Gradients for Potassium and Sodium Ions.

gradually losing energy. With the dam in place, even a small opening will release water under tremendous pressure. Similarly, any stimulus that increases the permeability of the plasma membrane to sodium or potassium ions will produce sudden and dramatic ion movement. For example, a stimulus that opens sodium ion channels will trigger an immediate rush of Na$^+$ into the cell. The nature of the stimulus does not determine the amount of ion movement; if the stimulus opens the door, the electrochemical gradient will do the rest.

Active Forces across the Membrane: The Sodium–Potassium Exchange Pump

We can compare a cell to a leaky fishing boat loaded with tiny fish. The hull represents the plasma membrane; the fish, K$^+$; and the ocean water, Na$^+$. As the boat rumbles and rolls, water comes in through the cracks, and fish swim out. If the boat

is to stay afloat and the catch kept, the water must be pumped out, and the lost fish recaptured.

Similarly, at the normal resting potential, the cell must bail out sodium ions that leak in and recapture potassium ions that leak out. The "bailing" occurs through the activity of an exchange pump powered by ATP. The ion pump involved is the carrier protein *sodium–potassium ATPase.* ∞ p. 95 This pump exchanges three intracellular sodium ions for two extracellular potassium ions. At the normal resting potential, this pump's primary significance is that it ejects sodium ions as quickly as they enter the cell. Thus, the activity of the exchange pump exactly balances the passive forces of diffusion, and the resting potential remains stable because the ionic concentration gradients are maintained.

Table 12–1 provides a summary of the important features of the resting potential.

SUMMARY TABLE 12–1 The Resting Potential
• Because the plasma membrane is highly permeable to potassium ions, the resting potential is fairly close to −90 mV, the equilibrium potential for K⁺.
• Although the electrochemical gradient for sodium ions is very large, the membrane's permeability to these ions is very low. Consequently, Na⁺ has only a small effect on the normal resting potential, making it just slightly less negative than it would be otherwise.
• The sodium–potassium exchange pump ejects 3 Na⁺ ions for every 2 K⁺ ions that it brings into the cell. It thus serves to stabilize the resting potential when the ratio of Na⁺ entry to K⁺ loss through passive channels is 3:2.
• At the normal resting potential, these passive and active mechanisms are in balance. The resting potential varies widely with the type of cell. A typical neuron has a resting potential of approximately −70 mV.

Changes in the Transmembrane Potential

As noted previously, the resting potential is the transmembrane potential of an "undisturbed" cell. Yet cells are dynamic structures that continually modify their activities, either in response to external stimuli or to perform specific functions. The transmembrane potential is equally dynamic, rising or falling in response to temporary changes in membrane permeability. Those changes result from the opening or closing of specific membrane channels.

Membrane channels control the movement of ions across the plasma membrane. Our discussion will focus on the permeability of the membrane to sodium and potassium ions, which are the primary determinants of the transmembrane potential of many cell types, including neurons. Sodium and potassium ion channels are either passive or active.

Passive channels, or **leak channels**, are always open. However, their permeability can vary from moment to moment as the proteins that make up the channel change shape in response to local conditions. As noted earlier, leak channels are important in establishing the normal resting potential of the cell (**Figure 12–9**).

Plasma membranes also contain **active channels**, often called **gated channels**, which open or close in response to specific stimuli. Each gated channel can be in one of three states: (1) closed but capable of opening, (2) open (**activated**), or (3) closed and incapable of opening (**inactivated**).

Three classes of gated channels exist: chemically gated channels, voltage-gated channels, and mechanically gated channels.

1. **Chemically gated channels** open or close when they bind specific chemicals (**Figure 12–11a**). The receptors that bind acetylcholine (ACh) at the neuromuscular junction are chemically gated channels. ∞ p. 304 Chemically gated channels are most abundant on the dendrites and cell body of a neuron, the areas where most synaptic communication occurs.

2. **Voltage-gated channels** are characteristic of areas of **excitable membrane**, a membrane capable of generating and conducting an action potential. Examples of excitable membranes are the axons of unipolar and multipolar neurons, and the sarcolemma (including T tubules) of skeletal muscle fibers and cardiac muscle cells. ∞ pp. 306, 327 Voltage-gated channels open or close in response to changes in the transmembrane potential. The most important voltage-gated channels, for our purposes, are voltage-gated sodium channels, potassium channels, and calcium channels. Sodium channels have two gates that function independently: an *activation gate* that opens on stimulation, letting sodium ions into the cell, and an *inactivation gate* that closes to stop the entry of sodium ions (**Figure 12–11b**).

3. **Mechanically gated channels** open or close in response to physical distortion of the membrane surface (**Figure 12–11c**). Such channels are important in sensory receptors that respond to touch, pressure, or vibration. We will discuss these receptors in more detail in Chapter 15.

At the resting potential, most gated channels are closed. The opening of gated channels increases the rate of ion movement across the plasma membrane and thus changes the transmembrane potential. The distribution of membrane channels can vary from one region of the plasma membrane to another, affecting how and where a cell responds to specific stimuli. For example, whereas chemically gated sodium channels are widespread on the surfaces of a neuron, voltage-gated sodium channels are most abundant on the axon, its branches, and the synaptic terminals, and mechanically gated channels are typically located only on the dendrites of sensory neurons. The functional implications of these differences in distribution will become apparent in later sections.

The A&P Top 100

#41 A transmembrane potential exists across the plasma membrane. It is there because (1) the cytosol differs from extracellular fluid in chemical and ionic composition and (2) the plasma membrane is selectively permeable. The transmembrane potential can change from moment to moment, as the plasma membrane changes its permeability in response to chemical or physical stimuli.

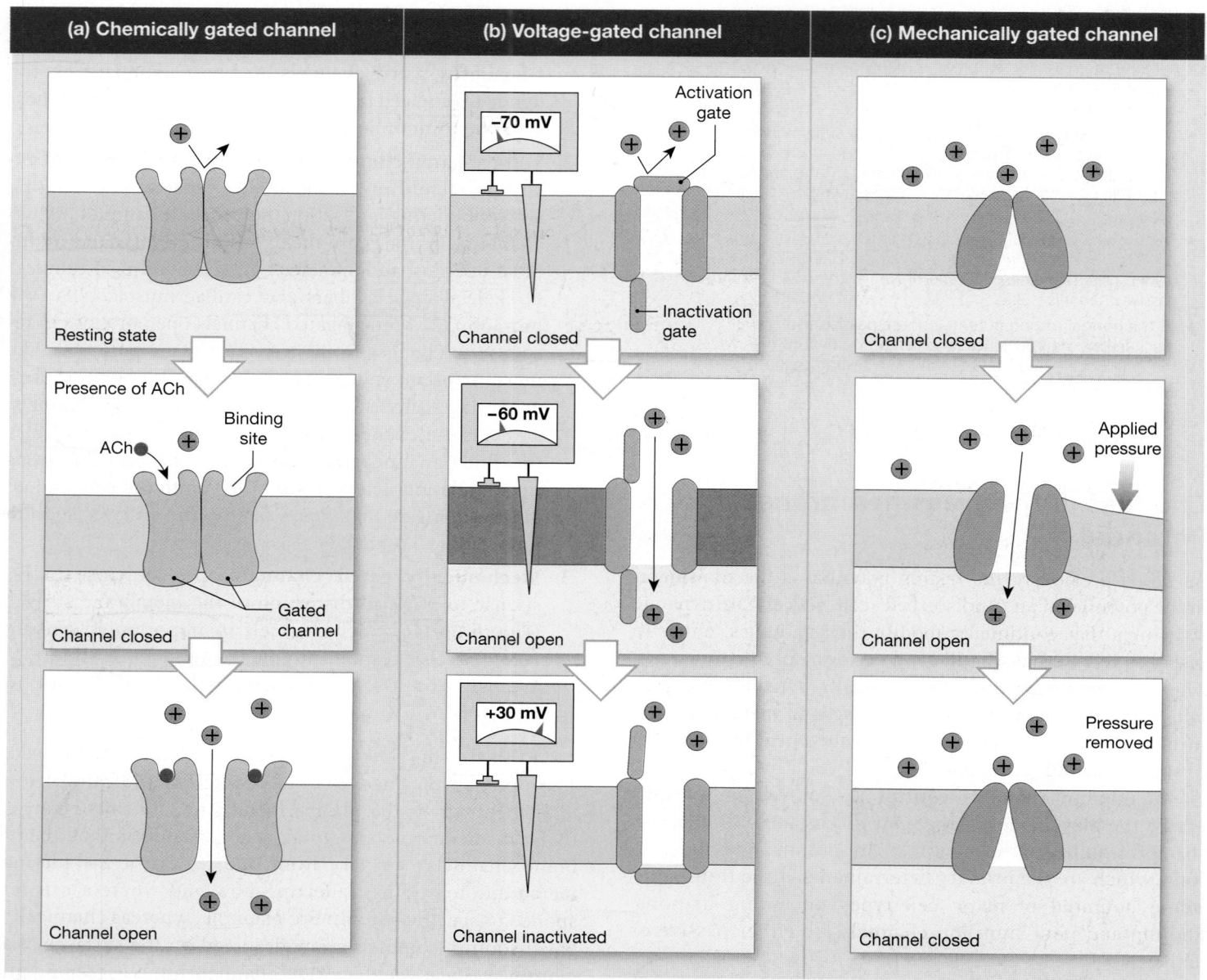

Figure 12–11 Gated Channels. Na$^+$ channels are shown here, but comparable gated channels regulate the movements of other cations and anions. **(a)** A chemically gated Na$^+$ channel that opens in response to the presence of ACh at a binding site. **(b)** A voltage-gated Na$^+$ channel that responds to changes in the transmembrane potential. At the normal resting potential, the channel is closed; at a membrane potential of −60 mV, the channel opens; at +30 mV, the channel is inactivated. **(c)** A mechanically gated channel, which opens in response to distortion of the membrane.

Graded Potentials

Graded potentials, or *local potentials*, are changes in the transmembrane potential that cannot spread far from the site of stimulation. Any stimulus that opens a gated channel will produce a graded potential. **Figure 12–12** shows what happens when a resting membrane is exposed to a chemical that opens chemically gated sodium channels. Sodium ions enter the cell and are attracted to the negative charges along the inner sur-

face of the membrane. The arrival and spreading out of additional positive charges shifts the transmembrane potential toward 0 mV (STEP 1). Any shift from the resting potential toward a more positive potential is called a **depolarization**, a term that applies to changes in potential from −70 mV to smaller negative values (−65 mV, −45 mV, −10 mV) as well as to membrane potentials above 0 mV (+10 mV, +30 mV)

At the resting potential, sodium ions are drawn to the outer surface of the plasma membrane, attracted by the excess

[handwritten annotation: cell becomes more ⊕]

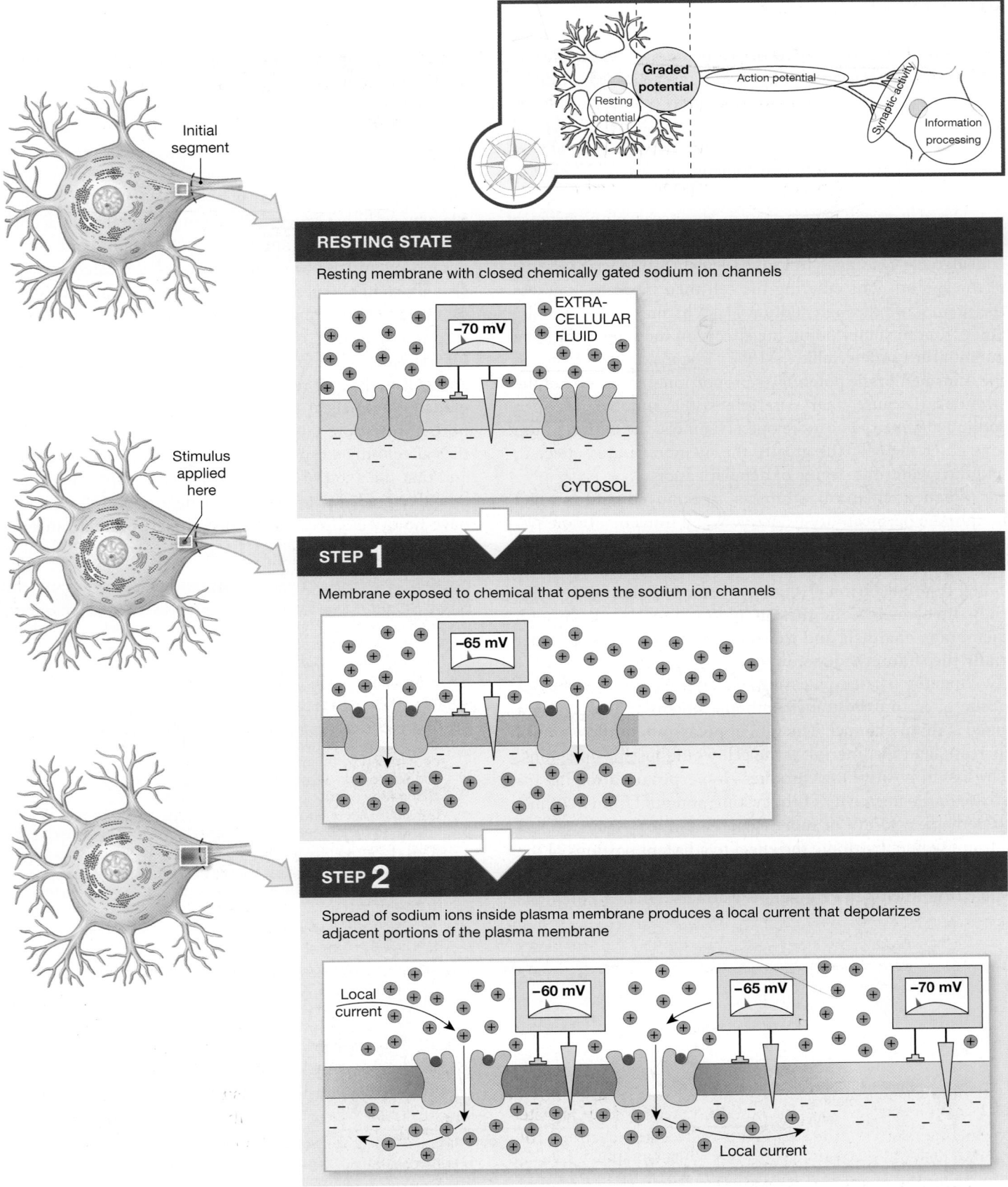

RESTING STATE

Resting membrane with closed chemically gated sodium ion channels

−70 mV

EXTRA-CELLULAR FLUID

CYTOSOL

Initial segment

STEP 1

Membrane exposed to chemical that opens the sodium ion channels

−65 mV

Stimulus applied here

STEP 2

Spread of sodium ions inside plasma membrane produces a local current that depolarizes adjacent portions of the plasma membrane

Local current

−60 mV   −65 mV   −70 mV

Local current

Graded potential

Resting potential

Action potential

Synaptic activity

Information processing

Figure 12–12 **Graded Potentials.** The depolarization radiates in all directions away from the source of stimulation. For clarity, only gated channels are shown; leak channels are present, but are not responsible for the production of graded potentials. Color changes in the phospholipid bilayer indicate that the resting potential has been disturbed and that the transmembrane potential is no longer −70 mV. Notice that the Navigator Icon now highlights the graded potential.

of negative ions on the inside of the membrane. As the plasma membrane depolarizes, sodium ions are released from its outer surface. These ions, accompanied by other extracellular sodium ions, then move toward the open channels, replacing ions that have already entered the cell. This movement of positive charges parallel to the inner and outer surfaces of a membrane is called a **local current** (STEP 2).

The degree of depolarization decreases with distance away from the stimulation site, because the cytosol offers considerable resistance to ion movement, and because some of the sodium ions entering the cell then move back across the membrane through sodium leak channels. At some distance from the entry point, the effects on the transmembrane potential are undetectable (STEP 2). The maximum change in the transmembrane potential is proportional to the size of the stimulus, because that determines the number of open sodium channels. The more open channels, the more sodium ions enter the cell, the greater the membrane area affected, and the greater the degree of depolarization.

When a chemical stimulus is removed and normal membrane permeability is restored, the transmembrane potential soon returns to resting levels. The process of restoring the normal resting potential after depolarization is called **repolarization** (Figure 12–13a). Repolarization typically involves a combination of ion movement through membrane channels and the activities of ion pumps, especially the sodium–potassium exchange pump.

Opening a gated potassium channel would have the opposite effect on the transmembrane potential as opening a gated sodium channel: The rate of potassium outflow would increase, and the interior of the cell would lose positive ions. The loss of positive ions produces **hyperpolarization**, an increase in the negativity of the resting potential from −70 mV to perhaps −80 mV or more (**Figure 12–13b**). Again, a local current would distribute the effect to adjacent portions of the plasma membrane, and the effect would decrease with distance from the open channel or channels. Graded potentials

occur in the membranes of many types of cells—not just nerve and muscle cells, but epithelial cells, gland cells, adipocytes, and a variety of sensory receptors.

Graded potentials are often the trigger for specific cell functions; for example, a graded potential at the surface of a gland cell may trigger the exocytosis of secretory vesicles. Similarly, it is the graded depolarization of the motor end plate by ACh that triggers an action potential in adjacent portions of the sarcolemma. The motor end plate supports graded potentials, whereas the rest of the sarcolemma consists of excitable membrane. Table 12–2 summarizes the basic characteristics of graded potentials.

An interesting observation—that each neuron receives information in the form of graded potentials on its dendrites and cell body, and releases neurotransmitter in response to graded potentials at synaptic terminals—brings us to an important question: Given that even the largest graded potentials affect only a tiny area (perhaps only 1 mm in diameter), and that the axon of a typical sensory or motor neuron is so long that graded potentials on the dendrites and cell body can have no direct effect on the distant synaptic terminals, how

SUMMARY TABLE 12–2 Graded Potentials
Graded potentials, whether depolarizing or hyperpolarizing, share four basic characteristics:
1. The transmembrane potential is most affected at the site of stimulation, and the effect decreases with distance.
2. The effect spreads passively, owing to local currents.
3. The graded change in membrane potential may involve either depolarization or hyperpolarization. The nature of the change is determined by the properties of the membrane channels involved. For example, in a resting membrane, the opening of sodium channels will cause depolarization, whereas the opening of potassium channels will cause hyperpolarization. That is, the change in membrane potential reflects whether positive charges enter or leave the cell.
4. The stronger the stimulus, the greater is the change in the transmembrane potential and the larger is the area affected.

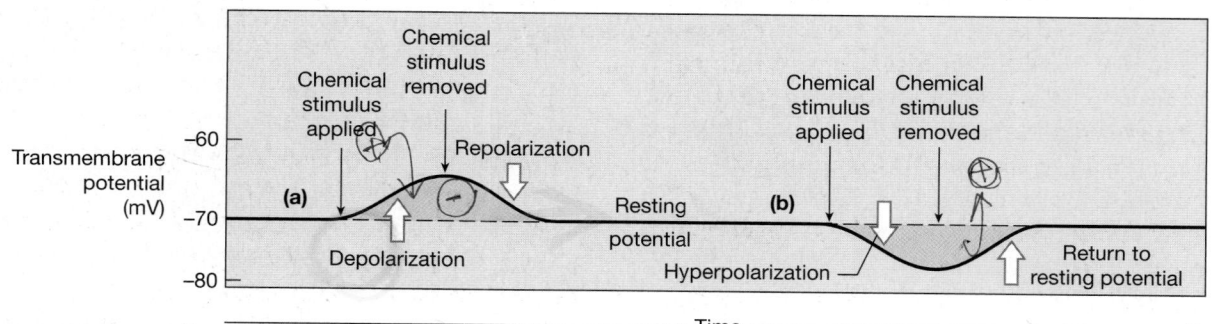

Figure 12–13 Depolarization, Repolarization, and Hyperpolarization. (a) Depolarization and repolarization in response to the application and removal of a stimulus that opens chemically gated sodium channels. **(b)** Hyperpolarization in response to the application of a stimulus that opens chemically gated potassium channels. When the stimulus is removed, the membrane potential returns to the resting level.

can graded potentials on the dendrites and cell body create a graded potential at the synaptic terminals to trigger the release of neurotransmitter? The answer is that the graded potentials at opposite ends of the cell are linked by an action potential, which we will study next.

[handwritten: doesn't matter how much/what as long as threshold met.]

CHECKPOINT

12. Define the resting potential. *[handwritten: transmembrane pot. of resting cell ⇒ homeostasis]*
13. What effect would a chemical that blocks the voltage-gated sodium channels in neuron plasma membranes have on a neuron's ability to depolarize? *[handwritten: Na+ wouldn't be able to flow in]*
14. What effect would decreasing the concentration of extracellular potassium ions have on the transmembrane potential of a neuron? *[handwritten: K+ would flow out ⇒ chemical gradient ⇒ hyperpolarization.]*

See the blue Answers tab at the end of the book.

12-5 An action potential is a nerve impulse

Action potentials are propagated changes in the transmembrane potential that, once initiated, affect an entire excitable membrane. The first step in the generation of an action potential is the opening of voltage-gated sodium ion channels at one site, usually the initial segment of the axon. The movement of sodium ions into the cell depolarizes adjacent sites, triggering the opening of additional voltage-gated channels. The result is a chain reaction that spreads across the surface of the membrane like a line of falling dominoes. In this way, the impulse is propagated along the length of the axon, ultimately reaching the synaptic terminals.

The All-or-None Principle

The stimulus that initiates an action potential is a depolarization large enough to open voltage-gated sodium channels. That opening occurs at a transmembrane potential known as the **threshold**. Threshold for an axon is typically between -60 mV and -55 mV corresponding to a depolarization of 10–15 mV. A stimulus that shifts the resting membrane potential from -70 mV to -62 mV will not produce an action potential, only a graded depolarization. When such a stimulus is removed, the transmembrane potential returns to the resting level. The depolarization of the initial segment of the axon is caused by local currents resulting from the graded depolarization of the axon hillock.

The initial depolarization acts like pressure on the trigger of a gun. If a slight pressure is applied, the gun will not fire. It will fire only when a certain minimum pressure is applied to the trigger. Once the pressure on the trigger reaches this threshold, the firing pin drops and the gun discharges. At that point, it no longer matters whether the pressure was applied gradually or suddenly or whether it was caused by the precise movement of just one finger or by the clenching of the entire hand. The speed and range of the bullet that leaves the gun do not change, regardless of the forces that were applied to the trigger.

In the case of an axon or another area of excitable membrane, a graded depolarization is analogous to the pressure on the trigger, and the action potential is like the firing of the gun. All stimuli that bring the membrane to threshold generate identical action potentials. In other words, the properties of the action potential are independent of the relative strength of the depolarizing stimulus, so long as that stimulus exceeds the threshold. This concept is called the **all-or-none principle**, because a given stimulus either triggers a typical action potential, or it does not produce one at all. The all-or-none principle applies to all excitable membranes.

We will now take a closer look at the mechanisms whereby action potentials are generated and propagated. Generation and propagation are closely related concepts, in terms of both time and space: An action potential must be generated at one site before it can be propagated away from that site.

Generation of Action Potentials

Figure 12–14 diagrams the steps involved in the generation of an action potential from the resting state. At the normal resting potential, the activation gates of the voltage-gated sodium channels are closed. The steps are as follows:

Step 1 *Depolarization to Threshold.* Before an action potential can begin, an area of excitable membrane must be depolarized to its threshold by local currents.

Step 2 *Activation of Sodium Channels and Rapid Depolarization.* At threshold, the sodium activation gates open, and the plasma membrane becomes much more permeable to Na^+. Driven by the large electrochemical gradient, sodium ions rush into the cytoplasm, and rapid depolarization occurs at the site. In less than a millisecond, the inner membrane surface has changed; it now contains more positive ions than negative ones, and the transmembrane potential has changed from -60 mV to positive values closer to the equilibrium potential for sodium ions.

Notice that the first two steps in the generation of an action potential are an example of positive feedback: A small depolarization triggers a larger depolarization.

Step 3 *Inactivation of Sodium Channels and the Activation of Potassium Channels.* As the transmembrane potential approaches $+30$ mV, the inactivation gates of the voltage-gated sodium channels begin closing. This step is known as **sodium channel inactivation**. While it is under way, voltage-gated potassium channels are opening. At a transmembrane potential of $+30$ mV, the cytosol along the

Axon hillock

Initial segment

[handwritten: Stimulation.]
[handwritten: if lower than threshold level, goes unnoticed]
[handwritten: must be stronger than threshold level.]

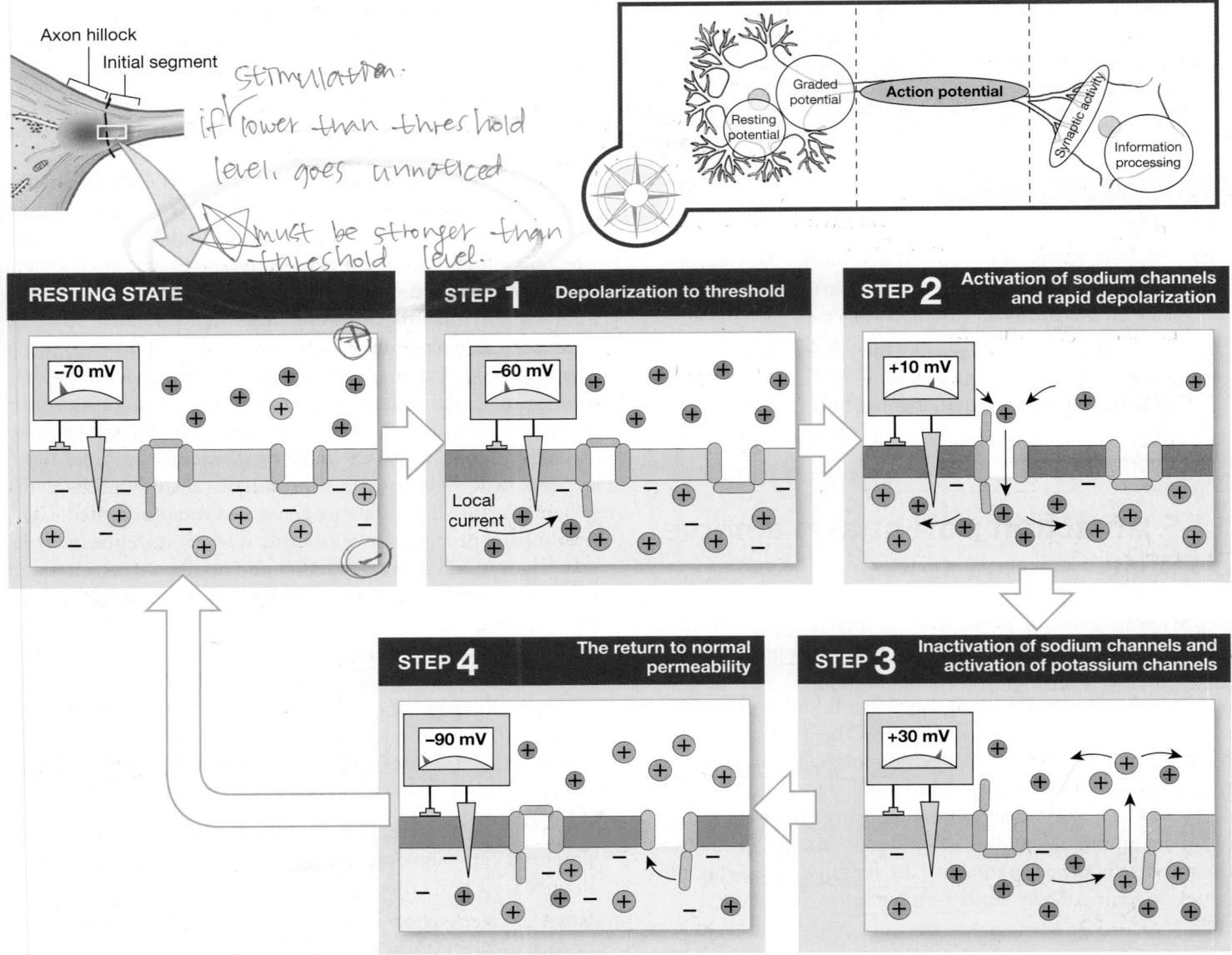

RESTING STATE	STEP **1** Depolarization to threshold	STEP **2** Activation of sodium channels and rapid depolarization
−70 mV	−60 mV	+10 mV
	Local current	

STEP **4** The return to normal permeability	STEP **3** Inactivation of sodium channels and activation of potassium channels
−90 mV	+30 mV

Figure 12–14 **The Generation of an Action Potential.** For clarity, only gated channels are shown.

interior of the membrane contains an excess of positive charges. Thus, in contrast to the situation in the resting membrane (p. 399), both the electrical *and* chemical gradients favor the movement of K^+ out of the cell. The sudden loss of positive charges then shifts the transmembrane potential back toward resting levels, and repolarization begins.

Step 4 *The Return to Normal Permeability.* The voltage-gated sodium channels remain inactivated until the membrane has repolarized to near threshold levels. At this time, they regain their normal status: closed but capable of opening. The voltage-gated potassium channels begin closing as the membrane reaches the normal resting potential (about −70 mV), but the process takes at least a millisecond. Over that period, potassium ions continue to move out

of the cell at a faster rate than when they are at rest, producing a brief hyperpolarization that brings the transmembrane potential very close to the equilibrium potential for potassium (−90 mV). As the voltage-gated potassium channels close, the transmembrane potential returns to normal resting levels. The membrane is now in a prestimulation condition, and the action potential is over.

The Refractory Period

From the time an action potential begins until the normal resting potential has stabilized, the membrane will not respond normally to additional depolarizing stimuli. This period is known as the **refractory period** of the membrane. From the moment the voltage-gated sodium channels open at threshold until sodium channel inactivation ends, the membrane cannot respond to further stimulation, because all the voltage-gated sodium channels either are already open or are inactivated. This portion of the refractory period, the **absolute refractory period**, lasts 0.4–1.0 msec; the smaller the axon diameter, the longer the duration. The **relative refractory period** begins when the sodium channels regain their normal resting condition, and continues until the transmembrane potential stabilizes at resting levels. Another action potential can occur over this period if the membrane is sufficiently depolarized. That depolarization, however, requires a larger-than-normal stimulus, because (1) the local current must deliver enough Na$^+$ to counteract the exit of positively charged K$^+$ through open voltage-gated K$^+$ channels, and (2) the membrane is hyperpolarized to some degree through most of the relative refractory period.

Tips&Tricks

Flushing a toilet provides a useful analogy for an action potential. Nothing happens while the handle is being pressed, until the water starts to flow (threshold is reached). Thereafter, the amount of water that is released is independent of how hard or quickly you pressed the handle (all-or-none principle). Finally, the toilet cannot be flushed again until the tank refills (refractory period).

The Role of the Sodium–Potassium Exchange Pump

In an action potential, depolarization results from the influx of Na$^+$, and repolarization involves the loss of K$^+$. Over time, the sodium–potassium exchange pump returns intracellular and extracellular ion concentrations to prestimulation levels. Compared with the total number of ions inside and outside the cell, however, the number involved in a single action potential is insignificant. Tens of thousands of action potentials can occur before intracellular ion concentrations change enough to disrupt the entire mechanism. Thus, the exchange pump is not essential to any single action potential.

However, a maximally stimulated neuron can generate action potentials at a rate of 1000 per second. Under these circumstances, the exchange pump is needed if ion concentrations are to remain within acceptable limits over a prolonged period. The sodium–potassium exchange pump requires energy in the form of ATP. Each time the pump exchanges two extracellular potassium ions for three intracellular sodium ions, one molecule of ATP must be broken down to ADP. The transmembrane protein of the exchange pump is called *sodium–potassium ATPase*, because it provides the energy to pump ions by splitting a phosphate group from a molecule of ATP, forming ADP. If the cell runs out of ATP, or if sodium–potassium ATPase is inactivated by a metabolic poison, a neuron will soon lose its ability to function.

Table 12–3 summarizes the generation of action potentials.

SUMMARY TABLE 12–3   Generation of Action Potentials

STEP 1: Depolarization to threshold
- A graded depolarization brings an area of excitable membrane to threshold (−60 mV).

STEP 2: Activation of sodium channels and rapid depolarization
- The voltage-gated sodium channels open (sodium channel activation).
- Sodium ions, driven by electrical attraction and the chemical gradient, flood into the cell.
- The transmembrane potential goes from −60 mV (the threshold level), toward +30 mV (the sodium equilibrium potential).

STEP 3: Inactivation of sodium channels and activation of potassium channels
- The voltage-gated sodium channels close (sodium channel inactivation occurs) at +30 mV.
- The voltage-gated potassium channels are now open, and potassium ions diffuse out of the cell.
- Repolarization begins.

STEP 4: Return to normal permeability
- The voltage-gated sodium channels regain their normal properties in 0.4–1.0 msec. The membrane is now capable of generating another action potential if a larger-than-normal stimulus is provided.
- The voltage-gated potassium channels begin closing at −70 mV. Because they do not all close at the same time, potassium loss continues and a temporary hyperpolarization to approximately −90 mV (the potassium equilibrium potential) occurs.
- At the end of relative refractory period, all voltage-gated channels have closed and the plasma membrane is back to its resting state.

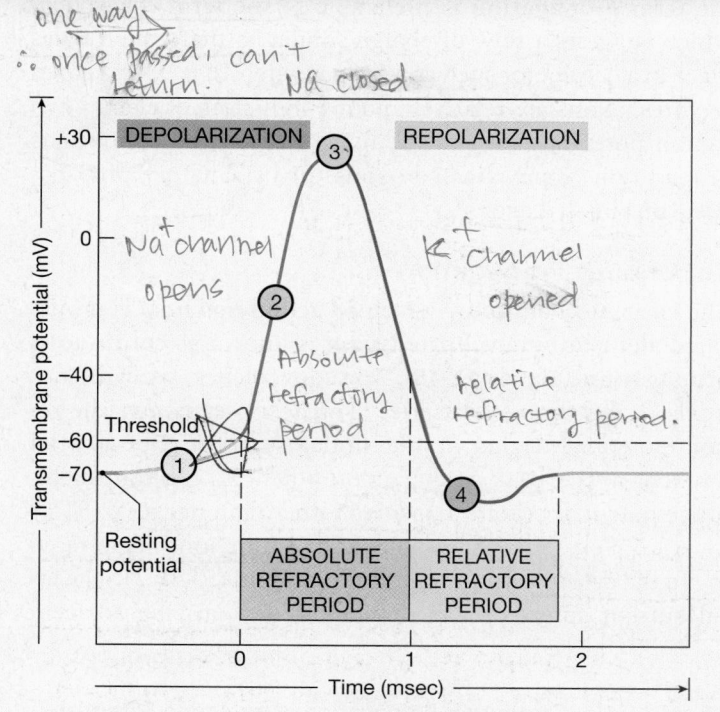

Propagation of Action Potentials

The sequence of events just described occurs in a relatively small portion of the total membrane surface. But we have already noted that, unlike graded potentials, which diminish rapidly with distance, action potentials spread and affect the entire excitable membrane. To understand the basic principle involved, imagine that you are standing by the doors of a movie theater at the start of a long line. Everyone is waiting for the doors to open. The manager steps outside and says to you, "Let everyone know that we're opening in 15 minutes." How would you spread the news?

If you treated the line as an inexcitable membrane, you would shout, "The doors open in 15 minutes!" as loudly as you could. The closest people in the line would hear the news very clearly, but those farther away might not hear the entire message, and those at the end of the line might not hear you at all. If, on the other hand, you treated the crowd as an excitable membrane, you would tell the message to the next person in line, with instructions to pass it on. In that way, the message would travel along the line undiminished, until everyone had heard the news. Such a message "moves" as each person repeats it to someone else. Distance is not a factor; the line can contain 50 people or 5000.

The situation just described is comparable to the way an action potential spreads along an excitable membrane. An action potential (message) is relayed from one location to another in a series of steps. At each step, the message is repeated. Because the same events take place over and over, the term **propagation** is preferable to the term *conduction*, which suggests a flow of charge similar to that which takes place in a conductor such as a copper wire. (In fact, compared to wires, axons are relatively poor conductors of electricity.) Action potentials may travel along an axon by continuous propagation (unmyelinated axons) or by saltatory propagation (myelinated axons).

Continuous Propagation

The basic mechanism by which an action potential is propagated along an unmyelinated axon is known as **continuous propagation** (Figure 12–15). For convenience, we will consider the membrane as a series of adjacent segments. The action potential begins at the initial segment. For a brief moment at the peak of the action potential, the transmembrane potential becomes positive rather than negative (STEP 1). A local current then develops as sodium ions begin moving in the cytosol and the extracellular fluid (STEP 2). The local current spreads in all directions, depolarizing adjacent

Figure 12–15 **Continuous Propagation of an Action Potential along an Unmyelinated Axon.** Events are best understood when the axon is viewed as a series of adjacent segments.

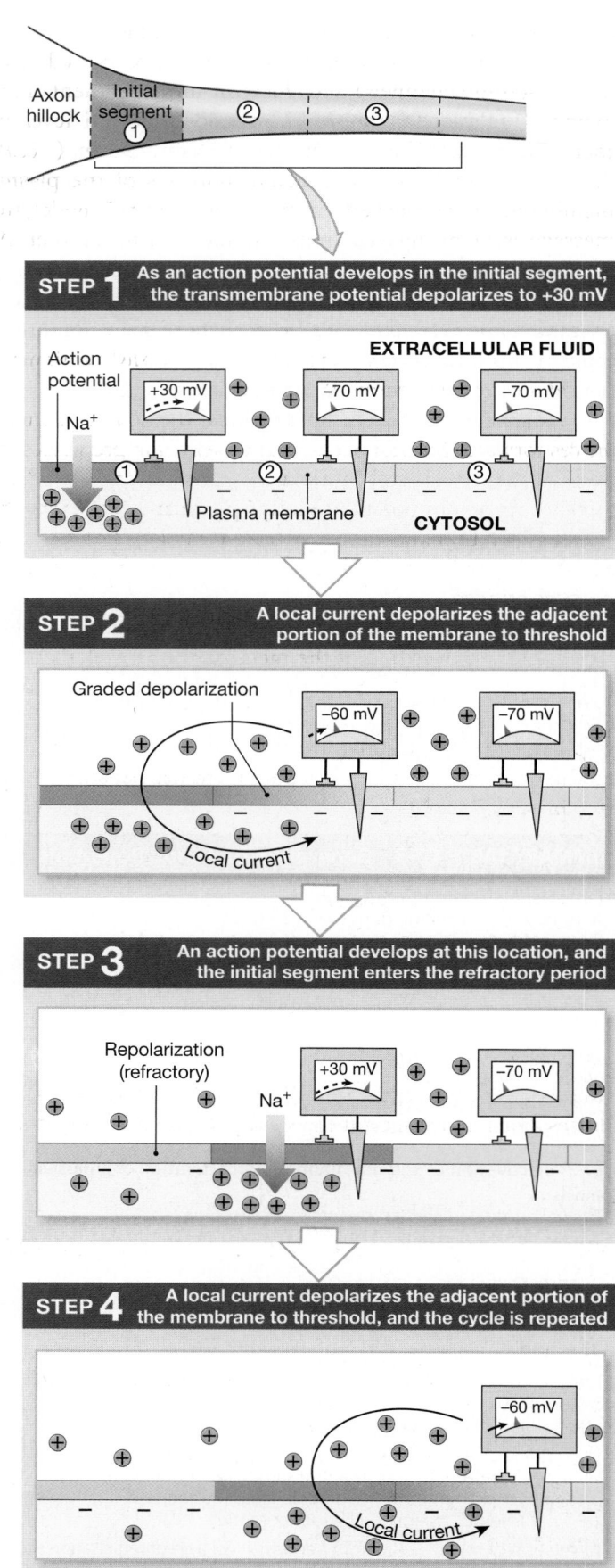

STEP 1 As an action potential develops in the initial segment, the transmembrane potential depolarizes to +30 mV

EXTRACELLULAR FLUID

Action potential

+30 mV   −70 mV   −70 mV

Na+

① ② ③

Plasma membrane   CYTOSOL

STEP 2 A local current depolarizes the adjacent portion of the membrane to threshold

Graded depolarization

−60 mV   −70 mV

Local current

STEP 3 An action potential develops at this location, and the initial segment enters the refractory period

Repolarization (refractory)

+30 mV   −70 mV

Na+

STEP 4 A local current depolarizes the adjacent portion of the membrane to threshold, and the cycle is repeated

−60 mV

Local current

portions of the membrane. The axon hillock cannot respond with an action potential (because like the rest of the cell body, it lacks sodium channels), but when the initial segment of the axon is depolarized to threshold, an action potential develops there. The process then continues in a chain reaction (STEPS 3, 4). Eventually, the most distant portions of the plasma membrane will be affected. As in our "movie line" model, the message is being relayed from one location to another. At each step along the way, the message is retold, so distance has no effect on the process. The action potential reaching the synaptic knob is identical to the one generated at the initial segment, and the net effect is the same as if a single action potential had traveled across the surface of the membrane.

Each time a local current develops, the action potential moves forward, but not backward, because the previous segment of the axon is still in the absolute refractory period. As a result, an action potential always proceeds away from the site of generation and cannot reverse direction. For a second action potential to occur at the same site, a second stimulus must be applied.

In continuous propagation, an action potential appears to move across the surface of the membrane in a series of tiny steps. Even though the events at any one location take only about a millisecond, the sequence must be repeated at each step along the way. Continuous propagation along unmyelinated axons occurs at a speed of about 1 meter per second (approximately 2 mph).

Tips&Tricks

The "wave" performed by fans in a football stadium illustrates the continuous propagation of an action potential. The "wave" moves, but the people remain in place.

Saltatory Propagation

Saltatory propagation in the CNS and PNS carries nerve impulses along an axon much more rapidly than does continu-

ous propagation. To get the general idea, let's return to the line in front of the movie theater, and assume that it takes 1 second to relay the message to another person. In a model of continuous propagation, the people are jammed together. In 4 seconds, four people would have heard the news, and the message would have moved perhaps 2 meters along the line. In a model of saltatory propagation, in contrast, the people in the line are spaced 5 meters apart. So after 4 seconds the same message would have moved 20 meters.

In a myelinated axon, the "people" are the nodes, and the spaces between them are the internodes wrapped in myelin (**Figures 12–5b** and **12–6a**). Continuous propagation cannot occur along a myelinated axon, because myelin increases resistance to the flow of ions across the membrane. Ions can readily cross the plasma membrane just at the nodes. As a result, only the nodes can respond to a depolarizing stimulus.

When an action potential appears at the initial segment of a myelinated axon, the local current skips the internodes and depolarizes the closest node to threshold (**Figure 12–16**). Because the nodes may be 1–2 mm apart in a large myelinated axon, the action potential "jumps" from node to node rather than moving along the axon in a series of tiny steps. This process is called **saltatory propagation** (*saltare*, leaping). In addition to being faster, saltatory propagation also uses proportionately less energy, because less surface area is involved and fewer sodium ions must be pumped out of the cytoplasm.

Table 12–4 reviews the key differences between graded potentials and action potentials.

CHECKPOINT

15. Define action potential.

16. List the steps involved in the generation and propagation of an action potential.

See the blue Answers tab at the end of the book.

TABLE 12–4 A Comparison of Graded Potentials and Action Potentials	
Graded Potentials	**Action Potentials**
Depolarizing or hyperpolarizing	Always depolarizing
No threshold value	Depolarization to threshold must occur before action potential begins
Amount of depolarization or hyperpolarization depends on intensity of stimulus	All-or-none phenomenon; all stimuli that exceed threshold will produce identical action potentials
Passive spread from site of stimulation	Action potential at one site depolarizes adjacent sites to threshold
Effect on membrane potential decreases with distance from stimulation site	Propagated along entire membrane surface without decrease in strength
No refractory period	Refractory period occurs
Occur in most plasma membranes	Occur only in excitable membranes of specialized cells such as neurons and muscle cells

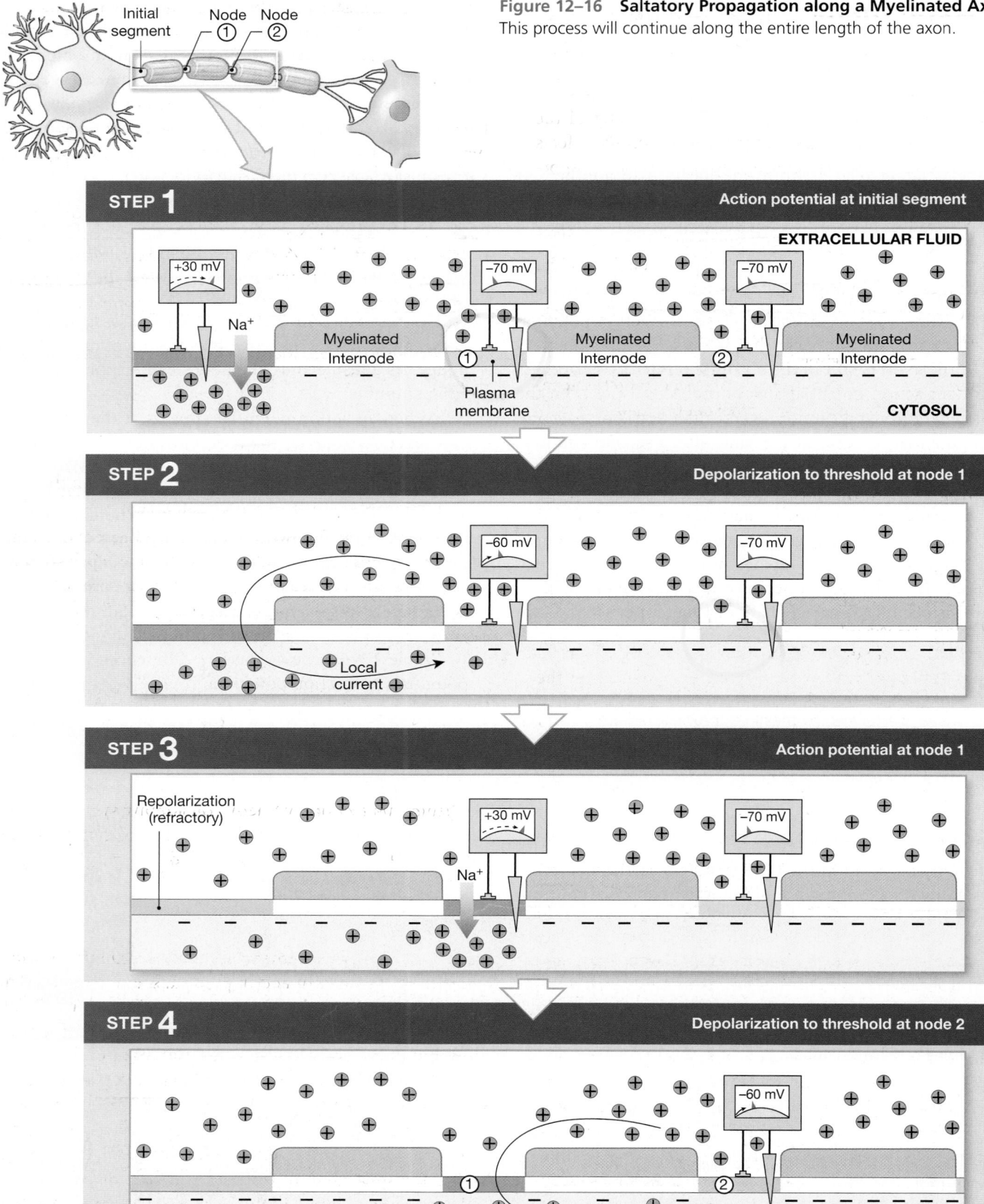

Figure 12–16 Saltatory Propagation along a Myelinated Axon. This process will continue along the entire length of the axon.

12-6 Axon diameter, in addition to myelin, affects propagation speed

As we have seen, the presence of myelin greatly increases the propagation speed of action potentials. The diameter of the axon also affects the propagation speed, although the effects are less dramatic. Axon diameter is important because in order to depolarize adjacent portions of the plasma membrane, ions must move through the cytoplasm. Cytoplasm offers resistance to ion movement, although much less resistance than the plasma membrane. In this instance, an axon behaves like an electrical cable: The larger the diameter, the lower the resistance. (That is why motors with large current demands, such as the starter on a car, an electric stove, or a big air conditioner, use such thick wires.)

Axons are classified into three groups according to the relationships among the diameter, myelination, and propagation speed:

1. **Type A fibers** are the largest axons, with diameters ranging from 4 to 20 μm. These fibers are myelinated axons that carry action potentials at speeds of up to 120 meters per second, or 268 mph.

2. **Type B fibers** are smaller myelinated axons, with diameters of 2–4 μm. Their propagation speeds average around 18 meters per second, or roughly 40 mph.

3. **Type C fibers** are unmyelinated and less than 2 μm in diameter. These axons propagate action potentials at the leisurely pace of 1 meter per second, or a mere 2 mph.

The relative importance of myelin becomes apparent by noting that in comparing Type C to Type A fibers, the diameter increases tenfold but the propagation speed increases by 120 times.

Type A fibers carry sensory information about position, balance, and delicate touch and pressure sensations from the skin surface to the CNS. The motor neurons that control skeletal muscles also send their commands over large, myelinated Type A axons. Type B fibers and Type C fibers carry information to the CNS; they deliver temperature, pain, and general touch and pressure sensations, and carry instructions to smooth muscle, cardiac muscle, glands, and other peripheral effectors.

Not every axon in the nervous system is large and myelinated, most likely because that would be physically impossible. If all sensory information were carried by large Type A fibers, your peripheral nerves would be the size of garden hoses, and your spinal cord would be the diameter of a garbage can. Instead, only about one-third of all axons carrying sensory information are myelinated, and most sensory information arrives over slender Type C fibers. In essence, information transfer in the nervous system represents a compromise between conduction time and available space. Messages are routed according to priority: Urgent news—sensory information about things that threaten survival and motor commands that prevent injury—travels over Type A fibers (the equivalent of Express Mail). Less urgent sensory information and motor commands are relayed by Type B fibers (Regular Mail) or Type C fibers (Bulk Mail).

The A&P Top 100

#42 "Information" travels within the nervous system primarily in the form of propagated electrical signals known as action potentials. The most important information—including vision and balance sensations, and the motor commands to skeletal muscles—is carried by large-diameter myelinated axons.

CHECKPOINT

17. What is the relationship between myelin and the propagation speed of action potentials?

18. Which of the following axons is myelinated: one that propagates action potentials at 50 meters per second, or one that carries them at 1 meter per second?

See the blue Answers tab at the end of the book.

12-7 At synapses, communication occurs among neurons or between neurons and other cells

We begin this section by briefly considering synaptic activity; then we examine the general properties of synapses before discussing the class of synapses called cholinergic synapses.

Synaptic Activity

In the nervous system, messages move from one location to another in the form of action potentials along axons. These electrical events are also known as **nerve impulses**. To be effective, a message must be not only propagated along an axon but also transferred in some way to another cell. At a synapse between two neurons, the impulse passes from the **presynaptic neuron** to the **postsynaptic neuron**. A synapse may also involve other types of postsynaptic cells. For example, the neuromuscular junction is a synapse where the postsynaptic cell is a skeletal muscle fiber. We will now take a closer look at the mechanisms involved in synaptic function.

General Properties of Synapses

A synapse may be *electrical*, with direct physical contact between the cells, or *chemical*, involving a neurotransmitter.

Electrical Synapses

At **electrical synapses**, the presynaptic and postsynaptic membranes are locked together at gap junctions (**Figure 4–2**, p. 116). The lipid portions of the opposing membranes, separated by only 2 nm, are held in position by binding between integral membrane proteins called *connexons*. These proteins contain pores that permit the passage of ions between the cells. Because the two cells are linked in this way, changes in the transmembrane potential of one cell will produce local currents that affect the other cell as if the two shared a common membrane. As a result, an electrical synapse propagates action potentials quickly and efficiently from one cell to the next.

Electrical synapses are located in both the CNS and PNS, but they are extremely rare. They are present in some areas of the brain, including the *vestibular nuclei*, in the eye, and in at least one pair of PNS ganglia (the *ciliary ganglia*).

Chemical Synapses

The situation at a **chemical synapse** is far more dynamic than that at an electrical synapse, because the cells are not directly coupled. For example, an action potential that reaches an electrical synapse will *always* be propagated to the next cell. But at a chemical synapse, an arriving action potential *may or may not* release enough neurotransmitter to bring the postsynaptic neuron to threshold. In addition, other factors may intervene and make the postsynaptic cell more or less sensitive to arriving stimuli. In essence, the postsynaptic cell at a chemical synapse is not a slave to the presynaptic neuron; its activity can be adjusted, or "tuned," by a variety of factors.

Chemical synapses are by far the most abundant type of synapse. Most synapses between neurons, and all communications between neurons and other types of cells, involve chemical synapses. Normally, communication across a chemical synapse can occur in only one direction: from the presynaptic membrane to the postsynaptic membrane.

Although acetylcholine is the neurotransmitter that has received the most attention, there are other important chemical transmitters. Based on their effects on postsynaptic membranes, neurotransmitters are often classified as excitatory or inhibitory. **Excitatory neurotransmitters** cause depolarization and promote the generation of action potentials, whereas **inhibitory neurotransmitters** cause hyperpolarization and suppress the generation of action potentials.

This classification is useful, but not always precise. For example, acetylcholine typically produces a depolarization in the postsynaptic membrane, but acetylcholine released at neuromuscular junctions in the heart has an inhibitory effect, producing a transient hyperpolarization of the postsynaptic membrane. This situation highlights an important aspect of neurotransmitter function: *The effect of a neurotransmitter on the postsynaptic membrane depends on the properties of the receptor, not on the nature of the neurotransmitter.*

We will continue our discussion of chemical synapses with a closer look at a synapse that releases the neurotransmitter **acetylcholine (ACh)**. In the next section we will briefly examine the activities of other important neurotransmitters that will be encountered in later chapters.

Cholinergic Synapses

Synapses that release ACh are known as **cholinergic synapses**. The neuromuscular junction is an example of a cholinergic synapse. 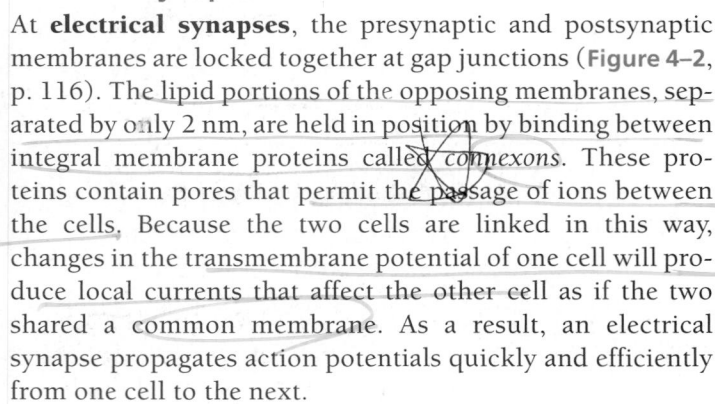 p. 304 ACh, the most widespread (and best-studied) neurotransmitter, is released (1) at all neuromuscular junctions involving skeletal muscle fibers, (2) at many synapses in the CNS, (3) at all neuron-to-neuron synapses in the PNS, and (4) at all neuromuscular and neuroglandular junctions within the parasympathetic division of the ANS.

At a cholinergic synapse between two neurons, the presynaptic and postsynaptic membranes are separated by a synaptic cleft that averages 20 nm (0.02 μm) in width. Most of the ACh in the synaptic knob is packaged in synaptic vesicles, each containing several thousand molecules of the neurotransmitter. A single synaptic knob may contain a million such vesicles.

Tips&Tricks

Cholinergic synapses are so named because the neurotransmitter involved is acetyl**cholin**e.

Events at a Cholinergic Synapse

Figure 12–17 diagrams the events that occur at a cholinergic synapse after an action potential arrives at a synaptic knob. For convenience, we will assume that this synapse is adjacent to the initial segment of the axon, an arrangement that is relatively easy to illustrate.

Step 1 *An Action Potential Arrives and Depolarizes the Synaptic Knob.* The normal stimulus for neurotransmitter release is the depolarization of the synaptic knob by the arrival of an action potential.

Step 2 *Extracellular Calcium Ions Enter the Synaptic Knob, Triggering the Exocytosis of ACh.* The depolarization of the synaptic knob opens voltage-gated calcium channels. In the brief period during which these channels remain open, calcium ions rush into the knob. Their arrival triggers exocytosis and the release of ACh into the synaptic cleft. The ACh is released in packets of roughly 3000 molecules,

STEP 1 — An action potential arrives and depolarizes the synaptic knob

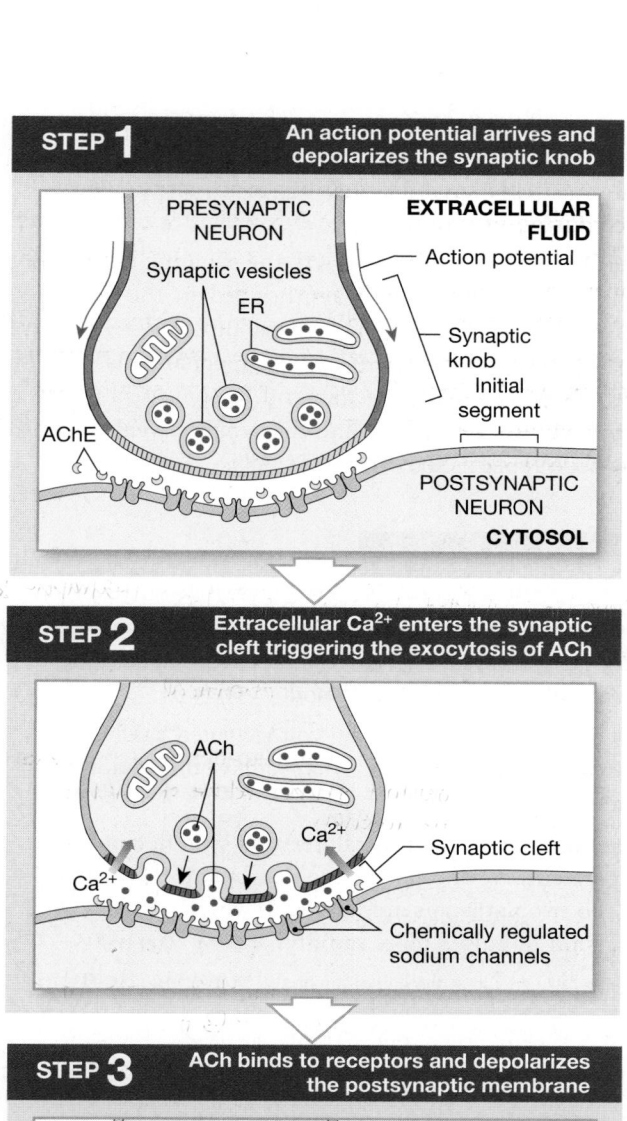

STEP 2 — Extracellular Ca²⁺ enters the synaptic cleft triggering the exocytosis of ACh

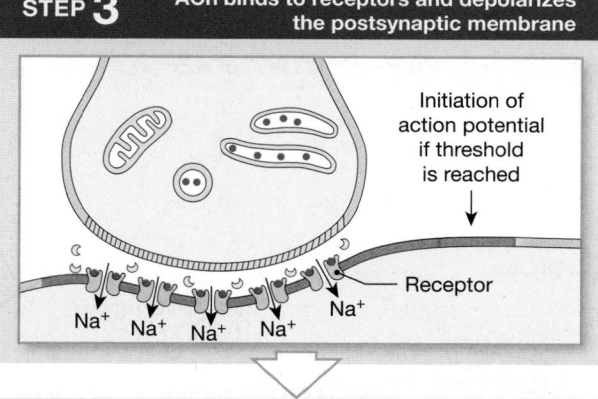

STEP 3 — ACh binds to receptors and depolarizes the postsynaptic membrane

STEP 4 — ACh is removed by AChE (acetylcholinesterase)

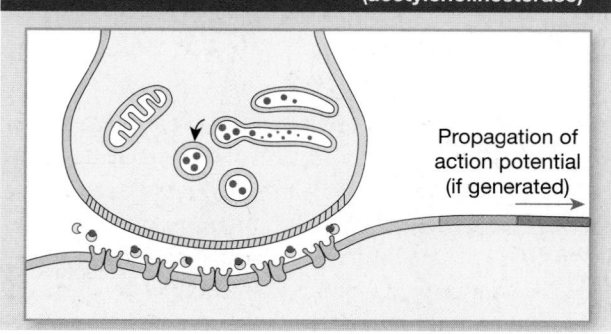

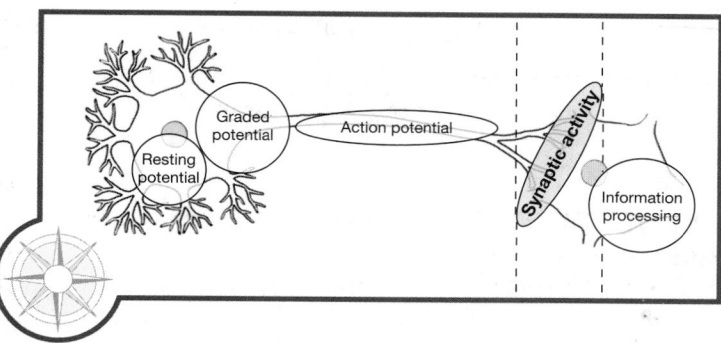

Figure 12–17 **Events in the Functioning of a Cholinergic Synapse.**

the average number of ACh molecules in a single vesicle. The release of ACh stops very soon, because the calcium ions that triggered exocytosis are rapidly removed from the cytoplasm by active transport mechanisms; they are either pumped out of the cell or transferred into mitochondria, vesicles, or the endoplasmic reticulum.

Step 3 *ACh Binds to Receptors and Depolarizes the Postsynaptic Membrane.* The ACh released through exocytosis diffuses across the synaptic cleft toward receptors on the postsynaptic membrane. These receptors are chemically gated ion channels. The primary response is an increased permeability to Na^+, producing a depolarization that lasts about 20 msec.[2]

This depolarization is a graded potential: The greater the amount of ACh released at the presynaptic membrane, the greater the number of open cation channels in the postsynaptic membrane, and so the larger the depolarization. If the depolarization brings the adjacent area of excitable membrane to threshold, an action potential will appear in the postsynaptic neuron.

Step 4 *ACh Is Removed by AChE.* The effects on the postsynaptic membrane are temporary, because the synaptic cleft and the postsynaptic membrane contain the enzyme *acetylcholinesterase* (*AChE*, or *cholinesterase*). Roughly half of the ACh released at the presynaptic membrane is broken down before it reaches receptors on the postsynaptic membrane. ACh molecules that succeed in binding to receptor sites are generally broken down within 20 msec of their arrival.

AChE breaks down (hydrolyzes) molecules of ACh into **acetate** and **choline**. The choline is actively absorbed by the

[2] These cation channels also let potassium ions out of the cell, but because sodium ions are driven by a much stronger electrochemical gradient, the net effect is a slight depolarization of the postsynaptic membrane.

synaptic knob and is used to synthesize more ACh, using acetate provided by *coenzyme A (CoA)*. (Recall from Chapter 2 that coenzymes derived from vitamins are required in many enzymatic reactions. ∞ p. 57) Acetate diffusing away from the synapse can be absorbed and metabolized by the postsynaptic cell or by other cells and tissues.

Table 12–5 summarizes the events that occur at a cholinergic synapse.

Synaptic Delay

A 0.2–0.5-msec **synaptic delay** occurs between the arrival of the action potential at the synaptic knob and the effect on the postsynaptic membrane. Most of that delay reflects the time involved in calcium influx and neurotransmitter release, not in the neurotransmitter's diffusion—the synaptic cleft is narrow, and neurotransmitters can diffuse across it in very little time.

Although a delay of 0.5 msec is not very long, in that time an action potential may travel more than 7 cm (about 3 in.) along a myelinated axon. When information is being passed along a chain of interneurons in the CNS, the cumulative synaptic delay may exceed the propagation time along the axons. This is why reflexes are important for survival—they involve only a few synapses and thus provide rapid and automatic responses to stimuli. The fewer synapses involved, the shorter the total synaptic delay and the faster the response. The fastest reflexes have just one synapse, with a sensory neuron directly controlling a motor neuron. The muscle spindle reflexes, discussed in Chapter 13, are arranged in this way.

Synaptic Fatigue

Because ACh molecules are recycled, the synaptic knob is not totally dependent on the ACh synthesized in the cell body and delivered by axoplasmic transport. But under intensive stimulation, resynthesis and transport mechanisms may be unable to keep pace with the demand for neurotransmitter. **Synaptic fatigue** then occurs, and the synapse weakens until ACh has been replenished.

CHECKPOINT

19. Describe the general structure of a synapse. *presynaptic & postsynaptic; divided by synaptic cleft*

20. If a synapse involves direct physical contact between cells, it is termed ___*electrical*___; if the synapse involves a neurotransmitter, it is termed ___*chemical*___.

21. What effect would blocking voltage-gated calcium channels at a cholinergic synapse have on synaptic communication? *wouldn't trigger release of ACh: Ca^{2+} no reaction.*

22. One pathway in the central nervous system consists of three neurons, another of five neurons. If the neurons in the two pathways are identical, which pathway will transmit impulses more rapidly? *fewer neurons → fewer synaptic delay → 3*

See the blue Answers tab at the end of the book.

SUMMARY TABLE 12–5 Synaptic Activity

THE SEQUENCE OF EVENTS AT A TYPICAL CHOLINERGIC SYNAPSE:

STEP 1
- An arriving action potential depolarizes the synaptic knob.

STEP 2
- Calcium ions enter the cytoplasm of the synaptic knob.
- ACh is released through exocytosis of neurotransmitter vesicles.

STEP 3
- ACh diffuses across the synaptic cleft and binds to receptors on the postsynaptic membrane.
- Chemically gated sodium channels on the postsynaptic surface are activated, producing a graded depolarization.
- ACh release ceases because calcium ions are removed from the cytoplasm of the synaptic knob.

STEP 4
- The depolarization ends as ACh is broken down into acetate and choline by AChE.
- The synaptic knob reabsorbs choline from the synaptic cleft and uses it to resynthesize ACh.

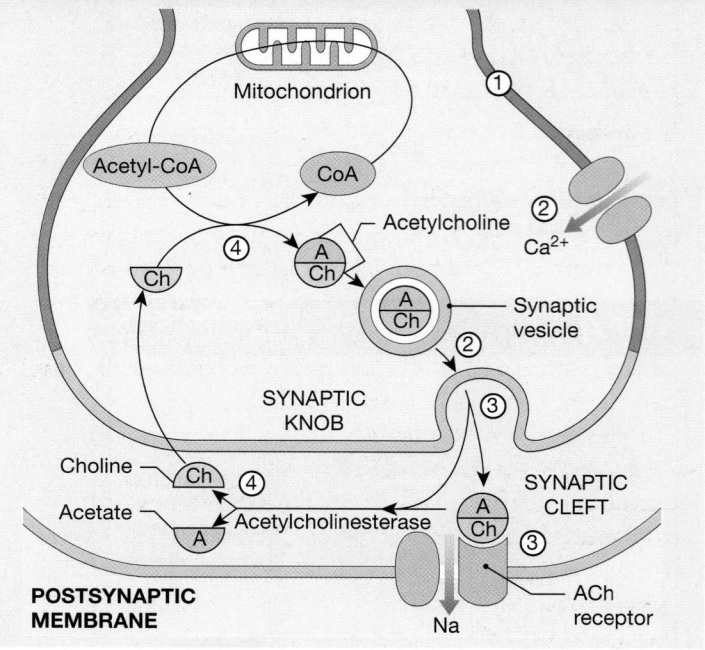

12-8 Neurotransmitters and neuromodulators have various functions

In the previous section we examined the actions of acetylcholine at cholinergic synapses; here we consider the actions of other neurotransmitters, and of neuromodulators, which change the cell's response to neurotransmitters.

The Activities of Other Neurotransmitters

causes depolarization & generates synapses [handwritten]

The nervous system relies on a complex form of chemical communication. Whereas it was once thought that neurons responded to a single neurotransmitter, we now realize that each neuron is continuously exposed to a variety of neurotransmitters. Some usually have excitatory effects, others usually have inhibitory effects. Yet in all cases, the observed effects depend on the nature of the receptor rather than the structure of the neurotransmitter. *causes hyperpolarization & suppresses action potential* [handwritten]

Major categories of neurotransmitters include *biogenic amines*, *amino acids*, *neuropeptides*, *dissolved gases*, and a variety of other compounds. Here we will consider only a few of the most important neurotransmitters; we will encounter additional examples in later chapters.

- **Norepinephrine** (nor-ep-i-NEF-rin), or **NE**, is a neurotransmitter that is widely distributed in the brain and in portions of the ANS. Norepinephrine is also called *noradrenaline*, and synapses that release NE are known as **adrenergic synapses**. Norepinephrine typically has an excitatory, depolarizing effect on the postsynaptic membrane, but the mechanism is quite distinct from that of ACh. We will consider specifics in Chapter 16.

- **Dopamine** (DŌ-puh-mēn), a CNS neurotransmitter released in many areas of the brain, may have either inhibitory or excitatory effects. Inhibitory effects play an important role in our precise control of movements. For example, dopamine release in one portion of the brain prevents the overstimulation of neurons that control skeletal muscle tone. If the neurons that produce dopamine are damaged or destroyed, the result can be the characteristic rigidity and stiffness of *Parkinson disease*, a condition we will describe in Chapter 14. At other sites, dopamine release has excitatory effects. Cocaine inhibits the removal of dopamine from synapses in specific areas of the brain. The resulting rise in dopamine concentrations at these synapses is responsible for the "high" experienced by cocaine users.

- **Serotonin** (ser-ō-TŌ-nin) is another important CNS neurotransmitter. Inadequate serotonin production can have widespread effects on a person's attention and emotional states and may be responsible for many cases of severe chronic depression. *Fluoxetine (Prozac)*, *paroxetine (Paxil)*, *sertraline (Zoloft)*, and related antidepressant drugs inhibit the reabsorption of serotonin by synaptic knobs (hence their classification as *selective serotonin reuptake inhibitors*, or *SSRIs*). This inhibition leads to increased serotonin concentrations at synapses; over time, the increase may relieve the symptoms of depression. Interactions among serotonin, norepinephrine, and other neurotransmitters are thought to be involved in the regulation of sleep and wake cycles.

- **Gamma-aminobutyric** (a-MĒ-nō-bū-TĒR-ik) **acid**, or **GABA**, generally has an inhibitory effect. Although roughly 20 percent of the synapses in the brain release GABA, its functions remain incompletely understood. In the CNS, GABA release appears to reduce anxiety, and some antianxiety drugs work by enhancing this effect.

The functions of many neurotransmitters are not well understood. In a clear demonstration of the principle "the more you look, the more you see," at least 50 neurotransmitters have been identified, including certain amino acids, peptides, polypeptides, prostaglandins, and ATP. In addition, two gases, nitric oxide and carbon monoxide, are now known to be important neurotransmitters. Nitric oxide (NO) is generated by synaptic terminals that innervate smooth muscle in the walls of blood vessels in the PNS, and at synapses in several regions of the brain. Carbon monoxide (CO), best known as a component of automobile exhaust, is also generated by specialized synaptic knobs in the brain, where it functions as a neurotransmitter. Our knowledge of the significance of these compounds and the mechanisms involved in their synthesis and release remains incomplete.

The A&P Top 100

#43 At a chemical synapse, a synaptic terminal releases a neurotransmitter that binds to the postsynaptic plasma membrane. The result is a temporary, localized change in the permeability of the postsynaptic cell. This change may have broader effects on the cell, depending on the nature and number of stimulated receptors. Many drugs affect the nervous system by stimulating receptors that otherwise respond only to neurotransmitters. These drugs can have complex effects on perception, motor control, and emotional states.

Neuromodulators

Although it is convenient to discuss each synapse as if it were releasing only one chemical, synaptic knobs may release a mixture of active compounds, either through diffusion across the membrane or via exocytosis, in the company of neurotransmitter molecules. These compounds may have a variety of functions. Those that alter the rate of neurotransmitter release by the presynaptic neuron or change the postsynaptic cell's response to neurotransmitters are called **neuromodulators** (noo-rō-MOD-ū-lā-torz). These substances are typically **neuropeptides**, small peptide chains synthesized and released by the synaptic knob. Most neuromodulators act by binding to receptors in the presynaptic or postsynaptic membranes and activating cytoplasmic enzymes.

Neuromodulators called **opioids** (Ō-pē-oydz) have effects similar to those of the drugs *opium* and *morphine*, because they bind to the same group of postsynaptic receptors. Four classes of opioids are identified in the CNS: (1) **endorphins** (en-DOR-finz), (2) **enkephalins** (en-KEF-a-linz), (3) **endomorphins**, and (4) **dynorphins** (DĪ-nor-finz). The primary function of opioids is probably the relief of pain—they inhibit the release of the neurotransmitter *substance P* at synapses that relay pain sensations. Dynorphins have far more powerful analgesic (pain-relieving) effects than morphine or the other opioids.

Tips&Tricks

Endorphins are so named because they act like *endo*genous (coming from within the body) m*orphin*e.

In general, neuromodulators (1) have long-term effects that are relatively slow to appear; (2) trigger responses that involve a number of steps and intermediary compounds; (3) may affect the presynaptic membrane, the postsynaptic membrane, or both; and (4) can be released alone or in the company of a neurotransmitter. Table 12–6 lists major neurotransmitters and neuromodulators of the brain and spinal cord, and their primary effects (if known). In practice, it can be very difficult to distinguish neurotransmitters from neuromodulators on either biochemical or functional grounds: A neuropeptide may function in one site as a neuromodulator and in another as a neurotransmitter. For this reason, Table 12–6 does not distinguish between neurotransmitters and neuromodulators.

How Neurotransmitters and Neuromodulators Work

Functionally, neurotransmitters and neuromodulators fall into one of three groups: (1) *compounds that have a direct effect on membrane potential*, (2) *compounds that have an indirect effect on membrane potential*, or (3) *lipid-soluble gases that exert their effects inside the cell*.

Compounds that have direct effects on membrane potential exert those effects by opening or closing gated ion channels (Figure 12–18a). Examples include ACh and the amino acids *glycine* and *aspartate*. Because these neurotransmitters alter ion movement across the membrane, they are said to have *ionotropic effects*. A few neurotransmitters, notably glutamate, GABA, NE, and serotonin, have both direct and indirect effects, because these compounds target two different classes of receptors. The direct effects are ionotropic; the indirect effects, which involve changes in the metabolic activity of the postsynaptic cell, are called *metabotropic*.

Compounds that have an indirect effect on membrane potential work through intermediaries known as *second messengers*. The neurotransmitter represents a *first messenger*, because it delivers the message to receptors on the plasma membrane or within the cell. Second messengers are ions or molecules that are produced or released inside the cell when a first messenger binds to one of these receptors.

Many neurotransmitters—including epinephrine, norepinephrine, dopamine, serotonin, histamine, and GABA, as well as many neuromodulators—bind to receptors in the plasma membrane. In these instances, the link between the first messenger and the second messenger involves a **G protein**, an enzyme complex coupled to a membrane receptor. The name *G protein* refers to the fact that these proteins bind GTP, a high-energy compound introduced in Chapter 2. ∞ p. 60 There are several types of G proteins, but each type includes an enzyme that is "turned on" when an extracellular compound binds to the associated receptor at the cell surface.

Figure 12–18b shows one possible result of this binding: the activation of the enzyme **adenylate cyclase**, also known as *adenylyl cyclase*. This enzyme converts ATP, the energy currency of the cell, to *cyclic-AMP*, a ring-shaped form of the compound AMP that was introduced in Chapter 2. ∞ p. 60 The conversion takes place at the inner surface of the plasma membrane. Cyclic-AMP (cAMP) is a second messenger that may open membrane channels, activate intracellular enzymes, or both, depending on the nature of the postsynaptic cell. This is only an overview of the function of one type of G protein; we will examine several types of G proteins more closely in later chapters.

Two lipid-soluble gases, nitric oxide (NO) and carbon monoxide (CO), are now known to be important neurotransmitters in specific regions of the brain. Because they can diffuse through lipid membranes, these gases can enter the cell and bind to enzymes on the inner surface of the membrane or elsewhere in the cytoplasm (Figure 12–18c). These enzymes then promote the appearance of second messengers that can affect cellular activity.

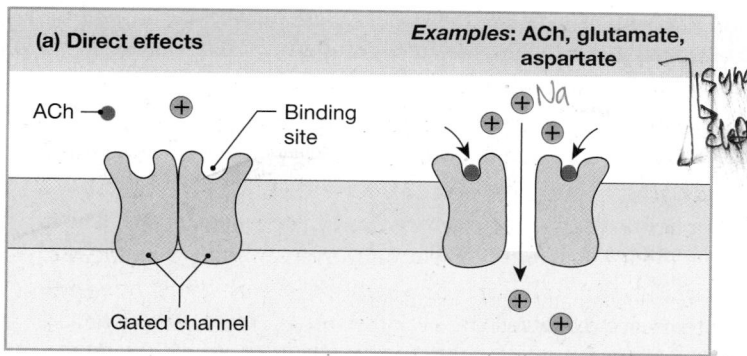

(a) Direct effects

Examples: ACh, glutamate, aspartate

ACh

Binding site

Gated channel

(handwritten:) Synapse cleft.

Na +

23. Differentiate between a neurotransmitter and a neuromodulator. *(handwritten: both released by presynaptic → neurotrans: transmits impulse released across synapse. mod: differs rate of neurotrans.)*

24. Identify the three functional groups into which neurotransmitters and neuromodulators fall. *(handwritten: Direct, indirect, lipid-soluble gas.)*

See the blue Answers tab at the end of the book.

12-9 Information processing by individual neurons involves integrating excitatory and inhibitory stimuli

A single neuron may receive information across thousands of synapses, and, as we have seen, some of the neurotransmitters arriving at the postsynaptic cell at any moment may be excitatory, whereas others may be inhibitory. The net effect on the transmembrane potential of the cell body—specifically, in the area of the axon hillock—determines how the neuron responds from moment to moment. If the net effect is a depolarization at the axon hillock, that depolarization will affect the transmembrane potential at the initial segment. If threshold is reached at the initial segment, an action potential will be generated and propagated along the axon.

Thus it is really the axon hillock that integrates the excitatory and inhibitory stimuli affecting the cell body and dendrites at any given moment. This integration process, which determines the rate of action potential generation at the initial segment, is the simplest level of **information processing** in the nervous system. The excitatory and inhibitory stimuli are integrated through interactions between *postsynaptic potentials*, which we discuss next. Higher levels of information processing involve interactions among neurons, and interactions among groups of neurons. These topics will be addressed in later chapters.

Postsynaptic Potentials

Postsynaptic potentials are graded potentials that develop in the postsynaptic membrane in response to a neurotransmitter. (**Figure 12–13** illustrated graded depolarizations and hyperpolarizations.) Two major types of postsynaptic potentials develop at neuron-to-neuron synapses: excitatory postsynaptic potentials and inhibitory postsynaptic potentials.

An **excitatory postsynaptic potential**, or **EPSP**, is a graded depolarization caused by the arrival of a neurotransmitter at the postsynaptic membrane. An EPSP results from the opening of chemically gated membrane channels that lead to depolarization of the plasma membrane. For example, the

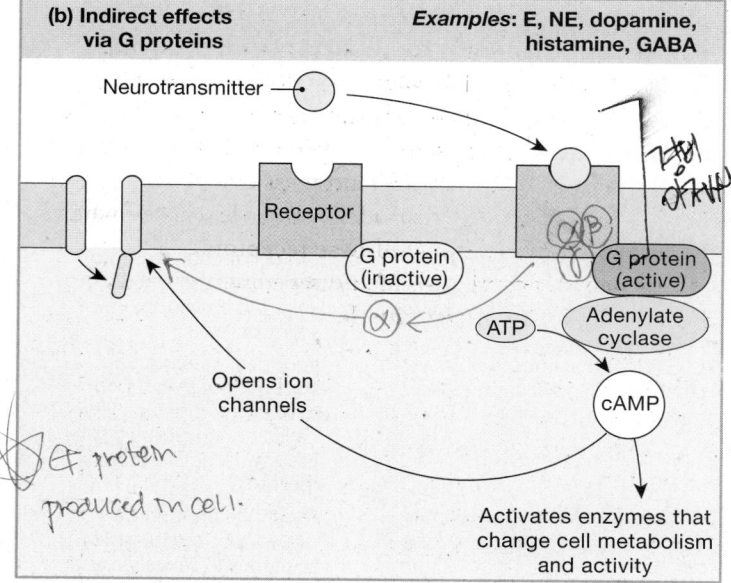

(b) Indirect effects via G proteins

Examples: E, NE, dopamine, histamine, GABA

Neurotransmitter

Receptor

G protein (inactive)

G protein (active)

ATP

Adenylate cyclase

Opens ion channels

cAMP

Activates enzymes that change cell metabolism and activity

(handwritten: G protein produced in cell.)

(handwritten: catecholamine → epinephrine, norepinephrine, dopamine.)

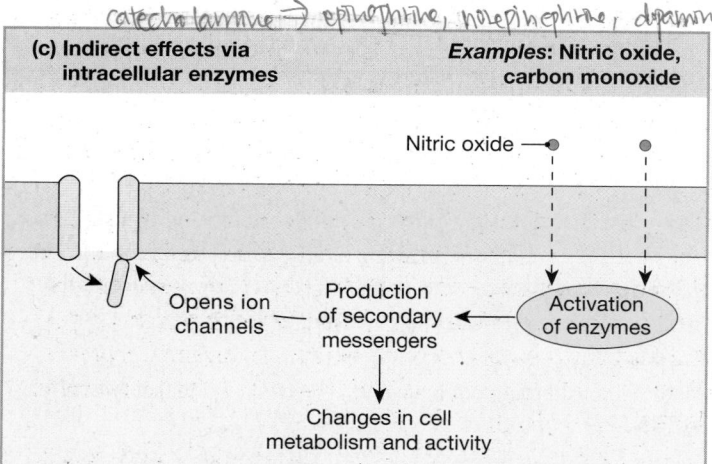

(c) Indirect effects via intracellular enzymes

Examples: Nitric oxide, carbon monoxide

Nitric oxide

Opens ion channels

Production of secondary messengers

Activation of enzymes

Changes in cell metabolism and activity

Figure 12–18 Mechanisms of Neurotransmitter Function.
(a) Direct effects on membrane channels. **(b)** Indirect effects mediated by G proteins. **(c)** Indirect effects mediated by intracellular enzymes.

(handwritten: esterase degrades acetylcholine.)

TABLE 12–6 Representative Neurotransmitters and Neuromodulators

Class and Neurotransmitter	Chemical Structure	Mechanism of Action	Location(s)	Comments
Acetylcholine		Primarily direct, through binding to chemically gated channels	CNS: Synapses throughout brain and spinal cord PNS: Neuromuscular junctions; preganglionic synapses of ANS; neuroglandular junctions of parasympathetic division and (rarely) sympathetic division of ANS; amacrine cells of retina	Widespread in CNS and PNS; best known and most studied of the neurotransmitters
BIOGENIC AMINES				
Norepinephrine		Indirect: G proteins and second messengers	CNS: Cerebral cortex, hypothalamus, brain stem, cerebellum, spinal cord PNS: Most neuromuscular and neuroglandular junctions of sympathetic division of ANS	Involved in attention and consciousness, control of body temperature, and regulation of pituitary gland secretion
Epinephrine		Indirect: G proteins and second messengers	CNS: Thalamus, hypothalamus, midbrain, spinal cord	Uncertain functions
Dopamine		Indirect: G proteins and second messengers	CNS: Hypothalamus, midbrain, limbic system, cerebral cortex, retina	Regulation of subconscious motor function; receptor abnormalities have been linked to development of schizophrenia
Serotonin		Primarily indirect: G proteins and second messengers	CNS: Hypothalamus, limbic system, cerebellum, spinal cord, retina	Important in emotional states, moods, and body temperature; several illicit hallucinogenic drugs, such as Ecstasy, target serotonin receptors
Histamine		Indirect: G proteins and second messengers	CNS: Neurons in hypothalamus, with axons projecting throughout the brain	Receptors are primarily on presynaptic membranes; functions in sexual arousal, pain threshold, pituitary hormone secretion, thirst, and blood pressure control
AMINO ACIDS				
Excitatory: Glutamate		Indirect: G proteins and second messengers Direct: opens calcium/sodium channels on pre- and postsynaptic membranes	CNS: Cerebral cortex and brain stem	Important in memory and learning; most important excitatory neurotransmitter in the brain
Aspartate		Direct or indirect (G proteins), depending on type of receptor	CNS: Cerebral cortex, retina, and spinal cord	Used by pyramidal cells that provide voluntary motor control over skeletal muscles
Inhibitory: Gamma-aminobutyric acid (GABA)		Direct or indirect (G proteins), depending on type of receptor	CNS: Cerebral cortex, cerebellum, interneurons throughout brain and spinal cord	Direct effects: open Cl^- channels; indirect effects: open K^+ channels and block entry of Ca^{2+}
Glycine		Direct: Opens Cl^- channels	CNS: Interneurons in brain stem, spinal cord, and retina	Produces postsynaptic inhibition; the poison *strychnine* produces fatal convulsions by blocking glycine receptors

TABLE 12–6   Continued

Class and Neurotransmitter	Chemical Structure	Mechanism of Action	Location(s)	Comments
NEUROPEPTIDES				
Substance P	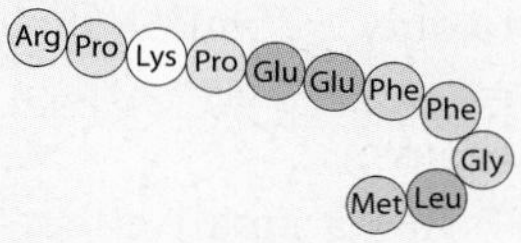 Arg–Pro–Lys–Pro–Glu–Glu–Phe–Phe–Gly–Leu–Met	Indirect: G proteins and second messengers	CNS: Synapses of pain receptors within spinal cord, hypothalamus, and other areas of the brain PNS: Enteric nervous system (network of neurons along the digestive tract)	Important in pain pathway, regulation of pituitary gland function, control of digestive tract reflexes
Neuropeptide Y	36-amino-acid peptide	As above	CNS: hypothalamus PNS: sympathetic neurons	Stimulates appetite and food intake
Opioids Endorphins	31-amino-acid peptide	Indirect: G proteins and second messengers	CNS: Thalamus, hypothalamus, brain stem, retina	Pain control; emotional and behavioral effects poorly understood
Enkephalins	 Tyr–Gly–Gly–Phe–Met	As above	CNS: Basal nuclei, hypothalamus, midbrain, pons, medulla oblongata, spinal cord	As above
Endomorphin	9- or 10-amino-acid peptide	As above	CNS: Thalamus, hypothalamus, basal nuclei	As above
Dynorphin	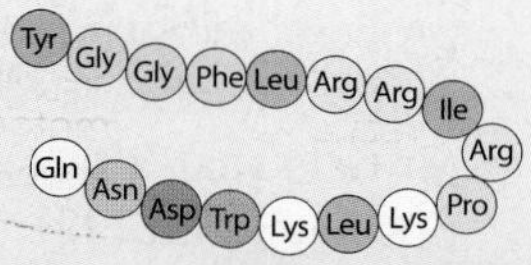 Tyr–Gly–Gly–Phe–Leu–Arg–Arg–Ile–Arg–Pro–Lys–Leu–Lys–Trp–Asp–Asn–Gln	As above	CNS: Hypothalamus, midbrain, medulla oblongata	As above
PURINES				
ATP, GTP	(see Figure 2–24)	Direct or indirect (G proteins), depending on type of receptor	CNS: Spinal cord PNS: Autonomic ganglia	
Adenosine	(see Figure 2–24)	Indirect: G proteins and second messengers	CNS: Cerebral cortex, hippocampus, cerebellum	Produces drowsiness; stimulatory effect of caffeine is due to inhibition of adenosine activity
HORMONES				
ADH, oxytocin, insulin, glucagon, secretin, CCK, GIP, VIP, inhibins, ANP, BNP, and many others	Peptide containing fewer than 200 amino acids	Typically indirect: G proteins and second messengers	CNS: Brain (widespread)	Numerous, complex, and incompletely understood
GASES				
Carbon monoxide (CO)	$C = O$	Indirect: By diffusion to enzymes activating second messengers	CNS: Brain PNS: Some neuromuscular and neuroglandular junctions	Localization and function poorly understood
Nitric oxide (NO)	$N = O$	As above	CNS: Brain, especially at blood vessels PNS: Some sympathetic neuromuscular and neuroglandular junctions	
LIPIDS				
Anandamide	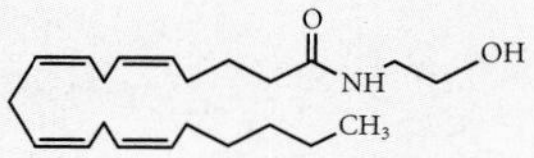 (structural diagram)	Indirect: G proteins and second messengers	CNS: cerebral cortex, hippocampus, cerebellum	Euphoria, drowsiness, appetite; receptors are targeted by the active ingredient in marijuana

12 NERVOUS

graded depolarization produced by the binding of ACh is an EPSP. Because it is a graded potential, an EPSP affects only the area immediately surrounding the synapse, as shown in Figure 12–12.

We have already noted that not all neurotransmitters have an excitatory (depolarizing) effect. An **inhibitory postsynaptic potential**, or **IPSP**, is a graded hyperpolarization of the postsynaptic membrane. For example, an IPSP may result from the opening of chemically gated potassium channels. While the hyperpolarization continues, the neuron is said to be **inhibited**, because a larger-than-usual depolarizing stimu-

lus must be provided to bring the membrane potential to threshold. A stimulus sufficient to shift the transmembrane potential by 10 mV (from −70 mV to −60 mV) would normally produce an action potential, but if the transmembrane potential were reset at −85 mV by an IPSP, the same stimulus would depolarize it to only −75 mV, which is below threshold.

Summation

An individual EPSP has a small effect on the transmembrane potential, typically producing a depolarization of about 0.5 mV at the postsynaptic membrane. Before an action potential will arise in the initial segment, local currents must depolarize that region by at least 10 mV. Therefore, a single EPSP will not result in an action potential, even if the synapse is on the axon hillock. But individual EPSPs combine through the process of **summation**, which integrates the effects of all the graded potentials that affect one portion of the plasma membrane. The graded potentials may be EPSPs, IPSPs, or both. We will consider EPSPs in our discussion. Two forms of summation exist: temporal summation and spatial summation (**Figure 12–19**).

Temporal summation (*tempus*, time) is the addition of stimuli occurring in rapid succession at a *single synapse* that is active *repeatedly*. This form of summation can be likened to using a bucket to fill up a bathtub; you can't fill the tub with a single bucket of water, but you will fill it eventually if you keep repeating the process. In the case of temporal summation, the water in a bucket corresponds to the sodium ions that enter the cytoplasm during an EPSP. A typical EPSP lasts about 20 msec, but under maximum stimulation an action potential can reach the synaptic knob each millisecond. **Figure 12–19a** shows what

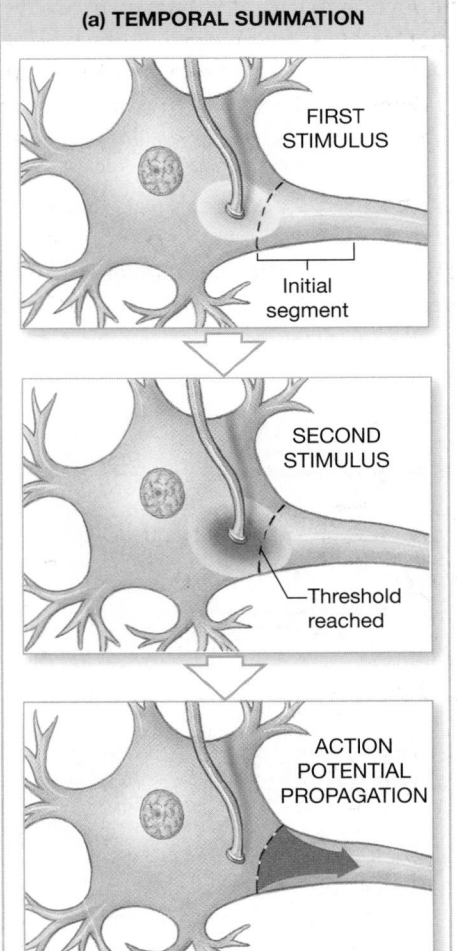

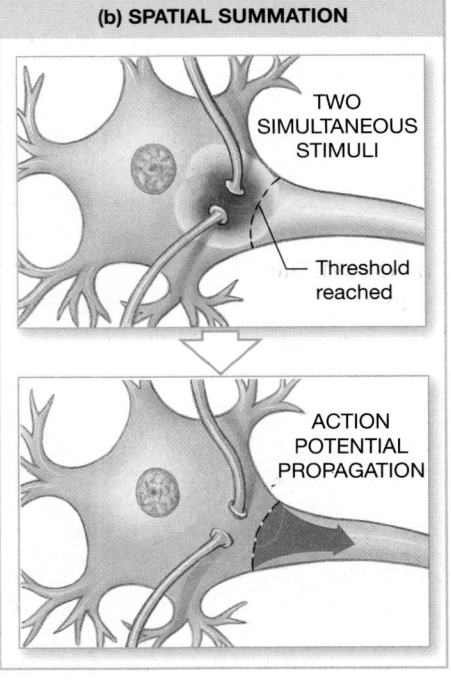

Figure 12–19 **Temporal and Spatial Summation. (a)** Temporal summation occurs on a membrane that receives two depolarizing stimuli from the same source in rapid succession. The effects of the second stimulus are added to those of the first. **(b)** Spatial summation occurs when sources of stimulation arrive simultaneously, but at different locations. Local currents spread the depolarizing effects, and areas of overlap experience the combined effects.

happens when a second EPSP arrives before the effects of the first EPSP have disappeared. Every time an action potential arrives, a group of vesicles discharges ACh into the synaptic cleft; every time more ACh molecules arrive at the postsynaptic membrane, more chemically gated channels open, and the degree of depolarization increases. In this way, a series of small steps can eventually bring the initial segment to threshold.

Spatial summation occurs when simultaneous stimuli applied at different locations have a cumulative effect on the transmembrane potential. In other words, spatial summation involves *multiple synapses* that are active *simultaneously*. In terms of our bucket analogy, you could fill the bathtub immediately if 50 friends emptied their buckets into it all at the same time.

In spatial summation, more than one synapse is active at the same time (**Figure 12–19b**), and each "pours" sodium ions across the postsynaptic membrane, producing a graded potential with localized effects. At each active synapse, the sodium ions that produce the EPSP spread out along the inner surface of the membrane and mingle with those entering at other synapses. As a result, the effects on the initial segment are cumulative. The degree of depolarization depends on how many synapses are active at any moment, and on their distance from the initial segment. As in temporal summation, an action potential results when the transmembrane potential at the initial segment reaches threshold.

Facilitation

Consider a situation in which summation of EPSPs is under way, but the initial segment has not been depolarized to threshold. The closer the initial segment is to threshold, the easier it will be for the *next* depolarizing stimulus to trigger an action potential. A neuron whose transmembrane potential

shifts closer to threshold is said to be **facilitated**. The larger the degree of facilitation, the smaller is the additional stimulus needed to trigger an action potential. In a highly facilitated neuron, even a small depolarizing stimulus produces an action potential.

Facilitation can result from the summation of EPSPs or from the exposure of a neuron to certain drugs in the extracellular fluid. For example, the nicotine in cigarettes stimulates postsynaptic ACh receptors, producing prolonged EPSPs that facilitate CNS neurons. Nicotine also increases the release of another neurotransmitter, dopamine, producing feelings of pleasure and reward, thereby leading to addiction.

Summation of EPSPs and IPSPs

Like EPSPs, IPSPs summate spatially and temporally. EPSPs and IPSPs reflect the activation of different types of chemically gated channels, producing opposing effects on the transmembrane potential. The antagonism between IPSPs and EPSPs is an important mechanism in cellular information processing. In terms of our bucket analogy, EPSPs put water into the bathtub, and IPSPs take water out. If more buckets add water than remove water, the water level in the tub will rise. If more buckets remove water, the level will fall. If a bucket of water is removed every time another bucket is dumped in, the level will remain stable. Comparable interactions between EPSPs and IPSPs (**Figure 12–20**) determine the transmembrane potential at the boundary between the axon hillock and the initial segment.

Neuromodulators, hormones, or both can change the postsynaptic membrane's sensitivity to excitatory or inhibitory neurotransmitters. By shifting the balance between EPSPs and IPSPs, these compounds promote facilitation or inhibition of CNS and PNS neurons.

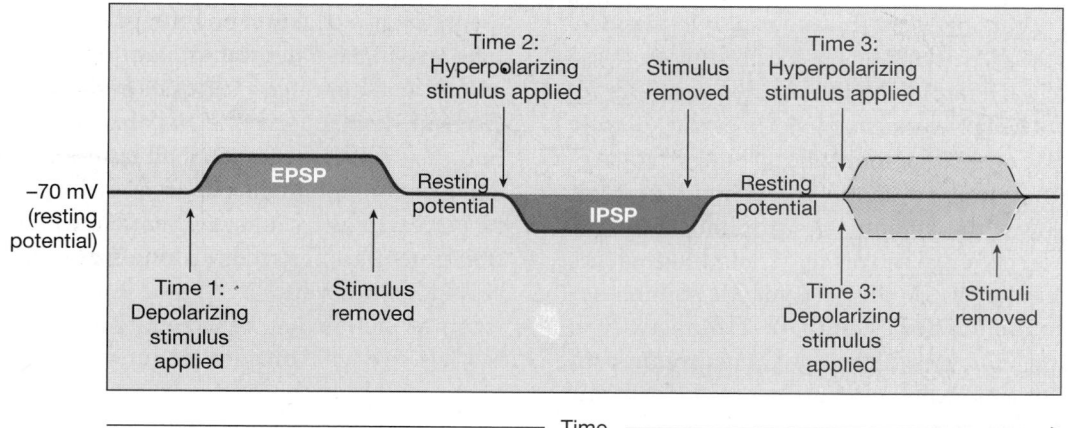

Figure 12–20 **Interactions between EPSPs and IPSPs.** At time 1, a small depolarizing stimulus produces an EPSP. At time 2, a small hyperpolarizing stimulus produces an IPSP of comparable magnitude. If the two stimuli are applied simultaneously, as they are at time 3, summation occurs. Because the two are equal in size but have opposite effects, the membrane potential remains at the resting level. If the EPSP were larger, a net depolarization would result; if the IPSP were larger, a net hyperpolarization would result instead.

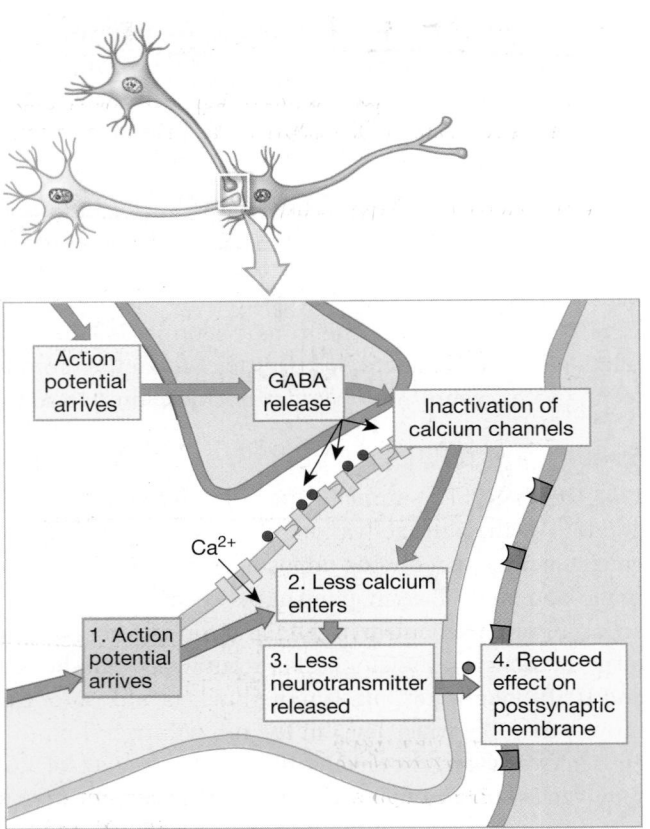

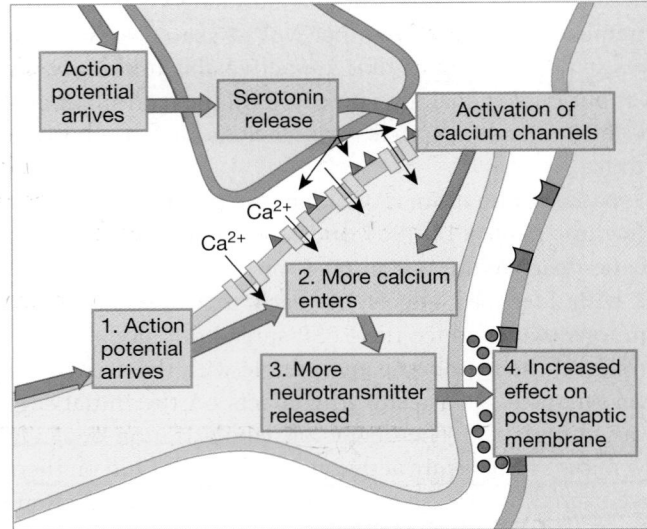

Figure 12–21 **Presynaptic Inhibition and Presynaptic Facilitation.** (a) Steps in presynaptic inhibition. (b) Steps in presynaptic facilitation.

(a) Presynaptic inhibition

(b) Presynaptic facilitation

Presynaptic Inhibition and Presynaptic Facilitation

Inhibitory or excitatory responses may occur not only at synapses involving the cell body and dendrites, but also at synapses found along an axon or its collaterals. At an *axoaxonic synapse*, a synapse occurs between the axons of two neurons. An axoaxonic synapse at the synaptic knob can either decrease (inhibit) or increase (facilitate) the rate of neurotransmitter release at the presynaptic membrane. In one form of **presynaptic inhibition**, the release of GABA inhibits the opening of voltage-gated calcium channels in the synaptic knob (**Figure 12–21a**). This inhibition reduces the amount of neurotransmitter released when an action potential arrives there, and thus reduces the effects of synaptic activity on the postsynaptic membrane.

In **presynaptic facilitation** (**Figure 12–21b**), activity at an axoaxonic synapse increases the amount of neurotransmitter released when an action potential arrives at the synaptic knob. This increase enhances and prolongs the neurotransmitter's effects on the postsynaptic membrane. The neurotransmitter *serotonin* is involved in presynaptic facilitation. In the presence of serotonin released at an axoaxonic synapse, voltage-gated calcium channels remain open longer.

The Rate of Generation of Action Potentials

In the nervous system, complex information is translated into action potentials that can be propagated along axons. On arrival, the message is often interpreted solely on the basis of the frequency of action potentials. For example, action potentials arriving at a neuromuscular junction at the rate of 1 per second may produce a series of isolated twitches in the associated skeletal muscle fiber, whereas at the rate of 100 per second they will cause a sustained tetanic contraction. Similarly, a few action potentials per second along a sensory fiber may be perceived as a featherlight touch, whereas hundreds of action potentials per second along that same axon may be perceived as unbearable pressure. In this section, we will examine factors that vary the rate of generation of action potentials. We will consider the functional significance of these changes in later chapters.

If a graded potential briefly depolarizes the axon hillock such that the initial segment reaches its threshold, an action potential will be propagated along the axon. Now consider what happens if the axon hillock *remains* depolarized past threshold for an extended period. The longer the initial segment remains above threshold, the more action potentials it will produce. The *frequency* of action potentials depends on the degree of depolarization above threshold: The greater the degree of depolarization, the higher the frequency of action potentials. The membrane can respond to a second stimulus as soon as the absolute refractory period ends. Holding the membrane above threshold has the same effect as applying a second, larger-than-normal stimulus.

The rate of generation of action potentials reaches a maximum when the relative refractory period has been completely eliminated. The maximum theoretical frequency is therefore established by the duration of the absolute refractory period. The absolute refractory period is shortest in large-diameter axons, in which the *theoretical* maximum frequency of action potentials is 2500 per second. However, the highest frequencies recorded from axons in the body range between 500 and 1000 per second.

Table 12–7 summarizes the basic principles of information processing.

The A&P Top 100

#44 In the nervous system, the changes in transmembrane potential that determine whether or not action potentials are generated represent the simplest form of information processing.

You are now familiar with the basic components of neural tissue, and the origin and significance of action potentials. In later chapters we will consider higher levels of anatomical and functional organization within the nervous system, examine information processing at these levels, and see how a single process—the generation of action potentials—can be

---

SUMMARY TABLE 12–7   Information Processing

- The neurotransmitters released at a synapse may have either excitatory or inhibitory effects. The effect on the axon's initial segment reflects a summation of the stimuli that arrive at any moment. The frequency of generation of action potentials is an indication of the degree of sustained depolarization at the axon hillock.

- Neuromodulators can alter either the rate of neurotransmitter release or the response of a postsynaptic neuron to specific neurotransmitters.

- Neurons may be facilitated or inhibited by extracellular chemicals other than neurotransmitters or neuromodulators.

- The response of a postsynaptic neuron to the activation of a presynaptic neuron can be altered by (1) the presence of neuromodulators or other chemicals that cause facilitation or inhibition at the synapse, (2) activity under way at other synapses affecting the postsynaptic cell, and (3) modification of the rate of neurotransmitter release through presynaptic facilitation or presynaptic inhibition. Information is relayed in the form of action potentials. In general, the degree of sensory stimulation or the strength of the motor response is proportional to the frequency of action potentials.

responsible for the incredible diversity of sensations and movements that we experience each day.

CHECKPOINT

25. One EPSP depolarizes the initial segment from a resting potential of −70 mV to −65 mV, and threshold is at −60 mV. Will an action potential be generated?

26. Given the situation in Checkpoint 25, if a second, identical EPSP occurs immediately after the first, will an action potential be generated?

27. If the two EPSPs in Checkpoint 26 occurred simultaneously, what form of summation would occur?

See the blue Answers tab at the end of the book.

Related Clinical Terms

anticholinesterase drug: A drug that blocks the breakdown of ACh by AChE.

atropine: A drug that prevents ACh from binding to the postsynaptic membrane of cardiac muscle and smooth muscle cells.

demyelination: The destruction of the myelin sheaths around axons in the CNS and PNS.

d-tubocurarine: A drug, derived from curare, that produces paralysis by preventing ACh from binding to the postsynaptic membrane of skeletal muscle fibers.

endorphins: Neuropeptides produced in the brain and spinal cord that appear to relieve pain and affect mood.

hyperkalemia: An abnormal physiological state resulting from a high extracellular concentration of potassium.

neurotoxin: A compound that disrupts normal nervous system function by interfering with the generation or propagation of action potentials. Examples include *tetrodotoxin (TTX)*, *saxitoxin (STX)*, and *ciguatoxin (CTX)*.

nicotine: A compound found in tobacco that binds to specific ACh receptor sites and stimulates the synaptic membrane.

rabies: A fatal disease caused by a virus that reaches the CNS via retrograde flow along peripheral axons.

Tay–Sachs disease: A genetic abnormality involving the metabolism of gangliosides, important components of neuron plasma membranes. The result is a gradual deterioration of neurons due to the buildup of metabolic by-products and the release of lysosomal enzymes.

Chapter Review

Study Outline

iP Use *Interactive Physiology*® (**IP**) —available on the IP CD packaged with your new textbook and online at **myA&P**™ (www.myaandp.com)—to help you understand difficult physiological concepts. Follow these navigation paths on IP for concepts in this chapter:

• Nervous I: The Membrane Potential
• Nervous I: Ion Channels
• Nervous I: The Action Potential
• Nervous II: Synaptic Transmission
• Nervous II: Synaptic Potentials and Cellular Integration

An Introduction to Neural Tissue p. 387

1. The nervous system includes all the neural tissue in the body. The basic functional unit is the **neuron**.

12-1 The nervous system has anatomical and functional divisions p. 387

2. The anatomical divisions of the nervous system are the **central nervous system (CNS)** (the brain and spinal cord) and the **peripheral nervous system (PNS)** (all the neural tissue outside the CNS). Bundles of **axons** (*nerve fibers*) in the PNS are called **nerves**.

3. Functionally, the PNS can be divided into an **afferent division**, which brings sensory information from **receptors** to the CNS, and an **efferent division**, which carries motor commands to muscles and glands called **effectors**.

4. The efferent division of the PNS includes the **somatic nervous system (SNS)**, which controls skeletal muscle contractions, and the **autonomic nervous system (ANS)**, which controls smooth muscle, cardiac muscle, adipose tissue, and glandular activity.

12-2 Neurons are nerve cells specialized for intercellular communication p. 388

5. The **perikaryon** of a multipolar neuron contains organelles, including **neurofilaments**, **neurotubules**, and **neurofibrils**. The **axon hillock** connects the **initial segment** of the **axon** to the **cell body**, or *soma*. The **axoplasm** contains numerous organelles. (*Figure 12–1*)

6. **Collaterals** may branch from an axon, with **telodendria** branching from the axon's tip.

7. A **synapse** is a site of intercellular communication. A **synaptic knob** is the most common type of synaptic terminal. **Neurotransmitters** released from the synaptic knob of the presynaptic cell affect the postsynaptic cell, which may be a neuron or another type of cell. (*Figures 12–1, 12–2*)

8. Neurons are structurally classified as **anaxonic**, **bipolar**, **unipolar**, or **multipolar**. (*Figure 12–3*)

9. The three functional categories of neurons are sensory neurons, motor neurons, and interneurons.

10. **Sensory neurons**, which form the afferent division of the PNS, deliver information received from **interoceptors**, **exteroceptors**, and **proprioceptors** to the CNS.

11. **Motor neurons**, which form the efferent division of the PNS, stimulate or modify the activity of a peripheral tissue, organ, or organ system.

12. **Interneurons** (*association neurons*) are always located in the CNS and may be situated between sensory and motor neurons. They distribute sensory inputs and coordinate motor outputs.

12-3 CNS and PNS neuroglia support and protect neurons p. 392

13. The four types of **neuroglia**, or *glial cells*, in the CNS are (1) **ependymal cells**, with functions related to the **cerebrospinal fluid (CSF)**; (2) **astrocytes**, the largest and most numerous neuroglia; (3) **oligodendrocytes**, which are responsible for the **myelination** of CNS axons; and (4) **microglia**, or phagocytic cells. (*Figures 12–4, 12–5*)

14. Neuron cell bodies in the PNS are clustered into **ganglia**.

15. **Satellite cells**, or *amphicytes*, surround neuron cell bodies within ganglia. **Schwann cells** ensheath axons in the PNS. A single Schwann cell may myelinate one segment of an axon or enclose segments of several unmyelinated axons. (*Figure 12–6*)

16. In the PNS, functional repair of axons may follow **Wallerian degeneration**. In the CNS, many factors complicate the repair process and reduce the chances of functional recovery. (*Figure 12–7*)

12-4 The transmembrane potential is the electrical potential of the cell's interior relative to its surroundings p. 398

17. All normal neural signaling depends on events that occur at the plasma membrane. (*Figure 12–8*)

18. The **electrochemical gradient** is the sum of all chemical and electrical forces acting across the plasma membrane. (*Figures 12–9, 12–10*)

19. The sodium–potassium exchange pump stabilizes the resting potential at approximately -70 mV. (*Summary Table 12–1*)

20. The plasma membrane contains **passive (leak) channels**, which are always open, and **active (gated) channels**, which open or close in response to specific stimuli. (*Figure 12–9*)

21. The three types of gated channels are **chemically gated channels**, **voltage-gated channels**, and **mechanically gated channels**. (*Figure 12–11*)

22. A localized **depolarization** or **hyperpolarization** is a **graded potential** (a change in potential that decreases with distance). (*Figures 12–12, 12–13; Summary Table 12–2*)

12-5 An action potential is a nerve impulse p. 407

23. An **action potential** arises when a region of excitable membrane depolarizes to its **threshold**. The steps involved, in order, are membrane depolarization to threshold, activation of sodium channels and rapid depolarization, inactivation of sodium channels and activation of potassium channels, and the return to normal permeability. (*Figure 12–14; Summary Table 12–3; Table 12–4*)

24. The generation of an action potential follows the **all-or-none principle**. The **refractory period** lasts from the time an action

potential begins until the normal resting potential has returned. (*Summary Table 12–3; Table 12–4*)

25. In **continuous propagation**, an action potential spreads across the entire excitable membrane surface in a series of small steps. (*Figure 12–15*)

26. In **saltatory propagation**, an action potential appears to leap from node to node, skipping the intervening membrane surface. Saltatory propagation carries nerve impulses many times more rapidly than does continuous propagation. (*Figure 12–16*)

12-6 Axon diameter, in addition to myelin, affects propagation speed p. 413

27. Axons are classified as **Type A fibers**, **Type B fibers**, or **Type C fibers** on the basis of their diameter, myelination, and propagation speed.

28. Compared with action potentials in neural tissue, those in muscle tissue have (1) larger resting potentials, (2) longer-lasting action potentials, and (3) slower propagation of action potentials.

12-7 At synapses, communication occurs among neurons or between neurons and other cells p. 413

29. An action potential traveling along an axon is a **nerve impulse**. At a synapse between two neurons, information passes from the **presynaptic neuron** to the **postsynaptic neuron**.

30. A synapse is either *electrical* (with direct physical contact between cells) or *chemical* (involving a neurotransmitter).

31. **Electrical synapses** occur in the CNS and PNS, but they are rare. At an electrical synapse, the presynaptic and postsynaptic plasma membranes are bound by interlocking membrane proteins at a gap junction. Pores within these proteins permit the passage of local currents, and the two neurons act as if they share a common plasma membrane.

32. **Chemical synapses** are far more common than electrical synapses. **Excitatory neurotransmitters** cause depolarization and promote the generation of action potentials, whereas **inhibitory neurotransmitters** cause hyperpolarization and suppress the generation of action potentials.

33. The effect of a neurotransmitter on the postsynaptic membrane depends on the properties of the receptor, not on the nature of the neurotransmitter.

34. **Cholinergic synapses** release the neurotransmitter **acetylcholine (ACh)**. Communication moves from the presynaptic neuron to the postsynaptic neuron across a synaptic cleft. A **synaptic delay** occurs because calcium influx and the release of the neurotransmitter takes an appreciable length of time. (*Figure 12–17*)

35. **Choline** released during the breakdown of ACh in the synaptic cleft is reabsorbed and recycled by the synaptic knob. If stores of ACh are exhausted, **synaptic fatigue** can occur. (*Summary Table 12–5*)

12-8 Neurotransmitters and neuromodulators have various functions p. 417

36. **Adrenergic synapses** release **norepinephrine (NE)**, also called *noradrenaline*. Other important neurotransmitters

include **dopamine**, **serotonin**, and **gamma aminobutyric acid (GABA)**. (*Table 12–6*)

37. **Neuromodulators** influence the postsynaptic cell's response to neurotransmitters.

38. Neurotransmitters can have a direct or indirect effect on membrane potential, or they can exert their effects via lipid-soluble gases that diffuse across the plasma membrane. (*Figure 12–18*)

12-9 Information processing by individual neurons involves integrating excitatory and inhibitory stimuli p. 419

39. Excitatory and inhibitory stimuli are integrated through interactions between **postsynaptic potentials**. This interaction is the simplest level of **information processing** in the nervous system.

40. A depolarization caused by a neurotransmitter is an **excitatory postsynaptic potential (EPSP)**. Individual EPSPs can combine through **summation**, which can be either **temporal** (occurring at a single synapse when a second EPSP arrives before the effects of the first have disappeared) or **spatial** (resulting from the cumulative effects of multiple synapses at various locations). (*Figure 12–19*)

41. Hyperpolarization of the postsynaptic membrane is an **inhibitory postsynaptic potential (IPSP)**.

42. The most important determinants of neural activity are EPSP–IPSP interactions. (*Figure 12–20*)

43. In **presynaptic inhibition**, GABA release at an *axoaxonic synapse* inhibits the opening of voltage-gated calcium channels in the synaptic knob. This inhibition reduces the amount of neurotransmitter released when an action potential arrives at the synaptic knob. (*Figure 12–21a*)

44. In **presynaptic facilitation**, activity at an axoaxonic synapse increases the amount of neurotransmitter released when an action potential arrives at the synaptic knob. This increase enhances and prolongs the effects of the neurotransmitter on the postsynaptic membrane. (*Figure 12–21b*)

45. The neurotransmitters released at a synapse have excitatory or inhibitory effects. The effect on the initial segment reflects an integration of the stimuli arriving at any moment. The frequency of generation of action potentials depends on the degree of depolarization above threshold at the axon hillock. (*Summary Table 12–7*)

46. Neuromodulators can alter either the rate of neurotransmitter release or the response of a postsynaptic neuron to specific neurotransmitters. Neurons may be facilitated or inhibited by extracellular chemicals other than neurotransmitters or neuromodulators. (*Summary Table 12–7*)

47. The effect of a presynaptic neuron's activation on a postsynaptic neuron may be altered by other neurons. (*Summary Table 12–7*)

48. The greater the degree of sustained depolarization at the axon hillock, the higher the frequency of generation of action potentials. At a frequency of about 1000 per second, the relative refractory period has been eliminated, and further depolarization will have no effect. (*Summary Table 12–7*)

Review Questions

See the blue Answers tab at the end of the book.

Access more review material online at
myA&P™ (**www.myaandp.com**). There,
you'll find chapter guides, chapter quizzes,
practice tests, animations, flashcards, a glossary with pronunciations,
Interactive Physiology® (IP) exercises and quizzes, and more to
help you succeed in the course.

LEVEL 1   Reviewing Facts and Terms

1. Label the structures in the following diagram of a neuron.

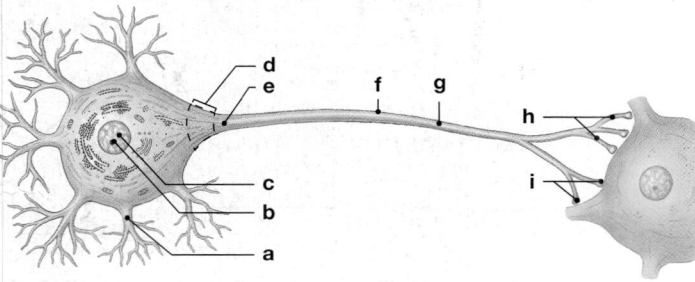

   (a) dendrite   (b) nucleolus
   (c) nucleus   (d) axon hillock
   (e) initial segment   (f) axolemma
   (g) axon   (h) telodendria
   (i) synapse knob

2. Regulation by the nervous system provides
   (a) relatively slow, but long-lasting, responses to stimuli.
   (b) swift, long-lasting responses to stimuli.
   (c) swift, but brief, responses to stimuli.
   (d) relatively slow, short-lived responses to stimuli.

3. In the CNS, a neuron typically receives information from other neurons at its
   (a) axon.   (b) Nissl bodies.
   (c) dendrites.   (d) nucleus.

4. Phagocytic cells in neural tissue of the CNS are
   (a) astrocytes.   (b) ependymal cells.
   (c) oligodendrocytes.   (d) microglia.

5. The neural cells responsible for the analysis of sensory inputs and coordination of motor outputs are
   (a) neuroglia.   (b) interneurons.
   (c) sensory neurons.   (d) motor neurons.

6. Depolarization of a neuron plasma membrane will shift the membrane potential toward
   (a) 0 mV.   (b) −70 mV.
   (c) −90 mV.   (d) all of these.

7. The primary determinant of the resting membrane potential is
   (a) the membrane permeability to sodium.
   (b) the membrane permeability to potassium.
   (c) intracellular negatively charged proteins.
   (d) negatively charged chloride ions in the ECF.

8. Receptors that bind acetylcholine at the postsynaptic membrane are
   (a) chemically gated channels.
   (b) voltage-gated channels.
   (c) passive channels.
   (d) mechanically gated channels.

9. What are the major components of (a) the central nervous system? (b) the peripheral nervous system?

10. Which two types of neuroglia insulate neuron cell bodies and axons in the PNS from their surroundings?

11. What three *functional* groups of neurons are found in the nervous system? What is the function of each type of neuron?

LEVEL 2   Reviewing Concepts

12. If the resting membrane potential is −70 mV and the threshold is −55 mV, a membrane potential of −60 mV will
   (a) produce an action potential.
   (b) make it easier to produce an action potential.
   (c) make it harder to produce an action potential.
   (d) hyperpolarize the membrane.

13. Why can't most neurons in the CNS be replaced when they are lost to injury or disease?

14. What is the difference between anterograde flow and retrograde flow?

15. What is the *functional* difference among voltage-gated, chemically gated, and mechanically gated channels?

16. State the all-or-none principle of action potentials.

17. Describe the steps involved in the generation of an action potential.

18. What is meant by saltatory propagation? How does it differ from continuous propagation?

19. What are the structural and functional differences among type A, B, and C fibers?

20. Describe the events that occur during nerve impulse transmission at a typical cholinergic synapse.

21. What is the difference between temporal summation and spatial summation?

LEVEL 3   Critical Thinking and Clinical Applications

22. Harry has a kidney condition that causes changes in his body's electrolyte levels (concentration of ions in the extracellular fluid). As a result, he is exhibiting tachycardia, an abnormally fast heart rate. Which ion is involved, and how does a change in its concentration cause Harry's symptoms?

23. Twenty neurons synapse with a single receptor neuron. Fifteen of the 20 neurons release neurotransmitters that produce EPSPs at the postsynaptic membrane, and the other five release neurotransmitters that produce IPSPs. Each time one of the neurons is stimulated, it releases enough neurotransmitter to produce a 2 mV change in potential at the postsynaptic membrane. If the threshold of the postsynaptic neuron is 10 mV, how many of the excitatory neurons must be stimulated to produce an action potential in the receptor neuron if all five inhibitory neurons are stimulated? (Assume that spatial summation occurs.)

24. In multiple sclerosis, there is intermittent and progressive damage to the myelin sheath of peripheral nerves. This results in poor motor control of the affected area. Why does destruction of the myelin sheath affect motor control?

25. What factor determines the maximum frequency of action potentials that could be conducted by an axon?

The Spinal Cord, Spinal Nerves, and Spinal Reflexes

13

Did you know...?

Loss of sensation and/or motor control in a specific part of the body due to trauma can indicate both the injury site and identity of the damaged nerve.

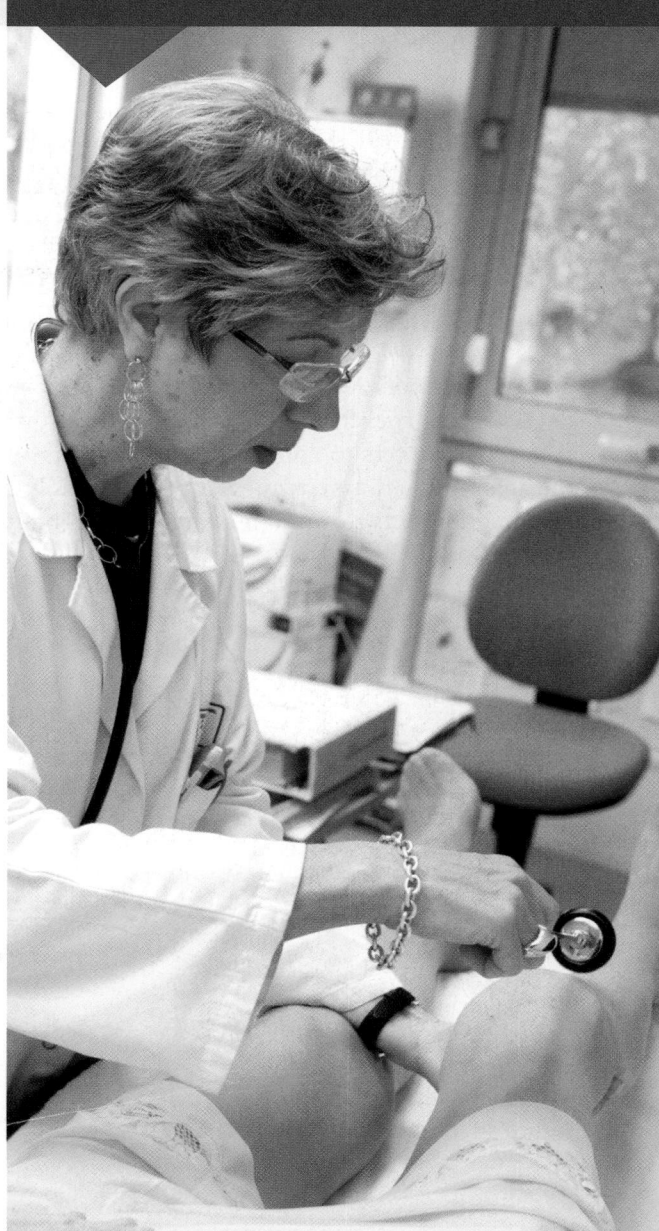

Learning Outcomes

After completing this chapter, you should be able to do the following:

13-1 Describe the basic structural and organizational characteristics of the nervous system.

13-2 Discuss the structure and functions of the spinal cord, and describe the three meningeal layers that surround the central nervous system.

13-3 Explain the roles of white matter and gray matter in processing and relaying sensory information and motor commands.

13-4 Describe the major components of a spinal nerve, and relate the distribution pattern of spinal nerves to the regions they innervate.

13-5 Discuss the significance of neuronal pools, and describe the major patterns of interaction among neurons within and among these pools.

13-6 Describe the steps in a neural reflex, and classify the types of reflexes.

13-7 Distinguish among the types of motor responses produced by various reflexes, and explain how reflexes interact to produce complex behaviors.

13-8 Explain how higher centers control and modify reflex responses.

Clinical Notes

Spinal Anesthesia p. 434
Shingles p. 440

An Introduction to the Spinal Cord, Spinal Nerves, and Spinal Reflexes

This chapter discusses the functional anatomy and organization of the spinal cord and spinal nerves, and describes simple spinal reflexes. Organization is usually the key to success in any complex environment. A large corporation, for example, has both a system to distribute messages on specific topics and executive assistants who decide whether an issue can be ignored or easily responded to; only the most complex and important problems reach the desk of the president. The nervous system works in much the same way: It has input pathways that route sensations, and processing centers that prioritize and distribute information. There are also several levels that issue motor responses. Your conscious mind (the president) gets involved only in a fraction of the day-to-day activities; the other decisions are handled at lower levels that operate outside your awareness. This very efficient system works only because it is so highly organized.

13-1 The brain and spinal cord make up the central nervous system, and the cranial nerves and spinal nerves constitute the peripheral nervous system

Because the nervous system has so many components and does so much, even a relatively superficial discussion will take four chapters to complete. If our primary interest were the anatomy of this system, we would probably start with an examination of the central nervous system (brain and spinal cord) and then consider the peripheral nervous system (cranial nerves and spinal nerves). But our primary interest is how the nervous system *functions*, so we will consider the system from a functional perspective. The basic approach has been diagrammed in **Figure 13–1**.

In the chapters that follow, we will look at increasing levels of structural and functional complexity. Chapter 12 provided the foundation by considering the function of individual neurons. In the current chapter, we consider the spinal cord and spinal nerves, and the basic wiring of relatively simple *spinal reflexes*—rapid, automatic responses triggered by specific stimuli. Spinal reflexes are controlled in the spinal cord; whether they involve a single spinal segment or multiple segments, they can function without any input from the brain. For example, a reflex controlled in the spinal cord makes you drop a frying pan you didn't realize was sizzling hot. Before the information reaches your brain and you become aware of the pain, you've already released the pan. Although there are much more complex spinal reflexes, this functional pattern still applies; a reflex provides a quick, automatic response to a specific stimulus.

Your spinal cord is structurally and functionally integrated with your brain. Chapter 14 provides an overview of the major components and functions of the brain and cranial nerves. It also discusses the *cranial reflexes*, relatively localized reflex responses comparable in organization and complexity to those of the spinal cord.

Chapters 15 and 16 consider the nervous system as an integrated functional unit. Chapter 15 deals with the interplay between centers in the brain and spinal cord that occurs in the processing of sensory information. It then examines the conscious and subconscious control of skeletal muscle activity provided by the *somatic nervous system* (SNS).

Chapter 16 continues with a discussion of the control of visceral functions by the *autonomic nervous system* (ANS).

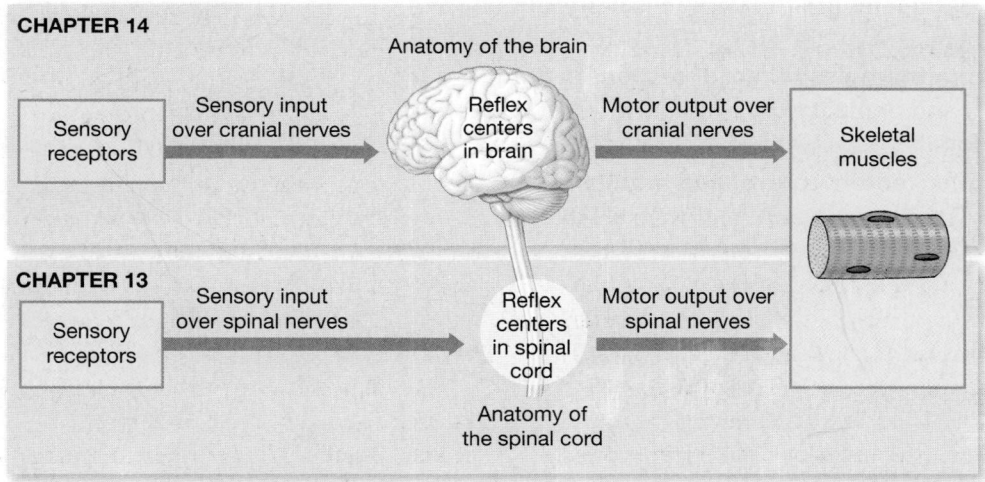

Figure 13–1 **An Overview of Chapters 13 and 14.** These chapters discuss basic functional neuroanatomy and simple reflexes.

The ANS, which has processing centers in the brain, spinal cord, and peripheral nervous system, is responsible for the control of visceral effectors, such as peripheral smooth muscles, cardiac muscle, and glands. We then conclude this section of the book by examining what are often called *higher-order functions*: memory, learning, consciousness, and personality. These fascinating topics are difficult to investigate, but they can affect activity along the sensory and motor pathways and alter our perception of those activities.

With these basic principles, definitions, and strategies in mind, we can begin our examination of the levels of functional organization in the nervous system.

CHECKPOINT

1. Name the components of the central nervous system and of the peripheral nervous system.
2. Define spinal reflex.

See the blue Answers tab at the end of the book.

13-2 The spinal cord is surrounded by three meninges and conveys sensory and motor information

We begin this section by considering the gross anatomy of the spinal cord. Then we examine the three layers that surround the spinal cord: the spinal meninges.

Gross Anatomy of the Spinal Cord

The adult spinal cord (**Figure 13–2a**) measures approximately 45 cm (18 in.) in length and has a maximum width of roughly 14 mm (0.55 in.). Note that the cord itself is not as long as the vertebral column—instead, the adult spinal cord ends between vertebrae L_1 and L_2. The posterior (dorsal) surface of the spinal cord bears a shallow longitudinal groove, the **posterior median sulcus** (**Figure 13–2b**). The **anterior median fissure** is a deeper groove along the anterior (ventral) surface.

The amount of gray matter is greatest in segments of the spinal cord dedicated to the sensory and motor control of the limbs. These segments are expanded, forming the **enlargements** of the spinal cord. The **cervical enlargement** supplies nerves to the shoulder and upper limbs; the **lumbar enlargement** provides innervation to structures of the pelvis and lower limbs. Inferior to the lumbar enlargement, the spinal cord becomes tapered and conical; this region is the **conus medullaris**. The **filum terminale** ("terminal thread"), a slender strand of fibrous tissue, extends from the inferior tip of the conus medullaris. It continues along the length of the vertebral canal as far as the second sacral vertebra, where it provides longitu-dinal support to the spinal cord as a component of the *coccygeal ligament*.

The series of sectional views in **Figure 13–2b** illustrates the variations in the relative mass of gray matter and white matter in the cervical, thoracic, lumbar, and sacral regions of the spinal cord. The entire spinal cord can be divided into 31 segments on the basis of the origins of the spinal nerves. Each segment is identified by a letter and number designation, the same method used to identify vertebrae. For example, C_3, the segment in the uppermost section of **Figure 13–2b**, is the third cervical segment.

Every spinal segment is associated with a pair of **dorsal root ganglia** (**Figure 13–2b**), situated near the spinal cord. These ganglia contain the cell bodies of sensory neurons. The axons of the neurons form the **dorsal roots**, which bring sensory information into the spinal cord. A pair of **ventral roots** contains the axons of motor neurons that extend into the periphery to control somatic and visceral effectors. On both sides, the dorsal and ventral roots of each segment pass between the vertebral canal and the periphery at the *intervertebral foramen* between successive vertebrae. The dorsal root ganglion lies between the pedicles of the adjacent vertebrae. (You can review vertebral anatomy in Chapter 7.  pp. 232–233)

Distal to each dorsal root ganglion, the sensory and motor roots are bound together into a single **spinal nerve**. Spinal nerves are classified as **mixed nerves**—that is, they contain both afferent (sensory) and efferent (motor) fibers. There are 31 pairs of spinal nerves, each identified by its association with adjacent vertebrae. For example, we may speak of "cervical spinal nerves" or even "cervical nerves" when we make a general reference to spinal nerves of the neck. However, when we indicate specific spinal nerves, it is customary to give them a regional number, as indicated in **Figure 13–2**. Each spinal nerve inferior to the first thoracic vertebra takes its name from the vertebra immediately superior to it. Thus, spinal nerve T_1 emerges immediately inferior to vertebra T_1, spinal nerve T_2 follows vertebra T_2, and so forth.

The arrangement differs in the cervical region, because the first pair of spinal nerves, C_1, passes between the skull and the first cervical vertebra. For this reason, each cervical nerve takes its name from the vertebra immediately inferior to it. In other words, cervical nerve C_2 *precedes* vertebra C_2, and the same system is used for the rest of the cervical series. The transition from one numbering system to another occurs between the last cervical vertebra and first thoracic vertebra. The spinal nerve found at this location has been designated C_8. Therefore, although there are only seven cervical vertebrae, there are *eight* cervical nerves.

The spinal cord continues to enlarge and elongate until an individual is approximately 4 years old. Up to that time, enlargement of the spinal cord keeps pace with the growth of

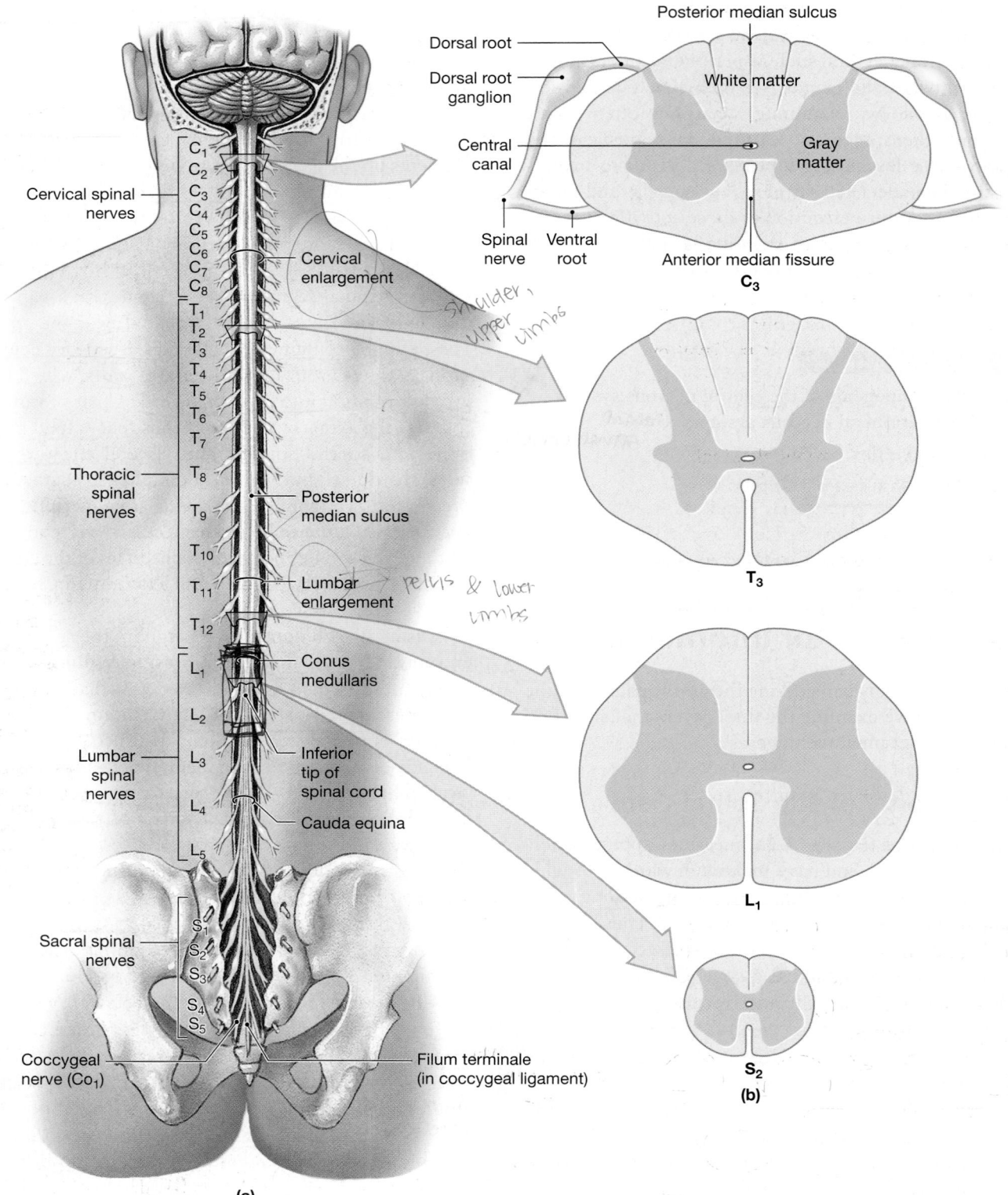

Figure 13–2 **Gross Anatomy of the Adult Spinal Cord.** (a) The superficial anatomy and orientation of the adult spinal cord. The numbers to the left identify the spinal nerves and indicate where the nerve roots leave the vertebral canal. The spinal cord, however, extends from the brain only to the level of vertebrae L_1–L_2; the spinal segments found at representative locations are indicated in the cross sections. (b) Inferior views of cross sections through representative segments of the spinal cord, showing the arrangement of gray matter and white matter.

ATLAS: Plates 2a; 20a,b, 24a–c

the vertebral column. Throughout this period, the ventral and dorsal roots are very short, and they enter the intervertebral foramina immediately adjacent to their spinal segment. After age 4, the vertebral column continues to elongate, but the spinal cord does not. This vertebral growth moves the intervertebral foramina, and thus the spinal nerves, farther and farther from their original positions relative to the spinal cord. As a result, the dorsal and ventral roots gradually elongate, and the correspondence between the spinal segment and the vertebral segment is lost. For example, in adults, the sacral segments of the spinal cord are at the level of vertebrae L_1–L_2.

Because the adult spinal cord extends only to the level of the first or second lumbar vertebra, the dorsal and ventral roots of spinal segments L_2 to S_5 extend inferiorly, past the inferior tip of the conus medullaris. When seen in gross dissection, the filum terminale and the long ventral and dorsal roots resemble a horse's tail. As a result, early anatomists called this complex the **cauda equina** (KAW-duh ek-WĪ-nuh; *cauda*, tail + *equus*, horse).

Spinal Meninges

The vertebral column and its surrounding ligaments, tendons, and muscles isolate the spinal cord from the rest of the body, and these structures also provide protection against bumps, shocks, and blows to the skin of the back. The delicate neural tissues must also be protected from damaging contacts with the surrounding bony walls of the vertebral canal. The **spinal meninges** (me-NIN-jēz; singular, *meninx*, membrane), a series of specialized membranes surrounding the spinal cord, provide the necessary physical stability and shock absorption. Blood vessels branching within these layers deliver oxygen and nutrients to the spinal cord.

The relationships among the spinal meninges are shown in **Figure 13–3a**. The spinal meninges consist of three layers: (1) the *dura mater*, (2) the *arachnoid mater*, and (3) the *pia mater*. At the foramen magnum of the skull, the spinal meninges are continuous with the **cranial meninges**, which surround the brain. (We will discuss the cranial meninges, which have the same three layers, in Chapter 14.)

Bacterial or viral infection can cause **meningitis**, or inflammation of the meningeal membranes. Meningitis is dan-

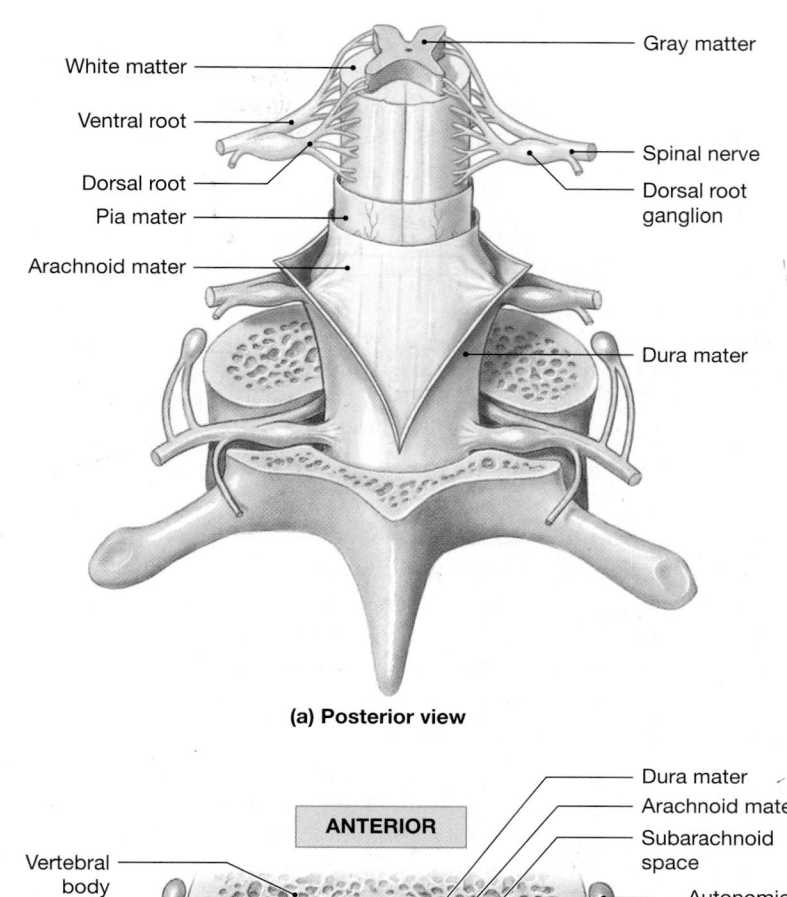

(a) Posterior view

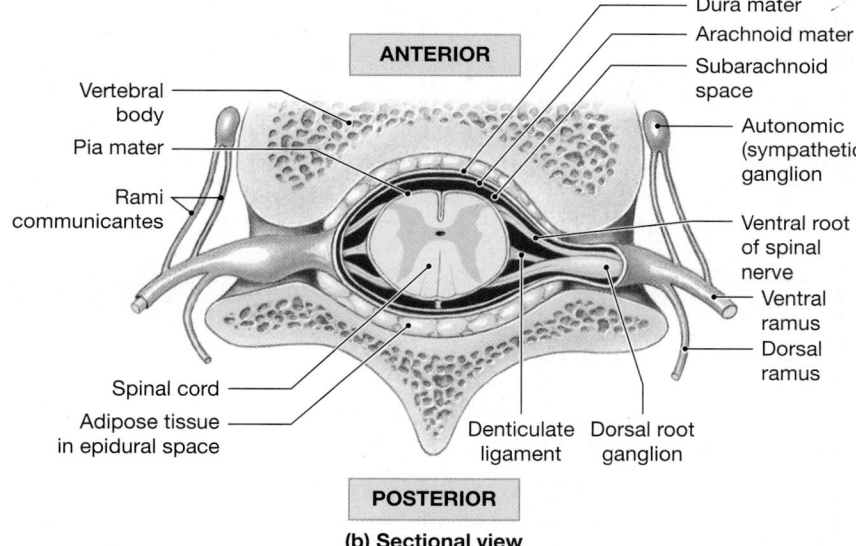

(b) Sectional view

Figure 13–3   The Spinal Cord and Spinal Meninges. (a) A posterior view of the spinal cord, showing the meningeal layers, superficial landmarks, and distribution of gray matter and white matter. **(b)** A sectional view through the spinal cord and meninges, showing the peripheral distribution of spinal nerves.

gerous because it can disrupt the normal circulatory and cerebrospinal fluid supplies, damaging or killing neurons and neuroglia in the affected areas. Although an initial diagnosis may specify the meninges of the spinal cord (*spinal meningitis*) or brain (*cerebral meningitis*), in later stages the entire meningeal system is usually affected.

The Dura Mater

The tough, fibrous **dura mater** (DOO-ruh MĀ-ter; *dura*, hard + *mater*, mother) is the layer that forms the outermost covering of the spinal cord (**Figure 13–3a**). This layer contains dense collagen fibers that are oriented along the longitudinal axis of the cord. Between the dura mater and the walls of the vertebral canal lies the **epidural space**, a region that contains areolar tissue, blood vessels, and a protective padding of adipose tissue (**Figure 13–3b**).

The spinal dura mater does not have extensive, firm connections to the surrounding vertebrae. Longitudinal stability is provided by localized attachment sites at either end of the vertebral canal. Cranially, the outer layer of the spinal dura mater fuses with the periosteum of the occipital bone around the margins of the foramen magnum. There, the spinal dura mater becomes continuous with the cranial dura mater. Within the sacral canal, the spinal dura mater tapers from a sheath to a dense cord of collagen fibers that blends with components of the filum terminale to form the **coccygeal ligament** (**Figure 13–2a**). The coccygeal ligament continues along the sacral canal, ultimately blending into the periosteum of the coccyx. Lateral support for the spinal dura mater is provided by loose connective tissue and adipose tissue within the epidural space. In addition, this dura mater extends between adjacent vertebrae at each intervertebral foramen, fusing with the connective tissues that surround the spinal nerves.

Anesthetics are often injected into the epidural space. Introduced in this way, a drug should affect only the spinal nerves in the immediate area of the injection. The result is an *epidural block*—a temporary sensory loss or a sensory and motor paralysis, depending on the anesthetic selected. Epidural blocks in the inferior lumbar or sacral regions may be used to control pain during childbirth.

The Arachnoid Mater

In most anatomical and histological preparations, a narrow **subdural space** separates the dura mater from deeper meningeal layers. It is likely, however, that in life no such space exists, and that the inner surface of the dura mater is in contact with the outer surface of the **arachnoid** (a-RAK-noyd; *arachne*, spider) **mater**, the middle meningeal layer (**Figure 13–3b**). The inner surface of the dura mater and the outer surface of the arachnoid mater are covered by simple squamous epithelia. The arachnoid mater includes this epithelium, called the *arachnoid membrane*, and the *arachnoid trabeculae*, a delicate network of collagen and elastic fibers that extends between the arachnoid membrane and the outer surface of the pia mater. The intervening region is called the **subarachnoid space**. It is filled with **cerebrospinal fluid (CSF)**, which acts as a shock absorber and a diffusion medium for dissolved gases, nutrients, chemical messengers, and waste products.

The spinal arachnoid mater extends inferiorly as far as the filum terminale, and the dorsal and ventral roots of the cauda equina lie within the fluid-filled subarachnoid space. In adults, the withdrawal of cerebrospinal fluid, a procedure known as a **lumbar puncture** or **spinal tap**, involves the insertion of a needle into the subarachnoid space in the inferior lumbar region.

CLINICAL NOTE

Spinal Anesthesia

Injecting a local anesthetic around a nerve produces a temporary blockage of sensory and motor nerve function. This procedure can be done either peripherally, as when skin lacerations are sewn up, or at sites around the spinal cord to obtain more widespread anesthetic effects. An *epidural block*—the injection of an anesthetic into the epidural space—has at least two advantages: (1) It affects only the spinal nerves in the immediate area of the injection, and (2) it provides mainly sensory anesthesia. If a catheter is left in place, continued injection allows sustained anesthesia. Epidural anesthesia can be difficult to achieve in the upper cervical and midthoracic regions, where the epidural space is extremely narrow. It is more effective in the lower lumbar region, inferior to the conus medullaris, because the epidural space is somewhat broader.

Caudal anesthesia involves the introduction of anesthetics into the epidural space of the sacrum. Injection at this site paralyzes and anesthetizes lower abdominal and perineal structures. Caudal anesthesia can be used to control pain during labor and delivery, but lumbar epidural anesthesia is often preferred.

Local anesthetics can also be introduced as a single dose into the subarachnoid space of the spinal cord. This procedure is commonly called "spinal anesthesia." However, the effects include both temporary muscle paralysis and sensory loss, which tend to spread as the movement of cerebrospinal fluid distributes the anesthetic along the spinal cord. Problems with overdosing are seldom serious, because controlling the patient's position during administration can limit the distribution of the drug to some degree. Moreover, because the diaphragmatic breathing muscles are controlled by upper cervical spinal nerves, respiration continues even if all thoracic and abdominal segments have been paralyzed.

The Pia Mater

The subarachnoid space extends between the arachnoid epithelium and the innermost meningeal layer, the **pia mater** (*pia*, delicate + *mater*, mother). The pia mater consists of a meshwork of elastic and collagen fibers that is firmly bound to the underlying neural tissue (**Figure 13–3**). These connective-tissue fibers are extensively interwoven with those that span the subarachnoid space, firmly binding the arachnoid to the pia mater. The blood vessels servicing the spinal cord run along the surface of the spinal pia mater, within the subarachnoid space (**Figure 13–4**).

Along the length of the spinal cord, paired **denticulate ligaments** extend from the pia mater through the arachnoid mater to the dura mater (**Figures 13–3b** and **13–4**). Denticulate ligaments, which originate along either side of the spinal cord, prevent lateral (side-to-side) movement. The dural connections at the foramen magnum and the coccygeal ligament prevent longitudinal (superior–inferior) movement.

The spinal meninges accompany the dorsal and ventral roots as these roots pass through the intervertebral foramina. As the sectional view in **Figure 13–3b** indicates, the meningeal membranes are continuous with the connective tissues that surround the spinal nerves and their peripheral branches.

ventral ⟹ CNS → out.
dorsal ⟹ out → CNS.

CHECKPOINT

3. Identify the three spinal meninges. *dura, arachnoid, pia*

4. Damage to which root of a spinal nerve would interfere with motor function? *Ventral root ⟹ composed of both visceral & somatic motor fibers*

5. Where is the cerebrospinal fluid that surrounds the spinal cord located? *subarachnoid space.*

See the blue Answers tab at the end of the book.

13-3 Gray matter is the region of integration and command initiation, and white matter carries information from place to place

To understand the functional organization of the spinal cord, you must become familiar with its sectional organization (**Figure 13–5**). Together, the anterior median fissure and the posterior median sulcus mark the division between the left and right sides of the spinal cord. The superficial white matter contains large numbers of myelinated and unmyelinated axons. The gray matter, dominated by the cell bodies of neurons, neuroglia, and unmyelinated axons, surrounds the narrow **central canal** and forms an H or butterfly shape. The projections of gray matter toward the outer surface of the spinal cord are called **horns**.

Organization of Gray Matter

The cell bodies of neurons in the gray matter of the spinal cord are organized into functional groups called *nuclei*. **Sensory nuclei** receive and relay sensory information from peripheral receptors. **Motor nuclei** issue motor commands to peripheral effectors. Although sensory and motor nuclei appear rather small in transverse section, they may extend for a considerable distance along the length of the spinal cord.

A frontal section along the length of the central canal of the spinal cord separates the sensory (posterior, or dorsal) nuclei from the motor (anterior, or ventral) nuclei. The

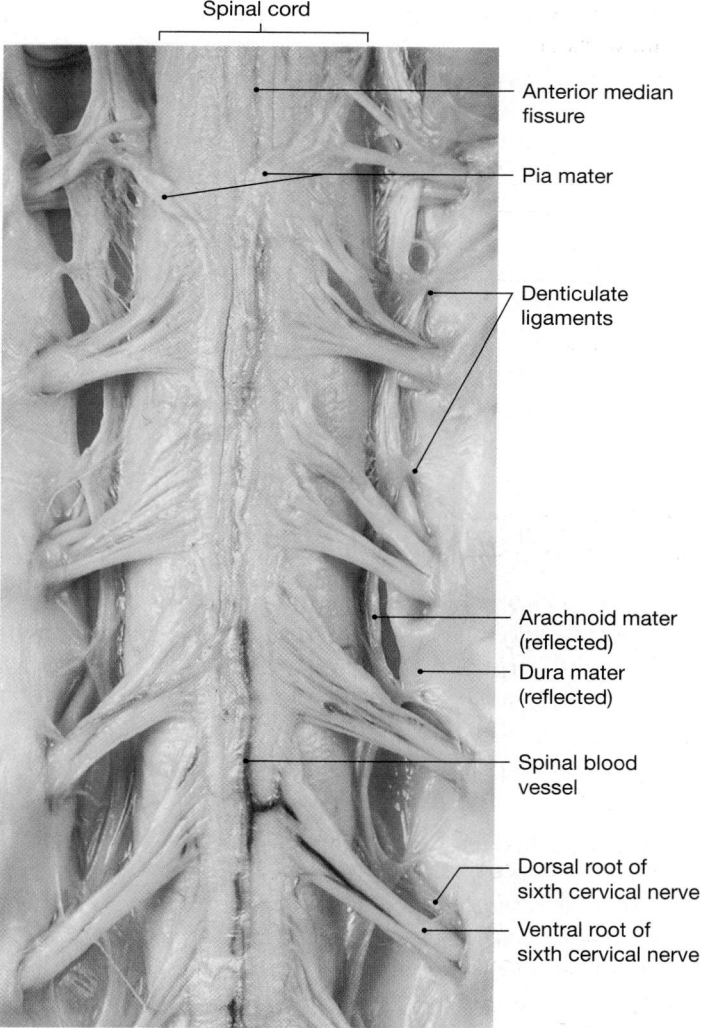

Spinal cord

Anterior median fissure

Pia mater

Denticulate ligaments

Arachnoid mater (reflected)

Dura mater (reflected)

Spinal blood vessel

Dorsal root of sixth cervical nerve

Ventral root of sixth cervical nerve

Anterior view

Figure 13–4   The Spinal Cord and Associated Structures. An anterior view of the cervical spinal cord and spinal nerve roots in the vertebral canal. The dura mater and arachnoid mater have been cut and reflected; notice the blood vessels that run in the subarachnoid space, bound to the outer surface of the delicate pia mater.

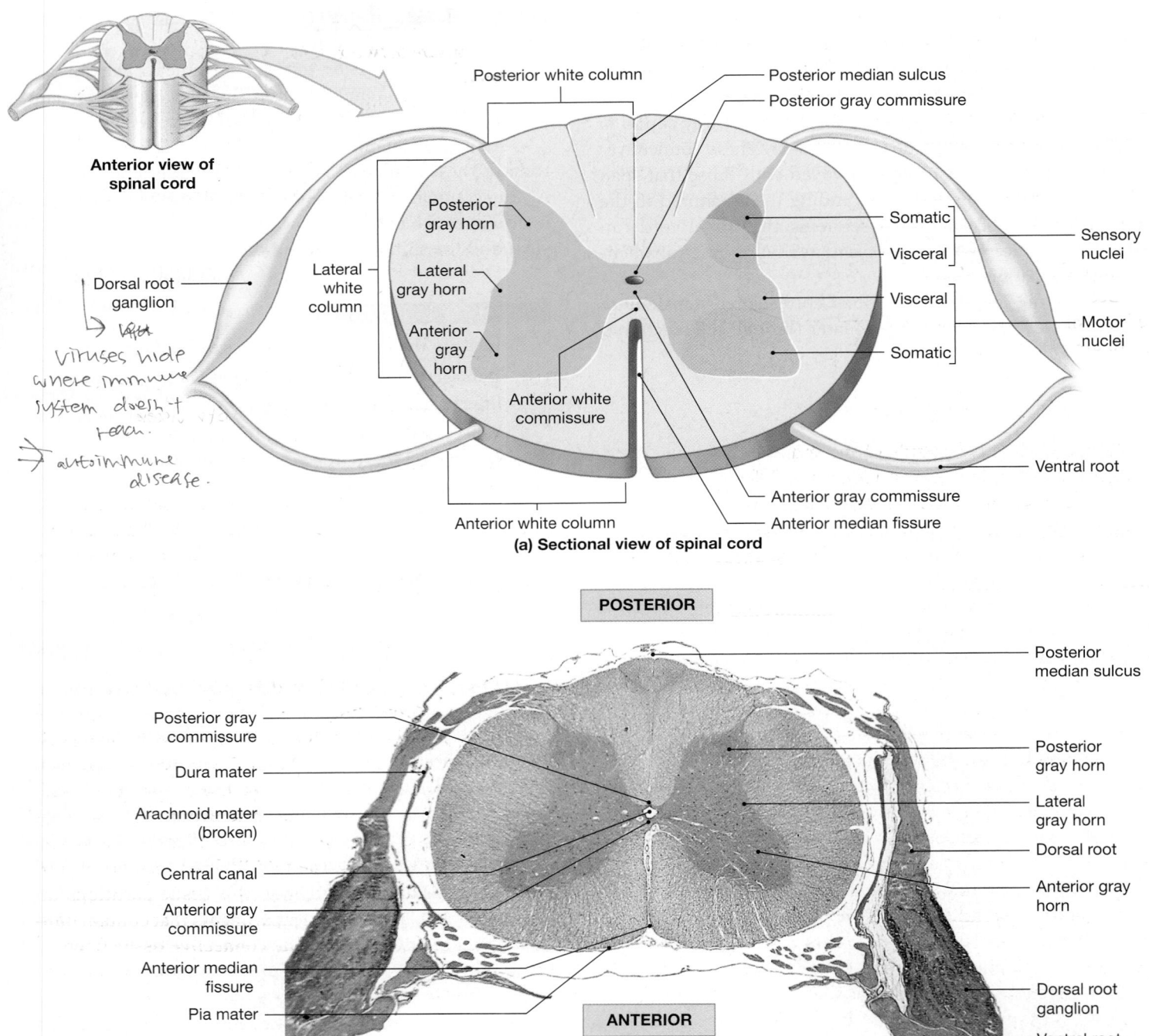

Handwritten note:
viruses hide
where immune
system doesn't
reach.
→ autoimmune
disease.

(a) Sectional view of spinal cord

POSTERIOR

ANTERIOR

(b) Micrograph of spinal cord section

Figure 13–5   The Sectional Organization of the Spinal Cord. (a) The left half of this sectional view shows important anatomical landmarks, including the three columns of white matter. The right half indicates the functional organization of the nuclei in the anterior, lateral, and posterior gray horns. **(b)** A micrograph of a section through the spinal cord, showing major landmarks in and surrounding the cord; compare with part (a).

posterior gray horns contain somatic and visceral sensory nuclei, whereas the **anterior gray horns** contain somatic motor nuclei. The **lateral gray horns**, located only in the thoracic and lumbar segments, contain visceral motor nuclei. The **gray commissures** (*commissura*, a joining together) pos-

terior to and anterior to the central canal contain axons that cross from one side of the cord to the other before they reach a destination in the gray matter.

Figure 13–5a shows the relationship between the function of a particular nucleus (sensory or motor) and its relative position in the gray matter of the spinal cord. The nuclei within each gray horn are also spatially organized. In the cervical enlargement, for example, the anterior gray horns contain nuclei whose motor neurons control the muscles of the upper limbs. On each side of the spinal cord, in medial to lateral sequence, are somatic motor nuclei that control (1) muscles that position the pectoral girdle, (2) muscles that move the arm, (3) muscles that move the forearm and hand, and (4) muscles that move the hand and fingers. Within each of these regions, the motor neurons that control flexor muscles are grouped separately from those that control extensor muscles. Because the spinal cord is so highly organized, we can predict which muscles will be affected by damage to a specific area of gray matter.

Organization of White Matter

The white matter on each side of the spinal cord can be divided into three regions called **columns**, or *funiculi* (**Figure 13–5a**). The **posterior white columns** lie between the posterior gray horns and the posterior median sulcus. The **anterior white columns** lie between the anterior gray horns and the anterior median fissure. The anterior white columns are interconnected by the **anterior white commissure**, a region where axons cross from one side of the spinal cord to the other. The white matter between the anterior and posterior columns on each side makes up the **lateral white column**.

Each column contains tracts whose axons share functional and structural characteristics. A **tract**, or *fasciculus* (fa-SIK-ū-lus; bundle), is a bundle of axons in the CNS that are relatively uniform with respect to diameter, myelination, and conduction speed. All the axons within a tract relay the same type of information (sensory or motor) in the same direction. Short tracts carry sensory or motor signals between segments of the spinal cord, and longer tracts connect the spinal cord with the brain. **Ascending tracts** carry *sensory* information toward the brain, and **descending tracts** convey *motor* commands to the spinal cord. We will describe the major tracts and their functions in Chapters 15 and 16. Because spinal tracts have very specific functions, damage to one produces a characteristic loss of sensation or motor control.

The A&P Top 100

#45 The spinal cord has a narrow central canal surrounded by gray matter containing sensory and motor nuclei. Sensory nuclei are dorsal; motor nuclei are ventral. A thick layer of white matter consisting of ascending and descending axons covers the gray matter. These axons are organized in columns that contain axon bundles with specific functions. Because the spinal cord is so highly organized, it is often possible to predict the results of injuries to localized areas.

CHECKPOINT

6. Differentiate between sensory nuclei and motor nuclei. *Gray matter ⇒ sensory: receives from peripheral motor: communicates to peripheral.*

7. A person with polio has lost the use of his leg muscles. In which area of his spinal cord would you expect the virus-infected motor neurons to be? *Anterior gray horn ⇒ somatic neurons located.*

8. A disease that damages myelin sheaths would affect which portion of the spinal cord? *white matter ⇒ columns b/c of myelinated axons.*

See the blue Answers tab at the end of the book.

13-4 Spinal nerves form plexuses that are named according to their level of emergence from the vertebral canal

In this section we consider the structure and function of spinal nerves. After we examine the anatomy and distribution of spinal nerves, we explore the interwoven networks of spinal nerves called nerve plexuses.

Anatomy of Spinal Nerves

Every segment of the spinal cord is connected to a pair of spinal nerves. Surrounding each spinal nerve is a series of connective tissue layers continuous with those of the associated peripheral nerves (**Figure 13–6**). These layers, best seen in sectional view, are comparable to those associated with skeletal muscles. ∞ p. 295 The **epineurium**, or outermost layer, consists of a dense network of collagen fibers. The fibers of the **perineurium**, the middle layer, extend inward from the epineurium. These connective tissue partitions divide the nerve into a series of compartments that contain bundles of axons, or *fascicles*. Delicate connective tissue fibers of the **endoneurium**, the innermost layer, extend from the perineurium and surround individual axons.

Arteries and veins penetrate the epineurium and branch within the perineurium. Capillaries leaving the perineurium branch in the endoneurium and supply the axons and Schwann cells of the nerve and the fibroblasts of the connective tissues.

As they extend into the periphery, the spinal nerves branch and interconnect, forming the peripheral nerves that innervate body tissues and organs. The connective tissue sheaths of peripheral nerves are the same as, and continuous with, those of spinal nerves.

If a peripheral axon is severed but not displaced, normal function may eventually return as the cut stump grows across the site of injury, away from the cell body and along its for

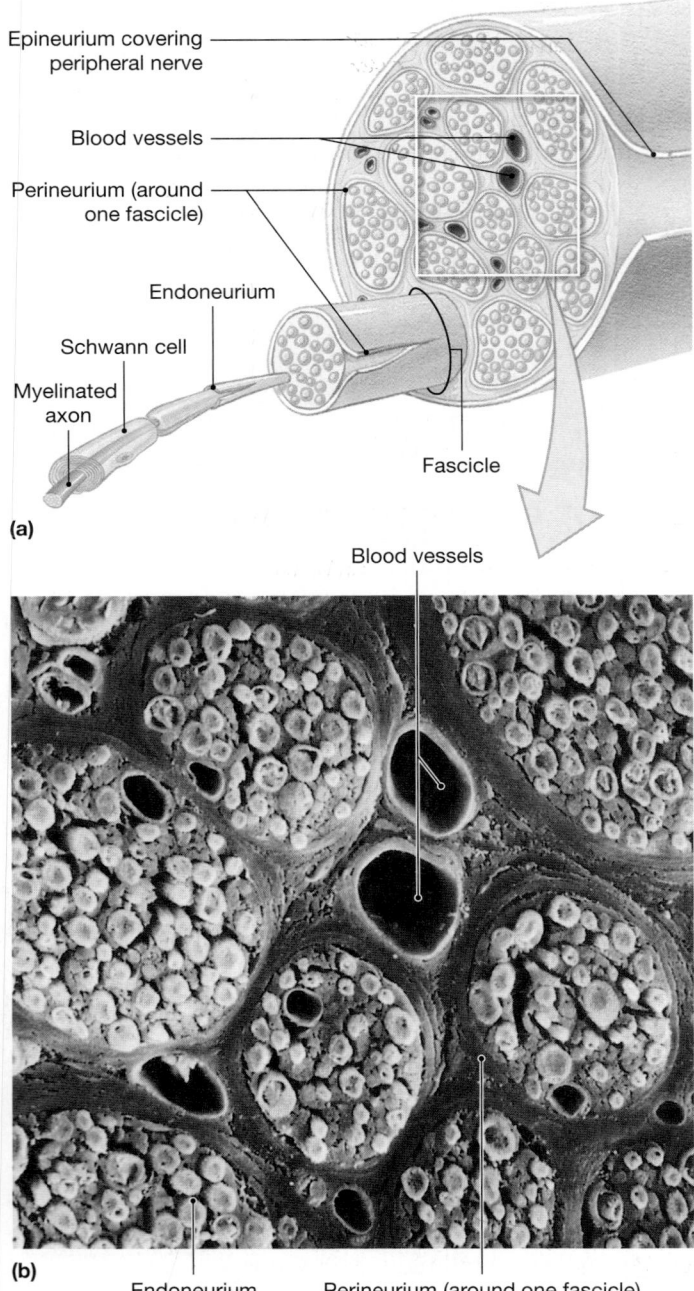

Epineurium covering peripheral nerve

Blood vessels

Perineurium (around one fascicle)

Endoneurium

Schwann cell

Myelinated axon

(a)

Fascicle

Blood vessels

(b)

Endoneurium   Perineurium (around one fascicle)

Figure 13–6   A Peripheral Nerve. (a) A diagrammatic view and **(b)** an electron micrograph of a typical peripheral nerve and the connective tissue wrappings (perineurium, endoneurium, and epineurium) continuous with those of the associated spinal nerve(s). (SEM × 340) [© R. G. Kessel and R. H. Kardon, *Tissues and Organs: A Text-Atlas of Scanning Electron Microscopy,* W. H. Freeman & Co., 1979. All Rights Reserved.]

path. ∞ p. 397 Repairs made after an entire peripheral *nerve* has been damaged are generally incomplete, primarily because of problems with axon alignment and regrowth. Various technologically sophisticated procedures designed to improve nerve regeneration and repair are currently under evaluation.

Peripheral Distribution of Spinal Nerves

Figure 13–7 shows the distribution, or pathway, of a typical spinal nerve that originates from the thoracic or superior lumbar segments of the spinal cord. The spinal nerve forms just lateral to the intervertebral foramen, where the dorsal and ventral roots unite. We will now consider the peripheral distribution of a representative spinal nerve from the thoracic region.

The ventral root of each spinal nerve contains the axons of somatic motor and visceral motor neurons (**Figure 13–7a**). Distally, the first branch from the spinal nerve carries visceral motor fibers to a nearby *sympathetic ganglion,* part of the *sympathetic division* of the autonomic nervous system. (Among its other functions, the sympathetic division is responsible for elevating metabolic rate and increasing heart rate.) Because preganglionic axons are myelinated, this branch has a light color and is therefore known as the **white ramus** ("branch"). Postganglionic fibers that innervate smooth muscles, glands, and organs in the thoracic cavity extend directly from the ganglion to their respective effector organs. These axons form a series of **sympathetic nerves**.

Postganglionic fibers innervating glands and smooth muscles in the body wall or limbs return from the ganglion to rejoin the spinal nerve. These fibers, which are unmyelinated and have a darker color, form the **gray ramus**. The gray ramus is typically proximal to the white ramus; together, they are known as the *rami communicantes* (RĀ-mī ko-mū-ni-KAN-tēz), or "communicating branches." The **dorsal ramus** of each spinal nerve contains somatic motor and visceral motor fibers that innervate the skin and skeletal muscles of the back. The axons in the relatively large **ventral ramus** supply the ventrolateral body surface, structures in the body wall, and the limbs.

The dorsal, ventral, and white rami also contain sensory (afferent) fibers (**Figure 13–7b**). Somatic sensory information arrives over the dorsal and ventral rami; visceral sensory information reaches the dorsal root through the dorsal, ventral, and white rami.

The specific bilateral region of the skin surface monitored by a single pair of spinal nerves is known as a **dermatome**. Each pair of spinal nerves services its own dermatome (**Figure 13–8**), although the boundaries of adjacent dermatomes overlap to some degree. Dermatomes are clinically important because damage or infection of a spinal nerve or dorsal root ganglion will produce a characteristic loss of sensation in the corresponding region of the skin. Additionally, characteristic signs may appear on the skin supplied by that specific nerve.

Peripheral *nerve palsies,* or **peripheral neuropathies**, are regional losses of sensory and motor function most often resulting from nerve trauma or compression. (You have experienced a mild, temporary palsy if your arm or leg has ever "fallen asleep" after you leaned or sat in an uncomfortable po-

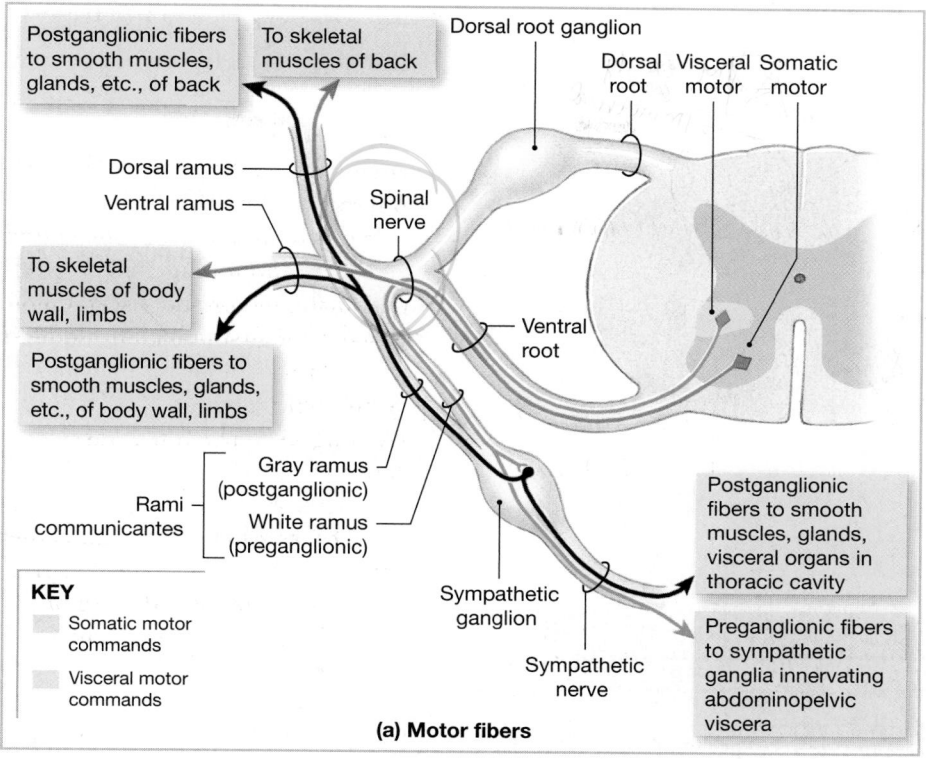

Postganglionic fibers to smooth muscles, glands, etc., of back

To skeletal muscles of back

Dorsal root ganglion

Dorsal root

Visceral motor

Somatic motor

Dorsal ramus

Ventral ramus

Spinal nerve

To skeletal muscles of body wall, limbs

Ventral root

Postganglionic fibers to smooth muscles, glands, etc., of body wall, limbs

Gray ramus (postganglionic)

White ramus (preganglionic)

Rami communicantes

Sympathetic ganglion

Postganglionic fibers to smooth muscles, glands, visceral organs in thoracic cavity

KEY

Somatic motor commands

Visceral motor commands

Sympathetic nerve

Postganglionic fibers to smooth muscles, glands, visceral organs in thoracic cavity

Preganglionic fibers to sympathetic ganglia innervating abdominopelvic viscera

(a) Motor fibers

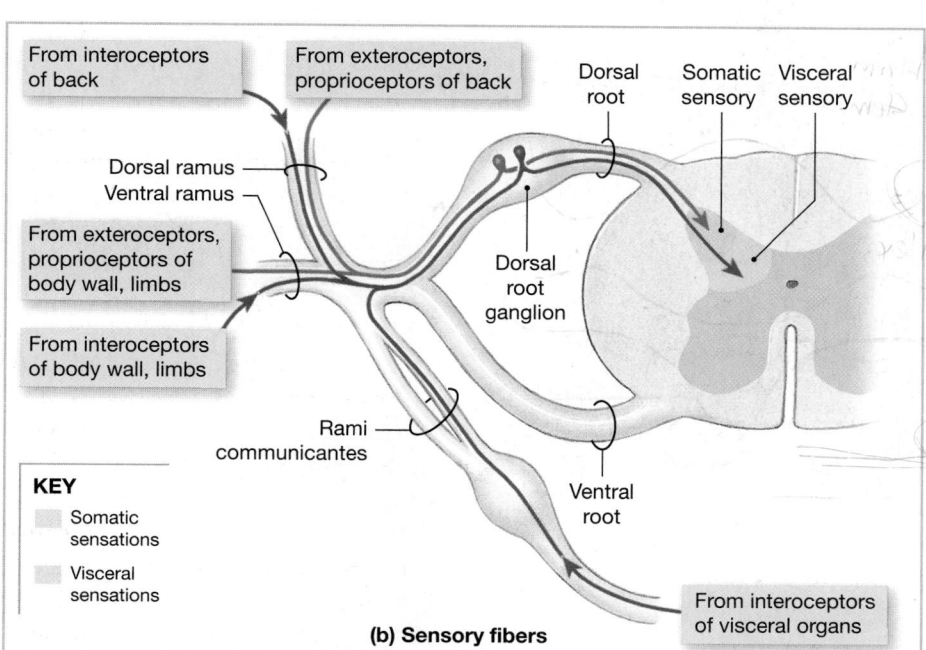

From interoceptors of back

From exteroceptors, proprioceptors of back

Dorsal root

Somatic sensory

Visceral sensory

Dorsal ramus

Ventral ramus

From exteroceptors, proprioceptors of body wall, limbs

Dorsal root ganglion

From interoceptors of body wall, limbs

Rami communicantes

Ventral root

KEY

Somatic sensations

Visceral sensations

From interoceptors of visceral organs

(b) Sensory fibers

Figure 13–7 Peripheral Distribution of Spinal Nerves. (a) The distribution of motor fibers in the major branches of a representative thoracic or superior lumbar spinal nerve. (Although the gray ramus is normally proximal to the white ramus, this diagrammatic view makes it easier to follow the relationships between preganglionic and postganglionic fibers.) **(b)** A comparable view of the distribution of sensory fibers.

sition.) The location of the affected dermatomes provides clues to the location of injuries along the spinal cord, but the information is not precise. More exact conclusions can be drawn if there is loss of motor control, on the basis of the origin and distribution of the peripheral nerves originating at nerve plexuses.

The A&P Top 100

#46 Each peripheral nerve provides sensory and/or motor innervation to specific structures.

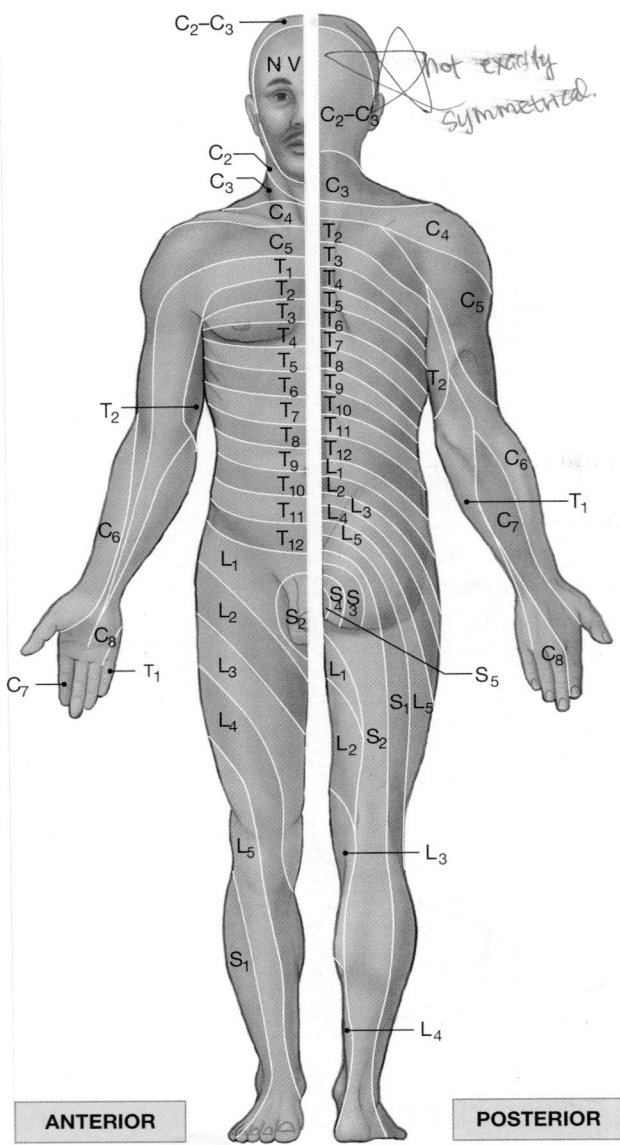

Figure 13–8 **Dermatomes.** Anterior and posterior distributions of dermatomes on the surface of the skin. N V = fifth cranial nerve (trigeminal nerve).

Nerve Plexuses

The simple distribution pattern of dorsal and ventral rami in **Figure 13–7** applies to spinal nerves T_1–T_{12}. But in segments controlling the skeletal musculature of the neck, upper limbs, or lower limbs, the situation is more complicated. During development, small skeletal muscles innervated by different ventral rami typically fuse to form larger muscles with compound origins. The anatomical distinctions between the component muscles may disappear, but separate ventral rami continue to provide sensory innervation and motor control to each part of the compound muscle. As they converge, the ventral rami of

CLINICAL NOTE

Shingles   **Shingles** (derived from the Latin *cingulum*, girdle) is caused by the varicella-zoster virus (VZV), the same virus that causes chickenpox. This herpes virus attacks neurons within the dorsal roots of spinal nerves and sensory ganglia of cranial nerves. This disorder produces a painful rash and blisters whose distribution corresponds to that of the affected sensory nerves and follows its dermatome (**Figure 13–9**). Any person who has had chickenpox is at risk of developing shingles. After the initial encounter, the virus remains dormant within neurons of the anterior gray horns of the spinal cord. It is not known what triggers the reactivation of this pathogen. Fortunately for those affected, attacks of shingles generally heal and leave behind only unpleasant memories.

   Most people who contract shingles suffer just a single episode in their adult lives. Postherpetic neuralgia is a painful condition experienced by some after a bout of shingles. Moreover, the problem can recur in people with weakened immune systems, including the elderly and individuals with AIDS or some forms of cancer. Treatment for shingles typically involves large doses of antiviral drugs. In 2006, the U.S. Food and Drug Administration approved a VZV vaccine (Zostavax) for use in people ages 60 and above who have had chickenpox.

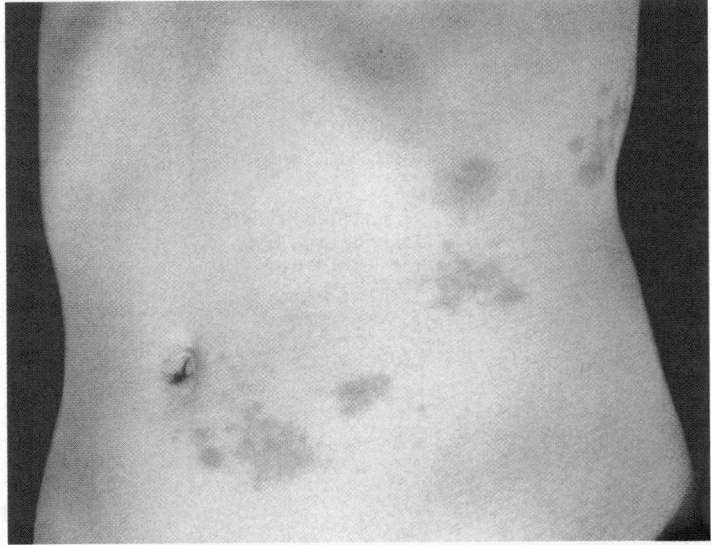

Figure 13–9 **Shingles.** The skin eruptions follow the distribution of the dermatomal innervation.

adjacent spinal nerves blend their fibers, producing a series of compound nerve trunks. Such a complex interwoven network of nerves is called a **nerve plexus** (PLEK-sus; *plexus*, braid). The ventral rami form four major plexuses: (1) the *cervical*

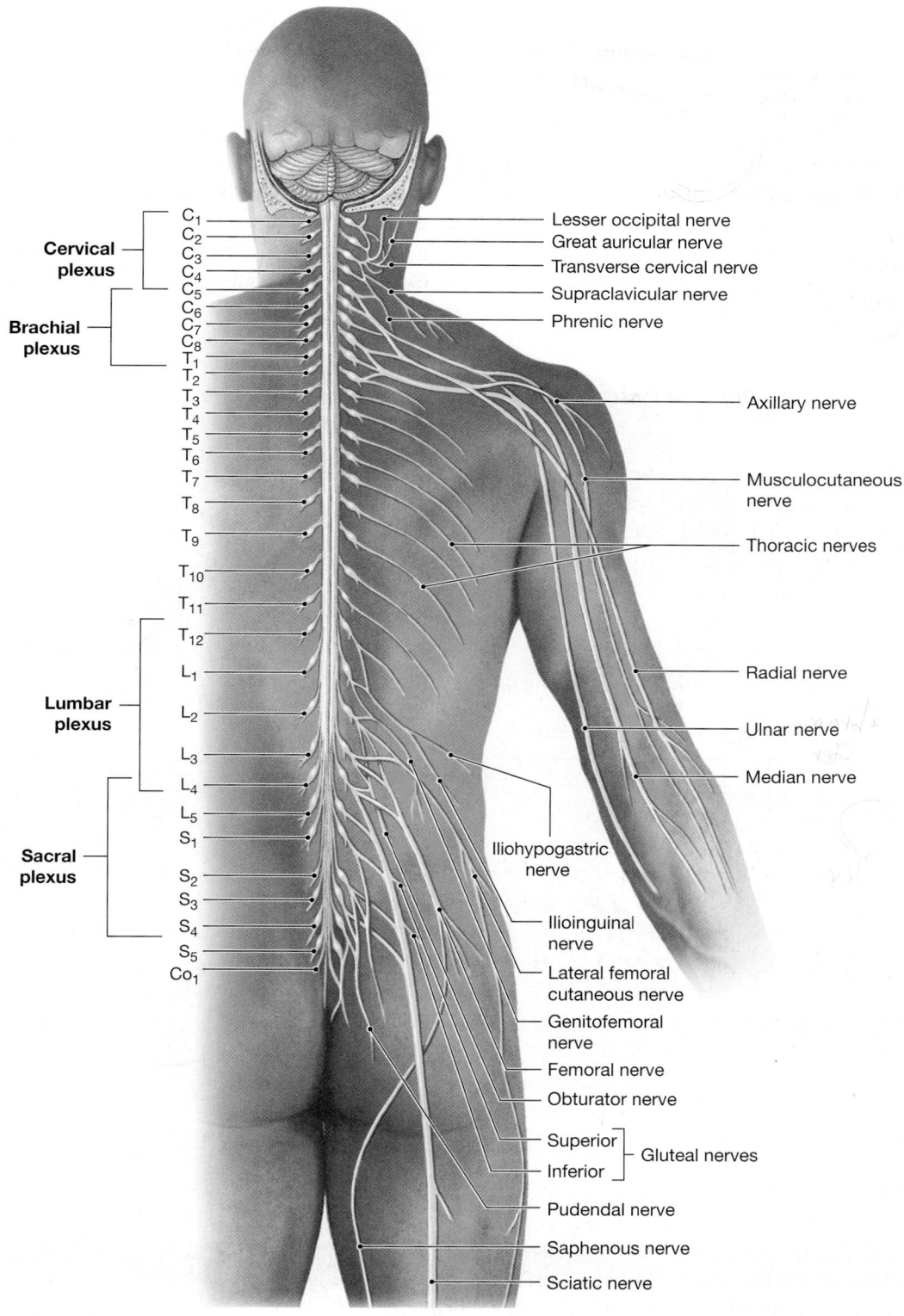

Figure 13–10 **Peripheral Nerves and Nerve Plexuses.** ATLAS: Plate 2a

plexus, (2) the *brachial plexus*, (3) the *lumbar plexus*, and (4) the *sacral plexus* (**Figure 13–10**). Because they form from the fusion of ventral rami, the nerves arising at these plexuses contain sensory as well as motor fibers (**Figure 13–7**).

In Chapter 11, we introduced the peripheral nerves that control the major axial and appendicular muscles. As we proceed, you may find it helpful to refer to the related tables in that chapter. ∞ pp. 347–380

The Cervical Plexus

The **cervical plexus** consists of the ventral rami of spinal nerves C_1–C_5 (**Figures 13–10, 13–11**; Table 13–1). The branches of the cervical plexus innervate the muscles of the neck and extend into the thoracic cavity, where they control the diaphragmatic muscles. The **phrenic nerve**, the major nerve of the cervical plexus, provides the entire nerve supply to the diaphragm, a key respiratory muscle. Other branches of this nerve plexus are distributed to the skin of the neck and the superior part of the chest.

The Brachial Plexus

The **brachial plexus** innervates the pectoral girdle and upper limb, with contributions from the ventral rami of spinal nerves C_5–T_1 (**Figures 13–10** and **13–12**; Table 13–2). The nerves that form this plexus originate from trunks and cords. **Trunks** are large bundles of axons contributed by several spinal nerves. **Cords** are smaller branches that originate at trunks. Both trunks and cords are named according to their location relative to the *axillary artery*, a large artery supplying the upper limb. Hence we have *superior*, *middle*, and *inferior*

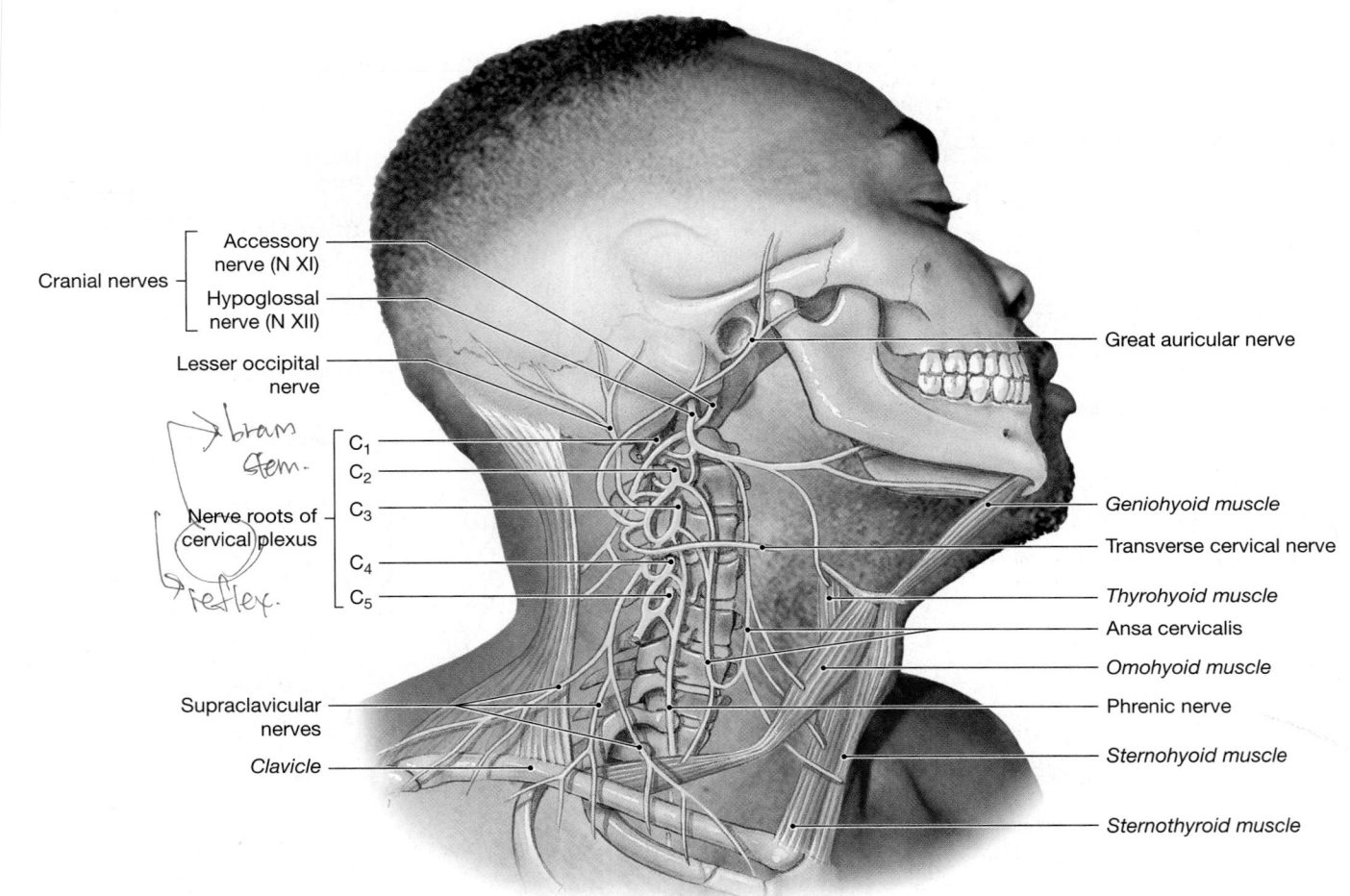

Figure 13–11   The Cervical Plexus. ATLAS: Plates 3c,d / 18a–c

TABLE 13–1 The Cervical Plexus		
Nerve(s)	**Spinal Segments**	**Distribution**
Ansa cervicalis (superior and inferior branches)	C_1–C_4	Five of the extrinsic laryngeal muscles: sternothyroid, sternohyoid, omohyoid, geniohyoid, and thyrohyoid muscles (via N XII)
Lesser occipital, transverse cervical, supraclavicular, and great auricular nerves	C_2–C_3	Skin of upper chest, shoulder, neck, and ear
Phrenic nerve	C_3–C_5	Diaphragm
Cervical nerves	C_1–C_5	Levator scapulae, scalene, sternocleidomastoid, and trapezius muscles (with N XI)

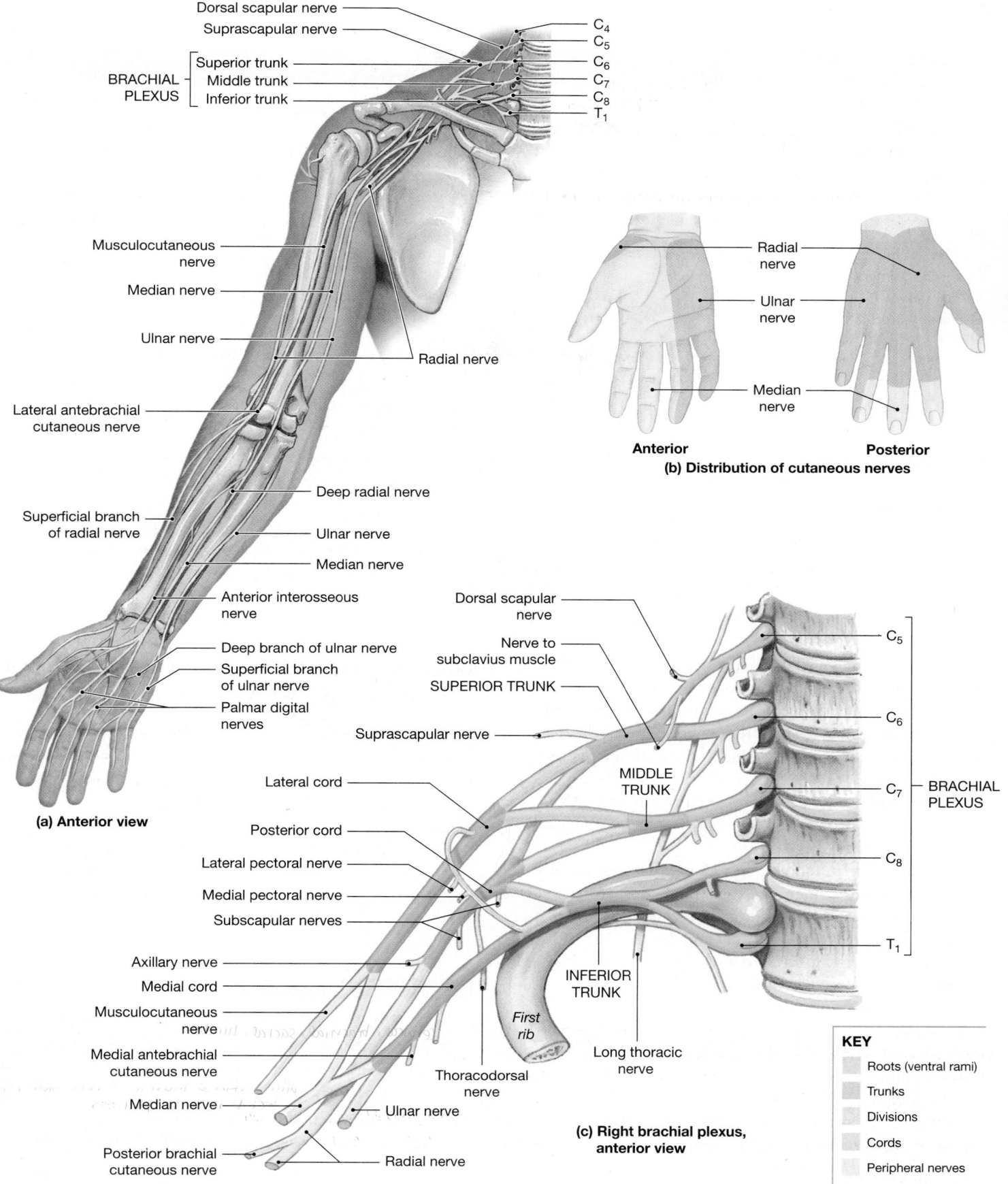

Dorsal scapular nerve
Suprascapular nerve
C4
C5
Superior trunk
Middle trunk
Inferior trunk
BRACHIAL PLEXUS
C6
C7
C8
T1

Musculocutaneous nerve
Median nerve
Ulnar nerve
Radial nerve

Lateral antebrachial cutaneous nerve

Superficial branch of radial nerve
Deep radial nerve
Ulnar nerve
Median nerve

Anterior interosseous nerve
Deep branch of ulnar nerve
Superficial branch of ulnar nerve
Palmar digital nerves

(a) Anterior view

Radial nerve
Ulnar nerve
Median nerve

Anterior   **Posterior**

(b) Distribution of cutaneous nerves

Dorsal scapular nerve
Nerve to subclavius muscle
SUPERIOR TRUNK
Suprascapular nerve

Lateral cord
Posterior cord
Lateral pectoral nerve
Medial pectoral nerve
Subscapular nerves
Axillary nerve
Medial cord
Musculocutaneous nerve
Medial antebrachial cutaneous nerve
Median nerve
Posterior brachial cutaneous nerve

MIDDLE TRUNK
INFERIOR TRUNK
First rib
Thoracodorsal nerve
Long thoracic nerve
Ulnar nerve
Radial nerve

C5
C6
C7
C8
T1
BRACHIAL PLEXUS

(c) Right brachial plexus, anterior view

KEY
Roots (ventral rami)
Trunks
Divisions
Cords
Peripheral nerves

Figure 13–12 The Brachial Plexus. (a) Major nerves originating at the right brachial plexus. **(b)** Areas of the hands serviced by nerves of the right brachial plexus. **(c)** The right brachial plexus. ATLAS: Plates 27a–c; 29b,c; 30

TABLE 13–2   The Brachial Plexus

Nerve(s)	Spinal Segments	Distribution
Nerve to subclavius	C_4–C_6	Subclavius muscle
Dorsal scapular nerve	C_5	Rhomboid and levator scapulae muscles
Long thoracic nerve	C_5–C_7	Serratus anterior muscle
Suprascapular nerve	C_5, C_6	Supraspinatus and infraspinatus muscles; sensory from shoulder joint and scapula
Pectoral nerves (medial and lateral)	C_5–T_1	Pectoralis muscles
Subscapular nerves	C_5, C_6	Subscapularis and teres major muscles
Thoracodorsal nerve	C_6–C_8	Latissimus dorsi muscle
Axillary nerve	C_5, C_6	Deltoid and teres minor muscles; sensory from the skin of the shoulder
Medial antebrachial cutaneous nerve	C_8, T_1	Sensory from skin over anterior, medial surface of arm and forearm
Radial nerve	C_5–T_1	Many extensor muscles on the arm and forearm (triceps brachii, anconeus, extensor carpi radialis, extensor carpi ulnaris, and brachioradialis muscles); supinator muscle, digital extensor muscles, and abductor pollicis muscle via the *deep branch*; sensory from skin over the posterolateral surface of the limb through the *posterior brachial cutaneous nerve* (arm), *posterior antebrachial cutaneous nerve* (forearm), and the *superficial branch* (radial half of hand)
Musculocutaneous nerve	C_5–T_1	Flexor muscles on the arm (biceps brachii, brachialis, and coracobrachialis muscles); sensory from skin over lateral surface of the forearm through the *lateral antebrachial cutaneous nerve*
Median nerve	C_6–T_1	Flexor muscles on the forearm (flexor carpi radialis and palmaris longus muscles); pronator quadratus and pronator teres muscles; digital flexors (through the *anterior interosseous nerve*); sensory from skin over anterolateral surface of the hand
Ulnar nerve	C_8, T_1	Flexor carpi ulnaris muscle, flexor digitorum profundus muscle, adductor pollicis muscle, and small digital muscles via the *deep branch*; sensory from skin over medial surface of the hand through the *superficial branch*

trunks, and *lateral*, *medial*, and *posterior cords*. The lateral cord forms the **musculocutaneous nerve** exclusively and, together with the medial cord, contributes to the **median nerve**. The **ulnar nerve** is the other major nerve of the medial cord. The posterior cord gives rise to the **axillary nerve** and the **radial nerve**. Table 13–2 provides further information about these and other major nerves of the brachial plexus.

The Lumbar and Sacral Plexuses

The **lumbar plexus** and the **sacral plexus** arise from the lumbar and sacral segments of the spinal cord, respectively. The nerves arising at these plexuses innervate the pelvic girdle and lower limbs (**Figures 13–10** and **13–13**). The individual nerves that form the lumbar and sacral plexuses are listed in Table 13–3.

The lumbar plexus contains axons from the ventral rami of spinal nerves T_{12}–L_4. The major nerves of this plexus are the **genitofemoral nerve**, the **lateral femoral cutaneous nerve**, and the **femoral nerve**. The sacral plexus contains axons from the ventral rami of spinal nerves L_4–S_4. Two major nerves arise at this plexus: the **sciatic nerve** and the **pudendal nerve**. The sciatic nerve passes posterior to the femur, deep to the long head of the biceps femoris muscle. As it approaches the knee, the sciatic nerve divides into two branches: the **fibular nerve** (or *peroneal nerve*) and the **tibial nerve**. The *sural nerve*, formed by branches of the fibular nerve, is a sensory nerve innervating the lateral portion of the foot. A section of this nerve is often removed for use in nerve grafts.

In discussions of motor performance, a distinction is usually made between the conscious ability to control motor function—something that requires communication and feedback between the brain and spinal cord—and automatic motor responses coordinated entirely within the spinal cord. These automatic responses, called *reflexes*, are relatively stereotyped motor responses to specific stimuli. The rest of this chapter looks at how sensory neurons, interneurons, and motor neurons interconnect, and how these interconnections produce both simple and complex reflexes. ATLAS: Embryology Summary 11: The Development of the Spinal Cord and Spinal Nerves

CHECKPOINT

9. Identify the major networks of nerves known as plexuses.

10. An anesthetic blocks the function of the dorsal rami of the cervical spinal nerves. Which areas of the body will be affected?

11. Injury to which of the nerve plexuses would interfere with the ability to breathe?

12. Compression of which nerve produces the sensation that your leg has "fallen asleep"?

See the blue Answers tab at the end of the book.

13 NERVOUS

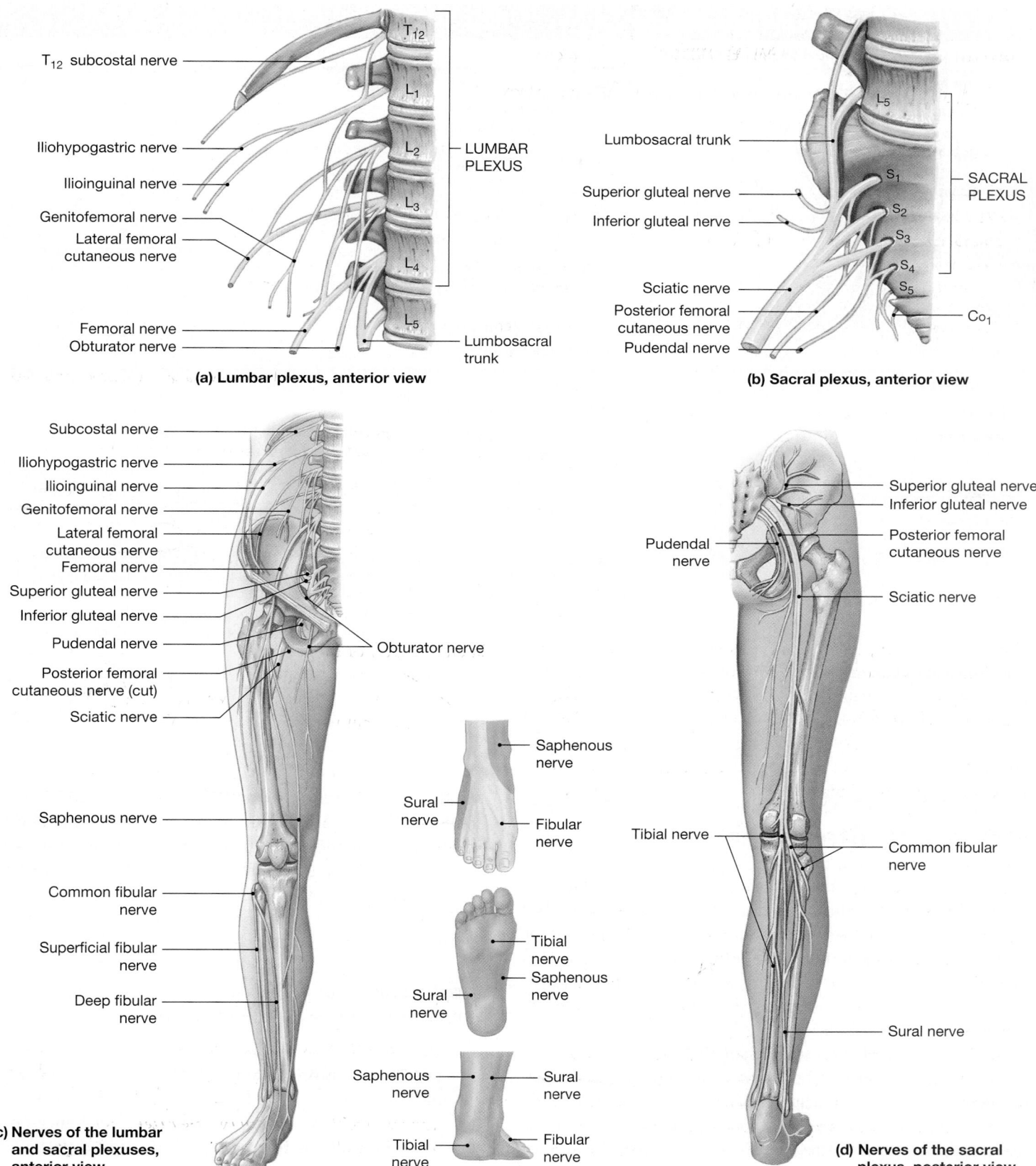

T₁₂ subcostal nerve

Iliohypogastric nerve

Ilioinguinal nerve

Genitofemoral nerve

Lateral femoral
cutaneous nerve

Femoral nerve

Obturator nerve

T₁₂

L₁

L₂

L₃

L₄

L₅

LUMBAR
PLEXUS

Lumbosacral
trunk

(a) Lumbar plexus, anterior view

Lumbosacral trunk

Superior gluteal nerve

Inferior gluteal nerve

Sciatic nerve

Posterior femoral
cutaneous nerve

Pudendal nerve

L₅

S₁

S₂

S₃

S₄

S₅

Co₁

SACRAL
PLEXUS

(b) Sacral plexus, anterior view

Subcostal nerve

Iliohypogastric nerve

Ilioinguinal nerve

Genitofemoral nerve

Lateral femoral
cutaneous nerve

Femoral nerve

Superior gluteal nerve

Inferior gluteal nerve

Pudendal nerve

Posterior femoral
cutaneous nerve (cut)

Sciatic nerve

Obturator nerve

Saphenous nerve

Common fibular
nerve

Superficial fibular
nerve

Deep fibular
nerve

Saphenous
nerve

Sural
nerve

Fibular
nerve

Tibial
nerve

Sural
nerve

Saphenous
nerve

Saphenous
nerve

Sural
nerve

Tibial
nerve

Fibular
nerve

**(c) Nerves of the lumbar
and sacral plexuses,
anterior view**

Superior gluteal nerve

Inferior gluteal nerve

Posterior femoral
cutaneous nerve

Pudendal
nerve

Sciatic nerve

Tibial nerve

Common fibular
nerve

Sural nerve

**(d) Nerves of the sacral
plexus, posterior view**

Figure 13–13   The Lumbar and Sacral Plexuses. (a) The right lumbar plexus. **(b)** The right sacral plexus. **(c)** The major branches of the right lumbar plexus. **(d)** The major branches of the right sacral plexus. ATLAS: Plates 70b; 76b; 82b

Nerve(s)	Spinal Segment(s)	Distribution
TABLE 13–3 **The Lumbar and Sacral Plexuses**		
LUMBAR PLEXUS		
Iliohypogastric nerve	T_{12}, L_1	Abdominal muscles (external and internal oblique muscles, transversus abdominis muscle); skin over inferior abdomen and buttocks
Ilioinguinal nerve	L_1	Abdominal muscles (with iliohypogastric nerve); skin over superior, medial thigh and portions of external genitalia
Genitofemoral nerve	L_1, L_2	Skin over anteromedial surface of thigh and portions of external genitalia
Lateral femoral cutaneous nerve	L_2, L_3	Skin over anterior, lateral, and posterior surfaces of thigh
Femoral nerve	L_2–L_4	Anterior muscles of thigh (sartorius muscle and quadriceps group); flexors and adductors of hip (pectineus and iliopsoas muscles); skin over anteromedial surface of thigh, medial surface of leg and foot
Obturator nerve	L_2–L_4	Adductors of hip (adductors magnus, brevis, and longus muscles); gracilis muscle; skin over medial surface of thigh
Saphenous nerve	L_2–L_4	Skin over medial surface of leg
SACRAL PLEXUS		
Gluteal nerves:	L_4–S_2	
Superior		Abductors of hip (gluteus minimus, gluteus medius, and tensor fasciae latae muscles)
Inferior		Extensor of hip (gluteus maximus muscle)
Posterior femoral cutaneous nerve	S_1–S_3	Skin of perineum and posterior surfaces of thigh and leg
Sciatic nerve:	L_4–S_3	Two of the hamstrings (semimembranosus and semitendinosus muscles); adductor magnus muscle (with obturator nerve)
Tibial nerve		Flexors of knee and extensors (plantar flexors) of ankle (popliteus, gastrocnemius, soleus, and tibialis posterior muscles and the long head of the biceps femoris muscle); flexors of toes; skin over posterior surface of leg, plantar surface of foot
Fibular nerve		Biceps femoris muscle (short head); fibularis muscles (brevis and longus) and tibialis anterior muscle; extensors of toes; skin over anterior surface of leg and dorsal surface of foot; skin over lateral portion of foot (through the *sural nerve*)
Pudendal nerve	S_2–S_4	Muscles of perineum, including urogenital diaphragm and external anal and urethral sphincter muscles; skin of external genitalia and related skeletal muscles (bulbospongiosus and ischiocavernosus muscles)

13-5 Neuronal pools are functional groups of interconnected neurons

The human body has about 10 million sensory neurons, one-half million motor neurons, and 20 *billion* interneurons. The sensory neurons deliver information to the CNS; the motor neurons distribute commands to peripheral effectors, such as skeletal muscles; and interactions among interneurons provide the interpretation, planning, and coordination of incoming and outgoing signals.

The billions of interneurons of the CNS are organized into a much smaller number of **neuronal pools**—functional groups of interconnected neurons. A neuronal pool may be diffuse, involving neurons in several regions of the brain, or localized, with neurons restricted to one specific location in the brain or spinal cord. Estimates of the actual number of neuronal pools range between a few hundred and a few thousand. Each has a limited number of input sources and output destinations, and each may contain both excitatory and inhibitory neurons. The output of the entire neuronal pool may stimulate or depress activity in other parts of the brain or spinal cord, affecting the interpretation of sensory information or the coordination of motor commands.

The pattern of interaction among neurons provides clues to the functional characteristics of a neuronal pool. It is customary to refer to the "wiring diagrams" in **Figure 13–14** as *neural circuits*, just as we refer to electrical circuits in the wiring of a house. We can distinguish five circuit patterns:

1. **Divergence** is the spread of information from one neuron to several neurons (**Figure 13–14a**), or from one pool to multiple pools. Divergence permits the broad distribution of a specific input. Considerable divergence occurs when sensory neurons bring information into the CNS, for the information is distributed to neuronal pools throughout the spinal cord and brain. Visual information arriving from the eyes, for example, reaches your con-

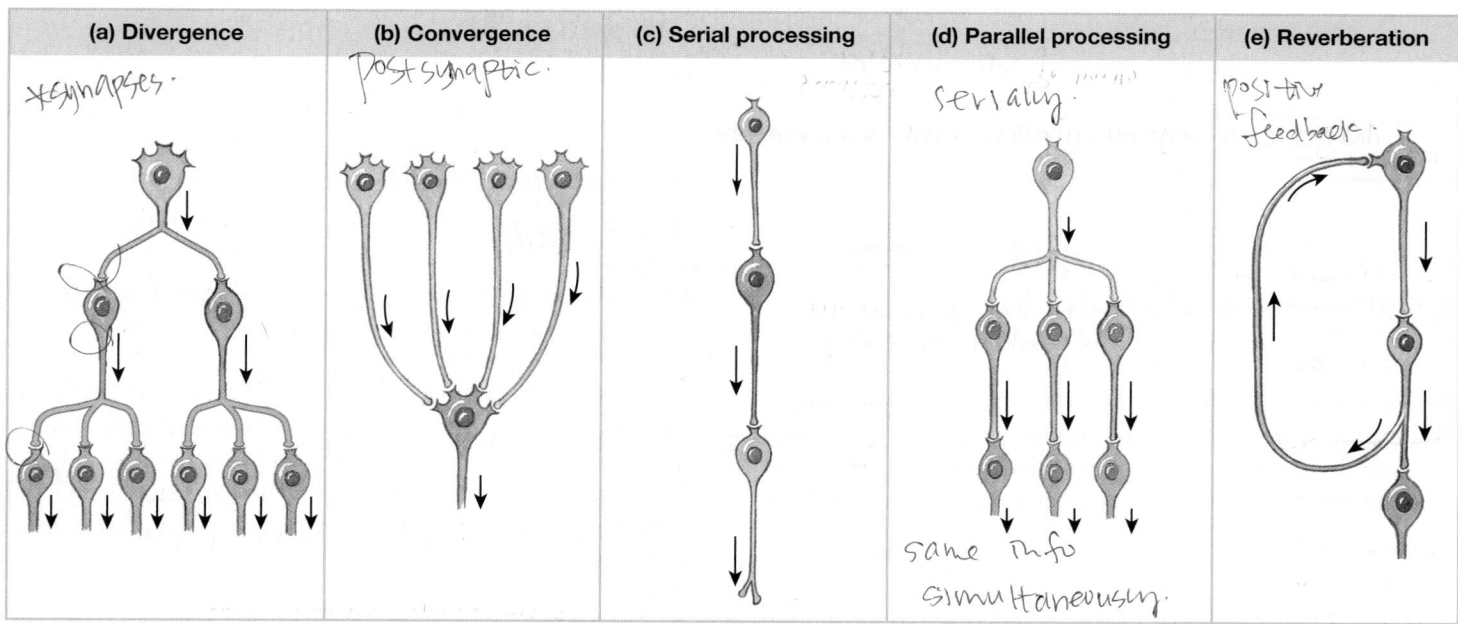

Figure 13–14   Neural Circuits: The Organization of Neuronal Pools. (a) Divergence, a mechanism for spreading stimulation to multiple neurons or neuronal pools in the CNS. **(b)** Convergence, a mechanism providing input to a single neuron from multiple sources. **(c)** Serial processing, in which neurons or pools work sequentially. **(d)** Parallel processing, in which neurons or pools process information simultaneously. **(e)** Reverberation, a positive feedback mechanism.

sciousness at the same time it is distributed to areas of the brain that control posture and balance at the subconscious level.

2. In **convergence**, several neurons synapse on a single postsynaptic neuron (**Figure 13–14b**). Several patterns of activity in the presynaptic neurons can therefore have the same effect on the postsynaptic neuron. Through convergence, the same motor neurons can be subject to both conscious and subconscious control. For example, the movements of your diaphragm and ribs are now being controlled by your brain at the subconscious level. But the same motor neurons can also be controlled consciously, as when you take a deep breath and hold it. Two neuronal pools are involved, both synapsing on the same motor neurons.

3. In **serial processing**, information is relayed in a stepwise fashion, from one neuron to another or from one neuronal pool to the next (**Figure 13–14c**). This pattern occurs as sensory information is relayed from one part of the brain to another. For example, pain sensations en route to your consciousness make two stops along the way, at neuronal pools along the pain pathway.

4. **Parallel processing** occurs when several neurons or neuronal pools process the same information simultaneously (**Figure 13–14d**). Divergence must take place before parallel processing can occur. Thanks to parallel processing, many responses can occur simultaneously. For example,

stepping on a sharp object stimulates sensory neurons that distribute the information to several neuronal pools. As a result of parallel processing, you might withdraw your foot, shift your weight, move your arms, feel the pain, and shout "Ouch!" at about the same time.

5. In **reverberation**, collateral branches of axons somewhere along the circuit extend back toward the source of an impulse and further stimulate the presynaptic neurons (**Figure 13–14e**). Reverberation is like a positive feedback loop involving neurons: Once a reverberating circuit has been activated, it will continue to function until synaptic fatigue or inhibitory stimuli break the cycle. Reverberation can occur within a single neuronal pool, or it may involve a series of interconnected pools. Highly complicated examples of reverberation among neuronal pools in the brain may help maintain consciousness, muscular coordination, and normal breathing.

The functions of the nervous system depend on the interactions among neurons organized in neuronal pools. The most complex neural processing steps occur in the spinal cord and brain. The simplest, which occur within the PNS and the spinal cord, control reflexes that are a bit like Legos: Individually, they are quite simple, but they can be combined in a great variety of ways to create very complex responses. Reflexes are thus the basic building blocks of neural function, as you will see in the next section.

13. Define neuronal pool. *group of interconnected neurons*

14. List the five circuit patterns found in neuronal pools. *divergence, convergence, parallel, serial, reverberation.*

See the blue Answers tab at the end of the book.

13-6 Reflexes are rapid, automatic responses to stimuli

not delivered any + to highest processes.

Conditions inside or outside the body can change rapidly and unexpectedly. **Reflexes** are rapid, automatic responses to specific stimuli. Reflexes preserve homeostasis by making rapid adjustments in the function of organs or organ systems. The response shows little variability: Each time a particular reflex is activated, it usually produces the same motor response. Chapter 1 introduced the basic functional components involved in all types of homeostatic regulation: a *receptor*, an *integration center*, and an *effector*. ∞ p. 12 Here we consider *neural reflexes*, in which sensory fibers deliver information from peripheral receptors to an integration center in the CNS, and motor fibers carry motor commands to peripheral effectors. We will examine *endocrine reflexes*, in which the commands to peripheral tissues and organs are delivered by hormones in the bloodstream, in Chapter 18.

The Reflex Arc

The "wiring" of a single reflex is called a **reflex arc**. A reflex arc begins at a receptor and ends at a peripheral effector, such as a muscle fiber or a gland cell. **Figure 13–15** diagrams the five steps in a simple neural reflex known as a *withdrawal reflex*:

Step 1 *The Arrival of a Stimulus and Activation of a Receptor.* A *receptor* is either a specialized cell or the dendrites of a sensory neuron. Receptors are sensitive to physical or chemical changes in the body or to changes in the external environment. The general categories of sensory receptors were introduced in Chapter 12. ∞ p. 391 If you lean on a tack, for example, pain receptors in the palm of your hand are activated. These receptors, the dendrites of sensory neurons, respond to stimuli that cause or accompany tissue damage. (We will discuss the link between receptor stimulation and sensory neuron activation further in Chapter 15.)

Step 2 *The Activation of a Sensory Neuron.* When the dendrites are stretched, there is a graded depolarization that leads to the formation and propagation of action potentials along the axons of the sensory neurons. This information reaches the spinal cord by way of a dorsal root. In our example, STEP 1 and STEP 2 involve the same cell. However, the two steps may involve different cells. For example, reflexes

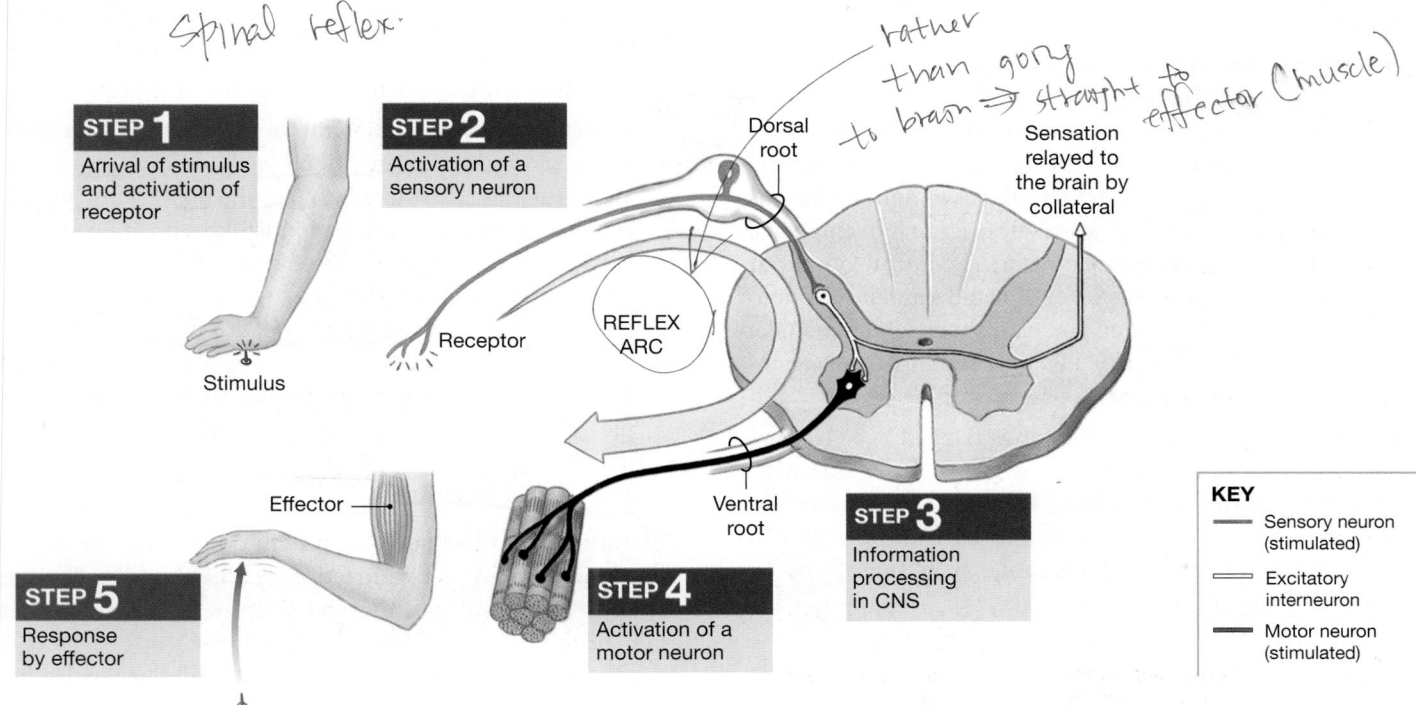

Spinal reflex.

rather than gory to brain ⇒ straight to effector (muscle)

STEP **1** Arrival of stimulus and activation of receptor

STEP **2** Activation of a sensory neuron

Dorsal root

Sensation relayed to the brain by collateral

Receptor

REFLEX ARC

Stimulus

Effector

Ventral root

STEP **3** Information processing in CNS

STEP **5** Response by effector

STEP **4** Activation of a motor neuron

KEY
— Sensory neuron (stimulated)
▭ Excitatory interneuron
— Motor neuron (stimulated)

Figure 13–15   Events in a Neural Reflex. A simple reflex arc, such as the withdrawal reflex, consists of a sensory neuron, an interneuron, and a motor neuron.

triggered by loud sounds begin when receptor cells in the inner ear release neurotransmitters that stimulate sensory neurons.

Step 3 *Information Processing.* In our example, information processing begins when excitatory neurotransmitter molecules, released by the synaptic knob of a sensory neuron, arrive at the postsynaptic membrane of an interneuron. The neurotransmitter produces an excitatory postsynaptic potential (EPSP), which is integrated with other stimuli arriving at the postsynaptic cell at that moment. ∞ p. 419 The information processing is thus performed by the interneuron. In the simplest reflexes, such as the *stretch reflex*, considered in a later section, the sensory neuron innervates a motor neuron directly. In that case, it is the motor neuron that performs the information processing. By contrast, complex reflexes introduced later in the chapter involve several interneurons, some releasing excitatory neurotransmitters (*excitatory interneurons*) and others releasing inhibitory neurotransmitters (*inhibitory interneurons*).

Step 4 *The Activation of a Motor Neuron.* The axons of the stimulated motor neurons carry action potentials into the periphery—in this example, through the ventral root of a spinal nerve.

Step 5 *The Response of a Peripheral Effector.* The release of neurotransmitters by the motor neurons at synaptic knobs then leads to a response by a peripheral effector—in this case, a skeletal muscle whose contraction pulls your hand away from the tack.

A reflex response generally removes or opposes the original stimulus; in this case, the contracting muscle pulls your hand away from a painful stimulus. This reflex arc is therefore an example of *negative feedback*. ∞ p. 14 By opposing potentially harmful changes in the internal or external environment, reflexes play an important role in homeostatic maintenance. The immediate reflex response is typically not the only response to a stimulus. The other responses, which are directed by your brain, involve multiple synapses and take longer to organize and coordinate.

Classification of Reflexes

Reflexes are classified on the basis of (1) their development, (2) the nature of the resulting motor response, (3) the complexity of the neural circuit involved, or (4) the site of information processing. These categories are not mutually exclusive—they represent different ways of describing a single reflex (**Figure 13–16**).

Development of Reflexes

Innate reflexes result from the connections that form between neurons during development. Such reflexes generally appear in a predictable sequence, from the simplest reflex responses (withdrawal from pain) to more complex motor patterns (chewing, suckling, or tracking objects with the eyes). The neural connections responsible for the basic motor patterns of an innate reflex are genetically or developmentally programmed. Examples include the reflexive removal of your

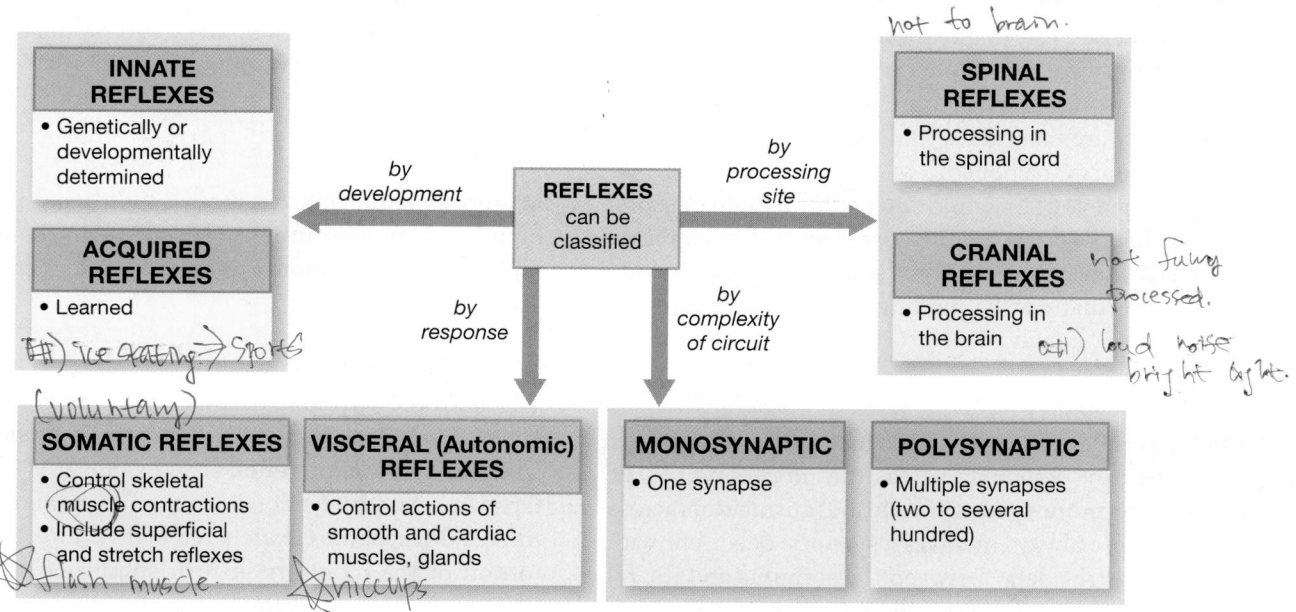

Figure 13–16   The Classification of Reflexes.

hand from a hot stove top and blinking when your eyelashes are touched.

More complex, learned motor patterns are called **acquired reflexes**. An experienced driver steps on the brake when trouble appears ahead; a professional skier must make equally quick adjustments in body position while racing. These motor responses are rapid and automatic, but they were learned rather than preestablished. Such reflexes are enhanced by repetition. The distinction between innate and acquired reflexes is not absolute: Some people can learn motor patterns more quickly than others, and the differences probably have a genetic basis.

Most reflexes, whether innate or acquired, can be modified over time or suppressed through conscious effort. For example, while walking a tightrope over the Grand Canyon, you might ignore a bee sting on your hand, although under other circumstances you would probably withdraw your hand immediately, while shouting and thrashing as well.

Nature of the Response

Somatic reflexes provide a mechanism for the involuntary control of the muscular system. *Superficial reflexes* are triggered by stimuli at the skin or mucous membranes. *Stretch reflexes* are triggered by the sudden elongation of a tendon, and thus of the muscle to which it attaches; a familiar example is the *patellar*, or *"knee-jerk,"* *reflex* that is usually tested during physical exams. These reflexes are also known as *deep tendon reflexes*, or *myotatic reflexes*. **Visceral reflexes**, or *autonomic reflexes*, control the activities of other systems. We will consider somatic reflexes in detail in this chapter and visceral reflexes in Chapter 16.

The movements directed by somatic reflexes are neither delicate nor precise. You might therefore wonder why they exist at all, because we have voluntary control over the same muscles. In fact, somatic reflexes are absolutely vital, primarily because they are *immediate*. Making decisions and coordinating voluntary responses take time, and in an emergency—when you slip while descending a flight of stairs, or press your hand against a knife edge—any delay increases the likelihood of severe injury. Thus, somatic reflexes provide a rapid response that can be modified later, if necessary, by voluntary motor commands.

Complexity of the Circuit

In the simplest reflex arc, a sensory neuron synapses directly on a motor neuron, which serves as the processing center. Such a reflex is a **monosynaptic reflex**. Transmission across a chemical synapse always involves a synaptic delay, but with only one synapse, the delay between the stimulus and the response is minimized. Most reflexes, however, have at least one interneuron between the sensory neuron and the motor neuron, as diagrammed in **Figure 13–15**. Such **polysynaptic**

reflexes have a longer delay between stimulus and response. The length of the delay is proportional to the number of synapses involved. Polysynaptic reflexes can produce far more complicated responses than monosynaptic reflexes, because the interneurons can control motor neurons that activate several muscle groups simultaneously.

Processing Sites

In **spinal reflexes**, the important interconnections and processing events occur in the spinal cord. We will discuss these reflexes further in the next section. Reflexes processed in the brain, called **cranial reflexes**, will be considered in Chapters 14, 16, and 17.

CHECKPOINT

15. Define reflex.
16. What is the minimum number of neurons in a reflex arc?
17. One of the first somatic reflexes to develop is the suckling reflex. Which type of reflex is this?

See the blue Answers tab at the end of the book.

13-7 Spinal reflexes vary in complexity

Spinal reflexes range in complexity from simple monosynaptic reflexes involving a single segment of the spinal cord to polysynaptic reflexes that involve many segments. In the most complicated spinal reflexes, called **intersegmental reflex arcs**, many segments interact to produce a coordinated, highly variable motor response.

Monosynaptic Reflexes

In monosynaptic reflexes, there is little delay between sensory input and motor output. These reflexes control the most rapid, stereotyped motor responses of the nervous system to specific stimuli.

The Stretch Reflex

The best-known monosynaptic reflex is the **stretch reflex**, which provides automatic regulation of skeletal muscle length. The **patellar reflex** is an example. When a physician taps your patellar tendon with a reflex hammer, receptors in the quadriceps muscle are stretched (**Figure 13–17**). The distortion of the receptors in turn stimulates sensory neurons that extend into the spinal cord and synapse on motor neurons that control the motor units in the stretched muscle. This leads to a reflexive contraction of the stretched muscle that ex-

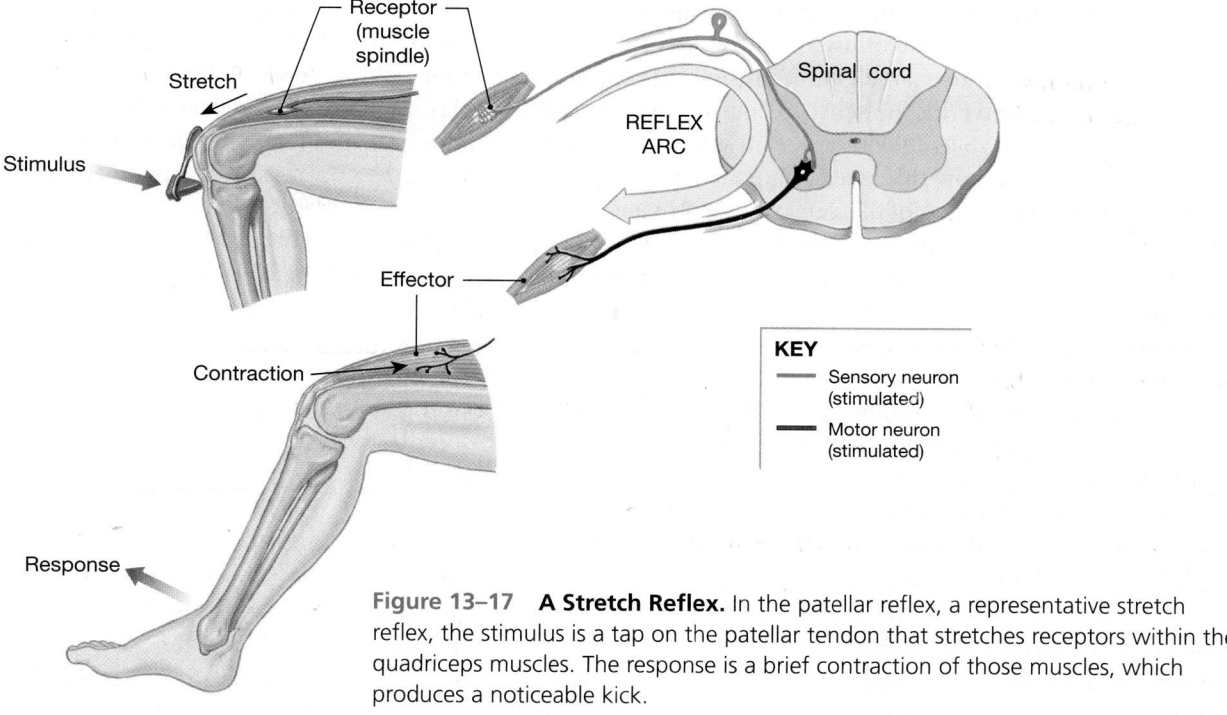

Figure 13–17 A Stretch Reflex. In the patellar reflex, a representative stretch reflex, the stimulus is a tap on the patellar tendon that stretches receptors within the quadriceps muscles. The response is a brief contraction of those muscles, which produces a noticeable kick.

tends the knee in a brief kick. To summarize: The stimulus (increasing muscle length) activates a sensory neuron, which triggers an immediate motor response (contraction of the stretched muscle) that counteracts the stimulus. Because the action potentials traveling toward and away from the spinal cord are conducted along large, myelinated Type A fibers, the entire reflex is completed within 20–40 msec.

The receptors in stretch reflexes are called *muscle spindles.* (The sensory mechanism will be described in the next section.) The stretching of muscle spindles produces a sudden burst of activity in the sensory neurons that monitor them. This in turn leads to stimulation of motor neurons that control the motor units in the stretched muscle. The result is rapid muscle shortening, and this returns the muscle spindles to their resting length. The rate of action potential generation in the sensory neurons then declines, causing a drop in muscle tone to resting levels.

Muscle Spindles

The sensory receptors involved in the stretch reflex are **muscle spindles**. Each consists of a bundle of small, specialized skeletal muscle fibers called **intrafusal muscle fibers** (**Figure 13–18**). The muscle spindle is surrounded by larger **extrafusal muscle fibers**, which are responsible for the resting muscle tone and, at greater levels of stimulation, for the contraction of the entire muscle.

Each intrafusal fiber is innervated by both sensory and motor neurons. The dendrites of the sensory neuron spiral around the intrafusal fiber in a central sensory region. Axons

from spinal motor neurons form neuromuscular junctions on either end of this fiber. Motor neurons innervating intrafusal fibers are called **gamma motor neurons**; their axons are called **gamma efferents**. An intrafusal fiber has one set of myofibrils

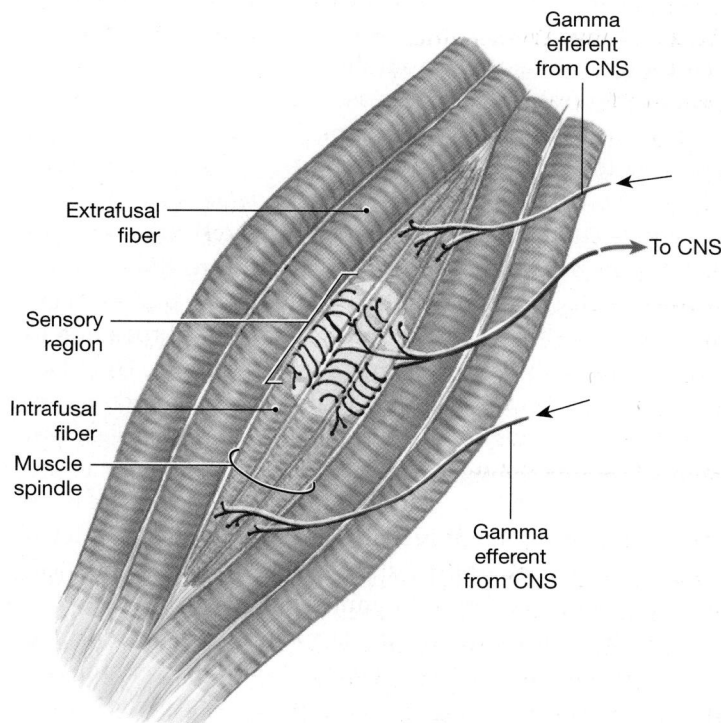

Figure 13–18 A Muscle Spindle. The location, structure, and innervation of a muscle spindle.

at each end. Instead of extending the length of the muscle fiber, as in extrafusal fibers, these myofibrils run from the end of the intrafusal fiber only to the sarcolemma in the central region that is closely monitored by the sensory neuron. The gamma efferents enable the CNS to adjust the sensitivity of the muscle spindle. Before seeing how this is accomplished, we will consider the normal functioning of this sensory receptor and its effects on the surrounding extrafusal fibers.

The sensory neuron is always active, conducting impulses to the CNS. The axon enters the CNS in a dorsal root and synapses on motor neurons in the anterior gray horn of the spinal cord. Collaterals distribute the information to the brain, providing information about the state of the muscle spindle. Stretching the central portion of the intrafusal fiber distorts the dendrites and stimulates the sensory neuron, increasing the frequency of action potential generation. Compressing the central portion inhibits the sensory neuron, decreasing the frequency of action potential generation.

The axon of the sensory neuron synapses on CNS motor neurons that control the extrafusal muscle fibers of the same muscle. An increase in stimulation of the sensory neuron, caused by stretching of the intrafusal fiber, will increase stimulation to the motor neuron controlling the surrounding extrafusal fibers, so muscle tone increases. This increase provides automatic resistance that reduces the chance of muscle damage due to overstretching. The patellar reflex and similar reflexes serve this function. A decrease in the stimulation of the sensory neuron, due to compression of the intrafusal fiber, will lead to a decrease in the stimulation of the motor neuron controlling the surrounding extrafusal fibers, so muscle tone decreases. This decrease reduces resistance to the movement under way. For example, if your elbow is flexed and you let gravity extend it, the triceps brachii muscle, which is compressed by this movement, relaxes.

Many stretch reflexes are **postural reflexes**—reflexes that help us maintain a normal upright posture. Standing, for example, involves a cooperative effort on the part of many muscle groups. Some of these muscles work in opposition to one another, exerting forces that keep the body's weight balanced over the feet. If the body leans forward, stretch receptors in the calf muscles are stimulated. Those muscles then respond by contracting, thereby returning the body to an upright position. If the muscles overcompensate and the body begins to lean back, the calf muscles relax. But then stretch receptors in muscles of the shins and thighs are stimulated, and the problem is corrected immediately.

Postural muscles generally have a firm muscle tone and extremely sensitive stretch receptors. As a result, very fine adjustments are continually being made, and you are not aware of the cycles of contraction and relaxation that occur. Stretch reflexes are only one type of postural reflex; there are many complex polysynaptic postural reflexes.

Now that you understand the basic stretch reflex, we can return to the role of the gamma efferents, which let the CNS adjust the sensitivity of muscle spindles. Gamma efferents play a vital role whenever voluntary contractions change the length of a muscle. Impulses arriving over gamma efferents cause the contraction of myofibrils in the intrafusal fibers as the biceps brachii muscle shortens. The myofibrils pull on the sarcolemma in the central portion of the intrafusal fiber—the region monitored by the sensory neuron—until that membrane is stretched to its normal resting length. As a result, the muscle spindles remain sensitive to any externally imposed changes in muscle length. Thus, if someone drops a ball into your palm when your elbow is partially flexed, the muscle spindles will automatically adjust the muscle tone to compensate for the increased load.

Polysynaptic Reflexes

Polysynaptic reflexes can produce far more complicated responses than can monosynaptic reflexes. One reason is that the interneurons involved can control several muscle groups. Moreover, these interneurons may produce either excitatory or inhibitory postsynaptic potentials (EPSPs or IPSPs) at CNS motor nuclei, so the response can involve the stimulation of some muscles and the inhibition of others.

The Tendon Reflex

The stretch reflex regulates the length of a skeletal muscle. The **tendon reflex** monitors the external tension produced during a muscular contraction and prevents tearing or breaking of the tendons. The sensory receptors for this reflex have not been identified, but they are distinct from both muscle spindles and proprioceptors in tendons. The receptors are stimulated when the collagen fibers are stretched to a dangerous degree. These receptors activate sensory neurons that stimulate inhibitory interneurons in the spinal cord. These interneurons in turn innervate the motor neurons controlling the skeletal muscle. The greater the tension in the tendon, the greater is the inhibitory effect on the motor neurons. As a result, a skeletal muscle generally cannot develop enough tension to break its tendons.

Withdrawal Reflexes

Withdrawal reflexes move affected parts of the body away from a stimulus. The strongest withdrawal reflexes are triggered by painful stimuli, but these reflexes are sometimes initiated by the stimulation of touch receptors or pressure receptors.

The **flexor reflex**, a representative withdrawal reflex, affects the muscles of a limb (**Figure 13–19**). Recall from Chapter 9 and Chapter 11 that flexion is a reduction in the angle between two articulating bones, and that the contrac-

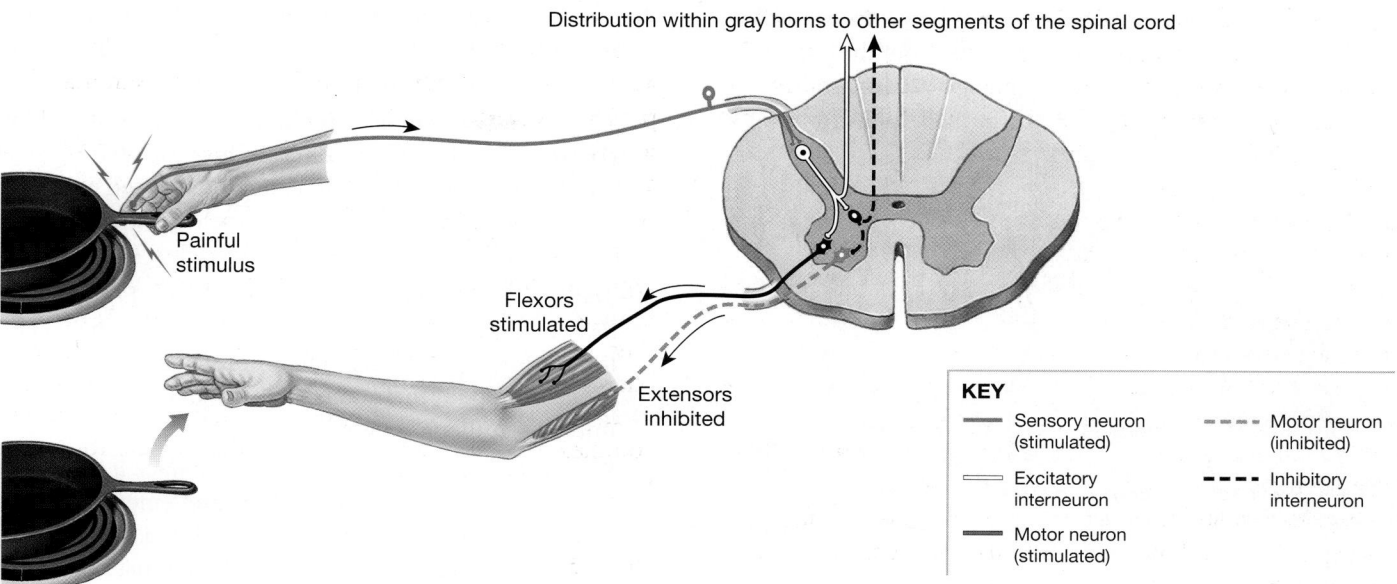

Distribution within gray horns to other segments of the spinal cord

Painful stimulus

Flexors stimulated

Extensors inhibited

KEY

— Sensory neuron (stimulated)

– – – – Motor neuron (inhibited)

═══ Excitatory interneuron

■■■■ Inhibitory interneuron

— Motor neuron (stimulated)

Figure 13–19　A Flexor Reflex. The withdrawal reflex is an example of a flexor reflex. In this example, the stimulus is the pain experienced when grabbing a hot frying pan. The response, contraction of the flexor muscles of the arm, yanks the forearm and hand away from the pan; the movement is sudden and powerful enough that the pan is released. This response occurs while pain sensations are ascending to the brain within the lateral column, as indicated in Figure 13–15.

tions of flexor muscles perform this movement. ∞ pp. 273, 340 If you grab an unexpectedly hot pan on the stove, a dramatic flexor reflex will occur. When the pain receptors in your hand are stimulated, the sensory neurons activate interneurons in the spinal cord that stimulate motor neurons in the anterior gray horns. The result is a contraction of flexor muscles that yanks your hand away from the stove.

When a specific muscle contracts, opposing muscles must relax to permit the movement. For example, flexor muscles that bend the elbow (such as the biceps brachii muscle) are opposed by extensor muscles (such as the triceps brachii muscle) that straighten it out. A potential conflict exists: In theory, the contraction of a flexor muscle should trigger a stretch reflex in the extensors that would cause them to contract, opposing the movement. Interneurons in the spinal cord prevent such competition through **reciprocal inhibition**. When one set of motor neurons is stimulated, those neurons that control antagonistic muscles are inhibited. The term *reciprocal* refers to the fact that the system works both ways: When the flexors contract, the extensors relax; when the extensors contract, the flexors relax.

Withdrawal reflexes are much more complex than any monosynaptic reflex. They also show tremendous versatility, because the sensory neurons activate many pools of interneurons. If the stimuli are strong, interneurons will carry excitatory and inhibitory impulses up and down the spinal cord, affecting motor neurons in many segments. The end result is always the same: a coordinated movement away from the stimulus. But the

distribution of the effects and the strength and character of the motor responses depend on the intensity and location of the stimulus. Mild discomfort might provoke a brief contraction in muscles of your hand and wrist. More powerful stimuli would produce coordinated muscular contractions affecting the positions of your hand, wrist, forearm, and arm. Severe pain would also stimulate contractions of your shoulder, trunk, and arm muscles. These contractions could persist for several seconds, owing to the activation of reverberating circuits. In contrast, monosynaptic reflexes are relatively invariable and brief; the patellar reflex is completed in roughly 20 msec.

Crossed Extensor Reflexes

The stretch, tendon, and withdrawal reflexes involve *ipsilateral* (*ipsi*, same + *lateral*, side) *reflex arcs*: The sensory stimulus and the motor response occur on the same side of the body. The **crossed extensor reflex** (**Figure 13–20**) involves a *contralateral reflex arc* (*contra*, opposite), because the motor response occurs on the side opposite the stimulus.

The crossed extensor reflex complements the flexor reflex, and the two occur simultaneously. When you step on a tack, while the flexor reflex pulls the affected foot away from the ground, the crossed extensor reflex straightens the other leg to support your body weight. In the crossed extensor reflex, the axons of interneurons responding to the pain cross to the other side of the spinal cord and stimulate motor neurons that control the extensor muscles of the uninjured leg. As a result, your opposite leg straightens to support the shifting

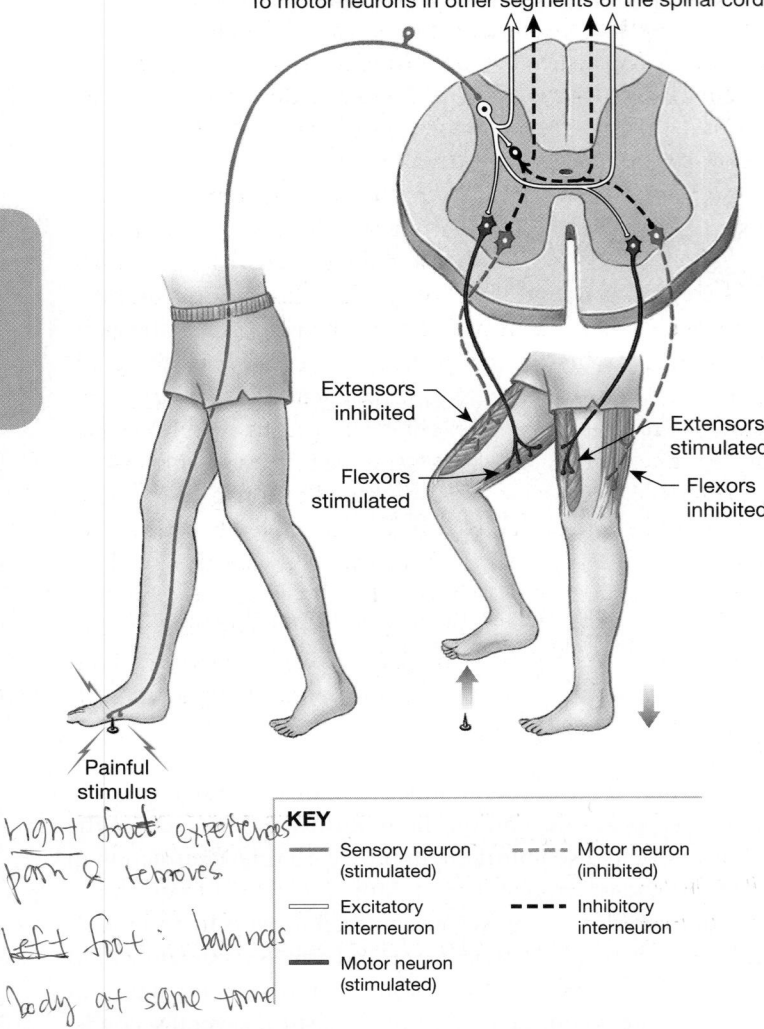

To motor neurons in other segments of the spinal cord

Extensors inhibited

Flexors stimulated

Extensors stimulated

Flexors inhibited

Painful stimulus

[handwritten:] right foot experiences pain & removes

[handwritten:] left foot : balances body at same time

KEY

—— Sensory neuron (stimulated)

– – – Motor neuron (inhibited)

☐ Excitatory interneuron

▬ ▬ Inhibitory interneuron

▬▬ Motor neuron (stimulated)

Figure 13–20   The Crossed Extensor Reflex. Pathways for sensations ascending to the brain are not shown.

weight. Reverberating circuits use positive feedback to ensure that the movement lasts long enough to be effective—all without motor commands from higher centers of the brain.

General Characteristics of Polysynaptic Reflexes

Polysynaptic reflexes range in complexity from a simple tendon reflex to the complex and variable reflexes associated with standing, walking, and running. Yet all polysynaptic reflexes share the following basic characteristics:

1. *They Involve Pools of Interneurons.* Processing occurs in pools of interneurons before motor neurons are activated. The result may be excitation or inhibition; the tendon reflex produces inhibition of motor neurons, whereas the flexor and crossed extensor reflexes direct specific muscle contractions.

2. *They Are Intersegmental in Distribution.* The interneuron pools extend across spinal segments and may activate muscle groups in many parts of the body.

3. *They Involve Reciprocal Inhibition.* Reciprocal inhibition coordinates muscular contractions and reduces resistance to movement. In the flexor and crossed extensor reflexes, the contraction of one muscle group is associated with the inhibition of opposing muscles.

4. *They Have Reverberating Circuits, Which Prolong the Reflexive Motor Response.* Positive feedback between interneurons that innervate motor neurons and the processing pool maintains the stimulation even after the initial stimulus has faded.

5. *Several Reflexes May Cooperate to Produce a Coordinated, Controlled Response.* As a reflex movement gets under way, antagonistic reflexes are inhibited. For example, during the stretch reflex, antagonistic muscles are inhibited; in the tendon reflex, antagonistic muscles are stimulated. In complex polysynaptic reflexes, commands may be distributed along the length of the spinal cord, producing a well-coordinated response.

CHECKPOINT

18. Identify the basic characteristics of polysynaptic reflexes.

19. For the patellar (knee-jerk) reflex, how would the stimulation of the muscle spindle by gamma motor neurons affect the speed of the reflex?

20. A weight lifter is straining to lift a 200-kg barbell above his head. Shortly after he lifts it to chest height, his muscles appear to relax and he drops the barbell. Which reflex has occurred?

21. During a withdrawal reflex of the foot, what happens to the limb on the side opposite the stimulus? What is this response called?

See the blue Answers tab at the end of the book.

13-8 The brain can affect spinal cord–based reflexes

Reflex motor behaviors occur automatically, without instructions from higher centers. However, higher centers can have a profound effect on the performance of a reflex. Processing centers in the brain can facilitate or inhibit reflex motor patterns based in the spinal cord. Descending tracts originating in the brain synapse on interneurons and motor neurons throughout the spinal cord. These synapses are continuously active, producing EPSPs or IPSPs at the postsynaptic membrane.

Voluntary Movements and Reflex Motor Patterns

Spinal reflexes produce consistent, stereotyped motor patterns that are triggered by specific external stimuli. However, the same motor patterns can also be activated as needed by centers in the brain. By making use of these preexisting patterns, relatively few descending fibers can control complex motor functions. For example, the motor patterns for walking, running, and jumping are directed primarily by neuronal pools in the spinal cord. The descending pathways from the brain provide appropriate facilitation, inhibition, or "fine-tuning" of the established patterns. This is a very efficient system that is similar to an order given in a military drill: A single command triggers a complex, predetermined sequence of events.

Motor control therefore involves a series of interacting levels. At the lowest level are monosynaptic reflexes that are rapid, but stereotyped and relatively inflexible. At the highest level are centers in the brain that can modulate or build on reflexive motor patterns.

The A&P Top 100

#47 Reflexes are rapid, automatic responses to stimuli that "buy time" for the planning and execution of more complex responses that are often consciously directed. The fastest reflexes are somatic motor reflexes.

Reinforcement and Inhibition

A single EPSP may not depolarize the postsynaptic neuron sufficiently to generate an action potential, but it does make that neuron more sensitive to other excitatory stimuli. This process of *facilitation* was introduced in Chapter 12. Alternatively, an IPSP will make the neuron less responsive to excitatory stimulation, through the process of *inhibition*. ∞ p. 414 By stimulating excitatory or inhibitory interneurons within the brain stem or spinal cord, higher centers can adjust the sensitivity of reflexes by creating EPSPs or IPSPs at the motor neurons involved in reflex responses.

When many of the excitatory synapses are chronically active, the postsynaptic neuron can enter a state of generalized facilitation. This facilitation of reflexes can result in **reinforcement**, an enhancement of spinal reflexes. For example, a voluntary effort to pull apart clasped hands elevates the general state of facilitation along the spinal cord, reinforcing all spinal reflexes. If a stimulus fails to elicit a particular reflex response during a clinical exam, there can be many reasons for the failure: The person may be consciously suppressing the response, the nerves involved may be damaged, or there may be underlying problems inside the CNS. The clinician may then ask the patient to perform an action designed to provide reinforcement. Reinforced reflexes are usually too strong to suppress consciously; if the reflex still fails to appear, the likelihood of nerve or CNS damage is increased, and more sophisticated tests, such as nerve conduction studies or scans, may be ordered.

Tips & Tricks

Facilitation and inhibition are similar to what happens when a symphony conductor raises or lowers one hand to control the music's volume while keeping the rhythm going with the baton hand: The basic pattern of beats doesn't change, but the loudness does.

Other descending fibers have an inhibitory effect on spinal reflexes. In adults, stroking the sole of the foot produces a curling of the toes, called a **plantar reflex**, or *negative Babinski reflex*, after about a 1-second delay (**Figure 13–21a**).

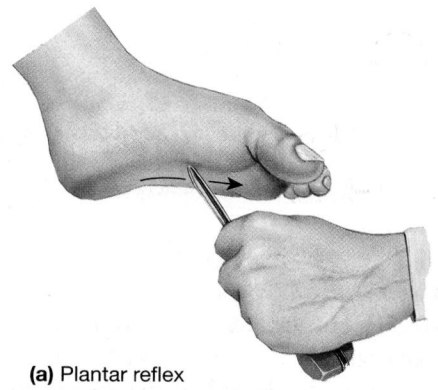

(a) Plantar reflex

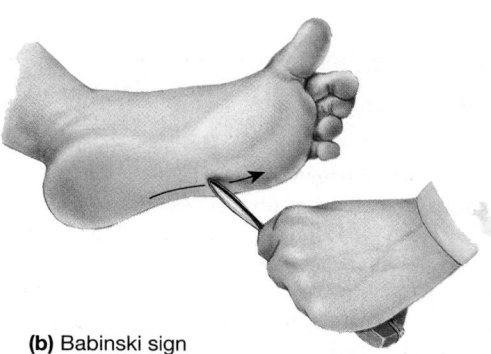

(b) Babinski sign

Figure 13–21   The Babinski Reflexes. (a) The plantar reflex (negative Babinski reflex), a curling of the toes, is seen in healthy adults. **(b)** The Babinski sign (positive Babinski reflex) occurs in the absence of descending inhibition. It is normal in infants, but pathological in adults.

Stroking an infant's foot on the side of the sole produces a fanning of the toes known as the **Babinski sign**, or *positive Babinski reflex*. This response disappears as descending motor pathways develop. If either the higher centers or the descending tracts are damaged, the Babinski sign will reappear in an adult (**Figure 13–21b**). As a result, this reflex is often tested if CNS injury is suspected.

CHECKPOINT

22. Define reinforcement as it pertains to spinal reflexes.

23. After injuring her back, Tina exhibits a positive Babinski reflex. What does this imply about Tina's injury?

See the blue Answers tab at the end of the book.

Related Clinical Terms

Babinski sign (*positive Babinski reflex*)**:** A spinal reflex in infants, consisting of a fanning of the toes in response to stroking the sole of the foot; in adults, a Babinski sign indicates CNS injury.

epidural block: The injection of anesthetic into the epidural space to eliminate sensory and motor innervation via spinal nerves in the area of injection.

lumbar puncture: A spinal tap performed between adjacent lumbar vertebrae inferior to the conus medullaris.

meningitis: An inflammation of the meninges involving either the spinal cord (*spinal meningitis*) or the brain (*cerebral meningitis*); generally caused by bacterial or viral pathogens.

myelography: A diagnostic procedure in which a radiopaque dye is introduced into the cerebrospinal fluid to obtain an x-ray image of the spinal cord and cauda equina.

nerve growth factor: A peptide factor that promotes the growth and maintenance of neurons. Other factors that are important to neuron growth and repair include BDNF, NT-3, NT-4, and GAP-43.

nerve palsies (peripheral neuropathies): Regional losses of sensory and motor function as a result of nerve trauma, disease, or compression. Common palsies include *radial nerve palsy, ulnar palsy, sciatica,* and *fibular palsy*.

paraplegia: Paralysis involving a loss of motor control of the lower, but not the upper, limbs.

patellar reflex (knee-jerk reflex): A stretch reflex resulting from the stimulation of stretch receptors in the quadriceps muscles.

plantar reflex (*negative Babinski reflex*)**:** A spinal reflex in adults, consisting of a curling of the toes in response to stroking the side of the foot sole.

quadriplegia: Paralysis involving the loss of sensation and motor control of the upper and lower limbs.

shingles: A condition caused by the infection of neurons in dorsal root ganglia by the varicella-zoster virus. The primary sign is a painful rash along the sensory distribution of the affected spinal nerves.

spinal tap: A procedure in which cerebrospinal fluid is removed from the subarachnoid space through a needle, generally inserted between the lumbar vertebrae.

Chapter Review

Study Outline

13-1 The brain and spinal cord make up the central nervous system, and the cranial nerves and spinal nerves constitute the peripheral nervous system p. 430

1. The CNS consists of the brain and spinal cord; the remainder of the nervous tissue forms the PNS. (*Figure 13–1*)

13-2 The spinal cord is surrounded by three meninges and conveys sensory and motor information p. 431

2. The adult spinal cord includes two localized **enlargements**, which provide innervation to the limbs. The spinal cord has 31 segments, each associated with a pair of **dorsal roots** and a pair of **ventral roots**. (*Figure 13–2*)

3. The **filum terminale** (a strand of fibrous tissue), which originates at the **conus medullaris**, ultimately becomes part of the **coccygeal ligament**. (*Figure 13–2*)

4. **Spinal nerves** are **mixed nerves**: They contain both afferent (sensory) and efferent (motor) fibers.

5. The **spinal meninges** provide physical stability and shock absorption for neural tissues of the spinal cord; the **cranial meninges** surround the brain. (*Figure 13–3*)

6. The **dura mater** covers the spinal cord; inferiorly, it tapers into the **coccygeal ligament**. The **epidural space** separates the dura mater from the walls of the vertebral canal. (*Figures 13–3, 13–4*)

7. Interior to the inner surface of the dura mater are the **subdural space**, the **arachnoid mater** (the second meningeal layer), and the **subarachnoid space**. The subarachnoid space contains **cerebrospinal fluid (CSF)**, which acts as a shock absorber and a diffusion medium for dissolved gases, nutrients, chemical messengers, and waste products. (*Figures 13–3, 13–4*)

8. The **pia mater**, a meshwork of elastic and collagen fibers, is the innermost meningeal layer. **Denticulate ligaments** extend from the pia mater to the dura mater. (*Figures 13–3, 13–4*)

13-3 Gray matter is the region of integration and command initiation, and white matter carries information from place to place p. 435

9. The white matter of the spinal cord contains myelinated and unmyelinated axons, whereas the gray matter contains cell bodies of neurons and neuroglia and unmyelinated axons. The

projections of gray matter toward the outer surface of the cord are called **horns**. *(Figure 13–5)*

10. The **posterior gray horns** contain somatic and visceral sensory nuclei; nuclei in the **anterior gray horns** function in somatic motor control. The **lateral gray horns** contain visceral motor neurons. The **gray commissures** contain axons that cross from one side of the spinal cord to the other. *(Figure 13–5)*

11. The white matter can be divided into six **columns** *(funiculi)*, each of which contains **tracts** *(fasciculi)*. **Ascending tracts** relay information from the spinal cord to the brain, and **descending tracts** carry information from the brain to the spinal cord. *(Figure 13–5)*

13-4 Spinal nerves form plexuses that are named according to their level of emergence from the vertebral canal p. 437

12. There are 31 pairs of spinal nerves. Each has an **epineurium** (outermost layer), a **perineurium**, and an **endoneurium** (innermost layer). *(Figure 13–6)*

13. A typical spinal nerve has a **white ramus** (containing myelinated axons), a **gray ramus** (containing unmyelinated fibers that innervate glands and smooth muscles in the body wall or limbs), a **dorsal ramus** (providing sensory and motor innervation to the skin and muscles of the back), and a **ventral ramus** (supplying the ventrolateral body surface, structures in the body wall, and the limbs). Each pair of nerves monitors a region of the body surface called a **dermatome**. *(Figures 13–7, 13–8, 13–9)*

14. A complex, interwoven network of nerves is a **nerve plexus**. The four large plexuses are the **cervical plexus**, the **brachial plexus**, the **lumbar plexus**, and the **sacral plexus**. *(Figures 13–10 to 13–13; Tables 13–1 to 13–3)*

13-5 Neuronal pools are functional groups of interconnected neurons p. 446

15. The body has sensory neurons, which deliver information to the CNS; motor neurons, which distribute commands to peripheral effectors; and interneurons, which interpret information and coordinate responses.

16. A functional group of interconnected neurons is a **neuronal pool**.

17. The neural circuit patterns are **divergence**, **convergence**, **serial processing**, **parallel processing**, and **reverberation**. *(Figure 13–14)*

13-6 Reflexes are rapid, automatic responses to stimuli p. 448

18. A *neural reflex* involves sensory fibers delivering information to the CNS, and motor fibers carrying commands to the effectors via the PNS.

19. A **reflex arc** is the neural "wiring" of a single reflex. *(Figure 13–15)*

20. The five steps involved in a neural reflex are (1) the arrival of a stimulus and activation of a receptor, (2) the activation of a sensory neuron, (3) information processing in the CNS, (4) the activation of a motor neuron, and (5) a response by an effector. *(Figure 13–15)*

21. Reflexes are classified according to (1) their development, (2) the nature of the resulting motor response, (3) the complexity of the neural circuit involved, and (4) the site of information processing. *(Figure 13–16)*

22. **Innate reflexes** result from the connections that form between neurons during development. **Acquired reflexes** are learned and typically are more complex.

23. **Somatic reflexes** control skeletal muscles; **visceral reflexes** *(autonomic reflexes)* control the activities of other systems.

24. In a **monosynaptic reflex**—the simplest reflex arc—a sensory neuron synapses directly on a motor neuron, which acts as the processing center. In a **polysynaptic reflex**, which has at least one interneuron between the sensory afferent and the motor efferent, there is a longer delay between stimulus and response.

25. Reflexes processed in the brain are **cranial reflexes**. In a **spinal reflex**, the important interconnections and processing events occur in the spinal cord.

13-7 Spinal reflexes vary in complexity p. 450

26. Spinal reflexes range from simple monosynaptic reflexes to more complex polysynaptic and **intersegmental reflexes**, in which many segments interact to produce a coordinated motor response.

27. The **stretch reflex** (such as the **patellar**, or **knee-jerk**, **reflex**) is a monosynaptic reflex that automatically regulates skeletal muscle length and muscle tone. The sensory receptors involved are **muscle spindles**. *(Figures 13–17, 13–18)*

28. A **postural reflex** maintains one's normal upright posture.

29. Polysynaptic reflexes can produce more complicated responses than can monosynaptic reflexes. Examples include the **tendon reflex** (which monitors the tension produced during muscular contractions and prevents damage to tendons) and **withdrawal reflexes** (which move affected portions of the body away from a source of stimulation). The **flexor reflex** is a withdrawal reflex affecting the muscles of a limb. The **crossed extensor reflex** complements withdrawal reflexes. *(Figures 13–19, 13–20)*

30. All polysynaptic reflexes (1) involve pools of interneurons, (2) are intersegmental in distribution, (3) involve reciprocal inhibition, and (4) have reverberating circuits, which prolong the reflexive motor response. Several reflexes may cooperate to produce a coordinated response.

13-8 The brain can affect spinal cord–based reflexes p. 454

31. The brain can facilitate or inhibit reflex motor patterns based in the spinal cord.

32. Motor control involves a series of interacting levels. Monosynaptic reflexes form the lowest level; at the highest level are the centers in the brain that can modulate or build on reflexive motor patterns.

33. Facilitation can produce an enhancement of spinal reflexes known as **reinforcement**. Spinal reflexes may also be inhibited, as when the **plantar reflex** in adults replaces the **Babinski sign** in infants. *(Figure 13–21)*

Review Questions

See the blue Answers tab at the end of the book.

 Access more review material online at **myA&P**™ (www.myaandp.com). There, you'll find chapter guides, chapter quizzes, practice tests, animations, flashcards, a glossary with pronunciations, *Interactive Physiology*® (IP) exercises and quizzes, and more to help you succeed in the course.

LEVEL 1 Reviewing Facts and Terms

1. Label the anatomical structures of the spinal cord in the following figure.

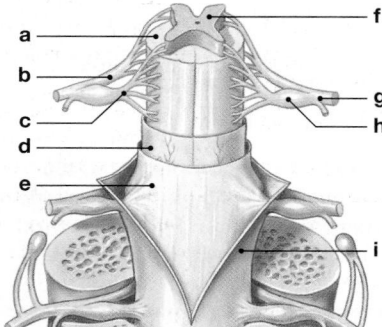

(a) white matter   (b) ventral root
(c) dorsal root   (d) ~~spinal cord~~ pia mate
(e) arachnoid matter (f) gray matter
(g) spinal nerve   (h) dorsal root ganglion
(i) dura mater

2. The ventral roots of each spinal segment
   (a) bring sensory information into the spinal cord.   dorsal : afferent
   (b) control peripheral effectors.   ventral : efferent
   (c) contain the axons of somatic motor and visceral motor neurons.
   (d) do both b and c.
3. Spinal nerves are called mixed nerves because they
   (a) contain sensory and motor fibers.
   (b) exit at intervertebral foramina.
   (c) are associated with a pair of dorsal root ganglia.
   (d) are associated with dorsal and ventral roots.
4. The adult spinal cord extends only to
   (a) the coccyx.
   (b) the sacrum.
   (c) the third or fourth lumbar vertebra.
   (d) the first or second lumbar vertebra.
   (e) the last thoracic vertebra.
5. Which of the following statements is *false* concerning the gray matter of the spinal cord?
   (a) It is located in the interior of the spinal cord around the central canal.
   (b) It functions in processing neural information.
   (c) It is primarily involved in relaying information to the brain.
   (d) It contains motor neurons.
   (e) It is divided into regions called horns.

6. The following are the steps involved in a reflex arc.
   1. activation of a sensory neuron
   2. activation of a motor neuron
   3. response by an effector
   4. arrival of a stimulus and activation of a receptor
   5. information processing
   The proper sequence of these steps is
   (a) 1, 3, 4, 5, 2.   (b) 4, 5, 3, 1, 2.
   (c) 4, 1, 5, 2, 3.   (d) 4, 3, 1, 5, 2.
   (e) 3, 1, 4, 5, 2.
7. A sensory region monitored by the dorsal rami of a single spinal segment is
   (a) a ganglion.   (b) a fascicle.
   (c) a dermatome.   (d) a ramus.
8. The major nerve of the cervical plexus that innervates the diaphragm is the
   (a) median nerve.   (b) axillary nerve.
   (c) phrenic nerve.   (d) fibular nerve.
9. The genitofemoral, femoral, and lateral femoral cutaneous nerves are major nerves of the
   (a) lumbar plexus.   (b) sacral plexus.
   (c) brachial plexus.   (d) cervical plexus.
10. The synapsing of several neurons on the same postsynaptic neuron is called
   (a) serial processing.   (b) reverberation.
   (c) divergence.   (d) convergence.
11. The reflexes that control the most rapid, stereotyped motor responses to stimuli are
   (a) monosynaptic reflexes. (b) polysynaptic reflexes.
   (c) tendon reflexes.   (d) extensor reflexes.
12. An example of a stretch reflex triggered by passive muscle movement is the
   (a) tendon reflex.   (b) patellar reflex.
   (c) flexor reflex.   (d) ipsilateral reflex.
13. The contraction of flexor muscles and the relaxation of extensor muscles illustrates the principle of
   (a) reverberating circuitry.
   (b) generalized facilitation.
   (c) reciprocal inhibition.
   (d) reinforcement.
14. Reflex arcs in which the sensory stimulus and the motor response occur on the same side of the body are
   (a) contralateral.   (b) paraesthetic.
   (c) ipsilateral.   (d) monosynaptic.
15. Proceeding deep from the most superficial layer, number the following in the correct sequence:
   (a) _____ walls of vertebral canal
   (b) _____ pia mater
   (c) _____ dura mater
   (d) _____ arachnoid membrane
   (e) _____ subdural space
   (f) _____ subarachnoid space
   (g) _____ epidural space
   (h) _____ spinal cord

LEVEL 2 Reviewing Concepts

16. Explain the anatomical significance of the fact that spinal cord growth ceases at age 4.
17. List, in sequence, the five steps involved in a neural reflex.
18. Polysynaptic reflexes can produce far more complicated responses than can monosynaptic reflexes because
    (a) the response time is quicker.
    (b) the response is initiated by highly sensitive receptors.
    (c) motor neurons carry impulses at a faster rate than do sensory neurons.
    (d) the interneurons involved can control several muscle groups.
19. Why do cervical nerves outnumber cervical vertebrae?
20. If the anterior gray horns of the spinal cord were damaged, what type of control would be affected?
21. List all of the CNS sites where cerebrospinal fluid (CSF) is located. What are the functions of CSF?
22. What five characteristics are common to all polysynaptic reflexes?
23. Predict the effects on the body of a spinal cord transection at C_7. How would these effects differ from those of a spinal cord transection at T_{10}?
24. The subarachnoid space contains
    (a) cerebrospinal fluid.
    (b) lymph.
    (c) air.
    (d) connective tissue and blood vessels.
    (e) denticulate ligaments.
25. Side-to-side movements of the spinal cord are prevented by the
    (a) filum terminale.   (b) denticulate ligaments.
    (c) dura mater.   (d) pia mater.
    (e) arachnoid mater.
26. Ascending tracts
    (a) carry sensory information to the brain.
    (b) carry motor information to the brain.
    (c) carry sensory information from the brain.
    (d) carry motor information from the brain.
    (e) connect perceptive areas with the brain.

LEVEL 3 Critical Thinking and Clinical Applications

27. What effect does the stimulation of a sensory neuron that innervates an intrafusal muscle fiber have on muscle tone?

28. Mary complains that when she wakes up in the morning, her thumb and forefinger are always "asleep." She mentions this condition to her physician, who asks Mary whether she sleeps with her wrists flexed. She replies that she does. The physician tells Mary that sleeping in that position may compress a portion of one of her peripheral nerves, producing her symptoms. Which nerve is involved?
29. The improper use of crutches can produce a condition known as "crutch paralysis," characterized by a lack of response by the extensor muscles of the arm, and a condition known as "wrist drop," consisting of an inability to extend the fingers and wrist. Which nerve is involved?
30. Bowel and bladder control involve spinal reflex arcs that are located in the sacral region of the spinal cord. In both instances two sphincter muscles, an inner sphincter of smooth muscle and an outer sphincter of skeletal muscle, control the passageway out of the body. How would a transection of the spinal cord at the L_1 level affect an individual's bowel and bladder control?
31. Karen falls down a flight of stairs and suffers spinal cord damage due to hyperextension of the cord during the fall. The injury results in edema of the central cord with resulting compression of the anterior horn cells of the lumbar region. What signs would you expect to observe as a result of this injury?

14

The Brain and Cranial Nerves

Did you know...?
A newborn baby's brain will triple in size by its first birthday.

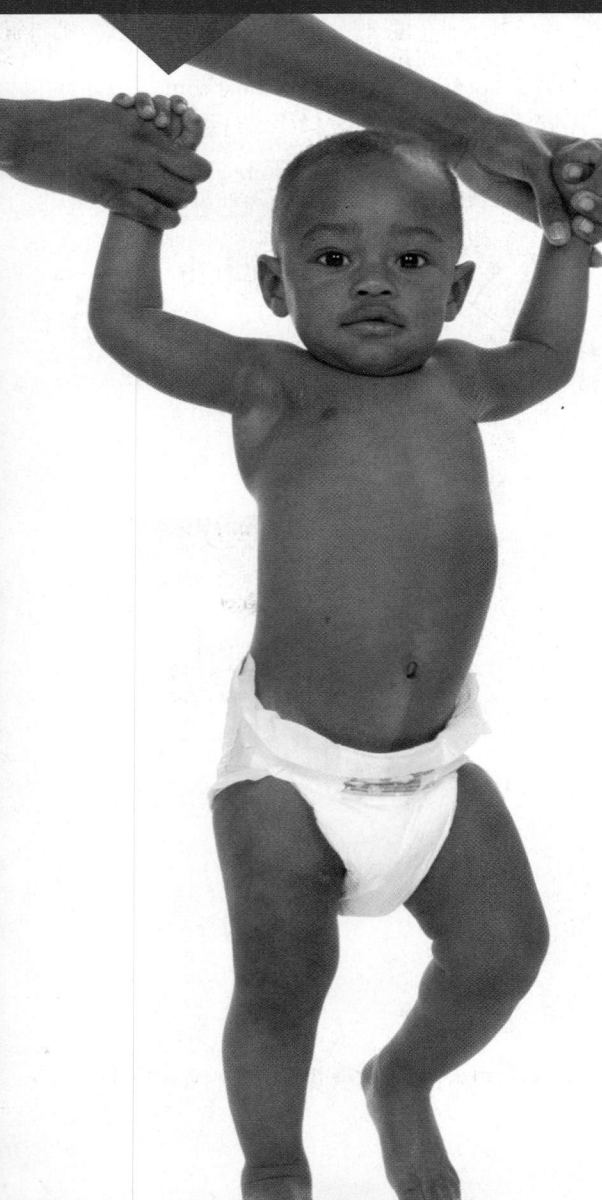

Learning Outcomes

After completing this chapter, you should be able to do the following:

14-1 Name the major brain regions, vesicles, and ventricles, and describe the locations and functions of each.

14-2 Explain how the brain is protected and supported, and discuss the formation, circulation, and function of cerebrospinal fluid.

14-3 Describe the anatomical differences between the medulla oblongata and the spinal cord, and identify the medulla oblongata's main components and functions.

14-4 List the main components of the pons, and specify the functions of each.

14-5 List the main components of the cerebellum, and specify the functions of each.

14-6 List the main components of the mesencephalon, and specify the functions of each.

14-7 List the main components of the diencephalon, and specify the functions of each.

14-8 Identify the main components of the limbic system, and specify the locations and functions of each.

14-9 Identify the major anatomical subdivisions and functions of the cerebrum, and discuss the origin and significance of the major types of brain waves seen in an electroencephalogram.

14-10 Describe representative examples of cranial reflexes that produce somatic responses or visceral responses to specific stimuli.

Clinical Notes

Epidural and Subdural Hemorrhages p. 466
Aphasia and Dyslexia p. 488
Disconnection Syndrome p. 488

An Introduction to the Brain and Cranial Nerves

This chapter introduces the functional organization of the brain and cranial nerves, and describes simple cranial reflexes. The adult human brain contains almost 97 percent of the body's neural tissue. A "typical" brain weighs 1.4 kg (3 lb) and has a volume of 1200 cc (71 in.^{3}). Brain size varies considerably among individuals. The brains of males are, on average, about 10 percent larger than those of females, owing to differences in average body size. No correlation exists between brain size and intelligence. Individuals with the smallest brains (750 cc) and the largest brains (2100 cc) are functionally normal.

14-1 The brain has several principal structures, each with specific functions

In this section we introduce the anatomical organization of the brain. We begin with an overview of the brain's major regions and landmarks; then we discuss the brain's embryological origins and some prominent internal cavities: the ventricles of the brain.

Major Brain Regions and Landmarks

The adult brain is dominated in size by the cerebrum (**Figure 14–1**). Viewed from the anterior and superior

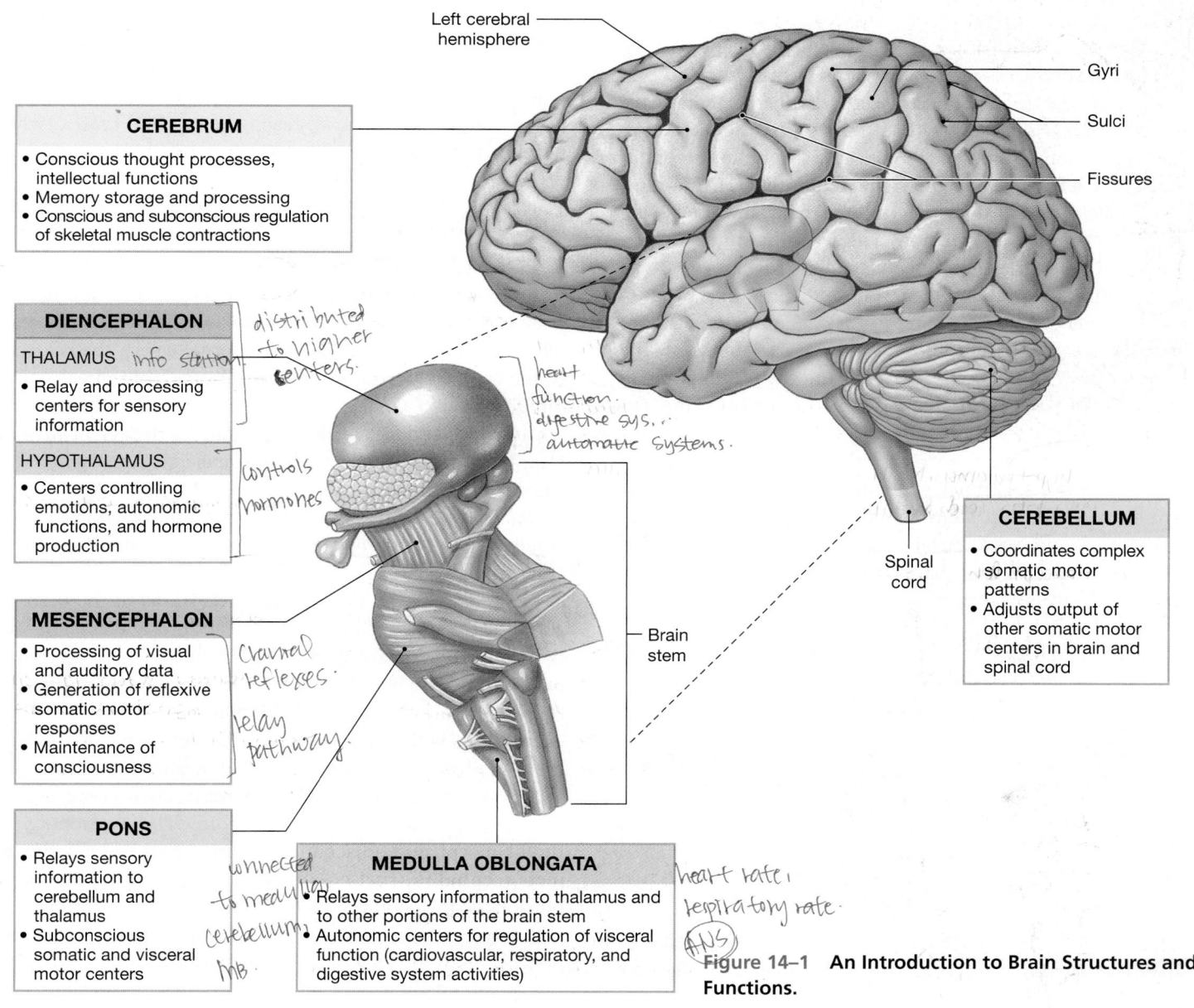

Left cerebral hemisphere

Gyri

Sulci

Fissures

CEREBRUM
- Conscious thought processes, intellectual functions
- Memory storage and processing
- Conscious and subconscious regulation of skeletal muscle contractions

DIENCEPHALON

THALAMUS   *info station*
- Relay and processing centers for sensory information

distributed to higher centers.

HYPOTHALAMUS
- Centers controlling emotions, autonomic functions, and hormone production

controls hormones

MESENCEPHALON
- Processing of visual and auditory data
- Generation of reflexive somatic motor responses
- Maintenance of consciousness

cranial reflexes.

relay pathway

PONS
- Relays sensory information to cerebellum and thalamus
- Subconscious somatic and visceral motor centers

connected to medulla, cerebellum, mb

MEDULLA OBLONGATA
- Relays sensory information to thalamus and to other portions of the brain stem
- Autonomic centers for regulation of visceral function (cardiovascular, respiratory, and digestive system activities)

heart rate, respiratory rate. (ANS)

heart function. digestive sys. automatic systems.

CEREBELLUM
- Coordinates complex somatic motor patterns
- Adjusts output of other somatic motor centers in brain and spinal cord

Spinal cord

Brain stem

Figure 14–1   An Introduction to Brain Structures and Functions.

surfaces, the **cerebrum** (se-RĒ-brum or SER-e-brum) of the adult brain can be divided into large, paired **cerebral hemispheres**. The surfaces of the cerebral hemispheres are highly folded and covered by **neural cortex** (*cortex*, rind or bark), a superficial layer of gray matter. This *cerebral cortex* forms a series of elevated ridges, or **gyri** (JĪ-rī singular, *gyrus*) that serve to increase its surface area. The gyri are separated by shallow depressions called **sulci** (SUL-sī) or by deeper grooves called **fissures**. The cerebrum is the seat of most higher mental functions. Conscious thoughts, sensations, intellect, memory, and complex movements all originate in the cerebrum.

The **cerebellum** (ser-e-BEL-um) is partially hidden by the cerebral hemispheres, but it is the second-largest part of the brain. Like the cerebrum, the cerebellum has hemispheres that are covered by a layer of gray matter, the *cerebellar cortex*. The cerebellum adjusts ongoing movements by comparing arriving sensations with previously experienced sensations, allowing you to perform the same movements over and over.

The other major anatomical regions of the brain can best be examined after the cerebral and cerebellar hemispheres have been removed (**Figure 14–1**). The walls of the **diencephalon** (dī-en-SEF-a-lon; *dia*, through + *encephalos*, brain) are composed of the **left thalamus** and **right thalamus** (THAL-a-mus; plural, *thalami*). Each thalamus contains relay and processing centers for sensory information. The **hypothalamus** (*hypo-*, below), or floor of the diencephalon, contains centers involved with emotions, autonomic function, and hormone production. The *infundibulum*, a narrow stalk, connects the hypothalamus to the **pituitary gland**, a component of the endocrine system. The hypothalamus and the pituitary gland are responsible for the integration of the nervous and endocrine systems.

The diencephalon is a structural and functional link between the cerebral hemispheres and the components of the brain stem. The **brain stem** contains a variety of important processing centers and nuclei that relay information headed to or from the cerebrum or cerebellum. The brain stem includes the *mesencephalon*, *pons*, and *medulla oblongata*.[1]

- The **mesencephalon** (*mesos*, middle), or midbrain, contains nuclei that process visual and auditory information and control reflexes triggered by these stimuli. For example, your immediate, reflexive responses to a loud, unexpected noise (eye movements and head turning) are directed by nuclei in the midbrain. This region also contains centers that help maintain consciousness.

- The **pons** of the brain connects the cerebellum to the brain stem (*pons* is Latin for "bridge"). In addition to tracts and relay centers, the pons also contains nuclei involved with somatic and visceral motor control.

- The spinal cord connects to the brain at the **medulla oblongata**. Near the pons, the posterior wall of the medulla oblongata is thin and membranous. The inferior portion of the medulla oblongata resembles the spinal cord in that it has a narrow central canal. The medulla oblongata relays sensory information to the thalamus and to centers in other portions of the brain stem. The medulla oblongata also contains major centers that regulate autonomic function, such as heart rate, blood pressure, and digestion.

The boundaries and general functions of the diencephalon and brain stem are indicated in **Figure 14–1**. In considering the individual components of the brain, we will begin at the inferior portion of the medulla oblongata. This region has the simplest organization found anywhere in the brain, and in many respects it resembles the spinal cord. We will then ascend to regions of increasing structural and functional complexity until we reach the cerebral cortex, whose functions and capabilities are as yet poorly understood.

Embryology of the Brain

To understand the internal organization of the adult brain, we must consider its embryological origins. The central nervous system (CNS) begins as a hollow cylinder known as the *neural tube*. This tube has a fluid-filled internal cavity, the *neurocoel*. In the cephalic portion of the neural tube, three areas enlarge rapidly through expansion of the neurocoel. This enlargement creates three prominent divisions called **primary brain vesicles**. The primary brain vesicles are named for their relative positions: the *prosencephalon* (prōz-en-SEF-a-lon; *proso*, forward + *enkephalos*, brain), or "forebrain"; the *mesencephalon*, or "midbrain"; and the *rhombencephalon* (rom-ben-SEF-a-lon), or "hindbrain."

The fates of the three primary divisions of the brain are summarized in Table 14–1. The prosencephalon and rhombencephalon are subdivided further, forming **secondary brain vesicles**. The prosencephalon forms the **telencephalon** (tel-en-SEF-a-lon; *telos*, end) and the diencephalon. The telencephalon will ultimately form the cerebrum of the adult brain. The walls of the mesencephalon thicken, and the neurocoel becomes a relatively narrow passageway, comparable to the central canal of the spinal cord. The portion of the rhombencephalon adjacent to the mesencephalon forms the **metencephalon** (met-en-SEF-a-lon; *meta*, after). The dorsal portion of the metencephalon will become the cerebellum, and the ventral portion will develop into the pons. The portion of the rhombencephalon closer to the spinal cord forms the **myelencephalon** (mī-el-en-SEF-a-lon; *myelon*, spinal

---

[1] Some sources consider the brain stem to include the diencephalon. We will use the more restrictive definition here.

cord), which will become the medulla oblongata. ATLAS: Embryology Summary 12: The Development of the Brain and Cranial Nerves

Ventricles of the Brain

During development, the neurocoel within the cerebral hemispheres, diencephalon, metencephalon, and medulla oblongata expands to form chambers called **ventricles** (VEN-tri-kls). The ventricles are lined by cells of the *ependyma.* ∞ p. 392

Each cerebral hemisphere contains a large **lateral ventricle** (Figure 14–2). The **septum pellucidum**, a thin medial partition, separates the two lateral ventricles. Because there are *two* lateral ventricles, the ventricle in the diencephalon is called the **third ventricle**. Although the two lateral ventricles

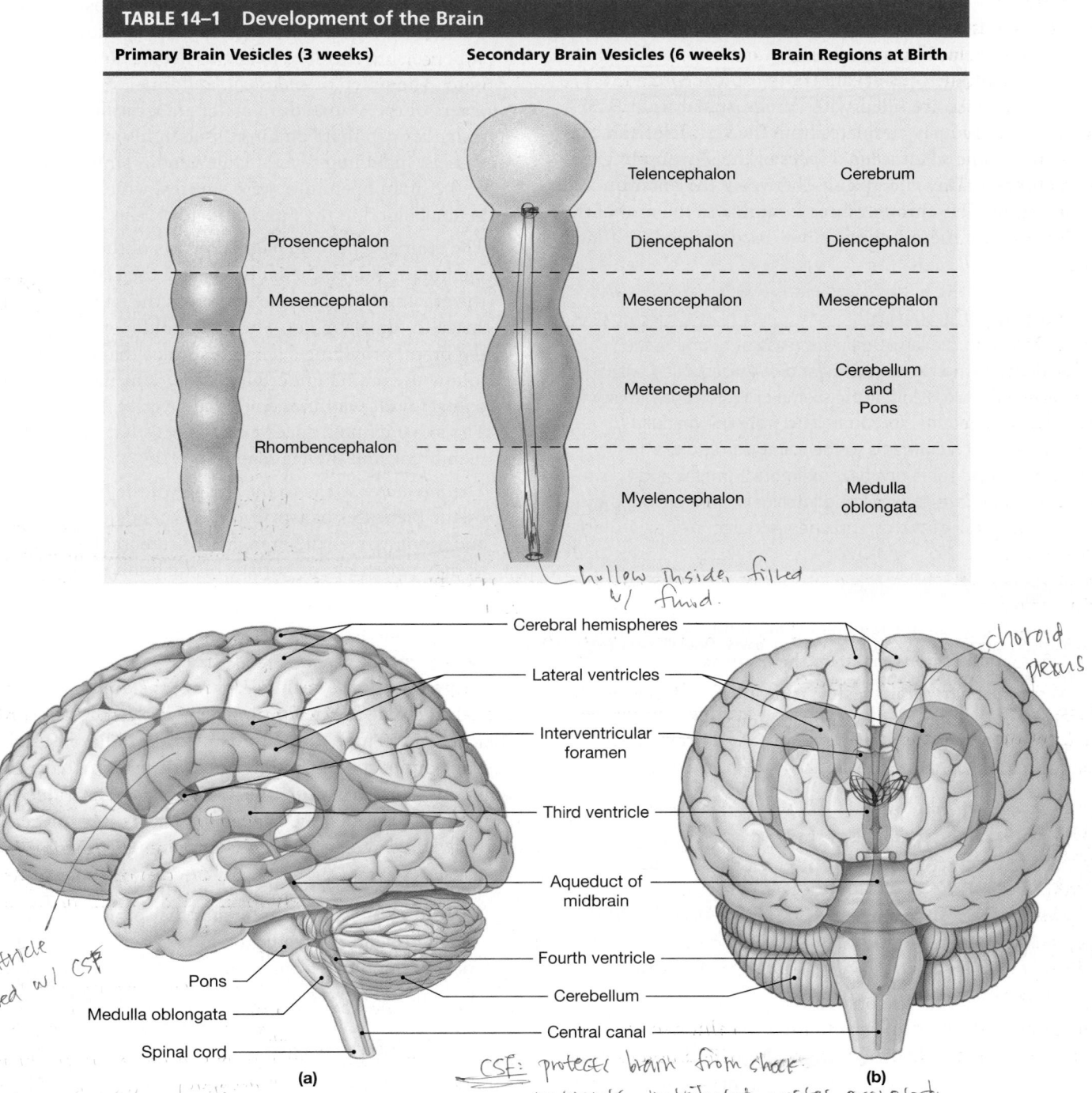

TABLE 14–1 Development of the Brain

Primary Brain Vesicles (3 weeks)	Secondary Brain Vesicles (6 weeks)	Brain Regions at Birth
	Telencephalon	Cerebrum
Prosencephalon	Diencephalon	Diencephalon
Mesencephalon	Mesencephalon	Mesencephalon
Rhombencephalon	Metencephalon	Cerebellum and Pons
	Myelencephalon	Medulla oblongata

[handwritten notes: hollow inside, filled w/ fluid.]

- Cerebral hemispheres
- Lateral ventricles
- Interventricular foramen
- Third ventricle
- Aqueduct of midbrain
- Fourth ventricle
- Cerebellum
- Central canal
- Pons
- Medulla oblongata
- Spinal cord

[handwritten notes: choroid plexus; ventricle filled w/ CSF; CSF: protects brain from shock. nutrients distributed, wastes excreted.]

(a) **(b)**

Figure 14–2 Ventricles of the Brain. The orientation and extent of the ventricles as they would appear if the brain were transparent. **(a)** A lateral view. **(b)** An anterior view. ATLAS: Plates 10; 12a–c; 13a–e

are not directly connected, each communicates with the third ventricle of the diencephalon through an **interventricular foramen** (*foramen of Monro*).

The mesencephalon has a slender canal known as the **aqueduct of midbrain** (or *the mesencephalic aqueduct, aqueduct of Sylvius,* or *cerebral aqueduct*). This passageway connects the third ventricle with the **fourth ventricle**. The superior portion of the fourth ventricle lies between the posterior surface of the pons and the anterior surface of the cerebellum. The fourth ventricle extends into the superior portion of the medulla oblongata. This ventricle then narrows and becomes continuous with the central canal of the spinal cord.

The ventricles are filled with cerebrospinal fluid (CSF). The CSF continuously circulates from the ventricles and central canal into the *subarachnoid space* of the surrounding cranial meninges. The CSF passes between the interior and exterior of the CNS through three foramina in the roof of the fourth ventricle; these foramina will be described in a later section.

The A&P Top 100

 # 48 The brain is a large, delicate mass of neural tissue containing internal passageways and chambers filled with cerebrospinal fluid. Each of the six major regions of the brain has specific functions. As you ascend from the medulla oblongata (which connects to the spinal cord) to the cerebrum, those functions become more complex and variable. Conscious thought and intelligence are provided by the neural cortex of the cerebral hemispheres.

CHECKPOINT

1. Name the six major regions of the brain.
2. What brain regions make up the brain stem?
3. Which primary brain vesicle is destined to form the cerebellum, pons, and medulla oblongata?

See the blue Answers tab at the end of the book.

14-2 The brain is protected and supported by the cranial meninges, cerebrospinal fluid, and the blood–brain barrier

The delicate tissues of the brain are protected from mechanical forces by the bones of the cranium, the *cranial meninges*, and cerebrospinal fluid. In addition, the neural tissue of the brain is biochemically isolated from the general circulation by the *blood–brain barrier*. Refer to **Figures 7–3** and **7–4** (pp. 215–217) for a review of the bones of the cranium. We will discuss the other protective factors here.

The Cranial Meninges

The layers that make up the cranial meninges—the cranial dura mater, arachnoid mater, and pia mater—are continuous with those of the spinal meninges. ∞ p. 433 However, the cranial meninges have distinctive anatomical and functional characteristics (**Figure 14–3a**):

- The cranial *dura mater* consists of outer and inner fibrous layers. The outer layer is fused to the periosteum of the cranial bones. As a result, there is no epidural space superficial to the dura mater, as occurs along the spinal cord. The outer, or *endosteal*, and inner, or *meningeal*, layers of the cranial dura mater are typically separated by a slender gap that contains tissue fluids and blood vessels, including several large venous sinuses. The veins of the brain open into these sinuses, which deliver the venous blood to the *internal jugular veins* of the neck.

- The cranial *arachnoid mater* consists of the arachnoid membrane (an epithelial layer) and the cells and fibers of the arachnoid trabeculae that cross the subarachnoid space to the pia mater. The arachnoid membrane covers the brain, providing a smooth surface that does not follow the brain's underlying folds. This membrane is in contact with the inner epithelial layer of the dura mater. The subarachnoid space extends between the arachnoid membrane and the pia mater.

- The *pia mater* sticks to the surface of the brain, anchored by the processes of astrocytes. It extends into every fold, and accompanies the branches of cerebral blood vessels as they penetrate the surface of the brain to reach internal structures.

Dural Folds

In several locations, the inner layer of the dura mater extends into the cranial cavity, forming a sheet that dips inward and then returns. These **dural folds** provide additional stabilization and support to the brain. **Dural sinuses** are large collecting veins located within the dural folds. The three largest dural folds are called the falx cerebri, the tentorium cerebelli, and the falx cerebelli (**Figure 14–3b**):

1. The **falx cerebri** (FALKS SER-e-brī; *falx*, curving or sickle-shaped) is a fold of dura mater that projects between the cerebral hemispheres in the longitudinal fissure. Its inferior portions attach anteriorly to the crista galli and posteriorly to the *internal occipital crest*, a ridge along the inner surface of the occipital bone. The **superior sagittal sinus** and the **inferior sagittal sinus**, two large venous sinuses, lie within this dural fold. The posterior margin of the falx cerebri intersects the tentorium cerebelli.

2. The **tentorium cerebelli** (ten-TO-ree-ē-um ser-e-BEL-ē; *tentorium,* a covering) protects the cerebellar hemi-

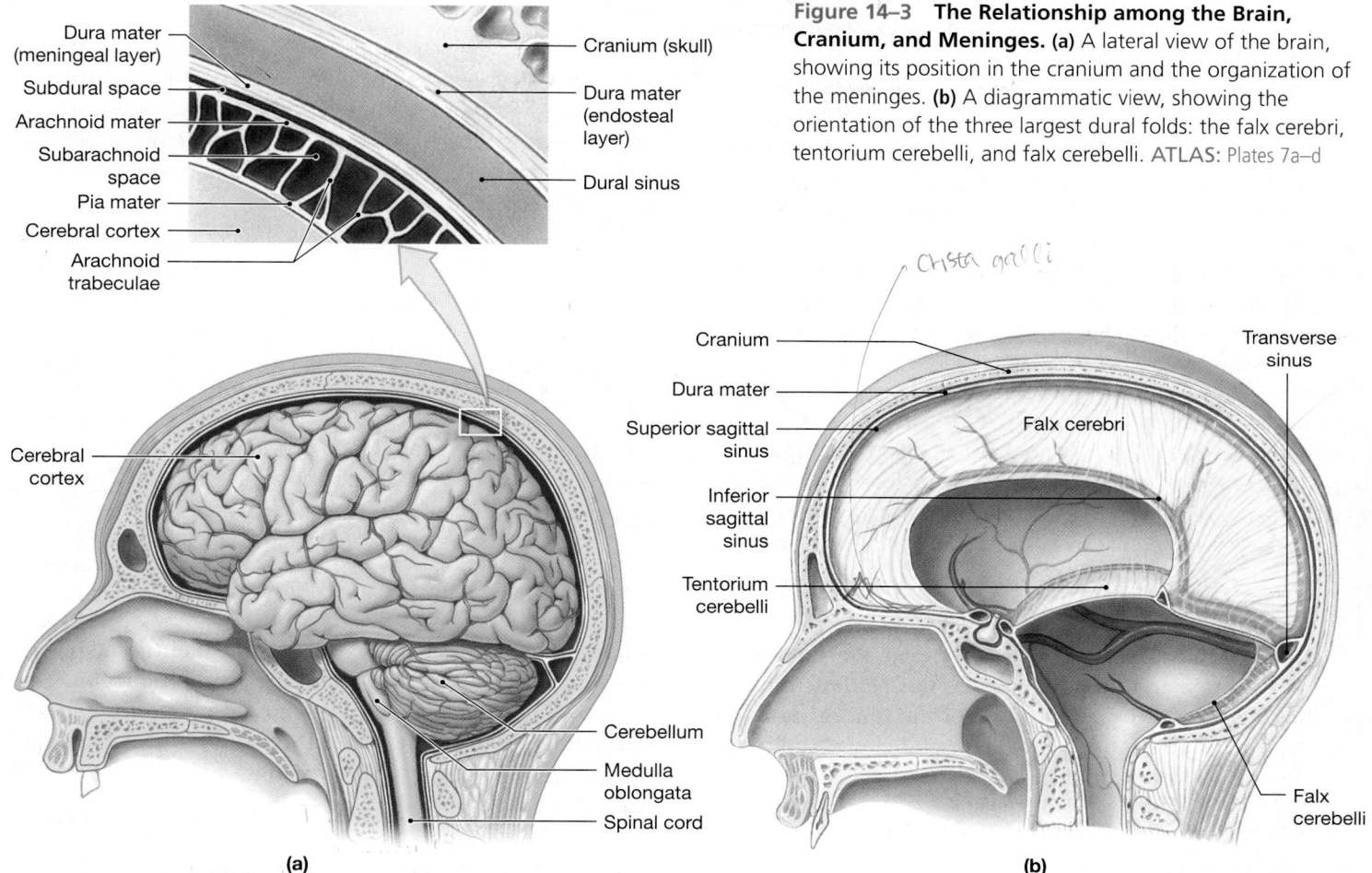

Dura mater (meningeal layer)
Subdural space
Arachnoid mater
Subarachnoid space
Pia mater
Cerebral cortex
Arachnoid trabeculae
Cranium (skull)
Dura mater (endosteal layer)
Dural sinus

Cerebral cortex

Cerebellum
Medulla oblongata
Spinal cord

(a)

Cranium
Dura mater
Superior sagittal sinus
Inferior sagittal sinus
Tentorium cerebelli
Crista galli
Transverse sinus
Falx cerebri
Falx cerebelli

(b)

Figure 14–3 The Relationship among the Brain, Cranium, and Meninges. (a) A lateral view of the brain, showing its position in the cranium and the organization of the meninges. **(b)** A diagrammatic view, showing the orientation of the three largest dural folds: the falx cerebri, tentorium cerebelli, and falx cerebelli. ATLAS: Plates 7a–d

spheres and separates them from those of the cerebrum. It extends across the cranium at right angles to the falx cerebri. The **transverse sinus** lies within the tentorium cerebelli.

3. The **falx cerebelli** divides the two cerebellar hemispheres along the midsagittal line inferior to the tentorium cerebelli.

The Protective Function of the Cranial Meninges

The overall shape of the brain roughly corresponds to that of the cranial cavity (**Figure 14–3a**). The massive cranial bones provide mechanical protection by cradling the brain, but they also pose a threat to safety that is countered by the cranial meninges and the CSF. The brain is like a person driving a car: If the car hits a tree, the car protects the driver from contact with the tree, but serious injury will occur unless a seat belt or airbag protects the driver from contact with the car. The tough, fibrous dural folds act like seat belts that hold the brain in position. The cerebrospinal fluid in the subarachnoid space acts like a bumper by cushioning against sudden jolts and shocks.

Cranial trauma is a head injury resulting from impact with another object. Each year in the United States, roughly

8 million cases of cranial trauma occur, but only 1 case in 8 results in serious brain damage. The percentage is relatively low because the cranial meninges and CSF are so effective in protecting the brain.

Cerebrospinal Fluid

Cerebrospinal fluid (CSF) completely surrounds and bathes the exposed surfaces of the CNS. The CSF has several important functions, including the following:

- *Cushioning Delicate Neural Structures.*
- *Supporting the Brain.* In essence, the brain is suspended inside the cranium and floats in the CSF. A human brain weighs about 1400 g (3.09 lb.) in air, but only about 50 g (1.8 oz.) when supported by CSF.
- *Transporting Nutrients, Chemical Messengers, and Waste Products.* Except at the choroid plexus, where CSF is produced, the ependymal lining is freely permeable and the CSF is in constant chemical communication with the interstitial fluid that surrounds the neurons and neuroglia of the CNS.

14 NERVOUS

CLINICAL NOTE

Epidural and Subdural Hemorrhages

A severe head injury may damage meningeal vessels and cause bleeding into the cranial cavity. The most serious cases involve an arterial break, because arterial blood pressure is relatively high. If blood is forced between the dura mater and the cranium, the condition is known as an **epidural hemorrhage**. The elevated fluid pressure then distorts the underlying tissues of the brain. The individual loses consciousness for a period lasting from minutes to hours after the injury, and death follows in untreated cases. An epidural hemorrhage involving a damaged vein does not produce massive symptoms immediately, and the individual may not have neurological problems for hours, days, or even weeks after the original injury. As a result, the problem may not be noticed until the nervous tissue has been severely damaged by distortion, compression, and

secondary hemorrhaging. Epidural hemorrhages are rare, occurring in fewer than 1 percent of head injuries. However, the mortality rate is 100 percent in untreated cases and over 50 percent even after the blood pool has been removed and the damaged vessels have been closed.

The term **subdural hemorrhage** may be misleading, because in many cases blood enters the meningeal layer of the dura mater, flowing deep to the epithelium that contacts the arachnoid membrane rather than between the dura mater and the arachnoid mater. Subdural hemorrhages are roughly twice as common as epidural hemorrhages. The most common source of blood is a small vein or one of the dural sinuses. Because the venous blood pressure in a subdural hemorrhage is lower than that in an arterial epidural hemorrhage, the distortion produced is gradual and the effects on brain function can be quite variable and difficult to diagnose.

Because free exchange occurs between the interstitial fluid of the brain and the CSF, changes in CNS function can produce changes in the composition of the CSF. As noted in Chapter 13, a *spinal tap* can provide useful clinical information about CNS injury, infection, or disease. ∞ p. 434

The Formation of CSF

The **choroid plexus** (*choroid*, vascular coat; *plexus*, network) consists of a combination of specialized ependymal cells and permeable capillaries involved in the production of cerebrospinal fluid. Two extensive folds of the choroid plexus originate in the roof of the third ventricle and extend through the interventricular foramina. These folds cover the floors of the lateral ventricles (**Figure 14–4a**). In the inferior brain stem, a region of the choroid plexus in the roof of the fourth ventricle projects between the cerebellum and the pons.

Specialized ependymal cells, interconnected by tight junctions, surround the capillaries of the choroid plexus. The ependymal cells secrete CSF into the ventricles; they also remove waste products from the CSF and adjust its composition over time. The differences in composition between CSF and blood plasma (blood with the cellular elements removed) are quite pronounced. For example, the blood contains high concentrations of soluble proteins, but the CSF does not. The concentrations of individual ions and the levels of amino acids, lipids, and waste products are also different.

Circulation of CSF

The choroid plexus produces CSF at a rate of about 500 mL/day. The total volume of CSF at any moment is approximately 150 mL; thus, the entire volume of CSF is replaced roughly every eight hours. Despite this rapid turnover, the

composition of CSF is closely regulated, and the rate of removal normally keeps pace with the rate of production.

The CSF circulates from the choroid plexus through the ventricles and fills the central canal of the spinal cord (**Figure 14–4a**). As the CSF circulates, diffusion between it and the interstitial fluid of the CNS is unrestricted between and across the ependymal cells. The CSF reaches the subarachnoid space through two **lateral apertures** and a single **median aperture**, openings in the roof of the fourth ventricle. Cerebrospinal fluid then flows through the subarachnoid space surrounding the brain, spinal cord, and cauda equina.

Fingerlike extensions of the arachnoid membrane, called the *arachnoid villi*, penetrate the meningeal layer of the dura mater and extend into the superior sagittal sinus. In adults, clusters of villi form large **arachnoid granulations** (**Figure 14–4b**). Cerebrospinal fluid is absorbed into the venous circulation at the arachnoid granulations.

If the normal circulation or reabsorption of CSF is interrupted, a variety of clinical problems may appear. For example, a problem with the reabsorption of CSF in infancy is responsible for *hydrocephalus*, or "water on the brain." Infants with this condition have enormously expanded skulls due to the presence of an abnormally large volume of CSF. In adults, a failure of reabsorption or a blockage of CSF circulation can distort and damage the brain.

The Blood Supply to the Brain

As noted in Chapter 12, neurons have a high demand for energy, but they have neither energy reserves in the form of carbohydrates or lipids, nor oxygen reserves in the form of myoglobin. Your brain, with billions of neurons, is an extremely

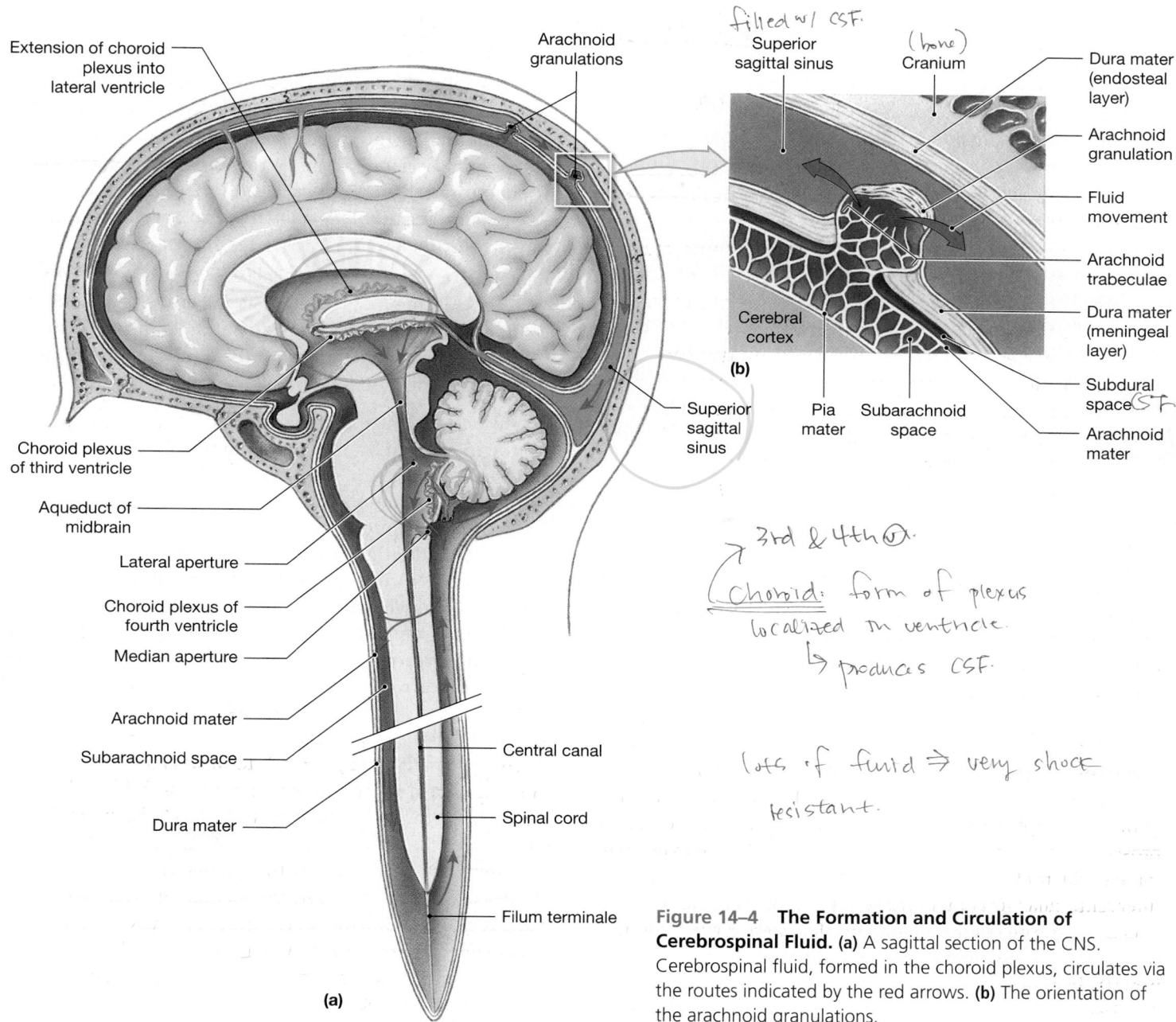

Extension of choroid plexus into lateral ventricle

Arachnoid granulations

filled w/ CSF.

Superior sagittal sinus

(bone) Cranium

Dura mater (endosteal layer)

Arachnoid granulation

Fluid movement

Arachnoid trabeculae

Dura mater (meningeal layer)

Cerebral cortex

(b)

Pia mater

Subarachnoid space

Subdural space *CSF*

Arachnoid mater

Choroid plexus of third ventricle

Superior sagittal sinus

Aqueduct of midbrain

Lateral aperture

Choroid plexus of fourth ventricle

Median aperture

Arachnoid mater

Subarachnoid space

Dura mater

Central canal

Spinal cord

Filum terminale

(a)

→ 3rd & 4th ⊙.

Choroid: form of plexus localized in ventricle.
↳ produces CSF.

lots of fluid ⇒ very shock resistant.

Figure 14–4 The Formation and Circulation of Cerebrospinal Fluid. (a) A sagittal section of the CNS. Cerebrospinal fluid, formed in the choroid plexus, circulates via the routes indicated by the red arrows. **(b)** The orientation of the arachnoid granulations.

active organ with a continuous demand for nutrients and oxygen. These demands are met by an extensive circulatory supply. Arterial blood reaches the brain through the *internal carotid arteries* and the *vertebral arteries*. Most of the venous blood from the brain leaves the cranium in the *internal jugular veins*, which drain the dural sinuses. A head injury that damages cerebral blood vessels may cause bleeding into the dura mater, either near the dural epithelium or between the outer layer of the dura mater and the bones of the skull. These are serious conditions, because the blood entering these spaces compresses and distorts the relatively soft tissues of the brain.

Cerebrovascular diseases are cardiovascular disorders that interfere with the normal blood supply to the brain. The particular distribution of the vessel involved determines the signs and symptoms, and the degree of oxygen or nutrient starvation determines their severity. A **cerebrovascular accident (CVA)**, or *stroke*, occurs when the blood supply to a portion of the brain is shut off. Affected neurons begin to die in a matter of minutes.

The Blood–Brain Barrier

Neural tissue in the CNS is isolated from the general circulation by the **blood–brain barrier (BBB)**. This barrier exists because the endothelial cells that line the capillaries of the CNS are extensively interconnected by tight junctions. These junctions prevent the diffusion of materials between adjacent endothelial cells. In general, only lipid-soluble compounds (including carbon dioxide; oxygen; ammonia; lipids, such as steroids or prostaglandins; and small alcohols) can diffuse across the membranes of endothelial cells into the interstitial fluid of the brain and spinal cord. Water and ions must pass through channels in the apical and basal plasma membranes. Larger, water-soluble compounds can cross the capillary walls only by active or passive transport.

The restricted permeability of the endothelial lining of brain capillaries is in some way dependent on chemicals secreted by astrocytes—cells that are in close contact with CNS capillaries. ∞ p. 392 The outer surfaces of the endothelial cells are covered by the processes of astrocytes. Because the astrocytes release chemicals that control the permeabilities of the endothelium to various substances, these cells play a key supporting role in the blood–brain barrier. If the astrocytes are damaged or stop stimulating the endothelial cells, the blood–brain barrier disappears.

The choroid plexus is not part of the neural tissue of the brain, so no astrocytes are in contact with the endothelial cells there. Substances do not have free access to the CNS, however, because a **blood–CSF barrier** is created by specialized ependymal cells. These cells, interconnected by tight junctions, surround the capillaries of the choroid plexus.

Transport across the blood–brain and blood–CSF barriers is selective and directional. Even the passage of small ions, such as sodium, hydrogen, potassium, or chloride, is controlled. As a result, the pH and concentrations of sodium, potassium, calcium, and magnesium ions in the blood and CSF are different. Some organic compounds are readily transported, and others cross only in minute amounts. Neurons have a constant need for glucose that must be met regardless of the relative concentrations in the blood and interstitial fluid. Even when circulating glucose levels are low, endothelial cells continue to transport glucose from the blood to the interstitial fluid of the brain. In contrast, only trace amounts of circulating norepinephrine, epinephrine, dopamine, and serotonin pass into the interstitial fluid or CSF of the brain. This limitation is important, because these compounds are neurotransmitters—their entry from the bloodstream (where concentrations can be relatively high) could result in the uncontrolled stimulation of neurons throughout the brain.

The blood–brain barrier remains intact throughout the CNS, with four noteworthy exceptions:

1. In portions of the hypothalamus, the capillary endothelium is extremely permeable. This permeability exposes hypothalamic nuclei to circulating hormones and permits the diffusion of hypothalamic hormones into the circulation.

2. Capillaries in the neurohypophysis, which is continuous with the floor of the hypothalamus, are highly permeable. At this site, the hormones *ADH* and *oxytocin*, produced by hypothalamic neurons, are released into the circulation.

3. Capillaries in the *pineal gland* are also very permeable. The pineal gland, an endocrine structure, is located on the posterior, superior surface of the diencephalon. The capillary permeability allows pineal secretions into the general circulation.

4. Capillaries at the choroid plexus are extremely permeable. Although the capillary characteristics of the blood–brain barrier are lost there, the transport activities of specialized ependymal cells in the choroid plexus maintain the blood–CSF barrier.

Physicians must sometimes get specific compounds into the interstitial fluid of the brain to fight CNS infections or to treat other neural disorders. To do this, they must understand the limitations of the blood–brain and blood–CSF barriers. For example, when considering possible treatments, the antibiotic *tetracycline* isn't used to treat meningitis or other CNS infections because this drug is excluded from the brain, whereas *sulfisoxazole* and *sulfadiazine* enter the CNS very rapidly. Sometimes, chemotherapeutic drugs or imaging dyes may be injected directly into the CSF by lumbar puncture.

The A&P Top 100

#49 The meninges stabilize the position of the brain within the cranial cavity, and cerebrospinal fluid provides protection against sudden jolts and shocks. CSF also provides nutrients and removes wastes generated by active neural tissues. The blood–brain barrier and the blood–CSF barrier selectively isolate the brain from chemicals in blood that might disrupt neural function.

CHECKPOINT

4. From superficial to deep, name the layers that constitute the cranial meninges.

5. What would happen if an interventricular foramen became blocked?

6. How would decreased diffusion across the arachnoid granulations affect the volume of cerebrospinal fluid in the ventricles?

7. Many water-soluble molecules that are relatively abundant in the blood occur in small amounts or not at all in the extracellular fluid of the brain. Why?

See the blue Answers tab at the end of the book.

14-3 The medulla oblongata, which is continuous with the spinal cord, contains vital centers

The medulla oblongata is the most inferior of the brain regions. **Figure 14–5** shows the position of the medulla oblongata in relation to the other components of the brain stem and the diencephalon. This figure includes the attachment sites for 11 of the 12 pairs of cranial nerves. The individual cranial nerves are identified by a capital N followed by a Roman numeral. (The full names and functions of these nerves will be introduced in a later section.)

In sectional view, the inferior portion of the medulla oblongata resembles the spinal cord, with a small central canal. However, the gray matter and white matter organization is

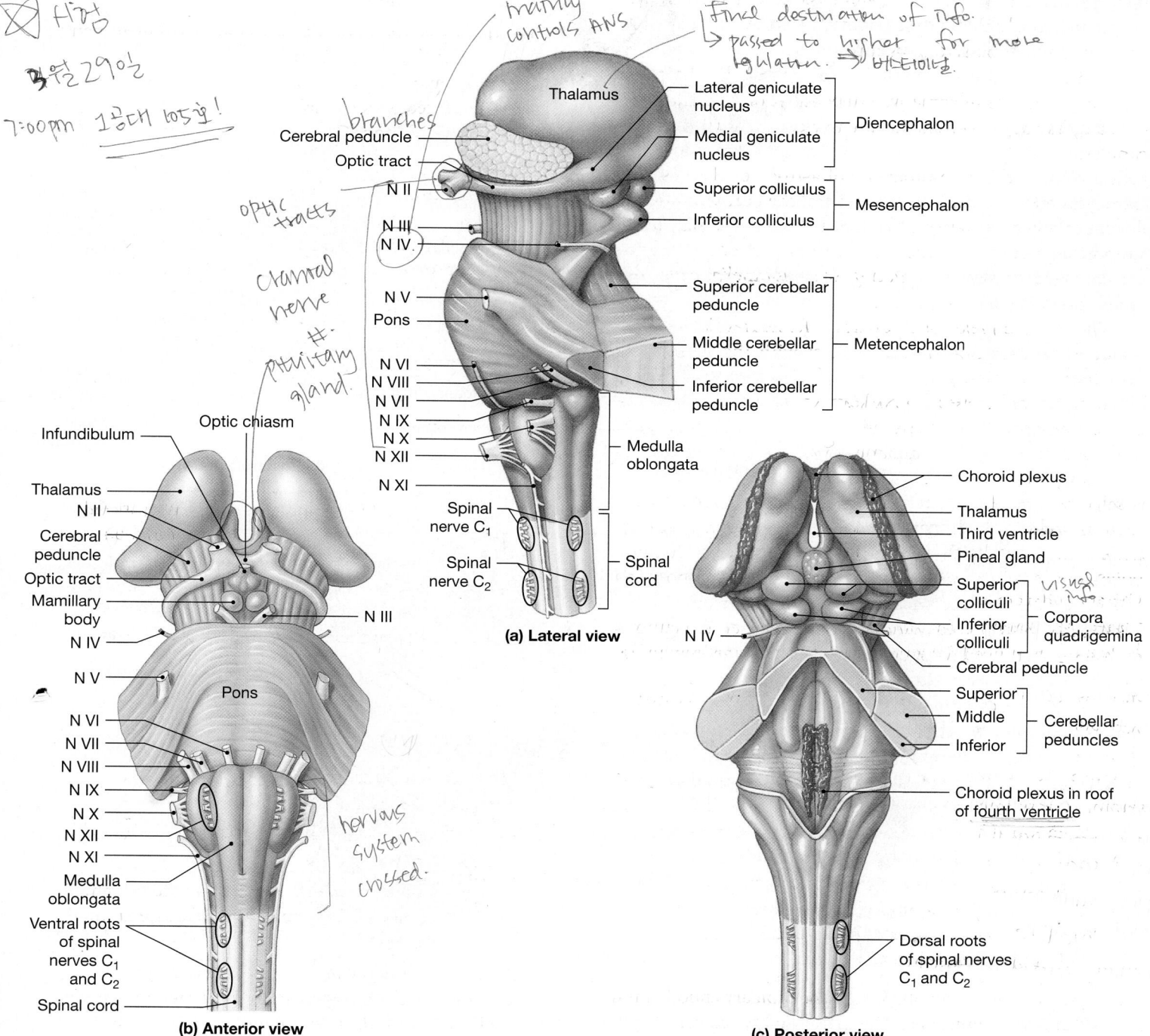

Figure 14–5   The Diencephalon and Brain Stem. (a) A lateral view, as seen from the left side. **(b)** An anterior view. **(c)** A posterior view.

more complex. As one ascends the medulla oblongata, the central canal opens into the fourth ventricle, and the similarity to the spinal cord all but disappears.

The medulla oblongata is a very busy place—all communication between the brain and spinal cord involves tracts that ascend or descend through the medulla oblongata. In addition, the medulla oblongata is a center for the coordination of relatively complex autonomic reflexes and the control of visceral functions.

The medulla oblongata (**Figure 14–6a,b**) includes three groups of nuclei that we will encounter in later chapters:

1. *Autonomic Nuclei Controlling Visceral Activities.* The **reticular formation** is a loosely organized mass of gray matter that contains embedded nuclei. It extends from the medulla oblongata to the mesencephalon. The portion of the reticular formation in the medulla oblongata contains nuclei and centers responsible for the regulation of vital autonomic functions. These **reflex centers** receive inputs from cranial nerves, the cerebral cortex, and the brain stem. Their output controls or adjusts the activities of one or more peripheral systems. There are two major groups of reflex centers. The **cardiovascular centers** adjust the

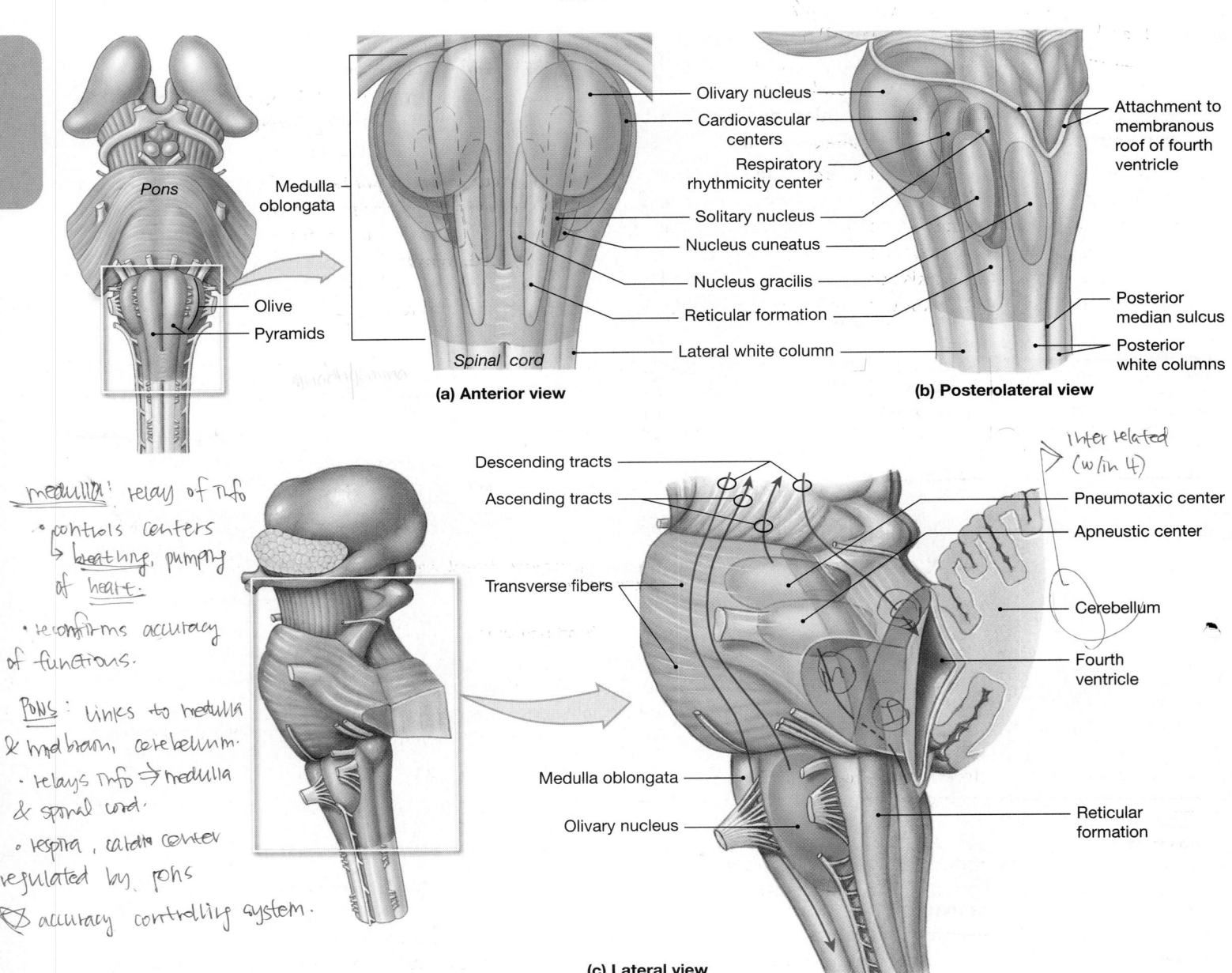

(a) Anterior view

(b) Posterolateral view

(c) Lateral view

Figure 14–6 **The Medulla Oblongata and Pons. (a, b)** Nuclei and tracts in the medulla oblongata. **(c)** Nuclei and tracts in the pons.
ATLAS: Plates 9a–c; 11c

heart rate, the strength of cardiac contractions, and the flow of blood through peripheral tissues. (In terms of function, the cardiovascular centers are subdivided into **cardiac** (*kardia*, heart) and **vasomotor** (*vas*, vessel) **centers**, but their anatomical boundaries are difficult to determine.) The **respiratory rhythmicity centers** set the basic pace for respiratory movements. Their activity is regulated by inputs from the apneustic and pneumotaxic centers of the pons.

2. *Sensory and Motor Nuclei of Cranial Nerves.* The medulla oblongata contains sensory and motor nuclei associated with five of the cranial nerves (VIII, IX, X, XI, and XII). These cranial nerves provide motor commands to muscles of the pharynx, neck, and back as well as to the visceral organs of the thoracic and peritoneal cavities. Cranial nerve VIII carries sensory information from receptors in the inner ear to the vestibular and cochlear nuclei, which extend from the pons into the medulla oblongata.

3. *Relay Stations along Sensory and Motor Pathways.* The **nucleus gracilis** and the **nucleus cuneatus** pass somatic sensory information to the thalamus. Tracts leaving these nuclei cross to the opposite side of the brain before reaching their destinations. This crossing over is called *decussation* (dē-kuh-SĀ-shun; *decussatio*, crossing over).

The **solitary nucleus** on either side receives visceral sensory information that reaches the CNS from the spinal nerves and cranial nerves. This information is integrated and forwarded to other autonomic centers in the medulla oblongata and elsewhere. The **olivary nuclei** relay information to the cerebellar cortex about somatic motor commands as they are issued by motor centers at higher levels. The bulk of the olivary nuclei create the **olives**, prominent olive-shaped bulges along the ventrolateral surface of the medulla oblongata.

Table 14–2 summarizes the major components of the medulla oblongata.

CHECKPOINT

8. Identify the components of the medulla oblongata that are responsible for relaying somatic sensory information to the thalamus.

9. The medulla oblongata is one of the smallest sections of the brain, yet damage there can cause death, whereas similar damage in the cerebrum might go unnoticed. Why?

See the blue Answers tab at the end of the book.

TABLE 14–2	Components and Functions of the Medulla Oblongata and Pons	
Region/Subdivision	**Component(s)**	**Function(s)**
MEDULLA OBLONGATA		
Gray matter	Nucleus gracilis Nucleus cuneatus	Relay somatic sensory information to the thalamus
	Olivary nuclei	Located within the olives; relay information from the red nucleus, other nuclei of the mesencephalon, and the cerebral cortex to the cerebellum
	Solitary nucleus	Integrates and relays visceral sensory information to autonomic processing centers
	Reflex centers	
	Cardiac centers	Regulate heart rate and force of contraction
	Vasomotor centers	Regulate distribution of blood flow
	Respiratory rhythmicity centers	Set the pace of respiratory movements
	Other nuclei/centers	Contain sensory and motor nuclei of cranial nerves VIII (in part), IX, X, XI (in part), and XII; relay ascending sensory information from the spinal cord or higher centers
White matter	Ascending and descending tracts	Link the brain with the spinal cord
PONS		
Gray matter	Nuclei associated with cranial nerves V, VI, VII, and VIII (in part)	Relay sensory information and issue somatic motor commands
	Apneustic and pneumotaxic centers	Adjust activities of the respiratory rhythmicity centers in the medulla oblongata
	Relay centers	Relay sensory and motor information to the cerebellum
White matter	Ascending tracts	Carry sensory information from the nucleus cuneatus and nucleus gracilis to the thalamus
	Descending tracts	Carry motor commands from higher centers to motor nuclei of cranial or spinal nerves

14-4 The pons contains nuclei and tracts that carry or relay sensory and motor information

The pons links the cerebellum with the mesencephalon, diencephalon, cerebrum, and spinal cord. Important features and regions of the pons are indicated in **Figures 14–5** and **14–6c** and Table 14–2. The pons contains four groups of components:

1. *Sensory and Motor Nuclei of Cranial Nerves.* These cranial nerves (V, VI, VII, and VIII) innervate the jaw muscles, the anterior surface of the face, one of the extra-ocular muscles (the lateral rectus), and the sense organs of the inner ear (the *vestibular* and *cochlear nuclei*).

2. *Nuclei Involved with the Control of Respiration.* On each side of the pons, the reticular formation in this region contains two respiratory centers: the *apneustic center* and the *pneumotaxic center*. These centers modify the activity of the *respiratory rhythmicity center* in the medulla oblongata.

3. *Nuclei and Tracts That Process and Relay Information Heading to or from the Cerebellum.* The pons links the cerebellum with the brain stem, cerebrum, and spinal cord.

4. *Ascending, Descending, and Transverse Tracts.* Longitudinal tracts interconnect other portions of the CNS. The middle cerebellar peduncles are connected to the **transverse fibers**, which cross the anterior surface of the pons. These fibers are axons that link nuclei of the pons (*pontine nuclei*) with the cerebellar hemisphere of the opposite side.

CHECKPOINT

10. Name the four groups of components found in the pons. *Sensory/motor nuclei in C5-C8, nuclei controlling respiratory rhythmic center, to/from cerebellum & Trans. tract.*

11. If the respiratory centers of the pons were damaged, what respiratory controls might be lost? *modification.*

See the blue Answers tab at the end of the book.

14-5 The cerebellum coordinates learned and reflexive patterns of muscular activity at the subconscious level

The cerebellum (**Figure 14–7** and Table 14–3) is an automatic processing center. It has two primary functions:

1. *Adjusting the Postural Muscles of the Body.* The cerebellum coordinates rapid, automatic adjustments that maintain balance and equilibrium. These alterations in muscle tone and position are made by modifying the activities of motor centers in the brain stem.

2. *Programming and Fine-Tuning Movements Controlled at the Conscious and Subconscious Levels.* The cerebellum refines learned movement patterns. This function is performed indirectly by regulating activity along motor pathways at the cerebral cortex, basal nuclei, and motor centers in the brain stem. The cerebellum compares the motor commands with proprioceptive information (position sense) and performs any adjustments needed to make the movement smooth.

The cerebellum has a complex, highly convoluted surface composed of neural cortex. The **folia** (FŌ-lē-uh; leaves), or folds of the cerebellum surface, are less prominent than the folds in the surfaces of the cerebral hemispheres (**Figure 14–7a**). The **anterior** and **posterior lobes** are separated by the **primary fissure**. Along the midline, a narrow band of cortex known as the **vermis** (VER-mis; worm) separates the **cerebellar hemispheres**. The slender **flocculonodular** (flok-ū-lō-NOD-ū-lar) **lobe** lies between the roof of the fourth ventricle and the cerebellar hemispheres and vermis (**Figure 14–7b**).

Like the cerebrum, the cerebellum has a superficial layer of neural cortex. The cerebellar cortex contains huge, highly branched **Purkinje** (pur-KIN-jē) **cells**. The extensive dendrites of each Purkinje cell receive input from up to 200,000 synapses. Internally, the white matter of the cerebellum forms a branching array that in sectional view resembles a tree. Anatomists call it the **arbor vitae**, or "tree of life."

The cerebellum receives proprioceptive information from the spinal cord and monitors all proprioceptive, visual, tactile, balance, and auditory sensations received by the brain. Most axons that carry sensory information do not synapse in the cerebellar nuclei but pass through the deeper layers of the cerebellum on their way to the Purkinje cells of the cerebellar cortex. Information about the motor commands issued at the conscious and subconscious levels reaches the Purkinje cells indirectly, after being relayed by nuclei in the pons or by the **cerebellar nuclei** embedded within the arbor vitae.

Tracts that link the cerebellum with the brain stem, cerebrum, and spinal cord leave the cerebellar hemispheres as the superior, middle, and inferior cerebellar peduncles. The **superior cerebellar peduncles** link the cerebellum with nuclei in the midbrain, diencephalon, and cerebrum. The **middle cerebellar peduncles** are connected to a broad band of fibers that cross the ventral surface of the pons at right angles to the axis of the brain stem. The middle cerebellar peduncles also connect the cerebellar hemispheres with sensory and motor nuclei in the pons. The **inferior cerebellar peduncles** permit communication between the cerebellum and nuclei in the medulla oblongata and carry ascending and descending cerebellar tracts from the spinal cord.

Figure 14–7 The Cerebellum. (a) The posterior, superior surface of the cerebellum, showing major anatomical landmarks and regions. **(b)** A sectional view of the cerebellum, showing the arrangement of gray matter and white matter.

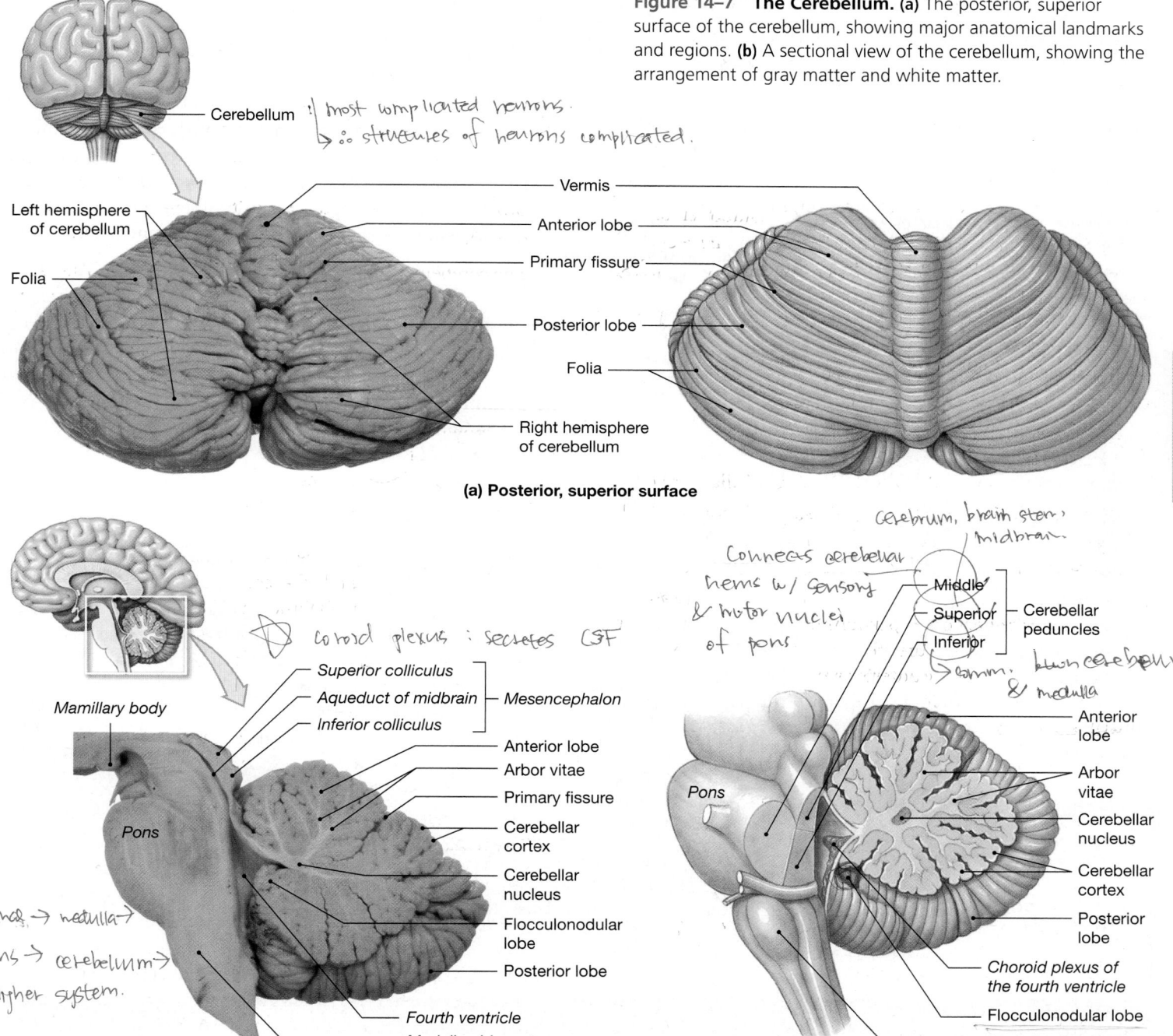

Cerebellum : *most complicated neurons.*
  ↳ ∴ structures of neurons complicated.

Left hemisphere of cerebellum

Folia

Vermis

Anterior lobe

Primary fissure

Posterior lobe

Folia

Right hemisphere of cerebellum

(a) Posterior, superior surface

14 NERVOUS

Mamillary body

Pons

choroid plexus : secretes CSF

Superior colliculus
Aqueduct of midbrain — Mesencephalon
Inferior colliculus

Anterior lobe
Arbor vitae
Primary fissure
Cerebellar cortex
Cerebellar nucleus
Flocculonodular lobe
Posterior lobe
Fourth ventricle
Medulla oblongata

Spinal → medulla →
pons → cerebellum →
higher system.

Connects cerebellar hems w/ sensory & motor nuclei of pons

cerebrum, brain stem, midbrain.

Middle
Superior — Cerebellar peduncles
Inferior

comm. btwn cerebellum & medulla

Anterior lobe
Arbor vitae
Cerebellar nucleus
Cerebellar cortex
Posterior lobe
Choroid plexus of the fourth ventricle
Flocculonodular lobe
Medulla oblongata

Pons

(b) Sagittal section

TABLE 14–3	Components of the Cerebellum	
Subdivision	**Region/Nuclei**	**Function(s)**
Gray matter	**Cerebellar cortex**	Involuntary coordination and control of ongoing body movements
	Cerebellar nuclei	As above
White matter	**Arbor vitae**	Connects cerebellar cortex and nuclei with cerebellar peduncles
	Cerebellar peduncles	
	Superior	Link the cerebellum with mesencephalon, diencephalon, and cerebrum
	Middle	Contain transverse fibers and carry communications between the cerebellum and pons
	Inferior	Link the cerebellum with the medulla oblongata and spinal cord
	Transverse fibers	Interconnect pontine nuclei with the cerebellar hemisphere on the opposite side

The cerebellum can be permanently damaged by trauma or stroke, or temporarily affected by drugs such as alcohol. The result is **ataxia** (a-TAK-sē-uh; *ataxia*, lack of order), a disturbance in muscular coordination. In severe ataxia, the individual cannot sit or stand without assistance.

CHECKPOINT

12. Identify the components of the cerebellar gray matter.

13. What part of the brain has the arbor vitae? What is its function?

See the blue Answers tab at the end of the book.

[handwritten notes:]
Cerebellar cortex & nuclei—involuntary control of body mov't.
Cerebellum → connects cerebellar cortex & nuclei to cerebellar peduncle. Responsible for info processing.
midbrain: cranial reflexes.
Integrates visual / auditory info.
superior coll. → Inferior colliculus.
subconscious control of upper limbs position of background of muscle tone → red nucleus.
maintains consciousness.

14-6 The mesencephalon regulates auditory and visual reflexes and controls alertness

The external anatomy of the mesencephalon, or midbrain, is shown in **Figure 14–5**, and the major nuclei are listed in Table 14–4 and shown in **Figure 14–8**. The **tectum**, or roof of the mesencephalon, is the region posterior to the aqueduct of midbrain. It contains two pairs of sensory nuclei known collectively as the **corpora quadrigemina** (KOR-pōr-uh qua-dri-JEM-i-nuh). These nuclei, the superior and inferior colliculi, process visual and auditory sensations. Each **superior colliculus**

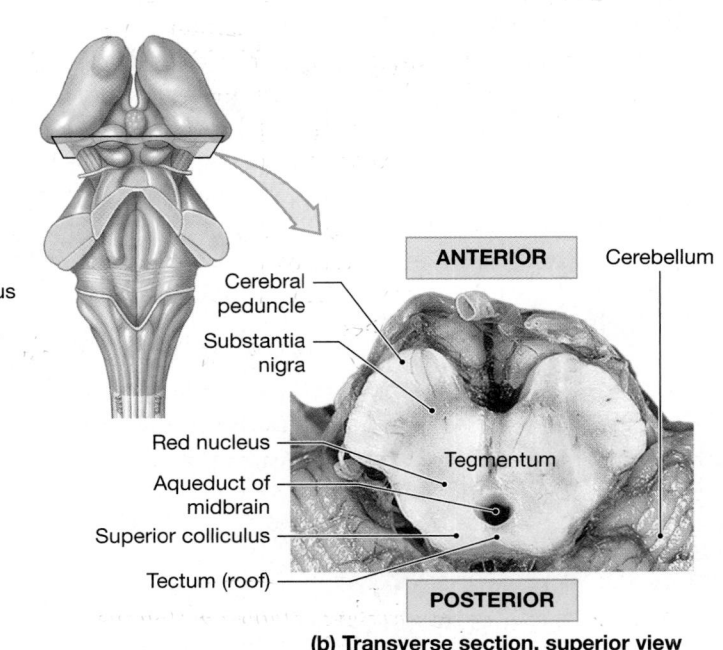

Thalamus

— Pineal gland
— Red nucleus
— Substantia nigra
— Superior colliculus
— Cerebral peduncle
— Inferior colliculus
— Reticular formation

corpora quadrigemina

(a) Posterior view

ANTERIOR — Cerebellum

Cerebral peduncle
Substantia nigra
Red nucleus — Tegmentum
Aqueduct of midbrain
Superior colliculus
Tectum (roof)

POSTERIOR

(b) Transverse section, superior view

Figure 14–8 **The Mesencephalon. (a)** A posterior view. The underlying nuclei are colored only on the right. **(b)** A superior view of a section at the level of the mesencephalon. ATLAS: Plates 7b; 9a–c; 11c; 12a,c

TABLE 14–4	**Components and Functions of the Mesencephalon**	
Subdivision	**Region/Nuclei**	**Functions**
GRAY MATTER		
Tectum (roof)	Superior colliculi	Integrate visual information with other sensory inputs; initiate reflex responses to visual stimuli
	Inferior colliculi	Relay auditory information to medial geniculate nuclei; initiate reflex responses to auditory stimuli
Walls and floor	Red nuclei	Subconscious control of upper limb position and background muscle tone
	Substantia nigra	Regulates activity in the basal nuclei
	Reticular formation (headquarters)	Automatic processing of incoming sensations and outgoing motor commands; can initiate involuntary motor responses to stimuli; helps maintain consciousness (RAS)
	Other nuclei/centers	Nuclei associated with two cranial nerves (III, IV)
WHITE MATTER	Cerebral peduncles	Connect primary motor cortex with motor neurons in brain and spinal cord; carry ascending sensory information to thalamus

(KO-LIK-ū-lus; *colliculus*, a small hill) receives visual inputs from the lateral geniculate nucleus of the thalamus on that side. The **inferior colliculus** receives auditory data from nuclei in the medulla oblongata and pons. Some of this information may be forwarded to the medial geniculate on the same side. The superior colliculi control the reflex movements of the eyes, head, and neck in response to visual stimuli, such as a bright light. The inferior colliculi control reflex movements of the head, neck, and trunk in response to auditory stimuli, such as a loud noise.

The area anterior to the aqueduct of midbrain is called the **tegmentum**. On each side, the tegmentum contains a red nucleus and the substantia nigra (**Figure 14–8a**). The **red nucleus** contains numerous blood vessels, which give it a rich red color. This nucleus, which receives information from the cerebrum and cerebellum, issues subconscious motor commands that affect upper limb position and background muscle tone. The **substantia nigra** (NĪ-gruh; *nigra*, black) is a nucleus that lies lateral to the red nucleus. The gray matter in this region contains darkly pigmented cells, giving it a black color.

The nerve fiber bundles on the ventrolateral surfaces of the mesencephalon (**Figures 14–5** and **14–8b**) are the **cerebral peduncles** (*peduncles*, little feet). They contain (1) descending fibers that go to the cerebellum by way of the pons and (2) descending fibers that carry voluntary motor commands issued by the cerebral hemispheres.

The mesencephalon also contains the headquarters of the *reticular activating system* (*RAS*), a specialized component of the reticular formation. Stimulation of this region makes you more alert and attentive. We will consider the role of the RAS in the maintenance of consciousness in Chapter 16.

CHECKPOINT

14. Identify the sensory nuclei contained within the corpora quadrigemina.

15. Which area(s) of the mesencephalon control reflexive movements of the eyes, head, and neck?

See the blue Answers tab at the end of the book.

14-7 The diencephalon integrates sensory information with motor output at the subconscious level

The diencephalon is a division of the brain that consists of the epithalamus, thalamus, and hypothalamus. **Figure 14–5** shows its position and its relationship to landmarks on the brain stem.

The *epithalamus* is the roof of the diencephalon superior to the third ventricle. The anterior portion of the epithalamus contains an extensive area of choroid plexus that extends through the interventricular foramina into the lateral ventri-

cles. The posterior portion of the epithalamus contains the **pineal gland** (**Figure 14–5c**), an endocrine structure that secretes the hormone **melatonin**. Melatonin is important in the regulation of day–night cycles and also in the regulation of reproductive functions. (We will describe the role of melatonin in Chapter 18.)

Most of the neural tissue in the diencephalon is concentrated in the *left thalamus* and *right thalamus*, which form the lateral walls, and the *hypothalamus*, which forms the floor. Ascending sensory information from the spinal cord and cranial nerves (other than the olfactory tract) synapses in a nucleus in the left or right thalamus before reaching the cerebral cortex and our conscious awareness. The hypothalamus contains centers involved with emotions and visceral processes that affect the cerebrum as well as other components of the brain stem. It also controls a variety of autonomic functions and forms the link between the nervous and endocrine systems.

The Thalamus

On each side of the diencephalon, the thalamus is the final relay point for ascending sensory information that will be projected to the primary sensory cortex. It acts as a filter, passing on only a small portion of the arriving sensory information. The thalamus also coordinates the activities of the basal nuclei and the cerebral cortex by relaying information between them.

The third ventricle separates the left thalamus and right thalamus. Each thalamus consists of a rounded mass of *thalamic nuclei* (**Figure 14–9**). Viewed in a midsagittal section through the brain (**Figure 14–10b**), each thalamus extends from the anterior commissure to the inferior base of the pineal gland. A projection of gray matter called an **interthalamic adhesion** or intermediate mass extends into the ventricle from the thalamus on either side (**Figures 14–10** and **14–11**), although no fibers cross the midline.

Functions of Thalamic Nuclei

The thalamic nuclei deal primarily with the relay of sensory information to the basal nuclei and cerebral cortex. The five major groups of thalamic nuclei, listed in Table 14–5 and shown in **Figure 14–9b**, are the anterior, medial, ventral, posterior, and lateral groups:

1. The *anterior group* includes the **anterior nuclei**, which are part of the *limbic system*. This system, which is involved with emotion and motivation, is discussed in a later section.

2. The nuclei of the *medial group* provide an awareness of emotional states by connecting emotional centers in the hypothalamus with the *frontal lobes* of the cerebral hemispheres. The medial group also receives and relays sensory information from other portions of the thalamus.

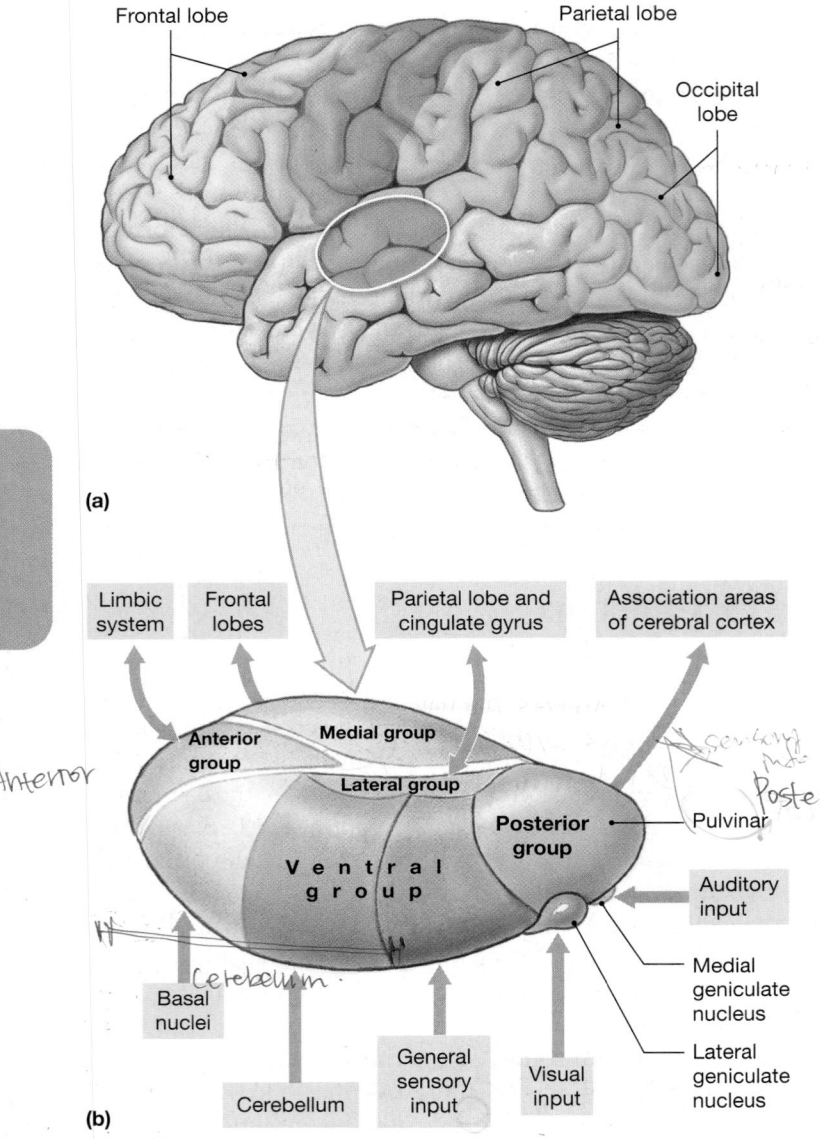

Figure 14–9 **The Thalamus. (a)** A lateral view of the brain, color coded to indicate the regions that receive input from the thalamic nuclei shown in part (b). **(b)** An enlarged view of the thalamic nuclei of the left side.

TABLE 14–5	The Thalamus
Group/Nuclei	**Function(s)**
ANTERIOR GROUP	Part of the limbic system
MEDIAL GROUP	Integrates sensory information for projection to the frontal lobes
VENTRAL GROUP	Projects sensory information to the primary sensory cortex; relays information from cerebellum and basal nuclei to motor area of cerebral cortex
POSTERIOR GROUP	
Pulvinar	Integrates sensory information for projection to association areas of cerebral cortex
Lateral geniculate nuclei	Project visual information to the visual cortex
Medial geniculate nuclei	Project auditory information to the auditory cortex
LATERAL GROUP	Integrates sensory information and influences emotional states

3. The nuclei of the *ventral group* relay information from the *basal nuclei* of the cerebrum and the cerebellum to somatic motor areas of the cerebral cortex. Ventral group nuclei also relay sensory information about touch, pressure, pain, temperature, and proprioception (position) to the sensory areas of the cerebral cortex.

4. The *posterior group* includes the pulvinar and the geniculate nuclei. The **pulvinar** integrates sensory information for projection to the cerebral cortex. The **lateral geniculate** (je-NIK-ū-lāt; *genicula*, little knee) **nucleus** of each thalamus receives visual information over the *optic tract*, which originates at the eyes. The output of the lateral geniculate nucleus goes to the *occipital lobes* of the cerebral hemispheres and to the mesencephalon. The **medial geniculate nucleus** relays auditory information to the appropriate area of the cerebral cortex from specialized receptors of the inner ear.

5. The nuclei of the *lateral group* form feedback loops with the limbic system, described in a later section, and the *parietal lobes* of the cerebral hemispheres. The lateral group affects emotional states and the integration of sensory information.

The Hypothalamus

The hypothalamus (**Figure 14–10a**) extends from the area superior to the *optic chiasm*, a crossover where the optic tracts from the eyes arrive at the brain, to the posterior margins of the **mamillary** (*mamilla*, little breast) **bodies**. The mamillary bodies process sensory information, including olfactory sensations. They also contain motor nuclei that control reflex movements associated with eating, such as chewing, licking, and swallowing.

Immediately posterior to the optic chiasm, a narrow stalk called the **infundibulum** (in-fun-DIB-ū-lum; *infundibulum*, funnel) extends inferiorly, connecting the floor of the hypothalamus to the pituitary gland (**Figure 14–10b**).

The floor of the hypothalamus between the infundibulum and the mamillary bodies is the **tuberal area** (*tuber,*

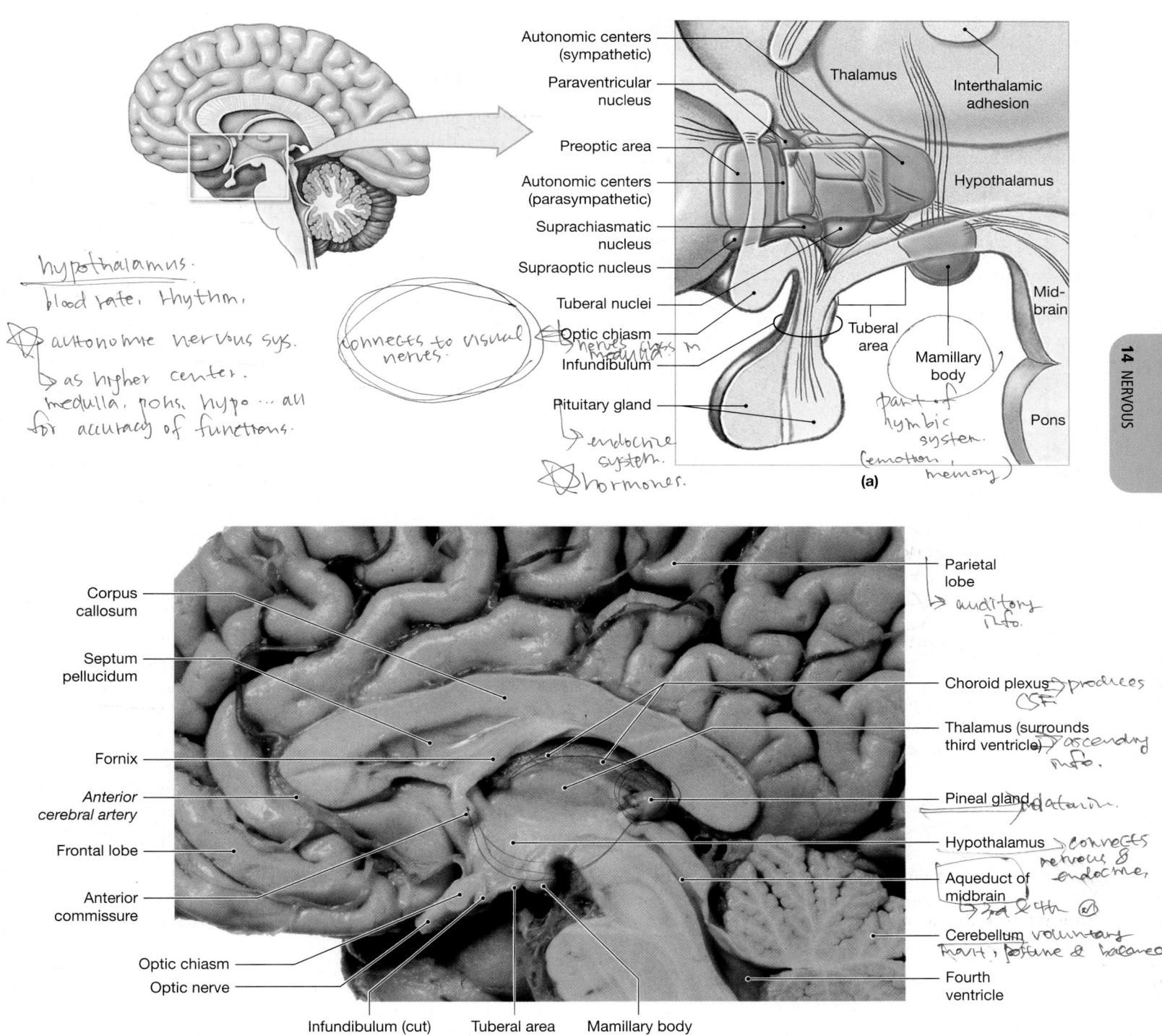

Figure 14–10   The Hypothalamus in Sagittal Section. (a) A diagrammatic view of the hypothalamus, showing the locations of major nuclei and centers. **(b)** The hypothalamus and adjacent portions of the brain.

swelling). The tuberal area contains nuclei that are involved with the control of pituitary gland function.

Functions of the Hypothalamus

The hypothalamus contains important control and integrative centers, in addition to those associated with the limbic system. These centers are shown in **Figure 14–10a**, and their functions are summarized in Table 14–6. Hypothalamic centers may be stimulated by (1) sensory information from the cerebrum, brain stem, and spinal cord; (2) changes in the compositions of the CSF and interstitial fluid; or (3) chemical stimuli in the circulating blood that move rapidly across

TABLE 14–6 **Components and Functions of the Hypothalamus**

Region/Nucleus	Function
Mamillary bodies	Control feeding reflexes (licking, swallowing, etc.)
Autonomic centers	Control medullary nuclei that regulate heart rate and blood pressure
Tuberal nuclei	Release hormones that control endocrine cells of the adenohypophysis
Supraoptic nucleus	Secretes ADH, restricting water loss at the kidneys
Paraventricular nucleus	Secretes oxytocin
Preoptic areas	Regulate body temperature
Suprachiasmatic nucleus	Coordinates day–night cycles of activity

highly permeable capillaries to enter the hypothalamus (where there is no blood–brain barrier).

The hypothalamus performs the following functions:

1. *The Subconscious Control of Skeletal Muscle Contractions.* The hypothalamus directs somatic motor patterns associated with rage, pleasure, pain, and sexual arousal by stimulating centers in other portions of the brain. For example, the changes in facial expression that accompany rage and the basic movements associated with sexual activity are controlled by hypothalamic centers.

2. *The Control of Autonomic Function.* The hypothalamus adjusts and coordinates the activities of autonomic centers in the pons and medulla oblongata that regulate heart rate, blood pressure, respiration, and digestive functions.

3. *The Coordination of Activities of the Nervous and Endocrine Systems.* The hypothalamus coordinates neural and endocrine activities by inhibiting or stimulating endocrine cells in the pituitary gland through the production of *regulatory hormones.* These hormones are produced at the tuberal area and are released into local capillaries for transport to the adenohypophysis.

4. *The Secretion of Two Hormones.* The hypothalamus secretes *antidiuretic hormone* (ADH, also called vasopressin), which is produced by the **supraoptic nucleus** and restricts water loss at the kidneys, and *oxytocin* (OT), which is produced by the **paraventricular nucleus** and stimulates smooth muscle contractions in the uterus and mammary glands of females and the prostate gland of males. These hormones are transported along axons that pass through the infundibulum to the neurohypophysis. There the hormones are released into the blood for distribution throughout the body.

5. *The Production of Emotions and Behavioral Drives.* Specific hypothalamic centers produce sensations that lead to conscious or subconscious changes in behavior. For example, stimulation of the **feeding center** produces the sensation of hunger, and stimulation of the **thirst center** produces the sensation of thirst. These unfocused "impressions" originating in the hypothalamus are called **drives**. The conscious sensations are only part of the hypothalamic response. For instance, the thirst center also orders the release of ADH by neurons in the supraoptic nucleus.

6. *Coordination between Voluntary and Autonomic Functions.* When you think about a dangerous or stressful situation, your heart rate and respiratory rate go up and your body prepares for an emergency. These autonomic adjustments are made by the hypothalamus.

7. *The Regulation of Body Temperature.* The **preoptic area** of the hypothalamus coordinates the activities of other CNS centers and regulates other physiological systems to maintain normal body temperature. If body temperature falls, the preoptic area sends instructions to the *vasomotor center*, an autonomic center in the medulla oblongata that controls blood flow by regulating the diameter of peripheral blood vessels. In response, the vasomotor center decreases the blood supply to the skin, reducing the rate of heat loss.

8. *The Control of Circadian Rhythms.* The **suprachiasmatic nucleus** coordinates daily cycles of activity that are linked to the 24-hour day–night cycle. This nucleus receives input from the retina of the eye, and its output adjusts the activities of other hypothalamic nuclei, the pineal gland, and the reticular formation.

CHECKPOINT

16. Name the main components of the diencephalon.

17. Damage to the lateral geniculate nuclei of the thalamus would interfere with the functions of which special sense?

18. Which component of the diencephalon is stimulated by changes in body temperature?

See the blue Answers tab at the end of the book.

14-8 The limbic system is a group of tracts and nuclei with various functions

The **limbic system** (*limbus*, border) includes nuclei and tracts along the border between the cerebrum and diencephalon. This system is a functional grouping rather than an anatomical one. Functions of the limbic system include (1) establishing emotional states; (2) linking the conscious, intellectual functions of the cerebral cortex with the unconscious and autonomic functions of the brain stem; and (3) facilitating memory storage and retrieval. Whereas the sensory cortex, motor cortex, and association areas of the cerebral cortex enable you to perform complex tasks, it is largely the limbic system that makes you *want* to do them. For this reason, the limbic system is also known as the *motivational system*.

Figure 14–11 focuses on major components of the limbic system. The **amygdaloid** (ah-MIG-da-loyd; *amygdale*, almond) **body** (**Figure 14–11b**), commonly referred to as the amygdala, appears to act as an interface between the limbic system, the cerebrum, and various sensory systems. It plays a role in the regulation of heart rate, in the control of the "fight or flight" response, and in linking emotions with specific memories. The **limbic lobe** of the cerebral hemisphere consists of the superficial folds, or *gyri*, and underlying structures adjacent to the diencephalon. The gyri curve along the *corpus callosum*, a fiber tract that links the two cerebral hemispheres, and continue onto the medial surface of the cerebrum lateral to the diencephalon (**Figure 14–11a**). There are three gyri in

the limbic lobe. The **cingulate** (SIN-gū-lāt) **gyrus** (*cingulum*, girdle or belt) sits superior to the corpus callosum. The **dentate gyrus** and the **parahippocampal** (pa-ra-hip-ō-KAM-pal) **gyrus** form the posterior and inferior portions of the limbic lobe. These gyri conceal the **hippocampus**, a nucleus inferior to the floor of the lateral ventricle. To early anatomists, this structure resembled a sea horse (*hippocampus*); it is important in learning, especially in the storage and retrieval of new long-term memories.

The **fornix** (FOR-niks, arch) is a tract of white matter that connects the hippocampus with the hypothalamus (**Figures 14–11 and 14–14b**). From the hippocampus, the fornix curves medially, meeting its counterpart from the opposing hemisphere. The fornix proceeds anteriorly, inferior to the corpus callosum, before curving toward the hypothalamus. Many fibers of the fornix end in the mamillary bodies of the hypothalamus.

Several other nuclei in the wall (thalamus) and floor (hypothalamus) of the diencephalon are components of the limbic

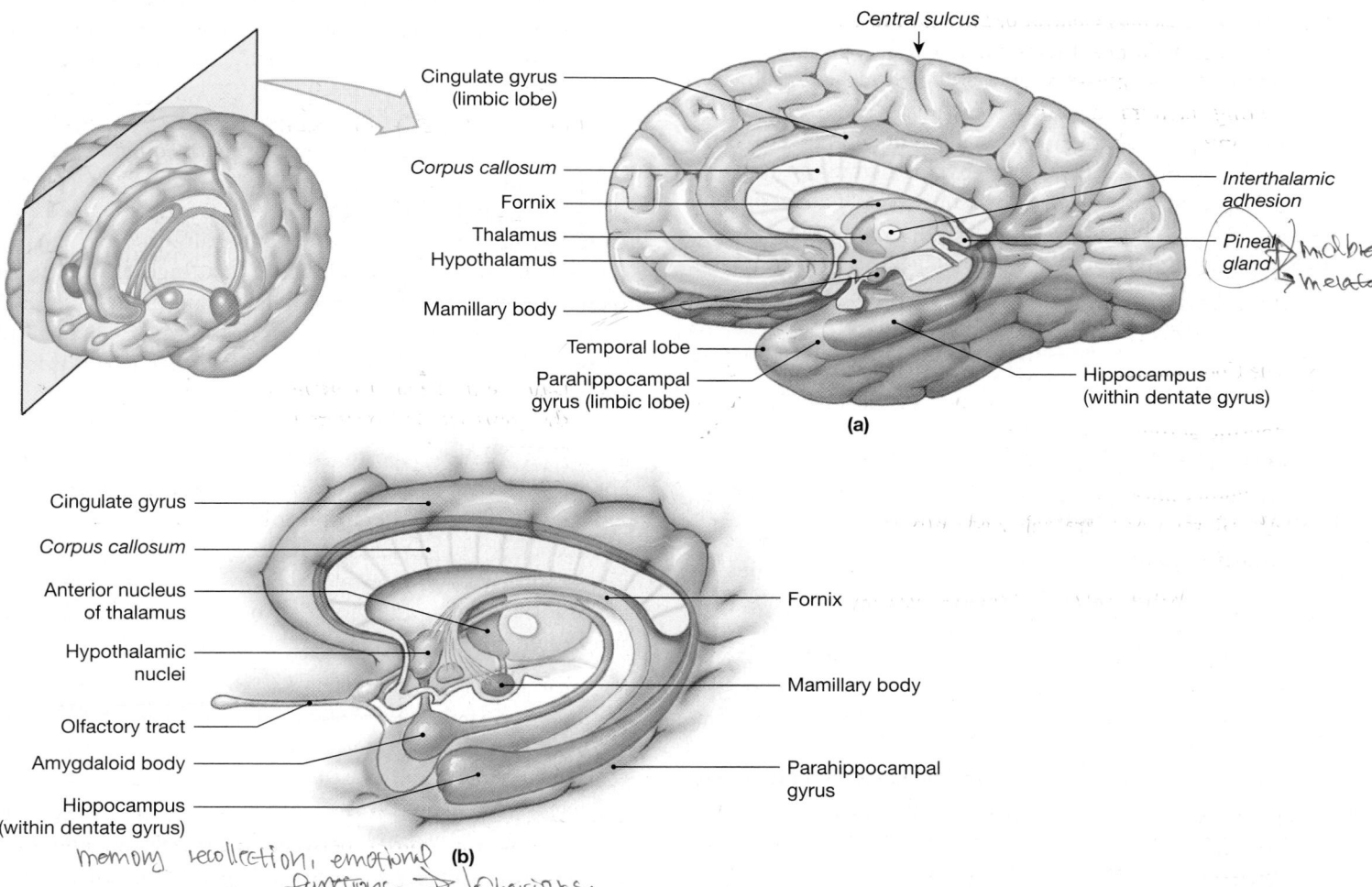

Figure 14–11 **The Limbic System. (a)** A diagrammatic sagittal section through the cerebrum, showing the cortical areas associated with the limbic system. The parahippocampal gyrus is shown as though transparent to make deeper limbic components visible. **(b)** A three-dimensional reconstruction of the limbic system, showing the relationships among the major components.

TABLE 14–7	The Limbic System

FUNCTIONS

Processing of memories; creation of emotional states, drives, and associated behaviors

CEREBRAL COMPONENTS

Cortical areas: limbic lobe (cingulate gyrus, dentate gyrus, and parahippocampal gyrus)
Nuclei: hippocampus, amygdaloid body
Tracts: fornix

DIENCEPHALIC COMPONENTS

Thalamus: anterior nuclear group
Hypothalamus: centers concerned with emotions, appetites (thirst, hunger), and related behaviors (*see Table 14–6*)

OTHER COMPONENTS

Reticular formation: network of interconnected nuclei throughout brain stem

system. The *anterior nucleus* of the thalamus (**Figure 14–11b**) relays information from the mamillary body (of the hypothalamus) to the cingulate gyrus on that side. The boundaries between the hypothalamic nuclei of the limbic system are often poorly defined, but experimental stimulation has outlined a number of important hypothalamic centers responsible for the emotions of rage, fear, pain, sexual arousal, and pleasure. The stimulation of specific regions of the hypothalamus can also produce heightened alertness and a generalized excitement or generalized lethargy and sleep. These responses are caused by the stimulation or inhibition of the reticular formation. Although the reticular formation extends the length of the brain stem, its headquarters reside in the mesencephalon.

Table 14–7 summarizes the organization and functions of the limbic system.

CHECKPOINT

19. What are the primary functions of the limbic system?
   emotional state, memory storage / retrieval.
20. Damage to the amygdaloid body would interfere with regulation of what division of the autonomic nervous system? *heart rate · cardiovascular.*

See the blue Answers tab at the end of the book.

14-9 The cerebrum, the largest region of the brain, contains motor, sensory, and association areas

The cerebrum is the largest region of the brain. Conscious thoughts and all intellectual functions originate in the cerebral hemispheres. Much of the cerebrum is involved in the processing of somatic sensory and motor information. Gray matter in the cerebrum is located in the *cerebral cortex* and in deeper *basal nuclei*. The white matter of the cerebrum lies deep to the neural cortex and around the basal nuclei.

The Cerebral Cortex

As previously noted, a blanket of neural cortex covers the paired cerebral hemispheres, which dominate the superior and lateral surfaces of the cerebrum. The gyri increase the surface area of the cerebral hemispheres, and thus the number of cortical neurons they contain; the total surface area of the cerebral hemispheres is roughly equivalent to 2200 cm^2 (2.5 ft^2) of flat surface. The entire brain has enlarged over the course of human evolution, but the cerebral hemispheres have enlarged at a much faster rate than has the rest of the brain, reflecting the large numbers of neurons needed for complex analytical and integrative functions. Since the neurons involved are in the superficial layer of cortex, it is there that the expansion has been most pronounced. The only solution available, other than an enlargement of the entire skull, was for the cortical layer to fold like a crumpled piece of paper.

Landmarks and features on the surface of one cerebral hemisphere are shown in **Figure 14–12a,b**. (The two cerebral hemispheres are almost completely separated by a deep **longitudinal fissure**, seen in **Figure 14–13b**). Each cerebral hemisphere can be divided into *lobes*, or regions, named after the overlying bones of the skull. Your brain has a unique pattern of sulci and gyri, as individual as a fingerprint, but the boundaries between lobes are reliable landmarks. On each hemisphere, the **central sulcus**, a deep groove, divides the anterior **frontal lobe** from the more posterior **parietal lobe**. The roughly horizontal **lateral sulcus** separates the frontal lobe from the **temporal lobe**. The **insula** (IN-sū-luh; *insula*, island), an "island" of cortex, lies medial to the lateral sulcus. The more posterior **parieto-occipital sulcus** separates the parietal lobe from the **occipital lobe**.

Each lobe contains functional regions whose boundaries are less clearly defined. Some of these regions deal with sensory information, and others with motor commands. Keep in mind three points about the cerebral lobes:

1. *Each Cerebral Hemisphere Receives Sensory Information from, and Sends Motor Commands to, the Opposite Side of the Body.* For example, the motor areas of the left cerebral hemisphere control muscles on the right side, and the right cerebral hemisphere controls muscles on the left side. This crossing over has no known functional significance.

2. *The Two Hemispheres Have Different Functions, Even Though They Look Almost Identical.* We will discuss these differences in a later section.

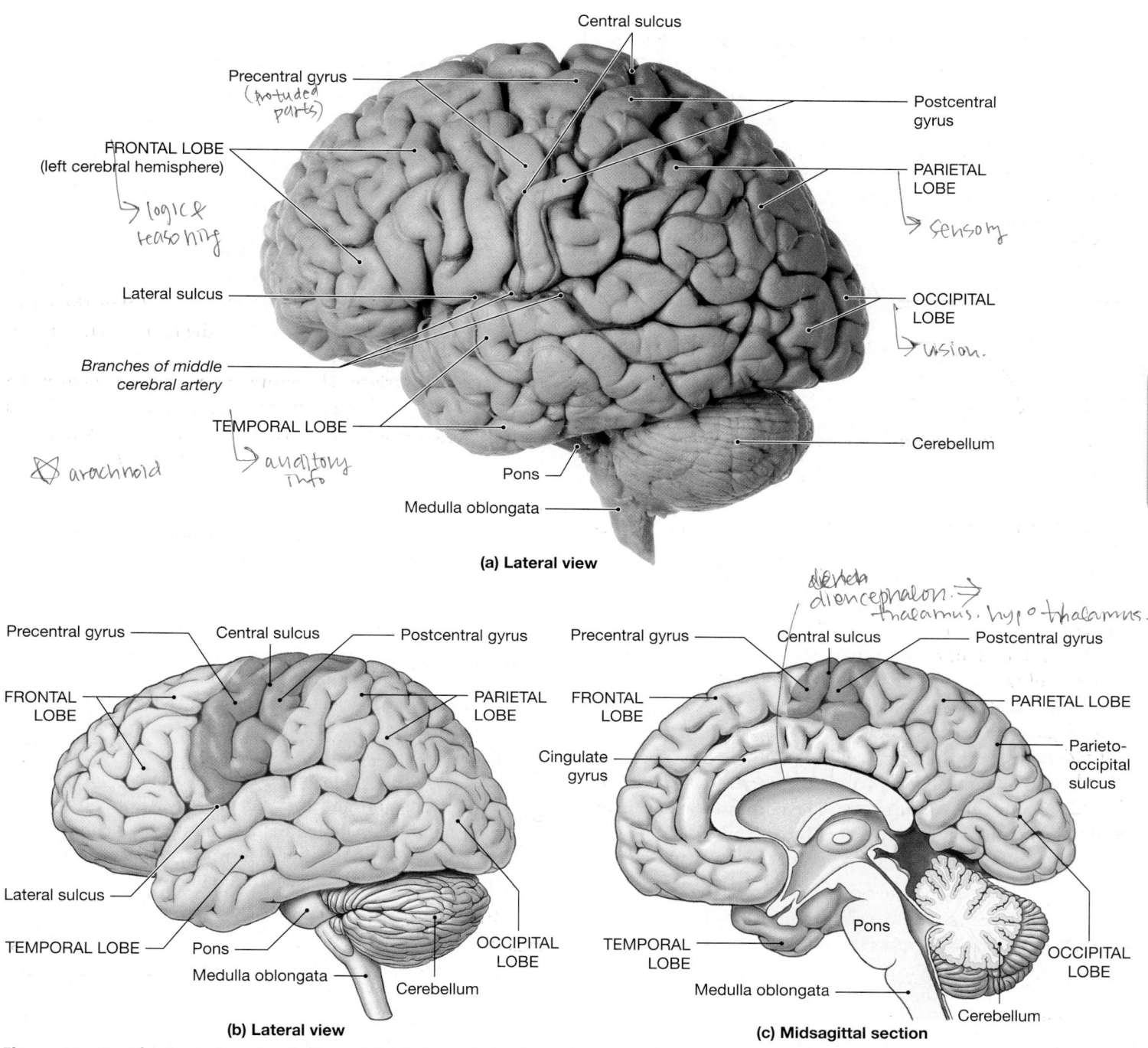

Figure 14–12   The Brain in Lateral View. ATLAS: Plates 11a,b; 13b–e; 14a–d

Handwritten annotations on figure (a):
- FRONTAL LOBE → logic & reasoning
- PARIETAL LOBE → sensory
- OCCIPITAL LOBE → vision
- TEMPORAL LOBE → auditory info
- arachnoid

Handwritten annotations on figure (c):
- diencephalon → thalamus, hypo-thalamus
- Precentral gyrus (protuded parts)

Figure labels, view (a) Lateral view:
Central sulcus, Precentral gyrus, Postcentral gyrus, FRONTAL LOBE (left cerebral hemisphere), PARIETAL LOBE, Lateral sulcus, OCCIPITAL LOBE, Branches of middle cerebral artery, TEMPORAL LOBE, Cerebellum, Pons, Medulla oblongata

(b) Lateral view:
Precentral gyrus, Central sulcus, Postcentral gyrus, FRONTAL LOBE, PARIETAL LOBE, Lateral sulcus, TEMPORAL LOBE, Pons, Medulla oblongata, Cerebellum, OCCIPITAL LOBE

(c) Midsagittal section:
Precentral gyrus, Central sulcus, Postcentral gyrus, FRONTAL LOBE, PARIETAL LOBE, Cingulate gyrus, Parieto-occipital sulcus, TEMPORAL LOBE, Pons, Medulla oblongata, Cerebellum, OCCIPITAL LOBE

3. *The Correspondence between a Specific Function and a Specific Region of the Cerebral Cortex Is Imprecise.* Because the boundaries are indistinct and have considerable overlap, one region may have several functions. Some aspects of cortical function, such as consciousness, cannot easily be assigned to any single region. However, we know that normal individuals use all portions of the brain.

The White Matter of the Cerebrum

The interior of the cerebrum consists primarily of white matter. The axons can be roughly classified as association fibers, commissural fibers, and projection fibers (**Figure 14–13**).

- **Association fibers** interconnect areas of neural cortex within a single cerebral hemisphere. Shorter association

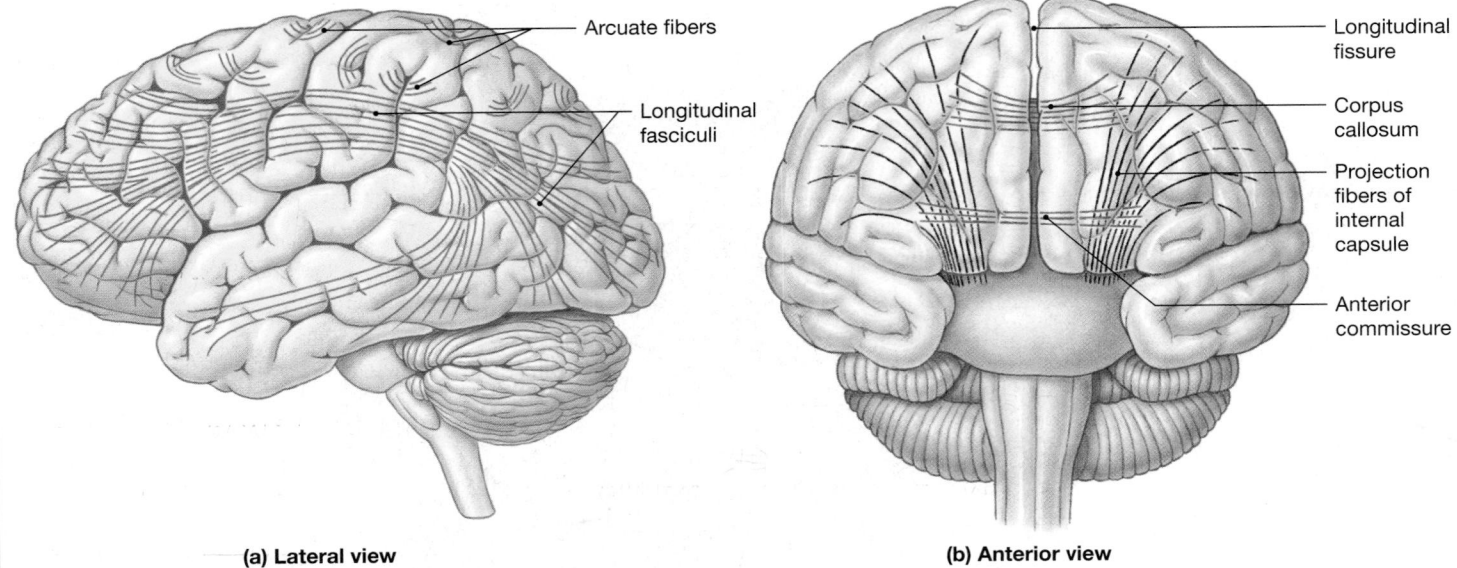

(a) Lateral view

(b) Anterior view

Figure 14–13   **Fibers of the White Matter of the Cerebrum.**

fibers are called **arcuate** (AR-kū-āt) **fibers**, because they curve in an arc to pass from one gyrus to another. Longer association fibers are organized into discrete bundles, or *fasciculi*. The **longitudinal fasciculi** connect the frontal lobe to the other lobes of the same hemisphere.

- **Commissural** (kom-i-SŪR-ul; *commissura*, crossing over) **fibers** interconnect and permit communication between the cerebral hemispheres. Bands of commissural fibers linking the hemispheres include the **corpus callosum** and the **anterior commissure**. The corpus callosum alone contains more than 200 million axons, carrying some 4 billion impulses per second!

- **Projection fibers** link the cerebral cortex to the diencephalon, brain stem, cerebellum, and spinal cord. All projection fibers must pass through the diencephalon, where axons heading to sensory areas of the cerebral cortex pass among the axons descending from motor areas of the cortex. In gross dissection, the ascending fibers and descending fibers look alike. The entire collection of projection fibers is known as the **internal capsule**.

The Basal Nuclei

While your cerebral cortex is consciously directing a complex movement or solving some intellectual puzzle, other centers of your cerebrum, diencephalon, and brain stem are processing sensory information and issuing motor commands outside your conscious awareness. Many of these activities, which occur at the subconscious level, are directed by the basal nuclei, or *cerebral nuclei*.

Anatomy of the Basal Nuclei

The **basal nuclei** are masses of gray matter that lie within each hemisphere deep to the floor of the lateral ventricle (**Figure 14–14**). They are embedded in the white matter of the cerebrum, and the radiating projection fibers and commissural fibers travel around or between these nuclei. Historically, the basal nuclei have been considered part of a larger functional group known as the *basal ganglia*. This group included the basal nuclei of the cerebrum and the associated motor nuclei in the diencephalon and mesencephalon. Although we will consider the functional interactions among these components in Chapter 15, we will avoid the term "basal ganglia" because ganglia are otherwise restricted to the PNS.

The **caudate nucleus** has a massive head and a slender, curving tail that follows the curve of the lateral ventricle. The head of the caudate nucleus lies anterior to the **lentiform** (lens-shaped) **nucleus**. The lentiform nucleus consists of a medial **globus pallidus** (GLŌ-bus PAL-i-dus; pale globe) and a lateral **putamen** (pū-TĀ-men). The term *corpus striatum* (striated body) has been used to refer to the caudate and lentiform nuclei, or to the caudate nucleus and putamen. The name refers to the striated (striped) appearance of the internal capsule as it passes among these nuclei. The amygdaloid body, part of the limbic system, lies anterior to the tail of the caudate nucleus and inferior to the lentiform nucleus.

Functions of the Basal Nuclei

The basal nuclei are involved with the subconscious control of skeletal muscle tone and the coordination of learned movement

14 NERVOUS

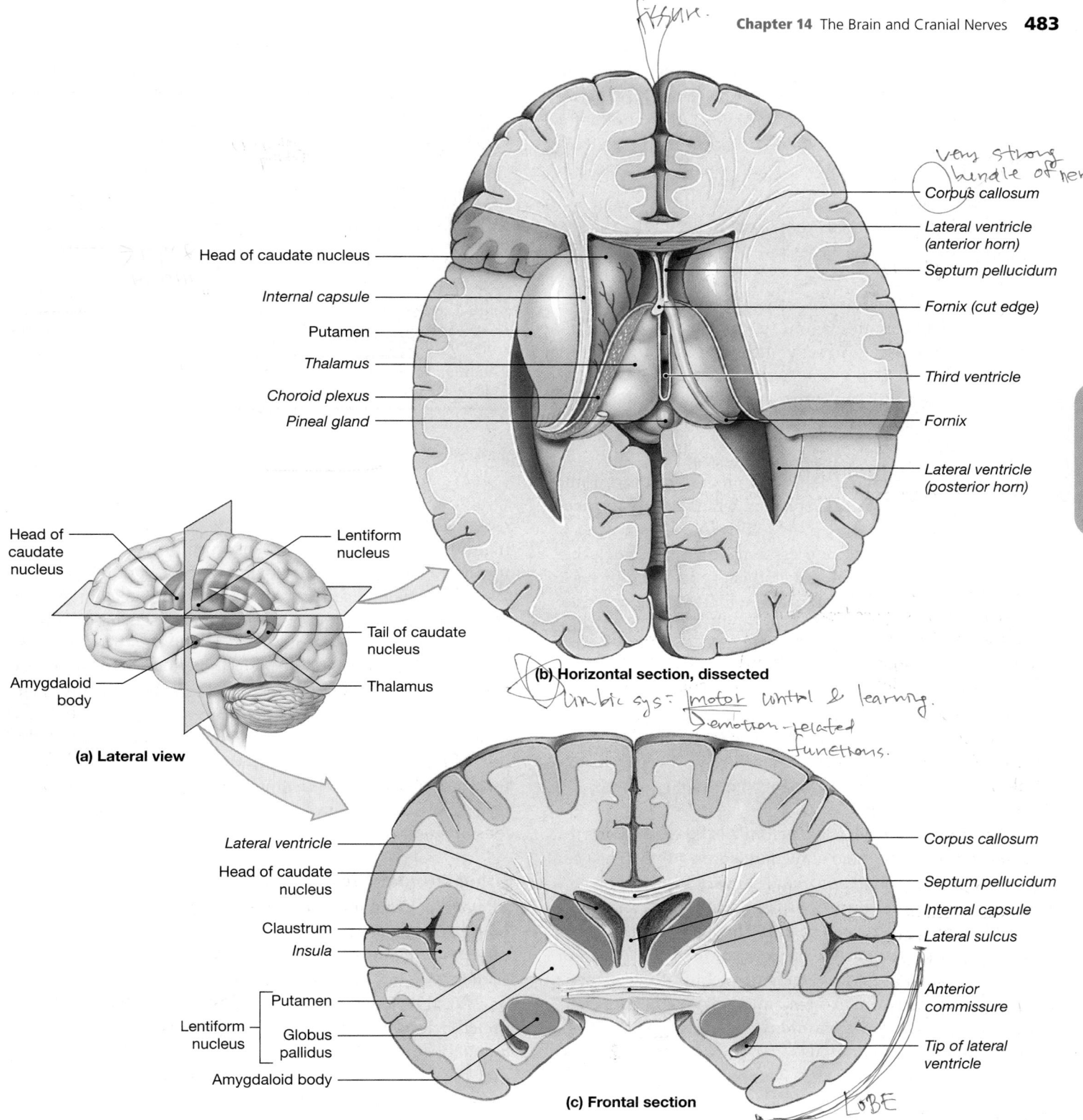

Corpus callosum — very strong bundle of nerves. *(handwritten)*

fissure (handwritten)

Head of caudate nucleus

Internal capsule

Putamen

Thalamus

Choroid plexus

Pineal gland

Corpus callosum

Lateral ventricle (anterior horn)

Septum pellucidum

Fornix (cut edge)

Third ventricle

Fornix

Lateral ventricle (posterior horn)

(b) Horizontal section, dissected

limbic sys = motor control & learning. → emotion-related functions. *(handwritten)*

Head of caudate nucleus

Lentiform nucleus

Tail of caudate nucleus

Amygdaloid body

Thalamus

(a) Lateral view

Lateral ventricle

Head of caudate nucleus

Claustrum

Insula

Putamen

Lentiform nucleus

Globus pallidus

Amygdaloid body

Corpus callosum

Septum pellucidum

Internal capsule

Lateral sulcus

Anterior commissure

Tip of lateral ventricle

LOBE *(handwritten)*

(c) Frontal section

Figure 14–14   The Basal Nuclei. (a) The relative positions of the basal nuclei in the intact brain. **(b)** A horizontal section as seen in a dissection. **(c)** Frontal section. ATLAS: Plates 11c; 13a,f

patterns. Under normal conditions, these nuclei do not initiate particular movements. But once a movement is under way, the basal nuclei provide the general pattern and rhythm, especially for movements of the trunk and proximal limb muscles.

Information arrives at the caudate nucleus and putamen from sensory, motor, and integrative areas of the cerebral cortex. Processing occurs in these nuclei and in the adjacent globus pallidus. Most of the output of the basal nuclei leaves the globus pallidus and synapses in the thalamus. Nuclei in the thalamus then project the information to appropriate areas of the cerebral cortex. The basal nuclei alter the motor commands issued by the cerebral cortex through this feedback loop. For example:

- When you walk, the basal nuclei control the cycles of arm and thigh movements that occur between the time you decide to "start" walking and the time you give the "stop" order.

- As you begin a voluntary movement, the basal nuclei control and adjust muscle tone, particularly in the appendicular muscles, to set your body position. When you decide to pick up a pencil, you consciously reach and grasp with your forearm, wrist, and hand while the basal nuclei operate at the subconscious level to position your shoulder and stabilize your arm.

Activity of the basal nuclei is inhibited by neurons in the substantia nigra of the mesencephalon, which release the neurotransmitter *dopamine*. ∞ p. 417 If the substantia nigra is damaged or the neurons secrete less dopamine, basal nuclei become more active. The result is a gradual, generalized increase in muscle tone and the appearance of symptoms characteristic of *Parkinson disease*. ∞ p. 107 People with Parkinson disease have difficulty starting voluntary movements, because opposing muscle groups do not relax; they must be overpowered. Once a movement is under way, every aspect must be voluntarily controlled through intense effort and concentration.

Motor and Sensory Areas of the Cortex

The major motor and sensory regions of the cerebral cortex are listed in Table 14–8 and shown in **Figure 14–15**. The central sulcus separates the motor and sensory areas of the cortex. The **precentral gyrus** of the frontal lobe forms the anterior border of the central sulcus. The surface of this gyrus is the **primary motor cortex**. Neurons of the primary motor cortex direct voluntary movements by controlling somatic motor neurons in the brain stem and spinal cord. These cortical neurons are called **pyramidal cells**, because their cell bodies resemble little pyramids.

The primary motor cortex is like the keyboard of a piano. If you strike a specific piano key, you produce a specific sound; if you stimulate a specific motor neuron in the primary motor cortex, you generate a contraction in a specific skeletal muscle.

Like the monitoring gauges in the dashboard of a car, the sensory areas of the cerebral cortex report key information. At each location, sensory information is reported in the pattern of neuron activity in the cortex. The **postcentral gyrus** of the parietal lobe forms the posterior border of the central sulcus, and its surface contains the **primary sensory cortex**. Neurons

TABLE 14–8	The Cerebral Cortex
Lobe/Region	**Function**
FRONTAL LOBE	
Primary motor cortex	Voluntary control of skeletal muscles
PARIETAL LOBE	
Primary sensory cortex	Conscious perception of touch, pressure, pain, vibration, taste, and temperature
OCCIPITAL LOBE	
Visual cortex	Conscious perception of visual stimuli
TEMPORAL LOBE	
Auditory cortex and olfactory cortex	Conscious perception of auditory and olfactory stimuli
ALL LOBES	
Association areas	Integration and processing of sensory data; processing and initiation of motor activities

in this region receive somatic sensory information from receptors for touch, pressure, pain, vibration, taste, or temperature. We are aware of these sensations only when nuclei in the thalamus relay the information to the primary sensory cortex.

Sensations of sight, sound, smell, and taste arrive at other portions of the cerebral cortex (**Figure 14–15a**). The **visual cortex** of the occipital lobe receives visual information, and the **auditory cortex** and **olfactory cortex** of the temporal lobe receive information about hearing and smell, respectively. The **gustatory cortex**, which receives information from taste receptors of the tongue and pharynx, lies in the anterior portion of the insula and adjacent portions of the frontal lobe.

Association Areas

The sensory and motor regions of the cortex are connected to nearby **association areas**, regions of the cortex that interpret incoming data or coordinate a motor response (**Figure 14–15a**). Like the information provided by the gauges in a car, the arriving information must be noticed and interpreted before the driver can take appropriate action. *Sensory association areas* are cortical regions that monitor and interpret the information that arrives at the sensory areas of the cortex. Examples include the somatic sensory association area, visual association area, and auditory association area.

The **somatic sensory association area** monitors activity in the primary sensory cortex. It is the somatic sensory association area that allows you to recognize a touch as light as the arrival of a mosquito on your arm (and gives you a chance to swat the mosquito before it bites).

The special senses of smell, sight, and hearing involve separate areas of the sensory cortex, and each has its own as-

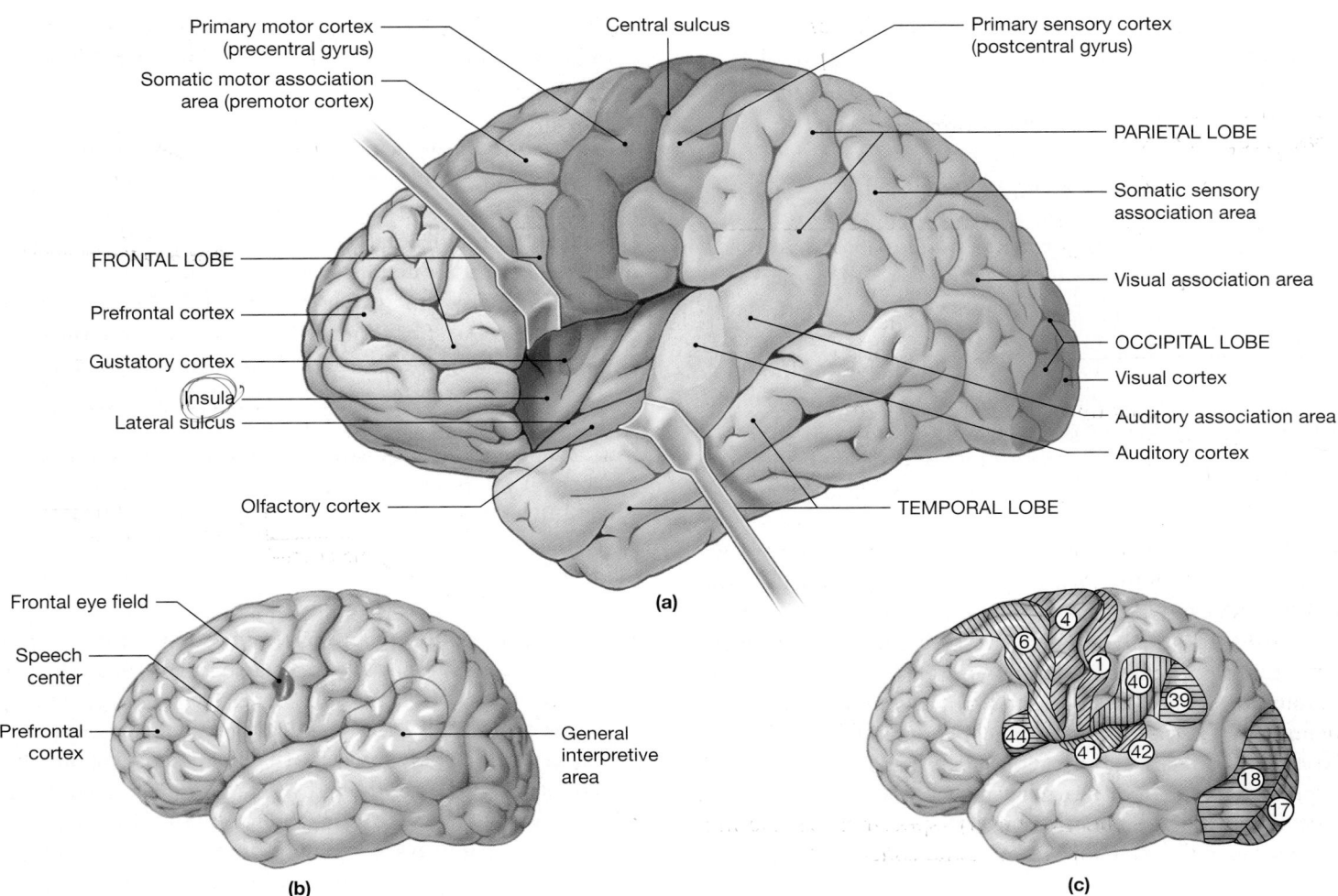

Figure 14–15 **Motor and Sensory Regions of the Cerebral Cortex.** (a) Major anatomical landmarks on the surface of the left cerebral hemisphere. The lateral sulcus has been pulled apart to expose the insula. (b) The left hemisphere generally contains the general interpretive area and the speech center. The prefrontal cortex of each hemisphere is involved with conscious intellectual functions. (c) Regions of the cerebral cortex as determined by histological analysis. Several of the 47 Brodmann areas (discussed in the text) are shown for comparison with the results of functional mapping.

sociation area. These areas monitor and interpret arriving sensations. For example, the **visual association area** monitors the patterns of activity in the visual cortex and interprets the results. You see the symbols *c*, *a*, and *r* when the stimulation of receptors in your eyes leads to the stimulation of neurons in your visual cortex. Your visual association area recognizes that these are letters and that *c* + *a* + *r* = *car*. An individual with a damaged visual association area could scan the lines of a printed page and see rows of symbols that are clear, but would perceive no meaning from the symbols. Similarly, the **auditory association area** monitors sensory activity in the auditory cortex; word recognition occurs in this association area.

The **somatic motor association area**, or **premotor cortex**, is responsible for the coordination of learned movements. The primary motor cortex does nothing on its own,

any more than a piano keyboard can play itself. The neurons in the primary motor cortex must be stimulated by neurons in other parts of the cerebrum. When you perform a voluntary movement, the instructions are relayed to the primary motor cortex by the premotor cortex. With repetition, the proper pattern of stimulation becomes stored in your premotor cortex. You can then perform the movement smoothly and easily by triggering the *pattern* rather than by controlling the individual neurons. This principle applies to any learned movement, from something as simple as picking up a glass to something as complex as playing the piano. One area of the premotor cortex, the *frontal eye field*, controls learned eye movements, such as when you scan these lines of type. Individuals with damage to the frontal eye field can understand written letters and words but cannot read, because their eyes cannot follow the lines on a printed page.

Integrative Centers

Integrative centers are areas that receive information from many association areas and direct extremely complex motor activities. These centers also perform complicated analytical functions. For example, the *prefrontal cortex* of the frontal lobe (**Figure 14–15b**) integrates information from sensory association areas and performs abstract intellectual functions, such as predicting the consequences of possible responses.

Integrative centers are located in the lobes and cortical areas of both cerebral hemispheres. Integrative centers concerned with the performance of complex processes, such as speech, writing, mathematical computation, and understanding spatial relationships, are restricted to either the left or the right hemisphere. These centers include the *general interpretive area* and the *speech center*. The corresponding regions on the opposite hemisphere are also active, but their functions are less well defined.

The General Interpretive Area. The **general interpretive area**, or the *Wernicke area* (**Figure 14–15b**), also called the *gnostic area*, receives information from all the sensory association areas. This analytical center is present in only one hemisphere (typically the left). This region plays an essential role in your personality by integrating sensory information and coordinating access to complex visual and auditory memories. Damage to the general interpretive area affects the ability to interpret what is seen or heard, even though the words are understood as individual entities. For example, if your general interpretive area were damaged, you might still understand the meaning of the spoken words *sit* and *here*, because word recognition occurs in the auditory association areas. But you would be totally bewildered by the request *sit here*. Damage to another portion of the general interpretive area might leave you able to see a chair clearly, and to know that you recognize it, but you would be unable to name it because the connection to your visual association area has been disrupted.

The Speech Center. Some of the neurons in the general interpretive area innervate the **speech center**, also called the *Broca area* or the *motor speech area* (**Figure 14–15b**). This center lies along the edge of the premotor cortex in the same hemisphere as the general interpretive area (usually the left). The speech center regulates the patterns of breathing and vocalization needed for normal speech. This regulation involves coordinating the activities of the respiratory muscles, the laryngeal and pharyngeal muscles, and the muscles of the tongue, cheeks, lips, and jaws. A person with a damaged speech center can make sounds but not words.

The motor commands issued by the speech center are adjusted by feedback from the auditory association area, also called the *receptive speech area*. Damage to the related sensory areas can cause a variety of speech-related problems. (See the discussion of *aphasia* on p. 488.) Some affected individuals have difficulty speaking although they know exactly which words to use; others talk constantly but use all the wrong words.

The Prefrontal Cortex. The **prefrontal cortex** of the frontal lobe (**Figure 14–15b**) coordinates information relayed from the association areas of the entire cortex. In doing so, it performs such abstract intellectual functions as predicting the consequences of events or actions. Damage to the prefrontal cortex leads to difficulties in estimating temporal relationships between events. Questions such as "How long ago did this happen?" or "What happened first?" become difficult to answer.

The prefrontal cortex has extensive connections with other cortical areas and with other portions of the brain. Feelings of frustration, tension, and anxiety are generated at the prefrontal cortex as it interprets ongoing events and makes predictions about future situations or consequences. If the connections between the prefrontal cortex and other brain regions are severed, the frustrations, tensions, and anxieties are removed. During the middle of the twentieth century, this rather drastic procedure, called **prefrontal lobotomy**, was used to "cure" a variety of mental illnesses, especially those associated with violent or antisocial behavior. After a lobotomy, the patient would no longer be concerned about what had previously been a major problem, whether psychological (hallucinations) or physical (severe pain). However, the individual was often equally unconcerned about tact, decorum, and toilet training. Drugs that target specific pathways and regions of the CNS have been developed, so lobotomies are no longer used to change behavior.

Brodmann Areas. Early in the 20th century, numerous researchers attempted to describe and classify regional differences in the histological organization of the cerebral cortex. They hoped to correlate the patterns of cellular organization with specific functions. By 1919, at least 200 patterns had been described, but most of the classification schemes have since been abandoned. However, the cortical map prepared by Korbinian Brodmann in 1909 has proved useful to neuroanatomists. Brodmann, a German neurologist, described 47 patterns of cellular organization in the cerebral cortex. Several of these *Brodmann areas* are shown in **Figure 14–15c**. Some correspond to known functional areas. For example, Brodmann area 44 corresponds to the speech center, and area 41 to the auditory cortex; area 4 follows the contours of the primary motor cortex. In other cases, the correspondence is less precise. For instance, Brodmann area 42 forms only a small portion of the auditory association area.

Hemispheric Lateralization

Each of the two cerebral hemispheres is responsible for specific functions that are not ordinarily performed by the opposite hemisphere. This regional specialization has been called

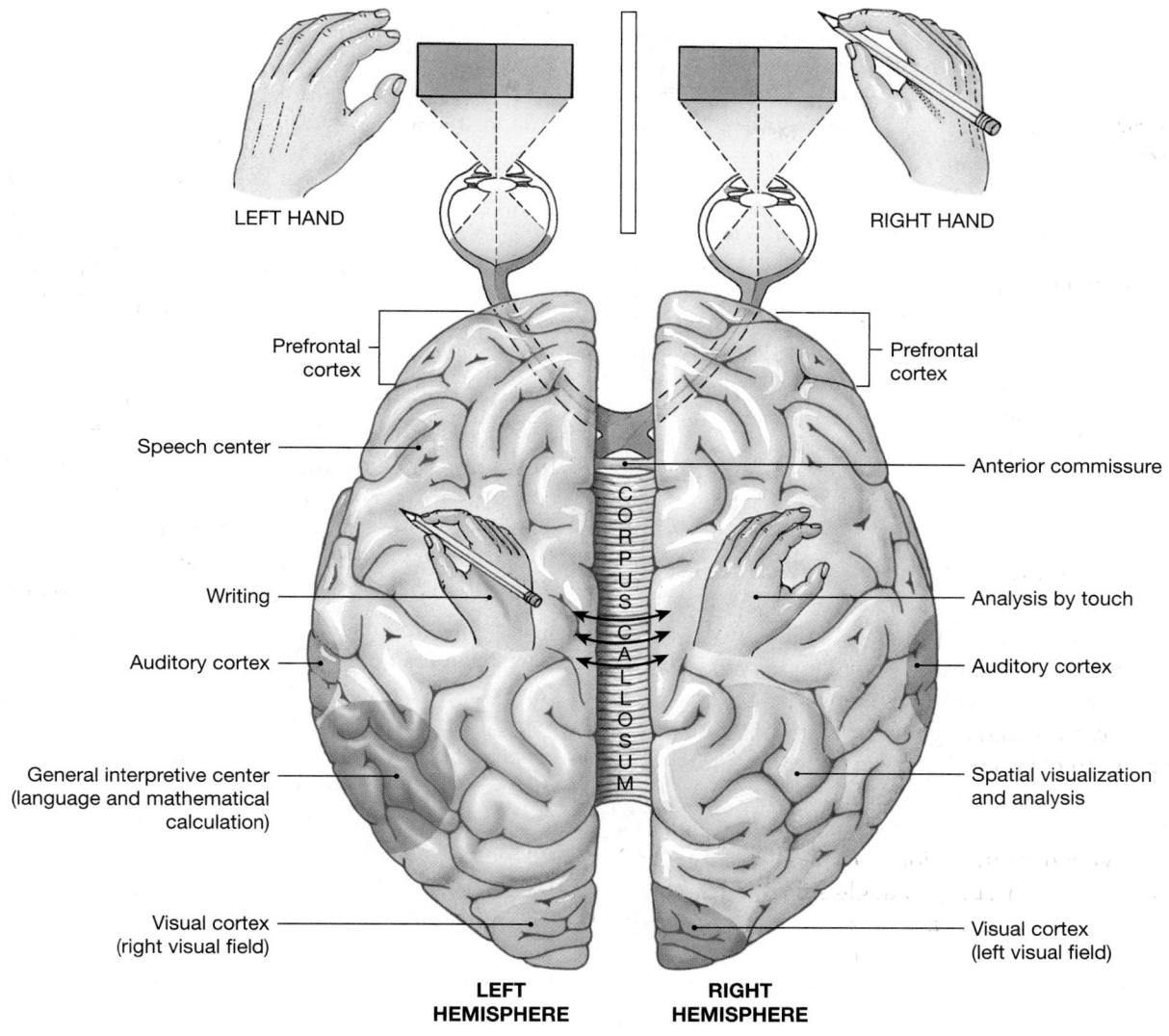

Figure 14–16 **Hemispheric Lateralization.** Functional differences between the left and right cerebral hemispheres.

hemispheric lateralization. **Figure 14–16** indicates the major functional differences between the hemispheres. In most people, the left hemisphere contains the general interpretive and speech centers and is responsible for language-based skills. For example, reading, writing, and speaking are dependent on processing done in the left cerebral hemisphere. In addition, the premotor cortex involved with the control of hand movements is larger on the left side for right-handed individuals than for left-handed ones. The left hemisphere is also important in performing analytical tasks, such as mathematical calculations and logical decision making. For these reasons, the left cerebral hemisphere has been called the *dominant hemisphere*, or the *categorical hemisphere*.

The right cerebral hemisphere analyzes sensory information and relates the body to the sensory environment. Interpretive centers in this hemisphere permit you to identify familiar objects by touch, smell, sight, taste, or feel. For ex-

ample, the right hemisphere plays a dominant role in recognizing faces and in understanding three-dimensional relationships. It is also important in analyzing the emotional context of a conversation—for instance, distinguishing between the threat "Get lost!" and the question "Get lost?" Individuals with a damaged right hemisphere may be unable to add emotional inflections to their own words.

Left-handed people represent 9 percent of the human population; in most cases, although the primary motor cortex of the right hemisphere controls motor function for the dominant hand, the centers involved with speech and analytical function are in the left hemisphere. Interestingly, an unusually high percentage of musicians and artists are left-handed. The complex motor activities performed by these individuals are directed by the primary motor cortex and association areas on the right cerebral hemisphere, near the association areas involved with spatial visualization and emotions.

CLINICAL NOTE

Aphasia and Dyslexia

Aphasia (*a-*, without + *phasia*, speech) is a disorder affecting the ability to speak or read. *Global aphasia* results from extensive damage to the general interpretive area or to the associated sensory tracts. Affected individuals are unable to speak, to read, or to understand the speech of others. Global aphasia often accompanies a severe stroke or tumor that affects a large area of cortex, including the speech and language areas. Recovery is possible when the condition results from edema or hemorrhage, but the process often takes months or even years.

Dyslexia (*dys-*, difficult, faulty + *lexis*, diction) is a disorder affecting the comprehension and use of written words. *Developmental dyslexia* affects children; estimates indicate that up to 15 percent of children in the United States have some degree of dyslexia. Children with dyslexia have difficulty reading and writing, although their other intellectual functions may be normal or above normal. Their writing looks uneven and disorganized; letters are typically written in the wrong order (*dig* becomes *gid*) or reversed (*E* becomes Ǝ). Recent evidence suggests that at least some forms of dyslexia result from problems in processing, sorting, and integrating visual or auditory information.

CLINICAL NOTE

Disconnection Syndrome The functional differences between the hemispheres become apparent if the corpus callosum is cut, a procedure sometimes performed to treat epileptic seizures that cannot be controlled by other methods. This surgery produces symptoms of **disconnection syndrome**. In this condition, the two hemispheres function independently, each "unaware" of stimuli or motor commands that involve its counterpart.

Individuals with this syndrome exhibit some rather interesting changes in their mental abilities. For example, objects touched by the left hand can be recognized but not verbally identified, because the sensory information arrives at the right hemisphere but the speech center is on the left. The object can be verbally identified if felt with the right hand, but the person cannot say whether it is the same object previously touched with the left hand. Sensory information from the left side of the body arrives at the right hemisphere and cannot reach the general interpretive area. Thus, conscious decisions are made without regard to sensations from the left side.

Two years after a surgical sectioning of the corpus callosum, the most striking behavioral abnormalities have disappeared and the person may test normally. In addition, individuals born without a functional corpus callosum do not have sensory, motor, or intellectual problems. In some way, the CNS adapts to these situations, probably by increasing the amount of information transferred across the anterior commissure.

Monitoring Brain Activity: The Electroencephalogram

The primary sensory cortex and the primary motor cortex have been mapped by direct stimulation in patients undergoing brain surgery. The functions of other regions of the cerebrum can be revealed by the behavioral changes that follow localized injuries or strokes, and the activities of specific regions can be examined by a PET scan or sequential MRI scans.

The electrical activity of the brain is commonly monitored to assess brain activity. Neural function depends on electrical events within the plasma membrane of neurons. The brain contains billions of neurons, and their activity generates an electrical field that can be measured by placing electrodes on the brain or on the outer surface of the skull. The electrical activity changes constantly, as nuclei and cortical areas are stimulated or they quiet down. A printed report of the electrical activity of the brain is called an **electroencephalogram (EEG)**. The electrical patterns observed are called *brain waves*.

Typical brain waves are shown in **Figure 14–17a–d**. **Alpha waves** occur in the brains of healthy, awake adults who are resting with their eyes closed. Alpha waves disappear during sleep, but they also vanish when the individual begins to concentrate on some specific task. During attention to stimuli or tasks, alpha waves are replaced by higher-frequency **beta waves**. Beta waves are typical of individuals who are either concentrating on a task, under stress, or in a state of psychological tension. **Theta waves** may appear transiently during sleep in normal adults but are most often observed in children and in intensely frustrated adults. The presence of theta waves under other circumstances may indicate the presence of a brain disorder, such as a tumor.

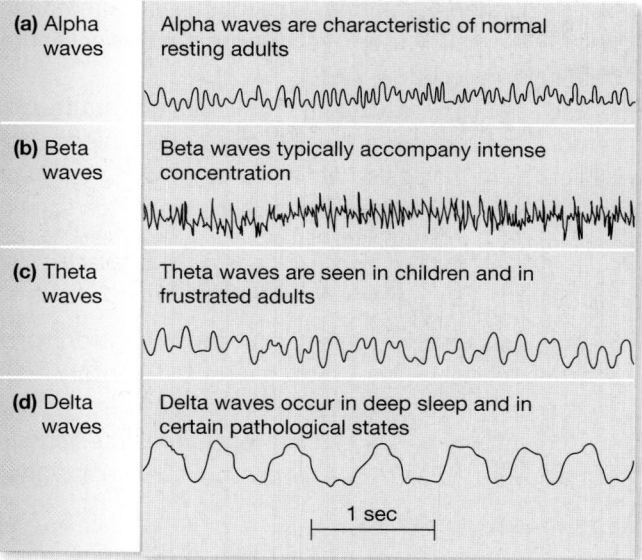

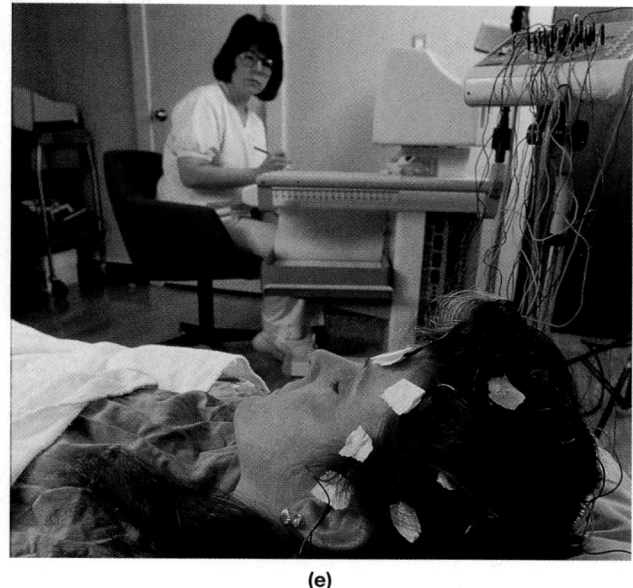

(e)

Figure 14–17   Brain Waves. (a)–(d) The four electrical patterns revealed by electroencephalograms (EEGs). The heights (amplitudes) of the four waves are not drawn to the same scale. **(e)** A patient wired for EEG monitoring.

Delta waves are very-large-amplitude, low-frequency waves. They are normally seen during deep sleep in individuals of all ages. Delta waves are also seen in the brains of infants (in whom cortical development is still incomplete) and in awake adults when a tumor, vascular blockage, or inflammation has damaged portions of the brain.

Electrical activity in the two hemispheres is generally synchronized by a "pacemaker" mechanism that appears to involve the thalamus. Asynchrony between the hemispheres can therefore indicate localized damage or other cerebral abnormalities. For example, a tumor or injury affecting one hemisphere typically changes the pattern in that hemisphere, and the patterns of the two hemispheres are no longer aligned. A **seizure** is a temporary cerebral disorder accompanied by abnormal movements, unusual sensations, inappropriate behavior, or some combination of these symptoms. Clinical conditions characterized by seizures are known as seizure disorders, or *epilepsies*. Seizures of all kinds are accompanied by a marked change in the pattern of electrical activity recorded in an electroencephalogram. The change begins in one portion of the cerebral cortex but may subsequently spread across the entire cortical surface, like a wave on the surface of a pond.

The nature of the signs and symptoms produced depends on the region of the cortex involved. If a seizure affects the primary motor cortex, movements will occur; if it affects the auditory cortex, the individual will hear strange sounds.

CHECKPOINT

21. What name is given to fibers carrying information between the brain and spinal cord, and through which brain regions do they pass?

22. What symptoms would you expect to observe in an individual who has damage to the basal nuclei?

23. A patient suffers a head injury that damages her primary motor cortex. Where is this area located?

24. Which senses would be affected by damage to the temporal lobes of the cerebrum?

25. After suffering a stroke, a patient is unable to speak. He can understand what is said to him, and he can understand written messages, but he cannot express himself verbally. Which part of his brain has been affected by the stroke?

26. A patient is having a difficult time remembering facts and recalling long-term memories. Which part of his cerebrum is probably involved?

See the blue Answers tab at the end of the book.

Cranial Nerves

Cranial nerves are PNS components that connect directly to the brain. The twelve pairs of cranial nerves are visible on the ventral surface of the brain (**Figure 14–18**); each has a name related to its appearance or its function.

The number assigned to a cranial nerve roughly corresponds to the nerve's position along the longitudinal axis of the brain,

beginning at the cerebrum. Roman numerals preceded by the letter N are usually used. (You may sometimes encounter these numerals preceded by the letters CN.)

Each cranial nerve attaches to the brain near the associated sensory or motor nuclei. The sensory nuclei act as switching centers, with the postsynaptic neurons relaying the information to other nuclei or to processing centers in the cerebral or cerebellar cortex. In a similar fashion, the motor nuclei receive convergent inputs from higher centers or from other nuclei along the brain stem.

In this section, we classify cranial nerves as primarily sensory, special sensory, motor, or mixed (sensory and motor). In this classification, sensory nerves carry somatic sensory information, including touch, pressure, vibration, temperature, or pain. Special sensory nerves carry the sensa-

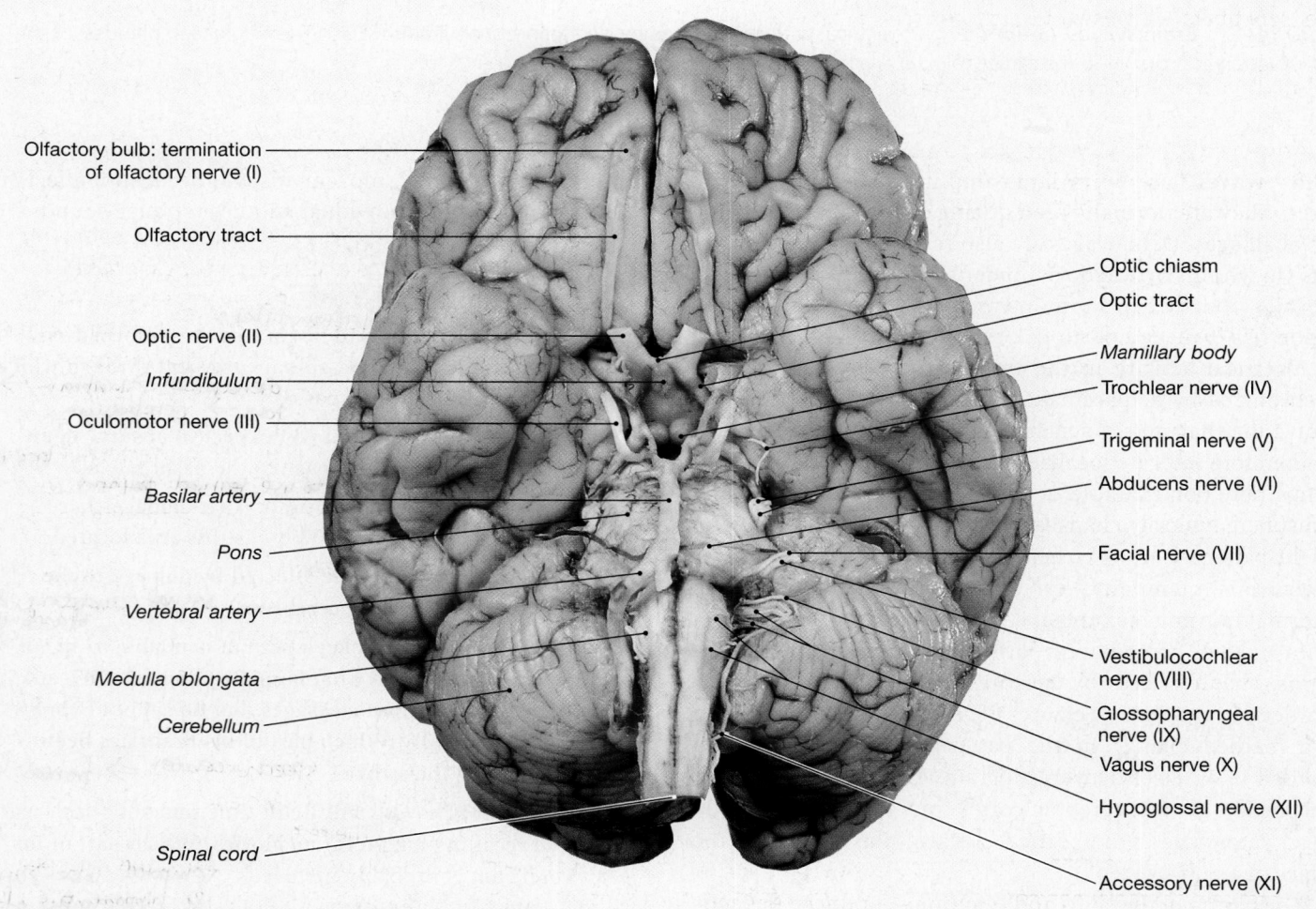

Figure 14–18 **Origins of the Cranial Nerves.** An inferior view of the brain.

- Olfactory bulb: termination of olfactory nerve (I)
- Olfactory tract
- Optic nerve (II)
- Infundibulum
- Oculomotor nerve (III)
- Basilar artery
- Pons
- Vertebral artery
- Medulla oblongata
- Cerebellum
- Spinal cord
- Optic chiasm
- Optic tract
- Mamillary body
- Trochlear nerve (IV)
- Trigeminal nerve (V)
- Abducens nerve (VI)
- Facial nerve (VII)
- Vestibulocochlear nerve (VIII)
- Glossopharyngeal nerve (IX)
- Vagus nerve (X)
- Hypoglossal nerve (XII)
- Accessory nerve (XI)

tions of smell, sight, hearing, or balance. Motor nerves are dominated by the axons of somatic motor neurons; mixed nerves have a mixture of sensory and motor fibers. This is a useful classification scheme, but it is based on the primary function, and a cranial nerve can have important secondary functions. Three examples are worth noting:

1. The olfactory receptors, the visual receptors, and the receptors of the inner ear are innervated by cranial nerves that are dedicated almost entirely to carrying special sensory information. The sensation of taste, considered to be one of the special senses, is carried by axons that form only a small part of large cranial nerves that have other primary functions.

2. As elsewhere in the PNS, a nerve containing tens of thousands of motor fibers that lead to a skeletal muscle will also contain sensory fibers from muscle spindles and tendon organs in that muscle. We assume that these sensory fibers are present but ignore them in the classification of the nerve.

3. Regardless of their other functions, several cranial nerves (III, VII, IX, and X) distribute autonomic fibers to peripheral ganglia, just as spinal nerves deliver them to ganglia along the spinal cord. We will note the presence of small numbers of autonomic fibers (and will discuss them further in Chapter 16) but ignore them in the classification of the nerve.

Tips&Tricks

Two useful mnemonics for remembering the names of the cranial nerves in order are "**O**h **O**h **O**h, **T**o **T**ouch **A**nd **F**eel **V**ery **G**reen **V**egetables, **A**h **H**eaven!" and "**O**h, **O**nce **O**ne **T**akes **T**he **A**natomy **F**inal, **V**ery **G**ood **V**acations **A**re **H**eavenly!" Another to assist with remembering cranial nerve function, is "N III, N IV, N V keep the diaphragm alive."

The Olfactory Nerves (I)

Primary function: Special sensory (smell)

Origin: Receptors of olfactory epithelium

Pass through: Olfactory foramina in cribriform plate of ethmoid ∞ pp. 217, 222

Destination: Olfactory bulbs

The first pair of cranial nerves (**Figure 14–19**) carries special sensory information responsible for the sense of smell. The

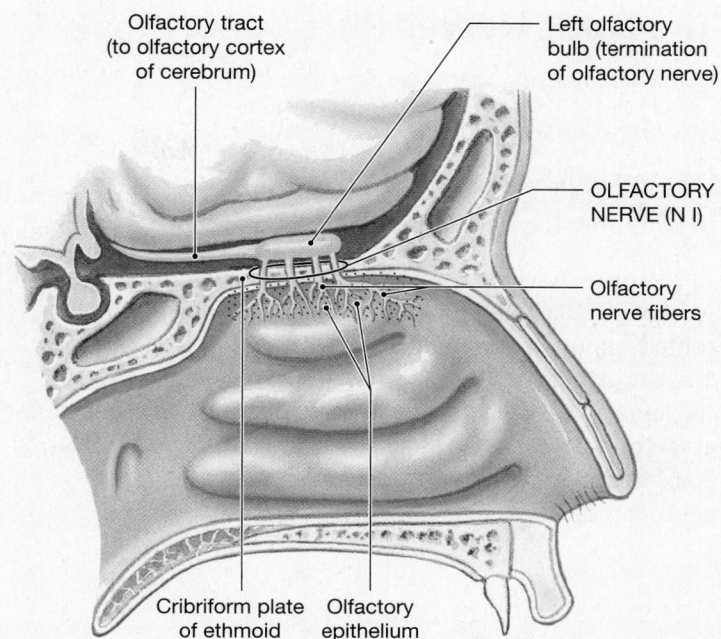

Figure 14–19   **The Olfactory Nerve.**

olfactory receptors are specialized neurons in the epithelium covering the roof of the nasal cavity, the superior nasal conchae, and the superior parts of the nasal septum. Axons from these sensory neurons collect to form 20 or more bundles that penetrate the cribriform plate of the ethmoid bone. These bundles are components of the **olfactory nerves** (I). Almost at once these bundles enter the **olfactory bulbs**, neural masses on either side of the crista galli. The olfactory afferents synapse within the olfactory bulbs. The axons of the post-synaptic neurons proceed to the cerebrum along the slender **olfactory tracts** (**Figures 14–18** and **14–19**).

Because the olfactory tracts look like typical peripheral nerves, anatomists about a century ago misidentified these tracts as the first cranial nerve. Later studies demonstrated that the olfactory tracts and bulbs are part of the cerebrum, but by then the numbering system was already firmly established. Anatomists were left with a forest of tiny olfactory nerve bundles lumped together as cranial nerve I.

The olfactory nerves are the only cranial nerves attached directly to the cerebrum. The rest originate or terminate within nuclei of the diencephalon or brain stem, and the ascending sensory information synapses in the thalamus before reaching the cerebrum.

The Optic Nerves (II)

Primary function: Special sensory (vision)

Origin: Retina of eye

Pass through: Optic canals of sphenoid ∞ p. 221

Destination: Diencephalon via the optic chiasm

The **optic nerves** (II) carry visual information from special sensory ganglia in the eyes. These nerves (**Figure 14–20**) contain about 1 million sensory nerve fibers. The optic nerves pass through the optic canals of the sphenoid. Then they converge at the ventral, anterior margin of the diencephalon, at the **optic chiasm** (*chiasma*, a crossing). At the optic chiasm, fibers from the nasal half of each retina cross over to the opposite side of the brain.

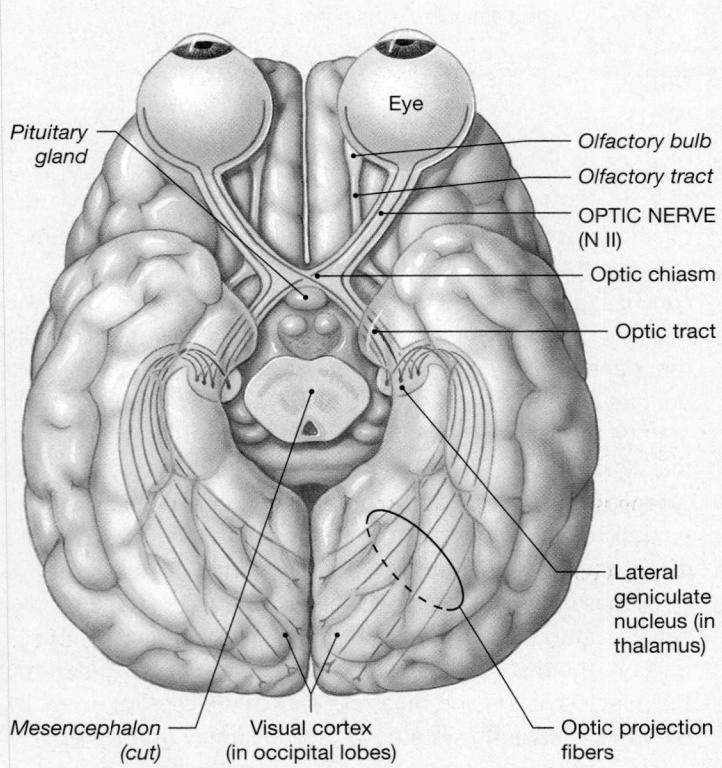

Eye
Pituitary gland
Olfactory bulb
Olfactory tract
OPTIC NERVE (N II)
Optic chiasm
Optic tract
Lateral geniculate nucleus (in thalamus)
Mesencephalon (cut)
Visual cortex (in occipital lobes)
Optic projection fibers

Figure 14–20 The Optic Nerve.

The reorganized axons continue toward the lateral geniculate nuclei of the thalamus as the **optic tracts** (**Figures 14–18 and 14–20**). After synapsing in the lateral geniculates, projection fibers deliver the information to the visual cortex of the occipital lobe. With this arrangement, each cerebral hemisphere receives visual information from the lateral half of the retina of the eye on that side and from the medial half of the retina of the eye of the opposite side (**Figure 14–16**). Relatively few axons in the optic tracts bypass the lateral geniculate nuclei and synapse in the superior colliculus of the midbrain. We will consider that pathway in Chapter 17.

The Oculomotor Nerves (III)

Primary function: Motor (eye movements)

Origin: Mesencephalon

Pass through: Superior orbital fissures of sphenoid ∞ pp. 216, 221, 224, 228

Destination: *Somatic motor*: superior, inferior, and medial rectus muscles; inferior oblique muscle; levator palpebrae superioris muscle. *Visceral motor*: intrinsic eye muscles

The mesencephalon contains the motor nuclei controlling the third and fourth cranial nerves. Each **oculomotor nerve** (III) innervates four of the six extra-ocular muscles that move the eye, and the levator palpebrae superioris muscle, which raises the upper eyelid (**Figure 14–21**). On each side of the brain, nerve III emerges from the ventral surface of the mesencephalon and penetrates the posterior wall of the orbit at the superior orbital fissure. Individuals with damage to this nerve often complain of pain over the eye, droopy eyelids, and double vision, because the movements of the left and right eyes cannot be coordinated properly.

The oculomotor nerve also delivers preganglionic autonomic fibers to neurons of the **ciliary ganglion**. The neurons of the ciliary ganglion control intrinsic eye muscles. These muscles change the diameter of the pupil, adjusting the amount of light entering the eye, and change the shape of the lens to focus images on the retina.

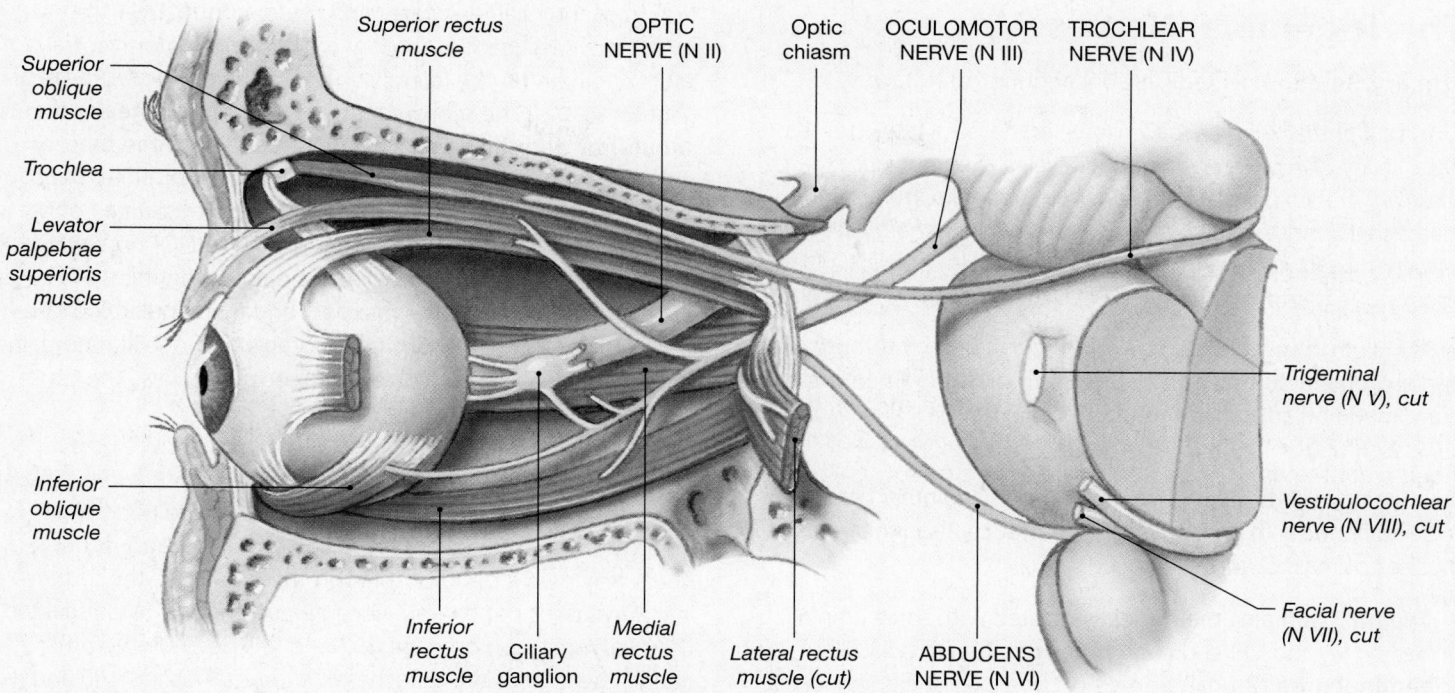

Figure 14–21 **Cranial Nerves Controlling the Extra-Ocular Muscles.** ATLAS: Plate 16a,b

Labels in figure:
Superior oblique muscle
Trochlea
Levator palpebrae superioris muscle
Inferior oblique muscle
Superior rectus muscle
OPTIC NERVE (N II)
Optic chiasm
OCULOMOTOR NERVE (N III)
TROCHLEAR NERVE (N IV)
Trigeminal nerve (N V), cut
Vestibulocochlear nerve (N VIII), cut
Facial nerve (N VII), cut
Inferior rectus muscle
Ciliary ganglion
Medial rectus muscle
Lateral rectus muscle (cut)
ABDUCENS NERVE (N VI)

The Trochlear Nerves (IV)

Primary function: Motor (eye movements)

Origin: Mesencephalon

Pass through: Superior orbital fissures of sphenoid ∞ pp. 216, 221, 224, 228

Destination: Superior oblique muscle

A **trochlear** (TRŌK-lē-ar; *trochlea*, a pulley) **nerve** (IV), the smallest cranial nerve, innervates the superior oblique muscle of each eye (**Figure 14–21**). The trochlea is a pulley-shaped, ligamentous sling. Each superior oblique muscle passes through a trochlea on its way to its insertion on the surface of the eye. An individual with damage to cranial nerve IV or to its nucleus will have difficulty looking down and to the side.

The Abducens Nerves (VI)

Primary function: Motor (eye movements)

Origin: Pons

Pass through: Superior orbital fissures of sphenoid ∞ pp. 216, 221, 224, 228

Destination: Lateral rectus muscle

The **abducens** (ab-DŪ-senz) **nerves** (VI) innervate the lateral rectus muscles, the sixth pair of extra-ocular muscles. Contraction of the lateral rectus muscle makes the eye look to the side; in essence, the *abducens* causes *abduction* of the eye. Each abducens nerve emerges from the inferior surface of the brain stem at the border between the pons and the medulla oblongata (**Figure 14–21**). Along with the oculomotor and trochlear nerves from that side, it reaches the orbit through the superior orbital fissure.

The Trigeminal Nerves (V)

Primary function: Mixed (sensory and motor) to face

Origin: *Ophthalmic branch* (sensory): orbital structures, cornea, nasal cavity, skin of forehead, upper eyelid, eyebrow, nose (part). *Maxillary branch* (sensory): lower eyelid, upper lip, gums, and teeth; cheek; nose, palate, and pharynx (part). *Mandibular branch* (mixed): sensory from lower gums, teeth, and lips; palate and tongue (part); motor from motor nuclei of pons

Pass through (on each side): Ophthalmic branch through superior orbital fissure, maxillary branch through foramen rotundum, mandibular branch through foramen ovale ∞ pp. 216, 217, 221, 228

Destination: Ophthalmic, maxillary, and mandibular branches to sensory nuclei in pons; mandibular branch also innervates muscles of mastication ∞ p. 349

The pons contains the nuclei associated with three cranial nerves (V, VI, and VII) and contributes to a fourth (VIII). The **trigeminal** (trī-JEM-i-nal) **nerves** (V), the largest cranial nerves, are mixed nerves. Each provides both somatic sensory informa-tion from the head and face, and motor control over the muscles of mastication. Sensory (dorsal) and motor (ventral) roots originate on the lateral surface of the pons (**Figure 14–22**). The sensory branch is larger, and the enormous **semilunar ganglion** contains the cell bodies of the sensory neurons. As the name implies, the trigeminal has three major branches; the relatively small motor root contributes to only one of the three. **Tic douloureux** (doo-luh-ROO; *douloureux*, painful), or *trigeminal neuralgia*, is a painful condition affecting the area innervated by the maxillary and mandibular branches of the trigeminal nerve. Sufferers complain of debilitating pain triggered by contact with the lip, tongue, or gums. The cause of the condition is unknown.

The trigeminal nerve branches are associated with the *ciliary*, *sphenopalatine*, *submandibular*, and *otic ganglia*. These are autonomic (parasympathetic) ganglia whose neurons innervate structures of the face. However, although its nerve fibers may pass around or through these ganglia, the trigeminal nerve does not contain visceral motor fibers. We discussed the ciliary ganglion on page 492 and will describe the other ganglia next, with the branches of the *facial nerves* (VII) and the *glossopharyngeal nerves* (IX).

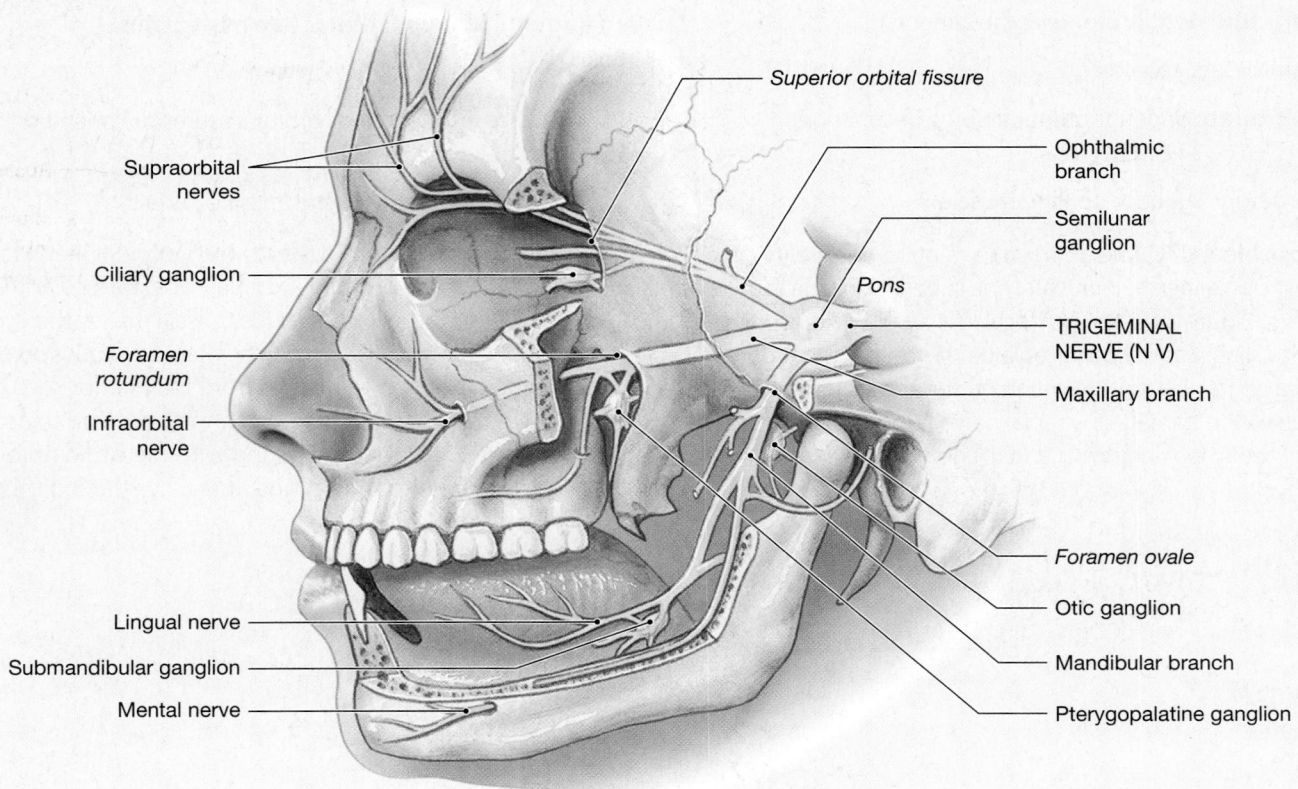

Figure 14–22   **The Trigeminal Nerve.**

The Facial Nerves (VII)

Primary function: Mixed (sensory and motor) to face

Origin: *Sensory*: taste receptors on anterior two-thirds of tongue. *Motor*: motor nuclei of pons

Pass through: Internal acoustic meatus to the *facial canals*, which end at the stylomastoid foramina ∞ pp. 216, 217, 221

Destination: *Sensory*: sensory nuclei of pons. *Somatic motor*: muscles of facial expression. ∞ p. 345 *Visceral motor*: lacrimal (tear) gland and nasal mucous glands by way of the pterygopalatine ganglion; submandibular and sublingual salivary glands by way of the submandibular ganglion

The **facial nerves** (VII) are mixed nerves. The cell bodies of the sensory neurons are located in the **geniculate ganglia**, and the motor nuclei are in the pons. On each side, the sensory and motor roots emerge from the pons and enter the internal acoustic meatus of the temporal bone. Each facial nerve then passes through the facial canal to reach the face by way of the stylomastoid foramen. The nerve then splits to form the temporal, zygomatic, buccal, mandibular, and cervical branches (**Figure 14–23**).

The sensory neurons monitor proprioceptors in the facial muscles, provide deep pressure sensations over the face, and receive taste information from receptors along the anterior two-thirds of the tongue. Somatic motor fibers control the superficial muscles of the scalp and face and deep muscles near the ear.

The facial nerves carry preganglionic autonomic fibers to the pterygopalatine and submandibular ganglia. Postganglionic fibers from the **pterygopalatine ganglia** innervate the lacrimal glands and small glands of the nasal cavity and pharynx. The **submandibular ganglia** innervate the *submandibular* and *sublingual* (*sub-*, under + *lingual*, pertaining to the tongue) *salivary glands*.

Bell palsy is a cranial nerve disorder that results from an inflammation of a facial nerve. The condition is probably due to a viral infection. Signs and symptoms include paralysis of facial muscles on the affected side and loss of taste sensations from the anterior two-thirds of the tongue. The condition is usually painless and in most cases the symptoms fade after a few weeks or months.

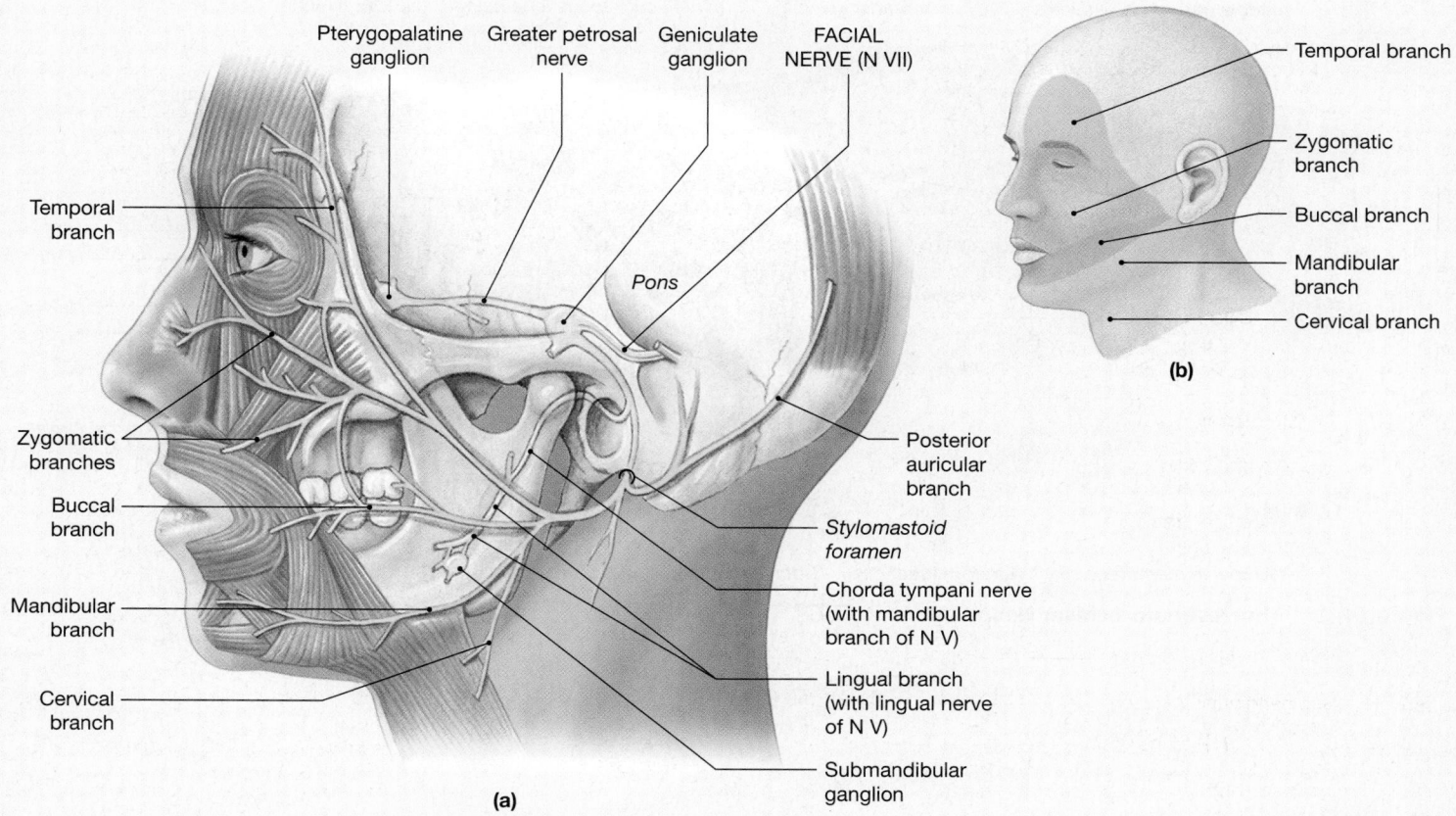

Figure 14–23 **The Facial Nerve.** (a) The origin and branches of the facial nerve. (b) The superficial distribution of the five major branches of the facial nerve.

The Vestibulocochlear Nerves (VIII)

Primary function: Special sensory: balance and equilibrium (vestibular branch) and hearing (cochlear branch)

Origin: Monitor receptors of the inner ear (vestibule and cochlea)

Pass through: Internal acoustic meatus of temporal bones
∞ pp. 217, 221

Destination: Vestibular and cochlear nuclei of pons and medulla oblongata

The **vestibulocochlear nerves** (VIII) are also known as the *acoustic nerves*, the *auditory nerves*, and the *stato-acoustic nerves*. We will use *vestibulocochlear*, because this term indicates the names of the two major branches: the vestibular branch and the cochlear branch. Each vestibulocochlear nerve lies posterior to the origin of the facial nerve, straddling the boundary between the pons and the medulla oblongata

(**Figure 14–24**). This nerve reaches the sensory receptors of the inner ear by entering the internal acoustic meatus in company with the facial nerve. Each vestibulocochlear nerve has two distinct bundles of sensory fibers. The **vestibular** (*vestibulum*, cavity) **branch** originates at the receptors of the *vestibule*, the portion of the inner ear concerned with balance sensations. The sensory neurons are located in an adjacent sensory ganglion, and their axons target the **vestibular nuclei** of the pons and medulla oblongata. These afferents convey information about the orientation and movement of the head. The **cochlear** (*cochlea*, snail shell) **branch** monitors the receptors in the *cochlea*, the portion of the inner ear that provides the sense of hearing. The cell bodies of the sensory neurons are located within a peripheral ganglion (the *spiral ganglion*), and their axons synapse within the **cochlear nuclei** of the pons and medulla oblongata. Axons leaving the vestibular and cochlear nuclei relay the sensory information to other centers or initiate reflexive motor responses. We will discuss balance and the sense of hearing in Chapter 17.

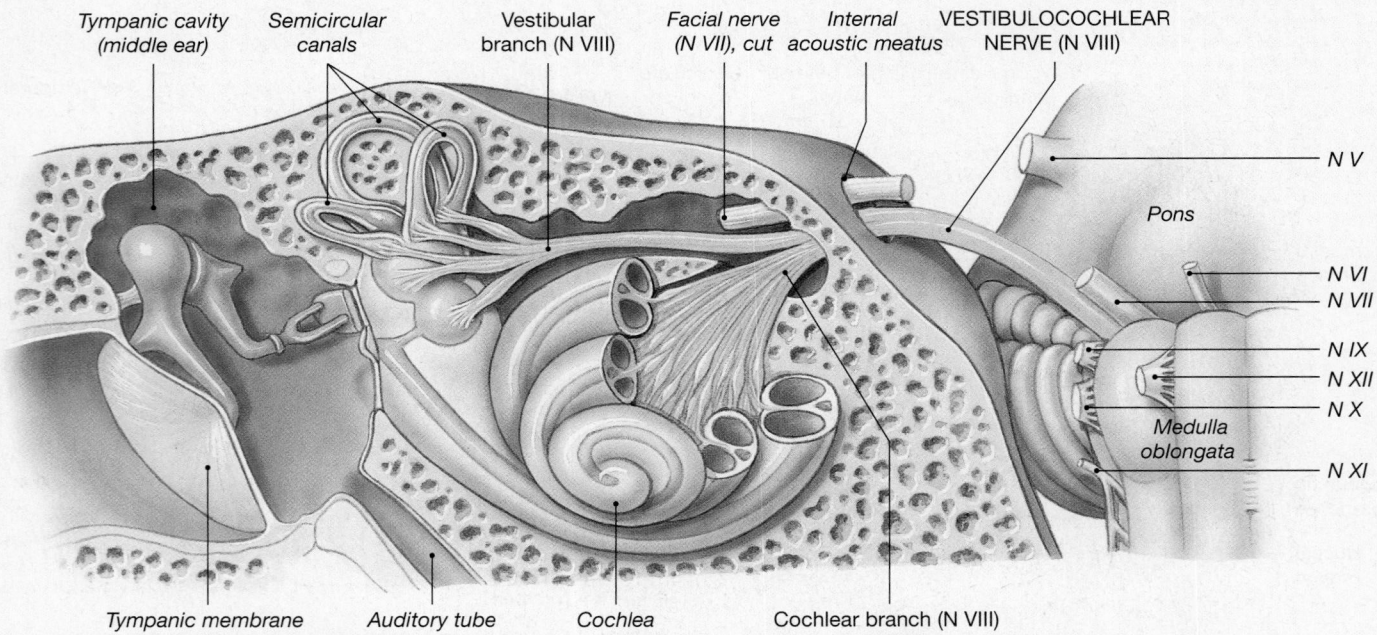

Figure 14–24 The Vestibulocochlear Nerve.

The Glossopharyngeal Nerves (IX)

Primary function: Mixed (sensory and motor) to head and neck

Origin: *Sensory*: posterior one-third of the tongue, part of the pharynx and palate, carotid arteries of the neck. *Motor*: motor nuclei of medulla oblongata

Pass through: Jugular foramina between the occipital bone and the temporal bones ∞ pp. 216, 217

Destination: *Sensory*: sensory nuclei of medulla oblongata. *Somatic motor*: pharyngeal muscles involved in swallowing. *Visceral motor*: parotid salivary gland by way of the otic ganglion

The medulla oblongata contains the sensory and motor nuclei of cranial nerves IX, X, XI, and XII, in addition to the vestibular nucleus of nerve VIII. The **glossopharyngeal** (glos-ō-fah-RIN-jē-al; *glossum*, tongue) **nerves** (IX) innervate the tongue and pharynx. Each glossopharyngeal nerve penetrates the cranium within the jugular foramen, with nerves X and XI.

The glossopharyngeal nerves are mixed nerves, but sensory fibers are most abundant. The sensory neurons on each side are in the **superior** (*jugular*) **ganglion** and **inferior** (*petrosal*) **ganglion** (**Figure 14–25**). The sensory fibers carry general sensory information from the lining of the pharynx and the soft palate to a nucleus in the medulla oblongata. These nerves also provide taste sensations from the posterior third of the tongue and have special receptors that monitor the blood pressure and dissolved gas concentrations in the carotid arteries, major blood vessels in the neck.

The somatic motor fibers control the pharyngeal muscles involved in swallowing. Visceral motor fibers synapse in the **otic ganglion**, and postganglionic fibers innervate the parotid salivary gland of the cheek.

The Vagus Nerves (X)

Primary function: Mixed (sensory and motor), widely distributed in the thorax and abdomen

Origin: *Sensory*: pharynx (part), auricle and external acoustic meatus, diaphragm, and visceral organs in thoracic and abdominopelvic cavities. *Motor*: motor nuclei in medulla oblongata

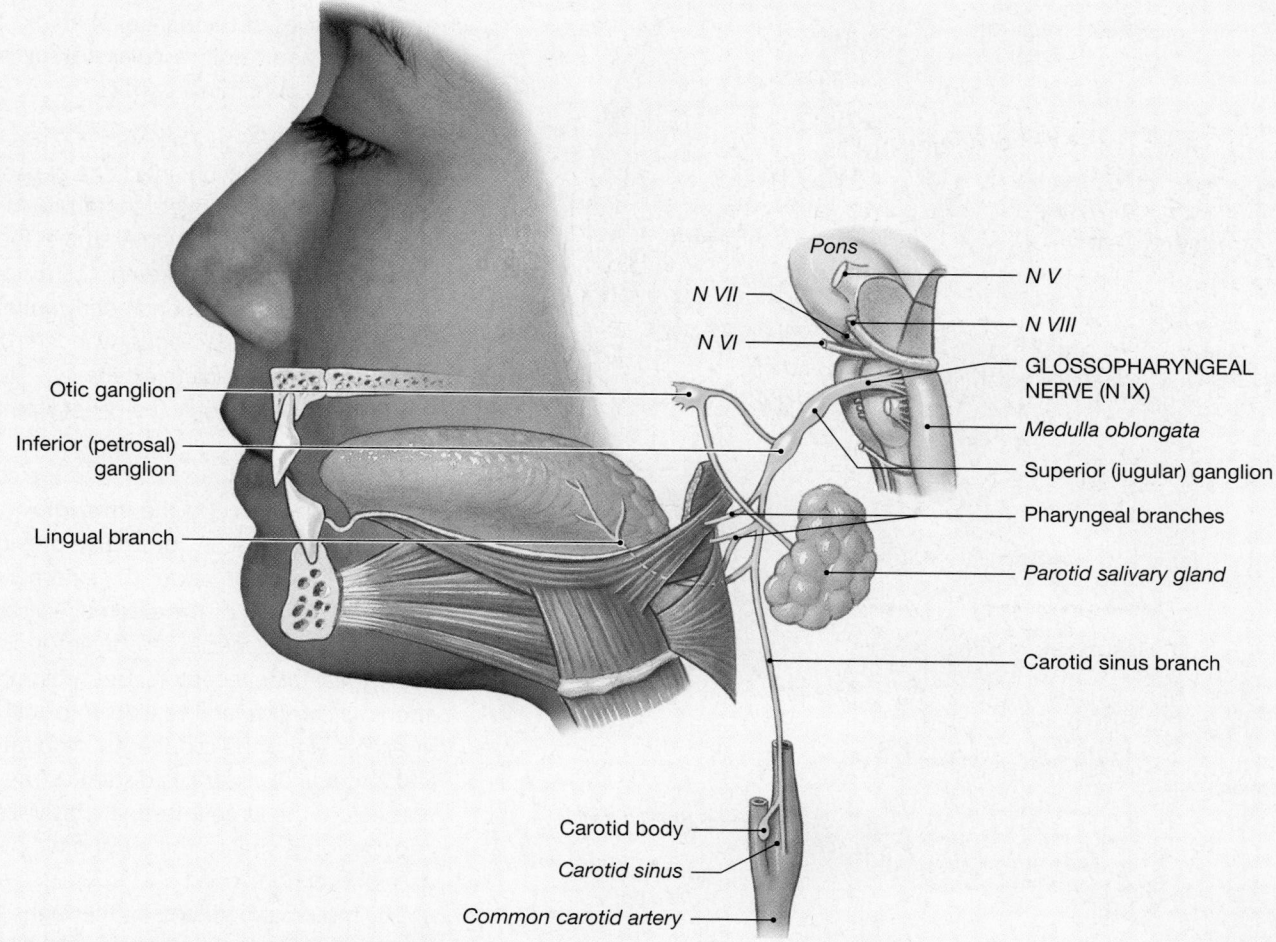

Figure 14–25 The Glossopharyngeal Nerve.

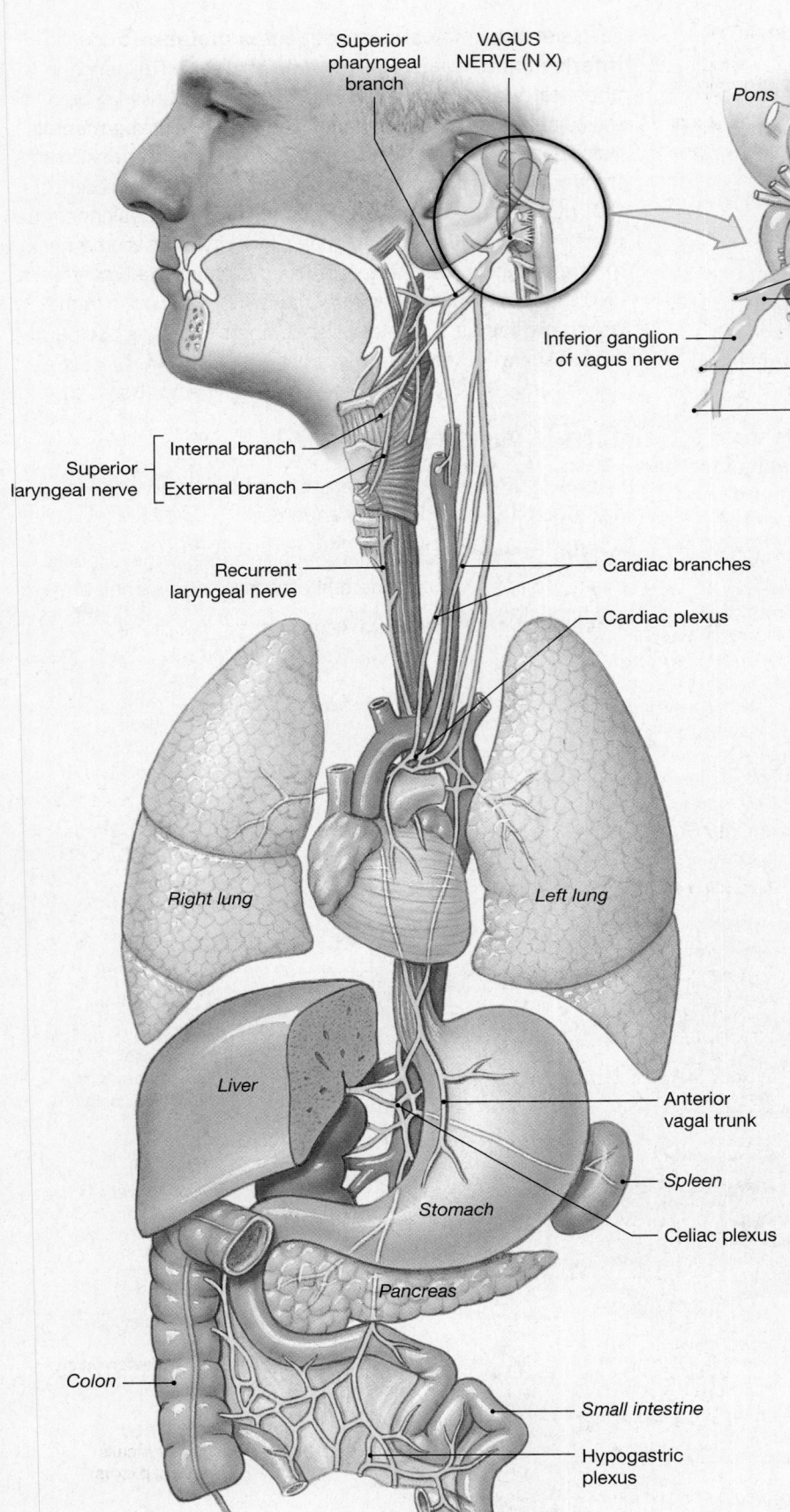

Superior pharyngeal branch

VAGUS NERVE (N X)

Pons

Medulla oblongata

Auricular branch to external ear

Superior ganglion of vagus nerve

Inferior ganglion of vagus nerve

Pharyngeal branch

Superior laryngeal nerve

Superior laryngeal nerve
— Internal branch
— External branch

Figure 14–26 **The Vagus Nerve.**

Recurrent laryngeal nerve

Cardiac branches

Cardiac plexus

Right lung

Left lung

Anterior vagal trunk

Liver

Spleen

Stomach

Celiac plexus

Pancreas

Colon

Small intestine

Hypogastric plexus

Pass through: Jugular foramina between the occipital bone and the temporal bones ∞ pp. 216, 217

Destination: *Sensory*: sensory nuclei and autonomic centers of medulla oblongata. *Visceral motor*: muscles of the palate, pharynx, digestive, respiratory, and cardiovascular systems in the thoracic and abdominal cavities

The **vagus** (VĀ-gus) **nerves** (X) arise immediately posterior to the attachment of the glossopharyngeal nerves. Many small rootlets contribute to their formation, and developmental studies indicate that these nerves probably represent the fusion of several smaller cranial nerves during our evolutionary history. As the name suggests (*vagus*, wanderer), the vagus nerves branch and radiate extensively. The adjacent **Figure 14–26** shows only the general pattern of distribution.

Sensory neurons are located in the **superior** (jugular) **ganglion** and the **inferior** (nodose; NŌ-dōs) **ganglion**. Each vagus nerve provides somatic sensory information about the external acoustic meatus (a portion of the external ear) and the diaphragm, and special sensory information from pharyngeal taste receptors. But most of the vagal afferents carry visceral sensory information from receptors along the esophagus, respiratory tract, and abdominal viscera as distant as the last portions of the large intestine. This visceral sensory information is vital to the autonomic control of visceral function.

The motor components of the vagus are equally diverse. Each vagus nerve carries preganglionic autonomic (parasympathetic) fibers that affect the heart and control smooth muscles and glands within the areas monitored by its sensory fibers, including the stomach, intestines, and gallbladder. Difficulty in swallowing is one of the most common signs of damage to either nerve IX or X, because damage to either one prevents the coordination of the swallowing reflex.

The Accessory Nerves (XI)

Primary function: Motor to muscles of the neck and upper back

Origin: Motor nuclei of spinal cord and medulla oblongata

Pass through: Jugular foramina between the occipital bone and the temporal bones ∞ pp. 216, 217

Destination: Internal branch innervates voluntary muscles of palate, pharynx, and larynx; external branch controls sternocleidomastoid and trapezius muscles

The **accessory nerves** (XI) are also known as the *spinal accessory nerves* or the *spinoaccessory nerves*. Unlike other cranial nerves, each accessory nerve has some motor fibers that originate in the lateral part of the anterior gray horns of the first five cervical segments of the spinal cord (**Figure 14–27**).

These somatic motor fibers form the **spinal root** of nerve XI. They enter the cranium through the foramen magnum. They then join the motor fibers of the **cranial root**, which originates at a nucleus in the medulla oblongata. The composite nerve leaves the cranium through the jugular foramen and divides into two branches.

The **internal branch** of nerve XI joins the vagus nerve and innervates the voluntary swallowing muscles of the soft palate and pharynx and the intrinsic muscles that control the vocal cords. The **external branch** of nerve XI controls the sternocleidomastoid and trapezius muscles of the neck and back. ∞ pp. 352, 362 The motor fibers of this branch originate in the lateral gray part of the anterior horns of cervical spinal nerves C_1 to C_5.

The Hypoglossal Nerves (XII)

Primary function: Motor (tongue movements)

Origin: Motor nuclei of medulla oblongata

Pass through: Hypoglossal canals of occipital bone ∞ pp. 216, 217, 218

Destination: Muscles of the tongue ∞ p. 350

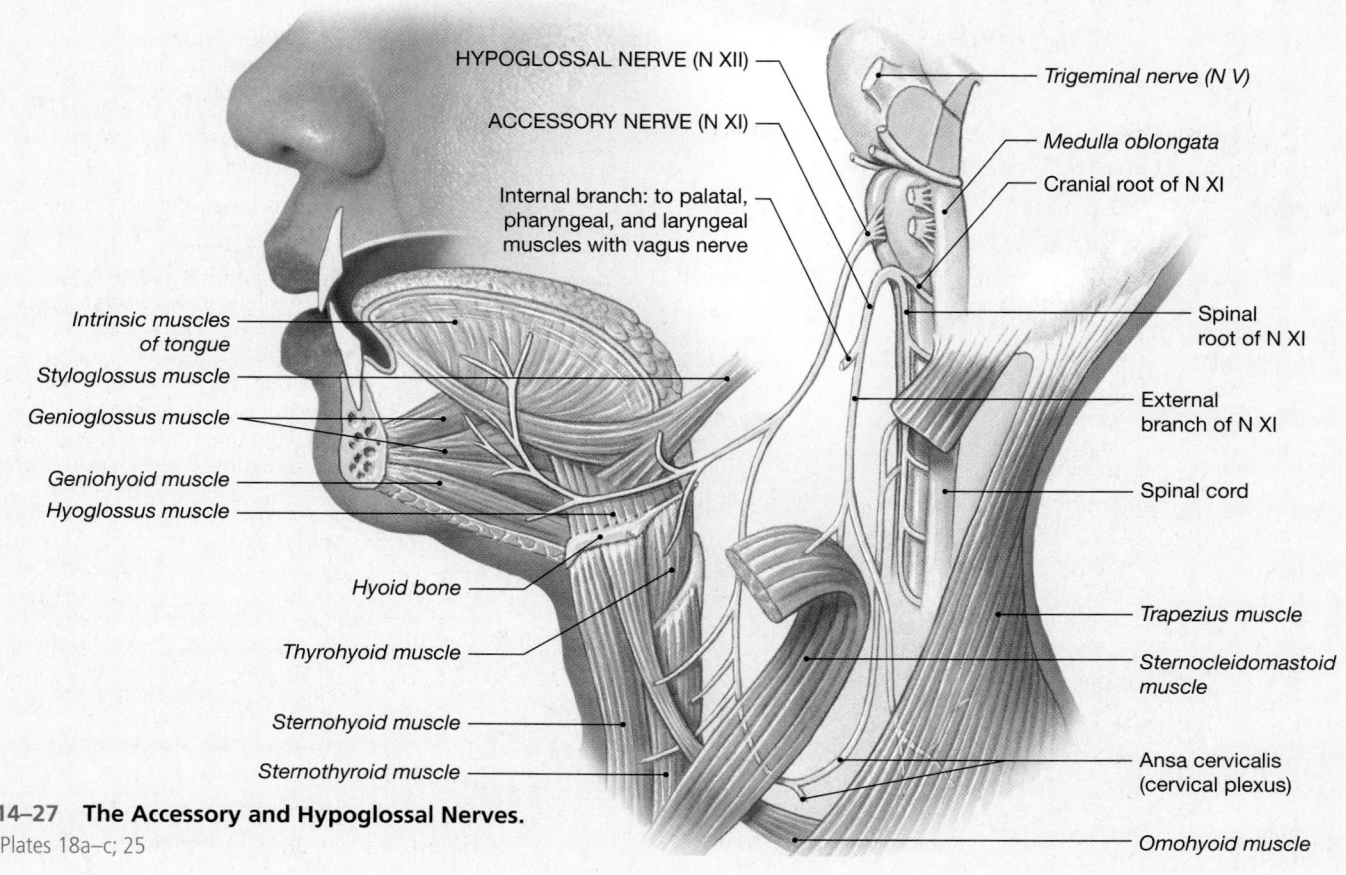

Figure 14–27   The Accessory and Hypoglossal Nerves.
ATLAS: Plates 18a–c; 25

Each **hypoglossal** (hī-pō-GLOS-al) **nerve** (XII) leaves the cranium through the hypoglossal canal. The nerve then curves to reach the skeletal muscles of the tongue (**Figure 14–27**). This cranial nerve provides voluntary motor control over movements of the tongue. Its condition is checked by having you stick out your tongue. Damage to one hypoglossal nerve or to its associ-ated nuclei causes the tongue to veer toward the affected side. Table 14–9 summarizes the basic distribution and function of each cranial nerve.

Cranial nerves are clinically important, in part because they can provide clues to underlying CNS problems. A number of standardized tests are used to assess cranial nerve function.

SUMMARY TABLE 14–9 Cranial Nerve Branches and Functions

Cranial Nerve (Number)	Sensory Ganglion	Branch	Primary Function	Foramen	Innervation
Olfactory (N I)			Special sensory	Olfactory foramina of ethmoid	Olfactory epithelium
Optic (N II)			Special sensory	Optic canal	Retina of eye
Oculomotor (N III)			Motor	Superior orbital fissure	Inferior, medial, superior rectus, inferior oblique and levator palpebrae superioris muscles; intrinsic eye muscles
Trochlear (N IV)			Motor	Superior orbital fissure	Superior oblique muscle
Trigeminal (N V)	Semilunar		Mixed	Superior orbital fissure	Areas associated with the jaws
		Ophthalmic	Sensory	Superior orbital fissure	Orbital structures, nasal cavity, skin of forehead, upper eyelid, eyebrows, nose (part)
		Maxillary		Foramen rotundum	Lower eyelid; superior lip, gums, and teeth; cheek, nose (part), palate, and pharynx (part)
		Mandibular		Foramen ovale	*Sensory*: inferior gums, teeth, lips, palate (part), and tongue (part) *Motor*: muscles of mastication
Abducens (N VI)			Motor	Superior orbital fissure	Lateral rectus muscle
Facial (N VII)	Geniculate		Mixed	Internal acoustic meatus to facial canal; exits at stylomastoid foramen	*Sensory*: taste receptors on anterior 2/3 of tongue *Motor*: muscles of facial expression, lacrimal gland, submandibular gland, sublingual salivary glands
Vestibulocochlear (Acoustic) (N VIII)		Cochlear Vestibular	Special sensory	Internal acoustic meatus	Cochlea (receptors for hearing) Vestibule (receptors for motion and balance)
Glossopharyngeal (N IX)	Superior (jugular) and inferior (petrosal)		Mixed	Jugular foramen	*Sensory*: posterior 1/3 of tongue; pharynx and palate (part); receptors for blood pressure, pH, oxygen, and carbon dioxide concentrations *Motor*: pharyngeal muscles and parotid salivary gland
Vagus (N X)	Superior (jugular) and inferior (nodose)		Mixed	Jugular foramen	*Sensory*: pharynx; auricle and external acoustic meatus; diaphragm; visceral organs in thoracic and abdominopelvic cavities *Motor*: palatal and pharyngeal muscles and visceral organs in thoracic and abdominopelvic cavities
Accessory (N XI)	Internal		Motor	Jugular foramen	Skeletal muscles of palate, pharynx, and larynx (with vagus nerve)
		External	Motor	Jugular foramen	Sternocleidomastoid and trapezius muscles
Hypoglossal (N XII)			Motor	Hypoglossal canal	Tongue musculature

14-10 Cranial reflexes involve sensory and motor fibers of cranial nerves

Cranial reflexes are monosynaptic and polysynaptic reflex arcs that involve the sensory and motor fibers of cranial nerves. Numerous examples of cranial reflexes will be encountered in later chapters, and this section will simply provide an overview and general introduction.

Table 14–10 lists representative examples of cranial reflexes and their functions. These reflexes are clinically important because they provide a quick and easy method for observing the condition of cranial nerves and specific nuclei and tracts in the brain. The somatic reflexes mediated by the cranial nerves are seldom more complex than the somatic reflexes of the spinal cord that were discussed in Chapter 13. ∞ p. 449 Table 14–10 includes four representative somatic reflexes: the corneal reflex, the tympanic reflex, the auditory reflexes, and the vestibulo-ocular reflexes. These cranial reflexes are often used to check for damage to the cranial nerves or the associated processing centers in the brain.

The brain stem contains many reflex centers that control visceral motor activity. The motor output of these reflexes is distributed by the autonomic nervous system. As you will see in Chapter 16, the cranial nerves carry most of the commands issued by the parasympathetic division of the ANS, whereas spinal nerves T_1–L_2 carry the sympathetic commands. Many of the centers that coordinate autonomic reflexes are located in the medulla oblongata. These centers can direct very complex visceral motor responses that are essential to the control of respiratory, digestive, and cardiovascular functions.

The A&P Top 100

#50 There are 12 pairs of cranial nerves. They are responsible for the special senses of smell, sight, and hearing/balance; for control over the muscles of the eye, jaw, face, and tongue; and for control over the superficial muscles of the neck, back, and shoulders. Cranial nerves also provide sensory information from the face, neck, and upper chest, and autonomic (parasympathetic) innervation to organs in the thoracic and abdominopelvic cavities.

CHECKPOINT

27. What are cranial reflexes?

See the blue Answers tab at the end of the book.

TABLE 14–10	Cranial Reflexes				
Reflex	**Stimulus**	**Afferents**	**Central Synapse**	**Efferents**	**Response**
SOMATIC REFLEXES					
Corneal reflex	Contact with corneal surface	N V (trigeminal)	Motor nucleus for N VII (facial)	N VII (facial)	Blinking of eyelids
Tympanic reflex	Loud noise	N VIII (vestibulocochlear)	Inferior colliculus	N VII	Reduced movement of auditory ossicles
Auditory reflexes	Loud noise	N VIII	Motor nuclei of brain stem and spinal cord	N III, IV, VI, N VII, X, and cervical nerves	Eye and/or head movements triggered by sudden sounds
Vestibulo-ocular reflexes	Rotation of head	N VIII	Motor nuclei controlling eye muscles	N III, IV, VI	Opposite movement of eyes to stabilize field of vision
VISCERAL REFLEXES					
Direct light reflex	Light striking photoreceptors	N II (optic)	Superior colliculus	N III (oculomotor)	Constriction of ipsilateral pupil
Consensual light reflex	Light striking photoreceptors	N II	Superior colliculus	N III	Constriction of contralateral pupil

Related Clinical Terms

aphasia: A disorder that impairs the ability to speak or read.

ataxia: A disturbance of muscular coordination that, in severe cases, leaves the individual unable to stand without assistance.

Bell palsy: A condition resulting from an inflammation of the facial nerve.

cerebrovascular accident (CVA), or *stroke:* A condition in which the blood supply to a portion of the brain is shut off.

cranial trauma: A head injury resulting from violent contact with another object.

disconnection syndrome: A condition in which the two cerebral hemispheres function independently.

dyslexia: A disorder affecting the comprehension and use of words.

electroencephalogram (EEG): A printed record of the brain's electrical activity over time.

epidural hemorrhage: A condition involving bleeding between the dura mater and the cranium, generally resulting from cranial trauma.

hydrocephalus, or "water on the brain": A condition resulting from interference with the normal circulation and/or reabsorption of cerebrospinal fluid.

Parkinson disease (*paralysis agitans*): A condition characterized by a pronounced increase in muscle tone, resulting from the excitation of neurons in the basal nuclei.

seizure: A temporary disorder of cerebral function, accompanied by abnormal movements, unusual sensations, and/or inappropriate behavior.

subdural hemorrhage: A condition in which blood accumulates under the dural epithelium in contact with the arachnoid membrane.

tic douloureux: A disorder of the maxillary and mandibular branches of nerve V characterized by almost total debilitating pain triggered by contact with the lip, tongue, or gums; also called trigeminal neuralgia.

Chapter Review

Study Outline

An Introduction to the Brain and Cranial Nerves p. 461

1. The adult human brain contains almost 97 percent of the body's neural tissue and averages 1.4 kg (3 lb) in weight and 1200 cc (71 in.^{3}) in volume.

14-1 The brain has several principal structures, each with specific functions p. 461

2. The six regions in the adult brain are the **cerebrum**, **cerebellum**, **diencephalon**, **mesencephalon** ("midbrain"), **pons**, and **medulla oblongata**. (*Figure 14–1*)

3. The brain contains extensive areas of **neural cortex**, a layer of gray matter on the surfaces of the cerebrum and cerebellum.

4. The brain forms from three swellings at the superior tip of the developing *neural tube*: the *prosencephalon*, the mesencephalon, and the *rhombencephalon*. The prosencephalon ("forebrain") forms the **telencephalon** (which becomes the cerebrum) and diencephalon; the rhombencephalon ("hindbrain") forms the **metencephalon** (cerebellum and pons) and **myelencephalon** (medulla oblongata). (*Table 14–1*)

5. The central passageway of the brain expands to form chambers called **ventricles**, which contain cerebrospinal fluid. (*Figure 14–2*)

14-2 The brain is protected and supported by the cranial meninges, cerebrospinal fluid, and the blood–brain barrier p. 464

6. The cranial meninges (the *dura mater*, *arachnoid mater*, and *pia mater*) are continuous with those of the spinal cord.

7. Folds of dura mater, including the **falx cerebri**, **tentorium cerebelli**, and **falx cerebelli**, stabilize the position of the brain. (*Figure 14–3*)

8. Cerebrospinal fluid (CSF) (1) protects delicate neural structures, (2) supports the brain, and (3) transports nutrients, chemical messengers, and waste products.

9. Cerebrospinal fluid is produced at the **choroid plexus**, reaches the subarachnoid space through the **lateral** and **median apertures**, and diffuses across the **arachnoid granulations** into the **superior sagittal sinus**. (*Figure 14–4*)

10. The **blood–brain barrier (BBB)** isolates neural tissue from the general circulation.

11. The blood–brain barrier is incomplete in parts of the **hypothalamus**, the **pituitary gland**, the **pineal gland**, and the **choroid plexus**.

14-3 The medulla oblongata, which is continuous with the spinal cord, contains vital centers p. 469

12. The medulla oblongata connects the brain and spinal cord. It contains relay stations such as the **olivary nuclei**, and **reflex centers**, including the **cardiovascular** and **respiratory rhythmicity centers**. The **reticular formation** begins in the medulla oblongata and extends into more superior portions of the brain stem. (*Figures 14–5, 14–6; Table 14–2*)

14-4 The pons contains nuclei and tracts that carry or relay sensory and motor information p. 472

13. The pons contains (1) sensory and motor nuclei for four cranial nerves; (2) nuclei that help control respiration; (3) nuclei and tracts linking the cerebellum with the brain stem, cerebrum, and spinal cord; and (4) ascending, descending, and transverse tracts. (*Figure 14–6; Table 14–2*)

14-5 The cerebellum coordinates learned and reflexive patterns of muscular activity at the subconscious level p. 472

14. The cerebellum adjusts postural muscles and programs and tunes ongoing movements. The **cerebellar hemispheres** consist of the **anterior** and **posterior lobes**, the **vermis**, and the **flocculonodular lobe**. (*Figure 14–7; Table 14–3*)

15. The **superior**, **middle**, and **inferior cerebellar peduncles** link the cerebellum with the brain stem, diencephalon, cerebrum, and spinal cord and interconnect the two cerebellar hemispheres.

14-6 The mesencephalon regulates auditory and visual reflexes and controls alertness p. 474

16. The **tectum** (roof of the mesencephalon) contains the **corpora quadrigemina** (**superior colliculi** and **inferior colliculi**). The tegmentum contains the **red nucleus**, the **substantia nigra**, the **cerebral peduncles**, and the headquarters of the reticular activating system (RAS). (*Figure 14–8; Table 14–4*)

14-7 The diencephalon integrates sensory information with motor output at the subconscious level p. 475

17. The diencephalon is composed of the epithalamus, the hypothalamus, and the thalamus. (*Figures 14–9, 14–10*)

18. The thalamus is the final relay point for ascending sensory information and coordinates the activities of the basal nuclei and cerebral cortex. *(Figures 14–9, 14–10; Table 14–5)*

19. The **hypothalamus** can (1) control somatic motor activities at the subconscious level, (2) control autonomic function, (3) coordinate activities of the nervous and endocrine systems, (4) secrete hormones, (5) produce emotions and behavioral **drives**, (6) coordinate voluntary and autonomic functions, (7) regulate body temperature, and (8) coordinate circadian cycles of activity. *(Figure 14–10; Table 14–6)*

14-8 The limbic system is a group of tracts and nuclei with various functions p. 478

20. The **limbic system**, or *motivational system*, includes the **amygdaloid body**, **cingulate gyrus**, **dentate gyrus**, **parahippocampal gyrus**, **hippocampus**, and **fornix**. The functions of the limbic system involve emotional states and related behavioral drives *(Figure 14–11; Table 14–7)*

14-9 The cerebrum, the largest region of the brain, contains motor, sensory, and association areas p. 480

21. The cortical surface contains **gyri** (elevated ridges) separated by **sulci** (shallow depressions) or **fissures** (deeper grooves). The **longitudinal fissure** separates the two **cerebral hemispheres**. The **central sulcus** separates the **frontal** and **parietal lobes**. Other sulci form the boundaries of the **temporal** and **occipital lobes**. *(Figure 14–12)*

22. The white matter of the cerebrum contains **association fibers**, **commissural fibers**, and **projection fibers**. *(Figure 14–13)*

23. The **basal nuclei** include the **caudate nucleus**, **globus pallidus**, and **putamen**; they control muscle tone and coordinate learned movement patterns and other somatic motor activities. *(Figure 14–14)*

24. The **primary motor cortex** of the **precentral gyrus** directs voluntary movements. The **primary sensory cortex** of the **postcentral gyrus** receives somatic sensory information from touch, pressure, pain, vibration, taste, and temperature receptors. *(Figure 14–15; Table 14–8)*

25. **Association areas**, such as the **somatic sensory association area**, **visual association area**, and **somatic motor association area (premotor cortex)**, control our ability to understand sensory information and coordinate a motor response. *(Figure 14–15)*

26. The **general interpretive area** receives information from all the sensory association areas. It is present in only one hemisphere—generally the left. *(Figure 14–15)*

27. The **speech center** regulates the patterns of breathing and vocalization needed for normal speech. *(Figure 14–15)*

28. The **prefrontal cortex** coordinates information from the secondary and special association areas of the entire cortex and performs abstract intellectual functions. *(Figure 14–15)*

29. The left hemisphere typically contains the general interpretive and speech centers and is responsible for language-based skills. The right hemisphere is typically responsible for spatial relationships and analyses. *(Figure 14–16)*

30. Brain activity is measured using an **electroencephalogram**. **Alpha waves** appear in healthy resting adults; **beta waves** occur when adults are concentrating; **theta waves** appear in children; and **delta waves** are normal during sleep. *(Figure 14–17)*

Focus: Cranial Nerves p. 490

31. We have 12 pairs of cranial nerves. Except for N I and N II, each nerve attaches to the ventrolateral surface of the brain stem near the associated sensory or motor nuclei. *(Figure 14–18)*

32. The **olfactory nerves** (N I) carry sensory information responsible for the sense of smell. The olfactory afferents synapse within the **olfactory bulbs**. *(Figures 14–18, 14–19)*

33. The **optic nerves** (N II) carry visual information from special sensory receptors in the eyes. *(Figures 14–18, 14–20)*

34. The **oculomotor nerves** (N III) are the primary source of innervation for four of the extra-ocular muscles. *(Figure 14–21)*

35. The **trochlear nerves** (N IV), the smallest cranial nerves, innervate the superior oblique muscles of the eyes. *(Figure 14–21)*

36. The **trigeminal nerves** (N V), the largest cranial nerves, are mixed nerves with *ophthalmic*, *maxillary*, and *mandibular* branches. *(Figure 14–22)*

37. The **abducens nerves** (N VI) innervate the lateral rectus muscles. *(Figure 14–21)*

38. The **facial nerves** (N VII) are mixed nerves that control muscles of the scalp and face. They provide pressure sensations over the face and receive taste information from the tongue. *(Figure 14–23)*

39. The **vestibulocochlear nerves** (N VIII) contain the **vestibular branch**, which monitors sensations of balance, position, and movement, and the **cochlear branch**, which monitors hearing receptors. *(Figure 14–24)*

40. The **glossopharyngeal nerves** (N IX) are mixed nerves that innervate the tongue and pharynx and control the action of swallowing. *(Figure 14–25)*

41. The **vagus nerves** (N X) are mixed nerves that are vital to the autonomic control of visceral function. *(Figure 14–26)*

42. The **accessory nerves** (N XI) have **internal branches**, which innervate voluntary swallowing muscles of the soft palate and pharynx, and **external branches**, which control muscles associated with the pectoral girdle. *(Figure 14–27)*

43. The **hypoglossal nerves** (N XII) provide voluntary motor control over tongue movements. *(Figure 14–27)*

44. The branches and functions of the cranial nerves are summarized in *Summary Table 14–9*.

14-10 Cranial reflexes involve sensory and motor fibers of cranial nerves p. 501

45. Cranial reflexes are monosynaptic and polysynaptic reflex arcs that involve sensory and motor fibers of cranial nerves. *(Table 14–10)*

Review Questions

See the blue Answers tab at the end of the book.

myA&P

Access more review material online at **myA&P**™ (www.myaandp.com). There, you'll find chapter guides, chapter quizzes, practice tests, animations, flashcards, a glossary with pronunciations, *Interactive Physiology*® (IP) exercises and quizzes, and more to help you succeed in the course.

LEVEL 1   Reviewing Facts and Terms

1. Identify the six principal parts of the brain in the following diagram.

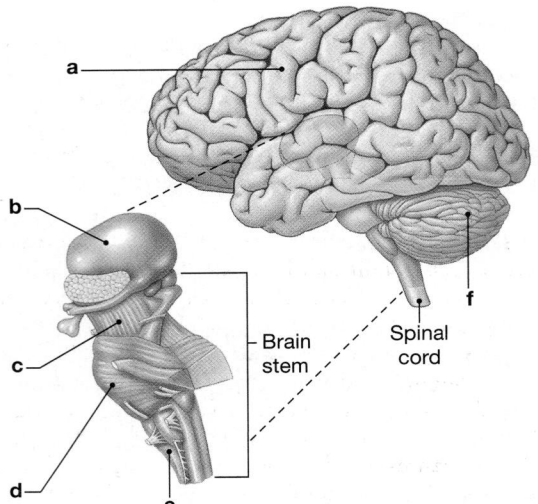

- Brain stem
- Spinal cord

(a) ___cortex___   (b) ___diencephalon___
(c) ___mesencephalon___   (d) ___pons___
(e) ___medulla oblongata___   (f) ___cerebellum___

2. Identify the layers of the cranial meninges in the following diagram.

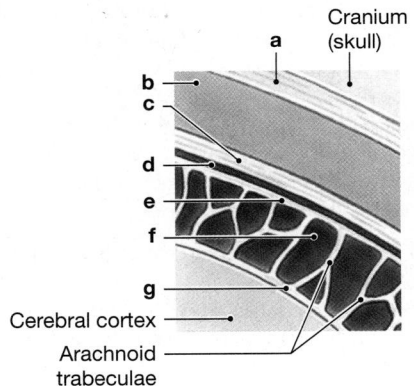

- Cranium (skull)
- Cerebral cortex
- Arachnoid trabeculae

(a) ___dura (endostal)___   (b) ___dura sinus___
(c) ___dura (meningeal)___   (d) ___subdural space___
(e) ___arachnoid mater___   (f) ___subarachnoid space___
(g) ___pia mater___

3. The term *higher brain centers* refers to those areas of the brain involved in higher-order functions. These centers would probably include nuclei, centers, and cortical areas of
   (a) the cerebrum.   (b) the cerebellum.
   (c) the diencephalon.   (d) all of these.
   (e) a and c only.

4. Which of the following is the site of cerebrospinal fluid production?
   (a) dural sinus   (b) choroid plexus
   (c) falx cerebri   (d) tentorium cerebelli
   (e) insula

5. The pons contains
   (a) sensory and motor nuclei for six cranial nerves.
   (b) nuclei concerned with control of blood pressure.
   (c) tracts that link the cerebellum with the brain stem.
   (d) no ascending or descending tracts.
   (e) both a and b.

6. The dural fold that divides the two cerebellar hemispheres is the
   (a) transverse sinus.   (b) falx cerebri.
   (c) tentorium cerebelli. (d) falx cerebelli.

7. Cerebrospinal fluid is produced and secreted by
   (a) neurons.   (b) ependymal cells.
   (c) Purkinje cells.   (d) basal nuclei.

8. The primary purpose of the blood–brain barrier (BBB) is to
   (a) provide the brain with oxygenated blood.
   (b) drain venous blood via the internal jugular veins.
   (c) isolate neural tissue in the CNS from the general circulation.
   (d) do all of these.

9. The centers in the pons that modify the activity of the respiratory rhythmicity centers in the medulla oblongata are the
   (a) apneustic and pneumotaxic centers.
   (b) inferior and superior peduncles.
   (c) cardiac and vasomotor centers.
   (d) nucleus gracilis and nucleus cuneatus.

10. The final relay point for ascending sensory information that will be projected to the primary sensory cortex is the
    (a) hypothalamus.   (b) thalamus.
    (c) spinal cord.   (d) medulla oblongata.

11. The establishment of emotional states is a function of the
    (a) limbic system.   (b) tectum.
    (c) mamillary bodies.   (d) thalamus.

12. Coordination of learned movement patterns at the subconscious level is performed by
    (a) the cerebellum.   (b) the substantia nigra.
    (c) association fibers. (d) the hypothalamus.

13. The two cerebral hemispheres are functionally different, even though anatomically they appear the same.
    (a) true
    (b) false

14. What are the three important functions of the CSF?

15. Which three areas in the brain are not isolated from the general circulation by the blood–brain barrier?

16. Using the mnemonic device "Oh, Once One Takes The Anatomy Final, Very Good Vacations Are Heavenly," list the names of the 12 pairs of cranial nerves.

LEVEL 2   Reviewing Concepts

17. Why can the brain respond to stimuli with greater versatility than the spinal cord?

18. Briefly summarize the overall function of the cerebellum.

19. The only cranial nerves that are attached to the cerebrum are the _____ nerves.
    - (a) optic
    - (b) oculomotor
    - (c) trochlear
    - (d) olfactory
    - (e) abducens

20. If symptoms characteristic of Parkinson disease appear, which part of the mesencephalon is inhibited from secreting a neurotransmitter? Which neurotransmitter is it?

21. What varied roles does the hypothalamus play in the body?

22. Stimulation of which part of the brain would produce sensations of hunger and thirst? *hypothalamus·*

23. Which structure in the brain would your A&P instructor be referring to when talking about a nucleus that resembles a sea horse and that appears to be important in the storage and retrieval of long-term memories? In which functional system of the brain is it located? *hippocampus → limbic system.*

24. What are the principal functional differences between the right and left hemispheres of the cerebrum?

25. Damage to the vestibular nucleus would lead to
    - (a) loss of sight.
    - (b) loss of hearing.
    - (c) inability to sense pain.
    - (d) difficulty in maintaining balance.
    - (e) inability to swallow.

26. A cerebrovascular accident occurs when *→ CVA = stroke*
    - (a) the reticular activating system fails to function.
    - (b) the prefrontal lobe is damaged.
    - (c) the blood supply to a portion of the brain is cut off.
    - (d) a descending tract in the spinal cord is severed.
    - (e) brain stem nuclei hypersecrete serotonin.

27. What kinds of problems are associated with the presence of lesions in the Wernicke area and the Broca area? *correlating v/a. speech*

LEVEL 3   Critical Thinking and Clinical Applications

28. Smelling salts can sometimes help restore consciousness after a person has fainted. The active ingredient of smelling salts is ammonia, and it acts by irritating the lining of the nasal cavity. Propose a mechanism by which smelling salts would raise a person from the unconscious state to the conscious state.

29. A police officer has just stopped Bill on suspicion of driving while intoxicated. The officer asks Bill to walk the yellow line on the road and then to place the tip of his index finger on the tip of his nose. How would these activities indicate Bill's level of sobriety? Which part of the brain is being tested by these activities?

30. Colleen falls down a flight of stairs and bumps her head several times. Soon after, she develops a headache and blurred vision. Diagnostic tests at the hospital reveal an epidural hematoma in the temporoparietal area. The hematoma is pressing against the brain stem. What other signs and symptoms might she experience as a result of the injury?

31. Cerebral meningitis is a condition in which the meninges of the brain become inflamed as the result of viral or bacterial infection. This condition can be life threatening. Why?

32. Infants have little to no control of the movements of their head. One of the consequences of this is that they are susceptible to shaken baby syndrome, caused by vigorous shaking of an infant or young child by the arms, legs, chest, or shoulders. Forceful shaking can cause brain damage leading to mental retardation, speech and learning disabilities, paralysis, seizures, hearing loss, and even death. Damage to which areas of the brain would account for the clinical signs observed in this syndrome?

---

(14) CSF: cushions brain & spinal cord
provides nutrients
supports brain

(15) portions of hypothalamus where capillary ependymal extremely permeable. capillaries of choroid plexus, capillaries in pineal gland. midbrain

(16) N1: olfactory. N2: optical N3: oculomotor N4: trochlear N5: trigeminal N6: abducen
N7: facial N8: vestibularcochlear N9: glossopharyngeal N10: vagus N11: accessory N12: hypoglossal

(17) more interneurons, pathways, connections than tracts of spinal cord

(18) Cerebellum: adjusts voluntary / involuntary responses based on sensory info & stored memories.

(19) only olfactory cranial nerve attached to cerebrum

(20) Parkinson's → substantia nigra produces less dopamine (inhibitor)

(21) hypothalamus.
- subconscious control of skeletal muscle contractions
- controls autonomic fun.
- coordinates nervous / endocrine systems
- secretes hormones
- produces emo & drives
- coordinates vol / invol
- body temp
- circadian rhythm.

(24) left: speech → language
logic/creative → logical
right: sensory interpret.

(28) RAS in mesencephalon
→ consciousness → activates neurons.

(29) cerebellum: maintains balance
& calculates speed & range of limb movnt

15

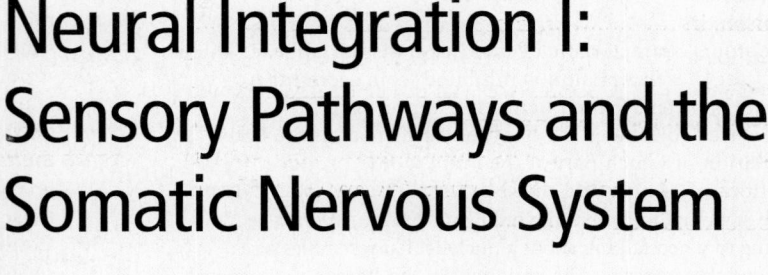

Neural Integration I: Sensory Pathways and the Somatic Nervous System

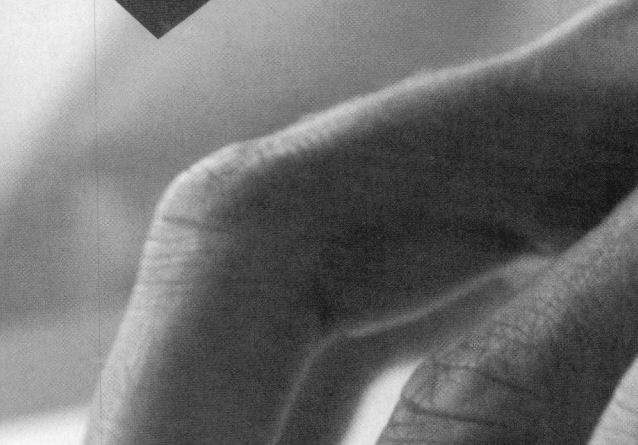

Learning Outcomes

After completing this chapter, you should be able to do the following:

15-1 Specify the components of the afferent and efferent divisions of the nervous system, and explain what is meant by the somatic nervous system.

15-2 Explain why receptors respond to specific stimuli, and how the organization of a receptor affects its sensitivity.

15-3 Identify the receptors for the general senses, and describe how they function.

15-4 Identify the major sensory pathways, and explain how it is possible to distinguish among sensations that originate in different areas of the body.

15-5 Describe the components, processes, and functions of the somatic motor pathways, and the levels of information processing involved in motor control.

Clinical Notes

Assessment of Tactile Sensitivities p. 513
Cerebral Palsy p. 522
Amyotrophic Lateral Sclerosis p. 524
Anencephaly p. 525

An Introduction to Sensory Pathways and the Somatic Nervous System

This chapter examines how the nervous system works as an integrated unit. It considers sensory receptors, sensory processing centers in the brain, and conscious and subconscious motor functions. The left-hand portion of **Figure 15–1** provides an overview of the topics we will cover in this chapter. Our discussion will focus on the "general senses" that provide information about the body and its environment. The "special senses"—smell, taste, sight, equilibrium (balance), and hearing—will be considered in Chapter 17.

15-1 Sensory information from all parts of the body is routed to the somatosensory cortex

Specialized cells called *sensory receptors* monitor specific conditions in the body or the external environment. When stimulated, a receptor passes information to the CNS in the form of action potentials along the axon of a sensory neuron. Such

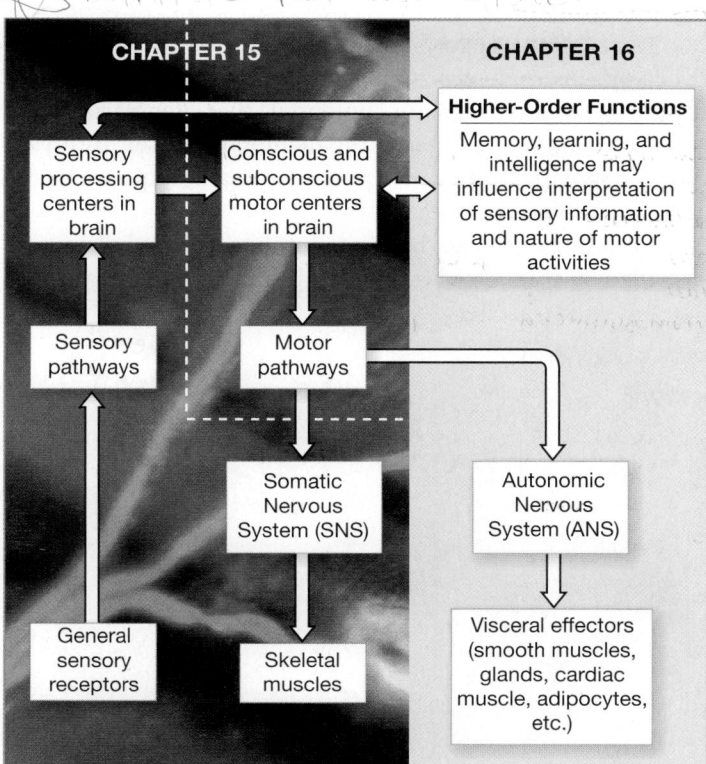

axons are parts of *sensory pathways*—the nerves, nuclei, and tracts that deliver somatic and visceral sensory information to their final destinations inside the CNS. Taken together, the receptors, sensory neurons, and sensory pathways constitute the afferent division of the nervous system. ∞ p. 387

Somatic and visceral sensory information often travels along the same pathway. Somatic sensory information is distributed to *sensory processing centers in the brain*—either the primary sensory cortex of the cerebral hemispheres or appropriate areas of the cerebellar hemispheres. Visceral sensory information is distributed primarily to reflex centers in the brain stem and diencephalon.

In this chapter we consider the somatic motor portion of the efferent division—the nuclei, motor tracts, and motor neurons that control peripheral effectors. Somatic motor commands—whether they arise at the conscious or subconscious levels—travel from motor centers in the brain along *somatic motor pathways*, which consist of motor nuclei, tracts, and nerves. The motor neurons and pathways that control skeletal muscles form the somatic nervous system (SNS).

Chapter 16 begins with a discussion of the visceral motor portion of the efferent division. All visceral motor commands are carried into the PNS by the autonomic nervous system (ANS). Both somatic and visceral motor commands may be issued in response to arriving sensory information, but these commands may be modified on the basis of planning, memories, and learning—the so-called *higher-order functions* of the brain that we will consider at the close of Chapter 16.

CHECKPOINT

1. What do we call the body's specialized cells that monitor specific internal or external conditions?

2. Is it possible for somatic motor commands to arise at the subconscious level?

See the blue Answers tab at the end of the book.

15-2 Sensory receptors connect our internal and external environments with the nervous system

Sensory receptors are specialized cells or cell processes that provide your central nervous system with information about conditions inside or outside the body. The term **general senses** is used to describe our sensitivity to temperature, pain, touch, pressure, vibration, and proprioception. General sensory receptors are distributed throughout the body, and they are relatively simple in structure. Some of the information they send to the CNS reaches the primary sensory cortex

Figure 15–1 An Overview of Neural Integration. This figure illustrates the relationships between Chapters 15 and 16 and indicates the major topics considered in this chapter.

and our awareness. As noted in Chapter 12, sensory information is interpreted on the basis of the frequency of arriving action potentials. ∞ p. 424 For example, when pressure sensations are arriving, the harder the pressure, the higher the frequency of action potentials. The arriving information is called a **sensation**. The conscious awareness of a sensation is called a **perception**.

The **special senses** are **olfaction** (smell), **vision** (sight), **gustation** (taste), **equilibrium** (balance), and **hearing**. These sensations are provided by receptors that are structurally more complex than those of the general senses. Special sensory receptors are located in **sense organs** such as the eye or ear, where the receptors are protected by surrounding tissues. The information these receptors provide is distributed to specific areas of the cerebral cortex (the auditory cortex, the visual cortex, and so forth) and to centers throughout the brain stem. We will consider the special senses in Chapter 17.

Sensory receptors represent the interface between the nervous system and the internal and external environments. A sensory receptor detects an arriving stimulus and translates it into an action potential that can be conducted to the CNS. This translation process is called *transduction*. If transduction does not occur, then as far as you are concerned, the stimulus doesn't exist. For example, bees can see ultraviolet light you can't see, and dogs can respond to sounds you can't hear. In each case the stimuli are there—but your receptors cannot detect them.

In the rest of this section we examine the basic concepts of receptor function and sensory processing. We begin by considering how receptors detect stimuli.

The Detection of Stimuli

Each receptor has a characteristic sensitivity. For example, a touch receptor is very sensitive to pressure but relatively insensitive to chemical stimuli, whereas a taste receptor is sensitive to dissolved chemicals but insensitive to pressure. This feature is called *receptor specificity*.

Specificity may result from the structure of the receptor cell, or from the presence of accessory cells or structures that shield the receptor cell from other stimuli. The simplest receptors are the dendrites of sensory neurons. The branching tips of these dendrites, called **free nerve endings**, are not protected by accessory structures. Free nerve endings extend through a tissue the way grass roots extend into the soil. They can be stimulated by many different stimuli and therefore exhibit little receptor specificity. For example, free nerve endings that respond to tissue damage by providing pain sensations may be stimulated by chemical stimulation, pressure, temperature changes, or trauma. Complex receptors, such as the eye's visual receptors, are protected by accessory

cells and connective tissue layers. These cells are seldom exposed to any stimulus other than light and so provide very specific information.

The area monitored by a single receptor cell is its *receptive field* (**Figure 15–2**). Whenever a sufficiently strong stimulus arrives in the receptive field, the CNS receives the information "stimulus arriving at receptor X." The larger the receptive field, the poorer your ability to localize a stimulus. A touch receptor on the general body surface, for example, may have a receptive field 7 cm (2.5 in.) in diameter. As a result, you can describe a light touch there as affecting only a general area, not an exact spot. On the tongue or fingertips, where the receptive fields are less than a millimeter in diameter, you can be very precise about the location of a stimulus.

An arriving stimulus can take many forms. It may be a physical force (such as pressure), a dissolved chemical, a sound, or light. Regardless of the nature of the stimulus, however, sensory information must be sent to the CNS in the form of action potentials, which are electrical events.

As noted earlier, transduction is the translation of an arriving stimulus into an action potential by a sensory receptor. Transduction begins when a stimulus changes the transmembrane potential of the receptor cell. This change, called a *receptor potential*, is either a graded depolarization or a graded hyperpolarization. The stronger the stimulus, the larger the receptor potential.

The typical receptors for the general senses are the dendrites of sensory neurons, and the sensory neuron is the receptor cell. Any receptor potential that depolarizes the plasma membrane will bring the membrane closer to threshold. A depolarizing receptor potential in a neural receptor is called a *generator potential*.

Sensations of taste, hearing, equilibrium, and vision are provided by specialized receptor cells that communicate with

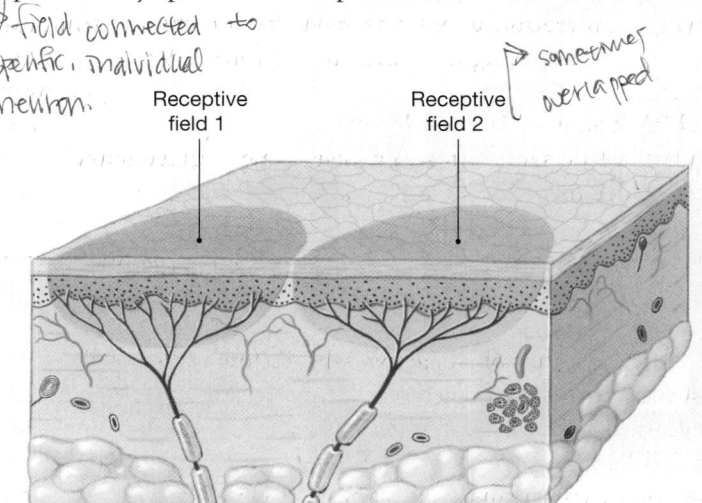

Receptive field 1   Receptive field 2

Figure 15–2   Receptors and Receptive Fields. Each receptor cell monitors a specific area known as the receptive field.

sensory neurons across chemical synapses. The receptor cells develop graded receptor potentials in response to stimulation, and the change in membrane potential alters the rate of neurotransmitter release at the synapse. The result is a depolarization or hyperpolarization of the sensory neuron. If sufficient depolarization occurs, an action potential appears in the sensory neuron. In this case, the receptor potential and the generator potential occur in different cells: The receptor potential develops in the receptor cell, and the generator potential appears later, in the sensory neuron.

Whenever a sufficiently large generator potential appears, action potentials develop in the axon of a sensory neuron. For reasons discussed in Chapter 12, the greater the degree of sustained depolarization at the axon hillock, the higher the frequency of action potentials in the afferent fiber. ∞ p. 424 The arriving information is then processed and interpreted by the CNS at the conscious and subconscious levels.

The Interpretation of Sensory Information

Sensory information that arrives at the CNS is routed according to the location and nature of the stimulus. Previous chapters emphasized the fact that axons in the CNS are organized in bundles with specific origins and destinations. Along sensory pathways, a series of neurons relays information from one point (the receptor) to another (a neuron at a specific site in the cerebral cortex). For example, sensations of touch, pressure, pain, and temperature arrive at the primary sensory cortex; visual, auditory, gustatory, and olfactory sensations reach the visual, auditory, gustatory, and olfactory regions of the cortex, respectively.

The link between peripheral receptor and cortical neuron is called a *labeled line*. Each labeled line consists of axons carrying information about one *modality*, or type of stimulus (touch, pressure, light, sound, and so forth). The CNS interprets the modality entirely on the basis of the labeled line over which it arrives. As a result, you cannot tell the difference between a true sensation and a false one generated somewhere along the line. For example, when you rub your eyes, you commonly see flashes of light. Although the stimulus is mechanical rather than visual, any activity along the optic nerve is projected to the visual cortex and experienced as a visual perception.

The identity of the active labeled line indicates the type of stimulus. Where it arrives within the sensory cortex determines its perceived location. For example, if activity in a labeled line that carries touch sensations stimulates the facial region of your primary sensory cortex, you perceive a touch on the face. All other characteristics of the stimulus—its strength, duration, and variation—are conveyed by the frequency and pattern of action potentials. The translation of complex sensory information into meaningful patterns of action potentials is called *sensory coding*.

Some sensory neurons, called **tonic receptors**, are always active. The frequency with which these receptors generate action potentials indicates the background level of stimulation. When the stimulus increases or decreases, the rate of action potential generation changes accordingly. Other receptors are normally inactive, but become active for a short time whenever a change occurs in the conditions they are monitoring. These receptors, called **phasic receptors**, provide information about the intensity and rate of change of a stimulus. Receptors that combine phasic and tonic coding can convey extremely complicated sensory information.

Adaptation

Adaptation is a reduction in sensitivity in the presence of a constant stimulus. You seldom notice the rumble of the tires when you ride in a car, or the background noise of the air conditioner, because your nervous system quickly adapts to stimuli that are painless and constant. *Peripheral adaptation* occurs when the level of receptor activity changes. The receptor responds strongly at first, but thereafter its activity gradually declines, in part because the size of the generator potential gradually decreases. This response is characteristic of phasic receptors, which are hence also called **fast-adapting receptors**. Temperature receptors (*thermoreceptors*) are phasic receptors; you seldom notice room temperature unless it changes suddenly. Tonic receptors show little peripheral adaptation and so are called **slow-adapting receptors**. Pain receptors (*nociceptors*) are slow-adapting receptors, which is one reason why pain sensations remind you of an injury long after the initial damage has occurred.

Adaptation also occurs along sensory pathways inside the CNS. For example, a few seconds after you have been exposed to a new smell, awareness of the stimulus virtually disappears, although the sensory neurons are still quite active. This process is known as *central adaptation*. Central adaptation generally involves the inhibition of nuclei along a sensory pathway.

Peripheral adaptation reduces the amount of information that reaches the CNS. Central adaptation at the subconscious level further restricts the amount of detail that arrives at the cerebral cortex. Most of the incoming sensory information is processed in centers along the spinal cord or brain stem at the subconscious level. Although this processing can produce reflexive motor responses, we are seldom consciously aware of either the stimuli or the responses.

The output from higher centers can increase receptor sensitivity or facilitate transmission along a sensory pathway. The reticular activating system in the mesencephalon helps focus our attention and thus heightens or reduces our awareness of arriving sensations. ∞ p. 475 This adjustment of sensitivity can occur under conscious or subconscious direction. When you "listen carefully," your sensitivity and awareness of auditory stimuli increase. Output from higher centers can also inhibit transmission along a sensory pathway. Such inhibition occurs when you enter a noisy factory or walk along a crowded city street, as you automatically tune out the high level of background noise.

Now that we have examined the basic concepts of receptor function and sensory processing, we consider how those concepts apply to the general senses.

The A&P Top 100

#51 Stimulation of a receptor produces action potentials along the axon of a sensory neuron. The frequency or pattern of action potentials contains information about the strength, duration, and variation of the stimulus. Your perception of the nature of that stimulus depends on the path it takes inside the CNS.

CHECKPOINT

3. Define adaptation.
4. Receptor A has a circular receptive field with a diameter of 2.5 cm. Receptor B has a circular receptive field 7.0 cm in diameter. Which receptor provides more precise sensory information?

See the blue Answers tab at the end of the book.

15-3 General sensory receptors can be classified by the type of stimulus that excites them

Receptors for the general senses are scattered throughout the body and are relatively simple in structure. The simple classification scheme introduced in Chapter 12 divides them into exteroceptors, proprioceptors, and interoceptors. ∞ p. 391 *Exteroceptors* provide information about the external environment; *proprioceptors* report the positions of skeletal muscles and joints; *interoceptors* monitor visceral organs and functions.

A more detailed classification system divides the general sensory receptors into four types by the nature of the stimulus that excites them: *nociceptors* (pain), *thermoreceptors* (temperature), *mechanoreceptors* (physical distortion), and *chemoreceptors* (chemical concentration). Each class of receptors has distinct structural and functional characteristics. The difference between a somatic receptor and a visceral receptor is its location, not its structure. A pain receptor in the gut looks and acts like a pain receptor in the skin, but the two sensations are delivered to separate locations in the CNS. However, proprioception is a purely somatic sensation—there are

no proprioceptors in the visceral organs of the thoracic and abdominopelvic cavities. Your mental map of your body doesn't include these organs; you cannot tell, for example, where your spleen, appendix, or pancreas is at the moment. The visceral organs also have fewer pain, temperature, and touch receptors than one finds elsewhere in the body, and the sensory information you receive is poorly localized because the receptive fields are very large and may be widely separated.

Although general sensations are widely distributed in the CNS, most of the processing occurs in centers along the sensory pathways in the spinal cord or brain stem. Only about 1 percent of the information provided by afferent fibers reaches the cerebral cortex and our awareness. For example, we usually do not feel the clothes we wear or hear the hum of the engine when riding in a car.

Nociceptors

Pain receptors, or **nociceptors** (*noxa*, harm), are especially common in the superficial portions of the skin, in joint capsules, within the periostea of bones, and around the walls of blood vessels. Other deep tissues and most visceral organs have few nociceptors. Pain receptors are free nerve endings with large receptive fields (**Figure 15–2**). As a result, it is often difficult to determine the exact source of a painful sensation.

Nociceptors may be sensitive to (1) extremes of temperature, (2) mechanical damage, and (3) dissolved chemicals, such as chemicals released by injured cells. Very strong stimuli, however, will excite all three receptor types. For that reason, people describing very painful sensations—whether caused by acids, heat, or a deep cut—use similar descriptive terms, such as "burning."

Stimulation of the dendrites of a nociceptor causes depolarization. When the initial segment of the axon reaches threshold, an action potential heads toward the CNS.

Two types of axons—Type A and Type C fibers—carry painful sensations. ∞ p. 413 Myelinated Type A fibers carry sensations of **fast pain**, or *prickling pain*. An injection or a deep cut produces this type of pain. These sensations very quickly reach the CNS, where they often trigger somatic reflexes. They are also relayed to the primary sensory cortex and so receive conscious attention. In most cases, the arriving information permits the stimulus to be localized to an area several inches in diameter.

Slower, Type C fibers carry sensations of **slow pain**, or *burning and aching pain*. These sensations cause a generalized activation of the reticular formation and thalamus. The individual becomes aware of the pain but has only a general idea of the area affected.

Pain receptors are tonic receptors. Significant peripheral adaptation does not occur, and the receptors continue to respond as long as the painful stimulus remains. Painful sensations cease only after tissue damage has ended. However, central adaptation may reduce the *perception* of the pain while pain receptors remain stimulated. This effect involves the inhibition of centers in the thalamus, reticular formation, lower brain stem, and spinal cord.

An understanding of the origins of pain sensations and an ability to control or reduce pain levels have always been among the most important aspects of medical treatment. After all, it is usually pain that induces someone to seek treatment; conditions that are not painful are typically ignored or tolerated. Although we often use the term *pain pathways*, it is becoming clear that pain distribution and perception are extremely complex—more so than had previously been imagined.

The sensory neurons that bring pain sensations into the CNS release *glutamate* and/or *substance P* as neurotransmitters. These neurotransmitters produce facilitation of neurons along the pain pathways. As a result, the level of pain experienced (especially chronic pain) can be out of proportion to the amount of painful stimuli or the apparent tissue damage. This effect may be one reason why people differ so widely in their perception of the pain associated with childbirth, headaches, or back pain. This facilitation is also presumed to play a role in phantom limb pain (discussed shortly); the sensory neurons may be inactive, but the hyperexcitable interneurons may continue to generate pain sensations.

The level of pain felt by an individual can be reduced by the release of endorphins and enkephalins within the CNS. As noted in Chapter 12, endorphins and enkephalins are neuromodulators whose release inhibits activity along pain pathways in the brain. ∞ p. 418 These compounds, structurally similar to morphine, are found in the limbic system, hypothalamus, and reticular formation. The pain centers in these areas also use substance P as a neurotransmitter. Endorphins bind to the presynaptic membrane and prevent the release of substance P, thereby reducing the conscious perception of pain, although the painful stimulus remains.

Tips&Tricks

The **P** in substance **P** stands for **p**eptide and is involved with **p**ain, which it transmits **p**eripherally.

Thermoreceptors

Temperature receptors, or **thermoreceptors**, are free nerve endings located in the dermis, in skeletal muscles, in the liver, and in the hypothalamus. Cold receptors are three or four times more numerous than warm receptors. No structural differences between warm and cold thermoreceptors have been identified.

Temperature sensations are conducted along the same pathways that carry pain sensations. They are sent to the reticular formation, the thalamus, and (to a lesser extent) the primary sensory cortex. Thermoreceptors are phasic receptors: They are very active when the temperature is changing, but they quickly adapt to a stable temperature. When you enter an air-conditioned classroom on a hot summer day or a warm lecture hall on a brisk fall evening, the temperature change seems extreme at first, but you quickly become comfortable as adaptation occurs.

Mechanoreceptors

Mechanoreceptors are sensitive to stimuli that distort their plasma membranes. These membranes contain *mechanically gated ion channels* whose gates open or close in response to stretching, compression, twisting, or other distortions of the membrane. There are three classes of mechanoreceptors:

1. **Tactile receptors** provide the closely related sensations of touch, pressure, and vibration. Touch sensations provide information about shape or texture, whereas pressure sensations indicate the degree of mechanical distortion. Vibration sensations indicate a pulsing or oscillating pressure. The receptors involved may be specialized in some way. For example, rapidly adapting tactile receptors are best suited for detecting vibration. But your interpretation of a sensation as touch rather than pressure is typically a matter of the degree of stimulation, and not of differences in the type of receptor stimulated.

2. **Baroreceptors** (bar-ō-rē-SEP-torz; *baro-*, pressure) detect pressure changes in the walls of blood vessels and in portions of the digestive, reproductive, and urinary tracts.

3. **Proprioceptors** monitor the positions of joints and muscles. They are the most structurally and functionally complex of the general sensory receptors.

Tactile Receptors

Fine touch and pressure receptors provide detailed information about a source of stimulation, including its exact location, shape, size, texture, and movement. These receptors are extremely sensitive and have relatively narrow receptive fields. **Crude touch and pressure receptors** provide poor localization and, because they have relatively large receptive fields, give little additional information about the stimulus.

Tactile receptors range in complexity from free nerve endings to specialized sensory complexes with accessory cells and supporting structures. **Figure 15–3** shows six types of tactile receptors in the skin:

1. Free nerve endings sensitive to touch and pressure are situated between epidermal cells (**Figure 15–3a**). There appear to be no structural differences between these receptors and the free nerve endings that provide temperature or pain sensations. These are the only sensory re-

ceptors on the corneal surface of the eye, but in other portions of the body surface, more specialized tactile receptors are probably more important. Free nerve endings that provide touch sensations are tonic receptors with small receptive fields.

2. Wherever hairs are located, the nerve endings of the **root hair plexus** monitor distortions and movements across the body surface (**Figure 15–3b**). When a hair is displaced, the movement of the follicle distorts the sensory dendrites and produces action potentials. These receptors adapt rapidly, so they are best at detecting initial contact and subsequent movements. Thus, you generally feel your clothing only when you move or when you consciously focus on tactile sensations from the skin.

3. **Tactile discs**, or *Merkel* (MER-kel) *discs*, are fine touch and pressure receptors (**Figure 15–3c**). They are extremely

sensitive tonic receptors, with very small receptive fields. The dendritic processes of a single myelinated afferent fiber make close contact with unusually large epithelial cells in the stratum germinativum of the skin; these *Merkel cells* were described in Chapter 5. ∞ p. 160

4. **Tactile corpuscles**, or *Meissner* (MĪS-ner) *corpuscles*, perceive sensations of fine touch and pressure and low-frequency vibration. They adapt to stimulation within a second after contact. Tactile corpuscles are fairly large structures, measuring roughly 100 μm in length and 50 μm in width. These receptors are most abundant in the eyelids, lips, fingertips, nipples, and external genitalia. The dendrites are highly coiled and interwoven, and they are surrounded by modified Schwann cells. A fibrous capsule surrounds the entire complex and anchors it within the dermis (**Figure 15–3d**).

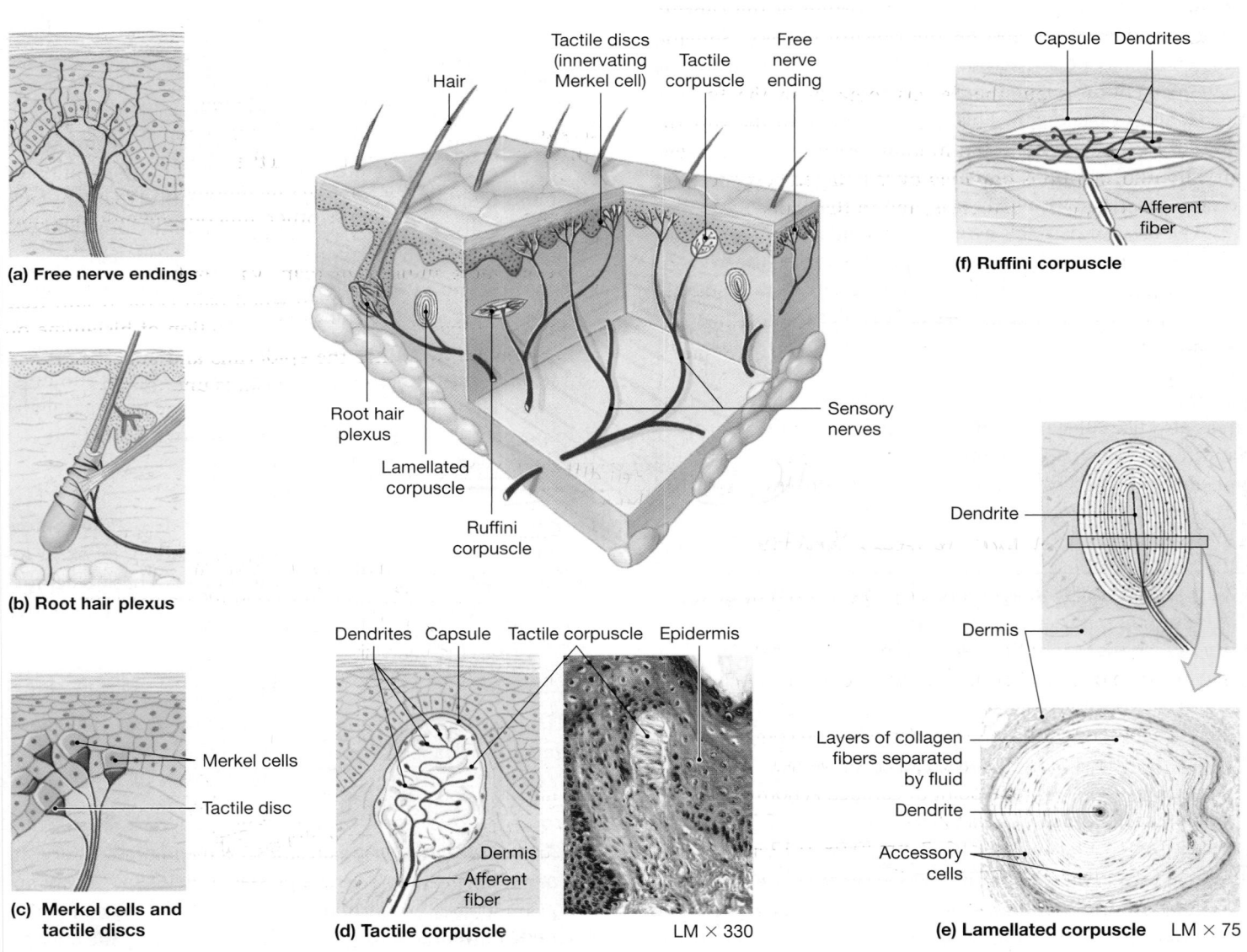

(a) Free nerve endings

(b) Root hair plexus

(c) Merkel cells and tactile discs

Hair
Tactile discs (innervating Merkel cell)
Tactile corpuscle
Free nerve ending
Root hair plexus
Lamellated corpuscle
Ruffini corpuscle
Sensory nerves

Capsule Dendrites
Afferent fiber
(f) Ruffini corpuscle

Merkel cells
Tactile disc

Dendrites Capsule Tactile corpuscle Epidermis
Dermis
Afferent fiber
(d) Tactile corpuscle   LM × 330

Dendrite
Dermis
Layers of collagen fibers separated by fluid
Dendrite
Accessory cells
(e) Lamellated corpuscle   LM × 75

Figure 15–3   **Tactile Receptors in the Skin.**

Tips&Tricks

To remember that Meissner corpuscles perceive pressure sensations, associate the *m* and *ss* in "**M**e**iss**ner" with **m**ass**age.

5. **Lamellated** (LAM-e-lāt-ed; *lamella*, a little thin plate) **corpuscles**, or *pacinian* (pa-SIN-ē-an) *corpuscles*, are sensitive to deep pressure. Because they are fast-adapting receptors, they are most sensitive to pulsing or high-frequency vibrating stimuli. A single dendrite lies within a series of concentric layers of collagen fibers and supporting cells (specialized fibroblasts) (**Figure 15–3e**). The entire corpuscle may reach 4 mm in length and 1 mm in diameter. The concentric layers, separated by interstitial fluid, shield the dendrite from virtually every source of stimulation other than direct pressure. Lamellated corpuscles adapt quickly because distortion of the capsule soon relieves pressure on the sensory process. Somatic sensory information is provided by lamellated corpuscles located throughout the dermis, notably in the fingers, mammary glands, and external genitalia; in the superficial and deep fasciae; and in joint capsules. Visceral sensory information is provided by lamellated corpuscles in mesenteries, in the pancreas, and in the walls of the urethra and urinary bladder.

6. **Ruffini** (roo-FĒ-nē) **corpuscles** are also sensitive to pressure and distortion of the skin, but they are located in the reticular (deep) dermis. These receptors are tonic and show little if any adaptation. The capsule surrounds a core of collagen fibers that are continuous with those of the surrounding dermis (**Figure 15–3f**). In the capsule, a network of dendrites is intertwined with the collagen fibers. Any tension or distortion of the dermis tugs or twists the capsular fibers, stretching or compressing the attached dendrites and altering the activity in the myelinated afferent fiber.

Our sensitivity to tactile sensations may be altered by infection, disease, or damage to sensory neurons or pathways. As a result, mapping tactile responses can sometimes aid clinical assessment. Sensory losses with clear regional boundaries indicate trauma to spinal nerves. For example, sensory loss within the boundaries of a dermatome can help identify the affected spinal nerve or nerves. ∞ p. 438

Tickle and itch sensations are closely related to the sensations of touch and pain. The receptors involved are free nerve endings, and the information is carried by unmyelinated Type C fibers. Tickle sensations, which are usually (but not always) described as pleasurable, are produced by a light touch that moves across the skin. Psychological factors are involved in the interpretation of tickle sensations, and tickle sensitivity differs greatly among individuals. Itching is probably produced by the stimulation of the same receptors. Specific "itch spots" can be mapped in the skin, the inner surfaces of the eyelids, and the mucous membrane of the nose. Itch sensations are absent from other mucous membranes and from deep tissues and viscera. Itching is extremely unpleasant, even more unpleasant than pain. Individuals with extreme itching will scratch even when pain is the result. Itch receptors can be stimulated by the injection of histamine or proteolytic enzymes into the epidermis and superficial dermis. The precise receptor mechanism is unknown.

CLINICAL NOTE

Assessment of Tactile Sensitivities

Regional sensitivity to light touch can be checked by gentle contact with a fingertip or a slender wisp of cotton. The **two-point discrimination test** provides a more detailed sensory map of tactile receptors. Two fine points of a bent paper clip or another object are applied to the skin surface simultaneously. The subject then describes the contact. When the points fall within a single receptive field, the individual will report only one point of contact. A normal individual loses two-point discrimination at 1 mm (0.04 in.) on the surface of the tongue, at 2–3 mm (0.08–0.12 in.) on the lips, at 3–5 mm (0.12–0.20 in.) on the backs of the hands and feet, and at 4–7 cm (1.6–2.75 in.) over the general body surface.

Applying the base of a tuning fork to the skin tests vibration receptors. Damage to an individual spinal nerve produces insensitivity to vibration along the paths of the related sensory nerves. If the sensory loss results from spinal cord damage, the injury site can typically be located by walking the tuning fork down the spinal column, resting its base on the vertebral spines.

Descriptive terms are used to indicate the degree of sensitivity in the area. *Anesthesia* implies a total loss of sensation; the individual cannot perceive touch, pressure, pain, or temperature sensations in that area. *Hypesthesia* is a reduction in sensitivity, and *paresthesia* is the presence of abnormal sensations such as the pins-and-needles sensation when an arm or leg "falls asleep" as a result of pressure on a peripheral nerve.

Baroreceptors

Baroreceptors monitor changes in pressure in an organ. A baroreceptor consists of free nerve endings that branch within the elastic tissues in the wall of a distensible organ, such as a blood vessel or a portion of the respiratory, digestive, or urinary tract. When the pressure changes, the elastic walls of the tract recoil or expand. This movement distorts the dendritic branches and alters the rate of action potential generation. Baroreceptors respond immediately to a change in pressure, but they adapt rapidly, and the output along the afferent fibers gradually returns to normal.

Baroreceptors monitor blood pressure in the walls of major vessels, including the carotid artery (at the *carotid sinus*) and the aorta (at the *aortic sinus*). The information plays a major role in regulating cardiac function and adjusting blood flow to vital tissues. Baroreceptors in the lungs monitor the degree of lung expansion. This information is relayed to the respiratory rhythmicity centers, which set the pace of respiration. Comparable stretch receptors at various sites in the digestive and urinary tracts trigger a variety of visceral reflexes, including those of urination and defecation. We will describe those baroreceptor reflexes in chapters that deal with specific physiological systems.

lung, heart, stomach, intestines, bladder wall.

Proprioceptors

Proprioceptors monitor the position of joints, the tension in tendons and ligaments, and the state of muscular contraction. There are three major groups of proprioceptors:

1. *Muscle Spindles.* Muscle spindles monitor skeletal muscle length and trigger stretch reflexes. ∞ p. 451

2. *Golgi Tendon Organs.* **Golgi tendon organs** are similar in function to Ruffini corpuscles but are located at the junction between a skeletal muscle and its tendon. In a Golgi tendon organ, dendrites branch repeatedly and wind around the densely packed collagen fibers of the tendon. These receptors are stimulated by tension in the tendon; they thus monitor the external tension developed during muscle contraction.

   external tension during muscle contraction

3. *Receptors in Joint Capsules.* Joint capsules are richly innervated by free nerve endings that detect pressure, tension, and movement at the joint. Your sense of body position results from the integration of information from these receptors with information provided by muscle spindles, Golgi tendon organs, and the receptors of the inner ear.

   integration of info received by receptors.
   provided by spindles.

Proprioceptors do not adapt to constant stimulation, and each receptor continuously sends information to the CNS. A relatively small proportion of the arriving proprioceptive information reaches your awareness; most proprioceptive information is processed at subconscious levels.

Chemoreceptors

Specialized chemoreceptive neurons can detect small changes in the concentration of specific chemicals or compounds. In general, **chemoreceptors** respond only to water-soluble and lipid-soluble substances that are dissolved in the surrounding fluid. These receptors exhibit peripheral adaptation over a period of seconds, and central adaptation may also occur.

The chemoreceptors included in the general senses do not send information to the primary sensory cortex, so we are not consciously aware of the sensations they provide. The arriving sensory information is routed to brain stem centers that deal with the autonomic control of respiratory and cardiovascular functions. Neurons in the respiratory centers of the brain respond to the concentration of hydrogen ions (pH) and levels of carbon dioxide molecules in the cerebrospinal fluid. Chemoreceptive neurons are also located in the **carotid bodies**, near the origin of the internal carotid arteries on each side of the neck, and in the **aortic bodies**, between the major branches of the aortic arch. These receptors monitor the pH and the carbon dioxide and oxygen levels in arterial blood. The afferent fibers leaving the carotid or aortic bodies reach the respiratory centers by traveling within cranial nerves IX (glossopharyngeal) and X (vagus).

carotid bodies → heart, arterial parts (pH, O2, CO2 levels) → sent to hypothalamus

> ### CHECKPOINT
>
> 5. List the four types of general sensory receptors, and identify the nature of the stimulus that excites each type.
>
> 6. Identify the three classes of mechanoreceptors.
>
> 7. What would happen to you if the information from proprioceptors in your legs were blocked from reaching the CNS?
>
> See the blue Answers tab at the end of the book.

individual receptors

15-4 Separate pathways carry somatic sensory and visceral sensory information

A sensory neuron that delivers sensations to the CNS is often called a **first-order neuron**. The cell body of a first-order general sensory neuron is located in a dorsal root ganglion or cranial nerve ganglion. In the CNS, the axon of that sensory neuron synapses on an interneuron known as a **second-order neuron**, which may be located in the spinal cord or brain stem. If the sensation is to reach our awareness, the second-order neuron synapses on a **third-order neuron** in the thalamus. Somewhere along its length, the axon of the second-order neuron crosses over to the opposite side of the CNS. As a result, the right side of the thalamus receives sensory information from the left side of the body, and the left side of the thalamus receives sensory information from the right side of the body.

The axons of the third-order neurons ascend without crossing over and synapse on neurons of the primary sensory cortex of the cerebral hemisphere. As a result, the right cerebral hemisphere receives sensory information from the left side of the body, and the left cerebral hemisphere receives sensations from the right side. The reason for this crossover is unknown. Although it has no apparent functional benefit, crossover occurs along sensory and motor pathways in all vertebrates.

Somatic Sensory Pathways

Somatic sensory pathways carry sensory information from the skin and musculature of the body wall, head, neck, and limbs. We will consider three major somatic sensory pathways: (1) the *posterior column pathway*, (2) the *spinothalamic pathway*, and (3) the *spinocerebellar pathway*. These pathways utilize pairs of spinal tracts, symmetrically arranged on opposite sides of the spinal cord. All the axons within a tract share a common origin and destination.

Figure 15–4 indicates the relative positions of the spinal tracts involved. Note that tract names often give clues to their function. For example, if the name of a tract begins with *spino-*, the tract must *start* in the spinal cord and *end* in the brain. It must therefore be an ascending tract that carries sen-

sory information. The rest of the name indicates the tract's destination. Thus, a *spinothalamic tract* begins in the spinal cord and carries sensory information to the thalamus.

If, on the other hand, the name of a tract ends in *-spinal*, the tract *ends* in the spinal cord and *starts* in a higher center of the brain. It must therefore be a descending tract that carries motor commands. The first part of the name indicates the nucleus or cortical area of the brain where the tract originates. For example, a *corticospinal tract* carries motor commands from the cerebral cortex to the spinal cord. Such tracts will be considered later in the chapter.

The Posterior Column Pathway

The **posterior column pathway** carries sensations of highly localized ("fine") touch, pressure, vibration, and proprioception (**Figure 15–5a**). This pathway, also known as the *dorsal column/medial lemniscus*, begins at a peripheral receptor and ends at the primary sensory cortex of the cerebral hemispheres. The spinal tracts involved are the left and right **fasciculus gracilis** (*gracilis*, slender) and the left and right **fasciculus cuneatus** (*cuneus*, wedge-shaped). On each side of the posterior median sulcus, the fasciculus gracilis is medial to the fasciculus cuneatus.

The axons of the first-order neurons reach the CNS within the dorsal roots of spinal nerves and the sensory roots of cranial nerves. The axons ascending within the posterior column are organized according to the region innervated. Axons carrying sensations from the inferior half of the body ascend within the fasciculus gracilis and synapse in the nucleus

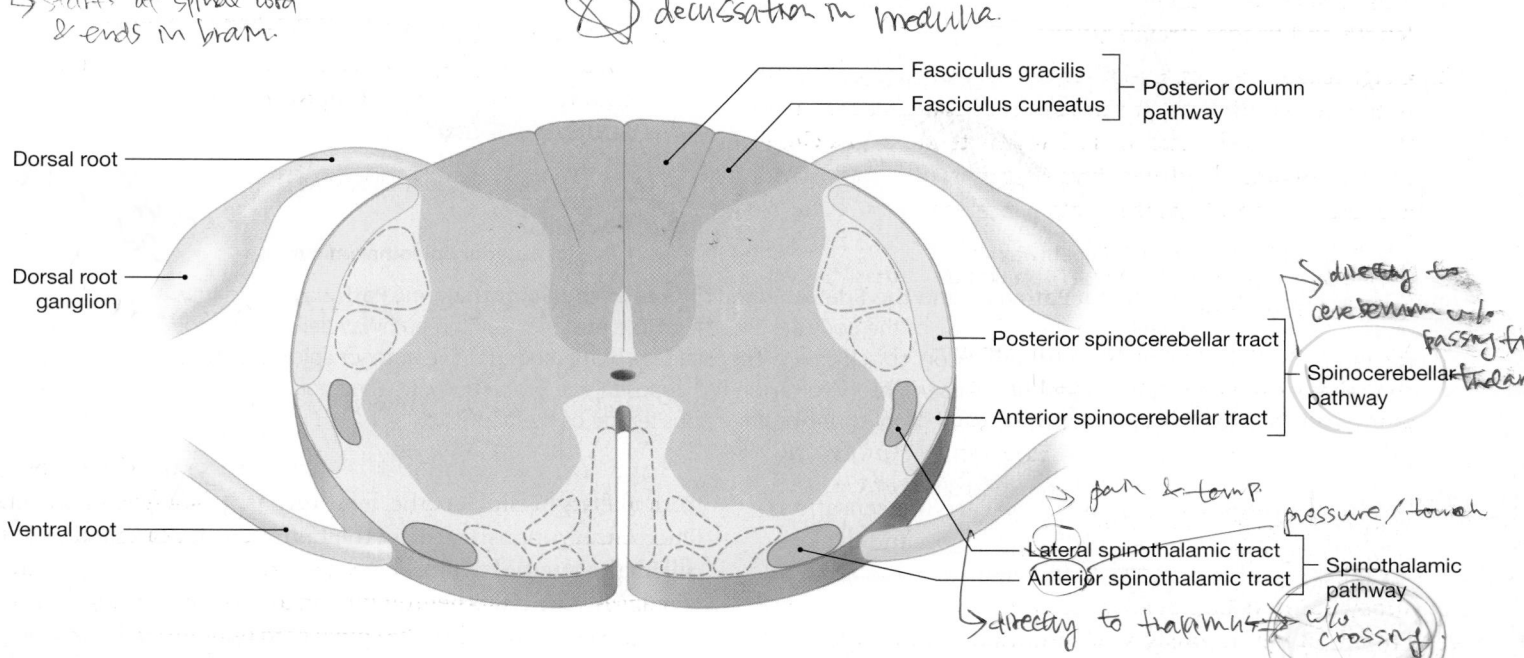

Figure 15–4 **Sensory Pathways and Ascending Tracts in the Spinal Cord.** A cross-sectional view of the spinal cord indicating the locations of the major ascending (sensory) tracts. For information about these tracts, see *Table 15–1*. Descending (motor) tracts (identified in *Figure 15–8*) are shown in dashed outline.

Labels in figure:
- Dorsal root
- Dorsal root ganglion
- Ventral root
- Fasciculus gracilis — Posterior column pathway
- Fasciculus cuneatus
- Posterior spinocerebellar tract — Spinocerebellar pathway
- Anterior spinocerebellar tract
- Lateral spinothalamic tract — Spinothalamic pathway
- Anterior spinothalamic tract

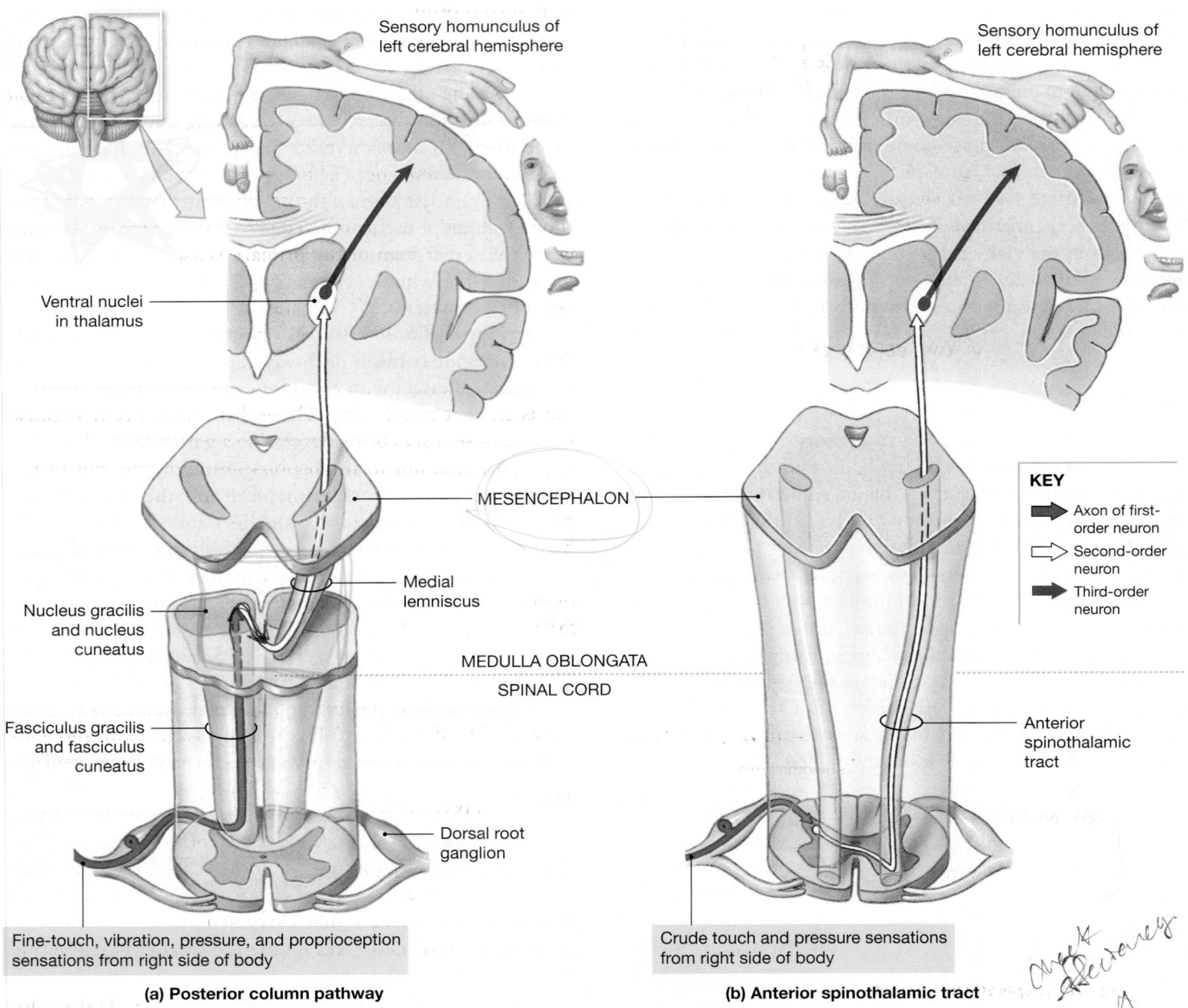

Sensory homunculus of
left cerebral hemisphere

Sensory homunculus of
left cerebral hemisphere

Ventral nuclei
in thalamus

MESENCEPHALON

KEY

➡ Axon of first-
order neuron

⇨ Second-order
neuron

➡ Third-order
neuron

Medial
lemniscus

Nucleus gracilis
and nucleus
cuneatus

MEDULLA OBLONGATA

SPINAL CORD

Fasciculus gracilis
and fasciculus
cuneatus

Anterior
spinothalamic
tract

Dorsal root
ganglion

Fine-touch, vibration, pressure, and proprioception
sensations from right side of body

Crude touch and pressure sensations
from right side of body

(a) Posterior column pathway

(b) Anterior spinothalamic tract

Figure 15–5 The Posterior Column Pathway, and the Spinothalamic Tracts of the Spinothalamic Pathway. For clarity, only the pathways for sensations originating on the right side of the body are shown. **(a)** The posterior column pathway delivers fine touch, vibration, and proprioception information to the primary sensory cortex on the opposite side of the body. **(b)** The anterior spinothalamic tracts carry sensations of crude touch and pressure to the primary sensory cortex on the opposite side of the body. **(c)** The lateral spinothalamic tracts carry sensations of pain and temperature to the primary sensory cortex on the opposite side of the body.

gracilis of the medulla oblongata. Axons carrying sensations from the superior half of the trunk, upper limbs, and neck ascend in the fasciculus cuneatus and synapse in the nucleus cuneatus. ∞ p. 471

Axons of the second-order neurons of the nucleus gracilis and nucleus cuneatus ascend to the thalamus. As they ascend, these axons cross over to the opposite side of the brain stem. The crossing of an axon from the left side to the right side, or from the right side to the left side, is called **decussation**. Once on the opposite side of the brain, the axons enter a tract called the **medial lemniscus** (*lemniskos*, ribbon). As it ascends, the medial lemniscus runs alongside a smaller tract that carries sensory information from the face, relayed from the sensory nuclei of the trigeminal nerve (N V).

The axons in these tracts synapse on third-order neurons in one of the ventral nuclei of the thalamus. ∞ p. 475 These

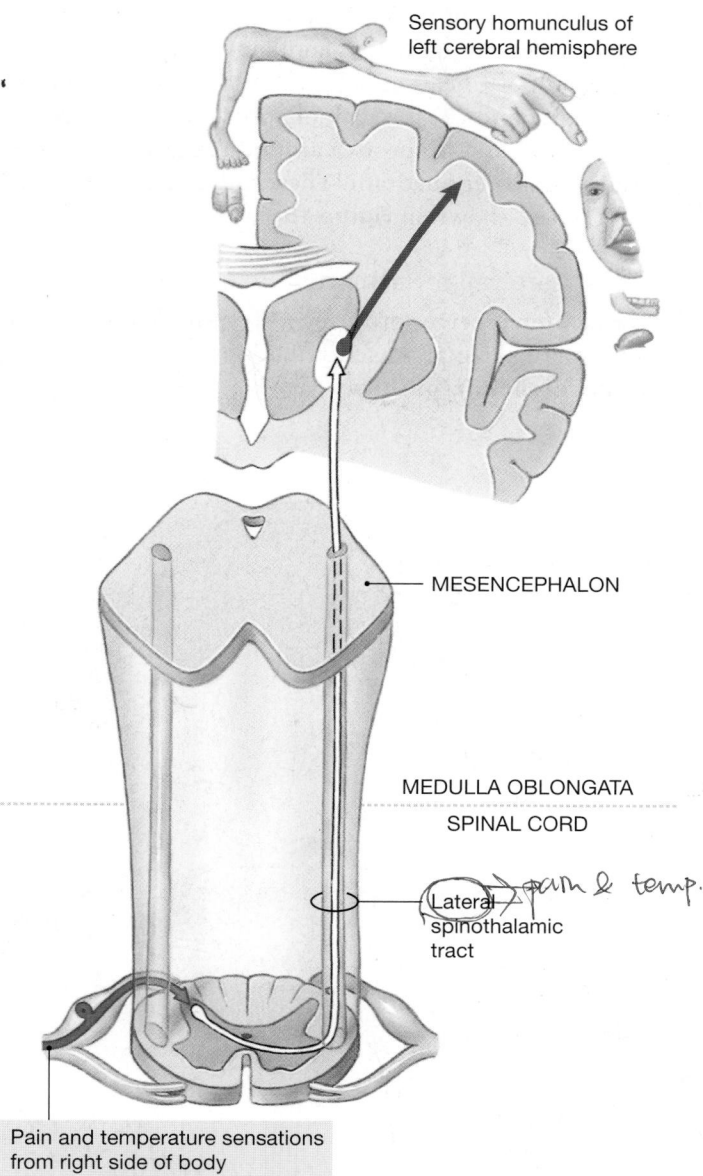

Sensory homunculus of
left cerebral hemisphere

MESENCEPHALON

MEDULLA OBLONGATA

SPINAL CORD

Lateral *pain & temp.*
spinothalamic
tract

Pain and temperature sensations
from right side of body

(c) Lateral spinothalamic tract

nuclei sort the arriving information according to (1) the nature of the stimulus and (2) the region of the body involved. Processing in the thalamus determines whether you perceive a given sensation as fine touch, or as pressure or vibration.

Our ability to localize the sensation—to determine precisely where on the body a specific stimulus originated—depends on the projection of information from the thalamus to the primary sensory cortex. Sensory information from the toes arrives at one end of the primary sensory cortex, and information from the head arrives at the other. When neurons in one portion of your primary sensory cortex are stimulated, you become aware of sensations originating at a specific loca-

tion. If your primary sensory cortex were damaged or the projection fibers were cut, you could detect a light touch but would be unable to determine its source.

The same sensations are reported whether the cortical neurons are activated by axons ascending from the thalamus or by direct electrical stimulation. Researchers have electrically stimulated the primary sensory cortex in awake individuals during brain surgery and asked the subjects where they thought the stimulus originated. The results were used to create a functional map of the primary sensory cortex. Such a map, three of which are shown in **Figure 15–5**, is called a **sensory homunculus** ("little man").

The proportions of the sensory homunculus are very different from those of any individual. For example, the face is huge and distorted, with enormous lips and tongue, whereas the back is relatively tiny. These distortions occur because the area of sensory cortex devoted to a particular body region is proportional not to the region's absolute size, but to the *number of sensory receptors* it contains. In other words, many more cortical neurons are required to process sensory information arriving from the tongue, which has tens of thousands of taste and touch receptors, than to analyze sensations originating on the back, where touch receptors are few and far between. ⟹ *distorted b/c some parts require processing of more nerves than others. #) hands & tongue complex*

The Spinothalamic Pathway

The **spinothalamic pathway** provides conscious sensations of poorly localized ("crude") touch, pressure, pain, and temperature. In this pathway, the axons of first-order sensory neurons enter the spinal cord and synapse on second-order neurons within the posterior gray horns. The axons of these interneurons cross to the opposite side of the spinal cord before ascending. This pathway includes relatively small tracts that deliver sensations to reflex centers in the brain stem as well as larger tracts that carry sensations destined for the cerebral cortex. We will ignore the smaller tracts in this discussion.

Sensations bound for the cerebral cortex ascend within the anterior or lateral spinothalamic tracts. The **anterior spinothalamic tracts** carry crude touch and pressure sensations (**Figure 15–5b**), whereas the **lateral spinothalamic tracts** carry pain and temperature sensations (**Figure 15–5c**). These tracts end at third-order neurons in the ventral nucleus group of the thalamus. After the sensations have been sorted and processed, they are relayed to the primary sensory cortex.

The perception that an arriving stimulus is painful rather than cold, hot, or vibrating depends on which second-order and third-order neurons are stimulated. The ability to localize that stimulus to a specific location in the body depends on the stimulation of an appropriate area of the primary sensory cortex. Any abnormality along the pathway can result in inappropriate sensations or inaccurate localization of the source. Consider these examples:

- An individual can experience painful sensations that are not real. For example, a person may continue to experience pain in an amputated limb. This *phantom limb pain* is caused by activity in the sensory neurons or interneurons along the spinothalamic pathway. The neurons involved were once part of the labeled line that monitored conditions in the intact limb. These labeled lines and pathways are developmentally programmed; even individuals born without limbs can have phantom limb pain.

- An individual can feel pain in an uninjured part of the body when the pain actually originates at another location. For example, strong visceral pain sensations arriving at a segment of the spinal cord can stimulate interneurons that are part of the spinothalamic pathway. Activity in these interneurons leads to the stimulation of

the primary sensory cortex, so the individual feels pain in a specific part of the body surface. This phenomenon is called **referred pain**. Two familiar examples are (1) the pain of a heart attack, which is frequently felt in the left arm, and (2) the pain of appendicitis, which is generally felt first in the area around the navel and then in the right lower quadrant. These and additional examples are shown in **Figure 15–6**.

The Spinocerebellar Pathway

The cerebellum receives proprioceptive information about the position of skeletal muscles, tendons, and joints along the **spinocerebellar pathway** (**Figure 15–7**). This infor-

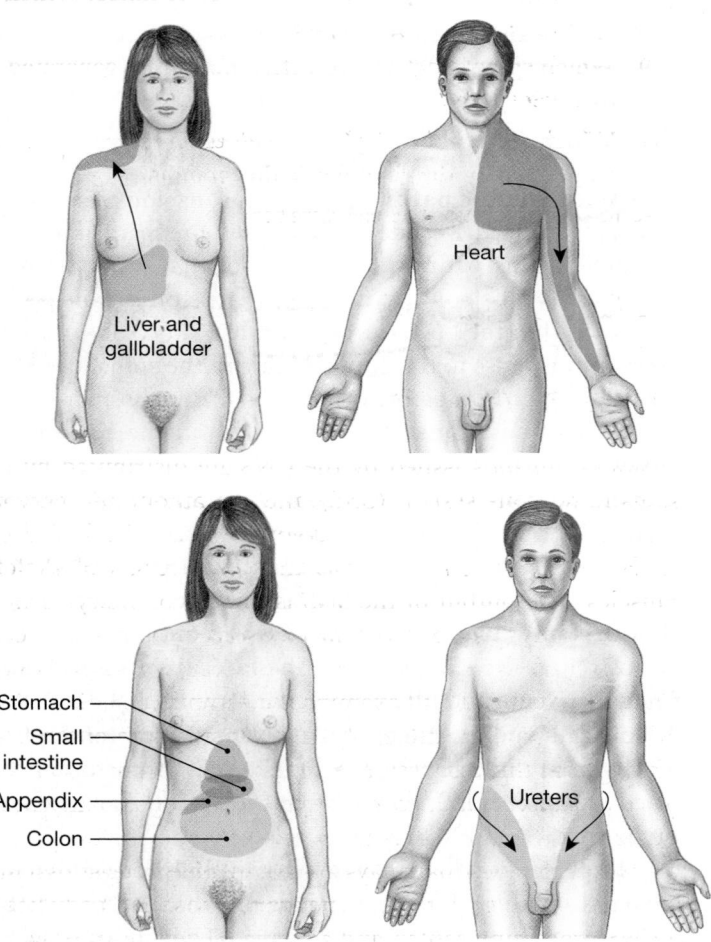

Figure 15–6 Referred Pain. Pain sensations from visceral organs are often perceived as involving specific regions of the body surface innervated by the same spinal segments. Each region of perceived pain is labeled according to the organ at which the pain originates.

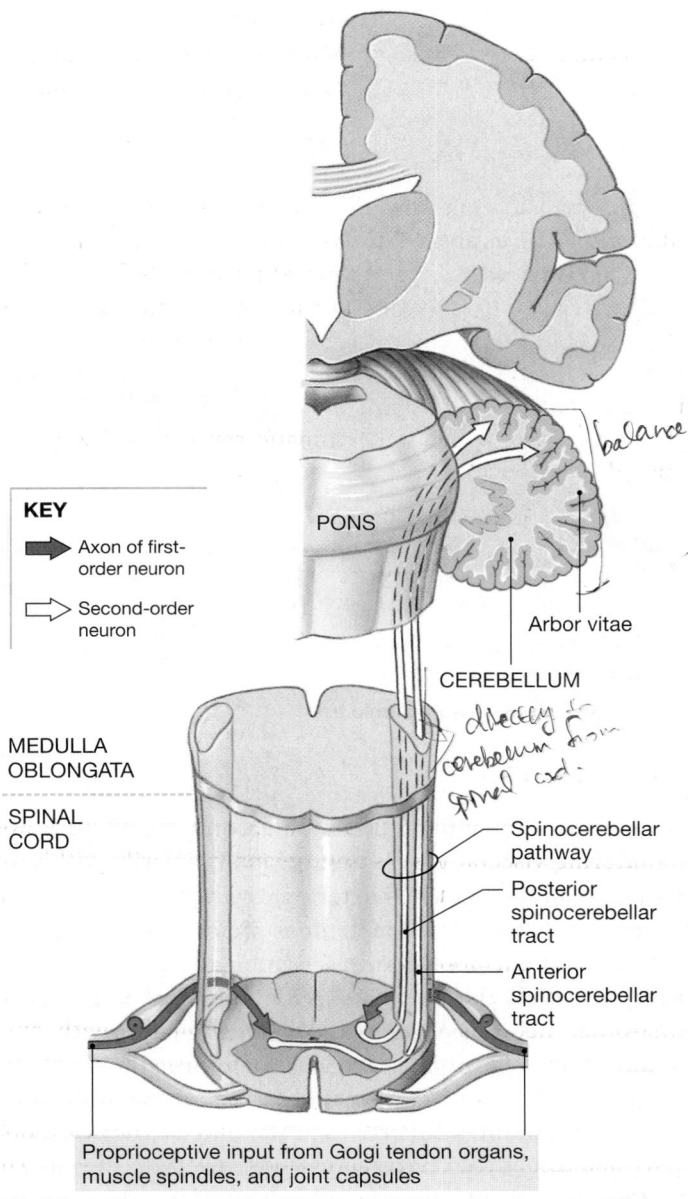

Figure 15–7 The Spinocerebellar Pathway.

mation does not reach our awareness. The axons of first-order sensory neurons synapse on interneurons in the dorsal gray horns of the spinal cord. The axons of these second-order neurons ascend in one of the spinocerebellar tracts:

- The **posterior spinocerebellar tracts** contain axons that do not cross over to the opposite side of the spinal cord. These axons reach the cerebellar cortex via the inferior cerebellar peduncle of that side.

- The **anterior spinocerebellar tracts** are dominated by axons that have crossed over to the opposite side of the spinal cord, although they do contain a significant number of uncrossed axons as well. The sensations carried by the anterior spinocerebellar tracts reach the cerebellar cortex via the superior cerebellar peduncle. Interestingly, many of the axons that cross over and ascend to the cerebellum then cross over again within the cerebellum, synapsing on the same side as the original stimulus. The functional significance of this "double cross" is unknown.

double cross → synapses on original side of stimulus.

The information carried by the spinocerebellar pathway ultimately arrives at the *Purkinje cells* of the cerebellar cortex. ∞ p. 472 Proprioceptive information from each part of the body is relayed to a specific portion of the cerebellar cortex. We will consider the integration of proprioceptive information and the role of the cerebellum in somatic motor control in a later section.

→ highly branched cells in cerebellum

Table 15–1 reviews the somatic sensory pathways discussed in this section.

→ superficial layer of cortex.

The A&P Top 100

#52 Most somatic sensory information is relayed to the thalamus for processing. A small fraction of the arriving information is projected to the cerebral cortex and reaches our awareness.

Visceral Sensory Pathways

Visceral sensory information is collected by interoceptors monitoring visceral tissues and organs, primarily within the thoracic and abdominopelvic cavities. These interoceptors include nociceptors, thermoreceptors, tactile receptors, baroreceptors, and chemoreceptors, although none of them are as numerous as they are in somatic tissues. The axons of the first-order neurons usually travel in company with autonomic motor fibers innervating the same visceral structures.

Cranial nerves V, VII, IX, and X carry visceral sensory information from the mouth, palate, pharynx, larynx, trachea, esophagus, and associated vessels and glands. ∞ pp. 494–498 This information is delivered to the **solitary nucleus**, a large nucleus in the medulla oblongata. The solitary nucleus is a major pro-cessing and sorting center for visceral sensory information; it has extensive connections with the various cardiovascular and respiratory centers as well as with the reticular formation.

The dorsal roots of spinal nerves T_1–L_2 carry visceral sensory information provided by receptors in organs located between the diaphragm and the pelvic cavity. The dorsal roots of spinal nerves S_2–S_4 carry visceral sensory information from organs in the inferior portion of the pelvic cavity, including the last portion of the large intestine, the urethra and base of the urinary bladder, and the prostate gland (males) or the cervix of the uterus and adjacent portions of the vagina (females).

The first-order neurons deliver the visceral sensory information to interneurons whose axons ascend within the spinothalamic pathway. Most of the sensory information is delivered to the solitary nucleus, and because it never reaches the primary sensory cortex we remain unaware of these sensations.

CHECKPOINT

8. **As a result of pressure on her spinal cord, Jill cannot feel fine touch or pressure on her lower limbs. Which spinal tract is being compressed?**

9. **Which spinal tract carries action potentials generated by nociceptors?**

10. **Which cerebral hemisphere receives impulses conducted by the right fasciculus gracilis?**

See the blue Answers tab at the end of the book.

15-5 The somatic nervous system is an efferent division that controls skeletal muscles

Motor commands issued by the CNS are distributed by the somatic nervous system (SNS) and the autonomic nervous system (ANS). The somatic nervous system, also called the *somatic motor system*, controls the contractions of skeletal muscles. The output of the SNS is under voluntary control. The autonomic nervous system, or *visceral motor system*, controls visceral effectors, such as smooth muscle, cardiac muscle, and glands. We will examine the organization of the ANS in Chapter 16; our interest here is the structure of the SNS. Throughout this discussion we will use the terms *motor neuron* and *motor control* to refer specifically to somatic motor neurons and pathways that control skeletal muscles.

Somatic motor pathways always involve at least two motor neurons: an **upper motor neuron**, whose cell body lies in a CNS processing center, and a **lower motor neuron**, whose cell body lies in a nucleus of the brain stem or spinal cord. The upper motor neuron synapses on the lower motor neuron, which in turn innervates a single motor unit in a skeletal muscle. Activity in the upper motor neuron may facilitate or

TABLE 15–1 Principal Ascending (Sensory) Pathways

Pathway/Tract	Sensation(s)	Location of Neuron Cell Bodies		
		First-Order	Second-Order	Third-Order
POSTERIOR COLUMN PATHWAY				
Fasciculus gracilis	Proprioception and fine touch, ventral pressure, and vibration from inferior half of body	Dorsal root ganglia of inferior half of body; axons enter CNS in dorsal roots and join fasciculus gracilis	Nucleus gracilis of medulla oblongata; axons cross over before entering medial lemniscus	Ventral nuclei of thalamus
Fasciculus cuneatus	Proprioception and fine touch, ventral pressure, and vibration from superior half of body	Dorsal root ganglia of superior half of body; axons enter CNS in dorsal roots and join fasciculus cuneatus	Nucleus cuneatus of medulla oblongata; axons cross over before entering medial lemniscus	As above
SPINOTHALAMIC PATHWAY				
Lateral spinothalamic tracts	Pain and temperature	Dorsal root ganglia; axons enter CNS in dorsal roots	Interneurons in posterior gray horn; axons enter lateral spinothalamic tract on opposite side	Ventral nuclei of thalamus
Anterior spinothalamic tracts	Crude touch and pressure	As above	Interneurons in posterior gray horn; axons enter anterior spinothalamic tract on opposite side	As above
SPINOCEREBELLAR PATHWAY				
Posterior spinocerebellar tracts	Proprioception	Dorsal root ganglia; axons enter CNS in dorsal roots	Interneurons in posterior gray horn; axons enter posterior spinothalamic tract on same side	Not present
Anterior spinocerebellar tracts	Proprioception	As above	Interneurons in same spinal section; axons enter anterior spinocerebellar tract on the same or opposite side	Not present

inhibit the lower motor neuron. Activation of the lower motor neuron triggers a contraction in the innervated muscle. Only the axon of the lower motor neuron extends outside the CNS. Destruction of or damage to a lower motor neuron eliminates voluntary and reflex control over the innervated motor unit.

Conscious and subconscious motor commands control skeletal muscles by traveling over three integrated motor pathways: the *corticospinal pathway*, the *medial pathway*, and the *lateral pathway*. **Figure 15–8** indicates the positions of the associated motor (descending) tracts in the spinal cord. Activity within these motor pathways is monitored and adjusted by the basal nuclei and cerebellum. The output of these centers stimulates or inhibits the activity of either (1) motor nuclei or (2) the primary motor cortex.

The Corticospinal Pathway

The **corticospinal pathway** (**Figure 15–9**), sometimes called the *pyramidal system*, provides voluntary control over skeletal muscles. This system begins at the *pyramidal cells* of the pri-

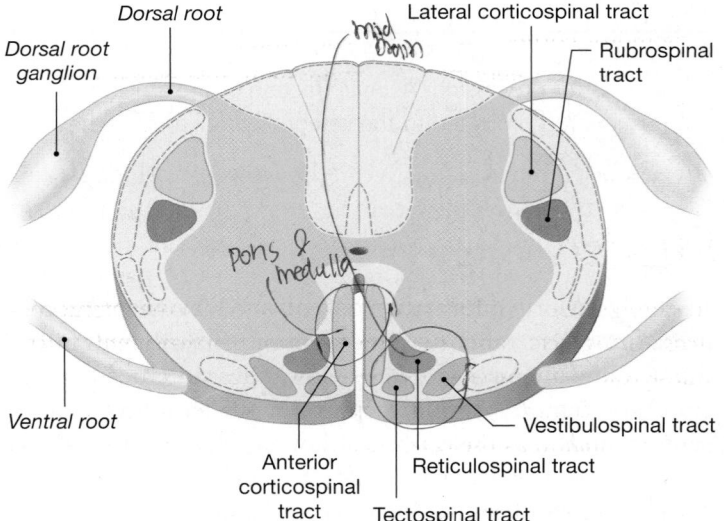

Figure 15–8 **Descending (Motor) Tracts in the Spinal Cord.** A cross-sectional view indicating the locations of the major descending (motor) tracts that contain the axons of upper motor neurons. The origins and destinations of these tracts are listed in *Table 15–2*. Sensory tracts (shown in *Figure 15–4*) appear in dashed outline.

Location of Neuron Cell Bodies	
Final Destination	**Site of Crossover**
Primary sensory cortex on side opposite stimulus	Axons of second-order neurons before entering the medial lemniscus
As above	As above
Primary sensory cortex on side opposite stimulus	Axons of second-order neurons at level of entry
As above	As above
Cerebellar cortex on side of stimulus	None
Cerebellar cortex on side opposite (and side of) stimulus	Axons of most second-order neurons cross over before entering tract; many re-cross at cerebellum

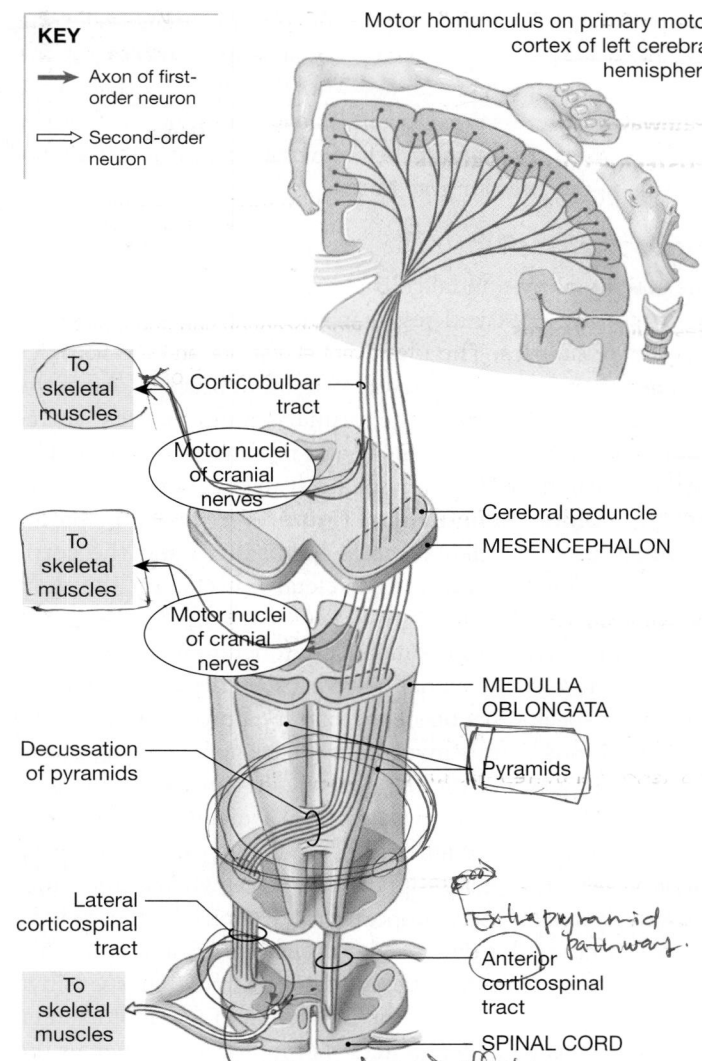

KEY

→ Axon of first-order neuron

⇨ Second-order neuron

Motor homunculus on primary motor cortex of left cerebral hemisphere

To skeletal muscles

Corticobulbar tract

Motor nuclei of cranial nerves

To skeletal muscles

Motor nuclei of cranial nerves

Cerebral peduncle

MESENCEPHALON

MEDULLA OBLONGATA

Decussation of pyramids

Pyramids

Lateral corticospinal tract

Extra pyramid pathway

Anterior corticospinal tract

To skeletal muscles

SPINAL CORD

anterior root → effector.

Figure 15–9   The Corticospinal Pathway. The corticospinal pathway originates at the primary motor cortex. The corticobulbar tracts end at the motor nuclei of cranial nerves on the opposite side of the brain. Most fibers in this pathway cross over in the medulla and enter the lateral corticospinal tracts; the rest descend in the anterior corticospinal tracts and cross over after reaching target segments in the spinal cord.

∞ p. 484 mary motor cortex. The axons of these upper motor neurons descend into the brain stem and spinal cord to synapse on lower motor neurons that control skeletal muscles. In general, the corticospinal pathway is direct: The upper motor neurons synapse directly on the lower motor neurons. However, the corticospinal pathway also works indirectly, as it innervates centers of the medial and lateral pathways.

The corticospinal pathway contains three pairs of descending tracts: (1) the *corticobulbar tracts*, (2) the *lateral corticospinal tracts*, and (3) the *anterior corticospinal tracts*. These tracts enter the white matter of the internal capsule, descend into the brain stem, and emerge on either side of the mesencephalon as the *cerebral peduncles*.

The Corticobulbar Tracts

Axons in the **corticobulbar** (kor-ti-kō-BUL-bar; *bulbar*, brain stem) **tracts** synapse on lower motor neurons in the motor nuclei of cranial nerves III, IV, V, VI, VII, IX, XI, and XII. The corticobulbar tracts provide conscious control over skeletal muscles that move the eye, jaw, and face, and some muscles of

⇒ conscious control of cranial nerves.

the neck and pharynx. The corticobulbar tracts also innervate the motor centers of the medial and lateral pathways.

The Corticospinal Tracts

Axons in the **corticospinal tracts** synapse on lower motor neurons in the anterior gray horns of the spinal cord. As they descend, the corticospinal tracts are visible along the ventral surface of the medulla oblongata as a pair of thick bands, the **pyramids**. Along the length of the pyramids, roughly 85 percent of the axons cross the midline (decussate) to enter the descending **lateral corticospinal tracts** on the opposite side of the spinal cord. The other 15 percent continue uncrossed along the spinal cord as the **anterior corticospinal tracts**. At the spinal segment it targets, an axon in the anterior corticospinal tract crosses over to the opposite side of the spinal cord in the anterior white commissure before synapsing on lower motor neurons in the anterior gray horns.

The Motor Homunculus

The activity of pyramidal cells in a specific portion of the primary motor cortex will result in the contraction of specific peripheral muscles. The identities of the stimulated muscles depend on the region of motor cortex that is active. As in the primary sensory cortex, the primary motor cortex corresponds point by point with specific regions of the body. The cortical areas have been mapped out in diagrammatic form, creating a **motor homunculus**. **Figure 15–9** shows the motor homunculus of the left cerebral hemisphere and the corticospinal pathway controlling skeletal muscles on the right side of the body.

The proportions of the motor homunculus are quite different from those of the actual body, because the motor area devoted to a specific region of the cortex is proportional to the number of motor units involved in the region's control, not to its actual size. As a result, the homunculus provides an indication of the degree of fine motor control available. For example, the hands, face, and tongue, all of which are capable of varied and complex movements, appear very large, whereas the trunk is relatively small. These proportions are similar to those of the sensory homunculus (**Figure 15–5**). The sensory and motor homunculi differ in other respects because some highly sensitive regions, such as the sole of the foot, contain few motor units, and some areas with an abundance of motor units, such as the eye muscles, are not particularly sensitive.

CLINICAL NOTE

Cerebral Palsy The term **cerebral palsy** refers to a number of disorders that affect voluntary motor performance; they appear during infancy or childhood and persist throughout the life of the affected individual. The cause may be trauma associated with premature or unusually stressful birth, maternal exposure to drugs (including alcohol), or a genetic defect that causes the improper development of motor pathways. Problems during labor and delivery may produce compression or interruption of placental circulation or oxygen supplies. If the oxygen concentration in fetal blood declines significantly for as little as 5–10 minutes, CNS function can be permanently impaired. The cerebral cortex, cerebellum, basal nuclei, hippocampus, and thalamus are likely targets, producing abnormalities in motor skills, posture and balance, memory, speech, and learning abilities.

The Medial and Lateral Pathways

Several centers in the cerebrum, diencephalon, and brain stem may issue somatic motor commands as a result of processing performed at a subconscious level. These centers and their associated tracts were long known as the *extrapyramidal system (EPS)*, because it was thought that they operated independently of, and in parallel with, the *pyramidal system* (corticospinal pathway). This classification scheme is both inaccurate and misleading, because motor control is integrated at all levels through extensive feedback loops and interconnections. It is more appropriate to group these nuclei and tracts in terms of their primary functions: The components of the **medial pathway** help control gross movements of the trunk and proximal limb muscles, whereas those of the **lateral pathway** help control the distal limb muscles that perform more precise movements.

The medial and lateral pathways can modify or direct skeletal muscle contractions by stimulating, facilitating, or inhibiting lower motor neurons. It is important to note that the axons of upper motor neurons in the medial and lateral pathways synapse on the same lower motor neurons innervated by the corticospinal pathway. This means that the various motor pathways interact not only within the brain, through interconnections between the primary motor cortex and motor centers in the brain stem, but also through excitatory or inhibitory interactions at the level of the lower motor neuron.

The Medial Pathway

The medial pathway is primarily concerned with the control of muscle tone and gross movements of the neck, trunk, and proximal limb muscles. The upper motor neurons of the medial pathway are located in the *vestibular nuclei*, the *superior* and *inferior colliculi*, and the *reticular formation*.

The vestibular nuclei receive information, over the vestibulocochlear nerve (N VIII), from receptors in the inner ear that monitor the position and movement of the head. These nuclei respond to changes in the orientation of the head, sending motor commands that alter the muscle tone, extension, and position of the neck, eyes, head, and limbs. The primary goal is to maintain posture and balance. The descending fibers in the spinal cord constitute the **vestibulospinal tracts**.

The superior and inferior colliculi are located in the *tectum*, or roof of the mesencephalon (**Figure 14–8**, p. 474). The colliculi receive visual (superior) and auditory (inferior) sensations. Axons of upper motor neurons in the colliculi descend in the **tectospinal tracts**. These axons cross to the opposite side immediately, before descending to synapse on lower motor neurons in the brain stem or spinal cord. Axons in the tectospinal tracts direct reflexive changes in the position of the head, neck, and upper limbs in response to bright lights, sudden movements, or loud noises.

The reticular formation is a loosely organized network of neurons that extends throughout the brain stem. ∞ p. 470 The reticular formation receives input from almost every ascending and descending pathway. It also has extensive interconnections with the cerebrum, the cerebellum, and brain stem nuclei. Axons of upper motor neurons in the reticular formation descend into the **reticulospinal tracts** without crossing to the opposite side. The effects of reticular formation stimulation are determined by the region stimulated. For example, the stimulation of upper motor neurons in one portion of the reticular formation produces eye movements, whereas the stimulation of another portion activates respiratory muscles.

The Lateral Pathway

The lateral pathway is primarily concerned with the control of muscle tone and the more precise movements of the distal parts of the limbs. The upper motor neurons of the lateral pathway lie within the red nuclei of the mesencephalon. ∞ p. 475 Axons of upper motor neurons in the red nuclei cross to the opposite side of the brain and descend into the spinal cord in the **rubrospinal tracts** (*ruber*, red). In humans, the rubrospinal tracts are small and extend only to the cervical spinal cord. There they provide motor control over distal muscles of the upper limbs; normally, their role is insignifi-

cant as compared with that of the lateral corticospinal tracts. However, the rubrospinal tracts can be important in maintaining motor control and muscle tone in the upper limbs if the lateral corticospinal tracts are damaged.

Table 15–2 reviews the major descending (motor) tracts discussed in this section.

The Basal Nuclei and Cerebellum

The basal nuclei and cerebellum are responsible for coordination and feedback control over muscle contractions, whether those contractions are consciously or subconsciously directed.

The Basal Nuclei

The basal nuclei provide the background patterns of movement involved in voluntary motor activities. For example, they may control muscles that determine the background position of the trunk or limbs, or they may direct rhythmic cycles of movement, as in walking or running. These nuclei do not exert direct control over lower motor neurons. Instead, they adjust the activities of upper motor neurons in the various motor pathways based on input from all portions of the cerebral cortex, as well as from the substantia nigra.

TABLE 15–2	Principal Descending (Motor) Pathways			
Tract	**Location of Upper Motor Neurons**	**Destination**	**Site of Crossover**	**Action**
CORTICOSPINAL PATHWAY				
Corticobulbar tracts	Primary motor cortex (cerebral hemisphere)	Lower motor neurons of cranial nerve nuclei in brain stem	Brain stem	Conscious motor control of skeletal muscles
Lateral corticospinal tracts	As above	Lower motor neurons of anterior gray horns of spinal cord	Pyramids of medulla oblongata	As above
Anterior corticospinal tracts	As above	As above	Level of lower motor neuron	As above
MEDIAL PATHWAY				
Vestibulospinal tracts	Vestibular nuclei (at border of pons and medulla oblongata)	As above	None (uncrossed)	Subconscious regulation of balance and muscle tone
Tectospinal tracts	Tectum (mesencephalon: superior and inferior colliculi)	Lower motor neurons of anterior gray horns (cervical spinal cord only)	Brain stem (mesencephalon)	Subconscious regulation of eye, head, neck, and upper limb position in response to visual and auditory stimuli
Reticulospinal tracts	Reticular formation (network of nuclei in brain stem)	Lower motor neurons of anterior gray horns of spinal cord	None (uncrossed)	Subconscious regulation of reflex activity
LATERAL PATHWAY				
Rubrospinal tracts	Red nuclei of mesencephalon	As above	Brain stem (mesencephalon)	Subconscious regulation of upper limb muscle tone and movement

The basal nuclei adjust or establish patterns of movement via two major pathways:

1. One group of axons synapses on thalamic neurons, whose axons extend to the premotor cortex, the motor association area that directs activities of the primary motor cortex. This arrangement creates a feedback loop that changes the sensitivity of the pyramidal cells and alters the pattern of instructions carried by the corticospinal tracts.

2. A second group of axons synapses in the reticular formation, altering the excitatory or inhibitory output of the reticulospinal tracts.

Two distinct populations of neurons exist: one that stimulates neurons by releasing acetylcholine (ACh), and another that inhibits neurons through the release of gamma aminobutyric acid (GABA). Under normal conditions, the excitatory interneurons are kept inactive, and the tracts leaving the basal nuclei have an inhibitory effect on upper motor neurons. In *Parkinson disease*, the excitatory neurons become more active, leading to problems with the voluntary control of movement. ∞ p. 484

If the primary motor cortex is damaged, the individual loses the ability to exert fine control over skeletal muscles. However, some voluntary movements can still be controlled by the basal nuclei. In effect, the medial and lateral pathways function as they usually do, but the corticospinal pathway cannot fine-tune the movements. For example, after damage to the primary motor cortex, the basal nuclei can still receive information about planned movements from the prefrontal cortex and can perform preparatory movements of the trunk and limbs. But because the corticospinal pathway is inoperative, precise movements of the forearms, wrists, and hands cannot occur. An individual in this condition can stand, maintain balance, and even walk, but all movements are hesitant, awkward, and poorly controlled.

> basic functions maintained, but not fine-tuned.

The Cerebellum

The cerebellum monitors proprioceptive (position) sensations, visual information from the eyes, and vestibular (balance) sensations from the inner ear as movements are under way. Axons within the spinocerebellar tracts deliver proprioceptive information to the cerebellar cortex. Visual information is relayed from the superior colliculi, and balance information is relayed from the vestibular nuclei. The output of the cerebellum affects upper motor neuron activity in the corticospinal, medial, and lateral pathways.

All motor pathways send information to the cerebellum when motor commands are issued. As the movement proceeds, the cerebellum monitors proprioceptive and vestibular information and compares the arriving sensations with those experienced during previous movements. It then adjusts the activities of the upper motor neurons involved. In general, any voluntary movement begins with the activation of far more motor units than are required—or even desirable. The cerebellum acts like a brake, providing the inhibition needed to minimize the number of motor commands used to perform the movement. The pattern and degree of inhibition changes from moment to moment, and this makes the movement efficient, smooth, and precisely controlled.

The patterns of cerebellar activity are learned by trial and error, over many repetitions. Many of the basic patterns are established early in life; examples include the fine balancing adjustments you make while standing and walking. The ability to fine-tune a complex pattern of movement improves with practice, until the movements become fluid and automatic. Consider the relaxed, smooth movements of acrobats, golfers, and sushi chefs. These people move without thinking about the details of their movements. This ability is important, because when you concentrate on voluntary control, the rhythm and pattern of the movement usually fall apart as your primary motor cortex starts overriding the commands of the basal nuclei and cerebellum.

CLINICAL NOTE

Amyotrophic Lateral Sclerosis **Amyotrophic lateral sclerosis (ALS)** is a progressive, degenerative disorder that affects motor neurons in the spinal cord, brain stem, and cerebral hemispheres. The degeneration affects both upper and lower motor neurons. A defect in axonal transport is thought to underlie the disease. Because a motor neuron and its dependent muscle fibers are so intimately related, the destruction of CNS neurons causes atrophy of the associated skeletal muscles. It is commonly known as Lou Gehrig disease, named after the famous New York Yankees player who died of the disorder. Noted physicist Stephen Hawking is also afflicted with this condition.

Levels of Processing and Motor Control

All sensory and motor pathways involve a series of synapses, one after the other. Along the way, the information is distributed to processing centers operating at the subconscious level. Consider what happens when you stumble—you often recover your balance even as you become aware that a problem exists. Long before your cerebral cortex could assess the situation, evaluate possible responses (shift weight *here*, move leg *there*, and so on), and issue appropriate motor commands, monosynaptic and polysynaptic reflexes, perhaps adjusted by the brain stem and cerebellum, successfully prevented a fall. This is a general pattern; spinal and cranial reflexes provide rapid, involuntary, preprogrammed responses that preserve homeostasis over the short term. Voluntary responses are more complex and require more time to prepare and execute.

CLINICAL NOTE

Anencephaly Although it may seem strange, physicians generally take newborn infants into a dark room and shine a light against the skull. They are checking for *anencephaly* (an-en-SEF-uh-lē) a rare condition in which the brain fails to develop at levels above the mesencephalon or lower diencephalon.

In most such cases, the cranium also fails to develop, and diagnosis is easy, but in some cases, a normal skull forms. In such instances, the cranium is empty and translucent enough to transmit light. Unless the condition is discovered right away, the parents may take the infant home, unaware of the problem. All the normal behavior patterns expected of a newborn are present, including suckling, stretching, yawning, crying, kicking, sticking fingers in the mouth, and tracking movements with the eyes. However, death will occur naturally within days or months.

This tragic condition provides a striking demonstration of the role of the brain stem in controlling complex motor patterns. During normal development, these patterns become incorporated into variable and versatile behaviors as control centers and analytical centers appear in the cerebral cortex.

Cranial and spinal reflexes control the most basic motor activities. Integrative centers in the brain perform more elaborate processing, and as we move from the medulla oblongata to the cerebral cortex, the motor patterns become increasingly complex and variable. The most complex and variable motor activities are directed by the primary motor cortex of the cerebral hemispheres.

During development, the spinal and cranial reflexes are the first to appear. More complex reflexes and motor patterns develop as CNS neurons multiply, enlarge, and interconnect. The process proceeds relatively slowly, as billions of neurons establish trillions of synaptic connections. At birth, neither the cerebral nor the cerebellar cortex is fully functional. The behavior of newborn infants is directed primarily by centers in the diencephalon and brain stem.

The A&P Top 100

#53 Neurons of the primary motor cortex innervate motor neurons in the brain and spinal cord responsible for stimulating skeletal muscles. Higher centers in the brain can suppress or facilitate reflex responses; reflexes can complement or increase the complexity of voluntary movements.

CHECKPOINT

11. What is the anatomical basis for the fact that the left side of the brain controls motor function on the right side of the body?

12. An injury involving the superior portion of the motor cortex affects which region of the body?

13. What effect would increased stimulation of the motor neurons of the red nucleus have on muscle tone?

See the blue Answers tab at the end of the book.

Related Clinical Terms

amyotrophic lateral sclerosis (ALS): A progressive, degenerative disorder affecting motor neurons of the spinal cord, brain stem, and cerebral hemispheres.

anencephaly: A rare condition in which the brain fails to develop at levels above the mesencephalon or inferior part of the diencephalon.

anesthesia: A total loss of sensation.

cerebral palsy: A disorder that affects voluntary motor performance and arises in infancy or childhood as a result of prenatal trauma, drug exposure, or a congenital defect.

hypesthesia: A reduction in sensitivity.

paresthesia: The presence of abnormal sensations, such as tingling, prickling, or burning.

two-point discrimination test: A diagnostic test in which two stimuli are applied simultaneously to the skin to assess tactile sensitivities.

Chapter Review
Study Outline

An Introduction to Sensory Pathways and the Somatic Nervous System p. 507

1. The nervous system works as an integrated unit. This chapter considers sensory receptors, sensory processing centers in the brain, and conscious and subconscious motor functions. (*Figure 15–1*)

15-1 Sensory information from all parts of the body is routed to the somatosensory cortex p. 507

2. The brain, spinal cord, and peripheral nerves continuously communicate with each other and with the internal and external environments. Information arrives via sensory receptors and ascends within the afferent division, while

motor commands descend and are distributed by the efferent division. (*Figure 15–1*)

15-2 Sensory receptors connect our internal and external environments with the nervous system p. 507

3. A sensory receptor is a specialized cell or cell process that monitors specific conditions within the body or in the external environment. Arriving information is called a **sensation**; awareness of a sensation is a **perception**.

4. The **general senses** are our sensitivity to pain, temperature, touch, pressure, vibration, and proprioception. Receptors for these senses are distributed throughout the body. **Special senses**, located in specific **sense organs**, are structurally more complex.

5. Each receptor cell monitors a specific receptive field. *Transduction* begins when a large enough stimulus depolarizes the *receptor potential* or *generator potential* to the point where action potentials are produced. (*Figure 15–2*)

6. **Tonic receptors** are always active. **Phasic receptors** provide information about the intensity and rate of change of a stimulus. **Adaptation** is a reduction in sensitivity in the presence of a constant stimulus. Tonic receptors are **slow-adapting receptors**, while phasic receptors are **fast-adapting receptors**.

15-3 General sensory receptors can be classified by the type of stimulus that excites them p. 510

7. Three types of **nociceptor** found in the body provide information on pain as related to extremes of temperature, mechanical damage, and dissolved chemicals. Myelinated Type A fibers carry **fast pain**. Slower Type C fibers carry **slow pain**.

8. **Thermoreceptors** are found in the dermis. **Mechanoreceptors** are sensitive to distortion of their membranes, and include **tactile receptors**, **baroreceptors**, and **proprioceptors**. There are six types of tactile receptors in the skin, and three kinds of proprioceptors. Chemoreceptors include **carotid bodies** and **aortic bodies**. (*Figure 15–3*)

15-4 Separate pathways carry somatic sensory and visceral sensory information p. 514

9. Sensory neurons that deliver sensation to the CNS are referred to as **first-order neurons**. These synapse on **second-order neurons** in the brain stem or spinal cord. The next neuron in this chain is a **third-order neuron**, found in the thalamus.

10. Three major somatic sensory pathways carry sensory information from the skin and musculature of the body wall, head, neck, and limbs: the *posterior column pathway*, the *spinothalamic pathway*, and the *spinocerebellar pathway*. (*Figure 15–4*)

11. The **posterior column pathway** carries fine touch, pressure, and proprioceptive sensations. The axons ascend within the **fasciculus gracilis** and **fasciculus cuneatus** and relay information to the thalamus via the **medial lemniscus**. Before the axons enter the medial lemniscus, they cross over to the opposite side of the brain stem. This crossing over is called **decussation**. (*Figure 15–5; Table 15–1*)

12. The **spinothalamic pathway** carries poorly localized sensations of touch, pressure, pain, and temperature. The axons involved decussate in the spinal cord and ascend within the **anterior** and **lateral spinothalamic tracts** to the ventral nuclei of the thalamus. Abnormalities along the spinothalamic pathway can lead to **referred pain**, inaccurate localizations of the source of pain. (*Figures 15–5, 15–6; Table 15–1*)

13. The **spinocerebellar pathway**, including the **posterior** and **anterior spinocerebellar tracts**, carries sensations to the cerebellum concerning the position of muscles, tendons, and joints. (*Figure 15–7; Table 15–1*)

14. Visceral sensory pathways carry information collected by interoceptors. Sensory information from cranial nerves V, VII, IX, and X is delivered to the **solitary nucleus** in the medulla oblongata. Dorsal roots of spinal nerves T_1–L_2 carry visceral sensory information from organs between the diaphragm and the pelvic cavity. Dorsal roots of spinal nerves S_2–S_4 carry sensory information from more inferior structures.

15-5 The somatic nervous system is an efferent division that controls skeletal muscles p. 519

15. Somatic motor (descending) pathways always involve an **upper motor neuron** (whose cell body lies in a CNS processing center) and a **lower motor neuron** (whose cell body is located in a nucleus of the brain stem or spinal cord). (*Figure 15–8*)

16. The neurons of the primary motor cortex are *pyramidal cells*. The **corticospinal pathway** provides voluntary skeletal muscle control. The **corticobulbar tracts** terminate at the cranial nerve nuclei; the **corticospinal tracts** synapse on lower motor neurons in the anterior gray horns of the spinal cord. The corticospinal tracts are visible along the medulla as a pair of thick bands, the **pyramids**, where most of the axons decussate to enter the descending **lateral corticospinal tracts**. Those that do not cross over enter the **anterior corticospinal tracts**. The corticospinal pathway provides a rapid, direct mechanism for controlling skeletal muscles. (*Figure 15–9; Table 15–2*)

17. The **medial** and **lateral pathways** include several other centers that issue motor commands as a result of processing performed at a subconscious level. (*Table 15–2*)

18. The medial pathway primarily controls gross movements of the neck, trunk, and proximal limbs. It includes the vestibulospinal, tectospinal, and reticulospinal tracts. The **vestibulospinal tracts** carry information related to maintaining balance and posture. Commands carried by the **tectospinal tracts** change the position of the head, neck, and upper limbs in response to bright lights, sudden movements, or loud noises. Motor commands carried by the **reticulospinal tracts** vary according to the region stimulated. (*Table 15–2*)

19. The lateral pathway consists of the **rubrospinal tracts**, which primarily control muscle tone and movements of the distal muscles of the upper limbs. (*Table 15–2*)

20. The basal nuclei adjust the motor commands issued in other processing centers and provide background patterns of movement involved in voluntary motor activities.

21. The cerebellum monitors proprioceptive sensations, visual information, and vestibular sensations. The integrative activities performed by neurons in the cortex and nuclei of the cerebellum are essential for the precise control of movements.

22. Spinal and cranial reflexes provide rapid, involuntary, preprogrammed responses that preserve homeostasis. Voluntary responses are more complex and require more time to prepare and execute.

23. During development, the spinal and cranial reflexes are first to appear. Complex reflexes develop over years, as the CNS matures and the brain grows in size and complexity.

Review Questions

See the blue Answers tab at the end of the book.

 Access more review material online at **myA&P™** (www.myaandp.com). There, you'll find chapter guides, chapter quizzes, practice tests, animations, flashcards, a glossary with pronunciations, *Interactive Physiology*® (IP) exercises and quizzes, and more to help you succeed in the course.

LEVEL 1   Reviewing Facts and Terms

1. Identify the tactile receptors of the skin in the following diagram.

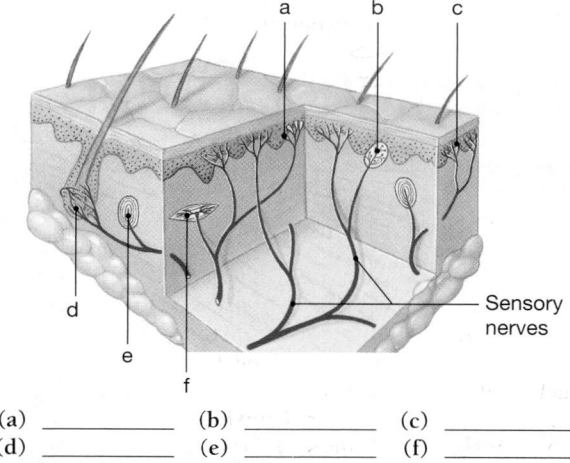

    (a) _____ (b) _____ (c) _____

    (d) _____ (e) _____ (f) _____

2. Shade in all the ascending sensory tracts in the following diagram of the spinal cord.

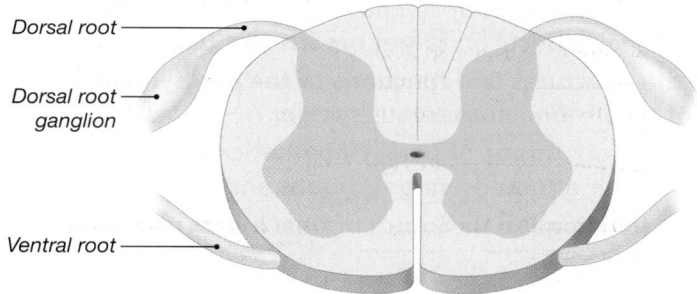

3. The larger the receptive field, the
    (a) larger the stimulus needed to stimulate a sensory receptor.
    (b) fewer sensory receptors there are.
    (c) harder it is to locate the exact point of stimulation.
    (d) larger the area of the somatosensory cortex in the brain that deals with the area.
    (e) closer together the receptor cells.
4. The CNS interprets information entirely on the basis of the
    (a) number of action potentials that it receives.
    (b) kind of action potentials that it receives.
    (c) line over which sensory information arrives.
    (d) intensity of the sensory stimulus.
    (e) number of sensory receptors that are stimulated.
5. The area of sensory cortex devoted to a body region is relative to the
    (a) size of the body area.
    (b) distance of the body area from the brain.
    (c) number of motor units in the area of the body.
    (d) number of sensory receptors in the area of the body.
    (e) size of the nerves that serve the area of the body.

6. _____ are receptors that are normally inactive, but become active for a short time whenever there is a change in the condition that they monitor.
7. Identify six types of tactile receptors located in the skin, and describe their sensitivities.
8. What three types of mechanoreceptors respond to stretching, compression, twisting, or other distortions of the plasma membrane?
9. What are the three major somatic sensory pathways, and what is the function of each pathway?
10. Which three pairs of descending tracts make up the corticospinal pathway?
11. Which three motor pathways make up the medial pathway?
12. What are the two primary functional roles of the cerebellum?
13. The corticospinal tract
    (a) carries motor commands from the cerebral cortex to the spinal cord.
    (b) carries sensory information from the spinal cord to the brain.
    (c) starts in the spinal cord and ends in the brain.
    (d) does all of these.
14. What three steps are necessary for transduction to occur?

LEVEL 2   Reviewing Concepts

15. Differentiate between a tonic receptor and a phasic receptor.
16. What is a motor homunculus? How does it differ from a sensory homunculus?
17. Describe the relationship among first-, second-, and third-order neurons.
18. Damage to the posterior spinocerebellar tract on the left side of the spinal cord at the L_1 level would interfere with the coordinated movement of which limb(s)?
19. What effect does injury to the primary motor cortex have on peripheral muscles?
20. By which structures and in which part of the brain is the level of muscle tone in the body's skeletal muscles controlled? How is this control exerted?
21. Explain the phenomenon of *referred pain* in terms of labeled lines and organization of sensory tracts and pathways.

LEVEL 3   Critical Thinking and Clinical Applications

22. Kelly is having difficulty controlling her eye movements and has lost some control of her facial muscles. After an examination and testing, Kelly's physician tells her that her cranial nerves are perfectly normal but that a small tumor is putting pressure on certain fiber tracts in her brain. This pressure is the cause of Kelly's symptoms. Where is the tumor most likely located?
23. Clarence, a construction worker, suffers a fractured skull when a beam falls on his head. Diagnostic tests indicate severe damage to the motor cortex. His wife is anxious to know if he will ever be able to move or walk again. What would you tell her?
24. Phil had to have his arm amputated at the elbow after an accident. He tells you that he can sometimes still feel pain in his fingers even though the hand is gone. He says this is especially true when he bumps the stub. How can this be?

16

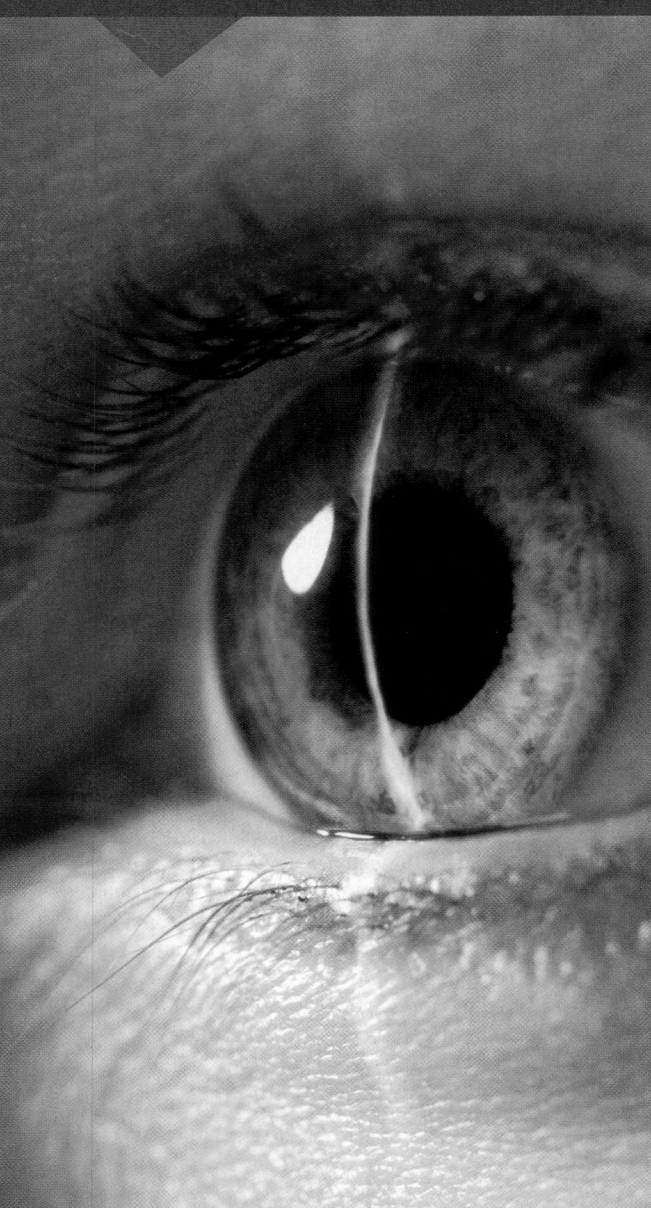

Neural Integration II: The Autonomic Nervous System and Higher-Order Functions

Did you know...?

When bright light is shone into one eye, the pupils of both eyes constrict. This is the "consensual light reflex;" loss of this reflex indicates brain, optic nerve, or eye damage.

Learning Outcomes

After completing this chapter, you should be able to do the following:

16-1 Compare the organization of the autonomic nervous system with that of the somatic nervous system.

16-2 Describe the structures and functions of the sympathetic division of the autonomic nervous system.

16-3 Describe the mechanisms of sympathetic neurotransmitter release and their effects on target organs and tissues.

16-4 Describe the structures and functions of the parasympathetic division of the autonomic nervous system.

16-5 Describe the mechanisms of parasympathetic neurotransmitter release and their effects on target organs and tissues.

16-6 Discuss the functional significance of dual innervation and autonomic tone.

16-7 Describe the hierarchy of interacting levels of control in the autonomic nervous system, including the significance of visceral reflexes.

16-8 Explain how memories are created, stored, and recalled, and distinguish among the levels of consciousness and unconsciousness.

16-9 Describe some of the ways in which the interactions of neurotransmitters influence brain function.

16-10 Summarize the effects of aging on the nervous system.

16-11 Give examples of interactions between the nervous system and each of the other organ systems.

Clinical Notes

Amnesia p. 551
Alzheimer Disease p. 555
Categorizing Nervous System Disorders p. 556

An Introduction to the Autonomic Nervous System and Higher-Order Functions

Chapter 16 examines the relationship between sensory pathways and the somatic nervous system with the autonomic nervous system (ANS). The ANS adjusts our basic life support systems without conscious control. This chapter also considers aspects of higher-order functions such as consciousness, learning, and intelligence.

Figure 16–1 provides an overview of the material covered in this chapter. We'll begin by completing our discussion of the efferent division of the nervous system with an examination of the ANS. The interpretation of arriving sensory information and the activities of the somatic nervous system (SNS) and the ANS can be influenced or modified in response to conscious planning, memories, and learning—the so-called higher-order functions of the brain, which we examine next. Finally, the chapter examines the effects of aging on the nervous system and concludes by discussing interactions between the nervous system and other body systems. We begin with an overview of the ANS.

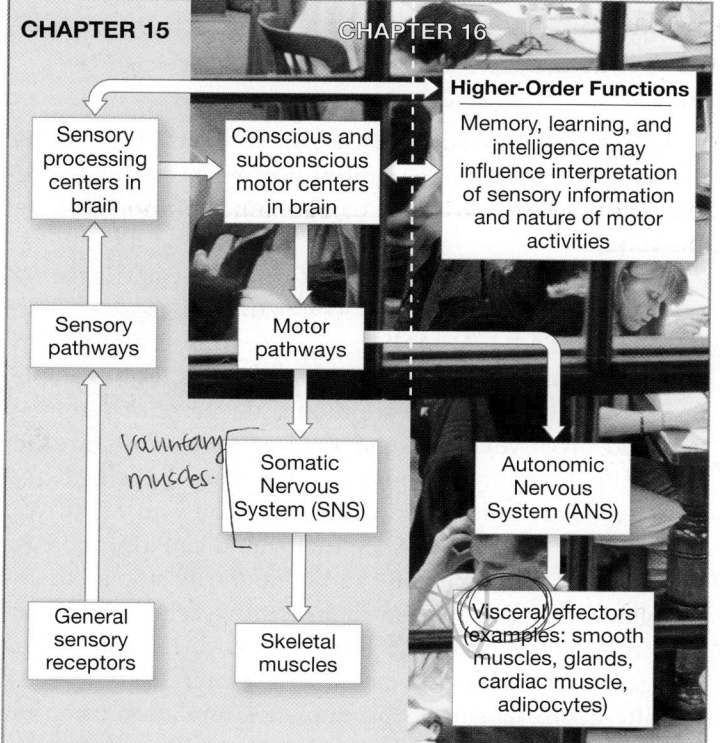

Figure 16–1   An Overview of Neural Integration. This figure illustrates the relationships between Chapters 15 and 16 and indicates the major topics considered in this chapter.

16-1 The autonomic nervous system, composed of the sympathetic and parasympathetic divisions, is involved in the unconscious regulation of visceral functions

Your conscious thoughts, plans, and actions represent a tiny fraction of the activities of the nervous system. In practical terms, your conscious thoughts and the somatic nervous system (SNS), which operates under conscious control, seldom have a direct effect on your long-term survival. Of course, the somatic nervous system can be important in moving you out of the way of a speeding bus or pulling your hand from a hot stove—but it was your conscious movements that put you in jeopardy in the first place. If all consciousness were eliminated, vital physiological processes would continue virtually unchanged; a night's sleep is not a life-threatening event. Longer, deeper states of unconsciousness are not necessarily more dangerous, as long as nourishment and other basic care is provided. People who have suffered severe brain injuries can survive in a coma for decades.

Survival under these conditions is possible because routine homeostatic adjustments in physiological systems are made by the **autonomic nervous system (ANS)**. The ANS coordinates cardiovascular, respiratory, digestive, urinary, and reproductive functions. In doing so, the ANS adjusts internal water, electrolyte, nutrient, and dissolved gas concentrations in body fluids—and it does so without instructions or interference from the conscious mind.

Our understanding of the ANS has had a profound effect on the practice of medicine. For example, in 1960, the five-year survival rate for patients surviving their first heart attack was very low, primarily because it was difficult and sometimes impossible to control high blood pressure. Forty-eight years later, many survivors of heart attacks lead normal lives. This dramatic change occurred as we learned to manipulate the ANS with specific drugs and clinical procedures.

It is useful to compare the organization of the ANS with that of the SNS, which controls our skeletal muscles. First we will focus on the neural interactions that direct motor output; then we will examine the subdivisions of the ANS, which are based on structural and functional patterns of peripheral innervation.

Organization of the ANS

Figure 16–2 compares the organization of the somatic and autonomic nervous systems. Both are efferent divisions that carry motor commands; the SNS controls skeletal muscles,

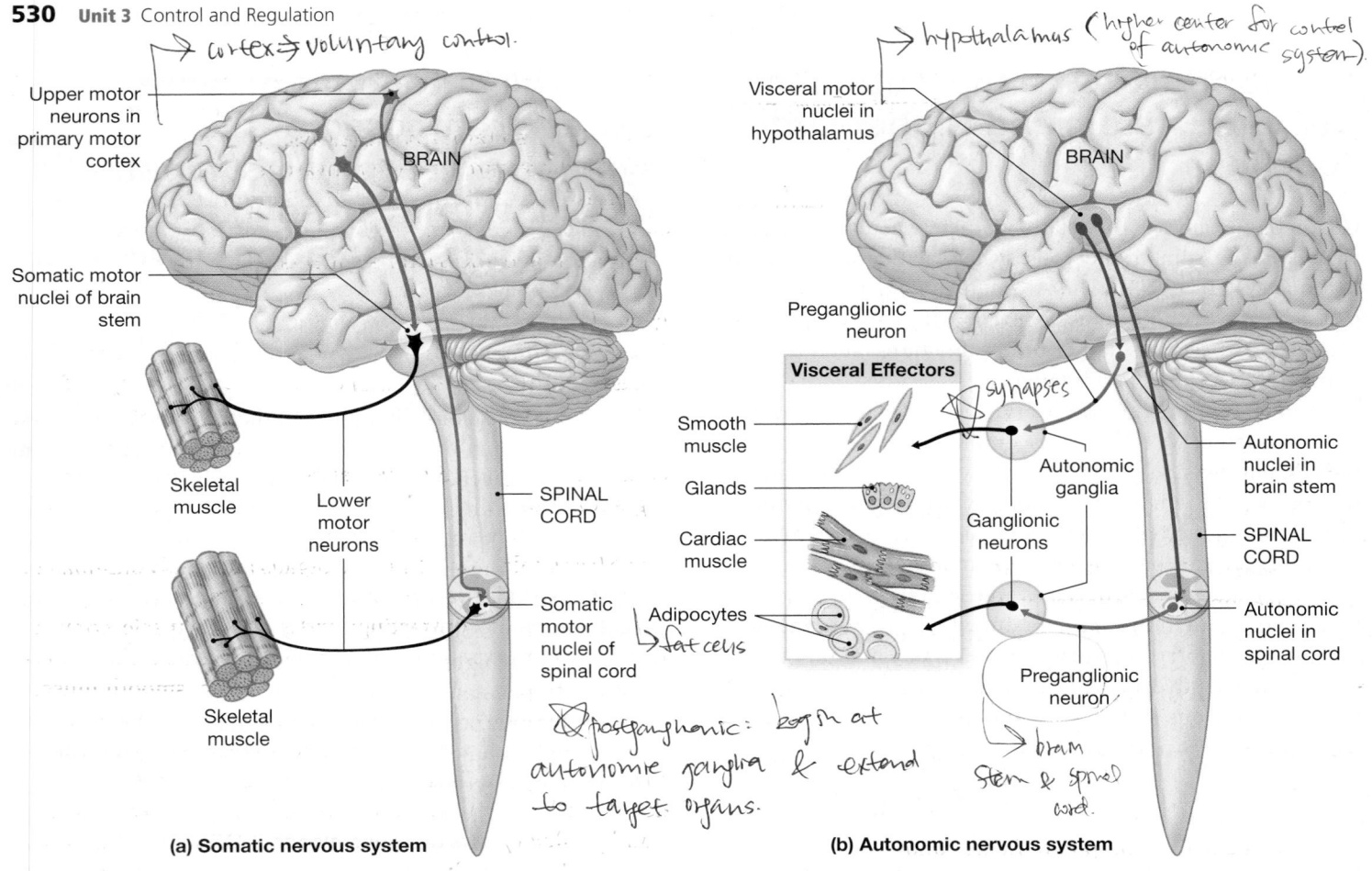

Handwritten annotations:
→ cortex → voluntary control.
→ hypothalamus (higher center for control of autonomic system).
synapses
⊗ postganglionic: begin at autonomic ganglia & extend to target organs.
Adipocytes → fat cells
→ brain stem & spinal cord.

Upper motor neurons in primary motor cortex

Somatic motor nuclei of brain stem

BRAIN

Skeletal muscle

Lower motor neurons

SPINAL CORD

Somatic motor nuclei of spinal cord

Skeletal muscle

(a) Somatic nervous system

Visceral motor nuclei in hypothalamus

BRAIN

Preganglionic neuron

Visceral Effectors

Smooth muscle

Glands

Cardiac muscle

Adipocytes

Autonomic ganglia

Autonomic nuclei in brain stem

Ganglionic neurons

SPINAL CORD

Preganglionic neuron

Autonomic nuclei in spinal cord

(b) Autonomic nervous system

Figure 16–2 The Organization of the Somatic and Autonomic Nervous Systems.

and the ANS controls visceral effectors. The primary structural difference between the two is that in the SNS, motor neurons of the central nervous system exert direct control over skeletal muscles (**Figure 16–2a**). In the ANS, by contrast, motor neurons of the central nervous system synapse on visceral motor neurons in autonomic ganglia, and these ganglionic neurons control visceral effectors (**Figure 16–2b**).

The integrative centers for autonomic activity are located in the hypothalamus. The neurons in these centers are comparable to the upper motor neurons in the SNS. Visceral motor neurons in the brain stem and spinal cord are known as **preganglionic neurons**. These neurons are part of visceral reflex arcs, and most of their activities represent direct reflex responses, rather than responses to commands from the hypothalamus. The axons of preganglionic neurons are called **preganglionic fibers**. Preganglionic fibers leave the CNS and synapse on **ganglionic neurons**—visceral motor neurons in peripheral ganglia. These ganglia, which contain hundreds to thousands of ganglionic neurons, are called **autonomic ganglia**. Ganglionic neurons innervate visceral effectors such as cardiac muscle, smooth

muscle, glands, and adipose tissue. The axons of ganglionic neurons are called **postganglionic fibers**, because they begin at the autonomic ganglia and extend to the peripheral target organs.

Somatic or visceral sensory information can trigger visceral reflexes, and the motor commands of those reflexes are distributed by the ANS. Sometimes those motor commands control the activities of target organs. For example, in cold weather, the ANS stimulates the arrector pili muscles and gives you "goosebumps." ∞ p. 169 In other cases, the motor commands may alter some ongoing activity. A sudden, loud noise can startle you and make you jump, but thanks to the ANS, that sound can also increase your heart rate dramatically and temporarily stop all digestive gland secretion. These changes in visceral activity occur in response to the release of neurotransmitters by postganglionic fibers. As noted in Chapter 12, whether a specific neurotransmitter produces a stimulation or an inhibition of activity depends on the response of the membrane receptors, and we will consider the major types of receptors later in the chapter.

We will now turn to the anatomy and physiology of the ANS. Then we'll consider the nature of *visceral reflexes*, polysynaptic reflexes that regulate visceral function.

Tips&Tricks

Each autonomic ganglion functions somewhat like a baton handoff in a relay race. Within the ganglion, one runner (the preganglionic fiber) hands off the baton (a neurotransmitter) to the next runner (the postganglionic fiber), who then carries on toward the finish line (the target effector).

Divisions of the ANS

The ANS contains two primary subdivisions whose names are probably already familiar to you: the *sympathetic division* and the *parasympathetic division*. Most often, these two divisions have opposing effects; if the sympathetic division causes excitation, the parasympathetic causes inhibition. However, this is not always the case, because (1) the two divisions may work independently, with some structures innervated by only one division and (2) the two divisions may work together, each controlling one stage of a complex process. In general, the sympathetic division "kicks in" only during periods of exertion, stress, or emergency, and the parasympathetic division predominates under resting conditions.

In the **sympathetic division**, or *thoracolumbar* (thor-a-kō-LUM-bar) *division*, preganglionic fibers from the thoracic and superior lumbar segments of the spinal cord synapse in ganglia near the spinal cord. In this division of the ANS, therefore, the preganglionic fibers are short, and the postganglionic fibers are long.

The sympathetic division prepares the body for heightened levels of somatic activity. When fully activated, this division produces what is known as the "fight or flight" response, which readies the body for a crisis that may require sudden, intense physical activity. An increase in sympathetic activity generally stimulates tissue metabolism and increases alertness. Imagine walking down a long, dark alley and hearing strange noises in the darkness ahead. Your body responds immediately, and you become more alert and aware of your surroundings. Your metabolic rate rises quickly, to as much as twice the resting level. Your digestive and urinary activities are suspended temporarily, and blood flow to your skeletal muscles increases. You begin breathing more quickly and more deeply. Both your heart rate and blood pressure increase, circulating your blood more rapidly. You feel warm and begin to perspire. The general pattern of responses to increased levels of sympathetic activity can be summarized as follows: (1) heightened mental alertness, (2) increased meta-

bolic rate, (3) reduced digestive and urinary functions, (4) activation of energy reserves, (5) increased respiratory rate and dilation of respiratory passageways, (6) increased heart rate and blood pressure, and (7) activation of sweat glands.

In the **parasympathetic division**, or *craniosacral* (krā-mē-ō-SĀ-krul) *division*, preganglionic fibers originate in the brain stem and the sacral segments of the spinal cord, and they synapse in ganglia very close to (or within) the target organs. Thus, in this division of the ANS, the preganglionic fibers are long, and the postganglionic fibers are short. The parasympathetic division stimulates visceral activity; for example, it is responsible for the state of "rest and digest" that follows a big dinner. General parasympathetic activation conserves energy and promotes sedentary activities, such as digestion. Your body relaxes, energy demands are minimal, and both your heart rate and blood pressure are relatively low. Meanwhile, your digestive organs are highly stimulated. Your salivary glands and other secretory glands are active; your stomach is contracting; and smooth muscle contractions move materials along your digestive tract. This movement promotes defecation; at the same time, smooth muscle contractions along your urinary tract promote urination. The overall pattern of responses to increased levels of parasympathetic activity is as follows: (1) decreased metabolic rate, (2) decreased heart rate and blood pressure, (3) increased secretion by salivary and digestive glands, (4) increased motility and blood flow in the digestive tract, and (5) stimulation of urination and defecation.

The ANS also includes a third division that most people have never heard of: the *enteric nervous system (ENS)*, an extensive network of neurons and nerve networks located in the walls of the digestive tract. Although the sympathetic and parasympathetic divisions influence the activities of the enteric nervous system, many complex visceral reflexes are initiated and coordinated locally, without instructions from the CNS. Altogether, the ENS has roughly 100 million neurons—at least as many as the spinal cord—and all of the neurotransmitters found in the brain. In this chapter, we focus on the sympathetic and parasympathetic divisions that integrate and coordinate visceral functions throughout the body. We will consider the activities of the enteric nervous system when we discuss visceral reflexes later in this chapter, and when we examine the control of digestive functions in Chapter 24.

The A&P Top 100

#54 The autonomic nervous system operates largely outside our awareness. It includes a sympathetic division that increases alertness, metabolic rate, and muscular abilities, and a parasympathetic division that reduces metabolic rate and promotes visceral activities such as digestion.

CHECKPOINT

1. Identify the two major divisions of the autonomic nervous system. *sympathetic: excitatory. parasympathetic: inhibitory.*

2. How many motor neurons are required to conduct an action potential from the spinal cord to smooth *2 synapse* muscles in the wall of the intestine? *preganglionic fibers → ganglia to / within target organ.*

3. While out for a walk, Julie is suddenly confronted by an angry dog. Which division of the autonomic *fight/flight* nervous system is responsible for the physiological changes that occur in Julie as she turns and runs? *sympathetic → heightened levels of somatic activity.*

4. On the basis of anatomy, how could you distinguish the sympathetic division from the parasympathetic division of the ANS? *Symp: T1 – L2. para: cranial & sacral segments of spinal cord.*

See the blue Answers tab at the end of the book.

Sympathetic: inhibitory T1 – L2. ⟹ thoracic.
⟹ antagonistic of parasympathetic.
homeostasis.

16-2 The sympathetic division consists of preganglionic neurons and ganglionic neurons involved in using energy and increasing metabolic rate

Figure 16–3 depicts the overall organization of the sympathetic division of the ANS. This division consists of preganglionic neurons that are located between segments T_1 and L_2 of the spinal cord, and ganglionic neurons that are located in ganglia near the vertebral column. The cell bodies of the preganglionic neurons are situated in the lateral gray horns, and their axons enter the ventral roots of these segments. The ganglionic neurons occur in three locations (**Figure 16–4**):

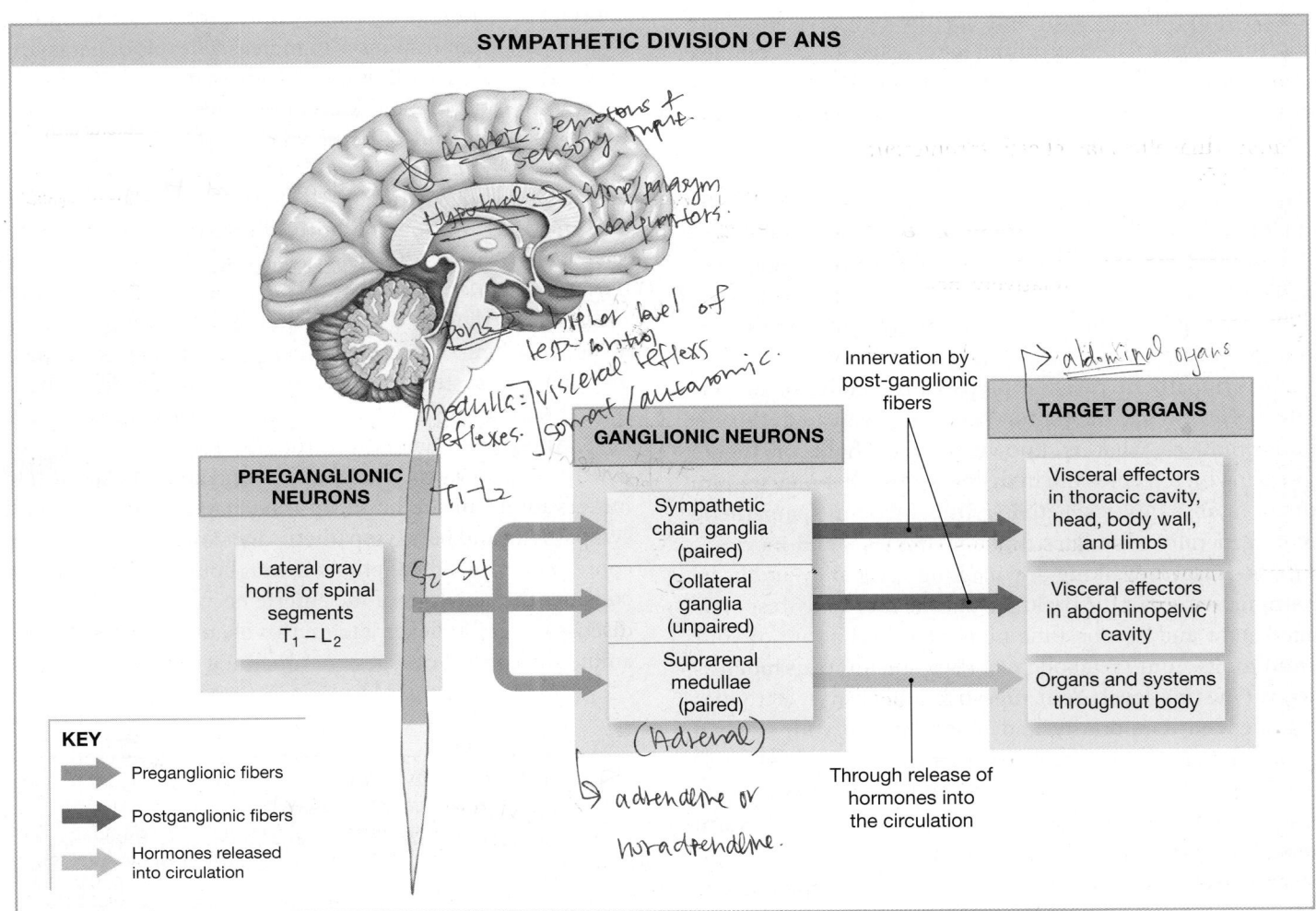

SYMPATHETIC DIVISION OF ANS

PREGANGLIONIC NEURONS

Lateral gray horns of spinal segments $T_1 - L_2$

GANGLIONIC NEURONS

Sympathetic chain ganglia (paired)

Collateral ganglia (unpaired)

Suprarenal medullae (paired)

Innervation by post-ganglionic fibers

TARGET ORGANS

Visceral effectors in thoracic cavity, head, body wall, and limbs

Visceral effectors in abdominopelvic cavity

Organs and systems throughout body

Through release of hormones into the circulation

KEY
- Preganglionic fibers
- Postganglionic fibers
- Hormones released into circulation

Figure 16–3   The Organization of the Sympathetic Division of the ANS.

1. *Sympathetic Chain Ganglia.* **Sympathetic chain ganglia**, also called *paravertebral ganglia* or *lateral ganglia*, lie on both sides of the vertebral column (**Figure 16–4a**). Neurons in these ganglia control effectors in the body wall, inside the thoracic cavity, and in the head and limbs.

2. *Collateral Ganglia.* **Collateral ganglia**, also known as *prevertebral ganglia*, are anterior to the vertebral bodies (**Figure 16–4b**). Collateral ganglia contain ganglionic neurons that innervate tissues and organs in the abdominopelvic cavity.

3. *The Suprarenal Medullae,* also called the *adrenal medullae.* The center of each (*supra-*, above + *renal*, kidney) gland, an area known as the *suprarenal medulla*, is a modified sympathetic ganglion (**Figure 16–4c**). The ganglionic neurons of the suprarenal medullae have very short axons; when stimulated, they release their neurotransmitters into the bloodstream. The release of neurotransmitters into a capillary, not at a synapse, allows them to function as hormones that affect target cells throughout the body.

In the sympathetic division, the preganglionic fibers are relatively short, because the ganglia are located relatively near the spinal cord. In contrast, postganglionic fibers are relatively long, except at the suprarenal medullae.

Organization and Anatomy of the Sympathetic Division

The ventral roots of spinal segments T_1 to L_2 contain sympathetic preganglionic fibers. The basic pattern of sympathetic innervation in these regions was described in **Figure 13–7a, p. 439**. After passing through the intervertebral foramen, each ventral root

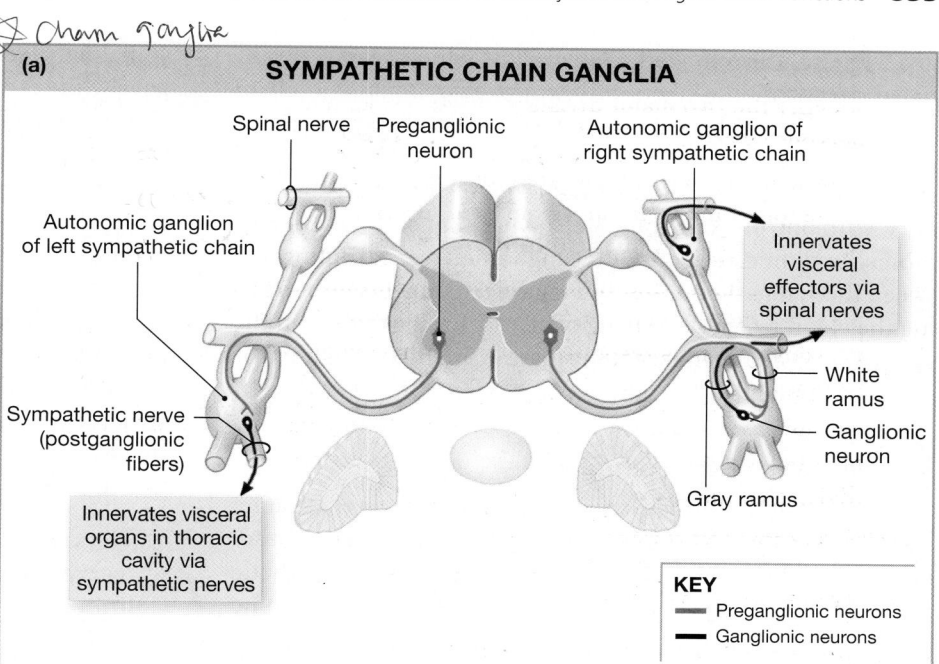

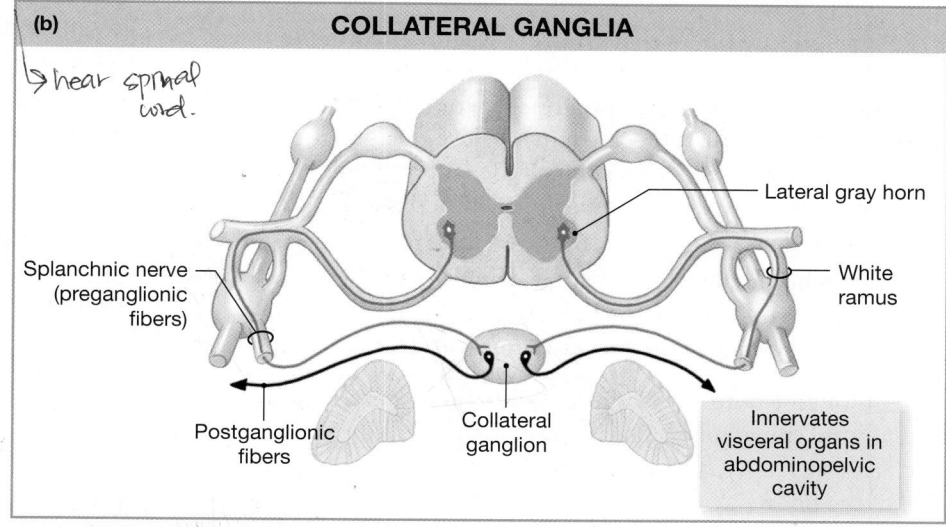

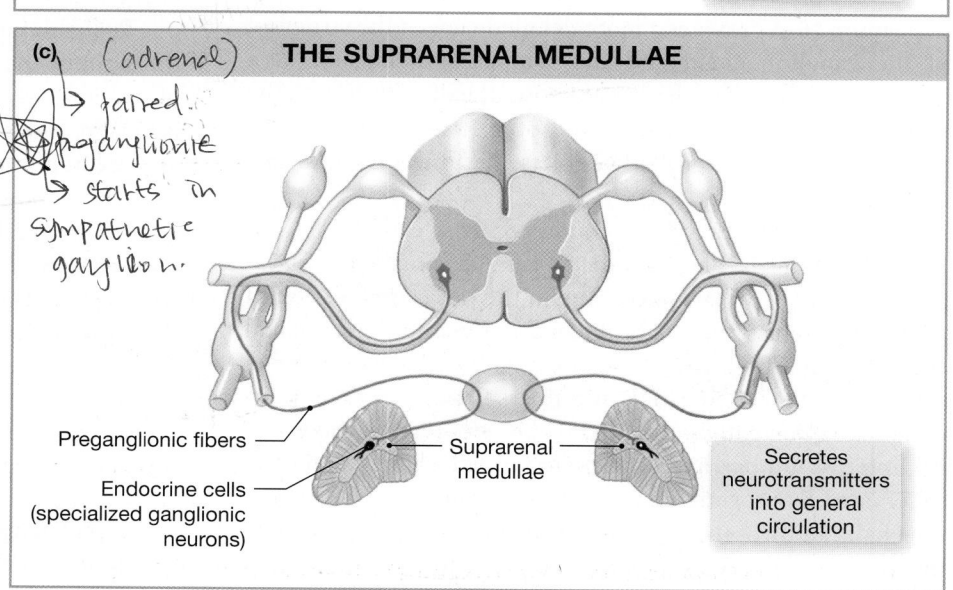

Figure 16–4 Sites of Ganglia in Sympathetic Pathways. Superior views of sections through the thoracic spinal cord, showing the three major patterns of distribution for preganglionic and postganglionic fibers.

gives rise to a myelinated *white ramus*, which carries myelinated preganglionic fibers into a nearby sympathetic chain ganglion. These fibers may synapse within the sympathetic chain ganglia, at one of the collateral ganglia, or in the suprarenal medullae (**Figure 16–4**). Extensive divergence occurs, with one preganglionic fiber synapsing on two dozen or more ganglionic neurons. Preganglionic fibers running between the sympathetic chain ganglia interconnect them, making the chain resemble a string of pearls. Each ganglion in the sympathetic chain innervates a particular body segment or group of segments.

Sympathetic Chain Ganglia

If a preganglionic fiber carries motor commands that target structures in the body wall or thoracic cavity, or in the head, neck, or limbs, it will synapse in one or more sympathetic chain ganglia. The paths of the unmyelinated postganglionic fibers differ, depending on whether their targets lie in the body wall or within the thoracic cavity:

- Postganglionic fibers that control visceral effectors in the body wall, head, neck, or limbs enter the *gray ramus* and return to the spinal nerve for subsequent distribution (**Figure 16–4a**, right). These postganglionic fibers innervate effectors such as the sweat glands of the skin and the smooth muscles in superficial blood vessels.

- Postganglionic fibers innervating structures in the thoracic cavity, such as the heart and lungs, form bundles known as **sympathetic nerves** (**Figure 16–4a**, left).

Although **Figure 16–4a** shows sympathetic nerves on the left side and spinal nerve distribution on the right for the sake of clarity, in reality *both* innervation patterns occur on *each* side of the body.

Figure 16–5 provides a more detailed view of the structure of the ganglion chain and the sympathetic division as a whole. The left side of the image shows the distribution to the skin (and to skeletal muscles and other tissues of the body wall), whereas the right side depicts the innervation of visceral organs.

Each sympathetic chain contains 3 cervical, 10–12 thoracic, 4–5 lumbar, and 4–5 sacral ganglia, plus 1 coccygeal ganglion. (The numbers vary due to the occasional fusion of adjacent ganglia.) Preganglionic neurons are limited to spinal cord segments $T_1–L_2$, and these spinal nerves have both white rami (myelinated preganglionic fibers) and gray rami (unmyelinated postganglionic fibers). The neurons in the cervical, inferior lumbar, and sacral sympathetic chain ganglia are innervated by preganglionic fibers that run along the axis of the chain. In turn, these chain ganglia provide postganglionic fibers, through gray rami, to the cervical, lumbar, and sacral

spinal nerves. As a result, although only spinal nerves $T_1–L_2$ have white rami, every spinal nerve has a gray ramus that carries sympathetic postganglionic fibers for distribution in the body wall.

About 8 percent of the axons in each spinal nerve are sympathetic postganglionic fibers. As a result, the spinal nerves, which provide somatic motor innervation to skeletal muscles of the body wall and limbs, also distribute sympathetic postganglionic fibers (**Figures 16–4a** and **16–5**). In the head and neck, postganglionic sympathetic fibers leaving the superior cervical sympathetic ganglia supply the regions and structures innervated by cranial nerves III, VII, IX, and X. ∞ pp. 492, 495, 497

In summary:

- The cervical, inferior lumbar, and sacral chain ganglia receive preganglionic innervation by preganglionic fibers from spinal segments $T_1–L_2$, and every spinal nerve receives a gray ramus from a ganglion of the sympathetic chain.

- Only the thoracic and superior lumbar ganglia ($T_1–L_2$) receive preganglionic fibers from white rami.

- Every spinal nerve receives a gray ramus from a ganglion of the sympathetic chain.

Collateral Ganglia

The abdominopelvic viscera receive sympathetic innervation via sympathetic preganglionic fibers that pass through the sympathetic chain without synapsing. They synapse in separate collateral ganglia (**Figures 16–3** and **16–4b**). Preganglionic fibers that innervate the collateral ganglia form the **splanchnic** (SPLANK-nik) **nerves**, which lie in the dorsal wall of the abdominal cavity. Although they originate as paired ganglia (left and right), the two usually fuse together, and in adults the collateral ganglia are typically single rather than paired.

Postganglionic fibers leaving the collateral ganglia extend throughout the abdominopelvic cavity, innervating a variety of visceral tissues and organs. The general functional pattern is (1) a reduction of blood flow and energy use by organs that are not important to short-term survival (such as the digestive tract) and (2) the release of stored energy reserves.

The splanchnic nerves innervate three collateral ganglia (**Figure 16–5**). Preganglionic fibers from the seven inferior thoracic segments end at either the **celiac** (SĒ-lē-ak) **ganglion** or the **superior mesenteric ganglion**. These ganglia are embedded in an extensive network of autonomic nerves. Preganglionic fibers from the lumbar segments form splanchnic nerves that end at the **inferior mesenteric ganglion**. All three ganglia are named for their association with adjacent arteries:

- The celiac ganglion is named after the *celiac trunk*, a major artery supplying the stomach, spleen, and liver.

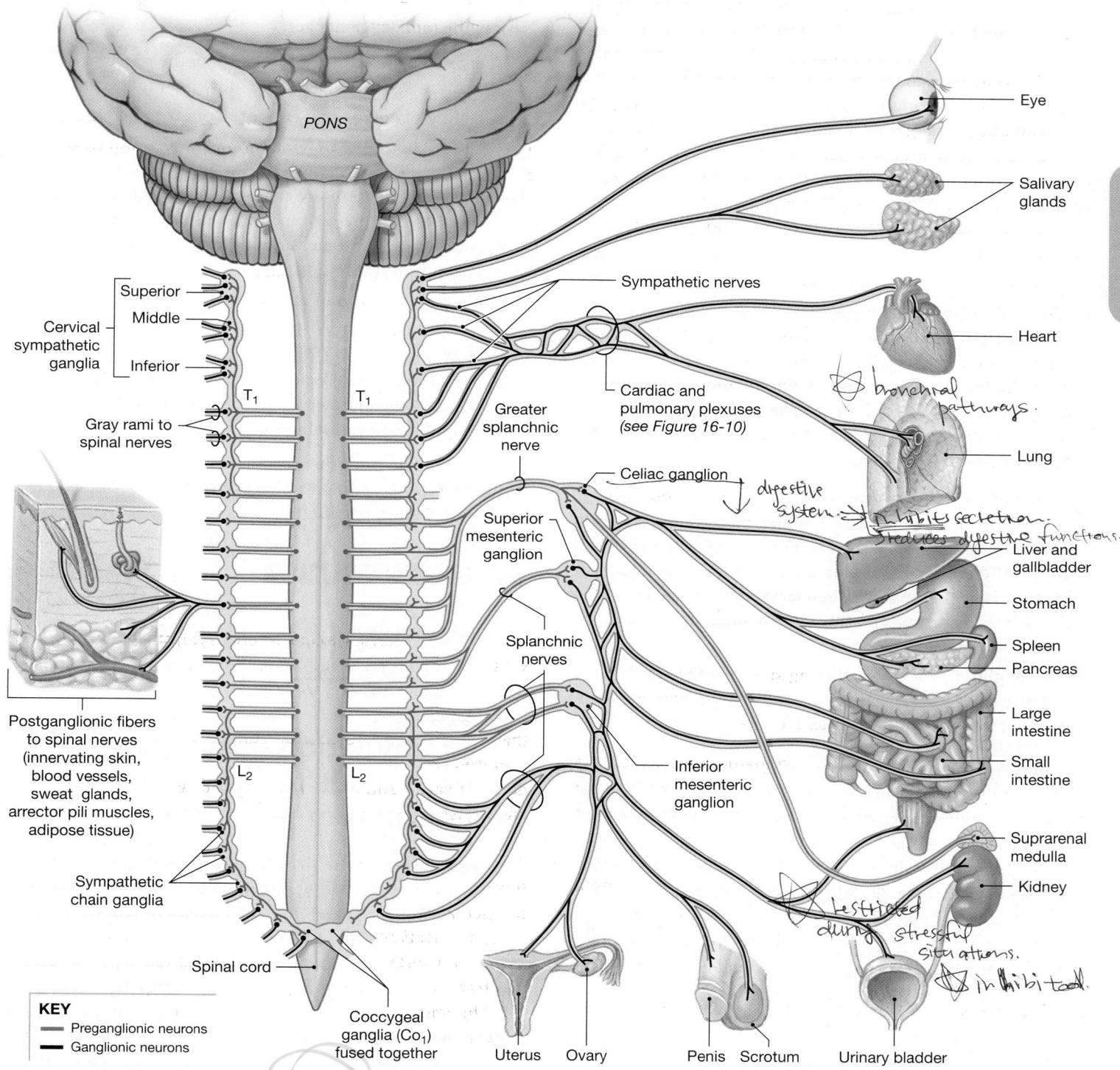

PONS

Superior ⎫
Middle   ⎬ Cervical sympathetic ganglia
Inferior ⎭

Gray rami to spinal nerves

T_1

Postganglionic fibers to spinal nerves (innervating skin, blood vessels, sweat glands, arrector pili muscles, adipose tissue)

L_2

Sympathetic chain ganglia

Spinal cord

KEY
— Preganglionic neurons
— Ganglionic neurons

Coccygeal ganglia (Co$_1$) fused together

T_1

L_2

Sympathetic nerves

Cardiac and pulmonary plexuses (see Figure 16-10)

Greater splanchnic nerve

Celiac ganglion

Superior mesenteric ganglion

Splanchnic nerves

Inferior mesenteric ganglion

Uterus   Ovary

Penis   Scrotum

Urinary bladder

Eye

Salivary glands

Heart

bronchial pathways.

Lung

digestive system ⇒ Inhibits secretion. reduces digestive functions-

Liver and gallbladder

Stomach

Spleen
Pancreas

Large intestine

Small intestine

Suprarenal medulla

Kidney

restricted during stressful situations.

inhibited.

Figure 16–5 The Distribution of Sympathetic Innervation. The distribution of sympathetic fibers is the same on both sides of the body. For clarity, the innervation of somatic structures is shown on the left, and the innervation of visceral structures on the right.

$T_1 - L_2$

The celiac ganglion most commonly consists of a pair of interconnected masses of gray matter situated at the base of that artery. The celiac ganglion may also form a single mass or many small, interwoven masses. Postganglionic fibers from this ganglion innervate the stomach, liver, gallbladder, pancreas, and spleen.

- The **superior mesenteric ganglion** is located near the base of the *superior mesenteric artery*, which provides blood to the stomach, small intestine, and pancreas. Postganglionic fibers leaving the superior mesenteric ganglion innervate the small intestine and the proximal two-thirds of the large intestine.

- The **inferior mesenteric ganglion** is located near the base of the *inferior mesenteric artery*, which supplies the large intestine and other organs in the inferior portion of the abdominopelvic cavity. Postganglionic fibers from this ganglion provide sympathetic innervation to the terminal portions of the large intestine, the kidney and urinary bladder, and the sex organs.

The Suprarenal Medullae

Preganglionic fibers entering a suprarenal gland proceed to its center, a region called the **suprarenal medulla** (**Figures 16–4c** and **16–5**). The suprarenal medulla is a modified sympathetic ganglion where preganglionic fibers synapse on *neuroendocrine cells*, specialized neurons that secrete hormones (chemical messengers) into the bloodstream. The neuroendocrine cells of the suprarenal medullae secrete the neurotransmitters *epinephrine* (*E*) and *norepinephrine* (*NE*). Epinephrine, or *adrenaline*, accounts for 75–80 percent of the secretory output; the rest is NE, or *noradrenaline*.

The bloodstream then carries the neurotransmitters throughout the body, causing changes in the metabolic activities of many different cells. These effects resemble those produced by the stimulation of sympathetic postganglionic fibers. They differ, however, in two respects: (1) Cells not innervated by sympathetic postganglionic fibers are affected as well; and (2) the effects last much longer than those produced by direct sympathetic innervation, because the hormones continue to diffuse out of the bloodstream for an extended period.

Sympathetic Activation

The sympathetic division can change the activities of tissues and organs by releasing NE at peripheral synapses, and by distributing E and NE throughout the body in the bloodstream. The visceral motor fibers that target specific effectors, such as smooth muscle fibers in blood vessels of the skin, can be activated in reflexes that do not involve other visceral effectors. In a crisis, however, the entire division responds. This event, called **sympathetic activation**, is controlled by sympathetic centers in the hypothalamus. The effects are not limited to peripheral tissues; sympathetic activation also alters CNS activity.

When sympathetic activation occurs, an individual experiences the following changes:

- Increased alertness via stimulation of the reticular activating system, causing the individual to feel "on edge."

- A feeling of energy and euphoria, often associated with a disregard for danger and a temporary insensitivity to painful stimuli.

- Increased activity in the cardiovascular and respiratory centers of the pons and medulla oblongata, leading to elevations in blood pressure, heart rate, breathing rate, and depth of respiration.

- A general elevation in muscle tone through stimulation of the medial and lateral pathways, so the person *looks* tense and may begin to shiver.

- The mobilization of energy reserves, through the accelerated breakdown of glycogen in muscle and liver cells and the release of lipids by adipose tissues.

These changes, plus the peripheral changes already noted, complete the preparations necessary for the individual to cope with a stressful situation.

CHECKPOINT

5. **Where do the nerves that synapse in collateral ganglia originate?** preganglionic fibers from splanchic nerves

See the blue Answers tab at the end of the book.

16-3 Stimulation of sympathetic neurons leads to the release of various neurotransmitters

We have examined the distribution of sympathetic impulses and the general effects of sympathetic activation. We now consider the cellular basis of these effects on peripheral organs.

The stimulation of sympathetic preganglionic neurons leads to the release of ACh at synapses with ganglionic neurons. Synapses that use ACh as a transmitter are called *cholinergic*. ∞ p. 414 The effect on the ganglionic neurons is always excitatory.

The stimulation of these ganglionic neurons leads to the release of neurotransmitters at specific target organs. The synaptic terminals are typically different from neuromuscular junctions of the somatic nervous system. Instead of forming synaptic knobs, telodendria form a branching network. Each branch resembles a string of pearls. Each "pearl," a swollen segment called a **varicosity**, is packed with neurotransmitter vesicles (**Figure 16–6**). Chains of varicosities pass along or near the surfaces of effector cells. There are no specialized

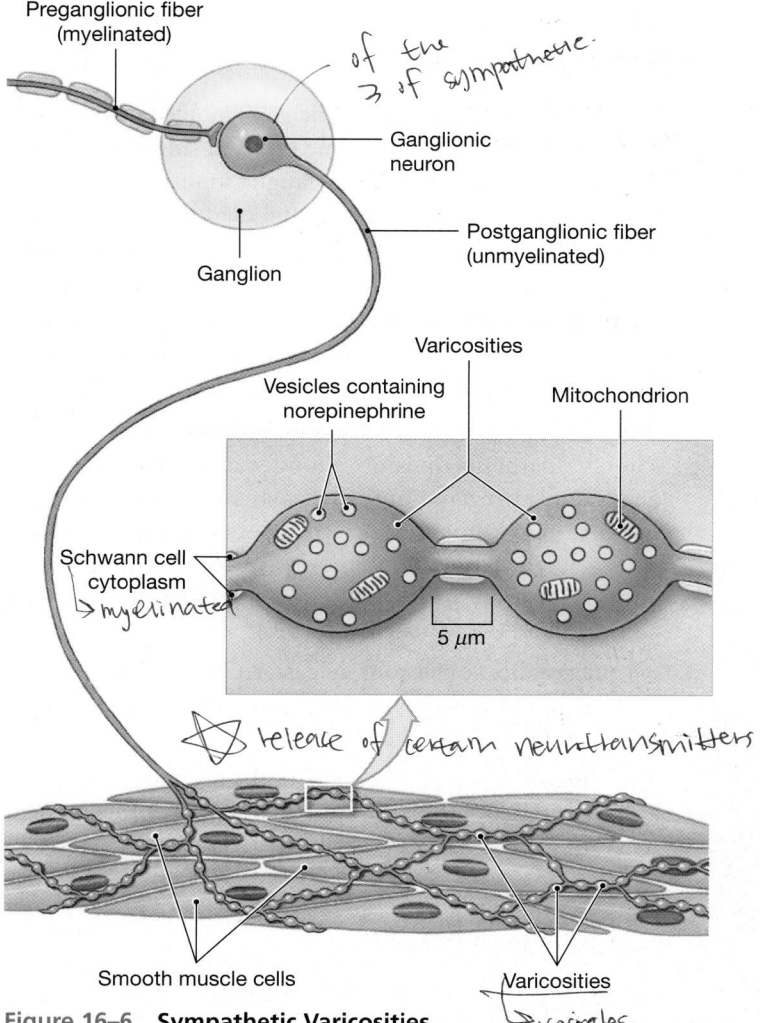

Preganglionic fiber (myelinated)

of the 3 of sympathetic. [handwritten]

Ganglionic neuron

Ganglion

Postganglionic fiber (unmyelinated)

Varicosities

Vesicles containing norepinephrine

Mitochondrion

Schwann cell cytoplasm
→ myelinated [handwritten]

5 μm

☆ release of certain neurotransmitters. [handwritten]

Smooth muscle cells

Varicosities
→ vesicles. [handwritten]

Figure 16–6 **Sympathetic Varicosities.**

postsynaptic membranes, but membrane receptors are scattered across the surfaces of the target cells.

Most sympathetic ganglionic neurons release NE at their varicosities. Neurons that use NE as a neurotransmitter are called *adrenergic*. ∞ p. 417 The sympathetic division also contains a small, but significant, number of ganglionic neurons that release ACh rather than NE. Varicosities releasing ACh are located in the body wall, the skin, the brain, and skeletal muscles.

The NE released by varicosities affects its targets until it is reabsorbed or inactivated by enzymes. From 50 to 80 percent of the NE is reabsorbed by varicosities and either reused or broken down by the enzyme *monoamine oxidase (MAO)*. The rest of the NE diffuses out of the area or is broken down by the enzyme *catechol-O-methyltransferase (COMT)* in surrounding tissues.

In general, the effects of NE on the postsynaptic membrane persist for a few seconds, significantly longer than the 20-msec duration of ACh effects. (As usual, the responses of the target cells vary with the nature of the receptor on the postsynaptic membrane.) The effects of NE or E released by the suprarenal medullae last even longer, because (1) the bloodstream does

not contain MAO or COMT, and (2) most tissues contain relatively low concentrations of those enzymes. After the suprarenal medullae are stimulated, tissue concentrations of NE and E throughout the body may remain elevated for as long as 30 seconds, and the effects may persist for several minutes.

Sympathetic Stimulation and the Release of NE and E

The effects of sympathetic stimulation result primarily from the interactions of NE and E with adrenergic membrane receptors. There are two classes of sympathetic receptors: *alpha receptors* and *beta receptors*. In general, norepinephrine stimulates alpha receptors to a greater degree than it does beta receptors, whereas epinephrine stimulates both classes of receptors. Thus, localized sympathetic activity, involving the release of NE at varicosities, primarily affects alpha receptors located near the active varicosities. By contrast, generalized sympathetic activation and the release of E by the suprarenal medulla affect alpha and beta receptors throughout the body.

Alpha receptors and beta receptors are *G proteins*. As we saw in Chapter 12, the effects of stimulating such a receptor depend on the production of *second messengers*, intracellular intermediaries with varied functions. ∞ p. 421

The stimulation of **alpha (α) receptors** activates enzymes on the inside of the plasma membrane. There are two types of alpha receptors: alpha-1 (α_1) and alpha-2 (α_2).

- The function of α_1, the more common type of alpha receptor, is the release of intracellular calcium ions into the cytosol from reserves in the endoplasmic reticulum. This action generally has an excitatory effect on the target cell. For example, the stimulation of α_1 receptors on the surfaces of smooth muscle cells is responsible for the constriction of peripheral blood vessels and the closure of sphincters along the digestive tract and urinary bladder.

- Stimulation of α_2 receptors results in a lowering of cyclic-AMP (cAMP) levels in the cytoplasm. Cyclic-AMP is an important second messenger that can activate or inactivate key enzymes. ∞ p. 421 This reduction generally has an inhibitory effect on the cell. The presence of α_2 receptors in the parasympathetic division helps coordinate sympathetic and parasympathetic activities. When the sympathetic division is active, the NE released binds to parasympathetic neuromuscular and neuroglandular junctions and inhibits their activity.

Beta (β) receptors are located on the membranes of cells in many organs, including skeletal muscles, the lungs, the heart, and the liver. The stimulation of beta receptors triggers changes in the metabolic activity of the target cell. These changes occur indirectly, as each beta receptor is a G protein whose stimulation results in an increase in intracellular cAMP levels. There are three major types of beta receptors: beta-1 (β_1) beta-2 (β_2), and beta-3 (β_3).

- The stimulation of β_1 receptors leads to an increase in metabolic activity. For example, the stimulation of β_1 receptors on skeletal muscles accelerates the metabolic activities of the muscles. The stimulation of β_1 receptors in the heart causes increases in heart rate and force of contraction.

- The stimulation of β_2 receptors causes inhibition, triggering a relaxation of smooth muscles along the respiratory tract. As a result, the diameters of the respiratory passageways increase, making breathing easier. This response accounts for the effectiveness of inhalers used to treat asthma.

- A third type of beta receptor, beta-3 (β_3), is found in adipose tissue. Stimulation of β_3 receptors leads to *lipolysis*, the breakdown of triglycerides stored within adipocytes. The fatty acids generated through lipolysis are released into the circulation for use by other tissues.

Sympathetic Stimulation and the Release of ACh and NO

Although the vast majority of sympathetic postganglionic fibers are adrenergic (release NE), a few are cholinergic (release ACh). These postganglionic fibers innervate sweat glands of the skin and the blood vessels to skeletal muscles and the brain. The activation of these sympathetic fibers stimulates sweat gland secretion and dilates the blood vessels.

In other regions of the body, ACh is released by the parasympathetic division rather than by the sympathetic division. However, neither the body wall nor skeletal muscles are innervated by the parasympathetic division, and in these areas both NE and ACh are needed to regulate visceral functions with precision. For example, whereas ACh causes dilation of most small peripheral arteries (*vasodilation*), NE causes their constriction (*vasoconstriction*). This means that the sympathetic division can increase blood flow to skeletal muscles, through activation of cholinergic terminals, at the same time that adrenergic terminals reduce the blood flow to other tissues in the body wall.

The sympathetic division also includes *nitroxidergic* synapses, which release *nitric oxide (NO)* as a neurotransmitter. Such synapses occur where neurons innervate smooth muscles in the walls of blood vessels in many regions, notably in skeletal muscles and the brain. The activity of these synapses produces vasodilation and increased blood flow through these regions.

Summary: The Sympathetic Division

To summarize our discussion of the sympathetic division:

1. The sympathetic division of the ANS includes two sets of sympathetic chain ganglia, one on each side of the vertebral column; three collateral ganglia anterior to the vertebral column; and two suprarenal medullae.

2. The preganglionic fibers are short, because the ganglia are close to the spinal cord. The postganglionic fibers are longer and extend a considerable distance before reaching their target organs. (In the case of the suprarenal medullae, very short axons end at capillaries that carry their secretions to the bloodstream.)

3. The sympathetic division shows extensive divergence, and a single preganglionic fiber may innervate two dozen or more ganglionic neurons in different ganglia. As a result, a single sympathetic motor neuron in the CNS can control a variety of visceral effectors and can produce a complex and coordinated response.

4. All preganglionic neurons release ACh at their synapses with ganglionic neurons. Most postganglionic fibers release NE, but a few release ACh or NO.

5. The effector response depends on the second messengers activated when NE or E binds to alpha receptors or beta receptors.

CHECKPOINT

6. How would a drug that stimulates acetylcholine receptors affect the sympathetic nervous system?

7. An individual with high blood pressure is given a medication that blocks beta receptors. How could this medication help correct that person's condition?

See the blue Answers tab at the end of the book.

16-4 The parasympathetic division consists of preganglionic neurons and ganglionic neurons involved in conserving energy and lowering metabolic rate

The parasympathetic division of the ANS (**Figure 16–7**) consists of:

1. *Preganglionic Neurons in the Brain Stem and in Sacral Segments of the Spinal Cord.* The mesencephalon, pons, and medulla oblongata contain autonomic nuclei associated with cranial nerves III, VII, IX, and X. In sacral segments of the spinal cord, the autonomic nuclei lie in the lateral gray horns of spinal segments S_2–S_4.

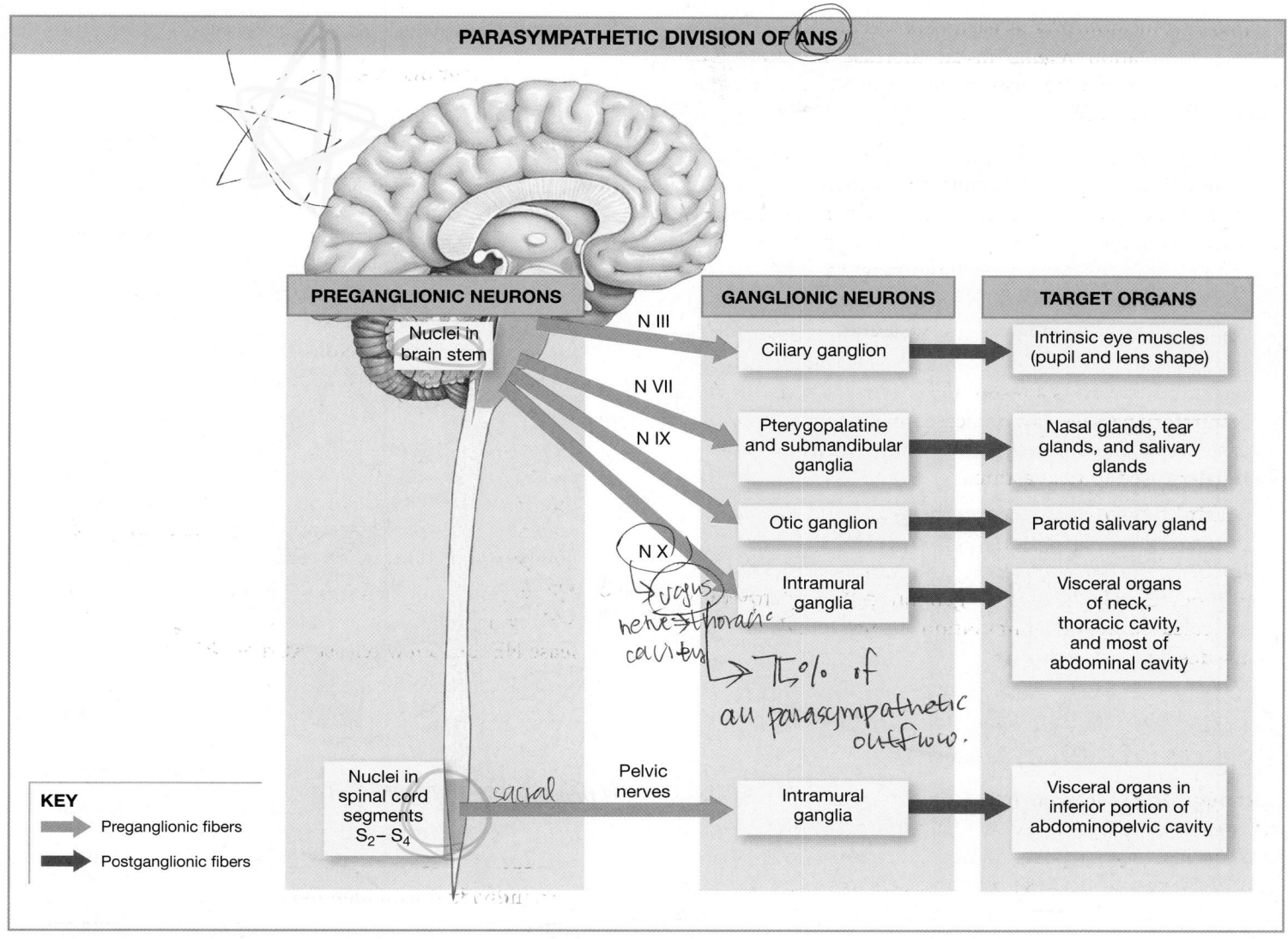

Figure 16–7   The Organization of the Parasympathetic Division of the ANS.

2. *Ganglionic Neurons in Peripheral Ganglia within or Adjacent to the Target Organs.* Preganglionic fibers of the parasympathetic division do not diverge as extensively as do those of the sympathetic division. A typical preganglionic fiber synapses on six to eight ganglionic neurons. These neurons may be situated in a **terminal ganglion**, located near the target organ, or in an **intramural** (*murus*, wall) **ganglion**, which is embedded in the tissues of the target organ. Terminal ganglia are usually paired; examples include the parasympathetic ganglia associated with the cranial nerves. Intramural ganglia typically consist of interconnected masses and clusters of ganglion cells.

In contrast to the pattern in the sympathetic division, all these ganglionic neurons are located in the same ganglion, and their postganglionic fibers influence the same target organ. Thus, the effects of parasympathetic stimulation are more specific and localized than are those of the sympathetic division.

Organization and Anatomy of the Parasympathetic Division

Parasympathetic preganglionic fibers leave the brain as components of cranial nerves III (oculomotor), VII (facial), IX (glossopharyngeal), and X (vagus) (**Figure 16–8**). These fibers carry the cranial parasympathetic output. Parasympathetic fibers in the oculomotor, facial, and glossopharyngeal nerves control visceral structures in the head. These fibers synapse in the *ciliary, pterygopalatine, submandibular,* and *otic ganglia.* ∞ pp. 494, 495 Short postganglionic fibers continue to their peripheral targets. The vagus nerve provides preganglionic parasympathetic innervation to structures in the neck and in the thoracic and abdominopelvic cavity as distant as the distal portion of the large intestine. The vagus nerve alone provides roughly 75 percent of all parasympathetic outflow. The numerous branches of the vagus nerve intermingle with preganglionic and postganglionic fibers of the sympathetic

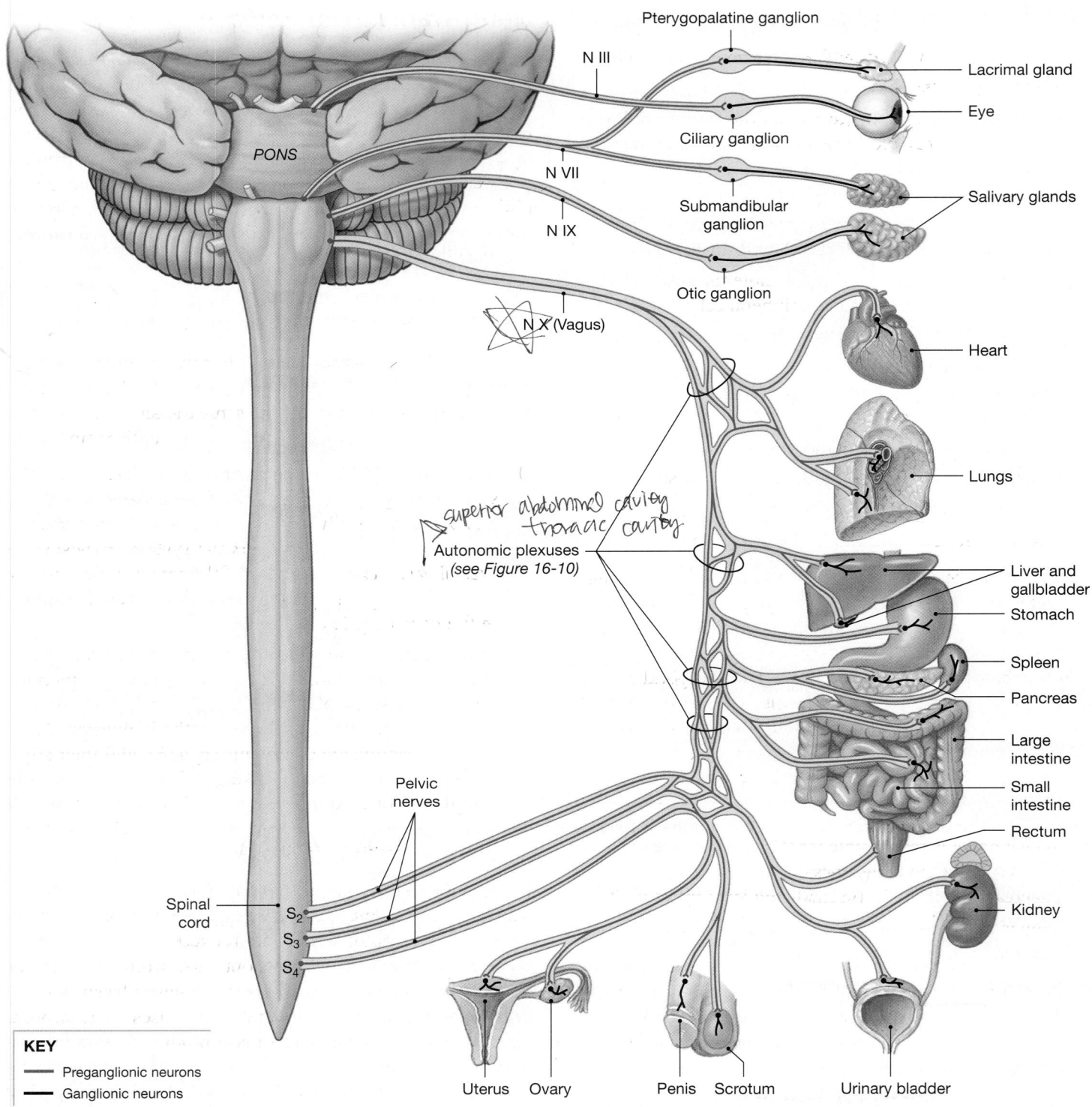

Figure 16–8 **The Distribution of Parasympathetic Innervation.**

division, forming plexuses comparable to those formed by spinal nerves innervating the limbs. These plexuses will be considered further in a later section.

The preganglionic fibers in the sacral segments of the spinal cord carry the sacral parasympathetic output. These fibers do not join the ventral roots of the spinal nerves. Instead, the preganglionic fibers form distinct **pelvic nerves**, which innervate intramural ganglia in the walls of the kidneys, urinary bladder, terminal portions of the large intestine, and sex organs.

Parasympathetic Activation

The major effects produced by the parasympathetic division include the following:

- Constriction of the pupils (to restrict the amount of light that enters the eyes) and focusing of the lenses of the eyes on nearby objects.

- Secretion by digestive glands, including salivary glands, gastric glands, duodenal glands, intestinal glands, the pancreas (exocrine and endocrine), and the liver.

- The secretion of hormones that promote the absorption and utilization of nutrients by peripheral cells.

- Changes in blood flow and glandular activity associated with sexual arousal.

- An increase in smooth muscle activity along the digestive tract.

- The stimulation and coordination of defecation.

- Contraction of the urinary bladder during urination.

- Constriction of the respiratory passageways.

- A reduction in heart rate and in the force of contraction.

These functions center on relaxation, food processing, and energy absorption. The parasympathetic division has been called the *anabolic system* (*anabole,* a rising up), because its stimulation leads to a general increase in the nutrient content of the blood. Cells throughout the body respond to this increase by absorbing nutrients and using them to support growth, cell division, and the creation of energy reserves in the form of lipids or glycogen when situations are favorable.

× plexus.

CHECKPOINT

8. **Which nerve is responsible for the parasympathetic innervation of the lungs, heart, stomach, liver, pancreas, and parts of the small and large intestines?**

9. **Why is the parasympathetic division sometimes referred to as the anabolic system?**

See the blue Answers tab at the end of the book.

16-5 Stimulation of parasympathetic neurons leads to the release of the neurotransmitter ACh

All parasympathetic neurons release ACh as a neurotransmitter. The effects on the postsynaptic cell can vary widely, however, due to variations in the type of receptor or the nature of the second messenger involved.

Neurotransmitter Release

The neuromuscular and neuroglandular junctions of the parasympathetic division are small and have narrow synaptic clefts. The effects of stimulation are short-lived, because most of the ACh released is inactivated by *acetylcholinesterase (AChE)* at the synapse. Any ACh diffusing into the surrounding tissues will be inactivated by the enzyme *tissue cholinesterase,* also called *pseudocholinesterase.* As a result, the effects of parasympathetic stimulation are quite localized, and they last a few seconds at most.

Membrane Receptors and Responses

Although all the synapses (neuron to neuron) and neuromuscular or neuroglandular junctions (neuron to effector) of the parasympathetic division use the same transmitter, ACh, two types of ACh receptors occur on the postsynaptic membranes:

1. **Nicotinic** (nik-ō-TIN-ik) **receptors** are located on the surfaces of ganglion cells of both the parasympathetic and sympathetic divisions, as well as at neuromuscular junctions of the somatic nervous system. Exposure to ACh always causes excitation of the ganglionic neuron or muscle fiber by the opening of chemically gated channels in the postsynaptic membrane.

2. **Muscarinic** (mus-ka-RIN-ik) **receptors** are located at cholinergic neuromuscular or neuroglandular junctions in the parasympathetic division, as well as at the few cholinergic junctions in the sympathetic division. Muscarinic receptors are G proteins (p. 418), and their stimulation produces longer-lasting effects than does the stimulation of nicotinic receptors. The response, which reflects the activation or inactivation of specific enzymes, can be excitatory or inhibitory.

The names *nicotinic* and *muscarinic* originated with researchers who found that dangerous environmental toxins bind to these receptor sites. Nicotinic receptors bind *nicotine,* a powerful toxin that can be obtained from a variety of sources, including tobacco leaves. The highest levels of nicotine are about 3 mg per gram of tobacco. Muscarinic receptors are stimulated by *muscarine,* a toxin produced by some poisonous mushrooms.

These compounds have discrete actions, targeting either the autonomic ganglia and skeletal neuromuscular junctions (nicotine) or the parasympathetic neuromuscular or neuroglandular junctions (muscarine). They produce dangerously exaggerated, uncontrolled responses due to abnormal stimulation of cholinergic or adrenergic receptors. Nicotine poisoning occurs if as little as 50 mg of the compound is ingested or absorbed through the skin. The signs and symptoms reflect widespread autonomic activation: vomiting, diarrhea, high blood pressure, rapid heart rate, sweating, and profuse salivation. Because the neuromuscular junctions of the somatic nervous system are stimulated, convulsions occur. In severe cases, the stimulation of nicotinic receptors inside the CNS can lead to coma and death within minutes. The signs and symptoms of muscarine poisoning are almost entirely restricted to the parasympathetic division: salivation, nausea, vomiting, diarrhea, constriction of respiratory passages, low blood pressure, and an abnormally slow heart rate (bradycardia). Sweating, a sympathetic response, is also prominent; hence, *Amanita muscaria*, commonly known as the fly-agaric mushroom, is used in some Native American sweat lodge rituals.

Table 16–1 summarizes details about the adrenergic and cholinergic receptors of the ANS.

Summary: The Parasympathetic Division

In summary:

- The parasympathetic division includes visceral motor nuclei associated with cranial nerves III, VII, IX, and X, and with sacral segments S_2–S_4.

- Ganglionic neurons are located within or next to their target organs.

- The parasympathetic division innervates areas serviced by the cranial nerves and organs in the thoracic and abdominopelvic cavities.

- All parasympathetic neurons are cholinergic. Ganglionic neurons have nicotinic receptors, which are excited by ACh. Muscarinic receptors at neuromuscular or neuroglandular junctions produce either excitation or inhibition, depending on the nature of the enzymes activated when ACh binds to the receptor.

- The effects of parasympathetic stimulation are generally brief and restricted to specific organs and sites.

The A&P Top 100

#55 The preganglionic neurons of the autonomic nervous system release acetylcholine (ACh) as a neurotransmitter. The ganglionic neurons of the sympathetic division primarily release norepinephrine as a neurotransmitter (and both NE and E as hormones at the suprarenal medulla). The ganglionic neurons of the parasympathetic division release ACh as a neurotransmitter.

CHECKPOINT

10. What neurotransmitter is released by all parasympathetic neurons?

11. Name the two types of ACh receptors on the postsynaptic membranes of parasympathetic neurons.

12. How would the stimulation of muscarinic receptors in cardiac muscle affect the heart?

See the blue Answers tab at the end of the book.

SUMMARY TABLE 16–1   Adrenergic and Cholinergic Receptors of the ANS

Receptor	Location(s)	Response(s)	Mechanism
ADRENERGIC			
α_1	Widespread, found in most tissues	Excitation, stimulation of metabolism	Enzyme activation; intracellular release of Ca^{2+}
α_2	Sympathetic neuromuscular or neuroglandular junctions	Inhibition of effector cell	Reduction of cAMP concentrations
	Parasympathetic neuromuscular or neuroglandular junctions	Inhibition of neurotransmitter release	Reduction of cAMP concentrations
β_1	Heart, kidneys, liver, adipose tissue*	Stimulation, increased energy consumption	Enzyme activation
β_2	Smooth muscle in vessels of heart and skeletal muscle; smooth muscle layers in intestines, lungs, bronchi	Inhibition, relaxation	Enzyme activation
CHOLINERGIC			
Nicotinic	All autonomic synapses between preganglionic and ganglionic neurons; neuromuscular junctions of SNS	Stimulation, excitation; muscular contraction	Opening of chemically gated Na^+ channels
Muscarinic	All parasympathetic and cholinergic sympathetic neuromuscular or neuroglandular junctions	Variable	Enzyme activation causing changes in membrane permeability to K^+

*Adipocytes also contain an additional receptor type, β_3, not found in other tissues. Stimulation of β_3 receptors causes lipolysis.

16-6 The sympathetic and parasympathetic divisions interact, creating dual innervation

Figure 16–9 and Table 16–2 compare key structural features of the sympathetic and parasympathetic divisions of the ANS. The distinctions have physiological and functional correlates. The sympathetic division has widespread impact, reaching organs and tissues throughout the body. The parasympathetic division innervates only visceral structures that are serviced by the cranial nerves or that lie within the abdominopelvic cavity. Although some organs are innervated by just one division, most vital organs receive **dual innervation**, receiving instructions from both the sympathetic and parasympathetic divisions. Where dual innervation exists, the two divisions commonly have opposing effects. Dual innervation with opposing effects is most evident in the digestive tract, heart, and lungs. At other sites, the responses may be separate or complementary. Table 16–3 provides a functional comparison of the two divisions, taking into account the effects of sympathetic or parasympathetic activity on specific organs and systems.

Anatomy of Dual Innervation

Parasympathetic postganglionic fibers from the ciliary, pterygopalatine, submandibular, and otic ganglia of the head accompany the cranial nerves to their peripheral destinations. The sympathetic innervation reaches the same structures by traveling directly from the superior cervical ganglia of the sympathetic chain.

In the thoracic and abdominopelvic cavities, the sympathetic postganglionic fibers mingle with parasympathetic preganglionic fibers, forming a series of nerve networks collectively called *autonomic plexuses*: the cardiac plexus, the pulmonary plexus, the esophageal plexus, the celiac plexus, the inferior mesenteric plexus, and the hypogastric plexus

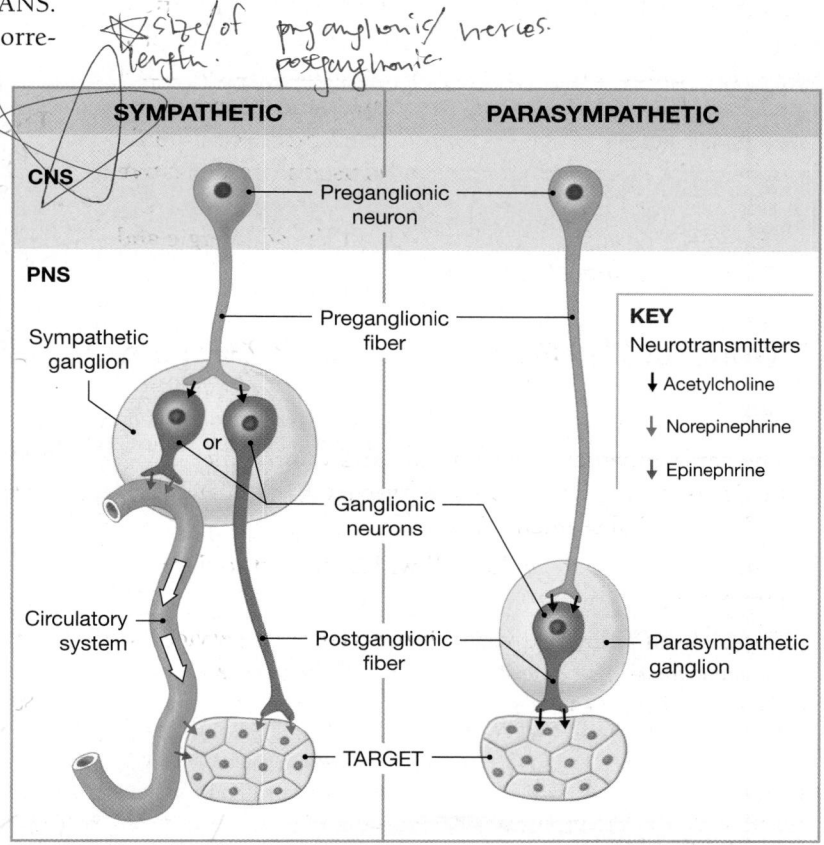

Figure 16–9 Summary: The Anatomical Differences between the Sympathetic and Parasympathetic Divisions.

SUMMARY TABLE 16–2	A Structural Comparison of the Sympathetic and Parasympathetic Divisions of the ANS	
Characteristic	**Sympathetic Division**	**Parasympathetic Division**
Location of CNS visceral motor neurons	Lateral gray horns of spinal segments T_1–L_2	Brain stem and spinal segments S_2–S_4
Location of PNS ganglia	Near vertebral column	Typically intramural
Preganglionic fibers		
Length	Relatively short	Relatively long
Neurotransmitter released	Acetylcholine	Acetylcholine
Postganglionic fibers		
Length	Relatively long	Relatively short
Neurotransmitter released	Normally NE; sometimes NO or ACh	Acetylcholine
Neuromuscular or neuroglandular junction	Varicosities and enlarged terminal knobs that release transmitter near target cells	Junctions that release transmitter to special receptor surface
Degree of divergence from CNS to ganglion cells	Approximately 1:32	Approximately 1:6
General function(s)	Stimulates metabolism; increases alertness; prepares for emergency ("fight or flight")	Promotes relaxation, nutrient uptake, energy storage ("rest and digest")

SUMMARY TABLE 16–3   A Functional Comparison of the Sympathetic and Parasympathetic Divisions of the ANS

Structure	Sympathetic Effects (receptor or synapse type)	Parasympathetic Effects (all muscarinic receptors)
EYE	Dilation of pupil (α_1); accommodation for distance vision (β_2)	Constriction of pupil; accommodation for close vision
Lacrimal glands	None (not innervated)	Secretion
SKIN		
Sweat glands	Increased secretion, palms and soles (α_1); generalized increase in secretion (cholinergic)	None (not innervated)
Arrector pili muscles	Contraction; erection of hairs (α_1)	None (not innervated)
CARDIOVASCULAR SYSTEM		
Blood vessels		None (not innervated)
To skin	Dilation (cholinergic); constriction (α_1)	
To skeletal muscles	Dilation (β_2 and cholinergic; nitroxidergic)	
To heart	Dilation (β_2); constriction (α_1, α_2)	
To lungs	Dilation (β_2); constriction (α_1)	
To digestive viscera	Constriction (α_1); dilation (α_2)	
To kidneys	Constriction, decreased urine production (α_1, α_2) dilation, increased urine production (β_1, β_2)	
To brain	Dilation (cholinergic and nitroxidergic)	
Veins	Constriction (α_1, β_2)	
Heart	Increased heart rate, force of contraction, and blood pressure (α_1, β_1)	Decreased heart rate, force of contraction, and blood pressure
ENDOCRINE SYSTEM		
Suprarenal gland	Secretion of epinephrine, norepinephrine by suprarenal medulla	None (not innervated)
Neurohypophysis	Secretion of ADH (β_1)	None (not innervated)
Pancreas	Decreased insulin secretion (α_2)	Increased insulin secretion
Pineal gland	Increased melatonin secretion (β)*	
RESPIRATORY SYSTEM		
Airways	Increased airway diameter (β_2)	Decreased airway diameter
Secretory glands	Mucous secretion (α_1)	None
DIGESTIVE SYSTEM		
Salivary glands	Production of viscous secretion (α_1, β_1) containing mucins and enzymes	Production of copious, watery secretion
Sphincters	Constriction (α_1)	Dilation
General level of activity	Decreased (α_2, β_2)	Increased
Secretory glands	Inhibition (α_2)	Stimulation
Liver	Glycogen breakdown, glucose synthesis and release (α_1, β_2)	Glycogen synthesis
Pancreas	Decreased exocrine secretion (α_1)	Increased exocrine secretion
SKELETAL MUSCLES	Increased force of contraction, glycogen breakdown (β_2)	None (not innervated)
	Facilitation of ACh release at neuromuscular junction (α_2)	None (not innervated)
ADIPOSE TISSUE	Lipolysis, fatty acid release (α_1, β_1, β_3)	

*The type of beta receptor has not yet been determined.

SUMMARY TABLE 16–3 (Continued)		
Structure	**Sympathetic Effects (receptor or synapse type)**	**Parasympathetic Effects (all muscarinic receptors)**
URINARY SYSTEM		
Kidneys	Secretion of renin (β_1)	Uncertain effects on urine production
Urinary bladder	Constriction of internal sphincter; relaxation of urinary bladder (α_1, β_2)	Tensing of urinary bladder, relaxation of internal sphincter to eliminate urine
MALE REPRODUCTIVE SYSTEM	Increased glandular secretion and ejaculation (α_1)	Erection
FEMALE REPRODUCTIVE SYSTEM	Increased glandular secretion; contraction of pregnant uterus (α_1)	Variable (depending on hormones present)
	Relaxation of nonpregnant uterus (β_2)	Variable (depending on hormones present)

(**Figure 16–10**). Nerves leaving these networks travel with the blood vessels and lymphatic vessels that supply visceral organs.

Autonomic fibers entering the thoracic cavity intersect at the **cardiac plexus** and the **pulmonary plexus**. These plexuses contain sympathetic and parasympathetic fibers bound for the heart and lungs, respectively, as well as the parasympathetic ganglia whose output affects those organs. The **esophageal plexus** contains descending branches of the vagus nerve and splanchnic nerves leaving the sympathetic chain on either side.

Parasympathetic preganglionic fibers of the vagus nerve enter the abdominopelvic cavity with the esophagus. There the fibers enter the **celiac plexus**, also known as the *solar plexus*. The celiac plexus and associated smaller plexuses, such as the **inferior mesenteric plexus**, innervate viscera within the abdominal cavity. The **hypogastric plexus** contains the parasympathetic outflow of the pelvic nerves, sympathetic postganglionic fibers from the inferior mesenteric ganglion, and splanchnic nerves from the sacral sympathetic chain. This plexus innervates the digestive, urinary, and reproductive organs of the pelvic cavity.

Autonomic Tone

Even in the absence of stimuli, autonomic motor neurons show a resting level of spontaneous activity. The background level of activation determines an individual's **autonomic tone**. Autonomic tone is an important aspect of ANS function, just as muscle tone is a key aspect of SNS function. If a nerve is absolutely inactive under normal conditions, then all it can do is increase its activity on demand. But if the nerve maintains a background level of activity, it can increase or decrease its activity, providing a greater range of control options.

Autonomic tone is significant where dual innervation occurs and the two ANS divisions have opposing effects. It is even more important in situations in which dual innervation does not occur. To demonstrate how autonomic tone affects ANS function, we will consider one example of each situation.

The heart is an organ that receives dual innervation. Recall that the heart consists of cardiac muscle tissue, and that its contractions are triggered by specialized pacemaker cells. ∞ p. 141 The two autonomic divisions have opposing effects on heart function. Acetylcholine released by postganglionic fibers of the parasympathetic division causes a reduction in heart rate, whereas NE released by varicosities of the sympathetic division accelerates heart rate. Because autonomic tone is present, small amounts of both of these neurotransmitters are released continuously. However, parasympathetic innervation dominates under resting conditions. Heart rate can be controlled very precisely to meet the demands of active tissues through small adjustments in the balance between parasympathetic stimulation and sympathetic stimulation. In a crisis, stimulation of the sympathetic innervation and inhibition of the parasympathetic innervation accelerate the heart rate to the maximum extent possible.

The sympathetic control of blood vessel diameter demonstrates how autonomic tone allows fine adjustment of peripheral activities when the target organ is not innervated by both ANS divisions. Blood flow to specific organs must be controlled to meet the tissue demands for oxygen and nutrients. When a blood vessel dilates, blood flow through it increases; when it constricts, blood flow is reduced. Sympathetic postganglionic fibers that release NE innervate the smooth muscle cells in the walls of peripheral vessels. The background sympathetic tone keeps these muscles partially contracted, so

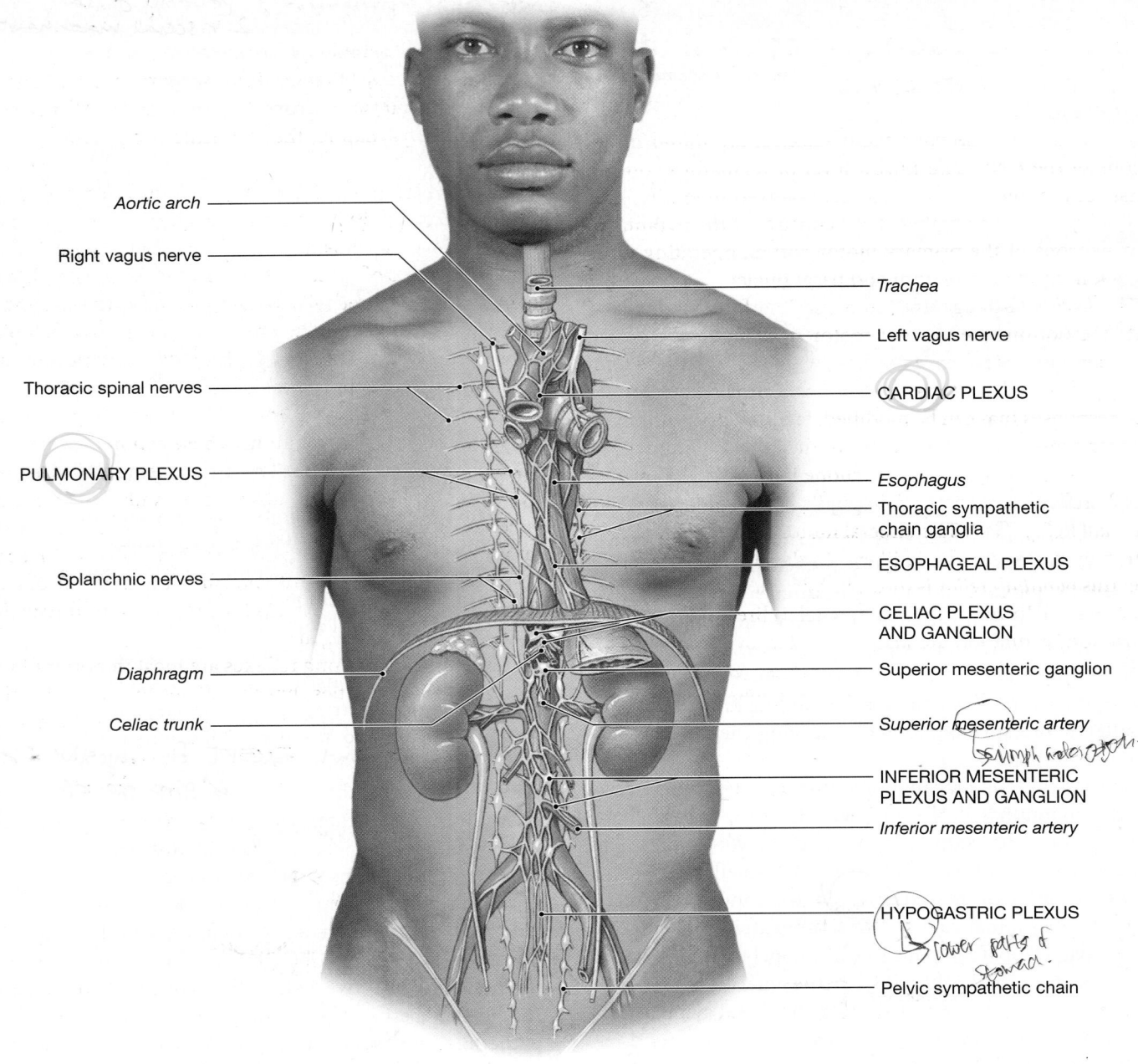

Figure 16–10 **The Autonomic Plexuses.**

the blood vessels are ordinarily at roughly half their maximum diameter. When increased blood flow is needed, the rate of NE release decreases and sympathetic cholinergic fibers are stimulated. As a result, the smooth muscle cells relax, the vessels dilate, and blood flow increases. By adjusting sympathetic tone and the activity of cholinergic fibers, the sympathetic division can exert precise control of vessel diameter over its entire range.

CHECKPOINT

13. What effect would the loss of sympathetic tone have on blood flow to a tissue?

14. What physiological changes would you expect in a patient who is about to undergo a root canal and is quite anxious about the procedure?

See the blue Answers tab at the end of the book.

16-7 Visceral reflexes play a role in the integration and control of autonomic functions

Centers involved in somatic motor control are found in all portions of the CNS. The lowest level of regulatory control consists of the lower motor neurons involved in cranial and spinal reflex arcs. The highest level consists of the pyramidal motor neurons of the primary motor cortex, operating with feedback from the cerebellum and basal nuclei.

The ANS is also organized into a series of interacting levels. At the bottom are visceral motor neurons in the lower brain stem and spinal cord that are involved in cranial and spinal visceral reflexes. **Visceral reflexes** provide automatic motor responses that can be modified, facilitated, or inhibited by higher centers, especially those of the hypothalamus.

For example, when a light is shone in one of your eyes, a visceral reflex constricts the pupils of *both* eyes (the *consensual light reflex*). The visceral motor commands are distributed by parasympathetic fibers. In darkness, your pupils dilate; this *pupillary reflex* is directed by sympathetic postganglionic fibers. However, the motor nuclei directing pupillary constriction or dilation are also controlled by hypothalamic centers concerned with emotional states. When you are queasy or nauseated, your pupils constrict; when you are sexually aroused, your pupils dilate.

Visceral Reflexes

Each **visceral reflex arc** consists of a receptor, a sensory neuron, a processing center (one or more interneurons), and two visceral motor neurons (**Figure 16–11**). All visceral reflexes are polysynaptic; they are either long reflexes or short reflexes.

Long reflexes are the autonomic equivalents of the polysynaptic reflexes introduced in Chapter 13. ∞ p. 452 Visceral sensory neurons deliver information to the CNS along the dorsal roots of spinal nerves, within the sensory branches of cranial nerves, and within the autonomic nerves that innervate visceral effectors. The processing steps involve interneurons within the CNS, and the ANS carries the motor commands to the appropriate visceral effectors.

Short reflexes bypass the CNS entirely; they involve sensory neurons and interneurons whose cell bodies are located within autonomic ganglia. The interneurons synapse on ganglionic neurons, and the motor commands are then distributed by postganglionic fibers. Short reflexes control very simple motor responses with localized effects. In general, short reflexes may control patterns of activity in one small part of a target organ, whereas long reflexes coordinate the activities of an entire organ.

In most organs, long reflexes are most important in regulating visceral activities, but this is not the case with the digestive tract and its associated glands. In these areas, short

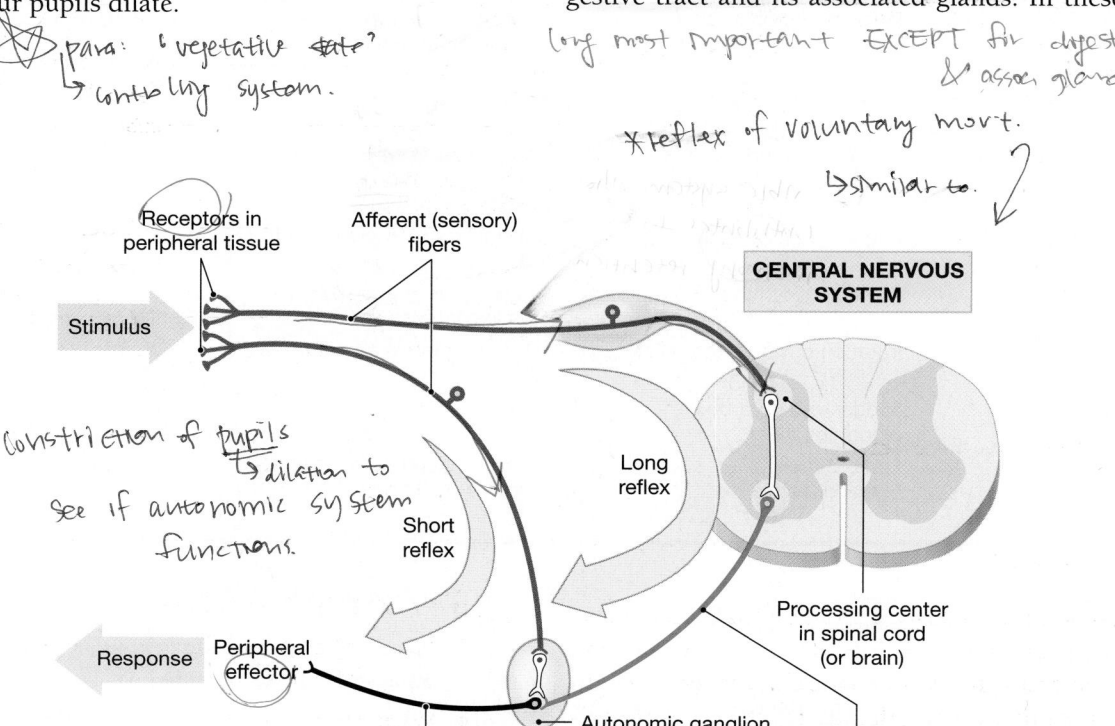

Figure 16–11 **Visceral Reflexes.** Visceral reflexes have the same basic components as somatic reflexes, but all visceral reflexes are polysynaptic. Note that short visceral reflexes bypass the CNS altogether.

reflexes provide most of the control and coordination required for normal functioning. The neurons involved form the *enteric nervous system*, introduced on p. 531. The ganglia in the walls of the digestive tract contain the cell bodies of visceral sensory neurons, interneurons, and visceral motor neurons, and their axons form extensive nerve nets. Although parasympathetic innervation of the visceral motor neurons can stimulate and coordinate various digestive activities, the enteric nervous system is quite capable of controlling digestive functions independent of the central nervous system. We will consider the functions of the enteric nervous system further in Chapter 24.

As we examine other body systems in later chapters, we will encounter many examples of autonomic reflexes involved in respiration, cardiovascular function, and other visceral activities. Some of the most important are previewed in Table 16–4. Notice that the parasympathetic division participates in a variety of reflexes that affect individual organs and systems. This specialization reflects the relatively specific and restricted pattern of innervation. In contrast, fewer sympathetic reflexes exist. The sympathetic division is typically activated as a whole, in part because it has such a high degree of divergence and in part because the release of hormones by the suprarenal medullae produces widespread peripheral effects.

Higher Levels of Autonomic Control

The levels of activity in the sympathetic and parasympathetic divisions of the ANS are controlled by centers in the brain stem that regulate specific visceral functions. As in the SNS, in the ANS simple reflexes based in the spinal cord provide relatively rapid and automatic responses to stimuli. More complex sympathetic and parasympathetic reflexes are coordinated by processing centers in the medulla oblongata. In addition to the cardiovascular and respiratory centers, the medulla oblongata contains centers and nuclei involved with salivation, swallowing, digestive secretions, peristalsis, and

TABLE 16–4	Representative Visceral Reflexes		
Reflex	**Stimulus**	**Response**	**Comments**
PARASYMPATHETIC REFLEXES			
Gastric and intestinal reflexes (*Chapter 24*)	Pressure and physical contact	Smooth muscle contractions that propel food materials and mix with secretions	Via vagus nerve
Defecation (*Chapter 24*)	Distention of rectum	Relaxation of internal anal sphincter	Requires voluntary relaxation of external anal sphincter
Urination (*Chapter 26*)	Distention of urinary bladder	Contraction of walls of urinary bladder; relaxation of internal urethral sphincter	Requires voluntary relaxation of external urethral sphincter
Direct light and consensual light reflexes (*Chapter 14*)	Bright light shining in eye(s)	Constriction of pupils of both eyes	
Swallowing reflex (*Chapter 24*)	Movement of food and liquids into pharynx	Smooth muscle and skeletal muscle contractions	Coordinated by medullary swallowing center
Coughing reflex (*Chapter 23*)	Irritation of respiratory tract	Sudden explosive ejection of air	Coordinated by medullary coughing center
Baroreceptor reflex (*Chapters 17, 20, 21*)	Sudden rise in carotid blood pressure	Reduction in heart rate and force of contraction	Coordinated in cardiac centers of medulla oblongata
Sexual arousal (*Chapter 28*)	Erotic stimuli (visual or tactile)	Increased glandular secretions, sensitivity, erection	
SYMPATHETIC REFLEXES			
Cardioacceleratory reflex (*Chapter 21*)	Sudden decline in blood pressure in carotid artery	Increase in heart rate and force of contraction	Coordinated in cardiac centers of medulla oblongata
Vasomotor reflexes (*Chapter 21*)	Changes in blood pressure in major arteries	Changes in diameter of peripheral vessels	Coordinated in vasomotor center in medulla oblongata
Pupillary reflex (*Chapter 17*)	Low light level reaching visual receptors	Dilation of pupil	
Ejaculation (in males) (*Chapter 28*)	Erotic stimuli (tactile)	Skeletal muscle contractions ejecting semen	

urinary function. These centers are in turn subject to regulation by the hypothalamus. ∞ p. 478

The term *autonomic* was originally applied because the regulatory centers involved with the control of visceral function were thought to operate autonomously—that is, independent of other CNS activities. This view has been drastically revised in light of subsequent research. Because the hypothalamus interacts with all other portions of the brain, activity in the limbic system, thalamus, or cerebral cortex can have dramatic effects on autonomic function. For example, when you become angry, your heart rate accelerates, your blood pressure rises, and your respiratory rate increases; when you consider your next meal, your stomach "growls" and your mouth waters.

The Integration of SNS and ANS Activities

Figure 16–12 and Table 16–5 indicate how the activities of the somatic nervous system (discussed in Chapter 15) and those of the autonomic nervous system are integrated. Although we have considered somatic and visceral motor pathways separately, the two have many parallels, in terms of both organization and function. Integration occurs at the level of the brain stem, and both systems are under the influence of higher centers.

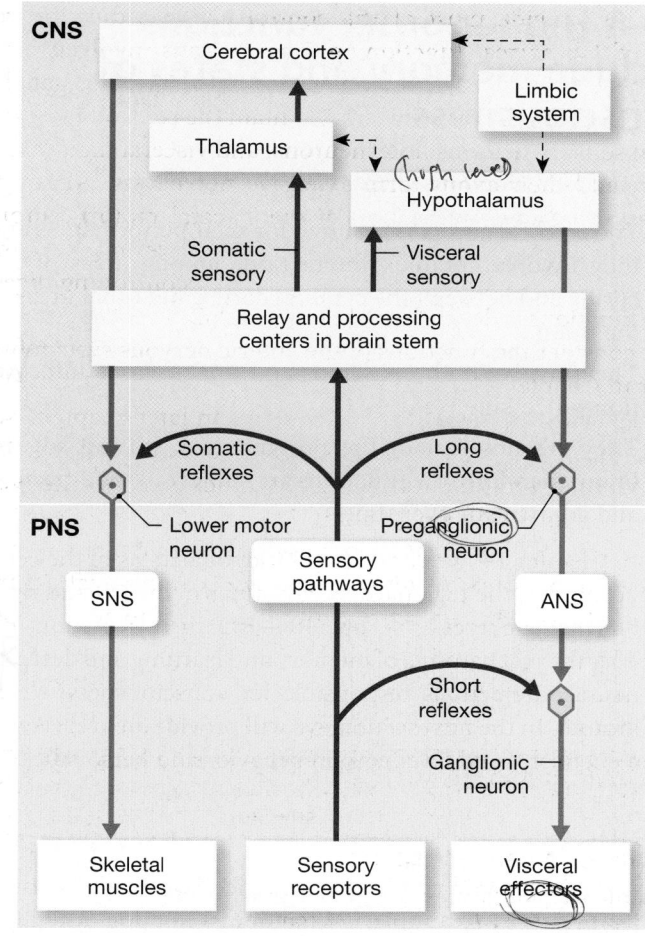

Figure 16–12   A Comparison of Somatic and Autonomic Function. The SNS and ANS are organized in parallel and are integrated at the level of the brain stem. Blue arrows indicate ascending sensory information; red arrows, descending motor commands; dashed lines indicate pathways of communication and feedback among higher centers.

CHECKPOINT

15. Define visceral reflex.

16. Harry has a brain tumor that is interfering with the function of his hypothalamus. Would you expect this tumor to interfere with autonomic function? Why or why not?

See the blue Answers tab at the end of the book.

SUMMARY TABLE 16–5	A Comparison of the ANS and SNS	
Characteristic	**ANS**	**SNS**
Innervation	Visceral effectors, including cardiac muscle, smooth muscle, glands, fat cells	Skeletal muscles
Activation	In response to sensory stimuli or from commands of higher centers	In response to sensory stimuli or from commands of higher centers
Relay and processing centers	Brain stem	Brain stem and thalamus
Headquarters	Hypothalamus	Cerebral cortex
Feedback received from	Limbic system and thalamus	Cerebellum and basal nuclei
Control method	Adjustment of activity in brain stem processing centers that innervate preganglionic neurons	Direct (corticospinal) and indirect (medial and lateral) pathways that innervate lower motor neurons
Reflexes	Polysynaptic (short and long)	Monosynaptic and polysynaptic (always long)

16-8 Higher-order functions include memory and states of consciousness

Higher-order functions share three characteristics:

1. The cerebral cortex is required for their performance, and they involve complex interactions among areas of the cortex and between the cerebral cortex and other areas of the brain.

2. They involve both conscious and unconscious information processing.

3. They are not part of the programmed "wiring" of the brain; therefore, the functions are subject to modification and adjustment over time.

In Chapter 14, we considered functional areas of the cerebral cortex and the regional specializations of the left and right cerebral hemispheres. ∞ pp. 484–485 In this section, we consider the mechanisms of memory and learning and describe the neural interactions responsible for consciousness, sleep, and arousal. In the next section, we will provide an overview of brain chemistry and its effects on behavior and personality.

Memory

What was the topic of the last sentence you read? What is your social security number? What does a hot dog taste like? Answering these questions involves accessing *memories*, stored bits of information gathered through experience. **Fact memories** are specific bits of information, such as the color

of a stop sign or the smell of a perfume. **Skill memories** are learned motor behaviors. You can probably remember how to light a match or open a screw-top jar, for example. With repetition, skill memories become incorporated at the unconscious level. Examples include the complex motor patterns involved in skiing, playing the violin, and similar activities. Skill memories related to programmed behaviors, such as eating, are stored in appropriate portions of the brain stem. Complex skill memories involve the integration of motor patterns in the basal nuclei, cerebral cortex, and cerebellum.

Two classes of memories are recognized. **Short-term memories**, or *primary memories*, do not last long, but while they persist the information can be recalled immediately. Primary memories contain small bits of information, such as a person's name or a telephone number. Repeating a phone number or other bit of information reinforces the original short-term memory and helps ensure its conversion to a long-term memory. **Long-term memories** last much longer, in some cases for an entire lifetime. The conversion from short-term to long-term memory is called **memory consolidation**. There are two types of long-term memory: (1) *Secondary memories* are long-term memories that fade with time and may require considerable effort to recall. (2) *Tertiary memories* are long-term memories that are with you for a lifetime, such as your name or the contours of your own body. Relationships among these memory classes are diagrammed in **Figure 16–13**.

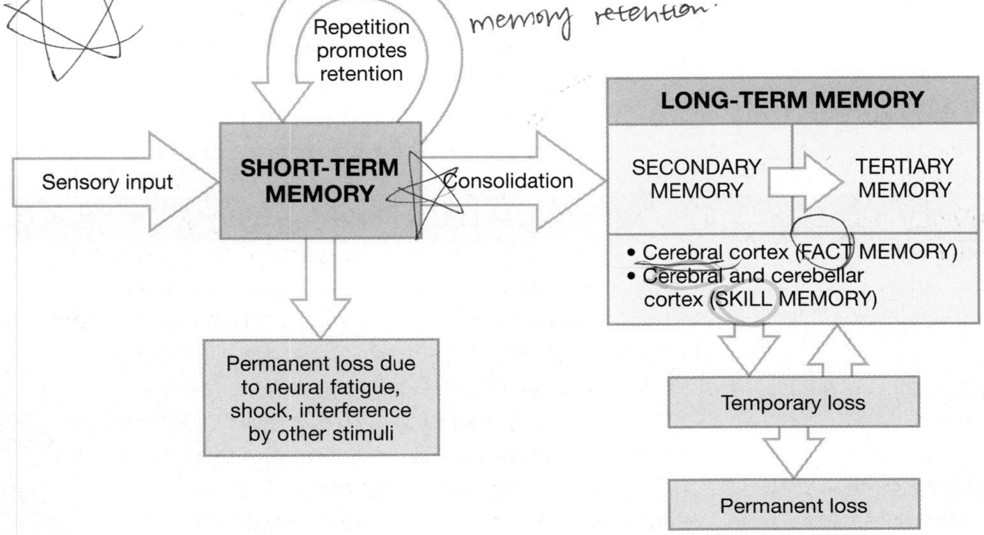

Brain Regions Involved in Memory Consolidation and Access

The amygdaloid body and the hippocampus, two components of the limbic system (**Figure 14–11**, p. 479), are essential to memory consolidation. Damage to the hippocampus leads to an inability to convert short-term memories to new long-term memories, although existing long-term memories remain intact and accessible. Tracts leading from the amygdaloid body to the hypothalamus may link memories to specific emotions.

The **nucleus basalis**, a cerebral nucleus near the diencephalon, plays an uncertain role in memory storage and retrieval. Tracts connect this nucleus with the hippocampus, amygdaloid body, and all areas of the cerebral cortex. Damage to this nucleus is associated with changes in emotional states, memory, and intellectual function (as we will see in the discussion of Alzheimer disease later in this chapter).

Most long-term memories are stored in the cerebral cortex. Conscious motor

Figure 16–13 **Memory Storage.** Steps in the storage of memories and the conversion from short-term to long-term memories.

and sensory memories are referred to the appropriate associ- ation areas. For example, visual memories are stored in the visual association area, and memories of voluntary motor activity are stored in the premotor cortex. Special portions of the occipital and temporal lobes are crucial to the memories of faces, voices, and words. In at least some cases, a specific memory probably depends on the activity of a single neuron. For example, in one portion of the temporal lobe an individual neuron responds to the sound of one word and ignores others. A specific neuron may also be activated by the proper combination of sensory stimuli associated with a particular individual, such as your grandmother. As a result, these neurons are called "grandmother cells."

Information on one subject is parceled out to many different regions of the brain. Your memories of cows are stored in the visual association area (what a cow looks like, that the letters *c-o-w* mean "cow"), the auditory association area (the "moo" sound and how the word *cow* sounds), the speech center (how to say the word *cow*), and the frontal lobes (how big cows are, what they eat). Related information, such as how you feel about cows and what milk tastes like, is stored in other locations. If one of those storage areas is damaged, your memory will be incomplete in some way. How these memories are accessed and assembled on demand remains a mystery.

Cellular Mechanisms of Memory Formation and Storage

Memory consolidation at the cellular level involves anatomical and physiological changes in neurons and synapses. For legal, ethical, and practical reasons, it is not possible to con- duct much research on these mechanisms with human subjects. Research on other animals, commonly those with relatively simple nervous systems, has indicated that the following mechanisms may be involved:

- *Increased Neurotransmitter Release.* A synapse that is frequently active increases the amount of neurotransmitter it stores, and it releases more on each stimulation. The more neurotransmitter released, the greater the effect on the postsynaptic neuron.

- *Facilitation at Synapses.* When a neural circuit is repeatedly activated, the synaptic terminals begin continuously releasing neurotransmitter in small quantities. The neurotransmitter binds to receptors on the postsynaptic membrane, producing a graded depolarization that brings the membrane closer to threshold. The facilitation that results affects all neurons in the circuit.

- *The Formation of Additional Synaptic Connections.* Evidence indicates that when one neuron repeatedly communicates with another, the axon tip branches and forms additional synapses on the postsynaptic neuron. As a result, the presynaptic neuron will have a greater effect on the transmembrane potential of the postsynaptic neuron.

These processes create anatomical changes that facilitate communication along a specific neural circuit. This facilitated communication is thought to be the basis of memory storage. A single circuit that corresponds to a single memory

---

CLINICAL NOTE

Amnesia

Amnesia is the loss of memory as a result of disease or trauma. The type of memory loss depends on the specific regions of the brain affected. For example, damage to the auditory association areas can make it difficult to remember sounds. Amnesia occurs suddenly or progressively, and recovery is complete, partial, or nonexistent, depending on the nature of the problem. In *retrograde amnesia* (retro-, behind), the individual loses memories of past events. Some degree of retrograde amnesia commonly follows a head injury; after a car wreck or a fall, many victims are unable to remember the moments preceding the accident. In *anterograde amnesia* (antero-, ahead), an individual may be unable to store additional memories, but earlier memories remain intact and accessible. The problem appears to involve an inability to generate long-term memories. Some degree of anterograde amnesia is a common sign of senility, a condition discussed further on p. 555. At least two drugs—*diazepam* (*Valium*) and *triazolam* (*Halcion*)—have been known to cause brief periods of anterograde amnesia. Brain injuries can cause more prolonged memory problems. A person with permanent anterograde amnesia lives in surroundings that are always new. Magazines can be read, chuckled over, and reread a few minutes later with equal pleasure, as if they had never been seen before. Clinicians must introduce themselves at every meeting, even if they have been treating the patient for years.

In *post-traumatic amnesia (PTA)*, a head injury produces a combination of retrograde and anterograde amnesias; the individual can neither remember the past nor consolidate memories of the present. The extent and duration of the amnesia varies with the severity of the injury.

has been called a **memory engram**. This definition is based on function rather than structure; we know too little about the organization and storage of memories to be able to describe the neural circuits involved. Memory engrams form as the result of experience and repetition. Repetition is crucial—that's why you probably need to read these chapters more than once before an exam.

Efficient conversion of a short-term memory into a memory engram takes time, usually at least an hour. Whether that conversion will occur depends on several factors, including the nature, intensity, and frequency of the original stimulus. Very strong, repeated, or exceedingly pleasant or unpleasant events are most likely to be converted to long-term memories. Drugs that stimulate the CNS, such as caffeine and nicotine, may enhance memory consolidation through facilitation; we discussed the membrane effects of those drugs in Chapter 12. ∞ pp. 423–424

The hippocampus plays a key role in the consolidation of memories. The mechanism, which remains unknown, is linked to the presence of *NMDA* (N-methyl D-aspartate) *receptors*, which are chemically gated calcium channels. When activated by the neurotransmitter *glutamate*, the gates open and calcium enters the cell. Blocking NMDA receptors in the hippocampus prevents long-term memory formation.

The A&P Top 100

#56 Memory storage involves anatomical as well as physiological changes in neurons. The hippocampus is involved in the conversion of temporary, short-term memories into durable, long-term memories.

States of Consciousness

The difference between a conscious individual and an unconscious one is obvious: A conscious individual is alert and attentive; an unconscious individual is not. However, there are many gradations of both states. Although *conscious* implies an awareness of and attention to external events and stimuli, a healthy conscious person can be nearly asleep, wide awake, or high-strung and jumpy. *Unconscious* can refer to conditions ranging from the deep, unresponsive state induced by anesthesia before major surgery, to deep sleep, to the light, drifting "nod" that occasionally plagues students who are reading anatomy and physiology textbooks.

The degree of wakefulness at any moment is an indication of the level of ongoing CNS activity. When you are asleep, you are unconscious but can still be awakened by normal sensory stimuli. Healthy individuals cycle between the alert, conscious state and sleep each day. When CNS function becomes abnormal or depressed, the state of wakefulness can be affected. An individual in a *coma*, for example, is uncon-

scious and cannot be awakened, even by strong stimuli. As a result, clinicians are quick to note any change in the responsiveness of comatose patients.

Sleep

Two general levels of sleep are recognized, each typified by characteristic patterns of brain wave activity (**Figure 16–14a**):

1. In **deep sleep**, also called *slow wave* or *non-REM (NREM) sleep*, your entire body relaxes, and activity at the cerebral cortex is at a minimum. Heart rate, blood pressure, respiratory rate, and energy utilization decline by up to 30 percent.

2. During **rapid eye movement (REM) sleep**, active dreaming occurs, accompanied by changes in blood pressure and respiratory rate. Although the EEG resembles that of the awake state, you become even less receptive to outside stimuli than in deep sleep, and muscle tone decreases markedly. Intense inhibition of somatic motor neurons probably prevents you from physically producing the responses you envision while dreaming. The neurons controlling the eye muscles escape this inhibitory influence, and your eyes move rapidly as dream events unfold.

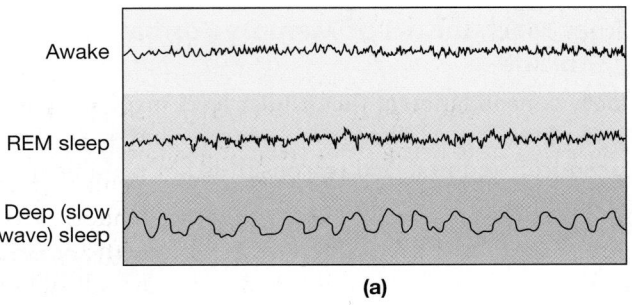

(a)

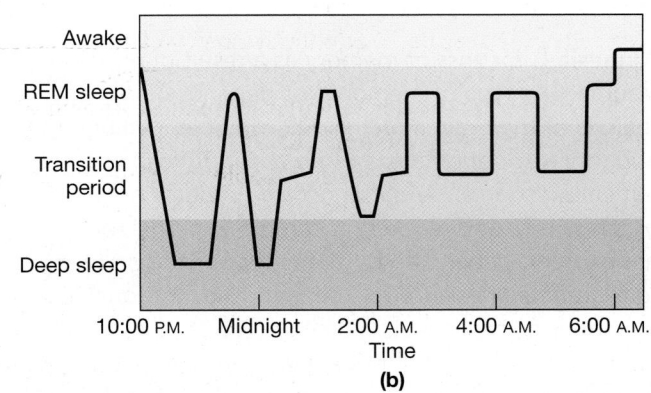

(b)

Figure 16–14 **Levels of Sleep.** **(a)** EEG from the awake, REM, and deep (slow wave) sleep states. The EEG pattern during REM sleep resembles the alpha waves typical of awake adults. **(b)** Typical pattern of sleep stages in a healthy young adult during a single night's sleep.

Periods of REM and deep sleep alternate throughout the night (**Figure 16–14b**), beginning with a period of deep sleep that lasts about an hour and a half. Rapid eye movement periods initially average about 5 minutes in length, but they gradually increase to about 20 minutes over an eight-hour night. Each night we probably spend less than two hours dreaming, but variation among individuals is significant. For example, children devote more time to REM sleep than do adults, and extremely tired individuals have very short and infrequent REM periods.

Sleep produces only minor changes in the physiological activities of other organs and systems, and none of these changes appear to be essential to normal function. The significance of sleep must lie in its impact on the CNS, but the physiological or biochemical basis remains to be determined. We do know that protein synthesis in neurons increases during sleep. Extended periods without sleep will lead to a variety of disturbances in mental function. Roughly 25 percent of the U.S. population experiences some form of *sleep disorder*. Examples of such disorders include abnormal patterns or duration of REM sleep or unusual behaviors performed while sleeping, such as sleepwalking. In some cases, these problems begin to affect the individual's conscious activities. Slowed reaction times, irritability, and behavioral changes may result.

Arousal and the Reticular Activating System

Arousal, or awakening from sleep, appears to be one of the functions of the reticular formation. The reticular formation is especially well suited for providing "watchdog" services, because it has extensive interconnections with the sensory, motor, and integrative nuclei and pathways all along the brain stem.

Your state of consciousness is determined by complex interactions between the reticular formation and the cerebral cortex. One of the most important brain stem components is a diffuse network in the reticular formation known as the **reticular activating system (RAS)**. ∞ p. 475 This network extends from the medulla oblongata to the mesencephalon (**Figure 16–15**). The output of the RAS projects to thalamic nuclei that influence large areas of the cerebral cortex. When the RAS is inactive, so is the cerebral cortex; stimulation of the RAS produces a widespread activation of the cerebral cortex.

The mesencephalic portion of the RAS appears to be the "headquarters" of the system. Stimulation of this area produces the most pronounced and long-lasting effects on the cerebral cortex. Stimulating other portions of the RAS seems to have an effect only to the degree that it changes the activity of the mesencephalic region. The greater the stimulation

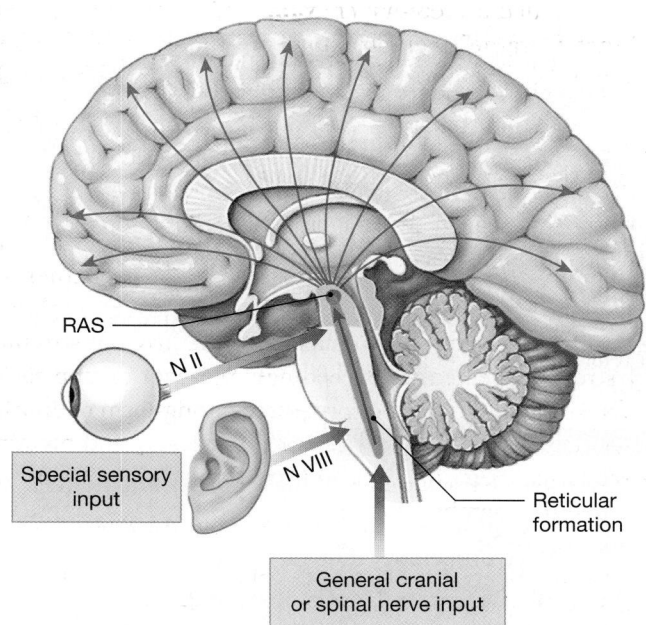

Figure 16–15 The Reticular Activating System. The mesencephalic "headquarters" of the reticular formation receives collateral inputs from a variety of sensory pathways. Stimulation of this region produces arousal and heightened states of attentiveness.

to the mesencephalic region of the RAS, the more alert and attentive the individual will be to incoming sensory information. The thalamic nuclei associated with the RAS may also play an important role in focusing attention on specific mental processes.

Sleep may be ended by any stimulus sufficient to activate the reticular formation and RAS. Arousal occurs rapidly, but the effects of a single stimulation of the RAS last less than a minute. Thereafter, consciousness can be maintained by positive feedback, because activity in the cerebral cortex, basal nuclei, and sensory and motor pathways will continue to stimulate the RAS.

After many hours of activity, the reticular formation becomes less responsive to stimulation. The individual becomes less alert and more lethargic. The precise mechanism remains unknown, but neural fatigue probably plays a relatively minor role in the reduction of RAS activity. Evidence suggests that the regulation of awake–asleep cycles involves an interplay between brain stem nuclei that use different neurotransmitters. One group of nuclei stimulates the RAS with norepinephrine and maintains the awake, alert state. The other group, which depresses RAS activity with serotonin, promotes deep sleep. These "dueling" nuclei are located in the brain stem.

The A&P Top 100

#57 An individual's state of consciousness is variable and complex, ranging from energized and "hyper" to unconscious and comatose. During deep sleep, all metabolic functions are significantly reduced; during REM sleep, muscular activities are inhibited while cerebral activity is similar to that seen in awake individuals. Sleep disorders result in abnormal reaction times, mood swings, and behaviors. Awakening occurs when the reticular activating system becomes active; the greater the level of activity, the more alert the individual.

CHECKPOINT

17. List three characteristics of higher-order functions.

18. As you recall facts while you take your A&P test, which type of memory are you using?

19. Name the two general levels of sleep that have characteristic patterns of brain wave activity.

20. You are asleep. What would happen if your reticular activating system (RAS) were suddenly stimulated?

See the blue Answers tab at the end of the book.

16-9 Neurotransmitters influence brain chemistry and behavior

Changes in the normal balance between two or more neurotransmitters can profoundly affect brain function. For example, the interplay between populations of neurons releasing serotonin and norepinephrine considered in the previous section appears to be involved in the regulation of awake–asleep cycles. Another example concerns *Huntington disease*. The primary problem in this inherited disease is the destruction of ACh-secreting and GABA-secreting neurons in the basal nuclei. The reason for this destruction is unknown. Symptoms appear as the basal nuclei and frontal lobes slowly degenerate. An individual with Huntington disease has difficulty controlling movements, and intellectual abilities gradually decline.

In many cases, the importance of a specific neurotransmitter has been revealed during the search for a mechanism for the effects of administered drugs. Here are two examples:

1. *Lysergic acid diethylamide (LSD)* is a powerful hallucinogenic drug that activates serotonin receptors in the brain stem, hypothalamus, and limbic system. Compounds that merely enhance the effects of serotonin also produce hallucinations, whereas compounds that inhibit serotonin production or block its action cause severe depression and anxiety. An effective antidepressive drug now in widespread use, *fluoxetine (Prozac)*, slows the removal of serotonin at synapses, causing an increase in serotonin concentrations at the postsynaptic membrane. Such drugs are classified as selective serotonin reuptake inhibitors (SSRIs). Other important SSRIs include *Celexa, Luvox, Paxil,* and *Zoloft.* It is now clear that an extensive network of tracts delivers serotonin to nuclei and higher centers throughout the brain, and variations in serotonin levels affect sensory interpretation and emotional states.

2. We have already seen that inadequate dopamine production causes the motor problems of Parkinson disease. ∞ p. 484 Amphetamines, or "speed," stimulate dopamine secretion and, in large doses, can produce symptoms resembling those of *schizophrenia,* a psychological disorder marked by pronounced disturbances of mood, thought patterns, and behavior. Dopamine is thus important not only in the nuclei involved in the control of intentional movements, but in many other centers of the diencephalon and cerebrum.

CHECKPOINT

21. What would be an effect of a drug that substantially increases the amount of serotonin released in the brain?

22. Identify the neurotransmitters thought to be involved with the regulation of awake–asleep cycles.

23. Amphetamines stimulate the secretion of which neurotransmitter?

See the blue Answers tab at the end of the book.

16-10 Aging produces various structural and functional changes in the nervous system

The aging process affects all body systems, and the nervous system is no exception. Anatomical and physiological changes begin shortly after maturity (probably by age 30) and accumulate over time. Although an estimated 85 percent of people above age 65 lead relatively normal lives, they exhibit noticeable changes in mental performance and in CNS function. Common age-related anatomical changes in the nervous system include the following:

- *A Reduction in Brain Size and Weight.* This reduction results primarily from a decrease in the volume of the cerebral cortex. The brains of elderly individuals have narrower gyri and wider sulci than do those of young people, and the subarachnoid space is larger.

- *A Reduction in the Number of Neurons.* Brain shrinkage has been linked to a loss of cortical neurons, although evidence indicates that neuronal loss does not occur (at least to the same degree) in brain stem nuclei.

- *A Decrease in Blood Flow to the Brain.* With age, fatty deposits gradually accumulate in the walls of blood vessels. Just as a clog in a drain reduces water flow, these deposits reduce the rate of blood flow through arteries. (This process, called *arteriosclerosis*, affects arteries throughout the body; we will discuss it further in Chapter 21.) Even if the reduction in blood flow is not sufficient to damage neurons, it increases the chances that the affected vessel wall will rupture, damaging the surrounding neural tissue and producing signs and symptoms of a *cerebrovascular accident (CVA)*, or stroke.

- *Changes in the Synaptic Organization of the Brain.* In many areas, the number of dendritic branches, spines, and interconnections appears to decrease. Synaptic connections are lost, and the rate of neurotransmitter production declines.

- *Intracellular and Extracellular Changes in CNS Neurons.* Many neurons in the brain accumulate abnormal intracellular deposits, including lipofuscin and neurofibrillary tangles. **Lipofuscin** is a granular pigment with no known function. **Neurofibrillary tangles** are masses of neurofibrils that form dense mats inside the cell body and axon. **Plaques** are extracellular accumulations of fibrillar proteins, surrounded by abnormal dendrites and axons. Both plaques and tangles contain deposits of several peptides—primarily two forms of **amyloid β (Aβ)** protein, fibrillar and soluble—and appear in brain regions such as the hippocampus, specifically associated with memory processing. The significance of these histological abnormalities is unknown. Evidence indicates that they appear in all aging brains, but when present in excess, they seem to be associated with clinical abnormalities.

These anatomical changes are linked to functional changes. In general, neural processing becomes less efficient with age. Memory consolidation typically becomes more difficult, and secondary memories, especially those of the recent past, become harder to access. The sensory systems of the elderly—notably, hearing, balance, vision, smell, and taste—become less sensitive. Lights must be brighter, sounds louder, and smells stronger before they are perceived. Reaction rates are slowed, and reflexes—even some withdrawal reflexes—weaken or disappear. The precision of motor control decreases, and it takes longer to perform a given motor pattern than it did 20 years earlier.

For roughly 85 percent of the elderly population, these changes do not interfere with their abilities to function in society. But for as yet unknown reasons, some elderly individuals become incapacitated by progressive CNS changes. These degenerative changes, which can include memory loss, anterograde amnesia, and emotional disturbances, are often lumped together under the general heading of *senile dementia*, or **senility**. By far the most common and incapacitating form of senile dementia is Alzheimer disease.

CLINICAL NOTE

Alzheimer Disease **Alzheimer disease (AD)** is a progressive disorder characterized by the loss of higher-order cerebral functions. It is the most common cause of **senile dementia**, or *senility*. Symptoms may appear at 50–60 years of age or later, although the disease occasionally affects younger individuals. Alzheimer disease has widespread impact. An estimated 2 million people in the United States—including roughly 15 percent of those over age 65, and nearly half of those over age 85—have some form of the condition, and it causes approximately 100,000 deaths each year.

The link is uncertain, but brain areas that on microscopic examination contain abnormal plaques and neurofibrillary tangles are the same regions involved with memory, emotions, and intellectual function. It remains to be determined whether these deposits cause Alzheimer disease or are secondary signs of ongoing metabolic alterations with an environmental, hereditary, or infectious basis.

Genetic factors certainly play a major role. The late-onset form of Alzheimer disease has been traced to a gene on chromosome 19 that codes for proteins involved in cholesterol transport. Fewer than 5 percent of people with Alzheimer disease have the early-onset form; these individuals may develop the condition before age 50. The early-onset form of Alzheimer disease has been linked to genes on chromosomes 1, 14, 19, and 21. Interestingly, the majority of individuals with *Down syndrome* develop Alzheimer disease relatively early in life. (Down syndrome results from an extra copy of chromosome 21; we will discuss this condition further in Chapter 29.) There is no cure for Alzheimer disease, but a few medications and supplements slow its progress in many patients. Diagnosis involves excluding metabolic disorders, chemical/drug interactions, nutritional deficiencies, and anatomical conditions that can mimic dementia; a detailed history and physical; and an evaluation of mental function and mood (depression may mimic dementia).

CHECKPOINT

24. One of the associated effects of aging is difficulty in recalling things, or even total loss of memory. What are some possible reasons for these changes?

25. Identify several common age-related anatomical changes in the nervous system.

26. Name the most common form of senile dementia.

See the blue Answers tab at the end of the book.

16-11 The nervous system is closely integrated with other body systems

Every moment of your life, billions of neurons in your nervous system are exchanging information across trillions of synapses and performing the most complex integrative functions in the body. As part of this process, the nervous system monitors all other systems and issues commands that adjust their activities. However, the significance and impact of these commands varies greatly from one system to another. The normal functions of the muscular system, for example, simply cannot be performed without instructions from the nervous system. By contrast, the cardiovascular system is relatively independent—the nervous system merely coordi-

nates and adjusts cardiovascular activities to meet the circulatory demands of other systems. In the final analysis, the nervous system is like the conductor of an orchestra, directing the rhythm and balancing the performances of each section to produce a symphony, instead of simply a very loud noise.

Figure 16–16 diagrams the relationships between the nervous system and other physiological systems. We will explore many of these relationships in greater detail in subsequent chapters.

CHECKPOINT

27. **Identify the role of the nervous system in the functioning of all other body systems.**

See the blue Answers tab at the end of the book.

CLINICAL NOTE

Categorizing Nervous System Disorders

Neural tissue is extremely delicate, and the characteristics of the extracellular environment must be kept within narrow homeostatic limits. When homeostatic regulatory mechanisms break down under the stress of genetic or environmental factors, infection, or trauma, signs and symptoms of neurological disorders appear.

Literally hundreds of disorders affect the nervous system. These disorders can be roughly categorized into the following groups:

- *Infections*, which include diseases such as rabies, meningitis, and polio

- *Congenital disorders*, such as spina bifida and hydrocephalus

- *Degenerative disorders*, such as Parkinson disease and Alzheimer disease

- *Tumors* of neural origin

- *Trauma*, such as spinal cord injuries and concussions

- *Toxins*, such as heavy metals and the neurotoxins found in certain seafoods

- *Secondary disorders*, which are problems resulting from dysfunction in other systems; examples include strokes and several demyelination disorders

A standard physical examination includes a neurological component, which the physician uses to check the general status of the CNS and PNS. In *neurological examinations*, physicians attempt to trace the location of a specific problem by evaluating the sensory, motor, behavioral, and cognitive functions of the nervous system.

Related Clinical Terms

alpha-blockers: Drugs that eliminate the peripheral vasoconstriction that accompanies sympathetic stimulation.

Alzheimer disease: A progressive CNS disorder marked by the loss of higher-order cerebral functions.

amnesia: A temporary or permanent loss of memory.

anterograde amnesia: Loss of the ability to store new memories.

beta-blockers: Drugs that decrease heart rate and force of contraction, lowering peripheral blood pressure.

Huntington disease: An inherited CNS disease marked by a progressive deterioration of mental abilities and by motor disturbances.

parasympathetic blocking agents: Drugs that target the muscarinic receptors at neuromuscular or neuroglandular junctions.

parasympathomimetic drugs: Drugs that mimic parasympathetic stimulation and increase the activity along the digestive tract.

post-traumatic amnesia (PTA): Loss of both past memories and the ability to consolidate new memories as a result of a head injury.

retrograde amnesia: Loss of memory of past events.

schizophrenia: A psychological disorder marked by pronounced disturbances of mood, thought patterns, and behavior.

senile dementia, or *senility*: A progressive loss of memory, spatial orientation, language, and personality as a consequence of aging.

sympathetic blocking agents: Drugs that bind to receptor sites, preventing a normal response to neurotransmitters or sympathomimetic drugs.

sympathomimetic drugs: Drugs that mimic the effects of sympathetic stimulation.

THE NERVOUS SYSTEM IN PERSPECTIVE

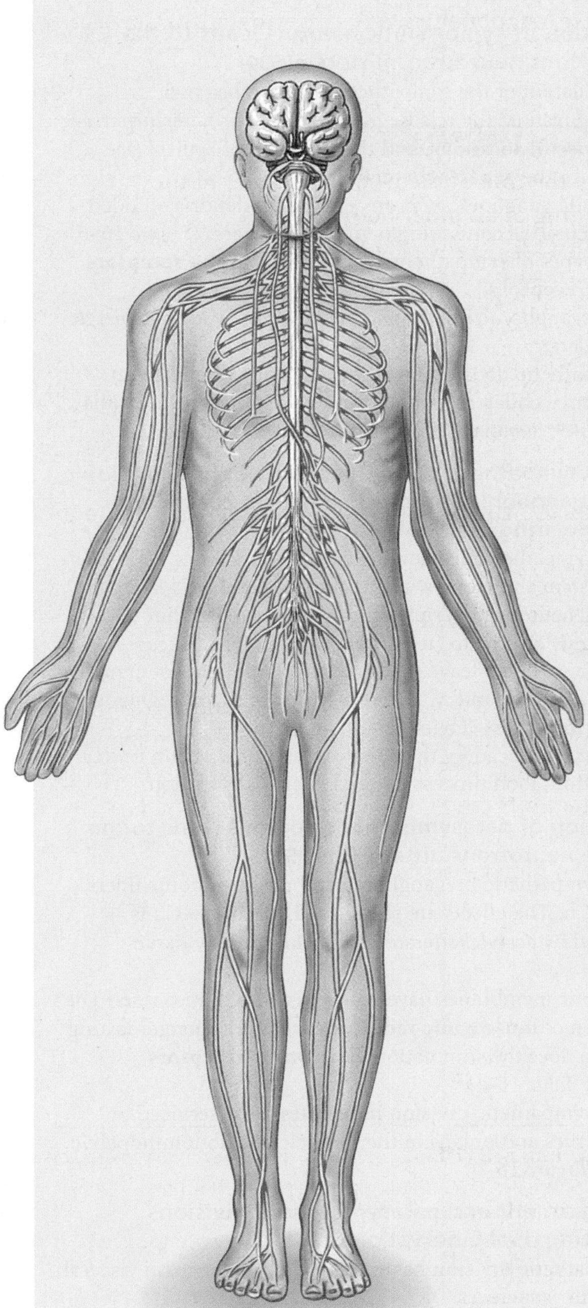

Integumentary System

- Provides sensations of touch, pressure, pain, vibration, and temperature; hair provides some protection and insulation for skull and brain; protects peripheral nerves
- Controls contraction of arrector pili muscles and secretion of sweat glands

Skeletal System

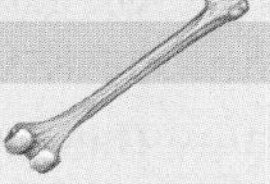

- Provides calcium for neural function; protects brain and spinal cord
- Controls skeletal muscle contractions that produce bone thickening and maintenance and determine bone position

Muscular System

- Facial muscles express emotional state; intrinsic laryngeal muscles permit communication; muscle spindles provide proprioceptive sensations
- Controls skeletal muscle contractions; coordinates respiratory and cardiovascular activities

Endocrine System

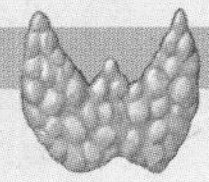

- Many hormones affect CNS neural metabolism; reproductive hormones and thyroid hormone influence CNS development and behavior
- Controls pituitary gland and many other endocrine organs; secretes ADH and oxytocin

Cardiovascular System

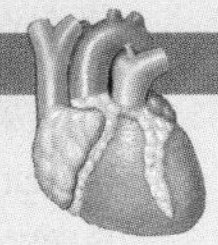

- Endothelial cells of capillaries maintain blood–brain barrier when stimulated by astrocytes; blood vessels (with ependymal cells) produce CSF
- Modifies heart rate and blood pressure; astrocytes stimulate maintenance of blood–brain barrier

Lymphoid System

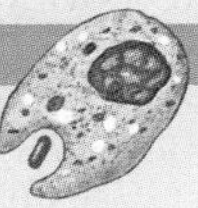

- Defends against infection and assists in tissue repairs
- Release of neurotransmitters and hormones affects sensitivity of immune response

Respiratory System

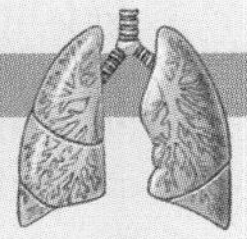

- Provides oxygen and eliminates carbon dioxide
- Controls pace and depth of respiration

Digestive System

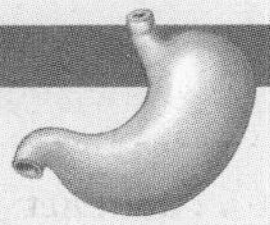

- Provides nutrients for energy production and neurotransmitter synthesis
- Regulates digestive tract movement and secretion

Urinary System

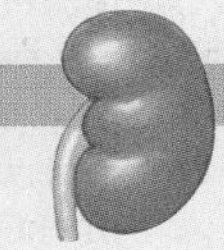

- Eliminates metabolic wastes; regulates body fluid pH and electrolyte concentrations
- Adjusts renal blood pressure; controls urination

Reproductive System

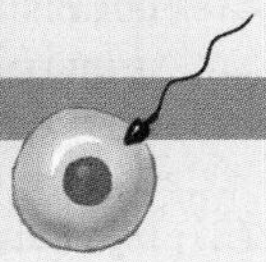

- Sex hormones affect CNS development and sexual behaviors
- Controls sexual behaviors and sexual function

FOR ALL SYSTEMS

Monitors pressure, pain, and temperature; adjusts tissue blood flow patterns

Figure 16–16 **Functional Relationships between the Nervous System and Other Systems.**

Chapter Review

Study Outline

An Introduction to the Autonomic Nervous System and Higher-Order Functions p. 529

1. The autonomic nervous system (ANS) adjusts our basic life support systems without conscious control.

16-1 The autonomic nervous system, composed of the sympathetic and parasympathetic divisions, is involved in the unconscious regulation of visceral functions p. 529

2. The **autonomic nervous system (ANS)** coordinates cardiovascular, respiratory, digestive, urinary, and reproductive functions. (*Figure 16–1*)
3. **Preganglionic neurons** in the CNS send axons to synapse on **ganglionic neurons** in **autonomic ganglia** outside the CNS. (*Figure 16–2*)

16-2 The sympathetic division consists of preganglionic neurons and ganglionic neurons involved in using energy and increasing metabolic rate p. 532

4. Preganglionic fibers from the thoracic and lumbar segments form the **sympathetic division**, or *thoracolumbar division* ("fight or flight" system), of the ANS. Preganglionic fibers leaving the brain and sacral segments form the **parasympathetic division**, or *craniosacral division* ("rest and digest" system).
5. The sympathetic division consists of preganglionic neurons between segments T_1 and L_2, ganglionic neurons in ganglia near the vertebral column, and specialized neurons in the suprarenal glands. (*Figure 16–3*)
6. The two types of sympathetic ganglia are **sympathetic chain ganglia** (*paravertebral ganglia*) and **collateral ganglia** (*prevertebral ganglia*). (*Figure 16–4*)
7. In spinal segments T_1–L_2, ventral roots give rise to the myelinated white ramus which, in turn, leads to the sympathetic chain ganglia. (*Figures 16–4, 16–5*)
8. **Postganglionic fibers** targeting structures in the body wall and limbs rejoin the spinal nerves and reach their destinations by way of the dorsal and ventral rami. (*Figures 16–4, 16–5*)
9. Postganglionic fibers targeting structures in the thoracic cavity form **sympathetic nerves**, which go directly to their visceral destinations. Preganglionic fibers run between the sympathetic chain ganglia and interconnect them. (*Figures 16–4, 16–5*)
10. The abdominopelvic viscera receive sympathetic innervation via preganglionic fibers that synapse within collateral ganglia. The preganglionic fibers that innervate the collateral ganglia form the **splanchnic nerves**. (*Figures 16–4, 16–5*)
11. The **celiac ganglion** innervates the stomach, liver, gallbladder, pancreas, and spleen; the **superior mesenteric ganglion** innervates the small intestine and initial segments of the large intestine; and the **inferior mesenteric ganglion** innervates the terminal portions of the large intestine, the kidneys and urinary bladder, and the sex organs. (*Figures 16–5, 16–10*)
12. Preganglionic fibers entering a suprarenal gland synapse within the **suprarenal medulla**. (*Figures 16–4, 16–5*)
13. In a crisis, the entire sympathetic division responds—an event called **sympathetic activation**. Its effects include increased alertness, a feeling of energy and euphoria, increased

cardiovascular and respiratory activities, a general elevation in muscle tone, and a mobilization of energy reserves.

16-3 Stimulation of sympathetic neurons leads to the release of various neurotransmitters p. 536

14. The stimulation of the sympathetic division has two distinctive results: the release of either ACh or *norepinephrine (NE)* at specific locations, and the secretion of *epinephrine (E)* and NE into the general circulation.
15. Sympathetic ganglionic neurons end in telodendria studded with **varicosities** containing neurotransmitters. (*Figure 16–6*)
16. The two types of sympathetic receptors are **alpha receptors** and **beta receptors**.
17. Most postganglionic fibers are *adrenergic*; a few are *cholinergic* or *nitroxidergic*.
18. The sympathetic division includes two sympathetic chain ganglia, three collateral ganglia, and two suprarenal medullae. (*Figure 16–9; Summary Tables 16–2, 16–3*)

16-4 The parasympathetic division consists of preganglionic neurons and ganglionic neurons involved in conserving energy and lowering metabolic rate p. 538

19. The parasympathetic division includes preganglionic neurons in the brain stem and sacral segments of the spinal cord, and ganglionic neurons in peripheral ganglia located within (**intramural**) or next to (**terminal**) target organs. (*Figure 16–7*)
20. Preganglionic fibers leave the brain as components of cranial nerves III, VII, IX, and X. Those leaving the sacral segments form **pelvic nerves**. (*Figure 16–8*)
21. The effects produced by the parasympathetic division center on relaxation, food processing, and energy absorption.

16-5 Stimulation of parasympathetic neurons leads to the release of the neurotransmitter ACh p. 541

22. All parasympathetic preganglionic and postganglionic fibers release ACh. The effects are short-lived, because ACh is inactivated by *acetylcholinesterase (AChE)* and by *tissue cholinesterase*.
23. Postsynaptic membranes have two types of ACh receptors. The stimulation of **muscarinic receptors** produces a longer-lasting effect than does the stimulation of **nicotinic receptors**. (*Summary Table 16–1*)
24. The parasympathetic division innervates areas serviced by cranial nerves and organs in the thoracic and abdominopelvic cavities. (*Figure 16–9*)

16-6 The sympathetic and parasympathetic divisions interact, creating dual innervation p. 543

25. The sympathetic division has widespread influence on visceral and somatic structures.
26. The parasympathetic division innervates only visceral structures that are serviced by cranial nerves or lying within the abdominopelvic cavity. Organs with **dual innervation** receive input from both divisions. (*Summary Tables 16–2, 16–3*)
27. In body cavities, the parasympathetic and sympathetic nerves intermingle to form a series of characteristic *autonomic plexuses* (nerve networks): the **cardiac**, **pulmonary**,

esophageal, **celiac**, **inferior mesenteric**, and **hypogastric plexuses**. (*Figure 16–10*)

28. Important physiological and functional differences exist between the sympathetic and parasympathetic divisions. (*Figure 16–9; Summary Tables 16–2, 16–3*)

29. Even when stimuli are absent, autonomic motor neurons show a resting level of activation, the **autonomic tone**.

16-7 Visceral reflexes play a role in the integration and control of autonomic functions p. 547

30. **Visceral reflex arcs** perform the simplest function of the ANS, and can be either **long reflexes** (with interneurons) or **short reflexes** (bypassing the CNS). (*Figure 16–11*)

31. Parasympathetic reflexes govern respiration, cardiovascular functions, and other visceral activities. (*Table 16–4*)

32. Levels of activity in the sympathetic and parasympathetic divisions of the ANS are controlled by centers in the brain stem that regulate specific visceral functions.

33. The SNS and ANS are organized in parallel. Integration occurs at the level of the brain stem and higher centers. (*Figure 16–12; Summary Table 16–5*)

16-8 Higher-order functions include memory and states of consciousness p. 550

34. Higher-order functions (1) are performed by the cerebral cortex and involve complex interactions among areas of the cerebral cortex and between the cortex and other areas of the brain, (2) involve conscious and unconscious information processing, and (3) are subject to modification and adjustment over time.

35. Memories can be classified as **short term** or **long term**.

36. The conversion from short-term to long-term memory is **memory consolidation**. (*Figure 16–13*)

37. **Amnesia** is the loss of memory as a result of disease or trauma.

38. In **deep sleep** (*slow wave* or *non-REM sleep*), the body relaxes and cerebral cortex activity is low. In **rapid eye movement (REM) sleep**, active dreaming occurs. (*Figure 16–14*)

39. The **reticular activating system (RAS)**, a network in the reticular formation, is most important to arousal and the maintenance of consciousness. (*Figure 16–15*)

16-9 Neurotransmitters influence brain chemistry and behavior p. 554

40. Changes in the normal balance between two or more neurotransmitters can profoundly affect brain function.

16-10 Aging produces various structural and functional changes in the nervous system p. 554

41. Age-related changes in the nervous system include a reduction in brain size and weight, a reduction in the number of neurons, a decrease in blood flow to the brain, changes in the synaptic organization of the brain, and intracellular and extracellular changes in CNS neurons.

16-11 The nervous system is closely integrated with other body systems p. 556

42. The nervous system monitors all other systems and issues commands that adjust their activities. (*Figure 16–16*)

Review Questions

See the blue Answers tab at the end of the book.

myA&P Access more review material online at **myA&P™** (www.myaandp.com). There, you'll find chapter guides, chapter quizzes, practice tests, animations, flashcards, a glossary with pronunciations, *Interactive Physiology*® (IP) exercises and quizzes, and more to help you succeed in the course.

LEVEL 1 Reviewing Facts and Terms

1. The preganglionic and ganglionic neurons are missing from the diagram at the right of the distribution of sympathetic innervation. Indicate their distribution using red for the preganglionic neurons and black for the ganglionic neurons.

2. The autonomic division of the nervous system directs
   (a) voluntary motor activity.
   (b) conscious control of skeletal muscles.
   (c) unconscious control of skeletal muscles.
   (d) processes that maintain homeostasis.
   (e) sensory input from the skin.

3. The division of the autonomic nervous system that prepares the body for activity and stress is the _____ division.
   (a) sympathetic   (b) parasympathetic
   (c) craniosacral   (d) intramural
   (e) somatomotor

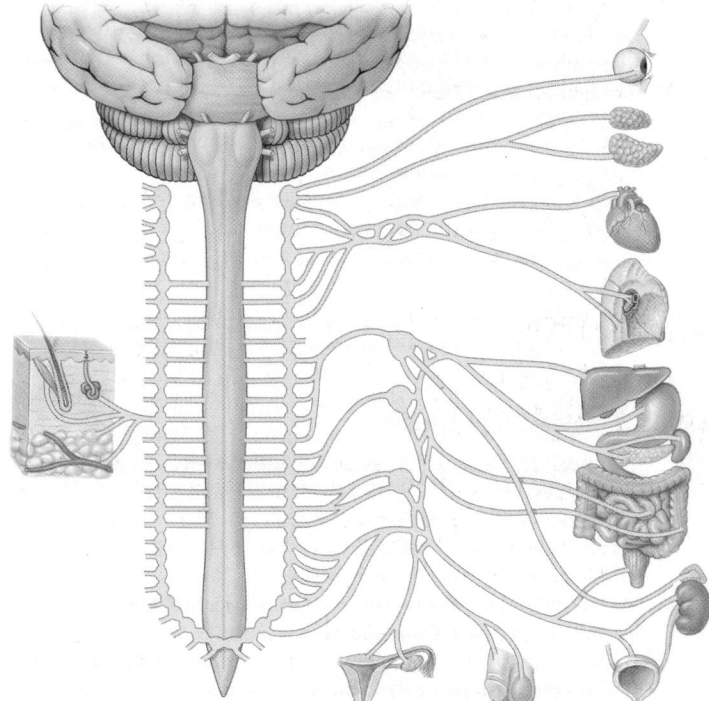

4. Effects produced by the parasympathetic branch of the autonomic nervous system include
   (a) dilation of the pupils.
   (b) increased secretion by digestive glands.
   (c) dilation of respiratory passages.
   (d) increased heart rate.
   (e) increased breakdown of glycogen by the liver.

5. A progressive disorder characterized by the loss of higher-order cerebral functions is
   (a) Parkinson disease. (b) parasomnia.
   (c) Huntington disease. (d) Alzheimer disease.

6. Starting in the spinal cord, trace an impulse through the sympathetic division of the ANS until it reaches a target organ in the abdominopelvic region.

7. Which four ganglia serve as origins for postganglionic fibers involved in control of visceral structures in the head?

8. What are the components of a visceral reflex arc?

9. What cellular mechanisms identified in animal studies are thought to be involved in memory formation and storage?

10. What physiological activities distinguish non-REM sleep from REM sleep?

11. What anatomical and functional changes in the brain are linked to alterations that occur with aging?

12. All preganglionic autonomic fibers release _____ at their synaptic terminals, and the effects are always

    _____.
    (a) norepinephrine; inhibitory
    (b) norepinephrine; excitatory
    (c) acetylcholine; excitatory
    (d) acetylcholine; inhibitory

13. The neurotransmitter at all synapses and neuromuscular or neuroglandular junctions in the parasympathetic division of the ANS is
    (a) epinephrine. (b) norepinephrine.
    (c) cyclic-AMP. (d) acetylcholine.

14. How does the emergence of sympathetic fibers from the spinal cord differ from the emergence of parasympathetic fibers?

15. Which three collateral ganglia serve as origins for ganglionic neurons that innervate organs or tissues in the abdominopelvic region?

16. What two distinctive results are produced by the stimulation of sympathetic ganglionic neurons?

17. Which four pairs of cranial nerves are associated with the cranial segment of the parasympathetic division of the ANS?

18. Which six plexuses in the thoracic and abdominopelvic cavities innervate visceral organs, and what are the effects of sympathetic versus parasympathetic stimulation?

19. What three characteristics are shared by higher-order functions?

LEVEL 2 Reviewing Concepts

20. Dual innervation refers to situations in which
    (a) vital organs receive instructions from both sympathetic and parasympathetic fibers.
    (b) the atria and ventricles of the heart receive autonomic stimulation from the same nerves.
    (c) sympathetic and parasympathetic fibers have similar effects.
    (d) a, b, and c are correct.

21. Damage to the hippocampus, a component of the limbic system, leads to
    (a) a loss of emotion due to forgetfulness.
    (b) a loss of consciousness.
    (c) a loss of long-term memory.
    (d) an immediate loss of short-term memory.

22. Why does sympathetic function remain intact even when the ventral roots of the cervical spinal nerves are damaged?

23. During sympathetic stimulation, a person may begin to feel "on edge"; this is the result of
    (a) increased energy metabolism by muscle tissue.
    (b) increased cardiovascular activity.
    (c) stimulation of the reticular activating system.
    (d) temporary insensitivity to painful stimuli.
    (e) decreased levels of epinephrine in the blood.

24. Under which of the following circumstances would the diameter of peripheral blood vessels be greatest?
    (a) increased sympathetic stimulation
    (b) decreased sympathetic stimulation
    (c) increased parasympathetic stimulation
    (d) decreased parasympathetic stimulation
    (e) both increased parasympathetic and sympathetic stimulation

25. A possible side effect of a drug used to open the airways of someone suffering from an asthma attack is
    (a) decreased activity of the digestive system.
    (b) diarrhea.
    (c) profuse urination.
    (d) increased blood pressure.
    (e) decreased heart rate.

26. You are home alone at night when you hear what sounds like breaking glass. What physiological effects would this experience probably produce, and what would be their cause?

27. Why is autonomic tone a significant part of ANS function?

28. Nicotine stimulates cholinergic receptors of the autonomic nervous system. Based on this information, how would cigarette smoking affect the cardiovascular system?

29. The condition known as shock is characterized in part by a decreased return of venous blood to the heart. How could an upsetting situation, such as the sight of a tragic accident or very bad news, produce some temporary symptoms of shock?

LEVEL 3 Critical Thinking and Clinical Applications

30. Phil is stung on his cheek by a wasp. Because Phil is allergic to wasp venom, his throat begins to swell and his respiratory passages constrict. Would acetylcholine or epinephrine be more helpful in relieving his condition? Why?

31. While studying the activity of smooth muscle in blood vessels, Shelly discovers that, when applied to a muscle plasma membrane, a molecule chemically similar to a neurotransmitter triggers an increase in intracellular calcium ions. Which neurotransmitter is the molecule mimicking, and to which receptors is it binding?

The Special Senses 17

Learning Outcomes

After completing this chapter, you should be able to do the following:

17-1 Describe the sensory organs of smell, trace the olfactory pathways to their destinations in the brain, and explain the physiological basis of olfactory discrimination.

17-2 Describe the sensory organs of taste, trace the gustatory pathways to their destinations in the brain, and explain the physiological basis of gustatory discrimination.

17-3 Identify the internal and accessory structures of the eye, and explain the functions of each.

17-4 Explain color and depth perception, describe how light stimulates the production of nerve impulses, and trace the visual pathways to their destinations in the brain.

17-5 Describe the structures of the external, middle, and inner ear, explain their roles in equilibrium and hearing, and trace the pathways for equilibrium and hearing to their destinations in the brain.

Clinical Notes

Diabetic Retinopathy p. 571
Detached Retina p. 573
Glaucoma p. 574
Accommodation Problems p. 577
Motion Sickness p. 591

An Introduction to the Special Senses

Our knowledge of the world around us is limited to those characteristics that stimulate our sensory receptors. Although we may not realize it, our picture of the environment is incomplete. Colors we cannot distinguish guide insects to flowers; sounds we cannot hear and smells we cannot detect provide dolphins, dogs, and cats with important information about their surroundings.

What we *do* perceive varies considerably with the state of our nervous systems. For example, during sympathetic activation, we experience a heightened awareness of sensory information and hear sounds that would normally escape our notice. Yet, when concentrating on a difficult problem, we may remain unaware of relatively loud noises. Finally, our perception of any stimulus reflects activity in the cerebral cortex, and that activity can be inappropriate. In cases of phantom limb pain, for example, a person feels pain in a missing limb, and during an epileptic seizure, an individual may experience sights, sounds, or smells that have no physical basis.

Our discussion of the general senses and sensory pathways in Chapter 15 introduced basic principles of receptor function and sensory processing. We now turn our attention to the five *special senses:* olfaction, gustation, vision, equilibrium, and hearing. Although the sense organs involved are structurally more complex than those of the general senses, the same basic principles of receptor function apply. ATLAS: Embryology Summary 13: The Development of Special Sense Organs

17-1 Olfaction, the sense of smell, involves olfactory receptors responding to chemical stimuli

The sense of smell, more precisely called *olfaction*, is provided by paired **olfactory organs**. These organs are located in the nasal cavity on either side of the nasal septum (**Figure 17–1a**). The olfactory organs are made up of two layers: the olfactory epithelium and the lamina propria. The **olfactory epithelium** (**Figure 17–1b**) contains the **olfactory receptor cells**, supporting cells, and regenerative **basal cells** (*stem cells*). It covers the inferior surface of the cribriform plate, the superior portion of the perpendicular plate, and the superior nasal conchae of the ethmoid. ∞ p. 222 The underlying lamina propria consists of areolar tissue, numerous blood vessels, and nerves. This layer also contains **olfactory glands**, or *Bowman glands*, whose secretions absorb water and form a thick, pigmented mucus.

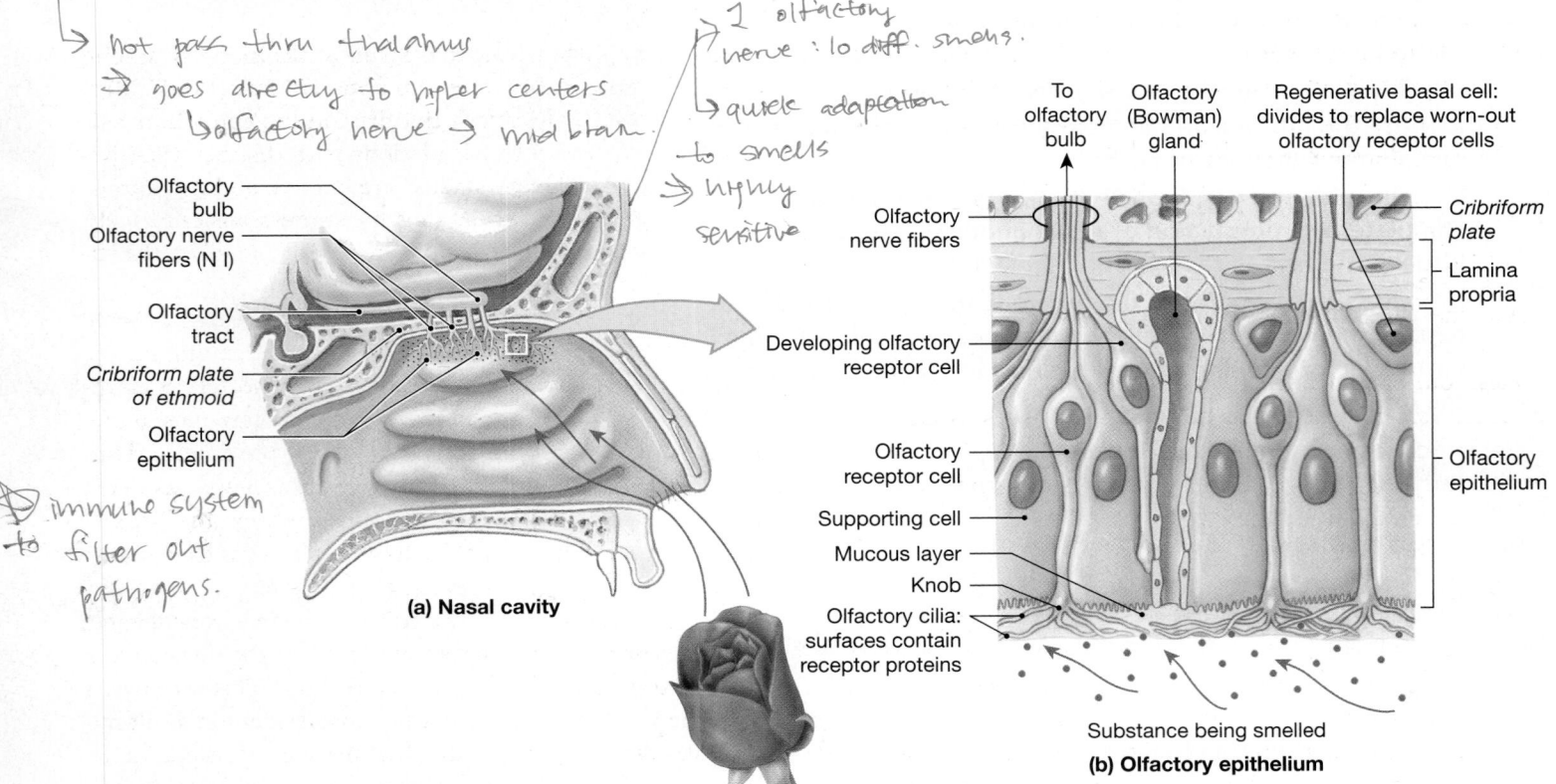

(a) Nasal cavity

(b) Olfactory epithelium

Figure 17–1   The Olfactory Organs. (a) The olfactory organ on the left side of the nasal septum. **(b)** An olfactory receptor is a modified neuron with multiple cilia extending from its free surface.

When you inhale through your nose, the air swirls and eddies within the nasal cavity, and this turbulence brings airborne compounds to your olfactory organs. A normal, relaxed inhalation carries a small sample of the inhaled air (about 2 percent) to the olfactory organs. Sniffing repeatedly increases the flow of air across the olfactory epithelium, intensifying the stimulation of the olfactory receptors. However, those receptors can be stimulated only by water-soluble and lipid-soluble materials that can diffuse into the overlying mucus.

Olfactory Receptors

Olfactory receptors are highly modified neurons. The exposed tip of each receptor cell forms a prominent knob that projects beyond the epithelial surface (**Figure 17–1b**). The knob provides a base for up to 20 cilia that extend into the surrounding mucus. These cilia lie parallel to the epithelial surface, exposing their considerable surface area to dissolved compounds.

Olfactory reception occurs on the surfaces of the olfactory cilia as dissolved chemicals interact with receptors, called *odorant-binding proteins*, on the membrane surface. *Odorants* are chemicals that stimulate olfactory receptors. In general, odorants are small organic molecules; the strongest smells are associated with molecules of high solubility both in water and in lipids. The receptors involved are G proteins; binding of an odorant to its receptor leads to the activation of adenylate cyclase, the enzyme that converts ATP to cyclic-AMP (cAMP). ∞ p. 418 The cAMP then opens sodium channels in the membrane, resulting in a localized depolarization. If sufficient depolarization occurs, an action potential is triggered in the axon, and the information is relayed to the CNS.

Between 10 and 20 million olfactory receptors are packed into an area of roughly 5 cm^2 (0.8 in.^{2}). If we take into account the exposed ciliary surfaces, the actual sensory area probably approaches that of the entire body surface. Nevertheless, our olfactory sensitivities cannot compare with those of other vertebrates such as dogs, cats, or fishes. A German shepherd dog sniffing for smuggled drugs or explosives has an olfactory receptor surface 72 times greater than that of the nearby customs inspector!

Olfactory Pathways

The olfactory system is very sensitive. As few as four odorant molecules can activate an olfactory receptor. However, the activation of an afferent fiber does not guarantee an awareness of the stimulus. Considerable convergence occurs along the olfactory pathway, and inhibition at the intervening synapses can prevent the sensations from reaching the *olfactory cortex* of the cerebral hemispheres. ∞ p. 484 The olfactory recep-

tors themselves adapt very little to a persistent stimulus. Rather, it is central adaptation that ensures that you quickly lose awareness of a new smell but retain sensitivity to others.

Axons leaving the olfactory epithelium collect into 20 or more bundles that penetrate the cribriform plate of the ethmoid to reach the *olfactory bulbs* of the cerebrum (**Figure 17–1**), where the first synapse occurs. Efferent fibers from nuclei elsewhere in the brain also innervate neurons of the olfactory bulbs. This arrangement provides a mechanism for central adaptation or facilitation of olfactory sensitivity. Axons leaving the olfactory bulb travel along the olfactory tract to reach the olfactory cortex, the hypothalamus, and portions of the limbic system.

Olfactory stimulation is the only type of sensory information that reaches the cerebral cortex directly; all other sensations are relayed from processing centers in the thalamus. The parallel distribution of olfactory information to the limbic system and hypothalamus explains the profound emotional and behavioral responses, as well as the memories, that can be triggered by certain smells. The perfume industry, which understands the practical implications of these connections, expends considerable effort to develop odors that trigger sexual responses.

Olfactory Discrimination

The olfactory system can make subtle distinctions among 2000–4000 chemical stimuli. No apparent structural differences exist among the olfactory cells, but the epithelium as a whole contains receptor populations with distinct sensitivities. At least 50 "primary smells" are known, and it is almost impossible to describe these sensory impressions effectively. It appears likely that the CNS interprets each smell on the basis of the overall pattern of receptor activity.

Although the human olfactory organs can discriminate among many smells, acuity varies widely, depending on the nature of the odorant. Many odorants are detected in amazingly small concentrations. One example is beta-mercaptan, an odorant commonly added to natural gas, propane, and butane, which are otherwise odorless. Because we can smell beta-mercaptan in extremely low concentrations (a few parts per billion), its addition enables us to detect a gas leak almost at once and take steps to prevent an explosion.

The olfactory receptor population undergoes considerable turnover; new receptor cells are produced by the division and differentiation of basal cells in the epithelium. This turnover is one of the few examples of neuronal replacement in adult humans. Despite this process, the total number of receptors declines with age, and the remaining receptors become less sensitive. As a result, elderly individuals have difficulty detecting odors in low concentrations. This decline in the number of receptors accounts for Grandmother's tendency to use too much perfume and explains why Grandfather's aftershave seems so strong: They must apply more to be able to smell it.

17-2 Gustation, the sense of taste, involves taste receptors responding to chemical stimuli

Gustation, or taste, provides information about the foods and liquids we consume. **Taste receptors**, or *gustatory* (GUS-ta-tor-ē) *receptors*, are distributed over the superior surface of the tongue and adjacent portions of the pharynx and larynx. The most important taste receptors are on the tongue; by the time we reach adulthood, the taste receptors on the pharynx, larynx, and epiglottis have decreased in importance and abundance. Taste receptors and specialized epithelial cells form sensory structures called **taste buds**. An adult has about 3000 taste buds.

The superior surface of the tongue bears epithelial projections called *lingual papillae* (pa-PIL-ē; *papilla*, a nipple-shaped mound). The human tongue bears three types of lingual papillae (**Figure 17–2**): (1) **filiform** (*filum*, thread) **papillae**, (2) **fungiform** (*fungus*, mushroom) **papillae**, and (3) **circumvallate** (sir-kum-VAL-āt; *circum-*, around + *vallum*, wall) **papillae** . The distribution of these lingual papillae varies by region. Filiform papillae provide friction that helps the tongue move objects around in the mouth, but do not contain taste buds. Each small fungiform papilla contains about five taste buds; each large circumvallate papilla contains as many as 100 taste buds. The circumvallate papillae form a V near the posterior margin of the tongue.

Taste Receptors

Taste buds are recessed into the surrounding epithelium, isolated from the relatively unprocessed contents of the mouth. Each taste bud (**Figure 17–2b,c**) contains about 40 slender, spindle-shaped cells of at least four different types. **Basal** cells appear to be stem cells. These cells divide to produce daughter cells that mature in stages; the cells of the last stage are called **gustatory cells**. Each gustatory cell extends slender microvilli, sometimes called *taste hairs*, into the surrounding fluids through the **taste pore**, a narrow opening.

Despite this relatively protected position, it's still a hard life: A typical gustatory cell survives for only about 10 days before it is replaced. Although everyone agrees that gustatory cells are taste receptors, it is not clear whether the cells at earlier stages of development also provide taste information. (Cells at all three stages are innervated by sensory neurons.)

Gustatory Pathways

Taste buds are monitored by cranial nerves VII (facial), IX (glossopharyngeal), and X (vagus). The facial nerve monitors all the taste buds located on the anterior two-thirds of the tongue, from the tip to the line of circumvallate papillae. The circumvallate papillae and the posterior one-third of the tongue are innervated by the glossopharyngeal nerve. The vagus nerve innervates taste buds scattered on the surface of the epiglottis. The sensory afferents carried by these cranial nerves synapse in the solitary nucleus of the medulla oblongata, and the axons of the postsynaptic neurons enter the medial lemniscus. There, the neurons join axons that carry somatic sensory information on touch, pressure, and proprioception. After another synapse in the thalamus, the information is projected to the appropriate portions of the primary sensory cortex.

A conscious perception of taste is produced as the information received from the taste buds is correlated with other sensory data. Information about the texture of food, along with taste-related sensations such as "peppery" or "burning hot," is provided by sensory afferents in the trigeminal cranial nerve (V). In addition, the level of stimulation from the olfactory receptors plays an overwhelming role in taste perception. Thus, you are several thousand times more sensitive to "tastes" when your olfactory organs are fully functional. By contrast, when you have a cold and your nose is stuffed up, airborne molecules cannot reach your olfactory receptors, so meals taste dull and unappealing. This reduction in taste perception occurs even though the taste buds may be responding normally.

Gustatory Discrimination

You are probably already familiar with the four **primary taste sensations**: sweet, salty, sour, and bitter. There is some evidence for differences in sensitivity to tastes along the axis of the tongue, with greatest sensitivity to salty–sweet anteriorly and sour–bitter posteriorly. However, there are no differences in the structure of the taste buds, and taste buds in all portions of the tongue provide all four primary taste sensations.

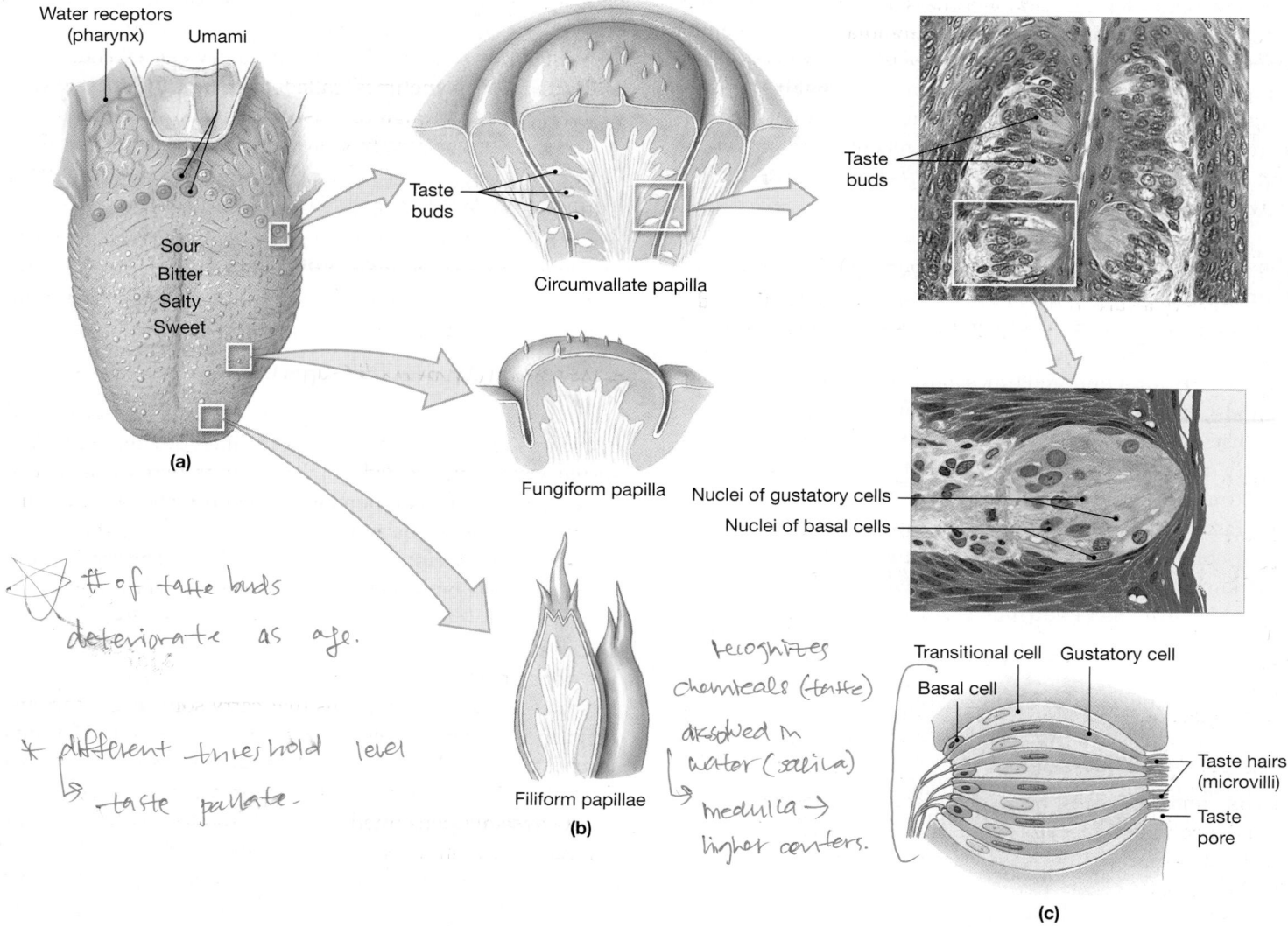

Figure 17–2 **Gustatory Receptors. (a)** Landmarks and receptors on the tongue. **(b)** The structure and representative locations of the three types of lingual papillae. Taste receptors are located in taste buds, which form pockets in the epithelium of fungiform or circumvallate papillae. **(c)** Taste buds in a circumvallate papilla. A diagrammatic view of a taste bud, showing receptor (gustatory) cells and supporting cells.

Humans have two additional taste sensations that are less widely known:

- *Umami.* **Umami** (oo-MAH-mē) is a pleasant taste that is characteristic of beef broth, chicken broth, and parmesan cheese. This taste is detected by receptors sensitive to the presence of amino acids (especially glutamate), small peptides, and nucleotides. The distribution of these receptors is not known in detail, but they are present in taste buds of the circumvallate papillae.

- *Water.* Most people say that water has no flavor. However, research on humans and other vertebrates has demonstrated the presence of **water receptors**, especially in the pharynx. The sensory output of these receptors is processed in the hypothalamus and affects several systems that affect water balance and the regulation of blood volume. For example, minor reductions in ADH secretion occur each time you take a long drink.

The mechanism behind gustatory reception resembles that of olfaction. Dissolved chemicals contacting the taste hairs bind to receptor proteins of the gustatory cell. The different tastes involve different receptor mechanisms. Salt receptors and sour receptors are chemically gated ion channels whose stimulation produces depolarization of the cell. Receptors responding to stimuli that produce sweet, bitter, and umami sensations are G proteins called **gustducins** (GUST-doos-inz)—protein complexes that use second messengers to produce their effects. The end result of taste receptor stimulation is the release of neurotransmitters by the receptor cell. The dendrites of the sensory afferents are tightly wrapped by folds of the receptor plasma membrane, and neurotransmitter release leads to the generation of action potentials in the afferent fiber. Taste receptors adapt slowly, but central adaptation quickly reduces your sensitivity to a new taste.

The threshold for receptor stimulation varies for each of the primary taste sensations, and the taste receptors respond more readily to unpleasant than to pleasant stimuli. For example, we are almost a thousand times more sensitive to acids, which taste sour, than to either sweet or salty chemicals, and we are a hundred times more sensitive to bitter compounds than to acids. This sensitivity has survival value, because acids can damage the mucous membranes of the mouth and pharynx, and many potent biological toxins have an extremely bitter taste.

Taste sensitivity differs significantly among individuals. Many conditions related to taste sensitivity are inherited. The best-known example involves sensitivity to the compound *phenylthiourea*, also known as *phenylthiocarbamide*, or **PTC**. This substance tastes bitter to some people, but is tasteless to others.

Our tasting abilities change with age. We begin life with more than 10,000 taste buds, but the number begins declining dramatically by age 50. The sensory loss becomes especially significant because, as we have already noted, aging individuals also experience a decline in the number of olfactory receptors. As a result, many elderly people find that their food tastes bland and unappetizing, whereas children tend to find the same foods too spicy.

The A&P Top 100

#58 Olfactory information is routed directly to the cerebrum, and olfactory stimuli have powerful effects on mood and behavior. Gustatory sensations are strongest and clearest when integrated with olfactory sensations.

CHECKPOINT

4. Define gustation.

5. If you completely dry the surface of your tongue and then place salt or sugar crystals on it, you can't taste them. Why not?

6. Your grandfather can't understand why foods he used to enjoy just don't taste the same anymore. How would you explain this to him?

See the blue Answers tab at the end of the book.

17-3 Internal eye structures contribute to vision, while accessory eye structures provide protection

We rely more on vision than on any other special sense. Our visual receptors are contained in the eyes, elaborate structures that enable us not only to detect light, but also to create detailed visual images. We will begin our discussion of these fascinating organs by considering the *accessory structures* of the eye, which provide protection, lubrication, and support.

> → *Protection, lubrication, & support.*

Accessory Structures of the Eye

The **accessory structures** of the eye include the eyelids and the superficial epithelium of the eye, and the structures associated with the production, secretion, and removal of tears. **Figure 17–3** shows the superficial anatomy of the eye and its accessory structures.

Eyelids and Superficial Epithelium of the Eye

The eyelids, or **palpebrae** (pal-PĒ-brē), are a continuation of the skin. Their continual blinking keeps the surface of the eye lubricated, and they act like windshield wipers, removing dust and debris. The eyelids can also close firmly to protect the delicate surface of the eye. The **palpebral fissure** is the gap that separates the free margins of the upper and lower eyelids. The two eyelids are connected, however, at the **medial canthus** (KAN-thus) and the **lateral canthus** (**Figure 17–3a**). The **eyelashes**, along the margins of the eyelids, are very robust hairs that help prevent foreign matter (including insects) from reaching the surface of the eye.

The eyelashes are associated with unusually large sebaceous glands. Along the inner margin of the lid, modified sebaceous glands called **tarsal glands**, or *Meibomian* (mī-BŌ-mē-an) *glands*, secrete a lipid-rich product that helps keep the eyelids from sticking together. (These glands are too small to be seen in **Figure 17–3**.) At the medial canthus, the **lacrimal caruncle** (KAR-ung-kul), a mass of soft tissue, contains glands producing the thick secretions that contribute to the gritty deposits that sometimes appear after a good night's sleep. These various glands are subject to occasional invasion and infection by bacteria. A *chalazion* (kah-LĀ-zē-on; small lump), or cyst, generally results from the infection of a tarsal gland. An infection in a sebaceous gland of one of the eyelashes, a tarsal gland, or one of the many sweat glands that open to the surface between the follicles produces a painful localized swelling known as a *sty*.

The skin covering the visible surface of the eyelid is very thin. Deep to the skin lie the muscle fibers of the *orbicularis*

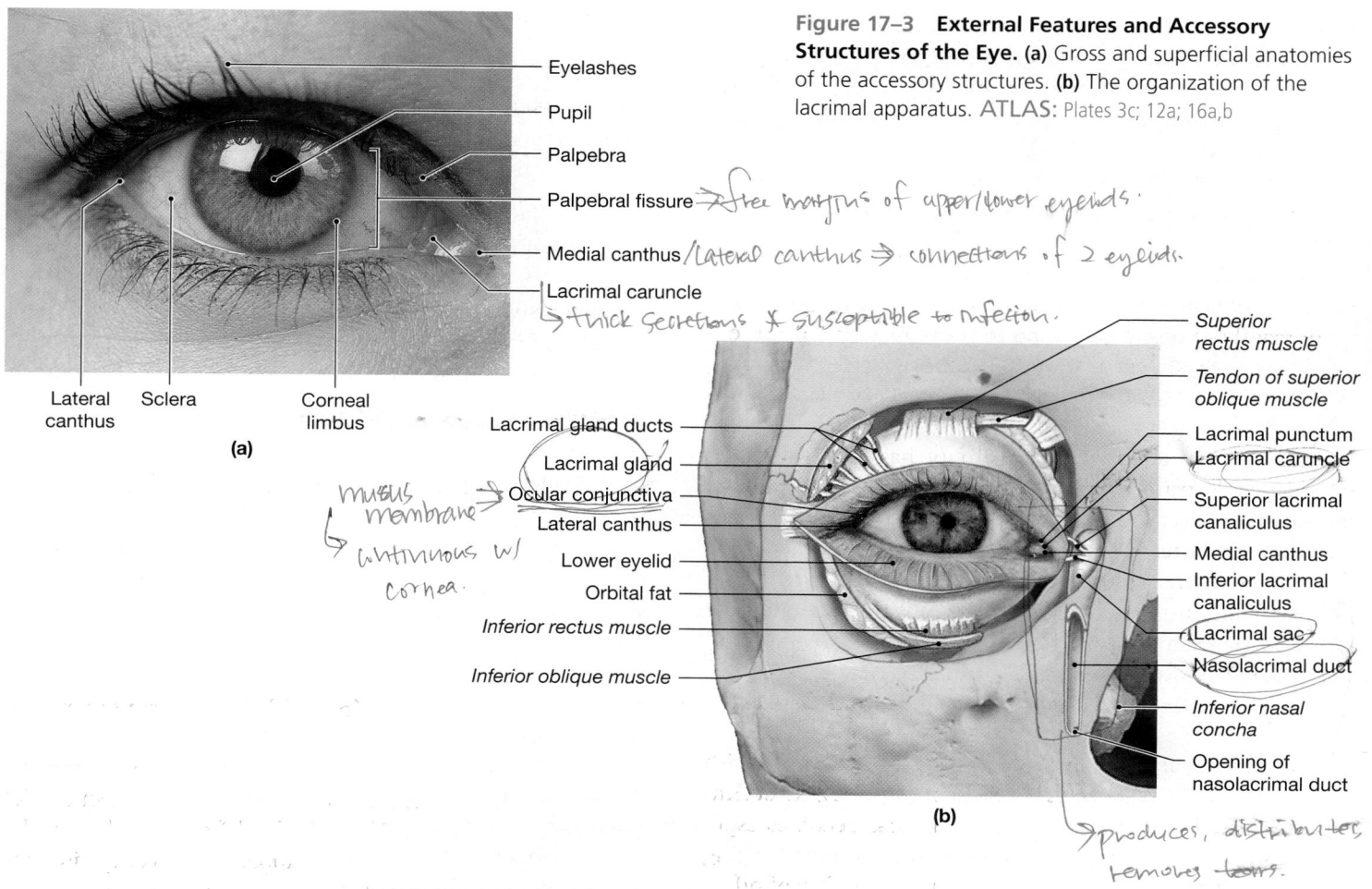

Figure 17–3 **External Features and Accessory Structures of the Eye.** (a) Gross and superficial anatomies of the accessory structures. (b) The organization of the lacrimal apparatus. ATLAS: Plates 3c; 12a; 16a,b

Handwritten annotations:
- Palpebral fissure → free margins of upper/lower eyelids.
- Medial canthus / Lateral canthus ⇒ connections of 2 eyelids.
- Lacrimal caruncle → thick secretions & susceptible to infection.
- mucous membrane → continuous w/ cornea.
- produces, distributes, removes tears.

(a) labels: Eyelashes, Pupil, Palpebra, Palpebral fissure, Medial canthus, Lacrimal caruncle, Lateral canthus, Sclera, Corneal limbus

(b) labels: Lacrimal gland ducts, Lacrimal gland, Ocular conjunctiva, Lateral canthus, Lower eyelid, Orbital fat, Inferior rectus muscle, Inferior oblique muscle, Superior rectus muscle, Tendon of superior oblique muscle, Lacrimal punctum, Lacrimal caruncle, Superior lacrimal canaliculus, Medial canthus, Inferior lacrimal canaliculus, Lacrimal sac, Nasolacrimal duct, Inferior nasal concha, Opening of nasolacrimal duct

---

oculi and *levator palpebrae superioris* muscles. ∞ p. 347 These skeletal muscles are responsible for closing the eyelids and raising the upper eyelid, respectively.

The epithelium covering the inner surfaces of the eyelids and the outer surface of the eye is called the **conjunctiva** (kon-junk-TĪ-vuh). It is a mucous membrane covered by a specialized stratified squamous epithelium. The **palpebral conjunctiva** covers the inner surface of the eyelids, and the **ocular conjunctiva**, or *bulbar conjunctiva*, covers the anterior surface of the eye (**Figure 17–3b**). The ocular conjunctiva extends to the edges of the **cornea** (KOR-nē-uh), a transparent part of the outer fibrous layer of the eye. The cornea is covered by a very delicate squamous *corneal epithelium*, five to seven cells thick, that is continuous with the ocular conjunctiva. A constant supply of fluid washes over the surface of the eyeball, keeping the ocular conjunctiva and cornea moist and clean. Goblet cells in the epithelium assist the accessory glands in providing a superficial lubricant that prevents friction and drying of the opposing conjunctival surfaces.

Conjunctivitis, or pinkeye, results from damage to, and irritation of, the conjunctival surface. The most obvious sign, redness, is due to the dilation of blood vessels deep to the conjunctival epithelium. This condition may be caused by pathogenic infection or by physical, allergic, or chemical irritation of the conjunctival surface.

The Lacrimal Apparatus

A constant flow of tears keeps conjunctival surfaces moist and clean. Tears reduce friction, remove debris, prevent bacterial infection, and provide nutrients and oxygen to portions of the conjunctival epithelium. The **lacrimal apparatus** produces, distributes, and removes tears. The lacrimal apparatus of each eye consists of (1) a *lacrimal gland* with associated ducts, (2) paired *lacrimal canaliculi*, (3) a *lacrimal sac*, and (4) a *nasolacrimal duct* (**Figure 17–3b**).

The pocket created where the palpebral conjunctiva becomes continuous with the ocular conjunctiva is known as the **fornix** of the eye (**Figure 17–4a**). The lateral portion of the superior fornix receives 10–12 ducts from the **lacrimal gland**, or tear gland (**Figure 17–3b**). This gland is about the size and shape of an almond, measuring roughly 12–20 mm (0.5–0.75 in.). It nestles within a depression in the frontal bone, just inside the orbit and superior and lateral to the eyeball. ∞ p. 219 The lacrimal gland normally provides the key

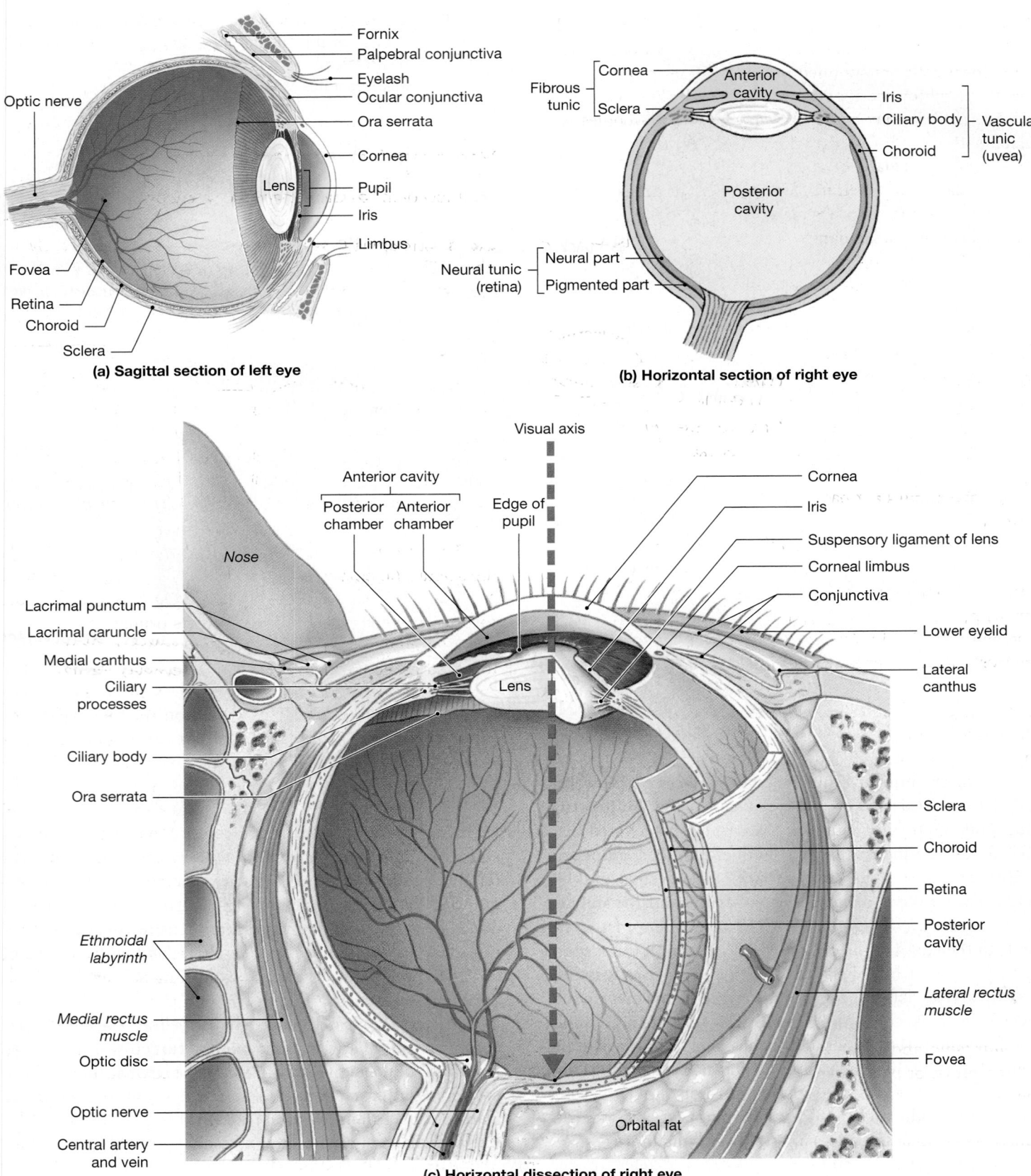

(a) Sagittal section of left eye

Fornix
Palpebral conjunctiva
Eyelash
Ocular conjunctiva
Ora serrata
Cornea
Pupil
Iris
Limbus
Optic nerve
Fovea
Retina
Choroid
Sclera
Lens

(b) Horizontal section of right eye

Cornea
Fibrous tunic
Sclera
Anterior cavity
Iris
Ciliary body
Choroid
Vascular tunic (uvea)
Posterior cavity
Neural tunic (retina)
Neural part
Pigmented part

(c) Horizontal dissection of right eye

Visual axis
Anterior cavity
Posterior chamber
Anterior chamber
Edge of pupil
Nose
Cornea
Iris
Suspensory ligament of lens
Corneal limbus
Conjunctiva
Lower eyelid
Lateral canthus
Lacrimal punctum
Lacrimal caruncle
Medial canthus
Ciliary processes
Ciliary body
Ora serrata
Lens
Ethmoidal labyrinth
Medial rectus muscle
Optic disc
Optic nerve
Central artery and vein
Sclera
Choroid
Retina
Posterior cavity
Lateral rectus muscle
Fovea
Orbital fat

Figure 17–4 **The Sectional Anatomy of the Eye. (a)** A sagittal section through the left eye, showing the position of the fornix. **(b)** Major landmarks in the eye. This horizontal section shows the anterior and posterior cavities and the three layers, or tunics, in the wall of the right eye. **(c)** A detailed horizontal section of the right eye. ATLAS: Plates 12a; 16a,b

ingredients and most of the volume of the tears that bathe the conjunctival surfaces. The nutrient and oxygen demands of the corneal cells are supplied by diffusion from the lacrimal secretions, which are watery and slightly alkaline. They contain the antibacterial enzyme **lysozyme** and antibodies that attack pathogens before they enter the body.

The lacrimal gland produces about 1 mL of tears each day. Once the lacrimal secretions have reached the ocular surface, they mix with the products of accessory glands and the oily secretions of the tarsal glands. The result is a superficial "oil slick" that assists in lubrication and slows evaporation.

Blinking sweeps the tears across the ocular surface, and they accumulate at the medial canthus in an area known as the *lacrimal lake* (*lacus lacrimalis*), or "lake of tears." The lacrimal lake covers the lacrimal caruncle, which bulges anteriorly. The **lacrimal puncta** (singular, *punctum*), two small pores, drain the lacrimal lake. They empty into the **lacrimal canaliculi**, small canals that in turn lead to the **lacrimal sac** (**Figure 17–3b**), which nestles within the lacrimal sulcus of the orbit. ∞ p. 225 From the inferior portion of the lacrimal sac, the **nasolacrimal duct** passes through the *nasolacrimal canal*, formed by the lacrimal bone and the maxillary bone. The nasolacrimal duct delivers tears to the nasal cavity on that side. The duct empties into the *inferior meatus*, a narrow passageway inferior and lateral to the inferior nasal concha. When a person cries, tears rushing into the nasal cavity produce a runny nose, and if the lacrimal puncta can't provide enough drainage, the lacrimal lake overflows and tears stream across the face.

The Eye

The eyes are extremely sophisticated visual instruments—more versatile and adaptable than the most expensive cameras, yet compact and durable. Each eye is a slightly irregular spheroid with an average diameter of 24 mm (almost 1 in., a little smaller than a Ping-Pong ball) and a weight of about 8 g (0.28 oz). Within the orbit, the eyeball shares space with the extrinsic eye muscles, the lacrimal gland, and the cranial nerves and blood vessels that supply the eye and adjacent portions of the orbit and face. **Orbital fat** cushions and insulates the eye (**Figures 17–3b** and **17–4c**).

The wall of the eye contains three distinct layers, or *tunics* (**Figure 17–4b**): (1) an outer *fibrous tunic*, (2) an intermediate *vascular tunic*, and (3) an inner *neural tunic* (retina). The visual receptors, or *photoreceptors*, are located in the neural tunic. The eyeball is hollow; its interior can be divided into two cavities (**Figure 17–4c**). The large **posterior cavity** is also called the *vitreous chamber*, because it contains the gelatinous *vitreous* (vitreo-, glassy) *body*. The smaller **anterior cavity** is subdivided into the *anterior* and *posterior chambers*. The shape of the eye is stabilized in part by the vitreous body and the clear *aqueous humor*, which fills the anterior cavity.

The Fibrous Tunic

The **fibrous tunic**, the outermost layer of the eye, consists of the *sclera* (SKLER-uh) and the *cornea*. The fibrous tunic (1) provides mechanical support and some degree of physical protection, (2) serves as an attachment site for the extrinsic eye muscles, and (3) contains structures that assist in the focusing process.

Most of the ocular surface is covered by the **sclera** (**Figure 17–4b,c**), or "white of the eye," which consists of a dense fibrous connective tissue containing both collagen and elastic fibers. This layer is thickest over the posterior surface of the eye, near the exit of the optic nerve, and thinnest over the anterior surface. The six extrinsic eye muscles insert on the sclera, blending their collagen fibers with those of the fibrous tunic. ∞ p. 348

The surface of the sclera contains small blood vessels and nerves that penetrate the sclera to reach internal structures. The network of small vessels interior to the ocular conjunctiva generally does not carry enough blood to lend an obvious color to the sclera, but on close inspection, the vessels are visible as red lines against the white background of collagen fibers.

The transparent cornea is structurally continuous with the sclera; the border between the two is called the **corneal limbus** (**Figures 17–3a** and **17–4a,c**). Deep to the delicate corneal epithelium, the cornea consists primarily of a dense matrix containing multiple layers of collagen fibers, organized so as not to interfere with the passage of light. The cornea has no blood vessels; the superficial epithelial cells must obtain oxygen and nutrients from the tears that flow across their free surfaces. The cornea also has numerous free nerve endings, and it is the most sensitive portion of the eye.

Corneal damage may cause blindness even though the functional components of the eye—including the photoreceptors—are perfectly normal. The cornea has a very restricted ability to repair itself, so corneal injuries must be treated immediately to prevent serious vision losses. Restoring vision after corneal scarring generally requires the replacement of the cornea through a corneal transplant. Corneal replacement is probably the most common form of transplant surgery. Such transplants can be performed between unrelated individuals, because there are no blood vessels to carry white blood cells, which attack foreign tissues, into the area. Corneal grafts are obtained from the eyes of donors who have died from illness or accident. For best results, the tissues must be removed within five hours after the donor's death.

The Vascular Tunic (Uvea)

The **vascular tunic**, or **uvea** (Ū-vē-uh), contains numerous blood vessels, lymphatic vessels, and the intrinsic (smooth) muscles of the eye (**Figure 17–4b,c**). The functions of this middle layer include (1) providing a route for blood vessels and lymphatics that supply tissues of the eye; (2) regulating the amount of light that enters the eye; (3) secreting and re-absorbing the *aqueous humor* that circulates within the chambers of the eye; and (4) controlling the shape of the *lens*, an essential part of the focusing process. The vascular tunic includes the *iris*, the *ciliary body*, and the *choroid*.

The Iris. The **iris**, which is visible through the transparent corneal surface, contains blood vessels, pigment cells, and two layers of smooth muscle fibers called *pupillary muscles* (**Figure 17–5**). When these muscles contract, they change the diameter of the **pupil**, or central opening of the iris. One group of smooth muscle fibers, the **pupillary constrictor muscles**, forms a series of concentric circles around the pupil. When these sphincter muscles contract, the diameter of the pupil decreases. A second group of smooth muscles, the **pupillary dilator muscles**, extends radially away from the edge of the pupil. Contraction of these muscles enlarges the pupil. Both muscle groups are controlled by the autonomic nervous system. For example, parasympathetic activation in response to bright light causes the pupils to constrict (the *consensual light reflex*), and sympathetic activation in response to dim light causes the pupils to dilate.

The body of the iris consists of a highly vascular, pigmented, loose connective tissue. The anterior surface has no epithelial covering; instead, it has an incomplete layer of fibrocytes and melanocytes. Melanocytes are also scattered within the body of the iris. The posterior surface is covered by a pigmented epithelium that is part of the neural tunic and contains melanin granules. Eye color is determined by genes that influence the density and distribution of melanocytes on the anterior surface and interior of the iris, as well as by the density of the pigmented epithelium. When the connective tissue of the iris contains few melanocytes, light passes through it and bounces off the pigmented epithelium. The eye then appears blue. Individuals with green, brown, or black eyes have increasing numbers of melanocytes in the body and surface of the iris. The eyes of human albinos appear a very pale gray or blue-gray.

The Ciliary Body. At its periphery, the iris attaches to the anterior portion of the **ciliary body**, a thickened region that begins deep to the junction between the cornea and the sclera. The ciliary body extends posteriorly to the level of the **ora serrata** (Ō-ra ser-RA-tuh; serrated mouth), the serrated anterior edge of the thick, inner portion of the neural tunic (**Figure 17–4a,c**). The bulk of the ciliary body consists of the **ciliary muscle**, a smooth muscular ring that projects into the interior of the eye. The epithelium covering this muscle is thrown into numerous folds called **ciliary processes**. The **suspensory ligaments** of the lens attach to the tips of these processes. The connective-tissue fibers of these ligaments hold the lens posterior to the iris and centered on the pupil. As a result, any light passing through the pupil will also pass through the lens.

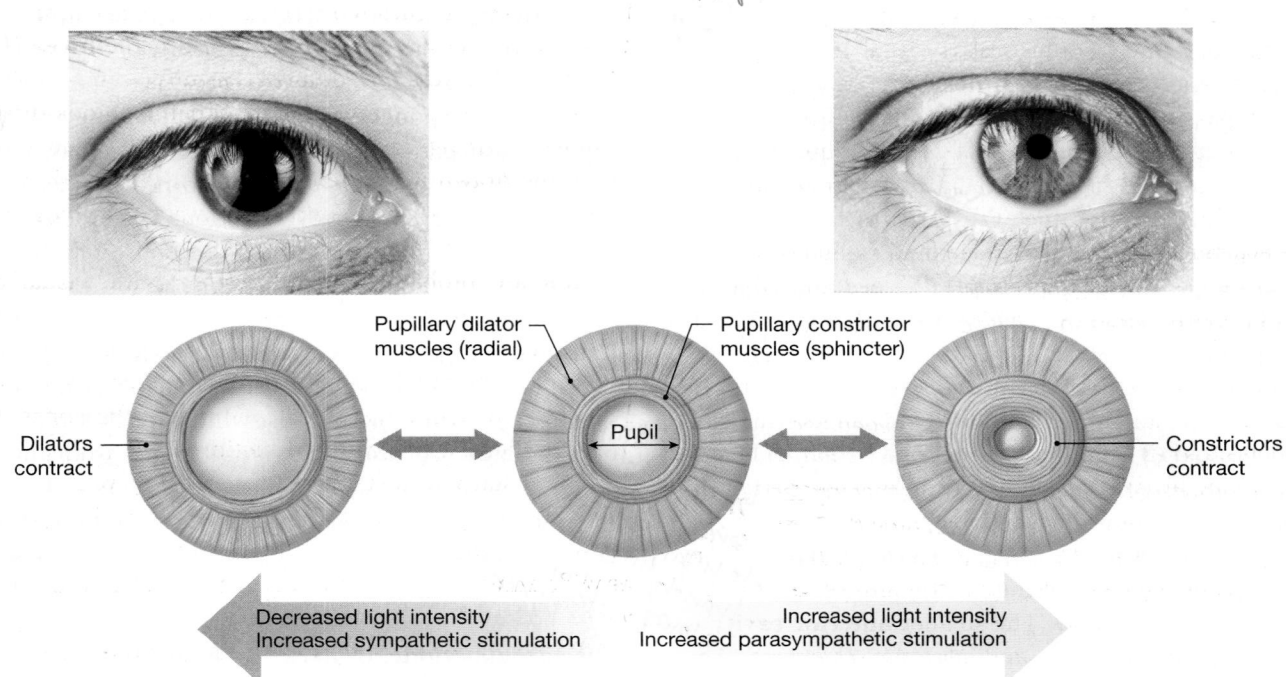

Figure 17–5 **The Pupillary Muscles.**

Pupillary dilator muscles (radial)

Pupillary constrictor muscles (sphincter)

Dilators contract

Pupil

Constrictors contract

Decreased light intensity
Increased sympathetic stimulation

Increased light intensity
Increased parasympathetic stimulation

The Choroid. The **choroid** is a vascular layer that separates the fibrous and neural tunics posterior to the ora serrata (**Figure 17–4c**). Covered by the sclera and attached to the outermost layer of the retina, the choroid contains an extensive capillary network that delivers oxygen and nutrients to the retina. The choroid also contains melanocytes, which are especially numerous near the sclera.

The Neural Tunic (Retina)

The **neural tunic**, or **retina**, is the innermost layer of the eye. It consists of a thin, outer layer called the *pigmented part*, and a thick inner layer called the *neural part*. The pigmented part of the retina absorbs light that passes through the neural part, preventing light from bouncing back through the neural part and producing visual "echoes." The pigment cells also have important biochemical interactions with the retina's light receptors, which are located in the neural part of the retina. In addition to light receptors, the neural part of the retina contains supporting cells and neurons that perform preliminary processing and integration of visual information.

The two layers of the retina are normally very close together, but not tightly interconnected. The pigmented part of the retina continues over the ciliary body and iris; the neural part extends anteriorly only as far as the ora serrata. The neural part of the retina thus forms a cup that establishes the posterior and lateral boundaries of the posterior cavity (**Figure 17–4b,c**).

CLINICAL NOTE

Diabetic Retinopathy A *retinopathy* is a disease of the retina. **Diabetic retinopathy** develops in many individuals with *diabetes mellitus*, an endocrine disorder that interferes primarily with glucose metabolism. Many systems are affected by diabetes, but serious cardiovascular problems are particularly common. Diabetic retinopathy, which develops over a period of years, results from the blockage of small retinal blood vessels followed by excessive growth of abnormal blood vessels that invade the retina and extend into the space between the pigment layer and the neural layer. Visual acuity is gradually lost through damage to photoreceptors (which are deprived of oxygen and nutrients), leakage of blood into the posterior chamber, and the overgrowth of blood vessels. Laser therapy can seal leaking vessels and block new vessel growth. The posterior chamber can be drained and the cloudy fluid replaced by a suitably clear substitute. This procedure is called a *vitrectomy*. These are only temporary, imperfect fixes, and diabetic retinopathy is a leading cause of blindness in the United States.

Organization of the Retina. In sectional view, the retina is seen to contain several layers of cells (**Figure 17–6a**). The outermost layer, closest to the pigmented part of the retina, contains the **photoreceptors**, or cells that detect light.

The eye has two types of photoreceptors: rods and cones. **Rods** do not discriminate among colors of light. Highly sensitive to light, they enable us to see in dimly lit rooms, at twilight, and in pale moonlight. **Cones** provide us with color vision. Three types of cones are present, and their stimulation in various combinations provides the perception of different colors. Cones give us sharper, clearer images than rods do, but cones require more intense light. If you sit outside at sunset with your textbook open to a colorful illustration, you can detect the gradual shift in your visual system from cone-based vision (a clear image in full color) to rod-based vision (a relatively grainy image in black and white).

Tips&Tricks

Associate the "r" in **r**od with the "r" in da**r**k, and associate the "c" in **c**ones with the "c" in **c**olor and in a**c**uity.

Rods and cones are not evenly distributed across the outer surface of the retina. Approximately 125 million rods form a broad band around the periphery of the retina; as you move away from the periphery, toward the center of the retina, the density of rods gradually decreases. In contrast, most of the roughly 6 million cones are concentrated in the area where a visual image arrives after it passes through the cornea and lens. This region, which is known as the **macula lutea** (MAK-ū-luh LOO-tē-uh; yellow spot), contains no rods. The very highest concentration of cones occurs at the center of the macula lutea, an area called the **fovea** (FŌ-vē-uh; shallow depression), or *fovea centralis* (**Figure 17–6c**). The fovea is the site of sharpest vision: When you look directly at an object, its image falls on this portion of the retina. An imaginary line drawn from the center of that object through the center of the lens to the fovea establishes the **visual axis** of the eye (**Figure 17–4c**).

You are probably already aware of the visual consequences of this distribution of photoreceptors. When you look directly at an object, you are placing its image on the fovea, the center of color vision. You see a very good image as long as there is enough light to stimulate the cones. But in very dim light, cones cannot function. That is why you can't see a dim star if you stare directly at it, but you can see it if you shift your gaze to one side or the other. Shifting your gaze moves the image of the star from the fovea, where it does not provide enough light to stimulate the cones, to the periphery, where it can affect the more sensitive rods.

Rods and cones synapse with roughly 6 million neurons called **bipolar cells** (**Figure 17–6a**), which in turn synapse within the layer of neurons called **ganglion cells** adjacent to

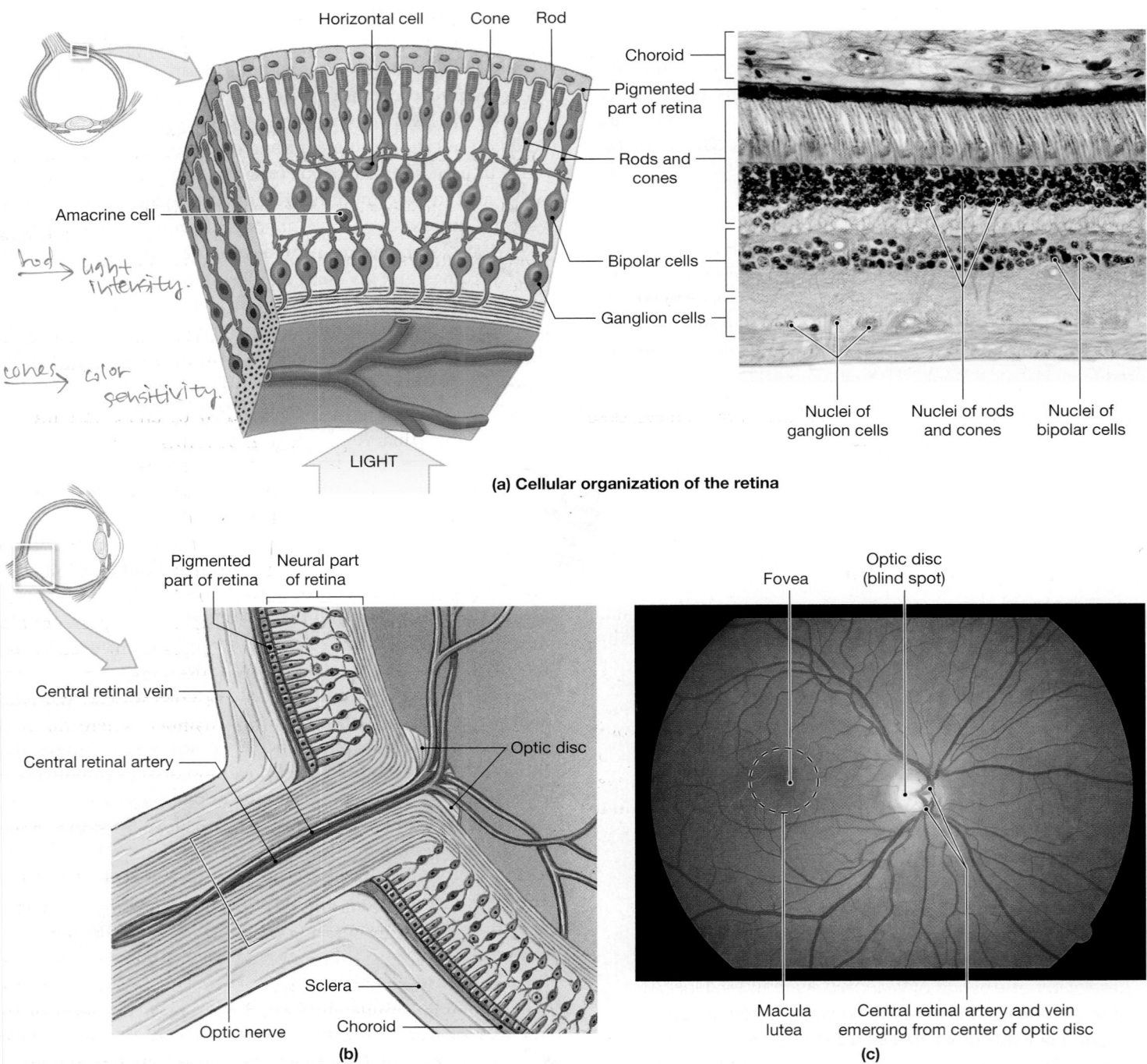

Horizontal cell · Cone · Rod

Choroid

Pigmented part of retina

Rods and cones

Amacrine cell

Bipolar cells

Ganglion cells

rod → light intensity.

cones → color sensitivity.

LIGHT

Nuclei of ganglion cells · Nuclei of rods and cones · Nuclei of bipolar cells

(a) Cellular organization of the retina

Pigmented part of retina · Neural part of retina

Central retinal vein

Central retinal artery

Optic disc

Sclera

Optic nerve · Choroid

(b)

Fovea · Optic disc (blind spot)

Macula lutea · Central retinal artery and vein emerging from center of optic disc

(c)

Figure 17–6 **The Organization of the Retina. (a)** The cellular organization of the retina. The photoreceptors are closest to the choroid, rather than near the posterior cavity (vitreous chamber). **(b)** The optic disc in diagrammatic sagittal section. **(c)** A photograph of the retina as seen through the pupil.

the posterior cavity. A network of **horizontal cells** extends across the outer portion of the retina at the level of the synapses between photoreceptors and bipolar cells. A comparable layer of **amacrine** (AM-a-krin) **cells** occurs where bipolar cells synapse with ganglion cells. Horizontal and amacrine cells can facilitate or inhibit communication between photoreceptors and ganglion cells, thereby altering the sensitivity

of the retina. The effect is comparable to adjusting the contrast on a television set. These cells play an important role in the eye's adjustment to dim or brightly lit environments.

The Optic Disc. Axons from an estimated 1 million ganglion cells converge on the **optic disc**, a circular region just medial to the fovea. The optic disc is the origin of the optic

Figure 17–7   A Demonstration of the Presence of a Blind Spot. Close your left eye and stare at the cross with your right eye, keeping the cross in the center of your field of vision. Begin with the page a few inches away from your eye, and gradually increase the distance. The dot will disappear when its image falls on the blind spot, at your optic disc. To check the blind spot in your left eye, close your right eye and repeat the sequence while you stare at the dot.

nerve (N II). From this point, the axons turn, penetrate the wall of the eye, and proceed toward the diencephalon (**Figure 17–6b**). The *central retinal artery* and *central retinal vein*, which supply the retina, pass through the center of the optic nerve and emerge on the surface of the optic disc (**Figure 17–6b,c**). The optic disc has no photoreceptors or other structures typical of the rest of the retina. Because light striking this area goes unnoticed, the optic disc is commonly called the **blind spot**. You do not notice a blank spot in your field of vision, primarily because involuntary eye movements keep the visual image moving and allow your brain to fill in the missing information. However, a simple activity using **Figure 17–7** will prove that a blind spot really exists in your field of vision.

CLINICAL NOTE

Detached Retina Photoreceptors are entirely dependent on the diffusion of oxygen and nutrients from blood vessels in the choroid. In a **detached retina**, the neural part of the retina becomes separated from the pigmented part. This condition can result from a variety of factors, including a sudden hard impact to the eye. Unless the two parts of the neural tunic are reattached, the photoreceptors will degenerate and vision will be lost. The reattachment is generally performed by "welding" the two layers together using laser beams focused through the cornea. These beams heat the layers, thereby fusing them together at several points around the retina. However, the procedure destroys the photoreceptors and other cells at the "welds," producing permanent blind spots.

The Chambers of the Eye

As noted earlier, the ciliary body and lens divide the interior of the eye into a large posterior cavity, or vitreous chamber, and a smaller anterior cavity (**Figure 17–4c**). The anterior cavity is subdivided into the **anterior chamber**, which extends from the cornea to the iris, and a **posterior chamber**, between the iris and the ciliary body and lens. The anterior and posterior chambers are filled with the fluid *aqueous humor*. The posterior cavity also contains aqueous humor, but most of its volume is taken up by a gelatinous substance known as the *vitreous body*, or *vitreous humor*.

Aqueous Humor. **Aqueous humor** is a fluid that circulates within the anterior cavity, passing from the posterior to the anterior chamber through the pupil (**Figure 17–8**). It also freely diffuses through the vitreous body and across the surface of the retina. Aqueous humor forms through active secretion by epithelial cells of the ciliary body's ciliary processes. The epithelial cells regulate its composition, which resembles that of cerebrospinal fluid. Because aqueous humor circulates, it provides an important route for nutrient and waste transport, in addition to forming a fluid cushion.

The eye is filled with fluid, and fluid pressure in the aqueous humor helps retain the eye's shape. Fluid pressure also stabilizes the position of the retina, pressing the neural part against the pigmented part. In effect, the aqueous humor acts like the air inside a balloon. The eye's **intraocular pressure** can be measured in the anterior chamber, where the fluid pushes against the inner surface of the cornea. Intraocular pressure is most often checked by applanation tonometry, in which a small, flat disk is placed on the anesthetized cornea to measure the tension. Normal intraocular pressure ranges from 12 to 21 mm Hg.

Aqueous humor is secreted into the posterior chamber at a rate of 1–2 μL per minute. It leaves the anterior chamber at the same rate. After filtering through a network of connective tissues located near the base of the iris, aqueous humor enters the **canal of Schlemm**, or *scleral venous sinus*, a passageway that extends completely around the eye at the level of the corneal limbus. Collecting channels deliver the aqueous humor from this canal to veins in the sclera. The rate of removal normally keeps pace with the rate of generation at the ciliary processes, and aqueous humor is removed and recycled within a few hours of its formation.

The Vitreous Body. The posterior cavity of the eye contains the **vitreous body**, a gelatinous mass. The vitreous body helps stabilize the shape of the eye, which might otherwise distort as the extraocular muscles change its position within the orbit. Specialized cells embedded in the vitreous body produce the collagen fibers and proteoglycans that account for the gelatinous consistency of this mass. Unlike the aqueous

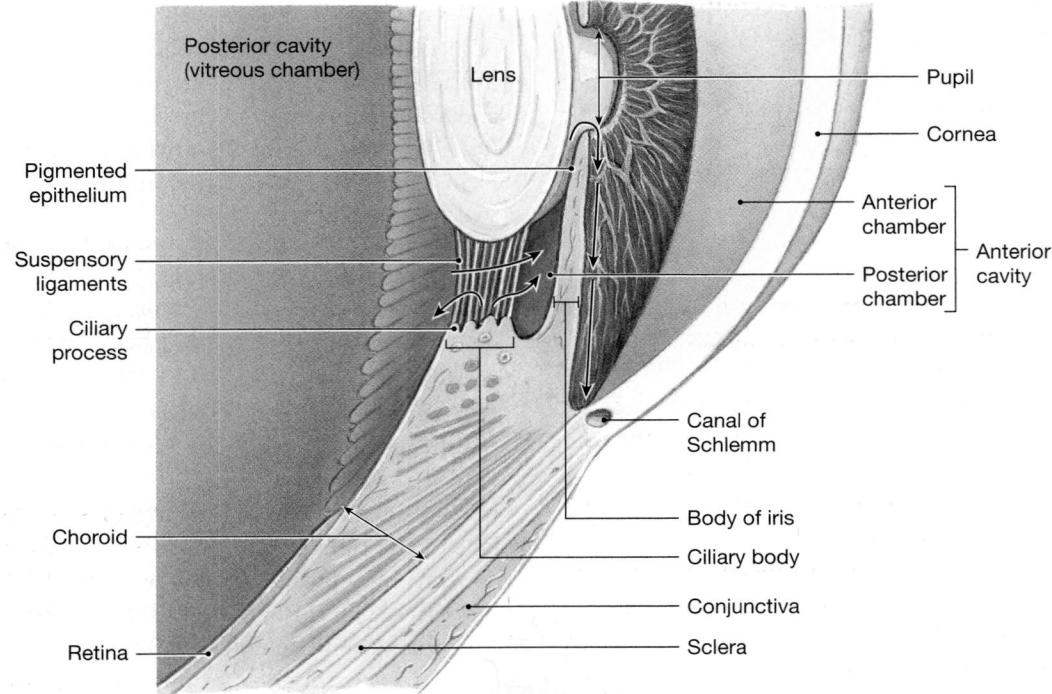

Figure 17–8 **The Circulation of Aqueous Humor.** Aqueous humor, which is secreted at the ciliary body, circulates through the posterior and anterior chambers before it is reabsorbed through the canal of Schlemm.

humor, the vitreous body is formed during development of the eye and is not replaced.

The Lens

The **lens** lies posterior to the cornea, held in place by the suspensory ligaments that originate on the ciliary body of the choroid (**Figures 17–4b** and **17–8**). The primary function of the lens is to focus the visual image on the photoreceptors. The lens does so by changing its shape.

The lens consists of concentric layers of cells that are precisely organized. A dense fibrous capsule covers the entire lens. Many of the capsular fibers are elastic. Unless an outside force is applied, they will contract and make the lens spherical. Around the edges of the lens, the capsular fibers intermingle with those of the suspensory ligaments. The cells in the interior of the lens are called **lens fibers**. These highly specialized cells have lost their nuclei and other organelles. They are slender and elongate and are

---

CLINICAL NOTE

Glaucoma

If aqueous humor cannot drain into the canal of Schlemm, intraocular pressure rises because of the continued production of aqueous humor, and **glaucoma** results. The sclera is a fibrous coat, so it cannot expand like an inflating balloon, but it does have one weak point—the optic disc, where the optic nerve penetrates the wall of the eye. Gradually the increasing pressure pushes the optic nerve outward, damaging its nerve fibers. When intraocular pressures have risen to roughly twice normal levels, the distortion of the optic nerve fibers begins to interfere with the propagation of action potentials, and peripheral vision

begins to deteriorate. If this condition is not corrected, tunnel vision and then complete blindness may result.

Glaucoma affects roughly 2 percent of individuals over age 35. In most cases, the primary factors responsible cannot be determined. Because glaucoma is a relatively common condition, with more than 2 million cases in the United States alone, most eye exams include a test of intraocular pressure. Glaucoma may be treated by the topical application of drugs that constrict the pupil and tense the edge of the iris, making the surface more permeable to aqueous humor. Surgical correction involves perforating the wall of the anterior chamber to encourage drainage. This procedure is now performed by laser surgery on an outpatient basis.

filled with transparent proteins called **crystallins**, which are responsible for both the clarity and the focusing power of the lens. Crystallins are extremely stable proteins that remain intact and functional for a lifetime without the need for replacement.

The transparency of the lens depends on a precise combination of structural and biochemical characteristics. When that balance becomes disturbed, the lens loses its transparency; this abnormality is known as a **cataract**. Cataracts can result from injuries, radiation, or reaction to drugs, but **senile cataracts**, a natural consequence of aging, are the most common form.

Over time, the lens turns yellowish and eventually begins to lose its transparency. As the lens becomes "cloudy," the individual needs brighter and brighter light for reading, and visual clarity begins to fade. If the lens becomes completely opaque, the person will be functionally blind, even though the photoreceptors are normal. Surgical procedures involve removal of the lens, either intact or after it has been shattered with high-frequency sound. The missing lens is replaced by an artificial substitute, and vision is then fine-tuned with glasses or contact lenses.

Refraction. The retina has about 130 million photoreceptors, each monitoring light striking a specific site on the retina. A visual image results from the processing of information from all the receptors. The eye is often compared to a camera. To provide useful information, the lens of the eye, like a camera lens, must focus the arriving image. To say that an image is "in focus" means that the rays of light arriving from an object strike the sensitive surface of the retina (or photographic film) in precise order so as to form a miniature image of the object. If the rays are not perfectly focused, the image is blurry. Focusing normally occurs in two steps, as light passes first through the cornea and then the lens.

Light is **refracted**, or bent, when it passes from one medium to another medium with a different density. You can demonstrate this effect by sticking a pencil into a glass of water. Because refraction occurs as the light passes into the air from the much denser water, the shaft of the pencil appears to bend sharply at the air–water interface.

In the human eye, the greatest amount of refraction occurs when light passes from the air into the corneal tissues, which have a density close to that of water. When you open your eyes under water, you cannot see clearly because refraction at the corneal surface has been largely eliminated; light passes unbent from one watery medium to another.

Additional refraction takes place when the light passes from the aqueous humor into the relatively dense lens. The lens provides the extra refraction needed to focus the light rays from an object toward a **focal point**—a specific point of intersection on the retina. The distance between the center of the lens and its focal point is the **focal distance** of the lens. Whether in the eye or in a camera, the focal distance is determined by two factors:

1. *The Distance from the Object to the Lens.* The closer an object is to the lens, the greater the focal distance (**Figure 17–9a,b**).
2. *The Shape of the Lens.* The rounder the lens, the more refraction occurs, so a very round lens has a shorter focal distance than a flatter one (**Figure 17–9b,c**).

Accommodation A camera focuses an image by moving the lens toward or away from the film. This method of focusing cannot work in our eyes, because the distance from the lens to the macula lutea cannot change. We focus images on the retina by changing the shape of the lens to keep the focal length constant, a process called **accommodation** (**Figure 17–10**). During accommodation, the lens becomes rounder to focus the image of a nearby object on the retina; the lens flattens when we focus on a distant object.

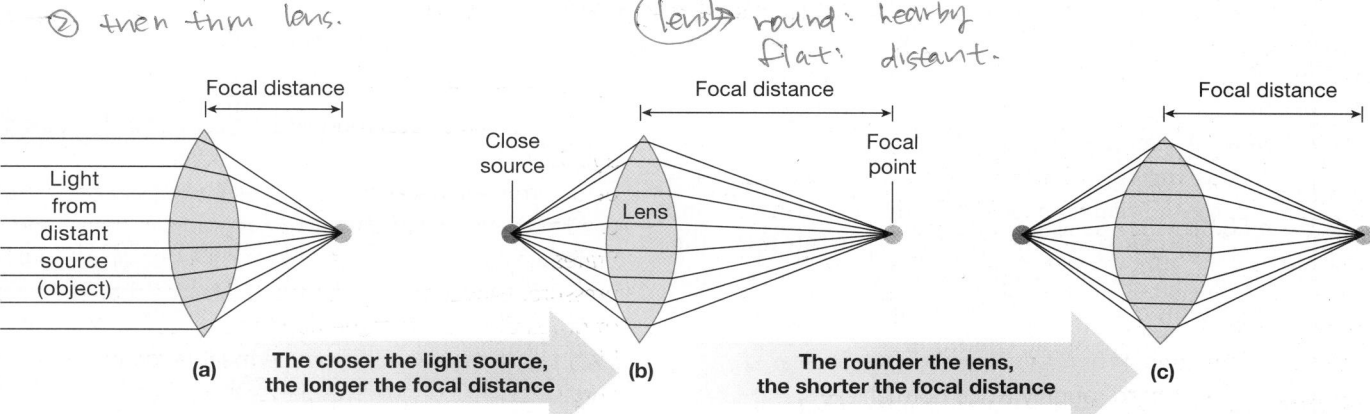

(a) **The closer the light source, the longer the focal distance**

(b) **The rounder the lens, the shorter the focal distance**

(c)

Figure 17–9 Factors Affecting Focal Distance. Light rays from a source are refracted when they reach the lens of the eye. The rays are then focused onto a single focal point.

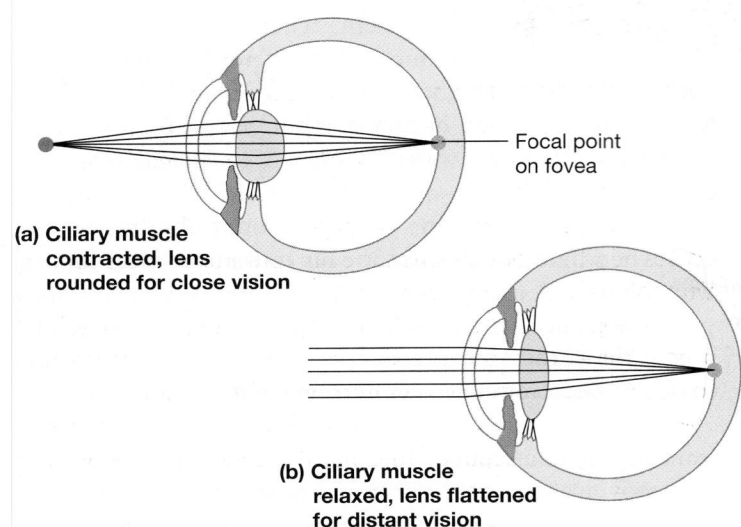

(a) Ciliary muscle contracted, lens rounded for close vision

Focal point on fovea

(b) Ciliary muscle relaxed, lens flattened for distant vision

Figure 17–10 **Accommodation.** For the eye to form a sharp image, the focal distance must equal the distance between the center of the lens and the retina.

The lens is held in place by the suspensory ligaments that originate at the ciliary body. Smooth muscle fibers in the ciliary body act like sphincter muscles. When the ciliary muscle contracts, the ciliary body moves toward the lens, thereby reducing the tension in the suspensory ligaments. The elastic capsule then pulls the lens into a more spherical shape that increases the refractive power of the lens, enabling it to bring light from nearby objects into focus on the retina (**Figure 17–10a**). When the ciliary muscle relaxes, the suspensory ligaments pull at the circumference of the lens, making the lens flatter (**Figure 17–10b**).

The greatest amount of refraction is required to view objects that are very close to the lens. The inner limit of clear vision, known as the *near point of vision*, is determined by the degree of elasticity in the lens. Children can usually focus on something 7–9 cm from the eye, but over time the lens tends to become stiffer and less responsive. A young adult can usually focus on objects 15–20 cm away. As aging proceeds, this distance gradually increases; the near point at age 60 is typically about 83 cm. (For more information on congenital and age-related changes in eye structure and function, see the Clinical Note "Accommodation Problems.")

If light passing through the cornea and lens is not refracted properly, the visual image will be distorted. In the condition called **astigmatism**, the degree of curvature in the cornea or lens varies from one axis to another. Minor astigmatism is very common; the image distortion may be so minimal that people are unaware of the condition.

Image Reversal. Thus far, we have considered light that originates at a single point, either near or far from the viewer. An object in view, however, is a complex light source that must be treated as a large number of individual points. Light from each point is focused on the retina as indicated in **Figure 17–12a,b**.

The result is the creation of a miniature image of the original, but the image arrives upside down and reversed.

To understand why the image is reversed in this fashion, consider **Figure 17–12c**, a sagittal section through an eye that is looking at a telephone pole. The image of the top of the pole lands at the bottom of the retina, and the image of the bottom hits the top of the retina. Now consider **Figure 17–12d**, a horizontal section through an eye that is looking at a picket fence. The image of the left edge of the fence falls on the right side of the retina, and the image of the right edge falls on the left side of the retina. The brain compensates for this image reversal, and we are not aware of any difference between the orientation of the image on the retina and that of the object.

Visual Acuity. Clarity of vision, or **visual acuity**, is rated against a "normal" standard. The standard vision rating of 20/20 is defined as the level of detail seen at a distance of 20 feet by an individual with normal vision. Vision rated as 20/15 is better than average, because at 20 feet the person is able to see details that would be clear to a normal eye only at a distance of 15 feet. Conversely, a person with 20/30 vision must be 20 feet from an object to discern details that a person with normal vision could make out at a distance of 30 feet.

When visual acuity falls below 20/200, even with the help of glasses or contact lenses, the individual is considered to be legally blind. There are probably fewer than 400,000 legally blind people in the United States; more than half are over 65 years old. The term *blindness* implies a total absence of vision due to damage to the eyes or to the optic pathways. Common causes of blindness include diabetes mellitus, cataracts, glaucoma, corneal scarring, detachment of the retina, accidental injuries, and hereditary factors that are as yet poorly understood.

Abnormal blind spots, or **scotomas** (skō-TŌ-muhz), may appear in the field of vision at positions other than at the optic disc. Scotomas are permanent abnormalities that are fixed in position. They may result from a compression of the optic nerve, damage to photoreceptors, or central damage along the visual pathway. *Floaters*, small spots that drift across the field of vision, are generally temporary phenomena that result from blood cells or cellular debris in the vitreous body. They can be detected by staring at a blank wall or a white sheet of paper.

The A&P Top 100

#59 Light passes through the conjunctiva and cornea, crosses the anterior cavity to reach the lens, transits the lens, crosses the posterior chamber, and then penetrates the neural tissue of the retina before reaching and stimulating the photoreceptors. Cones are most abundant at the fovea and macula lutea, and they provide high-resolution color vision in brightly lit environments. Rods dominate the peripheral areas of the retina, and they provide relatively low-resolution black-and-white vision in dimly lit environments.

CLINICAL NOTE

Accommodation Problems

In the healthy eye, when the ciliary muscle is relaxed and the lens is flattened, a distant image will be focused on the retina's surface. This condition is called **emmetropia** (*emmetro-*, proper + *opia*, vision), or normal vision (**Figure 17–11a**).

Figure 17–11b,c diagrams two common problems with the accommodation mechanism. If the eyeball is too deep or the resting curvature of the lens is too great, the image of a distant object is projected in front of the retina (**Figure 17–11b**). The individual will see distant objects as blurry and out of focus. Vision at close range will be normal, because the lens is able to round as needed to focus the image on the retina. Such individuals are said to be *nearsighted*. Their condition is more formally termed **myopia** (*myein*, to shut + *ops*, eye). Myopia can be treated by placing a *diverging* lens in front of the eye (**Figure 17–11d**). Diverging lenses have at least one concave surface and spread the light rays apart as if the object were closer to the viewer.

If the eyeball is too shallow or the lens is too flat, **hyperopia** results (**Figure 17–11c**). The ciliary muscle must contract to focus even a distant object on the retina, and at close range the lens cannot provide enough refraction to focus an image on the retina. Individuals with this problem are said to be *farsighted*, because they can see distant objects most clearly. Older individuals become farsighted as their lenses lose

elasticity; this form of hyperopia is called **presbyopia** (*presbys*, old man). Hyperopia can be corrected by placing a *converging* lens in front of the eye. Converging lenses have at least one convex surface and provide the additional refraction needed to bring nearby objects into focus (**Figure 17–11e**). The "reading glasses" sold without prescription have converging lenses sufficient to correct mild presbyopia.

Variable success at correcting myopia and hyperopia has been achieved by a form of refractive surgery that reshapes the cornea. In **photorefractive keratectomy (PRK)** a computer-guided laser shapes the cornea to exact specifications. Tissue is removed only to a depth of 10–20 μm—no more than about 10 percent of the cornea's thickness. The entire procedure can be done in less than a minute. A variation on PRK is called *LASIK (Laser-Assisted in-Situ Keratomileusis)*. In this procedure the interior layers of the cornea are reshaped and covered by a flap of normal corneal epithelium. Roughly 70 percent of LASIK patients achieve 20/20 vision; it is now the most common form of refractive surgery. Each year, an estimated 100,000 people undergo PRK therapy in the United States.

Corneal scarring is rare, and an estimated 10 million Americans have had corneal refractive surgery. However, many patients still need reading glasses, and both immediate and long-term visual problems can occur.

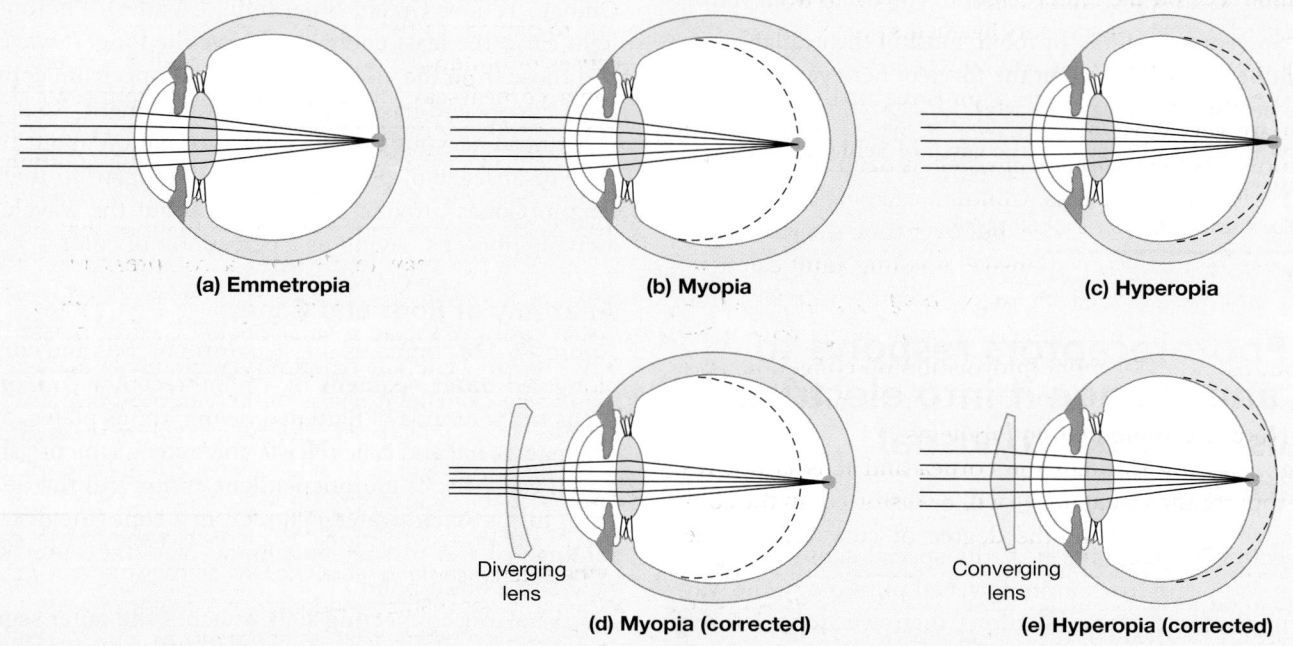

(a) Emmetropia (b) Myopia (c) Hyperopia

Diverging lens (d) Myopia (corrected) Converging lens (e) Hyperopia (corrected)

Figure 17–11 Visual Abnormalities. (a) In normal vision, or emmetropia, the lens focuses the visual image on the retina. Common problems with the accommodation mechanism involve **(b)** myopia, an inability to lengthen the focal distance enough to focus the image of a distant object on the retina, and **(c)** hyperopia, an inability to shorten the focal distance adequately for nearby objects. These conditions can be corrected by placing appropriately shaped lenses in front of the eyes. **(d)** A diverging lens is used to correct myopia, and **(e)** a converging lens is used to correct hyperopia.

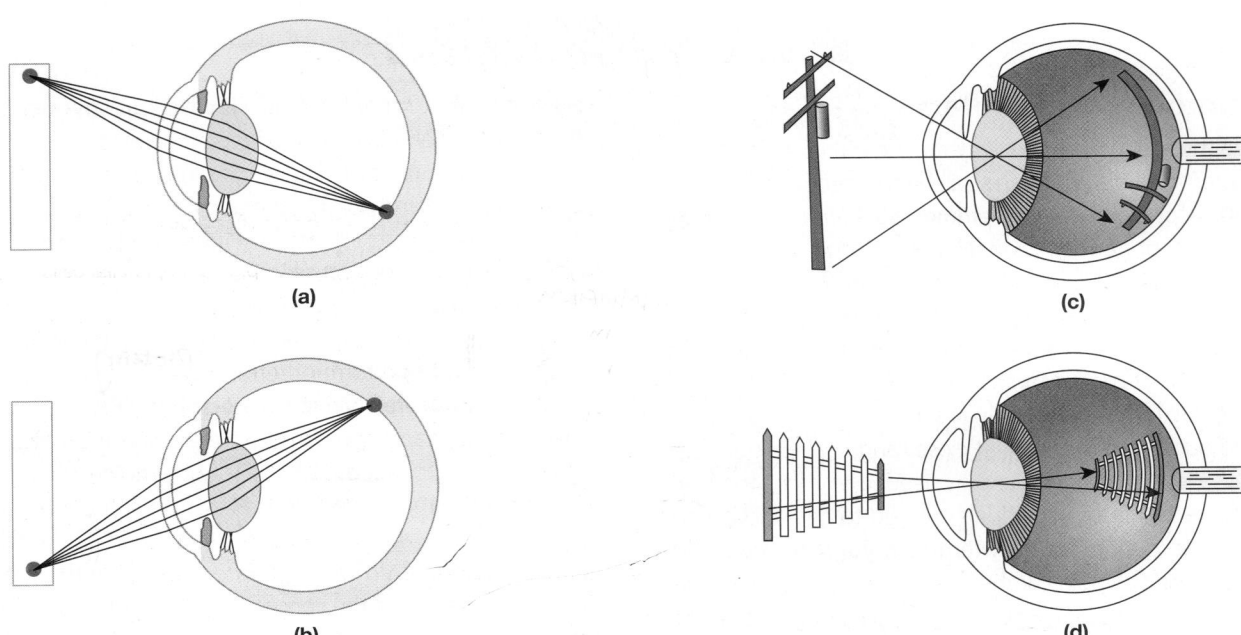

Figure 17–12   Image Formation. (a, b) Light from each portion of an object is focused on a different part of the retina. The resulting image arrives **(c)** upside down and **(d)** reversed.

17-4 Photoreceptors respond to light and change it into electrical signals essential to visual physiology

In this section we examine how the special sense of vision functions. We begin by examining visual physiology, the way in which photoreceptors function; then we consider the structure and function of the visual pathways.

Visual Physiology

The rods and cones of the retina are called *photoreceptors* because they detect *photons*, basic units of visible light. Light energy is a form of *radiant energy* that travels in waves with a characteristic *wavelength* (distance between wave peaks).

Our eyes are sensitive to wavelengths of 700–400 nm, the spectrum of visible light. This spectrum, seen in a rainbow, can be remembered by the acronym ROY G. BIV (*Red, Orange, Yellow, Green, Blue, Indigo, Violet*). Photons of red light carry the least energy and have the longest wavelength, and those from the violet portion of the spectrum contain the most energy and have the shortest wavelength. Rods provide the central nervous system with information about the presence or absence of photons, with little regard to their wavelength. Cones provide information about the wavelength of arriving photons, giving us a perception of color.

Anatomy of Rods and Cones

Figure 17–13a compares the structures of rods and cones. The elongated **outer segment** of a photoreceptor contains hundreds to thousands of flattened membranous plates, or **discs**. The names *rod* and *cone* refer to the outer segment's shape. In a rod, each disc is an independent entity, and the outer segment forms an elongated cylinder. In a cone, the discs are infoldings of the plasma membrane, and the outer segment tapers to a blunt point.

A narrow connecting stalk attaches the outer segment to the **inner segment**, a region that contains all the usual cellular organelles. The inner segment makes synaptic contact with other cells, and it is here that neurotransmitters are released.

Visual Pigments. The discs of the outer segment in both rods and cones contain special organic compounds called

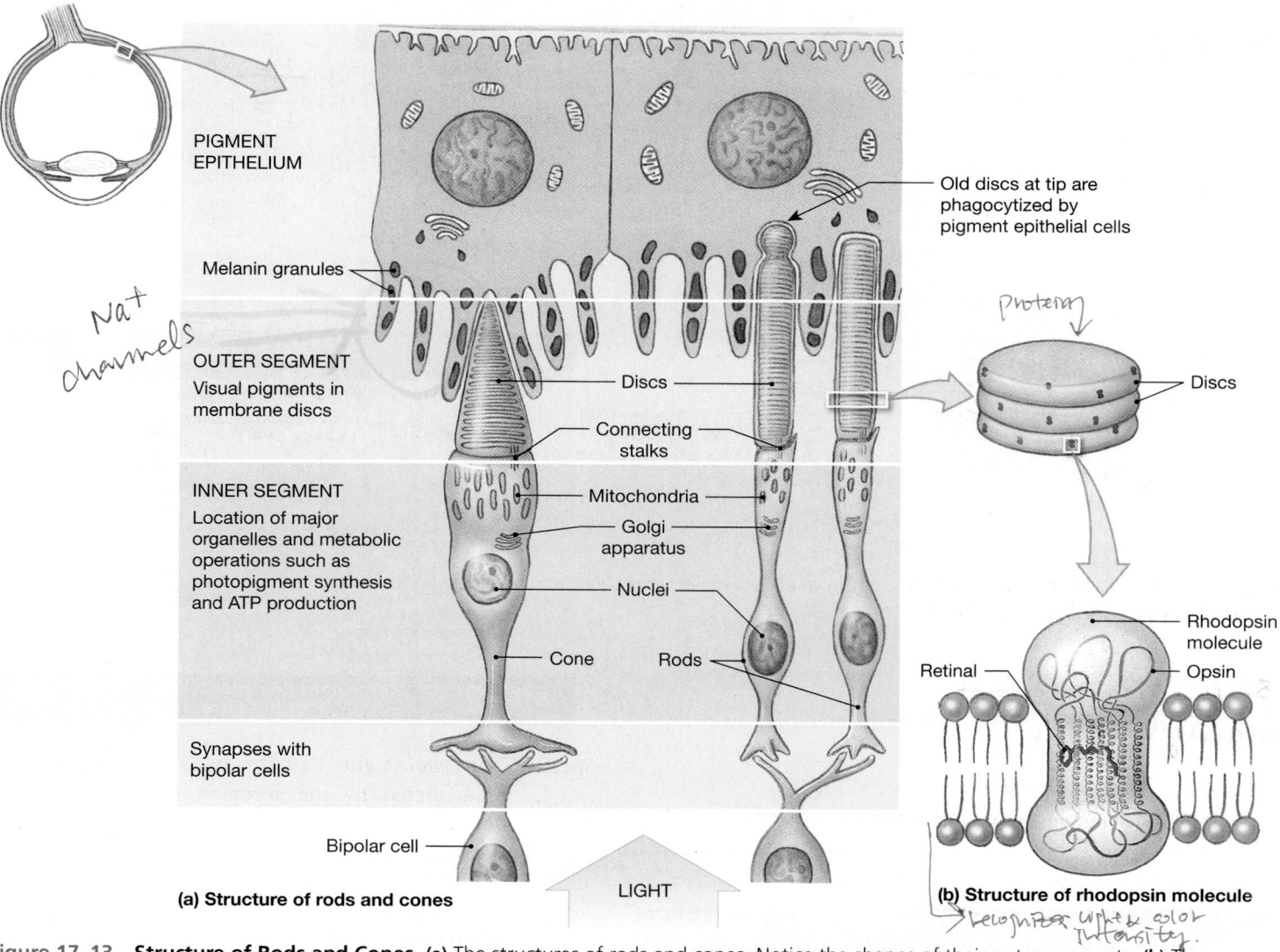

PIGMENT
EPITHELIUM

Old discs at tip are
phagocytized by
pigment epithelial cells

Melanin granules

Nat
Channels

Proteins

OUTER SEGMENT
Visual pigments in
membrane discs

Discs

Discs

Connecting
stalks

INNER SEGMENT
Location of major
organelles and metabolic
operations such as
photopigment synthesis
and ATP production

Mitochondria

Golgi
apparatus

Nuclei

Rhodopsin
molecule

Retinal

Opsin

Cone

Rods

Synapses with
bipolar cells

Bipolar cell

(a) Structure of rods and cones

LIGHT

(b) Structure of rhodopsin molecule

recognizes light & color
Intensity.

Figure 17–13   Structure of Rods and Cones. (a) The structures of rods and cones. Notice the shapes of their outer segments. **(b)** The structure of a rhodopsin molecule within the membrane of a disc.

visual pigments. The absorption of photons by visual pigments is the first key step in the process of *photoreception*—the detection of light. Visual pigments are derivatives of the compound **rhodopsin** (rō-DOP-sin), or *visual purple*, the visual pigment found in rods (**Figure 17–13b**). Rhodopsin consists of a protein, **opsin**, bound to the pigment **retinal** (RET-i-nal), or *retinene*, which is synthesized from **vitamin A**. One form of opsin is characteristic of all rods.

Cones contain the same retinal pigment that rods do but, in cones, retinal is attached to other forms of opsin. The type of opsin present determines the wavelength of light that can be absorbed by retinal. Differential stimulation of these cone populations is the basis of color vision.

New discs containing visual pigment are continuously assembled at the base of the outer segment. A completed disc then moves toward the tip of the segment. After about 10 days, the disc will be shed in a small droplet of cytoplasm. Droplets

with shed discs are absorbed by the pigment cells, which break down the membrane's components and reconvert the retinal to vitamin A. The vitamin A is then stored within the pigment cells for subsequent transfer to the photoreceptors.

The term **retinitis pigmentosa** (RP) refers to a collection of inherited retinopathies. Together, they are the most common inherited visual abnormality, affecting approximately 1 individual in 3000. The visual receptors gradually deteriorate, and blindness eventually results. The mutations that are responsible change the structure of the photoreceptors—specifically, the visual pigments of the membrane discs. It is not known how the altered pigments lead to the destruction of photoreceptors.

Photoreception

The plasma membrane in the outer segment of the photoreceptor contains chemically gated sodium ion channels. (Refer to the diagram of the resting state in **Figure 17–14**.)

Figure 17–14 **Photoreception.**

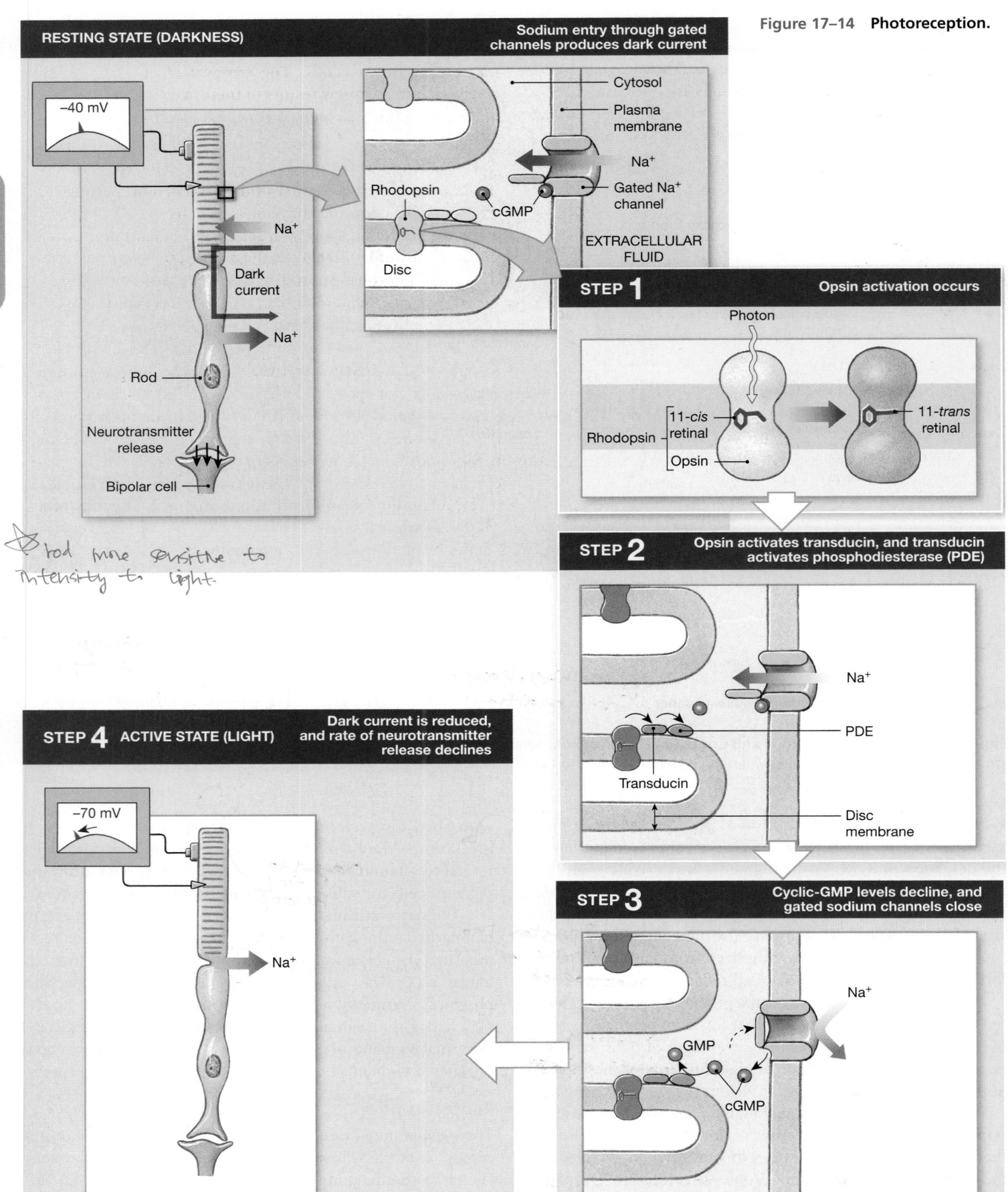

RESTING STATE (DARKNESS)

Sodium entry through gated channels produces dark current

−40 mV

Na⁺

Dark current

Na⁺

Rod

Neurotransmitter release

Bipolar cell

Cytosol

Plasma membrane

Rhodopsin

cGMP

Na⁺

Gated Na⁺ channel

EXTRACELLULAR FLUID

Disc

rod more sensitive to intensity to light.

STEP 1 **Opsin activation occurs**

Photon

Rhodopsin

11-*cis* retinal

Opsin

11-*trans* retinal

STEP 2 **Opsin activates transducin, and transducin activates phosphodiesterase (PDE)**

Na⁺

PDE

Transducin

Disc membrane

STEP 4 **ACTIVE STATE (LIGHT)** **Dark current is reduced, and rate of neurotransmitter release declines**

−70 mV

Na⁺

STEP 3 **Cyclic-GMP levels decline, and gated sodium channels close**

Na⁺

GMP

cGMP

In darkness, these gated channels are kept open in the presence of *cyclic-GMP* (*cyclic guanosine monophosphate*, or *cGMP*), a derivative of the high-energy compound *guanosine triphosphate* (GTP). Because the channels are open, the transmembrane potential is approximately −40 mV, rather than the −70 mV typical of resting neurons. At the −40 mV transmembrane potential, the photoreceptor is continuously releasing neurotransmitters (in this case, glutamate) across synapses at the inner segment. The inner segment also continuously pumps sodium ions out of the cytoplasm. The movement of sodium ions into the outer segment, on to the inner segment, and out of the cell is known as the *dark current*.

The process of rhodopsin-based photoreception begins when a photon strikes the retinal portion of a rhodopsin molecule embedded in the membrane of the disc (**Figure 17–14**):

Step 1   *Opsin Is Activated.*   The bound retinal molecule has two possible configurations: the **11-*cis*** form and the **11-*trans*** form. Normally, the molecule is in the 11-*cis* form; on absorbing light, it changes to the more linear 11-*trans* form. This change activates the opsin molecule.

Step 2   *Opsin Activates Transducin, Which in Turn Activates Phosphodiesterase.*   **Transducin** is a G protein—a membrane-bound enzyme complex. ∞ p. 421 In this case, transducin is activated by opsin, and transducin in turn activates **phosphodiesterase (PDE)**.

Step 3   *Cyclic-GMP (cGMP) Levels Decline, and Gated Sodium Channels Close.*   Phosphodiesterase is an enzyme that breaks down cGMP. The removal of cGMP from the gated sodium channels results in their inactivation. The rate of Na⁺ entry into the cytoplasm then decreases.

Step 4   *The Dark Current Is Reduced and the Rate of Neurotransmitter Release Declines.*   The reduction in the rate of Na⁺ entry reduces the dark current. Because active transport continues to remove Na⁺ from the cytoplasm, when the sodium channels close, the transmembrane potential drops toward −70 mV. As the membrane hyperpolarizes, the rate of neurotransmitter release decreases, indicating to the adjacent bipolar cell that the photoreceptor has absorbed a photon.

Recovery after Stimulation.   After absorbing a photon, retinal does not spontaneously revert to the 11-*cis* form. Instead, the entire rhodopsin molecule must be broken down and reassembled. Shortly after the change in shape occurs, the rhodopsin molecule begins to break down into retinal and opsin, a process known as **bleaching** (**Figure 17–15**). Before it can recombine with opsin, the retinal must be enzymatically converted to the 11-*cis* form. This conversion requires energy in the form of ATP (adenosine triphosphate), and it takes time.

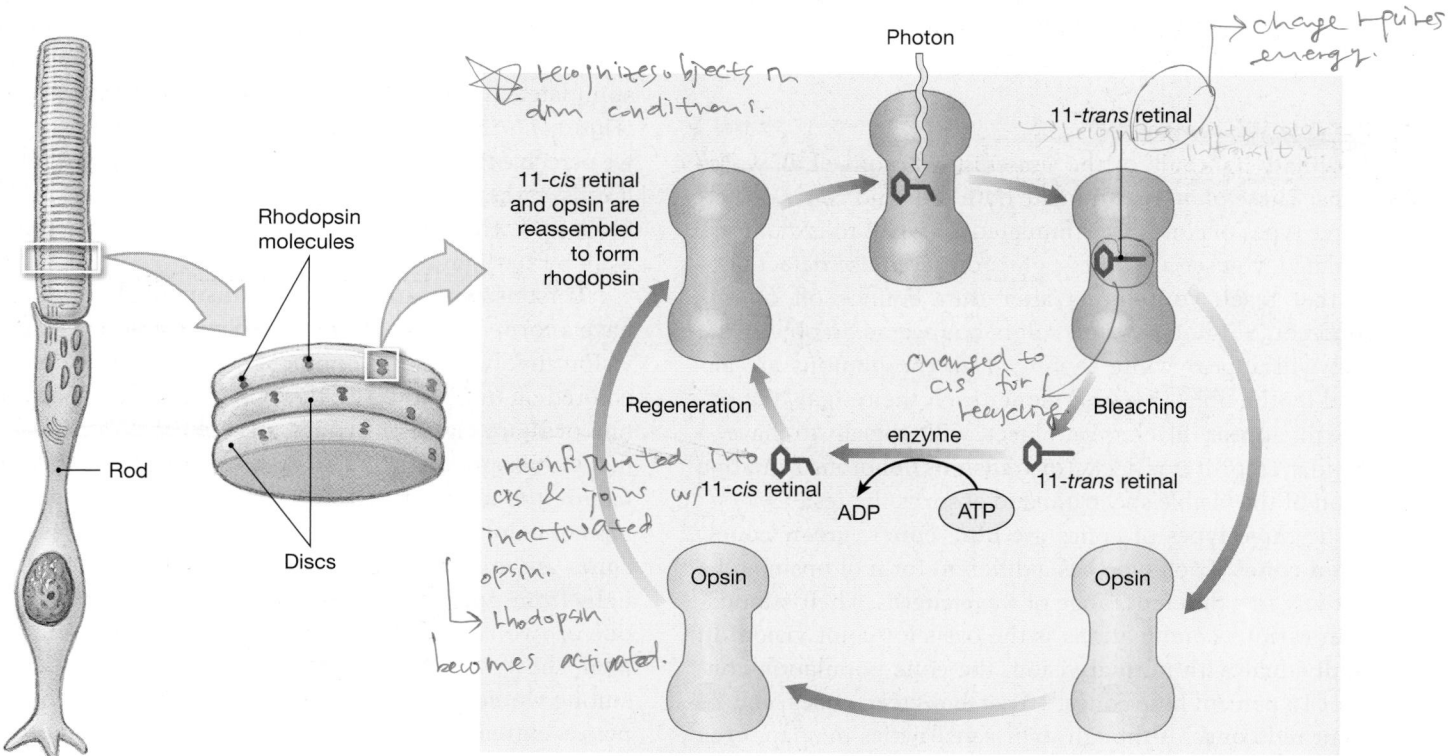

Figure 17–15   Bleaching and Regeneration of Visual Pigments.

Bleaching contributes to the lingering visual impression you have after you see a flashbulb go off. Following intense exposure to light, a photoreceptor cannot respond to further stimulation until its rhodopsin molecules have been regenerated. As a result, a "ghost" image remains on the retina. Bleaching is seldom noticeable under ordinary circumstances, because the eyes are constantly making small, involuntary changes in position that move the image across the retina's surface.

While the rhodopsin molecule is being reassembled, membrane permeability is returning to normal. Opsin is inactivated when bleaching occurs, and the breakdown of cGMP halts as a result. As other enzymes generate cGMP in the cytoplasm, the chemically gated sodium channels reopen.

As previously noted, the visual pigments of the photoreceptors are synthesized from vitamin A. The body contains vitamin A reserves sufficient for several months, and a significant amount is stored in the cells of the pigmented part of the retina. If dietary sources are inadequate, these reserves are gradually exhausted and the amount of visual pigment in the photoreceptors begins to decline. Daylight vision is affected, but in daytime the light is usually bright enough to stimulate any visual pigments that remain within the densely packed cone population of the fovea. As a result, the problem first becomes apparent at night, when the dim light proves insufficient to activate the rods. This condition, known as **night blindness**, can be treated by the administration of vitamin A. The body can convert the carotene pigments in many vegetables to vitamin A. Carrots are a particularly good source of carotene—hence the old adage that carrots are good for your eyes.

Color Vision

An ordinary lightbulb or the sun emits photons of all wavelengths. These photons stimulate both rods and cones. When all three types of cones are stimulated, or when rods alone are stimulated, you see a "white" light. Your eyes also detect photons that reach your retina after they bounce off objects around you. If photons of all colors bounce off an object, the object will appear white to you; if all the photons are absorbed by the object (so that none reach the retina), the object will appear black. An object will appear to have a particular color if it reflects (or transmits) photons from one portion of the visible spectrum and absorbs the rest.

The three types of cones are **blue cones**, **green cones**, and **red cones**. Each type has a different form of opsin and a sensitivity to a different range of wavelengths. Their stimulation in various combinations is the basis for color vision. In an individual with normal vision, the cone population consists of 16 percent blue cones, 10 percent green cones, and 74 percent red cones. Although their sensitivities overlap, each type is most sensitive to a specific portion of the visual spectrum (**Figure 17–16**).

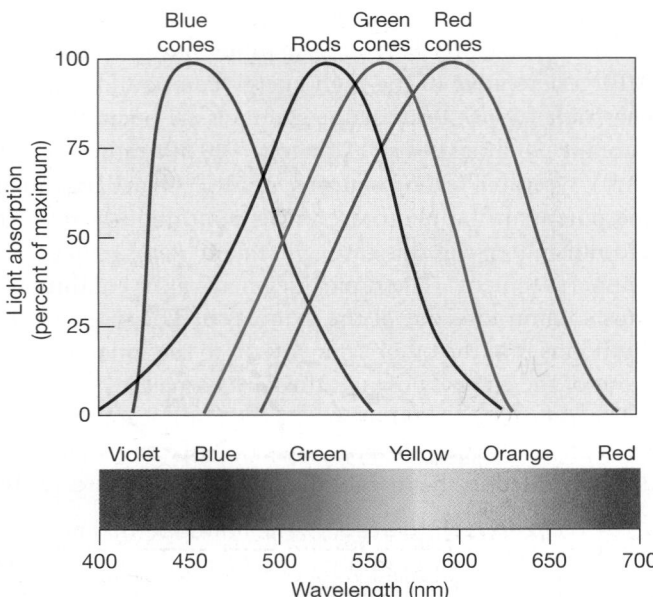

Figure 17–16   Cone Types and Sensitivity to Color. A graph comparing the absorptive characteristics of blue, green, and red cones with those of typical rods. Notice that the sensitivities of the rods overlap those of the cones, and that the three types of cones have overlapping sensitivity curves.

Color discrimination occurs through the integration of information arriving from all three types of cones. For example, the perception of yellow results from a combination of inputs from highly stimulated green cones, less strongly stimulated red cones, and relatively unaffected blue cones (**Figure 17–16**). If all three cone populations are stimulated, we perceive the color as white. Because we also perceive white if rods, rather than cones, are stimulated, everything appears black-and-white when we enter dimly lit surroundings or walk by starlight.

Persons who are unable to distinguish certain colors have a form of **color blindness**. The standard tests for color vision involve picking numbers or letters out of a complex colored picture (**Figure 17–17**). Color blindness occurs when one or more classes of cones are nonfunctional. The cones may be absent, or they may be present but unable to manufacture the necessary visual pigments. In the most common type of color blindness (red–green color blindness), the red cones are missing, so the individual cannot distinguish red light from green light. Inherited color blindness involving one or two cone pigments is not unusual. Ten percent of all men show some color blindness, whereas the incidence among women is only about 0.67 percent. Total color blindness is extremely rare; only 1 person in 300,000 fails to manufacture any cone pigments. We will consider the inheritance of color blindness in Chapter 29.

17 NERVOUS

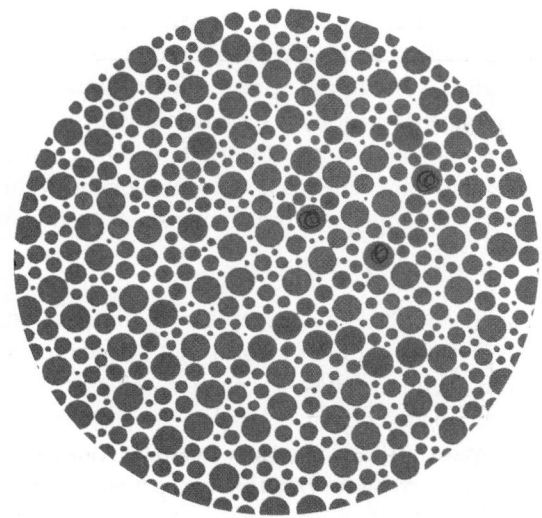

Figure 17–17 A Standard Test for Color Vision. Individuals lacking one or more populations of cones are unable to distinguish the patterned image (the number 12).

Light and Dark Adaptation

The sensitivity of your visual system varies with the intensity of illumination. After 30 minutes or more in the dark, almost all visual pigments will have recovered from photobleaching and be fully receptive to stimulation. This is the **dark-adapted state**. When dark-adapted, the visual system is extremely sensitive. For example, a single rod will hyperpolarize in response to a single photon of light. Even more remarkable, if as few as seven rods absorb photons at one time, you will see a flash of light.

When the lights come on, at first they seem almost unbearably bright, but over the next few minutes your sensitivity decreases as bleaching occurs. Eventually, the rate of bleaching is balanced by the rate at which the visual pigments re-form. This condition is the **light-adapted state**. If you moved from the depths of a cave to the full sunlight of midday, your receptor sensitivity would decrease by a factor of 25,000.

A variety of central responses further adjust light sensitivity. Constriction of the pupil, via the *pupillary constrictor reflex*, reduces the amount of light entering your eye to one-thirtieth the maximum dark-adapted level. Dilating the pupil fully can produce a thirtyfold increase in the amount of light entering the eye, and facilitating some of the synapses along the visual pathway can perhaps triple its sensitivity. Hence, the sensitivity of the entire system may increase by a factor of more than 1 million.

The Visual Pathways

The visual pathways begin at the photoreceptors and end at the visual cortex of the cerebral hemispheres. In other sensory pathways we have examined, at most one synapse lies between a receptor and a sensory neuron that delivers information to the CNS. In the visual pathways, the message must cross two synapses (photoreceptor to bipolar cell, and bipolar cell to ganglion cell) before it heads toward the brain. The extra synapse increases the synaptic delay, but it provides an opportunity for the processing and integration of visual information before it leaves the retina.

Processing by the Retina

Each photoreceptor in the retina monitors a specific receptive field. The retina contains about 130 million photoreceptors, 6 million bipolar cells, and 1 million ganglion cells. Thus, a considerable amount of convergence occurs at the start of the visual pathway. The degree of convergence differs between rods and cones. Regardless of the amount of convergence, each ganglion cell monitors a specific portion of the field of vision.

As many as a thousand rods may pass information via their bipolar cells to a single ganglion cell. The ganglion cells that monitor rods, called **M cells** (*magnocells*; *magnus*, great), are relatively large. They provide information about the general form of an object, motion, and shadows in dim lighting. Because so much convergence occurs, the activation of an M cell indicates that light has arrived in a general area rather than at a specific location.

The loss of specificity due to convergence is partially overcome by the fact that the activity of ganglion cells varies according to the pattern of activity in their receptive field, which is generally circular. Typically, a ganglion cell responds differently to stimuli that arrive in the center of its receptive field than to stimuli that arrive at the edges (**Figure 17–18**). Some ganglion cells (**on-center neurons**) are excited by light arriving in the center of their sensory field and are inhibited when light strikes the edges of their receptive field. Others (**off-center neurons**) are inhibited by light in the central zone, but are stimulated by illumination at the edges. On-center and off-center neurons provide information about which portion of their receptive field is illuminated. This kind of retinal processing within ganglion receptive fields improves the detection of the edges of objects within the visual field.

Cones typically show very little convergence; in the fovea, the ratio of cones to ganglion cells is 1:1. The ganglion cells that monitor cones, called **P cells** (*parvo cells*; *parvus*, small), are smaller and more numerous than M cells. P cells are active in bright light, and they provide information about edges, fine detail, and color. Because little convergence occurs, the activation of a P cell means that light has arrived at one specific location. As a result, cones provide more precise information about a visual image than do rods. In videographic terms, images formed by rods have a coarse, grainy, pixelated appearance that blurs details; by contrast, images produced by cones are fine-grained, of high density, sharp, and clear.

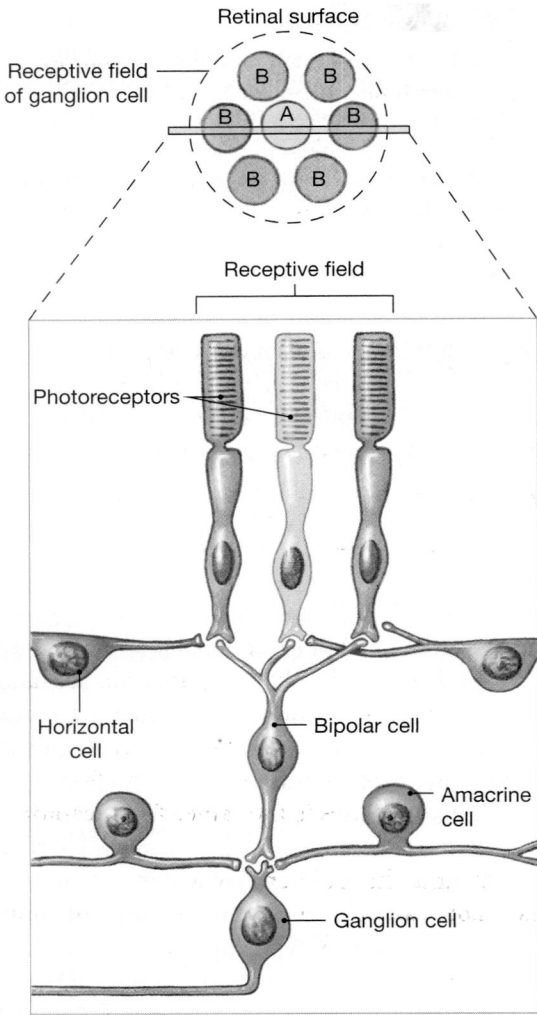

Figure 17–18 **Convergence and Ganglion Cell Function.**
Photoreceptors are organized in groups within a receptive field; each
ganglion cell monitors a well-defined portion of that field. Some
ganglion cells (on-center neurons, labeled A) respond strongly to
light arriving at the center of their receptive field. Others (off-center
neurons, labeled B) respond most strongly to illumination of the
edges of their receptive field.

Central Processing of Visual Information

Axons from the entire population of ganglion cells converge
on the optic disc, penetrate the wall of the eye, and proceed
toward the diencephalon as the optic nerve (II). The two optic nerves, one from each eye, reach the diencephalon at the
optic chiasm (**Figure 17–19**). From that point, approximately
half the fibers proceed toward the lateral geniculate nucleus
of the same side of the brain, whereas the other half cross over
to reach the lateral geniculate nucleus of the opposite side.
∞ p. 476 From each lateral geniculate nucleus, visual information travels to the occipital cortex of the cerebral hemi-

sphere on that side. The bundle of projection fibers linking
the lateral geniculates with the visual cortex is known as the
optic radiation. Collaterals from the fibers synapsing in the
lateral geniculate continue to subconscious processing centers in the diencephalon and brain stem. For example, the
pupillary reflexes and reflexes that control eye movement are
triggered by collaterals carrying information to the superior
colliculi.

The Field of Vision. The perception of a visual image reflects the integration of information that arrives at the visual
cortex of the occipital lobes. Each eye receives a slightly different visual image, because (1) the foveas are 5–7.5 cm
(2–3.0 in.) apart, and (2) the nose and eye socket block the
view of the opposite side. **Depth perception**, an interpretation of the three-dimensional relationships among objects in
view, is obtained by comparing the relative positions of objects within the images received by the two eyes.

When you look straight ahead, the visual images from
your left and right eyes overlap (**Figure 17–19**). The image received by the fovea of each eye lies in the center of the region
of overlap. A vertical line drawn through the center of this region marks the division of visual information at the optic chiasm. Visual information from the left half of the combined field
of vision reaches the visual cortex of your right occipital lobe;
visual information from the right half of the combined field of
vision arrives at the visual cortex of your left occipital lobe.

The cerebral hemispheres thus contain a map of the entire field of vision. As in the case of the primary sensory cortex, the map does not faithfully duplicate the relative areas
within the sensory field. For example, the area assigned to the
macula lutea and fovea covers about 35 times the surface it
would cover if the map were proportionally accurate. The
map is also upside down and backward, duplicating the orientation of the visual image at the retina.

The Brain Stem and Visual Processing. Many centers in
the brain stem receive visual information, either from the lateral geniculate nuclei or through collaterals from the optic
tracts. Collaterals that bypass the lateral geniculates synapse
in the superior colliculi or in the hypothalamus. The superior
colliculi of the mesencephalon issue motor commands that
control unconscious eye, head, or neck movements in response to visual stimuli. Visual inputs to the suprachiasmatic
nucleus of the hypothalamus affect the function of other brain
stem nuclei. ∞ p. 478 The suprachiasmatic nucleus and the
pineal gland of the epithalamus receive visual information and
use it to establish a daily pattern of visceral activity that is tied
to the day–night cycle. This **circadian rhythm** (*circa*, about
+ *dies*, day) affects your metabolic rate, endocrine function,
blood pressure, digestive activities, awake–asleep cycle, and
other physiological and behavioral processes.

LEFT SIDE | RIGHT SIDE

also relayed to limbic & memory system

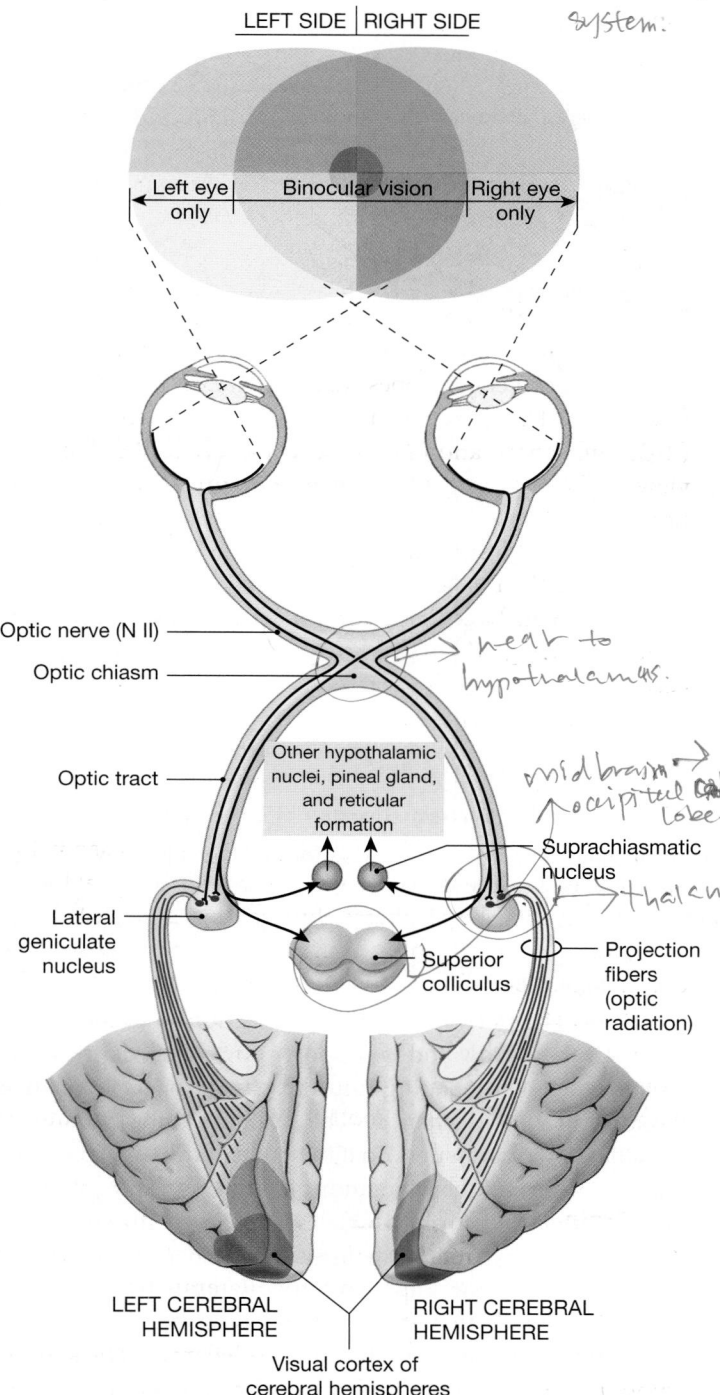

Left eye only | Binocular vision | Right eye only

Optic nerve (N II)

heat to hypothalamus

Optic chiasm

Optic tract

Other hypothalamic nuclei, pineal gland, and reticular formation

midbrain → occipital lobe

Suprachiasmatic nucleus

→ thalamus

Lateral geniculate nucleus

Superior colliculus

Projection fibers (optic radiation)

LEFT CEREBRAL HEMISPHERE

RIGHT CEREBRAL HEMISPHERE

Visual cortex of cerebral hemispheres

Figure 17–19 The Visual Pathways. The crossover of some nerve fibers occurs at the optic chiasm. As a result, each hemisphere receives visual information from the medial half of the field of vision of the eye on that side, and from the lateral half of the field of vision of the eye on the opposite side. Visual association areas integrate this information to develop a composite picture of the entire field of vision.

CHECKPOINT

11. **If you had been born without cones in your eyes, would you still be able to see? Explain.**

12. **How could a diet deficient in vitamin A affect vision?**

13. **What effect would a decrease in phosphodiesterase activity in photoreceptor cells have on vision?**

See the blue Answers tab at the end of the book.

17-5 Equilibrium sensations originate within the inner ear, while hearing involves the detection and interpretation of sound waves

The special senses of equilibrium and hearing are provided by the *inner ear*, a receptor complex located in the petrous part of the temporal bone of the skull. *Equilibrium* sensations inform us of the position of the head in space by monitoring gravity, linear acceleration, and rotation. *Hearing* enables us to detect and interpret sound waves. The basic receptor mechanism for both senses is the same. The receptors, called *hair cells*, are mechanoreceptors. The complex structure of the inner ear and the different arrangement of accessory structures enable hair cells to respond to different stimuli and thus to provide the input for both senses.

Anatomy of the Ear

The ear is divided into three anatomical regions: the external ear, the middle ear, and the inner ear (**Figure 17–20**). The *external ear*—the visible portion of the ear—collects and directs sound waves toward the *middle ear*, a chamber located within the petrous portion of the temporal bone. Structures of the middle ear collect sound waves and transmit them to an appropriate portion of the *inner ear*, which contains the sensory organs for hearing and equilibrium.

The External Ear

The **external ear** includes the fleshy and cartilaginous **auricle**, or *pinna*, which surrounds the **external acoustic meatus**, or *ear canal*. The auricle protects the opening of the canal and provides directional sensitivity; sounds coming from behind the head are blocked by the auricle, whereas sounds coming from the side or front are collected and channeled into the external acoustic meatus. (When you "cup" your ear with your hand to hear a faint sound more clearly, you are exaggerating this effect.) The external acoustic meatus is a passageway that ends at the **tympanic membrane**,

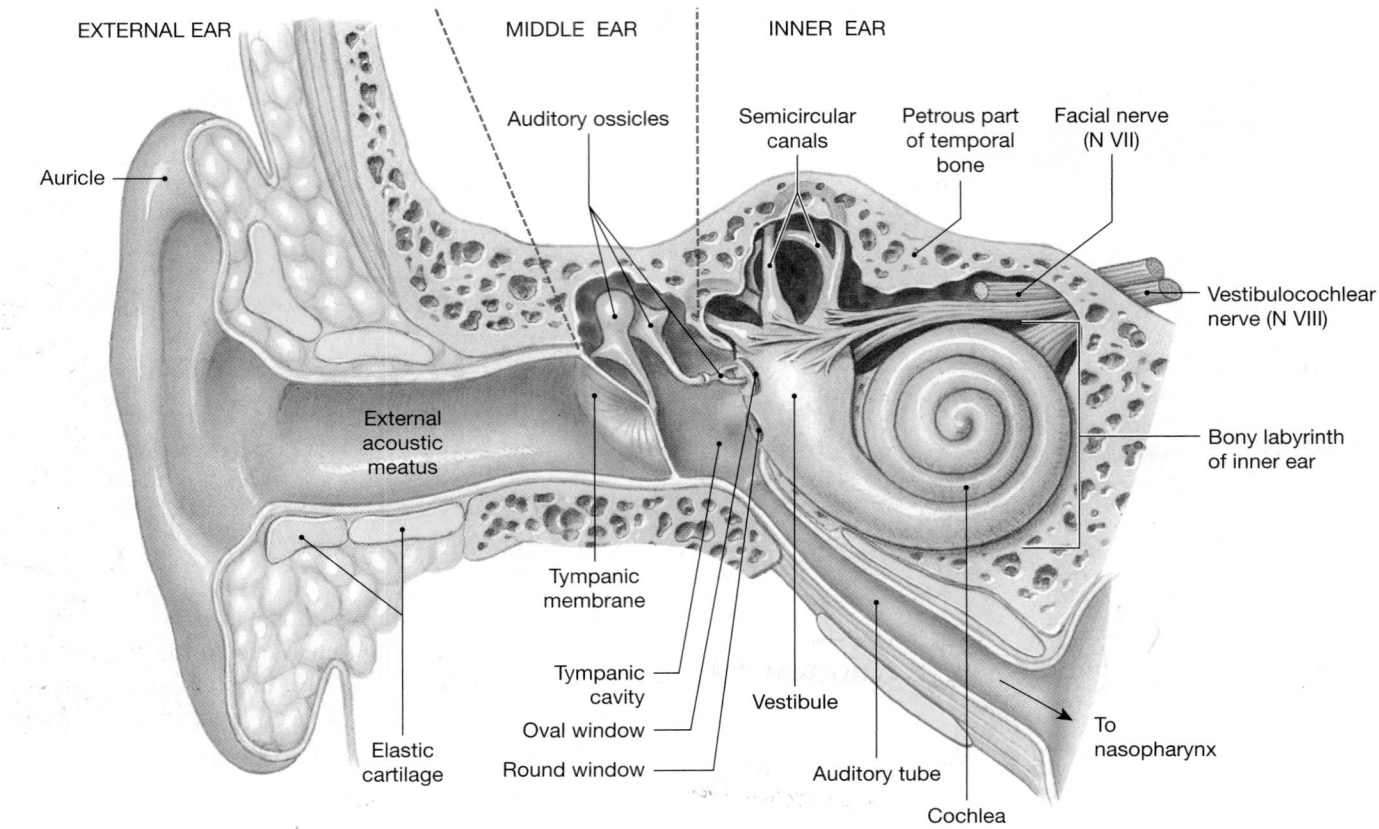

EXTERNAL EAR   MIDDLE EAR   INNER EAR

Auricle

Auditory ossicles

Semicircular canals

Petrous part of temporal bone

Facial nerve (N VII)

Vestibulocochlear nerve (N VIII)

External acoustic meatus

Bony labyrinth of inner ear

Tympanic membrane

Tympanic cavity

Oval window

Round window

Vestibule

Elastic cartilage

Auditory tube

To nasopharynx

Cochlea

Figure 17–20 **The Anatomy of the Ear.** The boundaries separating the three anatomical regions of the ear (external, middle, and inner) are indicated by the dashed lines. → processed form parietal cortex part of brain.

also called the *tympanum* or *eardrum*. The tympanic membrane is a thin, semitransparent sheet that separates the external ear from the middle ear.

The tympanic membrane is very delicate. The auricle and the narrow external acoustic meatus provide some protection from accidental injury. In addition, **ceruminous glands**—integumentary glands along the external acoustic meatus—secrete a waxy material that helps deny access to foreign objects or small insects, as do many small, outwardly projecting hairs. These hairs also provide increased tactile sensitivity through their root hair plexuses. The slightly waxy secretion of the ceruminous glands, called **cerumen**, also slows the growth of microorganisms in the external acoustic meatus and reduces the chances of infection.

The Middle Ear

The **middle ear**, or **tympanic cavity**, is an air-filled chamber separated from the external acoustic meatus by the tympanic membrane. The middle ear communicates both with the *nasopharynx* (the superior portion of the pharynx), through the **auditory tube**, and with the mastoid air cells, through a number of small connections (**Figures 17–20** and **17–21**). The auditory tube is also called the *pharyngotympanic tube* or the

Eustachian tube. About 4 cm (1.6 in.) long, it consists of two portions. The portion near the connection to the middle ear is relatively narrow and is supported by elastic cartilage. The portion near the opening into the nasopharynx is relatively broad and funnel shaped. The auditory tube permits the equalization of pressures on either side of the tympanic membrane. Unfortunately, the auditory tube can also allow microorganisms to travel from the nasopharynx into the middle ear. Invasion by microorganisms can lead to an unpleasant middle ear infection known as *otitis media*.

The Auditory Ossicles. The middle ear contains three tiny ear bones, collectively called **auditory ossicles**. The ear bones connect the tympanic membrane with one of the receptor complexes of the inner ear (**Figures 17–20** and **17–21**). The three auditory ossicles are the malleus, the incus, and the stapes. The **malleus** (*malleus*, hammer) attaches at three points to the interior surface of the tympanic membrane. The **incus** (*incus*, anvil), the middle ossicle, attaches the malleus to the **stapes** (*stapes*, stirrup), the inner ossicle. The edges of the base of the stapes are bound to the edges of the *oval window*, an opening in the bone that surrounds the inner ear. The articulations between the auditory ossicles are the smallest

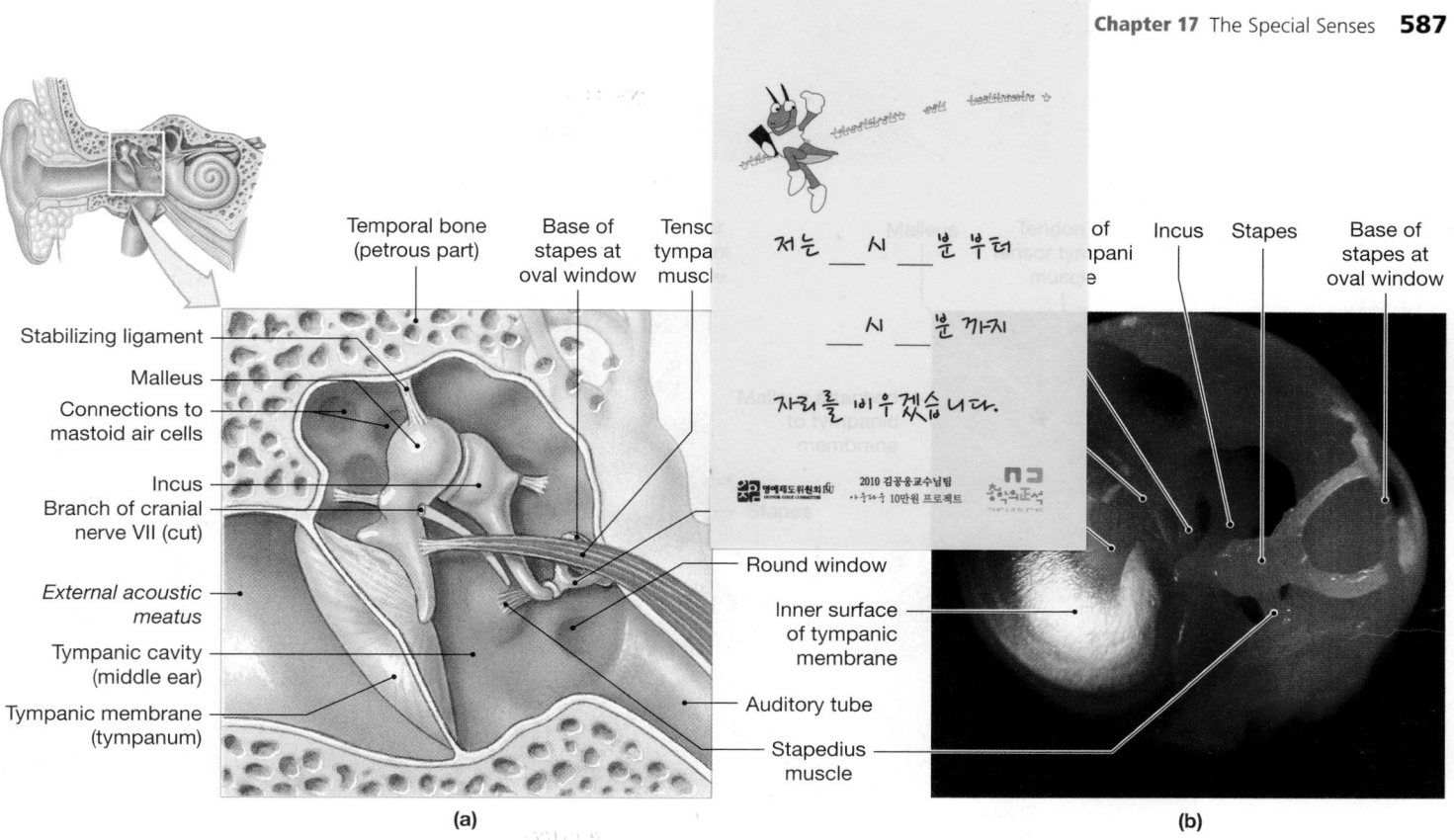

Figure 17–21   The Middle Ear. (a) The structures of the middle ear. **(b)** The tympanic membrane and auditory ossicles.

synovial joints in the body. Each has a tiny capsule and supporting extracapsular ligaments.

Vibration of the tympanic membrane converts arriving sound waves into mechanical movements. The auditory ossicles act as levers that conduct those vibrations to the inner ear. The ossicles are connected in such a way that an in–out movement of the tympanic membrane produces a rocking motion of the stapes. The ossicles thus function as a lever system that collects the force applied to the tympanic membrane and focuses it on the oval window. Because the tympanic membrane is 22 times larger and heavier than the oval window, considerable amplification occurs, so we can hear very faint sounds. But that degree of amplification can be a problem when we are exposed to very loud noises. In the middle ear, two small muscles protect the tympanic membrane and ossicles from violent movements under very noisy conditions:

1. The **tensor tympani** (TEN-sor tim-PAN-ē) **muscle** is a short ribbon of muscle whose origin is the petrous portion of the temporal bone and the auditory tube, and whose insertion is on the "handle" of the malleus. When the tensor tympani contracts, the malleus is pulled medially, stiffening the tympanic membrane. This increased stiffness reduces the amount of movement possible. The tensor tympani muscle is innervated by motor fibers of the mandibular branch of the trigeminal nerve (V).

2. The **stapedius** (sta-PĒ-dē-us) **muscle**, innervated by the facial nerve (VII), originates from the posterior wall of the middle ear and inserts on the stapes. Contraction of the stapedius pulls the stapes, reducing movement of the stapes at the oval window.

The Inner Ear

The senses of equilibrium and hearing are provided by receptors in the **inner ear** (**Figures 17–20** and **17–22a**).

The superficial contours of the inner ear are established by a layer of dense bone known as the **bony labyrinth** (*labyrinthos*, network of canals). The walls of the bony labyrinth are continuous with the surrounding temporal bone. The inner contours of the bony labyrinth closely follow the contours of the **membranous labyrinth**, a delicate, interconnected network of fluid-filled tubes. The receptors of the inner ear are found within those tubes. Between the bony and membranous labyrinths flows **perilymph** (PER-i-limf), a liquid whose properties closely resemble those of cerebrospinal fluid. The membranous labyrinth contains **endolymph** (EN-dō-limf), a fluid with electrolyte concentrations different from those of typical body fluids. The physical relationships are indicated in

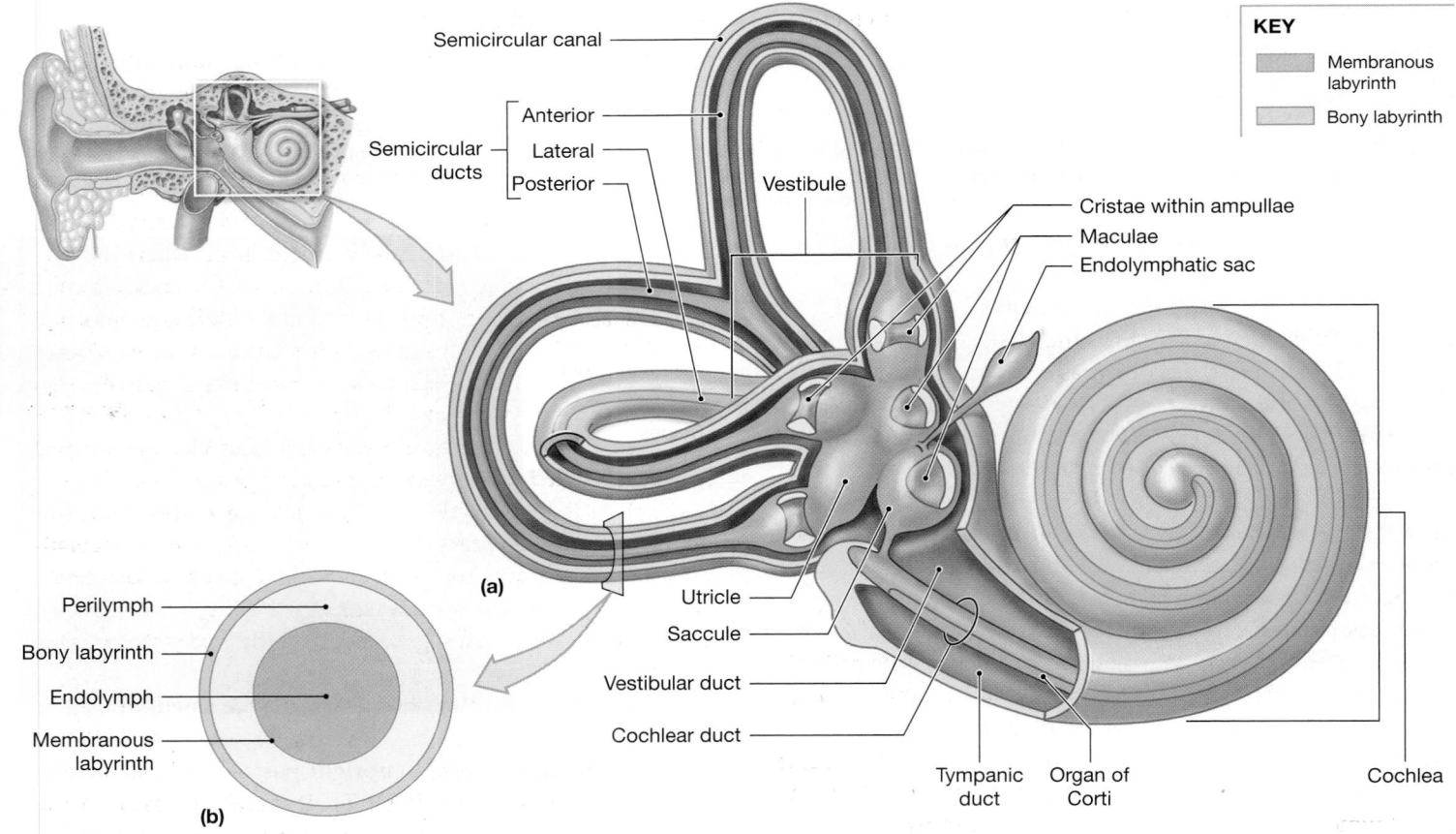

KEY

| Membranous labyrinth |
| Bony labyrinth |

Semicircular canal

Semicircular ducts
- Anterior
- Lateral
- Posterior

Vestibule

Cristae within ampullae
Maculae
Endolymphatic sac

(a)

Utricle
Saccule
Vestibular duct
Cochlear duct

Tympanic duct
Organ of Corti

Cochlea

Perilymph
Bony labyrinth
Endolymph
Membranous labyrinth

(b)

Figure 17–22 **The Inner Ear. (a)** The bony and membranous labyrinths. Areas of the membranous labyrinth containing sensory receptors (cristae, maculae, and the organ of Corti) are shown in purple. **(b)** A section through one of the semicircular canals, showing the relationship between the bony and membranous labyrinths, and the locations of perilymph and endolymph.

Figure 17–22b. (See Appendix IV for a chemical analysis of perilymph, endolymph, and other body fluids.)

The bony labyrinth can be subdivided into the *vestibule*, three *semicircular canals*, and the *cochlea* (**Figure 17–22a**). The **vestibule** (VES-ti-būl) consists of a pair of membranous sacs: the **saccule** (SAK-ūl) and the **utricle** (Ū-tri-kul), or *sacculus* and *utriculus*. Receptors in the saccule and utricle provide sensations of gravity and linear acceleration.

The **semicircular canals** enclose slender *semicircular ducts*. Receptors in the semicircular ducts are stimulated by rotation of the head. The combination of vestibule and semicircular canals is called the **vestibular complex**. The fluid-filled chambers within the vestibule are broadly continuous with those of the semicircular canals.

The **cochlea** (KOK-lē-uh; *cochlea*, a snail shell) is a spiral-shaped, bony chamber that contains the **cochlear duct** of the membranous labyrinth. Receptors within the cochlear duct provide the sense of hearing. The duct is sandwiched between a pair of perilymph-filled chambers. The entire complex makes turns around a central bony hub, much like a snail shell.

The walls of the bony labyrinth consist of dense bone everywhere except at two small areas near the base of the cochlear spiral (**Figure 17–20**). The **round window** is a thin, membranous partition that separates the perilymph of the cochlear chambers from the air-filled middle ear. Collagen fibers connect the bony margins of the opening known as the **oval window** to the base of the stapes.

Equilibrium

As just noted, equilibrium sensations are provided by receptors of the vestibular complex. The semicircular ducts convey information about rotational movements of the head. For example, when you turn your head to the left, receptors stimulated in the semicircular ducts tell you how rapid the movement is, and in which direction. The saccule and the utricle convey information about your position with respect to gravity. If you stand with your head tilted to one side, these receptors report the angle involved and whether your head tilts forward or backward. The saccule and the utricle are also stimulated by sudden acceleration. When your car accelerates from a stop, the saccular and utricular receptors give you the impression of increasing speed.

The Semicircular Ducts

Sensory receptors in the semicircular ducts respond to rotational movements of the head. These **hair cells** are active during a movement, but are quiet when the body is motionless. The **anterior**, **posterior**, and **lateral semicircular ducts** are continuous with the utricle (**Figure 17–23a**). Each semicircular duct contains an **ampulla**, an expanded region that contains the receptors. The region in the wall of the ampulla that contains the receptors is known as a crista (**Figure 17–23b**). Each crista is bound to a **cupula** (KŪ-pū-luh), a gelatinous structure that extends the full width of the ampulla. The receptors in the cristae are called hair cells (**Figure 17–23b,d**).

Hair cells are the receptors found in other portions of the membranous labyrinth as well. Regardless of location, they are always surrounded by supporting cells and monitored by the dendrites of sensory neurons. The free surface of each hair cell supports 80–100 long **stereocilia**, which resemble very long microvilli (**Figure 17–23d**). Each hair cell in the vestibule also contains a **kinocilium** (kī-nō-SIL-ē-um), a single large cilium. Hair cells do not actively move their kinocilia or stereocilia. However, when an external force pushes against these processes, the distortion of the plasma membrane alters the rate at which the hair cell releases chemical transmitters.

Hair cells provide information about the direction and strength of mechanical stimuli. The stimuli involved, however, depend on the hair cell's location: gravity or acceleration in the vestibule, rotation in the semicircular canals, and sound in the cochlea. The sensitivities of the hair cells differ, because each of these regions has different accessory structures that determine which stimulus will provide the force to deflect the kinocilia and stereocilia.

At a crista, the kinocilia and stereocilia of the hair cells are embedded in the cupula (**Figure 17–23b**). Because the cupula has a density very close to that of the surrounding endolymph, it essentially floats above the receptor surface. When your head rotates in the plane of the duct, the movement of endolymph along the length of the semicircular duct pushes the cupula to the side and distorts the receptor processes (**Figure 17–23c**). Movement of fluid in one direction stimulates the hair cells, and movement in the opposite direction inhibits them. When the endolymph stops moving, the elastic nature of the cupula makes it return to its normal position.

Even the most complex movement can be analyzed in terms of motion in three rotational planes. Each semicircular duct responds to one of these rotational movements. A horizontal rotation, as in shaking your head "no," stimulates the hair cells of the lateral semicircular duct. Nodding "yes" excites the anterior duct, and tilting your head from side to side activates receptors in the posterior duct.

The Utricle and Saccule

The utricle and saccule provide equilibrium sensations, whether the body is moving or is stationary. The two chambers are connected by a slender passageway that is continuous with the narrow endolymphatic duct. The **endolymphatic duct** ends in a blind pouch called the **endolymphatic sac** (**Figure 17–23a**). This sac projects through the dura mater that lines the temporal bone and into the subarachnoid space, where it is surrounded by a capillary network. Portions of the cochlear duct secrete endolymph continuously, and at the endolymphatic sac excess fluid returns to the general circulation as the capillaries absorb endolymph removed by a combination of active transport and vesicular transport.

The hair cells of the utricle and saccule are clustered in oval structures called **maculae** (MAK-ū-lē; *macula*, spot) (**Figure 17–24a**). As in the ampullae, the hair cell processes are embedded in a gelatinous mass. However, the surface of this gelatinous material contains densely packed calcium carbonate crystals known as **statoconia** (*statos*, standing + *conia*, dust). The complex as a whole (gelatinous matrix and statoconia) is called an **otolith** ("ear stone").

The macula of the saccule is diagrammed in **Figure 17–24b**, and its function is shown in **Figure 17–24c**. When your head is in the normal, upright position, the statoconia sit atop the macula (STEP 1). Their weight presses on the macular surface, pushing the hair cell processes down rather than to one side or another. When your head is tilted, the pull of gravity on the statoconia shifts them to the side, thereby distorting the hair cell processes (STEP 2). The change in receptor activity tells the CNS that your head is no longer level.

A similar mechanism accounts for your perception of linear acceleration when you are in a car that speeds up suddenly. The statoconia lag behind, and the effect on the hair cells is comparable to tilting your head back. Under normal circumstances, your nervous system distinguishes between the sensations of tilting and linear acceleration by integrating vestibular sensations with visual information. Many amusement park rides confuse your sense of equilibrium by combining rapid rotation with changes in position and acceleration while providing restricted or misleading visual information.

Pathways for Equilibrium Sensations

Hair cells of the vestibule and semicircular ducts are monitored by sensory neurons located in adjacent **vestibular ganglia**. Sensory fibers from these ganglia form the **vestibular branch** of the vestibulocochlear nerve (VIII). ∞ p. 496 These fibers innervate neurons within the pair of **vestibular nuclei** at the boundary between the pons and the medulla oblongata. The two vestibular nuclei have four functions:

1. Integrating sensory information about balance and equilibrium that arrives from both sides of the head.

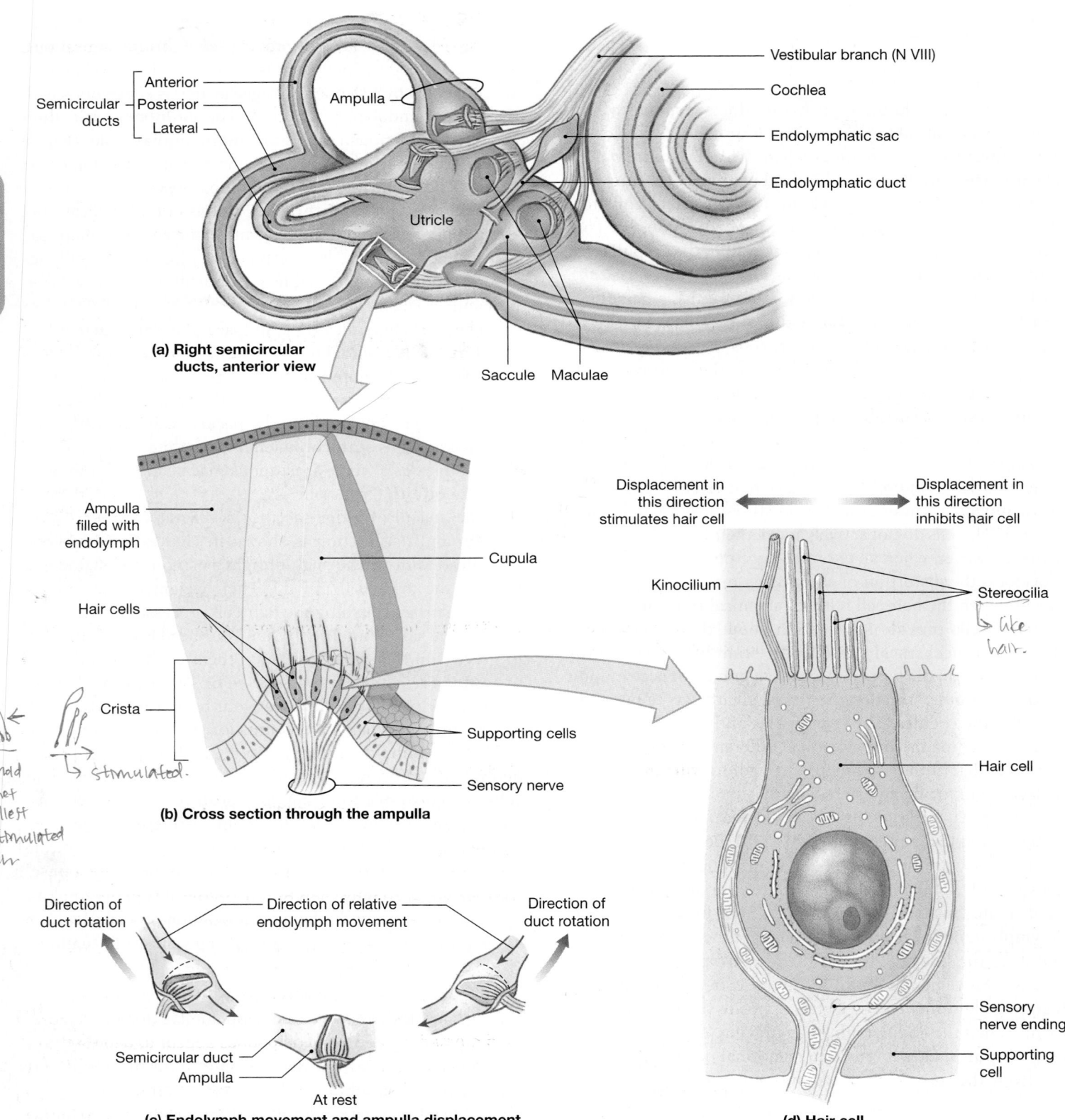

Figure 17–23 **The Semicircular Ducts.** **(a)** An anterior view of the right semicircular ducts, the utricle, and the saccule, showing the locations of sensory receptors. **(b)** A cross section through the ampulla of a semicircular duct. **(c)** Endolymph movement along the length of the duct moves the cupula and stimulates the hair cells. **(d)** A representative hair cell (receptor) from the vestibular complex. Bending the stereocilia toward the kinocilium depolarizes the cell and stimulates the sensory neuron. Displacement in the opposite direction inhibits the sensory neuron.

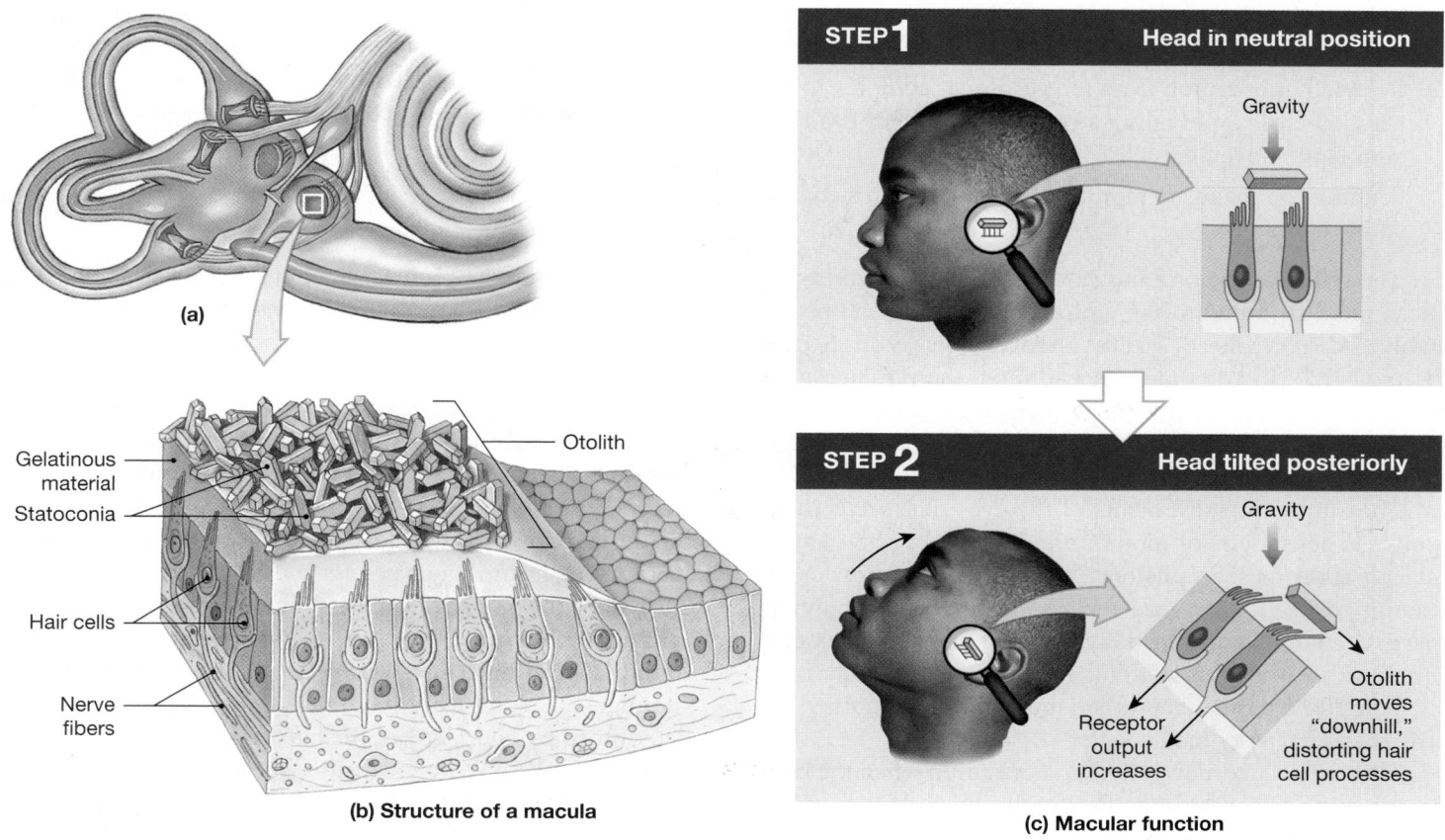

(a)

STEP 1 Head in neutral position

Gravity

STEP 2 Head tilted posteriorly

Gravity

Receptor output increases

Otolith moves "downhill," distorting hair cell processes

Gelatinous material

Statoconia

Hair cells

Nerve fibers

Otolith

(b) Structure of a macula

(c) Macular function

Figure 17–24   The Saccule and Utricle. (a) The location of the maculae. **(b)** The structure of an individual macula. **(c)** A diagrammatic view of macular function when the head is held horizontally (**STEP 1**) and then tilted back (**STEP 2**).

17 NERVOUS

CLINICAL NOTE

Motion Sickness

The exceedingly unpleasant signs and symptoms of **motion sickness** include headache, sweating, flushing of the face, nausea, vomiting, and various changes in mental perspective. (Sufferers may go from a state of giddy excitement to almost suicidal despair in a matter of moments.) It has been suggested that the condition results when central processing stations, such as the tectum of the mesencephalon, receive conflicting sensory information. Why and how these conflicting reports result in nausea, vomiting, and other symptoms are not known. Sitting belowdecks on a moving boat or reading in a car or airplane tends to provide the necessary conditions. Your eyes (which are tracking lines on a page) report that your

position in space is not changing, but your semicircular ducts report that your body is lurching and turning. To counter this effect, seasick sailors watch the horizon rather than their immediate surroundings, so that their eyes will provide visual confirmation of the movements detected by their inner ears. It is not known why some individuals are almost immune to motion sickness, whereas others find travel by boat or plane almost impossible.

Drugs commonly administered to prevent motion sickness include dimenhydrinate (*Dramamine*), scopolamine, and promethazine. These compounds appear to depress activity at the vestibular nuclei. Sedatives, such as prochlorperazine (*Compazine*), may also be effective. Scopolamine can be administered transdermally by using an adhesive patch (*Transderm Scop*).

2. Relaying information from the vestibular complex to the cerebellum.

3. Relaying information from the vestibular complex to the cerebral cortex, providing a conscious sense of head position and movement.

4. Sending commands to motor nuclei in the brain stem and spinal cord.

The reflexive motor commands issued by the vestibular nuclei are distributed to the motor nuclei for cranial nerves involved with eye, head, and neck movements (N III, N IV, N VI, and N XI). Instructions descending in the *vestibulospinal tracts* of the spinal cord adjust peripheral muscle tone and complement the reflexive movements of the head or neck. ∞ p. 522 These pathways are indicated in **Figure 17–25**.

The automatic movements of the eyes that occur in response to sensations of motion are directed by the *superior colliculi* of the mesencephalon. ∞ p. 474 These movements attempt to keep your gaze focused on a specific point in space, despite changes in body position and orientation. If your body is turning or spinning rapidly, your eyes will fix on one point for a moment and then jump ahead to another in a series of short, jerky movements.

This type of eye movement can occur even when the body is stationary if either the brain stem or the inner ear is damaged.

Individuals with this condition, which is called **nystagmus** (nis-TAG-mus), have trouble controlling their eye movements. Physicians commonly check for nystagmus by asking patients to watch a small penlight as it is moved across the field of vision.

Hearing

The receptors of the cochlear duct provide a sense of hearing that enables us to detect the quietest whisper, yet remain functional in a noisy room. The receptors responsible for auditory sensations are hair cells similar to those of the vestibular complex. However, their placement within the cochlear duct and the organization of the surrounding accessory structures shield them from stimuli other than sound.

In conveying vibrations from the tympanic membrane to the oval window, the auditory ossicles convert pressure fluctuations in air into much greater pressure fluctuations in the perilymph of the cochlea. These fluctuations stimulate hair cells along the cochlear spiral. The *frequency* of the perceived sound is determined by *which part* of the cochlear duct is stimulated. The *intensity* (volume) of the perceived sound is determined by *how many* of the hair cells at that location are stimulated. We will now consider the mechanics of this remarkably elegant process.

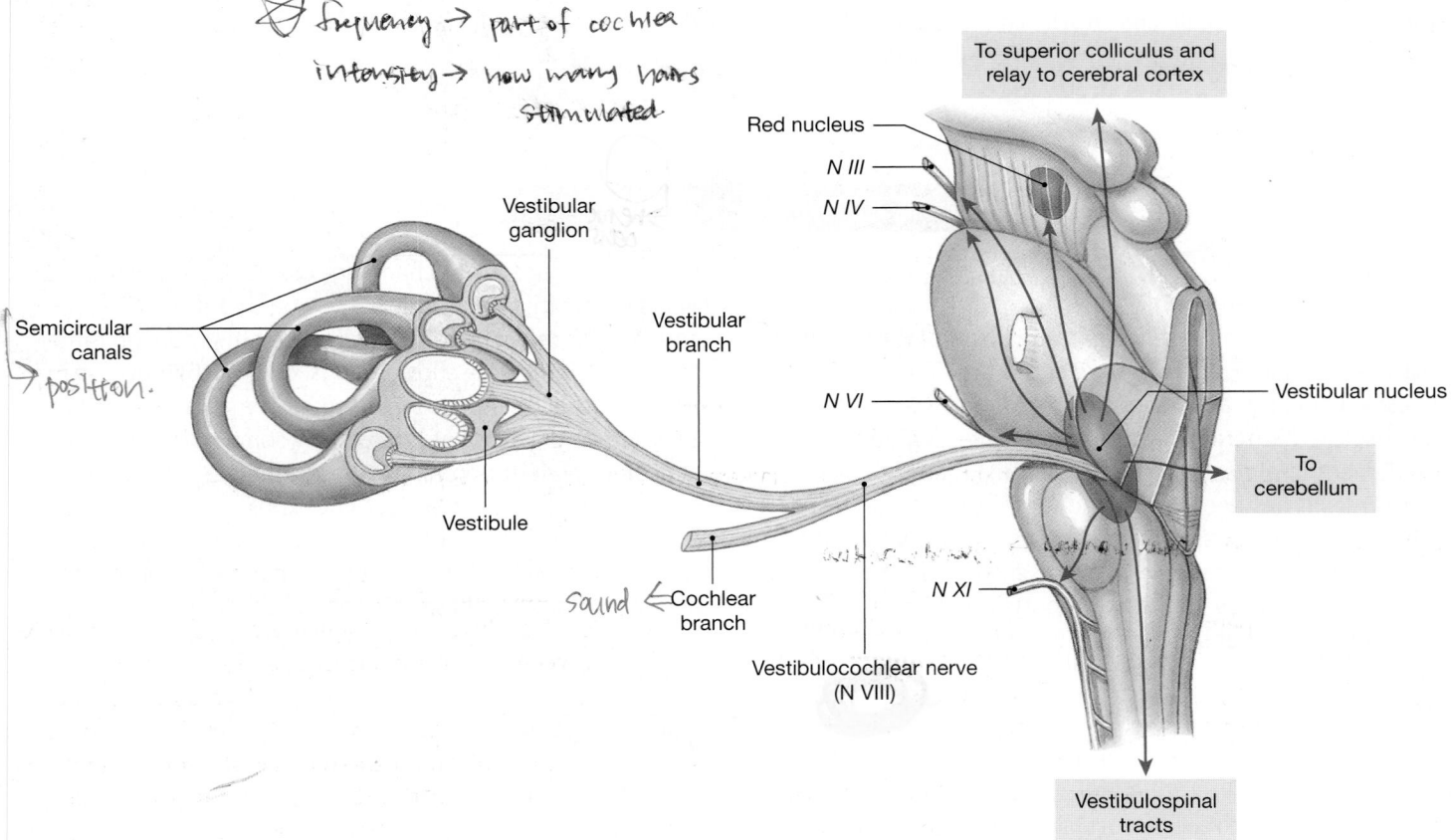

Figure 17–25 **Pathways for Equilibrium Sensations.**

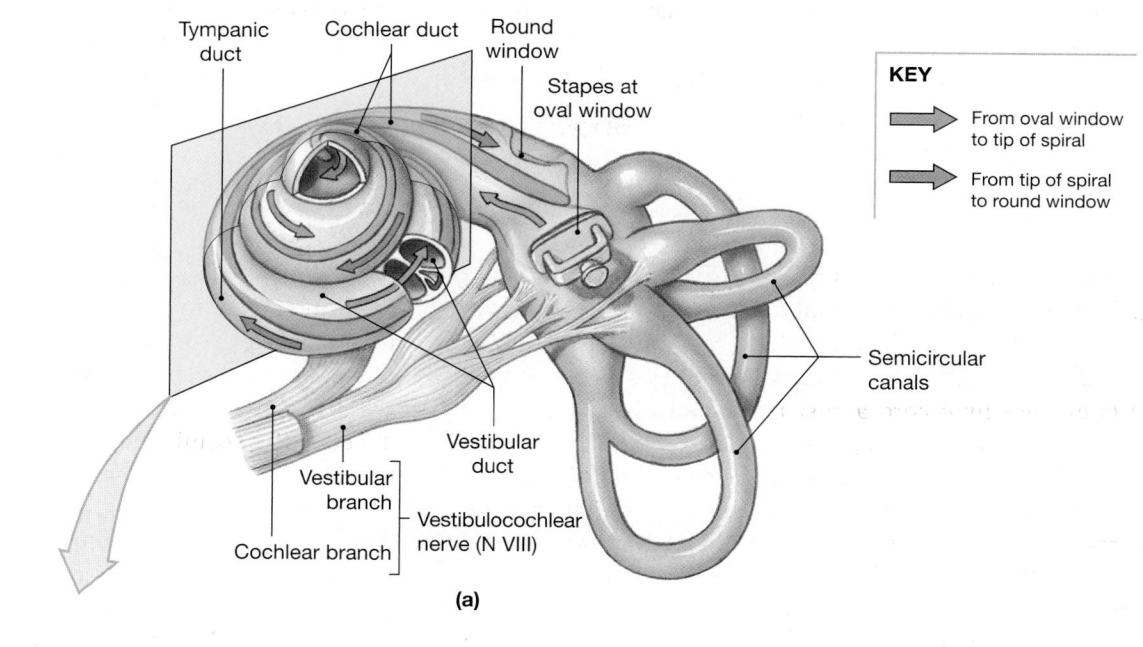

Tympanic duct

Cochlear duct

Round window

Stapes at oval window

KEY

From oval window to tip of spiral

From tip of spiral to round window

Semicircular canals

Vestibular duct

Vestibular branch

Cochlear branch

Vestibulocochlear nerve (N VIII)

(a)

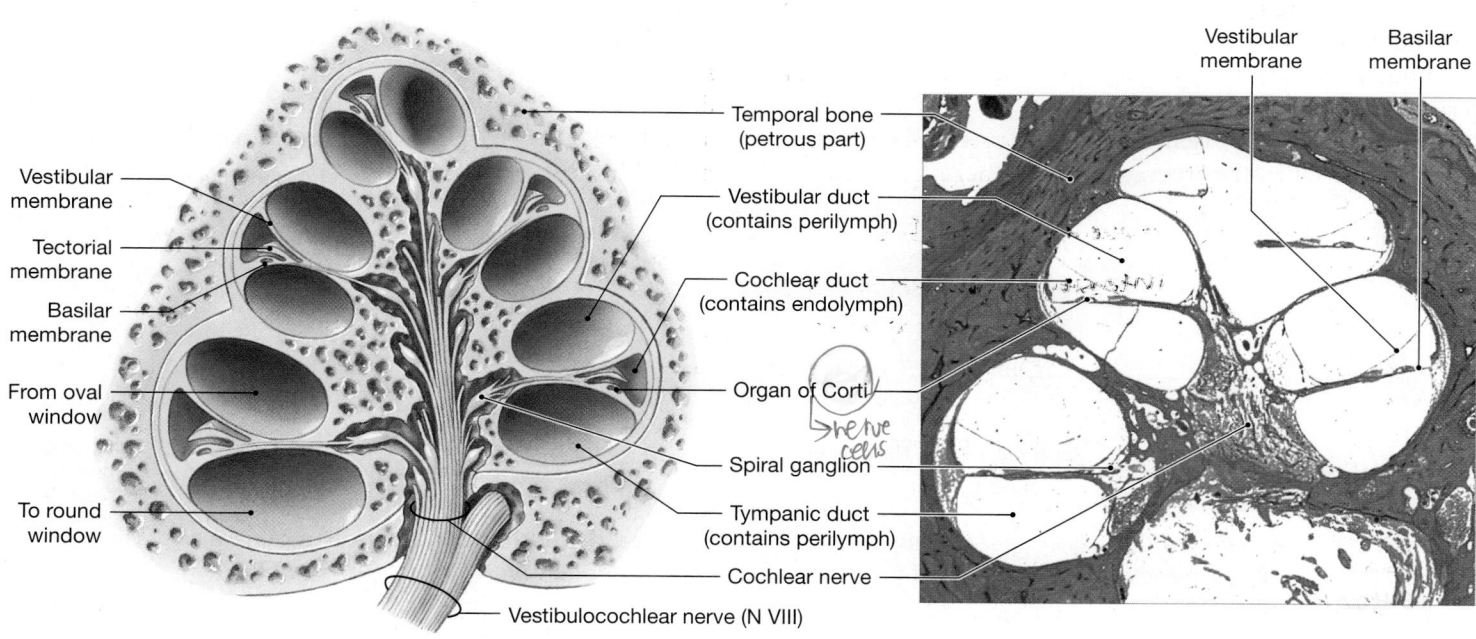

Vestibular membrane

Tectorial membrane

Basilar membrane

From oval window

To round window

Vestibular membrane

Basilar membrane

Temporal bone (petrous part)

Vestibular duct (contains perilymph)

Cochlear duct (contains endolymph)

Organ of Corti

Spiral ganglion

Tympanic duct (contains perilymph)

Cochlear nerve

Vestibulocochlear nerve (N VIII)

(b)

Figure 17–26   The Cochlea. (a) The structure of the cochlea. **(b)** Diagrammatic and sectional views of the cochlear spiral.

The Cochlear Duct

In sectional view (**Figures 17–26** and **17–27**), the cochlear duct, or *scala media*, lies between a pair of perilymphatic chambers: the **vestibular duct** (*scala vestibuli*) and the **tympanic duct** (*scala tympani*). The outer surfaces of these ducts are encased by the bony labyrinth everywhere except at the oval window (the base of the vestibular duct) and the round window (the base of the tympanic duct). Because the vestibular and tympanic ducts are interconnected at the tip of the cochlear spiral, they really form one long and continuous perilymphatic chamber. This chamber begins at the oval window; extends through the vestibular duct, around the top of the cochlea, and along the tympanic duct; and ends at the round window.

The cochlear duct is an elongated tubelike structure suspended between the vestibular duct and the tympanic duct. The hair cells of the cochlear duct are located in a structure called the **organ of Corti** (**Figures 17–26b** and **17–27a,b**). This sensory structure sits on the **basilar membrane**, a membrane that separates the cochlear duct from the tympanic duct. The hair cells are arranged in a series of longitudinal rows. They lack kinocilia, and their stereocilia are in contact with the overlying **tectorial** (tek-TOR-ē-al; *tectum*, roof) **membrane**.

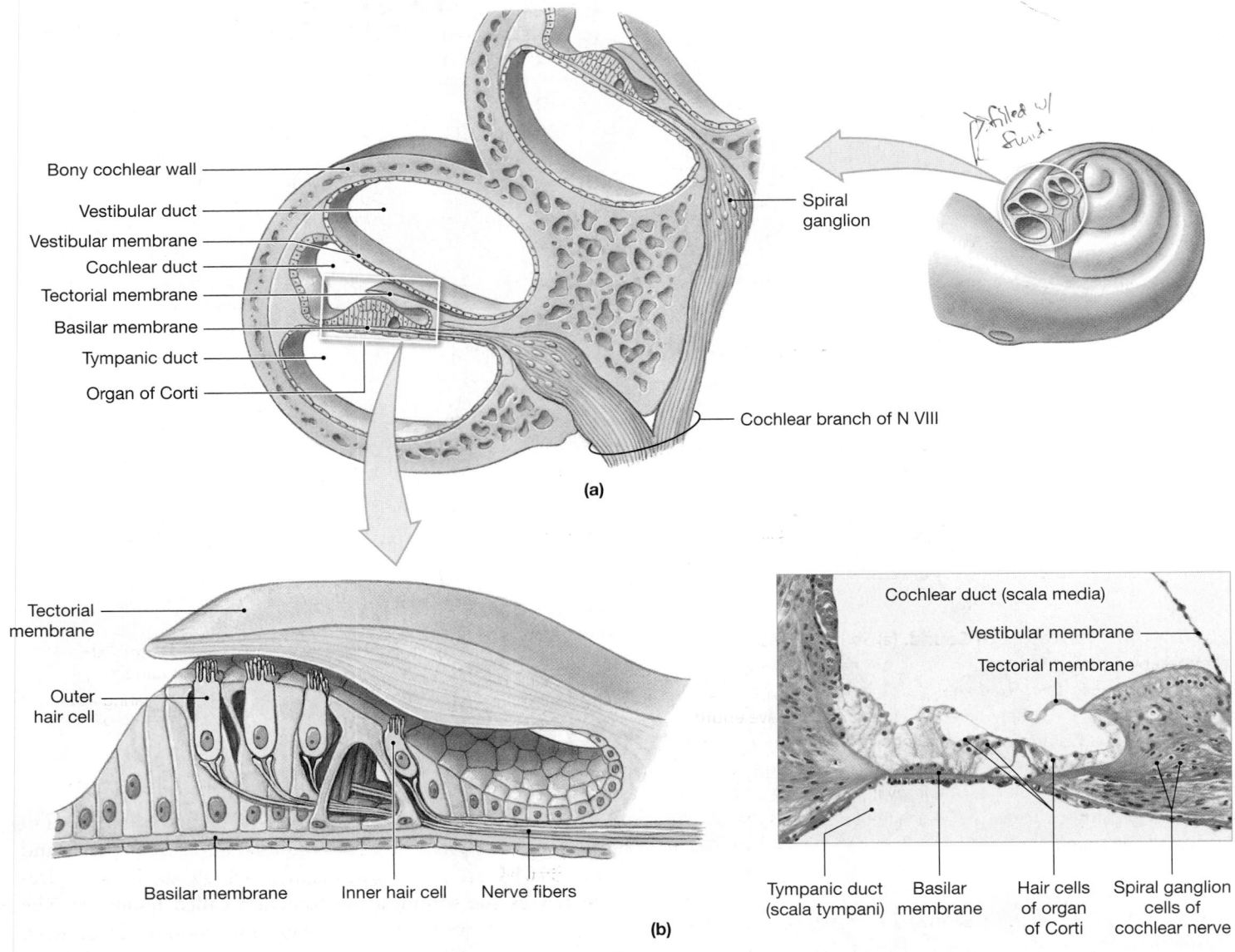

Figure 17–27 The Organ of Corti. (a) A three-dimensional section of the cochlea, showing the compartments, tectorial membrane, and organ of Corti. **(b)** Diagrammatic and sectional views of the receptor hair cell complex of the organ of Corti. (LM × 1233)

This membrane is firmly attached to the inner wall of the cochlear duct. When a portion of the basilar membrane bounces up and down, the stereocilia of the hair cells are pressed against the tectorial membrane and become distorted. The basilar membrane moves in response to pressure fluctuations within the perilymph. These pressure changes are triggered by sound waves arriving at the tympanic membrane. To understand this process, we must consider the basic properties of sound.

An Introduction to Sound

Hearing is the perception of sound, which consists of waves of pressure conducted through a medium such as air or water. In air, each *pressure wave* consists of a region where the air molecules are crowded together and an adjacent zone where they are farther apart (**Figure 17–28a**). These waves are sine waves—that is, S-shaped curves that repeat in a regular pattern—and travel through the air at about 1235 km/h (768 mph).

The *wavelength* of sound is the distance between two adjacent wave crests (peaks) or, equivalently, the distance between two adjacent wave troughs (**Figure 17–28b**). Wavelength is inversely related to **frequency**—the number of waves that pass a fixed reference point in a given time. Physicists use the term **cycles** rather than *waves*. Hence, the frequency of a sound is measured in terms of the number of cycles per second (cps), a unit called **hertz (Hz)**. What we perceive as the **pitch** of a sound is our sensory response to its

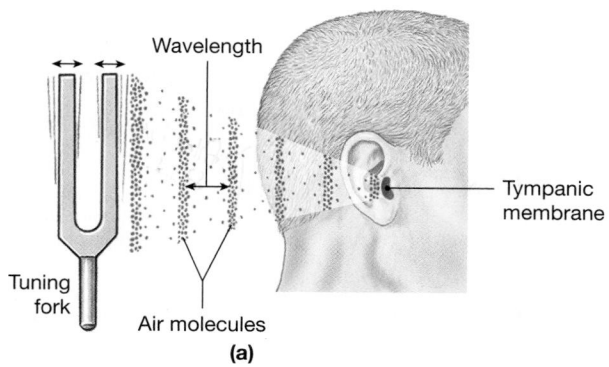

Wavelength

Tympanic membrane

Tuning fork

Air molecules

(a)

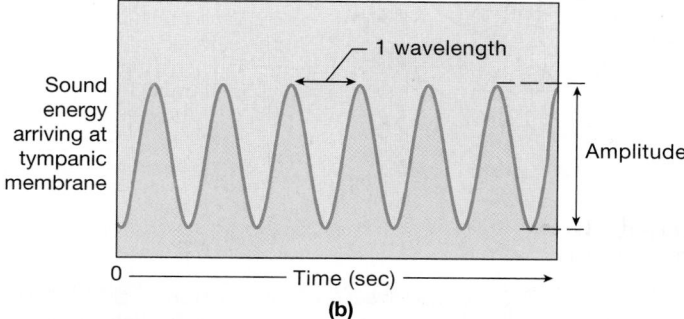

Sound energy arriving at tympanic membrane

1 wavelength

Amplitude

0 ——— Time (sec) ———→

(b)

Figure 17–28   The Nature of Sound. (a) Sound waves (here, generated by a tuning fork) travel through the air as pressure waves. **(b)** A graph showing the sound energy arriving at the tympanic membrane. The distance between wave peaks is the wavelength. The number of waves arriving each second is the frequency, which we perceive as pitch. Frequencies are reported in cycles per second (cps), or hertz (Hz). The amount of energy in each wave determines the wave's amplitude, or intensity, which we perceive as the loudness of the sound.

TABLE 17–1	The Power Content of Representative Sounds	
Typical Decibel Level	**Example**	**Dangerous Time Exposure**
0	Lowest audible sound	
30	Quiet library; soft whisper	
40	Quiet office; living room; bedroom away from traffic	
50	Light traffic at a distance; refrigerator; gentle breeze	
60	Air conditioner at 20 feet; conversation; sewing machine in operation	
70	Busy traffic; noisy restaurant	Some damage if continuous
80	Subway; heavy city traffic; alarm clock at 2 feet; factory noise	More than 8 hours
90	Truck traffic; noisy home appliances; shop tools; gas lawn mower	Less than 8 hours
100	Chain saw; boiler shop; pneumatic drill	2 hours
120	"Heavy metal" rock concert; sandblasting; thunderclap nearby	Immediate danger
140	Gunshot; jet plane	Immediate danger
160	Rocket launching pad	Hearing loss inevitable

frequency. A *high-frequency* sound (high pitch, short wavelength) might have a frequency of 15,000 Hz or more; a very *low-frequency* sound (low pitch, long wavelength) could have a frequency of 100 Hz or less.

It takes energy to produce sound waves. When you strike a tuning fork, it vibrates and pushes against the surrounding air, producing sound waves whose frequency depends on the instrument's frequency of vibration. The harder you strike the tuning fork, the more energy you provide; the energy increases the **amplitude** of the sound wave (**Figure 17–28b**). The measure of energy in a sound wave, or *intensity*, determines how loud it seems; the greater the energy content, the larger the amplitude, and the louder the sound. Sound energy is reported in **decibels** (DES-i-belz, dB). Table 17–1 indicates the decibel levels of familiar sounds.

When sound waves strike an object, their energy is a physical pressure. You may have seen windows move in a room in which a stereo is blasting. The more flexible the object, the more easily it will respond to sound pressure. Even

soft stereo music will vibrate a sheet of paper held in front of the speaker. Given the right combination of frequencies and amplitudes, an object will begin to vibrate at the same frequency as the sound, a phenomenon called *resonance*. The higher the decibel level, the greater the amount of vibration. For you to be able to hear any sound, your thin, flexible tympanic membrane must vibrate in resonance with the sound waves.

Probably more than 6 million people in the United States alone have at least a partial hearing deficit, due to problems with either the transfer of vibrations by the auditory ossicles or damage to the receptors or the auditory pathways.

The Hearing Process

The process of hearing can be divided into six basic steps (**Figure 17–29**):

Step 1   *Sound Waves Arrive at the Tympanic Membrane.* Sound waves enter the external acoustic meatus and travel toward the tympanic membrane. The orientation of the ear canal provides some directional sensitivity. Sound waves approaching a particular side of the head have direct access to the tympanic membrane on that side, whereas

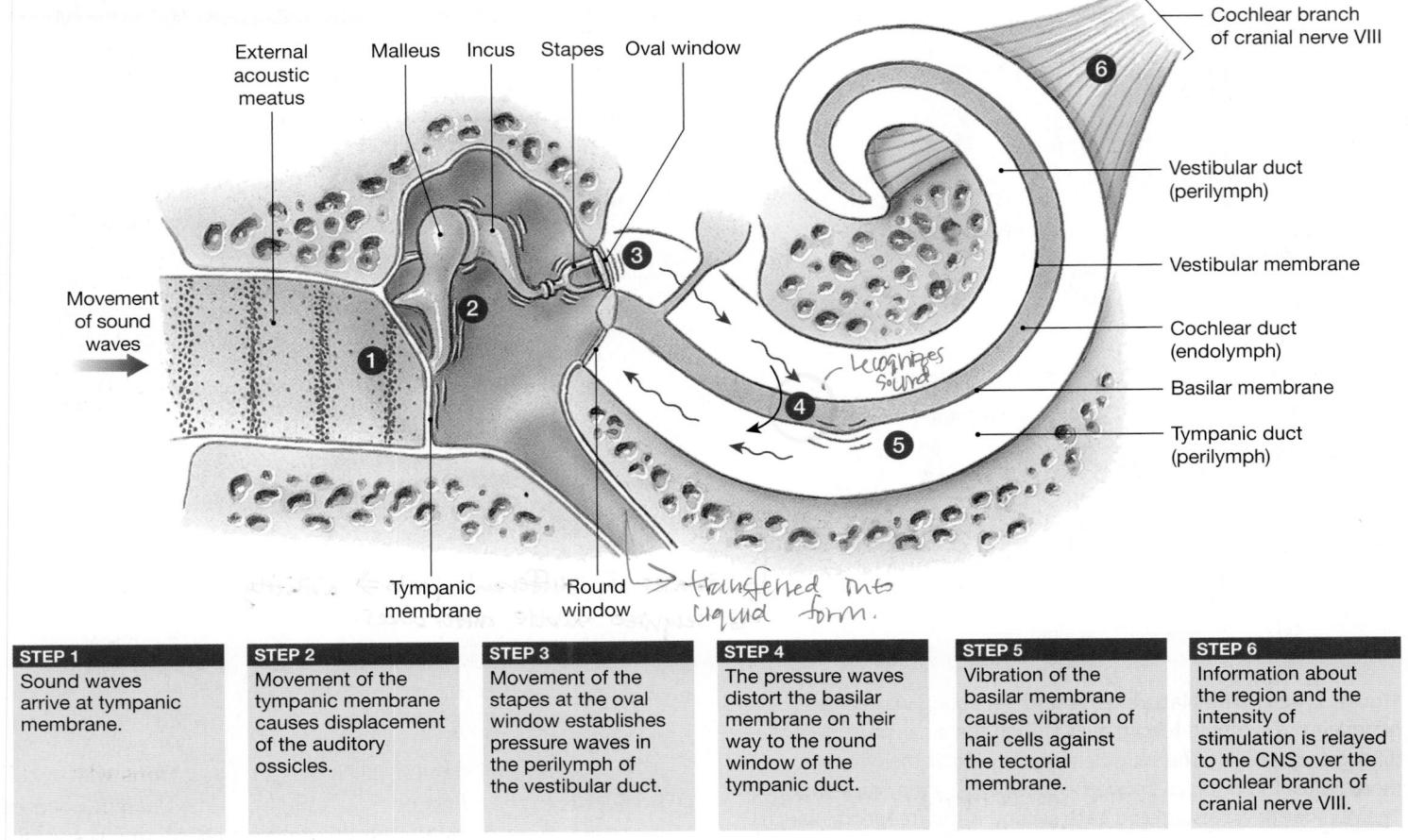

STEP 1	STEP 2	STEP 3	STEP 4	STEP 5	STEP 6
Sound waves arrive at tympanic membrane.	Movement of the tympanic membrane causes displacement of the auditory ossicles.	Movement of the stapes at the oval window establishes pressure waves in the perilymph of the vestibular duct.	The pressure waves distort the basilar membrane on their way to the round window of the tympanic duct.	Vibration of the basilar membrane causes vibration of hair cells against the tectorial membrane.	Information about the region and the intensity of stimulation is relayed to the CNS over the cochlear branch of cranial nerve VIII.

Figure 17–29 **Sound and Hearing.** Steps in the reception and transduction of sound energy.

sounds arriving from another direction must bend around corners or pass through the auricle or other body tissues.

Step 2 *Movement of the Tympanic Membrane Causes Displacement of the Auditory Ossicles.* The tympanic membrane provides a surface for the collection of sound, and it vibrates in resonance to sound waves with frequencies between approximately 20 and 20,000 Hz. When the tympanic membrane vibrates, so do the malleus and, through their articulations, the incus and stapes. In this way, the sound is amplified.

Step 3 *Movement of the Stapes at the Oval Window Establishes Pressure Waves in the Perilymph of the Vestibular Duct.* Liquids are incompressible: If you push down on one part of a water bed, the bed bulges somewhere else. Because the rest of the cochlea is sheathed in bone, pressure applied at the oval window can be relieved only at the round window. Although the stapes actually has a rocking movement, the in–out component is easiest to visualize and describe. Basically, when the stapes moves inward, the round window bulges outward, into the middle ear cavity.

As the stapes moves in and out, vibrating at the frequency of the sound arriving at the tympanic membrane, it creates pressure waves within the perilymph.

Step 4 *The Pressure Waves Distort the Basilar Membrane on Their Way to the Round Window of the Tympanic Duct.* The pressure waves established by the movement of the stapes travel through the perilymph of the vestibular and tympanic ducts to reach the round window. In doing so, the waves distort the basilar membrane. The location of maximum distortion varies with the frequency of the sound, owing to regional differences in the width and flexibility of the basilar membrane along its length. High-frequency sounds, which have a very short wavelength, vibrate the basilar membrane near the oval window. The lower the frequency of the sound, the longer the wavelength, and the farther from the oval window the area of maximum distortion will be (**Figure 17–30a–c**). Thus, information about frequency is translated into information about *position* along the basilar membrane.

The *amount* of movement at a given location depends on the amount of force applied by the stapes, which in turn is a

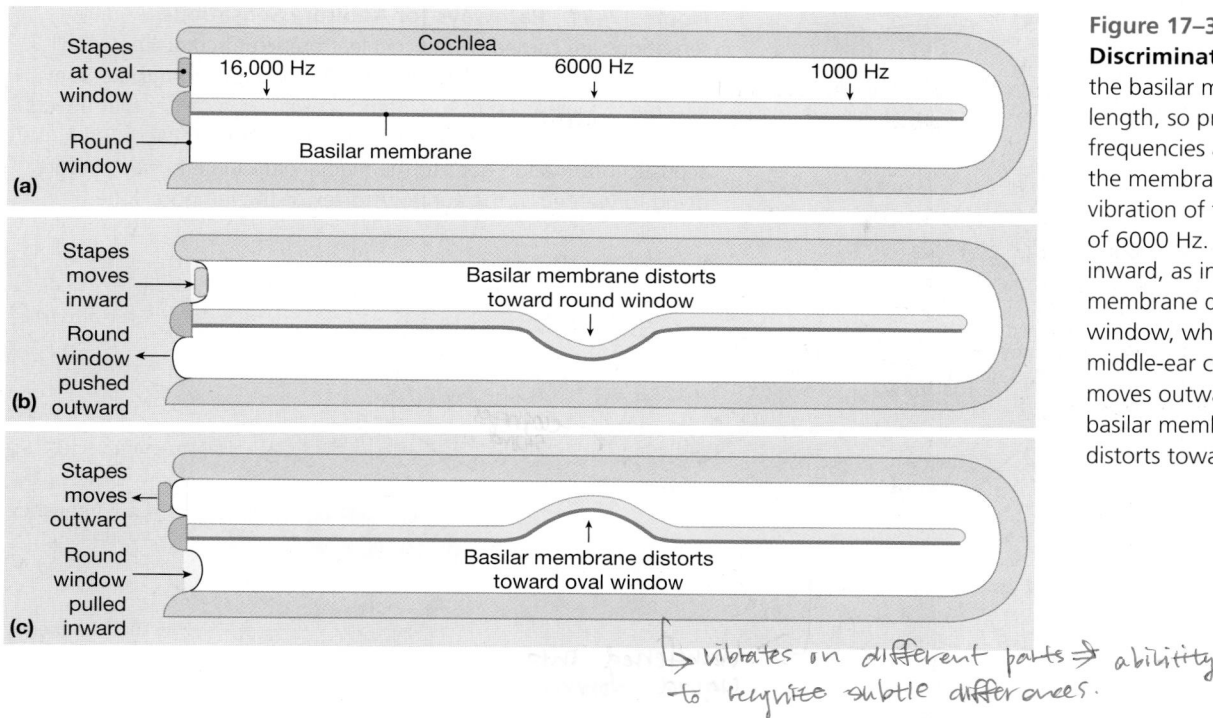

Figure 17–30 Frequency Discrimination. (a) The flexibility of the basilar membrane varies along its length, so pressure waves of different frequencies affect different parts of the membrane. **(b, c)** The effects of a vibration of the stapes at a frequency of 6000 Hz. When the stapes moves inward, as in part (b), the basilar membrane distorts toward the round window, which bulges into the middle-ear cavity. When the stapes moves outward, as in part (c), the basilar membrane rebounds and distorts toward the oval window.

function of energy content of the sound. The louder the sound, the more the basilar membrane moves.

Step 5 *Vibration of the Basilar Membrane Causes Vibration of Hair Cells against the Tectorial Membrane.* Vibration of the affected region of the basilar membrane moves hair cells against the tectorial membrane. This movement leads to the displacement of the stereocilia, which in turn opens ion channels in the hair cells' plasma membranes. The resulting inrush of ions depolarizes the hair cells, leading to the release of neurotransmitters and thus to the stimulation of sensory neurons.

The hair cells of the organ of Corti are arranged in several rows. A very soft sound may stimulate only a few hair cells in a portion of one row. As the intensity of a sound increases, not only do these hair cells become more active, but additional hair cells—at first in the same row and then in adjacent rows—are stimulated as well. The number of hair cells responding in a given region of the organ of Corti thus provides information on the intensity of the sound.

Step 6 *Information about the Region and the Intensity of Stimulation Is Relayed to the CNS over the Cochlear Branch of the Cranial Nerve (VIII).* The cell bodies of the bipolar sensory neurons that monitor the cochlear hair cells are located at the center of the bony cochlea, in the **spiral ganglion** (**Figure 17–27a**). From there, the information is carried by the cochlear branch of cranial nerve VIII to the cochlear nuclei of the medulla oblongata for subsequent distribution to other centers in the brain.

Auditory Pathways

Stimulation of hair cells activates sensory neurons whose cell bodies are in the adjacent spiral ganglion. The afferent fibers of those neurons form the **cochlear branch** of the vestibulocochlear nerve (VIII) (**Figure 17–31**). These axons enter the medulla oblongata, where they split into two groups and synapse at the left and right **dorsal** and **ventral cochlear nuclei**. From there, information ascends to the inferior colliculus of the mesencephalon. This processing center coordinates a number of responses to acoustic stimuli, including auditory reflexes that involve skeletal muscles of the head, face, and trunk. These reflexes automatically change the position of your head in response to a sudden loud noise; you usually turn your head and your eyes toward the source of the sound.

Before reaching the cerebral cortex and your awareness, ascending auditory sensations synapse in the medial geniculate nucleus of the thalamus. Projection fibers then deliver the information to the auditory cortex of the temporal lobe. Information travels to the cortex over labeled lines: High-frequency sounds activate one portion of the cortex, low-frequency sounds another. In effect, the auditory cortex contains a map of the organ of Corti. Thus, information about *frequency*, translated into information about *position* on the basilar membrane, is projected in that form onto the auditory cortex, where it is interpreted to produce your subjective sensation of pitch.

An individual whose auditory cortex is damaged will respond to sounds and have normal acoustic reflexes, but will find it difficult or impossible to interpret the sounds and

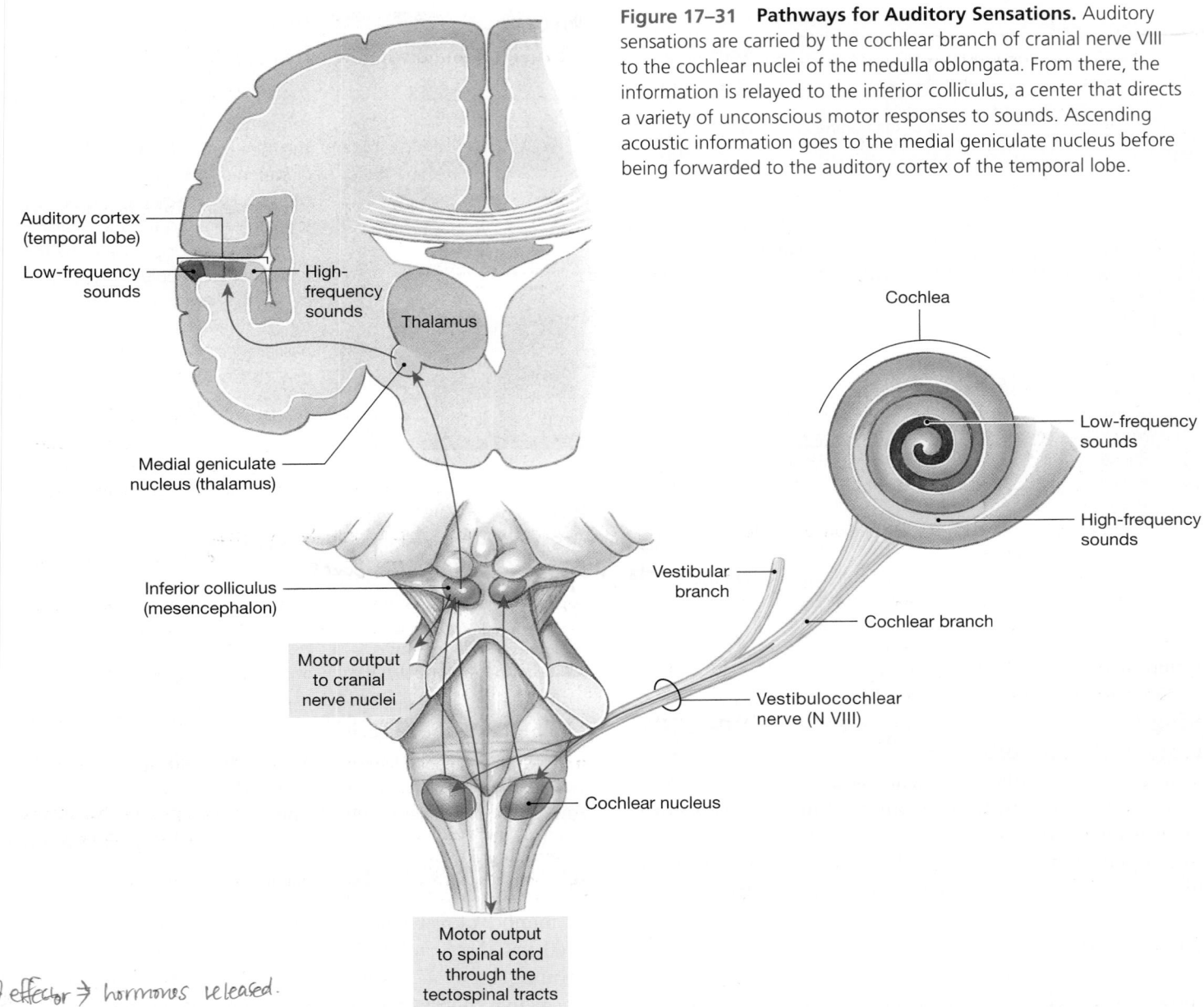

Figure 17–31 **Pathways for Auditory Sensations.** Auditory sensations are carried by the cochlear branch of cranial nerve VIII to the cochlear nuclei of the medulla oblongata. From there, the information is relayed to the inferior colliculus, a center that directs a variety of unconscious motor responses to sounds. Ascending acoustic information goes to the medial geniculate nucleus before being forwarded to the auditory cortex of the temporal lobe.

recognize a pattern in them. Damage to the adjacent association area leaves the ability to detect the tones and patterns intact, but produces an inability to comprehend their meaning.

Auditory Sensitivity

Our hearing abilities are remarkable, but it is difficult to assess the absolute sensitivity of the system. The range from the softest audible sound to the loudest tolerable blast represents a trillionfold increase in power. The receptor mechanism is so sensitive that, if we were to remove the stapes, we could, in theory, hear air molecules bouncing off the oval window. We never use the full potential of this system, because body movements and our internal organs produce squeaks, groans,

thumps, and other sounds that are tuned out by central and peripheral adaptation. When other environmental noises fade away, the level of adaptation drops and the system becomes increasingly sensitive. For example, when you relax in a quiet room, your heartbeat seems to get louder and louder as the auditory system adjusts to the level of background noise.

Young children have the greatest hearing range: They can detect sounds ranging from a 20-Hz buzz to a 20,000-Hz whine. With age, damage due to loud noises or other injuries accumulates. The tympanic membrane gets less flexible, the articulations between the ossicles stiffen, and the round window may begin to ossify. As a result, older individuals show some degree of hearing loss.

The A&P Top 100

#60 Balance and hearing rely on the same basic types of sensory receptors (hair cells). The anatomical structure of the associated sense organ determines what stimuli affect the hair cells. In the semicircular ducts, the stimulus is fluid movement caused by head rotation in the horizontal, sagittal, or frontal planes. In the utricle and saccule, the stimuli are gravity-induced shifts in the position of attached otoliths. In the cochlea, the stimulus is movement of the tectorial membrane as pressure waves distort the basilar membrane.

CHECKPOINT

14. If the round window were not able to bulge out with increased pressure in the perilymph, how would the perception of sound be affected?

15. How would the loss of stereocilia from hair cells of the organ of Corti affect hearing?

16. Why would blockage of the auditory tube produce an earache?

See the blue Answers tab at the end of the book.

Related Clinical Terms

astigmatism: Reduction in visual acuity due to a curvature irregularity in the cornea or lens.

cataract: An abnormal condition in which the lens has lost its transparency.

color blindness: A condition in which a person is unable to distinguish certain colors.

conductive deafness: Deafness resulting from conditions in the outer or middle ear that block the transfer of vibrations from the tympanic membrane to the oval window.

detached retina: Delamination of a portion of the neural retina, which separates the photoreceptor layer from the pigment layer. If untreated, blindness can result in the affected area.

diabetic retinopathy: Deterioration of the retinal photoreceptor layer due to vascular damage and the overgrowth and rupture of blood vessels on the retinal surface.

glaucoma: A condition characterized by increased intraocular pressure due to the impaired reabsorption of aqueous humor; can result in blindness.

hyperopia, or *farsightedness*: A condition in which nearby objects are blurry, but distant objects are clear.

motion sickness: A condition resulting from conflicting visual and equilibrium sensory stimuli. Signs and symptoms can include headache, sweating, nausea, vomiting, and changes in mental state.

myopia, or *nearsightedness*: A condition in which vision at close range is normal, but distant objects appear blurry.

nerve deafness: Deafness resulting from problems within the cochlea or along the auditory nerve pathway.

night blindness: Loss of visual acuity under dim light conditions due to inadequate visual pigment production, usually as a result of vitamin A deficiency.

nystagmus: Abnormal eye movements that may appear after damage to the brain stem or inner ear.

otitis media: Infection and tissue inflammation within the middle ear cavity.

presbyopia: A type of hyperopia that develops with age as lenses become less elastic.

retinitis pigmentosa: A group of inherited retinopathies characterized by the progressive deterioration of photoreceptors, eventually resulting in blindness.

scotomas: Abnormal blind spots that are fixed in position.

Chapter Review

Study Outline

17-1 Olfaction, the sense of smell, involves olfactory receptors responding to chemical stimuli p. 562

1. The **olfactory organs** contain the **olfactory epithelium** with **olfactory receptors**, supporting cells, and **basal (stem) cells**. The surfaces of the olfactory organs are coated with the secretions of the **olfactory glands**. (*Figure 17–1*)
2. The olfactory receptors are highly modified neurons.
3. Olfactory reception involves detecting dissolved chemicals as they interact with odorant-binding proteins.
4. In olfaction, the arriving information reaches the information centers without first synapsing in the thalamus. (*Figure 17–1*)

5. The olfactory system can distinguish thousands of chemical stimuli. The CNS interprets smells by the pattern of receptor activity.
6. The olfactory receptor population shows considerable turnover. The number of olfactory receptors declines with age.

17-2 Gustation, the sense of taste, involves taste receptors responding to chemical stimuli p. 564

7. **Taste** (gustatory) **receptors** are clustered in **taste buds**.
8. Taste buds are associated with epithelial projections (*lingual papillae*) on the dorsal surface of the tongue. (*Figure 17–2*)

9. Each taste bud contains **basal cells**, which appear to be stem cells, and **gustatory cells**, which extend *taste hairs* through a narrow **taste pore**. (*Figure 17–2*)

10. The taste buds are monitored by cranial nerves that synapse within the solitary nucleus of the medulla oblongata. Postsynaptic neurons carry the nerve impulses on to the thalamus, where third-order neurons project to the primary sensory cortex.

11. The **primary taste sensations** are sweet, salt, sour, and bitter. Receptors also exist for **umami** and **water**.

12. Taste sensitivity exhibits significant individual differences, some of which are inherited.

13. The number of taste buds declines with age.

17-3 Internal eye structures contribute to vision, while accessory eye structures provide protection p. 566

14. The **accessory structures** of the eye include the eyelids (**palpebrae**), separated by the **palpebral fissure**, the **eyelashes**, and the **tarsal glands**. (*Figures 17–3, 17–4*)

15. An epithelium called the **conjunctiva** covers most of the exposed surface of the eye. The **cornea** is transparent. (*Figures 17–3, 17–4*)

16. The secretions of the **lacrimal gland** contain **lysozyme**. Tears collect in the *lacrimal lake* and reach the inferior meatus of the nose after they pass through the **lacrimal puncta**, the **lacrimal canaliculi**, the **lacrimal sac**, and the **nasolacrimal duct**. (*Figure 17–3*)

17. The eye has three layers: an outer **fibrous tunic**, a middle **vascular tunic**, and an inner **neural tunic**. (*Figure 17–4*)

18. The fibrous tunic consists of the **sclera**, the cornea, and the **corneal limbus**. (*Figure 17–4*)

19. The vascular tunic, or **uvea**, includes the **iris**, the **ciliary body**, and the **choroid**. The iris contains muscle fibers that change the diameter of the **pupil**. The ciliary body contains the **ciliary muscle** and the **ciliary processes**, which attach to the **suspensory ligaments** of the *lens*. (*Figures 17–4, 17–5*)

20. The neural tunic, or **retina**, consists of an outer *pigmented part* and an inner *neural part*; the latter contains visual receptors and associated neurons. (*Figures 17–4, 17–6*)

21. The retina contains two types of **photoreceptors**: **rods** and **cones**.

22. Cones are densely clustered in the **fovea**, at the center of the **macula lutea**. (*Figure 17–6*)

23. The direct line to the CNS proceeds from the photoreceptors to **bipolar cells**, then to **ganglion cells**, and, finally, to the brain via the optic nerve. The axons of ganglion cells converge at the **optic disc**, or **blind spot**. **Horizontal cells** and **amacrine cells** modify the signals passed among other components of the retina. (*Figures 17–6, 17–7*)

24. The ciliary body and lens divide the interior of the eye into a large **posterior cavity**, or *vitreous chamber*, and a smaller **anterior cavity**. The anterior cavity is subdivided into the **anterior chamber**, which extends from the cornea to the iris, and a **posterior chamber**, between the iris and the ciliary body and lens. (*Figure 17–8*)

25. The fluid **aqueous humor** circulates within the eye and reenters the circulation after diffusing through the walls of the anterior chamber and into the **canal of Schlemm**. (*Figure 17–8*)

26. The **lens** lies posterior to the cornea and forms the anterior boundary of the posterior cavity. This cavity contains the **vitreous body**, a gelatinous mass that helps stabilize the shape of the eye and support the retina. (*Figure 17–8*)

27. The lens focuses a visual image on the photoreceptors. The condition in which a lens has lost its transparency is a **cataract**.

28. Light is **refracted** (bent) when it passes through the cornea and lens. During **accommodation**, the shape of the lens changes to focus an image on the retina. "Normal" **visual acuity** is rated 20/20. (*Figures 17–9 to 17–12*)

17-4 Photoreceptors respond to light and change it into electrical signals essential to visual physiology p. 578

29. The two types of photoreceptors are rods, which respond to almost any photon, regardless of its energy content, and cones, which have characteristic ranges of sensitivity. (*Figure 17–13*)

30. Each photoreceptor contains an **outer segment** with membranous **discs**. A narrow stalk connects the outer segment to the **inner segment**. Light absorption occurs in the **visual pigments**, which are derivatives of **rhodopsin** (opsin plus the pigment retinal, which is synthesized from vitamin A). (*Figures 17–13 to 17–15*)

31. Color sensitivity depends on the integration of information from **red**, **green**, and **blue** cones. **Color blindness** is the inability to detect certain colors. (*Figures 17–16, 17–17*)

32. In the **dark-adapted state**, most visual pigments are fully receptive to stimulation. In the **light-adapted state**, the pupil constricts and **bleaching** of the visual pigments occurs.

33. The ganglion cells that monitor rods, called **M cells** (*magnocells*), are relatively large. The ganglion cells that monitor cones, called **P cells** (*parvo cells*), are smaller and more numerous. (*Figure 17–18*)

34. Visual data from the left half of the combined field of vision arrive at the visual cortex of the right occipital lobe; data from the right half of the combined field of vision arrive at the visual cortex of the left occipital lobe. (*Figure 17–19*)

35. **Depth perception** is obtained by comparing relative positions of objects between the left- and right-eye images. (*Figure 17–19*)

36. Visual inputs to the suprachiasmatic nucleus of the hypothalamus affect the function of other brain stem nuclei. This nucleus establishes a visceral **circadian rhythm**, which is tied to the day–night cycle and affects other metabolic processes.

17-5 Equilibrium sensations originate within the inner ear, while hearing involves the detection and interpretation of sound waves p. 585

37. The senses of equilibrium and hearing are provided by the receptors of the inner ear.

38. The ear is divided into the **external ear**, the **middle ear**, and the **inner ear**. (*Figure 17–20*)

39. The external ear includes the **auricle**, or *pinna*, which surrounds the entrance to the **external acoustic meatus**, which ends at the **tympanic membrane** (*eardrum*). (*Figure 17–20*)

40. The middle ear communicates with the nasopharynx via the **auditory** (*pharyngotympanic*) **tube**. The middle ear encloses and protects the **auditory ossicles**. (*Figures 17–20, 17–21*)

41. The **membranous labyrinth** (the chambers and tubes) of the inner ear contains the fluid **endolymph**. The **bony labyrinth** surrounds and protects the membranous labyrinth and can be subdivided into the **vestibule**, the **semicircular canals**, and the **cochlea**. (*Figures 17–20, 17–22*)

42. The vestibule of the inner ear encloses the **saccule** and **utricle**. The semicircular canals contain the **semicircular ducts**. The cochlea contains the **cochlear duct**, an elongated portion of the membranous labyrinth. (*Figure 17–22*)

43. The **round window** separates the **perilymph** from the air spaces of the middle ear. The **oval window** is connected to the base of the stapes. (*Figure 17–20*)

44. The basic receptors of the inner ear are **hair cells**, which provide information about the direction and strength of mechanical stimuli. (*Figure 17–23*)

45. The **anterior**, **posterior**, and **lateral semicircular ducts** are continuous with the utricle. Each duct contains an **ampulla** with a gelatinous **cupula** and associated sensory receptors. (*Figures 17–22, 17–23*)

46. The saccule and utricle are connected by a passageway that is continuous with the **endolymphatic duct**, which terminates in the **endolymphatic sac**. In the saccule and utricle, hair cells cluster within **maculae**, where their cilia contact the **otolith** (densely packed mineral crystals, called **statoconia**, in a matrix). (*Figures 17–23, 17–24*)

47. The vestibular receptors activate sensory neurons of the **vestibular ganglia**. The axons form the **vestibular branch** of the vestibulocochlear nerve (N VIII), synapsing within the **vestibular nuclei**. (*Figure 17–25*)

48. The cochlear duct lies between the **vestibular duct** and the **tympanic duct**. The hair cells of the cochlear duct lie within the **organ of Corti**. (*Figures 17–26, 17–27*)

49. The energy content of a sound determines its *intensity*, measured in **decibels**. Sound waves travel toward the tympanic membrane, which vibrates; the auditory ossicles conduct these vibrations to the inner ear. Movement at the oval window applies pressure to the perilymph of the vestibular duct. (*Figures 17–28, 17–29; Table 17–1*)

50. Pressure waves distort the **basilar membrane** and push the hair cells of the organ of Corti against the **tectorial membrane**. The **tensor tympani** and **stapedius muscles** contract to reduce the amount of motion when very loud sounds arrive. (*Figures 17–29, 17–30*)

51. The sensory neurons are located in the **spiral ganglion** of the cochlea. The afferent fibers of these neurons form the **cochlear branch** of the vestibulocochlear nerve (VIII), synapsing at the **cochlear nuclei**. (*Figure 17–31*)

Review Questions

See the blue Answers tab at the end of the book.

myA&P Access more review material online at **myA&P**™ (www.myaandp.com). There, you'll find chapter guides, chapter quizzes, practice tests, animations, flashcards, a glossary with pronunciations, *Interactive Physiology*® (IP) exercises and quizzes, and more to help you succeed in the course.

LEVEL 1 Reviewing Facts and Terms

1. Identify the structures in the following horizontal section of the eye.

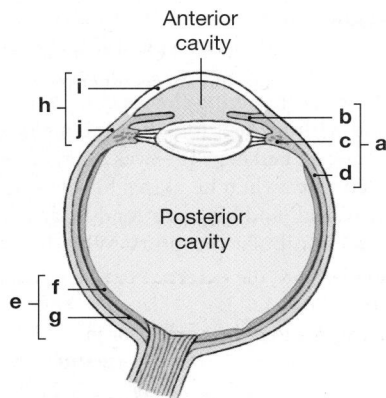

(b) Horizontal section of right eye

(a) _____ (b) _____
(c) _____ (d) _____
(e) _____ (f) _____
(g) _____ (h) _____
(i) _____ (j) _____

2. A reduction in sensitivity in the presence of a constant stimulus is
(a) transduction. (b) sensory coding.
(c) line labeling. (d) adaptation.

3. A blind spot occurs in the retina where
(a) the fovea is located.
(b) ganglion cells synapse with bipolar cells.
(c) the optic nerve attaches to the retina.
(d) rod cells are clustered to form the macula.
(e) amacrine cells are located.

4. Sound waves are converted into mechanical movements by the
(a) auditory ossicles. (b) cochlea.
(c) oval window. (d) round window.
(e) tympanic membrane.

5. The basic receptors in the inner ear are the
(a) utricles. (b) saccules.
(c) hair cells. (d) supporting cells.
(e) ampullae.

6. The retina is
(a) the vascular tunic. (b) the fibrous tunic.
(c) the neural tunic. (d) all of these.

7. At sunset, your visual system adapts to
(a) fovea vision. (b) rod-based vision.
(c) macular vision. (d) cone-based vision.

8. A better-than-average visual acuity rating is
(a) 20/20. (b) 20/30.
(c) 15/20. (d) 20/15.

9. The malleus, incus, and stapes are the tiny bones located in the
(a) outer ear. (b) middle ear.
(c) inner ear. (d) membranous labyrinth.

10. Identify the structures of the external, middle, and inner ear in the following figure.

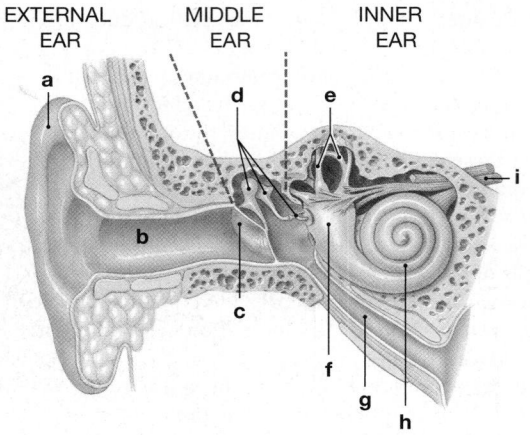

EXTERNAL EAR   MIDDLE EAR   INNER EAR

(a) _____   (b) _____
(c) _____   (d) _____
(e) _____   (f) _____
(g) _____   (h) _____
(i) _____

11. Receptors in the saccule and utricle provide sensations of
    (a) angular acceleration.
    (b) hearing.
    (c) vibration.
    (d) gravity and linear acceleration.

12. The organ of Corti is located in the _____ of the inner ear.
    (a) utricle   (b) bony labyrinth
    (c) vestibule   (d) cochlea

13. Auditory information about the frequency and intensity of stimulation is relayed to the CNS over the cochlear branch of cranial nerve
    (a) IV.   (b) VI.
    (c) VIII.   (d) X.

14. What are the three types of papillae on the human tongue?

15. (a) What structures make up the fibrous tunic of the eye?
    (b) What are the functions of the fibrous tunic?

16. What structures make up the vascular tunic of the eye?

17. What are the three auditory ossicles in the middle ear, and what are their functions?

LEVEL 2 Reviewing Concepts

18. Trace the olfactory pathway from the time an odor reaches the olfactory epithelium until nerve impulses reach their final destination in the brain.

19. Why are olfactory sensations long-lasting and an important part of our memories and emotions?

20. What is the usual result if a sebaceous gland of an eyelash or a tarsal gland becomes infected?

21. Displacement of sterocilia toward the kinocilium of a hair cell
    (a) produces a depolarization of the membrane.
    (b) produces a hyperpolarization of the membrane.
    (c) decreases the membrane permeability to sodium ions.
    (d) increases the membrane permeability to potassium ions.
    (e) does not affect the transmembrane potential of the cell.

22. Damage to the cupula of the lateral semicircular duct would interfere with the perception of
    (a) the direction of gravitational pull.
    (b) linear acceleration.
    (c) horizontal rotation of the head.
    (d) vertical rotation of the head.
    (e) angular rotation of the head.

23. When viewing an object *close* to you, your lens should be more _____.
    (a) rounded   (b) flattened
    (c) concave   (d) lateral
    (e) medial

LEVEL 3 Critical Thinking and Clinical Applications

24. You are at a park watching some deer 35 feet away from you. A friend taps you on the shoulder to ask a question. As you turn to look at your friend, who is standing just 2 feet away, what changes would your eyes undergo?

25. Your friend Shelly suffers from myopia (nearsightedness). You remember from your physics class that concave lenses cause light waves to spread or diverge and that convex lenses cause light waves to converge. What type of corrective lenses would you suggest to your friend?
    (a) concave lenses   (b) convex lenses

26. Tom has surgery to remove polyps (growths) from his sinuses. After he heals from the surgery, he notices that his sense of smell is not as keen as it was before the surgery. Can you suggest a reason for this?

27. For a few seconds after you ride the express elevator from the 20th floor to the ground floor, you still feel as if you are descending, even though you have come to a stop. Why?

28. Juan has a disorder involving the saccule and the utricle. He is asked to stand with his feet together and arms extended forward. As long as he keeps his eyes open, he exhibits very little movement. But when he closes his eyes, his body begins to sway a great deal, and his arms tend to drift in the direction of the impaired vestibular receptors. Why does this occur?

The Endocrine System

18

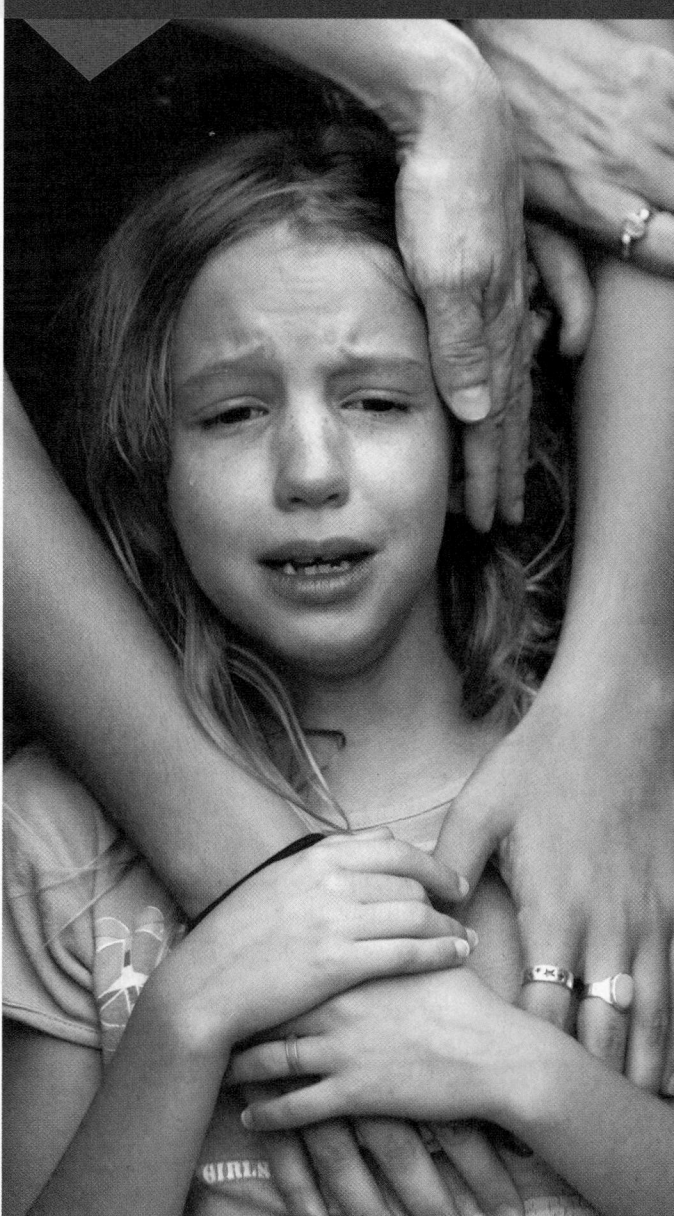

Learning Outcomes

After completing this chapter, you should be able to do the following:

18-1 Explain the importance of intercellular communication, describe the mechanisms involved, and compare the modes of intercellular communication that occur in the endocrine and nervous systems.

18-2 Compare the cellular components of the endocrine system with those of other systems, contrast the major structural classes of hormones, and explain the general mechanisms of hormonal action on target organs.

18-3 Describe the location, hormones, and functions of the pituitary gland, and discuss the effects of abnormal pituitary hormone production.

18-4 Describe the location, hormones, and functions of the thyroid gland, and discuss the effects of abnormal thyroid hormone production.

18-5 Describe the location, hormone, and functions of the parathyroid glands, and discuss the effects of abnormal parathyroid hormone production.

18-6 Describe the location, structure, hormones, and general functions of the suprarenal glands, and discuss the effects of abnormal adrenal hormone production.

18-7 Describe the location of the pineal gland, and discuss the functions of the hormone it produces.

18-8 Describe the location, structure, hormones, and functions of the pancreas, and discuss the effects of abnormal pancreatic hormone production.

18-9 Describe the functions of the hormones produced by the kidneys, heart, thymus, testes, ovaries, and adipose tissue.

18-10 Explain how hormones interact to produce coordinated physiological responses and influence behavior, describe the role of hormones in the general adaptation syndrome, and discuss how aging affects hormone production.

18-11 Give examples of interactions between the endocrine system and each of the other organ systems.

Clinical Notes

Diabetes Insipidus p. 618
Diabetes Mellitus p. 634
Endocrine Disorders p. 639
Hormones and Athletic Performance p. 643

An Introduction to the Endocrine System

This chapter examines the structural and functional organization of the endocrine system, in the process comparing it to what we have learned about the nervous system. After an overview of the endocrine system and the characteristics of the hormones it produces, we consider the structure and function of the body's various endocrine glands. Finally, we examine the mechanisms involved in the hormonal modification of metabolic operations, and the interactions between the endocrine system and other body systems. We begin by considering the role of intercellular communication in maintaining homeostasis.

18-1 Homeostasis is preserved through intercellular communication

To preserve homeostasis, cellular activities must be coordinated throughout the body. Neurons monitor or control specific cells or groups of cells. However, the number of cells innervated is only a small fraction of the total number of cells in the body, and the commands issued are very specific and of relatively brief duration. Many life processes are not short-lived; reaching adult stature takes decades. Maintenance of reproductive capabilities requires continual control for at least 30 years in a typical female, and even longer in males. There is no way that the nervous system can regulate such long-term processes as growth, development, or reproduction, which involve or affect metabolic activities in virtually every cell and tissue. This type of regulation is provided by the endocrine system, which uses chemical messengers to relay information and instructions between cells. To understand how these messages are generated and interpreted, we need to take a closer look at how cells communicate with one another.

In a few specialized cases, cellular activities are coordinated by the exchange of ions and molecules between adjacent cells across gap junctions. This **direct communication** occurs between two cells of the same type, and the cells must be in extensive physical contact. The two cells communicate so closely that they function as a single entity. Gap junctions (1) coordinate ciliary movement among epithelial cells, (2) coordinate the contractions of cardiac muscle cells, and (3) facilitate the propagation of action potentials from one neuron to the next at electrical synapses.

Direct communication is highly specialized and relatively rare. Most of the communication between cells involves the release and receipt of chemical messages. Each cell continuously "talks" to its neighbors by releasing chemicals into the extracellular fluid. These chemicals tell cells what their neighbors are doing at any moment. The result is the coordination of tissue function at the local level. The use of chemical messengers to transfer information from cell to cell within a single tissue is called **paracrine communication**. The chemicals involved are called paracrine factors, also known as *local hormones*. Examples of paracrine factors include the prostaglandins, introduced in Chapter 2, and the various growth factors, discussed in Chapter 3. ∞ pp. 49, 104

Paracrine factors enter the bloodstream, but their concentrations are usually so low that distant cells and tissues are not affected. However, some paracrine factors, including several of the prostaglandins and related chemicals, have primary effects in their tissues of origin and secondary effects in other tissues and organs. When secondary effects occur, the paracrine factors are also acting as **hormones**—chemical messengers that are released in one tissue and transported in the bloodstream to alter the activities of specific cells in other tissues. Whereas most cells release paracrine factors, typical hormones are produced only by specialized cells. Nevertheless, the difference between paracrine factors and hormones is mostly a matter of degree. Paracrine factors can diffuse out of their tissue of origin and have widespread effects, and hormones can affect their tissues of origin as well as distant cells. By convention, a substance with effects outside its tissue of origin is called a *hormone* if its chemical structure is known, and a *factor* if that structure remains to be determined.

In intercellular communication, hormones are like letters, and the cardiovascular system is the postal service. A hormone released into the bloodstream is distributed throughout the body. Each hormone has **target cells**, specific cells that possess the receptors needed to bind and "read" the hormonal message when it arrives. But hormones are really like bulk mail advertisements—cells throughout the body are exposed to them whether or not they have the necessary receptors. At any moment, each individual cell can respond to only a few of the hormones present. The other hormones are ignored, because the cell lacks the receptors to read the messages they contain. The activity of hormones in coordinating cellular activities in tissues in distant portions of the body is called **endocrine communication**.

Hormones alter the operations of target cells by changing the types, quantities, or activities of important enzymes and structural proteins. In other words, a hormone may

- stimulate the synthesis of an enzyme or a structural protein not already present in the cytoplasm by activating appropriate genes in the cell nucleus;

- increase or decrease the rate of synthesis of a particular enzyme or other protein by changing the rate of transcription or translation; or

- turn an existing enzyme or membrane channel "on" or "off" by changing its shape or structure.

Through one or more of these mechanisms, a hormone can modify the physical structure or biochemical properties of its target cells. Because the target cells can be anywhere in the body, a single hormone can alter the metabolic activities of multiple tissues and organs simultaneously. These effects may be slow to appear, but they typically persist for days. Consequently, hormones are effective in coordinating cell, tissue, and organ activities on a sustained, long-term basis. For example, circulating hormones keep body water content and levels of electrolytes and organic nutrients within normal limits 24 hours a day throughout our entire lives.

[handwritten: Single hormone alters metab. of tissues & organs simult.]

Cells can respond to several different hormones simultaneously. Gradual changes in the quantities and identities of circulating hormones can therefore produce complex changes in the body's physical structure and physiological capabilities. Examples include the processes of embryological and fetal development, growth, and puberty. Hormonal regulation is thus quite suitable for directing gradual, coordinated processes, but it is totally unable to handle situations requiring split-second responses. That kind of crisis management is the job of the nervous system.

Although the nervous system also relies primarily on chemical communication, it does not send messages through the bloodstream. Instead, neurons release a neurotransmitter at a synapse very close to target cells that bear the appropriate receptors. The command to release the neurotransmitter rapidly travels from one location to another in the form of action potentials propagated along axons. The nervous system thus acts like a telephone company, with a cable network carrying high-speed "messages" to specific destinations throughout the body. The effects of neural stimulation are generally short-lived, and they tend to be restricted to specific target cells—primarily because the neurotransmitter is rapidly broken down or recycled. This form of **synaptic communication** is ideal for crisis management: If you are in danger of being hit by a speeding bus, the nervous system can coordinate and direct your leap to safety. Once the crisis is over and the neural circuits quiet down, things soon return to normal.

Table 18–1 summarizes the ways cells and tissues communicate with one another. Viewed from a general perspective, the differences between the nervous and endocrine systems seem relatively clear. In fact, these broad organizational and

[side tab: 18 ENDOCRINE]

TABLE 18–1 Mechanisms of Intercellular Communication

Mechanism	Transmission	Chemical Mediators	Distribution of Effects
Direct communication *[handwritten: → physically interactive]*	Through gap junctions	Ions, small solutes, lipid-soluble materials	Usually limited to adjacent cells of the same type that are interconnected by connexons
Paracrine communication 	Through extracellular fluid	Paracrine factors	Primarily limited to local area, where concentrations are relatively high. Target cells must have appropriate receptors
Endocrine communication *[handwritten: → sends hormones thru vascular system to destination]*	Through the circulatory system	Hormones	Target cells are primarily in other tissues and organs and must have appropriate receptors
Synaptic communication *[handwritten: → very close to target ∴ specific, but short influence.]*	Across synaptic clefts	Neurotransmitters	Limited to very specific area. Target cells must have appropriate receptors

functional distinctions are the basis for treating them as two separate systems. Yet when we consider them in detail, we see that the two systems are similarly organized:

- Both systems rely on the release of chemicals that bind to specific receptors on their target cells.

- The two systems share many chemical messengers; for example, norepinephrine and epinephrine are called *hormones* when released into the bloodstream, but *neurotransmitters* when released across synapses.

- Both systems are regulated primarily by negative feedback control mechanisms.

- The two systems share a common goal: to preserve homeostasis by coordinating and regulating the activities of other cells, tissues, organs, and systems.

Next we introduce the components and functions of the endocrine system and further explore the interactions between the nervous and endocrine systems. We will consider specific endocrine organs, hormones, and functions in detail in later chapters.

CHECKPOINT

1. Define hormone.

2. Describe paracrine communication.

3. Identify four mechanisms of intercellular communication.

See the blue Answers tab at the end of the book.

18-2 The endocrine system regulates physiological processes through the binding of hormones to receptors

The **endocrine system** includes all the endocrine cells and tissues of the body that produce hormones or paracrine factors with effects beyond their tissues of origin. As noted in Chapter 4, *endocrine cells* are glandular secretory cells that release their secretions into the extracellular fluid. This characteristic distinguishes them from *exocrine cells*, which secrete their products onto epithelial surfaces, generally by way of ducts. ∞ p. 121 The chemicals released by endocrine cells may affect only adjacent cells, as in the case of most paracrine factors, or they may affect cells throughout the body.

The tissues, organs, and hormones of the endocrine system are introduced in **Figure 18–1**. Some of these organs, such as the pituitary gland, have endocrine secretion as a primary function. Others, such as the pancreas, have many other func-

tions in addition to endocrine secretion; chapters on other systems consider such endocrine organs in more detail.

Classes of Hormones

Hormones can be divided into three groups on the basis of their chemical structure: (1) *amino acid derivatives*, (2) *peptide hormones*, and (3) *lipid derivatives* (**Figure 18–2**).

Amino Acid Derivatives

Amino acid derivatives are relatively small molecules that are structurally related to amino acids, the building blocks of proteins. ∞ p. 53 This group of hormones, sometimes known as the *biogenic amines*, are synthesized from the amino acids *tyrosine* (TĪ-rō-sēn) and *tryptophan* (TRIP-tō-fan). Tyrosine derivatives include (1) thyroid hormones, produced by the thyroid gland, and (2) the compounds epinephrine (E), norepinephrine (NE), and dopamine, which are sometimes called *catecholamines* (kat-e-KŌ-la-mēnz). The primary hormone derivative of tryptophan is melatonin (mel-a-TŌ-nin), produced by the pineal gland.

Peptide Hormones

Peptide hormones are chains of amino acids. In general, peptide hormones are synthesized as prohormones—inactive molecules that are converted to active hormones either before or after they are secreted.

Peptide hormones can be divided into two groups. One group consists of glycoproteins. ∞ p. 57 These proteins are more than 200 amino acids long and have carbohydrate side chains. The glycoproteins include *thyroid-stimulating hormone* (*TSH*), *luteinizing hormone* (*LH*), and *follicle-stimulating hormone* (*FSH*) from the adenohypophysis (also called the anterior pituitary) as well as several hormones produced in other organs.

The second group of peptide hormones is large and diverse; it includes hormones that range from short polypeptide chains, such as *antidiuretic hormone* (*ADH*) and *oxytocin* (9 amino acids apiece), to small proteins, such as *growth hormone* (*GH*; 191 amino acids) and *prolactin* (*PRL*; 198 amino acids). This group includes all the hormones secreted by the hypothalamus, heart, thymus, digestive tract, pancreas, and neurohypophysis (also called the posterior pituitary) as well as most of the hormones secreted by the adenohypophysis.

Lipid Derivatives

There are two classes of *lipid derivatives*: (1) *eicosanoids*, derived from *arachidonic* (a-rak-i-DON-ik) *acid*, a 20-carbon fatty acid; and (2) *steroid hormones*, derived from cholesterol.

Eicosanoids. **Eicosanoids** (Ī-kō-sa-noydz) are small molecules with a five-carbon ring at one end. These compounds

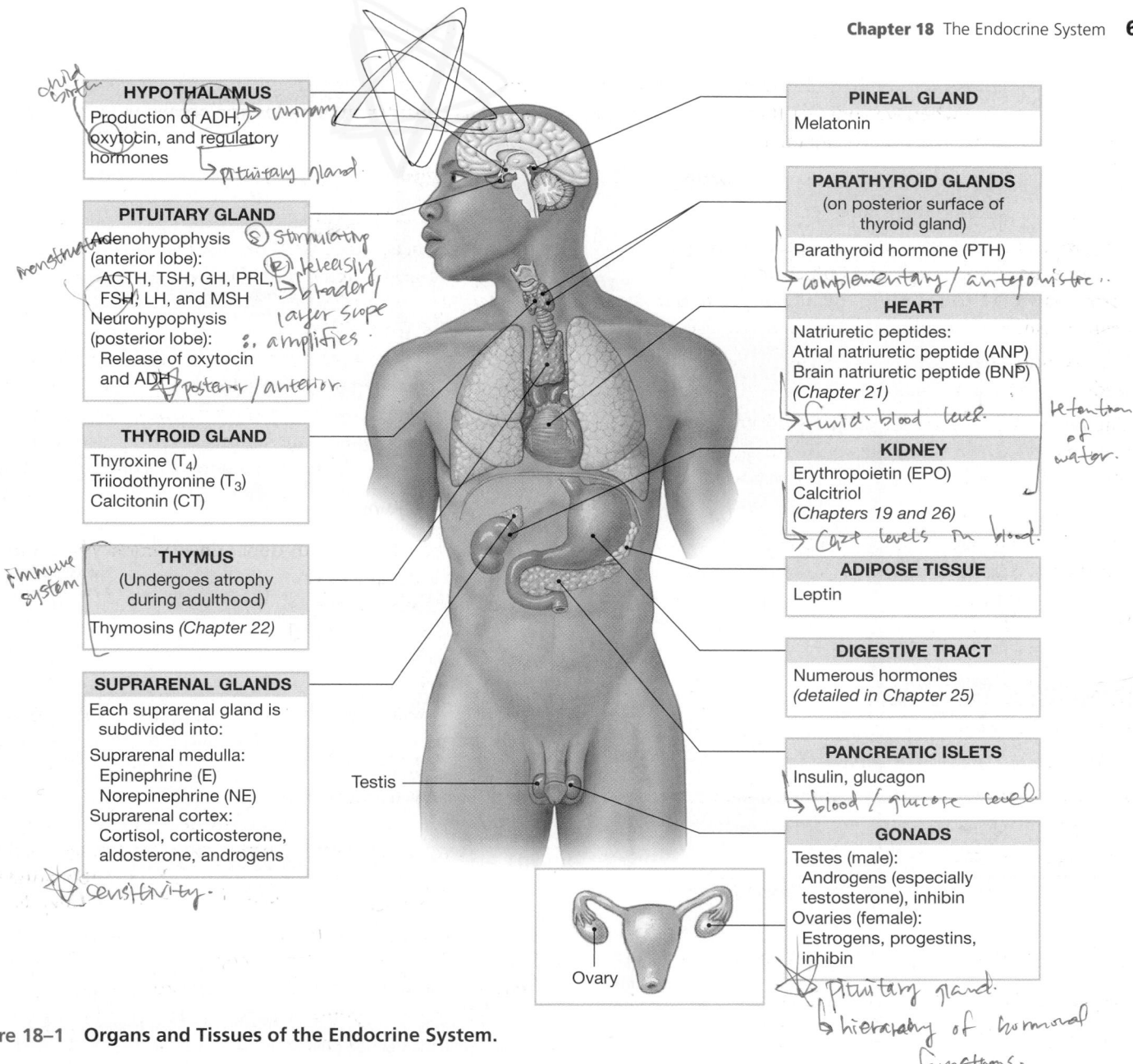

Figure 18–1 Organs and Tissues of the Endocrine System.

The following text appears within the figure:

HYPOTHALAMUS
Production of ADH, oxytocin, and regulatory hormones

[handwritten: child system, urinary, → pituitary gland]

PITUITARY GLAND
Adenohypophysis (anterior lobe): ACTH, TSH, GH, PRL, FSH, LH, and MSH
Neurohypophysis (posterior lobe): Release of oxytocin and ADH

[handwritten: menstruation, S stimulating, R releasing, bladder/layer slope, ∴ amplifies, posterior/anterior]

THYROID GLAND
Thyroxine (T₄)
Triiodothyronine (T₃)
Calcitonin (CT)

THYMUS
(Undergoes atrophy during adulthood)
Thymosins *(Chapter 22)*

[handwritten: immune system]

SUPRARENAL GLANDS
Each suprarenal gland is subdivided into:
Suprarenal medulla: Epinephrine (E) Norepinephrine (NE)
Suprarenal cortex: Cortisol, corticosterone, aldosterone, androgens

[handwritten: sensitivity]

PINEAL GLAND
Melatonin

PARATHYROID GLANDS
(on posterior surface of thyroid gland)
Parathyroid hormone (PTH)

[handwritten: complementary/antagonistic]

HEART
Natriuretic peptides:
Atrial natriuretic peptide (ANP)
Brain natriuretic peptide (BNP)
(Chapter 21)

[handwritten: fluid/blood level, retention of water]

KIDNEY
Erythropoietin (EPO)
Calcitriol
(Chapters 19 and 26)

[handwritten: Ca²⁺ levels in blood]

ADIPOSE TISSUE
Leptin

DIGESTIVE TRACT
Numerous hormones
(detailed in Chapter 25)

PANCREATIC ISLETS
Insulin, glucagon

[handwritten: blood/glucose level]

GONADS
Testes (male): Androgens (especially testosterone), inhibin
Ovaries (female): Estrogens, progestins, inhibin

[handwritten: pituitary gland, hierarchy of hormonal functions]

Testis

Ovary

[side tab: 18 ENDOCRINE]

---

are important paracrine factors that coordinate cellular activities and affect enzymatic processes (such as blood clotting) in extracellular fluids. Some of the eicosanoids also have secondary roles as hormones.

Leukotrienes (loo-kō-TRĪ-ēns) are eicosanoids released by activated white blood cells, or *leukocytes*. Leukotrienes are important in coordinating tissue responses to injury or disease. **Prostaglandins**, a second group of eicosanoids, are produced in most tissues of the body. Within each tissue, the prostaglandins released are involved primarily in coordinating local cellular activities. In some tissues, prostaglandins are converted to **thromboxanes** (throm-BOX-ānz) and **prostacyclins** (pros-ta-SĪ-klinz), which also have strong paracrine effects.

Steroid Hormones. **Steroid hormones** are lipids structurally similar to cholesterol (**Figure 2–16a,** p. 51). Steroid hormones are released by male and female reproductive organs (*androgens* by the testes, *estrogens* and *progestins* by the ovaries), by the cortex of the suprarenal glands (*corticosteroids*), and by the kidneys (*calcitriol*). The individual hormones differ in the side chains attached to the basic ring structure.

In the blood, steroid hormones are bound to specific transport proteins in the plasma. For this reason, they remain in circulation longer than do secreted peptide hormones. The liver gradually absorbs these steroids and converts them to a soluble form that can be excreted in the bile or urine.

Our focus in this chapter is on circulating hormones whose primary functions are the coordination of activities in

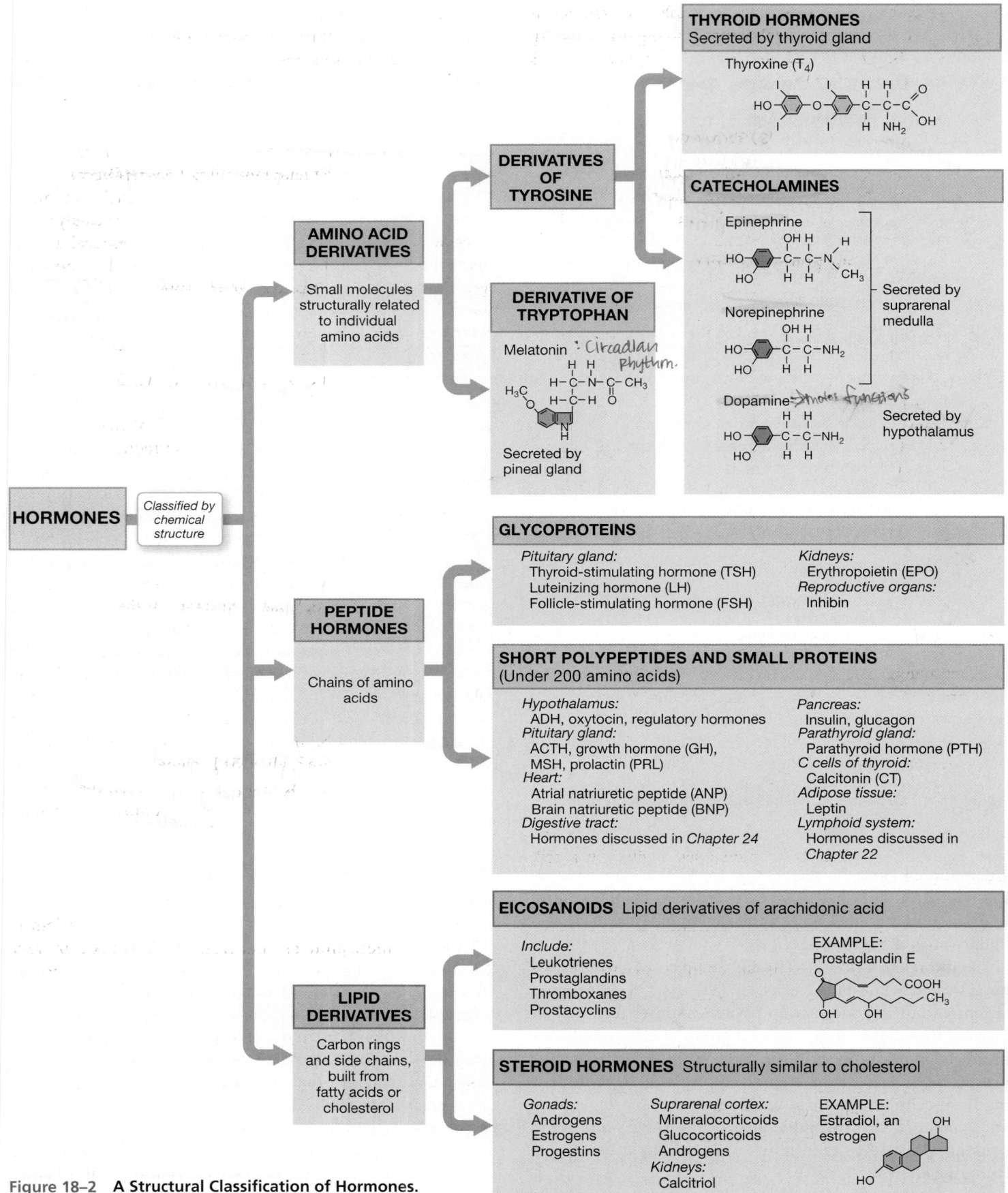

Figure 18–2   A Structural Classification of Hormones.

many tissues and organs. We will consider eicosanoids in chapters that discuss individual tissues and organs, including Chapters 19 (the blood), 22 (the lymphoid system), and 28 (the reproductive system).

Secretion and Distribution of Hormones

Hormone release typically occurs where capillaries are abundant, and the hormones quickly enter the bloodstream for distribution throughout the body. Within the blood, hormones may circulate freely or bound to special carrier proteins. A freely circulating hormone remains functional for less than one hour, and sometimes for as little as two minutes. It is inactivated when (1) it diffuses out of the bloodstream and binds to receptors on target cells, (2) it is absorbed and broken down by cells of the liver or kidneys, or (3) it is broken down by enzymes in the plasma or interstitial fluids.

Thyroid hormones and steroid hormones remain in circulation much longer, because when these hormones enter the bloodstream, more than 99 percent of them become attached to special transport proteins. For each hormone an equilibrium state exists between the free and bound forms: As the free hormones are removed and inactivated, they are replaced by the release of bound hormones. At any given time, the bloodstream contains a substantial reserve (several weeks' supply) of bound hormones.

Mechanisms of Hormone Action

To affect a target cell, a hormone must first interact with an appropriate receptor. A hormone receptor, like a neurotransmitter receptor, is a protein molecule to which a particular molecule binds strongly. Each cell has receptors for responding to several different hormones, but cells in different tissues have different combinations of receptors. This arrangement is one reason hormones have differential effects on specific tissues. For every cell, the presence or absence of a specific receptor determines the cell's hormonal sensitivities. If a cell has a receptor that can bind a particular hormone, that cell will respond to the hormone's presence. If a cell lacks the proper receptor for that hormone, the hormone will have no effect on that cell.

Hormone receptors are located either on the plasma membrane or inside the cell. Using a few specific examples, we will now introduce the basic mechanisms involved.

Hormones and Plasma Membrane Receptors

The receptors for catecholamines (E, NE, and dopamine), peptide hormones, and eicosanoids are in the plasma membranes of their respective target cells. Because catecholamines and peptide hormones are not lipid soluble, they are unable to penetrate a plasma membrane. Instead, these hormones

bind to receptor proteins at the *outer* surface of the plasma membrane (extracellular receptors). Eicosanoids, which *are* lipid soluble, diffuse across the membrane to reach receptor proteins on the *inner* surface of the membrane (intracellular receptors).

First and Second Messengers. A hormone that binds to receptors in the plasma membrane cannot have a direct effect on the activities under way inside the target cell. Such a hormone cannot, for example, begin building a protein or catalyzing a specific reaction. Instead, the hormone uses an intracellular intermediary to exert its effects. The hormone, or **first messenger**, does something that leads to the appearance of a **second messenger** in the cytoplasm. The second messenger may act as an enzyme activator, inhibitor, or cofactor, but the net result is a change in the rates of various metabolic reactions. The most important second messengers are (1) *cyclic-AMP (cAMP)*, a derivative of ATP; (2) *cyclic-GMP (cGMP)*, a derivative of GTP, another high-energy compound; and (3) calcium ions.

The binding of a small number of hormone molecules to membrane receptors may lead to the appearance of thousands of second messengers in a cell. This process, which magnifies the effect of a hormone on the target cell, is called *amplification*. Moreover, the arrival of a single hormone may promote the release of more than one type of second messenger, or the production of a linked sequence of enzymatic reactions known as a *receptor cascade*. Through such mechanisms, the hormone can alter many aspects of cell function simultaneously.

The presence or absence of a hormone can also affect the nature and number of hormone receptor proteins in the plasma membrane. **Down-regulation** is a process in which the presence of a hormone triggers a decrease in the number of hormone receptors. In down-regulation, when levels of a particular hormone are high, cells become less sensitive to it. Conversely, **up-regulation** is a process in which the absence of a hormone triggers an increase in the number of hormone receptors. In up-regulation, when levels of a particular hormone are low, cells become *more* sensitive to it.

The link between the first messenger and the second messenger generally involves a **G protein**, an enzyme complex coupled to a membrane receptor. The name *G protein* refers to the fact that these proteins bind GTP. ∞ p. 421 A G protein is activated when a hormone binds to its receptor at the membrane surface. What happens next depends on the nature of the G protein and its effects on second messengers in the cytoplasm; **Figure 18–3** diagrams the major patterns of response to G protein activation. Eighty percent of prescription drugs target G protein-coupled receptors.

G Proteins and cAMP. Many G proteins, once activated, exert their effects by changing the concentration of the second messenger *cyclic-AMP* (cAMP) within the cell. In most

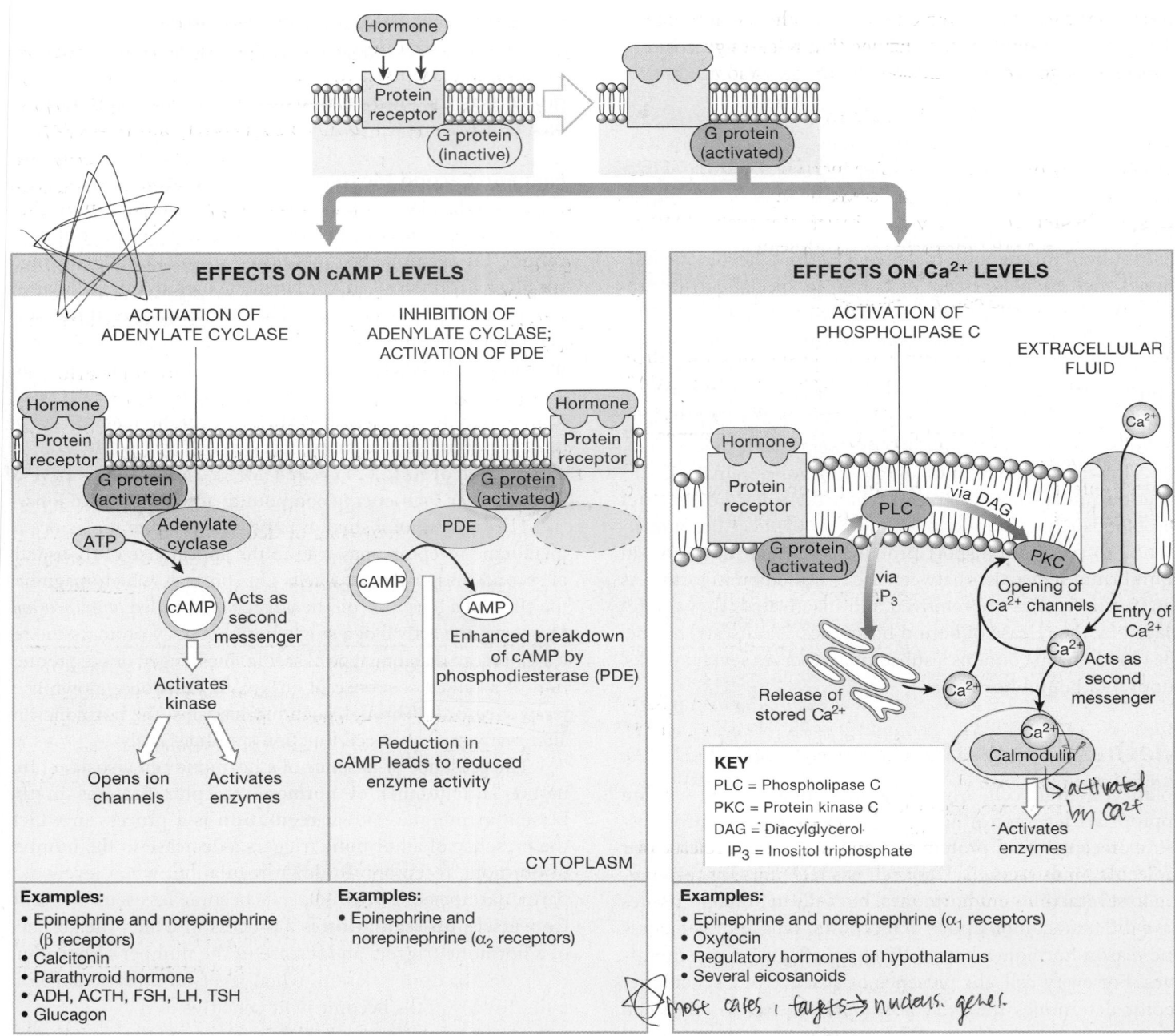

Figure 18–3 **G Proteins and Hormone Activity.** Peptide hormones, catecholamines, and eicosanoids bind to membrane receptors and activate G proteins. G protein activation may involve effects on cAMP levels (at left), or effects on Ca^{2+} levels (at right).

receptor / hormone relationship.

cases, the result is an increase in cAMP levels, and this accelerates metabolic activity within the cell.

The steps involved in increasing cAMP levels are diagrammed in **Figure 18–3** (left):

- The activated G protein activates the enzyme **adenylate cyclase**, also called *adenylyl cyclase.*
- Adenylate cyclase converts ATP to the ring-shaped molecule *cyclic-AMP.*

- Cyclic-AMP then functions as a second messenger, typically by activating a *kinase.* A kinase is an enzyme that performs *phosphorylation*, the attachment of a high-energy phosphate group ($\sim PO_4^{3-}$) to another molecule.

- Generally, the kinases activated by cyclic-AMP phosphorylate proteins. The effect on the target cell depends on the nature of the proteins affected. The phosphorylation of membrane proteins, for example, can open ion channels. In the cytoplasm, many important enzymes can be activated only by phosphorylation; one important example is the enzyme that releases glucose from glycogen reserves in skeletal muscles and the liver.

The hormones calcitonin, parathyroid hormone, ADH, ACTH, epinephrine, FSH, LH, TSH, and glucagon all produce their effects by this mechanism. The increase is usually short-lived, because the cytoplasm contains another enzyme, **phosphodiesterase (PDE)**, which inactivates cyclic-AMP by converting it to AMP (adenosine monophosphate).

Figure 18–3 (center) depicts one way the activation of a G protein can *lower* the concentration of cAMP within the cell. In this case, the activated G protein stimulates PDE activity and inhibits adenylate cyclase activity. Levels of cAMP then decline, because cAMP breakdown accelerates while cAMP synthesis is prevented. The decline has an inhibitory effect on the cell, because without phosphorylation, key enzymes remain inactive. This is the mechanism responsible for the inhibitory effects that follow the stimulation of α_2 adrenergic receptors by catecholamines, as discussed in Chapter 16. ∞ pp. 537–538

G Proteins and Calcium Ions. An activated G protein can trigger either the opening of calcium ion channels in the membrane or the release of calcium ions from intracellular stores. The steps involved are diagrammed in **Figure 18–3** (right panel). The G protein first activates the enzyme *phospholipase C (PLC)*. This enzyme triggers a receptor cascade that begins with the production of **diacylglycerol (DAG)** and **inositol triphosphate (IP₃)** from membrane phospholipids. The cascade then proceeds as follows:

- IP₃ diffuses into the cytoplasm and triggers the release of Ca^{2+} from intracellular reserves, such as those in the smooth endoplasmic reticulum of many cells.

- The combination of DAG and intracellular calcium ions activates another membrane protein: **protein kinase C (PKC)**. The activation of PKC leads to the phosphorylation of calcium channel proteins, a process that opens the channels and permits the entry of extracellular Ca^{2+}. This sets up a positive feedback loop that rapidly elevates intracellular calcium ion concentrations.

- The calcium ions themselves serve as messengers, generally in combination with an intracellular protein called **calmodulin**. Once it has bound calcium ions, calmodulin can activate specific cytoplasmic enzymes. This chain of events is responsible for the stimulatory effects that follow the activation of α_1 receptors by epinephrine or norepinephrine. ∞ p. 537 Calmodulin activation is also involved in the responses to oxytocin and to several regulatory hormones secreted by the hypothalamus.

Hormones and Intracellular Receptors

Steroid hormones diffuse across the lipid part of the plasma membrane and bind to receptors in the cytoplasm or nucleus. The hormone–receptor complexes then activate or deactivate specific genes (**Figure 18–4a**). By this mechanism, steroid hormones can alter the rate of DNA transcription in the nucleus, and thus change the pattern of protein synthesis. Alterations in the synthesis of enzymes or structural proteins will directly affect both the metabolic activity and the structure of the target cell. For example, the sex hormone *testosterone* stimulates the production of enzymes and structural proteins in skeletal muscle fibers, causing an increase in muscle size and strength.

Thyroid hormones cross the plasma membrane primarily by a transport mechanism. Once in the cytosol, these hormones bind to receptors within the nucleus and on mitochondria (**Figure 18–4b**). The hormone–receptor complexes in the nucleus activate specific genes or change the rate of transcription. The change in rate affects the metabolic activities of the cell by increasing or decreasing the concentration of specific enzymes. Thyroid hormones bound to mitochondria increase the mitochondrial rates of ATP production.

The A&P Top 100

#61 Hormones coordinate cell, tissue, and organ activities on a sustained basis. They circulate in the extracellular fluid and bind to specific receptors on or in target cells. They then modify cellular activities by altering membrane permeability, activating or inactivating key enzymes, or changing genetic activity.

Control of Endocrine Activity by Endocrine Reflexes

As noted earlier, the functional organization of the nervous system parallels that of the endocrine system in many ways. In Chapter 13, we considered the basic operation of neural reflex arcs, the simplest organizational units in the nervous system. ∞ p. 448 The most direct arrangement was a monosynaptic reflex, such as the stretch reflex. Polysynaptic reflexes provide more complex and variable responses to stimuli, and higher centers, which integrate multiple inputs, can facilitate or inhibit these reflexes as needed. **Endocrine reflexes** are the functional counterparts of neural reflexes. Endocrine reflexes can be triggered by (1) *humoral stimuli* (changes in the composition of the extracellular fluid), (2) *hormonal stimuli* (the arrival or removal of a specific hormone), or (3) *neural stimuli* (the arrival of neurotransmitters at neuroglandular junctions). In most cases, endocrine reflexes are controlled by negative feedback mechanisms: A stimulus triggers the production of a hormone whose direct or indirect effects reduce the intensity of the stimulus.

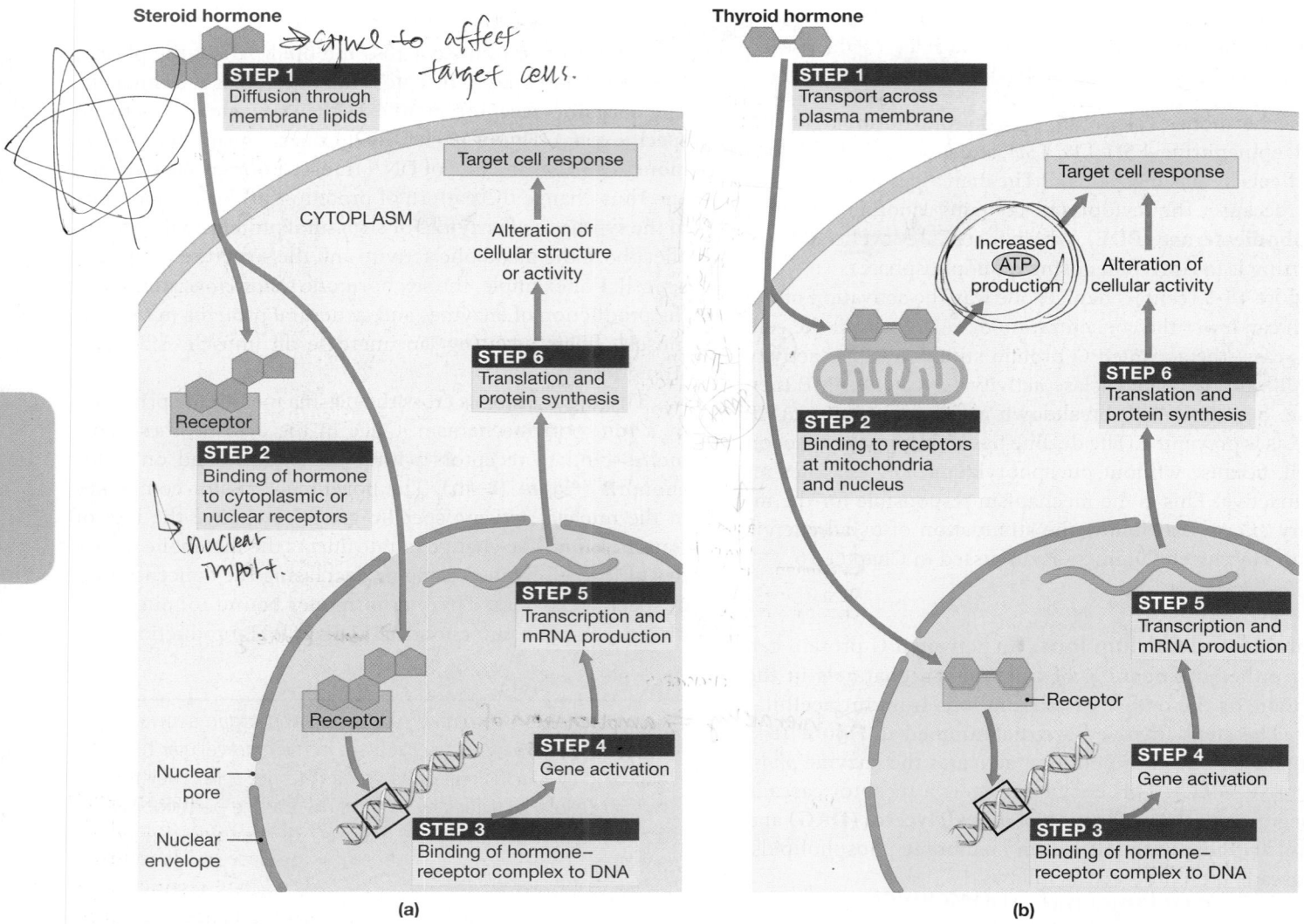

Figure 18–4 **Effects of Intracellular Hormone Binding.** **(a)** Steroid hormones diffuse through the membrane lipids and bind to receptors in the cytoplasm or nucleus. The complex then binds to DNA in the nucleus, activating specific genes. **(b)** Thyroid hormones enter the cytoplasm and bind to receptors in the nucleus to activate specific genes. They also bind to receptors on mitochondria and accelerate ATP production.

A simple endocrine reflex involves only one hormone. The endocrine cells involved respond directly to changes in the composition of the extracellular fluid. The secreted hormone adjusts the activities of target cells and restores homeostasis. Simple endocrine reflexes control hormone secretion by the heart, pancreas, parathyroid glands, and digestive tract.

More complex endocrine reflexes involve one or more intermediary steps and two or more hormones. The hypothalamus, which provides the highest level of endocrine control, integrates the activities of the nervous and endocrine systems in three ways (**Figure 18–5**):

1. The hypothalamus secretes **regulatory hormones**, special hormones that control endocrine cells in the pituitary gland. The hypothalamic regulatory hormones control the secretory activities of endocrine cells in the adenohypophysis. The hormones released by the adenohypophysis, in turn, control the activities of endocrine cells in the thyroid, suprarenal (adrenal) cortex, and reproductive organs.

2. The hypothalamus itself acts as an endocrine organ. Hypothalamic neurons synthesize hormones, transport them along axons within the infundibulum, and release them into the circulation at the neurohypophysis. We introduced two of these hormones, ADH and oxytocin, in Chapter 14. ∞ p. 478

3. The hypothalamus contains autonomic centers that exert direct neural control over the endocrine cells of the

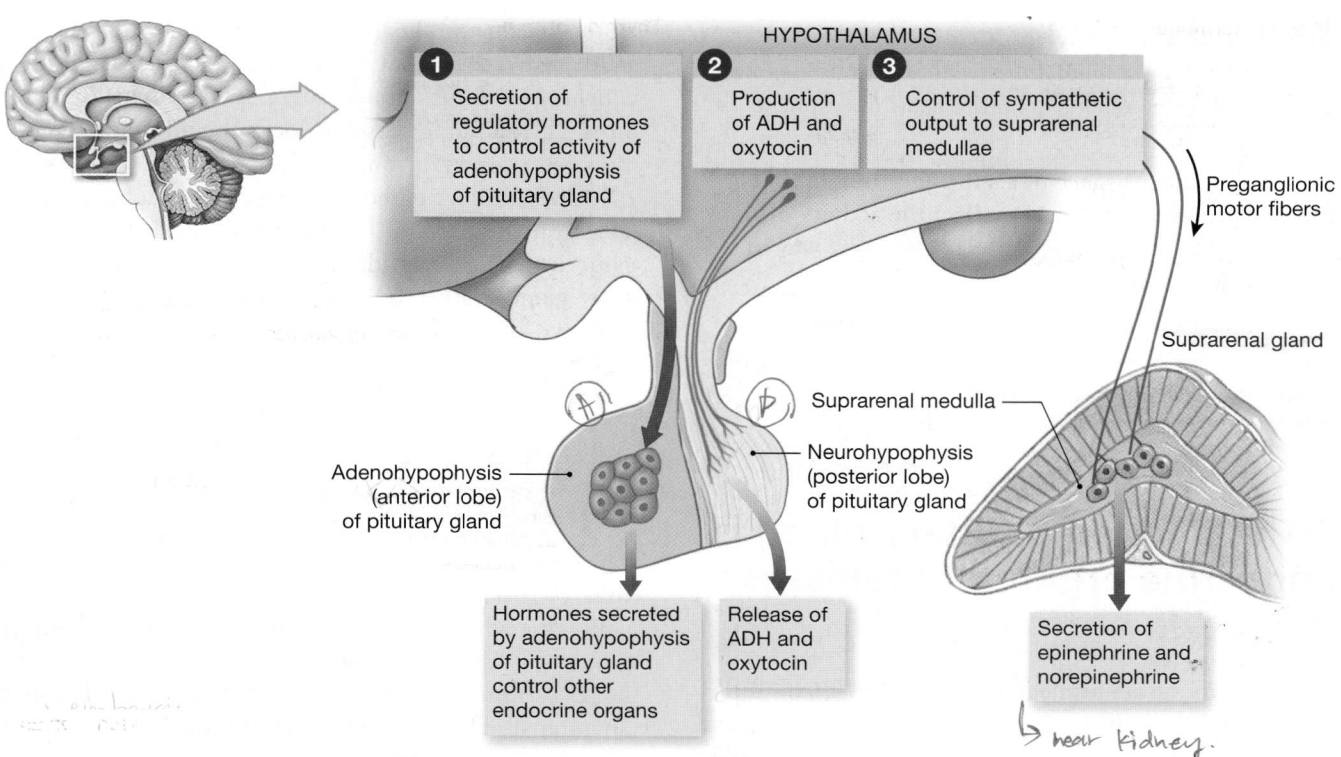

Figure 18–5 Three Mechanisms of Hypothalamic Control over Endocrine Function.

suprarenal medullae. When the sympathetic division is activated, the suprarenal medullae release hormones into the bloodstream.

The hypothalamus secretes regulatory hormones and ADH in response to changes in the composition of the circulating blood. The secretion of oxytocin (OXT), E, and NE involves both neural and hormonal mechanisms. For example, the suprarenal medullae secrete E and NE in response to action potentials rather than to circulating hormones. Such pathways are called *neuroendocrine reflexes*, because they include both neural and endocrine components. We will consider these reflex patterns in more detail as we examine specific endocrine tissues and organs.

In Chapter 15, we noted that sensory receptors provide complex information by varying the frequency and pattern of action potentials in a sensory neuron. In the endocrine system, complex commands are issued by changing the amount of hormone secreted and the pattern of hormone release. In a simple endocrine reflex, hormones are released continuously, but the rate of secretion rises and falls in response to humoral stimuli. For example, when blood glucose levels climb, the pancreas increases its secretion of *insulin*, a hormone that

stimulates glucose uptake and utilization. As insulin levels rise, glucose levels decline; in turn, the stimulation of the insulin-secreting cells is reduced. As glucose levels return to normal, the rate of insulin secretion reaches resting levels. (This regulatory pattern, called *negative feedback*, was introduced in Chapter 1 when we considered the control of body temperature. ∞ p. 14)

In this example, the responses of the target cells change over time, because the effect of insulin is proportional to its concentration. However, the relationship between hormone concentration and target cell response is not always predictable. For instance, a hormone can have one effect at low concentrations and more exaggerated effects—or even different effects—at high concentrations. (We will consider specific examples later in the chapter.)

Several hypothalamic and pituitary hormones are released in sudden bursts called *pulses*, rather than continuously. When hormones arrive in pulses, target cells may vary their response with the frequency of the pulses. For example, the target cell response to one pulse every three hours can differ from the response when pulses arrive every 30 minutes. The most complicated hormonal instructions issued by the hypothalamus involve changes in the frequency of pulses *and* in the amount secreted in each pulse.

4. How could you distinguish between a neural response and an endocrine response on the basis of response time and duration?

5. How would the presence of a substance that inhibits the enzyme adenylate cyclase affect the activity of a hormone that produces its cellular effects by way of the second messenger cAMP?

6. What primary factor determines each cell's hormonal sensitivities?

See the blue Answers tab at the end of the book.

[handwritten note: 9 peptide hormones, all 2nd messenger of cAMP]

18-3 The bilobed pituitary gland is an endocrine organ that releases nine peptide hormones

Figure 18–6 shows the anatomical organization of the pituitary gland, or **hypophysis** (hī-POF-i-sis). This small, oval gland lies nestled within the *sella turcica*, a depression in the sphenoid (**Figure 7–8**, p. 221). The pituitary gland hangs inferior to the hypothalamus, connected by the slender, funnel-shaped structure called the **infundibulum** (in-fun-DIB-ū-lum; funnel). The base of the infundibulum lies between the optic chiasm and the mamillary bodies. The pituitary gland is cradled by the sella turcica and held in position by the *diaphragma sellae*, a dural sheet that encircles the infundibulum. The diaphragma sellae locks the pituitary gland in position and isolates it from the cranial cavity.

The pituitary gland is divided into anterior and posterior lobes on the basis of function and developmental anatomy. Nine important peptide hormones are released by the pituitary gland—seven by the adenohypophysis (anterior lobe) and two by the neurohypophysis (posterior lobe). All nine hormones bind to membrane receptors, and all nine use cAMP as a second messenger. ATLAS: Embryology Summary 14: The Development of the Endocrine System

The Adenohypophysis

The **adenohypophysis** (ad-e-nō-hī-POF-i-sis), or **anterior lobe** of the pituitary gland, contains a variety of endocrine cells. The adenohypophysis can be subdivided into three regions: (1) the **pars distalis** (dis-TAL-is; distal part), the largest and most anterior portion of the pituitary gland; (2) an extension called the **pars tuberalis**, which wraps around the adjacent portion of the infundibulum; and (3) the slender **pars intermedia**, which forms a narrow band bordering the neurohypophysis (**Figure 18–6**). An extensive capillary network radiates through these regions, giving every endocrine cell immediate access to the circulatory system.

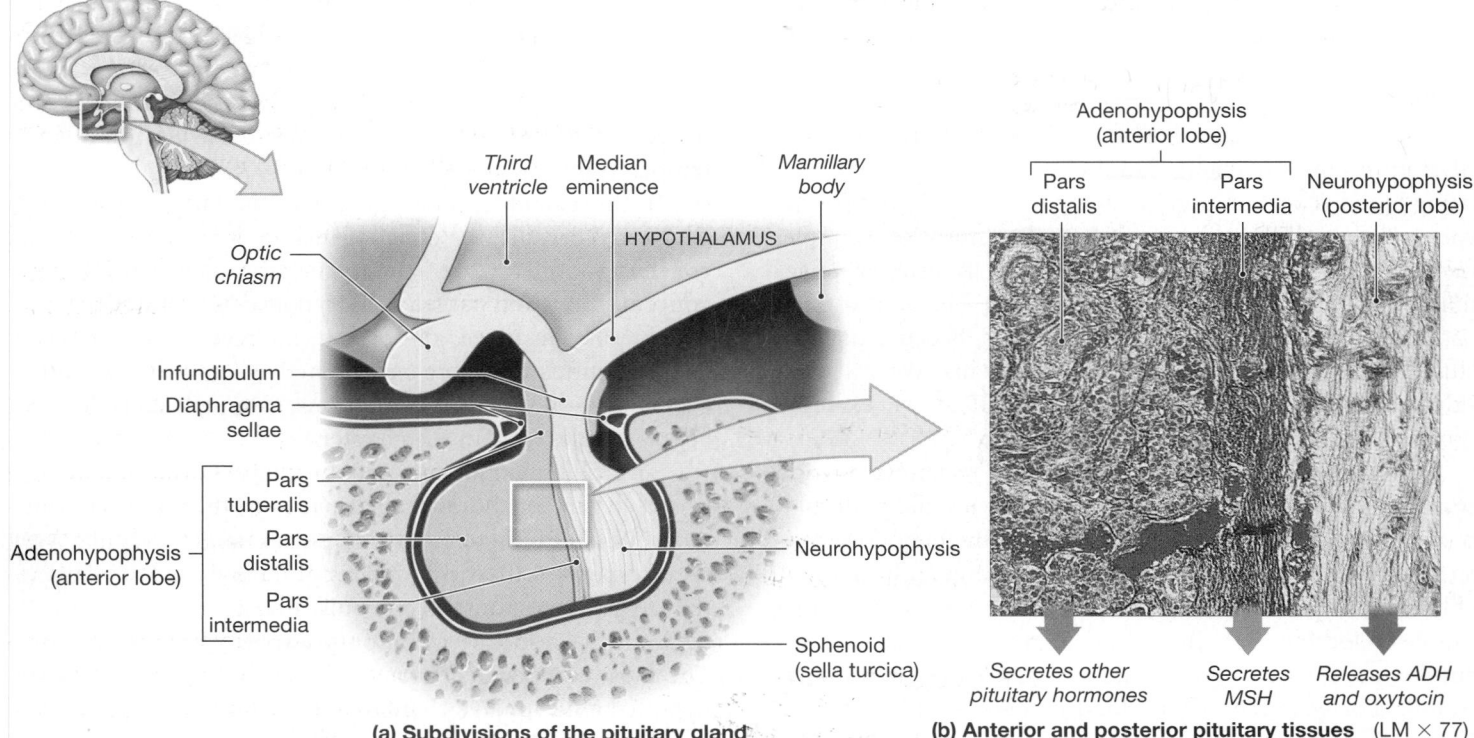

(a) Subdivisions of the pituitary gland

(b) Anterior and posterior pituitary tissues (LM × 77)

Figure 18–6   The Anatomy and Orientation of the Pituitary Gland.

The Hypophyseal Portal System

By secreting specific regulatory hormones, the hypothalamus controls the production of hormones in the adenohypophysis. At the *median eminence*, a swelling near the attachment of the infundibulum, hypothalamic neurons release regulatory factors into the surrounding interstitial fluids. Their secretions enter the bloodstream quite easily, because the endothelial cells lining the capillaries in this region are unusually permeable. These **fenestrated** (FEN-es-trā-ted; *fenestra*, window) **capillaries** allow relatively large molecules to enter or leave the circulatory system.

fenestrated caps.

The capillary networks in the median eminence are supplied by the *superior hypophyseal artery* (**Figure 18–7**). Before leaving the hypothalamus, the capillary networks unite to form a series of larger vessels that spiral around the infundibulum to reach the adenohypophysis. Once within the adenohypophysis, these vessels form a second capillary network that branches among the endocrine cells.

This vascular arrangement is unusual: A typical artery conducts blood from the heart to a capillary network, and a typical vein carries blood from a capillary network back to the heart. The vessels between the median eminence and the ade-

artery: heart to capillary
vein: capillary to heart

nohypophysis, however, carry blood from one capillary network to another. Blood vessels that link two capillary networks are called **portal vessels**; in this case, they have the histological structure of veins. The entire complex is a **portal system**. Portal systems are named after their destinations; hence, this particular network is known as the **hypophyseal** (hī-po-FI-sē-al) **portal system**.

capillary → capillary

Portal systems provide an efficient means of chemical communication by ensuring that all the hypothalamic hormones entering the portal vessels will reach the target cells in the adenohypophysis before being diluted through mixing with the general circulation. The communication is strictly one way, however, because any chemicals released by the cells "downstream" must do a complete circuit of the cardiovascular system before they reach the capillaries of the portal system.

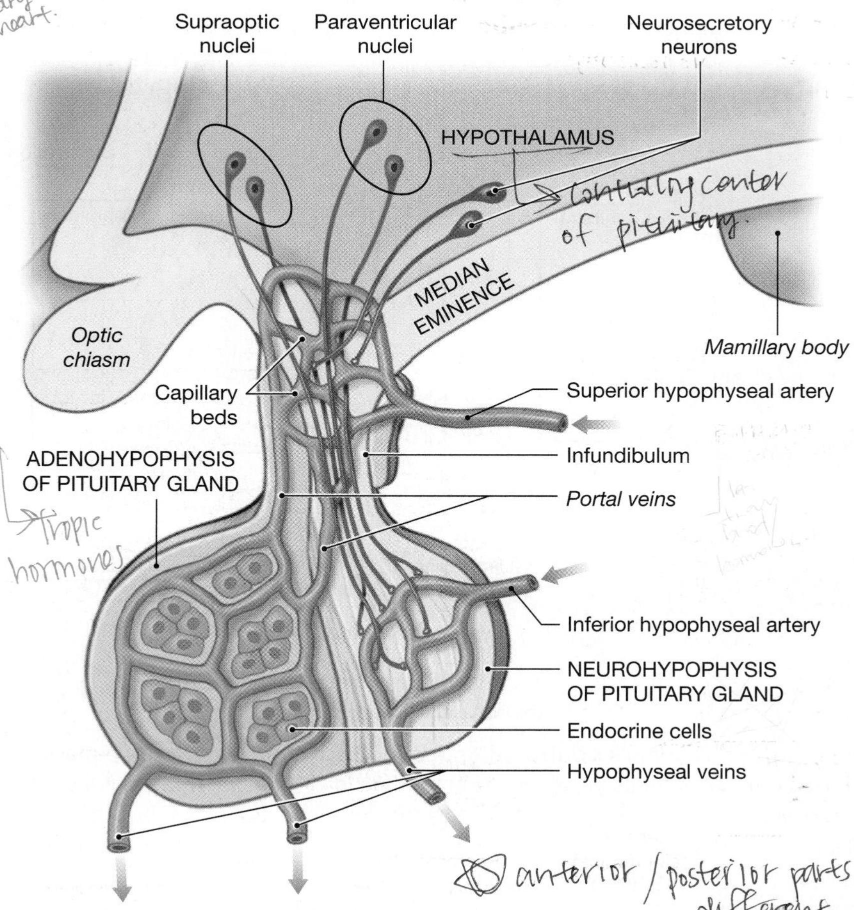

Hypothalamic Control of the Adenohypophysis

Releasing / inhibiting hormones.

Two classes of hypothalamic regulatory hormones exist: releasing hormones and inhibiting hormones. A **releasing hormone (RH)** stimulates the synthesis and secretion of one or more hormones at the adenohypophysis, whereas an **inhibiting hormone (IH)** prevents the synthesis and secretion of hormones from the adenohypophysis. An endocrine cell in the adenohypophysis may be controlled by releasing hormones, inhibiting hormones, or some combination of the two. The regulatory hormones released at the hypothalamus are transported directly to the adenohypophysis by the hypophyseal portal system.

The rate at which the hypothalamus secretes regulatory hormones is controlled by negative feedback. The primary regulatory patterns are diagrammed in **Figure 18–8**; we will refer to them as we examine specific pituitary hormones.

Hormones of the Adenohypophysis

We will discuss seven hormones whose functions and control mechanisms are reasonably well understood: *thyroid-stimulating hormone*; *adrenocorticotropic hormone*; two gonadotropins called *follicle-stimulating hormone* and *luteinizing hormone*; *prolactin*; *growth hormone*; and *melanocyte-stimulating hormone*. Of the six hormones produced by the pars distalis, four regulate the production of hormones by other endocrine glands. The names of these hormones indicate their activities, but many of the phrases are so long that abbreviations are often used instead.

The hormones of the adenohypophysis are also called *tropic hormones* (*trope*, a turning), because they "turn on" endocrine glands or support the functions of other organs. (Some sources call them *trophic hormones* [*trophe*, nourishment] instead.)

turn on endocrine glands.

Supraoptic nuclei
Paraventricular nuclei
Neurosecretory neurons
HYPOTHALAMUS
controlling center of pituitary.
MEDIAN EMINENCE
Optic chiasm
Mamillary body
Capillary beds
Superior hypophyseal artery
ADENOHYPOPHYSIS OF PITUITARY GLAND
tropic hormones
Infundibulum
Portal veins
Inferior hypophyseal artery
NEUROHYPOPHYSIS OF PITUITARY GLAND
Endocrine cells
Hypophyseal veins

anterior / posterior parts different.

Figure 18–7   The Hypophyseal Portal System and the Blood Supply to the Pituitary Gland.

Thyroid-Stimulating Hormone.

Thyroid-stimulating hormone (TSH), or *thyrotropin*, targets the thyroid gland and triggers the release of thyroid hormones. TSH is released in response to *thyrotropin-releasing hormone (TRH)* from the hypothalamus. As circulating concentrations of thyroid hormones rise, the rates of TRH and TSH production decline (**Figure 18–8a**).

Adrenocorticotropic Hormone.

Adrenocorticotropic hormone (ACTH), also known as *corticotropin*, stimulates the release of steroid hormones by the *suprarenal cortex*, the outer portion of the suprarenal gland. ACTH specifically targets cells that produce *glucocorticoids* (gloo-kō-KOR-ti-koydz), hormones that affect glucose metabolism. ACTH release occurs under the stimulation of **corticotropin-releasing hormone (CRH)** from the hypothalamus. As glucocorticoid levels increase, the rates of CRH release and ACTH release decline (**Figure 18–8a**).

The Gonadotropins.

The hormones called **gonadotropins** (gō-nad-ō-TRŌ-pinz) regulate the activities of the *gonads*. (These organs—the testes and ovaries in males and females, respectively—produce reproductive cells as well as hormones.) The production of gonadotropins occurs under stimulation by gonadotropin-releasing hormone (GnRH) from the hypothalamus. An abnormally low production of gonadotropins produces **hypogonadism**. Children with this condition will not undergo sexual maturation, and adults with hypogonadism cannot produce functional sperm (males) or oocytes (females). The two gonadotropins are follicle-stimulating hormone and luteinizing hormone.

- **Follicle-stimulating hormone (FSH)**, or *follitropin*, promotes follicle development in females and, in combination with luteinizing hormone, stimulates the secretion of *estrogens* (ES-trō-jenz) by ovarian cells. *Estradiol* is the most important estrogen. In males, FSH stimulates *nurse* (*sustentacular*) *cells*, specialized cells in the tubules where sperm differentiate. In response, the nurse cells promote the physical maturation of developing sperm. FSH production is inhibited by *inhibin*, a peptide hormone released by cells in the testes and ovaries (**Figure 18–8a**). (The role of inhibin in suppressing the release of GnRH as well as FSH is under debate.)

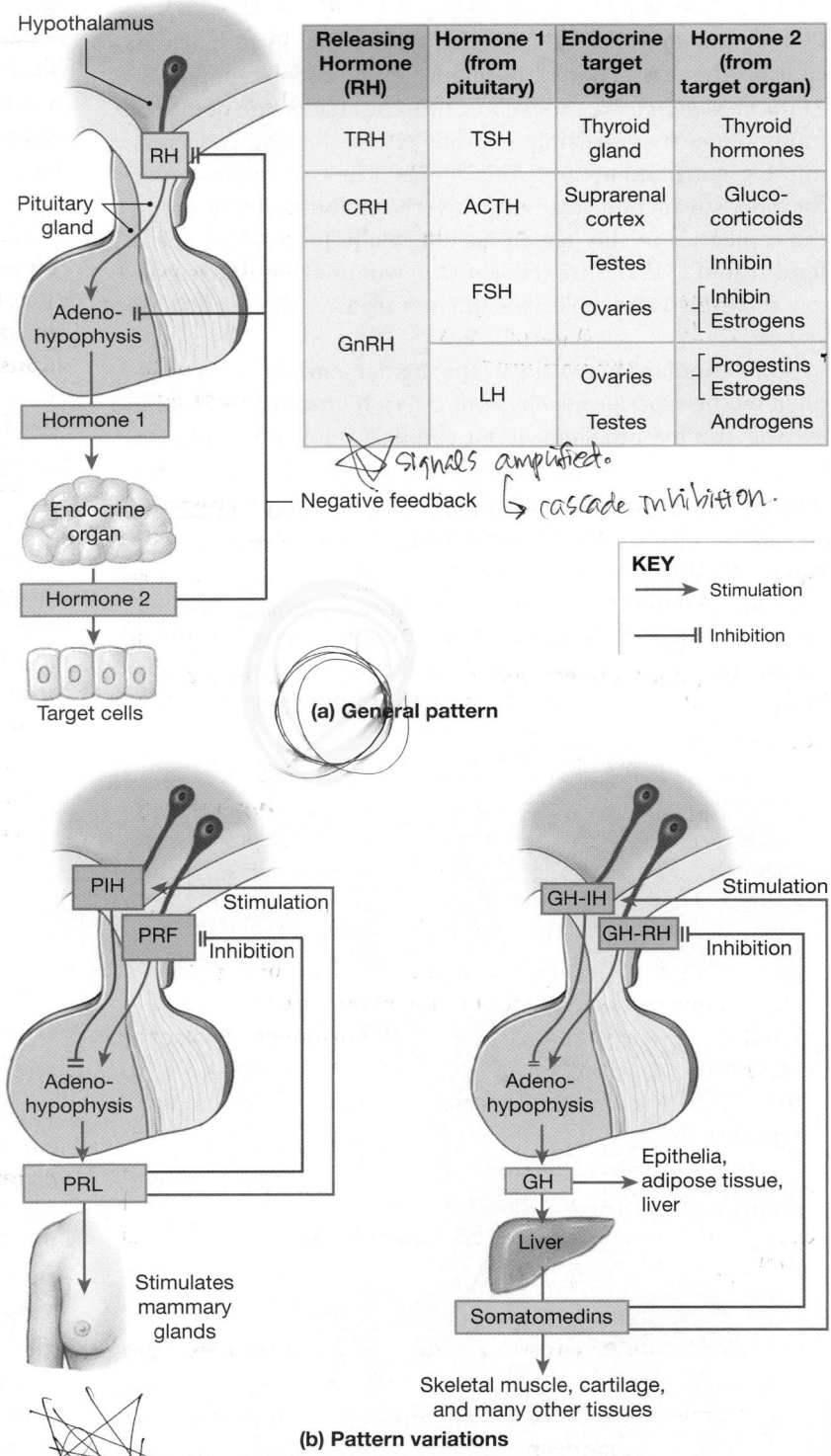

Releasing Hormone (RH)	Hormone 1 (from pituitary)	Endocrine target organ	Hormone 2 (from target organ)
TRH	TSH	Thyroid gland	Thyroid hormones
CRH	ACTH	Suprarenal cortex	Gluco-corticoids
GnRH	FSH	Testes	Inhibin
		Ovaries	Inhibin / Estrogens
	LH	Ovaries	Progestins / Estrogens
		Testes	Androgens

signals amplified. → *cascade inhibition.*

KEY
→ Stimulation
⊣ Inhibition

(a) General pattern

(b) Pattern variations

Figure 18–8 Feedback Control of Endocrine Secretion. (a) A typical pattern of regulation when multiple endocrine organs are involved. The hypothalamus produces a releasing hormone (RH) to stimulate hormone production by other glands; control occurs via negative feedback. **(b)** Variations on the theme outlined in part (a). Left: The regulation of prolactin (PRL) production by the adenohypophysis. In this case, the hypothalamus produces both a releasing factor (PRF) and an inhibiting hormone (PIH); when one is stimulated, the other is inhibited. Right: The regulation of growth hormone (GH) production by the adenohypophysis; when GH–RH release is inhibited, GH–IH release is stimulated.

- **Luteinizing** (LOO-tē-in-ī-zing) **hormone (LH)**, or *lutropin*, induces *ovulation*, the production of reproductive cells in females. It also promotes the secretion, by the ovaries, of estrogens and the *progestins* (such as *progesterone*), which prepare the body for possible pregnancy. In males, this gonadotropin is sometimes called *interstitial cell–stimulating hormone (ICSH)*, because it stimulates the production of sex hormones by the *interstitial cells* of the testes. These sex hormones are called **androgens** (AN-drō-jenz; *andros*, man), the most important of which is *testosterone*. LH production, like FSH production, is stimulated by GnRH from the hypothalamus. GnRH production is inhibited by estrogens, progestins, and androgens (**Figure 18–8a**).

Prolactin. **Prolactin** (*pro-*, before + *lac*, milk) (**PRL**), or *mammotropin*, works with other hormones to stimulate mammary gland development. In pregnancy and during the nursing period that follows delivery, PRL also stimulates milk production by the mammary glands. The functions of PRL in males are poorly understood, but evidence indicates that PRL helps regulate androgen production by making interstitial cells more sensitive to LH.

Prolactin production is inhibited by the neurotransmitter dopamine, also known as **prolactin-inhibiting hormone (PIH)**. The hypothalamus also secretes several *prolactin-releasing factors (PRF)*, few of which have been identified. Circulating PRL stimulates PIH release and inhibits the secretion of PRF (**Figure 18–8b**).

Although PRL exerts the dominant effect on the glandular cells, normal development of the mammary glands is regulated by the interaction of several hormones. Prolactin, estrogens, progesterone, glucocorticoids, pancreatic hormones, and hormones produced by the placenta cooperate in preparing the mammary glands for secretion, and milk ejection occurs only in response to oxytocin release at the neurohypophysis (posterior lobe of the pituitary gland). We will describe the functional development of the mammary glands in Chapter 28.

Growth Hormone. **Growth hormone (GH)**, or **somatotropin** (*soma*, body), stimulates cell growth and replication by accelerating the rate of protein synthesis. Although virtually every tissue responds to some degree, skeletal muscle cells and chondrocytes (cartilage cells) are particularly sensitive to GH.

The stimulation of growth by GH involves two mechanisms. The primary mechanism, which is indirect, is best understood. Liver cells respond to the presence of GH by synthesizing and releasing **somatomedins**, or **insulin-like growth factors (IGFs)**, which are peptide hormones that bind to receptor sites on a variety of plasma membranes (**Figure 18–8b**). In skeletal muscle fibers, cartilage cells, and other target cells, somatomedins increase the rate of uptake of amino acids and their incorporation into new proteins. These effects develop almost immediately after GH is released; they are particularly important after a meal, when the blood contains high concentrations of glucose and amino acids. In functional terms, cells can now obtain ATP easily through the aerobic metabolism of glucose, and amino acids are readily available for protein synthesis. Under these conditions, GH, acting through the somatomedins, stimulates protein synthesis and cell growth.

The direct actions of GH are more selective and tend not to appear until after blood glucose and amino acid concentrations have returned to normal levels:

- In epithelia and connective tissues, GH stimulates stem cell divisions and the differentiation of daughter cells. The subsequent growth of these daughter cells will be stimulated by somatomedins.

- In adipose tissue, GH stimulates the breakdown of stored triglycerides by adipocytes (fat cells), which then release fatty acids into the blood. As circulating fatty acid levels rise, many tissues stop breaking down glucose and start breaking down fatty acids to generate ATP. This process is termed a **glucose-sparing effect**.

- In the liver, GH stimulates the breakdown of glycogen reserves by liver cells, which then release glucose into the bloodstream. Because most other tissues are now metabolizing fatty acids rather than glucose, blood glucose concentrations begin to climb, perhaps to levels significantly higher than normal. The elevation of blood glucose levels by GH has been called a **diabetogenic effect**, because *diabetes mellitus*, an endocrine disorder we will consider later in the chapter, is characterized by abnormally high blood glucose concentrations.

The production of GH is regulated by **growth hormone–releasing hormone (GH–RH,** or *somatocrinin*) and **growth hormone–inhibiting hormone (GH–IH**, or *somatostatin*) from the hypothalamus. Somatomedins stimulate GH–IH and inhibit GH–RH (**Figure 18–8b**).

Melanocyte-Stimulating Hormone. The pars intermedia may secrete two forms of **melanocyte-stimulating hormone (MSH)**, or *melanotropin*. As the name indicates, MSH stimulates the melanocytes of the skin, increasing their production of melanin, a brown, black, or yellow-brown pigment. ∞ p. 162 The release of MSH is inhibited by dopamine.

Melanocyte-stimulating hormone from the pituitary gland is important in the control of skin pigmentation in fishes, amphibians, reptiles, and many mammals other than primates. In humans, MSH is produced locally, within sun-exposed skin. The pars intermedia in adult humans is virtually nonfunctional, and the circulating blood usually does not contain MSH. However, MSH is secreted by the human pars

18 ENDOCRINE

intermedia (1) during fetal development, (2) in very young children, (3) in pregnant women, and (4) in the course of some diseases. The functional significance of MSH secretion under these circumstances is not known. The administration of a synthetic form of MSH causes the skin to darken, so MSH has been suggested as a means of obtaining a "sunless tan."

The Neurohypophysis

The **neurohypophysis** (noo-rō-hī-POF-i-sis) is also called the **posterior lobe** of the pituitary gland, or *pars nervosa* (nervous part), because it contains the axons of hypothalamic neurons. Neurons of the **supraoptic** and **paraventricular nuclei** manufacture antidiuretic hormone (ADH) and oxytocin, respectively. These products move along axons in the infundibulum to axon terminals, which end on the basal laminae of capillaries in the neurohypophysis, by means of axoplasmic transport. ∞ p. 390

Antidiuretic Hormone

Antidiuretic hormone (ADH), also known as *arginine vasopressin (AVP)*, is released in response to a variety of stimuli, most notably a rise in the solute concentration in the blood or a fall in blood volume or blood pressure. A rise in the solute concentration stimulates specialized hypothalamic neurons. Because they respond to a change in the osmotic concentration of body fluids, these neurons are called *osmoreceptors*. These osmoreceptors then stimulate the neurosecretory neurons that release ADH.

The primary function of ADH is to decrease the amount of water lost at the kidneys. With losses minimized, any water absorbed from the digestive tract will be retained, reducing the concentrations of electrolytes in the extracellular fluid. In high concentrations, ADH also causes *vasoconstriction*, a constriction of peripheral blood vessels that helps elevate blood pressure. ADH release is inhibited by alcohol, which explains the

increased fluid excretion that follows the consumption of alcoholic beverages.

Oxytocin

In women, **oxytocin** (ɒxy-, quick + *tokos*, childbirth), or **OXT**, stimulates smooth muscle contraction in the wall of the uterus, promoting labor and delivery. After delivery, oxytocin stimulates the contraction of myoepithelial cells around the secretory alveoli and the ducts of the mammary glands, promoting the ejection of milk.

Until the last stages of pregnancy, the uterine smooth muscles are relatively insensitive to oxytocin, but sensitivity becomes more pronounced as the time of delivery approaches. The trigger for normal labor and delivery is probably a sudden rise in oxytocin levels at the uterus. There is good evidence, however, that oxytocin released at the neurohypophysis plays only a supporting role, and that most of the oxytocin involved is secreted by the uterus and fetus.

Oxytocin secretion and milk ejection are part of a neuroendocrine reflex. The normal stimulus is an infant suckling at the breast, and sensory nerves innervating the nipples relay the information to the hypothalamus. Oxytocin is then released into the circulation at the neurohypophysis, and the myoepithelial cells respond by squeezing milk from the secretory alveoli into large collecting ducts. This *milk let-down reflex* can be modified by any factor that affects the hypothalamus. For example, anxiety, stress, and other factors can prevent the flow of milk, even when the mammary glands are fully functional. In contrast, nursing mothers can become conditioned to associate a baby's crying with suckling. In these women, milk let-down may begin as soon as they hear a baby cry.

Although the functions of oxytocin in sexual activity remain uncertain, it is known that circulating concentrations of oxytocin rise during sexual arousal and peak at orgasm in both sexes. Evidence indicates that in men, oxytocin stimu-

---

CLINICAL NOTE

Diabetes Insipidus

Diabetes occurs in several forms, all characterized by excessive urine production (polyuria). Although diabetes can be caused by physical damage to the kidneys, most forms are the result of endocrine abnormalities. The two most prevalent forms are diabetes mellitus and diabetes insipidus. Diabetes mellitus is described on page 633. **Diabetes insipidus** generally develops because the neurohypophysis no longer releases adequate amounts of ADH.

Water conservation at the kidneys is impaired, and excessive amounts of water are lost in the urine. As a result, the individual is constantly thirsty, but the fluids consumed are not retained by the body.

Mild cases of diabetes insipidus may not require treatment if fluid and electrolyte intake keeps pace with urinary losses. In severe cases, the fluid losses can reach 10 liters per day, and dehydration and electrolyte imbalances are fatal without treatment. This condition can be effectively treated with desmopressin, a synthetic form of ADH.

lates smooth muscle contractions in the walls of the *ductus deferens* (sperm duct) and prostate gland. These actions may be important in *emission*—the ejection of sperm, secretions of the prostate gland, and the secretions of other glands into the male reproductive tract before ejaculation. Studies suggest that the oxytocin released in females during intercourse may stimulate smooth muscle contractions in the uterus and vagina that promote the transport of sperm toward the uterine tubes.

Summary: The Hormones of the Pituitary Gland

Figure 18–9 and Table 18–2 summarize important information about the hormonal products of the pituitary gland. Re-

view these carefully before considering the structure and function of other endocrine organs.

The A&P Top 100

#62 The hypothalamus produces regulatory factors that adjust the activities of the adenohypophysis, which produces seven hormones. Most of these hormones control other endocrine organs, including the thyroid gland, suprarenal gland, and gonads. The adenohypophysis also produces growth hormone, which stimulates cell growth and protein synthesis. The neurohypophysis releases two hormones produced in the hypothalamus; ADH restricts water loss and promotes thirst, and oxytocin stimulates smooth muscle contractions in the mammary glands and uterus (in females) and the prostate gland (in males).

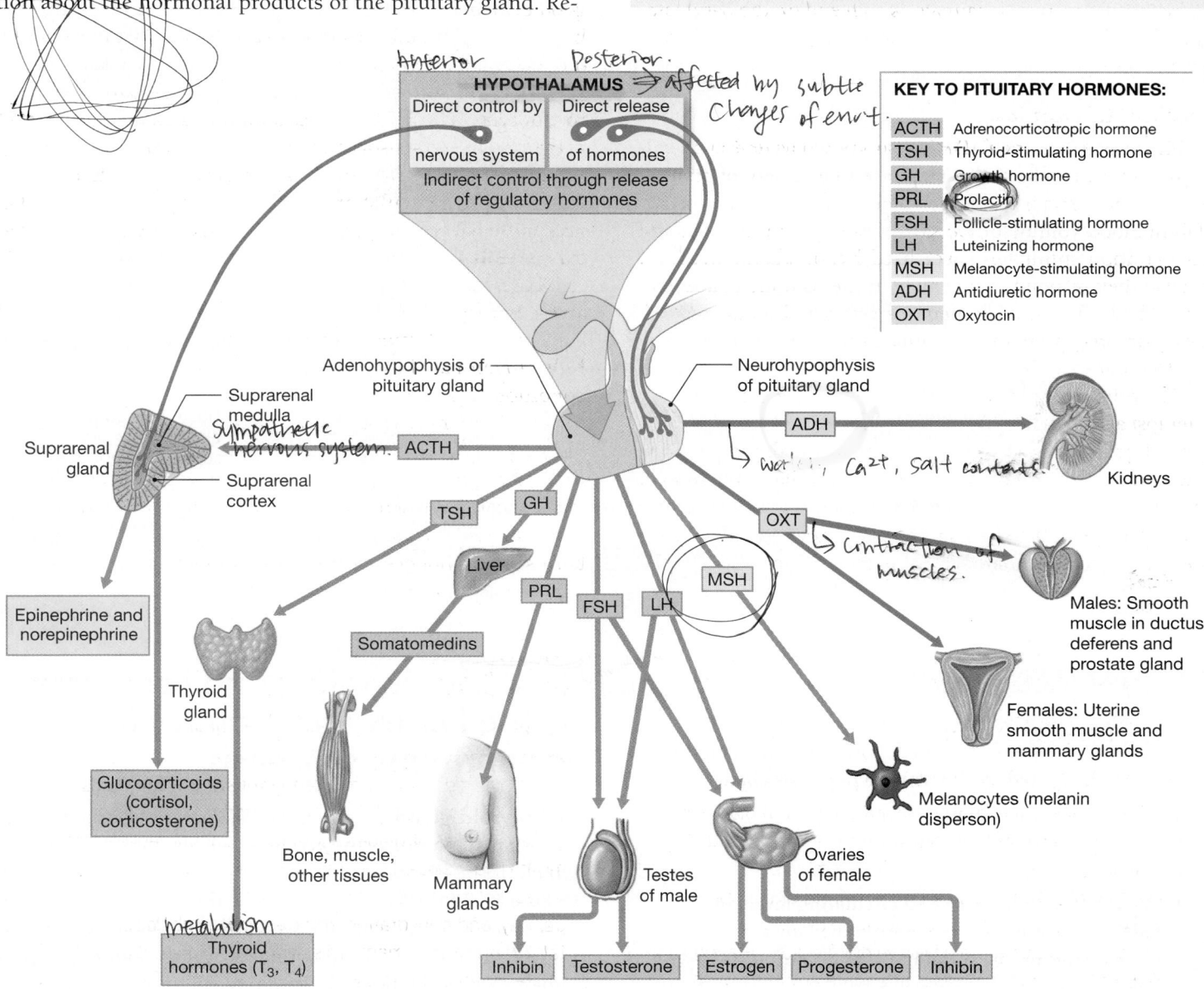

Figure 18–9   Pituitary Hormones and Their Targets.

SUMMARY TABLE 18–2   The Pituitary Hormones

Region/Area	Hormone(s)	Target(s)	Hormonal Effect(s)	Hypothalamic Regulatory Hormone
ADENOHYPOPHYSIS (ANTERIOR LOBE)				
Pars distalis	**Thyroid-stimulating hormone (TSH)**	Thyroid gland	Secretion of thyroid hormones	Thyrotropin-releasing hormone (TRH)
	Adrenocorticotropic hormone (ACTH)	Suprarenal cortex (zona fasciculata)	Secretion of glucocorticoids (cortisol, corticosterone)	Corticotropin-releasing hormone (CRH)
	Gonadotropins:			
	Follicle-stimulating hormone (FSH)	Follicle cells of ovaries	Secretion of estrogen, follicle development	Gonadotropin-releasing hormone (GnRH)
		Nurse (sustentacular) cells of testes	Stimulation of sperm maturation	Gonadotropin-releasing hormone (GnRH)
	Luteinizing hormone (LH)	Follicle cells of ovaries	Ovulation, formation of corpus luteum, secretion of progesterone	Gonadotropin-releasing hormone (GnRH)
		Interstitial cells of testes	Secretion of testosterone	Gonadotropin-releasing hormone (GnRH)
	Prolactin (PRL)	Mammary glands	Production of milk	Prolactin-releasing factor (PRF)
				Prolactin-inhibiting hormone (PIH)
	Growth hormone (GH)	All cells	Growth, protein synthesis, lipid mobilization and catabolism	Growth hormone–releasing hormone (GH–RH)
				Growth hormone–inhibiting hormone (GH–IH)
Pars intermedia (not active in normal adults)	**Melanocyte-stimulating hormone (MSH)**	Melanocytes	Increased melanin synthesis in epidermis	Melanocyte-stimulating hormone-inhibiting hormone (MSH–IH)
NEUROHYPOPHYSIS (POSTERIOR LOBE OR PARS NERVOSA)				
	Antidiuretic hormone (ADH)	Kidneys	Reabsorption of water, elevation of blood volume and pressure	None: Transported along axons from supraoptic nucleus to neurohypophysis
	Oxytocin (OXT)	Uterus, mammary glands (females)	Labor contractions, milk ejection	None: Transported along axons from paraventricular nucleus to neurohypophysis
		Ductus deferens and prostate gland (males)	Contractions of ductus deferens and prostate gland	

CHECKPOINT

7. Identify the two lobes of the pituitary gland.

8. If a person were dehydrated, how would the amount of ADH released by the neurohypophysis change?

9. A blood sample contains elevated levels of somatomedins. Which pituitary hormone would you also expect to be elevated?

10. What effect would elevated circulating levels of cortisol, a steroid hormone from the suprarenal cortex, have on the pituitary secretion of ACTH?

See the blue Answers tab at the end of the book.

18-4 The thyroid gland lies inferior to the larynx and requires iodine for hormone synthesis

The **thyroid gland** curves across the anterior surface of the trachea just inferior to the *thyroid* ("shield-shaped") *cartilage*, which forms most of the anterior surface of the larynx (**Figure 18–10a**). The two **lobes** of the thyroid gland are united by a slender connection, the **isthmus** (IS-mus). You can easily feel the gland with your fingers. When something goes wrong with it, the thyroid gland typically becomes visible as it enlarges and distorts the surface of the neck. The size of the

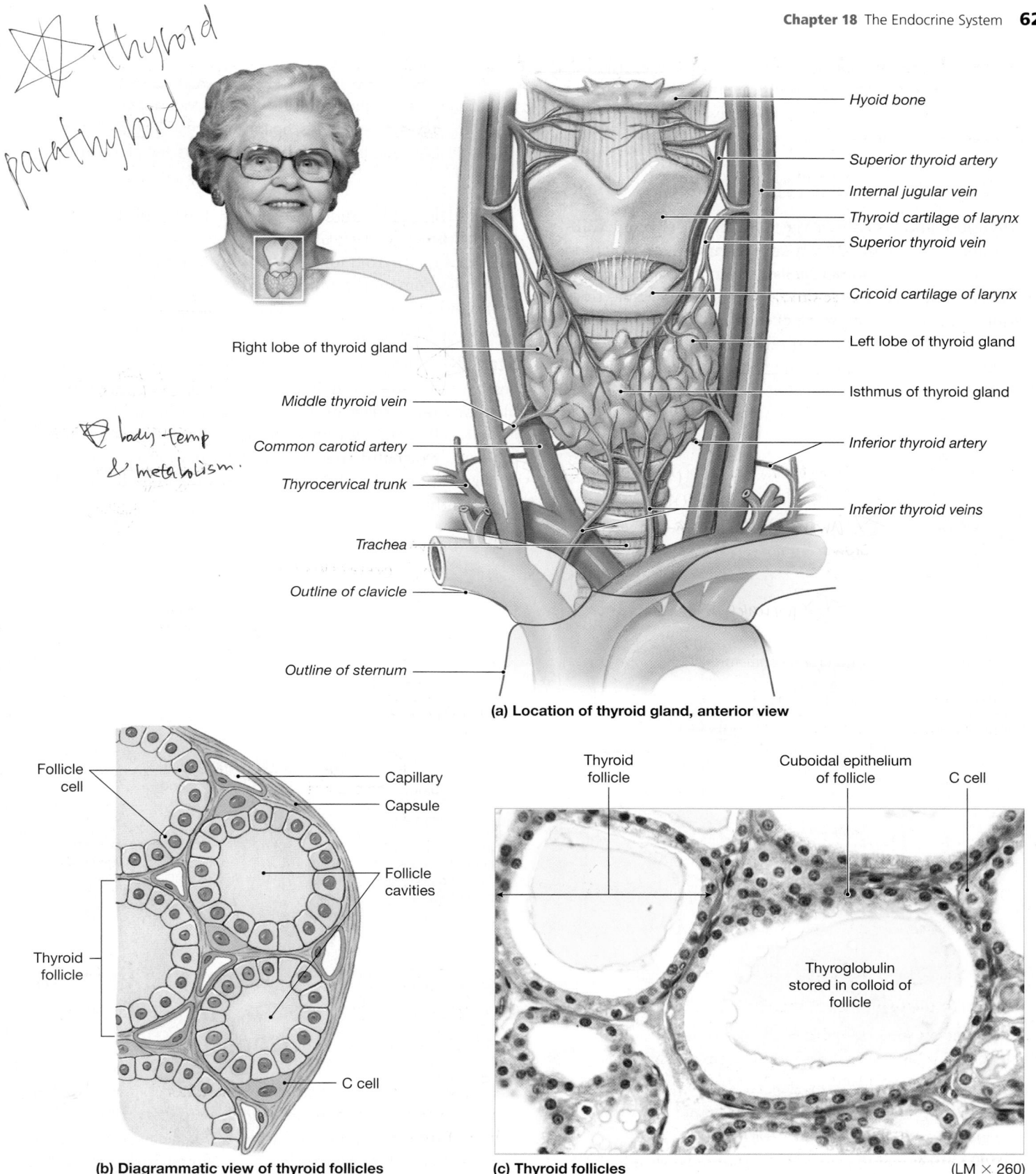

thyroid
parathyroid

body temp & metabolism.

(a) Location of thyroid gland, anterior view

Hyoid bone

Superior thyroid artery

Internal jugular vein

Thyroid cartilage of larynx

Superior thyroid vein

Cricoid cartilage of larynx

Left lobe of thyroid gland

Isthmus of thyroid gland

Inferior thyroid artery

Inferior thyroid veins

Right lobe of thyroid gland

Middle thyroid vein

Common carotid artery

Thyrocervical trunk

Trachea

Outline of clavicle

Outline of sternum

18 ENDOCRINE

(b) Diagrammatic view of thyroid follicles

Follicle cell

Capillary

Capsule

Follicle cavities

Thyroid follicle

C cell

(c) Thyroid follicles

Thyroid follicle

Cuboidal epithelium of follicle

C cell

Thyroglobulin stored in colloid of follicle

(LM × 260)

Figure 18–10   The Thyroid Gland. (a) The location, anatomy, and blood supply of the thyroid gland. **(b)** A diagrammatic view of a section through the wall of the thyroid gland. **(c)** Histological details, showing thyroid follicles. ATLAS: Plate 18c

gland is quite variable, depending on heredity and environmental and nutritional factors, but its average weight is about 34 g (1.2 oz). An extensive blood supply gives the thyroid gland a deep red color.

Thyroid Follicles and Thyroid Hormones

The thyroid gland contains large numbers of **thyroid follicles**, hollow spheres lined by a simple cuboidal epithelium (**Figure 18–11a,b**). The follicle cells surround a **follicle cavity** that holds a viscous *colloid*, a fluid containing large quantities of dissolved proteins. A network of capillaries surrounds each follicle, delivering nutrients and regulatory hormones to the glandular cells and accepting their secretory products and metabolic wastes.

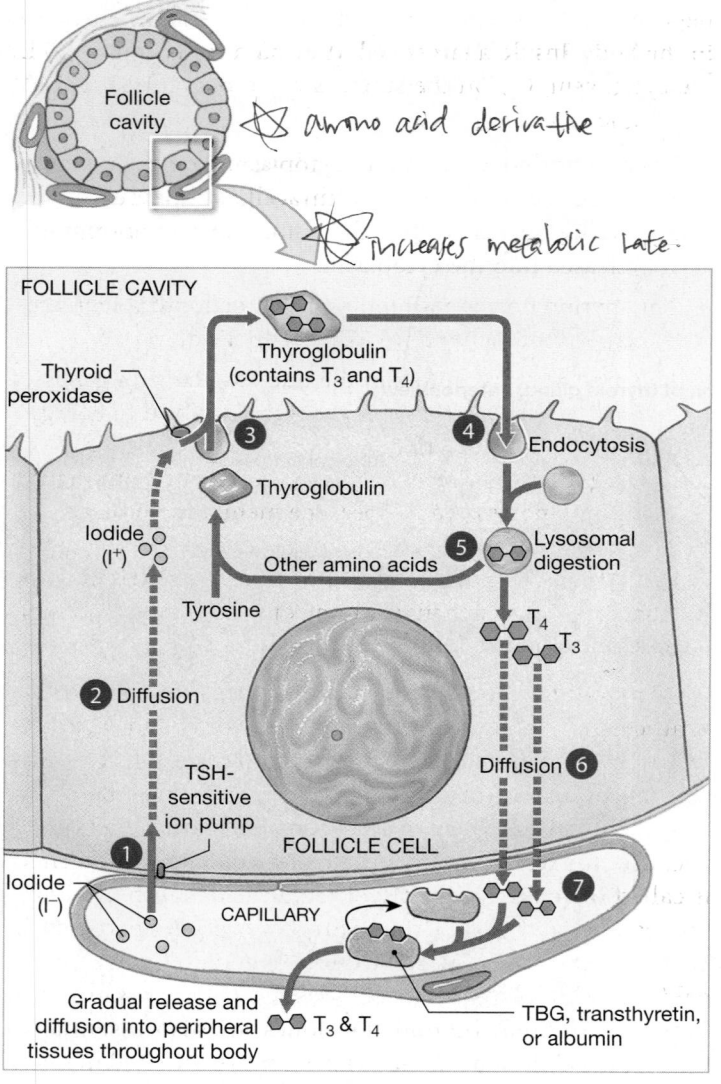

Follicle cells synthesize a globular protein called **thyroglobulin** (thī-rō-GLOB-ū-lin) and secrete it into the colloid of the thyroid follicles (**Figure 18–10c**). Thyroglobulin molecules contain the amino acid *tyrosine*, the building block of thyroid hormones. The formation of thyroid hormones involves the following basic steps (**Figure 18–11a**):

Step 1 Iodide ions are absorbed from the diet at the digestive tract and are delivered to the thyroid gland by the bloodstream. Carrier proteins in the basal membrane of the follicle cells actively transport iodide ions (I^-) into the cytoplasm. Normally, the follicle cells maintain intracellular concentrations of iodide that are many times higher than those in the extracellular fluid.

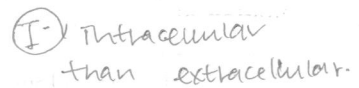

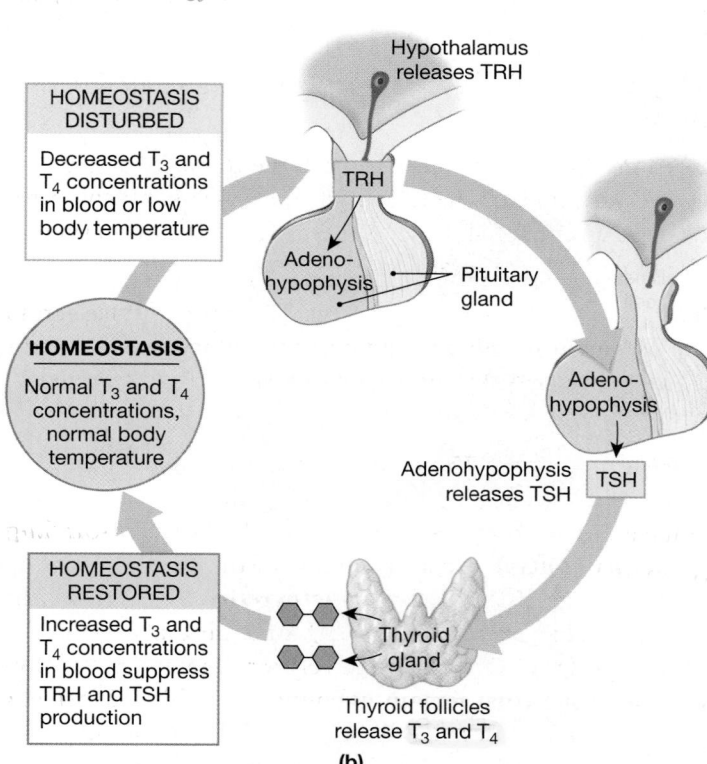

Figure 18–11 The Thyroid Follicles. (a) The synthesis, storage, and secretion of thyroid hormones. For a detailed explanation of the numbered events, see the text. **(b)** The regulation of thyroid secretion.

Step 2 The iodide ions diffuse to the apical surface of each [*activated*] follicle cell, where they are converted to an activated form of [*to*] iodide (I^+) by the enzyme *thyroid peroxidase*. This reaction se- [*I+*] quence, which occurs at the lumenal membrane surface, also attaches one or two iodide ions to the tyrosine portions of a thyroglobulin molecule within the lumen.

Step 3 Tyrosine molecules to which iodide ions have been attached become linked by covalent bonds, forming mole- cules of thyroid hormones that remain incorporated into thy- roglobulin. The pairing process is probably performed by thyroid peroxidase. The hormone **thyroxine** (thī-ROKS-ēn), also known as *tetraiodothyronine* or T_4, contains four iodide ions. A related molecule called **triiodothyronine**, or T_3, con- tains three iodide ions. Eventually, each molecule of thy- roglobulin contains four to eight molecules of T_3 or T_4 hormones or both. [→ *into capillaries → released into body.*]

The major factor controlling the rate of thyroid hormone release is the concentration of TSH in the circulating blood (**Figure 18–11b**). TSH stimulates iodide transport into the follicle cells and stimulates the production of thyroglobulin and thyroid peroxidase. TSH also stimulates the release of thyroid hormones. Under the influence of TSH, the follow- ing steps occur (**Figure 18–11a**):

Step 4 Follicle cells remove thyroglobulin from the folli- cles by endocytosis.

Step 5 Lysosomal enzymes break the thyroglobulin down, and the amino acids and thyroid hormones enter the cyto- plasm. The amino acids are then recycled and used to synthe- size more thyroglobulin.

Step 6 The released molecules of T_3 and T_4 diffuse across the basement membrane and enter the bloodstream. About 90 percent of all thyroid secretions is T_4; T_3 is secreted in comparatively small amounts.

Step 7 Roughly 75 percent of the T_4 molecules and 70 percent of the T_3 molecules entering the bloodstream be- come attached to transport proteins called **thyroid-binding globulins (TBGs)**. Most of the rest of the T_4 and T_3 in the circulation is attached to **transthyretin**, also known as *thyroid-binding prealbumin (TBPA)*, or to *albumin*, one of the plasma proteins. Only the relatively small quantities of thy- roid hormones that remain unbound—roughly 0.3 percent of the circulating T_3 and 0.03 percent of the circulating T_4— are free to diffuse into peripheral tissues.

An equilibrium exists between the bound and unbound thyroid hormones. At any moment, free thyroid hormones are being bound to carriers at the same rate at which bound hormones are being released. When unbound thyroid hor-

mones diffuse out of the bloodstream and into other tissues, the equilibrium is disturbed. The carrier proteins then re- lease additional thyroid hormones until a new equilibrium is reached. The bound thyroid hormones represent a sub- stantial reserve: The bloodstream normally contains more than a week's supply of thyroid hormones.

TSH plays a key role in both the synthesis and the release of thyroid hormones. In the absence of TSH, the thyroid fol- licles become inactive, and neither synthesis nor secretion occurs. TSH binds to membrane receptors and, by stimulat- ing adenylate cyclase, activates key enzymes involved in thyroid hormone production (**Figure 18–3**).

[→ *no TSH, follicles inactive.*]

Functions of Thyroid Hormones

Thyroid hormones enter target cells by means of an energy- dependent transport system, and they affect almost every cell in the body. Inside a target cell, they bind to receptors (1) in the cytoplasm, (2) on the surfaces of mitochondria, and (3) in the nucleus.

- Thyroid hormones bound to cytoplasmic receptors are essentially held in storage. If intracellular levels of thyroid hormones decline, the bound thyroid hormones are released into the cytoplasm.

- The thyroid hormones binding to mitochondria increase the rates of mitochondrial ATP production.

- The binding to receptors in the nucleus activates genes that control the synthesis of enzymes involved in energy transformation and utilization. One specific effect of binding to nuclear receptors is the accelerated production of sodium–potassium ATPase, the membrane protein responsible for the ejection of intracellular sodium and the recovery of extracellular potassium. As noted in Chapter 3, this exchange pump consumes large amounts of ATP. ∞ p. 95 [*ATPase → ejects intracellular Na+ & recovers K+ extracellular*]

Thyroid hormones also activate genes that code for the synthesis of enzymes involved in glycolysis and ATP pro- duction. This effect, coupled with the direct effect of thy- roid hormones on mitochondria, increases the metabolic rate of the cell. Because the cell consumes more energy and because this results in increased heat generation, the effect is called the **calorigenic effect** (*calor*, heat) of thyroid hor- mones. In young children, TSH production increases in cold weather; the calorigenic effect may help them adapt to cold climates. (This response does not occur in adults.) In growing children, thyroid hormones are also essential to normal development of the skeletal, muscular, and nervous systems.

The thyroid gland produces large amounts of T_4, but T_3 is primarily responsible for the observed effects of thyroid hormones: a strong, immediate, and short-lived increase in the rate of cellular metabolism. At any moment, T_3 released from the thyroid gland accounts for only 10–15 percent of the T_3 in peripheral tissues. However, enzymes in the liver, kidneys, and other tissues can convert T_4 to T_3. Roughly 85–90 percent of the T_3 that reaches the target cells is produced by the conversion of T_4 within peripheral tissues. Table 18–3 summarizes the effects of thyroid hormones on major organs and systems. *not much T3 produced & more converted*

Iodine and Thyroid Hormones

Iodine in the diet is absorbed at the digestive tract as I^-. Each day the follicle cells in the thyroid gland absorb 120–150 μg of I^-, the minimum dietary amount needed to maintain normal thyroid function. The iodine ions are actively transported into the thyroid follicle cells, so the concentration of I^- inside thyroid follicle cells is generally about 30 times higher than that in the blood plasma. If plasma I^- levels rise, so do levels inside the follicle cells. *more in follicle than blood*

release of CT depends on Ca2+ levels

The thyroid follicles contain most of the iodide reserves in the body. The active transport mechanism for iodide is stimulated by TSH. The resulting increase in the rate of iodide movement into the cytoplasm accelerates the formation of thyroid hormones.

The typical diet in the United States provides approximately 500 μg of iodide per day, roughly three times the minimum daily requirement. Much of the excess is due to the addition of I^- to the table salt sold in grocery stores as "iodized salt." Thus, iodide deficiency is seldom responsible for limiting the rate of thyroid hormone production. (This is not necessarily the case in other countries.) Excess I^- is removed from the blood at the kidneys, and each day a small amount of I^- (about 20 μg) is excreted by the liver into the *bile*, an exocrine product stored in the gallbladder. Iodide excreted at the kidneys is eliminated in urine; the I^- excreted in bile is eliminated in feces. The losses in the bile, which continue even if the diet contains less than the minimum iodide requirement, can gradually deplete the iodide reserves in the thyroid. Thyroid hormone production then declines, regardless of the circulating levels of TSH.

The C Cells of the Thyroid Gland and Calcitonin

A second population of endocrine cells lies sandwiched between the cuboidal follicle cells and their basement membrane. These cells, which are larger than those of the follicular epithelium and do not stain as clearly, are the **C (clear) cells**,

TABLE 18–3 Effects of Thyroid Hormones on Peripheral Tissues

1. Elevated rates of oxygen consumption and energy consumption; in children, may cause a rise in body temperature
2. Increased heart rate and force of contraction; generally results in a rise in blood pressure
3. Increased sensitivity to sympathetic stimulation
4. Maintenance of normal sensitivity of respiratory centers to changes in oxygen and carbon dioxide concentrations
5. Stimulation of red blood cell formation and thus enhanced oxygen delivery
6. Stimulation of activity in other endocrine tissues
7. Accelerated turnover of minerals in bone

or *parafollicular cells* (**Figure 18–10b,c**). C cells produce the hormone **calcitonin (CT)**, which aids in the regulation of Ca^{2+} concentrations in body fluids. The functions of this hormone were introduced in Chapter 6. ∞ p. 204 The net effect of calcitonin release is a drop in the Ca^{2+} concentration in body fluids, accomplished by (1) the inhibition of osteoclasts, which slows the rate of Ca^{2+} release from bone, and (2) the stimulation of Ca^{2+} excretion at the kidneys.

The control of calcitonin secretion is an example of direct endocrine regulation: Neither the hypothalamus nor the pituitary gland is involved. The C cells respond directly to an elevation in the Ca^{2+} concentration of blood. When the concentration rises, calcitonin secretion increases. The Ca^{2+} concentration then drops, eliminating the stimulus and "turning off" the C cells. *not regulated by hyp or pituitary*

Calcitonin is probably most important during childhood, when it stimulates bone growth and mineral deposition in the skeleton. It also appears to be important in reducing the loss of bone mass (1) during prolonged starvation and (2) in the late stages of pregnancy, when the maternal skeleton competes with the developing fetus for calcium ions absorbed by the digestive tract. The role of calcitonin in the healthy nonpregnant adult is unclear.

In several chapters, we have considered the importance of Ca^{2+} in controlling muscle cell and neuron activities. Calcium ion concentrations also affect the sodium permeabilities of excitable membranes. At high Ca^{2+} concentrations, sodium permeability decreases and membranes become less responsive. Such problems are relatively rare. Problems caused by lower-than-normal Ca^{2+} concentrations are equally dangerous and are much more common. When calcium ion concentrations decline, sodium permeabilities increase and cells become extremely excitable. If calcium levels fall too far, convulsions or muscular spasms can result. Maintenance of adequate calcium levels primarily involves the *parathyroid glands* and *parathyroid hormone*.

CHECKPOINT

11. Identify the hormones of the thyroid gland.

12. What signs and symptoms would you expect to see in an individual whose diet lacks iodine?

13. When a person's thyroid gland is removed, signs of decreased thyroid hormone concentration do not appear until about one week later. Why?

See the blue Answers tab at the end of the book.

18-5 The four parathyroid glands, embedded in the posterior surface of the thyroid gland, secrete parathyroid hormone to elevate plasma Ca²⁺

There are normally two pairs of **parathyroid glands** embedded in the posterior surfaces of the thyroid gland (**Figure 18–12a**). The cells of the two adjacent glands are separated by the dense

capsule that surrounds each parathyroid gland. Altogether, the four parathyroid glands weigh a mere 1.6 g (0.06 oz). The histological appearance of a single parathyroid gland is shown in **Figure 18–12b,c**. The parathyroid glands have at least two cell populations: The **parathyroid cells** (also termed **chief cells** or **principal cells**) produce parathyroid hormone; the functions of the other cells, called *oxyphils*, are unknown.

Like the C cells of the thyroid gland, the parathyroid cells monitor the circulating concentration of calcium ions. When the Ca^{2+} concentration of the blood falls below normal, the parathyroid cells secrete **parathyroid hormone (PTH)**, or *parathormone*. The net result of PTH secretion is an increase in Ca^{2+} concentration in body fluids. Parathyroid hormone has three major effects:

1. It mobilizes calcium from bone by affecting osteoblast and osteoclast activity. PTH probably inhibits osteoblasts, thereby reducing the rate of calcium deposition in bone, but it has other, more-significant effects. Although osteoclasts have no PTH receptors, PTH triggers the release of a growth factor known as RANKL, which increases osteoclast numbers. Because there are more osteoclasts, the

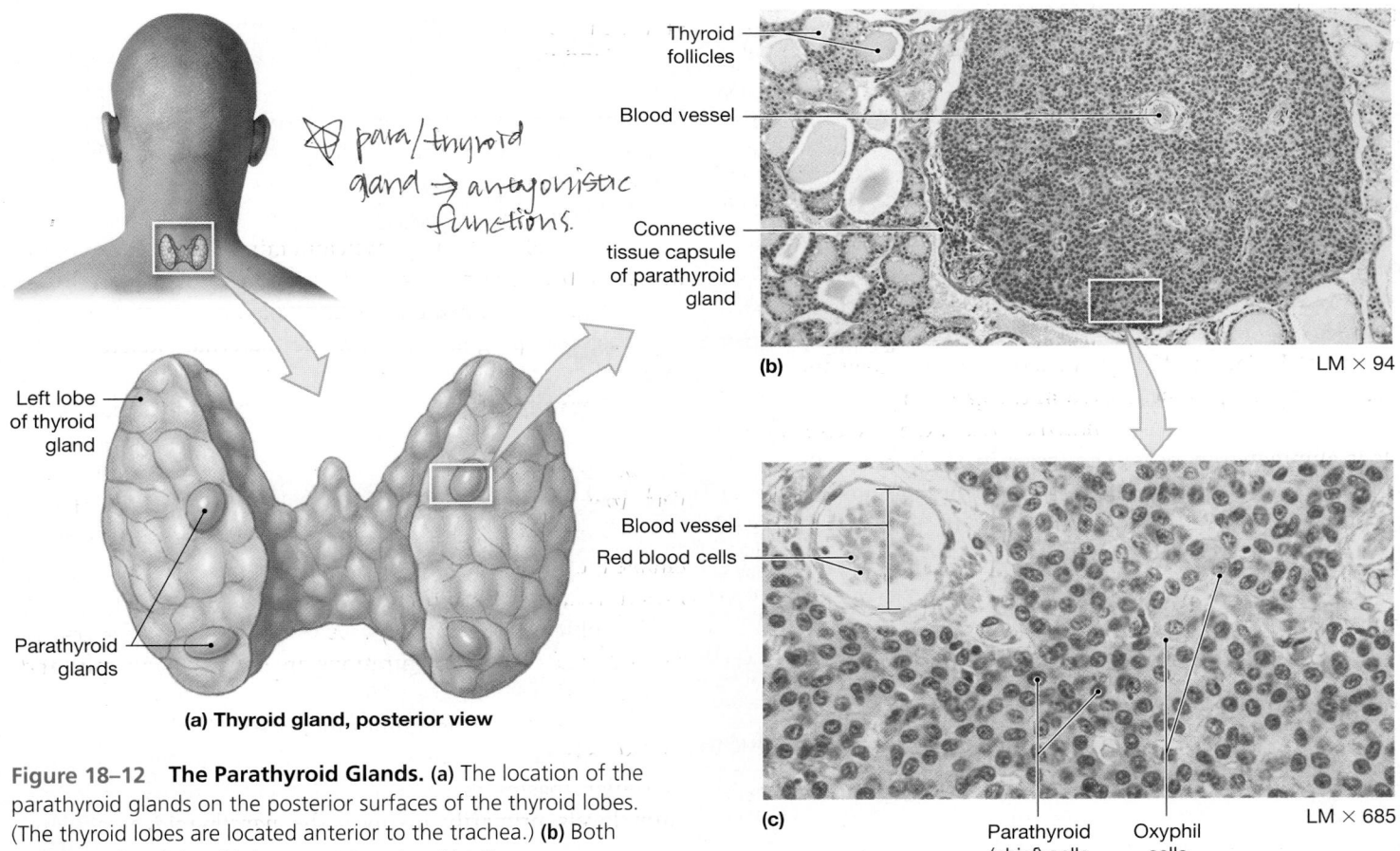

Figure 18–12 The Parathyroid Glands. (a) The location of the parathyroid glands on the posterior surfaces of the thyroid lobes. (The thyroid lobes are located anterior to the trachea.) **(b)** Both parathyroid and thyroid tissues. **(c)** Parathyroid cells.

rates of mineral turnover and Ca^{2+} release accelerate, and as bone matrix erodes, plasma Ca^{2+} rises.

2. It enhances the reabsorption of Ca^{2+} at the kidneys, reducing urinary losses.

3. It stimulates the formation and secretion of *calcitriol* at the kidneys. In general, the effects of calcitriol complement or enhance those of PTH, but one major effect of calcitriol is the enhancement of Ca^{2+} and PO_4^{3-} absorption by the digestive tract. ∞ p. 201

Figure 18–13 illustrates the roles of calcitonin and PTH in regulating Ca^{2+} concentrations. It is likely that PTH, aided by calcitriol, is the primary regulator of circulating calcium ion concentrations in healthy adults. Information about the hormones of the thyroid gland and parathyroid glands is summarized in Table 18–4.

The A&P Top 100

#63 The thyroid gland produces (1) hormones that adjust tissue metabolic rates and (2) a hormone that usually plays a minor role in calcium ion homeostasis by opposing the action of parathyroid hormone.

CHECKPOINT

14. Describe the location of the parathyroid glands.

15. Identify the hormone secreted by the parathyroid glands.

16. The removal of the parathyroid glands would result in a decrease in the blood concentration of which important mineral?

See the blue Answers tab at the end of the book.

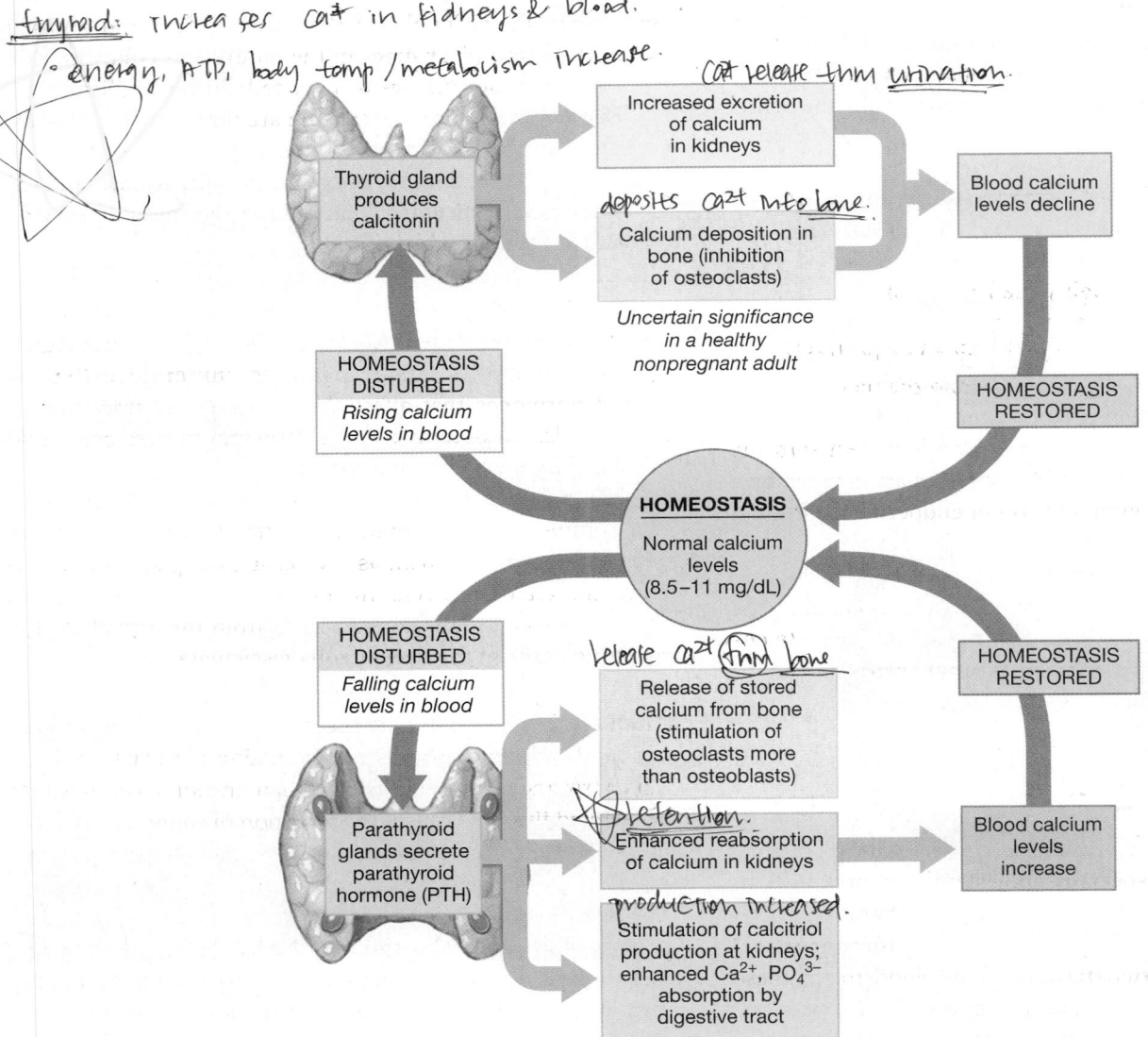

Figure 18–13   **The Homeostatic Regulation of Calcium Ion Concentrations.**

TABLE 18–4 Hormones of the Thyroid Gland and Parathyroid Glands

Gland/Cells	Hormone(s)	Targets	Hormonal Effects	Regulatory Control
THYROID GLAND				
Follicular epithelium	Thyroxine (T_4) Triiodothyronine (T_3)	Most cells	Increases energy utilization, oxygen consumption, growth, and development	Stimulated by TSH from adenohypophysis (see Table 18–3)
C cells	Calcitonin (CT)	Bone, kidneys	Decreases Ca^{2+} concentrations in body fluids	Stimulated by elevated blood Ca^{2+} levels; actions opposed by PTH
PARATHYROID GLANDS				
Parathyroid (chief) cells	Parathyroid hormone (PTH)	Bone, kidneys	Increases Ca^{2+} concentrations in body fluids	Stimulated by low blood Ca^{2+} levels; PTH effects enhanced by calcitriol and opposed by calcitonin

18-6 The suprarenal glands, consisting of a cortex and medulla, cap the superior borders of the kidneys and secrete several hormones

which genes transcribed, controls cellular metabolism

A yellow, pyramid-shaped **suprarenal** (soo-pra-RĒ-nal) **gland** (*supra-*, above + *renes*, kidneys), or *adrenal gland*, sits on the superior border of each kidney (**Figure 18–14**). Each suprarenal gland lies at roughly the level of the 12th rib and is firmly attached to the superior portion of each kidney by a dense fibrous capsule. The suprarenal gland on each side nestles among the kidney, the diaphragm, and the major arteries and veins that run along the posterior wall of the abdominopelvic cavity. The suprarenal glands project into the peritoneal cavity, and their anterior surfaces are covered by a layer of parietal peritoneum. Like other endocrine glands, the suprarenal glands are highly vascularized.

A typical suprarenal gland weighs about 5.0 g (0.18 oz), but its size can vary greatly as secretory demands change. The suprarenal gland is divided into two parts with separate endocrine functions: a superficial **suprarenal cortex** and an inner **suprarenal medulla** (**Figure 18–14b**).

The Suprarenal Cortex

The yellowish color of the suprarenal cortex is due to the presence of stored lipids, especially cholesterol and various fatty acids. The suprarenal cortex produces more than two dozen steroid hormones, collectively called **adrenocortical steroids**, or simply **corticosteroids**. In the bloodstream, these hormones are bound to transport proteins called *transcortins*. *>24 hormones, corticosteroids*

Corticosteroids are vital: If the suprarenal glands are destroyed or removed, the individual will die unless corticosteroids are administered. Corticosteroids, like other steroid hormones, exert their effects by determining which genes are transcribed in the nuclei of their target cells, and at what rates. The resulting changes in the nature and concentration of enzymes in the cytoplasm affect cellular metabolism.

Deep to the suprarenal capsule are three distinct regions, or zones, in the suprarenal cortex (**Figure 18–14c**): (1) an outer *zona glomerulosa*; (2) a middle *zona fasciculata*; and (3) an inner *zona reticularis*. Each zone synthesizes specific steroid hormones (Table 18–5).

The Zona Glomerulosa (outer)

The **zona glomerulosa** (glō-mer-ū-LŌ-suh), the outer region of the suprarenal cortex, produces **mineralocorticoids**, steroid hormones that affect the electrolyte composition of body fluids. **Aldosterone** is the principal mineralocorticoid produced by the suprarenal cortex.

The zona glomerulosa accounts for about 15 percent of the volume of the suprarenal cortex (**Figure 18–14c**). A *glomerulus* is a little ball; as the term *zona glomerulosa* implies, the endocrine cells in this region form small, dense knots or clusters. This zone extends from the capsule to the radiating cords of the deeper zona fasciculata.

keep Na+, lose K+

Aldosterone. Aldosterone secretion stimulates the conservation of sodium ions and the elimination of potassium ions. This hormone targets cells that regulate the ionic composition of excreted fluids. It causes the retention of sodium ions at the kidneys, sweat glands, salivary glands, and pancreas, preventing Na^+ loss in urine, sweat, saliva, and digestive secretions. The retention of Na^+ is accompanied by a loss of K^+. As a secondary effect, the reabsorption of Na^+ enhances the osmotic reabsorption of water at the kidneys, sweat glands, salivary glands, and pancreas. The effect at the kidneys is most dramatic when normal levels of ADH are present. In addition, aldosterone increases the sensitivity of salt receptors in the taste

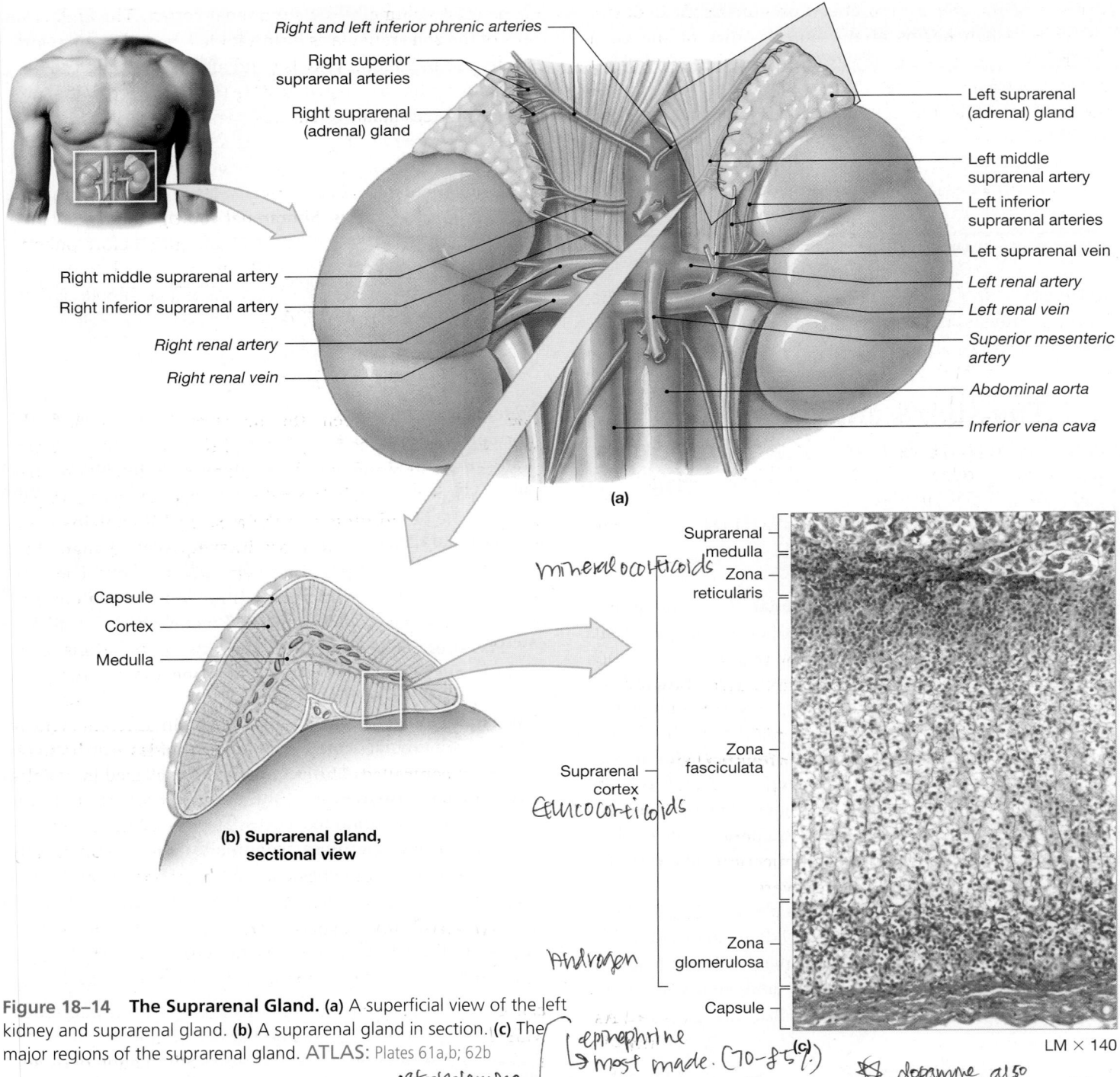

Right and left inferior phrenic arteries

Right superior suprarenal arteries

Right suprarenal (adrenal) gland

Left suprarenal (adrenal) gland

Left middle suprarenal artery

Left inferior suprarenal arteries

Left suprarenal vein

Right middle suprarenal artery

Right inferior suprarenal artery

Right renal artery

Right renal vein

Left renal artery

Left renal vein

Superior mesenteric artery

Abdominal aorta

Inferior vena cava

(a)

Capsule

Cortex

Medulla

(b) Suprarenal gland, sectional view

Suprarenal medulla

[handwritten: mineralocorticoids]

Zona reticularis

Suprarenal cortex

[handwritten: glucocorticoids]

Zona fasciculata

[handwritten: Androgen]

Zona glomerulosa

Capsule

LM × 140

Figure 18–14   The Suprarenal Gland. (a) A superficial view of the left kidney and suprarenal gland. **(b)** A suprarenal gland in section. **(c)** The major regions of the suprarenal gland. ATLAS: Plates 61a,b; 62b

[handwritten: catecholamine]

*[handwritten: epinephrine → most made. (70-85%) **(c)** / norepinephrine (20-25%) / dopamine also secreted → control of body most.]*

[handwritten left margin: fall in blood or rise in K+]

buds of the tongue. As a result, interest in (and consumption of) salty food increases.

Aldosterone secretion occurs in response to a drop in blood Na$^+$ content, blood volume, or blood pressure, or to a rise in blood K$^+$ concentration. Changes in either Na$^+$ or K$^+$ concentration have a direct effect on the zona glomerulosa, but the secretory cells are most sensitive to changes in potassium levels. A rise in potassium levels is very effective in

stimulating the release of aldosterone. Aldosterone release also occurs in response to *angiotensin II*. We will discuss this hormone, part of the *renin–angiotensin system*, later in this chapter.

The Zona Fasciculata

The **zona fasciculata** (fa-sik-ū-LA-tuh; *fasciculus*, little bundle) produces steroid hormones collectively known as

glucocorticoids, due to their effects on glucose metabolism. This zone, which begins at the inner border of the zona glomerulosa and extends toward the suprarenal medulla (**Figure 18–14c**), contributes about 78 percent of the cortical volume. The endocrine cells are larger and contain more lipids than those of the zona glomerulosa, and the lipid droplets give the cytoplasm a pale, foamy appearance. The cells of the zona fasciculata form individual cords composed of stacks of cells. Adjacent cords are separated by flattened blood vessels (sinusoids) with fenestrated walls.

The Glucocorticoids. When stimulated by ACTH from the adenohypophysis, the zona fasciculata secretes primarily **cortisol** (KOR-ti-sol), also called *hydrocortisone*, along with smaller amounts of the related steroid **corticosterone** (kor-ti-KOS-te-rōn). The liver converts some of the circulating cortisol to **cortisone**, another active glucocorticoid. Glucocorticoid secretion is regulated by negative feedback: The glucocorticoids released have an inhibitory effect on the production of corticotropin-releasing hormone (CRH) in the hypothalamus, and of ACTH in the adenohypophysis (**Figure 18–8a**). negative feedback → inhibits CRH

Effects of Glucocorticoids. Glucocorticoids accelerate the rates of glucose synthesis and glycogen formation, especially in the liver. Adipose tissue responds by releasing fatty acids into the blood, and other tissues begin to break down fatty acids and proteins instead of glucose. This process is another example of a glucose-sparing effect (p. 617).

Glucocorticoids also show **anti-inflammatory** effects; that is, they inhibit the activities of white blood cells and other components of the immune system. "Steroid creams" are commonly used to control irritating allergic rashes, such as those produced by poison ivy, and injections of glucocorticoids may be used to control more severe allergic reactions. Glucocorticoids slow the migration of phagocytic cells into an injury site and cause phagocytic cells already in the area to become less active. In addition, mast cells exposed to these steroids are less likely to release histamine and other chemicals that promote inflammation. ∞ pp. 141–142 As a result, swelling and further irritation are dramatically reduced. On the negative side, the rate of wound healing decreases, and the weakening of the region's defenses makes it more susceptible to infectious organisms. For that reason, topical steroids are used to treat superficial rashes, but should never be applied to open wounds. → steroids used for superficial rashes → not open wounds

The Zona Reticularis

The **zona reticularis** (re-tik-ū-LAR-is; *reticulum*, network) forms a narrow band bordering each suprarenal medulla (**Figure 18–14c**). This zone accounts for only about 7 percent of the total volume of the suprarenal cortex. The endocrine cells of the zona reticularis form a folded, branching network, and fenestrated blood vessels wind among the cells.

Under stimulation by ACTH, the zona reticularis normally produces small quantities of androgens, the sex hormones produced in large quantities by the testes in males. Once in the bloodstream, some of the androgens released by the zona reticularis are converted to estrogens, the dominant sex hormone in females. Suprarenal androgens stimulate the development of pubic hair in boys and girls before puberty. While not important in adult men, in adult women suprarenal androgens promote muscle mass, blood cell formation, and support the libido.

The Suprarenal Medulla

The boundary between the suprarenal cortex and the suprarenal medulla is irregular, and the supporting connective tissues and blood vessels are extensively interconnected. The suprarenal medulla is a pale gray or pink, owing in part to the many blood vessels in the area, and it contains large, rounded cells—similar to those in sympathetic ganglia that are innervated by preganglionic sympathetic fibers. The sympathetic division of the autonomic nervous system controls secretory activities of the suprarenal medullae. ∞ p. 533

The suprarenal medulla contains two populations of secretory cells: One produces epinephrine (adrenaline), the other norepinephrine (noradrenaline). Evidence suggests that the two types of cells are distributed in different areas of the medulla and that their secretory activities can be independently controlled. The secretions are packaged in vesicles that form dense clusters just inside cell membranes. The hormones in these vesicles are continuously released at low levels by exocytosis. Sympathetic stimulation dramatically accelerates the rate of exocytosis and hormone release.

Epinephrine and Norepinephrine

Epinephrine makes up 75–80 percent of the secretions from the suprarenal medullae, the rest being norepinephrine. The peripheral effects of these hormones, which result from interaction with alpha and beta receptors on plasma membranes, were described in Chapter 16. ∞ pp. 537–538 Stimulation of α_1 and β_1 receptors, the most common types, accelerates the utilization of cellular energy and the mobilization of energy reserves.

Activation of the suprarenal medullae has the following effects:

- In skeletal muscles, epinephrine and norepinephrine trigger a mobilization of glycogen reserves and accelerate the breakdown of glucose to provide ATP. This combination increases both muscular strength and endurance.

TABLE 18–5	The Suprarenal Hormones			
Region/Zone	**Hormone(s)**	**Primary Targets**	**Hormonal Effects**	**Regulatory Control**
CORTEX				
Zona glomerulosa	Mineralocorticoids (primarily aldosterone)	Kidneys	Increase renal reabsorption of Na^+ and water (especially in the presence of ADH) and accelerate urinary loss of K^+	Stimulated by angiotensin II, elevated plasma K^+ or a fall in plasma Na^+; inhibited by ANP and BNP
Zona fasciculate	Glucocorticoids (cortisol [hydrocortisone], corticosterone)	Most cells	Release amino acids from skeletal muscles and lipids from adipose tissues; promote liver formation of glucose and glycogen; promote peripheral utilization of lipids; anti-inflammatory effects	Stimulated by ACTH from adenohypophysis
Zona reticularis	Androgens		Not important in adult men; encourages bone growth, muscle growth, and blood formation in children and women	Stimulated by ACTH
MEDULLA	Epinephrine, norepinephrine	Most cells	Increases cardiac activity, blood pressure, glycogen breakdown, blood glucose levels; releases lipids by adipose tissue	Stimulated during sympathetic activation by sympathetic preganglionic fibers

- In adipose tissue, stored fats are broken down into fatty acids, which are released into the bloodstream for use by other tissues for ATP production.

- In the liver, glycogen molecules are broken down. The resulting glucose molecules are released into the bloodstream, primarily for use by neural tissues, which cannot shift to fatty acid metabolism.

- In the heart, the stimulation of β_1 receptors triggers an increase in the rate and force of cardiac muscle contraction.

The metabolic changes that follow the release of catecholamines such as E and NE are at their peak 30 seconds after suprarenal stimulation, and they persist for several minutes thereafter. As a result, the effects produced by the stimulation of the suprarenal medullae outlast the other signs of sympathetic activation.

The A&P Top 100

#64 The suprarenal glands produce hormones that adjust metabolic activities at specific sites, affecting either the pattern of nutrient utilization, mineral ion balance, or the rate of energy consumption by active tissues.

CHECKPOINT

17. Identify the two regions of the suprarenal gland, and cite the hormones secreted by each.

18. List the three zones of the suprarenal cortex.

19. What effect would elevated cortisol levels have on blood glucose levels?

See the blue Answers tab at the end of the book.

18-7 The pineal gland, attached to the third ventricle, secretes melatonin

The **pineal gland**, part of the epithalamus, lies in the posterior portion of the roof of the third ventricle. ∞ p. 475 The pineal gland contains neurons, neuroglia, and special secretory cells called **pinealocytes** (pin-Ē-al-ō-sīts). These cells synthesize the hormone **melatonin** from molecules of the neurotransmitter *serotonin*. Collaterals from the visual pathways enter the pineal gland and affect the rate of melatonin production, which is lowest during daylight hours and highest at night.

Among the functions suggested for melatonin in humans are the following:

- *Inhibiting Reproductive Functions.* In some mammals, melatonin slows the maturation of sperm, oocytes, and reproductive organs by reducing the rate of GnRH secretion. The significance of this effect in humans remains unclear, but circumstantial evidence suggests that melatonin may play a role in the timing of human sexual maturation. Melatonin levels in the blood decline at puberty, and pineal tumors that eliminate melatonin production cause premature puberty in young children.

- *Protecting against Damage by Free Radicals.* Melatonin is a very effective *antioxidant* that may protect CNS neurons from free radicals, such as nitric oxide (NO) or hydrogen peroxide (H_2O_2), that may be generated in active neural tissue.

- *Setting Circadian Rhythms.* Because pineal activity is cyclical, the pineal gland may also be involved with the maintenance of basic *circadian rhythms*—daily changes in physiological processes that follow a regular day–night pattern. ∞ p. 478 Increased melatonin secretion in darkness has been suggested as a primary cause of *seasonal affective disorder (SAD)*. This condition, characterized by changes in mood, eating habits, and sleeping patterns, can develop during the winter in people who live at high latitudes, where sunlight is scarce or lacking.

CHECKPOINT

20. Identify the hormone-secreting cells of the pineal gland.

21. Increased amounts of light would inhibit the production of which hormone?

22. List three functions of melatonin.

See the blue Answers tab at the end of the book.

18-8 The pancreas, located within the abdominopelvic cavity, is both an exocrine organ and endocrine gland

The **pancreas** lies within the abdominopelvic cavity in the loop formed between the inferior border of the stomach and the proximal portion of the small intestine (**Figure 18–1**). It is a slender, pale organ with a nodular (lumpy) consistency

(**Figure 18–15a**). The pancreas is 20–25 cm (8–10 in.) long and weighs about 80 g (2.8 oz) in adults. We will consider its anatomy further in Chapter 24, because it is primarily an exocrine organ that makes digestive enzymes.

The **exocrine pancreas**, roughly 99 percent of the pancreatic volume, consists of clusters of gland cells, called *pancreatic acini*, and their attached ducts. Together, the gland and duct cells secrete large quantities of an alkaline, enzyme-rich fluid that reaches the lumen of the digestive tract through a network of secretory ducts.

The **endocrine pancreas** consists of small groups of cells scattered among the exocrine cells. The endocrine clusters are known as **pancreatic islets**, or the *islets of Langerhans* (LAN-ger-hanz) (**Figure 18–15b**). Pancreatic islets account for only about 1 percent of all cells in the pancreas. Nevertheless, a typical pancreas contains roughly 2 million pancreatic islets.

The Pancreatic Islets

The pancreatic islets are surrounded by an extensive, fenestrated capillary network that carries pancreatic hormones into the bloodstream. Each islet contains four types of cells:

1. **Alpha cells** produce the hormone glucagon (GLOO-ka-gon). Glucagon raises blood glucose levels by increasing the rates of glycogen breakdown and glucose release by the liver.

2. **Beta cells** produce the hormone insulin (IN-suh-lin). Insulin lowers blood glucose levels by increasing the rate of

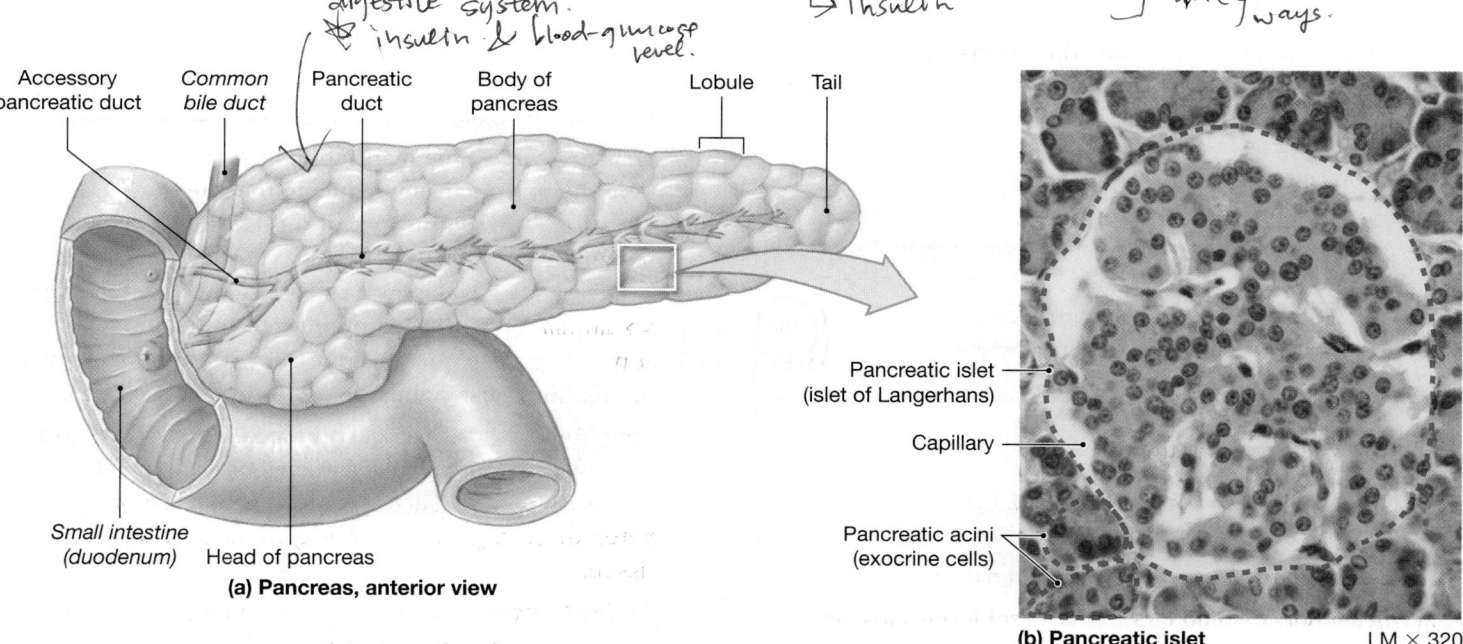

Accessory pancreatic duct | *Common bile duct* | **Pancreatic duct** | **Body of pancreas** | **Lobule** | **Tail**

Small intestine (duodenum) | **Head of pancreas**

(a) Pancreas, anterior view

Pancreatic islet (islet of Langerhans)

Capillary

Pancreatic acini (exocrine cells)

(b) Pancreatic islet LM × 320

Figure 18–15 The Endocrine Pancreas. (a) The gross anatomy of the pancreas. (b) A pancreatic islet surrounded by exocrine cells.

ATLAS: Plate 49e

glucose uptake and utilization by most body cells, and by increasing glycogen synthesis in skeletal muscles and the liver. Beta cells also secrete *amylin*, a recently discovered peptide hormone whose role is unclear.

3. **Delta cells** produce a peptide hormone identical to growth hormone–inhibiting hormone (GH–IH), a hypothalamic regulatory hormone. GH–IH suppresses the release of glucagon and insulin by other islet cells and slows the rates of food absorption and enzyme secretion along the digestive tract.

4. **F cells** produce the hormone **pancreatic polypeptide (PP)**. PP inhibits gallbladder contractions and regulates the production of some pancreatic enzymes, and it may also help control the rate of nutrient absorption by the digestive tract.

We will focus on insulin and glucagon, the hormones responsible for the regulation of blood glucose levels (**Figure 18–16**). When blood glucose levels rise, beta cells secrete insulin, which then stimulates the transport of glucose across plasma membranes. When blood glucose levels decline, alpha cells secrete glucagon, which stimulates glycogen breakdown and glucose release by the liver.

Tips&Tricks

To help in differentiating between the polysaccharide storage molecule "glycogen" and the pancreatic hormone "glucagon," remember that glyco**gen** literally means **gen**erates sugar; and associate gluca**gon** with the Penta**gon**, both of which "issue orders."

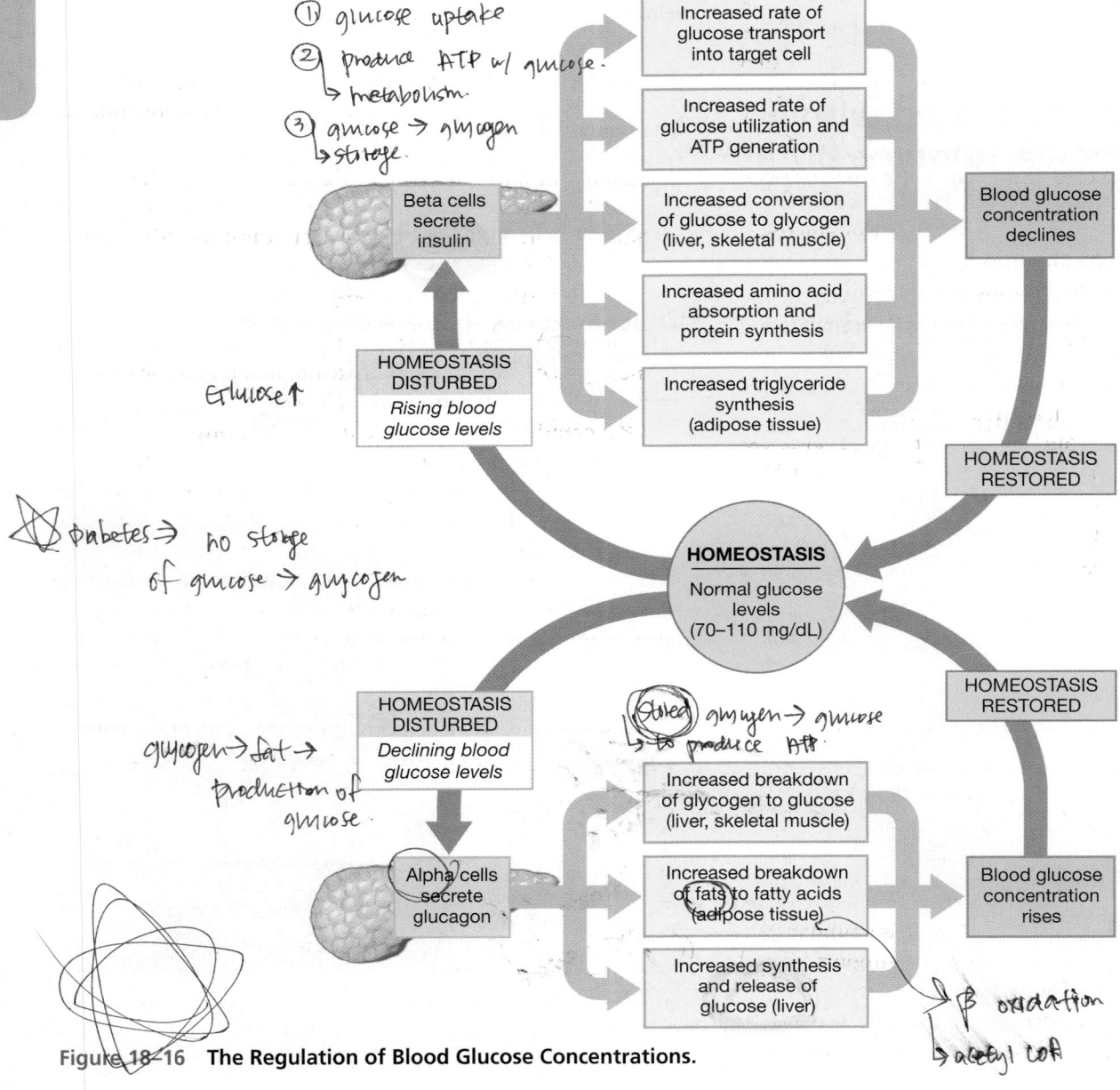

Figure 18–16 **The Regulation of Blood Glucose Concentrations.**

Insulin *[β cells]*

Insulin is a peptide hormone released by beta cells when glucose concentrations exceed normal levels (70–110 mg/dL). Secretion of this hormone is also stimulated by elevated levels of some amino acids, including arginine and leucine. Insulin exerts its effects on cellular metabolism in a series of steps that begins when insulin binds to receptor proteins on the plasma membrane. Binding leads to the activation of the receptor, which functions as a kinase, attaching phosphate groups to intracellular enzymes. The phosphorylation of enzymes then produces primary and secondary effects in the cell, the biochemical details of which remain unresolved.

One of the most important of these effects is the enhancement of glucose absorption and utilization. Insulin receptors are present in most plasma membranes; such cells are called *insulin dependent*. However, cells in the brain and kidneys, cells in the lining of the digestive tract, and red blood cells lack insulin receptors. These cells are called *insulin independent*, because they can absorb and utilize glucose without insulin stimulation.

The effects of insulin on its target cells include the following: *[Insulin secreted when too much glucose in circulation. So lowers glucose level thru increasy uptake.]*

- *The Acceleration of Glucose Uptake (All Target Cells).* This effect results from an increase in the number of glucose transport proteins in the plasma membrane. These proteins transport glucose into the cell by facilitated diffusion, a movement that follows the concentration gradient for glucose and for which ATP is not required.

- *The Acceleration of Glucose Utilization (All Target Cells) and Enhanced ATP Production.* This effect occurs for two reasons: (1) The rate of glucose use is proportional to its availability; when more glucose enters the cell, more is used. (2) Second messengers activate a key enzyme involved in the initial steps of glycolysis.

- *The Stimulation of Glycogen Formation (Skeletal Muscles and Liver Cells).* When excess glucose enters these cells, it is stored as glycogen.

- *The Stimulation of Amino Acid Absorption and Protein Synthesis.*

- *The Stimulation of Triglyceride Formation in Adipose Tissue.* Insulin stimulates the absorption of fatty acids and glycerol by adipocytes, which store these components as triglycerides. Adipocytes also increase their absorption of glucose; excess glucose is used in the synthesis of additional triglycerides.

In sum, insulin is secreted when glucose is abundant; the hormone stimulates glucose utilization to support growth and the establishment of carbohydrate (glycogen) and lipid (triglyceride) reserves. The accelerated use of glucose soon brings circulating glucose levels within normal limits.

Tips&Tricks

The function of ***in***sulin is to get glucose ***in***to cells.

Glucagon *[α cells.]*

When glucose concentrations fall below normal, alpha cells release glucagon and energy reserves are mobilized. When glucagon binds to a receptor in the target plasma membrane, the hormone activates adenylate cyclase. As we have seen, cAMP acts as a second messenger that activates cytoplasmic enzymes (p. 609). The primary effects of glucagon are as follows:

- *Stimulating the Breakdown of Glycogen in Skeletal Muscle and Liver Cells.* The glucose molecules released will be either metabolized for energy (in skeletal muscle fibers) or released into the bloodstream (by liver cells).

- *Stimulating the Breakdown of Triglycerides in Adipose Tissue.* The adipocytes then release the fatty acids into the bloodstream for use by other tissues.

- *Stimulating the Production of Glucose in the Liver.* Liver cells absorb amino acids from the bloodstream, convert them to glucose, and release the glucose into the circulation. This process of glucose synthesis in the liver is called *gluconeogenesis* (gloo-kō-nē-ō-JEN-e-sis).

[Glucagon secreted to use energy reserves.]

The results are a reduction in glucose use and the release of more glucose into the bloodstream. Blood glucose concentrations soon rise toward normal levels.

Pancreatic alpha cells and beta cells monitor blood glucose concentrations, and the secretion of glucagon and insulin occur without endocrine or nervous instructions. Yet because the alpha cells and beta cells are highly sensitive to changes in blood glucose levels, any hormone that affects blood glucose concentrations will indirectly affect the production of both insulin and glucagon. Insulin production is also influenced by autonomic activity: Parasympathetic stimulation enhances insulin release, and sympathetic stimulation inhibits it.

Information about insulin, glucagon, and other pancreatic hormones is summarized in Table 18–6.

The A&P Top 100

#65 The pancreatic islets release insulin and glucagon. Insulin is released when blood glucose levels rise, and it stimulates glucose transport into, and utilization by, peripheral tissues. Glucagon is released when blood glucose levels decline, and it stimulates glycogen breakdown, glucose synthesis, and fatty acid release.

18 ENDOCRINE

CLINICAL NOTE

Diabetes Mellitus

Whether glucose is absorbed across the digestive tract or manufactured and released by the liver, very little glucose leaves the body intact once it has entered the bloodstream. Glucose does not get filtered out of the blood at the kidneys; instead, virtually all of it is reabsorbed, so normally urinary glucose losses are negligible.

Diabetes mellitus (mel-Ī-tus; *mellitum*, honey) is characterized by glucose concentrations that are high enough to overwhelm the reabsorption capabilities of the kidneys. (The presence of abnormally high glucose levels in the blood in general is called *hyperglycemia* [hī-per-glī-SĒ-mē-ah].) Glucose appears in the urine (*glycosuria*; glī-kō-SOO-rē-a), and urine production generally becomes excessive (*polyuria*).

Diabetes mellitus can be caused by genetic abnormalities, and some of the genes responsible have been identified. Mutations that result in inadequate insulin production, the synthesis of abnormal insulin molecules, or the production of defective receptor proteins produce comparable symptoms. Under these conditions, obesity accelerates the onset and severity of the disease. Diabetes mellitus can also result from other pathological conditions, injuries, immune disorders, or hormonal imbalances. There are two major types of diabetes mellitus: **insulin-dependent (type 1) diabetes** and **non-insulin-dependent (type 2) diabetes**. Persons with type 1 diabetes require insulin to live and usually require multiple injections daily, or continuous infusion. Most diabetes patients have type 2 diabetes and initially produce sufficient insulin, but their bodies don't use it well. Weight loss through diet and exercise can be an effective treatment, but most persons require oral medicines, and some progress to needing insulin.

Probably because glucose levels cannot be stabilized adequately, even with treatment, individuals with diabetes mellitus commonly develop chronic medical problems. These problems arise because the tissues involved are experiencing an energy crisis—in essence, most of the tissues are responding as they would during chronic starvation, breaking down lipids and even proteins because they are unable to absorb glucose from their surroundings. Co-existing hypertension and high cholesterol contribute to many diabetes-related problems, and treating all three conditions can reduce these problems:

- The proliferation of capillaries and hemorrhaging at the retina may cause partial or complete blindness. This condition is called *diabetic retinopathy*.

- Changes occur in the clarity of the lens of the eye, producing cataracts.

- The degenerative changes in the kidneys found in diabetic nephropathy can lead to kidney failure.

- A variety of neural problems appear, including *peripheral neuropathies* and abnormal autonomic function. p. 438 These disorders, collectively termed *diabetic neuropathy*, are probably related to disturbances in the blood supply to neural tissues.

- Degenerative blockages in cardiac circulation can lead to early heart attacks. For a given age group, heart attacks are three to five times more likely in diabetic individuals than in nondiabetic people.

- Similar changes in the vascular system can disrupt normal blood flow to the distal portions of the limbs. For example, a reduction in blood flow to the feet can lead to tissue death, ulceration, infection, and loss of toes or a major portion of one or both feet.

CHECKPOINT

23. Identify the types of cells in the pancreatic islets and the hormones produced by each.

24. Why does a person with type 1 or type 2 diabetes urinate frequently and have increased thirst?

25. What effect would increased levels of glucagon have on the amount of glycogen stored in the liver?

See the blue Answers tab at the end of the book.

18-9 Many organs have secondary endocrine functions

As noted earlier, many organs of other body systems have secondary endocrine functions. Examples are the intestines (digestive system), the kidneys (urinary system), the heart (cardiovascular system), the thymus (lymphoid system), and the *gonads*—the testes in males and the ovaries in females (reproductive system).

TABLE 18–6 Hormones Produced by the Pancreatic Islets

Structure/Cells	Hormone	Primary Targets	Hormonal Effects	Regulatory Control
PANCREATIC ISLETS				
Alpha cells	Glucagon	Liver, adipose tissue	Mobilizes lipid reserves; promotes glucose synthesis and glycogen breakdown in liver; elevates blood glucose concentrations	Stimulated by low blood glucose concentrations; inhibited by GH–IH from delta cells
Beta cells	Insulin	Most cells	Facilitates uptake of glucose by target cells; stimulates formation and storage of lipids and glycogen	Stimulated by high blood glucose concentrations, parasympathetic stimulation, and high levels of some amino acids; inhibited by GH–IH from delta cells and by sympathetic activation
Delta cells	GH–IH (somatostatin)	Other islet cells, digestive epithelium	Inhibits insulin and glucagon secretion; slows rates of nutrient absorption and enzyme secretion along digestive tract	Stimulated by protein-rich meal; mechanism unclear
F cells	Pancreatic polypeptide (PP)	Digestive organs	Inhibits gallbladder contraction; regulates production of pancreatic enzymes; influences rate of nutrient absorption by digestive tract	Stimulated by protein-rich meal and by parasympathetic stimulation

Over the last decade, several new hormones from these endocrine tissues have been identified. In many cases, their structures and modes of action remain to be determined, and they have not been described in this chapter. However, in one instance, a significant new hormone was traced to an unexpected site of origin and led to the realization that the body's adipose tissue represents an important site of endocrine function. Although all of the details have yet to be worked out, we will consider the endocrine functions of adipose tissue in this section as well. Table 18–7 provides an overview of some of the hormones these organs produce.

The Intestines

The intestines, which process and absorb nutrients, release a variety of hormones that coordinate the activities of the digestive system. Although the pace of digestive activities can be affected by the autonomic nervous system, most digestive processes are hormonally controlled locally. These hormones will be described in Chapter 24.

The Kidneys

The kidneys release the steroid hormone *calcitriol*, the peptide hormone *erythropoietin*, and the enzyme *renin*. Calcitriol

TABLE 18–7 Representative Hormones Produced by Organs of Other Systems

Organ	Hormone(s)	Primary Target(s)	Hormonal Effects
Intestines	Many (secretin, gastrin, cholecystokinin, etc.)	Other regions and organs of the digestive system	Coordinate digestive activities
Kidneys	Erythropoietin (EPO) Calcitriol	Red bone marrow Intestinal lining, bone, kidneys	Stimulates red blood cell production Stimulates calcium and phosphate absorption; stimulates Ca^+ release from bone; inhibits PTH secretion
Heart	Natriuretic peptides (ANP and BNP)	Kidneys, hypothalamus, suprarenal gland	Increase water and salt loss at kidneys; decrease thirst; suppress secretion of ADH and aldosterone
Thymus	Thymosins (many)	Lymphocytes and other cells of the immune response	Coordinate and regulate immune response
Gonads	*See Table 18–8*		
Adipose tissues	Leptin	Hypothalamus	Suppression of appetite; permissive effects on GnRH and gonadotropin synthesis

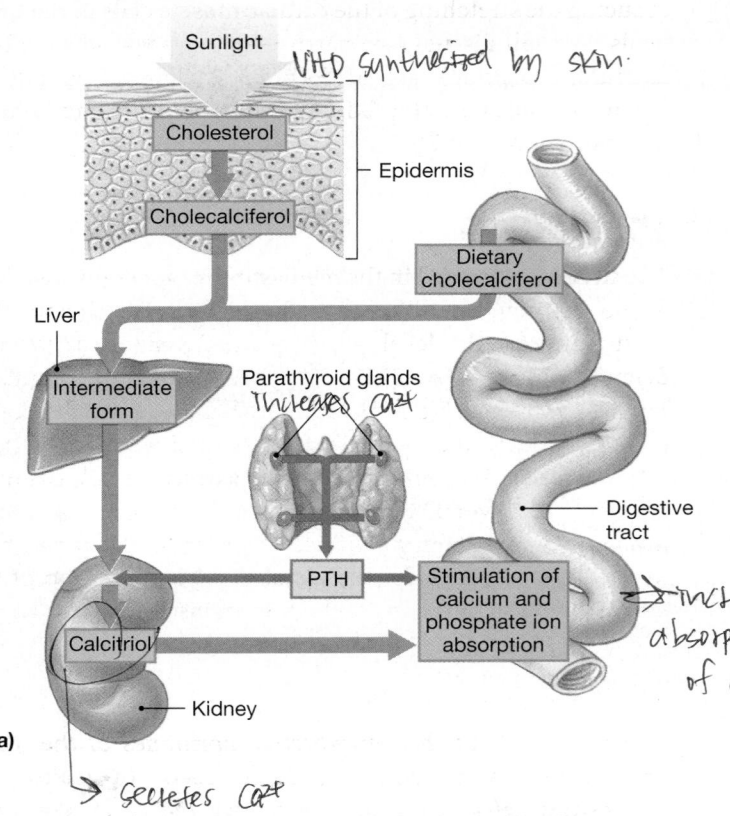

[handwritten annotations: VitD synthesized by skin; Increases Ca²⁺; → increased absorption of Ca²⁺; → secretes Ca²⁺]

(a)

is important for calcium ion homeostasis; erythropoietin and renin are involved in the regulation of blood volume and blood pressure.

Calcitriol

Calcitriol is a steroid hormone secreted by the kidneys in response to the presence of parathyroid hormone (PTH) (**Figure 18–17a**). *Cholecalciferol* (vitamin D_3) is a related steroid that is synthesized in the skin or absorbed from the diet. Cholecalciferol is converted to calcitriol, although not directly. The term *vitamin D* applies to the entire group of related steroids, including calcitriol, cholecalciferol, and various intermediate products.

The best-known function of calcitriol is the stimulation of calcium and phosphate ion absorption along the digestive tract. The effects of PTH on Ca^{2+} absorption result primarily from stimulation of calcitriol release. Calcitriol's other effects on calcium metabolism include (1) stimulating the formation

Figure 18–17 **Endocrine Functions of the Kidneys.** (a) The production of calcitriol. (b) The release of renin and erythropoietin, and an overview of the renin-angiotensin system.

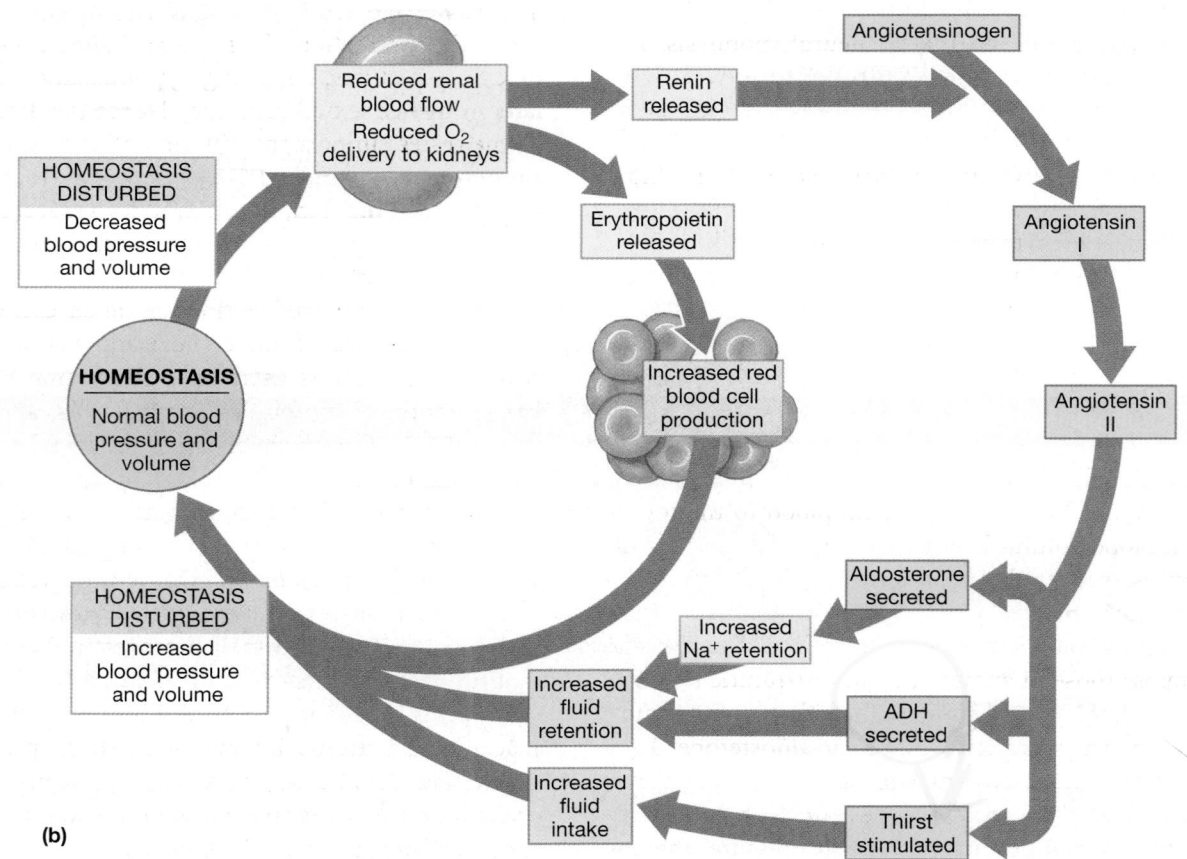

(b)

and differentiation of osteoprogenitor cells and osteoclasts, (2) stimulating bone resorption by osteoclasts, (3) stimulating Ca^{2+} reabsorption at the kidneys, and (4) suppressing PTH production. Evidence indicates that calcitriol also affects lymphocytes and keratinocytes in the skin; these effects have nothing to do with regulating calcium levels.

Erythropoietin

Erythropoietin (e-rith-rō-POY-e-tin; *erythros*, red + *poiesis*, making), or **EPO**, is a peptide hormone released by the kidneys in response to low oxygen levels in kidney tissues. EPO stimulates the production of red blood cells by bone marrow. The increase in the number of red blood cells elevates blood volume. Because these cells transport oxygen, the increase in their number improves oxygen delivery to peripheral tissues. We will consider EPO again in Chapter 19.

Renin

Renin is released by specialized kidney cells in response to (1) sympathetic stimulation or (2) a decline in renal blood flow. Once in the bloodstream, renin functions as an enzyme that starts an enzymatic cascade known as the *renin–angiotensin system* (**Figure 18–17b**). First, renin converts **angiotensinogen**, a plasma protein produced by the liver, to angiotensin I. In the capillaries of the lungs, **angiotensin I** is then modified to the hormone **angiotensin II**, which stimulates the secretion of aldosterone by the suprarenal cortex, and of ADH at the neurohypophysis. The combination of aldosterone and ADH restricts salt and water losses at the kidneys. Angiotensin II also stimulates thirst and elevates blood pressure.

Because renin plays such a key role in the renin–angiotensin system, many physiological and endocrinological references consider renin to be a hormone. We will take a closer look at the renin–angiotensin system when we examine the control of blood pressure and blood volume in Chapter 21.

The Heart

The endocrine cells in the heart are cardiac muscle cells in the walls of the *atria* (chambers that receive blood from the veins) and the *ventricles* (chambers that pump blood to the rest of the body). If blood volume becomes too great, these cells are stretched excessively, to the point at which they begin to secrete **natriuretic peptides** (nā-trē-ū-RET-ik; *natrium*, sodium + *ouresis*, making water). In general, the effects of natriuretic peptides oppose those of angiotensin II: Natriuretic peptides promote the loss of Na^+ and water at the kidneys, and inhibit renin release and the secretion of ADH and aldosterone. They also suppress thirst and prevent angiotensin II and norepinephrine from elevating blood pressure. The net result is a reduction in both blood volume and blood pressure, thereby

reducing the stretching of the cardiac muscle cells in the heart walls. We will discuss two natriuretic peptides—*ANP* (atrial natriuretic peptide) and *BNP* (brain natriuretic peptide)—when we consider the control of blood pressure and volume in Chapters 21 and 26.

The Thymus

The **thymus** is located in the mediastinum, generally just deep to the sternum. The thymus produces several hormones that are important to the development and maintenance of immune defenses. **Thymosin** (THĪ-mō-sin) is the name originally given to an extract from the thymus that promotes the development and maturation of *lymphocytes*, the white blood cells responsible for immunity. The thymic extract actually contains a blend of several complementary hormones; the term *thymosins* is now sometimes used to refer to all thymic hormones. We will consider the histological organization of the thymus and the functions of the thymosins in Chapter 22.

The Gonads

Information about the reproductive hormones of the testes and ovaries is presented in Table 18–8. In males, the **interstitial cells** of the testes produce the male hormones known as androgens. The most important of these androgens is **testosterone** (tes-TOS-ter-ōn). During embryonic development, the production of testosterone affects the development of CNS structures, including hypothalamic nuclei, that will later influence sexual behaviors. **Nurse (sustentacular) cells** in the testes support the differentiation and physical maturation of sperm. Under FSH stimulation, these cells secrete the hormone **inhibin**, which inhibits the secretion of FSH at the adenohypophysis and perhaps suppresses GnRH release at the hypothalamus.

In females, steroid hormones called **estrogens** are produced in the ovaries under FSH and LH stimulation. The principal estrogen is **estradiol**. Circulating FSH stimulates the secretion of inhibin by ovarian cells, and inhibin suppresses FSH release through a feedback mechanism comparable to that in males.

At ovulation, an immature gamete, or oocyte, is released by follicles in the ovary. The remaining follicle cells then reorganize into a *corpus luteum* (LOO-tē-um; "yellow body") that releases a mixture of estrogens and **progestins. Progesterone** (prō-JES-ter-ōn), the principal progestin, has several important functions, summarized in Table 18–8.

During pregnancy, additional hormones produced by the placenta and uterus interact with those produced by the ovaries and the pituitary gland to promote normal fetal development and delivery. We will consider the endocrinological aspects of pregnancy in Chapter 29.

18 ENDOCRINE

TABLE 18–8 Hormones of the Reproductive System

Structure/Cells	Hormone(s)	Primary Targets	Hormonal Effects	Regulatory Control
TESTES				
Interstitial cells	Androgens	Most cells	Support functional maturation of sperm, protein synthesis in skeletal muscles, male secondary sex characteristics, and associated behaviors	Stimulated by LH from the adenohypophysis (see Figure 18–8a)
Nurse cells	Inhibin	Adenohypophysis	Inhibits secretion of FSH	Stimulated by FSH from adenohypophysis (see Figure 18–8a)
OVARIES				
Follicular cells	Estrogens	Most cells	Support follicle maturation, female secondary sex characteristics, and associated behaviors	Stimulated by FSH and LH from adenohypophysis (see Figure 18–8a)
	Inhibin	Adenohypophysis	Inhibits secretion of FSH	Stimulated by FSH from adenohypophysis (see Figure 18–8a)
Corpus luteum	Progestins	Uterus, mammary glands	Prepare uterus for implantation; prepare mammary glands for secretory activity	Stimulated by LH from the adenohypophysis (see Figure 18–8a)

Adipose Tissue

Adipose tissue is a type of loose connective tissue introduced in Chapter 4. ∞ p. 127 Adipose tissue is known to produce a peptide hormone called **leptin**, which has several functions, the best known being the feedback control of appetite. When you eat, adipose tissue absorbs glucose and lipids and synthesizes triglycerides for storage. At the same time, it releases leptin into the bloodstream. Leptin binds to hypothalamic neurons involved with emotion and appetite control. The result is a sense of satiation and the suppression of appetite.

Leptin was first discovered in a strain of obese mice that had a defective leptin gene. The administration of leptin to these overweight mice quickly turned them into slim, athletic animals. The initial hope that leptin could be used to treat human obesity was soon dashed, however. Most obese people appear to have defective leptin receptors (or leptin pathways) in the appetite centers of the CNS. Their circulating leptin levels are already several times higher than those in individuals of normal body weight, so the administration of additional leptin would have no effect. Researchers are now investigating the structure of the receptor protein and the biochemistry of the pathway triggered by leptin binding. Another hormone, *resistin* (for resistance to insulin), has been associated with reduced insulin sensitivity. Experimental evidence from mice supports this linkage.

Leptin must be present if normal levels of GnRH and gonadotropin synthesis are to occur. This explains why (1) thin girls commonly enter puberty relatively late, (2) an increase in body fat content can improve fertility, and (3) women stop menstruating when their body fat content becomes very low.

CHECKPOINT

26. Identify two hormones secreted by the kidneys.

27. Identify a hormone released by adipose tissue.

28. Describe the action of renin in the bloodstream.

See the blue Answers tab at the end of the book.

18-10 Hormones interact to produce coordinated physiological responses

Although hormones are usually studied individually, the extracellular fluids contain a mixture of hormones whose concentrations change daily or even hourly. As a result, cells never respond to only one hormone; instead, they respond to multiple hormones simultaneously. When a cell receives instructions from two hormones at the same time, four outcomes are possible:

1. The two hormones may have **antagonistic** (opposing) **effects**, as in the case of PTH and calcitonin, or insulin and glucagon. The net result depends on the balance between the two hormones. In general, when antagonistic hormones are present, the observed effects are weaker than those produced by either hormone acting unopposed.

2. The two hormones may have additive effects, so that the net result is greater than the effect that each would produce acting alone. In some cases, the net result is greater than the *sum* of the hormones' individual effects. This

CLINICAL NOTE

Endocrine Disorders

Regulation of hormone levels often involves negative feedback control mechanisms involving the endocrine organ, neural regulatory factors, and the target tissues. Abnormalities may result from hormone overproduction (hypersecretion), or underproduction (hyposecretion), or from abnormal cellular sensitivity to the hormone.

Primary disorders result from problems within the endocrine organ. The underlying cause may be a metabolic factor; hypothyroidism due to a lack of dietary iodine is an example. An endocrine organ may also malfunction due to physical damage that destroys cells or disrupts the normal blood supply. Congenital problems may also affect the regulation, production, or release of hormones by endocrine cells.

Secondary disorders result from problems in other organs or target tissues. Such disorders often involve the hypothalamus or pituitary gland. For example, if the hypothalmus or pituitary gland don't produce enough TRH and/or TSH, then secondary hypothyroidism occurs.

Abnormalities in target cells can affect their sensitivity or responsiveness to a particular hormone. For example, type 2 diabetes results from a reduction in the target cell's sensitivity to insulin.

Endocrine disorders often reflect either abnormal hormone production or abnormal cellular sensitivity to hormones. The signs and symptoms highlight the significance of normally "silent" hormonal contributions. The characteristics of these disorders are summarized in Table 18–9.

TABLE 18–9 Clinical Implications of Endocrine Malfunctions

Hormone	Underproduction or Tissue Insensitivity	Principal Signs and Symptoms	Overproduction or Tissue Hypersensitivity	Principal Signs and Symptoms
Growth hormone (GH)	Pituitary growth failure (p. 202)	Retarded growth, abnormal fat distribution, low blood glucose hours after a meal	Gigantism, acromegaly (p. 202)	Excessive growth
Antidiuretic hormone (ADH) or arginine vasopressin (AVP)	Diabetes insipidus (p. 618)	Polyuria, dehydration, thirst	SIADH (syndrome of inappropriate ADH secretion)	Increased body weight and water content
Thyroxine (T_4), triiodothyronine (T_3)	Myxedema, cretinism	Low metabolic rate; low body temperature; impaired physical and mental development	Hyperthyroidism, Graves disease	High metabolic rate and body temperature
Parathyroid hormone (PTH)	Hypoparathyroidism	Muscular weakness, neurological problems, formation of dense bones, tetany due to low blood Ca^{2+} concentrations	Hyperparathyroidism	Neurological, mental, muscular problems due to high blood Ca^{2+} concentrations; weak and brittle bones
Insulin	Diabetes mellitus (type 1) (p. 634)	High blood glucose, impaired glucose utilization, dependence on lipids for energy; glycosuria	Excess insulin production or administration	Low blood glucose levels, possibly causing coma
Mineralocorticoids (MCs)	Hypoaldosteronism	Polyuria, low blood volume, high blood K^+, low blood Na^+ concentrations	Aldosteronism	Increased body weight due to Na^+ and water retention; low blood K^+ concentration
Glucocorticoids (GCs)	Addison disease	Inability to tolerate stress, mobilize energy reserves, or maintain normal blood glucose concentrations	Cushing disease	Excessive breakdown of tissue proteins and lipid reserves; impaired glucose metabolism
Epinephrine (E), norepinephrine (NE)	None identified		Pheochromocytoma	High metabolic rate, body temperature, and heart rate; elevated blood glucose levels
Estrogens (females)	Hypogonadism	Sterility, lack of secondary sex characteristics	Adrenogenital syndrome	Overproduction of androgens by zona reticularis of suprarenal cortex leads to masculinization
			Precocious puberty (p. 642)	Premature sexual maturation and related behavioral changes
Androgens (males)	Hypogonadism	Sterility, lack of secondary sex characteristics	Adrenogenital syndrome (gynecomastia)	Abnormal production of estrogen, sometimes due to suprarenal or interstitial cell tumors; leads to breast enlargement
			Precocious puberty (p. 642)	Premature sexual maturation and related behavioral changes

18 ENDOCRINE

phenomenon is called a **synergistic effect** (sin-er-JIS-tik; *synairesis*, a drawing together). An example is the glucose-sparing action of GH and glucocorticoids.

3. One hormone can have a **permissive effect** on another. In such cases, the first hormone is needed for the second to produce its effect. For example, epinephrine does not change energy consumption unless thyroid hormones are also present in normal concentrations.

4. Finally, hormones may produce different, but complementary, results in specific tissues and organs. These **integrative effects** are important in coordinating the activities of diverse physiological systems. The differing effects of calcitriol and parathyroid hormone on tissues involved in calcium metabolism are an example.

When multiple hormones are involved in the regulation of a complex process, it is very difficult to determine whether a hormone has synergistic, permissive, or integrative effects. Next we consider three examples of processes regulated by complex hormonal interactions: growth, the response to stress, and behavior.

Role of Hormones in Growth

Normal growth requires the cooperation of several endocrine organs. Several hormones—GH, thyroid hormones, insulin, PTH, calcitriol, and reproductive hormones—are especially important, although many others have secondary effects on growth. The circulating concentrations of these hormones are regulated independently. Every time the hormonal mixture changes, metabolic operations are modified to some degree. The modifications vary in duration and intensity, producing unique individual growth patterns.

- *Growth Hormone (GH).* The effects of GH on protein synthesis and cellular growth are most apparent in children, in whom GH supports muscular and skeletal development. In adults, growth hormone assists in the maintenance of normal blood glucose concentrations and in the mobilization of lipid reserves stored in adipose tissue. It is not the primary hormone involved, however, and an adult with a GH deficiency but normal levels of thyroxine (T_4), insulin, and glucocorticoids will have no physiological problems.

- *Thyroid Hormones.* Normal growth also requires appropriate levels of thyroid hormones. If these hormones are absent during fetal development or for the first year after birth, the nervous system will fail to develop normally, and mental retardation will result. If T_4 concentrations decline later in life but before puberty, normal skeletal development will not continue.

- *Insulin.* Growing cells need adequate supplies of energy and nutrients. Without insulin, the passage of glucose

and amino acids across plasma membranes will be drastically reduced or eliminated.

- *Parathyroid Hormone (PTH) and Calcitriol.* Parathyroid hormone and calcitriol promote the absorption of calcium salts for subsequent deposition in bone. Without adequate levels of both hormones, bones can still enlarge, but will be poorly mineralized, weak, and flexible. For example, in *rickets*, a condition typically caused by inadequate calcitriol production as a result of vitamin D deficiency in growing children, the lower limb bones are so weak that they bend under the body's weight. ∞ p. 204

- *Reproductive Hormones.* The activity of osteoblasts in key locations and the growth of specific cell populations are affected by the presence or absence of reproductive hormones (androgens in males, estrogens in females). These sex hormones stimulate cell growth and differentiation in their target tissues. The targets differ for androgens and estrogens, and the differential growth induced by each accounts for gender-related differences in skeletal proportions and secondary sex characteristics.

The Hormonal Responses to Stress

Any condition—physical or emotional—that threatens homeostasis is a form of **stress**. Many stresses are opposed by specific homeostatic adjustments. For example, a decline in body temperature leads to shivering or changes in the pattern of blood flow, which can restore normal body temperature.

In addition, the body has a *general* response to stress that can occur while other, more specific responses are under way. Exposure to a wide variety of stress-causing factors will produce the same general pattern of hormonal and physiological adjustments. These responses are part of the **general adaptation syndrome (GAS)**, also known as the **stress response**. The GAS, first described by Hans Selye in 1936, can be divided into three phases: the *alarm phase*, the *resistance phase*, and the *exhaustion phase* (**Figure 18–18**).

The Alarm Phase sympathetic of ANS

During the **alarm phase**, an immediate response to the stress occurs. This response is directed by the sympathetic division of the autonomic nervous system. In the alarm phase, (1) energy reserves are mobilized, mainly in the form of glucose, and (2) the body prepares to deal with the stress-causing factor by "fight or flight" responses. ∞ p. 531

Epinephrine is the dominant hormone of the alarm phase, and its secretion accompanies a generalized sympathetic activation. The characteristics of the alarm phase include the following:

- Increased mental alertness.
- Increased energy consumption by skeletal muscles and many other tissues.

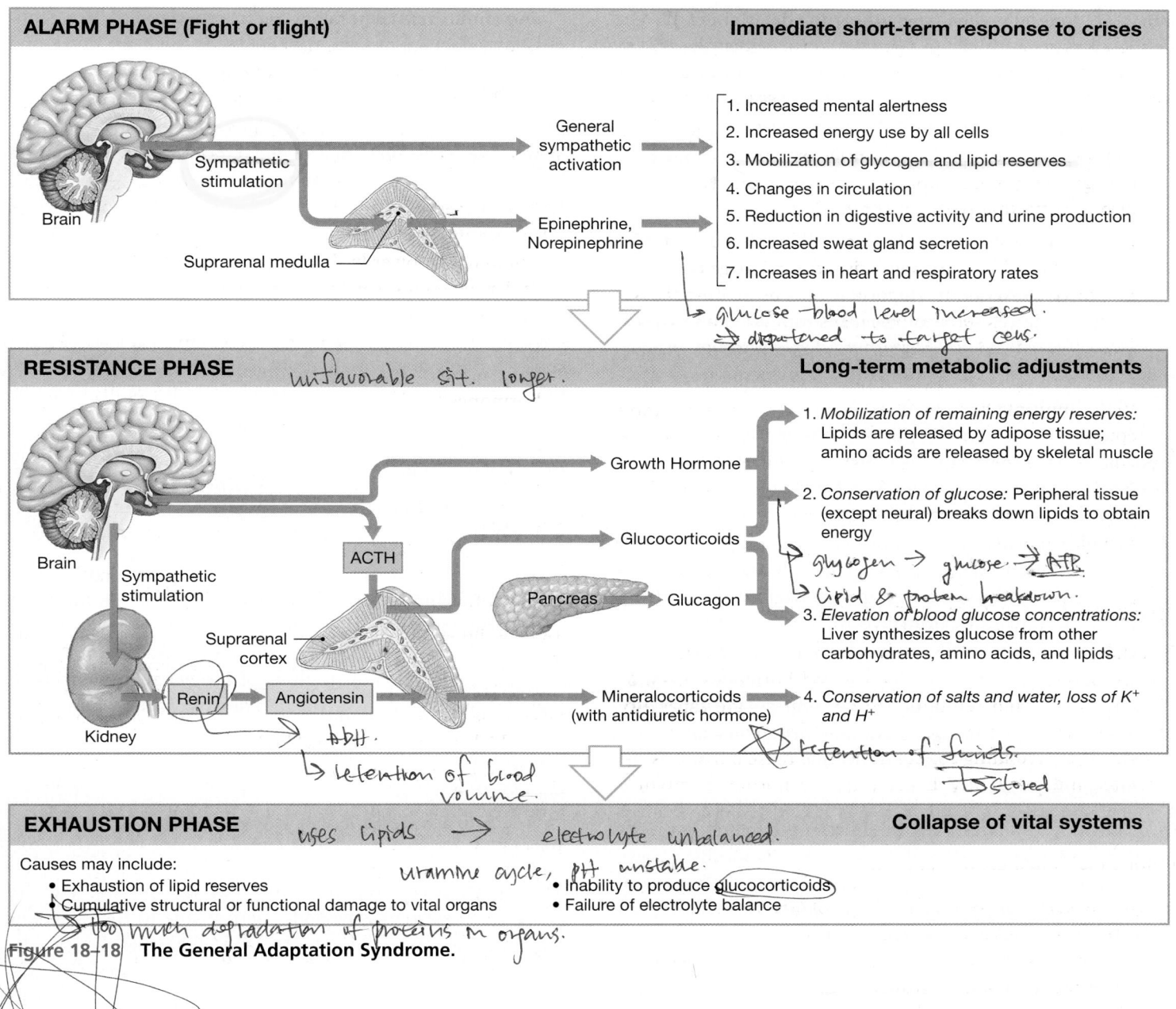

ALARM PHASE (Fight or flight) · **Immediate short-term response to crises**

Brain — Sympathetic stimulation → General sympathetic activation

Suprarenal medulla → Epinephrine, Norepinephrine

1. Increased mental alertness
2. Increased energy use by all cells
3. Mobilization of glycogen and lipid reserves
4. Changes in circulation
5. Reduction in digestive activity and urine production
6. Increased sweat gland secretion
7. Increases in heart and respiratory rates

glucose blood level increased.
⇒ distributed to target cells.

RESISTANCE PHASE · **Long-term metabolic adjustments**

unfavorable sit. longer.

Brain — Sympathetic stimulation

ACTH → Growth Hormone
Suprarenal cortex → Glucocorticoids
Pancreas → Glucagon

Kidney → Renin → Angiotensin → Mineralocorticoids (with antidiuretic hormone)

ADH.
↳ retention of blood volume.

1. *Mobilization of remaining energy reserves:* Lipids are released by adipose tissue; amino acids are released by skeletal muscle
2. *Conservation of glucose:* Peripheral tissue (except neural) breaks down lipids to obtain energy
3. *Elevation of blood glucose concentrations:* Liver synthesizes glucose from other carbohydrates, amino acids, and lipids
4. *Conservation of salts and water, loss of K^+ and H^+*

glycogen → glucose → ATP
lipid & protein breakdown.
retention of fluids.
stored

EXHAUSTION PHASE · **Collapse of vital systems**

uses lipids → electrolyte unbalanced.
uramme cycle, pH unstable.

Causes may include:
- Exhaustion of lipid reserves
- Cumulative structural or functional damage to vital organs
- Inability to produce glucocorticoids
- Failure of electrolyte balance

↳ too much degradation of proteins in organs.

Figure 18–18 **The General Adaptation Syndrome.**

- The mobilization of energy reserves (glycogen and lipids).
- Changes in circulation, including increased blood flow to skeletal muscles and decreased blood flow to the skin, kidneys, and digestive organs.
- A drastic reduction in digestion and urine production.
- Increased sweat gland secretion.
- Increases in blood pressure, heart rate, and respiratory rate.

Although the effects of epinephrine are most apparent during the alarm phase, other hormones play supporting roles. For example, the reduction of water losses resulting from ADH production and aldosterone secretion can be very important if the stress involves a loss of blood.

The Resistance Phase

The temporary adjustments of the alarm phase are often sufficient to remove or overcome a stress. But some stresses, including starvation, acute illness, or severe anxiety, can persist for hours, days, or even weeks. If a stress lasts longer than a few hours, the individual enters the **resistance phase** of the GAS.

Glucocorticoids are the dominant hormones of the resistance phase. Epinephrine, GH, and thyroid hormones are also involved. Energy demands in the resistance phase remain higher than normal, owing to the combined effects of these hormones.

Neural tissue has a high demand for energy, and neurons must have a reliable supply of glucose. If blood glucose concentrations fall too far, neural function deteriorates. Glycogen reserves are adequate to maintain normal glucose concentrations during the alarm phase, but are nearly exhausted after several hours. The endocrine secretions of the resistance phase are coordinated to achieve four integrated results:

1. *The Mobilization of Remaining Lipid and Protein Reserves.* The hypothalamus produces GH–RH and CRH, stimulating the release of GH and, by means of ACTH, the secretion of glucocorticoids. Adipose tissue responds to GH and glucocorticoids by releasing stored fatty acids. Skeletal muscles respond to glucocorticoids by breaking down proteins and releasing amino acids into the bloodstream.

2. *The Conservation of Glucose for Neural Tissues.* Glucocorticoids and GH from the adenohypophysis stimulate lipid metabolism in most tissues. These glucose-sparing effects maintain normal blood glucose levels even after long periods of starvation. Neural tissues do not alter their metabolic activities, however, and they continue to use glucose as an energy source.

3. *The Elevation and Stabilization of Blood Glucose Concentrations.* As blood glucose levels decline, glucagon and glucocorticoids stimulate the liver to manufacture glucose from other carbohydrates, from glycerol by way of triglycerides, and from amino acids provided by skeletal muscles. The glucose molecules are then released into the bloodstream, and blood glucose concentrations return to normal levels.

4. *The Conservation of Salts and Water, and the Loss of K^+ and H^+.* Blood volume is conserved through the actions of ADH and aldosterone. As Na^+ is conserved, K^+ and H^+ are lost.

The body's lipid reserves are sufficient to maintain the resistance phase for a period of weeks or even months. (These reserves account for the ability to endure lengthy periods of starvation.) But the resistance phase cannot be sustained indefinitely. If starvation is the primary stress, the resistance phase ends when lipid reserves are exhausted and structural proteins become the primary energy source. If another factor is the cause, the resistance phase ends due to complications brought about by hormonal side effects. Examples of hormone-related complications include the following:

- Although the metabolic effects of glucocorticoids are essential to normal resistance to stress, their anti-inflammatory action slows wound healing and increases the body's susceptibility to infection.

- The continued conservation of fluids under the influence of ADH and aldosterone stresses the cardiovascular system by producing elevated blood volumes and higher-than-normal blood pressures.

- The suprarenal cortex may become unable to continue producing glucocorticoids, quickly eliminating the ability to maintain acceptable blood glucose concentrations.

Poor nutrition, emotional or physical trauma, chronic illness, and damage to key organs such as the heart, liver, and kidneys hasten the end of the resistance phase.

The Exhaustion Phase

When the resistance phase ends, homeostatic regulation breaks down and the **exhaustion phase** begins. Unless corrective actions are taken almost immediately, the failure of one or more organ systems will prove fatal.

Mineral imbalances contribute to the existing problems with major systems. The production of aldosterone throughout the resistance phase results in a conservation of Na^+ at the expense of K^+. As the body's K^+ content declines, a variety of cells—notably neurons and muscle fibers—begin to malfunction. Although a single cause (such as heart failure) may be listed as the cause of death, the underlying problem is the inability to sustain the endocrine and metabolic adjustments of the resistance phase.

The Effects of Hormones on Behavior

As we have seen, many endocrine functions are regulated by the hypothalamus, and hypothalamic neurons monitor the levels of many circulating hormones. Other portions of the CNS are also quite sensitive to hormonal stimulation.

The clearest demonstrations of the behavioral effects of specific hormones involve individuals whose endocrine glands are oversecreting or undersecreting. But even normal changes in circulating hormone levels can cause behavioral changes. In *precocious* (premature) *puberty*, sex hormones are produced at an inappropriate time, perhaps as early as age 5 or 6. An affected child not only begins to develop adult secondary sex characteristics, but also undergoes significant behavioral changes. The "nice little kid" disappears, and the child becomes aggressive and assertive due to the effects of sex hormones on CNS function. Thus, behaviors that in normal teenagers are usually attributed to environmental stimuli, such as peer pressure, have a physiological basis as well. In adults, changes in the mixture of hormones reaching the CNS can have significant effects on intellectual capabilities, memory, learning, and emotional states.

To conclude this section, we briefly turn to another hormone-related topic: the effects of aging on hormone production.

Aging and Hormone Production

The endocrine system undergoes relatively few functional changes with age. The most dramatic exception is the decline in the concentrations of reproductive hormones. The effects of these hormonal changes on the skeletal system were noted in Chapter 6 (p. 207); we will continue the discussion in Chapter 29.

Blood and tissue concentrations of many other hormones, including TSH, thyroid hormones, ADH, PTH, prolactin, and glucocorticoids, remain unchanged with advancing age. Although circulating hormone levels may remain within normal limits, some endocrine tissues become less responsive to stimulation. For example, in elderly individuals, smaller amounts of GH and insulin are secreted after a carbohydrate-rich meal. The reduction in levels of GH and other tropic hormones affects tissues throughout the body; these hormonal effects are associated with the reductions in bone density and muscle mass noted in earlier chapters.

Finally, age-related changes in peripheral tissues may make them less responsive to some hormones. This loss of sensitivity has been documented in the case of glucocorticoids and ADH.

CLINICAL NOTE

Hormones and Athletic Performance

The use of hormones to improve athletic performance is banned by the International Olympic Committee, the U.S. Olympic Committee, the National Collegiate Athletic Association, Major League Baseball, and the National Football League, and it is condemned by the American Medical Association and the American College of Sports Medicine. A significant number of amateur and professional athletes, however, persist in this dangerous practice. Synthetic forms of testosterone are used most often, but athletes might use any combination of testosterone, GH, EPO, and a variety of synthetic hormones.

Androgen Abuse   The use of *anabolic steroids*, or androgens, has become popular with many amateur and professional athletes. The goal of steroid use is to increase muscle mass, endurance, and "competitive spirit." The use of steroids such as *androstenedione*, which the body can convert to testosterone, was highlighted during 2004 when several prominent sports trainers, including the trainer of baseball slugger Barry Bonds, were arrested for providing synthetic steroids to their clients. (Performance-enhancing drugs were banned by Major League Baseball in 1999.)

One supposed justification for this practice has been the unfounded opinion that compounds manufactured in the body are not only safe, but good for you. In reality, the administration of natural or synthetic androgens in abnormal amounts carries unacceptable health risks. Androgens are known to produce several complications, including (1) premature closure of epiphyseal cartilages, (2) various liver dysfunctions (including jaundice and liver tumors), (3) prostate gland enlargement and urinary tract obstruction, and (4) testicular atrophy and infertility. Links to heart attacks, impaired cardiac function, and strokes have also been suggested.

Moreover, the normal regulation of androgen production involves a feedback mechanism comparable to that described for suprarenal steroids earlier in this chapter. GnRH stimulates the production of LH, and LH stimulates the secretion of testosterone and other androgens by the interstitial cells of the testes. Circulating androgens, in turn, inhibit the production of both GnRH and LH. Thus, when synthetic androgens are administered in high doses, they can suppress the normal production of testosterone and depress the manufacture of GnRH by the hypothalamus. *This suppression of GnRH release can be permanent.*

The use of androgenic "bulking agents" by female bodybuilders not only may add muscle mass, but can alter muscular proportions and secondary sex characteristics as well. For example, women taking steroids can develop irregular menstrual periods and changes in body hair distribution (including baldness). Finally, androgen abuse can cause a generalized depression of the immune system.

EPO Abuse   Because it is now synthesized by recombinant DNA techniques, EPO is readily available. Endurance athletes, such as cyclists and marathon runners, may use it to boost the number of oxygen-carrying red blood cells in the bloodstream. Although this effect increases the oxygen content of blood, it also makes blood more viscous; thus, the heart must work harder to push it around the circulatory system. This can result in death due to heart failure or stroke in young and otherwise healthy individuals. The 2007 Tour de France bicycle race was tainted by competitors testing positive for banned substances including testosterone, erythropoietin, and an illegal blood transfusion.

Androgens and EPO are hormones with reasonably well-understood effects. Because drug testing is now widespread in amateur and professional sports, athletes interested in "getting an edge" are experimenting with drugs not easily detected by standard tests. The long-term and short-term effects of these drugs are difficult to predict.

GHB Use   One drug recently used by amateur athletes is *GHB*. **Gamma-hydroxybutyrate** (GHB) was tested for use as an anesthetic in the 1960s. It was rejected, in part because it was linked to seizures. In 1990, the drug appeared in health-food stores, where it was sold as an anabolic agent and diet aid. It has also been used as a "date rape" drug. According to the FDA, GHB and related compounds—sold or distributed under the names Renewtrient, Revivarant, Blue Nitro, Firewater, and Serenity—have recently been responsible for 145 serious illnesses and at least eight deaths. Signs and symptoms include reduced heart rate, lowered body temperature, confusion, hallucinations, seizures, and coma at doses from 0.25 teaspoon to 4 tablespoons.

18 ENDOCRINE

29. Insulin lowers blood glucose levels, and glucagon causes glucose levels to rise. What is this type of hormonal interaction called?

30. The lack of which hormones would inhibit skeletal formation?

31. Why do levels of GH–RH and CRH rise during the resistance phase of the general adaptation syndrome?

See the blue Answers tab at the end of the book.

18-11 Extensive integration occurs between the endocrine system and other body systems

The endocrine system provides long-term regulation and adjustments of homeostatic mechanisms that affect many body functions. For example, the endocrine system regulates fluid and electrolyte balance, cell and tissue metabolism, growth and development, and reproductive functions. It also assists the nervous system in responding to stressful stimuli through the general adaptation syndrome.

The relationships between the endocrine system and the other body systems are summarized in **Figure 18–19**. Not surprisingly, the most extensive interactions are with the nervous system. Although the hypothalamus modifies the activities of many endocrine organs via the adenohypophysis, the presence of so many complex feedback loops makes it difficult to determine whether the endocrine system or the nervous system is really in charge. Moreover, many hormones also serve as neurotransmitters in the brain, spinal cord, and/or enteric nervous system. As a result, circulating hormones that cross the blood–brain barrier can have direct and widespread effects on neural—and neuroendocrine—activity.

32. Discuss the general role of the endocrine system in the functioning of other body systems.

33. Discuss the functional relationship between the endocrine system and the muscular system.

34. Discuss the functional relationship between the endocrine system and the lymphoid system.

See the blue Answers tab at the end of the book.

Related Clinical Terms

Addison disease: A condition caused by the hyposecretion of glucocorticoids and mineralocorticoids; characterized by an inability to mobilize energy reserves and maintain normal blood glucose levels.

cretinism(*congenital hypothyroidism*)**:** A condition caused by hypothyroidism at birth or in infancy; marked by inadequate skeletal and nervous system development and a metabolic rate as much as 40 percent below normal levels.

Cushing disease: A condition caused by the hypersecretion of glucocorticoids; characterized by the excessive breakdown and relocation of lipid reserves and proteins.

diabetes insipidus: A disorder that develops when the neurohypophysis no longer releases adequate amounts of ADH, or when the kidneys cannot respond to ADH.

diabetes mellitus: A disorder that damages many organ systems; characterized by blood glucose concentrations high enough to overwhelm the kidneys' reabsorption capabilities.

diabetic retinopathy, nephropathy, neuropathy: Disorders of the retina, kidneys, and peripheral nerves, respectively, related to diabetes mellitus; the conditions most often afflict middle-aged or older diabetics.

general adaptation syndrome (GAS): The pattern of hormonal and physiological adjustments with which the body responds to all forms of stress.

glycosuria: The presence of glucose in the urine.

goiter: An abnormal enlargement of the thyroid gland.

hyperglycemia: Abnormally high blood glucose levels.

hypocalcemic tetany: Muscle spasms affecting the face and upper extremities; caused by low Ca^{2+} concentrations in body fluids.

insulin-dependent diabetes or *type 1 diabetes*, or *juvenile-onset diabetes*: A type of diabetes mellitus; the primary cause is inadequate insulin production by the beta cells of the pancreatic islets.

myxedema: Condition resulting from severe hyposecretion of thyroid hormones; characterized by subcutaneous swelling, hair loss, dry skin, low body temperature, muscle weakness, and slowed reflexes.

non-insulin-dependent diabetes or *type 2 diabetes*, or *maturity-onset diabetes*: A type of diabetes mellitus in which insulin levels are normal or elevated, but peripheral tissues no longer respond normally.

polyuria: The production of excessive amounts of urine; a sign of diabetes.

seasonal affective disorder (SAD): A condition characterized by depression, lethargy, an inability to concentrate, and altered sleep and eating habits; linked to elevated melatonin levels in individuals exposed to only short periods of daylight.

thyrotoxicosis: A condition caused by the oversecretion of thyroid hormones (hyperthyroidism). Signs and symptoms include increases in metabolic rate, blood pressure, and heart rate; excitability and emotional instability; and lowered energy reserves.

THE ENDOCRINE SYSTEM IN PERSPECTIVE

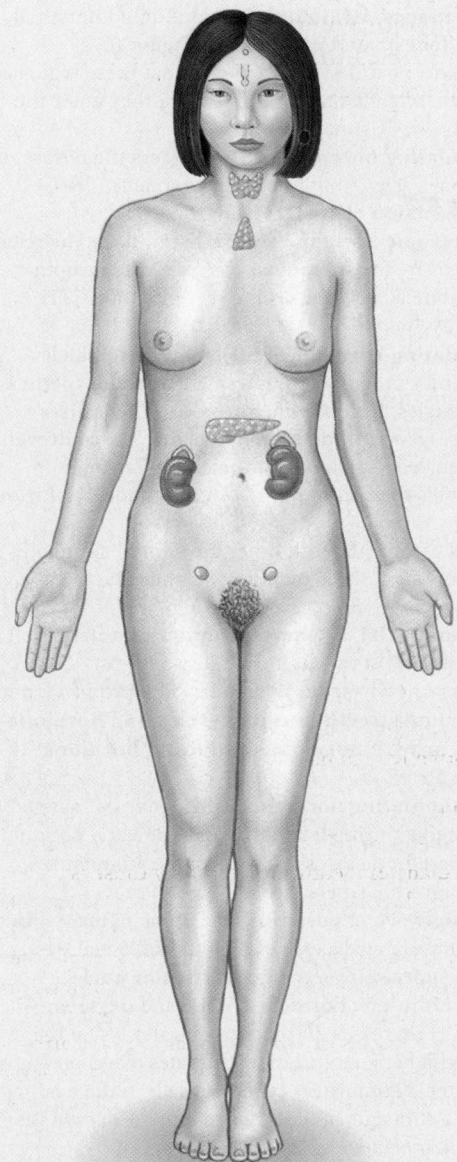

Integumentary System

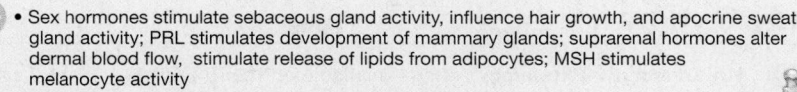

- Protects superficial endocrine organs; epidermis synthesizes cholecalciferol

- Sex hormones stimulate sebaceous gland activity, influence hair growth, and apocrine sweat gland activity; PRL stimulates development of mammary glands; suprarenal hormones alter dermal blood flow, stimulate release of lipids from adipocytes; MSH stimulates melanocyte activity

Skeletal System

- Protects endocrine organs, especially in brain, chest, and pelvic cavity

- Skeletal growth regulated by several hormones; calcium mobilization regulated by parathyroid hormone and calcitonin; sex hormones speed growth and closure of epiphyseal cartilages at puberty and help maintain bone mass in adults

Muscular System

- Skeletal muscles protect some endocrine organs

- Hormones adjust muscle metabolism, energy production, and growth; regulate calcium and phosphate levels in body fluids; speed skeletal muscle growth

Nervous System

- Hypothalamic hormones directly control pituitary and indirectly control secretions of other endocrine organs; controls suprarenal medullae; secretes ADH and oxytocin

- Several hormones affect neural metabolism and brain development; hormones help regulate fluid and electrolyte balance; reproductive hormones influence CNS development and behaviors

Cardiovascular System

- Circulatory system distributes hormones; heart secretes ANP and BNP; endothelium secretes nitric oxide

- Erythropoietin regulates production of RBCs; several hormones elevate blood pressure; epinephrine elevates heart rate and contractile force

Lymphoid System

- Lymphocytes provide defense against infection and, with other WBCs, assist in repair after injury

- Glucocorticoids have anti-inflammatory effects; thymosins stimulate development of lymphocytes; many hormones affect immune function

Respiratory System

- Provides oxygen and eliminates carbon dioxide generated by endocrine cells; converting enzyme in lung capillaries converts angiotensin I to angiotensin II

- Epinephrine and norepinephrine stimulate respiratory activity and dilate respiratory passageways

Digestive System

- Provides nutrients and substrates to endocrine cells; endocrine cells of pancreas secrete insulin and glucagon; liver produces and releases angiotensinogen

- E and NE stimulate constriction of sphincters and depress activity along digestive tract; hormones coordinate secretory activities along tract

Urinary System

- Kidney cells (1) release renin and erythropoietin when local blood pressure and blood flow declines and (2) produce calcitriol

- Aldosterone, ADH, ANP, and BNP adjust rates of fluid and electrolyte reabsorption in kidneys

Reproductive System

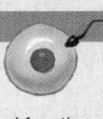

- Steroid sex hormones and inhibin suppress secretory activities in hypothalamus and pituitary gland

- Hypothalamic factors and pituitary hormones regulate sexual development and function; oxytocin stimulates smooth muscle contractions in uterus, prostate, and mammary glands

FOR ALL SYSTEMS

Adjusts metabolic rates and substrate utilization; regulates growth and development

Figure 18–19 Functional Relationships between the Endocrine System and Other Systems.

Chapter Review

Study Outline

iP Use *Interactive Physiology®* (**IP**) —available on the IP CD packaged with your new textbook and online at **myA&P™** (www.myaandp.com)—to help you understand difficult physiological concepts. Follow these navigation paths on IP for concepts in this chapter:

- Endocrine System/The Hypothalamic-Pituitary Axis
- Endocrine System/Endocrine System Review
- Endocrine System/Response to Stress

18-1 Homeostasis is preserved through intercellular communication p. 604

1. In general, the nervous system performs short-term "crisis management," whereas the endocrine system regulates longer-term, ongoing metabolic processes.

2. **Paracrine communication** involves the use of chemical signals to transfer information from cell to cell within a single tissue.

3. **Endocrine communication** results when chemicals, called hormones, are released into the circulation by *endocrine cells*. The hormones alter the metabolic activities of many tissues and organs simultaneously by modifying the activities of **target cells**. (*Table 18–1*)

18-2 The endocrine system regulates physiological processes through the binding of hormones to receptors p. 606

4. The endocrine system includes all the cells and endocrine tissues of the body that produce hormones or paracrine factors. (Figure 18–1)

5. Hormones can be divided into three groups: *amino acid derivatives*; *peptide hormones*; and *lipid derivatives*, including **steroid hormones** and **eicosanoids**. (*Figure 18–2*)

6. Hormones may circulate freely or bound to transport proteins. Free hormones are rapidly removed from the bloodstream.

7. Receptors for *catecholamines*, peptide hormones, and eicosanoids are in the plasma membranes of target cells. Thyroid and steroid hormones cross the plasma membrane and bind to receptors in the cytoplasm or nucleus, activating or inactivating specific genes. (*Figures 18–3, 18–4*)

8. **Endocrine reflexes** are the functional counterparts of neural reflexes. (*Figure 18–5*)

9. The hypothalamus regulates the activities of the nervous and endocrine systems by (1) secreting **regulatory hormones**, which control the activities of endocrine cells in the adenohypophysis; (2) acting as an endocrine organ by releasing hormones into the bloodstream at the neurohypophysis; and (3) exerting direct neural control over the endocrine cells of the suprarenal medullae. (*Figure 18–5*)

18-3 The bilobed pituitary gland is an endocrine organ that releases nine peptide hormones p. 614

10. The **pituitary gland**, or **hypophysis**, releases nine important peptide hormones; all bind to membrane receptors and use cyclic-AMP as a second messenger. (*Figures 18–6 through 18–9; Table 18–2*)

11. The **adenohypophysis**, or **anterior lobe**, of the pituitary gland, can be subdivided into the **pars distalis**, the **pars intermedia**, and the **pars tuberalis**. (*Figure 18–6*)

12. At the median eminence of the hypothalamus, neurons release regulatory factors (either **releasing hormones**, **RH**, or **inhibiting hormones**, **IH**) into the surrounding interstitial fluids through **fenestrated capillaries**. (*Figure 18–7*)

13. The **hypophyseal portal system** ensures that these regulatory factors reach the intended target cells before they enter the general circulation. (*Figure 18–7*)

14. **Thyroid-stimulating hormone (TSH)** triggers the release of thyroid hormones. *Thyrotropin-releasing hormone (TRH)* promotes the secretion of TSH. (*Figure 18–8*)

15. **Adrenocorticotropic hormone (ACTH)** stimulates the release of *glucocorticoids* by the suprarenal cortex. Corticotropin-releasing hormone (CRH) causes the secretion of ACTH. (*Figure 18–8*)

16. **Follicle-stimulating hormone (FSH)** stimulates follicle development and estrogen secretion in females and sperm production in males. **Luteinizing hormone (LH)** causes *ovulation* and *progestin* production in females, and androgen production in males. Gonadotropin-releasing hormone (GnRH) promotes the secretion of both FSH and LH. (*Figure 18–8*)

17. **Prolactin (PRL)**, together with other hormones, stimulates both the development of the mammary glands and milk production. (*Figure 18–8*)

18. **Growth hormone (GH**, or **somatotropin)** stimulates cell growth and replication through the release of **somatomedins** or **IGFs** from liver cells. The production of GH is regulated by **growth hormone–releasing hormone (GH–RH)** and **growth hormone–inhibiting hormone (GH–IH)**. (*Figure 18–8*)

19. **Melanocyte-stimulating hormone (MSH)** may be secreted by the pars intermedia during fetal development, early childhood, pregnancy, or certain diseases. This hormone stimulates melanocytes to produce melanin.

20. The **neurohypophysis**, or **posterior lobe** of the pituitary gland, contains the unmyelinated axons of hypothalamic neurons. Neurons of the **supraoptic** and **paraventricular nuclei** manufacture **antidiuretic hormone (ADH)** and **oxytocin**, respectively. ADH decreases the amount of water lost at the kidneys and, in higher concentrations, elevates blood pressure. In women, oxytocin stimulates contractile cells in the mammary glands and has a stimulatory effect on smooth muscles in the uterus. (*Figure 18–9; Summary Table 18–2*)

18-4 The thyroid gland lies inferior to the larynx and requires iodine for hormone synthesis p. 620

21. The thyroid gland lies anterior to the *thyroid cartilage* of the larynx and consists of two **lobes** connected by a narrow **isthmus**. (*Figure 18–10*)

22. The thyroid gland contains numerous **thyroid follicles**. Thyroid follicles release several hormones, including

thyroxine and **triiodothyronine** (*Figures 18–10, 18–11; Table 18–4*)

23. Most of the thyroid hormones entering the bloodstream are attached to special **thyroid-binding globulins (TBGs)**; the rest are attached to either **transthyretin** or albumin. (*Figure 18–11*)

24. Thyroid hormones are held in storage, bound to mitochondria (thereby increasing ATP production), or bound to receptors activating genes that control energy utilization. They also exert a **calorigenic effect**. (*Table 18–3*)

25. The **C cells** of the thyroid follicles produce **calcitonin (CT)**, which helps regulate Ca^{2+} concentrations in body fluids, especially during childhood and pregnancy. (*Figure 18–10, Table 18–4*)

18-5 The four parathyroid glands, embedded in the posterior surface of the thyroid, secrete parathyroid hormone to elevate plasma Ca^{2+} p. 625

26. Four **parathyroid glands** are embedded in the posterior surface of the thyroid gland. **Parathyroid chief cells** produce **parathyroid hormone (PTH)** in response to lower-than-normal concentrations of Ca^{2+}. The parathyroid glands, aided by *calcitriol*, are the primary regulators of blood calcium levels in healthy adults. (*Figures 18–12, 18–13; Table 18–4*)

18-6 The suprarenal glands, consisting of a cortex and medulla, cap the superior borders of the kidneys and secrete several hormones p. 627

27. One **suprarenal** (*adrenal*) **gland** lies along the superior border of each kidney. The gland is subdivided into the superficial **suprarenal cortex** and the inner **suprarenal medulla**. (*Figure 18–14*)

28. The suprarenal cortex manufactures steroid hormones called **adrenocortical steroids (corticosteroids)**. The cortex can be subdivided into three areas: (1) the **zona glomerulosa**, which releases **mineralocorticoids**, principally **aldosterone**; (2) the **zona fasciculata**, which produces **glucocorticoids**, notably **cortisol** and **corticosterone**; and (3) the **zona reticularis**, which produces androgens under ACTH stimulation. (*Figure 18–14; Table 18–5*)

29. The suprarenal medulla produces epinephrine (75–80 percent of medullary secretion) and norepinephrine (20–25 percent). (*Figure 18–14; Table 18–5*)

18-7 The pineal gland, attached to the third ventricle, secretes melatonin p. 630

30. The **pineal gland** contains **pinealocytes**, which synthesize **melatonin**. Suggested functions include inhibiting reproductive functions, protecting against damage by free radicals, and setting circadian rhythms.

18-8 The pancreas, located within the abdominopelvic cavity, is both an exocrine organ and endocrine gland p. 631

31. The pancreas contains both exocrine and endocrine cells. Cells of the endocrine pancreas form clusters called **pancreatic islets** (*islets of Langerhans*). These islets contain **alpha cells**, which secrete the hormone glucagon; **beta cells**, which secrete **insulin**; **delta cells**, which secrete **somatostatin (GH–IH)**; and **F cells**, which secrete **pancreatic polypeptide**. (*Figure 18–15; Table 18–6*)

32. Insulin lowers blood glucose by increasing the rate of glucose uptake and utilization; glucagon raises blood glucose by increasing the rates of glycogen breakdown and glucose manufacture in the liver. (*Figure 18–16, Table 18–6*)

18-9 Many organs have secondary endocrine functions p. 634

33. The intestines produce hormones important to the coordination of digestive activities. (*Table 18–7*)

34. Endocrine cells in the kidneys produce the hormones *calcitriol* and *erythropoietin* and the enzyme *renin*. (*Table 18–7*)

35. **Calcitriol** stimulates calcium and phosphate ion absorption along the digestive tract. (*Figure 18–17*)

36. **Erythropoietin (EPO)** stimulates red blood cell production by the bone marrow. (*Figure 18–17*)

37. **Renin** converts **angiotensinogen** to **angiotensin I**. In the capillaries of the lungs, the latter compound is converted to **angiotensin II**, a hormone that (1) stimulates the suprarenal production of aldosterone, (2) stimulates the pituitary release of ADH, (3) promotes thirst, and (4) elevates blood pressure. (*Figure 18–17*)

38. Specialized muscle cells in the heart produce **natriuretic peptides** (*ANP* and *BNP*) when blood volume becomes excessive. In general, their actions oppose those of angiotensin II. (*Table 18–7*)

39. The thymus produces several hormones, collectively known as **thymosins**, which play a role in developing and maintaining normal immune defenses. (*Table 18–7*)

40. The **interstitial cells** of the testes produce androgens. **Testosterone** is the most important sex hormone in males. (*Table 18–8*)

41. In females, *oocytes* develop in follicles; follicle cells produce **estrogens**, especially **estradiol**. After ovulation, the remaining follicle cells reorganize into a *corpus luteum*. Those cells release a mixture of estrogens and **progestins**, especially **progesterone**. (*Table 18–8*)

42. Adipose tissue secretes **leptin** (a feedback control for appetite).

18-10 Hormones interact to produce coordinated physiological responses p. 638

43. Endocrine system hormones often interact, producing (1) **antagonistic** (opposing) **effects**; (2) **synergistic** (additive) **effects**; (3) **permissive effects**, in which one hormone is necessary for another to produce its effect; or (4) **integrative effects**, in which hormones produce different, but complementary, results.

44. Normal growth requires the cooperation of several endocrine organs. Several hormones are especially important: GH, thyroid hormones, insulin, PTH, calcitriol, and reproductive hormones.

45. Any condition that threatens homeostasis is a **stress**. Our bodies respond to a variety of stress-causing factors through the **general adaptation syndrome (GAS)**, or stress response.

46. The GAS can be divided into three phases: (1) the **alarm phase** (an immediate, "fight or flight" response, under the direction of the sympathetic division of the ANS); (2) the **resistance phase**, dominated by glucocorticoids; and (3) the **exhaustion phase**, the eventual breakdown of homeostatic regulation and failure of one or more organ systems. (*Figure 18–18*)

18 ENDOCRINE

47. Many hormones affect the CNS; changes in the normal mixture of hormones can significantly alter intellectual capabilities, memory, learning, and emotional states.

48. The endocrine system undergoes few functional changes with advanced age. The major changes include a decline in the concentration of growth hormone and reproductive hormones.

18-11 Extensive integration occurs between the endocrine system and other body systems p. 644

49. The endocrine system provides long-term regulation and homeostatic adjustments that affect many body systems. (*Figure 18–19*)

Review Questions

See the blue Answers tab at the end of the book.

myA&P Access more review material online at **myA&P™** (www.myaandp.com). There, you'll find chapter guides, chapter quizzes, practice tests, animations, flashcards, a glossary with pronunciations, *Interactive Physiology*® (IP) exercises and quizzes, and more to help you succeed in the course.

LEVEL 1 Reviewing Facts and Terms

1. Identify the endocrine glands and tissues in the following diagram.

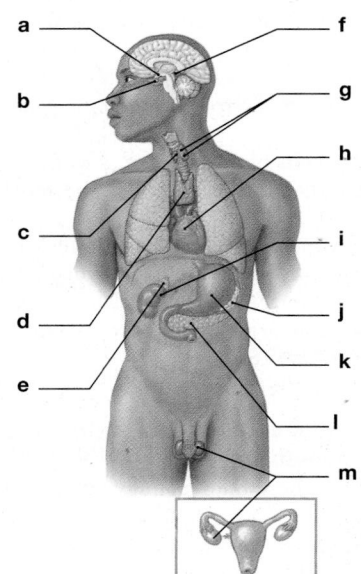

(a) _____
(b) _____
(c) _____
(d) _____
(e) _____
(f) _____
(g) _____
(h) _____
(i) _____
(j) _____
(k) _____
(l) _____
(m) _____

2. The use of a chemical messenger to transfer information from cell to cell within a single tissue is referred to as _____ communication.
 (a) direct
 (b) paracrine
 (c) hormonal
 (d) endocrine

3. Cyclic-AMP functions as a second messenger to
 (a) build proteins and catalyze specific reactions.
 (b) activate adenylate cyclase.
 (c) open ion channels and activate key enzymes in the cytoplasm.
 (d) bind the hormone–receptor complex to DNA segments.

4. Adrenocorticotropic hormone (ACTH) stimulates the release of
 (a) thyroid hormones by the hypothalamus.
 (b) gonadotropins by the suprarenal glands.
 (c) growth hormones by the hypothalamus.
 (d) steroid hormones by the suprarenal glands.

5. FSH production in males supports
 (a) the maturation of sperm by stimulating nurse cells.
 (b) the development of muscles and strength.
 (c) the production of male sex hormones.
 (d) an increased desire for sexual activity.

6. The two hormones released by the neurohypophysis are
 (a) GH and gonadotropin.
 (b) estrogen and progesterone.
 (c) GH and prolactin.
 (d) ADH and oxytocin.

7. All of the following are true of the endocrine system, *except* that it
 (a) releases chemicals into the bloodstream for distribution throughout the body.
 (b) releases hormones that simultaneously alter the metabolic activities of many different tissues and organs.
 (c) produces effects that can last for hours, days, and even longer.
 (d) produces rapid, local, brief-duration responses to specific stimuli.
 (e) functions to control ongoing metabolic processes.

8. A cell's hormonal sensitivities are determined by the
 (a) chemical nature of the hormone.
 (b) quantity of circulating hormone.
 (c) shape of the hormone molecules.
 (d) presence or absence of appropriate receptors.
 (e) thickness of its plasma membrane.

9. Endocrine organs can be regulated by all of the following, *except*
 (a) hormones from other endocrine glands.
 (b) changes in the genetic makeup of certain hypothalamic cells.
 (c) direct neural stimulation.
 (d) changes in the composition of extracellular fluid.
 (e) releasing hormones from the hypothalamus.

10. What three higher-level mechanisms are involved in integrating the activities of the nervous and endocrine systems?

11. Which seven hormones are released by the adenohypophysis?

12. What six hormones primarily affect growth?

13. What five primary effects result from the action of thyroid hormones?

14. What effects do calcitonin and parathyroid hormone have on blood calcium levels?

15. What three zones make up the suprarenal cortex, and what kind of hormones does each zone produce?

16. Which two hormones are released by the kidneys, and what is the importance of each hormone?

17. What are the four opposing effects of atrial natriuretic peptide and angiotensin II?
18. What four cell populations make up the endocrine pancreas? Which hormone does each type of cell produce?

LEVEL 2   Reviewing Concepts

19. What is the primary difference in the way the nervous and endocrine systems communicate with their target cells?
20. In what ways can a hormone modify the activities of its target cells?
21. What is an endocrine reflex? Compare endocrine reflexes and neural reflexes.
22. How would blocking the activity of phosphodiesterase affect a cell that responds to hormonal stimulation by the cAMP second-messenger system?
23. How does control of the suprarenal medulla differ from control of the suprarenal cortex?
24. A researcher observes that stimulation by a particular hormone induces a marked increase in the activity of G proteins in the target plasma membrane. The hormone being studied is probably
    (a) a steroid.   (b) a peptide.
    (c) testosterone.   (d) estrogen.
    (e) aldosterone.
25. Increased blood calcium levels would result in *increased*
    (a) secretion of calcitonin.
    (b) secretion of PTH.
    (c) retention of calcium by the kidneys.
    (d) osteoclast activity.
    (e) excitability of neural membranes.
26. In type 2 diabetes mellitus, insulin levels are frequently normal, yet the target cells are less sensitive to the effects of insulin. This suggests that the target cells
    (a) are impermeable to insulin.
    (b) may lack enough insulin receptors.
    (c) cannot convert insulin to an active form.
    (d) have adequate internal supplies of glucose.
    (e) are both b and c.

LEVEL 3   Critical Thinking and Clinical Applications

27. Roger has been extremely thirsty; he drinks numerous glasses of water every day and urinates a great deal. Name two disorders that could produce these signs and symptoms. What test could a clinician perform to determine which disorder Roger has?
28. Julie is pregnant but is not receiving prenatal care. She has a poor diet consisting mostly of fast food. She drinks no milk, preferring colas instead. How would this situation affect Julie's level of parathyroid hormone?
29. Sherry tells her physician that she has been restless and irritable lately. She has a hard time sleeping and complains of diarrhea and weight loss. During the examination, her physician notices a higher-than-normal heart rate and a fine tremor in her outstretched fingers. What tests could the physician perform to make a positive diagnosis of Sherry's condition?
30. What are two benefits of having a portal system connect the median eminence of the hypothalamus with the adenohypophysis?
31. Pamela and her teammates are considering taking testosterone supplements (anabolic steroids) to enhance their competitive skills. What natural effects of this hormone are they hoping to gain? What additional side effects might these women expect should they begin an anabolic steroid regime?

19

Blood

Did you know...?
Platelet function and the coagulation cascade are impaired by hypothermia, a significantly low core body temperature.

Learning Outcomes

After completing this chapter, you should be able to do the following:

19-1 Describe the components and major functions of blood, identify blood collection sites, and list the physical characteristics of blood.

19-2 Specify the composition and functions of plasma.

19-3 List the characteristics and functions of red blood cells, describe the structure and functions of hemoglobin, describe how red blood cell components are recycled, and explain erythropoiesis.

19-4 Explain the importance of blood typing, and the basis for ABO and Rh incompatibilities.

19-5 Categorize white blood cell types based on their structures and functions, and discuss the factors that regulate the production of each type.

19-6 Describe the structure, function, and production of platelets.

19-7 Discuss the mechanisms that control blood loss after an injury, and describe the reaction sequences responsible for blood clotting.

Clinical Notes

Collecting Blood for Analysis p. 653
Plasma Expanders p. 654
Abnormal Hemoglobin p. 658
Hemolytic Disease of the Newborn p. 664

An Introduction to the Cardiovascular System

This chapter discusses the nature of blood, the fluid component of the **cardiovascular system**. This body system also includes a pump (the heart) that circulates the fluid and a series of conducting hoses (the blood vessels) that carry it throughout the body. In Chapter 18, we noted the importance of this system as the transport medium for hormones, but that is only one of its many vital roles.

In adults, circulating blood provides each of the body's roughly 75 trillion cells a source of nutrients, oxygen, and chemical instructions, and a way of removing wastes. The blood also transports specialized cells that defend tissues from infection and disease. These services are essential—so much so that cells in a body region deprived of circulation may die in a matter of minutes. This chapter takes a close look at the structure and functions of blood, a fluid connective tissue with remarkable properties.

19-1 Blood has several important functions and unique physical characteristics

In this chapter, we examine the structure and functions of **blood**, a specialized fluid connective tissue that contains cells suspended in a fluid matrix. As you may recall, Chapter 4 introduced the components and properties of this connective tissue. ∞ p. 130

The functions of blood include the following:

- *The Transportation of Dissolved Gases, Nutrients, Hormones, and Metabolic Wastes*. Blood carries oxygen from the lungs to peripheral tissues, and carbon dioxide from those tissues to the lungs. Blood distributes nutrients absorbed at the digestive tract or released from storage in adipose tissue or in the liver. It carries hormones from endocrine glands toward their target cells, and it absorbs and carries the wastes produced by tissue cells to the kidneys for excretion.

- *The Regulation of the pH and Ion Composition of Interstitial Fluids*. Diffusion between interstitial fluids and blood eliminates local deficiencies or excesses of ions such as calcium or potassium. Blood also absorbs and neutralizes acids generated by active tissues, such as lactic acid produced by skeletal muscles.

- *The Restriction of Fluid Losses at Injury Sites*. Blood contains enzymes and other substances that respond to breaks in vessel walls by initiating the process of *clotting*.

A blood clot acts as a temporary patch that prevents further blood loss.

- *Defense against Toxins and Pathogens*. Blood transports *white blood cells*, specialized cells that migrate into other tissues to fight infections or remove debris. Blood also delivers *antibodies*, proteins that specifically attack invading organisms or foreign compounds.

- *The Stabilization of Body Temperature*. Blood absorbs the heat generated by active skeletal muscles and redistributes it to other tissues. If body temperature is already high, that heat will be lost across the surface of the skin. If body temperature is too low, the warm blood is directed to the brain and to other temperature-sensitive organs.

Blood has a unique composition (**Figure 19–1**). It is a fluid connective tissue with a matrix called **plasma** (PLAZ-muh). Plasma proteins are in solution rather than forming insoluble fibers like those in other connective tissues, such as loose connective tissue or cartilage. Because these proteins are in solution, plasma is slightly denser than water. Plasma is similar to interstitial fluid, although it contains a unique assortment of suspended proteins. A continuous exchange of fluid between the tissues and the blood is driven by a combination of hydrostatic pressure, concentration gradients, and osmosis. These relationships will be considered further in Chapter 21.

Formed elements are blood cells and cell fragments that are suspended in plasma. Three types of formed elements exist: red blood cells, white blood cells, and platelets. **Red blood cells (RBCs)**, or **erythrocytes** (e-RITH-rō-sīts; *erythros*, red + *-cyte*, cell), are the most abundant blood cells. These specialized cells are essential for the transport of oxygen in the blood. The less numerous **white blood cells (WBCs)**, or **leukocytes** (LOO-kō-sīts; *leukos*, white + *-cyte*, cell), participate in the body's defense mechanisms. There are five classes of leukocytes, each with slightly different functions. **Platelets** are small, membrane-bound cell fragments that contain enzymes and other substances important to the process of clotting.

Formed elements are produced through the process of **hemopoiesis** (hē-mō-poy-Ē-sis), or *hematopoiesis*. Two populations of stem cells—*myeloid stem cells* and *lymphoid stem cells*—are responsible for the production of all the kinds of formed elements. We will consider the fates of the myeloid and lymphoid stem cells as we discuss the formation of each type of formed element.

Together, the plasma and the formed elements constitute **whole blood**. The components of whole blood can be **fractionated**, or separated, for analytical or clinical purposes. We will encounter examples of uses for fractionated blood later in the chapter.

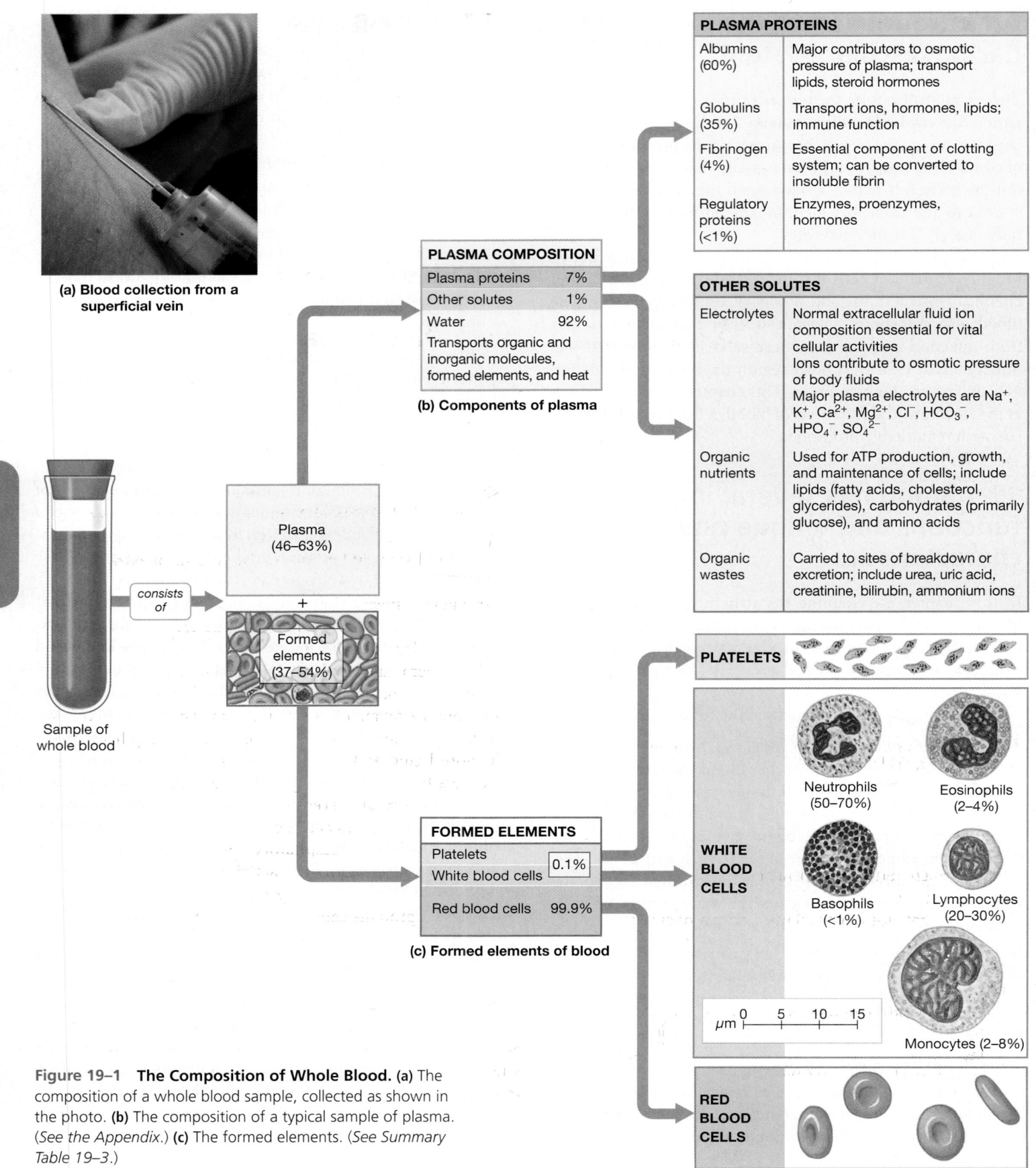

(a) Blood collection from a superficial vein

PLASMA PROTEINS

Albumins (60%)	Major contributors to osmotic pressure of plasma; transport lipids, steroid hormones
Globulins (35%)	Transport ions, hormones, lipids; immune function
Fibrinogen (4%)	Essential component of clotting system; can be converted to insoluble fibrin
Regulatory proteins (<1%)	Enzymes, proenzymes, hormones

PLASMA COMPOSITION

Plasma proteins	7%
Other solutes	1%
Water	92%
Transports organic and inorganic molecules, formed elements, and heat	

(b) Components of plasma

OTHER SOLUTES

Electrolytes	Normal extracellular fluid ion composition essential for vital cellular activities. Ions contribute to osmotic pressure of body fluids. Major plasma electrolytes are Na^+, K^+, Ca^{2+}, Mg^{2+}, Cl^-, HCO_3^-, HPO_4^-, SO_4^{2-}
Organic nutrients	Used for ATP production, growth, and maintenance of cells; include lipids (fatty acids, cholesterol, glycerides), carbohydrates (primarily glucose), and amino acids
Organic wastes	Carried to sites of breakdown or excretion; include urea, uric acid, creatinine, bilirubin, ammonium ions

Plasma (46–63%)

+

Formed elements (37–54%)

Sample of whole blood

consists of

PLATELETS

WHITE BLOOD CELLS

Neutrophils (50–70%)   Eosinophils (2–4%)

Basophils (<1%)   Lymphocytes (20–30%)

μm   0   5   10   15

Monocytes (2–8%)

FORMED ELEMENTS

Platelets	0.1%
White blood cells	
Red blood cells	99.9%

(c) Formed elements of blood

RED BLOOD CELLS

Figure 19–1 **The Composition of Whole Blood. (a)** The composition of a whole blood sample, collected as shown in the photo. **(b)** The composition of a typical sample of plasma. (*See the Appendix.*) **(c)** The formed elements. (*See Summary Table 19–3.*)

Whole blood from any source—venous blood, blood from peripheral capillaries, or arterial blood—has the same basic physical characteristics:

- Blood temperature is roughly 38°C (100.4°F), slightly above normal body temperature.

- Blood is five times as viscous as water—that is, five times as sticky, five times as cohesive, and five times as resistant to flow as water. The high viscosity results from interactions among dissolved proteins, formed elements, and water molecules in plasma.

- Blood is slightly alkaline, with a pH between 7.35 and 7.45 (average: 7.4).

The cardiovascular system of an adult male contains 5–6 liters (5.3–6.4 quarts) of whole blood; that of an adult female contains 4–5 liters (4.2–5.3 quarts). The sex differences in blood volume primarily reflect differences in average body size. Blood volume in liters can be estimated for an individual of either sex by calculating 7 percent of the body weight in kilograms. For example, a 75-kg (165-lb) individual would have a blood volume of approximately 5.25 liters (5.4 quarts).

CLINICAL NOTE

Collecting Blood for Analysis Fresh whole blood is generally collected from a superficial vein, such as the *median cubital vein* on the anterior surface of the elbow (**Figure 19–1a**). The procedure is called **venipuncture** (VĒN-i-punk-chur; *vena*, vein + *punctura*, a piercing). It is a common sampling technique because (1) superficial veins are easy to locate, (2) the walls of veins are thinner than those of comparably sized arteries, and (3) blood pressure in the venous system is relatively low, so the puncture wound seals quickly. The most common clinical procedures examine venous blood.

A small drop of blood can be used to prepare a *blood smear*, a thin film of blood on a microscope slide. The blood smear is then stained with special dyes to show each type of formed element. Blood from peripheral capillaries can be obtained by puncturing the tip of a finger, an earlobe, or (in infants) the great toe or heel. Small amounts of capillary blood can also be used to test (among other items) glucose, cholesterol, and hemoglobin levels. This technical feat is valuable when venous access is difficult.

An **arterial puncture**, or "arterial stick," can be used for checking the efficiency of gas exchange at the lungs. Samples are generally drawn from the *radial artery* at the wrist or the *brachial artery* at the elbow.

CHECKPOINT

1. List five major functions of blood.
2. Identify the composition of the formed elements in blood.
3. What two components constitute whole blood?
4. Why is venipuncture a common technique for obtaining a blood sample?

See the blue Answers tab at the end of the book.

19-2 Plasma, the fluid portion of blood, contains significant quantities of plasma proteins

In this section we consider the composition of plasma and the kinds of proteins it contains.

The Composition of Plasma

As shown in **Figure 19–1a**, plasma makes up 46–63 percent of the volume of whole blood. Water accounts for 92 percent of the plasma volume (**Figure 19–1b**). Together, plasma and interstitial fluid account for most of the volume of extracellular fluid (ECF) in the body.

In many respects, the composition of plasma resembles that of interstitial fluid. The concentrations of the major plasma ions, for example, are similar to those of interstitial fluid but differ markedly from those inside cells. This similarity is understandable, as water, ions, and small solutes are continuously exchanged between plasma and interstitial fluids across the walls of capillaries. Normally, the capillaries deliver more liquid and solutes to a tissue than they remove. The excess fluid flows through the tissue, into vessels of the lymphoid system, and eventually back to the bloodstream. The primary differences between plasma and interstitial fluid involve (1) the levels of respiratory gases (oxygen and carbon dioxide, due to the respiratory activities of tissue cells), and (2) the concentrations and types of dissolved proteins (because plasma proteins cannot cross capillary walls).

Plasma Proteins

Plasma contains significant quantities of dissolved proteins. On average, each 100 mL of plasma contains 7.6 g (0.3 oz) of protein, almost five times the concentration in interstitial fluid. The large size and globular shapes of most blood proteins prevent them from crossing capillary walls, so they remain trapped within the circulatory system. Three major types of plasma proteins exist: *albumins* (al-BŪ-minz), *globulins* (GLOB-ū-linz), and *fibrinogen* (fī-BRIN-ō-jen). These three types make up more than 99 percent of the plasma proteins. The remainder consists of circulating enzymes, hormones, and prohormones.

Albumins

Albumins constitute roughly 60 percent of the plasma proteins. As the most abundant plasma proteins, they are major contributors to the osmotic pressure of plasma. Albumins are also important in the transport of fatty acids, thyroid hormones, some steroid hormones, and other substances.

Globulins

Globulins account for approximately 35 percent of the proteins in plasma. Important plasma globulins include antibodies and transport globulins. **Antibodies**, also called **immunoglobulins** (i-mū-nō-GLOB-ū-linz), attack foreign proteins and pathogens. We will examine several classes of immunoglobulins in Chapter 22. **Transport globulins** bind small ions, hormones, and compounds that might otherwise be lost at the kidneys or that have very low solubility in water. Important examples of transport globulins include the following:

- *Hormone-binding proteins*, which provide a reserve of hormones in the bloodstream. Examples include *thyroid-binding globulin* and *transthyretin*, which transport thyroid hormones, and *transcortin*, which transports ACTH. ∞ pp. 623, 627

- *Metalloproteins*, which transport metal ions. *Transferrin*, for example, is a metalloprotein that transports iron (Fe^{2+}).

- *Apolipoproteins* (ap-ō-lip-ō-PRŌ-tenz), which carry triglycerides and other lipids in blood. When bound to lipids, an apolipoprotein becomes a **lipoprotein** (LĪ-pō-prō-tēn).

- *Steroid-binding proteins*, which transport steroid hormones in blood. For example, *testosterone-binding globulin* (*TeBG*) binds and transports testosterone.

Fibrinogen

The third major type of plasma protein, **fibrinogen**, functions in clotting. Fibrinogen normally accounts for roughly 4 percent of plasma proteins. Under certain conditions, fibrinogen molecules interact, forming large, insoluble strands of **fibrin** (FĪ-brin). These fibers provide the basic framework for a blood clot. If steps are not taken to prevent clotting in a blood sample, the conversion of fibrinogen to fibrin will occur. This conversion removes the clotting proteins, leaving a fluid known as **serum**. The clotting process also removes calcium ions and other materials from solution, so plasma and serum differ in several significant ways. (See Appendix.) Thus, the results of a blood test generally indicate whether the sample was plasma or serum.

Other Plasma Proteins

The remaining 1 percent of plasma proteins is composed of specialized proteins whose levels vary widely. Peptide hormones—including insulin, prolactin (PRL), and the glycoproteins thyroid-stimulating hormone (TSH), follicle-stimulating hormone (FSH), and luteinizing hormone (LH)—are normally present in circulating blood. Their plasma concentrations rise and fall from day to day or even hour to hour.

CLINICAL NOTE

Plasma Expanders

Plasma expanders can be used to increase blood volume temporarily, over a period of hours. They are often used to buy time for lab work to determine a person's blood type. (Transfusion of the wrong blood type can kill the recipient.) Isotonic electrolyte solutions such as normal (physiological) saline can be used as a plasma expander, but their effects are short-lived due to diffusion into interstitial fluid and cells. This fluid loss is slowed by the addition of solutes that cannot freely diffuse across plasma membranes. One example is lactated *Ringer's solution*, an isotonic saline also containing lactate, potassium chloride, and calcium chloride ions. The effects of Ringer's solution fade gradually as the liver, skeletal muscles, and other tissues absorb and metabolize the lactate ions. Another option is the

administration of isotonic saline solution containing purified human albumin. However, the plasma expanders in clinical use often contain large carbohydrate molecules, rather than proteins, to maintain proper osmotic concentration. (The emergency use of the carbohydrate dextran in sodium chloride solutions was noted in Chapter 3. ∞ p. 93) Although these carbohydrates are not metabolized, they are gradually removed from the bloodstream by phagocytes, and blood volume slowly declines. Plasma expanders are easily stored, and their sterile preparation avoids viral or bacterial contamination, which can be a problem with donated plasma. Note that although they provide a temporary solution to low blood volume, plasma expanders do not increase the amount of oxygen carried by the blood; that function is performed by red blood cells.

Origins of the Plasma Proteins

The liver synthesizes and releases more than 90 percent of the plasma proteins, including all albumins and fibrinogen, most globulins, and various prohormones. Antibodies are produced by *plasma cells*. Plasma cells are derived from *lymphocytes*, the primary cells of the lymphoid system. Peptide hormones are produced in a variety of endocrine organs.

Because the liver is the primary source of plasma proteins, liver disorders can alter the composition and functional properties of blood. For example, some forms of liver disease can lead to uncontrolled bleeding due to the inadequate synthesis of fibrinogen and other proteins involved in clotting.

The A&P Top 100

#66 Your total blood volume, in liters, is roughly equal to 7 percent of your body weight in kilograms. Approximately half the volume of whole blood consists of cells and cell products. Plasma resembles interstitial fluid, but it contains a unique mixture of proteins not found in other extracellular fluids.

CHECKPOINT

5. List the three major types of plasma proteins.

6. What would be the effects of a decrease in the amount of plasma proteins?

7. Which specific plasma protein would you expect to be elevated during a viral infection?

See the blue Answers tab at the end of the book.

19-3 Red blood cells, formed by erythropoiesis, contain hemoglobin that can be recycled

The most abundant blood cells are the red blood cells (RBCs), which account for 99.9 percent of the formed elements. These cells give whole blood its deep red color because they contain the red pigment *hemoglobin* (HĒ-mō-glō-bin), which binds and transports oxygen and carbon dioxide.

Abundance of RBCs

A standard blood test reports the number of RBCs per microliter (μL) of whole blood as the *red blood cell count*. In adult males, 1 microliter, or 1 *cubic millimeter* (mm^3), of whole blood contains 4.5–6.3 million RBCs; in adult females, 1 microliter contains 4.2–5.5 million. A single drop of whole blood contains approximately 260 million RBCs, and the blood of an average adult has 25 trillion RBCs. RBCs thus account for roughly one-third of all cells in the human body.

The **hematocrit** (he-MAT-ō-krit) is the percentage of whole blood volume contributed by formed elements, 99.9 percent of which are red blood cells (**Figure 19–1a**). The normal hematocrit in adult males averages 46 (range: 40–54); the average for adult females is 42 (range: 37–47). The sex difference in hematocrit primarily reflects the fact that androgens (male hormones) stimulate red blood cell production, whereas estrogens (female hormones) do not.

The hematocrit is determined by centrifuging a blood sample so that all the formed elements come out of suspension. Whole blood contains roughly 1000 red blood cells for each white blood cell. After centrifugation, the white blood cells and platelets form a very thin *buffy coat* above a thick layer of RBCs. Because the hematocrit value is due almost entirely to the volume of RBCs, hematocrit is commonly reported as the *volume of packed red cells* (VPRC), or simply the *packed cell volume* (PCV).

Many conditions can affect the hematocrit. For example, the hematocrit increases during dehydration, owing to a reduction in plasma volume, or after *erythropoietin* (*EPO*) stimulation. ∞ p. 637 The hematocrit can decrease as a result of internal bleeding or problems with RBC formation. As a result, the hematocrit alone does not provide specific diagnostic information. Still, an abnormal hematocrit is an indication that other, more specific tests are needed. (We will consider some of those tests later in the chapter.)

Structure of RBCs

Red blood cells are among the most specialized cells of the body. A red blood cell is very different from the "typical cell" we discussed in Chapter 3. Each RBC is a biconcave disc with a thin central region and a thicker outer margin (**Figure 19–2**). An average RBC has a diameter of 7.8 μm and a maximum thickness of 2.85 μm, although the center narrows to about 0.8 μm.

This unusual shape has three important effects on RBC function:

1. *It Gives Each RBC a Large Surface Area-to-Volume Ratio.* Each RBC carries oxygen bound to intracellular proteins. That oxygen must be absorbed or released quickly as the RBC passes through the capillaries of the lungs or peripheral tissues. The greater the surface area per unit volume, the faster the exchange between the RBC's interior and the surrounding plasma. The total surface area of all the RBCs in the blood of a typical adult is about 3800 square meters, (nearly 4600 square yards), some 2000 times the total surface area of the body.

2. *It Enables RBCs to Form Stacks, Like Dinner Plates, That Smooth the Flow through Narrow Blood Vessels.* These stacks, known as *rouleaux* (roo-LŌ; *rouleau*, singular), form and dissociate repeatedly without affecting the cells involved. An entire stack can pass along a blood vessel

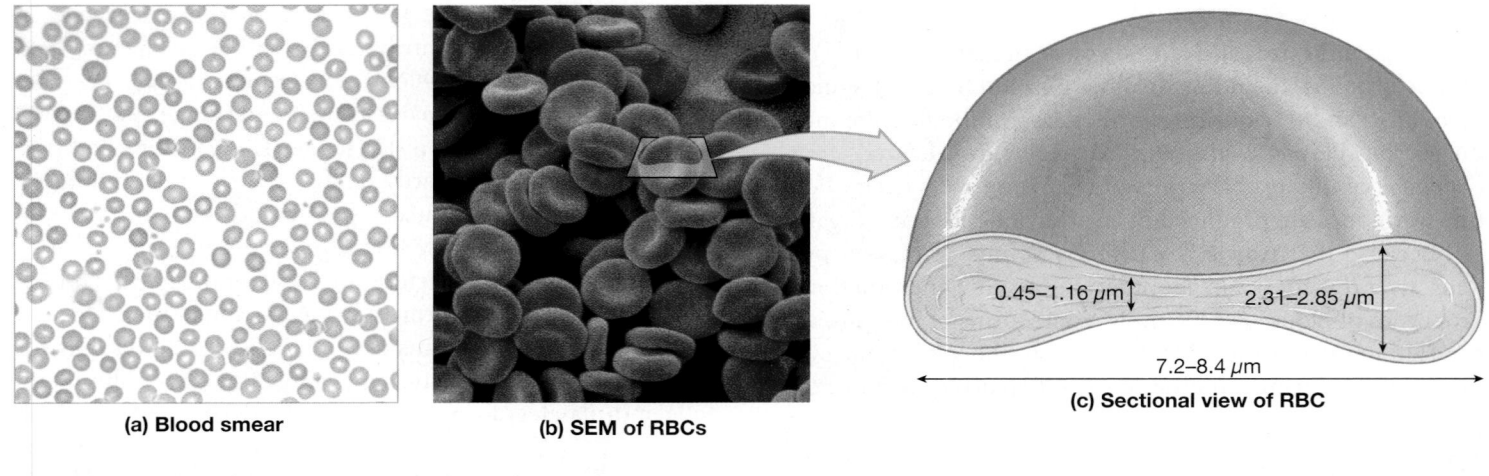

(a) Blood smear

(b) SEM of RBCs

0.45–1.16 µm   2.31–2.85 µm

7.2–8.4 µm

(c) Sectional view of RBC

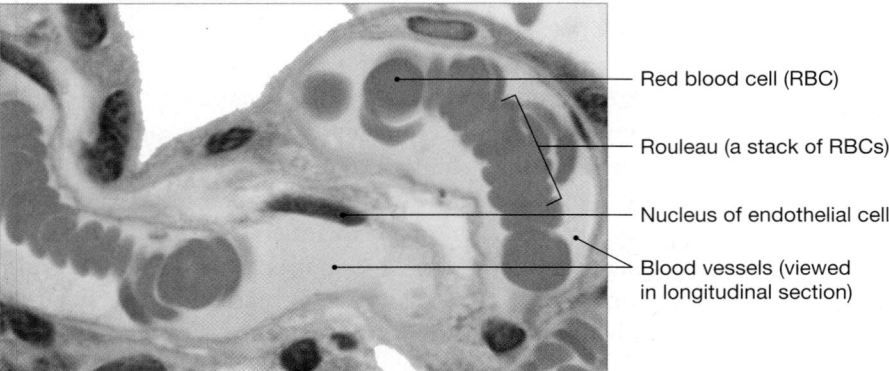

— Red blood cell (RBC)

— Rouleau (a stack of RBCs)

— Nucleus of endothelial cell

— Blood vessels (viewed in longitudinal section)

(d) Sectional view of capillaries

Figure 19–2   The Anatomy of Red Blood Cells. (a) When viewed in a standard blood smear, red blood cells appear as two-dimensional objects, because they are flattened against the surface of the slide. **(b)** The three-dimensional structure of red blood cells. **(c)** A sectional view of a mature red blood cell, showing the normal ranges for its dimensions. **(d)** When traveling through relatively narrow capillaries, RBCs may stack like dinner plates.

only slightly larger than the diameter of a single RBC, whereas individual cells would bump the walls, bang together, and form logjams that could restrict or prevent blood flow. Such stacks are shown in **Figure 19–2d**.

3. *It Enables RBCs to Bend and Flex When Entering Small Capillaries and Branches.* Red blood cells are very flexible. By changing shape, individual RBCs can squeeze through capillaries as narrow as 4 µm.

During their differentiation, the RBCs of humans and other mammals lose most of their organelles, including nuclei; the cells retain only the cytoskeleton. (The RBCs of vertebrates other than mammals have nuclei.) Because they lack nuclei and ribosomes, circulating mammalian RBCs cannot divide or synthesize structural proteins or enzymes. As a result, the RBCs cannot perform repairs, so their life span is relatively short—normally less than 120 days. With few organelles and no ability to synthesize proteins, their energy demands are low. In the absence of mitochondria, they obtain the energy they need through the anaerobic metabolism of glucose absorbed from the surrounding plasma. The lack of mitochondria ensures that absorbed oxygen will be carried to peripheral tissues, not "stolen" by mitochondria in the RBC.

Hemoglobin

In effect, a developing red blood cell loses any organelle not directly associated with the cell's primary function: the transport of respiratory gases. Molecules of **hemoglobin (Hb)** account for more than 95 percent of its intracellular proteins. The hemoglobin content of whole blood is reported in grams of Hb per deciliter (100 mL) of whole blood (g/dL). Normal ranges are 14–18 g/dL in males and 12–16 g/dL in females. Hemoglobin is responsible for the cell's ability to transport oxygen and carbon dioxide.

Hemoglobin Structure

Hb molecules have complex quaternary structures. ∞ p. 54 Each Hb molecule has two *alpha* (α) *chains* and two *beta* (β) *chains* of polypeptides (**Figure 19–3**). Each chain is a globular protein subunit that resembles the myoglobin in skeletal and cardiac muscle cells. Like myoglobin, each Hb chain contains a single molecule of **heme**, a non-protein pigment complex. Each heme unit holds an iron ion in such a way that the iron can interact with an oxygen molecule, forming **oxyhemoglobin**, HbO_2. Blood containing RBCs filled with oxyhemoglobin is bright red. The iron–oxygen interaction is very

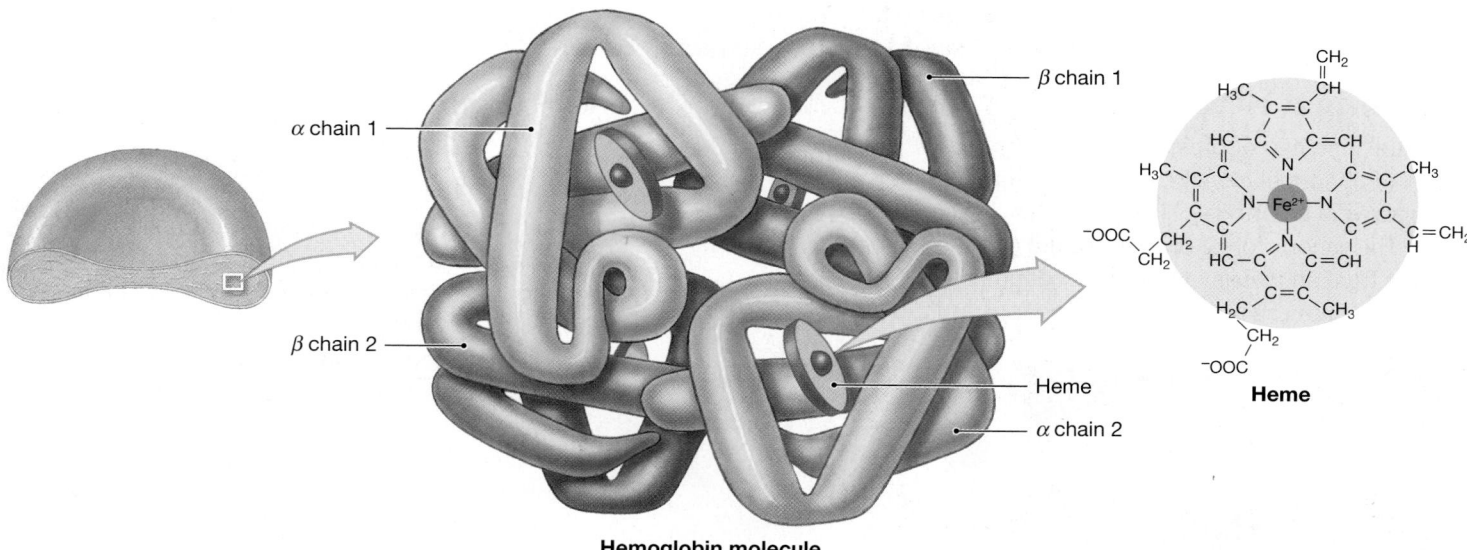

Figure 19–3 **The Structure of Hemoglobin.** Hemoglobin consists of four globular protein subunits. Each subunit contains a single molecule of heme—a nonprotein ring surrounding a single ion of iron.

weak; the two can easily dissociate without damaging the heme unit or the oxygen molecule. The binding of an oxygen molecule to the iron in a heme unit is therefore completely reversible. A hemoglobin molecule whose iron is not bound to oxygen is called **deoxyhemoglobin**. Blood containing RBCs filled with deoxyhemoglobin is dark red—almost burgundy.

The RBCs of an embryo or a fetus contain a different form of hemoglobin, known as *fetal hemoglobin*, which binds oxygen more readily than does the hemoglobin of adults. For this reason, a developing fetus can "steal" oxygen from the maternal bloodstream at the placenta. The conversion from fetal hemoglobin to the adult form begins shortly before birth and continues over the next year. The production of fetal hemoglobin can be stimulated in adults by the administration of drugs such as *hydroxyurea* or *butyrate*. This is one method of treatment for conditions, such as *sickle cell anemia* or *thalassemia*, that result from the production of abnormal forms of adult hemoglobin.

Hemoglobin Function

Each red blood cell contains about 280 million Hb molecules. Because a Hb molecule contains four heme units, each RBC can potentially carry more than a billion molecules of oxygen at a time. Roughly 98.5 percent of the oxygen carried by the blood travels through the bloodstream bound to Hb molecules inside RBCs.

The amount of oxygen bound to hemoglobin depends primarily on the oxygen content of the plasma. When plasma oxygen levels are low, hemoglobin releases oxygen. Under these conditions, typical of peripheral capillaries, plasma carbon dioxide levels are elevated. The alpha and beta chains of

hemoglobin then bind carbon dioxide, forming **carbamino-hemoglobin**. In the capillaries of the lungs, plasma oxygen levels are high and carbon dioxide levels are low. Upon reaching these capillaries, RBCs absorb oxygen (which is then bound to hemoglobin) and release carbon dioxide. We will revisit these processes in Chapter 23.

Normal activity levels can be sustained only when tissue oxygen levels are kept within normal limits. If the hematocrit is low or the Hb content of the RBCs is reduced, the condition called **anemia** exists. Anemia interferes with oxygen delivery to peripheral tissues. Every system is affected as organ function deteriorates owing to oxygen starvation. Anemic individuals become weak, lethargic, and often confused, because the brain is affected as well.

RBC Formation and Turnover

A red blood cell is exposed to severe mechanical stresses. A single round trip from the heart, through the peripheral tissues, and back to the heart usually takes less than a minute. In that time, an RBC gets pumped out of the heart and forced along vessels, where it bounces off the walls and collides with other RBCs. It forms stacks, contorts and squeezes through tiny capillaries, and then is rushed back to the heart to make yet another round trip.

With all this wear and tear and no repair mechanisms, a typical RBC has a relatively short life span. After it travels about 700 miles in 120 days, either its plasma membrane ruptures or some other damage is detected by phagocytes, which engulf the RBC. The continuous elimination of RBCs usually goes unnoticed, because new ones enter the blood-

CLINICAL NOTE

Abnormal Hemoglobin

Several inherited disorders are characterized by the production of abnormal hemoglobin. Two of the best known are thalassemia and sickle cell anemia (SCA).

The various forms of **thalassemia** (thal-ah-SĒ-mē-uh) result from an inability to produce adequate amounts of alpha or beta chains of hemoglobin. As a result, the rate of

RBC production is slowed, and mature RBCs are fragile and short lived. The scarcity of healthy RBCs reduces the oxygen-carrying capacity of the blood and leads to problems with the development and growth of systems throughout the body. Individuals with severe thalassemia must periodically undergo *transfusions*—the administration of blood components—to keep adequate numbers of RBCs in the bloodstream.

Sickle cell anemia results from a mutation affecting the amino acid sequence of the beta chains of the Hb molecule. When blood contains abundant oxygen, the Hb molecules and the RBCs that carry them appear normal (**Figure 19–4a**). However, when the defective hemoglobin gives up enough of its bound oxygen, the adjacent Hb molecules interact and the cells become stiff and curved (**Figure 19–4b**). This "sickling" makes the RBCs fragile and easily damaged. Moreover, an RBC that has folded to squeeze into a narrow capillary delivers its oxygen to the surrounding tissue, but the cell can become stuck as sickling occurs. A circulatory blockage results, and nearby tissues become starved for oxygen.

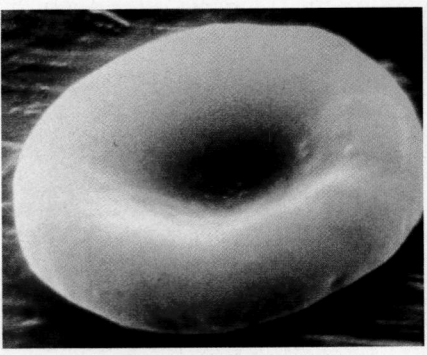

(a) Normal RBC

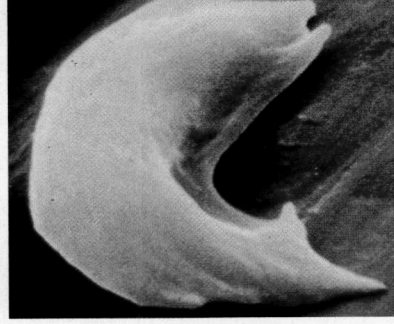

(b) Sickled RBC

Figure 19–4 **"Sickling" in Red Blood Cells. (a)** When fully oxygenated, the red blood cells of an individual with the sickling trait appear relatively normal. **(b)** At lower oxygen concentrations, the RBCs change shape, becoming more rigid and sharply curved.

stream at a comparable rate. About 1 percent of the circulating RBCs are replaced each day, and in the process approximately 3 million new RBCs enter the bloodstream *each second!*

Hemoglobin Conservation and Recycling

Macrophages of the liver, spleen, and bone marrow monitor the condition of circulating RBCs, generally recognizing and engulfing them before they **hemolyze**, or rupture. These phagocytes also detect and remove Hb molecules and cell fragments from the relatively small proportion of RBCs that hemolyze in the bloodstream (about 10 percent of the total recycled each day).

If the Hb released by hemolysis is not phagocytized, its components will not be recycled. Hemoglobin remains intact only under the conditions inside RBCs. When hemolysis occurs, the Hb breaks down, and the alpha and beta chains are filtered by the kidneys and eliminated in urine. When abnormally large numbers of RBCs break down in the bloodstream, urine may turn red or brown. This condition is called **hemoglobinuria**. The presence of intact RBCs in urine—a sign called **hematuria** (hē-ma-TOO-rē-uh)—occurs only after kidney damage or damage to vessels along the urinary tract.

Once an RBC has been engulfed and broken down by a phagocytic cell, each component of the Hb molecule has a different fate (**Figure 19–5**). The globular proteins are disassembled into their component amino acids, which are then either metabolized by the cell or released into the bloodstream for use by other cells. Each heme unit is stripped of its iron and converted to **biliverdin** (bil-i-VER-din), an organic compound with a green color. (Bad bruises commonly develop a greenish tint due to biliverdin formation in the blood-filled tissues.) Biliverdin is then converted to **bilirubin** (bil-i-ROO-bin), an orange-yellow pigment, and released into the bloodstream. There, the bilirubin binds to albumin and is transported to the liver for excretion in bile.

If the bile ducts are blocked or the liver cannot absorb or excrete bilirubin, circulating levels of the compound climb rapidly. Bilirubin then diffuses into peripheral tissues, giving them a yellow color that is most apparent in the skin and over the sclera of the eyes. This combination of signs (yellow skin and eyes) is called **jaundice** (JAWN-dis).

In the large intestine, bacteria convert bilirubin to related pigments called *urobilinogens* (ūr-ō-bī-LIN-ō-jens) and *stercobilinogens* (ster-kō-bī-LIN-ō-jens). Some of the urobilinogens are absorbed into the bloodstream and are subsequently excreted into urine. On exposure to oxygen, some of the uro-

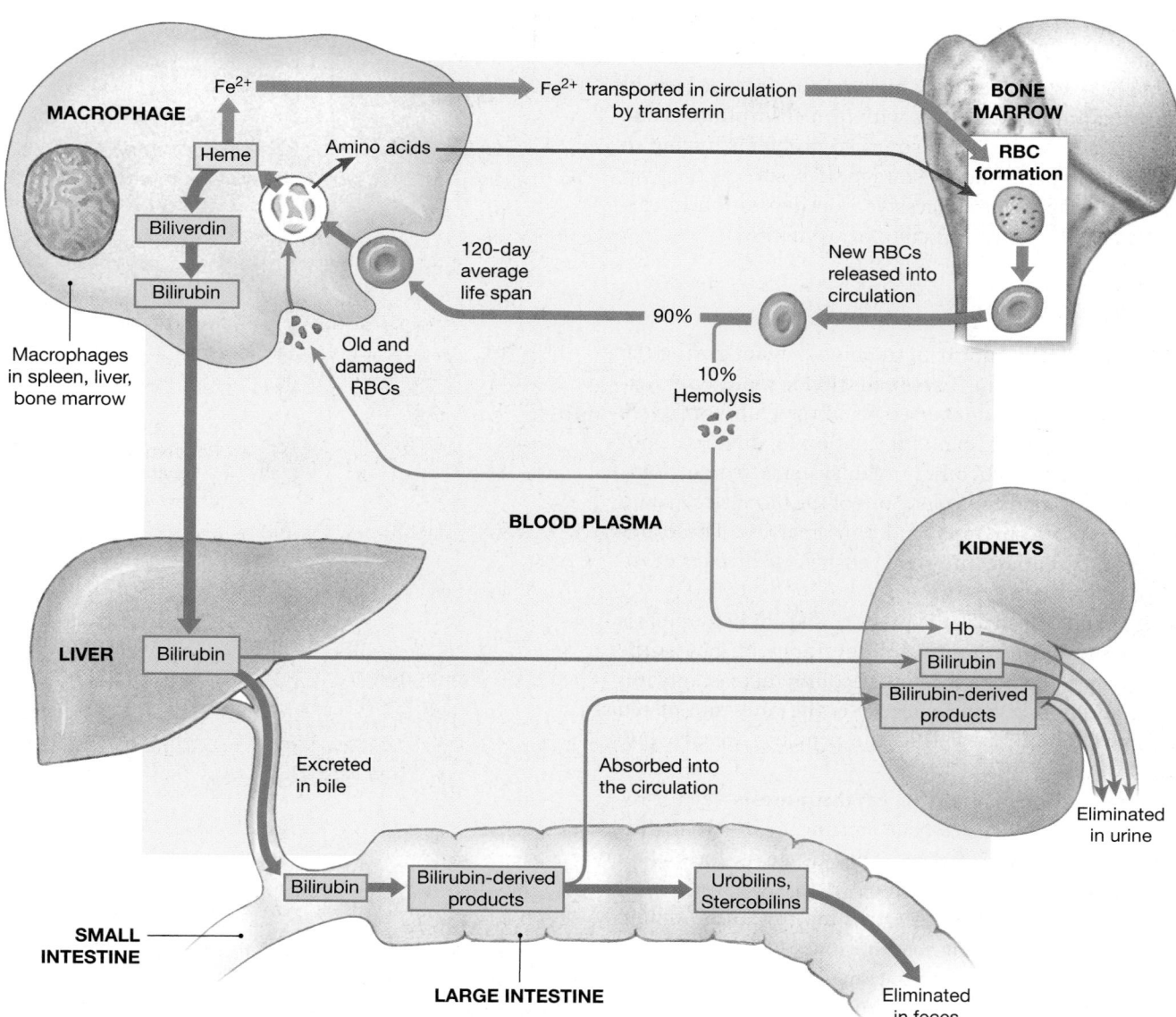

Figure 19–5 **Recycling of Red Blood Cell Components.** The normal pathways for recycling amino acids and iron from aging or damaged RBCs, broken down by macrophages. The amino acids are absorbed, especially by developing cells in bone marrow. The iron is stored in many sites. The rings of the heme units are converted to bilirubin, absorbed by the liver, and excreted in bile or urine; some of the breakdown products produced in the large intestine are reabsorbed and recirculated.

bilinogens and stercobilinogens are converted to **urobilins** (ūr-ō-BĪ-lins) and **stercobilins** (ster-kō-BĪ-lins). Urine is usually yellow because it contains urobilins; feces are yellow-brown or brown owing to the presence of urobilins and stercobilins in varying proportions.

Iron

Large quantities of free iron are toxic to cells, so in the body iron is generally bound to transport or storage proteins. Iron extracted from heme molecules may be bound and stored in a phagocytic cell or released into the bloodstream, where it

binds to **transferrin** (trans-FER-in), a plasma protein. Red blood cells developing in the bone marrow absorb the amino acids and transferrins from the bloodstream and use them to synthesize new Hb molecules. Excess transferrins are removed in the liver and spleen, and the iron is stored in two special protein–iron complexes: **ferritin** (FER-i-tin) and **hemosiderin** (hē-mō-SID-e-rin).

This recycling system is remarkably efficient. Although roughly 26 mg of iron is incorporated into new Hb molecules each day, a dietary supply of 1–2 mg can keep pace with the incidental losses that occur at the kidneys and digestive tract.

Any impairment in iron uptake or metabolism can cause serious clinical problems, because RBC formation will be affected. *Iron-deficiency anemia*, which results from a lack of iron in the diet or from problems with iron absorption, is one example. Too much iron can also cause problems, owing to excessive buildup in secondary storage sites, such as the liver and cardiac muscle tissue. Excessive iron deposition in cardiac muscle cells has been linked to heart disease.

RBC Production

Embryonic blood cells appear in the bloodstream during the third week of development. These cells divide repeatedly, rapidly increasing in number. The vessels of the embryonic *yolk sac* are the primary site of blood formation for the first eight weeks of development. As other organ systems appear, some of the embryonic blood cells move out of the bloodstream and into the liver, spleen, thymus, and bone marrow. These embryonic cells differentiate into stem cells whose divisions produce blood cells.

The liver and spleen are the primary sites of hemopoiesis from the second to fifth months of development, but as the skeleton enlarges, the bone marrow becomes increasingly important. In adults, red bone marrow is the only site of red blood cell production, as well as the primary site of white blood cell formation.

Red blood cell formation, or **erythropoiesis** (e-rith-rō-poy-Ē-sis), occurs only in *red bone marrow*, or **myeloid** (MĪ-e-loyd; *myelos*, marrow) **tissue**. This tissue is located in portions of the vertebrae, sternum, ribs, skull, scapulae, pelvis, and proximal limb bones. Other marrow areas contain a fatty tissue known as *yellow bone marrow*. 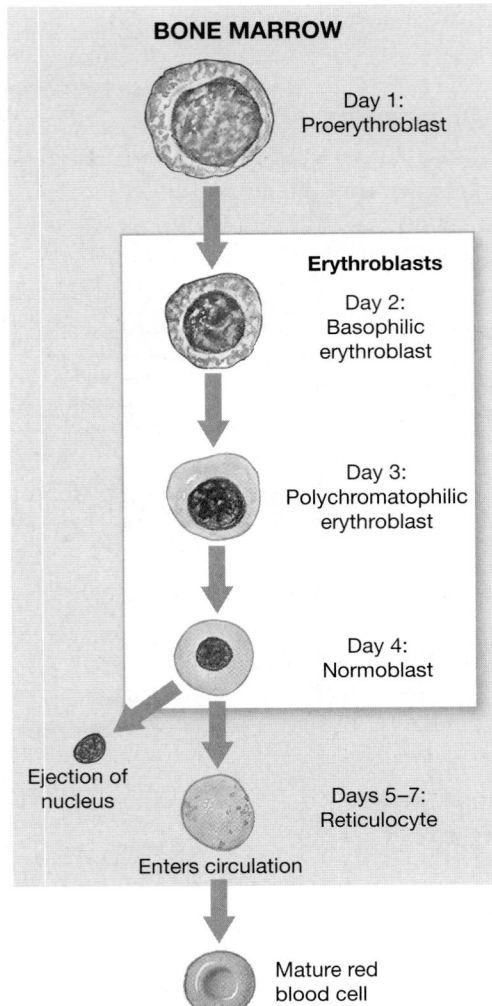 p. 192 Under extreme stimulation, such as severe and sustained blood loss, areas of yellow marrow can convert to red marrow, increasing the rate of RBC formation.

Stages in RBC Maturation

During its maturation, a red blood cell passes through a series of stages. **Hematologists** (hē-ma-TOL-o-jists), specialists in blood formation and function, have given specific names to key stages. Divisions of **hemocytoblasts** (*hemo-*, blood + *cyte*, cell + *blastos*, precursor), or *multipotent stem cells*, in bone marrow produce (1) **myeloid stem cells**, which in turn divide to produce red blood cells and several classes of white blood cells, and (2) **lymphoid stem cells**, which divide to produce the various classes of lymphocytes. Cells destined to become RBCs first differentiate into **proerythroblasts** and then proceed through various **erythroblast** stages (**Figure 19–6**). Erythroblasts, which actively synthesize hemoglobin, are named based on total size, amount of hemoglobin present, and size and appearance of the nucleus.

BONE MARROW

Day 1: Proerythroblast

Erythroblasts

Day 2: Basophilic erythroblast

Day 3: Polychromatophilic erythroblast

Day 4: Normoblast

Ejection of nucleus

Days 5–7: Reticulocyte

Enters circulation

Mature red blood cell

Figure 19–6 Stages of RBC Maturation. Red blood cells are produced in the red bone marrow. The color density in the cytoplasm indicates the abundance of hemoglobin. Note the reductions in the sizes of the cell and nucleus leading up to the formation of a reticulocyte.

After roughly four days of differentiation, the erythroblast, now called a *normoblast*, sheds its nucleus and becomes a **reticulocyte** (re-TIK-ū-lō-sīt), which contains 80 percent of the Hb of a mature RBC. Hb synthesis then continues for two to three more days. During this period, while the cells are synthesizing hemoglobin and other proteins, their cytoplasm still contains RNA which can be seen under the microscope with certain stains. After two days in the bone marrow, reticulocytes enter the bloodstream. At this time, reticulocytes normally account for about 0.8 percent of the RBC population in the blood and can still be detected by staining. After 24 hours in circulation, the reticulocytes complete their maturation and become indistinguishable from other mature RBCs.

Regulation of Erythropoiesis

For erythropoiesis to proceed normally, the red bone marrow must receive adequate supplies of amino acids, iron, and vitamins (including B_{12}, B_6, and folic acid) required for protein synthesis. We obtain **vitamin B_{12}** from dairy products and meat, and its absorption requires the presence of *intrinsic factor* produced in the stomach. If vitamin B_{12} is not obtained from the diet, normal stem cell divisions cannot occur and *pernicious anemia* results. Thus, pernicious anemia is caused by either a vitamin B_{12} deficiency, a problem with the production of intrinsic factor, or a problem with the absorption of vitamin B_{12} bound to intrinsic factor.

Erythropoiesis is stimulated directly by the peptide hormone erythropoietin (∞ p. 637) and indirectly by several hormones, including thyroxine, androgens, and growth hormone. As previously noted, estrogens do not stimulate erythropoiesis, a fact that accounts for the differences in hematocrit values between males and females.

Erythropoietin (EPO), also called **erythropoiesis-stimulating hormone**, is a glycoprotein, formed by the kidneys and liver, that appears in the plasma when peripheral tissues, especially the kidneys, are exposed to low oxygen concentrations. The state of low tissue oxygen levels is called **hypoxia** (hī-POKS-ē-uh; *hypo-*, below + *ox-*, presence of oxygen). Erythropoietin is released (1) during anemia; (2) when blood flow to the kidneys declines; (3) when the oxygen content of air in the lungs declines, owing to disease or high altitude; and (4) when the respiratory surfaces of the lungs are damaged. Once in the bloodstream, EPO travels to areas of red bone marrow, where it stimulates stem cells and developing RBCs.

Erythropoietin has two major effects: (1) It stimulates increased cell division rates in erythroblasts and in the stem cells that produce erythroblasts, and (2) it speeds up the maturation of RBCs, mainly by accelerating the rate of Hb synthesis. Under maximum EPO stimulation, bone marrow can increase the rate of RBC formation tenfold, to about 30 million cells per second.

The ability to increase the rate of blood formation quickly and dramatically is important to a person recovering from a severe blood loss. But if EPO is administered to a healthy individual, as in the case of the cyclists mentioned in Chapter 18, the hematocrit may rise to 65 or more. ∞ p. 643 Such an increase can place an intolerable strain on the heart. Comparable problems can occur after **blood doping**, a practice in which athletes attempt to elevate their hematocrits by reinfusing packed RBCs that were removed and stored at an earlier date. The goal is to improve oxygen delivery to muscles, thereby enhancing performance. The strategy can be dangerous, however, because it elevates blood viscosity and increases the workload on the heart.

Blood tests provide information about the general health of an individual, usually with a minimum of trouble and expense. Several common blood tests focus on red blood cells, the most abundant formed elements. These *RBC tests* assess the number, size, shape, and maturity of circulating RBCs, providing an indication of the erythropoietic activities under way. The tests can also be useful in detecting problems, such as internal bleeding, that may not produce other obvious signs or symptoms. Table 19–1 lists examples of important blood tests and related terms.

The A&P Top 100

#67 Red blood cells (RBCs) are the most numerous cells in the body. They remain in circulation for approximately 4 months before being recycled; several million are produced each second. The hemoglobin inside RBCs transports oxygen from the lungs to peripheral tissues; it also carries carbon dioxide from those tissues to the lungs.

TABLE 19–1	RBC Tests and Related Terminology		
		Terms Associated with Abnormal Values	
Test	**Determines**	**Elevated**	**Depressed**
Hematocrit (Hct)	Percentage of formed elements in whole blood Normal = 37–54%	Polycythemia (may reflect erythrocytosis or leukocytosis)	Anemia
Reticulocyte count (Retic.)	Percentage of circulating reticulocytes Normal = 0.8%	Reticulocytosis	
Hemoglobin concentration (Hb)	Concentration of hemoglobin in blood Normal = 12–18 g/dL		Anemia
RBC count	Number of RBCs per μL of whole blood Normal = 4.2–6.3 million/μL	Erythrocytosis/polycythemia	Anemia
Mean corpuscular volume (MCV)	Average volume of single RBC Normal = 82–101 μm^3 (normocytic)	Macrocytic	Microcytic
Mean corpuscular hemoglobin concentration (MCHC)	Average amount of Hb in one RBC Normal = 27–34 pg/μL (normochromic)	Hyperchromic	Hypochromic

19 CARDIOVASCULAR

8. Describe hemoglobin.

9. How would the hematocrit change after an individual suffered a significant blood loss?

10. Dave develops a blockage in his renal arteries that restricts blood flow to the kidneys. Will his hematocrit change?

11. In what way would a disease that causes damage to the liver affect the level of bilirubin in the blood?

See the blue Answers tab at the end of the book.

19-4 The ABO blood types and Rh system are based on antigen–antibody responses

Antigens are substances that can trigger a protective defense mechanism called an *immune response*. Most antigens are proteins, although some other types of organic molecules are antigens as well. Your plasma membranes contain **surface antigens**, substances that your immune system recognizes as "normal." In other words, your immune system ignores these substances rather than attacking them as "foreign."

Your **blood type** is a classification determined by the presence or absence of specific surface antigens in RBC plasma membranes. The surface antigens involved are integral membrane glycoproteins or glycolipids whose characteristics are genetically determined. Although red blood cells have at least 50 kinds of surface antigens, three surface antigens are of particular importance: **A**, **B**, and **Rh** (or **D**).

Based on RBC surface antigens, there are four blood types (**Figure 19–7a**): **Type A** blood has surface antigen A only, **Type B** has surface antigen B only, **Type AB** has both A and B, and **Type O** has neither A nor B. Individuals with these blood types are not evenly distributed throughout the population. The average values for the U.S. population are as follows: Type O, 46 percent; Type A, 40 percent; Type B, 10 percent; and Type AB, 4 percent (Table 19–2).

The term **Rh positive (Rh⁺)** indicates the presence of the Rh surface antigen, sometimes called the *Rh factor*. The absence of this antigen is indicated as **Rh negative (Rh⁻)**. When the complete blood type is recorded, the term *Rh* is usually omitted and the data are reported as O negative (O⁻), A positive (A⁺), and so on. As in the distribution of A and B surface antigens, Rh type differs by ethnic group and by region (Table 19–2).

Your immune system ignores the surface antigens—called **agglutinogens** (a-gloo-TIN-ō-jenz)—on your own RBCs. However, your plasma contains antibodies, sometimes called *agglutinins* (a-GLOO-ti-ninz), that will attack the antigens on "foreign" RBCs. When these antibodies attack, the foreign cells **agglutinate**, or clump together; this process is called **agglutination**. If you have Type A blood, your plasma contains anti-B antibodies, which will attack Type B surface antigens. If you have Type B blood, your plasma contains anti-A antibodies. The RBCs of an individual with Type O blood have neither A nor B surface antigens, and that person's plasma contains both anti-A and anti-B antibodies. A Type AB individual has RBCs with both A and B surface antigens, and the plasma does not contain anti-A or anti-B antibodies. The presence of anti-A and/or anti-B antibodies is genetically determined and they are present throughout life, regardless of whether the individual has ever been exposed to foreign RBCs.

In contrast, the plasma of an Rh-negative individual does not necessarily contain anti-Rh antibodies. These antibodies are present only if the individual has been **sensitized** by previous exposure to Rh-positive RBCs. Such exposure can occur accidentally during a transfusion, but it can also accompany a seemingly normal pregnancy involving an Rh-negative mother and an Rh-positive fetus. (See the Clinical Note "Hemolytic Disease of the Newborn" on pages 664–665.)

Cross-Reactions in Transfusions

When an antibody meets its specific surface antigen, the RBCs agglutinate and may also hemolyze. This reaction is called a **cross-reaction** (**Figure 19–7b**). For instance, an anti-A antibody that encounters A surface antigens will cause the RBCs bearing the surface antigens to clump or even break up. Clumps and fragments of RBCs under attack form drifting masses that can plug small blood vessels in the kidneys, lungs, heart, or brain, damaging or destroying affected tissues. Such cross-reactions, or *transfusion reactions*, can be prevented by ensuring that the blood types of the donor and the recipient are **compatible**—that is, that the donor's blood cells and the recipient's plasma will not cross-react.

In practice, the surface antigens on the donor's cells are more important in determining compatibility than are the antibodies in the donor's plasma. Unless large volumes of whole blood or plasma are transferred, cross-reactions between the donor's plasma and the recipient's blood cells will fail to produce significant agglutination. This is because the donated plasma is diluted quickly through mixing with the relatively large plasma volume of the recipient. (One unit of whole blood, 500 mL, contains roughly 275 mL of plasma, only about 10 percent of normal plasma volume.) Nonetheless, when increasing the blood's oxygen-carrying capacity rather than its plasma volume is the primary goal, packed RBCs, with a minimal amount of plasma, are often transfused. This practice minimizes the risk of a reaction between the donated plasma and the blood cells of the recipient.

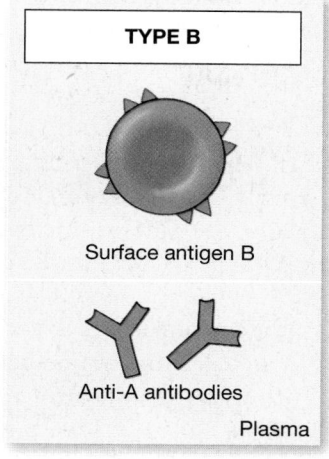

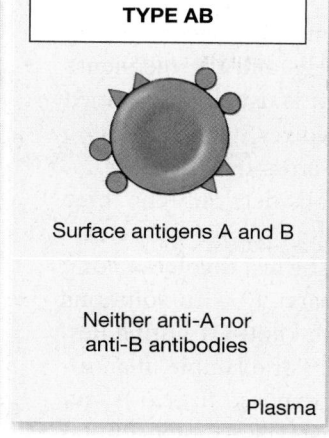

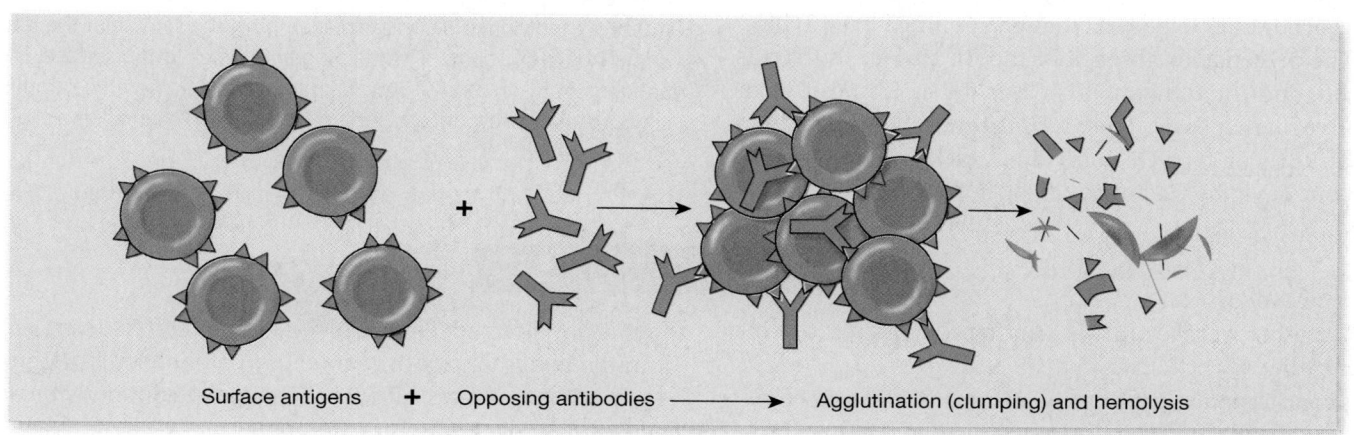

Figure 19–7 **Blood Types and Cross-Reactions.** Blood type depends on the presence of surface antigens (agglutinogens) on RBC surfaces. **(a)** The plasma contains antibodies (agglutinins) that will react with foreign surface antigens. The relative frequencies of each blood type in the U.S. population are listed in Table 19–2. **(b)** In a cross-reaction, antibodies that encounter their target antigens lead to agglutination and hemolysis of the affected RBCs.

TABLE 19–2   Differences in Blood Group Distribution

Population	Percentage with Each Blood Type				
	O	A	B	AB	Rh⁺
U.S. (AVERAGE)	46	40	10	4	85
African American	49	27	20	4	95
Caucasian	45	40	11	4	85
Chinese American	42	27	25	6	100
Filipino American	44	22	29	6	100
Hawaiian	46	46	5	3	100
Japanese American	31	39	21	10	100
Korean American	32	28	30	10	100
NATIVE NORTH AMERICAN	79	16	4	1	100
NATIVE SOUTH AMERICAN	100	0	0	0	100
AUSTRALIAN ABORIGINE	44	56	0	0	100

Testing For Transfusion Compatibility

Extra care must be taken to avoid potentially life-threatening cross-reactions between the donor's cells and the recipient's plasma. As a result, a compatibility test is usually performed in advance. This process normally involves two steps: (1) a determination of blood type and (2) a cross-match test.

The standard test for blood type considers only the three surface antigens most likely to produce dangerous cross-reactions: A, B, and Rh (**Figure 19–8**). The test involves taking drops of blood and mixing them separately with solutions containing anti-A, anti-B, and anti-Rh (anti-D) antibodies. Any cross-reactions are then recorded. For example, if an individual's RBCs clump together when exposed to anti-A and to anti-B antibodies, the individual has Type AB blood. If no reactions occur after exposure, that person must have Type O blood. The presence or absence of the Rh surface antigen is also noted, and the individual is classified as Rh positive or Rh negative on that basis. Type O$^+$ is the most common blood type. The RBCs of Type O$^+$ individuals lack surface antigens A and B but have the Rh antigen.

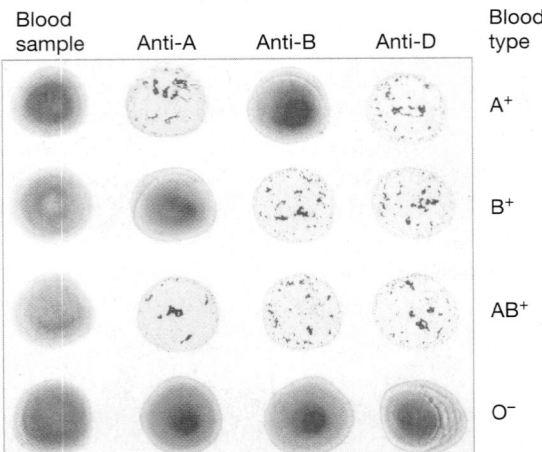

Figure 19–8 **Blood Type Testing.** Test results for blood samples from four individuals. Drops are taken from the sample at the left and mixed with solutions containing antibodies to the surface antigens A, B, AB, and D (Rh). Clumping occurs when the sample contains the corresponding surface antigen(s). The individuals' blood types are shown at right.

CLINICAL NOTE

Hemolytic Disease of the Newborn

Genes controlling the presence or absence of any surface antigen in the plasma membrane of a red blood cell are provided by both parents, so a child can have a blood type different from that of either parent. During pregnancy, when fetal and maternal circulatory systems are closely intertwined, the mother's antibodies may cross the placenta, attacking and destroying fetal RBCs. The resulting condition is called **hemolytic disease of the newborn (HDN)**.

This disease has many forms, some so mild as to remain undetected. Those involving the Rh surface antigen are quite dangerous, because unlike anti-A and anti-B antibodies, anti-Rh antibodies are able to cross the placenta and enter the fetal bloodstream. An Rh-positive mother (who lacks anti-Rh antibodies) can carry an Rh-negative fetus without difficulty. Because maternal antibodies can cross the placenta, problems may appear when an Rh-negative woman carries an Rh-positive fetus. Exposure to fetal red blood cell antigens generally occurs at delivery, when bleeding takes place at the placenta and uterus. Such mixing of fetal and maternal blood can stimulate the mother's immune system to produce anti-Rh antibodies, leading to sensitization. Roughly 20 percent of Rh-negative mothers who carried Rh-positive children become sensitized within 6 months of delivery.

Because the anti-Rh antibodies are not produced in significant amounts until after delivery, a woman's first in-

fant is not affected (**Figure 19–9**). (Some fetal RBCs cross into the maternal bloodstream during pregnancy, but generally not in numbers sufficient to stimulate antibody production.) But if a subsequent pregnancy involves an Rh-positive fetus, maternal anti-Rh antibodies produced after the first delivery cross the placenta and enter the fetal bloodstream. These anti-bodies destroy fetal RBCs and produce a dangerous anemia. The fetal demand for blood cells increases, and they begin leaving the bone marrow and entering the bloodstream before completing their development. Because these immature RBCs are erythroblasts, HDN is also known as **erythroblastosis fetalis** (e-rith-rō-blas-TŌ-sis fē-TAL-is).

Without treatment, the fetus will probably die before delivery or shortly thereafter. A newborn with severe HDN is anemic, and the high concentration of circulating bilirubin produces jaundice. Because the maternal antibodies will remain active for one to two months after delivery, the entire blood volume of the infant may have to be replaced. Replacing the blood removes most of the maternal anti-Rh antibodies, as well as the affected RBCs, reducing complications and the chance of the infant dying.

If the fetus is in danger of not surviving to full term, delivery may be induced after seven to eight months of development. In severe cases affecting a fetus at an earlier stage, one or more transfusions can be given while the fetus continues to develop in the uterus.

The maternal production of anti-Rh antibodies can be prevented by administering such antibodies (available under the name *RhoGam*) to the mother in the last three months of pregnancy and during and after delivery. (Anti-Rh antibodies are also given after a miscarriage or an abortion.) These antibodies will destroy any fetal RBCs that cross the placenta before they can stimulate an immune response in the mother. Because sensitization does not occur, no anti-Rh antibodies are produced. This relatively simple procedure has almost entirely prevented HDN mortality caused by Rh incompatibilities.

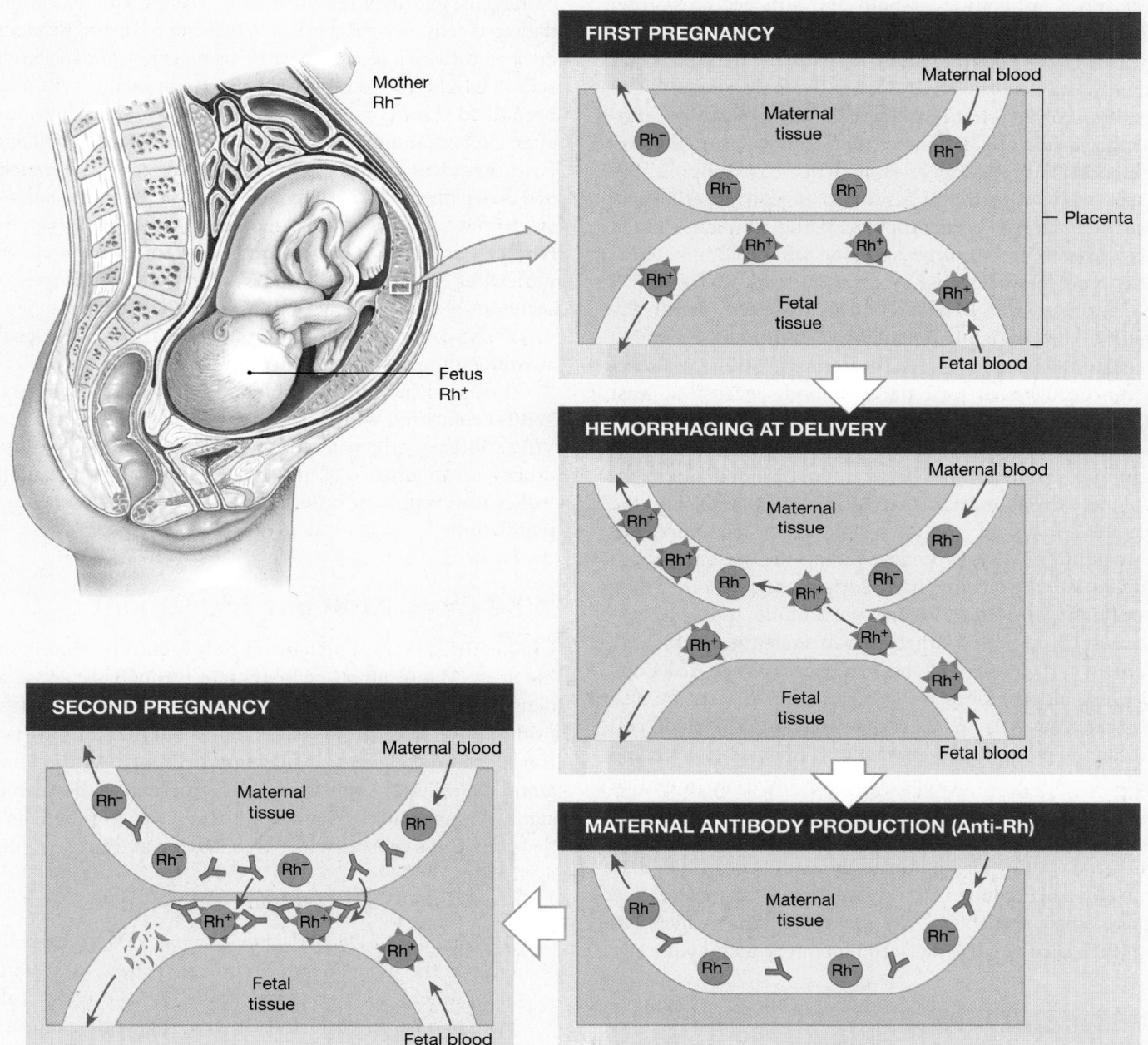

Figure 19–9   Rh Factors and Pregnancy. When an Rh-negative woman delivers her first Rh-positive child, fetal and maternal blood mix when the placenta breaks down. The presence of Rh-positive blood cells in the maternal bloodstream sensitizes the mother, stimulating the production of anti-Rh antibodies. If another pregnancy involves an Rh-positive fetus, maternal anti-Rh antibodies can cross the placenta and attack fetal blood cells, producing hemolytic disease of the newborn.

Standard blood-typing of both donor and recipient can be completed in a matter of minutes. However, in an emergency, there may not be time for preliminary testing. For example, a person with a severe gunshot wound may require 5 *liters* or more of blood before the damage can be repaired. Under these circumstances, Type O blood (preferably O⁻) will be administered. Because the donated RBCs lack both A and B surface antigens, the recipient's blood can have anti-A antibodies, anti-B antibodies, or both and still not cross-react with the donor's blood. Because cross-reactions with Type O blood are very unlikely, Type O individuals are sometimes called *universal donors*. Type AB individuals were once called *universal recipients*, because they lack anti-A or anti-B antibodies that would attack donated RBCs, and so can safely receive blood of any type. However, now that blood supplies are adequate and compatibility testing is regularly performed, the term has largely been dropped. If the recipient's blood type is known to be AB, Type AB blood will be administered.

It is now possible to use enzymes to strip off the A or B surface antigens from RBCs and create Type O blood in the laboratory. The procedure is expensive and time-consuming and has limited use in emergency treatment. Still, cross-reactions can occur, even to Type O⁻ blood, because at least 48 other surface antigens are present. As a result, whenever time and facilities permit, further testing is performed to ensure complete compatibility between donor blood and recipient blood. **Cross-match testing** involves exposing the donor's RBCs to a sample of the recipient's plasma under controlled conditions. This procedure reveals the presence of significant cross-reactions involving surface antigens other than A, B, or Rh. Another way to avoid compatibility problems is to replace lost blood with synthetic blood substitutes, which do not contain surface antigens that can trigger a cross-reaction.

Because blood groups are inherited, blood tests are also used as paternity tests and in crime detection. The blood collected cannot prove that a particular individual *is* a certain child's father or *is* guilty of a specific crime, but it can prove that the individual is *not* involved. It is impossible, for example, for an adult with Type AB blood to be the parent of an infant with Type O blood. Testing for additional surface antigens, other than the standard ABO groups, can increase the accuracy of the conclusions. When available, DNA identity testing, which has nearly 100% accuracy, has replaced blood type identity testing.

CHECKPOINT

12. What is the function of surface antigens on RBCs?

13. Which blood type(s) can be safely transfused into a person with Type O blood?

14. Why can't a person with Type A blood safely receive blood from a person with Type B blood?

See the blue Answers tab at the end of the book.

19-5 The various types of white blood cells contribute to the body's defenses

Unlike red blood cells, white blood cells (WBCs) have nuclei and other organelles, but they lack hemoglobin. White blood cells, or leukocytes, help defend the body against invasion by pathogens, and they remove toxins, wastes, and abnormal or damaged cells. Several types of WBCs can be distinguished microscopically in a blood smear by using either of two standard stains: *Wright stain* or *Giemsa stain*. Traditionally, WBCs have been divided into two groups on the basis of their appearance after such staining: (1) *granular leukocytes*, or *granulocytes* (with abundant stained granules)—the *neutrophils, eosinophils,* and *basophils;* and (2) *agranular leukocytes*, or *agranulocytes* (with few, if any, stained granules)—the *monocytes* and *lymphocytes*. This categorization is convenient but somewhat misleading, because the granules in "granular leukocytes" are secretory vesicles and lysosomes, and the "agranular leukocytes" also contain vesicles and lysosomes; they are just smaller and difficult to see with the light microscope.

A typical microliter of blood contains 5000 to 10,000 WBCs, compared with 4.2 to 6.3 million RBCs. Most of the WBCs in the body at any moment are in connective tissue proper or in organs of the lymphoid system. Circulating WBCs thus represent only a small fraction of the total WBC population.

WBC Circulation and Movement

Unlike RBCs, WBCs circulate for only a short portion of their life span. White blood cells migrate through the loose and dense connective tissues of the body, using the bloodstream primarily to travel from one organ to another and for rapid transportation to areas of infection or injury. As they travel along the miles of capillaries, WBCs can detect the chemical signs of damage to surrounding tissues. When problems are detected, these cells leave the bloodstream and enter the damaged area.

Circulating WBCs have four characteristics:

1. *All Can Migrate Out of the Bloodstream.* When white blood cells in the bloodstream become activated, they contact and adhere to the vessel walls in a process called *margination.* After further interaction with endothelial cells, the activated WBCs squeeze between adjacent endothelial cells and enter the surrounding tissue. This process is called *emigration,* or *diapedesis* (*dia,* through + *pedesis,* a leaping).

2. *All Are Capable of Amoeboid Movement.* Amoeboid movement is a gliding motion accomplished by the flow of cytoplasm into slender cellular processes extended in front

of the cell. (The movement is so named because it is similar to that of an *amoeba*, a type of protozoan.) The mechanism is not fully understood, but it involves the continuous rearrangement of bonds between actin filaments in the cytoskeleton, and it requires calcium ions and ATP. This mobility allows WBCs to move through the endothelial lining and into peripheral tissues.

3. *All Are Attracted to Specific Chemical Stimuli.* This characteristic, called **positive chemotaxis** (kē-mō-TAK-sis), guides WBCs to invading pathogens, damaged tissues, and other active WBCs.

4. *Neutrophils, Eosinophils, and Monocytes Are Capable of Phagocytosis.* These cells may engulf pathogens, cell debris, or other materials. Neutrophils and eosinophils are sometimes called *microphages*, to distinguish them from the larger macrophages in connective tissues. Macrophages are monocytes that have moved out of the bloodstream and have become actively phagocytic. ∞ p. 125

Types of WBCs

Neutrophils, eosinophils, basophils, and monocytes contribute to the body's *nonspecific defenses.* Such defenses are activated by a variety of stimuli, but they do not discriminate between one type of threat and another. Lymphocytes, in contrast, are responsible for *specific defenses:* the mounting of a counterattack against specific types of invading pathogens or foreign proteins. We will discuss the interactions among WBCs and the relationships between specific and nonspecific defenses in Chapter 22.

Neutrophils

Fifty to 70 percent of the circulating WBCs are **neutrophils** (NOO-trō-filz). This name reflects the fact that the granules of these WBCs are chemically neutral and thus are difficult to stain with either acidic or basic dyes. A mature neutrophil has a very dense, segmented nucleus with two to five lobes resembling beads on a string (**Figure 19–10a**). This structure has given neutrophils another name: **polymorphonuclear** (pol-ē-mor-fō-NOO-klē-ar) **leukocytes** (*poly,* many + *morphe,* form), or *PMNs.* "Polymorphs," or "polys," as they are often called, are roughly 12 μm in diameter. Their cytoplasm is packed with pale granules containing lysosomal enzymes and bactericidal (bacteria-killing) compounds.

Neutrophils are highly mobile, and consequently are generally the first of the WBCs to arrive at the site of an injury. These very active cells specialize in attacking and digesting bacteria that have been "marked" with antibodies or with *complement proteins*—plasma proteins involved in tissue defenses. (We will discuss the complement system in Chapter 22.)

Upon encountering a bacterium, a neutrophil quickly engulfs it, and the metabolic rate of the neutrophil increases dramatically. This *respiratory burst* accompanies the production of

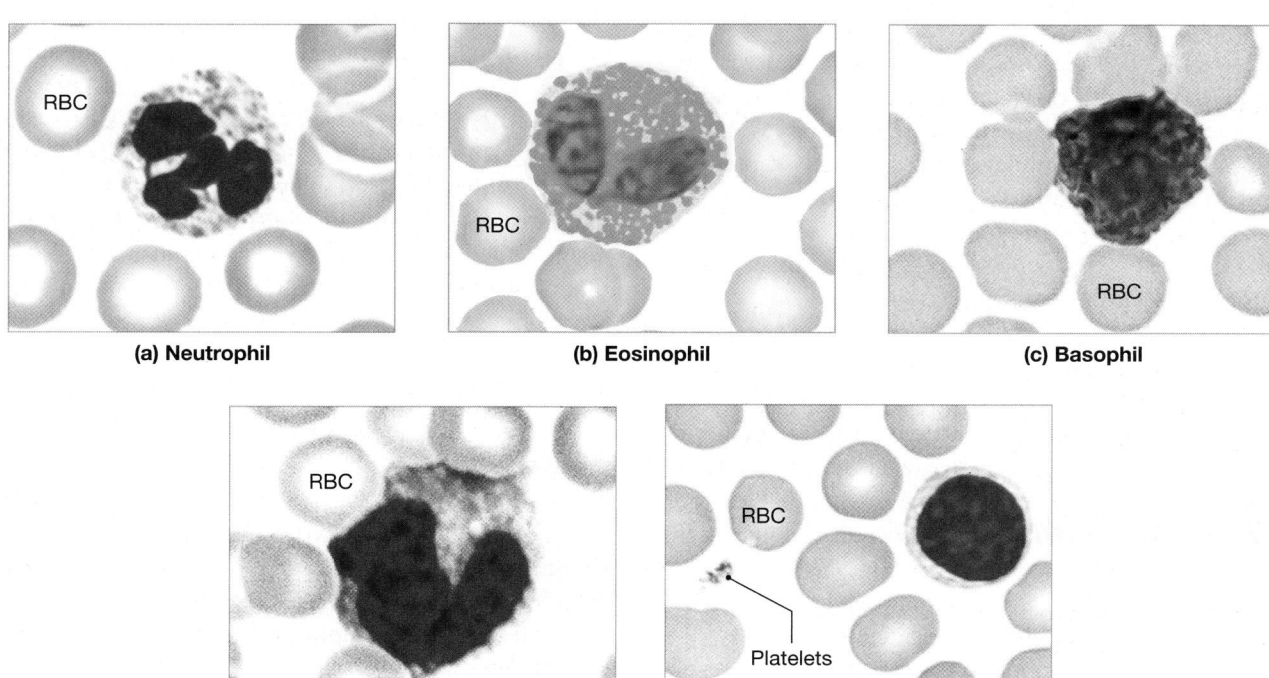

(a) **Neutrophil** (b) **Eosinophil** (c) **Basophil**

(d) **Monocyte** (e) **Lymphocyte**

Figure 19–10 White Blood Cells. (LM×1500)

highly reactive, destructive chemical agents, including *hydrogen peroxide* (H_2O_2) and *superoxide anions* (O_2^-), which can kill bacteria.

Meanwhile, the vesicle containing the engulfed pathogen fuses with lysosomes that contain digestive enzymes and small peptides called **defensins**. This process, which reduces the number of granules in the cytoplasm, is called **degranulation**. Defensins kill a variety of pathogens, including bacteria, fungi, and enveloped viruses, by combining to form large channels in their plasma membranes. The digestive enzymes then break down the bacterial remains. While actively engaged in attacking bacteria, a neutrophil releases prostaglandins and leukotrienes. ∞ p. 607 The prostaglandins increase capillary permeability in the affected region, thereby contributing to local inflammation and restricting the spread of injury and infection. Leukotrienes are hormones that attract other phagocytes and help coordinate the immune response.

Most neutrophils have a short life span, surviving in the bloodstream for only about 10 hours. When actively engulfing debris or pathogens, they may last 30 minutes or less. A neutrophil dies after engulfing one to two dozen bacteria, but its breakdown releases chemicals that attract other neutrophils to the site. A mixture of dead neutrophils, cellular debris, and other waste products form the *pus* associated with infected wounds.

Tips&Tricks

Remember that **neu**trophils are the most **nu**merous of the white blood cells.

Eosinophils

Eosinophils (ē-ō-SIN-ō-filz) were so named because their granules stain darkly with *eosin*, a red dye. The granules also stain with other acid dyes, so the name **acidophils** (a-SID-ō-filz) applies as well. Eosinophils, which generally represent 2–4 percent of the circulating WBCs, are similar in size to neutrophils. However, the combination of deep red granules and a bilobed (two-lobed) nucleus makes eosinophils easy to identify (**Figure 19–10b**).

Eosinophils attack objects that are coated with antibodies. Although they will engulf antibody-marked bacteria, protozoa, or cellular debris, their primary mode of attack is the exocytosis of toxic compounds, including nitric oxide and cytotoxic enzymes. This is particularly effective against multicellular parasites, such as flukes or parasitic roundworms, that are too big to engulf. The number of circulating eosinophils increases dramatically during a parasitic infection.

Because they are sensitive to circulating *allergens* (materials that trigger allergies), eosinophils increase in number during allergic reactions as well. Eosinophils are also attracted to sites of injury, where they release enzymes that reduce the degree of inflammation produced by mast cells and neutrophils, thereby controlling the spread of inflammation to adjacent tissues.

Basophils

Basophils (BĀ-sō-filz; *baso-*, base+ *phileo*, to love) have numerous granules that stain darkly with basic dyes. In a standard blood smear, the inclusions are deep purple or blue (**Figure 19–10c**). Measuring 8–10 μm in diameter, basophils are smaller than neutrophils or eosinophils. They are also relatively rare, accounting for less than 1 percent of the circulating WBC population.

Basophils migrate to injury sites and cross the capillary endothelium to accumulate in the damaged tissues, where they discharge their granules into the interstitial fluids. The granules contain *histamine*, which dilates blood vessels, and *heparin*, a compound that prevents blood clotting. Stimulated basophils release these chemicals into the interstitial fluids, and their arrival enhances the local inflammation initiated by mast cells. ∞ p. 142 Although the same compounds are released by mast cells in damaged connective tissues, mast cells and basophils are distinct populations with separate origins. Other chemicals released by stimulated basophils attract eosinophils and other basophils to the area.

Monocytes

Monocytes (MON-ō-sīts) in blood are spherical cells that may exceed 15 μm in diameter, nearly twice the diameter of a typical red blood cell. When flattened in a blood smear, they look even larger, so monocytes are relatively easy to identify. The nucleus is large and tends to be oval or kidney bean–shaped rather than lobed (**Figure 19–10d**). Monocytes normally account for 2–8 percent of circulating WBCs.

An individual monocyte uses the bloodstream for transportation, remaining in circulation for only about 24 hours before entering peripheral tissues to become a tissue macrophage. Macrophages are aggressive phagocytes, often attempting to engulf items as large as or larger than themselves. While phagocytically active, they release chemicals that attract and stimulate neutrophils, monocytes, and other phagocytic cells. Active macrophages also secrete substances that draw fibrocytes into the region. The fibrocytes then begin producing scar tissue, which will wall off the injured area.

Tips&Tricks

A **mon**ocyte is the **mon**ster cell that engulfs debris and pathogens.

Lymphocytes

Typical **lymphocytes** (LIM-fō-sīts) are slightly larger than RBCs and lack abundant, deeply stained granules. In blood smears, lymphocytes typically have a relatively large, round nucleus surrounded by a thin halo of cytoplasm (**Figure 19–10e**).

Lymphocytes account for 20–30 percent of the circulating WBC population. Lymphocytes continuously migrate from the bloodstream, through peripheral tissues, and back to the bloodstream. Circulating lymphocytes represent only a minute fraction of all lymphocytes, for at any moment most of your body's lymphocytes are in other connective tissues and in organs of the lymphoid system.

The circulating blood contains three functional classes of lymphocytes, which cannot be distinguished with a light microscope:

1. **T cells** are responsible for *cell-mediated immunity*, a specific defense mechanism against invading foreign cells and tissues, and for the coordination of the immune response. T cells either enter peripheral tissues and attack foreign cells directly or control the activities of other lymphocytes.

2. **B cells** are responsible for *humoral immunity*, a specific defense mechanism that involves the production of antibodies, which after distribution by blood, lymph, and interstitial fluid attack foreign antigens throughout the body. Activated B cells differentiate into **plasma cells**, which are specialized to synthesize and secrete antibodies. Whereas the T cells responsible for cellular immunity must migrate to their targets, the antibodies produced by plasma cells in one location can destroy antigens almost anywhere in the body.

3. **Natural killer (NK) cells** are responsible for *immune surveillance*—the detection and subsequent destruction of abnormal tissue cells. NK cells, sometimes known as *large granular lymphocytes*, are important in preventing cancer.

Tips&Tricks

To remember the various white blood cell populations, think "Never let monkeys eat bananas" for neutrophils, lymphocytes, monocytes, eosinophils, and basophils.

The Differential Count and Changes in WBC Profiles

A variety of conditions, including pathogenic infection, inflammation, and allergic reactions, cause characteristic changes in circulating populations of WBCs. By examining a stained blood smear, we can obtain a **differential count** of the WBC population. The values reported indicate the number of each type of cell in a sample of 100 WBCs.

The normal range for each type of WBC is indicated in Table 19–3. The term **leukopenia** (loo-kō-PĒ-nē-uh; *penia*, poverty) indicates inadequate numbers of WBCs. **Leukocytosis** (loo-kō-sī-TŌ-sis) refers to excessive numbers of WBCs. A modest leukocytosis is normal during an infection. Extreme leukocytosis (100,000/μl or more) generally indicates the presence of some form of **leukemia** (loo-KĒ-mē-uh). The endings *-penia* and *-osis* can also indicate low or

high numbers of specific types of WBCs. For example, *lymphopenia* means too few lymphocytes, and *lymphocytosis* means too many.

The A&P Top 100

#68 Red blood cells (RBCs) outnumber white blood cells (WBCs) by a ratio of 1000:1. WBCs are responsible for defending the body against infection, foreign cells, or toxins, and for assisting in the cleanup and repair of damaged tissues. The most numerous are neutrophils, which engulf bacteria, and lymphocytes, which are responsible for the specific defenses of the immune response.

WBC Production

Stem cells responsible for the production of WBCs originate in the bone marrow, with the divisions of hemocytoblasts (**Figure 19–11**). As previously noted, hemocytoblast divisions produce myeloid stem cells and lymphoid stem cells. Myeloid stem cell division creates **progenitor cells**, which give rise to all the formed elements except lymphocytes. One type of progenitor cell produces daughter cells that mature into RBCs; a second type produces cells that manufacture platelets. Neutrophils, eosinophils, basophils, and monocytes develop from daughter cells produced by a third type of progenitor cell.

Granulocytes (basophils, eosinophils, and neutrophils) complete their development in the bone marrow. These WBCs go through a characteristic series of maturational stages, proceeding from *blast cells* to *myelocytes* to *band cells* before becoming mature WBCs. For example, a cell differentiating into a neutrophil goes from a myeloblast to a *neutrophilic myelocyte* and then becomes a *neutrophilic band cell*. Some band cells enter the bloodstream before completing their maturation; normally, 3–5 percent of all circulating WBCs are band cells.

Monocytes begin their differentiation in the bone marrow, enter the bloodstream, and complete development when they become free macrophages in peripheral tissues. Some lymphocytes are derived from lymphoid stem cells that remain in bone marrow; these lymphocytes differentiate into either B cells or natural killer cells. Many of the lymphoid stem cells responsible for the production of lymphocytes migrate from the bone marrow to peripheral **lymphoid tissues**, including the thymus, spleen, and lymph nodes. As a result, lymphocytes are produced in these organs as well as in the bone marrow. Lymphoid stem cells migrating to the thymus mature into T cells. The process of lymphocyte production is called **lymphopoiesis**. Lymphocytes will be further discussed in Chapter 22.

Regulation of WBC Production

Factors that regulate lymphocyte maturation remain incompletely understood. Until adulthood, hormones produced

SUMMARY TABLE 19–3 Formed Elements of the Blood

Cell	Abundance (average number per μL)	Appearance in a Stained Blood Smear	Functions	Remarks
RED BLOOD CELLS	5.2 million (range: 4.4–6.0 million)	Flattened, circular cell; no nucleus, mitochondria, or ribosomes; red	Transport oxygen from lungs to tissues and carbon dioxide from tissues to lungs	Remain in bloodstream; 120-day life expectancy; amino acids and iron recycled; produced in bone marrow
WHITE BLOOD CELLS	7000 (range: 5000–10,000)			
Neutrophils	4150 (range: 1800–7300) Differential count: 50–70%	Round cell; nucleus lobed and may resemble a string of beads; cytoplasm contains large, pale inclusions	Phagocytic: Engulf pathogens or debris in tissues, release cytotoxic enzymes and chemicals	Move into tissues after several hours; may survive minutes to days, depending on tissue activity; produced in bone marrow
Eosinophils	165 (range: 0–700) Differential count: 2–4%	Round cell; nucleus generally in two lobes; cytoplasm contains large granules that generally stain bright red	Phagocytic: Engulf antibody-labeled materials, release cytotoxic enzymes, reduce inflammation; increase in allergic and parasitic situations	Move into tissues after several hours; survive minutes to days, depending on tissue activity; produced in bone marrow
Basophils	44 (range: 0–150) Differential count: <1%	Round cell; nucleus generally cannot be seen through dense, blue-stained granules in cytoplasm	Enter damaged tissues and release histamine and other chemicals that promote inflammation	Survival time unknown; assist mast cells of tissues in producing inflammation; produced in bone marrow
Monocytes	456 (range: 200–950) Differential count: 2–8%	Very large cell; kidney bean–shaped nucleus; abundant pale cytoplasm	Enter tissues to become macrophages; engulf pathogens or debris	Move into tissues after 1–2 days; survive for months or longer; produced primarily in bone marrow
Lymphocytes	2185 (range: 1500–4000) Differential count: 20–30%	Generally round cell, slightly larger than RBC; round nucleus; very little cytoplasm	Cells of lymphoid system, providing defense against specific pathogens or toxins	Survive for months to decades; circulate from blood to tissues and back; produced in bone marrow and lymphoid tissues
PLATELETS	350,000 (range: 150,000–500,000)	Round to spindle-shaped cytoplasmic fragment; contain enzymes, proenzymes, actin, and myosin; no nucleus	Hemostasis: Clump together and stick to vessel wall (platelet phase); activate intrinsic pathway of coagulation phase	Remain in bloodstream or in vascular organs; remain intact for 7–12 days; produced by megakaryocytes in bone marrow

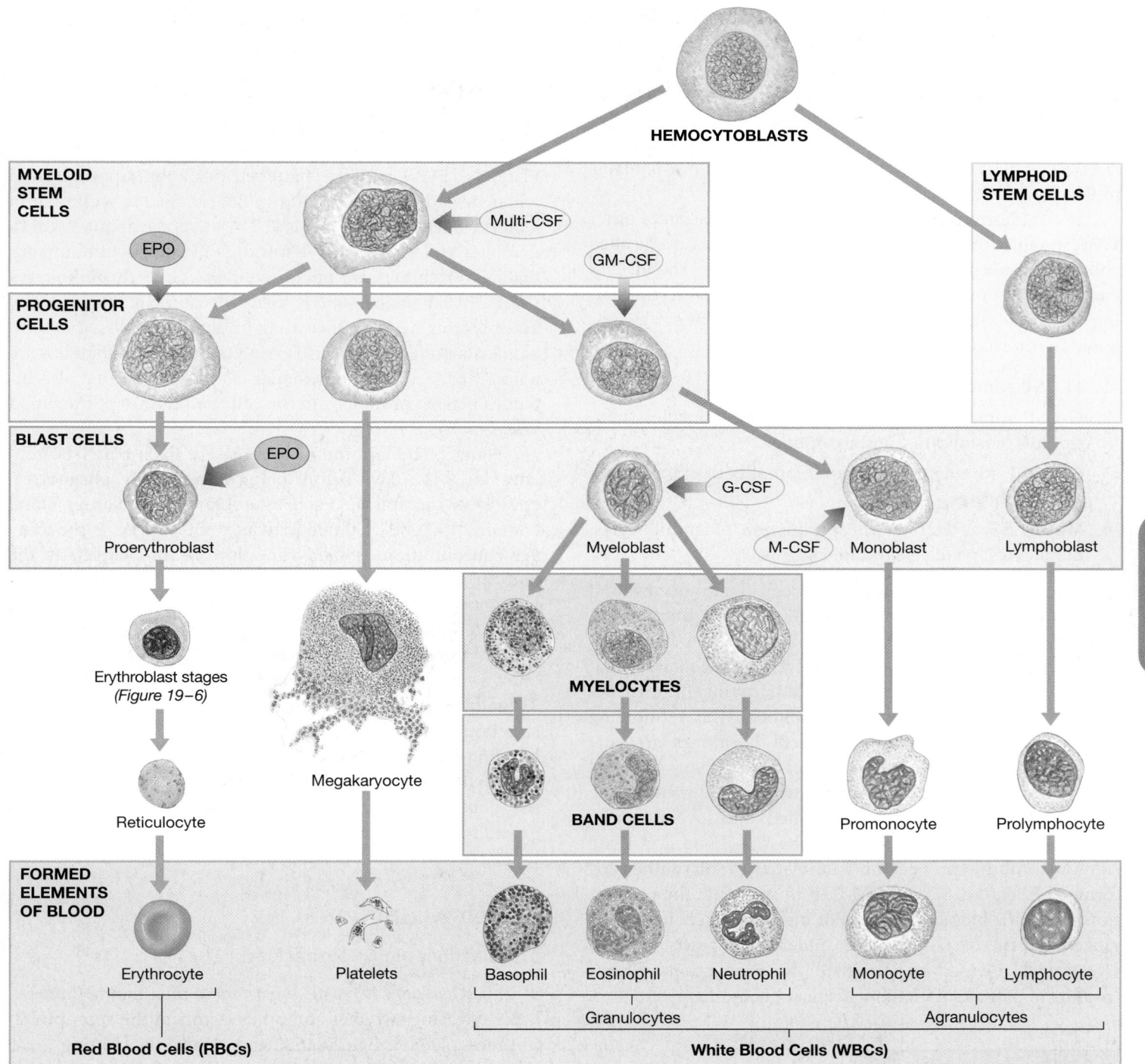

Figure 19–11 **The Origins and Differentiation of Formed Elements.** Hemocytoblast divisions give rise to myeloid stem cells or lymphoid stem cells. Lymphoid stem cells produce the various lymphocytes. Myeloid stem cells produce progenitor cells that divide to produce the other classes of formed elements. The targets of EPO and the four colony-stimulating factors (CSFs) are indicated.

by the thymus promote the differentiation and maintenance of T cell populations. The importance of the thymus in adults, especially with respect to aging, remains controversial. In adults, the production of B and T lymphocytes is regulated primarily by exposure to antigens (foreign proteins, cells, or toxins). When antigens appear, lymphocyte production escalates. We will describe the control mechanisms in Chapter 22.

Several hormones are involved in the regulation of other WBC populations. The targets of these hormones, called **colony-stimulating factors (CSFs)**, are shown in Figure 19–11. Four CSFs have been identified, each stimulating the formation of WBCs or both WBCs and RBCs. The designation for each factor indicates its target:

1. **M-CSF** stimulates the production of monocytes.

2. **G-CSF** stimulates the production of granulocytes (neutrophils, eosinophils, and basophils).

3. **GM-CSF** stimulates the production of both granulocytes and monocytes.

4. **Multi-CSF** accelerates the production of granulocytes, monocytes, platelets, and RBCs.

Chemical communication between lymphocytes and other WBCs assists in the coordination of the immune response. For example, active macrophages release chemicals that make lymphocytes more sensitive to antigens and that accelerate the development of specific immunity. In turn, active lymphocytes release multi-CSF and GM-CSF, reinforcing nonspecific defenses. Immune system hormones are currently being studied intensively because of their potential clinical importance. The molecular structures of many of the stimulating factors have been identified, and several can be produced by genetic engineering. The U.S. Food and Drug Administration approved the administration of synthesized forms of EPO, G-CSF, and GM-CSF to stimulate the production of specific blood cell lines. For instance, a genetically engineered form of G-CSF, sold under the name *filgrastim* (*Neupogen*), is used to stimulate the production of neutrophils in patients undergoing cancer chemotherapy.

CHECKPOINT

15. Identify the five types of white blood cells.

16. Which type of white blood cell would you find in the greatest numbers in an infected cut?

17. Which type of cell would you find in elevated numbers in a person who is producing large amounts of circulating antibodies to combat a virus?

18. How do basophils respond during inflammation?

See the blue Answers tab at the end of the book.

19-6 Platelets, disc-shaped structures formed from megakaryocytes, function in the clotting process

Platelets (PLĀT-lets) are flattened discs that appear round when viewed from above, and spindle shaped in section or in a blood smear (**Figure 19–10e**). They average about 4 μm in diameter and are roughly 1 μm thick. Platelets in nonmammalian vertebrates are nucleated cells called **thrombocytes** (THROM-bō-sīts; *thrombos*, clot). Because in humans they are cell fragments rather than individual cells, the term *platelet* is preferred when referring to our blood. Platelets are a major participant in a vascular *clotting system* that also includes plasma proteins and the cells and tissues of the blood vessels.

Platelets are continuously replaced. Each platelet circulates for 9–12 days before being removed by phagocytes, mainly in the spleen. Each microliter of circulating blood contains 150,000–500,000 platelets; 350,000/μL is the average concentration. Roughly one-third of the platelets in the body at any moment are held in the spleen and other vascular organs, rather than in the bloodstream. These reserves are mobilized during a circulatory crisis, such as severe bleeding.

An abnormally low platelet count (80,000/μL or less) is known as **thrombocytopenia** (throm-bō-sī-tō-PĒ-nē-uh). Thrombocytopenia generally indicates excessive platelet destruction or inadequate platelet production. Signs include bleeding along the digestive tract, within the skin, and occasionally inside the CNS. In **thrombocytosis** (throm-bō-sī-TŌ-sis), platelet counts can exceed 1,000,000/μL. Thrombocytosis generally results from accelerated platelet formation in response to infection, inflammation, or cancer.

Platelet Functions

The functions of platelets include:

- *The Release of Chemicals Important to the Clotting Process.* By releasing enzymes and other factors at the appropriate times, platelets help initiate and control the clotting process.

- *The Formation of a Temporary Patch in the Walls of Damaged Blood Vessels.* Platelets clump together at an injury site, forming a *platelet plug*, which can slow the rate of blood loss while clotting occurs.

- *Active Contraction after Clot Formation Has Occurred.* Platelets contain filaments of actin and myosin. After a blood clot has formed, the contraction of platelet filaments shrinks the clot and reduces the size of the break in the vessel wall.

Platelet Production

Platelet production, or **thrombocytopoiesis**, occurs in the bone marrow. Normal bone marrow contains a number of **megakaryocytes** (meg-a-KAR-ē-ō-sīts; *mega-*, big + *karyon*, nucleus + *-cyte*, cell), enormous cells (up to 160 μm in diameter) with large nuclei (**Figure 19–11**). During their development and growth, megakaryocytes manufacture structural proteins, enzymes, and membranes. They then begin shedding cytoplasm in small membrane-enclosed packets. These packets are the platelets that enter the bloodstream. A mature megakaryocyte gradually loses all of its cytoplasm, producing about 4000 platelets before the nucleus is engulfed by phagocytes and broken down for recycling.

The rate of megakaryocyte activity and platelet formation is stimulated by (1) *thrombopoietin* (TPO), or *thrombocyte-stimulating factor*, a peptide hormone produced in the kidneys (and perhaps other sites) that accelerates platelet formation and stimulates the production of megakaryocytes; (2) *interleukin-6* (IL-6), a hormone that stimulates platelet formation; and (3) *multi-CSF*, which stimulates platelet production by promoting the formation and growth of megakaryocytes.

C H E C K P O I N T

19. Define thrombocytopoiesis.

20. Explain the difference between platelets and thrombocytes.

21. List the three primary functions of platelets.

See the blue Answers tab at the end of the book.

19-7 Hemostasis involves vascular spasm, platelet plug formation, and blood coagulation

The process of **hemostasis** (*haima*, blood + *stasis*, halt), the cessation of bleeding, halts the loss of blood through the walls of damaged vessels. At the same time, it establishes a framework for tissue repairs. Hemostasis consists of three phases: the *vascular phase*, the *platelet phase*, and the *coagulation phase*. However, the boundaries of these phases are somewhat arbitrary. In reality, hemostasis is a complex cascade in which many things happen at once, and all of them interact to some degree.

The Vascular Phase

Cutting the wall of a blood vessel triggers a contraction in the smooth muscle fibers of the vessel wall (**Figure 19–12**). This local contraction of the vessel is a **vascular spasm**, which decreases the diameter of the vessel at the site of injury. Such a constriction can slow or even stop the loss of blood through the wall of a small vessel. The vascular spasm lasts about 30 minutes, a period called the **vascular phase** of hemostasis.

During the vascular phase, changes occur in the endothelium of the vessel at the injury site:

- *The Endothelial Cells Contract and Expose the Underlying Basal Lamina to the Bloodstream.*

- *The Endothelial Cells Begin Releasing Chemical Factors and Local Hormones.* We will discuss several of these factors, including *ADP*, *tissue factor*, and *prostacyclin*, in later sections. Endothelial cells also release **endothelins**, peptide hormones that (1) stimulate smooth muscle

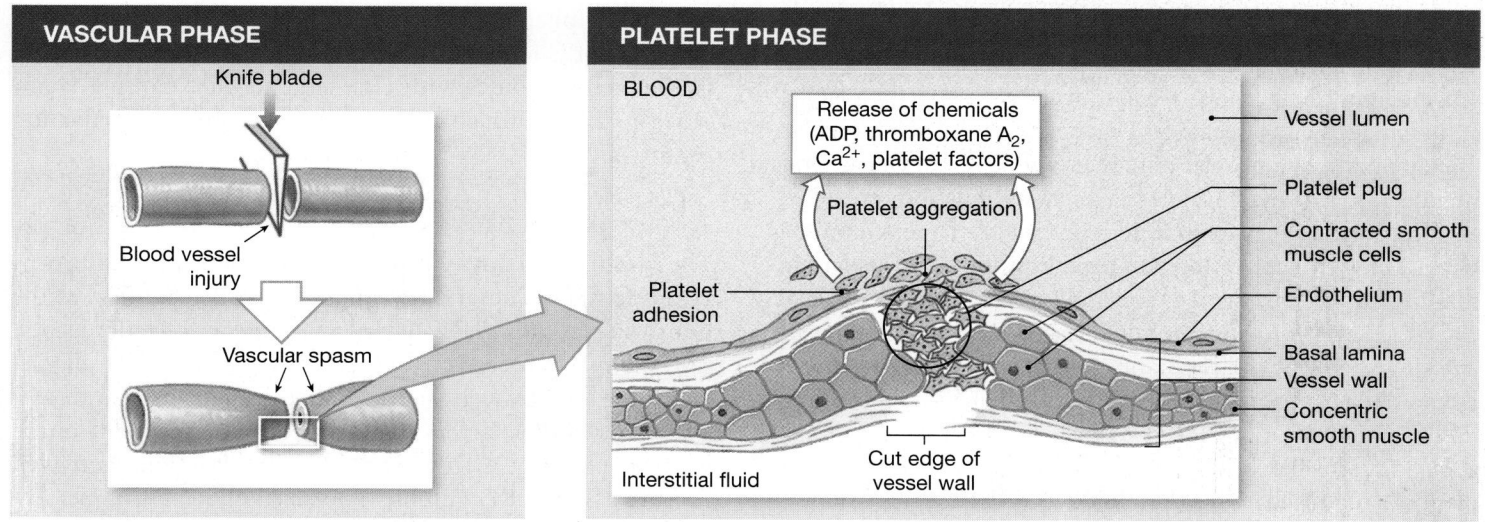

Figure 19–12 **The Vascular and Platelet Phases of Hemostasis.**

contraction and promote vascular spasms and (2) stimulate the division of endothelial cells, smooth muscle cells, and fibrocytes to accelerate the repair process.

- *The Endothelial Plasma Membranes Become "Sticky."* A tear in the wall of a small artery or vein may be partially sealed off by the attachment of endothelial cells on either side of the break. In small capillaries, endothelial cells on opposite sides of the vessel may stick together and prevent blood flow along the damaged vessel. The stickiness is also important because it facilitates the attachment of platelets as the platelet phase gets under way.

The Platelet Phase

The attachment of platelets to sticky endothelial surfaces, to the basal lamina, and to exposed collagen fibers marks the start of the **platelet phase** of hemostasis (**Figure 19–12**). The attachment of platelets to exposed surfaces is called **platelet adhesion**. As more and more platelets arrive, they begin sticking to one another as well. This process, called **platelet aggregation**, forms a **platelet plug** that may close the break in the vessel wall if the damage is not severe or the vessel is relatively small. Platelet aggregation begins within 15 seconds after an injury occurs.

As they arrive at the injury site, platelets become activated. The first sign of activation is that they become more spherical and develop cytoplasmic processes that extend toward adjacent platelets. At this time, the platelets begin releasing a wide variety of compounds, including (1) *adenosine diphosphate (ADP)*, which stimulates platelet aggregation and secretion; (2) *thromboxane A₂* and *serotonin*, which stimulate vascular spasms; (3) *clotting factors*, proteins that play a role in blood clotting; (4) *platelet-derived growth factor (PDGF)*, a peptide that promotes vessel repair; and (5) calcium ions, which are required for platelet aggregation and in several steps in the clotting process.

The platelet phase proceeds rapidly, because ADP, thromboxane, and calcium ions released from each arriving platelet stimulate further aggregation. This positive feedback loop ultimately produces a platelet plug that will be reinforced as clotting occurs. However, platelet aggregation must be controlled and restricted to the site of injury. Several key factors limit the growth of the platelet plug: (1) **prostacyclin**, a prostaglandin that inhibits platelet aggregation and is released by endothelial cells; (2) inhibitory compounds released by white blood cells entering the area; (3) circulating plasma enzymes that break down ADP near the plug; (4) compounds that, when abundant, inhibit plug formation (for example, serotonin, which at high concentrations blocks the action of ADP); and (5) the development of a blood clot, which reinforces the platelet plug, but isolates it from the general circulation.

The Coagulation Phase

The vascular and platelet phases begin within a few seconds after the injury. The **coagulation** (cō-ag-ū-LĀ-shun) **phase** does not start until 30 seconds or more after the vessel has been damaged. **Coagulation**, or *blood clotting*, involves a complex sequence of steps leading to the conversion of circulating fibrinogen into the insoluble protein fibrin. As the fibrin network grows, it covers the surface of the platelet plug. Passing blood cells and additional platelets are trapped in the fibrous tangle, forming a **blood clot** that effectively seals off the damaged portion of the vessel. **Figure 19–13** shows the formation and structure of a blood clot.

Clotting Factors

Normal blood clotting depends on the presence of **clotting factors**, or **procoagulants**, in the plasma. Important clotting factors include Ca^{2+} and 11 different proteins (Table 19–4).

Many of the proteins are **proenzymes**, which, when converted to active enzymes, direct essential reactions in the clotting response. The activation of one proenzyme commonly creates an enzyme that activates a second proenzyme, and so on in a chain reaction, or *cascade*. During the coagulation phase, enzymes and proenzymes interact.

Figure 19–13a depicts the cascades involved in the *extrinsic*, *intrinsic*, and *common pathways*. The extrinsic pathway begins outside the bloodstream, in the vessel wall; the intrinsic pathway begins inside the bloodstream, with the activation of a circulating proenzyme. These two pathways converge at the common pathway.

The Extrinsic Pathway

The **extrinsic pathway** begins with the release of **Factor III**, also known as **tissue factor (TF)**, by damaged endothelial cells or peripheral tissues. The greater the damage, the more tissue factor is released and the faster clotting occurs. Tissue factor then combines with Ca^{2+} and another clotting factor (Factor VII) to form an enzyme complex capable of activating Factor X, the first step in the common pathway.

The Intrinsic Pathway

The **intrinsic pathway** begins with the activation of proenzymes (usually Factor XII) exposed to collagen fibers at the injury site (or to a glass surface of a slide or collection tube). This pathway proceeds with the assistance of **PF-3**, a platelet factor released by aggregating platelets. Platelets also release a variety of other factors that accelerate the reactions of the intrinsic pathway. After a series of linked reactions, activated Factors VIII and IX combine to form an enzyme complex capable of activating Factor X.

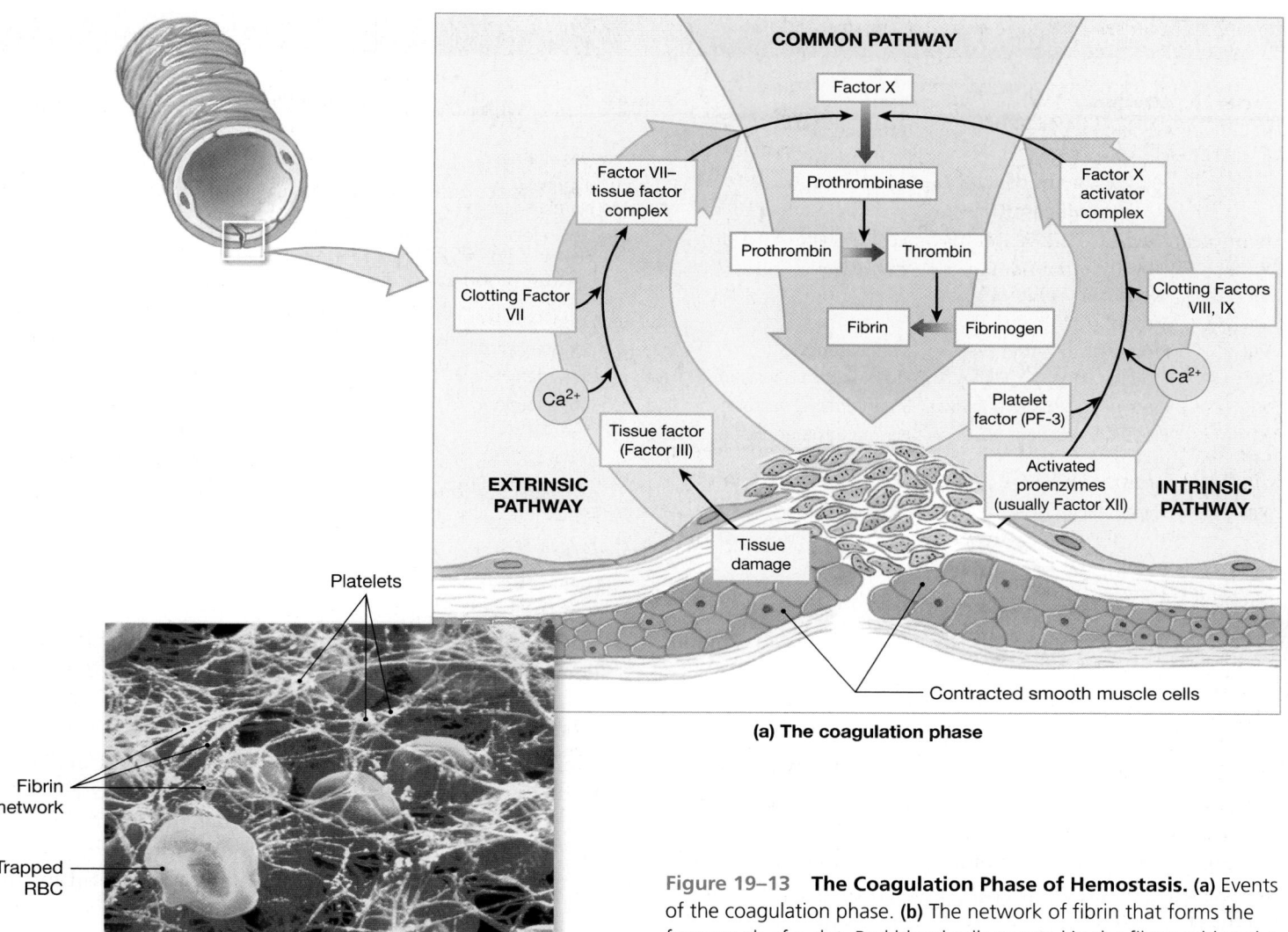

Platelets

Fibrin
network

Trapped
RBC

(b) A blood clot

(a) The coagulation phase

Figure 19–13 **The Coagulation Phase of Hemostasis. (a)** Events of the coagulation phase. **(b)** The network of fibrin that forms the framework of a clot. Red blood cells trapped in the fibers add to the mass of the blood clot and give it a red color.

The Common Pathway

The **common pathway** begins when enzymes from either the extrinsic or intrinsic pathway activate Factor X, forming the enzyme **prothrombinase**. Prothrombinase converts the proenzyme prothrombin into the enzyme **thrombin** (THROM-bin). Thrombin then completes the clotting process by converting fibrinogen, a soluble plasma protein, to insoluble strands of fibrin.

Interactions among the Pathways

When a blood vessel is damaged, both the extrinsic and the intrinsic pathways respond. The extrinsic pathway is shorter and faster than the intrinsic pathway, and it is usually the first to initiate clotting. In essence, the extrinsic pathway produces a small amount of thrombin very quickly. This quick patch is reinforced by the intrinsic pathway, which later produces more thrombin.

The time required to complete clot formation varies with the site and the nature of the injury. In tests of the clotting system, blood held in fine glass tubes normally clots in 8–18 minutes (the *coagulation time*), and a small puncture wound typically stops bleeding in 1–4 minutes (the *bleeding time*).

Feedback Control of Blood Clotting

Thrombin generated in the common pathway stimulates blood clotting by (1) stimulating the formation of tissue factor and (2) stimulating the release of PF-3 by platelets. Thus, the activity of the common pathway stimulates both the intrinsic and extrinsic pathways. This positive feedback loop

				Concentration	
Factor	Structure	Name	Source	in Plasma(μg/mL)	Pathway
I	Protein	Fibrinogen	Liver	2500–3500	Common
II	Protein	Prothrombin	Liver, requires vitamin K	100	Common
III	Lipoprotein	Tissue factor (TF)	Damaged tissue, activated platelets	0	Extrinsic
IV	Ion	Calcium ions	Bone, diet, platelets	100	Entire process
V	Protein	Proaccelerin	Liver, platelets	10	Extrinsic and intrinsic
VI	(No longer used)				
VII	Protein	Proconvertin	Liver, requires vitamin K	0.5	Extrinsic
VIII	Protein factor (AHF)	Antihemophilic	Platelets, endothelial cells	15	Intrinsic
IX	Protein factor	Plasma thromboplastin	Liver, requires vitamin K	3	Intrinsic
X	Protein	Stuart–Prower factor	Liver, requires vitamin K	10	Extrinsic and intrinsic
XI	Protein antecedent (PTA)	Plasma thromboplastin	Liver	<5	Intrinsic
XII	Protein	Hageman factor	Liver	<5	Intrinsic; also activates plasmin
XIII	Protein factor (FSF)	Fibrin-stabilizing	Liver, platelets	20	Stabilizes fibrin, slows fibrinolysis

accelerates the clotting process, and speed can be very important in reducing blood loss after a severe injury.

Blood clotting is restricted by substances that either deactivate or remove clotting factors and other stimulatory agents from the blood. Examples include the following:

- Normal plasma contains several **anticoagulants**—enzymes that inhibit clotting. One, **antithrombin-III**, inhibits several clotting factors, including thrombin.

- **Heparin**, a compound released by basophils and mast cells, is a cofactor that accelerates the activation of antithrombin-III. Heparin is used clinically to impede or prevent clotting.

- **Thrombomodulin** is released by endothelial cells. This protein binds to thrombin and converts it to an enzyme that activates protein C. **Protein C** is a plasma protein that inactivates several clotting factors and stimulates the formation of *plasmin*, an enzyme that gradually breaks down fibrin strands.

- Prostacyclin released during the platelet phase inhibits platelet aggregation and opposes the stimulatory action of thrombin, ADP, and other factors.

- Other plasma proteins with anticoagulant properties include *alpha-2-macroglobulin*, which inhibits thrombin, and C_1 *inactivator*, which inhibits several clotting factors involved in the intrinsic pathway.

The clotting process involves a complex chain of events, and disorders that affect any individual clotting factor can disrupt the entire process. As a result, managing many clinical conditions involves controlling or manipulating the clotting response.

Calcium Ions, Vitamin K, and Blood Clotting

Calcium ions and **vitamin K** affect almost every aspect of the clotting process. All three pathways (intrinsic, extrinsic, and common) require Ca^{2+}, so any disorder that lowers plasma Ca^{2+} concentrations will impair blood clotting.

Adequate amounts of vitamin K must be present for the liver to be able to synthesize four of the clotting factors, including prothrombin. Vitamin K is a fat-soluble vitamin, present in green vegetables, grain, and organ meats, that is absorbed with dietary lipids. Roughly half of the daily requirement is obtained from the diet, and the other half is manufactured by bacteria in the large intestine. A diet inadequate in fats or in vitamin K, or a disorder that affects fat digestion and absorption (such as problems with bile production), or prolonged use of antibiotics that kill normal intestinal bacteria may lead to a vitamin K deficiency. This condition will cause the eventual breakdown of the common pathway due to a lack of clotting factors and, ultimately, deactivation of the entire clotting system.

Clot Retraction

Once the fibrin meshwork has formed, platelets and red blood cells stick to the fibrin strands. The platelets then contract, and the entire clot begins to undergo **clot retraction**, or *syneresis* (si-NER-e-sis; "a drawing together"). Clot retraction, which occurs over a period of 30–60 minutes, (1) pulls the torn edges of the vessel closer together, reducing residual bleeding and stabilizing the injury site, and (2) reduces the size of the damaged area, making it easier for fibrocytes, smooth muscle cells, and endothelial cells to complete repairs.

Fibrinolysis

As the repairs proceed, the clot gradually dissolves. This process, called **fibrinolysis** (fī-bri-NOL-i-sis), begins with the activation of the proenzyme **plasminogen** by two enzymes: thrombin, produced by the common pathway, and **tissue plasminogen activator** (t-PA), released by damaged tissues at the site of injury. The activation of plasminogen produces the enzyme **plasmin** (PLAZ-min), which begins digesting the fibrin strands and eroding the foundation of the clot.

The A&P Top 100

#69 Platelets are involved in the coordination of hemostasis (blood clotting). When platelets are activated by abnormal changes in their local environment, they release clotting factors and other chemicals. Hemostasis is a complex cascade that establishes a fibrous patch that can subsequently be remodeled and then removed as the damaged area is repaired.

To perform its vital functions, blood must be kept in motion. On average, an RBC completes two circuits around the cardiovascular system each minute. The circulation of blood begins in the third week of embryonic development and continues throughout life. If the blood supply is cut off, dependent tissues may die in a matter of minutes. In Chapter 20, we will examine the structure and function of the heart—the pump that maintains this vital blood flow.

CHECKPOINT

22. A sample of bone marrow has unusually few megakaryocytes. What body process would you expect to be impaired as a result?

23. Vitamin K is fat soluble, and some dietary fat is required for its absorption. How could a diet of fruit juice and water have an effect on blood clotting?

24. Unless chemically treated, blood will coagulate in a test tube. This clotting process begins when Factor XII becomes activated. Which clotting pathway is involved in this process?

See the blue Answers tab at the end of the book.

Related Clinical Terms

anemia: A condition in which the oxygen-carrying capacity of blood is reduced, owing to low hematocrit or low blood hemoglobin concentrations.

embolism: A condition in which a drifting blood clot (an embolus) becomes stuck in a blood vessel, blocking circulation to the area downstream.

hematocrit: The value that indicates the percentage of whole blood occupied by formed elements.

hematuria: The presence of red blood cells in urine.

hemoglobinuria: The presence of hemoglobin in urine.

hemolytic disease of the newborn (HDN): A condition in which fetal red blood cells have been destroyed by maternal antibodies.

hemophilia: Inherited disorders characterized by the inadequate production of clotting factors.

hypervolemic: Having an excessive blood volume.

hypovolemic: Having a low blood volume.

hypoxia: Low tissue oxygen levels.

jaundice: A condition characterized by yellow skin and eyes, caused by abnormally high levels of plasma bilirubin; examples include *hemolytic jaundice* and *obstructive jaundice*.

leukemia: A condition characterized by extremely elevated levels of circulating white blood cells; includes both *myeloid* and *lymphoid* forms.

leukocytosis: Abnormally high numbers of white blood cells in the bloodstream.

leukopenia: Abnormally low numbers of white blood cells in the bloodstream.

normochromic: Referring to red blood cells that contain normal amounts of hemoglobin.

normocytic: Referring to red blood cells of normal size.

normovolemic: Referring to a normal blood volume.

plaque: An abnormal accumulation of large quantities of lipids within a blood vessel.

RBC tests: These tests include a *reticulocyte count, hematocrit, hemoglobin concentration, RBC count, mean corpuscular volume,* and *mean corpuscular hemoglobin concentration.*

sickle cell anemia: An anemia resulting from the production of an abnormal form of hemoglobin; causes red blood cells to become sickle shaped at low oxygen levels.

thalassemia: A disorder resulting from the production of an abnormal form of hemoglobin.

thrombus: A blood clot attached to the luminal surface of a blood vessel.

transfusion: A procedure in which blood components are given to someone to restore blood volume or to remedy a deficiency in blood composition.

venipuncture: The puncturing of a vein for any purpose, including the withdrawal of blood or the administration of medication.

Chapter Review
Study Outline

An Introduction to the Cardiovascular System p. 651

1. The **cardiovascular system** enables the rapid transport of nutrients, respiratory gases, waste products, and cells within the body.

19-1 Blood has several important functions and unique physical characteristics p. 651

2. **Blood** is a specialized fluid connective tissue. Its functions include (1) transporting dissolved gases, nutrients, hormones, and metabolic wastes; (2) regulating the pH and ion composition of interstitial fluids; (3) restricting fluid losses at injury sites; (4) defending the body against toxins and pathogens; and (5) regulating body temperature by absorbing and redistributing heat.

3. Blood contains **plasma** and **formed elements—red blood cells (RBCs)**, **white blood cells (WBCs)**, and **platelets**. The plasma and formed elements constitute **whole blood**, which can be **fractionated** for analytical or clinical purposes. (*Figure 19–1*)

4. **Hemopoiesis** is the process of blood cell formation. Circulating stem cells divide to form all types of blood cells.

5. Whole blood from any region of the body has roughly the same temperature, viscosity, and pH.

19-2 Plasma, the fluid portion of blood, contains significant quantities of plasma proteins p. 653

6. Plasma accounts for 46–63 percent of the volume of blood; roughly 92 percent of plasma is water. (*Figure 19–1*)

7. Plasma differs from interstitial fluid in terms of its oxygen and carbon dioxide levels and the concentrations and types of dissolved proteins.

8. The three major types of plasma proteins are *albumins*, *globulins*, and *fibrinogen*.

9. **Albumins** constitute about 60 percent of plasma proteins. **Globulins** constitute roughly 35 percent of plasma proteins; they include **antibodies (immunoglobulins)**, which attack foreign proteins and pathogens, and **transport globulins**, which bind ions, hormones, and other compounds. **Fibrinogen** molecules are converted to **fibrin** in the clotting process. The removal of fibrinogen from plasma leaves a fluid called **serum**.

10. The liver synthesizes and releases more than 90 percent of the plasma proteins.

19-3 Red blood cells, formed by erythropoiesis, contain hemoglobin that can be recycled p. 655

11. Red blood cells account for slightly less than half of the blood volume and 99.9 percent of the formed elements. The **hematocrit** value indicates the percentage of whole blood occupied by formed elements and is commonly reported as the *volume of packed red cells (VPRC)*. (*Figure 19–1; Table 19–1*)

12. Each RBC is a biconcave disc, providing a large surface-to-volume ratio. This shape allows RBCs to stack, bend, and flex. (*Figure 19–2*)

13. Red blood cells lack most organelles, including mitochondria and nuclei, retaining only the cytoskeleton. They typically degenerate after about 120 days in the bloodstream.

14. Molecules of **hemoglobin (Hb)** account for more than 95 percent of the proteins in RBCs. Hemoglobin is a globular protein formed from two pairs of polypeptide subunits. Each subunit contains a single molecule of **heme**, which also has an iron atom that can reversibly bind an oxygen molecule. Damaged or dead RBCs are recycled by phagocytes. (*Figures 19–3, 19–5*)

15. Damaged RBCs are continuously replaced at a rate of approximately 3 million new RBCs entering the bloodstream per second. They are replaced before they **hemolyze**.

16. The components of hemoglobin are individually recycled; the heme is stripped of its iron and converted to **biliverdin**, which is converted to **bilirubin**. If bile ducts are blocked, bilirubin builds up in skin and eyes, resulting in **jaundice**. (*Figure 19–5*)

17. Iron is also recycled by being stored in phagocytic cells or transported through the bloodstream, bound to **transferrin**.

18. **Erythropoiesis**, the formation of red blood cells, occurs only in *red bone marrow* (**myeloid tissue**). The process speeds up under stimulation by erythropoietin (EPO or **erythropoiesis-stimulating hormone**). Stages in RBC development include **erythroblasts** and **reticulocytes**. (*Figure 19–6*)

19-4 The ABO blood types and Rh system are based on antigen–antibody responses p. 662

19. **Blood type** is determined by the presence or absence of specific **surface antigens** (*agglutinogens*) in the RBC plasma membranes: antigens **A**, **B**, and **Rh (D)**. Antibodies (*agglutinins*) in the plasma will react with RBCs bearing different surface antigens. When an antibody meets its specific surface antigen, the resulting reaction is a **cross-reaction**. (*Figures 19–7 to 19–9; Table 19–2*)

19-5 The various types of white blood cells contribute to the body's defenses p. 666

20. White blood cells (**leukocytes**) have nuclei and other organelles. They defend the body against pathogens and remove toxins, wastes, and abnormal or damaged cells.

21. White blood cells are capable of *margination*, amoeboid movement, and **positive chemotaxis**. Some WBCs are also capable of *phagocytosis*.

22. *Granular leukocytes (granulocytes)* are subdivided into **neutrophils**, **eosinophils**, and **basophils**. Fifty to 70 percent of circulating WBCs are neutrophils, which are highly mobile phagocytes. The much less common eosinophils are phagocytes attracted to foreign compounds that have reacted with circulating antibodies. The relatively rare basophils migrate to damaged tissues and release *histamine* and *heparin*, aiding the inflammatory response. (*Figure 19–10*)

23. *Agranular leukocytes (agranulocytes)* include **monocytes** and **lymphocytes**. Monocytes that migrate into peripheral tissues become tissue macrophages. Lymphocytes, the primary cells of the lymphoid system, include **T cells** (which enter peripheral tissues and attack foreign cells directly, or affect the activities of other lymphocytes), **B cells** (which produce antibodies), and **natural killer (NK) cells** (which destroy abnormal tissue cells). (*Figure 19–10; Summary Table 19–3*)

24. A **differential count** of the WBC population can indicate a variety of disorders. **Leukemia** is indicated by extreme **leukocytosis**—that is, excessive numbers of WBCs. (*Summary Table 19–3*)

25. Granulocytes and monocytes are produced by myeloid stem cells in the bone marrow that divide to create **progenitor cells**. Lymphoid stem cells also originate in the bone marrow, but many migrate to peripheral **lymphoid tissues**. (*Figure 19–11*)

26. Factors that regulate lymphocyte maturation are not completely understood. Several **colony-stimulating factors** (CSFs) are involved in regulating other WBC populations and in coordinating RBC and WBC production. (*Figure 19–11*)

19-6 Platelets, disc-shaped structures formed from megakaryocytes, function in the clotting process p. 672

27. Platelets are flattened discs that appear round from above and spindle shaped in section. They circulate for 9–12 days before being removed by phagocytes. (*Figure 19–10*)

28. The functions of platelets include (1) transporting and releasing chemicals important to the clotting process, (2) forming a temporary patch in the walls of damaged blood vessels, and (3) contracting after a clot has formed, to reduce the size of the break in the vessel wall.

29. During **thrombocytopoiesis**, **megakaryocytes** in the bone marrow release packets of cytoplasm (platelets) into the circulating blood. The rate of platelet formation is stimulated by thrombopoietin or thrombocyte-stimulating factor, interleukin-6, and multi-CSF.

19-7 Hemostasis involves vascular spasm, platelet plug formation, and blood coagulation p. 673

30. **Hemostasis** halts the loss of blood through the walls of damaged vessels. It consists of three phases: the *vascular phase*, the *platelet phase*, and the *coagulation phase*.

31. The **vascular phase** is a period of local blood vessel constriction, or **vascular spasm**, at the injury site. (*Figure 19–12*)

32. The **platelet phase** follows as platelets are activated, aggregate at the site, and adhere to the damaged surfaces. (*Figure 19–12*)

33. The **coagulation phase** occurs as factors released by platelets and endothelial cells interact with **clotting factors** (through either the **extrinsic pathway**, the **intrinsic pathway**, or the **common pathway**) to form a **blood clot**. In this reaction sequence, suspended fibrinogen is converted to large, insoluble fibers of fibrin. (*Figure 19–13; Table 19–4*)

34. During **clot retraction**, platelets contract and pull the torn edges of the damaged vessel closer together.

35. During **fibrinolysis**, the clot gradually dissolves through the action of **plasmin**, the activated form of circulating **plasminogen**.

19 CARDIOVASCULAR

Review Questions

See the blue Answers tab at the end of the book.

Access more review material online at **myA&P**™ (www.myaandp.com). There, you'll find chapter guides, chapter quizzes, practice tests, animations, flashcards, a glossary with pronunciations, *Interactive Physiology*® (IP) exercises and quizzes, and more to help you succeed in the course.

LEVEL 1 Reviewing Facts and Terms

1. Identify the five types of white blood cells in the following photographs.

(a) _____   (b) _____
(c) _____   (d) _____
(e) _____

2. The formed elements of the blood include
   (a) plasma, fibrin, and serum.
   (b) albumins, globulins, and fibrinogen.
   (c) WBCs, RBCs, and platelets.
   (d) a, b, and c.

3. Blood temperature is approximately _____, and blood pH averages _____.
   (a) 36°C, 7.0   (b) 39°C, 7.8
   (c) 38°C, 7.4   (d) 37°C, 7.0

4. Plasma contributes approximately _____ percent of the volume of whole blood, and water accounts for _____ percent of the plasma volume.
   (a) 55, 92   (b) 25, 55
   (c) 92, 55   (d) 35, 72

5. Serum is
   (a) the same as blood plasma.
   (b) plasma minus the formed elements.
   (c) plasma minus the proteins.
   (d) plasma minus fibrinogen.
   (e) plasma minus the electrolytes.

6. A hemoglobin molecule is composed of
   (a) two protein chains.
   (b) three protein chains.
   (c) four protein chains and nothing else.
   (d) four protein chains and four heme groups.
   (e) four heme groups but no protein.

7. The following is a list of the steps involved in the process of hemostasis.

   1. coagulation   2. fibrinolysis
   3. vascular spasm   4. retraction
   5. platelet phase

   The correct sequence of these steps is
   (a) 5, 1, 4, 2, 3.   (b) 3, 5, 1, 4, 2.
   (c) 2, 3, 5, 1, 4.   (d) 3, 5, 4, 1, 2.
   (e) 4, 3, 5, 2, 1.

8. Stem cells responsible for lymphopoiesis are located in
   (a) the thymus and spleen.
   (b) the lymph nodes.
   (c) the red bone marrow.
   (d) all of these structures.

9. _____ and _____ affect almost every aspect of the clotting process.
   (a) Calcium and vitamin K
   (b) Calcium and vitamin B_{12}
   (c) Sodium and vitamin K
   (d) Sodium and vitamin B_{12}

10. What five major functions are performed by blood?

11. What three major types of plasma proteins are in the blood? What is the major function of each?

12. Which type of antibodies does plasma contain for each of the following blood types?
    (a) Type A   (b) Type B
    (c) Type AB   (d) Type O

13. What four characteristics of WBCs are important to their response to tissue invasion or injury?

14. Which kinds of WBCs contribute to the body's nonspecific defenses?

15. Which three classes of lymphocytes are the primary cells of the lymphoid system? What are the functions of each class?

16. What are the three functions of platelets during the clotting process?

17. What four conditions cause the release of erythropoietin?

18. What contribution from the intrinsic and the extrinsic pathways is necessary for the common pathway to begin?

LEVEL 2 Reviewing Concepts

19. Dehydration would
    (a) cause an increase in the hematocrit.
    (b) cause a decrease in the hematocrit.
    (c) have no effect on the hematocrit.
    (d) cause an increase in plasma volume.

20. Erythropoietin directly stimulates RBC formation by
    (a) increasing rates of mitotic divisions in erythroblasts.
    (b) speeding up the maturation of red blood cells.
    (c) accelerating the rate of hemoglobin synthesis.
    (d) a, b, and c.

21. The waste product bilirubin is formed from
    (a) transferrin.   (b) globin.
    (c) heme.   (d) hemosiderin.
    (e) ferritin.

22. A difference between the A, B, and O blood types and the Rh factor is
    (a) Rh agglutinogens are not found on the surface of red blood cells.
    (b) Rh agglutinogens do not produce a cross-reaction.
    (c) individuals who are Rh^- do not carry agglutinins to Rh factor unless they have been previously sensitized.
    (d) Rh agglutinogens are found free in the plasma.
    (e) Rh agglutinogens are found bound to plasma proteins.

23. How do red blood cells differ from white blood cells in both form and function?

24. How do elements of blood defend against toxins and pathogens in the body?

25. What is the role of blood in the stabilization and maintenance of body temperature?

26. Describe the structure of hemoglobin. How does the structure relate to its function?

27. Why is aspirin sometimes prescribed for the prevention of vascular problems?

LEVEL 3 Critical Thinking and Clinical Applications

28. A test for prothrombin time is used to identify deficiencies in the extrinsic clotting pathway; prothrombin time is prolonged if any of the factors are deficient. A test for activated partial thromboplastin time is used in a similar fashion to detect deficiencies in the intrinsic clotting pathway. Which factor would be deficient if a person had a prolonged prothrombin time but a normal partial thromboplastin time?

29. In the disease mononucleosis ("mono"), the spleen enlarges because of increased numbers of phagocytes and other cells. Common signs and symptoms of this disease include pale complexion, a tired feeling, and a lack of energy sometimes to the point of not being able to get out of bed. What might cause these signs and symptoms?

30. Almost half of our vitamin K is synthesized by bacteria that inhabit the large intestine. Based on this information, how could taking a broad-spectrum antibiotic for a long time cause frequent nosebleeds?

31. After Randy was diagnosed with stomach cancer, nearly all of his stomach had to be removed. Postoperative treatment included regular injections of vitamin B_{12}. Why was this vitamin prescribed, and why were injections specified?

The Heart

20

Learning Outcomes

After completing this chapter, you should be able to do the following:

20-1 Describe the anatomy of the heart, including vascular supply and pericardium structure, and trace the flow of blood through the heart, identifying the major blood vessels, chambers, and heart valves.

20-2 Explain the events of an action potential in cardiac muscle, indicate the importance of calcium ions to the contractile process, describe the conducting system of the heart, and identify the electrical events associated with a normal electrocardiogram.

20-3 Explain the events of the cardiac cycle, including atrial and ventricular systole and diastole, and relate the heart sounds to specific events in the cycle.

20-4 Define cardiac output, describe the factors that influence heart rate and stroke volume, and explain how adjustments in stroke volume and cardiac output are coordinated at different levels of physical activity.

Clinical Notes

Coronary Artery Disease p. 694
Myocardial Infarction p. 702
Abnormal Conditions Affecting Cardiac Output p. 709

An Introduction to the Cardiovascular System

This chapter considers the structure and function of the heart. Blood flows through a network of blood vessels that extend between the heart and peripheral tissues. Those blood vessels can be organized into a **pulmonary circuit**, which carries blood to and from the gas exchange surfaces of the lungs, and a **systemic circuit**, which transports blood to and from the rest of the body (**Figure 20–1**). Each circuit begins and ends at the heart, and blood travels through these circuits in sequence. Thus, blood returning to the heart from the systemic circuit must complete the pulmonary circuit before reentering the systemic circuit. Blood is carried away from the heart by **arteries**, or *efferent vessels*, and returns to the heart by way of **veins**, or *afferent vessels*. Microscopic thin-walled vessels called **capillaries** interconnect the smallest arteries and the smallest veins. Capillaries are called **exchange vessels**, because their thin walls permit the exchange of nutrients, dissolved gases, and waste products between the blood and surrounding tissues.

Unlike most other muscles, the heart never rests. This extraordinary organ beats approximately 100,000 times each day, pumping roughly 8000 liters of blood—enough to fill forty 55-gallon drums, or 8800 quart-sized milk cartons. Try transferring a gallon of water by using a squeeze pump, and you'll appreciate just how hard the heart has to work to keep you alive. Despite its impressive workload, the heart is a small organ, roughly the size of a clenched fist.

The heart contains four muscular chambers, two associated with each circuit. The **right atrium** (Ā-trē-um; entry chamber; plural, *atria*) receives blood from the systemic circuit and passes it to the **right ventricle** (VEN-tri-kl; little belly), which pumps blood into the pulmonary circuit. The **left atrium** collects blood from the pulmonary circuit and empties it into the **left ventricle**, which pumps blood into the systemic circuit. When the heart beats, first the atria contract, and then the ventricles contract. The two ventricles contract at the same time and eject equal volumes of blood into the pulmonary and systemic circuits.

20-1 The heart is a four-chambered organ, supplied by the coronary circulation, that pumps oxygen-poor blood to the lungs and oxygen-rich blood to the rest of the body

The heart is located near the anterior chest wall, directly posterior to the sternum (**Figure 20–2a**). The great veins and arteries are connected to the superior end of the heart at the attached base. The base sits posterior to the sternum at the level of the third costal cartilage, centered about 1.2 cm (0.5 in.) to the left side. The inferior, pointed tip of the heart is the **apex** (Ā-peks). A typical adult heart measures approximately 12.5 cm (5 in.) from the base to the apex, which reaches the fifth intercostal space approximately 7.5 cm (3 in.) to the left of the midline. A midsagittal section through the trunk does not divide the heart into two equal halves, because (1) the center of the base lies slightly to the left of the midline, (2) a line drawn between the center of the base and the apex points further to the left, and (3) the entire heart is rotated to the left around this line, so that the right atrium and right ventricle dominate an anterior view of the heart.

The heart, surrounded by the *pericardial* (per-i-KAR-dē-al) *sac*, sits in the anterior portion of the mediastinum. The **mediastinum**, the region between the two pleural cavities, also contains the great vessels (the large arteries and veins linked to the heart), thymus, esophagus, and trachea. **Figure 20–2b** is a sectional view that illustrates the position of the heart relative to other structures in the mediastinum.

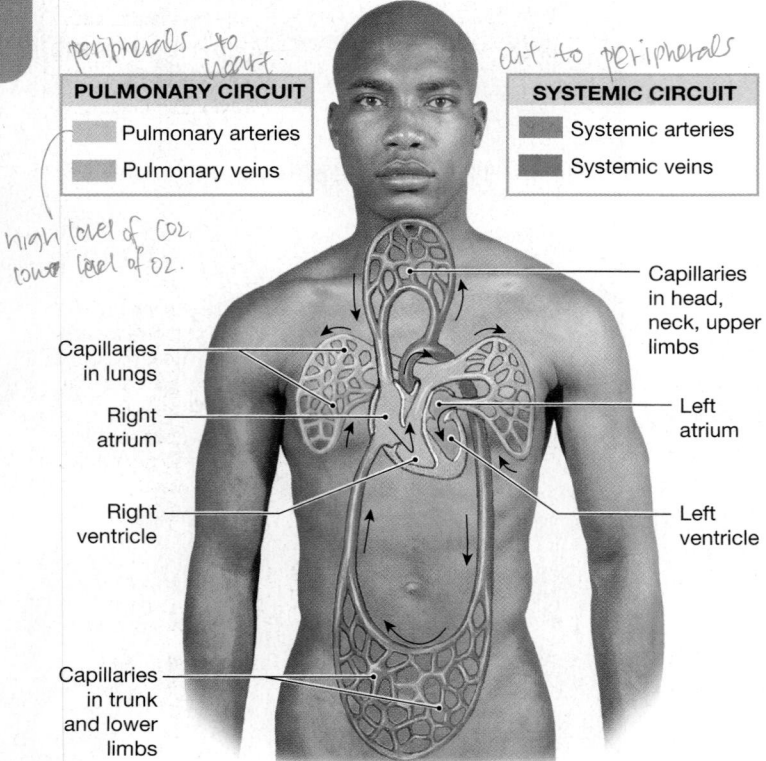

PULMONARY CIRCUIT	SYSTEMIC CIRCUIT
Pulmonary arteries	Systemic arteries
Pulmonary veins	Systemic veins

Capillaries in head, neck, upper limbs

Capillaries in lungs

Right atrium

Left atrium

Right ventricle

Left ventricle

Capillaries in trunk and lower limbs

Figure 20–1 An Overview of the Cardiovascular System.
Driven by the pumping of the heart, blood flows through the pulmonary and systemic circuits in sequence. Each circuit begins and ends at the heart and contains arteries, capillaries, and veins.

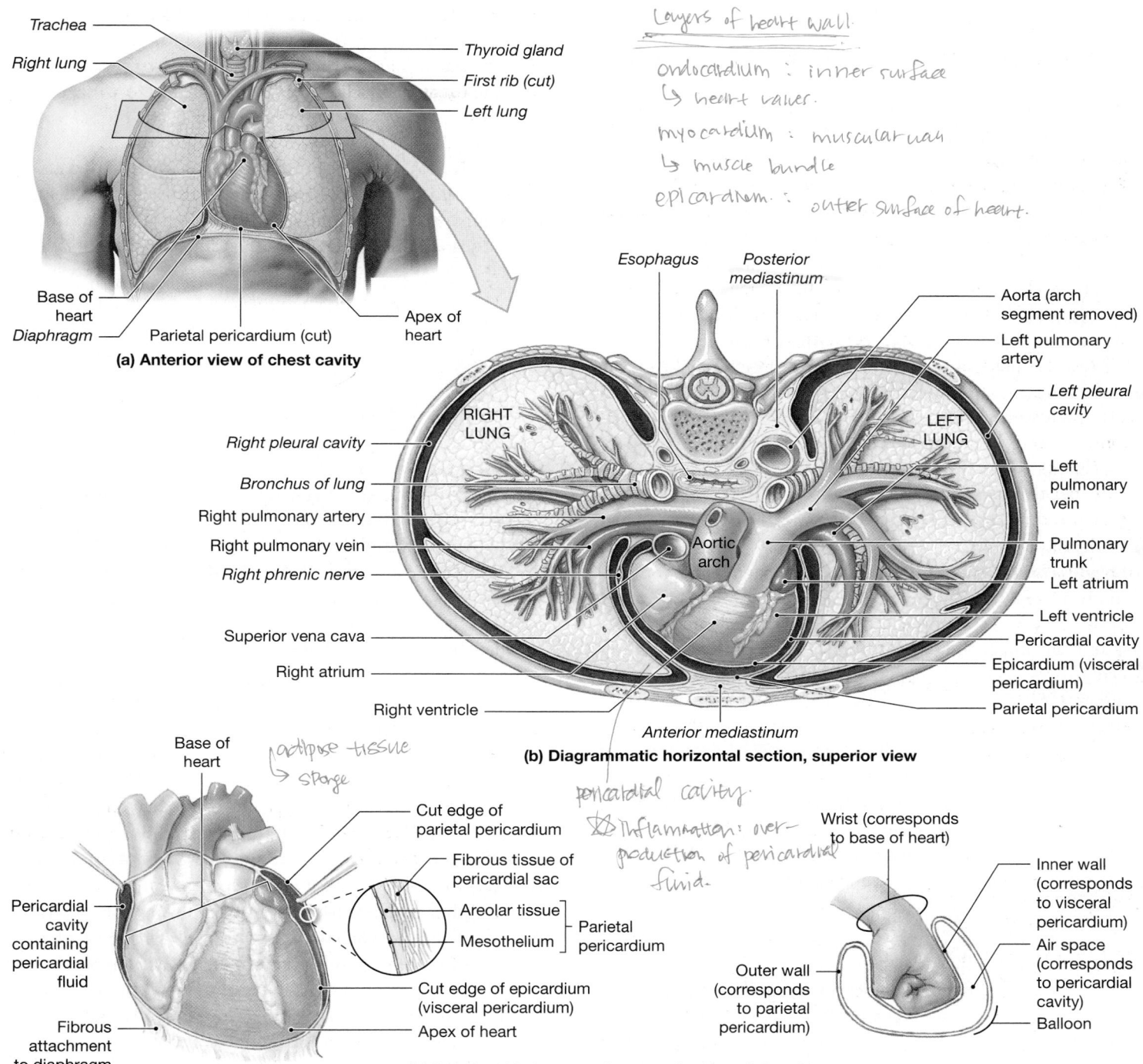

Layers of heart wall.

ondocardium : inner surface
⤷ heart valves.
myocardium : muscular way
⤷ muscle bundle
epicardium : outer surface of heart.

(a) Anterior view of chest cavity

adipose tissue
⤷ sponge

pericardial cavity.
✵ Inflammation: over-
production of pericardial
fluid.

(b) Diagrammatic horizontal section, superior view

20 CARDIOVASCULAR

(c) Relationship between heart and pericardial cavity

Figure 20–2   The Location of the Heart in the Thoracic Cavity. (a) An anterior view of the chest, showing the position of the heart and major vessels relative to the ribs, sternum, and lungs. **(b)** A superior view of the organs in the mediastinum; portions of the lungs have been removed to reveal the blood vessels and airways. The heart is situated in the anterior part of the mediastinum, immediately posterior to the sternum. **(c)** The relationship between the heart and the pericardial cavity; compare with the fist-and-balloon example. ATLAS: Plate 47a,b

✵ aorta: art
veins : m.

The Pericardium

The lining of the pericardial cavity is called the **pericardium**. To visualize the relationship between the heart and the pericardial cavity, imagine pushing your fist toward the center of a large, partially inflated balloon (**Figure 20–2c**). The balloon represents the pericardium, and your fist is the heart. Your wrist, where the balloon folds back on itself, corresponds to the *base* of the heart, to which the *great vessels*, the largest veins and arteries in the body, are attached. The air space inside the balloon corresponds to the pericardial cavity.

The pericardium is lined by a delicate serous membrane that can be subdivided into the visceral pericardium and the parietal pericardium. The **visceral pericardium**, or *epicardium*, covers and adheres closely to the outer surface of the heart; the **parietal pericardium** lines the inner surface of the **pericardial sac**, which surrounds the heart (**Figure 20–2c**). The pericardial sac, or *fibrous pericardium*, which consists of a dense network of collagen fibers, stabilizes the position of the heart and associated vessels within the mediastinum.

The small space between the parietal and visceral surfaces is the pericardial cavity. It normally contains 15–50 mL of **pericardial fluid**, secreted by the pericardial membranes. This fluid acts as a lubricant, reducing friction between the opposing surfaces as the heart beats. Pathogens and inflammation can infect the pericardium, producing the condition **pericarditis**. The inflamed pericardial surfaces rub against one another, producing a distinctive scratching sound that can be heard through a stethoscope. The pericardial inflammation also commonly results in an increased production of pericardial fluid. Fluid then collects in the pericardial cavity, restricting the movement of the heart. This condition, called *cardiac tamponade* (tam-po-NĀD; *tampon*, plug), can also be caused by traumatic injuries (such as stab wounds) that produce bleeding into the pericardial cavity.

Superficial Anatomy of the Heart

The four cardiac chambers can easily be identified in a superficial view of the heart (**Figure 20–3**). The two atria have relatively thin muscular walls and are highly expandable. When not filled with blood, the outer portion of each atrium deflates and becomes a lumpy, wrinkled flap. This expandable extension of an atrium is called an *atrial appendage*, or an **auricle** (AW-ri-kl; *auris*, ear), because it reminded early anatomists of the external ear (**Figure 20–3a**). The **coronary sulcus**, a deep groove, marks the border between the atria and the ventricles. The **anterior interventricular sulcus** and the **posterior interventricular sulcus**, shallower depressions, mark the boundary between the left and right ventricles (**Figure 20–3a,b**).

The connective tissue of the epicardium at the coronary and interventricular sulci generally contains substantial amounts of fat. In fresh or preserved hearts, this fat must be stripped away to expose the underlying grooves. These sulci also contain the arteries and veins that carry blood to and from the cardiac muscle.

The Heart Wall

A section through the wall of the heart reveals three distinct layers: an outer epicardium, a middle myocardium, and an inner endocardium. **Figure 20–4a** illustrates these three layers:

1. The **epicardium** is the visceral pericardium that covers the outer surface of the heart. This serous membrane consists of an exposed mesothelium and an underlying layer of loose areolar connective tissue that is attached to the myocardium.

2. The **myocardium**, or muscular wall of the heart, forms both atria and ventricles. This layer contains cardiac muscle tissue, blood vessels, and nerves. The myocardium consists of concentric layers of cardiac muscle tissue. The atrial myocardium contains muscle bundles that wrap around the atria and form figure eights that encircle the great vessels (**Figure 20–4b**). Superficial ventricular muscles wrap around both ventricles; deeper muscle layers spiral around and between the ventricles toward the apex in a figure-eight pattern.

3. The inner surfaces of the heart, including those of the heart valves, are covered by the **endocardium**, a simple squamous epithelium that is continuous with the endothelium of the attached great vessels.

Cardiac Muscle Tissue

As noted in Chapter 10, **cardiac muscle cells** are interconnected by **intercalated discs** (**Figure 20–5a,c**). At an intercalated disc, the interlocking membranes of adjacent cells are held together by desmosomes and linked by gap junctions (**Figure 20–5b**). Intercalated discs transfer the force of contraction from cell to cell and propagate action potentials. Table 20–1 provides a quick review of the structural and functional differences between cardiac muscle cells and skeletal muscle fibers. Among the histological characteristics of cardiac muscle cells that distinguish them from skeletal muscle fibers are (1) small size; (2) a single, centrally located nucleus; (3) branching interconnections between cells; and (4) the presence of intercalated discs.

Tips&Tricks

The term *intercalated* means "inserted between other elements"; thus, intercalated discs appear to have been inserted between cardiac muscle cells.

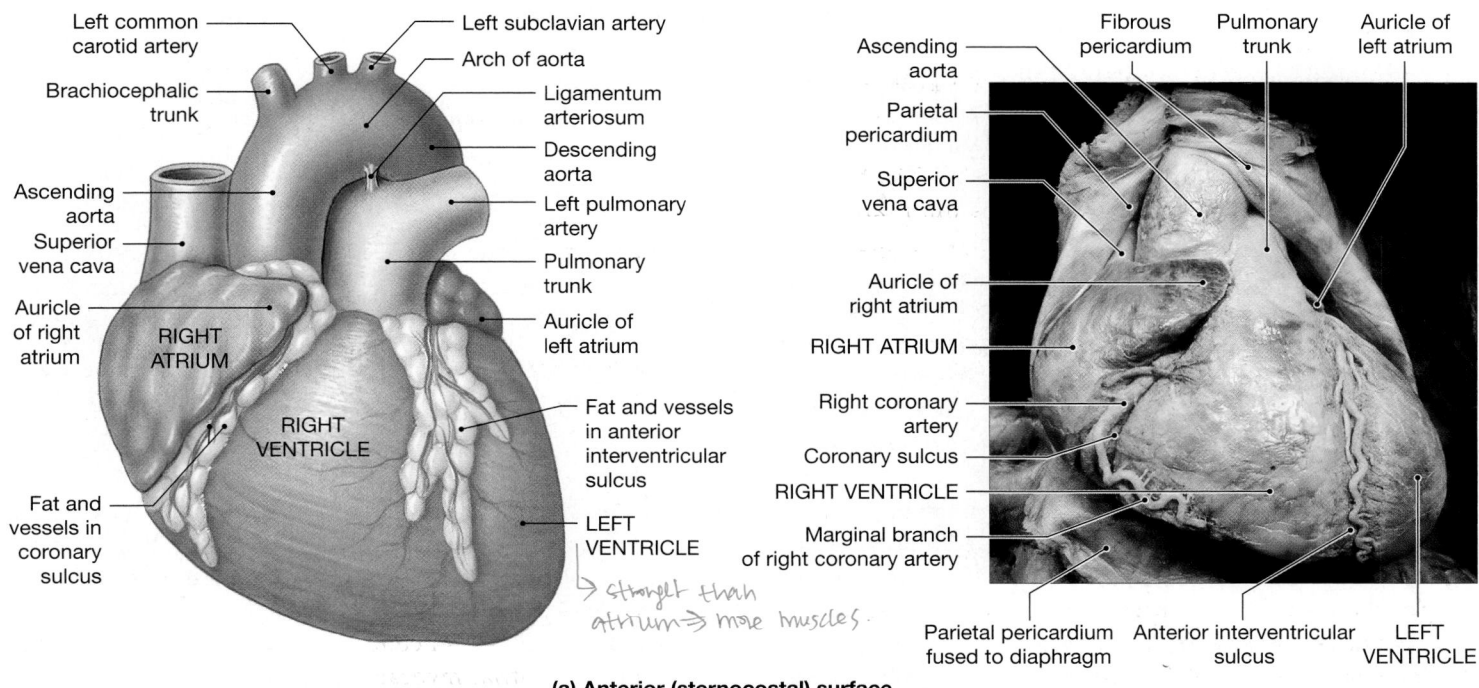

(a) Anterior (sternocostal) surface

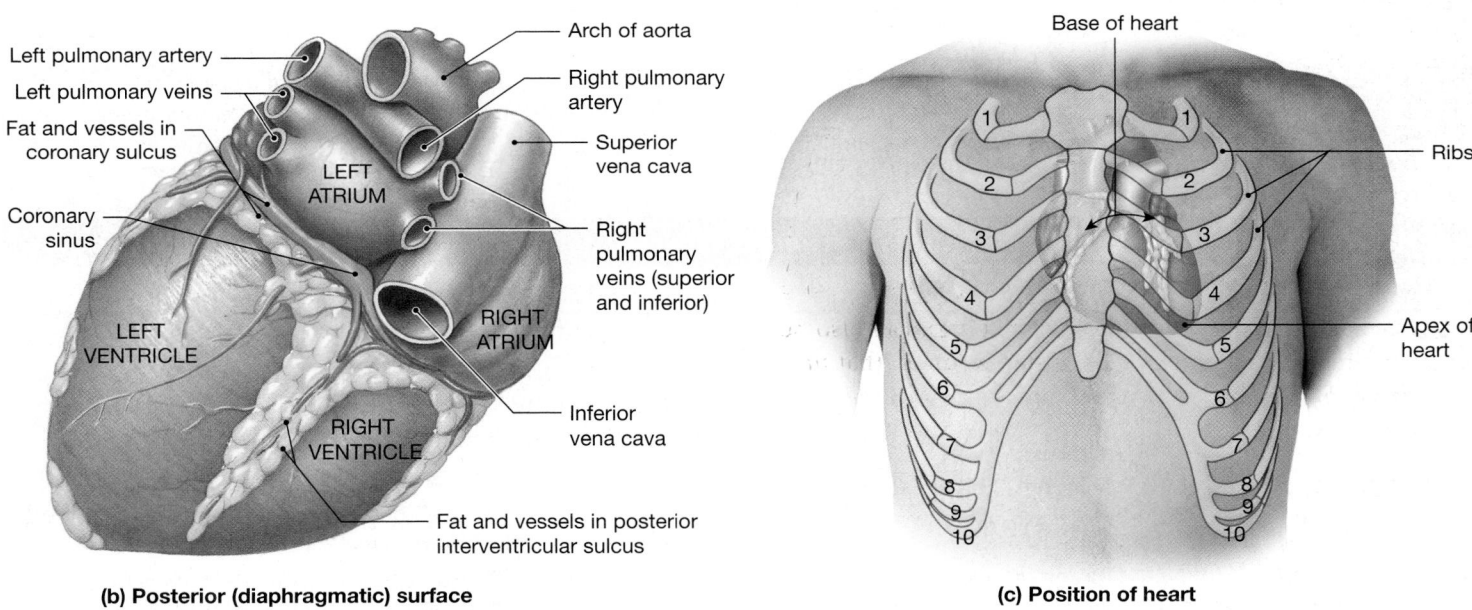

(b) Posterior (diaphragmatic) surface

(c) Position of heart

Figure 20–3 **The Superficial Anatomy of the Heart.** (a) Major anatomical features on the anterior surface. (b) Major landmarks on the posterior surface. Coronary arteries (which supply the heart itself) are shown in red; coronary veins are shown in blue. (c) Heart position relative to the rib cage.

Internal Anatomy and Organization

Next we examine the major landmarks and structures visible on the interior surface of the heart. In a sectional view, you can see that the right atrium communicates with the right ventricle, and the left atrium with the left ventricle (**Figure 20–6a,c**). The atria are separated by the **interatrial septum**

(*septum*, wall); the ventricles are separated by the much thicker **interventricular septum**. Each septum is a muscular partition. **Atrioventricular (AV) valves**, folds of fibrous tissue, extend into the openings between the atria and ventricles. These valves permit blood flow in one direction only: from the atria to the ventricles.

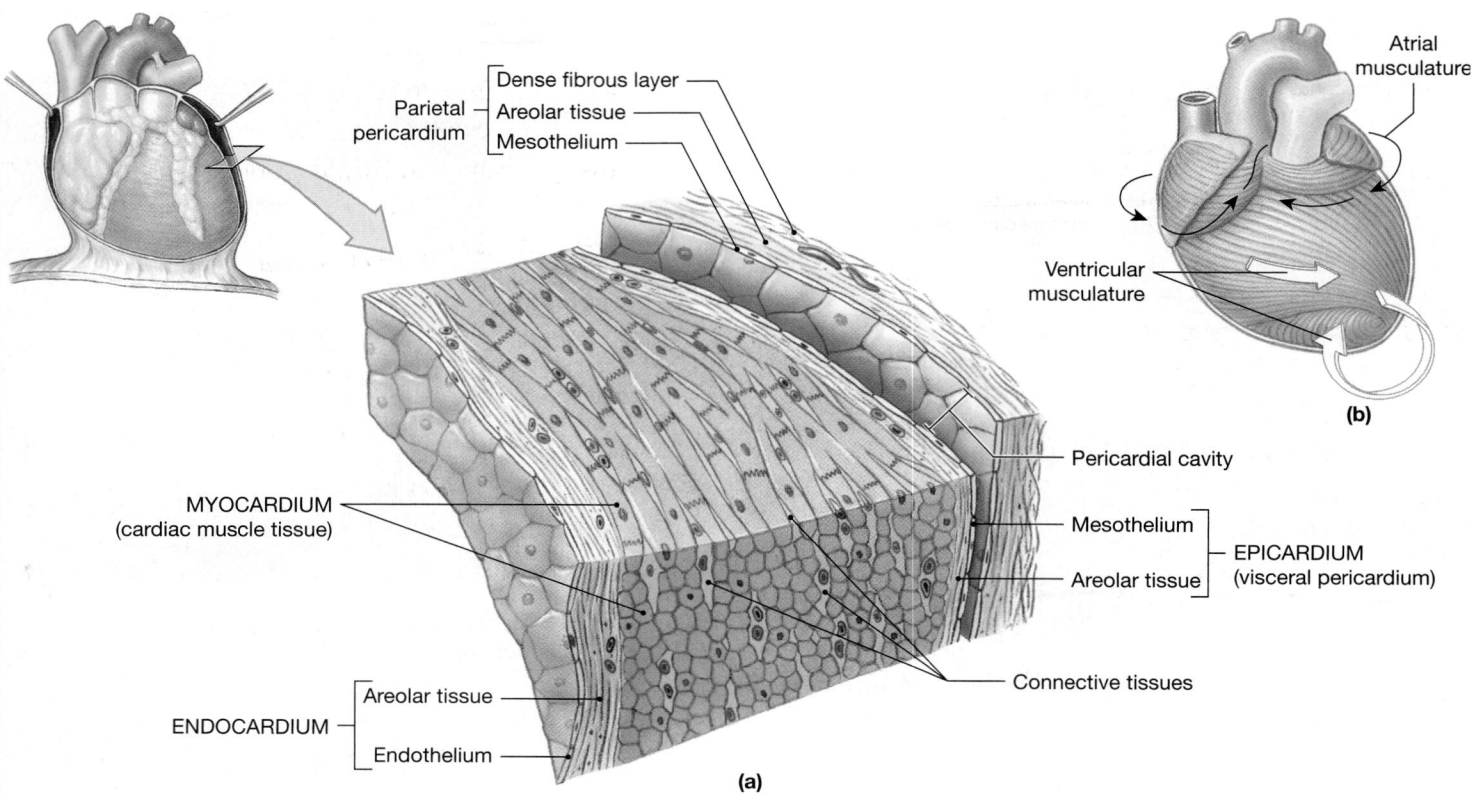

Figure 20–4 **The Heart Wall.** **(a)** A diagrammatic section through the heart wall, showing the relative positions of the epicardium, myocardium, and endocardium. The proportions are not to scale; the relative thickness of the myocardial wall has been greatly reduced. **(b)** Cardiac muscle tissue forms concentric layers that wrap around the atria or spiral within the walls of the ventricles.

SUMMARY TABLE 20–1	Structural and Functional Differences between Cardiac Muscle Cells and Skeletal Muscle Fibers	
Feature	**Cardiac Muscle Cells**	**Skeletal Muscle Fibers**
Size	10–$20\ \mu m \times 50$–$100\ \mu m$	$100\ \mu m \times$ up to 40 cm
Nuclei	Typically 1 (rarely 2–5)	Multiple (hundreds)
Contractile proteins	Sarcomeres along myofibrils	Sarcomeres along myofibrils
Internal membranes	Short T tubules; no triads formed with sarcoplasmic reticulum	Long T tubules form triads with cisternae of the sarcoplasmic reticulum
Mitochondria	Abundant (25% of cell volume)	Much less abundant
Inclusions	Myoglobin, lipids, glycogen	Little myoglobin, few lipids, but extensive glycogen reserves
Blood supply	Very extensive	More extensive than in most connective tissues, but sparse compared with supply to cardiac muscle cells
Metabolism (resting)	Not applicable	Aerobic, primarily lipid-based
Metabolism (active)	Aerobic, primarily using lipids and carbohydrates	Anaerobic, through breakdown of glycogen reserves
Contractions	Twitches with brief relaxation periods; long refractory period prevents tetanic contractions	Usually sustained contractions
Stimulus for contraction	Autorhythmicity of pacemaker cells generates action potentials	Activity of somatic motor neuron generates action potentials in sarcolemma
Trigger for contraction	Calcium entry from the ECF and calcium release from the sarcoplasmic reticulum	Calcium release from the sarcoplasmic reticulum
Intercellular connections	Branching network with plasma membranes locked together at intercalated discs; connective tissue fibers tie adjacent layers together	Adjacent fibers tied together by connective tissue fibers

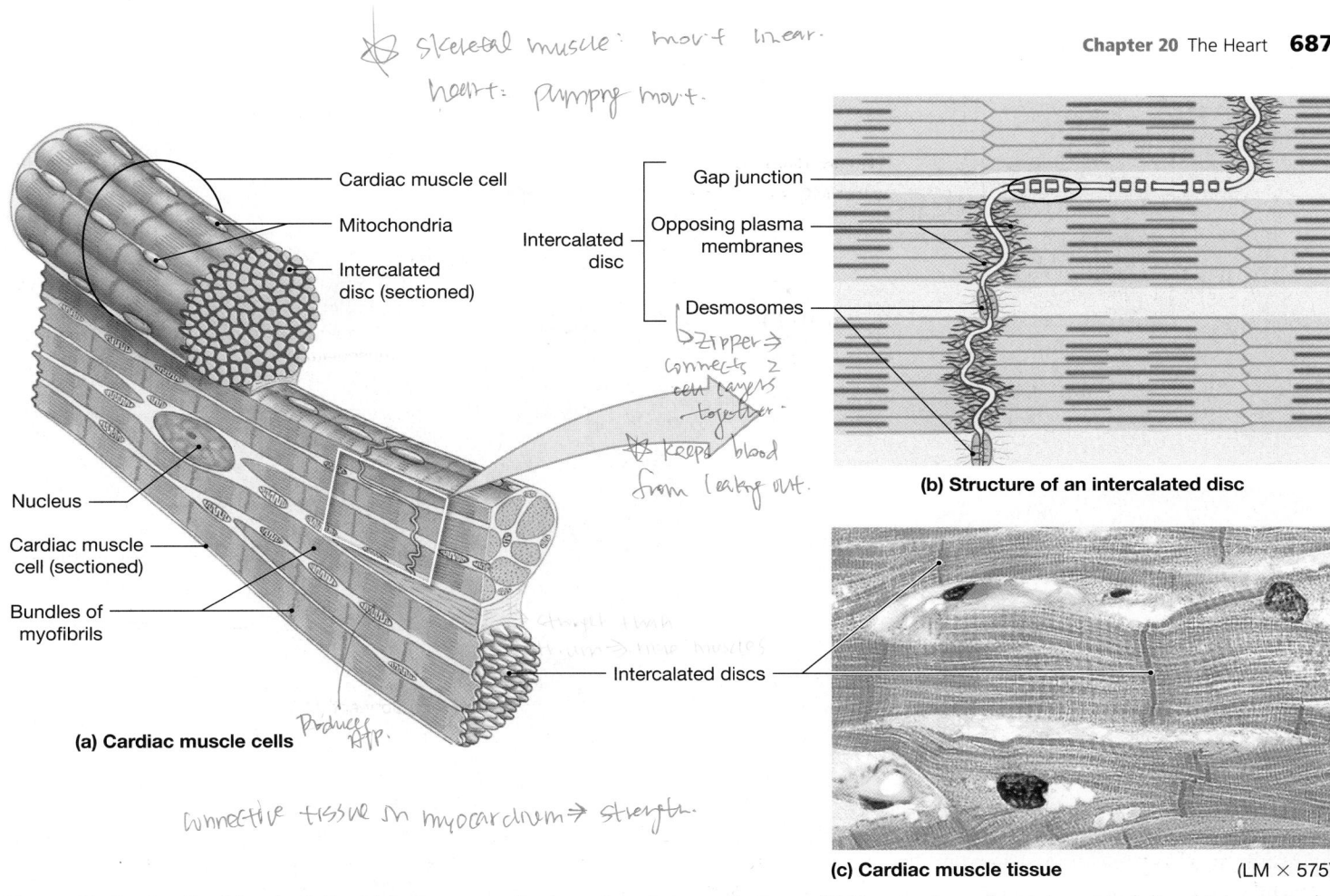

(handwritten notes) skeletal muscle: mov't linear.
heart: pumping mov't.

Cardiac muscle cell
Mitochondria
Intercalated disc (sectioned)

Gap junction
Opposing plasma membranes
Intercalated disc
Desmosomes

(handwritten) zipper → connects 2 cell layers together. keeps blood from leaking out.

(b) Structure of an intercalated disc

Nucleus
Cardiac muscle cell (sectioned)
Bundles of myofibrils

(handwritten) Produces ATP.

Intercalated discs

(a) Cardiac muscle cells

(handwritten) connective tissue in myocardium → strength.

(c) Cardiac muscle tissue (LM × 575)

Figure 20–5 **Cardiac Muscle Cells. (a)** A diagrammatic view of cardiac muscle tissue. **(b)** The structure of an intercalated disc. **(c)** A sectional view of cardiac muscle tissue.

(handwritten) nucleus located in peripheral surface layer → pericardial.

The Right Atrium

The right atrium receives blood from the systemic circuit through the two great veins: the **superior vena cava** (VĒ-na KĀ-vuh; *venae cavae,* plural) and the **inferior vena cava**. The superior vena cava, which opens into the posterior and superior portion of the right atrium, delivers blood to the right atrium from the head, neck, upper limbs, and chest. The inferior vena cava, which opens into the posterior and inferior portion of the right atrium, carries blood to the right atrium from the rest of the trunk, the viscera, and the lower limbs. The *cardiac veins* of the heart return blood to the **coronary sinus**, a large, thin-walled vein that opens into the right atrium inferior to the connection with the superior vena cava.

The opening of the coronary sinus lies near the posterior edge of the interatrial septum. From the fifth week of embryonic development until birth, the **foramen ovale**, an oval opening, penetrates the interatrial septum and connects the two atria of the fetal heart. Before birth, the foramen ovale permits blood flow from the right atrium to the left atrium while the lungs are developing. At birth, the foramen ovale closes, and the opening is permanently sealed off within three

months of delivery. (If the foramen ovale does not close, serious cardiovascular problems may result; these are considered in Chapter 21.) The **fossa ovalis**, a small, shallow depression, persists at this site in the adult heart (**Figure 20–6a,c**). ATLAS: Embryology Summary 15: The Development of the Heart

The posterior wall of the right atrium and the interatrial septum have smooth surfaces. In contrast, the anterior atrial wall and the inner surface of the auricle contain prominent muscular ridges called the **pectinate muscles** (*pectin,* comb), or *musculi pectinati* (**Figure 20–6a,c**).

The Right Ventricle

Blood travels from the right atrium into the right ventricle through a broad opening bounded by three fibrous flaps. These flaps, called **cusps** or *leaflets,* are part of the **right atrioventricular (AV) valve**, also known as the **tricuspid** (trī-KUS-pid; *tri,* three) **valve**. The free edge of each cusp is attached to connective tissue fibers called the **chordae tendineae** (KOR-dē TEN-di-nē-ē; tendinous cords). The fibers originate at the **papillary** (PAP-i-ler-ē) **muscles**, conical muscular projections that arise from the inner surface of

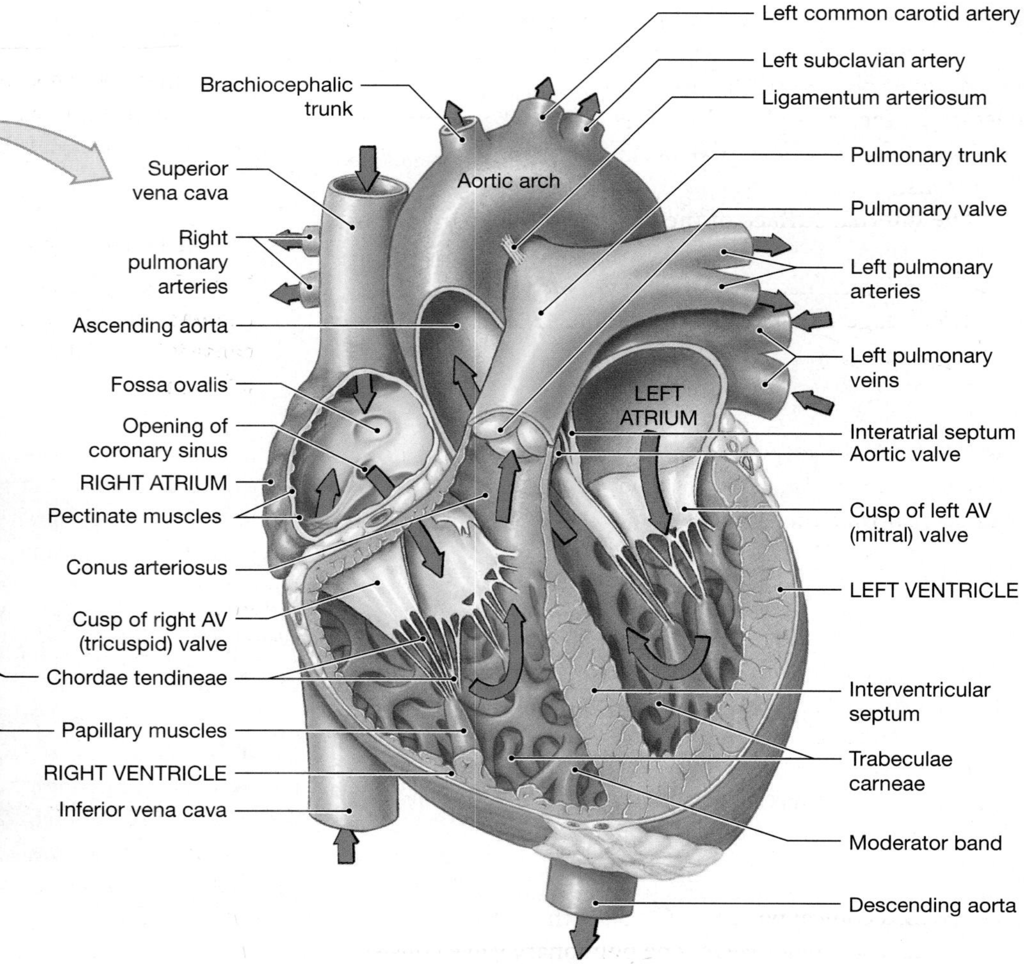

Left common carotid artery

Left subclavian artery

Ligamentum arteriosum

Pulmonary trunk

Pulmonary valve

Left pulmonary arteries

Left pulmonary veins

Interatrial septum

Aortic valve

Cusp of left AV (mitral) valve

LEFT VENTRICLE

Interventricular septum

Trabeculae carneae

Moderator band

Descending aorta

Brachiocephalic trunk

Aortic arch

LEFT ATRIUM

Superior vena cava

Right pulmonary arteries

Ascending aorta

Fossa ovalis

Opening of coronary sinus

RIGHT ATRIUM

Pectinate muscles

Conus arteriosus

Cusp of right AV (tricuspid) valve

Chordae tendineae

Papillary muscles

RIGHT VENTRICLE

Inferior vena cava

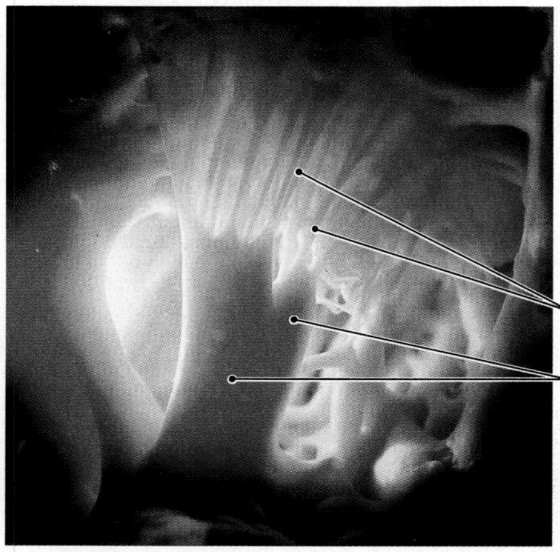

(b) Papillary muscles and chordae tendineae

(a) Frontal section through the heart

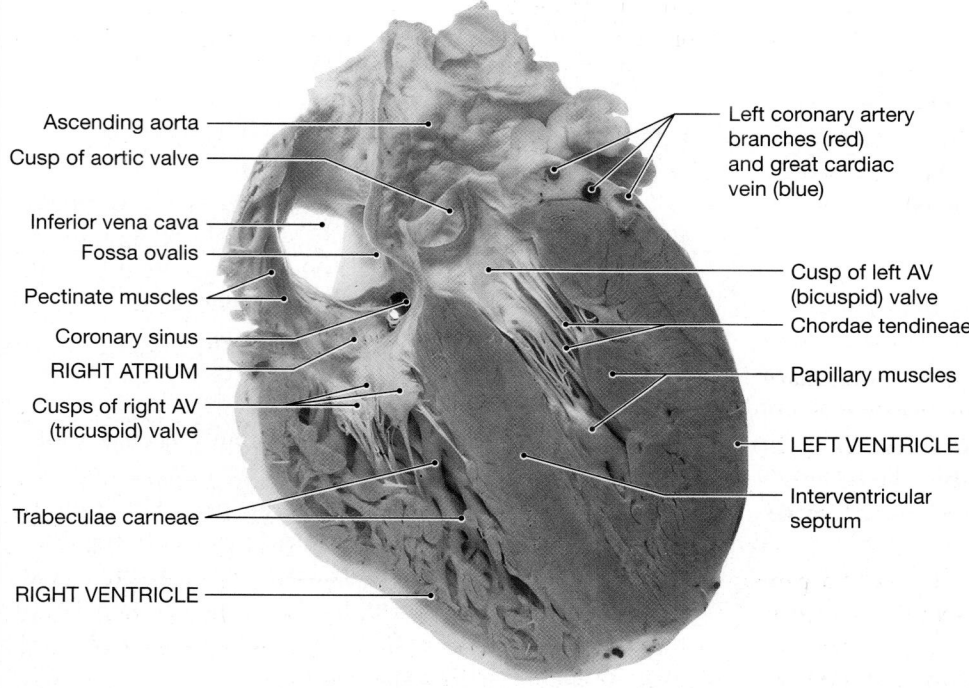

Ascending aorta

Cusp of aortic valve

Inferior vena cava

Fossa ovalis

Pectinate muscles

Coronary sinus

RIGHT ATRIUM

Cusps of right AV (tricuspid) valve

Trabeculae carneae

RIGHT VENTRICLE

Left coronary artery branches (red) and great cardiac vein (blue)

Cusp of left AV (bicuspid) valve

Chordae tendineae

Papillary muscles

LEFT VENTRICLE

Interventricular septum

(c) Frontal section, anterior view

Figure 20–6 The Sectional Anatomy of the Heart.
(a) A diagrammatic frontal section through the heart, showing major landmarks and the path of blood flow (marked by arrows) through the atria, ventricles, and associated vessels.
(b) The papillary muscles and chordae tendineae supporting the right AV (tricuspid) valve. The photograph was taken from inside the right ventricle, looking toward a light shining from the right atrium. **(c)** A sectional view of the heart.

the right ventricle (**Figure 20–6a,b**). The right AV valve closes when the right ventricle contracts, preventing the backflow of blood into the right atrium. Without the chordae tendineae to anchor their free edges, the cusps would be like swinging doors that permitted blood flow in both directions.

The internal surface of the ventricle also contains a series of muscular ridges: the **trabeculae carneae** (tra-BEK-ū-lē KAR-nē-ē; *carneus*, fleshy). The *moderator band* is a muscular ridge that extends horizontally from the inferior portion of the interventricular septum and connects to the anterior papillary muscle. This ridge contains a portion of the *conducting system*, an internal network that coordinates the contractions of cardiac muscle cells. The moderator band delivers the stimulus for contraction to the papillary muscles, so that they begin tensing the chordae tendineae before the rest of the ventricle contracts.

Tips&Tricks

The saying "To tug on your heartstrings" may help you remember the functions of the papillary muscles and the chordae tendineae: Contractions of the papillary muscles pull on the chordae tendineae, which "tug" on your heart's valves.

The superior end of the right ventricle tapers to the **conus arteriosus**, a conical pouch that ends at the **pulmonary valve**, or *pulmonary semilunar valve*. The pulmonary valve consists of three semilunar (half-moon–shaped) cusps of thick connective tissue. Blood flowing from the right ventricle passes through this valve to enter the **pulmonary trunk**, the start of the pulmonary circuit. The arrangement of cusps prevents backflow as the right ventricle relaxes. Once in the pulmonary trunk, blood flows into the **left pulmonary arteries** and the **right pulmonary arteries**. These vessels branch repeatedly within the lungs before supplying the capillaries, where gas exchange occurs.

The Left Atrium

From the respiratory capillaries, blood collects into small veins that ultimately unite to form the four pulmonary veins. The posterior wall of the left atrium receives blood from two **left** and two **right pulmonary veins**. Like the right atrium, the left atrium has an auricle. A valve, the **left atrioventricular (AV) valve**, or **bicuspid** (bī-KUS-pid) **valve**, guards the entrance to the left ventricle (**Figure 20–6a,c**). As the name *bicuspid* implies, the left AV valve contains a pair, not a trio, of cusps. Clinicians often call this valve the **mitral** (MĪ-tral; *mitre*, a bishop's hat) **valve**. The left AV valve permits the flow of blood from the left atrium into the left ventricle but prevents backflow during ventricular contraction.

Tips&Tricks

To remember the locations of the tricuspid and bicuspid (mitral) valves, think **"try to be right"** for the *tri*cuspid, and associate the *l* in mitra*l* with the *l* in *l*eft.

The Left Ventricle

Even though the two ventricles hold equal amounts of blood, the left ventricle is much larger than the right ventricle because it has thicker walls. Its thick, muscular walls enable the left ventricle to develop pressure sufficient to push blood through the large systemic circuit, whereas the right ventricle needs to pump blood, at lower pressure, only about 15 cm (6 in.) to and from the lungs. The internal organization of the left ventricle generally resembles that of the right ventricle, except for the absence of a moderator band (**Figure 20–6a,c**). The trabeculae carneae are prominent, and a pair of large papillary muscles tense the chordae tendineae that anchor the cusps of the AV valve and prevent the backflow of blood into the left atrium.

Blood leaves the left ventricle by passing through the **aortic valve**, or *aortic semilunar valve*, into the **ascending aorta**. The arrangement of cusps in the aortic valve is the same as that in the pulmonary valve. Once the blood has been pumped out of the heart and into the systemic circuit, the aortic valve prevents backflow into the left ventricle. From the ascending aorta, blood flows through the **aortic arch** and into the **descending aorta** (**Figure 20–6a**). The pulmonary trunk is attached to the aortic arch by the *ligamentum arteriosum*, a fibrous band that is a remnant of an important fetal blood vessel that once linked the pulmonary and systemic circuits.

Structural Differences between the Left and Right Ventricles

The function of an atrium is to collect blood that is returning to the heart and convey it to the attached ventricle. The functional demands on the right and left atria are similar, and the two chambers look almost identical. The demands on the right and left ventricles, however, are very different, and the two have significant structural differences.

Anatomical differences between the left and right ventricles are best seen in a three-dimensional view (**Figure 20–7a**). The lungs are close to the heart, and the pulmonary blood vessels are relatively short and wide. Thus, the right ventricle normally does not need to work very hard to push blood through the pulmonary circuit. Accordingly, the wall of the right ventricle is relatively thin. In sectional view, it resembles a pouch attached to the massive wall of the left ventricle. When it contracts, the right ventricle acts like a bellows, squeezing the blood against the thick wall of the left ventricle. This mechanism moves blood very efficiently with minimal effort, but it develops relatively low pressures.

A comparable pumping arrangement would not be suitable for the left ventricle, because four to six times as much pressure must be exerted to push blood around the systemic circuit as around the pulmonary circuit. The left ventricle has an extremely thick muscular wall and is round in cross section (**Figure 20–7a**). When this ventricle contracts, (1) the distance between the base and apex decreases, and (2) the diameter of the ventricular chamber decreases. The effect is similar to simultaneously squeezing and rolling up the end of a toothpaste tube. The pressure generated is more than enough to open the aortic valve and eject blood into the ascending aorta.

As the powerful left ventricle contracts, it also bulges into the right ventricular cavity (**Figure 20–7b**). This action im-

proves the efficiency of the right ventricle's efforts. Individuals whose right ventricular musculature has been severely damaged may survive, because the contraction of the left ventricle helps push blood into the pulmonary circuit. We will return to this topic in Chapter 21, where we consider the integrated functioning of the cardiovascular system.

The Heart Valves

The heart has two pairs of one-way valves that prevent the backflow of blood as the chambers contract. We will now consider the structure and function of these heart valves.

The Atrioventricular Valves. The atrioventricular (AV) valves prevent the backflow of blood from the ventricles to the atria when the ventricles are contracting. The chordae tendineae and papillary muscles play important roles in the normal function of the AV valves. When the ventricles are relaxed, the chordae tendineae are loose, and the AV valves offer no resistance to the flow of blood from the atria into the ventricles (**Figure 20–8a**). When the ventricles contract, blood moving back toward the atria swings the cusps together, closing the valves (**Figure 20–8b**). At the same time, the contraction of the papillary muscles tenses the chordae tendineae, stopping the cusps before they swing into the atria. If the chordae tendineae are cut or the papillary muscles are damaged, backflow (**regurgitation**) of blood into the atria occurs each time the ventricles contract.

The Semilunar Valves. The pulmonary and aortic valves prevent the backflow of blood from the pulmonary trunk and aorta into the right and left ventricles, respectively. Unlike the AV valves, the semilunar valves do not require muscular braces, because the arterial walls do not contract and the relative positions of the cusps are stable. When the semilunar valves close, the three symmetrical cusps support one another like the legs of a tripod (**Figure 20–8a,c**).

Saclike dilations of the base of the ascending aorta are adjacent to each cusp of the aortic valve. These sacs, called **aortic sinuses**, prevent the individual cusps from sticking to the wall of the aorta when the valve opens. The *right* and *left coronary arteries*, which deliver blood to the myocardium, originate at the aortic sinuses.

Serious valve problems can interfere with cardiac function. If valve function deteriorates to the point at which the heart cannot maintain adequate circulatory flow, symptoms of **valvular heart disease (VHD)** appear. Congenital malformations may be responsible, but in many cases the condition develops after **carditis**, an inflammation of the heart, occurs. One important cause of carditis is **rheumatic** (roo-MAT-ik) **fever**, an inflammatory autoimmune response to an infection by streptococcal bacteria that most often occurs in children.

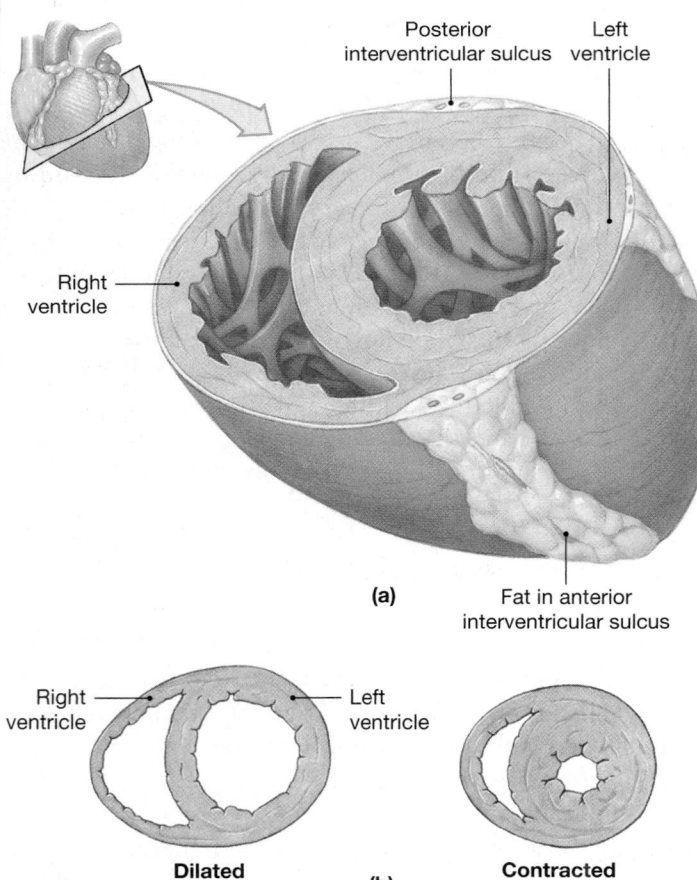

Figure 20–7 Structural Differences between the Left and Right Ventricles. (a) A diagrammatic sectional view through the heart, showing the relative thicknesses of the two ventricles. Notice the pouchlike shape of the right ventricle and the thickness of the left ventricle. **(b)** Diagrammatic views of the ventricles just before a contraction (dilated) and just after a contraction (contracted).
ATLAS: Plate 45d

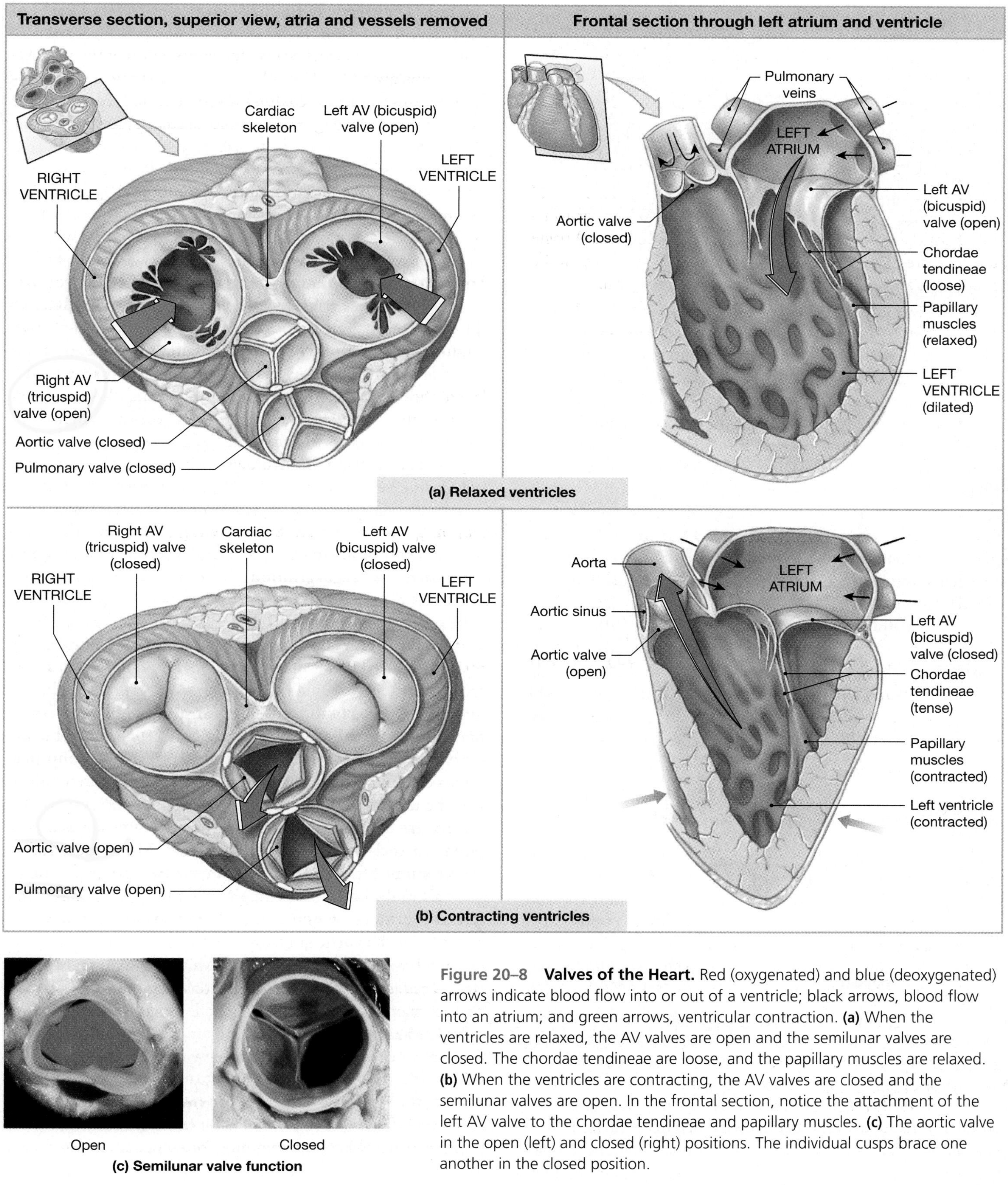

Transverse section, superior view, atria and vessels removed

Cardiac skeleton

Left AV (bicuspid) valve (open)

LEFT VENTRICLE

RIGHT VENTRICLE

Right AV (tricuspid) valve (open)

Aortic valve (closed)

Pulmonary valve (closed)

(a) Relaxed ventricles

Right AV (tricuspid) valve (closed)

Cardiac skeleton

Left AV (bicuspid) valve (closed)

RIGHT VENTRICLE

LEFT VENTRICLE

Aortic valve (open)

Pulmonary valve (open)

(b) Contracting ventricles

Frontal section through left atrium and ventricle

Pulmonary veins

LEFT ATRIUM

Aortic valve (closed)

Left AV (bicuspid) valve (open)

Chordae tendineae (loose)

Papillary muscles (relaxed)

LEFT VENTRICLE (dilated)

Aorta

LEFT ATRIUM

Aortic sinus

Aortic valve (open)

Left AV (bicuspid) valve (closed)

Chordae tendineae (tense)

Papillary muscles (contracted)

Left ventricle (contracted)

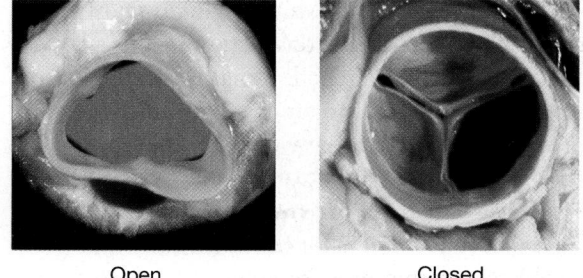

Open   Closed

(c) Semilunar valve function

Figure 20–8 Valves of the Heart. Red (oxygenated) and blue (deoxygenated) arrows indicate blood flow into or out of a ventricle; black arrows, blood flow into an atrium; and green arrows, ventricular contraction. **(a)** When the ventricles are relaxed, the AV valves are open and the semilunar valves are closed. The chordae tendineae are loose, and the papillary muscles are relaxed. **(b)** When the ventricles are contracting, the AV valves are closed and the semilunar valves are open. In the frontal section, notice the attachment of the left AV valve to the chordae tendineae and papillary muscles. **(c)** The aortic valve in the open (left) and closed (right) positions. The individual cusps brace one another in the closed position.

The A&P Top 100

#70 The heart has four chambers, two associated with the pulmonary circuit (right atrium and right ventricle) and two with the systemic circuit (left atrium and left ventricle). The left ventricle has a greater workload and is much more massive than the right ventricle, but these two chambers pump equal amounts of blood. AV valves prevent backflow from the ventricles into the atria, and semilunar valves prevent backflow from the aortic and pulmonary trunks into the ventricles.

Connective Tissues and the Cardiac Skeleton

The connective tissues of the heart include large numbers of collagen and elastic fibers. Each cardiac muscle cell is wrapped in a strong, but elastic, sheath, and adjacent cells are tied together by fibrous cross-links, or "struts." These fibers are, in turn, interwoven into sheets that separate the superficial and deep muscle layers. The connective tissue fibers (1) provide physical support for the cardiac muscle fibers, blood vessels, and nerves of the myocardium; (2) help distribute the forces of contraction; (3) add strength and prevent overexpansion of the heart; and (4) provide elasticity that helps return the heart to its original size and shape after a contraction.

The **cardiac skeleton** (sometimes called the *fibrous skeleton*) of the heart consists of four dense bands of tough elastic tissue that encircle the heart valves and the bases of the pulmonary trunk and aorta (**Figure 20–8**). These bands stabilize the positions of the heart valves and ventricular muscle cells and electrically insulate the ventricular cells from the atrial cells.

The Blood Supply to the Heart

The heart works continuously, so cardiac muscle cells require reliable supplies of oxygen and nutrients. Although a great volume of blood flows through the chambers of the heart, the myocardium needs its own, separate blood supply. The **coronary circulation** supplies blood to the muscle tissue of the heart. During maximum exertion, the demand for oxygen rises considerably. The blood flow to the myocardium may then increase to nine times that of resting levels. The coronary circulation includes an extensive network of coronary blood vessels (**Figure 20–9**).

The Coronary Arteries

The left and right **coronary arteries** originate at the base of the ascending aorta, at the aortic sinuses (**Figure 20–9a**). Blood pressure here is the highest in the systemic circuit. Each

time the left ventricle contracts, it forces blood into the aorta. The arrival of additional blood at elevated pressures stretches the elastic walls of the aorta. When the left ventricle relaxes, blood no longer flows into the aorta, pressure declines, and the walls of the aorta recoil. This recoil, called *elastic rebound*, pushes blood both forward, into the systemic circuit, and backward, through the aortic sinuses and then into the coronary arteries. Thus, the combination of elevated blood pressure and elastic rebound ensures a continuous flow of blood to meet the demands of active cardiac muscle tissue. However, myocardial blood flow is not steady; it peaks while the heart muscle is relaxed, and almost ceases while it contracts.

The **right coronary artery**, which follows the coronary sulcus around the heart, supplies blood to (1) the right atrium, (2) portions of both ventricles, and (3) portions of the conducting system of the heart, including the *sinoatrial (SA) node* and the *atrioventricular (AV) node*. The cells of these nodes are essential to establishing the normal heart rate. We will focus on their functions and their part in regulating the heart rate in a later section.

Inferior to the right atrium, the right coronary artery generally gives rise to one or more **marginal arteries**, which extend across the surface of the right ventricle (**Figure 20–9a,b**). The right coronary artery then continues across the posterior surface of the heart, supplying the **posterior interventricular artery**, or *posterior descending artery*, which runs toward the apex within the posterior interventricular sulcus (**Figure 20–9b,c**). The posterior interventricular artery supplies blood to the interventricular septum and adjacent portions of the ventricles.

The **left coronary artery** supplies blood to the left ventricle, left atrium, and interventricular septum. As it reaches the anterior surface of the heart, it gives rise to a circumflex branch and an anterior interventricular branch. The **circumflex artery** curves to the left around the coronary sulcus, eventually meeting and fusing with small branches of the **right coronary artery** (**Figure 20–9a–c**). The much larger **anterior interventricular artery**, or *left anterior descending artery* (LAD), swings around the pulmonary trunk and runs along the surface within the anterior interventricular sulcus (**Figure 20–9a**).

The anterior interventricular artery supplies small tributaries continuous with those of the posterior interventricular artery. Such interconnections between arteries are called **arterial anastomoses** (a-nas-tō-MŌ-sēz; *anastomosis*, outlet). Because the arteries are interconnected in this way, the blood supply to the cardiac muscle remains relatively constant despite pressure fluctuations in the left and right coronary arteries as the heart beats.

The Cardiac Veins

The various cardiac veins are shown in **Figure 20–9**. The **great cardiac vein** begins on the anterior surface of the ventricles, along the interventricular sulcus. This vein drains blood from

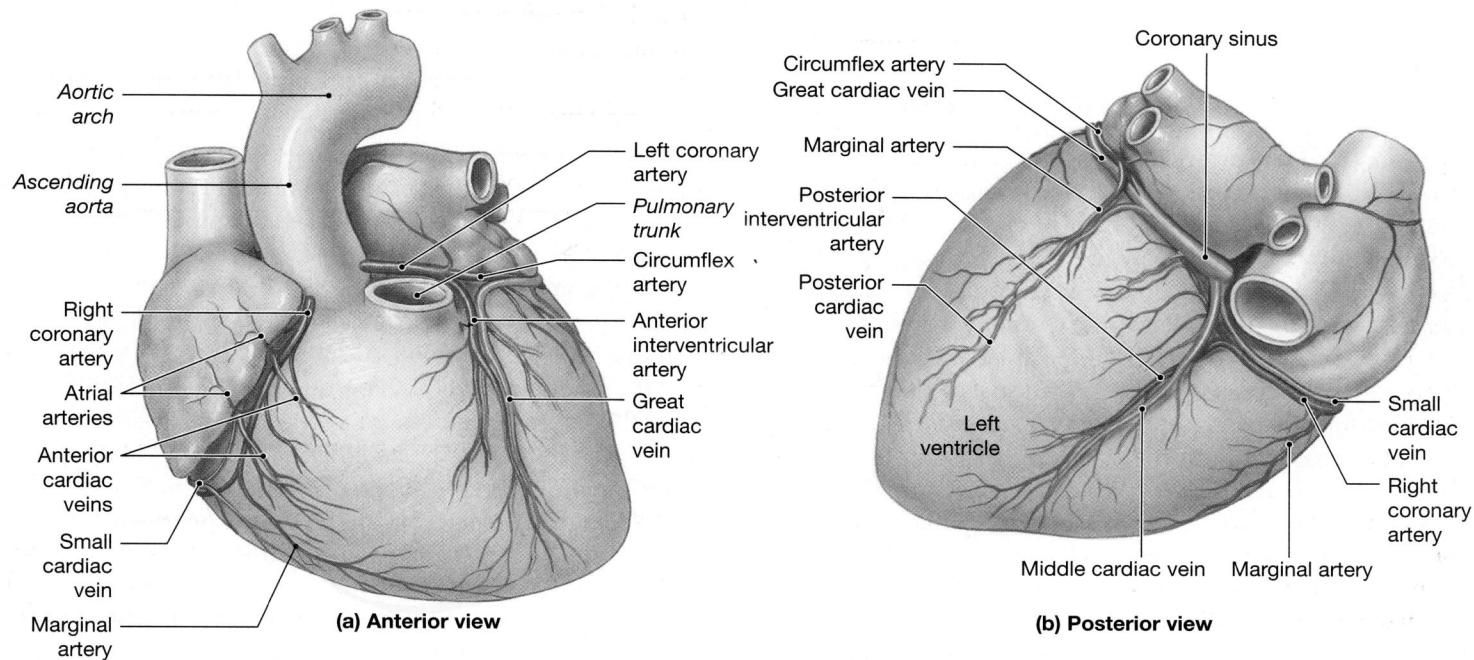

Aortic arch

Ascending aorta

Right coronary artery

Atrial arteries

Anterior cardiac veins

Small cardiac vein

Marginal artery

Left coronary artery

Pulmonary trunk

Circumflex artery

Anterior interventricular artery

Great cardiac vein

(a) Anterior view

Circumflex artery

Great cardiac vein

Marginal artery

Posterior interventricular artery

Posterior cardiac vein

Left ventricle

Coronary sinus

Small cardiac vein

Right coronary artery

Middle cardiac vein   Marginal artery

(b) Posterior view

Figure 20–9   Coronary Circulation. (a) Coronary vessels supplying and draining the anterior surface of the heart. **(b)** Coronary vessels supplying and draining the posterior surface of the heart. **(c)** A posterior view of the heart; the vessels have been injected with colored latex (liquid rubber). ATLAS: Plate 45b,c

artery & veins together.

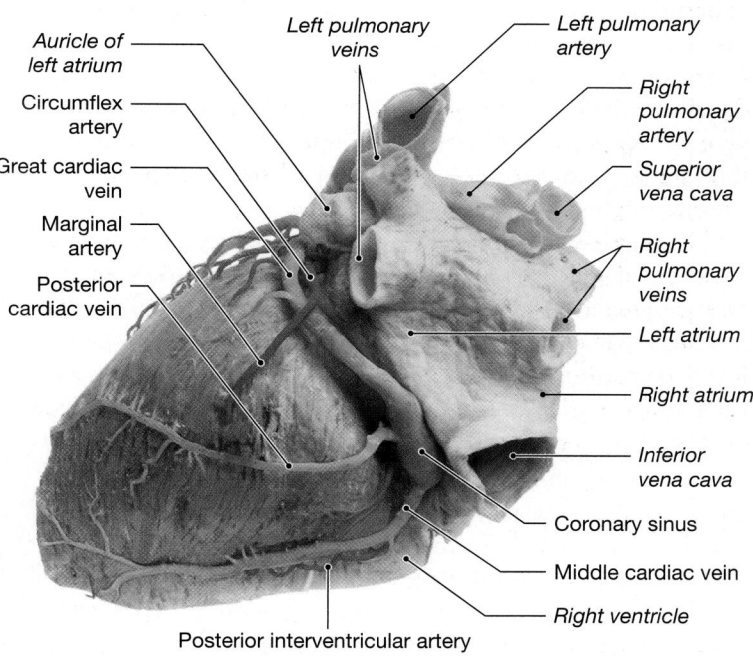

Auricle of left atrium

Circumflex artery

Great cardiac vein

Marginal artery

Posterior cardiac vein

Left pulmonary veins

Left pulmonary artery

Right pulmonary artery

Superior vena cava

Right pulmonary veins

Left atrium

Right atrium

Inferior vena cava

Coronary sinus

Middle cardiac vein

Right ventricle

Posterior interventricular artery

(c) Posterior view

the region supplied by the anterior interventricular artery, a branch of the left coronary artery. The great cardiac vein reaches the level of the atria and then curves around the left side of the heart within the coronary sulcus. The vein empties into the coronary sinus, which lies in the posterior portion of the coronary sulcus. The coronary sinus opens into the right atrium near the base of the inferior vena cava.

Other cardiac veins that empty into the great cardiac vein or the coronary sinus include (1) the **posterior cardiac vein**, draining the area served by the circumflex artery, (2) the **middle cardiac vein**, draining the area supplied by the posterior interventricular artery, and (3) the **small cardiac vein**, which receives blood from the posterior surfaces of the right atrium and ventricle. The **anterior cardiac veins**, which drain the anterior surface of the right ventricle, empty directly into the right atrium.

CHECKPOINT

1. **Damage to the semilunar valve on the right side of the heart would affect blood flow to which vessel?**

2. **What prevents the AV valves from swinging into the atria?**

systemic

3. **Why is the left ventricle more muscular than the right ventricle?**

pulmonary

See the blue Answers tab at the end of the book.

CLINICAL NOTE

Coronary Artery Disease

The term **coronary artery disease (CAD)** refers to areas of partial or complete blockage of coronary circulation. Cardiac muscle cells need a constant supply of oxygen and nutrients, so any reduction in blood flow to the heart muscle produces a corresponding reduction in cardiac performance. Such reduced circulatory supply, known as **coronary ischemia** (is-KĒ-mē-uh), generally results from partial or complete blockage of the coronary arteries. The usual cause is the formation of a fatty deposit, or *atherosclerotic plaque*, in the wall of a coronary vessel. The plaque, or an associated *thrombus* (clot), then narrows the passageway and reduces blood flow. Spasms in the smooth muscles of the vessel wall can further decrease or even stop blood flow. Plaques may be visible by *angiography* or high-resolution ultrasound, and the effects on coronary blood flow can be detected in digital subtraction angiography (DSA) scans of the heart (**Figure 20–10a,b**). We will consider the development and growth of plaques in Chapter 21.

One of the first symptoms of CAD is commonly **angina pectoris** (an-JĪ-nuh PEK-tor-is; *angina*, pain spasm + *pectoris*, of the chest). In its most common form, a temporary ischemia develops when the workload of the heart increases. Although the individual may feel comfortable at rest, exertion or emotional stress can produce a sensation of pressure, chest constriction, and pain that may radiate from the sternal area to the arms, back, and neck.

Coronary artery disease is treated by a combination of drug therapy and changes in lifestyle, including (1) limiting activities known to trigger angina attacks (such as strenuous exercise and stress), (2) stopping smoking, (3) lowering fat consumption, and (4) attending to associated problems such as hypertension, diabetes, and high cholesterol. Among the medications used to control angina are drugs that block sympathetic stimulation (*propranolol* or *metoprolol*), vasodilators such as nitroglycerin (nī-trō-GLIS-er-in), and drugs that block calcium movement into the cardiac and vascular smooth muscle cells (*calcium channel blockers*).

CAD can also be treated surgically. Blockage by a single, soft plaque may be reduced with the aid of a long, slender **catheter** (KATH-e-ter), a small-diameter tube. The catheter is inserted into a large peripheral artery and guided into a coronary artery to the plaque. A variety of surgical tools can be slid into the catheter, and the plaque can then be removed using laser beams or chewed to pieces by a device that resembles a miniature version of a Roto-Rooter drain cleaner. Debris created during the destruction of a plaque could block smaller vessels, so the pieces are sucked up through the catheter (an *atherectomy*).

In **balloon angioplasty** (AN-jē-ō-plas-tē; *angeion*, vessel), the tip of the catheter contains an inflatable balloon. Once in position, the balloon is inflated, pressing the plaque against the vessel walls. This procedure works best for small (under 10 mm), soft plaques. Several factors make angioplasty a highly attractive treatment: (1) The mortality rate during angioplasty is only about 1 percent; (2) the success rate is more than 90 percent; and (3) the procedure can be performed on an outpatient basis. Because plaques commonly redevelop after angioplasty, a fine tubular wire mesh called a *stent* may be inserted into the vessel, holding it open. Stents are now part of the standard protocol for many cardiac specialists, as the long-term success rate and incidence of complications are significantly lower than those of balloon angioplasty alone. If the circulatory blockage is too large for a single stent, mutiple stents can be inserted along the length of the vessel.

When angioplasty and stents are not advisable, coronary bypass surgery may be done. In a **coronary artery bypass graft (CABG)**, a small section is removed from either a small artery (commonly the *internal thoracic artery*) or a peripheral vein (such as the *great saphenous vein* of the leg) and is used to create a detour around the obstructed portion of a coronary artery. As many as four coronary arteries can be rerouted this way during a single operation. The procedures are named according to the number of vessels repaired, so we speak of single, double, triple, or quadruple coronary bypasses. The mortality rate during surgery for such operations—when they are performed before significant heart damage has occurred—is 1–2 percent. Under these conditions, the procedure eliminates angina symptoms in 70 percent of the cases and provides partial relief in another 20 percent.

Although coronary bypass surgery offers certain advantages, recent studies have shown that for mild angina, it does not yield significantly better results than drug therapy. Current guidelines recommend that coronary bypass surgery be reserved for cases of CAD and angina that are unlikely to respond to other treatment.

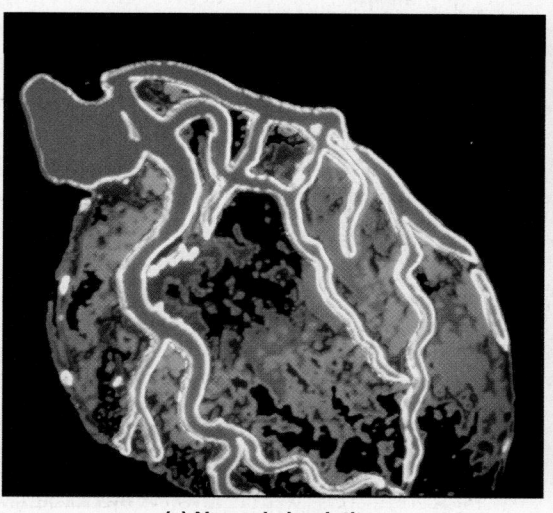

(a) Normal circulation   **(b) Restricted circulation**

Figure 20–10   Coronary Circulation and Clinical Testing. (a) A color-enhanced digital subtraction angiography (DSA) image of a healthy heart. The ventricular walls have an extensive circulatory supply. (The atria are not shown.) **(b)** A color-enhanced DSA image of a damaged heart. Most of the ventricular myocardium is deprived of circulation.

20-2 The conducting system distributes electrical impulses through the heart, and an electrocardiogram records the associated electrical events

Next we examine several aspects of each contraction of the heart, including the structure and function of the conducting system, electrical events as recorded in an electrocardiogram, and the functioning of contractile cells. We begin with an overview of cardiac physiology.

Cardiac Physiology

atria → ventricles

Figure 20–11 presents an overview of the aspects of cardiac physiology we will consider in this chapter. In a single cardiac contraction, or heartbeat, the entire heart contracts in series— first the atria and then the ventricles. Two types of cardiac muscle cells are involved in a normal heartbeat: (1) specialized muscle cells of the *conducting system*, which control and coordinate the heartbeat, and (2) *contractile cells*, which produce the powerful contractions that propel blood. Each heartbeat begins with an action potential generated at a pacemaker called the *SA node*, which is part of the conducting system. This electrical impulse is then propagated by the conducting

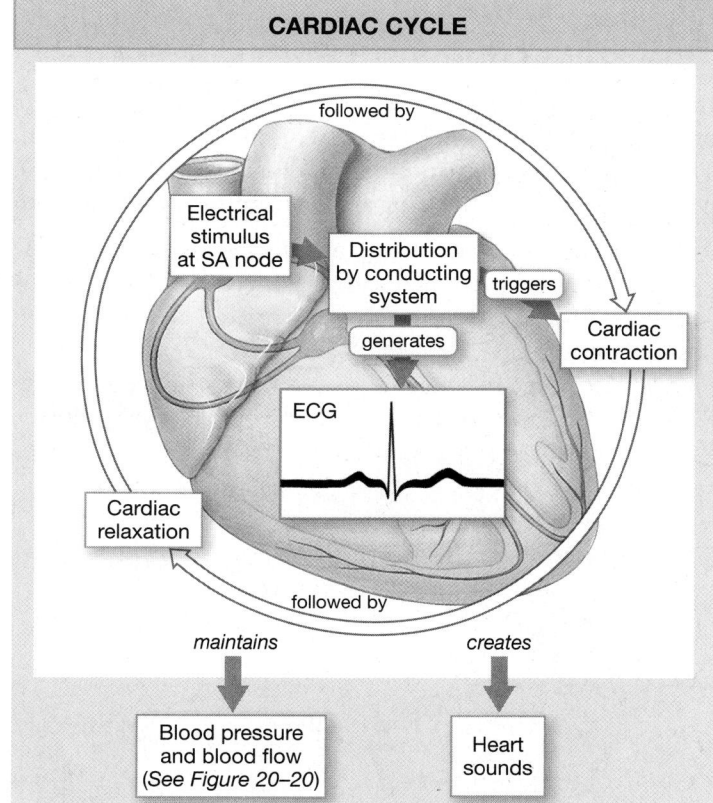

CARDIAC CYCLE

Figure 20–11   An Overview of Cardiac Physiology. The major events and relationships are indicated.

20 CARDIOVASCULAR

system and distributed so that the stimulated contractile cells will push blood in the right direction at the proper time. The electrical events under way in the conducting system can be monitored from the surface of the body through a procedure known as *electrocardiography*; the printed record of the result is called an *electrocardiogram* (*ECG* or *EKG*).

The arrival of an impulse at a cardiac muscle plasma membrane produces an action potential that is comparable to an action potential in a skeletal muscle fiber. As in a skeletal muscle fiber, this action potential triggers the contraction of the cardiac muscle cell. Thanks to the coordination provided by the conducting system, the atria contract first, driving blood into the ventricles through the AV valves, and the ventricles contract next, driving blood out of the heart through the semilunar valves.

The SA node generates impulses at regular intervals, and one heartbeat follows another throughout your life. After each heartbeat there is a brief pause—less than half a second—before the next heartbeat begins. The period between the start of one heartbeat and the start of the next is called the *cardiac cycle*.

A heartbeat lasts only about 370 msec. Although brief, it is a very busy period! We will begin our analysis of cardiac function by following the steps that produce a single heartbeat, from the generation of an action potential at the SA node through the contractions of the atria and ventricles.

The Conducting System

In contrast to skeletal muscle, cardiac muscle tissue contracts on its own, in the absence of neural or hormonal stimulation. This property is called **automaticity**, or *autorhythmicity*. The cells responsible for initiating and distributing the stimulus to

contract are part of the heart's **conducting system**, also known as the *cardiac conduction system* or the *nodal system*. This system is a network of specialized cardiac muscle cells that initiates and distributes electrical impulses. The actual contraction lags behind the beginning of an electrical impulse (the action potential), with the delay representing the time it takes for calcium ions to enter the sarcoplasm and activate the contraction process, as described in Chapter 10. ∞ p. 313

The conducting system (**Figure 20–12a**) includes the following elements:

- The *sinoatrial (SA) node*, located in the wall of the right atrium.
- The *atrioventricular (AV) node*, located at the junction between the atria and ventricles.
- *Conducting cells*, which interconnect the two nodes and distribute the contractile stimulus throughout the myocardium. In the atria, conducting cells are found in **internodal pathways**, which distribute the contractile stimulus to atrial muscle cells as the impulse travels from the SA node to the AV node. (The importance of these pathways in relaying the signal to the AV node remains in dispute, because an impulse can also spread from contractile cell to contractile cell, reaching the AV node at about the same time as an impulse that traverses an internodal pathway.) The ventricular conducting cells include those in the *AV bundle* and the *bundle branches*, as well as the *Purkinje* (pur-KIN-jē) *fibers*, which distribute the stimulus to the ventricular myocardium.

Most of the cells of the conducting system are smaller than the contractile cells of the myocardium and contain very few

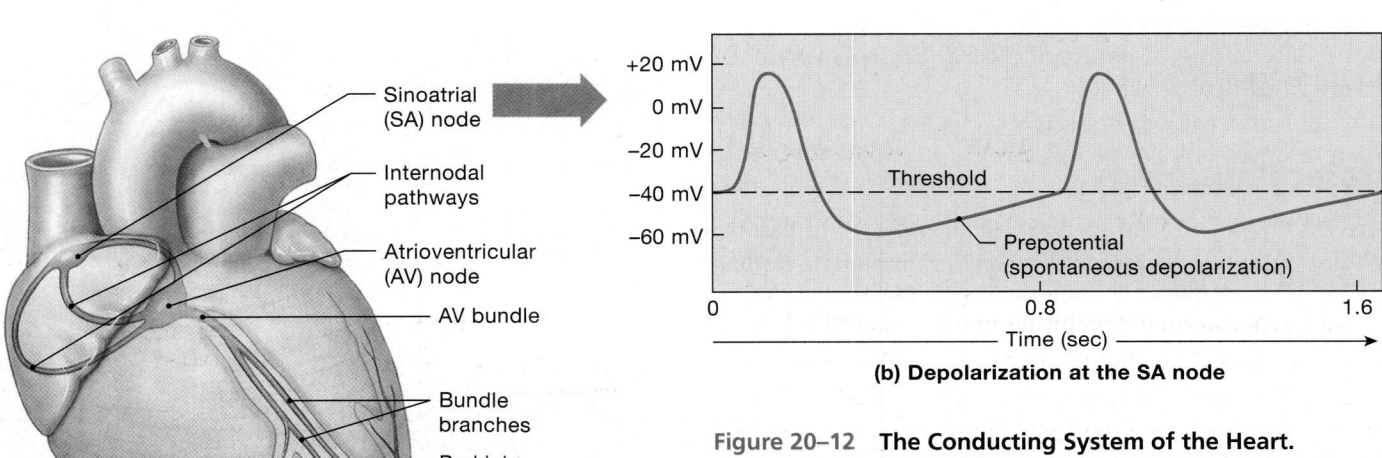

(a) The conducting system

- Sinoatrial (SA) node
- Internodal pathways
- Atrioventricular (AV) node
- AV bundle
- Bundle branches
- Purkinje fibers

(b) Depolarization at the SA node

+20 mV, 0 mV, −20 mV, −40 mV, −60 mV

Threshold

Prepotential (spontaneous depolarization)

Time (sec) — 0, 0.8, 1.6

Figure 20–12 **The Conducting System of the Heart.**
(a) Components of the conducting system. **(b)** Changes in the membrane potential of a pacemaker cell in the SA node that is establishing a heart rate of 72 beats per minute. Note the presence of a prepotential, a gradual spontaneous depolarization.

myofibrils. Purkinje cells, however, are much larger in diameter than the contractile cells; as a result, they conduct action potentials more quickly than other conducting cells. Conducting cells of the SA and AV nodes share an important characteristic: Their excitable membranes cannot maintain a stable resting potential. After each repolarization, the membrane gradually drifts toward threshold. This gradual depolarization is called a **prepotential** or *pacemaker potential* (**Figure 20–12b**).

The rate of spontaneous depolarization differs in various portions of the conducting system. It is fastest at the SA node, which in the absence of neural or hormonal stimulation generates action potentials at a rate of 80–100 per minute. Isolated cells of the AV node depolarize more slowly, generating 40–60 action potentials per minute. Because the SA node reaches threshold first, it establishes the heart rate—the impulse generated by the SA node brings the AV nodal cells to threshold faster than does the prepotential of the AV nodal cells. The normal resting heart rate is somewhat slower than 80–100 beats per minute, however, due to the effects of parasympathetic innervation. (The influence of autonomic innervation on heart rate is discussed in a later section.)

If any of the atrial pathways or the SA node becomes damaged, the heart will continue to beat, but at a slower rate, usually 40–60 beats per minute, as dictated by the AV node. Certain cells in the Purkinje fiber network depolarize spontaneously at an even slower rate, and if the rest of the conducting system is damaged, they can stimulate a heart rate of 20–40 beats per minute. Under normal conditions, cells of the AV bundle, the bundle branches, and most Purkinje fibers do not depolarize spontaneously. If, due to damage or disease, these cells *do* begin depolarizing spontaneously, the heart may no longer pump blood effectively, and death can result if the problem persists.

We will now trace the path of an impulse from its initiation at the SA node, examining its effects on the surrounding myocardium as we proceed.

The Sinoatrial (SA) Node

The **sinoatrial** (sī-nō-Ā-trē-al) **node (SA node)** is embedded in the posterior wall of the right atrium, near the entrance of the superior vena cava (**Figure 20–13** [STEP 1]). The SA node contains **pacemaker cells**, which establish the heart rate. As a result, the SA node is also known as the *cardiac pacemaker* or the *natural pacemaker*. The SA node is connected to the larger AV node by the internodal pathways in the atrial walls. It takes roughly 50 msec for an action potential to travel from the SA node to the AV node along these pathways. Along the way, the conducting cells pass the stimulus to contractile cells of both atria. The action potential then spreads across the atrial surfaces by cell-to-cell contact (**Figure 20–13** [STEP 2]). The stimulus affects only the atria, because the cardiac skeleton isolates the atrial myocardium from the ventricular myocardium.

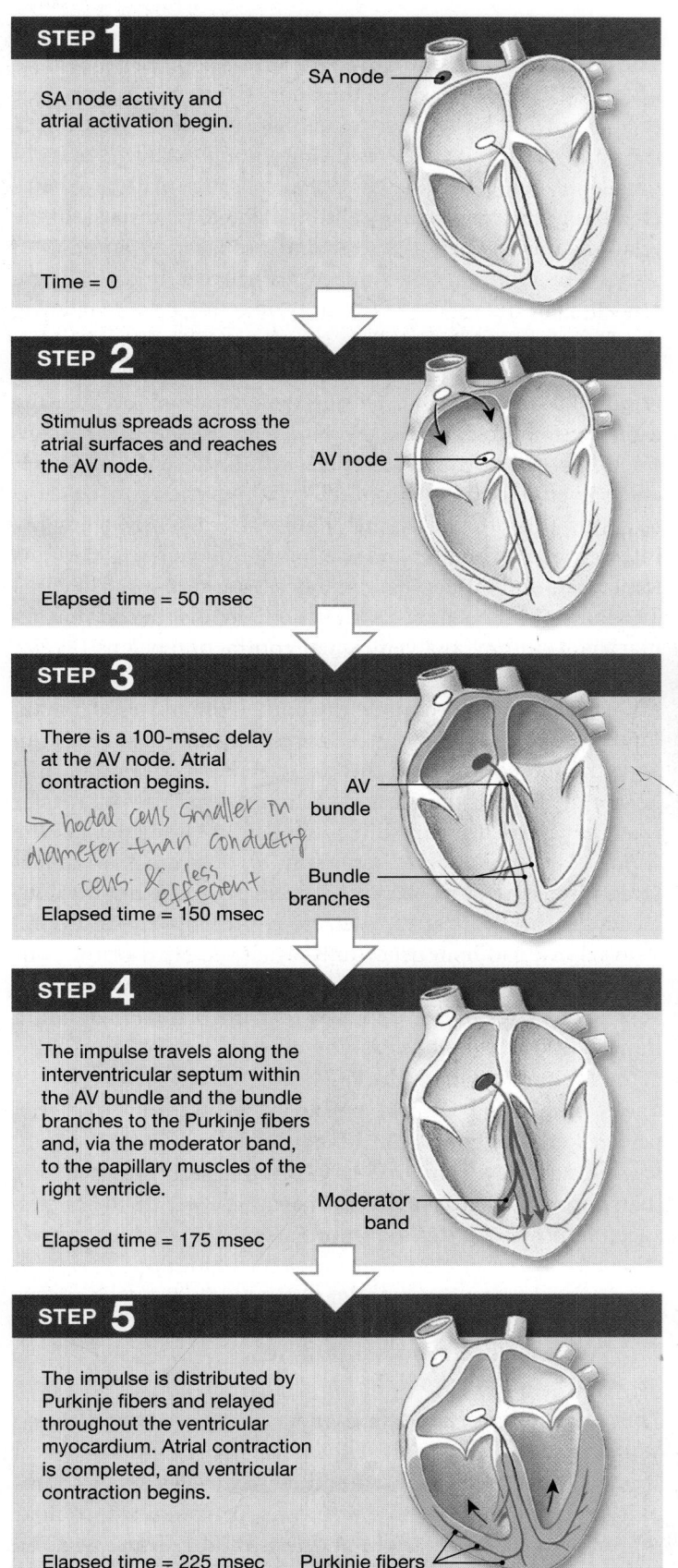

STEP 1

SA node activity and atrial activation begin.

SA node

Time = 0

STEP 2

Stimulus spreads across the atrial surfaces and reaches the AV node.

AV node

Elapsed time = 50 msec

STEP 3

There is a 100-msec delay at the AV node. Atrial contraction begins.

AV bundle

Bundle branches

→ *nodal cells smaller in diameter than conducting cells & less efficient*

Elapsed time = 150 msec

STEP 4

The impulse travels along the interventricular septum within the AV bundle and the bundle branches to the Purkinje fibers and, via the moderator band, to the papillary muscles of the right ventricle.

Moderator band

Elapsed time = 175 msec

STEP 5

The impulse is distributed by Purkinje fibers and relayed throughout the ventricular myocardium. Atrial contraction is completed, and ventricular contraction begins.

Elapsed time = 225 msec

Purkinje fibers

Figure 20–13 **Impulse Conduction through the Heart.**

Purkinje cells & fibers.

The Atrioventricular (AV) Node

The relatively large **atrioventricular (AV) node** (Figure 20–13 [STEP 2]) sits within the floor of the right atrium near the opening of the coronary sinus. The rate of propagation of the impulse slows as it leaves the internodal pathways and enters the AV node, because the nodal cells are smaller in diameter than the conducting cells. (Chapter 12 discussed the relationship between diameter and propagation speed. ∞ p. 413) In addition, the connections between nodal cells are less efficient than those between conducting cells at relaying the impulse from one cell to another. As a result, it takes about 100 msec for the impulse to pass through the AV node (Figure 20–13 [STEP 3]). This delay is important because the atria must contract before the ventricles do. Otherwise, contraction of the powerful ventricles would close the AV valves and prevent blood flow from the atria into the ventricles.

After this brief delay, the impulse is conducted along the interventricular bundle and the bundle branches to the Purkinje fibers and the papillary muscles (Figure 20–13 [STEP 4]). The Purkinje fibers then distribute the impulse to the ventricular myocardium, and ventricular contraction begins (Figure 20–13 [STEP 5]).

The cells of the AV node can conduct impulses at a maximum rate of 230 per minute. Because each impulse results in a ventricular contraction, this value is the maximum normal heart rate. Even if the SA node generates impulses at a faster rate, the ventricles will still contract at 230 beats per minute (bpm). This limitation is important, because mechanical factors (discussed later) begin to decrease the pumping efficiency of the heart at rates above approximately 180 bpm. Rates above 230 bpm occur only when the heart or the conducting system has been damaged or stimulated by drugs. As ventricular rates increase toward their theoretical maximum limit of 300–400 bpm, pumping effectiveness becomes dangerously, if not fatally, reduced.

A number of clinical problems result from abnormal pacemaker function. **Bradycardia** (brād-ē-KAR-dĕ-uh; *bradys*, slow) is a condition in which the heart rate is slower than normal, whereas **tachycardia** (tak-ē-KAR-dĕ-uh; *tachys*, swift) indicates a faster-than-normal heart rate. These are relative terms, and in clinical practice the definitions vary with the normal resting heart rate of the individual.

The AV Bundle, Bundle Branches, and Purkinje Fibers

The connection between the AV node and the **AV bundle**, also called the *bundle of His* (hiss), is normally the only electrical connection between the atria and the ventricles. Once an impulse enters the AV bundle, it travels to the interventricular septum and enters the **right** and **left bundle branches**. The left bundle branch, which supplies the massive left ventricle, is much larger than the right bundle branch. Both branches extend toward the apex of the heart, turn, and fan out deep to the endocardial surface. As the branches diverge, they conduct the impulse to **Purkinje fibers** and, through the moderator band, to the papillary muscles of the right ventricle.

Purkinje fibers conduct action potentials very rapidly—as fast as small myelinated axons. Within about 75 msec, the signal to begin a contraction has reached all the ventricular cardiac muscle cells. The entire process, from the generation of an impulse at the SA node to the complete depolarization of the ventricular myocardium, normally takes around 225 msec. By this time, the atria have completed their contractions and ventricular contraction can safely occur.

Because the bundle branches deliver the impulse across the moderator band to the papillary muscles directly, rather than by way of Purkinje fibers, the papillary muscles begin contracting before the rest of the ventricular musculature does. Contraction of the papillary muscles applies tension to the chordae tendineae, bracing the AV valves. By limiting the movement of the cusps, tension in the chordae tendineae prevents the backflow of blood into the atria when the ventricles contract. The Purkinje fibers radiate from the apex toward the base of the heart. As a result, ventricular contraction proceeds in a wave that begins at the apex and spreads toward the base. Blood is therefore pushed toward the base of the heart, into the aorta and pulmonary trunk.

If the conducting pathways are damaged, the normal rhythm of the heart will be disturbed. The resulting problems are called *conduction deficits*. If the SA node or internodal pathways are damaged, the AV node will assume command. The heart will continue beating normally, although at a slower rate. If an abnormal conducting cell or ventricular muscle cell begins generating action potentials at a higher rate, the impulses can override those of the SA or AV node. The origin of these abnormal signals is called an **ectopic** (ek-TOP-ik; out of place) **pacemaker**. The activity of an ectopic pacemaker partially or completely bypasses the conducting system, disrupting the timing of ventricular contraction. The result may be a dangerous reduction in the pumping efficiency of the heart. Such conditions are commonly diagnosed with the aid of an *electrocardiogram*.

The Electrocardiogram

The electrical events occurring in the heart are powerful enough to be detected by electrodes on the surface of the body. A recording of these events is an **electrocardiogram** (ē-lek-trō-KAR-dē-ō-gram), also called an **ECG** or **EKG**. Each time the heart beats, a wave of depolarization radiates through the atria, pauses at the AV node, then travels down the interventricular septum to the apex, turns, and spreads through the ventricular myocardium toward the base (Figure 20–13).

An ECG integrates electrical information obtained by placing electrodes at different locations on the body surface. Clinicians can use an ECG to assess the performance of specific nodal, conducting, and contractile components. When a portion of the heart has been damaged by a heart attack, for example, the ECG will reveal an abnormal pattern of impulse conduction.

The appearance of the ECG varies with the placement of the monitoring electrodes, or *leads*. **Figure 20–14a** shows the leads in one of the standard configurations. **Figure 20–14b** depicts the important features of an ECG obtained with that configuration. Note the following ECG features:

- The small **P wave**, which accompanies the depolarization of the atria. The atria begin contracting about 25 msec after the start of the P wave.

- The **QRS complex**, which appears as the ventricles depolarize. This is a relatively strong electrical signal, because the ventricular muscle is much more massive than that of the atria. It is also a complex signal, largely because of the complex pathway that the spread of depolarization takes through the ventricles. The ventricles begin contracting shortly after the peak of the **R wave**.

- The smaller **T wave**, which indicates ventricular repolarization. A deflection corresponding to atrial repolarization is not apparent, because it occurs while the ventricles are depolarizing, and the electrical events are masked by the QRS complex.

To analyze an ECG, you must measure the size of the voltage changes and determine the durations and temporal relationships of the various components. Of particular diagnostic importance is the amount of depolarization occurring during the P wave and the QRS complex. For example, an excessively large QRS complex often indicates that the heart has become enlarged. A smaller-than-normal electrical signal may mean that the mass of the heart muscle has decreased (although monitoring problems are more often responsible). The size and shape of the T wave may also be affected by any condition that slows ventricular repolarization. For example, starvation and low cardiac energy reserves, coronary ischemia, or abnormal ion concentrations will reduce the size of the T wave.

The times between waves are reported as *segments* or *intervals*. Segments generally extend from the end of one wave to the start of another; intervals are more variable, but always include at least one entire wave. Commonly used segments

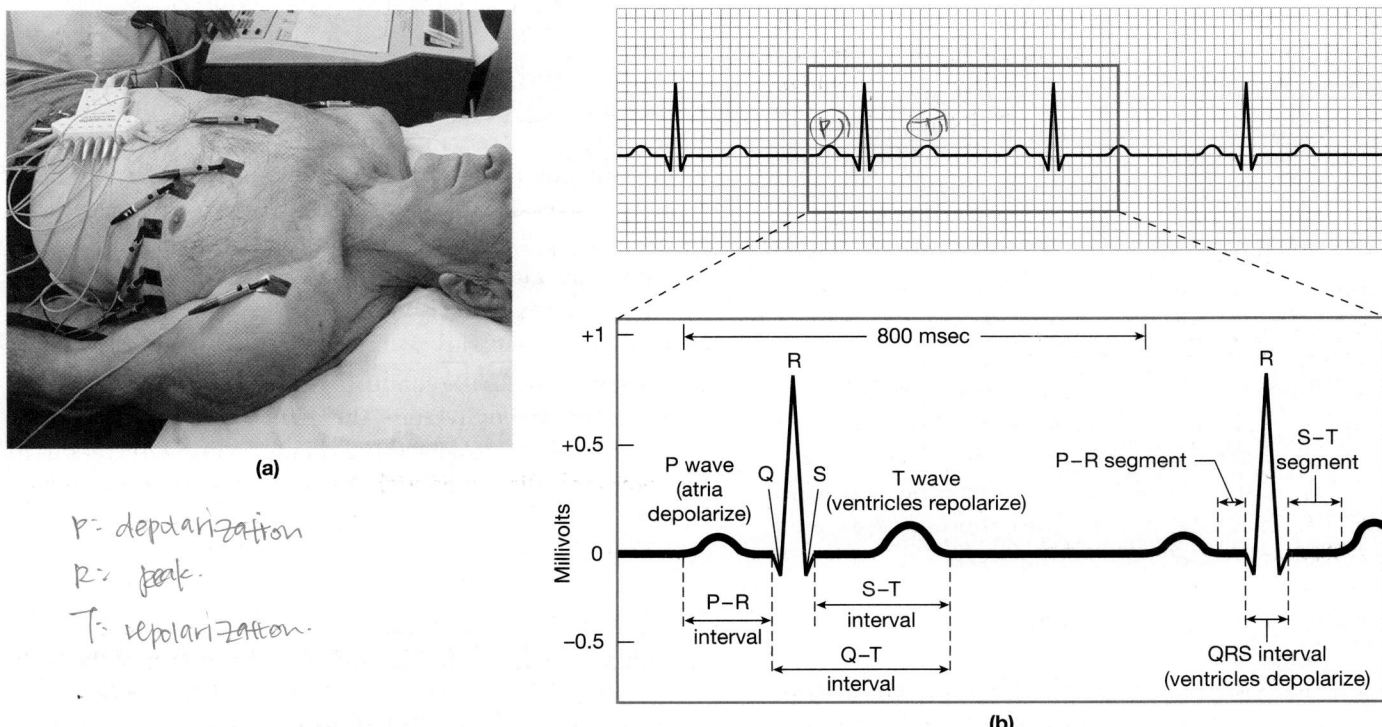

Figure 20–14 An Electrocardiogram. (a) Electrode placement for recording a standard ECG. **(b)** An ECG printout is a strip of graph paper containing a record of the electrical events monitored by the electrodes. The placement of electrodes on the body surface affects the size and shape of the waves recorded. This example is a normal ECG; the enlarged section indicates the major components of the ECG and the measurements most often taken during clinical analysis.

and intervals are indicated in **Figure 20–14b**. The names, however, can be somewhat misleading. For example:

- The **P–R interval** extends from the start of atrial depolarization to the start of the QRS complex (ventricular depolarization) rather than to R, because in abnormal ECGs the peak at R can be difficult to determine. Extension of the P–R interval to more than 200 msec can indicate damage to the conducting pathways or AV node.

- The **Q–T interval** indicates the time required for the ventricles to undergo a single cycle of depolarization and repolarization. It is usually measured from the end of the P–R interval rather than from the bottom of the Q wave. The Q–T interval can be lengthened by electrolyte disturbances, some medications, conduction problems, coronary ischemia, or myocardial damage. A congenital heart defect that can cause sudden death without warning may be detectable as a prolonged Q–T interval.

Despite the variety of sophisticated equipment available to assess or visualize cardiac function, in the majority of cases the ECG provides the most important diagnostic information. ECG analysis is especially useful in detecting and diagnosing **cardiac arrhythmias** (ā-RITH-mē-az)—abnormal patterns of cardiac electrical activity. Momentary arrhythmias are not inherently dangerous; about 5 percent of healthy individuals experience a few abnormal heartbeats each day. Clinical problems appear when arrhythmias reduce the pumping efficiency of the heart. Serious arrhythmias may indicate damage to the myocardium, injuries to the pacemakers or conduction pathways, exposure to drugs, or abnormalities in the electrolyte composition of extracellular fluids.

The A&P Top 100

#71 The heart rate is normally established by cells of the SA node, but that rate can be modified by autonomic activity, hormones, and other factors. From the SA node, the stimulus is conducted to the AV node, the AV bundle, the bundle branches, and Purkinje fibers before reaching the ventricular muscle cells. The electrical events associated with the heartbeat can be monitored in an electrocardiogram (ECG).

Contractile Cells

The Purkinje fibers distribute the stimulus to the **contractile cells**, which form the bulk of the atrial and ventricular walls. In the discussions of cardiac muscle tissue in previous chapters, we considered only the structure of contractile cells, which account for roughly 99 percent of the muscle cells in the heart. In both cardiac muscle cells and skeletal muscle fibers, (1) an action potential leads to the appearance of Ca^{2+} among the myofibrils, and (2) the binding of Ca^{2+} to troponin on the thin filaments initiates the contraction. But skeletal and cardiac muscle cells differ in terms of the nature of the action potential, the source of the Ca^{2+}, and the duration of the resulting contraction. ⚭ p. 329

The Action Potential in Cardiac Muscle Cells

The resting potential of a ventricular contractile cell is approximately −90 mV, comparable to that of a resting skeletal muscle fiber (−85 mV). (The resting potential of an atrial contractile cell is about −80 mV, but the basic principles described here apply to atrial cells as well.) An action potential begins when the membrane of the ventricular muscle cell is brought to threshold, usually at about −75 mV. Threshold is normally reached in a portion of the membrane next to an intercalated disc. The typical stimulus is the excitation of an adjacent muscle cell. Once threshold has been reached, the action potential proceeds in three basic steps (**Figure 20–15a**):

Step 1 *Rapid Depolarization.* The stage of rapid depolarization in a cardiac muscle cell resembles that in a skeletal muscle fiber. At threshold, voltage-gated sodium channels open, and the membrane suddenly becomes permeable to Na^+. The result is a massive influx of sodium ions and the rapid depolarization of the sarcolemma. The channels involved are called **fast sodium channels**, because they open quickly and remain open for only a few milliseconds.

Step 2 *The Plateau.* As the transmembrane potential approaches +30 mV, the voltage-gated sodium channels close. They will remain closed and inactivated until the membrane potential reaches −60 mV. The cell now begins actively pumping Na^+ out of the cell. However, a net loss of positive charges does not continue, because as the sodium channels are closing, voltage-gated calcium channels are opening. These channels are called **slow calcium channels**, because they open slowly and remain open for a relatively long period—roughly 175 msec. While the slow calcium channels are open, calcium ions enter the sarcoplasm. The entry of positive charges through the calcium channels, in the form of Ca^{2+}, roughly balances the loss of positive ions through the active transport of Na^+, and the transmembrane potential remains near 0 mV for an extended period. This portion of the action potential curve is called the plateau. The presence of a plateau is the major difference between action potentials in cardiac muscle cells and in skeletal muscle fibers. In a skeletal muscle fiber, rapid depolarization is immediately followed by rapid repolarization.

Step 3 *Repolarization.* As the plateau continues, slow calcium channels begin closing, and **slow potassium channels** begin opening. As the channels open, potassium ions (K^+) rush out of the cell, and the net result is a period of rapid repolarization that restores the resting potential.

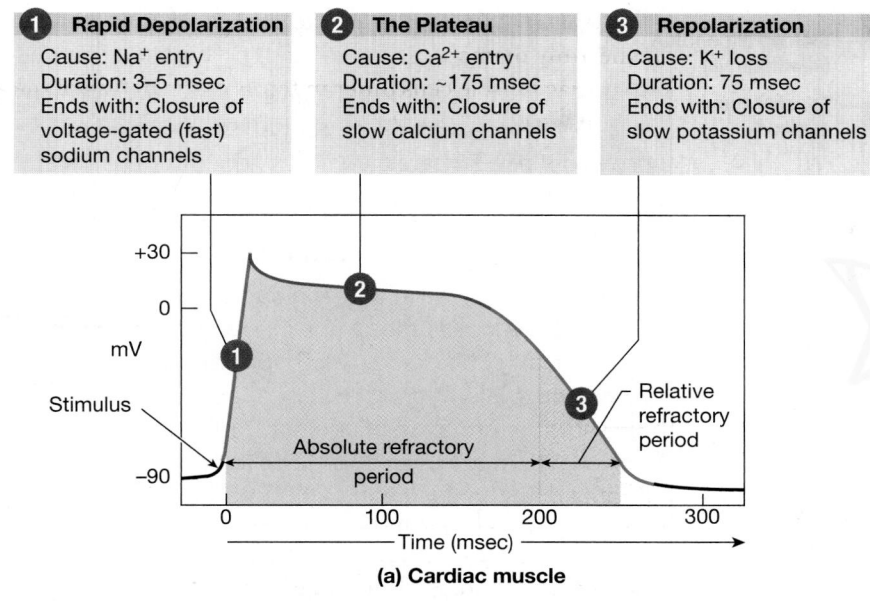

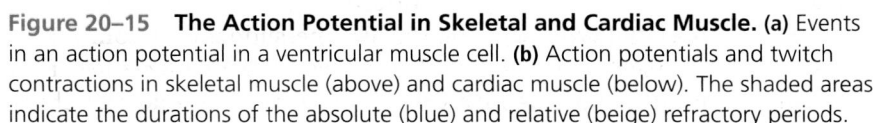

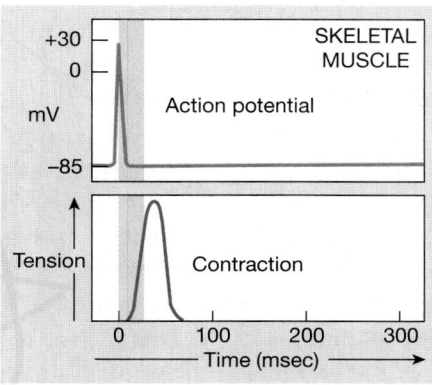

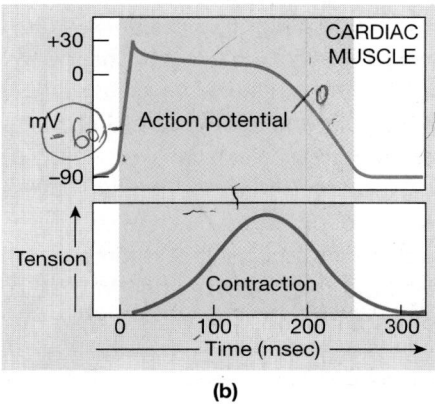

(b)

Figure 20–15   The Action Potential in Skeletal and Cardiac Muscle. (a) Events in an action potential in a ventricular muscle cell. **(b)** Action potentials and twitch contractions in skeletal muscle (above) and cardiac muscle (below). The shaded areas indicate the durations of the absolute (blue) and relative (beige) refractory periods.

The Refractory Period. As with skeletal muscle contractions, for some time after an action potential begins, the membrane will not respond normally to a second stimulus. This time is called the refractory period. Initially, in the absolute refractory period, the membrane cannot respond at all, because the sodium channels either are already open or are closed and inactivated. In a ventricular muscle cell, the absolute refractory period lasts approximately 200 msec, spanning the duration of the plateau and the initial period of rapid repolarization.

The absolute refractory period is followed by a shorter (50 msec) relative refractory period. During this period, the voltage-gated sodium channels are closed, but can open. The membrane will respond to a stronger-than-normal stimulus by initiating another action potential. In total, an action potential in a ventricular contractile cell lasts 250–300 msec, roughly 30 times as long as a typical action potential in a skeletal muscle fiber.

The Role of Calcium Ions in Cardiac Contractions

The appearance of an action potential in the cardiac muscle plasma membrane produces a contraction by causing an increase in the concentration of Ca^{2+} around the myofibrils. This process occurs in two steps:

1. Calcium ions crossing the plasma membrane during the plateau phase of the action potential provide roughly 20 percent of the Ca^{2+} required for a contraction.

2. The arrival of extracellular Ca^{2+} is the trigger for the release of additional Ca^{2+} from reserves in the sarcoplasmic reticulum (SR).

Extracellular calcium ions thus have both direct and indirect effects on cardiac muscle cell contraction. For this reason, cardiac muscle tissue is highly sensitive to changes in the Ca^{2+} concentration of the extracellular fluid.

In a skeletal muscle fiber, the action potential is relatively brief and ends as the related twitch contraction begins (**Figure 20–15b**). The twitch contraction is short and ends as the SR reclaims the Ca^{2+} it released. In a cardiac muscle cell, as we have seen, the action potential is prolonged, and calcium ions continue to enter the cell throughout the plateau. As a result, the period of active muscle cell contraction continues until the plateau ends. As the slow calcium channels close, the intracellular calcium ions are absorbed by the SR or are pumped out of the cell, and the muscle cell relaxes.

In skeletal muscle fibers, the refractory period ends before peak tension develops. As a result, twitches can summate and tetanus can occur. In cardiac muscle cells, the absolute refractory period continues until relaxation is under

CLINICAL NOTE

Myocardial Infarction

In a **myocardial** (mī-ō-KAR-dē-al) **infarction (MI)**, or *heart attack*, part of the coronary circulation becomes blocked, and cardiac muscle cells die from lack of oxygen. The death of affected tissue creates a nonfunctional area known as an *infarct*. Heart attacks most commonly result from severe coronary artery disease (CAD). The consequences depend on the site and nature of the circulatory blockage. If it occurs near the start of one of the coronary arteries, the damage will be widespread and the heart may stop beating. If the blockage involves one of the smaller arterial branches, the individual may survive the immediate crisis but may have many complications, all unpleasant. As scar tissue forms in the damaged area, the heartbeat may become irregular and less effective, and other vessels can become constricted, creating additional circulatory problems.

Myocardial infarctions are generally associated with fixed, partial blockages, such as those seen in CAD. When a crisis develops as a result of thrombus formation at a plaque (the most common cause of an MI), the condition is called **coronary thrombosis**. A vessel already narrowed by plaque formation may also become blocked by a sudden spasm in the smooth muscles of the vascular wall. The individual then may experience intense pain, similar to that felt in an angina attack, but persisting even at rest. However, pain does not always accompany a heart attack, and *silent heart attacks* may be even more dangerous than more apparent attacks, because the condition may go undiagnosed and may not be treated before a fatal MI occurs. Roughly 25 percent of heart attacks are not recognized when they occur.

The cytoplasm of a damaged cardiac muscle cell differs from that of a normal muscle cell. As the supply of oxygen decreases, the cells become more dependent on anaerobic metabolism to meet their energy needs. ∞ p. 320 Over time, the cytoplasm accumulates large quantities of enzymes involved with anaerobic energy production. As the membranes of damaged cardiac muscle cells deteriorate, these enzymes enter the surrounding intercellular fluids. The appearance of such enzymes in the circulation thus indicates that infarction has occurred. The most common enzymes that may be measured in a diagnostic blood test include **cardiac troponin T**, **cardiac troponin I**, and the MB isomer of a special form of creatine phosphokinase that occurs only in cardiac muscle, called **CK-MB**.

About 25 percent of MI patients die before obtaining medical assistance, and 65 percent of MI deaths among those under age 50 occur within an hour after the initial infarction. The goals of treatment are to limit the size of the infarct and to avoid additional complications by preventing irregular contractions, improving circulation with vasodilators, providing pain relief and additional oxygen, reducing the cardiac workload, and, if possible, eliminating the cause of the circulatory blockage. Anticoagulants (even aspirin chewed and swallowed at the start of an MI) are essential to prevent the formation of additional thrombi, and clot-dissolving enzymes may reduce the extent of the damage if they are administered within six hours after the MI occurred. Current evidence suggests that tissue plasminogen activator (t-PA), which is relatively expensive, is more beneficial than other fibrinolytic agents, such as *urokinase* or *streptokinase*. ∞ p. 677 Follow-up treatment with coumadin (an oral anticoagulant), aspirin, or both is recommended; without further treatment, circulatory blockages will reappear in about 20 percent of patients.

Roughly 1.3 million MIs occur in the United States each year, and half the victims die within a year of the incident. The following factors appear to increase the risk of a heart attack: (1) smoking, (2) high blood pressure, (3) high blood cholesterol levels, (4) high circulating levels of low-density lipoproteins (LDLs), (5) diabetes, (6) male gender (below age 70), (7) severe emotional stress, (8) obesity, (9) genetic predisposition, and (10) a sedentary lifestyle. We will consider the role of lipoproteins and cholesterol in plaque formation and heart disease in Chapter 21. Although the heart attack rate in women under age 70 is lower than that in men, the mortality rate for women is higher—perhaps because heart disease in women is neither diagnosed as early nor treated as aggressively as that in men.

The presence of any two risk factors more than doubles the risk of heart attack, so eliminating as many risk factors as possible improves the chances of preventing or surviving a heart attack. Limiting cholesterol in the diet, exercising to reduce weight, and seeking treatment for high blood pressure are steps in the right direction. It has been estimated that a reduction in coronary risk factors could prevent 150,000 deaths each year in the United States alone.

---

way. Because summation is not possible, tetanic contractions cannot occur in a normal cardiac muscle cell, regardless of the frequency or intensity of stimulation. This feature is vital: A heart in tetany could not pump blood. With a single twitch lasting 250 msec or longer, a normal cardiac muscle cell could reach 300–400 contractions per minute under maximum stimulation. This rate is not reached in a normal heart, due to limitations imposed by the conducting system.

The Energy for Cardiac Contractions

When a normal heart is beating, the energy required is obtained by the mitochondrial breakdown of fatty acids (stored as lipid droplets) and glucose (stored as glycogen). These aerobic reactions can occur only when oxygen is readily available. ∞ p. 320

In addition to obtaining oxygen from the coronary circulation, cardiac muscle cells maintain their own sizable reserves of oxygen. In these cells, oxygen molecules are bound to the heme units of myoglobin molecules. (We discussed this globular protein, which reversibly binds oxygen molecules, and its function in muscle fibers in Chapter 10.) ∞ p. 323 Normally, the combination of circulatory supplies plus myoglobin reserves is enough to meet the oxygen demands of the heart, even when it is working at maximum capacity.

> ### CHECKPOINT
>
> 4. Define automaticity.
>
> 5. Which structure of the heart is known as the cardiac pacemaker or the natural pacemaker?
>
> 6. If the cells of the SA node failed to function, how would the heart rate be affected?
>
> 7. Why is it important for impulses from the atria to be delayed at the AV node before they pass into the ventricles?
>
> See the blue Answers tab at the end of the book.

20-3 Events during a complete heartbeat constitute a cardiac cycle

Each heartbeat is followed by a brief resting phase, allowing time for the chambers to relax and prepare for the next heartbeat. The period between the start of one heartbeat and the beginning of the next is a single **cardiac cycle**. The cardiac cycle, therefore, includes alternating periods of contraction and relaxation. For any one chamber in the heart, the cardiac cycle can be divided into two phases: (1) systole and (2) diastole. During **systole** (SIS-tō-lē), or contraction, the chamber contracts and pushes blood into an adjacent chamber or into an arterial trunk. Systole is followed by **diastole** (dī-AS-tō-lē), or relaxation. During diastole, the chamber fills with blood and prepares for the next cardiac cycle.

Tips&Tricks

To remember that systole is contraction, relate "**syst**" to "**syst**em," as in during systole, a contraction sends blood out of the heart and into the circulatory system. The word part "**di**" can mean two, so during **di**astole, the heart "puts two feet up" and relaxes.

Fluids move from an area of higher pressure to one of lower pressure. In the course of the cardiac cycle, the pressure within each chamber rises during systole and falls during diastole. Valves between adjacent chambers help ensure that blood flows in the required direction, but blood will flow from one chamber to another only if the pressure in the first chamber exceeds that in the second. This basic principle governs the movement of blood between atria and ventricles, between ventricles and arterial trunks, and between major veins and atria.

The correct pressure relationships are dependent on the careful timing of contractions. For example, blood could not move in the proper direction if an atrium and its attached ventricle contracted at precisely the same moment. The elaborate pacemaking and conducting systems normally provide the required spacing between atrial and ventricular systoles. At a representative heart rate of 75 bpm, a sequence of systole and diastole in either the atria or the ventricles lasts 800 msec. For convenience, we will assume that the cardiac cycle is determined by the atria, and that it includes one cycle of atrial systole and atrial diastole. This convention follows the previous description of the conducting system and the propagation of the stimulus for contraction.

Phases of the Cardiac Cycle

The phases of the cardiac cycle—atrial systole, atrial diastole, ventricular systole, and ventricular diastole—are diagrammed in **Figure 20–16** for a heart rate of 75 bpm. When the cardiac cycle begins, all four chambers are relaxed, and the ventricles are partially filled with blood. During atrial systole, the atria contract, filling the ventricles completely with blood (**Figure 20–16a,b**). Atrial systole lasts 100 msec. Over this period, blood cannot flow into the atria because atrial pressure exceeds venous pressure. Yet there is very little backflow into the veins, even though the connections with the venous system lack valves, because blood takes the path of least resistance. Resistance to blood flow through the broad AV connections and into the ventricles is less than that through the smaller, angled openings of the large veins.

The atria next enter atrial diastole, which continues until the start of the next cardiac cycle. Atrial diastole and ventricular systole begin at the same time. Ventricular systole lasts 270 msec. During this period, blood is pushed through the systemic and pulmonary circuits and toward the atria (**Figure 20–16c,d**). The heart then enters ventricular diastole (**Figure 20–16e,f**), which lasts 530 msec (the 430 msec remaining in this cardiac cycle, plus the first 100 msec of the next). For the rest of this cycle, filling occurs passively, and both the atria and the ventricles are relaxed. The next cardiac cycle begins with atrial systole and the completion of ventricular filling.

When the heart rate increases, all the phases of the cardiac cycle are shortened. The greatest reduction occurs in the

[handwritten marginal note: atrial diastole & vent. systole start at same time.]

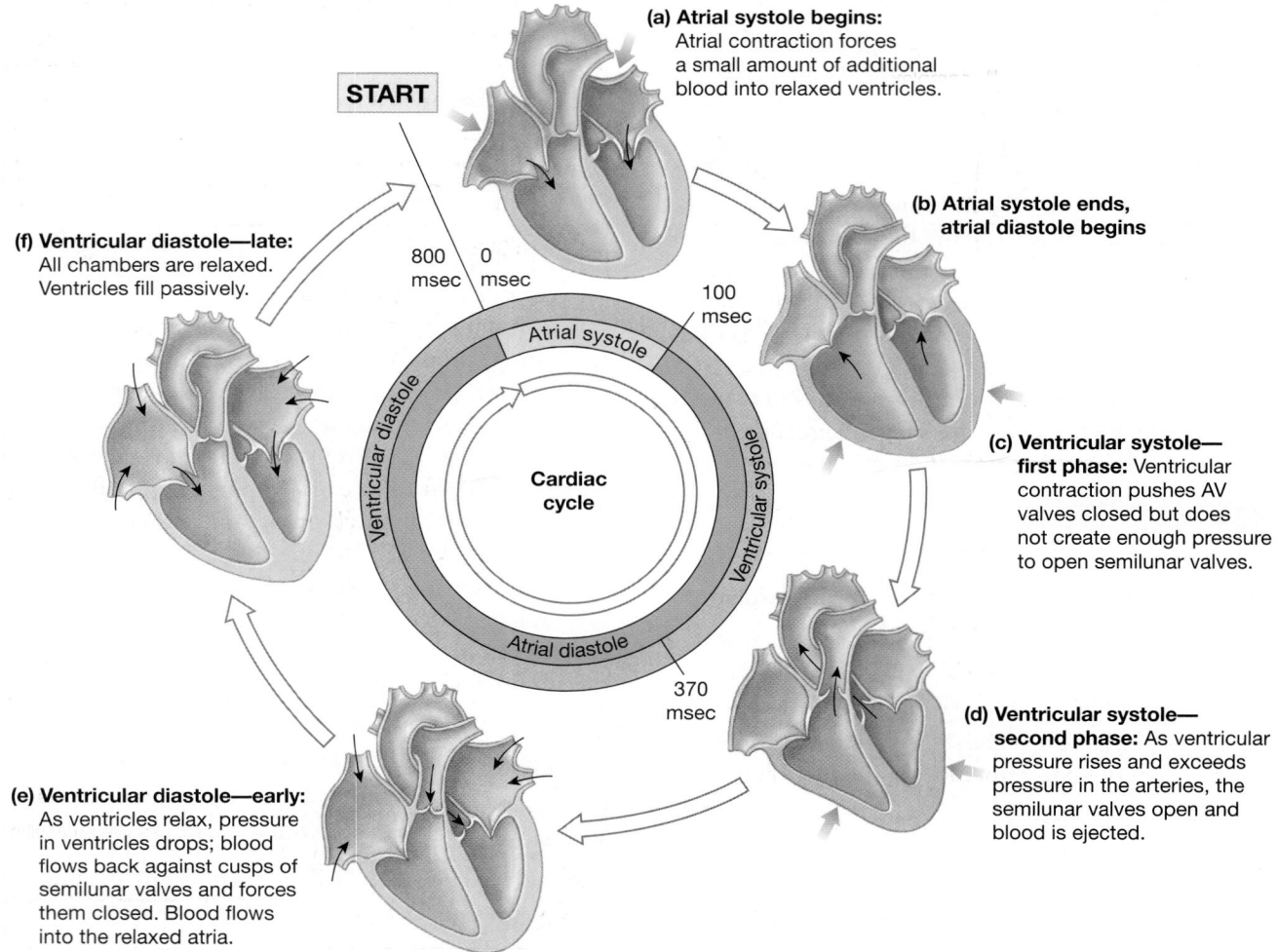

START

(a) Atrial systole begins: Atrial contraction forces a small amount of additional blood into relaxed ventricles.

(f) Ventricular diastole—late: All chambers are relaxed. Ventricles fill passively.

800 msec   0 msec

100 msec

Atrial systole

Ventricular diastole

Cardiac cycle

Ventricular systole

Atrial diastole

(b) Atrial systole ends, atrial diastole begins

(c) Ventricular systole— first phase: Ventricular contraction pushes AV valves closed but does not create enough pressure to open semilunar valves.

370 msec

(d) Ventricular systole— second phase: As ventricular pressure rises and exceeds pressure in the arteries, the semilunar valves open and blood is ejected.

(e) Ventricular diastole—early: As ventricles relax, pressure in ventricles drops; blood flows back against cusps of semilunar valves and forces them closed. Blood flows into the relaxed atria.

Figure 20–16 **Phases of the Cardiac Cycle.** Thin black arrows indicate blood flow, and green arrows indicate contractions.

length of time spent in diastole. When the heart rate climbs from 75 bpm to 200 bpm, the time spent in systole drops by less than 40 percent, but the duration of diastole is reduced by almost 75 percent.

Pressure and Volume Changes in the Cardiac Cycle

Figure 20–17 plots the pressure and volume changes during the cardiac cycle; the circled numbers in the figure correspond to numbered items in the text. The figure shows pressure and volume within the left atrium and left ventricle, but the discussion that follows applies to both sides of the heart. Although pressures are lower in the right atrium and right ventricle, both sides of the heart contract at the same time, and they eject equal volumes of blood.

Atrial Systole

The cardiac cycle begins with atrial systole, which lasts about 100 msec in a resting adult:

1. As the atria contract, rising atrial pressures push blood into the ventricles through the open right and left AV valves.

2. At the start of atrial systole, the ventricles are already filled to about 70 percent of their normal capacity, due to passive blood flow during the end of the previous cardiac cycle. As the atria contract, rising atrial pressures provide the remaining 30 percent by pushing blood through the open AV valves. Atrial systole essentially "tops off" the ventricles.

3. At the end of atrial systole, each ventricle contains the maximum amount of blood that it will hold in this cardiac cycle. That quantity is called the **end-diastolic volume (EDV)**. In an adult who is standing at rest, the end-diastolic volume is typically about 130 mL (about 4.4 oz).

Ventricular Systole

As atrial systole ends, ventricular systole begins. This period lasts approximately 270 msec in a resting adult. As the pressures in the ventricles rise above those in the atria, the AV valves are pushed closed.

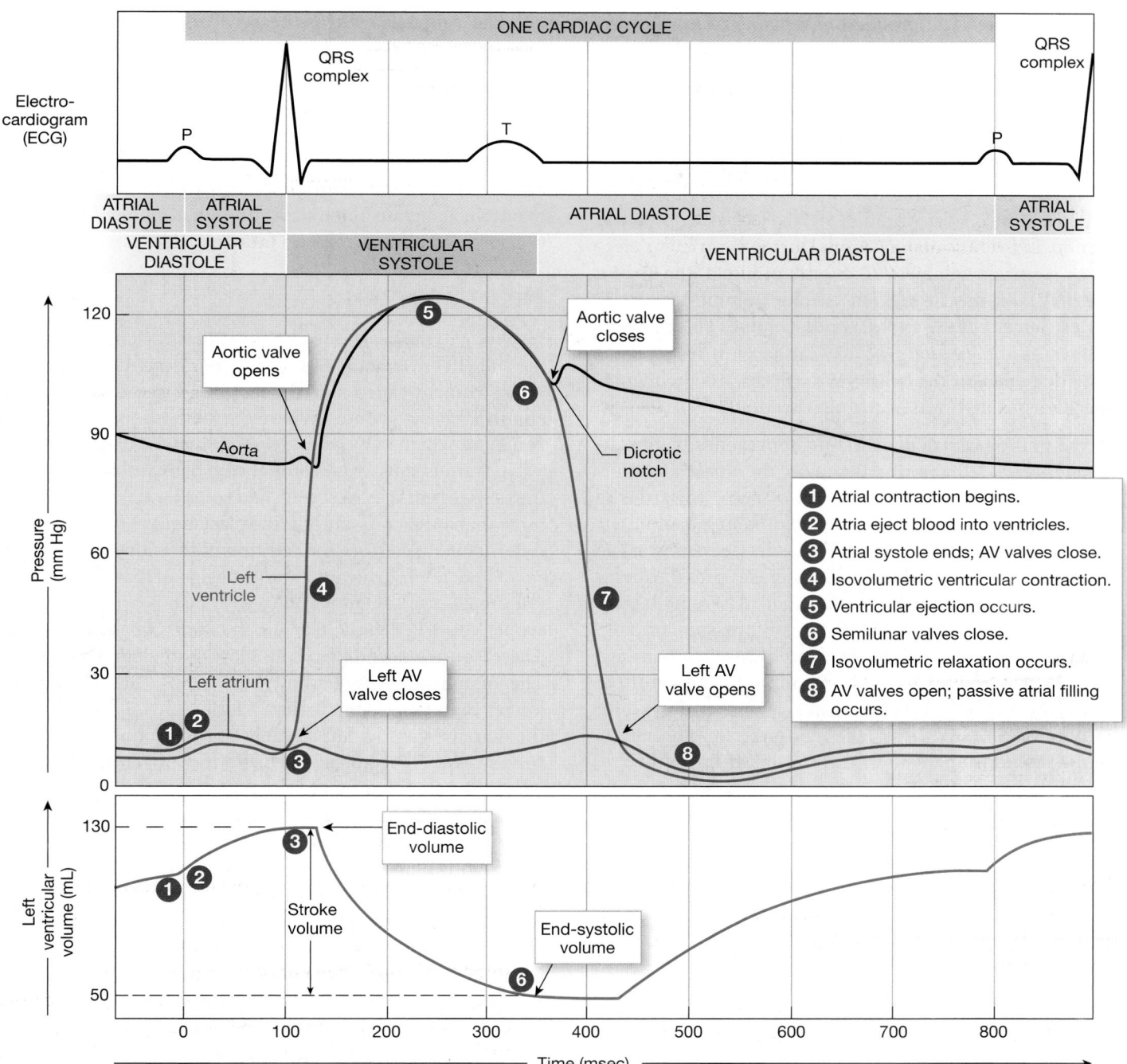

Figure 20–17 **Pressure and Volume Relationships in the Cardiac Cycle.** Major features of the cardiac cycle are shown for a heart rate of 75 bpm. The circled numbers correspond to those in the associated box; for further details, see the numbered list in the text.

4. During this stage of ventricular systole, the ventricles are contracting. Blood flow has yet to occur, however, because ventricular pressures are not high enough to force open the semilunar valves and push blood into the pulmonary or aortic trunk. Over this period, the ventricles contract isometrically: They generate tension and ventricular pressures rise, but blood flow does not occur. The ventricles are now in the period of **isovolumetric**

contraction: All the heart valves are closed, the volumes of the ventricles remain constant, and ventricular pressures rise.

5. Once pressure in the ventricles exceeds that in the arterial trunks, the semilunar valves open and blood flows into the pulmonary and aortic trunks. This point marks the beginning of the period of **ventricular ejection**. The ventricles now contract isotonically: The muscle cells

shorten, and tension production remains relatively constant. (To review isotonic versus isometric contractions, see **Figure 10–18**, p. 317.)

After reaching a peak, ventricular pressures gradually decline near the end of ventricular systole. **Figure 20–17** shows values for the left ventricle and aorta. Although pressures in the right ventricle and pulmonary trunk are much lower, the right ventricle also goes through periods of isovolumetric contraction and ventricular ejection. During ventricular ejection, each ventricle will eject 70–80 mL of blood, the **stroke volume (SV)** of the heart. The stroke volume at rest is roughly 60 percent of the end-diastolic volume. This percentage, known as the *ejection fraction*, can vary in response to changing demands on the heart. (We will discuss the regulatory mechanisms involved in the next section.)

6. As the end of ventricular systole approaches, ventricular pressures fall rapidly. Blood in the aorta and pulmonary trunk now starts to flow back toward the ventricles, and this movement closes the semilunar valves. As the backflow begins, pressure decreases in the aorta. When the semilunar valves close, pressure rises again as the elastic arterial walls recoil. This small, temporary rise produces a valley in the pressure tracing, called a *dicrotic* (dī-KROT-ik; *dikrotos*, double beating) *notch*. The amount of blood remaining in the ventricle when the semilunar valve closes is the **end-systolic volume (ESV)**. At rest, the end-systolic volume is 50 mL, about 40 percent of the end-diastolic volume.

Ventricular Diastole

The period of ventricular diastole lasts for the 430 msec remaining in the current cardiac cycle and continues through atrial systole in the next cycle.

7. All the heart valves are now closed, and the ventricular myocardium is relaxing. Because ventricular pressures are still higher than atrial pressures, blood cannot flow into the ventricles. This is the period of **isovolumetric relaxation**. Ventricular pressures drop rapidly over this period, because the elasticity of the connective tissues of the heart and cardiac skeleton helps reexpand the ventricles toward their resting dimensions.

8. When ventricular pressures fall below those of the atria, the atrial pressures force the AV valves open. Blood now flows from the atria into the ventricles. Both the atria and the ventricles are in diastole, but the ventricular pressures continue to fall as the ventricular chambers expand. Throughout this period, pressures in the ventricles are so far below those in the major veins that blood pours through the relaxed atria and on through the open AV valves into the ventricles. This passive mechanism is the primary method of ventricular filling. → passive

The ventricles will be nearly three-quarters full before the cardiac cycle ends.

The relatively minor contribution that atrial systole makes to ventricular volume explains why individuals can survive quite normally when their atria have been so severely damaged that they can no longer function. In contrast, damage to one or both ventricles can leave the heart unable to maintain adequate blood flow through peripheral tissues and organs. A condition of **heart failure** then exists.

Heart Sounds

Listening to the heart, a technique called *auscultation*, is a simple and effective method of cardiac assessment. Clinicians use an instrument called a **stethoscope** to listen for normal and abnormal heart sounds. Where the stethoscope is placed depends on which valve is under examination (**Figure 20–18a**). Valve sounds must pass through the pericardium, surrounding tissues, and the chest wall, and some tissues muffle sounds more than others. As a result, the placement of the stethoscope differs somewhat from the position of the valve under review.

There are four heart sounds, designated as S_1 through S_4 (**Figure 20–18b**). If you listen to your own heart with a stethoscope, you will clearly hear the *first and second heart sounds*. These sounds accompany the closing of your heart valves. The first heart sound, known as "lubb" (S_1), lasts a little longer than the second, called "dubb" (S_2). S_1, which marks the start of ventricular contraction, is produced as the AV valves close; S_2 occurs at the beginning of ventricular filling, when the semilunar valves close.

Third and *fourth heart sounds* are usually very faint and seldom are audible in healthy adults. These sounds are associated with blood flowing into the ventricles (S_3) and atrial contraction (S_4), rather than with valve action.

If the valve cusps are malformed or there are problems with the papillary muscles or chordae tendineae, the heart valves may not close properly. AV valve regurgitation then occurs during ventricular systole. The surges, swirls, and eddies that accompany regurgitation create a rushing, gurgling sound known as a **heart murmur**. Minor heart murmurs are common and inconsequential.

CHECKPOINT

8. Provide the alternate terms for heart contraction and heart relaxation. → systole → diastole

9. List the phases of the cardiac cycle.

10. Is the heart always pumping blood when pressure in the left ventricle is rising? Explain.

11. What factor or factors could cause an increase in the size of the QRS complex in an electrocardiogram?

See the blue Answers tab at the end of the book.

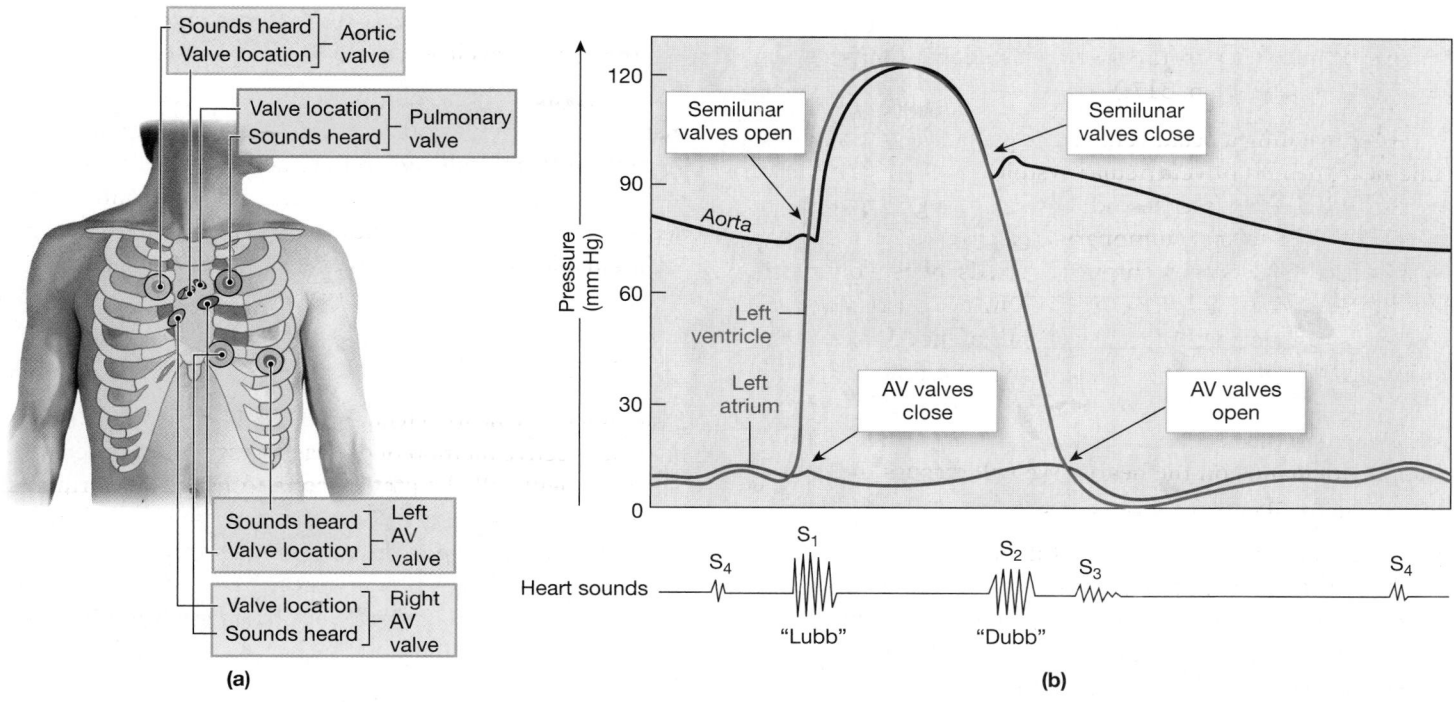

Figure 20–18   Heart Sounds. **(a)** Placements of a stethoscope for listening to the different sounds produced by individual valves. **(b)** The relationship between heart sounds and key events in the cardiac cycle.

[handwritten annotation] ⭐ Stroke volume → when EDV largest, ESV smallest.

20-4 Cardiodynamics examines the factors that affect cardiac output

The term **cardiodynamics** refers to the movements and forces generated during cardiac contractions. Each time the heart beats, the two ventricles eject equal amounts of blood. Earlier we introduced these terms:

- *End-Diastolic Volume (EDV):* The amount of blood in each ventricle at the end of ventricular diastole (the start of ventricular systole).

- *End-Systolic Volume (ESV):* The amount of blood remaining in each ventricle at the end of ventricular systole (the start of ventricular diastole).

- *Stroke Volume (SV):* The amount of blood pumped out of each ventricle during a single beat; it can be expressed as SV = EDV − ESV.

- *Ejection Fraction:* The percentage of the EDV represented by the SV.

Stroke volume is the most important factor in an examination of a single cardiac cycle. If the heart were an old-fashioned bicycle pump, the stroke volume would be the amount of air pumped in one up–down cycle of the handle (**Figure 20–19**). Where you stop when you lift the handle determines how much air the pump contains—the end-diastolic volume. How far down you push the handle determines how much air remains in the pump at the end of the cycle—the end-systolic volume. You pump the maximum amount of air when the handle is pulled all the way to the top and then pushed all the way to the bottom (**Figure 20–19b,d**). In other words, you get the largest stroke volume when the EDV is as large as it can be and the ESV is as small as it can be.

When considering cardiac function over time, physicians generally are most interested in the **cardiac output (CO)**, the amount of blood pumped by the left ventricle in one minute. In essence, cardiac output is an indication of the blood flow through peripheral tissues—without adequate blood flow, homeostasis cannot be maintained. The cardiac output provides a useful indication of ventricular efficiency over time. We can calculate it by multiplying the heart rate (HR) by the average stroke volume (SV):

CO	=	HR	×	SV
cardiac output (mL/min)		heart rate (beats/min)		stroke volume (mL/beat)

For example, if the heart rate is 75 bpm and the stroke volume is 80 mL per beat, the cardiac output will be

CO = 75 bpm × 80 mL/beat = 6000 mL/min (6 L/min)

The body precisely adjusts cardiac output such that peripheral tissues receive an adequate circulatory supply under

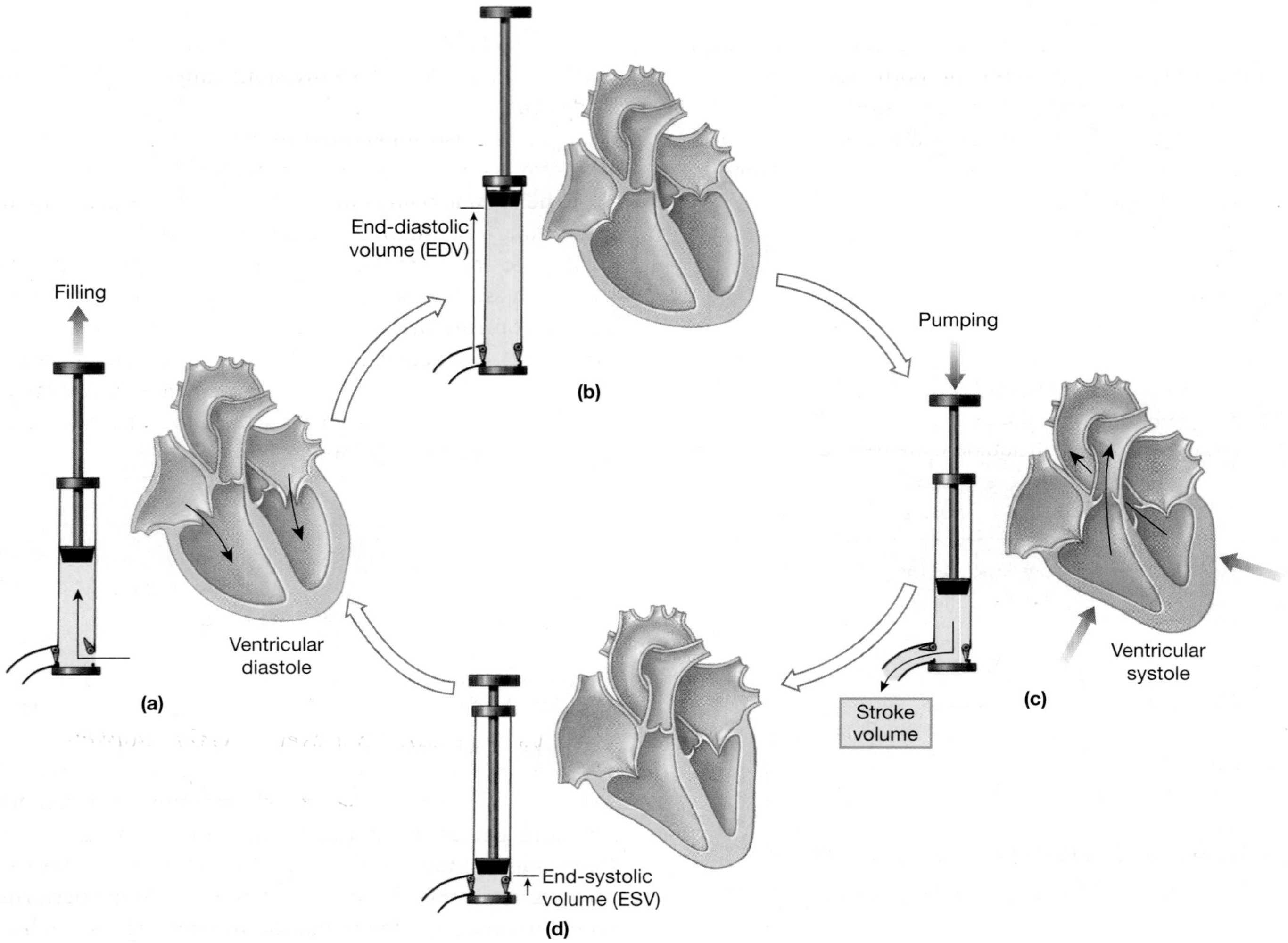

Filling

End-diastolic
volume (EDV)

(b)

Pumping

(a)

Ventricular
diastole

Ventricular
systole

Stroke
volume

(c)

End-systolic
volume (ESV)

(d)

Figure 20–19   A Simple Model of Stroke Volume. The stroke volume of the heart can be compared to the amount of air pumped from an old-fashioned bicycle pump. The amount pumped varies with the amount of movement of the pump handle **(a, c)**. The extent of upward movement determines the EDV **(b)**; the extent of downward movement determines the ESV **(d)**. The stroke volume is equal to the difference between the EDV and the ESV.

a variety of conditions. When necessary, the heart rate can increase by 250 percent, and stroke volume in a normal heart can almost double.

Overview: Factors Affecting Cardiac Output

Figure 20–20 summarizes the factors involved in the normal regulation of cardiac output. Cardiac output can be adjusted by changes in either heart rate or stroke volume. For convenience, we will consider these independently as we discuss the individual factors involved. However, changes in cardiac output generally reflect changes in both heart rate and stroke volume.

The heart rate can be adjusted by the activities of the autonomic nervous system or by circulating hormones. The stroke volume can be adjusted by changing the end-diastolic

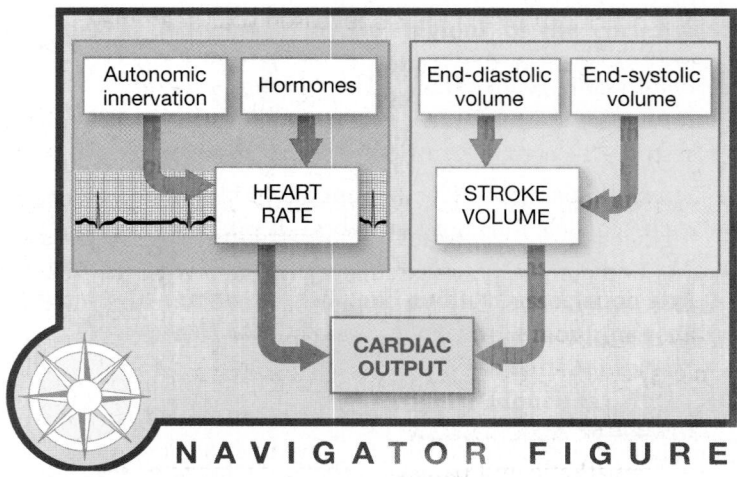

NAVIGATOR FIGURE

Figure 20–20   Factors Affecting Cardiac Output. A simplified version of this figure will appear as a Navigator icon in key figures as we move from one topic to the next.

volume (how full the ventricles are when they start to contract), the end-systolic volume (how much blood remains in the ventricle after it contracts), or both. As **Figure 20–19** shows, stroke volume peaks when EDV is highest and ESV is lowest. A variety of other factors can influence cardiac output under abnormal circumstances; we will consider several examples in the Clinical Note.

CLINICAL NOTE

Abnormal Conditions Affecting Cardiac Output
Various drugs, abnormal variations in concentrations, and changes in body temperature can alter the basic rhythm of contraction established by the SA node. In Chapter 12, we noted that several drugs, including caffeine and nicotine, have a stimulatory effect on the excitable membranes in the nervous system. ∞ p. 423 These drugs also cause an increase in heart rate. Caffeine acts directly on the conducting system and increases the rate of depolarization at the SA node. Nicotine acts directly by stimulating the activity of sympathetic neurons that innervate the heart.

Disorders affecting ion concentrations or body temperature can have direct effects on cardiac output by changing the stroke volume, the heart rate, or both. Abnormal ion concentrations can change both the contractility of the heart, by affecting the cardiac muscle cells, and the heart rate, by affecting the SA nodal cells. The most obvious and clinically important examples of problems with ion concentrations involve K^+ and Ca^{2+}.

Temperature changes affect metabolic operations throughout the body. For example, a reduction in temperature slows the rate of depolarization at the SA node, lowers the heart rate, and reduces the strength of cardiac contractions. (In open-heart surgery, the exposed heart may be deliberately chilled until it stops beating.) An elevated body temperature accelerates the heart rate and the contractile force. That is one reason your heart may seem to race and pound when you have a fever.

Factors Affecting the Heart Rate

Under normal circumstances, autonomic activity and circulating hormones are responsible for making delicate adjustments to the heart rate as circulatory demands change. These factors act by modifying the natural rhythm of the heart. Even a heart removed for a heart transplant will continue to beat unless steps are taken to prevent it from doing so.

Autonomic Innervation

The sympathetic and parasympathetic divisions of the autonomic nervous system innervate the heart by means of the *cardiac plexus* (**Figure 16–10**, p. 546, and **Figure 20–21**). Postganglionic sympathetic neurons are located in the cervical

and upper thoracic ganglia. The vagus nerves (N X) carry parasympathetic preganglionic fibers to small ganglia in the cardiac plexus. Both ANS divisions innervate the SA and AV nodes and the atrial muscle cells. Although ventricular muscle cells are also innervated by both divisions, sympathetic fibers far outnumber parasympathetic fibers there.

The *cardiac centers* of the medulla oblongata contain the autonomic headquarters for cardiac control. ∞ p. 471 The **cardioacceleratory center** controls sympathetic neurons that increase the heart rate; the adjacent **cardioinhibitory center** controls the parasympathetic neurons that slow the heart rate. The activities of the cardiac centers are regulated by reflex pathways and through input from higher centers, especially from the parasympathetic and sympathetic headquarters in the hypothalamus.

Tips&Tricks

To remember the effect of the sympathetic nervous system on cardiac performance, remember that **s**ympathetic input **s**peeds and **s**trengthens the heartbeat.

Cardiac Reflexes. Information about the status of the cardiovascular system arrives over visceral sensory fibers accompanying the vagus nerve and the sympathetic nerves of the cardiac plexus. The cardiac centers monitor baroreceptors and chemoreceptors innervated by the glossopharyngeal (N IX) and vagus (N X) nerves. ∞ p. 497, 498 On the basis of the information received, the centers adjust cardiac performance to maintain adequate circulation to vital organs, such as the brain. The centers respond to changes in blood pressure as reported by baroreceptors, and to changes in arterial concentrations of dissolved oxygen and carbon dioxide as reported by chemoreceptors. For example, a decline in blood pressure or oxygen concentrations or an increase in carbon dioxide levels generally indicates that the heart must work harder to meet the demands of peripheral tissues. The cardiac centers then call for an increase in cardiac activity. We will detail these reflexes and their effects on the heart and peripheral vessels in Chapter 21.

Autonomic Tone. As is the case in other organs with dual innervation, the heart has a resting autonomic tone. Both autonomic divisions are normally active at a steady background level, releasing ACh and NE at the nodes and into the myocardium. Thus, cutting the vagus nerves increases the heart rate, and sympathetic blocking agents slow the heart rate.

In a healthy, resting individual, parasympathetic effects dominate. In the absence of autonomic innervation, the heart rate is established by the pacemaker cells of the SA node. Such a heart beats at a rate of 80–100 bpm. At rest, a typical adult heart with normal innervation beats at 70–80 bpm due to activity in the parasympathetic nerves innervating the SA node.

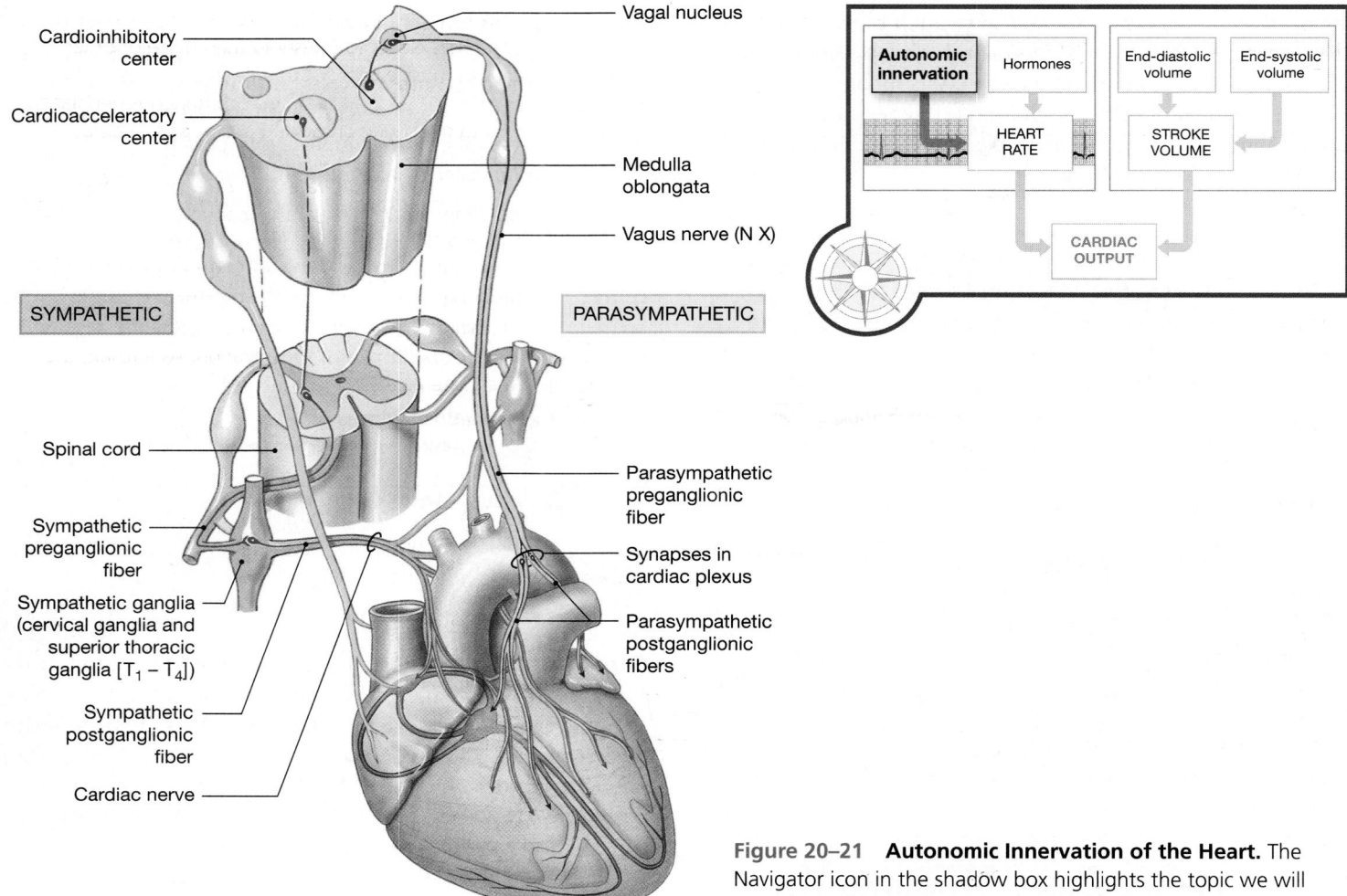

Figure 20–21 **Autonomic Innervation of the Heart.** The Navigator icon in the shadow box highlights the topic we will consider in this section.

If parasympathetic activity increases, the heart rate declines further. Conversely, the heart rate will increase if either parasympathetic activity decreases or sympathetic activation occurs. Through dual innervation and adjustments in autonomic tone, the ANS can make very delicate adjustments in cardiovascular function to meet the demands of other systems.

Effects on the SA Node. The sympathetic and parasympathetic divisions alter the heart rate by changing the ionic permeabilities of cells in the conducting system. The most dramatic effects are seen at the SA node, where changes in the rate at which impulses are generated affect the heart rate.

Consider the SA node of a resting individual whose heart is beating at 75 bpm (**Figure 20–22a**). Any factor that changes the rate of spontaneous depolarization or the duration of repolarization will alter the heart rate by changing the time required to reach threshold. Acetylcholine released by parasympathetic neurons opens chemically gated K^+ chan-

nels in the plasma membrane, thereby dramatically slowing the rate of spontaneous depolarization and also slightly extending the duration of repolarization (**Figure 20–22b**). The result is a decline in heart rate.

NE released by sympathetic neurons binds to beta-1 receptors, leading to the opening of sodium-calcium ion channels. The subsequent influx of positively charged ions increases the rate of depolarization and shortens the period of repolarization. The nodal cells reach threshold more quickly, and the heart rate increases (**Figure 20–22c**).

The Atrial Reflex. The **atrial reflex**, or *Bainbridge reflex*, involves adjustments in heart rate in response to an increase in the venous return. When the walls of the right atrium are stretched, the stimulation of stretch receptors in the atrial walls triggers a reflexive increase in heart rate caused by increased sympathetic activity (**Figure 20–22**). Thus, when the rate of venous return to the heart increases, the heart rate, and hence the cardiac output, rises as well.

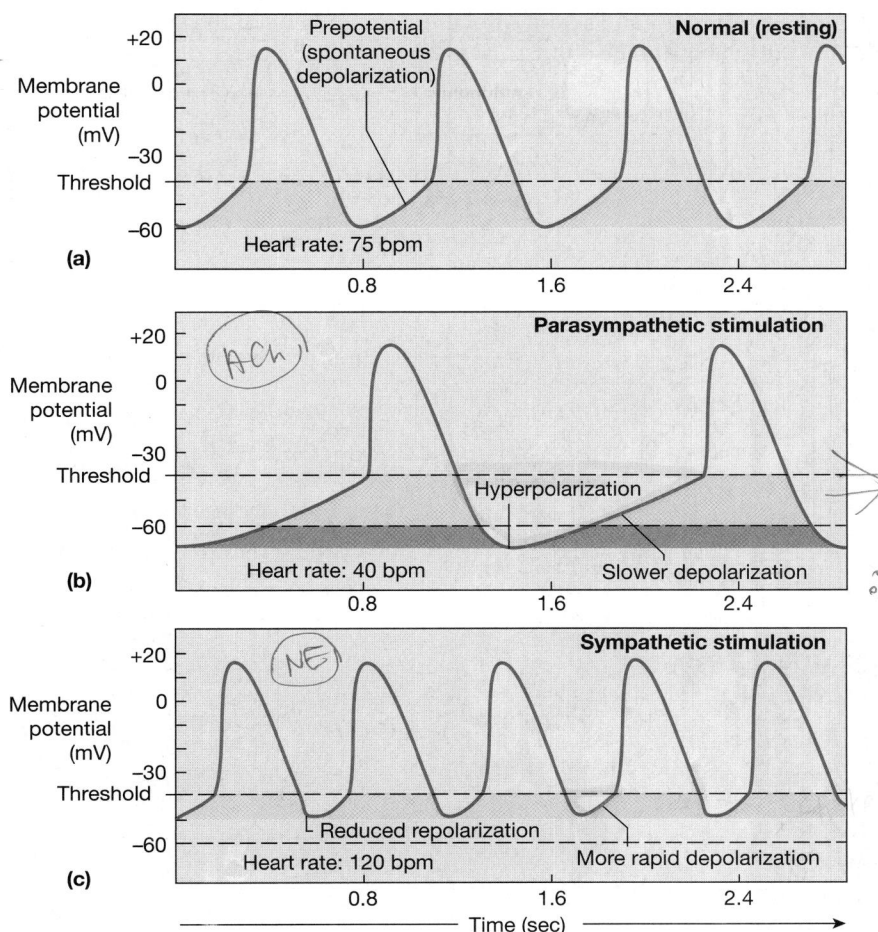

(a)

Membrane potential (mV)
+20
0
−30 Threshold
−60

Prepotential (spontaneous depolarization)
Normal (resting)
Heart rate: 75 bpm

0.8   1.6   2.4

(b)

Membrane potential (mV)
+20
0
−30 Threshold
−60

(ACh)
Parasympathetic stimulation
Hyperpolarization
Heart rate: 40 bpm
Slower depolarization

0.8   1.6   2.4

(c)

Membrane potential (mV)
+20
0
−30 Threshold
−60

(NE)
Sympathetic stimulation
Reduced repolarization
Heart rate: 120 bpm
More rapid depolarization

0.8   1.6   2.4
Time (sec)

(handwritten annotation): → decreases rate of spontaneous depolarization. ∴ heart rate slows.

Figure 20–22 Autonomic Regulation of Pacemaker Function. (a) Pacemaker cells have membrane potentials closer to threshold than those of other cardiac muscle cells (−60 mV versus −90 mV). Their plasma membranes undergo spontaneous depolarization to threshold, producing action potentials at a frequency determined by (1) the resting-membrane potential and (2) the rate of depolarization (slope of the prepotential). **(b)** Parasympathetic stimulation releases ACh, which extends repolarization and decreases the rate of spontaneous depolarization. The heart rate slows. **(c)** Sympathetic stimulation releases NE, which shortens repolarization and accelerates the rate of spontaneous depolarization. As a result, the heart rate increases.

Hormones

(handwritten annotation): E: sympathetic stimulation of suprarenal medullae → myocardium.

Epinephrine, norepinephrine, and thyroid hormone increase heart rate by their effect on the SA node. The effects of epinephrine on the SA node are similar to those of norepinephrine. Epinephrine also affects the contractile cells; after massive sympathetic stimulation of the suprarenal medullae, the myocardium may become so excitable that abnormal contractions occur.

Venous Return

In addition to its indirect effect on heart rate via the atrial reflex, venous return also has direct effects on nodal cells. When venous return increases, the atria receive more blood and the walls are stretched. Stretching of the cells of the SA node leads to more rapid depolarization and an increase in the heart rate.

Factors Affecting the Stroke Volume

(handwritten annotation): → changes in EDV & ESV.

The stroke volume is the difference between the end-diastolic volume and the end-systolic volume. Thus, changes in either EDV or ESV can change the stroke volume, and thus cardiac output. The factors involved in the regulation of stroke volume are indicated in **Figure 20–23**.

The EDV

The EDV is the amount of blood a ventricle contains at the end of diastole, just before a contraction begins. This volume is affected by two factors: the filling time and the venous return. **Filling time** is the duration of ventricular diastole. As such, it depends entirely on the heart rate: The faster the heart rate, the shorter is the available filling time. **Venous return** is the rate of blood flow over this period. Venous return changes in response to alterations in cardiac output, blood volume, patterns of peripheral circulation, skeletal muscle activity, and other factors that affect the rate of blood flow through the venae cavae. (We will explore these factors in Chapter 21.)

Preload. The degree of stretching experienced by ventricular muscle cells during ventricular diastole is called the **preload**. The preload is directly proportional to the EDV: The greater the EDV, the larger the preload. Preload is significant

20 CARDIOVASCULAR

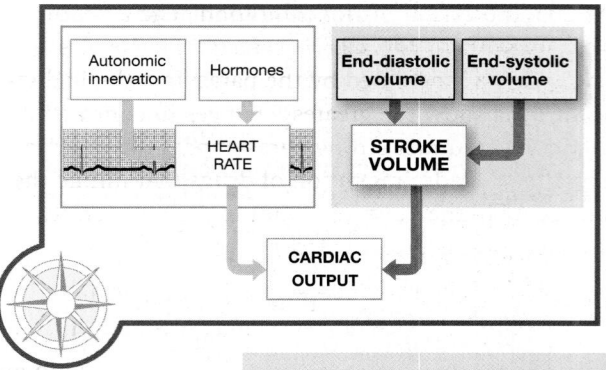

Figure 20–23 **Factors Affecting Stroke Volume.** The arrows indicate the nature of the effects: ↑ = increases, ↓ = decreases.

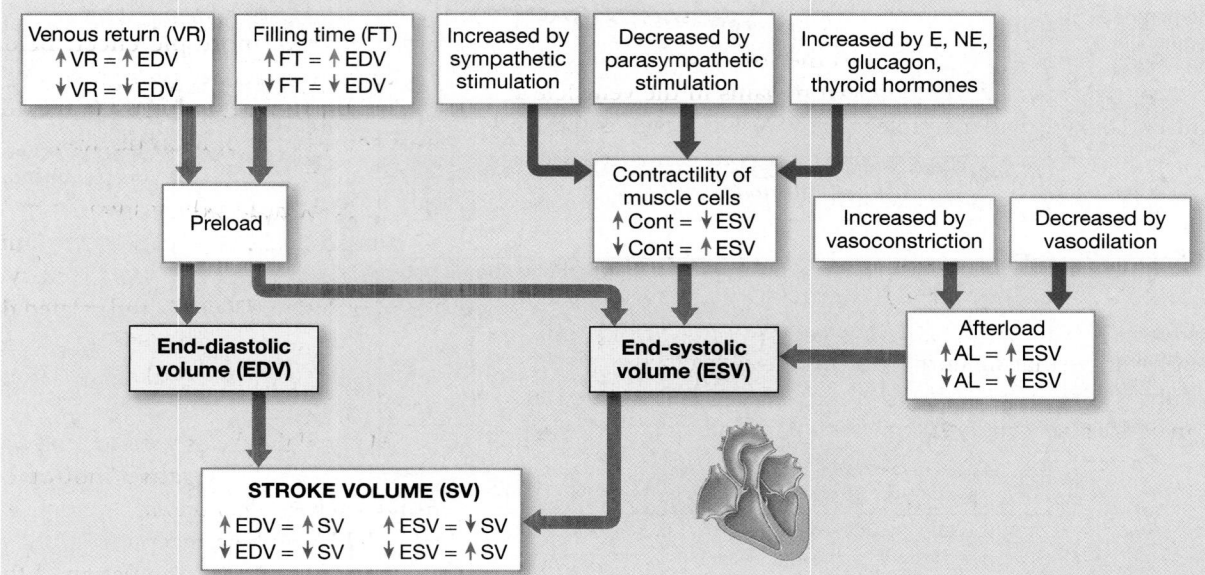

because it affects the ability of muscle cells to produce tension. As sarcomere length increases past resting length, the amount of force produced during systole increases.

The amount of preload, and hence the degree of myocardial stretching, varies with the demands on the heart. When you are standing at rest, your EDV is low; the ventricular muscle is stretched very little, and the sarcomeres are relatively short. During ventricular systole, the cardiac muscle cells develop little power, and the ESV (the amount of blood remaining in the ventricle after contraction) is relatively high because the muscle cells contract only a short distance. If you begin exercising, venous return increases and more blood flows into your heart. Your EDV increases, and the myocardium stretches further. As the sarcomeres approach optimal lengths, the ventricular muscle cells can contract more efficiently and produce more forceful contractions. They also shorten more, and more blood is pumped out of your heart.

The EDV and Stroke Volume. In general, the greater the EDV, the larger the stroke volume. Stretching *past* the optimal length, which would reduce the force of contraction, does not

normally occur, because ventricular expansion is limited by myocardial connective tissues, the cardiac skeleton, and the pericardial sac.

The relationship between the amount of ventricular stretching and the contractile force means that, within normal physiological limits, increasing the EDV results in a corresponding increase in the stroke volume. This general rule of "more in = more out" was first proposed by Ernest H. Starling based on his studies and research by Otto Frank. The relationship is therefore known as the **Frank–Starling principle**, or *Starling law of the heart*.

Autonomic adjustments to cardiac output make the effects of the Frank–Starling principle difficult to see. However, it can be demonstrated effectively in individuals who have received a heart transplant, because the implanted heart is not innervated by the ANS. The most obvious effect of the Frank–Starling principle in these hearts is that the outputs of the left and right ventricles remain balanced under a variety of conditions.

Consider, for example, an individual at rest, with the two ventricles ejecting equal volumes of blood. Although the ventricles contract together, they function in series: When the heart contracts, blood leaving the right ventricle heads to the lungs; during the next ventricular diastole, that volume of blood will pass through the left atrium, to be ejected by the left ventricle at the next contraction. If the venous return decreases, the EDV of the right ventricle will decline. During ventricular systole, it will then pump less blood into the pulmonary circuit. In the next cardiac cycle, the EDV of the left ventricle will be reduced, and that ventricle will eject a smaller volume of blood. The output of the two ventricles will again be in balance, but both will have smaller stroke volumes than they did initially.

The ESV

(+) NE, E, thyroid hormones, & glucagon.

After the ventricle contracts and the stroke volume has been ejected, the amount of blood that remains in the ventricle at the end of ventricular systole is the ESV. Three factors that influence the ESV are the *preload* (discussed earlier), the *contractility* of the ventricle, and the *afterload*.

Contractility. **Contractility** is the amount of force produced during a contraction, at a given preload. Under normal circumstances, contractility can be altered by autonomic innervation or circulating hormones. Under special circumstances, contractility can be altered by drugs or as a result of abnormal ion concentrations in the extracellular fluid.

Factors that increase contractility are said to have a *positive inotropic action*; those that decrease contractility have a *negative inotropic action*. Positive inotropic (*ino-*, fiber) agents typically stimulate Ca^{2+} entry into cardiac muscle cells, thus increasing the force and duration of ventricular contractions. Negative inotropic agents may block Ca^{2+} movement or depress cardiac muscle metabolism. Positive and negative inotropic factors include ANS activity, hormones, and changes in extracellular ion concentrations.

Effects of Autonomic Activity on Contractility. Autonomic activity alters the degree of contraction and changes the ESV in the following ways:

- Sympathetic stimulation has a positive inotropic effect, causing the release of norepinephrine (NE) by postganglionic fibers of the cardiac nerves and the secretion of epinephrine (E) and NE by the suprarenal medullae. In addition to their effects on heart rate, discussed shortly, these hormones stimulate cardiac muscle cell metabolism and increase the force and degree of contraction by stimulating alpha and beta receptors in cardiac muscle plasma membranes. The net effect is that the ventricles contract more forcefully, increasing the ejection fraction and decreasing the ESV.

- Parasympathetic stimulation from the vagus nerves has a negative inotropic effect. The primary effect of acetylcholine (ACh) is at the membrane surface, where it pro-

duces hyperpolarization and inhibition. The force of cardiac contractions is reduced; because the ventricles are not extensively innervated by the parasympathetic division, the atria show the greatest changes in contractile force. However, under strong parasympathetic stimulation or after the administration of drugs that mimic the actions of ACh, the ventricles contract less forcefully, the ejection fraction decreases, and the ESV enlarges.

Hormones. Many hormones affect the contractility of the heart. For example, epinephrine, norepinephrine, and thyroid hormones all have positive inotropic effects. Glucagon also has a positive inotropic effect. Before synthetic inotropic agents were available, glucagon was widely used to stimulate cardiac function. It is still used in cardiac emergencies and to treat some forms of heart disease.

The drugs *isoproterenol*, *dopamine*, and *dobutamine* mimic the action of E and NE by stimulating beta-1 receptors on cardiac muscle cells. ∞ p. 538 Dopamine (at high doses) and dobutamine also stimulate Ca^{2+} entry through alpha-1 receptor stimulation. *Digitalis* and related drugs elevate intracellular Ca^{2+} concentrations, but by a different mechanism: They interfere with the removal of Ca^{2+} from the sarcoplasm of cardiac muscle cells.

Many of the drugs used to treat hypertension (high blood pressure) have a negative inotropic action. Beta-blocking drugs such as *propranolol*, *timolol*, *metoprolol*, *atenolol*, and *labetalol* block beta receptors, alpha receptors, or both, and prevent sympathetic stimulation of the heart. Calcium channel blockers such as *nifedipine* or *verapamil* also have a negative inotropic effect.

Afterload. The **afterload** is the amount of tension the contracting ventricle must produce to force open the semilunar valve and eject blood. The greater the afterload, the longer the period of isovolumetric contraction, the shorter the duration of ventricular ejection, and the larger the ESV. In other words, as the afterload increases, the stroke volume decreases.

Afterload is increased by any factor that restricts blood flow through the arterial system. For example, the constriction of peripheral blood vessels or a circulatory blockage will elevate arterial blood pressure and increase the afterload. If the afterload is too great, the ventricle cannot eject blood. Such a high afterload is rare in a normal heart, but damage to the heart muscle can weaken the myocardium enough that even a modest rise in arterial blood pressure can reduce stroke volume to dangerously low levels, producing symptoms of heart failure.

Summary: The Control of Cardiac Output

Figure 20–24 summarizes the factors involved in the regulation of heart rate and stroke volume, the two factors that interact to determine cardiac output under normal conditions.

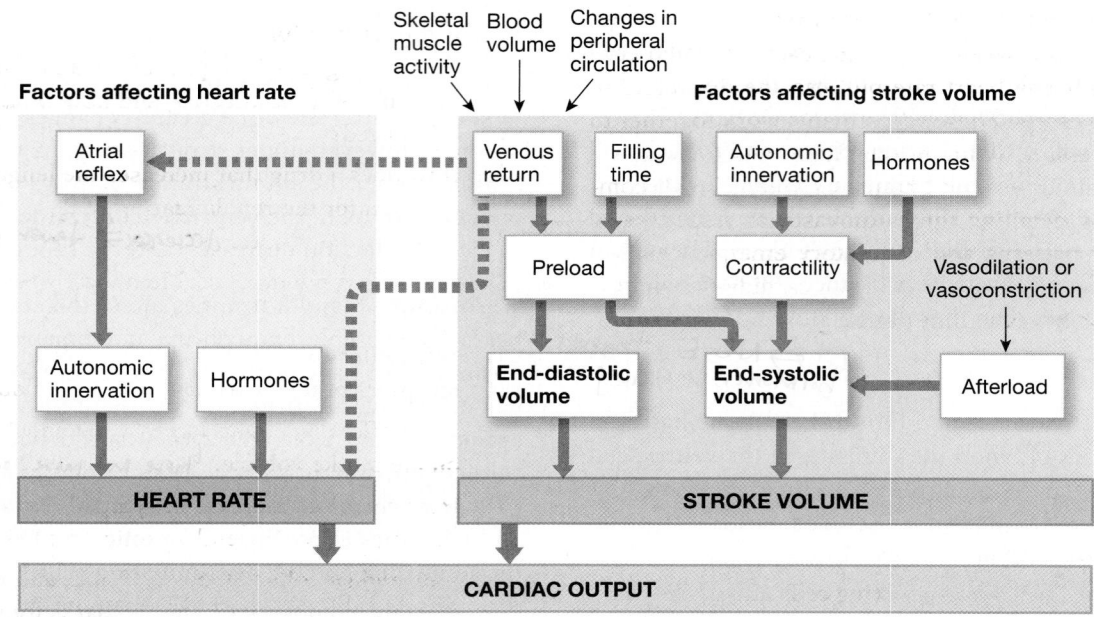

Figure 20–24   A Summary of the Factors Affecting Cardiac Output.

The heart rate is influenced by the autonomic nervous system, circulating hormones, and the venous return.

- Sympathetic stimulation increases the heart rate; parasympathetic stimulation decreases it. Under resting conditions, parasympathetic tone dominates, and the heart rate is slightly slower than the intrinsic heart rate. When activity levels rise, venous return increases and triggers the atrial reflex. The result is an increase in sympathetic tone and an increase in heart rate.
- Circulating hormones, specifically E, NE, and T_3, accelerate heart rate.
- An increase in venous return stretches the nodal cells and increases heart rate.
- The stroke volume is the difference between the end-diastolic volume (EDV) and the end-systolic volume (ESV).
- The EDV is determined by the available filling time and the rate of venous return.
- The ESV is determined by the amount of preload (the degree of myocardial stretching), the degree of contractility (adjusted by hormones and autonomic innervation), and the afterload (the amount of arterial resistance).

In most healthy people, increasing both the stroke volume and the heart rate, such as occurs during heavy exercise, can raise the cardiac output by 300–500 percent, to 18–30 L/min. The difference between resting and maximal cardiac outputs is the **cardiac reserve**. Trained athletes exercising at maximal levels may increase cardiac output by nearly 700 percent, to 40 L/min.

Cardiac output cannot increase indefinitely, primarily because the available filling time shortens as the heart rate increases. At heart rates up to 160–180 bpm, the combination of increased rate of venous return and increased contractility compensates for the reduction in filling time. Over this range, cardiac output and heart rate increase together. But if the heart rate continues to climb, the stroke volume begins to drop. Cardiac output first plateaus and then declines.

The A&P Top 100

#72 Cardiac output is the amount of blood pumped by the left ventricle each minute. It is adjusted on a moment-to-moment basis by the ANS, and in response to circulating hormones, changes in blood volume, and alterations in venous return. Most healthy people can increase cardiac output by 300–500 percent.

The Heart and the Cardiovascular System

The purpose of cardiovascular regulation is maintaining adequate blood flow to all body tissues. The heart cannot accomplish this by itself, and it does not work in isolation. For example, when blood pressure changes, the cardiovascular centers adjust not only the heart rate but also the diameters of peripheral blood vessels. These adjustments work together to keep the blood pressure within normal limits and to maintain circulation to vital tissues and organs. Chapter 21 will complete this story by detailing the cardiovascular responses to changing activity patterns and circulatory emergencies. We will then conclude our discussion of the cardiovascular system by examining the anatomy of the pulmonary and systemic circuits.

CHECKPOINT

12. Define cardiac output.

13. Caffeine has effects on conducting cells and contractile cells that are similar to those of NE. What effect would drinking large amounts of caffeinated drinks have on the heart?

14. If the cardioinhibitory center of the medulla oblongata were damaged, which part of the autonomic nervous system would be affected, and how would the heart be influenced?

15. How does a drug that increases the length of time required for the repolarization of pacemaker cells affect the heart rate? *decreases → fewer action pot/min*

16. Why is it a potential problem if the heart beats too rapidly?

17. What effect would stimulating the acetylcholine receptors of the heart have on cardiac output? *inhibitory*

18. What effect would an increase in venous return have on the stroke volume? *more in more out'.*

19. How would an increase in sympathetic stimulation of the heart affect the end-systolic volume?

20. Joe's end-systolic volume is 40 mL, and his end-diastolic volume is 125 mL. What is Joe's stroke volume?

See the blue Answers tab at the end of the book.

Related Clinical Terms

angina pectoris: A condition in which exertion or stress produces severe chest pain, resulting from temporary ischemia when the heart's workload increases.

balloon angioplasty: A technique for reducing the size of a coronary plaque by compressing it against the arterial walls, using a catheter with an inflatable collar.

bradycardia: A heart rate that is slower than normal.

calcium channel blockers: Drugs that reduce the contractility of the heart by slowing the influx of calcium ions during the plateau phase of the cardiac muscle action potential.

cardiac arrhythmias: Abnormal patterns of cardiac electrical activity, indicating abnormal contractions.

cardiac tamponade: A condition, resulting from pericardial irritation and inflammation, in which fluid collects in the pericardial sac and restricts cardiac output.

carditis: A general term indicating inflammation of the heart.

conduction deficit: An abnormality in the conducting system of the heart that affects the timing and pacing of cardiac contractions.

coronary artery bypass graft (CABG): The routing of blood around an obstructed coronary artery (or one of its branches) by a vessel transplanted from another part of the body.

coronary artery disease (CAD): The obstruction of coronary circulation.

coronary ischemia: The restriction of the circulatory supply to the heart, potentially causing tissue damage and a reduction in cardiac efficiency.

coronary thrombosis: A blockage due to the formation of a clot (thrombus) at a plaque in a coronary artery.

electrocardiogram (ECG or EKG): A recording of the electrical activities of the heart over time.

heart failure: A condition in which the heart weakens and peripheral tissues suffer from oxygen and nutrient deprivation.

heart murmur: The sound produced by regurgitation or turbulent flow through an incompletely closed heart valve.

mitral valve prolapse: A condition in which the mitral valve cusps do not close properly and are pushed back toward the left atrium.

myocardial infarction (MI): A condition in which the coronary circulation becomes blocked and cardiac muscle cells die from oxygen starvation; also called a *heart attack*.

pericarditis: Inflammation of the pericardium.

rheumatic heart disease (RHD): A disorder in which the heart valves become thickened and stiffen in a partially closed position, affecting the efficiency of the heart.

tachycardia: A heart rate that is faster than normal.

valvular heart disease (VHD): A condition caused by abnormal functioning of one of the cardiac valves. The severity of the condition depends on the degree of damage and the valve involved.

20 CARDIOVASCULAR

Chapter Review

Study Outline

iP Use *Interactive Physiology*® (**IP**)—available on the IP CD packaged with your new textbook and online at myA&P™ (www.myaandp.com)—to help you understand difficult physiological concepts. Follow these navigation paths on IP for concepts in this chapter:

- Cardiovascular System/Anatomy Review: the Heart
- Cardiovascular System/Intrinsic Conduction System
- Cardiovascular System/Cardiac Action Potential
- Cardiovascular System/Cardiac Cycle
- Cardiovascular System/Cardiac Output

An Introduction to the Cardiovascular System p. 682

1. The blood vessels can be subdivided into the **pulmonary circuit** (which carries blood to and from the lungs) and the **systemic circuit** (which transports blood to and from the rest of the body).
2. **Arteries** carry blood away from the heart; **veins** return blood to the heart. **Capillaries**, or *exchange vessels*, are thin-walled, narrow-diameter vessels that connect the smallest arteries and veins. (*Figure 20–1*)
3. The heart has four chambers: the **right atrium** and **right ventricle**, and the **left atrium** and **left ventricle**.

20-1 The heart is a four-chambered organ, supplied by the coronary circulation, that pumps oxygen-poor blood to the lungs and oxygen-rich blood to the rest of the body p. 682

4. The heart is surrounded by the **pericardial cavity** and lies within the anterior portion of the **mediastinum**, which separates the two pleural cavities. (*Figure 20–2*)
5. The pericardial cavity is lined by the **pericardium**. The **visceral pericardium (epicardium)** covers the heart's outer surface, and the **parietal pericardium** lines the inner surface of the **pericardial sac**, which surrounds the heart. (*Figure 20–2*)
6. The **coronary sulcus**, a deep groove, marks the boundary between the atria and the ventricles. Other surface markings also provide useful reference points in describing the heart and associated structures. (*Figure 20–3*)
7. The bulk of the heart consists of the muscular **myocardium**. The **endocardium** lines the inner surfaces of the heart, and the **epicardium** covers the outer surface. (*Figure 20–4*)
8. **Cardiac muscle cells** are interconnected by **intercalated discs**, which convey the force of contraction from cell to cell and conduct action potentials. (*Figure 20–5; Summary Table 20–1*)
9. The atria are separated by the **interatrial septum**, and the ventricles are divided by the **interventricular septum**. The right atrium receives blood from the systemic circuit via two large veins, the **superior vena cava** and the **inferior vena cava**. (The atrial walls contain the **pectinate muscles**, prominent muscular ridges.) (*Figure 20–6*)
10. Blood flows from the right atrium into the right ventricle via the **right atrioventricular (AV) valve (tricuspid valve)**. This opening is bounded by three **cusps** of fibrous tissue braced by the **chordae tendineae**, which are connected to **papillary muscles**. (*Figure 20–6*)

11. Blood leaving the right ventricle enters the **pulmonary trunk** after passing through the **pulmonary valve**. The pulmonary trunk divides to form the **left** and **right pulmonary arteries**. The **left** and **right pulmonary veins** return blood from the lungs to the left atrium. Blood leaving the left atrium flows into the left ventricle via the **left atrioventricular (AV) valve** (**bicuspid**, or **mitral**, **valve**). Blood leaving the left ventricle passes through the **aortic valve** and into the systemic circuit via the **ascending aorta**. (*Figure 20–6*)
12. Anatomical differences between the ventricles reflect the functional demands placed on them. The wall of the right ventricle is relatively thin, whereas the left ventricle has a massive muscular wall. (*Figure 20–7*)
13. Valves normally permit blood flow in only one direction, preventing the **regurgitation** (backflow) of blood. (*Figure 20–8*)
14. The connective tissues of the heart (mainly collagen and elastic fibers) and the **cardiac skeleton** support the heart's contractile cells and valves. (*Figure 20–8*)
15. The **coronary circulation** meets the high oxygen and nutrient demands of cardiac muscle cells. The **coronary arteries** originate at the base of the ascending aorta. Interconnections between arteries, called **arterial anastomoses**, ensure a constant blood supply. The **great**, **posterior**, **small**, **anterior**, and **middle cardiac veins** are epicardial vessels that carry blood from the coronary capillaries to the **coronary sinus**. (*Figure 20–9*)
16. In **coronary artery disease (CAD)**, portions of the coronary circulation undergo partial or complete blockage. (*Figure 20–10*)

20-2 The conducting system distributes electrical impulses through the heart, and an electrocardiogram records the associated electrical events p. 695

17. Two general classes of cardiac muscle cells are involved in the normal **heartbeat: contractile cells** and cells of the **conducting system**. (*Figure 20–11*)
18. The **conducting system** is composed of the *sinoatrial node*, the *atrioventricular node*, and *conducting cells*. The conducting system initiates and distributes electrical impulses within the heart. Nodal cells establish the rate of cardiac contraction, and conducting cells distribute the contractile stimulus from the SA node to the atrial myocardium and the AV node (along *internodal pathways*), and from the AV node to the ventricular myocardium. (*Figure 20–12*)
19. Unlike skeletal muscle, cardiac muscle contracts without neural or hormonal stimulation. **Pacemaker cells** in the **sinoatrial (SA) node** (*cardiac pacemaker*) normally establish the rate of contraction. From the SA node, the stimulus travels to the **atrioventricular (AV) node**, and then to the **AV bundle**, which divides into **bundle branches**. From there, **Purkinje fibers** convey the impulses to the ventricular myocardium. (*Figures 20–12, 20–13*)
20. A recording of electrical activities in the heart is an **electrocardiogram** (**ECG** or **EKG**). Important landmarks of an ECG include the **P wave** (atrial depolarization), the **QRS complex** (ventricular depolarization), and the **T wave** (ventricular repolarization). (*Figure 20–14*)

21. **Contractile cells** form the bulk of the atrial and ventricular walls. Cardiac muscle cells have a long refractory period, so rapid stimulation produces twitches rather than tetanic contractions. (*Figure 20–15*)

20-3 Events during a complete heartbeat constitute a cardiac cycle p. 703

22. The **cardiac cycle** contains periods of **atrial** and **ventricular systole** (contraction) and **atrial** and **ventricular diastole** (relaxation). (*Figure 20–16*)
23. When the heart beats, the two ventricles eject equal volumes of blood. (*Figure 20–17*)
24. The closing of valves and rushing of blood through the heart cause characteristic heart sounds, which can be heard during *auscultation*. (*Figure 20–18*)

20-4 Cardiodynamics examines the factors that affect cardiac output p. 707

25. The amount of blood ejected by a ventricle during a single beat is the **stroke volume (SV)**. The amount of blood pumped by a ventricle each minute is the **cardiac output (CO)**. (*Figure 20–19*)
26. Cardiac output can be adjusted by changes in either stroke volume or heart rate. (*Figure 20–20*)
27. The **cardioacceleratory center** in the medulla oblongata activates sympathetic neurons; the **cardioinhibitory center** controls the parasympathetic neurons that slow the heart rate. These cardiac centers receive inputs from higher centers and from receptors monitoring blood pressure and the concentrations of dissolved gases. (*Figure 20–21*)

28. The basic heart rate is established by the pacemaker cells of the SA node, but it can be modified by the autonomic nervous system. The **atrial reflex** accelerates the heart rate when the walls of the right atrium are stretched. (*Figure 20–22*)
29. Sympathetic activity produces more powerful contractions that reduce the ESV. Parasympathetic stimulation slows the heart rate, reduces the contractile strength, and raises the ESV.
30. Cardiac output is affected by various factors, including autonomic innervation and hormones. (*Figure 20–22*)
31. The stroke volume is the difference between the **end-diastolic volume (EDV)** and the **end-systolic volume (ESV)**. The **filling time** and **venous return** interact to determine the EDV. Normally, the greater the EDV, the more powerful is the succeeding contraction (the **Frank–Starling principle**). (*Figure 20–23*)
32. The difference between resting and maximal cardiac outputs is the **cardiac reserve**. (*Figure 20–24*)
33. The heart does not work in isolation in maintaining adequate blood flow to all tissues.

Review Questions

See the blue Answers tab at the end of the book.

Access more review material online at **myA&P™** (www.myaandp.com). There, you'll find chapter guides, chapter quizzes, practice tests, animations, flashcards, a glossary with pronunciations, *Interactive Physiology*® (IP) exercises and quizzes, and more to help you succeed in the course.

LEVEL 1 Reviewing Facts and Terms

1. Identify the superficial structures in the following diagram of the heart.

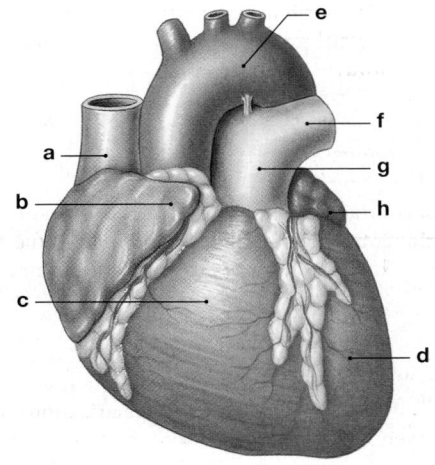

(a) superior venal cavae   (b) right atrium
(c) right ventricle   (d) left ventricle
(e) arch of aorta   (f) left pulm. artery.
(g) pulmonary trunk   (h) auricle of left atrium

2. The great cardiac vein drains blood from the heart muscle to the
   (a) left ventricle.   (b) right ventricle.
   (c) right atrium.   (d) left atrium.

3. The autonomic centers for cardiac function are located in
   (a) the myocardial tissue of the heart.
   (b) the cardiac centers of the medulla oblongata.
   (c) the cerebral cortex.
   (d) all of these structures.

4. The serous membrane covering the inner surface of the heart is the
   (a) parietal pericardium.   (b) endocardium.
   (c) myocardium.   (d) visceral pericardium.

5. The simple squamous epithelium covering the valves of the heart constitutes the
   (a) epicardium.   (b) endocardium.
   (c) myocardium.   (d) cardiac skeleton.

6. The heart lies in the
   (a) pleural cavity.   (b) peritoneal cavity.
   (c) abdominopelvic cavity.   (d) mediastinum.
   (e) abdominal cavity.

7. Identify the structures in the following diagram of a sectional view of the heart.

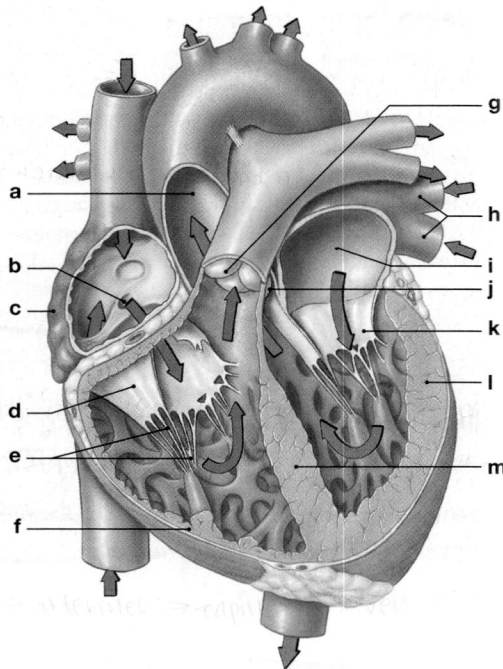

(a) ascending aorta   (b) fossa ovalis
(c) right atrium   (d) cusp of right AV
(e) chordae tendinae   (f) right ventricle
(g) pulmonary valve   (h) left pulm. veins
(i) left atrium   (j) aortic valve
(k) cusp of left AV   (l) left ventricle.
(m) intervent. septum

8. The cardiac skeleton of the heart has which *two* of the following functions?
   (a) It physically isolates the muscle fibers of the atria from those of the ventricles.
   (b) It maintains the normal shape of the heart.
   (c) It helps distribute the forces of cardiac contraction.
   (d) It allows more rapid contraction of the ventricles.
   (e) It strengthens and helps prevent overexpansion of the heart.

9. Cardiac output is equal to the
   (a) difference between the end-diastolic volume and the end-systolic volume.
   (b) product of heart rate and stroke volume.
   (c) difference between the stroke volume at rest and the stroke volume during exercise.
   (d) stroke volume less the end-systolic volume.
   (e) product of heart rate and blood pressure.

10. During diastole, a chamber of the heart
    (a) relaxes and fills with blood.
    (b) contracts and pushes blood into an adjacent chamber.
    (c) experiences a sharp increase in pressure.
    (d) reaches a pressure of approximately 120 mm Hg.

11. During the cardiac cycle, the amount of blood ejected from the left ventricle when the semilunar valve opens is the
    (a) stroke volume (SV). ⟶ amount of blood pumped out of ventricle.
    (b) end-diastolic volume (EDV).
    (c) end-systolic volume (ESV).
    (d) cardiac output (CO).

12. What role do the chordae tendineae and papillary muscles play in the normal function of the AV valves? *lowers stroke volume*

13. Describe the three distinct layers that make up the heart wall.
14. What are the valves in the heart, and what is the function of each?
15. Trace the normal pathway of an electrical impulse through the conducting system of the heart.
16. What is the cardiac cycle? What phases and events are necessary to complete a cardiac cycle?
17. What three factors regulate stroke volume to ensure that the left and right ventricles pump equal volumes of blood?

LEVEL 2 Reviewing Concepts

18. The cells of the conducting system differ from the contractile cells of the heart in that
    (a) conducting cells are larger and contain more myofibrils.
    (b) contractile cells exhibit prepotentials.
    (c) contractile cells do not normally exhibit automaticity.
    (d) both a and b are correct.

19. Which of the following is *longer*?
    (a) the refractory period of cardiac muscle
    (b) the refractory period of skeletal muscle

20. If the papillary muscles fail to contract,
    (a) the ventricles will not pump blood.
    (b) the atria will not pump blood.
    (c) the semilunar valves will not open.
    (d) the AV valves will not close properly.
    (e) none of these happen.

21. Cardiac output cannot increase indefinitely because
    (a) the available filling time becomes shorter as the heart rate increases. *increases both cardiac output & heart rate.*
    (b) the cardiovascular centers adjust the heart rate.
    (c) the rate of spontaneous depolarization decreases.
    (d) the ion concentrations of pacemaker plasma membranes decrease.

22. Describe the function of the SA node in the cardiac cycle. How does this function differ from that of the AV node?
23. What are the sources and significance of the four heart sounds?
24. Differentiate between stroke volume and cardiac output. How is cardiac output calculated?
25. What factors influence cardiac output?
26. What effect does sympathetic stimulation have on the heart? What effect does parasympathetic stimulation have on the heart?
27. Describe the effects of epinephrine, norepinephrine, glucagon, and thyroid hormones on the contractility of the heart.

LEVEL 3 Critical Thinking and Clinical Applications

28. Vern is suffering from cardiac arrhythmias and is brought into the emergency room of a hospital. In the emergency room he begins to exhibit tachycardia and as a result loses consciousness. Explain why Vern lost consciousness.
29. Harvey has a heart murmur in his left ventricle that produces a loud "gurgling" sound at the beginning of systole. Which valve is probably faulty?
30. The following measurements were made on two individuals (the values recorded remained stable for one hour):
    Person 1: heart rate, 75 bpm; stroke volume, 60 mL
    Person 2: heart rate, 90 bpm; stroke volume, 95 mL
    Which person has the greater venous return? Which person has the longer ventricular filling time?
31. Karen is taking the medication *verapamil*, a drug that blocks the calcium channels in cardiac muscle cells. What effect should this medication have on Karen's stroke volume?

Blood Vessels and Circulation

21

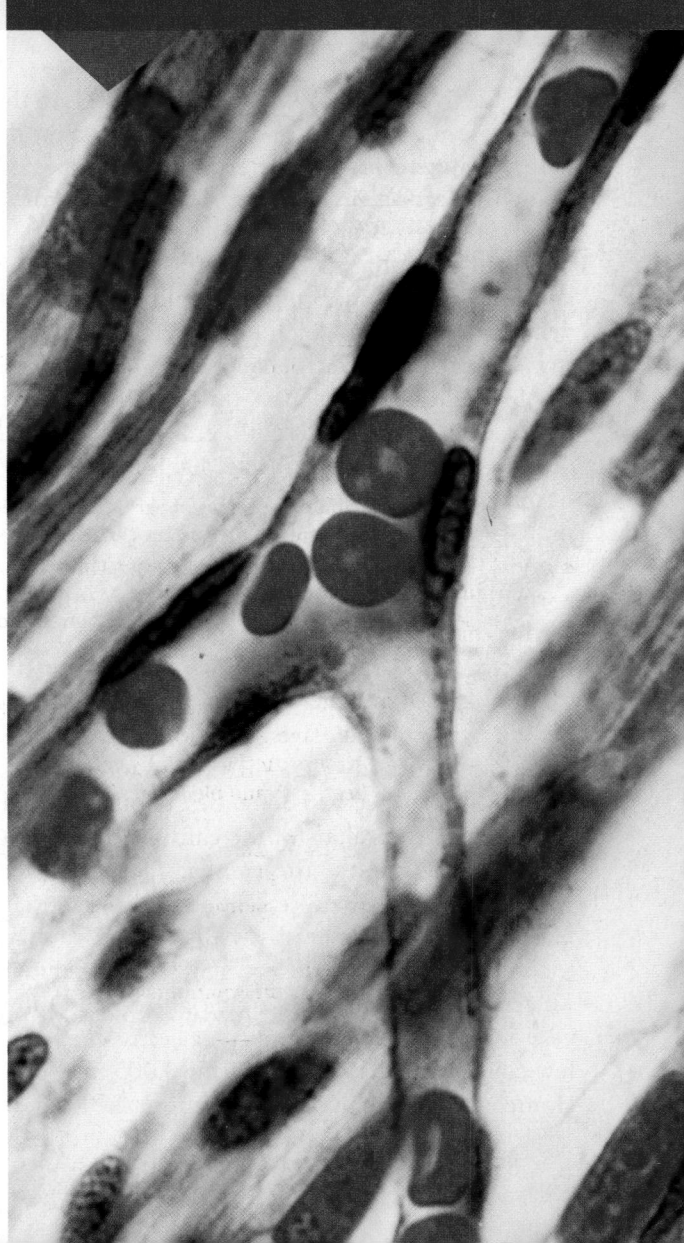

Learning Outcomes

After completing this chapter, you should be able to do the following:

21-1 Distinguish among the types of blood vessels based on their structure and function, and describe how and where fluid and dissolved materials enter and leave the cardiovascular system.

21-2 Explain the mechanisms that regulate blood flow through vessels, describe the factors that influence blood pressure, and discuss the mechanisms that regulate movement of fluids between capillaries and interstitial spaces.

21-3 Describe the control mechanisms that regulate blood flow and pressure in tissues, and explain how the activities of the cardiac, vasomotor, and respiratory centers are coordinated to control blood flow through the tissues.

21-4 Explain the cardiovascular system's homeostatic response to exercise and hemorrhaging, and identify the principal blood vessels and functional characteristics of the special circulation to the brain, heart, and lungs.

21-5 Describe the three general functional patterns seen in the pulmonary and systemic circuits of the cardiovascular system.

21-6 Identify the major arteries and veins of the pulmonary circuit.

21-7 Identify the major arteries and veins of the systemic circuit.

21-8 Identify the differences between fetal and adult circulation patterns, and describe the changes in the patterns of blood flow that occur at birth.

21-9 Discuss the effects of aging on the cardiovascular system.

21-10 Give examples of interactions between the cardiovascular system and each of the other organ systems.

Clinical Notes

Arteriosclerosis p. 724
Edema p. 737
Congenital Cardiovascular Problems p. 768

An Introduction to Blood Vessels and Circulation

This chapter examines the organization of blood vessels and considers the integrated functions of the circulatory system as a whole. We begin with a description of the histological organization of arteries, capillaries, and veins. Then we explore the functions of these vessels, the basic principles of cardiovascular regulation, and the distribution of major blood vessels in the body. We will then be ready to consider the organization and function of the lymphoid system, the focus of Chapter 22.

21-1 Arteries, arterioles, capillaries, venules, and veins differ in size, structure, and functional properties

arteries → arterioles → capillaries → venules → veins.

There are five general classes of blood vessels in the cardiovascular system. **Arteries** carry blood away from the heart. As they enter peripheral tissues, arteries branch repeatedly, and the branches decrease in diameter. The smallest arterial branches are called **arterioles** (ar-TER-ē-ōls). From the arterioles, blood moves into **capillaries**, where diffusion occurs between blood and interstitial fluid. From the capillaries, blood enters small **venules** (VEN-ūls), which unite to form larger **veins** that return blood to the heart.

Blood leaves the heart through the pulmonary trunk, which originates at the right ventricle, and the aorta, which originates at the left ventricle. Each of these arterial trunks has an internal diameter of about 2.5 cm (1 in.). The pulmonary arteries that branch from the pulmonary trunk carry blood to the lungs. The systemic arteries that branch from the aorta distribute blood to all other organs. Within these organs, further branching occurs, creating several hundred million tiny arterioles that provide blood to more than 10 billion capillaries. These capillaries, barely the diameter of a single red blood cell, form extensive branching networks. If all the capillaries in your body were placed end to end, their combined length would exceed 25,000 miles, enough to circle the globe.

The vital functions of the cardiovascular system depend entirely on events at the capillary level: All chemical and gaseous exchange between blood and interstitial fluid takes place across capillary walls. Cells rely on capillary diffusion to obtain nutrients and oxygen and to remove metabolic wastes, such as carbon dioxide and urea. Diffusion occurs very rapidly, because the distances involved are very short; few cells lie farther than 125 μm (0.005 in.) from a capillary. Homeostatic mechanisms operating at the local, regional, and systemic levels adjust blood flow through the capillaries to meet the demands of peripheral tissues.

Blood vessels must be resilient enough to withstand changes in pressure, and flexible enough to move with underlying tissues and organs. The pressures experienced by vessels vary with distance from the heart, and their structures reflect this fact. Moreover, arteries, veins, and capillaries differ in function, and these functional differences are associated with distinctive anatomical features.

The Structure of Vessel Walls

The walls of arteries and veins contain three distinct layers—the tunica intima, tunica media, and tunica externa (**Figure 21–1**):

1. The **tunica intima** (IN-ti-muh), or *tunica interna*, is the innermost layer of a blood vessel. This layer includes the endothelial lining and an underlying layer of connective tissue containing a variable number of elastic fibers. In arteries, the outer margin of the tunica intima contains a thick layer of elastic fibers called the **internal elastic membrane**. *esp-artery* → *pressure requires expansion.*

2. The **tunica media**, the middle layer, contains concentric sheets of smooth muscle tissue in a framework of loose connective tissue. The collagen fibers bind the tunica media to the tunica intima and tunica externa. Commonly the thickest layer in the wall of a small artery, the tunica media is separated from the surrounding tunica externa by a thin band of elastic fibers called the **external elastic membrane**. The smooth muscle cells of the tunica media encircle the endothelium lining the lumen of the blood vessel. When these smooth muscles contract, the vessel decreases in diameter; when they relax, the diameter increases. Large arteries also contain layers of longitudinally arranged smooth muscle cells. *keeps vessels intact.*

3. The **tunica externa** (eks-TER-nuh) or *tunica adventitia* (ad-ven-TISH-a), the outermost layer of a blood vessel, is a connective tissue sheath. In arteries, this layer contains collagen fibers with scattered bands of elastic fibers. In veins, it is generally thicker than the tunica media and contains networks of elastic fibers and bundles of smooth muscle cells. The connective tissue fibers of the tunica externa typically blend into those of adjacent tissues, stabilizing and anchoring the blood vessel. *keeps in original position.*

Their layered walls give arteries and veins considerable strength. The muscular and elastic components also permit controlled alterations in diameter as blood pressure or blood volume changes. However, the walls of arteries and veins are too thick to allow diffusion between the bloodstream and surrounding tissues, or even between the blood and the tissues

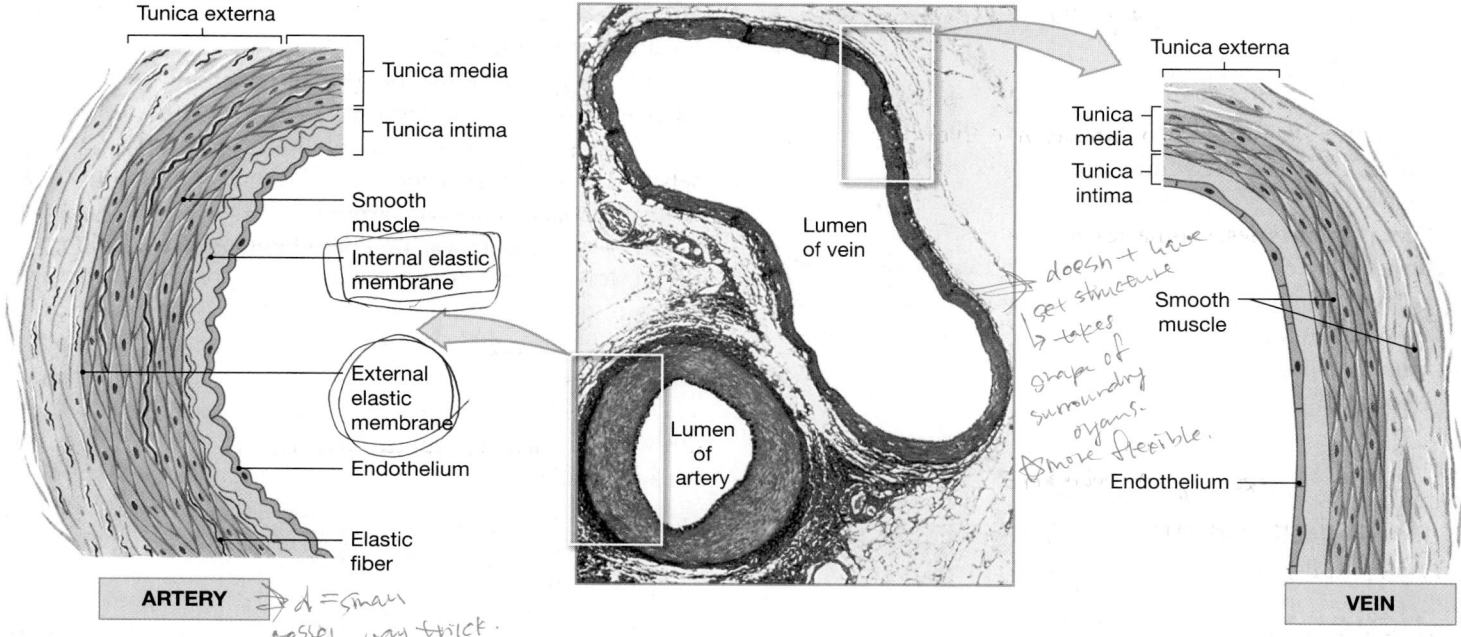

[Handwritten annotations near vein diagram: "doesn't have set structure → takes shape of surrounding organs. ↑more flexible."]

[Handwritten near ARTERY label: "→ d = small vessel way thick."]

Figure 21–1 Comparisons of a Typical Artery and a Typical Vein.

Feature	Typical Artery	Typical Vein
GENERAL APPEARANCE IN SECTIONAL VIEW	Usually round, with relatively thick wall	Usually flattened or collapsed, with relatively thin wall
TUNICA INTIMA Endothelium Internal elastic membrane	Usually rippled, due to vessel constriction Present	Often smooth Absent
TUNICA MEDIA	Thick, dominated by smooth muscle cells and elastic fibers	Thin, dominated by smooth muscle cells and collagen fibers
External elastic membrane	Present	Absent
TUNICA EXTERNA	Collagen and elastic fibers	Collagen and elastic fibers and smooth muscle cells

[Handwritten notes below table: "Artery: blood away from heart.   Venules: small veins.   arteriol: branch of artery.   veins: small veins   capillary: diffusion into interstitial fluid"]

of the vessel itself. For this reason, the walls of large vessels contain small arteries and veins that supply the smooth muscle cells and fibrocytes of the tunica media and tunica externa. These blood vessels are called the *vasa vasorum* ("vessels of vessels").

Differences between Arteries and Veins

Arteries and veins servicing the same region typically lie side by side (**Figure 21–1**). In sectional view, arteries and veins may be distinguished by the following characteristics:

- In general, the walls of arteries are thicker than those of veins. The tunica media of an artery contains more smooth muscle and elastic fibers than does that of a vein. These contractile and elastic components resist the pressure generated by the heart as it forces blood into the circuit.

[Handwritten: "bp higher in vein than artery."]

- When not opposed by blood pressure, the elastic fibers in the arterial walls recoil, constricting the lumen. Thus, seen on dissection or in sectional view, the lumen of an artery often looks smaller than that of the corresponding vein. Because the walls of arteries are relatively thick and strong, they retain their circular shape in section. Cut veins tend to collapse, and in section they often look flattened or grossly distorted. *[Handwritten: "blood flow faster in artery & volume doesn't matter."]*

- The endothelial lining of an artery cannot contract, so when an artery constricts, its endothelium is thrown into folds that give sectioned arteries a pleated appearance. The lining of a vein lacks these folds.

In gross dissection, arteries and veins can generally be distinguished because:

- The thicker walls of arteries can be felt if the vessels are compressed.

- Arteries usually retain their cylindrical shape, whereas veins often collapse.

- Arteries are more resilient: When stretched, they keep their shape and elongate, and when released, they snap back. A small vein cannot tolerate as much distortion without collapsing or tearing.

- Veins typically contain *valves*—internal structures that prevent the backflow of blood toward the capillaries. In an intact vein, the location of each valve is marked by a slight distension of the vessel wall. (We will consider valve structure in a later section.)

Arteries

Their relatively thick, muscular walls make arteries elastic and contractile. Elasticity permits passive changes in vessel diameter in response to changes in blood pressure. For example, it allows arteries to absorb the surging pressure waves that accompany the contractions of the ventricles.

The contractility of the arterial walls enables them to change in diameter actively, primarily under the control of the sympathetic division of the autonomic nervous system. When stimulated, arterial smooth muscles contract, thereby constricting the artery—a process called **vasoconstriction**. Relaxation of these smooth muscles increases the diameter of the lumen—a process called **vasodilation**. Vasoconstriction and vasodilation affect (1) the afterload on the heart, (2) peripheral blood pressure, and (3) capillary blood flow. We will explore these effects in a later section. Contractility is also important during the vascular phase of hemostasis, when the contraction of a damaged vessel wall helps reduce bleeding. ∞ p. 673

In traveling from the heart to peripheral capillaries, blood passes through *elastic arteries, muscular arteries*, and *arterioles* (Figure 21–2).

Elastic Arteries

Elastic arteries, or *conducting arteries*, are large vessels with diameters up to 2.5 cm (1 in.). These vessels transport large volumes of blood away from the heart. The pulmonary trunk and aorta, as well as their major arterial branches (the *pulmonary, common carotid, subclavian*, and *common iliac arteries*), are elastic arteries.

The walls of elastic arteries (**Figure 21–2**) are extremely resilient because the tunica media contains a high density of elastic fibers and relatively few smooth muscle cells. As a result, elastic arteries can tolerate the pressure changes that occur during the cardiac cycle. We have already considered the role played by elastic rebound in the aorta in maintaining blood flow in the coronary arteries. ∞ p. 692 However, elastic rebound occurs to some degree in all elastic arteries. During ventricular systole, pressures rise rapidly and the elastic arteries expand as the stroke volume is ejected. During ventricular diastole, blood pressure within the arterial system falls and the elastic fibers recoil to their original dimensions. Their expansion cushions the sudden rise in pressure during ventricular systole, and their recoil slows the drop in pressure during ventricular diastole. This feature is important because blood pressure is the driving force behind blood flow: The greater the pressure oscillations, the greater the changes in blood flow. The elasticity of the arterial system dampens the pressure peaks and valleys that accompany the heartbeat. By the time blood reaches the arterioles, the pressure oscillations have disappeared, and blood flow is continuous.

Muscular Arteries

Muscular arteries, also known as *medium-sized arteries* or *distribution arteries*, distribute blood to the body's skeletal muscles and internal organs. Most of the vessels of the arterial system are muscular arteries. These arteries are characterized by a thick tunica media that contains more smooth muscle cells than does the tunica media of elastic arteries (**Figures 21–1** and **21–2**). A typical muscular artery has a lumen diameter of approximately 4.0 mm (0.16 in.), but some have diameters as small as 0.5 mm. The *external carotid arteries* of the neck, the *brachial arteries* of the arms, the *mesenteric arteries* of the abdomen, and the *femoral arteries* of the thighs are examples of muscular arteries. Superficial muscular arteries are important as *pressure points*—places in the body where muscular arteries can be pressed against deeper bones to reduce blood flow and control severe bleeding.

Arterioles

Arterioles, with an internal diameter of 30 μm or less, are considerably smaller than muscular arteries. Arterioles have a poorly defined tunica externa, and the tunica media in the larger arterioles consists of one or two layers of smooth muscle cells (**Figure 21–2**). The tunica media of the smallest arterioles contains scattered smooth muscle cells that do not form a complete layer.

The diameters of smaller muscular arteries and arterioles change in response to local conditions or to sympathetic or endocrine stimulation. For example, arterioles in most tissues vasodilate when oxygen levels are low and, as we saw in Chapter 16, vasoconstrict under sympathetic stimulation. ∞ p. 537 Changes in their diameter affect the amount of force required to push blood around the cardiovascular system: More pressure is required to push blood through a constricted vessel than through a dilated one. The force opposing blood flow is called **resistance (R)**, so arterioles are also called **resistance vessels**.

Vessel characteristics change gradually with distance from the heart. Each type of vessel described here actually represents the midpoint in a portion of a continuum. Thus, the largest

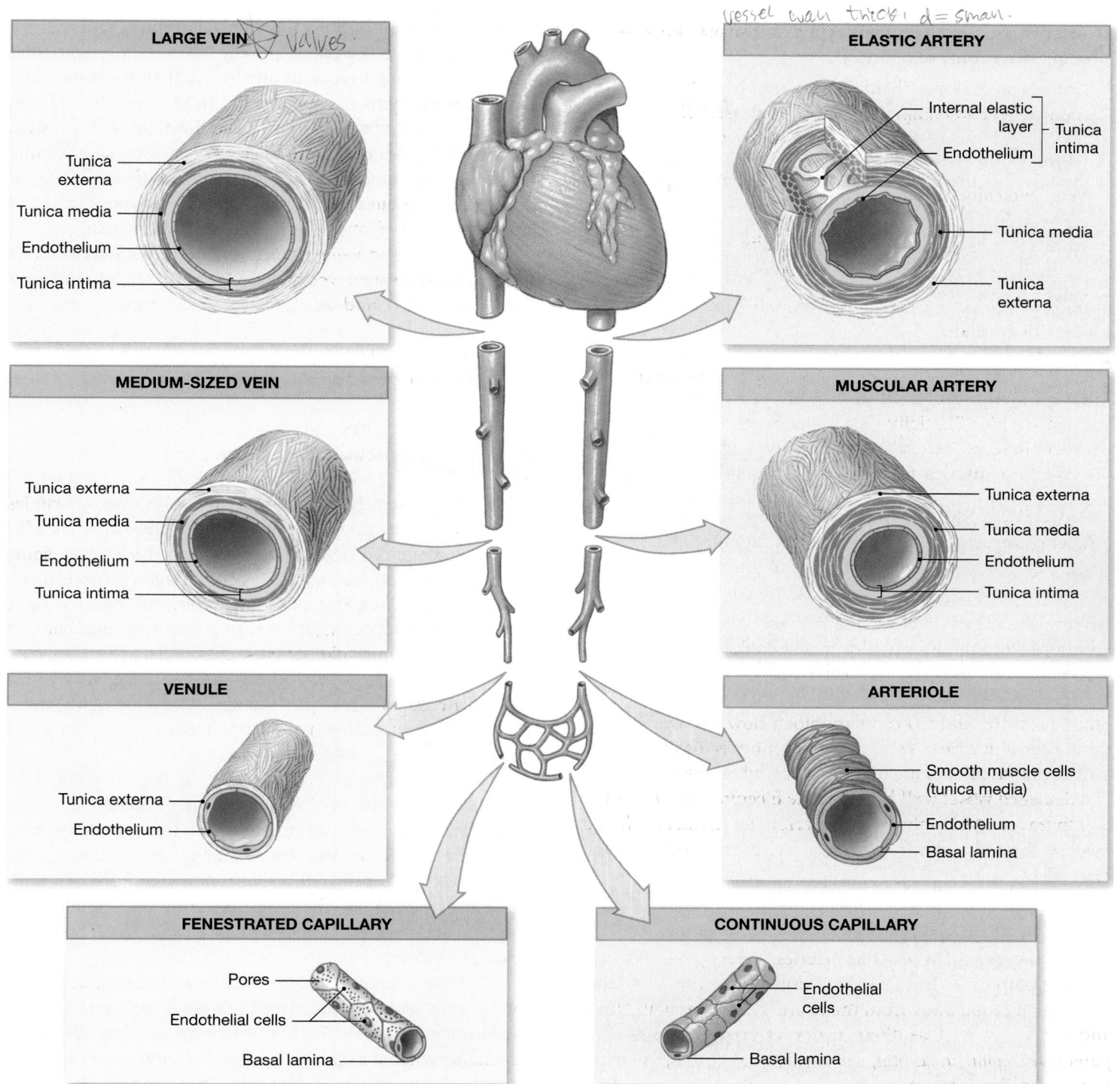

Figure 21–2 Histological Structure of Blood Vessels. Representative diagrammatic cross-sectional views of the walls of arteries, capillaries, and veins. Notice the relative sizes of the layers in these vessels.

muscular arteries contain a considerable amount of elastic tissue, whereas the smallest resemble heavily muscled arterioles.

Arteries carry blood under great pressure, and their walls are adapted to handle that pressure. Occasionally, local arterial pressure exceeds the capacity of the elastic components of

the tunics, producing an **aneurysm** (AN-ū-rizm), or bulge in the weakened wall of an artery. The bulge resembles a bubble in the wall of a tire—and like a bad tire, the affected artery can suffer a catastrophic blowout. The most dangerous aneurysms occur in arteries of the brain (where they cause strokes) or in the aorta (where a rupture will cause fatal bleeding in a matter of minutes).

Capillaries → *[handwritten: reaches nearly all organs.]*

When we think of the cardiovascular system, we think first of the heart or the great blood vessels connected to it. But the real work of the cardiovascular system is done in the microscopic capillaries that permeate most tissues. These delicate vessels weave throughout active tissues, forming intricate networks that surround muscle fibers, radiate through connective tissues, and branch beneath the basal laminae of epithelia.

[handwritten: → exchange of O2 & nutrients btwn blood & interstitial fluid.]

Capillaries are the *only* blood vessels whose walls permit exchange between the blood and the surrounding interstitial fluids. Because capillary walls are thin, diffusion distances are short, so exchange can occur quickly. In addition, blood flows through capillaries relatively slowly, allowing sufficient time for the diffusion or active transport of materials across the capillary walls. Thus, the histological structure of capillaries permits a two-way exchange of substances between blood and interstitial fluid.

A typical capillary consists of an endothelial tube inside a delicate basal lamina; neither a tunica media nor a tunica externa is present (**Figure 21–2**). The average diameter of a capillary is a mere 8 μm, very close to that of a single red blood

[handwritten: → slightus layer ... → single RBC.]

CLINICAL NOTE

Arteriosclerosis

Arteriosclerosis (ar-tēr-ē-ō-skler-Ō-sis; *arterio-*, artery + *sklerosis*, hardness) is a thickening and toughening of arterial walls. This condition may not sound life-threatening, but complications related to arteriosclerosis account for roughly half of all deaths in the United States. The effects of arteriosclerosis are varied; for example, arteriosclerosis of coronary vessels is responsible for *coronary artery disease (CAD)*, and arteriosclerosis of arteries supplying the brain can lead to strokes. ∞ p. 694 Arteriosclerosis takes two major forms:

1. **Focal calcification** is the deposition of calcium salts following the gradual degeneration of smooth muscle in the tunica media. Some focal calcification occurs as part of the aging process, and it may develop in association with atherosclerosis (described next). Rapid and severe calcification may occur as a complication of diabetes mellitus, an endocrine disorder. ∞ p. 634

   [handwritten: → LDL attaches low density lipoprotein.]

2. **Atherosclerosis** (ath-er-ō-skler-Ō-sis; *athero-*, fatty degeneration) is the formation of lipid deposits in the tunica media associated with damage to the endothelial lining. Atherosclerosis is the most common form of arteriosclerosis. *[handwritten: → consuming more fat than needed.]*

   [handwritten: → growth of LDL.]

   [handwritten left margin: diameter becomes smaller]

   Many factors may be involved in the development of atherosclerosis. One major factor is lipid levels in the blood. Atherosclerosis tends to develop in people whose blood contains elevated levels of plasma lipids—specifically, cholesterol. Circulating cholesterol is transported to peripheral tissues in

lipoproteins, which are protein–lipid complexes. (We will discuss the various types of lipoproteins in Chapter 25.)

When plasma cholesterol levels are chronically elevated, cholesterol-rich lipoproteins remain in circulation for an extended period. Circulating monocytes then begin removing them from the bloodstream. Eventually, the monocytes become filled with lipid droplets. Now called *foam cells*, they attach themselves to the endothelial walls of blood vessels, where they release cytokines. These growth factors stimulate the divisions of smooth muscle cells near the tunica intima, thickening the vessel wall. *[handwritten: keeps growing, vessel becomes narrow.]*

Other monocytes then invade the area, migrating between the endothelial cells. As these changes occur, the monocytes, smooth muscle cells, and endothelial cells begin phagocytizing lipids as well. The result is an atherosclerotic **plaque**, a fatty mass of tissue that projects into the lumen of the vessel. At this point, the plaque has a relatively simple structure, and evidence suggests that the process can be reversed if appropriate dietary adjustments are made.

If the conditions persist, the endothelial cells become swollen with lipids, and gaps appear in the endothelial lining. Platelets now begin sticking to the exposed collagen fibers. The combination of platelet adhesion and aggregation leads to the formation of a localized blood clot, which further restricts blood flow through the artery. The structure of the plaque is now relatively complex.

A typical plaque is shown in **Figure 21–3**. Elderly individuals—especially elderly men—are most likely to develop atherosclerotic plaques. Estrogens may slow plaque formation, which may account for the lower incidence of CAD, myocardial infarctions (MIs), and strokes in women. After menopause, when estrogen production declines, the risks of CAD, MIs, and strokes in women increase markedly.

In addition to advanced age and male gender, other important risk factors for atherosclerosis include high blood cholesterol levels, high blood pressure, and cigarette

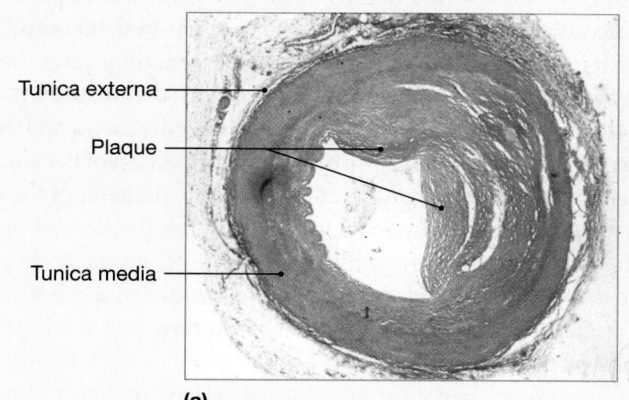

(a)

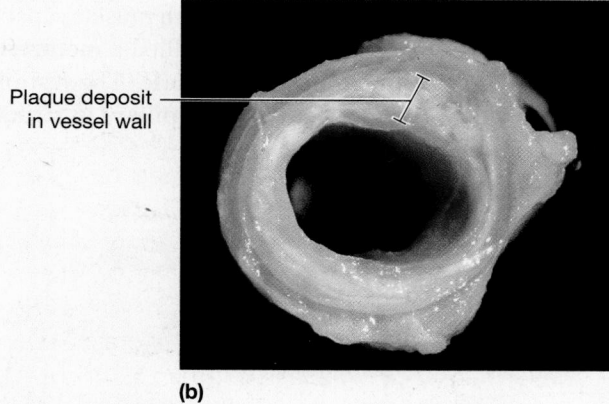

(b)

Figure 21–3 A Plaque within an Artery.
(a) A cross-sectional view of a large plaque. **(b)** A section of a coronary artery narrowed by plaque formation.

smoking. Roughly 20 percent of middle-aged men have all three of these risk factors; these individuals are four times as likely to experience an MI or a cardiac arrest as other men in their age group. Although fewer women develop atherosclerotic plaques, elderly female smokers with high blood cholesterol and high blood pressure are at much greater risk than other women. Factors that can promote the development of atherosclerosis in both men and women include diabetes mellitus, obesity, and stress. Evidence also indicates that at least some forms of atherosclerosis may be linked to chronic infection with *Chlamydia pneumoniae*, a bacterium responsible for several types of respiratory infections, including some forms of pneumonia.

We discussed potential treatments for atherosclerotic plaques, such as catheterization with balloon angioplasty and stenting, and bypass surgery, in Chapter 20. ∞ p. 694 In the many cases where dietary modifications do not lower circulating LDL levels sufficiently, drug therapies can bring them under control. Genetic engineering techniques have recently been used to treat an inherited form of *hypercholesterolemia* (high blood cholesterol) linked to extensive plaque formation. (Individuals with this condition are unable to absorb and recycle cholesterol in the liver.) In this experimental procedure, circulating cholesterol levels declined after copies of appropriate genes were inserted into some of the individual's liver cells.

Without question, the best approach to atherosclerosis is avoiding it by eliminating or reducing associated risk factors. Suggestions include (1) reducing the intake of dietary cholesterol, saturated fats, and trans fatty acids by restricting consumption of fatty meats (such as beef, lamb, and pork), egg yolks, and cream; (2) not smoking; (3) checking your blood pressure and taking steps to lower it if necessary; (4) having your blood cholesterol levels checked annually; (5) controlling your weight; and (6) exercising regularly.

Continuous Capillaries

cell. There are two major types of capillaries: *continuous capillaries* and *fenestrated capillaries*.

Continuous Capillaries

Most regions of the body are supplied by continuous capillaries. In a **continuous capillary**, the endothelium is a complete lining. A cross section through a large continuous capillary cuts across several endothelial cells (**Figure 21–4a**). In a small continuous capillary, a single endothelial cell may completely encircle the lumen.

Continuous capillaries are located in all tissues except epithelia and cartilage. Continuous capillaries permit the dif-

fusion of water, small solutes, and lipid-soluble materials into the surrounding interstitial fluid, but prevent the loss of blood cells and plasma proteins. In addition, some exchange may occur between blood and interstitial fluid by bulk transport—the movement of vesicles that form through endocytosis at the inner endothelial surface. ∞ p. 96

In specialized continuous capillaries throughout most of the central nervous system and in the thymus, the endothelial cells are bound together by tight junctions. These capillaries have very restricted permeability characteristics. We discussed one example—the capillaries responsible for the *blood–brain barrier*—in Chapters 12 and 14. ∞ pp. 392, 467

Fenestrated Capillaries *↳ much exchange, even small peptides*

Fenestrated (FEN-es-trā-ted; *fenestra*, window) **capillaries** are capillaries that contain "windows," or pores, that penetrate the endothelial lining (**Figure 21–4b**). The pores permit the rapid exchange of water and solutes as large as small peptides between plasma and interstitial fluid. Examples of fenestrated capillaries include the *choroid plexus* of the brain and the blood vessels in a variety of endocrine organs, such as the hypothalamus and the pituitary, pineal, and thyroid glands. Fenestrated capillaries are also located along absorptive areas of the intestinal tract and at filtration sites in the kidneys. Both the number of pores and their permeability characteristics may vary from one region of the capillary to another.

Sinusoids (SĪ-nuh-soydz), also called **sinusoidal capillaries**, resemble fenestrated capillaries that are flattened and irregularly shaped. In contrast to fenestrated capillaries, sinusoids commonly have gaps between adjacent endothelial cells, and the basal lamina is either thinner or absent. As a result, sinusoids permit the free exchange of water and solutes as large as plasma proteins between blood and interstitial fluid.

Blood moves through sinusoids relatively slowly, maximizing the time available for exchange across the sinusoidal walls. Sinusoids occur in the liver, bone marrow, spleen, and many endocrine organs, including the pituitary and suprarenal glands. At liver sinusoids, plasma proteins secreted by liver cells enter the bloodstream. Along sinusoids of the liver, spleen, and bone marrow, phagocytic cells monitor the passing blood, engulfing damaged red blood cells, pathogens, and cellular debris.

Capillary Beds *branches & direct connection of arteriole & venous system → venules*

Capillaries do not function as individual units but as part of an interconnected network called a **capillary bed**, or **capillary plexus** (**Figure 21–5**). A single arteriole generally gives rise to dozens of capillaries that empty into several *venules*, the smallest vessels of the venous system. The entrance to each capillary is guarded by a **precapillary sphincter**. Contraction of the smooth muscle cell(s) of this sphincter narrows the diameter of the capillary entrance, thereby reducing or stopping the flow of blood. When one precapillary sphincter constricts, blood is diverted into other branches of the plexus. Relaxation of a precapillary sphincter dilates the opening, and blood flow into the capillary accelerates as a result.

A capillary bed contains several relatively direct connections between arterioles and venules. The wall in the initial part of such a passageway possesses smooth muscle capable of changing its diameter. This segment is called a **metarteriole** (met-ar-TĒR-ē-ōl) or **precapillary arteriole**. The rest of the passageway, which resembles a typical capillary in structure, is called a **thoroughfare channel**.

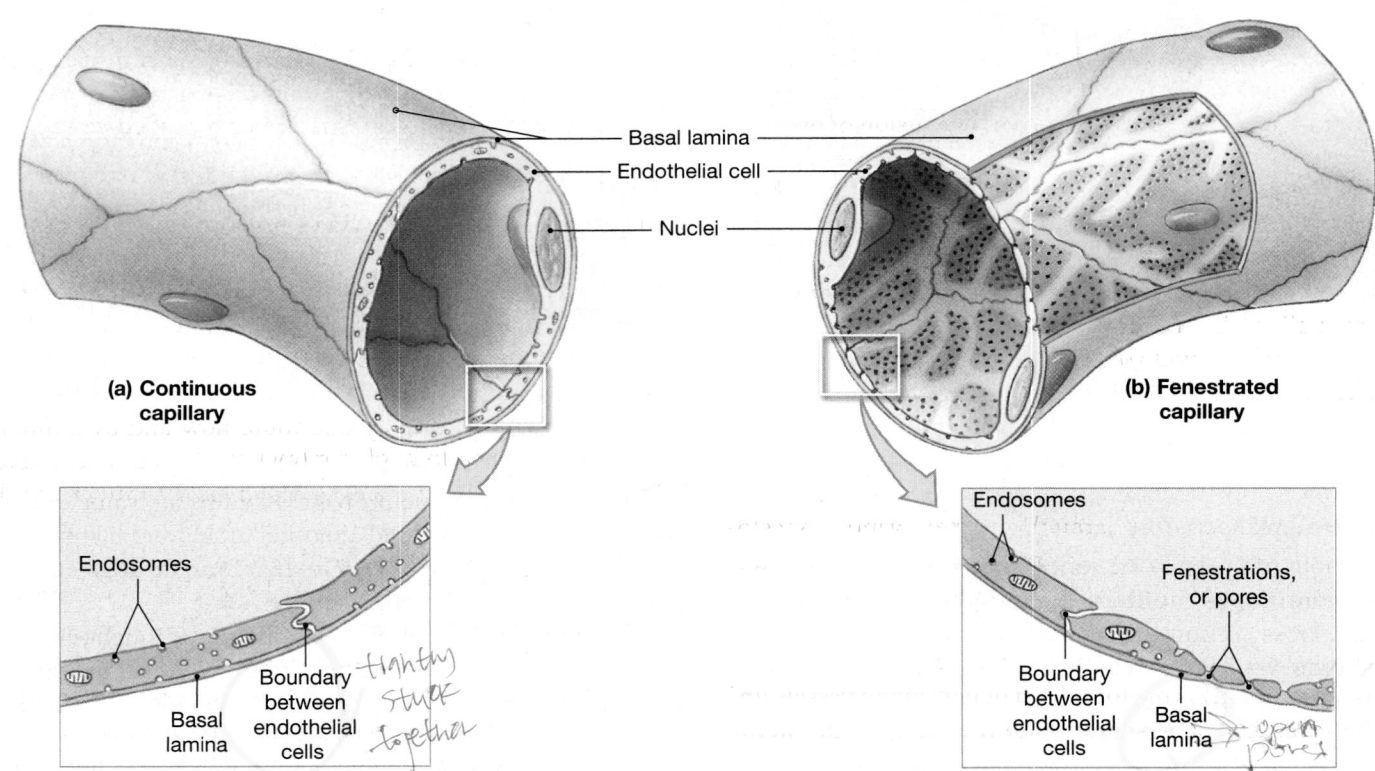

Figure 21–4 **Capillary Structure.** **(a)** A continuous capillary. The enlargement shows routes for the diffusion of water and solutes. **(b)** A fenestrated capillary. Note the pores, which facilitate diffusion across the endothelial lining.

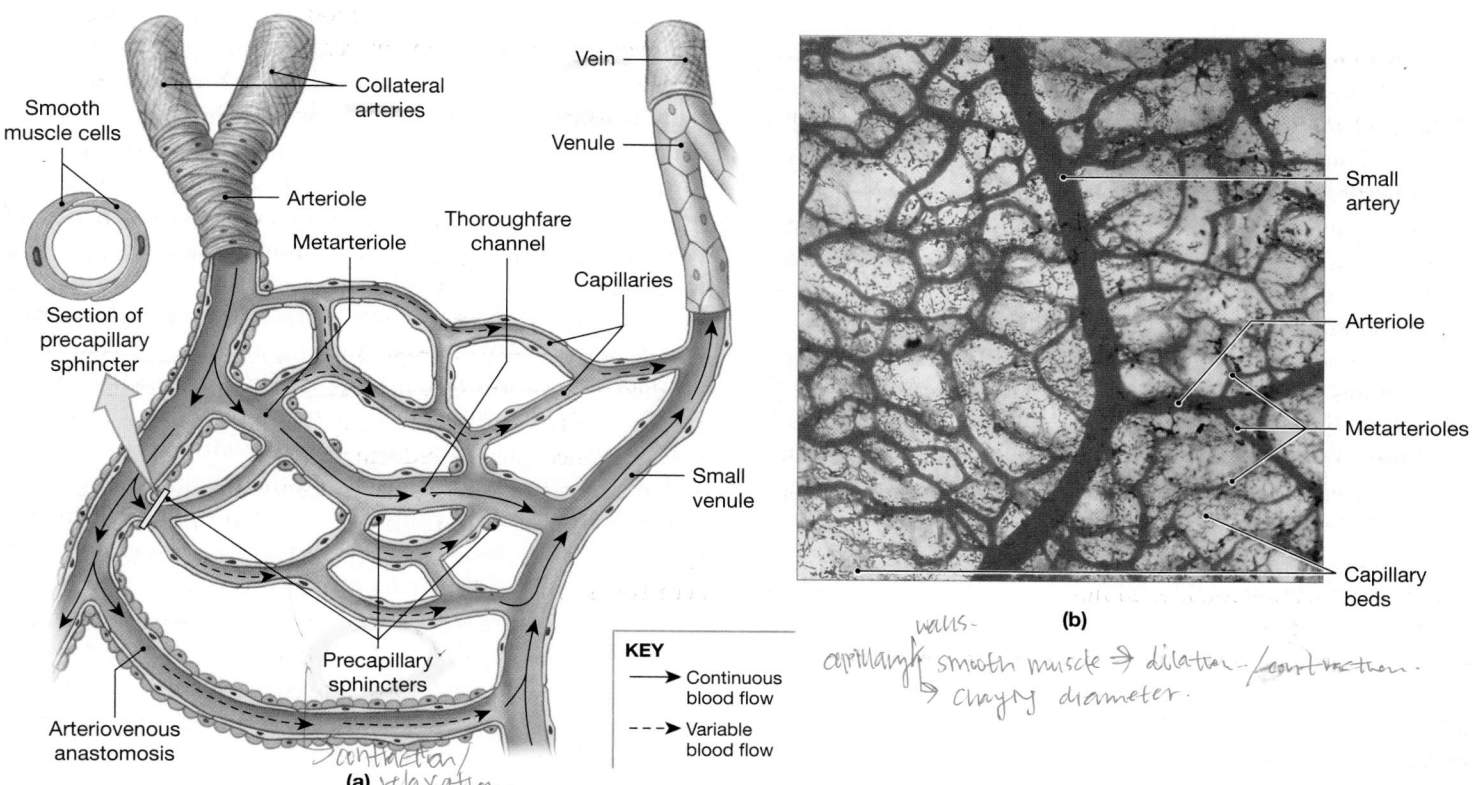

Figure 21–5   The Organization of a Capillary Bed. (a) A typical capillary bed. Solid arrows indicate consistent blood flow; dashed arrows indicate variable or pulsating blood flow. **(b)** A micrograph of a number of capillary beds.

A capillary bed may receive blood from more than one artery. The multiple arteries, called **collaterals**, enter the region and fuse before giving rise to arterioles. The fusion of two collateral arteries that supply a capillary bed is an example of an **arterial anastomosis**. (An *anastomosis* is the joining of blood vessels.) The interconnections between the *anterior* and *posterior interventricular arteries* of the heart are arterial anastomoses. ∞ p. 692 An arterial anastomosis acts like an insurance policy: If one artery is compressed or blocked, capillary circulation will continue.

Arteriovenous (ar-tēr-ē-ō-VĒ-nus) **anastomoses** are direct connections between arterioles and venules. When an arteriovenous anastomosis is dilated, blood will bypass the capillary bed and flow directly into the venous circulation. The pattern of blood flow through an arteriovenous anastomosis is regulated primarily by sympathetic innervation under the control of the cardiovascular centers of the medulla oblongata.

Angiogenesis (an-jē-ō-JEN-e-sis; *angio-*, blood vessel + *genesis*, production) is the formation of new blood vessels under the direction of **vascular endothelial growth factor (VEGF)**. Angiogenesis occurs embryologically during the development of tissues and organs. It may also occur at other times in any body tissue in response to factors released by hy-poxic (oxygen-starved) cells. Clinically, it is probably most important in cardiac muscle, where angiogenesis occurs in response to a chronically constricted or occluded vessel.

Vasomotion

Although blood normally flows from arterioles to venules at a constant rate, the flow within each capillary is quite variable. Each precapillary sphincter alternately contracts and relaxes, perhaps a dozen times per minute. As a result, the blood flow within any capillary occurs in pulses rather than as a steady and constant stream. The net effect is that blood may reach the venules by one route now and by a different route later. The cycling of contraction and relaxation of smooth muscles that changes blood flow through capillary beds is called **vasomotion**.

Vasomotion is controlled locally by changes in the concentrations of chemicals and dissolved gases in the interstitial fluids. For example, when dissolved oxygen concentrations decline within a tissue, the capillary sphincters relax, so blood flow to the area increases. This process, an example of capillary *autoregulation*, will be the focus of a later section.

When you are at rest, blood flows through roughly 25 percent of the vessels within a typical capillary bed in your body. Your cardiovascular system does not contain enough blood to maintain adequate blood flow to all the capillaries in all the capillary beds in your body at the same time. As a result, when many tissues become active, the blood flow through capillary beds must be coordinated. We will describe the mechanisms by which the cardiovascular centers perform this coordination later in the chapter.

Veins

Veins collect blood from all tissues and organs and return it to the heart. The walls of veins can be thinner than those of corresponding arteries because the blood pressure in veins is lower than that in arteries. Veins are classified on the basis of their size. Even though their walls are thinner, in general veins are larger in diameter than their corresponding arteries. (Review **Figure 21–2** to compare typical arteries and veins.)

Venules *larger than artery*

Venules, which collect blood from capillary beds, are the smallest venous vessels. They vary widely in size and structure. An average venule has an internal diameter of roughly 20 μm. Venules smaller than 50 μm lack a tunica media, and the smallest venules resemble expanded capillaries.

Medium-Sized Veins

Medium-sized veins range from 2 to 9 mm in internal diameter, comparable in size to muscular arteries. In these veins, the tunica media is thin and contains relatively few smooth muscle cells. The thickest layer of a medium-sized vein is the tunica externa, which contains longitudinal bundles of elastic and collagen fibers.

Large Veins

Large veins include the superior and inferior venae cavae and their tributaries within the abdominopelvic and thoracic cavities. All the tunica layers are present in all large veins. The slender tunica media is surrounded by a thick tunica externa composed of a mixture of elastic and collagen fibers.

Venous Valves *prevents backflow.*

The arterial system is a high-pressure system: Almost all the force developed by the heart is required to push blood along the network of arteries and through miles of capillaries. Blood pressure in a peripheral venule is only about 10 percent of that in the ascending aorta, and pressures continue to fall along the venous system.

The blood pressure in venules and medium-sized veins is so low that it cannot overcome the force of gravity. In the limbs, veins of this size contain **valves**, folds of the tunica intima that project from the vessel wall and point in the direction of blood flow. These valves, like those in the heart, permit

blood flow in one direction only. Venous valves prevent the backflow of blood toward the capillaries (**Figure 21–6**).

As long as the valves function normally, any movement that distorts or compresses a vein will push blood toward the heart. This effect improves *venous return*, the rate of blood flow to the heart. ∞ p. 711 The mechanism is particularly important when you are standing, because blood returning from your feet must overcome the pull of gravity to ascend to the heart. Valves compartmentalize the blood within the veins, thereby dividing the weight of the blood between the compartments. Any contraction of the surrounding skeletal muscles squeezes the blood toward the heart. Although you are probably not aware of it, when you stand, rapid cycles of contraction and relaxation are occurring within your leg muscles, helping to push blood toward the trunk. When you lie down, venous valves contribute less to venous return, because your heart and major vessels are at the same level.

If the walls of the veins near the valves weaken or become stretched and distorted, the valves may not work properly. Blood then pools in the veins, and the vessels become

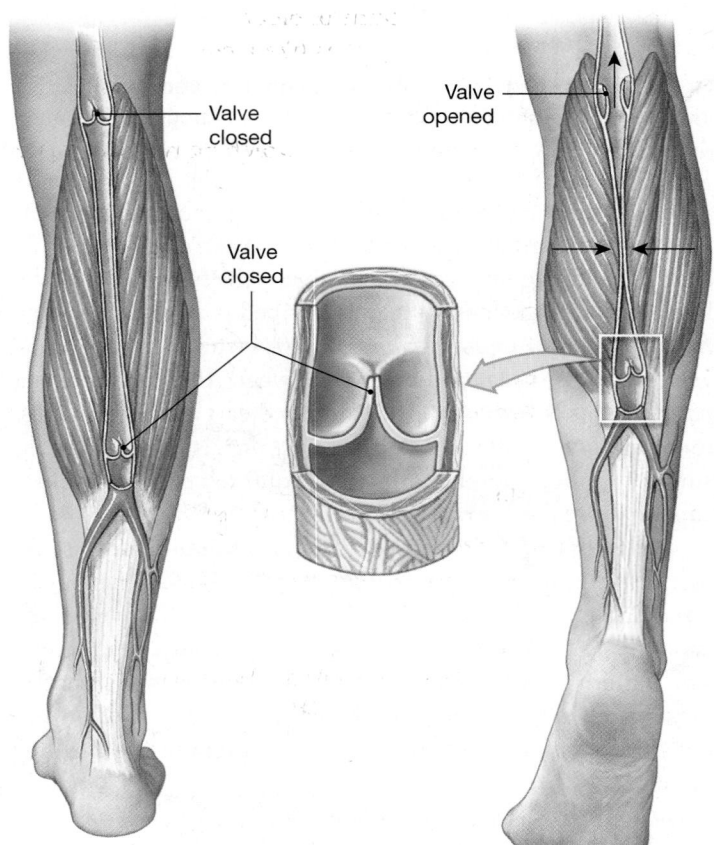

Figure 21–6   The Function of Valves in the Venous System.
Valves in the walls of medium-sized veins prevent the backflow of blood. Venous compression caused by the contraction of adjacent skeletal muscles assists in maintaining venous blood flow.

grossly distended. The effects range from mild discomfort and a cosmetic problem, as in superficial **varicose veins** in the thighs and legs, to painful distortion of adjacent tissues, as in **hemorrhoids**.

The Distribution of Blood

The total blood volume is unevenly distributed among arteries, veins, and capillaries (**Figure 21–7**). The heart, arteries, and capillaries in the pulmonary and systemic circuits normally contain 30–35 percent of the blood volume (roughly 1.5 liters of whole blood), and the venous system contains the rest (65–70 percent, or about 3.5 liters). Roughly one-third of the blood in the venous system (about a liter) is circulating within the liver, bone marrow, and skin. These organs have extensive venous networks that at any moment contain large volumes of blood.

Because their walls are thinner, with a lower proportion of smooth muscle, veins are much more distensible (expandable) than arteries. For a given rise in blood pressure, a typical vein will stretch about eight times as much as a corresponding artery. The *capacitance* of a blood vessel is the relationship between the volume of blood it contains and the blood pressure. If a vessel behaves like a child's balloon, expanding easily at low pressures, it has high capacitance. If it behaves more like a truck tire, expanding only at high pressures, it has low capacitance. Veins, which expand easily, are

veins ⇒ larger diameter, volume, blood flow rate slower than arteries..

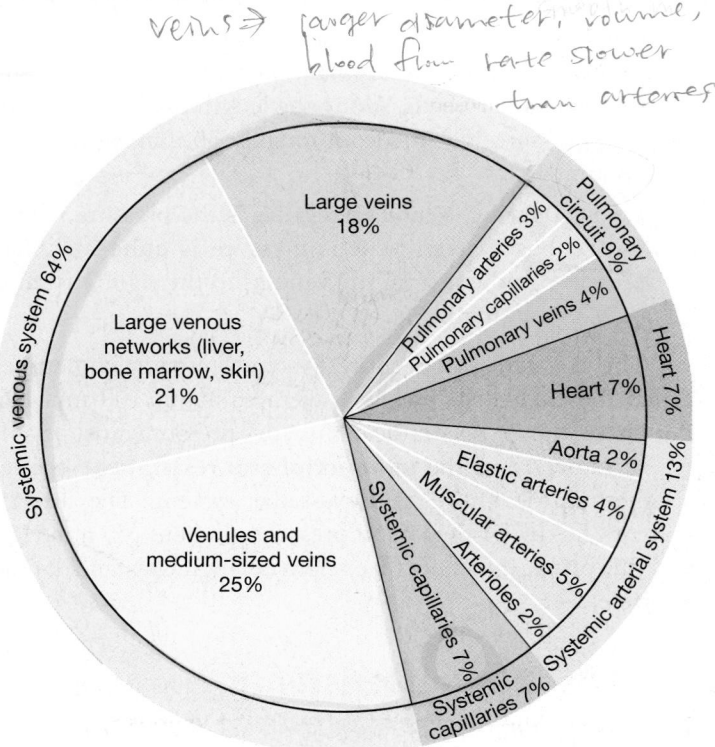

Figure 21–7 **The Distribution of Blood in the Cardiovascular System.**

called **capacitance vessels**. Because veins have high capacitance, they can accommodate large changes in blood volume. If the blood volume rises or falls, the elastic walls stretch or recoil, changing the volume of blood in the venous system.

If serious hemorrhaging occurs, the *vasomotor center* of the medulla oblongata stimulates sympathetic nerves that innervate smooth muscle cells in the walls of medium-sized veins. This activity has two major effects:

1. *Systemic Veins Constrict.* This process, called **venoconstriction** (vē-nō-kon-STRIK-shun), reduces the amount of blood within the venous system, thereby increasing the volume within the arterial system and capillaries. Venoconstriction can keep the blood volume within the arteries and capillaries at near-normal levels despite a significant blood loss.

2. *The Constriction of Veins in the Liver, Skin, and Lungs Redistributes a Significant Proportion of the Total Blood Volume.* As a result, blood flow to delicate organs (such as the brain) and to active skeletal muscles can be increased or maintained after blood loss. The amount of blood that can be shifted from veins in the liver, skin, and lungs to the general circulation, called the **venous reserve**, is normally about 20 percent of total blood volume.

CHECKPOINT

1. List the five general classes of blood vessels.

2. A cross section of tissue shows several small, thin-walled vessels with very little smooth muscle tissue in the tunica media. Which type of vessel are these?

3. Why are valves located in veins, but not in arteries?

4. Where in the body would you find fenestrated capillaries?

See the blue Answers tab at the end of the book.

21-2 Pressure and resistance determine blood flow and affect rates of capillary exchange

Figure 21–8 provides an overview of the discussion of cardiovascular physiology that follows. The purpose of cardiovascular regulation is the maintenance of adequate blood flow through the capillaries in peripheral tissues and organs. Under normal circumstances, blood flow is equal to cardiac output. When cardiac output goes up, so does the blood flow through capillary beds; when cardiac output declines, capillary blood flow is reduced. Capillary blood flow is determined by the interplay between *pressure* (P) and *resistance* (R) in the cardiovascular

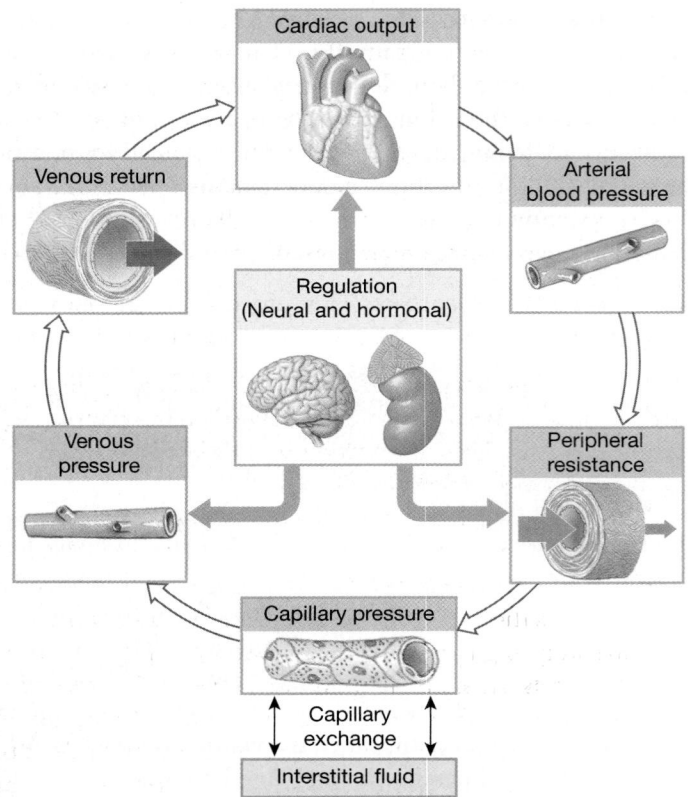

Figure 21–8 **An Overview of Cardiovascular Physiology.**
Neural and hormonal activities influence cardiac output, peripheral
resistance, and venous pressure (through venoconstriction). Capillary
pressure is the primary drive for exchange between blood and
interstitial fluid.

network. To keep blood moving, the heart must generate sufficient pressure to overcome the resistance to blood flow in the
pulmonary and systemic circuits. In general terms, flow (*F*) is
directly proportional to the pressure (increased pressure → increased flow), and inversely proportional to resistance (increased resistance → decreased flow). However, the absolute
pressure is less important than the pressure *gradient*—the difference in pressure from one end of the vessel to the other. This relationship can be summarized as

$$F \propto \frac{\Delta P}{R}$$

where the symbol ∝ means "is proportional to" and Δ means
"the difference in." The largest pressure gradient is found between the base of the aorta and the proximal ends of peripheral
capillary beds. Cardiovascular control centers can alter this
pressure gradient, and thereby change the rate of capillary blood
flow, by adjusting cardiac output and peripheral resistance.

Blood leaving the peripheral capillaries enters the venous
system. Although the pressure gradient across the venous sys-

tem is relatively small, venous resistance is very low. This low
venous blood pressure—aided by valves, skeletal muscle contraction, gravity, and other factors—is sufficient to return the
blood to the heart. When necessary, cardiovascular control centers can elevate venous pressure (through venoconstriction) to
enhance venous return and maintain adequate cardiac output.

We will begin this section by examining blood pressure
and resistance more closely. We will then consider the mechanisms of *capillary exchange*, the transfer of liquid and solutes
between the blood and interstitial fluid. Capillary exchange
provides tissues with oxygen and nutrients and removes the
carbon dioxide and waste products generated by active cells.

Active tissues require more blood flow than inactive ones;
even something as simple as a change in position—going from
sitting to standing, for instance—triggers a number of cardiovascular changes. We will end this section with a discussion
of what those changes are and how they are coordinated.

Pressure

When talking about cardiovascular pressures, three values are
usually reported:

1. *Blood Pressure.* The term **blood pressure (BP)** refers to arterial pressure, usually reported in millimeters of mercury
   (mm Hg). Average systemic arterial pressures range from
   an average of 100 mm Hg at the entrance to the aorta to
   roughly 35 mm Hg at the start of a capillary network.

2. *Capillary Hydrostatic Pressure.* **Capillary hydrostatic
   pressure (CHP)**, or *capillary pressure*, is the pressure
   within capillary beds. Along the length of a typical capillary, pressures decline from roughly 35 mm Hg to about
   18 mm Hg.

3. *Venous Pressure.* **Venous pressure** is the pressure within
   the venous system. Venous pressure is quite low: The
   pressure gradient from the venules to the right atrium is
   only about 18 mm Hg.

The ΔP across the entire systemic circuit, sometimes
called the *circulatory pressure*, averages about 100 mm Hg.
For circulation to occur, the circulatory pressure must be sufficient to overcome the **total peripheral resistance**—the resistance of the entire cardiovascular system. The arterial
network has by far the largest pressure gradient (65 mm Hg),
and this primarily reflects the relatively high resistance of the
arterioles.

Total Peripheral Resistance

The total peripheral resistance of the cardiovascular system
reflects a combination of factors: *vascular resistance*, *blood
viscosity*, and *turbulence*.

Vascular Resistance

Vascular resistance, the resistance of the blood vessels, is the largest component. The most important factor in vascular resistance is friction between blood and the vessel walls. The amount of friction depends on two factors: vessel length and vessel diameter.

Vessel Length. Increasing the length of a blood vessel increases friction: The longer the vessel, the larger the surface area in contact with blood. You can easily blow the water out of a snorkel that is 2.5 cm (1 in.) in diameter and 25 cm (10 in.) long, but you cannot blow the water out of a 15 m (16 yard)-long garden hose, because the total friction is too great. The most dramatic changes in blood vessel length occur between birth and maturity, as individuals grow to adult size. In adults, vessel length can increase or decrease gradually when individuals gain or lose weight, but on a day-to-day basis this component of vascular resistance can be considered constant.

Vessel Diameter. The effects of friction on blood act primarily in a narrow zone closest to the vessel wall. In a small-diameter vessel, nearly all the blood is slowed down by friction with the walls. Resistance is therefore relatively high. Blood near the center of a large-diameter vessel does not encounter friction with the walls, so the resistance in large vessels is relatively low.

Differences in diameter have much more significant effects on resistance than do differences in length. If two vessels are equal in diameter but one is twice as long as the other, the longer vessel offers twice as much resistance to blood flow. But for two vessels of equal length, one twice the diameter of the other, the narrower one offers 16 times as much resistance to blood flow. This relationship, expressed in terms of the vessel radius r and resistance R, can be summarized as $R \propto 1/r^4$.

More significantly, there is no way to control vessel length, but vessel diameter can change quickly through vasoconstriction or vasodilation. Most of the peripheral resistance occurs in arterioles, the smallest vessels of the arterial system. As noted earlier in the chapter, arterioles are extremely muscular: The wall of an arteriole with a luminal diameter of 30 μm can have a 20-μm-thick layer of smooth muscle. When these smooth muscles contract or relax, peripheral resistance increases or decreases. Because a small change in diameter produces a large change in resistance, mechanisms that alter the diameters of arterioles provide control over peripheral resistance and blood flow.

Blood Viscosity

Viscosity is the resistance to flow caused by interactions among molecules and suspended materials in a liquid. Liquids of low viscosity, such as water (viscosity 1.0), flow at low pressures; thick, syrupy fluids, such as molasses (viscosity 300), flow only under higher pressures. Whole blood has a viscosity about five times that of water, owing to the presence of plasma proteins and blood cells. Under normal conditions, the viscosity of blood remains stable, but anemia, polycythemia, and other disorders that affect the hematocrit also change blood viscosity, and thus peripheral resistance.

Turbulence

High flow rates, irregular surfaces, and sudden changes in vessel diameter upset the smooth flow of blood, creating eddies and swirls. This phenomenon, called **turbulence**, increases resistance and slows blood flow.

Turbulence normally occurs when blood flows between the atria and the ventricles, and between the ventricles and the aortic and pulmonary trunks. It also develops in large arteries, such as the aorta, when cardiac output and arterial flow rates are very high. However, turbulence seldom occurs in smaller vessels unless their walls are damaged. For example, the development of an atherosclerotic plaque creates abnormal turbulence and restricts blood flow. Because of the distinctive sound, or *bruit* (broo-Ē), produced by turbulence, the presence of plaques in large blood vessels can often be detected with a stethoscope.

Table 21–1 provides a quick review of the terms and relationships discussed in this section.

An Overview of Cardiovascular Pressures

The graphs in **Figure 21–9** provide an overview of the vessel diameters, cross-sectional areas, pressures, and velocity of blood flow in the systemic circuit.

- As you proceed from the aorta toward the capillaries, divergence occurs; the arteries branch repeatedly, and each branch is smaller in diameter than the preceding one (**Figure 21–9a**). As you proceed from the capillaries toward the venae cavae, convergence occurs; vessel diameters increase as venules combine to form small and medium-sized veins.

- Although the arterioles, capillaries, and venules are small in diameter, the body has a great many of them. All the blood flowing through the aorta also flows through peripheral capillaries. Blood pressure and the speed of blood flow are proportional to the cross-sectional area of the vessels involved. What is important is not the cross-sectional area of each individual vessel, but the *combined* cross-sectional area of *all* the vessels (**Figure 21–9b**). Even

SUMMARY TABLE 21–1   Key Terms and Relationships Pertaining to Blood Circulation

Blood Flow (*F*):	The volume of blood flowing per unit of time through a vessel or a group of vessels; may refer to circulation through a capillary, a tissue, an organ, or the entire vascular network. Total blood flow is equal to cardiac output.
Blood Pressure (*BP*):	The hydrostatic pressure in the arterial system that pushes blood through capillary beds.
Circulatory Pressure:	The pressure difference between the base of the ascending aorta and the entrance to the right atrium.
Hydrostatic Pressure:	A pressure exerted by a liquid in response to an applied force.
Peripheral Resistance (*PR*):	The resistance of the arterial system; affected by such factors as vascular resistance, viscosity, and turbulence.
Resistance (*R*):	A force that opposes movement (in this case, blood flow).
Total Peripheral Resistance:	The resistance of the entire cardiovascular system.
Turbulence:	A resistance due to the irregular, swirling movement of blood at high flow rates or exposure to irregular surfaces.
Vascular Resistance:	A resistance due to friction within a blood vessel, primarily between the blood and the vessel walls. Increases with increasing length or decreasing diameter; vessel length is constant, but vessel diameter can change.
Venous Pressure:	The hydrostatic pressure in the venous system.
Viscosity:	A resistance to flow due to interactions among molecules within a liquid.

RELATIONSHIPS AMONG THE PRECEDING TERMS

$F \propto P$	Flow is proportional to the pressure gradient.
$F \propto 1/R$	Flow is inversely proportional to resistance.
$F \propto P/R$	Flow is directly proportional to the pressure gradient, and inversely proportional to resistance.
$F \propto BP/PR$	Flow is directly proportional to blood pressure, and inversely proportional to peripheral resistance.
$R \propto 1/r^4$	Resistance is inversely proportional to the fourth power of the vessel radius.

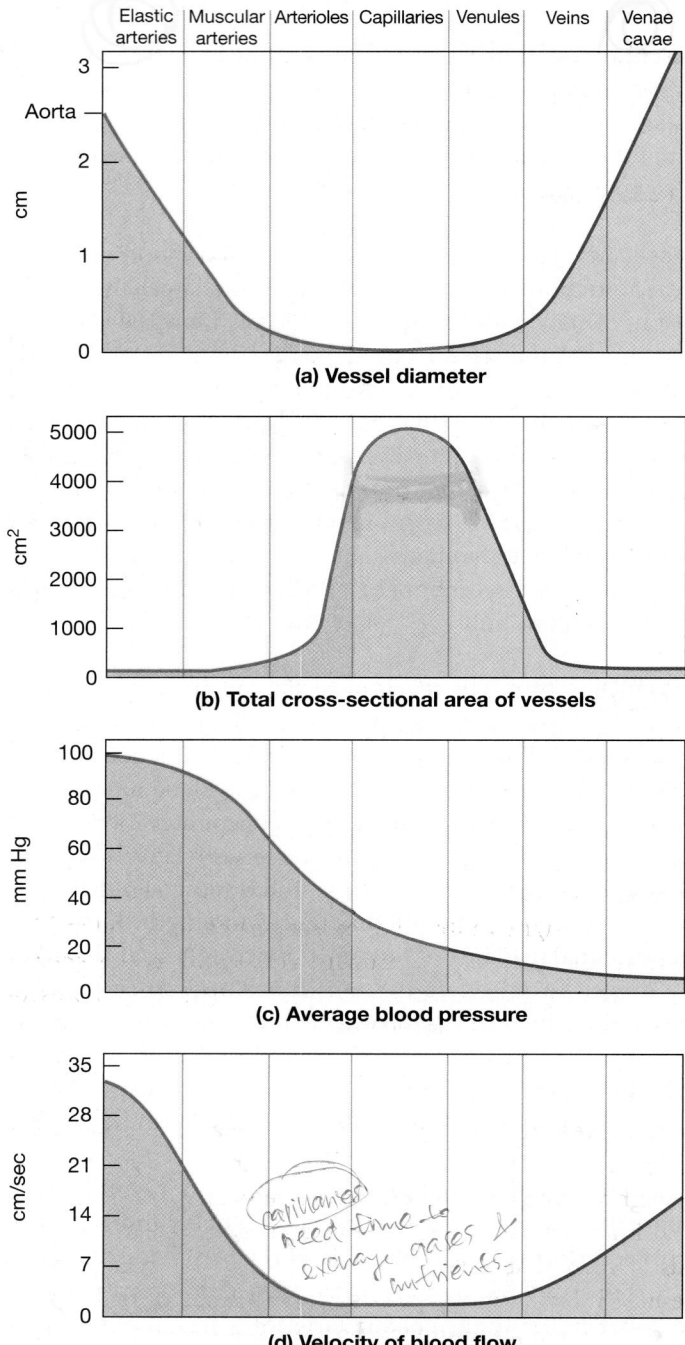

Figure 21–9   Relationships among Vessel Diameter, Cross-Sectional Area, Blood Pressure, and Blood Velocity.

though the arterioles, capillaries, and venules are small in diameter, the body has a great many of them. In effect, your blood moves from one big pipe (the aorta, with a cross-sectional area of 4.5 cm²) into countless tiny ones (the peripheral capillaries, with a total cross-sectional area of 5000 cm²), and then back to the heart through two large venae cavae.

- As arterial branching occurs, the cross-sectional area increases, and blood pressure falls rapidly (**Figure 21–9c**). Most of the decline occurs in the small arteries and arterioles of the arterial system; venous pressures are relatively low.

- As the total cross-sectional area of the vessels increases from the aorta toward the capillaries, the velocity of blood flow decreases (**Figure 21–9d**). Blood flow velocity then rises as the cross-sectional area drops from the capillaries toward the venae cavae.

Figure 21–10 graphs the blood pressure throughout the cardiovascular system. Systemic pressures are highest in the

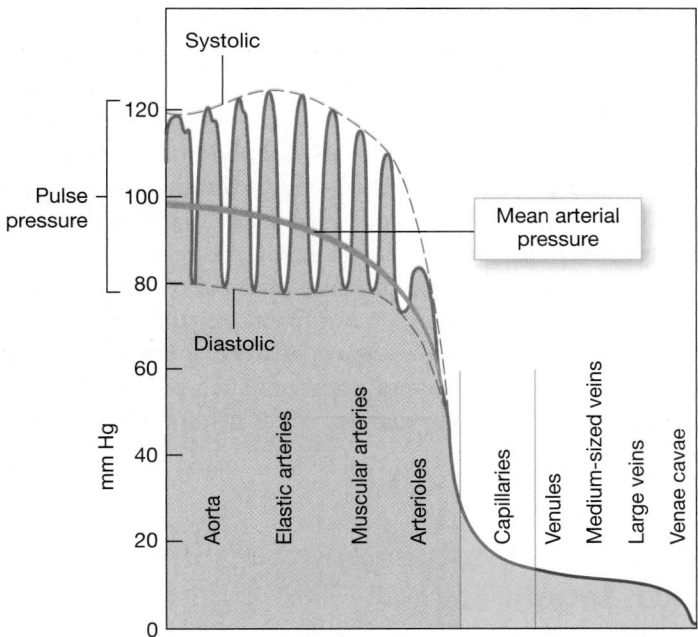

Figure 21–10 Pressures within the Systemic Circuit. Notice the general reduction in circulatory pressure within the systemic circuit and the elimination of the pulse pressure within the arterioles.

aorta, peaking at about 120 mm Hg, and reach a minimum of 2 mm Hg at the entrance to the right atrium. Pressures in the pulmonary circuit are much lower than those in the systemic circuit. The right ventricle does not ordinarily develop high pressures because the pulmonary vessels are much shorter and more distensible than the systemic vessels, thus providing less resistance to blood flow.

Arterial Blood Pressure

Arterial pressure is important because it maintains blood flow through capillary beds. To do this, it must always be high enough to overcome the peripheral resistance. Arterial pressure is not constant; it rises during ventricular systole and falls during ventricular diastole. The peak blood pressure measured during ventricular systole is called **systolic pressure**, and the minimum blood pressure at the end of ventricular diastole is called **diastolic pressure**. In recording blood pressure, we separate systolic and diastolic pressures by a slash, as in "120/80" ("one-twenty over eighty") or "110/75."

A *pulse* is a rhythmic pressure oscillation that accompanies each heartbeat. The difference between the systolic and diastolic pressures is the **pulse pressure** (**Figure 21–10**). To report a single blood pressure value, we use the **mean arterial pressure (MAP)**, which is calculated by adding one-third of the pulse pressure to the diastolic pressure:

$$MAP = \text{diastolic pressure} + \frac{\text{pulse pressure}}{3}$$

For a systolic pressure of 120 mm Hg and a diastolic pressure of 90 mm Hg, MAP can be calculated as follows:

$$MAP = 90 + \frac{(120 - 90)}{3} = 90 + 10 = 100 \text{ mm Hg}$$

A normal range of systolic and diastolic pressures occurs in healthy individuals. When pressures shift outside of the normal range, clinical problems develop. Abnormally high blood pressure is termed **hypertension**; abnormally low blood pressure, **hypotension**. Hypertension is much more common, and in fact many cases of hypotension result from overly aggressive drug treatment for hypertension.

The usual criterion established by the American Heart Association for hypertension in adults is a blood pressure greater than 140/90. Blood pressure at or below 120/80 is normal; values between 121/81 and 139/89 indicate a state of pre-hypertension. Cardiologists often recommend some combination of diet modification and drug therapy for people whose blood pressures are consistently pre-hypertensive. Hypertension significantly increases the workload on the heart, and the left ventricle gradually enlarges. More muscle mass means a greater oxygen demand. When the coronary circulation cannot keep pace, symptoms of coronary ischemia appear. ∞ p. 694 Increased arterial pressures also place a physical stress on the walls of blood vessels throughout the body. This stress promotes or accelerates the development of arteriosclerosis and increases the risk of aneurysms, heart attacks, and strokes.

Elastic Rebound

As systolic pressure climbs, the arterial walls stretch, just as an extra puff of air expands a partially inflated balloon. This expansion allows the arterial system to accommodate some of the blood provided by ventricular systole. When diastole begins and blood pressures fall, the arteries recoil to their original dimensions. This phenomenon is called **elastic rebound**. Some blood is forced back toward the left ventricle, closing the aortic valve and helping to drive additional blood into the coronary arteries. However, most of the push generated by elastic rebound forces blood toward the capillaries. This maintains blood flow along the arterial network while the left ventricle is in diastole.

Pressures in Small Arteries and Arterioles

The mean arterial pressure and the pulse pressure become smaller as the distance from the heart increases (**Figure 21–10**):

- The mean arterial pressure declines as the arterial branches become smaller and more numerous. In essence, blood pressure decreases as it overcomes friction and produces blood flow.

- The pulse pressure lessens as a result of the cumulative effects of elastic rebound along the arterial system. The

21 CARDIOVASCULAR

effect can be likened to a series of ever-softer echoes following a loud shout. Each time an echo is produced, the reflecting surface absorbs some of the sound energy. Eventually, the echo disappears. The pressure surge accompanying ventricular ejection is analogous to the shout, and it is reflected by the wall of the aorta, echoing down the arterial system until it finally disappears at the level of the small arterioles. By the time blood reaches a precapillary sphincter, no pressure oscillations remain, and the blood pressure remains steady at approximately 35 mm Hg.

Venous Pressure and Venous Return

Venous pressure, although low, determines venous return— the amount of blood arriving at the right atrium each minute. Venous return has a direct impact on cardiac output. ∞ p. 711 Although blood pressure at the start of the venous system is only about one-tenth that at the start of the arterial system, the blood must still travel through a vascular network as complex as the arterial system before returning to the heart.

Pressures at the entrance to the right atrium fluctuate, but they average about 2 mm Hg. Thus, the effective pressure in the venous system is roughly 16 mm Hg (from 18 mm Hg in the venules to 2 mm Hg in the venae cavae), compared with 65 mm Hg in the arterial system (from 100 mm Hg at the aorta to 35 mm Hg at the capillaries). Yet, although venous pressures are low, veins offer comparatively little resistance, so pressure declines very slowly as blood moves through the venous system. As blood continues toward the heart, the veins become larger, resistance drops, and the velocity of blood flow increases (**Figure 21–9**).

When you stand, the venous blood returning from your body inferior to the heart must overcome gravity as it ascends within the inferior vena cava. Two factors assist the low venous pressures in propelling blood toward your heart: *muscular compression* of peripheral veins and the *respiratory pump*.

Muscular Compression.

The contractions of skeletal muscles near a vein compress it, helping to push blood toward the heart. The valves in small and medium-sized veins ensure that blood flows in one direction only (**Figure 21–6**). When standing and walking, the cycles of contraction and relaxation that accompany normal movements assist venous return. If you stand at attention, with knees locked and leg muscles immobilized, that assistance is lost. The reduction in venous return then leads to a fall in cardiac output, which reduces the blood supply to the brain. This decline is sometimes enough to cause **fainting**, a temporary loss of consciousness. You would then collapse, but while you were in the horizontal position, both venous return and cardiac output would return to normal.

The Respiratory Pump. As you inhale, your thoracic cavity expands, reducing the pressure within the pleural cavities. This drop in pressure pulls air into your lungs. At the same time, blood is pulled into the inferior vena cava and right atrium from the smaller veins of your abdominal cavity and lower body. The effect on venous return through the superior vena cava is less pronounced, as blood in that vessel is normally assisted by gravity. As you exhale, your thoracic cavity decreases in size. Internal pressure then rises, forcing air out of your lungs and pushing venous blood into the right atrium. This mechanism is called the **respiratory pump**, or *thoracoabdominal pump*. The importance of such pumping action increases during heavy exercise, when respirations are deep and frequent.

Capillary Pressures and Capillary Exchange

Because capillary exchange plays such an important role in homeostasis, we will now consider the factors and mechanisms involved. The most important processes that move materials across typical capillary walls are *diffusion*, *filtration*, and *reabsorption*.

Diffusion

As we saw in Chapter 3, *diffusion* is the net movement of ions or molecules from an area where their concentration is higher to an area where their concentration is lower. ∞ p. 89 The difference between the high and low concentrations represents a *concentration gradient*, and diffusion tends to eliminate that gradient. Diffusion occurs most rapidly when (1) the distances involved are short, (2) the concentration gradient is large, and (3) the ions or molecules involved are small.

Different substances diffuse across capillary walls by different routes:

1. Water, ions, and small organic molecules, such as glucose, amino acids, and urea, can usually enter or leave the bloodstream by diffusion between adjacent endothelial cells or through the pores of fenestrated capillaries.

2. Many ions, including sodium, potassium, calcium, and chloride, can diffuse across endothelial cells by passing through channels in plasma membranes.

3. Large water-soluble compounds are unable to enter or leave the bloodstream except at fenestrated capillaries, such as those of the hypothalamus, the kidneys, many endocrine organs, and the intestinal tract.

4. Lipids, such as fatty acids and steroids, and lipid-soluble materials, including soluble gases such as oxygen and carbon dioxide, can cross capillary walls by diffusion through the endothelial plasma membranes.

5. Plasma proteins are normally unable to cross the endothelial lining anywhere except in sinusoids, such as those of the liver, where plasma proteins enter the bloodstream.

Filtration → large particles filtered out.

Filtration is the removal of solutes as a solution flows across a porous membrane; solutes too large to pass through the pores are filtered out of the solution. The driving force for filtration is hydrostatic pressure, which, as we saw earlier, pushes water from an area of higher pressure to an area of lower pressure.

In *capillary filtration*, water and small solutes are forced across a capillary wall, leaving larger solutes and suspended proteins in the bloodstream (**Figure 21–11**). The solute molecules leaving the bloodstream are those small enough to pass between adjacent endothelial cells or through the pores in a fenestrated capillary. Filtration occurs primarily at the arterial end of a capillary, where capillary hydrostatic pressure (CHP) is highest.

Reabsorption

Reabsorption occurs as the result of osmosis. *Osmosis* is a special term used to refer to the diffusion of water across a selectively permeable membrane separating two solutions of differing solute concentrations. Water molecules tend to diffuse across a membrane toward the solution containing the higher solute concentration (**Figure 3–16**, p. 92).

The **osmotic pressure (OP)** of a solution is an indication of the force of osmotic water movement—in other words, the pressure that must be applied to prevent osmotic movement across a membrane. The higher the solute concentration of a solution, the greater the solution's osmotic pressure. The presence of suspended proteins unable to cross capillary walls creates an osmotic pressure called *blood colloid osmotic pressure (BCOP)*. Clinicians often use the term *oncotic pressure* (*onkos*, a swelling) when referring to the colloid osmotic pressure of body fluids. The two terms are equivalent. Osmotic water movement will continue until either the solute concentrations are equalized or the movement is prevented by an opposing hydrostatic pressure.

We will now consider the interplay between filtration and reabsorption along the length of a typical capillary. As the discussion proceeds, remember that hydrostatic pressure forces water *out of* a solution, whereas osmotic pressure draws water *into* a solution.

CHP → diffuses arterial → venous end.

The Interplay between Filtration and Reabsorption

Capillary blood pressure declines as blood flows from the arterial end to the venous end. As a result, the rates of filtration and reabsorption gradually change as blood passes along the length of a capillary. The factors involved are diagrammed in **Figure 21–12**.

The *net hydrostatic pressure* tends to push water and solutes out of capillaries and into the interstitial fluid. The net hydrostatic pressure is the difference between

1. the *capillary hydrostatic pressure* (CHP), which ranges from 35 mm Hg at the arterial end of a capillary to 18 mm Hg at the venous end, and

2. the *hydrostatic pressure of the interstitial fluid (IHP)*. Measurements of IHP have yielded very small values that differ from tissue to tissue—from +6 mm Hg in the brain to −6 mm Hg in subcutaneous tissues. A positive IHP opposes CHP, and the tissue hydrostatic pressure must be overcome before fluid can move out of a capillary. A negative IHP assists CHP, and additional fluid will be pulled out of the capillary. However, under normal circumstances the average IHP is 0 mm Hg, and we can assume that the net hydrostatic pressure is equal to CHP. (For this reason, IHP is not included in **Figure 21–12**.)

The *net colloid osmotic pressure* tends to pull water and solutes into a capillary from the interstitial fluid. The net colloid osmotic pressure is the difference between

1. the *blood colloid osmotic pressure (BCOP)*, which is roughly 25 mm Hg, and

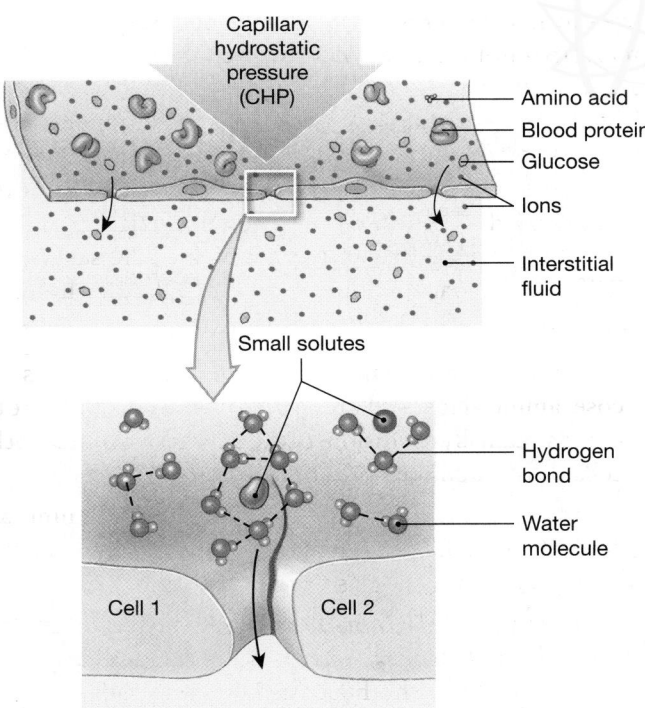

Figure 21–11 Capillary Filtration. Capillary hydrostatic pressure (CHP) forces water and solutes through the gaps between adjacent endothelial cells in continuous capillaries. The sizes of solutes that move across the capillary wall are determined primarily by the dimensions of the gaps.

Labels in figure:
- Capillary hydrostatic pressure (CHP)
- Amino acid
- Blood protein
- Glucose
- Ions
- Interstitial fluid
- Small solutes
- Hydrogen bond
- Water molecule
- Cell 1
- Cell 2

21 CARDIOVASCULAR

2. the *interstitial fluid colloid osmotic pressure (ICOP)*. The ICOP is as variable and low as the IHP, because the interstitial fluid in most tissues contains negligible quantities of suspended proteins. Reported values of ICOP are from 0 to 5 mm Hg, within the range of pressures recorded for the IHP. It is thus safe to assume that under normal circumstances the net colloid osmotic pressure is equal to the BCOP. (For this reason, ICOP is not included in **Figure 21–12**.)

The **net filtration pressure (NFP)** is the difference between the net hydrostatic pressure and the net osmotic pressure. In terms of the factors just listed, this means that

net filtration = net hydrostatic − net colloid
pressure   pressure   osmotic pressure
NFP   = (CHP − IHP) − (BCOP − ICOP)

At the arterial end of a capillary, the net filtration pressure can be calculated as follows:

NFP = (35 − 0) − (25 − 0) = 35 − 25 = 10 mm Hg

Because this value is positive, it indicates that fluid will tend to move *out of* the capillary and into the interstitial fluid. At the venous end of the capillary, the net filtration pressure will be:

NFP = (18 − 0) − (25 − 0) = 18 − 25 = −7 mm Hg

The minus sign indicates that fluid tends to move *into* the capillary; that is, reabsorption is occurring.

The transition between filtration and reabsorption occurs where the CHP is 25 mm Hg, because at that point the hydrostatic and osmotic forces are equal—that is, the NFP is 0 mm Hg. If the maximum filtration pressure at the arterial end of the capillary were equal to the maximum reabsorption pressure at the venous end, this transition point would lie midway along the length of the capillary. Under these circumstances, filtration would occur along the first half of the capillary, and an identical amount of reabsorption would occur along the second half. However, the maximum filtration pressure is higher than the maximum reabsorption pressure, so the transition point between filtration and reabsorption normally lies closer to the venous end of the capillary than to the arterial end. As a result, more filtration than reabsorption occurs along the capillary. Of the roughly 24 liters of fluid that moves out of the plasma and into the interstitial fluid each day, 20.4 liters (85 percent) is reabsorbed. The remainder (3.6 liters) flows through the tissues and into lymphatic vessels, for eventual return to the venous system.

This continuous movement of water out of the capillaries, through peripheral tissues, and then back to the bloodstream by way of the lymphoid system has four important functions:

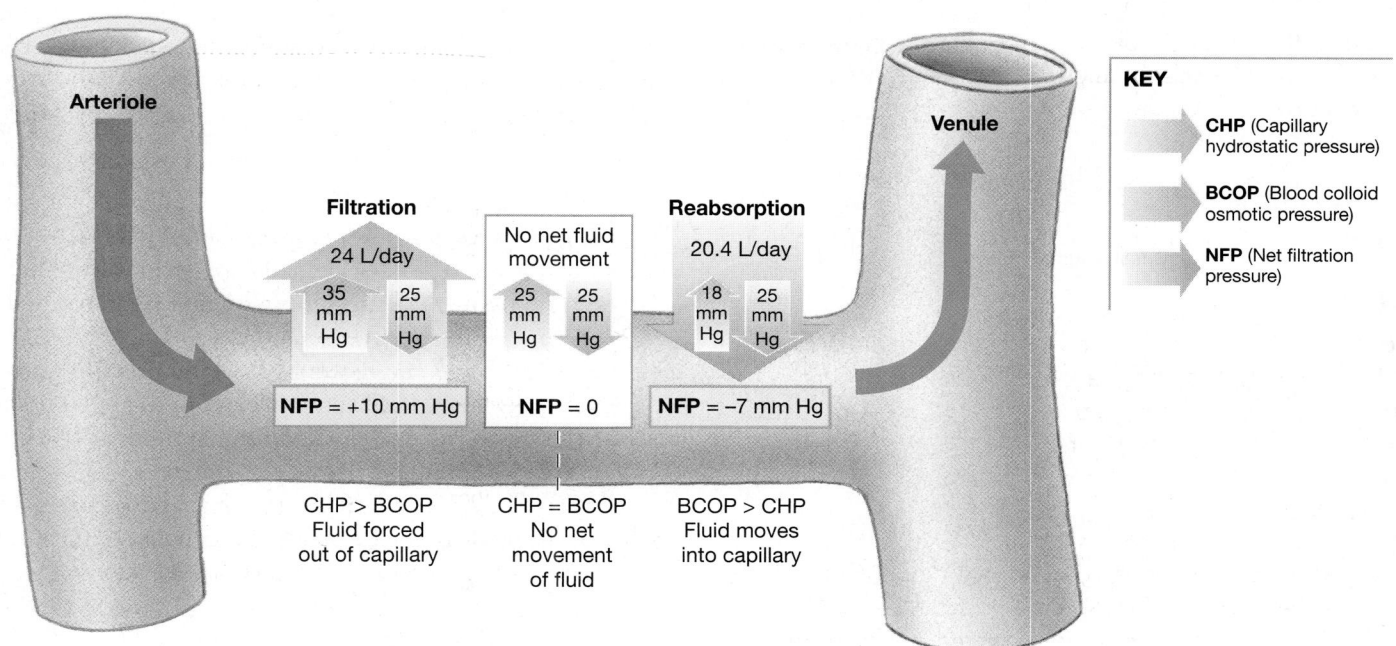

Figure 21–12 **Forces Acting across Capillary Walls.** At the arterial end of the capillary, capillary hydrostatic pressure (CHP) is greater than blood colloid osmotic pressure (BCOP), so fluid moves out of the capillary (filtration). Near the venule, CHP is lower than BCOP, so fluid moves into the capillary (reabsorption). In this model, interstitial fluid colloid osmotic pressure (ICOP) and interstitial fluid hydrostatic pressure (IHP) are assumed to be 0 mm Hg and so are not shown.

1. It ensures that plasma and interstitial fluid, two major components of extracellular fluid, are in constant communication and mutual exchange.

2. It accelerates the distribution of nutrients, hormones, and dissolved gases throughout tissues.

3. It assists in the transport of insoluble lipids and tissue proteins that cannot enter the bloodstream by crossing the capillary walls.

4. It has a flushing action that carries bacterial toxins and other chemical stimuli to lymphoid tissues and organs responsible for providing immunity to disease.

Any condition that affects hydrostatic or osmotic pressures in the blood or tissues will shift the balance between hydrostatic and osmotic forces. We can then predict the effects on the basis of an understanding of capillary dynamics. For example,

- If hemorrhaging occurs, both blood volume and blood pressure decline. This reduction in CHP lowers the NFP and increases the amount of reabsorption. The result is a reduction in the volume of interstitial fluid and an increase in the circulating plasma volume. This process is known as a *recall of fluids*.

- If dehydration occurs, the plasma volume decreases owing to water loss, and the concentration of plasma proteins increases. The increase in BCOP accelerates reabsorption and a recall of fluids that delays the onset and severity of clinical signs and symptoms.

- If the CHP rises or the BCOP declines, fluid moves out of the blood and builds up in peripheral tissues, a condition called *edema*.

The A&P Top 100

#73 The goal of cardiovascular regulation is the maintenance of blood flow, and total peripheral blood flow is equal to cardiac output. Blood pressure is needed to overcome friction and elastic forces and to sustain blood flow. If blood pressure is too low, vessels collapse, blood flow stops, and tissues die; if blood pressure is too high, vessel walls stiffen and capillary beds may rupture.

CHECKPOINT

5. Identify the factors that contribute to total peripheral resistance.

6. In a healthy individual, where is blood pressure greater: at the aorta or at the inferior vena cava? Explain.

7. While standing in the hot sun, Sally begins to feel light-headed and faints. Explain what happened.

8. Mike's blood pressure is 125/70. What is his mean arterial pressure?

See the blue Answers tab at the end of the book.

21-3 Cardiovascular regulatory mechanisms involve autoregulation, neural mechanisms, and endocrine responses

Homeostatic mechanisms regulate cardiovascular activity to ensure that **tissue perfusion**, or blood flow through tissues, meets the demand for oxygen and nutrients. The factors that

---

CLINICAL NOTE

Edema

Edema (e-DĒ-muh) is an abnormal accumulation of interstitial fluid. Edema has many causes, and we will encounter specific examples in later chapters. The underlying problem in all types of edema is a disturbance in the normal balance between hydrostatic and osmotic forces at the capillary level. For instance:

- When a capillary is damaged, plasma proteins can cross the capillary wall and enter the interstitial fluid. The resulting elevation of the ICOP reduces the rate of capillary reabsorption and produces a localized edema. This is why you usually have swelling at a bruise.

- In starvation, the liver cannot synthesize enough plasma proteins to maintain normal concentrations in the blood. BCOP declines, and fluids begin moving from the blood into peripheral tissues. In children, fluid accumulates in the abdominopelvic cavity, producing the swollen bellies typical of starvation victims. A reduction in BCOP is also seen after severe burns and in several types of liver and kidney diseases.

- In the U.S. population, most serious cases of edema result from increases in arterial blood pressure, venous pressure, or total circulatory pressure. The increase may result from heart problems such as heart failure, venous blood clots that elevate venous pressures, or other circulatory abnormalities. The net result is an increase in CHP that accelerates fluid movement into the tissues.

affect tissue perfusion are (1) cardiac output, (2) peripheral resistance, and (3) blood pressure. We discussed cardiac output in Chapter 20 (p. 708) and considered peripheral resistance and blood pressure earlier in this chapter.

Most cells are relatively close to capillaries. When a group of cells becomes active, the circulation to that region must increase to deliver the necessary oxygen and nutrients, and to carry away the waste products and carbon dioxide they generate. The purpose of cardiovascular regulation is to ensure that these blood flow changes occur (1) at an appropriate time, (2) in the right area, and (3) without drastically changing blood pressure and blood flow to vital organs.

The regulatory mechanisms focus on controlling cardiac output and blood pressure to restore adequate blood flow after a fall in blood pressure. These mechanisms can be broadly categorized as follows:

- *Autoregulation.* Local factors change the pattern of blood flow within capillary beds as precapillary sphincters open and close in response to chemical changes in interstitial fluids. This is an example of autoregulation at the tissue level. Autoregulation causes immediate, localized homeostatic adjustments. If autoregulation fails to

normalize conditions at the tissue level, neural mechanisms and endocrine factors are activated.

- *Neural Mechanisms.* Neural mechanisms respond to changes in arterial pressure or blood gas levels sensed at specific sites. When those changes occur, the cardiovascular centers of the autonomic nervous system adjust cardiac output and peripheral resistance to maintain blood pressure and ensure adequate blood flow.

- *Endocrine Mechanisms.* The endocrine system releases hormones that enhance short-term adjustments and that direct long-term changes in cardiovascular performance.

We will next consider each of these regulatory mechanisms individually by examining regulatory responses to inadequate perfusion of skeletal muscles. The regulatory relationships are diagrammed in **Figure 21–13**.

Autoregulation of Blood Flow within Tissues

Under normal resting conditions, cardiac output remains stable, and peripheral resistance within individual tissues is adjusted to control local blood flow.

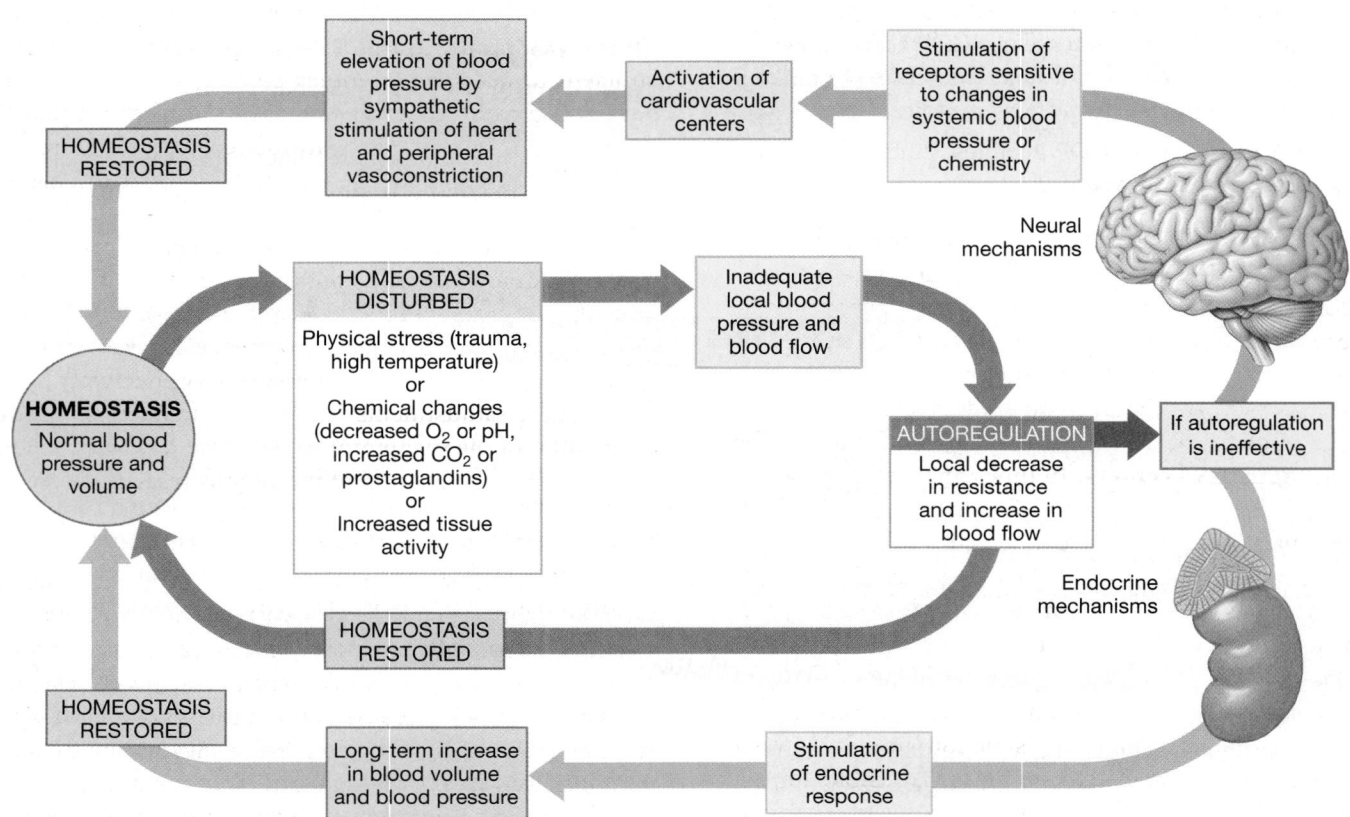

Figure 21–13 **Short-Term and Long-Term Cardiovascular Responses.** This diagram indicates general mechanisms that compensate for a reduction in blood pressure and blood flow.

Factors that promote the dilation of precapillary sphincters are called **vasodilators**. **Local vasodilators** act at the tissue level to accelerate blood flow through their tissue of origin. Examples of local vasodilators include the following:

- Decreased tissue oxygen levels or increased CO_2 levels.

- Lactic acid or other acids generated by tissue cells.

- Nitric oxide (NO) released from endothelial cells.

- Rising concentrations of potassium ions or hydrogen ions in the interstitial fluid.

- Chemicals released during local inflammation, including histamine and NO. ∞ p. 142

- Elevated local temperature.

These factors work by relaxing the smooth muscle cells of the precapillary sphincters. All of them indicate that tissue conditions are in some way abnormal. An increase in blood flow, which will bring oxygen, nutrients, and buffers, may be sufficient to restore homeostasis.

As noted in Chapter 19, aggregating platelets and damaged tissues produce compounds that stimulate the constriction of precapillary sphincters. These compounds are **local vasoconstrictors**. Examples include prostaglandins and thromboxanes released by activated platelets and white blood cells, and the endothelins released by damaged endothelial cells.

Local vasodilators and vasoconstrictors control blood flow within a single capillary bed (**Figure 21–5**). In high concentrations, these factors also affect arterioles, increasing or decreasing blood flow to all the capillary beds in a given area.

Neural Mechanisms

The nervous system is responsible for adjusting cardiac output and peripheral resistance in order to maintain adequate blood flow to vital tissues and organs. Centers responsible for these regulatory activities include the *cardiac centers* and the *vasomotor center* of the medulla oblongata. ∞ p. 471 It is difficult to distinguish the cardiac and vasomotor centers anatomically, and they are often considered to form complex **cardiovascular (CV) centers**. In functional terms, however, the cardiac and vasomotor centers often act independently. As noted in Chapter 20, each cardiac center consists of a *cardioacceleratory center*, which increases cardiac output through sympathetic innervation, and a *cardioinhibitory center*, which reduces cardiac output through parasympathetic innervation. ∞ p. 709 large: vasoconstriction. small: dilation.

The vasomotor center contains two populations of neurons: (1) a very large group responsible for widespread vasoconstriction and (2) a smaller group responsible for the vasodilation of arterioles in skeletal muscles and the brain. The vasomotor center exerts its effects by controlling the activity of sympathetic motor neurons:

1. *Control of Vasoconstriction.* The neurons innervating peripheral blood vessels in most tissues are *adrenergic*; that is, they release the neurotransmitter norepinephrine (NE). The response to NE release is the stimulation of smooth muscle in the walls of arterioles, producing vasoconstriction.

2. *Control of Vasodilation.* Vasodilator neurons innervate blood vessels in skeletal muscles and in the brain. The stimulation of these neurons relaxes smooth muscle cells in the walls of arterioles, producing vasodilation. The relaxation of smooth muscle cells is triggered by the appearance of NO in their surroundings. The vasomotor center may control NO release indirectly or directly. The most common vasodilator synapses are *cholinergic*—their synaptic knobs release ACh. In turn, ACh stimulates endothelial cells in the area to release NO, which causes local vasodilation. Another population of vasodilator synapses is *nitroxidergic*—the synaptic knobs release NO as a neurotransmitter. Nitric oxide has an immediate and direct relaxing effect on the vascular smooth muscle cells in the area.

Vasomotor Tone

In Chapter 16, we discussed the significance of autonomic tone in setting a background level of neural activity that can increase or decrease on demand. ∞ p. 545 The sympathetic vasoconstrictor nerves are chronically active, producing a significant **vasomotor tone**. Vasoconstrictor activity is normally sufficient to keep the arterioles partially constricted. Under maximal stimulation, arterioles constrict to about half their resting diameter, whereas a fully dilated arteriole increases its resting diameter by roughly 1.5 times. Constriction has a significant effect on resistance, because, as we saw earlier, resistance increases sharply as luminal diameter decreases. The resistance of a maximally constricted arteriole is roughly 80 *times* that of a fully dilated arteriole. Because blood pressure varies directly with peripheral resistance, the vasomotor center can control arterial blood pressure very effectively by making modest adjustments in vessel diameters. Extreme stimulation of the vasomotor centers also produces venoconstriction and mobilization of the venous reserve.

Reflex Control of Cardiovascular Function

The cardiovascular centers detect changes in tissue demand by monitoring arterial blood, with particular attention to blood pressure, pH, and the concentrations of dissolved gases. The *baroreceptor reflexes* (baro-, pressure) respond to changes in blood pressure, and the *chemoreceptor reflexes* monitor changes in the chemical composition of arterial blood. These reflexes are regulated through a negative feedback loop: The stimulation of a receptor by an abnormal condition leads to a response that counteracts the stimulus and restores normal conditions.

Baroreceptor Reflexes. Baroreceptors are specialized receptors that monitor the degree of stretch in the walls of expandable organs. ∞ p. 514 The baroreceptors involved in cardiovascular regulation are located in the walls of (1) the **carotid sinuses**, expanded chambers near the bases of the *internal carotid arteries* of the neck (**Figure 21–22**); (2) the **aortic sinuses**, pockets in the walls of the ascending aorta adjacent to the heart (**Figure 20–8b**, p. 691); and (3) the wall of the right atrium. These receptors are components of the **baroreceptor reflexes**, which adjust cardiac output and peripheral resistance to maintain normal arterial pressures.

Aortic baroreceptors monitor blood pressure within the ascending aorta. Any changes trigger the **aortic reflex**, which adjusts blood pressure to maintain adequate blood pressure and blood flow through the systemic circuit. In response to changes in blood pressure at the carotid sinus, carotid sinus baroreceptors trigger reflexes that maintain adequate blood flow to the brain. Because blood flow to the brain must remain constant, the carotid sinus receptors are extremely sensitive. **Figure 21–14** presents the basic organization of the baroreceptor reflexes triggered by changes in blood pressure at the carotid and aortic sinuses.

When blood pressure climbs, the increased output from the baroreceptors alters activity in the CV centers and produces two major effects (**Figure 21–14a**):

1. *A decrease in cardiac output*, due to parasympathetic stimulation and the inhibition of sympathetic activity.

2. *Widespread peripheral vasodilation*, due to the inhibition of excitatory neurons in the vasomotor center.

The decrease in cardiac output reflects primarily a reduction in heart rate due to the release of acetylcholine at the sinoatrial (SA) node. ∞ p. 710 The widespread vasodilation lowers peripheral resistance, and this effect, combined with a reduction in cardiac output, leads to a decline in blood pressure to normal levels.

When blood pressure falls below normal, baroreceptor output is reduced accordingly (**Figure 21–14b**). This change has two major effects:

1. *An increase in cardiac output*, through the stimulation of sympathetic innervation to the heart. This results from the stimulation of the cardioacceleratory center and is accompanied by an inhibition of the cardioinhibitory center.

2. *Widespread peripheral vasoconstriction*, caused by the stimulation of sympathetic vasoconstrictor neurons by the vasomotor center.

The effects on the heart result from the release of NE by sympathetic neurons innervating the SA node, the atrioventric-

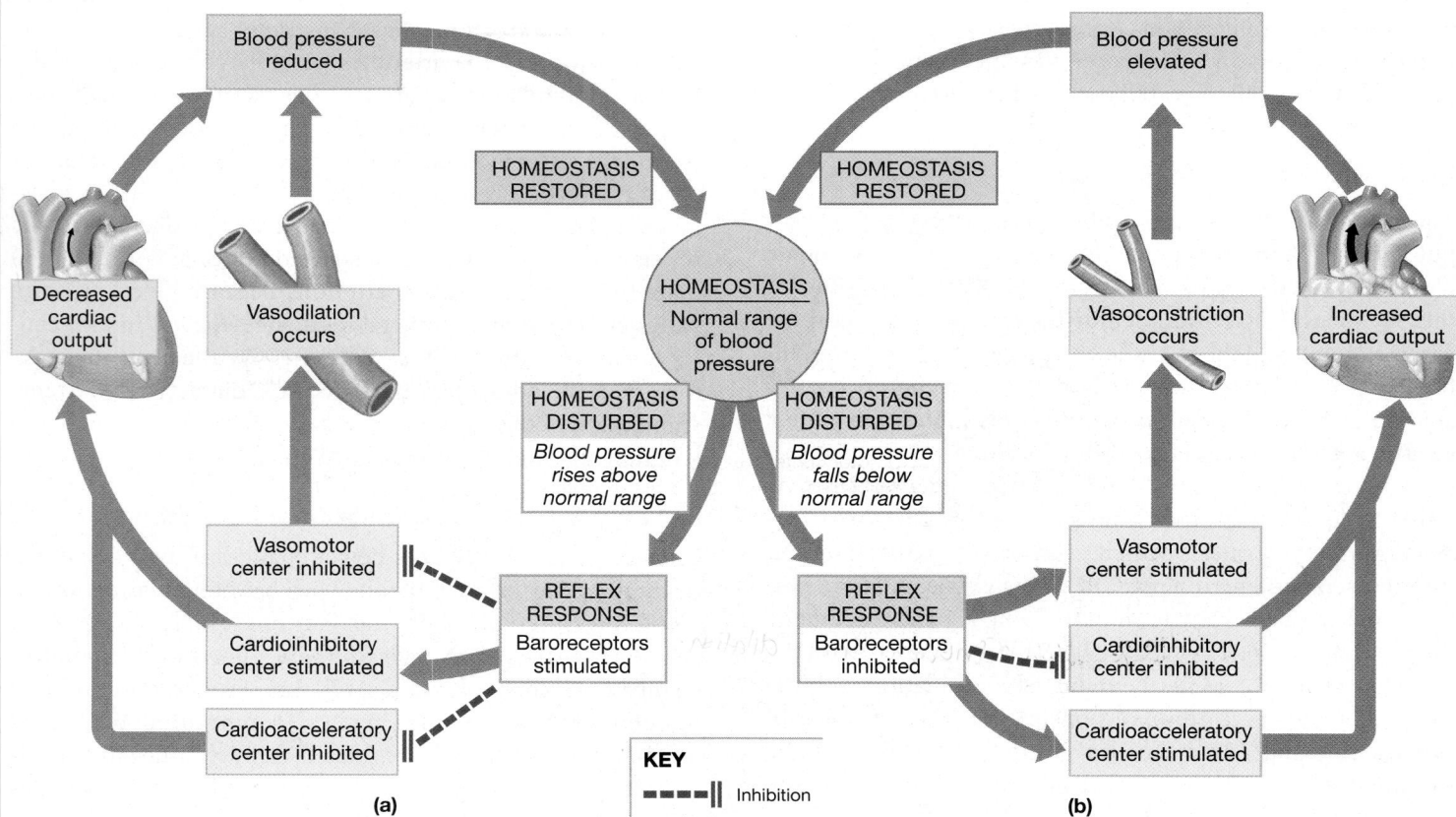

Figure 21–14 **Baroreceptor Reflexes of the Carotid and Aortic Sinuses.**

ular (AV) node, and the general myocardium. In a crisis, sympathetic activation occurs, and its effects are enhanced by the release of both NE and epinephrine (E) from the suprarenal medullae. The net effect is an immediate increase in heart rate and stroke volume, and a corresponding rise in cardiac output. The vasoconstriction, which also results from the release of NE by sympathetic neurons, increases peripheral resistance. These adjustments—increased cardiac output and increased peripheral resistance—work together to elevate blood pressure.

Atrial baroreceptors are receptors that monitor blood pressure at the end of the systemic circuit—at the venae cavae and the right atrium. The **atrial reflex** responds to a stretching of the wall of the right atrium. ∞ p. 710

Under normal circumstances, the heart pumps blood into the aorta at the same rate at which blood arrives at the right atrium. When blood pressure rises at the right atrium, blood is arriving at the heart faster than it is being pumped out. The atrial baroreceptors correct the situation by stimulating the CV centers and increasing cardiac output until the backlog of venous blood is removed. Atrial pressure then returns to normal.

Exhaling forcefully against a closed glottis, a procedure known as the *Valsalva maneuver*, causes reflexive changes in blood pressure and cardiac output due to increased intrathoracic pressure, which impedes venous return to the right atrium. When internal pressures rise, the venae cavae collapse, and the venous return decreases. The resulting fall in cardiac output and blood pressure stimulates the aortic and carotid baroreceptors, causing reflexive increase in heart rate and peripheral vasoconstriction. When the glottis opens and pressures return to normal, venous return increases suddenly and so does cardiac output. Because vasoconstriction has occurred, blood pressure rises sharply, and this inhibits the baroreceptors. As a result, cardiac output, heart rate, and blood pressure quickly return to normal levels. The Valsalva maneuver is thus a simple way to check for normal cardiovascular responses to changes in arterial pressure and venous return.

Chemoreceptor Reflexes. The **chemoreceptor reflexes** respond to changes in carbon dioxide, oxygen, or pH levels in blood and cerebrospinal fluid (CSF) (**Figure 21–15**). The chemoreceptors involved are sensory neurons located in the **carotid bodies**, situated in the neck near the carotid sinus, and the **aortic bodies**, near the arch of the aorta. ∞ p. 514 These receptors monitor the composition of arterial blood.

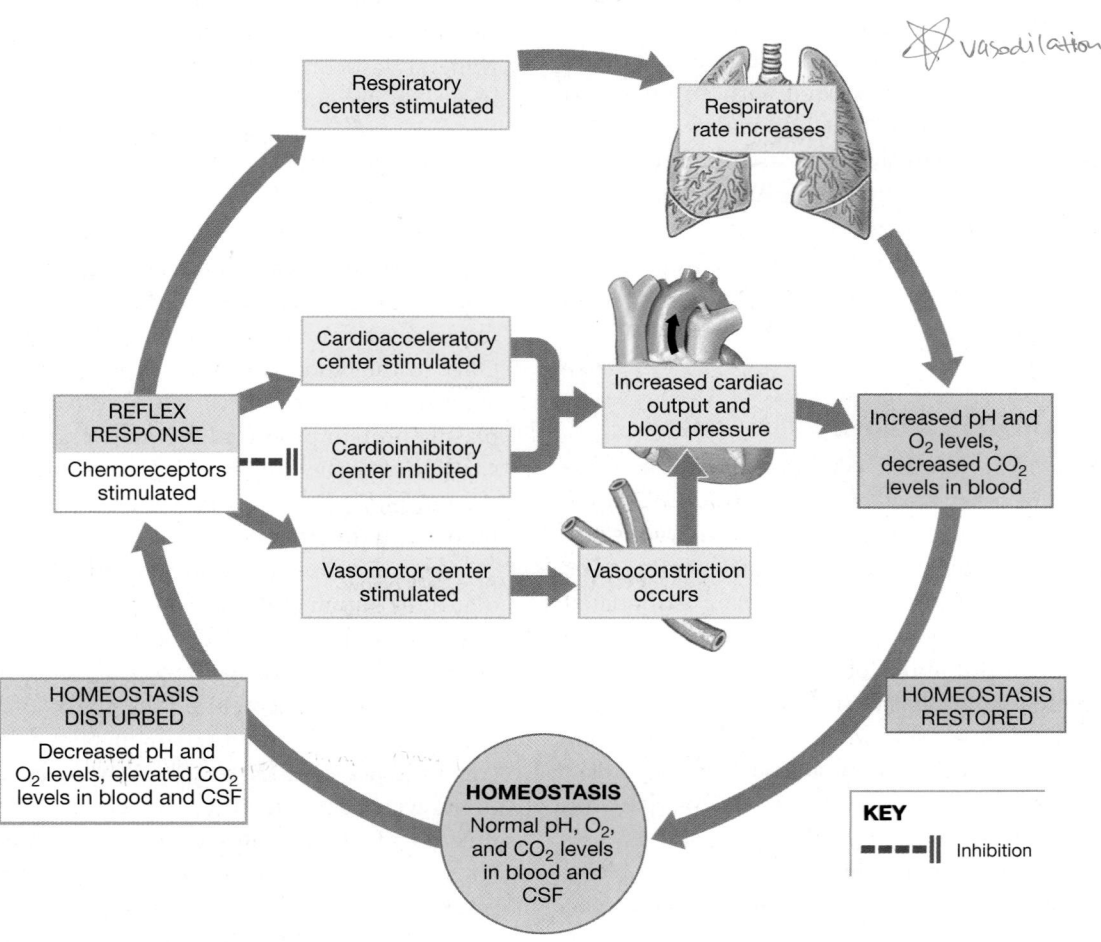

vasodilation.

Figure 21–15 The Chemoreceptor Reflexes.

Additional chemoreceptors located on the ventrolateral surfaces of the medulla oblongata monitor the composition of CSF.

When chemoreceptors in the carotid bodies or aortic bodies detect either a rise in the carbon dioxide content or a fall in the pH of the arterial blood, the cardioacceleratory and vasomotor centers are stimulated, and the cardioinhibitory center is inhibited. This dual effect causes an increase in cardiac output, peripheral vasoconstriction, and an elevation in arterial blood pressure. A drop in the oxygen level at the aortic bodies has the same effects. Strong stimulation of the carotid or aortic chemoreceptors causes widespread sympathetic activation, with more dramatic increases in heart rate and cardiac output.

The chemoreceptors of the medulla oblongata are involved primarily with the control of respiratory function, and secondarily with regulating blood flow to the brain. For example, a steep rise in CSF carbon dioxide levels will trigger the vasodilation of cerebral vessels, but will produce vasoconstriction in most other organs. The result is increased blood flow—and hence increased oxygen delivery—to the brain.

Arterial CO_2 levels can be reduced and O_2 levels increased most effectively by coordinating cardiovascular and respiratory activities. Chemoreceptor stimulation also stimulates the respiratory centers, and the rise in cardiac output and blood pressure is associated with an increased respiratory rate. Coordination of cardiovascular and respiratory activities is vital, because accelerating blood flow in the tissues is useful only if the circulating blood contains an adequate amount of oxygen. In addition, a rise in the respiratory rate accelerates venous return through the action of the respiratory pump. (We will consider other aspects of chemoreceptor activity and respiratory control in Chapter 23.)

CNS Activities and the Cardiovascular Centers

The output of the cardiovascular centers can also be influenced by activities in other areas of the brain. For example, the activation of either division of the autonomic nervous system will affect output from the cardiovascular centers. The cardioacceleratory and vasomotor centers are stimulated when a general sympathetic activation occurs. The result is an increase in cardiac output and blood pressure. In contrast, when the parasympathetic division is activated, the cardioinhibitory center is stimulated, producing a reduction in cardiac output. Parasympathetic activity does not directly affect the vasomotor center, but vasodilation occurs as sympathetic activity declines.

The activities of higher brain centers can also affect blood pressure. Our thought processes and emotional states can produce significant changes in blood pressure by influencing cardiac output and vasomotor tone. For example, strong emotions of anxiety, fear, and rage are accompanied by an elevation in blood pressure, caused by cardiac stimulation and vasoconstriction.

Hormones and Cardiovascular Regulation

The endocrine system provides both short-term and long-term regulation of cardiovascular performance. As we have seen, E and NE from the suprarenal medullae stimulate cardiac output and peripheral vasoconstriction. Other hormones important in regulating cardiovascular function include (1) antidiuretic hormone (ADH), (2) angiotensin II, (3) erythropoietin (EPO), and (4) the natriuretic peptides (ANP and BNP). ∞ p. 637 Although ADH and angiotensin II also affect blood pressure, all four are concerned primarily with the long-term regulation of blood volume (**Figure 21–16**).

Antidiuretic Hormone

Antidiuretic hormone (ADH) is released at the neurohypophysis in response to a decrease in blood volume, to an increase in the osmotic concentration of the plasma, or (secondarily) to circulating angiotensin II. The immediate result is a peripheral vasoconstriction that elevates blood pressure. This hormone also stimulates the conservation of water at the kidneys, thus preventing a reduction in blood volume that would further reduce blood pressure (**Figure 21–16a**).

Angiotensin II

Angiotensin II appears in the blood after the release of the enzyme renin by *juxtaglomerular cells*, specialized kidney cells, in response to a fall in renal blood pressure (**Figure 21–16a**). Once in the bloodstream, renin starts an enzymatic chain reaction. In the first step, renin converts *angiotensinogen*, a plasma protein produced by the liver, to *angiotensin I*. In the capillaries of the lungs, *angiotensin-converting enzyme* (ACE) then modifies angiotensin I to angiotensin II, an active hormone with diverse effects.

Angiotensin II has four important functions: (1) It stimulates the suprarenal production of aldosterone, causing Na^+ retention and K^+ loss at the kidneys; (2) it stimulates the secretion of ADH, in turn stimulating water reabsorption at the kidneys and complementing the effects of aldosterone; (3) it stimulates thirst, resulting in increased fluid consumption (the presence of ADH and aldosterone ensures that the additional water consumed will be retained, elevating blood volume); and (4) it stimulates cardiac output and triggers the constriction of arterioles, in turn elevating the systemic blood pressure. The effect of angiotensin II on blood pressure is four to eight times greater than that produced by norepinephrine.

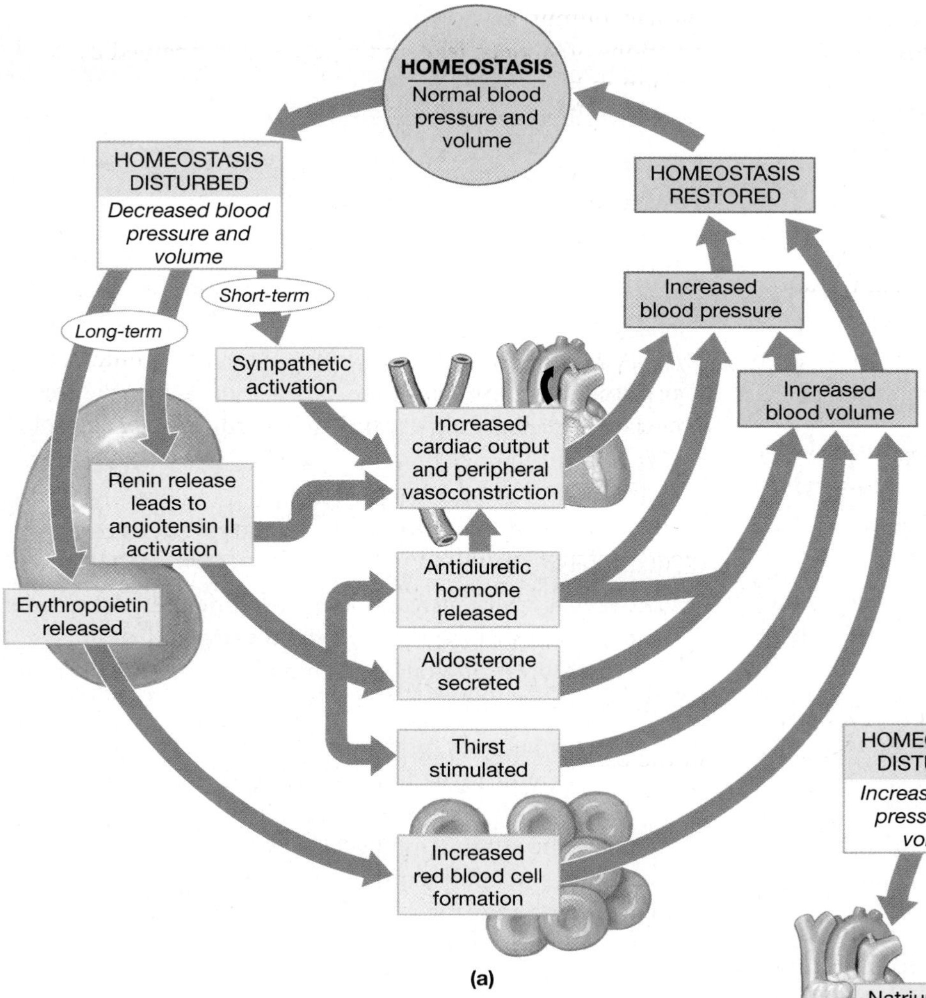

Figure 21–16 **The Hormonal Regulation of Blood Pressure and Blood Volume.** Shown are factors that compensate for **(a)** decreased blood pressure and volume and for **(b)** increased blood pressure and volume.

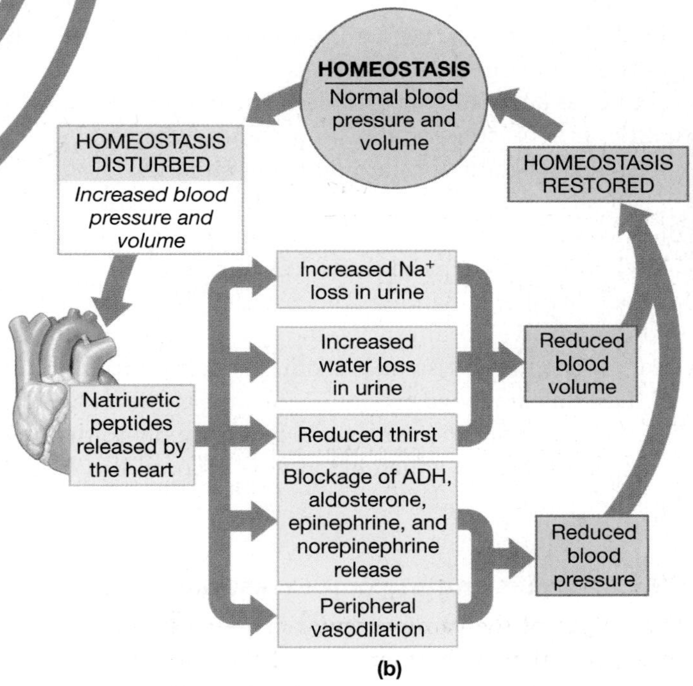

Erythropoietin

Erythropoietin (EPO) is released at the kidneys if blood pressure falls or if the oxygen content of the blood becomes abnormally low (**Figure 21–16a**). EPO stimulates the production and maturation of red blood cells, thereby increasing the volume and viscosity of the blood and improving its oxygen-carrying capacity.

Natriuretic Peptides

Atrial natriuretic peptide (nā-trē-ū-RET-ik; *natrium*, sodium + *ouresis*, making water), or *ANP*, is produced by cardiac muscle cells in the wall of the right atrium in response to excessive stretching during diastole. A related hormone called *brain natriuretic peptide*, or *BNP*, is produced by ventricular muscle cells exposed to comparable stimuli. These peptide hormones reduce blood volume and blood pressure by (1) increasing sodium ion excretion at the kidneys; (2) promoting water losses by increasing the volume of urine produced; (3)

reducing thirst; (4) blocking the release of ADH, aldosterone, epinephrine, and norepinephrine; and (5) stimulating peripheral vasodilation (**Figure 21–16b**). As blood volume and blood pressure decline, the stresses on the walls of the heart are removed, and natriuretic peptide production ceases.

The A&P Top 100

#74 Cardiac output cannot increase indefinitely, and blood flow to active versus inactive tissues must be differentially controlled. This is accomplished by a combination of autoregulation, neural regulation, and hormone release.

CHECKPOINT

9. Describe the actions of vasodilators and local vasodilators.

10. How would applying slight pressure to the common carotid artery affect your heart rate?

11. What effect would the vasoconstriction of the renal artery have on blood pressure and blood volume?

See the blue Answers tab at the end of the book.

21-4 The cardiovascular system adapts to physiological stress and maintains a special vascular supply to the brain, heart, and lungs

In this and the two previous chapters, we have considered the blood, the heart, and the cardiovascular system as individual entities. Yet in our day-to-day lives, the cardiovascular system operates as an integrated complex. The interactions are fascinating and of considerable importance when physical or physiological conditions are changing rapidly.

In this section we begin by examining the patterns of cardiovascular responses to two common stresses, exercise and blood loss, which provide two examples of that system's adaptability in maintaining homeostasis. The homeostatic responses involve an interplay among the cardiovascular system, the endocrine system, and other systems, and the central mechanisms are aided by autoregulation at the tissue level. Then we consider the patterns of blood supply to the brain, heart, and lungs, in which blood flow is controlled by separate mechanisms.

The Cardiovascular Response to Exercise

At rest, cardiac output averages about 5.8 liters per minute. That value changes dramatically during exercise. In addition, the pattern of blood distribution changes markedly, as detailed in Table 21–2.

Light Exercise

Before you begin to exercise, your heart rate increases slightly due to a general rise in sympathetic activity as you think about the workout ahead. As you begin light exercise, three interrelated changes take place:

- *Extensive vasodilation occurs* as the rate of oxygen consumption in skeletal muscles increases. Peripheral resistance drops, blood flow through the capillaries

TABLE 21–2 Changes in Blood Distribution during Exercise

Organ	Rest	Light Exercise	Strenuous Exercise
		Tissue Blood Flow (mL/min)	
Skeletal muscles	1200	4500	12,500
Heart	250	350	750
Brain	750	750	750
Skin	500	1500	1900
Kidney	1100	900	600
Abdominal viscera	1400	1100	600
Miscellaneous	600	400	400
Total cardiac output	5800	9500	17,500

increases, and blood enters the venous system at an accelerated rate.

- *The venous return increases* as skeletal muscle contractions squeeze blood along the peripheral veins and an increased breathing rate pulls blood into the venae cavae via the respiratory pump.

- *Cardiac output rises*, primarily in response to (1) the rise in venous return (the Frank–Starling principle ∞ p. 712) and (2) atrial stretching (the atrial reflex). Some sympathetic stimulation occurs, leading to increases in heart rate and contractility, but there is no massive sympathetic activation. The increased cardiac output keeps pace with the elevated demand, and arterial pressures are maintained despite the drop in peripheral resistance.

This regulation by venous feedback produces a gradual increase in cardiac output to about double resting levels. The increase supports accelerated blood flow to skeletal muscles, cardiac muscle, and the skin. The increased flow to skeletal and cardiac muscles reflects the dilation of arterioles and precapillary sphincters in response to local factors; the increased flow to the skin occurs in response to the rise in body temperature.

Heavy Exercise

At higher levels of exertion, other physiological adjustments occur as the cardiac and vasomotor centers call for the general activation of the sympathetic nervous system. Cardiac output increases toward maximal levels, and major changes in the peripheral distribution of blood occur, facilitating the blood flow to active skeletal muscles.

Under massive sympathetic stimulation, the cardioacceleratory center can increase cardiac output to levels as high as 20–25 liters per minute. But that is still not enough to meet the demands of active skeletal muscles unless the vasomotor center severely restricts the blood flow to "nonessential" organs, such as those of the digestive system. During exercise at maximal levels, your blood essentially races between the skeletal muscles and the lungs and heart. Although blood flow to most tissues is diminished, skin perfusion increases further, because body temperature continues to climb. Only the blood supply to the brain remains unaffected.

Exercise, Cardiovascular Fitness, and Health

Cardiovascular performance improves significantly with training. Table 21–3 compares the cardiac performance of athletes with that of nonathletes. Trained athletes have bigger hearts and larger stroke volumes than do nonathletes, and these are important functional differences.

Recall that cardiac output is equal to the stroke volume times the heart rate; thus, for the same cardiac output, the person with a larger stroke volume has a slower heart rate. An athlete at rest can maintain normal blood flow to peripheral tissues at a heart rate as low as 32 bpm (beats per minute), and, when necessary, the athlete's cardiac output can increase to levels 50 percent higher than those of nonathletes. Thus, a trained athlete can tolerate sustained levels of activity that are well beyond the capabilities of nonathletes.

Exercise and Cardiovascular Disease

Regular exercise has several beneficial effects. Even a moderate exercise routine (jogging 5 miles a week, for example) can lower total blood cholesterol levels. A high cholesterol level is one of the major risk factors for atherosclerosis, which leads to cardiovascular disease and strokes. In addition, a healthy lifestyle—regular exercise, a balanced diet, weight control, and not smoking—reduces stress, lowers blood pressure, and slows the formation of plaques.

Regular moderate exercise may cut the incidence of heart attacks almost in half. However, only an estimated 8 percent of adults in the United States currently exercise at recommended levels. Exercise is also beneficial in accelerating recovery after a heart attack. Regular light-to-moderate exercise (such as walking, jogging, or bicycling), coupled with a low-fat diet and a low-stress lifestyle, not only reduces symptoms of coronary artery disease (such as angina), but also improves one's mood and overall quality of life. However, exercise does not remove any underlying medical problem, and atherosclerotic plaques, described on p. 724, do not disappear and seldom grow smaller with exercise.

There is no evidence that *intense* athletic training lowers the incidence of cardiovascular disease. On the contrary, the strains placed on all physiological systems—including the cardiovascular system—during an ultramarathon, iron-man triathlon, or other extreme athletic event can be severe. Individuals with congenital aneurysms, cardiomyopathy, or cardiovascular disease risk fatal circulatory problems, such as an arrhythmia or heart attack, during severe exercise. Even healthy individuals can develop acute physiological disorders, such as kidney failure, after extreme exercise. We will discuss the effects of exercise on other systems in later chapters.

The Cardiovascular Response to Hemorrhaging

In Chapter 19, we considered the local circulatory reaction to a break in the wall of a blood vessel. ∞ p. 673 When hemostasis fails to prevent significant blood loss, the entire cardiovascular system makes adjustments to maintain blood pressure and restore blood volume (**Figure 21–17**). The immediate problem is the maintenance of adequate blood pressure and peripheral blood flow. The long-term problem is the restoration of normal blood volume.

Short-Term Elevation of Blood Pressure

Almost as soon as the pressures start to decline, several short-term responses appear:

- The initial neural response occurs as carotid and aortic reflexes increase cardiac output and cause peripheral vasoconstriction (pp. 739–741). With blood volume reduced, cardiac output is maintained by increasing the heart rate, typically to 180–200 bpm.

- The combination of stress and anxiety stimulates the sympathetic nervous system headquarters in the

	Heart Weight (g)	Stroke Volume (mL)	Heart Rate (bpm)	Cardiac Output (L/min)	Blood Pressure (systolic/diastolic)
TABLE 21–3 Effects of Training on Cardiovascular Performance					
Subject	**Heart Weight (g)**	**Stroke Volume (mL)**	**Heart Rate (bpm)**	**Cardiac Output (L/min)**	**Blood Pressure (systolic/diastolic)**
Nonathlete (rest)	300	60	83	5.0	120/80
Nonathlete (maximum)		104	192	19.9	187/75
Trained athlete (rest)	500	100	53	5.3	120/80
Trained athlete (maximum)		167	182	30.4	200/90*

*Diastolic pressures in athletes during maximal activity have not been accurately measured.

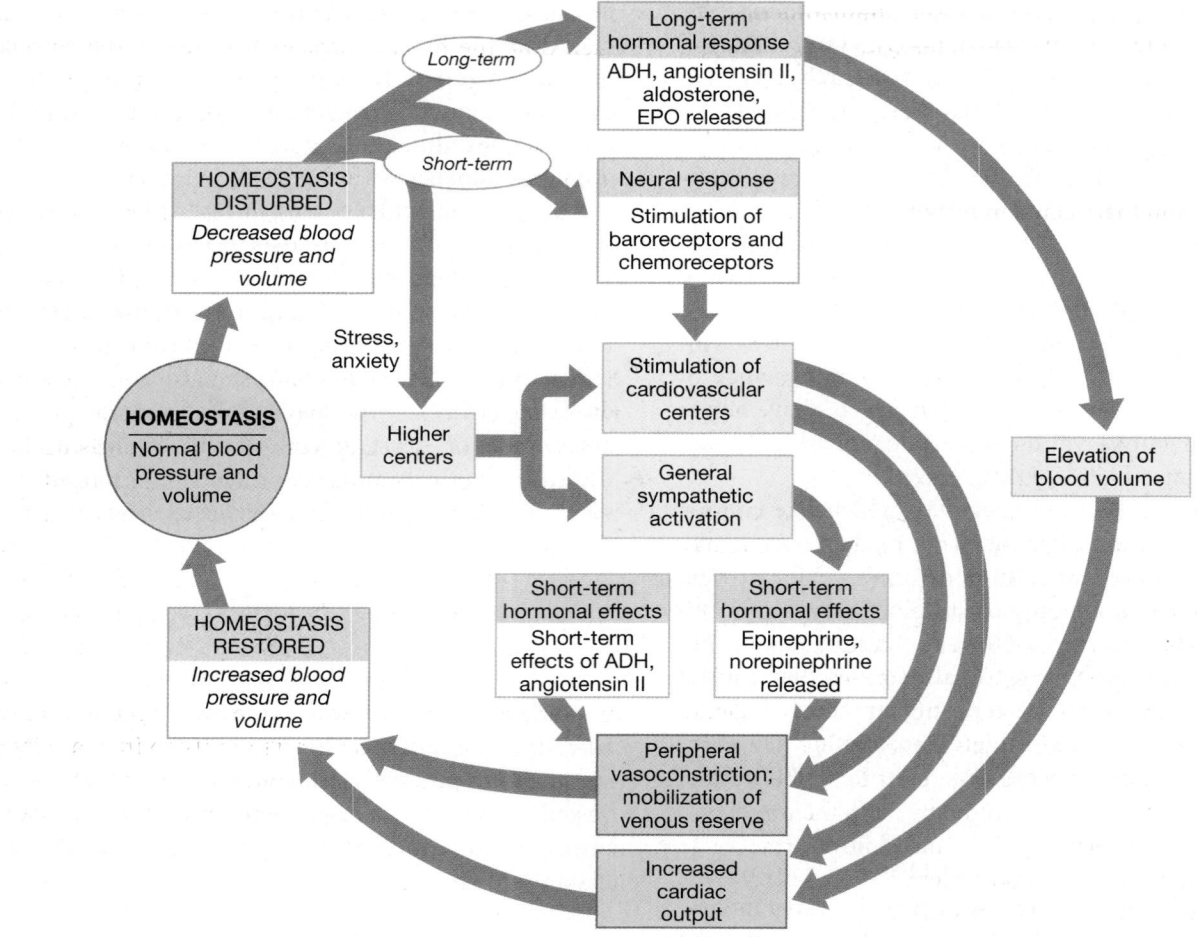

Figure 21–17 **Cardiovascular Responses to Hemorrhaging and Blood Loss.** These mechanisms can cope with blood losses equivalent to approximately 30 percent of total blood volume.

hypothalamus, which in turn triggers a further increase in vasomotor tone, constricting the arterioles and elevating blood pressure. At the same time, venoconstriction mobilizes the venous reserve and quickly improves venous return (p. 729).

- Short-term hormonal effects also occur. For instance, sympathetic activation causes the secretion of E and NE by the suprarenal medullae, increasing cardiac output and extending peripheral vasoconstriction. In addition, the release of ADH by the neurohypophysis and the production of angiotensin II enhance vasoconstriction while participating in the long-term response.

This combination of short-term responses elevates blood pressure and improves peripheral blood flow, often restoring normal arterial pressures and peripheral circulation after blood losses of up to 20 percent of total blood volume. Such adjustments are more than sufficient to compensate for the blood loss experienced when you donate blood. (Most blood

banks collect 500 mL of whole blood, roughly 10 percent of your total blood volume.) If compensatory mechanisms fail, the individual develops signs of *shock*.

Long-Term Restoration of Blood Volume

Short-term responses temporarily compensate for a reduction in blood volume. Long-term responses are geared to restoring normal blood volume, a process that can take several days after a serious hemorrhage. The steps include the following:

- The decline in capillary blood pressure triggers a recall of fluids from the interstitial spaces (p. 737).

- Aldosterone and ADH promote fluid retention and reabsorption at the kidneys, preventing further reductions in blood volume.

- Thirst increases, and additional water is obtained by absorption across the digestive tract. This intake of fluid elevates the plasma volume and ultimately replaces the interstitial fluids "borrowed" at the capillaries.

- Erythropoietin targets the bone marrow, stimulating the maturation of red blood cells, which increase blood volume and improve oxygen delivery to peripheral tissues.

Vascular Supply to Special Regions

The vasoconstriction that occurs in response to a fall in blood pressure or a rise in CO_2 levels affects multiple tissues and organs simultaneously. The term *special circulation* refers to the vascular supply through organs in which blood flow is controlled by separate mechanisms. We will consider three important examples: the blood flow to the brain, the heart, and the lungs.

Blood Flow to the Brain

In Chapter 14, we noted the existence of the blood–brain barrier, which isolates most CNS tissue from the general circulation. ⚬⚬ p. 467 The brain has a very high demand for oxygen and receives a substantial supply of blood. Under a variety of conditions, blood flow to the brain remains steady at about 750 mL/min—roughly 12 percent of the cardiac output delivered to an organ that represents less than 2 percent of body weight. Neurons do not maintain significant energy reserves, and in functional terms most of the adjustments made by the cardiovascular system treat blood flow to the brain as the top priority. Even during a cardiovascular crisis, blood flow through the brain remains as near normal as possible: While the cardiovascular centers are calling for widespread peripheral vasoconstriction, the cerebral vessels are instructed to dilate.

Although total blood flow to the brain remains relatively constant, blood flow to specific regions of the brain changes from moment to moment. These changes occur in response to local changes in the composition of interstitial fluid that accompany neural activity. When you read, write, speak, or walk, specific regions of your brain become active. Blood flow to those regions increases almost instantaneously, ensuring that the active neurons will continue to receive the oxygen and nutrients they require.

The brain receives arterial blood through four arteries. Because these arteries form anastomoses inside the cranium, an interruption of flow in any one of these large vessels will not significantly reduce blood flow to the brain as a whole. However, a plaque or a blood clot may still block a small artery, and weakened arteries may rupture. Such incidents temporarily or permanently shut off blood flow to a localized area of the brain, damaging or killing the dependent neurons. Signs and symptoms of a *stroke*, or *cerebrovascular accident (CVA)*, then appear.

Blood Flow to the Heart

The anatomy of the coronary circulation was described in Chapter 20. ⚬⚬ p. 692 The coronary arteries arise at the base of the ascending aorta, where systemic pressures are highest. Each time the heart contracts, it squeezes the coronary vessels, so blood flow is reduced. In the left ventricle, systolic pressures are high enough that blood can flow into the myocardium only during diastole; over this period, elastic rebound helps drive blood along the coronary vessels. Normal cardiac muscle cells can tolerate these brief circulatory interruptions because they have substantial oxygen reserves.

When you are at rest, coronary blood flow is about 250 mL/min. When the workload on your heart increases, local factors, such as reduced O_2 levels and lactic acid production, dilate the coronary vessels and increase blood flow. Epinephrine released during sympathetic stimulation promotes the vasodilation of coronary vessels while increasing heart rate and the strength of cardiac contractions. As a result, coronary blood flow increases while vasoconstriction occurs in other tissues.

For unclear reasons, some individuals experience *coronary spasms*, which can temporarily restrict coronary circulation and produce symptoms of angina. A permanent restriction or blockage of coronary vessels (as in coronary artery disease) and tissue damage (as caused by a myocardial infarction) can limit the heart's ability to increase its output, even under maximal stimulation. Individuals with these conditions experience signs and symptoms of heart failure when the cardiac workload increases much above resting levels.

Blood Flow to the Lungs

The lungs contain roughly 300 million *alveoli* (al-VĒ-ō-lī; *alveolus*, sac), delicate epithelial pockets where gas exchange occurs. Each alveolus is surrounded by an extensive capillary network. Blood flow through the lungs is regulated primarily by local responses to oxygen levels within individual alveoli. When an alveolus contains oxygen in abundance, the associated vessels dilate, so blood flow increases, promoting the absorption of oxygen from the alveolar air. When the oxygen content of the air is very low, the vessels constrict, so blood is shunted to alveoli that still contain significant levels of oxygen. This mechanism maximizes the efficiency of the respiratory system, because the circulation of blood through the capillaries of an alveolus has no benefit unless that alveolus contains oxygen.

This mechanism is precisely the opposite of that in other tissues, where a decline in oxygen levels causes local vasodilation rather than vasoconstriction. The difference makes functional sense, but its physiological basis remains a mystery.

Blood pressure in pulmonary capillaries (average: 10 mm Hg) is lower than that in systemic capillaries. The BCOP (25 mm Hg) is the same as elsewhere in the bloodstream. As a result, reabsorption exceeds filtration in pulmonary capillaries. Fluid moves continuously into the pulmonary capillaries

21 CARDIOVASCULAR

across the alveolar surfaces, thereby preventing a buildup of fluid in the alveoli that could interfere with the diffusion of respiratory gases. If the blood pressure in pulmonary capillaries rises above 25 mm Hg, fluid enters the alveoli, causing *pulmonary edema*.

CHECKPOINT

12. Why does blood pressure increase during exercise?

13. Name the immediate and long-term problems related to the cardiovascular response to hemorrhaging.

14. Explain the role of aldosterone and ADH in long-term restoration of blood volume.

See the blue Answers tab at the end of the book.

21-5 The pulmonary and systemic circuits of the cardiovascular system exhibit three general functional patterns

Pulmonary & Systemic system.

You already know that the cardiovascular system consists of the *pulmonary circuit* and the *systemic circuit*. The pulmonary circuit is composed of arteries and veins that transport blood between the heart and the lungs. This circuit begins at the right ventricle and ends at the left atrium. From the left ventricle, the arteries of the systemic circuit transport oxygenated blood and nutrients to all organs and tissues, ultimately returning deoxygenated blood to the right atrium. **Figure 21–18** summarizes the primary distribution routes within the pulmonary and systemic circuits.

In the pages that follow, we will examine the vessels of the pulmonary and systemic circuits further. Three general functional patterns are worth noting at the outset:

1. The peripheral distributions of arteries and veins on the body's left and right sides are generally identical, except near the heart, where the largest vessels connect to the atria or ventricles. Corresponding arteries and veins usually follow the same path. For example, the distributions of the left and right subclavian *arteries* parallel those of the left and right subclavian *veins*.

2. A single vessel may have several names as it crosses specific anatomical boundaries, making accurate anatomical descriptions possible when the vessel extends far into the periphery. For example, the *external iliac artery* becomes the *femoral artery* as it leaves the trunk and enters the lower limb.

3. Tissues and organs are usually serviced by several arteries and veins. Often, anastomoses between adjacent arteries or veins reduce the impact of a temporary or even permanent occlusion (blockage) of a single blood vessel.

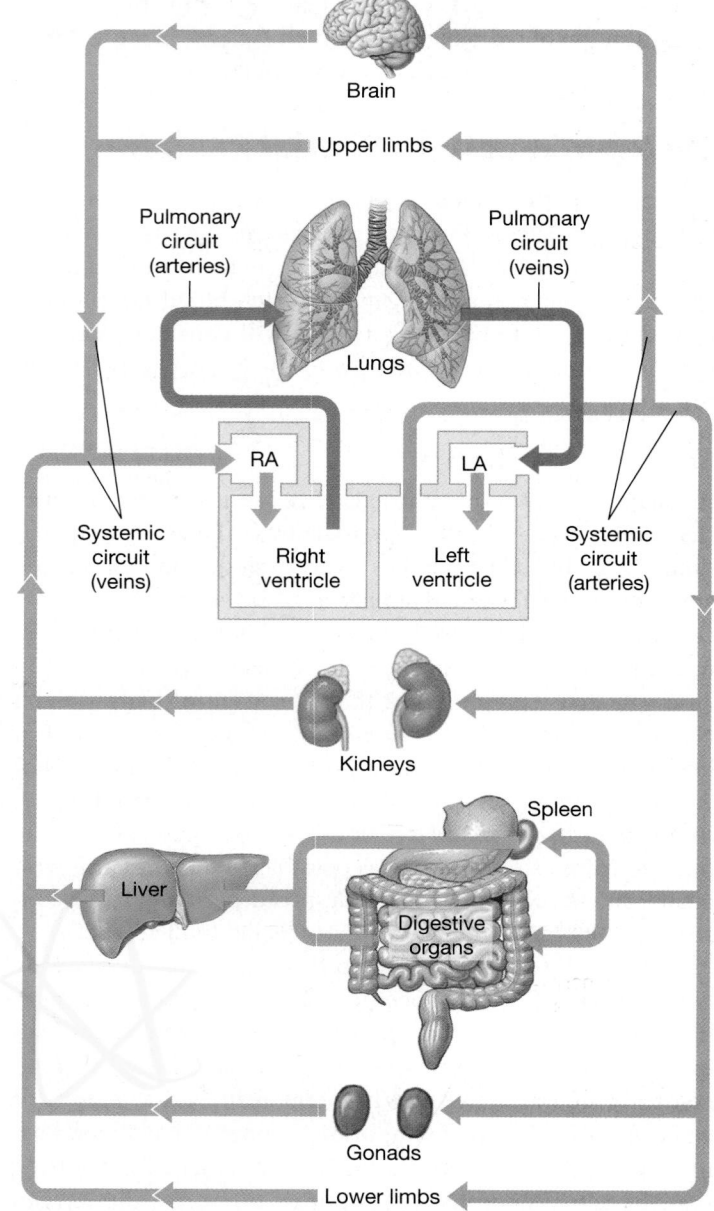

Figure 21–18 **A Schematic Overview of the Pattern of Circulation.** *RA* stands for right atrium, *LA* for left atrium.

pulmonary = lungs → gas exchange.

CHECKPOINT

15. Identify the two circulatory circuits of the cardiovascular system.

16. Identify the three general functional patterns seen in the pulmonary and systemic circuits of the cardiovascular system.

See the blue Answers tab at the end of the book.

21-6 In the pulmonary circuit, deoxygenated blood enters the lungs in arteries, and oxygenated blood leaves the lungs via veins

Blood entering the right atrium has just returned from the peripheral capillary beds, where oxygen was released and carbon dioxide absorbed. After traveling through the right atrium and ventricle, this deoxygenated blood enters the pulmonary trunk, the start of the pulmonary circuit (**Figure 21–19**). At the lungs, oxygen is replenished, carbon dioxide is released, and the oxygenated blood is returned to the heart for distribution via the systemic circuit. Compared with the systemic circuit, the pulmonary circuit is short: The base of the pulmonary trunk and the lungs are only about 15 cm (6 in.) apart.

The arteries of the pulmonary circuit differ from those of the systemic circuit in that they carry deoxygenated blood. (For this reason, most color-coded diagrams show the pulmonary arteries in blue, the same color as systemic veins.) As the pulmonary trunk curves over the superior border of the heart, it gives rise to the **left** and **right pulmonary arteries**. These large arteries enter the lungs before branching repeatedly, giving rise to smaller and smaller arteries. The smallest branches, the *pulmonary arterioles*, provide blood to capillary networks that surround *alveoli*. The walls of these small air pockets are thin enough for gas to be exchanged between the capillary blood and inspired air. As it leaves the alveolar capillaries, oxygenated blood enters venules that in turn unite to form larger vessels carrying blood toward the **pulmonary veins**. These four veins, two from each lung, empty into the left atrium, completing the pulmonary circuit.

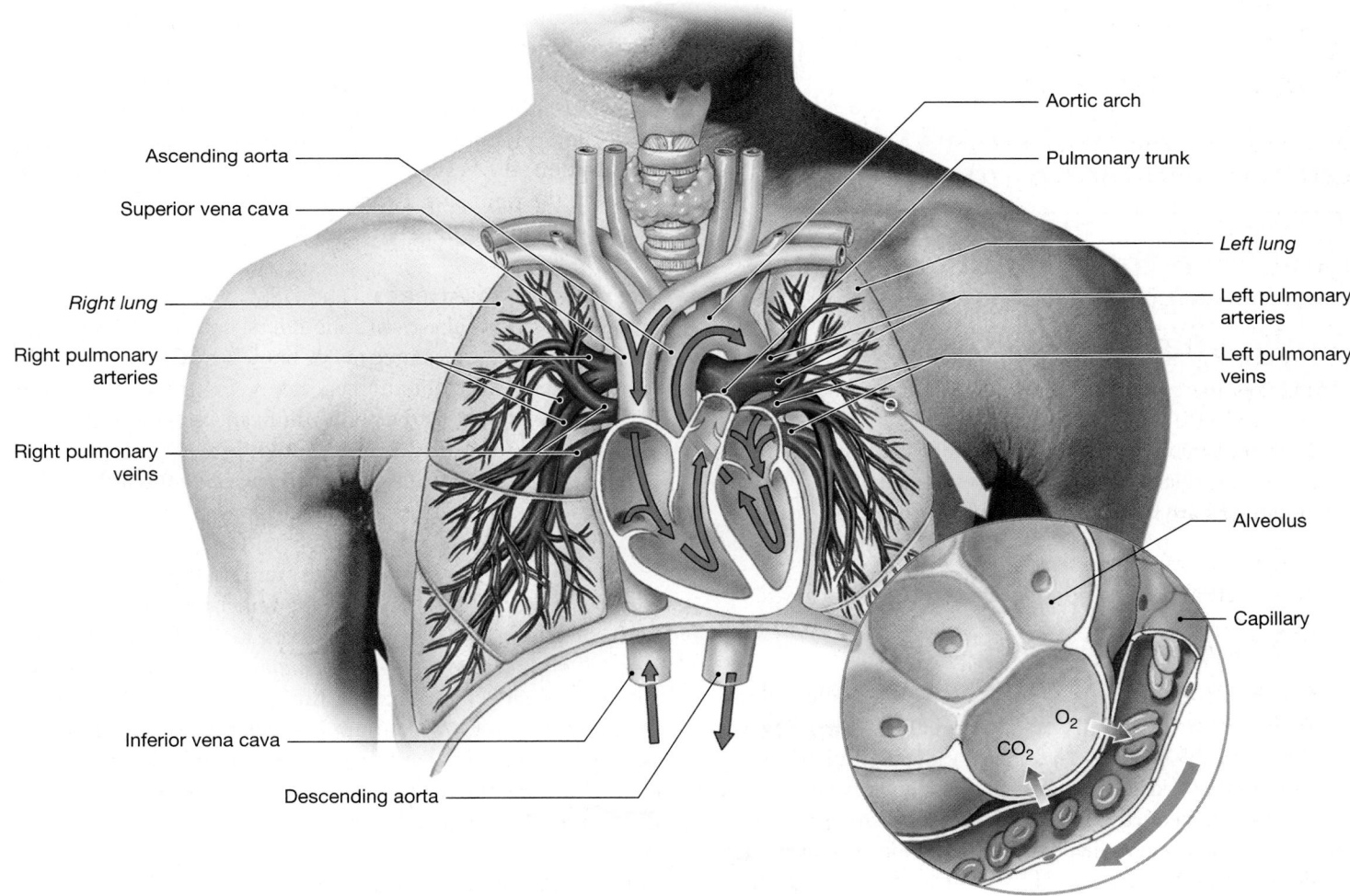

Figure 21–19   The Pulmonary Circuit. The pulmonary circuit consists of pulmonary arteries, which deliver deoxygenated blood from the right ventricle to the lungs; pulmonary capillaries, where gas exchange occurs; and pulmonary veins, which deliver oxygenated blood to the left atrium. As the enlarged view shows, diffusion across the capillary walls at alveoli removes carbon dioxide and provides oxygen to the blood. ATLAS: Plates 42a; 44c; 47b

Tips&Tricks

Arteries and veins are defined by the direction of blood flow relative to the heart, not the oxygen content of the blood they carry. So if you remember that **a**rteries carry blood **a**way from the heart, and veins carry blood toward the heart, you can remember that the pulmonary **a**rteries carry oxygen-poor blood **a**way from the heart to the lungs, and the pulmonary veins deliver oxygen-rich blood to the heart.

CHECKPOINT

17. Name the blood vessels that enter and exit the lungs, and indicate the relative oxygen content of the blood in each.

18. Trace a drop of blood through the lungs, beginning at the right ventricle and ending at the left atrium.

See the blue Answers tab at the end of the book.

21-7 The systemic circuit carries oxygenated blood from the left ventricle to tissues and organs other than the pulmonary exchange surfaces, and returns deoxygenated blood to the right atrium

The systemic circuit supplies the capillary beds in all parts of the body not serviced by the pulmonary circuit. The systemic circuit, which at any moment contains about 84 percent of total blood volume, begins at the left ventricle and ends at the right atrium.

Systemic Arteries

Figure 21–20 provides an overview of the systemic arterial system, indicating the relative locations of major systemic arteries. **Figures 21–21 to 21–26** show the detailed distribution of these vessels and their branches. By convention, several large arteries are called *trunks*; examples are the *pulmonary, brachiocephalic, thyrocervical,* and *celiac trunks.* Because most of the major arteries are paired, with one artery of each pair on either side of the body, the terms *right* and *left* will appear in figures only when the arteries on both sides are labeled.

The Ascending Aorta

The **ascending aorta** (**Figure 21–21**) begins at the aortic valve of the left ventricle. The left and right coronary arteries originate in the aortic sinus at the base of the ascending aorta, just superior to the aortic valve. The distribution of coronary vessels was described in Chapter 20 and illustrated in **Figure 20–9**, p. 693.

The Aortic Arch

The **aortic arch** curves like the handle of a cane across the superior surface of the heart, connecting the ascending aorta with the *descending aorta* (**Figure 21–20**). Three elastic arteries originate along the aortic arch and deliver blood to the head, neck, shoulders, and upper limbs: (1) the **brachiocephalic** (brā-kē-ō-se-FAL-ik) **trunk**, (2) the **left common carotid artery**, and (3) the **left subclavian artery** (**Figures 21–21** and **21–22**). The brachiocephalic trunk, also called the *innominate artery* (i-NOM-i-nat; unnamed), ascends for a short distance before branching to form the **right subclavian artery** and the **right common carotid artery**.

We have only one brachiocephalic trunk, with the left common carotid and left subclavian arteries arising separately from the aortic arch. However, in terms of their peripheral distribution, the vessels on the left side are mirror images of those on the right side. **Figures 21–21** and **21–22** illustrate the major branches of these arteries.

The Subclavian Arteries. The subclavian arteries supply blood to the arms, chest wall, shoulders, back, and CNS (**Figures 21–20** and **21–21**). Three major branches arise before a subclavian artery leaves the thoracic cavity: (1) the **internal thoracic artery**, supplying the pericardium and anterior wall of the chest; (2) the **vertebral artery**, which provides blood to the brain and spinal cord; and (3) the **thyrocervical trunk**, which provides blood to muscles and other tissues of the neck, shoulder, and upper back.

After leaving the thoracic cavity and passing across the superior border of the first rib, the subclavian is called the **axillary artery**. This artery crosses the axilla to enter the arm, where it gives rise to *humeral circumflex arteries,* which supply structures near the head of the humerus. Distally, it becomes the **brachial artery**, which supplies blood to the rest of the upper limb. The brachial artery gives rise to the *deep brachial artery,* which supplies deep structures on the posterior aspect of the arm, and the *ulnar collateral arteries,* which supply the area around the elbow. As it approaches the coronoid fossa of the humerus, the brachial artery divides into the **radial artery**, which follows the radius, and the **ulnar artery**, which follows the ulna to the wrist. These arteries supply blood to the forearm

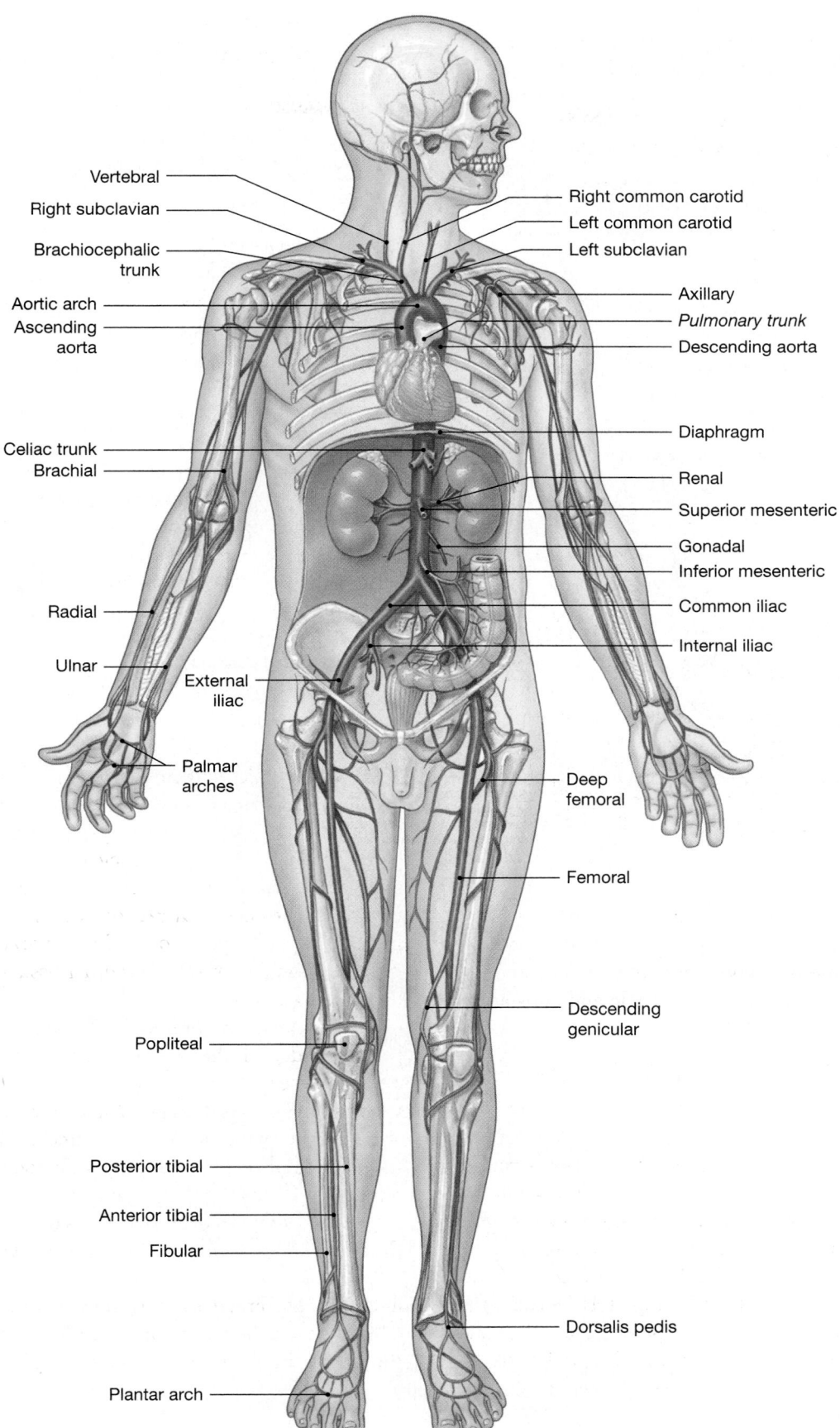

Vertebral

Right subclavian

Brachiocephalic trunk

Aortic arch

Ascending aorta

Celiac trunk

Brachial

Radial

Ulnar

External iliac

Palmar arches

Popliteal

Posterior tibial

Anterior tibial

Fibular

Plantar arch

Right common carotid

Left common carotid

Left subclavian

Axillary

Pulmonary trunk

Descending aorta

Diaphragm

Renal

Superior mesenteric

Gonadal

Inferior mesenteric

Common iliac

Internal iliac

Deep femoral

Femoral

Descending genicular

Dorsalis pedis

Figure 21–20   An Overview of the Major Systemic Arteries.

21 CARDIOVASCULAR

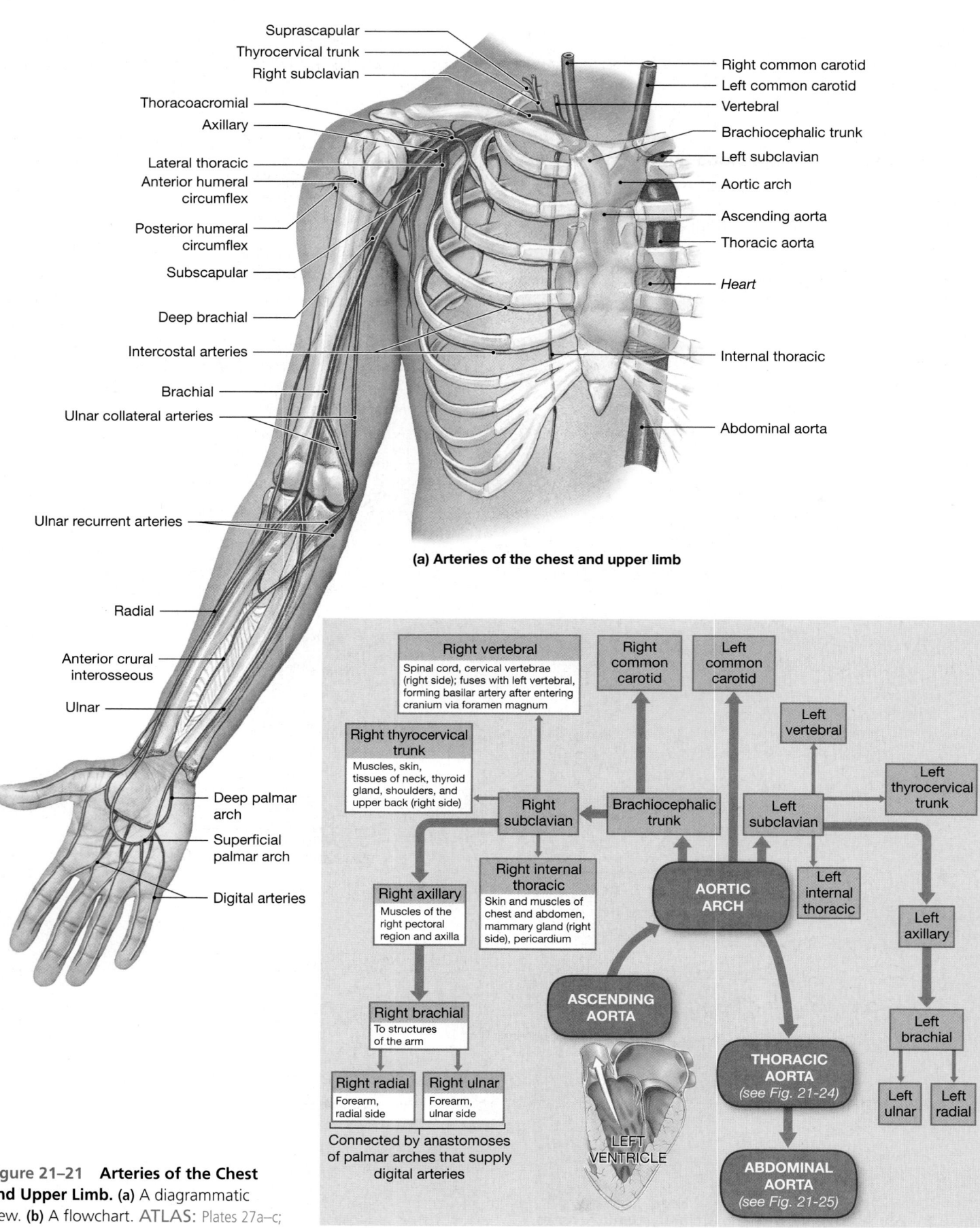

(a) Arteries of the chest and upper limb

Figure 21–21 Arteries of the Chest and Upper Limb. (a) A diagrammatic view. **(b)** A flowchart. ATLAS: Plates 27a–c; 29c; 30; 45a

(b) Flowchart

and, through the *ulnar recurrent arteries*, the region around the elbow. At the wrist, the radial and ulnar arteries fuse to form the **superficial** and **deep palmar arches**, which supply blood to the hand and to the **digital arteries** of the thumb and fingers.

The Carotid Artery and the Blood Supply to the Brain. The common carotid arteries ascend deep in the tissues of the neck. You can usually locate the carotid artery by pressing gently along either side of the windpipe (trachea) until you feel a strong pulse.

Each common carotid artery divides into an **external carotid artery** and an **internal carotid artery** (**Figure 21–22**). The **carotid sinus**, located at the base of the internal carotid artery, may extend along a portion of the common carotid. The external carotid arteries supply blood to the structures of the neck, esophagus, pharynx, larynx, lower jaw,

and face. The internal carotid arteries enter the skull through the carotid canals of the temporal bones, delivering blood to the brain. (**Figures 7–3** and **7–4**, pp. 215–217.)

Tips&Tricks

Remember that the *external* carotid artery supplies the face (which is external), and that the *internal* carotid supplies the brain (which is internal).

The internal carotid arteries ascend to the level of the optic nerves, where each artery divides into three branches: (1) an **ophthalmic artery**, which supplies the eyes; (2) an **anterior cerebral artery**, which supplies the frontal and parietal lobes of the brain; and (3) a **middle cerebral artery**, which supplies the mesencephalon and lateral surfaces of the cerebral hemispheres (**Figures 21–22** and **21–23**).

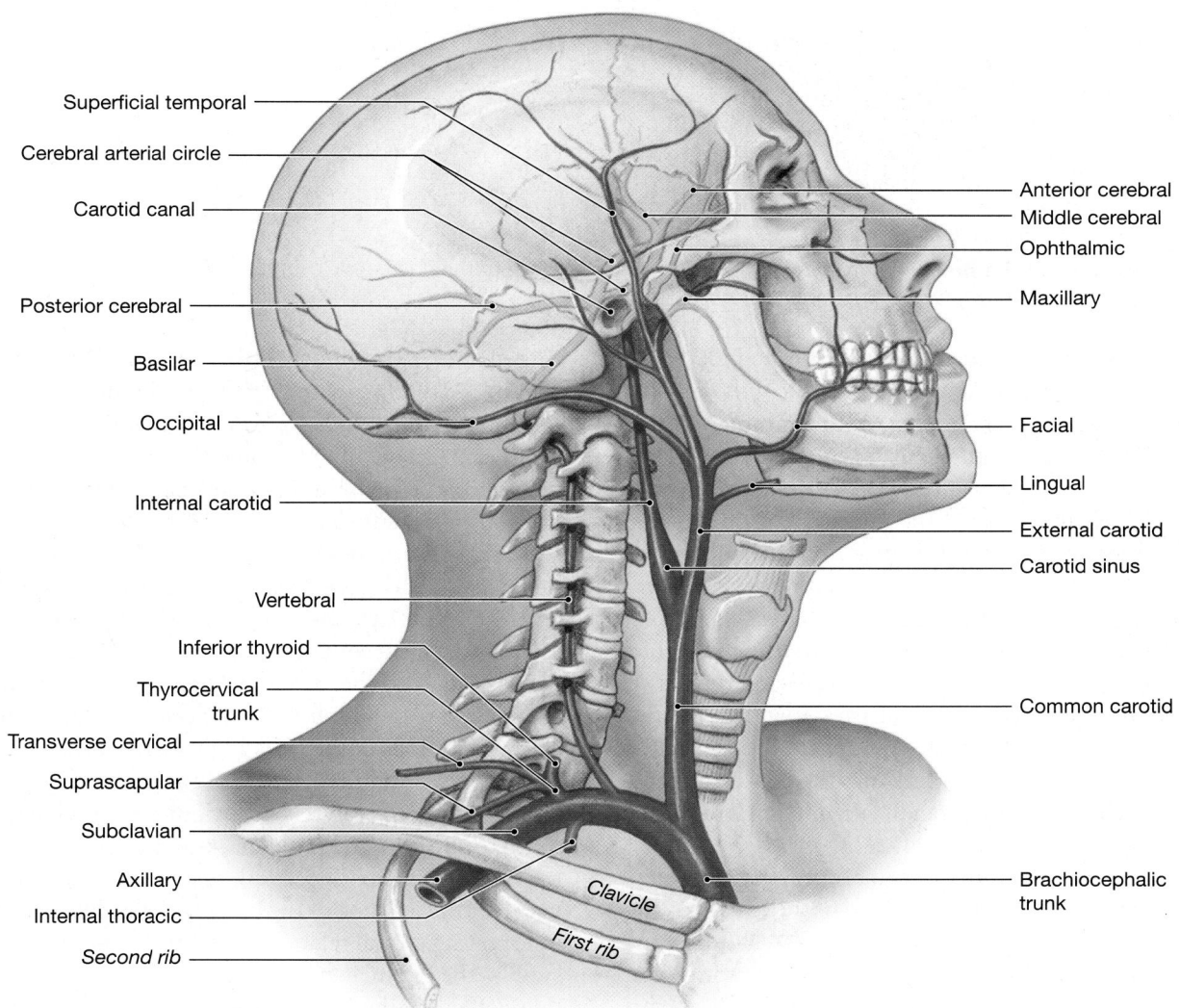

Figure 21–22 **Arteries of the Neck and Head.** Shown as seen from the right side. ATLAS: Plates 3c,d; 15b; 18a–c; 45a

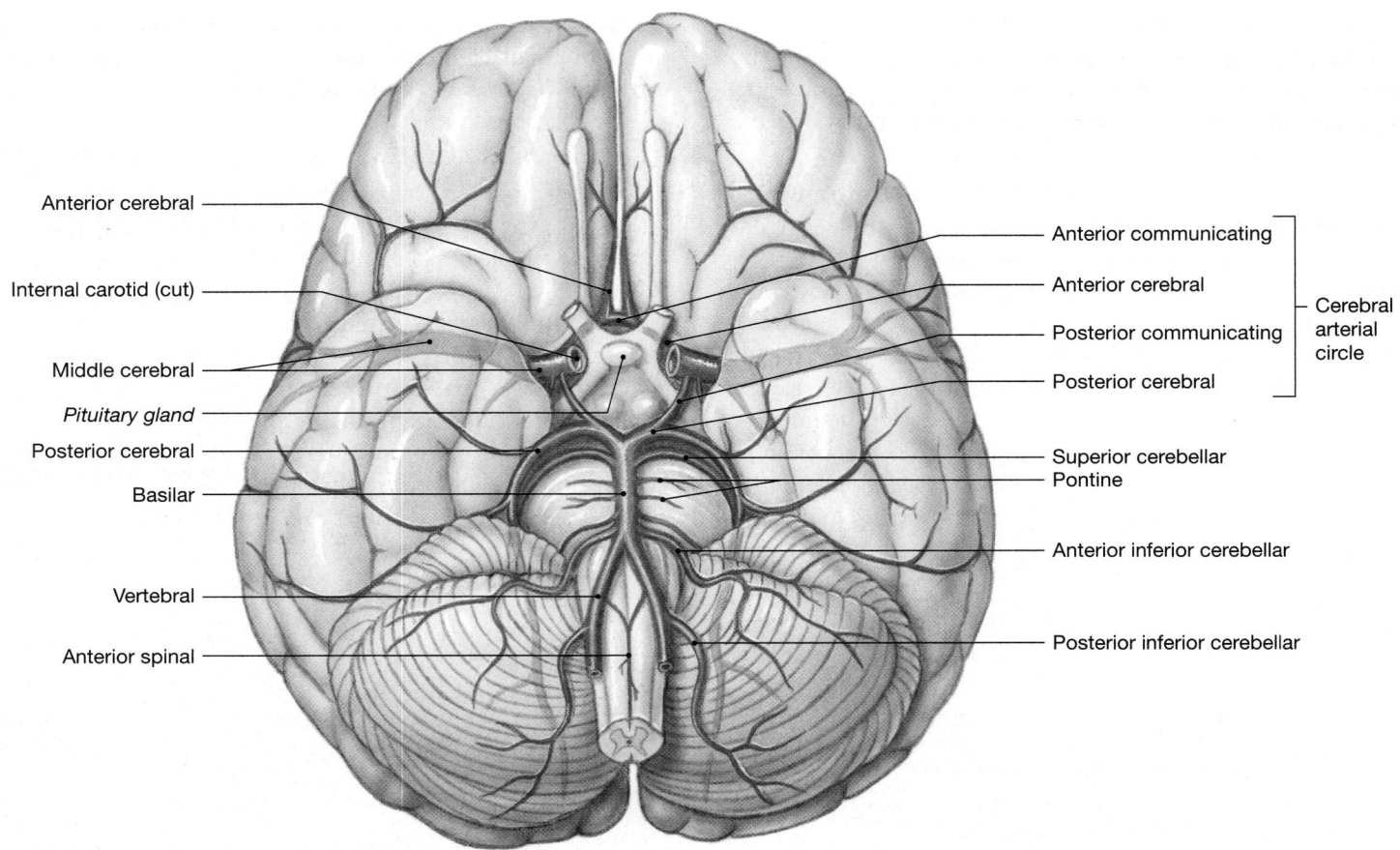

Anterior cerebral

Internal carotid (cut)

Middle cerebral

Pituitary gland

Posterior cerebral

Basilar

Vertebral

Anterior spinal

Anterior communicating

Anterior cerebral

Posterior communicating

Posterior cerebral

Cerebral arterial circle

Superior cerebellar

Pontine

Anterior inferior cerebellar

Posterior inferior cerebellar

Figure 21–23 **Arteries of the Brain.** The major arteries on the inferior surface of the brain. ATLAS: Plate 15a–c

The brain is extremely sensitive to changes in blood supply. An interruption of blood flow for several seconds will produce unconsciousness, and after four minutes some permanent neural damage can occur. Such circulatory crises are rare, because blood reaches the brain through the vertebral arteries as well as by way of the internal carotid arteries. The left and right vertebral arteries arise from the subclavian arteries and ascend within the transverse foramina of the cervical vertebrae. (**Figure 7–19b,c**, p. 235.) The vertebral arteries enter the cranium at the foramen magnum, where they fuse along the ventral surface of the medulla oblongata to form the **basilar artery**. The vertebral arteries and the basilar artery supply blood to the spinal cord, medulla oblongata, pons, and cerebellum before dividing into the **posterior cerebral arteries**, which in turn branch off into the **posterior communicating arteries** (**Figure 21–23**).

The internal carotid arteries normally supply the arteries of the anterior half of the cerebrum, and the rest of the brain receives blood from the vertebral arteries. But this circulatory pattern can easily change, because the internal carotid arteries and the basilar artery are interconnected in a

ring-shaped anastomosis called the **cerebral arterial circle**, or *circle of Willis*, which encircles the infundibulum of the pituitary gland (**Figure 21–23**). With this arrangement, the brain can receive blood from either the carotid or the vertebral arteries, so the likelihood of a serious interruption of circulation is reduced.

Strokes, or *cerebrovascular accidents (CVAs)*, are interruptions of the vascular supply to a portion of the brain. The *middle cerebral artery*, a major branch of the cerebral arterial circle, is the most common site of a stroke. Superficial branches deliver blood to the temporal lobe and large portions of the frontal and parietal lobes; deep branches supply the basal nuclei and portions of the thalamus. If a stroke blocks the middle cerebral artery on the left side of the brain, aphasia and a sensory and motor paralysis of the right side of the body result. In a stroke affecting the middle cerebral artery on the right side, the individual experiences a loss of sensation and motor control over the left side of the body and has difficulty drawing or interpreting spatial relationships. Strokes affecting vessels that supply the brain stem also produce distinctive symptoms; those affecting the lower brain stem are commonly fatal.

The Descending Aorta

The **descending aorta** is continuous with the aortic arch. The diaphragm divides the descending aorta into a superior **thoracic aorta** and an inferior **abdominal aorta** (**Figures 21–24** and **21–25**).

The Thoracic Aorta. The thoracic aorta begins at the level of vertebra T_5 and penetrates the diaphragm at the level of vertebra T_{12}. It travels within the mediastinum, on the posterior thoracic wall, slightly to the left of the vertebral column. This vessel supplies blood to branches that service the tissues and organs of the mediastinum, the muscles of the chest and the diaphragm, and the thoracic spinal cord.

The branches of the thoracic aorta are anatomically grouped as either visceral or parietal:

- *Visceral branches* supply the organs of the chest. The **bronchial arteries** supply the tissues of the lungs not involved in gas exchange, the **pericardial arteries** supply

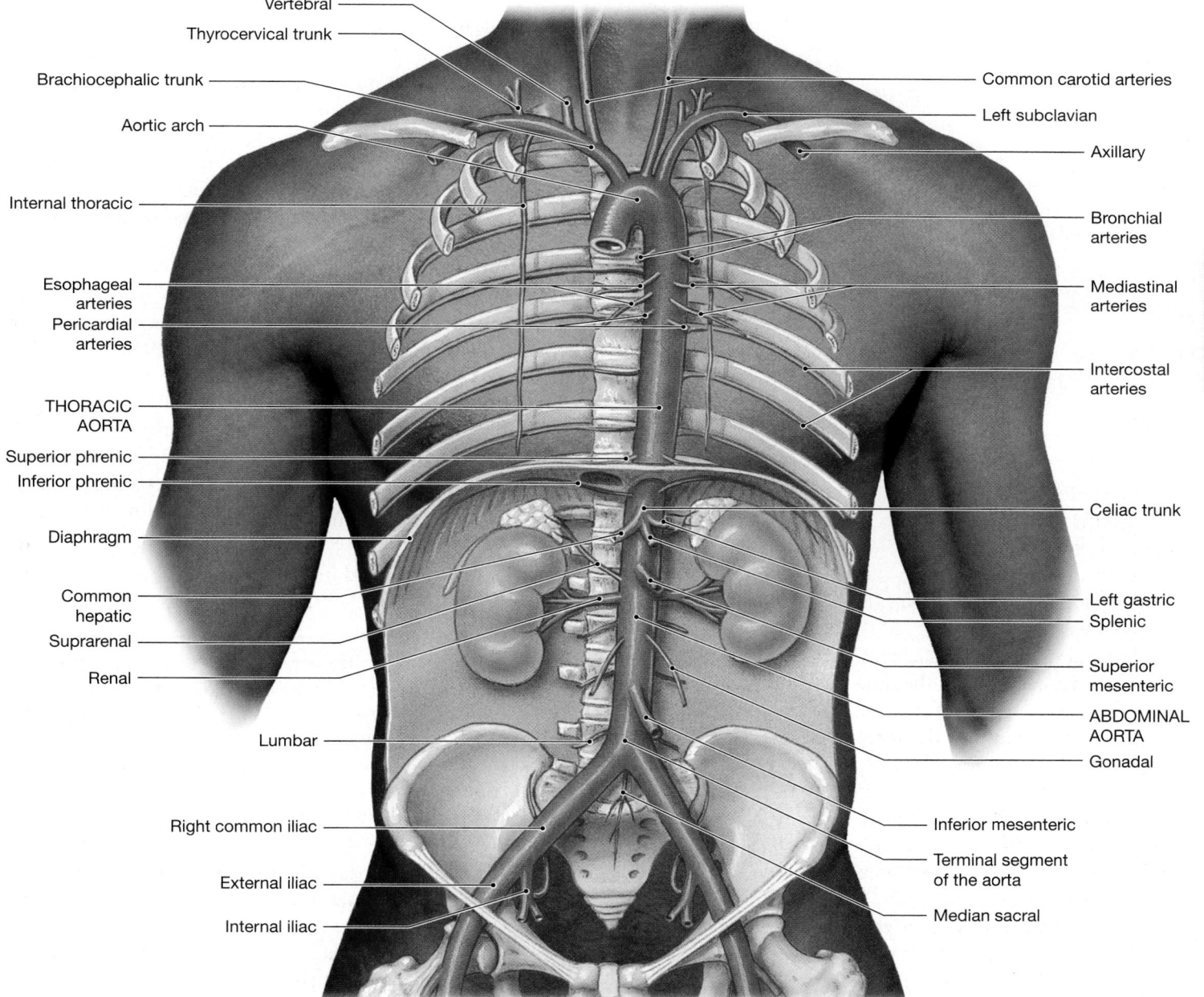

(a)

Figure 21–24   Major Arteries of the Trunk. (a) A diagrammatic view, with most of the thoracic and abdominal organs removed.
ATLAS: Plates 47d; 53c,e; 62a,b.

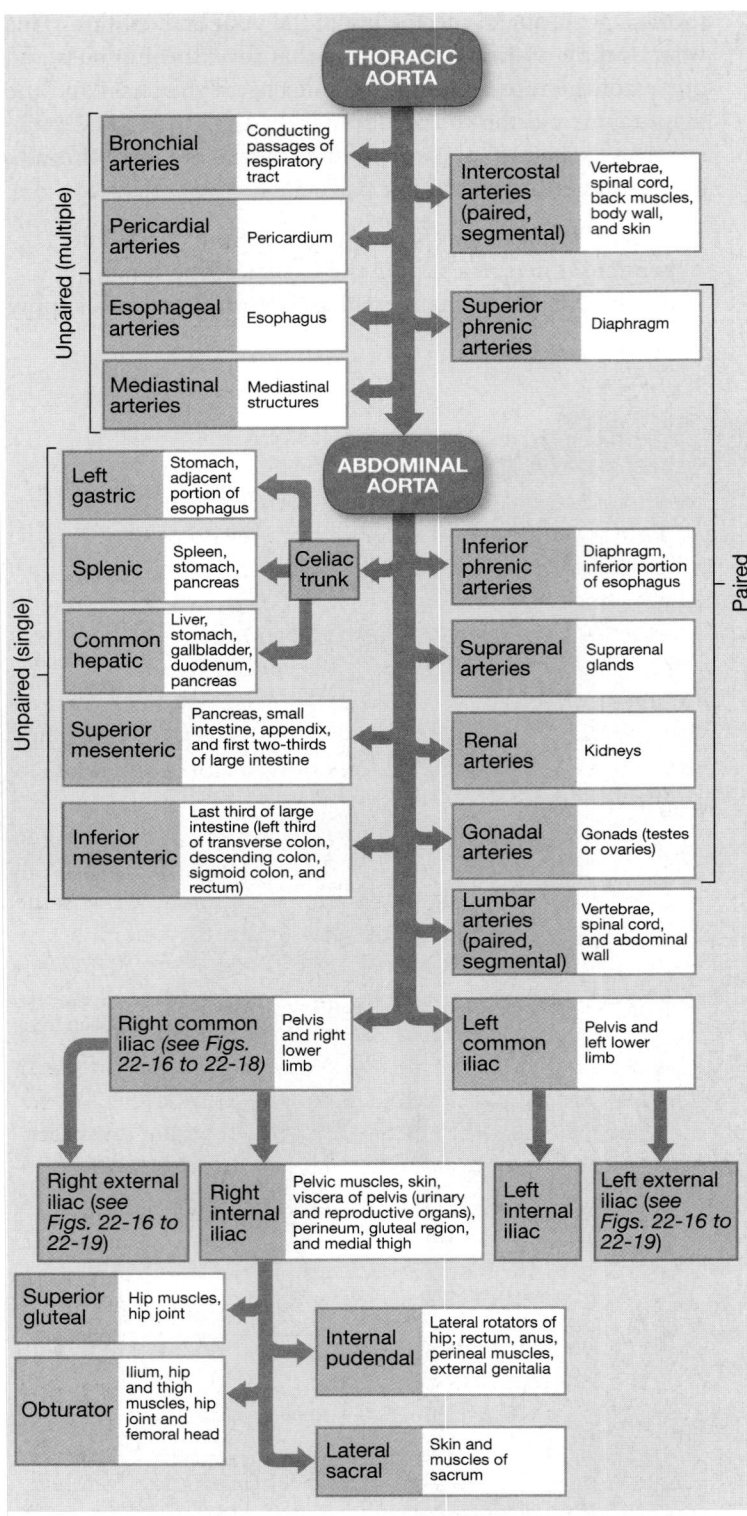

Figure 21–24   Major Arteries of the Trunk *(continued)*.
(b) A flowchart.

the pericardium, the **esophageal arteries** supply the esophagus, and the **mediastinal arteries** supply the tissues of the mediastinum.

- *Parietal branches* supply the chest wall. The **intercostal arteries** supply the chest muscles and the vertebral column area, and the **superior phrenic** (FREN-ik) **arteries** deliver blood to the superior surface of the diaphragm, which separates the thoracic and abdominopelvic cavities.

The branches of the thoracic aorta are shown in **Figure 21–24**.

The Abdominal Aorta. The abdominal aorta, which begins immediately inferior to the diaphragm, is a continuation of the thoracic aorta (**Figure 21–24a**). Descending slightly to the left of the vertebral column but posterior to the peritoneal cavity, the abdominal aorta is commonly surrounded by a cushion of adipose tissue. At the level of vertebra L_4, it splits into two major arteries—the *left* and *right common iliac arteries*—that supply deep pelvic structures and the lower limbs. The region where the abdominal aorta splits is called the *terminal segment of the aorta.*

The abdominal aorta delivers blood to all the abdominopelvic organs and structures. The major branches to visceral organs are unpaired; they arise on the anterior surface of the abdominal aorta and extend into the mesenteries. By contrast, branches to the body wall, the kidneys, the urinary bladder, and other structures outside the peritoneal cavity are paired, and originate along the lateral surfaces of the abdominal aorta. **Figure 21–24a** shows the major arteries of the trunk after the removal of most thoracic and abdominal organs. **Figure 21–25** shows the distribution of those arteries to abdominopelvic organs.

Tips&Tricks

The aorta resembles a walking cane: the ascending aorta, aortic arch, and the start of the descending aorta form the cane's handle, while the thoracic and abdominal segments of the descending aorta form the cane's shaft.

The abdominal aorta gives rise to three unpaired arteries (**Figures 21–24** and **21–25**).

1. The **celiac** (SĒ-lē-ak) **trunk** delivers blood to the liver, stomach, and spleen. The celiac trunk divides into three branches: (a) the **left gastric artery**, which supplies the stomach and the inferior portion of the esophagus, (b) the **splenic artery**, which supplies the spleen and arteries to the stomach (*left gastroepiploic artery*) and pancreas (*pancreatic arteries*), and (c) the **common hepatic**

artery, which supplies arteries to the liver (*hepatic artery proper*), stomach (*right gastric artery*), gallbladder (*cystic artery*), and duodenal area (*gastroduodenal, right gastroepiploic,* and *superior pancreaticoduodenal arteries*).

2. The **superior mesenteric** (mez-en-TER-ik) **artery** arises about 2.5 cm (1 in.) inferior to the celiac trunk to supply arteries to the pancreas and duodenum (*inferior pancreaticoduodenal artery*), small intestine (*intestinal arteries*), and most of the large intestine (*right* and *middle colic* and the *ileocolic arteries*).

3. The **inferior mesenteric artery** arises about 5 cm (2 in.) superior to the terminal aorta and delivers blood to the terminal portions of the colon (*left colic* and *sigmoid arteries*) and the rectum (*rectal arteries*).

The abdominal aorta also gives rise to five paired arteries (**Figure 21–24**):

1. The **inferior phrenic arteries**, which supply the inferior surface of the diaphragm and the inferior portion of the esophagus.

2. The **suprarenal arteries**, which originate on either side of the aorta near the base of the superior mesenteric artery. Each suprarenal artery supplies one suprarenal gland, which caps the superior part of a kidney.

3. The short (about 7.5 cm) **renal arteries**, which arise along the posterolateral surface of the abdominal aorta, about 2.5 cm (1 in.) inferior to the superior mesenteric artery, and travel posterior to the peritoneal lining to reach the suprarenal glands and kidneys. We will consider the branches of the renal arteries in Chapter 26.

4. The **gonadal** (gō-NAD-al) **arteries**, which originate between the superior and inferior mesenteric arteries. In males, they are called *testicular arteries* and are long, thin arteries that supply blood to the testes and scrotum. In females, they are termed *ovarian arteries* and supply blood to the ovaries, uterine tubes, and uterus. The distribution of gonadal vessels (both arteries and veins) differs by gender; we will describe the differences in Chapter 28.

5. Small **lumbar arteries**, which arise on the posterior surface of the aorta and supply the vertebrae, spinal cord, and abdominal wall.

Arteries of the Pelvis and Lower Limbs

Near the level of vertebra L$_4$, the terminal segment of the abdominal aorta divides to form a pair of elastic arteries—the **right** and **left common iliac** (IL-ē-ak) **arteries**—plus the small **median sacral artery** (**Figure 21–24**). The common iliac arteries, which carry blood to the pelvis and lower limbs, descend posterior to the cecum and sigmoid colon along the inner surface of the ilium. At the level of the lumbosacral joint, each common iliac divides to form an **internal iliac artery** and an **external iliac artery** (**Figure 21–25**). The internal iliac arteries enter the pelvic cavity to supply the urinary bladder, the internal and external walls of the pelvis, the external genitalia, the medial side of the thigh, and, in females, the uterus and vagina. The major tributaries of the internal iliac artery are the *gluteal, internal pudendal, obturator,* and *lateral sacral arteries*. The external iliac arteries supply blood to the lower limbs and are much larger in diameter than the internal iliac arteries.

Arteries of the Thigh and Leg. Each external iliac artery crosses the surface of an iliopsoas muscle and penetrates the abdominal wall midway between the anterior superior iliac spine and the pubic symphysis on that side. It emerges on the anterior, medial surface of the thigh as the **femoral artery** (**Figure 21–26a,b**). Roughly 5 cm (2 in.) distal to the emergence of the femoral artery, the **deep femoral artery** branches off its lateral surface. The deep femoral artery, which gives rise to the *femoral circumflex arteries*, supplies blood to the ventral and lateral regions of the skin and deep muscles of the thigh.

The femoral artery continues inferiorly and posterior to the femur. As it approaches the knee, it gives rise to the *descending genicular artery*, which supplies the area around the knee. At the popliteal fossa, posterior to the knee joint, the femoral artery becomes the **popliteal** (pop-LIT-ē-al) **artery**, which then branches to form the **posterior** and **anterior tibial arteries**. The posterior tibial artery gives rise to the **fibular artery**, or *peroneal* (perone, fibula) *artery*, before continuing inferiorly along the posterior surface of the tibia. The anterior tibial artery passes between the tibia and fibula, emerging on the anterior surface of the tibia. As it descends toward the foot, the anterior tibial artery provides blood to the skin and muscles of the anterior portion of the leg.

Arteries of the Foot. When it reaches the ankle, the anterior tibial artery becomes the **dorsalis pedis artery**, which then branches repeatedly, supplying the ankle and dorsal portion of the foot (**Figure 21–26a,b**).

As it reaches the ankle, the posterior tibial artery divides to form the **medial** and **lateral plantar arteries**, which supply blood to the plantar surface of the foot. These arteries are connected to the dorsalis pedis artery through a pair of anastomoses. The arrangement produces a **dorsal arch** (*arcuate arch*) and a **plantar arch**; small arteries branching off these arches supply the distal portions of the foot and the toes.

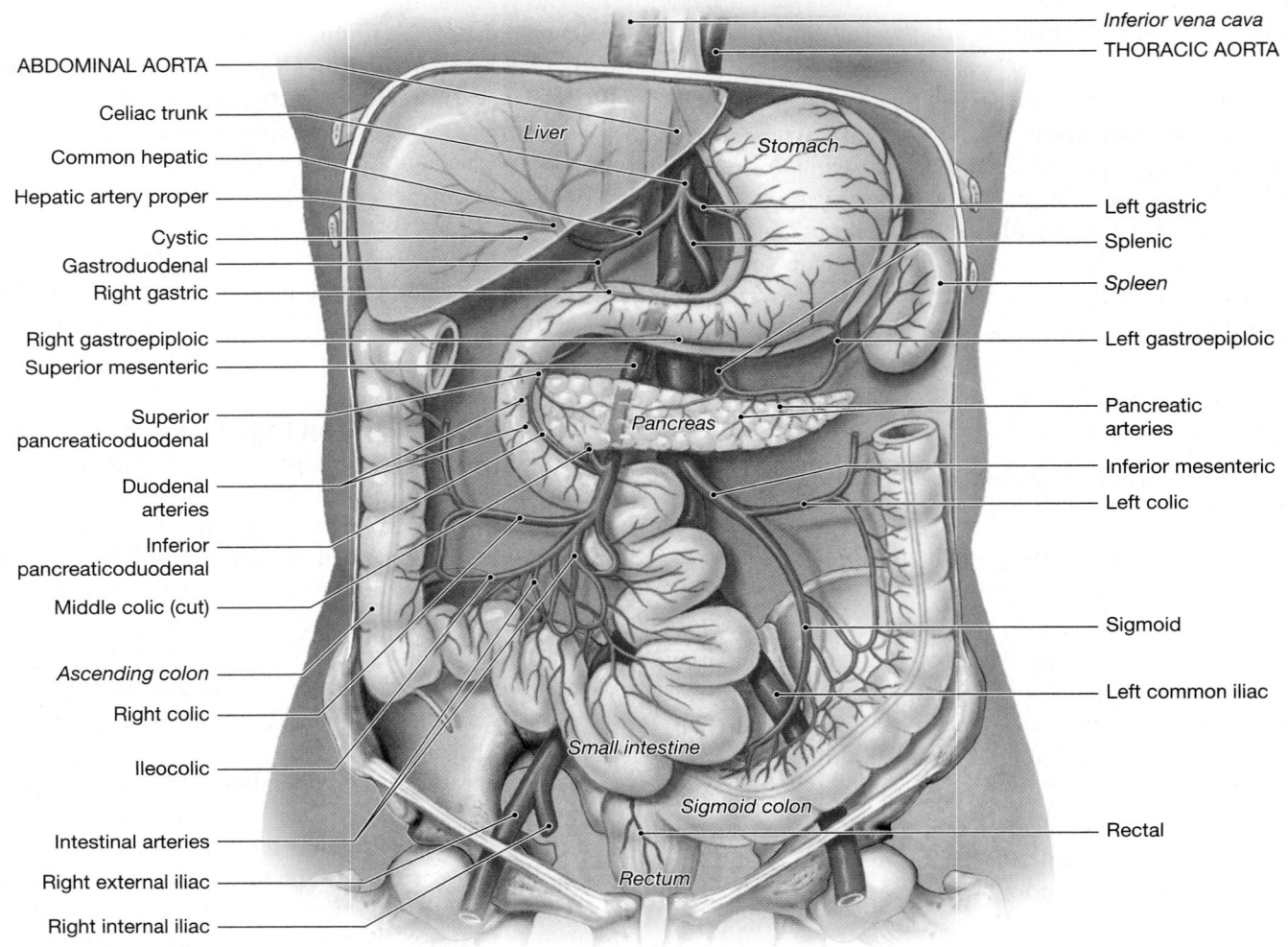

ABDOMINAL AORTA
Celiac trunk
Common hepatic
Hepatic artery proper
Cystic
Gastroduodenal
Right gastric
Right gastroepiploic
Superior mesenteric
Superior pancreaticoduodenal
Duodenal arteries
Inferior pancreaticoduodenal
Middle colic (cut)
Ascending colon
Right colic
Ileocolic
Intestinal arteries
Right external iliac
Right internal iliac

Liver
Stomach
Pancreas
Ascending colon
Small intestine
Sigmoid colon
Rectum

Inferior vena cava
THORACIC AORTA
Left gastric
Splenic
Spleen
Left gastroepiploic
Pancreatic arteries
Inferior mesenteric
Left colic
Sigmoid
Left common iliac
Rectal

Figure 21–25 **Arteries Supplying the Abdominopelvic Organs.** *(See also Figure 24–23, p. 912.)* ATLAS: Plates 53a–e; 54c; 55

Systemic Veins

Veins collect blood from each of the tissues and organs of the body by means of an elaborate venous network that drains into the right atrium of the heart via the superior and inferior venae cavae (**Figure 21–27**). The branching pattern of peripheral veins is much more variable than is the branching pattern of arteries. The discussion that follows is based on the most common arrangement of veins. Complementary arteries and veins commonly run side by side, and in many cases they have comparable names.

One significant difference between the arterial and venous systems concerns the distribution of major veins in the neck and limbs. Arteries in these areas are located deep beneath the skin, protected by bones and surrounding soft tissues. In contrast, the neck and limbs generally have two sets of peripheral veins, one superficial and the other deep. This

dual venous drainage is important for controlling body temperature. In hot weather, venous blood flows through superficial veins, where heat loss can occur; in cold weather, blood is routed to the deep veins to minimize heat loss.

The Superior Vena Cava

All the body's systemic veins (except the cardiac veins) ultimately drain into either the superior vena cava or the inferior vena cava. The **superior vena cava (SVC)** receives blood from the tissues and organs of the head, neck, chest, shoulders, and upper limbs.

Venous Return from the Cranium. Numerous veins drain the cerebral hemispheres. The *superficial cerebral veins* and small veins of the brain stem empty into a network of dural sinuses (**Figure 21–28a**), including the *superior* and *inferior sagittal sinuses*, the *petrosal sinuses*, the *occipital sinus*, the *left*

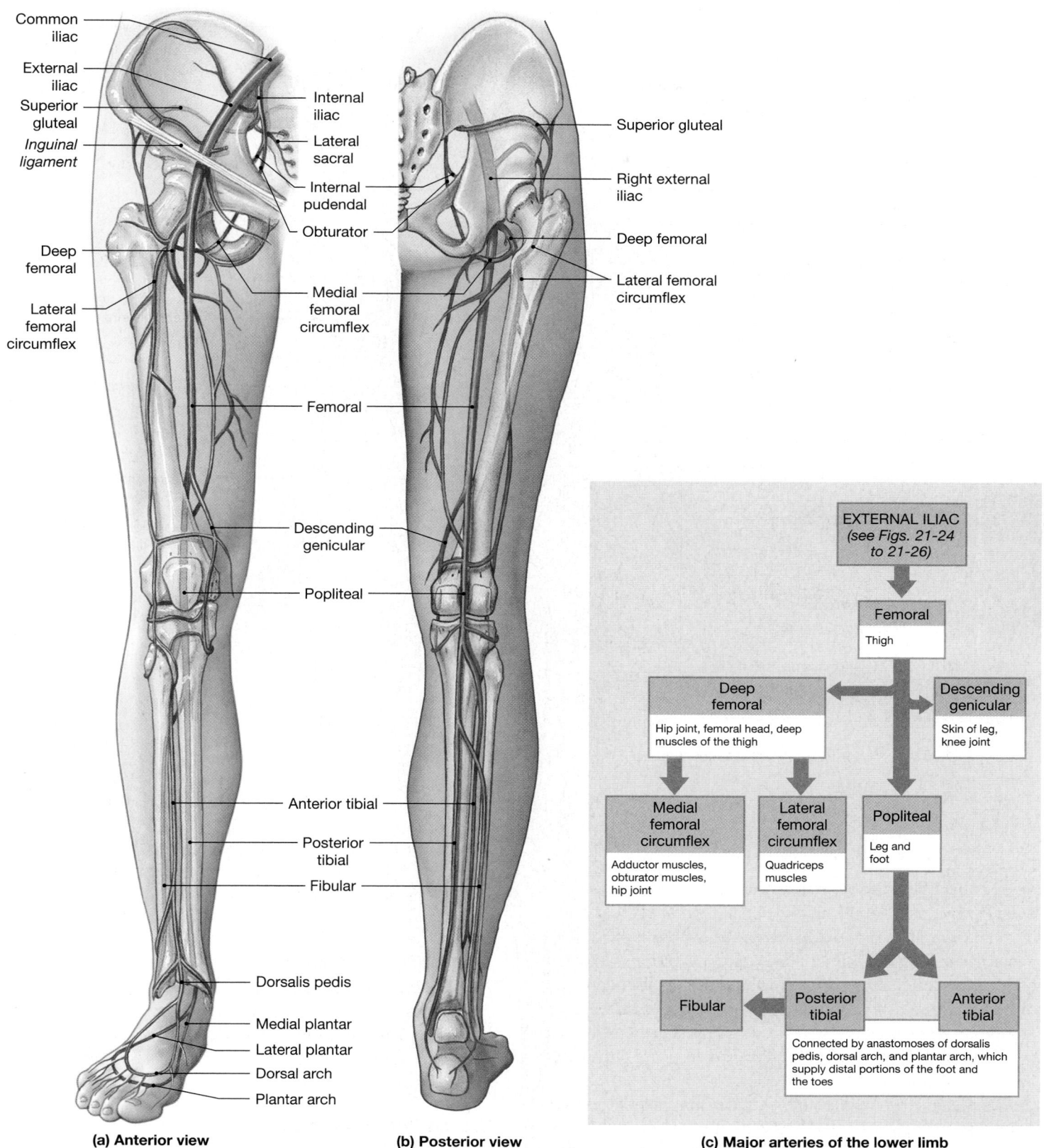

(a) Anterior view

(b) Posterior view

(c) Major arteries of the lower limb

Figure 21–26 **Arteries of the Lower Limb.** **(a)** An anterior view. **(b)** A posterior view. **(c)** A flowchart of blood flow to a lower limb.
ATLAS: Plates 68c; 70b; 78b–g

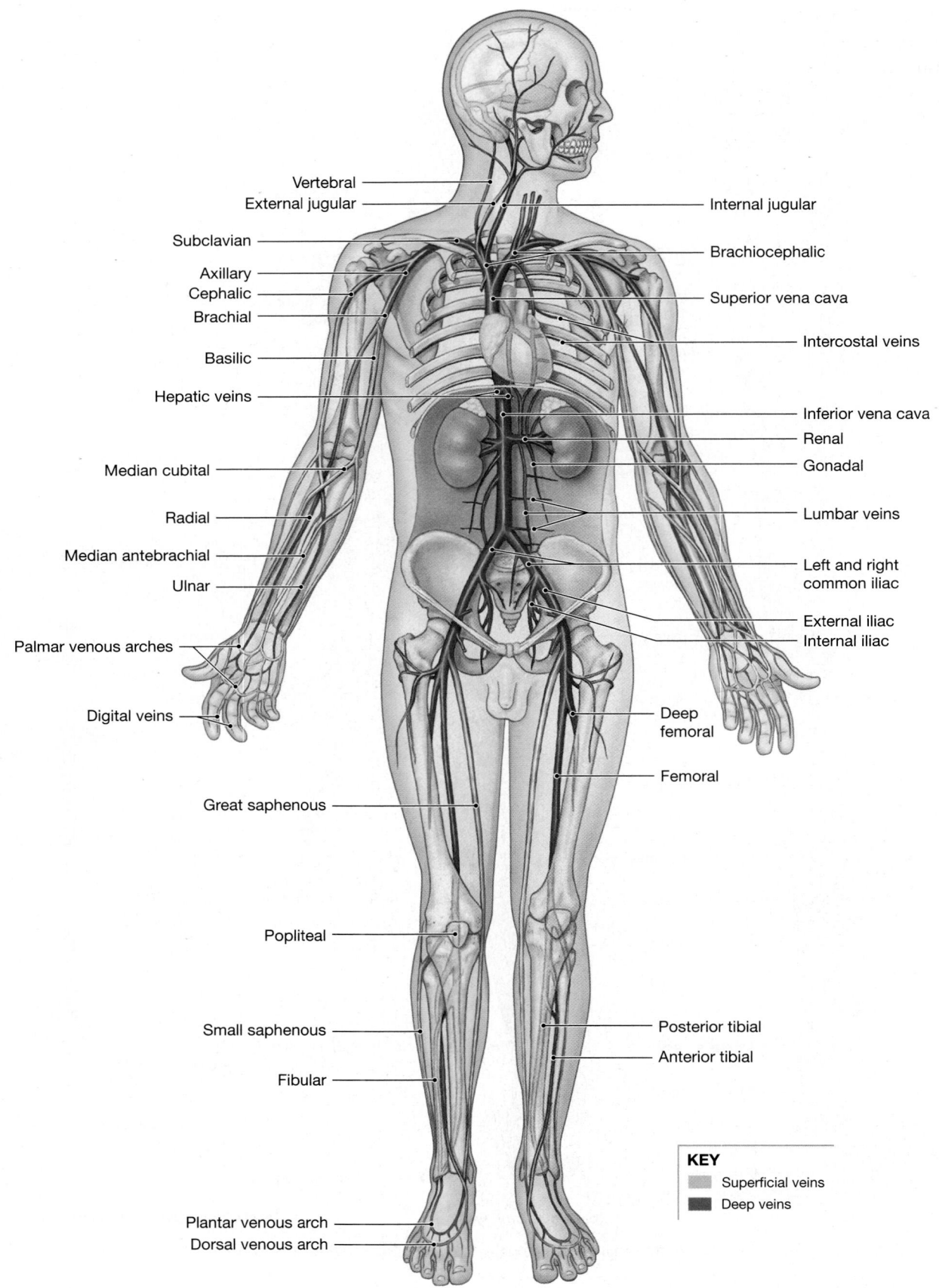

Vertebral
External jugular
Subclavian
Axillary
Cephalic
Brachial
Basilic
Hepatic veins
Median cubital
Radial
Median antebrachial
Ulnar
Palmar venous arches
Digital veins
Great saphenous
Popliteal
Small saphenous
Fibular
Plantar venous arch
Dorsal venous arch

Internal jugular
Brachiocephalic
Superior vena cava
Intercostal veins
Inferior vena cava
Renal
Gonadal
Lumbar veins
Left and right
common iliac
External iliac
Internal iliac
Deep
femoral
Femoral
Posterior tibial
Anterior tibial

KEY
Superficial veins
Deep veins

Figure 21–27   **An Overview of the Major Systemic Veins.**

Figure 21–28   Major Veins of the Head, Neck, and Brain.
(a) An inferior view of the brain, showing the venous distribution. For the relationship of these veins to meningeal layers, *see Figure 14–4, p. 467.* **(b)** Veins draining the brain and the superficial and deep portions of the head and neck. ATLAS: Plates 3c,d; 18a–c

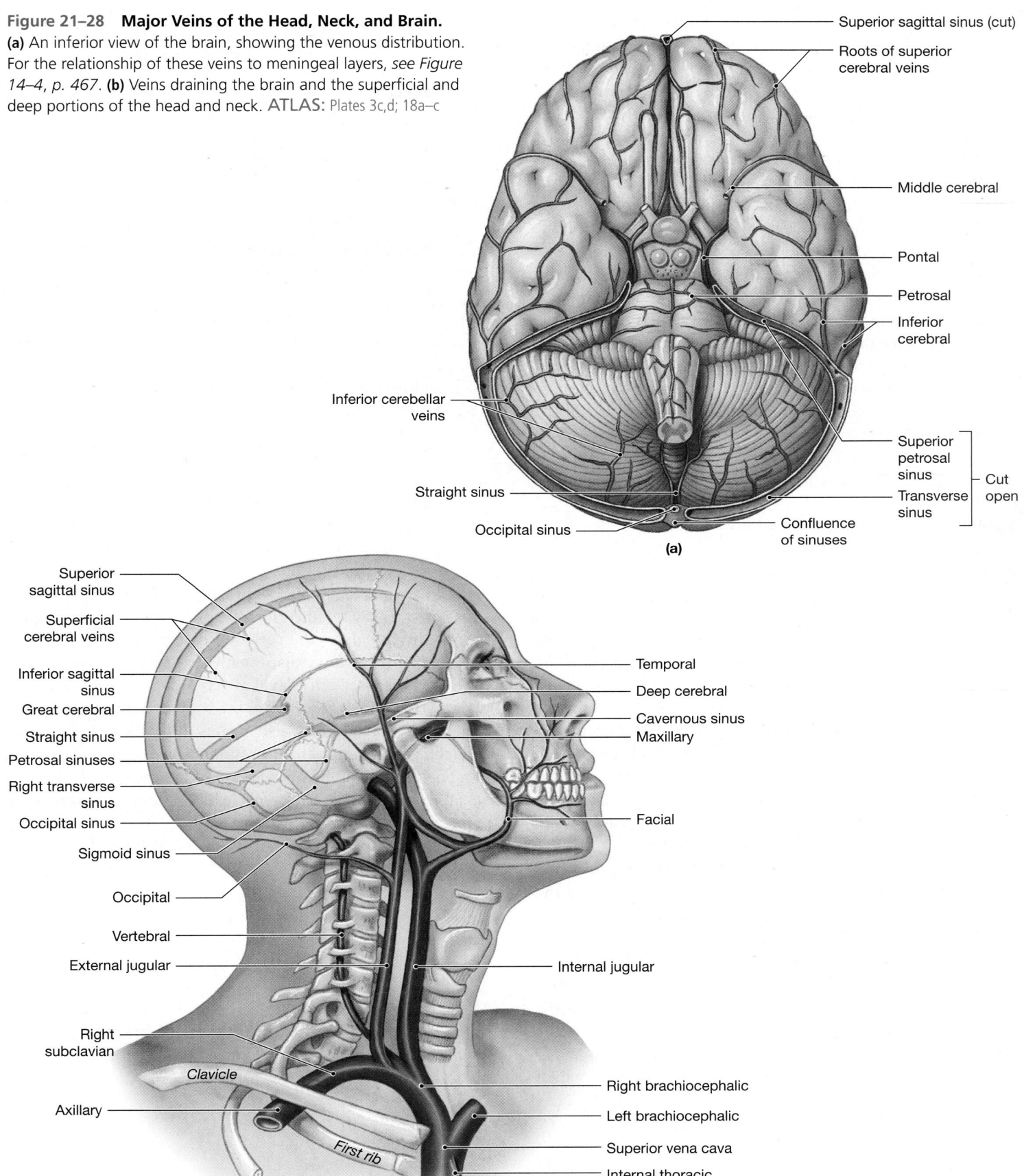

Superior sagittal sinus (cut)

Roots of superior cerebral veins

Middle cerebral

Pontal

Petrosal

Inferior cerebral

Inferior cerebellar veins

Superior petrosal sinus — Cut open

Straight sinus

Transverse sinus

Occipital sinus

Confluence of sinuses

(a)

Superior sagittal sinus

Superficial cerebral veins

Inferior sagittal sinus

Great cerebral

Straight sinus

Petrosal sinuses

Right transverse sinus

Occipital sinus

Sigmoid sinus

Occipital

Vertebral

External jugular

Right subclavian

Clavicle

Axillary

First rib

Temporal

Deep cerebral

Cavernous sinus

Maxillary

Facial

Internal jugular

Right brachiocephalic

Left brachiocephalic

Superior vena cava

Internal thoracic

(b)

21 CARDIOVASCULAR

and *right transverse sinuses*, and the *straight sinus* (**Figure 21–28b**). The largest, the **superior sagittal sinus**, is in the falx cerebri (**Figure 14–4**, p. 467). Most of the *inferior cerebral veins* converge within the brain to form the **great cerebral vein**, which delivers blood from the interior of the cerebral hemispheres and the choroid plexus to the **straight sinus**. Other cerebral veins drain into the **cavernous sinus** with numerous small veins from the orbit. Blood from the cavernous sinus reaches the internal jugular vein through the petrosal sinuses.

The venous sinuses converge within the dura mater in the region of the lambdoid suture. The left and right transverse sinuses begin at the confluence of the occipital, sagittal, and straight sinuses. Each transverse sinus drains into a **sigmoid sinus**, which penetrates the jugular foramen and leaves the skull as the **internal jugular vein**, descending parallel to the common carotid artery in the neck (p. 752).

Vertebral veins drain the cervical spinal cord and the posterior surface of the skull. These vessels descend within the transverse foramina of the cervical vertebrae, in company with the vertebral arteries. The vertebral veins empty into the *brachiocephalic veins* of the chest (discussed later in the chapter).

Superficial Veins of the Head and Neck. The superficial veins of the head converge to form the **temporal**, **facial**, and **maxillary veins** (**Figure 21–28b**). The temporal vein and the maxillary vein drain into the **external jugular vein**. The facial vein drains into the internal jugular vein. A broad anastomosis between the external and internal jugular veins at the angle of the mandible provides dual venous drainage of the face, scalp, and cranium. The external jugular vein descends toward the chest just deep to the skin on the anterior surface of the sternocleidomastoid muscle. Posterior to the clavicle, the external jugular vein empties into the *subclavian vein*. In healthy individuals, the external jugular vein is easily palpable, and a *jugular venous pulse* (*JVP*) is sometimes detectable at the base of the neck.

Venous Return from the Upper Limbs. The **digital veins** empty into the **superficial** and **deep palmar veins** of the hand, which are interconnected to form the **palmar venous arches** (**Figure 21–29**). The superficial arch empties into the **cephalic vein**, which ascends along the radial side of the forearm; the **median antebrachial vein**; and the **basilic vein**, which ascends on the ulnar side. Anterior to the elbow is the superficial **median cubital vein**, which passes from the cephalic vein, medially and at an oblique angle, to connect to the basilic vein. (The median cubital is the vein from which

venous blood samples are typically collected.) From the elbow, the basilic vein passes superiorly along the medial surface of the biceps brachii muscle.

The deep palmar veins drain into the **radial vein** and the **ulnar vein**. After crossing the elbow, these veins fuse to form the **brachial vein**, running parallel to the brachial artery. As the brachial vein continues toward the trunk, it merges with the basilic vein and becomes the **axillary vein**, which enters the axilla.

Formation of the Superior Vena Cava. The cephalic vein joins the axillary vein on the lateral surface of the first rib, forming the **subclavian vein**, which continues into the chest. The subclavian vein passes superior to the first rib and along the superior margin of the clavicle, to merge with the external and internal jugular veins of that side. This fusion creates the **brachiocephalic vein**, or *innominate vein*, which penetrates the body wall and enters the thoracic cavity.

Each brachiocephalic vein receives blood from the **vertebral vein** of the same side, which drains the back of the skull and spinal cord. Near the heart, at the level of the first and second ribs, the left and right brachiocephalic veins combine, creating the superior vena cava. Close to the point of fusion, the **internal thoracic vein** empties into the brachiocephalic vein.

The **azygos** (AZ-i-gos) **vein** is the major tributary of the superior vena cava. This vein ascends from the lumbar region over the right side of the vertebral column to enter the thoracic cavity through the diaphragm. The azygos vein joins the superior vena cava at the level of vertebra T_2. On the left side, the azygos receives blood from the smaller **hemiazygos vein**, which in many people also drains into the left brachiocephalic vein through the *highest intercostal vein*.

The azygos and hemiazygos veins are the chief collecting vessels of the thorax. They receive blood from (1) **intercostal veins**, which in turn receive blood from the chest muscles; (2) **esophageal veins**, which drain blood from the inferior portion of the esophagus; and (3) smaller veins draining other mediastinal structures.

Figure 21–30a diagrams the venous tributaries of the superior vena cava.

The Inferior Vena Cava

The **inferior vena cava (IVC)** collects most of the venous blood from organs inferior to the diaphragm. (A small amount reaches the superior vena cava via the azygos and hemiazygos veins.)

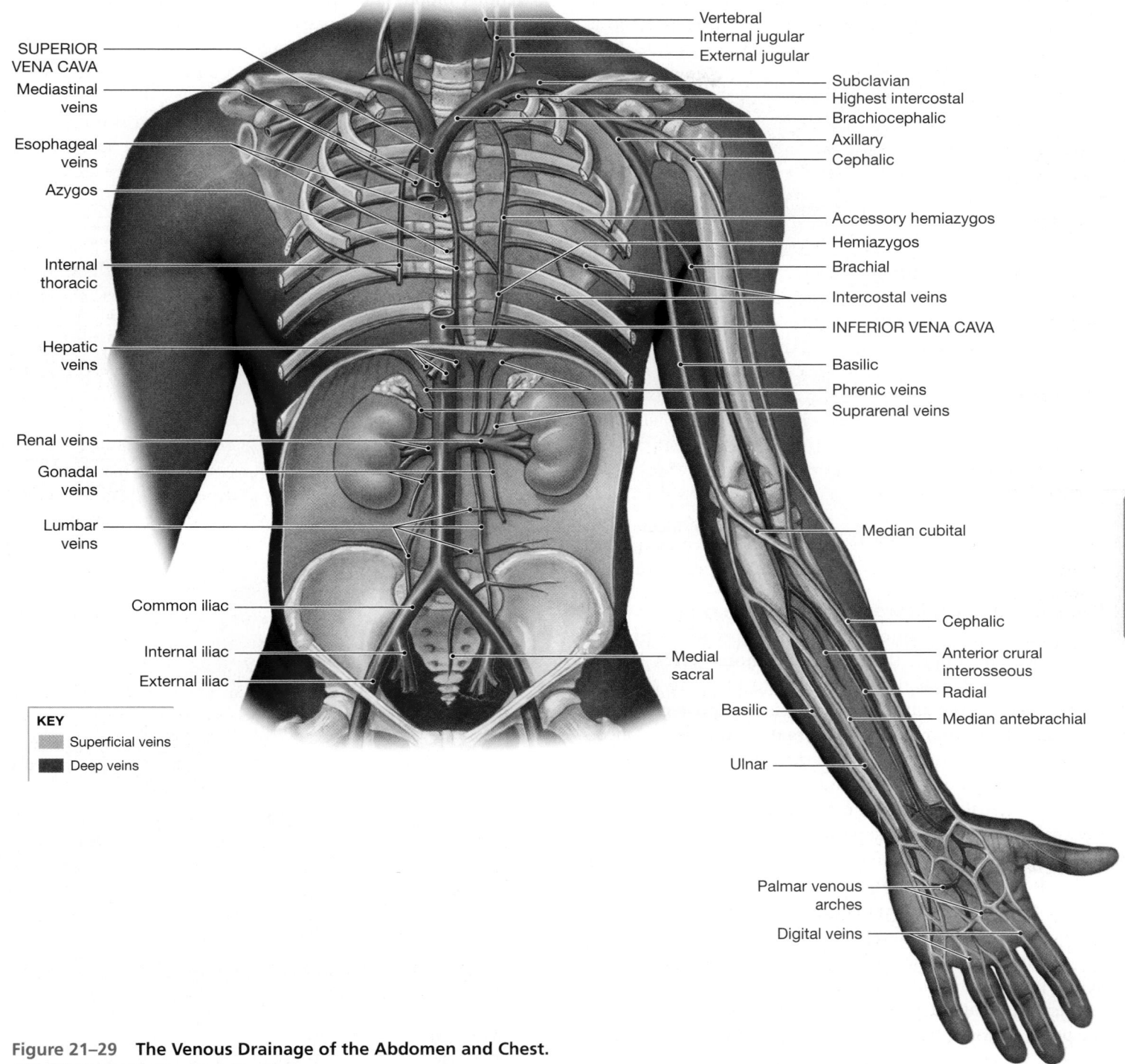

Figure 21–29   The Venous Drainage of the Abdomen and Chest.
ATLAS: Plates 27c; 29c; 47b,d; 61a; 62a,b

21 CARDIOVASCULAR

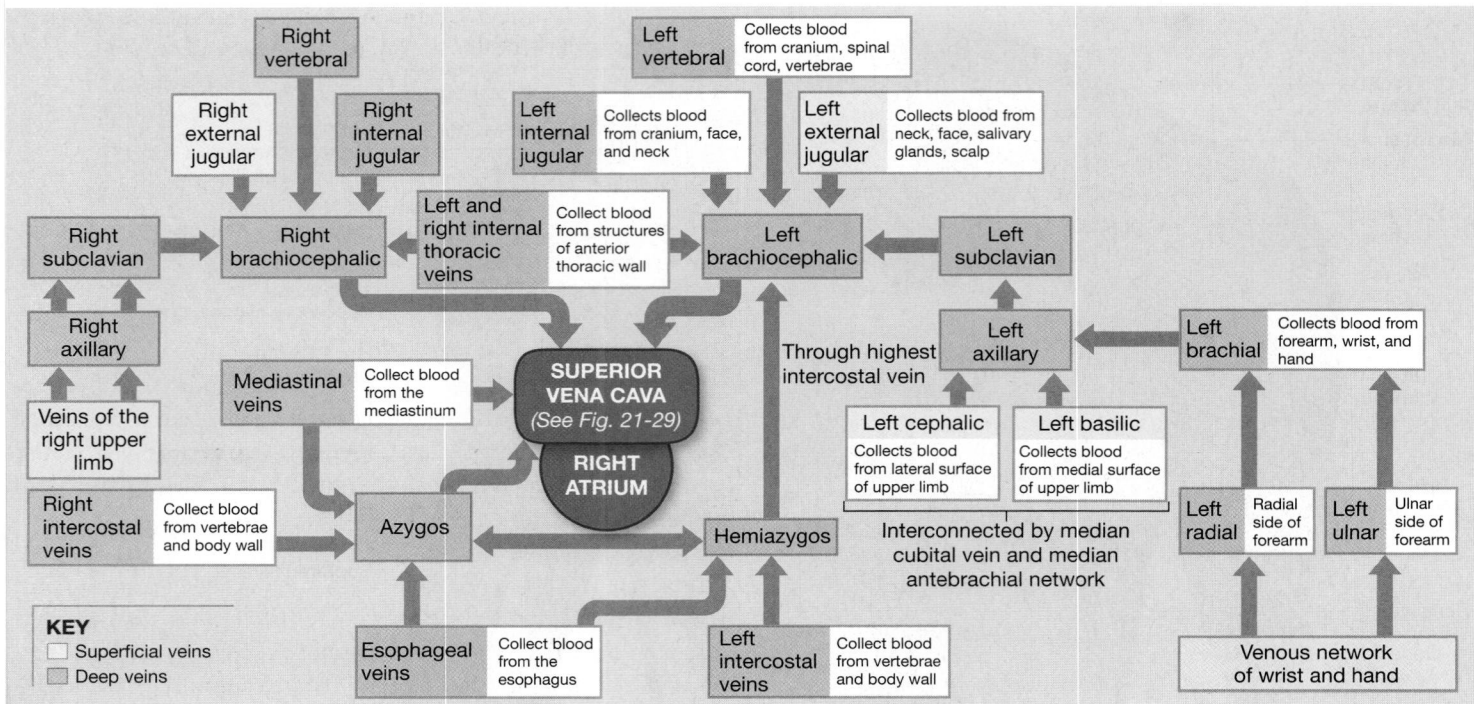

(a) Tributaries of the superior vena cava

Veins Draining the Lower Limbs. Blood leaving capillaries in the sole of each foot collects into a network of **plantar veins**, which supply the **plantar venous arch** (**Figure 21–31a**). The plantar network provides blood to the deep veins of the leg: the **anterior tibial vein**, the **posterior tibial vein**, and the **fibular vein**. The **dorsal venous arch** collects blood from capillaries on the superior surface of the foot and the **digital veins** of the toes. The plantar arch and the dorsal arch are extensively interconnected, and the path of blood flow can easily shift from superficial to deep veins.

The dorsal venous arch is drained by two superficial veins: the **great saphenous** (sa-FĒ-nus; *saphenes*, prominent) **vein** and the **small saphenous vein**. The great saphenous vein ascends along the medial aspect of the leg and thigh, draining into the *femoral vein* near the hip joint. The small saphenous vein arises from the dorsal venous arch and ascends along the posterior and lateral aspect of the calf. This vein then enters the popliteal fossa, where it meets the **popliteal vein**, formed by the union of the fibular and both tibial veins (**Figure 21–31b**). The popliteal vein is easily palpated in the popliteal fossa adjacent to the adductor magnus muscle. At the femur, the popliteal vein becomes the **femoral vein**, which ascends along the thigh, next to the femoral artery. Immediately before penetrating the abdominal wall, the femoral vein receives blood from (1) the great saphenous vein; (2) the **deep femoral vein**, which collects blood from deeper structures in the thigh; and (3) the **femoral circumflex vein**, which drains the region

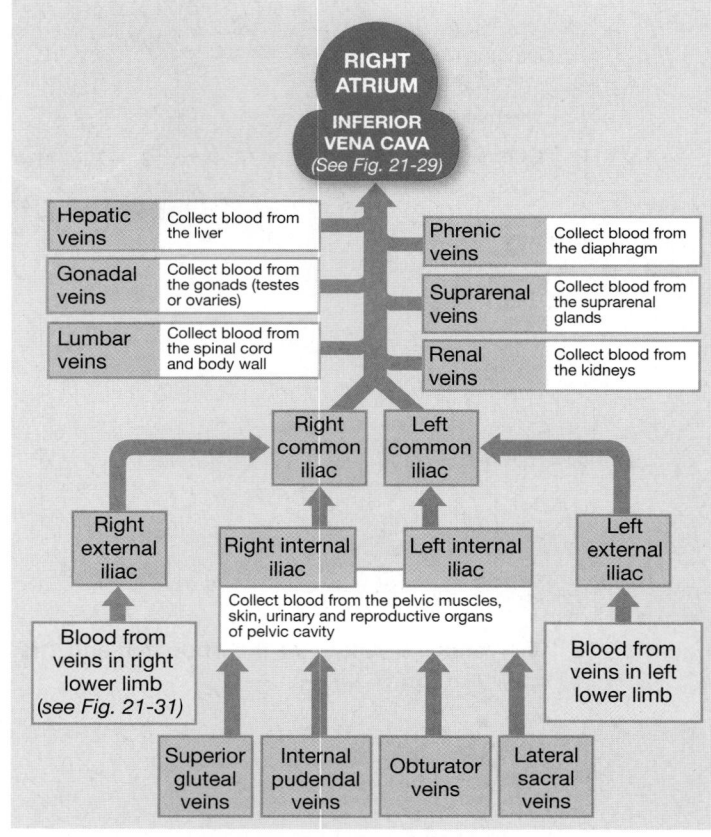

(b) Tributaries of the inferior vena cava

Figure 21–30 Flowcharts of Circulation to the Superior and Inferior Venae Cavae.

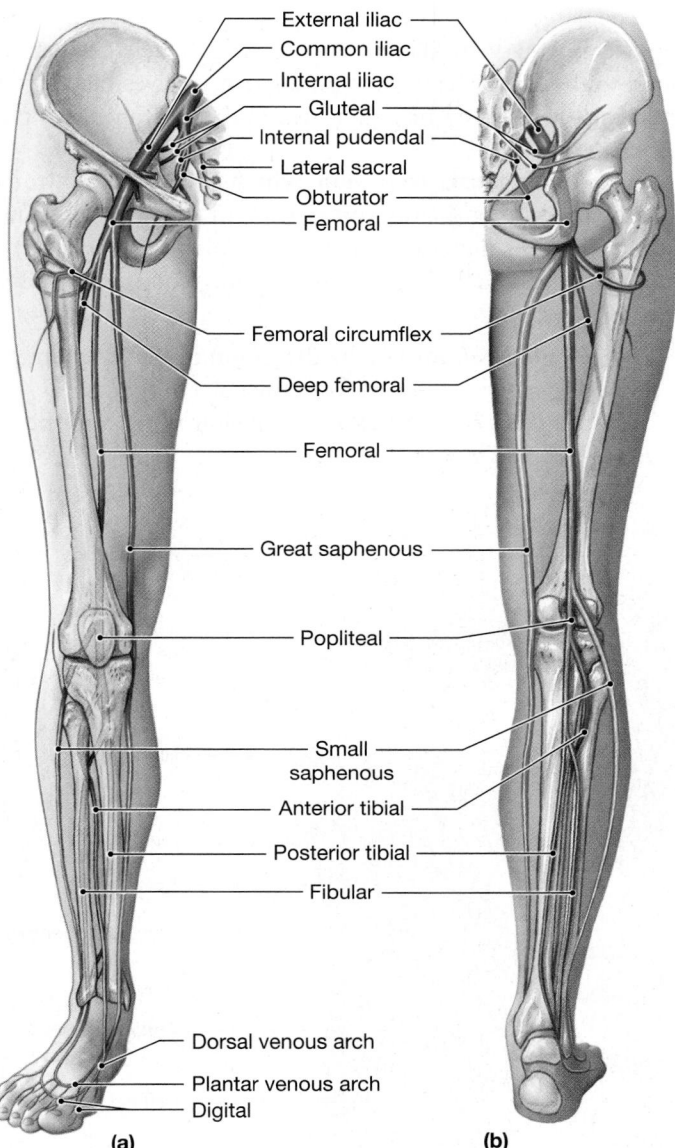

(a)

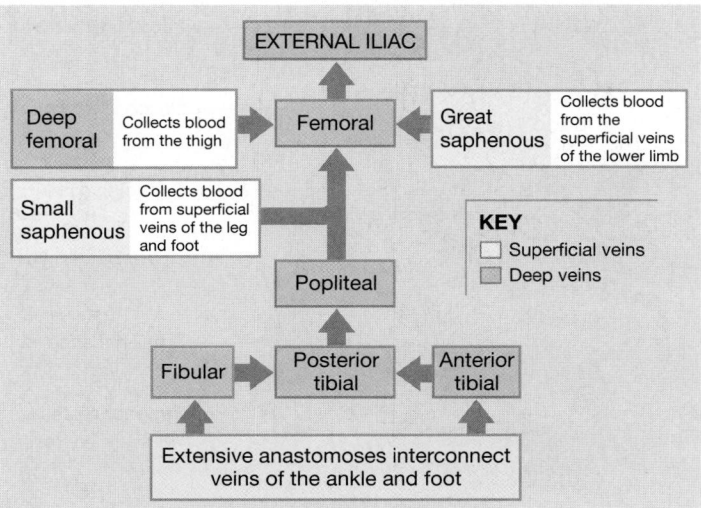

(c)

Figure 21–31 Venous Drainage from the Lower Limb. (a) An anterior view. **(b)** A posterior view. **(c)** A flowchart of venous circulation to a lower limb. ATLAS: Plates 70b; 74; 78a–g

around the neck and head of the femur. The femoral vein penetrates the body wall and emerges in the pelvic cavity as the **external iliac vein**.

Veins Draining the Pelvis. The external iliac veins receive blood from the lower limbs, the pelvis, and the lower abdomen. As the left and right external iliac veins cross the inner surface of the ilium, they are joined by the **internal iliac veins**, which drain the pelvic organs (**Figure 21–30**). The internal iliac veins are formed by the fusion of the *gluteal*, *internal pudendal*, *obturator*, and *lateral sacral veins* (**Figure 21–31a**). The union of external and internal iliac veins forms the **common iliac vein**, the right and left branches of which ascend at an oblique angle. The left common iliac vein receives blood from the *median sacral vein*, which drains the area supplied by the middle sacral artery (**Figure 21–29**). Anterior to vertebra L_5, the common iliac veins unite to form the inferior vena cava.

Veins Draining the Abdomen. The inferior vena cava ascends posterior to the peritoneal cavity, parallel to the aorta. The abdominal portion of the inferior vena cava collects blood from six major veins (**Figures 21–29** and **21–30b**):

1. **Lumbar veins** drain the lumbar portion of the abdomen, including the spinal cord and body wall muscles. Superior branches of these veins are connected to the azygos vein (right side) and hemiazygos vein (left side), which empty into the superior vena cava.

2. **Gonadal** (*ovarian* or *testicular*) **veins** drain the ovaries or testes. The right gonadal vein empties into the inferior vena cava; the left gonadal vein generally drains into the left renal vein.

3. **Hepatic veins** from the liver empty into the inferior vena cava at the level of vertebra T_{10}.

4. **Renal veins**, the largest tributaries of the inferior vena cava, collect blood from the kidneys.

5. **Suprarenal veins** drain the suprarenal glands. In most people, only the right suprarenal vein drains into the inferior vena cava; the left suprarenal vein drains into the left renal vein.

6. **Phrenic veins** drain the diaphragm. Only the right phrenic vein drains into the inferior vena cava; the left drains into the left renal vein.

Figure 21–30b diagrams the tributaries of the inferior vena cava.

The Hepatic Portal System

The **hepatic portal system** (**Figure 21–32**) begins in the capillaries of the digestive organs and ends in the liver sinusoids. (As you may recall from Chapter 18, a blood vessel connecting two capillary beds is called a *portal vessel*; the network is a *portal system*.) Blood flowing in the hepatic portal system is quite different from that in other systemic veins, because hepatic portal vessels contain substances absorbed by the stomach and intestines. For example, levels of blood glucose and amino acids in the hepatic portal vein often exceed those found anywhere else in the cardiovascular system. The hepatic portal system delivers these and other absorbed compounds directly to the liver for storage, metabolic conversion, or excretion.

The largest vessel of the hepatic portal system is the **hepatic portal vein** (**Figure 21–32**), which delivers venous blood to the liver. The hepatic portal vein receives blood from three large veins draining organs within the peritoneal cavity:

- The **inferior mesenteric vein**, which collects blood from capillaries along the inferior portion of the large intestine. Its tributaries include the *left colic vein* and the *superior rectal veins*, which drain the descending colon, sigmoid colon, and rectum.

- The **splenic vein**, formed by the union of the inferior mesenteric vein and veins from the spleen, the lateral border of the stomach (*left gastroepiploic vein*), and the pancreas (*pancreatic veins*).

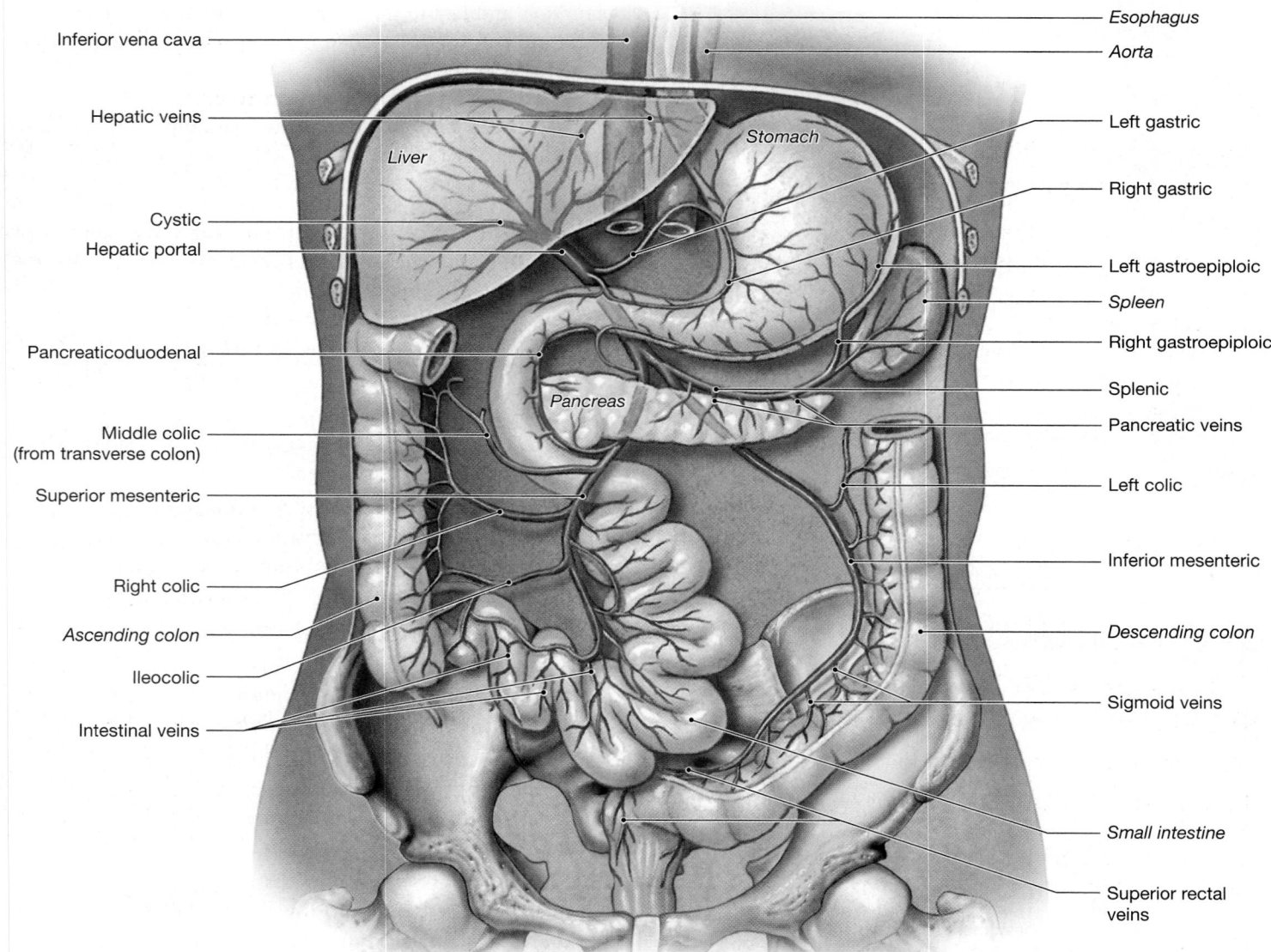

Figure 21–32 The Hepatic Portal System. *See also Figure 24–23, p. 912.* ATLAS: Plates 53b; 54a–c; 55; 57a,b

- The **superior mesenteric vein**, which collects blood from veins draining the stomach (*right gastroepiploic vein*), the small intestine (*intestinal* and *pancreaticoduodenal veins*), and two-thirds of the large intestine (*ileocolic, right colic,* and *middle colic veins*).

The hepatic portal vein forms through the fusion of the superior mesenteric and splenic veins. The superior mesenteric vein normally contributes the greater volume of blood and most of the nutrients. As it proceeds, the hepatic portal vein receives blood from the left and right **gastric veins**, which drain the medial border of the stomach, and from the **cystic vein**, emanating from the gallbladder.

After passing through liver sinusoids, blood collects in the hepatic veins, which empty into the inferior vena cava. Because blood from the intestines goes to the liver first, and because the liver regulates the nutrient content of the blood before it enters the inferior vena cava, the composition of the blood in the systemic circuit is relatively stable despite changes in diet and digestive activity.

CHECKPOINT

19. A blockage of which branch from the aortic arch would interfere with blood flow to the left arm?

20. Why would compression of the common carotid arteries cause a person to lose consciousness?

21. Grace is in an automobile accident, and her celiac trunk is ruptured. Which organs will be affected most directly by this injury?

22. Whenever Tim gets angry, a large vein bulges in the lateral region of his neck. Which vein is this?

23. A thrombus that blocks the popliteal vein would interfere with blood flow in which other veins?

See the blue Answers tab at the end of the book.

21-8 Modifications of fetal and maternal cardiovascular systems promote the exchange of materials, and independence is achieved at birth

The fetal and adult cardiovascular systems exhibit significant differences, reflecting different sources of respiratory and nutritional support. Most strikingly, the embryonic lungs are collapsed and nonfunctional, and the digestive tract has nothing to digest. The nutritional and respiratory needs of the fetus are provided by diffusion across the placenta.

Placental Blood Supply

Fetal patterns of blood flow are diagrammed in **Figure 21–33a**. Blood flow to the placenta is provided by a pair of **umbilical arteries**, which arise from the internal iliac arteries and enter the umbilical cord. Blood returns from the placenta in the single **umbilical vein**, bringing oxygen and nutrients to the developing fetus. The umbilical vein drains into the **ductus venosus**, a vascular connection to an intricate network of veins within the developing liver. The ductus venosus collects blood from the veins of the liver and from the umbilical vein, and empties into the inferior vena cava. When the placental connection is broken at birth, blood flow ceases along the umbilical vessels, and they soon degenerate. However, remnants of these vessels persist throughout life as fibrous cords.

Fetal Circulation in the Heart and Great Vessels

One of the most interesting aspects of circulatory development reflects the differences between the life of an embryo or fetus and that of an infant. Throughout embryonic and fetal life, the lungs are collapsed; yet after delivery, the newborn infant must be able to extract oxygen from inspired air rather than across the placenta. ATLAS: Embryology Summary 16: The Development of the Cardiovascular System

Although the interatrial and interventricular septa develop early in fetal life, the interatrial partition remains functionally incomplete until birth. The **foramen ovale**, or *interatrial opening*, is associated with a long flap that acts as a valve. Blood can flow freely from the right atrium to the left atrium, but any backflow will close the valve and isolate the two chambers from one another. Thus, blood entering the heart at the right atrium can bypass the pulmonary circuit. A second short-circuit exists between the pulmonary and aortic trunks. This connection, the **ductus arteriosus**, consists of a short, muscular vessel.

With the lungs collapsed, the capillaries are compressed and little blood flows through the lungs. During diastole, blood enters the right atrium and flows into the right ventricle, but it also passes into the left atrium through the foramen ovale. About 25 percent of the blood arriving at the right atrium bypasses the pulmonary circuit in this way. In addition, more than 90 percent of the blood leaving the right ventricle passes through the ductus arteriosus and enters the systemic circuit rather than continuing to the lungs.

Cardiovascular Changes at Birth

At birth, dramatic changes occur. When an infant takes the first breath, the lungs expand, and so do the pulmonary vessels. The resistance in the pulmonary circuit declines suddenly, and

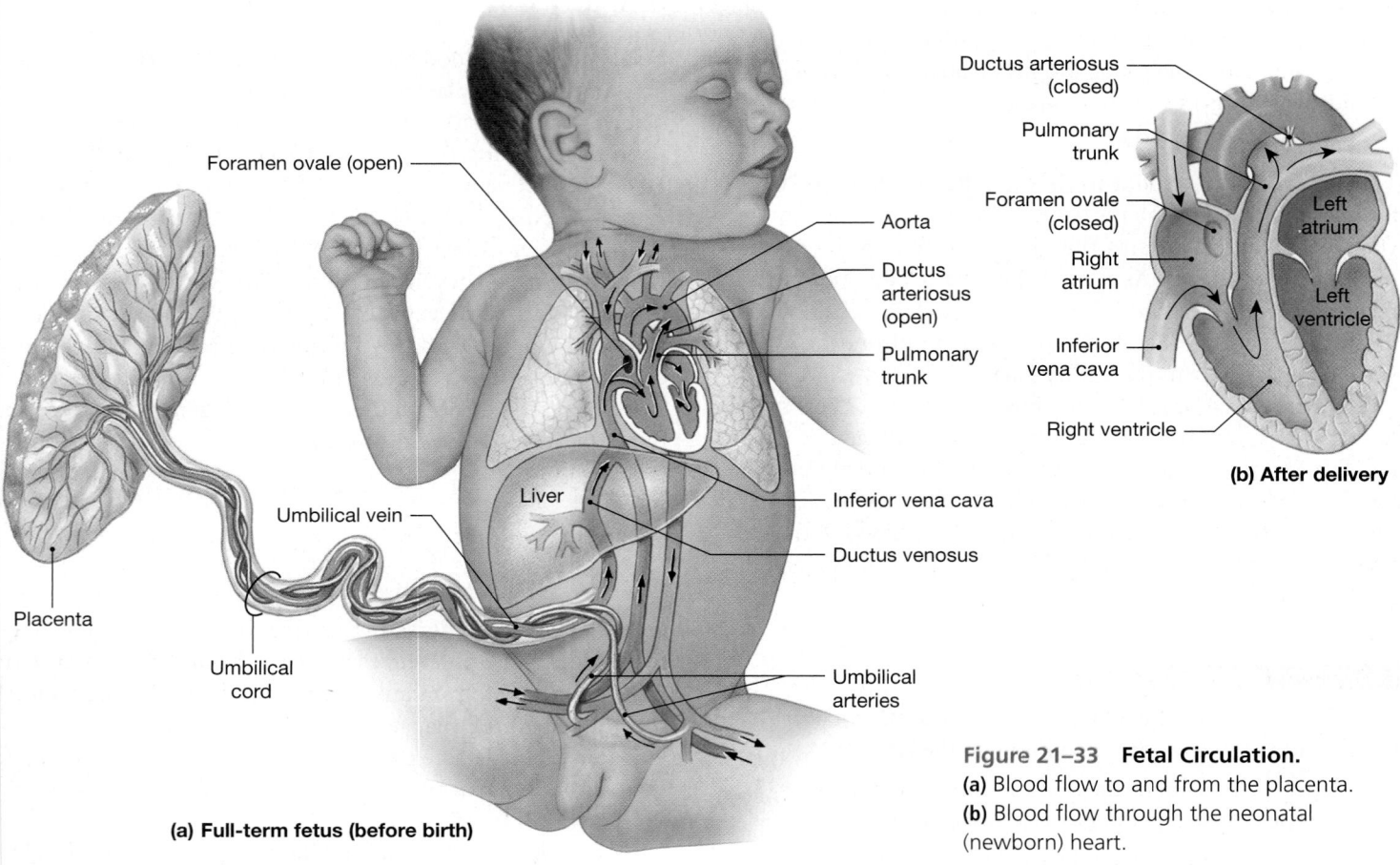

Figure 21–33 labels (a) Full-term fetus (before birth):
- Foramen ovale (open)
- Aorta
- Ductus arteriosus (open)
- Pulmonary trunk
- Inferior vena cava
- Ductus venosus
- Umbilical arteries
- Liver
- Umbilical vein
- Placenta
- Umbilical cord

Figure 21–33 labels (b) After delivery:
- Ductus arteriosus (closed)
- Pulmonary trunk
- Foramen ovale (closed)
- Right atrium
- Left atrium
- Inferior vena cava
- Left ventricle
- Right ventricle

(a) Full-term fetus (before birth)

Figure 21–33   Fetal Circulation.
(a) Blood flow to and from the placenta.
(b) Blood flow through the neonatal (newborn) heart.

blood rushes into the pulmonary vessels. Within a few seconds, rising O_2 levels stimulate the constriction of the ductus arteriosus, isolating the pulmonary and aortic trunks from one another. As pressures rise in the left atrium, the valvular flap closes the foramen ovale. In adults, the interatrial septum bears the *fossa ovalis*, a shallow depression that marks the site of the foramen ovale. (**Figure 20–6a,c**, p. 688.) The remnants of the ductus arteriosus persist throughout life as the *ligamentum arteriosum*, a fibrous cord.

If the proper circulatory changes do not occur at birth or shortly thereafter, problems will eventually develop. The severity of the problems depends on which connection remains open and on the size of the opening. Treatment may involve surgical closure of the foramen ovale, the ductus arteriosus, or both. Other forms of congenital heart defects result from abnormal cardiac development or inappropriate connections between the heart and major arteries and veins. See the following Clinical Note.

CLINICAL NOTE

Congenital Cardiovascular Problems

Minor individual variations in the vascular network are quite common. For example, very few individuals have identical patterns of venous distribution. Congenital cardiovascular problems serious enough to threaten homeostasis are relatively rare. They generally reflect abnormal formation of the heart or problems with the interconnections between the heart and the great vessels. Several examples of congenital cardiovascular defects are illustrated in **Figure 21–34**. All

these conditions can be surgically corrected, although multiple surgeries may be required.

If the foramen ovale remains open, or *patent*, blood may pass, or shunt, between the atria either right-to-left or left-to-right depending on relative atrial pressures (**Figure 21–34a**). A right-to-left shunt may allow clots in the systemic venous blood, which are normally trapped and removed at the lungs, to enter the systemic arteries and cause blockages, such as strokes. A left-to-right shunt recirculates blood through the lungs, increasing the workload on the heart, elevating pulmonary pressures, and potentially reducing lung function. Up to one in four persons may have a patent foramen ovale, but because shunting is intermittent or small, most have no problems, and the condition is not diagnosed. The abnormality may not be immediately apparent, but pulmonary hypertension, pulmonary edema, and cardiac enlargement are eventual results. If the ductus arteriosus remains open, higher systemic pressures create a left-to-right shunt which, if large enough, overloads the pulmonary circuit and the heart. If the ductus arteriosus remains open but valve defects, constricted pulmonary vessels, or other abnormalities occur as well, pulmonary pressures can rise enough to force blood into the systemic circuit. This sends poorly oxygenated blood back into the systemic circulation. Because normal blood oxygenation does not occur, the circulating blood develops a deep red color. The skin and lips then develop the blue tones typical of *cyanosis,* and the infant is known as a "blue baby."

Ventricular septal defects are openings in the interventricular septum (**Figure 21–34b**). These are the most common congenital heart problems, affecting 0.12 percent of newborn infants. When the more powerful left ventricle beats, it ejects blood into the right ventricle

and pulmonary circuit. The end result is a left-to-right shunt, which may lead to pulmonary hypertension, pulmonary edema, and cardiac enlargement.

The *tetralogy of Fallot* (fa-LŌ) is a complex group of heart and circulatory defects that affect 0.10 percent of newborn infants. In this condition, (1) the pulmonary trunk is abnormally narrow (pulmonary *stenosis*), (2) the interventricular septum is incomplete, (3) the aorta originates where the interventricular septum normally ends, and (4) the right ventricle is enlarged and both ventricles are thickened owing to increased workloads (**Figure 21–34c**).

In the *transposition of great vessels*, the aorta is connected to the right ventricle, and the pulmonary artery is connected to the left ventricle (**Figure 21–34d**). This malformation affects 0.05 percent of newborn infants.

In an *atrioventricular septal defect*, both the atria and ventricles are incompletely separated (**Figure 21–34e**). The results are quite variable, depending on the extent of the defect and the effects on the atrioventricular valves. This type of defect most commonly affects infants with *Down syndrome*, a disorder caused by the presence of an extra copy of chromosome 21.

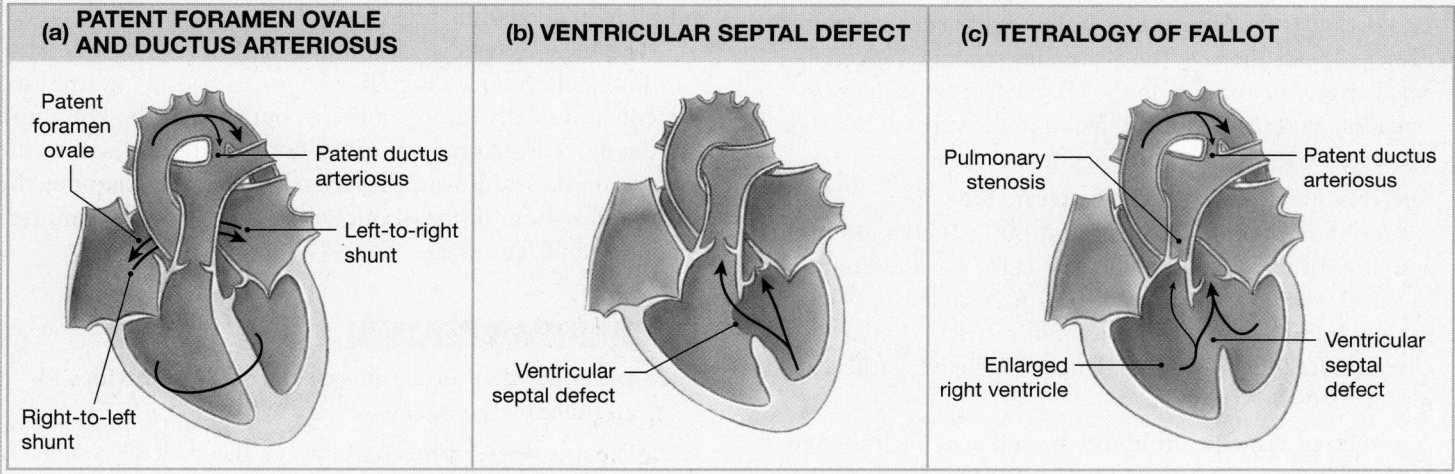

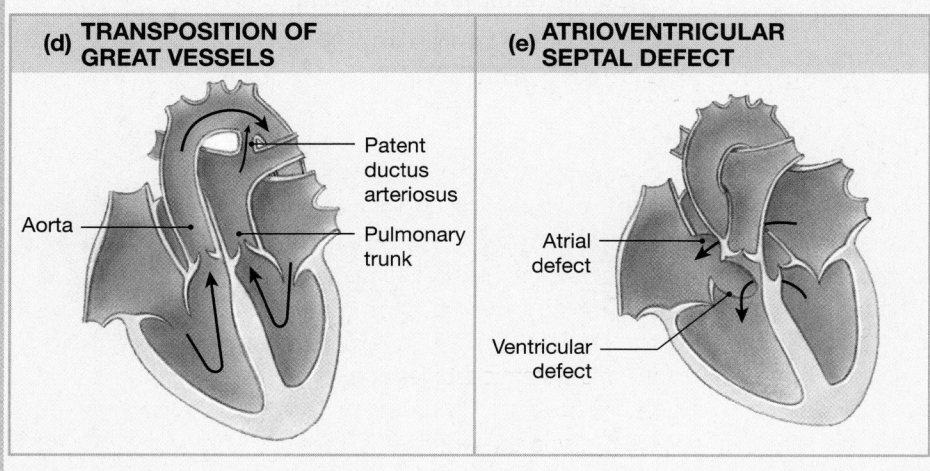

Figure 21–34 Congenital Cardiovascular Problems.

24. Name the three vessels that carry blood to and from the placenta.

25. A blood sample taken from the umbilical cord contains high levels of oxygen and nutrients, and low levels of carbon dioxide and waste products. Is this sample from an umbilical artery or from the umbilical vein? Explain.

26. Name the structures that are necessary in the fetal circulation but cease to function at birth. What becomes of these structures?

See the blue Answers tab at the end of the book.

21-9 Aging affects the blood, heart, and blood vessels

The capabilities of the cardiovascular system gradually decline. As you age, your cardiovascular system undergoes the following major changes:

- **Age-related changes in blood** may include (1) a decreased hematocrit; (2) constriction or blockage of peripheral veins by a *thrombus* (stationary blood clot), which can become detached, pass through the heart, and become wedged in a small artery (commonly in the lungs), causing *pulmonary embolism*; and (3) pooling of blood in the veins of the legs because valves are not working effectively.

- **Age-related changes in the heart** include (1) a reduction in maximum cardiac output, (2) changes in the activities of nodal and conducting cells, (3) a reduction in the elasticity of the cardiac (fibrous) skeleton, (4) progressive atherosclerosis that can restrict coronary circulation, and (5) replacement of damaged cardiac muscle cells by scar tissue.

- **Age-related changes in blood vessels** may be linked to arteriosclerosis: (1) The inelastic walls of arteries become less tolerant of sudden pressure increases, which can lead to an *aneurysm*, whose rupture may (depending on the vessel) cause a stroke, myocardial infarction, or massive blood loss; (2) calcium salts can be deposited on weakened vascular walls, increasing the risk of a stroke or myocardial infarction; and (3) thrombi can form at atherosclerotic plaques.

CHECKPOINT

27. Identify components of the cardiovascular system that are affected by age.

28. Define thrombus.

29. Define aneurysm.

See the blue Answers tab at the end of the book.

21-10 The cardiovascular system is both anatomically and functionally linked to all other systems

The section on vessel distribution demonstrated the extent of the anatomical connections between the cardiovascular system and other organ systems; **Figure 21–35** summarizes the physiological relationships involved.

The most extensive communication occurs between the cardiovascular and lymphoid systems. Not only are the two systems physically interconnected, but cells of the lymphoid system move from one part of the body to another within the vessels of the cardiovascular system. We will examine the lymphoid system in detail, including its role in the immune response, in Chapter 22.

CHECKPOINT

30. Describe what the cardiovascular system provides for all other body systems.

31. What is the relationship between the skeletal system and the cardiovascular system?

See the blue Answers tab at the end of the book.

THE CARDIOVASCULAR SYSTEM IN PERSPECTIVE

Integumentary System

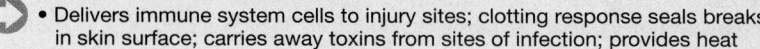

- Stimulation of mast cells produces localized changes in blood flow and capillary permeability
- Delivers immune system cells to injury sites; clotting response seals breaks in skin surface; carries away toxins from sites of infection; provides heat

Skeletal System

- Provides calcium needed for normal cardiac muscle contraction; protects blood cells developing in bone marrow
- Provides calcium and phosphate for bone deposition; delivers EPO to bone marrow, parathyroid hormone and calcitonin to osteoblasts and osteoclasts

Muscular System

- Skeletal muscle contractions assist in moving blood through veins; protects superficial blood vessels, especially in neck and limbs
- Delivers oxygen and nutrients, removes carbon dioxide, lactic acid, and heat during skeletal muscle activity

Nervous System

- Controls patterns of circulation in peripheral tissues; modifies heart rate and regulates blood pressure; releases ADH
- Endothelial cells maintain blood–brain barrier; helps generate CSF

Endocrine System

- Erythropoietin regulates production of RBCs; several hormones elevate blood pressure; epinephrine stimulates cardiac muscle, elevating heart rate and contractile force
- Distributes hormones throughout the body; heart secretes ANP and BNP

Lymphoid System

- Defends against pathogens or toxins in blood; fights infections of cardiovascular organs; returns tissue fluid to bloodstream
- Distributes WBCs; carries antibodies that attack pathogens; clotting response assists in restricting spread of pathogens; granulocytes and lymphocytes produced in bone marrow

Respiratory System

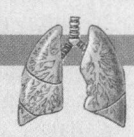

- Provides oxygen to cardiovascular organs and removes carbon dioxide
- RBCs transport oxygen and carbon dioxide between lungs and peripheral tissues

Digestive System

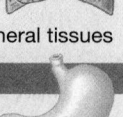

- Provides nutrients to cardiovascular organs; absorbs water and ions essential to maintenance of normal blood volume
- Distributes digestive tract hormones; carries nutrients, water, and ions away from sites of absorption; delivers nutrients and toxins to liver

Urinary System

- Releases renin to elevate blood pressure and erythropoietin to accelerate red blood cell production
- Delivers blood to capillaries, where filtration occurs; accepts fluids and solutes reabsorbed during urine production

Reproductive System

- Sex hormones maintain healthy vessels; estrogen slows development of atherosclerosis
- Distributes reproductive hormones; provides nutrients, oxygen, and waste removal for developing fetus; local blood pressure changes responsible for physical changes during sexual arousal

FOR ALL SYSTEMS

Delivers oxygen, hormones, nutrients, and WBCs; removes carbon dioxide and metabolic wastes; transfers heat

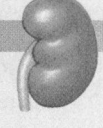

21 CARDIOVASCULAR

Figure 21–35   Functional Relationships between the Cardiovascular System and Other Systems.

Related Clinical Terms

aneurysm: A bulge in the weakened wall of a blood vessel, generally an artery.

angiogenesis: Formation of new blood vessels; a common response to tumors or tissue hypoxia from occlusion of a vessel.

arteriosclerosis: A thickening and toughening of arterial walls.

atherosclerosis: A type of arteriosclerosis characterized by changes in the endothelial lining and the formation of a plaque.

edema: An abnormal accumulation of fluid in peripheral tissues.

hemorrhoids: Varicose veins in the walls of the rectum, the anus, or both; commonly associated with frequent straining to force bowel movements.

hypertension: Abnormally high blood pressure; usually defined in adults as blood pressure higher than 140/90.

hypotension: Blood pressure so low that circulation to vital organs may be impaired.

pressure points: Locations where muscular arteries can be compressed against skeletal elements to restrict or stop the flow of blood in an emergency.

pulmonary embolism: Blockage of a pulmonary artery caused by an embolus (often a detached thrombus).

shock: An acute cardiovascular crisis marked by hypotension and inadequate peripheral blood flow.

sounds of Korotkoff: Distinctive sounds, caused by turbulent arterial blood flow, heard through the stethoscope while measuring blood pressure.

sphygmomanometer: A device that measures blood pressure using an inflatable cuff placed around a limb.

stroke, or *cerebrovascular accident (CVA)*: An interruption of the vascular supply to a portion of the brain.

thrombus: A stationary blood clot within a blood vessel.

varicose veins: Sagging, swollen veins distorted by gravity and pooling of blood due to the failure of venous valves.

Chapter Review

Study Outline

21-1 Arteries, arterioles, capillaries, venules, and veins differ in size, structure, and functional properties p. 720

1. Blood flows through a network of arteries, veins, and capillaries. All chemical and gaseous exchange between blood and interstitial fluid takes place across capillary walls.

2. **Arteries** and **veins** form an internal distribution system through which the heart propels blood. Arteries branch repeatedly, decreasing in size until they become **arterioles**. From the arterioles, blood enters **capillary** networks. Blood flowing from the capillaries enters small **venules** before entering larger veins.

3. The walls of arteries and veins contain three layers: the innermost **tunica intima**, the **tunica media**, and the outermost **tunica externa**. (*Figure 21–1*)

4. In general, the walls of arteries are thicker than those of veins. Arteries constrict when blood pressure does not distend them; veins constrict very little. The endothelial lining cannot contract, so when constriction occurs, the lining of an artery is thrown into folds. (*Figure 21–1*)

5. The arterial system includes the large **elastic arteries**, medium-sized **muscular arteries**, and smaller arterioles. As we proceed toward the capillaries, the number of vessels increases, but the diameters of the individual vessels decrease and the walls become thinner. (*Figure 21–2*)

6. **Atherosclerosis**, a type of **arteriosclerosis**, is associated with changes in the endothelial lining of arteries. Fatty masses of tissue called **plaques** typically develop during atherosclerosis. (*Figure 21–3*)

7. Capillaries are the only blood vessels whose walls are thin enough to permit an exchange between blood and interstitial fluid. Capillaries are **continuous** or **fenestrated**. **Sinusoids** have fenestrated walls and form elaborate networks that allow very slow blood flow. Sinusoids are located in the liver and in various endocrine organs. (*Figure 21–4*)

8. Capillaries form interconnected networks called **capillary beds (capillary plexuses)**. A **precapillary sphincter** (a band of smooth muscle) adjusts the blood flow into each capillary. Blood flow in a capillary changes as **vasomotion** occurs. The entire capillary bed may be bypassed by blood flow through **arteriovenous anastomoses**. (*Figure 21–5*)

9. Venules collect blood from the capillaries and merge into **medium-sized veins** and then **large veins**. The arterial system is a high-pressure system; blood pressure in veins is much lower. **Valves** in veins prevent the backflow of blood. (*Figures 21–1, 21–2, 21–6*)

10. Peripheral **venoconstriction** helps maintain adequate blood volume in the arterial system after a hemorrhage. The **venous reserve** normally accounts for about 20 percent of total blood volume. (*Figure 21–7*)

21-2 Pressure and resistance determine blood flow and affect rates of capillary exchange p. 729

11. Cardiovascular regulation involves the manipulation of blood pressure and resistance to control the rates of blood flow and capillary exchange. (*Figure 21–8*)

12. Blood flow occurs from an area of higher pressure to one of lower pressure and is proportional to the pressure gradient. The *circulatory pressure* is the pressure gradient across the systemic circuit. It is reported as three values: arterial **blood pressure (BP)**, **capillary hydrostatic pressure (CHP)**, and **venous pressure**.

13. The **resistance (R)** determines the rate of blood flow through the systemic circuit. The major determinant of blood flow rate is the **peripheral resistance**—the resistance of the arterial

system. Neural and hormonal control mechanisms regulate blood pressure and peripheral resistance.

14. **Vascular resistance** is the resistance of blood vessels. It is the largest component of peripheral resistance and depends on vessel length and vessel diameter.

15. **Viscosity** and **turbulence** also contribute to peripheral resistance. (*Summary Table 21–1*)

16. The high arterial pressures overcome peripheral resistance and maintain blood flow through peripheral tissues. Capillary pressures are normally low; small changes in capillary pressure determine the rate of movement of fluid into or out of the bloodstream. Venous pressure, normally low, determines *venous return* and affects cardiac output and peripheral blood flow. (*Figures 21–9, 21–10; Summary Table 21–1*)

17. Arterial blood pressure rises during ventricular systole and falls during ventricular diastole. The difference between these two blood pressures is the pulse pressure. Blood pressure is measured at the brachial artery with the use of a sphygmomanometer. (*Figures 21–9, 21–10*)

18. Valves, muscular compression, and the **respiratory pump** (*thoracoabdominal pump*) help the relatively low venous pressures propel blood toward the heart. (*Figures 21–6, 21–9*)

19. At the capillaries, blood pressure forces water and solutes out of the plasma, across capillary walls. Water moves out of the capillaries, through the peripheral tissues, and back to the bloodstream by way of the lymphoid system. Water movement across capillary walls is determined by the interplay between osmotic pressures and hydrostatic pressures. (*Figure 21–11*)

20. **Osmotic pressure (OP)** is a measure of the pressure that must be applied to prevent osmotic movement across a membrane. Osmotic water movement continues until either solute concentrations are equalized or the movement is prevented by an opposing hydrostatic pressure.

21. The rates of filtration and reabsorption gradually change as blood passes along the length of a capillary, as determined by the **net filtration pressure** (the difference between the net hydrostatic pressure and the net osmotic pressure). (*Figure 21–12*)

21-3 Cardiovascular regulatory mechanisms involve autoregulation, neural mechanisms, and endocrine responses p. 737

22. Homeostatic mechanisms ensure that **tissue perfusion** (blood flow) delivers adequate oxygen and nutrients.

23. Autoregulation, neural mechanisms, and endocrine mechanisms influence the coordinated regulation of cardiovascular function. Autoregulation involves local factors changing the pattern of blood flow within capillary beds in response to chemical changes in interstitial fluids. Neural mechanisms respond to changes in arterial pressure or blood gas levels. Hormones can assist in short-term adjustments (changes in cardiac output and peripheral resistance) and long-term adjustments (changes in blood volume that affect cardiac output and gas transport). (*Figure 21–13*)

24. Peripheral resistance is adjusted at the tissues by local factors that result in the dilation or constriction of precapillary sphincters. (*Figure 21–5*)

25. **Cardiovascular (CV) centers** of the medulla oblongata are responsible for adjusting cardiac output and peripheral resistance to maintain adequate blood flow. The vasomotor center contains one group of neurons responsible for controlling vasoconstriction, and another group responsible for controlling vasodilation.

26. **Baroreceptor reflexes** monitor the degree of stretch within expandable organs. Baroreceptors are located in the **carotid sinuses**, the **aortic sinuses**, and the right atrium. (*Figure 21–14*)

27. **Chemoreceptor reflexes** respond to changes in the oxygen or CO_2 levels in the blood. They are triggered by sensory neurons located in the **carotid bodies** and the **aortic bodies**. (*Figure 21–15*)

28. The endocrine system provides short-term regulation of cardiac output and peripheral resistance with epinephrine and norepinephrine from the suprarenal medullae. Hormones involved in the long-term regulation of blood pressure and volume are *antidiuretic hormone (ADH), angiotensin II, erythropoietin (EPO),* and *natriuretic peptides (ANP and BNP).* (*Figure 21–16*)

21-4 The cardiovascular system adapts to physiological stress and maintains a special vascular supply to the brain, heart, and lungs p. 744

29. During exercise, blood flow to skeletal muscles increases at the expense of blood flow to nonessential organs, and cardiac output rises. Cardiovascular performance improves with training. Athletes have larger stroke volumes, slower resting heart rates, and larger cardiac reserves than do nonathletes. (*Tables 21–2, 21–3*)

30. Blood loss lowers blood volume and venous return and decreases cardiac output. Compensatory mechanisms include an increase in cardiac output, a mobilization of venous reserves, peripheral vasoconstriction, and the release of hormones that promote the retention of fluids and the manufacture of erythrocytes. (*Figure 21–17*)

31. The blood–brain barrier, the coronary circulation, and the circulation to alveolar capillaries in the lungs are examples of special circulations, in which cardiovascular dynamics and regulatory mechanisms differ from those in other tissues.

21-5 The pulmonary and systemic circuits of the cardiovascular system exhibit three general functional patterns p. 748

32. The peripheral distributions of arteries and veins are generally identical on both sides of the body, except near the heart. (*Figure 21–18*)

21-6 In the pulmonary circuit, deoxygenated blood enters the lungs in arteries, and oxygenated blood leaves the lungs via veins p. 749

33. The pulmonary circuit includes the pulmonary trunk, the **left** and **right pulmonary arteries**, and the **pulmonary veins**, which empty into the left atrium. (*Figure 21–19*)

21-7 The systemic circuit carries oxygenated blood from the left ventricle to tissues and organs other than the pulmonary exchange surfaces, and returns deoxygenated blood to the right atrium p. 750

34. The **ascending aorta** gives rise to the coronary circulation. The **aortic arch** communicates with the **descending aorta**. (*Figures 21–20 to 21–26*)

35. Three elastic arteries originate along the aortic arch: the **left common carotid artery**, the **left subclavian artery**, and the **brachiocephalic trunk**. (*Figures 21–21, 21–22, 21–23*)

21 CARDIOVASCULAR

36. The remaining major arteries of the body originate from the **descending aorta**. (*Figures 21–24, 21–25, 21–26*)
37. Arteries in the neck and limbs are deep beneath the skin; in contrast, there are generally two sets of peripheral veins, one superficial and one deep. This dual venous drainage is important for controlling body temperature. (*Figure 21–27*)
38. The **superior vena cava** receives blood from the head, neck, chest, shoulders, and arms. (*Figures 21–27 to 21–30*)
39. The **inferior vena cava** collects most of the venous blood from organs inferior to the diaphragm. (*Figures 21–29 to 21–31*)
40. The **hepatic portal system** directs blood from the other digestive organs to the liver before the blood returns to the heart. (*Figure 21–32*)

21-8 Modifications of fetal and maternal cardiovascular systems promote the exchange of materials, and independence is achieved at birth p. 767

41. Blood flow to the placenta is provided by a pair of **umbilical arteries** and is drained by a single **umbilical vein**. (*Figure 21–33*)
42. The interatrial partition remains functionally incomplete until birth. The **foramen ovale** allows blood to flow freely from the right to the left atrium, and the **ductus arteriosus** short-circuits the pulmonary trunk.
43. The foramen ovale closes, leaving the fossa ovalis. The ductus arteriosus constricts, leaving the ligamentum arteriosum. (*Figure 21–33*)

44. Congenital cardiovascular problems generally reflect abnormalities of the heart or of interconnections between the heart and great vessels. (*Figure 21–34*)

21-9 Aging affects the blood, heart, and blood vessels p. 770

45. Age-related changes in the blood include (1) a decreased hematocrit, (2) constriction or blockage of peripheral veins by a *thrombus* (stationary blood clot), and (3) pooling of blood in the veins of the legs because valves are not working effectively.
46. Age-related changes in the heart include (1) a reduction in the maximum cardiac output, (2) changes in the activities of nodal and conducting cells, (3) a reduction in the elasticity of the fibrous skeleton, (4) progressive atherosclerosis that can restrict coronary circulation, and (5) the replacement of damaged cardiac muscle cells by scar tissue.
47. Age-related changes in blood vessels, commonly related to arteriosclerosis, include (1) a weakening in the walls of arteries, potentially leading to the formation of an *aneurysm*; (2) deposition of calcium salts on weakened vascular walls, increasing the risk of a stroke or myocardial infarction; and (3) the formation of a thrombus at atherosclerotic plaques.

21-10 The cardiovascular system is both anatomically and functionally linked to all other systems p. 770

48. The cardiovascular system is anatomically and functionally connected to all other body systems. (*Figure 21–35*)

Review Questions

See the blue Answers tab at the end of the book.

myA&P Access more review material online at **myA&P™** (**www.myaandp.com**). There, you'll find chapter guides, chapter quizzes, practice tests, animations, flashcards, a glossary with pronunciations, *Interactive Physiology*® (IP) exercises and quizzes, and more to help you succeed in the course.

LEVEL 1 Reviewing Facts and Terms

1. Identify the major arteries in the following diagram.
   (a) brachiocephalic trunk
   (b) brachial
   (c) radial
   (d) external iliac
   (e) anterior tibial
   (f) right common carotid
   (g) left subclavian
   (h) common iliac
   (i) femoral

2. The blood vessels that play the most important role in regulating blood pressure and blood flow to a tissue are the
   (a) arteries.   (b) arterioles.
   (c) veins.   (d) venules.
   (e) capillaries.
3. Cardiovascular function is regulated by all of the following *except*
   (a) local factors.   (b) neural factors.
   (c) endocrine factors. (d) venous return.
   (e) conscious control.
4. Baroreceptors that function in the regulation of blood pressure are located in the
   (a) left ventricle.   (b) brain stem.
   (c) carotid sinus.   (d) common iliac artery.
   (e) pulmonary trunk.
5. The two-way exchange of substances between blood and body cells occurs only through
   (a) arterioles.   (b) capillaries.
   (c) venules.   (d) a, b, and c.
6. Large molecules such as peptides and proteins move into and out of the bloodstream by way of
   (a) continuous capillaries.   (b) fenestrated capillaries.
   (c) thoroughfare channels.   (d) metarterioles.
7. The local control of blood flow due to the action of precapillary sphincters is
   (a) vasomotion.   (b) autoregulation.
   (c) selective resistance. (d) turbulence.
8. Blood is transported through the venous system by means of
   (a) muscular contractions.
   (b) increasing blood pressure.
   (c) the respiratory pump.
   (d) a and c.

9. Identify the major veins in the following diagram.
   (a) external jugular
   (b) axillary
   (c) median cubital
   (d) radial
   (e) great saphenous
   (f) internal jugular
   (g) sup. vena cava
   (h) left & right common iliac
   (i) femoral

10. The most important factor in vascular resistance is
    (a) the viscosity of the blood.
    (b) the diameter of the lumen of blood vessels.
    (c) turbulence due to irregular surfaces of blood vessels.
    (d) the length of the blood vessels.

11. Net hydrostatic pressure forces water _____ a capillary; net osmotic pressure forces water _____ a capillary.
    (a) into, out of   (b) out of, into
    (c) out of, out of   (d) into, into

12. The two arteries formed by the division of the brachiocephalic trunk are the
    (a) aorta and internal carotid.
    (b) axillary and brachial.
    (c) external and internal carotid.
    (d) common carotid and subclavian.

13. The unpaired arteries supplying blood to the visceral organs include
    (a) the suprarenal, renal, and lumbar arteries.
    (b) the iliac, gonadal, and femoral arteries.
    (c) the celiac and superior and inferior mesenteric arteries.
    (d) a, b, and c.

14. The paired arteries supplying blood to the body wall and other structures outside the abdominopelvic cavity include the
    (a) left gastric, hepatic, splenic, and phrenic arteries.
    (b) suprarenal, colic, lumbar, and gonadal arteries.
    (c) iliac, femoral, and lumbar arteries.
    (d) celiac, left gastric, and superior and inferior mesenteric arteries.

15. The vein that drains the dural sinuses of the brain is the
    (a) cephalic vein.   (b) great saphenous vein.
    (c) internal jugular vein. (d) superior vena cava.

16. The vein that collects most of the venous blood from below the diaphragm is the
    (a) superior vena cava.   (b) great saphenous vein.
    (c) inferior vena cava.   (d) azygos vein.

17. What are the primary forces that cause fluid to move
    (a) out of a capillary at its arterial end and into the interstitial fluid? capillary hydrostatic pressure.
    (b) into a capillary at its venous end from the interstitial fluid? osmotic

18. What cardiovascular changes occur at birth?

LEVEL 2   Reviewing Concepts

19. A major difference between the arterial and venous systems is that
    (a) arteries are usually more superficial than veins.
    (b) in the limbs there is dual venous drainage.
    (c) veins are usually less branched compared to arteries.
    (d) veins exhibit a much more orderly pattern of branching in the limbs.
    (e) veins are not found in the abdominal cavity.

20. Which of the following conditions would have the *greatest* effect on peripheral resistance?
    (a) doubling the length of a vessel
    (b) doubling the diameter of a vessel → greatest factor.
    (c) doubling the viscosity of the blood
    (d) doubling the turbulence of the blood
    (e) doubling the number of white cells in the blood

21. Which of the following is *greater*?
    (a) the osmotic pressure of the interstitial fluid during inflammation
    (b) the osmotic pressure of the interstitial fluid during normal conditions
    (c) neither is greater

22. Relate the anatomical differences between arteries and veins to their functions.

23. Why do capillaries permit the diffusion of materials, whereas arteries and veins do not? others several layers thick.

24. How is blood pressure maintained in veins to counter the force of gravity? valves & muscular pumps.

25. How do pressure and resistance affect cardiac output and peripheral blood flow? cardiac output & perip. blood ∝ bp.

26. Why is blood flow to the brain relatively continuous and constant? 4 arteries forming anastomos to brain

27. Compare the effects of the cardioacceleratory and cardioinhibitory centers on cardiac output and blood pressure.

LEVEL 3   Critical Thinking and Clinical Applications

28. Bob is sitting outside on a warm day and is sweating profusely. Mary wants to practice taking blood pressures, and he agrees to play the patient. Mary finds that Bob's blood pressure is elevated, even though he is resting and has lost fluid from sweating. (She reasons that fluid loss should lower blood volume and, thus, blood pressure.) Why is Bob's blood pressure high instead of low? high bp compensates fluid loss.

29. People with allergies commonly take antihistamines with decongestants to relieve their symptoms. The container warns that individuals who are being treated for high blood pressure should not take the medication. Why not? stimulates symp.

30. Jolene awakens suddenly to the sound of her alarm clock. Realizing that she is late for class, she jumps to her feet, feels light-headed, and falls back on her bed. What probably caused this reaction? Why doesn't this happen all the time?

22

The Lymphoid System and Immunity

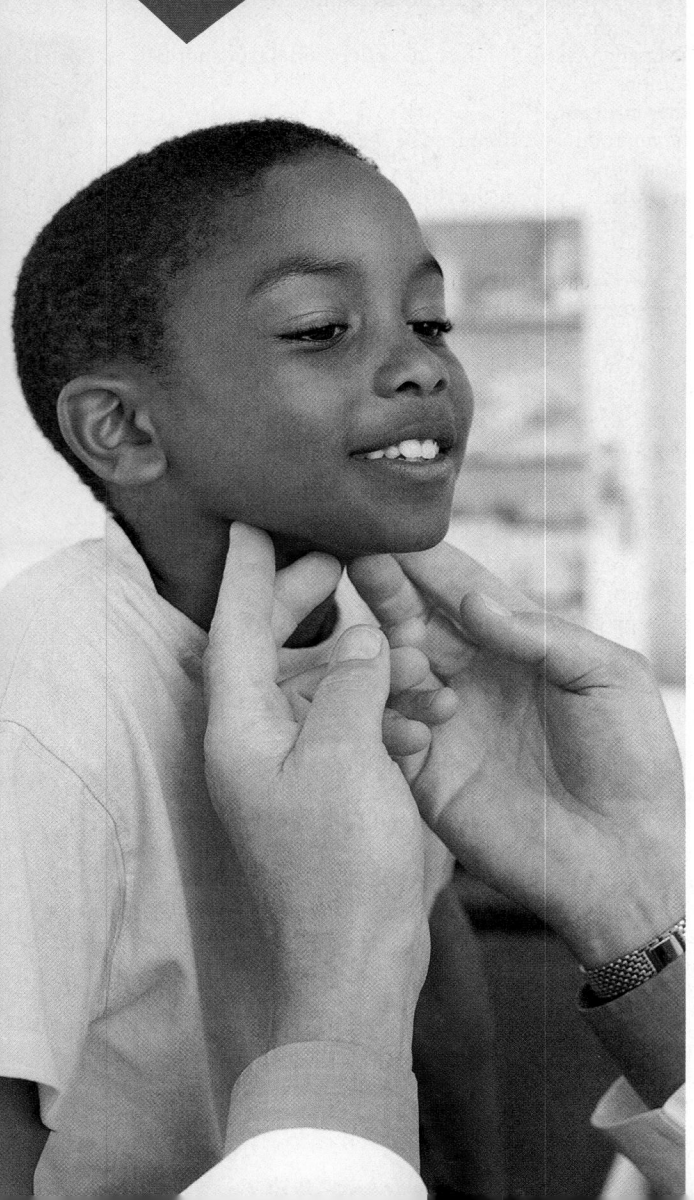

Learning Outcomes

After completing this chapter, you should be able to do the following:

22-1 Distinguish between nonspecific and specific defense, and explain the role of lymphocytes in the immune response.

22-2 Identify the major components of the lymphoid system, describe the structure and functions of each component, and discuss the importance of lymphocytes.

22-3 List the body's nonspecific defenses, and describe the components, mechanisms, and functions of each.

22-4 Define specific defenses, identify the forms and properties of immunity, and distinguish between cell-mediated (cellular) immunity and antibody-mediated (humoral) immunity.

22-5 Discuss the types of T cells and their roles in the immune response, and describe the mechanisms of T cell activation and differentiation.

22-6 Discuss the mechanisms of B cell activation and differentiation, describe the structure and function of antibodies, and explain the primary and secondary responses to antigen exposure.

22-7 Describe the development of immunological competence, list and explain examples of immune disorders and allergies, and discuss the effects of stress on immune function.

22-8 Describe the effects of aging on the lymphoid system and the immune response.

22-9 Give examples of interactions between the lymphoid system and each of the other organ systems.

Clinical Notes

Cancer and the Lymphoid System p. 786
Graft Rejection and Immunosuppression p. 803
AIDS p. 819

An Introduction to the Lymphoid System and Immunity

Many organs and systems work together to keep us alive and healthy. In the ongoing struggle to maintain health, the lymphoid system plays a central role. This chapter discusses the components of the lymphoid system and the ways those components interact.

22-1 Anatomical barriers and defense mechanisms constitute nonspecific defense, and lymphocytes provide specific defense

The world is not always kind to the human body. Accidental bumps, cuts, and scrapes; chemical and thermal burns; extreme cold; and ultraviolet radiation are just a few of the hazards in our physical environment. Making matters worse, the world around us contains an assortment of viruses, bacteria, fungi, and parasites capable of not only surviving but thriving inside our bodies—and potentially causing us great harm. These organisms, called **pathogens,** are responsible for many diseases in humans. Each pathogen has a different lifestyle and attacks the body in a specific way. For example, viruses spend most of their time hidden within cells, which they often eventually destroy, whereas some of the largest parasites actually burrow through internal organs. Many bacteria multiply in interstitial fluids, where they release foreign proteins—enzymes or toxins—that can damage cells, tissues, even entire organ systems. And as if that were not enough, we are constantly at risk from renegade body cells that have the potential to produce lethal cancers. ∞ p. 105

The **lymphoid system** includes the cells, tissues, and organs responsible for defending the body against both environmental hazards, such as various pathogens, and internal threats, such as cancer cells. *Lymphocytes*, the primary cells of the lymphoid system, were introduced in Chapters 4 and 19. ∞ pp. 131, 668 These cells are vital to the body's ability to resist or overcome infection and disease. Lymphocytes respond to the presence of invading pathogens (such as bacteria or viruses), abnormal body cells (such as virus-infected cells or cancer cells), and foreign proteins (such as the toxins released by some bacteria). They act to eliminate these threats or render them harmless through a combination of physical and chemical attacks.

The body has several anatomical barriers and defense mechanisms that either prevent or slow the entry of infectious organisms, or attack them if they do succeed in gaining entry. These

mechanisms are called *nonspecific defenses*, because they do not distinguish one potential threat from another. In contrast, lymphocytes respond specifically: If a bacterial pathogen invades peripheral tissues, lymphocytes organize a defense against that particular type of bacterium. For this reason, lymphocytes are said to provide a *specific defense*, known as the **immune response**. The ability to resist infection and disease through the activation of specific defenses constitutes **immunity**.

All the cells and tissues involved in the production of immunity are sometimes considered part of an *immune system*—a physiological system that includes not only the lymphoid system, but also components of the integumentary, cardiovascular, respiratory, digestive, and other systems. For example, interactions between lymphocytes and dendritic (Langerhans) cells of the skin are important in mobilizing specific defenses against skin infections.

We begin this chapter by examining the organization of the lymphoid system. We will then consider the body's nonspecific defenses. Finally, we will see how the lymphoid system interacts with cells and tissues of other systems to defend the body against infection and disease.

> ### CHECKPOINT
>
> 1. Define pathogen.
> 2. Explain the difference between nonspecific defense and specific defense.
>
> See the blue Answers tab at the end of the book.

22-2 Lymphatic vessels, lymphocytes, lymphoid tissues, and lymphoid organs function in body defenses

The lymphoid system consists of (1) **lymph,** a fluid that resembles plasma but contains a much lower concentration of suspended proteins; (2) a network of **lymphatic vessels,** often called **lymphatics,** which begin in peripheral tissues and end at connections to veins; (3) an array of **lymphoid tissues** and **lymphoid organs** scattered throughout the body; and (4) lymphocytes and smaller numbers of phagocytes and other cells. **Figure 22–1** provides a general overview of the primary tissues, vessels, and organs of this system.

Functions of the Lymphoid System

The primary function of the lymphoid system is the production, maintenance, and distribution of lymphocytes that provide de-

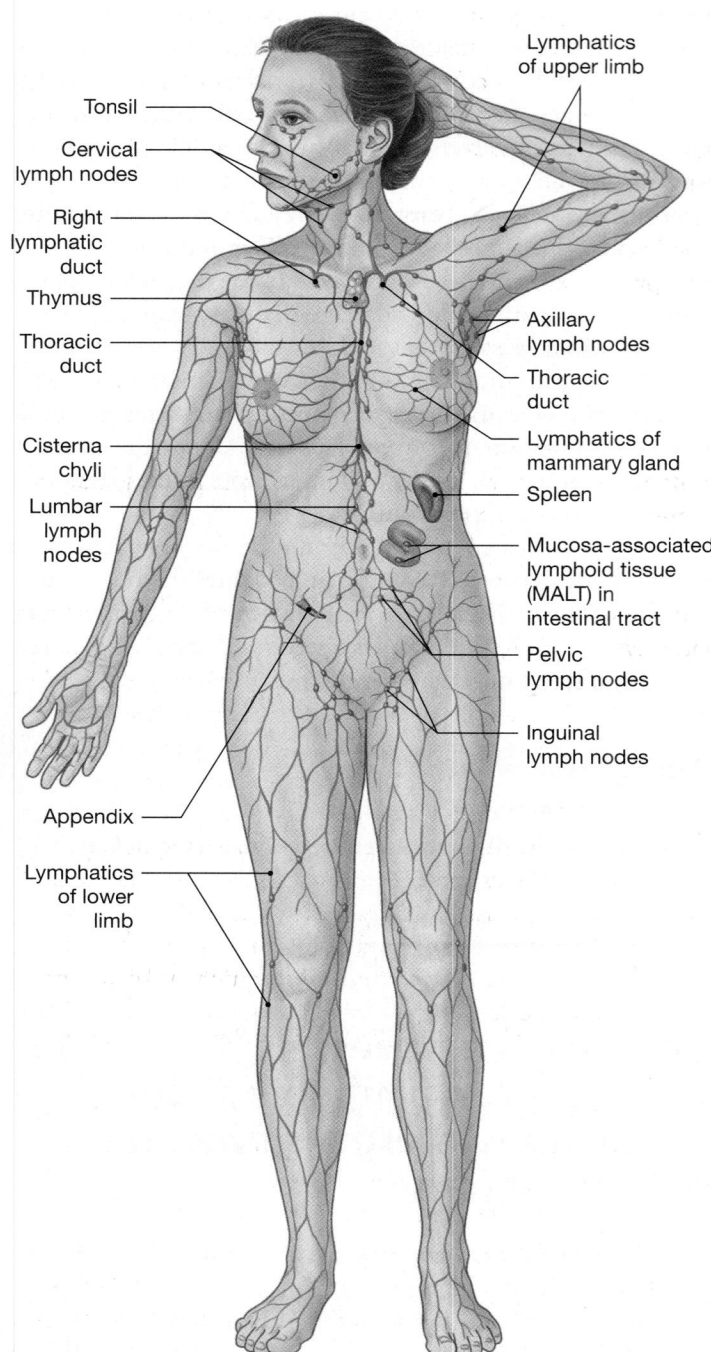

Tonsil

Cervical
lymph nodes

Right
lymphatic
duct

Thymus

Thoracic
duct

Cisterna
chyli

Lumbar
lymph
nodes

Appendix

Lymphatics
of lower
limb

Lymphatics
of upper limb

Axillary
lymph nodes

Thoracic
duct

Lymphatics of
mammary gland

Spleen

Mucosa-associated
lymphoid tissue
(MALT) in
intestinal tract

Pelvic
lymph nodes

Inguinal
lymph nodes

Figure 22–1   An Overview of the Lymphoid System: The Lymphatic Vessels, Lymphoid Tissues, and Lymphoid Organs.

fense against infections and other environmental hazards. Most of the body's lymphocytes are produced and stored within lymphoid tissues (such as the tonsils) and lymphoid organs (such as the spleen and thymus). However, lymphocytes are also produced in areas of red bone marrow, along with other defense cells, such as monocytes and macrophages.

To provide an effective defense, lymphocytes must be able to detect problems, and they must be able to reach the site of injury or infection. Lymphocytes, macrophages, and microphages circulate within the blood and are able to enter or leave the capillaries that supply most of the tissues of the body. As noted in Chapter 21, capillaries normally deliver more fluid to peripheral tissues than they carry away. ∞ p. 736 The excess fluid returns to the bloodstream through lymphatic vessels. This continuous circulation of extracellular fluid helps transport lymphocytes and other defense cells from one organ to another. In the process, it maintains normal blood volume and eliminates local variations in the composition of the interstitial fluid by distributing hormones, nutrients, and waste products from their tissues of origin to the general circulation.

Lymphatic Vessels

Lymphatic vessels carry lymph from peripheral tissues to the venous system. The smallest lymphatic vessels are called *lymphatic capillaries*.

Lymphatic Capillaries

[handwritten: blind pocket.]   *[handwritten: lower pressure.]*

The lymphatic network begins with **lymphatic capillaries,** or *terminal lymphatics,* which branch through peripheral tissues. Lymphatic capillaries differ from blood capillaries in that they (1) originate as pockets rather than forming continuous tubes, (2) have larger diameters, (3) have thinner walls, and (4) typically have a flattened or irregular outline in sectional view (**Figure 22–2**). Although lymphatic capillaries are lined by endothelial cells, the basal lamina is incomplete or absent. The endothelial cells of a lymphatic capillary are not bound tightly together, but they do overlap. The region of overlap acts as a one-way valve, permitting the entry of fluids and solutes (including those as large as proteins), as well as viruses, bacteria, and cell debris, but preventing their return to the intercellular spaces.

Lymphatic capillaries are present in almost every tissue and organ in the body. Prominent lymphatic capillaries in the small intestine called *lacteals* are important in the transport of lipids absorbed by the digestive tract. Lymphatic capillaries are absent in areas that lack a blood supply, such as the cornea of the eye. The bone marrow and the central nervous system also lack lymphatic vessels.

Small Lymphatic Vessels

From the lymphatic capillaries, lymph flows into larger lymphatic vessels that lead toward the body's trunk. The walls of these vessels contain layers comparable to those of veins, and, like veins, the larger lymphatic vessels contain valves (**Figure 22–3**). The valves are quite close together, and at each the

plasma → lymphatic vessel.

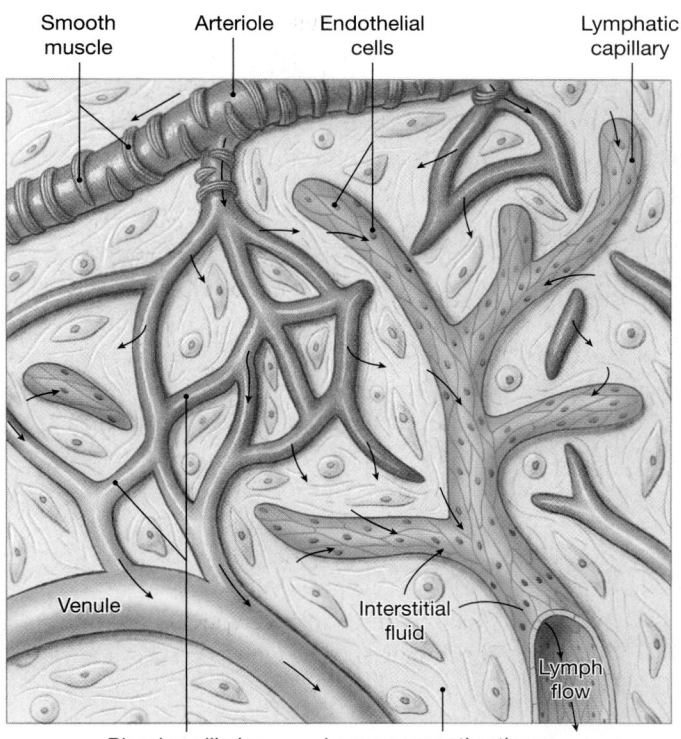

Smooth muscle — Arteriole — Endothelial cells — Lymphatic capillary

Venule — Interstitial fluid — Lymph flow

Blood capillaries — Loose connective tissue

(a) Association of blood capillaries, tissue, and lymphatic capillaries

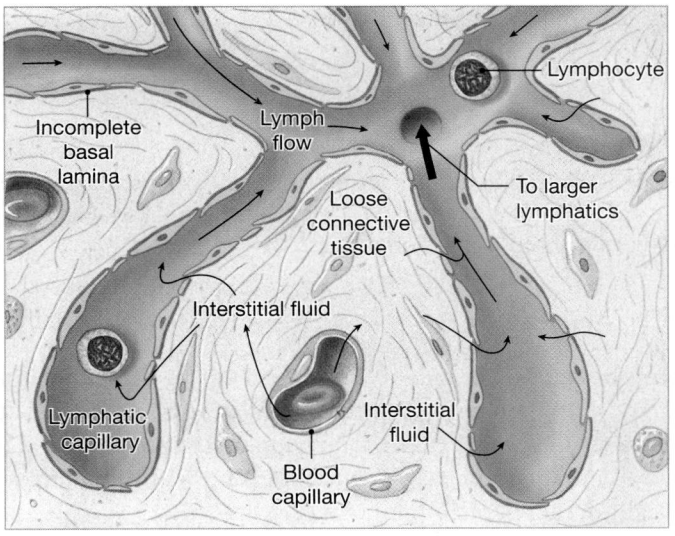

Incomplete basal lamina — Lymph flow — Lymphocyte

Loose connective tissue — To larger lymphatics

Interstitial fluid

Lymphatic capillary — Interstitial fluid

Blood capillary

(b) Sectional view

Figure 22–2 Lymphatic Capillaries. (a) The interwoven network formed by blood capillaries and lymphatic capillaries. Arrows indicate the movement of fluid out of blood vessels and the net flow of interstitial fluid and lymph. **(b)** A sectional view indicating the movement of fluid from the plasma, through the interstitial fluid, and into the lymphoid system.

lymphatic vessel bulges noticeably. As a result, large lymphatic vessels have a beaded appearance (**Figure 22–3a**). The valves prevent the backflow of lymph within lymph vessels, especially those of the limbs. Pressures within the lymphoid system are minimal, and the valves are essential to maintaining normal lymph flow toward the thoracic cavity.

Lymphatic vessels commonly occur in association with blood vessels (**Figure 22–3a**). Differences in relative size, general appearance, and branching pattern distinguish lymphatic vessels from arteries and veins. Characteristic color differences are also apparent on examining living tissues. Most arteries are bright red, veins are dark red (although usually illustrated as blue to distinguish them from arteries), and lymphatic vessels are a pale golden color. In general, a tissue contains many more lymphatic vessels than veins, but the lymphatic vessels are much smaller.

Major Lymph-Collecting Vessels
collects lymphs from capillaries.

Two sets of lymphatic vessels collect lymph from the lymphatic capillaries: superficial lymphatics and deep lymphatics. **Superficial lymphatics** are located in the subcutaneous layer deep to the skin; in the areolar tissues of the mucous membranes lining the digestive, respiratory, urinary, and reproductive tracts; and in the areolar tissues of the serous membranes lining the pleural, pericardial, and peritoneal cavities. **Deep lymphatics** are larger lymphatic vessels that accompany deep arteries and veins supplying skeletal muscles and other organs of the neck, limbs, and trunk, and the walls of visceral organs.

Superficial and deep lymphatics converge to form even larger vessels called *lymphatic trunks*, which in turn empty into two large collecting vessels: the thoracic duct and the right lymphatic duct. The *thoracic duct* collects lymph from the body inferior to the diaphragm and from the left side of the body superior to the diaphragm. The smaller *right lymphatic duct* collects lymph from the right side of the body superior to the diaphragm (**Figure 22–4a**).

The **thoracic duct** begins inferior to the diaphragm at the level of vertebra L_2 (**Figure 22–4b**). The base of the thoracic duct is an expanded, saclike chamber called the **cisterna chyli** (KĪ-lī; *chylos*, juice). The cisterna chyli receives lymph from the inferior part of the abdomen, the pelvis, and the lower limbs by way of the *right* and *left lumbar trunks* and the *intestinal trunk*.

The inferior segment of the thoracic duct lies anterior to the vertebral column. From the second lumbar vertebra, it passes posterior to the diaphragm alongside the aorta and ascends along the left side of the vertebral column to the level of the left clavicle. After collecting lymph from the *left bronchomediastinal trunk*, the *left subclavian trunk*, and the *left jugular trunk*, it empties into the left subclavian vein near the left internal jugular vein (**Figure 22–4b**). Lymph collected

22 LYMPHOID

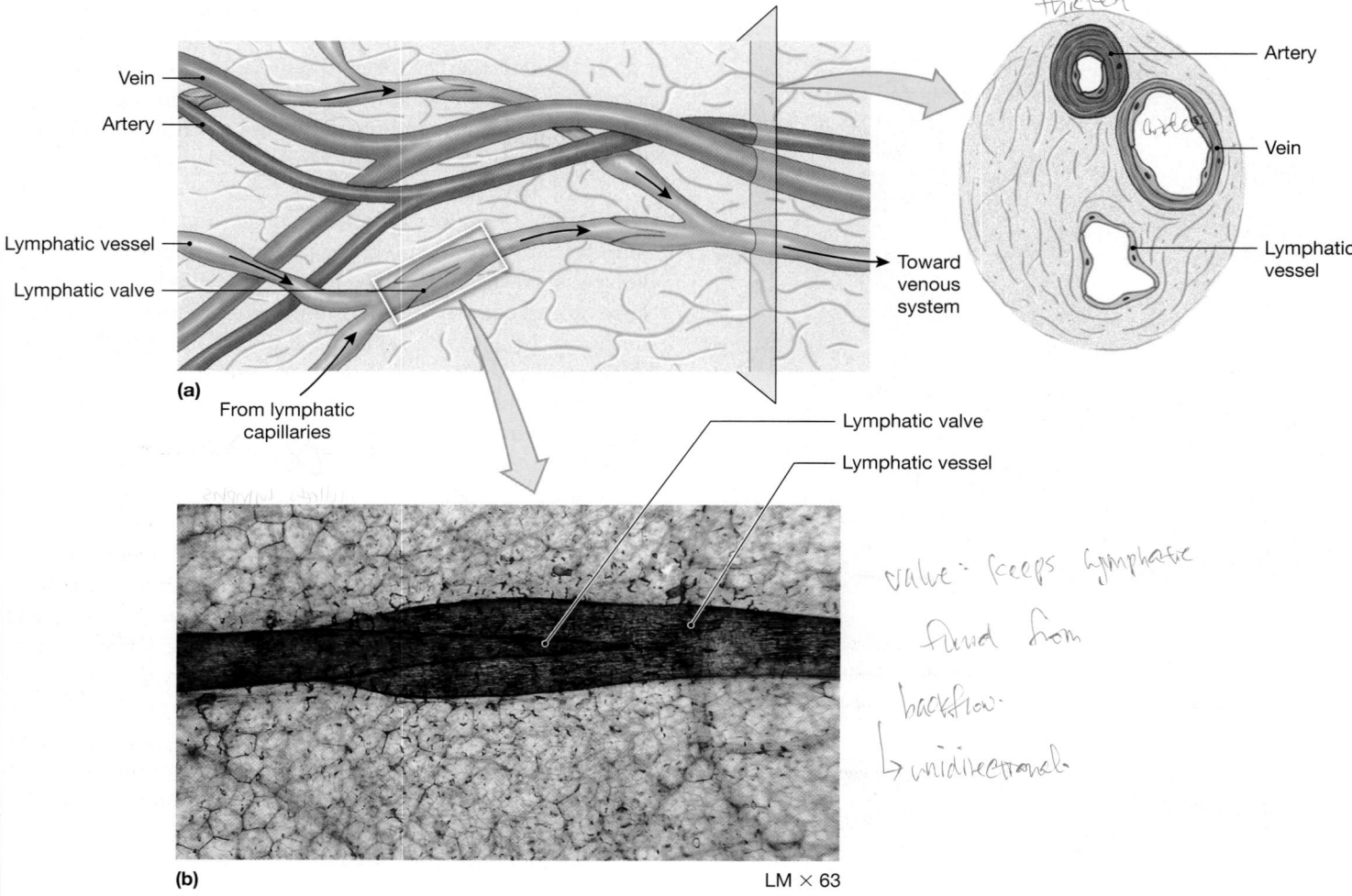

thickest (handwritten)

valve: keeps lymphatic fluid from backflow. (handwritten)
↳ unidirectional (handwritten)

Figure 22–3 **Lymphatic Vessels and Valves. (a)** A diagrammatic view of loose connective tissue containing small blood vessels and a lymphatic vessel. The cross-sectional view emphasizes the structural differences among these structures. **(b)** A lymphatic valve. Like valves in veins, each lymphatic valve consists of a pair of flaps that permit movement of fluid in only one direction.

from the left side of the head, neck, and thorax, as well as lymph from the entire body inferior to the diaphragm, reenters the venous circulation in this way.

The **right lymphatic duct** is formed by the merging of the *right jugular*, *right subclavian*, and *right bronchomediastinal trunks* in the area near the right clavicle. This duct empties into the right subclavian vein, delivering lymph from the right side of the body superior to the diaphragm.

Blockage of the lymphatic drainage from a limb produces **lymphedema** (limf-e-DĒ-muh), a condition in which interstitial fluids accumulate and the limb gradually becomes swollen and grossly distended. If the condition persists, the connective tissues lose their elasticity and the swelling becomes permanent. Lymphedema by itself does not pose a major threat to life. The danger comes from the constant risk that an uncontrolled infection will develop in the affected area. Because the interstitial fluids are essentially stagnant, toxins and pathogens can accumulate and overwhelm local defenses without fully activating the immune system.

Lymphocytes

Lymphocytes account for 20–30 percent of the circulating leukocyte population. However, circulating lymphocytes are only a small fraction of the total lymphocyte population. The body contains some 10^{12} lymphocytes, with a combined weight of more than a kilogram (2.2 lbs).

Types of Lymphocytes

Three classes of lymphocytes circulate in blood: (1) **T** (thymus-dependent) **cells,** (2) **B** (bone marrow–derived) **cells,** and (3) **NK** (natural killer) **cells**. Each type has distinctive biochemical and functional characteristics.

Approximately 80 percent of circulating lymphocytes are classified as T cells. The primary types of T cells are the following:

- **Cytotoxic T cells,** which attack foreign cells or body cells infected by viruses. Their attack commonly involves

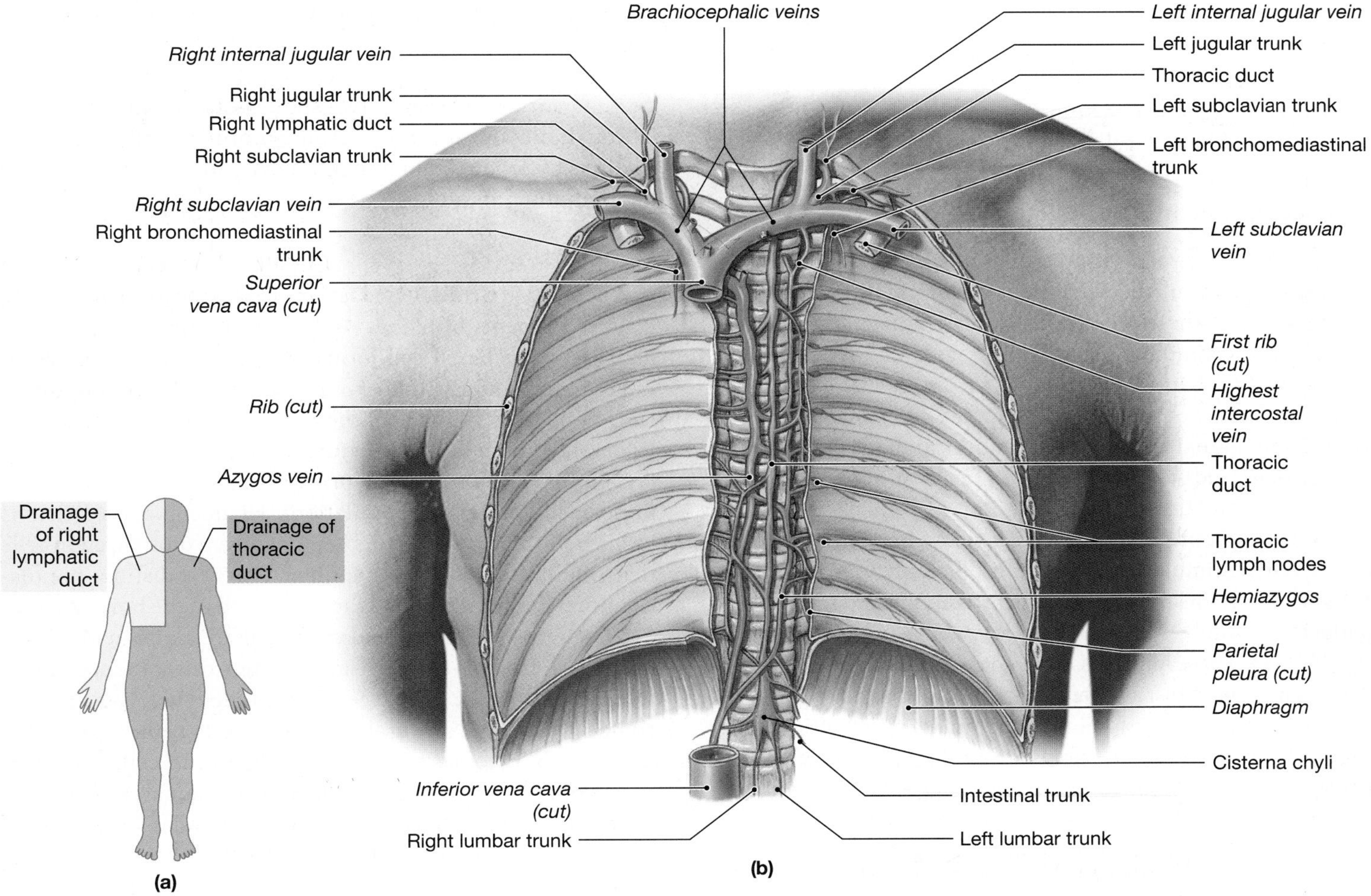

Brachiocephalic veins

Right internal jugular vein

Right jugular trunk

Right lymphatic duct

Right subclavian trunk

Right subclavian vein

Right bronchomediastinal trunk

Superior vena cava (cut)

Rib (cut)

Azygos vein

Left internal jugular vein

Left jugular trunk

Thoracic duct

Left subclavian trunk

Left bronchomediastinal trunk

Left subclavian vein

First rib (cut)

Highest intercostal vein

Thoracic duct

Thoracic lymph nodes

Hemiazygos vein

Parietal pleura (cut)

Diaphragm

Cisterna chyli

Drainage of right lymphatic duct

Drainage of thoracic duct

Inferior vena cava (cut)

Right lumbar trunk

Intestinal trunk

Left lumbar trunk

(a)

(b)

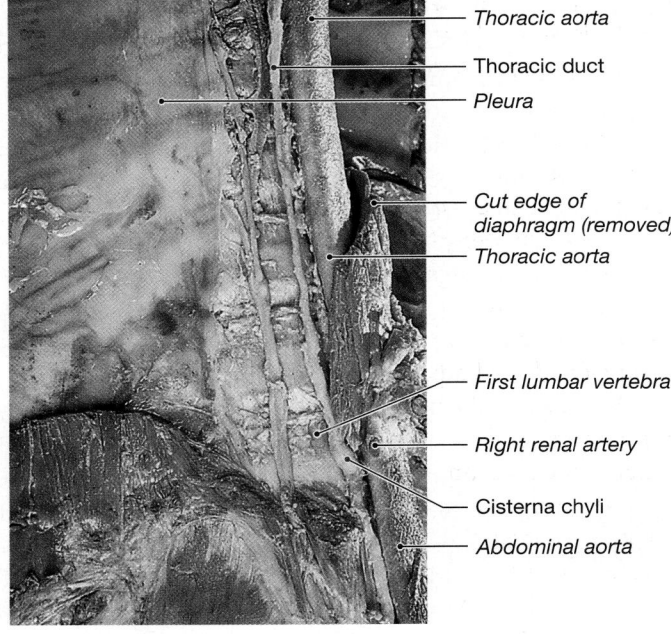

Thoracic aorta

Thoracic duct

Pleura

Cut edge of diaphragm (removed)

Thoracic aorta

First lumbar vertebra

Right renal artery

Cisterna chyli

Abdominal aorta

(c)

22 LYMPHOID

Figure 22–4 **The Relationship between the Lymphatic Ducts and the Venous System. (a)** The thoracic duct carries lymph originating in tissues inferior to the diaphragm and from the left side of the upper body. The smaller right lymphatic duct delivers lymph from the rest of the body. **(b)** The thoracic duct empties into the left subclavian vein. The right lymphatic duct drains into the right subclavian vein. **(c)** Anterior view of a dissection of the thoracic duct, cisterna chyli, and adjacent blood vessels. ATLAS: Plate 48a,b

direct contact. These lymphocytes are the primary cells involved in the production of *cell-mediated immunity*, or *cellular immunity*.

- **Helper T cells**, which stimulate the activation and function of both T cells and B cells. → *helps produce antibodies.*
- **Suppressor T cells**, which inhibit the activation and function of both T cells and B cells.

The interplay between suppressor and helper T cells helps establish and control the sensitivity of the immune response. For this reason, these cells are also known as *regulatory T cells*.

We will examine cytotoxic and regulatory T cells in the course of this chapter. Other types of T cells also participate in the immune response. For example, *inflammatory T cells* stimulate regional inflammation and local defenses in an injured tissue, and *suppressor/inducer T cells* suppress B cell activity but stimulate other T cells.

B cells account for 10–15 percent of circulating lymphocytes. When stimulated, B cells can differentiate into **plasma cells**, which are responsible for the production and secretion of *antibodies*—soluble proteins also known as *immunoglobulins*. ∞ p. 654 These proteins bind to specific chemical targets called **antigens**. Most antigens are pathogens, parts or products of pathogens, or other foreign compounds. Most antigens are proteins, but some lipids, polysaccharides, and nucleic acids can also stimulate antibody production. The binding of an antibody to its target antigen starts a chain reaction leading to the destruction of the target compound or organism. B cells are responsible for *antibody-mediated immunity*, which is also known as *humoral* ("liquid") *immunity* because antibodies occur in body fluids.

The remaining 5–10 percent of circulating lymphocytes are NK cells, also known as **large granular lymphocytes**. These lymphocytes attack foreign cells, normal cells infected with viruses, and cancer cells that appear in normal tissues. Their continuous "policing" of peripheral tissues has been called *immunological surveillance*.

not evenly distributed.

Life Span and Circulation of Lymphocytes

The various types of lymphocytes are not evenly distributed in the blood, bone marrow, spleen, thymus, and peripheral lymphoid tissues. The ratio of B cells to T cells varies among tissues and organs. For example, B cells are seldom found in the thymus, whereas in blood, T cells outnumber B cells by a ratio of 8:1.

The lymphocytes in these organs are visitors, not residents. All types of lymphocytes move throughout the body, wandering through tissues and then entering blood vessels or lymphatic vessels for transport.

T cells move relatively quickly. For example, a wandering T cell may spend about 30 minutes in the blood, 5–6 hours in the spleen, and 15–20 hours in a lymph node. B cells, which are responsible for antibody production, move more slowly. A typical B cell spends about 30 hours in a lymph node before moving on.

Lymphocytes have relatively long life spans. Roughly 80 percent survive 4 years, and some last 20 years or more. Throughout your life, you maintain normal lymphocyte populations by producing new lymphocytes in your bone marrow and lymphoid tissues.

Lymphocyte Production   *erythro → bone marrow*

In Chapter 19, we discussed *hemopoiesis*—the formation of the cellular elements of blood. ∞ pp. 660, 669 In adults, *erythropoiesis* (red blood cell formation) is normally confined to bone marrow, but lymphocyte production, or **lymphopoiesis** (lim-fō-poy-Ē-sis), involves the bone marrow, thymus, and peripheral lymphoid tissues (**Figure 22–5**).   *(T cell)*

Bone marrow plays the primary role in the maintenance of normal lymphocyte populations. Hemocytoblast divisions in the bone marrow of adults generate the lymphoid stem cells that produce all types of lymphocytes. Two distinct populations of lymphoid stem cells are produced in the bone marrow.

One group of lymphoid stem cells remains in the bone marrow (**Figure 22–5a**). Divisions of these cells produce immature B cells and NK cells. B cell development involves intimate contact with large **stromal** (*stroma*, a bed) **cells** in the bone marrow. The cytoplasmic extensions of stromal cells contact or even wrap around the developing B cells. Stromal cells produce an immune system hormone, or *cytokine*, called *interleukin-7*, that promotes the differentiation of B cells. (We will consider cytokines and their varied effects in a later section.)

As they mature, B cells and NK cells enter the bloodstream and migrate to peripheral tissues (**Figure 22–5c**). Most of the B cells move into lymph nodes, the spleen, or other lymphoid tissues. The NK cells migrate throughout the body, moving through peripheral tissues in search of abnormal cells.

The second group of lymphoid stem cells migrates to the thymus (**Figure 22–5b**). There, these cells and their descendants develop further in an environment that is isolated from the general circulation by the **blood–thymus barrier**. Under the influence of thymic hormones, the lymphoid stem cells divide repeatedly, producing the various kinds of T cells. At least seven thymic hormones have been identified, although their precise functions and interactions have yet to be determined.

When their development is nearing completion, T cells reenter the bloodstream and return to the bone marrow. They also travel to peripheral tissues, including lymphoid tissues and organs, such as the spleen (**Figure 22–5c**).

The T cells and B cells that migrate from their sites of origin retain the ability to divide. Their divisions produce daughter cells of the same type; for example, a dividing B cell produces other B cells, not T cells or NK cells. As we will see, the ability of specific types of lymphocytes to increase in number is crucial to the success of the immune response.

Figure 22–5   The Derivation and Distribution of Lymphocytes. (a) Multipotent stem cell divisions in bone marrow produce stem cells with two fates. One group remains in the bone marrow, producing daughter cells that mature into B cells or NK cells. **(b)** The other group migrates to the thymus, where subsequent divisions produce daughter cells that mature into T cells. The mature B cells, NK cells, and T cells circulate throughout the body in the bloodstream, reaching and (if necessary) defending peripheral tissues from infection and disease **(c)**.

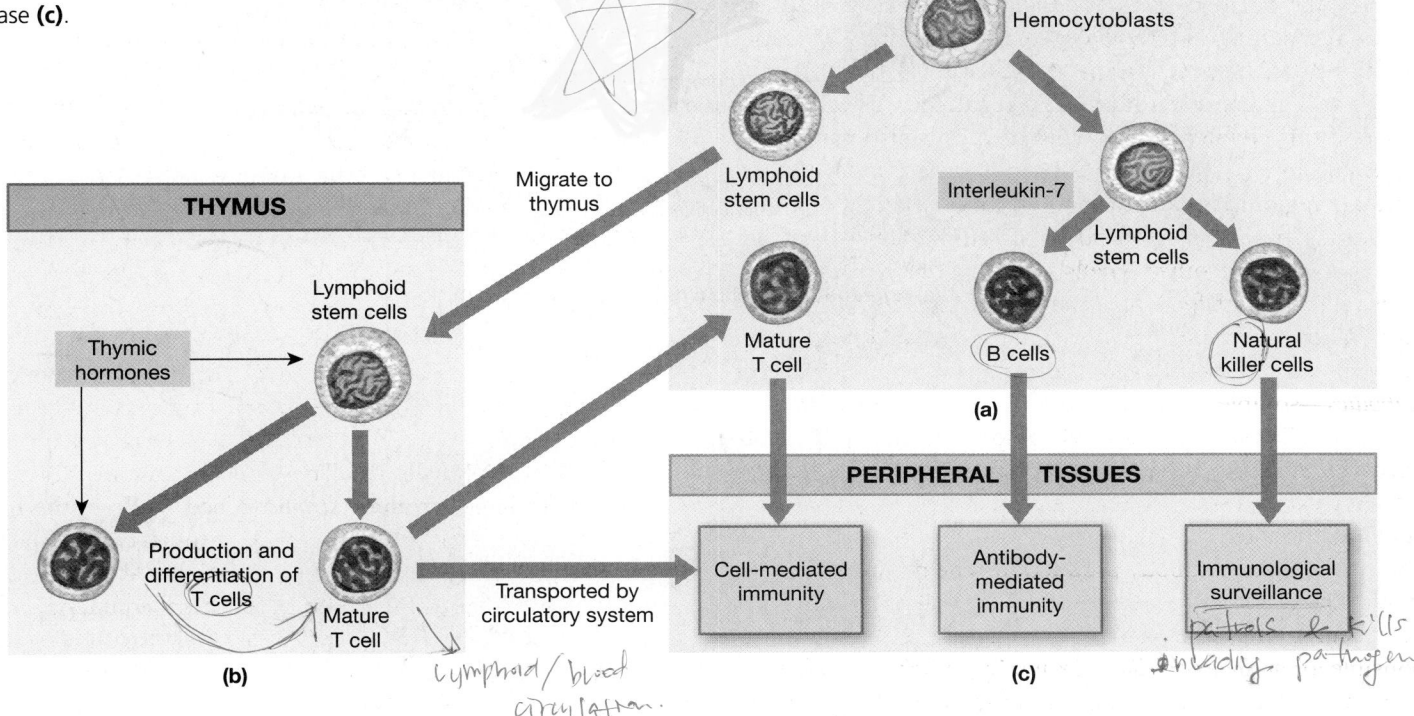

Lymphoid Tissues

Lymphoid tissues are connective tissues dominated by lymphocytes. In a **lymphoid nodule**, or *lymphatic nodule*, the lymphocytes are densely packed in an area of areolar tissue. In many areas, lymphoid nodules form large clusters (**Figure 22–6**). Lymphoid nodules occur in the connective tissue deep to the epithelia lining the respiratory tract, where they are known as *tonsils*, and along the digestive and urinary tracts. They are also found within more complex lymphoid organs, such as lymph nodes or the spleen. A single nodule averages about a millimeter in diameter, but the boundaries are not distinct, because no fibrous capsule surrounds them. Each nodule often has a central zone called a **germinal center**, which contains dividing lymphocytes (**Figure 22–6b**).

MALT

The collection of lymphoid tissues linked with the digestive system is called the **mucosa-associated lymphoid tissue (MALT)**. Clusters of lymphoid nodules deep to the epithelial lining of the intestine are known as **aggregated lymphoid nodules**, or *Peyer patches* (**Figure 22–6a**). In addition, the walls of the *appendix*, or *vermiform* ("worm-shaped") *appendix*—a blind pouch that originates near the junction between the small and large intestines—contain a mass of fused lymphoid nodules.

Tonsils

The **tonsils** are large lymphoid nodules in the walls of the pharynx (**Figure 22–6b**). Most people have five tonsils. Left and right **palatine tonsils** are located at the posterior, inferior margin of the oral cavity, along the boundary with the pharynx. A single **pharyngeal tonsil**, often called the *adenoid*, lies in the posterior superior wall of the nasopharynx, and a pair of **lingual tonsils** lie deep to the mucous epithelium covering the base (pharyngeal portion) of the tongue. Because of their location, the latter are usually not visible unless they become infected and swollen, a condition known as **tonsillitis**.

Lymphoid Organs

A fibrous connective tissue capsule separates lymphoid organs—the *lymph nodes*, the *thymus*, and the *spleen*—from surrounding tissues.

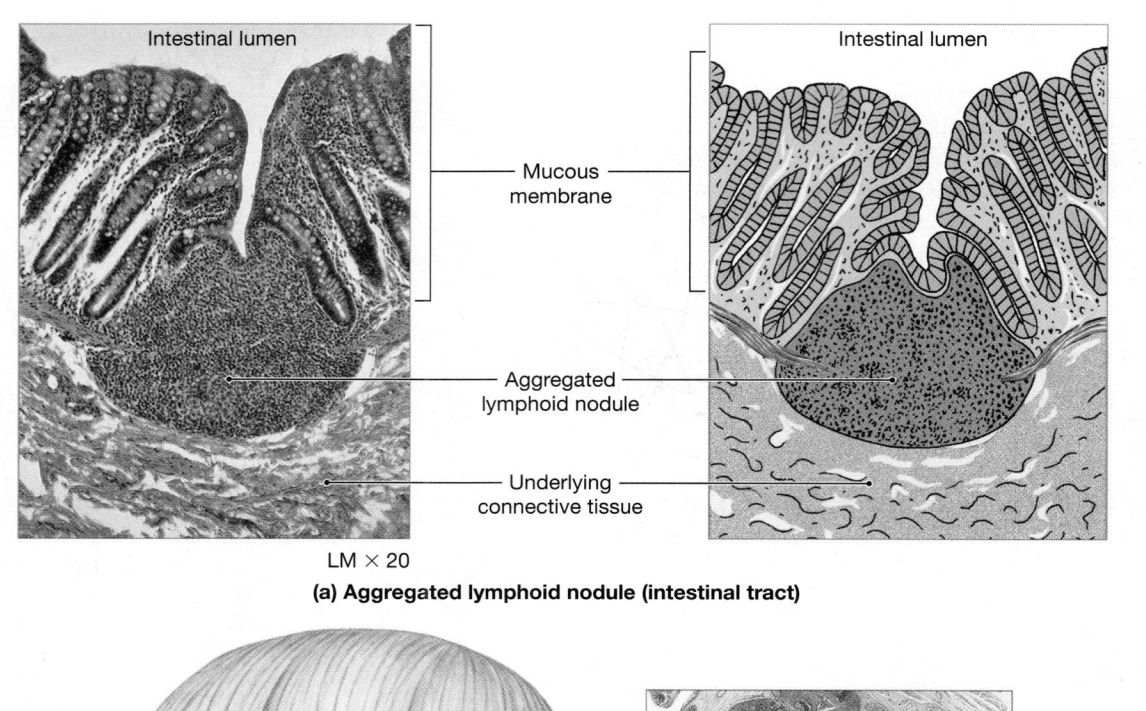

LM × 20

(a) Aggregated lymphoid nodule (intestinal tract)

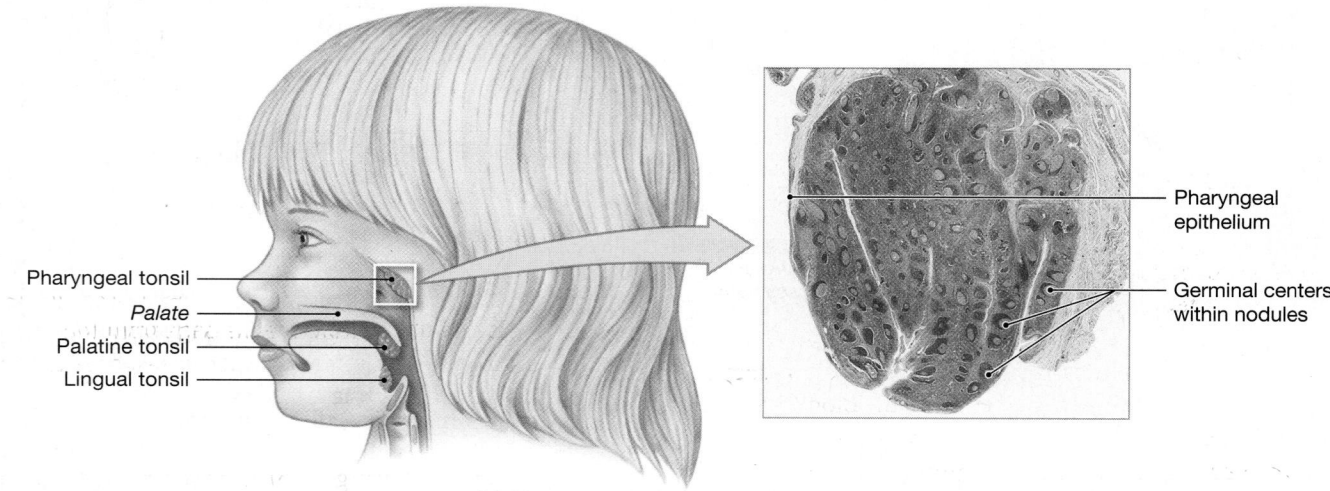

(b) Pharyngeal tonsil (respiratory tract)

Figure 22–6   Lymphoid Nodules. (a) A representative lymphoid nodule in section. **(b)** The positions of the tonsils and a tonsil in section. Notice the relatively pale germinal centers, where lymphocyte cell divisions occur.

Lymph Nodes

Lymph nodes are small lymphoid organs ranging in diameter from 1 mm to 25 mm (to about 1 in.). **Figure 22–1** shows the general pattern of lymph node distribution in the body. Each lymph node is covered by a dense connective tissue capsule (**Figure 22–7**). Bundles of collagen fibers extend from the capsule into the interior of the node. These fibrous partitions are called **trabeculae** (*trabecula*, a beam).

The shape of a typical lymph node resembles that of a kidney bean (**Figure 22–7**). Blood vessels and nerves reach the lymph node at the **hilum**, a shallow indentation. Two sets of lymphatic vessels, afferent lymphatics and efferent lymphatics, are connected to each lymph node. **Afferent** (*afferens*, to bring to) **lymphatics** carry lymph to the lymph node from pe-

ripheral tissues. The afferent lymphatics penetrate the capsule of the lymph node on the side opposite the hilum. **Efferent** (*efferens*, to bring out) **lymphatics** leave the lymph node at the hilum. These vessels carry lymph away from the lymph node and toward the venous circulation.

Lymph Flow. Lymph delivered by the afferent lymphatics flows through the lymph node within a network of sinuses, open passageways with incomplete walls (**Figure 22–7**). Lymph first enters a *subcapsular space* (formerly called the *subcapsular sinus*), which contains a meshwork of branching reticular fibers, macrophages, and dendritic cells. **Dendritic cells** are involved in the initiation of the immune response; we will consider their role in a later section. After passing through

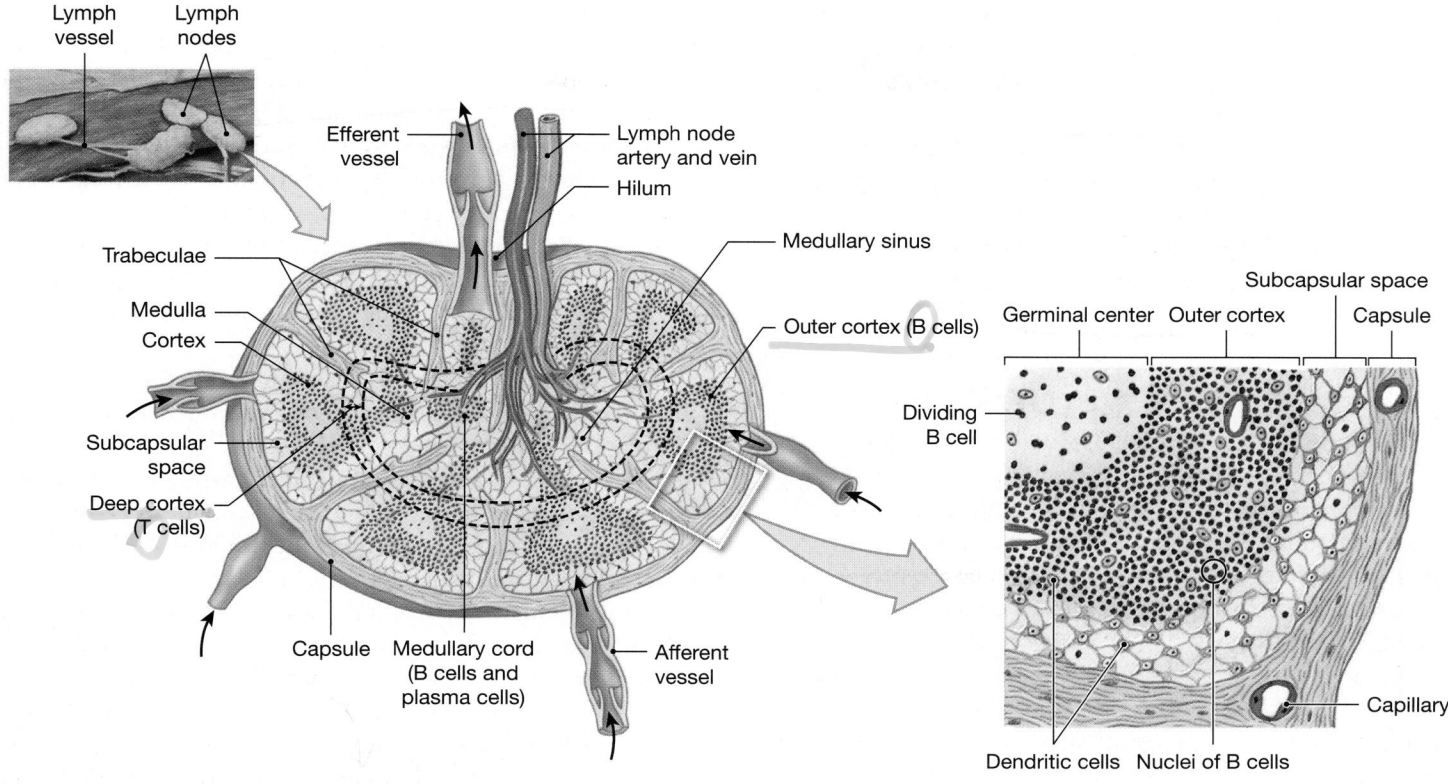

Figure 22–7 **The Structure of a Lymph Node.** ATLAS: Plate 70a

the subcapsular space, lymph flows through the **outer cortex** of the node. The outer cortex contains B cells within germinal centers that resemble those of lymphoid nodules.

Lymph then continues through lymph sinuses in the **deep cortex** (*paracortical area*). Lymphocytes leave the bloodstream and enter the lymph node by crossing the walls of blood vessels within the deep cortex, which is dominated by T cells.

After flowing through the sinuses of the deep cortex, lymph continues into the core, or **medulla**, of the lymph node. The medulla contains B cells and plasma cells organized into elongate masses known as **medullary cords**. After passing through a network of sinuses in the medulla, lymph enters the efferent lymphatics at the hilum.

Lymph Node Function. A lymph node functions like a kitchen water filter, purifying lymph before it reaches the venous circulation. As lymph flows through a lymph node, at least 99 percent of the antigens in the lymph are removed. Fixed macrophages in the walls of the lymphatic sinuses engulf debris or pathogens in lymph as it flows past. Antigens removed in this way are then processed by the macrophages and "presented" to nearby lymphocytes. Other antigens bind to receptors on the surfaces of dendritic cells, where they can stimulate lymphocyte activity. This process—*antigen presentation*—is generally the first step in the activation of the immune response.

Tips&Tricks

Helper T cells "help" translate the message from the antigen-presenting cells of the nonspecific response to the cells of the specific immune responses.

In addition to filtering, lymph nodes provide an early-warning system. Any infection or other abnormality in a peripheral tissue will introduce abnormal antigens into the interstitial fluid, and thus into the lymph leaving the area. These antigens then stimulate macrophages and lymphocytes in nearby lymph nodes.

To protect a house against intrusion, you might guard all the entrances and exits or place traps by the windows and doors. The distribution of lymphoid tissues and lymph nodes follows such a pattern. The largest lymph nodes are located where peripheral lymphatics connect with the trunk, such as in the groin, the axillae, and the base of the neck. These nodes are often called *lymph glands*. Because lymph is monitored in the cervical, inguinal, or axillary lymph nodes, potential problems can be detected and dealt with before they affect the vital organs of the trunk. Aggregations of lymph nodes also exist in the mesenteries of the gut, near the trachea and passageways leading to the lungs, and in association with the thoracic duct. These lymph nodes protect against pathogens and other antigens within the digestive and respiratory systems.

A minor injury commonly produces a slight enlargement of the nodes along the lymphatic vessels draining the region. This sign, often called "swollen glands," typically indicates inflammation in peripheral structures. The enlargement generally results from an increase in the number of lymphocytes and phagocytes in the node in response to a minor, localized infection. Chronic or excessive enlargement of lymph nodes constitutes **lymphadenopathy** (lim-fad-e-NOP-a-thē), a condition that may occur in response to bacterial or viral infections, endocrine disorders, or cancer.

CLINICAL NOTE

Cancer and the Lymphoid System Lymphatic vessels are located in almost all portions of the body except the central nervous system, and lymphatic capillaries offer little resistance to the passage of cancer cells. As a result, metastasizing cancer cells commonly spread along lymphatic vessels. Under these circumstances, the lymph nodes serve as "way stations" for migrating cancer cells. Thus, an analysis of lymph nodes can provide information on the spread of the cancer cells, and such information helps determine the appropriate therapies. In Chapter 29, we will discuss one example: identifying the stages of breast cancer by the degree of nodal involvement. *Lymphomas* are one group of cancers originating in the cells of the lymphoid system.

The Thymus

The **thymus** is a pink, grainy organ located in the mediastinum, generally just posterior to the sternum (**Figure 22–8a,b**). In newborn infants and young children, the thymus is relatively large, commonly extending from the base of the neck to the superior border of the heart. The thymus reaches its greatest size relative to body size in the first year or two after birth. (Although the organ continues to increase in mass throughout childhood, the body as a whole grows even faster, so the size of the thymus relative to that of the other organs in the mediastinum gradually decreases.) The thymus reaches its maximum absolute size, at a weight of about 40 g (1.4 oz), just before puberty. After puberty, it gradually diminishes in size and becomes increasingly fibrous, a process called *involution*. By the time an individual reaches age 50, the thymus may weigh less than 12 g (0.3 oz). The gradual decrease in the size and secretory abilities of the thymus may make elderly individuals more susceptible to disease.

The capsule that covers the thymus divides it into two **thymic lobes** (**Figure 22–8b**). Fibrous partitions called **septa** (singular, *septum*) originate at the capsule and divide the lobes into **lobules** averaging 2 mm in diameter (**Figure 22–8b,c**). Each lobule consists of a densely packed outer cortex and a paler, central **medulla**. Lymphocytes in the cortex are dividing; as the T cells mature, they migrate into the medulla. After roughly three weeks, these T cells leave the thymus by entering one of the medullary blood vessels.

Lymphocytes in the cortex are arranged in clusters that are completely surrounded by **reticular epithelial cells**. These cells, which developed from epithelial cells of the embryo, also encircle the blood vessels of the cortex. The reticular epithelial cells maintain the blood–thymus barrier and secrete the thymic hormones that stimulate stem cell divisions and T cell differentiation.

As they mature, T cells leave the cortex and enter the medulla of the thymus. The medulla has no blood–thymus barrier. The reticular epithelial cells in the medulla cluster together in concentric layers, forming distinctive structures known as **thymic (Hassall) corpuscles** (**Figure 22–8d**). Despite their imposing appearance, the function of thymic corpuscles remains unknown. T cells in the medulla can enter or leave the bloodstream across the walls of blood vessels in this region or within one of the efferent lymphatics that collect lymph from the thymus.

The thymus produces several hormones that are important to the development and maintenance of normal immunological defenses. *Thymosin* (THĪ-mō-sin) is the name originally given to an extract from the thymus that promotes the development and maturation of lymphocytes. This thymic extract actually contains several complementary hormones: *thymosin-a*, *thymosin-b*, *thymosin V*, *thymopoietin*, *thymulin*, and others. The term *thymosins* is now sometimes used to refer to all thymic hormones.

The Spleen

The adult **spleen** contains the largest collection of lymphoid tissue in the body. In essence, the spleen performs the same functions for blood that lymph nodes perform for lymph. Functions of the spleen can be summarized as (1) the removal of abnormal blood cells and other blood components by phagocytosis, (2) the storage of iron recycled from red blood cells, and (3) the initiation of immune responses by B cells and T cells in response to antigens in circulating blood.

Anatomy of the Spleen. The spleen is about 12 cm (5 in.) long and weighs, on average, nearly 160 g (5.6 oz). In gross dissection, the spleen is deep red, owing to the blood it contains. The spleen lies along the curving lateral border of the stomach, extending between the 9th and 11th ribs on the left side. It is attached to the lateral border of the stomach by the **gastrosplenic ligament**, a broad band of mesentery (**Figure 22–9a**).

The spleen has a soft consistency, so its shape primarily reflects its association with the structures around it. The spleen is in contact with the stomach, the left kidney, and the muscu-

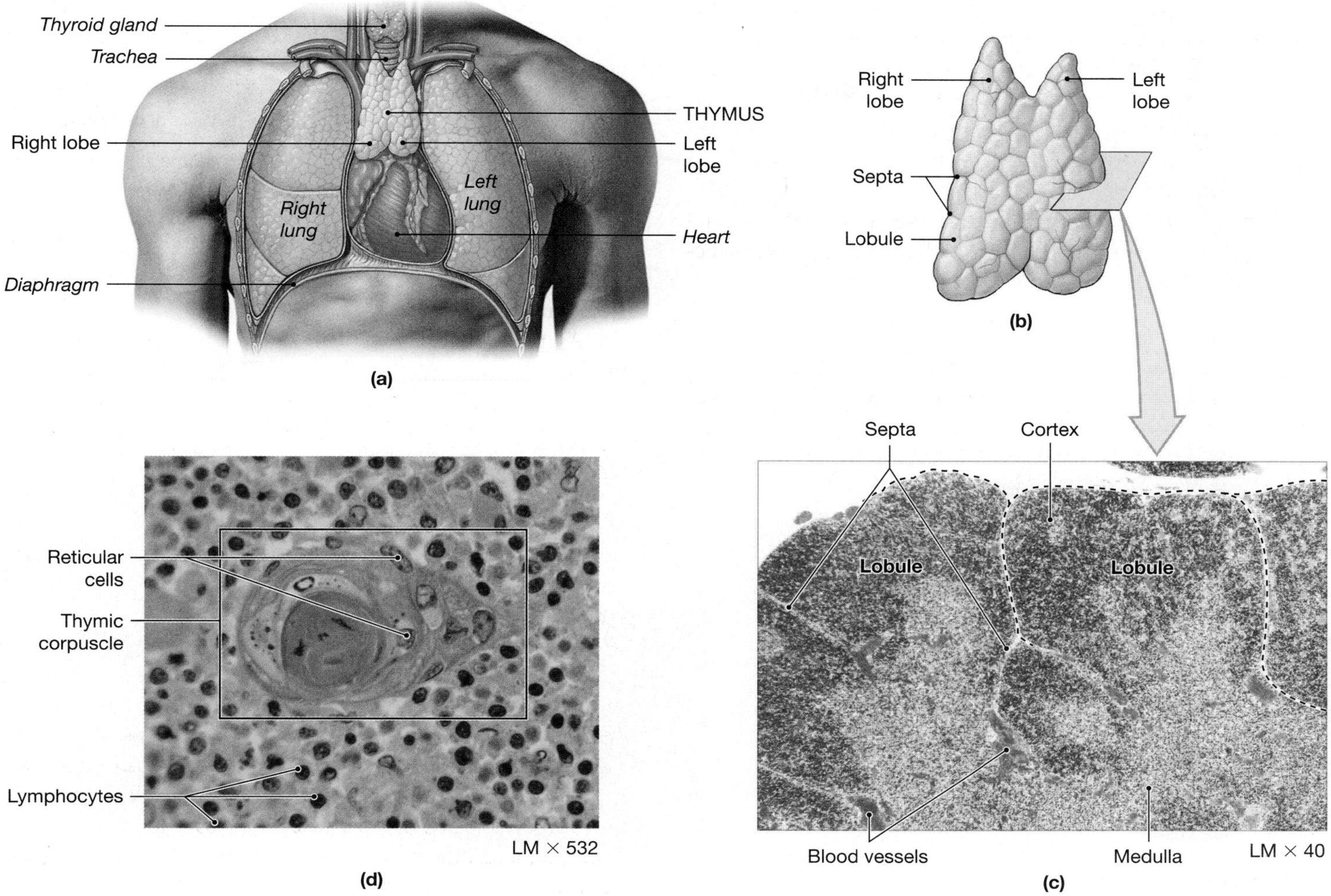

Thyroid gland

Trachea

Right lobe

Right lung

Left lung

Diaphragm

THYMUS

Left lobe

Heart

(a)

Right lobe

Left lobe

Septa

Lobule

(b)

Septa   Cortex

Lobule   Lobule

Blood vessels   Medulla   LM × 40

(c)

Reticular cells

Thymic corpuscle

Lymphocytes

LM × 532

(d)

Figure 22–8   The Thymus. (a) The appearance and position of the thymus in relation to other organs in the chest. **(b)** Anatomical landmarks on the thymus. **(c)** Fibrous septa divide the tissue of the thymus into lobules resembling interconnected lymphoid nodules. **(d)** Higher magnification reveals the unusual structure of thymic corpuscles. The small cells are lymphocytes in various stages of development.
ATLAS: Plate 47a

lar diaphragm. The *diaphragmatic surface* is smooth and convex, conforming to the shape of the diaphragm and body wall. The *visceral surface* contains indentations that conform to the shape of the stomach (the *gastric area*) and the kidney (the *renal area*) (**Figure 22–9b**). Splenic blood vessels (the *splenic artery* and *splenic vein*) and lymphatic vessels communicate with the spleen on the visceral surface at the **hilum**, a groove marking the border between the gastric and renal areas.

Histology of the Spleen. The spleen is surrounded by a capsule containing collagen and elastic fibers.[1] The cellular components within constitute the **pulp** of the spleen

[1]The spleens of dogs, cats, and other mammals of the order *Carnivora* have extensive layers of smooth muscle that can contract to eject blood into the bloodstream. The human spleen lacks those muscle layers and cannot contract.

(**Figure 22–9c**). **Red pulp** contains large quantities of red blood cells, whereas **white pulp** resembles lymphoid nodules.

The splenic artery enters at the hilum and branches to produce a number of arteries that radiate outward toward the capsule. These **trabecular arteries** in turn branch extensively, and their finer branches are surrounded by areas of white pulp. Capillaries then discharge the blood into the red pulp.

The cell population of the red pulp includes all the normal components of circulating blood, plus fixed and free macrophages. The structural framework of the red pulp consists of a network of reticular fibers. The blood passes through this meshwork and enters large sinusoids, also lined by fixed macrophages. The sinusoids empty into small veins, which ultimately collect into **trabecular veins** that continue toward the hilum.

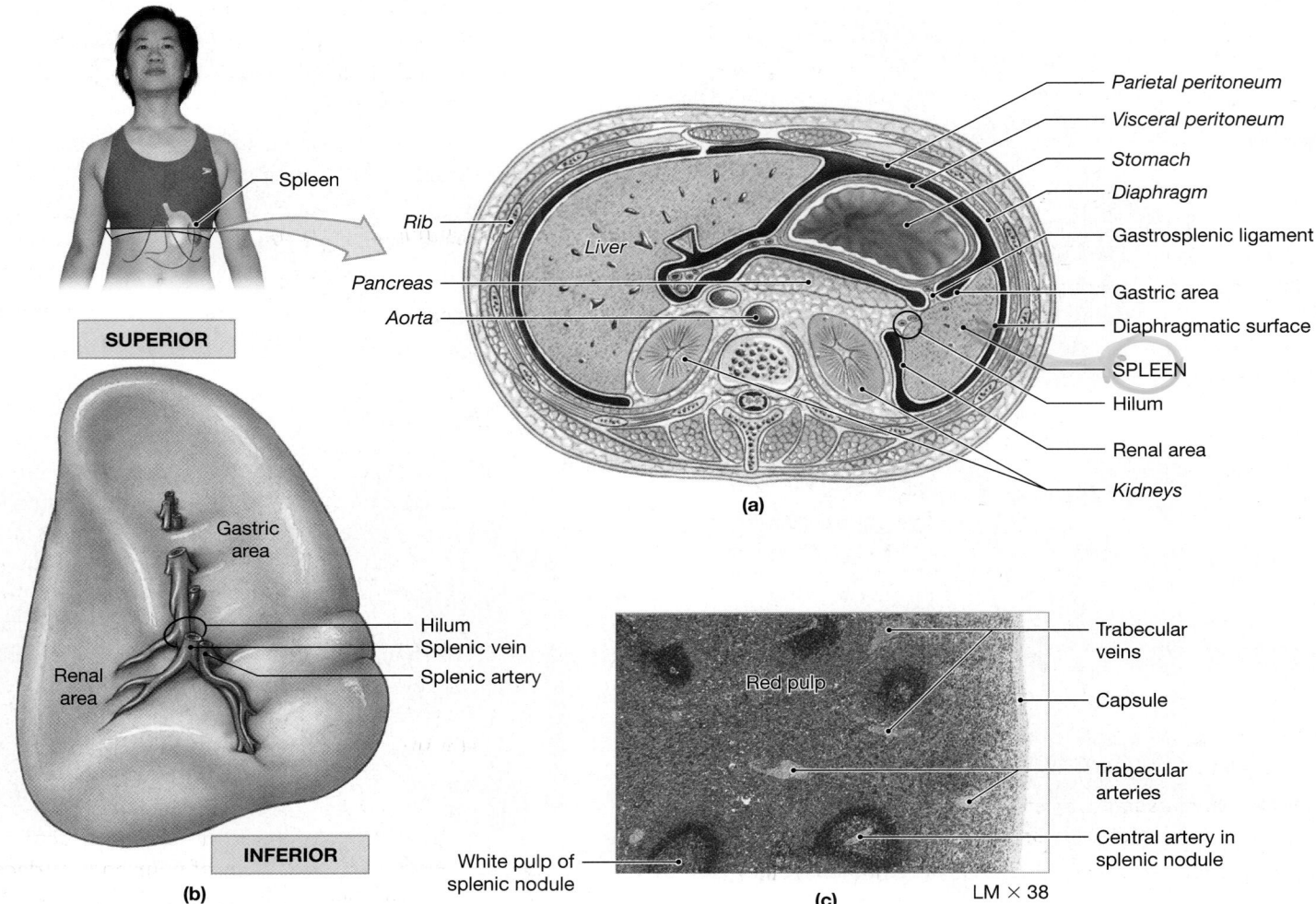

Figure 22–9   The Spleen. (a) A transverse section through the trunk, showing the typical position of the spleen within the abdominopelvic cavity. The shape of the spleen roughly conforms to the shapes of adjacent organs. **(b)** The external appearance of the intact spleen, showing major anatomical landmarks. Compare this view with that of part (a). **(c)** The histological appearance of the spleen. White pulp is dominated by lymphocytes; it appears purple because the nuclei of lymphocytes stain very darkly. Red pulp contains a preponderance of red blood cells.
ATLAS: Plates 49e; 55; 56e,f; 57a,b

This circulatory arrangement gives the phagocytes of the spleen an opportunity to identify and engulf any damaged or infected cells in circulating blood. Lymphocytes are scattered throughout the red pulp, and the area surrounding the white pulp has a high concentration of macrophages and dendritic cells. Thus, any microorganism or other antigen in the blood will quickly come to the attention of the splenic lymphocytes.

The spleen tears so easily that a seemingly minor blow to the left side of the abdomen can rupture the capsule. The result is serious internal bleeding and eventual circulatory shock. Such an injury is a known risk of contact sports (such as football and hockey) and of more solitary athletic activities, such as skiing and sledding.

Because the spleen is relatively fragile, it is very difficult to repair surgically. (Sutures typically tear out before they

have been tensed enough to stop the bleeding.) A severely ruptured spleen is removed, a process called a **splenectomy** (splē-NEK-tō-mē). A person without a spleen survives but has a greater risk of bacterial infection (particularly involving pneumococcal bacteria) than do individuals with a functional spleen.

The Lymphoid System and Body Defenses

The human body has multiple defense mechanisms that together provide resistance—the ability to fight infection, illness, and disease. Body defenses can be sorted into two general categories:

1. **Nonspecific defenses** do not distinguish one type of threat from another. Their response is the same, regardless of the type of invading agent. These defenses, which are present at birth, include *physical barriers, phagocytic cells, immunological surveillance, interferons, complement, inflammation,* and *fever.* They provide a defensive capability known as **nonspecific resistance**.

2. **Specific defenses** protect against particular threats. For example, a specific defense may protect against infection by one type of bacterium, but be ineffective against other bacteria and viruses. Many specific defenses develop after birth as a result of accidental or deliberate exposure to environmental hazards. *Specific defenses depend on the activities of specific lymphocytes.* The body's specific defenses produce a state of protection known as immunity, or **specific resistance**.

Nonspecific and specific resistances are complementary. Both must function normally to provide adequate resistance to infection and disease.

CHECKPOINT

3. List the components of the lymphoid system.

4. How would blockage of the thoracic duct affect the circulation of lymph?

5. If the thymus failed to produce thymic hormones, which population of lymphocytes would be affected?

6. Why do lymph nodes enlarge during some infections?

See the blue Answers tab at the end of the book.

22-3 Nonspecific defenses do not discriminate between potential threats and respond the same regardless of the invader

Nonspecific defenses prevent the approach, deny the entry, or limit the spread of microorganisms or other environmental hazards. Seven major categories of nonspecific defenses are summarized in **Figure 22–10**.

1. *Physical barriers* keep hazardous organisms and materials outside the body. For example, a mosquito that lands on your head may be unable to reach the surface of the scalp if you have a full head of hair.

2. *Phagocytes* are cells that engulf pathogens and cell debris. Examples of phagocytes are the macrophages of peripheral tissues and the microphages of blood.

3. *Immunological surveillance* is the destruction of abnormal cells by NK cells in peripheral tissues.

4. *Interferons* are chemical messengers that coordinate the defenses against viral infections.

5. *Complement* is a system of circulating proteins that assists antibodies in the destruction of pathogens.

6. The *inflammatory response* is a localized, tissue-level response that tends to limit the spread of an injury or infection.

7. *Fever* is an elevation of body temperature that accelerates tissue metabolism and the activity of defenses.

Physical Barriers

To cause trouble, an antigenic compound or a pathogen must enter body tissues, which requires crossing an epithelium—either at the skin or across a mucous membrane. The epithelial covering of the skin has multiple layers, a keratin coating, and a network of desmosomes that lock adjacent cells together. ∞ pp. 160–161 These barriers provide very effective protection for underlying tissues. Even along the more delicate internal passageways of the respiratory, digestive, and urinary tracts, epithelial cells are tied together by tight junctions and generally are supported by a dense and fibrous basal lamina.

In addition to the barriers posed by epithelial cells, most epithelia are protected by specialized accessory structures and secretions. The hairs on most areas of your body surface provide some protection against mechanical abrasion (especially on the scalp), and they often prevent hazardous materials or insects from contacting your skin surface. The epidermal surface also receives the secretions of sebaceous and sweat glands. These secretions, which flush the surface to wash away microorganisms and chemical agents, may also contain bactericidal chemicals, destructive enzymes (*lysozymes*), and antibodies.

The epithelia lining the digestive, respiratory, urinary, and reproductive tracts are more delicate, but they are equally well defended. Mucus bathes most surfaces of your digestive tract, and your stomach contains a powerful acid that can destroy many pathogens. Mucus moves across the respiratory tract lining, urine flushes the urinary passageways, and glandular secretions do the same for the reproductive tract. Special enzymes, antibodies, and an acidic pH add to the effectiveness of these secretions.

Phagocytes

Phagocytes perform janitorial and police services in peripheral tissues, removing cellular debris and responding to invasion by

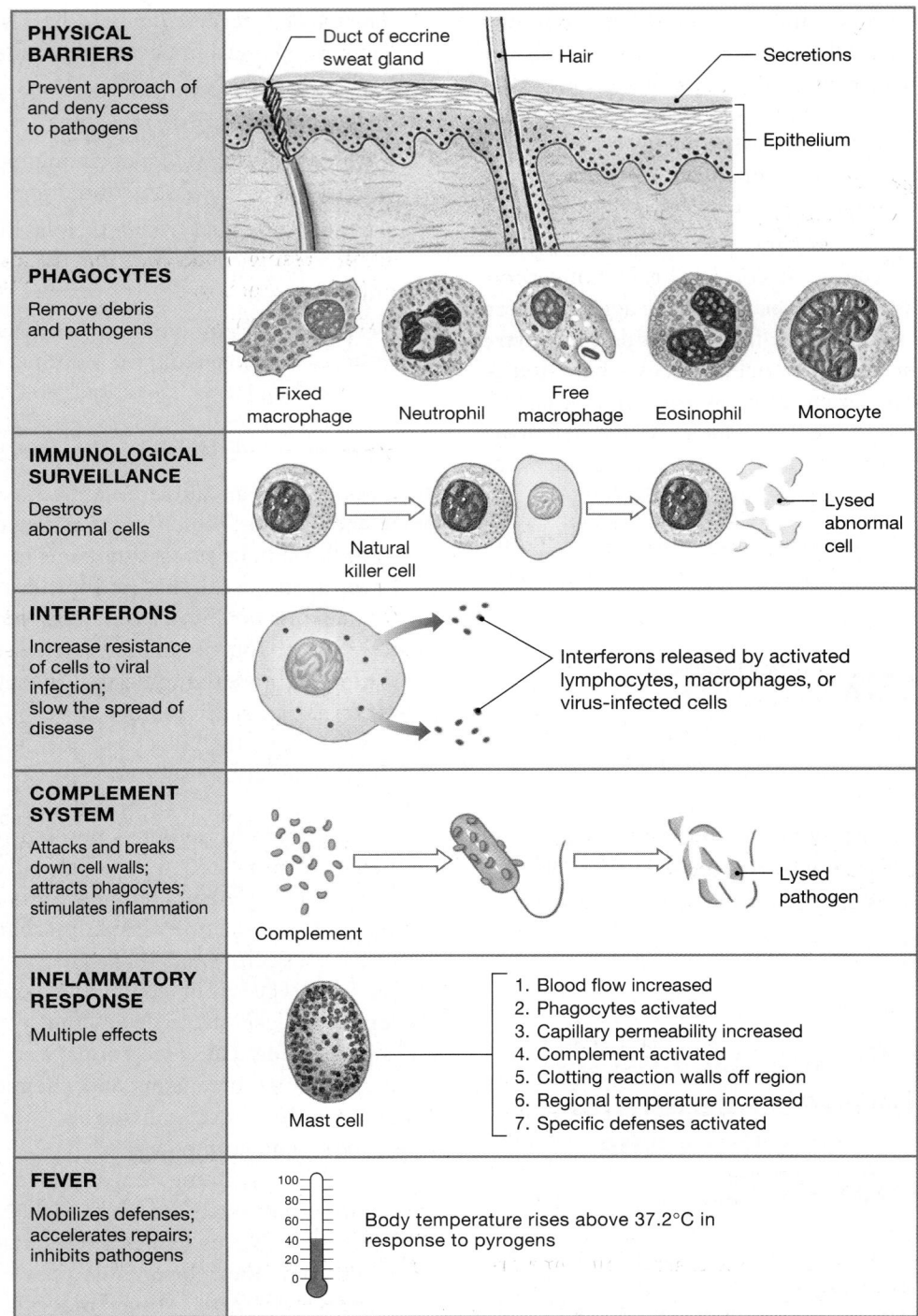

PHYSICAL BARRIERS Prevent approach of and deny access to pathogens	Duct of eccrine sweat gland — Hair — Secretions — Epithelium
PHAGOCYTES Remove debris and pathogens	Fixed macrophage — Neutrophil — Free macrophage — Eosinophil — Monocyte
IMMUNOLOGICAL SURVEILLANCE Destroys abnormal cells	Natural killer cell — Lysed abnormal cell
INTERFERONS Increase resistance of cells to viral infection; slow the spread of disease	Interferons released by activated lymphocytes, macrophages, or virus-infected cells
COMPLEMENT SYSTEM Attacks and breaks down cell walls; attracts phagocytes; stimulates inflammation	Complement — Lysed pathogen
INFLAMMATORY RESPONSE Multiple effects	Mast cell — 1. Blood flow increased 2. Phagocytes activated 3. Capillary permeability increased 4. Complement activated 5. Clotting reaction walls off region 6. Regional temperature increased 7. Specific defenses activated
FEVER Mobilizes defenses; accelerates repairs; inhibits pathogens	Body temperature rises above 37.2°C in response to pyrogens

Figure 22–10   Nonspecific Defenses. Nonspecific defenses deny pathogens access to the body or destroy them without distinguishing among specific types.

foreign compounds or pathogens. Phagocytes represent the "first line of cellular defense" against pathogenic invasion. Many phagocytes attack and remove microorganisms even before lymphocytes detect their presence. The human body has two general classes of phagocytic cells: *microphages* and *macrophages*.

Microphages

Microphages are the neutrophils and eosinophils that normally circulate in the blood. These phagocytic cells leave the bloodstream and enter peripheral tissues that have been subjected to injury or infection. As noted in Chapter 19, neutrophils are abundant, mobile, and quick to phagocytize cellular debris or invading bacteria. ∞ p. 667 Eosinophils, which are less abundant, target foreign compounds or pathogens that have been coated with antibodies.

Macrophages

Macrophages are large, actively phagocytic cells. Your body contains several types of macrophages, and most are derived from the monocytes of the circulating blood. Typically, macrophages are either fixed in position or freely mobile, and as a result they are usually classified as fixed macrophages or free macrophages. The distinction is not absolute, however; during an infection, fixed macrophages may lose their attachments and begin roaming around the damaged tissue.

Although no organs or tissues are dominated by phagocytes, almost every tissue in the body shelters resident or visiting macrophages. This relatively diffuse collection of phagocytic cells has been called the **monocyte–macrophage system**, or the *reticuloendothelial system*.

An activated macrophage may respond to a pathogen in several ways:

- It may engulf a pathogen or other foreign object and destroy it with lysosomal enzymes.
- It may bind to or remove a pathogen from the interstitial fluid, but be unable to destroy the invader until assisted by other cells.
- It may destroy its target by releasing toxic chemicals, such as *tumor necrosis factor*, nitric oxide, or hydrogen peroxide, into the interstitial fluid.

We will consider those responses further in a later section.

Fixed Macrophages. **Fixed macrophages**, or *histiocytes*, are permanent residents of specific tissues and organs. These cells are normally incapable of movement, so their targets must diffuse or otherwise move through the surrounding tissue until they are within range. Fixed macrophages are scattered among connective tissues, usually in close association with collagen or reticular fibers. Their presence has already been noted in the papillary and reticular layers of the dermis, in the subarachnoid space of the meninges, and in bone marrow. In some organs, the fixed macrophages have special names: **Microglia** are macrophages in the central nervous system, and **Kupffer cells** are macrophages located in and around the liver sinusoids.

Free Macrophages. **Free macrophages**, or *wandering macrophages*, travel throughout the body, arriving at the site of an injury by migrating through adjacent tissues or by recruitment from the circulating blood. Some tissues contain free macrophages with distinctive characteristics; for example, the exchange surfaces of the lungs are monitored by **alveolar macrophages**, also known as *phagocytic dust cells*.

Movement and Phagocytosis

Free macrophages and microphages share a number of functional characteristics:

- Both can move through capillary walls by squeezing between adjacent endothelial cells, a process known as *emigration*, or *diapedesis*. ∞ p. 666 The endothelial cells in an injured area develop membrane "markers" that signal passing blood cells that something is wrong. The cells then attach to the endothelial lining and migrate into the surrounding tissues.

- Both may be attracted to or repelled by chemicals in the surrounding fluids, a phenomenon called **chemotaxis**. They are particularly sensitive to cytokines released by other body cells and to chemicals released by pathogens.

- For both, phagocytosis begins with **adhesion**, the attachment of the phagocyte to its target. In adhesion, receptors on the plasma membrane of the phagocyte bind to the surface of the target. Adhesion is followed by the formation of a vesicle containing the bound target (**Figure 3–22**, p. 98). The contents of the vesicle are digested once the vesicle fuses with lysosomes or peroxisomes.

All phagocytic cells function in much the same way, although the target of phagocytosis may differ from one type of phagocyte to another. The life span of an actively phagocytic cell can be rather brief. For example, most neutrophils die before they have engulfed more than 25 bacteria, and in an infection a neutrophil may attack that many in an hour.

Tips & Tricks

Membrane markers and chemotaxis are like putting up the flag on your mailbox: They signal the need for awareness and action.

Immunological Surveillance

The immune system generally ignores the body's own cells unless they become abnormal in some way. Natural killer (NK) cells are responsible for recognizing and destroying abnormal cells when they appear in peripheral tissues. The constant monitoring of normal tissues by NK cells is called **immunological surveillance**.

The plasma membrane of an abnormal cell generally contains antigens that are not found on the membranes of normal cells. NK cells recognize an abnormal cell by detecting the presence of those antigens. NK cells are much less selective about their targets than are other lymphocytes: They respond to a *variety* of abnormal antigens that may appear anywhere on a plasma membrane, and *any* membrane containing abnormal antigens will be attacked. As a result, NK cells are highly versatile: A single NK cell can attack bacteria in the interstitial fluid, body cells infected with viruses, or cancer cells.

NK cells also respond much more rapidly than T cells or B cells. The activation of T cells and B cells involves a relatively complex and time-consuming sequence of events; NK cells respond immediately on contact with an abnormal cell.

NK Cell Activation → attacks abnormal (cancer) cells, cells infected by virus.

Activated NK cells react in a predictable way (**Figure 22–11**):

Step 1 If a cell has unusual components in its plasma membrane, an NK cell recognizes that other cell as abnormal. Such recognition activates the NK cell, which then adheres to its target cell.

Step 2 The Golgi apparatus moves around the nucleus until the maturing face points directly toward the abnormal cell. The process might be compared to the rotation of a tank turret to point the cannon toward the enemy. A flood of secretory vesicles is then produced at the Golgi apparatus. These vesicles, which contain proteins called **perforins**, travel through the cytoplasm toward the cell surface. *creates network of pores*

Step 3 The perforins are released at the cell surface by exocytosis and diffuse across the narrow gap separating the NK cell from its target.

Step 4 On reaching the opposing plasma membrane, perforin molecules interact with one another and with the membrane to create a network of pores in it. These pores are large enough to permit the free passage of ions, proteins,

and other intracellular materials. As a result, the target cell can no longer maintain its internal environment, and it quickly disintegrates.

Tips&Tricks

Perforin gets its name because it **perfor**ates the target cell.

It is not clear why perforin does not affect the membrane of the NK cell itself. NK cell membranes contain a second protein, called *protectin*, which may be responsible for binding and inactivating perforin.

NK cells attack cancer cells and cells infected with viruses. Cancer cells probably appear throughout life, but their plasma membranes generally contain unusual proteins called **tumor-specific antigens**, which NK cells recognize as abnormal. The affected cells are then destroyed, preserving tissue integrity. Unfortunately, some cancer cells avoid detection, perhaps because they lack tumor-specific antigens or because these antigens are covered in some way. Other cancer cells are able to destroy the NK cells that detect them. This process of avoiding detection or neutralizing body defenses is called **immunological escape**. Once immunological escape has occurred, cancer cells can multiply and spread without interference by NK cells.

In viral infections, the viruses replicate inside cells, beyond the reach of circulating antibodies. However, infected cells incorporate viral antigens into their plasma membranes, and NK cells recognize these infected cells as abnormal. By destroying them, NK cells can slow or prevent the spread of a viral infection.

→ some cancer cells destroy NK cells

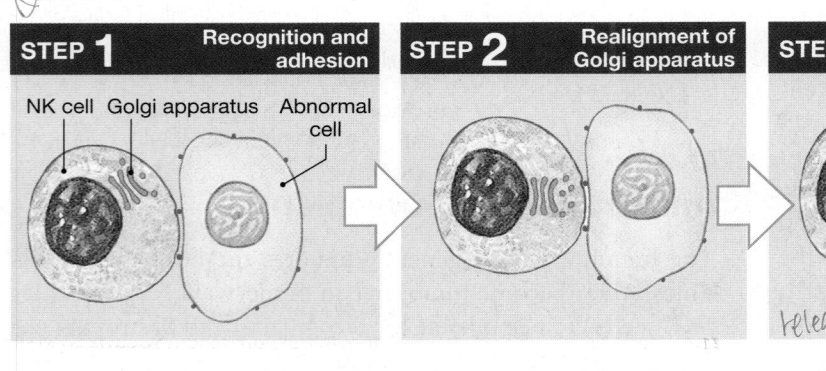

STEP 1 — Recognition and adhesion — NK cell — Golgi apparatus — Abnormal cell

STEP 2 — Realignment of Golgi apparatus

STEP 3 — Secretion of perforin — *released by exocytosis*

STEP 4 — Lysis of abnormal cell

Figure 22–11 How Natural Killer Cells Kill Cellular Targets. NK cell activity involves a series of overlapping steps. STEP 1: The NK cell recognizes another cell as abnormal if that cell's plasma membrane contains unusual proteins or other components. The NK cell then attaches to the target cell. STEP 2: The Golgi apparatus of the NK cell faces the target and secretory activity begins. STEP 3: Vesicles containing perforin are released by exocytosis. STEP 4: Perforin lyses the target cell by creating large pores in the plasma membrane.

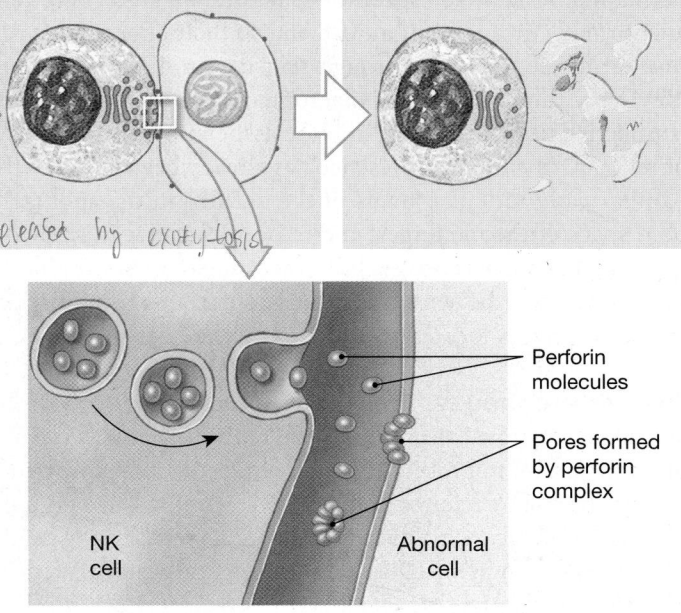

Perforin molecules

Pores formed by perforin complex

NK cell

Abnormal cell

Interferons

Interferons (in-ter-FĒR-onz) are small proteins released by activated lymphocytes and macrophages, and by tissue cells infected with viruses. On reaching the membrane of a normal cell, an interferon binds to surface receptors on the cell and, via second messengers, triggers the production of **antiviral proteins** in the cytoplasm. Antiviral proteins do not interfere with the entry of viruses, but they do interfere with viral replication inside the cell. In addition to their role in slowing the spread of viral infections, interferons stimulate the activities of macrophages and NK cells.

At least three types of interferons exist, each of which has additional specialized functions: (1) **Alpha-(α) interferons**, produced by several types of leukocytes, attract and stimulate NK cells; (2) **beta-(β) interferons**, secreted by fibrocytes slow inflammation in a damaged area; and (3) **gamma-(γ) interferons**, secreted by T cells and NK cells, stimulate macrophage activity. Most cells other than lymphocytes and macrophages respond to viral infection by secreting beta-interferon.

Interferons are examples of **cytokines** (SĪ-tō-kīnz)—chemical messengers released by tissue cells to coordinate local activities. Cytokines produced by most cells are used only for paracrine communication—that is, cell-to-cell communication within one tissue. However, cytokines released by defense cells also act as hormones, affecting cells and tissues throughout the body. We will discuss their role in the regulation of specific defenses in a later section.

Complement

Plasma contains 11 special **complement (C) proteins** that form the **complement system**. The term *complement* refers to the fact that this system complements the action of antibodies.

The complement proteins interact with one another in chain reactions, or *cascades*, reminiscent of those of the clotting system. **Figure 22–12** provides an overview of the complement system.

The activation of complement can occur by two different routes: the *classical pathway* and the *alternative pathway*.

Complement Activation: The Classical Pathway

The most rapid and effective activation of the complement system occurs through the **classical pathway** (**Figure 22–12**). The process begins when one of the complement proteins (*C1*) binds to an antibody molecule already attached to its specific antigen—in this case, a bacterial cell wall. The bound complement protein then acts as an enzyme, catalyzing a series of reactions involving other complement proteins. The classical pathway ends with the conversion of an inactive complement protein, *C3*, to an active form, *C3b*.

Complement Activation: The Alternative Pathway

A less effective, slower activation of the complement system occurs in the absence of antibody molecules. This **alternative pathway**, or *properdin pathway*, is important in the defense against bacteria, some parasites, and virus-infected cells. The pathway begins when several complement proteins—including **properdin** (or *factor P*), *factor B*, and *factor D*—interact in the plasma (**Figure 22–12**). This interaction can be triggered by exposure to foreign materials, such as the capsule of a bacterium. As does the classical pathway, the alternative pathway ends with the conversion of C3 to C3b.

Effects of Complement Activation

Known effects of complement activation include the following:

- *Stimulation of Inflammation*. Activated complement proteins enhance the release of histamine by mast cells and basophils. Histamine increases the degree of local inflammation and accelerates blood flow to the region.

- *Attraction of Phagocytes*. Activated complement proteins attract neutrophils and macrophages to the area, improving the likelihood that phagocytic cells will be able to cope with the injury or infection.

- *Enhancement of Phagocytosis*. A coating of complement proteins and antibodies both attracts phagocytes and makes the target easier to engulf. Macrophage membranes contain receptors that can detect and bind to complement proteins and bound antibodies. After binding, the pathogens are easily engulfed. The antibodies involved are called **opsonins**, and the effect is called **opsonization**.

- *Destruction of Target Plasma Membranes*. In the presence of C3b, five of the interacting complement proteins (C5–C9) bind to the plasma membrane, forming a functional unit called a **membrane attack complex (MAC)**. The MACs create pores in the membrane that are comparable to those produced by perforin and have the same effect: The target cell is soon destroyed.

Inflammation

Inflammation, or the *inflammatory response*, is a localized tissue response to injury. ∞ p. 142 Inflammation produces local swelling (*tumor*), redness (*rubor*), heat (*calor*), and pain (*dolor*); these are known as the *cardinal signs and symptoms of inflammation*. Many stimuli, including impact, abrasion, distortion, chemical irritation, infection by pathogens, and extreme temperatures (hot or cold), can produce inflammation. Each of these stimuli kills cells, damages connective tissue fibers, or injures the tissue in some other way. The changes alter the chemical composition of the interstitial fluid. Damaged cells release prostaglandins, proteins, and potassium ions, and

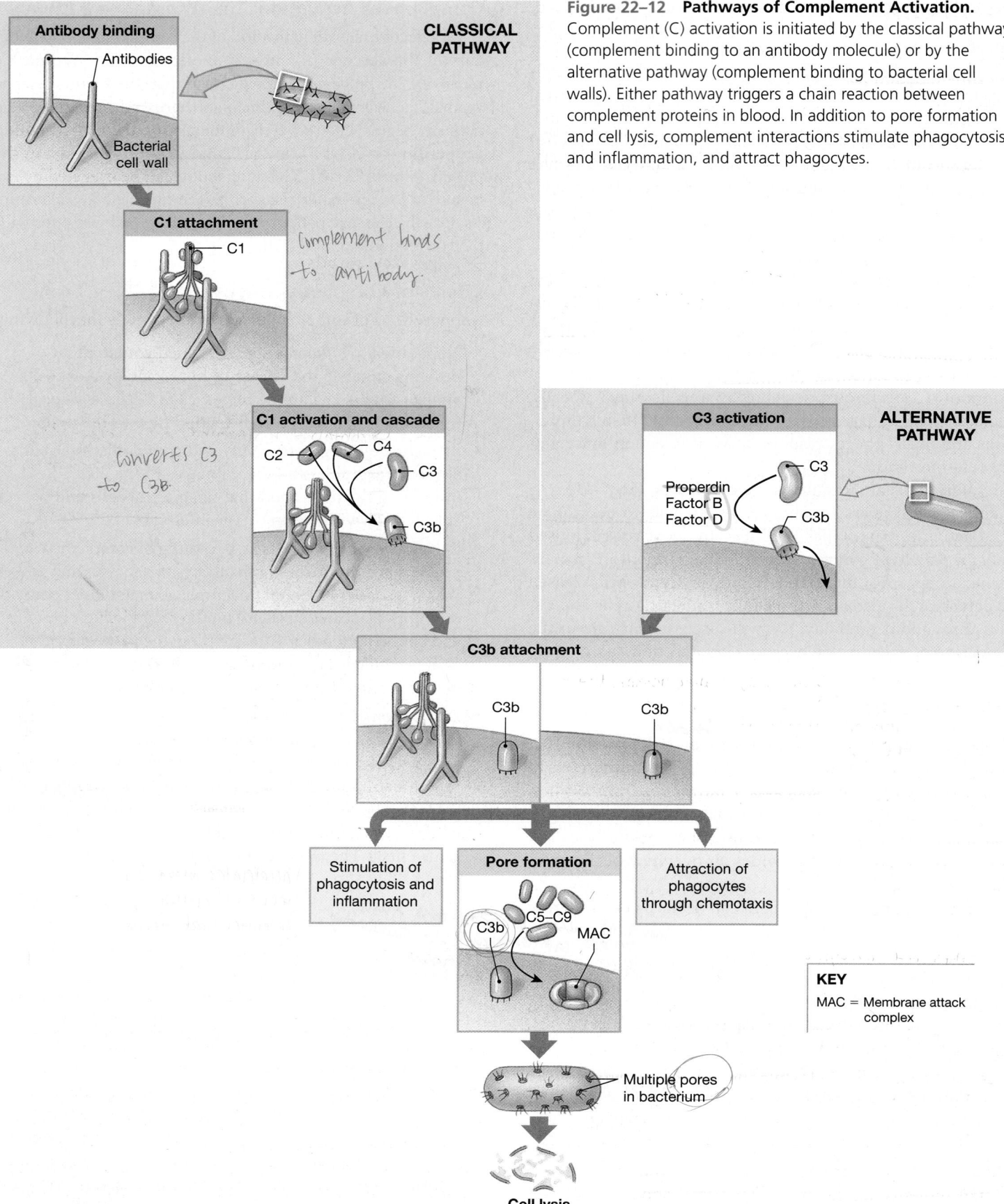

Figure 22–12 **Pathways of Complement Activation.**
Complement (C) activation is initiated by the classical pathway
(complement binding to an antibody molecule) or by the
alternative pathway (complement binding to bacterial cell
walls). Either pathway triggers a chain reaction between
complement proteins in blood. In addition to pore formation
and cell lysis, complement interactions stimulate phagocytosis
and inflammation, and attract phagocytes.

CLASSICAL PATHWAY

Antibody binding
Antibodies
Bacterial cell wall

C1 attachment
C1

Complement binds to antibody.

C1 activation and cascade
C2 — C4
C3
C3b

Converts C3 to C3b.

C3 activation
Properdin
Factor B
Factor D
C3
C3b

ALTERNATIVE PATHWAY

C3b attachment
C3b
C3b

Stimulation of phagocytosis and inflammation

Pore formation
C3b
C5–C9
MAC

Attraction of phagocytes through chemotaxis

Multiple pores in bacterium

Cell lysis

KEY

MAC = Membrane attack complex

the injury itself may have introduced foreign proteins or pathogens. The changes in the interstitial environment trigger the complex process of inflammation.

Inflammation has several effects:

- The injury is temporarily repaired, and additional pathogens are prevented from entering the wound.

- The spread of pathogens away from the injury is slowed.

- Local, regional, and systemic defenses are mobilized to overcome the pathogens and facilitate permanent repairs. This repair process is called *regeneration*.

The Response to Injury *releases histamine, heparin, prostaglandins*

Mast cells play a pivotal role in the inflammatory response. **Figure 22–13** summarizes the events of inflammation in the skin; comparable events take place in almost any tissue subjected to physical damage or infection.

When stimulated by mechanical stress or chemical changes in the local environment, mast cells release histamine, heparin, prostaglandins, and other chemicals into interstitial fluid. The released histamine increases capillary permeability and accelerates blood flow through the area. The combination of abnormal tissue conditions and chemicals released by mast cells stimulates local sensory neurons, producing sensations of pain. The individual then becomes aware of these sensations and may take steps to limit the damage they signal, such as removing a splinter or cleaning a wound.

The increased blood flow reddens the area and elevates the local temperature, increasing the rate of enzymatic reactions and accelerating the activity of phagocytes. The rise in temperature may also denature foreign proteins or vital enzymes of invading microorganisms. *tender.*

Because vessel permeability has increased, clotting factors and complement proteins can leave the bloodstream and enter the injured or infected area. Clotting does not occur at the actual site of injury, due to the presence of heparin. However, a clot soon forms around the damaged area, both isolating the region and slowing the spread of the chemical or pathogen into healthy tissues. Meanwhile, complement activation through the alternative pathway breaks down bacterial cell walls and attracts phagocytes.

Debris and bacteria are attacked by neutrophils drawn to the area by chemotaxis. As they circulate through a blood vessel in an injured area, neutrophils undergo *activation*, a process in which (1) they stick to the side of the vessel and move into the tissue by *diapedesis*; (2) their metabolic rate goes up dramatically, and while this *respiratory burst* continues, they generate reactive compounds, such as nitric oxide and hydrogen peroxide, that can destroy engulfed pathogens; and (3) they secrete cytokines that attract other neutrophils and macrophages to the area. As inflammation proceeds, the foreign proteins, toxins, microorganisms, and active phagocytes in the area activate the body's specific defenses.

movt of passage of blood cells thru intact capillary walls to surrounding body tissue.

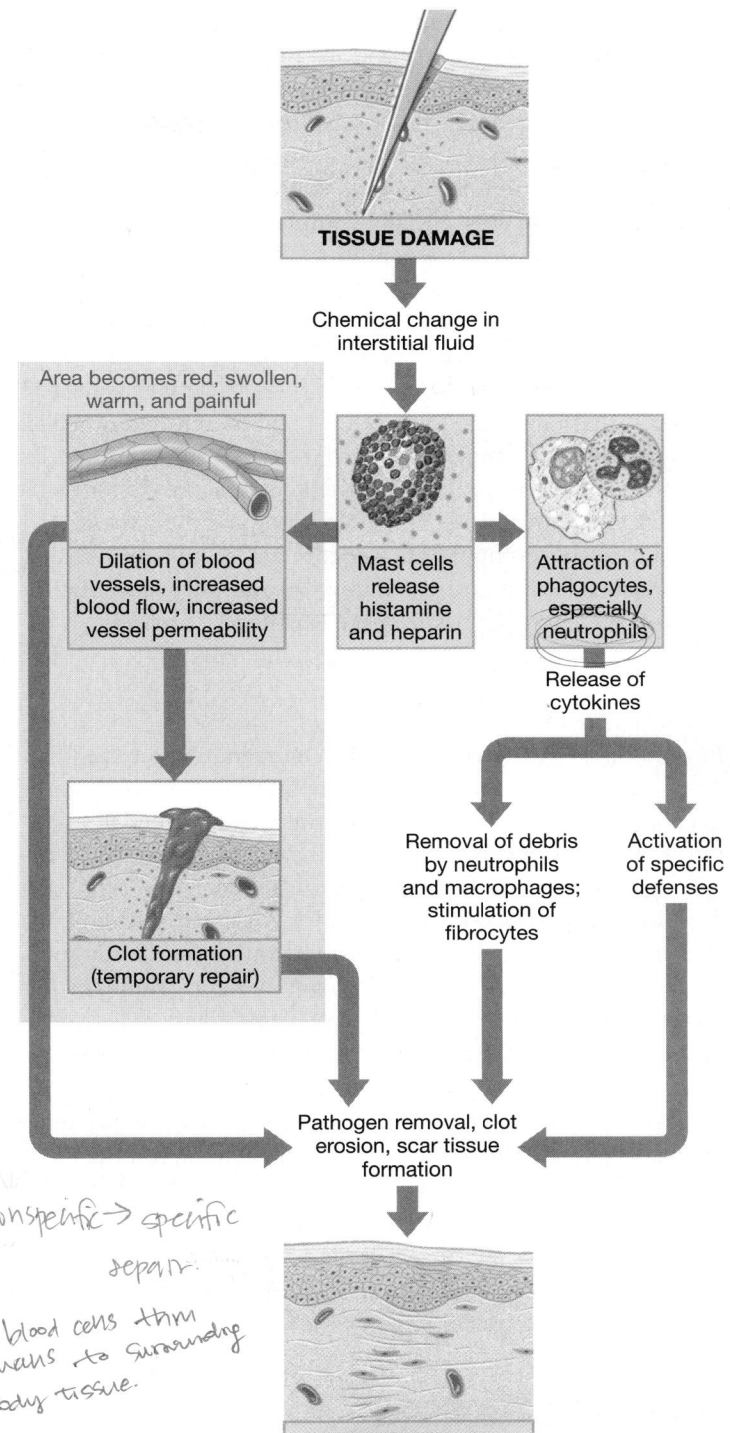

nonspecific → specific repair.

Figure 22–13 Inflammation and the Steps in Tissue Repair.

Fixed and free macrophages engulf pathogens and cell debris. At first, these cells are outnumbered by neutrophils, but as the macrophages and neutrophils continue to secrete cytokines, the number of macrophages increases rapidly. Eosinophils may get involved if the foreign materials become coated with antibodies.

The cytokines released by active phagocytes stimulate fibrocytes in the area. The fibrocytes then begin forming scar tissue that reinforces the clot and slows the invasion of adjacent tissues. Over time, the clot is broken down and the injured tissues are either repaired or replaced by scar tissue. Although subsequent remodeling may occur over a period of years, the process is essentially complete.

After an injury, tissue conditions generally become even more abnormal before they begin to improve. The tissue destruction that occurs after cells have been injured or destroyed is called **necrosis** (ne-KRŌ-sis). The process begins several hours after the initial event, and the damage is caused by lysosomal enzymes. Lysosomes break down by autolysis, releasing digestive enzymes that first destroy the injured cells and then attack surrounding tissues. ∞ p. 79 As local inflammation continues, debris, fluid, dead and dying cells, and necrotic tissue components accumulate at the injury site. This viscous fluid mixture is known as **pus**. An accumulation of pus in an enclosed tissue space is called an **abscess**.

Fever

Fever is the maintenance of body temperature greater than 37.2°C (99°F). The presence of a temperature-regulating center in the preoptic area of the hypothalamus was described in Chapter 14. ∞ p. 478 Circulating proteins called **pyrogens** (PĪ-rō-jenz; *pyro-*, fever or heat + *-gen*, substance) can reset this thermostat and raise body temperature. A variety of stimuli, including pathogens, bacterial toxins, and antigen–antibody complexes, either act as pyrogens themselves or stimulate the release of pyrogens by macrophages. The pyrogen released by active macrophages is a cytokine called **endogenous pyrogen**, or **interleukin-1** (in-ter-LOO-kin), abbreviated **IL-1**.

Within limits, a fever can be beneficial. High body temperatures may inhibit some viruses and bacteria, but the most likely beneficial effect is on body metabolism. For each 1°C rise in body temperature, metabolic rate jumps by 10 percent. Cells can move faster, and enzymatic reactions occur faster. The net results may be the quicker mobilization of tissue defenses and an accelerated repair process.

CHECKPOINT

7. List the body's nonspecific defenses.

8. What types of cells would be affected by a decrease in the number of monocyte-forming cells in bone marrow?

9. A rise in the level of interferon in the body suggests what kind of infection?

10. What effects do pyrogens have in the body?

See the blue Answers tab at the end of the book.

22-4 Specific defenses (immunity) respond to individual threats and are either cell-mediated or antibody-mediated

Specific defense, or immunity, is provided by the coordinated activities of T cells and B cells, which respond to the presence of specific antigens. In general, T cells are responsible for **cell-mediated immunity** (or *cellular immunity*), which defends against abnormal cells and pathogens inside cells, and B cells provide **antibody-mediated immunity** (or *humoral immunity*), which defends against antigens and pathogens in body fluids.

Both mechanisms are important, because they come into play under different circumstances. Activated T cells do not respond to antigenic materials in solution, and antibodies (produced by activated B cells) cannot cross plasma membranes. Moreover, helper T cells play a crucial role in antibody-mediated immunity by stimulating the activity of B cells.

Our understanding of immunity has greatly improved in the past two decades, and a comprehensive discussion would involve hundreds of pages and thousands of details. The discussion that follows emphasizes important patterns and introduces general principles that will provide a foundation for future courses in microbiology and immunology.

Forms of Immunity

Another way of classifying immunity is to categorize it as either *innate* or *acquired* (**Figure 22–14**). **Innate immunity** is genetically determined; it is present at birth and has no relationship to previous exposure to the antigen involved. For example, people do not get the same diseases that goldfish do. Innate immunity breaks down only in the case of *AIDS* or other conditions that depress all aspects of specific resistance. **Acquired immunity** is not present at birth; you acquire immunity to a specific antigen only when you have been exposed to that antigen. Acquired immunity can be *active* or *passive*, and these forms can be further divided, depending on whether the immunity is naturally acquired or induced.

Active immunity develops after exposure to an antigen, as a consequence of the immune response. In active immunity, the body responds to the antigen by making its own antibody. The immune system is *capable* of defending against a huge number of antigens. However, the appropriate defenses are mobilized only after you encounter a particular antigen. Active immunity can develop as a result of natural exposure to an antigen in the environment (*naturally acquired active immunity*) or from deliberate exposure to an antigen (*induced active immunity*).

[handwritten notes: nonspecific: epithelium, macrophage, & NK cells.]

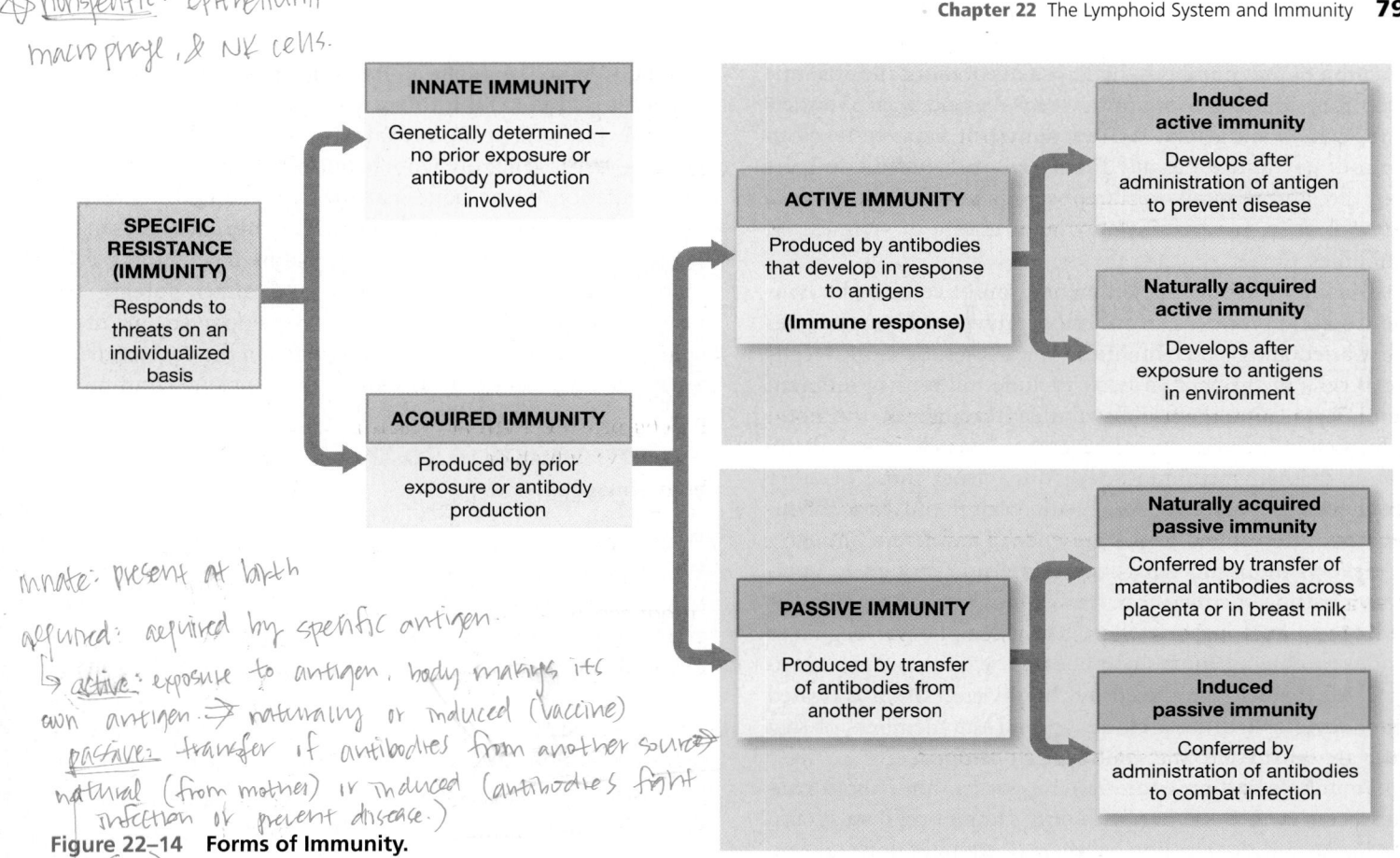

*[handwritten notes: innate: present at birth

acquired: acquired by specific antigen.
↳ active: exposure to antigen, body making its own antigen → naturally or induced (vaccine)
passive: transfer if antibodies from another source → natural (from mother) or Induced (antibodies fight infection or prevent disease.)]*

Figure 22–14   Forms of Immunity.

[handwritten note: (IgG)]

- **Naturally acquired active immunity** normally begins to develop after birth, and it is continually enhanced as you encounter "new" pathogens or other antigens. You might compare this process to the development of a child's vocabulary: The child begins with a few basic common words and learns new ones as they are encountered.

- The purpose of **induced active immunity** is to stimulate the production of antibodies under controlled conditions so that you will be able to overcome natural exposure to the pathogen some time in the future. This is the basic principle behind *immunization*, or *vaccination*, to prevent disease. A **vaccine** is a preparation designed to induce an immune response. It contains either a dead or an inactive pathogen, or antigens derived from that pathogen.

   Passive immunity is produced by the transfer of antibodies from another source.

- In **naturally acquired passive immunity**, a mother's antibodies protect her baby against infections, either during gestation (by crossing the placenta) or in early infancy (through breast milk).

- In **induced passive immunity**, antibodies are administered to fight infection or prevent disease. For

example, antibodies against the rabies virus are injected into a person bitten by a rabid animal.

Properties of Immunity

Regardless of the form, immunity exhibits four general properties: (1) *specificity*, (2) *versatility*, (3) *memory*, and (4) *tolerance*.

Specificity *→ [handwritten: specific antigen activates specific defense.]*

A specific defense is activated by a specific antigen, and the immune response targets that particular antigen and no others. **Specificity** results from the activation of appropriate lymphocytes and the production of antibodies with targeted effects. Specificity occurs because T cells and B cells respond to the molecular structure of an antigen. The shape and size of the antigen determine which lymphocytes will respond to its presence. Each T cell or B cell has receptors that will bind to one specific antigen, ignoring all others. The response of an activated T cell or B cell is equally specific. Either lymphocyte will destroy or inactivate that antigen without affecting other antigens or normal tissues.

[handwritten: Ⓑ style specificity.]

Versatility

Millions of antigens in the environment can pose a threat to health. Over a normal lifetime, an individual encounters only a fraction of that number—perhaps tens of thousands of antigens. Your immune system, however, has no way of anticipating which antigens it will encounter. It must be ready to confront *any* antigen at *any* time. **Versatility** results in part from the large diversity of lymphocytes present in the body, and in part from variability in the structure of synthesized antibodies.

During development, differentiation of cells in the lymphoid system produces an enormous number of lymphocytes with varied antigen sensitivities. The trillion or more T cells and B cells in the human body include millions of different lymphocyte populations, distributed throughout the body. Each population consists of several thousand cells with receptors in their membranes that differ from those of other lymphocyte populations. As a result, each population of lymphocytes will respond to the presence of a different antigen.

Several thousand lymphocytes are not enough to overcome a pathogenic invasion. However, when activated in the presence of an appropriate antigen, a lymphocyte begins to divide, producing more lymphocytes with the same specificity. All the cells produced by the division of an activated lymphocyte constitute a **clone**, and all the members of that clone are sensitive to the same specific antigen.

To understand how this system works, think about running a commercial kitchen with only samples on display. You can display a wide selection because the samples don't take up much space, and you don't have to expend energy now preparing food that might never be eaten. When a customer selects one of your samples and places an order for several dozen, you prepare them on the spot.

The same principle applies to the lymphoid system: Your body contains a small number of many different kinds of lymphocytes. When an antigen arrives, lymphocytes sensitive to its presence are "selected," and these lymphocytes divide to generate a large number of additional lymphocytes of the same type.

Memory *may be dormant up to 20 years.*

As we just saw, during the initial response to an antigen, lymphocytes that are sensitive to its presence undergo repeated cycles of cell division. Immunologic **memory** exists because those cell divisions produce two groups of cells: one group that attacks the invader immediately, and another that remains inactive unless it is exposed to the same antigen at a later date. These inactive *memory cells* enable your immune system to "remember" antigens it has previously encountered, and to launch a faster, stronger, and longer-lasting counterattack if such an antigen reappears.

Tolerance

The immune system does not respond to all antigens. All cells and tissues in the body, for example, contain antigens that normally do not stimulate an immune response. The immune system is said to exhibit **tolerance** toward such antigens.

The immune response targets foreign cells and compounds, but it generally ignores normal tissues. During their differentiation in the bone marrow (B cells) and thymus (T cells), cells that react to antigens that are normally present in the body are destroyed. As a result, mature B cells and T cells ignore normal (or *self*) antigens, and attack foreign (or *nonself*) antigens. Tolerance can also develop over time in response to chronic exposure to an antigen in the environment. Such tolerance lasts only as long as the exposure continues.

An Introduction to the Immune Response

Figure 22–15 provides an overview of the immune response. When an antigen triggers an immune response, it usually activates both T cells and B cells. The activation of T cells generally occurs first, but only after phagocytes have been exposed to the antigen. Once activated, T cells attack the antigen and stimulate the activation of B cells. Activated B cells mature into cells that produce antibodies; antibodies distributed in the bloodstream bind to and attack the antigen. We will examine each of these processes more closely in the sections that follow.

> **CHECKPOINT**
>
> 11. **Distinguish between cell-mediated (cellular) immunity and antibody-mediated (humoral) immunity.**
>
> 12. **Identify the two forms of active immunity and the two types of passive immunity.**
>
> 13. **List the four general properties of immunity.**
>
> See the blue Answers tab at the end of the book.

22-5 T cells play a role in the initiation, maintenance, and control of the immune response

The role of T cells in the immune response is varied and important in resisting pathogens. We have already noted three major types of T cells: *directly destroys antigen.*

1. *Cytotoxic T cells* (T_C cells) are responsible for cell-mediated immunity. These cells enter peripheral tissues and directly attack antigens physically and chemically.

2. *Helper T cells* (T_H cells) stimulate the responses of both T cells and B cells. Helper T cells are absolutely vital to the immune response, because B cells must be activated by helper T cells before the B cells can produce antibodies.

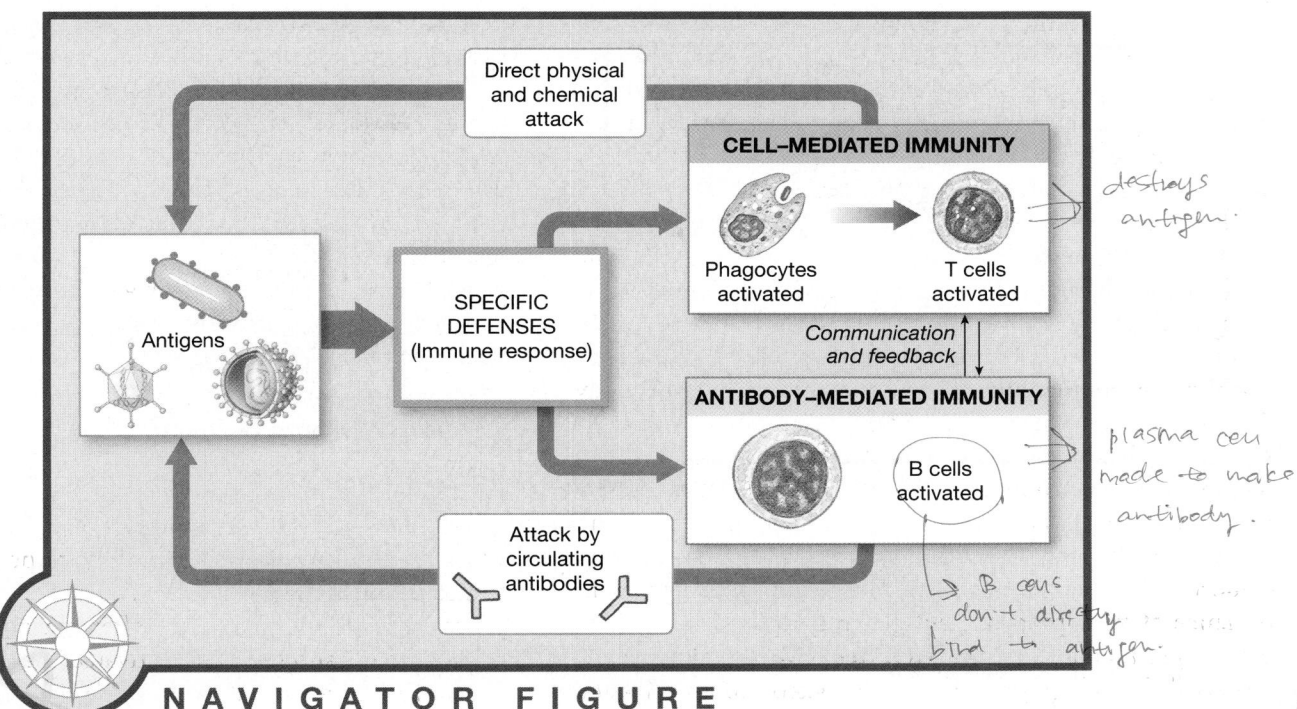

destroys antigen.

plasma cell made to make antibody.

B cells don't directly bind to antigen.

NAVIGATOR FIGURE

Figure 22–15   An Overview of the Immune Response. This figure will be repeated, in reduced and simplified form as Navigator icons, in key figures throughout the chapter as we change topics.

(T): initiation. maintenance, & control of immune system.

The reduction in the helper T cell population that occurs in AIDS is largely responsible for the loss of immunity. (We will discuss AIDS on p. 819.)

3. *Suppressor T cells* (T_S cells) inhibit T cell and B cell activities and moderate the immune response.

→ involved in regulation. → negative inhibition.

Before an immune response can begin, T cells must be activated by exposure to an antigen. This activation seldom occurs through direct lymphocyte–antigen interaction, and foreign compounds or pathogens entering a tissue commonly fail to stimulate an immediate immune response.

Antigen Presentation
→ large molecule of lipids + proteins.
→ part of antigen presented to T cell.

T cells recognize antigens when the antigens are bound to glycoproteins in plasma membranes. Glycoproteins are integral membrane components. ∞ p. 71 **Antigen presentation** occurs when an antigen–glycoprotein combination capable of activating T cells appears in a plasma membrane. The structure of these glycoproteins is genetically determined. The genes controlling their synthesis are located along one portion of chromosome 6, in a region called the **major histocompatibility complex (MHC).** These membrane glycoproteins are called **MHC proteins,** or *human leukocyte antigens (HLAs).*

The amino acid sequences and the shapes of MHC proteins differ among individuals. Each MHC molecule has a dis-

tinct three-dimensional shape with a relatively narrow central groove. An antigen that fits into this groove can be held in position by hydrogen bonding.

Two major classes of MHC proteins are known: *Class I* and *Class II.* An antigen bound to a Class I MHC protein acts like a red flag that in effect tells the immune system "Hey, I'm an abnormal cell—kill me!" An antigen bound to a Class II MHC protein tells the immune system "Hey, this antigen is dangerous—get rid of it!" *chopped & made into small peptides)*

Class I MHC proteins are in the membranes of all nucle- *class I* ated cells. These proteins are continuously synthesized and *picks up small peptide.* exported to the plasma membrane in vesicles created at the Golgi apparatus. As they form, Class I proteins pick up small peptides from the surrounding cytoplasm and carry them to the cell surface. If the cell is healthy and the peptides are normal, T cells will ignore them. If the cytoplasm contains abnormal (nonself) peptides or viral proteins (**Figure 22–16a**), they will soon appear in the plasma membrane, and T cells will be activated. Ultimately, their activation leads to the destruction of the abnormal cells. This is the primary reason that donated organs are commonly rejected by the recipient; despite preliminary cross-match testing, the recipient's T cells recognize the transplanted tissue as foreign. *dendrite*

Class II MHC proteins are present only in the membranes of antigen-presenting cells and lymphocytes. **Antigen-presenting cells (APCs)** are specialized cells responsible for

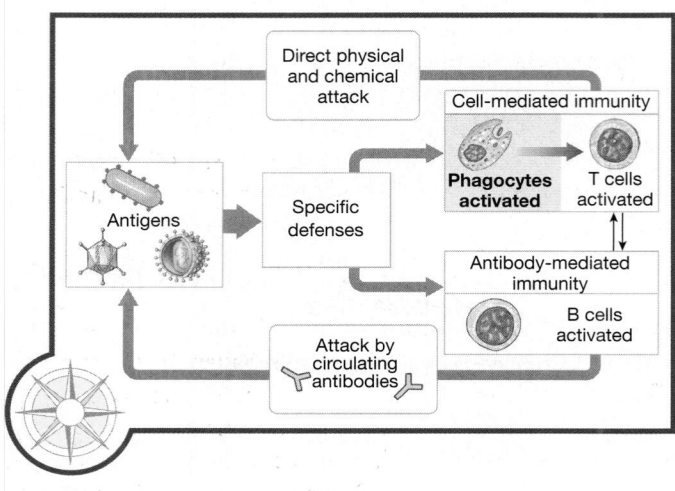

(a) Infected cell

Figure 22–16 **Antigens and MHC Proteins.** The Navigator icon in the shadow box highlights how phagocyte activation relates to the rest of the immune response. **(a)** Viral or other foreign antigens appear in plasma membranes bound to Class I MHC proteins. **(b)** Processed antigens appear on the surfaces of antigen-presenting cells bound to Class II MHC proteins.

(b) Phagocytic antigen-presenting cell

activating T cell defenses against foreign cells (including bacteria) and foreign proteins. Antigen-presenting cells include all the phagocytic cells of the monocyte–macrophage group discussed in other chapters, including (1) free and fixed macrophages in connective tissues, (2) the Kupffer cells of the liver, and (3) the microglia in the central nervous system (Chapter 12). ∞ pp. 126, 395 The dendritic (Langerhans) cells of the skin and the dendritic cells of the lymph nodes and spleen are APCs that are not phagocytic. ∞ p. 161

Phagocytic APCs engulf and break down pathogens or foreign antigens. Such **antigen processing** creates antigenic fragments, which are then bound to Class II MHC proteins and inserted into the plasma membrane (**Figure 22–16b**). *Class II*

MHC proteins appear in the plasma membrane only when the cell is processing antigens. Exposure to an APC membrane containing a processed antigen can stimulate appropriate T cells.

The dendritic cells remove antigenic materials from their surroundings via pinocytosis rather than phagocytosis. However, their plasma membranes still present antigens bound to Class II MHC proteins.

Antigen Recognition

Inactive T cells have receptors that recognize Class I or Class II MHC proteins. The receptors also have binding sites that detect the presence of specific bound antigens. If an MHC protein contains any antigen other than the specific target of a particular kind of T cell, the T cell remains inactive. If the MHC protein contains the antigen that the T cell is programmed to detect, binding will occur. This process is called **antigen recognition**, because the T cell recognizes that it has found an appropriate target.

Some T cells can recognize antigens bound to Class I MHC proteins, whereas others can recognize antigens bound to Class II MHC proteins. Whether a T cell responds to antigens held by Class I or Class II proteins depends on the structure of the T cell plasma membrane. The membrane proteins involved are members of a larger class of proteins called **CD** (*cluster of differentiation*) **markers**.

Lymphocytes, macrophages, and other, related cells have CD markers. Each of the more than 70 types of CD markers is designated by an identifying number. All T cells have a **CD3 receptor complex** in their membranes. Two other CD markers are of particular importance in specific groups of T cells:

1. **CD8** markers are found on cytotoxic T cells and suppressor T cells, which together are often called *CD8 T cells* or *CD8+ T cells*. CD8 T cells respond to antigens presented by Class I MHC proteins.

2. **CD4** markers are found on helper T cells, often called *CD4 T cells* or *CD4+ T cells*. CD4 T cells respond to antigens presented by Class II MHC proteins.

Costimulation

CD8 or CD4 markers are bound to the CD3 receptor complex, which ultimately activates the T cell. However, such activation usually does not occur upon the first encounter with the antigen. Antigen recognition simply prepares the cell for activation. Before activation can occur, a T cell must bind to the stimulating cell at a second site. This vital secondary binding process, called *costimulation*, essentially confirms the initial activation signal. Appropriate costimulation proteins appear in the presenting cell only if that cell has engulfed antigens or is infected by viruses. Many costimulation proteins are structurally related to the cytokines released by activated lymphocytes. The effects of these proteins on the exposed T cell vary, but they typically include the stimulation of transcription at the nucleus and the promotion of cell division and differentiation.

Costimulation is like the safety on a gun: It helps prevent T cells from mistakenly attacking normal (self) tissues. If a cell displays an unusual antigen but does not display the "I am an active phagocyte" or "I am infected" signal, T cell activation will not occur. Costimulation is important only in determining whether a T cell will become activated. Once activation has occurred, the "safety" is off and the T cell will attack any cells that carry the target antigens.

Activation of CD8 T Cells

Two different classes of CD8 T cells are activated by exposure to antigens bound to Class I MHC proteins. One type of CD8 T cell responds quickly, giving rise to large numbers of *cytotoxic T cells* and cytotoxic *memory T cells* (**Figure 22–17**). The other type of CD8 T cell responds more slowly and produces relatively small numbers of *suppressor T cells*.

Cytotoxic T Cells

Cytotoxic T cells, also called T_C *cells* or *killer T cells*, seek out and destroy abnormal and infected cells. Killer T cells are highly mobile cells that roam throughout injured tissues. When a cytotoxic T cell encounters its target antigens bound to Class I MHC proteins of another cell, it immediately destroys the target cell (**Figure 22–17**). The T cell may (1) destroy the antigenic plasma membrane through the release of perforin, (2) kill the target cell by secreting a poisonous **lymphotoxin** (lim-fō-TOK-sin), or (3) activate genes in the target cell's nucleus that tell that cell to die. (We introduced genetically programmed cell death, called *apoptosis*, in Chapter 3.) ∞ p. 100

The entire sequence of events, from the appearance of the antigen in a tissue to cell destruction by cytotoxic T cells, takes a significant amount of time. After the first exposure to an antigen, two days or more may pass before the concentration of cytotoxic T cells reaches effective levels at the site of injury or infection. Over this period, the damage or infection may spread, making it more difficult to control.

Memory T_C Cells

Memory T_C cells are produced by the same cell divisions that produce cytotoxic T cells. Thousands of these cells are produced, but they do not differentiate further the first time the antigen triggers an immune response. However, if the same antigen appears a second time, memory T cells will *immediately* differentiate into cytotoxic T cells, producing a prompt, effective cellular response that can overwhelm an invading organism before it becomes well established in the tissues.

Suppressor T Cells

Suppressor T cells (T_S *cells*) suppress the responses of other T cells and of B cells by secreting *suppression factors*—inhibitory cytokines of unknown structure. Suppression does not occur immediately, because suppressor T cell activation takes much longer than the activation of other types of T cells. In addition, upon activation, most of the CD8 T cells in the bloodstream produce cytotoxic T cells rather than suppressor T cells. As a result, suppressor T cells act *after* the initial immune response. In effect, these cells limit the degree of immune system activation from a single stimulus.

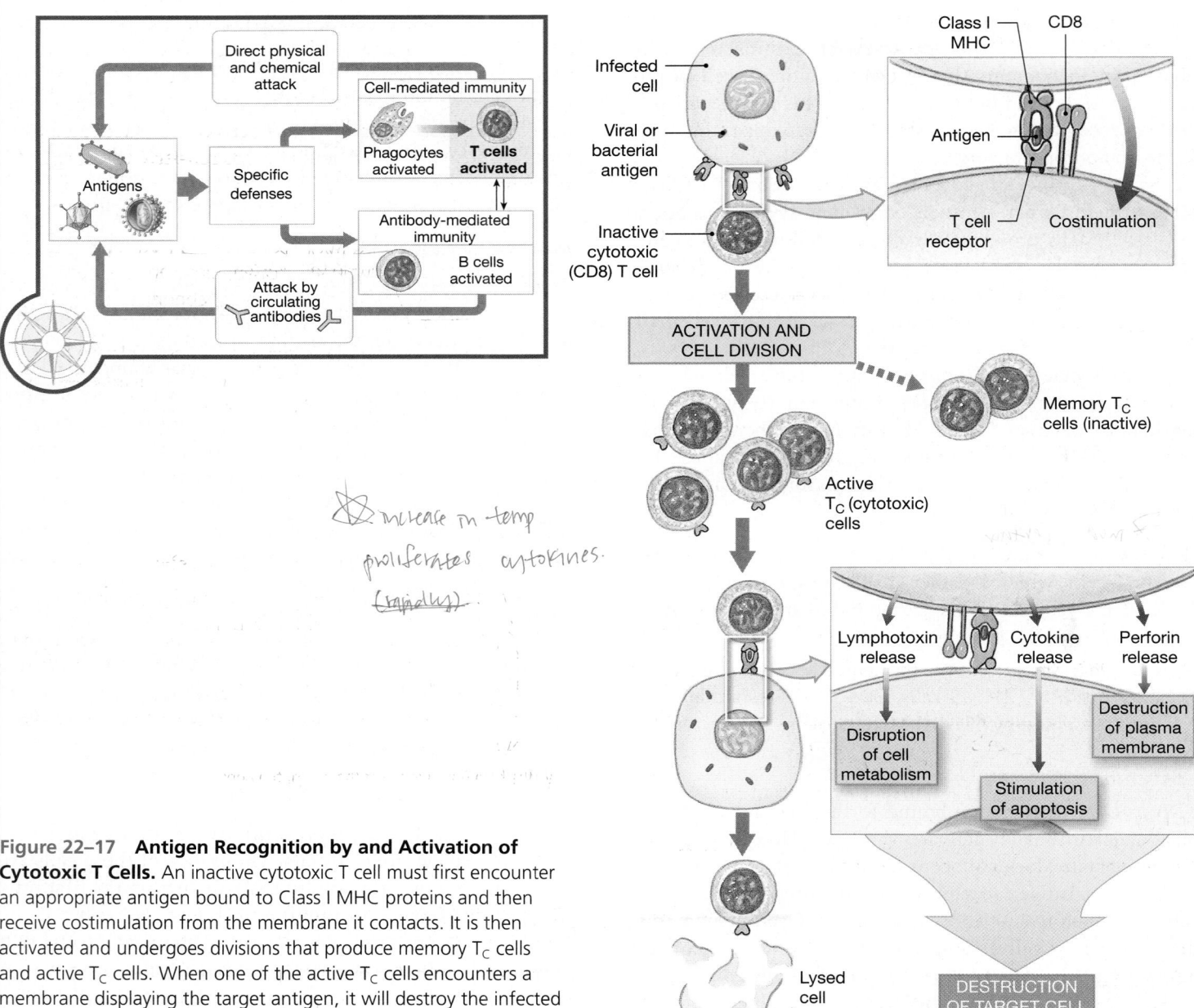

increase in temp
proliferates cytokines.
(rapidly).

Figure 22–17 **Antigen Recognition by and Activation of Cytotoxic T Cells.** An inactive cytotoxic T cell must first encounter an appropriate antigen bound to Class I MHC proteins and then receive costimulation from the membrane it contacts. It is then activated and undergoes divisions that produce memory T_C cells and active T_C cells. When one of the active T_C cells encounters a membrane displaying the target antigen, it will destroy the infected cell by one of several methods.

Activation of CD4 T Cells

HIV directly targets (CD4) T cells.

Upon activation, CD4 T cells undergo a series of divisions that produce active helper T cells (T_H *cells*) and **memory T_H cells** (**Figure 22–18**). The memory T_H cells remain in reserve, whereas the active helper T cells secrete a variety of cytokines that coordinate specific and nonspecific defenses and stimulate cell-mediated and antibody-mediated immunities. Activated helper T cells secrete cytokines that do the following:

1. stimulate the T cell divisions that produce memory T_H cells and accelerate the maturation of cytotoxic T cells;

2. enhance nonspecific defenses by attracting macrophages to the affected area, preventing their departure, and stimulating their phagocytic activity and effectiveness;

3. attract and stimulate the activity of NK cells, providing another mechanism for the destruction of abnormal cells and pathogens; and

4. promote the activation of B cells, leading to B cell division, plasma cell maturation, and antibody production.

Figure 22–19 provides a review of the methods of antigen presentation and T cell stimulation. The plasma membranes of infected or otherwise abnormal cells trigger an immune re-

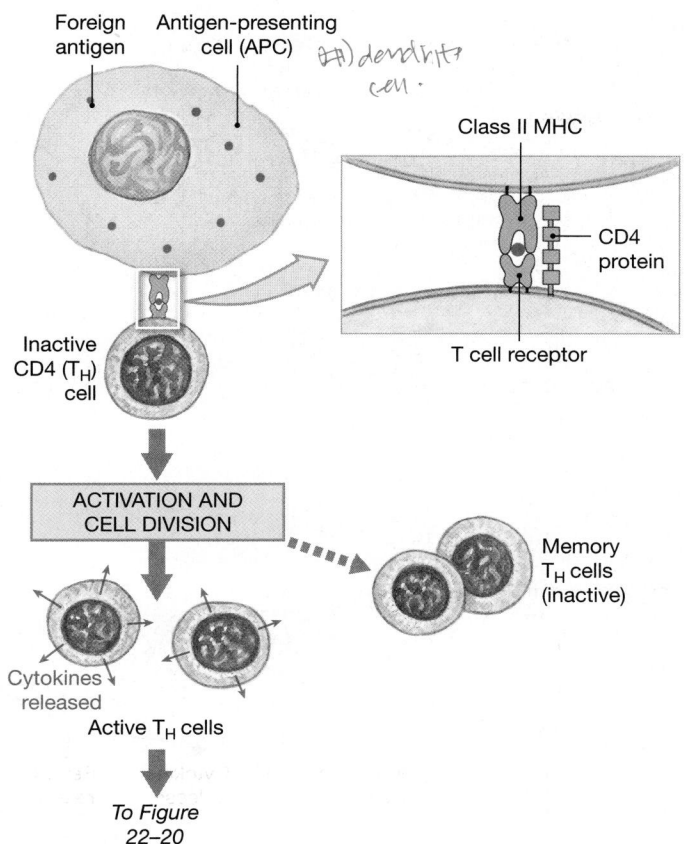

Figure 22–18 **Antigen Recognition and Activation of Helper T Cells.** Inactive CD4 T cells (T_H cells) must be exposed to appropriate antigens bound to Class II MHC proteins. The T_H cells then undergo activation, dividing to produce active T_H cells and memory T_H cells. Active T_H cells secrete cytokines that stimulate cell-mediated and antibody-mediated immunities. They also interact with sensitized B cells, as shown in *Figure 22–20*.

sponse when CD8 T cells recognize antigens bound to Class I MHC proteins. Extracellular pathogens or foreign proteins trigger an immune response when CD4 T cells recognize antigens displayed by Class II MHC proteins. In the next section, we will consider how the T_H cells derived from activated CD4 T cells in turn activate B cells that are sensitive to the specific antigen involved.

The A&P Top 100

#75 Cell-mediated immunity involves close physical contact between activated T_C cells and foreign, abnormal, or infected cells. T cell activation usually involves (1) antigen presentation by a phagocytic cell and (2) costimulation by cytokines released by active phagocytes. T_C cells may destroy target cells through the local release of cytokines, lymphotoxins, or perforin.

CHECKPOINT

14. Identify the three major types of T cells.

15. How can the presence of an abnormal peptide in the cytoplasm of a cell initiate an immune response?

16. A decrease in the number of cytotoxic T cells would affect which type of immunity?

17. How would a lack of helper T cells affect the antibody-mediated immune response?

See the blue Answers tab at the end of the book.

CLINICAL NOTE

Graft Rejection and Immunosuppression

Organ transplantation may be a treatment option for patients with severe disorders of the bone marrow, kidneys, liver, heart, lungs, or pancreas. Finding a suitable donor is the first major problem. In the United States, many people die each day while awaiting an organ transplant, and dozens are added to the transplant waiting list. After surgery has been performed, the major problem is **graft rejection**. In graft rejection, T cells are activated by contact with MHC proteins on plasma membranes in the donated tissues. The cytotoxic T cells that develop then attack and destroy the foreign cells.

Significant improvements in transplant success can be made by immunosuppression, a reduction in the sensitivity of the immune system. Until recently, the drugs used to produce immunosuppression did not selectively target the

immune response. For example, **prednisone** (PRED-ni-sōn), a corticosteroid, was used because it has anti-inflammatory effects that reduce the number of circulating white blood cells and depress the immune response. However, corticosteroid use also causes undesirable side effects in systems other than the immune system.

An understanding of the communication among T cells, macrophages, and B cells has now led to the development of drugs with more selective effects. **Cyclosporin A (CsA)**, a compound derived from a fungus, was the most important *immunosuppressive drug* developed in the 1980s. This compound suppresses all aspects of the immune response primarily by suppressing helper T cell activity while leaving suppressor T cells relatively unaffected. Even more narrowly focused drugs are available, including monoclonal antibodies that prevent antigen recognition by T cell receptors.

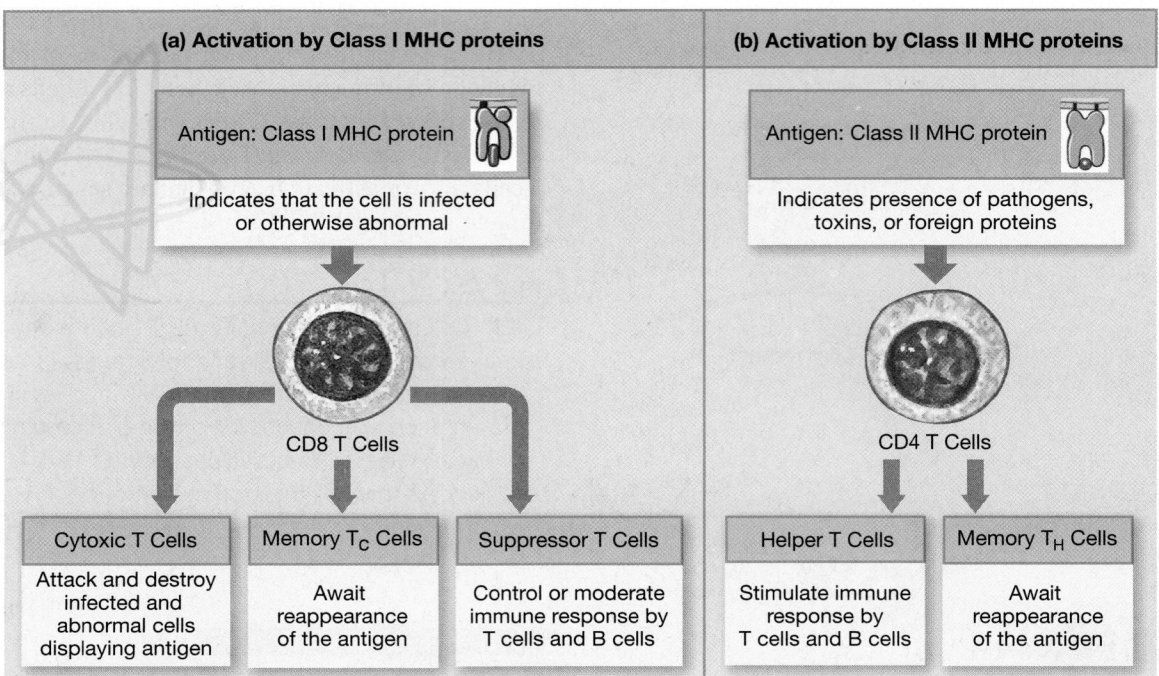

Figure 22–19   **A Summary of the Pathways of T Cell Activation.**

22-6 B cells respond to antigens by producing specific antibodies

B cells are responsible for launching a chemical attack on antigens by producing appropriate, specific *antibodies*.

B Cell Sensitization and Activation

As noted earlier, the body has millions of B cell populations. Each kind of B cell carries its own particular antibody molecules in its plasma membrane. If corresponding antigens appear in the interstitial fluid, they will interact with these superficial antibodies. When binding occurs, the B cell prepares to undergo activation. This preparatory process is called **sensitization**. Because B cells migrate throughout the body, pausing briefly in one lymphoid tissue or another, sensitization typically occurs within the lymph node nearest the site of infection or injury.

As noted earlier, B cell plasma membranes contain Class II MHC proteins. During sensitization, antigens are brought into the cell by endocytosis. The antigens subsequently appear on the surface of the B cell, bound to Class II MHC proteins. (The mechanism is comparable to that shown in **Figure 22–16b**). Once this happens, the sensitized B cell is on "standby" but generally will not undergo activation unless it receives the "OK" from a helper T cell (**Figure 22–20**). The

need for activation by a helper T cell helps prevent inappropriate activation, the same way that costimulation acts as a "safety" in cell-mediated immunity.

When a sensitized B cell encounters a helper T cell already activated by antigen presentation, the helper T cell binds to the MHC complex, recognizes the presence of the antigen, and begins secreting cytokines that promote B cell activation. After activation has occurred, these same cytokines stimulate B cell division, accelerate plasma cell formation, and enhance antibody production.

The activated B cell typically divides several times, producing daughter cells that differentiate into plasma cells and *memory B cells*. The plasma cells begin synthesizing and secreting large quantities of antibodies into the interstitial fluid. These antibodies have the same target as the antibodies on the surface of the sensitized B cell. When stimulated by cytokines from helper T cells, a plasma cell can secrete up to 100 million antibody molecules each hour.

Memory B cells perform the same role in antibody-mediated immunity that memory T cells perform in cell-mediated immunity. Memory B cells do not respond to a threat on first exposure. Instead, they remain in reserve to deal with subsequent injuries or infections that involve the same antigens. On subsequent exposure, the memory B cells respond by dividing and differentiating into plasma cells that secrete antibodies in massive quantities.

Figure 22–20   The Sensitization and Activation of B Cells.
A B cell is sensitized by exposure to antigens. Once antigens are bound to antibodies in the B cell membrane, the B cell displays those antigens in its plasma membrane. Activated helper T cells encountering the antigens release cytokines that trigger the activation of the B cell. The activated B cell then divides, producing memory B cells and plasma cells that secrete antibodies.

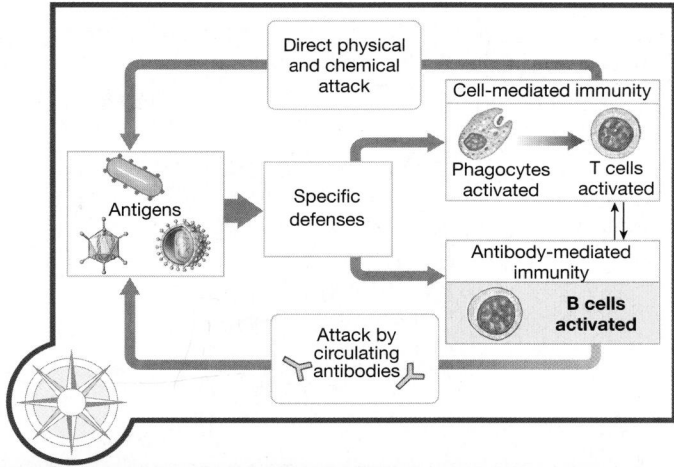

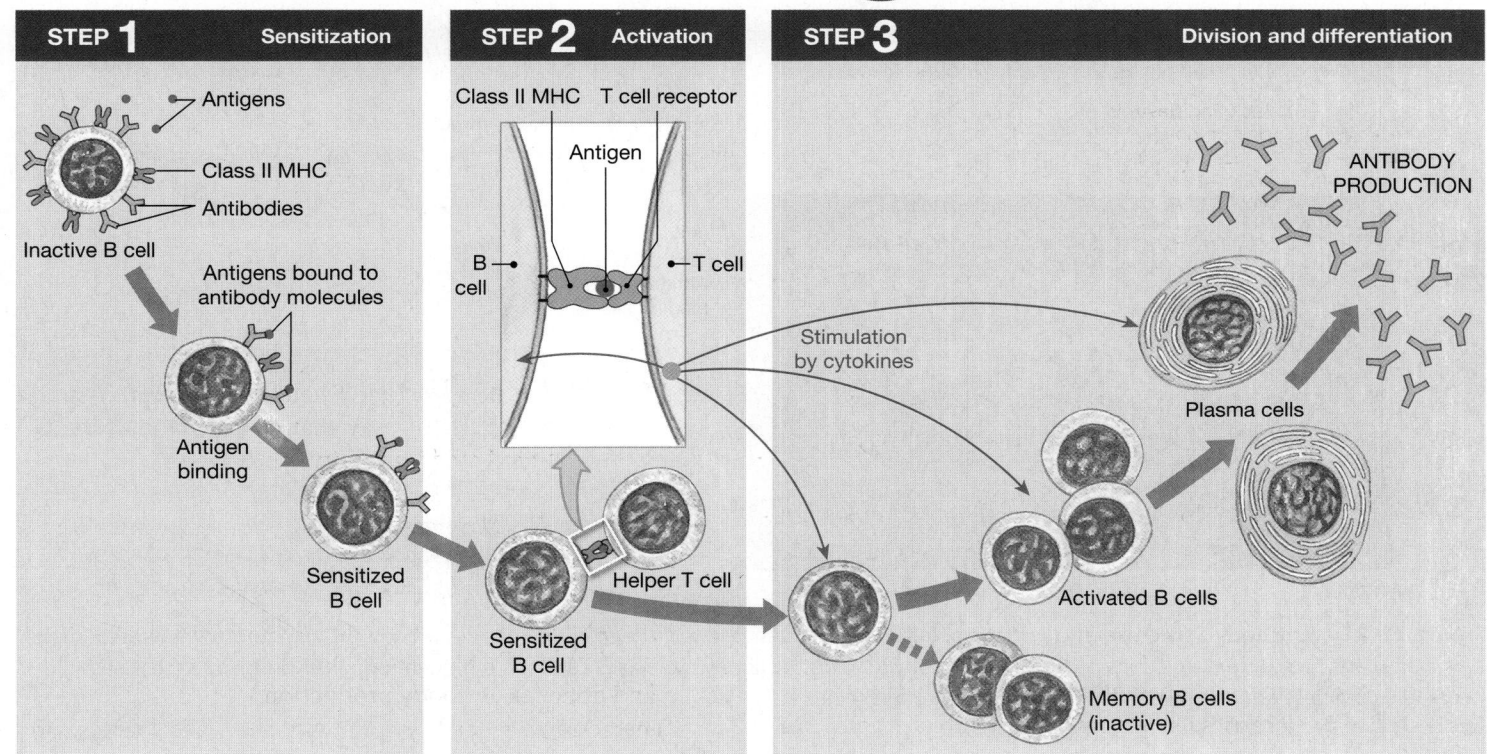

Antibody Structure

An antibody molecule consists of two parallel pairs of polypeptide chains: one pair of **heavy chains** and one pair of **light chains** (**Figure 22–21**). Each chain contains both *constant segments* and *variable segments*.

The constant segments of the heavy chains form the base of the antibody molecule (**Figure 22–21a,b**). B cells produce only five types of constant segments. These form the basis of a classification scheme that identifies antibodies as *IgG, IgE, IgD, IgM,* or *IgA,* as we will discuss in the next section. The structure of the constant segments of the heavy chains determines the way the antibody is secreted and how it is distributed within the body. For example, antibodies in one class circulate in body fluids, whereas those of another class bind to the membranes of basophils and mast cells.

The heavy-chain constant segments, which are bound to constant segments of the light chains, also contain binding sites that can activate the complement system. These binding sites are covered when the antibody is secreted but become exposed when the antibody binds to an antigen.

The specificity of an antibody molecule depends on the structure of the variable segments of the light and heavy chains. The free tips of the two variable segments form the

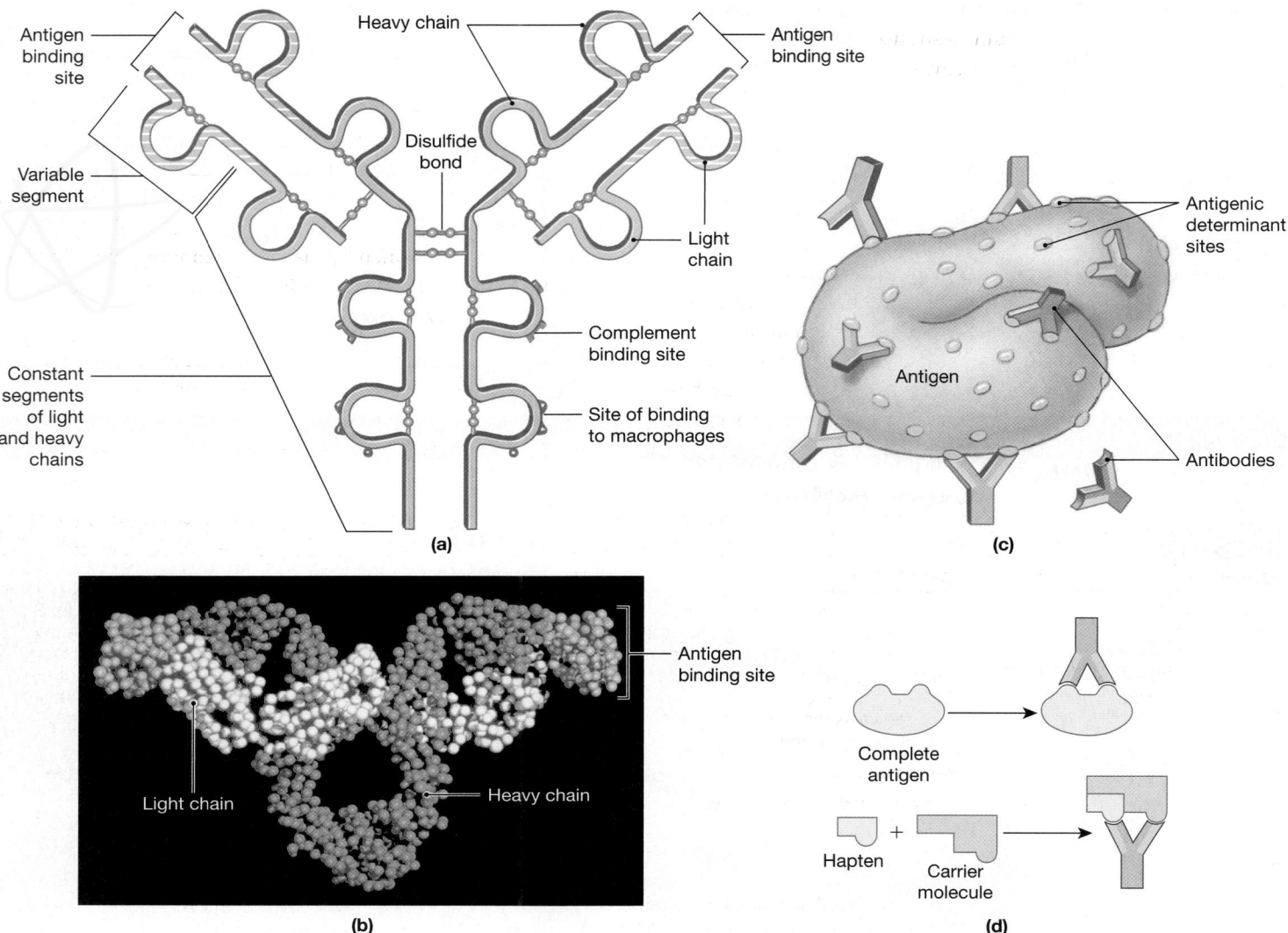

Figure 22–21 Antibody Structure and Function. (a) A diagrammatic view of the structure of an antibody. **(b)** A computer-generated image of a typical antibody. **(c, d)** Depictions of antigen–antibody binding. Antibody molecules can bind a hapten, or partial antigen, once it has become a complete antigen by combining with a carrier molecule.

antigen binding sites of the antibody molecule (**Figure 22–21a**). These sites can interact with an antigen in the same way that the active site of an enzyme interacts with a substrate molecule. ∞ p. 56

Small differences in the amino acid sequence of the variable segments affect the precise shape of the antigen binding site. These differences account for differences in specificity among the antibodies produced by different B cells. The distinctions are the result of minor genetic variations that occur during the production, division, and differentiation of B cells. A normal adult body contains roughly 10 trillion B cells, which can produce an estimated 100 million types of antibodies, each with a different specificity.

The Antigen–Antibody Complex

When an antibody molecule binds to its corresponding antigen molecule, an **antigen–antibody complex** is formed. Once the two molecules are in position, hydrogen bonding and other weak chemical forces lock them together.

Antibodies bind not to the entire antigen, but to specific portions of its exposed surface—regions called **antigenic determinant sites** (**Figure 22–21c**). The specificity of the binding depends initially on the three-dimensional "fit" between the variable segments of the antibody molecule and the corresponding antigenic determinant sites. A **complete antigen** is an antigen with at least two antigenic determinant sites, one for each of the antigen binding sites on an antibody molecule.

Exposure to a complete antigen can lead to B cell sensitization and a subsequent immune response. Most environmental antigens have multiple antigenic determinant sites; entire microorganisms may have thousands.

Haptens, or *partial antigens*, do not ordinarily cause B cell activation and antibody production. Haptens include short peptide chains, steroids and other lipids, and several drugs, including antibiotics such as *penicillin*. However, haptens may become attached to carrier molecules, forming combinations that can function as complete antigens (**Figure 22–21d**). In some cases, the carrier contributes an antigenic determinant site. The antibodies produced will attack both the hapten and the carrier molecule. If the carrier molecule is normally present in the tissues, the antibodies may begin attacking and destroying normal cells. This process is

the basis for several drug reactions, including allergies to penicillin.

Classes and Actions of Antibodies

Body fluids have five classes of antibodies, or **immunoglobulins (Igs)**: *IgG, IgE, IgD, IgM,* and *IgA* (Table 22–1). The classes are determined by differences in the structure of the heavy-chain constant segments and so have no effect on the antibody's specificity, which is determined by the antigen binding sites. The formation of an antigen–antibody complex may cause the elimination of the antigen in seven ways:

1. *Neutralization.* Both viruses and bacterial toxins have specific sites that must bind to target regions on body cells before they can enter or injure those cells. Antibodies may bind to

Hapten → Incomplete ∴ combination to function properly.

TABLE 22–1	Classes of Antibodies
Structure	**Description**
IgG	**IgG** is the largest and most diverse class of antibodies. There are several types of IgG, but each type occurs as an individual molecule. Together, they account for 80 percent of all antibodies. IgG antibodies are responsible for resistance against many viruses, bacteria, and bacterial toxins. These antibodies can cross the placenta, and maternal IgG provides passive immunity to the fetus during embryological development. However, the anti-Rh (anti-D) antibodies produced by Rh-negative mothers sensitized to Rh surface antigens are also IgG antibodies that can cross the placenta and attack fetal Rh-positive red blood cells, producing *hemolytic disease of the newborn.* ∞ p. 664
IgE	**IgE** attaches as an individual molecule to the exposed surfaces of basophils and mast cells. When a suitable antigen is bound by IgE molecules, the cell is stimulated to release histamine and other chemicals that accelerate inflammation in the immediate area. IgE is also important in the allergic response.
IgD	**IgD** is an individual molecule on the surfaces of B cells, where it can bind antigens in the extracellular fluid. This binding can play a role in the activation of the B cell involved.
IgM	**IgM** is the first class of antibody secreted after an antigen arrives. The concentration of IgM declines as IgG production accelerates. Although plasma cells secrete individual IgM molecules, IgM circulates as a five-antibody starburst. This configuration makes these antibodies particularly effective in forming immune complexes. The anti-A and anti-B antibodies responsible for the agglutination of incompatible blood types are IgM antibodies. ∞ p. 662 IgM antibodies may also attack bacteria that are insensitive to IgG.
IgA (Secretory piece)	**IgA** is found primarily in glandular secretions such as mucus, tears, saliva, and semen. These antibodies attack pathogens before they gain access to internal tissues. IgA antibodies circulate in blood as individual molecules or in pairs. Epithelial cells absorb them from the blood and attach a *secretory piece,* which confers solubility, before secreting the IgA molecules onto the epithelial surface.

those sites, making the virus or toxin incapable of attaching itself to a cell. This mechanism is known as **neutralization**.

2. *Precipitation and Agglutination.* Each antibody molecule has two antigen binding sites, and most antigens have many antigenic determinant sites. If separate antigens (such as macromolecules or bacterial cells) are far apart, an antibody molecule will necessarily bind to two antigenic sites on the same antigen. However, if antigens are close together, an antibody can bind to antigenic determinant sites on two different antigens. In this way, antibodies can form extensive "bridges" that tie large numbers of antigens together. The three-dimensional structure created by such binding is known as an **immune complex**. When the antigen is a soluble molecule, such as a toxin, this process may create complexes that are too large to remain in solution. The formation of insoluble immune complexes is called **precipitation**. When the target antigen is on the surface of a cell or virus, the formation of large complexes is called **agglutination**. The clumping of erythrocytes that occurs when incompatible blood types are mixed is an agglutination reaction. ∞ p. 664

3. *Activation of Complement.* On binding to an antigen, portions of the antibody molecule change shape, exposing areas that bind complement proteins. The bound complement molecules then activate the complement system, destroying the antigen (as discussed previously).

4. *Attraction of Phagocytes.* Antigens covered with antibodies attract eosinophils, neutrophils, and macrophages—cells that phagocytize pathogens and destroy foreign or abnormal plasma membranes.

5. *Opsonization.* A coating of antibodies and complement proteins increases the effectiveness of phagocytosis. This effect is called *opsonization* (p. 793). Some bacteria have slick plasma membranes or capsules, and phagocytes must be able to hang onto their prey before they can engulf it. Phagocytes can bind more easily to antibodies and complement proteins on the surface of a pathogen than they can to the bare surface.

6. *Stimulation of Inflammation.* Antibodies may promote inflammation through the stimulation of basophils and mast cells.

7. *Prevention of Bacterial and Viral Adhesion.* Antibodies dissolved in saliva, mucus, and perspiration coat epithelia, providing an additional layer of defense. A covering of antibodies makes it difficult for pathogens to attach to and penetrate body surfaces.

Tips&Tricks

The classes of immunoglobulins—Ig**M**, Ig**A**, Ig**D**, Ig**G**, Ig**E**—spell **MADGE**.

Primary and Secondary Responses to Antigen Exposure

The initial response to exposure to an antigen is called the **primary response**. When the antigen appears again, it triggers a more extensive and prolonged **secondary response**. This response reflects the presence of large numbers of memory cells that are primed for the arrival of the antigen. Primary and secondary responses are characteristic of both cell-mediated and antibody-mediated immunities. The differences between the responses are most easily demonstrated by following the production of antibodies over time.

The Primary Response

Because the antigen must activate the appropriate B cells, which must then differentiate into plasma cells, the primary response takes time to develop (**Figure 22–22a**). As plasma cells differentiate and begin secreting, the concentration of circulating antibodies undergoes a gradual, sustained rise.

During the primary response, the **antibody titer,** or level of antibody activity, in the plasma does not peak until one to two weeks after the initial exposure. If the individual is no longer exposed to the antigen, the antibody concentration declines thereafter. This reduction in antibody production occurs because (1) plasma cells have very high metabolic rates and survive for only a short time, and (2) further production of plasma cells is inhibited by suppression factors released by suppressor T cells. However, suppressor T cell activity does not begin immediately after antigen exposure, and under normal conditions helper cells outnumber suppressors by more than 3 to 1. As a result, many B cells are activated before suppressor T cell activity has a noticeable effect.

Activated B cells start dividing immediately. At each cycle of division, some of the daughter cells differentiate into plasma cells, while others continue to divide. Molecules of *immunoglobulin M*, or *IgM*, are the first to appear in the bloodstream. The plasma cells responsible for IgM production differentiate after only a few cycles of B cell division. Levels of *immunoglobulin G*, or *IgG*, rise more slowly, because the plasma cells responsible differentiate only after repeated cell divisions that also generate large numbers of

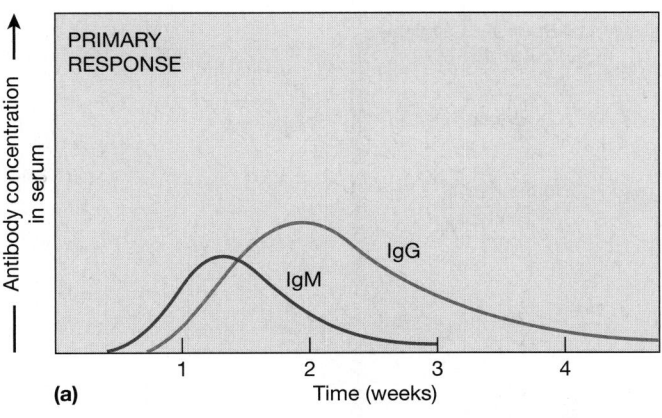

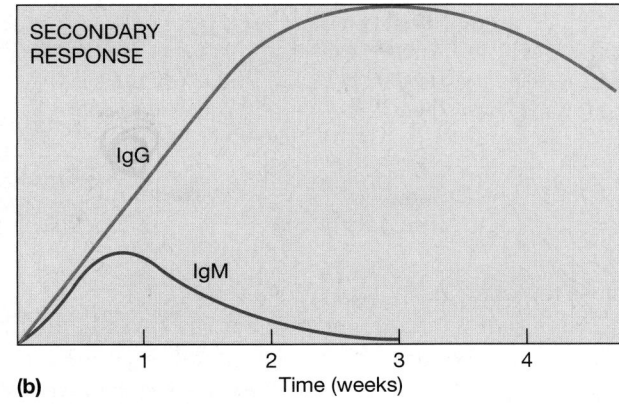

Figure 22–22 **The Primary and Secondary Responses in Antibody-Mediated Immunity. (a)** The primary response, which takes about two weeks to develop peak antibody levels and activities (titers). IgM and IgG antibody concentrations do not remain elevated. **(b)** The secondary response, which is characterized by a very rapid increase in IgG antibody titer, rises to levels much higher than those of the primary response. Antibody activity remains elevated for an extended period after the second exposure to the antigen.

memory B cells. In general, IgM is less effective as a defense mechanism than IgG. However, IgM provides an immediate defense that can fight the infection until massive quantities of IgG can be produced.

The Secondary Response

Unless they are exposed to the same antigen a second time, memory B cells do not differentiate into plasma cells. If and when that exposure occurs, the memory B cells respond immediately—faster than the B cells stimulated during the initial exposure. This response is immediate in part because memory B cells are activated at relatively low antigen concentrations, and in part because they synthesize more effective and destructive antibodies. Activated memory B cells divide and differentiate into plasma cells that secrete these antibodies in massive quantities. This secretion is the secondary response to antigen exposure.

During the secondary response, antibody titers increase more rapidly and reach levels many times higher than they did in the primary response (**Figure 22–22b**). The secondary response appears even if the second exposure occurs years after the first, because memory cells may survive for 20 years or more.

Because the primary response develops slowly and antibodies are not produced in massive quantities, it may not prevent an infection the first time a pathogen appears in the body. However, a person who survives that first infection will probably be resistant to that pathogen in the future, because the secondary response will be so rapid and overwhelming that the pathogens won't be able to survive in body tissues. The effectiveness of the secondary response is one of the basic principles behind the use of immunization to prevent disease.

The A&P Top 100

#77 Immunization produces a primary response to a specific antigen under controlled conditions. If the same antigen is encountered at a later date, it triggers a powerful secondary response that is usually sufficient to prevent infection and disease.

Tips&Tricks

The antibody response is like ordering a custom suit. The first suit (the primary antibody response) takes time to make because the tailor (an activated B cell) must first make a pattern (a clone of memory cells). Subsequent suits (secondary responses) are made much more quickly because the pattern already exists.

Summary of the Immune Response

We have now examined the basic cellular and chemical interactions that follow the appearance of a foreign antigen in the body. **Figure 22–23** presents an integrated view of the immune response and its relationship to nonspecific defenses.

Figure 22–24 provides an overview of the course of events responsible for overcoming a bacterial infection. In the early stages of infection, before antigen processing has occurred, neutrophils and NK cells migrate into the threatened area and destroy bacteria. Over time, cytokines draw increasing numbers of phagocytes into the region. Cytotoxic T cells appear as arriving T cells are activated by antigen presentation. Last of all, the population of plasma cells rises as activated B cells differentiate. This rise is followed by a gradual, sustained increase in the level of circulating antibodies.

22 LYMPHOID

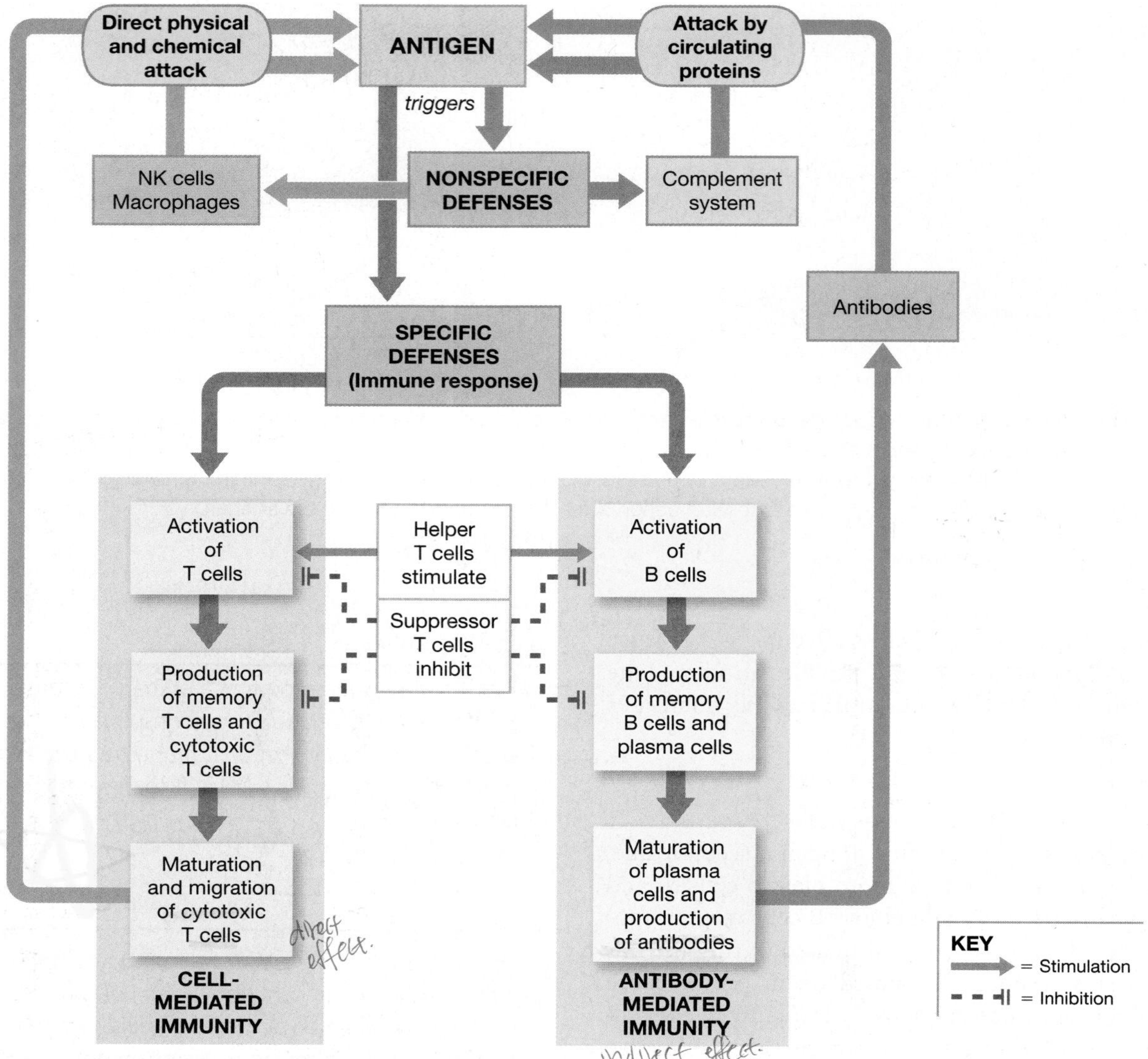

Direct physical and chemical attack → ANTIGEN ← Attack by circulating proteins

triggers

NK cells / Macrophages ← → NONSPECIFIC DEFENSES → Complement system

Antibodies

SPECIFIC DEFENSES (Immune response)

CELL-MEDIATED IMMUNITY

Activation of T cells

Production of memory T cells and cytotoxic T cells

Maturation and migration of cytotoxic T cells

direct effect.

Helper T cells stimulate

Suppressor T cells inhibit

ANTIBODY-MEDIATED IMMUNITY

Activation of B cells

Production of memory B cells and plasma cells

Maturation of plasma cells and production of antibodies

indirect effect.

KEY
→ = Stimulation
- - - ⊣| = Inhibition

Figure 22–23 **An Integrated Summary of the Immune Response.**

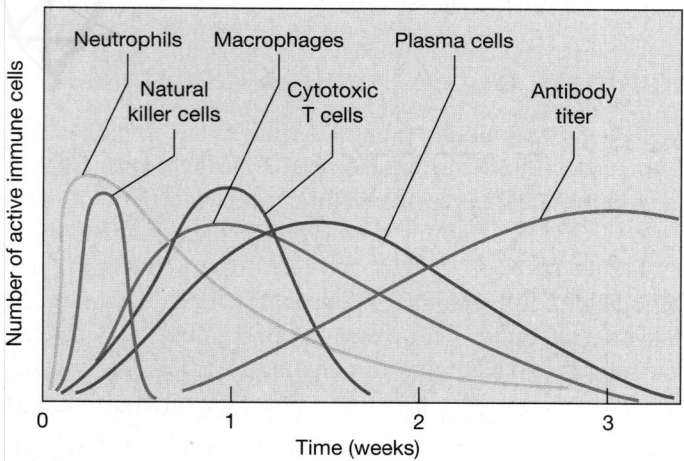

Neutrophils   Macrophages   Plasma cells

Natural killer cells   Cytotoxic T cells   Antibody titer

(Number of active immune cells vs. Time (weeks): 0, 1, 2, 3)

Figure 22–24 **The Course of the Body's Response to a Bacterial Infection.** The basic sequence of events, which begins with the appearance of bacteria in peripheral tissues at time 0.

The basic sequence of events is similar when a viral infection occurs. The initial steps are different, however, because cytotoxic T cells and NK cells can be activated by contact with virus-infected cells. **Figure 22–25** contrasts the events involved in defending against bacterial infection with those involved in defending against viral infection. Table 22–2 reviews the cells that participate in tissue defenses.

The A&P Top 100

#78 Viruses replicate inside cells, whereas bacteria may live extracellularly. Antibodies (and administered antibiotics) work outside cells, so they are primarily effective against bacteria, but not viruses. (That's why antibiotics can't fight the common cold or flu.) T cells, NK cells, and interferons are the primary defenses against viral infection.

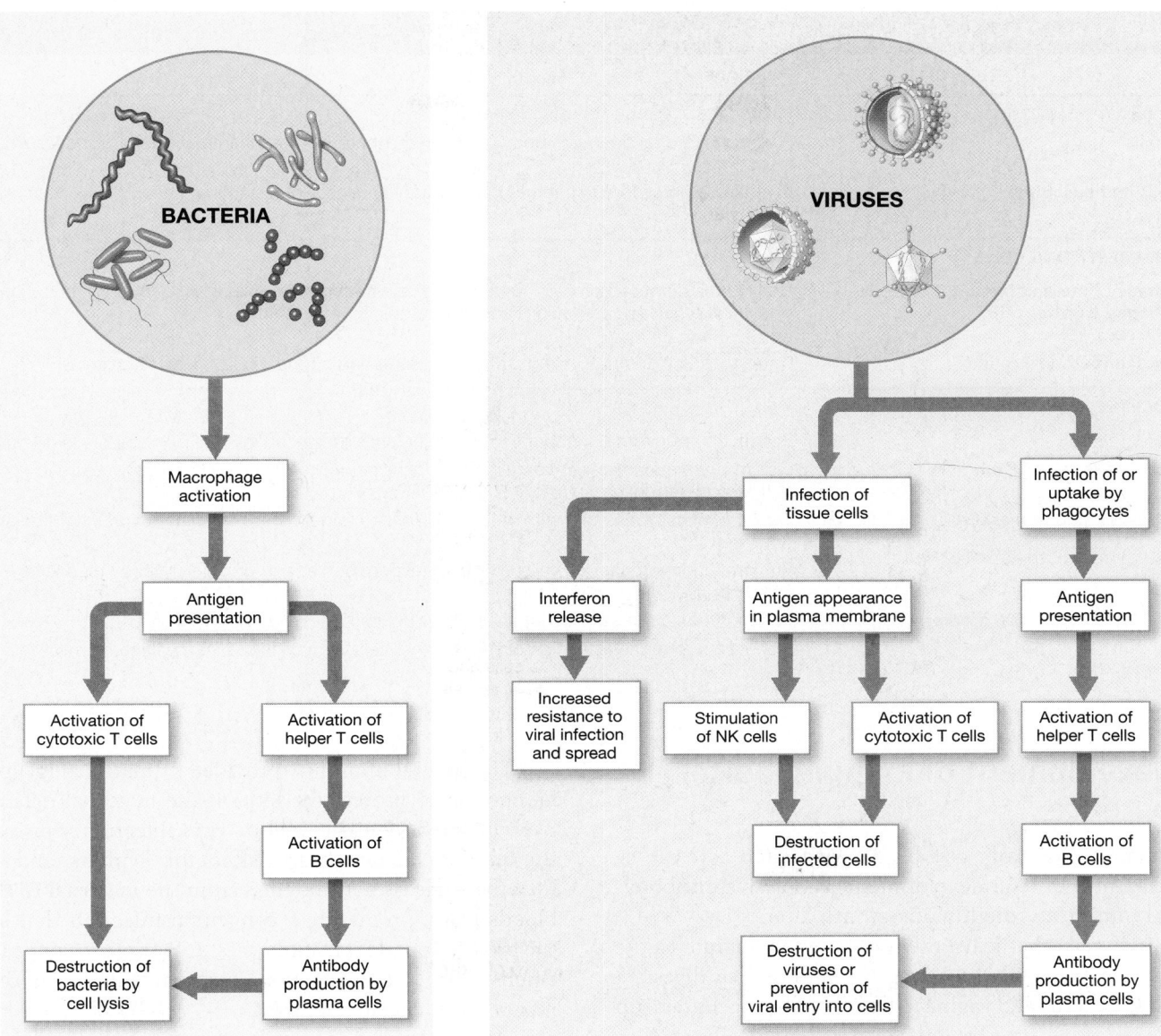

Figure 22–25 **Defenses against Bacterial and Viral Pathogens. (a)** Defenses against bacteria are usually initiated by active macrophages. **(b)** Defenses against viruses are usually activated after the infection of normal cells.

22 LYMPHOID

CHECKPOINT

18. Define sensitization.

19. Describe the structure of an antibody.

20. A sample of lymph contains an elevated number of plasma cells. Would you expect the number of antibodies in the blood to be increasing or decreasing? Why?

21. Would the primary response or the secondary response be more affected by a lack of memory B cells for a particular antigen?

See the blue Answers tab at the end of the book.

22-7 Immunological competence enables a normal immune response; abnormal responses result in immune disorders

We begin this section by exploring **immunological competence**, the ability to produce an immune response after exposure to an antigen. Then we consider some of the disorders that result when immune responses go awry. Finally, we examine the role of stress in immune responses.

SUMMARY TABLE 22–2   Cells That Participate in Tissue Defenses

Cell	Functions
Neutrophils	Phagocytosis; stimulation of inflammation
Eosinophils	Phagocytosis of antigen–antibody complexes; suppression of inflammation; participation in allergic response
Mast cells and basophils	Stimulation and coordination of inflammation by release of histamine, heparin, leukotrienes, prostaglandins
ANTIGEN-PRESENTING CELLS	
Macrophages (free and fixed macrophages, Kupffer cells, microglia, etc.)	Phagocytosis; antigen processing; antigen presentation with Class II MHC proteins; secretion of cytokines, especially interleukins and interferons
Dendritic (Langerhans) cells	Pinocytosis; antigen processing; antigen presentation bound to Class II MHC proteins
LYMPHOCYTES	
NK cells	Destruction of plasma membranes containing abnormal antigens
Cytotoxic T cells (T_C, CD8 marker)	Lysis of plasma membranes containing antigens bound to Class I MHC proteins; secretion of perforins, defensins, lymphotoxins, and other cytokines
Helper T cells (T_H, CD4 marker)	Secretion of cytokines that stimulate cell-mediated and antibody-mediated immunity; activation of sensitized B cells
B cells	Differentiation into plasma cells, which secrete antibodies and provide antibody-mediated immunity
Suppressor T cells (T_S, CD8 marker)	Secretion of suppression factors that inhibit the immune response
Memory cells (T_S, T_H, B)	Produced during the activation of T cells and B cells; remain in tissues awaiting rearrival of antigens

The Development of Immunological Competence

Cell-mediated immunity can be demonstrated as early as the third month of fetal development, and active antibody-mediated immunity roughly one month later.

The first cells that leave the fetal thymus migrate to the skin and into the epithelia lining the mouth, the digestive tract, and the uterus and vagina in females. These cells take up residence in these tissues as antigen-presenting cells, such as the dendritic cells of the skin, whose primary function will be the activation of T cells. T cells that leave the thymus later in development populate lymphoid organs throughout the body.

The plasma membranes of the first B cells produced in the liver and bone marrow carry IgM antibodies. Sometime after the fourth month *in utero* the fetus may, if exposed to specific pathogens, produce IgM antibodies. Fetal antibody production is uncommon, however, because the developing fetus has naturally acquired passive immunity due to the transfer of IgG antibodies from the maternal bloodstream. These are the only antibodies that can cross the placenta, and they include the antibodies responsible for the clinical problems that accompany fetal–maternal Rh incompatibility, discussed in Chapter 19. ∞ p. 664 Because the anti-A and anti-B antibodies are IgM antibodies, which cannot cross the placenta, problems with maternal–fetal incompatibilities involving the ABO blood groups rarely occur.

The natural immunity provided by maternal IgG may not be enough to protect the fetus if the maternal defenses are overwhelmed by a bacterial or viral infection. For example, the microorganisms responsible for syphilis and rubella ("German measles") can cross from the maternal to the fetal bloodstream, producing a congenital infection that leads to the production of fetal antibodies. IgM provides only a partial defense, and these infections can result in severe developmental problems for the fetus.

Delivery eliminates the maternal supply of IgG. Although the mother provides IgA antibodies in breast milk, the infant gradually loses its passive immunity. The amount of maternal IgG in the infant's bloodstream declines rapidly over the first two months after birth. During this period, the infant becomes vulnerable to infection by bacteria or viruses that were previously overcome by maternal antibodies. The infant also begins producing its own IgG, as its immune system begins to respond to infections, environmental antigens, and vaccinations. It has been estimated that, from birth to age 12, children encounter a "new" antigen every six weeks. (This fact explains why most parents, exposed to the same antigens when they were children, remain healthy while their children develop runny noses and colds.) Over this period, the concentration of circulating antibodies gradually rises toward normal adult levels, and the populations of memory B cells and T cells continue to increase.

Focus

Hormones of the Immune System

The specific and nonspecific defenses of the body are coordinated by both physical interaction and the release of chemical messengers. One example of physical interaction is antigen presentation by activated macrophages and helper T cells. An example of the release of chemical messengers is the secretion of cytokines by many cell types involved in the immune response. Cytokines are often classified according to their origins: *Lymphokines* are produced by lymphocytes, *monokines* by active macrophages and other antigen-presenting cells. These terms are misleading, however, because lymphocytes and macrophages may secrete the same cytokines, and cytokines can also be secreted by cells involved in nonspecific defenses and tissue repair.

Table 22–3 summarizes the cytokines identified to date. Six subgroups merit special attention: (1) *interleukins*, (2) *interferons*, (3) *tumor necrosis factors*, (4) chemicals that regulate phagocytic activities, (5) *colony-stimulating factors*, and (6) miscellaneous cytokines.

Interleukins

Interleukins may be the most diverse and important chemical messengers in the immune system. Nearly 20 types of interleukins have been identified; several are listed in Table 22–3. Lymphocytes and macrophages are the primary sources of interleukins, but certain interleukins, such as interleukin-1 (IL-1), are also produced by endothelial cells, fibrocytes, and astrocytes. Interleukins have the following general functions:

1. *Increasing T Cell Sensitivity to Antigens Exposed on Macrophage Membranes.* Heightened sensitivity accelerates the production of cytotoxic and regulatory T cells.

2. *Stimulating B Cell Activity, Plasma Cell Formation, and Antibody Production.* These events promote the production of antibodies and the development of antibody-mediated immunity.

3. *Enhancing Nonspecific Defenses.* Known effects of interleukin production include (a) stimulation of inflammation, (b) formation of scar tissue by fibrocytes, (c) elevation of body temperature via the preoptic nucleus of the hypothalamus, (d) stimulation of mast cell formation,

and (e) promotion of adrenocorticotropic hormone (ACTH) secretion by the adenohypophysis.

4. *Moderating the Immune Response.* Some interleukins help suppress immune function and shorten the duration of an immune response.

Two interleukins, IL-1 and IL-2, are important in stimulation and maintenance of the immune response. When released by activated macrophages and lymphocytes, these cytokines not only stimulate the activities of other immune cells but also further stimulate the secreting cell. The result is a positive feedback loop that promotes the recruitment of additional immune cells. Although mechanisms exist to control the degree of stimulation, the regulatory process sometimes breaks down, and massive production of interleukins can cause problems at least as severe as those of the primary infection. For example, in *Lyme disease* the release of IL-1 by activated macrophages in response to a localized bacterial infection produces fever, pain, skin rash, and arthritis throughout the entire body.

Some interleukins enhance the immune response, whereas others suppress it. The relative quantities secreted at any moment therefore have significant effects on the nature and intensity of the response to an antigen. In the course of a typical infection, the pattern of interleukin secretion is constantly changing. Whether the individual succeeds in overcoming the infection is determined in part by whether stimulatory or suppressive interleukins predominate. As a result, interleukins and their interactions are now the focus of an intensive research effort.

Interferons

Interferons make the cell that synthesizes them, and that cell's neighbors, resistant to viral infection, thereby slowing the spread of the virus. These compounds may have other beneficial effects in addition to their antiviral activity. For example, alpha-interferons and gamma-interferons attract and stimulate NK cells, and beta-interferons slow the progress of inflammation associated with viral infection. Gamma-interferons also stimulate macrophages, making them more effective at killing bacterial or fungal pathogens.

Because they stimulate NK cell activity, interferons can be used to fight some cancers. For example, alpha-interferons have been used in the treatment of malignant melanoma, bladder cancer, ovarian cancer, and two forms of leukemia. Alpha- or gamma-interferons may be used to treat Kaposi sarcoma, a cancer that typically develops in individuals with AIDS.

Tumor Necrosis Factors

Tumor necrosis factors (TNFs) slow the growth of a tumor and kill sensitive tumor cells. Activated macrophages secrete one type of TNF and carry the molecules in their plasma membranes. Cytotoxic T cells produce a different type of TNF. In addition to their effects on tumor cells, tumor necrosis factors stimulate granular leukocyte production, promote eosinophil activity, cause fever, and increase T cell sensitivity to interleukins.

Chemicals Regulating Phagocytic Activities

Several cytokines coordinate immune defenses by adjusting the activities of phagocytic cells. These cytokines include factors that attract free macrophages and microphages and prevent their premature departure from the site of an injury.

Colony-Stimulating Factors

Colony-stimulating factors (CSFs) were introduced in Chapter 19. ∞ p. 672 These factors are produced by active T cells, cells of the monocyte–macrophage group, endothelial cells, and fibrocytes. CSFs stimulate the production of blood cells in bone marrow and lymphocytes in lymphoid tissues and organs.

Miscellaneous Cytokines

This general category includes many chemicals that have been discussed in earlier chapters. Examples include leukotrienes, lymphotoxins, perforin, hemopoiesis-stimulating factor, and suppression factors.

TABLE 22–3 Chemical Mediators (Cytokines) of the Immune Response	
Compound	**Functions**
INTERLEUKINS	
Interleukin-1 (IL-1)	Stimulates T cells to produce IL-2, promotes inflammation; causes fever
IL-2	Stimulates growth and activation of other T cells and NK cells
IL-3	Stimulates production of mast cells and other blood cells
IL-4 (B cell differentiating factor); IL-5 (B cell growth factor); IL-6, IL-7 (B cell stimulating factors); IL-10; IL-11	Promote differentiation and growth of B cells and stimulate plasma cell formation and antibody production; each has somewhat different effects on macrophages and their activities
IL-8	Stimulates blood vessel formation (angiogenesis)
IL-9	Stimulates myeloid cell production (RBCs, platelets, granulocytes, monocytes)
IL-12	Stimulates T cell activity and cell-mediated immunity
IL-13	Suppresses production of several other cytokines (IL-1, IL-8, TNF); stimulates Class II MHC antigen presentation
INTERFERONS (ALPHA, BETA, GAMMA)	Activate other cells to prevent viral entry; inhibit viral replication; stimulate NK cells and macrophages
TUMOR NECROSIS FACTORS (TNFS)	Kill tumor cells; slow tumor growth; stimulate activities of T cells and eosinophils; inhibit parasites and viruses
PHAGOCYTE-ACTIVATING CHEMICALS	
Monocyte-chemotactic factor (MCF)	Attracts monocytes; transforms them into macrophages
Migration-inhibitory factor (MIF)	Prevents migration of macrophages from the area
Macrophage-activating factor (MAF)	Makes macrophages more active and aggressive
Microphage-chemotactic factor	Attracts microphages from the blood
COLONY-STIMULATING FACTORS (CSFS)	Stimulate RBC and WBC production
Miscellaneous cytokines	
Leukotrienes	Stimulate regional inflammation
Lymphotoxins	Kill cells; damage tissue; promote inflammation
Perforin	Destroys plasma membranes by creating large pores
Defensins	Kill cells by piercing plasma membrane
Hemopoiesis-stimulating factor	Promotes blood cell production in bone marrow
Suppression factors	Inhibit T_C cell and B cell activity; depress immune response
Growth-inhibitory factor (GIF)	Reduces or inhibits replication of target cells

Skin tests can sometimes determine whether an individual has been exposed to a particular antigen. In this procedure, small quantities of antigen are injected into the skin, generally on the anterior surface of the forearm. If resistance has developed, the region will become inflamed over the next two to four days. Many states require a tuberculosis test, called a *tuberculin skin test*, before children enter public school, and when adults apply for a food-service or health-service job. If the test is positive, further tests must then be performed to determine whether an infection is currently under way. Skin tests are also used to check for allergies to environmental antigens.

Immune Disorders

[handwritten: AIDS: helper T cells.]

Because the immune response is so complex, many opportunities exist for things to go wrong. A variety of clinical conditions result from disorders of the immune function. **Autoimmune disorders** develop when the immune response inappropriately targets normal body cells and tissues. In an **immunodeficiency disease,** either the immune system fails to develop normally or the immune response is blocked in some way. Autoimmune disorders and immunodeficiency diseases are relatively rare—clear *[handwritten: can't produce neither B nor T cells.]* evidence of the effectiveness of the immune system's control mechanisms. A far more common (and generally far less dangerous) class of immune disorders is **allergies**. We next consider examples of each type of immune disorder.

Autoimmune Disorders

Autoimmune disorders affect an estimated 5 percent of adults in North America and Europe. Previous chapters cited many examples of the effects of autoimmune disorders on the function of major systems.

The immune system usually recognizes but ignores antigens normally found in the body—self-antigens. When the recognition system malfunctions, however, activated B cells make antibodies against other body cells and tissues. These "misguided" antibodies are called **autoantibodies**. The trigger may be a reduction in suppressor T cell activity, the excessive stimulation of helper T cells, tissue damage that releases large quantities of antigenic fragments, haptens bound to compounds normally ignored, viral or bacterial toxins, or a combination of factors.

The condition produced depends on the specific antigen attacked by autoantibodies. For example,

- The inflammation of *thyroiditis* results from the release of autoantibodies against thyroglobulin;

- *Rheumatoid arthritis* occurs when autoantibodies form immune complexes within connective tissues around the joints; and

- *Insulin-dependent diabetes mellitus* (IDDM) is generally caused by autoantibodies that attack cells in the pancreatic islets.

Many autoimmune disorders appear to be cases of mistaken identity. For example, proteins associated with the measles, Epstein–Barr, influenza, and other viruses contain amino acid sequences that are similar to those of myelin proteins. As a result, antibodies that target these viruses may also attack myelin sheaths. This mechanism accounts for the neurological complications that sometimes follow a vaccination or a viral infection. It is also the mechanism that is likely responsible for *multiple sclerosis*.

For unknown reasons, the risk of autoimmune problems increases if an individual has an unusual type of MHC protein. At least 50 clinical conditions have been linked to specific variations in MHC structure.

Immunodeficiency Diseases

Immunodeficiency diseases result from (1) problems with the embryological development of lymphoid organs and tissues; (2) an infection with a virus, such as HIV, that depresses immune function; or (3) treatment with, or exposure to, immunosuppressive agents, such as radiation or drugs.

Individuals born with **severe combined immunodeficiency disease (SCID)** fail to develop either cell- or antibody-mediated immunity. Their lymphocyte populations are low, and normal B cells and T cells are absent. Such infants cannot produce an immune response, so even a mild infection can prove fatal. Total isolation offers protection but at great cost—extreme restrictions on lifestyle. Bone marrow transplants from compatible donors, normally a close relative, have been used to colonize lymphoid tissues with functional lymphocytes. Gene-splicing techniques have led to therapies that can treat at least one form of SCID.

AIDS, an immunodeficiency disease that we consider on page 819, is the result of a viral infection that targets primarily helper T cells. As the number of T cells declines, the normal immune control mechanism breaks down. When a subsequent infection occurs, suppressor factors released by suppressor T cells inhibit an immune response before the few surviving helper T cells can stimulate the formation of cytotoxic T cells or plasma cells in adequate numbers.

Immunosuppressive drugs have been used for many years to prevent graft rejection after transplant surgery. But immunosuppressive agents can destroy stem cells and lymphocytes, leading to a complete immunological failure. This outcome is one of the potentially fatal consequences of radiation exposure.

Allergies

Allergies are inappropriate or excessive immune responses to antigens. The sudden increase in cellular activity or antibody titers can have several unpleasant side effects. For example, neutrophils or cytotoxic T cells may destroy normal cells while attacking the antigen, or the antigen–antibody complex

[vertical margin text: 22 LYMPHOID]

may trigger a massive inflammatory response. Antigens that trigger allergic reactions are often called **allergens**.

There are several types of allergies. A complete classification recognizes four categories: *immediate hypersensitivity (Type I)*, *cytotoxic reactions (Type II)*, *immune complex disorders (Type III)*, and *delayed hypersensitivity (Type IV)*. In Chapter 19 we discussed one example of a cytotoxic (Type II) reaction: the cross-reaction that follows the transfusion of an incompatible blood type. ∞ p. 662 Here we concentrate on immediate (Type I) hypersensitivity, probably the most common type of allergy.

Immediate hypersensitivity is a rapid and especially severe response to the presence of an antigen. One form, *allergic rhinitis*, includes hay fever and environmental allergies and may affect 15 percent of the U.S. population. Sensitization to an allergen during the initial exposure leads to the production of large quantities of IgE. The tendency to produce IgE antibodies in response to specific allergens may be genetically determined.

Due to the lag time needed to activate B cells, produce plasma cells, and synthesize antibodies, the first exposure to an allergen does not produce symptoms, but merely sets the stage for the next encounter. After sensitization, the IgE molecules become attached to the plasma membranes of basophils and mast cells throughout the body. When the individual is subsequently exposed to the same allergen, the bound antibodies stimulate these cells to release histamine, heparin, several cytokines, prostaglandins, and other chemicals into the surrounding tissues. A sudden, massive inflammation of the affected tissues results.

The cytokines and other mast cell secretions draw basophils, eosinophils, T cells, and macrophages into the area. These cells release their own chemicals, extending and intensifying the responses initiated by mast cells. The severity of the allergic reaction depends on the individual's sensitivity and on the location involved. If allergen exposure occurs at the body surface, the response may be restricted to that area. If the allergen enters the bloodstream, the response could be lethal.

In **anaphylaxis** (an-a-fi-LAK-sis; *ana-*, again + *phylaxis*, protection), a circulating allergen affects mast cells throughout the body (**Figure 22–26**). (In drug reactions, such as allergies to penicillin, IgE antibodies are produced in response to a hapten [partial antigen] bound to a larger molecule that is widely distributed within the body; the combination acts as an allergen.) A wide range of signs and symptoms can develop within minutes. Changes in capillary permeabilities produce swelling and edema in the dermis, and raised welts, or *hives*, appear on the surface of the skin. Smooth muscles along the respiratory passageways contract; the narrowed passages make breathing extremely difficult. In severe cases, an extensive peripheral vasodilation occurs, producing a fall in blood pressure that can lead to a circulatory collapse. This response is **anaphylactic shock**.

Many of the signs and symptoms of immediate hypersensitivity can be prevented by the prompt administration of **anti-**

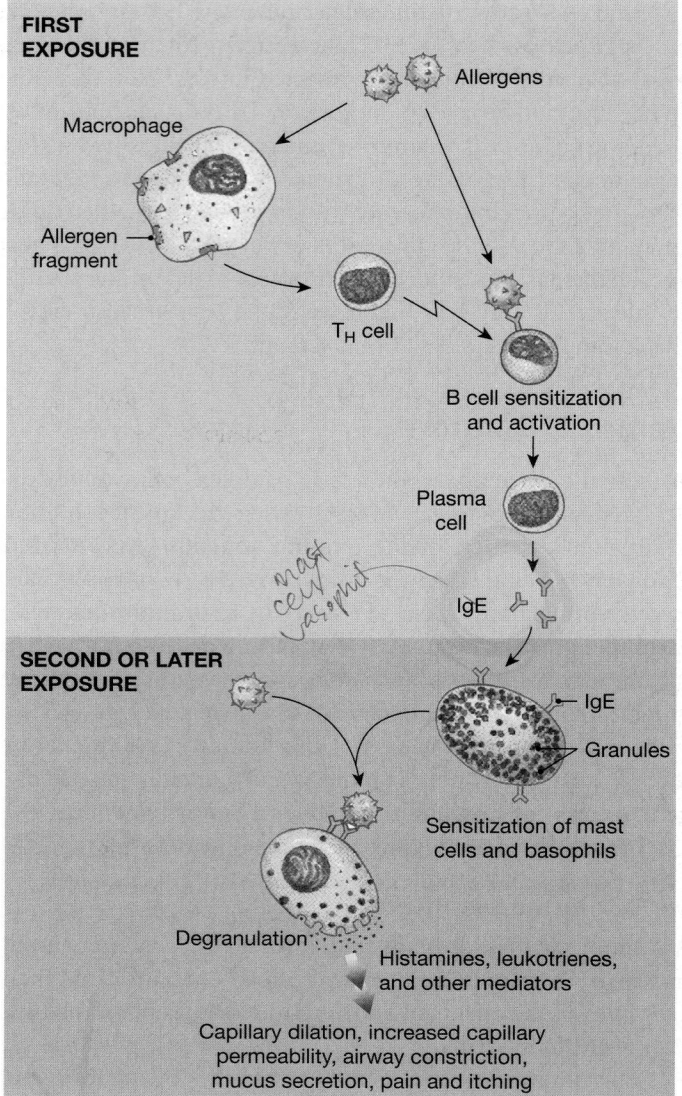

Figure 22–26   The Mechanism of Anaphylaxis.

histamines (an-tē-HIS-ta-mēnz)—drugs that block the action of histamine. *Benadryl* (*diphenhydramine hydrochloride*) is a popular antihistamine that is available over the counter. The treatment of severe anaphylaxis involves antihistamine, corticosteroid, and epinephrine injections.

Stress and the Immune Response

One of the normal effects of interleukin-1 secretion is the stimulation of adrenocorticotropic hormone (ACTH) production by the adenohypophysis. The production of ACTH in turn leads to the secretion of glucocorticoids by the suprarenal cortex. ∞ p. 629 The anti-inflammatory effects of the glucocorticoids may help control the extent of the immune response. However, the long-term secretion of glucocorticoids, as in the resistance phase of the *general adaptation syndrome*, can inhibit the immune response and lower resistance to disease. ∞ p. 640 The effects of glucocorticoids that alter the effectiveness of specific and nonspecific defenses include the following:

- *Depression of the Inflammatory Response.* Glucocorticoids inhibit mast cells and reduce the permeability of capillaries. Inflammation is therefore less likely. When it does occur, the reduced permeability of the capillaries slows the entry of fibrinogen, complement proteins, and cellular defenders into tissues.

- *Reduction in the Abundance and Activity of Phagocytes in Peripheral Tissues.* This reduction further impairs nonspecific defense mechanisms and interferes with the processing and presentation of antigens to lymphocytes.

- *Inhibition of Interleukin Secretion.* A reduction in interleukin production depresses the response of lymphocytes, even to antigens bound to MHC proteins.

The mechanisms responsible for these changes are still under investigation. It is clear, however, that depression of the immune system due to chronic stress can be a serious threat to health.

CHECKPOINT

22. Which kind of immunity protects a developing fetus, and how is that immunity produced?

23. What is an autoimmune disorder?

24. How does increased stress reduce the effectiveness of the immune response?

See the blue Answers tab at the end of the book.

22-8 The immune response diminishes with advancing age

With advancing age, the immune system becomes less effective at combating disease. T cells become less responsive to antigens, so fewer cytotoxic T cells respond to an infection. This effect may, at least in part, be associated with the gradual involution of the thymus and a reduction in circulating levels of thymic hormones. Because the number of helper T cells is also reduced, B cells are less responsive, so antibody levels do not rise as quickly after antigen exposure. The net result is an increased susceptibility to viral and bacterial infections. For this reason, vaccinations for acute viral diseases such as the flu (influenza), and for pneumococcal pneumonia, are strongly recommended for elderly individuals. The increased incidence of cancer in the elderly reflects the fact that immune surveillance declines, so tumor cells are not eliminated as effectively.

CHECKPOINT

25. Why are the elderly more susceptible to viral and bacterial infections?

26. What may account for the increased incidence of cancer among the elderly?

See the blue Answers tab at the end of the book.

22-9 For all body systems, the lymphoid system provides defenses against infection and returns tissue fluid to the circulation

Figure 22–27 summarizes the interactions between the lymphoid system and other physiological systems. Of all the system interactions, those among elements of the immune response and the nervous and endocrine systems are now the focus of intense research. For example,

- The thymus secretes oxytocin, ADH, and endorphins as well as thymic hormones. The effects on the CNS are not known, but removal of the thymus lowers brain endorphin levels.

- Both thymic hormones and cytokines help establish the normal levels of CRH and TRH produced by the hypothalamus.

- Other thymic hormones affect the adenohypophysis directly, stimulating the secretion of prolactin and GH.

Conversely, the nervous system can apparently adjust the sensitivity of the immune response:

- The CNS innervates dendritic cells in the lymph nodes, spleen, skin, and other antigen-presenting cells. The nerve endings release neurotransmitters that heighten local immune responses. For this reason, some skin conditions, such as *psoriasis*, worsen when a person is under stress.

- Neuroglia in the CNS produce cytokines that promote an immune response.

- A sudden decline in immune function can occur after even a brief period of emotional distress.

CHECKPOINT

27. What is the relationship between the endocrine system and the lymphoid system?

28. Identify the role of the lymphoid system for all body systems.

See the blue Answers tab at the end of the book.

THE LYMPHOID SYSTEM IN PERSPECTIVE

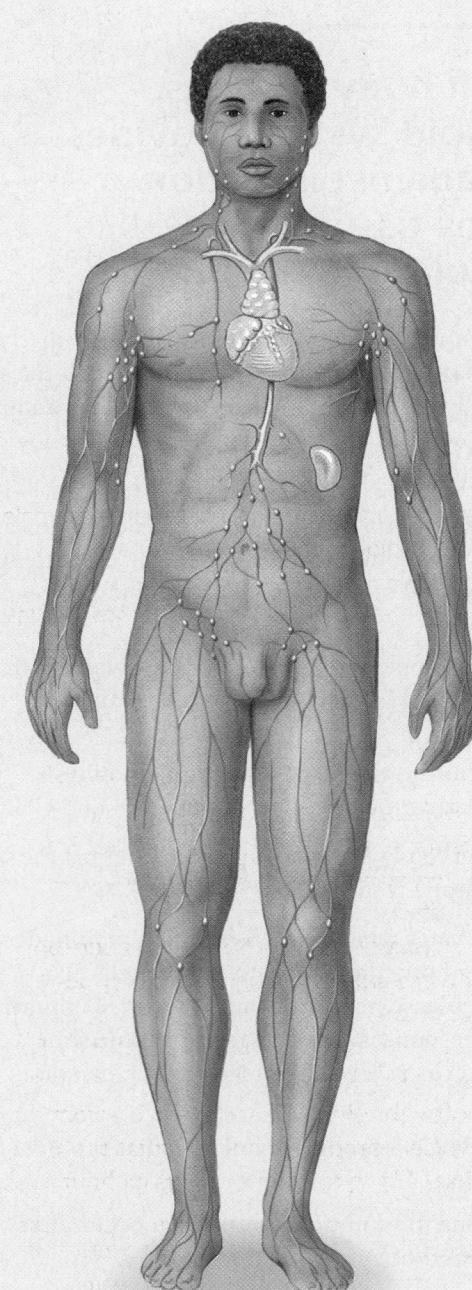

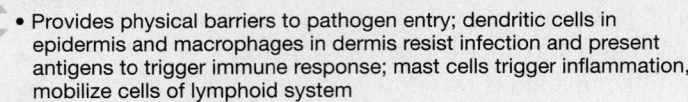

Integumentary System

- Provides physical barriers to pathogen entry; dendritic cells in epidermis and macrophages in dermis resist infection and present antigens to trigger immune response; mast cells trigger inflammation, mobilize cells of lymphoid system
- Provides IgA for secretion onto integumentary surfaces

Skeletal System

- Lymphocytes and other cells involved in the immune response are produced and stored in bone marrow
- Assists in repair of bone after injuries; osteoclasts differentiate from monocyte–macrophage cell line

Muscular System

- Protects superficial lymph nodes and the lymphatic vessels in the abdominopelvic cavity; muscle contractions help propel lymph along lymphatic vessels
- Assists in repair after injury

Nervous System

- Microglia present antigens that stimulate specific defenses; neuroglia secrete cytokines; innervation stimulates antigen-presenting cells
- Cytokines affect hypothalamic production of CRH and TRH

Endocrine System

- Glucocorticoids have anti-inflammatory effects; thymic hormones stimulate development and maturation of lymphocytes; many hormones affect immune function
- Thymus secretes thymic hormones; cytokines affect cells throughout the body

Cardiovascular System

- Distributes WBCs; carries antibodies that attack pathogens; clotting response helps restrict spread of pathogens; granulocytes and lymphocytes produced in bone marrow
- Fights infections of cardiovascular organs; returns tissue fluid to circulation

Respiratory System

- Alveolar phagocytes present antigens and trigger specific defenses; provides oxygen required by lymphocytes and eliminates carbon dioxide generated during their metabolic activities
- Tonsils protect against infection at entrance to respiratory tract

Digestive System

- Provides nutrients required by lymphoid tissues, digestive acids, and enzymes; provides nonspecific defense against pathogens
- Tonsils and MALT defend against infection and toxins absorbed from digestive tract; lymphatic vessels carry absorbed lipids to venous system

Urinary System

- Eliminates metabolic wastes generated by cellular activity
- Acid pH of urine provides nonspecific defense against urinary tract infection; provides specific defenses against urinary tract infection

Reproductive System

- Lysosomes and bactericidal chemicals in secretions provide nonspecific defense against reproductive tract infection
- Provides IgA for secretion by epithelial glands

FOR ALL SYSTEMS
Provides specific defenses against infection; immune surveillance eliminates cancer cells; returns tissue fluid to circulation

Figure 22–27   Functional Relationships between the Lymphoid System and Other Systems.

CLINICAL NOTE

AIDS

Acquired immune deficiency syndrome (AIDS), or *late-stage HIV disease*, is caused by the **human immunodeficiency virus (HIV)**. This virus is a *retrovirus*: It carries its genetic information in RNA rather than in DNA. The virus enters human leukocytes by receptor-mediated endocytosis. ∞ p. 97 Specifically, the virus binds to CD4, the membrane protein characteristic of helper T cells. Several types of antigen-presenting cells, including those of the monocyte–macrophage line, also are infected by HIV, but it is the infection of helper T cells that leads to clinical problems.

Once the virus is inside a cell, the viral enzyme *reverse transcriptase* synthesizes a complementary strand of DNA, which is then incorporated into the cell's genetic material. When these inserted viral genes are activated, the infected cell begins synthesizing viral proteins. In effect, the introduced viral genes take over the cell's synthetic machinery and force the cell to produce additional viruses. These new viruses are then shed at the cell surface.

Cells infected with HIV are ultimately destroyed by (1) formation of pores in the plasma membrane as the viruses are shed, (2) cessation of cell maintenance due to the continuing synthesis of viral components, (3) autolysis, or (4) stimulation of apoptosis.

The gradual destruction of helper T cells impairs the immune response, because these cells play a central role in coordinating cell-mediated and antibody-mediated responses to antigens. To make matters worse, suppressor T cells are relatively unaffected by the virus, and over time the excess of suppressing factors "turns off" the normal immune response. Circulating antibody levels decline, cell-mediated immunity is reduced, and the body is left with impaired defenses against a wide variety of microbial invaders. With the affected person's immune function reduced, ordinarily harmless microorganisms can initiate lethal *opportunistic infections*. Because immune surveillance is also depressed, the risk of cancer increases.

Infection with HIV occurs through intimate contact with the body fluids of infected individuals. Although all body fluids may contain the virus, the major routes of transmission involve contact with blood, semen, or vaginal secretions. Worldwide, most individuals with AIDS become infected through sexual contact with an HIV-infected person (who may *not* necessarily be exhibiting the clinical signs of AIDS). The next largest group of infected individuals consists of intravenous drug users who shared contaminated needles. Relatively few individuals have become infected with the virus after receiving a transfusion of contaminated blood or blood products. Finally, an increasing number of infants are born with AIDS, having acquired it from infected mothers.

AIDS is a public health problem of massive proportions. Over half a million people have already died of AIDS in the United States alone. The estimated number of individuals infected with HIV in the United States as of 2005 was 956,000. In 2005, an estimated 16,000 of these people died, and more than 44,000 new cases of AIDS were diagnosed. The numbers worldwide are even more frightening. The World Health Organization (WHO) estimates that, as of 2006, 40 million individuals were infected. Every 6 seconds another person becomes infected with HIV, resulting in well over 4 million new cases in 2006. Every 11 seconds someone dies of AIDS, and the total death toll for 2006 was estimated to be 3 million people, including 400,000 children under the age of 15. (Data compiled by the WHO are estimates based on the best available information.)

The most effective ways to prevent HIV infection include abstinence, mutual monogamy of uninfected partners, male circumcision, and the avoidance of needle sharing. Circumcised men are up to 70% less likely to become infected than uncircumcised men, and are also less likely to infect their female partners. The use of latex condoms labeled "for the prevention of disease" has been shown to prevent the passage of HIV and the hepatitis and herpes viruses. Natural condoms, made from lambskin or gut tissue, are not effective against disease transmission due to the presence of naturally occurring pores. Synthetic condoms, usually made of urethane, are considered to offer similar protection as latex, but differences in their manufacture cause variable reliability.

Clinical signs and symptoms of AIDS may not appear until 5–10 years or more after infection. When they do appear, they are commonly mild, consisting of lymphadenopathy and chronic, but nonfatal, infections. So far as is known, however, AIDS is almost always fatal, and most people who carry the virus will eventually die from complications of the disease. (A handful of infected individuals have been able to tolerate the virus without apparent illness for many years.)

Despite intensive efforts, a vaccine has yet to be developed that prevents HIV infection in an uninfected person exposed to the virus. The survival rate for AIDS patients has been steadily increasing, because new drugs and drug combinations that slow the progression of the disease are available, and because improved antibiotic therapies help combat secondary infections.

Related Clinical Terms

acquired immune deficiency syndrome (AIDS): A disorder that develops following HIV infection and is characterized by reduced circulating antibody levels and depressed cell-mediated immunity.

allergen: An antigen capable of triggering an allergic reaction.

allergy: An inappropriate or excessive immune response to antigens, triggered by the stimulation of mast cells bound to IgE.

anaphylactic shock: A drop in blood pressure that may lead to circulatory collapse, resulting from a severe case of anaphylaxis.

anaphylaxis: A type of allergy in which a circulating allergen affects mast cells throughout the body, producing numerous signs and symptoms very quickly.

appendicitis: An infection and inflammation of the aggregated lymphoid nodules in the appendix.

autoimmune disorder: A disorder that develops when the immune response inappropriately targets normal body cells and tissues.

bacteria: Prokaryotic cells (cells lacking nuclei and other membranous organelles) that may be extracellular or intracellular pathogens.

filariasis: An infection by parasitic roundworms; the adult worms may scar and block lymphatic vessels, causing acute lymphedema, commonly in the external genitalia and lower limbs (**elephantiasis**).

fungi (singular, *fungus*): Eukaryotic organisms that absorb organic materials from the remains of dead cells; some fungi are pathogenic.

human immunodeficiency virus (HIV): The virus responsible for AIDS and related immunodeficiency disorders.

immune complex disorder: A disorder caused by the precipitation of immune complexes at sites such as the kidneys, where their presence disrupts normal tissue function.

immunodeficiency disease: A disease in which either the immune system fails to develop normally or the immune response is blocked.

immunosuppression: A reduction in the sensitivity of the immune system.

immunosuppressive drugs: Drugs administered to inhibit the immune response; examples include prednisone, cyclophosphamide, azathioprine, cyclosporin, and tacrolimus (FK506).

lymphadenopathy: A chronic or excessive enlargement of lymph nodes.

lymphedema: An accumulation of lymph in a region whose lymphatic drainage has been blocked.

lymphomas: Cancers consisting of abnormal lymphocytes or lymphoid stem cells; examples include *Hodgkin disease* and *non-Hodgkin lymphoma*.

severe combined immunodeficiency disease (SCID): A congenital disorder resulting from the failure of both cell-mediated and antibody-mediated immunity to develop.

tonsillitis: An infection of one or more tonsils; signs and symptoms include a sore throat, high fever, and leukocytosis (an abnormally high white blood cell count).

vaccine: A preparation of antigens derived from a specific pathogen; administered during *immunization*, or *vaccination*.

viruses: Noncellular pathogens that replicate by directing the synthesis of virus-specific proteins and nucleic acids inside tissue cells.

Chapter Review
Study Outline

iP Use *Interactive Physiology*® **(IP)** —available on the IP CD packaged with your new textbook and online at **myA&P**™ (www.myaandp.com)—to help you understand difficult physiological concepts. Follow these navigation paths on IP for concepts in this chapter:

- Immune System/Orientation
- Immune System/Anatomy Review
- Immune System/Innate Host Defenses
- Immune System/Common Characteristics of B and T Lymphocytes
- Immune System/Humoral Immunity
- Immune System/Cellular Immunity

22-1 Anatomical barriers and defense mechanisms constitute nonspecific defense, and lymphocytes provide specific defense p. 777

1. The cells, tissues, and organs of the **lymphoid system** play a central role in the body's defenses against a variety of **pathogens**, or disease-causing organisms.

2. *Lymphocytes*, the primary cells of the lymphoid system, are central to an **immune response** against specific threats to the body. **Immunity** is the ability to resist infection and disease through the activation of specific defenses.

22-2 Lymphatic vessels, lymphocytes, lymphoid tissues, and lymphoid organs function in body defenses p. 777

3. The lymphoid system includes a network of **lymphatic vessels**, or **lymphatics**, that carries **lymph** (a fluid similar to plasma, but with a lower concentration of proteins). An array of **lymphoid tissues** and **lymphoid organs** is connected to the lymphatic vessels. (*Figure 22–1*)

4. The lymphoid system produces, maintains, and distributes lymphocytes (which attack invading organisms, abnormal cells, and foreign proteins); it also helps maintain blood volume and eliminate local variations in the composition of interstitial fluid.

5. Lymph flows along a network of lymphatic vessels, the smallest of which are the **lymphatic capillaries** (*terminal lymphatics*). The lymphatic vessels empty into the **thoracic duct** and the **right lymphatic duct**. (*Figures 22–2 to 22–4*)

6. The three classes of lymphocytes are **T** (**t**hymus-dependent) **cells**, **B** (**b**one marrow–derived) **cells**, and **NK** (**n**atural **k**iller) **cells**.

7. **Cytotoxic T cells** attack foreign cells or body cells infected by viruses and provide **cell-mediated (cellular) immunity**. *Regulatory T cells* (**helper T cells** and **suppressor T cells**) regulate and coordinate the immune response.

8. B cells can differentiate into **plasma cells**, which produce and secrete *antibodies* that react with specific chemical targets called **antigens**. Antibodies in body fluids are called *immunoglobulins*. B cells are responsible for **antibody-mediated (humoral) immunity**.

9. NK cells (also called **large granular lymphocytes**) attack foreign cells, normal cells infected with viruses, and cancer cells. NK cells provide *immunological surveillance*.

10. Lymphocytes continuously migrate into and out of the blood through the lymphoid tissues and organs. **Lymphopoiesis** (lymphocyte production) involves the bone marrow, thymus, and peripheral lymphoid tissues. (*Figure 22–5*)

11. **Lymphoid tissues** are connective tissues dominated by lymphocytes. In a **lymphoid nodule**, the lymphocytes are densely packed in an area of loose connective tissue. The lymphoid tissue embedded within the organs of the digestive system is called **mucosa-associated lymphoid tissue (MALT)**. (*Figure 22–6*)

12. Important lymphoid organs include the **lymph nodes**, the **thymus**, and the **spleen**. Lymphoid tissues and organs are distributed in areas that are especially vulnerable to injury or invasion.

13. Lymph nodes are encapsulated masses of lymphoid tissue. The **deep cortex** is dominated by T cells; the **outer cortex** and **medulla** contain B cells. (*Figure 22–7*)

14. The thymus lies behind the sternum, in the anterior mediastinum. **Reticular epithelial cells** scattered among the lymphocytes maintain the blood–thymus barrier and secrete thymic hormones. (*Figure 22–8*)

15. The adult spleen contains the largest mass of lymphoid tissue in the body. The cellular components form the **pulp** of the spleen. **Red pulp** contains large numbers of red blood cells, and **white pulp** resembles lymphoid nodules. (*Figure 22–9*)

16. The lymphoid system is a major component of the body's defenses, which are classified as either (1) **nonspecific defenses**, which protect without distinguishing one threat from another, or (2) **specific defenses**, which protect against particular threats only.

22-3 Nonspecific defenses do not discriminate between potential threats and respond the same regardless of the invader p. 789

17. Nonspecific defenses prevent the approach, deny the entry, or limit the spread of living or nonliving hazards. (*Figure 22–10*)

18. Physical barriers include skin, mucous membranes, hair, epithelia, and various secretions of the integumentary and digestive systems.

19. The two types of phagocytic cells are **microphages** and **macrophages** (cells of the **monocyte–macrophage system**). Microphages are neutrophils and eosinophils in circulating blood.

20. **Phagocytes** leave the bloodstream by *emigration*, or *diapedesis* (migration between adjacent endothelial cells), and exhibit **chemotaxis** (sensitivity and orientation to chemical stimuli).

21. **Immunological surveillance** involves constant monitoring of normal tissues by NK cells that are sensitive to abnormal antigens on the surfaces of otherwise normal cells. NK cells kill cancer cells that have **tumor-specific antigens** on their surfaces. (*Figure 22–11*)

22. **Interferons**—small proteins released by cells infected with viruses—trigger the production of **antiviral proteins**, which interfere with viral replication inside the cell. Interferons are **cytokines**—chemical messengers released by tissue cells to coordinate local activities.

23. At least 11 **complement proteins** make up the **complement system**. These proteins interact with each other in cascades to destroy target plasma membranes, stimulate inflammation, attract phagocytes, or enhance phagocytosis. The complement system can be activated by either the **classical pathway** or the **alternative pathway**. (*Figure 22–12*)

24. **Inflammation** is a localized tissue response to injury. (*Figure 22–13*)

25. A **fever** (body temperature greater than 37.2°C [99°F]) can inhibit pathogens and accelerate metabolic processes. **Pyrogens** can reset the body's thermostat and raise the temperature.

22-4 Specific defenses (immunity) respond to individual threats and are either cell mediated or antibody mediated p. 796

26. T cells are responsible for **cell-mediated (cellular) immunity**. B cells provide **antibody-mediated (humoral) immunity**.

27. Specific resistance or immunity involves **innate immunity** (genetically determined and present at birth) or **acquired immunity**. The two types of acquired immunity are **active immunity** (which appears after exposure to an antigen) and **passive immunity** (produced by the transfer of antibodies from another source). (*Figure 22–14*)

28. Immunity exhibits four general properties: **specificity**, **versatility**, **memory**, and **tolerance**. *Memory cells* enable the immune system to "remember" previous target antigens. Tolerance is the ability of the immune system to ignore some antigens, such as those of normal body cells.

29. The immune response is triggered by the presence of an antigen and includes cell-mediated and antibody-mediated defenses. (*Figure 22–15*)

22-5 T cells play a role in the initiation, maintenance, and control of the immune response p. 798

30. **Antigen presentation** occurs when an antigen–glycoprotein combination appears in a plasma membrane (typically, that of a macrophage). T cells sensitive to this antigen are activated if they contact the membrane of the antigen-presenting cell.

31. All body cells have membrane glycoproteins. The genes controlling their synthesis make up a chromosomal region called the **major histocompatibility complex (MHC)**. The membrane glycoproteins are called **MHC proteins**. **APCs (antigen-presenting cells)** are involved in antigen stimulation.

22 LYMPHOID

32. Lymphocytes are not activated by lone antigens, but will respond to an antigen bound to either a **Class I** or a **Class II** MHC protein in a process called **antigen recognition**. *(Figure 22–16)*

33. Class I MHC proteins are in all nucleated body cells. Class II MHC proteins are only in antigen-presenting cells (APCs) and lymphocytes.

34. Whether a T cell responds to antigens held in Class I or Class II MHC proteins depends on the structure of the T cell plasma membrane. T cell plasma membranes contain proteins called **CD** (**c**luster of **d**ifferentiation) **markers**. **CD3 markers** are present on all T cells. **CD8 markers** are on cytotoxic and suppressor T cells. **CD4 markers** are on all helper T cells.

35. One type of CD8 cell responds quickly, giving rise to large numbers of cytotoxic T cells and memory T cells. The other type of CD8 cell responds more slowly, giving rise to small numbers of suppressor T cells.

36. Cytotoxic T cells seek out and destroy abnormal and infected cells, using three different methods, including the secretion of **lymphotoxin**. *(Figure 22–17)*

37. Cell-mediated immunity (cellular immunity) results from the activation of CD8 T cells by antigens bound to Class I MHCs. When activated, most of these T cells divide to generate cytotoxic T cells and **memory T$_C$ cells**, which remain in reserve to guard against future such attacks. Suppressor T cells depress the responses of other T cells and of B cells. *(Figures 22–17, 22–19)*

38. Helper, or CD4, T cells respond to antigens presented by Class II MHC proteins. When activated, helper T cells secrete lymphokines that aid in coordinating specific and nonspecific defenses, and regulate cell-mediated and antibody-mediated immunity. *(Figures 22–18, 22–19)*

22-6 B cells respond to antigens by producing specific antibodies p. 804

39. B cells become **sensitized** when antibody molecules in their membranes bind antigens. The antigens are then displayed on the Class II MHC proteins of the B cells, which become activated by helper T cells activated by the same antigen. *(Figure 22–20)*

40. An active B cell may differentiate into a plasma cell or produce daughter cells that differentiate into plasma cells and **memory B cells**. Antibodies are produced by plasma cells. *(Figure 22–20)*

41. An antibody molecule consists of two parallel pairs of polypeptide chains containing *constant* and *variable segments*. *(Figure 22–21)*

42. When antibody molecules bind to an antigen, they form an **antigen–antibody complex**. Effects that appear after binding include **neutralization** (antibody binding such that viruses or bacterial toxins cannot bind to body cells); **precipitation** (formation of an insoluble **immune complex**) and **agglutination** (formation of large complexes); *opsonization* (coating of pathogens with antibodies and complement proteins to enhance phagocytosis); stimulation of inflammation; and prevention of bacterial or viral adhesion. *(Figure 22–21)*

43. The five classes of antibodies (**immunoglobulins, Ig**) in body fluids are (1) **IgG**, responsible for resistance against many viruses, bacteria, and bacterial toxins; (2) **IgE**, which releases chemicals that accelerate local inflammation; (3) **IgD**, located

on the surfaces of B cells; (4) **IgM**, the first type of antibody secreted after an antigen arrives; and (5) **IgA**, found in glandular secretions. *(Table 22–1)*

44. In humoral immunity, the antibodies first produced by plasma cells are the agents of the **primary response**. The maximum **antibody titer** appears during the **secondary response** to antigen exposure. *(Figure 22–22)*

45. The initial steps in the immune responses to viral and bacterial infections differ. *(Figures 22–23 to 22–25; Summary Table 22–2)*

FOCUS: Hormones of the Immune System p. 813

46. **Interleukins** increase T cell sensitivity to antigens exposed on macrophage membranes; stimulate B cell activity, plasma cell formation, and antibody production; enhance nonspecific defenses; and moderate the immune response. *(Table 22–3)*

47. Interferons slow the spread of a virus by making the synthesizing cell and its neighbors resistant to viral infections. *(Table 22–3)*

48. **Tumor necrosis factors (TNFs)** slow tumor growth and kill tumor cells. *(Table 22–3)*

49. Several cytokines adjust the activities of phagocytic cells to coordinate specific and nonspecific defenses. *(Table 22–3)*

50. **Colony-stimulating factors (CSFs)** are factors produced by active T cells, cells of the monocyte-macrophage group, endothelial cells, and fibrocytes. *(Table 22–3)*

22-7 Immunological competence enables a normal immune response; abnormal responses result in immune disorders p. 811

51. **Immunological competence** is the ability to produce an immune response after exposure to an antigen. A developing fetus receives passive immunity from the maternal bloodstream. After delivery, the infant begins developing active immunity following exposure to environmental antigens.

52. **Autoimmune disorders** develop when the immune response inappropriately targets normal body cells and tissues.

53. In an **immunodeficiency disease**, either the immune system does not develop normally or the immune response is blocked.

54. **Allergies** are inappropriate or excessive immune responses to **allergens** (antigens that trigger allergic reactions). The four types of allergies are *immediate hypersensitivity (Type I)*, *cytotoxic reactions (Type II)*, *immune complex disorders (Type III)*, and *delayed hypersensitivity (Type IV)*.

55. In **anaphylaxis**, a circulating allergen affects mast cells throughout the body. *(Figure 22–26)*

56. Interleukin-1 released by active macrophages triggers the release of ACTH by the adenohypophysis. Glucocorticoids produced by the suprarenal cortex moderate the immune response, but their long-term secretion can lower a person's resistance to disease.

22-8 The immune response diminishes with advancing age p. 817

57. With aging, the immune system becomes less effective at combating disease.

22-9 For all body systems, the lymphoid system provides defenses against infection and returns tissue fluid to the circulation p. 817

58. The lymphoid system has extensive interactions with the neural and endocrine systems *(Figure 22–27)*.

Review Questions

See the blue Answers tab at the end of the book.

 Access more review material online at **myA&P™** (www.myaandp.com). There, you'll find chapter guides, chapter quizzes, practice tests, animations, flashcards, a glossary with pronunciations, *Interactive Physiology*® (IP) exercises and quizzes, and more to help you succeed in the course.

LEVEL 1 Reviewing Facts and Terms

1. Identify the structures of the lymphoid system in the following diagram.

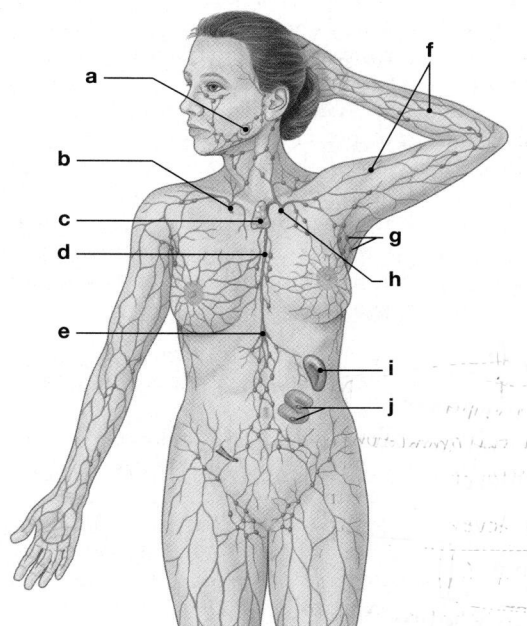

  (a) tonsil   (b) cervical lymph nodes
  (c) thymus   (d) thoracic duct
  (e) cisterna chyli   (f) upper limb lymphatics
  (g) axillary lymph nodes   (h) thoracic duct
  (i) MALT in intestine ←→   (j) spleen

2. Lymph from the right arm, the right half of the head, and the right chest is received by the
  (a) cisterna chyli.   (b) right lymphatic duct.
  (c) right thoracic duct.   (d) aorta.
3. Anatomically, lymph vessels resemble
  (a) elastic arteries.   (b) muscular arteries.
  (c) arterioles.   (d) medium veins.
  (e) the venae cavae.
4. The specificity of an antibody is determined by the
  (a) fixed segment.   (b) antigenic determinants.
  (c) variable region.   (d) size of the antibody.
  (e) antibody class.
5. The major histocompatibility complex (MHC)
  (a) is responsible for forming lymphocytes.
  (b) produces antibodies in lymph glands.
  (c) is a group of genes that codes for human leukocyte antigens.
  (d) is a membrane protein that can recognize foreign antigens.
  (e) is the antigen found on bacteria that stimulates an immune response.

6. Red blood cells that are damaged or defective are removed from the bloodstream by the
  (a) thymus.   (b) lymph nodes.
  (c) spleen.   (d) tonsils.
7. Phagocytes move through capillary walls by squeezing between adjacent endothelial cells, a process known as
  (a) diapedesis.   (b) chemotaxis.
  (c) adhesion.   (d) perforation.
8. Perforins are proteins associated with the activity of
  (a) T cells.   (b) B cells.
  (c) NK cells.   (d) plasma cells.
9. Complement activation
  (a) stimulates inflammation.
  (b) attracts phagocytes.
  (c) enhances phagocytosis.
  (d) achieves a, b, and c.
10. The most beneficial effect of fever is that it
  (a) inhibits the spread of some bacteria and viruses.
  (b) increases the metabolic rate by up to 10 percent.
  (c) stimulates the release of pyrogens.
  (d) achieves a and b.
11. CD4 markers are associated with
  (a) cytotoxic T cells.   (b) suppressor T cells.
  (c) helper T cells.   (d) a, b, and c.
12. List the specific functions of each of the body's lymphoid tissues and organs.
13. Give a function for each of the following:
  (a) cytotoxic T cells   (b) helper T cells
  (c) suppressor T cells   (d) plasma cells
  (e) NK cells   (f) stromal cells
  (g) reticular epithelial cells   (h) interferons
  (i) pyrogens   (j) T cells
  (k) B cells   (l) interleukins
  (m) tumor necrosis factor   (n) colony-stimulating factors
14. What are the three classes of lymphocytes, and where does each class originate?
15. What seven defenses, present at birth, provide the body with the defensive capability known as nonspecific resistance?

LEVEL 2 Reviewing Concepts

16. Compared with nonspecific defenses, specific defenses
  (a) do not distinguish between one threat and another.
  (b) are always present at birth.
  (c) protect against threats on an individual basis.
  (d) deny the entry of pathogens to the body.
17. Blocking the antigen receptors on the surface of lymphocytes would interfere with
  (a) phagocytosis of the antigen.
  (b) that lymphocyte's ability to produce antibodies.
  (c) antigen recognition.
  (d) the ability of the lymphocyte to present antigen.
  (e) opsonization of the antigen.
18. A *decrease* in which population of lymphocytes would impair all aspects of an immune response?
  (a) cytoxic T cells   (b) helper T cells
  (c) suppressor T cells   (d) B cells
  (e) plasma cells

19. Skin tests are used to determine if a person
    (a) has an active infection.
    (b) has been exposed to a particular antigen.
    (c) carries a particular antigen.
    (d) has measles.
    (e) can produce antibodies.

20. Compare and contrast the effects of complement with those of interferon.

21. How does a cytotoxic T cell destroy another cell displaying antigens bound to Class I MHC proteins?

22. How does the formation of an antigen–antibody complex cause the elimination of an antigen?

23. Give one example of each type of immunity: innate immunity, naturally acquired active immunity, induced active immunity, induced passive immunity, and naturally acquired passive immunity.

24. An anesthesia technician is advised that she should be vaccinated against hepatitis B, which is caused by a virus. She is given one injection and is told to come back for a second injection in a month and a third injection after six months. Why is this series of injections necessary?

LEVEL 3 Critical Thinking and Clinical Applications

25. An investigator at a crime scene discovers some body fluid on the victim's clothing. The investigator carefully takes a sample and sends it to the crime lab for analysis. On the basis of the

analysis of antibodies, could the crime lab determine whether the sample is blood plasma or semen? Explain.

26. Ted finds out that he has been exposed to measles. He is concerned that he might have contracted the disease, so he goes to see his physician. The physician takes a blood sample and sends it to a lab for antibody titers. The results show an elevated level of IgM antibodies to rubella (measles) virus but very few IgG antibodies to the virus. Has Ted contracted the disease?

27. While walking along the street, you and your friend see an elderly woman whose left arm appears to be swollen to several times its normal size. Your friend remarks that the woman must have been in the tropics and contracted a form of filariasis that produces elephantiasis. You disagree, saying that it is more likely that the woman had a radical mastectomy (the removal of a breast because of cancer). Explain the rationale behind your answer.

28. Paula's grandfather is diagnosed as having lung cancer. His physician takes biopsies of several lymph nodes from neighboring regions of the body, and Paula wonders why, since his cancer is in the lungs. What would you tell her?

29. Willy is allergic to ragweed pollen and tells you that he read about a medication that can help his condition by blocking certain antibodies. Do you think that this treatment could help Willy? Explain.

The Respiratory System

23

Learning Outcomes

After completing this chapter, you should be able to do the following:

23-1 Describe the primary functions of the respiratory system, and explain how the delicate respiratory exchange surfaces are protected from pathogens, debris, and other hazards.

23-2 Identify the organs of the upper respiratory system, and describe their functions.

23-3 Describe the structure of the larynx, and discuss its roles in normal breathing and in the production of sound.

23-4 Discuss the structure of the extrapulmonary airways.

23-5 Describe the superficial anatomy of the lungs, the structure of a pulmonary lobule, and the functional anatomy of alveoli.

23-6 Define and compare the processes of external respiration and internal respiration.

23-7 Summarize the physical principles governing the movement of air into the lungs, and describe the origins and actions of the muscles responsible for respiratory movements.

23-8 Summarize the physical principles governing the diffusion of gases into and out of the blood and body tissues.

23-9 Describe the structure and function of hemoglobin, and the transport of oxygen and carbon dioxide in the blood.

23-10 List the factors that influence respiration rate, and discuss reflex respiratory activity and the brain centers involved in the control of respiration.

23-11 Describe age-related changes in the respiratory system.

23-12 Give examples of interactions between the respiratory system and each of the other organ systems.

Clinical Notes

Breakdown of the Respiratory Defense System p. 829
Pneumothorax p. 847
Decompression Sickness p. 853
Blood Gas Analysis p. 854
Carbon Monoxide Poisoning p. 856
Emphysema and Lung Cancer p. 867

An Introduction to the Respiratory System

When we think of the respiratory system, we generally think of the mechanics of breathing—pulling air into and out of our bodies. However, the requirements for an efficient respiratory system go beyond merely moving air. Cells need energy for maintenance, growth, defense, and division. Our cells obtain that energy primarily through aerobic mechanisms that require oxygen and produce carbon dioxide.

Many aquatic organisms can obtain oxygen and excrete carbon dioxide through diffusion across the surface of the skin or specialized structures, such as the gills of a fish. But such arrangements are poorly suited for life on land, because the exchange surfaces must be very thin and relatively delicate to permit rapid diffusion. In air, the exposed membranes collapse, evaporation and dehydration reduce blood volume, and the delicate surfaces become vulnerable to attack by pathogens. The respiratory exchange surfaces of humans are just as delicate as those of an aquatic organism, but they are confined to the inside of the *lungs*—a warm, moist, protected environment. Under these conditions, diffusion can occur between the air and the blood. The cardiovascular system provides the link between the interstitial fluids and the exchange surfaces of your lungs. Circulating blood carries oxygen from the lungs to peripheral tissues; it also accepts and transports the carbon dioxide generated by those tissues, delivering it to the lungs. This chapter describes how air enters the lungs as a result of the actions of respiratory muscles, and how oxygen and carbon dioxide are exchanged across delicate epithelial surfaces within the lungs.

23-1 The respiratory system, organized into an upper respiratory system and a lower respiratory system, has several basic functions

The **respiratory system** is composed of structures involved in ventilation and gas exchange. In this section we consider this body system's functions and structural organization.

Functions of the Respiratory System

The respiratory system has five basic functions:

1. Providing an extensive surface area for gas exchange between air and circulating blood.

2. Moving air to and from the exchange surfaces of the lungs along the respiratory passageways.

3. Protecting respiratory surfaces from dehydration, temperature changes, or other environmental variations, and defending the respiratory system and other tissues from invasion by pathogens.

4. Producing sounds involved in speaking, singing, and other forms of communication.

5. Facilitating the detection of olfactory stimuli by olfactory receptors in the superior portions of the nasal cavity.

In addition, the capillaries of the lungs indirectly assist in the regulation of blood volume and blood pressure, through the conversion of angiotensin I to angiotensin II. ∞ p. 637

Organization of the Respiratory System

The respiratory system can be organized from either an anatomical or a functional perspective. Anatomically, we can divide the system into an upper respiratory system and a lower respiratory system (**Figure 23–1**). The **upper respiratory system** consists of the nose, nasal cavity, paranasal sinuses, and pharynx. These passageways filter, warm, and humidify incoming air—protecting the more delicate surfaces of the lower respiratory system—and cool and dehumidify outgoing air. The **lower respiratory system** includes the larynx (voice box), trachea (windpipe), bronchi, bronchioles, and alveoli of the lungs.

Tips&Tricks

To recall the boundary between the upper and lower respiratory systems, remember that the *l*ower respiratory system begins at the *l*arynx.

The term **respiratory tract** refers to the passageways that carry air to and from the exchange surfaces of the lungs. The *conducting portion* of the respiratory tract begins at the entrance to the nasal cavity and extends through the pharynx, larynx, trachea, bronchi, and larger bronchioles. The *respiratory portion* of the tract includes the smallest, most delicate *bronchioles* and the associated **alveoli** (al-VĒ-ō-lī), air-filled pockets within the lungs where all gas exchange between air and blood occurs.

Gas exchange can occur quickly and efficiently because the distance between the blood in an alveolar capillary and the air inside an alveolus is generally less than 1 μm and in some cases as small as 0.1 μm. To meet the metabolic requirements of peripheral tissues, the surface area devoted to gas exchange in the lungs must be very large; it is roughly 35 times the sur-

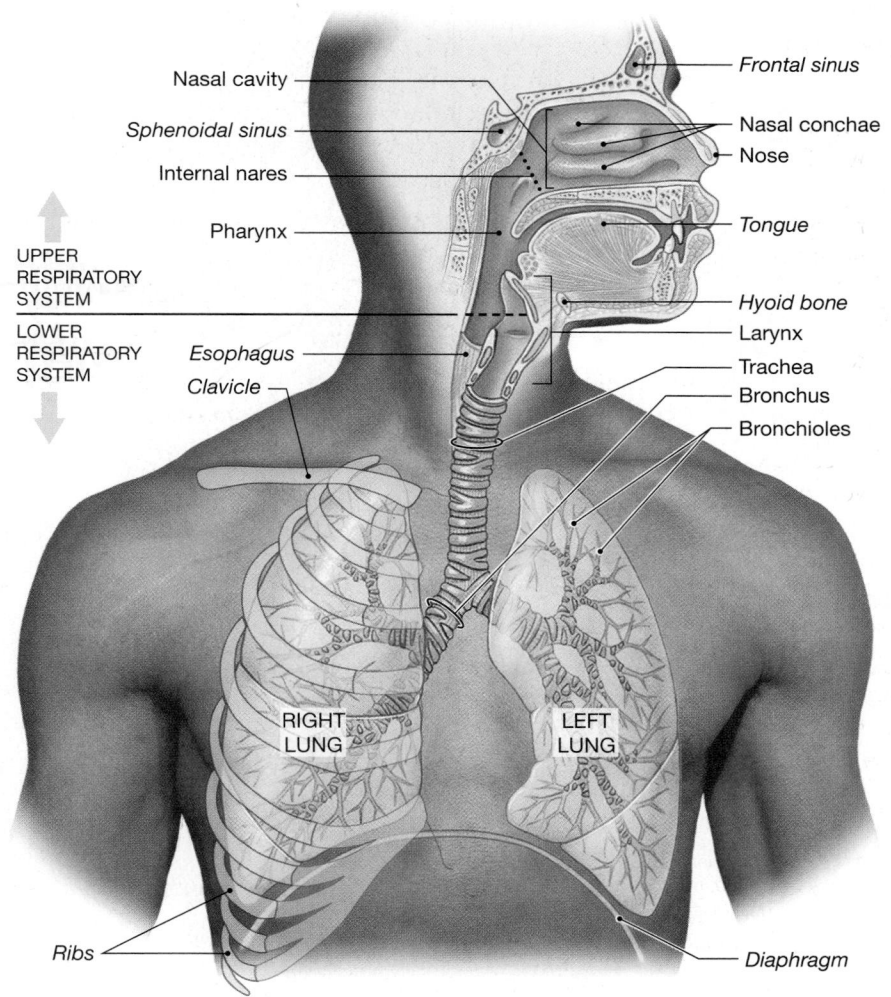

Nasal cavity

Sphenoidal sinus

Internal nares

Pharynx

UPPER RESPIRATORY SYSTEM

LOWER RESPIRATORY SYSTEM

Esophagus

Clavicle

Ribs

Frontal sinus

Nasal conchae

Nose

Tongue

Hyoid bone

Larynx

Trachea

Bronchus

Bronchioles

RIGHT LUNG

LEFT LUNG

Diaphragm

Figure 23–1   The Components of the Respiratory System. Only the conducting portion of the respiratory system is shown; the smaller bronchioles and alveoli have been omitted. ATLAS: Plate 47a,b

face area of the body. Estimates of the surface area involved in gas exchange range from 70 m² to 140 m² (753 ft² to 1506 ft²).

Filtering, warming, and humidification of the inhaled air begin at the entrance to the respiratory tract and continue as air passes along the conducting portion. By the time air reaches the alveoli, most foreign particles and pathogens have been removed, and the humidity and temperature are within acceptable limits. The success of this "conditioning process" is due primarily to the properties of the respiratory mucosa.

The Respiratory Mucosa

The **respiratory mucosa** (mū-KŌ-suh) lines the conducting portion of the respiratory system. A *mucosa* is a *mucous membrane*, one of the four types of membranes introduced in Chapter 4. It consists of an epithelium and an underlying layer of areolar tissue. ∞ p. 136 The **lamina propria** (LAM-

i-nuh PRŌ-prē-uh) is the underlying layer of areolar tissue that supports the respiratory epithelium. In the upper respiratory system, trachea, and bronchi, the lamina propria contains mucous glands that discharge their secretions onto the epithelial surface. The lamina propria in the conducting portions of the lower respiratory system contains bundles of smooth muscle cells. At the bronchioles, the smooth muscles form relatively thick bands that encircle or spiral around the lumen.

The structure of the respiratory epithelium changes along the respiratory tract. A pseudostratified ciliated columnar epithelium with numerous mucous cells lines the nasal cavity and the superior portion of the pharynx. ∞ p. 119 The epithelium lining inferior portions of the pharynx is a stratified squamous epithelium similar to that of the oral cavity. These portions of the pharynx, which conduct air to the larynx, also convey food to the esophagus. The pharyngeal epithelium must therefore protect against abrasion and chemical attack.

A pseudostratified ciliated columnar epithelium comparable to that of the nasal cavity lines the superior portion of the lower respiratory system. In the smaller bronchioles, this pseudostratified epithelium is replaced by a cuboidal epithelium with scattered cilia. The exchange surfaces of the alveoli are lined by a very delicate simple squamous epithelium. Other, more specialized cells are scattered among the squamous cells; together they form the *alveolar epithelium*.

[handwritten annotation: Goblet cell: mucus gland ⟹ filters dust, particles, pathogens...]

The Respiratory Defense System

The delicate exchange surfaces of the respiratory system can be severely damaged if inhaled air becomes contaminated with debris or pathogens. Such contamination is prevented by a series of filtration mechanisms that constitute the **respiratory defense system**.

Along much of the length of the respiratory tract, mucous cells in the epithelium and mucous glands in the lamina propria produce a sticky mucus that bathes exposed surfaces. In the nasal cavity, cilia sweep that mucus and any trapped debris or microorganisms toward the pharynx, where it will be swallowed and exposed to the acids and enzymes of the stomach. In the lower respiratory system, the cilia also beat toward the pharynx, moving a carpet of mucus in that direction and cleaning the respiratory surfaces. This process is often described as a *mucus escalator* (**Figure 23–2c**).

23 RESPIRATORY

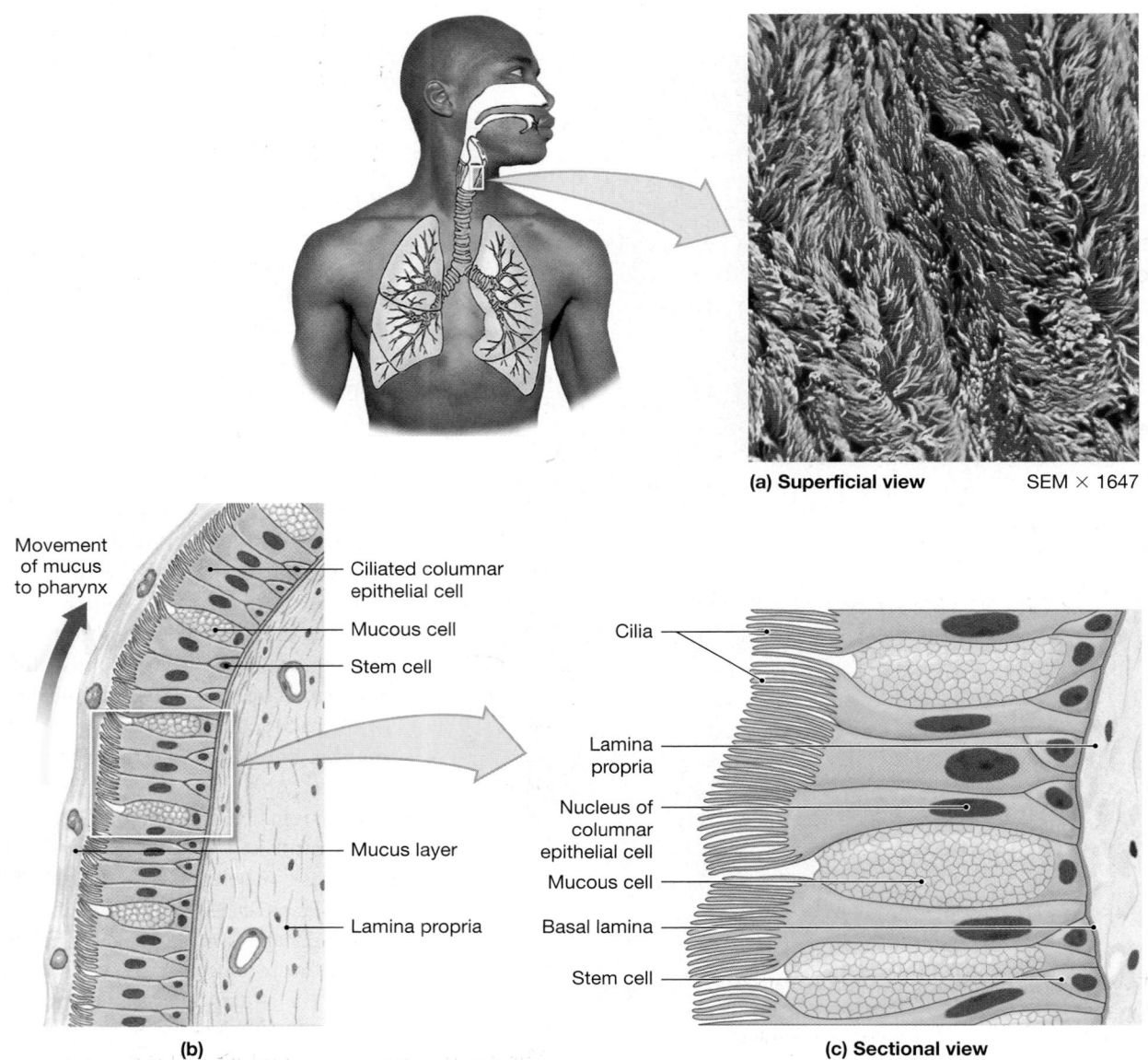

(a) **Superficial view**   SEM × 1647

Movement of mucus to pharynx

Ciliated columnar epithelial cell

Mucous cell

Stem cell

Mucus layer

Lamina propria

(b)

Cilia

Lamina propria

Nucleus of columnar epithelial cell

Mucous cell

Basal lamina

Stem cell

(c) **Sectional view**

Figure 23–2   The Respiratory Epithelium of the Nasal Cavity and Conducting System. (a) A surface view of the epithelium. The cilia of the epithelial cells form a dense layer that resembles a shag carpet. The movement of these cilia propels mucus across the epithelial surface. **(b)** A diagrammatic view of the respiratory epithelium of the trachea, indicating the mechanism of mucus transport. **(c)** The sectional appearance of the respiratory epithelium, a pseudostratified ciliated columnar epithelium.

Tips&Tricks

The mucus layer of the respiratory epithelium functions like sticky flypaper, but instead of trapping flies, it traps particles and debris in the air moving past it.

Filtration in the nasal cavity removes virtually all particles larger than about 10 μm from the inhaled air. Smaller particles may be trapped by the mucus of the nasopharynx or by secretions of the pharynx before proceeding along the conducting system. Exposure to unpleasant stimuli, such as noxious vapors,

large quantities of dust and debris, allergens, or pathogens, generally causes a rapid increase in the rate of mucus production in the nasal cavity and paranasal sinuses. (The familiar signs and symptoms of the "common cold" result from the invasion of the respiratory epithelium by any of more than 200 types of viruses.)

Most particles 1–5 μm in diameter are trapped in the mucus coating the respiratory bronchioles or in the liquid covering the alveolar surfaces. These areas are outside the boundaries of the mucus escalator, but the foreign particles can be engulfed by alveolar macrophages. Most particles smaller than about 0.5 μm remain suspended in the air.

CLINICAL NOTE

Breakdown of the Respiratory Defense System Large quantities of airborne particles may overload the respiratory defenses and produce a variety of illnesses. For example, the presence of irritants in the lining of the conducting passageways can provoke the formation of mucous plugs that block airflow and reduce pulmonary function, and damage to the epithelium in the affected area may allow irritants to enter the surrounding tissues of the lung. The irritants then produce local inflammation. Airborne irritants—such as those in cigarette smoke—are known to promote the development of lung cancer (p. 867).

Respiratory defenses can also be overwhelmed by aggressive pathogens. **Tuberculosis** (tū-ber-kū-LŌ-sis), or TB, results from an infection of the lungs by the bacterium *Mycobacterium tuberculosis;* other organs may be invaded as well. Bacteria may colonize the respiratory passageways, the interstitial spaces, the alveoli, or a combination of the three. Signs and symptoms are variable, but generally include coughing and chest pain, plus fever, night sweats, fatigue, and weight loss. In 1900, TB, then known as "consumption," was the leading cause of death. It remains among the most common and serious infectious diseases, killing 1.6 million people in 2005, primarily in Southeast Asia and Africa.

The respiratory defense system can also fail due to inherited congenital defects affecting mucus production or transport. One example, **cystic fibrosis (CF),** is the most common lethal inherited disease affecting individuals of Northern European descent, occurring at a frequency of 1 birth in 2500. The respiratory mucosa in these individuals produces dense, viscous mucus that cannot be transported by the respiratory defense system. The mucus escalator stops working, leading to frequent infections, and mucus blocks the smaller respiratory passageways, making breathing difficult. Individuals with CF seldom survive past age 30; death is generally the result of heart failure associated with a massive chronic bacterial infection of the lungs.

CHECKPOINT

1. Identify several functions of the respiratory system.
2. List the two anatomical divisions of the respiratory system.
3. What membrane lines the conducting portion of the respiratory system?

See the blue Answers tab at the end of the book.

23-2 Located outside the thoracic cavity, the upper respiratory system consists of the nose, nasal cavity, paranasal sinuses, and pharynx

As previously noted, the upper respiratory system consists of the nose, nasal cavity, paranasal sinuses, and pharynx (**Figures 23–1** and **23–3**).

The Nose, Nasal Cavity, and Paranasal Sinuses

The nose is the primary passageway for air entering the respiratory system. Air normally enters through the paired **external nares** (NĀ-res), or *nostrils* (**Figure 23–3a**), which open into the *nasal cavity.* The **nasal vestibule** is the space contained within the flexible tissues of the nose (**Figure 23–3c**). The epithelium of the vestibule contains coarse hairs that extend across the external nares. Large airborne particles, such as sand, sawdust, or even insects, are trapped in these hairs and are thereby prevented from entering the nasal cavity.

The **nasal septum** divides the nasal cavity into left and right portions (**Figure 23–3b**). The bony portion of the nasal septum is formed by the fusion of the perpendicular plate of the ethmoid bone and the plate of the vomer (**Figure 7–3d**, p. 216). The anterior portion of the nasal septum is formed of hyaline cartilage. This cartilaginous plate supports the *dorsum nasi* (DOR-sum NĀ-zī), or bridge, and *apex* (tip) of the nose.

The maxillary, nasal, frontal, ethmoid, and sphenoid bones form the lateral and superior walls of the nasal cavity. The mucous secretions produced in the associated *paranasal sinuses* (**Figure 7–14**, p. 229), aided by the tears draining through the nasolacrimal ducts, help keep the surfaces of the nasal cavity moist and clean. The *olfactory region*, or superior portion of the nasal cavity, includes the areas lined by olfactory epithelium: (1) the inferior surface of the cribriform plate, (2) the superior portion of the nasal septum, and (3) the superior nasal conchae. Receptors in the olfactory epithelium provide your sense of smell. ∞ p. 562

The *superior, middle,* and *inferior nasal conchae* project toward the nasal septum from the lateral walls of the nasal cavity. ∞ pp. 222, 224 To pass from the vestibule to the internal nares, air tends to flow between adjacent conchae, through the **superior, middle,** and **inferior meatuses** (mē-Ā-tus-ez; *meatus,* a passage) (**Figure 23–3b**). These are narrow grooves rather than open passageways; the incoming air bounces off the conchal surfaces and churns like a stream flowing over

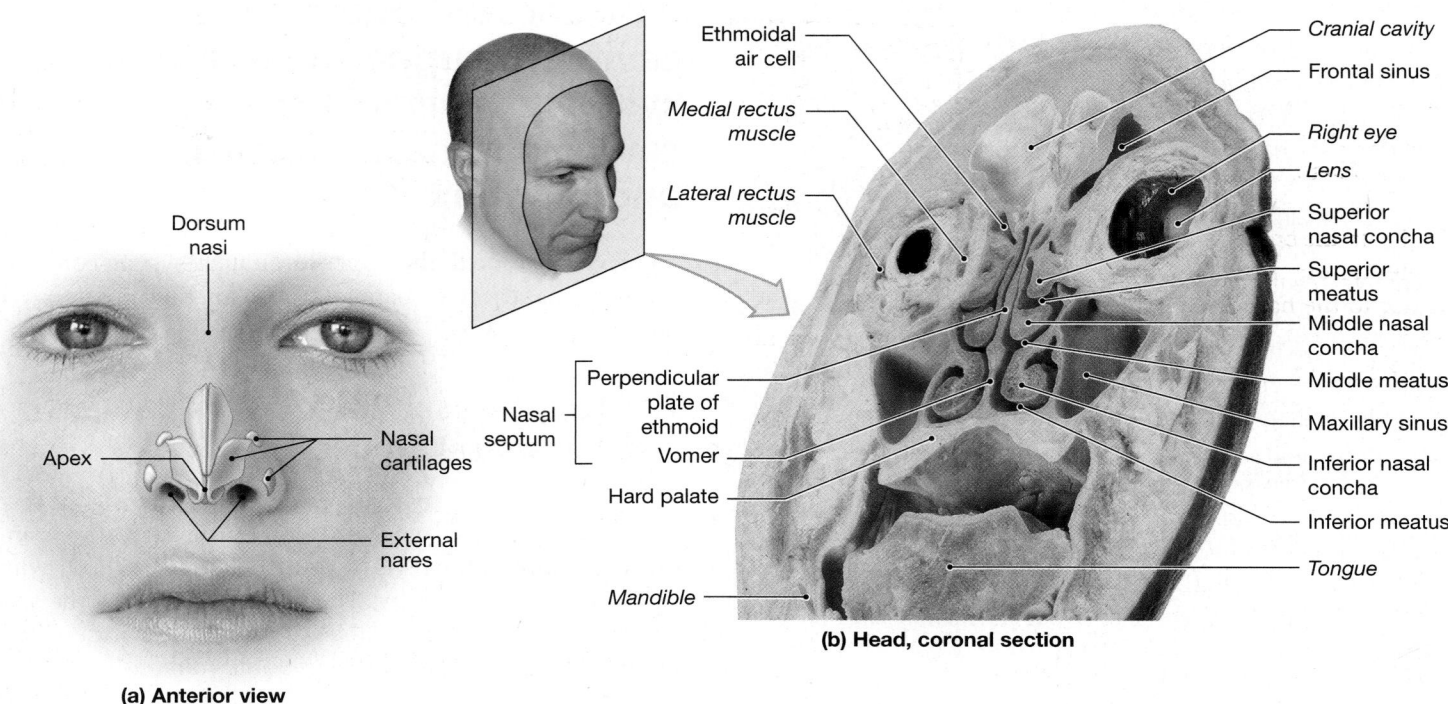

Dorsum nasi

Apex

Nasal cartilages

External nares

(a) Anterior view

Ethmoidal air cell

Medial rectus muscle

Lateral rectus muscle

Nasal septum

Perpendicular plate of ethmoid

Vomer

Hard palate

Mandible

Cranial cavity

Frontal sinus

Right eye

Lens

Superior nasal concha

Superior meatus

Middle nasal concha

Middle meatus

Maxillary sinus

Inferior nasal concha

Inferior meatus

Tongue

(b) Head, coronal section

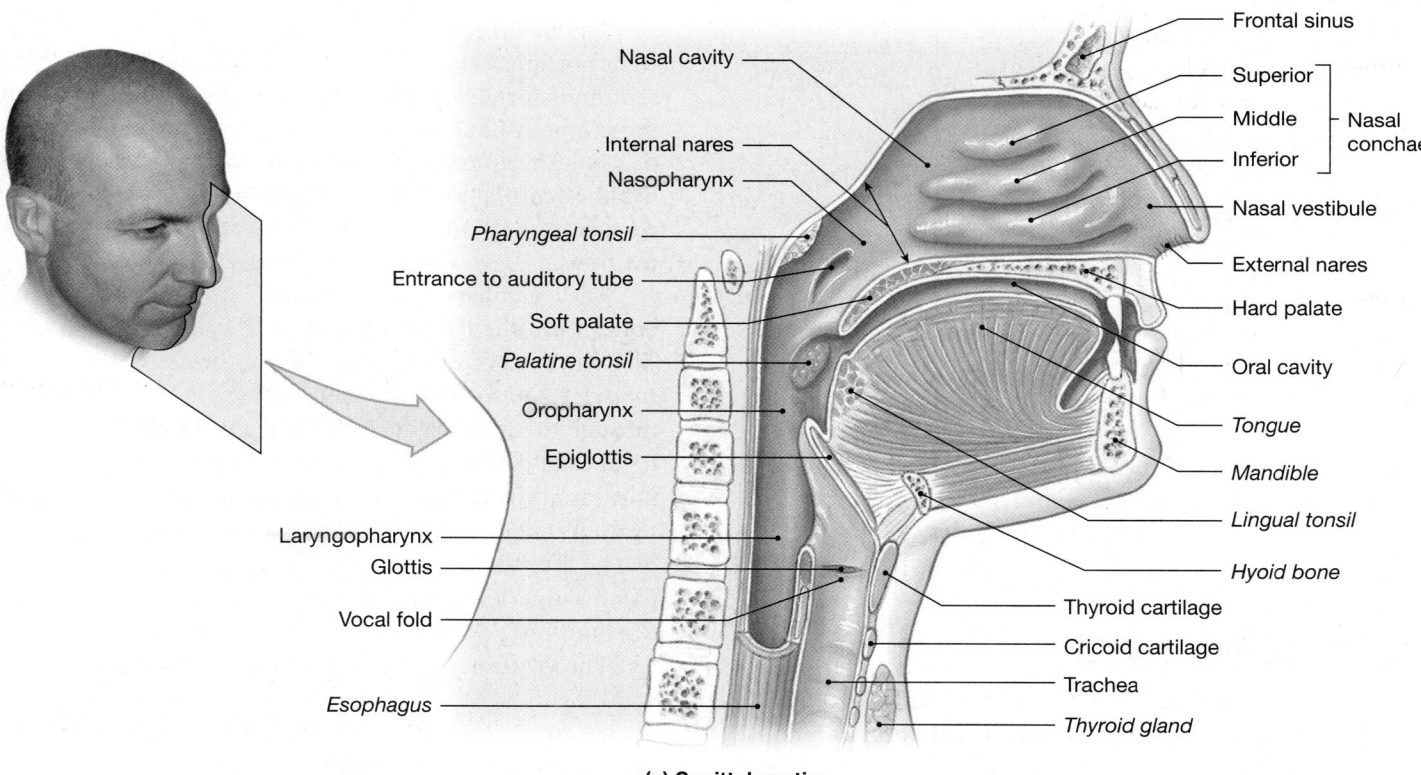

Nasal cavity

Internal nares

Nasopharynx

Pharyngeal tonsil

Entrance to auditory tube

Soft palate

Palatine tonsil

Oropharynx

Epiglottis

Laryngopharynx

Glottis

Vocal fold

Esophagus

Frontal sinus

Superior

Middle — Nasal conchae

Inferior

Nasal vestibule

External nares

Hard palate

Oral cavity

Tongue

Mandible

Lingual tonsil

Hyoid bone

Thyroid cartilage

Cricoid cartilage

Trachea

Thyroid gland

(c) Sagittal section

Figure 23–3 Structures of the Upper Respiratory System. (a) The nasal cartilages and external landmarks on the nose. **(b)** A frontal section through the head, showing the meatuses and the maxillary sinuses and air cells of the ethmoidal labyrinth. **(c)** The nasal cavity and pharynx, as seen in sagittal section with the nasal septum removed. ATLAS: Plate 19

rocks. This turbulence serves several purposes: As the air swirls, small airborne particles are likely to come into contact with the mucus that coats the lining of the nasal cavity. In addition, the turbulence provides extra time for warming and humidifying incoming air, and it creates eddy currents that bring olfactory stimuli to the olfactory receptors.

A bony **hard palate,** made up of portions of the maxillary and palatine bones, forms the floor of the nasal cavity and separates it from the oral cavity. A fleshy **soft palate** extends posterior to the hard palate, marking the boundary between the superior *nasopharynx* (nā-zō-FAR-ingks) and the rest of the pharynx. The nasal cavity opens into the nasopharynx through a connection known as the **internal nares**.

The Nasal Mucosa

The mucosa of the nasal cavity prepares inhaled air for arrival at the lower respiratory system. Throughout much of the nasal cavity, the lamina propria contains an abundance of arteries, veins, and capillaries that bring nutrients and water to the secretory cells. The lamina propria of the nasal conchae also contains an extensive network of large and highly expandable veins. This vascularization provides a mechanism for warming and humidifying the incoming air (as well as for cooling and dehumidifying the outgoing air). As cool, dry air passes inward over the exposed surfaces of the nasal cavity, the warm epithelium radiates heat, and water in the mucus evaporates. Air moving from your nasal cavity to your lungs is thus heated almost to body temperature, and it is nearly saturated with water vapor. This mechanism protects more delicate respiratory surfaces from chilling or drying out—two potentially disastrous events. Breathing through your mouth eliminates much of the preliminary filtration, heating, and humidifying of the inhaled air. To avoid alveolar damage, patients breathing on a respirator (mechanical ventilator), which utilizes a tube to conduct air directly into the trachea, must receive air that has been externally filtered and humidified.

As air moves out of the respiratory tract, it again passes over the epithelium of the nasal cavity. This air is warmer and more humid than the air that enters; it warms the nasal mucosa, and moisture condenses on the epithelial surfaces. Thus, breathing through your nose also helps prevent heat loss and water loss.

The extensive vascularization of the nasal cavity and the relatively vulnerable position of the nose make a nosebleed, or *epistaxis* (ep-i-STAK-sis), a fairly common event. This bleeding generally involves vessels of the mucosa covering the cartilaginous portion of the septum. A variety of factors may be responsible, including trauma (such as a punch in the nose), drying, infections, allergies, or clotting disorders. Hypertension can also provoke epistaxis by rupturing small vessels of the lamina propria.

The Pharynx

The **pharynx** (FAR-ingks) is a chamber shared by the digestive and respiratory systems. It extends between the internal nares and the entrances to the larynx and esophagus. The curving superior and posterior walls of the pharynx are closely bound to the axial skeleton, but the lateral walls are flexible and muscular.

The pharynx is divided into the nasopharynx, the oropharynx, and the laryngopharynx (**Figure 23–3c**):

1. The **nasopharynx** is the superior portion of the pharynx. It is connected to the posterior portion of the nasal cavity through the internal nares and is separated from the oral cavity by the soft palate. The nasopharynx is lined by the same pseudostratified ciliated columnar epithelium as that in the nasal cavity. The *pharyngeal tonsil* is located on the posterior wall of the nasopharynx; the left and right *auditory tubes* open into the nasopharynx on either side of this tonsil. ∞ pp. 586, 783

2. The **oropharynx** (*oris*, mouth) extends between the soft palate and the base of the tongue at the level of the hyoid bone. The posterior portion of the oral cavity communicates directly with the oropharynx, as does the posterior inferior portion of the nasopharynx. At the boundary between the nasopharynx and the oropharynx, the epithelium changes from pseudostratified columnar to stratified squamous.

3. The narrow **laryngopharynx** (la-rin-gō-FAR-ingks), the inferior part of the pharynx, includes that portion of the pharynx between the hyoid bone and the entrance to the larynx and esophagus. Like the oropharynx, the laryngopharynx is lined with a stratified squamous epithelium that resists abrasion, chemical attack, and invasion by pathogens.

CHECKPOINT

4. Name the structures of the upper respiratory system.

5. Why is the vascularization of the nasal cavity important?

6. Why is the lining of the nasopharynx different from that of the oropharynx and the laryngopharynx?

See the blue Answers tab at the end of the book.

23-3 Composed of cartilages, ligaments, and muscles, the larynx produces sound

Inhaled air leaves the pharynx and enters the larynx through a narrow opening called the **glottis** (GLOT-is). The **larynx** (LAR-ingks) is a cartilaginous structure that surrounds and protects the glottis. The larynx begins at the level of vertebra C_4 or C_5 and ends at the level of vertebra C_6 Essentially a cylinder, the larynx has incomplete cartilaginous walls that are stabilized by ligaments and skeletal muscles (**Figure 23–4**).

Cartilages and Ligaments of the Larynx

Three large, unpaired cartilages form the larynx: (1) the thyroid cartilage, (2) the cricoid cartilage, and (3) the epiglottis (**Figure 23–4**). The **thyroid cartilage** (*thyroid*, shield shaped) is the largest laryngeal cartilage. Consisting of hyaline cartilage, it forms most of the anterior and lateral walls of the larynx. In section, this cartilage is U-shaped; posteriorly, it is incomplete. The prominent anterior surface of the thyroid cartilage, which you can easily see and feel, is called the *laryngeal prominence* or *Adam's apple*. The inferior surface articulates with the cricoid cartilage. The superior surface has ligamentous attachments to the hyoid bone and to the epiglottis and smaller laryngeal cartilages.

The thyroid cartilage sits superior to the **cricoid** (KRĪ-koyd; ring shaped) **cartilage**, another hyaline cartilage. The posterior portion of the cricoid is greatly expanded, providing support in the absence of the thyroid cartilage. The

cricoid and thyroid cartilages protect the glottis and the entrance to the trachea, and their broad surfaces provide sites for the attachment of important laryngeal muscles and ligaments. Ligaments attach the inferior surface of the cricoid cartilage to the first tracheal cartilage. The superior surface of the cricoid cartilage articulates with the small, paired *arytenoid cartilages*.

The shoehorn-shaped **epiglottis** (ep-i-GLOT-is) projects superior to the glottis and forms a lid over it. Composed of elastic cartilage, the epiglottis has ligamentous attachments to the anterior and superior borders of the thyroid cartilage and the hyoid bone. During swallowing, the larynx is elevated and the epiglottis folds back over the glottis, preventing the entry of both liquids and solid food into the respiratory tract.

The larynx also contains three pairs of smaller hyaline cartilages: (1) The **arytenoid** (ar-i-TĒ-noyd; ladle shaped) **cartilages** articulate with the superior border of the enlarged portion of the cricoid cartilage. (2) The **corniculate** (kor-NIK-ū-lāt; horn shaped) **cartilages** articulate with the arytenoid cartilages. The corniculate and arytenoid cartilages function in the opening and closing of the glottis and the production of sound. (3) Elongated, curving **cuneiform** (kū-NĒ-i-form; wedge shaped) **cartilages** lie within folds of tissue (the *aryepiglottic*

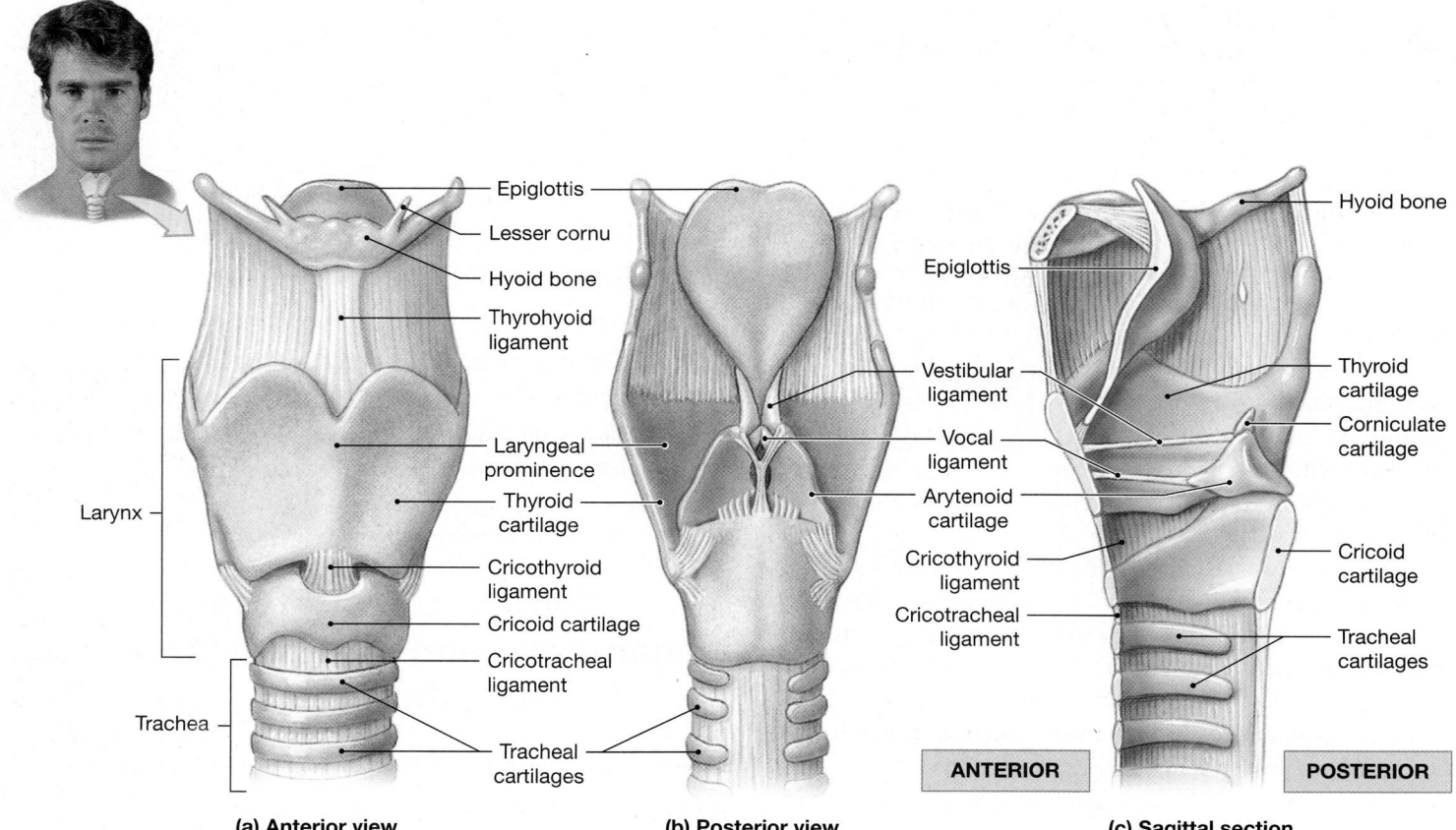

(a) Anterior view   **(b) Posterior view**   **(c) Sagittal section**

Figure 23–4 **The Anatomy of the Larynx. (a)** An anterior view. **(b)** A posterior view. **(c)** A sagittal section through the larynx.

folds) that extend between the lateral surface of each arytenoid cartilage and the epiglottis (**Figures 23–4c** and **23–5**).

The various laryngeal cartilages are bound together by ligaments; additional ligaments attach the thyroid cartilage to the hyoid bone, and the cricoid cartilage to the trachea (cricotracheal ligament) (**Figure 23–4a,b**). The median cricothyroid ligament attaches the thyroid cartilage to the cricoid cartilage and is the common placement site for a tracheostomy, a tracheal incision to bypass an airway obstruction. The **vestibular ligaments** and the **vocal ligaments** extend between the thyroid cartilage and the arytenoid cartilages.

The vestibular and vocal ligaments are covered by folds of laryngeal epithelium that project into the glottis. The vestibular ligaments lie within the superior pair of folds, known as the **vestibular folds** (**Figure 23–5**). These folds, which are relatively inelastic, help prevent foreign objects from entering the glottis and protect the more delicate **vocal folds**.

The vocal folds, inferior to the vestibular folds, guard the entrance to the glottis. The vocal folds are highly elastic, because the vocal ligaments consist of elastic tissue. The vocal folds are involved with the production of sound, and for this reason they are known as the **vocal cords**.

Sound Production

Air passing through the glottis vibrates the vocal folds and produces sound waves. The pitch of the sound produced depends on the diameter, length, and tension in the vocal folds. The diameter and length are directly related to the size of the larynx. The tension is controlled by the contraction of voluntary muscles that reposition the arytenoid cartilages relative to the thyroid cartilage. When the distance increases, the vocal folds tense and the pitch rises; when the distance decreases, the vocal folds relax and the pitch falls.

Children have slender, short vocal folds; their voices tend to be high-pitched. At puberty, the larynx of males enlarges much more than does that of females. The vocal cords of an adult male are thicker and longer, and produce lower tones, than those of an adult female.

Sound production at the larynx is called *phonation* (fō-NĀ-shun; *phone*, voice). Phonation is one component of speech production. However, clear speech also requires *articulation*, the modification of those sounds by other structures, such as the tongue, teeth, and lips. In a stringed instrument, such as a guitar, the quality of the sound produced does not depend solely on the nature of the vibrating string. Rather, the entire instrument becomes involved as the walls vibrate and the composite sound echoes within the hollow body. Similar amplification and resonance occur within your pharynx, oral cavity, nasal cavity, and paranasal sinuses. The combination determines the particular and distinctive sound of your voice.

Tips&Tricks

Intelligible sound requires both phonation and articulation. Saying "ahhhh" while your tongue is depressed during a tonsil examination is an example of phonation; saying "hot" adds articulation to that sound.

When the nasal cavity and paranasal sinuses are filled with mucus rather than air, as in sinus infections, the sound changes. The final production of distinct words depends further on voluntary movements of the tongue, lips, and cheeks. An infection or inflammation of the larynx is known as *laryngitis* (lar-in-JĪ-tis). It commonly affects the vibrational qualities of the vocal folds; hoarseness is the most familiar manifestation. Mild cases are temporary and seldom serious. However, bacterial or viral infections of the epiglottis can be very dangerous; the resulting swelling may close the glottis

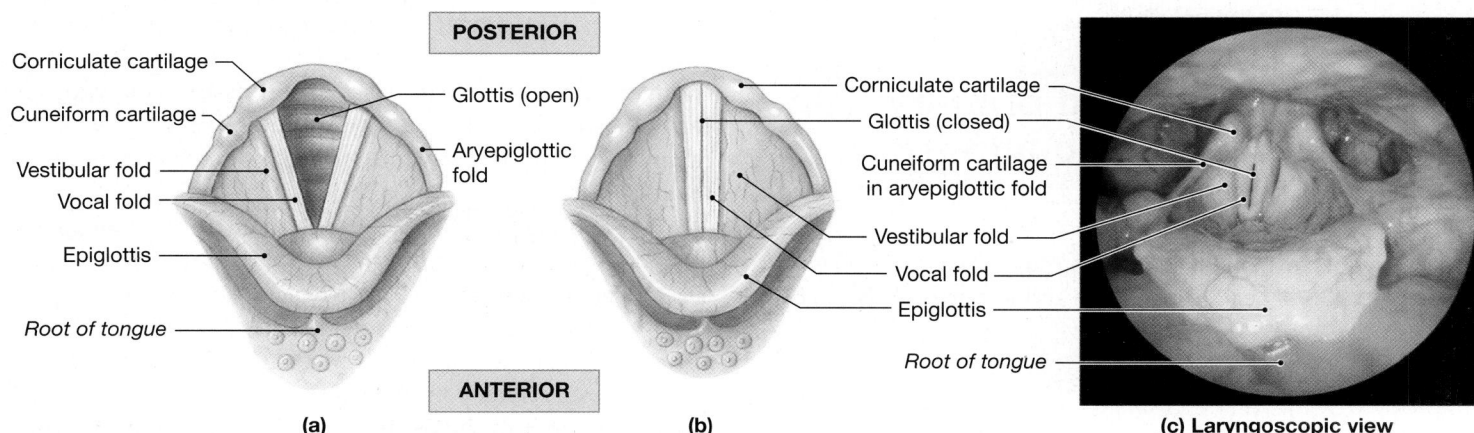

Figure 23–5 **The Glottis and Surrounding Structures.** (a, b) A diagrammatic superior view of the entrance to the larynx, with the glottis open (a) and closed (b). **(c)** A laryngoscopic view of the entrance to the larynx, corresponding to the image in part (b). Note that the glottis is almost completely closed by the vocal folds.

and cause suffocation. This condition, *acute epiglottitis* (ep-i-glot-TĪ-tis), can develop rapidly after a bacterial infection of the throat. Young children are most likely to be affected.

The Laryngeal Musculature

The larynx is associated with (1) muscles of the neck and pharynx, which position and stabilize the larynx (∞ pp. 350–352), and (2) smaller intrinsic muscles that control tension in the vocal folds or open and close the glottis. These latter muscles insert on the thyroid, arytenoid, and corniculate cartilages. The opening or closing of the glottis involves rotational movements of the arytenoid cartilages that move the vocal folds.

When you swallow, both sets of muscles cooperate to prevent food or drink from entering the glottis. Before food is swallowed, it is crushed and chewed into a pasty mass known as a *bolus*. Muscles of the neck and pharynx then elevate the larynx, bending the epiglottis over the glottis, so that the bolus can glide across the epiglottis rather than falling into the larynx. While this movement is under way, the glottis is closed.

Food or liquids that touch the vestibular or vocal folds trigger the *coughing reflex*. In a cough, the glottis is kept closed while the chest and abdominal muscles contract, compressing the lungs. When the glottis is opened suddenly, the resulting blast of air from the trachea ejects material that blocks the entrance to the glottis.

CHECKPOINT

7. Identify the paired and unpaired cartilages associated with the larynx.

8. What are the highly elastic vocal folds of the larynx better known as?

9. When the tension in your vocal folds increases, what happens to the pitch of your voice?

See the blue Answers tab at the end of the book.

23-4 The trachea and primary bronchi convey air to and from the lungs

In this section we consider three large airways associated with the lungs: the trachea and the right and left primary bronchi.

The Trachea

The epithelium of the larynx is continuous with that of the **trachea** (TRĀ-kē-uh), or windpipe, a tough, flexible tube with a diameter of about 2.5 cm (1 in.) and a length of about 11 cm (4.33 in.) (**Figure 23–6**). The trachea begins anterior to

vertebra C_6 in a ligamentous attachment to the cricoid cartilage. It ends in the mediastinum, at the level of vertebra T_5, where it branches to form the *right* and *left primary bronchi*.

The mucosa of the trachea resembles that of the nasal cavity and nasopharynx (**Figure 23–2a**). The **submucosa** (sub-mū-KŌ-suh), a thick layer of connective tissue, surrounds the mucosa. The submucosa contains mucous glands that communicate with the epithelial surface through a number of secretory ducts. The trachea contains 15–20 **tracheal cartilages** (**Figure 23–6a**), which serve to stiffen the tracheal walls and protect the airway. They also prevent its collapse or overexpansion as pressures change in the respiratory system.

Each tracheal cartilage is C-shaped. The closed portion of the C protects the anterior and lateral surfaces of the trachea. The open portion of the C faces posteriorly, toward the esophagus (**Figure 23–6b**). Because these cartilages are not continuous, the posterior tracheal wall can easily distort when you swallow, permitting large masses of food to pass through the esophagus.

An elastic ligament and the **trachealis muscle**, a band of smooth muscle, connect the ends of each tracheal cartilage (**Figure 23–6b**). Contraction of the trachealis muscle reduces the diameter of the trachea, increasing the vessel's resistance to airflow. The normal diameter of the trachea changes from moment to moment, primarily under the control of the sympathetic division of the ANS. Sympathetic stimulation increases the diameter of the trachea and makes it easier to move large volumes of air along the respiratory passageways.

The Primary Bronchi

The trachea branches within the mediastinum, giving rise to the **right** and **left primary bronchi** (BRONG-kī; singular, *bronchus*). An internal ridge called the **carina** (ka-RĪ-nuh) separates the two bronchi (**Figure 23–6a**). Like the trachea, the primary bronchi have cartilaginous C-shaped supporting rings. The right primary bronchus supplies the right lung, and the left supplies the left lung. The right primary bronchus is larger in diameter than the left, and descends toward the lung at a steeper angle. Thus, most foreign objects that enter the trachea find their way into the right bronchus rather than the left.

Before branching further, each primary bronchus travels to a groove along the medial surface of its lung. This groove, the **hilum** of the lung, also provides access for entry to pulmonary vessels, nerves, and lymphatics (**Figure 23–7**). The entire array is firmly anchored in a meshwork of dense connective tissue. This complex, the **root** of the lung (**Figure 23–6a**), attaches to the mediastinum and fixes the positions of the major nerves, blood vessels, and lymphatic vessels. The roots of the lungs are anterior to vertebrae T_5 (right) and T_6 (left).

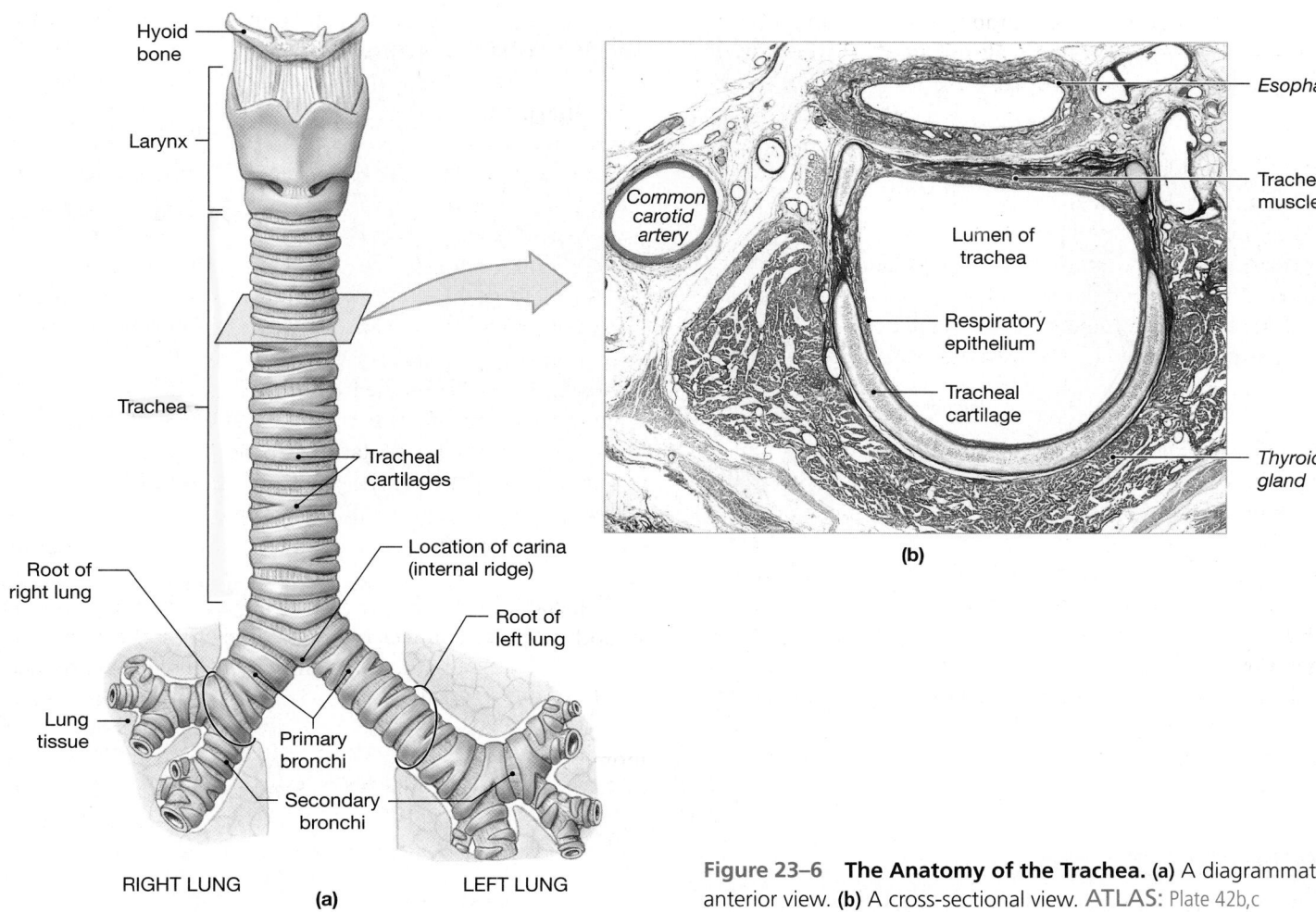

Figure 23–6 The Anatomy of the Trachea. (a) A diagrammatic anterior view. **(b)** A cross-sectional view. ATLAS: Plate 42b,c

---

CHECKPOINT

10. List functions of the trachea.

11. Why are the cartilages that reinforce the trachea C-shaped?

12. If food accidentally enters the bronchi, in which bronchus (left or right) is it more likely to lodge?

See the blue Answers tab at the end of the book.

23-5 Enclosed by a pleural membrane, the lungs are paired organs containing alveoli, which permit gaseous exchange

The left and right lungs (**Figure 23–7**) are in the left and right pleural cavities, respectively. Each lung is a blunt cone, the tip (or apex) of which points superiorly. The apex on each side extends superior to the first rib. The broad concave inferior portion (or base) of each lung rests on the superior surface of the diaphragm.

Lobes and Surfaces of the Lungs

The lungs have distinct **lobes** that are separated by deep fissures (**Figure 23–7**). The right lung has three lobes—*superior, middle,* and *inferior*—separated by the *horizontal* and *oblique fissures*. The left lung has only two lobes—*superior* and *inferior*—separated by the *oblique fissure*. The right lung is broader than the left, because most of the heart and great vessels project into the left thoracic cavity. However, the left lung is longer than the right lung, because the diaphragm rises on the right side to accommodate the mass of the liver.

The curving anterior and lateral surfaces of each lung follow the inner contours of the rib cage. The medial surface, which contains the hilum, has a more irregular shape. The medial surfaces of both lungs bear grooves that mark the positions of the great vessels and the heart (**Figures 23–7** and **23–8**). The heart is located to the left of the midline, so the corresponding impression is larger in the left lung than in the

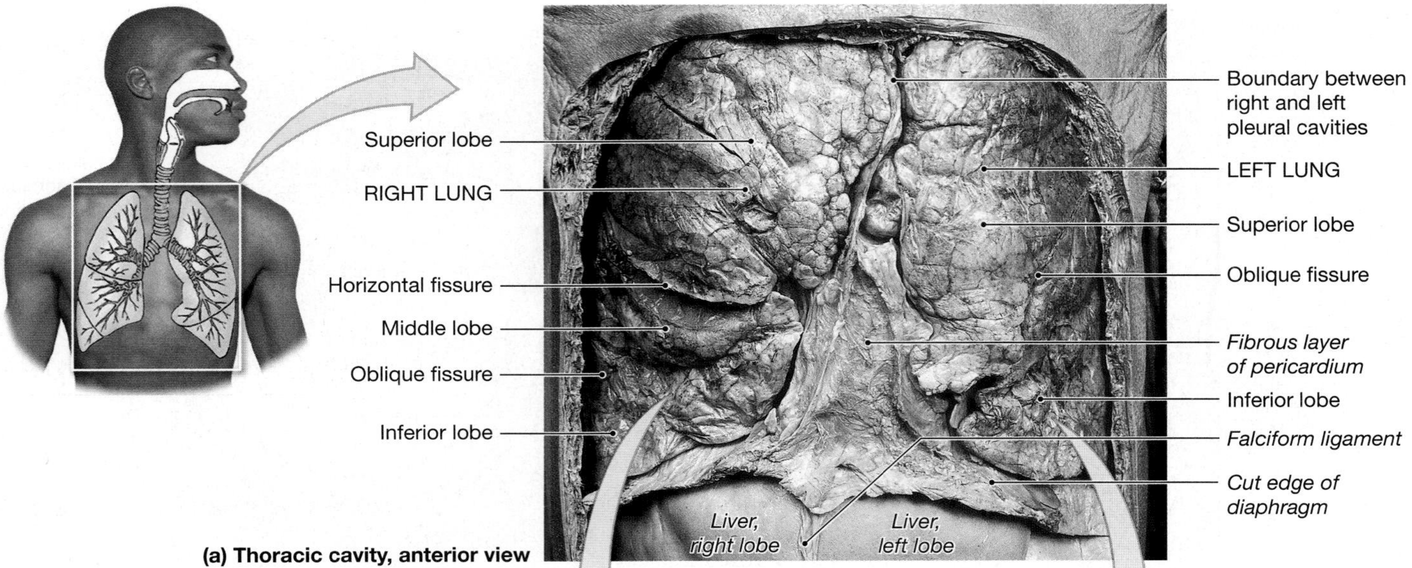

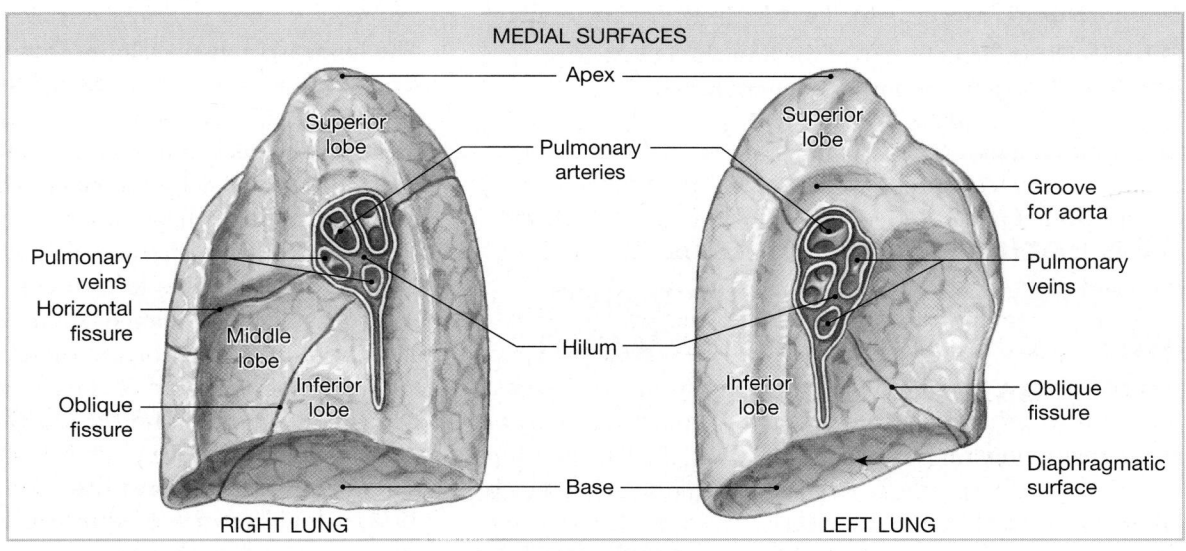

Superior lobe

RIGHT LUNG

Horizontal fissure

Middle lobe

Oblique fissure

Inferior lobe

Boundary between right and left pleural cavities

LEFT LUNG

Superior lobe

Oblique fissure

Fibrous layer of pericardium

Inferior lobe

Falciform ligament

Cut edge of diaphragm

Liver, right lobe   *Liver, left lobe*

(a) Thoracic cavity, anterior view

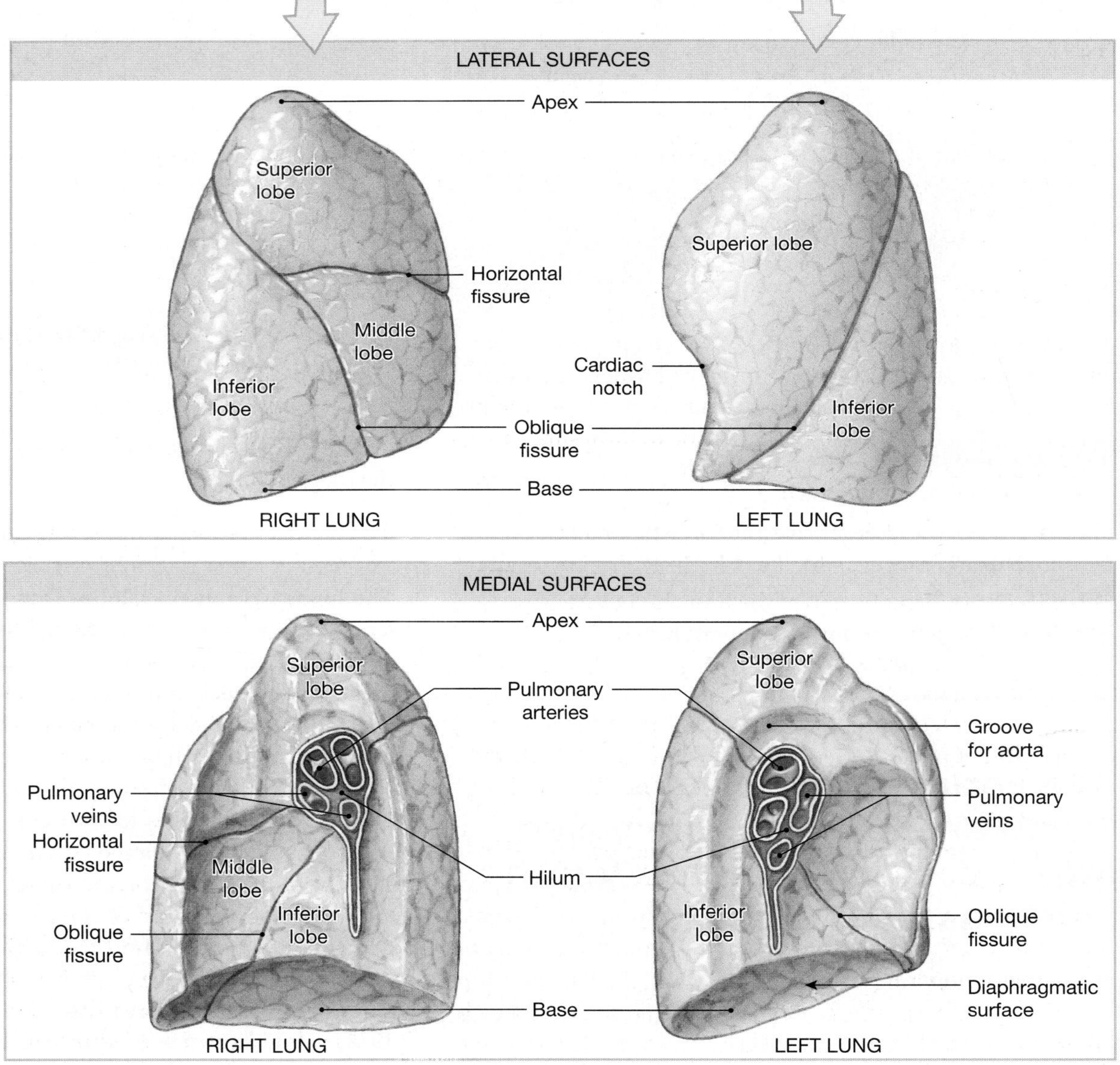

LATERAL SURFACES

Apex

Superior lobe

Horizontal fissure

Middle lobe

Inferior lobe

Oblique fissure

Base

RIGHT LUNG

Superior lobe

Cardiac notch

Inferior lobe

LEFT LUNG

MEDIAL SURFACES

Apex

Superior lobe

Pulmonary arteries

Superior lobe

Groove for aorta

Pulmonary veins

Horizontal fissure

Middle lobe

Inferior lobe

Hilum

Pulmonary veins

Inferior lobe

Oblique fissure

Oblique fissure

Base

Diaphragmatic surface

RIGHT LUNG

LEFT LUNG

Figure 23–7   The Gross Anatomy of the Lungs.

ATLAS: Plates 42–47

(b) The right and left lungs

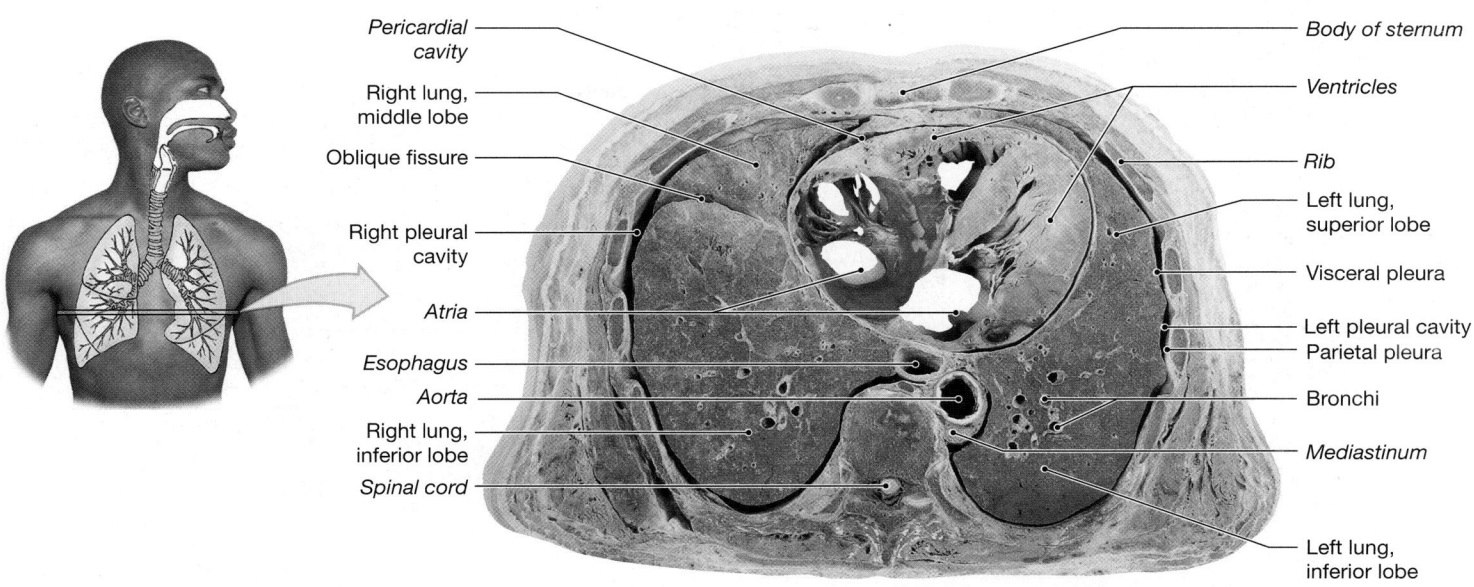

Figure 23–8 **The Relationship between the Lungs and Heart.** This transverse section was taken at the level of the cardiac notch.

right. In anterior view, the medial edge of the right lung forms a vertical line, whereas the medial margin of the left lung is indented at the **cardiac notch** (**Figure 23–7**).

The Bronchi

The primary bronchi and their branches form the **bronchial tree**. Because the left and right primary bronchi are outside the lungs, they are called *extrapulmonary bronchi*. As the primary bronchi enter the lungs, they divide to form smaller passageways (**Figure 23–6a**). The branches within the lungs are collectively called the *intrapulmonary bronchi*.

Each primary bronchus divides to form **secondary bronchi**, also known as *lobar bronchi*. In each lung, one secondary bronchus goes to each lobe, so the right lung has three secondary bronchi, and the left lung has two.

Figure 23–9 depicts the branching pattern of the left primary bronchus as it enters the lung. (The number of branches has been reduced for clarity.) In each lung, the secondary bronchi branch to form **tertiary bronchi**, or *segmental bronchi*. The branching pattern differs between the two lungs, but each tertiary bronchus ultimately supplies air to a single **bronchopulmonary segment**, a specific region of one lung (**Figure 23–9a**). The right lung has 10 bronchopulmonary segments. During development, the left lung also has 10 segments, but subsequent fusion of adjacent tertiary bronchi generally reduces that number to eight or nine.

The walls of the primary, secondary, and tertiary bronchi contain progressively less cartilage. In the secondary and tertiary bronchi, the cartilages form plates arranged around the lumen. These cartilages serve the same purpose as the rings of cartilage in the trachea and primary bronchi. As the amount of cartilage decreases, the relative amount of smooth muscle increases. With less cartilaginous support, the amount of tension in those smooth muscles has a greater effect on bronchial diameter and the resistance to airflow. During a respiratory infection, the bronchi and bronchioles can become inflamed and constricted, increasing resistance. In this condition, called **bronchitis**, the individual has difficulty breathing.

The Bronchioles

Each tertiary bronchus branches several times within the bronchopulmonary segment, giving rise to multiple **bronchioles**. These passageways further branch into the finest conducting branches, called **terminal bronchioles**. Roughly 6500 terminal bronchioles arise from each tertiary bronchus. The lumen of each terminal bronchiole has a diameter of 0.3–0.5 mm.

The walls of bronchioles, which lack cartilaginous supports, are dominated by smooth muscle tissue (**Figure 23–9b**). In functional terms, bronchioles are to the respiratory system what arterioles are to the cardiovascular system. Varying the diameter of the bronchioles controls the resistance to airflow and the distribution of air in the lungs.

The autonomic nervous system regulates the activity in this smooth muscle layer and thereby controls the diameter of the bronchioles. Sympathetic activation leads to **bronchodilation**, the enlargement of airway diameter. Parasympathetic stimulation leads to **bronchoconstriction**, a reduction in the diameter of the airway. Bronchoconstriction also occurs during allergic

Figure 23–9 **The Bronchi and Lobules of the Lung. (a)** The branching pattern of bronchi in the left lung, simplified. **(b)** The structure of a single pulmonary lobule, part of a bronchopulmonary segment.
ATLAS: Plates 42b,c; 47b–d

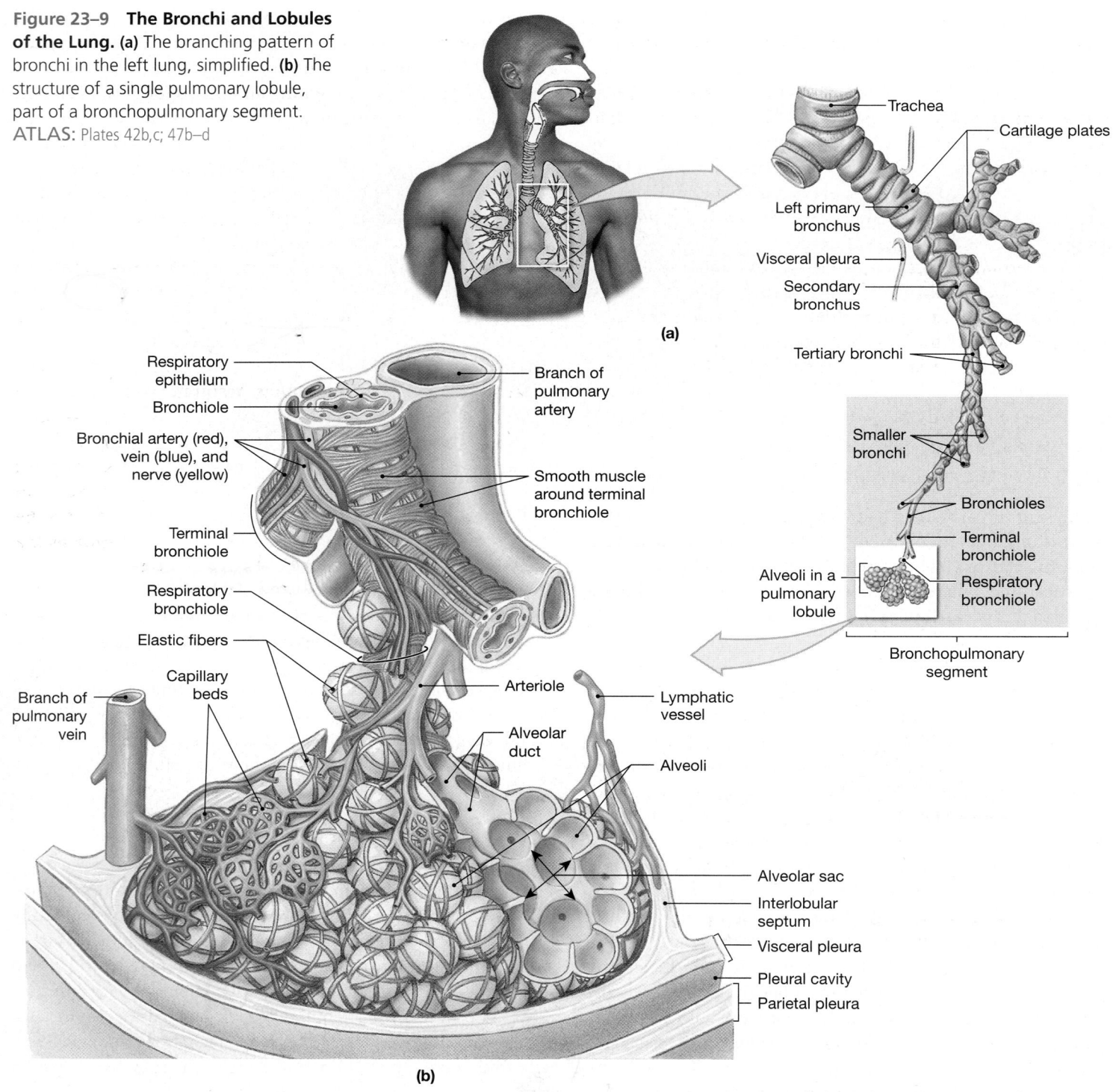

(a)

(b)

reactions such as anaphylaxis, in response to histamine released by activated mast cells and basophils. ∞ p. 816

By adjusting the resistance to airflow, bronchodilation and bronchoconstriction direct airflow toward or away from specific portions of the respiratory exchange surfaces. Tension in the smooth muscles commonly throws the bronchiolar mucosa into a series of folds, limiting airflow; excessive stimulation, as in **asthma** (AZ-muh), can almost completely prevent airflow along the terminal bronchioles.

Pulmonary Lobules

The connective tissues of the root of each lung extend into the lung's parenchyma. These fibrous partitions, or *trabeculae*, contain elastic fibers, smooth muscles, and lymphatic vessels.

The trabeculae branch repeatedly, dividing the lobes into ever-smaller compartments. The branches of the conducting passageways, pulmonary vessels, and nerves of the lungs follow these trabeculae. The finest partitions, or **interlobular septa** (*septum*, a wall), divide the lung into **pulmonary lobules** (LOB-ūlz), each of which is supplied by branches of the pulmonary arteries, pulmonary veins, and respiratory passageways (**Figure 23–9b**). The connective tissues of the septa are, in turn, continuous with those of the *visceral pleura*, the serous membrane covering the lungs.

Each terminal bronchiole delivers air to a single pulmonary lobule. Within the lobule, the terminal bronchiole branches to form several **respiratory bronchioles**. The thinnest and most delicate branches of the bronchial tree, the respiratory bronchioles deliver air to the gas exchange surfaces of the lungs.

The preliminary filtration and humidification of incoming air are completed before the air moves beyond the terminal bronchioles. A cuboidal epithelium lines the terminal bronchioles and respiratory bronchioles. There are only scattered cilia, and no mucous cells or underlying mucous glands are present. If particulate matter or pathogens reach this part of the respiratory tract, there is little to prevent them from damaging the delicate exchange surfaces of the lungs.

Alveolar Ducts and Alveoli

Respiratory bronchioles are connected to individual alveoli and to multiple alveoli along regions called **alveolar ducts** (**Figures 23–9b** and **23–10**). Alveolar ducts end at **alveolar sacs**, common chambers connected to multiple individual alveoli. Each lung contains about 150 million alveoli, and their abundance gives the lung an open, spongy appearance. An extensive network of capillaries is associated with each alveolus (**Figure 23–11a**). The capillaries are surrounded by a network of elastic fibers, which help maintain the relative positions of the alveoli and respiratory bronchioles. Recoil of these fibers during exhalation reduces the size of the alveoli and helps push air out of the lungs.

The alveolar epithelium consists primarily of simple squamous epithelium (**Figure 23–11b**). The squamous epithelial cells, called **pneumocytes type I**, are unusually thin and delicate. Roaming **alveolar macrophages** (*dust cells*) patrol the epithelial surface, phagocytizing any particulate matter that has eluded other respiratory defenses and reached the alveolar surfaces. **Pneumocytes type II**, also called *septal cells*, are scattered among the squamous cells. The large pneumocytes type II cells produce **surfactant** (sur-FAK-tant), an oily secretion containing a mixture of phospholipids and proteins. Surfactant is secreted onto the alveolar surfaces, where it forms a superficial coating over a thin layer of water. *reduces tension when alveoli contracts.*

Tips&Tricks

The term **surfactant** is derived from its purpose as a **surf**ace **act**ive age**nt**.

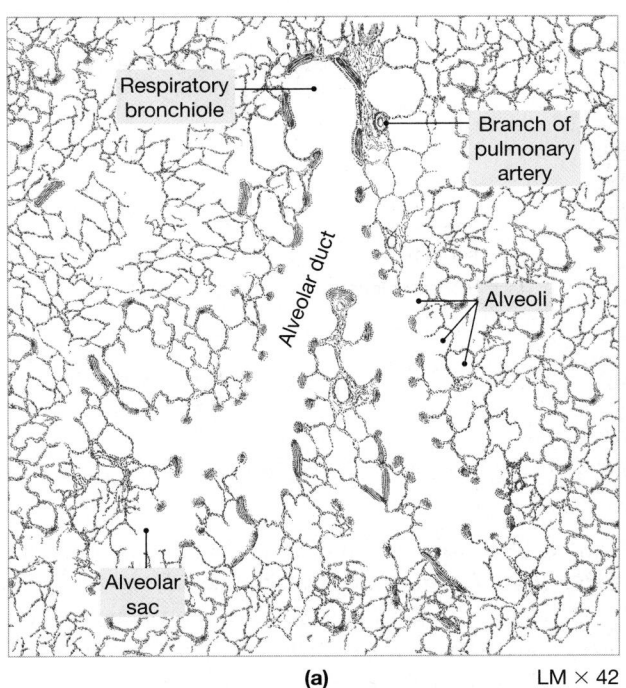

(a)   LM × 42

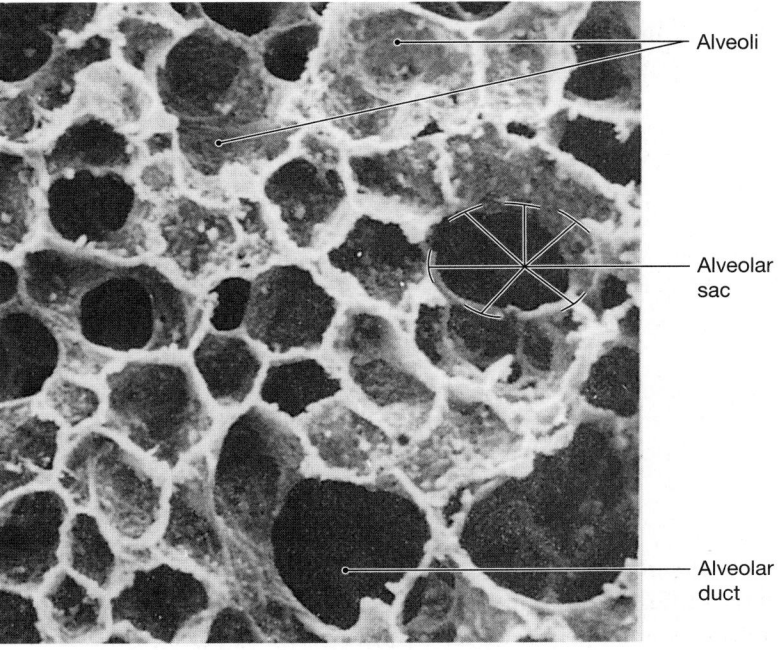

(b)

Figure 23–10 Respiratory Tissue. (a) The distribution of a respiratory bronchiole supplying a portion of a lobule. **(b)** A scanning electron micrograph (SEM) of the lung. Notice the open, spongy appearance of the lung tissue; *compare with Figure 23–9b.*

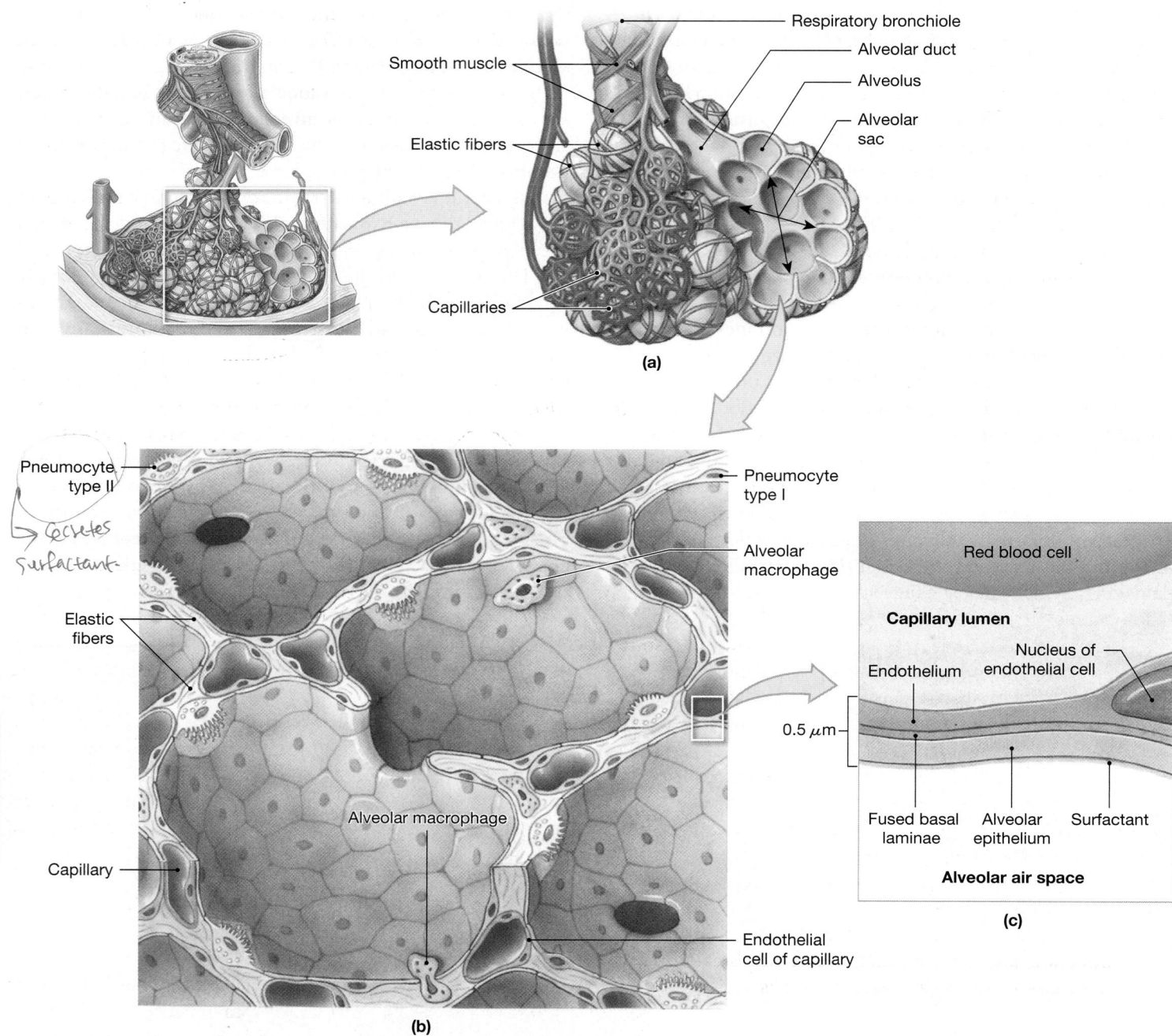

Figure 23–11 **Alveolar Organization.** (a) The basic structure of a portion of a single lobule. A network of capillaries, supported by elastic fibers, surrounds each alveolus. Respiratory bronchioles are also wrapped by smooth muscle cells that can change the diameter of these airways. (b) A diagrammatic view of alveolar structure. A single capillary may be involved in gas exchange with several alveoli simultaneously. (c) The respiratory membrane, which consists of an alveolar epithelial cell, a capillary endothelial cell, and their fused basal laminae.

Surfactant reduces surface tension in the liquid coating the alveolar surface. Recall from Chapter 2 that *surface tension* results from the attraction between water molecules at an air–water boundary. ∞ p. 35 Surface tension creates a barrier that keeps small objects from entering the water, but it also tends to collapse small bubbles. Alveolar walls, like air bubbles, are very

delicate; without surfactant, the surface tension would be so high that the alveoli would collapse. Surfactant forms a thin surface layer that interacts with the water molecules, reducing the surface tension and keeping the alveoli open.

If pneumocytes type II produce inadequate amounts of surfactant due to injury or genetic abnormalities, the alveoli will collapse after each exhalation, and respiration will become difficult. With each breath, the inhalation must be forceful enough to pop open the alveoli. A person who does not produce enough surfactant is soon exhausted by the effort required to keep inflating and deflating the lungs. This condition is called *respiratory distress syndrome.*

Gas exchange occurs across the **respiratory membrane** of the alveoli. The respiratory membrane is a composite structure consisting of three layers (**Figure 23–11c**): (1) the squamous epithelial cells lining the alveolus, (2) the endothelial cells lining an adjacent capillary, and (3) the fused basal laminae that lie between the alveolar and endothelial cells.

At the respiratory membrane, the total distance separating alveolar air from blood can be as little as 0.1 μm; it averages about 0.5 μm. Diffusion across the respiratory membrane proceeds very rapidly, because the distance is short and both oxygen and carbon dioxide are lipid soluble. The membranes of the epithelial and endothelial cells thus do not pose a barrier to the movement of oxygen and carbon dioxide between blood and alveolar air spaces.

In certain disease states, function of the respiratory membrane can be compromised. **Pneumonia** (noo-MŌ-nē-uh) develops from an infection or any other stimulus that causes inflammation of the lobules of the lung. As inflammation occurs, fluids leak into the alveoli and the respiratory bronchioles swell and constrict. Respiratory function deteriorates as a result. When bacteria are involved, they are generally types that normally inhabit the mouth and pharynx but have managed to evade the respiratory defenses. Pneumonia becomes more likely when the respiratory defenses have already been compromised by other factors, such as epithelial damage from smoking or the breakdown of the immune system in AIDS. The most common pneumonia that develops in individuals with AIDS results from infection by the fungus *Pneumocystis carinii.* The respiratory defenses of healthy individuals are able to prevent infection and tissue damage, but the breakdown of those defenses in AIDS can result in a massive, potentially fatal lung infection.

The Blood Supply to the Lungs ⊗ *gas exchange*

Lung tissue is nourished by two circuits, one suppling the *respiratory* portion of the lungs and the other perfusing the *conducting* portion.

The respiratory exchange surfaces receive blood from arteries of the pulmonary circuit. The pulmonary arteries enter the lungs at the hilum and branch with the bronchi as they approach the lobules. Each lobule receives an arteriole and a venule, and a network of capillaries surrounds each alveolus as part of the respiratory membrane. In addition to providing a mechanism for gas exchange, the endothelial cells of the alveolar capillaries are the primary source of *angiotensin-converting enzyme (ACE)*, which converts circulating angiotensin I to angiotensin II. This enzyme plays an important role in the regulation of blood volume and blood pressure. ∞ p. 742 Blood from the alveolar capillaries passes through the pulmonary venules and then enters the pulmonary veins, which deliver it to the left atrium.

The capillaries supplied by the bronchial arteries provide oxygen and nutrients to the tissues of conducting passageways of your lungs. The venous blood from these bronchial capillaries empties into bronchial veins or anastomoses and then into pulmonary veins. Blood flow outside the pulmonary vein bypasses the rest of the systemic circuit and dilutes the oxygenated blood leaving the alveoli.

Blood pressure in the pulmonary circuit is usually relatively low, with systemic pressures of 30 mm Hg or less. With such pressures, pulmonary vessels can easily become blocked by small blood clots, fat masses, or air bubbles in the pulmonary arteries. Because the lungs receive the entire cardiac output, any such objects drifting in blood are likely to be trapped in the pulmonary arterial or capillary networks. Very small blood clots occasionally form in the venous system. These are usually trapped in the pulmonary capillary network, where they soon dissolve. Larger emboli are much more dangerous. The blockage of a branch of a pulmonary artery will stop blood flow to a group of lobules or alveoli. This condition is called **pulmonary embolism**. If a pulmonary embolus is in place for several hours, the alveoli will permanently collapse. If the blockage occurs in a major pulmonary vessel rather than a minor tributary, pulmonary resistance increases. The resistance places extra strain on the right ventricle, which may be unable to maintain cardiac output, and congestive heart failure can result.

The Pleural Cavities and Pleural Membranes

The thoracic cavity has the shape of a broad cone. Its walls are the rib cage, and the muscular diaphragm forms its floor. The two **pleural cavities** are separated by the mediastinum (**Figure 23–8**). Each lung occupies a single pleural cavity, which is lined by a serous membrane called the **pleura** (PLOOR-uh; plural, *pleurae*). The pleura consists of two layers: the parietal pleura and the visceral pleura. The **parietal pleura** covers the inner surface of the thoracic wall and extends over the diaphragm and mediastinum. The **visceral pleura** covers the outer surfaces of the lungs, extending into the fissures between the lobes. Each pleural cavity actually represents a potential space rather than an open chamber, because the parietal and visceral pleurae are usually in close contact.

Both pleurae secrete a small amount of transudate (fluid that has passed through a membrane) referred to as **pleural fluid**.

23 RESPIRATORY

Pleural fluid forms a moist, slippery coating that provides lubrication, thereby reducing friction between the parietal and visceral surfaces as you breathe. Samples of pleural fluid, obtained through a long needle inserted between the ribs, are sometimes obtained for diagnostic purposes. This sampling procedure is called *thoracentesis* (thōr-a-sen-TĒ-sis; *thora-*, thoracic + *centesis*, puncture). The extracted fluid is examined for the presence of bacteria, blood cells, or other abnormal components.

In some disease states, the normal coating of pleural fluid is unable to prevent friction between the opposing pleural surfaces. The result is pain and pleural inflammation, a condition called *pleurisy*. When pleurisy develops, the secretion of pleural fluid may be excessive, or the inflamed pleurae may adhere to one another, limiting relative movement. In either case, breathing becomes difficult, and prompt medical attention is required.

CHECKPOINT

13. What would happen to the alveoli if surfactant were not produced?

14. Trace the path air takes in flowing from the glottis to the respiratory membrane.

15. Which arteries supply blood to the conducting portions and respiratory exchange surfaces of the lungs?

16. List the functions of the pleura. What does it secrete?

See the blue Answers tab at the end of the book.

23-6 External respiration and internal respiration allow gaseous exchange within the body

The general term *respiration* refers to two integrated processes: *external respiration* and *internal respiration*. The precise definitions of these terms vary among references. In this discussion, **external respiration** includes all the processes involved in the exchange of oxygen and carbon dioxide between the body's interstitial fluids and the external environment. The purpose of external respiration, and the primary function of the respiratory system, is meeting the respiratory demands of cells. **Internal respiration** is the absorption of oxygen and the release of carbon dioxide by those cells. We will consider the biochemical pathways responsible for oxygen consumption and for the generation of carbon dioxide by mitochondria—pathways known collectively as *cellular respiration*—in Chapter 25.

Our discussion here focuses on three integrated steps involved in external respiration (**Figure 23–12**):

1. *Pulmonary ventilation*, or breathing, which involves the physical movement of air into and out of the lungs.

2. *Gas diffusion* across the respiratory membrane between alveolar air spaces and alveolar capillaries, and across capillary walls between blood and other tissues.

3. *The transport of oxygen and carbon dioxide* between alveolar capillaries and capillary beds in other tissues.

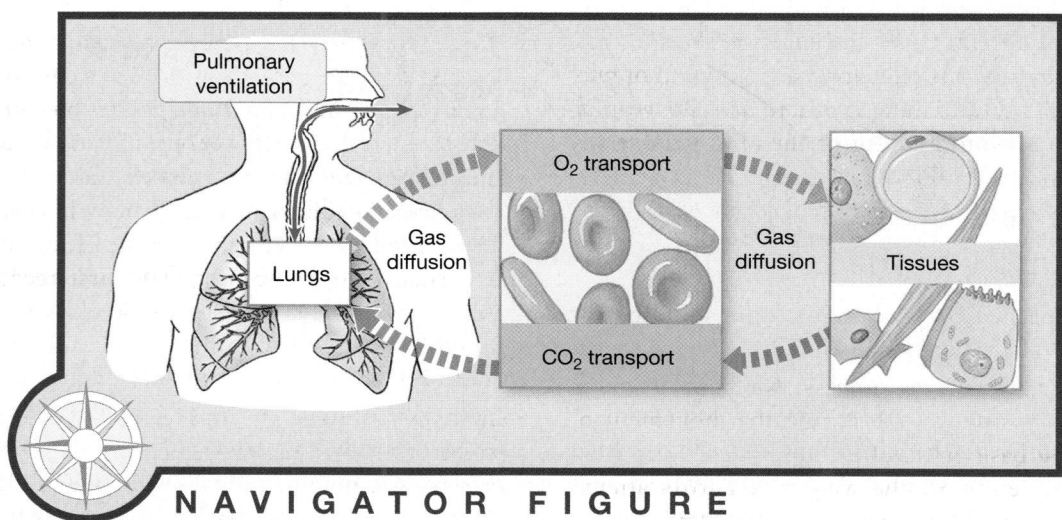

NAVIGATOR FIGURE

Figure 23–12   An Overview of the Key Steps in External Respiration. This figure will be repeated, in reduced and simplified form, as Navigator icons in key figures throughout this chapter as we move from one topic to the next.

Abnormalities affecting any of the steps involved in external respiration will ultimately affect the gas concentrations of interstitial fluids, and thus cellular activities as well. If the oxygen content declines, the affected tissues will become starved for oxygen. **Hypoxia**, or low tissue oxygen levels, places severe limits on the metabolic activities of the affected area. For example, the effects of coronary ischemia result from chronic hypoxia affecting cardiac muscle cells. ∞ p. 694 If the supply of oxygen is cut off completely, the condition called **anoxia** (an-OK-sē-uh; *a*-, without + *ox*-, oxygen) results. Anoxia kills cells very quickly. Much of the damage caused by strokes and heart attacks is the result of localized anoxia.

In the sections that follow, we examine each of the processes involved in external respiration in greater detail.

CHECKPOINT

17. Define external respiration and internal respiration.

18. Name the integrated steps involved in external respiration.

See the blue Answers tab at the end of the book.

23-7 Pulmonary ventilation—the exchange of air between the atmosphere and the lungs—involves pressure changes, muscle movement, and respiratory rates and volumes

Pulmonary ventilation is the physical movement of air into and out of the respiratory tract. The primary function of pulmonary ventilation is to maintain adequate *alveolar ventilation*—movement of air into and out of the alveoli. Alveolar ventilation prevents the buildup of carbon dioxide in the alveoli and ensures a continuous supply of oxygen that keeps pace with absorption by the bloodstream.

The Movement of Air

To understand this mechanical process, we need to grasp some basic physical principles governing the movement of air. One of the most basic is that our bodies and everything around us are compressed by the weight of Earth's atmosphere. Although we are seldom aware of it, this **atmospheric pressure** has several important physiological effects. For example, air moves into and out of the respiratory tract as the air pressure in the lungs cycles between below atmospheric pressure and above atmospheric pressure.

Gas Pressure and Volume (Boyle's Law)

The primary differences between liquids and gases reflect the interactions among individual molecules. Although the molecules in a liquid are in constant motion, they are held closely together by weak interactions, such as the hydrogen bonding between adjacent water molecules. ∞ p. 35 Yet because the electrons of adjacent atoms tend to repel one another, liquids tend to resist compression. If you squeeze a balloon filled with water, it will distort into a different shape, but the volumes of the two shapes will be the same.

In a gas, such as air, the molecules bounce around as independent entities. At normal atmospheric pressures, gas molecules are much farther apart than the molecules in a liquid, so the density of air is relatively low. The forces acting between gas molecules are minimal (the molecules are too far apart for weak interactions to occur), so an applied pressure can push them closer together. Consider a sealed container of air at atmospheric pressure. The pressure exerted by the enclosed gas results from the collision of gas molecules with the walls of the container. The greater the number of collisions, the higher the pressure.

You can change the gas pressure within a sealed container by changing the volume of the container, thereby giving the gas molecules more or less room in which to bounce around. If you decrease the volume of the container, collisions occur more frequently over a given period, elevating the pressure of the gas (**Figure 23–13a**). If you increase the volume, fewer collisions occur per unit time, because it takes longer for a gas molecule to travel from one wall to another. As a result, the gas pressure inside the container declines (**Figure 23–13b**).

For a gas in a closed container and at a constant temperature, pressure (*P*) is inversely proportional to volume (*V*). That is, *if you decrease the volume of a gas, its pressure will rise; if you increase the volume of a gas, its pressure will fall*. In particular, the relationship between pressure and volume is reciprocal: If you double the external pressure on a flexible container, its volume will drop by half; if you reduce the external pressure by half, the volume of the container will double. This relationship, $P = 1/V$, first recognized by Robert Boyle in the 1600s, is called **Boyle's law**.

Pressure and Airflow to the Lungs

Air will flow from an area of higher pressure to an area of lower pressure. This tendency for directed airflow, plus the pressure–volume relationship of Boyle's law, provides the basis for pulmonary ventilation. A single respiratory cycle consists of an *inspiration*, or inhalation, and an *expiration*, or exhalation. Inhalation and exhalation involve changes in the

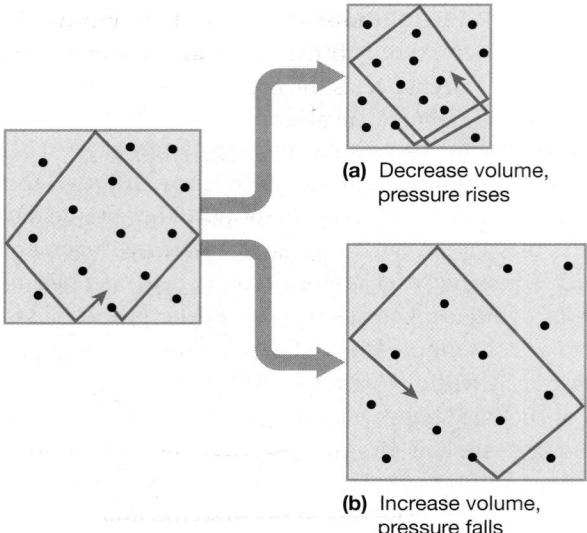

(a) Decrease volume, pressure rises

(b) Increase volume, pressure falls

Figure 23–13   Gas Pressure and Volume Relationships. Gas molecules in a sealed container bounce off the walls and off one another, traveling the distance indicated in a given period of time. **(a)** If the volume decreases, each molecule travels the same distance in that same period, but strikes the walls more frequently. The pressure exerted by the gas thus increases. **(b)** If the volume of the container increases, each molecule strikes the walls less often, lowering the pressure in the container.

volume of the lungs. These volume changes create pressure gradients that move air into or out of the respiratory tract.

Each lung lies within a pleural cavity. The parietal and visceral pleurae are separated by only a thin film of pleural fluid. The two membranes can slide across one another, but they are held together by that fluid film. You encounter the same principle when you set a wet glass on a smooth table-top. You can slide the glass easily, but when you try to lift it, you experience considerable resistance. As you pull the glass away from the tabletop, you create a powerful suction. The only way to overcome it is to tilt the glass so that air is pulled between the glass and the table, breaking the fluid bond.

A comparable fluid bond exists between the parietal pleura and the visceral pleura covering the lungs. As a result, the surface of each lung sticks to the inner wall of the chest and to the superior surface of the diaphragm. As a result, movements of the diaphragm or rib cage that change the volume of the thoracic cavity also change the volume of the lungs.

Changes in the volume of the thoracic cavity result from movements of the diaphragm or rib cage (**Figure 23–14a**):

- The diaphragm forms the floor of the thoracic cavity. The relaxed diaphragm has the shape of a dome that projects superiorly into the thoracic cavity. When the diaphragm contracts, it tenses and moves inferiorly. This movement increases the volume of the thoracic cavity, reducing the pressure within it. When the diaphragm relaxes, it

returns to its original position, and the volume of the thoracic cavity decreases.

- Owing to the nature of the articulations between the ribs and the vertebrae, superior movement of the rib cage increases the depth and width of the thoracic cavity, increasing its volume. Inferior movement of the rib cage reverses the process, reducing the volume of the thoracic cavity.

At the start of a breath, pressures inside and outside the thoracic cavity are identical, and no air moves into or out of the lungs (**Figure 23–14b**). When the thoracic cavity enlarges, the pleural cavities and lungs expand to fill the additional space (**Figure 23–14c**). This increase in volume lowers the pressure inside the lungs. Air then enters the respiratory passageways, because the pressure inside the lungs (P_{inside}) is lower than atmospheric pressure ($P_{outside}$). Air continues to enter the lungs until their volume stops increasing and the internal pressure is the same as that outside. When the thoracic cavity decreases in volume, pressures rise inside the lungs, forcing air out of the respiratory tract (**Figure 23–14d**).

Compliance

The **compliance** of the lungs is an indication of their expandability, or how easily the lungs expand and contract. The lower the compliance, the greater the force required to fill and empty the lungs. The greater the compliance, the easier it is to fill and empty the lungs. Factors affecting compliance include the following:

- *The Connective Tissue Structure of the Lungs.* The loss of supporting tissues resulting from alveolar damage, as in *emphysema*, increases compliance.

- *The Level of Surfactant Production.* On exhalation, the collapse of alveoli as a result of inadequate surfactant, as in respiratory distress syndrome, reduces the lungs' compliance.

- *The Mobility of the Thoracic Cage.* Arthritis or other skeletal disorders that affect the articulations of the ribs or spinal column also reduce compliance.

At rest, the muscular activity involved in pulmonary ventilation accounts for 3–5 percent of the resting energy demand. If compliance is reduced, that figure climbs dramatically, and an individual may become exhausted simply trying to continue breathing.

Pressure Changes during Inhalation and Exhalation

To understand the mechanics of respiration and the principles of gas exchange, we must know the pressures existing inside and outside the respiratory tract. We can report pres-

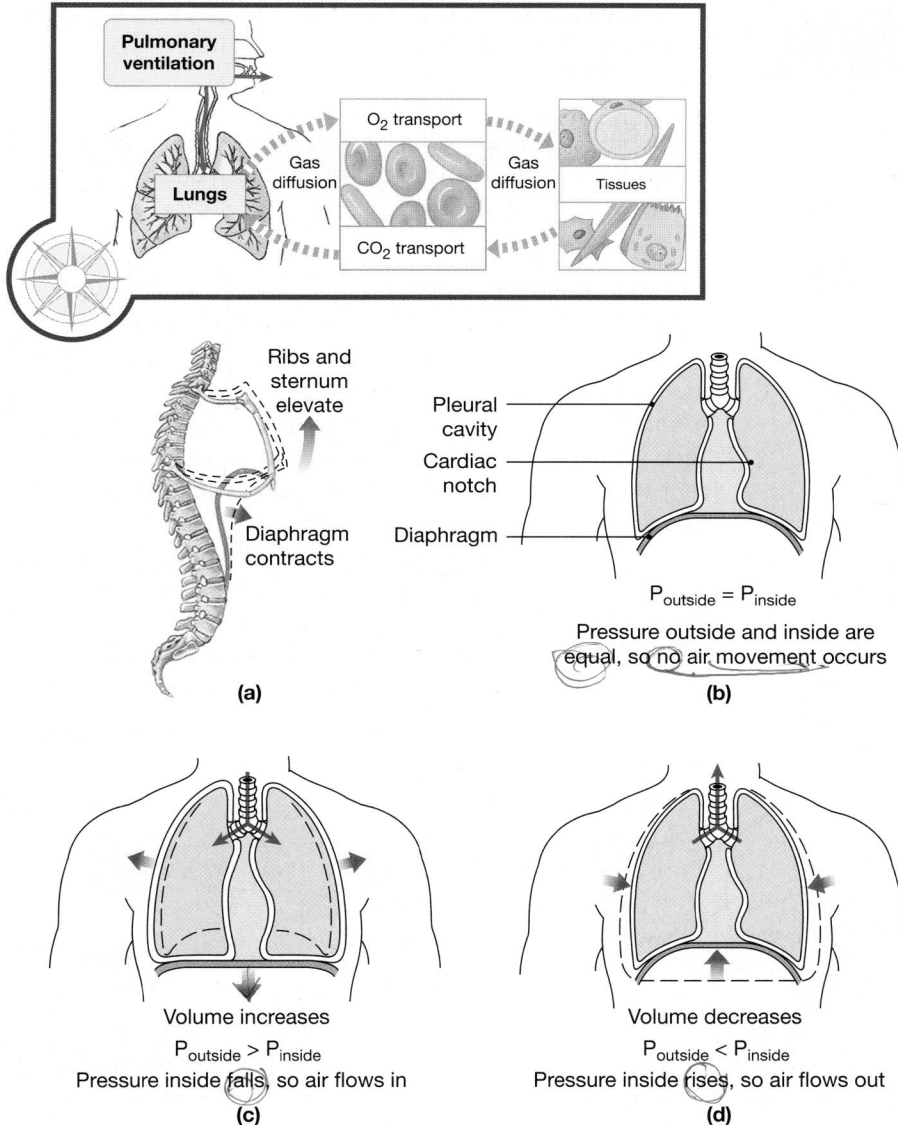

Figure 23–14 Mechanisms of Pulmonary Ventilation. The Navigator icon in the shadow box highlights the topic of the current figure. **(a)** As the rib cage is elevated or the diaphragm is depressed, the volume of the thoracic cavity increases. **(b)** An anterior view with the diaphragm at rest; no air movement occurs. **(c)** *Inhalation*: Elevation of the rib cage and contraction of the diaphragm increase the size of the thoracic cavity. Pressure within the thoracic cavity decreases, and air flows into the lungs. **(d)** *Exhalation*: When the rib cage returns to its original position and the diaphragm relaxes, the volume of the thoracic cavity decreases. Pressure rises, and air moves out of the lungs.

sure readings in several ways (Table 23–1); in this text, we will use millimeters of mercury (mm Hg), as we did for blood pressure. Atmospheric pressure is also measured in *atmospheres*; one atmosphere of pressure (1 *atm*) is equivalent to 760 mm Hg.

The Intrapulmonary Pressure

The direction of airflow is determined by the relationship between atmospheric pressure and intrapulmonary pressure.

Intrapulmonary (in-tra-PUL-mo-nār-ē) **pressure,** or **intra-alveolar** (in-tra-al-VĒ-ō-lar) **pressure,** is the pressure inside the respiratory tract, at the alveoli.

When you are relaxed and breathing quietly, the difference between atmospheric pressure and intrapulmonary pressure is relatively small. On inhalation, your lungs expand, and the intrapulmonary pressure drops to about 759 mm Hg. Because the intrapulmonary pressure is 1 mm Hg below atmospheric pressure, it is generally reported as −1 mm Hg. On exhalation, your lungs recoil, and intrapulmonary pressure rises to 761 mm Hg, or +1 mm Hg (**Figure 23–15a**).

The size of the pressure gradient increases when you breathe heavily. When a trained athlete breathes at maximum capacity, the pressure differentials can reach −30 mm Hg during inhalation and +100 mm Hg if the individual is straining with the glottis kept closed. This is one reason you are told to exhale while lifting weights; exhaling keeps your intrapulmonary pressures and peritoneal pressure from climbing so high that an alveolar rupture or hernia could occur.

The Intrapleural Pressure

Intrapleural pressure is the pressure in the space between the parietal and visceral pleurae. Intrapleural pressure averages about −4 mm Hg (**Figure 23–15b**) but can reach −18 mm Hg during a powerful inhalation. This pressure is below atmospheric pressure, due to the relationship between the lungs and the body wall. Earlier, we noted that the lungs are highly elastic. In fact, they would collapse to about 5 percent of their normal resting volume if the elastic fibers could recoil completely. The elastic fibers cannot recoil significantly, however, because they are not strong enough to overcome the fluid bond between the parietal and visceral pleurae.

The elastic fibers continuously oppose that fluid bond and pull the lungs away from the chest wall and diaphragm, lowering the intrapleural pressure slightly. Because the elastic fibers remain stretched even after a full exhalation, intrapleural pressure remains below atmospheric pressure throughout normal cycles of inhalation and exhalation. The cyclical changes in the intrapleural pressure are responsible for the *respiratory pump*—the mechanism that assists the venous return to the heart. ∞ p. 734

TABLE 23–1 The Four Most Common Methods of Reporting Gas Pressures

millimeters of mercury (mm Hg): This is the most common method of reporting blood pressure and gas pressures. Normal atmospheric pressure is approximately 760 mm Hg.

torr: This unit of measurement is preferred by many respiratory therapists; it is also commonly used in Europe and in some technical journals. One torr is equivalent to 1 mm Hg; in other words, normal atmospheric pressure is equal to 760 torr.

centimeters of water (cm H_2O): In a hospital setting, anesthetic gas pressures and oxygen pressures are commonly measured in centimeters of water. One cm H_2O is equivalent to 0.735 mm Hg; normal atmospheric pressure is 1033.6 cm H_2O.

pounds per square inch (psi): Pressures in compressed gas cylinders and other industrial applications are generally reported in psi. Normal atmospheric pressure at sea level is approximately 15 psi.

The Respiratory Cycle

A **respiratory cycle** is a single cycle of inhalation and exhalation. The curves in **Figure 23–15a,b** indicate the intrapulmonary and intrapleural pressures during a single respiratory cycle of an individual at rest. The graph in **Figure 23–15c** plots

the **tidal volume**, the amount of air you move into or out of your lungs during a single respiratory cycle.

Tips&Tricks

Tidal volume floods and ebbs like the ocean tides.

At the start of the respiratory cycle, the intrapulmonary and atmospheric pressures are equal, and no air is moving. Inhalation begins with the fall of intrapleural pressure that accompanies the expansion of the thoracic cavity. This pressure gradually falls to approximately −6 mm Hg. Over the period, intrapulmonary pressure drops to just under −1 mm Hg; it then begins to rise as air flows into the lungs. When exhalation begins, intrapleural and intrapulmonary pressures rise rapidly, forcing air out of the lungs. At the end of exhalation, air movement again ceases when the difference between intrapulmonary and atmospheric pressures has been eliminated. The amount of air moved into the lungs during inhalation is equal to the amount moved out of the lungs during exhalation. That amount is the tidal volume.

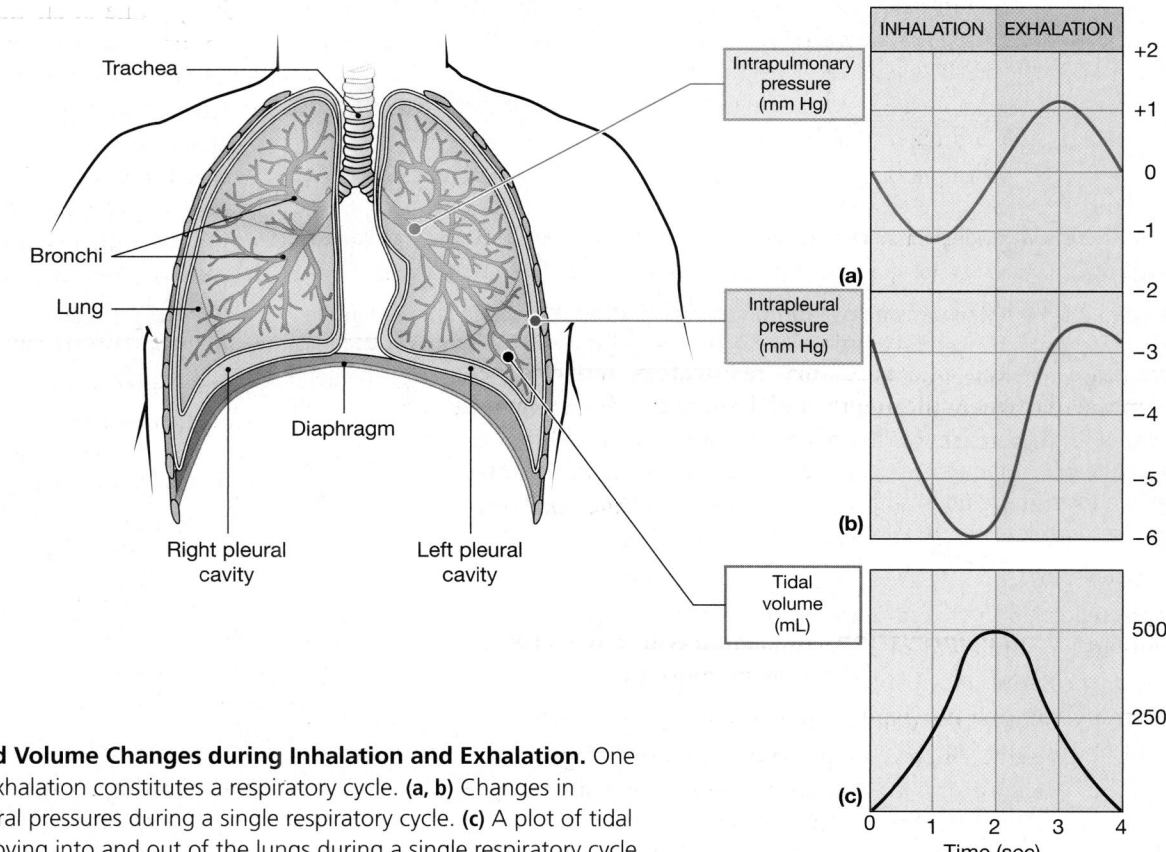

Figure 23–15 **Pressure and Volume Changes during Inhalation and Exhalation.** One sequence of inhalation and exhalation constitutes a respiratory cycle. **(a, b)** Changes in intrapulmonary and intrapleural pressures during a single respiratory cycle. **(c)** A plot of tidal volume, the amount of air moving into and out of the lungs during a single respiratory cycle.

CLINICAL NOTE

Pneumothorax An injury to the chest wall that penetrates the parietal pleura, or a rupture of the alveoli that breaks through the visceral pleura, can allow air into the pleural cavity. This condition, called **pneumothorax** (noo-mō-THOR-aks; *pneumo-*, air), breaks the fluid bond between the pleurae and allows the elastic fibers to recoil, resulting in a "collapsed lung," or **atelectasis** (at-e-LEK-ta-sis; *atel-*, imperfect or incomplete + *ectasia*, distention). The opposite lung would not be affected due to compartmentalization. The treatment for a collapsed lung involves the removal of as much of the air as possible from the affected pleural cavity before the opening is sealed. This treatment lowers the intrapleural pressure and reinflates the lung.

The Mechanics of Breathing

As we have just seen, you move air into and out of the respiratory system by changing the volume of the lungs. Those changes alter the pressure relationships, producing air movement. The changes of volume in the lungs occur through the contraction of skeletal muscles—specifically, those that insert on the rib cage—and the *diaphragm*, which separates the thoracic and abdominopelvic cavities. Because of the nature of their articulations with the vertebrae, when the ribs are elevated they swing outward, increasing the depth of the thoracic cavity. ∞ p. 241 This movement can be compared to the elevation of a bucket handle (**Figure 23–16a**).

The Respiratory Muscles

The skeletal muscles involved in respiratory movements were introduced in Chapter 11. Of those muscles, the most important are the *diaphragm* and the *external intercostal muscles*. ∞ pp. 355–356 These muscles are active during normal breathing at rest. The **accessory respiratory muscles** become active when the depth and frequency of respiration must be increased markedly. These muscles include the *internal intercostal, sternocleidomastoid, serratus anterior, pectoralis minor, scalene, transversus thoracis, transversus abdominis, external* and *internal oblique,* and *rectus abdominis muscles* (**Figure 23–16b–d**). ∞ pp. 351, 355, 361

Muscles Used in Inhalation.
Inhalation is an active process involving one or more of the following muscles:

- Contraction of the diaphragm flattens the floor of the thoracic cavity, increasing its volume and drawing air into the lungs. Diaphragmatic contraction is responsible for roughly 75 percent of the air movement in normal breathing at rest.

- Contraction of the external intercostal muscles assists in inhalation by elevating the ribs. This action contributes roughly 25 percent to the volume of air in the lungs at rest.

- Contraction of accessory muscles, including the sternocleidomastoid, serratus anterior, pectoralis minor, and scalene muscles, can assist the external intercostal muscles in elevating the ribs. These muscles increase the speed and amount of rib movement.

Muscles Used in Exhalation.
Exhalation is either passive or active, depending on the level of respiratory activity. When exhalation is active, it may involve one or more of the following muscles:

- The internal intercostal and transversus thoracis muscles depress the ribs and reduce the width and depth of the thoracic cavity.

- The abdominal muscles, including the external and internal oblique, transversus abdominis, and rectus abdominis muscles, can assist the internal intercostal muscles in exhalation by compressing the abdomen and forcing the diaphragm upward.

Modes of Breathing

The respiratory muscles are used in various combinations, depending on the volume of air that must be moved into or out of the system. Respiratory movements are usually classified as *quiet breathing* or *forced breathing* according to the pattern of muscle activity during a single respiratory cycle.

Quiet Breathing. In **quiet breathing**, or **eupnea** (ŪP-nē-uh; *eu-*, true or normal + *-pnea*, respiration), inhalation involves muscular contractions, but exhalation is a passive process. Inhalation usually involves the contraction of both the diaphragm and the external intercostal muscles. The relative contributions of these muscles can vary:

- During **diaphragmatic breathing**, or **deep breathing**, contraction of the diaphragm provides the necessary change in thoracic volume. Air is drawn into the lungs as the diaphragm contracts, and is exhaled passively when the diaphragm relaxes.

- In **costal breathing**, or **shallow breathing**, the thoracic volume changes because the rib cage alters its shape. Inhalation occurs when contractions of the external intercostal muscles elevate the ribs and enlarge the thoracic cavity. Exhalation occurs passively when these muscles relax.

During quiet breathing, expansion of the lungs stretches their elastic fibers. In addition, elevation of the rib cage stretches opposing skeletal muscles and elastic fibers in the

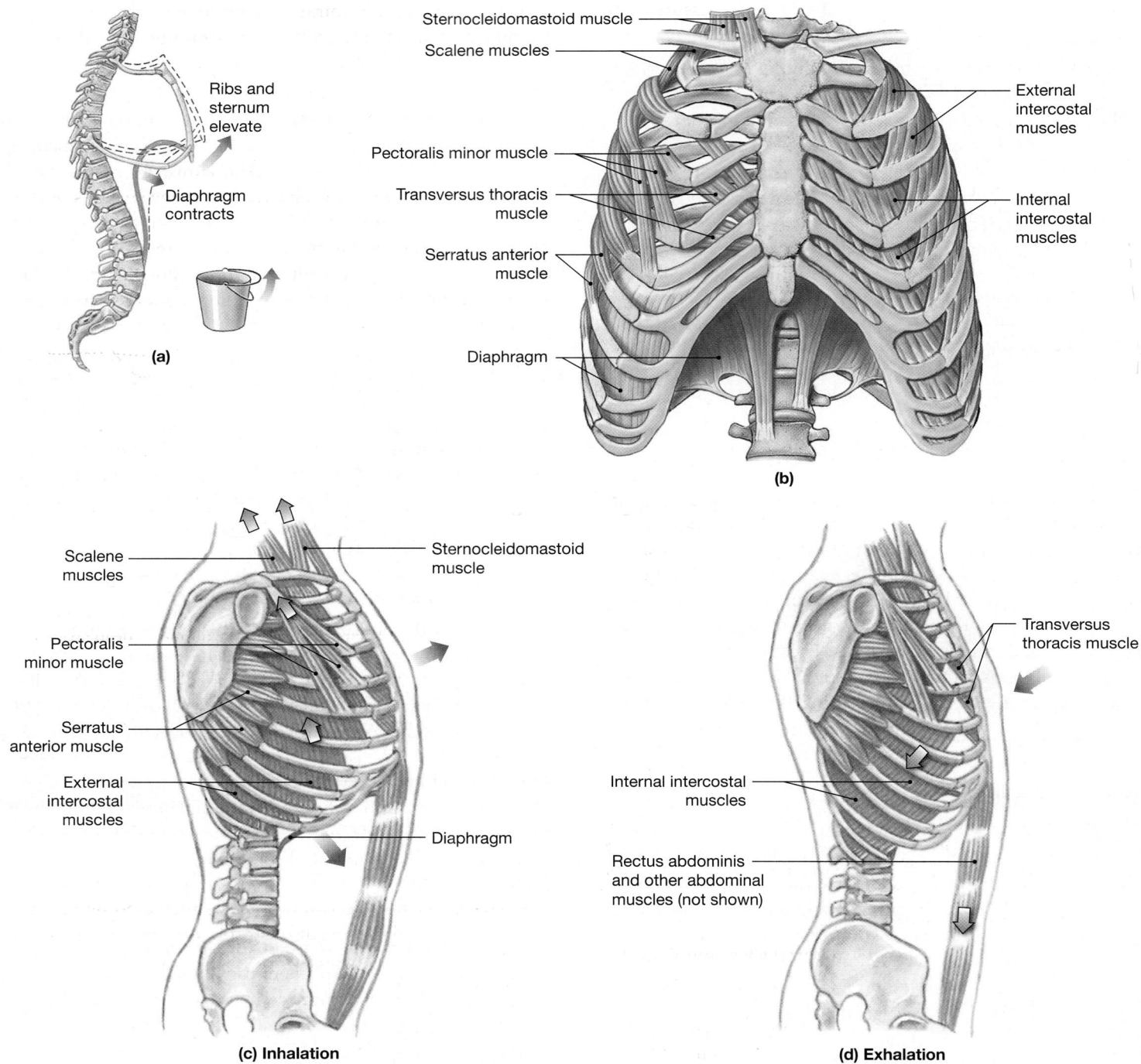

Figure 23–16   The Respiratory Muscles. (a) Movements of the ribs and diaphragm that increase the volume of the thoracic cavity. Diaphragmatic movements were also illustrated in *Figure 23–14*. **(b)** An anterior view at rest (with no air movement), showing the primary and accessory respiratory muscles. **(c)** A lateral view during inhalation, showing the muscles that elevate the ribs. **(d)** A lateral view during exhalation, showing the muscles that depress the ribs. The abdominal muscles that assist in exhalation are represented by a single muscle (the rectus abdominis).

connective tissues of the body wall. When the muscles of inhalation relax, these elastic components recoil, returning the diaphragm, the rib cage, or both to their original positions. This phenomenon is called **elastic rebound**.

Diaphragmatic breathing typically occurs at minimal levels of activity. As increased volumes of air are required, inspiratory movements become larger and the contribution of rib movement increases. Even when you are at rest, costal breathing can predominate when abdominal pressures, fluids, or masses restrict diaphragmatic movements. For example, pregnant women increasingly rely on costal breathing as the enlarging uterus pushes the abdominal viscera against the diaphragm.

Forced Breathing. Forced breathing, or **hyperpnea** (hī-PERP-nē-uh), involves active inspiratory and expiratory movements. Forced breathing calls on the accessory muscles to assist with inhalation, and exhalation involves contraction of the internal intercostal muscles. At absolute maximum levels of forced breathing, the abdominal muscles are involved in exhalation. Their contraction compresses the abdominal contents, pushing them up against the diaphragm and further reducing the volume of the thoracic cavity.

Respiratory Rates and Volumes

Your respiratory system is extremely adaptable. You can be breathing slowly and quietly one moment, rapidly and deeply the next. The respiratory system adapts to meet the oxygen demands of the body by varying both the number of breaths per minute and the amount of air moved per breath. When you are exercising at peak levels, the amount of air moving into and out of the respiratory tract can be 50 times the amount moved at rest.

Respiratory Rate

Your **respiratory rate** is the number of breaths you take each minute. As you read this, you are probably breathing quietly, with a low respiratory rate. The normal respiratory rate of a resting adult ranges from 12 to 18 breaths each minute, roughly one for every four heartbeats. Children breathe more rapidly, at rates of about 18–20 breaths per minute.

The Respiratory Minute Volume

We can calculate the amount of air moved each minute, symbolized $\dot{V}_E$, by multiplying the respiratory rate f by the tidal volume V_T. This value is called the **respiratory minute volume**. The respiratory rate at rest averages 12 breaths per minute, and the tidal volume at rest averages around 500 mL per breath. On that basis, we can calculate the respiratory minute volume as follows:

$$\dot{V}_E = f \times V_T$$

$$\begin{pmatrix} \text{volume of air moved} \\ \text{each minute} \end{pmatrix} = \begin{pmatrix} \text{breaths per} \\ \text{minute} \end{pmatrix} \times \begin{pmatrix} \text{tidal} \\ \text{volume} \end{pmatrix}$$

$$= 12 \times 500 \text{ mL per minute}$$
$$= 6000 \text{ mL per minute}$$
$$= 6.0 \text{ liters per minute}$$

In other words, the respiratory minute volume at rest is approximately 6 liters (1.6 gallons) per minute.

Alveolar Ventilation

The respiratory minute volume measures pulmonary ventilation and provides an indication of how much air is moving into and out of the respiratory tract. However, only some of the inhaled air reaches the alveolar exchange surfaces. A typical inhalation pulls about 500 mL of air into the respiratory system. The first 350 mL of inhaled air travels along the conducting passageways and enters the alveolar spaces. The last 150 mL of inhaled air never gets farther than the conducting passageways and thus does not participate in gas exchange with blood. The volume of air in the conducting passages is known as the **anatomic dead space**, denoted V_D. **Alveolar ventilation**, symbolized $\dot{V}_A$ is the amount of air reaching the alveoli each minute. The alveolar ventilation is less than the respiratory minute volume, because some of the air never reaches the alveoli, but remains in the dead space of the lungs. We can calculate alveolar ventilation by subtracting the dead space from the tidal volume:

$$\dot{V}_A = f \times (V_T - V_D)$$

$$\begin{pmatrix} \text{alveolar} \\ \text{ventilation} \end{pmatrix} = \begin{pmatrix} \text{breaths} \\ \text{per minute} \end{pmatrix} \times \begin{pmatrix} \text{tidal} \\ \text{volume} - \frac{\text{anatomic}}{\text{dead space}} \end{pmatrix}$$

At rest, alveolar ventilation rates are approximately 4.2 liters per minute (12×350 mL) However, the gas arriving in the alveoli is significantly different from that of the surrounding atmosphere, because inhaled air always mixes with "used" air in the conducting passageways (the anatomic dead space) on its way to the exchange surfaces. The air in alveoli thus contains less oxygen and more carbon dioxide than does atmospheric air.

Relationships among V_T, $\dot{V}_E$, and $\dot{V}_A$

The respiratory minute volume can be increased by increasing either the tidal volume or the respiratory rate. Under maximum stimulation, the tidal volume can increase to roughly 4.8 liters. At peak respiratory rates of 40–50 breaths per minute and maximum cycles of inhalation and exhalation, the respiratory minute volume can approach 200 liters (about 55 gal) per minute.

In functional terms, the alveolar ventilation rate is more important than the respiratory minute volume, because it determines the rate of delivery of oxygen to the alveoli. The respiratory rate and the tidal volume together determine the alveolar ventilation rate:

- For a given respiratory rate, increasing the tidal volume increases the alveolar ventilation rate.

- For a given tidal volume, increasing the respiratory rate increases the alveolar ventilation rate.

The alveolar ventilation rate can change independently of the respiratory minute volume. In our previous example, the respiratory minute volume at rest was 6 liters and the alveolar ventilation rate was 4.2 L/min. If the respiratory rate rises to 20 breaths per minute, but the tidal volume drops to 300 mL, the respiratory minute volume remains the same (20 × 300 = 6000). However, the alveolar ventilation rate drops to only 3 L/min (20 × [300 − 150] = 3000). Thus, whenever the demand for oxygen increases, both the tidal volume *and* the respiratory rate must be regulated closely. (The mechanisms involved are the focus of a later section.)

Respiratory Performance and Volume Relationships

Only a small proportion of the air in the lungs is exchanged during a single quiet respiratory cycle; the tidal volume can be increased by inhaling more vigorously and exhaling more completely. We can divide the total volume of the lungs into a series of *volumes* and *capacities* (each the sum of various volumes), as indicated in **Figure 23–17**. The values obtained are useful in diagnosing problems with pulmonary ventilation. Adult females, on average, have smaller bodies and thus smaller lung volumes than do males. As a result, there are sex-related differences in respiratory volumes and capacities. Representative values for both sexes are indicated in the figure.

Pulmonary volumes include the following:

- The **resting tidal volume** is the amount of air you move into or out of your lungs during a single respiratory cycle under resting conditions. The resting tidal volume averages about 500 mL in both males and females.

- The **expiratory reserve volume (ERV)** is the amount of air that you can voluntarily expel after you have completed a normal, quiet respiratory cycle. As an example, with maximum use of the accessory muscles, males can expel an additional 1000 mL of air, on average. Female expiratory reserve volume averages 700 mL.

- The **residual volume** is the amount of air that remains in your lungs even after a maximal exhalation—typically about 1200 mL in males and 1100 mL in females. The **minimal volume**, a component of the residual volume, is the amount of air that would remain in your lungs if they were allowed to collapse. The minimal volume ranges from 30 to 120 mL, but, unlike other volumes, it cannot

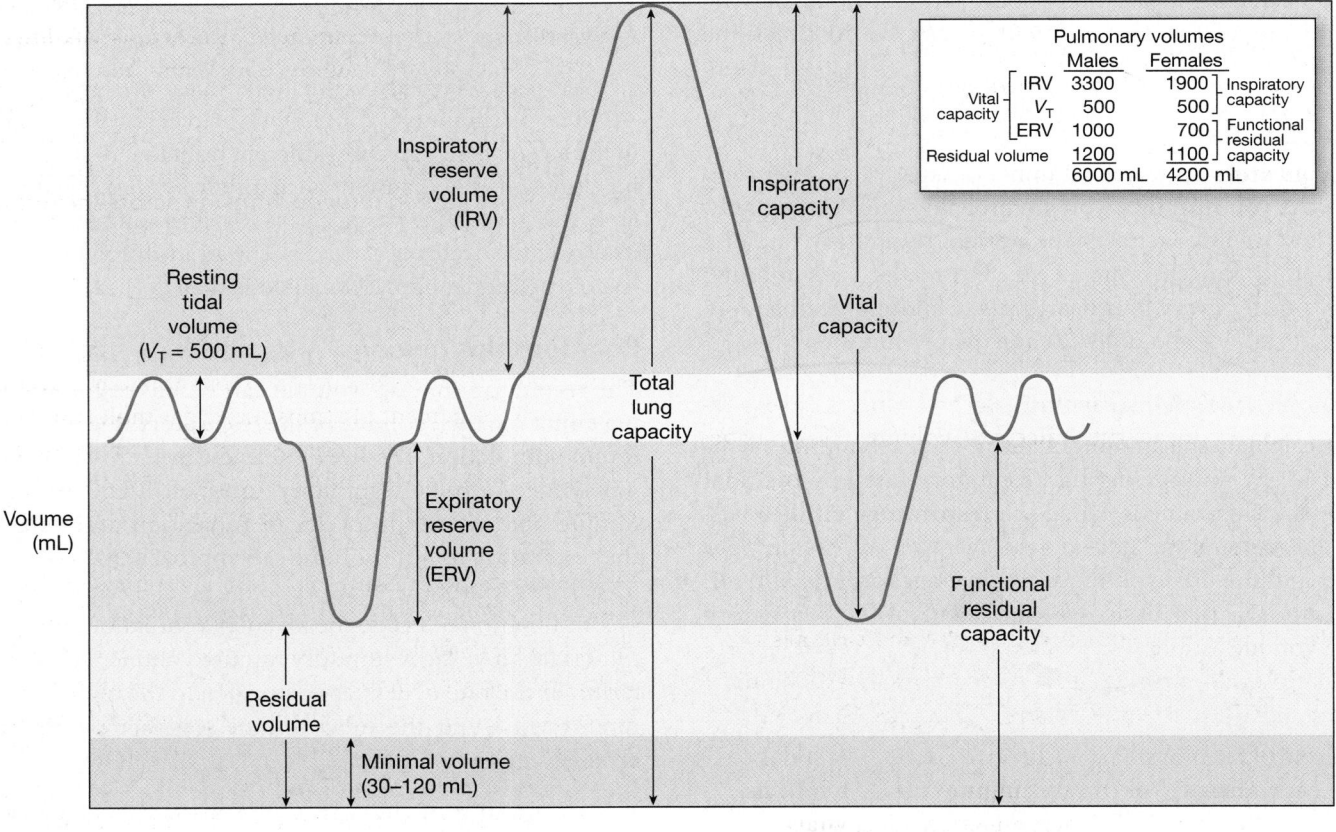

Figure 23–17   Pulmonary Volumes and Capacities. The red line indicates the volume of air within the lung as respiratory movements are performed.

be measured in a healthy person. The minimal volume and the residual volume are very different, because the fluid bond between the lungs and the chest wall prevents the recoil of the elastic fibers. Some air remains in the lungs, even at minimal volume, because the surfactant coating the alveolar surfaces prevents their collapse.

- The **inspiratory reserve volume (IRV)** is the amount of air that you can take in over and above the tidal volume. On average, the lungs of males are larger than those of females, and as a result, the inspiratory reserve volume of males averages 3300 mL, compared with 1900 mL in females.

We can calculate respiratory capacities by adding the values of various volumes. Examples include the following:

- The **inspiratory capacity** is the amount of air that you can draw into your lungs after you have completed a quiet respiratory cycle. The inspiratory capacity is the sum of the tidal volume and the inspiratory reserve volume.

- The **functional residual capacity (FRC)** is the amount of air remaining in your lungs after you have completed a quiet respiratory cycle. The FRC is the sum of the expiratory reserve volume and the residual volume.

- The **vital capacity** is the maximum amount of air that you can move into or out of your lungs in a single respiratory cycle. The vital capacity is the sum of the expiratory reserve volume, the tidal volume, and the inspiratory reserve volume and averages around 4800 mL in males and 3400 mL in females.

- The **total lung capacity** is the total volume of your lungs, calculated by adding the vital capacity and the residual volume. The total lung capacity averages around 6000 mL in males and 4200 mL in females.

Pulmonary function tests monitor several aspects of respiratory function by measuring rates and volumes of air movement.

CHECKPOINT

19. **Define compliance and identify the factors that affect it.**

20. **Name the various measurable pulmonary volumes.**

21. **Mark breaks a rib that punctures the chest wall on his left side. What do you expect will happen to his left lung as a result?**

22. **In pneumonia, fluid accumulates in the alveoli of the lungs. How would this accumulation affect vital capacity?**

See the blue Answers tab at the end of the book.

23-8 Gas exchange depends on the partial pressures of gases and the diffusion of molecules

Pulmonary ventilation both ensures that your alveoli are supplied with oxygen and that the carbon dioxide arriving from your bloodstream is removed. The actual process of gas exchange occurs between blood and alveolar air across the respiratory membrane. To understand these events, we will first consider (1) the *partial pressures* of the gases involved and (2) the diffusion of molecules between a gas and a liquid. We can then proceed to discuss the movement of oxygen and carbon dioxide across the respiratory membrane.

The Gas Laws

Gases are exchanged between the alveolar air and the blood through diffusion, which occurs in response to concentration gradients. As we saw in Chapter 3, the rate of diffusion varies in response to a variety of factors, including the size of the concentration gradient and the temperature. ∞ p. 90 The principles that govern the movement and diffusion of gas molecules, such as those in the atmosphere, are relatively straightforward. These principles, known as *gas laws*, have been understood for roughly 250 years. You have already heard about Boyle's law, which determines the direction of air movement in pulmonary ventilation. In this section, you will learn about gas laws and other factors that determine the rate of oxygen and carbon dioxide diffusion across the respiratory membrane.

Dalton's Law and Partial Pressures

The air we breathe is not a single gas but a mixture of gases. Nitrogen molecules (N_2) are the most abundant, accounting for about 78.6 percent of atmospheric gas molecules. Oxygen molecules (O_2), the second most abundant, make up roughly 20.9 percent of air. Most of the remaining 0.5 percent consists of water molecules, with carbon dioxide (CO_2) contributing a mere 0.04 percent.

Atmospheric pressure, 760 mm Hg, represents the combined effects of collisions involving each type of molecule in air. At any moment, 78.6 percent of those collisions will involve nitrogen molecules, 20.9 percent oxygen molecules, and so on. Thus, each of the gases contributes to the total pressure in proportion to its relative abundance. This relationship is known as **Dalton's law**.

The **partial pressure** of a gas is the pressure contributed by a single gas in a mixture of gases. The partial pressure is abbreviated by the symbol P or p. All the partial pressures added together equal the total pressure exerted by the gas

TABLE 23–2 Partial Pressures (mm Hg) and Normal Gas Concentrations (%) in Air

Source of Sample	Nitrogen (N₂)	Oxygen (O₂)	Carbon Dioxide (CO₂)	Water Vapor (H₂O)
Inhaled air (dry)	597 (78.6%)	159 (20.9%)	0.3 (0.04%)	3.7 (0.5%)
Alveolar air (saturated)	573 (75.4%)	100 (13.2%)	40 (5.2%)	47 (6.2%)
Exhaled air (saturated)	569 (74.8%)	116 (15.3%)	28 (3.7%)	47 (6.2%)

mixture. For the atmosphere, this relationship can be summarized as follows:

$$P_{N_2} + P_{O_2} + P_{H_2O} + P_{CO_2} = 760 \text{ mm Hg}$$

Because we know the individual percentages in air, we can easily calculate the partial pressure of each gas. For example, the partial pressure of oxygen, P_{O_2}, is 20.9 percent of 760 mm Hg, or roughly 159 mm Hg. The partial pressures for other atmospheric gases are included in Table 23–2.

Diffusion between Liquids and Gases (Henry's Law)

Differences in pressure, which move gas molecules from one place to another, also affect the movement of gas molecules into and out of solution. At a given temperature, the amount of a particular gas in solution is directly proportional to the partial pressure of that gas. This principle is known as **Henry's law**.

When a gas under pressure contacts a liquid, the pressure tends to force gas molecules into solution. At a given pressure, the number of dissolved gas molecules will rise until an equilibrium is established. At equilibrium, gas molecules diffuse out of the liquid as quickly as they enter it, so the total number of gas molecules in solution remains constant. If the partial pressure goes up, more gas molecules will go into solution; if the partial pressure goes down, gas molecules will come out of solution.

You see Henry's law in action whenever you open a can of soda. The soda was put into the can under pressure, and the gas (carbon dioxide) is in solution (**Figure 23–18a**). When you open the can, the pressure falls and the gas molecules begin coming out of solution (**Figure 23–18b**). Theoretically, the process will continue until an equilibrium develops between the surrounding air and the gas in solution. In fact, the volume of the can is so small, and the volume of the atmosphere so great, that within a half hour or so virtually all the carbon dioxide comes out of solution. You are then left with "flat" soda.

The actual *amount* of a gas in solution at a given partial pressure and temperature depends on the solubility of the gas in that particular liquid. In body fluids, carbon dioxide is highly soluble, oxygen is somewhat less soluble, and nitrogen has very limited solubility. The dissolved gas content is usually reported in milliliters of gas per 100 mL (1 dL) of solution. To see the differences in relative solubility, we can compare the gas content of blood in the pulmonary veins with the partial pressure of each gas in the alveoli. In a pulmonary vein, plasma generally contains 2.62 mL/dL of dissolved CO_2 (P_{CO_2} = 40 mm Hg),

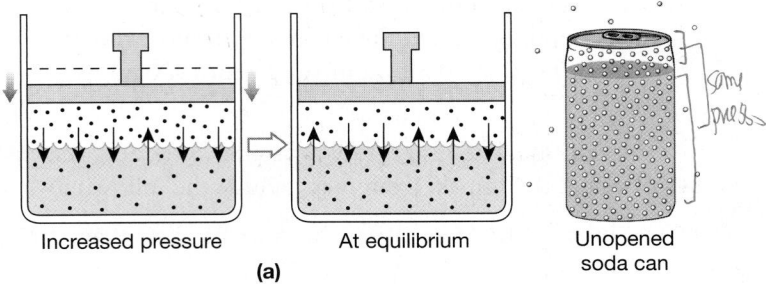

Increased pressure   At equilibrium   Unopened soda can

(a)

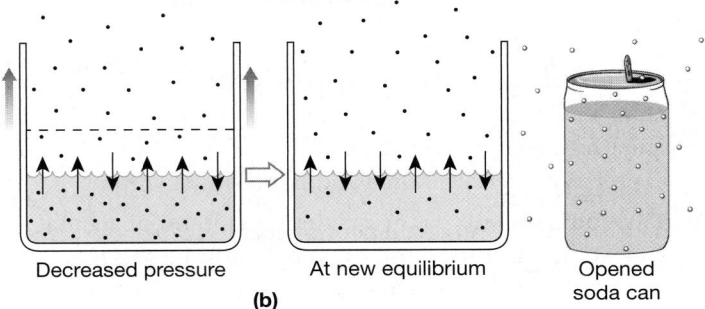

Decreased pressure   At new equilibrium   Opened soda can

(b)

Figure 23–18 **Henry's Law and the Relationship between Solubility and Pressure.** **(a)** Increasing the pressure drives gas molecules into solution until an equilibrium is established. A sealed can of carbonated soda is under higher-than-atmospheric pressure. **(b)** When the gas pressure decreases, dissolved gas molecules leave the solution until a new equilibrium is reached. Opening the soda can relieves the pressure, and bubbles form as dissolved gases leave the solution.

0.29 mL/dL of dissolved O_2 (P_{O_2} = 100 mm Hg), and 1.25 mL/dL of dissolved N_2 (P_{N_2} = 573 mm Hg.)

Diffusion and Respiratory Function

The gas laws apply to the diffusion of oxygen, carbon dioxide, and nitrogen between a gas and a liquid. We will now consider how differing partial pressures and solubilities determine the direction and rate of diffusion across the respiratory membrane that separates the air within the alveoli from the blood in alveolar capillaries.

The Composition of Alveolar Air

As soon as air enters the respiratory tract, its characteristics begin to change. In passing through the nasal cavity, inhaled

CLINICAL NOTE

Decompression Sickness

Decompression sickness is a painful condition that develops when a person is exposed to a sudden drop in atmospheric pressure. Nitrogen is the gas responsible for the problems experienced, owing to its high partial pressure in air. When the pressure drops, nitrogen comes out of solution, forming bubbles like those in a shaken can of soda. The bubbles may form

in joint cavities, in the bloodstream, and in the cerebrospinal fluid. Individuals with decompression sickness typically curl up from the pain in affected joints; this reaction accounts for the condition's common name: *the bends*. Decompression sickness most commonly affects scuba divers who return to the surface too quickly after breathing air under greater-than-normal pressure while submerged. It can also develop in airline passengers subject to sudden losses of cabin pressure.

air becomes warmer, and the amount of water vapor increases. Humidification and filtration continue as the air travels through the pharynx, trachea, and bronchial passageways. On reaching the alveoli, the incoming air mixes with air remaining in the alveoli from the previous respiratory cycle. Alveolar air thus contains more carbon dioxide and less oxygen than does atmospheric air.

The last 150 mL of inhaled air never gets farther than the conducting passageways and remains in the anatomic dead space of the lungs. During the subsequent exhalation, the departing alveolar air mixes with air in the dead space, producing yet another mixture that differs from both atmospheric and alveolar samples. The differences in composition between atmospheric (inhaled) and alveolar air are given in Table 23–2.

Efficiency of Diffusion at the Respiratory Membrane

Gas exchange at the respiratory membrane is efficient for the following five reasons:

1. *The Differences in Partial Pressure across the Respiratory Membrane Are Substantial.* This fact is important, because the greater the difference in partial pressure, the faster the rate of gas diffusion. Conversely, if P_{O_2} in alveoli decreases, the rate of oxygen diffusion into blood will drop. This is why many people feel light-headed at altitudes of 3000 m or more; the partial pressure of oxygen in their alveoli has dropped low enough that the rate of oxygen absorption is significantly reduced.

2. *The Distances Involved in Gas Exchange Are Short.* The fusion of capillary and alveolar basal laminae reduces the distance for gas exchange to an average of 0.5 μm. Inflammation of the lung tissue or a buildup of fluid in alveoli increases the diffusion distance and impairs alveolar gas exchange.

3. *The Gases Are Lipid Soluble.* Both oxygen and carbon dioxide diffuse readily through the surfactant layer and the alveolar and endothelial plasma membranes.

4. *The Total Surface Area Is Large.* The combined alveolar surface area at peak inhalation may approach 140 m² (1506 ft²). Damage to alveolar surfaces, which occurs in

emphysema, reduces the available surface area and the efficiency of gas transfer.

5. *Blood Flow and Airflow Are Coordinated.* This arrangement improves the efficiency of both pulmonary ventilation and pulmonary circulation. For example, blood flow is greatest around alveoli with the highest P_{O_2} values, where oxygen uptake can proceed with maximum efficiency. If the normal blood flow is impaired (as it is in *pulmonary embolism*), or if the normal airflow is interrupted (as it is in various forms of *pulmonary obstruction*), this coordination is lost and respiratory efficiency decreases.

Partial Pressures in Alveolar Air and Alveolar Capillaries

Figure 23–19 illustrates the partial pressures of oxygen and carbon dioxide in the pulmonary and systemic circuits. Blood arriving in the pulmonary arteries has a lower P_{O_2} and a higher P_{CO_2} than does alveolar air (**Figure 23–19a**). Diffusion between the alveolar gas mixture and the alveolar capillaries thus elevates the P_{O_2} of blood while lowering its P_{CO_2}. By the time the blood enters the pulmonary venules, it has reached equilibrium with the alveolar air. Hence, blood departs the alveoli with a P_{O_2} of about 100 mm Hg and a P_{CO_2} of roughly 40 mm Hg.

Diffusion between alveolar air and blood in the alveolar capillaries occurs very rapidly. When you are at rest, a red blood cell moves through one of your alveolar capillaries in about 0.75 second; when you exercise, the passage takes less than 0.3 second. This amount of time is usually sufficient to reach an equilibrium between the alveolar air and the blood.

Partial Pressures in the Systemic Circuit

The oxygenated blood now leaves the alveolar capillaries and returns to the heart, to be discharged into the systemic circuit. As this blood enters the pulmonary veins, it mixes with blood that flowed through capillaries around conducting passageways. Because gas exchange occurs only at alveoli, the blood leaving the conducting passageways carries relatively little oxygen. The partial pressure of oxygen in the pulmonary veins therefore drops to about 95 mm Hg. This is the P_{O_2} in the blood that enters the systemic circuit, and no further

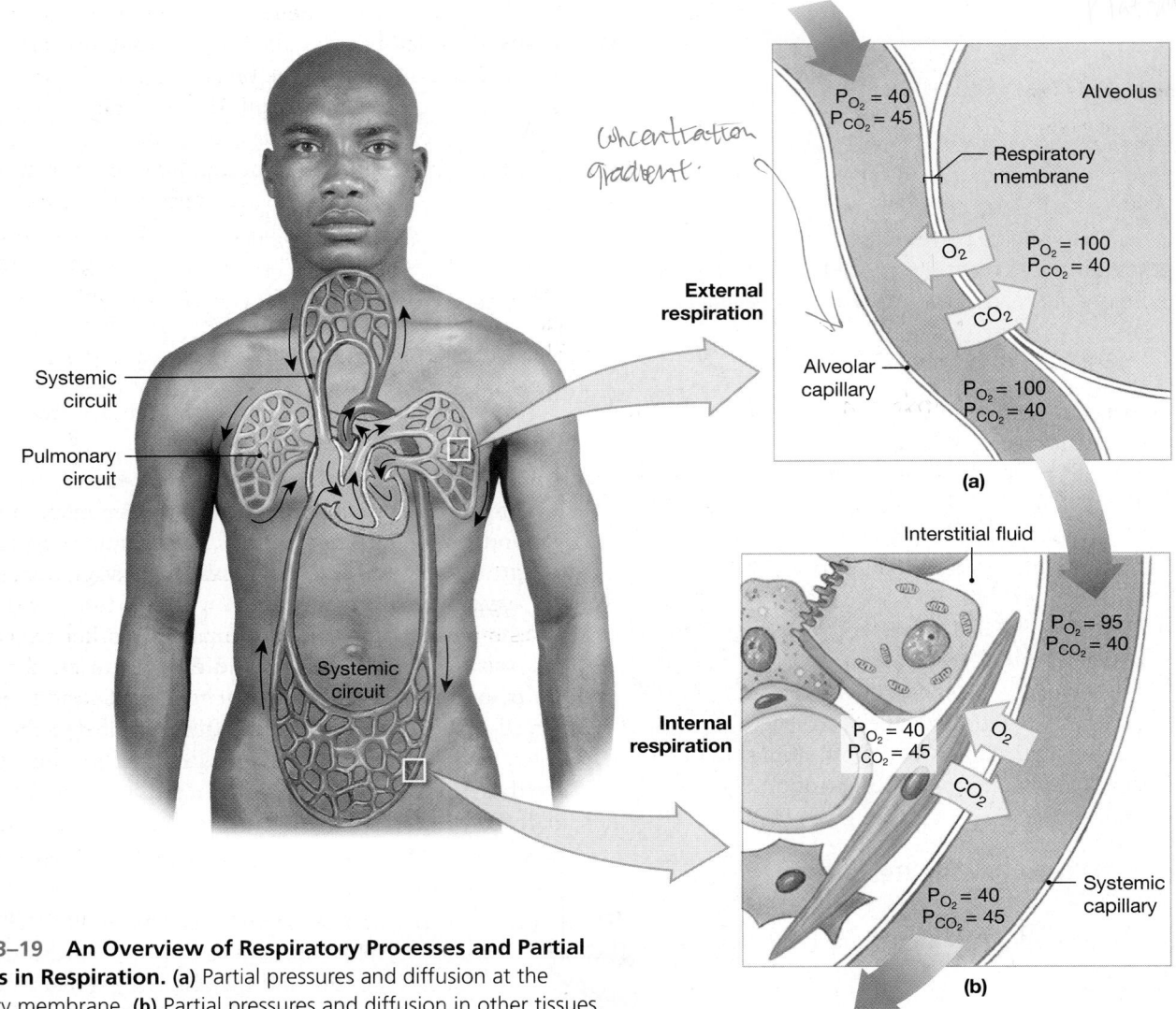

Figure 23–19   **An Overview of Respiratory Processes and Partial Pressures in Respiration. (a)** Partial pressures and diffusion at the respiratory membrane. **(b)** Partial pressures and diffusion in other tissues.

changes in partial pressure occur until the blood reaches the peripheral capillaries (**Figure 23–19b**).

Normal interstitial fluid has a P_{O_2} of 40 mm Hg. As a result, oxygen diffuses out of the capillaries and carbon dioxide diffuses in, until the capillary partial pressures are the same as those in the adjacent tissues. Inactive peripheral tissues normally have a P_{CO_2} of about 45 mm Hg, whereas blood entering peripheral capillaries normally has a P_{CO_2} of 40 mm Hg. As a result, carbon dioxide diffuses into the blood as oxygen diffuses out (**Figure 23–19b**).

CLINICAL NOTE

Blood Gas Analysis

Blood samples can be analyzed to determine their concentrations of dissolved gases. The usual tests include the determination of pH, P_{CO_2}, and P_{O_2} in an arterial sample. Such samples provide information about the degree of oxygenation in peripheral tissues. For example, if the arterial P_{CO_2} is very high and the P_{O_2} is very low, tissues are not receiving adequate oxygen. This problem may be solved by providing a gas mixture that has a high P_{O_2} (or even pure oxygen, with a P_{O_2} of 760 mm Hg).

Blood gas measurements also provide information on the efficiency of gas exchange at the lungs. If the arterial P_{O_2} remains low despite the administration of oxygen, or if the P_{CO_2} is very high, pulmonary exchange problems, such as pulmonary edema, asthma, or pneumonia, may exist.

CHECKPOINT

23. Define Dalton's law.

24. Define Henry's law.

See the blue Answers tab at the end of the book.

23-9 Most oxygen is transported bound to hemoglobin; and carbon dioxide is transported in three ways: as carbonic acid, bound to hemoglobin, or dissolved in plasma

Oxygen and carbon dioxide have limited solubilities in blood plasma. For example, at the normal P_{O_2} of alveoli, 100 mL of plasma will absorb only about 0.3 mL of oxygen. The limited solubilities of these gases are a problem, because peripheral tissues need more oxygen and generate more carbon dioxide than the plasma can absorb and transport.

The problem is solved by red blood cells (RBCs), which remove dissolved oxygen and CO_2 molecules from plasma and bind them (in the case of oxygen) or use them to manufacture soluble compounds (in the case of carbon dioxide). Because these reactions remove dissolved gases from blood plasma, gases continue to diffuse into the blood, but never reach equilibrium.

The important thing about these reactions is that they are both temporary and completely reversible. When plasma oxygen or carbon dioxide concentrations are high, the excess molecules are removed by RBCs. When plasma concentrations are falling, the RBCs release their stored reserves.

Oxygen Transport

Each 100 mL of blood leaving the alveolar capillaries carries away roughly 20 mL of oxygen. Of this amount, only about 0.3 mL (1.5 percent) consists of oxygen molecules in solution. The rest of the oxygen molecules are bound to *hemoglobin (Hb) molecules*—specifically, to the iron ions in the center of heme units. ∞ p. 656 Recall that the hemoglobin molecule consists of four globular protein subunits, each containing a heme unit. Thus, each hemoglobin molecule can bind four molecules of oxygen, forming oxyhemoglobin (HbO_2). This is a reversible reaction that can be summarized as

$$Hb + O_2 \rightleftharpoons HbO_2.$$

Each red blood cell has approximately 280 million molecules of hemoglobin. Because a hemoglobin molecule contains four heme units, each RBC potentially can carry more than a billion molecules of oxygen.

The percentage of heme units containing bound oxygen at any given moment is called the **hemoglobin saturation**. If all the Hb molecules in the blood are fully loaded with oxygen, saturation is 100 percent. If, on average, each Hb molecule carries two O_2 molecules, saturation is 50 percent.

In Chapter 2, we saw that the shape and functional properties of a protein change in response to changes in its environment. ∞ p. 54 Hemoglobin is no exception: Any changes in shape that occur can affect oxygen binding. Under normal conditions, the most important environmental factors affecting hemoglobin are (1) the P_{O_2} of blood, (2) blood pH, (3) temperature, and (4) ongoing metabolic activity within RBCs.

Hemoglobin and P_{O_2}

An **oxygen–hemoglobin saturation curve**, or *oxygen–hemoglobin dissociation curve*, is a graph that relates the saturation of hemoglobin to the partial pressure of oxygen (**Figure 23–20**). The binding and dissociation of oxygen to hemoglobin is a typical reversible reaction. At equilibrium, oxygen molecules bind to heme at the same rate that other oxygen molecules are being released. If the P_{O_2} increases, the reaction shifts to the right, and more oxygen gets bound to hemoglobin. If the P_{O_2} decreases, the reaction shifts to the left, and more oxygen is released by hemoglobin. The graph of this relationship is a curve rather than a straight line, because the shape of the Hb molecule changes slightly each time it binds an oxygen molecule, in a way that enhances its ability to bind *another* oxygen molecule. In other words, the attachment of the first oxygen molecule makes it easier to bind the second; binding the second promotes binding of the third; and binding of the third enhances binding of the fourth.

Because each arriving oxygen molecule increases the affinity of hemoglobin for the *next* oxygen molecule, the saturation curve takes the form shown in **Figure 23–20**. Once the first oxygen molecule binds to the hemoglobin, the slope rises rapidly until reaching a plateau near 100 percent saturation. While the slope is steep, a very small change in plasma P_{O_2} will result in a large change in the amount of oxygen bound to Hb or released from HbO_2. Because the curve rises quickly, hemoglobin will be more than 90 percent saturated if exposed to an alveolar P_{O_2} above 60 mm Hg. Thus, near-normal oxygen transport can continue despite a decrease in the oxygen content of alveolar air. Without this ability, you could not survive at high altitudes, and conditions significantly reducing pulmonary ventilation would be immediately fatal.

At normal alveolar pressures ($P_{O_2} = 100$ mm Hg) the hemoglobin saturation is very high (97.5 percent), although complete saturation does not occur until the P_{O_2} reaches excessively high levels (about 250 mm Hg). In functional terms, the maximum saturation is not as important as the ability of hemoglobin to provide oxygen over the normal P_{O_2} range in

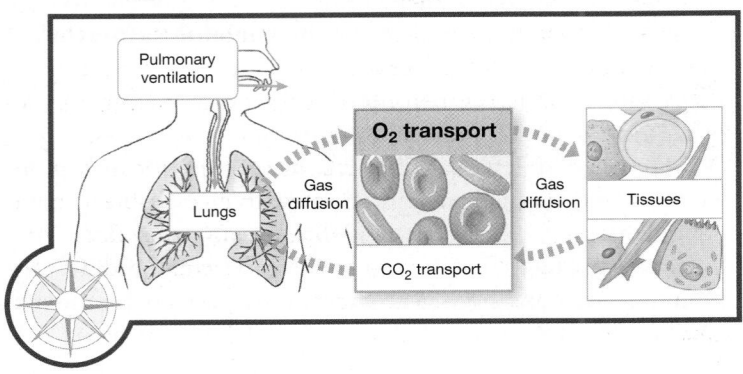

Figure 23–20 **An Oxygen–Hemoglobin Saturation Curve.** The saturation characteristics of hemoglobin at various partial pressures of oxygen under normal conditions (body temperature of 37°C and blood pH of 7.4).

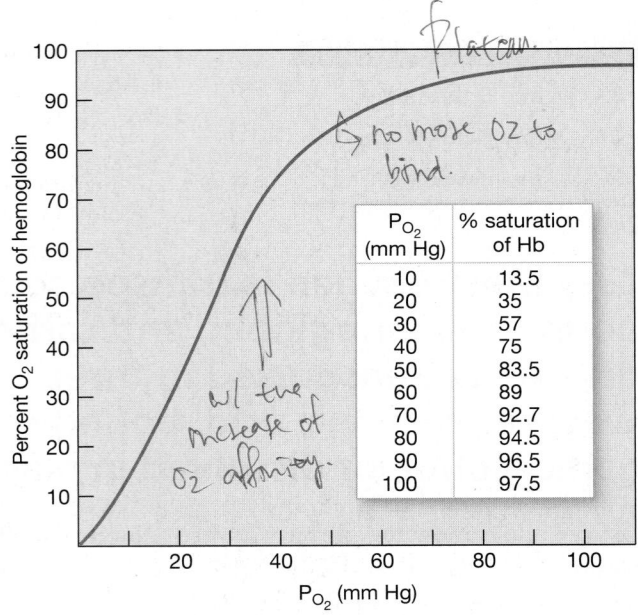

P_{O_2} (mm Hg)	% saturation of Hb
10	13.5
20	35
30	57
40	75
50	83.5
60	89
70	92.7
80	94.5
90	96.5
100	97.5

body tissues. Over that range—from 100 mm Hg at the alveoli to perhaps 15 mm Hg in active tissues—the saturation drops from 97.5 percent to less than 20 percent, and a small change in P_{O_2} makes a big difference in terms of the amount of oxygen bound to hemoglobin.

Note that the relationship between P_{O_2} and hemoglobin saturation remains valid whether the P_{O_2} is rising or falling. If the P_{O_2} increases, the saturation goes up and hemoglobin stores oxygen. If the P_{O_2} decreases, hemoglobin releases oxygen into its surroundings. When oxygenated blood arrives in the peripheral capillaries, the blood P_{O_2} declines rapidly as a result of gas exchange with the interstitial fluid. As the P_{O_2} falls, hemoglobin gives up its oxygen.

The relationship between the P_{O_2} and hemoglobin saturation provides a mechanism for automatic regulation of oxygen delivery. Inactive tissues have little demand for oxygen, and the local P_{O_2} is usually about 40 mm Hg. Under these conditions, hemoglobin will not release much oxygen. As it passes through the capillaries, it will go from 97 percent saturation (P_{O_2} = 95 mm Hg) to 75 percent saturation (P_{O_2} = 40 mm Hg). Because it still retains three-quarters of its oxygen, venous blood has a relatively large oxygen reserve. This reserve is important, because it can be mobilized if tissue oxygen demands increase.

Active tissues consume oxygen at an accelerated rate, so the P_{O_2} may drop to 15–20 mm Hg. Hemoglobin passing through these capillaries will then go from 97 percent saturation to about 20 percent saturation. In practical terms, this means that as blood circulates through peripheral capillaries, active tissues will receive 3.5 times as much oxygen as will inactive tissues.

CLINICAL NOTE

Carbon Monoxide Poisoning The exhaust of automobiles and other petroleum-burning engines, of oil lamps, and of fuel-fired space heaters contains *carbon monoxide* (CO), and each winter entire families die from **carbon monoxide poisoning**. Carbon monoxide competes with oxygen molecules for the binding sites on heme units. Unfortunately, the carbon monoxide usually wins the competition, because at very low partial pressures it has a much stronger affinity for hemoglobin than does oxygen. The bond formed between CO and heme is extremely durable, so the attachment of a CO molecule essentially makes that heme unit inactive for respiratory purposes.

If CO molecules make up just 0.1 percent of the components of inhaled air, enough hemoglobin will be affected that survival will become impossible without medical assistance. Treatment includes: (1) preventing further CO exposure, (2) the administration of pure oxygen, because at sufficiently high partial pressures, the oxygen molecules will gradually replace CO at the hemoglobin molecules; and, if necessary, (3) the transfusion of compatible red blood cells.

Hemoglobin and pH

The oxygen–hemoglobin saturation curve in **Figure 23–20** was determined in normal blood, with a pH of 7.4 and a temperature of 37°C. In addition to consuming oxygen, active tissues generate acids that lower the pH of the interstitial fluid. When the pH drops, the shape of hemoglobin molecules changes. As a result of this change, the molecules release their oxygen reserves more readily, so the slope of the hemoglobin saturation curve changes (**Figure 23–21a**). In other words, as pH drops, the saturation declines. At a tissue P_{O_2} of 40 mm Hg, for example, a pH drop from 7.4 to 7.2 changes hemoglobin saturation from 75 percent to 60 percent. This means that hemoglobin molecules will release 20 percent more oxygen in peripheral tissues at a pH of 7.2 than they will at a pH of 7.4. This effect of pH on the hemoglobin saturation curve is called the **Bohr effect**.

Carbon dioxide is the primary compound responsible for the Bohr effect. When CO_2 diffuses into the blood, it rapidly diffuses into red blood cells. There, an enzyme called **carbonic anhydrase** catalyzes the reaction of CO_2 with water molecules:

$$CO_2 + H_2O \overset{\text{carbonic anhydrase}}{\rightleftharpoons} H_2CO_3 \rightleftharpoons H^+ + HCO_3^-$$

The product of this enzymatic reaction, H_2CO_3, is called *carbonic acid*, because it dissociates into a hydrogen ion (H^+) and a bicarbonate ion (HCO_3^-). The rate of carbonic acid formation depends on the amount of carbon dioxide in solution, which, as noted earlier, depends on the P_{CO_2}. When the P_{CO_2} rises, the reaction proceeds from left to right, and the rate of carbonic acid formation accelerates. The hydrogen ions that are generated diffuse out of the RBCs, and the pH of the plasma drops. When the P_{CO_2} declines, the reaction proceeds from right to left; hydrogen ions then diffuse into the RBCs, so the pH of the plasma rises.

Hemoglobin and Temperature

Changes in temperature also affect the slope of the hemoglobin saturation curve (**Figure 23–21b**). As the temperature rises, hemoglobin releases more oxygen; as the temperature declines, hemoglobin holds oxygen more tightly. Temperature effects are significant only in active tissues in which large amounts of heat are being generated. For example, active skeletal muscles generate heat, and the heat warms blood that flows through these organs. As the blood warms, the Hb molecules release more oxygen than can be used by the active muscle fibers.

Hemoglobin and BPG

Red blood cells, which lack mitochondria, produce adenosine triphosphate (ATP) only by glycolysis, in which, as we saw in Chapter 10, lactic acid is formed. ∞ p. 320 The metabolic pathways involved in glycolysis in an RBC also generate the compound **2,3-bisphosphoglycerate** (biz-fos-fō-GLIS-er-āt), or **BPG**. Normal RBCs always contain BPG, which has a direct effect on oxygen binding and release. For any partial pressure of oxygen, the higher the concentration of BPG, the greater the release of oxygen by Hb molecules.

The concentration of BPG can be increased by thyroid hormones, growth hormone, epinephrine, androgens, and high blood pH. These factors improve oxygen delivery to the tissues, because when BPG levels are elevated, hemoglobin releases

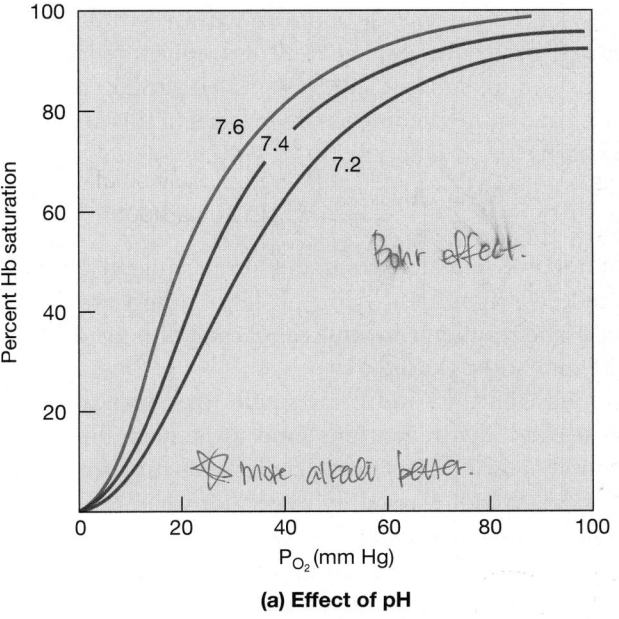

(a) Effect of pH

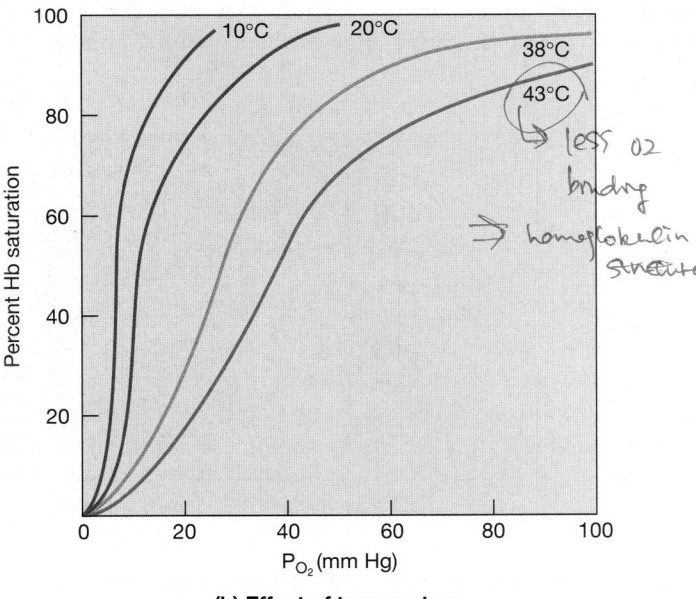

(b) Effect of temperature

Figure 23–21   The Effects of pH and Temperature on Hemoglobin Saturation. (a) When the pH drops below normal levels, more oxygen is released; the hemoglobin saturation curve shifts to the right. When the pH increases, less oxygen is released; the curve shifts to the left. **(b)** When the temperature rises, the saturation curve shifts to the right.

about 10 percent more oxygen at a given P_{O_2} than it would otherwise. Both BPG synthesis and the Bohr effect improve oxygen delivery when the pH changes: BPG levels rise when the pH increases, and the Bohr effect appears when the pH decreases.

The production of BPG decreases as RBCs age. Thus, the level of BPG can determine how long a blood bank can store fresh whole blood. When BPG levels get too low, hemoglobin becomes firmly bound to the available oxygen. The blood is then useless for transfusions, because the RBCs will no longer release oxygen to peripheral tissues, even at a disastrously low P_{O_2}.

Fetal Hemoglobin

→ higher affinity of O_2.

The RBCs of a developing fetus contain **fetal hemoglobin**. The structure of fetal hemoglobin differs from that of adult hemoglobin, giving it a much higher affinity for oxygen. At the same P_{O_2}, fetal hemoglobin binds more oxygen than does adult hemoglobin (**Figure 23–22**). This trait is important in transferring oxygen across the placenta.

A fetus obtains oxygen from the maternal bloodstream. At the placenta, maternal blood has a relatively low P_{O_2}, ranging from 35 to 50 mm Hg. If maternal blood arrives at the placenta with a P_{O_2} of 40 mm Hg, hemoglobin saturation will be roughly 75 percent. The fetal blood arriving at the placenta has a P_{O_2} close to 20 mm Hg. However, because fetal hemoglobin has a higher affinity for oxygen, it is still 58 percent saturated.

As diffusion occurs between fetal blood and maternal blood, oxygen enters the fetal bloodstream until the P_{O_2} equilibrates at 30 mm Hg. At this P_{O_2}, the maternal hemoglobin is less than 60 percent saturated, but the fetal hemoglobin is

over 80 percent saturated. The steep slope of the saturation curve for fetal hemoglobin means that when fetal RBCs reach peripheral tissues, the Hb molecules will release a large amount of oxygen in response to a very small change in P_{O_2}.

The A&P Top 100

#79 Hemoglobin within RBCs carries most of the oxygen in the bloodstream and releases it in response to changes in the oxygen partial pressure in the surrounding plasma. If the P_{O_2} increases, hemoglobin binds oxygen; if the P_{O_2} decreases, hemoglobin releases oxygen. At a given P_{O_2} hemoglobin will release additional oxygen if the pH decreases or the temperature increases.

Carbon Dioxide Transport

Carbon dioxide is generated by aerobic metabolism in peripheral tissues. After entering the bloodstream, a CO_2 molecule is either (1) converted to a molecule of carbonic acid, (2) bound to the protein portion of hemoglobin molecules within red blood cells, or (3) dissolved in plasma. All three reactions are completely reversible. We will consider the events that occur as blood enters peripheral tissues in which the P_{CO_2} is 45 mm Hg.

Carbonic Acid Formation

Most of the carbon dioxide absorbed by blood (roughly 70 percent of the total) is transported as molecules of carbonic acid. Carbon dioxide is converted to carbonic acid through the activity of the enzyme carbonic anhydrase in RBCs. The carbonic acid molecules immediately dissociate into a hydrogen ion and a bicarbonate ion, as described earlier (p. 857). Hence, we can ignore the intermediate steps in this sequence and summarize the reaction as

$$CO_2 + H_2O \overset{\text{carbonic anhydrase}}{\rightleftharpoons} H^+ + HCO_3^-$$

This reaction is completely reversible. In peripheral capillaries, it proceeds vigorously, tying up large numbers of CO_2 molecules. The reaction continues as carbon dioxide diffuses out of the interstitial fluids.

The hydrogen ions and bicarbonate ions have different fates. Most of the hydrogen ions bind to hemoglobin molecules, forming HbH^+. The Hb molecules thus function as pH buffers, tying up the released hydrogen ions before the ions can leave the RBCs and lower the plasma pH. The bicarbonate ions move into the surrounding plasma with the aid of a countertransport mechanism that exchanges intracellular bi-

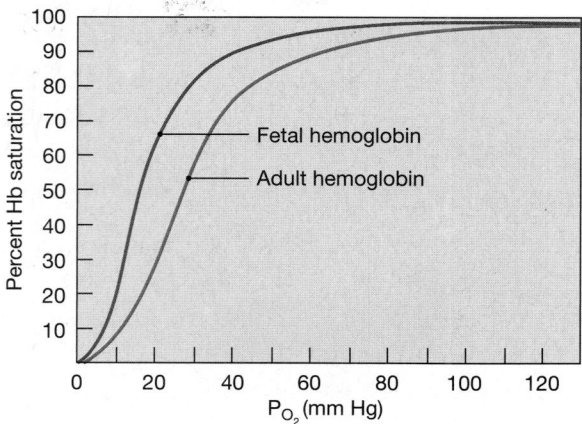

Figure 23–22 **A Functional Comparison of Fetal and Adult Hemoglobin.**

carbonate ions (HCO_3^-) for extracellular chloride ions (Cl^-). This exchange, which trades one anion for another, does not require ATP. The result is a mass movement of chloride ions into the RBCs, an event known as the **chloride shift**.

Binding to Hemoglobin

Roughly 23 percent of the carbon dioxide carried by blood is bound to the globular protein portions of Hb molecules inside RBCs. These CO_2 molecules are attached to exposed amino groups ($—NH_2$) of the Hb molecules. The resulting compound is called **carbaminohemoglobin** (kar-BAM-i-nō-hē-mō-glō-bin), $HbCO_2$. The reversible reaction is summarized as follows:

$$CO_2 + HbNH_2 \rightleftharpoons HbNHCOOH$$

This reaction can be abbreviated without the amino groups as

$$CO_2 + Hb \rightleftharpoons HbCO_2.$$

Transport in Plasma

Plasma becomes saturated with carbon dioxide quite rapidly, and only about 7 percent of the carbon dioxide absorbed by peripheral capillaries is transported as dissolved gas molecules. The rest is absorbed by the RBCs for conversion by carbonic anhydrase or storage as carbaminohemoglobin.

Mechanisms of carbon dioxide transport are summarized in **Figure 23–23**.

The A&P Top 100

#80 Carbon dioxide travels in the bloodstream primarily as bicarbonate ions, which form through dissociation of carbonic acid produced by carbonic anhydrase inside RBCs. Lesser amounts of CO_2 are bound to hemoglobin or dissolved in plasma.

Summary: Gas Transport

Figure 23–24 summarizes the transportation of oxygen and carbon dioxide in the respiratory and cardiovascular systems. The process is dynamic, capable of varying its responses to meet changing circumstances. Some of the responses are automatic and result from the basic chemistry of the transport mechanisms. Other responses require coordinated adjustments in the activities of the cardiovascular and respiratory systems. We will consider those levels of control and regulation next.

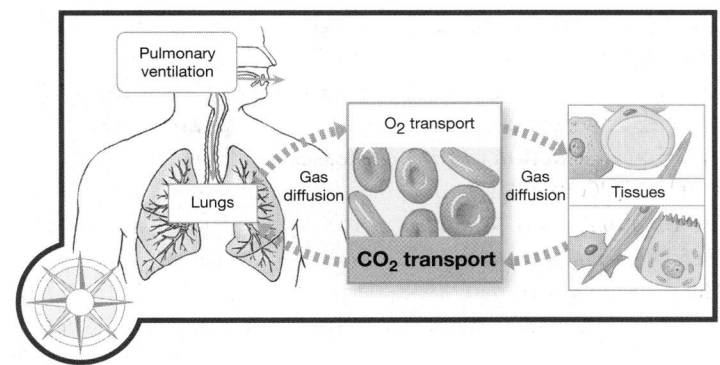

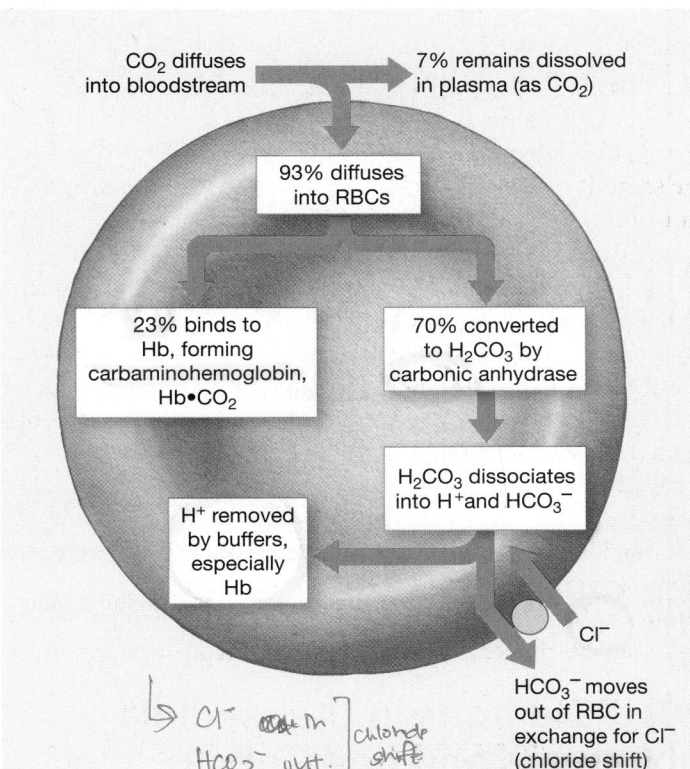

Figure 23–23   Carbon Dioxide Transport in Blood.

CHECKPOINT

25. Identify the three ways that carbon dioxide is transported in the bloodstream.

26. As you exercise, hemoglobin releases more oxygen to active skeletal muscles than it does when those muscles are at rest. Why?

27. How would blockage of the trachea affect blood pH?

See the blue Answers tab at the end of the book.

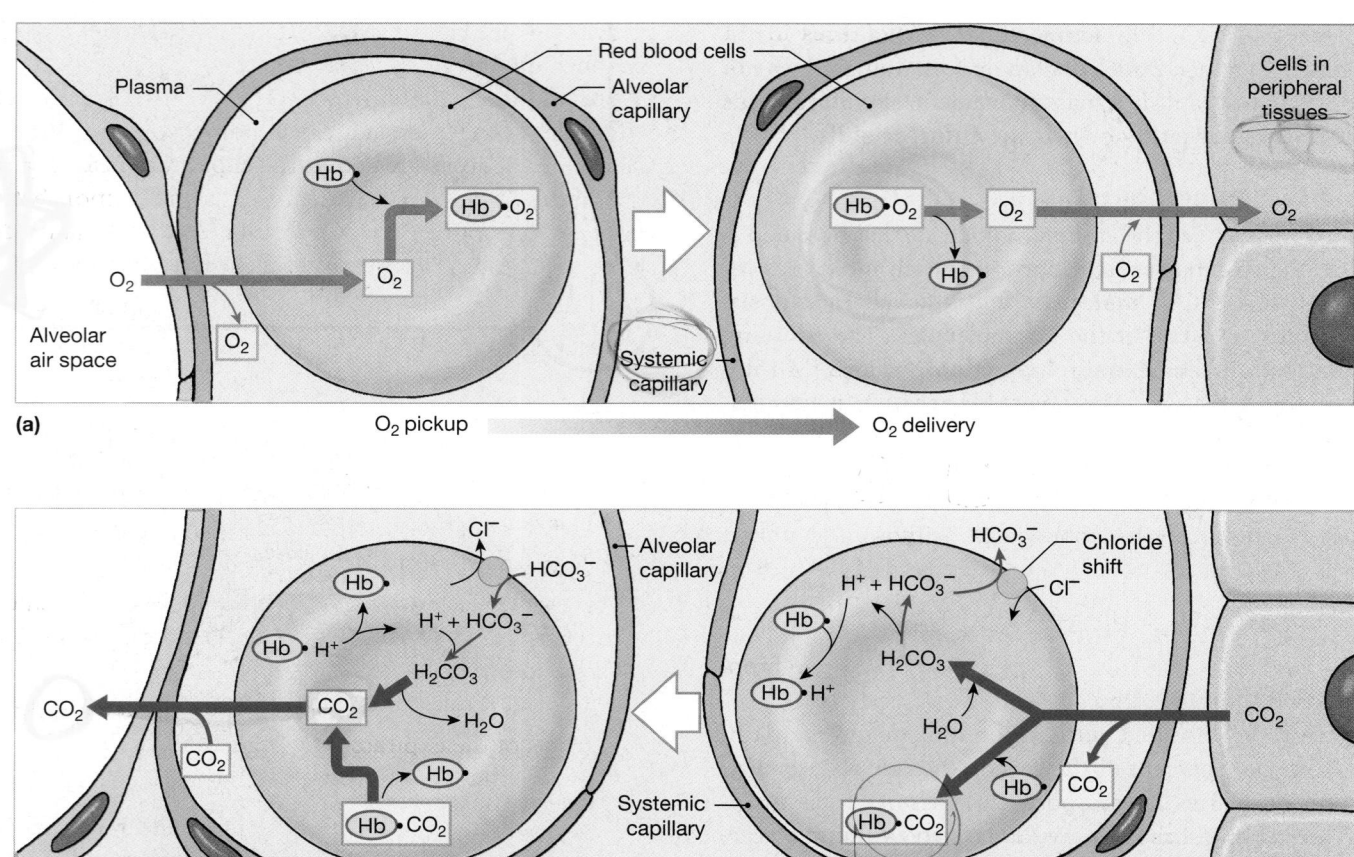

Figure 23–24 A Summary of the Primary Gas Transport Mechanisms. (a) Oxygen transport. **(b)** Carbon dioxide transport.

23-10 Neurons in the medulla oblongata and pons, along with respiratory reflexes, control respiration

Peripheral cells continuously absorb oxygen from interstitial fluids and generate carbon dioxide. Under normal conditions, the cellular rates of absorption and generation are matched by the capillary rates of delivery and removal. Both rates are identical to those of oxygen absorption and carbon dioxide excretion at the lungs. If diffusion rates at the peripheral and alveolar capillaries become unbalanced, homeostatic mechanisms intervene to restore equilibrium. Such mechanisms involve (1) changes in blood flow and oxygen delivery that are regulated at the local level and (2) changes in the depth and rate of respiration under the control of the brain's respiratory centers. The activities of the respiratory centers are coordinated with changes in cardiovascular function, such as fluctuations in blood pressure and cardiac output.

Local Regulation of Gas Transport and Alveolar Function

The rate of oxygen delivery in each tissue and the efficiency of oxygen pickup at the lungs are largely regulated at the local level. For example, when a peripheral tissue becomes more active, the interstitial P_{O_2} falls and the P_{CO_2} rises. This change increases the difference between the partial pressures in the tissues and in the arriving blood, so more oxygen is delivered and more carbon dioxide is carried away. In addition, the rising P_{CO_2} levels cause the relaxation of smooth muscles in the walls of arterioles and capillaries in the area, increasing local blood flow.

Local factors also coordinate (1) *lung perfusion*, or blood flow to the alveoli, with (2) *alveolar ventilation*, or airflow, over a wide range of conditions and activity levels. As blood flows toward the alveolar capillaries, it is directed toward lobules in which the P_{O_2} is relatively high. This movement occurs because alveolar capillaries constrict when the local P_{O_2} is low. (This response, the opposite of that seen in peripheral tissues, was noted in Chapter 21. ∞ p. 747) Such a shift in blood flow tends to eliminate temporary differences in the oxygen and carbon dioxide contents of alveoli, lobules, or groups of lobules that could otherwise result from minor variations in local blood flow.

Smooth muscles in the walls of bronchioles are sensitive to the P_{CO_2} of the air they contain. When the P_{CO_2} goes up, the bronchioles increase in diameter (bronchodilation). When the P_{CO_2} declines, the bronchioles constrict (bronchoconstriction). Airflow is therefore directed to lobules in which the P_{CO_2} is high. Because these lobules obtain carbon dioxide from blood, they are actively engaged in gas exchange. This response is especially important, because the improvement of airflow to functional alveoli can at least partially compensate for damage to pulmonary lobules.

By directing blood flow to alveoli with low CO_2 levels and increasing airflow to alveoli with high CO_2 levels, local adjustments improve the efficiency of gas transport. When activity levels increase and the demand for oxygen rises, the cardiac output and respiratory rates increase under neural control, but the adjustments in alveolar blood flow and bronchiole diameter occur automatically.

The Respiratory Centers of the Brain

Respiratory control has both involuntary and voluntary components. Your brain's involuntary centers regulate the activities of the respiratory muscles and control the respiratory minute volume by adjusting the frequency and depth of pulmonary ventilation. They do so in response to sensory information arriving from your lungs and other portions of the respiratory tract, as well as from a variety of other sites.

The voluntary control of respiration reflects activity in the cerebral cortex that affects either the output of the respiratory centers in the medulla oblongata and pons or of motor neurons in the spinal cord that control respiratory muscles. The **respiratory centers** are three pairs of nuclei in the reticular formation of the medulla oblongata and pons. The motor neurons in the spinal cord are generally controlled by *respiratory reflexes*, but they can also be controlled voluntarily through commands delivered by the corticospinal pathway. ∞ p. 520

Respiratory Centers in the Medulla Oblongata

The *respiratory rhythmicity centers* of the medulla oblongata were introduced in Chapter 14. ∞ p. 471 These paired centers set the pace of respiration. Each center can be subdivided into a **dorsal respiratory group (DRG)** and a **ventral respiratory group (VRG)**. The DRG's *inspiratory center* contains neurons that control lower motor neurons innervating the external intercostal muscles and the diaphragm. The DRG functions in every respiratory cycle, whether quiet or forced. The VRG functions only during forced breathing. It includes neurons that innervate lower motor neurons controlling accessory respiratory

muscles involved in active exhalation (an *expiratory center*) and maximal inhalation (an *inspiratory center*).

Reciprocal inhibition occurs between the neurons involved with inhalation and exhalation. ∞ p. 453 When the inspiratory neurons are active, the expiratory neurons are inhibited, and vice versa. The pattern of interaction between these groups differs between quiet breathing and forced breathing. During quiet breathing (**Figure 23–25a**):

- Activity in the DRG increases over a period of about 2 seconds, providing stimulation to the inspiratory muscles. Over this period, inhalation occurs.

- After 2 seconds, the DRG neurons become inactive. They remain quiet for the next 3 seconds and allow the inspiratory muscles to relax. Over this period, passive exhalation occurs.

During forced breathing (**Figure 23–25b**):

- Increases in the level of activity in the DRG stimulate neurons of the VRG that activate the accessory muscles involved in inhalation.

- After each inhalation, active exhalation occurs as the neurons of the expiratory center stimulate the appropriate accessory muscles.

The basic pattern of respiration thus reflects a cyclic interaction between the DRG and the VRG. The pace of this interaction is thought to be established by pacemaker cells that spontaneously undergo rhythmic patterns of activity. Attempts to locate the pacemaker, however, have been unsuccessful.

Central nervous system stimulants, such as amphetamines or even caffeine, increase the respiratory rate by facilitating the respiratory centers. These actions are opposed by CNS depressants, such as barbiturates or opiates.

The Apneustic and Pneumotaxic Centers of the Pons

The **apneustic** (ap-NOO-stik) **centers** and the **pneumotaxic** (noo-mō-TAKS-ik) **centers** of the pons are paired nuclei that adjust the output of the respiratory rhythmicity centers. Their activities regulate the respiratory rate and the depth of respiration in response to sensory stimuli or input from other centers in the brain. Each apneustic center provides continuous stimulation to the DRG on that side of the brain stem. During quiet breathing, stimulation from the apneustic center helps increase the intensity of inhalation over the next 2 seconds. Under normal conditions, after 2 seconds the apneustic center is inhibited by signals from the pneumotaxic center on that side. During forced breathing, the apneustic centers also respond to sensory input from the vagus nerves regarding the amount of lung inflation.

The pneumotaxic centers inhibit the apneustic centers and promote passive or active exhalation. Centers in the hypothalamus and cerebrum can alter the activity of the pneumotaxic

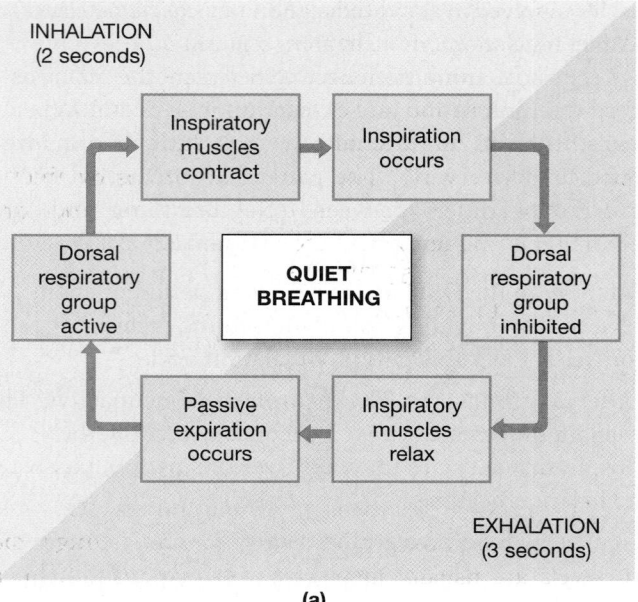

INHALATION
(2 seconds)

Inspiratory muscles contract → Inspiration occurs

Dorsal respiratory group active

QUIET BREATHING

Dorsal respiratory group inhibited

Passive expiration occurs ← Inspiratory muscles relax

EXHALATION
(3 seconds)

(a)

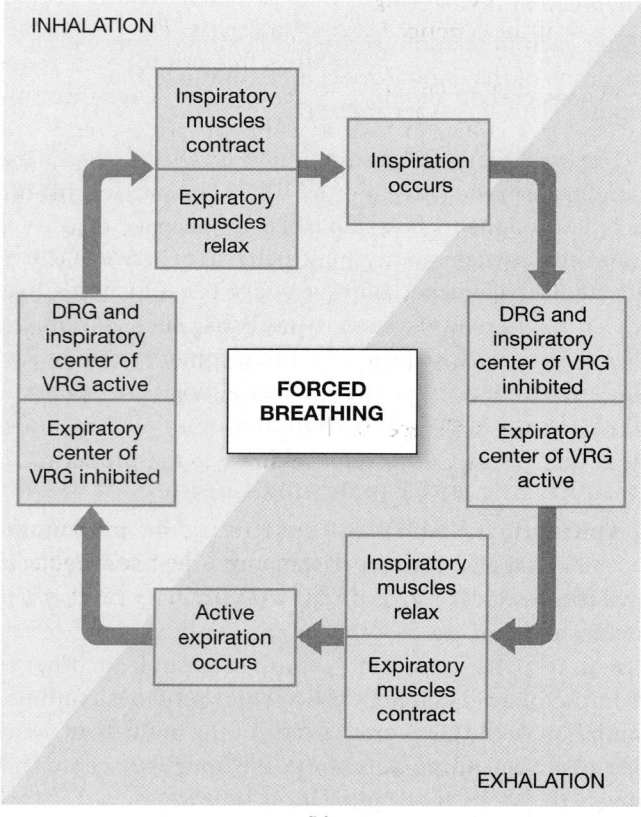

INHALATION

Inspiratory muscles contract / Expiratory muscles relax → Inspiration occurs

DRG and inspiratory center of VRG active

Expiratory center of VRG inhibited

FORCED BREATHING

DRG and inspiratory center of VRG inhibited

Expiratory center of VRG active

Active expiration occurs ← Inspiratory muscles relax / Expiratory muscles contract

EXHALATION

(b)

Figure 23–25 Basic Regulatory Patterns of Respiration.
(a) Quiet breathing. **(b)** Forced breathing.

centers, as well as the respiratory rate and depth. However, essentially normal respiratory cycles continue even if the brain stem superior to the pons has been severely damaged. If the inhibitory output of the pneumotaxic centers is cut off by a stroke or other damage to the brain stem, and if sensory inner-

vation from the lungs is eliminated due to damage to the vagus nerves, the person inhales to maximum capacity and maintains that state for 10–20 seconds at a time. Intervening exhalations are brief, and little pulmonary ventilation occurs.

The CNS regions involved with respiratory control are diagrammed in **Figure 23–26**. Interactions between the DRG and the VRG establish the basic pace and depth of respiration. The pneumotaxic centers modify that pace: An increase in pneumotaxic output quickens the pace of respiration by shortening the duration of each inhalation; a decrease in pneumotaxic output slows the respiratory pace, but increases the depth of respiration, because the apneustic centers are more active.

Sudden infant death syndrome (SIDS), also known as *crib death*, kills an estimated 10,000 infants each year in the United States alone. Most crib deaths occur between midnight and 9:00 A.M., in the late fall or winter, and involve infants 2 to 4 months old. Eyewitness accounts indicate that the sleeping infant suddenly stops breathing, turns blue, and relaxes. Genetic factors appear to be involved, but controversy remains as to the relative importance of other factors. The age at the time of death corresponds with a period when the pacemaker complex and respiratory centers are establishing connections with other portions of the brain. It has been suggested that SIDS results from a problem in the interconnection process that disrupts the reflexive respiratory pattern.

Respiratory Reflexes

The activities of the respiratory centers are modified by sensory information from several sources:

- Chemoreceptors sensitive to the P_{CO_2}, pH, or P_{O_2} of the blood or cerebrospinal fluid.

- Baroreceptors in the aortic or carotid sinuses sensitive to changes in blood pressure.

- Stretch receptors that respond to changes in the volume of the lungs.

- Irritating physical or chemical stimuli in the nasal cavity, larynx, or bronchial tree.

- Other sensations, including pain, changes in body temperature, and abnormal visceral sensations.

Information from these receptors alters the pattern of respiration. The induced changes have been called *respiratory reflexes*.

The Chemoreceptor Reflexes

The respiratory centers are strongly influenced by chemoreceptor inputs from cranial nerves IX and X, and from receptors that monitor the composition of the cerebrospinal fluid (CSF):

- The glossopharyngeal nerves (N IX) carry chemoreceptive information from the carotid bodies, adjacent to the carotid sinus. ∞ pp. 514, 753 The carotid bodies are stimulated

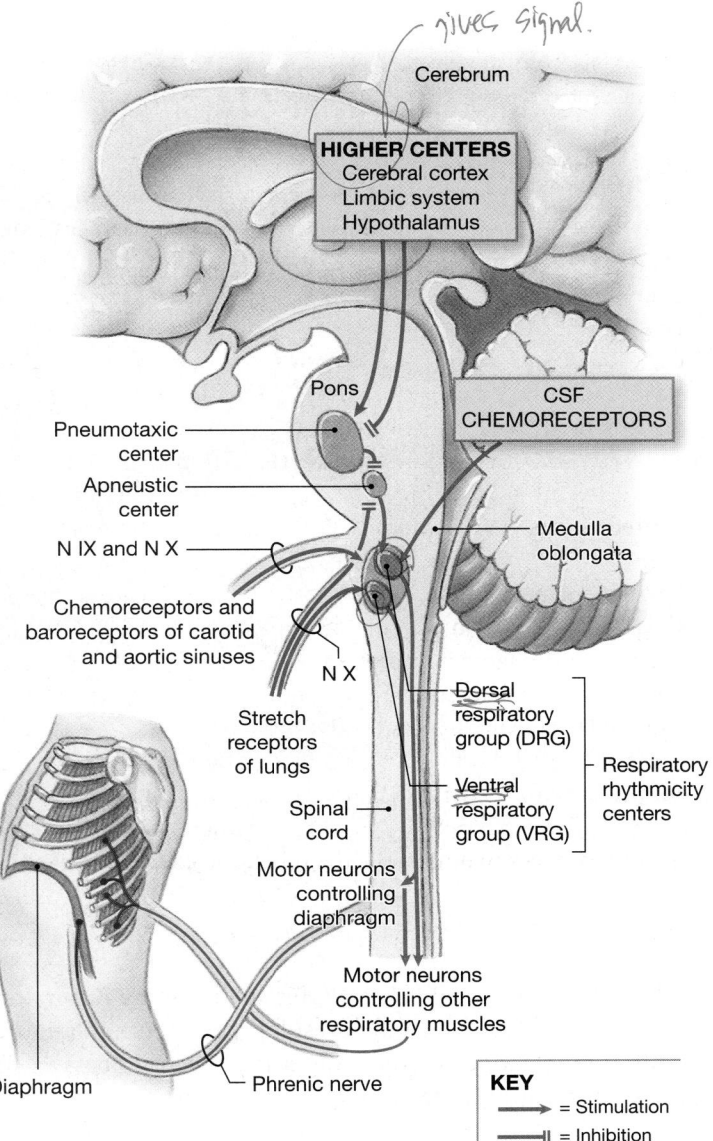

gives signal.

Figure 23–26 Respiratory Centers and Reflex Controls. The locations of the major respiratory centers and other structures important to the reflex control of respiration. Pathways for conscious control over respiratory muscles are not shown.

by a decrease in the pH or P_{O_2} of blood. Because changes in P_{CO_2} affect pH, these receptors are indirectly stimulated by a rise in the P_{CO_2}.

- The vagus nerves (N X) monitor chemoreceptors in the aortic bodies, near the aortic arch. ∞ pp. 514, 750 These receptors are sensitive to the same stimuli as the carotid bodies. Carotid and aortic body receptors are often called *peripheral chemoreceptors.*

- Chemoreceptors are located on the ventrolateral surface of the medulla oblongata in a region known as the *chemosensitive area.* The neurons in that area respond only to the P_{CO_2} and pH of the CSF and are often called *central chemoreceptors.*

Chemoreceptors and their effects on cardiovascular function were discussed in Chapters 15 and 21. ∞ pp. 514, 741–742 Stimulation of these chemoreceptors leads to an increase in the depth and rate of respiration. Under normal conditions, a drop in arterial P_{O_2} has little effect on the respiratory centers, until the arterial P_{O_2} drops by about 40 percent, to below 60 mm Hg. If the P_{O_2} of arterial blood drops to 40 mm Hg (the level in peripheral tissues), the respiratory rate increases by only 50–70 percent. In contrast, a rise of just 10 percent in the arterial P_{CO_2} causes the respiratory rate to double, even if the P_{O_2} remains completely normal. Carbon dioxide levels are therefore responsible for regulating respiratory activity under normal conditions.

Although the receptors monitoring CO_2 levels are more sensitive, oxygen and carbon dioxide receptors work together in a crisis. Carbon dioxide is generated during oxygen consumption, so when oxygen concentrations are falling rapidly, CO_2 levels are usually increasing. This cooperation breaks down only under unusual circumstances. For example, you can hold your breath longer than normal by taking deep, full breaths, but the practice is very dangerous. The danger lies in the fact that the increased ability is due not to extra oxygen, but to less carbon dioxide. If the P_{CO_2} is driven down far enough, your ability to hold your breath can increase to the point at which you become unconscious from oxygen starvation in the brain without ever feeling the urge to breathe.

The chemoreceptors are subject to adaptation—a decrease in sensitivity after chronic stimulation—if the P_{O_2} or P_{CO_2} remains abnormal for an extended period. This adaptation can complicate the treatment of chronic respiratory disorders. For example, if the P_{O_2} remains low for an extended period while the P_{CO_2} remains chronically elevated, the chemoreceptors will reset to those values and will oppose any attempts to return the partial pressures to the proper range.

Because the chemoreceptors monitoring CO_2 levels are also sensitive to pH, any condition altering the pH of blood or CSF will affect respiratory performance. The rise in lactic acid levels after exercise, for example, causes a drop in pH that helps stimulate respiratory activity.

Hypercapnia and Hypocapnia. An increase in the P_{CO_2} of arterial blood constitutes **hypercapnia**. **Figure 23–27a** diagrams the central response to hypercapnia, which is triggered by the stimulation of chemoreceptors in the carotid and aortic bodies and is reinforced by the stimulation of CNS chemoreceptors. Carbon dioxide crosses the blood–brain barrier quite rapidly, so a rise in arterial P_{CO_2} almost immediately elevates CO_2 levels in the CSF, lowering the pH of the CSF and stimulating the chemoreceptive neurons of the medulla oblongata.

These receptors stimulate the respiratory centers to increase the rate and depth of respiration. Your breathing becomes more rapid, and more air moves into and out of your lungs with each breath. Because more air moves into and out

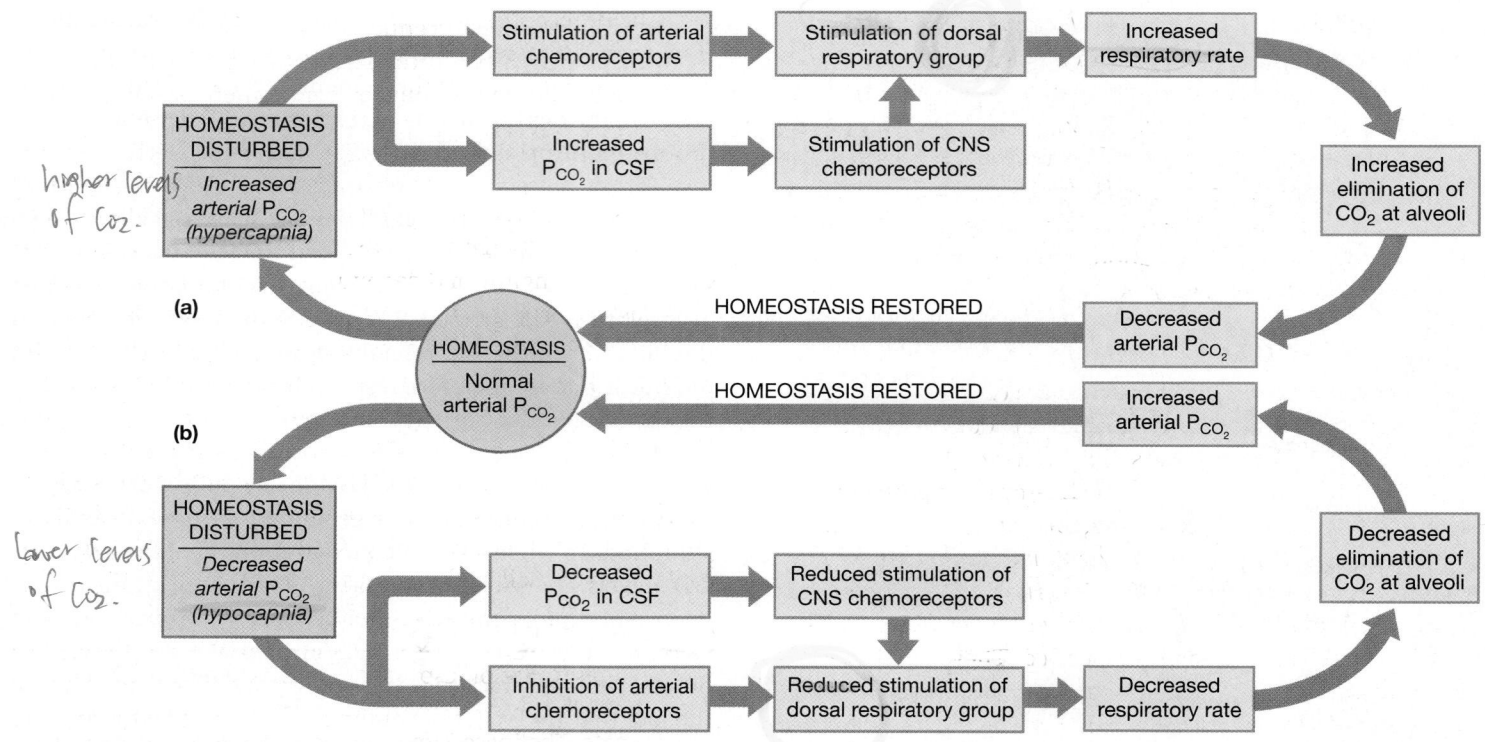

higher levels of CO₂.

lower levels of CO₂.

Figure 23–27 **The Chemoreceptor Response to Changes in P_{CO_2}.** **(a)** A rise in arterial P_{CO_2} stimulates chemoreceptors that accelerate breathing cycles at the inspiratory center. This change increases the respiratory rate, encourages CO_2 loss at the lungs, and lowers arterial P_{CO_2}. **(b)** A drop in arterial P_{CO_2} inhibits these chemoreceptors. In the absence of stimulation, the rate of respiration decreases, slowing the rate of CO_2 loss and elevating arterial P_{CO_2}.

of the alveoli each minute, alveolar concentrations of carbon dioxide decline, accelerating the diffusion of carbon dioxide out of alveolar capillaries. Thus, homeostasis is restored.

The most common cause of hypercapnia is hypoventilation. In **hypoventilation**, the respiratory rate remains abnormally low and is insufficient to meet the demands for normal oxygen delivery and carbon dioxide removal. Carbon dioxide then accumulates in the blood.

If the rate and depth of respiration exceed the demands for oxygen delivery and carbon dioxide removal, the condition called **hyperventilation** exists. Hyperventilation gradually leads to **hypocapnia**, an abnormally low P_{CO_2}. If the arterial P_{CO_2} drops below normal levels, chemoreceptor activity decreases and the respiratory rate falls (**Figure 23–27b**). This situation continues until the P_{CO_2} returns to normal and homeostasis is restored.

The Baroreceptor Reflexes

The effects of carotid and aortic baroreceptor stimulation on systemic blood pressure were described in Chapter 21. ∞ p. 740 The output from these baroreceptors also affects the respiratory centers. When blood pressure falls, the respira-

tory rate increases; when blood pressure rises, the respiratory rate declines. This adjustment results from the stimulation or inhibition of the respiratory centers by sensory fibers in the glossopharyngeal (N IX) and vagus (N X) nerves.

The Hering–Breuer Reflexes

The **Hering–Breuer reflexes** are named after the physiologists who described them in 1865. The sensory information from these reflexes is distributed to the apneustic centers and the ventral respiratory group. The Hering–Breuer reflexes are not involved in normal quiet breathing (eupnea) or in tidal volumes under 1000 mL. There are two such reflexes:

1. The **inflation reflex** prevents overexpansion of the lungs during forced breathing. Stretch receptors located in the smooth muscle tissue around bronchioles are stimulated by lung expansion. Sensory fibers leaving the stretch receptors of each lung reach the respiratory rhythmicity center on the same side via the vagus nerve. As lung volume increases, the dorsal respiratory group is gradually inhibited, and the expiratory center of the VRG is stimulated. Inhalation stops as the lungs near maximum volume, and active exhalation then begins.

2. The **deflation reflex** inhibits the expiratory centers and stimulates the inspiratory centers when the lungs are deflating. These receptors, which are distinct from those of the inflation reflex, are located in the alveolar wall near the alveolar capillary network. The smaller the volume of the lungs, the greater the degree of inhibition, until exhalation stops and inhalation begins. This reflex normally functions only during forced exhalation, when both the inspiratory and expiratory centers are active.

Protective Reflexes

Protective reflexes operate when you are exposed to toxic vapors, chemical irritants, or mechanical stimulation of the respiratory tract. The receptors involved are located in the epithelium of the respiratory tract. Examples of protective reflexes include sneezing, coughing, and laryngeal spasms.

Sneezing is triggered by an irritation of the nasal cavity wall. Coughing is triggered by an irritation of the larynx, trachea, or bronchi. Both reflexes involve **apnea** (AP-nē-uh), a period in which respiration is suspended. They are usually followed by a forceful expulsion of air to remove the offending stimulus. The glottis is forcibly closed while the lungs are still relatively full. The abdominal and internal intercostal muscles then contract suddenly, creating pressures that blast air out of the respiratory passageways when the glottis reopens. Air leaving the larynx can travel at 160 kph (99 mph), carrying mucus, foreign particles, and irritating gases out of the respiratory tract via the nose or mouth.

Laryngeal spasms result from the entry of chemical irritants, foreign objects, or fluids into the area around the glottis. This reflex generally closes the airway temporarily. A very strong stimulus, such as a toxic gas, could close the glottis so powerfully that you could lose consciousness and die without taking another breath. Fine chicken bones or fish bones that pierce the laryngeal walls can also stimulate laryngeal spasms, swelling, or both, restricting the airway.

Voluntary Control of Respiration

Activity of the cerebral cortex has an indirect effect on the respiratory centers, as the following examples show:

- Conscious thought processes tied to strong emotions, such as rage or fear, affect the respiratory rate by stimulating centers in the hypothalamus.

- Emotional states can affect respiration through the activation of the sympathetic or parasympathetic division of the autonomic nervous system. Sympathetic activation causes bronchodilation and increases the respiratory rate; parasympathetic stimulation has the opposite effect.

- An anticipation of strenuous exercise can trigger an automatic increase in the respiratory rate, along with increased cardiac output, by sympathetic stimulation.

Conscious control over respiratory activities may bypass the respiratory centers completely, using pyramidal fibers that innervate the same lower motor neurons that are controlled by the DRG and VRG. This control mechanism is an essential part of speaking, singing, and swimming, when respiratory activities must be precisely timed. Higher centers can also have an inhibitory effect on the apneustic centers and on the DRG and VRG; this effect is important when you hold your breath.

Your abilities to override the respiratory centers have limits, however. The chemoreceptor reflexes are extremely powerful respiratory stimulators, and they cannot be consciously suppressed. For example, you cannot kill yourself by holding your breath "till you turn blue." Once the P_{CO_2} rises to critical levels, you will be forced to take a breath.

The A&P Top 100

#81 A basic pace of respiration is established by the interplay between respiratory centers in the pons and medulla oblongata. That pace is modified in response to input from chemoreceptors, baroreceptors, and stretch receptors. In general, carbon dioxide levels, rather than oxygen levels, are the primary drivers of respiratory activity. Respiratory activity can also be interrupted by protective reflexes and adjusted by the conscious control of respiratory muscles.

Changes in the Respiratory System at Birth

The respiratory systems of fetuses and newborns differ in several important ways. Before delivery, pulmonary arterial resistance is high, because the pulmonary vessels are collapsed. The rib cage is compressed, and the lungs and conducting passageways contain only small amounts of fluid and no air. During delivery, the lungs are compressed further, and as the placental connection is lost, blood oxygen levels fall and carbon dioxide levels climb rapidly. At birth, the newborn infant takes a truly heroic first breath through powerful contractions of the diaphragmatic and external intercostal muscles. The inhaled air must enter the respiratory passageways with enough force to overcome surface tension and inflate the bronchial tree and most of the alveoli. The same drop in pressure that pulls air into the lungs pulls blood into the pulmonary circulation. The changes in blood flow and rise in oxygen levels that occur lead to the closure of the *foramen ovale*, an interatrial connection, and the *ductus arteriosus*, the fetal connection between the pulmonary trunk and the aorta. ∞ p. 767

ATLAS: Embryology Summary 18: The Development of the Respiratory System

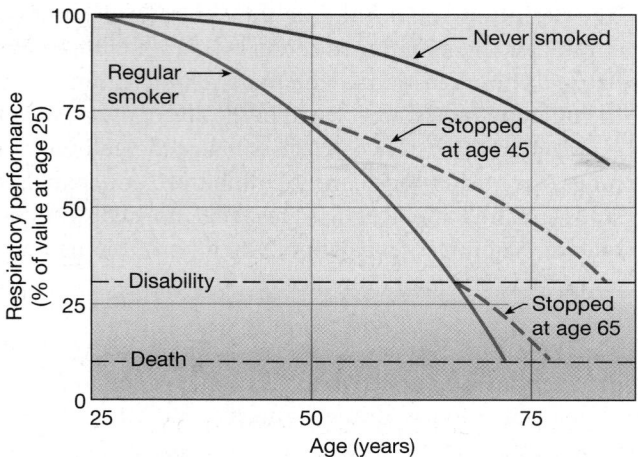

The exhalation that follows fails to empty the lungs completely, because the rib cage does not return to its former, fully compressed state. Cartilages and connective tissues keep the conducting passageways open, and surfactant covering the alveolar surfaces prevents their collapse. Subsequent breaths complete the inflation of the alveoli. Pathologists sometimes use these physical changes to determine whether a newborn infant died before delivery or shortly thereafter. Before the first breath, the lungs are completely filled with amniotic fluid, and extracted lungs will sink if placed in water. After the infant's first breath, even the collapsed lungs contain enough air to keep them afloat.

CHECKPOINT

28. What effect does exciting the pneumotaxic centers have on respiration?

29. Are peripheral chemoreceptors as sensitive to levels of carbon dioxide as they are to levels of oxygen?

30. Little Johnny is angry with his mother, so he tells her that he will hold his breath until he turns blue and dies. Should Johnny's mother worry?

See the blue Answers tab at the end of the book.

23-11 Respiratory performance declines with age

Many factors interact to reduce the efficiency of the respiratory system in elderly individuals. Three examples are particularly noteworthy:

1. As one's age increases, elastic tissue deteriorates throughout the body, altering the compliance of the lungs and lowering their vital capacity.

2. Chest movements are restricted by arthritic changes in the rib articulations and by decreased flexibility at the costal cartilages. Along with the changes in item 1, the stiffening and reduction in chest movement effectively limit the respiratory minute volume. This restriction contributes to the reduction in exercise performance and capabilities with increasing age.

3. Some degree of emphysema is normal in individuals over age 50. However, the extent varies widely with the lifetime exposure to cigarette smoke and other respiratory irritants. **Figure 23–28** compares the respiratory performance of individuals who have never smoked with individuals who have smoked for various periods of time. The message is quite clear: Although some decrease in respiratory

Figure 23–28 **Decline in Respiratory Performance with Age and Smoking.** The relative respiratory performances of individuals who have never smoked, individuals who quit smoking at age 45, individuals who quit smoking at age 65, and lifelong smokers.

performance is inevitable, you can prevent serious respiratory deterioration by stopping smoking or never starting.

CHECKPOINT

31. Name several age-related factors that affect the respiratory system.

See the blue Answers tab at the end of the book.

23-12 The respiratory system provides oxygen to, and eliminates carbon dioxide from, other organ systems

The goal of respiratory activity is to maintain homeostatic oxygen and carbon dioxide levels in peripheral tissues. Changes in respiratory activity alone are seldom sufficient to accomplish this; coordinated changes in cardiovascular activity must also occur.

Consider these examples of the integration between the respiratory and cardiovascular systems:

• At the local level, changes in lung perfusion in response to changes in alveolar P_{O_2} improve the efficiency of gas exchange within or among lobules.

• Chemoreceptor stimulation not only increases the respiratory drive, it also causes an elevation in blood pressure and increased cardiac output.

CLINICAL NOTE

Emphysema and Lung Cancer

Emphysema and lung cancer are two relatively common disorders that are often associated with cigarette smoking. **Emphysema** (em-fi-ZĒ-muh) is a chronic, progressive condition characterized by shortness of breath and an inability to tolerate physical exertion. The underlying problem is the destruction of alveolar surfaces and inadequate surface area for oxygen and carbon dioxide exchange. In essence, respiratory bronchioles and alveoli are functionally eliminated. The alveoli gradually expand, and adjacent alveoli merge to form larger air spaces supported by fibrous tissue without alveolar capillary networks. As connective tissues are eliminated, compliance increases; air moves into and out of the lungs more easily than before. However, the loss of respiratory surface area restricts oxygen absorption, so the individual becomes short of breath.

Emphysema has been linked to the inhalation of air that contains fine particulate matter or toxic vapors, such as those in cigarette smoke. Genetic factors also predispose individuals to the condition. Some degree of emphysema is a normal consequence of aging, however. An estimated 66 percent of adult males and 25 percent of adult females have detectable areas of emphysema in their lungs.

Lung cancer, or *bronchopulmonary carcinoma*, is an aggressive class of malignancies originating in the bronchial passageways or alveoli. These cancers affect the epithelial cells that line conducting passageways, mucous glands, or alveoli. Signs and symptoms generally do not appear until the condition has progressed to the point at which the tumor masses are restricting airflow or compressing adjacent structures. Chest pain, shortness of breath, a cough or a wheeze, and weight loss commonly occur. Treatment programs vary with the cellular organization of the tumor and whether metastasis (cancer cell migration) has occurred, but surgery, radiation therapy, or chemotherapy may be involved.

Deaths from lung cancer were rare at the turn of the 20th century, but 29,000 such deaths occurred in 1956, 105,000 in 1978, and 163,510 in 2005 in the United States. This rise coincides with an increased rate of smoking in the population. The number of diagnosed cases is also rising, doubling every 15 years. In 2005, 12.6% percent of *new* cancers detected were lung cancers, affecting an estimated 93,010 men and 79,560 women. Compared to rates in 1989, these numbers reflect a marked decrease among men (fewer men are smoking) and an increase among women (more women are smoking).

- The stimulation of baroreceptors in the lungs has secondary effects on cardiovascular function. For example, the stimulation of airway stretch receptors not only triggers the inflation reflex, but also increases heart rate. Thus, as the lungs fill, cardiac output rises and more blood flows through the alveolar capillaries.

The adaptations that occur at high altitudes provide an excellent example of the functional interplay between the respiratory and cardiovascular systems. Atmospheric pressure decreases with increasing altitude, and so do the partial pressures of the component gases, including oxygen. People living in Denver or Mexico City function normally with alveolar oxygen pressures in the 80–90 mm Hg range. At higher elevations, alveolar P_{O_2} is even lower. At 3300 meters (10,826 ft), an altitude many hikers and skiers have experienced, alveolar P_{O_2} is about 60 mm Hg.

Despite the low alveolar P_{O_2} millions of people live and work at altitudes this high or higher. Important physiological adjustments include increased respiratory rate, increased heart rate, increased BPG levels, and elevated hematocrit. Thus, even though the hemoglobin is not fully saturated, the bloodstream holds more of it, and the round trip between the lungs and the peripheral tissues takes less time. However, most such adaptations take days to weeks to develop. As a result, athletes planning to compete in events held at high altitude must begin training under such conditions well in advance.

The respiratory system is functionally linked to all other systems as well. **Figure 23–29** illustrates these interrelationships.

CHECKPOINT

32. **Identify the functional relationship between the respiratory system and all other organ systems.**

33. **Describe the functional relationship between the respiratory system and the urinary system.**

See the blue Answers tab at the end of the book.

THE RESPIRATORY SYSTEM IN PERSPECTIVE

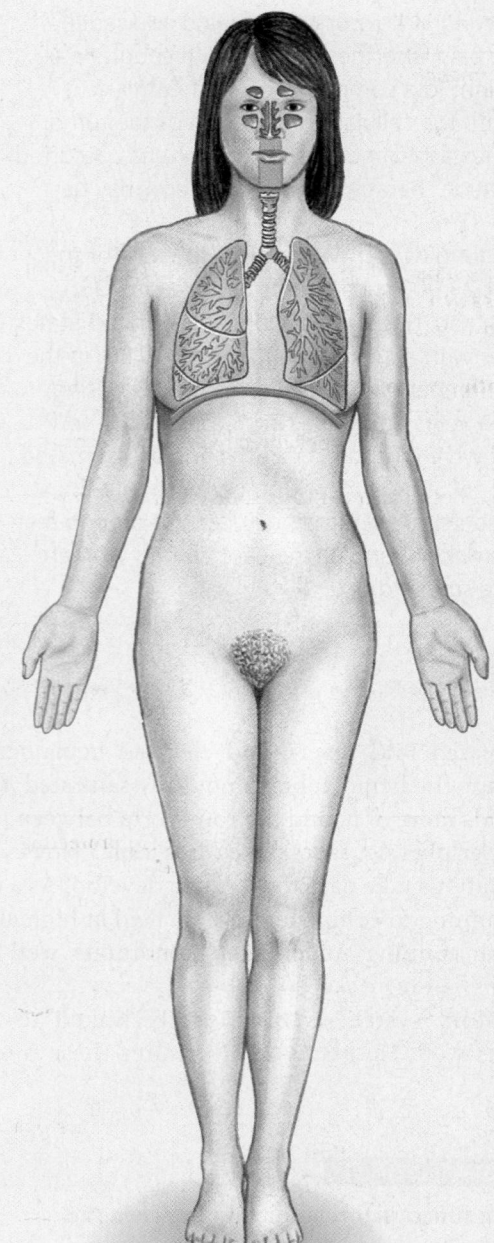

Integumentary System

 • Protects portions of upper respiratory system; hairs guard entry to external nares

Skeletal System

 • Movements of the ribs important in breathing; axial skeleton surrounds and protects lungs

Muscular System

 • Muscular activity generates carbon dioxide; respiratory muscles fill and empty lungs; other muscles control entrances to respiratory tract; intrinsic laryngeal muscles control airflow through larynx and produce sound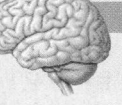

Nervous System

• Monitors respiratory volume and blood gas levels; controls pace and depth of respiration

Endocrine System

• Epinephrine and norepinephrine stimulate respiratory activity and dilate respiratory passageways

 • Converting enzyme along capillaries of lung converts angiotensin I to angiotensin II

Cardiovascular System

• Red blood cells transport oxygen and carbon dioxide between the lungs and peripheral tissues

 • Activation of angiotensin II by converting enzyme important in regulation of blood pressure and volume; bicarbonate ions contribute to buffering capacity of blood

Lymphoid System

• Tonsils protect against infection at entrance to respiratory tract; lymphatic vessels monitor lymph drainage from lungs and mobilize specific defenses when infection occurs

 • Alveolar phagocytes present antigens to trigger specific defenses; respiratory defense system traps pathogens, protects deeper tissues

Digestive System

• Provides substrates, vitamins, water, and ions that are necessary to all cells of the respiratory system

 • Increased thoracic and abdominal pressure through contraction of respiratory muscles can assist in defecation

Urinary System

• Eliminates organic wastes generated by cells of the respiratory system; maintains normal fluid and ion balance in the blood

 • Assists in the regulation of pH by eliminating carbon dioxide

Reproductive System

• Changes in respiratory rate and depth occur during sexual arousal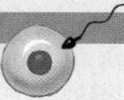

FOR ALL SYSTEMS
Provides oxygen and eliminates carbon dioxide

Figure 23–29 **Functional Relationships between the Respiratory System and Other Systems.**

Related Clinical Terms

anoxia: A condition of tissue oxygen starvation caused by (1) circulatory blockage, (2) respiratory problems, or (3) cardiovascular problems.

asthma: An acute respiratory disorder characterized by unusually reactive, constricting airways.

atelectasis: A collapsed lobe or lung.

bronchitis: An inflammation of the bronchial lining.

bronchodilation: An enlargement of the respiratory passageways.

bronchography: A procedure in which radiopaque materials are introduced into the airways to improve x-ray imaging of the bronchial tree.

bronchoscope: A fiber-optic bundle small enough to be inserted into the trachea and finer airways; the procedure is called *bronchoscopy*.

carbon monoxide poisoning: Toxic condition resulting from the strong affinity of carbon monoxide for hemoglobin, which reduces oxygen transport and blocks oxygen utilization.

cardiopulmonary resuscitation (CPR): The application of cycles of compression to the rib cage and mouth-to-mouth breathing to maintain cardiovascular and respiratory function.

cystic fibrosis (CF): A lethal inherited disease caused by an abnormal plasma membrane channel protein; mucous secretions become too thick to be transported easily, leading to respiratory problems.

decompression sickness, or *the bends*: A condition caused by a rapid drop in atmospheric pressure and the resulting formation of nitrogen gas bubbles in body fluids, tissues, and organs.

emphysema: A chronic, progressive condition characterized by shortness of breath and an inability to tolerate physical exertion.

epistaxis: A nosebleed.

Heimlich maneuver, or *abdominal thrust*: Sudden compression applied to the abdomen just inferior to the diaphragm, to force air out of the lungs to clear a blocked trachea or larynx.

hypercapnia: An increase in the P_{CO_2} of arterial blood.

hypocapnia: An abnormally low arterial P_{CO_2}.

hypoxia: A condition of reduced tissue P_{O_2}.

lung cancer (*bronchopulmonary carcinoma*): A class of aggressive malignancies originating in the bronchial passageways or alveoli.

mountain sickness: An acute disorder resulting from CNS effects due to the low gas partial pressures that occur at high altitudes.

pleurisy: An inflammation of the pleurae, often accompanied by the secretion of excess amounts of pleural fluid.

pneumonia: A respiratory disorder characterized by fluid leakage into the alveoli or swelling and constriction of the respiratory bronchioles.

pneumothorax: The entry of air into the pleural cavity.

pulmonary embolism: Blockage of a branch of a pulmonary artery, with interruption of blood flow to a group of lobules or alveoli.

respiratory distress syndrome: A condition resulting from the production of inadequate surfactant and associated alveolar collapse.

sudden infant death syndrome (SIDS), or *crib death*: The death of an infant due to respiratory arrest; the causes remain obscure.

tracheostomy: The insertion of a tube directly into the trachea to bypass a blocked or damaged larynx.

tuberculosis: A respiratory disorder caused by infection of the lungs by the bacterium *Mycobacterium tuberculosis*.

Chapter Review

Study Outline

iP Use *Interactive Physiology*® (IP) —available on the IP CD packaged with your new textbook and online at **myA&P**™ (www.myaandp.com)—to help you understand difficult physiological concepts. Follow these navigation paths on IP for concepts in this chapter:

- Respiratory System/Pulmonary Ventilation
- Respiratory System/Gas Transport
- Respiratory System/Gas Exchange
- Respiratory System/Control of Respiration

An introduction to the respiratory system p. 826

1. Body cells must obtain oxygen and eliminate carbon dioxide. The respiratory surfaces where gas exchange occurs are inside the lungs.

23-1 The respiratory system, organized into an upper respiratory system and a lower respiratory system, has several basic functions p. 826

2. The functions of the **respiratory system** include (1) providing an area for gas exchange between air and circulating blood;

(2) moving air to and from exchange surfaces; (3) protecting respiratory surfaces from environmental variations and defending the respiratory system and other tissues from invasion by pathogens; (4) producing sounds; and (5) facilitating the detection of olfactory stimuli.

3. The respiratory system includes the **upper respiratory system**, composed of the nose, nasal cavity, paranasal sinuses, and pharynx, and the **lower respiratory system**, which includes the larynx, trachea, bronchi, bronchioles, and alveoli of the lungs. (*Figure 23–1*)

4. The **respiratory tract** consists of the conducting airways that carry air to and from the **alveoli**. The passageways of the upper respiratory tract filter and humidify incoming air. The lower respiratory tract includes delicate conduction passages and the alveolar exchange surfaces.

5. The **respiratory mucosa** (respiratory epithelium and underlying connective tissue) lines the conducting portion of the respiratory tract.

6. The respiratory epithelium changes in structure along the respiratory tract. It is supported by the **lamina propria**, a layer of areolar tissue. (*Figure 23–2*)

7. Contamination of the respiratory system is prevented by the **respiratory defense system**. (*Figure 23–2*)

23-2 Located outside the thoracic cavity, the upper respiratory system consists of the nose, nasal cavity, paranasal sinuses, and pharynx p. 829

8. The components of the upper respiratory system consist of the nose, nasal cavity, paranasal sinuses, and pharynx. (*Figures 23–1, 23–3*)
9. Air normally enters the respiratory system through the **external nares**, which open into the *nasal cavity*. The **nasal vestibule** (entryway) is guarded by hairs that screen out large particles. (*Figure 23–3*)
10. Incoming air flows through the **superior**, **middle**, and **inferior meatuses** (narrow grooves) and bounces off the conchal surfaces. (*Figure 23–3*)
11. The **hard palate** separates the oral and nasal cavities. The **soft palate** separates the superior nasopharynx from the rest of the pharynx. The connections between the nasal cavity and nasopharynx are the **internal nares**.
12. The nasal mucosa traps particles, warms and humidifies incoming air, and cools and dehumidifies outgoing air.
13. The **pharynx** is a chamber shared by the digestive and respiratory systems. The **nasopharynx** is the superior part of the pharynx. The **oropharynx** is continuous with the oral cavity. The **laryngopharynx** includes the narrow zone between the hyoid bone and the entrance to the esophagus. (*Figure 23–3*)

23-3 Composed of cartilages, ligaments, and muscles, the larynx produces sound p. 831

14. Inhaled air passes through the **glottis** en route to the lungs; the **larynx** surrounds and protects the glottis. (*Figure 23–4*)
15. The cylindrical larynx is composed of three large cartilages (the **thyroid cartilage**, **cricoid cartilage**, and **epiglottis**) and three smaller pairs of cartilages (the **arytenoid**, **corniculate**, and **cuneiform cartilages**). The epiglottis projects into the pharynx. (*Figures 23–4, 23–5*)
16. Two pairs of folds span the glottis: the inelastic **vestibular folds** and the more delicate **vocal folds**. (*Figure 23–5*)
17. Air passing through the glottis vibrates the vocal folds, producing sound. The pitch of the sound depends on the diameter, length, and tension of the vocal folds.
18. The muscles of the neck and pharynx position and stabilize the larynx. The smaller intrinsic muscles regulate tension in the vocal folds or open and close the glottis. During swallowing, both sets of muscles help prevent particles from entering the glottis.

23-4 The trachea and primary bronchi convey air to and from the lungs p. 834

19. The **trachea** extends from the sixth cervical vertebra to the fifth thoracic vertebra. The **submucosa** contains C-shaped **tracheal cartilages**, which stiffen the tracheal walls and protect the airway. The posterior tracheal wall can distort to permit large masses of food to pass through the esophagus. (*Figure 23–6*)
20. The trachea branches within the mediastinum to form the **right** and **left primary bronchi**. Each bronchus enters a lung at the **hilum** (a groove). The **root** is a connective tissue mass

that includes the bronchus, pulmonary vessels, and nerves. (*Figures 23–6, 23–7*)

23-5 Enclosed by a pleural membrane, the lungs are paired organs containing alveoli, which permit gaseous exchange p. 835

21. The **lobes** of the lungs are separated by fissures. The right lung has three lobes, the left lung two. (*Figure 23–7*)
22. The anterior and lateral surfaces of the lungs follow the inner contours of the rib cage. The concavity of the medial surface of the left lung is the **cardiac notch**, which conforms to the shape of the pericardium. (*Figures 23–7, 23–8*)
23. The primary bronchi and their branches form the **bronchial tree**. The **secondary** and **tertiary bronchi** are branches within the lungs. As they branch, the amount of cartilage in their walls decreases and the amount of smooth muscle increases. (*Figure 23–9*)
24. Each tertiary bronchus supplies air to a single **bronchopulmonary segment**. (*Figure 23–9*)
25. **Bronchioles** within the bronchopulmonary segments ultimately branch into **terminal bronchioles**. Each terminal bronchiole delivers air to a single **pulmonary lobule** in which the terminal bronchiole branches into **respiratory bronchioles**. The connective tissues of the root of the lung extend into the parenchyma of the lung as a series of *trabeculae* (partitions) that branch to form **interlobular septa**, which divide the lung into lobules. (*Figure 23–9*)
26. The respiratory bronchioles open into **alveolar ducts**, at each of which many alveoli are interconnected. The respiratory exchange surfaces are extensively connected to the circulatory system via the vessels of the pulmonary circuit. (*Figure 23–10*)
27. The **respiratory membrane** consists of a simple squamous epithelium, the endothelial cell lining an adjacent capillary, and the fused basal laminae; **pneumocytes type II** (septal cells) scattered in the respiratory membrane produce **surfactant** that keeps the alveoli from collapsing. **Alveolar macrophages** patrol the epithelium and engulf foreign particles. (*Figure 23–11*)
28. The conducting portions of the respiratory tract receive blood from the external carotid arteries, the thyrocervical trunks, and the bronchial arteries. Venous blood flows into the pulmonary veins, bypassing the rest of the systemic circuit and diluting the oxygenated blood leaving the alveoli.
29. Each lung occupies a single **pleural cavity** lined by a **pleura** (serous membrane). The two types of pleurae are the **parietal pleura**, covering the inner surface of the thoracic wall, and the **visceral pleura**, covering the lungs.

23-6 External respiration and internal respiration allow gaseous exchange within the body p. 842

30. Respiratory physiology focuses on a series of integrated processes. **External respiration** (the exchange of oxygen and carbon dioxide between interstitial fluid and the external environment) includes **pulmonary ventilation** (breathing). **Internal respiration** is the exchange of oxygen and carbon dioxide between interstitial fluid and cells. If the oxygen content declines, the affected tissues suffer from **hypoxia**; if the oxygen supply is completely shut off, **anoxia** and tissue death result. (*Figure 23–12*)

23-7 Pulmonary ventilation—the exchange of air between the atmosphere and the lungs—involves pressure changes, muscle movement, and respiratory rates and volumes p. 843

31. **Pulmonary ventilation** is the physical movement of air into and out of the respiratory tract.

32. As pressure on a gas decreases, its volume expands; as pressure increases, gas volume contracts. This inverse relationship is **Boyle's law**. (*Figure 23–13; Table 23–1*)

33. Lung volume is directly affected by movement of the diaphragm and ribs.

34. The relationship between **intrapulmonary pressure** (the pressure inside the respiratory tract) and **atmospheric pressure** (**atm**) determines the direction of airflow. **Intrapleural pressure** is the pressure in the space between the parietal and visceral pleurae. (*Figures 23–14, 23–15*)

35. A **respiratory cycle** is a single cycle of inhalation and exhalation. The amount of air moved in one respiratory cycle is the **tidal volume**. (*Figure 23–15*)

36. The diaphragm and the external and internal intercostal muscles are involved in normal **quiet breathing**, or **eupnea**. Accessory muscles become active during the active inspiratory and expiratory movements of **forced breathing**, or **hyperpnea**. (*Figure 23–16*)

37. **Alveolar ventilation** is the amount of air reaching the alveoli each minute. The **vital capacity** includes the **tidal volume** plus the **expiratory** and **inspiratory reserve volumes**. The air left in the lungs at the end of maximum exhalation is the **residual volume**. (*Figure 23–17*)

23-8 Gas exchange depends on the partial pressures of gases and the diffusion of molecules p. 851

38. In a mixed gas, the individual gases exert a pressure proportional to their abundance in the mixture (**Dalton's law**). The pressure contributed by a single gas is its **partial pressure**. (Table 23–2)

39. The amount of a gas in solution is directly proportional to the partial pressure of that gas (**Henry's law**). (*Figure 23–18*)

40. Alveolar air and atmospheric air differ in composition. Gas exchange across the respiratory membrane is efficient due to differences in partial pressures, the short diffusion distance, lipid-soluble gases, the large surface area of all the alveoli combined, and the coordination of blood flow and airflow. (*Figure 23–19*)

23-9 Most oxygen is transported bound to hemoglobin; and carbon dioxide is transported in three ways: as carbonic acid, bound to hemoglobin, or dissolved in plasma p. 855

41. Blood entering peripheral capillaries delivers oxygen and absorbs carbon dioxide. The transport of oxygen and carbon dioxide in blood involves reactions that are completely reversible.

42. Oxygen is carried mainly by RBCs, reversibly bound to hemoglobin. At alveolar P_{O_2} hemoglobin is almost fully saturated; at the P_{O_2} of peripheral tissues, it retains a substantial oxygen reserve. The effect of pH on the hemoglobin saturation curve is called the **Bohr effect**. When low plasma P_{O_2} continues for extended periods, red blood cells generate more **2,3-bisphosphoglycerate** (**BPG**), which reduces hemoglobin's affinity for oxygen. (*Figures 23–20, 23–21*)

43. **Fetal hemoglobin** has a stronger affinity for oxygen than does adult hemoglobin, aiding the removal of oxygen from maternal blood. (*Figure 23–22*)

44. Aerobic metabolism in peripheral tissues generates CO_2. About 7 percent of the CO_2 transported in blood is dissolved in the plasma, 23 percent is bound as **carbaminohemoglobin**, and 70 percent is converted to carbonic acid, which dissociates into H^+; and HCO_3^- (*Figure 23–23*)

45. Driven by differences in partial pressure, oxygen enters the blood at the lungs and leaves it in peripheral tissues; similar forces drive carbon dioxide into the blood at the tissues and into the alveoli at the lungs. (*Figure 23–24*)

23-10 Neurons in the medulla oblongata and pons, along with respiratory reflexes, control respiration p. 860

46. Normally, the cellular rates of gas absorption and generation are matched by the capillary rates of delivery and removal and are identical to the rates of oxygen absorption and carbon dioxide removal at the lungs. When these rates are unbalanced, homeostatic mechanisms restore equilibrium.

47. Local factors regulate alveolar blood flow (*lung perfusion*) and airflow (*alveolar ventilation*). Alveolar capillaries constrict under conditions of low oxygen, and bronchioles dilate under conditions of high carbon dioxide.

48. The **respiratory centers** include three pairs of nuclei in the reticular formation of the pons and medulla oblongata. The *respiratory rhythmicity centers* set the pace for respiration; the **apneustic centers** cause strong, sustained inspiratory movements; and the **pneumotaxic centers** inhibit the apneustic centers and promote exhalation. (*Figures 23–25, 23–26*)

49. Stimulation of the chemoreceptor reflexes is based on the level of carbon dioxide in the blood and CSF. The **inflation reflex** prevents overexpansion of the lungs during forced breathing. The **deflation reflex** stimulates inhalation when the lungs are collapsing. (*Figures 23–26, 23–27*)

50. Conscious and unconscious thought processes can affect respiration by affecting the respiratory centers.

51. Before delivery, the fetal lungs are filled with body fluids and collapsed. At the first breath, the lungs inflate and do not collapse completely thereafter.

23-11 Respiratory performance declines with age p. 866

52. The respiratory system is generally less efficient in the elderly because (1) elastic tissue deteriorates, lowering the vital capacity of the lungs; (2) movements of the chest are restricted by arthritic changes and decreased flexibility of costal cartilages; and (3) some degree of emphysema is generally present. (*Figure 23–28*)

23-12 The respiratory system provides oxygen to, and eliminates carbon dioxide from, other organ systems p. 866

53. The respiratory system has extensive anatomical and physiological connections to the cardiovascular system. (*Figure 23–29*)

Review Questions

See the blue Answers tab at the end of the book.

 Access more review material online at **myA&P**™ (**www.myaandp.com**). There, you'll find chapter guides, chapter quizzes, practice tests, animations, flashcards, a glossary with pronunciations, **Interactive Physiology**® (IP) exercises and quizzes, and more to help you succeed in the course.

LEVEL 1 Reviewing Facts and Terms

1. Identify the structures of the respiratory system in the following figure:

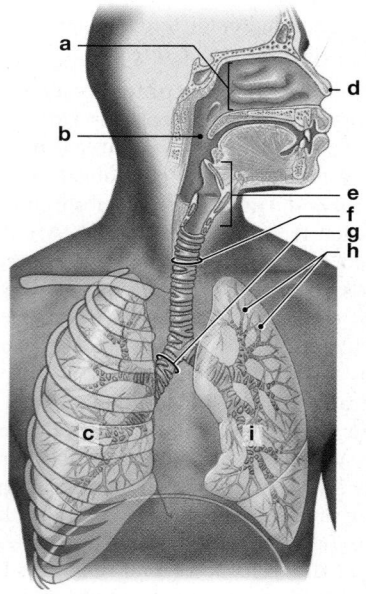

    (a) nasal cavity   (b) pharynx
    (c) right lung   (d) nose
    (e) larynx   (f) trachea
    (g) bronchus   (h) bronchioles
    (i) left lung

2. Surfactant
    (a) protects the surface of the lungs.
    (b) phagocytizes small particulates.
    (c) replaces mucus in the alveoli.
    (d) helps prevent the alveoli from collapsing.
    (e) is not found in healthy lung tissue.

3. The hard palate separates the
    (a) nasal cavity from the larynx.
    (b) left and right sides of the nasal cavity.
    (c) nasal cavity and the oral cavity.
    (d) external nares from the internal nares.
    (e) soft palate from the nasal cavity.

4. Air moves into the lungs because
    (a) the gas pressure in the lungs is less than atmospheric pressure.
    (b) the volume of the lungs decreases with inspiration.
    (c) the thorax is muscular.
    (d) contraction of the diaphragm decreases the volume of the pleural cavity.
    (e) the respiratory control center initiates active expansion of the thorax.

5. The glottis closes partway through an exhalation. The abdominal and internal intercostal muscles then contract suddenly, creating pressure that blasts the air out of the respiratory passages. This describes a
    (a) sneeze.   (b) hiccough.
    (c) cough.   (d) laryngeal spasm.
    (e) gag.

6. When the diaphragm and external intercostal muscles contract,
    (a) expiration occurs. → *Inhalation*
    (b) intrapulmonary pressure increases.
    (c) intrapleural pressure decreases.
    (d) the volume of the lungs decreases. → *increases*
    (e) the size of the pleural cavity increases. → *decreases*

7. During the winter, Brad sleeps in a dorm room that lacks any humidifier for the heated air. In the mornings he notices that his nose is "stuffy" similar to when he has a cold, but after showering and drinking some water, the stuffiness disappears until the next morning. What might be the cause of Brad's nasal condition?

8. Distinguish the structures of the upper respiratory system from those of the lower respiratory system.

9. Name the three regions of the pharynx. Where is each region located?

10. List the cartilages of the larynx. What are the functions of each?

11. What three integrated steps are involved in external respiration?

12. What important physiological differences exist between fetal hemoglobin and maternal hemoglobin?

13. By what three ways is carbon dioxide transported in the bloodstream?

LEVEL 2 Reviewing Concepts

14. Which of the following does *not* occur in internal respiration?
    (a) Oxygen diffuses from the blood to the interstitial spaces.
    (b) Carbon dioxide diffuses from the interstitial spaces to the blood.
    (c) Hemoglobin binds more oxygen.
    (d) Bicarbonate ions are formed in red blood cells.
    (e) Chloride ions diffuse into red blood cells as bicarbonate ions diffuse out.

15. Gas exchange at the respiratory membrane is efficient because
    (a) the differences in partial pressure are substantial.
    (b) the gases are lipid soluble.
    (c) the total surface area is large.
    (d) of a, b, and c.

16. For any partial pressure of oxygen, if the concentration of 2,3-bisphosphoglycerate (BPG) increases,
    (a) the amount of oxygen released by hemoglobin will decrease.
    (b) the oxygen levels in hemoglobin will be unaffected.
    (c) the amount of oxygen released by hemoglobin will increase.
    (d) the amount of carbon dioxide carried by hemoglobin will increase.

 *BPG↑ direct effect of O2 binding & release.*

17. An increase in the partial pressure of carbon dioxide in arterial blood causes chemoreceptors to stimulate the respiratory centers, resulting in
    (a) a decreased respiratory rate.
    (b) an increased respiratory rate.
    (c) hypocapnia.
    (d) hypercapnia.
18. Why is breathing through the nasal cavity more desirable than breathing through the mouth?
19. How would you justify the statement "The bronchioles are to the respiratory system what the arterioles are to the cardiovascular system"?
20. How are pneumocytes type II involved with keeping the alveoli from collapsing?
21. How does pulmonary ventilation differ from alveolar ventilation, and what is the function of each type of ventilation?
22. What is the significance of (a) Boyle's law, (b) Dalton's law, and (c) Henry's law to the process of respiration?
23. What happens to the process of respiration when a person is sneezing or coughing?
24. What are the differences between pulmonary volumes and respiratory capacities? How are pulmonary volumes and respiratory capacities determined?
25. What is the functional difference between the dorsal respiratory group (DRG) and the ventral respiratory group (VRG) of the medulla oblongata?

LEVEL 3 Critical Thinking and Clinical Applications

26. Billy's normal alveolar ventilation rate (AVR) during mild exercise is 6.0 L/min. While at the beach on a warm summer day, he goes snorkeling. The snorkel has a volume of 50 mL. Assuming that the water is not too cold and that snorkeling is mild exercise for Billy, what would his respiratory rate have to be for him to maintain an AVR of 6.0 L/min.while snorkeling? (Assume a constant tidal volume of 500 mL and an anatomic dead space of 150 mL.)
27. Mr. B. has had chronic advanced emphysema for 15 years. While hospitalized with a respiratory infection, he goes into respiratory distress. Without thinking, his nurse immediately administers pure oxygen, which causes Mr. B. to stop breathing. Why?
28. Cary hyperventilates for several minutes before diving into a swimming pool. Shortly after he enters the water and begins swimming, he blacks out and almost drowns. What caused this to happen?
29. Why do individuals who are anemic generally not exhibit an increase in respiratory rate or tidal volume, even though their blood is not carrying enough oxygen?
30. Doris has an obstruction of her right primary bronchus. As a result, how would you expect the oxygen–hemoglobin saturation curve for her right lung to compare with that for her left?

24

The Digestive System

Did you know...?
The stomach can stretch up to 50 times its empty size. The 2007 world record for eating hot dogs is 59½!

Learning Outcomes

After completing this chapter, you should be able to do the following:

24-1 Identify the organs of the digestive system, list their major functions, describe the functional histology of the digestive tract, and outline the mechanisms that regulate digestion.

24-2 Discuss the anatomy of the oral cavity, and list the functions of its major structures and regions.

24-3 Describe the structure and functions of the pharynx.

24-4 Describe the structure and functions of the esophagus.

24-5 Describe the anatomy of the stomach, including its histological features, and discuss its roles in digestion and absorption.

24-6 Describe the anatomical and histological characteristics of the small intestine, explain the functions and regulation of intestinal secretions, and describe the structure, functions, and regulation of the accessory digestive organs.

24-7 Describe the gross and histological structure of the large intestine, including its regional specializations and role in nutrient absorption.

24-8 List the nutrients required by the body, describe the chemical events responsible for the digestion of organic nutrients, and describe the mechanisms involved in the absorption of organic and inorganic nutrients.

24-9 Summarize the effects of aging on the digestive system.

24-10 Give examples of interactions between the digestive system and each of the other organ systems.

Clinical Notes

Peritonitis p. 877
Epithelial Renewal and Repair p. 879
Mumps Virus p. 885
Gastritis and Peptic Ulcers p. 894
Pancreatitis p. 903

Cirrhosis p. 907
Colorectal Cancer p. 913
Inflammatory and Infectious Disorders of the Digestive System p. 922

An Introduction to the Digestive System

The digestive system may not have the visibility of the integumentary system or the glamour of the reproductive system, but it is certainly just as important. All living organisms must obtain nutrients from their environment to sustain life. These substances are used as raw materials for synthesizing essential compounds (anabolism) or are decomposed to provide the energy that cells need to continue functioning (catabolism). ∞ pp. 36–39, 319–320 Catabolic reactions require two essential ingredients: oxygen and organic molecules (such as carbohydrates, fats, or proteins) that can be broken down by intracellular enzymes. Obtaining oxygen and organic molecules can be relatively straightforward for a single-celled organism like an amoeba, but the situation is much more complicated for animals as large and complex as humans. Along with increasing size and complexity comes a division of labor and the need for the coordination of organ system activities.

This chapter discusses the structure and function of the elongate digestive tract and several digestive glands (notably the liver and pancreas), how the process of digestion breaks down large and complex organic molecules to smaller fragments that can be absorbed by the digestive epithelium, and how a few organic wastes are removed from the body.

24-1 The digestive system, consisting of the digestive tract and accessory organs, has overlapping food utilization functions

In our bodies, the respiratory system works in concert with the cardiovascular system to supply the oxygen required for catabolism. The **digestive system**, working with the cardiovascular and lymphoid systems, provides the needed organic molecules. In effect, the digestive system provides both the fuel that keeps all the body's cells running and the building blocks needed for cell growth and repair.

The digestive system consists of a muscular tube, the **digestive tract**, also called the *gastrointestinal (GI) tract* or *alimentary canal*, and various **accessory organs**. The *oral cavity* (mouth), *pharynx*, *esophagus*, *stomach*, *small intestine*, and *large intestine* make up the digestive tract. Accessory digestive organs include the teeth, tongue, and various *glandular organs*, such as the salivary glands, liver, and pancreas, which secrete their products into ducts emptying into the digestive tract. Food enters the digestive tract and passes along its length. On the way, the secretions of the glandular organs,

which contain water, enzymes, buffers, and other components, assist in preparing organic and inorganic nutrients for absorption across the epithelium of the digestive tract.

Figure 24–1 shows the major components of the digestive system. The digestive tract begins at the oral cavity and continues through the pharynx, esophagus, stomach, small intestine, and large intestine, which opens to the exterior at the anus. These structures have overlapping functions, but each has certain areas of specialization and shows distinctive histological characteristics.

Functions of the Digestive System

We can regard digestive functions as a series of integrated steps:

1. **Ingestion** occurs when materials enter the digestive tract via the mouth. Ingestion is an active process involving conscious choice and decision making.

2. **Mechanical processing** is crushing and shearing that makes materials easier to propel along the digestive tract. It also increases their surface area, making them more susceptible to enzymatic attack. Mechanical processing may or may not be required before ingestion; you can swallow liquids immediately, but must process most solids first. Tearing and mashing with the teeth, followed by squashing and compaction by the tongue, are examples of preliminary mechanical processing. Swirling, mixing, and churning motions of the stomach and intestines provide mechanical processing after ingestion.

3. **Digestion** refers to the chemical breakdown of food into small organic fragments suitable for absorption by the digestive epithelium. Simple molecules in food, such as glucose, can be absorbed intact, but epithelial cells have no way to absorb molecules the size and complexity of proteins, polysaccharides, or triglycerides. Prior to absorption, digestive enzymes disassemble these molecules. For example, the starches in a potato are of no nutritional value until enzymes have broken them down to simple sugars that the digestive epithelium can absorb for distribution to body cells. hydrolytic enzymes

4. **Secretion** is the release of water, acids, enzymes, buffers, and salts by the epithelium of the digestive tract and by glandular organs.

5. **Absorption** is the movement of organic substrates, electrolytes (inorganic ions), vitamins, and water across the digestive epithelium and into the interstitial fluid of the digestive tract. mostly in fat (nutrients), int (water).

6. **Excretion** is the removal of waste products from body fluids. The digestive tract and glandular organs discharge waste products in secretions that enter the lumen of the tract. Most of these waste products, after mixing with the

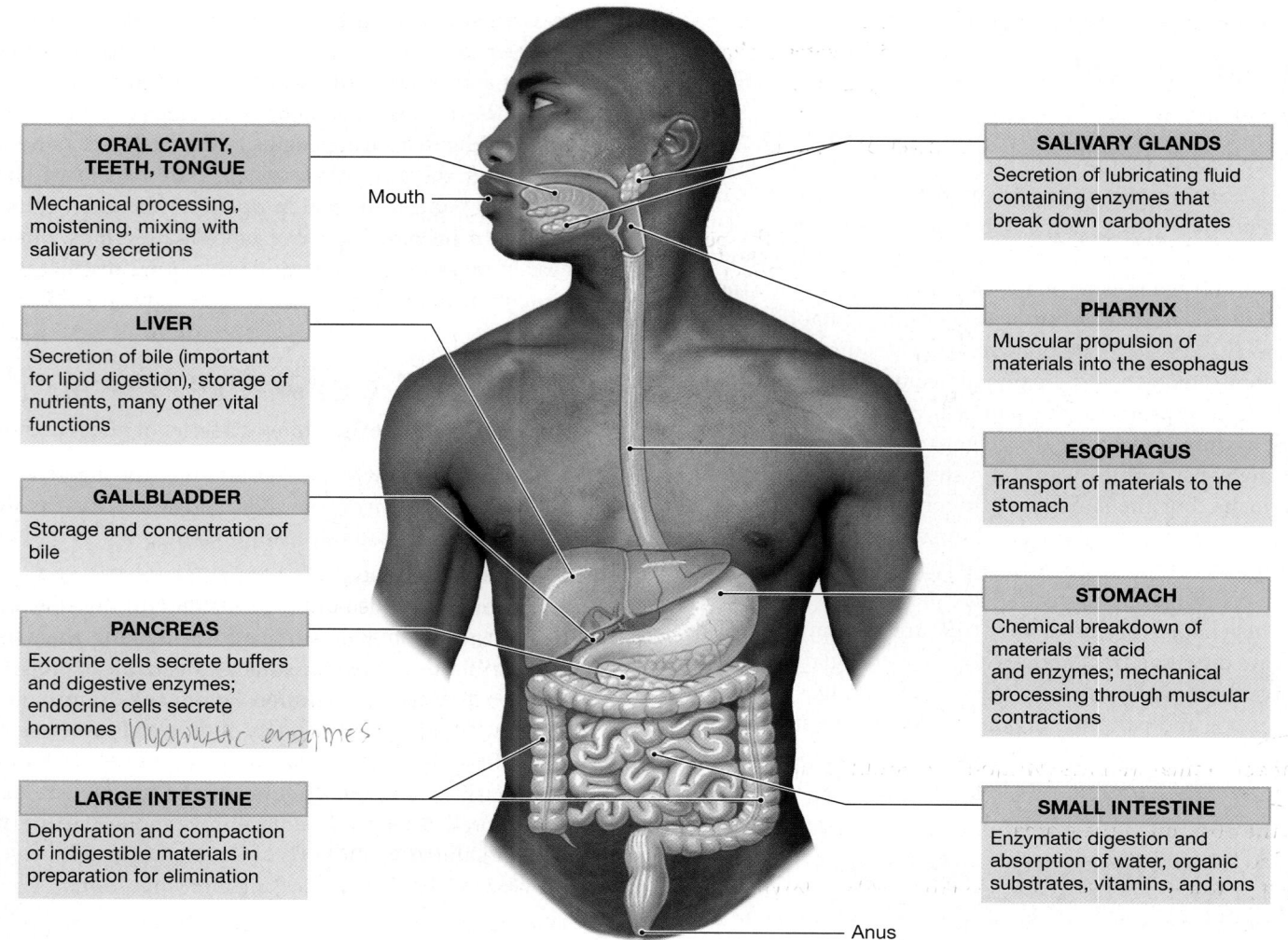

ORAL CAVITY, TEETH, TONGUE
Mechanical processing, moistening, mixing with salivary secretions

LIVER
Secretion of bile (important for lipid digestion), storage of nutrients, many other vital functions

GALLBLADDER
Storage and concentration of bile

PANCREAS
Exocrine cells secrete buffers and digestive enzymes; endocrine cells secrete hormones *hydrolytic enzymes*

LARGE INTESTINE
Dehydration and compaction of indigestible materials in preparation for elimination

SALIVARY GLANDS
Secretion of lubricating fluid containing enzymes that break down carbohydrates

PHARYNX
Muscular propulsion of materials into the esophagus

ESOPHAGUS
Transport of materials to the stomach

STOMACH
Chemical breakdown of materials via acid and enzymes; mechanical processing through muscular contractions

SMALL INTESTINE
Enzymatic digestion and absorption of water, organic substrates, vitamins, and ions

Mouth

Anus

Figure 24–1 **The Components of the Digestive System.** The major regions and accessory organs of the digestive tract, together with their primary functions.

indigestible residue of the digestive process, will leave the body. The ejection of materials from the digestive tract, a process called **defecation** (def-e-KĀ-shun), or *egestion*, eliminates materials as **feces**.

The lining of the digestive tract also plays a protective role by safeguarding surrounding tissues against (1) the corrosive effects of digestive acids and enzymes; (2) mechanical stresses, such as abrasion; and (3) bacteria that either are swallowed with food or reside in the digestive tract. The digestive epithelium and its secretions provide a nonspecific defense against these bacteria. When bacteria reach the underlying layer of areolar tissue, the *lamina propria*, macrophages and other cells of the immune system attack them.

We will explore specific functions in more detail as we proceed through the individual regions and components of the system. First, however, we consider several structural and functional characteristics of the system as a whole.

The Digestive Organs and the Peritoneum

The abdominopelvic cavity contains the *peritoneal cavity*, which is lined by a serous membrane consisting of a superficial mesothelium covering a layer of areolar tissue. ∞ pp. 24, 136 We can divide the serous membrane into the serosa or *visceral peritoneum*, which covers organs within the peritoneal cavity, and the *parietal peritoneum*, which lines the inner surfaces of the body wall.

The serous membrane lining the peritoneal cavity continuously produces peritoneal fluid, which provides essential lubrication. Because a thin layer of peritoneal fluid separates the parietal and visceral surfaces, sliding movement can occur without friction and resulting irritation. About 7 liters (7.4 quarts) of fluid are secreted and reabsorbed each day, although the volume within the peritoneal cavity at any one time is very small. Liver disease, kidney disease, and heart failure can cause an increase in the rate at which fluids move into the peritoneal cavity. The accumulation of fluid creates a characteristic abdominal swelling called **ascites** (a-SĪ-tēz). The distortion of internal organs by this fluid can result in symptoms such as heartburn, indigestion, and lower back pain.

Mesenteries

Portions of the digestive tract are suspended within the peritoneal cavity by sheets of serous membrane that connect the parietal peritoneum with the visceral peritoneum. These **mesenteries** (MEZ-en-ter-ēz) are double sheets of peritoneal membrane. The areolar tissue between the mesothelial surfaces provides an access route for the passage of blood vessels, nerves, and lymphatic vessels to and from the digestive tract. Mesenteries also stabilize the positions of the attached organs and prevent the intestines from becoming entangled during digestive movements or sudden changes in body position.

During embryonic development, the digestive tract and accessory organs are suspended within the peritoneal cavity by *dorsal* and *ventral mesenteries* (**Figure 24–2a**). The ventral mesentery later disappears along most of the digestive tract, persisting in adults in only two places: on the ventral surface of the stomach, between the stomach and the liver (the *lesser omentum*; **Figure 24–2b,d**); and between the liver and the anterior abdominal wall (the *falciform ligament*; **Figure 24–2c,d**). The **lesser omentum** (ō-MEN-tum; *omentum*, fat skin) stabilizes the position of the stomach and provides an access route for blood vessels and other structures entering or leaving the liver. The **falciform** (FAL-si-form; *falx*, sickle + *forma*, form) **ligament** helps stabilize the position of the liver relative to the diaphragm and abdominal wall. ATLAS: Embryology Summary 19: The Development of the Digestive System.

As the digestive tract elongates, it twists and turns within the crowded peritoneal cavity. The dorsal mesentery of the stomach becomes greatly enlarged and forms an enormous pouch that extends inferiorly between the body wall and the anterior surface of the small intestine. This pouch, the **greater omentum** (**Figure 24–2b,d**), hangs like an apron from the lateral and inferior borders of the stomach. Adipose tissue in the greater omentum conforms to the shapes of the surrounding organs, providing padding and protection across the anterior and lateral surfaces of the abdomen. The lipids in the adipose tissue are an important energy reserve. The greater omentum also provides insulation that reduces heat loss across the anterior abdominal wall.

All but the first 25 cm (10 in.) of the small intestine is suspended by the **mesentery proper**, a thick mesenterial sheet that provides stability, but permits some independent movement. The mesentery associated with the initial portion of the small intestine (the *duodenum*) and the pancreas fuses with the posterior abdominal wall, locking those structures in position. After this fusion is completed, only their anterior surfaces remain covered by peritoneum. Because their mass lies posterior to, rather than within, the peritoneal cavity, these organs are **retroperitoneal** (*retro*, behind).

A **mesocolon** is a mesentery associated with a portion of the large intestine. During normal development, the mesocolon of the *ascending colon*, the *descending colon*, and the *rectum* of the large intestine fuse to the dorsal body wall. These regions become locked in place. Thereafter, these organs are retroperitoneal, with the visceral peritoneum covering only their anterior surfaces and portions of their lateral surfaces (**Figure 24–2b,c,d**). The **transverse mesocolon**, which supports the transverse colon, and the **sigmoid mesocolon**, which supports the sigmoid colon, are all that remain of the original embryonic mesocolon.

CLINICAL NOTE

Peritonitis An inflammation of the peritoneal membrane is called **peritonitis** (per-i-tō-NĪ-tis), a painful condition that interferes with the normal functioning of the affected organs. Physical damage, chemical irritation, and bacterial invasion of the peritoneum can lead to severe and even fatal cases of peritonitis. In untreated appendicitis, the rupturing of the appendix and the subsequent release of bacteria into the peritoneal cavity may cause peritonitis. Peritonitis is a potential complication of any surgery in which the peritoneal cavity is opened, or of any disease or injury that perforates the walls of the stomach or intestines.

Histological Organization of the Digestive Tract

The major layers of the digestive tract include (1) the *mucosa*, (2) the *submucosa*, (3) the *muscularis externa*, and (4) the *serosa*. The structure of these layers varies by region; **Figure 24–3** is a composite view that most closely resembles the appearance of the small intestine, the longest segment of the digestive tract.

The Mucosa

The inner lining, or **mucosa**, of the digestive tract is a *mucous membrane* consisting of an epithelium, moistened by glandular secretions, and a *lamina propria* of areolar tissue.

The Digestive Epithelium. The mucosal epithelium is either simple or stratified, depending on its location and the stresses to which it is most often subjected. The oral cavity, pharynx, and esophagus (where mechanical stresses are most severe) are lined by a stratified squamous epithelium, whereas the stomach, the small intestine, and almost the entire length

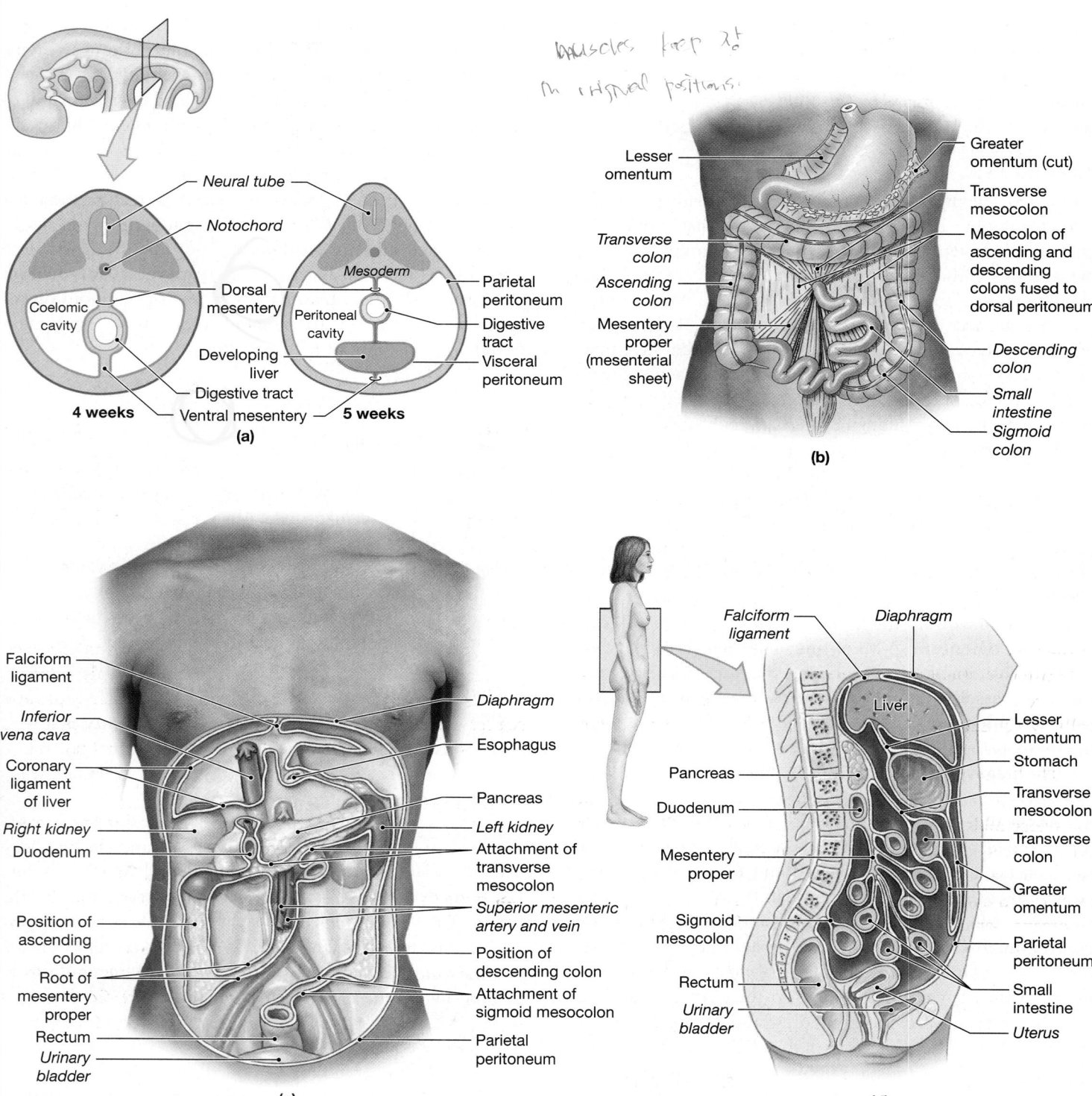

Figure 24–2 **Mesenteries.** **(a)** During embryonic development, the digestive tube is initially suspended by dorsal and ventral mesenteries. In adults, the ventral mesentery is lost, except where it connects the stomach to the liver (at the lesser omentum) and the liver to the anterior body wall and diaphragm (at the falciform ligament). **(b)** A diagrammatic view of the organization of mesenteries in an adult. As the digestive tract enlarges, mesenteries associated with the proximal portion of the small intestine, the pancreas, and the ascending and descending portions of the colon fuse to the body wall. **(c)** An anterior view of the empty peritoneal cavity, showing attachment sites where fusion occurs. **(d)** A sagittal section showing the mesenteries of an adult. Notice that the pancreas, duodenum, and rectum are retroperitoneal.

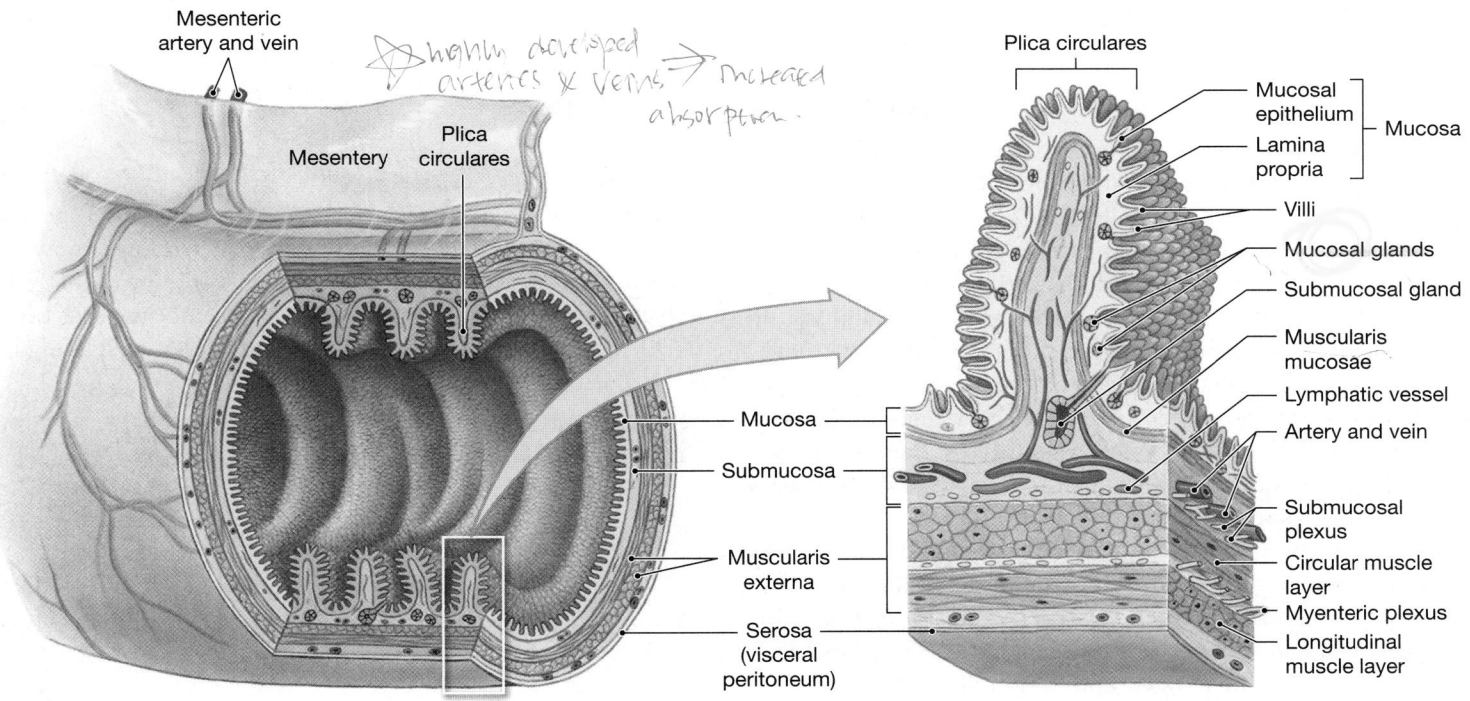

highly developed arteries & veins → increased absorption.

microvilli → increases general surface area of intestines.

Figure 24–3   The Structure of the Digestive Tract. A diagrammatic view of a representative portion of the digestive tract. The features illustrated are typical of those of the small intestine.

of the large intestine (where absorption occurs) have a simple columnar epithelium that contains mucous (goblet) cells. Scattered among the columnar cells are **enteroendocrine cells**, which secrete hormones that coordinate the activities of the digestive tract and the accessory glands.

The lining of the digestive tract is often thrown into longitudinal folds, which disappear as the tract fills, and permanent transverse folds, or *plicae* (PLĪ-sē; singular, *plica* [PLĪ-ka]) *circulares* (**Figure 24–3**). The folding dramatically increases the surface area available for absorption. The secretions of gland cells located in the mucosa and submucosa—or in accessory glandular organs—are carried to the epithelial surfaces by ducts.

The Lamina Propria. The lamina propria consists of a layer of areolar tissue that also contains blood vessels, sensory nerve endings, lymphatic vessels, smooth muscle cells, and scattered areas of lymphoid tissue. In the oral cavity, pharynx, esophagus, stomach, and *duodenum* (the proximal portion of the small intestine), the lamina propria also contains the secretory cells of mucous glands.

In most areas of the digestive tract, the lamina propria contains a narrow sheet of smooth muscle and elastic fibers. This sheet is called the **muscularis** (mus-kū-LĀ-ris) **mucosae** (mū-KŌ-sē) (**Figure 24–3**). The smooth muscle cells in the muscularis mucosae are arranged in two concentric layers. The inner layer encircles the lumen (the *circular muscle*), and the outer layer contains muscle cells oriented parallel to the long axis of the tract (the *longitudinal layer*). Contractions in these layers alter the shape of the lumen and move the epithelial pleats and folds.

The Submucosa

The **submucosa** is a layer of dense irregular connective tissue that surrounds the muscularis mucosae (**Figure 24–3**). The submucosa has large blood vessels and lymphatic vessels, and in some regions it also contains exocrine glands that secrete buffers and enzymes into the lumen of the digestive tract. Along its outer margin, the submucosa contains a network of intrinsic nerve fibers and scattered neurons. This **submucosal plexus**, or *plexus of Meissner*, contains sensory neurons, parasympathetic ganglionic neurons, and sympathetic postganglionic fibers that innervate the mucosa and submucosa.

The Muscularis Externa

The submucosal plexus lies along the inner border of the **muscularis externa,** a region dominated by smooth muscle cells. Like the smooth muscle cells in the muscularis mucosae, those in the muscularis externa are arranged in an inner circular layer and an outer longitudinal layer. These layers play an essential role in mechanical processing and in the movement of materials along the digestive tract. The movements are coordinated primarily by the sensory neurons, interneurons, and motor neurons of the enteric nervous system (ENS). The ENS is innervated primarily by the parasympathetic division of the ANS. Sympathetic postganglionic fibers also synapse here, although many continue onward to innervate the mucosa and the **myenteric** (mī-en-TER-ik) **plexus** (*mys*, muscle + *enteron*, intestine), or *plexus of Auerbach*. This network of parasympathetic ganglia, sensory neurons, interneurons, and sympathetic postganglionic fibers lies sandwiched between the circular and longitudinal muscle layers. In general, parasympathetic stimulation increases muscle tone and activity; sympathetic stimulation promotes muscular inhibition and relaxation.

Tips&Tricks

Because the parasympathetic nervous system plays a role in the digestive process, it is often referred to as the "rest and digest" division.

The Serosa

Along most portions of the digestive tract inside the peritoneal cavity, the muscularis externa is covered by a serous membrane known as the **serosa** (**Figure 24–3**). There is no serosa covering the muscularis externa of the oral cavity, pharynx, esophagus, and rectum, where a dense network of collagen fibers firmly attaches the digestive tract to adjacent structures. This fibrous sheath is called an *adventitia* (adven-TISH-uh).

The Movement of Digestive Materials

The muscular layers of the digestive tract consist of *visceral smooth muscle tissue*, a type of smooth muscle introduced in Chapter 10. p. 330 The smooth muscle along the digestive tract has rhythmic cycles of activity due to the presence of *pacesetter cells*. These smooth muscle cells undergo spontaneous depolarization, triggering a wave of contraction that spreads throughout the entire muscular sheet. Pacesetter cells are located in the muscularis mucosae and muscularis externa, the layers of which surround the lumen of the digestive

tract. The coordinated contractions of the muscularis externa play a vital role in the movement of materials along the tract, through *peristalsis*, and in mechanical processing, through *segmentation.*

Peristalsis

The muscularis externa propels materials from one portion of the digestive tract to another by contractions known as **peristalsis** (per-i-STAL-sis). Peristalsis (**Figure 24–4**) consists of waves of muscular contractions that move a **bolus** (BŌ-lus), or small oval mass of digestive contents, along the

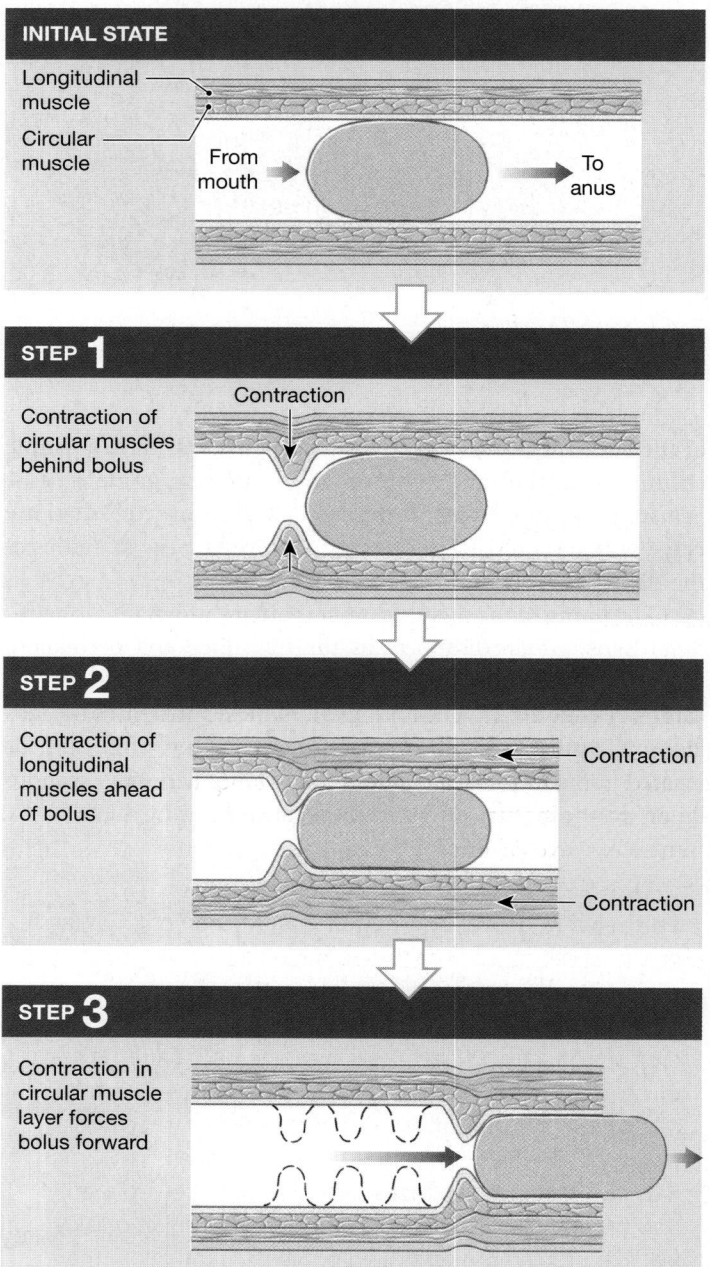

Figure 24–4 Peristalsis. Peristalsis propels materials along the length of the digestive tract.

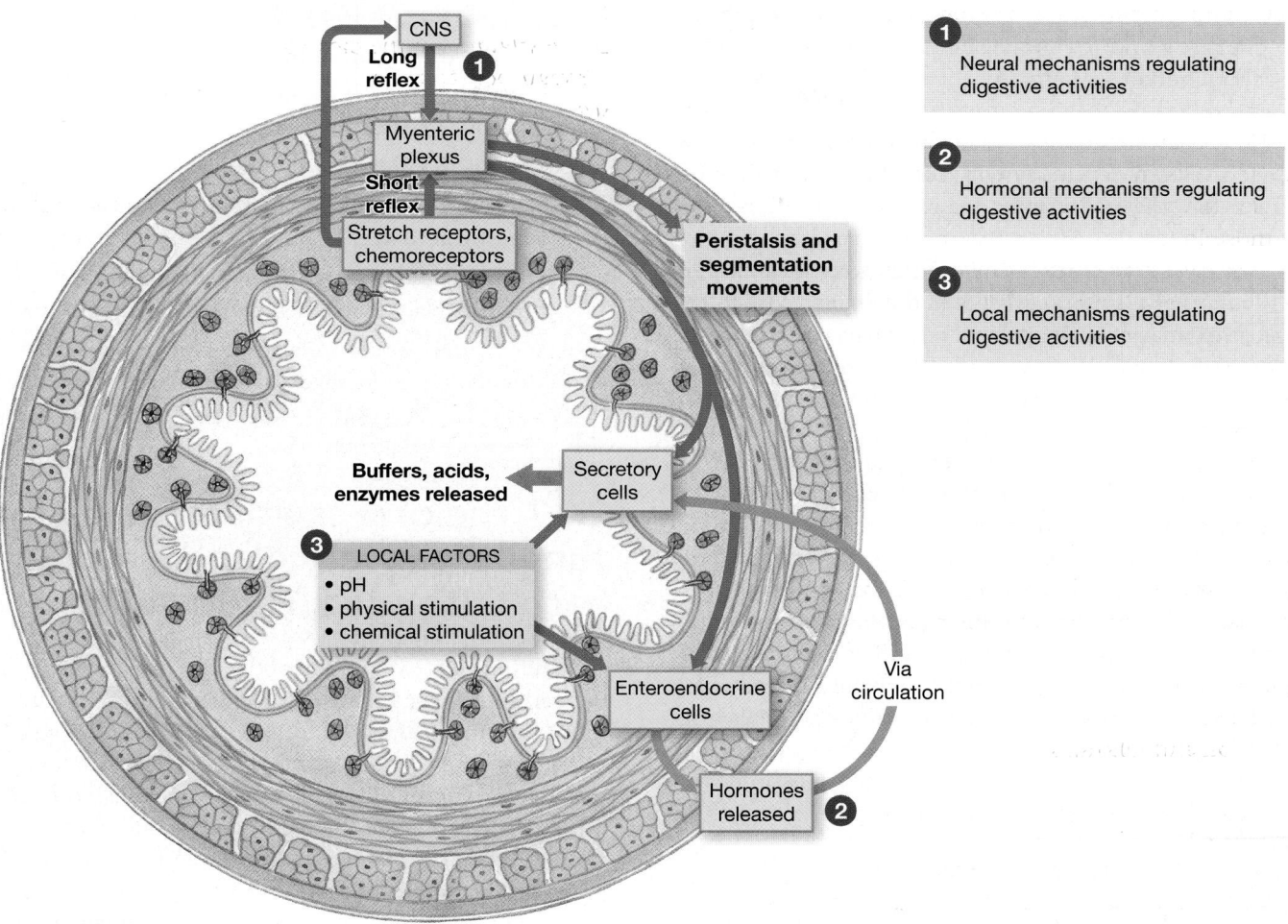

Figure 24–5 **The Regulation of Digestive Activities.** The major factors responsible for regulating digestive activities are neural mechanisms, hormonal mechanisms, and local mechanisms.

length of the digestive tract. During a peristaltic movement, the circular muscles contract behind the bolus while circular muscles ahead of the bolus relax. Longitudinal muscles ahead of the bolus then contract, shortening adjacent segments. A wave of contraction in the circular muscles then forces the bolus forward.

Tips&Tricks

Squeezing toothpaste out of a tube is similar to peristalsis: Your squeezing hand (contracting circular muscles) forces toothpaste (the bolus) along and out of the tube (the digestive tract).

Segmentation

Most areas of the small intestine and some portions of the large intestine undergo cycles of contraction that churn and fragment the bolus, mixing the contents with intestinal secretions. This activity, called **segmentation**, does not follow a set pattern, and thus does not push materials along the tract in any one direction.

Control of Digestive Functions

The activities of the digestive system are regulated by neural, hormonal, and local mechanisms (**Figure 24–5**).

Neural Mechanisms

The movement of materials along your digestive tract, as well as many secretory functions, is controlled primarily by neural mechanisms. For example, peristaltic movements limited to a few centimeters of the digestive tract are triggered by sensory receptors in the walls of the digestive tract. The motor neurons that control smooth muscle contraction and glandular secretion are located in the myenteric plexus. These neurons are usually considered parasympathetic, because some of them are innervated by parasympathetic preganglionic fibers. However, the plexus also contains sensory neurons, motor neurons, and interneurons responsible for local reflexes whose internal workings operate entirely outside the control of the central nervous system. As noted in Chapter 16, the reflexes controlled by these neurons are called *short reflexes.* ∞ p. 547 These reflexes are also called *myenteric reflexes*, and the term *enteric nervous system* is often used to refer to the neural network that coordinates the myenteric reflexes along the digestive tract.

In general, short reflexes control relatively localized activities that involve small segments of the digestive tract. For example, they may coordinate local peristalsis and trigger the secretion of digestive glands in response to the arrival of a bolus. Many neurons are involved: The enteric nervous system has roughly as many neurons as the spinal cord, and as many neurotransmitters as the brain. The specific functions and interactions of these neurotransmitters in the enteric nervous system remain largely unknown.

Sensory information from receptors in the digestive tract is also distributed to the CNS, where it can trigger *long reflexes* ∞ p. 547, which involve interneurons and motor neurons in the CNS. Long reflexes provide a higher level of control over digestive and glandular activities, generally controlling large-scale peristaltic waves that move materials from one region of the digestive tract to another. Long reflexes may involve parasympathetic motor fibers in the glossopharyngeal, vagus, or pelvic nerves that synapse in the myenteric plexus.

Hormonal Mechanisms ~secretes 18 hormones =endocrine

The sensitivity of the smooth muscle cells to neural commands can be enhanced or inhibited by digestive hormones. The digestive tract produces at least 18 hormones that affect almost every aspect of digestive function, and some of them also affect the activities of other systems. The hormones (*gastrin*, *secretin*, and others), which are peptides produced by enteroendocrine cells in the digestive tract, reach their target organs after their distribution in the bloodstream. We will consider each of these hormones further as we proceed down the length of the digestive tract.

Local Mechanisms

Prostaglandins, histamine, and other chemicals released into interstitial fluid may affect adjacent cells within a small segment of the digestive tract. These local messengers are important in coordinating a response to changing conditions (such as variations in the local pH or certain chemical or physical stimuli) that affect only a portion of the tract. For example, the release of histamine in the lamina propria of the stomach stimulates the secretion of acid by cells in the adjacent epithelium.

↳ vasdilation of smooth muscle.

CHECKPOINT

1. **Identify the organs of the digestive system.**

2. **List and define the six primary functions of the digestive system.**

3. **What is the importance of the mesenteries?**

4. **Name the layers of the gastrointestinal tract from superficial to deep.**

5. **Which is more efficient in propelling intestinal contents from one place to another: peristalsis or segmentation?**

6. **What effect would a drug that blocks parasympathetic stimulation of the digestive tract have on peristalsis?**

See the blue Answers tab at the end of the book.

24-2 The oral cavity contains the tongue, salivary glands, and teeth, each with specific functions

We continue our exploration of the digestive tract by following the path of ingested materials. We begin at the mouth, which opens into the **oral cavity**, or **buccal** (BUK-ul) **cavity** (**Figure 24–6**). The functions of the oral cavity include (1) *sensory analysis* of material before swallowing; (2) *mechanical processing* through the actions of the teeth, tongue, and palatal surfaces; (3) *lubrication* by mixing with mucus and salivary gland secretions; and (4) limited *digestion* of carbohydrates and lipids.

The oral cavity is lined by the **oral mucosa**, which has a stratified squamous epithelium. A layer of keratinized cells covers regions exposed to severe abrasion, such as the superior surface of the tongue and the opposing surface of the hard palate (part of the roof of the mouth). The epithelial lining of the cheeks, lips, and inferior surface of the tongue is relatively thin, nonkeratinized, and delicate. Although nutrients are not absorbed in the oral cavity, the mucosa inferior to the tongue is thin enough and vascular enough to permit the rapid absorption of lipid-soluble drugs. *Nitroglycerin* may be administered via this route to treat acute angina attacks. ∞ p. 694

The mucosae of the **cheeks**, or lateral walls of the oral cavity, are supported by pads of fat and the buccinator muscles. Anteriorly, the mucosa of each cheek is continuous with that of the lips, or **labia** (LĀ-bē-uh; singular, *labium*). The **vestibule** is the space between the cheeks (or lips) and the teeth. The **gingivae** (JIN-ji-vē), or *gums*, are ridges of oral mucosa that surround the base of each tooth on the alveolar processes of the maxillary bones and mandible. In most regions, the gingivae are firmly bound to the periostea of the underlying bones.

The hard and soft palates form the roof of the oral cavity; the tongue dominates its floor (**Figure 24–6b**). The floor of the mouth inferior to the tongue receives extra support from the

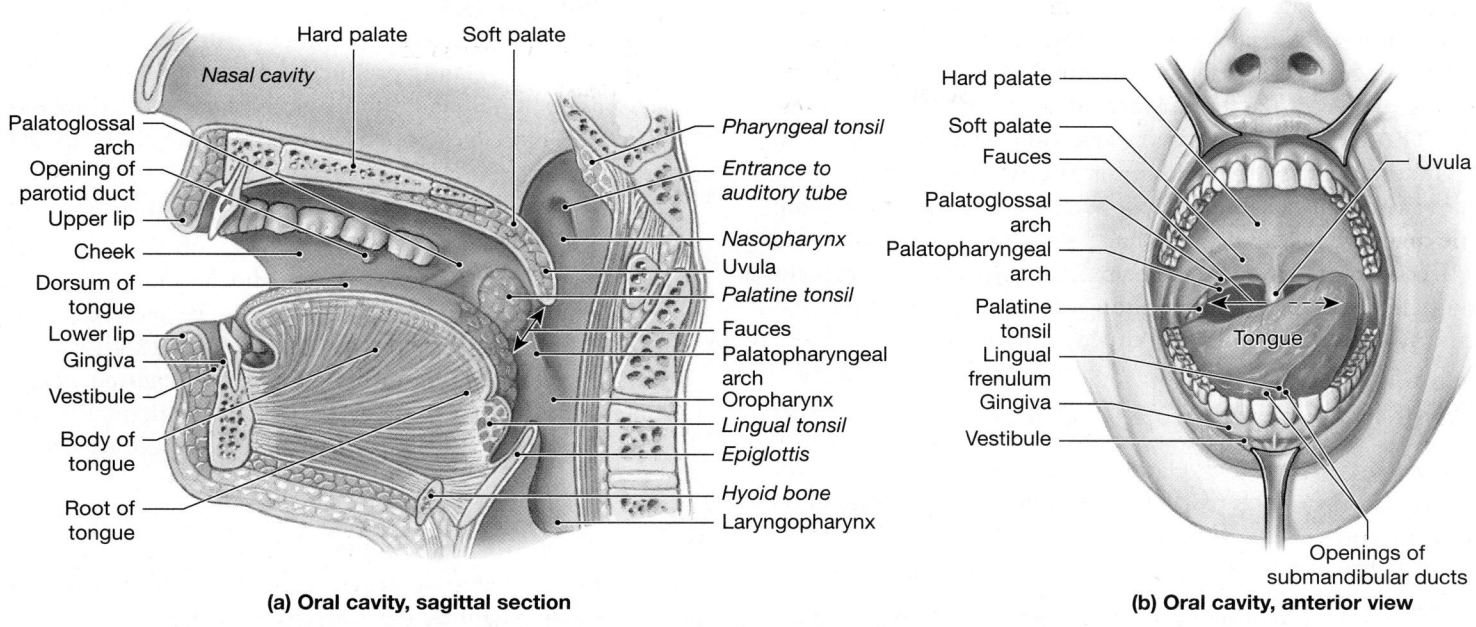

(a) Oral cavity, sagittal section

(b) Oral cavity, anterior view

Figure 24–6 **The Oral Cavity. (a)** A sagittal section of the oral cavity. **(b)** An anterior view of the oral cavity, as seen through the open mouth.
ATLAS: Plates 11a; 19

geniohyoid and mylohyoid muscles. ∞ p. 351 The palatine processes of the maxillary bones and the horizontal plates of the palatine bones form the *hard palate*. A prominent central ridge, or *raphe* (RĀ-fee), extends along the midline of the hard palate. The mucosa lateral and anterior to the raphe is thick, with complex ridges. When your tongue compresses food against the hard palate, these ridges provide traction. The *soft palate* lies posterior to the hard palate. A thinner and more delicate mucosa covers the posterior margin of the hard palate and extends onto the soft palate.

The posterior margin of the soft palate supports the **uvula** (Ū-vū-luh), a dangling process that helps prevent food from entering the pharynx prematurely (**Figure 24–6a**). On either side of the uvula are two pairs of muscular *pharyngeal arches* (**Figure 24–6b**). The more anterior **palatoglossal** (pal-a-tō-GLOS-al) **arch** extends between the soft palate and the base of the tongue. A curving line that connects the palatoglossal arches and uvula forms the boundaries of the **fauces** (FAW-sēz), the passageway between the oral cavity and the oropharynx. The more posterior **palatopharyngeal** (pal-a-tō-fa-RIN-jē-al) **arch** extends from the soft palate to the pharyngeal wall. A palatine tonsil lies between the palatoglossal and palatopharyngeal arches on either side.

The Tongue

The **tongue** (**Figure 24–6**) manipulates materials inside the mouth and is occasionally used to bring foods (such as ice cream on a cone) into the oral cavity. The primary functions of the tongue are (1) mechanical processing by compression, abrasion, and distortion; (2) manipulation to assist in chewing and to prepare material for swallowing; (3) sensory analysis by touch, temperature, and taste receptors; and (4) secretion of mucins and the enzyme *lingual lipase*.

We can divide the tongue into an anterior **body,** or *oral portion*, and a posterior **root,** or *pharyngeal portion*. The superior surface, or *dorsum*, of the body contains a forest of fine projections, the *lingual papillae*. ∞ p. 564 The thickened epithelium covering each papilla assists the tongue in moving materials. A V-shaped line of circumvallate papillae roughly demarcates the boundary between the body and the root of the tongue, which is situated in the oropharynx (**Figure 24–6a**).

The epithelium covering the inferior surface of the tongue is thinner and more delicate than that of the dorsum. Along the inferior midline is the **lingual frenulum** (FREN-ū-lum; *frenulum*, a small bridle), a thin fold of mucous membrane that connects the body of the tongue to the mucosa covering the floor of the oral cavity (**Figure 24–6a**). Ducts from two pairs of salivary glands open on either side of the lingual frenulum, which serves to prevent extreme movements of the tongue. However, an overly restrictive lingual frenulum prevents normal eating or speech. Properly diagnosed, this condition, called *ankyloglossia* (ang-ki-lō-GLOS-ē-uh), can be corrected surgically.

The tongue's epithelium is flushed by the secretions of small glands that extend into the underlying lamina propria. These secretions contain water, mucins, and the enzyme **lingual lipase,** which works over a broad pH range (3.0–6.0), enabling it to start lipid digestion immediately. Because lingual lipase tolerates an acid environment, it can continue to break down lipids—specifically, triglycerides—for a considerable time after the food reaches the stomach.

The tongue contains two groups of skeletal muscles. The relatively large **extrinsic tongue muscles** perform all gross movements of the tongue. ∞ p. 350 The smaller **intrinsic tongue muscles** change the shape of the tongue and assist the extrinsic muscles during precise movements, as in speech. Both intrinsic and extrinsic tongue muscles are under the control of the hypoglossal cranial nerves (N XII).

Salivary Glands

Three pairs of salivary glands secrete into the oral cavity (**Figure 24–7a**). Each pair has a distinctive cellular organization and produces *saliva*, a mixture of glandular secretions, with slightly different properties:

1. The large **parotid** (pa-ROT-id) **salivary glands** lie inferior to the zygomatic arch deep to the skin covering the lateral and posterior surface of the mandible. Each gland has an irregular shape, extending from the mastoid process of the temporal bone across the outer surface of the masseter muscle. The parotid salivary glands produce a serous secretion containing large amounts of *salivary amylase*, an enzyme that breaks down starches (complex carbohydrates). The secretions of each parotid gland are drained by a **parotid duct** (*Stensen duct*), which empties into the vestibule at the level of the second upper molar.

2. The **sublingual** (sub-LING-gwal) **salivary glands** are covered by the mucous membrane of the floor of the mouth. These glands produce a mucous secretion that acts as a buffer and lubricant. Numerous **sublingual ducts** (*Rivinus ducts*) open along either side of the lingual frenulum.

3. The **submandibular salivary glands** are in the floor of the mouth along the inner surfaces of the mandible within a depression called the *mandibular groove*. Cells of the submandibular glands (**Figure 24–7b**) secrete a mixture of buffers, glycoproteins called *mucins*, and salivary amylase. The **submandibular ducts** (*Wharton ducts*) open into the mouth on either side of the lingual frenulum immediately posterior to the teeth (**Figure 24–6b**).

Saliva

The salivary glands produce 1.0–1.5 liters of saliva each day. Saliva is 99.4 percent water; the remaining 0.6 percent includes an assortment of electrolytes (principally Na^+, Cl^-, and HCO_3^-), buffers, glycoproteins, antibodies, enzymes, and waste products. The glycoproteins, called **mucins**, are primarily responsible for the lubricating action of saliva. About 70 percent of saliva originates in the submandibular salivary glands, 25 percent in the parotids, and the remaining 5 percent in the sublingual salivary glands.

A continuous background level of saliva secretion flushes the oral surfaces, helping keep them clean. Buffers in the saliva keep the pH of your mouth near 7.0 and prevent the buildup of acids produced by bacterial action. In addition,

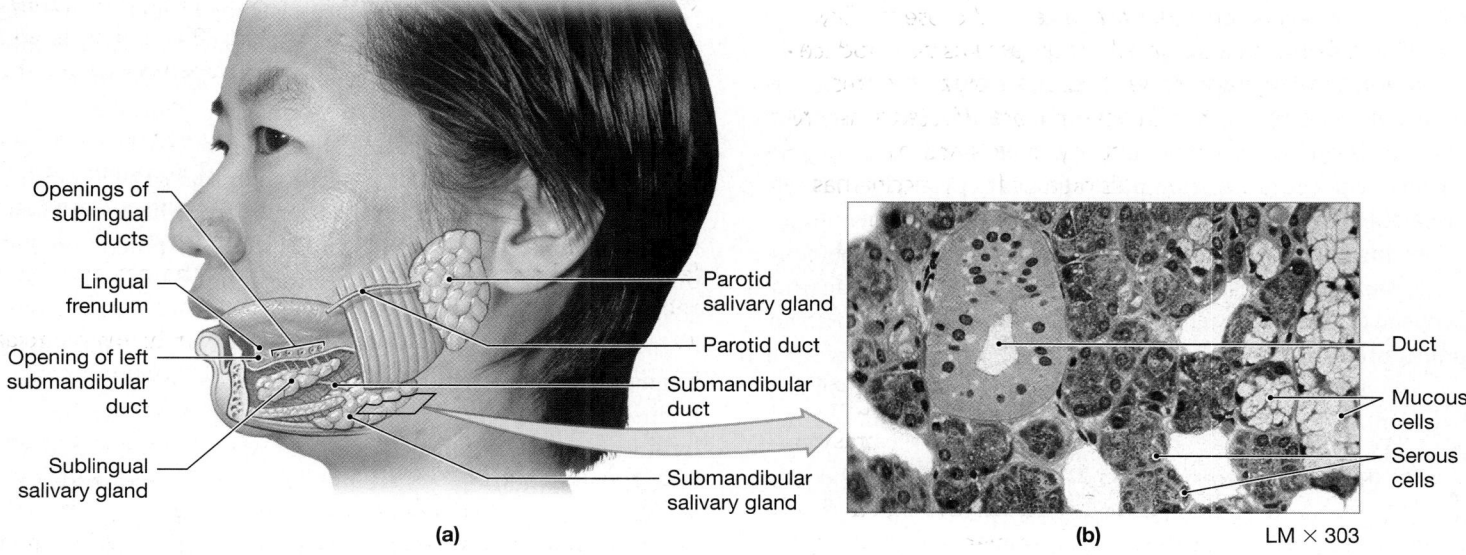

Openings of sublingual ducts

Lingual frenulum

Opening of left submandibular duct

Sublingual salivary gland

Parotid salivary gland

Parotid duct

Submandibular duct

Submandibular salivary gland

Duct

Mucous cells

Serous cells

(a)

(b)   LM × 303

Figure 24–7   The Salivary Glands. (a) A lateral view, showing the relative positions of the salivary glands and ducts on the left side of the head. For clarity, the left ramus and body of the mandible have been removed. For the positions of the parotid and submandibular ducts in the oral cavity, see *Figure 24–6*. **(b)** The submandibular gland secretes a mixture of mucins, produced by mucous cells, and enzymes, produced by serous cells. ATLAS: Plates 3c,d; 18a,b

saliva contains antibodies (IgA) and *lysozyme,* which help control populations of oral bacteria. A reduction in or elimination of salivary secretions—caused by radiation exposure, emotional distress, certain drugs, sleep, or other factors—triggers a bacterial population explosion in the oral cavity. This proliferation rapidly leads to recurring infections and progressive erosion of the teeth and gums.

The saliva produced when you eat has a variety of functions, including the following:

- Lubricating the mouth.
- Moistening and lubricating materials in the mouth.
- Dissolving chemicals that can stimulate the taste buds and provide sensory information about the material.
- Initiating the digestion of complex carbohydrates before the material is swallowed. The enzyme involved is **salivary amylase**, also known as *ptyalin* or *alpha-amylase*. Although the digestive process begins in the oral cavity, it is not completed there, and no absorption of nutrients occurs across the lining of the oral cavity. Saliva also contains a small amount of lingual lipase that is secreted by the glands of the tongue.

CLINICAL NOTE

Mumps The *mumps virus* most often targets the salivary glands, especially the parotid salivary glands, although other organs can also become infected. Infection typically occurs at 5–9 years of age. The first exposure stimulates the production of antibodies and, in most cases, confers permanent immunity. In post-adolescent males, the mumps virus can also infect the testes and cause sterility. Infection of the pancreas by the mumps virus can produce temporary or permanent diabetes; other organ systems, including the central nervous system, are affected in severe cases. A mumps vaccine effectively confers active immunity. Widespread administration of that vaccine has almost eliminated the incidence of the disease in the United States.

Control of Salivary Secretions

Salivary secretions are normally controlled by the autonomic nervous system. Each salivary gland receives parasympathetic and sympathetic innervation. The parasympathetic outflow originates in the **salivatory nuclei** of the medulla oblongata and synapses in the submandibular and otic ganglia. ∞ pp. 495, 497 Any object in your mouth can trigger a salivary reflex by stimulating receptors monitored by the trigeminal cranial nerve (N V) or taste buds innervated by cranial nerves VII, IX, or X. Parasympathetic stimulation accelerates

secretion by all the salivary glands, resulting in the production of large amounts of saliva. The role of sympathetic innervation remains unclear; evidence suggests that it provokes the secretion of small amounts of very thick saliva.

The salivatory nuclei are also influenced by other brain stem nuclei, as well as by the activities of higher centers. For example, chewing with an empty mouth, smelling food, or even thinking about food initiates an increase in salivary secretion rates; that is why chewing gum is so effective at keeping your mouth moist. The presence of irritating stimuli in the esophagus, stomach, or intestines also accelerates the production of saliva, as does nausea. Increased saliva production in response to unpleasant stimuli helps reduce the magnitude of the stimulus by dilution, by a rinsing action, or by buffering strong acids or bases.

The Teeth

Movements of the tongue are important in passing food across the opposing surfaces, or occlusal surfaces, of the **teeth**. These surfaces perform chewing, or **mastication** (mas-ti-KĀ-shun), of food. Mastication breaks down tough connective tissues in meat and the plant fibers in vegetable matter, and it helps saturate the materials with salivary secretions and enzymes.

Figure 24–8a is a sectional view through an adult tooth. The bulk of each tooth consists of a mineralized matrix similar to that of bone. This material, called **dentin**, differs from bone in that it does not contain cells. Instead, cytoplasmic processes extend into the dentin from cells in the central **pulp cavity**, an interior chamber. The pulp cavity receives blood vessels and nerves through the **root canal**, a narrow tunnel located at the **root**, or base, of the tooth. Blood vessels and nerves enter the root canal through an opening called the **apical foramen** to supply the pulp cavity.

The root of each tooth sits in a bony socket called an *alveolus*. Collagen fibers of the **periodontal ligament** extend from the dentin of the root to the bone of the alveolus, creating a strong articulation known as a gomphosis. ∞ p. 269 A layer of **cementum** (se-MEN-tum) covers the dentin of the root, providing protection and firmly anchoring the periodontal ligament. Cementum is very similar in histological structure to bone and is less resistant to erosion than is dentin.

The **neck** of the tooth marks the boundary between the root and the **crown**, the exposed portion of the tooth that projects beyond the soft tissue of the gingiva. A shallow groove called the **gingival sulcus** surrounds the neck of each tooth. The mucosa of the gingival sulcus is very thin and is not tightly bound to the periosteum. The epithelium is bound to the tooth at the base of the sulcus. This epithelial attachment prevents bacterial access to the lamina propria of the

24 DIGESTIVE

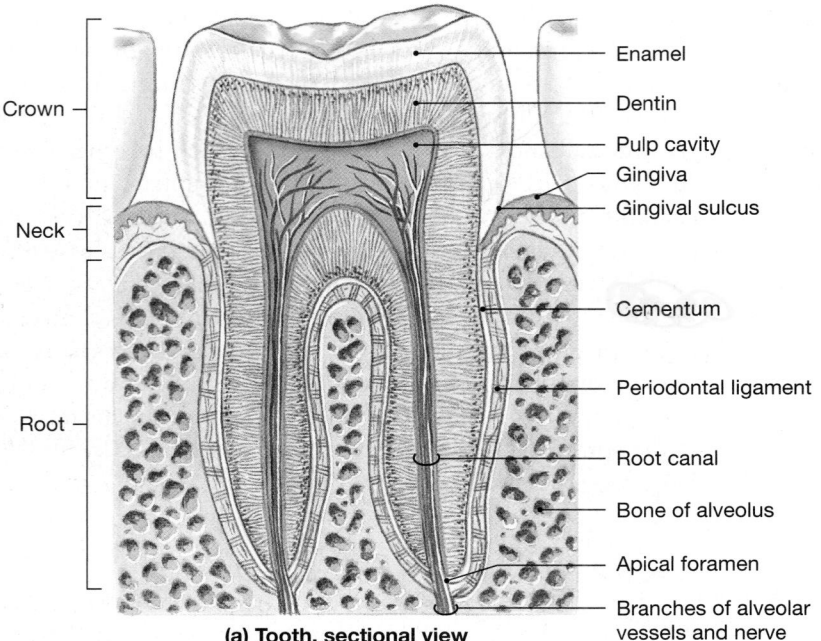

Enamel

Dentin

Pulp cavity

Gingiva

Gingival sulcus

Cementum

Periodontal ligament

Root canal

Bone of alveolus

Apical foramen

Branches of alveolar vessels and nerve

Crown

Neck

Root

(a) Tooth, sectional view

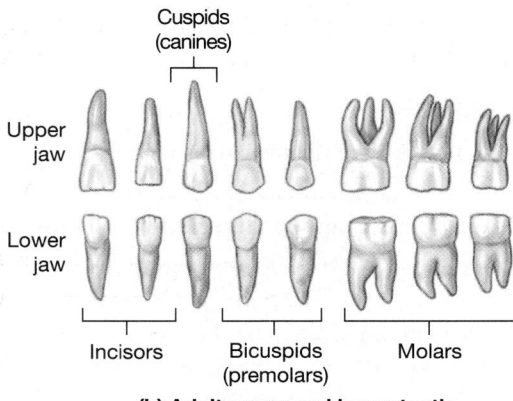

Figure 24–8   Teeth. (a) A diagrammatic section through a typical adult tooth. **(b)** The adult teeth from the right side of the upper and lower jaws. *Figure 24–9* provides a view of the occlusal surfaces.

Cuspids (canines)

Upper jaw

Lower jaw

Incisors

Bicuspids (premolars)

Molars

(b) Adult upper and lower teeth

gingiva and the relatively soft cementum of the root. When you brush and massage your gums, you stimulate the epithelial cells and strengthen the attachment. A condition called *gingivitis*, a bacterial infection of the gingivae, can occur if the attachment breaks down.

A layer of **enamel** covers the dentin of the crown. Enamel, which contains calcium phosphate in a crystalline form, is the hardest biologically manufactured substance. Adequate amounts of calcium, phosphates, and vitamin D during childhood are essential if the enamel coating is to be complete and resistant to decay.

Tooth decay generally results from the action of bacteria that inhabit your mouth. Bacteria adhering to the surfaces of the teeth produce a sticky matrix that traps food particles and creates deposits known as *dental plaque*. Over time, this organic material can become calcified, forming a hard layer of *tartar*, or *dental calculus*, which can be difficult to remove. Tartar deposits most commonly develop at or near the gingival sulcus, where brushing cannot remove the relatively soft plaque deposits.

Types of Teeth

The alveolar processes of the maxillae and the mandible form the *maxillary* and *mandibular arcades*, or upper and lower dental arches, respectively. These arcades contain four types of teeth, each with specific functions (**Figure 24–8b**):

1. **Incisors** (in-SĪ-zerz) are blade-shaped teeth located at the front of the mouth. Incisors are useful for clipping or cutting, as when you nip off the tip of a carrot stick. These teeth have a single root.

2. The **cuspids** (KUS-pidz), or *canines*, are conical, with a sharp ridgeline and a pointed tip. They are used for tearing or slashing. You might weaken a tough piece of celery using the clipping action of the incisors and then take advantage of the shearing action provided by the cuspids. Cuspids have a single root.

3. **Bicuspids** (bī-KUS-pidz), or *premolars*, have flattened crowns with prominent ridges. They crush, mash, and grind. Bicuspids have one or two roots.

4. **Molars** have very large, flattened crowns with prominent ridges adapted for crushing and grinding. You can usually shift a tough nut to your bicuspids and molars for successful crunching. Molars typically have three or more roots.

Dental Succession

Two sets of teeth form during development. The first to appear are the **deciduous teeth** (de-SID-ū-us; *deciduus*, falling off), the temporary teeth of the **primary dentition**. Deciduous teeth are also called *primary teeth, milk teeth*, or *baby teeth*. Most children have 20 deciduous teeth (**Figure 24–9a**). On each side of the upper or lower jaw, the primary dentition consists of two incisors, one cuspid, and a pair of deciduous molars. These teeth will later be replaced by the **secondary dentition**, or *permanent dentition* of the larger adult jaws (**Figure 24–9b**). Three additional molars appear on each side of the upper and lower jaws as the individual ages, extending the length of the tooth rows posteriorly and bringing the permanent tooth count to 32.

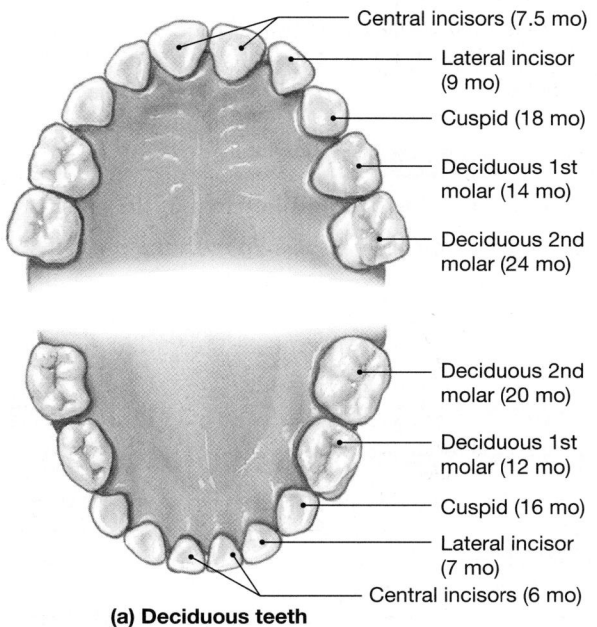

Central incisors (7.5 mo)
Lateral incisor (9 mo)
Cuspid (18 mo)
Deciduous 1st molar (14 mo)
Deciduous 2nd molar (24 mo)

Deciduous 2nd molar (20 mo)
Deciduous 1st molar (12 mo)
Cuspid (16 mo)
Lateral incisor (7 mo)
Central incisors (6 mo)

(a) Deciduous teeth

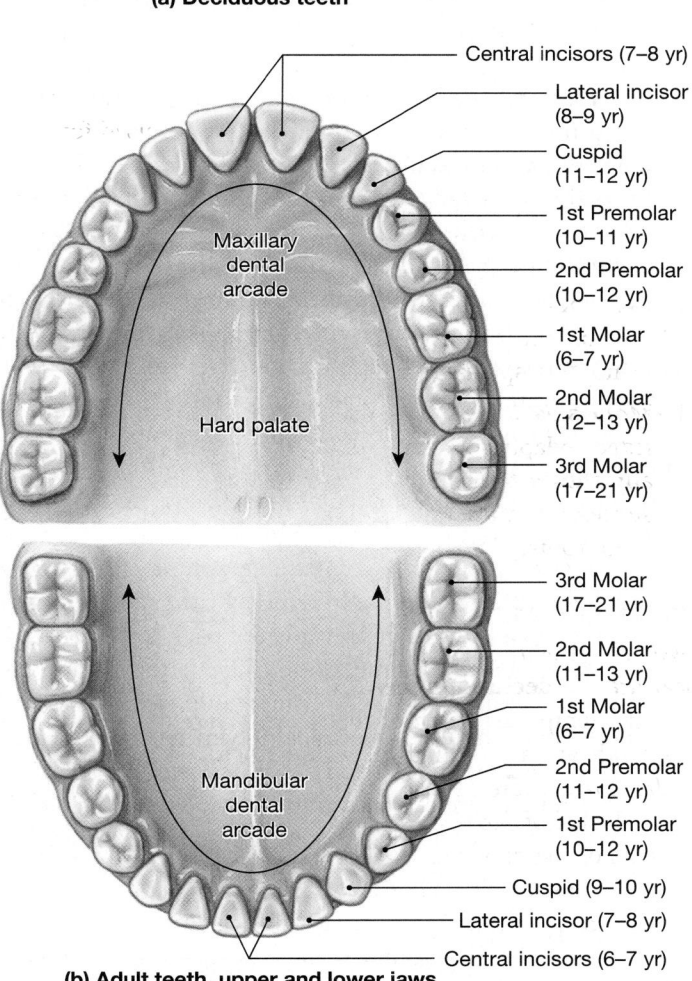

Central incisors (7–8 yr)
Lateral incisor (8–9 yr)
Cuspid (11–12 yr)
1st Premolar (10–11 yr)
2nd Premolar (10–12 yr)
1st Molar (6–7 yr)
2nd Molar (12–13 yr)
3rd Molar (17–21 yr)

Maxillary dental arcade
Hard palate

3rd Molar (17–21 yr)
2nd Molar (11–13 yr)
1st Molar (6–7 yr)
2nd Premolar (11–12 yr)
1st Premolar (10–12 yr)
Cuspid (9–10 yr)
Lateral incisor (7–8 yr)
Central incisors (6–7 yr)

Mandibular dental arcade

(b) Adult teeth, upper and lower jaws

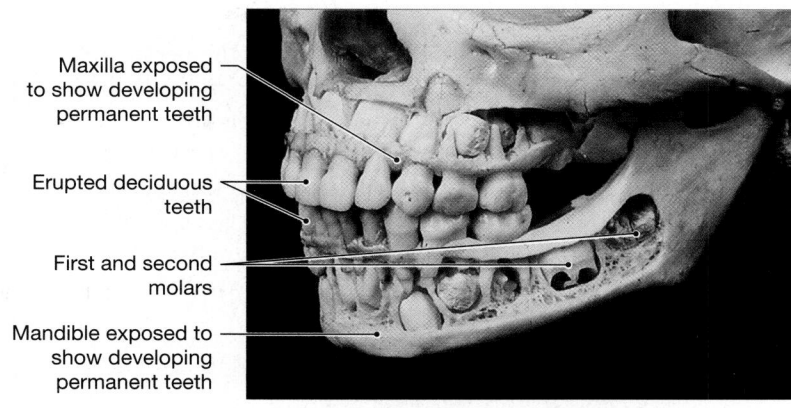

Maxilla exposed to show developing permanent teeth
Erupted deciduous teeth
First and second molars
Mandible exposed to show developing permanent teeth

(c) Exposed unerupted teeth

Figure 24–9   Primary and Secondary Dentitions. (a) The primary teeth, with the age at eruption given in months. **(b)** The adult teeth, with the age at eruption given in years. **(c)** Maxilla and mandible with unerupted teeth exposed.

Tips&Tricks

Here's how to remember the timing of dental succession: The deciduous (primary) teeth are present during the primary grades, and the secondary (permanent) teeth are present during and after the secondary grades.

As replacement proceeds, the periodontal ligaments and roots of the primary teeth erode until the deciduous teeth either fall out or are pushed aside by the **eruption**, or emergence, of the secondary teeth (**Figure 24–9c**). The adult premolars take the place of the deciduous molars, and the adult molars extend the tooth row as the jaw enlarges. The third molars, or *wisdom teeth*, may not erupt before age 21. Wisdom teeth may fail to erupt because they develop in inappropriate positions or because space on the dental arch is inadequate. Any teeth that develop in locations that do not permit their eruption are called *impacted teeth*. Impacted wisdom teeth can be surgically removed to prevent the formation of abscesses.

Mastication

The *muscles of mastication* close your jaws and slide or rock your lower jaw from side to side. p. 348 Chewing is not a simple process; it can involve any combination of mandibular elevation/depression, protraction/retraction, and medial/lateral movement. (Try classifying the movements involved the next time you eat.)

During mastication, you force food from the oral cavity to the vestibule and back, crossing and recrossing the **occlusal surfaces**. This movement results in part from the action of the muscles of mastication, but control would be impossible without the aid of the muscles of the cheeks, lips, and tongue. Once you have shredded or torn the material to a satisfactory consistency and have moistened it with salivary secretions, your tongue begins compacting the debris into a moist, cohesive bolus that can be swallowed relatively easily.

CHECKPOINT

7. Name the structures associated with the oral cavity.

8. Which type of epithelium lines the oral cavity?

9. The digestion of which nutrient would be affected by damage to the parotid salivary glands?

10. Which type of tooth is most useful for chopping off bits of relatively rigid foods?

11. Where exactly in the human body is the fauces?

See the blue Answers tab at the end of the book.

24-3 The pharynx is a passageway between the oral cavity and esophagus

The **pharynx** (FAR-ingks), an anatomical space, serves as a common passageway for solid food, liquids, and air. The epithelial lining and regions of the pharynx—the nasopharynx, the oropharynx, and the laryngopharynx—were described in Chapter 23. ∞ p. 831 Food normally passes through the oropharynx and laryngopharynx on its way to the esophagus. Both of these regions have a stratified squamous epithelium similar to that of the oral cavity. The lamina propria contains scattered mucous glands and the lymphoid tissue of the pharyngeal, palatal, and lingual tonsils. Deep to the lamina propria lies a dense layer of elastic fibers, bound to the underlying skeletal muscles.

The specific pharyngeal muscles involved in swallowing were described in Chapter 11. ∞ p. 350

- The *pharyngeal constrictor muscles* push the bolus toward the esophagus.

- The *palatopharyngeus* and *stylopharyngeus muscles* elevate the larynx.

- The *palatal muscles* elevate the soft palate and adjacent portions of the pharyngeal wall.

These muscles cooperate with muscles of the oral cavity and esophagus to initiate swallowing, which pushes the bolus along the esophagus and into the stomach.

CHECKPOINT

12. Describe the structure and function of the pharynx.

13. Identify the muscles associated with the pharynx.

See the blue Answers tab at the end of the book.

24-4 The esophagus is a muscular tube that transports solids and liquids from the pharynx to the stomach

The **esophagus** (Figure 24–10) is a hollow muscular tube with a length of approximately 25 cm (10 in.) and a diameter of about 2 cm (0.80 in.) at its widest point. The primary function of the esophagus is to convey solid food and liquids to the stomach.

The esophagus begins posterior to the cricoid cartilage, at the level of vertebra C_6. From this point, where it is at its narrowest, the esophagus descends toward the thoracic cavity posterior to the trachea. It passes inferiorly along the dorsal wall of the mediastinum and enters the abdominopelvic cavity through the **esophageal hiatus** (hī-Ā-tus), an opening in the diaphragm. The esophagus then empties into the stomach anterior to vertebra T_7.

The esophagus is innervated by parasympathetic and sympathetic fibers from the esophageal plexus. ∞ p. 545 Resting muscle tone in the circular muscle layer in the superior 3 cm (1.2 in.) of the esophagus normally prevents air from entering the esophagus. A comparable zone at the inferior end of the esophagus normally remains in a state of active contraction. This condition prevents the backflow of materials from the stomach into the esophagus. Neither region has a well-defined sphincter muscle comparable to those located elsewhere along the digestive tract. Nevertheless, the terms *upper esophageal sphincter* and *lower esophageal sphincter (cardiac sphincter)* are often used to describe these regions at either end of the esophagus, because they are similar in function to other sphincters.

Tips&Tricks

Just as security gates control the passage of people by opening and closing, sphincters control the passage of material through them by dilating and constricting.

Figure 24–10　**The Esophagus. (a)** A transverse section through an empty esophagus. **(b)** The esophageal mucosa.

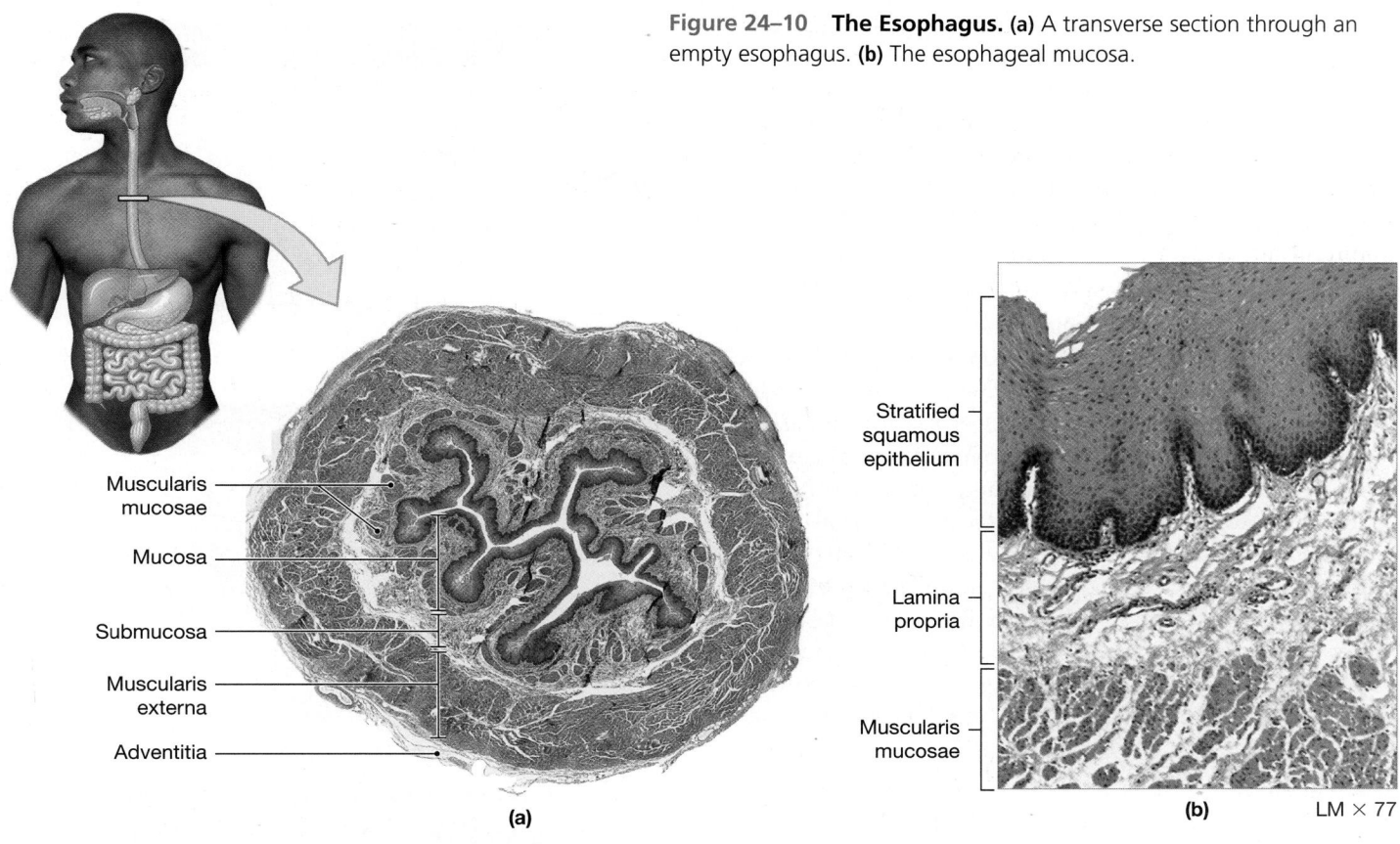

Muscularis mucosae

Mucosa

Submucosa

Muscularis externa

Adventitia

(a)

Stratified squamous epithelium

Lamina propria

Muscularis mucosae

(b)　LM × 77

Ⓧsmooth muscle → involuntary

24 DIGESTIVE

Histology of the Esophagus

The wall of the esophagus contains mucosal, submucosal, and muscularis layers comparable to those depicted in **Figure 24–3**. Distinctive features of the esophageal wall (**Figure 24–10**) include the following:

- The mucosa of the esophagus contains a nonkeratinized, stratified squamous epithelium similar to that of the pharynx and oral cavity.

- The mucosa and submucosa are thrown into large folds that extend the length of the esophagus. These folds allow for expansion during the passage of a large bolus; muscle tone in the walls keeps the lumen closed, except when you swallow.

- The muscularis mucosae consists of an irregular layer of smooth muscle.

- The submucosa contains scattered *esophageal glands*, which produce a mucous secretion that reduces friction between the bolus and the esophageal lining.

- The muscularis externa has the usual inner circular and outer longitudinal layers. However, in the superior third of the esophagus, these layers contain skeletal muscle fibers; the middle third contains a mixture of skeletal and smooth muscle tissue; along the inferior third, only smooth muscle occurs.

- There is no serosa, but an adventitia of connective tissue outside the muscularis externa anchors the esophagus to the dorsal body wall. Over the 1–2 cm (0.4–0.8 in.) between the diaphragm and stomach, the esophagus is retroperitoneal, with peritoneum covering the anterior and left lateral surfaces.

Swallowing

Swallowing, or **deglutition** (dē-gloo-TISH-un), is a complex process that can be initiated voluntarily but proceeds automatically once it begins. Although you take conscious control over swallowing when you eat or drink, swallowing is also controlled at the subconscious level. For example, swallowing occurs at regular intervals as saliva collects at the back of the mouth. Each day you swallow approximately 2400 times.

We can divide swallowing into buccal, pharyngeal, and esophageal phases:

1. The **buccal phase** begins with the compression of the bolus against the hard palate. Subsequent retraction of the

tongue then forces the bolus into the oropharynx and assists in the elevation of the soft palate, thereby sealing off the nasopharynx (**Figure 24–11**, STEP 1). The buccal phase is strictly voluntary. Once the bolus enters the oropharynx, reflex responses are initiated and the bolus is moved toward the stomach.

2. The **pharyngeal phase** begins as the bolus comes into contact with the palatoglossal and palatopharyngeal arches and the posterior pharyngeal wall (**Figure 24–11**, STEP 2). The **swallowing reflex** begins when tactile receptors on the palatal arches and uvula are stimulated by the passage of the bolus. The information is relayed to the **swallowing center** of the medulla oblongata over the trigeminal and glossopharyngeal nerves. Motor commands originating at this center then signal the pharyngeal musculature, producing a coordinated and stereotyped pattern of muscle contraction. Elevation of the larynx and folding of the epiglottis direct the bolus past the closed glottis while the uvula and soft palate block passage back to the nasopharynx. It takes less than a second for the pharyngeal muscles to propel the bolus into the esophagus. During this period, the respiratory centers are inhibited and breathing stops.

3. The **esophageal phase** of swallowing begins as the contraction of pharyngeal muscles forces the bolus through the entrance to the esophagus (**Figure 24–11**, STEP 3). Once in the esophagus, the bolus is pushed toward the stomach by a peristaltic wave. The approach of the bolus triggers the opening of the lower esophageal sphincter, and the bolus then continues into the stomach (**Figure 24–11**, STEP 4).

Primary peristaltic waves are peristaltic movements coordinated by afferent and efferent fibers in the glossopharyngeal and vagus nerves. For a typical bolus, the entire trip takes about 9 seconds. Liquids may make the journey in a few seconds, flowing ahead of the peristaltic contractions with the assistance of gravity. A dry or poorly lubricated bolus travels much more slowly, and a series of *secondary peristaltic waves* may be required to push it all the way to the stomach. Secondary peristaltic waves are local reflexes triggered by the stimulation of sensory receptors in the esophageal walls.

CHECKPOINT

14. Name the structure connecting the pharynx to the stomach.

15. Compared to other segments of the digestive tract, what is unusual about the muscularis externa of the esophagus?

16. What is occurring when the soft palate and larynx elevate and the glottis closes?

See the blue Answers tab at the end of the book.

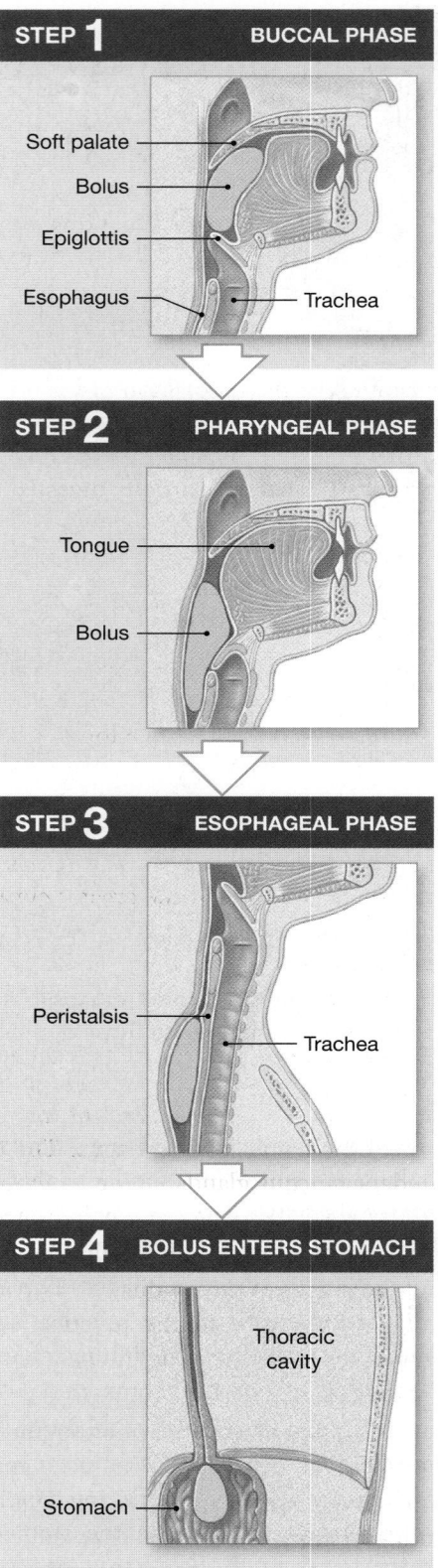

Figure 24–11 The Swallowing Process. This sequence, based on a series of x-rays, shows the phases of swallowing and the movement of a bolus from the mouth to the stomach.

24-5 The stomach is a J-shaped organ that receives the bolus from the esophagus and aids in chemical and mechanical digestion

The **stomach** performs four major functions: (1) storage of ingested food, (2) mechanical breakdown of ingested food, (3) disruption of chemical bonds in food material through the action of acid and enzymes, and (4) production of *intrinsic factor*, a glycoprotein whose presence in the digestive tract is required for the absorption of vitamin B_{12} in the small intestine. Ingested substances combine with the secretions of the glands of the stomach, producing a viscous, highly acidic, soupy mixture of partially digested food called **chyme** (KĪM).

Anatomy of the Stomach

The stomach has the shape of an expanded J (**Figure 24–12**). A short **lesser curvature** forms the medial surface of the organ, and a long **greater curvature** forms the lateral surface. The anterior and posterior surfaces are smoothly rounded. The shape and size of the stomach are extremely variable from individual to individual and even from one meal to the next. In an "average" stomach, the lesser curvature has a length of approximately 10 cm (4 in.), and the greater curvature measures about 40 cm (16 in.). The stomach typically extends between the levels of vertebrae T_7 and L_3.

We can divide the stomach into four regions (**Figure 24–12**):

1. *The Cardia.* The **cardia** (KAR-dē-uh) is the smallest part of the stomach. It consists of the superior, medial portion of the stomach within 3 cm (1.2 in.) of the junction between the stomach and the esophagus. The cardia contains abundant mucous glands whose secretions coat the connection with the esophagus and help protect that tube from the acid and enzymes of the stomach.

2. *The Fundus.* The **fundus** (FUN-dus) is the portion of the stomach that is superior to the junction between the stomach and the esophagus. The fundus contacts the inferior, posterior surface of the diaphragm (**Figure 24–12a**).

3. *The Body.* The area of the stomach between the fundus and the curve of the J is the **body**, the largest region of the stomach. The body functions as a mixing tank for ingested food and secretions produced in the stomach. *Gastric* (*gaster*, stomach) *glands* in the fundus and body secrete most of the acid and enzymes involved in gastric digestion.

4. *The Pylorus.* The **pylorus** (pī-LOR-us) forms the sharp curve of the J. The pylorus is divided into a **pyloric antrum** (*antron*, cavity), which is connected to the body, and a **pyloric canal,** which empties into the *duodenum*, the proximal segment of the small intestine. As mixing movements occur during digestion, the pylorus frequently changes shape. A muscular **pyloric sphincter** regulates the release of chyme into the duodenum. Glands in the pylorus secrete mucus and important digestive hormones, including *gastrin*, a hormone that stimulates the activity of gastric glands.

Tips&Tricks

The stomach squeezes chyme into the small intestine just as you squeeze cake frosting out of a pastry bag.

The stomach's volume increases while you eat and then decreases as chyme enters the small intestine. When the stomach is relaxed (empty), the mucosa is thrown into prominent folds called **rugae** (ROO-gē; wrinkles) (**Figure 24–12b**)—temporary features that let the gastric lumen expand. As the stomach fills, the rugae gradually flatten out until, at maximum distension, they almost disappear. When empty, the stomach resembles a muscular tube with a narrow, constricted lumen. When full, it can contain 1–1.5 liters of material.

The muscularis mucosae and muscularis externa of the stomach contain extra layers of smooth muscle cells in addition to the usual circular and longitudinal layers. The muscularis mucosae generally contain an outer, circular layer of muscle cells. The muscularis externa has an inner, **oblique layer** of smooth muscle (**Figure 24–12b**). The extra layers of smooth muscle strengthen the stomach wall and assist in the mixing and churning activities essential to the formation of chyme.

Tips&Tricks

Both plicae circulares and rugae allow a large surface area to fit within a small volume, just as crumpling a sheet of paper rearranges its surface area so that it occupies a small space. When the stomach fills, its rugae allow its volume to expand, so that the stomach's lining can hold a larger volume of chyme.

Histology of the Stomach

A simple columnar epithelium lines all portions of the stomach (**Figure 24–13a**). The epithelium is a *secretory sheet*, which produces a carpet of mucus that covers the interior surfaces of the stomach. The alkaline mucous layer protects epithelial cells against the acid and enzymes in the gastric lumen.

Shallow depressions called **gastric pits** open onto the gastric surface (**Figure 24–13b**). The mucous cells at the base, or *neck*, of each gastric pit actively divide, replacing superficial cells that are shed into the chyme. A typical gastric epithelial cell has a life span of three to seven days, but exposure

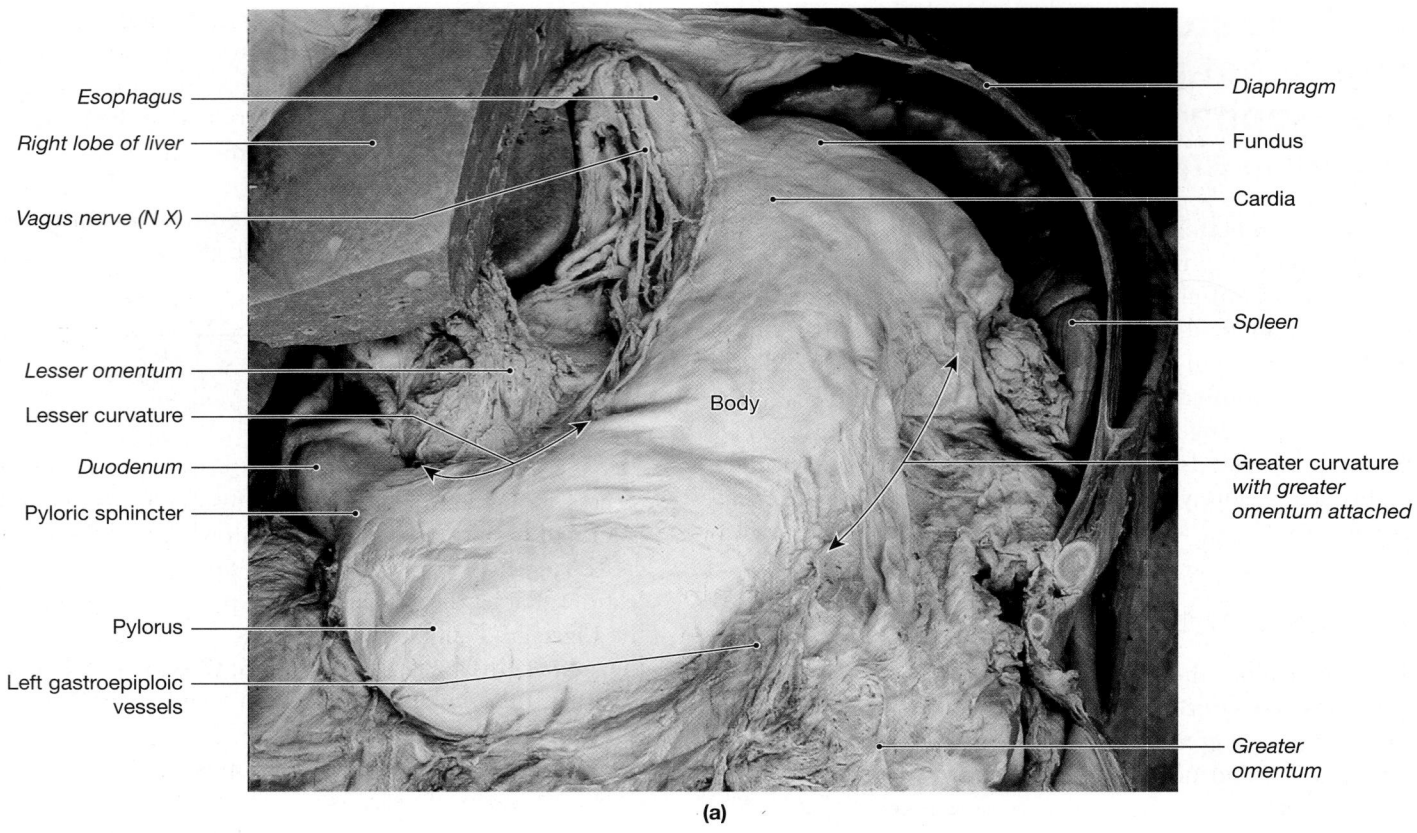

(a)

Figure 24–12 The Stomach. (a) The position and external appearance of the stomach, showing superficial landmarks. **(b)** The structure of the stomach wall. ATLAS: Plates 49a–c; 50a–c

(b)

Figure 24–13 The Stomach Lining. (a) The organization of the stomach wall. **(b)** A gastric gland.

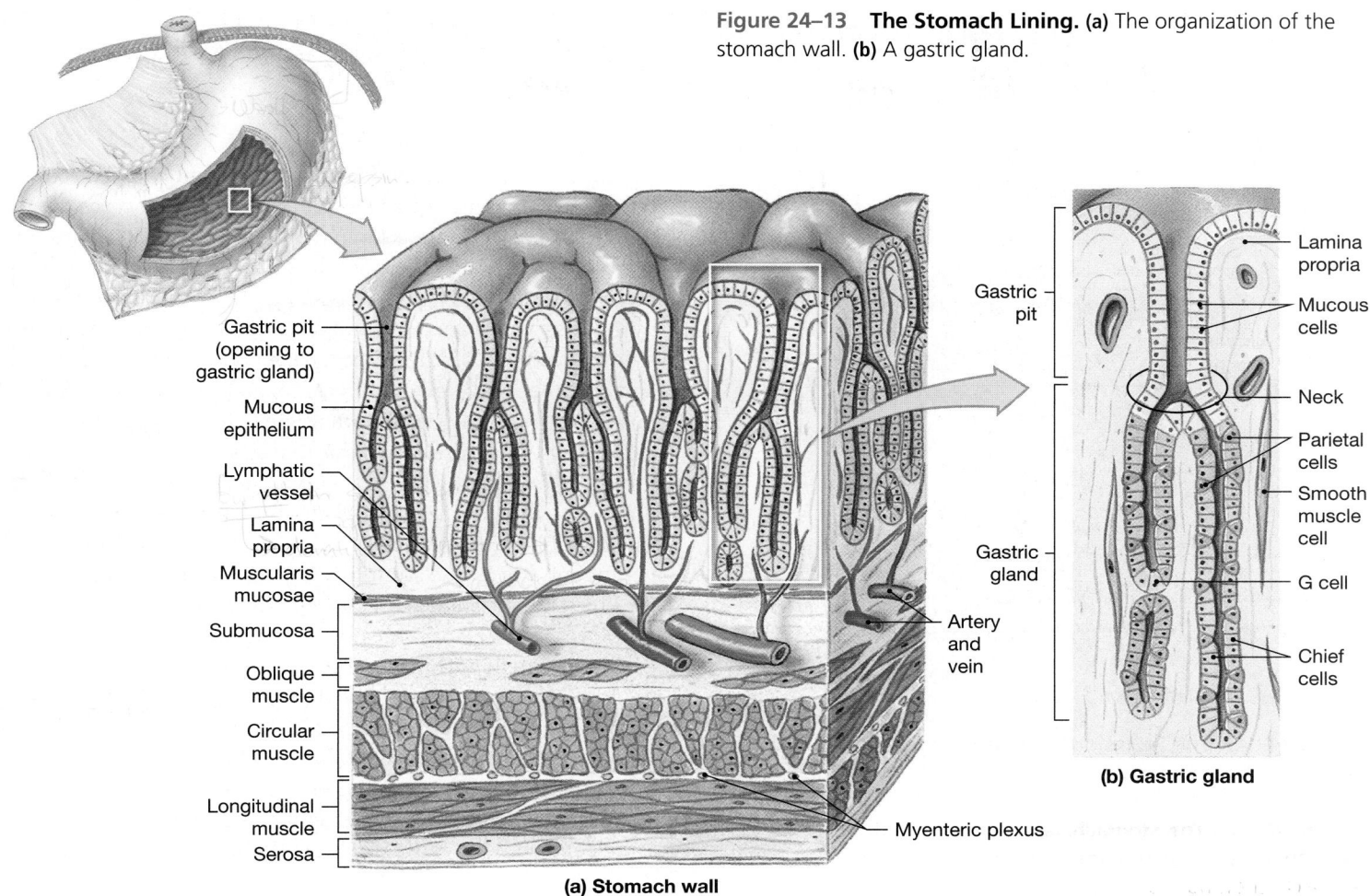

Gastric pit (opening to gastric gland)

Mucous epithelium

Lymphatic vessel

Lamina propria

Muscularis mucosae

Submucosa

Oblique muscle

Circular muscle

Longitudinal muscle

Serosa

Artery and vein

Myenteric plexus

(a) Stomach wall

Gastric pit

Gastric gland

Lamina propria

Mucous cells

Neck

Parietal cells

Smooth muscle cell

G cell

Chief cells

(b) Gastric gland

to alcohol or other chemicals that damage or kill epithelial cells increases the rate of cell turnover.

Gastric Glands

In the fundus and body of the stomach, each gastric pit communicates with several **gastric glands**, which extend deep into the underlying lamina propria (**Figure 24–13b**). Gastric glands are dominated by two types of secretory cells: *parietal cells* and *chief cells*. Together, they secrete about 1500 mL (1.6 qt.) of **gastric juice** each day.

Parietal cells are especially common along the proximal portions of each gastric gland (**Figure 24–13b**). These cells secrete **intrinsic factor**, a glycoprotein that facilitates the absorption of **vitamin B₁₂** across the intestinal lining. (Recall from Chapter 19 that this vitamin is essential for normal erythropoiesis.) ∞ p. 661

Parietal cells also secrete *hydrochloric acid* (HCl). They do not produce HCl in the cytoplasm, however, because it is such a strong acid that it would erode a secretory vesicle and destroy the cell. Instead, H⁺ and Cl⁻ the two ions that form

kills most microorganisms.

HCl, are transported independently by different mechanisms (**Figure 24–14**). Hydrogen ions are generated inside a parietal cell as the enzyme carbonic anhydrase converts carbon dioxide and water to carbonic acid (H_2CO_3). The carbonic acid promptly dissociates into hydrogen ions and bicarbonate ions (HCO_3^-). The hydrogen ions are actively transported into the lumen of the gastric gland. The bicarbonate ions are ejected into the interstitial fluid by a countertransport mechanism that exchanges intracellular bicarbonate ions for extracellular chloride ions. The chloride ions then diffuse across the cell and through open chloride channels in the plasma membrane into the lumen of the gastric gland.

The bicarbonate ions released by parietal cells diffuse through the interstitial fluid into the bloodstream. When gastric glands are actively secreting, enough bicarbonate ions enter the bloodstream to increase the pH of the blood significantly. This sudden influx of bicarbonate ions has been called the *alkaline tide*.

The secretory activities of the parietal cells can keep the stomach contents at pH 1.5–2.0. Although this highly acidic

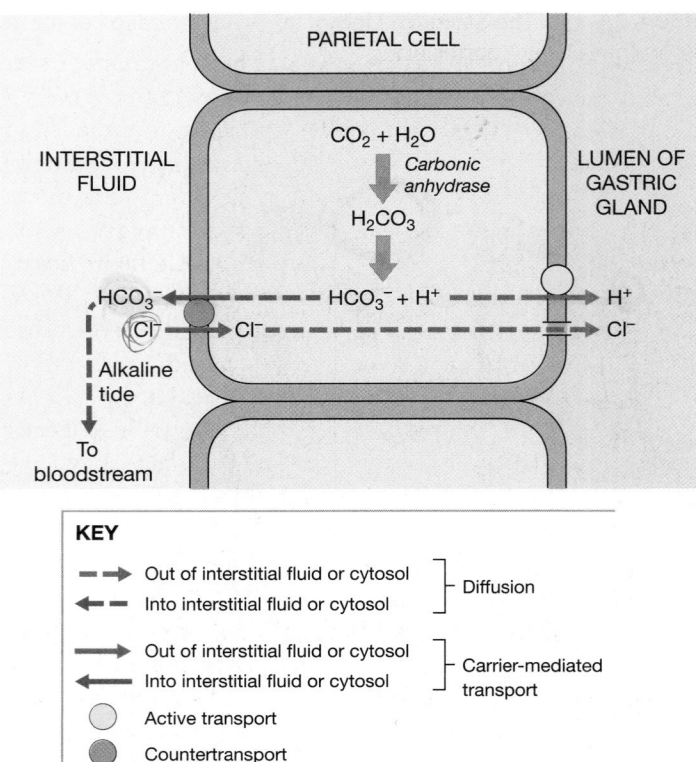

PARIETAL CELL

INTERSTITIAL FLUID

LUMEN OF GASTRIC GLAND

$CO_2 + H_2O$

Carbonic anhydrase

H_2CO_3

HCO_3^-

$HCO_3^- + H^+$

H^+

Cl^-

Cl^-

Cl^-

Alkaline tide

To bloodstream

KEY

- - -> Out of interstitial fluid or cytosol
<- - - Into interstitial fluid or cytosol
⎫ Diffusion

——> Out of interstitial fluid or cytosol
<—— Into interstitial fluid or cytosol
⎫ Carrier-mediated transport

○ Active transport

● Countertransport

Figure 24–14 The Secretion of Hydrochloric Acid. An active parietal cell generates H^+ by the dissociation of carbonic acid within the cell. The bicarbonate is exchanged for Cl^- in the interstitial fluid; the chloride ions diffuse into the lumen of the gastric gland as the hydrogen ions are transported out of the cell.

environment does not by itself digest chyme, it has four important functions:

1. The acidity of gastric juice kills most of the microorganisms ingested with food.

2. The acidity denatures proteins and inactivates most of the enzymes in food.

3. The acidity helps break down plant cell walls and the connective tissues in meat.

4. An acidic environment is essential for the activation and function of *pepsin*, a protein-digesting enzyme secreted by chief cells.

Chief cells are most abundant near the base of a gastric gland (**Figure 24–13b**). These cells secrete **pepsinogen** (pep-SIN-ō-jen), an inactive proenzyme. Acid in the gastric lumen converts pepsinogen to **pepsin**, an active *proteolytic* (protein-digesting) enzyme. Pepsin functions most effectively at a strongly acidic pH of 1.5–2.0. In addition, the stomachs of newborn infants (but not of adults) produce **rennin** (also known as *chymosin*) and **gastric lipase**,

enzymes important for the digestion of milk. Rennin coagulates milk proteins; gastric lipase initiates the digestion of milk fats.

Pyloric Glands

Glands in the pylorus produce primarily a mucous secretion, rather than enzymes or acid. In addition, several types of enteroendocrine cells are scattered among the mucus-secreting cells. These enteroendocrine cells produce at least seven hormones, most notably **gastrin** (GAS-trin). Gastrin is produced by *G cells*, which are most abundant in the gastric pits of the pyloric antrum. Gastrin stimulates secretion by both parietal and chief cells, as well as contractions of the gastric wall that mix and stir the gastric contents. The pyloric glands also contain *D cells*, which release **somatostatin**, a hormone that inhibits the release of gastrin. D cells continuously release their secretions into the interstitial fluid adjacent to the G cells. This inhibition of gastrin production can be overridden by neural and hormonal stimuli when the stomach is preparing for digestion or is already engaged in digestion.

Several other hormones play a role in hunger and satiety. Levels of *ghrelin*, a hormone produced by *P/D1 cells*, lining the fundic region of the stomach, rise before meals to initiate hunger and decline shortly after eating to curb appetite. Ghrelin is also antagonistic to *leptin*, a fat tissue-derived hormone that induces satiety. Another hormone from the stomach and small intestine, *obestatin*, decreases appetite. The same gene encodes both ghrelin and obestatin.

CLINICAL NOTE

Gastritis and Peptic Ulcers A superficial inflammation of the gastric mucosa is called *gastritis* (gas-TRĪ-tis). The condition can develop after a person has swallowed drugs, including beverage alcohol and aspirin. Gastritis is also associated with smoking, severe emotional or physical stress, bacterial infection of the gastric wall, or ingestion of strongly acidic or alkaline chemicals. Over time, gastritis can lead to the erosion of the gastric lining and the development of *gastric* or *peptic ulcers*.

Regulation of Gastric Activity

The production of acid and enzymes by the gastric mucosa can be (1) controlled by the CNS; (2) regulated by short reflexes of the enteric nervous system, coordinated in the wall of the stomach; and (3) regulated by hormones of the digestive tract. Gastric control proceeds in three overlapping phases, named according to the location of the control center: the *cephalic phase*, the *gastric phase*, and the *intestinal phase* (**Figure 24–15** and Table 24–1).

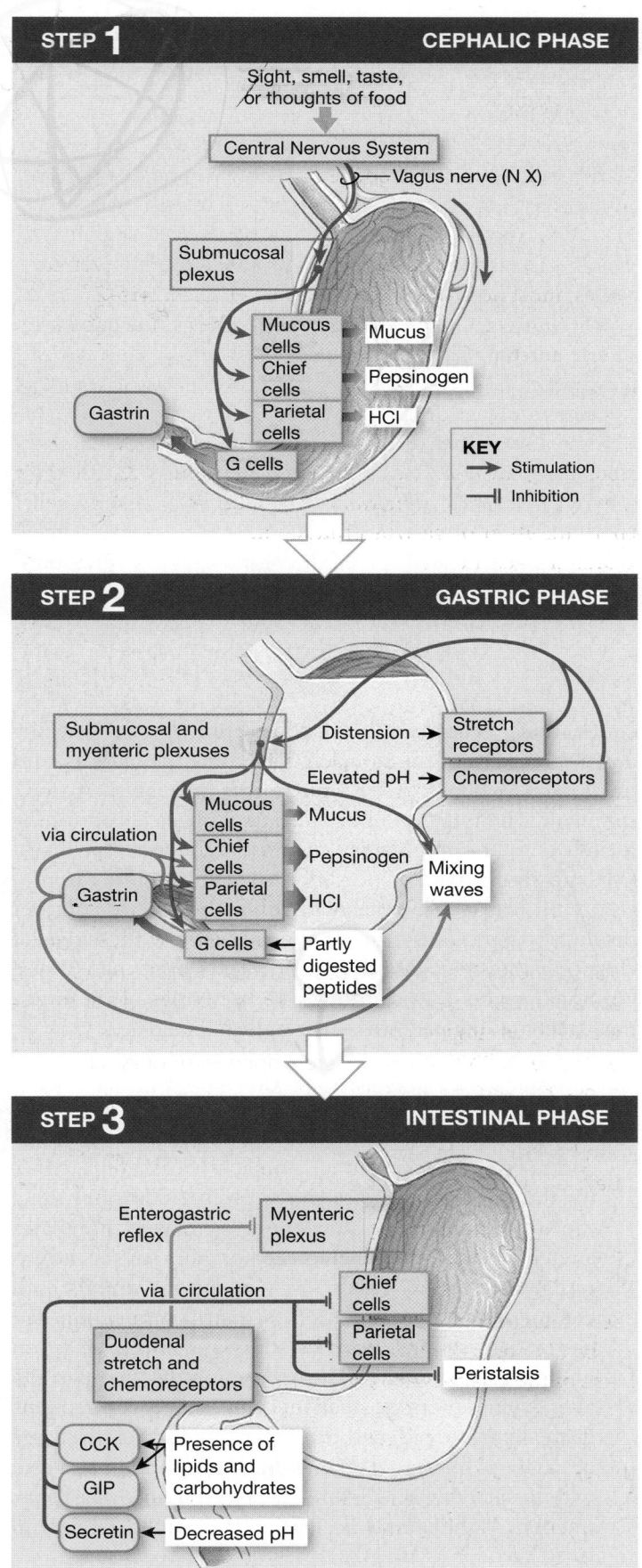

The Cephalic Phase

The **cephalic phase** of gastric secretion begins when you see, smell, taste, or think of food (**Figure 24–15**, STEP 1). This stage, which is directed by the CNS, prepares the stomach to receive food. The neural output proceeds by way of the parasympathetic division of the autonomic nervous system, and the vagus nerves innervate the submucosal plexus of the stomach. Next, postganglionic parasympathetic fibers innervate mucous cells, chief cells, parietal cells, and G cells of the stomach. In response to stimulation, the production of gastric juice accelerates, reaching rates of about 500 mL/h. This phase generally lasts only minutes. Emotional states can exaggerate or inhibit the cephalic phase. For example, anger or hostility leads to excessive gastric secretion, whereas anxiety, stress, or fear decreases gastric secretion and gastric contractions, or *motility*.

The Gastric Phase

The **gastric phase** begins with the arrival of food in the stomach and builds on the stimulation provided during the cephalic phase (**Figure 24–15**, STEP 2). The stimuli that initiate the gastric phase are (1) distension of the stomach, (2) an increase in the pH of the gastric contents, and (3) the presence of undigested materials in the stomach, especially proteins and peptides. The gastric phase consists of the following mechanisms:

1. *A Neural Response.* The stimulation of stretch receptors in the stomach wall and chemoreceptors in the mucosa triggers short reflexes coordinated in the submucosal and myenteric plexuses. The postganglionic fibers leaving the submucosal plexus innervate parietal cells and chief cells, and the release of ACh stimulates their secretion. Proteins, beverage alcohol in small doses, and caffeine enhance gastric secretion markedly by stimulating chemoreceptors in the gastric lining. The stimulation of the myenteric plexus produces mixing waves in the muscularis externa.

2. *A Hormonal Response.* Neural stimulation and the presence of peptides and amino acids in chyme stimulate the secretion of gastrin, primarily by G cells of the pyloric antrum. Gastrin entering the interstitial fluid of the stomach must penetrate capillaries and complete a round trip of the bloodstream before the hormone stimulates parietal and chief cells of the fundus and body. Both parietal and chief cells respond to the presence of gastrin by accelerating their rates of secretion. The effect on the parietal cells is the most pronounced, and the pH of the gastric juice declines as a result. In addition, gastrin stimulates gastric motility.

3. *A Local Response.* Distention of the gastric wall also stimulates the release of histamine in the lamina propria. The source of the histamine is thought to be mast cells in the

**Figure 24–15 The Phases of Gastric Secretion. ** ATLAS: Plate 50c

TABLE 24–1 **The Phases of Gastric Secretion**

Phase	Functions	Duration	Mechanism	Actions
Cephalic phase	Prepares stomach for arrival of food	Short (minutes)	*Neural:* preganglionic fibers in vagus nerves and synapses in submucosal plexus	*Primary:* stimulation of mucus, enzyme, and acid production, leading to increased volume of gastric juice *Secondary:* stimulation of gastrin release by G cells
Gastric phase	Enhances secretion started in cephalic stage; homogenizes and acidifies chyme; initiates digestion of proteins by pepsin	Long (3–4 hr)	*Neural:* short reflexes triggered by stretch receptors and chemoreceptors *Hormonal:* stimulation of gastrin release by G cells *Local:* release of histamine by mast cells as stomach fills	Increased acid and pepsinogen production; increased motility and initiation of mixing waves
Intestinal phase	Controls rate of chyme entry into duodenum	Long (hours)	*Neural:* short reflexes triggered by distension of duodenum *Hormonal:* Primary—stimulation of CCK, GIP, and secretin by acids, carbohydrates, and lipids Secondary—release of gastrin in response to presence of undigested proteins	Feedback inhibition of gastric acid and pepsinogen production; reduction in gastric motility

connective tissue of that layer. Histamine binds to receptors on the parietal cells and stimulates acid secretion.

The gastric phase may continue for three to four hours while the acid and enzymes process the ingested materials. During this period, gastrin stimulates contractions in the muscularis externa of the stomach and intestinal tract. The effects are strongest in the stomach, where stretch receptors are stimulated as well. The initial contractions are weak pulsations in the gastric walls. These *mixing waves* occur several times per minute and gradually increase in intensity. After an hour, the material in the stomach is churning like clothing in a washing machine.

When the contractions begin, the pH of the gastric contents is high; only the material in contact with the gastric epithelium is exposed to undiluted digestive acid and enzymes. As mixing occurs, the acid is diluted, and the pH remains elevated until a large volume of gastric juice has been secreted and the contents are thoroughly mixed. This process generally takes several hours. As the pH throughout the chyme reaches 1.5–2.0 and the amount of undigested protein decreases, gastrin production declines, as do the rates of acid and enzyme secretion by parietal and chief cells.

The Intestinal Phase

The **intestinal phase** of gastric secretion begins when chyme first enters the small intestine (**Figure 24–15**, STEP 3). The intestinal phase generally starts after several hours of mixing

contractions, when waves of contraction begin sweeping down the length of the stomach. Each time the pylorus contracts, a small quantity of chyme squirts through the pyloric sphincter. The function of the intestinal phase is controlling the rate of gastric emptying to ensure that the secretory, digestive, and absorptive functions of the small intestine can proceed with reasonable efficiency. Although here we consider the intestinal phase as it affects stomach activity, the arrival of chyme in the small intestine also triggers other neural and hormonal events that coordinate the activities of the intestinal tract and the pancreas, liver, and gallbladder.

The intestinal phase involves a combination of neural and hormonal responses:

1. *Neural Responses.* Chyme leaving the stomach relieves some of the distension in the stomach wall, thereby reducing the stimulation of stretch receptors. At the same time, the distension of the duodenum by chyme stimulates stretch receptors and chemoreceptors that trigger the **enterogastric reflex**. This reflex temporarily inhibits both central and local stimulation of gastrin production and gastric contractions, as well as stimulating the contraction of the pyloric sphincter. The net result is that immediately after chyme enters the small intestine, gastric contractions decrease in strength and frequency, and further discharge of chyme is prevented, giving the duodenum time to deal with the arriving acid before the next wave of gastric contraction. At the same time, local reflexes at the duodenum

stimulate mucus production, which helps protect the intestinal lining from the arriving acid and enzymes.

2. *Hormonal Responses.* Several hormonal responses are triggered by the arrival of chyme in the duodenum:

- The arrival of lipids (especially triglycerides and fatty acids) and carbohydrates in the duodenum stimulates the secretion of the hormones *cholecystokinin* (kō-lē-sis-tō-KĪ-nin), or CCK, and *gastric inhibitory peptide* (GIP). CCK inhibits gastric secretion of acid and enzymes; GIP, which also targets the pancreas, inhibits gastric secretion and reduces the rate and force of gastric contractions. As a result, a meal high in fats stays in the stomach longer, and enters the duodenum at a more leisurely pace, than does a low-fat meal. This delay allows more time for lipids to be digested and absorbed in the small intestine.

- A drop in pH below 4.5 stimulates the secretion of the hormone *secretin* (sē-KRĒ-tin) by enteroendocrine cells of the duodenum. Secretin inhibits parietal cell and chief cell activity in the stomach. It also targets two accessory organs: the pancreas, where it stimulates the production of bicarbonate ions that will protect the duodenum by neutralizing the acid in chyme, and the liver, where it stimulates the secretion of bile.

- The arrival of partially digested proteins in the duodenum stimulates G cells in the duodenal wall. These cells secrete gastrin, which circulates to the stomach and accelerates acid and enzyme production. In effect, this is a feedback mechanism that regulates the amount of gastric processing to meet the requirements of a specific meal.

In general, the rate of movement of chyme into the small intestine is highest when the stomach is greatly distended and the meal contains relatively little protein. A large meal containing small amounts of protein, large amounts of carbohydrates (such as rice or pasta), wine (alcohol), and after-dinner coffee (caffeine) will leave your stomach extremely quickly because both alcohol and caffeine stimulate gastric secretion and motility.

Digestion and Absorption in the Stomach

The stomach performs preliminary digestion of proteins by pepsin and, for a variable period, permits the digestion of carbohydrates and lipids by salivary amylase and lingual lipase. Until the pH throughout the contents of the stomach falls below 4.5, salivary amylase and lingual lipase continue to digest carbohydrates and lipids in the meal. These enzymes generally remain active one to two hours after a meal.

As the stomach contents become more fluid and the pH approaches 2.0, pepsin activity increases and protein disassembly begins. Protein digestion is not completed in the stomach, because time is limited and pepsin attacks only specific types of peptide bonds, not all of them. However, pepsin generally has enough time to break down complex proteins into smaller peptide and polypeptide chains before the chyme enters the duodenum.

Although digestion occurs in the stomach, nutrients are not absorbed there, for several reasons: (1) The epithelial cells are covered by a blanket of alkaline mucus and are not directly exposed to chyme, (2) the epithelial cells lack the specialized transport mechanisms of cells that line the small intestine, (3) the gastric lining is relatively impermeable to water, and (4) digestion has not been completed by the time chyme leaves the stomach. At this stage, most carbohydrates, lipids, and proteins are only partially broken down.

Some drugs can be absorbed in the stomach. For example, ethyl alcohol can diffuse through the mucous barrier and penetrate the lipid membranes of the epithelial cells. As a result, beverage alcohol is absorbed in your stomach before any nutrients in a meal reach the bloodstream. Meals containing large amounts of fat slow the rate of alcohol absorption, because alcohol is lipid soluble, and some of it will be dissolved in fat droplets in the chyme. Aspirin is another lipid-soluble drug that can enter the bloodstream across the gastric mucosa. Such drugs alter the properties of the mucous layer and can promote epithelial damage by stomach acid and enzymes. Prolonged use of aspirin can cause gastric bleeding, so individuals with stomach ulcers usually avoid aspirin.

The A&P Top 100

#82 Carbohydrate digestion begins in the mouth through the action of salivary amylase before swallowing, and continues after substances arrive in the stomach. Food is stored in the stomach while it is physically broken down. Pepsin in the stomach starts the chemical digestion of proteins.

CHECKPOINT

17. Name the four major regions of the stomach.

18. Discuss the significance of the low pH in the stomach.

19. How does a large meal affect the pH of blood leaving the stomach?

20. When a person suffers from chronic gastric ulcers, the branches of the vagus nerves that serve the stomach are sometimes cut in an attempt to provide relief. Why might this be an effective treatment?

See the blue Answers tab at the end of the book.

24 DIGESTIVE

24-6 The small intestine digests and absorbs nutrients, and associated glandular organs assist with the digestive process

The stomach is a holding tank in which food is saturated with gastric juices and exposed to stomach acid and the digestive effects of pepsin. These are preliminary steps; most of the important digestive and absorptive functions occur in the small intestine, where chemical digestion is completed and the products of digestion are absorbed. The mucosa of the small intestine produces only a few of the enzymes involved. The pancreas provides digestive enzymes, as well as buffers that help neutralize chyme. The liver secretes *bile*, a solution stored in the gallbladder for subsequent discharge into the small intestine. Bile contains buffers and *bile salts*, compounds that facilitate the digestion and absorption of lipids.

The Small Intestine

The **small intestine** plays the key role in the digestion and absorption of nutrients. Ninety percent of nutrient absorption occurs in the small intestine; most of the rest occurs in the large intestine. The small intestine averages 6 m (19.7 ft)

in length (range: 4.5–7.5 m; 14.8–24.6 ft) and has a diameter ranging from 4 cm (1.6 in.) at the stomach to about 2.5 cm (1 in.) at the junction with the large intestine. It occupies all abdominal regions except the right and left hypochondriac and epigastric regions (see **Figure 1–7b**, p. 19). The small intestine has three segments: the duodenum, the jejunum, and the ileum (**Figure 24–16a**).

The **duodenum** (doo-ō-DĒ-num), 25 cm (10 in.) in length, is the segment closest to the stomach. This portion of the small intestine is a "mixing bowl" that receives chyme from the stomach and digestive secretions from the pancreas and liver. From its connection with the stomach, the duodenum curves in a C that encloses the pancreas. Except for the proximal 2.5 cm (1 in.), the duodenum is in a retroperitoneal position between vertebrae L$_1$ and L$_4$ (**Figure 24–2d**).

A rather abrupt bend marks the boundary between the duodenum and the **jejunum** (je-JOO-num). At this junction, the small intestine reenters the peritoneal cavity, supported by a sheet of mesentery. The jejunum is about 2.5 meters (8.2 ft) long. The bulk of chemical digestion and nutrient absorption occurs in the jejunum.

The **ileum** (IL-ē-um), the final segment of the small intestine, is also the longest, averaging 3.5 meters (11.5 ft) in length. The ileum ends at the **ileocecal** (il-ē-ō-SĒ-kal) **valve,** a sphincter that controls the flow of material from the ileum into the *cecum* of the large intestine.

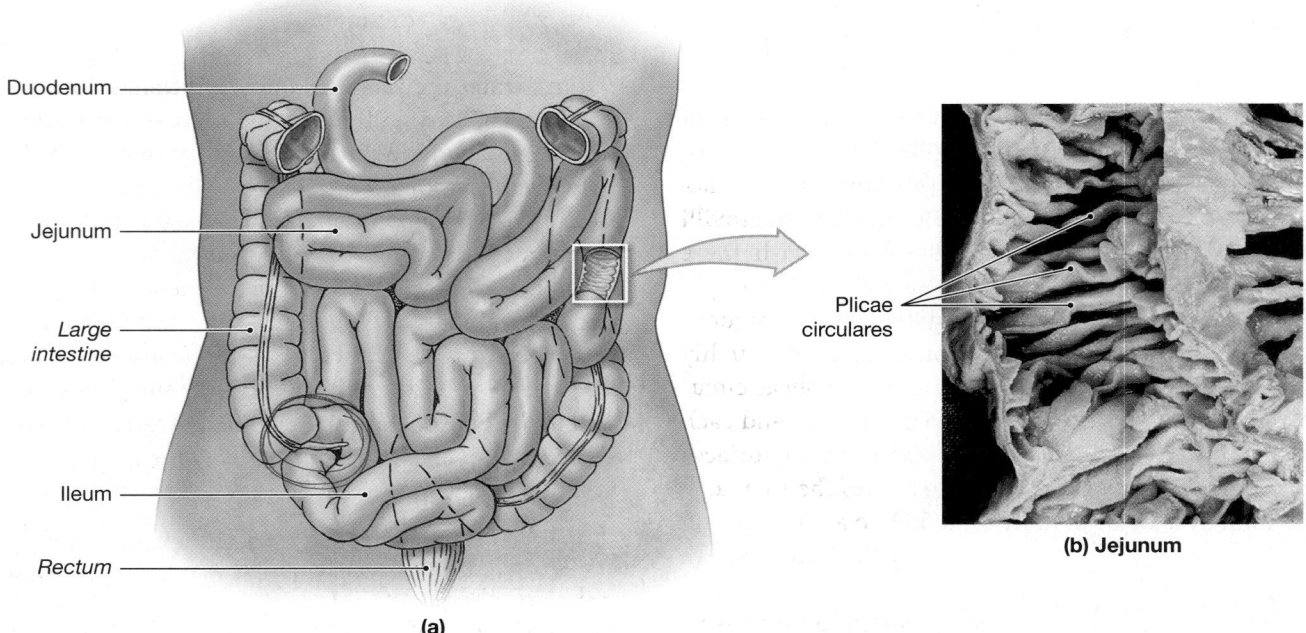

(a)

(b) Jejunum

Figure 24–16 **Segments of the Intestine. (a)** The positions of the duodenum, jejunum, and ileum in the abdominopelvic cavity. **(b)** A representative view of the jejunum. ATLAS: Plates 49a,b,d; 51a,b

Tips & Tricks

To remember the order of the small intestine segments, beginning at the stomach, use this mnemonic: **D**on't **j**ump **i**n—**d**uodenum, **j**ejunum, and **i**leum. Also, do not confuse ileum, the last segment of the small intestine, with ilium, which is a bone.

The small intestine fills much of the peritoneal cavity, and its position is stabilized by the mesentery proper, a broad mesentery attached to the dorsal body wall (**Figure 24–2c,d**). The stomach, large intestine, abdominal wall, and pelvic girdle restrict movement of the small intestine during digestion. Blood vessels, lymphatic vessels, and nerves reach the segments of the small intestine within the connective tissue of the mesentery. The primary blood vessels involved are branches of the superior mesenteric artery and the superior mesenteric vein. ∞ pp. 757, 767

The segments of the small intestine—the duodenum, jejunum, and ileum—are distinguished by both histological specialization and primary function.

Histology of the Small Intestine

The intestinal lining bears a series of transverse folds called **plicae circulares** (**Figure 24–16b**). Unlike the rugae in the stomach, the plicae circulares are permanent features that do not disappear when the small intestine fills. The small intestine contains roughly 800 plicae circulares—roughly 2 per cm. Their presence greatly increases the surface area available for absorption.

Intestinal Villi

The mucosa of the small intestine is thrown into a series of fingerlike projections, the **intestinal villi** (**Figure 24–17a,b**). These structures are covered by a simple columnar epithelium that is carpeted with microvilli. Because the microvilli project from the epithelium like the bristles on a brush, these cells are said to have a *brush border* (**Figure 24–17d**).

If the small intestine were a simple tube with smooth walls, it would have a total absorptive area of roughly 3300 cm^2 (3.6 ft^2). Instead, the mucosa contains plicae circulares; each plica circulares supports a forest of villi, and each villus is covered by epithelial cells whose exposed surfaces contain microvilli. This arrangement increases the total area for absorption by a factor of more than 600, to approximately 2 million cm^2 (more than 2200 ft^2, roughly the floor space of a spacious four-bedroom home).

The lamina propria of each villus contains an extensive network of capillaries that originate in a vascular network within the submucosa (**Figure 24–17c**). These capillaries carry absorbed nutrients to the hepatic portal circulation for delivery to the liver, which adjusts the nutrient concentrations of blood before the blood reaches the general systemic circulation.

In addition to capillaries and nerve endings, each villus contains a lymphatic capillary called a **lacteal** (LAK-tē-ul; *lacteus*, milky) (**Figure 24–17b,c**). Lacteals transport materials that cannot enter blood capillaries. For example, absorbed fatty acids are assembled into protein–lipid packages that are too large to diffuse into the bloodstream. These packets, called *chylomicrons*, reach the venous circulation via the thoracic duct, which delivers lymph into the left subclavian vein. The name *lacteal* refers to the pale, milky appearance of lymph that contains large quantities of lipids.

Contractions of the muscularis mucosae and smooth muscle cells within the intestinal villi move the villi back and forth, exposing the epithelial surfaces to the liquefied intestinal contents. This movement improves the efficiency of absorption by quickly eliminating local differences in nutrient concentration. Movements of the villi also squeeze the lacteals, thereby assisting in the movement of lymph out of the villi.

Intestinal Glands

Mucous cells between the columnar epithelial cells eject mucins onto the intestinal surfaces (**Figure 24–17c,d**). At the bases of the villi are the entrances to the **intestinal glands**, or *crypts of Lieberkühn*. These glandular pockets extend deep into the underlying lamina propria. Near the base of each intestinal gland, stem cell divisions produce new generations of epithelial cells, which are continuously displaced toward the intestinal surface. In a few days the new cells will have reached the tip of a villus, where they are shed into the intestinal lumen. This ongoing process renews the epithelial surface, and the subsequent disintegration of the shed cells adds enzymes to the lumen.

Several important brush border enzymes enter the intestinal lumen in this way. *Brush border enzymes* are integral membrane proteins located on the surfaces of intestinal microvilli. These enzymes perform the important digestive function of breaking down materials that come in contact with the brush border. The epithelial cells then absorb the breakdown products. Once the epithelial cells are shed, they disintegrate within the lumen, releasing intracellular and brush border enzymes. *Enteropeptidase* (previously called *enterokinase*), one brush border enzyme that enters the lumen in this way, does not directly participate in digestion, but it activates a key pancreatic proenzyme, trypsinogen. (We will consider the functions of enteropeptidase and other brush border enzymes in a later section.) Intestinal glands also contain enteroendocrine cells responsible for the production of several intestinal hormones, including gastrin, cholecystokinin, and secretin.

globular cells ⇒ produces mucin

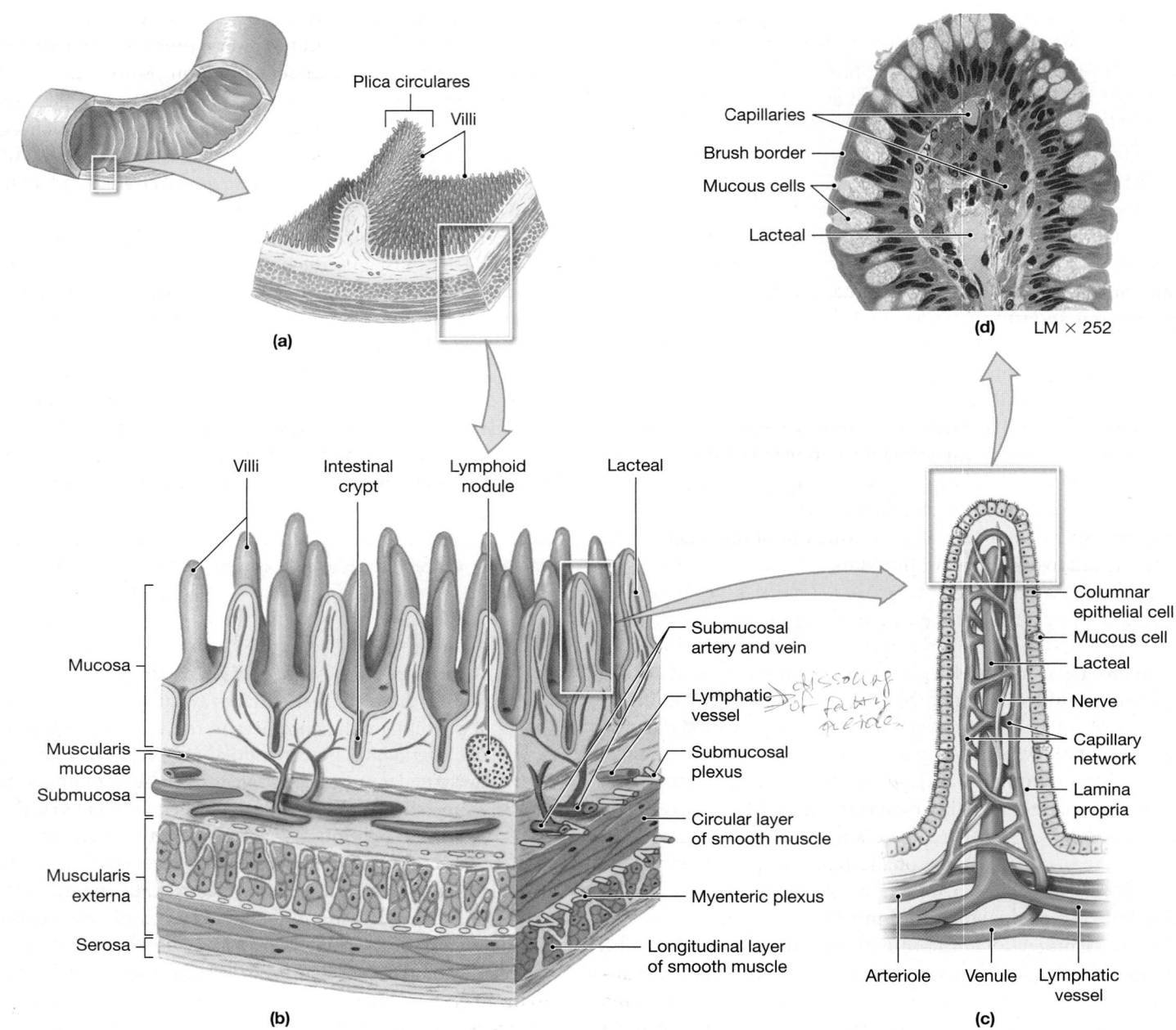

Figure 24–17 **The Intestinal Wall. (a)** A single plica circulares and multiple villi. **(b)** The organization of the intestinal wall. **(c)** Internal structures in a single villus, showing the capillary and lymphatic supplies. **(d)** A villus in sectional view. ATLAS: Plate 51a–d

The duodenum has numerous mucous glands, both in the epithelium and deep to it. In addition to intestinal glands, its submucosa contains **duodenal glands**, also called *submucosal glands* or *Brunner glands*, which produce copious quantities of mucus when chyme arrives from the stomach. The mucus protects the epithelium from the acidity of chyme and also contains bicarbonate ions that help elevate the pH of the chyme. Along the length of the duodenum, the pH of chyme goes from 1–2 to 7–8. The duodenal glands also secrete the hormone *urogastrone*, which inhibits gastric acid production and stimulates the division of epithelial stem cells along the digestive tract.

Regional Specializations

The duodenum has few plicae circulares, and their villi are small. The primary function of the duodenum is to receive chyme from the stomach and neutralize its acids before they can damage the absorptive surfaces of the small intestine. Over the proximal half of the jejunum, however, plicae circulares and villi are very prominent. Thereafter, the plicae circulares and villi gradually decrease in size. This reduction parallels a reduction in absorptive activity: Most nutrient absorption has occurred before ingested materials reach the ileum. One rather drastic surgical method of promoting weight loss is the removal of a significant portion of the jejunum. The resulting reduction in absorptive area causes a marked weight loss and may not interfere with adequate nutrition, but the side effects can be very serious.

The distal portions of the ileum lack plicae circulares, and the lamina propria there contains 20–30 masses of lymphoid tissue called aggregated lymphoid nodules (*Peyer patches*). These lymphoid tissues are most abundant in the terminal portion of the ileum, near the entrance to the large intestine. The lymphocytes in the aggregated lymphoid nodules protect the small intestine from bacteria that are normal inhabitants of the large intestine.

Intestinal Secretions

Roughly 1.8 liters of watery **intestinal juice** enters the intestinal lumen each day. Intestinal juice moistens chyme, assists in buffering acids, and keeps both the digestive enzymes and the products of digestion in solution. Much of this fluid volume arrives by osmosis, as water flows out of the mucosa and into the relatively concentrated chyme. The rest is secreted by intestinal glands, stimulated by the activation of touch receptors and stretch receptors in the intestinal walls.

The duodenal glands help protect the duodenal epithelium from gastric acids and enzymes. These glands increase their secretory activities in response to (1) local reflexes, (2) the release of the hormone *enterocrinin* by enteroendocrine cells of the duodenum, and (3) parasympathetic stimulation via the vagus nerves. The first two mechanisms operate only after chyme arrives in the duodenum. However, because vagus nerve activity triggers their secretion, the duodenal glands begin secreting during the cephalic phase of gastric secretion, long before chyme reaches the pyloric sphincter. Thus, the duodenal lining has protection in advance.

Sympathetic stimulation inhibits the activation of the duodenal glands, leaving the duodenal lining relatively unprepared for the arrival of chyme. This fact probably accounts for the common observation that duodenal ulcers can be caused by chronic stress or other factors that promote sympathetic activation.

Intestinal Movements

After chyme has arrived in the duodenum, weak peristaltic contractions move it slowly toward the jejunum. The contractions are myenteric reflexes that are not under CNS control. Their effects are limited to within a few centimeters of the site of the original stimulus. These short reflexes are controlled by motor neurons in the submucosal and myenteric plexuses. In addition, some of the smooth muscle cells contract periodically, even without stimulation, establishing a basic contractile rhythm that then spreads from cell to cell.

The stimulation of the parasympathetic system increases the sensitivity of the weak myenteric reflexes and accelerates both local peristalsis and segmentation. More elaborate reflexes coordinate activities along the entire length of the small intestine. Two reflexes are triggered by the stimulation of stretch receptors in the stomach as it fills. The **gastroenteric reflex** stimulates motility and secretion along the entire small intestine; the **gastroileal** (gas-trō-IL-ē-al) **reflex** triggers the relaxation of the ileocecal valve. The net result is that materials pass from the small intestine into the large intestine. Thus, the gastroenteric and gastroileal reflexes accelerate movement along the small intestine—the opposite effect of the enterogastric reflex.

Hormones released by the digestive tract can enhance or suppress reflexes. For example, the gastroileal reflex is triggered by stretch receptor stimulation, but the degree of ileocecal valve relaxation is enhanced by gastrin, which is secreted in large quantities when food enters the stomach.

The A&P Top 100

#83 The small intestine receives chyme from the stomach and decreases its acidity by raising the pH. It then absorbs water, ions, vitamins, and the chemical products released by the action of digestive enzymes produced by intestinal glands and exocrine glands of the pancreas.

The Pancreas

The **pancreas** lies posterior to the stomach, extending laterally from the duodenum toward the spleen. The pancreas is an elongate, pinkish-gray organ about 15 cm (6 in.) long and weighing about 80 g (3 oz) (**Figure 24–18a**). The broad **head** of the pancreas lies within the loop formed by the duodenum as it leaves the pylorus. The slender **body** of the pancreas extends toward the spleen, and the **tail** is short and bluntly rounded. The pancreas is retroperitoneal and is firmly bound to the posterior wall of the abdominal cavity. The surface of the pancreas has a lumpy, lobular texture. A thin, transparent capsule of connective tissue wraps the entire organ. The pancreatic lobules, associated blood vessels, and excretory ducts are visible through the anterior capsule and the overlying layer of peritoneum. Arterial blood reaches the pancreas by way of branches of the splenic, superior mesenteric, and common hepatic arteries. The pancreatic arteries and pancreaticoduodenal arteries are the major branches from these vessels. The splenic vein and its branches drain the pancreas.

The pancreas is primarily an exocrine organ, producing digestive enzymes and buffers. The large **pancreatic duct** (*duct of Wirsung*) delivers these secretions to the duodenum. (In 3–10 percent of the population, a small **accessory pancreatic duct** [*duct of Santorini*] branches from the pancreatic duct.) The pancreatic duct extends within the attached

24 DIGESTIVE

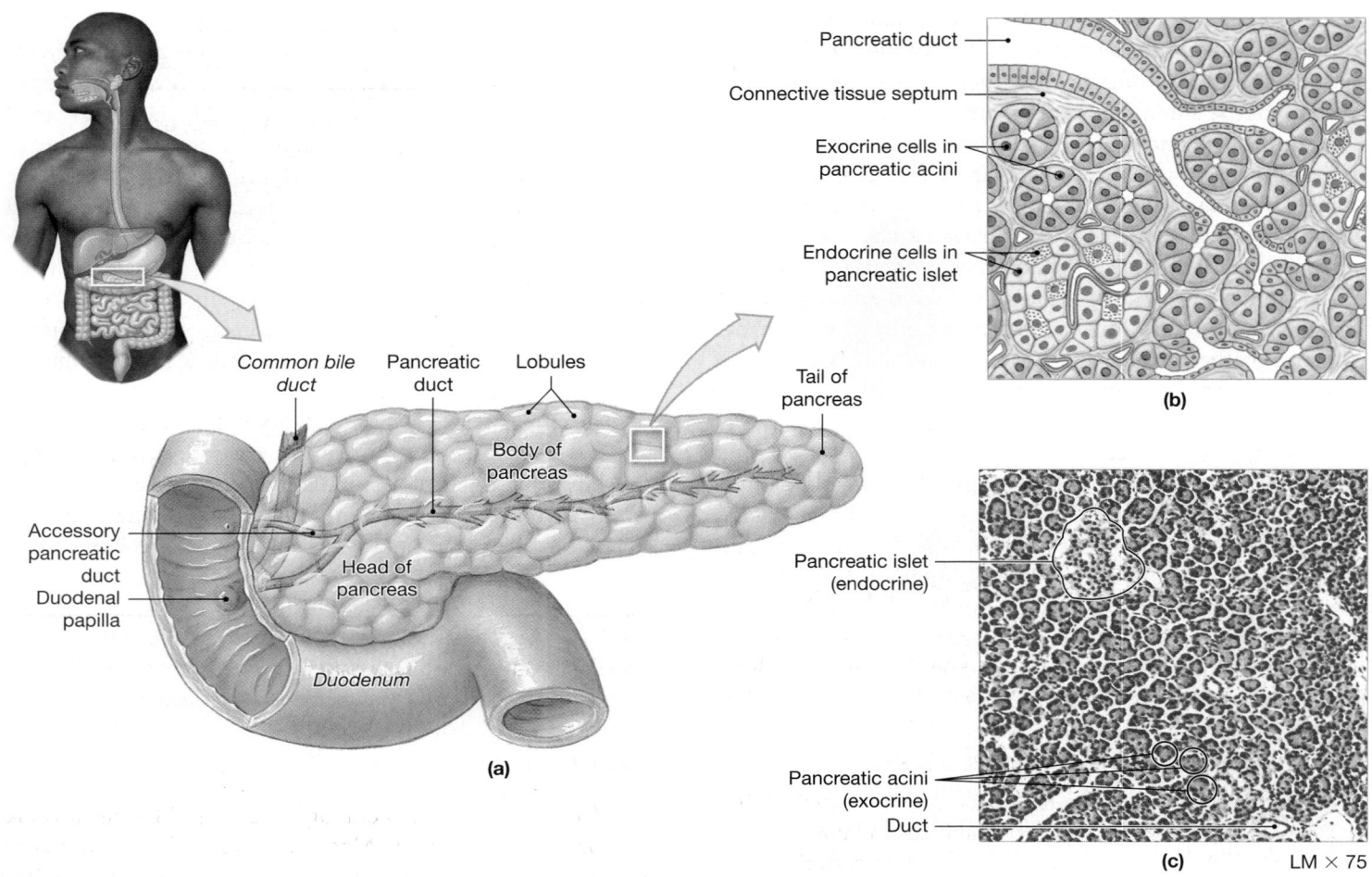

Pancreatic duct

Connective tissue septum

Exocrine cells in pancreatic acini

Endocrine cells in pancreatic islet

(b)

Common bile duct

Pancreatic duct

Lobules

Tail of pancreas

Body of pancreas

Accessory pancreatic duct

Duodenal papilla

Head of pancreas

Duodenum

(a)

Pancreatic islet (endocrine)

Pancreatic acini (exocrine)

Duct

(c)   LM × 75

Figure 24–18   The Pancreas. (a) The gross anatomy of the pancreas. The head of the pancreas is tucked into a C-shaped curve of the duodenum that begins at the pylorus of the stomach. The cellular organization of the pancreas is shown **(b)** diagrammatically and **(c)** in a micrograph. ATLAS: Plates 54d; 55; 57a

mesentery to reach the duodenum, where it meets the *common bile duct* from the liver and gallbladder (**Figure 24–21b**). The two ducts then empty into the *duodenal ampulla* (*ampulla of Vater*), a chamber located roughly halfway along the length of the duodenum. When present, the accessory pancreatic duct generally empties into the duodenum independently, outside the duodenal ampulla.

Histological Organization

Partitions of connective tissue divide the interior of the pancreas into distinct lobules. The blood vessels and tributaries of the pancreatic ducts are situated within these connective tissue septa (**Figure 24–18b**). The pancreas is an example of a *compound tubuloalveolar gland*, a structure described in Chapter 4. ∞ p. 123 In each lobule, the ducts branch repeatedly before ending in blind pockets called **pancreatic acini** (AS-i-nī). Each pancreatic acinus is lined with a simple cuboidal epithelium. *Pancreatic islets*, the endocrine tissues of the pancreas, are scattered among the acini (**Figure 24–18b,c**).

The islets account for only about 1 percent of the cell population of the pancreas.

The pancreas has two distinct functions, one endocrine and the other exocrine. The endocrine cells of the pancreatic islets secrete insulin and glucagon into the bloodstream. The exocrine cells include the acinar cells and the epithelial cells that line the duct system. Together, the acinar cells and the epithelial cells secrete **pancreatic juice**—an alkaline mixture of digestive enzymes, water, and ions—into the small intestine. Acinar cells secrete pancreatic enzymes, which do most of the digestive work in the small intestine, breaking down ingested materials into small molecules suitable for absorption. The water and ions, secreted primarily by the cells lining the pancreatic ducts, assist in diluting and neutralizing acid in the chyme.

Physiology of the Pancreas

Each day, the pancreas secretes about 1000 mL (1 qt) of pancreatic juice. The secretory activities are controlled primarily by hormones from the duodenum. When chyme arrives in the duodenum, secretin is released. This hormone triggers the pancreatic secretion of a watery buffer solution with a pH of 7.5–8.8. Among its other components, the secretion contains bicarbonate and phosphate buffers that help elevate the pH of the chyme. A different duodenal hormone, cholecystokinin, stimulates the production and secretion of pancreatic enzymes. Pancreatic enzyme secretion also increases under stimulation by the vagus nerves. As noted earlier, this stimulation occurs during the cephalic phase of gastric regulation, so the pancreas starts to synthesize enzymes before food even reaches the stomach. Such a head start is important, because enzyme synthesis takes much longer than the production of buffers. By starting early, the pancreatic cells are ready to meet the demand when chyme arrives in the duodenum.

The specific pancreatic enzymes involved include the following:

- **Pancreatic alpha-amylase**, a **carbohydrase** (kar-bō-HĪ-drās)—an enzyme that breaks down certain starches. Pancreatic alpha-amylase is almost identical to salivary amylase.

- **Pancreatic lipase**, which breaks down certain complex lipids, releasing products (such as fatty acids) that can be easily absorbed.

- **Nucleases**, which break down RNA or DNA.

- **Proteolytic enzymes**, which break apart certain proteins. The proteolytic enzymes of the pancreas include **proteases**, which break apart large protein complexes, and **peptidases**, which break small peptide chains into individual amino acids.

Proteolytic enzymes account for about 70 percent of total pancreatic enzyme production. The enzymes are secreted as inactive proenzymes and are activated only after they reach the small intestine. Proenzymes discussed earlier in the text include pepsinogen, angiotensinogen, plasminogen, fibrinogen, and many of the clotting factors and enzymes of the complement system. ∞ pp. 635, 654, 677, 793 As in the stomach, the release of a proenzyme rather than an active enzyme protects the secretory cells in the pancreas from the destructive effects of their own products. Among the proenzymes secreted by the pancreas are **trypsinogen** (trip-SIN-ō-jen), **chymotrypsinogen** (kī-mō-trip-SIN-ō-jen), **procarboxypeptidase** (prō-kar-bok-sē-PEP-ti-dās), and **proelastase** (pro-ē-LAS-tās).

Once inside the duodenum, enteropeptidase located in the brush border and in the lumen triggers the conversion of trypsinogen to **trypsin**, an active protease. Trypsin then activates the other proenzymes, producing **chymotrypsin, carboxypeptidase**, and **elastase**. Each enzyme attacks peptide bonds linking specific amino acids and ignores others. Together, they break down proteins into a mixture of dipeptides, tripeptides, and amino acids.

CLINICAL NOTE

Pancreatitis *Pancreatitis* (pan-krē-a-TĪ-tis) is an inflammation of the pancreas. A blockage of the excretory ducts, bacterial or viral infections, ischemia, and drug reactions, especially those involving alcohol, are among the factors that may produce this extremely painful condition. These stimuli provoke a crisis by injuring exocrine cells in at least a portion of the organ. Lysosomes in the damaged cells then activate the proenzymes, and autolysis begins. The proteolytic enzymes digest the surrounding, undamaged cells, activating their enzymes and starting a chain reaction. In most cases, only a portion of the pancreas is affected, and the condition subsides in a few days. In 10–15 percent of pancreatitis cases, the process does not subside; the enzymes can then ultimately destroy the pancreas. If the islet cells are damaged, diabetes mellitus may result. ∞ p. 634

The A&P Top 100

#84 The exocrine pancreas produces essential digestive buffers and enzymes in response to the release of regulatory hormones (CCK and secretin) by the duodenum.

The Liver

The **liver**, the largest visceral organ, is one of the most versatile organs in the body. Most of its mass lies in the right hypochondriac and epigastric regions, but it may extend into the left hypochondriac and umbilical regions as well. The liver weighs about 1.5 kg (3.3 lb). This large, firm, reddish-brown organ performs essential metabolic and synthetic functions.

Anatomy of the Liver

The liver is wrapped in a tough fibrous capsule and is covered by a layer of visceral peritoneum. On the anterior surface, the **falciform ligament** marks the division between the organ's left lobe and the right lobe (**Figure 24–19a,b**). A thickening in the posterior margin of the falciform ligament is the **round ligament**, or *ligamentum teres*, a fibrous band that marks the path of the fetal umbilical vein.

On the posterior surface of the liver, the impression left by the inferior vena cava marks the division between the right lobe and the small **caudate** (KAW-dāt) **lobe** (**Figure 24–19a,c**). Inferior to the caudate lobe lies the **quadrate lobe**, sandwiched between the left lobe and the gallbladder. Afferent blood vessels and other structures reach the liver by traveling within the connective tissue of the lesser omentum. They converge at a region called the **porta hepatis** ("doorway to the liver").

The circulation to the liver was discussed in Chapter 21 and summarized in **Figures 21–25** and **21–32**, pp. 758, 766. Roughly

24 DIGESTIVE

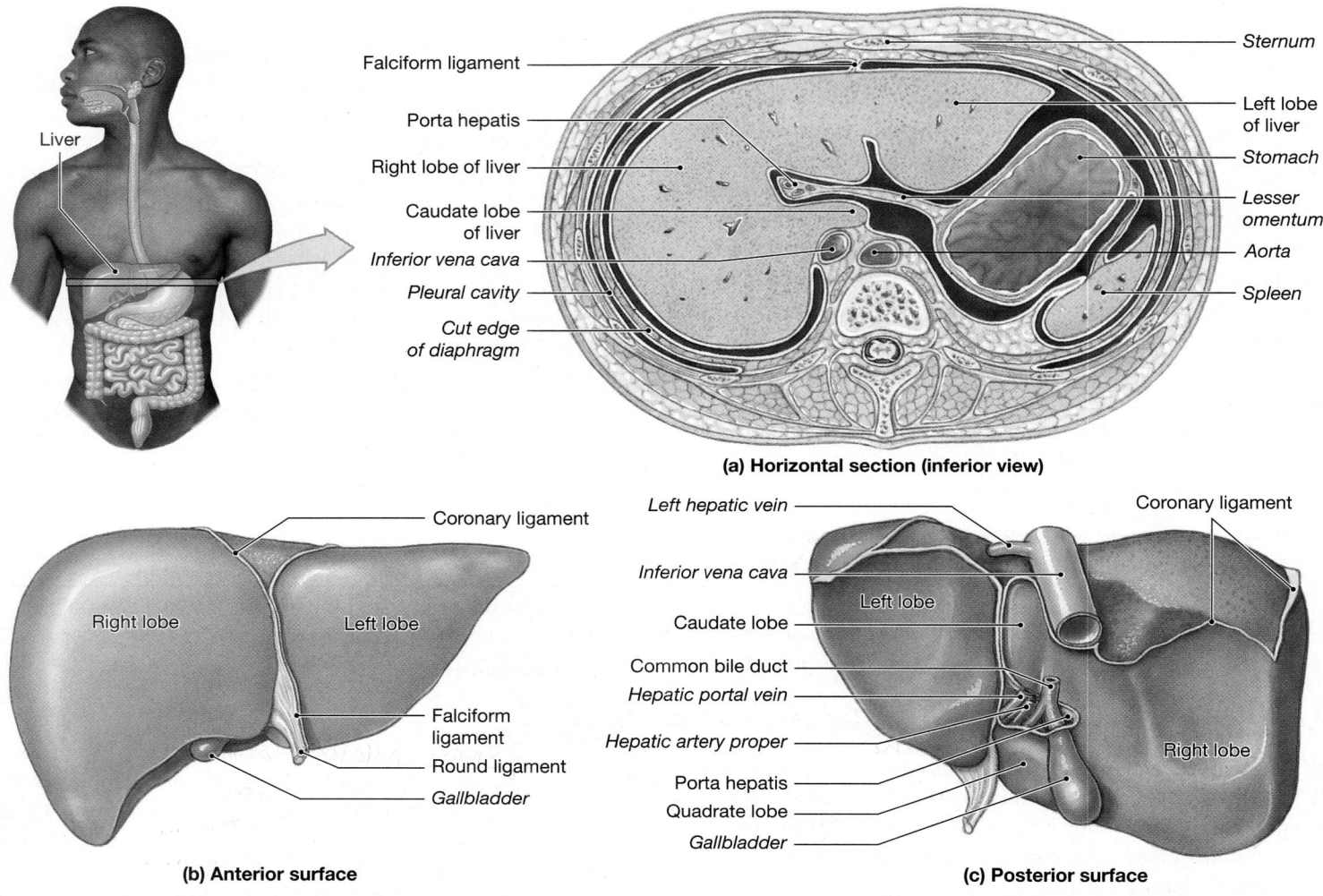

(a) Horizontal section (inferior view)

Labels for (a): Falciform ligament, Porta hepatis, Right lobe of liver, Caudate lobe of liver, Inferior vena cava, Pleural cavity, Cut edge of diaphragm, Sternum, Left lobe of liver, Stomach, Lesser omentum, Aorta, Spleen

Liver

(b) Anterior surface

Labels for (b): Coronary ligament, Right lobe, Left lobe, Falciform ligament, Round ligament, Gallbladder

(c) Posterior surface

Labels for (c): Left hepatic vein, Inferior vena cava, Caudate lobe, Common bile duct, Hepatic portal vein, Hepatic artery proper, Porta hepatis, Quadrate lobe, Gallbladder, Coronary ligament, Left lobe, Right lobe

Figure 24–19   The Anatomy of the Liver. **(a)** A horizontal sectional view through the superior abdomen. **(b)** The anterior surface of the liver. **(c)** The posterior surface of the liver. ATLAS: Plates 49a,b,e; 54a–c; 57a,b

one-third of the blood supply to the liver is arterial blood from the hepatic artery proper. The rest is venous blood from the hepatic portal vein, which begins in the capillaries of the esophagus, stomach, small intestine, and most of the large intestine. Liver cells, called **hepatocytes** (HEP-a-tō-sīts), adjust circulating levels of nutrients through selective absorption and secretion. The blood leaving the liver returns to the systemic circuit via the hepatic veins, which open into the inferior vena cava.

Histological Organization of the Liver

Each lobe of the liver is divided by connective tissue into approximately 100,000 **liver lobules**, the basic functional units of the liver. The histological organization and structure of a typical liver lobule are shown in **Figure 24–20.**

Each lobule is roughly 1 mm in diameter. Adjacent lobules are separated from each other by an *interlobular septum.* The hepatocytes in a liver lobule form a series of irregular plates arranged like the spokes of a wheel (**Figure 24–20a,b**).

The plates are only one cell thick, and exposed hepatocyte surfaces are covered with short microvilli. Within a lobule, sinusoids between adjacent plates empty into the **central vein**. (Sinusoids were introduced in Chapter 21. ∞ p. 726) The liver sinusoids lack a basal lamina, so large openings between the endothelial cells allow solutes—even those as large as plasma proteins—to pass out of the bloodstream and into the spaces surrounding the hepatocytes.

In addition to containing typical endothelial cells, the sinusoidal lining includes a large number of **Kupffer** (KOOP-fer) **cells**, also known as *stellate reticuloendothelial cells.* ∞ p. 791 These phagocytic cells, part of the monocyte–macrophage system, engulf pathogens, cell debris, and damaged blood cells. Kupffer cells are also responsible for storing iron, some lipids, and heavy metals (such as tin or mercury) that are absorbed by the digestive tract.

Blood enters the liver sinusoids from small branches of the hepatic portal vein and hepatic artery proper. A typical

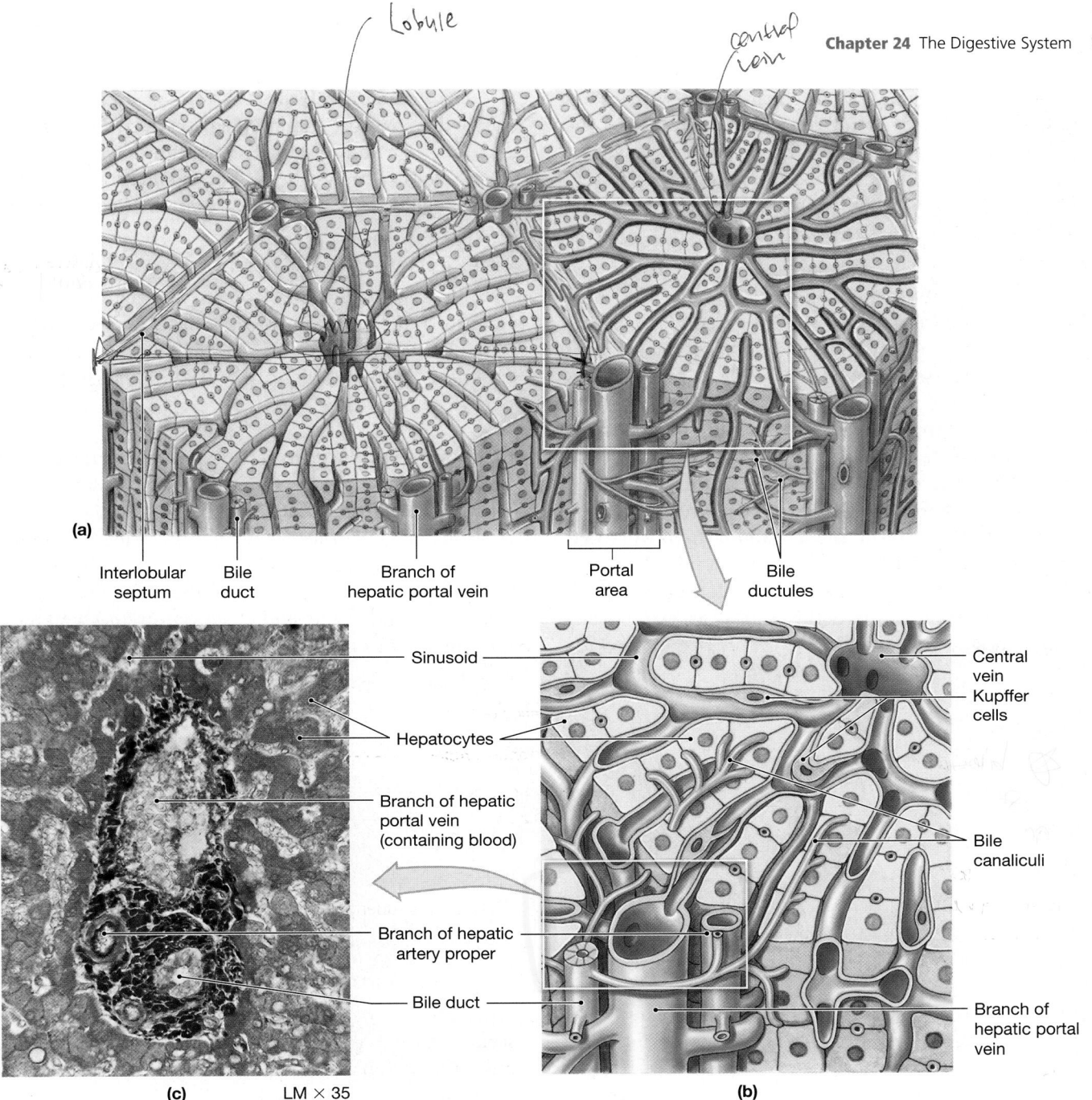

Figure 24–20 **Liver Histology.** **(a)** A diagrammatic view of liver structure, showing relationships among lobules. **(b)** A single liver lobule and its cellular components. **(c)** A portal area.

Labels in figure:
(a) Interlobular septum · Bile duct · Branch of hepatic portal vein · Portal area · Bile ductules
(c) LM × 35 — Sinusoid · Hepatocytes · Branch of hepatic portal vein (containing blood) · Branch of hepatic artery proper · Bile duct
(b) Central vein · Kupffer cells · Bile canaliculi · Branch of hepatic portal vein

liver lobule has a hexagonal shape in cross section (**Figure 24–20a**). There are six **portal areas**, or *hepatic triads*, one at each corner of the lobule. A portal area contains three structures: (1) a branch of the hepatic portal vein, (2) a branch of the hepatic artery proper, and (3) a small branch of the bile duct (**Figure 24–20a–c**).

Branches from the arteries and veins deliver blood to the sinusoids of adjacent liver lobules (**Figure 24–20a,b**). As blood flows through the sinusoids, hepatocytes absorb solutes from the plasma and secrete materials such as plasma proteins.

Blood then leaves the sinusoids and enters the central vein of the lobule. The central veins ultimately merge to form the hepatic veins, which then empty into the inferior vena cava. Liver diseases, such as the various forms of *hepatitis*, and conditions such as alcoholism, can lead to degenerative changes in the liver tissue and constriction of the circulatory supply.

Pressures in the hepatic portal system are usually low, averaging 10 mm Hg or less. This pressure can increase markedly, however, if blood flow through the liver becomes restricted as a result of a blood clot or damage to the organ.

Such a rise in portal pressure is called *portal hypertension*. As pressures rise, small peripheral veins and capillaries in the portal system become distended; if they rupture, extensive bleeding can occur. Portal hypertension can also force fluid into the peritoneal cavity across the serosal surfaces of the liver and viscera, producing ascites (p. 877).

The Bile Duct System

The liver secretes a fluid called **bile** into a network of narrow channels between the opposing membranes of adjacent liver cells. These passageways, called **bile canaliculi**, extend outward, away from the central vein (**Figure 24–20b**). Eventually, they connect with fine **bile ductules** (DUK-tūlz), which carry bile to bile ducts in the nearest portal area (**Figure 24–20a**). The **right** and **left hepatic ducts** (**Figure 24–21a**) collect bile from all the bile ducts of the liver lobes. These ducts unite to form the **common hepatic duct**, which leaves the liver. The bile in the common hepatic duct either flows into the *common bile duct*, which empties into the duodenal ampulla, or enters the *cystic duct*, which leads to the gallbladder.

The **common bile duct** is formed by the union of the **cystic duct** and the common hepatic duct. The common bile duct passes within the lesser omentum toward the stomach, turns, and penetrates the wall of the duodenum to meet the pancreatic duct at the duodenal ampulla (**Figure 24–21b**).

The Physiology of the Liver

The liver is responsible for three general categories of functions: (1) *metabolic regulation*, (2) *hematological regulation*, and (3) *bile production*. The liver has more than 200 functions; this discussion will provide only a general overview.

Metabolic Regulation. The liver is the primary organ involved in regulating the composition of circulating blood. All blood leaving the absorptive surfaces of the digestive tract enters the hepatic portal system and flows into the liver. Liver cells extract nutrients or toxins from the blood before it reaches the systemic circulation through the hepatic veins. The liver removes and stores excess nutrients, and it corrects nutrient deficiencies by mobilizing stored reserves or

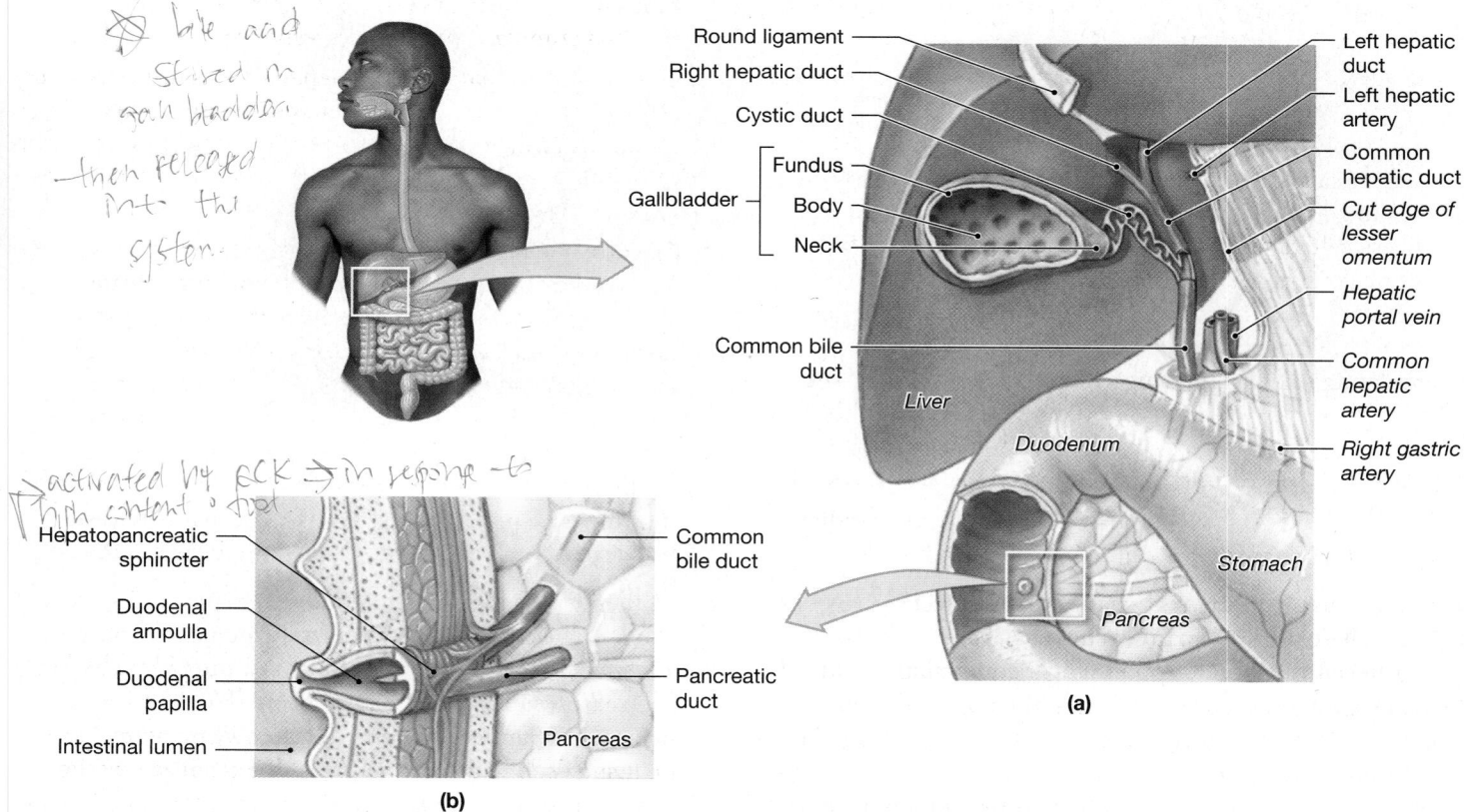

Figure 24–21 **The Gallbladder and Bile Ducts.** (a) A view of the inferior surface of the liver, showing the position of the gallbladder and ducts that transport bile from the liver to the gallbladder and duodenum. A portion of the lesser omentum has been cut away. (b) An interior view of the duodenum, showing the duodenal ampulla and related structures. ATLAS: Plates 49c,e; 51a; 54b–d

performing synthetic activities. The liver's regulatory activities affect the following:

- *Carbohydrate Metabolism.* The liver stabilizes blood glucose levels at about 90 mg/dL. If blood glucose levels drop, hepatocytes break down glycogen reserves and release glucose into the bloodstream. They also synthesize glucose from other carbohydrates or from available amino acids. The synthesis of glucose from other compounds is a process called *gluconeogenesis.* If blood glucose levels climb, hepatocytes remove glucose from the bloodstream and either store it as glycogen or use it to synthesize lipids that can be stored in the liver or other tissues. Circulating hormones, such as insulin and glucagon, regulate these metabolic activities. ∞ pp. 631–633

- *Lipid Metabolism.* The liver regulates circulating levels of triglycerides, fatty acids, and cholesterol. When those levels decline, the liver breaks down its lipid reserves and releases the breakdown products into the bloodstream. When the levels are high, the lipids are removed for storage. However, because most lipids absorbed by the digestive tract bypass the hepatic portal circulation, this regulation occurs only after lipid levels have risen within the general circulation.

- *Amino Acid Metabolism.* The liver removes excess amino acids from the bloodstream. These amino acids can be used to synthesize proteins or can be converted to lipids or glucose for energy storage.

- *Waste Product Removal.* When converting amino acids to lipids or carbohydrates, or when breaking down amino acids to get energy, the liver strips off the amino groups, a process called *deamination.* Ammonia, a toxic waste product, is formed. The liver neutralizes ammonia by converting it to *urea,* a fairly harmless compound excreted by the kidneys. Other waste products, circulating toxins, and drugs are also removed from the blood for inactivation, storage, or excretion.

- *Vitamin Storage.* Fat-soluble vitamins (A, D, E, and K) and vitamin B_{12} are absorbed from the blood and stored in the liver. These reserves are called on when your diet contains inadequate amounts of those vitamins.

- *Mineral Storage.* The liver converts iron reserves to ferritin and stores this protein–iron complex. ∞ p. 659

- *Drug Inactivation.* The liver removes and breaks down circulating drugs, thereby limiting the duration of their effects. When physicians prescribe a particular drug, they must take into account the rate at which the liver removes that drug from the bloodstream. For example, a drug that is absorbed relatively quickly must be administered every few hours to keep plasma concentrations at therapeutic levels.

CLINICAL NOTE

Cirrhosis Any condition that severely damages the liver represents a serious threat to life. The liver has a limited ability to regenerate itself after injury, but liver function will not fully recover unless the normal vascular pattern is restored. Examples of important types of liver disease include **cirrhosis,** which is characterized by the replacement of lobules by fibrous tissue, and various forms of *hepatitis* caused by viral infections. In some cases, liver transplants are used to treat liver failure, but the supply of suitable donor tissue is limited, and the success rate is highest in young, otherwise healthy individuals. Clinical trials are now under way to test an artificial liver known as *ELAD* (extracorporeal *liver assist device*) that may prove suitable for the long-term support of individuals with chronic liver disease.

Hematological Regulation. The liver, the largest blood reservoir in your body, receives about 25 percent of cardiac output. As blood passes through it, the liver performs the following functions:

- *Phagocytosis and Antigen Presentation.* Kupffer cells in the liver sinusoids engulf old or damaged red blood cells, cellular debris, and pathogens, removing them from the bloodstream. Kupffer cells are antigen-presenting cells that can stimulate an immune response. ∞ p. 791

- *Synthesis of Plasma Proteins.* Hepatocytes synthesize and release most of the plasma proteins, including the albumins (which contribute to the osmotic concentration of the blood), the various types of transport proteins, clotting proteins, and complement proteins.

- *Removal of Circulating Hormones.* The liver is the primary site for the absorption and recycling of epinephrine, norepinephrine, insulin, thyroid hormones, and steroid hormones, such as the sex hormones (estrogens and androgens) and corticosteroids. The liver also absorbs cholecalciferol (vitamin D_3) from the blood. Liver cells then convert the cholecalciferol, which may be synthesized in the skin or absorbed in the diet, into an intermediary product, 25-hydroxy-D_3, that is released back into the bloodstream. This intermediary is absorbed by the kidneys and used to generate calcitriol, a hormone important to Ca^{2+} metabolism. ∞ p. 636

- *Removal of Antibodies.* The liver absorbs and breaks down antibodies, releasing amino acids for recycling.

- *Removal or Storage of Toxins.* Lipid-soluble toxins in the diet, such as the insecticide DDT, are absorbed by the liver and stored in lipid deposits, where they do not disrupt cellular functions. Other toxins are removed from the bloodstream and are either broken down or excreted in the bile.

- *The Synthesis and Secretion of Bile.* Bile is synthesized in the liver and excreted into the lumen of the duodenum. Bile secretion is regulated by hormonal and neural mechanisms. Bile consists mostly of water, with minor amounts of ions, *bilirubin* (a pigment derived from hemoglobin), cholesterol, and an assortment of lipids collectively known as **bile salts**. (Bile salts play a role in the digestion of lipids, as discussed in the next section.) The water and ions assist in the dilution and buffering of acids in chyme as it enters the small intestine. Bile salts are synthesized from cholesterol in the liver. Several related compounds are involved; the most abundant are derivatives of the steroids *cholate* and *chenodeoxycholate*.

The Functions of Bile. Most dietary lipids are not water soluble. Mechanical processing in the stomach creates large drops containing a variety of lipids. Pancreatic lipase is not lipid soluble, so the enzymes can interact with lipids only at the surface of a lipid droplet. The larger the droplet, the more lipids are inside, isolated and protected from these enzymes. Bile salts break the droplets apart in a process called **emulsification** (ē-mul-si-fi-KĀ-shun), which dramatically increases the surface area accessible to enzymatic attack.

Emulsification creates tiny *emulsion droplets* with a superficial coating of bile salts. The formation of tiny droplets increases the surface area available for enzymatic attack. In addition, the layer of bile salts facilitates interaction between the lipids and lipid-digesting enzymes supplied by the pancreas. After lipid digestion has been completed, bile salts promote the absorption of lipids by the intestinal epithelium. More than 90 percent of the bile salts are themselves reabsorbed, primarily in the ileum, as lipid digestion is completed. The reabsorbed bile salts enter the hepatic portal circulation and are collected and recycled by the liver. The cycling of bile salts from the liver to the small intestine and back is called the **enterohepatic circulation of bile**.

Tips&Tricks

Oil and water don't mix. The emulsification process helps mix oils and water-soluble enzymes so that fats can be broken down.

The Gallbladder

The **gallbladder** is a hollow, pear-shaped organ that stores and concentrates bile prior to its excretion into the small intestine. This muscular sac is located in a fossa, or recess, in the posterior surface of the liver's right lobe (**Figure 24–21a**). The gallbladder is divided into three regions: (1) the **fundus**, (2) the **body**, and (3) the **neck**. The cystic duct extends from the gallbladder to the point where its union with the common hepatic duct forms the common bile duct. At the duodenum, the common bile duct meets the pancreatic duct before emptying into a chamber called the **duodenal ampulla** (am-PUL-luh) (**Figure 24–21b**), which receives buffers and enzymes from the pancreas and bile from the liver and gallbladder. The duodenal ampulla opens into the duodenum at the **duodenal papilla**, a small mound.

The muscular **hepatopancreatic sphincter** (*sphincter of Oddi*) encircles the lumen of the common bile duct and, generally, the pancreatic duct and duodenal ampulla as well.

Physiology of the Gallbladder

A major function of the gallbladder is *bile storage*. Bile is secreted continuously—roughly 1 liter is produced each day—but it is released into the duodenum only under the stimulation of the intestinal hormone CCK. In the absence of CCK, the hepatopancreatic sphincter remains closed, so bile exiting the liver in the common hepatic duct cannot flow through the common bile duct and into the duodenum. Instead, it enters the cystic duct and is stored within the expandable gallbladder. Whenever chyme enters the duodenum, CCK is released, relaxing the hepatopancreatic sphincter and stimulating contractions in the walls of the gallbladder that push bile into the small intestine. The amount of CCK secreted increases markedly when the chyme contains large amounts of lipids.

The gallbladder also functions in *bile modification*. When full, the gallbladder contains 40–70 mL of bile. The composition of bile gradually changes as it remains in the gallbladder: Much of the water is absorbed, and the bile salts and other components of bile become increasingly concentrated.

If bile becomes too concentrated, crystals of insoluble minerals and salts begin to form. These deposits are called *gallstones*. Small gallstones are not a problem so long as they can be flushed down the bile duct and excreted. In *cholecystitis* (kō-lē-sis-TĪ-tis; *chole*, bile + *kystis*, bladder + *itis*, inflammation), the gallstones are so large that they can damage the wall of the gallbladder or block the cystic duct or common bile duct. In that case, the gallbladder may need to be surgically removed. This does not seriously impair digestion, because bile production continues at normal levels. However, the bile is more dilute, and its entry into the small intestine is not as closely tied to the arrival of food in the duodenum.

The A&P Top 100

#85 The liver is the center for metabolic regulation in the body. It also produces bile that is stored in the gallbladder and ejected into the duodenum in response to the release of the hormone, CCK. Bile is needed for the efficient digestion of lipids because it breaks down large lipid droplets so that digestive enzymes can work on individual lipid molecules.

The Coordination of Secretion and Absorption

A combination of neural and hormonal mechanisms coordinates the activities of the digestive glands. These regulatory mechanisms are centered around the duodenum, where acids must be neutralized and the appropriate enzymes added.

Neural mechanisms involving the CNS (1) prepare the digestive tract for activity (parasympathetic innervation) or inhibit gastrointestinal activity (sympathetic innervation) and (2) coordinate the movement of materials along the length of the digestive tract (the enterogastric, gastroenteric, and gastroileal reflexes).

In addition, motor neurons synapsing in the digestive tract release a variety of neurotransmitters. Many of these chemicals are also released in the CNS, but in general, their functions are poorly understood. Examples of potentially important neurotransmitters include substance P, enkephalins, and endorphins.

We will now summarize the information presented thus far on the regulation of intestinal and glandular function and consider some additional details about the regulatory mechanisms involved.

Intestinal Hormones

The intestinal tract secretes a variety of peptide hormones with similar chemical structures. Many of these hormones have multiple effects in several regions of the digestive tract, and in the accessory glandular organs as well.

Duodenal enteroendocrine cells produce the following hormones known to coordinate digestive functions:

- **Secretin** is released when chyme arrives in the duodenum. Secretin's primary effect is an increase in the secretion of bile and buffers by the liver and pancreas. Among its secondary effects, secretin reduces gastric motility and secretory rates.

- **Cholecystokinin (CCK)** is secreted when chyme arrives in the duodenum, especially when the chyme contains lipids and partially digested proteins. In the pancreas, CCK accelerates the production and secretion of all types of digestive enzymes. It also causes a relaxation of the hepatopancreatic sphincter and contraction of the gallbladder, resulting in the ejection of bile and pancreatic juice into the duodenum. Thus, the net effects of CCK are to increase the secretion of pancreatic enzymes and to push pancreatic secretions and bile into the duodenum. The presence of CCK in high concentrations has two additional effects: It inhibits gastric activity, and it appears to have CNS effects that reduce the sensation of hunger.

- **Gastric inhibitory peptide (GIP)** is secreted when fats and carbohydrates—especially glucose—enter the small intestine. The inhibition of gastric activity is accompanied by the stimulation of insulin release at the pancreatic islets, so GIP is also known as *glucose-dependent insulinotropic peptide*. This hormone has several secondary effects; for instance, it stimulates the activity of the duodenal glands, stimulates lipid synthesis in adipose tissue, and increases glucose use by skeletal muscles.

- **Vasoactive intestinal peptide (VIP)** stimulates the secretion of intestinal glands, dilates regional capillaries, and inhibits acid production in the stomach. By dilating capillaries in active areas of the intestinal tract, VIP provides an efficient mechanism for removing absorbed nutrients.

- **Gastrin** is secreted by G cells in the duodenum when they are exposed to large quantities of incompletely digested proteins. The functions of gastrin include promoting increased stomach motility and stimulating the production of acids and enzymes. (Gastrin is also secreted by the stomach, as detailed on p. 894.)

- **Enterocrinin**, a hormone released when chyme enters the small intestine, stimulates mucin production by the submucosal glands of the duodenum.

- **Other intestinal hormones** are produced in relatively small quantities. Examples include *motilin*, which stimulates intestinal contractions; *villikinin*, which promotes the movement of villi and the associated lymph flow; and *somatostatin*, which inhibits gastric secretion.

Table 24–2 summarizes the origins and primary effects of these important digestive hormones. Functional interactions among gastrin, secretin, CCK, GIP, and VIP are diagrammed in **Figure 24–22**.

Intestinal Absorption

On average, it takes about five hours for materials to pass from the duodenum to the end of the ileum, so the first of the materials to enter the duodenum after you eat breakfast may leave the small intestine at lunchtime. Along the way, the organ's absorptive effectiveness is enhanced by the fact that so much of the mucosa is movable. The microvilli can be moved by their supporting microfilaments, the individual villi by smooth muscle cells, groups of villi by the muscularis mucosae, and the plicae circulares by the muscularis mucosae and the muscularis externa. These movements stir and mix the intestinal contents, changing the environment around each epithelial cell from moment to moment.

TABLE 24–2   Major Digestive Hormones and Their Primary Effects

Hormone	Stimulus	Origin	Target	Effects
Cholecystokinin (CCK)	Arrival of chyme containing lipids and partially digested proteins	Duodenum	Pancreas	Stimulates production of pancreatic enzymes
			Gallbladder	Stimulates contraction of gallbladder
			Duodenum	Causes relaxation of hepatopancreatic sphincter
			Stomach	Inhibits gastric secretion and motion
			CNS	May reduce hunger
Enterocrinin	Arrival of chyme in duodenum	Duodenum	Duodenum	Stimulates mucin production
Gastric inhibitory peptide (GIP)	Arrival of chyme containing large quantities of fats and glucose	Duodenum	Pancreas	Stimulates release of insulin by pancreatic islets
			Stomach	Inhibits gastric secretion and motility
			Adipose tissue	Stimulates lipid synthesis
			Skeletal muscle	Stimulates glucose use
Gastrin	Vagus nerve stimulation or arrival of food in the stomach	Stomach	Stomach	Stimulates production of acids and enzymes; increases motility
	Arrival of chyme containing large quantities of undigested proteins	Duodenum	Stomach	As above
Secretin	Arrival of chyme in the duodenum	Duodenum	Pancreas	Stimulates production of alkaline buffers
			Stomach	Inhibits gastric secretion and motility
			Liver	Increases rate of bile secretion
Vasoactive intestinal peptide (VIP)	Arrival of chyme in the duodenum	Duodenum	Duodenal glands, stomach	Stimulates buffer secretion; inhibits acid production; dilates intestinal capillaries

CHECKPOINT

21. **Name the three regions of the small intestine from proximal to distal.**

22. **How is the small intestine adapted for the absorption of nutrients?**

23. **Does a high-fat meal raise or lower the level of cholecystokinin in the blood?**

24. **How would the pH of the intestinal contents be affected if the small intestine did not produce secretin?**

25. **The digestion of which nutrient would be most impaired by damage to the exocrine pancreas?**

See the blue Answers tab at the end of the book.

24-7 The large intestine is divided into three parts with regional specialization

The horseshoe-shaped **large intestine** begins at the end of the ileum and ends at the anus. The large intestine lies inferior to the stomach and liver and almost completely frames the small intestine (**Figure 24–1**). The major functions of the large intestine include (1) the reabsorption of water and the compaction of the intestinal contents into feces, (2) the absorption of important vitamins liberated by bacterial action, and (3) the storage of fecal material prior to defecation.

The large intestine, also known as the *large bowel*, has an average length of about 1.5 meters (4.9 ft) and a width of 7.5 cm (3 in.). We can divide it into three parts: (1) the pouch-like *cecum*, the first portion of the large intestine; (2) the *colon*, the largest portion; and (3) the *rectum*, the last 15 cm (6 in.) of the large intestine and the end of the digestive tract (**Figure 24–23a**).

The Cecum

Material arriving from the ileum first enters an expanded pouch called the **cecum** (SĒ-kum). The ileum attaches to the medial surface of the cecum and opens into the cecum at the *ileocecal valve* (**Figure 24–23a,b**). The cecum collects and stores materials from the ileum and begins the process of compaction.

The slender, hollow **appendix**, or *vermiform appendix* (*vermis*, worm), is attached to the posteromedial surface of the cecum (**Figure 24–23a,b**). The appendix is generally

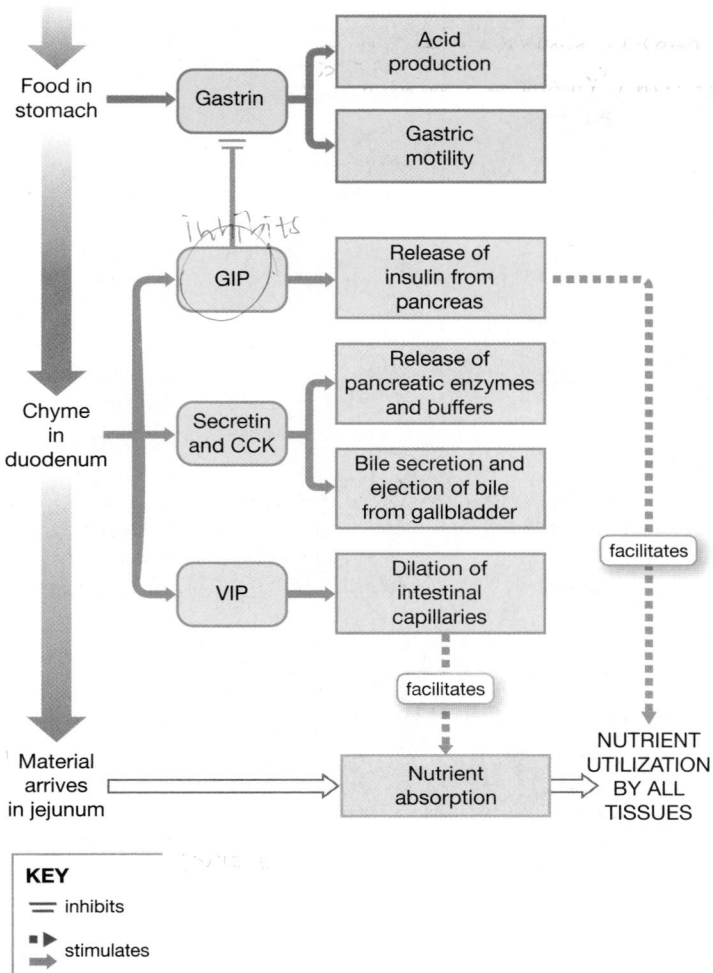

Figure 24–22 The Activities of Major Digestive Tract Hormones. The primary actions of gastrin, secretin, CCK, GIP, and VIP are depicted.

about 9 cm (3.6 in.) long, but its size and shape are quite variable. A small mesentery called the **mesoappendix** connects the appendix to the ileum and cecum. The mucosa and submucosa of the appendix are dominated by lymphoid nodules, and the primary function of the appendix is as an organ of the lymphoid system. Inflammation of the appendix is known as *appendicitis*.

The Colon

The **colon** has a larger diameter and a thinner wall than the small intestine. Distinctive features of the colon include the following (**Figure 24–23a**):

- The wall of the colon forms a series of pouches, or **haustra** (HAWS-truh; singular, *haustrum*). Cutting into the intestinal lumen reveals that the creases between the

haustra affect the mucosal lining as well, producing a series of internal folds. Haustra permit the expansion and elongation of the colon, rather like the bellows that allow an accordion to lengthen.

- Three separate longitudinal bands of smooth muscle—called the **taeniae coli** (TĒ-nē-ē KŌ-lē; singular, *taenia*)—run along the outer surfaces of the colon just deep to the serosa. These bands correspond to the outer layer of the muscularis externa in other portions of the digestive tract. Muscle tone within the taeniae coli is what creates the haustra.

- The serosa of the colon contains numerous teardrop-shaped sacs of fat called **fatty appendices**, or *epiploic* (ep-i-PLŌ-ik; *epiploon*, omentum) *appendages*.

We can subdivide the colon into four regions: the ascending colon, transverse colon, descending colon, and sigmoid colon (**Figure 24–23a**).

1. The **ascending colon** begins at the superior border of the cecum and ascends along the right lateral and posterior wall of the peritoneal cavity to the inferior surface of the liver. There, the colon bends sharply to the left at the **right colic flexure**, or *hepatic flexure*, which marks the end of the ascending colon and the beginning of the transverse colon.

2. The **transverse colon** curves anteriorly from the right colic flexure and crosses the abdomen from right to left. The transverse colon is supported by the transverse mesocolon and is separated from the anterior abdominal wall by the layers of the greater omentum. As the transverse colon reaches the left side of the body, it passes inferior to the greater curvature of the stomach. Near the spleen, the colon makes a 90° turn at the **left colic flexure**, or *splenic flexure*, and becomes the descending colon.

3. The **descending colon** proceeds inferiorly along the person's left side until reaching the iliac fossa formed by the inner surface of the left ilium. The descending colon is retroperitoneal and firmly attached to the abdominal wall. At the iliac fossa, the descending colon curves at the **sigmoid flexure** and becomes the sigmoid colon.

4. The sigmoid flexure is the start of the **sigmoid** (SIG-moyd) **colon** (*sigmeidos*, the Greek letter *S*), an S-shaped segment that is only about 15 cm (6 in.) long. The sigmoid colon lies posterior to the urinary bladder, suspended from the sigmoid mesocolon. The sigmoid colon empties into the *rectum*.

The large intestine receives blood from tributaries of the superior mesenteric and inferior mesenteric arteries. Venous blood is collected from the large intestine by the superior mesenteric and inferior mesenteric veins. ∞ pp. 766, 767

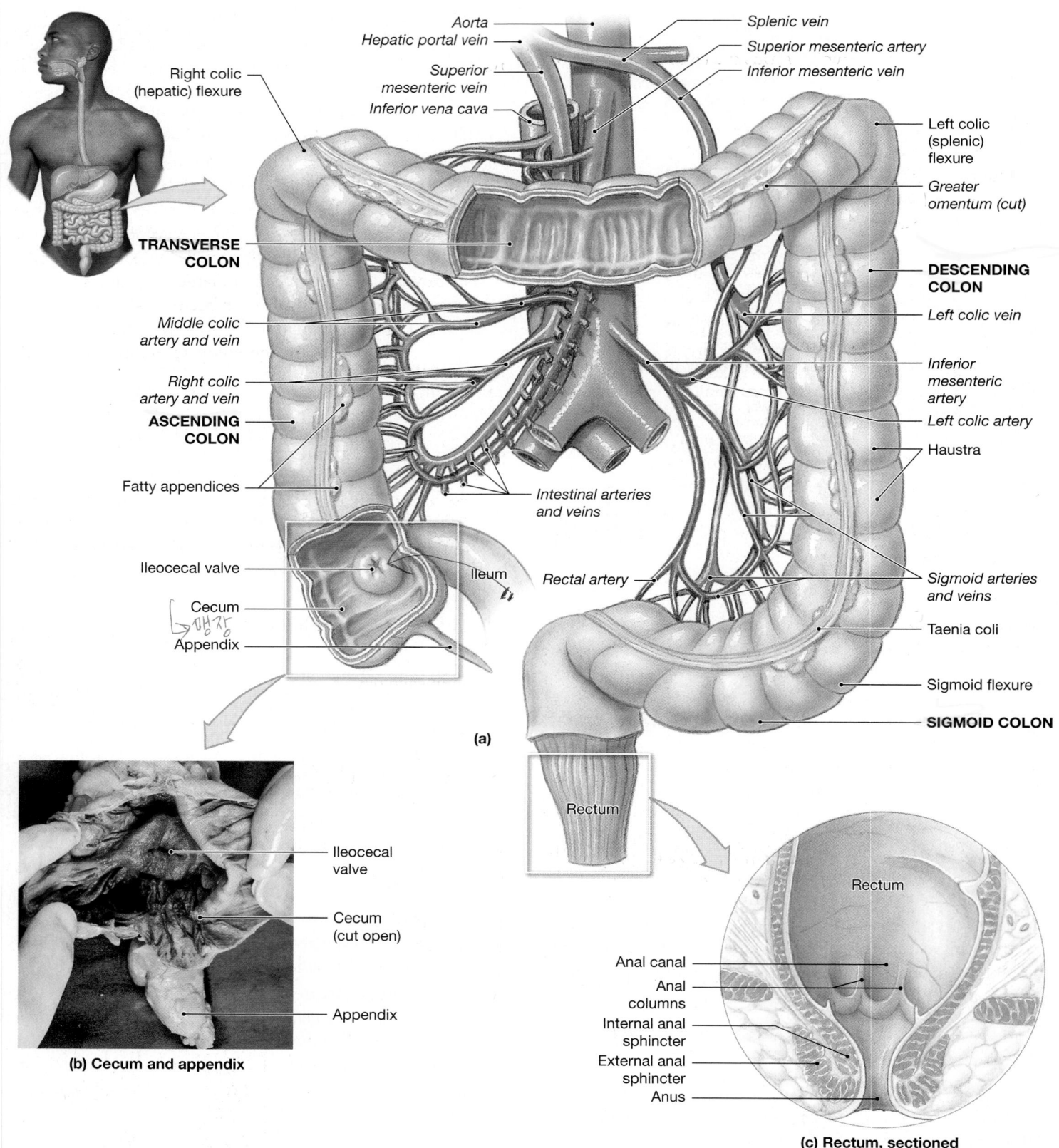

TRANSVERSE COLON

Right colic (hepatic) flexure

Aorta
Hepatic portal vein
Superior mesenteric vein
Inferior vena cava

Splenic vein
Superior mesenteric artery
Inferior mesenteric vein

Left colic (splenic) flexure

Greater omentum (cut)

DESCENDING COLON

Left colic vein

Middle colic artery and vein

Right colic artery and vein

ASCENDING COLON

Fatty appendices

Ileocecal valve

Cecum

Appendix

Ileum

Intestinal arteries and veins

Rectal artery

Inferior mesenteric artery

Left colic artery

Haustra

Sigmoid arteries and veins

Taenia coli

Sigmoid flexure

SIGMOID COLON

(a)

Rectum

Ileocecal valve

Cecum (cut open)

Appendix

(b) Cecum and appendix

Rectum

Anal canal

Anal columns

Internal anal sphincter

External anal sphincter

Anus

(c) Rectum, sectioned

Figure 24–23 The Large Intestine. (a) The gross anatomy and regions of the large intestine. **(b)** The cecum and appendix. **(c)** The rectum and anus. ATLAS: Plates 49a–c; 58a–c; 59; 64; 65

The Rectum

The **rectum** (REK-tum), which forms the last 15 cm (6 in.) of the digestive tract (**Figure 24–23a,c**), is an expandable organ for the temporary storage of feces. The movement of fecal material into the rectum triggers the urge to defecate.

The last portion of the rectum, the **anal canal**, contains small longitudinal folds called **anal columns**. The distal margins of these columns are joined by transverse folds that mark the boundary between the columnar epithelium of the proximal rectum and a stratified squamous epithelium like that in the oral cavity. The **anus**, or *anal orifice*, is the exit of the anal canal. There, the epidermis becomes keratinized and identical to the surface of the skin. The circular muscle layer of the muscularis externa in this region forms the **internal anal sphincter** (**Figure 24–23c**), the smooth muscle cells of which are not under voluntary control. The **external anal sphincter**, which guards the anus, consists of a ring of skeletal muscle fibers that encircles the distal portion of the anal canal. This sphincter consists of skeletal muscle and is under voluntary control.

The lamina propria and submucosa of the anal canal bear a network of veins. If venous pressures there rise too high due to straining during defecation, the veins can become distended, producing *hemorrhoids*.

CLINICAL NOTE

Colorectal Cancer Colorectal cancer is relatively common in the United States. Aside from skin cancers, colorectal cancer is the third most common cancer, affecting both men and women. In 2006, approximately 106,680 new cases of colon cancer and 41,930 cases of rectal cancers were reported. Colorectal cancer is the second leading cause of cancer-related deaths; however, the death rate has declined over the past 15 years. The best defense appears to be early detection and prompt treatment. Standard screening involves checking the feces for blood. This is a simple procedure that can be performed easily on a stool (fecal) sample as part of a routine physical. For those individuals at increased risk because of family history, associated disease, or older age, visual inspection of the lumen by fiberoptic colonoscopy to discover polyps before they develop into cancers is prudent. The 5-year survival rate for people whose cancer is found at an early stage and treated immediately is greater than 90%.

Histology of the Large Intestine

Although the diameter of the colon is roughly three times that of the small intestine, its wall is much thinner. The major characteristics of the colon are the lack of villi, the abundance of mucous cells, and the presence of distinctive intestinal glands (**Figure 24–24**). The glands in the large intestine are deeper than those of the small intestine and are dominated by mucous cells. The mucosa of the large intestine does not produce enzymes; any digestion that occurs results from enzymes introduced in the small intestine or from bacterial action. The mucus provides lubrication as the fecal material becomes drier and more compact. Mucus is secreted as local stimuli, such as friction or exposure to harsh chemicals, trigger short reflexes involving local nerve plexuses. Large lymphoid nodules are scattered throughout the lamina propria and submucosa.

The muscularis externa of the large intestine is unusual, because the longitudinal layer has been reduced to the muscular bands of the taeniae coli. However, the mixing and propulsive contractions of the colon resemble those of the small intestine.

Physiology of the Large Intestine

Less than 10 percent of the nutrient absorption under way in the digestive tract occurs in the large intestine. Nevertheless, the absorptive operations in this segment of the digestive tract are important. The large intestine also prepares fecal material for ejection from the body.

Absorption in the Large Intestine

The reabsorption of water is an important function of the large intestine. Although roughly 1500 mL of material enters the colon each day, only about 200 mL of feces is ejected. The remarkable efficiency of digestion can best be appreciated by considering the average composition of feces: 75 percent water, 5 percent bacteria, and the rest a mixture of indigestible materials, small quantities of inorganic matter, and the remains of epithelial cells.

In addition to reabsorbing water, the large intestine absorbs a number of other substances that remain in the feces or were secreted into the digestive tract along its length. Examples include useful compounds such as bile salts and vitamins, organic waste products such as urobilinogen, and various toxins generated by bacterial action. Most of the bile salts entering the large intestine are promptly reabsorbed in the cecum and transported in blood to the liver for secretion into bile.

Vitamins. Vitamins are organic molecules that are important as cofactors or coenzymes in many metabolic pathways. The normal bacterial residents of the colon generate three vitamins that supplement our dietary supply:

1. *Vitamin K*, a fat-soluble vitamin the liver requires for synthesizing four clotting factors, including prothrombin. Intestinal bacteria produce roughly half of your daily vitamin K requirements.

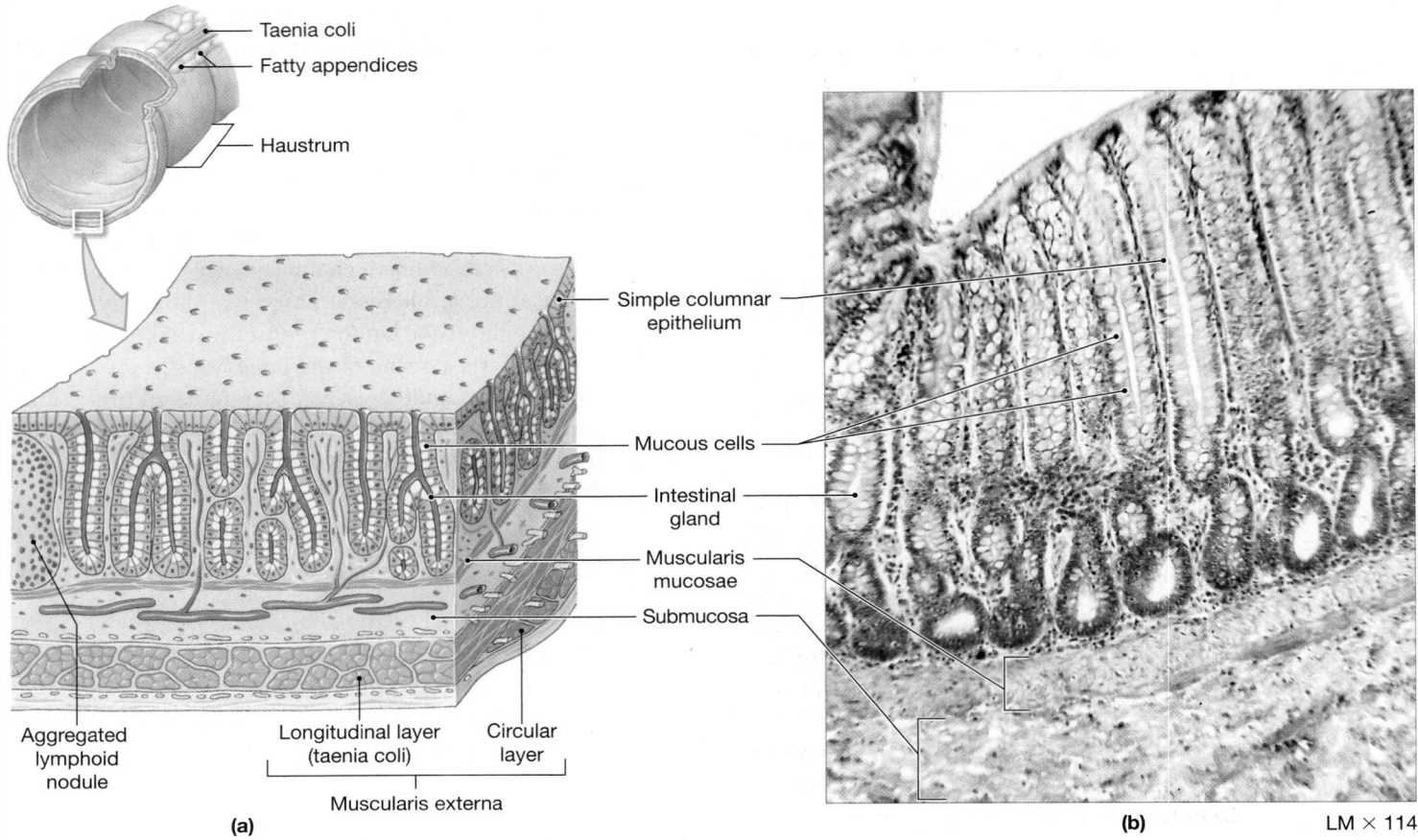

Figure 24–24 **The Mucosa and Glands of the Colon.**

2. *Biotin*, a water-soluble vitamin important in various reactions, notably those of glucose metabolism.

3. *Vitamin B₅* (pantothenic acid), a water-soluble vitamin required in the manufacture of steroid hormones and some neurotransmitters.

Vitamin K deficiencies, which lead to impaired blood clotting, result from either (1) a deficiency of lipids in the diet, which impairs the absorption of all fat-soluble vitamins, or (2) problems affecting lipid processing and absorption, such as inadequate bile production or chronic diarrhea. Disorders resulting from deficiencies of biotin or vitamin B₅ are extremely rare after infancy, because the intestinal bacteria produce sufficient amounts to supplement any dietary shortage.

Organic Wastes. The fate of bilirubin, a breakdown product of heme, was discussed in Chapter 19. ∞ p. 658 In the large intestine, bacteria convert bilirubin to *urobilinogens* and *stercobilinogens*. Some urobilinogens are absorbed into the bloodstream and then excreted in urine. The urobilinogens and stercobilinogens remaining within the colon are con-

verted to **urobilins** and **stercobilins** by exposure to oxygen. These pigments in various proportions give feces a yellow-brown or brown coloration. Bacterial action breaks down peptides that remain in the feces and generates (1) ammonia, in the form of soluble *ammonium ions* (NH_4^+); (2) *indole* and *skatole*, two nitrogen-containing compounds that are primarily responsible for the odor of feces; and (3) hydrogen sulfide (H_2S), a gas that produces a "rotten egg" odor. Significant amounts of ammonia and smaller amounts of other toxins cross the colonic epithelium and enter the hepatic portal circulation. These toxins are removed by the liver and converted to relatively nontoxic compounds that can be released into the blood and excreted at the kidneys.

Intestinal enzymes do not alter indigestible carbohydrates, so they arrive in the colon virtually intact. These complex polysaccharides provide a reliable nutrient source for colonic bacteria, whose metabolic activities are responsible for the small quantities of **flatus**, or intestinal gas, in the large intestine. Foods containing large amounts of indigestible carbohydrates (such as beans) stimulate bacterial gas production, leading to distension of the colon, cramps, and the frequent discharge of intestinal gases.

Movements of the Large Intestine

The gastroileal and gastroenteric reflexes move materials into the cecum while you eat. Movement from the cecum to the transverse colon is very slow, allowing hours for water absorption to convert the already thick material into a sludgy paste. Peristaltic waves move material along the length of the colon, and segmentation movements, called *haustral churning*, mix the contents of adjacent haustra. Movement from the transverse colon through the rest of the large intestine results from powerful peristaltic contractions called **mass movements**, which occur a few times each day. The stimulus is distension of the stomach and duodenum; the commands are relayed over the intestinal nerve plexuses. The contractions force feces into the rectum and produce the conscious urge to defecate.

The rectal chamber is usually empty, except when a powerful peristaltic contraction forces feces out of the sigmoid colon. Distension of the rectal wall then triggers the **defecation reflex**, which involves two positive feedback loops (**Figure 24–25**). Both loops are triggered by the stimulation of stretch receptors in the walls of the rectum. The first loop is a short reflex that triggers a series of peristaltic contractions in the rectum that move feces toward the anus (STEPS 1, 2, and 3). The second loop is a long reflex coordinated by the sacral parasympathetic system. This reflex stimulates mass movements that push feces toward the rectum from the descending colon and sigmoid colon (STEPS 1, 2a, and 3).

Rectal stretch receptors also trigger two reflexes important to the *voluntary* control of defecation. One is a long reflex mediated by parasympathetic innervation within the pelvic nerves. This reflex causes the relaxation of the internal anal sphincter, the smooth muscle sphincter that controls the movement of feces into the anal canal. The second (involving STEP 2b in **Figure 24–25**) is a somatic reflex that stimulates the immediate contraction of the external anal sphincter, a skeletal muscle. ∞ pp. 356–358 The motor commands are carried by the pudendal nerves.

The elimination of feces requires that both the internal and external anal sphincters be relaxed, but the two reflexes just mentioned open the internal sphincter and close the external sphincter. The actual release of feces requires a conscious effort to open the external sphincter. In addition to opening the external sphincter, consciously directed activities such as tensing the abdominal muscles or making expiratory movements while closing the glottis elevate intra-abdominal pressures and help force fecal material out of the rectum.

If the external anal sphincter remains constricted, the peristaltic contractions cease until additional rectal expansion triggers the defecation reflex a second time. The urge to defecate usually develops when rectal pressure reaches about 15 mm Hg. If pressure inside the rectum exceeds 55 mm Hg, the external anal sphincter will involuntarily relax and defe-

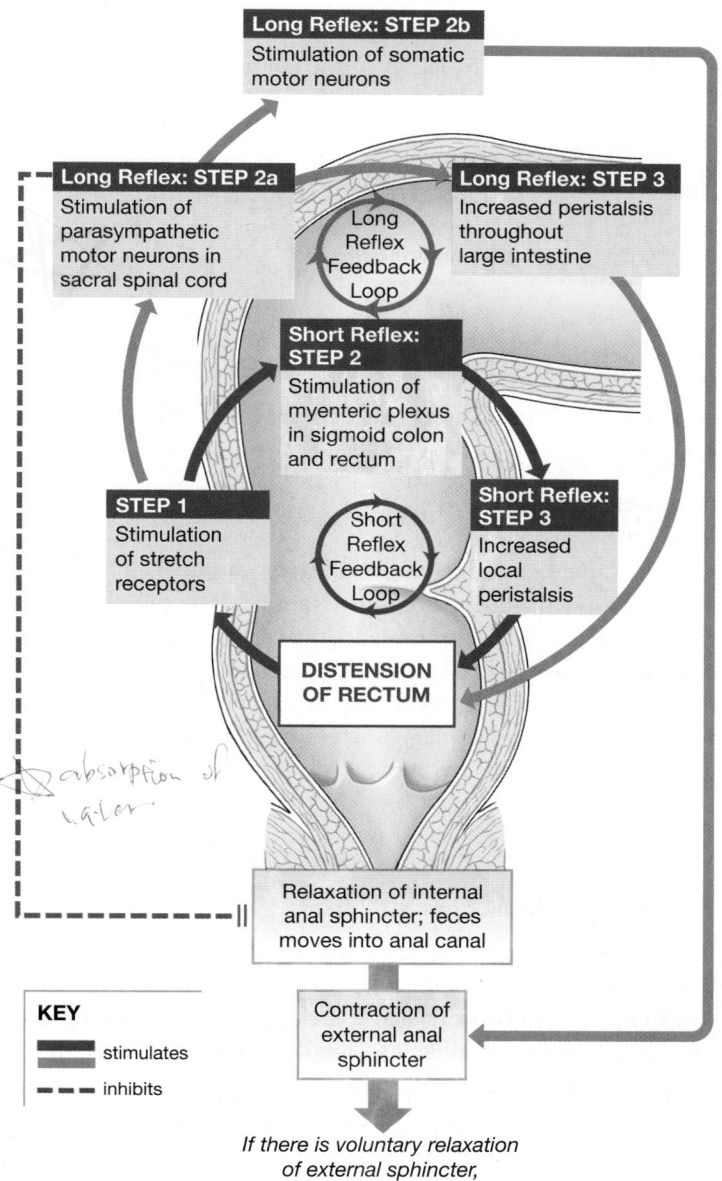

Long Reflex: STEP 2b
Stimulation of somatic motor neurons

Long Reflex: STEP 2a
Stimulation of parasympathetic motor neurons in sacral spinal cord

Long Reflex Feedback Loop

Long Reflex: STEP 3
Increased peristalsis throughout large intestine

Short Reflex: STEP 2
Stimulation of myenteric plexus in sigmoid colon and rectum

STEP 1
Stimulation of stretch receptors

Short Reflex Feedback Loop

Short Reflex: STEP 3
Increased local peristalsis

DISTENSION OF RECTUM

Relaxation of internal anal sphincter; feces moves into anal canal

KEY
━━━ stimulates
╌╌╌ inhibits

Contraction of external anal sphincter

If there is voluntary relaxation of external sphincter,
DEFECATION OCCURS

Figure 24–25 The Defecation Reflex. Short and long reflexes promote movement of fecal material toward the anus. Another long reflex triggered by rectal stretch receptor stimulation (STEPS 2a and 2b) prevents involuntary defecation.

cation will occur. This mechanism regulates defecation in infants and in adults with severe spinal cord injuries.

The A&P Top 100

#86 The large intestine stores digestive wastes and reabsorbs water. Bacterial residents of the large intestine are an important source of vitamins, especially vitamin K, biotin, and vitamin B_5.

26. Identify the four regions of the colon.

27. What are some major histological differences between the large intestine and the small intestine?

28. Differentiate between haustral churning and mass movements.

See the blue Answers tab at the end of the book.

24-8 Digestion is the mechanical and chemical alteration of food that allows the absorption and use of nutrients

A typical meal contains carbohydrates, proteins, lipids, water, electrolytes (minerals), and vitamins. The digestive system handles each component differently. Large organic molecules must be broken down by digestion before absorption can occur. Water, electrolytes, and vitamins can be absorbed without preliminary processing, but special transport mechanisms may be involved.

The Processing and Absorption of Nutrients

Food contains large organic molecules, many of them insoluble. The digestive system first breaks down the physical structure of the ingested material and then proceeds to disassemble the component molecules into smaller fragments. This disassembly eliminates any antigenic properties, so that the fragments do not trigger an immune response after absorption. Cells absorb the molecules released into the bloodstream and either (1) break them down to provide energy for the synthesis of ATP or (2) use these molecules to synthesize carbohydrates, proteins, and lipids. This section focuses on the mechanics of digestion and absorption; the fates of the compounds inside cells are the focus in Chapter 25.

Most ingested organic materials are complex chains of simpler molecules. In a typical dietary carbohydrate, the basic molecules are simple sugars; in a protein, the building blocks are amino acids; in lipids, they are generally fatty acids; and in nucleic acids, they are nucleotides. Digestive enzymes break the bonds between the component molecules of carbohydrates, proteins, lipids, and nucleic acids in a process called *hydrolysis*. ∞ p. 38

The classes of digestive enzymes differ with respect to their targets. *Carbohydrases* break the bonds between simple sugars, *proteases* split the linkages between amino acids, and *lipases* separate fatty acids from glycerides. Some enzymes in each class are even more selective, breaking bonds between specific molecules. For example, a particular carbohydrase might break the bond between two glucose molecules, but not those between glucose and another simple sugar.

Digestive enzymes secreted by the salivary glands, tongue, stomach, and pancreas are mixed into the ingested material as it passes along the digestive tract. These enzymes break down large carbohydrates, proteins, lipids, and nucleic acids into smaller fragments, which in turn must typically be broken down further before absorption can occur. The final enzymatic steps involve brush border enzymes, which are attached to the exposed surfaces of microvilli.

Nucleic acids are broken down into their component nucleotides. Brush border enzymes digest these nucleotides into sugars, phosphates, and nitrogenous bases that are absorbed by active transport. However, nucleic acids represent only a small fraction of all the nutrients absorbed each day. The digestive fates of carbohydrates, lipids, and proteins, the major dietary components, are depicted in **Figure 24–26**. Table 24–3 summarizes the major digestive enzymes and their functions. Next we take a closer look at the digestion and absorption of carbohydrates, lipids, and proteins.

Carbohydrate Digestion and Absorption

The digestion of complex carbohydrates (simple polysaccharides and starches) proceeds in two steps. One step involves carbohydrases produced by the salivary glands and pancreas; the other, brush border enzymes.

The Actions of Salivary and Pancreatic Enzymes

The digestion of complex carbohydrates involves two enzymes—salivary amylase and pancreatic alpha-amylase (**Figure 24–26a**)—that function effectively at a pH of 6.7–7.5. Carbohydrate digestion begins in the mouth during mastication, through the action of salivary amylase from the parotid and submandibular salivary glands. Salivary amylase breaks down starches (complex carbohydrates), producing a mixture composed primarily of *disaccharides* (two simple sugars) and *trisaccharides* (three simple sugars). Salivary amylase continues to digest the starches and glycogen in the food for 1–2 hours before stomach acids render the enzyme inactive. Because the enzymatic content of saliva is not high, only a small amount of digestion occurs over this period.

In the duodenum, the remaining complex carbohydrates are broken down by the action of pancreatic alpha-amylase. Any disaccharides or trisaccharides produced, and any present in the food, are not broken down further by salivary and pancreatic amylases. Additional hydrolysis does not occur until these molecules contact the intestinal mucosa.

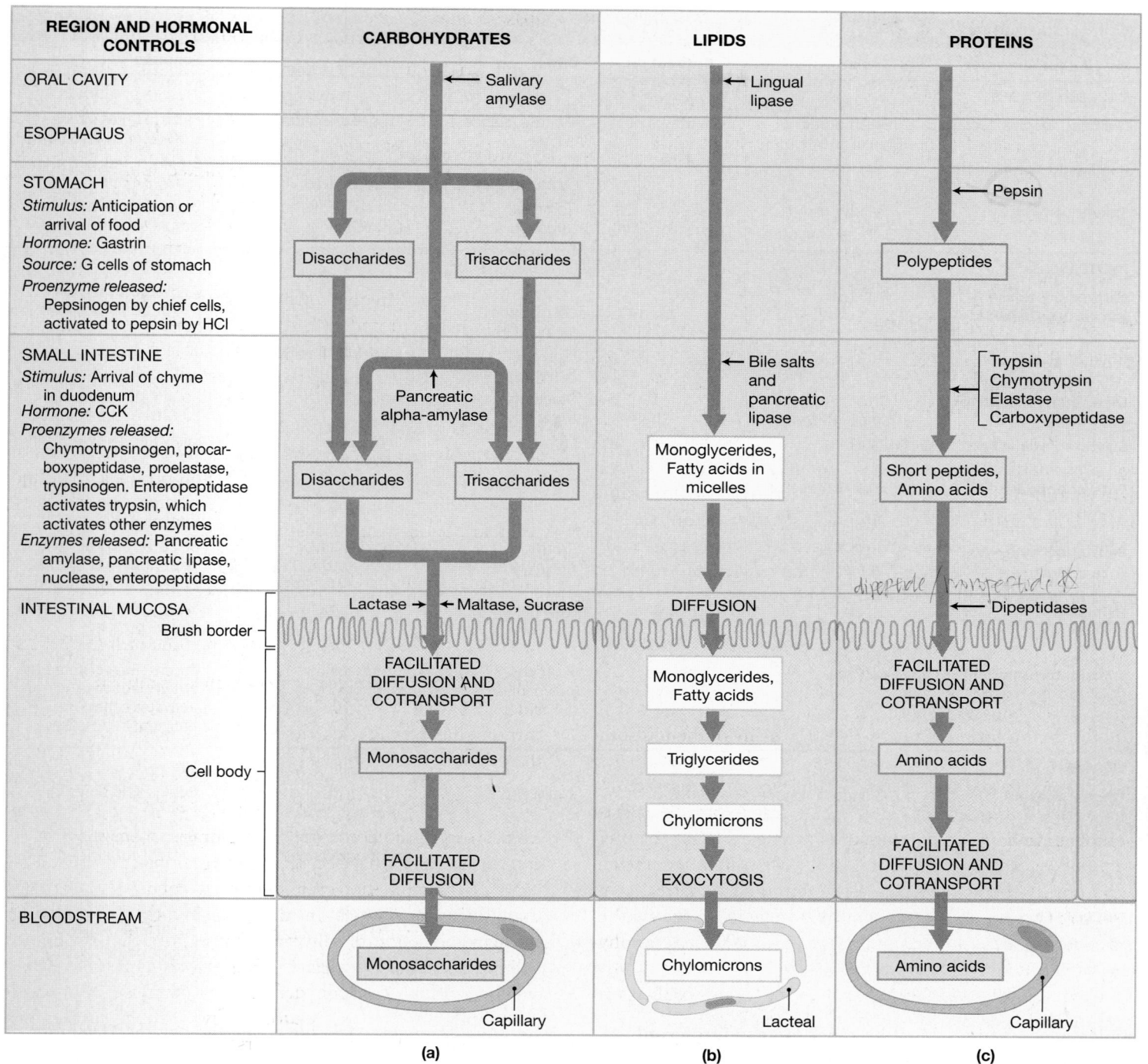

Figure 24–26 **A Summary of the Chemical Events in Digestion.** For further details on the enzymes involved, see *Table 24–3*.

Actions of Brush Border Enzymes

Prior to absorption, disaccharides and trisaccharides are fragmented into *monosaccharides* (simple sugars) by brush border enzymes of the intestinal microvilli. The enzyme **maltase** splits bonds between the two glucose molecules of the disaccharide **maltose**. **Sucrase** breaks the disaccharide **sucrose** into glucose and *fructose*, another six-carbon sugar. **Lactase** hydrolyzes the disaccharide **lactose** into a molecule of glucose and one of *galactose*. Lactose is the primary carbohydrate in milk, so by breaking down lactose, lactase provides an essential function throughout infancy and early childhood. If the intestinal mucosa stops producing lactase by the time of

SUMMARY TABLE 24–3	Digestive Enzymes and Their Functions				
Enzyme (proenzyme)	**Source**	**Optimal pH**	**Target**	**Products**	**Remarks**
CARBOHYDRASES					
Maltase, sucrase, lactase	Brush border of small intestine	7–8	Maltose, sucrose, lactose	Monosaccharides	Found in membrane surface of microvilli
Pancreatic alpha-amylase	Pancreas	6.7–7.5	Complex carbohydrates	Disaccharides and trisaccharides	Breaks bonds between simple sugars
Salivary amylase	Salivary glands	6.7–7.5	Complex carbohydrates	Disaccharides and trisaccharides	Breaks bonds between simple sugars
PROTEASES					
Carboxypeptidase (procarboxypeptidase)	Pancreas	7–8	Proteins, polypeptides, amino acids	Short-chain peptides	Activated by trypsin
Chymotrypsin (chymotrypsinogen)	Pancreas	7–8	Proteins, polypeptides	Short-chain peptides	Activated by trypsin
Dipeptidases, peptidases	Brush border of small intestine	7–8	Dipeptides, tripeptides	Amino acids	Found in membrane surface of brush border
Elastase (proelastase)	Pancreas	7–8	Elastin	Short-chain peptides	Activated by trypsin
Enteropeptidase	Brush border and lumen of small intestine	7–8	Trypsinogen	Trypsin	Reaches lumen through disintegration of shed epithelial cells
Pepsin (pepsinogen)	Chief cells of stomach	1.5–2.0	Proteins, polypeptides	Short-chain polypeptides	Secreted as proenzyme pepsinogen; activated by H^+ in stomach acid
Rennin	Stomach	3.5–4.0	Milk proteins		Secreted only in infants; causes protein coagulation
Trypsin (trypsinogen)	Pancreas	7–8	Proteins, polypeptides	Short-chain peptides	Proenzyme activated by enteropeptidase; activates other pancreatic proteases
LIPASES					
Lingual lipase	Glands of tongue	3.0–6.0	Triglycerides	Fatty acids and monoglycerides	Begins lipid digestion
Pancreatic lipase	Pancreas	7–8	Triglycerides	Fatty acids and monoglycerides	Bile salts must be present for efficient action
NUCLEASES	Pancreas	7–8	Nucleic acids	Nitrogenous bases and simple sugars	Includes ribonuclease for RNA and deoxyribonuclease for DNA

adolescence, the individual becomes **lactose intolerant**. After ingesting milk and other dairy products, lactose-intolerant individuals can experience a variety of unpleasant digestive problems.

Absorption of Monosaccharides

The intestinal epithelium then absorbs the monosaccharides by facilitated diffusion and cotransport mechanisms (see **Figure 3–18**, p. 95). Both methods involve a carrier protein. Facilitated diffusion and cotransport differ in three major ways:

1. *Facilitated Diffusion Moves Only One Molecule or Ion through the Plasma Membrane, Whereas Cotransport Moves More Than One Molecule or Ion through the Membrane at the Same Time.* In cotransport, the transported materials move in the same direction: down the concentration gradient for at least one of the transported substances.

2. *Facilitated Diffusion Does Not Require ATP.* Although cotransport by itself does not consume ATP, the cell must often expend ATP to preserve homeostasis. For example, the process may introduce sodium ions that must later be pumped out of the cell.

3. *Facilitated Diffusion Will Not Occur if There Is an Opposing Concentration Gradient for the Particular Molecule or Ion.* By contrast, cotransport can occur despite an opposing

concentration gradient for one of the transported substances. For example, cells lining the small intestine will continue to absorb glucose when glucose concentrations inside the cells are much higher than they are in the intestinal contents.

The cotransport system responsible for the uptake of glucose also brings sodium ions into the cell. This passive process resembles facilitated diffusion, except that both a sodium ion and a glucose molecule must bind to the carrier protein before they can move into the cell. Glucose cotransport is an example of sodium-linked cotransport. Comparable cotransport mechanisms exist for other simple sugars and for some amino acids. Although these mechanisms deliver valuable nutrients to the cytoplasm, they also bring in sodium ions that must be ejected by the sodium–potassium exchange pump.

The simple sugars that are transported into the cell at its apical surface diffuse through the cytoplasm and reach the interstitial fluid by facilitated diffusion across the basolateral surfaces. These monosaccharides then diffuse into the capillaries of the villus for eventual transport to the liver in the hepatic portal vein.

Lipid Digestion and Absorption

Lipid digestion involves lingual lipase from glands of the tongue, and pancreatic lipase from the pancreas (**Figure 24–26b**). The most important and abundant dietary lipids are triglycerides, which consist of three fatty acids attached to a single molecule of glycerol (see **Figure 2–15,** p. 50). The lingual and pancreatic lipases break off two of the fatty acids, leaving monoglycerides.

Lipases are water-soluble enzymes, and lipids tend to form large drops that exclude water molecules. As a result, lipases can attack only the exposed surfaces of the lipid drops. Lingual lipase begins breaking down triglycerides in the mouth and continues for a variable time within the stomach, but the lipid drops are so large, and the available time so short, that only about 20 percent of the lipids have been digested by the time the chyme enters the duodenum.

Bile salts improve chemical digestion by emulsifying the lipid drops into tiny emulsion droplets, thereby providing better access for pancreatic lipase. The emulsification occurs only after the chyme has been mixed with bile in the duodenum. Pancreatic lipase then breaks apart the triglycerides to form a mixture of fatty acids and monoglycerides. As these molecules are released, they interact with bile salts in the surrounding chyme to form small lipid–bile salt complexes called **micelles** (mī-SELZ). A micelle is only about 2.5 nm (0.0025 μm) in diameter.

When a micelle contacts the intestinal epithelium, the lipids diffuse across the plasma membrane and enter the cytoplasm. The intestinal cells synthesize new triglycerides from the monoglycerides and fatty acids. These triglycerides, in company with absorbed steroids, phospholipids, and fat-soluble vitamins, are then coated with proteins, creating complexes known as **chylomicrons** (kī-lō-MĪ-kronz; *chylos,* juice + *mikros,* small).

The intestinal cells then secrete the chylomicrons into interstitial fluid by exocytosis. The superficial protein coating of the chylomicrons keeps them suspended in the interstitial fluid, but their size generally prevents them from diffusing into capillaries. Most of the chylomicrons released diffuse into the intestinal lacteals, which lack basal laminae and have large gaps between adjacent endothelial cells. From the lacteals, the chylomicrons proceed along the lymphatic vessels and through the thoracic duct, finally entering the bloodstream at the left subclavian vein.

Most of the bile salts within micelles are reabsorbed by sodium-linked cotransport. Only about 5 percent of the bile salts secreted by the liver enters the colon, and only about 1 percent is lost in feces.

Protein Digestion and Absorption

Proteins have very complex structures, so protein digestion is both complex and time-consuming. The first task is to disrupt the three-dimensional organization of the food so that proteolytic enzymes can attack individual proteins. This step involves mechanical processing in the oral cavity, through mastication, and chemical processing in the stomach, through the action of hydrochloric acid. Exposure of the bolus to a strongly acidic environment kills pathogens and breaks down plant cell walls and the connective tissues in animal products. Acid also disrupts tertiary and secondary protein structure, exposing peptide bonds to enzymatic attack.

The acidic contents of the stomach also provide the proper environment for the activity of pepsin, the proteolytic enzyme secreted by chief cells of the stomach (**Figure 24–26c**). Pepsin, which works effectively at a pH of 1.5–2.0, breaks the peptide bonds within a polypeptide chain. When chyme enters the duodenum, enteropeptidase produced in the small intestine triggers the conversion of trypsinogen to trypsin, and the pH is adjusted to 7–8. Pancreatic proteases can now begin working. Trypsin, chymotrypsin, and elastase are like pepsin in that they break specific peptide bonds within a polypeptide. For example, trypsin breaks peptide bonds involving the amino acids *arginine* or *lysine,* whereas chymotrypsin targets peptide bonds involving *tyrosine* or *phenylalanine.*

Carboxypeptidase also acts in the small intestine. This enzyme chops off the last amino acid of a polypeptide chain, ignoring the identities of the amino acids involved. Thus, while the other peptidases generate a variety of short peptides, carboxypeptidase produces free amino acids.

24

DIGESTIVE

The epithelial surfaces of the small intestine contain several peptidases, notably **dipeptidases**—enzymes that break short peptide chains into individual amino acids. (Dipeptidases break apart *dipeptides*.) These amino acids, as well as those produced by the pancreatic enzymes, are absorbed through both facilitated diffusion and cotransport mechanisms.

After diffusing to the basal surface of the cell, the amino acids are released into interstitial fluid by facilitated diffusion and cotransport. Once in the interstitial fluid, the amino acids diffuse into intestinal capillaries for transport to the liver by means of the hepatic portal vein.

Water Absorption

Cells cannot actively absorb or secrete water. All movement of water across the lining of the digestive tract, as well as the production of glandular secretions, involves passive water flow down osmotic gradients. When two solutions are separated by a selectively permeable membrane, water tends to flow into the solution that has the higher concentration of solutes. ∞ p. 91 Osmotic movements are rapid, so interstitial fluid and the fluids in the intestinal lumen always have the same osmolarity (osmotic concentration of solutes).

Intestinal epithelial cells continuously absorb nutrients and ions, and these activities gradually lower the solute concentration in the lumen. As the solute concentration drops, water moves into the surrounding tissues, maintaining osmotic equilibrium.

Each day, roughly 2000 mL of water enters the digestive tract in the form of food or drink. The salivary, gastric, intestinal, pancreatic, and bile secretions provide an additional 7000 mL. Of that total, only about 150 mL is lost in feces. The sites of secretion and absorption of water are indicated in **Figure 24–27**.

Ion Absorption

Osmosis does not distinguish among solutes; all that matters is the total concentration of solutes. To maintain homeostasis, however, the concentrations of specific ions must be closely regulated. Thus, each ion must be handled individually, and the rate of intestinal absorption of each must be tightly controlled (Table 24–4). Many of the regulatory mechanisms controlling the rates of absorption are poorly understood.

Sodium ions (Na^+), usually the most abundant cations in food, may enter intestinal cells by diffusion, by cotransport with another nutrient, or by active transport. These ions are then pumped into interstitial fluid across the base of the cell. The rate of Na^+ uptake from the lumen is generally proportional to the concentration of Na^+ in the intestinal contents. As a result, eating heavily salted foods leads to increased

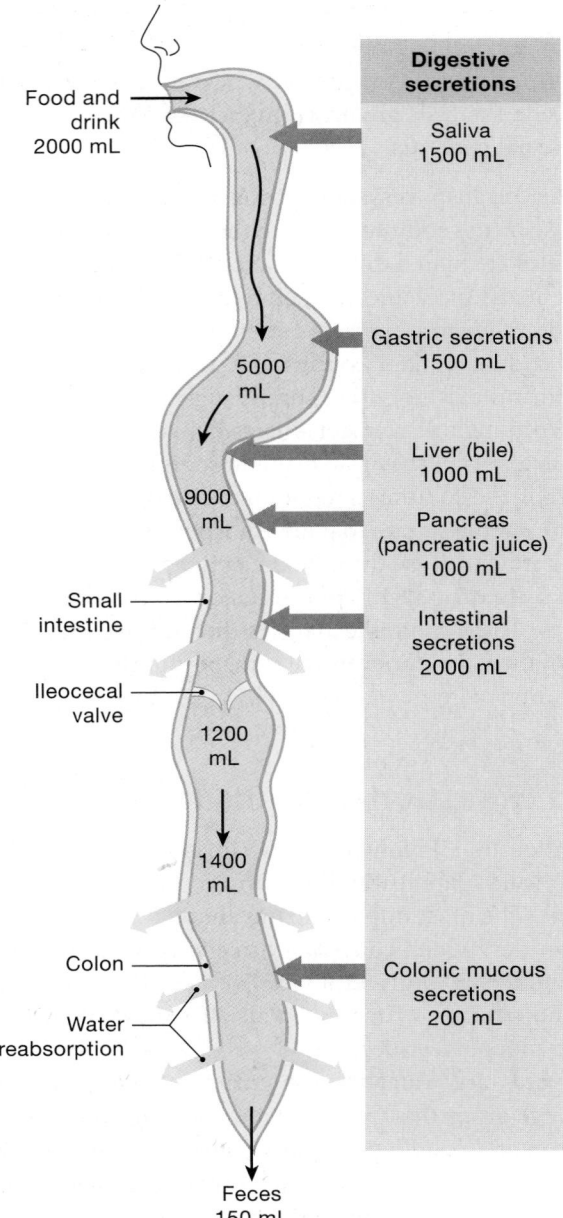

Figure 24–27 Digestive Secretion and Absorption of Water. The gray arrows indicate secretion, the blue arrows water reabsorption.

sodium ion absorption and an associated gain of water through osmosis. The rate of sodium ion absorption by the digestive tract is increased by aldosterone, a steroid hormone from the suprarenal cortex. ∞ p. 627

Calcium ion (Ca^{2+}) absorption involves active transport at the epithelial surface. The rate of transport is accelerated by calcitriol. ∞ p. 625

As other solutes move out of the lumen, the concentration of potassium ions (K^+) increases. These ions can diffuse into the epithelial cells along the concentration gradient. The

TABLE 24–4 The Absorption of Ions and Vitamins

Ion or Vitamin	Transport Mechanism	Regulatory Factors
Na^+	Channel-mediated diffusion, cotransport, or active transport	Increased when sodium-linked cotransport is under way; stimulated by aldosterone
Ca^{2+}	Active transport	Stimulated by calcitriol and PTH
K^+	Channel-mediated diffusion	Follows concentration gradient
Mg^{2+}	Active transport	
Fe^{2+}	Active transport	
Cl^-	Channel-mediated diffusion or carrier-mediated transport	
I^-	Channel-mediated diffusion or carrier-mediated transport	
HCO_3^-	Channel-mediated diffusion or carrier-mediated transport	
NO_3^-	Channel-mediated diffusion or carrier-mediated transport	
PO_4^{3-}	Active transport	
SO_4^{2-}	Active transport	
Water-soluble vitamins (except B_{12})	Channel-mediated diffusion	Follows concentration gradient
Vitamin B_{12}	Active transport	Must be bound to intrinsic factor prior to absorption
Fat-soluble vitamins	Diffusion	Absorbed from micelles along with dietary lipids

absorption of magnesium (Mg^{2+}), iron (Fe^{2+}), and other cations involves specific carrier proteins; the cell must use ATP to obtain and transport these ions to interstitial fluid. Regulatory factors controlling their absorption are poorly understood.

The anions chloride (Cl^-), iodide (I^-), bicarbonate (HCO_3^-), and nitrate (NO_3^-) are absorbed by diffusion or carrier-mediated transport. Phosphate (PO_4^{3-}) and sulfate (SO_4^{2-}) ions enter epithelial cells only by active transport.

Vitamin Absorption

Vitamins are organic compounds required in very small quantities. There are two major groups of vitamins: fat-soluble vitamins and water-soluble vitamins. Vitamins A, D, E, and K are **fat-soluble vitamins**; their structure allows them to dissolve in lipids. The nine **water-soluble vitamins** include the B vitamins, common in milk and meats, and vitamin C, found in citrus fruits. We will consider the functions of vitamins and associated nutritional problems in Chapter 25.

All but one of the water-soluble vitamins are easily absorbed by diffusion across the digestive epithelium. Vitamin B_{12} cannot be absorbed by the intestinal mucosa in normal amounts, unless this vitamin has been bound to *intrinsic factor*, a glycoprotein secreted by the parietal cells of the stomach (p. 893). The combination is then absorbed through active transport.

Fat-soluble vitamins in the diet enter the duodenum in fat droplets, mixed with triglycerides. The vitamins remain in association with these lipids as they form emulsion droplets and, after further digestion, micelles. The fat-soluble vitamins are then absorbed from the micelles along with the fatty acids and monoglycerides. Vitamin K produced in the colon is absorbed with other lipids released through bacterial action. Taking supplements of fat-soluble vitamins while you have an empty stomach, are fasting, or are on a low-fat diet is relatively ineffective, because proper absorption of these vitamins requires the presence of other lipids.

CHECKPOINT

29. **What kinds of nutrients does the body require?**

30. **What component of food would increase the number of chylomicrons in the lacteals?**

31. **The absorption of which vitamin would be impaired by the removal of the stomach?**

32. **Why is it that diarrhea is potentially life threatening, but constipation is not?**

See the blue Answers tab at the end of the book.

24-9 Many age-related changes affect digestion and absorption

Essentially normal digestion and absorption occur in elderly individuals. However, many changes in the digestive system parallel age-related changes we have already discussed in connection with other systems:

- *The Division Rate of Epithelial Stem Cells Declines.* The digestive epithelium becomes more susceptible to damage by abrasion, acids, or enzymes. Peptic ulcers therefore become more likely. Stem cells in the epithelium divide less frequently with age, so tissue repair is less efficient. In the mouth, esophagus, and anus, the stratified epithelium becomes thinner and more fragile.

- *Smooth Muscle Tone Decreases.* General motility decreases, and peristaltic contractions are weaker as a result of a decrease in smooth muscle tone. These changes slow the rate of fecal movement and promote constipation. Sagging and inflammation of the haustra in the colon can occur. Straining to eliminate compacted feces can stress the less resilient walls of blood vessels, producing hemorrhoids. Problems are not restricted to the lower digestive tract; for example, weakening of muscular sphincters can lead to esophageal reflux and frequent bouts of "heartburn."

- *The Effects of Cumulative Damage Become Apparent.* A familiar example is the gradual loss of teeth due to *dental caries* (cavities) or gingivitis. Cumulative damage can involve internal organs as well. Toxins such as alcohol and other injurious chemicals that are absorbed by the digestive tract are transported to the liver for processing. The cells of the liver are not immune to these toxic compounds, and chronic exposure can lead to cirrhosis or other types of liver disease.

- *Cancer Rates Increase.* Cancers are most common in organs in which stem cells divide to maintain epithelial cell populations. Rates of colon cancer and stomach cancer rise with age; oral, esophageal, and pharyngeal cancers are particularly common among elderly smokers.

- *Dehydration Is Common among the Elderly.* Because osmoreceptor sensitivity declines with age, dehydration is common among the elderly.

- *Changes in Other Systems Have Direct or Indirect Effects on the Digestive System.* For example, reduction in bone mass and calcium content in the skeleton is associated with erosion of the tooth sockets and eventual tooth loss. The decline in olfactory and gustatory sensitivities with age can lead to dietary changes that affect the entire body.

CHECKPOINT

33. **Identify general digestive system changes that occur with aging.**

See the blue Answers tab at the end of the book.

24-10 The digestive system is extensively integrated with other body systems

Figure 24–28 summarizes the physiological relationships between the digestive system and other organ systems. The digestive system has particularly extensive anatomical and physiological connections to the nervous, cardiovascular, endocrine, and lymphoid systems. As we have seen, the digestive tract is also an endocrine organ that produces a variety of hormones. Many of these hormones, and some of the neurotransmitters produced by the digestive system, can enter the circulation, cross the blood–brain barrier, and alter CNS activity. Thus, a continual exchange of chemical information occurs among these systems.

CLINICAL NOTE

Inflammatory and Infectious Disorders of the Digestive System Because the digestive system has so many components, and those components have so many functions, digestive system disorders are both very diverse and relatively common.

The largest category of digestive disorders includes those resulting from inflammation or infection of the digestive tract. In part, this is because the epithelium lining most of the digestive tract has two properties that are difficult to reconcile: (1) It must be thin and delicate enough to permit the rapid and efficient absorption of nutrients, and (2) it must resist damage by ingested materials and the enzymes secreted to promote digestion.

The relative delicacy of the epithelium makes it susceptible to damage from chemical attack or abrasion. For example, *peptic ulcers* develop if acids and enzymes contact and erode the gastric or duodenal lining. Pathogens in food, including bacteria, viruses, and multicellular parasites, may also penetrate the epithelial barriers and cause infections. Small battles are continually being fought; the fact that 80 percent of the body's plasma cells are normally located within the lamina propria of the digestive tract indicates how often antigens of one kind or another somehow cross the epithelial barriers.

High rates of cell division and exposure to strong chemical agents are both correlated with an increased risk of cancer. As a result, cancers of the digestive tract are relatively common. Predictably, most of these are epithelial cancers that develop in the stem cell populations responsible for epithelial cell renewal.

CHECKPOINT

34. **Identify the functional relationships between the digestive system and other body systems.**

35. **What body systems may be affected by inadequate calcium absorption?**

See the blue Answers tab at the end of the book.

THE DIGESTIVE SYSTEM IN PERSPECTIVE

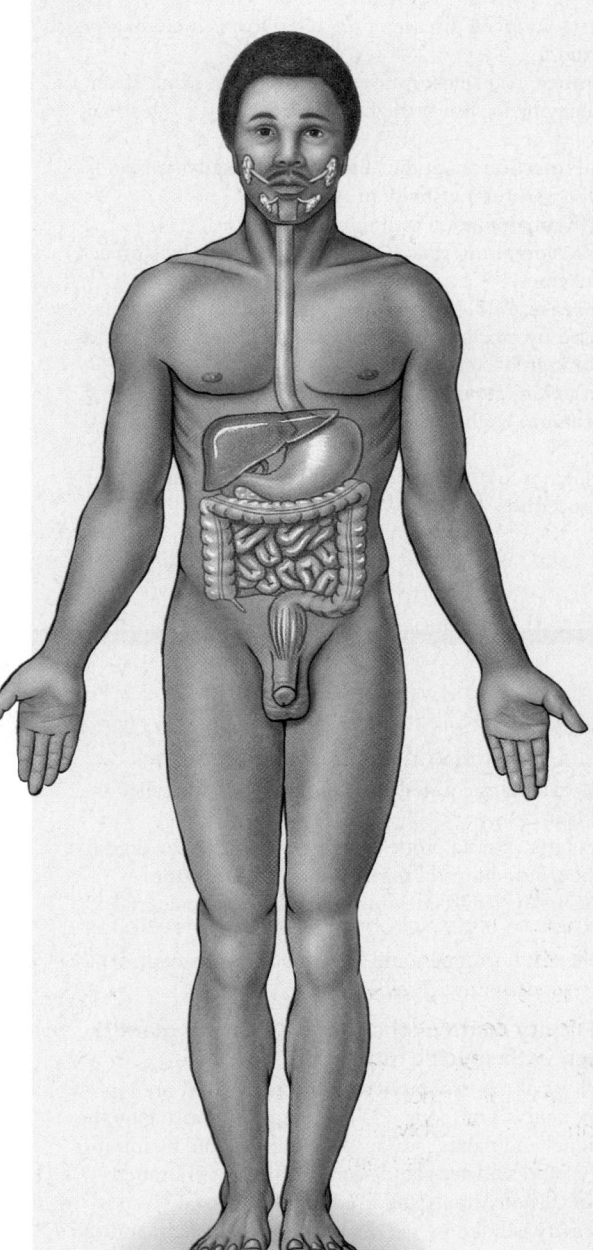

Integumentary System

↪ • Provides vitamin D_3 needed for the absorption of calcium and phosphorus

↩ • Provides lipids for storage by adipocytes in subcutaneous layer

Skeletal System

↪ • Skull, ribs, vertebrae, and pelvic girdle support and protect parts of digestive tract; teeth important in mechanical processing of food

↩ • Absorbs calcium and phosphate ions for incorporation into bone matrix; provides lipids for storage in yellow marrow

Muscular System

↪ • Protects and supports digestive organs in abdominal cavity; controls entrances and exits to digestive tract

↩ • Liver regulates blood glucose and fatty acid levels, metabolizes lactic acid from active muscles

Nervous System

↪ • ANS regulates movement and secretion; reflexes coordinate passage of materials along tract; control over skeletal muscles regulates ingestion and defecation; hypothalamic centers control hunger, satiation, and feeding behaviors

↩ • Provides substrates essential for neurotransmitter synthesis

Endocrine System

↪ • Epinephrine and norepinephrine stimulate constriction of sphincters and depress digestive activity; hormones coordinate activity along tract

↩ • Provides nutrients and substrates to endocrine cells; endocrine cells of pancreas secrete insulin and glucagon; liver produces angiotensinogen

Cardiovascular System

↪ • Distributes hormones of the digestive tract; carries nutrients, water, and ions from sites of absorption; delivers nutrients and toxins to liver

↩ • Absorbs fluid to maintain normal blood volume; absorbs vitamin K; liver excretes heme (as bilirubin), synthesizes coagulation proteins

Lymphoid System

↪ • Tonsils and other lymphoid nodules along digestive tract defend against infection and toxins absorbed from the tract; lymphatic vessels carry absorbed lipids to venous system

↩ • Secretions of digestive tract (acids and enzymes) provide nonspecific defense against pathogens

Respiratory System

↪ • Increased thoracic and abdominal pressure through contraction of respiratory muscles can assist in defecation

↩ • Pressure of digestive organs against the diaphragm can assist in exhalation and limit inhalation

Urinary System

↪ • Excretes toxins absorbed by the digestive epithelium; excretes some bilirubin produced by liver

↩ • Absorbs water needed to excrete waste products at the kidneys; absorbs ions needed to maintain normal body fluid concentrations

Reproductive System

↩ • Provides additional nutrients required to support gamete production and (in pregnant women) embryonic and fetal development

FOR ALL SYSTEMS
Absorbs organic substrates, vitamins, ions, and water required by all cells

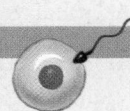

Figure 24–28 **Functional Relationships between the Digestive System and Other Systems.**

Related Clinical Terms

ascites: Fluid leakage into the peritoneal cavity across the serous membranes of the liver and viscera.

cathartics: Drugs that promote defecation.

cholecystitis: An inflammation of the gallbladder usually due to a blockage of the cystic duct or common bile duct by gallstones.

cholelithiasis: The presence of gallstones in the gallbladder.

cholera: A bacterial infection of the digestive tract that causes massive fluid losses through diarrhea.

cirrhosis: A disease characterized by the widespread destruction of hepatocytes due to exposure to drugs (especially beverage alcohol), viral infection, ischemia, or blockage of the hepatic ducts.

colitis: A general term for a condition characterized by inflammation of the colon.

constipation: Infrequent defecation, generally involving dry, hard feces.

diarrhea: Frequent, watery bowel movements.

diverticulitis: An infection and inflammation of mucosal pockets of the large intestine (diverticula).

diverticulosis: The formation of diverticula, generally along the sigmoid colon.

esophageal varices: Swollen and fragile esophageal veins that result from portal hypertension.

gallstones: Deposits of minerals, bile salts, and cholesterol that form if bile becomes too concentrated.

gastrectomy: The surgical removal of the stomach, generally to treat advanced stomach cancer.

gastritis: An inflammation of the gastric mucosa.

gastroscope: A fiber-optic instrument inserted into the mouth and directed along the esophagus and into the stomach; used to examine the interior of the stomach and to perform minor surgical procedures.

hepatitis: A virus-induced disease of the liver; forms include *hepatitis A, B, C, D,* and *E.*

lactose intolerance: A malabsorption syndrome that results from the lack of the enzyme *lactase* at the brush border of the intestinal epithelium.

mumps: A viral infection that typically targets the salivary glands (primarily the parotids), usually in children.

pancreatitis: An inflammation of the pancreas.

peptic ulcer: Erosion of the gastric or duodenal lining by stomach acids and enzymes.

periodontal disease: A loosening of the teeth within the alveolar sockets caused by erosion of the periodontal ligaments by acids produced through bacterial action.

peritonitis: An inflammation of the peritoneal membrane.

portal hypertension: High venous pressures in the hepatic portal system.

pulpitis: An infection of the pulp of a tooth; treatment may involve a *root canal* procedure.

Chapter Review

Study Outline

iP Use *Interactive Physiology*® **(IP)** —available on the IP CD packaged with your new textbook and online at **myA&P**™ (www.myaandp.com)—to help you understand difficult physiological concepts. Follow these navigation paths on IP for concepts in this chapter:

- Digestive System/Orientation
- Digestive System/Motility
- Digestive System/Anatomy Review
- Digestive System/Digestion and Absorption

24-1 The digestive system, consisting of the digestive tract and accessory organs, has overlapping food utilization functions p. 875

1. The **digestive system** consists of the muscular **digestive tract** and various **accessory organs**. (*Figure 24–1*)

2. Double sheets of peritoneal membrane called **mesenteries** suspend the digestive tract. The **greater omentum** lies anterior to the abdominal viscera. Its adipose tissue provides padding, protection, insulation, and an energy reserve. (*Figure 24–2*)

3. The *lamina propria* and epithelium form the **mucosa** (mucous membrane) of the digestive tract. Proceeding outward is the **submucosa**, the **muscularis externa**, and a layer of areolar tissue called the *adventitia*. Within the peritoneal cavity, the muscularis externa is covered by a serous membrane called the **serosa**. (*Figure 24–3*)

4. The muscularis externa propels materials through the digestive tract by the contractions of **peristalsis**. **Segmentation** movements in the small intestine churn digestive materials. (*Figure 24–4*)

5. Neural reflexes, hormones, and local mechanisms control digestive tract activities. (*Figure 24–5*)

24-2 The oral cavity contains the tongue, salivary glands, and teeth, each with specific functions p. 882

6. The functions of the **oral cavity**, or **buccal cavity**, are (1) *sensory analysis* of foods; (2) *mechanical processing* by the teeth, tongue, and palatal surfaces; (3) *lubrication*, by mixing with mucus and salivary gland secretions; and (4) limited *digestion* of carbohydrates and lipids.

7. The oral cavity is lined by the **oral mucosa**. The *hard* and *soft palates* form the roof of the oral cavity, and the *tongue* forms its floor. (*Figure 24–6*)

8. **Intrinsic** and **extrinsic tongue muscles** are controlled by the hypoglossal nerves. (*Figure 24–6*)

9. The **parotid**, **sublingual**, and **submandibular salivary glands** discharge their secretions into the oral cavity. (*Figure 24–7*)

10. **Mastication** (chewing) of the **bolus** occurs through the contact of the **occlusal** (opposing) **surfaces** of the **teeth**. The **periodontal ligament** anchors each tooth in an *alveolus*, or bony socket. **Dentin** forms the basic structure of a tooth. The

crown is coated with **enamel**, the **root** with **cementum**. (*Figure 24–8*)

11. During childhood and early adulthood, the 20 primary teeth, or **deciduous teeth**, are replaced by the 32 teeth of the **secondary dentition** (*Figure 24–9*)

24-3 The pharynx is a passageway between the oral cavity and esophagus p. 888

12. Propulsion of the bolus through the **pharynx** results from contractions of the *pharyngeal constrictor muscles* and the *palatal muscles*, and from elevation of the larynx.

24-4 The esophagus is a muscular tube that transports solids and liquids from the pharynx to the stomach p. 888

13. The **esophagus** carries solids and liquids from the pharynx to the stomach through the **esophageal hiatus**, an opening in the diaphragm. (*Figure 24–10*)

14. The esophageal mucosa consists of a stratified epithelium. Mucous secretion by esophageal glands of the submucosa reduces friction during the passage of foods. The proportions of skeletal and smooth muscle of the muscularis externa change from the pharynx to the stomach. (*Figure 24–10*)

15. Swallowing (**deglutition**) can be divided into **buccal**, **pharyngeal**, and **esophageal phases**. Swallowing begins with the compaction of a bolus and its movement into the pharynx, followed by the elevation of the larynx, reflection of the *epiglottis*, and closure of the *glottis*. After the *upper esophageal sphincter* is opened, peristalsis moves the bolus down the esophagus to the *lower esophageal sphincter*. (*Figure 24–11*)

24-5 The stomach is a J-shaped organ that receives the bolus from the esophagus and aids in chemical and mechanical digestion p. 891

16. The **stomach** has four major functions: (1) storage of ingested food, (2) mechanical breakdown of food, (3) disruption of chemical bonds by acid and enzymes, and (4) production of *intrinsic factor*.

17. The four regions of the stomach are the **cardia**, **fundus**, **body**, and **pylorus**. The **pyloric sphincter** guards the exit out of the stomach. In a relaxed state, the stomach lining contains numerous **rugae** (ridges and folds). (*Figure 24–12*)

18. Within the **gastric glands**, *parietal cells* secrete *intrinsic factor* and *hydrochloric acid*. **Chief cells** secrete **pepsinogen**, which is converted by acids in the gastric lumen to the enzyme **pepsin**. **Enteroendocrine** cells of the stomach secrete several compounds, notably the hormone **gastrin**. (*Figures 24–13, 24–14*)

19. Gastric control involves (1) the **cephalic phase**, which prepares the stomach to receive ingested materials; (2) the **gastric phase**, which begins with the arrival of food in the stomach; and (3) the **intestinal phase**, which controls the rate of gastric emptying. (*Figure 24–15; Table 24–1*)

24-6 The small intestine digests and absorbs nutrients, and associated glandular organs assist with the digestive process p. 898

20. Most of the important digestive and absorptive functions occur in the **small intestine**. The pancreas, liver, and gallbladder provide digestive secretions and buffers.

21. The small intestine consists of the **duodenum**, the **jejunum**, and the **ileum**. A sphincter, the **ileocecal valve**, marks the transition between the small and large intestines. (*Figure 24–16*)

22. The intestinal mucosa bears transverse folds called **plicae circulares** and small projections called **intestinal villi**. These folds and projections increase the surface area for absorption. Each villus contains a terminal lymphatic called a **lacteal**. Pockets called **intestinal glands** are lined by enteroendocrine, mucous, and stem cells. (*Figures 24–16, 24–17*)

23. **Intestinal juice** moistens chyme, helps buffer acids, and holds digestive enzymes and digestive products in solution.

24. The **duodenal** (*submucosal* or *Brunner*) **glands** of the duodenum produce mucus, bicarbonate ions, and the hormone **urogastrone**. The ileum contains masses of lymphoid tissue called *aggregated lymphoid nodules*, or *Peyer patches*, near the entrance to the large intestine.

25. The **gastroenteric reflex**, initiated by stretch receptors in the stomach, stimulates motility and secretion along the entire small intestine. The **gastroileal reflex** triggers the relaxation of the ileocecal valve.

26. The **pancreatic duct** penetrates the wall of the duodenum. Within each lobule of the **pancreas**, ducts branch repeatedly before ending in the **pancreatic acini** (blind pockets). (*Figure 24–18*)

27. The pancreas has two functions: endocrine (secreting insulin and glucagon into the blood) and exocrine (secreting **pancreatic juice** into the small intestine). Pancreatic enzymes include **carbohydrases**, **lipases**, **nucleases**, and **proteolytic enzymes**.

28. The **liver** performs metabolic and hematological regulation and produces **bile**. The bile ducts from all the **liver lobules** unite to form the **common hepatic duct**. That duct meets the **cystic duct** to form the **common bile duct**, which empties into the duodenum. (*Figures 24–19 to 24–21*)

29. The liver lobule is the organ's basic functional unit. **Hepatocytes** form irregular plates arranged in the form of spokes of a wheel. **Bile canaliculi** carry bile to the **bile ductules**, which lead to **portal areas**. (*Figure 24–20*)

30. In **emulsification**, **bile salts** break apart large drops of lipids, making the lipids accessible to lipases secreted by the pancreas.

31. The liver is the primary organ involved in regulating the composition of circulating blood because all the blood leaving the absorptive surfaces of the digestive tract flows into the liver before it enters the systemic circulation. The liver regulates metabolism as it removes and stores excess nutrients, vitamins, and minerals from the blood; mobilizes stored reserves; synthesizes needed nutrients; and removes waste products.

32. The liver's hematological activities include the monitoring of circulating blood by phagocytes and antigen-presenting cells; the synthesis of plasma proteins; the removal of circulating hormones and antibodies; and the removal or storage of toxins.

33. The liver synthesizes bile, which is composed of water, ions, bilirubin, cholesterol, and bile salts (an assortment of lipids).

34. The **gallbladder** stores, modifies, and concentrates bile. (*Figure 24–21*)

35. Neural and hormonal mechanisms coordinate the activities of the digestive glands. Gastrointestinal activity is stimulated by parasympathetic innervation and inhibited by sympathetic innervation. The **enterogastric**, **gastroenteric**, and **gastroileal**

24 DIGESTIVE

reflexes coordinate movement from the stomach to the large intestine.

36. Intestinal hormones include **secretin**, **cholecystokinin (CCK)**, **gastric inhibitory peptide (GIP)**, **vasoactive intestinal peptide (VIP)**, **gastrin**, and **enterocrinin**. (*Figure 24–22*; *Table 24–2*)

24-7 The large intestine is divided into three parts with regional specialization p. 910

37. The main functions of the **large intestine** are to (1) reabsorb water and compact materials into feces, (2) absorb vitamins produced by bacteria, and (3) store fecal material prior to defecation. The large intestine consists of the *cecum*, *colon*, and *rectum*. (*Figure 24–23*)

38. The **cecum** collects and stores material from the ileum and begins the process of compaction. The **appendix** is attached to the cecum. (*Figure 24–23*)

39. The **colon** has a larger diameter and a thinner wall than the small intestine. The colon bears **haustra** (pouches), **taeniae coli** (longitudinal bands of muscle), and sacs of fat (**fatty appendices**). (*Figure 24–23*)

40. The four regions of the colon are the **ascending colon**, **transverse colon**, **descending colon**, and **sigmoid colon**. (*Figure 24–23*)

41. The **rectum** terminates in the **anal canal**, leading to the **anus**. (*Figure 24–23*)

42. Histological characteristics of the colon include the absence of villi and the presence of mucous cells and deep intestinal glands. (*Figure 24–24*)

43. The large intestine reabsorbs water and other substances such as vitamins, urobilinogen, bile salts, and toxins. Bacteria are responsible for intestinal gas, or **flatus**.

44. The gastroileal reflex moves materials from the ileum into the cecum while you eat. Distension of the stomach and duodenum stimulates **mass movements** of materials from the transverse colon through the rest of the large intestine and into the rectum. Muscular sphincters control the passage of fecal material to the anus. Distension of the rectal wall triggers the **defecation reflex**. (*Figure 24–25*)

24-8 Digestion is the mechanical and chemical alteration of food that allows the absorption and use of nutrients p. 916

45. The digestive system first breaks down the physical structure of the ingested material and then disassembles the component molecules into smaller fragments by *hydrolysis*. (*Figure 24–26*; *Summary Table 24–3*)

46. Salivary and pancreatic amylases break down complex carbohydrates into *disaccharides* and *trisaccharides*. These in turn are broken down into *monosaccharides* by enzymes at the epithelial surface. The monosaccharides are then absorbed by the intestinal epithelium by facilitated diffusion or cotransport. (*Figure 24–26*)

47. *Triglycerides* are emulsified into lipid droplets that interact with bile salts to form **micelles**. The fatty acids and monoglycerides resulting from the action of pancreatic lipase diffuse from the micelles across the intestinal epithelium. Triglycerides are then synthesized and released into the interstitial fluid, for transport to the general circulation by way of the lymphoid system. (*Figure 24–26*)

48. Protein digestion involves a low pH (which destroys tertiary and quaternary structure), the gastric enzyme pepsin, and various pancreatic proteases. Peptidases liberate amino acids that are absorbed and exported to interstitial fluid. (*Figure 24–26*)

49. About 2000 mL of water is ingested each day, and digestive secretions provide another 7000 mL. Nearly all is reabsorbed by osmosis. (*Figure 24–27*)

50. Various processes, including diffusion, cotransport, and carrier-mediated and active transport, are responsible for the movements of cations (sodium, calcium, potassium, and so on) and anions (chloride, iodide, bicarbonate, and so on) into epithelial cells. (*Table 24–4*)

51. The **water-soluble vitamins** (except B_{12}) diffuse easily across the digestive epithelium. **Fat-soluble vitamins** are enclosed within fat droplets and absorbed with the products of lipid digestion. (*Table 24–4*)

24-9 Many age-related changes affect digestion and absorption p. 921

52. Age-related changes include a thinner and more fragile epithelium due to a reduction in epithelial stem cell divisions, weaker peristaltic contractions as smooth muscle tone decreases, the effects of cumulative damage, increased cancer rates, and increased dehydration.

24-10 The digestive system is extensively integrated with other body systems p. 922

53. The digestive system has extensive anatomical and physiological connections to the nervous, endocrine, cardiovascular, and lymphoid systems. (*Figure 24–28*)

Review Questions

See the blue Answers tab at the end of the book.

myA&P Access more review material online at **myA&P**™ (www.myaandp.com). There, you'll find chapter guides, chapter quizzes, practice tests, animations, flashcards, a glossary with pronunciations, *Interactive Physiology*® (IP) exercises and quizzes, and more to help you succeed in the course.

LEVEL 1 Reviewing Facts and Terms

1. The enzymatic breakdown of large molecules into their basic building blocks is called
   (a) absorption.　　(b) secretion.
   (c) mechanical digestion.　　(d) chemical digestion.

2. The outer layer of the digestive tract is known as the
   (a) serosa.　　(b) mucosa.
   (c) submucosa.　　(d) muscularis.

3. Double sheets of peritoneum that provide support and stability for the organs of the peritoneal cavity are the
   (a) mediastina.　　(b) mucous membranes.
   (c) omenta.　　(d) mesenteries.

4. A branch of the portal vein, hepatic artery, and tributary of the bile duct form
   (a) a liver lobule.　　(b) the sinusoids.
   (c) a portal area.　　(d) the hepatic duct.
   (e) the pancreatic duct.

5. Label the digestive system structures in the following figure.

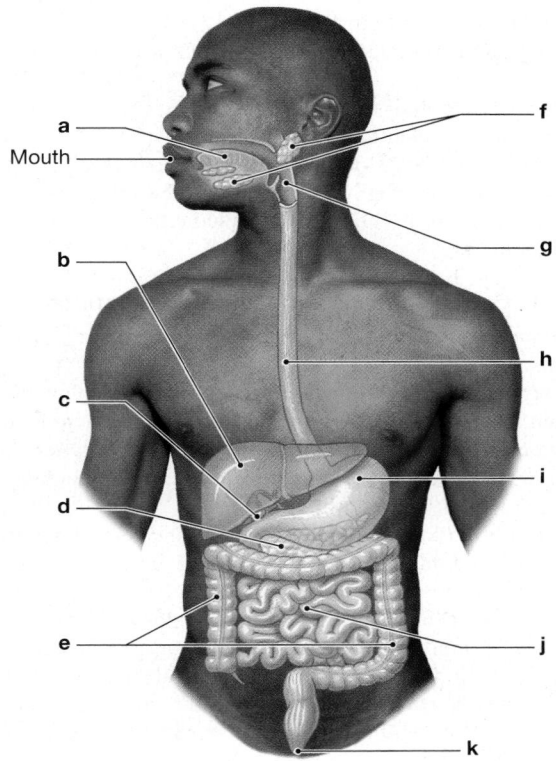

Mouth

(a) oral cavity, teeth, tongue   (b) liver
(c) gallbladder   (d) pancreas
(e) 대장   (f) Salivary glands
(g) pharynx 인두   (h) esophagus 식도
(i) stomach 위장   (j) small int. 소장
(k) anus 항문

6. Label the four layers of the digestive tract in the following figure.

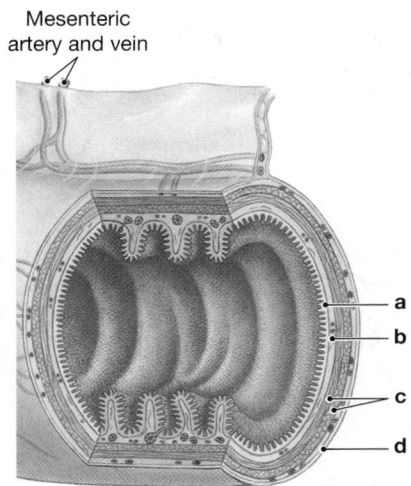

Mesenteric artery and vein

(a) mucosa   (b) Muscularis externa
(c) submucosa   (d) serosa (visceral peritoneum)

7. Most of the digestive tract is lined by _____ epithelium.
   (a) pseudostratified ciliated columnar
   (b) cuboidal
   (c) stratified squamous
   (d) simple
   (e) simple columnar

8. Regional movements that occur in the small intestine and function to churn and fragment the digestive material are called
   (a) segmentation.   (b) pendular movements.
   (c) peristalsis.   (d) mass movements.
   (e) mastication.

9. Bile release from the gallbladder into the duodenum occurs only under the stimulation of
   (a) cholecystokinin.   (b) secretin.
   (c) gastrin.   (d) enteropeptidase.

10. Label the three segments of the small intestine in the following figure.

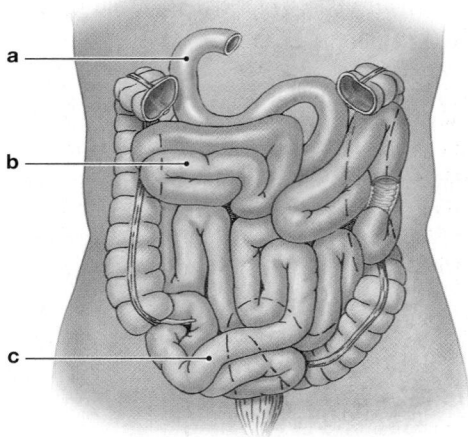

(a) 십이지장   (b) 공장
(c) 회장

11. The major function(s) of the large intestine is (are)
    (a) reabsorption of water and compaction of feces.
    (b) absorption of vitamins liberated by bacterial action.
    (c) storage of fecal material prior to defecation.
    (d) a, b, and c.

12. Vitamins generated by bacteria in the colon are
    (a) vitamins A, D, and E.
    (b) B complex vitamins and vitamin C.
    (c) vitamin K, biotin, and pantothenic acid.
    (d) niacin, thiamine, and riboflavin.

13. The final enzymatic steps in the digestive process are accomplished by
    (a) brush border enzymes of the microvilli.
    (b) enzymes secreted by the stomach.
    (c) enzymes secreted by the pancreas.
    (d) the action of bile from the gallbladder.

14. What are the six steps of digestion?

15. Name and describe the layers of the digestive tract, proceeding from the innermost layer to the outermost layer.

16. What three basic mechanisms regulate the activities of the digestive tract?

neural, hormonal, local semulations

17. What are the three phases of swallowing, and how are they controlled? buccal pharyngeal oesphageal

18. What are the primary digestive functions of the pancreas, liver, and gallbladder?

19. Which hormones produced by duodenal enteroendocrine cells effectively coordinate digestive functions?

20. What are the three primary functions of the large intestine?

21. What two positive feedback loops are involved in the defecation reflex?

LEVEL 2 Reviewing Concepts

22. During defecation,
    (a) stretch receptors in the rectal wall initiate a series of peristaltic contractions in the colon and rectum.
    (b) stretch receptors in the rectal wall activate parasympathetic centers in the sacral region of the spinal cord.
    (c) the internal anal sphincter relaxes while the external anal sphincter contracts.
    (d) all of these occur.
    (e) only a and b occur.

23. *Increased* parasympathetic stimulation of the intestine would result in
    (a) decreased motility.
    (b) decreased secretion.
    (c) decreased sensitivity of local reflexes.
    (d) decreased segmentation.
    (e) none of these.

24. A drop in pH below 4.5 in the duodenum stimulates the secretion of
    (a) secretin.   (b) cholecystokinin.
    (c) gastrin.   (d) a, b, and c.

25. Through which layers of a molar would an oral surgeon drill to perform a root canal (removal of the alveolar nerve in a severely damaged tooth)?

26. How is the epithelium of the stomach protected from digestion?

27. How does each of the three phases of gastric secretion promote and facilitate gastric control?

28. Nutritionists have found that after a heavy meal, the pH of blood increases slightly, especially in the veins that carry blood away from the stomach. What causes this "postenteric alkaline tide"?

LEVEL 3 Critical Thinking and Clinical Applications

29. Some people with gallstones develop pancreatitis. How could this occur?

30. Harry is suffering from an obstruction in his colon. He notices that when he urinates, the color of his urine is much darker than normal, and he wonders if there is any relationship between the color of his urine and his intestinal obstruction. What would you tell him?

31. A condition known as lactose intolerance is characterized by painful abdominal cramping, gas, and diarrhea. The cause of the problem is an inability to digest the milk sugar, lactose. How would this cause the observed signs and symptoms?

32. Recently, more people have turned to surgery to help them lose weight. One form of weight control surgery involves stapling a portion of the stomach shut, creating a smaller volume. How would such a surgery result in weight loss?

Metabolism and Energetics

25

Did you know...?
Basal metabolic rate represents the calories that we use when at rest. This energy supports ongoing maintenance, repair, and other vital functions.

Learning Outcomes

After completing this chapter, you should be able to do the following:

25-1 Define metabolism, and explain why cells must synthesize new organic components.

25-2 Describe the basic steps in glycolysis, the TCA cycle, and the electron transport system, and summarize the energy yields of glycolysis and cellular respiration.

25-3 Describe the pathways involved in lipid metabolism, and summarize the mechanisms of lipid transport and distribution.

25-4 Summarize the main processes of protein metabolism, and discuss the use of proteins as an energy source.

25-5 Differentiate between the absorptive and postabsorptive metabolic states, and summarize the characteristics of each.

25-6 Explain what constitutes a balanced diet and why such a diet is important.

25-7 Define metabolic rate, discuss the factors involved in determining an individual's BMR, and discuss the homeostatic mechanisms that maintain a constant body temperature.

Clinical Notes

Carbohydrate Loading p. 940
Dietary Fats and Cholesterol p. 943
Vitamins p. 954
Alcohol—A Risky Diversion p. 956
Induced Hypothermia p. 960
Thermoregulatory Disorders p. 961

An Introduction to Metabolism and Energetics

The amount and type of nutrients you obtain in meals can vary widely. Your body stores energy reserves when nutrients are abundant, and mobilizes them when nutrients are in short supply. The neuroendocrine system adjusts and coordinates the metabolic activities of the body's tissues and controls the storage and mobilization of these energy reserves. This chapter examines how the body obtains energy released by the breakdown of organic molecules, stores it as ATP, and uses it to support intracellular operations such as the construction of new organic molecules.

25-1 Metabolism refers to all the chemical reactions that occur in the body

Cells are chemical factories that break down organic molecules to obtain energy, which can then be used to generate ATP. Reactions within mitochondria provide most of the energy a typical cell needs. ∞ p. 80 To carry out these metabolic reactions, cells must have a reliable supply of oxygen and nutrients, including water, vitamins, mineral ions, and organic substrates (the reactants in enzymatic reactions). Oxygen is absorbed at the lungs; the other substances are obtained through absorption by the digestive tract. The cardiovascular system then carries these substances throughout the body. They diffuse from the bloodstream into the tissues, where they can be absorbed and used by our cells.

Mitochondria break down the organic nutrients to provide energy for cell growth, cell division, contraction, secretion, and other functions. Each tissue contains a unique mixture of various kinds of cells. As a result, the energy and nutrient requirements of any two tissues—loose connective tissue and cardiac muscle, for instance—can be quite different. Moreover, activity levels can change rapidly within a tissue, and such changes affect the metabolic requirements of the body. For example, when skeletal muscles start contracting, the tissue demand for oxygen skyrockets. Thus, the energy and nutrient requirements of the body vary from moment to moment (resting versus exercising), hour to hour (asleep versus awake), and year to year (growing child versus adult).

The term **metabolism** (me-TAB-ō-lizm) refers to all the chemical reactions that occur in an organism. Chemical reactions within cells, collectively known as *cellular metabolism*, provide the energy needed to maintain homeostasis and to perform essential functions. Such functions include (1) *metabolic turnover*, the periodic breakdown and replacement of the organic components of a cell; (2) growth and cell divi-

sion; and (3) special processes, such as secretion, contraction, and the propagation of action potentials.

Figure 25–1 provides a broad overview of the processes involved in cellular metabolism. The cell absorbs organic molecules from the surrounding interstitial fluids. Amino acids, lipids, and simple sugars cross the plasma membrane and join nutrients already in the cytoplasm. All the cell's organic building blocks collectively form a *nutrient pool* that the cell relies on to provide energy and to create new intracellular components.

The breakdown of organic substrates is called **catabolism**. This process releases energy that can be used to synthesize ATP or other high-energy compounds. Catabolism proceeds in a series of steps. In general, the initial steps occur in the cytosol, where enzymes break down large organic molecules previously assembled by the cell (such as glycogen, triglycerides, or proteins) into smaller fragments that join the nutrient pool. For example, carbohydrates are broken down into simple sugars, triacylglycerols are split into fatty acids and glycerol, and proteins are broken down to individual amino acids.

Relatively little ATP is produced during these preparatory steps. However, further catabolic activity produces smaller organic molecules that can be absorbed and processed by mitochondria. It is that mitochondrial activity that releases significant amounts of energy. As mitochondrial enzymes break the covalent bonds that hold these molecules together, they capture roughly 40 percent of the energy released and use it to convert ADP to ATP. The other 60 percent escapes as heat that warms the interior of the cell and the surrounding tissues.

The ATP produced by mitochondria provides energy to support both **anabolism**—the synthesis of new organic molecules—and other cell functions. Those additional functions, such as ciliary or cell movement, contraction, active transport, and cell division, vary among cell types. For example, muscle fibers need ATP to provide energy for contraction, whereas gland cells need ATP to synthesize and transport their secretions. We have considered such specialized functions in other chapters, so here we will restrict our focus to anabolic processes.

In terms of energy, anabolism is an "uphill" process that involves the formation of new chemical bonds. Cells synthesize new organic components for four basic reasons:

1. *To Perform Structural Maintenance or Repairs.* All cells must expend energy to perform ongoing maintenance and repairs, because most structures in the cell are temporary rather than permanent. Their removal and replacement are part of the process of *metabolic turnover.* ∞ p. 61

2. *To Support Growth.* Cells preparing to divide increase in size and synthesize extra proteins and organelles.

3. *To Produce Secretions.* Secretory cells must synthesize their products and deliver them to the interstitial fluid.

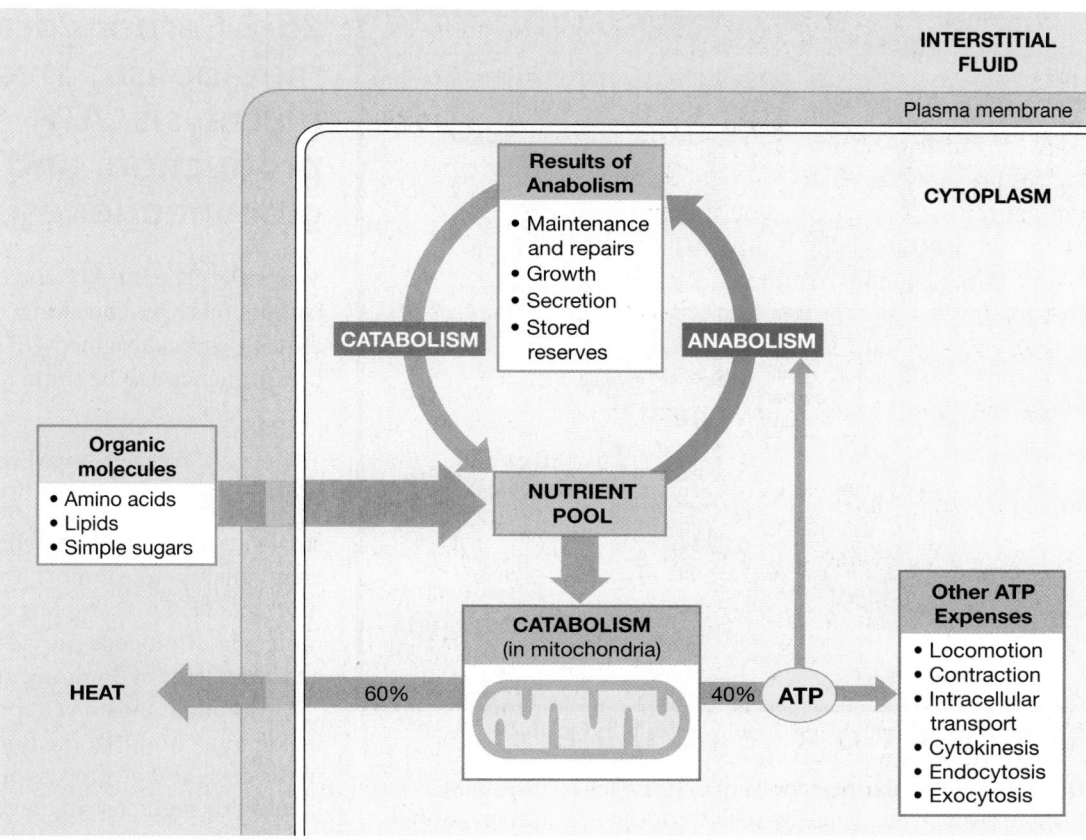

Figure 25–1   An Introduction to Cellular Metabolism. Cells obtain organic molecules from the interstitial fluid and break them down to produce ATP. Only about 40 percent of the energy released by catabolism is captured in the form of ATP; the rest is lost as heat. The ATP generated by catabolism provides energy for all vital cellular activities, including anabolism.

25 DIGESTIVE

4. *To Store Nutrient Reserves.* Most cells "prepare for a rainy day"—a period of emergency, an interval of extreme activity, or a time when the supply of nutrients in the blood-stream is inadequate. Cells prepare for such times by building up reserves—nutrients stored in a form that can be mobilized as needed. The most abundant storage form of carbohydrate is glycogen, a branched chain of glucose molecules; the most abundant storage lipids are triglycerides, consisting primarily of fatty acids. Thus, muscle cells and liver cells, for example, store glucose in the form of glycogen, whereas adipocytes and liver cells store triglycerides. Proteins, the most abundant organic components in the body, perform a variety of vital functions for the cell, and when energy is available, cells synthesize additional proteins. However, when glucose or fatty acids are unavailable, proteins can be broken down into their component amino acids, and the amino acids catabolized as an energy source. So, although their primary function is not to serve as an energy source, proteins are so abundant and accessible that they represent an important "last-ditch" nutrient reserve.

The nutrient pool is the source of the substrates for both catabolism and anabolism. As you might expect, cells tend to conserve the materials needed to build new compounds and break down the rest. Cells continuously replace membranes, organelles, enzymes, and structural proteins. These anabolic activities require more amino acids than lipids, and few carbohydrates. In general, when a cell with excess carbohydrates, lipids, and amino acids needs energy, it will break down carbohydrates first. Lipids are the second choice, and amino acids are seldom broken down if other energy sources are available.

Mitochondria provide the energy that supports cellular operations. The cell feeds its mitochondria from its nutrient pool, and in return, the cell gets the ATP it needs. However, mitochondria are picky eaters: They will accept only specific organic molecules for processing and energy production. Thus, chemical reactions in the cytoplasm take whichever organic nutrients are available and break them down into smaller fragments that the mitochondria can process. The mitochondria then break the fragments down further, generating carbon dioxide, water, and ATP (**Figure 25–2**). This mitochondrial activity involves two pathways: the *TCA cycle* and the *electron transport system.* We will describe these important catabolic and anabolic cellular reactions in the next section.

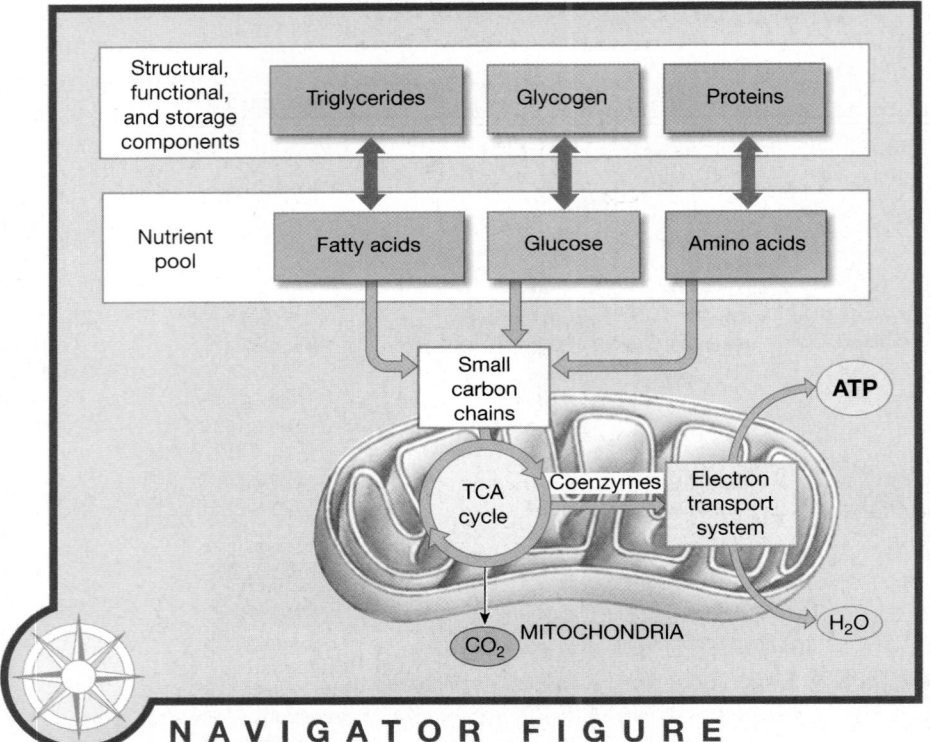

N A V I G A T O R   F I G U R E

Figure 25–2   Nutrient Use in Cellular Metabolism. Cells use the contents of the nutrient pool to build up reserves and to synthesize cellular structures. Catabolism within mitochondria provides the ATP needed to sustain cell functions. Mitochondria are "fed" small carbon chains produced by the breakdown of carbohydrates (primarily glucose, stored as glycogen), lipids (especially fatty acids from triglycerides), and proteins (amino acids). The mitochondria absorb these breakdown products for further catabolism by means of the tricarboxylic acid (TCA) cycle and the electron transport system. This figure will be repeated, in reduced and simplified form as Navigator icons, as the text changes topics.

The A&P Top 100

#87 There is an energy cost to staying alive, even at rest. All cells must expend ATP to perform routine maintenance, like removing and replacing intracellular and extracellular structures and components. In addition, cells must spend additional energy performing other vital functions, such as growth, secretion, and contraction.

C H E C K P O I N T

1. Define metabolism.

2. Define catabolism.

3. Define anabolism.

See the blue Answers tab at the end of the book.

25-2 Carbohydrate metabolism involves glycolysis, ATP production, and gluconeogenesis

Most cells generate ATP and other high-energy compounds by breaking down carbohydrates—especially glucose. The complete reaction sequence can be summarized as follows:

$$C_6H_{12}O_6 + 6\ O_2 \longrightarrow 6\ CO_2 + 6\ H_2O$$
glucose oxygen   carbon dioxide water

The breakdown occurs in a series of small steps, several of which release sufficient energy to support the conversion of ADP to ATP. The complete catabolism of one molecule of glucose provides a typical body cell a net gain of 36 molecules of ATP.

Although most ATP production occurs inside mitochondria, the first steps take place in the cytosol. The process of *glycolysis*, which breaks down glucose in the cytosol and generates smaller molecules that can be absorbed and utilized by mitochondria, was introduced in Chapter 10. ∞ p. 319 Because glycolysis does not require oxygen, the reactions are said to be *anaerobic*. The subsequent reactions, which occur in mitochondria, consume oxygen and are considered *aerobic*. The mitochondrial activity responsible for ATP production is called **aerobic metabolism**, or *cellular respiration*.

Glycolysis

Glycolysis (glī-KOL-i-sis; *glykus*, sweet + *lysis*, dissolution) is the breakdown of glucose to **pyruvic acid**. In this process, a series of enzymatic steps breaks the six-carbon glucose molecule ($C_6H_{12}O_6$) into two three-carbon molecules of pyruvic acid (CH_3—CO—COOH). At the normal pH inside cells, each pyruvic acid molecule loses a hydrogen ion and exists as the negatively charged ion CH_3−CO—COO⁻. This ionized form is usually called *pyruvate*, rather than pyruvic acid.

Glycolysis requires (1) glucose molecules, (2) appropriate cytoplasmic enzymes, (3) ATP and ADP, (4) inorganic phosphates, and (5) **NAD** (*nicotinamide adenine dinucleotide*), a coenzyme that removes hydrogen atoms during one of the enzymatic reactions. (Recall from Chapter 2 that coenzymes are organic molecules that are essential to enzyme function. ∞ p. 57) If any of these participants is missing, glycolysis cannot occur.

Figure 25–3 provides an overview of the steps in glycolysis. Glycolysis begins when an enzyme *phosphorylates*—that is, attaches a phosphate group—to the last (sixth) carbon atom of a glucose molecule, creating **glucose-6-phosphate**. This step, which "costs" the cell one ATP molecule, has two important results: (1) It traps the glucose molecule within the cell, because phosphorylated glucose cannot cross the plasma membrane; and (2) it prepares the glucose molecule for further biochemical reactions.

A second phosphorylation occurs in the cytosol before the six-carbon chain is broken into two three-carbon frag-

ments. Energy benefits begin to appear as these fragments are converted to pyruvic acid. Two of the steps release enough energy to generate ATP from ADP and inorganic phosphate (PO_4^{3-} or P_i). In addition, two molecules of NAD are converted to NADH. The net reaction looks like this:

$$\text{Glucose} + 2\,\text{NAD} + 2\,\text{ADP} + 2\,P_i \longrightarrow$$
$$2\,\text{Pyruvic acid} + 2\,\text{NADH} + 2\,\text{ATP}$$

This anaerobic reaction sequence provides the cell a net gain of two molecules of ATP for each glucose molecule converted to two pyruvic acid molecules. A few highly specialized

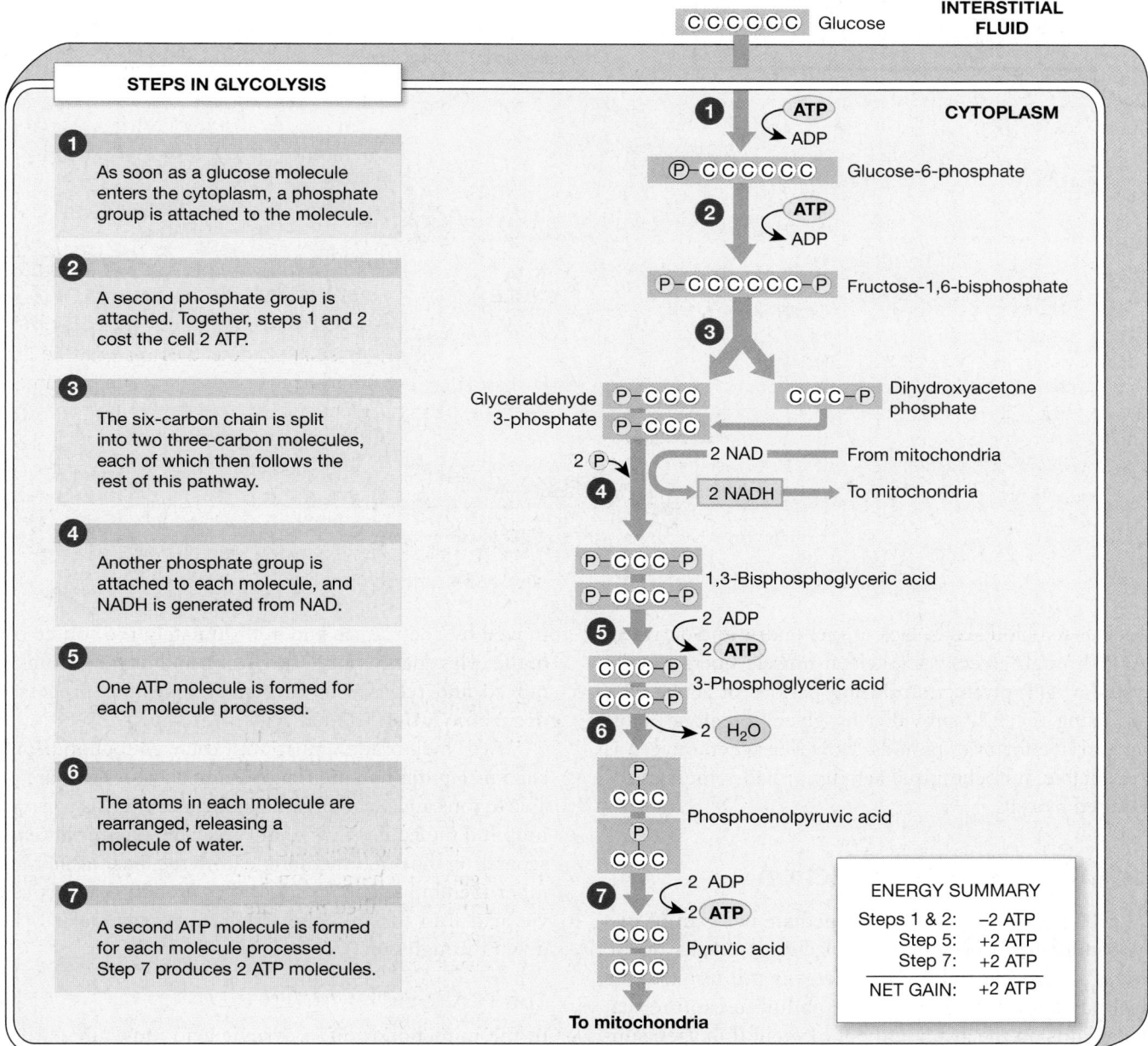

Figure 25–3 **Glycolysis.** Glycolysis breaks down a six-carbon glucose molecule into two three-carbon molecules of pyruvic acid through a series of enzymatic steps. The further catabolism of pyruvic acid begins with its entry into a mitochondrion. *(See Figure 25–4.)*

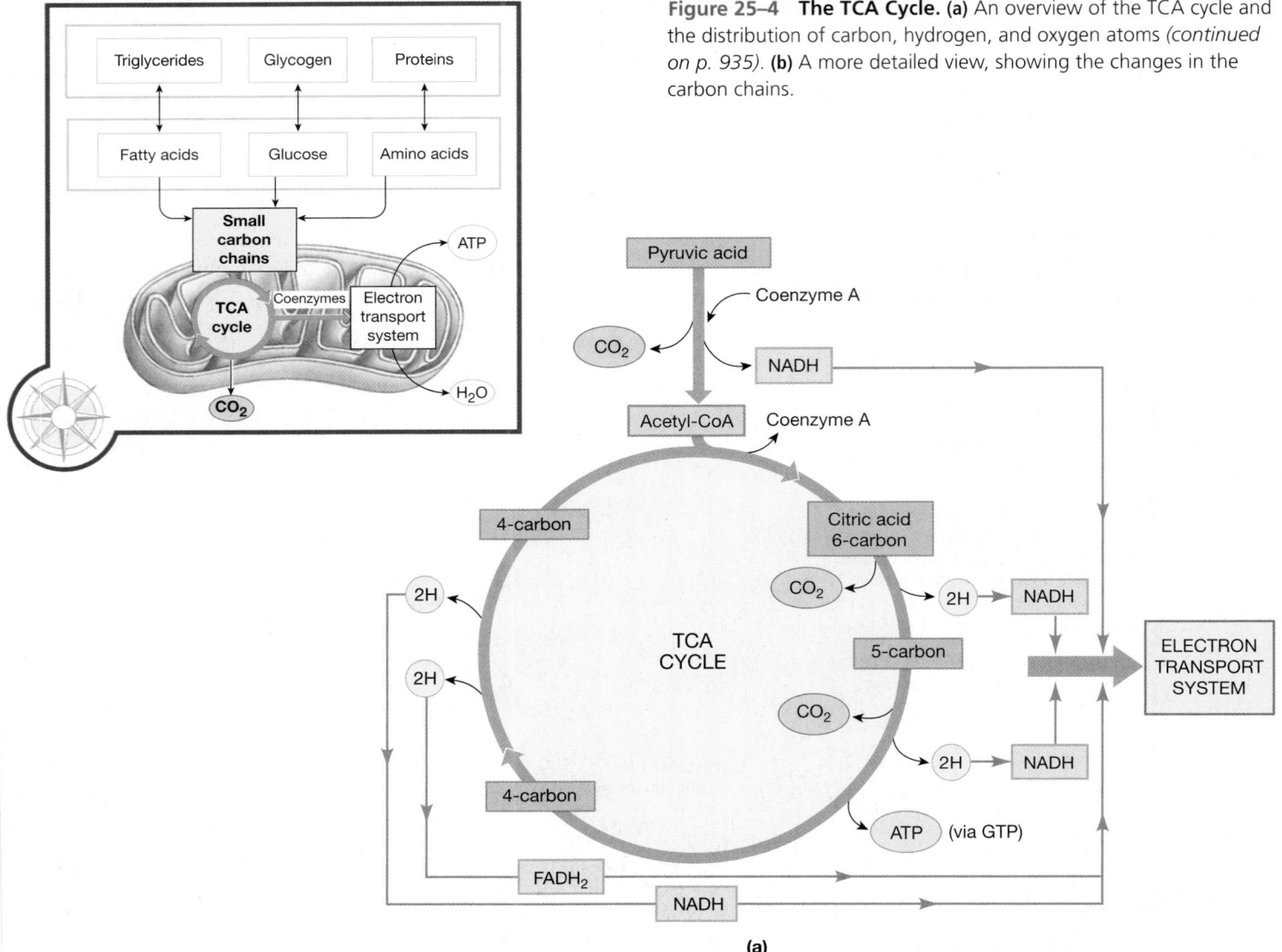

Figure 25–4 **The TCA Cycle. (a)** An overview of the TCA cycle and the distribution of carbon, hydrogen, and oxygen atoms *(continued on p. 935)*. **(b)** A more detailed view, showing the changes in the carbon chains.

(a)

cells, such as red blood cells, lack mitochondria and derive all their ATP through glycolysis. Skeletal muscle fibers rely on glycolysis for ATP production during periods of active contraction; using the ATP provided by glycolysis alone, most cells can survive for brief periods. However, when oxygen is readily available, mitochondrial activity provides most of the ATP required by cells.

Mitochondrial ATP Production

For the cell, glycolysis yields an immediate net gain of two ATP molecules for each glucose molecule it breaks down. However, a great deal of additional energy is still stored in the chemical bonds of pyruvic acid. The ability to capture that energy depends on the availability of oxygen. If oxygen supplies are adequate, mitochondria absorb the pyruvic acid molecules and break them down. The hydrogen atoms of each pyruvic acid molecule ($CH_3-CO-COOH$) are re-

moved by coenzymes and are ultimately the source of most of the cell's energy gain. The carbon and oxygen atoms are removed and released as carbon dioxide in a process called **decarboxylation** (dē-kar-boks-i-LĀ-shun).

Two membranes surround each mitochondrion. The *outer membrane* contains large-diameter pores that are permeable to ions and small organic molecules such as pyruvic acid. Ions and molecules thus easily enter the *intermembrane space* separating the outer membrane from the *inner membrane*. The inner membrane contains a carrier protein that moves pyruvic acid into the mitochondrial matrix, where it is broken down through the TCA cycle (**Figure 25–4**).

The TCA Cycle

In the mitochondrion, a pyruvic acid molecule participates in a complex reaction involving NAD and another coenzyme, **coenzyme A**, or **CoA**. This reaction yields one molecule of carbon dioxide, one of NADH, and one of **acetyl-CoA**

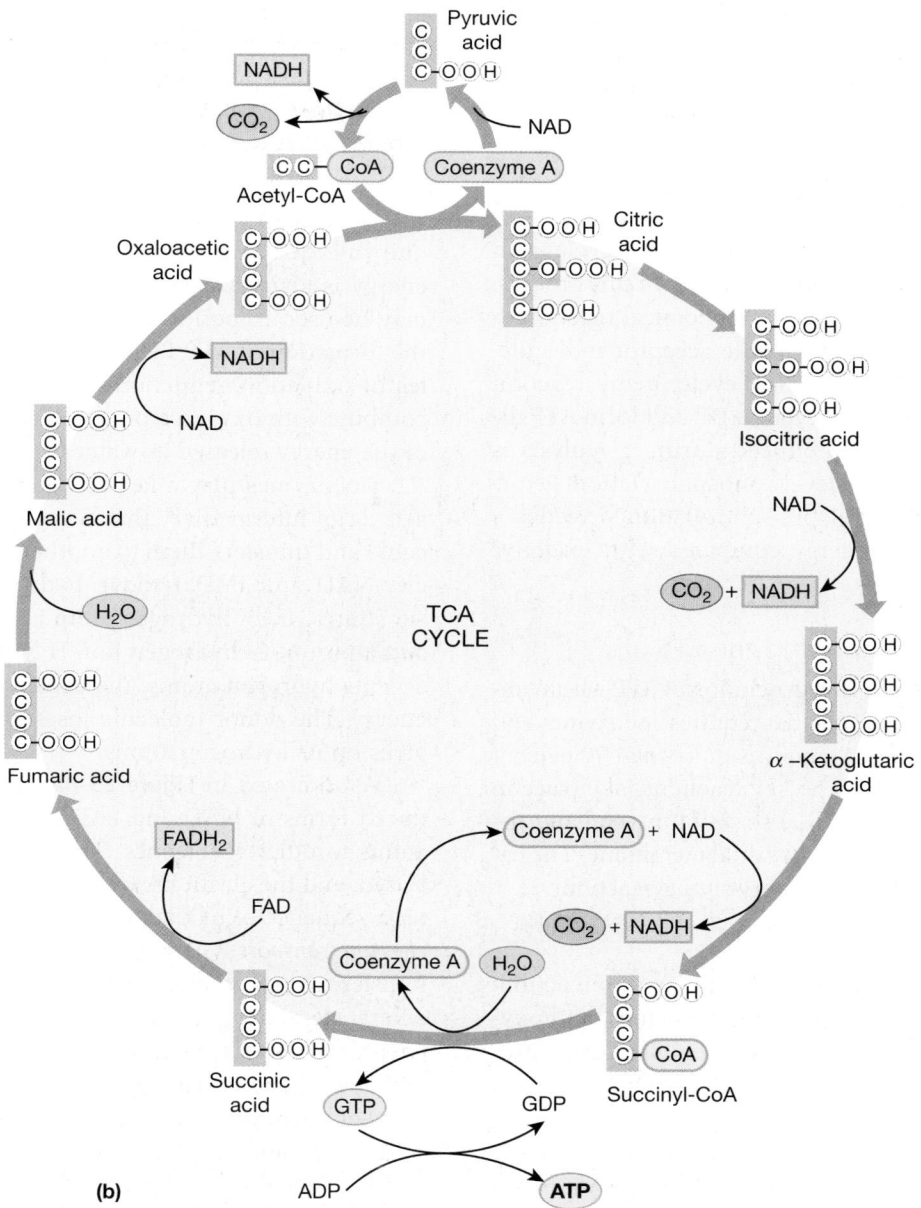

(b)

(AS-e-til-KŌ-ā)—a two-carbon **acetyl group** (CH_3CO) bound to coenzyme A. This sets the stage for a sequence of enzymatic reactions called the **tricarboxylic** (trī-kar-bok-SIL-ik) **acid (TCA) cycle**, or **citric acid cycle**. In the first step of that cycle, the acetyl group is transferred from acetyl-CoA to a four-carbon molecule of *oxaloacetic acid*, producing *citric acid*.

The function of the citric acid cycle is to remove hydrogen atoms from organic molecules and transfer them to coenzymes. The overall pattern of the TCA cycle is shown in **Figure 25–4a**.

At the start of the TCA cycle, the two-carbon acetyl group carried by CoA is attached to the four-carbon oxaloacetic acid molecule to make the six-carbon compound citric acid. Coenzyme A is released intact and can thus bind another acetyl group. A complete revolution of the TCA cycle removes two carbon atoms, regenerating the four-carbon chain. (This

is why the reaction sequence is called a *cycle*.) We can summarize the fate of the atoms in the acetyl group as follows:

- The two carbon atoms are removed in enzymatic reactions that incorporate four oxygen atoms and form two molecules of carbon dioxide, a waste product.

- The hydrogen atoms are removed by the coenzyme NAD or a related coenzyme called **FAD** (*flavin adenine dinucleotide*) (**Figure 25–4b**).

Several of the steps involved in a revolution of the TCA cycle involve more than one reaction and require more than one enzyme. Water molecules are tied up in two of those steps. The entire sequence can be summarized as follows:

$$CH_3CO - CoA + 3\,NAD + FAD + GDP + P_i + 2\,H_2O \longrightarrow$$
$$CoA + 2\,CO_2 + 3\,NADH + FADH_2 + 2\,H^+ + GTP$$

The only immediate energy benefit of one revolution of the TCA cycle is the formation of a single molecule of GTP (*guanosine **tri**phosphate*) from GDP (*guanosine **di**phosphate*) and P_i. In practical terms, GTP is the equivalent of ATP, because GTP readily transfers a phosphate group to ADP, producing ATP:

$$GTP + ADP \longrightarrow GDP + ATP$$

The formation of GTP from GDP in the TCA cycle is an example of **substrate-level phosphorylation**. In this process, an enzyme uses the energy released by a chemical reaction to transfer a phosphate group to a suitable acceptor molecule. Although GTP is formed in the TCA cycle, many reaction pathways in the cytosol phosphorylate ADP and form ATP directly. For example, the ATP produced during glycolysis is generated through substrate-level phosphorylation. Normally, however, substrate-level phosphorylation provides a relatively small amount of energy compared with *oxidative phosphorylation*.

Oxidative Phosphorylation and the ETS

Oxidative phosphorylation is the generation of ATP within mitochondria in a reaction sequence that requires coenzymes and consumes oxygen. The process produces more than 90 percent of the ATP used by body cells. The key reactions take place in the *electron transport system (ETS)*, a series of integral and peripheral proteins in the inner mitochondrial membrane. The basis of oxidative phosphorylation is a very simple reaction:

$$2 H_2 + O_2 \longrightarrow 2 H_2O$$

Cells can easily obtain the ingredients for this reaction: Hydrogen is a component of all organic molecules, and oxygen is an atmospheric gas. The only problem is that the reaction releases a tremendous amount of energy all at once. In fact, this reaction releases so much energy that it is used to launch space shuttles into orbit. Cells cannot handle energy explosions; energy release must be gradual, as it is in oxidative phosphorylation. This powerful reaction proceeds in a series of small, enzymatically controlled steps. Under these conditions, energy can be captured safely, and ATP generated.

Oxidation, Reduction, and Energy Transfer. The enzymatic steps of oxidative phosphorylation involve oxidation and reduction. (There are different types of oxidation and reduction reactions, but for our purposes the most important are those involving the transfer of electrons.) The loss of electrons is a form of **oxidation**; the gain of electrons is a form of **reduction**. The two reactions are always paired. When electrons pass from one molecule to another, the electron donor is oxidized and the electron recipient reduced. Oxidation and reduction are important because electrons carry chemical energy. In a typical oxidation–reduction reaction, *the reduced molecule gains energy at the expense of the oxidized molecule.*

In such an exchange, the reduced molecule does not acquire all the energy released by the oxidized molecule. Some energy is always released as heat, but the remaining energy may be used to perform physical or chemical work, such as the formation of ATP. By sending the electrons through a series of oxidation–reduction reactions before they ultimately combine with oxygen atoms, cells can capture and use much of the energy released as water is formed.

Coenzymes play a key role in this process. A coenzyme acts as an intermediary that accepts electrons from one molecule and transfers them to another molecule. In the TCA cycle, NAD and FAD remove hydrogen atoms from organic substrates. Each hydrogen atom consists of an electron (e^-) and a proton (a hydrogen ion, H^+). Thus, when a coenzyme accepts hydrogen atoms, the coenzyme is reduced and gains energy. The donor molecule loses electrons and energy as it gives up its hydrogen atoms.

As indicated in **Figure 25–4a**, NADH and $FADH_2$, the reduced forms of NAD and FAD, then transfer their hydrogen atoms to other coenzymes. The protons are subsequently released, and the electrons, which carry the chemical energy, enter a sequence of oxidation–reduction reactions known as the *electron transport system*. This sequence ends with the electrons' transfer to oxygen and the formation of a water molecule. At several steps along the oxidation–reduction sequence, enough energy is released to support the synthesis of ATP from ADP. We will now consider that reaction sequence in greater detail.

The coenzyme FAD accepts two hydrogen atoms from the TCA cycle and in doing so gains two electrons, forming $FADH_2$. The oxidized form of the coenzyme NAD has a positive charge (NAD^+). This coenzyme also gains two electrons as two hydrogen atoms are removed from the donor molecule, resulting in the formation of NADH and the release of a proton (H^+). For this reason, the reduced form of NAD is often described as "$NADH + H^+$."

The Electron Transport System. The **electron transport system (ETS)**, or *respiratory chain*, is a sequence of proteins called **cytochromes** (SĪ-tō-krōmz; *cyto-*, cell + *chroma*, color). Each cytochrome has two components: a protein and a pigment. The protein, embedded in the inner membrane of a mitochondrion, surrounds the pigment complex, which contains a metal ion—either iron (Fe^{3+}) or copper (Cu^{2+}). We will consider four cytochromes: b, c, a, and a_3.

Figure 25–5 summarizes the basic steps in oxidative phosphorylation. We will first consider the general path taken by

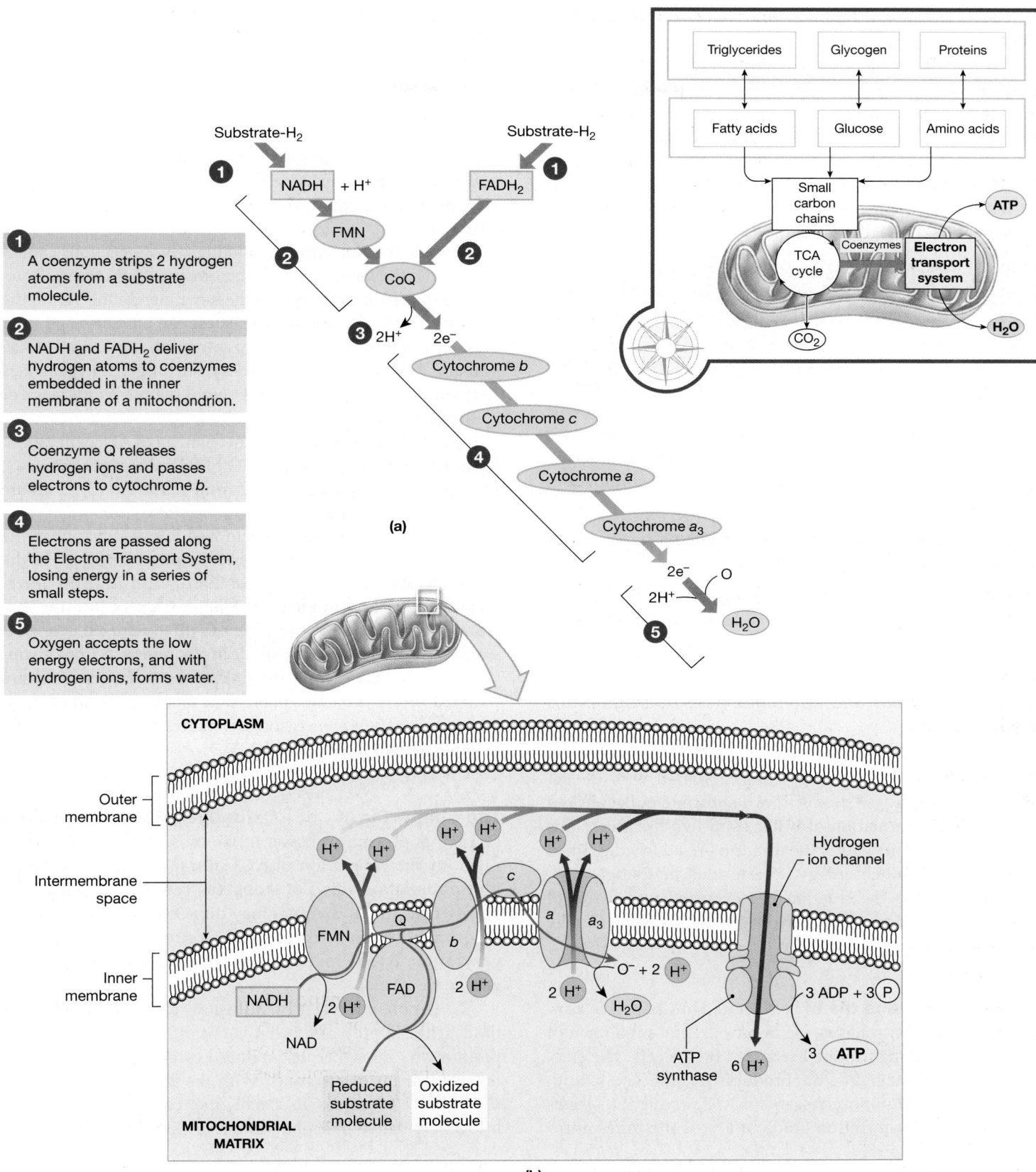

1 A coenzyme strips 2 hydrogen atoms from a substrate molecule.

2 NADH and $FADH_2$ deliver hydrogen atoms to coenzymes embedded in the inner membrane of a mitochondrion.

3 Coenzyme Q releases hydrogen ions and passes electrons to cytochrome *b*.

4 Electrons are passed along the Electron Transport System, losing energy in a series of small steps.

5 Oxygen accepts the low energy electrons, and with hydrogen ions, forms water.

(a)

(b)

Figure 25–5 Oxidative Phosphorylation. (a) The sequence of oxidation–reduction reactions involved in oxidative phosphorylation. **(b)** The locations of the coenzymes and the electron transport system. Notice the sites where hydrogen ions are pumped into the intermembrane space, providing the concentration gradient essential to the generation of ATP. The red line indicates the path taken by the electrons.

the electrons that are captured and delivered by coenzymes (**Figure 25–5a**):

Step 1 *A Coenzyme Strips a Pair of Hydrogen Atoms from a Substrate Molecule.* As we have seen, different coenzymes are used for different substrate molecules. During glycolysis, which occurs in the cytoplasm, NAD is reduced to NADH. Within mitochondria, both NAD and FAD are reduced through reactions of the TCA cycle, producing NADH and $FADH_2$, respectively.

Step 2 *NADH and $FADH_2$ Deliver Hydrogen Atoms to Coenzymes Embedded in the Inner Mitochondrial Membrane.* The electrons carry the energy, and the protons that accompany them are released before the electrons are transferred to the ETS. As indicated in **Figure 25–5a**, one of two paths is taken to the ETS; which one depends on whether the donor is NADH or $FADH_2$. The path from NADH involves the coenzyme **FMN** (*flavin mononucleotide*), whereas the path from $FADH_2$ proceeds directly to **coenzyme Q** (*ubiquinone*). Both FMN and coenzyme Q are bound to the inner mitochondrial membrane.

Step 3 *Coenzyme Q Releases Hydrogen Ions and Passes Electrons to Cytochrome b.*

Step 4 *Electrons Are Passed along the Electron Transport System, Losing Energy in a Series of Small Steps.* The sequence is cytochrome *b* to *c* to *a* to a_3.

Step 5 *At the End of the ETS, an Oxygen Atom Accepts the Electrons and Combines with Hydrogen Ions to Form Water.*

Notice that this reaction sequence started with the removal of two hydrogen atoms from a substrate molecule and ended with the formation of water from two hydrogen ions and one oxygen ion. This is the reaction previously described as releasing too much energy if performed in a single step. However, because the reaction has occurred in a series of small steps, the combining of hydrogen and oxygen can take place safely rather than explosively.

ATP Generation and the ETS. As noted in Chapter 3, concentration gradients across membranes represent a form of potential energy that can be harnessed by the cell. The electron transport system does not produce ATP directly. Instead, it creates the conditions necessary for ATP production by creating a steep concentration gradient across the inner mito-

chondrial membrane. The electrons that travel along the ETS release energy as they pass from coenzyme to cytochrome and from cytochrome to cytochrome. The energy released at each of several steps drives hydrogen ion pumps that move hydrogen ions from the mitochondrial matrix into the intermembrane space between the inner and outer mitochondrial membranes. These pumps create a large concentration gradient for hydrogen ions across the inner membrane. It is this concentration gradient that provides the energy to convert ADP to ATP.

Despite the concentration gradient, hydrogen ions cannot diffuse into the matrix because they are not lipid soluble. However, hydrogen ion channels in the inner membrane permit the diffusion of H^+ into the matrix. These ion channels and their attached *coupling factors* make up a protein complex called *ATP synthase*. The kinetic energy of passing hydrogen ions generates ATP in a process known as *chemiosmosis* (kem-ē-oz-MŌ-sis), or *chemiosmotic phosphorylation*.

Figure 25–5b diagrams the mechanism of ATP generation. Hydrogen ions are pumped as (1) FMN reduces coenzyme Q, (2) cytochrome *b* reduces cytochrome *c*, and (3) electrons are passed from cytochrome *a* to cytochrome a_3.

For each pair of electrons removed from a substrate in the TCA cycle by NAD, six hydrogen ions are pumped across the inner membrane of the mitochondrion and into the intermembrane space (**Figure 25–5b**). Their reentry into the matrix provides the energy to generate three molecules of ATP. Alternatively, for each pair of electrons removed from a substrate in the TCA cycle by FAD, four hydrogen ions are pumped across the inner membrane and into the intermembrane space. Their reentry into the matrix provides the energy to generate two molecules of ATP.

The Importance of Oxidative Phosphorylation. Oxidative phosphorylation is the most important mechanism for the generation of ATP. In most cases, if oxidative phosphorylation slows or stops, the cell dies. If many cells are affected, the individual may die. Oxidative phosphorylation requires both oxygen and electrons; the rate of ATP generation is thus limited by the availability of either oxygen or electrons.

Cells obtain oxygen by diffusion from the extracellular fluid. If the supply of oxygen is cut off, mitochondrial ATP production ceases, because reduced cytochrome a_3 will have no acceptor for its electrons. With the last reaction stopped, the entire ETS comes to a halt, like cars at a washed-out bridge. Oxidative phosphorylation can no longer take place, and cells quickly succumb to energy starvation. Because neurons have a high demand for energy, the brain is one of the first organs to be affected.

Hydrogen cyanide gas is sometimes used as a pesticide to kill rats or mice; in some states where capital punishment is legal, it is used to execute criminals. The cyanide ion (CN^-) binds to cytochrome a_3 and prevents the transfer of electrons to oxygen. As a result, cells die from energy starvation.

Energy Yield of Glycolysis and Cellular Respiration

For most cells, the complete reaction pathway that begins with glucose and ends with carbon dioxide and water is the main method of generating ATP. **Figure 25–6** summarizes the process in terms of energy production:

- *Glycolysis.* During glycolysis, the cell gains a net two molecules of ATP for each glucose molecule broken down

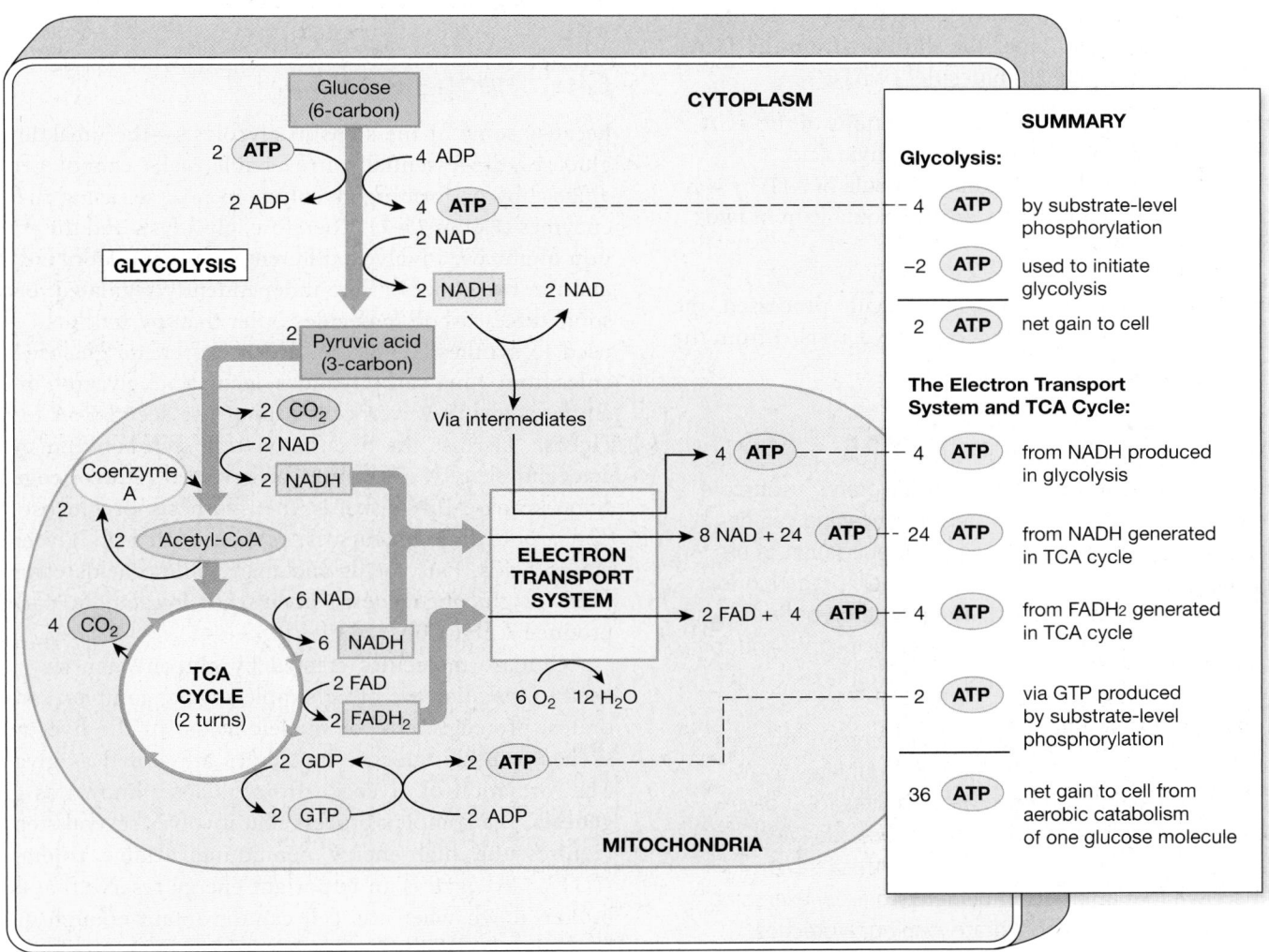

Figure 25–6   A Summary of the Energy Yield of Aerobic Metabolism. For each glucose molecule broken down by glycolysis, only two molecules of ATP (net) are produced. However, glycolysis, the formation of acetyl-CoA, and the TCA cycle all yield molecules of reduced coenzymes (NADH or $FADH_2$). Many additional ATP molecules are produced when electrons from these coenzymes pass through the electron transport system. In most cells, each of the two NADH molecules produced in glycolysis provides another two ATP molecules. Each of the eight NADH molecules produced in the mitochondria yields three ATP molecules, for a total of 24. Another two ATP molecules are gained from each of the two $FADH_2$ molecules generated in the mitochondria. The TCA cycle generates an additional two ATP molecules in the form of GTP.

anaerobically to pyruvic acid. Two molecules of NADH are also produced. In most cells, electrons are passed from NADH to FAD via an intermediate in the intermembrane space, and then to CoQ and the electron transport system. This sequence of events ultimately provides an additional four ATP molecules.

- *The Electron Transport System.* The TCA cycle breaks down the 2 pyruvic acid molecules, transferring hydrogen atoms to NADH and FADH$_2$. These coenzymes provide electrons to the ETS; each of the 8 molecules of NADH yields 3 molecules of ATP and 1 water molecule; each of the 2 FADH$_2$ molecules yields 2 ATP molecules and 1 water molecule. Thus, the shuffling from the TCA cycle to the ETS yields 28 molecules of ATP.

- *The TCA Cycle.* Each of the two revolutions of the TCA cycle required to break down both pyruvic acid molecules completely yields one molecule of ATP by way of GTP. This cycling provides an additional gain of two molecules of ATP.

Summing up, for each glucose molecule processed, the cell gains 36 molecules of ATP: 2 from glycolysis, 4 from the NADH generated in glycolysis, 2 from the TCA cycle (by means of GTP), and 28 from the ETS.

Cardiac muscle cells and liver cells are able to gain an additional two ATP molecules for each glucose molecule broken down. This gain is accomplished by increasing the energy yield from the NADH generated during glycolysis from four to six ATP molecules. In these cells, each NADH molecule passes electrons to an intermediate that generates NADH, not FADH$_2$ in the mitochondrial matrix. The subsequent transfer of electrons to FMN, CoQ, and the ETS results in the production of six ATP molecules, just as if the two NADH molecules had been generated in the TCA cycle.

Gluconeogenesis

Because some of the steps in glycolysis—the breakdown of glucose—are essentially irreversible, cells cannot generate glucose by performing glycolysis in reverse, using the same enzymes (**Figure 25–7**). Therefore, glycolysis and the production of glucose involve a different set of regulatory enzymes, and the two processes are independently regulated. Because some three-carbon molecules other than pyruvic acid can be used to synthesize glucose, a cell can create glucose molecules from other carbohydrates, lactic acid, glycerol, or some amino acids. However, cells cannot use acetyl-CoA to make glucose, because the decarboxylation step between pyruvic acid and acetyl-CoA cannot be reversed. **Gluconeogenesis** (gloo-kō-nē-ō-JEN-e-sis) is the synthesis of glucose from noncarbohydrate precursors, such as lactic acid, glycerol, or amino acids. Fatty acids and many amino acids cannot be used for gluconeogenesis, because their catabolic pathways produce acetyl-CoA.

Glucose molecules created by gluconeogenesis can be used to manufacture other simple sugars, complex carbohydrates, proteoglycans, or nucleic acids. In the liver and in skeletal muscle, glucose molecules are stored as glycogen. The formation of glycogen from glucose, known as **glycogenesis,** is a complex process that involves several steps and requires the high-energy compound **uridine triphosphate** (UTP). Glycogen is an important energy reserve that can be broken down when the cell cannot obtain enough glucose from interstitial fluid. The breakdown of glycogen, called **glycogenolysis**, occurs quickly and involves a single enzymatic step. Although glycogen molecules are large, glycogen reserves take up very little space because they form compact, insoluble granules.

CLINICAL NOTE

Carbohydrate Loading. Eating carbohydrates just before exercise does not improve your performance and can decrease your endurance by slowing the mobilization of existing energy reserves. Runners or swimmers preparing for lengthy endurance events, such as a marathon or 5-km swim, do not eat immediately before participating, and for two hours before the event they limit their intake to drinking water.

Performance in endurance sports improves if muscles have large stores of glycogen. Endurance athletes try to achieve this by eating carbohydrate-rich diets for 3 days before competing. This is called *carbohydrate loading.* Studies in Sweden, Australia, and South Africa have recently shown that attempts to deplete stores by exercising to exhaustion before carbohydrate loading, a practice called *carbohydrate depletion/loading* are less effective than 3 days of rest or minimal exercise during carbohydrate loading. This less intense approach improves mood and reduces the risks of muscle and kidney damage.

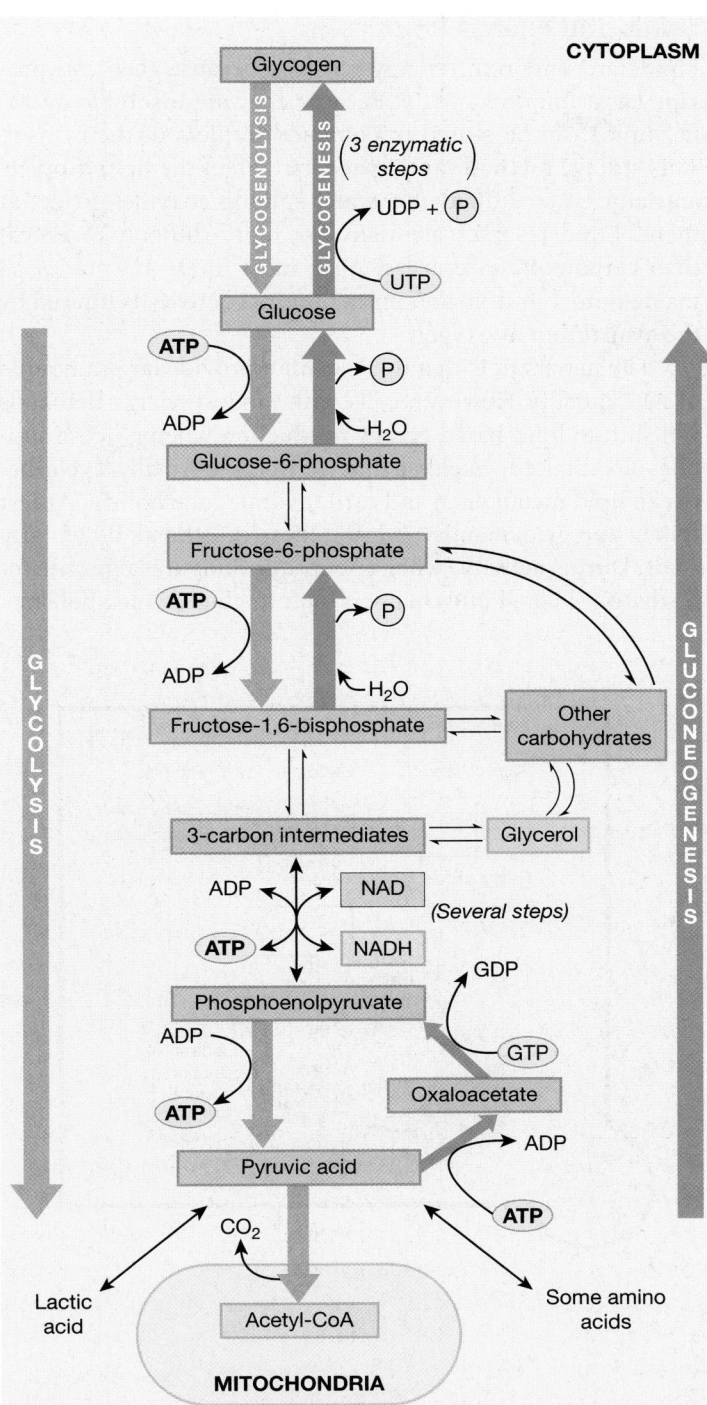

Figure 25–7 Carbohydrate Breakdown and Synthesis. The pathways for glycolysis and gluconeogenesis. Many of the reactions are freely reversible, but separate regulatory enzymes control the key steps, which are indicated by colored arrows. Some amino acids, carbohydrates (including lactic acid), and glycerol can be converted to glucose. The enzymatic reaction that converts pyruvic acid to acetyl-CoA cannot be reversed.

Tips&Tricks

To make several similar-sounding terms easier to tell apart, learn these word parts: *genesis* means "the formation of," *lysis* means "the breakdown of," *glyco* and *gluco* refer to glucose, and *neo* means "new." So, **gluconeogenesis** is the formation of new glucose, **glycogenesis** is the formation of glycogen (the storage form of glucose), **glycogenolysis** is the breakdown of glycogen to glucose, and **glycolysis** is the breakdown of glucose to pyruvic acid.

CHECKPOINT

4. **What is the primary role of the TCA cycle in the production of ATP?**

5. **NADH produced by glycolysis in skeletal muscle fibers leads to the production of two ATP molecules in the mitochondria, but NADH produced by glycolysis in cardiac muscle cells leads to the production of three ATP molecules. Why?**

6. **How would a decrease in the level of cytoplasmic NAD affect ATP production in mitochondria?**

See the blue Answers tab at the end of the book.

25-3 Lipid metabolism involves lipolysis, beta-oxidation, and the transport and distribution of lipids as free fatty acids and lipoproteins

Like carbohydrates, lipid molecules contain carbon, hydrogen, and oxygen, but the atoms are present in different proportions. Triglycerides are the most abundant lipid in the body, so our discussion will focus on pathways for triglyceride breakdown and synthesis.

Lipid Catabolism

During lipid catabolism, or **lipolysis**, lipids are broken down into pieces that can be either converted to pyruvic acid or channeled directly into the TCA cycle. A triglyceride is first split into its component parts by hydrolysis, yielding one molecule of glycerol and three fatty acid molecules. Glycerol enters the TCA cycle after enzymes in the cytosol convert it to pyruvic acid. The catabolism of fatty acids involves a completely different set of enzymes that generate acetyl-CoA directly.

Beta-Oxidation

Fatty acid molecules are broken down into two-carbon fragments in a sequence of reactions known as **beta-oxidation**. This process occurs inside mitochondria, so the carbon chains can enter the TCA cycle immediately as acetyl-CoA. **Figure 25–8** diagrams one step in the process of beta-oxidation. Each step generates molecules of acetyl-CoA, NADH, and $FADH_2$, leaving a shorter carbon chain bound to coenzyme A.

Beta-oxidation provides substantial energy benefits. For each two-carbon fragment removed from the fatty acid, the cell gains 12 ATP molecules from the processing of acetyl-CoA in the TCA cycle, plus 5 ATP molecules from the NADH and $FADH_2$. The cell can therefore gain 144 ATP molecules from the breakdown of one 18-carbon fatty acid molecule. This number of ATP molecules yields almost 1.5 times the energy obtained by the complete breakdown of three 6-carbon glucose molecules. The catabolism of other lipids follows similar patterns, generally ending with the formation of acetyl-CoA.

Lipids and Energy Production

Lipids are important energy reserves because they can provide large amounts of ATP. Because they are insoluble in water, lipids can be stored in compact droplets in the cytosol. This storage method saves space, but when the lipid droplets are large, it is difficult for water-soluble enzymes to get at them. Lipid reserves are therefore more difficult to access than carbohydrate reserves. Also, most lipids are processed inside mitochondria, and mitochondrial activity is limited by the availability of oxygen.

The net result is that lipids cannot provide large amounts of ATP quickly. However, cells with modest energy demands can shift to lipid-based energy production when glucose supplies are limited. Skeletal muscle fibers normally cycle between lipid metabolism and carbohydrate metabolism. At rest (when energy demands are low), these cells break down fatty acids. During activity (when energy demands are large and immediate), skeletal muscle fibers shift to glucose metabolism.

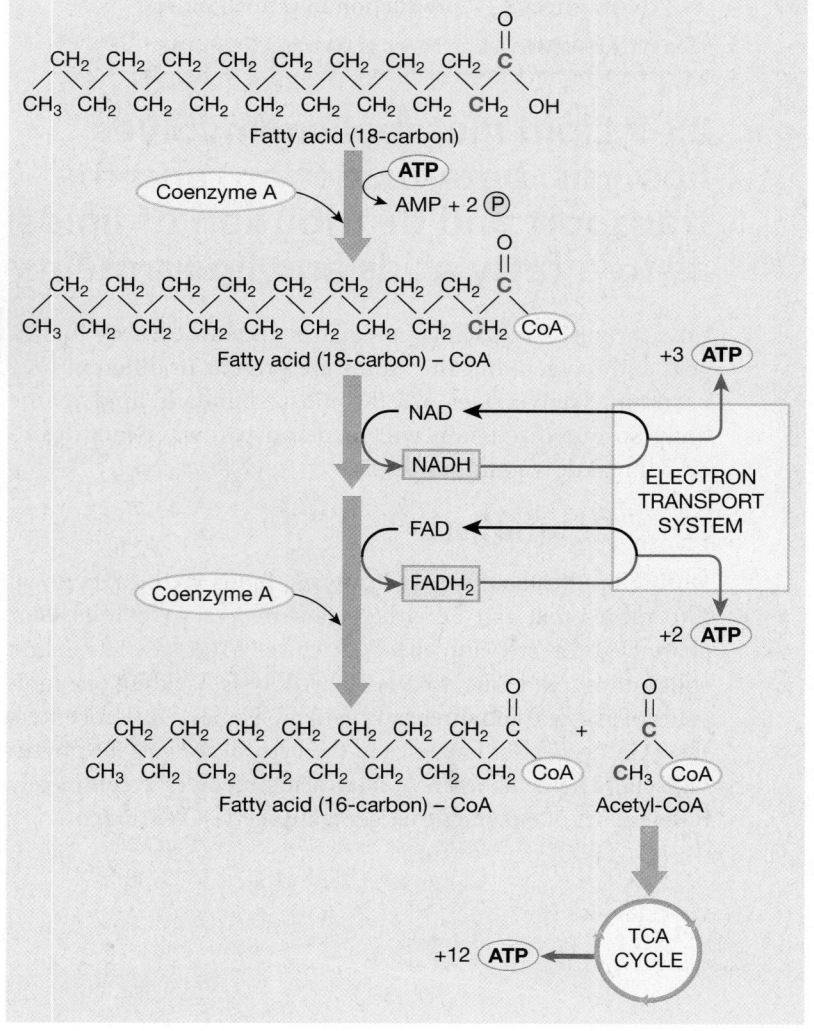

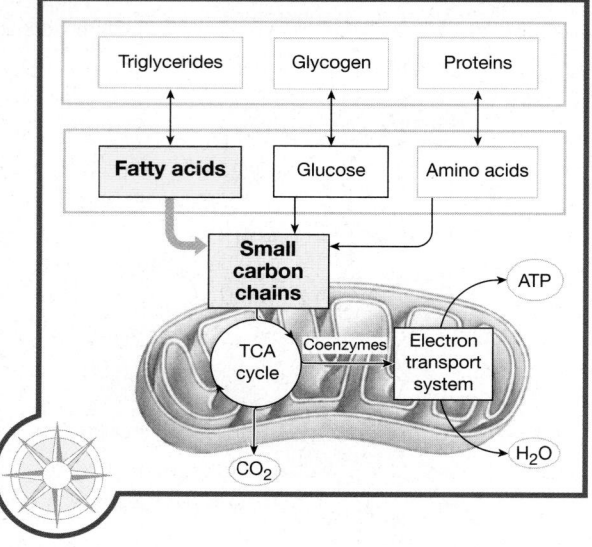

Figure 25–8 **Beta-Oxidation.** During beta-oxidation, the carbon chains of fatty acids are broken down to yield molecules of acetyl-CoA, which can be used in the TCA cycle. The reaction also donates hydrogen atoms to coenzymes, which then deliver them to the electron transport system.

CLINICAL NOTE

Dietary Fats and Cholesterol

Elevated cholesterol levels are associated with the development of atherosclerosis (see Chapter 21) and coronary artery disease (see Chapter 20). ∞ pp. 724, 694 Nutritionists currently recommend that you reduce cholesterol intake to under 300 mg per day. This amount represents a 40 percent reduction for the average American adult. As a result of rising concerns about cholesterol, such phrases as "low cholesterol," "contains no cholesterol," and "cholesterol free" are now widely used in the advertising and packaging of foods. Cholesterol content alone, however, does not tell the entire story; consider the following basic information about cholesterol and about lipid metabolism in general:

- **Cholesterol Has Many Vital Functions in the Human Body.** Cholesterol serves as a waterproofing for the epidermis, a lipid component of all plasma membranes, a key constituent of bile, and the precursor of several steroid hormones and one vitamin (vitamin D_3). Because cholesterol is so important, the goal of dietary restrictions is not eliminating it from the diet or from the circulating blood. The goal is to keep cholesterol levels within acceptable limits.

- **The Cholesterol Content of the Diet Is Not the Only Source of Circulating Cholesterol.** The human body can manufacture cholesterol from acetyl-CoA obtained through glycolysis or the beta-oxidation of other lipids. That cholesterol probably accounts for only about 20 percent of the cholesterol in the bloodstream; the rest results from metabolism of saturated fats in the diet. If the diet contains an abundance of saturated fats, cholesterol levels in the blood will rise because excess lipids are broken down to acetyl-CoA and used to synthesize cholesterol. Consequently, individuals trying to lower serum cholesterol levels by dietary control must also restrict other lipids—especially saturated fats.

- **Genetic Factors Affect Each Individual's Cholesterol Level.** If you reduce your dietary supply of cholesterol, your body will synthesize more to maintain "acceptable" concentrations in the blood. What an "acceptable" level is depends on your genetic makeup. Because individuals have different genes, their cholesterol levels can vary, even on similar diets. In virtually all instances, however, dietary restrictions can lower blood cholesterol significantly.

- **Cholesterol Levels Vary with Age and Physical Condition.** In general, as we age, cholesterol levels gradually climb. At age 19, three out of four males have fasting cholesterol levels (levels measured 8–12 hours after a meal) below 170 mg/dL. Cholesterol levels in females of this age are slightly higher, typically at or below 175 mg/dL. As age increases, the cholesterol values gradually climb; over age 70, the values are 230 mg/dL (males) and 250 mg/dL (females). Cholesterol levels are considered unhealthy if they are higher than those of 90 percent of the population in that age group. For males, this value ranges from 185 mg/dL at age 19 to 250 mg/dL at age 70. For females, the comparable values are 190 mg/dL and 275 mg/dL respectively.

What levels are considered optimal to reduce risk of atherosclerosis and coronary artery disease depend on the presence or absence of associated risk factors, including but not limited to hypertension, diabetes mellitus, personal and family history of heart disease, and smoking.

The recommendations and guidelines vary depending on the reference consulted. However, everyone should be screened for high blood cholesterol levels as they age, and men should be screened at a younger age than women.

When ordering a blood test for cholesterol, most physicians also request information about circulating triglycerides. In fasting individuals, triglycerides are usually present at levels of 40–150 mg/dL. (After a person has consumed a fatty meal, triglyceride levels may be temporarily elevated.)

When cholesterol levels are high, or when an individual has a family history of atherosclerosis or CAD, further tests and calculations may be performed. The HDL level is measured, and the LDL level is calculated as:

$$\text{LDL} = \text{Cholesterol} - \text{HDL} - \frac{\text{Triglycerides}}{5}$$

A high total cholesterol value linked to a high LDL level spells trouble. In effect, an excessive amount of cholesterol is being exported to peripheral tissues. Problems can also exist in individuals with high total cholesterol—or even normal total cholesterol—but low HDL levels (below 35 mg/dL). In such cases, excess cholesterol delivered to the tissues cannot easily be returned to the liver for excretion. In either event, the amount of cholesterol in peripheral tissues—and especially in arterial walls—is likely to increase. For years, LDL–HDL ratios were considered valid predictors of the risk of developing atherosclerosis. Risk-factor analysis and LDL levels are now thought to be more accurate indicators. For males with more than one risk factor, many clinicians recommend dietary changes and drug therapy if LDL levels exceed 130 mg/dL, or even 100 mg/dL, regardless of the total cholesterol or HDL levels.

25 DIGESTIVE

Lipid Synthesis

The synthesis of lipids is known as **lipogenesis** (lip-ō-JEN-e-sis). Glycerol is synthesized from *dihydroxyacetone phosphate*, one of the 3-carbon intermediate products shared by the pathways of glycolysis and gluconeogenesis (**Figure 25–9**). The synthesis of most types of lipids, including nonessential fatty acids and steroids, begins with acetyl-CoA. Lipogenesis can use almost any organic substrate, because lipids, amino acids, and carbohydrates can be converted to acetyl-CoA.

Fatty acid synthesis involves a reaction sequence quite distinct from that of beta-oxidation. Body cells cannot *build* every fatty acid they can break down. For example, **linoleic acid** and **linolenic acid**, both 18-carbon unsaturated fatty acids synthesized by plants, cannot be synthesized in the body. They are thus considered **essential fatty acids**, because they must be included in your diet. These fatty acids are also needed to synthesize prostaglandins and some of the phospholipids incorporated into plasma membranes throughout the body.

Lipid Transport and Distribution

Like glucose, lipids are needed throughout the body. All cells require lipids to maintain their plasma membranes, and important steroid hormones must reach target cells in many different tissues. Free fatty acids constitute a relatively small percentage of the total circulating lipids. Because most lipids are not soluble in water, special transport mechanisms carry them from one region of the body to another. Most lipids circulate through the bloodstream as lipoproteins.

Free Fatty Acids

Free fatty acids (FFAs) are lipids that can diffuse easily across plasma membranes. In the blood, free fatty acids are generally bound to albumin, the most abundant plasma protein. Sources of free fatty acids in the blood include the following:

- Fatty acids that are not used in the synthesis of triglycerides, but diffuse out of the intestinal epithelium and into the blood.

- Fatty acids that diffuse out of lipid stores (such as those in the liver and adipose tissue) when triglycerides are broken down.

Liver cells, cardiac muscle cells, skeletal muscle fibers, and many other body cells can metabolize free fatty acids, which are an important energy source during periods of starvation, when glucose supplies are limited.

Lipoproteins

Lipoproteins are lipid–protein complexes that contain large insoluble glycerides and cholesterol. A superficial coating of phospholipids and proteins makes the entire complex soluble.

Exposed proteins, which can bind to specific membrane receptors, determine which cells absorb the associated lipids.

Lipoproteins are usually classified into five major groups according to size and the relative proportions of lipid and protein:

1. *Chylomicrons.* Roughly 95 percent of the weight of a chylomicron consists of triglycerides. The largest lipoproteins, ranging in diameter from 0.03 to 0.5 μm, chylomicrons are produced by intestinal epithelial cells from the fats in food. ∞ p. 899 Chylomicrons carry absorbed lipids from the intestinal tract to the bloodstream. The liver is the primary source of all the other groups of lipoproteins, which shuttle lipids among various tissues.

2. *Very Low-Density Lipoproteins (VLDLs).* Very low-density lipoproteins contain triglycerides manufactured by the liver, plus small amounts of phospholipids and cholesterol. The primary function of VLDLs is to transport these triglycerides to peripheral tissues. The VLDLs range in diameter from 25 to 75 nm (0.025–0.075 μm).

3. *Intermediate-Density Lipoproteins (IDLs).* Intermediate-density lipoproteins are intermediate in size and lipid composition between VLDLs and low-density lipoproteins (LDLs). IDLs contain smaller amounts of triglycerides than do VLDLs, and relatively more phospholipids and cholesterol than do LDLs.

4. *Low-Density Lipoproteins (LDLs).* Low-density lipoproteins contain cholesterol, lesser amounts of phospholipids, and very few triglycerides. These lipoproteins, which are about 25 nm in diameter, deliver cholesterol to peripheral tissues. Because the cholesterol may wind up in arterial plaques, LDL cholesterol is often called "bad cholesterol."

5. *High-Density Lipoproteins (HDLs).* High-density lipoproteins, about 10 nm in diameter, have roughly equal amounts of lipid and protein. The lipids are largely cholesterol and phospholipids. The primary function of HDLs is to transport excess cholesterol from peripheral tissues back to the liver for storage or excretion in the bile. Because HDL cholesterol is returning from peripheral tissues and does not cause circulatory problems, it is called "good cholesterol." Actually, applying the terms *good* and *bad* to cholesterol can be misleading, for cholesterol metabolism is complex and variable. (For more details, see the Clinical Note "Dietary Fats and Cholesterol" on p. 943.)

Figure 25–9 depicts the relationships among these lipoproteins. Chylomicrons produced in the intestinal tract enter lymphatic capillaries and travel through the thoracic duct to reach the venous circulation and then the systemic arteries (**Figure 25–9a**). Chylomicrons are too large to diffuse across a capillary wall. However, when they contact the endothelial walls of capillaries in skeletal muscle, cardiac muscle, adipose tissue, and the liver, the enzyme **lipoprotein lipase** breaks down the complex

Figure 25–9 **Lipid Transport and Utilization.**

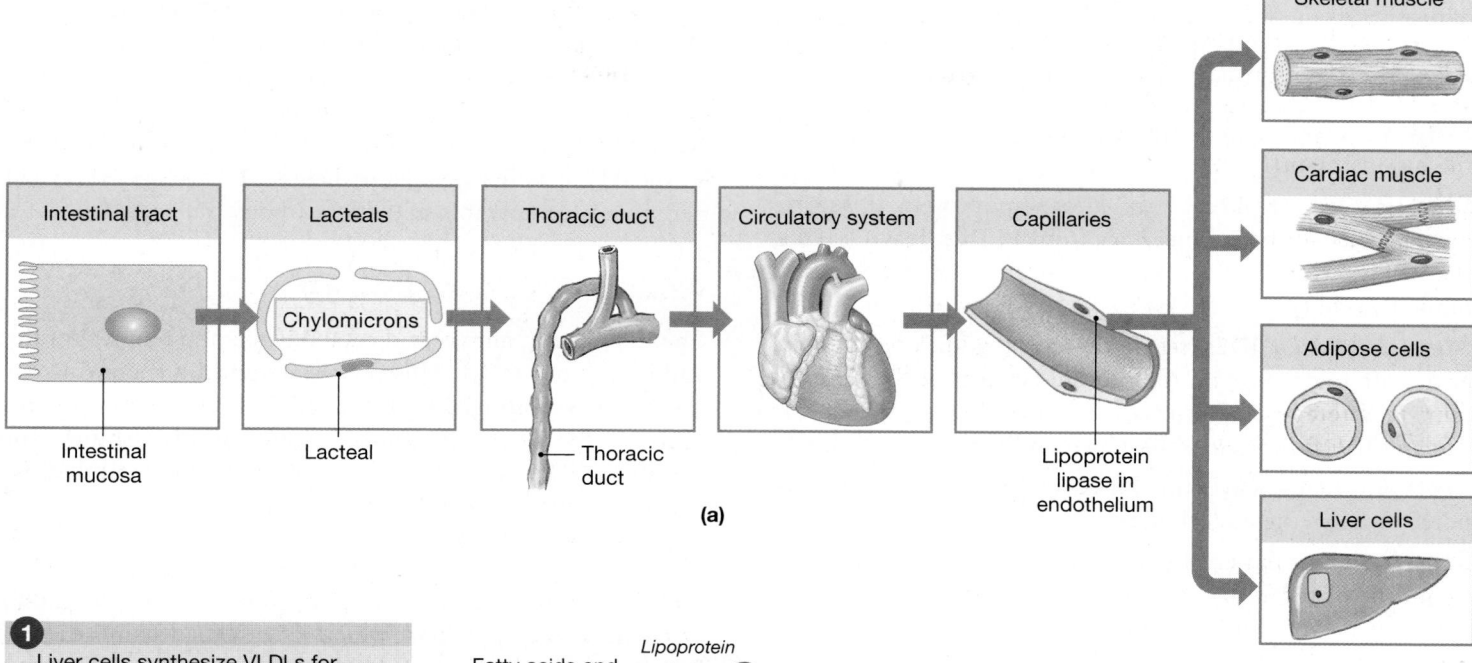

(a)

1. Liver cells synthesize VLDLs for discharge into the bloodstream.

2. In peripheral capillaries, lipoprotein lipase removes many of the triglycerides from VLDLs, leaving IDLs.

3. When IDLs reach the liver, additional triglycerides are removed and the protein content is altered. This process creates LDLs.

4. LDLs leave the bloodstream through capillary pores or cross the endothelium by vesicular transport.

5. Once in peripheral tissues, the LDLs are absorbed by means of receptor-mediated endocytosis.

6. The cholesterol not used diffuses out of the cell.

7. The cholesterol then reenters the bloodstream, where it is absorbed by HDLs and returned to the liver.

8. In the liver, the HDLs are absorbed and the cholesterol is extracted. Some of the recovered cholesterol is used in the synthesis of LDLs.

9. The HDLs stripped of their cholesterol are released into the bloodstream to travel to peripheral tissues and absorb additional cholesterol.

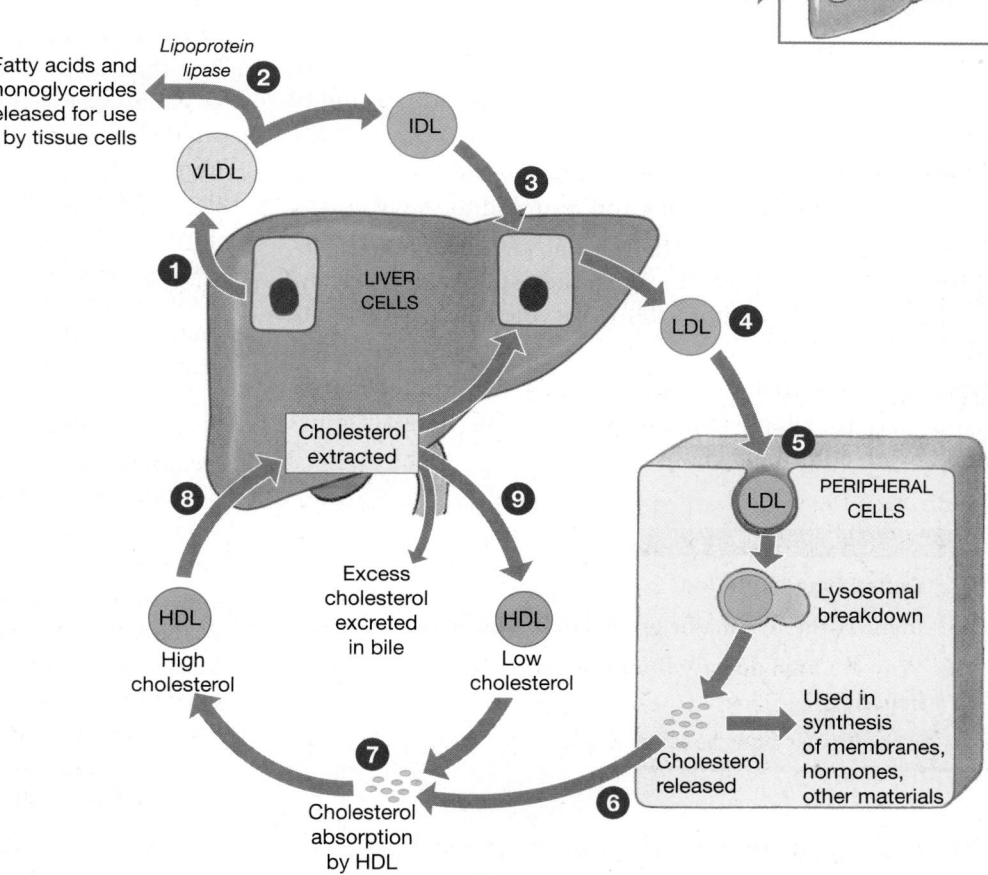

(b)

25 DIGESTIVE

lipids in them, releasing fatty acids and monoglycerides that can diffuse across the endothelium and into interstitial fluid.

The liver controls the distribution of other lipoproteins through a series of steps (**Figure 25–9b**):

Step 1 Liver cells synthesize VLDLs for discharge into the bloodstream.

Step 2 In peripheral capillaries, lipoprotein lipase removes many of the triglycerides from VLDLs, leaving IDLs; the triglycerides are broken down into fatty acids and monoglycerides.

Step 3 When IDLs reach the liver, additional triglycerides are removed, and the protein content of the lipoprotein is altered. This process creates LDLs, which are transported to peripheral tissues to deliver cholesterol.

Step 4 LDLs leave the bloodstream through capillary pores or cross the endothelium by vesicular transport.

Step 5 Once in peripheral tissues, the LDLs are absorbed by means of receptor-mediated endocytosis. ∞ p. 97 The amino acids and cholesterol then enter the cytoplasm.

Step 6 The cholesterol not used by the cell (in the synthesis of lipid membranes or other products) diffuses out of the cell.

Step 7 The cholesterol then reenters the bloodstream, where it is absorbed by HDLs and returned to the liver.

Step 8 In the liver, the HDLs are absorbed and their cholesterol is extracted. Some of the cholesterol that is recovered is used in the synthesis of LDLs; the rest is excreted in bile salts.

Step 9 The HDLs stripped of their cholesterol are released into the bloodstream to travel into peripheral tissues and absorb additional cholesterol.

CHECKPOINT

7. Define beta-oxidation.

8. Identify the five major groups of lipoproteins.

9. Why are high-density lipoproteins (HDLs) considered beneficial?

See the blue Answers tab at the end of the book.

25-4 Protein catabolism involves transamination and deamination, whereas protein synthesis involves amination and transamination

The body can synthesize 100,000 to 140,000 different proteins with various forms, functions, and structures. Yet, each protein contains some combination of the same 20 amino acids. Under normal conditions, cellular proteins are continuously recycled in the cytosol. Peptide bonds are broken, and the free amino acids are used in new proteins.

If other energy sources are inadequate, mitochondria can generate ATP by breaking down amino acids in the TCA cycle. Not all amino acids enter the cycle at the same point, however, so the ATP benefits vary. Nonetheless, the average ATP yield per gram is comparable to that of carbohydrate catabolism.

Amino Acid Catabolism

The first step in amino acid catabolism is the removal of the amino group ($-NH_2$). This process requires a coenzyme derivative of **vitamin B$_6$** (*pyridoxine*). The amino group is removed by *transamination* (tranz-am-i-NĀ-shun) or *deamination* (dē-am-i-NĀ-shun). We will consider other aspects of amino acid catabolism in a later section.

Transamination

Transamination attaches the amino group of an amino acid to a **keto acid** (**Figure 25–10a**), which resembles an amino acid except that the second carbon binds an oxygen atom rather than an amino group. Transamination converts the keto acid into an amino acid that can leave the mitochondrion and enter the cytosol, where it can be used for protein synthesis. In the process, the original amino acid becomes a keto acid that can be broken down in the TCA cycle. Cells in many different tissues perform transaminations, enabling cells to synthesize many of the amino acids needed for protein synthesis. Cells of the liver, skeletal muscles, heart, lung, kidney, and brain, which are particularly active in protein synthesis, perform many transaminations.

Deamination

Deamination (**Figure 25–10b**) prepares an amino acid for breakdown in the TCA cycle. Deamination is the removal of an amino group and a hydrogen atom in a reaction that generates an ammonium ion (NH_4^+). Ammonium ions are highly toxic, even in low concentrations. Liver cells, the primary sites of deamination, have enzymes that use ammonium ions to synthesize **urea**, a relatively harmless water-soluble compound excreted in urine. The **urea cycle** is the reaction sequence responsible for the production of urea (**Figure 25–10c**).

When glucose supplies are low and lipid reserves are inadequate, liver cells break down internal proteins and absorb additional amino acids from the blood. The amino acids are deaminated, and the carbon chains are broken down to provide ATP.

Proteins and ATP Production

Three factors make protein catabolism an impractical source of quick energy:

1. Proteins are more difficult to break apart than are complex carbohydrates or lipids.

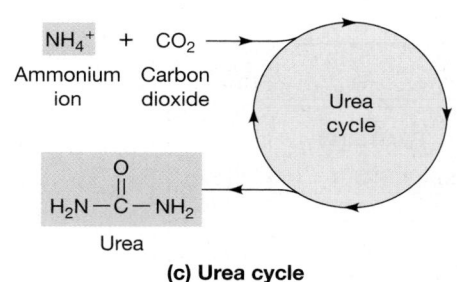

Glutamic acid Keto acid 1

(a) Transamination

Keto acid 2 Tyrosine

Transaminase

Glutamic acid

Deaminase

H_2O NAD **NADH**

Keto acid $+$ NH_4^+

Ammonium ion

(b) Deamination

NH_4^+ $+$ CO_2

Ammonium ion Carbon dioxide

Urea cycle

$H_2N-\overset{\overset{\displaystyle O}{\|}}{C}-NH_2$

Urea

(c) Urea cycle

Triglycerides Glycogen Proteins

Fatty acids Glucose **Amino acids**

Small carbon chains

ATP

TCA cycle Coenzymes Electron transport system

CO_2 H_2O

Figure 25–10 Amino Acid Catabolism. (a) In transamination, an enzyme removes the amino group (–NH_2) from one molecule and attaches it to a keto acid. **(b)** In deamination, an enzyme strips the amino group and a hydrogen atom from an amino acid and produces a keto acid and an ammonium ion. **(c)** The urea cycle takes two metabolic waste products—ammonium ions and carbon dioxide—and produces urea, a relatively harmless, soluble compound that is excreted in the urine.

2. One of the by-products, ammonium ions, is toxic to cells.

3. Proteins form the most important structural and functional components of any cell. Thus extensive protein catabolism threatens homeostasis at the cellular and systems levels.

Several inherited metabolic disorders result from an inability to produce specific enzymes involved in amino acid metabolism. Individuals with *phenylketonuria* (fen-il-kē-tō-NOO-rē-uh), or PKU, for example, cannot convert phenylalanine to tyrosine, due to a defect in the enzyme *phenylalanine hydroxylase*. This reaction is an essential step in the synthesis of norepinephrine, epinephrine, dopamine, and melanin. If PKU is not detected in infancy, central nervous system development is inhibited, and severe brain damage results. The condition is common enough that a warning is printed on the packaging of products, such as diet drinks, that contain phenylalanine.

Protein Synthesis

The basic mechanism for protein synthesis was discussed in Chapter 3 (**Figures 3–12** and **3–13**, pp. 85, 86–87). Your body can synthesize roughly half of the various amino acids needed to build proteins. There are 10 **essential amino acids**. Your body cannot synthesize eight of them (*isoleucine, leucine, lysine, threonine, tryptophan, phenylalanine, valine,* and *methionine*); the other two (*arginine* and *histidine*) can be synthesized, but in amounts that are insufficient for growing children. Because the body can make other amino acids on demand, they are called **nonessential amino acids**. Your body cells can readily synthesize the carbon frameworks of the nonessential amino acids, and a nitrogen group can be added by **amination**—the attachment of an amino group (**Figure 25–11**)—or by transamination.

Protein deficiency diseases develop when individuals do not consume adequate amounts of all essential amino acids. All amino acids must be available if protein synthesis is to occur. Every transfer RNA molecule must appear at the active ribosome in the proper sequence, bearing its individual amino acid. If that does not happen, the entire process comes to a halt. Regardless of its energy content, if the diet is deficient in essential amino acids, the individual will be malnourished to some degree. Examples of protein deficiency diseases include *marasmus* and *kwashiorkor*. More than 100 million children worldwide have symptoms of these disorders, although none of these conditions are common in the United States today.

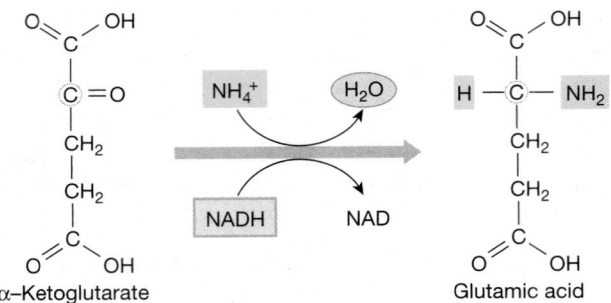

Figure 25–11 **Amination.** Amination attaches an amino group to a keto acid. This important step in the synthesis of nonessential amino acids attaches an amino group to a keto acid. Amino groups can also be attached through transamination *(Figure 25–10a)*.

Figure 25–12 summarizes the metabolic pathways for lipids, carbohydrates, and proteins. Although the diagram presents the reactions in a "typical" cell, no one cell can perform all the anabolic and catabolic operations and interconversions required by the body as a whole. As differentiation proceeds, each type of cell develops its own complement of enzymes that determines the cell's metabolic capabilities. In the presence of such cellular diversity, homeostasis can be preserved only when the metabolic activities of tissues, organs, and organ systems are coordinated.

CHECKPOINT

10. Define transamination.

11. Define deamination.

12. How would a diet that is deficient in pyridoxine (vitamin B_6) affect protein metabolism?

See the blue Answers tab at the end of the book.

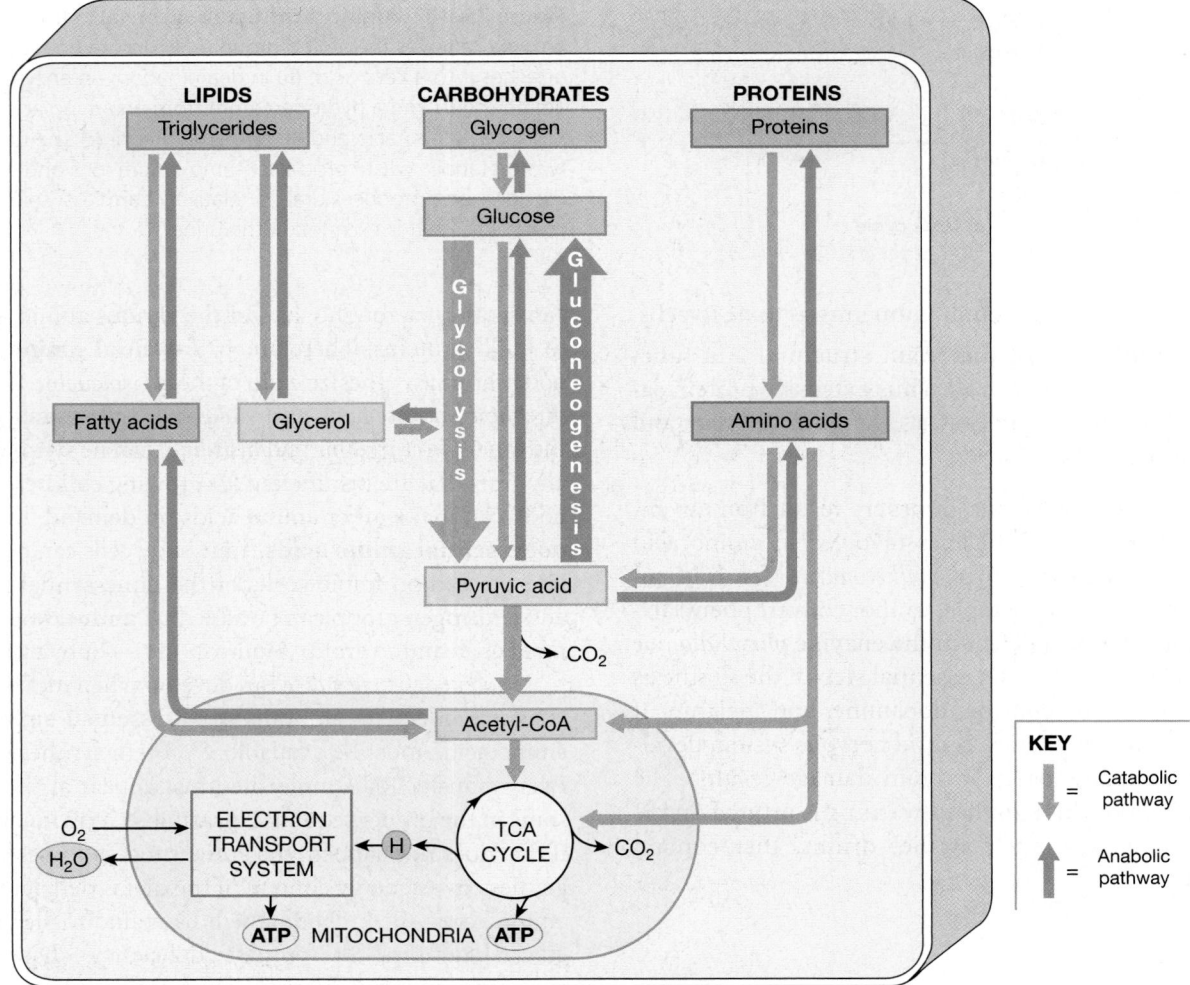

Figure 25–12 **A Summary of the Pathways of Catabolism and Anabolism.**

25-5 The body experiences two patterns of metabolic activity: the absorptive and postabsorptive states

The nutrient requirements of each tissue vary with the types and quantities of enzymes present in the cytoplasm of cells. From a metabolic standpoint, we can consider the body to have five distinctive components: the liver, adipose tissue, skeletal muscle, neural tissue, and other peripheral tissues:

1. *The Liver.* The liver is the focal point of metabolic regulation and control. Liver cells contain a great diversity of enzymes, so they can break down or synthesize most of the carbohydrates, lipids, and amino acids needed by other body cells. Liver cells have an extensive blood supply, so they are in an excellent position to monitor and adjust the nutrient composition of circulating blood. The liver also contains significant energy reserves in the form of glycogen deposits.

2. *Adipose Tissue.* Adipose tissue stores lipids, primarily as triglycerides. Adipocytes are located in many areas: in areolar tissue, in mesenteries, within red and yellow marrows, in the epicardium, and around the eyes and the kidneys.

3. *Skeletal Muscle.* Skeletal muscle accounts for almost half of a healthy individual's body weight, and skeletal muscle fibers maintain substantial glycogen reserves. In addition, if other nutrients are unavailable, their contractile proteins can be broken down and the amino acids used as an energy source.

4. *Neural Tissue.* Neural tissue has a high demand for energy, but the cells do not maintain reserves of carbohydrates, lipids, or proteins. Neurons must be provided with a reliable supply of glucose, because they are generally unable to metabolize other molecules. If blood glucose levels become too low, neural tissue in the central nervous system cannot continue to function, and the individual falls unconscious.

5. *Other Peripheral Tissues.* Other peripheral tissues do not maintain large metabolic reserves, but they are able to metabolize glucose, fatty acids, or other substrates. Their preferred source of energy varies according to instructions from the endocrine system.

The interrelationships among these five components can best be understood by considering events over a typical 24-hour period. During this time, the body experiences two broad patterns of metabolic activity: the *absorptive state* and the *postabsorptive state.*

The **absorptive state** is the period following a meal, when nutrient absorption is under way. After a typical meal, the absorptive state continues for about 4 hours. If you are fortunate enough to eat three meals a day, you spend 12 out of every 24 hours in the absorptive state. Insulin is the primary hormone of the absorptive state, although various other hormones stimulate amino acid uptake (growth hormone, or GH) and protein synthesis (GH, androgens, and estrogens) (Table 25–1).

The **postabsorptive state** is the period when nutrient absorption is not under way and your body must rely on internal energy reserves to continue meeting its energy demands. You spend roughly 12 hours each day in the postabsorptive state, although a person who is skipping meals can extend that time considerably. Metabolic activity in the postabsorptive state is

TABLE 25–1	Regulatory Hormones and Their Effects on Peripheral Metabolism	
Hormone	**Effect on General Peripheral Tissues**	**Selective Effects on Target Tissues**
ABSORPTIVE STATE		
Insulin	Increased glucose uptake and utilization	*Liver*: Glycogenesis *Adipose tissue*: Lipogenesis *Skeletal muscle*: Glycogenesis
Insulin and growth hormone	Increased amino acid uptake and protein synthesis	*Skeletal muscle*: Fatty acid catabolism
Androgens, estrogens	Increased amino acid use in protein synthesis	*Skeletal muscle*: Muscle hypertrophy (especially androgens)
POSTABSORPTIVE STATE		
Glucagon		*Liver*: Glycogenolysis
Epinephrine		*Liver*: Glycogenolysis *Adipose tissue*: Lipolysis
Glucocorticoids	Decreased use of glucose; increased reliance on ketone bodies and fatty acids	*Liver*: Glycogenolysis *Adipose tissue*: Lipolysis, gluconeogenesis *Skeletal muscle*: Glycogenolysis, protein breakdown, amino acid release
Growth hormone	Complements effects of glucocorticoids	Acts with glucocorticoids

25 DIGESTIVE

focused on the mobilization of energy reserves and the maintenance of normal blood glucose levels. These activities are coordinated by several hormones, including glucagon, epinephrine, glucocorticoids, and growth hormone (Table 25–1).

Tips&Tricks

The absorptive state is like harvest time: Food is being gathered and stored. The postabsorptive state corresponds to the time between harvests, when stored food is used for sustenance.

During the postabsorptive state, liver cells conserve glucose and break down lipids and amino acids. Both lipid catabolism and amino acid catabolism generate acetyl-CoA. As the concentration of acetyl-CoA rises, compounds called **ketone bodies** begin to form. There are three such compounds: (1) **acetoacetate** (as-e-tō-AS-e-tāt), (2) **acetone** (AS-e-tōn), and (3) **betahydroxybutyrate** (bā-ta-hī-droks-ē-BŪ-te-rāt). Liver cells do not catabolize any of the ketone bodies, and these compounds diffuse through the cytoplasm and into the general circulation. Cells in peripheral tissues then absorb the ketone bodies and reconvert them to acetyl-CoA for introduction into the TCA cycle.

The normal concentration of ketone bodies in the blood is about 30 mg/dL, and very few of these compounds appear in urine. During even a brief period of fasting, the increased production of ketone bodies results in *ketosis* (kē-TŌ-sis), a high concentration of ketone bodies in body fluids. A ketone body is an organic compound, produced by fatty acid metabolism, that dissociates in solution, releasing a hydrogen ion. As a result, the appearance of ketone bodies in the bloodstream—*ketonemia*—lowers plasma pH, which must be controlled by buffers. During prolonged starvation, ketone levels continue to rise. Eventually, buffering capacities are exceeded and a dangerous drop in pH occurs. This acidification of the blood by ketone bodies is called *ketoacidosis* (kē-tō-as-i-DŌ-sis). In severe ketoacidosis, the circulating concentration of ketone bodies can reach 200 mg/dL, and the pH may fall below 7.05. A pH that low can disrupt tissue activities and cause coma, cardiac arrhythmias, and death.

In summary, during the postabsorptive state, the liver acts to stabilize blood glucose concentrations, first through the breakdown of glycogen reserves and later by gluconeogenesis. Over the remainder of the postabsorptive state, the combination of lipid and amino acid catabolism provides the necessary ATP and generates large quantities of ketone bodies that diffuse into the bloodstream.

The changes in activity of the liver, adipose tissue, skeletal muscle, and other peripheral tissues ensure that the supply of glucose to the nervous system continues unaffected, despite daily or even weekly changes in nutrient availability. Only after a prolonged period of starvation will neural tissue begin to metabolize ketone bodies and lactic acid molecules, as well as glucose.

The A&P Top 100

#88 In the absorption state that follows a meal, cells absorb nutrients to be used for growth, maintenance, and energy reserves. Hours later, in the postabsorptive state, blood glucose levels are maintained by gluconeogenesis within the liver, but most cells begin conserving energy and shifting from glucose-based to lipid-based metabolism. If necessary, ketone bodies become the preferred energy source. This metabolic shift reserves the circulating glucose for use by neurons.

CHECKPOINT

13. What process in the liver increases after you have eaten a high-carbohydrate meal?

14. Why do blood levels of urea increase during the postabsorptive state?

15. If a cell accumulates more acetyl-CoA than it can metabolize in the TCA cycle, what products are likely to form?

See the blue Answers tab at the end of the book.

25-6 Adequate nutrition is necessary to prevent deficiency disorders and ensure physiological functioning

The postabsorptive state can be maintained for a considerable period, but homeostasis can be maintained indefinitely only if the digestive tract regularly absorbs enough fluids, organic substrates, minerals, and vitamins to keep pace with cellular demands. The absorption of nutrients from food is called **nutrition**.

The body's requirement for each nutrient varies from day to day and from person to person. *Nutritionists* attempt to analyze a diet in terms of its ability to meet the needs of a specific individual. A **balanced diet** contains all the ingredients needed to maintain homeostasis, including adequate substrates for energy generation, essential amino acids and fatty acids, minerals, and vitamins. In addition, the diet must include enough water to replace losses in urine, feces, and evaporation. A balanced diet prevents **malnutrition**, an unhealthy state resulting from inadequate or excessive absorption of one or more nutrients.

Food Groups and the MyPyramid Plan

One method of avoiding malnutrition is to consume a diet based on the updated food pyramid, called the MyPyramid Plan (**Figure 25–13** and Table 25-2). After twelve years, the United States Department of Agriculture updated its food pyramid to reflect the balance between food consumption and physical activity. The widths of the color-coded vertical food group bands running from the apex to the base indicate the proportions of food we should consume from each of the **five basic food groups**: grains (orange), vegetables (green), fruits (red), milk products (blue), and meat and beans (purple). Oils (yellow) are to be used sparingly in addition to the five basic food groups. Twelve different pyramids have been developed based on level of physical activity and general health; all aim to increase healthful eating habits. For more information, visit www.mypyramid.gov.

What is important concerning the food you consume is that you obtain nutrients in sufficient *quantity* (adequate to meet your energy needs) and *quality* (including essential amino acids, fatty acids, vitamins, and minerals). There is nothing magical about five groups; at various times since 1940, the U.S.

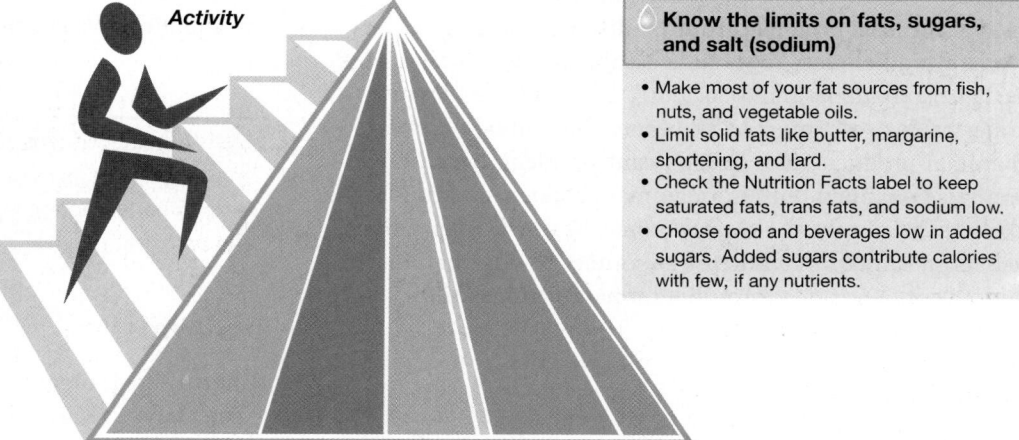

Find your balance between food and physical activity

- Be sure to stay within your daily calorie needs.
- Be physically active for at least 30 minutes most days of the week.
- About 60 minutes a day of physical activity may be needed to prevent weight gain.
- For sustaining weight loss, at least 60 to 90 minutes a day of physical activity may be required.
- Children and teenagers should be physically active for 60 minutes most days.

Activity

Know the limits on fats, sugars, and salt (sodium)

- Make most of your fat sources from fish, nuts, and vegetable oils.
- Limit solid fats like butter, margarine, shortening, and lard.
- Check the Nutrition Facts label to keep saturated fats, trans fats, and sodium low.
- Choose food and beverages low in added sugars. Added sugars contribute calories with few, if any nutrients.

GRAINS	VEGETABLES	FRUITS	OILS	MILK	MEAT & BEANS
Make half your grains whole	Vary your veggies	Focus on fruits		Get your calcium-rich foods	Go lean with proteins

Figure 25–13   The MyPyramid Plan. The proportion of each food group is indicated by the widths of the color-coded bands. Foods should be consumed in proportions based on both the food group and the individual's level of activity.

TABLE 25–2   Basic Food Groups of the 2005 Dietary Guidelines and Their General Effects on Health

Nutrient Group	Provides	Health Effects
Grains (recommended: at least half of the total eaten as whole grains)	Carbohydrates; vitamins E, thiamine, niacin, folate; calcium; phosphorus; iron; sodium; dietary fiber	Whole grains prevent rapid rise in blood glucose levels, and consequent rapid rise in insulin levels
Vegetables (recommended: especially dark-green and orange vegetables)	Carbohydrates; vitamins A, C, E, folate; dietary fiber; potassium	Reduce risk of cardiovascular disease; protect against colon cancer (folate) and prostate cancer (lycopene in tomatoes)
Fruits (recommended: a variety of fruit each day)	Carbohydrates; vitamins A, C, E, folate; dietary fiber; potassium	Reduce risk of cardiovascular disease; protect against colon cancer (folate)
Milk (recommended: low-fat or fat-free milk, yogurt, and cheese)	Complete proteins; fats; carbohydrates; calcium; potassium; magnesium; sodium; phosphorus; vitamins A, B_{12}, pantothenic acid, thiamine, riboflavin	Whole milk: High in calories, may cause weight gain; saturated fats correlated with heart disease
Meat and Beans (recommended: lean meats, fish, poultry, eggs, dry beans, nuts, legumes)	Complete proteins; fats; calcium; potassium; phosphorus; iron; zinc; vitamins E, thiamine, B_6	Fish and poultry lower risk of heart disease and colon cancer (compared to red meat). Consumption of up to one egg per day does not appear to increase incidence of heart disease; nuts and legumes improve blood cholesterol ratios, lower risk of heart disease and diabetes

government has advocated 11, 7, 4, or 6 food groups. The key is making intelligent choices about what you eat. Poor choices can lead to malnutrition even if all five groups are represented.

For example, consider the essential amino acids. The liver cannot synthesize any of these amino acids, so you must obtain them from your diet. Some members of the milk products and meat and beans groups—specifically, beef, fish, poultry, eggs, and milk—provide all the essential amino acids in sufficient quantities. They are said to contain **complete proteins**. Many plants also supply adequate *amounts* of protein but contain **incomplete proteins**, which are deficient in one or more of the essential amino acids. Vegetarians, who largely restrict themselves to the grains, vegetables, and fruits groups (with or without the milk products group), must become adept at varying their food choices to include combinations of ingredients that meet all their amino acid requirements. Even with a proper balance of amino acids, vegans, individuals who avoid all animal products, face a significant problem, because vitamin B_{12} is obtained only from animal products or from fortified cereals or tofu. (Although some health-food products, such as *Spirulina*, are marketed as sources of this vitamin, the B_{12} they contain is in a form that humans cannot utilize.)

Nitrogen Balance

A variety of important compounds in the body contain nitrogen atoms. These **N compounds** include:

- Amino acids, which are part of the framework of all proteins and protein derivatives, such as glycoproteins and lipoproteins.

- Purines and pyrimidines, the nitrogenous bases of RNA and DNA.

- *Creatine*, important in energy storage in muscle tissue (as creatine phosphate).

- *Porphyrins*, complex ring-shaped molecules that bind metal ions and are essential to the function of hemoglobin, myoglobin, and the cytochromes.

Despite the importance of nitrogen to these compounds, your body neither stores nitrogen nor maintains reserves of N compounds, as it does carbohydrates (glycogen) and lipids (triglycerides). Your body can synthesize the carbon chains of the N compounds, but you must obtain nitrogen atoms either by recycling N compounds already in the body or by absorbing nitrogen from your diet. You are in **nitrogen balance** when the amount of nitrogen you absorb from the diet balances the amount you lose in urine and feces. This is the normal condition, and it means that the rates of synthesis and breakdown of N compounds are equivalent.

Growing children, athletes, people recovering from an illness or injury, and pregnant or lactating women actively synthesize N compounds, so these individuals must absorb more nitrogen than they excrete. Such individuals are in a state of **positive nitrogen balance**. When excretion exceeds ingestion, a **negative nitrogen balance** exists. This is an extremely unsatisfactory situation: The body contains only about a kilogram of nitrogen tied up in N compounds, and a decrease of one-third can be fatal. Even when energy reserves are mobilized (as during starvation), carbohydrates and lipid reserves are broken down first and N compounds are conserved.

Like N compounds, minerals and vitamins are essential components of the diet. Your body cannot synthesize minerals, and your cells can generate only a small quantity of a very few vitamins. We consider minerals and vitamins next.

The A&P Top 100

#89 A balanced diet contains all the ingredients needed to maintain homeostasis, including adequate substrates for energy generation, essential amino acids and fatty acids, minerals, vitamins, and water.

Minerals

Minerals are inorganic ions released through the dissociation of electrolytes. Minerals are important for three reasons:

1. *Ions Such as Sodium and Chloride Determine the Osmotic Concentrations of Body Fluids.* Potassium is important in maintaining the osmotic concentration of the cytoplasm inside body cells.

2. *Ions in Various Combinations Play Major Roles in Important Physiological Processes.* As we have seen, these processes include the maintenance of transmembrane potentials, the construction and maintenance of the skeleton, muscle contraction, the generation of action potentials, the release of neurotransmitters, hormone production, blood clotting, the transport of respiratory gases, buffer systems, fluid absorption, and waste removal.

3. *Ions Are Essential Cofactors in a Variety of Enzymatic Reactions.* For example, calcium-dependent ATPase in skeletal muscle also requires the presence of magnesium ions, and another ATPase required for the conversion of glucose to pyruvic acid needs both potassium and magnesium ions. Carbonic anhydrase, important in CO_2 transport, buffering systems, and gastric acid secretion, requires the presence of zinc ions. Finally, each component of the electron transport system requires an iron atom, and the final cytochrome (a_3) of the ETS must bind a copper ion as well.

The major minerals and a summary of their functional roles are listed in Table 25–3. Your body contains significant reserves of several important minerals; these reserves help re-

TABLE 25–3 Minerals and Mineral Reserves*

Mineral	Significance	Total Body Content	Primary Route of Excretion	Recommended Daily Intake (RDA or AI)
BULK MINERALS				
Sodium	Major cation in body fluids; essential for normal membrane function	110 g, primarily in body fluids	Urine, sweat, feces	1.5 g
Potassium	Major cation in cytoplasm; essential for normal membrane function	140 g, primarily in cytoplasm	Urine	4.7 g
Chloride	Major anion in body fluids	89 g, primarily in body fluids	Urine, sweat	2.3 g
Calcium	Essential for normal muscle and neuron function and normal bone structure	1.36 kg, primarily in skeleton	Urine, feces	1000–1200 mg
Phosphorus	In high-energy compounds, nucleic acids, and bone matrix (as phosphate)	744 g, primarily in skeleton	Urine, feces	700 mg
Magnesium	Cofactor of enzymes, required for normal membrane functions	29 g (skeleton, 17 g; cytoplasm and body fluids, 12 g)	Urine	310–400 mg
TRACE MINERALS				
Iron	Component of hemoglobin, myoglobin, cytochromes	3.9 g (1.6 g stored as ferritin or hemosiderin)	Urine (traces)	8–18 mg
Zinc	Cofactor of enzyme systems, notably carbonic anhydrase	2 g	Urine, hair (traces)	8–11 mg
Copper	Required as cofactor for hemoglobin synthesis	127 mg	Urine, feces (traces)	900 μg
Manganese	Cofactor for some enzymes	11 mg	Feces, urine (traces)	1.8–2.3 mg
Cobalt	Cofactor for transaminations; mineral in vitamin B_{12} (cobalamin)	1.1 g	Feces, urine	0.0001 mg
Selenium	Antioxidant	Variable	Feces, urine	55 μg
Chromium	Cofactor for glucose metabolism	0.0006 mg	Feces, urine	20–35 μg

*For information on the effects of deficiencies and excesses, see Table 27–2, p. 1021. The recommended intakes are reported as either RDA (recommended dietary allowance for most people) or AI (adequate intake) when RDA data is lacking. Ranges indicate the differences between males and females and/or ages.

duce the effects of variations in the dietary supply. However, the reserves are often small, and chronic dietary reductions can lead to a variety of clinical problems. Alternatively, because storage capabilities are limited, a dietary excess of mineral ions can be equally dangerous.

Problems involving iron are particularly common. The body of a healthy man contains about 3.5 g of iron in the ionic form Fe^{2+}. Of that amount, 2.5 g is bound to the hemoglobin of circulating red blood cells, and the rest is stored in the liver and bone marrow. In women, the total body iron content averages 2.4 g, with roughly 1.9 g incorporated into red blood cells. Thus, a woman's iron reserves consist of only 0.5 g, half that of a typical man. If the diet contains inadequate amounts of iron, premenopausal women are therefore more likely to develop signs of iron deficiency than are men.

Vitamins

Vitamins are organic molecules that are assigned to either of two groups based on their chemical structure and characteristics: *fat-soluble vitamins* or *water-soluble vitamins*.

Fat-Soluble Vitamins

Vitamins A, D, E, and K are the **fat-soluble vitamins**. These vitamins are absorbed primarily from the digestive tract along with the lipid contents of micelles. However, when exposed to sunlight, your skin can synthesize small amounts of vitamin D, and intestinal bacteria produce some vitamin K.

The modes of action of these vitamins require further study. Vitamin A has long been recognized as a structural component of the visual pigment retinal, but its more general metabolic effects are not well understood. Vitamin D is ultimately converted to calcitriol, which binds to cytoplasmic receptors within the intestinal epithelium and promotes an increase in the rate of intestinal calcium and phosphorus absorption. Vitamin E is thought to stabilize intracellular membranes. Vitamin K is a necessary participant in a reaction essential to the synthesis of several proteins, including at least three of the clotting factors. Current information about the fat-soluble vitamins is summarized in Table 25–4.

Because fat-soluble vitamins dissolve in lipids, they normally diffuse into plasma membranes and other lipids in the body, including the lipid inclusions in the liver and adipose

TABLE 25–4	The Fat-Soluble Vitamins				
Vitamin	**Significance**	**Sources**	**Recommended Daily Intake (RDA or AI)**	**Effects of Deficiency**	**Effects of Excess**
A	Maintains epithelia; required for synthesis of visual pigments; supports immune system; promotes growth and bone remodeling	Leafy green and yellow vegetables	700–900 μg	Retarded growth, night blindness, deterioration of epithelial membranes	Liver damage, skin paling, CNS effects (nausea, anorexia)
D (steroids, including cholecalciferol, or D$_3$)	Required for normal bone growth, calcium and phosphorus absorption at gut and retention at kidneys	Synthesized in skin exposed to sunlight	5–15 μg*	Rickets, skeletal deterioration	Calcium deposits in many tissues, disrupting functions
E (tocopherols)	Prevents breakdown of vitamin A and fatty acids	Meat, milk, vegetables	15 mg	Anemia, other problems suspected	Nausea, stomach cramps, blurred vision, fatigue
K	Essential for liver synthesis of prothrombin and other clotting factors	Vegetables; production by intestinal bacteria	90–120 μg	Bleeding disorders	Liver dysfunction, jaundice

*Unless exposure to sunlight is inadequate for extended periods and alternative sources (fortified milk products) are unavailable.

tissue. Your body therefore contains a significant reserve of these vitamins, and normal metabolic operations can continue for several months after dietary sources have been cut off. For this reason, signs and symptoms of **avitaminosis** (ā-vī-tuh-min-Ō-sis), or **vitamin deficiency disease,** rarely result from a dietary insufficiency of fat-soluble vitamins. However, avitaminosis involving either fat-soluble or water-soluble vitamins can be caused by a variety of factors other than dietary deficiencies. An inability to absorb a vitamin from the digestive tract, inadequate storage, or excessive demand may each produce the same result.

Too much of a vitamin can also produce harmful effects. **Hypervitaminosis** (hī-per-vī-tuh-mi-NŌ-sis) occurs when dietary intake exceeds the body's abilities to store, utilize, or excrete a particular vitamin. This condition most commonly involves one of the fat-soluble vitamins, because the excess is retained and stored in body lipids.

Water-Soluble Vitamins

Most of the **water-soluble vitamins** (Table 25–5) are components of coenzymes. For example, NAD is derived from niacin, FAD from vitamin B$_2$ (riboflavin), and coenzyme A from vitamin B$_5$ (pantothenic acid).

Water-soluble vitamins are rapidly exchanged between the fluid compartments of the digestive tract and the circulating blood, and excessive amounts are readily excreted in urine. For this reason, hypervitaminosis involving water-soluble vitamins is relatively uncommon, except among individuals taking large doses of vitamin supplements. However, only vitamins B$_{12}$ and C are stored in significant quantities, so

insufficient intake of other water-soluble vitamins can lead to initial signs and symptoms of vitamin deficiency within a period of days to weeks.

The bacterial inhabitants of the intestines help prevent deficiency diseases by producing small amounts of five of the nine water-soluble vitamins, in addition to fat-soluble vitamin K. The intestinal epithelium can easily absorb all the water-soluble vitamins except B$_{12}$. The B$_{12}$ molecule is large, and as discussed in Chapter 24, it must be bound to *intrinsic factor* from the gastric mucosa before absorption can occur. ∞ p. 893

CLINICAL NOTE

Vitamins "If a little is good, a lot must be better" is a common—but dangerously incorrect—attitude about vitamins. When the dietary supply of fat-soluble vitamins is excessive, tissue lipids absorb the additional vitamins. Because these vitamins later diffuse back into the bloodstream, the signs and symptoms of hypervitaminosis, once apparent, are likely to persist. When absorbed in massive amounts (from ten to thousands of times the recommended daily allowance), fat-soluble vitamins can produce acute symptoms of *vitamin toxicity*. Vitamin A toxicity is the most common condition, afflicting some children whose parents are overanxious about proper nutrition and vitamins. A single enormous overdose can produce nausea, vomiting, headache, dizziness, lethargy, and even death. Chronic overdose can lead to hair loss, joint pain, hypertension, weight loss, and liver enlargement.

TABLE 25–5 The Water-Soluble Vitamins

Vitamin	Significance	Sources	Recommended Daily Intake (RDA or AI)	Effects of Deficiency	Effects of Excess
B$_1$ (thiamine)	Coenzyme in decarboxylations	Milk, meat, bread	1.1–1.2 mg	Muscle weakness, CNS and cardiovascular problems, including heart disease; called *beriberi*	Hypotension
B$_2$ (riboflavin)	Part of FMN and FAD	Milk, meat	1.1–1.3 mg	Epithelial and mucosal deterioration	Itching, tingling
Niacin (nicotinic acid)	Part of NAD	Meat, bread, potatoes	14–16 mg	CNS, GI, epithelial, and mucosal deterioration; called *pellagra*	Itching, burning; vasodilation; death after large dose
B$_5$ (pantothenic acid)	Part of acetyl-CoA	Milk, meat	5 mg	Retarded growth, CNS disturbances	None reported
B$_6$ (pyridoxine)	Coenzyme in amino acid and lipid metabolisms	Meat	1.3–1.7 mg	Retarded growth, anemia, convulsions, epithelial changes	CNS alterations, perhaps fatal
Folate (folic acid)	Coenzyme in amino acid and nucleic acid metabolisms	Vegetables, cereal, bread	400 μg	Retarded growth, anemia, gastrointestinal disorders, developmental abnormalities	Few noted, except at massive doses
B$_{12}$ (cobalamin)	Coenzyme in nucleic acid metabolism	Milk, meat	2.4 μg	Impaired RBC production, causing *pernicious anemia*	Polycythemia
Biotin	Coenzyme in decarboxylations	Eggs, meat, vegetables	30 μg	Fatigue, muscular pain, nausea, dermatitis	None reported
C (ascorbic acid)	Coenzyme; delivers hydrogen ions, antioxidant	Citrus fruits	75–90 mg Smokers: add 35 mg	Epithelial and mucosal deterioration; called *scurvy*	Kidney stones

Diet and Disease

Diet has a profound influence on a person's general health. We have already considered the effects of too many or too few nutrients, above-normal or below-normal concentrations of minerals, and hypervitaminosis or avitaminosis. More-subtle long-term problems can occur when the diet includes the wrong proportions or combinations of nutrients. The average diet in the United States contains too much sodium and too many calories, and lipids—particularly saturated fats—provide too great a proportion of those calories. This diet increases the incidence of obesity, heart disease, atherosclerosis, hypertension, and diabetes in the U.S. population.

CHECKPOINT

16. Identify the two classes of vitamins.

17. Would an athlete in intensive training try to maintain a positive or a negative nitrogen balance?

18. How would a decrease in the amount of bile salts in the bile affect the amount of vitamin A in the body?

See the blue Answers tab at the end of the book.

25-7 Metabolic rate is the average caloric expenditure, and thermoregulation involves balancing heat-producing and heat-losing mechanisms

In this section we consider topics that relate to energy gains and losses. First we consider aspects of energy intake and expenditure; then we turn to the topic of thermoregulation.

Energy Gains and Losses

To understand energy dynamics within the body, we first need to know something about the measurement of energy. When chemical bonds are broken, energy is released. Inside cells, a significant amount of energy may be used to synthesize ATP, but much of it is lost to the environment as heat. The process of *calorimetry* (kal-ō-RIM-e-trē) measures the total amount of energy released when the bonds of organic molecules are broken. The unit of measurement is the **calorie** (KAL-ō-rē) (cal), defined as the amount of energy required to raise the temperature of 1 g of water 1 degree Celsius. One gram of water is not a very practical measure when you are interested in the

Alcohol—A Risky Diversion

Alcohol production and sales are big business throughout the world. Beer commercials on television, billboards advertising various brands of liquor, and TV or movie characters enjoying a drink—all demonstrate the prominence of alcohol in many societies. Many people are unaware of the medical consequences of this cultural fondness for alcohol. Problems with alcohol are usually divided into those stemming from alcohol abuse and those involving alcoholism. The boundary between these conditions is hazy. *Alcohol abuse* is the general term for overuse and the resulting behavioral and physical effects of overindulgence. *Alcoholism* is chronic alcohol abuse accompanied by the physiological changes associated with addiction to other CNS-active drugs. Alcoholism has received the most attention in recent years, although alcohol abuse—especially when combined with driving an automobile—is also in the limelight.

Consider these statistics:

- Alcoholism affects more than 10 million people in the United States alone. The lifetime risk of developing alcoholism for those who drink alcohol is estimated at 10 percent.

- Alcoholism is probably society's most expensive health problem, with an annual estimated direct cost of more than $136 billion. Indirect costs, in terms of damage to automobiles, property, and innocent accident victims, are unknown.

- An estimated 25–40 percent of U.S. hospital patients are undergoing treatment related to alcohol consumption. Approximately 200,000 deaths occur annually from alcohol-related medical conditions. Some major clinical conditions are caused almost entirely by alcohol consumption. For example, alcohol is responsible for 60–90 percent of all liver disease in the United States.

- Alcohol affects all physiological systems. Major clinical symptoms of alcoholism include (1) disorientation and confusion (nervous system); (2) ulcers, diarrhea, and cirrhosis (digestive system); (3) cardiac arrhythmias, cardiomyopathy, and anemia (cardiovascular system); (4) depressed sexual drive and testosterone levels (reproductive system); and (5) itching and angiomas (integumentary system).

- The toll on newborn infants has risen steadily since the 1960s as the number of female drinkers has increased. Women consuming 1 ounce of alcohol per day during pregnancy have a higher rate of spontaneous abortion and produce children with lower birth weights than do women who consume no alcohol. Women who drink heavily may bear children with *fetal alcohol syndrome (FAS)*, a condition marked by characteristic facial abnormalities, a small head, slow growth, and mental retardation.

- Alcohol abuse is considerably more widespread than alcoholism. Although the medical effects are less well documented, they are clearly significant.

Several factors interact to produce alcoholism. The primary risk factors are gender (males are more likely to become alcoholics than are females) and a family history of alcoholism. There does appear to be a genetic component: A gene on chromosome 11 has been implicated in some inherited forms of alcoholism. The relative importance of genes versus social environment has been difficult to assess. It is likely that alcohol abuse and alcoholism result from a variety of factors. Treatment may consist of counseling and behavior modification. To be successful, treatment must include total avoidance of alcohol. Support groups, such as Alcoholics Anonymous (AA), can be very helpful in providing a social framework for abstinence. The drug *disulfiram (Antabuse)*—which sensitizes the individual to alcohol such that a drink produces intense nausea—has not proved to be as effective a deterrent as originally anticipated. Clinical tests indicated that it could increase the time between drinks but could not prevent drinking altogether.

Another drug, naltrexone (ReVia, Depade, and injectable Vivitrol), has shown promise in reducing the frequency and severity of drinking relapses in alcoholics. Naltrexone, an opioid receptor antagonist, blocks the part of the brain that elicits pleasurable feelings after the consumption of alcohol. The result is a decreased craving for alcohol, enabling individuals to stop drinking more readily. Unlike Antabuse, there is no associated nausea.

metabolic operations that keep a 70-kg human alive, so the **kilocalorie** (KIL-ō-kal-o-rē) (kcal), or **Calorie** (with a capital C), also known as "large calorie," is used instead. One Calorie is the amount of energy needed to raise the temperature of 1 *kilogram* of water 1 degree Celsius. Calorie-counting guides that indicate the caloric value of various foods list Calories, not calories.

The Energy Content of Food

In cells, organic molecules are oxidized to carbon dioxide and water. Oxidation also occurs when something burns, and this process can be experimentally controlled. A known amount of food is placed in a chamber called a **calorimeter** (kal-ō-RIM-e-ter), which is filled with oxygen and surrounded by a known volume of water. Once the food is inside, the chamber is sealed and the contents are electrically ignited. When the material has completely oxidized and only ash remains in the chamber, the number of Calories released can be determined by comparing the water temperatures before and after the test.

The energy potential of food is usually expressed in Calories per gram (Cal/g). The catabolism of lipids entails the release of a considerable amount of energy, roughly 9.46 Cal/g. The catabolism of carbohydrates or proteins is not as productive, because many of the carbon and hydrogen atoms are already bound to oxygen. Their average yields are comparable: 4.18 Cal/g for carbohydrates and 4.32 Cal/g for proteins. Most foods are mixtures of fats, proteins, and carbohydrates, so the calculated values listed in a "calorie counter" vary.

Energy Expenditure: Metabolic Rate

Clinicians can examine your metabolic state and determine how many Calories you are utilizing. The result can be expressed as Calories per hour, Calories per day, or Calories per unit of body weight per day. What is actually measured is the sum of all the varied anabolic and catabolic processes occurring in your body—your **metabolic rate** at that time. Metabolic rate changes according to the activity under way; for instance, measurements taken while one is sprinting are quite different from those taken while one is sleeping. In an attempt to reduce the variations, clinicians standardize the testing conditions so as to determine the **basal metabolic rate (BMR)**. Ideally, the BMR is the minimum resting energy expenditure of an awake, alert person.

A direct method of determining the BMR involves monitoring respiratory activity, because in resting individuals energy utilization is proportional to oxygen consumption. If we assume that average amounts of carbohydrates, lipids, and proteins are being catabolized, 4.825 Calories are expended per liter of oxygen consumed.

An average individual has a BMR of 70 Cal per hour, or about 1680 Cal per day. Although the test conditions are standardized, many uncontrollable factors, including age, gender, physical condition, body weight, and genetic differences, can influence the BMR.

Because the BMR is technically difficult to measure, and because circulating thyroid hormone levels have a profound effect on the BMR, clinicians usually monitor the concentration of thyroid hormones rather than the actual metabolic rate. The results are then compared with normal values to obtain an index of metabolic activity. One such test, the T_4 **assay**, measures the amount of thyroxine in the blood.

Daily energy expenditures for a given individual vary widely with activity. For example, a person leading a sedentary life may have near-basal energy demands, but a single hour of swimming can increase the daily caloric requirements by 500 Cal or more. If your daily energy intake exceeds your total energy demands, you will store the excess energy, primarily as triglycerides in adipose tissue. If your daily caloric expenditures exceed your dietary supply, the result is a net reduction in your body's energy reserves and a corresponding loss in weight. This relationship accounts for the significance of both Calorie counting and exercise in a weight-control program.

The control of appetite is poorly understood. Stretch receptors along the digestive tract, especially in the stomach, play a role, but other factors are probably more significant. Social factors, psychological pressures, and dietary habits are important. Evidence also indicates that complex hormonal stimuli interact to affect appetite. For example, the hormones *cholecystokinin (CCK)* and *adrenocorticotropic hormone (ACTH)* suppress the appetite. The hormone *leptin*, released by adipose tissues, also plays a role. During the absorptive state, adipose tissues release leptin into the bloodstream as they synthesize triglycerides. When leptin binds to CNS neurons that function in emotion and the control of appetite, the result is a sense of satiation and the suppression of appetite. *Ghrelin* (GREL-in), a hormone secreted from the gastric mucosa, stimulates appetite. Ghrelin levels decline when the stomach is full, and increase during the fasting state. Blood ghrelin levels tend to be low in obese individuals, suggesting that it plays a role in obesity. Other effects of ghrelin are currently being studied.

Thermoregulation

The BMR estimates the rate of energy use by the body. The energy not captured and harnessed by cells is released as heat and serves an important homeostatic purpose. Humans are subject to vast changes in environmental temperatures, but our complex biochemical systems have a major limitation: Enzymes operate over only a relatively narrow temperature range. Accordingly, our bodies have anatomical and physiological mechanisms that keep body temperatures within acceptable limits, regardless of environmental conditions. This homeostatic process is called **thermoregulation**. Failure to control body temperature can result in a series of physiological changes. For example, a body temperature below 36°C (97°F) or above 40°C (104°F) can cause disorientation, and a temperature above 42°C (108°F) can cause convulsions and permanent cell damage.

We continuously produce heat as a by-product of metabolism. When energy use increases due to physical activity, or when our cells are more active metabolically (as they are during the absorptive state), additional heat is generated. The heat produced by biochemical reactions is retained by water, which accounts for nearly two-thirds of body weight. Water is an excellent conductor of heat, so the heat produced in one region of the body is rapidly distributed by diffusion, as well as through the bloodstream. If body temperature is to remain constant, that heat must be lost to the environment at the same rate it is generated. When environmental conditions rise above or fall below "ideal," the body must control the gains or losses to maintain homeostasis.

Mechanisms of Heat Transfer

Heat exchange with the environment involves four basic processes: (1) *radiation*, (2) *conduction*, (3) *convection*, and (4) *evaporation.*

Warm objects lose heat energy as infrared **radiation.** When you feel the heat from the sun, you are experiencing radiant heat. Your body loses heat the same way, but in proportionately smaller amounts. More than 50 percent of the heat you lose indoors is attributable to radiation; the exact amount varies with both body temperature and skin temperature.

Conduction is the direct transfer of energy through physical contact. When you arrive in an air-conditioned classroom and sit on a cold plastic chair, you are immediately aware of this process. Conduction is generally not an effective mechanism for gaining or losing heat. We cannot estimate the value of its contribution, because it varies with the temperature of the object and with the amount of surface area involved. When you are lying on cool sand in the shade, conductive losses can be considerable; when you are standing, conductive losses are negligible.

Convection is the result of conductive heat loss to the air that overlies the surface of the body. Warm air rises, because it is lighter than cool air. As your body conducts heat to the air next to your skin, that air warms and rises, moving away from the surface of the skin. Cooler air replaces it, and as this air in turn becomes warmed, the pattern repeats. Convection accounts for roughly 15 percent of the body's heat loss indoors but is insignificant as a mechanism of heat gain.

When water evaporates, it changes from a liquid to a vapor. **Evaporation** absorbs energy—roughly 0.58 Cal per gram of water evaporated—and cools the surface where evaporation occurs. The rate of evaporation occurring at your skin is highly variable. Each hour, 20–25 mL of water crosses epithelia and evaporates from the alveolar surfaces and the surface of the skin. This insensible water loss remains relatively constant; at rest, it accounts for roughly 20 percent of your body's average indoor heat loss. The sweat glands responsible for sensible perspiration have a tremendous scope of activity, ranging from virtual inactivity to secretory rates of 2–4 liters (2.1–4.2 quarts) per hour.

The Regulation of Heat Gain and Heat Loss

Heat loss and heat gain involve the activities of many systems. Those activities are coordinated by the **heat-loss center** and **heat-gain center**, respectively, in the preoptic area of the anterior hypothalamus. ∞ p. 478 These centers modify the activities of other hypothalamic nuclei. The overall effect is to control temperature by influencing two events: the rate of heat production and the rate of heat loss to the environment. These events may be further supported by behavioral modifications.

Mechanisms for Increasing Heat Loss. When the temperature at the preoptic nucleus exceeds its set point, the heat-loss center is stimulated, producing three major effects:

1. *The Inhibition of the Vasomotor Center Causes Peripheral Vasodilation, and Warm Blood Flows to the Surface of the Body.* The skin takes on a reddish color, skin temperatures rise, and radiational and convective losses increase.

2. *As Blood Flow to the Skin Increases, Sweat Glands Are Stimulated to Increase Their Secretory Output.* The perspiration flows across the body surface, and evaporative heat losses accelerate. Maximal secretion, if completely evaporated, would remove 2320 Cal per hour.

3. *The Respiratory Centers Are Stimulated, and the Depth of Respiration Increases.* Often, the individual begins respiring through an open mouth rather than through the nasal passageways, increasing evaporative heat losses through the lungs.

Mechanisms for Promoting Heat Gain. The function of the heat-gain center of the brain is to prevent **hypothermia** (hī-pō-THER-mē-uh), or below-normal body temperature. When the temperature at the preoptic nucleus drops below acceptable levels, the heat-loss center is inhibited and the heat-gain center is activated.

Heat Conservation. The sympathetic vasomotor center decreases blood flow to the dermis, thereby reducing losses by radiation, convection, and conduction. The skin cools, and with blood flow restricted, it may take on a bluish or pale color. The epithelial cells are not damaged, because they can tolerate extended periods at temperatures as low as 25°C (77°F) or as high as 49°C (120°F).

In addition, blood returning from the limbs is shunted into a network of deep veins. ∞ p. 758 Under warm conditions, blood flows in a superficial venous network (**Figure 25–14a**). In cold conditions, blood is diverted to a network of veins that lie deep to an insulating layer of subcutaneous fat (**Figure 25–14b**). This venous network wraps around the deep arteries, and heat is conducted from the warm blood flowing outward to the limbs to the cooler blood returning from the periphery (**Figure 25–14c**). This arrangement traps the heat close to the body core and restricts heat loss. Such exchange between fluids that are moving in opposite directions is called *countercurrent exchange.* (We will return to this topic in Chapter 26.)

Heat Generation. The mechanisms for generating heat can be divided into two broad categories: shivering thermogenesis and nonshivering thermogenesis. In **shivering thermogenesis** (ther-mō-JEN-e-sis), a gradual increase in muscle tone increases the energy consumption of skeletal muscle tissue throughout your body, and the more energy consumed, the

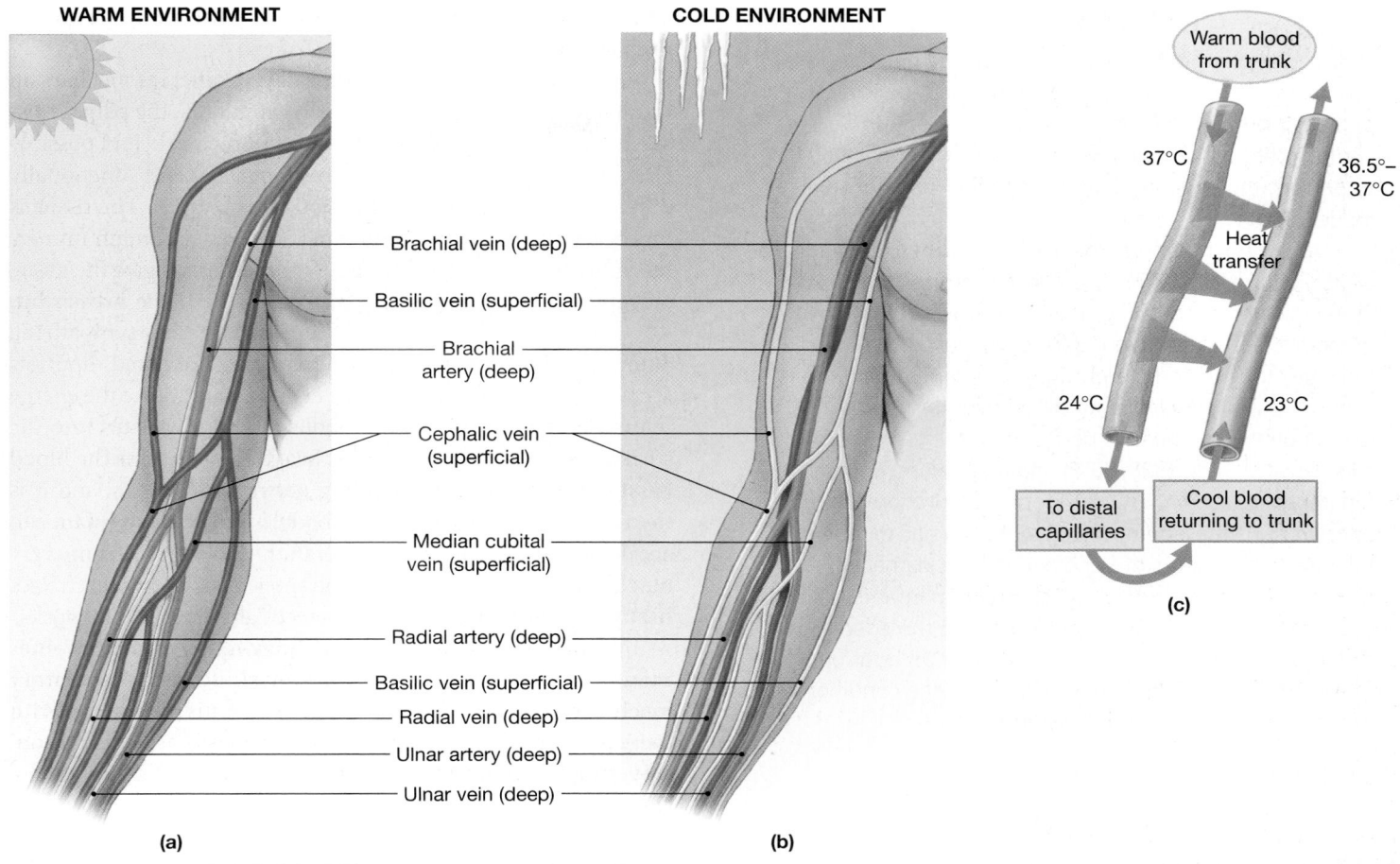

Figure 25–14 Vascular Adaptations for Heat Loss and Conservation. (a) Circulation through the blood vessels of the forearm in a warm environment. Blood enters the limb in a deep artery and returns to the trunk in a network of superficial veins that radiate heat into the environment through the overlying skin. **(b)** Circulation through the blood vessels of the forearm in a cold environment. Blood now returns to the trunk via a network of deep veins that flow around the artery. The amount of heat loss is reduced, as indicated in part (c). **(c)** Countercurrent heat exchange occurs as heat radiates from the warm arterial blood into the cooler venous blood flowing in the opposite direction. By the time the arterial blood reaches distal capillaries, where most heat loss to the environment occurs, it is already 13°C cooler than it was when it left the trunk. This mechanism conserves body heat by trapping heat near the trunk and thereby reducing the rate of heat loss.

more heat is produced. Both agonists and antagonists are involved, and the degree of stimulation varies with the demand.

If the heat-gain center is extremely active, muscle tone increases to the point at which stretch receptor stimulation will produce brief, oscillatory contractions of antagonistic muscles. In other words, you begin to **shiver**. Shivering increases the workload of the muscles and further elevates oxygen and energy consumption. The heat that is produced warms the deep vessels, to which blood has been shunted by the sympathetic vasomotor center. Shivering can elevate body temperature quite effectively, increasing the rate of heat generation by as much as 400 percent.

Nonshivering thermogenesis involves the release of hormones that increase the metabolic activity of all tissues:

- The heat-gain center stimulates the suprarenal medullae via the sympathetic division of the autonomic nervous

system, and epinephrine is released. Epinephrine increases the rates of glycogenolysis in liver and skeletal muscle, and the metabolic rate of most tissues. The effects are immediate.

- The preoptic nucleus regulates the production of thyrotropin-releasing hormone (TRH) by the hypothalamus. In children, when body temperatures are below normal, additional TRH is released, stimulating the release of thyroid-stimulating hormone (TSH) by the adenohypophysis. The thyroid gland responds to this release of TSH by increasing the rate at which thyroxine is released into the blood. Thyroxine increases not only the rate of carbohydrate catabolism, but also the rate of catabolism of all other nutrients. These effects develop gradually, over a period of days to weeks.

CLINICAL NOTE

Induced Hypothermia Hypothermia may be intentionally produced during surgery to reduce the metabolic rate of a particular organ or of the patient's entire body. In controlled hypothermia, the individual is first anesthetized to prevent the shivering that would otherwise fight the induction of hypothermia.

During open-heart surgery, the body is typically cooled to 25°–32°C (79°–89°F). This cooling reduces the metabolic demands of the body, which will be receiving blood from an external pump or oxygenator. The heart must be stopped completely during the operation, and it cannot be well supplied with blood over this period. Consequently, the heart is perfused with an *arresting solution* at 0°–4°C (32°–39°F) and maintained at a temperature below 15°C (60°F) for the duration of the operation. At these temperatures, the cardiac muscle can tolerate several hours of ischemia without damage.

When cardiac surgery is performed on infants, a deep hypothermia may be produced by cooling the entire body to temperatures as low as 11°C (52°F) for an hour or more. In effect, this procedure duplicates the conditions experienced by accidental drowning victims.

Sources of Individual Variation in Thermoregulation

The timing of thermoregulatory responses differs from individual to individual. A person may undergo **acclimatization** (a-klī-ma-ti-ZĀ-shun)—a physiological adjustment to a particular environment over time. For example, natives of Tierra del Fuego once lived naked in the snow, but Hawaii residents often unpack their sweaters when the temperature drops below 22°C (72°F).

Another interesting source of variation is body size. Although heat *production* occurs within the mass of the body, heat *loss* occurs across a body surface. As an object (or person) gets larger, its surface area increases at a much slower rate than does its total volume. This relationship affects thermoregulation, because heat generated by the "volume" (that is, by internal tissues) is lost at the body surface. Hence, small individuals lose heat more readily than do large individuals.

Thermoregulatory Problems of Infants.

During embryonic development, the maternal surroundings are at normal body temperature. At birth, the infant's temperature-regulating mechanisms are not fully functional; also, infants lose heat quickly as a result of their small size. Consequently, newborns must be dried promptly and kept bundled up; for those born prematurely, a thermally-regulated incubator is required. Infants' body temperatures are also less stable than those of adults. Their metabolic rates decline while they sleep and then rise after they awaken.

Infants cannot shiver, but they have a different mechanism for raising body temperature rapidly. In infants, the adipose tissue between the shoulder blades, around the neck, and possibly elsewhere in the upper body, is histologically and functionally different from most of the adipose tissue in adults. The tissue is highly vascularized, and individual adipocytes contain numerous mitochondria. Together, these characteristics give the tissue a deep, rich color that is responsible for the name **brown fat**. ∞ p. 127 Individual adipocytes are innervated by sympathetic autonomic fibers. When these nerves are stimulated, lipolysis accelerates in the adipocytes. The cells do not capture the energy released through fatty acid catabolism, and it radiates into the surrounding tissues as heat. This heat quickly warms the blood passing through the surrounding network of vessels, and it is then distributed throughout the body. In this way, an infant can accelerate metabolic heat generation by 100 percent very quickly, whereas nonshivering thermogenesis in an adult raises heat production by only 10–15 percent after a period of weeks.

With increasing age and size, body temperature becomes more stable, so the importance of this thermoregulatory mechanism declines. Adults have little, if any, brown fat; with increased body size, skeletal muscle mass, and insulation, shivering thermogenesis is significantly more effective in elevating body temperature.

Thermoregulatory Variations among Adults.

Adults of a given body weight may differ in their thermal responses if their weight is distributed differently. Which tissues account for their weight is also a factor. Adipose tissue is an excellent insulator, conducting heat at only about one-third the rate of other tissues. As a result, individuals with a more substantial layer of subcutaneous fat may not begin to shiver until long after their thinner companions.

Two otherwise similar individuals may also differ in their response to temperature changes because their hypothalamic "thermostats" are at different settings. We experience daily oscillations in body temperature, with temperatures falling 1°–2°C (1.8°–3.6°F) at night and peaking during the day or early evening. The ovulatory cycle also causes temperature fluctuations, as you will see in Chapter 28.

Individuals vary in terms of the timing of their maximum temperature setting; some have a series of peaks, with an afternoon low. The origin of these patterns is unclear. It is not the result of daily activity regimens: The temperatures of people who work at night still peak during the same range of times as the rest of the population.

Fevers

Any elevated body temperature is called **pyrexia** (pī-REK-sē-uh). Pyrexia is usually temporary. A **fever** is a body tem-

CLINICAL NOTE

Thermoregulatory Disorders

Heat exhaustion and heat stroke are malfunctions of thermoregulatory mechanisms. In **heat exhaustion**, also known as *heat prostration*, the individual has difficulty maintaining blood volume. The heat-loss center stimulates sweat glands, whose secretions moisten the surface of the skin to provide evaporative cooling. As fluid losses mount, blood volume decreases. The resulting decline in blood pressure is not countered by peripheral vasoconstriction, because the heat-loss center is actively stimulating peripheral vasodilation. As blood flow to the brain declines, headache, nausea, and eventual collapse follow. Treatment is straightforward: Provide fluids, salts, and a cooler environment.

Heat stroke is more serious and can follow an untreated case of heat exhaustion. Predisposing factors include any preexisting condition that affects peripheral circulation, such as heart disease or diabetes. The thermoregulatory center ceases to function, the sweat glands are inactive, and the skin becomes hot and dry. Unless the situation is recognized in time, body temperature may climb to 41°–45°C (106°–113°F). Temperatures in this range quickly disrupt a variety of vital physiological systems and destroy brain, liver, skeletal muscle, and kidney cells. Effective treatment involves lowering the body temperature as rapidly as possible.

If body temperature drops significantly below normal levels, thermoregulatory mechanisms lose sensitivity and effectiveness. Cardiac output and respiratory rate decrease, and if the core temperature falls below 28°C (82°F), cardiac arrest is likely. The individual then has no heartbeat, no respiratory rate, and no response to external stimuli—even painful ones. The body temperature continues to decline, and the skin turns blue or pale and cold.

At this point, people commonly assume that the individual has died. But because metabolic activities have decreased systemwide, the victim may still be saved, even after several hours have elapsed. Treatment consists of cardiopulmonary support and gradual rewarming, both external and internal. The skin can be warmed up to 45°C (113°F) without damage; warm baths or blankets can be used. One effective method of raising internal temperatures involves the introduction of warm saline solution into the peritoneal cavity.

Hypothermia is a significant risk for those engaged in water sports, and it may complicate the treatment of a drowning victim. Water absorbs heat roughly 27 times as fast as air does, and the body's heat-gain mechanisms are unable to keep pace over long periods or when faced with a large temperature gradient. But hypothermia in cold water does have a positive side: On several occasions, small children who have drowned in cold water have been successfully revived after periods of up to four hours. Children lose body heat quickly, and their systems soon stop functioning as their body temperature declines. This rapid drop in temperature prevents the oxygen starvation and tissue damage that would otherwise occur when breathing ceases.

Resuscitation is not attempted if the individual has actually frozen. Water expands roughly 7 percent during ordinary freezing, and plasma membranes throughout the body are destroyed in the process. Very small organisms can be frozen and subsequently thawed without ill effects, because their surface-to-volume ratio is enormous, and the freezing process occurs so rapidly that ice crystals never form.

---

perature maintained at greater than 37.2°C (99°F). We discussed fevers when we examined nonspecific defenses. p. 796 Fevers occur for a variety of reasons, not all of them pathological. In young children, transient fevers with no ill effects can result from exercise in warm weather. Similar exercise-related elevations were rarely encountered in adults until running marathons became popular. Temperatures from 39° to 41°C (103° to 109°F) may result. For this reason, competitions are usually held when the air temperature is below 28°C (82°F).

Fevers can also result from the following factors:

- Abnormalities affecting the entire thermoregulatory mechanism, such as heat exhaustion or heat stroke.

- Clinical problems that restrict blood flow, such as congestive heart failure.

- Conditions that impair sweat gland activity, such as drug reactions and some skin conditions.

- The resetting of the hypothalamic "thermostat" by circulating *pyrogens*—most notably, interleukin-1.

CHECKPOINT

19. **How would the BMR of a pregnant woman compare with her own BMR before she became pregnant?**

20. **What effect does peripheral vasoconstriction on a hot day have on an individual's body temperature?**

21. **Why do infants have greater problems with thermoregulation than adults do?**

See the blue Answers tab at the end of the book.

Related Clinical Terms

avitaminosis: A disease caused by a deficiency in one or more vitamins.

carbohydrate loading: Eating large quantities of carbohydrates in the days preceding an athletic event in order to increase one's endurance.

eating disorders: Psychological problems that result in inadequate or excessive food consumption. Examples include anorexia nervosa and bulimia.

heat exhaustion: A malfunction of thermoregulatory mechanisms, caused by excessive fluid loss in perspiration.

heat stroke: A condition in which the thermoregulatory center stops functioning and body temperature rises uncontrollably.

hyperuricemia: A condition in which the plasma uric acid level exceeds 7.4 mg/dL; can result in the condition called *gout*.

hypervitaminosis: A disorder caused by the ingestion of excessive quantities of one or more vitamins.

hypothermia: Below-normal body temperature.

ketoacidosis: The acidification of blood due to the presence of ketone bodies.

ketonemia: Elevated levels of ketone bodies in blood.

ketonuria: The presence of ketone bodies in urine.

ketosis: Abnormally high concentration of ketone bodies in body fluids.

obesity: A body weight more than 20 percent above the ideal weight for a given individual.

phenylketonuria (PKU): An inherited metabolic disorder resulting from an inability to convert phenylalanine to tyrosine.

protein deficiency diseases: Nutritional disorders resulting from a lack of one or more essential amino acids.

Chapter Review

Study Outline

25-1 Metabolism refers to all the chemical reactions that occur in the body p. 930

1. In general, during *cellular metabolism*, cells break down excess carbohydrates first and then lipids, while conserving amino acids. Only about 40 percent of the energy released through **catabolism** is captured in ATP; the rest is released as heat. (*Figure 25–1*)

2. Cells synthesize new compounds (**anabolism**) to (1) perform structural maintenance or repairs, (2) support growth, (3) produce secretions, and (4) build and store nutrient reserves.

3. Cells "feed" small organic molecules to their mitochondria; in return, the cells get the ATP they need to perform cellular functions. (*Figure 25–2*)

25-2 Carbohydrate metabolism involves glycolysis, ATP production, and gluconeogenesis p. 932

4. Most cells generate ATP and other high-energy compounds through the breakdown of carbohydrates.

5. **Glycolysis** and **aerobic metabolism**, or *cellular respiration*, provide most of the ATP used by typical cells. Glycogen can be broken down to glucose molecules. In glycolysis, each molecule of glucose yields two molecules of **pyruvic acid** (as pyruvate ions), a net two molecules of ATP, and two **NADH** molecules. (*Figure 25–3*)

6. In the presence of oxygen, pyruvic acid molecules enter mitochondria, where they are broken down completely in the **tricarboxylic acid (TCA) cycle**. Carbon and oxygen atoms are lost as carbon dioxide (**decarboxylation**); hydrogen atoms are passed to coenzymes, which initiate the oxygen-consuming and ATP-generating reaction **oxidative phosphorylation**. (*Figure 25–4*)

7. **Cytochromes** of the **electron transport system (ETS)** pass electrons to oxygen, resulting in the formation of water. As this transfer occurs, the ETS generates ATP. (*Figure 25–5*)

8. For each glucose molecule processed through glycolysis, the TCA cycle, and the ETS, most cells gain 36 molecules of ATP. (*Figure 25–6*)

9. Cells can break down other nutrients to provide substrates for the TCA cycle if supplies of glucose are limited.

10. **Gluconeogenesis**, the synthesis of glucose from noncarbohydrate precursors such as lactic acid, glycerol, or amino acids, enables a liver cell to synthesize glucose molecules when carbohydrate reserves are depleted. **Glycogenesis** is the process of glycogen formation. Glycogen is an important energy reserve when cells cannot obtain enough glucose from interstitial fluid. (*Figure 25–7*)

25-3 Lipid metabolism involves lipolysis, beta-oxidation, and the transport and distribution of lipids as free fatty acids and lipoproteins p. 941

11. During **lipolysis** (lipid catabolism), lipids are broken down into pieces that can be converted into pyruvic acid or channeled into the TCA cycle.

12. Triglycerides, the most abundant lipids in the body, are split into glycerol and fatty acids. The glycerol enters the glycolytic pathways, and the fatty acids enter the mitochondria.

13. **Beta-oxidation** is the breakdown of a fatty acid molecule into two-carbon fragments that can be used in the TCA cycle. The steps of beta-oxidation cannot be reversed, and the body cannot manufacture all the fatty acids needed for normal metabolic operations. (*Figure 25–8*)

14. Lipids cannot provide large amounts of ATP quickly. However, cells can shift to lipid-based energy production when glucose reserves are limited.

15. In **lipogenesis** (the synthesis of lipids), almost any organic substrate can be used to form glycerol. **Essential fatty acids** cannot be synthesized and must be included in the diet.

16. Most lipids circulate as **lipoproteins** (lipid–protein complexes that contain large glycerides and cholesterol). The largest lipoproteins, chylomicrons, carry absorbed lipids from the intestinal tract to the bloodstream. All other lipoproteins are derived from the liver and carry lipids to and from various tissues of the body. (*Figure 25–9*)

17. Capillary walls of adipose tissue, skeletal muscle, cardiac muscle, and the liver contain **lipoprotein lipase**, an enzyme that breaks down complex lipids, releasing a mixture of fatty acids and monoglycerides. *(Figure 25–9)*

25-4 Protein catabolism involves transamination and deamination, whereas protein synthesis involves amination and transamination p. 946

18. If other energy sources are inadequate, mitochondria can break down amino acids in the TCA cycle to generate ATP. In the mitochondria, the amino group can be removed by either **transamination** or **deamination**. *(Figure 25–10)*
19. Protein catabolism is impractical as a source of quick energy.
20. Roughly half the amino acids needed to build proteins can be synthesized. There are 10 **essential amino acids**, which must be acquired through the diet. **Amination**, the attachment of an amino acid group to a carbon framework, is an important step in the synthesis of **nonessential amino acids**. *(Figures 25–11, 25–12)*

25-5 The body experiences two patterns of metabolic activity: the absorptive and postabsorptive states p. 949

21. No one cell of a human can perform all the anabolic and catabolic operations necessary to support life. Homeostasis can be preserved only when metabolic activities of different tissues are coordinated.
22. The body has five metabolic components: the liver, adipose tissue, skeletal muscle, neural tissue, and other peripheral tissues. The liver is the focal point for metabolic regulation and control. Adipose tissue stores lipids, primarily in the form of triglycerides. Skeletal muscle contains substantial glycogen reserves, and the contractile proteins can be degraded and the amino acids used as an energy source. Neural tissue does not contain energy reserves; glucose must be supplied to it for energy. Other peripheral tissues are able to metabolize glucose, fatty acids, or other substrates under the direction of the endocrine system.
23. For about four hours after a meal, nutrients enter the blood as intestinal absorption proceeds. *(Table 25–1)*
24. The liver closely regulates the glucose content of blood and the circulating levels of amino acids.
25. The **absorptive state** exists when nutrients are being absorbed by the digestive tract. Adipocytes remove fatty acids and glycerol from the bloodstream and synthesize new triglycerides to be stored for later use.
26. During the absorptive state, glucose molecules are catabolized and amino acids are used to build proteins. Skeletal muscles may also catabolize circulating fatty acids, and the energy obtained is used to increase glycogen reserves.
27. The **postabsorptive state** extends from the end of the absorptive state to the next meal.
28. When blood glucose levels fall, the liver begins breaking down glycogen reserves and releasing glucose into the bloodstream. As the time between meals increases, liver cells synthesize glucose molecules from smaller carbon fragments and from glycerol molecules. Fatty acids undergo beta-oxidation; the fragments enter the TCA cycle or combine to form **ketone bodies**. *(Table 25–1)*
29. Some amino acids can be converted to pyruvic acid and used for gluconeogenesis; others, including most of the essential amino acids, are converted to acetyl-CoA and are either catabolized or converted to ketone bodies.
30. Neural tissue continues to be supplied with glucose as an energy source until blood glucose levels become very low.

25-6 Adequate nutrition is necessary to prevent deficiency disorders and ensure physiological functioning p. 950

31. **Nutrition** is the absorption of nutrients from food. A **balanced diet** contains all the ingredients needed to maintain homeostasis; a balanced diet prevents **malnutrition**.
32. The five **basic food groups** are grains, vegetables, fruits, milk, and meat and beans. These are arranged in a *food pyramid* to reflect the recommended daily food consumption balanced with daily physical activity. *(Figure 25–13; Table 25–2)*
33. Amino acids, purines, pyrimidines, creatine, and porphyrins are **N compounds**, which contain nitrogen atoms. An adequate dietary supply of nitrogen is essential, because the body does not maintain large nitrogen reserves. **Nitrogen balance** is a state in which the amount of nitrogen absorbed equals that lost in urine and feces.
34. **Minerals** act as cofactors in a variety of enzymatic reactions. They also contribute to the osmotic concentration of body fluids and play a role in transmembrane potentials, action potentials, the release of neurotransmitters, muscle contraction, skeletal construction and maintenance, gas transport, buffer systems, fluid absorption, and waste removal. *(Table 25–3)*
35. Vitamins are needed in very small amounts. Vitamins A, D, E, and K are **fat-soluble vitamins**; taken in excess, they can lead to **hypervitaminosis**. **Water-soluble vitamins** are not stored in the body; a lack of adequate dietary supplies may lead to **vitamin deficiency disease** (**avitaminosis**). *(Tables 25–4, 25-5)*
36. A balanced diet can improve one's general health.

25-7 Metabolic rate is the average caloric expenditure, and thermoregulation involves balancing heat-producing and heat-losing mechanisms p. 955

37. The energy content of food is usually expressed in **kilocalories (kcal)** or as **Calories** per gram (Cal/g).
38. The catabolism of lipids releases 9.46 Cal/g, about twice the amount released by equivalent weights of carbohydrates or proteins.
39. The sum of all the anabolic and catabolic processes under way is an individual's **metabolic rate**. The **basal metabolic rate (BMR)** is the rate of energy utilization by a person at rest.
40. The homeostatic regulation of body temperature is **thermoregulation**. Heat exchange with the environment involves four processes: **radiation**, **conduction**, **convection**, and **evaporation**.
41. The *preoptic area* of the hypothalamus acts as the body's thermostat, affecting the **heat-loss center** and the **heat-gain center**.
42. Mechanisms for increasing heat loss include both physiological mechanisms (peripheral vasodilation, increased perspiration, and increased respiration) and behavioral modifications.
43. Responses that conserve heat include a decreased blood flow to the dermis and *countercurrent exchange*. *(Figure 25–14)*
44. Heat is generated by **shivering thermogenesis** and **nonshivering thermogenesis**.

25 DIGESTIVE

45. Thermoregulatory responses differ among individuals. One important source of variation is **acclimatization** (a physiological adjustment to an environment over time).

46. **Pyrexia** is an elevated body temperature. **Fever**, a body temperature above 37.2°C (99°F), can result from problems with the thermoregulatory mechanism, circulation, or sweat gland activity, or from the resetting of the hypothalamic "thermostat" by circulating pyrogens.

Review Questions

See the blue Answers tab at the end of the book.

myA&P Access more review material online at myA&P™ (www.myaandp.com). There, you'll find chapter guides, chapter quizzes, practice tests, animations, flashcards, a glossary with pronunciations, *Interactive Physiology*® (IP) exercises and quizzes, and more to help you succeed in the course.

LEVEL 1   Reviewing Facts and Terms

1. Catabolism refers to
   (a) the creation of a nutrient pool.
   (b) the sum total of all chemical reactions in the body.
   (c) the production of organic compounds.
   (d) the breakdown of organic substrates.

2. The breakdown of glucose to pyruvic acid is an _____ process.
   (a) anaerobic   (b) aerobic
   (c) anabolic   (d) oxidative

3. The process that produces more than 90 percent of the ATP used by our cells is
   (a) glycolysis.
   (b) the TCA cycle.
   (c) substrate-level phosphorylation.
   (d) oxidative phosphorylation.

4. The sequence of reactions responsible for the breakdown of fatty acid molecules is
   (a) beta-oxidation.   (b) the TCA cycle.
   (c) lipogenesis.   (d) all of these.

5. The TCA cycle must turn _____ times to completely metabolize the pyruvic acid produced from one glucose molecule.
   (a) 1   (b) 2
   (c) 3   (d) 4
   (e) 5

6. Use the following word bank to fill in the missing terms in the Nutrient Use in Cellular Metabolism diagram below.

Word Bank: TCA cycle, amino acids, glucose, electron transport system, fatty acids, ATP, CO_2, small carbon chains, H_2O.
   (a) _____   (b) _____   (c) _____
   (d) _____   (e) _____   (f) _____
   (g) _____   (h) _____   (i) _____

7. The largest metabolic reserves for the average adult are stored as
   (a) carbohydrates.   (b) proteins.
   (c) amino acids.   (d) triglycerides.
   (e) fatty acids.

8. The vitamins generally associated with vitamin toxicity are
   (a) fat-soluble vitamins.
   (b) water-soluble vitamins.
   (c) the B complex vitamins.
   (d) vitamins C and B_{12}.

9. What is a lipoprotein? What are the major groups of lipoproteins, and how do they differ?

LEVEL 2 Reviewing Concepts

10. What is oxidative phosphorylation? Explain how the electron transport system is involved in this process.

11. How are lipids catabolized in the body? How is beta-oxidation involved in lipid catabolism?

12. How do the absorptive and postabsorptive states maintain normal blood glucose levels?

13. Why is the liver the focal point for metabolic regulation and control?

14. Why are vitamins and minerals essential components of the diet?

15. Some articles in the popular media refer to "good cholesterol" and "bad cholesterol." To which types and functions of cholesterol might these terms refer? Explain your answer.

LEVEL 3   Critical Thinking and Clinical Applications

16. When blood levels of glucose, amino acids, and insulin are high, and glycogenesis is occurring in the liver, the body is in the _____ state.
   (a) fasting   (b) postabsorptive
   (c) absorptive   (d) stress
   (e) bulimic

17. Charlie has a blood test that shows a normal level of LDLs but an elevated level of HDLs in his blood. Given that his family has a history of cardiovascular disease, he wonders if he should modify his lifestyle. What would you tell him?

18. Jill suffers from anorexia nervosa. One afternoon she is rushed to the emergency room because of cardiac arrhythmias. Her breath has the smell of an aromatic hydrocarbon, and blood and urine samples contained high levels of ketone bodies. Why do you think she is having the arrhythmias?

The Urinary System

26

Learning Outcomes

After completing this chapter, you should be able to do the following:

26-1 Identify the components of the urinary system, and describe the functions it performs.

26-2 Describe the location and structural features of the kidneys, identify major blood vessels associated with each kidney, trace the path of blood flow through a kidney, describe the structure of a nephron, and outline the processes involved in urine formation.

26-3 Discuss the major functions of each portion of the nephron and collecting system, and describe the primary factors responsible for urine production.

26-4 Describe the factors that influence glomerular filtration pressure and the rate of filtrate formation.

26-5 Identify the types and functions of transport mechanisms found along each segment of the nephron, explain the role of countercurrent multiplication, describe hormonal influence on the volume and concentration of urine, and describe the characteristics of a normal urine sample.

26-6 Describe the structures and functions of the ureters, urinary bladder, and urethra, discuss the voluntary and involuntary regulation of urination, and describe the micturition reflex.

26-7 Describe the effects of aging on the urinary system.

26-8 Give examples of interactions between the urinary system and each of the other organ systems.

Clinical Notes

Analysis of Renal Blood Flow p. 969
Glomerulonephritis p. 975
Diuretics p. 991
Urinary Obstruction p. 998
Renal Failure and Kidney Transplantation p. 998

An Introduction to the Urinary System

Most physiological wastes are removed by the urinary system. In this chapter, we will consider the functional organization of the urinary system and describe how the kidneys remove metabolic waste products from the circulation to produce urine. We also explain the major regulatory mechanisms controlling urine production and concentration, and identify how urine is transported to the urinary bladder and released from the body through the urinary tract passageways.

26-1 Consisting of the kidneys, ureters, urinary bladder, and urethra, the urinary system has three primary functions

The **urinary system** (**Figure 26–1**) has three major functions: (1) *excretion*, the removal of organic waste products from body fluids, (2) *elimination*, the discharge of these waste products into the environment, and (3) homeostatic regulation of the volume and solute concentration of blood plasma. The excretory functions of the urinary system are performed by the two **kidneys**—organs that produce **urine**, a fluid containing water, ions, and small soluble compounds. Urine leaving the kidneys flows along the **urinary tract**, which consists of paired tubes called **ureters** (ū-RĒ-terz), to the **urinary bladder**, a muscular sac for temporary storage of urine. On leaving the urinary bladder, urine passes through the **urethra** (ū-RĒ-thra), which conducts the urine to the exterior. The urinary bladder and the urethra are responsible for the elimination of urine, a process called **urination** or **micturition** (mik-choo-RISH-un). In this process, contraction of the muscular urinary bladder forces urine through the urethra and out of the body. ATLAS: Embryology Summary 20: The Development of the Urinary System

In addition to removing waste products generated by cells throughout the body, the urinary system has several other essential homeostatic functions that are often overlooked, including the following:

- *Regulating blood volume and blood pressure*, by adjusting the volume of water lost in urine, releasing erythropoietin, and releasing renin.

- *Regulating plasma concentrations of sodium, potassium, chloride, and other ions*, by influencing the quantities lost in urine and controlling calcium ion levels through the synthesis of calcitriol.

- *Helping to stabilize blood pH*, by controlling the loss of hydrogen ions and bicarbonate ions in urine.

- *Conserving valuable nutrients*, by preventing their excretion in urine while excreting organic waste products—especially nitrogenous wastes such as *urea* and *uric acid*.

- *Assisting the liver* in detoxifying poisons and, during starvation, deaminating amino acids so that other tissues can metabolize them down.

These activities are carefully regulated to keep the composition of blood within acceptable limits. A disruption of any one of them has immediate and potentially fatal consequences.

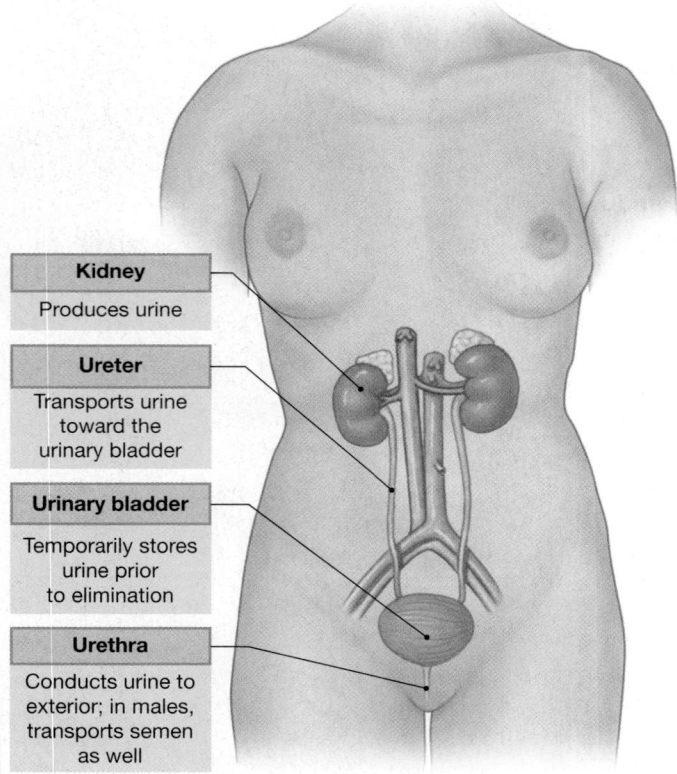

| **Kidney** |
| Produces urine |

| **Ureter** |
| Transports urine toward the urinary bladder |

| **Urinary bladder** |
| Temporarily stores urine prior to elimination |

| **Urethra** |
| Conducts urine to exterior; in males, transports semen as well |

Anterior view

Figure 26–1   An Introduction to the Urinary System. An anterior view of the urinary system, showing the positions of its components.

CHECKPOINT

1. Name the three primary functions of the urinary system.

2. Identify the components of the urinary system.

3. Define micturition.

See the blue Answers tab at the end of the book.

26-2 Kidneys are highly vascular structures containing functional units called nephrons, which perform filtration, reabsorption, and secretion

The kidneys are located on either side of the vertebral column, between vertebrae T_{12} and L_3 (**Figure 26–2a**). The left kidney lies slightly superior to the right kidney.

The superior surface of each kidney is capped by a suprarenal gland. The kidneys and suprarenal glands lie between the muscles of the dorsal body wall and the parietal peritoneum, in a retroperitoneal position (**Figure 26–2b**).

Tips&Tricks

To visualize the kidneys' retroperitoneal positions, think of each kidney as a picture on the body wall that got covered over by wallpaper (the parietal peritoneum).

The position of the kidneys in the abdominal cavity is maintained by (1) the overlying peritoneum, (2) contact with adjacent visceral organs, and (3) supporting connective tissues. Each kidney is protected and stabilized by three concentric layers of connective tissue (**Figure 26–2b**):

1. The **fibrous capsule**, a layer of collagen fibers that covers the outer surface of the entire organ.

2. The **perinephric fat capsule**, a thick layer of adipose tissue that surrounds the fibrous capsule.

3. The **renal fascia**, a dense, fibrous outer layer that anchors the kidney to surrounding structures. Collagen fibers extend outward from the fibrous capsule through the perinephric fat to this layer. Posteriorly, the renal fascia fuses with the deep fascia surrounding the muscles of the body wall. Anteriorly, the renal fascia forms a thick layer that fuses with the peritoneum.

In effect, each kidney hangs suspended by collagen fibers from the renal fascia and is packed in a soft cushion of adipose tissue. This arrangement prevents the jolts and shocks of day-to-day living from disturbing normal kidney function. If the suspensory fibers break or become detached, a slight bump or jar can displace the kidney and stress the attached vessels and ureter. This condition, called a *floating kidney*, may cause pain or other problems from the distortion of the ureter or blood vessels during movement.

A typical adult kidney (**Figures 26–3** and **26–4**) is reddish-brown and about 10 cm (4 in.) long, 5.5 cm (2.2 in.) wide, and 3 cm (1.2 in.) thick. Each kidney weighs about 150 g

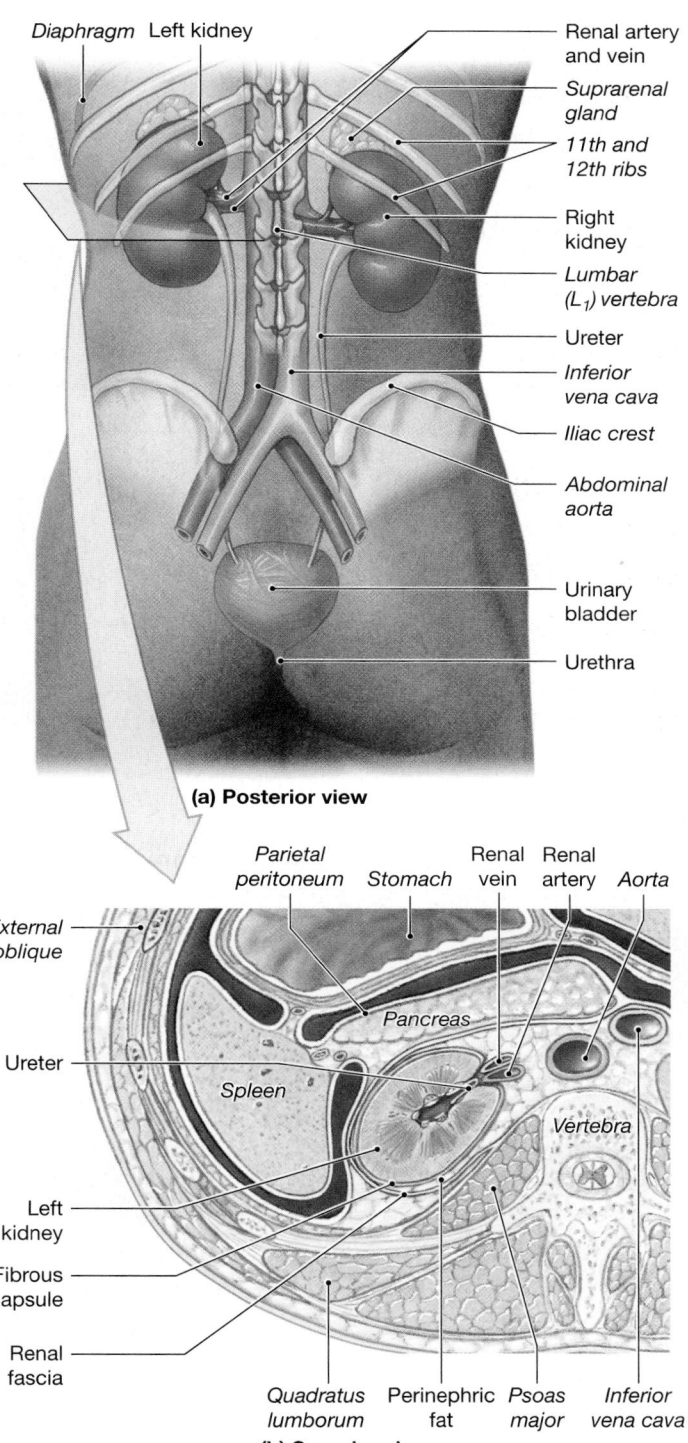

(a) Posterior view

(b) Superior view

Figure 26–2   The Position of the Kidneys. (a) A posterior view of the trunk. **(b)** A superior view of a transverse section at the level indicated in part (a). ATLAS: Plate 57a,b

(5.25 oz). The **hilum**, a prominent medial indentation, is the point of entry for the *renal artery* and *renal nerves*, and the point of exit for the *renal vein* and the ureter.

26 URINARY

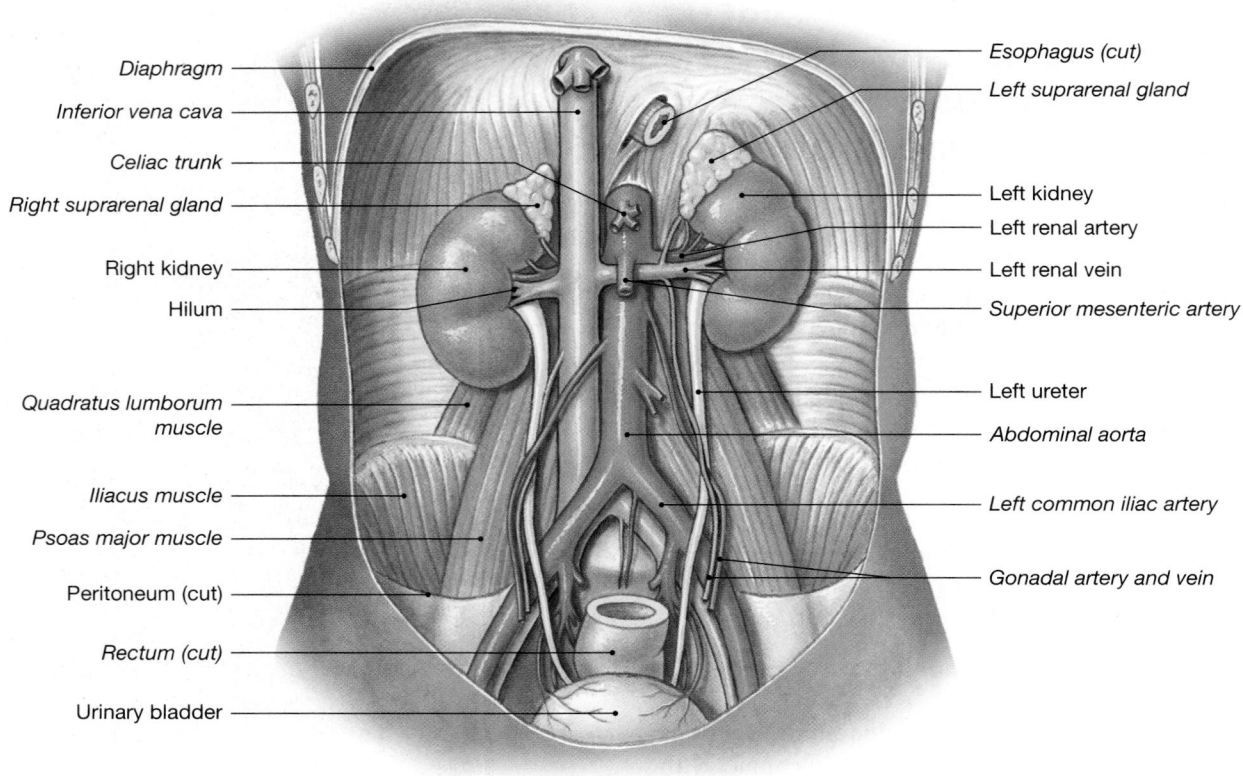

Diaphragm
Inferior vena cava
Celiac trunk
Right suprarenal gland
Right kidney
Hilum
Quadratus lumborum
muscle
Iliacus muscle
Psoas major muscle
Peritoneum (cut)
Rectum (cut)
Urinary bladder

Esophagus (cut)
Left suprarenal gland
Left kidney
Left renal artery
Left renal vein
Superior mesenteric artery
Left ureter
Abdominal aorta
Left common iliac artery
Gonadal artery and vein

Anterior view

Figure 26–3   The Gross Anatomy of the Urinary System. The abdominopelvic cavity (with the digestive organs removed), showing the kidneys, ureters, urinary bladder, and blood supply to the urinary structures. ATLAS: Plates 61a; 62a,b

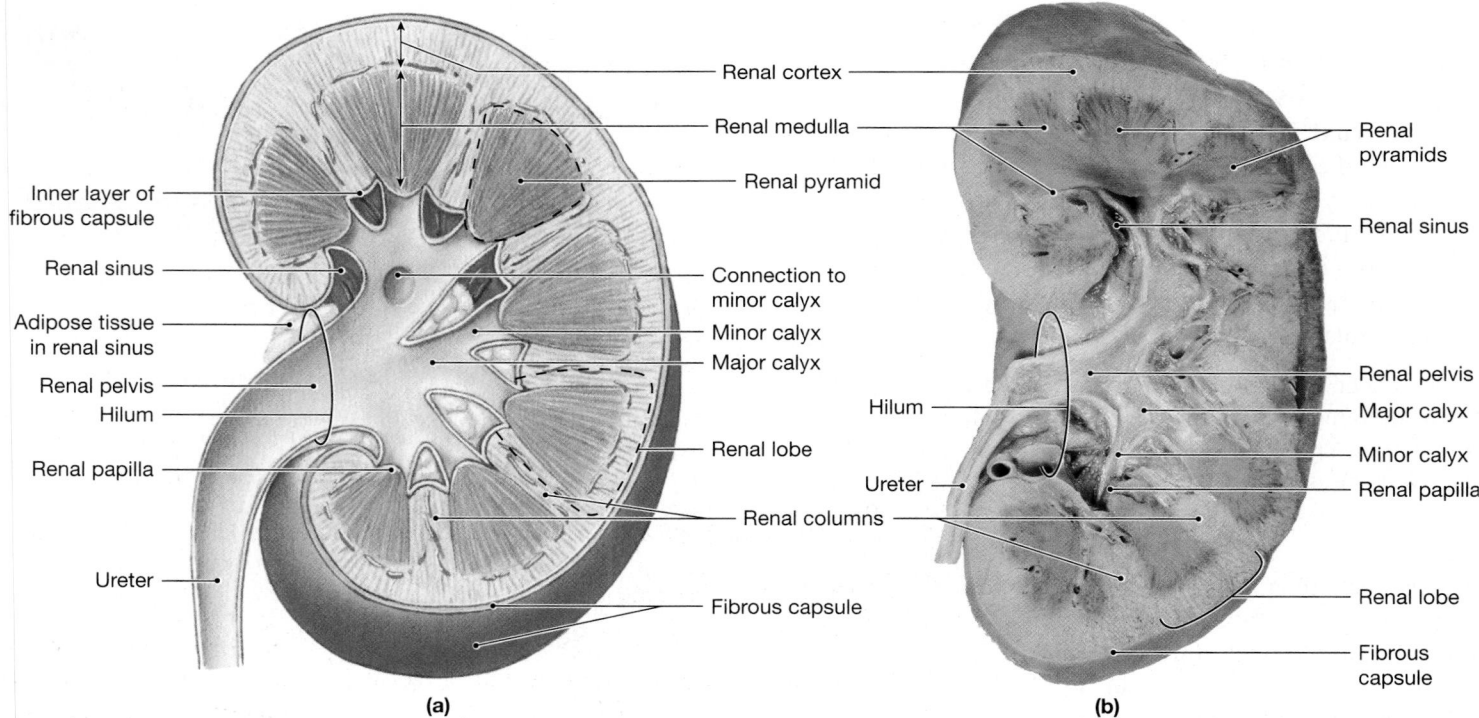

Renal cortex
Renal medulla
Renal pyramid

Inner layer of
fibrous capsule
Renal sinus
Adipose tissue
in renal sinus
Renal pelvis
Hilum
Renal papilla
Ureter

Connection to
minor calyx
Minor calyx
Major calyx

Renal lobe

Renal columns

Fibrous capsule

Renal pyramids
Renal sinus

Hilum
Ureter

Renal pelvis
Major calyx
Minor calyx
Renal papilla

Renal lobe

Fibrous capsule

(a)   **(b)**

Figure 26–4   The Structure of the Kidney. (a) A diagrammatic view of a frontal section through the left kidney. **(b)** A frontal section of the left kidney. ATLAS: Plates 57a,b; 61b

Sectional Anatomy of the Kidneys

The fibrous capsule covering the outer surface of the kidney also lines the **renal sinus**, an internal cavity within the kidney (**Figure 26–4a**). The fibrous capsule is bound to the outer surfaces of the structures within the renal sinus, stabilizing the positions of the ureter and of the renal blood vessels and nerves.

The kidney itself has an outer cortex and an inner medulla. The **renal cortex** is the superficial portion of the kidney, in contact with the fibrous capsule. The cortex is reddish brown and granular. The **renal medulla** consists of 6 to 18 distinct conical or triangular structures called **renal pyramids**. The base of each pyramid abuts the cortex, and the tip of each pyramid—a region known as the **renal papilla**—projects into the renal sinus. Each pyramid has a series of fine grooves that converge at the papilla. Adjacent renal pyramids are separated by bands of cortical tissue called **renal columns**, which extend into the medulla. The columns have a distinctly granular texture, similar to that of the cortex. A **renal lobe** consists of a renal pyramid, the overlying area of renal cortex, and adjacent tissues of the renal columns.

Urine production occurs in the renal lobes. Ducts within each renal papilla discharge urine into a cup-shaped drain called a **minor calyx** (KĀ-liks). Four or five minor calyces (KA-li-sēz) merge to form a **major calyx**, and two or three major calyces combine to form the **renal pelvis**, a large, funnel-shaped chamber. The renal pelvis, which fills most of the renal sinus, is connected to the ureter, which drains the kidney.

Urine production begins in microscopic, tubular structures called **nephrons** (NEF-ronz) in the cortex of each renal lobe. Each kidney has roughly 1.25 million nephrons, with a combined length of about 145 km (85 miles).

Blood Supply and Innervation of the Kidneys

Your kidneys receive 20–25 percent of your total cardiac output. In normal, healthy individuals, about 1200 mL of blood flow through the kidneys each minute—a phenomenal amount of blood for organs with a combined weight of less than 300 g (10.5 oz)!

Each kidney receives blood through a **renal artery**, which originates along the lateral surface of the abdominal aorta near the level of the superior mesenteric artery (**Figure 21–24a**, pp. 755–756). As it enters the renal sinus, the renal artery provides blood to the **segmental arteries** (**Figure 26–5a**). Segmental arteries further divide into a series of **interlobar arteries**, which radiate outward through the renal columns between the renal pyramids. The interlobar arteries supply blood to the **arcuate** (AR-kū-āt) **arteries**, which arch along the boundary

between the cortex and medulla of the kidney. Each arcuate artery gives rise to a number of **cortical radiate arteries**, also called *interlobular arteries*, which supply the cortical portions of the adjacent renal lobes. Branching from each cortical radiate artery are numerous **afferent arterioles**, which deliver blood to the capillaries supplying individual nephrons (**Figure 26–5b,c**).

From the capillaries of the nephrons, blood enters a network of venules and small veins that converge on the **cortical radiate veins**, also called *interlobular veins* (**Figure 26–5a,c**). The cortical radiate veins deliver blood to **arcuate veins**; these in turn empty into **interlobar veins**, which drain directly into the **renal vein**; there are no segmental veins.

The kidneys and ureters are innervated by **renal nerves**. Most of the nerve fibers involved are sympathetic postganglionic fibers from the celiac plexus and the inferior splanchnic nerves. ∞ pp. 534, 545 A renal nerve enters each kidney at the hilum and follows the tributaries of the renal arteries to reach individual nephrons. The sympathetic innervation (1) adjusts rates of urine formation by changing blood flow and blood pressure at the nephron and (2) stimulates the release of renin, which ultimately restricts losses of water and salt in the urine by stimulating reabsorption at the nephron.

CLINICAL NOTE

Analysis of Renal Blood Flow The rate of blood flow through the kidneys can be estimated by administering the compound *para-aminohippuric acid (PAH)*, which is removed at the nephrons and eliminated in urine. Virtually all the PAH contained in the blood that arrives at the kidneys is removed before the blood departs in the renal veins. Renal blood flow can thus be approximated by comparing plasma concentrations of PAH with the amount secreted in urine. In practice, however, it is usually easier to measure the glomerular filtration rate (p. 981).

The Nephron

Each nephron (**Figure 26–6**) consists of a renal tubule and a renal corpuscle. The **renal tubule** is a long tubular passageway which may be 50 mm (1.97 in.) in length. It begins at the **renal corpuscle** (KOR-pus-ul), a spherical structure consisting of the *glomerular* (Bowman) *capsule*, a cup-shaped chamber approximately 200 μm in diameter, and a capillary network known as the *glomerulus*.

Blood arrives at the renal corpuscle by way of an afferent arteriole. This arteriole delivers blood to the **glomerulus** (glo-MER-ū-lus; plural, *glomeruli*), which consists of about 50 intertwining capillaries. The glomerulus projects into the glomerular capsule much as the heart projects into the

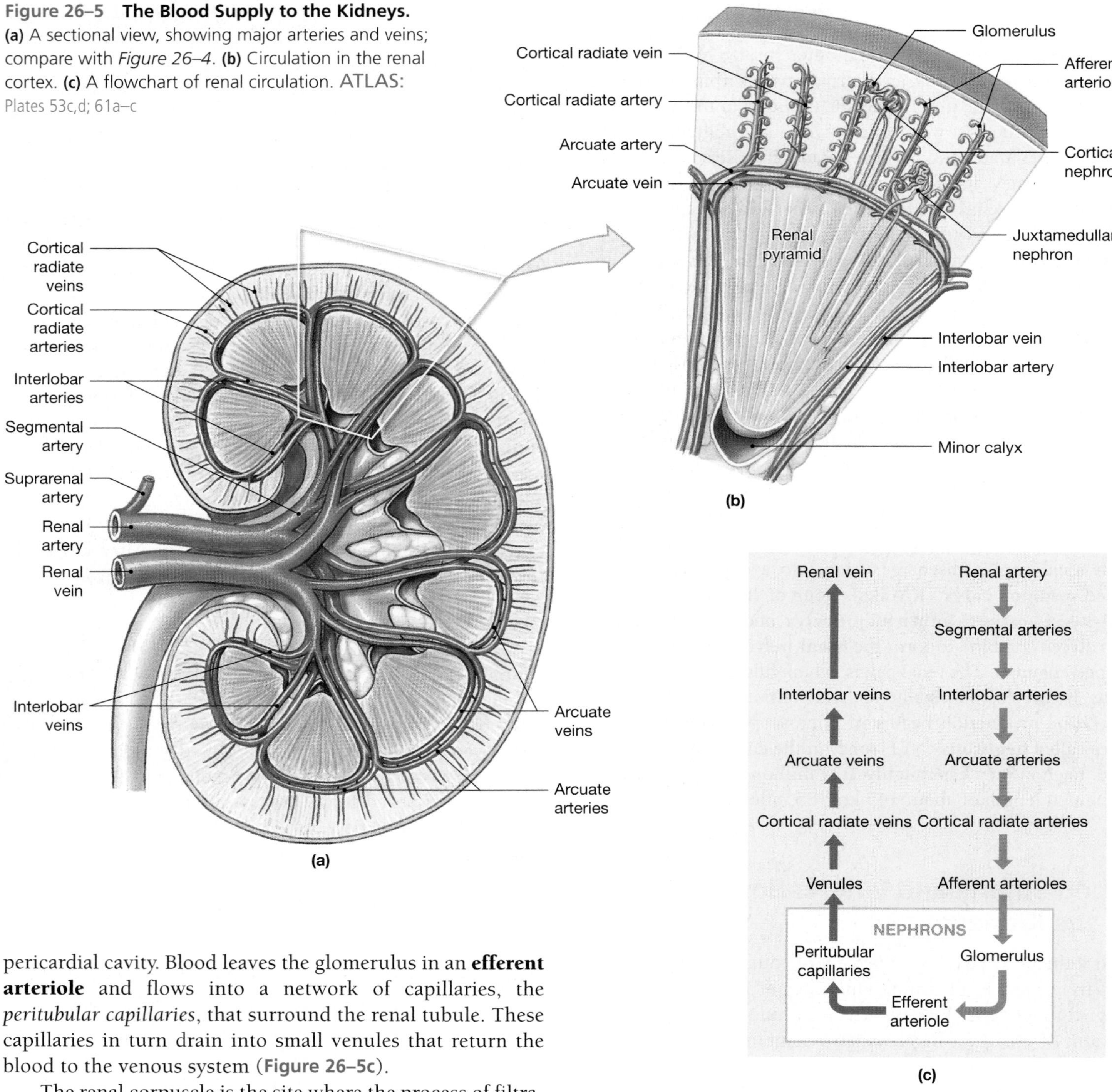

Figure 26–5 **The Blood Supply to the Kidneys.**
(a) A sectional view, showing major arteries and veins; compare with *Figure 26–4*. (b) Circulation in the renal cortex. (c) A flowchart of renal circulation. ATLAS: Plates 53c,d; 61a–c

(a)

(b)

(c)

pericardial cavity. Blood leaves the glomerulus in an **efferent arteriole** and flows into a network of capillaries, the *peritubular capillaries*, that surround the renal tubule. These capillaries in turn drain into small venules that return the blood to the venous system (**Figure 26–5c**).

The renal corpuscle is the site where the process of filtration occurs. In this process, blood pressure forces water and dissolved solutes out of the glomerular capillaries and into a chamber—the *capsular space*—that is continuous with the lumen of the renal tubule (**Figure 26–6**). Filtration produces an essentially protein-free solution, known as a **filtrate**, that is otherwise similar to blood plasma.

From the renal corpuscle, filtrate enters the renal tubule, which is responsible for three crucial functions: (1) reabsorbing all the useful organic nutrients that enter the filtrate, (2) re-

absorbing more than 90 percent of the water in the filtrate, and (3) secreting into the tubule any waste products that failed to enter the renal corpuscle through filtration at the glomerulus.

The renal tubule has two convoluted (coiled or twisted) segments—the *proximal convoluted tubule* (PCT) and the *distal convoluted tubule* (DCT)—separated by a simple U-shaped tube, the *nephron loop*, also called the *loop of Henle*

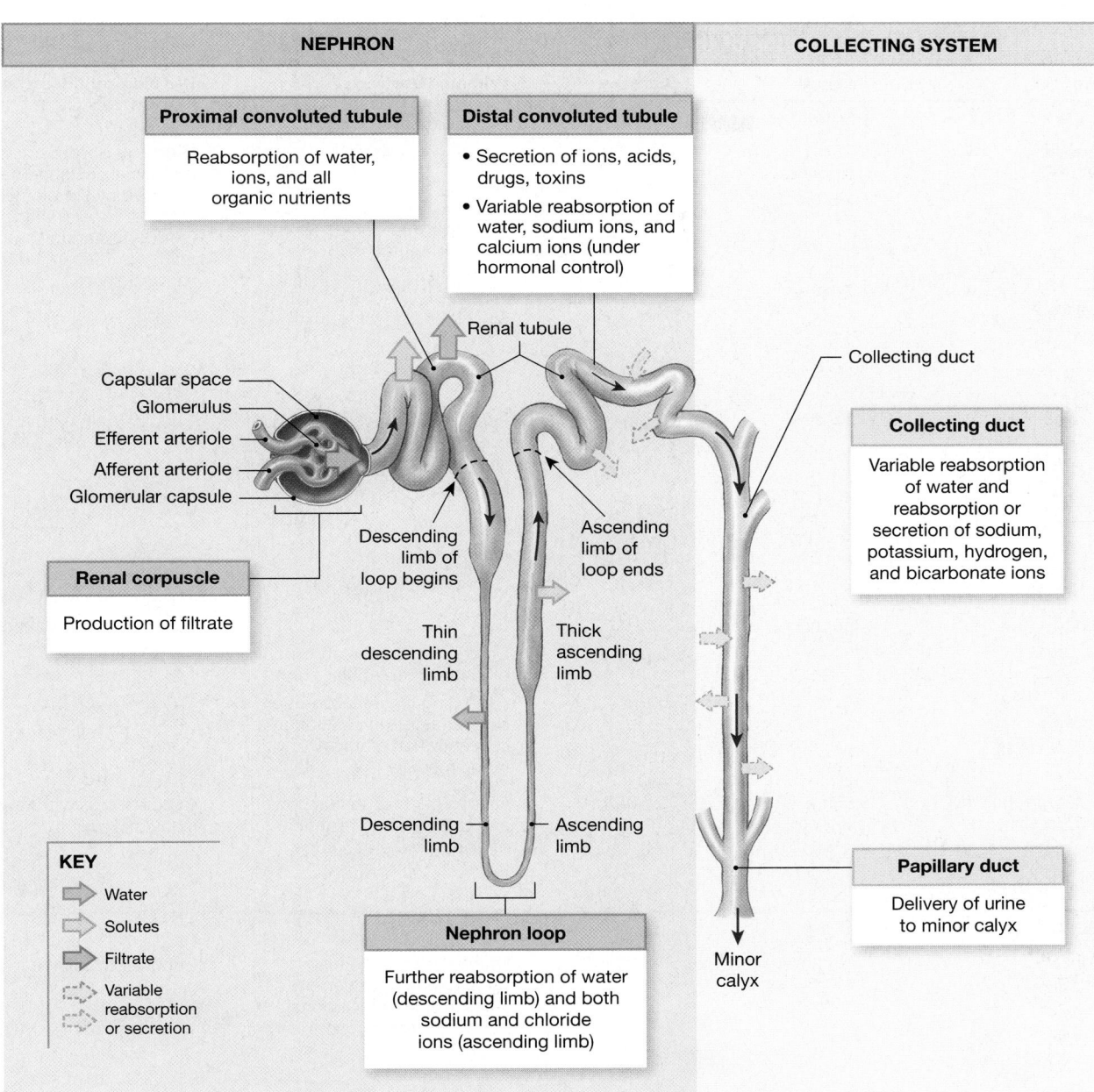

| NEPHRON | COLLECTING SYSTEM |

Proximal convoluted tubule

Reabsorption of water, ions, and all organic nutrients

Distal convoluted tubule

- Secretion of ions, acids, drugs, toxins
- Variable reabsorption of water, sodium ions, and calcium ions (under hormonal control)

Renal tubule

Collecting duct

Capsular space
Glomerulus
Efferent arteriole
Afferent arteriole
Glomerular capsule

Renal corpuscle

Production of filtrate

Descending limb of loop begins

Ascending limb of loop ends

Thin descending limb

Thick ascending limb

Collecting duct

Variable reabsorption of water and reabsorption or secretion of sodium, potassium, hydrogen, and bicarbonate ions

KEY

⇨ Water
⇨ Solutes
⇨ Filtrate
⇢ Variable reabsorption or secretion

Descending limb

Ascending limb

Nephron loop

Further reabsorption of water (descending limb) and both sodium and chloride ions (ascending limb)

Papillary duct

Delivery of urine to minor calyx

Minor calyx

Figure 26–6   The Functional Anatomy of a Representative Nephron and the Collecting System. The major functions of each segment of the nephron (purple) and the collecting system (tan) are noted. The nephron has been significantly shortened, and some components rearranged in space, in this representational figure.

(HEN-lē). The convoluted segments are in the cortex, and the nephron loop extends at least partially into the medulla. For clarity, the nephron shown in **Figure 26–6** has been shortened and straightened. The regions of the nephron vary by structure and function. As it travels along the tubule, the filtrate, now called **tubular fluid**, gradually changes in composition. The changes that occur and the characteristics of the urine that result vary with the activities under way in each segment of the nephron. **Figure 26–6** and Table 26–1 survey the regional specializations.

Tips&Tricks

In the nephron, the terms *proximal tubule* and *distal tubule* refer to how far along the renal tubule from the renal corpuscle these structures are situated. Think of *proximal* (nearer the renal corpuscle) as being first, and *distal* (farther from the renal corpuscle) as being second.

Each nephron empties into the **collecting system**, a series of tubes that carry tubular fluid away from the nephron.

SUMMARY TABLE 26–1   The Organization of the Nephron and Collecting System

Region	Length	Diameter	Primary Function	Histological Characteristics
NEPHRON				
Renal corpuscle	150–250 μm (spherical)	150–250 μm	Filtration of plasma	Glomerulus (capillary knot), mesangial cells, and dense layer, enclosed by the glomerular capsule; visceral epithelium (podocytes) and capsular epithelium separated by capsular space
Renal tubule				
Proximal convoluted tubule (PCT)	14 mm	60 μm	Reabsorption of ions, organic molecules, vitamins, water; secretion of drugs, toxins, acids	Cuboidal cells with microvilli
Nephron loop	30 mm			Squamous or low cuboidal cells
		15 μm	Descending limb: reabsorption of water from tubular fluid	
		30 μm	Ascending limb: reabsorption of ions; assists in creation of a concentration gradient in the medulla	
Distal convoluted tubule (DCT)	5 mm	30–50 μm	Reabsorption of sodium ions and calcium ions; secretion of acids, ammonia, drugs, toxins	Cuboidal cells with few if any microvilli
COLLECTING SYSTEM				
Collecting duct	15 mm	50–100 μm	Reabsorption of water, sodium ions; secretion or reabsorption of bicarbonate ions or hydrogen ions	Cuboidal to columnar cells
Papillary duct	5 mm	100–200 μm	Conduction of tubular fluid to minor calyx; contributes to concentration gradient of the medulla	Columnar cells

Renal corpuscle diagram labels: Capsular space, Capsular epithelium, Capillaries of glomerulus, Visceral epithelium. PCT diagram label: Microvilli.

Collecting ducts receive this fluid from many nephrons. Each collecting duct begins in the cortex and descends into the medulla, carrying fluid to a *papillary duct* that drains into a minor calyx.

Nephrons from different locations differ slightly in structure. Roughly 85 percent of all nephrons are **cortical nephrons**, located almost entirely within the superficial cortex of the kid-

ney (**Figure 26–7a,b**). In a cortical nephron, the nephron loop is relatively short, and the efferent arteriole delivers blood to a network of **peritubular capillaries**, which surround the entire renal tubule. These capillaries drain into small venules that carry blood to the coritical radiate veins (**Figure 26–5c**).

The remaining 15 percent of nephrons, termed **juxtamedullary** (juks-tuh-MED-u-lār-e; *juxta*, near) **nephrons**,

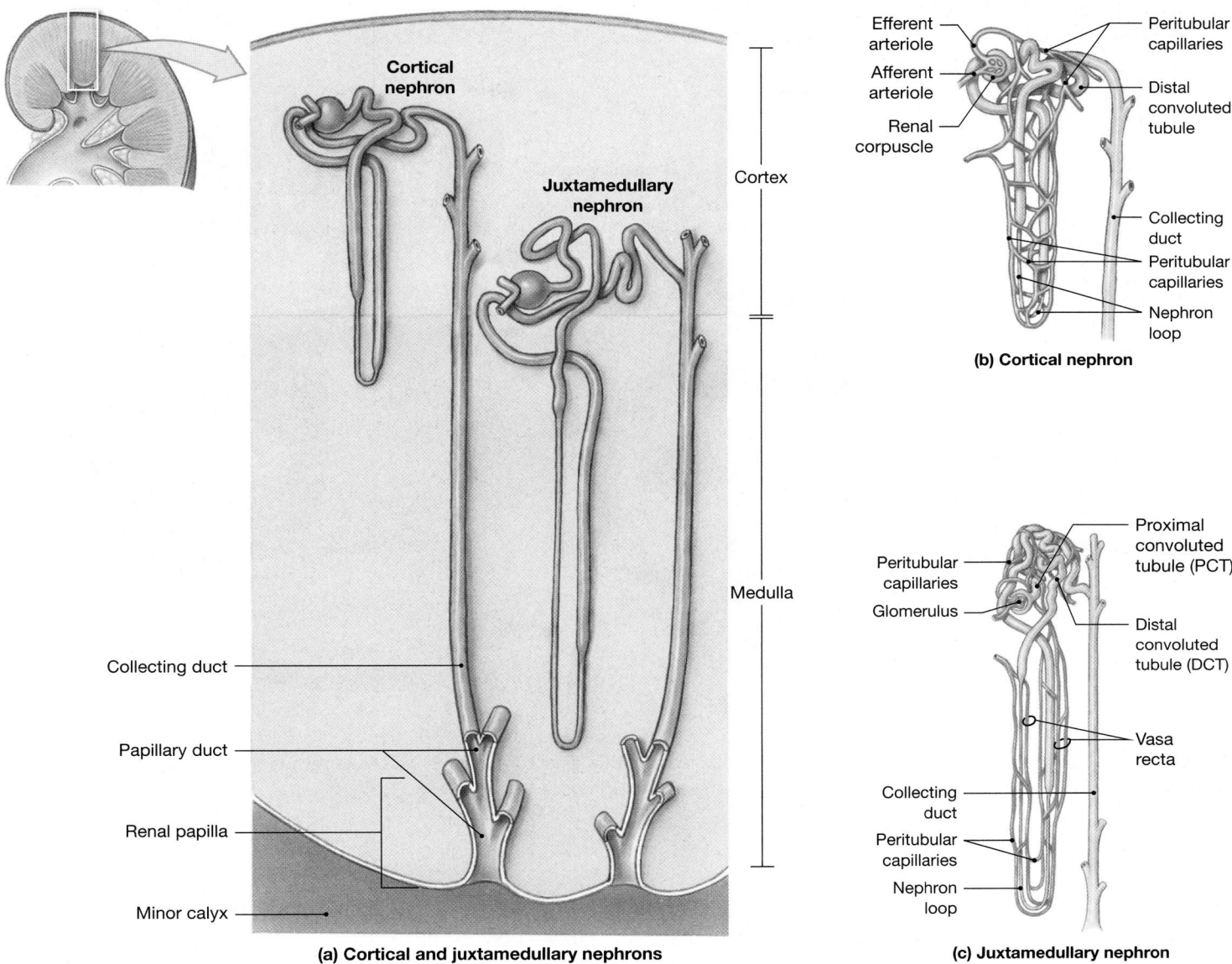

Cortical nephron

Juxtamedullary nephron

Cortex

Medulla

Collecting duct

Papillary duct

Renal papilla

Minor calyx

(a) Cortical and juxtamedullary nephrons

Efferent arteriole

Afferent arteriole

Renal corpuscle

Peritubular capillaries

Distal convoluted tubule

Collecting duct

Peritubular capillaries

Nephron loop

(b) Cortical nephron

Peritubular capillaries

Glomerulus

Proximal convoluted tubule (PCT)

Distal convoluted tubule (DCT)

Vasa recta

Collecting duct

Peritubular capillaries

Nephron loop

(c) Juxtamedullary nephron

Figure 26–7 The Locations and Structures of Cortical and Juxtamedullary Nephrons. (a) The general appearance and location of nephrons in the kidneys. The circulation to **(b)** a cortical nephron and **(c)** a juxtamedullary nephron. The lengths of the nephron loop are not drawn to scale.

have long nephron loops that extend deep into the medulla (**Figure 26–7a,c**). In these nephrons, the peritubular capillaries are connected to the **vasa recta** (*vasa*, vessel + *recta*, straight)—long, straight capillaries that parallel the nephron loop.

Because they are more numerous than juxtamedullary nephrons, cortical nephrons perform most of the reabsorptive and secretory functions of the kidneys. However, as you will see later in the chapter, it is the juxtamedullary nephrons that enable the kidneys to produce concentrated urine.

Next we examine the structure of each segment of a representative nephron.

The A&P Top 100

#90 The kidneys remove waste products from the blood; they also assist in the regulation of blood volume and blood pressure, ion levels, and blood pH. In the kidneys, the functional unit—the smallest structure that can carry out all the functions of a system—is the nephron.

The Renal Corpuscle

Each renal corpuscle (**Figure 26–8**) is 150–250 μm in diameter. It includes both a structure known as the **glomerular capsule** and the capillary network of the **glomerulus**

26 URINARY

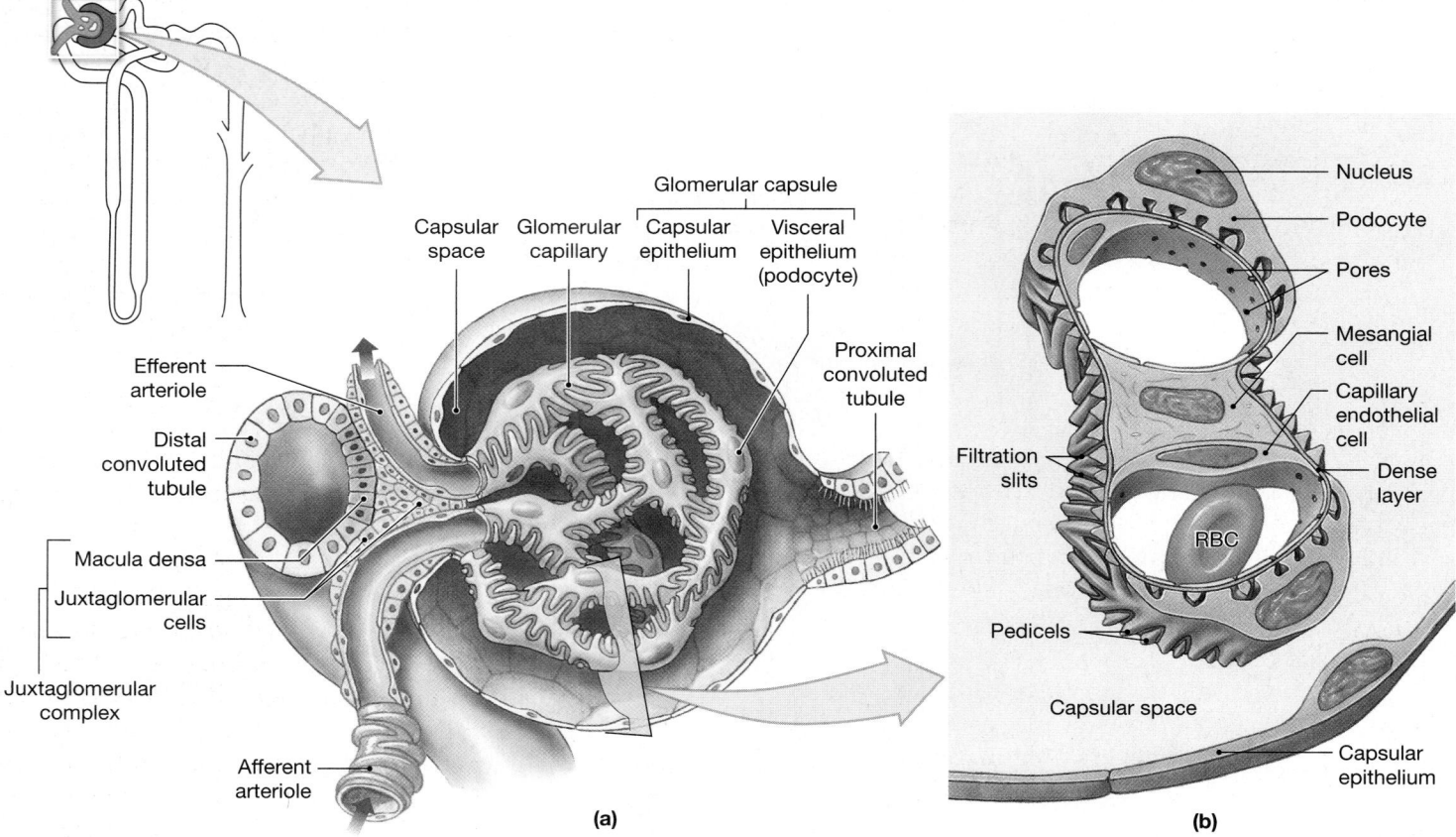

Figure 26–8   **The Renal Corpuscle. (a)** Important structural features of a renal corpuscle. **(b)** This cross section through a segment of the glomerulus shows the components of the filtration membrane of the nephron.

(**Figure 26–8a**). Connected to the initial segment of the renal tubule, the glomerular capsule forms the outer wall of the renal corpuscle and encapsulates the glomerular capillaries.

The outer wall of the capsule is lined by a simple squamous **capsular epithelium** (**Figure 26–8a**). This layer is continuous with the **visceral epithelium**, which covers the glomerular capillaries. The **capsular space** separates the capsular and visceral epithelia. The two epithelial layers are continuous where the glomerular capillaries are connected to the afferent arteriole and efferent arteriole.

The visceral epithelium consists of large cells with complex processes, or "feet," that wrap around the specialized dense layer of the glomerular capillaries. These unusual cells are called **podocytes** (PŌ-do-sīts; *podos*, foot + *-cyte*, cell), and their feet are known as **pedicels** (**Figure 26–8b**). Materials passing out of the blood at the glomerulus must be small enough to pass between the narrow gaps, or **filtration slits**, between adjacent pedicels. **Mesangial cells**, special supporting cells, lie between adjacent capillaries. Actin-like filaments in these cells enable them to contract, thereby

controlling capillary diameter and the rate of capillary blood flow. Several substances, including angiotensin II, vasopressin, and histamine, have been shown to affect mesangial cell contraction. ∞ pp. 637 Furthermore, some evidence suggests that these cells also make renin.

The glomerular capillaries are fenestrated capillaries— that is, their endothelium contains large-diameter pores (**Figure 26–8b**). The dense layer differs from that found in the basal lamina of other capillary networks in that it may encircle more than one capillary. Together, the fenestrated endothelium, the dense layer, and the filtration slits form the *filtration membrane*. During filtration, blood pressure forces water and small solutes across this membrane and into the capsular space. The larger solutes, especially plasma proteins, are excluded. Filtration at the renal corpuscle is both effective and passive, but it has one major limitation: In addition to metabolic wastes and excess ions, compounds such as glucose, free fatty acids, amino acids, vitamins, and other solutes also enter the capsular space. These potentially useful materials are recaptured before filtrate leaves the kidneys; much of the reabsorption occurs in the proximal convoluted tubule.

CLINICAL NOTE

Glomerulonephritis Glomerulonephritis (glo-mer-ū-lō-nef-RĪ-tis) is an inflammation of the glomeruli that affects the filtration mechanism of the kidneys. The condition is often an *immune complex disorder*, which may develop after an infection involving *Streptococcus* bacteria. ∞ p. 815 The kidneys are not the site of infection, but as the immune system responds, the number of circulating antigen–antibody complexes skyrockets. These complexes are small enough to pass through the dense layer, but too large to fit between the filtration slits of the filtration membrane. As a result, the complexes clog up the filtration mechanism, and filtrate production drops. Any condition that leads to a massive immune response, including viral infections and autoimmune disorders, can cause glomerulonephritis.

The Proximal Convoluted Tubule

The **proximal convoluted tubule (PCT)** is the first segment of the renal tubule (**Figure 26–6**). The entrance to the PCT lies almost directly opposite the point where the afferent and efferent arterioles connect to the glomerulus. The lining of the PCT is a simple cuboidal epithelium whose apical surfaces bear microvilli (Table 26–1). The tubular cells absorb organic nutrients, ions, water, and plasma proteins (if present) from the tubular fluid and release them into the **peritubular fluid**, the interstitial fluid surrounding the renal tubule. Reabsorption is the primary function of the PCT, but the epithelial cells can also secrete substances into the lumen.

The Nephron Loop (Loop of Henle)

The PCT makes an acute bend that turns the renal tubule toward the renal medulla. This turn leads to the **nephron loop**, or *loop of Henle* (**Figure 26–7**). The nephron loop can be divided into a **descending limb** and an **ascending limb**. Fluid in the descending limb flows toward the renal pelvis, and that in the ascending limb flows toward the renal cortex. Each limb contains a **thick segment** and a **thin segment**. The terms *thick* and *thin* refer to the height of the epithelium, not to the diameter of the lumen: Thick segments have a cuboidal epithelium, whereas a squamous epithelium lines the thin segments (Table 26–1).

The thick descending limb has functions similar to those of the PCT: It pumps sodium and chloride ions out of the tubular fluid. The effect of this pumping is most noticeable in the medulla, where the long ascending limbs of juxtamedullary nephrons create unusually high solute concentrations in peritubular fluid. The thin segments are freely permeable to water, but not to solutes; water movement out of these segments helps concentrate the tubular fluid.

The Distal Convoluted Tubule

The thick ascending limb of the nephron loop ends where it forms a sharp angle near the renal corpuscle. The **distal convoluted tubule (DCT)**, the third segment of the renal tubule, begins there. The initial portion of the DCT passes between the afferent and efferent arterioles (**Figure 26–8a**).

In sectional view, the DCT differs from the PCT in that the DCT has a smaller diameter and its epithelial cells lack microvilli (Table 26–1). The DCT is an important site for three vital processes: (1) the active secretion of ions, acids, drugs, and toxins; (2) the selective reabsorption of sodium ions and calcium ions from tubular fluid; and (3) the selective reabsorption of water, which assists in concentrating the tubular fluid.

The Juxtaglomerular Complex. The epithelial cells of the DCT near the renal corpuscle are taller than those elsewhere along the DCT, and their nuclei are clustered together. This region is called the **macula densa** (MAK-ū-la DEN-sa) (**Figure 26–8a**). The cells of the macula densa are closely associated with unusual smooth muscle fibers in the wall of the afferent arteriole. These fibers are known as **juxtaglomerular cells**. Together, the macula densa and juxtaglomerular cells form the **juxtaglomerular complex (JGC)**, an endocrine structure that secretes the hormone *erythropoietin* and the enzyme *renin*. ∞ p. 637

The Collecting System

The distal convoluted tubule, the last segment of the nephron, opens into the collecting system (**Figure 26–6**). Individual nephrons drain into a nearby **collecting duct**. Several collecting ducts then converge into a larger **papillary duct**, which in turn empties into a minor calyx. The epithelium lining the collecting system is typically columnar (Table 26–1).

In addition to transporting tubular fluid from the nephron to the renal pelvis, the collecting system adjusts the fluid's composition and determines the final osmotic concentration and volume of urine.

CHECKPOINT

4. Which portions of the nephron are in the renal cortex?

5. Why don't plasma proteins pass into the capsular space under normal circumstances?

6. Damage to which part of the nephron would interfere with the hormonal control of blood pressure?

See the blue Answers tab at the end of the book.

26 URINARY

26-3 Different segments of the nephron form urine by filtration, reabsorption, and secretion

The goal of urine production is to maintain homeostasis by regulating the volume and composition of blood. This process involves the excretion of solutes—specifically, metabolic waste products. Three organic waste products are noteworthy:

1. *Urea.* Urea is the most abundant organic waste. You generate roughly 21 g of urea each day, most of it through the breakdown of amino acids.

2. *Creatinine.* Creatinine is generated in skeletal muscle tissue through the breakdown of *creatine phosphate*, a high-energy compound that plays an important role in muscle contraction. ∞ p. 319 Your body generates roughly 1.8 g of creatinine each day, and virtually all of it is excreted in urine.

3. *Uric Acid.* **Uric acid** is a waste product formed during the recycling of the nitrogenous bases from RNA molecules. You produce approximately 480 mg of uric acid each day.

These waste products are dissolved in the bloodstream and can be eliminated only when dissolved in urine. As a result, their removal is accompanied by an unavoidable water loss. The kidneys are usually capable of producing concentrated urine with an osmotic concentration of 1200–1400 mOsm/L, more than four times that of plasma. (Methods of reporting solute concentrations are discussed in a later section.) If the kidneys were unable to concentrate the filtrate produced by glomerular filtration, fluid losses would lead to fatal dehydration in a matter of hours. The kidneys also ensure that the fluid that *is* lost does not contain potentially useful organic substrates that are present in blood plasma, such as sugars or amino acids. These valuable materials must be reabsorbed and retained for use by other tissues.

Basic Processes of Urine Formation

To perform their functions, the kidneys rely on three distinct processes:

1. *Filtration.* In **filtration**, blood pressure forces water and solutes across the wall of the glomerular capillaries and into the capsular space. Solute molecules small enough to pass through the filtration membrane are carried by the surrounding water molecules.

2. *Reabsorption.* **Reabsorption** is the removal of water and solutes from the filtrate, and their movement across the tubular epithelium and into the peritubular fluid. Reabsorption occurs after filtrate has left the renal corpuscle. Most of the reabsorbed materials are nutrients the body can use. Whereas filtration occurs solely based on size, reabsorption is a selective process involving either simple diffusion or the activity of carrier proteins in the tubular epithelium. The reabsorbed substances in the peritubular fluid eventually reenter the blood. Water reabsorption occurs passively, through osmosis.

3. *Secretion.* **Secretion** is the transport of solutes from the peritubular fluid, across the tubular epithelium, and into the tubular fluid. Secretion is necessary because filtration does not force all the dissolved materials out of the plasma. Tubular secretion, which provides a backup process for filtration, can further lower the plasma concentration of undesirable materials. Secretion is often the primary method of excretion for some compounds, including many drugs.

Together, these processes produce a fluid that is very different from other body fluids. Table 26–2 indicates the efficiency of the renal system by comparing the concentrations of some substances in urine and plasma. Before considering the functions of the individual portions of the nephron, we will briefly examine each of the three major processes involved in urine formation.

Tips&Tricks

Secretion occurs when cells produce and then discharge substances into the urine, whereas *excretion* is the elimination of wastes from the body in the form of urine, sweat, and feces.

Filtration

In filtration, hydrostatic pressure forces water through membrane pores, and solute molecules small enough to pass through those pores are carried along. Filtration occurs as larger solutes and suspended materials are left behind. We

TABLE 26–2 Significant Differences between Solute Concentrations in Urine and Plasma

Solute	Urine	Plasma
IONS (mEq/L)*		
Sodium (Na^+)	90	140
Potassium (K^+)	47.5	4.6
Chloride (Cl^-)	153.3	99
Bicarbonate (HCO_3^-)	1.9	24.8
METABOLITES AND NUTRIENTS (mg/dL)		
Glucose	0	100
Lipids	0.002	600
Amino acids	100	4.2
Proteins	62	7.5 g/dL
NITROGENOUS WASTES (mg/dL)		
Urea	900	10–20
Creatinine	150	1–1.5
Ammonia	60	<0.1
Uric acid	40	3

*See the discussion of solute concentrations on page 43.

can see filtration in action in a drip coffee machine. Gravity forces hot water through the filter, and the water carries a variety of dissolved compounds into the pot. The large coffee grounds never reach the pot, because they cannot fit through the pores of the filter. In other words, they are "filtered out" of the solution; the coffee we drink is the filtrate.

In the body, the heart pushes blood around the cardiovascular system and generates hydrostatic pressure. Filtration occurs across the walls of capillaries as water and dissolved materials are pushed into the interstitial fluids of the body (see **Figure 21–11**, p. 735). In some sites (for example, the liver), the pores are so large that even plasma proteins can enter the interstitial fluids. At the renal corpuscle, however, a specialized filtration membrane restricts the passage of even the smallest circulating proteins.

Reabsorption and Secretion

The processes of reabsorption and secretion at the kidneys involve a combination of diffusion, osmosis, channel-mediated diffusion, and carrier-mediated transport. Diffusion and osmosis were considered in several other chapters, so here we will briefly review carrier-mediated transport mechanisms.

Types of Carrier-Mediated Transport.   In previous chapters, we considered four major types of *carrier-mediated transport:*

- In *facilitated diffusion*, a carrier protein transports a molecule across the plasma membrane without expending energy (see **Figure 3–18**, p. 95). Such transport always follows the concentration gradient for the ion or molecule transported.

- *Active transport* is driven by the hydrolysis of ATP to ADP on the inner membrane surface (see **Figure 3–19**, p. 96). Exchange pumps and other carrier proteins are active along the kidney tubules. Active transport mechanisms can operate despite an opposing concentration gradient.

- In *cotransport*, carrier protein activity is not directly linked to the hydrolysis of ATP (see **Figure 3–20**, p. 96). Instead, two substrates (ions, molecules, or both) cross the membrane while bound to the carrier protein. The movement of the substrates always follows the concentration gradient of at least one of the transported substances. Cotransport mechanisms are responsible for the reabsorption of organic and inorganic compounds from the tubular fluid.

- *Countertransport* resembles cotransport in all respects, except that the two transported ions move in *opposite* directions (see **Figures 23–24**, p. 860, and **24–14**, p. 894). Countertransport mechanisms operate in the PCT, DCT, and collecting system.

Characteristics of Carrier-Mediated Transport.   All carrier-mediated processes share five features that are important for an understanding of kidney function:

1. *A Specific Substrate Binds to a Carrier Protein That Facilitates Movement across the Membrane.*

2. *A Given Carrier Protein Typically Works in One Direction Only.* In facilitated diffusion, that direction is determined by the concentration gradient of the substance being transported. In active transport, cotransport, and countertransport, the location and orientation of the carrier proteins determine whether a particular substance is reabsorbed or secreted. The carrier protein that transports amino acids from the tubular fluid to the cytoplasm, for example, will not carry amino acids back into the tubular fluid.

3. *The Distribution of Carrier Proteins Can Vary among Portions of the Cell Surface.* Transport between tubular fluid and interstitial fluid involves two steps—the material must enter the cell at its apical surface and then leave the cell and enter the peritubular fluid at the cell's basolateral surface. Each step involves a different carrier protein. For example, the apical surfaces of cells along the proximal convoluted tubule contain carrier proteins that bring amino acids, glucose, and many other nutrients into these cells by sodium-linked cotransport. In contrast, the basolateral surfaces contain carrier proteins that move those nutrients out of the cell by facilitated diffusion.

4. *The Membrane of a Single Tubular Cell Contains Many Types of Carrier Proteins.* Each cell can have multiple functions, and a cell that reabsorbs one compound can secrete another.

5. *Carrier Proteins, Like Enzymes, Can Be Saturated.* When an enzyme is *saturated*, further increases in substrate concentration have no effect on the rate of reaction. ∞ p. 56 When a carrier protein is saturated, further increases in substrate concentration have no effect on the rate of transport across the plasma membrane. For any substance, the concentration at saturation is called the **transport maximum (T_m)** or *tubular maximum.* The saturation of carrier proteins involved in tubular secretion seldom occurs in healthy individuals, but carriers involved in tubular reabsorption are often at risk of saturation, especially during the absorptive state following a meal.

T_m and the Renal Threshold.   Normally, any plasma proteins and nutrients, such as amino acids and glucose, are removed from the tubular fluid by cotransport or facilitated diffusion. If the concentrations of these nutrients rise in the tubular fluid, the rates of reabsorption increase until the carrier proteins are saturated. A concentration higher than the transport maximum will exceed the reabsorptive abilities of the nephron, so some of the material will remain in the tubular fluid and appear in the urine. The transport maximum thus determines the **renal threshold**—the plasma concentration at which a specific compound or ion begins to appear in the urine.

26 URINARY

The renal threshold varies with the substance involved. The renal threshold for glucose is approximately 180 mg/dL. When plasma glucose concentrations exceed 180 mg/dL, glucose concentrations in tubular fluid exceed the T_m of the tubular cells, so glucose appears in urine. The presence of glucose in urine is a condition called *glycosuria*. After you have eaten a meal rich in carbohydrates, your plasma glucose levels may exceed the T_m for a brief period. The liver will quickly lower circulating glucose levels, and very little glucose will be lost in your urine. However, chronically elevated plasma and urinary glucose concentrations are highly abnormal. (Glycosuria is one of the key signs of diabetes mellitus. ∞ p. 634)

The renal threshold for amino acids is lower than that for glucose; amino acids appear in urine when plasma concentrations exceed 65 mg/dL. Plasma amino acid levels commonly exceed the renal threshold after you have eaten a protein-rich meal, causing some amino acids to appear in your urine. This condition is termed **aminoaciduria** (am-i-nō-as-i-DOO-rē-uh).

T_m values for water-soluble vitamins are relatively low; as a result, excess quantities of these vitamins are excreted in urine. (This is typically the fate of water-soluble vitamins in daily supplements.) Cells of the renal tubule ignore a number of other compounds in the tubular fluid. As water and other compounds are removed, the concentrations of the ignored materials in the tubular fluid gradually rise. Table 26–3 lists some substances that are actively reabsorbed or secreted by the renal tubules, as well as several that are not transported at all.

An Overview of Renal Function

Figure 26–9 summarizes the general functions of the various segments of the nephron and collecting system in the formation of urine. Most regions perform a combination of reabsorption and secretion, but the balance between the two processes shifts from one region to another:

- Filtration occurs exclusively in the renal corpuscle, across the filtration membrane.

- Water and solute reabsorption occurs primarily along the proximal convoluted tubules, but also elsewhere along the renal tubule and within the collecting system.

- Active secretion occurs primarily at the proximal and distal convoluted tubules.

- Regulation of the final volume and solute concentration of the urine results from the interaction between the collecting system and the nephron loops—especially the long loops of the juxtamedullary nephrons.

TABLE 26–3	Tubular Reabsorption and Secretion	
Reabsorbed	**Secreted**	**No Transport Mechanism**
Ions	**Ions**	Urea
Na^+, Cl^-, K^+	K^+, H^+, Ca^{2+},	Water
Ca^{2+}, Mg^{2+},	PO_4^{3-}	Urobilinogen
SO_4^{2-}, HCO_3^-		Bilirubin
	Wastes	
	Creatinine	
Metabolites	Ammonia	
Glucose	Organic acids and bases	
Amino acids		
Proteins		
Vitamins	**Miscellaneous**	
	Neurotransmitters (ACh, NE, E, dopamine)	
	Histamine	
	Drugs (penicillin, atropine, morphine, many others)	

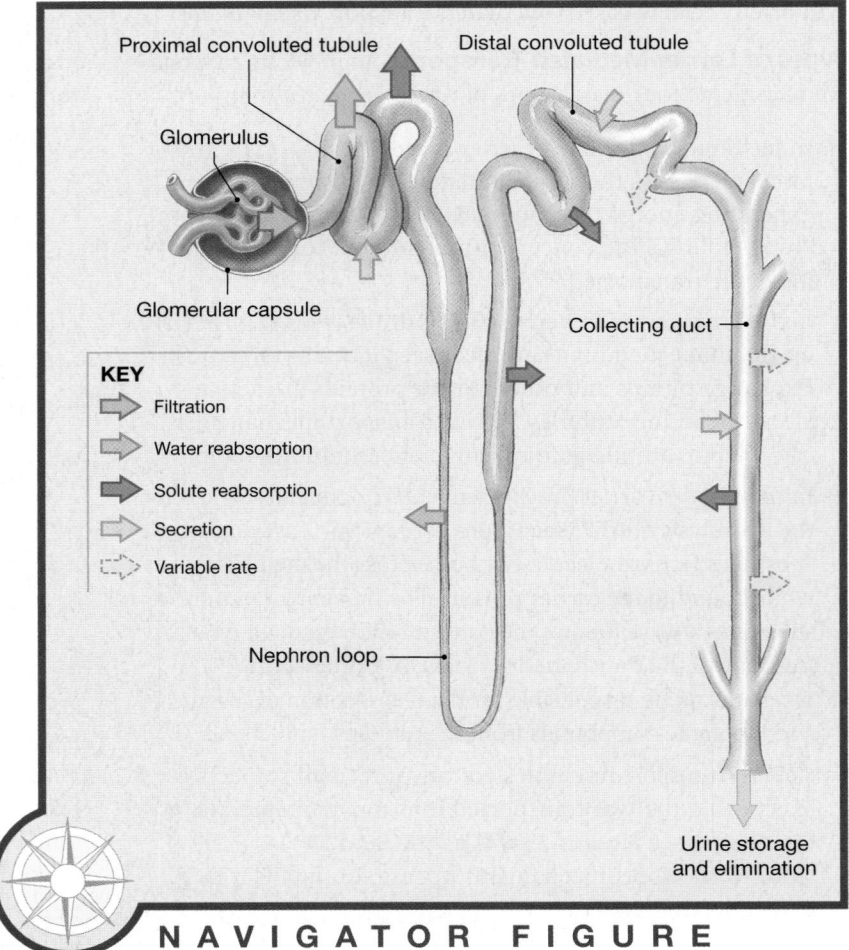

NAVIGATOR FIGURE

Figure 26–9 **An Overview of Urine Formation.** This figure will be repeated, in reduced and simplified form as Navigator icons, in key figures as the text changes topics.

Normal kidney function can continue only as long as filtration, reabsorption, and secretion function within relatively narrow limits. A disruption in kidney function has immediate effects on the composition of the circulating blood. If both kidneys are affected, death will occur within a few days unless medical assistance is provided.

Next we will proceed along the nephron to consider the formation of filtrate and the changes in the composition and concentration of the filtrate as it passes along the renal tubule. Most of what follows applies equally to cortical and juxtamedullary nephrons. The major differences between the two types of nephron are that the nephron loop of a cortical nephron is shorter and does not extend as far into the medulla as does the nephron loop of a juxtamedullary nephron (**Figure 26–7a**). The long nephron loop in a juxtamedullary nephron extends deep into the renal pyramids, where it plays a vital role in water conservation and the formation of concentrated urine. Because this process is so important, affecting the tubular fluid produced by every nephron in the kidney, and because the functions of the renal corpuscle and of the proximal and distal convoluted tubules are the same in all nephrons, we will use a juxtamedullary nephron as our example. Table 26–4 summarizes the functions of the various parts of the nephron.

Summary Table 26–4 Renal Structures and Their Function

Segment	General Functions	Specific Functions	Mechanisms
Renal corpuscle	*Filtration* of plasma; generates approximately 180 L/day of filtrate similar in composition to blood plasma without plasma proteins	*Filtration* of water, inorganic and organic solutes from plasma; retention of plasma proteins and blood cells	Glomerular hydrostatic (blood) pressure working across capillary endothelium, dense layer, and filtration slits
Proximal convoluted tubule (PCT)	*Reabsorption* of 60–70% of the water (108–116 L/day), 99–100% of the organic substrates, and 60–70% of the sodium and chloride ions in the original filtrate	*Active reabsorption:* glucose, other simple sugars, amino acids, vitamins, ions (including sodium, potassium, calcium, magnesium, phosphate, and bicarbonate) *Passive reabsorption:* urea, chloride ions, lipid-soluble materials, water *Secretion:* Hydrogen ions, ammonium ions, creatinine, drugs, and toxins (as at DCT)	Carrier-mediated transport, including facilitated transport (glucose, amino acids), cotransport (glucose, ions), and countertransport (with secretion of H^+) Diffusion (solutes) or osmosis (water) Countertransport with sodium ions
Nephron loop	*Reabsorption* of 25% of the water (45 L/day) and 20–25% of the sodium and chloride ions present in the original filtrate; creation of the concentration gradient in the medulla	*Reabsorption:* Sodium and chloride ions Water	Active transport via $Na^+–K^+/2\ Cl^-$ transporter Osmosis
Distal convoluted tubule (DCT)	*Reabsorption* of a variable amount of water (usually 5%, or 9 L/day), under ADH stimulation, and a variable amount of sodium ions, under aldosterone stimulation	*Reabsorption:* Sodium and chloride ions Sodium ions (variable) Calcium ions (variable) Water (variable) *Secretion:* Hydrogen ions, ammonium ions Creatinine, drugs, toxins	Cotransport Countertransport with potassium ions; aldosterone-regulated Carrier-mediated transport stimulated by parathyroid hormone and calcitriol Osmosis; ADH regulated Countertransport with sodium ions Carrier-mediated transport
Collecting system	*Reabsorption* of a variable amount of water (usually 9.3%, or 16.8 L/day) under ADH stimulation, and a variable amount of sodium ions, under aldosterone stimulation	*Reabsorption:* Sodium ions (variable) Bicarbonate ions (variable) Water (variable) Urea (distal portions only) *Secretion:* Potassium and hydrogen ions (variable)	Countertransport with potassium or hydrogen ions; aldosterone-regulated Diffusion, generated within tubular cells Osmosis; ADH-regulated Diffusion Carrier-mediated transport
Peritubular capillaries	*Redistribution* of water and solutes reabsorbed in the cortex	Return of water and solutes to the general circulation	Osmosis and diffusion
Vasa recta	*Redistribution* of water and solutes reabsorbed in the medulla and stabilization of the concentration gradient of the medulla	Return of water and solutes to the general circulation	Osmosis and diffusion

The osmotic concentration, or *osmolarity*, of a solution is the total number of solute particles in each liter. ∞ p. 93 Osmolarity is usually expressed in **osmoles** per liter (Osm/L) or **milliosmoles** per liter (mOsm/L). If each liter of a fluid contains 1 mole of dissolved particles, the solute concentration is 1 Osm/L, or 1000 mOsm/L. Body fluids have an osmotic concentration of about 300 mOsm/L. In comparison, that of seawater is about 1000 mOsm/L, and that of fresh water about 5 mOsm/L. Ion concentrations are often reported in *milliequivalents* per liter (mEq/L), whereas the concentrations of large organic molecules are usually reported in grams or milligrams per unit volume of solution (typically, mg or g per dL).

CHECKPOINT

7. **Identify the three distinct processes of urine formation in the kidney.**

8. **What occurs when the plasma concentration of a substance exceeds its tubular maximum?**

See the blue Answers tab at the end of the book.

26-4 Hydrostatic and colloid osmotic pressures influence glomerular filtration pressure, which in turn affects the glomerular filtration rate

Filtration occurs in the renal corpuscle as fluids move across the wall of the glomerulus and into the capsular space. The process of **glomerular filtration** involves passage across a filtration membrane, which has three components: (1) the capillary endothelium, (2) the dense layer, and (3) the filtration slits (**Figures 26–8b** and **26–10**).

Glomerular capillaries are fenestrated capillaries with pores ranging from 60 to 100 nm (0.06 to 0.1 μm) in diameter. These openings are small enough to prevent the passage of blood cells, but they are too large to restrict the diffusion of solutes, even those the size of plasma proteins. The dense layer is more selective: Only small plasma proteins, nutrients, and ions can cross it. The filtration slits are the finest filters of all. Their gaps are only 6–9 nm wide, which is small enough to prevent the passage of most small plasma proteins. As a result, under normal circumstances only a few plasma proteins—such as albumin molecules, with an average diameter of 7 nm—can cross the filtration membrane and enter the capsular space. However, plasma proteins are all that stay behind, so the filtrate contains dissolved ions and small organic molecules in roughly the same concentrations as in plasma.

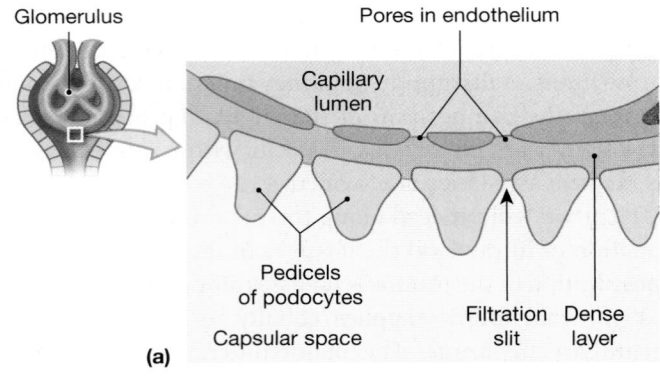

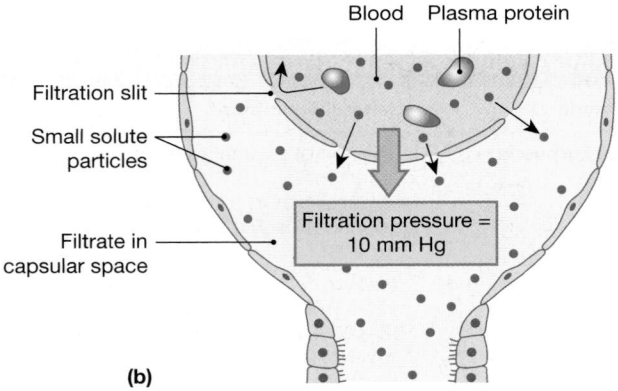

Figure 26–10 **Glomerular Filtration.** (a) The filtration membrane. (b) Filtration pressure.

Filtration Pressures

The major forces that act across capillary walls were discussed in Chapters 21 and 22. (You may find it helpful to review **Figures 21–11** and **21–12**, pp. 735, 736, before you proceed.) The primary factor involved in glomerular filtration is basically the same as that governing fluid and solute movement across capillaries throughout the body: the balance between hydrostatic pressure (fluid pressure) and colloid osmotic pressure (pressure due to materials in solution) on either side of the capillary walls.

Hydrostatic Pressure

The **glomerular hydrostatic pressure (GHP)** is the blood pressure in the glomerular capillaries. This pressure tends to push water and solute molecules out of the plasma and into the filtrate. The GHP is significantly higher than capillary pressures elsewhere in the systemic circuit, due to the arrangement of vessels at the glomerulus.

Blood pressure is low in typical systemic capillaries because capillary blood flows into the venous system, where resistance is relatively low. However, at the glomerulus, blood leaving the glomerular capillaries flows into an efferent arteriole, whose diameter is *smaller* than that of the afferent arteriole. The efferent arteriole thus offers considerable resistance, so relatively high pressures are needed to force blood into it. As a result, glomerular pressures are similar to those of small arteries, averaging about 50 mm Hg instead of the 35 mm Hg typical of peripheral capillaries.

Glomerular hydrostatic pressure is opposed by the **capsular hydrostatic pressure (CsHP)**, which tends to push water and solutes out of the filtrate and into the plasma. This pressure results from the resistance to flow along the nephron and the conducting system. (Before additional filtrate can enter the capsule, some of the filtrate already present must be forced into the PCT.) The CsHP averages about 15 mm Hg.

The *net hydrostatic pressure (NHP)* is the difference between the glomerular hydrostatic pressure, which tends to push water and solutes out of the bloodstream, and the capsular hydrostatic pressure, which tends to push water and solutes into the bloodstream. Net hydrostatic pressure can be calculated as follows:

$$NHP = (GHP - CsHP) = (50 \text{ mm Hg} - 15 \text{ mm Hg}) = 35 \text{ mm Hg}$$

Colloid Osmotic Pressure

The colloid osmotic pressure of a solution is the osmotic pressure resulting from the presence of suspended proteins. The **blood colloid osmotic pressure (BCOP)** tends to draw water out of the filtrate and into the plasma; it thus opposes filtration. Over the entire length of the glomerular capillary bed, the BCOP averages about 25 mm Hg. Under normal conditions, very few plasma proteins enter the capsular space, so no opposing colloid osmotic pressure exists within the capsule. However, if the glomeruli are damaged by disease or injury, and plasma proteins begin passing into the capsular space, a *capsular colloid osmotic pressure* is created that promotes filtration and increases fluid losses in urine.

Filtration Pressure

The **filtration pressure (FP)** at the glomerulus is the difference between the hydrostatic pressure and the colloid osmotic pressure acting across the glomerular capillaries. Under normal circumstances, this relationship can be summarized as

$$FP = NHP - BCOP$$

or

$$FP = 35 \text{ mm Hg} - 25 \text{ mm Hg} = 10 \text{ mm Hg}$$

This is the average pressure forcing water and dissolved materials out of the glomerular capillaries and into the capsular spaces (**Figure 26–10b**). Problems that affect filtration pressure can seriously disrupt kidney function and cause a variety of clinical signs and symptoms.

The Glomerular Filtration Rate

The **glomerular filtration rate (GFR)** is the amount of filtrate the kidneys produce each minute. Each kidney contains about 6 m²—some 64 square feet—of filtration surface, and the GFR averages an astounding *125 mL per minute*. This means that roughly 10 percent of the fluid delivered to the kidneys by the renal arteries leaves the bloodstream and enters the capsular spaces.

A *creatinine clearance test* is often used to estimate the GFR. Creatinine, which results from the breakdown of creatine phosphate in muscle tissue, is normally eliminated in urine. Creatinine enters the filtrate at the glomerulus and is not reabsorbed in significant amounts. By monitoring the creatinine concentrations in blood and the amount excreted in urine in a 24-hour period, a clinician can easily estimate the GFR. Consider, for example, a person who eliminates 84 mg of creatinine each hour and has a plasma creatinine concentration of 1.4 mg/dL. Because the GFR is equal to the amount secreted divided by the plasma concentration, this person's GFR is

$$\frac{84 \text{ mg/h}}{1.4 \text{ mg/dL}} = 60 \text{ dL/h} = 100 \text{ mL/min}.$$

The GFR is usually reported in milliliters per minute.

The value 100 mL/min is only an approximation of the GFR, because up to 15 percent of creatinine in the urine enters by means of active tubular secretion. When necessary, a more accurate GFR determination can be performed by using the complex carbohydrate *inulin*, which is not metabolized in the body and is neither reabsorbed nor secreted by the kidney tubules.

In the course of a single day, the glomeruli generate about 180 liters (48 gal) of filtrate, roughly 70 times the total plasma volume. But as filtrate passes through the renal tubules, about 99 percent of it is reabsorbed. You should now appreciate the significance of tubular reabsorption!

The glomerular filtration rate depends on the filtration pressure across glomerular capillaries. Any factor that alters the filtration pressure therefore alters the GFR, thereby affecting kidney function. One of the most significant factors is a drop in renal blood pressure. If blood pressure at the glomeruli drops by 20 percent (from 50 mm Hg to 40 mm Hg), kidney filtration will cease, because the filtration pressure will be 0 mm Hg. The kidneys are therefore sensitive to changes in blood pressure that have little or no effect on other organs. Hemorrhaging, shock, and dehydration are relatively common clinical conditions that can cause a dangerous decline in the GFR and lead to acute renal failure (p. 998).

Control of the GFR

Glomerular filtration is the vital first step essential to all other kidney functions. If filtration does not occur, waste products are not excreted, pH control is jeopardized, and an important mechanism for regulating blood volume is lost. It should be no surprise that a variety of regulatory mechanisms ensure that GFR remains within normal limits.

Filtration depends on adequate blood flow to the glomerulus and on the maintenance of normal filtration pressures. Three interacting levels of control stabilize GFR: (1) *autoregulation*, at the local level; (2) *hormonal regulation*, initiated by the kidneys; and (3) *autonomic regulation*, primarily by the sympathetic division of the autonomic nervous system.

Autoregulation of the GFR

Autoregulation maintains an adequate GFR despite changes in local blood pressure and blood flow. Maintenance of the GFR is accomplished by changing the diameters of afferent arterioles, efferent arterioles, and glomerular capillaries. The most important regulatory mechanisms stabilize the GFR when systemic blood pressure declines. A reduction in blood flow and a decline in glomerular blood pressure trigger (1) dilation of the afferent arteriole, (2) relaxation of supporting cells and dilation of the glomerular capillaries, and (3) constriction of the efferent arteriole. This combination increases blood flow and elevates glomerular blood pressure to normal levels. As a result, filtration rates remain relatively constant. The GFR also remains relatively constant when systemic blood pressure rises. A rise in renal blood pressure stretches the walls of afferent arterioles, and the smooth muscle cells respond by contracting. The reduction in the diameter of afferent arterioles decreases glomerular blood flow and keeps the GFR within normal limits.

Hormonal Regulation of the GFR

The GFR is regulated by the hormones of the renin–angiotensin system and the natriuretic peptides (ANP and BNP). These hormones and their actions were introduced in Chapters 18 and 21. ∞ pp. 635–637, 742–743 There are three triggers for the release of renin by the juxtaglomerular complex (JGC): (1) a decline in blood pressure at the glomerulus as the result of a decrease in blood volume, a fall in systemic pressures, or a blockage in the renal artery or its tributaries; (2) stimulation of juxtaglomerular cells by sympathetic innervation; or (3) a decline in the osmotic concentration of the tubular fluid at the macula densa. These stimuli are often interrelated. For example, a decline in systemic blood pressure reduces the glomerular filtration rate while baroreceptor reflexes cause sympathetic activation. Meanwhile, a reduction in the GFR slows the movement of tubular fluid along the nephron. Because the tubular fluid is then in the ascending limb of the nephron loop longer, the concentration of sodium and chloride ions in the tubular fluid reaching the macula densa and DCT becomes abnormally low.

Figure 26–11a provides a general overview of the response of the renin–angiotensin system to a decline in GFR. A fall in GFR leads to the release of renin, which activates angiotensin. Angiotensin II then elevates the blood volume and blood pressure, increasing the GFR. Figure 26–11b presents a more detailed view of the mechanisms involved. Once released into the bloodstream by the juxtaglomerular complex, renin converts the inactive protein angiotensinogen to angiotensin I. Angiotensin I, which is also inactive, is then converted to angiotensin II by **angiotensin-converting enzyme (ACE)**, primarily in the capillaries of the lungs. Angiotensin II is an active hormone whose primary effects include the following:

- *At the nephron*, angiotensin II causes the constriction of the efferent arteriole, further elevating glomerular pressures and filtration rates. Angiotensin II also directly stimulates the reabsorption of sodium ions and water at the PCT.

- *At the suprarenal glands*, angiotensin II stimulates the secretion of aldosterone by the suprarenal cortex. The aldosterone then accelerates sodium reabsorption in the DCT and cortical portion of the collecting system.

- *In the CNS*, angiotensin II (1) causes the sensation of thirst; (2) triggers the release of antidiuretic hormone (ADH), stimulating the reabsorption of water in the distal portion of the DCT and the collecting system; and (3) increases sympathetic motor tone, mobilizing the venous reserve, increasing cardiac output, and stimulating peripheral vasoconstriction.

- *In peripheral capillary beds*, angiotensin II causes a brief but powerful vasoconstriction of arterioles and precapillary sphincters, elevating arterial pressures throughout the body.

The combined effect is an increase in systemic blood volume and blood pressure and the restoration of normal GFR.

If blood volume rises, the GFR increases automatically, and this promotes fluid losses that help return blood volume to normal levels. If the elevation in blood volume is severe, hormonal factors further increase the GFR and accelerate fluid losses in the urine. As noted in Chapter 18, the *natriuretic peptides* are released in response to the stretching of the walls of the heart by increased blood volume or blood pressure. These hormones are released by the heart; atrial natriuretic peptide (ANP) is released by the atria, and brain natriuretic peptide (BNP) is released by the ventricles. ∞ pp. 637, 743 Among their other effects, the natriuretic peptides trigger the dilation of afferent arterioles and

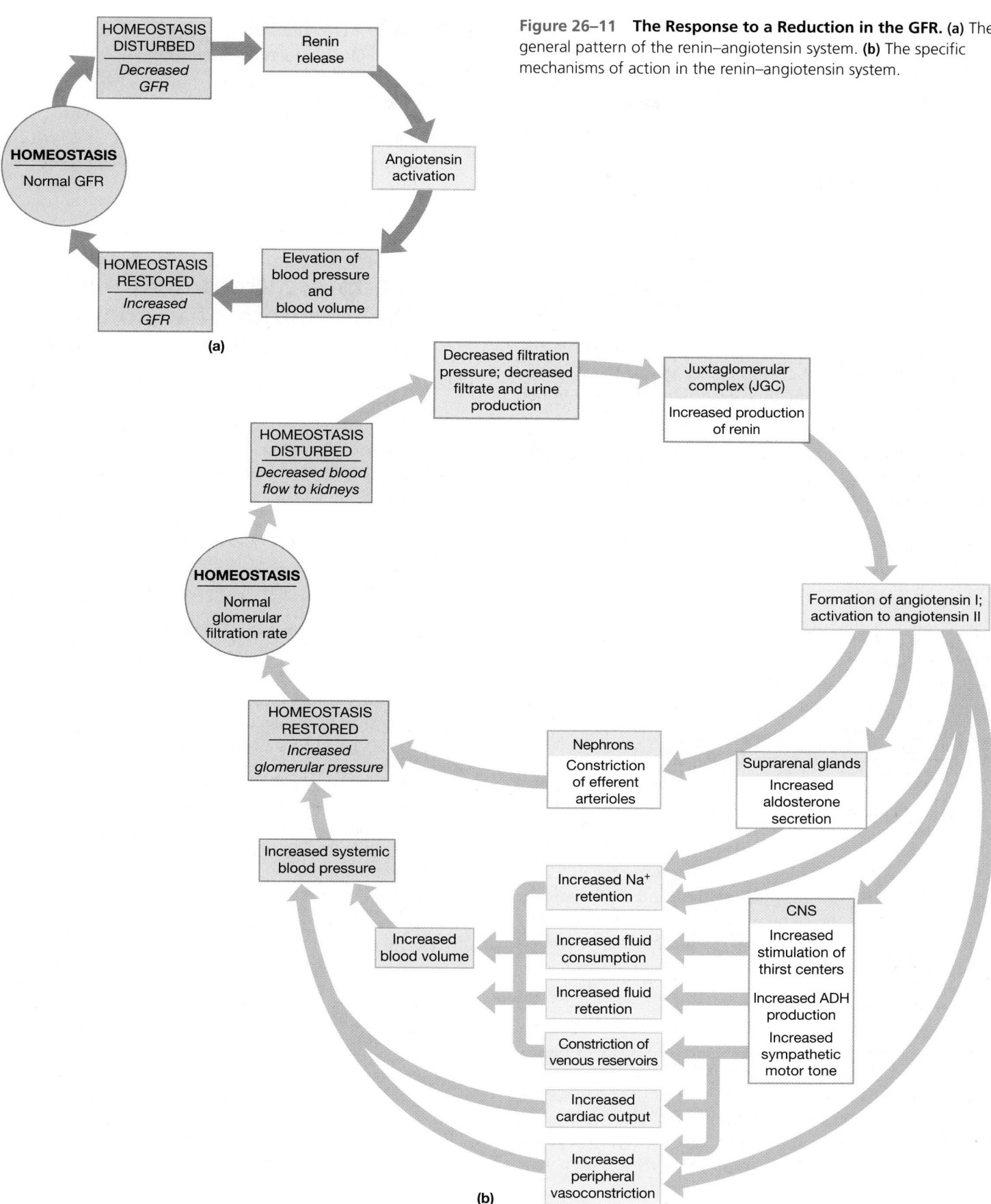

Figure 26–11   The Response to a Reduction in the GFR. (a) The general pattern of the renin–angiotensin system. **(b)** The specific mechanisms of action in the renin–angiotensin system.

(a)

(b)

26 URINARY

the constriction of efferent arterioles. This mechanism elevates glomerular pressures and increases the GFR. The natriuretic peptides also increase tubular reabsorption of sodium ions, and the net result is increased urine production and decreased blood volume and pressure.

Autonomic Regulation of the GFR

Most of the autonomic innervation of the kidneys consists of sympathetic postganglionic fibers. (The role of the few parasympathetic fibers in regulating kidney function is not known.) Sympathetic activation has one direct effect on the GFR: It produces a powerful vasoconstriction of afferent arterioles, decreasing the GFR and slowing the production of filtrate. Thus the sympathetic activation triggered by an acute fall in blood pressure or a heart attack overrides the local regulatory mechanisms that act to stabilize the GFR. As the crisis passes and sympathetic tone decreases, the filtration rate gradually returns to normal levels.

When the sympathetic division alters regional patterns of blood circulation, blood flow to the kidneys is often affected. For example, the dilation of superficial vessels in warm weather shunts blood away from the kidneys, so glomerular filtration declines temporarily. The effect becomes especially pronounced during periods of strenuous exercise. As the blood flow to your skin and skeletal muscles increases, kidney perfusion gradually decreases. These changes may be opposed, with variable success, by autoregulation at the local level.

At maximal levels of exertion, renal blood flow may be less than 25 percent of normal resting levels. This reduction can create problems for endurance athletes, because metabolic wastes build up over the course of a long event. *Proteinuria* (protein in the urine) commonly occurs after such events because the glomerular cells have been injured by prolonged hypoxia (low oxygen levels). If the damage is substantial, *hematuria* (blood in the urine) occurs. Hematuria develops in roughly 18 percent of marathon runners as a result of trauma to the bladder epithelium from the shocks of running. Proteinuria and hematuria generally disappear within 48 hours as the glomerular tissues are repaired. However, a small number of marathon and ultramarathon runners experience *acute renal failure*, with permanent impairment of kidney function.

The A&P Top 100

#91 Roughly 180 L (about 48 gal) of filtrate is produced at the glomeruli each day, which represents 70 times the total plasma volume. Almost all of that fluid volume must be reabsorbed to avoid fatal dehydration.

CHECKPOINT

9. What nephron structures are involved in filtration?

10. List the factors that influence filtration pressure.

11. List the factors that influence the rate of filtrate formation.

12. How would a decrease in blood pressure affect the GFR?

See the blue Answers tab at the end of the book.

26-5 Countercurrent multiplication and the influence of antidiuretic hormone and aldosterone affect reabsorption and secretion

Reabsorption recovers useful materials that have entered the filtrate, and secretion ejects waste products, toxins, or other undesirable solutes that did not leave the bloodstream at the glomerulus. Both processes occur in every segment of the nephron except the renal corpuscle, but their relative importance changes from segment to segment.

Reabsorption and Secretion at the PCT

The cells of the proximal convoluted tubule normally reabsorb 60–70 percent of the volume of the filtrate produced in the renal corpuscle. The reabsorbed materials enter the peritubular fluid, diffuse into peritubular capillaries, and are quickly returned to the circulation.

The PCT has five major functions:

1. *Reabsorption of Organic Nutrients.* Under normal circumstances, before the tubular fluid enters the nephron loop, the PCT reabsorbs more than 99 percent of the glucose, amino acids, and other organic nutrients in the fluid. This reabsorption involves a combination of facilitated transport and cotransport.

2. *Active Reabsorption of Ions.* The PCT actively transports several ions, including sodium, potassium, and bicarbonate ions (**Figure 26–12**), plus magnesium, phosphate, and sulfate ions. The ion pumps involved are individually regulated and may be influenced by circulating ion or hormone levels. For example, angiotensin II stimulates Na^+ reabsorption along the PCT. By absorbing carbon dioxide, the PCT indirectly recaptures roughly 90 percent of the bicarbonate ions from tubular fluid. Bicarbonate is important in stabilizing blood pH, a process we will examine further in Chapter 27.

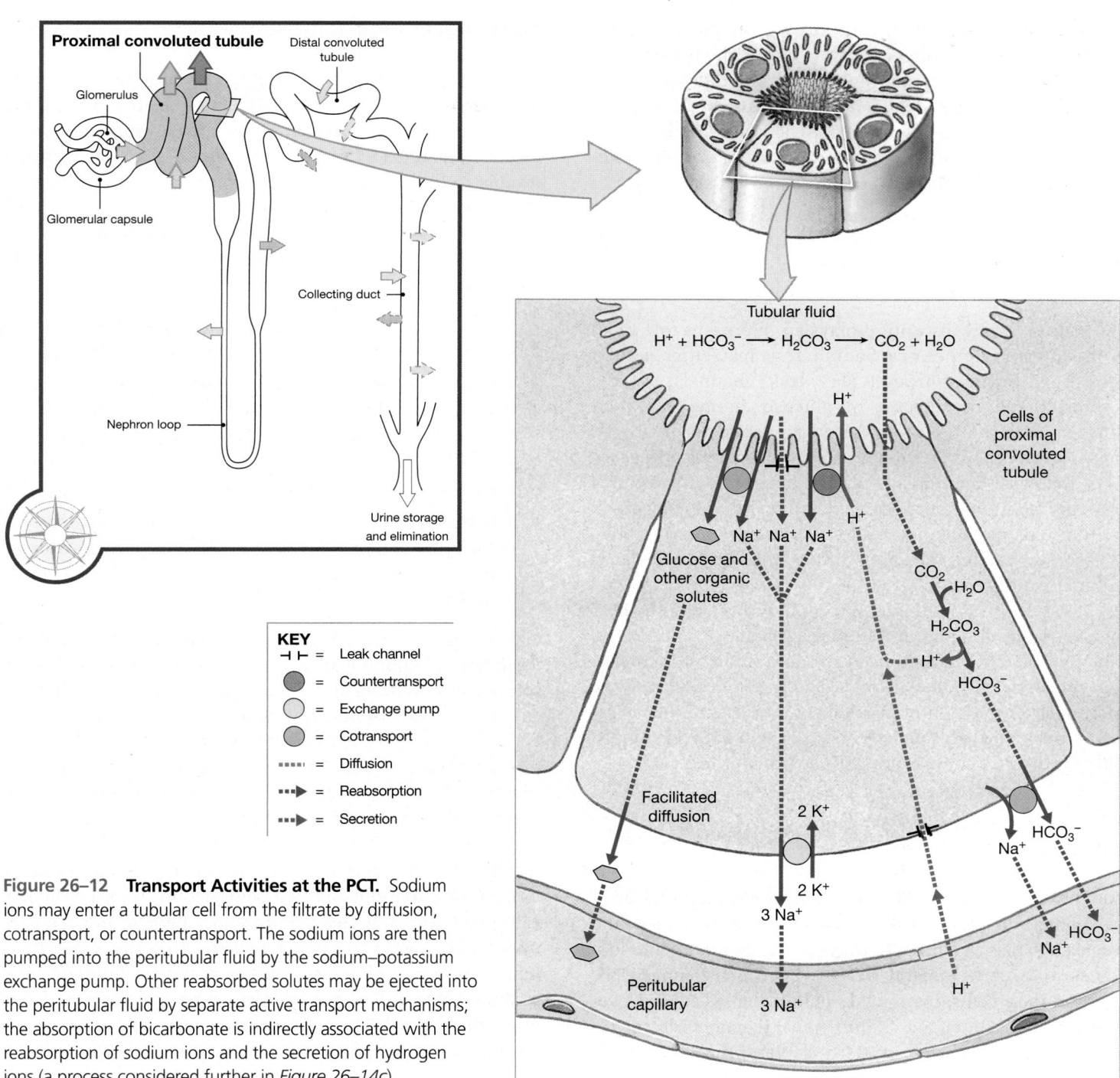

Figure 26–12 Transport Activities at the PCT. Sodium ions may enter a tubular cell from the filtrate by diffusion, cotransport, or countertransport. The sodium ions are then pumped into the peritubular fluid by the sodium–potassium exchange pump. Other reabsorbed solutes may be ejected into the peritubular fluid by separate active transport mechanisms; the absorption of bicarbonate is indirectly associated with the reabsorption of sodium ions and the secretion of hydrogen ions (a process considered further in *Figure 26–14c*).

3. *Reabsorption of Water.* The reabsorptive processes have a direct effect on the solute concentrations inside and outside the tubules. The filtrate entering the PCT has the same osmotic concentration as that of the surrounding peritubular fluid. As transport activities proceed, the solute concentration of tubular fluid decreases, and that of peritubular fluid and adjacent capillaries increases. Os-

mosis then pulls water out of the tubular fluid and into the peritubular fluid. Along the PCT, this mechanism results in the reabsorption of roughly 108 liters of water each day.

4. *Passive Reabsorption of Ions.* As active reabsorption of ions occurs and water leaves tubular fluid by osmosis, the concentration of other solutes in tubular fluid increases above

that in peritubular fluid. If the tubular cells are permeable to them, those solutes will move across the tubular cells and into the peritubular fluid by passive diffusion. Urea, chloride ions, and lipid-soluble materials may diffuse out of the PCT in this way. Such diffusion further reduces the solute concentration of the tubular fluid and promotes additional water reabsorption by osmosis.

5. *Secretion.* Active secretion also occurs along the PCT. Because the DCT performs comparatively little reabsorption, we will consider secretory mechanisms when we discuss the DCT.

Sodium ion reabsorption plays an important role in all of the foregoing processes. Sodium ions may enter tubular cells by diffusion through Na^+ leak channels; by the sodium-linked cotransport of glucose, amino acids, or other organic solutes; or by countertransport for hydrogen ions (**Figure 26–12**). In tubular cells, sodium ions diffuse toward the basal lamina. The plasma membrane in this area contains sodium–potassium exchange pumps that eject sodium ions in exchange for extracellular potassium ions. Reabsorbed sodium ions then diffuse into the adjacent peritubular capillaries.

The reabsorption of ions and compounds along the PCT involves many different carrier proteins. Some people have an inherited inability to manufacture one or more of these carrier proteins and are therefore unable to recover specific solutes from tubular fluid. In *renal glycosuria* (glī-kō-SOO-rē-uh), for example, a defective carrier protein makes it impossible for the PCT to reabsorb glucose from tubular fluid.

The Nephron Loop and Countercurrent Multiplication

Roughly 60–70 percent of the volume of filtrate produced at the glomerulus has been reabsorbed before the tubular fluid reaches the nephron loop. In the process, useful organic substrates and many mineral ions have been reclaimed. The nephron loop reabsorbs roughly half of the water, and two-thirds of the sodium and chloride ions, remaining in the tubular fluid. This reabsorption is performed efficiently according to the principle of countercurrent exchange, which was introduced in Chapter 25 in our discussion of heat conservation mechanisms. ∞ p. 958

The thin descending limb and the thick ascending limb of the nephron loop are very close together, separated only by peritubular fluid. The exchange that occurs between these segments is called **countercurrent multiplication**. *Countercurrent* refers to the fact that the exchange occurs between fluids moving in opposite directions: Tubular fluid in the descending limb flows toward the renal pelvis, whereas tubular fluid in the ascending limb flows toward the cortex. *Multiplication* refers to the fact that the effect of the exchange increases as movement of the fluid continues.

The two parallel segments of the nephron loop have very different permeability characteristics. The thin descending limb is permeable to water but relatively impermeable to solutes. The thick ascending limb, which is relatively impermeable to both water and solutes, contains active transport mechanisms that pump sodium and chloride ions from the tubular fluid into the peritubular fluid of the medulla.

A quick overview of countercurrent multiplication will help you make sense of the details:

- Sodium and chloride are pumped out of the thick ascending limb and into the peritubular fluid.
- This pumping action elevates the osmotic concentration in the peritubular fluid around the thin descending limb.
- The result is an osmotic flow of water out of the thin descending limb and into the peritubular fluid, increasing the solute concentration in the thin descending limb.
- The arrival of the highly concentrated solution in the thick ascending limb accelerates the transport of sodium and chloride ions into the peritubular fluid of the medulla.

Notice that this arrangement is a simple positive feedback loop: Solute pumping at the ascending limb leads to higher solute concentrations in the descending limb, which then result in accelerated solute pumping in the ascending limb.

We can now take a closer look at the mechanics of the process. **Figure 26–13a** diagrams ion transport across the epithelium of the thick ascending limb. Active transport at the apical surface moves sodium, potassium, and chloride ions out of the tubular fluid. The carrier is called a **Na^+–K^+/2 Cl^- transporter**, because each cycle of the pump carries a sodium ion, a potassium ion, and two chloride ions into the tubular cell. Potassium and chloride ions are pumped into the peritubular fluid by cotransport carriers. However, potassium ions are removed from the peritubular fluid as the sodium–potassium exchange pump pumps sodium ions out of the tubular cell. The potassium ions then diffuse back into the lumen of the tubule through potassium leak channels. The net result is that Na^+ and Cl^- enter the peritubular fluid of the renal medulla.

The removal of sodium and chloride ions from the tubular fluid in the ascending limb elevates the osmotic concentration of the peritubular fluid around the thin descending limb (**Figure 26–13b**). Because the thin descending limb is permeable to water but impermeable to solutes, as tubular fluid travels deeper into the medulla along the thin descending limb, osmosis moves water into the peritubular fluid. Solutes remain behind, so the tubular fluid reaching the turn of the nephron loop has a higher osmotic concentration than it did at the start.

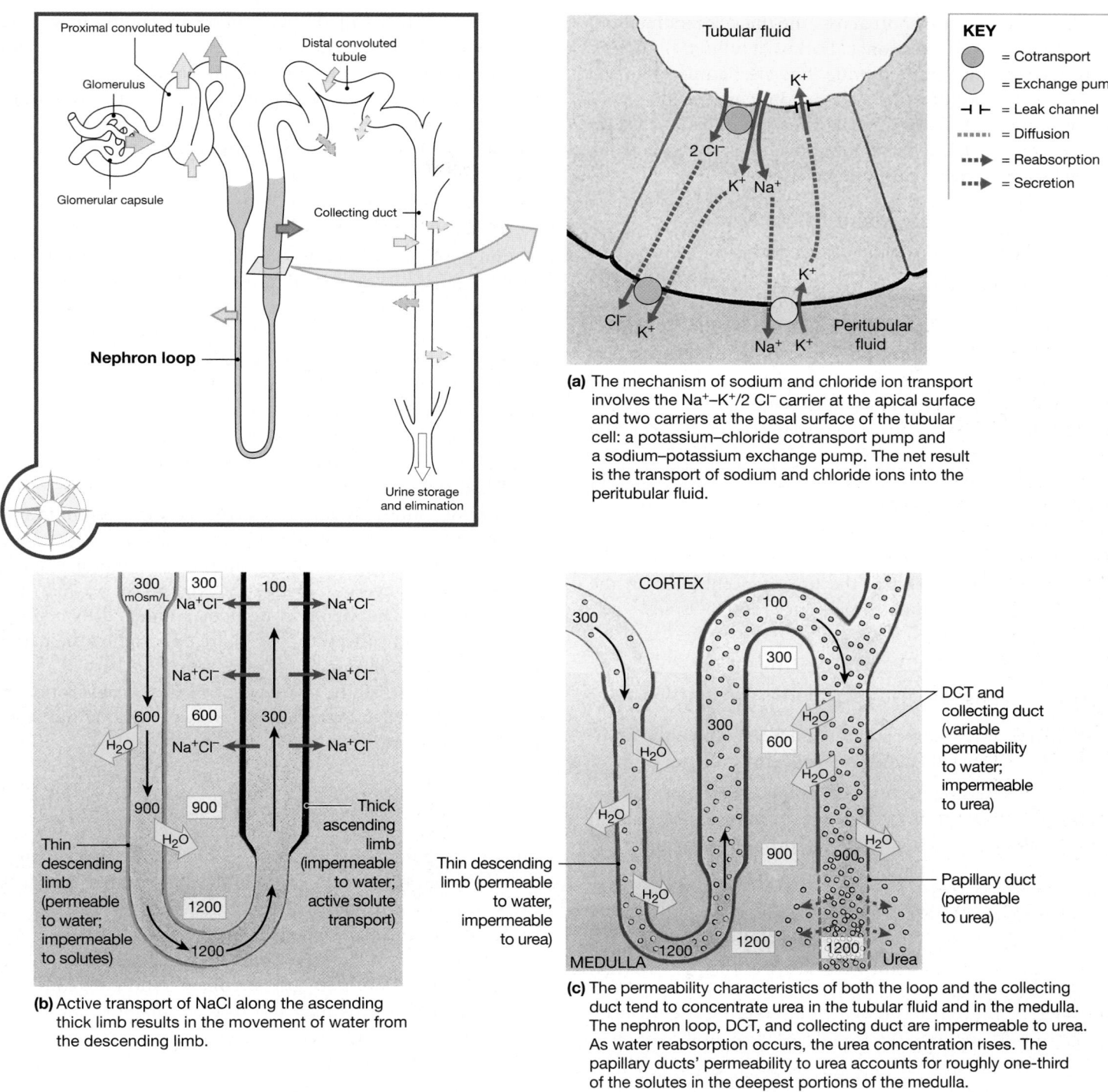

(a) The mechanism of sodium and chloride ion transport involves the Na^+–K^+/2 Cl^- carrier at the apical surface and two carriers at the basal surface of the tubular cell: a potassium–chloride cotransport pump and a sodium–potassium exchange pump. The net result is the transport of sodium and chloride ions into the peritubular fluid.

KEY
= Cotransport
= Exchange pump
= Leak channel
= Diffusion
= Reabsorption
= Secretion

(b) Active transport of NaCl along the ascending thick limb results in the movement of water from the descending limb.

(c) The permeability characteristics of both the loop and the collecting duct tend to concentrate urea in the tubular fluid and in the medulla. The nephron loop, DCT, and collecting duct are impermeable to urea. As water reabsorption occurs, the urea concentration rises. The papillary ducts' permeability to urea accounts for roughly one-third of the solutes in the deepest portions of the medulla.

Figure 26–13 Countercurrent Multiplication and Concentration of Urine.

The pumping mechanism of the thick ascending limb is highly effective: Almost two-thirds of the sodium and chloride ions that enter it are pumped out of the tubular fluid before that fluid reaches the DCT. In other tissues, differences in solute concentration are quickly resolved by osmosis. However, osmosis cannot occur across the wall of the thick ascending limb, because the tubular epithelium there is impermeable to water. Thus, as Na^+ and Cl^- are removed, the solute concentration in the tubular fluid declines. Tubular fluid arrives at the DCT with an osmotic concentration of only about 100 mOsm/L, one-third the concentration of the peritubular fluid of the renal cortex.

The rate of ion transport across the thick ascending limb is proportional to an ion's concentration in tubular fluid. As a result, more sodium and chloride ions are pumped into the medulla at the start of the thick ascending limb, where NaCl concentrations are highest, than near the cortex. This regional difference in the rate of ion transport is the basis of the concentration gradient within the medulla.

The Concentration Gradient of the Medulla

Normally, the maximum solute concentration of the peritubular fluid near the turn of the nephron loop is about 1200 mOsm/L (**Figure 26–13b**). Sodium and chloride ions pumped out of the loop's ascending limb account for roughly two-thirds of that gradient (750 mOsm/L). The rest of the concentration gradient results from the presence of urea. To understand how urea arrived in the medulla, we must look ahead to events in the last segments of the collecting system (**Figure 26–13c**). The thick ascending limb of the nephron loop, the DCT, and the collecting ducts are impermeable to urea. As water is reabsorbed, the concentration of urea gradually rises. The tubular fluid reaching the papillary duct typically contains urea at a concentration of about 450 mOsm/L. Because the papillary ducts are permeable to urea, the urea concentration in the deepest parts of the medulla also averages 450 mOsm/L.

Benefits of Countercurrent Multiplication

The countercurrent mechanism performs two functions:

1. It efficiently reabsorbs solutes and water before the tubular fluid reaches the DCT and collecting system.

2. It establishes a concentration gradient that permits the passive reabsorption of water from the tubular fluid in the collecting system. This reabsorption is regulated by circulating levels of antidiuretic hormone (ADH).

The tubular fluid arriving at the descending limb of the nephron loop has an osmotic concentration of roughly 300 mOsm/L, due primarily to the presence of ions such as Na^+ and Cl^-. The concentration of organic wastes, such as urea, is low. Roughly half of the tubular fluid entering the nephron loop is reabsorbed along the thin descending limb, and two-thirds of the Na^+ and Cl^- is reabsorbed along the thick ascending limb. As a result, the DCT receives a reduced volume of tubular fluid with an osmotic concentration of about 100 mOsm/L. Urea and other organic wastes, which were not pumped out of the thick ascending limb, now represent a significant proportion of the dissolved solutes.

Reabsorption and Secretion at the DCT

The composition and volume of tubular fluid change dramatically as it flows from the capsular space to the distal convoluted tubule. Only 15–20 percent of the initial filtrate volume reaches the DCT, and the concentrations of electrolytes and organic wastes in the arriving tubular fluid no longer resemble the concentrations in blood plasma. Selective reabsorption or secretion, primarily along the DCT, makes the final adjustments in the solute composition and volume of the tubular fluid.

Reabsorption at the DCT

Throughout most of the DCT, the tubular cells actively transport Na^+ and Cl^- out of the tubular fluid (**Figure 26–14a**). Tubular cells along the distal portions of the DCT also contain ion pumps that reabsorb tubular Na^+ in exchange for another cation (usually K^+) (**Figure 26–14b**). The ion pump and the Na^+ channels involved are controlled by the hormone *aldosterone*, produced by the suprarenal cortex. Aldosterone stimulates the synthesis and incorporation of sodium ion pumps and sodium channels in plasma membranes along the DCT and collecting duct. The net result is a reduction in the number of sodium ions lost in urine. However, sodium ion conservation is associated with potassium ion loss. Prolonged aldosterone stimulation can therefore produce *hypokalemia*, a dangerous reduction in the plasma K^+ concentration. The secretion of aldosterone and its actions on the DCT and collecting system are opposed by the natriuretic peptides (ANP and BNP).

The DCT is also the primary site of Ca^{2+} reabsorption, a process regulated by circulating levels of parathyroid hormone and calcitriol. ∞ pp. 625–626

Secretion at the DCT

The blood entering peritubular capillaries still contains a number of potentially undesirable substances that did not cross the filtration membrane at the glomerulus. In most cases, the concentrations of these materials are too low to cause physiological problems. However, any ions or compounds that are present in peritubular capillaries will diffuse into the peritubular fluid. If those concentrations become too high, the tubular cells may absorb these materials from the peritubular fluid and secrete them into the tubular fluid. Table 26–3 lists some of the substances secreted into tubular fluid by the distal and proximal convoluted tubules.

The rate of K^+ and H^+ secretion rises or falls in response to changes in their concentrations in peritubular fluid. The higher their concentration in the peritubular fluid, the higher the rate of secretion. Potassium and hydrogen ions merit special attention, because their concentrations in body fluids must be maintained within relatively narrow limits.

Potassium Ion Secretion. Figure 26–14a,b diagrams the mechanism of K^+ secretion. Potassium ions are removed from the peritubular fluid in exchange for sodium ions obtained from the tubular fluid. These potassium ions diffuse into the lumen through potassium channels at the apical surfaces of the tubular cells. In effect, tubular cells trade sodium ions in the tubular fluid for excess potassium ions in body fluids.

Hydrogen Ion Secretion. Hydrogen ion secretion is also associated with the reabsorption of sodium. **Figure 26–14c** depicts two routes of secretion. Both involve the generation of carbonic acid by the enzyme *carbonic anhydrase.* ∞ pp. 858, 893 Hydrogen ions generated by the dissociation of the carbonic acid are secreted by sodium-linked countertransport in exchange for Na^+ in the tubular fluid. The bicarbonate ions diffuse into the peritubular fluid and then into the bloodstream, where they help prevent changes in plasma pH.

Hydrogen ion secretion acidifies the tubular fluid while elevating the pH of the blood. Hydrogen ion secretion accelerates when the pH of the blood falls—as in *lactic acidosis,* which can develop after exhaustive muscle activity, or *ketoacidosis,* which can develop in starvation or diabetes mellitus. ∞ p. 950 The combination of H^+ removal and HCO_3^- production at the kidneys plays an important role in the control of blood pH. Because one of the secretory pathways is aldosterone sensitive, aldosterone stimulates H^+ secretion. Prolonged aldosterone stimulation can cause *alkalosis,* or abnormally high blood pH.

In Chapter 25, we noted that the production of lactic acid and ketone bodies during the postabsorptive state can cause acidosis. Under these conditions, the PCT and DCT deaminate amino acids in reactions that strip off the amino groups ($—NH_2$). The reaction sequence ties up H^+ and yields both **ammonium ions** (NH_4^+) and HCO_3^-. As indicated in **Figure 26–14c**, the ammonium ions are then pumped into the tubular fluid by sodium-linked countertransport, and the bicarbonate ions enter the bloodstream by way of the peritubular fluid.

Tubular deamination thus has two major benefits: It provides carbon chains suitable for catabolism, and it generates bicarbonate ions that add to the buffering capabilities of plasma.

Reabsorption and Secretion along the Collecting System

The collecting ducts receive tubular fluid from many nephrons and carry it toward the renal sinus, through the concentration gradient in the medulla. The normal amount of water and solute loss in the collecting system is regulated in two ways:

- By aldosterone, which controls sodium ion pumps along most of the DCT and the proximal portion of the collecting system. As noted above, these actions are opposed by the natriuretic peptides.

- By ADH, which controls the permeability of the DCT and collecting system to water. The secretion of ADH is suppressed by the natriuretic peptides, and this—combined with its effects on aldosterone secretion and action—can dramatically increase urinary water losses.

The collecting system also has other reabsorptive and secretory functions, many of which are important to the control of body fluid pH.

Reabsorption in the Collecting System

Important examples of solute reabsorption in the collecting system include the following:

- *Sodium Ion Reabsorption.* The collecting system contains aldosterone-sensitive ion pumps that exchange Na^+ in tubular fluid for K^+ in peritubular fluid (**Figure 26–14b**).

- *Bicarbonate Reabsorption.* Bicarbonate ions are reabsorbed in exchange for chloride ions in the peritubular fluid (**Figure 26–14c**).

- *Urea Reabsorption.* The concentration of urea in the tubular fluid entering the collecting duct is relatively high. The fluid entering the papillary duct generally has the same osmotic concentration as that of interstitial fluid of the medulla—about 1200 mOsm/L—but contains a much higher concentration of urea. As a result, urea tends to diffuse out of the tubular fluid and into the peritubular fluid in the deepest portion of the medulla.

Secretion in the Collecting System

The collecting system is an important site for the control of body fluid pH by means of the secretion of hydrogen or bicarbonate ions. If the pH of the peritubular fluid declines, carrier proteins pump hydrogen ions into the tubular fluid and reabsorb bicarbonate ions that help restore normal pH. If the pH of the peritubular fluid rises (a much less common event), the collecting system secretes bicarbonate ions and pumps hydrogen ions into the peritubular fluid. The net result is that the body eliminates a buffer and gains hydrogen ions that lower the pH. We will examine these responses in more detail in Chapter 27, when we consider acid–base balance.

The A&P Top 100

#92 Reabsorption involves a combination of diffusion, osmosis, channel-mediated diffusion, and active transport. Many of these processes are independently regulated by local or hormonal mechanisms. The primary mechanism governing water reabsorption can be described as "water follows salt." Secretion is a selective, carrier-mediated process.

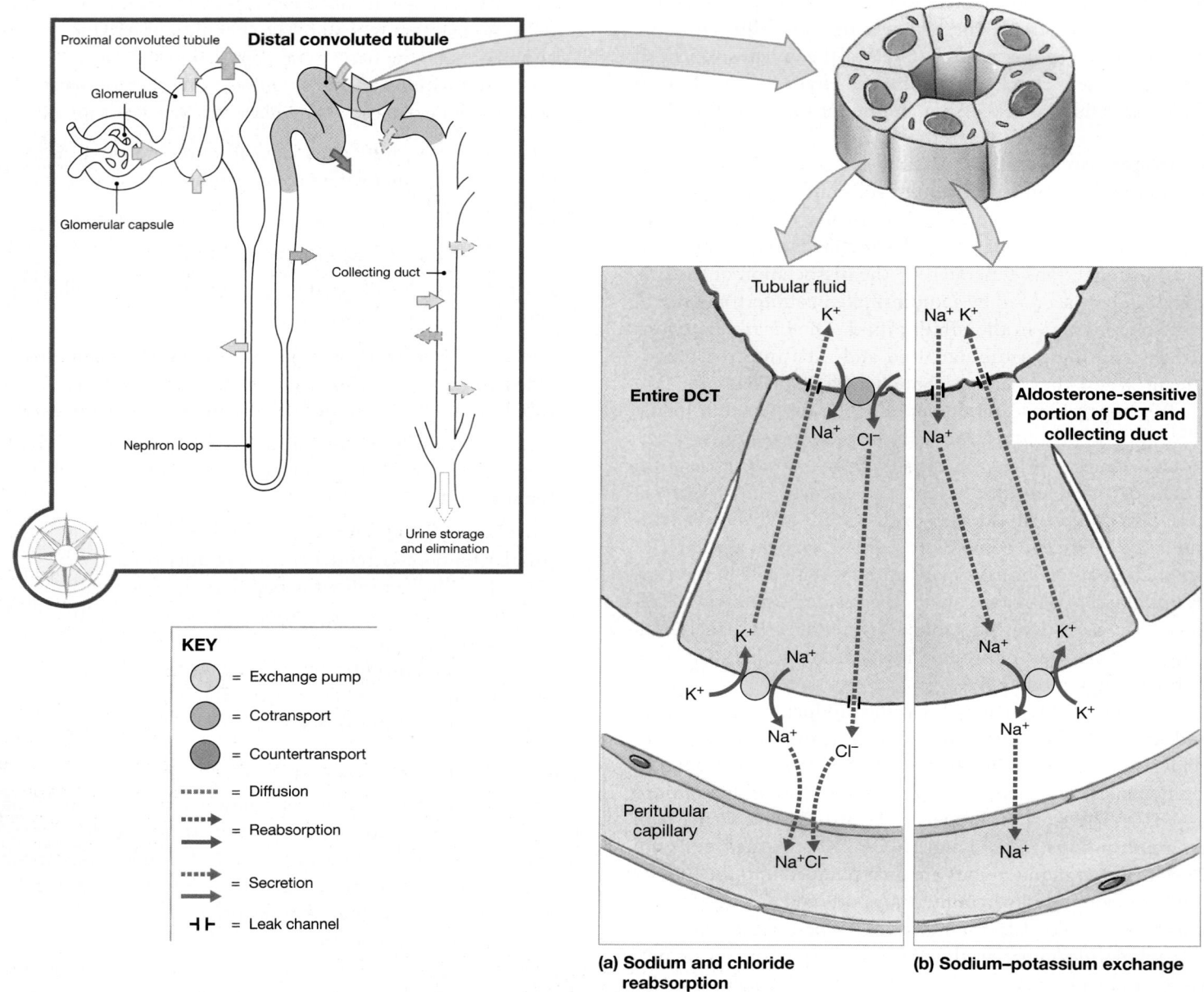

Figure 26–14 **Tubular Secretion and Solute Reabsorption at the DCT.** **(a)** The basic pattern of the absorption of sodium and chloride ions and the secretion of potassium ions. **(b)** Aldosterone-regulated absorption of sodium ions, linked to the passive loss of potassium ions. **(c)** Hydrogen ion secretion and the acidification of urine occur by two routes. The central theme is the exchange of hydrogen ions in the cytoplasm for sodium ions in the tubular fluid, and the reabsorption of the bicarbonate ions generated in the process.

The Control of Urine Volume and Osmotic Concentration

Urine volume and osmotic concentration are regulated through the control of water reabsorption. Water is reabsorbed by osmosis in the proximal convoluted tubule and the descending limb of the nephron loop. The water permeabilities of these regions cannot be adjusted, and water reabsorption occurs whenever the osmotic concentration of the peritubular fluid exceeds that of the tubular fluid. The ascending limb of the nephron loop is impermeable to water, but 1–2 percent of the volume of water in the original filtrate is recovered during sodium ion reabsorption in the distal convoluted tubule and collecting system. Because these water movements cannot be prevented, they represent *obligatory water reabsorption*, which usually recovers 85 percent of the volume of filtrate produced.

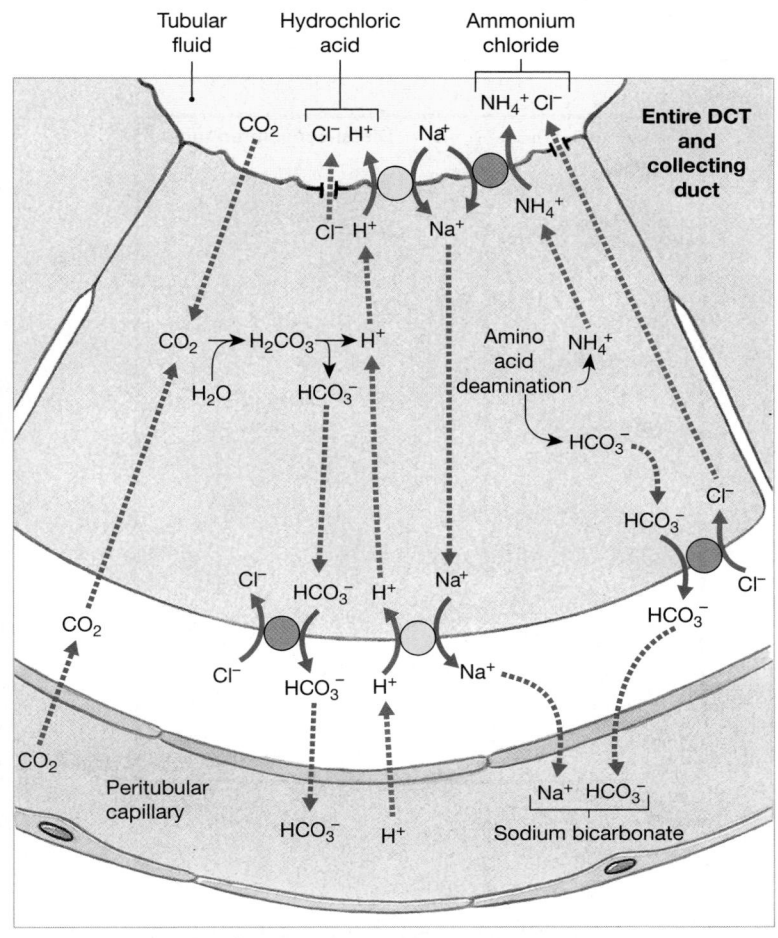

Tubular fluid | Hydrochloric acid | Ammonium chloride

(c) H⁺ secretion and HCO₃⁻ reabsorption

The volume of water lost in urine depends on how much of the water in the remaining tubular fluid (15 percent of the filtrate volume, or roughly 27 liters per day) is reabsorbed along the DCT and collecting system. The amount can be precisely controlled by a process called *facultative water reabsorption*. Precise control is possible because these segments are relatively impermeable to water except in the presence of ADH. This hormone causes the appearance of special *water channels* (*aquaporins*) in the apical plasma membranes, dramatically enhancing the rate of osmotic water movement. The higher the circulating levels of ADH, the greater the number of water channels, and the greater the water permeability of these segments.

As noted earlier in this chapter, the tubular fluid arriving at the DCT has an osmotic concentration of only about 100 mOsm/L. In the presence of ADH, osmosis occurs, and water moves out of the DCT until the osmotic concentration of the tubular fluid equals that of the surrounding cortex (roughly 300 mOsm/L). The tubular fluid then flows along the collecting duct, which passes through the concentration gradient of the medulla. Additional water is then reabsorbed, and the urine reaching the minor calyx has an osmotic concentration closer to 1200 mOsm/L. Just how closely the osmotic concentration approaches 1200 mOsm/L depends on how much ADH is present.

Figure 26–15 diagrams the effects of ADH on the DCT and collecting system. In the absence of ADH (**Figure 26–15a**), water is not reabsorbed in these segments, so all the fluid reaching the DCT is lost in the urine. The individual then produces large amounts of very dilute urine. That is just what happens in cases of *diabetes insipidus* ∞ p. 618, in which urinary water losses may reach 24 liters per day and the urine osmotic

CLINICAL NOTE

Diuretics

Diuresis (dī-ū-RĒ-sis; *dia*, through + *ouresis*, urination) is the elimination of urine. Whereas *urination* is an equivalent term in a general sense, *diuresis* typically indicates the production of a large volume of urine. **Diuretics** (dī-ū-RET-iks) are drugs that promote the loss of water in urine. The usual goal in diuretic therapy is a reduction in blood volume, blood pressure, extracellular fluid volume, or all three. The ability to control renal water losses with relatively safe and effective diuretics has saved the lives of many individuals, especially those with high blood pressure or congestive heart failure.

Diuretics have many mechanisms of action, but all such drugs affect transport activities or water reabsorption along the nephron and collecting system. For example, a class of diuretics called *thiazides* (THĪ-uh-zīdz) promotes water loss by reducing sodium and chloride ion transport in the proximal and distal convoluted tubules.

Diuretic use for nonclinical reasons is on the rise. For example, large doses of diuretics are taken by some bodybuilders to improve muscle definition temporarily, and by some fashion models or horse jockeys to reduce body weight for brief periods. This practice of "cosmetic dehydration" is extremely dangerous and has caused several deaths due to electrolyte imbalances and consequent cardiac arrest.

concentration is 30–400 mOsm/L. As ADH levels rise (**Figure 26–15b**), the DCT and collecting system become more permeable to water, the amount of water reabsorbed increases, and the urine osmotic concentration climbs. Under maximum ADH stimulation, the DCT and collecting system become so permeable to water that the osmotic concentration of the urine is equal to that of the deepest portion of the medulla. Note that the concentration of urine can never *exceed* that of the medulla, because the concentrating mechanism relies on osmosis.

The hypothalamus continuously secretes ADH at low levels, and the DCT and collecting system always have a significant degree of water permeability. The DCT normally reabsorbs roughly 9 liters of water per day, or about 5 percent of the original volume of filtrate produced by the glomeruli. At normal ADH levels, the collecting system reabsorbs roughly 16.8 liters per day, or about 9.3 percent of the original volume of filtrate. A healthy adult typically produces 1200 mL of urine per day (about 0.6 percent of the filtrate volume), with an osmotic concentration of 800–1000 mOsm/L.

The effects of ADH are opposed by those of the natriuretic peptides, ANP and BNP. These hormones stimulate the production of a large volume of relatively dilute urine, soon restoring plasma volume to normal.

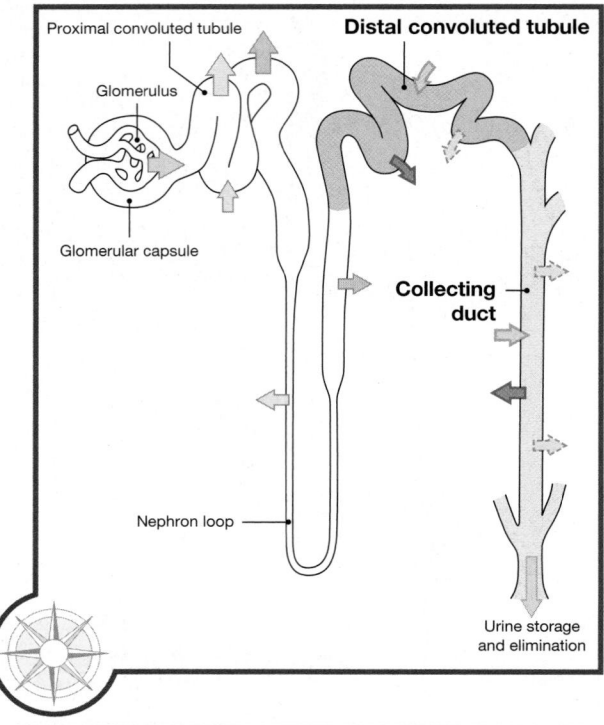

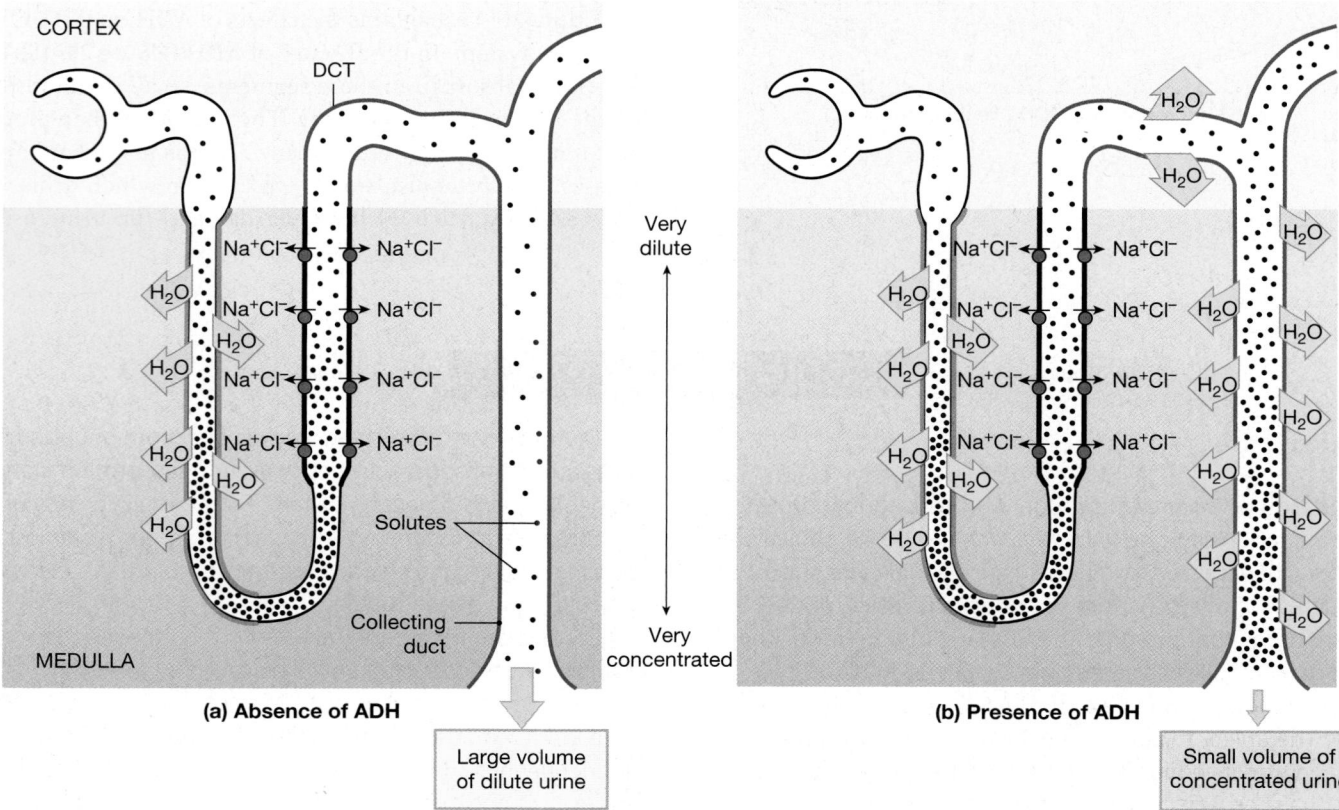

(a) **Absence of ADH**

Large volume of dilute urine

(b) **Presence of ADH**

Small volume of concentrated urine

Figure 26–15 **The Effects of ADH on the DCT and Collecting Duct.** (a) Tubule permeabilities and the osmotic concentration of urine in the absence of ADH. (b) Tubule permeabilities and the osmotic concentration of urine at maximal ADH levels.

The Function of the Vasa Recta

The solutes and water reabsorbed in the medulla must be returned to the general circulation without disrupting the concentration gradient. This return is the function of the vasa recta. Blood entering the vasa recta has an osmotic concentration of approximately 300 mOsm/L. The blood descending into the medulla gradually increases in osmotic concentration as the solute concentration in the peritubular fluid rises. This increase in blood osmotic concentration involves both solute absorption and water loss, but solute absorption predominates, because the plasma proteins limit osmotic flow out of the blood. ∞ p. 735 Blood flowing toward the cortex gradually decreases in osmotic concentration as the solute concentration of the peritubular fluid declines. Again, this decrease involves a combination of solute diffusion and osmosis, but in this case osmosis predominates, because the presence of plasma proteins does not oppose the osmotic flow of water into the blood.

The net results are that (1) some of the solutes absorbed in the descending portion of the vasa recta do not diffuse out in the ascending portion and (2) more water moves into the ascending portion of the vasa recta than is moved out in the descending portion. Thus, the vasa recta carries both water and solutes out of the medulla. Under normal conditions, the removal of solutes and water by the vasa recta precisely balances the rates of solute reabsorption and osmosis in the medulla.

The Composition of Normal Urine

More than 99 percent of the 180 liters of filtrate produced each day by the glomeruli is reabsorbed and never reaches the renal pelvis. General characteristics of the remaining filtrate—normal urine—are listed in Table 26–5. However, the composition of the urine produced each day varies with the metabolic and hormonal events under way.

The composition and concentration of urine are two related but distinct properties. The *composition* of urine reflects the filtration, absorption, and secretion activities of the nephrons. Some compounds (such as urea) are neither ac-

tively excreted nor reabsorbed along the nephron. In contrast, organic nutrients are completely reabsorbed, and other compounds, such as creatinine, that are missed by the filtration process are actively secreted into the tubular fluid.

The processes mentioned in the previous paragraph determine the identities and amounts of materials excreted in urine. The *concentration* of these components in a given urine sample depends on the osmotic movement of water across the walls of the tubules and collecting ducts. Because the composition and concentration of urine vary independently, you can produce a small volume of concentrated urine or a large volume of dilute urine and still excrete the same amount of dissolved materials. As a result, physicians who are interested in a detailed assessment of renal function commonly analyze the urine produced over a 24-hour period rather than testing a single urine sample.

Normal urine is a clear, sterile solution. Its yellow color results from the presence of the pigment urobilin, generated in the kidneys from the urobilinogens produced by intestinal bacteria and absorbed in the colon (see **Figure 19–5**, p. 659). The evaporation of small molecules, such as ammonia, accounts for the characteristic odor of urine. Other substances not normally present, such as acetone or other ketone bodies, can also impart a distinctive smell. **Urinalysis**, the analysis of a urine sample, is a diagnostic tool of considerable importance, even in high-technology medicine. A standard urinalysis includes an assessment of the color and appearance of urine, two characteristics that can be determined without specialized equipment. In the 17th century, physicians classified the taste of the urine as sweet, salty, and so on, but quantitative analytical tests have long since replaced the taste-bud assay. Table 26–6 gives some typical values obtained from urinalysis.

Summary: Renal Function

Table 26–4 lists the functions of each segment of the nephron and collecting system. **Figure 26–16** provides a functional overview that summarizes the major steps involved in the reabsorption of water and the production of concentrated urine:

Step 1 The filtrate produced at the renal corpuscle has the same osmotic concentration as does plasma—about 300 mOsm/L. It has the composition of blood plasma, minus the plasma proteins.

Step 2 In the PCT, the active removal of ions and organic substrates produces a continuous osmotic flow of water out of the tubular fluid. This process reduces the volume of filtrate but keeps the solutions inside and outside the tubule isotonic. Between 60 and 70 percent of the filtrate volume has been reabsorbed before the tubular fluid reaches the descending limb of the nephron loop.

TABLE 26–5 General Characteristics of Normal Urine	
Characteristic	**Normal Range**
pH	4.5–8 (average: 6.0)
Specific gravity	1.003–1.030
Osmotic concentration (osmolarity)	855–1335 mOsm/L
Water content	93–97%
Volume	700–2000 mL/day
Color	Clear yellow
Odor	Varies with composition
Bacterial content	None (sterile)

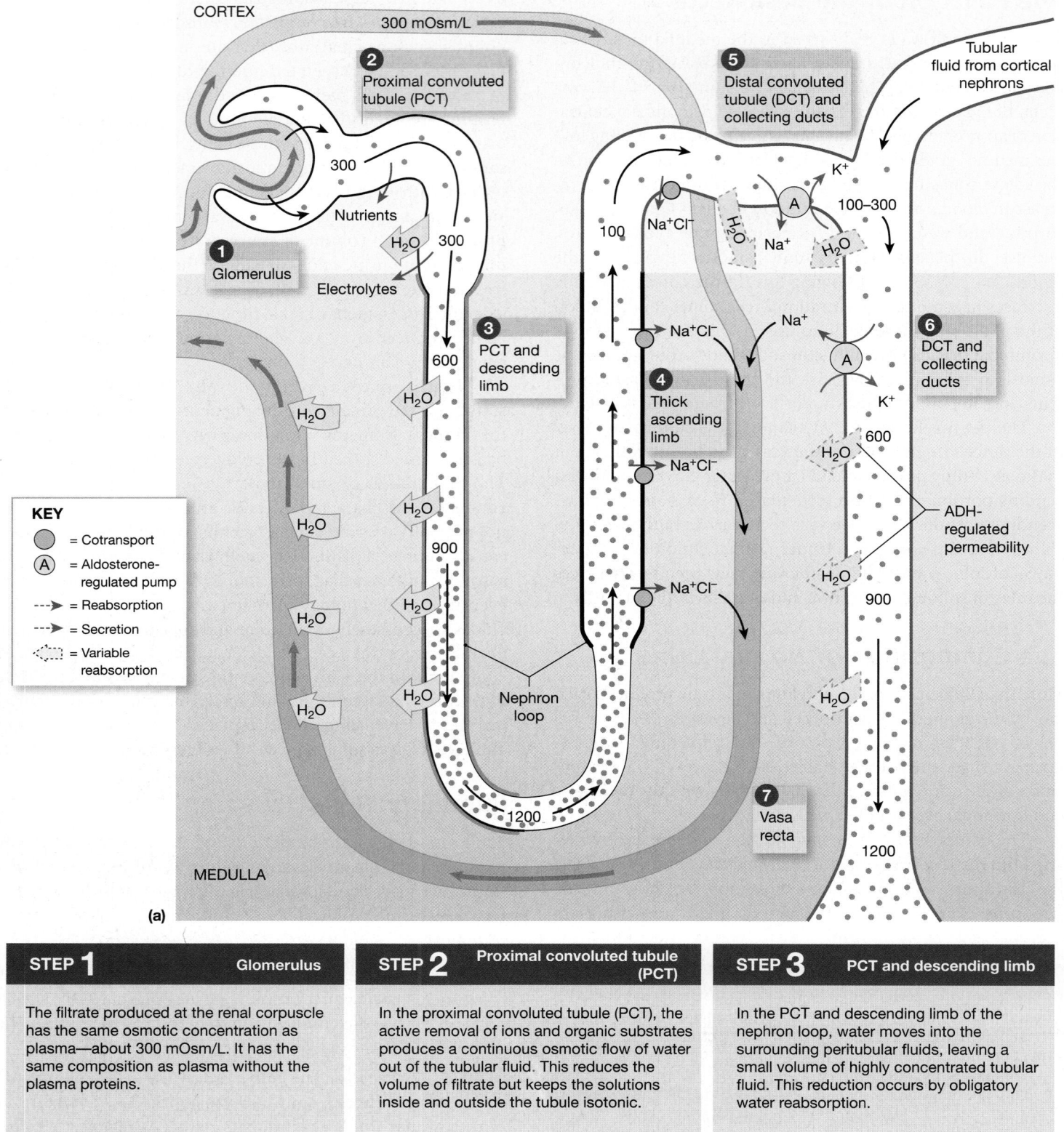

(a)

KEY

- ⬤ = Cotransport
- Ⓐ = Aldosterone-regulated pump
- --→ = Reabsorption
- --→ = Secretion
- ⬠ = Variable reabsorption

STEP **1** Glomerulus	STEP **2** Proximal convoluted tubule (PCT)	STEP **3** PCT and descending limb
The filtrate produced at the renal corpuscle has the same osmotic concentration as plasma—about 300 mOsm/L. It has the same composition as plasma without the plasma proteins.	In the proximal convoluted tubule (PCT), the active removal of ions and organic substrates produces a continuous osmotic flow of water out of the tubular fluid. This reduces the volume of filtrate but keeps the solutions inside and outside the tubule isotonic.	In the PCT and descending limb of the nephron loop, water moves into the surrounding peritubular fluids, leaving a small volume of highly concentrated tubular fluid. This reduction occurs by obligatory water reabsorption.

Figure 26–16   A Summary of Renal Function. (a) A general overview of steps and events. **(b)** Specific changes in the composition and concentration of the tubular fluid as it flows along the nephron and collecting duct.

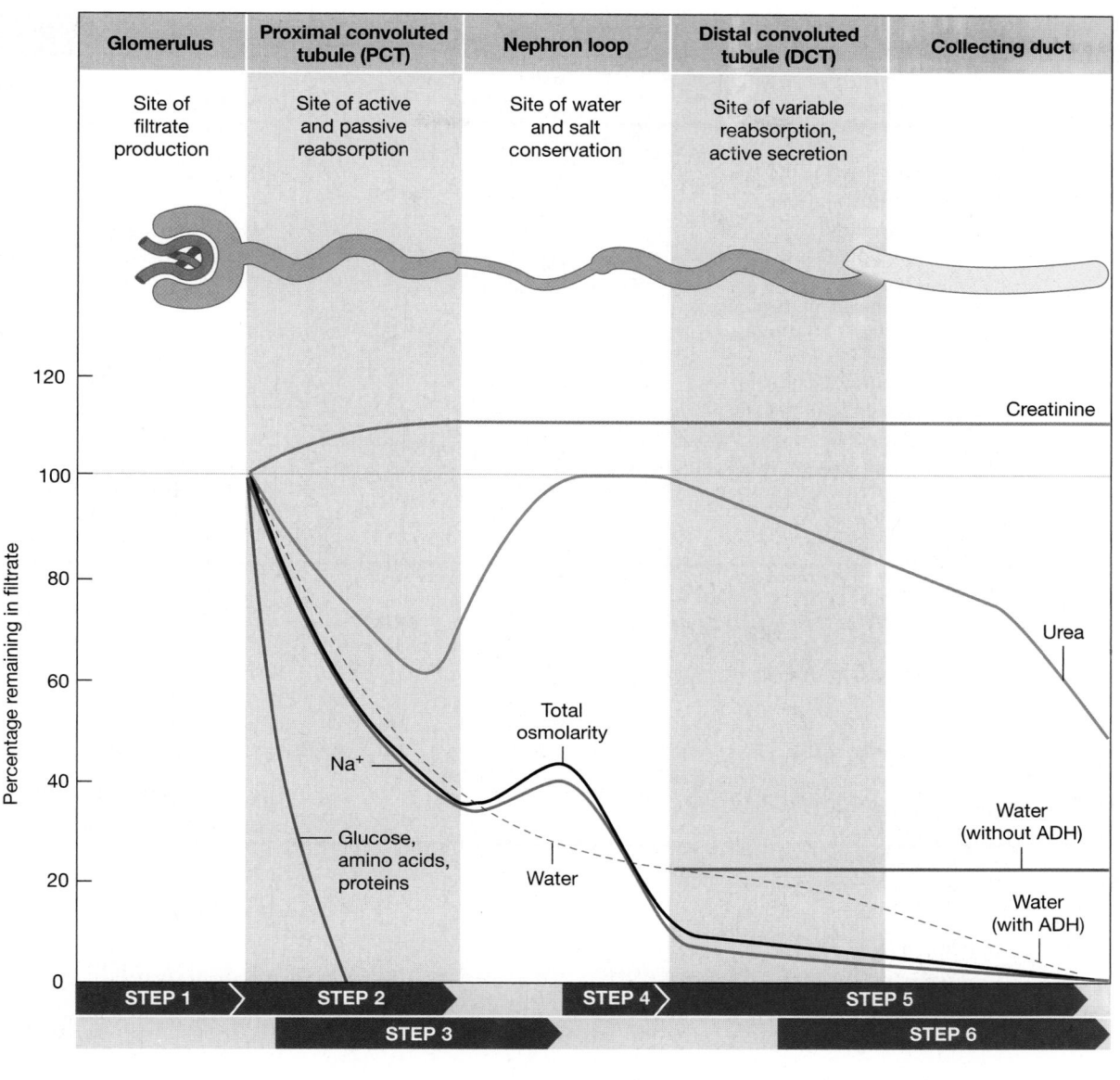

Glomerulus	Proximal convoluted tubule (PCT)	Nephron loop	Distal convoluted tubule (DCT)	Collecting duct
Site of filtrate production	Site of active and passive reabsorption	Site of water and salt conservation	Site of variable reabsorption, active secretion	

(b)

STEP **4**	Thick ascending limb	STEP **5**	DCT and collecting ducts	STEP **6**	DCT and collecting ducts	STEP **7**	Vasa recta
The thick ascending limb is impermeable to water and solutes. The tubular cells actively transport Na⁺ and Cl⁻ out of the tubule, thereby lowering the osmotic concentration of the tubular fluid. Because just Na⁺ and Cl⁻ are removed, urea accounts for a higher proportion of the total osmotic concentration at the end of the loop.		The final adjustments in the composition of the tubular fluid occur in the DCT and the collecting system. The osmotic concentration of the tubular fluid can be adjusted through active transport (reabsorption or secretion).		The final adjustments in the volume and osmotic concentration of the tubular fluid are made by controlling the water permeabilities of the distal portions of the DCT and the collecting system. The level of exposure to ADH determines the final urine concentration.		The vasa recta absorbs the solutes and water reabsorbed by the nephron loop and the collecting ducts. By removing these solutes and water into the main circulatory system, the vasa recta maintains the concentration gradient of the medulla.	

TABLE 26–6 Typical Values Obtained from Standard Urinalysis				
Compound	**Primary Source**	**Daily Elimination***	**Concentration**	**Remarks**
NITROGENOUS WASTES				
Urea	Deamination of amino acids by liver and kidneys	21 g	1.8 g/dL	Rises if negative nitrogen balance exists
Creatinine	Breakdown of creatine phosphate in skeletal muscle	1.8 g	150 mg/dL	Proportional to muscle mass; decreases during atrophy or muscle disease
Ammonia	Deamination by liver and kidney, absorption from intestinal tract	0.68 g	60 mg/dL	
Uric acid	Breakdown of purines	0.53 g	40 mg/dL	Increases in gout, liver diseases
Hippuric acid	Breakdown of dietary toxins	4.2 mg	350 μg/dL	
Urobilin	Urobilinogens absorbed at colon	1.5 mg	125 μg/dL	Gives urine its yellow color
Bilirubin	Hemoglobin breakdown product	0.3 mg	20 μg/dL	Increase may indicate problem with liver elimination or excess production; causes yellowing of skin and mucous membranes in jaundice
NUTRIENTS AND METABOLITES				
Carbohydrates		0.11 g	9 μg/dL	Primarily glucose; *glycosuria* develops if T_m is exceeded
Ketone bodies		0.21 g	17 μg/dL	Ketonuria may occur during postabsorptive state
Lipids		0.02 g	1.6 μg/dL	May increase in some kidney diseases
Amino acids		2.25 g	287.5 μg/dL	Note relatively high loss compared with other metabolites due to low T_m; excess *(aminoaciduria)* indicates T_m problem
IONS				
Sodium		4.0 g	333 mg/dL	Varies with diet, urine pH, hormones, etc.
Chloride		6.4 g	533 mg/dL	
Potassium		2.0 g	166 mg/dL	Varies with diet, urine pH, hormones, etc.
Calcium		0.2 g	17 mg/dL	Hormonally regulated (PTH/CT)
Magnesium		0.15 g	13 mg/dL	
BLOOD CELLS†				
RBCs		130,000/day	100/mL	Excess *(hematuria)* indicates vascular damage in urinary system
WBCs		650,000/day	500/mL	Excess *(pyuria)* indicates renal infection or inflammation

*Representative values for a 70-kg (154-lb) male.

†Usually estimated by counting the cells in a sample of sediment after urine centrifugation.

Step 3 In the PCT and descending limb of the nephron loop, water moves into the surrounding peritubular fluids, leaving a small volume (15–20 percent of the original filtrate) of highly concentrated tubular fluid. The reduction in volume has occurred by obligatory water reabsorption.

Step 4 The thick ascending limb is impermeable to water and solutes. The tubular cells actively transport Na$^+$ and Cl$^-$ out of the tubular fluid, thereby lowering the osmotic concentration of tubular fluid without affecting its volume. The tubular fluid reaching the distal convoluted tubule is hypotonic relative to the peritubular fluid, with an osmotic concentration of only about 100 mOsm/L. Because only Na$^+$ and Cl$^-$ are removed, urea accounts for a significantly higher proportion of the total osmotic concentration at the end of the loop than it did at the start of it.

Step 5 The final adjustments in the composition of the tubular fluid are made in the DCT and the collecting system.

Although the DCT and collecting duct are generally impermeable to solutes, the osmotic concentration of tubular fluid can be adjusted through active transport (reabsorption or secretion). Some of these transport activities are hormonally regulated.

Step 6 The final adjustments in the volume and osmotic concentration of the tubular fluid are made by controlling the water permeabilities of the distal portions of the DCT and the collecting system. These segments are relatively impermeable to water unless exposed to ADH. Under maximum ADH stimulation, urine volume is at a minimum, and urine osmotic concentration is equal to that of the peritubular fluid in the deepest portion of the medulla (roughly 1200 mOsm/L).

Step 7 The vasa recta absorbs solutes and water reabsorbed by the nephron loop and the collecting ducts, thereby maintaining the concentration gradient of the medulla.

CHECKPOINT

13. What effect would increased amounts of aldosterone have on the K$^+$ concentration in urine?

14. What effect would a decrease in the Na$^+$ concentration of filtrate have on the pH of tubular fluid?

15. How would the lack of juxtamedullary nephrons affect the volume and osmotic concentration of urine?

16. Why does a decrease in the amount of Na$^+$ in the distal convoluted tubule lead to an increase in blood pressure?

See the blue Answers tab at the end of the book.

26-6 Urine is transported via the ureters, stored in the bladder, and eliminated through the urethra, aided by the micturition reflex

Filtrate modification and urine production end when the fluid enters the renal pelvis. The urinary tract (the ureters, urinary bladder, and urethra) is responsible for the transport, storage, and elimination of urine. A **pyelogram** (PĪ-el-ō-gram) is an image of the urinary system, obtained by taking an x-ray of the kidneys after a radiopaque compound has been administered intravenously. Such an image provides an orientation to the relative sizes and positions of the main structures (**Figure 26–17**). The sizes of the minor and major calyces, the renal pelvis, the ureters, the urinary bladder, and

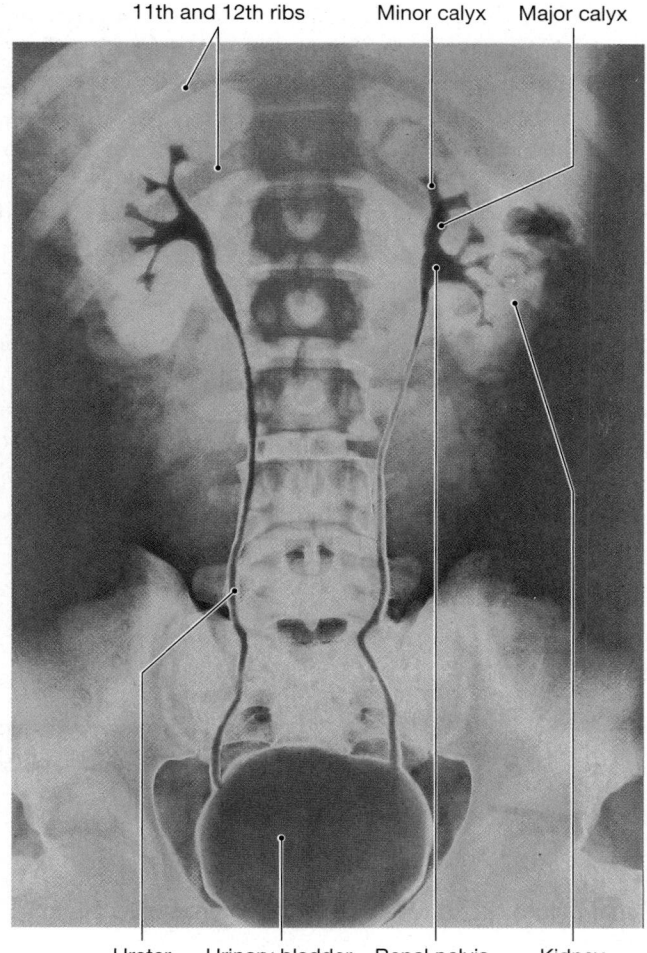

11th and 12th ribs   Minor calyx   Major calyx

Ureter   Urinary bladder   Renal pelvis   Kidney

Figure 26–17 **A Pyelogram.** This posterior view of urinary system structures was color-enhanced. ATLAS: Plate 62b

26 URINARY

the proximal portion of the urethra are somewhat variable, because these regions are lined by a *transitional epithelium* that can tolerate cycles of distension and contraction without damage. ∞ p. 119

The Ureters

The ureters are a pair of muscular tubes that extend from the kidneys to the urinary bladder—a distance of about 30 cm (12 in.). Each ureter begins at the funnel-shaped renal pelvis (**Figure 26–4**). The ureters extend inferiorly and medially, passing over the anterior surfaces of the *psoas major muscles* (**Figure 26–3**). The ureters are retroperitoneal and are firmly attached to the posterior abdominal wall. The paths taken by the ureters in men and women are different, due to variations in the nature, size, and position of the reproductive organs. In males, the base of the urinary bladder lies between the rectum and the pubic symphysis (**Figure 26–18a**);

in females, the base of the urinary bladder sits inferior to the uterus and anterior to the vagina (**Figure 26–18b**).

The ureters penetrate the posterior wall of the urinary bladder without entering the peritoneal cavity. They pass through the bladder wall at an oblique angle, and the **ureteral openings** are slitlike rather than rounded (**Figure 26–18c**). This shape helps prevent the backflow of urine toward the ureter and kidneys when the urinary bladder contracts.

Histology of the Ureters

The wall of each ureter consists of three layers (**Figure 26–19a**): (1) an inner mucosa, comprising a transitional epithelium and the surrounding lamina propria; (2) a middle muscular layer made up of longitudinal and circular bands of smooth muscle; and (3) an outer connective-tissue layer that is continuous with the fibrous capsule and peritoneum. About every 30 seconds, a peristaltic contraction begins at the renal pelvis and sweeps along the ureter, forcing urine toward the urinary bladder.

The Urinary Bladder

The urinary bladder is a hollow, muscular organ that functions as a temporary reservoir for the storage of urine (**Figure 26–18c**). The dimensions of the urinary bladder vary with its state of distension, but a full urinary bladder can contain as much as a liter of urine.

The superior surfaces of the urinary bladder are covered by a layer of peritoneum; several peritoneal folds assist in stabilizing its position. The **median umbilical ligament** extends from the anterior, superior border toward the umbilicus (navel). The **lateral umbilical ligaments** pass along the sides of the bladder to the umbilicus. These fibrous cords are the vestiges of the two *umbilical arteries*, which supplied blood to

CLINICAL NOTE

Renal Failure and Kidney Transplantation

Renal failure occurs when the kidneys become unable to perform the excretory functions needed to maintain homeostasis. When kidney filtration slows for any reason, urine production declines. As the decline continues, signs and symptoms of renal failure appear because water, ions, and metabolic wastes are retained. Virtually all systems in the body are affected. Fluid balance, pH, muscular contraction, metabolism, and digestive function are disturbed. The individual generally becomes hypertensive; anemia develops due to a decline in erythropoietin production; and CNS problems can lead to sleeplessness, seizures, delirium, and even coma.

Acute renal failure occurs when renal ischemia, urinary obstruction, trauma, or exposure to nephrotoxic drugs causes filtration to slow suddenly or stop. The reduction in kidney function occurs over a few days and may persist for weeks. Sensitized individuals can also develop acute renal failure after an allergic response to antibiotics or anesthetics. Most deaths associated with acute renal failure are caused by the underlying non-renal disease. With supportive treatment, the kidneys may then regain partial or complete function. (With supportive treatment, the mortality rate is approximately 50 percent.) In *chronic renal failure*, kidney function deteriorates gradually, and the associated problems accumulate over years. This condition generally cannot be reversed; its progression can only be slowed, and symptoms of *end-stage renal failure* eventually develop. The symptoms of end-stage and acute renal failure can be relieved by *kidney dialysis*, but this treatment is not a cure.

Probably the most satisfactory solution to the problem of end-stage renal failure, in terms of overall quality of life, is *kidney transplantation*. This procedure involves the implantation of a new kidney obtained from a living donor or a cadaver. Of the 17,090 kidneys transplanted in 2006, 6434 came from living donors; the rest came from cadavers. In 2007, 73,551 people were on the kidney waiting list. The recipient's nonfunctioning kidney(s) may be removed, especially if an infection is present. The transplanted kidney and ureter is usually placed retroperitoneally in the pelvic cavity (within the iliac fossa) and the ureter connected to the recipient's urinary bladder.

The success rate for kidney transplantation varies, depending on how aggressively the recipient's T cells attack the donated organ and whether infection develops. The one-year success rate for transplantation is now 89% when a cadaver kidney is used, and 95.1% when the kidney comes from a living donor. The use of kidneys taken from close relatives significantly improves the chances that the transplant will succeed. Immunosuppressive drugs are administered to reduce tissue rejection, but unfortunately, this treatment also lowers the individual's resistance to infection or cancer.

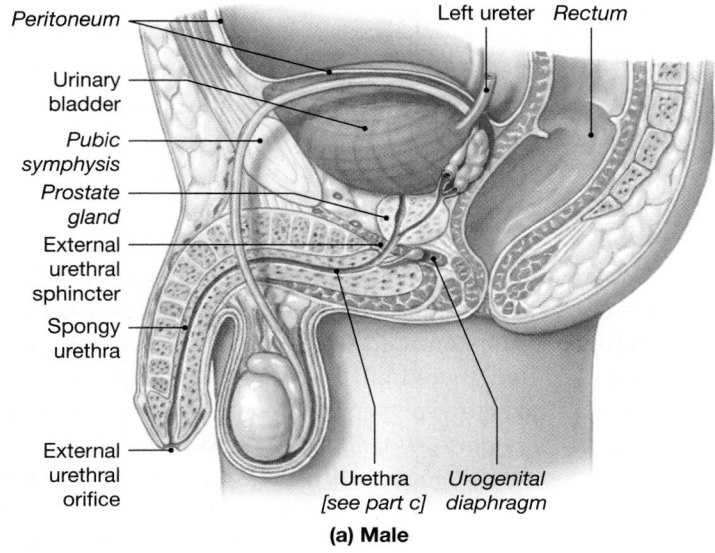

Peritoneum
Urinary bladder
Pubic symphysis
Prostate gland
External urethral sphincter
Spongy urethra
External urethral orifice
Left ureter
Rectum
Urethra [see part c]
Urogenital diaphragm

(a) Male

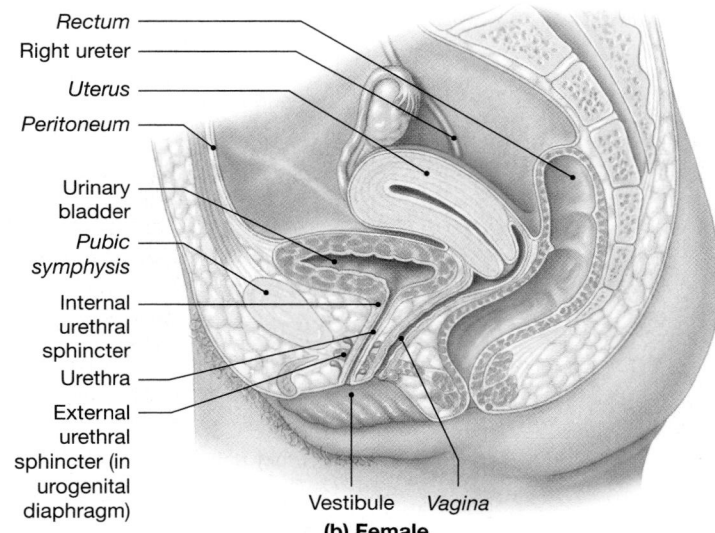

Rectum
Right ureter
Uterus
Peritoneum
Urinary bladder
Pubic symphysis
Internal urethral sphincter
Urethra
External urethral sphincter (in urogenital diaphragm)
Vestibule
Vagina

(b) Female

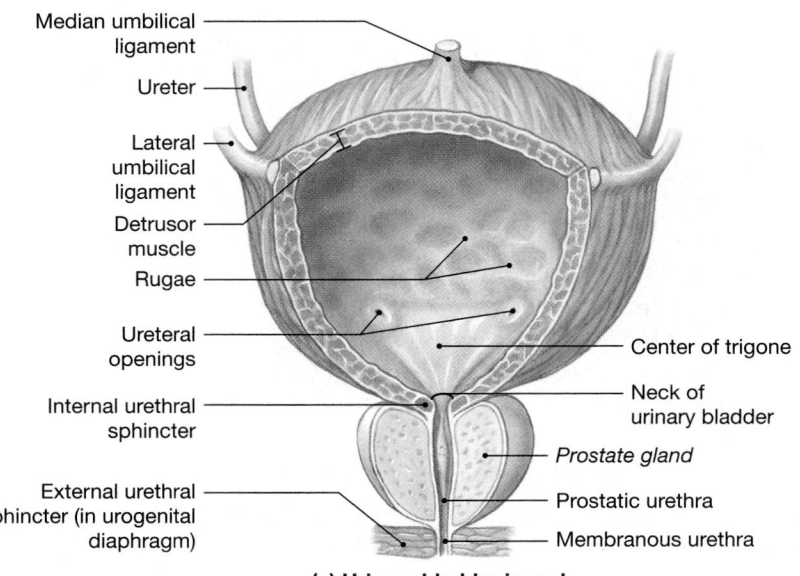

Median umbilical ligament
Ureter
Lateral umbilical ligament
Detrusor muscle
Rugae
Ureteral openings
Internal urethral sphincter
External urethral sphincter (in urogenital diaphragm)
Center of trigone
Neck of urinary bladder
Prostate gland
Prostatic urethra
Membranous urethra

(c) Urinary bladder in male

Figure 26–18 **Organs for the Conduction and Storage of Urine.** The ureter, urinary bladder, and urethra **(a)** in the male and **(b)** in the female. **(c)** The urinary bladder of a male. ATLAS: Plates 62b; 64; 65

the placenta during embryonic and fetal development. ∞ p. 767 The urinary bladder's posterior, inferior, and anterior surfaces lie outside the peritoneal cavity. In these areas, tough ligamentous bands anchor the urinary bladder to the pelvic and pubic bones.

In sectional view, the mucosa lining the urinary bladder is usually thrown into folds, or **rugae**, that disappear as the bladder fills. The triangular area bounded by the openings of the ureters and the entrance to the urethra constitutes a region called the **trigone** (TRĪ-gōn) of the urinary bladder. There, the mucosa is smooth and very thick. The trigone acts as a funnel that channels urine into the urethra when the urinary bladder contracts.

The urethral entrance lies at the apex of the trigone, at the most inferior point in the urinary bladder. The region surrounding the urethral opening, known as the **neck** of the urinary bladder, contains a muscular **internal urethral sphincter**, or *sphincter vesicae*. The smooth muscle fibers of this sphincter provide involuntary control over the discharge of urine from the bladder. The urinary bladder is innervated by postganglionic fibers from ganglia in the hypogastric plexus and by parasympathetic fibers from intramural ganglia that are controlled by branches of the pelvic nerves.

Histology of the Urinary Bladder

The wall of the urinary bladder contains mucosa, submucosa, and muscularis layers (**Figure 26–19b**). The muscularis layer consists of inner and outer layers of longitudinal smooth muscle, with a circular layer between the two. Collectively, these layers form the powerful **detrusor** (de-TROO-sor) **muscle** of the urinary bladder. Contraction of this muscle compresses the urinary bladder and expels its contents into the urethra.

The Urethra

The urethra extends from the neck of the urinary bladder and transports urine to the exterior of the body. The urethrae of males and females differ in length and in function. In males, the urethra extends from the neck of the urinary bladder to the tip of the penis (**Figure 26–18a,c**), a distance that may be 18–20 cm (7–8 in.). The male urethra can be subdivided into three portions: the prostatic urethra, the membranous urethra, and the

26 URINARY

spongy urethra. The **prostatic urethra** passes through the center of the prostate gland. The **membranous urethra** includes the short segment that penetrates the *urogenital diaphragm*, the muscular floor of the pelvic cavity. The **spongy urethra**, or *penile* (PĒ-nīl) *urethra*, extends from the distal border of the urogenital diaphragm to the external opening, or **external urethral orifice**, at the tip of the penis. In females, the urethra is very short, extending 3–5 cm (1–2 in.) from the bladder to the vestibule (**Figure 26–18b**). The external urethral orifice is near the anterior wall of the vagina.

In both sexes, where the urethra passes through the urogenital diaphragm, a circular band of skeletal muscle forms the **external urethral sphincter**. This muscular band acts as a valve. The external urethral sphincter, which is under voluntary control via the perineal branch of the pudendal nerve, has a resting muscle tone and must be voluntarily relaxed to permit micturition.

Histology of the Urethra

The urethral lining consists of a stratified epithelium that varies from transitional at the neck of the urinary bladder, to stratified columnar at the midpoint, to stratified squamous near the external urethral orifice. The lamina propria is thick and elastic, and the mucous membrane is thrown into longitudinal folds (**Figure 26–19c**). Mucin-secreting cells are located in the epithelial pockets. In males, the epithelial mucous glands may form tubules that extend into the lamina propria. Connective tissues of the lamina propria anchor the urethra to

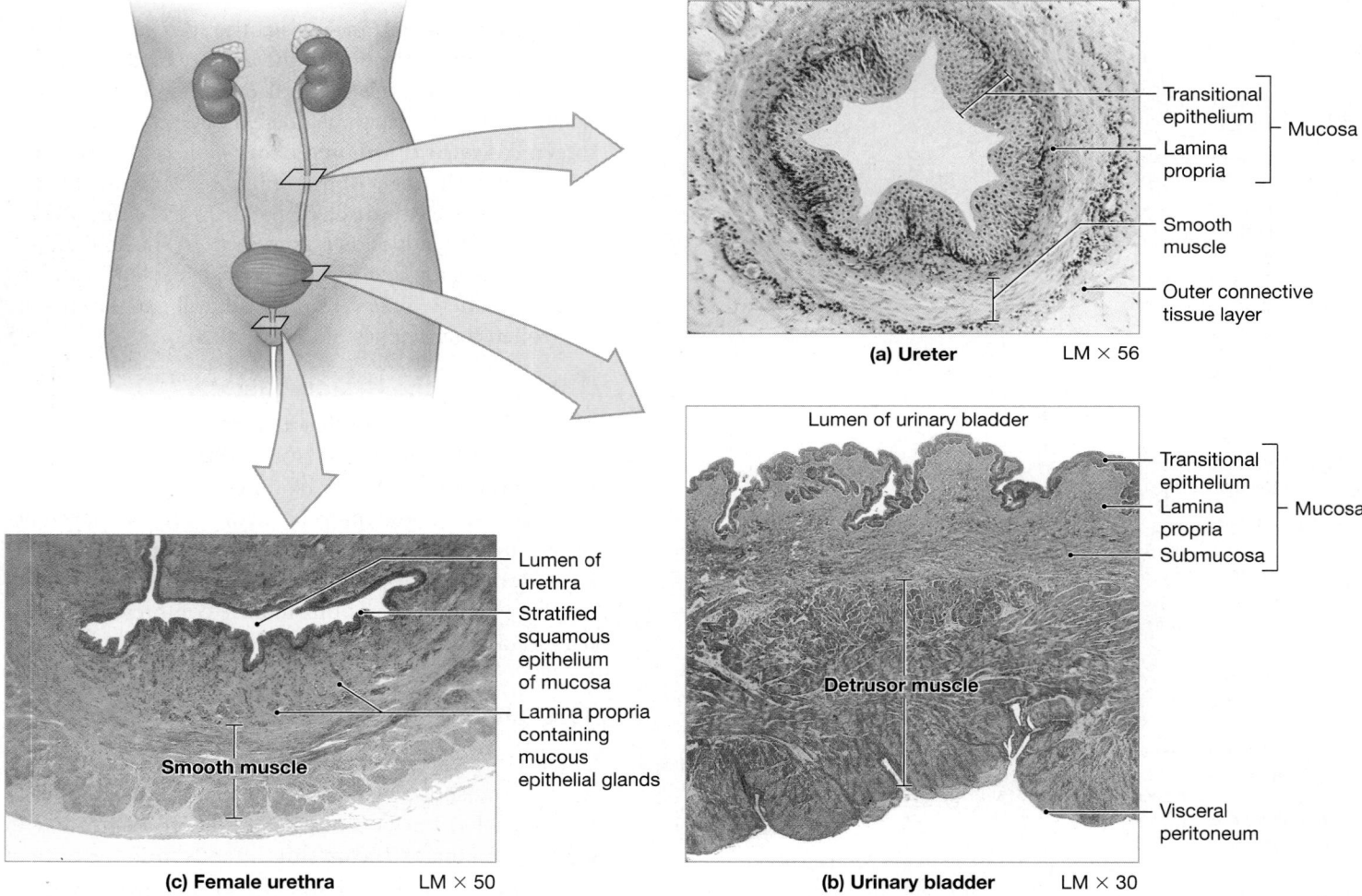

(a) Ureter LM × 56

(c) Female urethra LM × 50

(b) Urinary bladder LM × 30

Figure 26–19 **The Histology of the Organs That Collect and Transport Urine. (a)** A transverse section through the ureter. A thick layer of smooth muscle surrounds the lumen. (For a close-up of transitional epithelium, review *Figure 4–4c*, p. 120.) **(b)** The wall of the urinary bladder. **(c)** A transverse section through the female urethra.

surrounding structures. In females, the lamina propria contains an extensive network of veins, and the entire complex is surrounded by concentric layers of smooth muscle.

The Micturition Reflex and Urination

Urine reaches the urinary bladder by peristaltic contractions of the ureters. The process of urination is coordinated by the **micturition reflex** (**Figure 26–20**).

Stretch receptors in the wall of the urinary bladder are stimulated as the bladder fills with urine. Afferent fibers in the pelvic nerves carry the impulses generated to the sacral spinal cord. The increased level of activity in the fibers (1) facilitates parasympathetic motor neurons in the sacral spinal cord and (2) stimulates interneurons that relay sensations to the thalamus and then, through projection fibers, to the cerebral cortex. As a result, you become aware of the fluid pressure in your urinary bladder.

The urge to urinate generally appears when the bladder contains about 200 mL of urine. The micturition reflex begins to function when the stretch receptors have provided adequate

stimulation to parasympathetic preganglionic motor neurons (STEP 1). Action potentials carried by efferent fibers within the pelvic nerves then stimulate ganglionic neurons in the wall of the urinary bladder (STEP 2a). These neurons in turn stimulate sustained contraction of the detrusor muscle (STEP 3a).

This contraction elevates fluid pressure in the urinary bladder, but urine ejection does not occur unless both the internal and external urethral sphincters are relaxed. An awareness of the fullness of the urinary bladder depends on interneurons that relay information from stretch receptors to the thalamus (STEP 2b) and from the thalamus to the cerebral cortex (STEP 3b). The relaxation of the external urethral sphincter occurs under voluntary control; when the external urethral sphincter relaxes, so does the internal urethral sphincter, and urination occurs (STEP 4). If the external urethral sphincter does not relax, the internal urethral sphincter remains closed, and the urinary bladder gradually relaxes.

A further increase in bladder volume begins the cycle again, usually within an hour. Each increase in urinary volume leads to an increase in stretch receptor stimulation that makes the sensation more acute. Once the volume exceeds

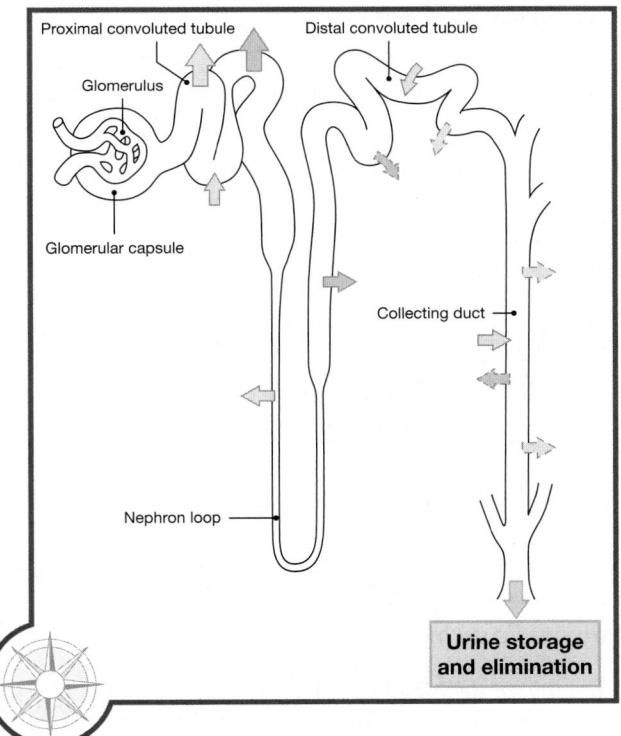

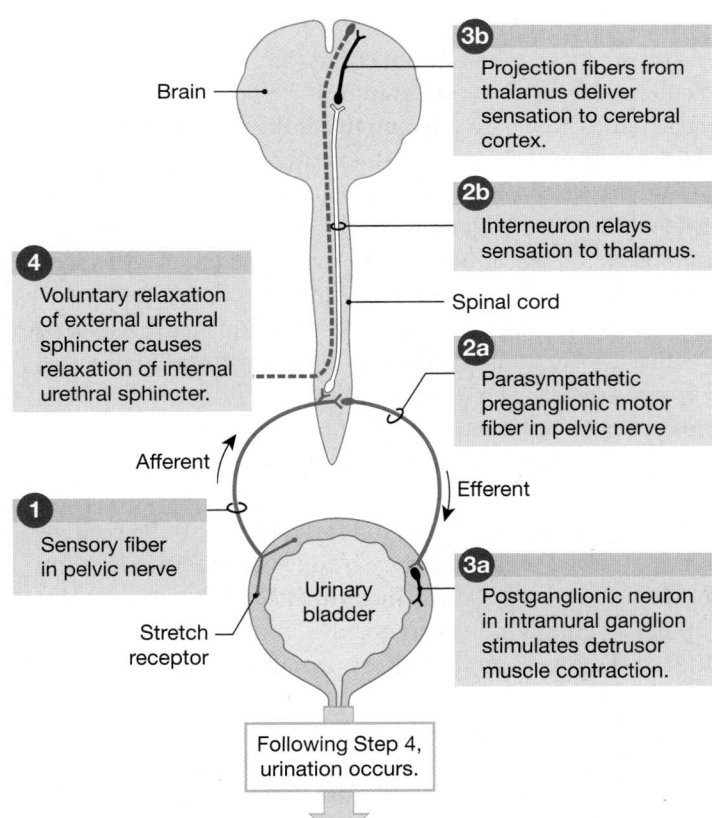

Figure 26–20 The Micturition Reflex. The components of the reflex arc that stimulates smooth muscle contractions in the urinary bladder. Micturition occurs after voluntary relaxation of the external urethral sphincter.

CLINICAL NOTE

Urinary Obstruction

Local blockages of the collecting ducts or ureters can result from the formation of *casts*—small blood clots, epithelial cells, lipids, or other materials that form in the collecting ducts. Casts are commonly eliminated in urine and are visible in microscopic analyses of urine samples. **Renal calculi** (KAL-kū-lī), or *kidney stones*, form within the urinary tract from calcium deposits, magnesium salts, or crystals of uric acid. The condition is called *nephrolithiasis* (nef-rō-li-THĪ-uh-sis; *nephros*, kidney; *lithos*, stone). The blockage of the ureter by a stone or by other means (such as external compression) results in **urinary obstruction**, a serious problem because, in

addition to causing pain, it reduces or prevents filtration in the affected kidney by elevating the capsular hydrostatic pressure.

Calculi are generally visible on an x-ray. If peristalsis and fluid pressures are insufficient to dislodge them, they must be either surgically removed or destroyed. One interesting nonsurgical procedure involves disintegrating the stones with a *lithotripter*, a device originally developed from machines used to de-ice airplane wings. Lithotripters focus sound waves on the stones, breaking them into smaller fragments that can be passed in the urine. Another nonsurgical approach is the insertion of a catheter armed with a laser that can shatter calculi with intense light beams.

---

500 mL, the bladder contractions triggered by the micturition reflex may generate enough pressure to force open the internal urethral sphincter. This opening leads to a reflexive relaxation of the external urethral sphincter, and urination occurs despite voluntary opposition or potential inconvenience. At the end of a typical micturition, less than 10 mL of urine remains in the bladder.

Infants lack voluntary control over urination, because the necessary corticospinal connections have yet to be established. Accordingly, "toilet training" before age 2 often involves training the parent to anticipate the timing of the reflex rather than training the child to exert conscious control.

Incontinence (in-KON-ti-nens) is the inability to control urination voluntarily. Trauma to the internal or external urethral sphincter can contribute to incontinence in otherwise healthy adults. For example, some mothers develop *stress incontinence* if childbirth overstretches and damages the sphincter muscles. In this condition, elevated intra-abdominal pressures—caused, for example, by a cough or sneeze—can overwhelm the sphincter muscles, causing urine to leak out. Incontinence can also develop in older individuals due to a general loss of muscle tone.

Damage to the central nervous system, the spinal cord, or the nerve supply to the urinary bladder or external urethral sphincter can also produce incontinence. For example, incontinence commonly accompanies Alzheimer disease or spinal cord damage. In most cases, the affected individual develops an *automatic bladder*. The micturition reflex remains intact, but voluntary control of the external urethral sphincter is lost, so the person cannot prevent the reflexive emptying of the urinary bladder. Damage to the pelvic nerves can abolish the micturition reflex entirely, because those nerves carry both afferent and efferent fibers of this reflex arc. The urinary bladder then becomes greatly distended with urine

and remains filled to capacity while the excess urine flows into the urethra in an uncontrolled stream. The insertion of a catheter is often needed to facilitate the discharge of urine.

CHECKPOINT

17. **What effect would a high-protein diet have on the composition of urine?**

18. **Obstruction of a ureter by a kidney stone would interfere with the flow of urine between which two points?**

19. **The ability to control the micturition reflex depends on your ability to control which muscle?**

See the blue Answers tab at the end of the book.

26-7 Age-related changes affect kidney function and the micturition reflex

In general, aging is associated with an increased incidence of kidney problems. One example—*nephrolithiasis*, the formation of calculi, or kidney stones—is described in the above Urinary Obstruction Clinical Note. Other age-related changes in the urinary system include the following:

* *A Decline in the Number of Functional Nephrons.* The total number of kidney nephrons drops by 30–40 percent between ages 25 and 85.

* *A Reduction in the GFR.* This reduction results from fewer glomeruli, cumulative damage to the filtration apparatus in the remaining glomeruli, and diminished renal blood flow.

- *A Reduced Sensitivity to ADH.* With age, the distal portions of the nephron and collecting system become less responsive to ADH. Reabsorption of water and sodium ions occurs at a reduced rate, and more sodium ions are lost in urine.

- *Problems with the Micturition Reflex.* Three factors are involved in such problems: (1) The sphincter muscles lose muscle tone and become less effective at voluntarily retaining urine. This leads to incontinence, often involving a slow leakage of urine. (2) The ability to control micturition can be lost after a stroke, Alzheimer disease, or other CNS problems affecting the cerebral cortex or hypothalamus. (3) In males, *urinary retention* may develop if enlargement of the prostate gland compresses the urethra and restricts the flow of urine.

> **CHECKPOINT**
>
> 20. List four age-related changes in the urinary system.
> 21. Define nephrolithiasis.
> 22. Describe how incontinence may develop in an elderly person.
>
> See the blue Answers tab at the end of the book.

26-8 The urinary system is one of several body systems involved in waste excretion

Although the urinary system excretes wastes produced by other body systems, it is not the only organ system involved in excretion. Indeed, the urinary, integumentary, respiratory, and digestive systems are together regarded as an anatomically diverse **excretory system** whose components perform all the excretory functions that affect the composition of body fluids:

- *Integumentary System.* Water losses and electrolyte losses in sensible perspiration can affect the volume and composition of the plasma. The effects are most apparent when losses are extreme, such as during peak sweat production. Small amounts of metabolic wastes, including urea, also are eliminated in perspiration.

- *Respiratory System.* The lungs remove the carbon dioxide generated by cells. Small amounts of other compounds, such as acetone and water, evaporate into the alveoli and are eliminated when you exhale.

- *Digestive System.* The liver excretes small amounts of metabolic waste products in bile, and you lose a variable amount of water in feces.

These excretory activities have an impact on the composition of body fluids. The respiratory system, for example, is the primary site of carbon dioxide excretion. Even though these excretory activities have an effect on the composition of body fluids, the excretory functions of these systems are not regulated as closely as are those of the kidneys. Under normal circumstances, the effects of integumentary and digestive excretory activities are minor compared with those of the urinary system.

Figure 26–21 summarizes the functional relationships between the urinary system and other systems. We will explore many of these relationships further in Chapter 27 when we consider major aspects of fluid, pH, and electrolyte balance.

> **CHECKPOINT**
>
> 23. Identify the role the urinary system plays for all other body systems.
> 24. Name the components of the body's excretory system.
>
> See the blue Answers tab at the end of the book.

Related Clinical Terms

aminoaciduria: Amino acid loss in urine; the most common form is *cystinuria.*

calculi: Insoluble deposits that form within the urinary tract from calcium salts, magnesium salts, or uric acid.

clearance test: A procedure that estimates the GFR by comparing plasma and renal concentrations of a specific solute, such as creatinine.

diuretics: Drugs that promote fluid loss in urine.

glomerulonephritis: An inflammation of the renal cortex.

hematuria: The presence of blood in urine.

hemodialysis: A technique in which an artificial membrane is used to regulate the composition of blood when kidney function is seriously impacted.

incontinence: An inability to control urination voluntarily.

polycystic kidney disease: An inherited abnormality that affects the development and structure of kidney tubules.

proteinuria: The presence of protein in urine.

pyelogram: An x-ray image of the kidneys after a radiopaque compound has been administered.

renal failure: An inability of the kidneys to excrete wastes in sufficient quantities to maintain homeostasis.

urinalysis: A physical and chemical assessment of urine.

urinary obstruction: A blockage of the urinary tract.

THE URINARY SYSTEM IN PERSPECTIVE

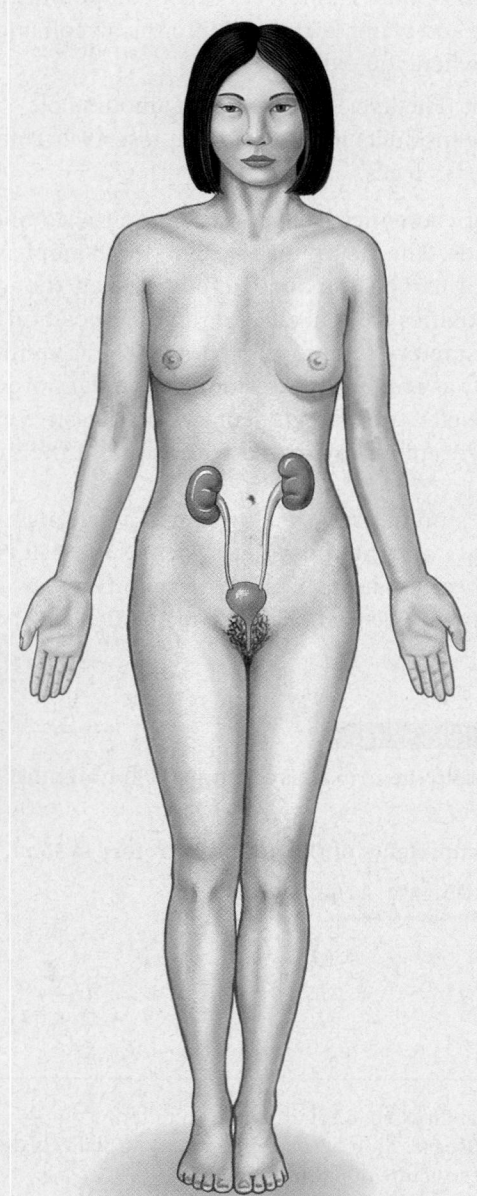

FOR ALL SYSTEMS
Excretes waste products; maintains normal body fluid pH and ion composition

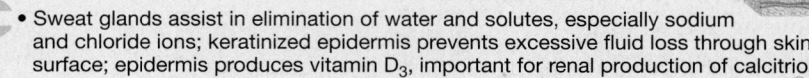

Integumentary System

• Sweat glands assist in elimination of water and solutes, especially sodium and chloride ions; keratinized epidermis prevents excessive fluid loss through skin surface; epidermis produces vitamin D_3, important for renal production of calcitriol

Skeletal System

• Axial skeleton provides some protection for kidneys and ureters; pelvis protects urinary bladder and proximal portion of urethra

• Conserves calcium and phosphate needed for bone growth

Muscular System

• Sphincter controls urination by closing urethral opening; muscle layers of trunk provide some protection for urinary organs

• Removes waste products of protein metabolism; assists in regulation of calcium and phosphate concentrations

Nervous System

• Adjusts renal blood pressure; monitors distension of urinary bladder and controls urination

Endocrine System

• Aldosterone, ADH, ANP, and BNP adjust rates of fluid and electrolyte reabsorption in kidneys

• Kidney cells release renin when local blood pressure declines and erythropoietin (EPO) when renal oxygen levels decline

Cardiovascular System

• Delivers blood to capillaries, where filtration occurs; accepts fluids and solutes reabsorbed during urine production

• Releases renin to elevate blood pressure and erythropoietin to accelerate red blood cell production

Lymphoid System

• Provides specific defenses against urinary tract infections

• Eliminates toxins and wastes generated by cellular activities; acid pH of urine provides nonspecific defense against urinary tract infections

Respiratory System

• Assists in the regulation of pH by eliminating carbon dioxide

• Assists in the elimination of carbon dioxide; provides bicarbonate buffers that assist in pH regulation

Digestive System

• Absorbs water needed to excrete wastes at kidneys; absorbs ions needed to maintain normal body fluid concentrations; liver removes bilirubin

• Excretes toxins absorbed by the digestive epithelium; excretes bilirubin and nitrogenous wastes produced by the liver; calcitriol production by kidneys aids calcium and phosphate absorption along digestive tract

Reproductive System

• Accessory organ secretions may have antibacterial action that helps prevent urethral infections in males

• Urethra in males carries semen to the exterior

Figure 26–21 **Functional Relationships between the Urinary System and Other Systems.**

Chapter Review

Study Outline

iP Use *Interactive Physiology®* (**IP**)—available on the IP CD packaged with your new textbook and online at **myA&P™** (www.myaandp.com)—to help you understand difficult physiological concepts. Follow these navigation paths on IP for concepts in this chapter:

• Urinary System/Glomerular Filtration

• Urinary System/Early Filtrate Processing

• Urinary System/Late Filtrate Processing

26-1 Consisting of the kidneys, ureters, urinary bladder, and urethra, the urinary system has three primary functions p. 966

1. The three major functions of the **urinary system** are *excretion*, the removal of organic waste products from body fluids; *elimination*, the discharge of these waste products into the environment; and homeostatic regulation of the volume and solute concentration of blood plasma. Other homeostatic functions include regulating blood volume and pressure by adjusting the volume of water lost and releasing hormones; regulating plasma concentrations of ions; helping to stabilize blood pH; conserving nutrients; assisting the liver in detoxifying poisons; and, during starvation, deaminating amino acids so that they can be catabolized by other tissues.

2. The urinary system includes the **kidneys**, the **ureters**, the **urinary bladder**, and the **urethra**. The kidneys produce **urine**, a fluid containing water, ions, and soluble compounds. During **urination** (**micturition**), urine is forced out of the body. (*Figure 26–1*)

26-2 Kidneys are highly vascular structures containing functional units called nephrons, which perform filtration, reabsorption, and secretion p. 967

3. The left kidney extends superiorly slightly more than the right kidney. Both kidneys and the suprarenal gland that overlies each are retroperitoneal. (*Figures 26–1, 26–2, 26–3*)

4. The **hilum**, a medial indentation, provides entry for the *renal artery* and *renal nerves* and exit for the *renal vein* and the ureter. (*Figures 26–3, 26–4*)

5. The superficial portion of the kidney, the cortex, surrounds the **medulla**. The ureter communicates with the **renal pelvis**, a chamber that branches into two **major calyces**. Each major calyx is connected to four or five **minor calyces**, which enclose the **renal papillae**. (*Figure 26–4*)

6. The blood supply to the kidneys includes the **renal**, **segmental**, **interlobar**, **arcuate**, and **cortical radiate arteries**. (*Figure 26–5*)

7. The **renal nerves**, which innervate the kidneys and ureters, are dominated by sympathetic postganglionic fibers.

8. The **nephron** (the basic functional unit in the kidney) consists of the **renal corpuscle** and **renal tubule**. The renal tubule is long and narrow and divided into the *proximal convoluted tubule*, the *nephron loop*, and the *distal convoluted tubule*. **Filtrate** is produced at the renal corpuscle. The nephron empties **tubular fluid** into the **collecting system**, consisting of **collecting ducts** and **papillary ducts**. (*Figures 26–6, 26–7*)

9. Nephrons are responsible for the production of filtrate, the reabsorption of organic nutrients, the reabsorption of water and ions, and the secretion into the tubular fluid of waste products missed by filtration. (*Summary Table 26–1*)

10. Roughly 85 percent of the nephrons are **cortical nephrons**, located in the renal cortex. **Juxtamedullary nephrons** are closer to the renal medulla, with their *nephron loops* extending deep into the **renal pyramids**. (*Figure 26–7*)

11. Blood travels from the efferent arteriole to the **peritubular capillaries** and the **vasa recta**. (*Figure 26–7*)

12. The renal tubule begins at the renal corpuscle and includes a knot of intertwined capillaries called the **glomerulus**, surrounded by the **glomerular capsule**. Blood arrives at the glomerulus via the **afferent arteriole** and departs in the **efferent arteriole**. (*Figure 26–8*)

13. At the glomerulus, **podocytes** cover the **dense layer** of the capillaries that project into the **capsular space**. The **secondary processes** of the podocytes are separated by narrow **filtration slits**. **Mesangial cells** lie between adjacent capillaries and control capillary diameter. (*Figure 26–8*)

14. The **proximal convoluted tubule** (**PCT**) actively reabsorbs nutrients, plasma proteins, and ions from the filtrate. These substances are released into the **peritubular fluid**, which surrounds the nephron. (*Figure 26–6*)

15. The **nephron loop**, also called the **loop of Henle**, includes a **descending limb** and an **ascending limb**; each limb contains a **thick segment** and a **thin segment**. The ascending limb delivers fluid to the **distal convoluted tubule** (**DCT**), which actively secretes ions, toxins, and drugs, and reabsorbs sodium ions from the tubular fluid. (*Figures 26–6, 26–7*)

26-3 Different segments of the nephron form urine by filtration, reabsorption, and secretion p. 976

16. Urine production maintains homeostasis by regulating blood volume and composition. In the process, organic waste products—notably urea, creatinine, and uric acid—are excreted.

17. Urine formation involves **filtration**, **reabsorption**, and **secretion**.

18. Four types of *carrier-mediated transport* (*facilitated diffusion*, *active transport*, *cotransport*, and *countertransport*) are involved in modifying filtrate. The saturation limit of a carrier protein is its **transport maximum**, which determines the **renal threshold**—the plasma concentration at which various compounds will appear in urine. (*Table 26–2*)

19. The transport maximum determines the renal threshold for the reabsorption of substances in tubular fluid. (*Table 26–3*)

20. Most regions of the nephron perform a combination of reabsorption and secretion. (*Figure 26–9, Summary Table 26–4*)

26-4 Hydrostatic and colloid osmotic pressures influence glomerular filtration pressure, which in turn affects the glomerular filtration rate p. 980

21. **Glomerular filtration** occurs as fluids move across the wall of the glomerulus into the capsular space in response to the **glomerular hydrostatic pressure (GHP)**—the hydrostatic (blood) pressure in the glomerular capillaries. This movement is opposed by the **capsular hydrostatic pressure (CsHP)** and by the **blood colloid osmotic pressure (BCOP)**. The **filtration pressure (FP)** at the glomerulus is the difference between the blood pressure and the opposing capsular and osmotic pressures. (*Figure 26–10*)

22. The **glomerular filtration rate (GFR)** is the amount of filtrate produced in the kidneys each minute. Any factor that alters the filtration pressure acting across the glomerular capillaries will change the GFR and affect kidney function.

23. A drop in filtration pressures stimulates the **juxtaglomerular complex (JGC)** to release renin and erythropoietin. (*Figure 26–11*)

24. Sympathetic activation (1) produces a powerful vasoconstriction of the afferent arterioles, decreasing the GFR and slowing the production of filtrate; (2) alters the GFR by changing the regional pattern of blood circulation; and (3) stimulates the release of renin by the juxtaglomerular complex. (*Figure 26–11*)

26-5 Countercurrent multiplication and the influence of antidiuretic hormone and aldosterone affect reabsorption and secretion p. 984

25. Glomerular filtration produces a filtrate with a composition similar to blood plasma, but with few, if any, plasma proteins.

26. The cells of the PCT normally reabsorb sodium and other ions, water, and almost all the organic nutrients that enter the filtrate. These cells also secrete various substances into the tubular fluid. (*Figure 26–12*)

27. Water and ions are reclaimed from tubular fluid by the nephron loop. A concentration gradient in the renal medulla encourages the osmotic flow of water out of the tubular fluid. The **countercurrent multiplication** between the ascending and descending limbs of the nephron loop helps create the osmotic gradient in the medulla. As water is lost by osmosis and the volume of tubular fluid decreases, the urea concentration rises. (*Figure 26–13*)

28. The DCT performs final adjustments by actively secreting or absorbing materials. Sodium ions are actively absorbed, in exchange for potassium or hydrogen ions discharged into tubular fluid. Aldosterone secretion increases the rate of sodium reabsorption and potassium loss. (*Figure 26–14*)

29. The amount of water and solutes in the tubular fluid of the collecting ducts is further regulated by aldosterone and ADH secretions. (*Figure 26–14*)

30. Urine volume and osmotic concentration are regulated by controlling water reabsorption. Precise control over this occurs via *facultative water reabsorption*. (*Figure 26–15*)

31. Normally, the removal of solutes and water by the vasa recta precisely balances the rates of reabsorption and osmosis in the renal medulla.

32. More than 99 percent of the filtrate produced each day is reabsorbed before reaching the renal pelvis. (*Table 26–5*)

33. *Urinalysis* is the chemical and physical analysis of a urine sample. (*Table 26–6*)

34. Each segment of the nephron and collecting system contributes to the production of concentrated urine. (*Figure 26–16, Summary Table 26–4*)

26-6 Urine is transported via the ureters, stored in the bladder, and eliminated through the urethra, aided by the micturition reflex p. 997

35. Urine production ends when tubular fluid enters the renal pelvis. The rest of the urinary system is responsible for transporting, storing, and eliminating urine. (*Figure 26–17*)

36. The ureters extend from the renal pelvis to the urinary bladder. Peristaltic contractions by smooth muscles move the urine along the tract. (*Figures 26–18, 26–19*)

37. The urinary bladder is stabilized by the **middle umbilical ligament** and the **lateral umbilical ligaments**. Internal features include the **trigone**, the **neck**, and the **internal urethral sphincter**. The mucosal lining contains prominent **rugae** (folds). Contraction of the **detrusor muscle** compresses the urinary bladder and expels urine into the urethra. (*Figures 26–18, 26–19*)

38. In both sexes, where the urethra passes through the *urogenital diaphragm*, a circular band of skeletal muscles forms the **external urethral sphincter**, which is under voluntary control. (*Figure 26–18*)

39. Urination is coordinated by the **micturition reflex**, which is initiated by stretch receptors in the wall of the urinary bladder. Voluntary urination involves coupling this reflex with the voluntary relaxation of the external urethral sphincter, which allows the opening of the **internal urethral sphincter**. (*Figure 26–20*)

26-7 Age-related changes affect kidney function and the micturition reflex p. 1002

40. Aging is generally associated with increased kidney problems. Age-related changes in the urinary system include (1) declining numbers of functional nephrons, (2) reduced GFR, (3) reduced sensitivity to ADH, and (4) problems with the micturition reflex. (**Urinary retention** may develop in men whose prostate gland is enlarged.)

26-8 The urinary system is one of several body systems involved in waste excretion p. 1003

41. The urinary system is the major component of an anatomically diverse *excretory system* that includes the integumentary system, the respiratory system, and the digestive system.

Review Questions

See the blue Answers tab at the end of the book.

myA&P Access more review material online at **myA&P™** (www.myaandp.com). There, you'll find chapter guides, chapter quizzes, practice tests, animations, flashcards, a glossary with pronunciations, *Interactive Physiology*® (IP) exercises and quizzes, and more to help you succeed in the course.

LEVEL 1   Reviewing Facts and Terms

1. Identify the structures of the kidney in the following diagram.

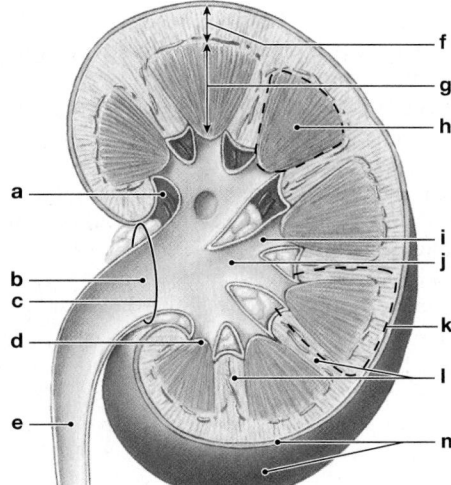

(a) _____	(b) _____
(c) _____	(d) _____
(e) _____	(f) _____
(g) _____	(h) _____
(i) _____	(j) _____
(k) _____	(l) _____
(m) _____	

2. The basic functional unit of the kidney is the
   (a) nephron.   (b) renal corpuscle.
   (c) glomerulus.   (d) nephron loop.
   (e) filtration unit.

3. The process of urine formation involves all of the following, *except*
   (a) filtration of plasma.
   (b) reabsorption of water.
   (c) reabsorption of certain solutes.
   (d) secretion of wastes.
   (e) secretion of excess lipoprotein and glucose molecules.

4. The glomerular filtration rate is regulated by all of the following, *except*
   (a) autoregulation.
   (b) sympathetic neural control.
   (c) cardiac output.
   (d) angiotensin II.
   (e) the hormone ADH.

5. The distal convoluted tubule is an important site for
   (a) active secretion of ions.
   (b) active secretion of acids and other materials.
   (c) selective reabsorption of sodium ions from the tubular fluid.
   (d) all of these.

6. Changing the diameter of the afferent and efferent arterioles to alter the GFR can be an example of _____ regulation.
   (a) hormonal   (b) autonomic
   (c) autoregulation   (d) a, b, and c

7. What is the primary function of the urinary system?

8. What structures are components of the urinary system?

9. Trace the pathway of the protein-free filtrate from where it is produced in the renal corpuscle until it drains into the renal pelvis in the form of urine. (Use arrows to indicate the direction of flow.)

10. Name the segments of the nephron distal to the renal corpuscle, and state the function(s) of each.

11. What is the function of the juxtaglomerular complex?

12. Using arrows, trace a drop of blood from its entry into the renal artery until its exit via a renal vein.

13. Name and define the three distinct processes involved in the production of urine.

14. What are the primary effects of angiotensin II on kidney function and regulation?

15. Which parts of the urinary system are responsible for the transport, storage, and elimination of urine?

LEVEL 2   Reviewing Concepts

16. When the renal threshold for a substance exceeds its tubular maximum,
   (a) more of the substance will be filtered.
   (b) more of the substance will be reabsorbed.
   (c) more of the substance will be secreted.
   (d) the amount of the substance that exceeds the tubular maximum will be found in the urine.
   (e) both a and d occur.

17. Sympathetic activation of nerve fibers in the nephron causes
   (a) the regulation of glomerular blood flow and pressure.
   (b) the stimulation of renin release from the juxtaglomerular complex.
   (c) the direct stimulation of water and Na^+ reabsorption.
   (d) all of these.

18. Sodium reabsorption in the DCT and in the cortical portion of the collecting system is accelerated by the secretion of
   (a) ADH.   (b) rennin.
   (c) aldosterone.   (d) erythropoietin.

19. When ADH levels rise,
   (a) the amount of water reabsorbed increases.
   (b) the DCT becomes impermeable to water.
   (c) the amount of water reabsorbed decreases.
   (d) sodium ions are exchanged for potassium ions.

20. The control of blood pH by the kidneys during acidosis involves
    (a) the secretion of hydrogen ions and reabsorption of bicarbonate ions from the tubular fluid.
    (b) a decrease in the amount of water reabsorbed.
    (c) hydrogen ion reabsorption and bicarbonate ion loss.
    (d) potassium ion secretion.
21. How are proteins excluded from filtrate? Why is this important?
22. What interacting controls stabilize the glomerular filtration rate (GFR)?
23. What primary changes occur in the composition and concentration of filtrate as a result of activity in the proximal convoluted tubule?
24. Describe two functions of countercurrent multiplication in the kidney.
25. Describe the micturition reflex.

LEVEL 3 Critical Thinking and Clinical Applications

26. In a normal kidney, which of the following conditions would cause an increase in the glomerular filtration rate (GFR)?
    (a) constriction of the afferent arteriole
    (b) a decrease in the pressure of the glomerulus
    (c) an increase in the capsular hydrostatic pressure
    (d) a decrease in the concentration of plasma proteins in the blood
    (e) a decrease in the net glomerular filtration process

27. In response to *excess* water in the body,
    (a) antidiuretic hormone is secreted by the adenohypophysis.
    (b) the active transport mechanisms in the ascending limb of the nephron loop cease functioning.
    (c) the permeability of the distal convoluted tubules and collecting ducts to water is decreased.
    (d) the permeability of the ascending limb of the nephron loop is increased.
    (e) the glomerular filtration rate is reduced.
28. Sylvia is suffering from severe edema in her arms and legs. Her physician prescribes a diuretic (a substance that increases the volume of urine produced). Why might this help alleviate Sylvia's problem?
29. David's grandfather suffers from hypertension. His doctor tells him that part of his problem stems from renal arteriosclerosis. Why would this cause hypertension?
30. *Mannitol* is a sugar that is filtered, but not reabsorbed, by the kidneys. What effect would drinking a solution of mannitol have on the volume of urine produced?
31. The drug *Diamox* is sometimes used to treat mountain sickness. Diamox inhibits the action of carbonic anhydrase in the proximal convoluted tubule. Polyuria (the elimination of an unusually large volume of urine) is a side effect associated with the medication. Why does polyuria occur?

27

Fluid, Electrolyte, and Acid–Base Balance

Did you know...?
Electrolyte levels often change when water levels in the body change. Exhaustive exercise can cause potentially dangerous disruptions of fluid and electrolyte balance.

Learning Outcomes

After completing this chapter, you should be able to do the following:

27-1 Explain what is meant by the terms fluid balance, electrolyte balance, and acid–base balance, and discuss their importance for homeostasis.

27-2 Compare the composition of intracellular and extracellular fluids, explain the basic concepts involved in the regulation of fluids and electrolytes, and identify the hormones that play important roles in fluid and electrolyte regulation.

27-3 Describe the movement of fluid within the ECF, between the ECF and the ICF, and between the ECF and the environment.

27-4 Discuss the mechanisms by which sodium, potassium, calcium, and chloride ion concentrations are regulated to maintain electrolyte balance.

27-5 Explain the buffering systems that balance the pH of the intracellular and extracellular fluids, and describe the compensatory mechanisms involved in the maintenance of acid–base balance.

27-6 Identify the most frequent disturbances of acid–base balance, and explain how the body responds when the pH of body fluids varies outside normal limits.

27-7 Describe the effects of aging on fluid, electrolyte, and acid–base balance.

Clinical Notes

Water and Weight Loss p. 1017
Athletes and Salt Loss p. 1020

An Introduction to Fluid, Electrolyte, and Acid-Base Balance

The next time you see a small pond, think about the fish it contains. They live out their lives totally dependent on the quality of that isolated environment. Severe water pollution will kill them, but even subtle changes can have equally grave effects. Changes in the volume of the pond, for example, can be quite important. If evaporation removes too much of the water, the fish become overcrowded; oxygen and food supplies run out, and the fish suffocate or starve. The ionic concentration of the water is also crucial. Most of the fish in a freshwater pond will die if the water becomes too salty; those in a saltwater pond will die if their environment becomes too dilute. The pH of the pond water, too, is a vital factor; that is one reason acid rain is such a problem.

Your cells live in a pond whose shores are the exposed surfaces of your skin. Most of your body weight is water. Water accounts for up to 99 percent of the volume of the fluid outside cells, and it is an essential ingredient of cytoplasm. All of a cell's operations rely on water as a diffusion medium for the distribution of gases, nutrients, and waste products. If the water content of the body changes, cellular activities are jeopardized. For example, when the water content reaches very low levels, proteins denature, enzymes cease functioning, and cells die. This chapter discusses the homeostatic mechanisms that regulate ion concentrations, volume, and pH in the fluid surrounding cells.

27-1 Fluid balance, electrolyte balance, and acid–base balance are interrelated and essential to homeostasis

To survive, we must maintain a normal volume and composition of both the **extracellular fluid** or **ECF** (the interstitial fluid, plasma, and other body fluids) and the **intracellular fluid** or **ICF** (the cytosol). The ionic concentrations and pH (hydrogen ion concentration) of these fluids are as important as their absolute quantities. If concentrations of calcium or potassium ions in the ECF become too high, cardiac arrhythmias develop and death can result. A pH outside the normal range can also lead to a variety of serious problems. Low pH is especially dangerous, because hydrogen ions break chemical bonds, change the shapes of complex molecules, disrupt plasma membranes, and impair tissue functions.

Tips&Tricks

The "p" in pH refers to power. Hence, pH refers to the **p**ower of **H**ydrogen.

In this chapter, we will consider the dynamics of exchange among the various body fluids, and between the body and the external environment. Stabilizing the volumes, solute concentrations, and pH of the ECF and the ICF involves three interrelated processes:

1. *Fluid Balance.* You are in **fluid balance** when the amount of water you gain each day is equal to the amount you lose to the environment. The maintenance of normal fluid balance involves regulating the content and distribution of body water in the ECF and the ICF. The digestive system is the primary source of water gains; a small amount of additional water is generated by metabolic activity. The urinary system is the primary route for water loss under normal conditions, but as we saw in Chapter 25, sweat gland activity can become important when body temperature is elevated. ∞ p. 958 Although cells and tissues cannot transport water, they can transport ions and create concentration gradients that are then eliminated by osmosis.

2. *Electrolyte Balance.* **Electrolytes** are ions released through the dissociation of inorganic compounds; they are so named because they can conduct an electrical current in a solution. ∞ p. 41 Each day, your body fluids gain electrolytes from the food and drink you consume, and lose electrolytes in urine, sweat, and feces. For each ion, daily gains must balance daily losses. For example, if you lose 500 mg of Na^+ in urine and insensible perspiration, you need to gain 500 mg of Na^+ from food and drink to remain in sodium balance. If the gains and losses for every electrolyte are in balance, you are said to be in **electrolyte balance**. Electrolyte balance primarily involves balancing the rates of absorption across the digestive tract with rates of loss at the kidneys, although losses at sweat glands and other sites can play a secondary role.

3. *Acid–Base Balance.* You are in **acid–base balance** when the production of hydrogen ions in your body is precisely offset by their loss. When acid–base balance exists, the pH of body fluids remains within normal limits. ∞ p. 43 Preventing a reduction in pH is the primary problem, because your body generates a variety of acids during normal metabolic operations. The kidneys play a major role by secreting hydrogen ions into the urine and generating buffers that enter the bloodstream. Such secretion occurs primarily in the distal segments of the distal convoluted tubule (DCT) and along the collecting system. ∞ p. 989 The lungs also play a key role through the elimination of carbon dioxide.

Much of the material in this chapter was introduced in earlier chapters, in discussions considering aspects of fluid, electrolyte, or acid–base balance that affect specific systems. This chapter provides an overview that integrates those discussions to highlight important functional patterns. Few other chapters have such wide-ranging clinical importance: The treatment of any serious illness affecting the nervous, cardiovascular, respiratory, urinary, or digestive system must include steps to restore normal fluid, electrolyte, and acid–base balances. Because this chapter builds on information presented in earlier chapters, you will encounter many references to relevant discussions and figures that can provide a quick review.

1. **Identify the three interrelated processes essential to stabilizing body fluid volumes.**

2. **List the components of extracellular fluid (ECF) and intracellular fluid (ICF), respectively.**

See the blue Answers tab at the end of the book.

27-2 The ECF and ICF make up the fluid compartments, which also contain cations and anions

Figure 27–1a presents an overview of the body composition of a 70-kg (154-pound) individual with a minimum of body fat. The distribution was obtained by averaging values for males and females ages 18–40 years. Water accounts for roughly 60 percent of the total body weight of an adult male, and 50 percent of that of an adult female (**Figure 27–1b**). This difference between the sexes primarily reflects the proportionately larger mass of adipose tissue in adult females, and the greater average muscle mass in adult males. (Adipose tissue is

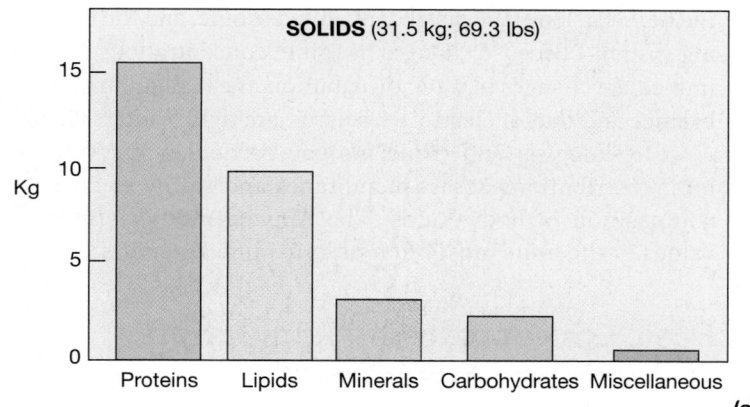

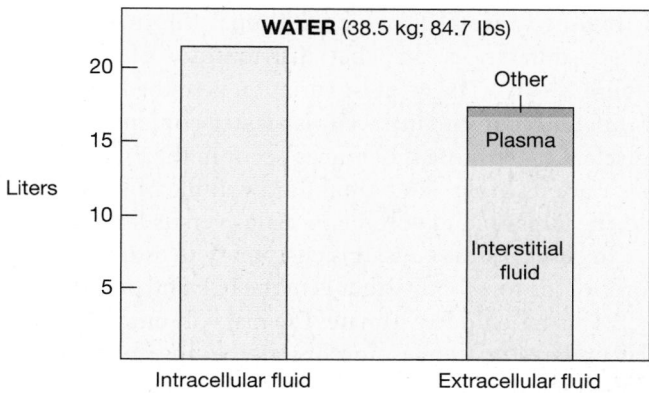

(a)

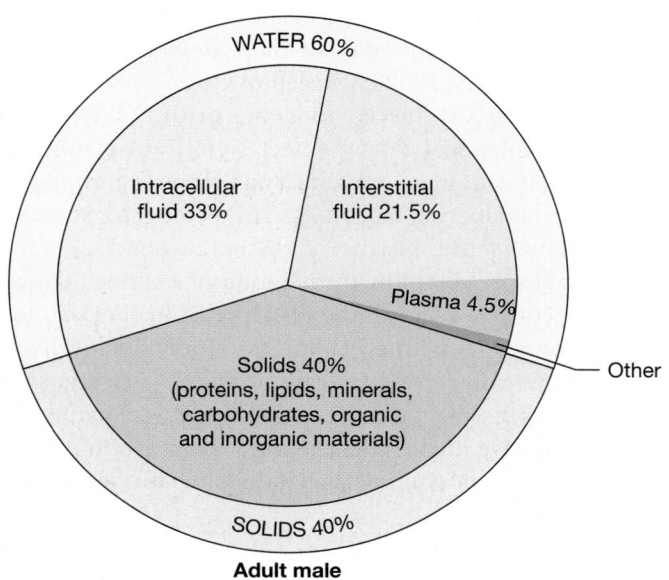

Adult male

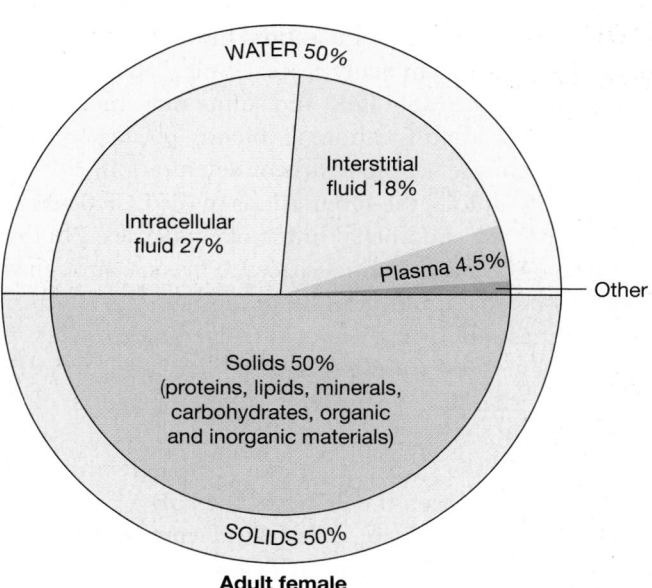

Adult female

(b)

Figure 27–1 The Composition of the Human Body. (a) The body composition (by weight, averaged for both sexes) and major body fluid compartments of a 70-kg individual. For technical reasons, it is extremely difficult to determine the precise size of any of these compartments; estimates of their relative sizes vary widely. **(b)** A comparison of the body compositions of adult males and females, ages 18–40 years.

only 10 percent water, whereas skeletal muscle is 75 percent water.) In both sexes, intracellular fluid contains a greater proportion of total body water than does extracellular fluid. Exchange between the ICF and the ECF occurs across plasma membranes by osmosis, diffusion, and carrier-mediated transport. (To review the mechanisms involved, see Table 3–3, p. 99.)

The ECF and the ICF

The largest subdivisions of the ECF are the interstitial fluid of peripheral tissues and the plasma of circulating blood (**Figure 27–1a**). Minor components of the ECF include lymph, cerebrospinal fluid (CSF), synovial fluid, serous fluids (pleural, pericardial, and peritoneal fluids), aqueous humor, perilymph, and endolymph. More precise measurements of total body water provide additional information on sex-linked differences in the distribution of body water (**Figure 27–1b**). The greatest variation is in the ICF, as a result of differences in the intracellular water content of fat versus muscle. Less striking differences occur in the ECF values, due to variations in the interstitial fluid volume of various tissues and the larger blood volume in males versus females.

In clinical situations, it is customary to estimate that two-thirds of the total body water is in the ICF and one-third in the ECF. This ratio underestimates the real volume of the ECF, but that underestimation is appropriate because portions of the ECF—including the water in bone, in many dense connective tissues, and in many of the minor ECF components—are relatively isolated. Exchange between these fluid volumes and the rest of the ECF occurs more slowly than does exchange between plasma and other interstitial fluids. As a result, they can be safely ignored in many cases. Clinical attention is usually focused on the rapid fluid and solute movements associated with the administration of blood, plasma, or saline solutions to counteract blood loss or dehydration.

Exchange among the subdivisions of the ECF occurs primarily across the endothelial lining of capillaries. Fluid may also travel from the interstitial spaces to plasma through lymphatic vessels that drain into the venous system. ∞ p. 779 The identities and quantities of dissolved electrolytes, proteins, nutrients, and waste products in the ECF vary regionally. (For a chemical analysis of the composition of ECF compartments, see the Appendix.) Still, the variations among the segments of the ECF seem minor compared with the major differences between the ECF and the ICF.

The ECF and ICF are called **fluid compartments**, because they commonly behave as distinct entities. The presence of a plasma membrane and active transport at the membrane surface enable cells to maintain internal environments with a composition that differs from their surroundings. The principal ions in the ECF are sodium, chloride, and

bicarbonate. The ICF contains an abundance of potassium, magnesium, and phosphate ions, plus large numbers of negatively charged proteins. **Figure 27–2** compares the ICF with the two major subdivisions of the ECF.

If the plasma membrane were freely permeable, diffusion would continue until these ions were evenly distributed across the membrane. But it does not, because plasma membranes are selectively permeable: Ions can enter or leave the cell only via specific membrane channels. In addition, carrier mechanisms move specific ions into or out of the cell.

Despite the differences in the concentration of specific substances, the osmotic concentrations of the ICF and ECF are identical. Osmosis eliminates minor differences in concentration almost at once, because most plasma membranes are freely permeable to water. (The only noteworthy exceptions are the apical surfaces of epithelial cells along the ascending limb of the nephron loop, the distal convoluted tubule, and the collecting system.) Because changes in solute concentrations lead to immediate changes in water distribution, the regulation of fluid balance and that of electrolyte balance are tightly intertwined.

Physiologists and clinicians pay particular attention to ionic distributions across membranes and to the electrolyte composition of body fluids. The Appendix reports normal values in the units most often used in clinical reports.

Basic Concepts in the Regulation of Fluids and Electrolytes

Before we can proceed to a discussion of fluid balance and electrolyte balance, you must understand four basic principles:

1. *All the Homeostatic Mechanisms That Monitor and Adjust the Composition of Body Fluids Respond to Changes in the ECF, Not in the ICF.* Receptors monitoring the composition of two key components of the ECF—plasma and cerebrospinal fluid—detect significant changes in their composition or volume and trigger appropriate neural and endocrine responses. This arrangement makes functional sense, because a change in one ECF component will spread rapidly throughout the extracellular compartment and affect all the body's cells. In contrast, the ICF is contained within trillions of individual cells that are physically and chemically isolated from one another by their plasma membranes. Thus, changes in the ICF in one cell have no direct effect on the composition of the ICF in distant cells and tissues, unless those changes also affect the ECF.

2. *No Receptors Directly Monitor Fluid or Electrolyte Balance.* In other words, receptors cannot detect how many liters of water or grams of sodium, chloride, or potassium the body contains, or count how many liters or grams we gain or lose in the course of a day. But receptors *can* mon-

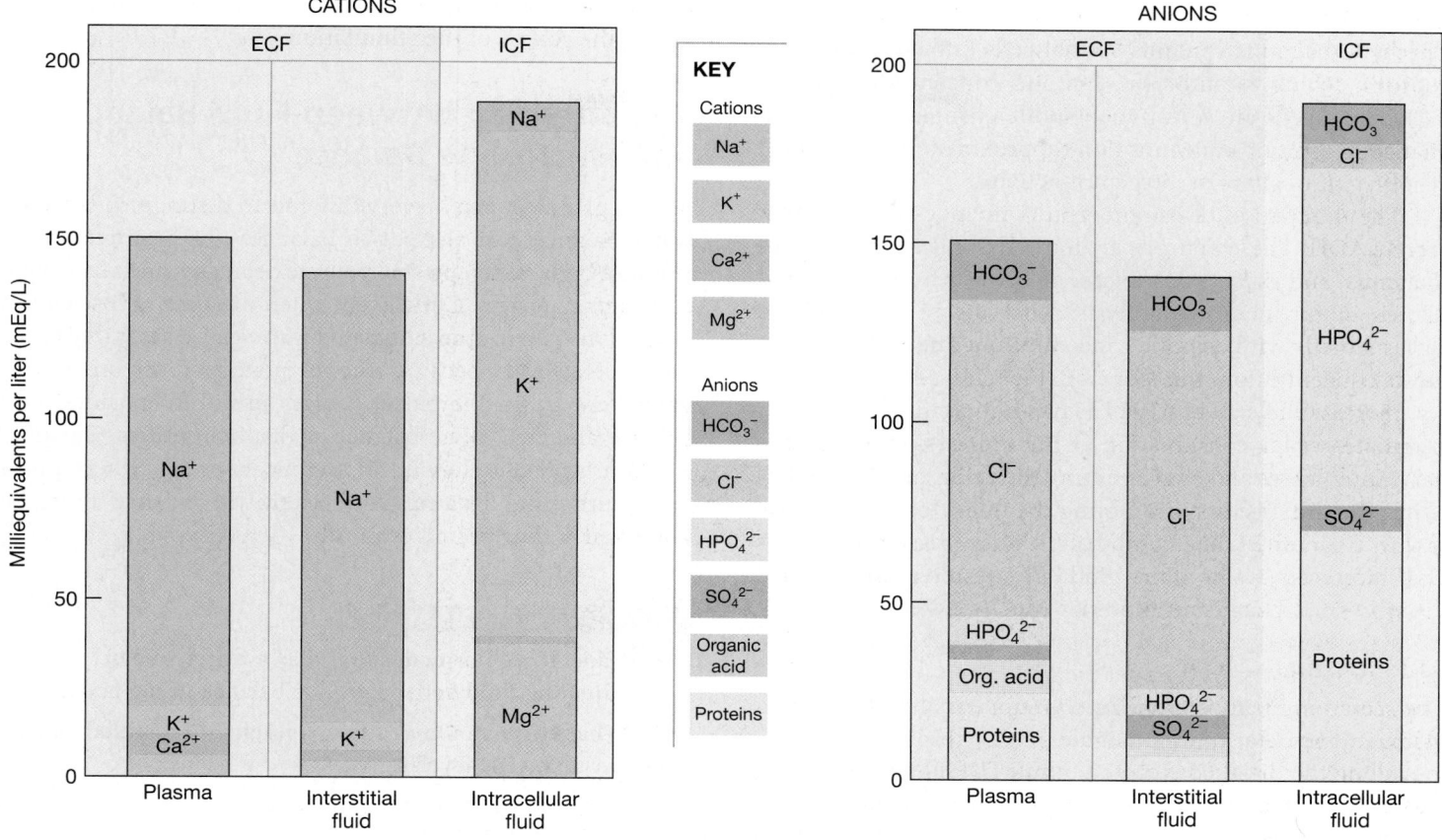

Figure 27–2 Cations and Anions in Body Fluids. Notice the differences in cation and anion concentrations in the various body fluid compartments. For information about the composition of other body fluids, see the Appendix.

itor *plasma volume* and *osmotic concentration*. Because fluid continuously circulates between interstitial fluid and plasma, and because exchange occurs between the ECF and the ICF, the plasma volume and osmotic concentration are good indicators of the state of fluid balance and electrolyte balance for the body as a whole.

3. *Cells Cannot Move Water Molecules by Active Transport.* All movement of water across plasma membranes and epithelia occurs passively, in response to osmotic gradients established by the active transport of specific ions, such as sodium and chloride. You may find it useful to remember that *"water follows salt."* As we saw in earlier chapters, when sodium and chloride ions (or other solutes) are actively transported across a membrane or epithelium, water follows by osmosis. ∞ p. 986 This basic principle accounts for water absorption across the digestive epithelium, and for water conservation in the kidneys.

4. *The Body's Content of Water or Electrolytes Will Rise if Dietary Gains Exceed Losses to the Environment, and Will Fall if Losses Exceed Gains.* This basic rule is important when you consider the mechanics of fluid balance and electrolyte balance. Homeostatic adjustments generally affect

the balance between urinary excretion and dietary absorption. As we saw in Chapter 26, the physiological adjustments in renal function are regulated primarily by circulating hormones. These hormones can also produce complementary changes in behavior. For example, the combination of angiotensin II and aldosterone can give you a sensation of thirst—which stimulates you to drink fluids—and a taste for heavily salted foods.

An Overview of the Primary Regulatory Hormones

Major physiological adjustments affecting fluid balance and electrolyte balance are mediated by three hormones: (1) *antidiuretic hormone (ADH)*, (2) *aldosterone*, and (3) the *natriuretic peptides (ANP and BNP)*. These hormones were introduced and discussed in earlier chapters; we will summarize their effects next. Those interested in a more detailed review should refer to the appropriate sections of Chapters 18, 21, and 26. The interactions among these hormones were illustrated in **Figures 18–17b, 21–16, 21–17,** and **26–11** (pp. 636, 743, 746, 983).

Antidiuretic Hormone

The hypothalamus contains special cells known as **osmoreceptors**, which monitor the osmotic concentration of the ECF. These cells are sensitive to subtle changes: A 2 percent change in osmotic concentration (approximately 6 mOsm/L) is sufficient to alter osmoreceptor activity.

The population of osmoreceptors includes neurons that secrete ADH. These neurons are located in the anterior hypothalamus, and their axons release ADH near fenestrated capillaries in the neurohypophysis. The rate of ADH release varies directly with osmotic concentration: The higher the osmotic concentration, the more ADH is released.

Increased release of ADH has two important effects: (1) It stimulates water conservation at the kidneys, reducing urinary water losses and concentrating the urine; and (2) it stimulates the thirst center, promoting the intake of fluids. As we saw in Chapter 21, the combination of decreased water loss and increased water gain gradually restores the normal plasma osmotic concentration. ∞ pp. 742–743

Aldosterone

The secretion of aldosterone by the suprarenal cortex plays a major role in determining the rate of Na^+ absorption and K^+ loss along the distal convoluted tubule (DCT) and collecting system of the kidneys. ∞ p. 988 The higher the plasma concentration of aldosterone, the more efficiently the kidneys conserve Na^+. Because "water follows salt," the conservation of Na^+ also stimulates water retention: As Na^+ is reabsorbed, Cl^- follows (see **Figure 26–14a**, p. 990), and as sodium and chloride ions move out of the tubular fluid, water follows by osmosis. Aldosterone also increases the sensitivity of salt receptors on the tongue. This effect may increase your interest in, and consumption of, salty foods.

Aldosterone is secreted in response to rising K^+ or falling Na^+ levels in the blood reaching the suprarenal cortex, or in response to the activation of the renin–angiotensin system. As we saw in earlier chapters, renin release occurs in response to (1) a drop in plasma volume or blood pressure at the juxtaglomerular complex of the nephron, (2) a decline in filtrate osmotic concentration at the DCT, or, as we will soon see, (3) falling Na^+ or rising K^+ concentrations in the renal circulation.

Natriuretic Peptides

The natriuretic (*natrium*, sodium; *ouron*, urine) peptides ANP and BNP are released by cardiac muscle cells in response to abnormal stretching of the heart walls, caused by elevated blood pressure or an increase in blood volume. Among their other effects, they reduce thirst and block the release of ADH and aldosterone that might otherwise lead to the conservation of water and salt. The resulting diuresis (fluid loss at the kidneys) lowers both blood pressure and plasma volume, eliminating the source of the stimulation.

The Interplay between Fluid Balance and Electrolyte Balance

At first glance, it can be very difficult to distinguish between water balance and electrolyte balance. For example, when you lose body water, plasma volume decreases and electrolyte concentrations rise. Conversely, when you gain or lose excess electrolytes, there is an associated water gain or loss due to osmosis. However, because the regulatory mechanisms involved are quite different, it is often useful to consider fluid balance and electrolyte balance as distinct entities. This distinction is absolutely vital in a clinical setting, where problems with fluid balance and electrolyte balance must be identified and corrected promptly.

CHECKPOINT

3. Name three hormones that play a major role in adjusting fluid and electrolyte balance in the body.

4. What effect would drinking a pitcher of distilled water have on ADH levels?

See the blue Answers tab at the end of the book.

27-3 Hydrostatic and osmotic pressures regulate the movement of water and electrolytes to maintain fluid balance

Water circulates freely within the ECF compartment. At capillary beds throughout the body, hydrostatic pressure forces water out of plasma and into interstitial spaces. Some of that water is reabsorbed along the distal portion of the capillary bed, and the rest enters lymphatic vessels for transport to the venous circulation. There is also a continuous movement of fluid among the minor components of the ECF:

1. Water moves back and forth across the mesothelial surfaces that line the peritoneal, pleural, and pericardial cavities, and through the synovial membranes that line joint capsules. The flow rate is significant; for example, roughly 7 liters (1.8 gal) of peritoneal fluid is produced and reabsorbed each day. However, the actual volume present at any time in the peritoneal cavity is very small—less than 35 mL (1.2 oz).

2. Water also moves between blood and cerebrospinal fluid (CSF), between the aqueous humor and vitreous humor

of the eye, and between the perilymph and endolymph of the inner ear. The volumes involved in these water movements are very small, and the volume and composition of the fluids are closely regulated. For those reasons, we will largely ignore them in the discussion that follows.

Water movement can also occur between the ECF and the ICF, but under normal circumstances the two are in osmotic equilibrium, and no large-scale circulation occurs between the two compartments. (A small amount of water moves from the ICF to the ECF each day, as the result of mitochondrial water generation; this will be considered in a separate section.)

The body's water content cannot easily be determined. However, the concentration of Na^+, the most abundant ion in the ECF, provides useful clues to the state of water balance. When the body's water content rises, the Na^+ concentration of the ECF becomes abnormally low; when the body water content declines, the Na^+ concentration becomes abnormally high.

Fluid Movement within the ECF

In the discussion of capillary dynamics in Chapter 21, we considered the basic principles that determine fluid movement among the divisions of the ECF. ∞ p. 735 The exchange between plasma and interstitial fluid, by far the largest components of the ECF, is determined by the relationship between the net hydrostatic pressure, which tends to push water out of the plasma and into the interstitial fluid, and the net colloid osmotic pressure, which tends to draw water out of the interstitial fluid and into the plasma. The interaction between these opposing forces, diagrammed in **Figure 21–12** (p. 736), results in the continuous filtration of fluid from the capillaries into the interstitial fluid. This volume of fluid is then redistributed: After passing through the channels of the lymphoid system, the fluid returns to the venous system. At any moment, interstitial fluid and minor fluid compartments contain roughly 80 percent of the ECF volume, and plasma contains the other 20 percent.

Any factor that affects the net hydrostatic pressure or the net colloid osmotic pressure will alter the distribution of fluid within the ECF. The movement of abnormal amounts of water from plasma into interstitial fluid is called *edema*. Pulmonary edema, for example, can result from an increase in the blood pressure in pulmonary capillaries, and generalized edema can result from a decrease in blood colloid osmotic pressure (as in advanced starvation, when plasma protein concentrations decline). Localized edema can result from damage to capillary walls (as in bruising), the constriction of regional venous circulation, or a blockage of the lymphatic drainage (as in *lymphedema,* introduced in Chapter 22). ∞ p. 780

Tips&Tricks

Water movement between compartments, driven by osmotic pressure, is like water movements between compartments in a waterbed mattress: The total amount of fluid doesn't change; fluid merely moves from one compartment to another, driven by pressure differences.

Fluid Gains and Losses

Figure 27–3 and Table 27–1 indicate the major factors involved in fluid balance and highlight the routes of fluid exchange with the environment:

• *Water Losses.* You lose roughly 2500 mL of water each day through urine, feces, and *insensible perspiration*—the

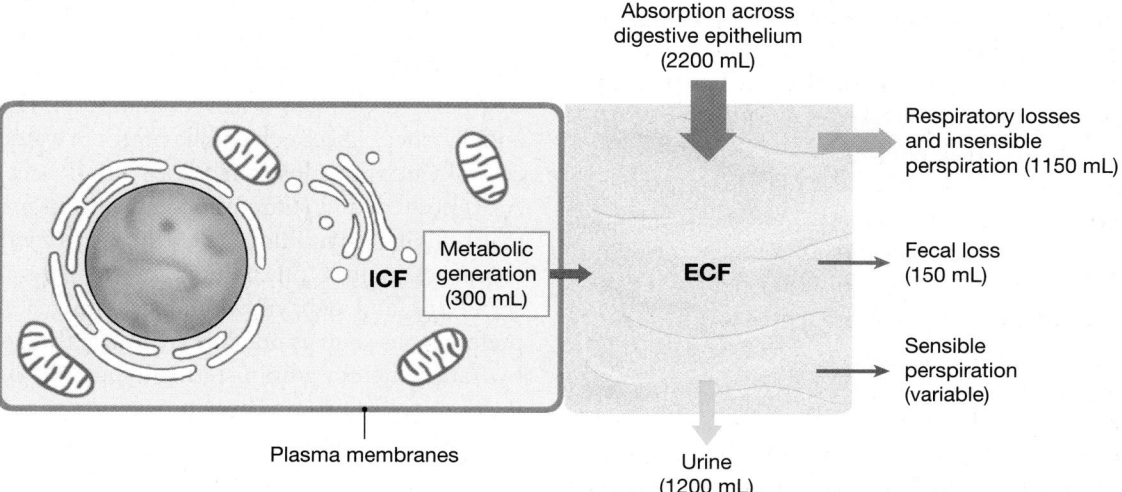

Figure 27–3 Fluid Gains and Losses. Fluid movements that maintain fluid balance in a normal individual. The volumes are drawn to scale; the ICF is roughly twice as large as the ECF.

TABLE 27–1 Water Balance

Source	Daily Input (mL)
Water content of food	1000
Water consumed as liquid	1200
Metabolic water produced during catabolism	300
Total	**2500**

Method of Elimination	Daily Output (mL)
Urination	1200
Evaporation at skin	750
Evaporation at lungs	400
Loss in feces	150
Total	**2500**

gradual movement of water across the epithelia of the skin and respiratory tract. The losses due to *sensible perspiration*—the secretory activities of the sweat glands—vary with the activities you undertake. Sensible perspiration can cause significant water deficits, with maximum perspiration rates reaching 4 liters per hour. ∞ p. 958 Fever can also increase water losses. For each degree that body temperature rises above normal, daily insensible water losses increase by 200 mL. The advice "Drink plenty of fluids" for anyone who is sick has a definite physiological basis.

- *Water Gains.* A water gain of roughly 2500 mL/day is required to balance your average water losses. This value amounts to roughly 40 mL/kg of body weight per day. You obtain water through eating (1000 mL), drinking (1200 mL), and *metabolic generation* (300 mL).

Metabolic generation of water is the production of water within cells, primarily as a result of oxidative phosphorylation in mitochondria. (The synthesis of water at the end of the electron transport system was described in Chapter 25. ∞ p. 938) When a cell breaks down 1 g of lipid, 1.7 mL of water is generated. Breaking down proteins or carbohydrates yields much lower values (0.41 mL/g and 0.55 mL/g, respectively). A typical diet in the United States contains 46 percent carbohydrates, 40 percent lipids, and 14 percent protein. Such a diet produces roughly 300 mL of water per day, about 12 percent of your average daily requirement.

Fluid Shifts

A rapid water movement between the ECF and the ICF in response to an osmotic gradient is called a **fluid shift**. Fluid shifts occur rapidly in response to changes in the osmotic concentration of the ECF and reach equilibrium within minutes to hours.

- *If the Osmotic Concentration of the ECF Increases, that Fluid Will Become Hypertonic with Respect to the ICF.* Water will then move from the cells into the ECF until osmotic equilibrium is restored. The osmotic concentration of the ECF will increase if you lose water but retain electrolytes.

- *If the Osmotic Concentration of the ECF Decreases, that Fluid Will Become Hypotonic with Respect to the ICF.* Water will then move from the ECF into the cells, and the ICF volume will increase. The osmotic concentration of the ECF will decrease if you gain water but do not gain electrolytes.

In sum, if the osmotic concentration of the ECF changes, a fluid shift between the ICF and ECF will tend to oppose the change. Because the volume of the ICF is much greater than that of the ECF, the ICF acts as a water reserve. In effect, instead of a large change in the osmotic concentration of the ECF, smaller changes occur in both the ECF and ICF. Two examples will demonstrate the dynamic exchange of water between the ECF and ICF.

Allocation of Water Losses

Dehydration, or *water depletion*, develops when water losses outpace water gains. When you lose water but retain electrolytes, the osmotic concentration of the ECF rises. Osmosis then moves water out of the ICF and into the ECF until the two solutions are again isotonic. At that point, both the ECF and ICF are somewhat more concentrated than normal, and both volumes are lower than they were before the fluid loss. Because the ICF has roughly twice the functional volume of the ECF, the net change in the ECF is relatively small. However, if the fluid imbalance continues unchecked, the loss of body water will produce severe thirst, dryness, and wrinkling of the skin. Eventually a significant fall in plasma volume and blood pressure occurs, and shock may develop.

Conditions that cause severe water losses include excessive perspiration (brought about by exercising in hot weather), inadequate water consumption, repeated vomiting, and diarrhea. These conditions promote water losses far in excess of electrolyte losses, so body fluids become increasingly concentrated, and sodium ion concentrations become abnormally high (a condition called *hypernatremia*). Homeostatic responses include physiologic mechanisms (ADH and renin secretion) and behavioral changes (increasing fluid intake, preferably as soon as possible). Clinical therapies for acute dehydration include administering hypotonic fluids by mouth or intravenous infusion. These procedures rapidly increase ECF volume and promote the shift of water back into the ICF.

Distribution of Water Gains

When you drink a glass of pure water or when you are given hypotonic solutions intravenously, your body's water content increases without a corresponding increase in the concentration of electrolytes. As a result, the ECF increases in volume but becomes hypotonic with respect to the ICF. A fluid shift then occurs, and the volume of the ICF increases at the expense of the ECF. Once again, the larger volume of the ICF limits the amount of osmotic change. After the fluid shift, the ECF and ICF have slightly larger volumes and slightly lower osmotic concentrations than they did originally.

Normally, this situation will be promptly corrected. The reduced plasma osmotic concentration depresses the secretion of ADH, discouraging fluid intake and increasing water losses in urine. If the situation is *not* corrected, a variety of clinical problems will develop as water shifts into the intracellular fluid, distorting cells, changing the solute concentrations around enzymes, and disrupting normal cell functions. This condition is called **overhydration**, or *water excess*. It can be caused by (1) the ingestion of a large volume of fresh water or the infusion (injection into the bloodstream) of a hypotonic solution; (2) an inability to eliminate excess water in urine, due to chronic renal failure, heart failure, cirrhosis, or some other disorder; and (3) endocrine disorders, such as excessive ADH production.

The most obvious sign of overhydration is abnormally low sodium ion concentrations (*hyponatremia*), and the reduction in Na^+ concentrations in the ECF leads to a fluid shift into the ICF. The first signs are the effects on central nervous system function. The individual initially behaves as if drunk on alcohol. This condition, called *water intoxication*, may sound odd, but is extremely dangerous. Untreated cases can rapidly progress from confusion to hallucinations, convulsions, coma, and then death. Treatment of severe overhydration generally involves administering diuretics and infusing a concentrated salt solution that promotes a fluid shift from the ICF to the ECF and returns Na^+ concentrations to near-normal levels.

CHECKPOINT

5. Define edema.

6. Describe a fluid shift.

7. Define dehydration.

8. What effect would being in the desert without water for a day have on your plasma osmotic concentration?

See the blue Answers tab at the end of the book.

27-4 Balance of the electrolytes sodium, potassium, calcium, and chloride is essential for maintaining homeostasis

You are in electrolyte balance when the rates of gain and loss are equal for each electrolyte in your body. Electrolyte balance is important because:

- *Total electrolyte concentrations directly affect water balance*, as previously described, and

- *The concentrations of individual electrolytes can affect cell functions*. We saw many examples in earlier chapters, including the effect of abnormal Na^+ concentrations on neuron activity and the effects of high or low Ca^{2+} and K^+ concentrations on cardiac muscle tissue.

Two cations, Na^+ and K^+ merit particular attention, because (1) they are major contributors to the osmotic concentrations of the ECF and the ICF, respectively, and (2) they directly affect the normal functioning of all cells. Sodium is the dominant cation in the ECF. More than 90 percent of the

27 URINARY

CLINICAL NOTE

Water and Weight Loss

The safest way to lose weight is to reduce the intake of food while ensuring that all nutrient requirements are met. Water must be included on the list of nutrients, along with carbohydrates, fats, proteins, vitamins, and minerals. Because nearly half of our normal water intake comes from food, a person who eats less becomes more dependent on drinking fluids and on whatever water is generated metabolically.

At the start of a diet, the body conserves water and catabolizes lipids. That is why the first week of dieting may seem rather unproductive. Over that week, the level of circulating ketone bodies gradually increases. In the weeks that follow, the rate of water loss at the kidneys increases in order to excrete waste products, such as the hydrogen ions released by ketone bodies and the urea and ammonia generated during protein catabolism. The rate of fluid intake must also increase, or else the individual risks dehydration. While a person is dieting, dehydration is especially serious, because as water is lost, the concentration of solute in the extracellular fluid (ECF) climbs, further increasing the concentration of waste products and acids generated during the catabolism of energy reserves. This soon becomes a positive feedback loop: These waste products enter the filtrate at the kidneys, and their excretion accelerates urinary water losses.

osmotic concentration of the ECF results from the presence of sodium salts, mainly sodium chloride (NaCl) and sodium bicarbonate ($NaHCO_3$), so changes in the osmotic concentration of body fluids generally reflect changes in Na^+ concentration. Normal Na^+ concentrations in the ECF average about 140 mEq/L, versus 10 mEq/L or less in the ICF. Potassium is the dominant cation in the ICF, where concentrations reach 160 mEq/L. Extracellular K^+ concentrations are generally very low, from 3.8 to 5.0 mEq/L.

Two general rules concerning sodium balance and potassium balance are worth noting:

1. *The Most Common Problems with Electrolyte Balance Are Caused by an Imbalance between Gains and Losses of Sodium Ions.*

2. *Problems with Potassium Balance Are Less Common, but Significantly More Dangerous than Are Those Related to Sodium Balance.*

Sodium Balance

The total amount of sodium in the ECF represents a balance between two factors:

1. *Sodium Ion Uptake across the Digestive Epithelium.* Sodium ions enter the ECF by crossing the digestive epithelium through diffusion and carrier-mediated transport. The rate of absorption varies directly with the amount of sodium in the diet.

2. *Sodium Ion Excretion at the Kidneys and Other Sites.* Sodium losses occur primarily by excretion in urine and through perspiration. The kidneys are the most important sites of Na^+ regulation. The mechanisms for sodium reabsorption at the kidneys were discussed in Chapter 26. ∞ pp. 984, 988

A person in sodium balance typically gains and loses 48–144 mEq (1.1–3.3 g) of Na^+ each day. When sodium gains exceed sodium losses, the total Na^+ content of the ECF goes up; when losses exceed gains, the Na^+ content declines. However, a change in the Na^+ content of the ECF does not produce a change in the Na^+ concentration. When sodium intake or output changes, a corresponding gain or loss of water tends to keep the Na^+ concentration constant. For example, if you eat a very salty meal, the osmotic concentration of the ECF will not increase. When sodium ions are pumped across the digestive epithelium, the solute concentration in that portion of the ECF increases, whereas that of the intestinal contents decreases. Osmosis then occurs. Additional water enters the ECF from the digestive tract, elevating the blood volume and blood pressure. For this reason, people with high blood pressure or a renal salt sensitivity are advised to restrict the amount of salt in their diets.

Sodium Balance and ECF Volume

The sodium regulatory mechanism, diagrammed in **Figure 27–4**, changes the ECF volume but keeps the Na^+ concentration relatively stable. If you consume large amounts of salt *without* adequate fluid, as when you eat salty potato chips without taking a drink, the plasma Na^+ concentration rises temporarily. A change in ECF volume soon follows, however. Fluid will exit the ICF, increasing ECF volume and lowering Na^+ concentrations somewhat. The secretion of ADH restricts water loss and stimulates thirst, promoting additional water consumption. Due to the inhibition of water receptors in the pharynx, ADH secretion begins even before Na^+ absorption occurs; the secretion rate rises further after Na^+ absorption, due to osmoreceptor stimulation. ∞ pp. 565, 618

When sodium losses exceed gains, the volume of the ECF decreases. This reduction occurs without a significant change in the osmotic concentration of the ECF. Thus, if you perspire heavily but consume only pure water, you will lose sodium,

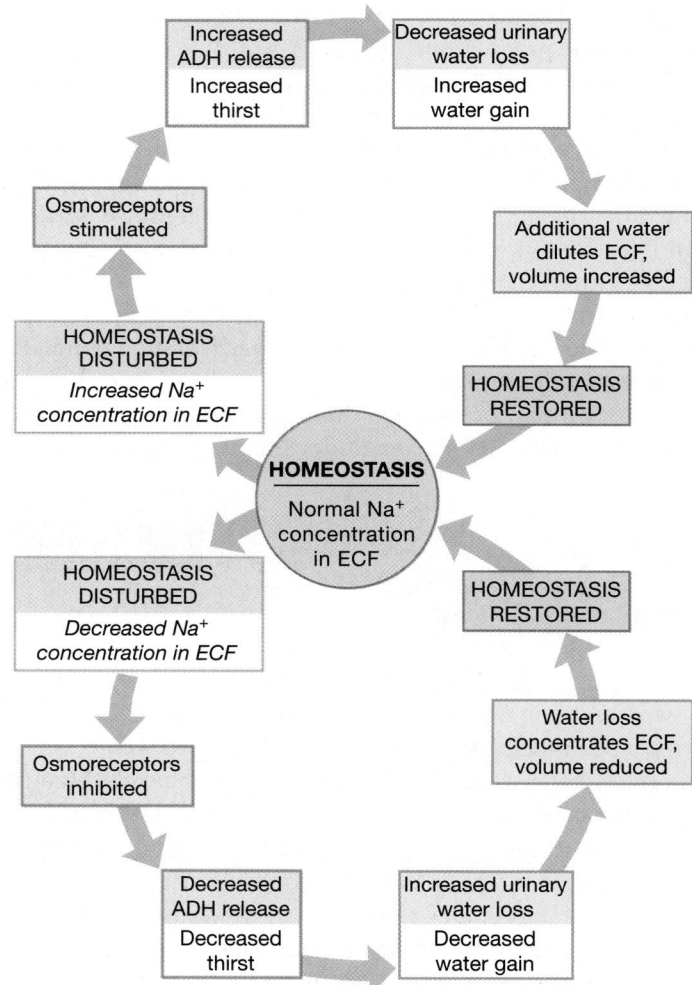

Figure 27–4 The Homeostatic Regulation of Normal Sodium Ion Concentrations in Body Fluids.

and the osmotic concentration of the ECF will drop briefly. However, as soon as the osmotic concentration drops by 2 percent or more, ADH secretion decreases, so water losses at your kidneys increase. As water leaves the ECF, the osmotic concentration returns to normal.

Minor changes in ECF volume do not matter, because they do not cause adverse physiological effects. If, however, regulation of Na+ concentrations results in a large change in ECF volume, the situation will be corrected by the same homeostatic mechanisms responsible for regulating blood volume and blood pressure. This is the case because when ECF volume changes, so does plasma volume and, in turn, blood volume. If ECF volume rises, blood volume goes up; if ECF volume drops, blood volume goes down. As we saw in Chapter 21, blood volume has a direct effect on blood pressure. A rise in blood volume elevates blood pressure; a drop

lowers blood pressure. The net result is that homeostatic mechanisms can monitor ECF volume indirectly by monitoring blood pressure. The receptors involved are baroreceptors at the carotid sinus, the aortic sinus, and the right atrium. The regulatory steps involved are reviewed in **Figure 27–5**.

Sustained abnormalities in the Na+ concentration in the ECF occur only when there are severe problems with fluid balance, such as dehydration or overhydration. When the body's water content rises enough to reduce the Na+ concentration of the ECF below 136 mEq/L, a state of *hyponatremia* (*natrium*, sodium) exists. When body water content declines, the Na+ concentration rises; when that concentration exceeds 145 mEq/L, *hypernatremia* exists.

If the ECF volume is inadequate, both blood volume and blood pressure decline, and the renin–angiotensin system is activated. In response, losses of water and Na+ are reduced,

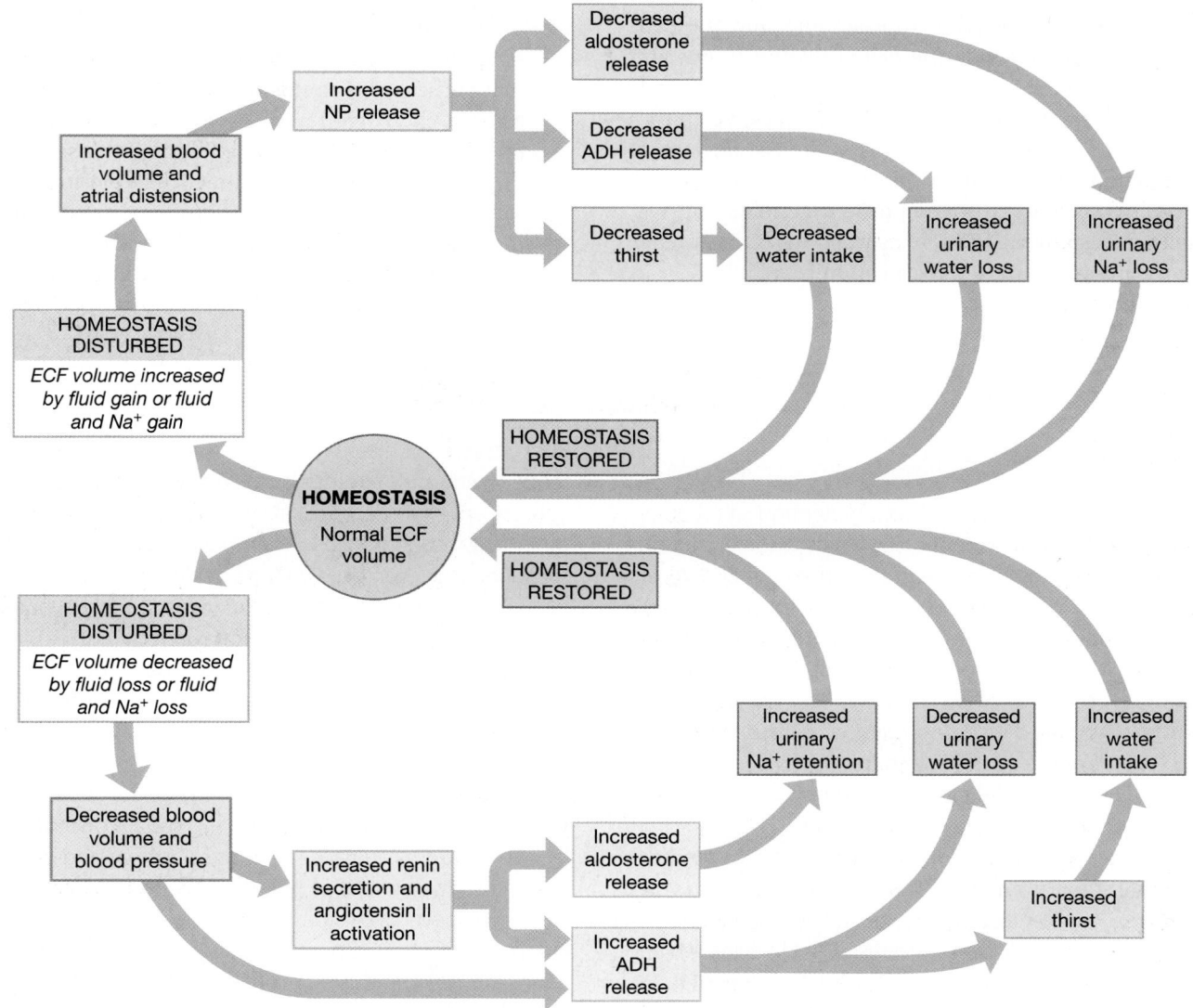

Figure 27–5 **The Integration of Fluid Volume Regulation and Sodium Ion Concentrations in Body Fluids.** NP = Natriuretic peptides.

and gains of water and Na^+ are increased. The net result is that ECF volume increases. Although the total amount of Na^+ in the ECF is increasing (gains exceed losses), the Na^+ concentration in the ECF remains unchanged, because absorption is accompanied by osmotic water movement.

If the plasma volume becomes abnormally large, venous return increases, stretching the atrial and ventricular walls and stimulating the release of natriuretic peptides (ANP and BNP). This in turn reduces thirst and blocks the secretion of ADH and aldosterone, which together promote water or salt conservation. As a result, salt and water loss at the kidneys increases and the volume of the ECF declines.

Potassium Balance

Roughly 98 percent of the potassium content of the human body is in the ICF. Cells expend energy to recover potassium ions as they diffuse out of the cytoplasm and into the ECF. The K^+ concentration outside the cell is relatively low, and the concentration in the ECF at any moment represents a balance between (1) the rate of gain across the digestive epithelium and (2) the rate of loss into urine. Potassium loss in urine is regulated by controlling the activities of ion pumps along the distal portions of the nephron and collecting system. Whenever a sodium ion is reabsorbed from the tubular fluid, it generally is exchanged for a cation (typically K^+) in the peritubular fluid.

Urinary K^+ losses are usually limited to the amount gained by absorption across the digestive epithelium, typically 50–150 mEq (1.9–5.8 g) per day. (Potassium losses in feces and perspiration are negligible.) The K^+ concentration in the ECF is controlled by adjustments in the rate of active secretion along the distal convoluted tubule and collecting system of the nephron.

The rate of tubular secretion of K^+ varies in response to three factors:

1. *Changes in the K^+ Concentration of the ECF.* In general, the higher the extracellular concentration of potassium, the higher the rate of secretion.

2. *Changes in pH.* When the pH of the ECF falls, so does the pH of peritubular fluid. The rate of potassium secretion then declines, because hydrogen ions, rather than potassium ions, are secreted in exchange for sodium ions in tubular fluid. The mechanisms for H^+ secretion were summarized in **Figure 26–14c** (p. 991).

3. *Aldosterone Levels.* The rate at which K^+ is lost in urine is strongly affected by aldosterone, because the ion pumps that are sensitive to this hormone reabsorb Na^+ from filtrate in exchange for K^+ from peritubular fluid. Aldosterone secretion is stimulated by angiotensin II as part of the regulation of blood volume. High plasma K^+ concentrations also stimulate aldosterone secretion directly. Either way, under the influence of aldosterone the amount of sodium conserved and the amount of potassium excreted in urine are directly related.

CLINICAL NOTE

Athletes and Salt Loss

Unfounded notions and rumors about water and salt requirements during exercise abound. Sweat is a hypotonic solution that contains Na^+ in lower concentration than the ECF. As a result, a person who is sweating profusely loses more water than salt, and this loss leads to a rise in the Na^+ concentration of the ECF. The water content of the ECF decreases as the water loss occurs, so blood volume drops. Clinically, this condition is often called *volume depletion*. Because volume depletion occurs at the same time that blood is being shunted away from the kidneys, kidney function is impaired and waste products accumulate in the blood.

To prevent volume depletion, exercising athletes should drink liquids at regular intervals. The primary problem in volume depletion is water loss, and research has revealed no basis for the rumor that cramps will result if you drink while exercising. Salt pills and the various sports beverages that claim "faster absorption" and "better electrolyte balance" have no apparent benefits, despite their relatively high cost.

Body reserves of electrolytes are sufficient to tolerate extended periods of strenuous activity, and problems with Na^+ balance are extremely unlikely except during marathons or other activities that involve maximal exertion for more than six hours. However, both volume depletion (causing acute renal failure) and water intoxication (causing fatal hyponatremia) have occurred in marathon runners.

Some sports beverages contain sugars and vitamins as well as electrolytes. During endurance events (marathons, ultramarathons, and distance cycling), solutions containing less than 10 g/dL of glucose may improve one's performance if consumed late in the event, when metabolic reserves are exhausted. However, high sugar concentrations (above 10 g/dL) can cause cramps, diarrhea, and other problems. The benefit of "glucose polymers" (often corn starch) in sports drinks has yet to be proved. Drinking beverages "fortified" with vitamins is actively discouraged: Vitamins are not lost during exercise, and the consumption of these beverages in large volumes could, over time, cause hypervitaminosis.

When the plasma concentration of potassium falls below 3.5 mEq/L, extensive muscular weakness develops, followed by eventual paralysis. This condition, called *hypokalemia* (*hypo-*, below + *kalium*, potassium), has potentially lethal effects on cardiac function.

Tips&Tricks

The chemical symbols for sodium (Na) and potassium (K) are derived from their Latin names, **Na**trium and **K**alium.

Balance of Other Electrolytes

The ECF concentrations of other electrolytes are regulated as well. Here we will consider the most important ions involved. Additional information about sodium, potassium, and these other ions is listed in Table 27–2.

Calcium Balance

Calcium is the most abundant mineral in the body. A typical individual has 1–2 kg (2.2–4.4 lb) of this element, 99 percent

---

TABLE 27–2 Electrolyte Balance for Average Adult

Ion and Normal ECF Range (mEq/L)	Disorder (mEq/L)	Signs and Symptoms	Causes	Treatments
Sodium (136–145)	Hypernatremia (>145)	Thirst, dryness and wrinkling of skin, reduced blood volume and pressure, eventual circulatory collapse	Dehydration; loss of hypotonic fluid	Ingestion of water or intravenous infusion of hypotonic solution
	Hyponatremia (<136)	Disturbed CNS function (water intoxication): confusion, hallucinations, convulsions, coma; death in severe cases	Infusion or ingestion of large volumes of hypotonic solution	Diuretic use and infusion of hypertonic salt solution
Potassium (3.5–5.5)	Hyperkalemia (>5.5)	Severe cardiac arrhythmias; muscle spasms	Renal failure; use of diuretics; chronic acidosis	Infusion of hypotonic solution; selection of different diuretics; infusion of buffers; dietary restrictions
	Hypokalemia (<3.5)	Muscular weakness and paralysis	Low-potassium diet; diuretics; hypersecretion of aldosterone; chronic alkalosis	Increase in dietary K^+ content; ingestion of K^+ tablets or solutions; infusion of potassium solution
Calcium (4.5–5.5)	Hypercalcemia (>5.5)	Confusion, muscle pain, cardiac arrhythmias, kidney stones, calcification of soft tissues	Hyperparathyroidism; cancer; vitamin D toxicity; calcium supplement overdose	Infusion of hypotonic fluid to lower Ca^{2+} levels; surgery to remove parathyroid gland; administration of calcitonin
	Hypocalcemia (<4.5)	Muscle spasms, convulsions, intestinal cramps, weak heartbeats, cardiac arrhythmias, osteoporosis	Poor diet; lack of vitamin D; renal failure; hypoparathyroidism; hypomagnesemia	Calcium supplements; administration of vitamin D
Magnesium (1.4–2.1)	Hypermagnesemia (>2.1)	Confusion, lethargy, respiratory depression, hypotension	Overdose of magnesium supplements or antacids (rare)	Infusion of hypotonic solution to lower plasma concentration
	Hypomagnesemia (<1.4)	Hypocalcemia, muscle weakness, cramps, cardiac arrhythmias, hypertension	Poor diet; alcoholism; severe diarrhea; kidney disease; malabsorption syndrome; ketoacidosis	Intravenous infusion of solution high in Mg^{2+}
Phosphate (1.8–2.9)	Hyperphosphatemia (>2.9)	No immediate symptoms; chronic elevation leads to calcification of soft tissues	High dietary phosphate intake; hypoparathyroidism	Dietary reduction; PTH supplementation
	Hypophosphatemia (<1.8)	Anorexia, dizziness, muscle weakness, cardiomyopathy, osteoporosis	Poor diet; kidney disease; malabsorption syndrome; hyperparathyroidism; vitamin D deficiency	Dietary improvement; vitamin D and/or calcitriol supplementation
Chloride (97–107)	Hyperchloremia (>107)	Acidosis, hyperkalemia	Dietary excess; increased chloride retention	Infusion of hypotonic solution to lower plasma concentration
	Hypochloremia (<97)	Alkalosis, anorexia, muscle cramps, apathy	Vomiting; hypokalemia	Diuretic use and infusion of hypertonic salt solution

27 URINARY

of which is deposited in the skeleton. In addition to forming the crystalline component of bone, calcium ions play key roles in the control of muscular and neural activities, in blood clotting, as cofactors for enzymatic reactions, and as second messengers.

As noted in Chapters 6 and 18, calcium homeostasis primarily reflects an interplay between the reserves in bone, the rate of absorption across the digestive tract, and the rate of loss at the kidneys. The hormones parathyroid hormone (PTH), calcitriol, and (to a lesser degree) calcitonin, maintain calcium homeostasis in the ECF. Parathyroid hormone and calcitriol raise Ca^{2+} concentrations; their actions are opposed by calcitonin. ∞ pp. 624–626

A small amount of Ca^{2+} is lost in the bile, and under normal circumstances very little Ca^{2+} escapes in urine or feces. To keep pace with biliary, urinary, and fecal Ca^{2+} losses, an adult must absorb only 0.8–1.2 g/day of Ca^{2+}. That amount represents only about 0.03 percent of the calcium reserve in the skeleton. Calcium absorption at the digestive tract and reabsorption along the distal convoluted tubule are stimulated by PTH from the parathyroid glands and by calcitriol from the kidneys.

Hypercalcemia exists when the Ca^{2+} concentration of the ECF exceeds 5.5 mEq/L. The primary cause of hypercalcemia in adults is *hyperparathyroidism*, a condition resulting from oversecretion of PTH. Less common causes include malignant cancers of the breast, lung, kidney, and bone marrow, and excessive use of calcium or vitamin D supplements. Severe hypercalcemia (12–13 mEq/L) causes such signs and symptoms as fatigue, confusion, cardiac arrhythmias, and calcification of the kidneys and soft tissues throughout the body.

Hypocalcemia (a Ca^{2+} concentration under 4.5 mEq/L) is much less common than hypercalcemia. *Hypoparathyroidism* (undersecretion of PTH), vitamin D deficiency, or chronic renal failure is typically responsible for hypocalcemia. Signs and symptoms include muscle spasms, sometimes with generalized convulsions, weak heartbeats, cardiac arrhythmias, and osteoporosis.

Magnesium Balance

The adult body contains about 29 g of magnesium; almost 60 percent of it is deposited in the skeleton. The magnesium in body fluids is contained primarily in the ICF, where the concentration of Mg^{2+} averages about 26 mEq/L. Magnesium is required as a cofactor for several important enzymatic reactions, including the phosphorylation of glucose within cells and the use of ATP by contracting muscle fibers. Magnesium is also important as a structural component of bone.

The typical range of Mg^{2+} concentrations in the ECF is 1.4–2.1 mEq/L, considerably lower than levels in the ICF. The proximal convoluted tubule reabsorbs magnesium very effectively. Keeping pace with the daily urinary loss requires a minimum dietary intake of only 24–32 mEq (0.3–0.4 g) per day.

Phosphate Balance

Phosphate ions are required for bone mineralization, and roughly 740 g of PO_4^{3-} is bound up in the mineral salts of the skeleton. In body fluids, the most important functions of PO_4^{3-} involve the ICF, where the ions are required for the formation of high-energy compounds, the activation of enzymes, and the synthesis of nucleic acids.

The PO_4^{3-} concentration of the plasma is usually 1.8–2.9 mEq/L. Phosphate ions are reabsorbed from tubular fluid along the proximal convoluted tubule; urinary and fecal losses of PO_4^{3-} amount to 30–45 mEq (0.8–1.2 g) per day. Phosphate ion reabsorption along the PCT is stimulated by calcitriol.

Chloride Balance

Chloride ions are the most abundant anions in the ECF. The normal plasma concentration is 97–107 mEq/L. In the ICF, Cl^- concentrations are usually low (3 mEq/L). Chloride ions are absorbed across the digestive tract together with sodium ions; several carrier proteins along the renal tubules reabsorb Cl^- with Na^+ ∞ pp. 986, 988 The rate of Cl^- loss is small; a gain of 48–146 mEq (1.7–5.1 g) per day will keep pace with losses in urine and perspiration.

The A&P Top 100

#93 Fluid balance and electrolyte balance are interrelated. Small water gains or losses affect electrolyte concentrations only temporarily. The impacts are reduced by fluid shifts between the ECF and ICF, and by hormonal responses that adjust the rates of water intake and excretion. Similarly, electrolyte gains or losses produce only temporary changes in solute concentration. These changes are opposed by fluid shifts, adjustments in the rates of ion absorption and secretion, and adjustments to the rates of water gain and loss.

CHECKPOINT

9. Identify four physiologically important cations and two important anions in the extracellular fluid.

10. Why does prolonged sweating increase plasma sodium ion levels?

11. Which is more dangerous, disturbances of sodium balance or disturbances of potassium balance?

See the blue Answers tab at the end of the book.

27-5 In acid-base balance, regulation of hydrogen ions in body fluids involves buffer systems and renal and respiratory compensatory mechanisms

The topic of pH and the chemical nature of acids, bases, and buffers was introduced in Chapter 2. Table 27–3 reviews key terms important to the discussion that follows. If you need a more detailed review, refer to the appropriate sections of Chapter 2 before you proceed. ∞ pp. 41–45

The pH of body fluids can be altered by the introduction of either acids or bases. In general, acids and bases can be categorized as either *strong* or *weak*.

- *Strong acids* and *strong bases* dissociate completely in solution. For example, hydrochloric acid (HCl), a strong acid, dissociates in solution via the reaction

$$HCl \longrightarrow H^+ + Cl^-$$

- When *weak acids* or *weak bases* enter a solution, a significant number of molecules remain intact; dissociation is not complete. Thus, if you place molecules of a weak acid in one solution and the same number of molecules of a strong acid in another solution, the weak acid will liberate fewer hydrogen ions and have less effect on the pH of the solution than will the strong acid. Carbonic acid is a weak acid. At the normal pH of the ECF, an equilibrium state exists, and the reaction can be diagrammed as follows:

$$\underset{\text{carbonic acid}}{H_2CO_3} \rightleftharpoons H^+ + \underset{\text{bicarbonate ion}}{HCO_3^-}$$

The Importance of pH Control

The pH of body fluids reflects interactions among all the acids, bases, and salts in solution in the body. The pH of the ECF normally remains within relatively narrow limits, usually 7.35–7.45. Any deviation from the normal range is extremely dangerous, because changes in H^+ concentrations disrupt the stability of plasma membranes, alter the structure of proteins, and change the activities of important enzymes. You could not survive for long with an ECF pH below 6.8 or above 7.7.

When the pH of plasma falls below 7.35, *acidemia* exists. The physiological state that results is called **acidosis**. When the pH of plasma rises above 7.45, *alkalemia* exists. The physiological state that results is called **alkalosis**. Acidosis and alkalosis affect virtually all body systems, but the nervous and cardiovascular systems are particularly sensitive to pH fluctuations. For example, severe acidosis (pH below 7.0) can be deadly, because (1) central nervous system function deteriorates, and the individual may become comatose; (2) cardiac contractions grow weak and irregular, and signs and symptoms of heart failure may develop; and (3) peripheral vasodilation produces a dramatic drop in blood pressure, potentially producing circulatory collapse.

The control of pH is therefore a homeostatic process of great physiological and clinical significance. Although both acidosis and alkalosis are dangerous, in practice, problems with acidosis are much more common. This is so because several acids, including carbonic acid, are generated by normal cellular activities.

Types of Acids in the Body

The body contains three general categories of acids: (1) *volatile acids*, (2) *fixed acids*, and (3) *organic acids*.

A **volatile acid** is an acid that can leave solution and enter the atmosphere. Carbonic acid (H_2CO_3)is an important volatile acid in body fluids. At the lungs, carbonic acid breaks down into carbon dioxide and water; the carbon dioxide diffuses into the alveoli. In peripheral tissues, carbon dioxide in solution interacts with water to form carbonic acid, which dissociates to release hydrogen ions and bicarbonate ions. The complete reaction sequence is:

$$\underset{\substack{\text{carbon} \\ \text{dioxide}}}{CO_2} + \underset{\text{water}}{H_2O} \rightleftharpoons \underset{\substack{\text{carbonic} \\ \text{acid}}}{H_2CO_3} \rightleftharpoons H^+ + \underset{\substack{\text{biocarbonate} \\ \text{ion}}}{HCO_3^-}$$

TABLE 27–3 A Review of Important Terms Relating to Acid–Base Balance	
pH	The negative exponent (negative logarithm) of the hydrogen ion concentration [H^+]
Neutral	A solution with a pH of 7; the solution contains equal numbers of hydrogen ions and hydroxide ions
Acidic	A solution with a pH below 7; in this solution, hydrogen ions [H^+] predominate
Basic, or alkaline	A solution with a pH above 7; in this solution, hydroxide ions [OH^-] predominate
Acid	A substance that dissociates to release hydrogen ions, decreasing pH
Base	A substance that dissociates to release hydroxide ions or to tie up hydrogen ions, increasing pH
Salt	An ionic compound consisting of a cation other than hydrogen and an anion other than a hydroxide ion
Buffer	A substance that tends to oppose changes in the pH of a solution by removing or replacing hydrogen ions; in body fluids, buffers maintain blood pH within normal limits (7.35–7.45)

27

URINARY

This reaction occurs spontaneously in body fluids, but it proceeds much more rapidly in the presence of *carbonic anhydrase (CA)*, an enzyme found in the cytoplasm of red blood cells, liver and kidney cells, parietal cells of the stomach, and many other types of cells.

Because most of the carbon dioxide in solution is converted to carbonic acid, and most of the carbonic acid dissociates, the partial pressure of carbon dioxide and the pH are inversely related (**Figure 27–6**). When carbon dioxide levels rise, additional hydrogen ions and bicarbonate ions are released, so the pH goes down. (Recall that the pH is a *negative exponent*, so when the concentration of hydrogen ions goes up, the pH goes down.) The P_{CO_2} is the most important factor affecting the pH in body tissues.

At the alveoli, carbon dioxide diffuses into the atmosphere, the number of hydrogen ions and bicarbonate ions in the alveolar capillaries drops, and blood pH rises. We will consider this process, which effectively removes hydrogen ions from solution, in more detail later in the chapter.

Fixed acids are acids that do not leave solution; once produced, they remain in body fluids until they are eliminated at the kidneys. Sulfuric acid and phosphoric acid are the most important fixed acids in the body. They are generated in small amounts during the catabolism of amino acids and compounds that contain phosphate groups, including phospholipids and nucleic acids.

Organic acids are acid participants in or by-products of aerobic metabolism. Important organic acids include lactic acid, produced by the anaerobic metabolism of pyruvate, and ketone bodies, synthesized from acetyl-CoA. Under normal conditions, most organic acids are metabolized rapidly, so significant accumulations do not occur. But relatively large amounts of organic acids are produced (1) during periods of anaerobic metabolism, because lactic acid builds up rapidly, and (2) during starvation or excessive lipid catabolism, because ketone bodies accumulate.

Mechanisms of pH Control

To maintain acid–base balance over long periods of time, your body must balance gains and losses of hydrogen ions. Hydrogen ions are gained at the digestive tract and through metabolic activities within cells. Your body must eliminate these ions at the kidneys, by secreting H^+ into urine, and at the lungs, by forming water and carbon dioxide from H^+ and HCO_3^-. The sites of elimination are far removed from the sites of acid production. As the hydrogen ions travel through the body, they must be neutralized to avoid tissue damage.

Buffers and buffer systems in body fluids temporarily neutralize the acids produced in the course of normal metabolic operations. *Buffers* are dissolved compounds that stabilize the pH of a solution by providing or removing H^+. Buffers include weak acids that can donate H^+, and weak bases that can absorb H^+. A **buffer system** in body fluids generally consists of a combination of a weak acid and the anion released by its dissociation. The anion functions as a weak base. In solution, molecules of the weak acid exist in equilibrium with its dissociation products. In chemical notation, this relationship is represented as:

$$HY \rightleftharpoons H^+ + Y^-$$

Adding H^+ to the solution upsets the equilibrium, and the resulting formation of additional molecules of the weak acid removes some of the H^+ from the solution.

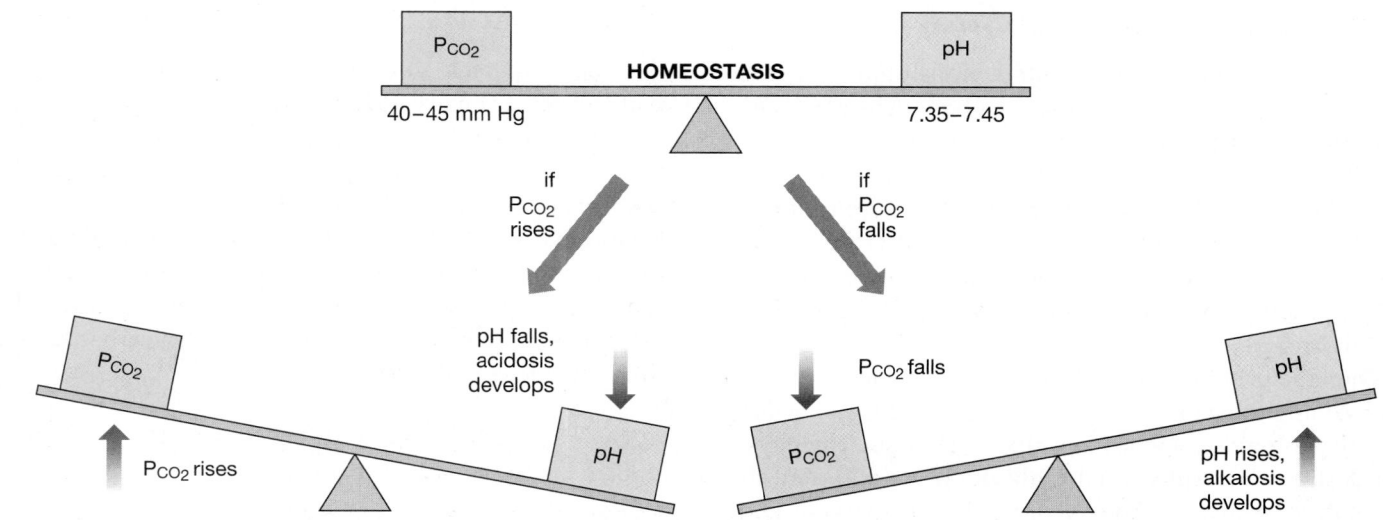

Figure 27–6 **The Basic Relationship between P_{CO_2} and Plasma pH.** The P_{CO_2} is inversely related to the pH.

The body has three major buffer systems, each with slightly different characteristics and distributions (**Figure 27–7**):

1. *Protein buffer systems* contribute to the regulation of pH in the ECF and ICF. These buffer systems interact extensively with the other two buffer systems.

2. *The carbonic acid–bicarbonate buffer system* is most important in the ECF.

3. *The phosphate buffer system* has an important role in buffering the pH of the ICF and of urine.

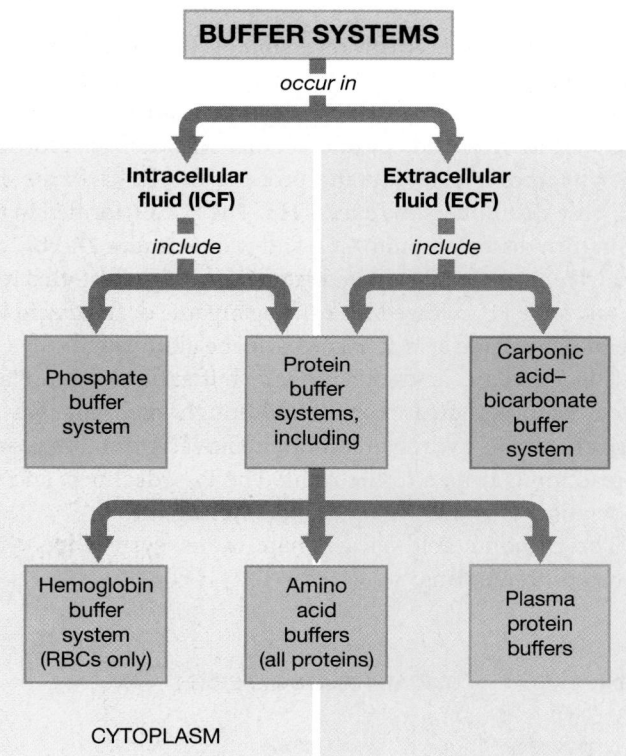

BUFFER SYSTEMS

occur in

Intracellular fluid (ICF)	Extracellular fluid (ECF)
include	*include*

Phosphate buffer system

Protein buffer systems, including

Carbonic acid–bicarbonate buffer system

Hemoglobin buffer system (RBCs only)

Amino acid buffers (all proteins)

Plasma protein buffers

CYTOPLASM

Figure 27–7   Buffer Systems in Body Fluids. Phosphate buffers occur primarily in the ICF, whereas the carbonic acid–bicarbonate buffer system occurs primarily in the ECF. Protein buffer systems are in both the ICF and the ECF. Extensive interactions take place among these systems.

Protein Buffer Systems

Protein buffer systems depend on the ability of amino acids to respond to pH changes by accepting or releasing H^+. The underlying mechanism is shown in **Figure 27–8**:

- If pH climbs, the carboxyl group (—COOH) of the amino acid can dissociate, acting as a weak acid and releasing a hydrogen ion. The carboxyl group then becomes a carboxylate ion (—COO^-). At the normal pH of body fluids (7.35–7.45), the carboxyl groups of most amino acids have already given up their hydrogen ions. (Proteins carry negative charges primarily for that reason.) However, some amino acids, notably *histidine* and *cysteine*, have R groups (side chains) that will donate hydrogen ions if the pH climbs outside the normal range. Their buffering effects are very important in both the ECF and the ICF.

- If pH drops, the carboxylate ion and the amino group (—NH_2) can act as weak bases and accept additional hydrogen ions, forming a carboxyl group (—COOH) and an amino ion (—NH_3^+), respectively. In free amino acids, both the main structural chain and the side chain can act as buffers. In a protein, most of the carboxyl and amino groups in the main chain are tied up in peptide bonds, leaving only the —NH_2 of the first amino acid and the —COOH of the last as available buffers. So most of the buffering capacity of proteins is provided by the R-groups.

Plasma proteins contribute to the buffering capabilities of blood. Interstitial fluid contains extracellular protein fibers and dissolved amino acids that also assist in regulating pH. In the ICF of active cells, structural and other proteins provide an extensive buffering capability that prevents destructive changes in pH when organic acids, such as lactic acid or pyruvic acid, are produced by cellular metabolism.

Because exchange occurs between the ECF and the ICF, the protein buffer system can help stabilize the pH of the ECF. For example, when the pH of the ECF decreases, cells pump H^+ out of the ECF and into the ICF, where they can be buffered by intracellular proteins. When the pH of the ECF rises, pumps in plasma membranes exchange H^+ in the ICF for H^+ in the ECF.

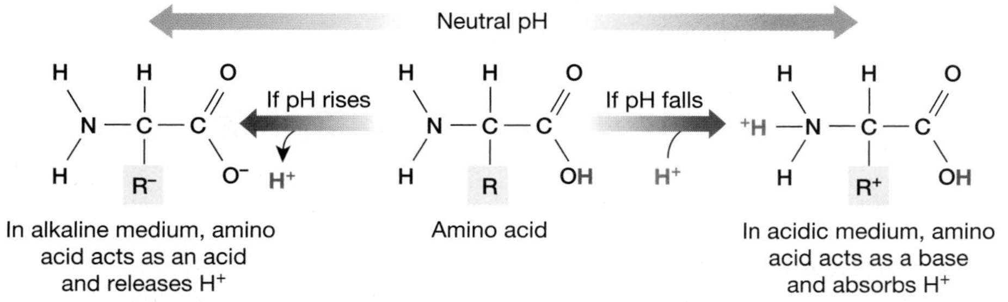

27 URINARY

Neutral pH

In alkaline medium, amino acid acts as an acid and releases H^+

If pH rises

Amino acid

If pH falls

In acidic medium, amino acid acts as a base and absorbs H^+

Figure 27–8   The Role of Amino Acids in Protein Buffer Systems. Depending on the pH of their surroundings, amino acids either donate a hydrogen ion (as at left) or accept a hydrogen ion (as at right). Additionally, several R groups may release or absorb H^+.

These mechanisms can assist in buffering the pH of the ECF. The process is slow, however, because hydrogen ions must be individually transported across the plasma membrane. As a result, the protein buffer system in most cells cannot make rapid, large-scale adjustments in the pH of the ECF.

The Hemoglobin Buffer System. The situation is somewhat different for red blood cells. These cells, which contain roughly 5.5 percent of the ICF, are normally suspended in the plasma. They are densely packed with hemoglobin, and their cytoplasm contains large amounts of carbonic anhydrase. Red blood cells have a significant effect on the pH of the ECF, because they absorb carbon dioxide from the plasma and convert it to carbonic acid. Carbon dioxide can diffuse across the RBC membrane very quickly, so no transport mechanism is needed. As the carbonic acid dissociates, the bicarbonate ions diffuse into the plasma in exchange for chloride ions, a swap known as the *chloride shift.* ∞ p. 859 The hydrogen ions are buffered by hemoglobin molecules. At the lungs, the entire reaction sequence diagrammed in **Figure 23–23** (p. 859) proceeds in reverse. This mechanism is known as the **hemoglobin buffer system**.

The hemoglobin buffer system is the only intracellular buffer system that can have an immediate effect on the pH of the ECF. *The hemoglobin buffer system helps prevent drastic changes in pH when the plasma P_{CO_2} is rising or falling.*

The Carbonic Acid–Bicarbonate Buffer System

With the exception of red blood cells, some cancer cells, and tissues temporarily deprived of oxygen, body cells generate carbon dioxide virtually 24 hours a day. As we have seen, most of the carbon dioxide is converted to carbonic acid, which then dissociates into a hydrogen ion and a bicarbonate ion. The carbonic acid and its dissociation products form the **carbonic acid–bicarbonate buffer system**. The primary role of the carbonic acid–bicarbonate buffer system is to prevent changes in pH caused by organic acids and fixed acids in the ECF.

This buffer system consists of the reaction introduced in our discussion of volatile acids (**Figure 27–9a**):

$$CO_2 + H_2O \rightleftharpoons H_2CO_3 \rightleftharpoons H^+ + HCO_3^-$$

carbon dioxide   water   carbonic acid   biocarbonate ion

Because the reaction is freely reversible, a change in the concentration of any participant affects the concentrations of all other participants. For example, if hydrogen ions are added, most of them will be removed by interactions with HCO_3^-, forming H_2CO_3 (carbonic acid). In the process, the HCO_3^- acts as a weak base that buffers the excess H^+. The H_2CO_3 formed in this way in turn dissociates into CO_2 and water (**Figure 27–9b**). The extra CO_2 can then be excreted at the lungs. In effect, this reaction takes the H^+ released by a strong organic or fixed acid and generates a volatile acid that can easily be eliminated.

The carbonic acid–bicarbonate buffer system can also protect against increases in pH, although such changes are relatively rare. If hydrogen ions are removed from the plasma, the reaction is driven to the right: The P_{CO_2} declines, and the dissociation of H_2CO_3 replaces the missing H^+.

The carbonic acid–bicarbonate buffer system has three important limitations:

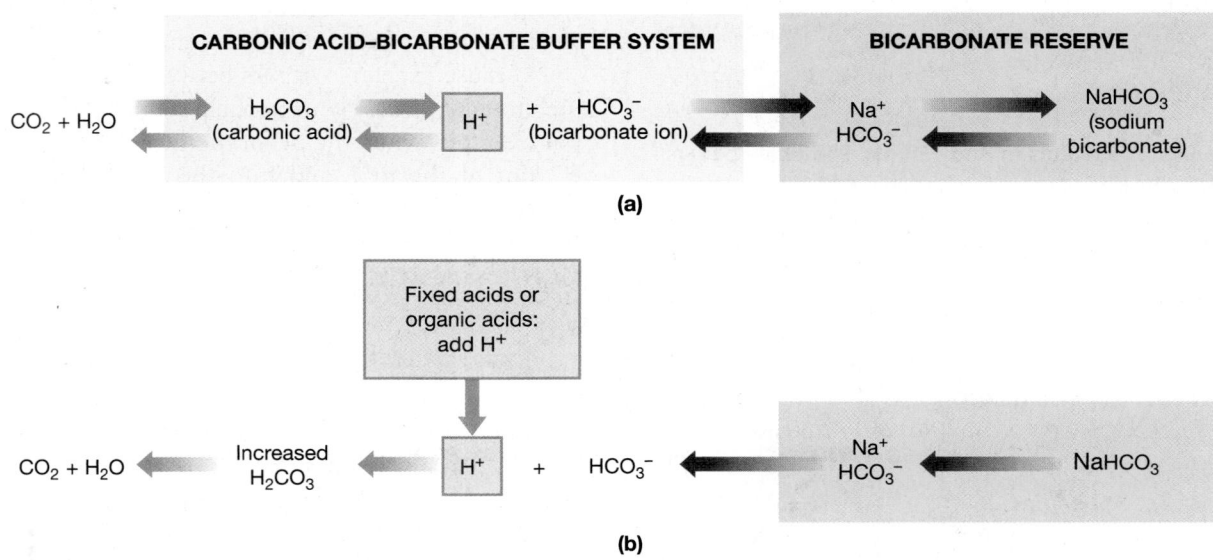

(a)

(b)

Figure 27–9 **The Carbonic Acid–Bicarbonate Buffer System.** **(a)** Basic components of the carbonic acid–bicarbonate buffer system, and their relationships to carbon dioxide and the bicarbonate reserve. **(b)** The response of the carbonic acid–bicarbonate buffer system to hydrogen ions generated by fixed or organic acids in body fluids.

1. *It Cannot Protect the ECF from Changes in pH that Result from Elevated or Depressed Levels of CO_2.* A buffer system cannot protect against changes in the concentration of its own weak acid. As **Figure 27–9a** indicates, an equilibrium exists among the components of this buffer system. Thus, in this system, the addition of excess H^+ from an outside source would drive the reaction to the left. But the addition of excess CO_2 would form H_2CO_3 and drive the reaction to the right. The dissociation of H_2CO_3 would release H^+ and HCO_3^-, reducing the pH of the plasma.

2. *It Can Function Only When the Respiratory System and the Respiratory Control Centers Are Working Normally.* Normally, the elevation in P_{CO_2} that occurs when fixed or organic acids are buffered stimulates an increase in the respiratory rate. This increase accelerates the removal of CO_2 at the lungs. If the respiratory passageways are blocked, or blood flow to the lungs is impaired, or the respiratory centers do not respond normally, the efficiency of the buffer system will be reduced. This buffer system cannot eliminate H^+ and remove the threat to homeostasis unless the respiratory system is functioning normally.

3. *The Ability to Buffer Acids Is Limited by the Availability of Bicarbonate Ions.* Every time a hydrogen ion is removed from the plasma, a bicarbonate ion goes with it. When all the bicarbonate ions have been tied up, buffering capabilities are lost.

Problems due to a lack of bicarbonate ions are rare, for several reasons. First, body fluids contain a large reserve of HCO_3^-, primarily in the form of dissolved molecules of the weak base *sodium bicarbonate* ($NaHCO_3$). This readily available supply of HCO_3^- is known as the **bicarbonate reserve**. The reaction involved (**Figure 27–9a**) is

$$Na^+ \;+\; HCO_3^- \;\rightleftharpoons\; NaHCO_3$$
$$\text{biocarbonate} \qquad\qquad \text{sodium}$$
$$\text{ion} \qquad\qquad\qquad \text{bicarbonate}$$

When hydrogen ions enter the ECF, the bicarbonate ions tied up in H_2CO_3 molecules are replaced by HCO_3^- from the bicarbonate reserve (**Figure 27–9b**).

Second, additional HCO_3^- can be generated at the kidneys, through mechanisms described in Chapter 26 (**Figure 26–14c**, p. 991). In the distal convoluted tubule and collecting system, carbonic anhydrase converts CO_2 within tubular cells into H_2CO_3, which then dissociates. The hydrogen ion is pumped into tubular fluid in exchange for a sodium ion, and the bicarbonate ion is transported into peritubular fluid in exchange for a chloride ion. In effect, tubular cells remove HCl from peritubular fluid in exchange for $NaHCO_3$.

The Phosphate Buffer System

The **phosphate buffer system** consists of the anion $H_2PO_4^-$, which is a weak acid. The operation of the phosphate buffer system resembles that of the carbonic acid–bicarbonate buffer system. The reversible reaction involved is

$$H_2PO_4^- \;\rightleftharpoons\; H^+ \;+\; HPO_4^{2-}$$
$$\text{dihydrogen} \qquad\qquad \text{monohydrogen}$$
$$\text{phosphate} \qquad\qquad\qquad \text{phosphate}$$

The weak acid is *dihydrogen phosphate* ($H_2PO_4^-$), and the anion released is *monohydrogen phosphate* (HPO_4^{2-}). In the ECF, the phosphate buffer system plays only a supporting role in the regulation of pH, primarily because the concentration of HCO_3^- far exceeds that of HPO_4^{2-}. However, the phosphate buffer system is quite important in buffering the pH of the ICF. In addition, cells contain a *phosphate reserve* in the form of the weak base *sodium monohydrogen phosphate* (Na_2HPO_4). The phosphate buffer system is also important in stabilizing the pH of urine. The dissociation of Na_2HPO_4 provides additional HPO_4^{2-} for use by this buffer system:

$$2\,Na^+ \;+\; HPO_4^{2-} \;\rightleftharpoons\; Na_2HPO_4$$
$$\text{monohydrogen} \qquad\qquad \text{sodium}$$
$$\text{phosphate} \qquad\qquad\qquad \text{monohydrogen}$$
$$\text{phosphate}$$

Maintenance of Acid–Base Balance

Although buffer systems can tie up excess H^+, they provide only a temporary solution to an acid–base imbalance. The hydrogen ions are not eliminated, but merely rendered harmless. For homeostasis to be preserved, the captured H^+ must ultimately be either permanently tied up in water molecules, through the removal of carbon dioxide at the lungs, or removed from body fluids, through secretion at the kidneys. The underlying problem is that the body's supply of buffer molecules is limited. Suppose that a buffer molecule prevents a change in pH by binding a hydrogen ion that enters the ECF. That buffer molecule is then tied up, reducing the capacity of the ECF to cope with any additional H^+. Eventually, all the buffer molecules are bound to H^+, and further pH control becomes impossible.

The situation can be resolved only by either removing H^+ from the ECF (thereby freeing the buffer molecules) or replacing the buffer molecules. Similarly, if a buffer provides a hydrogen ion to maintain normal pH, homeostatic conditions will return only when either another hydrogen ion has been obtained or the buffer has been replaced.

The maintenance of acid–base balance thus includes balancing H^+ gains and losses. This "balancing act" involves coordinating the actions of buffer systems with respiratory mechanisms and renal mechanisms. These mechanisms support the buffer systems by (1) secreting or absorbing H^+, (2) controlling the excretion of acids and bases, and, when necessary, (3) generating additional buffers. It is the *combination* of buffer systems and these respiratory and renal mechanisms that maintains body pH within narrow limits.

Respiratory Compensation

Respiratory compensation is a change in the respiratory rate that helps stabilize the pH of the ECF. Respiratory compensation occurs whenever body pH strays outside normal limits. Such compensation is effective because respiratory activity has a direct effect on the carbonic acid–bicarbonate buffer system. Increasing or decreasing the rate of respiration alters pH by lowering or raising the P_{CO_2}. When the P_{CO_2} rises, the pH falls, because the addition of CO_2 drives the carbonic acid–bicarbonate buffer system to the right. When the P_{CO_2} falls, the pH rises because the removal of CO_2 drives that buffer system to the left.

The mechanisms responsible for the control of respiratory rate were described in Chapter 23; hence, only a brief summary is presented here. (If necessary, review **Figures 23–26** and **23–27**, pp. 863, 864.)

Chemoreceptors of the carotid and aortic bodies are sensitive to the P_{CO_2} of circulating blood; other receptors, located on the ventrolateral surfaces of the medulla oblongata, monitor the P_{CO_2} of the CSF. A rise in P_{CO_2} stimulates the chemoreceptors, leading to an increase in the respiratory rate. As the rate of respiration increases, more CO_2 is lost at the lungs, so the P_{CO_2} returns to normal levels. Conversely, when the P_{CO_2} of the blood or CSF declines, the chemoreceptors are inhibited. Respiratory activity becomes depressed and the breathing rate decreases, causing an elevation of the P_{CO_2} in the ECF.

Renal Compensation

Renal compensation is a change in the rates of H^+ and HCO_3^- secretion or reabsorption by the kidneys in response to changes in plasma pH. Under normal conditions, the body generates enough organic and fixed acids to add about 100 mEq of H^+ to the ECF each day. An equivalent number of hydrogen ions must therefore be excreted in urine to maintain acid–base balance. In addition, the kidneys assist the lungs by eliminating any CO_2 that either enters the renal tubules during filtration or diffuses into the tubular fluid as it travels toward the renal pelvis.

Hydrogen ions are secreted into the tubular fluid along the proximal convoluted tubule (PCT), the distal convoluted tubule (DCT), and the collecting system. The basic mechanisms involved are depicted in **Figures 26–12** and **26–14c** (pp. 985, 991). The ability to eliminate a large number of hydrogen ions in a normal volume of urine depends on the presence of buffers in the urine. The secretion of H^+ can continue only until the pH of the tubular fluid reaches

4.0–4.5. (At that point, the H^+ concentration gradient is so great that hydrogen ions leak out of the tubule as fast as they are pumped in.)

If the tubular fluid lacked buffers to absorb the hydrogen ions, the kidneys could secrete less than 1 percent of the acid produced each day before the pH reached this limit. To maintain acid balance under these conditions, the kidneys would have to produce about 1000 liters of urine each day just to keep pace with the generation of H^+ in the body. Buffers in tubular fluid are therefore extremely important, because they keep the pH high enough for H^+ secretion to continue. Metabolic acids are being generated continuously; without these buffering mechanisms, the kidneys would be unable to maintain homeostasis.

Figure 27–10 diagrams the primary routes of H^+ secretion and the buffering mechanisms that stabilize the pH of tubular fluid. The three major buffers involved are the carbonic acid–bicarbonate buffer system, the phosphate buffer system, and the ammonia buffer system (**Figure 27–10a**). Glomerular filtration puts components of the carbonic acid–bicarbonate and phosphate buffer systems into the filtrate. The ammonia is generated by tubule cells (primarily those of the PCT).

Figure 27–10a shows the secretion of H^+, which relies on carbonic anhydrase (CA) activity within tubular cells. The hydrogen ions generated may be pumped into the lumen in exchange for sodium ions, individually or together with chloride ions. The net result is the secretion of H^+, accompanied by the removal of CO_2 (from the tubular fluid, the tubule cells, and the ECF), and the release of sodium bicarbonate into the ECF.

Figure 27–10b shows the generation of ammonia within the tubules. As tubule cells use the enzyme *glutaminase* to break down the amino acid *glutamine,* amino groups are released as either ammonium ions (NH_4^+) or ammonia (NH_3). The ammonium ions are transported into the lumen in exchange for Na^+ in the tubular fluid. The NH_3, which is highly volatile and also toxic to cells, diffuses rapidly into the tubular fluid. There it reacts with a hydrogen ion, forming NH_4^+.

This reaction buffers the tubular fluid and removes a potentially dangerous compound from body fluids. The carbon chains of the glutamine molecules are ultimately converted to HCO_3^-, which is cotransported with Na^+ into the ECF. The generation of ammonia by tubule cells thus ties up H^+ in the tubular fluid and releases sodium bicarbonate into the ECF, where it contributes to the bicarbonate reserve. These mechanisms of H^+ secretion and buffering are always functioning, but their levels of activity vary widely with the pH of the ECF.

The Renal Responses to Acidosis and Alkalosis.

Acidosis (low body fluid pH) develops when the normal plasma buffer mechanisms are stressed by excessive numbers of hydrogen ions. The kidney tubules do not distinguish among the various acids that may cause acidosis. Whether the fall in pH results from the production of

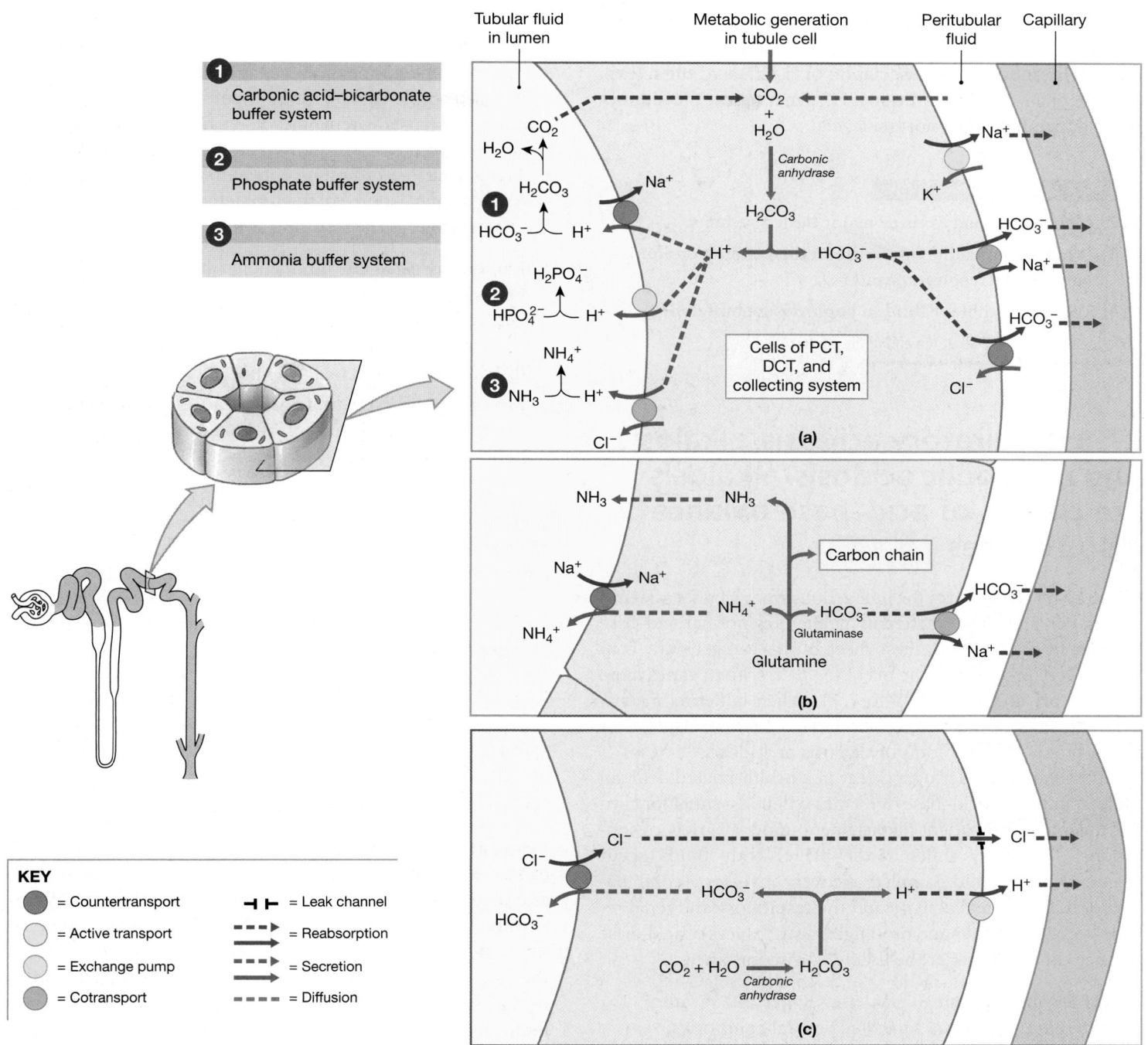

Figure 27–10 Kidney Tubules and pH Regulation. (a) The three major buffering mechanisms in tubular fluid, which are essential to the secretion of hydrogen ions. **(b)** The production of ammonium ions and ammonia by the breakdown of glutamine. **(c)** The response of the kidney tubules to alkalosis.

volatile, fixed, or organic acids, the renal contribution remains limited to (1) the secretion of H^+, (2) the activity of buffers in the tubular fluid, (3) the removal of CO_2, and (4) the reabsorption of $NaHCO_3$.

Tubule cells thus bolster the capabilities of the carbonic acid–bicarbonate buffer system. They do so by increasing the concentration of bicarbonate ions in the ECF, replacing those already used to remove hydrogen ions from solution. In a

starving individual, tubule cells break down amino acids, yielding ammonium ions that are pumped into the tubular fluid, bicarbonates to help buffer ketone bodies in the blood, and carbon chains for catabolism (**Figure 27–10b**).

When alkalosis (high body fluid pH) develops, (1) the rate of H^+ secretion at the kidneys declines, (2) tubule cells do not reclaim the bicarbonates in tubular fluid, and (3) the collecting system transports HCO_3^- into tubular fluid while releasing a strong acid (HCl) into peritubular fluid (**Figure 27–10c**). The concentration of HCO_3^- in plasma decreases, promoting the dissociation of H_2CO_3 and the release of hydrogen ions. The additional H^+ generated at the kidneys helps return the pH to normal levels.

CHECKPOINT

12. Identify the body's three major buffer systems.

13. What effect would a decrease in the pH of body fluids have on the respiratory rate?

14. Why must tubular fluid in nephrons be buffered?

See the blue Answers tab at the end of the book.

27-6 Respiratory acidosis/alkalosis and metabolic acidosis/alkalosis are classes of acid–base balance disturbances

Figure 27–11 summarizes the interactions among buffer systems, respiration, and renal function in maintaining normal acid–base balance. In combination, these mechanisms can generally control pH very precisely, so the pH of the ECF seldom varies more than 0.1 pH unit, from 7.35 to 7.45. When buffering mechanisms are severely stressed, however, the pH drifts outside these limits, producing symptoms of alkalosis or acidosis.

If you are considering a career in a health-related field, an understanding of acid–base dynamics will be essential for clinical diagnosis and patient management under a variety of conditions. Temporary shifts in the pH of body fluids occur frequently. Rapid and complete recovery involves a combination of buffer system activity and the respiratory and renal responses. More serious and prolonged disturbances of acid–base balance can result under the following circumstances:

- *Any Disorder Affecting Circulating Buffers, Respiratory Performance, or Renal Function.* Several conditions, including *emphysema* and *renal failure*, are associated with dangerous changes in pH. ∞ pp. 867, 998

- *Cardiovascular Conditions.* Conditions such as *heart failure* or *hypotension* can affect the pH of internal fluids by causing fluid shifts and by changing glomerular filtration rates and respiratory efficiency. ∞ pp. 706, 733

- *Conditions Affecting the Central Nervous System.* Neural damage or disease that affects the CNS can affect the respiratory and cardiovascular reflexes that are essential to normal pH regulation.

Serious abnormalities in acid–base balance generally have an initial *acute phase*, in which the pH moves rapidly out of the normal range. If the condition persists, physiological adjustments occur; the individual then enters the *compensated phase*. Unless the underlying problem is corrected, compensation cannot be completed, and blood chemistry will remain abnormal. The pH typically remains outside normal limits even after compensation has occurred. Even if the pH is within the normal range, the P_{CO_2} or HCO_3^- concentrations can be abnormal.

The primary source of the problem is usually indicated by the name given to the resulting condition:

- *Respiratory acid–base disorders* result from a mismatch between carbon dioxide generation in peripheral tissues and carbon dioxide excretion at the lungs. When a respiratory acid–base disorder is present, the carbon dioxide level of the ECF is abnormal.

- *Metabolic acid–base disorders* are caused by the generation of organic or fixed acids or by conditions affecting the concentration of HCO_3^- in the ECF.

Respiratory compensation alone may restore normal acid–base balance in individuals with respiratory acid–base disorders. In contrast, compensation mechanisms for metabolic acid–base disorders may be able to stabilize pH, but other aspects of acid–base balance (buffer system function, bicarbonate and P_{CO_2} levels) remain abnormal until the underlying metabolic cause is corrected.

We can subdivide the respiratory and metabolic categories to create four major classes of acid–base disturbances: (1) *respiratory acidosis*, (2) *respiratory alkalosis*, (3) *metabolic acidosis*, and (4) *metabolic alkalosis*.

Respiratory Acidosis

Respiratory acidosis develops when the respiratory system cannot eliminate all the carbon dioxide generated by peripheral tissues. The primary sign is low plasma pH due to **hypercapnia**, an elevated plasma P_{CO_2}. The usual cause is hypoventilation, an abnormally low respiratory rate. When the P_{CO_2} in the ECF rises, H^+ and HCO_3^- concentrations also begin rising as H_2CO_3 forms and dissociates. Other buffer systems can tie up some of the H^+, but once the combined buffering capacity has been exceeded, the pH begins to fall rapidly. The effects are diagrammed in **Figure 27–12a**.

Respiratory acidosis is the most common challenge to acid–base equilibrium. Body tissues generate carbon dioxide rapidly. Even a few minutes of hypoventilation can cause acidosis, reducing the pH of the ECF to as low as 7.0. Under normal circumstances, the chemoreceptors that monitor the P_{CO_2} of plasma and of cerebrospinal fluid (CSF) eliminate the problem by calling for an increase in breathing (pulmonary ventilation) rates.

If the chemoreceptors fail to respond, if the breathing rate cannot be increased, or if the circulatory supply to the lungs is inadequate, pH will continue to decline. If the decline is severe,

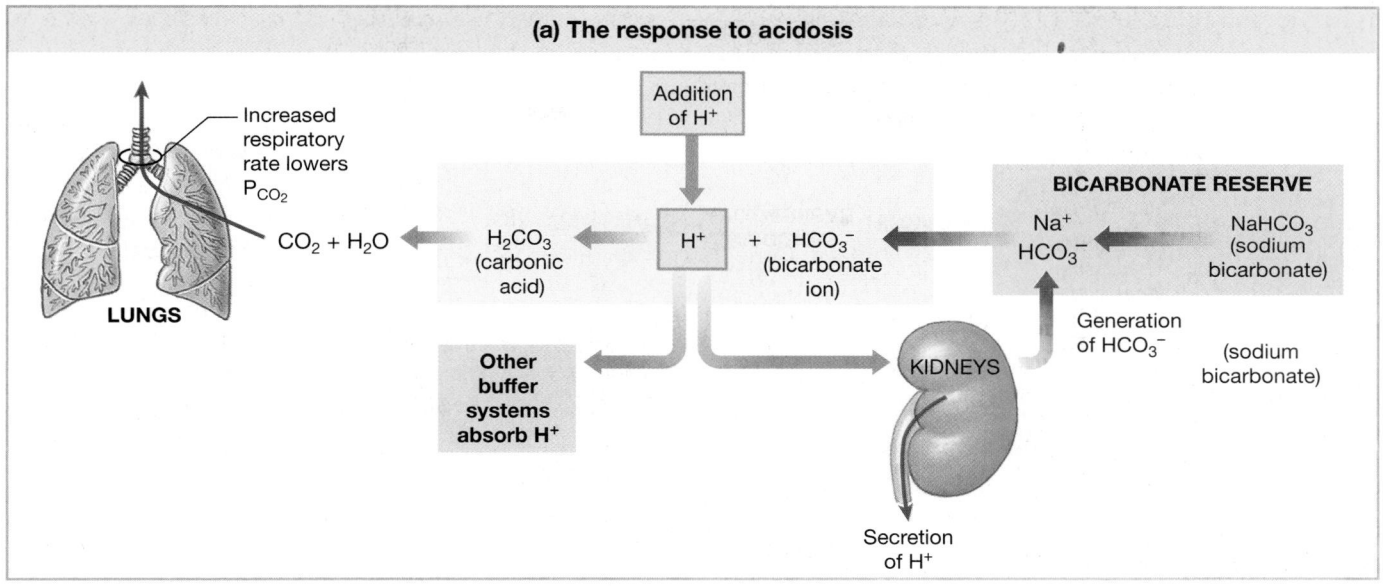

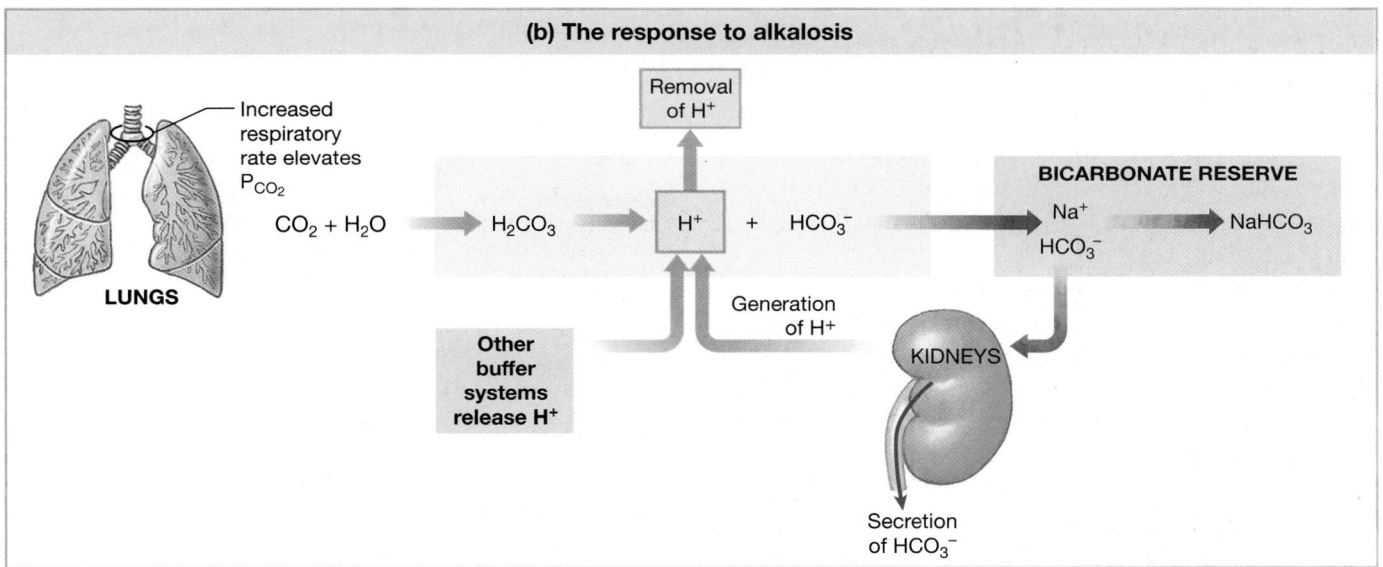

Figure 27–11   Interactions among the Carbonic Acid–Bicarbonate Buffer System and Compensatory Mechanisms in the Regulation of Plasma pH. The central role of the carbonic acid–bicarbonate buffer system is highlighted. **(a)** The response to acidosis caused by the addition of H^+. **(b)** The response to alkalosis caused by the removal of H^+.

acute respiratory acidosis develops. Acute respiratory acidosis is an immediate, life-threatening condition. It is especially dangerous in people whose tissues are generating large amounts of carbon dioxide, or in individuals who are incapable of normal respiratory activity. For this reason, the reversal of acute respiratory acidosis is probably the major goal in the resuscitation of cardiac arrest or drowning victims. Thus, first-aid, CPR, and lifesaving courses always stress the "ABCs" of emergency care: *A*irway, *B*reathing, and *C*irculation.

Chronic respiratory acidosis develops when normal respiratory function has been compromised, but the compensatory mechanisms have not failed completely. For example, normal respiratory compensation may not occur in response to chemoreceptor stimulation in individuals with CNS injuries and those whose respiratory centers have been desensitized by drugs such as alcohol or barbiturates. As a result, these people are prone to developing acidosis due to chronic hypoventilation.

Even when respiratory centers are intact and functional, damage to some respiratory system components can prevent increased pulmonary exchange. Examples of conditions fostering chronic respiratory acidosis include emphysema, congestive heart failure, and pneumonia (in which alveolar damage or blockage typically occurs). Pneumothorax and respiratory muscle paralysis have a similar effect, because they, too, limit the ability to maintain adequate breathing rates.

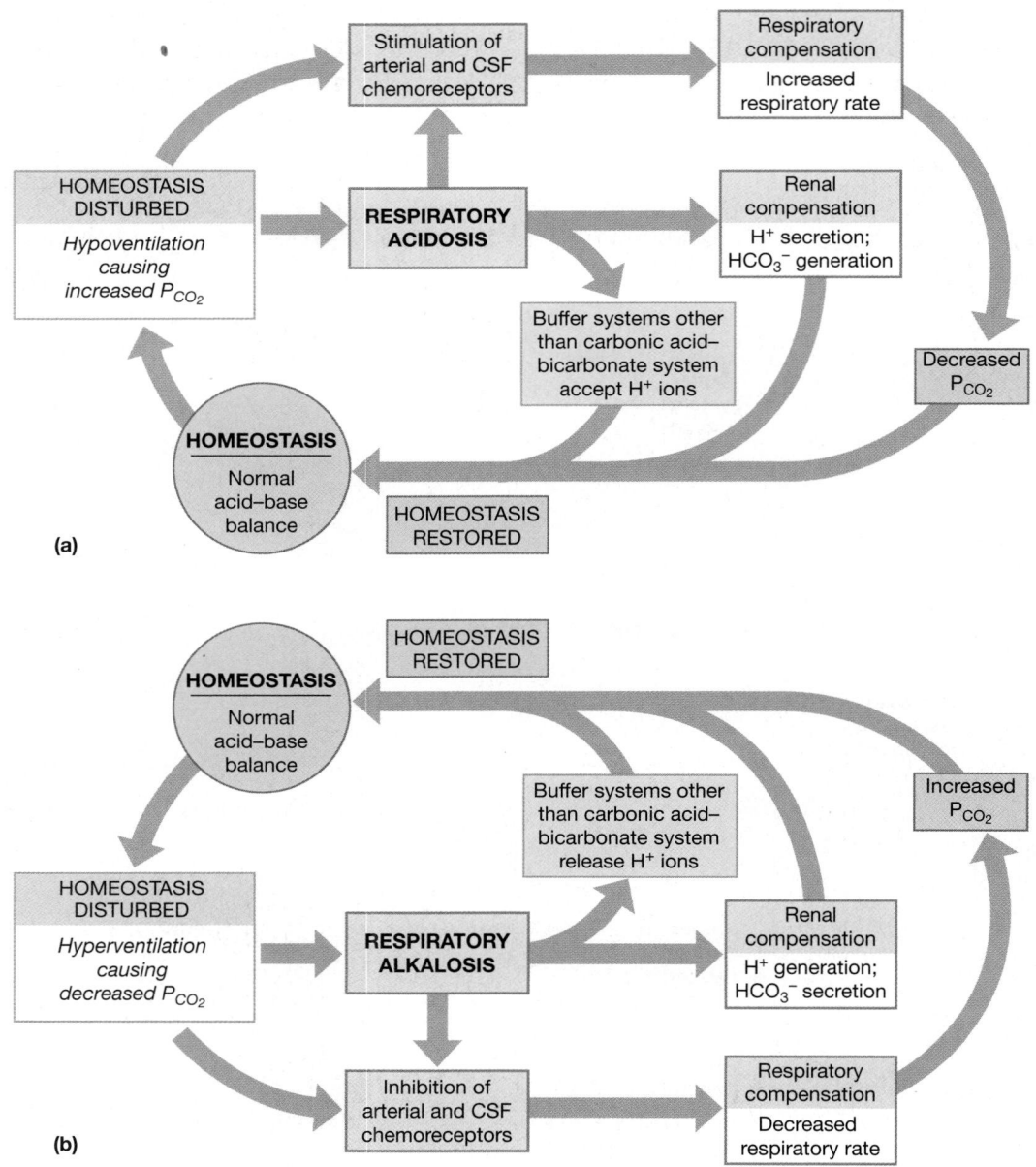

Figure 27–12 **Respiratory Acid–Base Regulation** Respiratory acidosis **(a)** and respiratory alkalosis **(b)**, which result from inadequate and excessive breathing, respectively. In healthy individuals, respiratory responses, combined with renal responses, can generally restore normal acid–base balance.

When a normal pulmonary response does not occur, the kidneys respond by increasing the rate of H^+ secretion into tubular fluid. This response slows the rate of pH change. However, renal mechanisms alone cannot return the pH to normal until the underlying respiratory or circulatory problems are corrected.

The primary problem in respiratory acidosis is that the rate of pulmonary exchange is inadequate to keep the arterial P_{CO_2} within normal limits. Breathing efficiency can typically be improved temporarily by inducing bronchodilation or by using mechanical aids that provide air under positive pressure. If breathing has ceased, artificial respiration or a mechanical ventilator is required. These measures may restore normal pH if the

respiratory acidosis was neither severe nor prolonged. Treatment of acute respiratory acidosis is complicated by the fact that, as we will soon see, it causes a complementary *metabolic acidosis* due to the generation of lactic acid in oxygen-starved tissues.

Respiratory Alkalosis

Problems resulting from **respiratory alkalosis** (**Figure 27–12b**) are relatively uncommon. Respiratory alkalosis develops when respiratory activity lowers plasma P_{CO_2} to below-normal levels, a condition called **hypocapnia** (*-capnia*, presence of carbon dioxide). A temporary hypocapnia can be produced by *hyperventilation* when increased respiratory activity leads to a reduction in the arterial P_{CO_2}. Continued hyperventilation can elevate the pH to levels as high as 8.0. This condition generally corrects itself, because the reduction in P_{CO_2} halts the stimulation of the chemoreceptors, so the urge to breathe fades until carbon dioxide levels have returned to normal. *Respiratory alkalosis caused by hyperventilation seldom persists long enough to cause a clinical emergency.*

Common causes of hyperventilation include physical stresses such as pain, or psychological stresses such as extreme anxiety. Hyperventilation gradually elevates the pH of the cerebrospinal fluid, and central nervous system function is affected. The initial symptoms involve tingling sensations in the hands, feet, and lips. A light-headed feeling may also be noted. If hyperventilation continues, the individual may lose consciousness. When unconsciousness occurs, any contributing psychological stimuli are removed, and the breathing rate declines. The P_{CO_2} then rises until pH returns to normal.

A simple treatment for respiratory alkalosis caused by hyperventilation consists of having the individual rebreathe air exhaled into a small paper bag. As the P_{CO_2} in the bag rises, so do the person's alveolar and arterial CO_2 concentrations. This change eliminates the problem and restores the pH to normal levels. Other problems with respiratory alkalosis are rare and involve primarily (1) individuals adapting to high altitudes, where the low P_{O_2} promotes hyperventilation, (2) patients on mechanical respirators, or (3) individuals whose brain stem injuries render them incapable of responding to shifts in plasma CO_2 concentrations.

Metabolic Acidosis

Metabolic acidosis is the second most common type of acid–base imbalance. It has three major causes:

1. The most widespread cause of metabolic acidosis is the production of a large number of fixed or organic acids. The hydrogen ions released by these acids overload the carbonic acid–bicarbonate buffer system, so pH begins to decline (**Figure 27–13a**). We considered two examples of metabolic acidosis earlier:

   • **Lactic acidosis** can develop after strenuous exercise or prolonged tissue hypoxia (oxygen starvation) as active cells rely on anaerobic respiration (see **Figure 10–20c**, p. 321).

   • **Ketoacidosis** results from the generation of large quantities of ketone bodies during the postabsorptive state of metabolism. Ketoacidosis is a problem in starvation, and a potentially lethal complication of poorly controlled diabetes mellitus. In either case, peripheral tissues are unable to obtain adequate glucose from the bloodstream and they begin metabolizing lipids and ketone bodies.

2. A less common cause of metabolic acidosis is an impaired ability to excrete H^+ at the kidneys

(**Figure 27–13a**). For example, conditions marked by severe kidney damage, such as *glomerulonephritis*, typically result in severe metabolic acidosis. ∞ p. 975 Metabolic acidosis is also caused by diuretics that "turn off" the sodium–hydrogen transport system in the kidney tubules. The secretion of H^+ is directly or indirectly linked to the reabsorption of Na^+. When Na^+ reabsorption stops, so does H^+ secretion.

3. Metabolic acidosis occurs after severe bicarbonate loss (**Figure 27–13b**). The carbonic acid–bicarbonate buffer system relies on bicarbonate ions to balance hydrogen ions that threaten pH balance. A drop in the HCO_3^- concentration in the ECF thus reduces the effectiveness of this buffer system, and acidosis soon develops. The

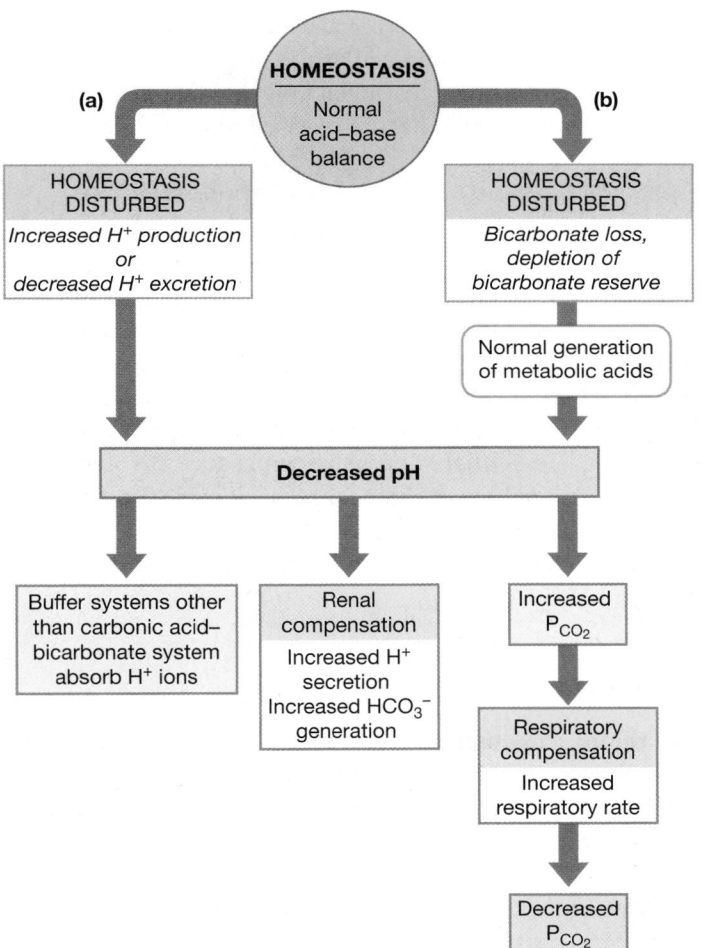

Figure 27–13 Responses to Metabolic Acidosis. Metabolic acidosis can result from either **(a)** increased acid production or decreased acid secretion, leading to a buildup of H^+ in body fluids, or **(b)** a loss of bicarbonate ions that makes the carbonic acid–bicarbonate buffer system incapable of preventing a fall in pH. Respiratory and renal compensation mechanisms can stabilize pH, but blood chemistry remains abnormal until the levels of acid production, acid secretion, and bicarbonate ions return to normal.

most common cause of HCO_3^- depletion is chronic diarrhea. Under normal conditions, most of the bicarbonate ions secreted into the digestive tract in pancreatic, hepatic, and mucous secretions are reabsorbed before the feces are eliminated. In diarrhea, these bicarbonates are lost, and thus the HCO_3^- concentration of the ECF drops.

The nature of the problem must be understood before treatment can begin. Potential causes are so varied that clinicians must piece together relevant clues to make a diagnosis. In some cases, the diagnosis is straightforward; for example, a patient with metabolic acidosis after a bicycle race probably has lactic acidosis. In other cases, clinicians must be detectives.

Compensation for metabolic acidosis generally involves a combination of respiratory and renal mechanisms. Hydrogen ions interacting with bicarbonate ions form carbon dioxide molecules that are eliminated at the lungs, whereas the kidneys excrete additional hydrogen ions into the urine and generate bicarbonate ions that are released into the ECF.

Combined Respiratory and Metabolic Acidosis

Respiratory acidosis and metabolic acidosis are typically linked, because oxygen-starved tissues generate large quantities of lactic acid, and because sustained hypoventilation leads to decreased arterial P_{O_2}. The problem can be especially serious in cases of near drowning, in which body fluids have high P_{CO_2}, low P_{O_2}, and large amounts of lactic acid generated by the muscles of the struggling person. Prompt emergency treatment is essential. The usual procedure involves some form of artificial or mechanical respiratory assistance, coupled with intravenous infusion of an isotonic solution that contains sodium lactate, sodium gluconate, or sodium bicarbonate.

Metabolic Alkalosis

Metabolic alkalosis occurs when HCO_3^- concentrations become elevated (**Figure 27–14**). The bicarbonate ions then interact with hydrogen ions in solution, forming H_2CO_3. The resulting reduction in H^+ concentrations causes signs of alkalosis.

Metabolic alkalosis is relatively rare, but we noted one interesting cause in Chapter 24. ∞ p. 893 The phenomenon known as the *alkaline tide*—produced by the influx into the ECF of large numbers of bicarbonate ions associated with the secretion of hydrochloric acid (HCl) by the gastric mucosa—temporarily elevates the HCO_3^- concentration in the ECF during meals. But serious metabolic alkalosis may result from bouts of repeated vomiting, because the stomach continues to generate stomach acids to replace those

that are lost. As a result, the HCO_3^- concentration of the ECF continues to rise.

Compensation for metabolic alkalosis involves a reduction in the breathing rate, coupled with an increased loss of HCO_3^- in urine. Treatment of mild cases typically addresses the primary cause—generally by controlling the vomiting—and may involve the administration of solutions that contain NaCl or KCl.

Treatment of acute cases of metabolic alkalosis may involve the administration of ammonium chloride (NH_4Cl). Metabolism of the ammonium ion in the liver liberates a hydrogen ion, so in effect the introduction of NH_4Cl leads to the internal generation of HCl, a strong acid. As the HCl diffuses into the bloodstream, pH falls toward normal levels.

The Detection of Acidosis and Alkalosis

Virtually anyone who has a problem that affects the cardiovascular, respiratory, urinary, digestive, or nervous system

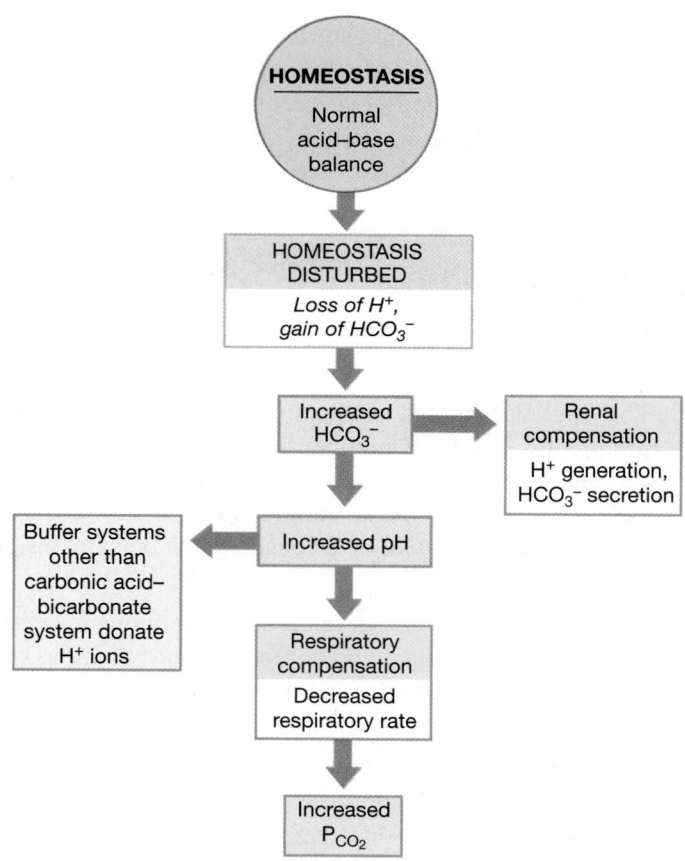

Figure 27–14 **Metabolic Alkalosis.** Metabolic alkalosis most commonly results from the loss of acids, especially stomach acid lost through vomiting. As replacement gastric acids are produced, the alkaline tide introduces a great many bicarbonate ions into the bloodstream, so pH increases.

may develop potentially dangerous acid–base imbalances. For this reason, most diagnostic blood tests include several screens designed to provide information about pH and buffer function. Standard tests monitor blood pH, P_{CO_2}, and HCO_3^- levels. These measurements make recognition of acidosis or alkalosis, and the classification of a particular condition as respiratory or metabolic, relatively straightforward. **Figure 27–15** and Table 27–4 indicate the patterns that characterize the four major categories of acid–base disorders. Additional steps, such as determining the *anion gap*, plotting blood test results on a numerical graph called a *nomogram*, or using a di-

agnostic chart, can help in identifying possible causes of the problem and in distinguishing compensated from uncompensated conditions.

The A&P Top 100

#94 The most common and acute acid–base disorder is respiratory acidosis, which develops when respiratory activity cannot keep pace with the rate of carbon dioxide generation in peripheral tissues.

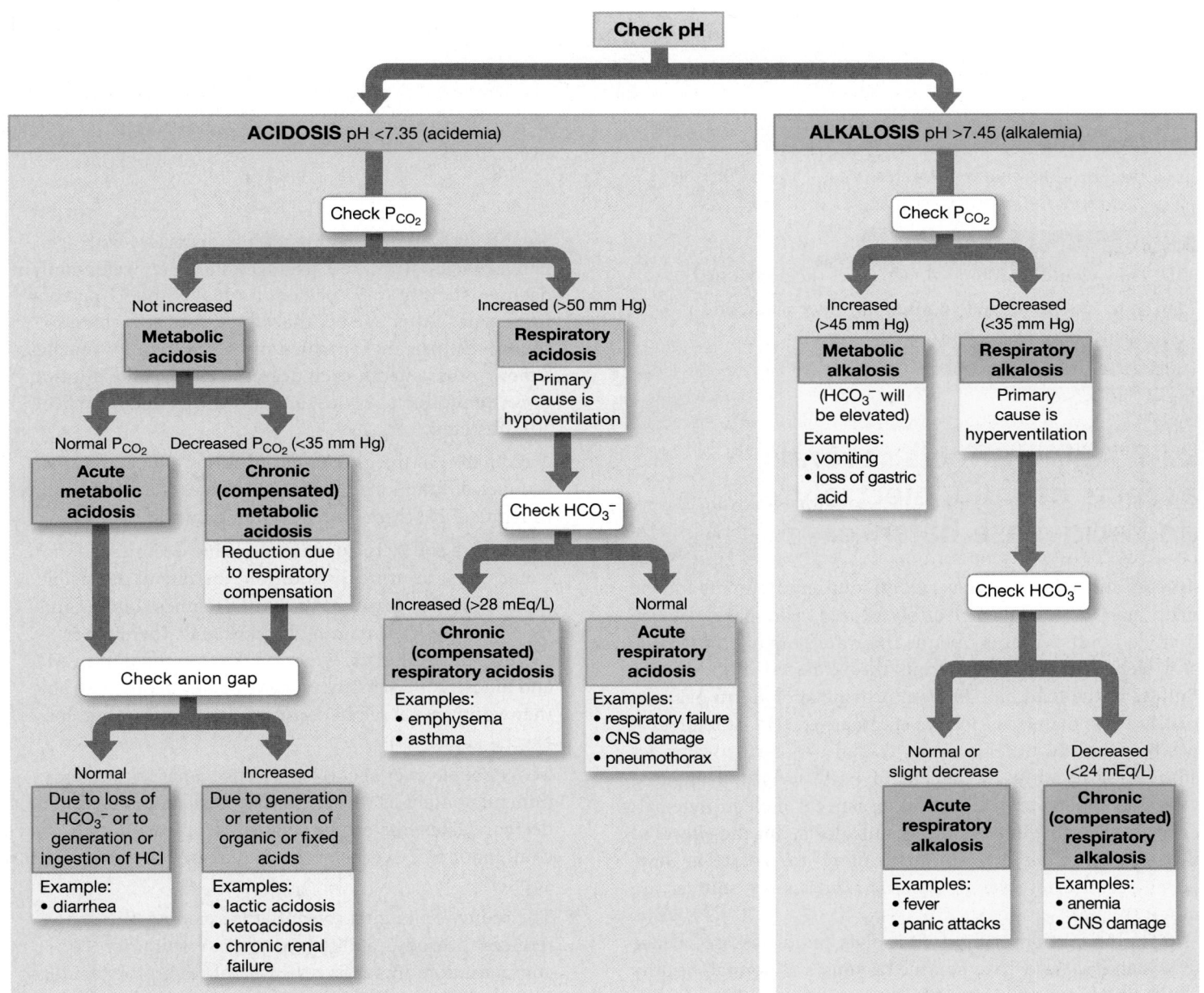

Figure 27–15   A Diagnostic Chart for Acid–Base Disorders. The anion gap is defined as: Na^+ concentration − (HCO_3^- concentration + Cl^- concentration).

27 URINARY

TABLE 27–4 Changes in Blood Chemistry Associated with the Major Classes of Acid–Base Disorders

Disorder	pH (normal 7.35–7.45)	HCO$_3^-$ (normal 24–28 mEq/L)	P$_{CO_2}$ (mm Hg) (normal = 35–45)	Remarks	Treatments
Respiratory acidosis	Decreased (below 7.35)	Acute: normal Compensated: increased (above 28)	Increased (above 45)	Generally caused by hypoventilation and CO$_2$ buildup in tissues and blood	Improve ventilation; in some cases, with bronchodilation and mechanical assistance
Metabolic acidosis	Decreased (below 7.35)	Decreased (below 24)	Acute: normal Compensated: decreased (below 35)	Caused by buildup of organic or fixed acid, impaired H$^+$ elimination at kidneys, or HCO$_3^-$ loss in urine or feces	Administration of bicarbonate (gradual), with other steps as needed to correct primary cause
Respiratory alkalosis	Increased (above 7.45)	Acute: normal Compensated: decreased (below 24)	Decreased (below 35)	Generally caused by hyperventilation and reduction in plasma CO$_2$ levels	Reduce respiratory rate, allow rise in P$_{CO_2}$
Metabolic alkalosis	Increased (above 7.45)	Increased (above 28)	Increased (above 45)	Generally caused by prolonged vomiting and associated acid loss	pH below 7.55: no treatment; pH above 7.55: may require administration of NH$_4$Cl

CHECKPOINT

15. How would a prolonged fast affect the body's pH?

16. Why can prolonged vomiting produce metabolic alkalosis?

See the blue Answers tab at the end of the book.

27-7 Aging affects several aspects of fluid, electrolyte, and acid–base balance

Fetuses and infants have very different requirements for the maintenance of fluid, electrolyte, and acid–base balance than do adults. A fetus obtains the water, organic nutrients, and electrolytes it needs from the maternal bloodstream. Buffers in the fetal bloodstream provide short-term pH control, and the maternal kidneys eliminate the H$^+$ generated. A newborn's body water content is high: At birth, water accounts for roughly 75 percent of body weight, compared with 50–60 percent in adults. Basic aspects of electrolyte balance are the same in newborns as in adults, but the effects of fluctuations in the diet are much more immediate in newborns because reserves of minerals and energy sources are much smaller.

The descriptions of fluid, electrolyte, and acid–base balance in this chapter were based on the responses of normal, healthy adults under age 40. Aging affects many aspects of fluid, electrolyte, and acid–base balance, including the following:

- Total body water content gradually decreases with age. Between ages 40 and 60, average total body water content declines slightly, to 55 percent for males and 47 percent for females. After age 60, the values decline further, to roughly 50 percent for males and 45 percent for females. Among other effects, each decrease reduces the dilution of waste products, toxins, and any drugs that have been administered.

- A reduction in the glomerular filtration rate and in the number of functional nephrons reduces the body's ability to regulate pH through renal compensation.

- The body's ability to concentrate urine declines, so more water is lost in urine. In addition, the rate of insensible perspiration increases as the skin becomes thinner and more delicate. Maintaining fluid balance therefore requires a higher daily water intake. A reduction in ADH and aldosterone sensitivity makes older people less able than younger people to conserve body water when losses exceed gains.

- Many people over age 60 experience a net loss in body mineral content as muscle mass and skeletal mass decrease. This loss can be prevented, at least in part, by a combination of exercise and an increased dietary mineral supply.

- The reduction in vital capacity that accompanies aging reduces the body's ability to perform respiratory compensation, increasing the risk of respiratory acidosis. This problem can be compounded by arthritis, which can reduce vital capacity by limiting rib movements, and by emphysema, another condition that, to some degree, develops with aging.

- Disorders affecting major systems become more common with increasing age. Most, if not all, of these disorders have some effect on fluid, electrolyte, and/or acid–base balance.

17. As one ages, the glomerular filtration rate and the number of functional nephrons declines. What effect would these changes have on pH regulation?

18. After the age of 40, does the total body water content increase or decrease?

See the blue Answers tab at the end of the book.

Related Clinical Terms

hyperkalemia: Plasma K^+ levels above 5.5 mEq/L.
hypernatremia: Plasma Na^+ levels above 145 mEq/L.
hypokalemia: Plasma K^+ levels below 3.5 mEq/L.
hyponatremia: Plasma Na^+ levels below 136 mEq/L.
metabolic acidosis: Acidosis caused by the kidneys' inability to excrete hydrogen ions, the production of numerous fixed or organic acids, or a severe bicarbonate loss.

metabolic alkalosis: A rare form of alkalosis resulting from high concentrations of bicarbonate ions in body fluids.
respiratory acidosis: Acidosis resulting from inadequate respiratory activity, characterized by elevated levels of carbon dioxide (hypercapnia) in body fluids.
respiratory alkalosis: Alkalosis due to excessive respiratory activity, which depresses carbon dioxide levels and elevates the pH of body fluids.

Chapter Review

Study Outline

iP Use *Interactive Physiology®* **(IP)** —available on the IP CD packaged with your new textbook and online at **myA&P**™ (**www.myaandp.com**)—to help you understand difficult physiological concepts. Follow these navigation paths on IP for concepts in this chapter:

- Fluids and Electrolytes/Water Homeostasis
- Fluids and Electrolytes/Electrolyte Homestasis
- Fluids and Electrolytes/Acid–Base Homestasis

27-1 Fluid balance, electrolyte balance, and acid-base balance are interrelated and essential to homeostasis p. 1010

1. The maintenance of normal volume and composition of extracellular and intracellular fluids is vital to life. Three types of homeostasis are involved: **fluid balance**, **electrolyte balance**, and **acid–base balance**.

27-2 The ECF and ICF make up the fluid compartments, which also contain cations and anions p. 1011

2. The **intracellular fluid (ICF)** contains nearly two-thirds of the total body water; the **extracellular fluid (ECF)** contains the rest. Exchange occurs between the ICF and the ECF, but the two **fluid compartments** retain their distinctive characteristics. (*Figures 27–1, 27–2*)

3. Homeostatic mechanisms that monitor and adjust the composition of body fluids respond to changes in the ECF.

4. No receptors directly monitor fluid or electrolyte balance; receptors involved in fluid balance and in electrolyte balance

respond to changes in plasma volume and osmotic concentration.

5. Body cells cannot move water molecules by active transport; all movements of water across plasma membranes and epithelia occur passively, in response to osmotic gradients.

6. The body's content of water or electrolytes will rise if intake exceeds outflow and will fall if losses exceed gains.

7. ADH encourages water reabsorption at the kidneys and stimulates thirst. Aldosterone increases the rate of sodium reabsorption at the kidneys. Natriuretic peptides (ANP and BNP) oppose those actions and promote fluid and electrolyte losses in urine.

8. The regulatory mechanisms of fluid balance and electrolyte balance are quite different, and the distinction is clinically important.

27-3 Hydrostatic and osmotic pressures regulate the movement of water and electrolytes to maintain fluid balance p. 1014

9. Water circulates freely within the ECF compartment.

10. Water losses are normally balanced by gains through eating, drinking, and **metabolic generation**. (*Figure 27–3; Table 27–1*)

11. Water movement between the ECF and ICF is called a **fluid shift**. If the ECF becomes hypertonic relative to the ICF, water will move from the ICF into the ECF until osmotic equilibrium has been restored. If the ECF becomes hypotonic relative to the ICF, water will move from the ECF into the cells, and the volume of the ICF will increase.

27 URINARY

27-4 Balance of the electrolytes sodium, potassium, calcium, and chloride is essential for maintaining homeostasis p. 1017

12. Problems with electrolyte balance generally result from a mismatch between gains and losses of sodium. Problems with potassium balance are less common, but more dangerous.

13. The rate of sodium uptake across the digestive epithelium is directly proportional to the amount of sodium in the diet. Sodium losses occur mainly in urine and through perspiration. (*Figure 27–4*)

14. Shifts in sodium balance result in expansion or contraction of the ECF. Large variations in ECF volume are corrected by homeostatic mechanisms triggered by changes in blood volume. If the volume becomes too low, ADH and aldosterone are secreted; if the volume becomes too high, ANP is secreted. (*Figure 27–5*)

15. Potassium ion concentrations in the ECF are very low and not as closely regulated as are sodium ion concentrations. Potassium excretion increases as ECF concentrations rise, under aldosterone stimulation, and when the pH rises. Potassium retention occurs when the pH falls.

16. ECF concentrations of other electrolytes, such as calcium, magnesium, phosphate, and chloride, are also regulated. (*Table 27–2*)

27-5 In acid–base balance, regulation of hydrogen ions in body fluids involves buffer systems and renal and respiratory compensatory mechanisms p. 1023

17. Acids and bases are either **strong** or **weak**. (*Table 27–3*)

18. The pH of normal body fluids ranges from 7.35 to 7.45; variations outside this relatively narrow range produce symptoms of **acidosis** or **alkalosis**.

19. **Volatile acids** can leave solution and enter the atmosphere; **fixed acids** remain in body fluids until excreted at the kidneys; **organic acids** are participants in, or are by-products of, aerobic metabolism.

20. Carbonic acid is the most important factor affecting the pH of the ECF. In solution, CO_2 reacts with water to form carbonic acid. An inverse relationship exists between pH and the concentration of CO_2. (*Figure 27–6*)

21. Sulfuric acid and phosphoric acid, the most important fixed acids, are generated during the catabolism of amino acids and compounds containing phosphate groups.

22. Organic acids include metabolic products such as lactic acid and ketone bodies.

23. A **buffer system** typically consists of a weak acid and the anion released by its dissociation. The anion functions as a weak base that can absorb H^+. The three major buffer systems are (1) **protein buffer systems** in the ECF and ICF; (2) the **carbonic acid–bicarbonate buffer system**, most important in the ECF; and (3) the **phosphate buffer system** in the ICF and urine. (*Figure 27–7*)

24. In protein buffer systems, the initial, final, and R groups of component amino acids respond to changes in pH by accepting or releasing hydrogen ions. The **hemoglobin buffer system** is a protein buffer system that helps prevent drastic changes in pH when the P_{CO_2} is rising or falling. (*Figure 27–8*)

25. The carbonic acid–bicarbonate buffer system prevents pH changes caused by organic acids and fixed acids in the ECF. The readily available supply of bicarbonate ions is the **bicarbonate reserve**. (*Figure 27–9*)

26. The phosphate buffer system plays a supporting role in regulating the pH of the ECF, but it is important in buffering the pH of the ICF and of urine.

27. The lungs help regulate pH by affecting the carbonic acid–bicarbonate buffer system. A change in respiratory rate can raise or lower the P_{CO_2} of body fluids, affecting the body's buffering capacity. This process is called **respiratory compensation**.

28. In **renal compensation**, the kidneys vary their rates of hydrogen ion secretion and bicarbonate ion reabsorption, depending on the pH of the ECF. (*Figure 27–10*)

27-6 Respiratory acidosis/alkalosis and metabolic acidosis/alkalosis are classes of acid–base balance disturbances p. 1030

29. Interactions among buffer systems, respiration, and renal function normally maintain tight control of the pH of the ECF, generally within a range of 7.35–7.45. (*Figure 27–11*)

30. *Respiratory acid–base disorders* result when abnormal respiratory function causes an extreme rise or fall in CO_2 levels in the ECF. *Metabolic acid–base disorders* are caused by the generation of organic or fixed acids or by conditions affecting the concentration of bicarbonate ions in the ECF.

31. **Respiratory acidosis** results from excessive levels of CO_2 in body fluids. (*Figure 27–12*)

32. **Respiratory alkalosis** is a relatively rare condition associated with hyperventilation. (*Figure 27–12*)

33. **Metabolic acidosis** results from the depletion of the bicarbonate reserve, caused by an inability to excrete hydrogen ions at the kidneys, the production of large numbers of fixed and organic acids, or bicarbonate loss that accompanies chronic diarrhea. (*Figure 27–13*)

34. **Metabolic alkalosis** results when bicarbonate ion concentrations become elevated, as occurs during extended periods of vomiting. (*Figure 27–14*)

35. Standard diagnostic blood tests such as blood pH, P_{CO_2}, and bicarbonate levels are used to recognize and classify acidosis and alkalosis as either respiratory or metabolic in nature. (*Figure 27–15; Table 27–4*)

27-7 Aging affects several aspects of fluid, electrolyte, and acid–base balance p. 1036

36. Changes affecting fluid, electrolyte, and acid–base balance in the elderly include (1) reduced total body water content, (2) impaired ability to perform renal compensation, (3) increased water demands due to reduced ability to concentrate urine and reduced sensitivity to ADH and aldosterone, (4) a net loss of minerals, (5) reductions in respiratory efficiency that affect the ability to perform respiratory compensation, and (6) increased incidence of conditions that secondarily affect fluid, electrolyte, or acid–base balance.

Review Questions

See the blue Answers tab at the end of the book.

myA&P Access more review material online at **myA&P™** (www.myaandp.com). There, you'll find chapter guides, chapter quizzes, practice tests, animations, flashcards, a glossary with pronunciations, *Interactive Physiology*® (IP) exercises and quizzes, and more to help you succeed in the course.

LEVEL 1 Reviewing Facts and Terms

1. The primary components of the extracellular fluid are
   (a) lymph and cerebrospinal fluid.
   (b) plasma and serous fluids
   (c) interstitial fluid and plasma.
   (d) all of these.

2. The principal anions in the ICF are
   (a) phosphate and proteins (Pr^-).
   (b) phosphate and bicarbonate.
   (c) sodium and chloride.
   (d) sodium and potassium.

3. Osmoreceptors in the hypothalamus monitor the osmotic concentration of the ECF and secrete _____ in response to higher osmotic concentrations.
   (a) BNP   (b) ANP
   (c) aldosterone   (d) ADH

4. Write the missing names and molecular formulas for the following reactions between the carbonic acid–bicarbonate buffer system and the bicarbonate reserve.

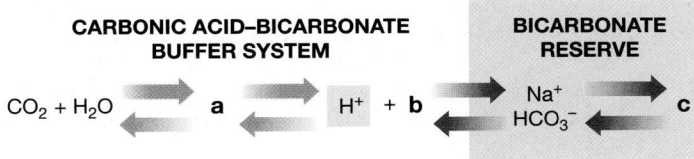

   (a) _____ (b) _____ (c) _____

5. Calcium homeostasis primarily reflects
   (a) a balance between absorption in the gut and excretion by the kidneys.
   (b) careful regulation of blood calcium levels by the kidneys.
   (c) an interplay between parathormone and aldosterone
   (d) an interplay among reserves in the bone, the rate of absorption, and the rate of excretion.
   (e) hormonal control of calcium reserves in the bones.

6. The *most* important factor affecting the pH of body tissues is the concentration of
   (a) lactic acid.   (b) ketone bodies.
   (c) organic acids.   (d) carbon dioxide.
   (e) hydrochloric acid.

7. Changes in the pH of body fluids are compensated for by all of the following *except*
   (a) an increase in urine output.
   (b) the carbonic acid–bicarbonate buffer system.
   (c) the phosphate buffer system
   (d) changes in the rate and depth of breathing.
   (e) protein buffers.

8. Respiratory acidosis develops when the plasma pH is
   (a) elevated due to a decreased plasma P_{CO_2} level.
   (b) decreased due to an elevated plasma P_{CO_2} level.
   (c) elevated due to an elevated plasma P_{CO_2} level.
   (d) decreased due to a decreased plasma P_{CO_2} level.

9. Metabolic alkalosis occurs when
   (a) bicarbonate ion concentrations become elevated.
   (b) a severe bicarbonate loss occurs.
   (c) the kidneys fail to excrete hydrogen ions.
   (d) ketone bodies are generated in abnormally large quantities.

10. Identify four hormones that mediate major physiological adjustments affecting fluid and electrolyte balance. What are the primary effects of each hormone?

LEVEL 2 Reviewing Concepts

11. Drinking a hypotonic solution causes the ECF to
    (a) increase in volume and become hypertonic with respect to the ICF.
    (b) decrease in volume and become hypertonic with respect to the ICF.
    (c) decrease in volume and become hypotonic with respect to the ICF.
    (d) increase in volume and become hypotonic with respect to the ICF.

12. The osmotic concentration of the ECF decreases if an individual gains water without a corresponding
    (a) gain of electrolytes.
    (b) loss of water.
    (c) fluid shift from the ECF to the ICF.
    (d) a, b, and c.

13. When the pH of body fluids begins to *fall*, free amino acids and proteins will
    (a) release a hydrogen from the carboxyl group.
    (b) release a hydrogen from the amino group.
    (c) release a hydrogen at the carboxyl group.
    (d) bind a hydrogen at the amino group.

14. In a protein buffer system, if the pH rises,
    (a) the protein acquires a hydrogen ion from carbonic acid.
    (b) hydrogen ions are buffered by hemoglobin molecules.
    (c) a hydrogen ion is released and a carboxyl ion is formed.
    (d) a chloride shift occurs.

15. Differentiate among fluid balance, electrolyte balance, and acid–base balance, and explain why each is important to homeostasis.

16. What are fluid shifts? What is their function, and what factors can cause them?

17. Why should a person with a fever drink plenty of fluids?

18. Define and give an example of (a) a volatile acid, (b) a fixed acid, and (c) an organic acid. Which represents the greatest threat to acid–base balance? Why?

19. What are the three major buffer systems in body fluids? How does each system work?

20. How do respiratory and renal mechanisms support the buffer systems?

21. Differentiate between respiratory compensation and renal compensation.

22. Distinguish between respiratory and metabolic disorders that disturb acid–base balance.

23. What is the difference between metabolic acidosis and respiratory acidosis? What can cause these conditions?

24. The most recent advice from medical and nutritional experts is to monitor one's intake of salt so that it does not exceed the amount needed to maintain a constant ECF volume. What effect does excessive salt ingestion have on blood pressure?

25. Exercise physiologists recommend that adequate amounts of fluid be ingested before, during, and after exercise. Why is fluid replacement during extensive sweating important?

LEVEL 3   Critical Thinking and Clinical Applications

26. After falling into an abandoned stone quarry filled with water and nearly drowning, a young boy is rescued. In assessing his condition, rescuers find that his body fluids have high P_{CO_2} and low P_{O_2} levels, and that the boy's muscles generated large amounts of lactic acid as he struggled in the water. As a clinician, diagnose the boy's condition and recommend treatment to restore his body to homeostasis.

27. Dan has been lost in the desert for 2 days with very little water. As a result of this exposure, you would expect to observe which of the following?
    (a) elevated ADH levels
    (b) decreased blood osmolarity
    (c) normal urine production
    (d) increased blood volume
    (e) cells enlarged with fluid

28. Mary, a nursing student, has been caring for burn patients. She notices that they consistently show elevated levels of potassium in their urine and wonders why. What would you tell her?

29. While visiting a foreign country, Milly inadvertently drinks some water, even though she had been advised not to. She contracts an intestinal disease that causes severe diarrhea. How would you expect her condition to affect her blood pH, urine pH, and pattern of ventilation?

30. Yuka is dehydrated, so her physician prescribes intravenous fluids. The attending nurse becomes distracted and erroneously gives Yuka a hypertonic glucose solution instead of normal saline. What effect will this mistake have on Yuka's plasma ADH levels and urine volume?

31. Refer to the diagnostic flowchart in **Figure 27–15**. Use information from the blood test results in the accompanying table to categorize the acid–base disorders that affect the patients represented in the table.

Results	Patient 1	Patient 2	Patient 3	Patient 4
pH	7.5	7.2	7.0	7.7
P_{CO_2} (mmHg)	32	45	60	50
Na^+ (mEq/L)	138	140	140	136
HCO_3^- (mEq/L)	22	20	28	34
Cl^- (mEq/L)	106	102	101	91
Anion gap* (mEq/L)	10	18	12	11

*Anion gap = Na^+ concentration − (HCO_3^- concentration + Cl^- concentration).

The Reproductive System

28

Did you know...?
There are approximately 6.6 billion people currently inhabiting the Earth.

Learning Outcomes

After completing this chapter, you should be able to do the following:

28-1 **List the basic components of the human reproductive system, and summarize the functions of each.**

28-2 **Describe the components of the male reproductive system and the roles played by the reproductive tract and accessory glands in producing spermatozoa; specify the composition of semen; and summarize the hormonal mechanisms that regulate male reproductive functions.**

28-3 **Describe the components of the female reproductive system and the ovarian roles in oogenesis; explain the complete ovarian and uterine cycles; outline the histology, anatomy, and functions of the vagina; and summarize all aspects of the female reproductive cycle.**

28-4 **Discuss the physiology of sexual intercourse in males and females.**

28-5 **Describe the reproductive system changes that occur with aging.**

28-6 **Give examples of interactions between the reproductive system and each of the other organ systems.**

Clinical Notes

Dehydroepiandrosterone (DHEA) p. 1058
Prostatic Hypertrophy and Prostate Cancer p. 1059
Ovarian Cancer p. 1065
Breast Cancer p. 1073

An Introduction to the Reproductive System

The reproductive system is the only system that is not essential to the life of the individual (although its activities do affect other systems). This chapter discusses how the male and female reproductive organs produce and store specialized reproductive cells that combine to form new individuals, and how various reproductive organs also secrete hormones that play major roles in the maintenance of normal sexual function.

28-1 Basic reproductive system structures are gonads, ducts, accessory glands and organs, and external genitalia

In this chapter we examine the anatomy and physiology of the human **reproductive system**. This system ensures the continued existence of the human species—by producing, storing, nourishing, and transporting functional male and female reproductive cells, or **gametes** (GAM-ēts).

The reproductive system includes the following basic components:

- **Gonads** (GŌ-nadz; *gone*, seed, generation), or reproductive organs that produce gametes and hormones.
- Ducts that receive and transport the gametes.
- Accessory glands and organs that secrete fluids into the ducts of the reproductive system or into other excretory ducts.
- Perineal structures that are collectively known as the **external genitalia** (jen-i-TĀ-lē-uh).

In both males and females, the ducts are connected to chambers and passageways that open to the exterior of the body. The structures involved constitute the *reproductive tract*. The male and female reproductive systems are functionally quite different, however. In adult males, the **testes** (TES-tēz; singular, *testis*), or male gonads, secrete sex hormones called *androgens* (principally *testosterone*, which, together with other sex hormones, was introduced in Chapter 18). ∞ p. 637 The testes also produce the male gametes, called **spermatozoa** (sper-ma-tō-ZŌ-uh; singular, *spermatozoon*), or *sperm*—one-half billion each day. During *emission*, mature spermatozoa travel along a lengthy duct system, where they are mixed with the secretions of accessory glands. The mixture created is known as **semen** (SĒ-men). During *ejaculation*, semen is expelled from the body.

In adult females, the **ovaries**, or female gonads, typically release only one immature gamete, an **oocyte**, per month. This immature gamete travels along one of two short *uterine tubes*, which end in the muscular organ called the *uterus* (Ū-ter-us). If a sperm reaches the oocyte and initiates the process of *fertilization*, the oocyte matures into an **ovum** (plural, *ova*). A short passageway, the *vagina* (va-JĪ-nuh), connects the uterus with the exterior. Ejaculation introduces semen into the vagina during *sexual intercourse*, and the spermatozoa then ascend the female reproductive tract. If fertilization occurs, the uterus will enclose and support a developing *embryo* as it grows into a *fetus* and prepares for birth.

Next we will examine the anatomy of the male and female reproductive systems further, and will consider the physiological and hormonal mechanisms responsible for the regulation of reproductive function. Earlier chapters introduced the anatomical reference points used in the discussions that follow; you may find it helpful to review the figures on the pelvic girdle (**Figures 8–7** and **8–8**, pp. 255, 256), perineal musculature (**Figure 11–12**, p. 357), pelvic innervation (**Figure 13–13**, p. 445), and regional blood supply (**Figures 21–25** and **21–29**, pp. 758, 763).

CHECKPOINT

1. **Define gamete.**
2. **List the basic components of the reproductive system.**
3. **Define gonads.**

See the blue Answers tab at the end of the book.

28-2 Spermatogenesis occurs in the testes, and hormones from the hypothalamus, adenohypophysis, and testes control male reproductive functions

The principal structures of the male reproductive system are shown in **Figure 28–1**. Proceeding from a testis, the spermatozoa travel within the *epididymis* (ep-i-DID-i-mus); the *ductus deferens* (DUK-tus DEF-e-renz), or *vas deferens*; the *ejaculatory duct*; and the *urethra* before leaving the body. Accessory organs—the *seminal* (SEM-i-nal) *glands* (vesicles), the *prostate* (PROS-tāt) *gland*, and the *bulbo-urethral* (bul-bō-ū-RĒ-thral) *glands*—secrete various fluids into the ejaculatory ducts and urethra. The external genitalia consist of the *scrotum* (SKRŌ-tum), which encloses the testes, and the *penis* (PĒ-nis), an erectile organ through which the distal portion of the urethra passes.

The Testes

Each testis has the shape of a flattened egg that is roughly 5 cm (2 in.) long, 3 cm (1.2 in.) wide, and 2.5 cm (1 in.) thick. Each has a weight of 10–15 g (0.35–0.53 oz). The testes hang within the **scrotum**, a fleshy pouch suspended inferior to the perineum, anterior to the anus and posterior to the base of the penis (**Figure 28–1**). ATLAS: Embryology Summary 21: The Development of the Reproductive System

Descent of the Testes

During development of the fetus, the testes form inside the body cavity adjacent to the kidneys. A bundle of connective tissue fibers—called the **gubernaculum testis** (goo-bur-NAK-ū-lum TES-tis)—extends from each testis to the posterior wall of a small anterior and inferior pocket of the peritoneum (**Figure 28–2a**). As the fetus grows, the gubernacula do not get any longer, so they lock the testes in position. As a result, the relative position of each testis changes as the body enlarges: The testis gradually moves inferiorly and anteriorly toward the anterior abdominal wall. During the seventh developmental month, fetal growth continues at a rapid pace, and circulating hormones stimulate a contraction of the gubernaculum testis. Over this period, each testis moves through the abdominal musculature, accompanied by small pockets of the peritoneal cavity. This process is called the **descent of the testes**.

In *cryptorchidism* (krip-TOR-ki-dizm; *crypto*, hidden + *orchis*, testis), one or both of the testes have not descended into the scrotum by the time of birth. Typically, the cryptorchid (abdominal) testes are lodged in the abdominal cavity or within the inguinal canal. Cryptorchidism occurs in about 3 percent of full-term deliveries and in roughly 30 percent of premature births. In most instances, normal descent occurs a few weeks later, but the condition can be surgically corrected if it persists. Corrective measures should be taken before *puberty* (sexual maturation), because a cryptorchid testis will not produce spermatozoa. If both testes are cryptorchid, the individual will be *sterile* (*infertile*) and unable to father children. If the testes cannot be moved into the scrotum, in most cases they will be removed, because about 10 percent of males with uncorrected cryptorchid testes eventually develop testicular cancer. This surgical procedure is called an *orchiectomy* (or-kē-EK-tō-mē).

As each testis moves through the body wall, it is accompanied by the ductus deferens and the testicular blood vessels,

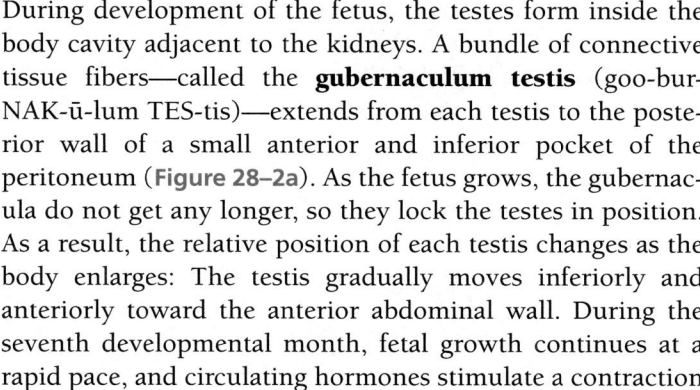

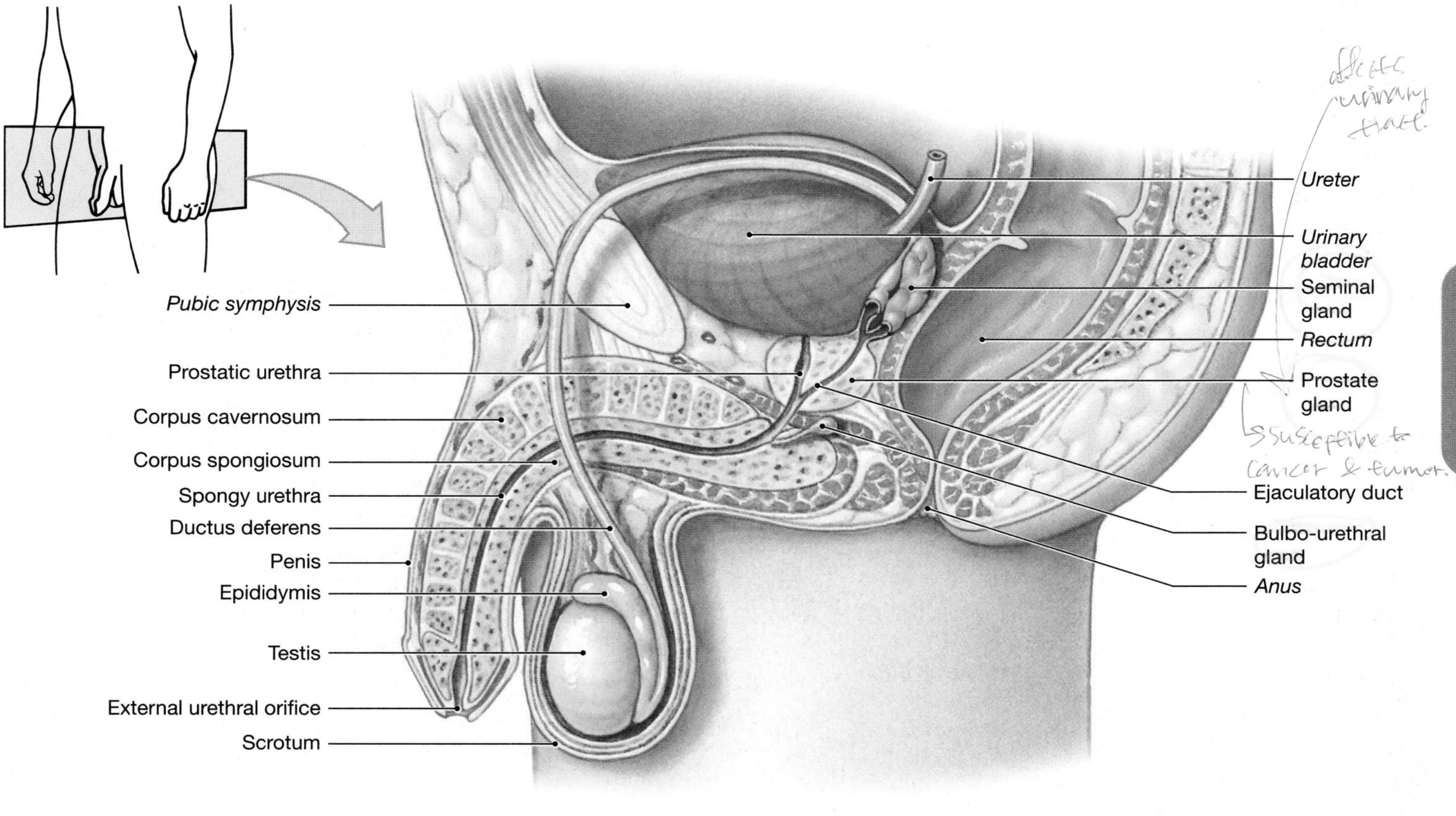

Pubic symphysis	Ureter
Prostatic urethra	Urinary bladder
Corpus cavernosum	Seminal gland
Corpus spongiosum	Rectum
Spongy urethra	Prostate gland
Ductus deferens	Ejaculatory duct
Penis	Bulbo-urethral gland
Epididymis	Anus
Testis	
External urethral orifice	
Scrotum	

Figure 28–1   The Male Reproductive System. A sagittal section of the male reproductive organs. ATLAS: Plate 64

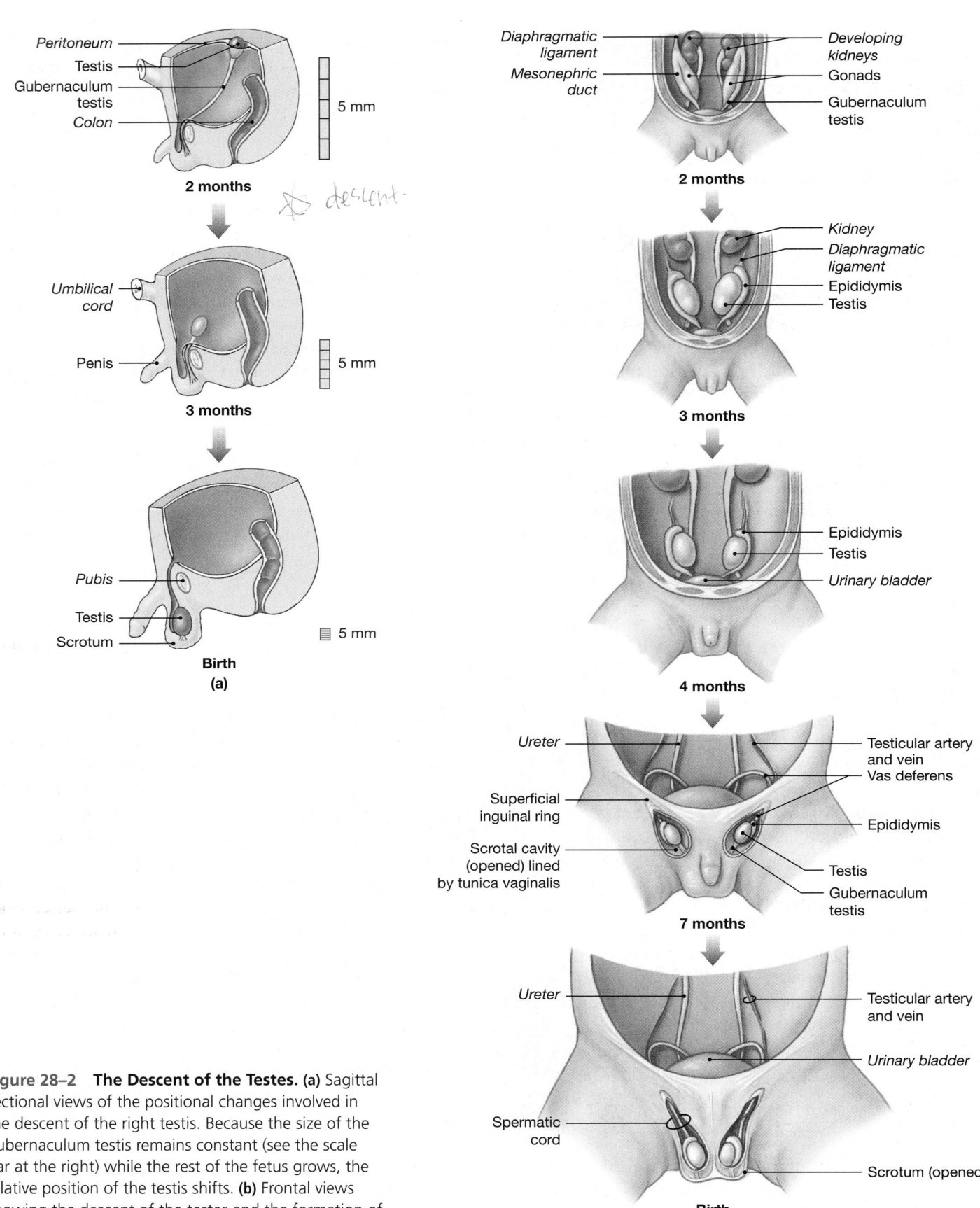

Figure 28–2 **The Descent of the Testes. (a)** Sagittal sectional views of the positional changes involved in the descent of the right testis. Because the size of the gubernaculum testis remains constant (see the scale bar at the right) while the rest of the fetus grows, the relative position of the testis shifts. **(b)** Frontal views showing the descent of the testes and the formation of the spermatic cords.

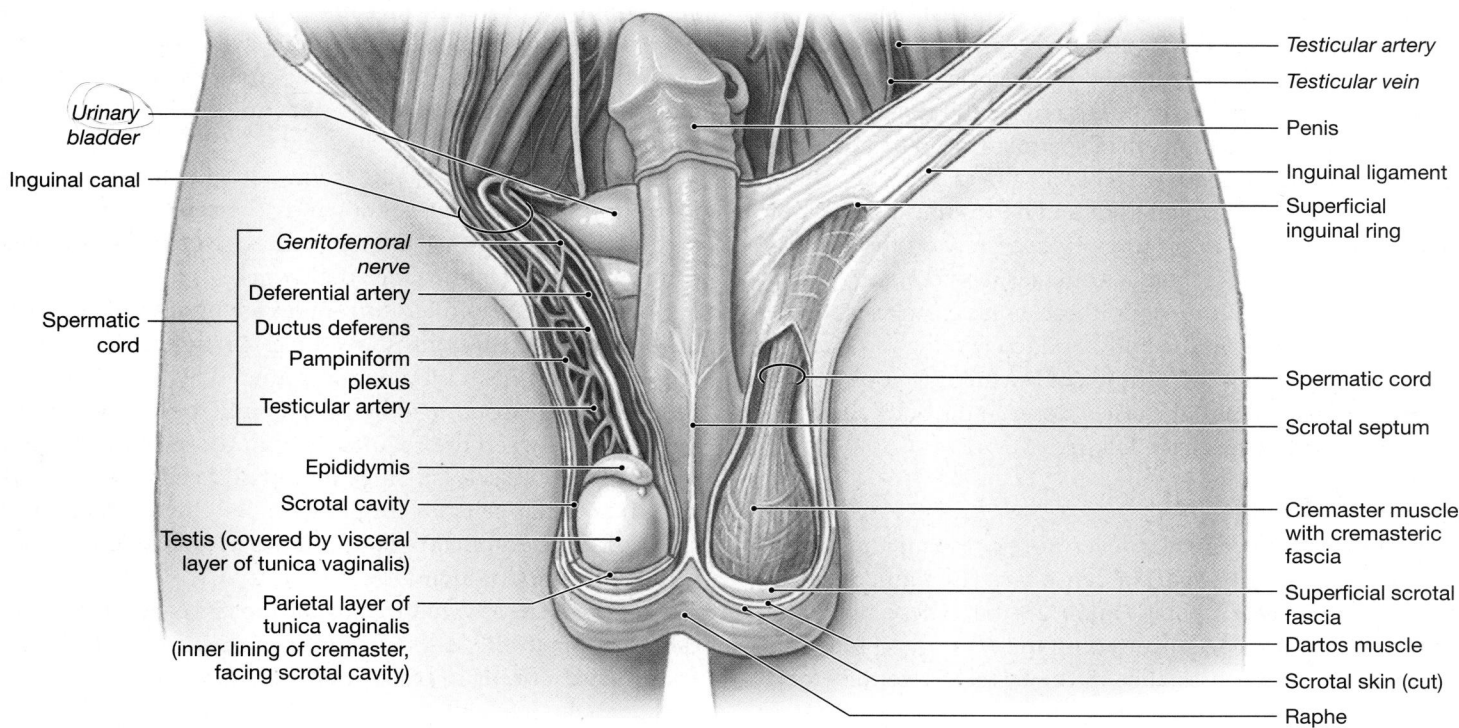

Figure 28–3 **The Male Reproductive System in Anterior View.**

Labels (left, top to bottom): Urinary bladder · Inguinal canal · Spermatic cord · Genitofemoral nerve · Deferential artery · Ductus deferens · Pampiniform plexus · Testicular artery · Epididymis · Scrotal cavity · Testis (covered by visceral layer of tunica vaginalis) · Parietal layer of tunica vaginalis (inner lining of cremaster, facing scrotal cavity)

Labels (right, top to bottom): Testicular artery · Testicular vein · Penis · Inguinal ligament · Superficial inguinal ring · Spermatic cord · Scrotal septum · Cremaster muscle with cremasteric fascia · Superficial scrotal fascia · Dartos muscle · Scrotal skin (cut) · Raphe

nerves, and lymphatic vessels. Together, these structures form the body of the spermatic cord, which we will discuss next.

The Spermatic Cords

The **spermatic cords** are paired structures extending between the abdominopelvic cavity and the testes (**Figure 28–3**). Each spermatic cord consists of layers of fascia and muscle enclosing the ductus deferens and the blood vessels, nerves, and lymphatic vessels that supply the testes. The blood vessels include the *deferential artery*, a *testicular artery*, and the **pampiniform** (pam-PIN-i-form; *pampinus*, tendril + *forma*, form) **plexus** of a testicular vein. Branches of the *genitofemoral nerve* from the lumbar plexus provide innervation. Each spermatic cord begins at the entrance to the *inguinal canal* (a passageway through the abdominal musculature). After passing through the inguinal canal, the spermatic cord descends into the scrotum.

The inguinal canals form during development as the testes descend into the scrotum; at that time, these canals link the scrotal cavities with the peritoneal cavity. In normal adult males, the inguinal canals are closed, but the presence of the spermatic cords creates weak points in the abdominal wall that remain throughout life. As a result, *inguinal hernias*— protrusions of visceral tissues or organs into the inguinal canal—are relatively common in males. Note that the inguinal canals in females are very small, containing only the *ilioinguinal nerves* and the *round ligaments* of the uterus; the abdominal wall is nearly intact, so inguinal hernias in women are very rare.

The Scrotum and the Position of the Testes

The scrotum is divided internally into two chambers. The partition between the two is marked by a raised thickening in the scrotal surface known as the **raphe** (RĀ-fē) (**Figure 28–3**). Each testis lies in a separate chamber, or **scrotal cavity**. Because the scrotal cavities are separated by a partition, infection or inflammation of one testis does not ordinarily spread to the other. A narrow space separates the inner surface of the scrotum from the outer surface of the testis. The **tunica vaginalis** (TOO-ni-ka vaj-i-NAL-is), a serous membrane, lines the scrotal cavity and reduces friction between the opposing parietal (scrotal) and visceral (testicular) surfaces. The tunica vaginalis is an isolated portion of the peritoneum that lost its connection with the peritoneal cavity after the testes descended, when the inguinal canal closed.

The scrotum consists of a thin layer of skin and the underlying superficial fascia. The dermis contains a layer of smooth muscle, the **dartos** (DAR-tōs) **muscle**. Resting muscle tone in the dartos muscle elevates the testes and causes the characteristic wrinkling of the scrotal surface. A layer of skeletal muscle, the **cremaster** (krē-MAS-ter) **muscle**, lies deep to the dermis. Contraction of the cremaster muscle during sexual arousal or in response to decreased testicular temperature tenses the scrotum and pulls the testes closer to the body. The cremasteric reflex can be initiated by stroking the skin on the upper thigh, causing the scrotum to move the testes closer to the body. Normal development of spermatozoa in the testes requires temperatures about 1.1°C (2°F) lower than those elsewhere in the body. The cremaster and dartos muscles relax or contract to move the testes away from or toward the body as needed to maintain acceptable testicular temperatures. When air or body temperature rises, these muscles relax and the testes move away from the body. Sudden cooling of the scrotum, as occurs during entry into a cold swimming pool, results in contractions that pull the testes closer to the body and keep testicular temperatures from falling.

Structure of the Testes

Deep to the tunica vaginalis covering the testis is the **tunica albuginea** (al-bū-JIN-ē-uh), a dense layer of connective tissue rich in collagen fibers (**Figure 28–4a**). These fibers are continuous with those surrounding the adjacent epididymis and extend into the substance of the testis. There they form fibrous partitions, or *septa*, that converge toward the region nearest the entrance to the epididymis. The connective tissues in this region support the blood vessels and lymphatic vessels that supply and drain the testis, and the *efferent ductules*, which transport spermatozoa to the epididymis.

Histology of the Testes

The septa subdivide the testis into a series of **lobules** (**Figure 28–4a**). Distributed among the lobules are roughly 800 slen-

der, tightly coiled **seminiferous** (sem-i-NIF-er-us) **tubules** (**Figures 28–4** and **28–5**). Each tubule averages about 80 cm (32 in.) in length, and a typical testis contains nearly one-half mile of seminiferous tubules. Sperm production occurs within these tubules.

Each seminiferous tubule forms a loop that is connected to a maze of passageways known as the **rete** (RĒ-tē; *rete*, a net) **testis** (**Figure 28–4**). Fifteen to 20 large **efferent ductules** connect the rete testis to the epididymis.

Because the seminiferous tubules are tightly coiled, most histological preparations show them in transverse section (**Figure 28–5a**). Each tubule is surrounded by a delicate connective tissue capsule (**Figure 28–5b**), and areolar tissue fills the spaces between the tubules. Within those spaces are numerous blood vessels and large **interstitial cells** (*cells of Leydig*). Interstitial cells are responsible for the production of *androgens*, the dominant sex hormones in males. *Testosterone* is the most important androgen.

Spermatozoa are produced by the process of **spermatogenesis** (sper-ma-tō-JEN-e-sis). Spermatogenesis begins at the outermost layer of cells in the seminiferous tubules and proceeds toward the lumen (**Figure 28–5c,d**). At each step in this process, the daughter cells move closer to the lumen. First, stem cells called **spermatogonia** (sper-ma-tō-GŌ-nē-uh) divide by mitosis to produce two daughter cells, one of which remains at that location as a spermatogonium while the other differentiates into a primary spermatocyte. **Primary spermatocytes** (sper-MA-tō-sīts) are the cells that begin *meiosis*, a specialized form of cell division involved only in the production of gametes (spermatozoa in males, ova in females). Primary spermatocytes give rise to

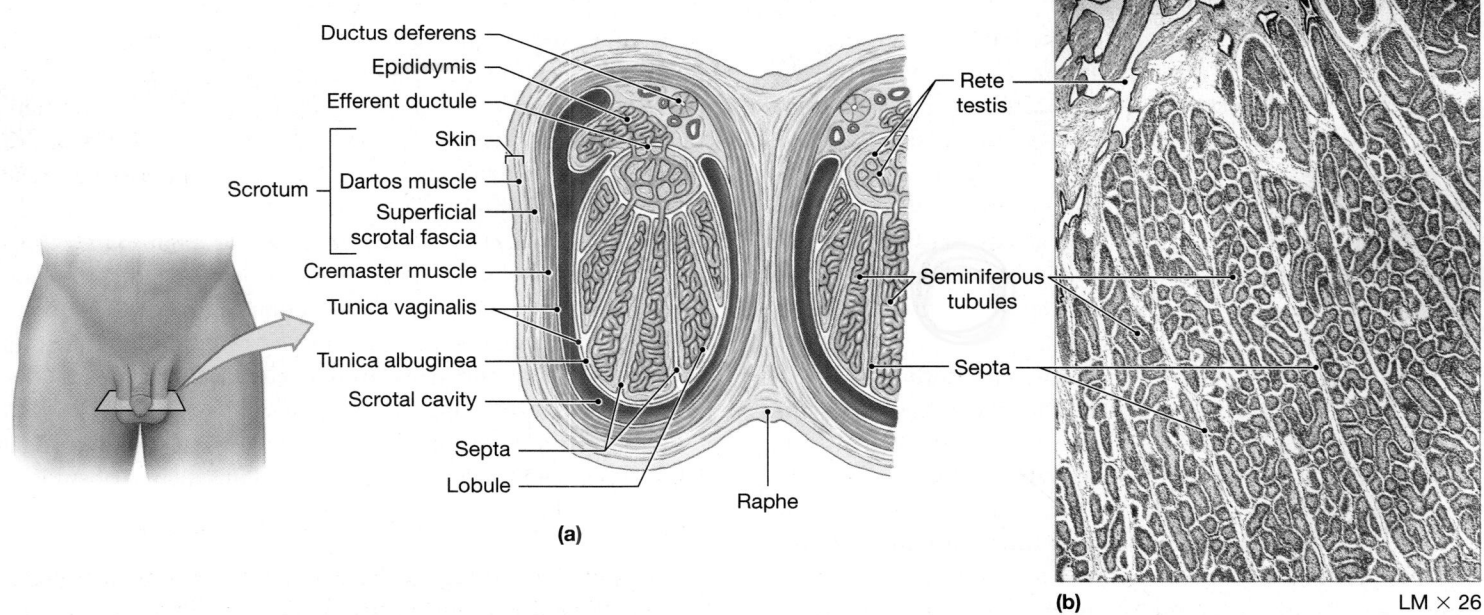

Ductus deferens
Epididymis
Efferent ductule
Skin
Scrotum {
Dartos muscle
Superficial scrotal fascia
Cremaster muscle
Tunica vaginalis
Tunica albuginea
Scrotal cavity
Septa
Lobule
Raphe

Rete testis
Seminiferous tubules
Septa

(a)

(b) LM × 26

Figure 28–4   The Structure of the Testes. (a) A transverse section. **(b)** A section through a testis.

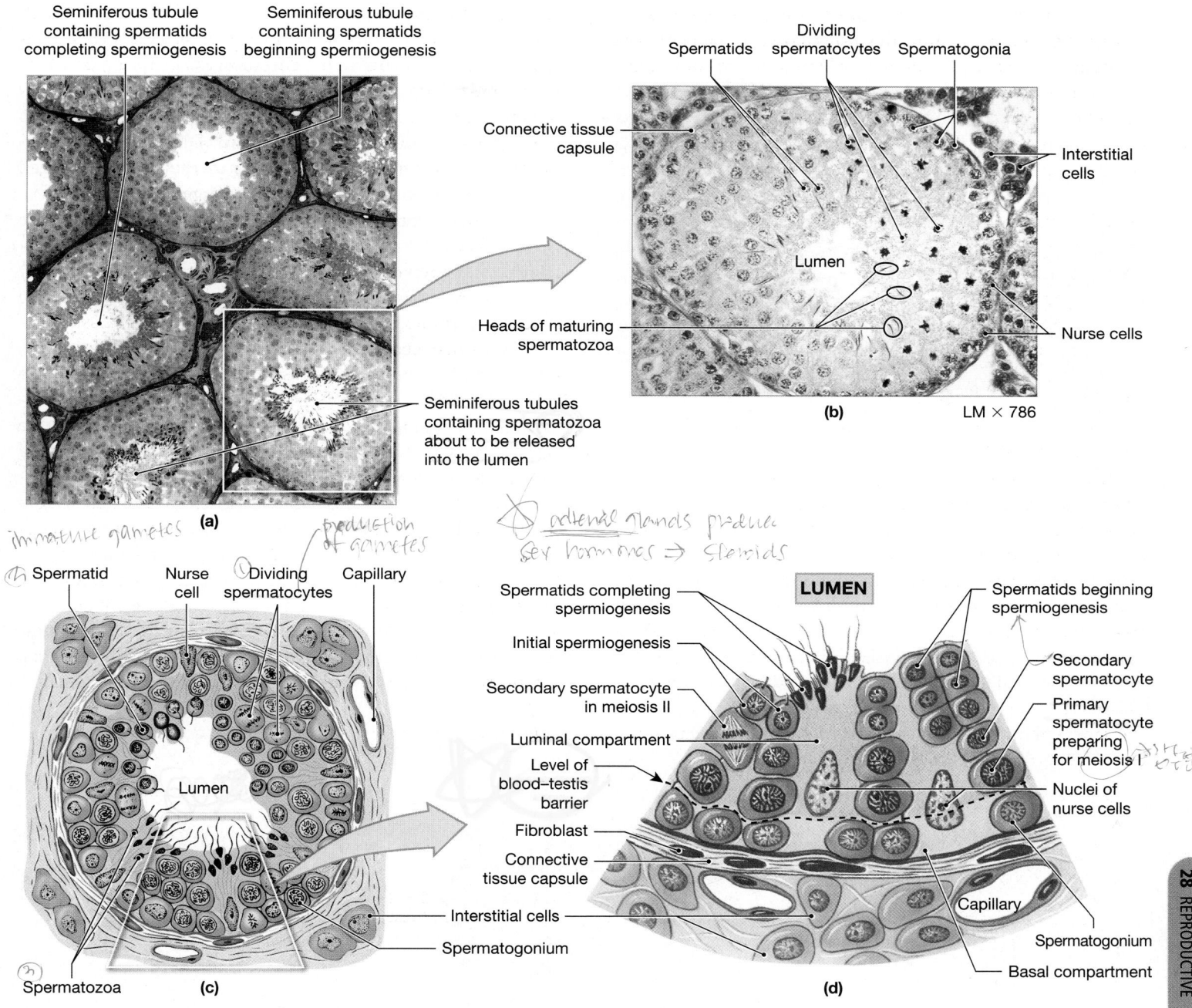

Figure 28–5   The Seminiferous Tubules. (a) A section through a coiled seminiferous tubule. **(b)** A cross section through a single tubule. **(c,d)** Stages in spermatogenesis in the wall of a seminiferous tubule. Nurse cells surround the stem cells of the tubule and support the developing spermatocytes and spermatids.

secondary spermatocytes that divide and differentiate into **spermatids** (SPER-ma-tidz)—immature gametes that subsequently differentiate into spermatozoa. The spermatozoa lose contact with the wall of the seminiferous tubule and enter the fluid in the lumen.

Each seminiferous tubule contains spermatogonia, spermatocytes at various stages of meiosis, spermatids, spermatozoa, and large **nurse cells**, or *sustentacular* (sus-ten-TAK-ū-lar) (or *Sertoli*) cells. Nurse cells are attached to

the tubular capsule and extend to the lumen between the other types of cells (**Figure 28–5b,c**).

Spermatogenesis

Spermatogenesis involves three integrated processes:

1. *Mitosis.* Spermatogonia undergo cell divisions throughout adult life. (You can review the description of mitosis and cell division in Chapter 3. ∞ pp. 102–103) One

daughter cell from each division remains in place while the other is pushed toward the lumen of the seminiferous tubule. The displaced cells differentiate into primary spermatocytes, which prepare to begin meiosis.

2. *Meiosis.* Meiosis (mī-Ō-sis) is a special form of cell division involved in gamete production. In humans, gametes contain 23 chromosomes, half the normal set. As a result, the fusion of the nuclei of a male gamete and a female gamete produces a cell that has the normal number of chromosomes (46), rather than twice that number. In the seminiferous tubules, meiotic divisions that begin with primary spermatocytes produce spermatids, the undifferentiated male gametes.

3. *Spermiogenesis.* Spermatids are small, relatively unspecialized cells. In *spermiogenesis,* spermatids differentiate into physically mature spermatozoa, which are among the most highly specialized cells in the body. Spermiogenesis involves major changes in a spermatid's internal and external structures.

Mitosis and Meiosis

In both males and females, mitosis and meiosis differ significantly in terms of the events occurring in the nucleus. As you may recall from Chapter 3, somatic cells contain 23 pairs of chromosomes. Each pair consists of one chromosome provided by the father, and another provided by the mother, at the time of fertilization. Mitosis is part of the process of somatic cell division, producing two daughter cells each containing identical pairs of chromosomes; the pattern is illustrated in **Figure 28–6a.** Because daughter cells contain both members of each chromosome pair (for a total of 46 chromosomes), they are called **diploid** (DIP-loyd; *diplo,* double) cells. Meiosis (**Figure 28–6b**) involves two cycles of cell division (*meiosis I* and *meiosis II*) and produces four cells, each of which contains 23 *individual* chromosomes. Because these cells contain only one member of each pair of chromosomes, they are called **haploid** (HAP-loyd; *haplo,* single) cells. The events in the nucleus shown in **Figure 28–6b** are the same for the formation of spermatozoa or ova.

As a cell prepares to begin meiosis, DNA replication occurs within the nucleus just as it does in a cell preparing to undergo mitosis. This similarity continues as prophase I arrives; the chromosomes condense and become visible with a light microscope. As in mitosis, each chromosome consists of two duplicate *chromatids.*

At this point, the close similarities between meiosis and mitosis end. In meiosis, the corresponding maternal (inherited from the mother) and paternal (inherited from the father) chromosomes now come together, an event known as **synapsis** (si-NAP-sis). Synapsis involves 23 pairs of chromosomes; each member of each pair consists of two chromatids. A matched set of four chromatids is called a **tetrad** (TET-rad; *tetras,* four) (**Figure 28–6b**). Some exchange of genetic mater-

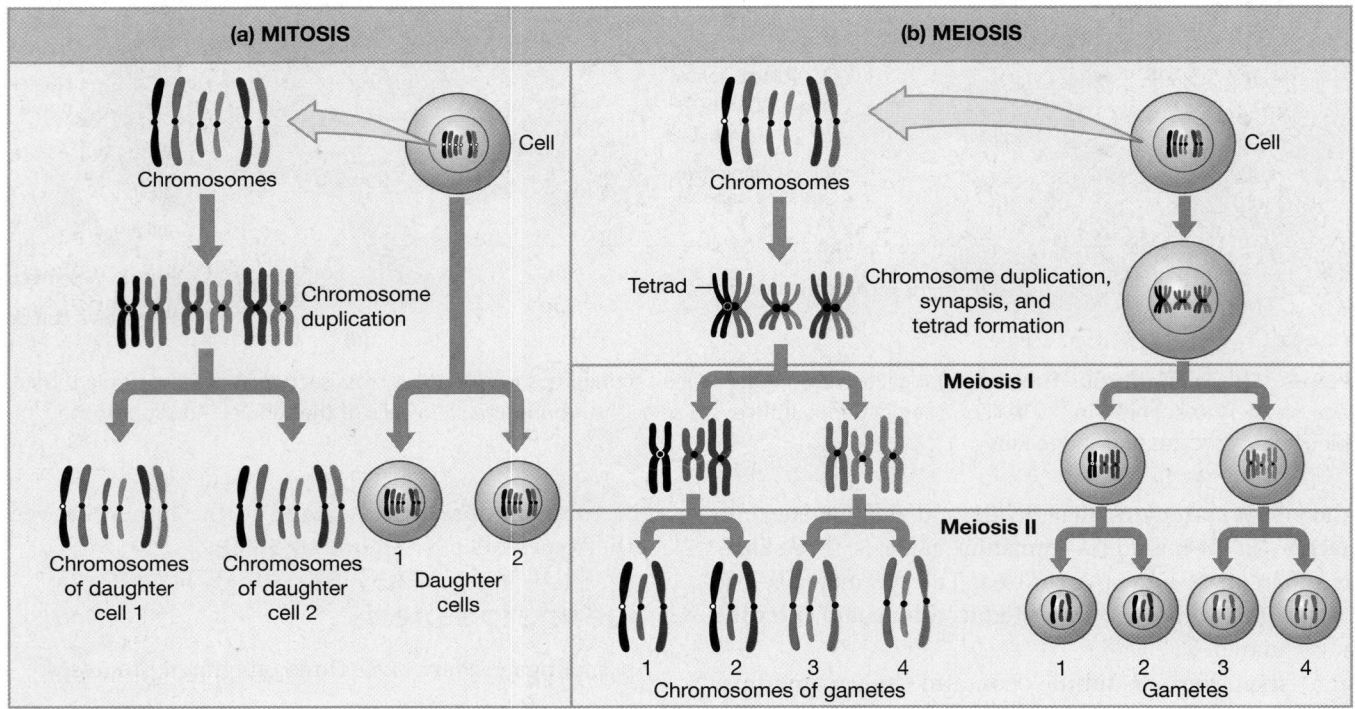

Figure 28–6 Chromosomes in Mitosis and Meiosis. (a) The fates of three representative chromosomes during mitosis. (See *Figure 3–25,* pp. 102–103.) **(b)** The fates of three representative chromosomes during the two stages of meiosis.

ial can occur between the chromatids of a chromosome pair at this stage of meiosis. Such an exchange, called *crossing over*, increases genetic variation among offspring; we will discuss it in Chapter 29.

Meiosis includes two division cycles, referred to as **meiosis I** and **meiosis II**. The stages within each phase are identified as prophase I, metaphase II, and so on. The nuclear envelope disappears at the end of prophase I. As metaphase I begins, the tetrads line up along the metaphase plate. As anaphase I begins, the tetrads break up—the maternal and paternal chromosomes separate. This is a major difference between mitosis and meiosis: In mitosis, each daughter cell receives one of the two copies of every chromosome, maternal and paternal; in meiosis I, each daughter cell receives both copies of *either* the maternal chromosome *or* the paternal chromosome from each tetrad. (Compare the two parts of **Figure 28–6**.)

As anaphase proceeds, the maternal and paternal components are randomly and independently distributed. That is, as each tetrad splits, one cannot predict which daughter cell will receive copies of the maternal chromosome, and which will receive copies of the paternal chromosome. As a result, telophase I ends with the formation of two daughter cells containing unique combinations of maternal and paternal chromosomes. Both cells contain 23 chromosomes. Because the first meiotic division reduces the number of chromosomes from 46 to 23, it is called a **reductional division**. Each of these chromosomes still consists of two duplicate chromatids. The duplicates will separate during meiosis II.

The interphase separating meiosis I and meiosis II is very brief, and no DNA is replicated during that period. Each cell proceeds through prophase II, metaphase II, and anaphase II. During anaphase II, the duplicate chromatids separate. Telophase II thus yields *four cells*, each containing 23 chromosomes. Because the number of chromosomes has not changed, meiosis II is an **equational division**. Although chromosomes are evenly distributed among these four cells, the cytoplasm may not be. In males, meiosis produces four immature gametes that are identical in size; each will develop into a functional sperm. In females, meiosis produces one relatively huge ovum and three tiny, nonfunctional polar bodies. We will examine the details of spermatogenesis here and will consider oogenesis in a later section.

In spermatogenesis (**Figure 28–7**), the mitotic division of each diploid spermatogonium produces two daughter cells. One is a spermatogonium that remains in contact with the basal lamina, and the other is a primary spermatocyte that is displaced toward the lumen. As meiosis begins, each primary spermatocyte contains 46 individual chromosomes. At the end of meiosis I, the daughter cells are called **secondary spermatocytes**. Every secondary spermatocyte contains 23 chromosomes, each of which consists of a pair of duplicate chromatids. The secondary spermatocytes soon enter prophase II. The completion of

metaphase II, anaphase II, and telophase II yields four haploid spermatids, each containing 23 chromosomes.

For each primary spermatocyte that enters meiosis, four spermatids are produced. Because cytokinesis (cytoplasmic division) is not completed in meiosis I or meiosis II, the four spermatids initially remain interconnected by bridges of cytoplasm. These connections assist in the transfer of nutrients and hormones between the cells, helping ensure that the cells develop in synchrony. The bridges are not broken until the last stages of physical maturation.

The A&P Top 100

#95 Meiosis produces gametes that contain half the number of chromosomes found in somatic cells. For each cell entering meiosis, the testes produce four spermatozoa, whereas the ovaries produce a single ovum.

Spermiogenesis

In **spermiogenesis**, the last step of spermatogenesis, each spermatid matures into a single spermatozoon, or sperm (**Figure 28–7**). Developing spermatocytes undergoing meiosis, and spermatids undergoing spermiogenesis, are not free in the seminiferous tubules. Instead, they are surrounded by the cytoplasm of the nurse cells. As spermiogenesis proceeds, the spermatids gradually develop the appearance of mature spermatozoa. At *spermiation*, a spermatozoon loses its attachment to the nurse cell and enters the lumen of the seminiferous tubule. The entire process, from spermatogonial division to spermiation, takes approximately nine weeks.

Nurse Cells. Nurse cells play a key role in spermatogenesis. These cells have six important functions that directly or indirectly affect mitosis, meiosis, and spermiogenesis within the seminiferous tubules:

1. *Maintenance of the Blood–Testis Barrier.* The seminiferous tubules are isolated from the general circulation by a **blood–testis barrier**, comparable in function to the blood–brain barrier. ∞ p. 467 Nurse cells are joined by tight junctions, forming a layer that divides the seminiferous tubule into an outer *basal compartment*, which contains the spermatogonia, and an inner *luminal compartment* (or *adluminal compartment*), where meiosis and spermiogenesis occur (**Figure 28–5d**). Transport across the nurse cells is tightly regulated, so conditions in the luminal compartment remain very stable. The fluid in the lumen of a seminiferous tubule is produced by the nurse cells, which also regulate the fluid's composition. This fluid is very different from the surrounding interstitial fluid; it is high in androgens, estrogens, potassium, and amino acids. The blood–testis barrier is essential to preserving the differences between the tubular fluid and

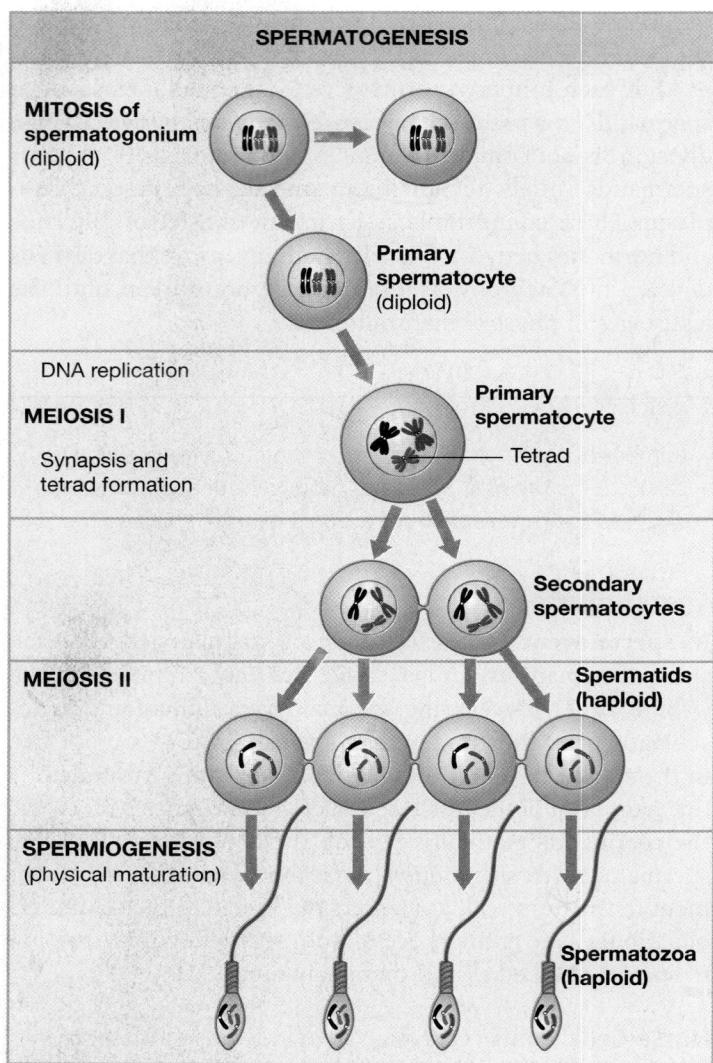

SPERMATOGENESIS

MITOSIS of spermatogonium (diploid)

Primary spermatocyte (diploid)

DNA replication

MEIOSIS I

Synapsis and tetrad formation

Primary spermatocyte

Tetrad

Secondary spermatocytes

MEIOSIS II

Spermatids (haploid)

SPERMIOGENESIS (physical maturation)

Spermatozoa (haploid)

Figure 28–7   Spermatogenesis. The events depicted occur in the seminiferous tubules. The fates of three representative chromosomes are shown; compare with *Figure 28–6b.*

4. *Secretion of Inhibin.* Nurse cells secrete the peptide hormone *inhibin* (in-HIB-in) in response to factors released by developing spermatozoa. Inhibin depresses the pituitary production of FSH, and perhaps the hypothalamic secretion of gonadotropin-releasing hormone (GnRH). The faster the rate of sperm production, the more inhibin is secreted. By regulating FSH and GnRH secretion, nurse cells provide feedback control of spermatogenesis.

5. *Secretion of Androgen-Binding Protein. Androgen-binding protein (ABP)* binds androgens (primarily testosterone) in the fluid contents of the seminiferous tubules. This protein is thought to be important in both elevating the concentration of androgens within the seminiferous tubules and stimulating spermiogenesis. The production of ABP is stimulated by FSH.

6. *Secretion of Müllerian-Inhibiting Factor. Müllerian-inhibiting factor (MIF)* is secreted by nurse cells in the developing testes. This hormone causes regression of the fetal *Müllerian (paramesonephric) ducts,* passageways that participate in the formation of the uterine tubes and the uterus in females. In males, inadequate MIF production during fetal development leads to the retention of these ducts and the failure of the testes to descend into the scrotum.

The Anatomy of a Spermatozoon

Each spermatozoon has three distinct regions: the head, the middle piece, and the tail (**Figure 28–8a**). In humans, the **head** is a flattened ellipse containing a nucleus with densely packed chromosomes. At the tip of the head is the **acrosomal** (ak-rō-SŌ-mal) **cap**, a membranous compartment containing enzymes essential to fertilization. During spermiogenesis, saccules of the spermatid's Golgi apparatus fuse and flatten into an *acrosomal vesicle,* which ultimately forms the acrosomal cap of the spermatozoon.

A short **neck** attaches the head to the **middle piece**. The neck contains both centrioles of the original spermatid. The microtubules of the distal centriole are continuous with those of the middle piece and tail. Mitochondria in the middle piece are arranged in a spiral around the microtubules. Mitochondrial activity provides the ATP required to move the tail.

The **tail** is the only flagellum in the human body. A *flagellum,* a whiplike organelle, moves a cell from one place to another. Whereas cilia beat in a predictable, wavelike fashion, the flagellum of a spermatozoon has a complex, corkscrew motion.

Unlike other, less specialized cells, a mature spermatozoon lacks an endoplasmic reticulum, a Golgi apparatus, lysosomes, peroxisomes, inclusions, and many other intracellular structures. The loss of these organelles reduces the cell's size and mass; it is essentially a mobile carrier for the enclosed chromosomes, and extra weight would slow it down.

the interstitial fluid. In addition, this barrier prevents immune system cells from detecting and attacking the developing spermatozoa, which have in their plasma membranes sperm-specific antigens not found in somatic cell membranes and thus might be identified as "foreign."

2. *Support of Mitosis and Meiosis.* Spermatogenesis depends on the stimulation of nurse cells by circulating follicle-stimulating hormone (FSH) and testosterone. Stimulated nurse cells then promote the division of spermatogonia and the meiotic divisions of spermatocytes.

3. *Support of Spermiogenesis.* Spermiogenesis requires the presence of nurse cells. These cells surround and enfold the spermatids, providing nutrients and chemical stimuli that promote their development. Nurse cells also phagocytize cytoplasm that is shed by spermatids as they develop into spermatozoa.

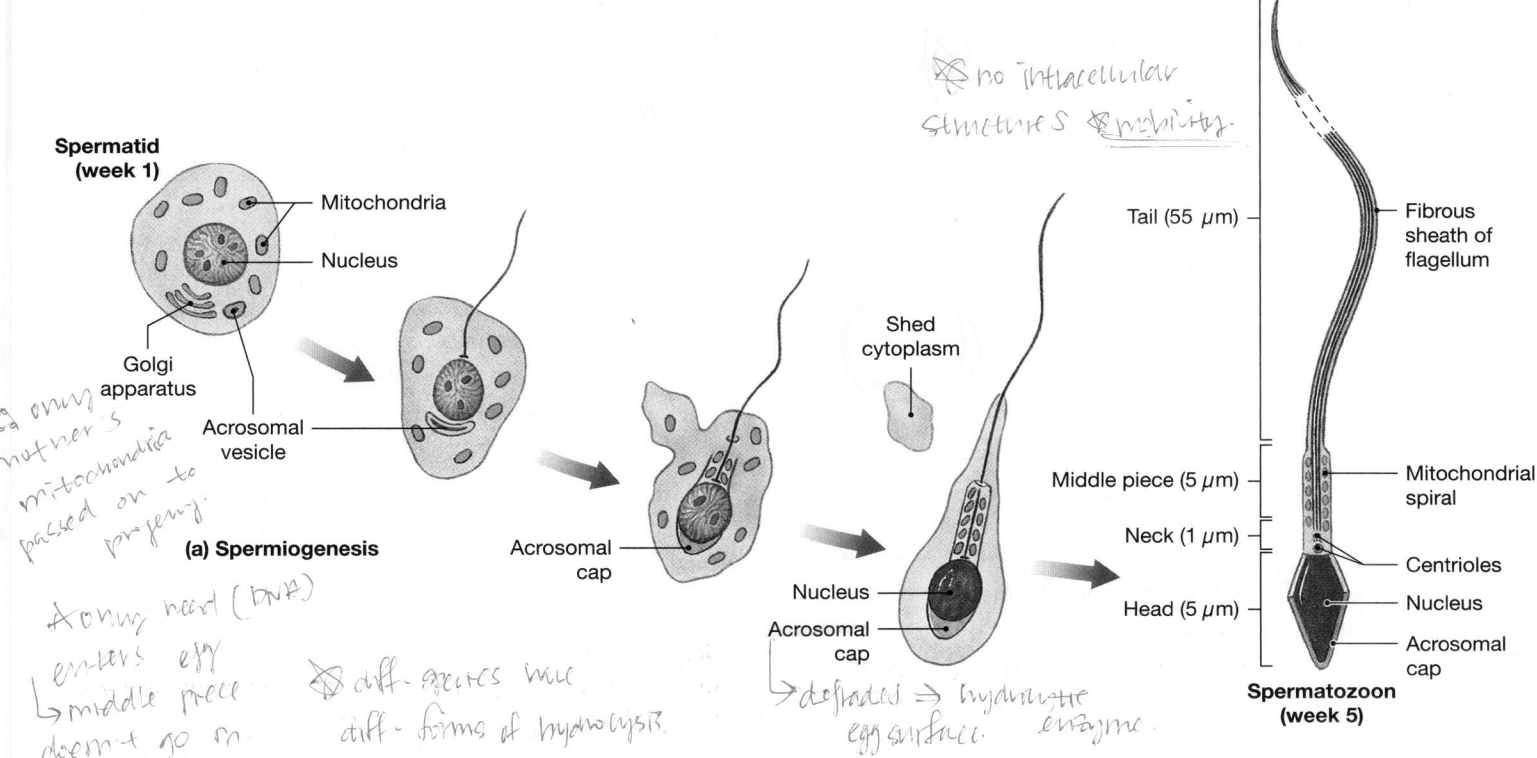

Handwritten notes:
- no intracellular structures & mobility.
- only mother's mitochondria passed on to progeny.
- only head (DNA) enters egg
  - middle piece doesn't go on
- diff species have diff. forms of hyaluronidase
- degraded → hyaluronidase egg surface enzyme

Spermatid (week 1)
— Mitochondria
— Nucleus
— Golgi apparatus
— Acrosomal vesicle

(a) Spermiogenesis

— Acrosomal cap

Shed cytoplasm

Nucleus
Acrosomal cap

Tail (55 μm)

Fibrous sheath of flagellum

Middle piece (5 μm)
Neck (1 μm)
Head (5 μm)

Mitochondrial spiral
Centrioles
Nucleus
Acrosomal cap

Spermatozoon (week 5)

Because a spermatozoon lacks glycogen or other energy reserves, it must absorb nutrients (primarily fructose) from the surrounding fluid.

The A&P Top 100

#96 Spermatogenesis begins at puberty and continues until relatively late in life (past age 70). It is a continuous process, and all stages of meiosis can be observed within the seminiferous tubules.

The Male Reproductive Tract

The testes produce physically mature spermatozoa that are incapable of successfully fertilizing an oocyte. The other portions of the male reproductive system are responsible for the functional maturation, nourishment, storage, and transport of spermatozoa.

The Epididymis

Late in their development, spermatozoa detach from the nurse cells and lie within the lumen of the seminiferous tubule. They have most of the physical characteristics of mature spermatozoa, but are functionally immature and incapable of coordinated locomotion or fertilization. Fluid currents, created by cilia lining the efferent ductules, transport the immobile gametes into the epididymis (**Figure 28–4a**). The **epididymis** (ep-i-DID-i-mis; *epi*, on + *didymos*, twin), the start of the male reproductive tract, is a coiled tube bound to the posterior border of each testis.

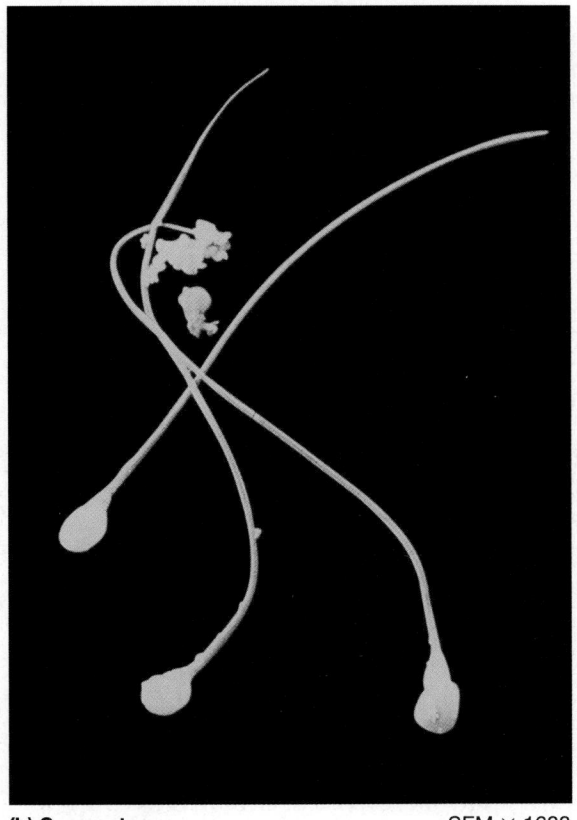

(b) Spermatozoa　　　　　　　SEM × 1688

Figure 28–8 Spermiogenesis and Spermatozoon Structure.
(a) The differentiation of a spermatid into a spermatozoon. This differentiation process is completed in approximately five weeks. **(b)** Human spermatozoa.

The epididymides (ep-i-DID-i-mi-dēz) can be felt through the skin of the scrotum. A tubule almost 7 m (23 ft) long, the epididymis is coiled and twisted so as to take up very little space. It has a head, a body, and a tail (**Figure 28–9a**). The superior **head** is the portion of the epididymis proximal to the testis. The head receives spermatozoa from the efferent ductules.

The **body** begins distal to the last efferent ductule and extends inferiorly along the posterior margin of the testis. Near the inferior border of the testis, the number of coils decreases, marking the start of the **tail**. The tail recurves and ascends to its connection with the ductus deferens. Spermatozoa are stored primarily within the tail of the epididymis.

The epididymis has three functions:

1. *It Monitors and Adjusts the Composition of the Fluid Produced by the Seminiferous Tubules.* The pseudostratified columnar epithelial lining of the epididymis bears distinctive stereocilia (**Figure 28–9b**), which increase the surface area available for absorption from, and secretion into, the fluid in the tubule.

2. *It Acts as a Recycling Center for Damaged Spermatozoa.* Cellular debris and damaged spermatozoa are absorbed in the epididymis, and the products of enzymatic breakdown are released into the surrounding interstitial fluids for pickup by the epididymal blood vessels.

3. *It Stores and Protects Spermatozoa and Facilitates Their Functional Maturation.* A spermatozoon passes through the epididymis in about two weeks and completes its functional maturation at that time. Over this period, spermatozoa exist in a sheltered environment that is precisely regulated by the surrounding epithelial cells. Although spermatozoa leaving the epididymis are mature, they remain immobile. To become *motile* (actively swimming) and fully functional, spermatozoa must undergo a process called **capacitation**. Capacitation normally occurs in two steps: (1) Spermatozoa become motile when they are mixed with secretions of the seminal glands, and (2) they become capable of successful fertilization when exposed to conditions in the female reproductive tract. The epididymis secretes a substance (as yet unidentified) that prevents premature capacitation.

Transport along the epididymis involves a combination of fluid movement and peristaltic contractions of smooth muscle in the walls of the epididymis. After passing along the tail of the epididymis, the spermatozoa enter the ductus deferens.

The Ductus Deferens

Each **ductus deferens**, or *vas deferens*, is 40–45 cm (16–18 in.) long. It begins at the tail of the epididymis (**Figure 28–9a**) and, as part of the spermatic cord, ascends through the inguinal canal (**Figure 28–3**). Inside the abdominal cavity, the ductus deferens passes posteriorly, curving inferiorly along

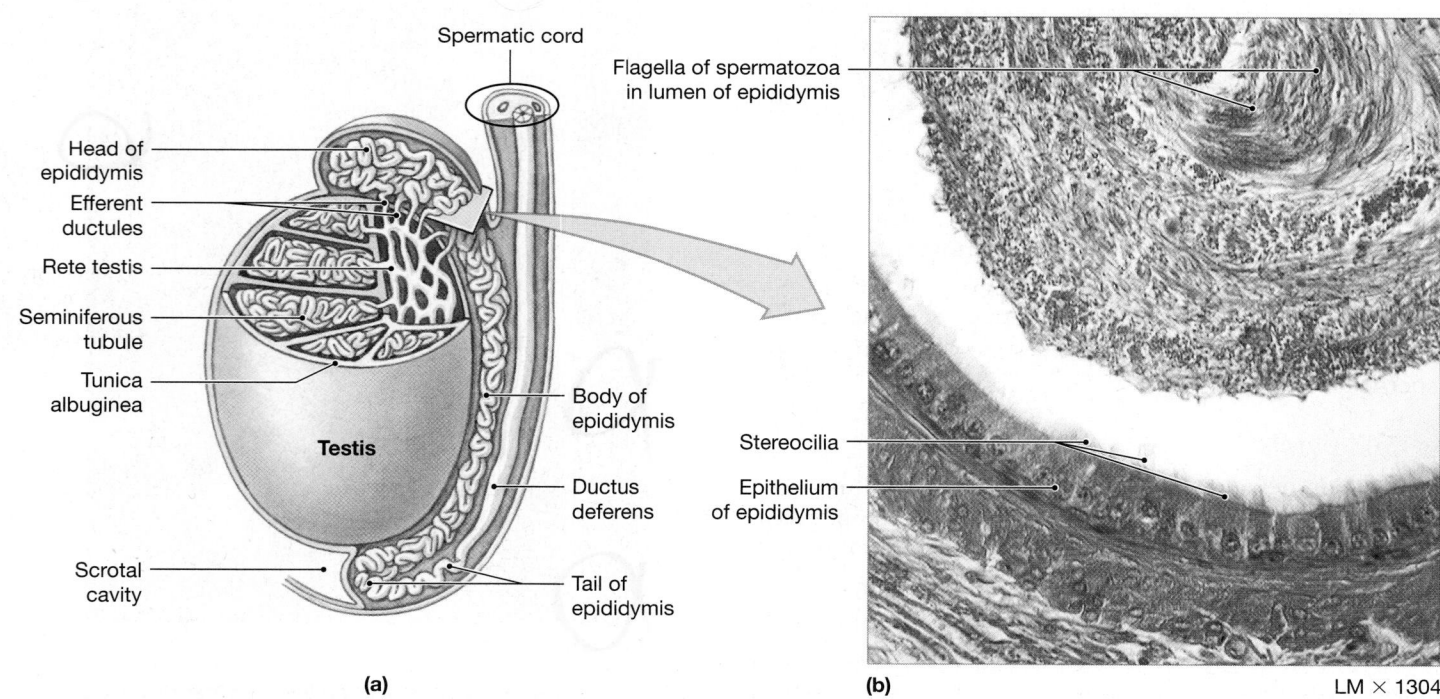

(a)

(b) LM × 1304

Figure 28–9 **The Epididymis. (a)** A diagrammatic view. **(b)** Epithelial features, especially the elongate stereocilia characteristic of the epididymis. ATLAS: Plate 60a

the lateral surface of the urinary bladder toward the superior and posterior margin of the prostate gland (**Figure 28–1**). Just before the ductus deferens reaches the prostate gland and seminal glands, its lumen enlarges. This expanded portion is known as the **ampulla** (am-PUL-luh) of the ductus deferens (**Figure 28–10a**).

The wall of the ductus deferens contains a thick layer of smooth muscle (**Figure 28–10b**). Peristaltic contractions in this layer propel spermatozoa and fluid along the duct, which is lined by a pseudostratified ciliated columnar epithelium. In addition to transporting spermatozoa, the ductus deferens can store spermatozoa for several months. During this time,

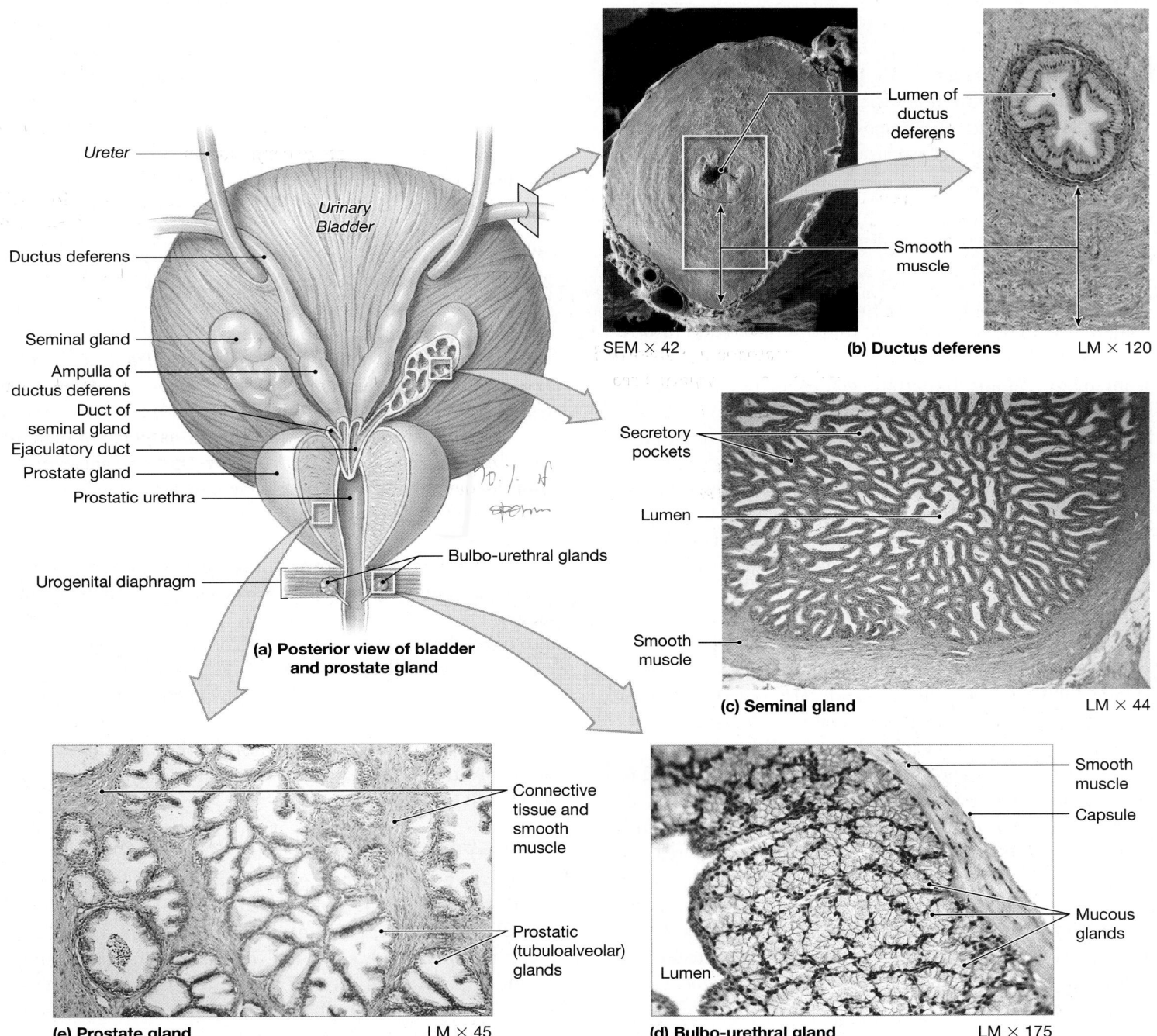

Figure 28–10 **The Ductus Deferens and Accessory Glands.** (a) A posterior view of the prostate gland, showing subdivisions of the ductus deferens in relation to surrounding structures. (b) The ductus deferens, showing the smooth muscle around the lumen. [© R. G. Kessel and R. H. Kardon, *Tissues and Organs: A Text-Atlas of Scanning Electron Microscopy*, W. H. Freeman & Co., 1979. All Rights Reserved.] Sections of (c) the seminal gland, (d) a bulbo-urethral gland, and (e) the prostate gland.

the spermatozoa remain in a state of suspended animation and have low metabolic rates.

The junction of the ampulla with the duct of the seminal gland marks the start of the **ejaculatory duct**. This short passageway (2 cm, or less than 1 in.) penetrates the muscular wall of the prostate gland and empties into the urethra near the opening of the ejaculatory duct from the opposite side (**Figures 28–1** and **28–10a**).

The Urethra

In males, the **urethra** extends 18–20 cm (7–8 in.) from the urinary bladder to the tip of the penis (**Figure 28–1**). It is divided into *prostatic*, *membranous*, and *spongy* regions. The male urethra is a passageway used by both the urinary and reproductive systems.

female ovan- acidic (4.0 pH).

The Accessory Glands

The fluids contributed by the seminiferous tubules and the epididymis account for only about 5 percent of the volume of semen. The fluid component of semen is a mixture of secretions—each with distinctive biochemical characteristics—from many glands. Important glands include the *seminal glands*, the *prostate gland*, and the *bulbo-urethral glands*, all of which occur only in males. Among the major functions of these glands are (1) activating spermatozoa; (2) providing the nutrients spermatozoa need for motility; (3) propelling spermatozoa and fluids along the reproductive tract, mainly by peristaltic contractions; and (4) producing buffers that counteract the acidity of the urethral and vaginal environments.

The Seminal Glands (Seminal Vesicles)

The ductus deferens on each side ends at the junction between the ampulla and the duct that drains the seminal gland (**Figure 28–10a**). The **seminal glands**, also called the **seminal vesicles**, are glands embedded in connective tissue on either side of the midline, sandwiched between the posterior wall of the urinary bladder and the rectum. Each seminal gland is a tubular gland with a total length of about 15 cm (6 in.). The body of the gland has many short side branches. The entire assemblage is coiled and folded into a compact, tapered mass roughly 5 cm × 2.5 cm (2 in. × 1 in.).

Seminal glands are extremely active secretory glands with an epithelial lining that contains extensive folds (**Figure 28–10c**). The seminal glands contribute about 60 percent of the volume of semen. Although the glandular fluid generally has the same osmotic concentration as that of blood plasma, the compositions of the two fluids are quite different. In particular, the secretion of the seminal glands contains (1) higher concentrations of fructose, which is easily metabolized by spermatozoa; (2) prostaglandins, which can stimulate smooth muscle contractions along the male and female reproductive tracts;

and (3) fibrinogen, which after ejaculation forms a temporary clot within the vagina. The secretions of the seminal glands are slightly alkaline, helping to neutralize acids in the secretions of the prostate gland and within the vagina. When mixed with the secretions of the seminal glands, previously inactive but functional spermatozoa undergo the first step in capacitation and begin beating their flagella, becoming highly motile.

The secretions of the seminal glands are discharged into the ejaculatory duct at *emission*, when peristaltic contractions are under way in the ductus deferens, seminal glands, and prostate gland. These contractions are under the control of the sympathetic nervous system.

The Prostate Gland → slightly acidic

The **prostate gland** is a small, muscular, rounded organ about 4 cm (1.6 in.) in diameter. The prostate gland encircles the proximal portion of the urethra as it leaves the urinary bladder (**Figure 28–10a**). The glandular tissue of the prostate (**Figure 28–10e**) consists of a cluster of 30–50 compound tubuloalveolar glands. ∞ p. 124 These glands are surrounded by and wrapped in a thick blanket of smooth muscle fibers.

The prostate gland produces **prostatic fluid**, a slightly acidic solution that contributes 20–30 percent of the volume of semen. In addition to several other compounds of uncertain significance, prostatic secretions contain **seminalplasmin** (sem-i-nal-PLAZ-min), an antibiotic that may help prevent urinary tract infections in males. These secretions are ejected into the prostatic urethra by peristaltic contractions of the muscular prostate wall.

Prostatic inflammation, or **prostatitis** (pros-ta-TĪ-tis), can occur in males at any age, but it most commonly afflicts older men. Prostatitis can result from bacterial infections but also occurs in the apparent absence of pathogens. Symptoms can resemble those of prostate cancer. Individuals with prostatitis may complain of pain in the lower back, perineum, or rectum, in some cases accompanied by painful urination and the discharge of mucous secretions from the external urethral orifice. Antibiotic therapy is effective in treating most cases that result from bacterial infection.

The Bulbo-urethral Glands *lubrication*

The paired **bulbo-urethral glands**, or *Cowper glands*, are situated at the base of the penis, covered by the fascia of the urogenital diaphragm (**Figures 28–10a** and **28–11a**). The bulbo-urethral glands are round, with diameters approaching 10 mm (less than 0.5 in.). The duct of each gland travels alongside the penile urethra for 3–4 cm (1.2–1.6 in.) before emptying into the urethral lumen. The bulbo-urethral glands are compound tubular mucous glands (**Figure 28–10d**) that secrete a thick, alkaline mucus. The secretion helps neutralize any urinary acids that may remain in the urethra, and it lubricates the *glans*, or tip of the penis.

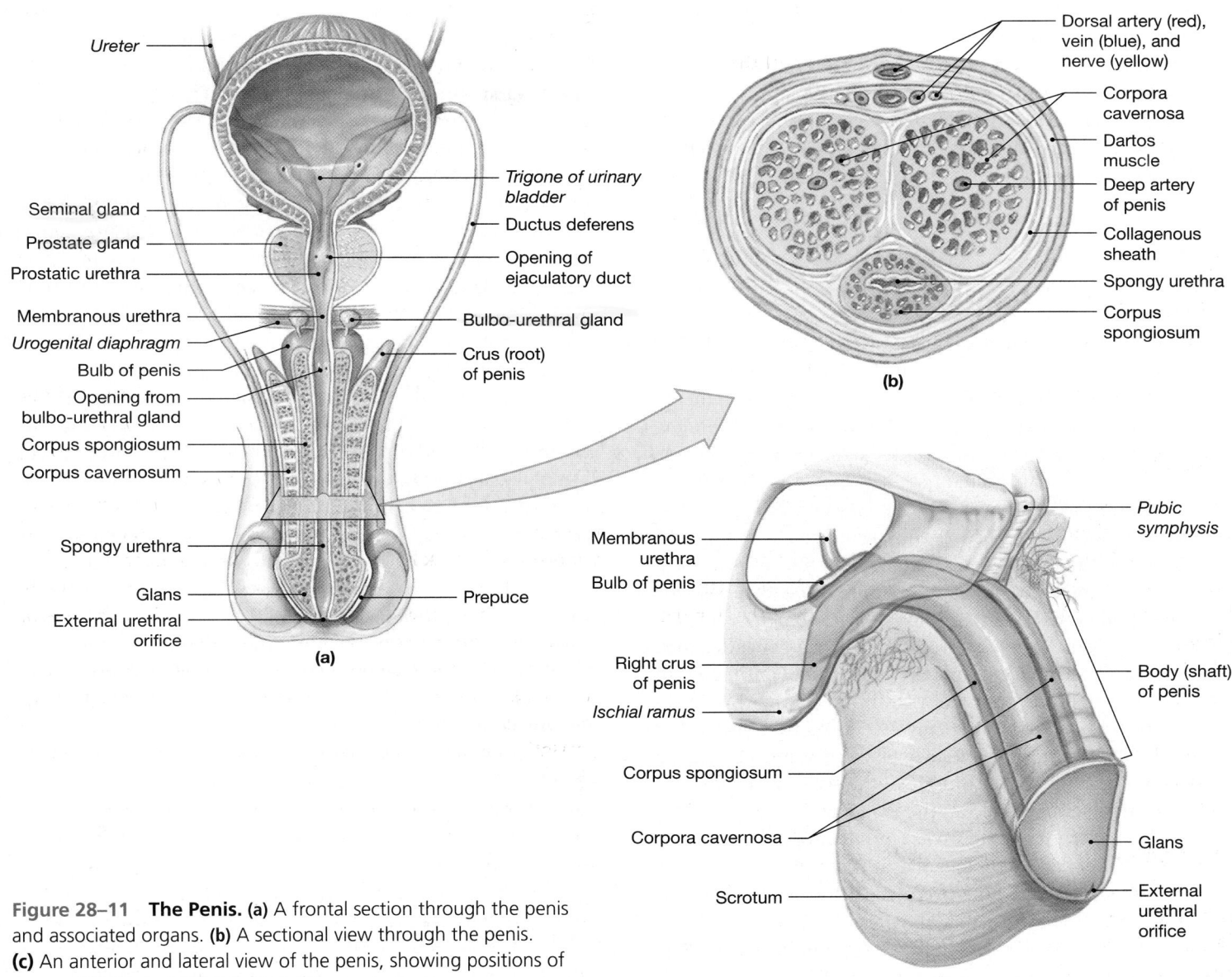

Figure 28–11 **The Penis. (a)** A frontal section through the penis and associated organs. **(b)** A sectional view through the penis. **(c)** An anterior and lateral view of the penis, showing positions of the erectile tissues. ATLAS: Plate 60b

Semen

A typical ejaculation releases 2–5 mL of semen; an abnormally low volume may indicate problems with the prostate gland or seminal glands. When sampled for analysis, semen is collected after a 36-hour period of sexual abstinence. The volume of fluid produced by an ejaculation, called the **ejaculate**, typically contains the following:

- *Spermatozoa.* The normal **sperm count** ranges from 20 million to 100 million spermatozoa per milliliter of semen. Most individuals with lower sperm counts are infertile, because too few spermatozoa survive the ascent of the female reproductive tract to perform fertilization. A low sperm count may reflect inflammation of the epididymis,

ductus deferens, or prostate gland. In a fertile male, at least 60 percent of the spermatozoa in the sample are normal in appearance; common abnormalities are malformed heads and "twin" spermatozoa that did not separate at the time of spermiation. The normal sperm will be swimming actively.

- *Seminal Fluid.* **Seminal fluid**, the fluid component of semen, is a mixture of glandular secretions with a distinct ionic and nutrient composition. A typical sample of seminal fluid contains the combined secretions of the seminal glands (60 percent), the prostate gland (30 percent), the nurse cells and epididymis (5 percent), and the bulbo-urethral glands (less than 5 percent).

- *Enzymes.* Several important enzymes are present in seminal fluid, including (1) a protease that may help dissolve mucous secretions in the vagina; (2) seminalplasmin, an antibiotic prostatic enzyme that kills a variety of bacteria, including *Escherichia coli*; (3) a prostatic enzyme that coagulates the semen within a few minutes after ejaculation by converting fibrinogen to fibrin; and (4) *fibrinolysin*, which liquefies the clotted semen after 15–30 minutes.

A complete chemical analysis of semen appears in the Appendix.

The External Genitalia

The male external genitalia consist of the scrotum and penis. The structure of the scrotum has already been described (p. 1045). The **penis** is a tubular organ through which the distal portion of the urethra passes (**Figure 28–11a**). It conducts urine to the exterior and introduces semen into the female's vagina during sexual intercourse. The penis is divided into three regions: the root, the body, and the glans (**Figure 28–11c**). The **crus (root)** of the penis is the fixed portion that attaches the penis to the body wall. This connection occurs within the urogenital triangle immediately inferior to the pubic symphysis. The **body (shaft)** of the penis is the tubular, movable portion of the organ. The **glans** of the penis is the expanded distal end that surrounds the external urethral orifice. The *neck* is the narrow portion of the penis between the shaft and the glans.

The skin overlying the penis resembles that of the scrotum. The dermis contains a layer of smooth muscle that is a continuation of the dartos muscle of the scrotum, and the underlying areolar tissue allows the thin skin to move without distorting underlying structures. The subcutaneous layer also contains superficial arteries, veins, and lymphatic vessels.

A fold of skin called the **prepuce** (PRĒ-poos), or *foreskin*, surrounds the tip of the penis. The prepuce attaches to the relatively narrow neck of the penis and continues over the glans. **Preputial** (prē-PŪ-shē-al) **glands** in the skin of the neck and the inner surface of the prepuce secrete a waxy material known as **smegma** (SMEG-ma). Unfortunately, smegma can be an excellent nutrient source for bacteria. Mild inflammation and infections in this locale are common, especially if the area is not washed thoroughly and frequently. One way to avoid trouble is **circumcision** (ser-kum-SIZH-un), the surgical removal of the prepuce. In Western societies (especially the United States), this procedure is generally performed shortly after birth. Circumcision lowers the risks of developing urinary tract infections, HIV infection, and penile cancer. Because it is a surgical procedure with risk of bleeding, infection, and other complications, the practice remains controversial.

Deep to the areolar tissue, a dense network of elastic fibers encircles the internal structures of the penis. Most of the body of the penis consists of three cylindrical columns of **erectile tissue** (**Figure 28–11b**). Erectile tissue consists of a three-dimensional maze of vascular channels incompletely separated by partitions of elastic connective tissue and smooth muscle fibers. In the resting state, the arterial branches are constricted and the muscular partitions are tense. This combination restricts blood flow into the erectile tissue. The parasympathetic innervation of the penile arteries involves neurons that release nitric oxide (NO) at their synaptic knobs. The smooth muscles in the arterial walls relax when NO is released, at which time the vessels dilate, blood flow increases, the vascular channels become engorged with blood, and **erection** of the penis occurs. The flaccid (nonerect) penis hangs inferior to the pubic symphysis and anterior to the scrotum, but during erection the penis stiffens and assumes a more upright position.

The anterior surface of the flaccid penis covers two cylindrical masses of erectile tissue: the **corpora cavernosa** (KOR-por-a ka-ver-NŌ-suh; singular, *corpus cavernosum*). The two are separated by a thin septum and encircled by a dense collagenous sheath (**Figure 28–11b**). The corpora cavernosa diverge at their bases, forming the **crura** (*crura*, legs; singular, *crus*) of the penis (**Figure 28–11a**). Each crus is bound to the ramus of the ischium and pubis by tough connective tissue ligaments. The corpora cavernosa extend along the length of the penis as far as its neck. The erectile tissue within each corpus cavernosum surrounds a central artery (**Figure 28–11b**).

The relatively slender **corpus spongiosum** (spon-jē-Ō-sum) surrounds the penile urethra (**Figure 28–11a,b**). This erectile body extends from the superficial fascia of the urogenital diaphragm to the tip of the penis, where it expands to form the glans. The sheath surrounding the corpus spongiosum contains more elastic fibers than does that of the corpora cavernosa, and the erectile tissue contains a pair of small arteries.

Hormones and Male Reproductive Function

The hormonal interactions that regulate male reproductive function are diagrammed in **Figure 28–12**. The major reproductive hormones were introduced in Chapter 18. ∞ p. 637 The adenohypophysis releases *follicle-stimulating hormone* (**FSH**) and *luteinizing hormone* (**LH**). The pituitary release of these hormones occurs in response to *gonadotropin-releasing hormone* (**GnRH**), a peptide synthesized in the hypothalamus and carried to the adenohypophysis by the hypophyseal portal system.

The hormone GnRH is secreted in pulses rather than continuously. In adult males, small pulses occur at 60–90-minute intervals. As levels of GnRH change, so do the rates of secretion of FSH and LH (and testosterone, which is released in response to LH). Unlike the situation in women, which we will consider later in the chapter, the GnRH pulse frequency in adult males remains relatively steady from hour to hour, day to day, and year to year. As a result, plasma levels of FSH, LH,

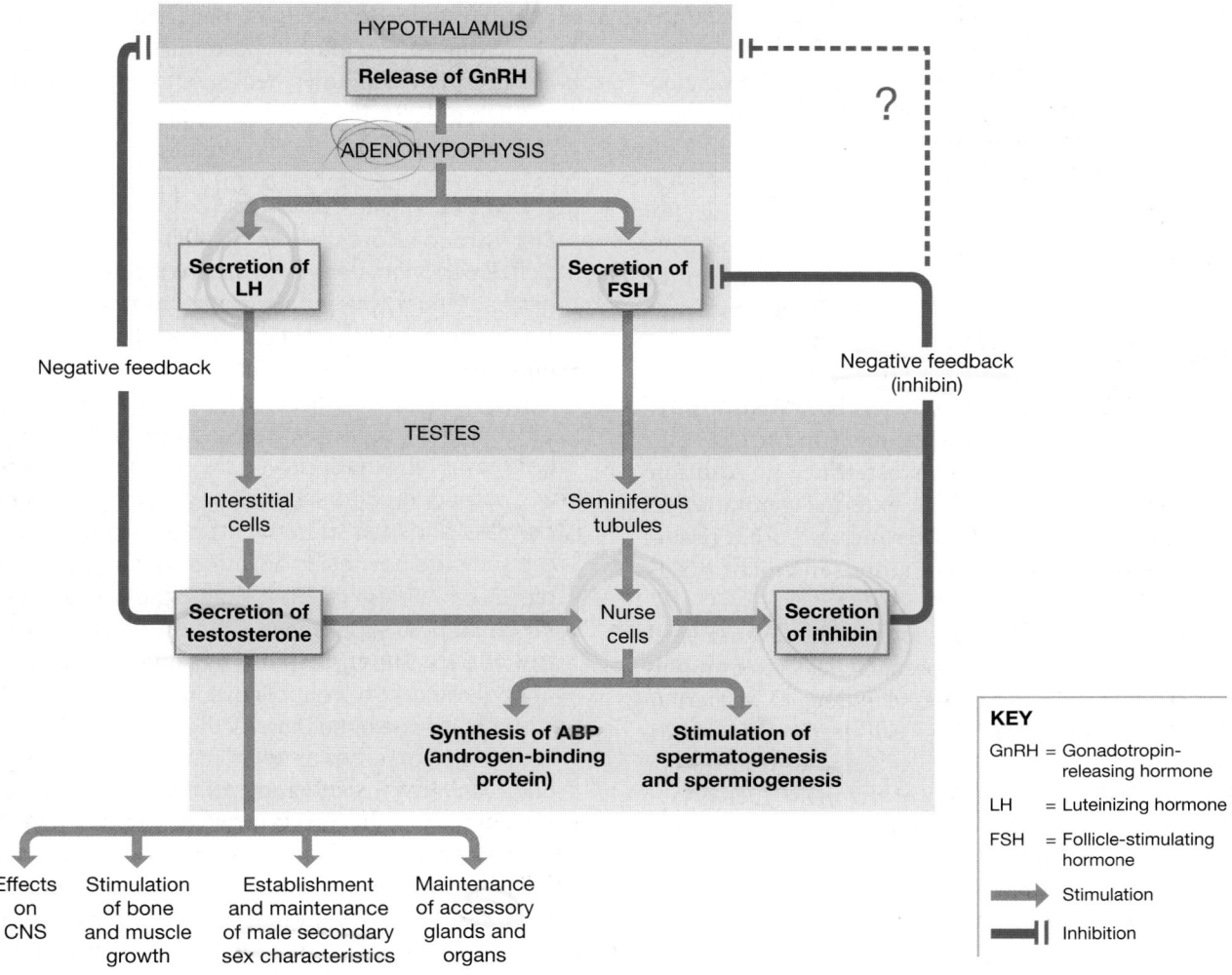

Figure 28–12   Hormonal Feedback and the Regulation of Male Reproductive Function.

and testosterone remain within a relatively narrow range until fairly late in life (see p. 1079).

FSH and Spermatogenesis

In males, FSH targets primarily the nurse cells of the seminiferous tubules. Under FSH stimulation, and in the presence of testosterone from the interstitial cells, nurse cells (1) promote spermatogenesis and spermiogenesis and (2) secrete androgen-binding protein (ABP).

The rate of spermatogenesis is regulated by a negative feedback mechanism involving GnRH, FSH, and inhibin. Under GnRH stimulation, FSH promotes spermatogenesis along the seminiferous tubules. As spermatogenesis accelerates, however, so does the rate of inhibin secretion by the nurse cells of the testes (**Figure 28–12**). Inhibin inhibits FSH production in the adenohypophysis and may also suppress the secretion of GnRH at the hypothalamus.

The net effect is that when FSH levels become elevated, inhibin production increases until FSH levels return to nor-

mal. If FSH levels decline, inhibin production falls, so the rate of FSH production accelerates.

LH and Androgen Production

In males, LH induces the secretion of testosterone and other androgens by the interstitial cells of the testes. Testosterone, the most important androgen, has numerous functions: (1) stimulating spermatogenesis and promoting the functional maturation of spermatozoa, through its effects on nurse cells; (2) affecting central nervous system (CNS) function, including the libido (sexual drive) and related behaviors; (3) stimulating metabolism throughout the body, especially pathways concerned with protein synthesis, blood cell formation, and muscle growth; (4) establishing and maintaining male secondary sex characteristics, such as the distribution of facial hair, increased muscle mass and body size, and the quantity and location of characteristic adipose tissue deposits; and (5) maintaining the accessory glands and organs of the male reproductive tract.

Testosterone functions like other steroid hormones, circulating in the bloodstream while bound to one of two types of transport proteins: (1) *gonadal steroid-binding globulin* (GBG), which carries roughly two-thirds of the circulating testosterone, and (2) the albumins, which bind the remaining one-third. Testosterone diffuses across the plasma membrane of target cells and binds to an intracellular receptor. The hormone–receptor complex then binds to the DNA in the nucleus. In many target tissues, some of the arriving testosterone is converted to **dihydrotestosterone (DHT)**. A small amount of DHT diffuses back out of the cell and into the bloodstream, and DHT levels are usually about 10 percent of circulating testosterone levels. Dihydrotestosterone can also enter peripheral cells and bind to the same hormone receptors targeted by testosterone. In addition, some tissues (notably those of the external genitalia) respond to DHT rather than to testosterone, and other tissues (including the prostate gland) are more sensitive to DHT than to testosterone.

Testosterone production begins around the seventh week of fetal development and reaches a prenatal peak after roughly six months. Over this period, the secretion of Müllerian-inhibiting factor by developing nurse cells leads to the regression of the Müllerian ducts. The early surge in testosterone levels stimulates the differentiation of the male duct system and accessory organs and affects CNS development. The best-known CNS effects occur in the developing hypothalamus. There, testosterone apparently programs the hypothalamic centers that are involved with (1) GnRH production and the regulation of pituitary FSH and LH secretion, (2) sexual behaviors, and (3) sexual drive. As a result of this prenatal exposure to testosterone, the hypothalamic centers will respond appropriately when the individual becomes sexually mature. The factors responsible for regulating the fetal production of testosterone are not known.

Testosterone levels are low at birth. Until puberty, background testosterone levels, although still relatively low, are higher in males than in females. Testosterone secretion accelerates markedly at puberty, initiating sexual maturation and the appearance of secondary sex characteristics. In adult males, negative feedback controls the level of testosterone production. Above-normal testosterone levels inhibit the release of GnRH by the hypothalamus, causing a reduction in LH secretion and lowering testosterone levels (**Figure 28–12**).

The plasma of adult males also contains relatively small amounts of estradiol (2 ng/dL versus 525 ng/dL of testosterone). Seventy percent of the estradiol is formed from circulating testosterone; the rest is secreted, primarily by the interstitial and nurse cells of the testes. The conversion of testosterone to estradiol is performed by an enzyme called *aromatase*. For unknown reasons, estradiol production increases in older men.

CLINICAL NOTE

Dehydroepiandrosterone (DHEA)

Dehydroepiandrosterone, or **DHEA**, is the primary androgen secreted by the zona reticularis of the suprarenal cortex. ∞ p. 629 As noted in Chapter 18, these androgens, which are secreted in small amounts, are converted to testosterone (or estrogens) by other tissues. The significance of this small suprarenal androgen contribution in both sexes remains unclear, but DHEA is being promoted as a wonder drug for increasing vitality, strength, and muscle mass, and food supplements prepared from wild Mexican yams are now being advertised as containing "DHEA precursors." These claims are false; the compounds contained in these supplements have no effect on circulating DHEA levels. The current recommendations are that DHEA use be restricted to controlled, supervised clinical trials, and that no one under age 40 use the drug. The effects of long-term high doses of DHEA remain largely unknown; however, recall from Chapter 18 that the long-term effects of androgen abuse can be quite serious. ∞ p. 643 High levels of DHEA in women have been linked to an increased risk of ovarian cancer as well as to masculinization, due to the conversion of DHEA to testosterone. The IOC (International Olympic Committee), NCAA, and NFL have banned the use of DHEA for muscle enhancement; and it is detectable via urinalysis.

CHECKPOINT

4. Name the male reproductive structures.

5. On a warm day, would the cremaster muscle be contracted or relaxed? Why?

6. What happens when the arteries within the penis dilate?

7. What effect would low FSH levels have on sperm production?

8. Trace the pathway that a sperm travels from the site of its production to outside the body.

See the blue Answers tab at the end of the book.

CLINICAL NOTE

Prostatic Hypertrophy and Prostate Cancer

Enlargement of the prostate gland, or **benign prostatic hypertrophy**, typically occurs spontaneously in men over age 50. The increase in size occurs as testosterone production by the interstitial cells decreases. For unknown reasons, small masses called *prostatic concretions* may form within the glands. At the same time, the interstitial cells begin releasing small quantities of estrogens into the bloodstream. The combination of lower testosterone levels and the presence of estrogens probably stimulates prostatic growth. In severe cases, prostatic swelling constricts and blocks the urethra and constricts the rectum. If not corrected, the urinary obstruction can cause permanent kidney damage.[1] Partial surgical removal is the most effective treatment. In the procedure known as a **TURP** (*transurethral prostatectomy*), an instrument pushed along the urethra restores normal function by cutting away the swollen prostatic tissue. Most of the prostate gland remains in place, and no external scars result.

Prostate cancer, a malignancy of the prostate gland, is the second most common cancer and the second most common cause of cancer deaths in males. The American Cancer Society estimates that approximately 219,000 new prostate cancer cases in 2007 will result in about 27,000 deaths. Most patients are elderly. (The average age at diagnosis is 72.) For reasons that are poorly understood, prostate cancer rates for Asian-American males are relatively low compared with those of either Caucasian-Americans or African-Americans. For all age and ethnic groups, the rates of prostate cancer are rising sharply. The reason for the increase is not known. Aggressive diagnosis and treatment of localized prostate cancer in elderly patients is controversial because many of these men have nonmetastatic tumors, and even if untreated are more likely to die of some other disease.

Prostate cancer normally originates in one of the secretory glands. As the cancer progresses, it produces a nodular lump or swelling on the surface of the prostate gland. Palpation of this gland through the rectal wall—a procedure

known as a *digital rectal exam* (DRE)—is the easiest diagnostic screening procedure. *Transrectal prostatic ultrasound* (TRUS) can be used to obtain more detailed information about the status of the prostate. Blood tests are also used for screening purposes. The most sensitive is a blood test for **prostate-specific antigen (PSA)**. Elevated levels of this antigen, normally present in low concentrations, may indicate the presence of prostate cancer. Once prostrate cancer is detected, treatment decisions vary depending on how rapidly PSA levels are rising. Screening with periodic PSA tests is now being recommended for men over age 50.

If cancer is detected before it has spread to other organs, and the patient is elderly or has other serious health problems, "watchful waiting" is an option. In other cases, the usual treatment is localized radiation or surgical removal of the prostate gland. This operation, a **prostatectomy** (prosta-TEK-tō-mē), can be effective in controlling the condition, but both surgery and radiation can have undesirable side effects, including urinary incontinence and loss of sexual function. Modified treatment procedures along with medications such as Viagra can reduce the risks and maintain normal sexual function in perhaps three out of four patients.

The prognosis is much worse for prostate cancer diagnosed after metastasis has occurred, because metastasis rapidly involves the lymphoid system, lungs, bone marrow, liver, or suprarenal glands. Survival rates at this stage are relatively low. Treatments for metastasized prostate cancer include widespread irradiation, hormonal manipulation, lymph node removal, and aggressive chemotherapy. Because the cancer cells are stimulated by testosterone, treatment may involve castration or administering hormones that depress GnRH or LH production. There are three treatment options: (1) an estrogen, typically diethylstilbestrol (DES); (2) drugs that mimic GnRH, which when given in high doses produce a surge in LH production followed by a sharp decline to very low levels (presumably as the endocrine cells adapt to the excessive stimulation); and (3) drugs that block the binding of androgens to the receptors on target cells (including the drugs flutamide and bicalutamide), which prevent the stimulation of cancer cells by testosterone. The death rate for prostate cancer may be falling in some countries, perhaps from earlier detection and more effective treatment.

[1]Symptoms are improved by drug therapy with alpha-blockers that relax smooth muscle, or with *finasteride*, which inhibits production of DHT, an active derivative of testosterone.

28-3 Oogenesis occurs in the ovaries, and hormones from the pituitary gland and gonads control female reproductive functions

A woman's reproductive system produces sex hormones and functional gametes, and it must also be able to protect and support a developing embryo and nourish a newborn infant. The principal organs of the female reproductive system are the *ovaries*, the *uterine tubes*, the *uterus*, the *vagina*, and the components of the external genitalia (**Figure 28–13**). As in males, a variety of accessory glands release secretions into the female reproductive tract.

The ovaries, uterine tubes, and uterus are enclosed within an extensive mesentery known as the **broad ligament**. The uterine tubes run along the superior border of the broad ligament and open into the pelvic cavity lateral to the ovaries. The **mesovarium** (mez-ō-VĀ-rē-um), a thickened fold of mesentery, supports and stabilizes the position of each ovary (**Figure 28–14a**). The broad ligament attaches to the sides and floor of the pelvic cavity, where it becomes continuous with the parietal peritoneum. The broad liga-

ment thus subdivides this part of the peritoneal cavity. The pocket formed between the posterior wall of the uterus and the anterior surface of the colon is the **rectouterine** (rek-tō-Ū-ter-in) **pouch** (**Figure 28–13**); the pocket formed between the uterus and the posterior wall of the bladder is the **vesico-uterine** (ves-i-kō-Ū-ter-in) **pouch**. These subdivisions are most apparent in sagittal section.

Several other ligaments assist the broad ligament in supporting and stabilizing the position of the uterus and associated reproductive organs. These ligaments lie within the mesentery sheet of the broad ligament and are connected to the ovaries or uterus. The broad ligament limits side-to-side movement and rotation, and the other ligaments (described in our discussion of the ovaries and uterus) prevent superior–inferior movement.

The Ovaries

The paired ovaries are small, lumpy, almond-shaped organs near the lateral walls of the pelvic cavity (**Figure 28–14**). The ovaries perform three main functions: (1) production of immature female gametes, or oocytes; (2) secretion of female sex hormones, including estrogens and progestins; and (3) secre-

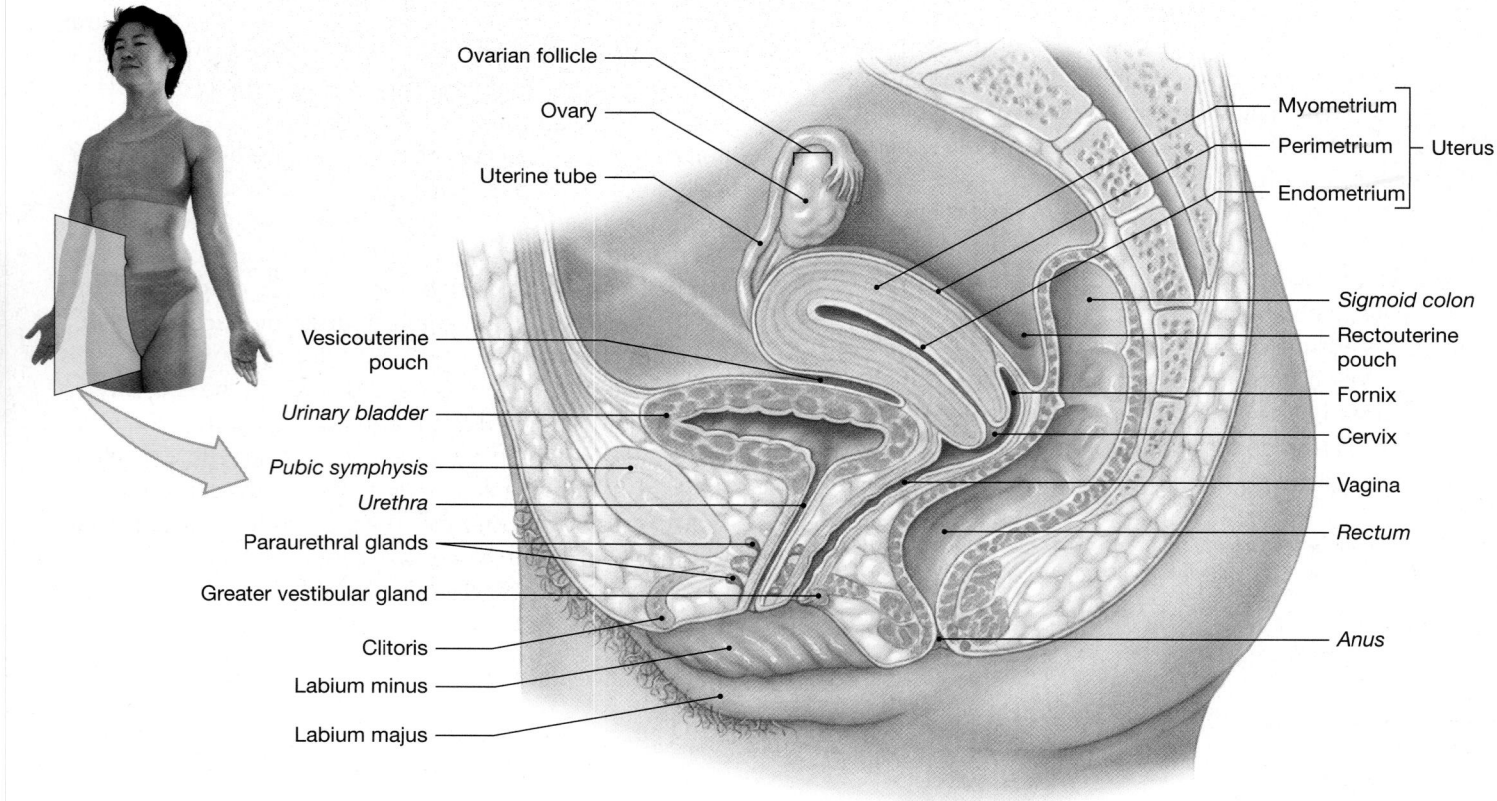

Figure 28–13 **The Female Reproductive System.** A sagittal section of the female reproductive organs. ATLAS: Plate 65

tion of inhibin, involved in the feedback control of pituitary FSH production.

The position of each ovary is stabilized by the mesovarium and by a pair of supporting ligaments: the ovarian ligament and the infundibulopelvic (suspensory) ligament (**Figure 28–14a**). The **ovarian ligament** extends from the uterus, near the attachment of the uterine tube, to the medial surface of the ovary. The **infundibulopelvic (suspensory) ligament** extends from the lateral surface of the ovary past the open end of the uterine tube to the pelvic wall. The infundibulopelvic ligament contains the major blood vessels of the ovary: the **ovarian artery** and **ovarian vein**. These vessels are connected to the ovary at the **ovarian hilum**, where the ovary attaches to the mesovarium (**Figure 28–14b**).

A typical ovary is about 5 cm long, 2.5 cm wide, and 8 mm thick (2 in. × 1 in. × 0.33 in.) and weighs 6–8 g (roughly 0.25 oz). An ovary is pink or yellowish and has a nodular con-

sistency. The visceral peritoneum, or *germinal epithelium*, covering the surface of each ovary consists of a layer of columnar epithelial cells that overlies a dense connective tissue layer called the **tunica albuginea** (**Figure 28–14b**). We can divide the interior tissues, or **stroma**, of the ovary into a superficial *cortex* and a deeper *medulla*. Gametes are produced in the cortex.

Oogenesis

Ovum production, or **oogenesis** (ō-ō-JEN-e-sis; *oon*, egg), begins before a woman's birth, accelerates at puberty, and ends at *menopause*. Between puberty and menopause, oogenesis occurs on a monthly basis as part of the *ovarian cycle*.

Oogenesis is summarized in **Figure 28–15**. Unlike spermatogonia, the **oogonia** (ō-ō-GŌ-nē-uh), or stem cells of females, complete their mitotic divisions before birth. Between the third and seventh months of fetal development, the

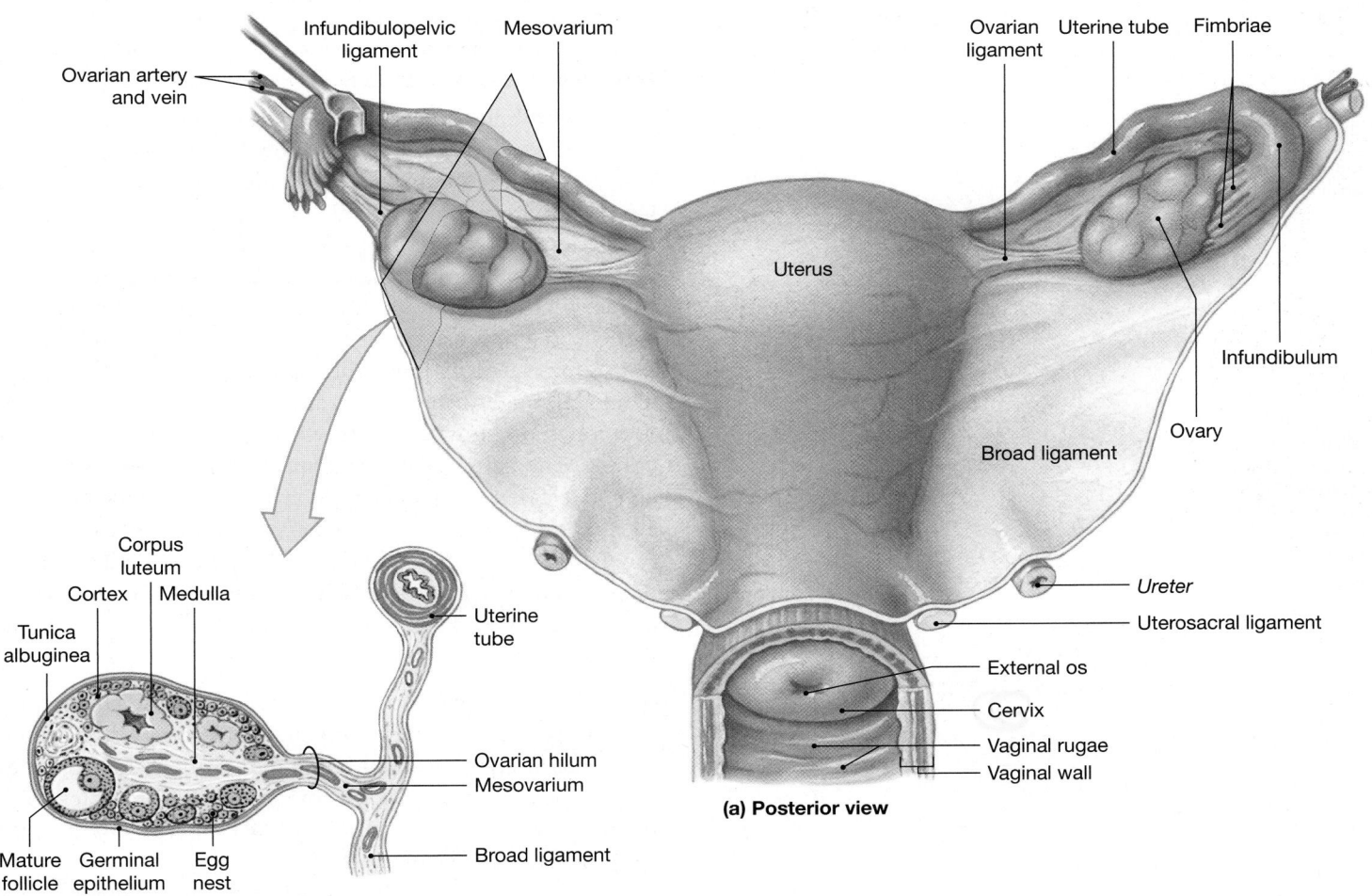

(a) **Posterior view**

(b) **Ovary and mesenteries, sectional view**

Figure 28–14 **The Ovaries and Their Relationships to the Uterine Tube and Uterus.** (a) A posterior view of the uterus, uterine tubes, and ovaries. (b) A sectional view of the ovary, uterine tube, and associated mesenteries. ATLAS: Plate 67

28 REPRODUCTIVE

daughter cells, or **primary oocytes** (Ō-ō-sīts), prepare to undergo meiosis. They proceed as far as the prophase of meiosis I, but then the process comes to a halt. The primary oocytes remain in a state of suspended development until the individual reaches puberty, when rising levels of FSH trigger the start of the ovarian cycle. Each month thereafter, some of the primary oocytes are stimulated to undergo further development. Not all primary oocytes produced during development survive until puberty. The ovaries have roughly 2 million *primordial follicles* at birth, each containing a primary oocyte. Primordial follicles exist at the earliest stages of development. By the time of puberty, the number has dropped to about 400,000. The rest of the primordial follicles degenerate in a process called **atresia** (a-TRĒ-zē-uh).

Although the nuclear events in the ovaries during meiosis are the same as those in the testes, the process differs in two important details:

1. The cytoplasm of the primary oocyte is unevenly distributed during the two meiotic divisions. Oogenesis produces one functional ovum, which contains most of the original cytoplasm, and two or three **polar bodies**, nonfunctional cells that later disintegrate (**Figure 28–15**).

2. The ovary releases a **secondary oocyte** rather than a mature ovum. The secondary oocyte is suspended in metaphase of meiosis II; meiosis will not be completed unless and until fertilization occurs.

The Ovarian Cycle

Ovarian follicles are specialized structures in the cortex of the ovaries where both oocyte growth and meiosis I occur. Primary oocytes are located in the outer portion of the ovarian cortex, near the tunica albuginea, in clusters called *egg nests* (**Figure 28–16**). Each primary oocyte within an egg nest is surrounded by a single squamous layer of *follicle cells*. The primary oocyte and its follicle cells form a **primordial follicle**. Beginning at puberty, primordial follicles are continuously activated to join other follicles already in development. Although the activating mechanism is unknown, local hormones or growth factors within the ovary may be involved. The activated primordial follicle will either mature and be released as a secondary oocyte or degenerate (atresia). This monthly process is known as the **ovarian cycle**.

The ovarian cycle can be divided into a **follicular phase**, or *preovulatory phase*, and a **luteal phase**, or *postovulatory phase*. Important steps in the ovarian cycle can be summarized as follows (**Figure 28–16**):

Step 1 *The Formation of Primary Follicles.* The preliminary steps in follicle development are of variable length but may take almost a year to complete. Follicle development begins with the activation of primordial follicles into **primary follicles**. The follicular cells enlarge, divide, and form several layers of cells around the growing primary oocyte. These follicle cells, which become rounded in appearance, are now called **granulosa cells**. Microvilli from the surrounding follicular cells intermingle with microvilli

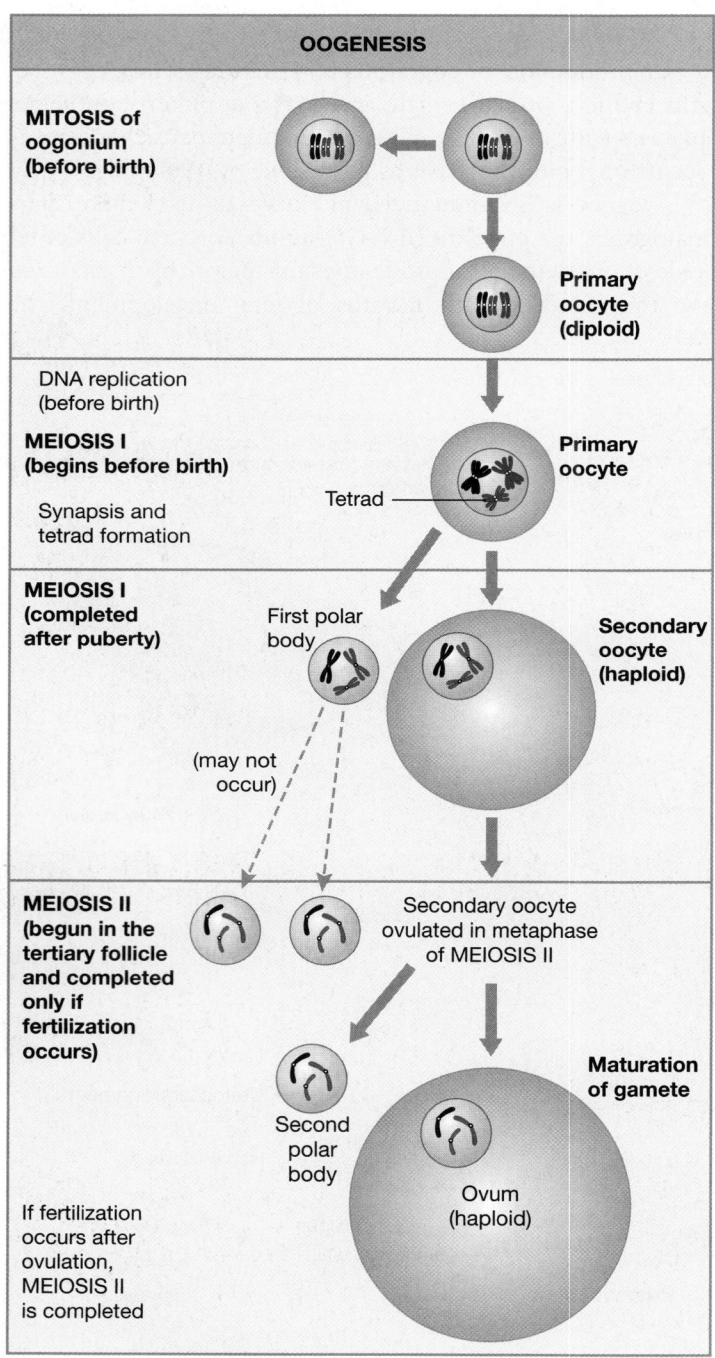

Figure 28–15 **Oogenesis.** In oogenesis, a single primary oocyte produces an ovum and two or three nonfunctional polar bodies. Compare this schematic diagram with *Figure 28–7.*

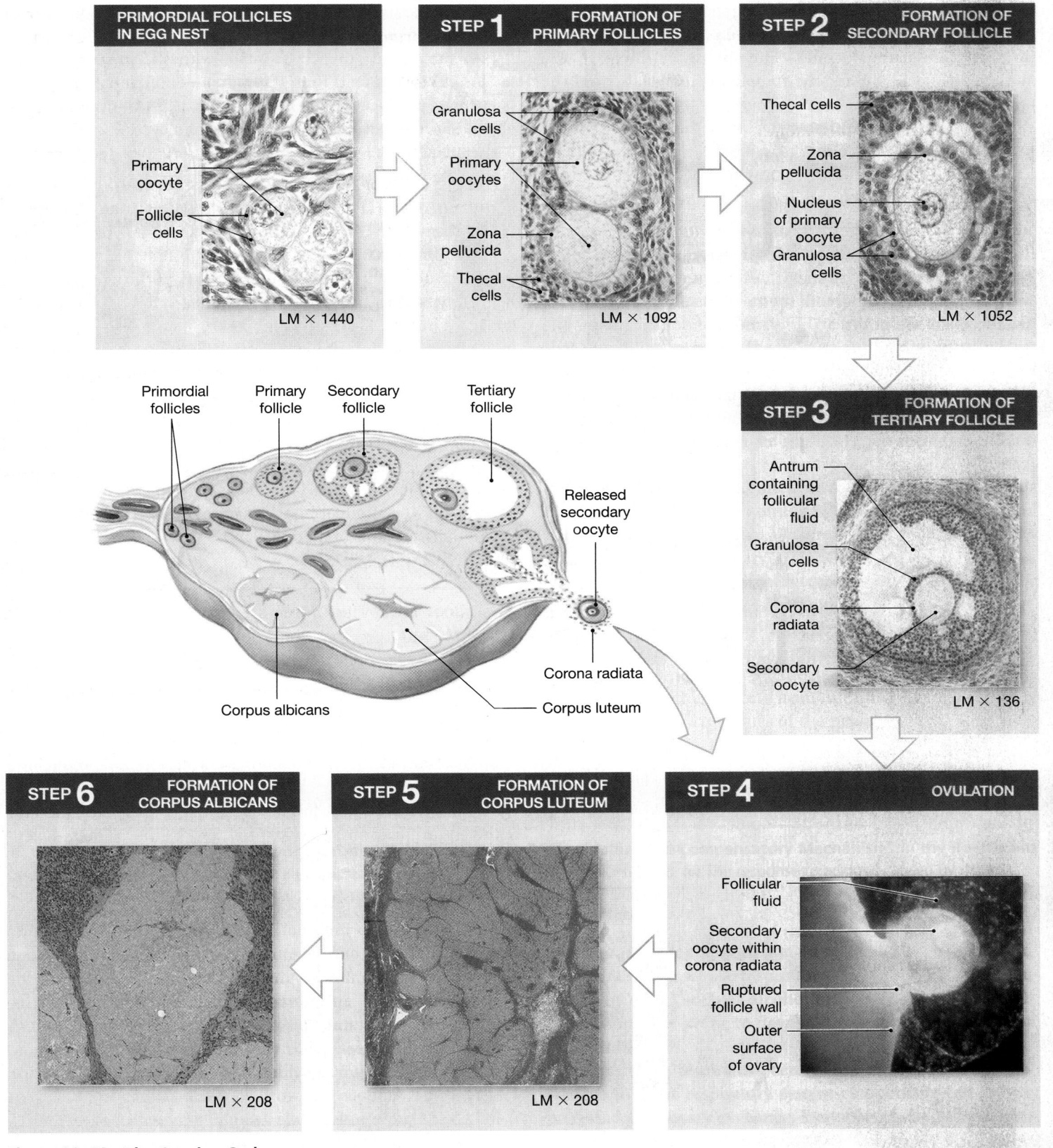

Figure 28–16 **The Ovarian Cycle.**

originating at the surface of the oocyte. This glycoprotein-rich region is called the **zona pellucida** (ZŌ-na pe-LOO-sid-uh; *pellucidus,* translucent). The microvilli increase the surface area available for the transfer of materials from the follicular cells to the growing oocyte. As the granulosa cells enlarge and multiply, adjacent cells in the ovarian stroma form a layer of **thecal cells** (*theca,* a box) around the follicle. Thecal cells and granulosa cells work together to produce sex hormones called estrogens.

Step 2 *The Formation of Secondary Follicles.* Although many primordial follicles develop into primary follicles, only a few of the primary follicles mature further. This process is apparently under the control of a growth factor produced by the oocyte. The transformation begins as the wall of the follicle thickens and the deeper follicular cells begin secreting small amounts of fluid. This **follicular fluid**, or *liquor folliculi,* accumulates in small pockets that gradually expand and separate the inner and outer layers of the follicle. At this stage, the complex is known as a **secondary follicle**. Although the primary oocyte continues to grow slowly, the follicle as a whole enlarges rapidly because follicular fluid accumulates.

Step 3 *The Formation of a Tertiary Follicle.* Eight to 10 days after the start of the ovarian cycle, the ovaries generally contain only a single secondary follicle destined for further development. By the 10th to the 14th day of the cycle, that follicle has become a **tertiary follicle**, or *mature graafian* (GRAF-ē-an) *follicle,* roughly 15 mm in diameter. This complex spans the entire width of the ovarian cortex and distorts the ovarian capsule, creating a prominent bulge in the surface of the ovary. The oocyte projects into the **antrum** (AN-trum), or expanded central chamber of the follicle. The antrum is surrounded by a mass of granulosa cells.

Until this time, the primary oocyte has been suspended in prophase of meiosis I. As the development of the tertiary follicle ends, LH levels begin rising, prompting the primary oocyte to complete meiosis I. Instead of producing two secondary oocytes, the first meiotic division yields a secondary oocyte and a small, nonfunctional polar body. The secondary oocyte then enters meiosis II, but stops once again upon reaching metaphase. Meiosis II will not be completed unless fertilization occurs.

Generally, on day 14 of a 28-day cycle, the secondary oocyte and the attached granulosa cells lose their connections with the follicular wall and drift free within the antrum. The granulosa cells still associated with the secondary oocyte form a protective layer known as the **corona radiata** (kō-RŌ-nuh rā-dē-AH-tuh).

Step 4 *Ovulation.* At **ovulation**, the tertiary follicle releases the secondary oocyte. The distended follicular wall suddenly ruptures, ejecting the follicular contents, including the secondary oocyte and corona radiata, into the pelvic cavity. The sticky follicular fluid keeps the corona radiata (and the oocyte) attached to the surface of the ovary. The oocyte is then moved into the uterine tube by contact with the fimbriae (**Figure 28–14a**) that extend from its funnel-like opening, or by fluid currents produced by the cilia that line it. Ovulation marks the end of the follicular phase of the ovarian cycle and the start of the luteal phase.

Step 5 *The Formation and Degeneration of the Corpus Luteum.* The empty tertiary follicle initially collapses, and ruptured vessels bleed into the antrum. The remaining granulosa cells then invade the area, proliferating to create an endocrine structure known as the **corpus luteum** (LOO-tē-um), named for its yellow color (*lutea,* yellow). This process occurs under LH stimulation.

The cholesterol contained in the corpus luteum is used to manufacture steroid hormones known as **progestins** (prō-JES-tinz), principally the steroid **progesterone** (prō-JES-ter-ōn). Although the corpus luteum also secretes moderate amounts of estrogens, levels are not as high as they were at ovulation, and progesterone is the principal hormone in the luteal phase. Progesterone's primary function is to prepare the uterus for pregnancy by stimulating the maturation of the uterine lining and the secretions of uterine glands.

Tips&Tricks:

Progesterone literally means a steroid (**-one**) that favors (**pro-**) gestation (**-gest**).

Step 6 *Degeneration of the Corpus Luteum Beginning Roughly 12 Days after Ovulation (Unless Fertilization Occurs).* Progesterone and estrogen levels then fall markedly. Fibroblasts invade the nonfunctional corpus luteum, producing a knot of pale scar tissue called a **corpus albicans** (AL-bi-kanz). The disintegration, or *involution,* of the corpus luteum marks the end of the ovarian cycle. A new ovarian cycle then begins with the activation of another group of primordial follicles.

Age and Oogenesis

Although many primordial follicles may have developed into primary follicles, and several primary follicles may have been converted to secondary follicles, generally only a single oocyte is released into the pelvic cavity at ovulation. The rest undergo atresia. At puberty, each ovary contains about 200,000 primordial follicles. Forty years later, few if any follicles remain, although only about 500 will have been ovulated in the interim.

Tips&Tricks

The maturation of an ovum takes several cycles to complete; that is why at any given time, oocytes are in various stages of development within the ovary.

The A&P Top 100

#97 Oogenesis begins during embryonic development, and primary oocyte production is completed before birth. Each month after puberty, the ovarian cycle produces one or more secondary oocytes from the pre-existing population of primary oocytes. The number of viable and responsive primary oocytes declines markedly over time, until ovarian cycles typically end around age 45–55.

CLINICAL NOTE

Ovarian Cancer A woman in the United States has a 1-in-70 chance of developing **ovarian cancer** in her lifetime. In 2007, an estimated 22,000 ovarian cancer cases will be diagnosed, and about 15,000 deaths will occur as a result of this condition. Although ovarian cancer is the third most common reproductive cancer among women, it is the most dangerous because it is seldom diagnosed in its early stages. The prognosis is relatively good for cancers that originate in the general ovarian tissues or from abnormal oocytes. These cancers respond well to some combination of chemotherapy, radiation, and surgery. However, 85 percent of ovarian cancers develop from epithelial cells, and sustained remission can be obtained in only about one-third of the cases of this type.

The Uterine Tubes

Each **uterine tube** (*Fallopian tube* or *oviduct*) is a hollow, muscular tube measuring roughly 13 cm (5.2 in.) in length (**Figures 28–13** and **28–14**). Each uterine tube is divided into three segments (**Figure 28–17a**):

1. *The Infundibulum.* The end closest to the ovary forms an expanded funnel, or **infundibulum**, with numerous fingerlike projections that extend into the pelvic cavity. The projections are called **fimbriae** (FIM-brē-ē). Fimbriae drape over the surface of the ovary, but there is no physical connection between the two structures. The inner surfaces of the infundibulum are lined with cilia that beat toward the middle segment of the uterine tube, called the *ampulla.*

2. *The Ampulla.* The thickness of the smooth muscle layers in the wall of the **ampulla**, the middle segment of the uterine tube, gradually increases as the tube approaches the uterus.

3. *The Isthmus.* The ampulla leads to the **isthmus** (IS-mus), a short segment connected to the uterine wall.

Histology of the Uterine Tube

The epithelium lining the uterine tube is composed of ciliated columnar epithelial cells with scattered mucin-secreting cells (**Figure 28–17c**). Concentric layers of smooth muscle surround the mucosa (**Figure 28–17b**). Oocyte transport involves a combination of ciliary movement and peristaltic contractions in the walls of the uterine tube. A few hours before ovulation, sympathetic and parasympathetic nerves from the hypogastric plexus "turn on" this beating pattern and initiate peristalsis. It normally takes three to four days for an oocyte to travel from the infundibulum to the *uterine cavity*. If fertilization is to occur, the secondary oocyte must encounter spermatozoa during the first 12–24 hours of its passage. Fertilization typically occurs near the boundary between the ampulla and isthmus of the uterine tube.

In addition to its transport function, the uterine tube provides a nutrient-rich environment that contains lipids and glycogen. This mixture supplies nutrients to both spermatozoa and a developing *pre-embryo* (the cell cluster produced by the initial mitotic divisions following fertilization). Unfertilized oocytes degenerate in the terminal portions of the uterine tubes or within the uterus without completing meiosis.

In addition to ciliated cells, the epithelium lining the uterine tubes contains *peg cells* and scattered mucin-secreting cells. The peg cells project into the lumen of the uterine tube, and they secrete a fluid that both completes the capacitation of spermatozoa and supplies nutrients to spermatozoa and the developing pre-embryo.

The Uterus

The **uterus** provides mechanical protection, nutritional support, and waste removal for the developing *embryo* (weeks 1–8) and *fetus* (week 9 through delivery). In addition, contractions in the muscular wall of the uterus are important in ejecting the fetus at birth.

The uterus is a small, pear-shaped organ (**Figure 28–18a**) about 7.5 cm (3 in.) long and with a maximum diameter of 5 cm (2 in.). It weighs 30–40 g (1–1.4 oz). In its normal position, the uterus bends anteriorly near its base (**Figure 28–13**), a condition known as *anteflexion* (an-tē-FLEK-shun). In this position, the uterus covers the superior and posterior surfaces of the urinary bladder. If the uterus bends backward toward the sacrum, the condition is termed *retroflexion* (re-trō-FLEK-shun). Retroflexion, which occurs in about 20 percent of adult women, has no clinical significance. (A retroflexed uterus generally becomes anteflexed spontaneously during the third month of pregnancy.)

Suspensory Ligaments of the Uterus

In addition to the broad ligament, three pairs of suspensory ligaments stabilize the position of the uterus and limit its range of movement (**Figure 28–18b**). The **uterosacral** (ū-te-rō-SĀ-krul) **ligaments** extend from the lateral surfaces of the uterus to the anterior face of the sacrum, keeping the body of

28 REPRODUCTIVE

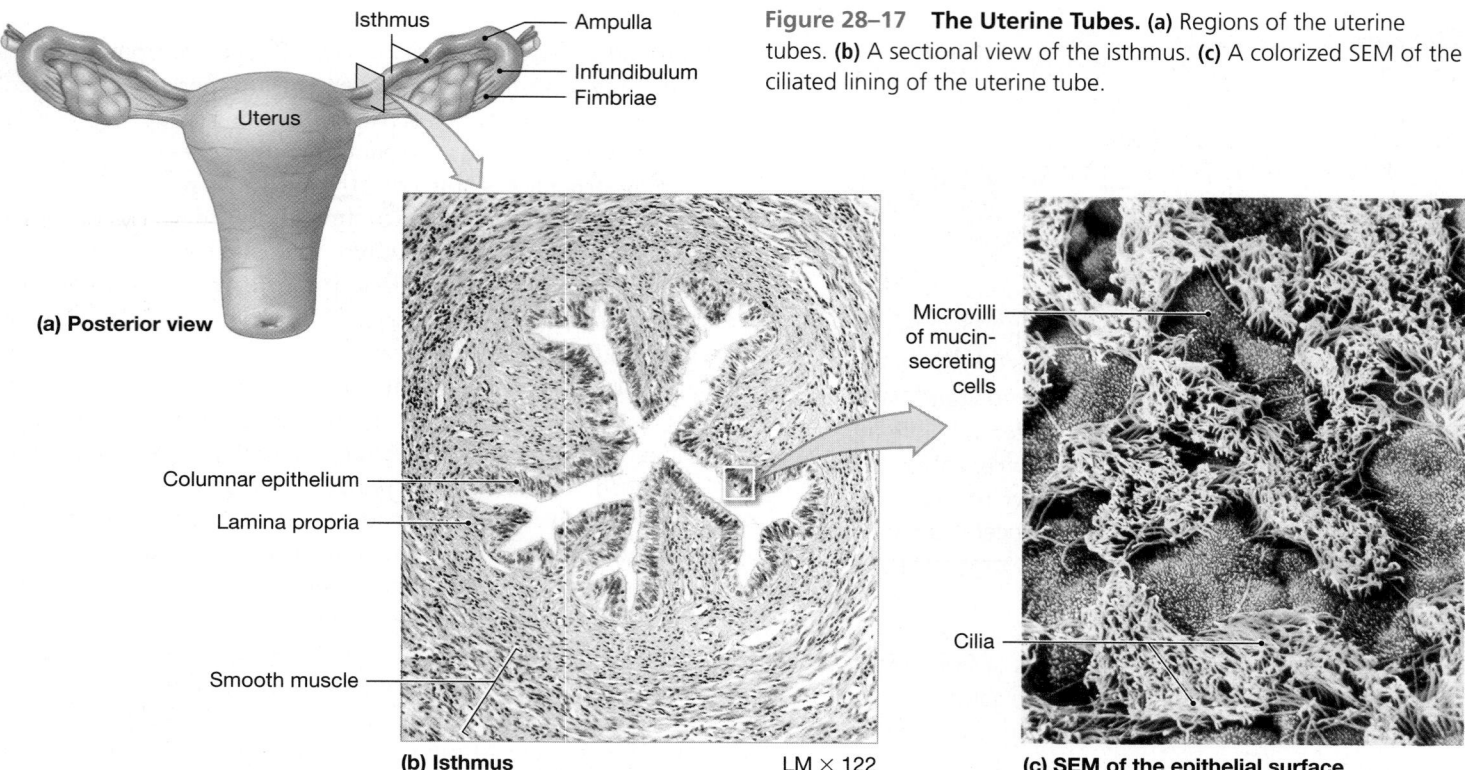

Figure 28–17 **The Uterine Tubes. (a)** Regions of the uterine tubes. **(b)** A sectional view of the isthmus. **(c)** A colorized SEM of the ciliated lining of the uterine tube.

Isthmus — Ampulla
Infundibulum
Fimbriae
Uterus

(a) Posterior view

Columnar epithelium
Lamina propria
Smooth muscle

(b) Isthmus   LM × 122

Microvilli of mucin-secreting cells

Cilia

(c) SEM of the epithelial surface

the uterus from moving inferiorly and anteriorly. The **round ligaments** arise on the lateral margins of the uterus just posterior and inferior to the attachments of the uterine tubes. These ligaments extend through the inguinal canal and end in the connective tissues of the external genitalia. The round ligaments restrict primarily posterior movement of the uterus. The **cardinal** (*lateral*) **ligaments** extend from the base of the uterus and vagina to the lateral walls of the pelvis. These ligaments tend to prevent inferior movement of the uterus. The muscles and fascia of the pelvic floor provide additional mechanical support.

Internal Anatomy of the Uterus

We can divide the uterus into two anatomical regions (**Figure 28–18a**): the body and the cervix. The uterine **body**, or *corpus*, is the largest portion of the uterus. The **fundus** is the rounded portion of the body superior to the attachment of the uterine tubes. The body ends at a constriction known as the **isthmus** of the uterus. The **cervix** (SER-viks) is the inferior portion of the uterus that extends from the isthmus to the vagina.

The tubular cervix projects about 1.25 cm (0.5 in.) into the vagina. Within the vagina, the distal end of the cervix forms a curving surface that surrounds the **external os** (*os*, an opening or mouth), or *external orifice* of the uterus. The external os leads into the **cervical canal**, a constricted passageway that opens into the **uterine cavity** of the body at the **internal os**, or *internal orifice*.

The uterus receives blood from branches of the **uterine arteries**, which arise from branches of the *internal iliac arteries*, and from the *ovarian arteries*, which arise from the abdominal aorta inferior to the renal arteries. The arteries to the uterus are extensively interconnected, ensuring a reliable flow of blood to the organ despite changes in its position and shape during pregnancy. Numerous veins and lymphatic vessels also drain each portion of the uterus. The organ is innervated by autonomic fibers from the hypogastric plexus (sympathetic) and from sacral segments S_3 and S_4 (parasympathetic). Sensory information reaches the central nervous system within the dorsal roots of spinal nerves T_{11} and T_{12}. The most delicate anesthetic procedures used during labor and delivery, known as *segmental blocks*, target only spinal nerves $T_{10} - L_1$.

The Uterine Wall

The dimensions of the uterus are highly variable. In women of reproductive age who have not given birth, the uterine wall is about 1.5 cm (0.6 in.) thick. The wall has a thick, outer, muscular **myometrium** (mī-ō-MĒ-trē-um; *myo-*, muscle + *metra*, uterus) and a thin, inner, glandular **endometrium** (en-dō-MĒ-trē-um), or *mucosa* (**Figure 28–19**). The fundus and the posterior surface of the uterine body and isthmus are covered by a serous membrane that is continuous with the peritoneal lining. This incomplete serosa is called the **perimetrium**.

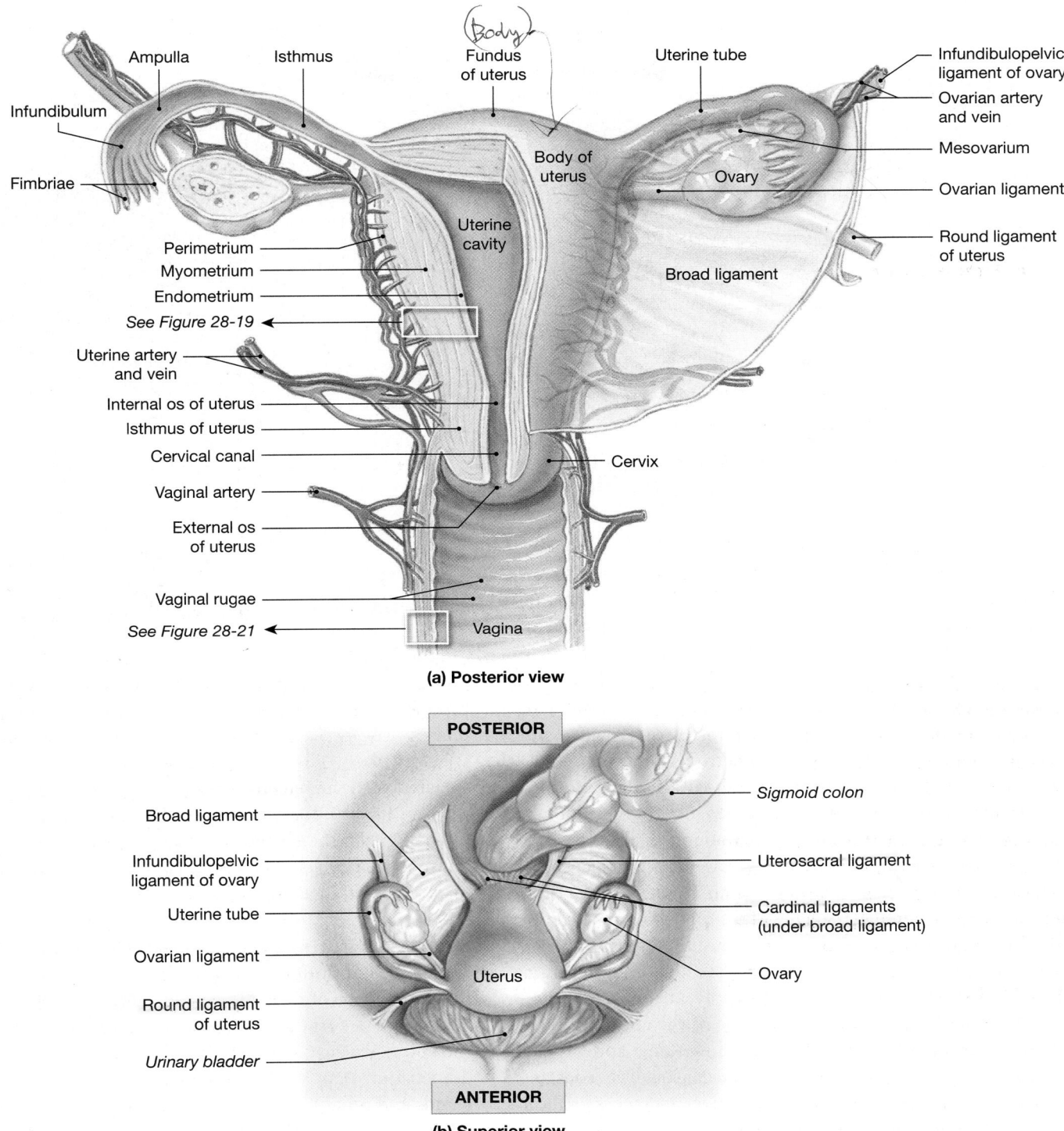

(a) **Posterior view**

(b) **Superior view**

Figure 28–18   The Uterus. (a) A posterior view with the left portion of the uterus, left uterine tube, and left ovary shown in section. **(b)** The ligaments that stabilize the position of the uterus in the pelvic cavity. ATLAS: Plates 66; 67

The endometrium contributes about 10 percent to the mass of the uterus. The glandular and vascular tissues of the endometrium support the physiological demands of the growing fetus. Vast numbers of uterine glands open onto the endometrial surface and extend deep into the lamina propria, almost to the myometrium. Under the influence of estrogen, the uterine glands, blood vessels, and epithelium change with the phases of the monthly *uterine cycle*.

The myometrium, the thickest portion of the uterine wall, constitutes almost 90 percent of the mass of the uterus.

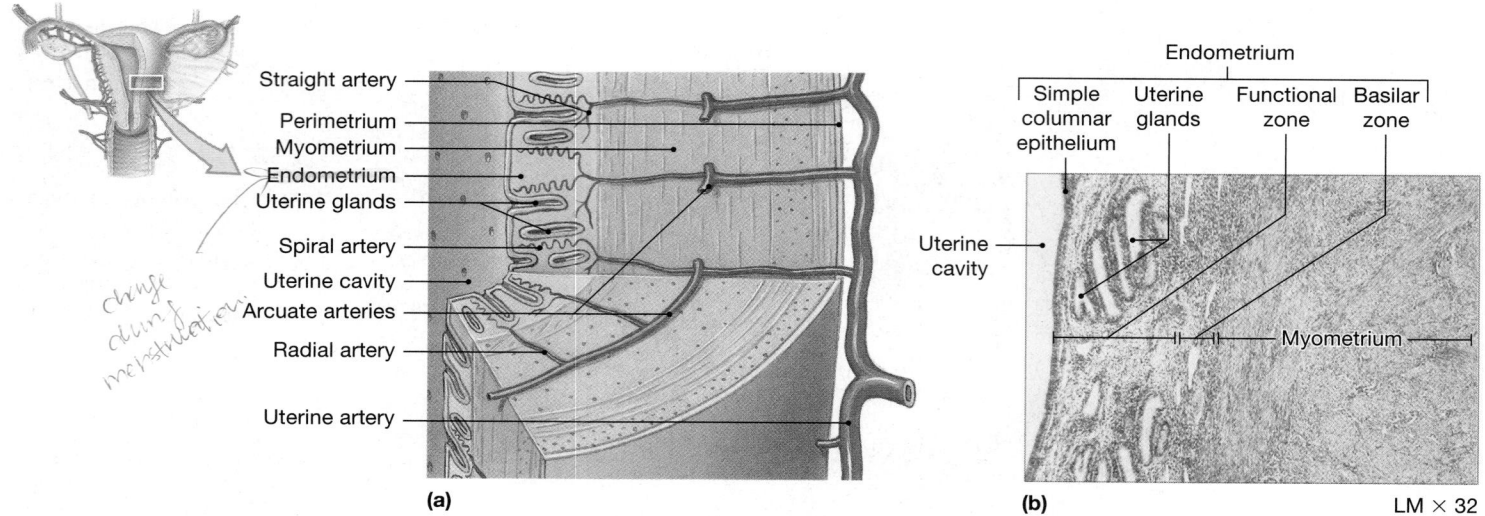

Figure 28–19 **The Uterine Wall. (a)** A diagrammatic sectional view of the uterine wall, showing the endometrial regions and the circulatory supply to the endometrium. **(b)** The basic histological structure of the endometrium.

Smooth muscle in the myometrium is arranged into longitudinal, circular, and oblique layers. The smooth muscle tissue of the myometrium provides much of the force needed to move a fetus out of the uterus and into the vagina.

Histology of the Uterus. We can divide the endometrium into a **functional zone**—the layer closest to the uterine cavity—and a **basilar zone**, adjacent to the myometrium (**Figure 28–19b**). The functional zone contains most of the **uterine glands**, also called *endometrial glands*, and contributes most of the endometrial thickness. It is this zone that undergoes the dramatic changes in thickness and structure during the menstrual cycle. The basilar zone attaches the endometrium to the myometrium and contains the terminal branches of the tubular uterine glands.

Within the myometrium, branches of the uterine arteries form **arcuate arteries**, which encircle the endometrium (**Figure 28–19a**). **Radial arteries** supply **straight arteries**, which deliver blood to the basilar zone of the endometrium, and **spiral arteries**, which supply the functional zone.

The structure of the basilar zone remains relatively constant over time, but that of the functional zone undergoes cyclical changes in response to sex hormone levels. These cyclical changes produce the characteristic histological features of the uterine cycle.

Cervical cancer is the most common cancer of the reproductive system in women ages 15–34. Each year roughly 13,000 U.S. women are diagnosed with invasive cervical cancer, and approximately one-third of them eventually die from the condition. Another 35,000 women are diagnosed with a less aggressive form of cervical cancer. *Gardisil* is a new vaccine against two types of human papillomavirus (HPV) that cause most cervical cancers.

The Uterine Cycle

The **uterine cycle**, or *menstrual* (MEN-stroo-ul) *cycle*, is a repeating series of changes in the structure of the endometrium (**Figure 28–20**). The uterine cycle averages 28 days in length, but it can range from 21 to 35 days in healthy women of reproductive age. We can divide the uterine cycle into three phases: (1) *menses*, (2) the *proliferative phase*, and (3) the *secretory phase*. The phases occur in response to hormones associated with the regulation of the ovarian cycle. Menses and the proliferative phase occur during the follicular phase of the ovarian cycle; the secretory phase corresponds to the luteal phase of the cycle. We will consider the regulatory mechanism involved in a later section.

Menses. The uterine cycle begins with the onset of **menses** (MEN-sēz), an interval marked by the degeneration of the functional zone of the endometrium (**Figure 28–20a**). This degeneration occurs in patches and is caused by constriction of the spiral arteries, which reduces blood flow to areas of endometrium. Deprived of oxygen and nutrients, the secretory glands and other tissues in the functional zone begin to deteriorate. Eventually, the weakened arterial walls rupture, and blood pours into the connective tissues of the functional zone. Blood cells and degenerating tissues then break away and enter the uterine lumen, to be lost by passage through the external os and into the vagina. Only the functional zone is affected, because the deeper, basilar layer is provided with blood from the straight arteries, which remain unconstricted.

The sloughing off (shedding) of tissue is gradual, and at each site repairs begin almost at once. Nevertheless, before menses has ended, the entire functional zone has been lost.

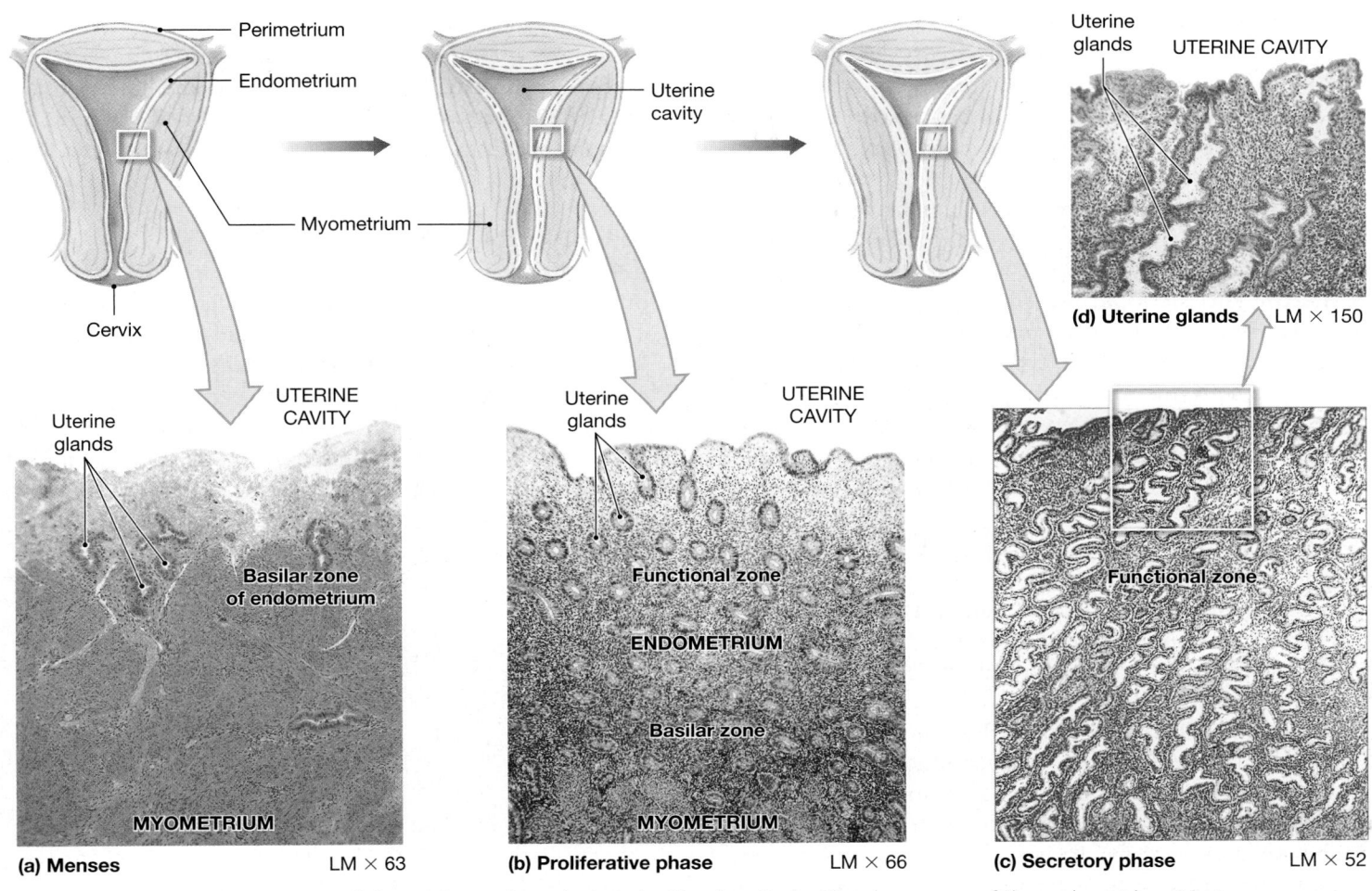

Figure 28–20 **The Appearance of the Endometrium during the Uterine Cycle.** The appearance of the endometrium **(a)** at menses, **(b)** during the proliferative phase, and **(c)** during the secretory phase of the uterine cycle. **(d)** Detail of uterine (endometrial) glands.

The process of endometrial sloughing, called **menstruation** (men-stroo-Ā-shun), generally lasts from one to seven days. Over this period roughly 35 to 50 mL of blood are lost. The process can be relatively painless. Painful menstruation, or **dysmenorrhea**, can result from uterine inflammation, myometrial contractions ("cramps"), or from conditions involving adjacent pelvic structures.

The Proliferative Phase. The basilar zone, including the basal parts of the uterine glands, survives menses intact. In the days after menses, the epithelial cells of the uterine glands multiply and spread across the endometrial surface, restoring the integrity of the uterine epithelium (**Figure 28–20b**). Further growth and vascularization result in the complete restoration of the functional zone. As this reorganization proceeds, the endometrium is in the **proliferative phase**. Restoration occurs at the same time as the enlargement of primary and secondary follicles in the ovary. The proliferative phase is stimulated and sustained by estrogens secreted by the developing ovarian follicles.

By the time ovulation occurs, the functional zone is several millimeters thick, and prominent mucous glands extend to the border with the basilar zone. At this time, the endometrial glands are manufacturing a mucus rich in glycogen. This specialized mucus appears to be essential for the survival of the fertilized egg through its earliest developmental stages. (These stages will be considered in Chapter 29.) The entire functional zone is highly vascularized, with small arteries spiraling toward the endometrial surface from larger arteries in the myometrium.

The Secretory Phase. During the **secretory phase** of the uterine cycle, the endometrial glands enlarge, accelerating their rates of secretion, and the arteries that supply the uterine wall elongate and spiral through the tissues of the functional zone (**Figure 28–20c**). This activity occurs under the combined stimulatory effects of progestins and estrogens from the corpus luteum. The secretory phase begins at the time of ovulation and persists as long as the corpus luteum remains intact.

Secretory activities peak about 12 days after ovulation. Over the next day or two, glandular activity declines, and the uterine cycle ends as the corpus luteum stops producing stimulatory hormones. A new cycle then begins with the onset of menses and the disintegration of the functional zone. The secretory phase generally lasts 14 days. As a result, you can identify the date of ovulation by counting backward 14 days from the first day of menses.

Menarche and Menopause. The uterine cycle begins at puberty. The first cycle, known as **menarche** (me-NAR-kē; *men*, month + *arche*, beginning), typically occurs at age 11–12. The cycles continue until **menopause** (MEN-ō-pawz), the termination of the uterine cycle, at age 45–55. Over the interim, the regular appearance of uterine cycles is interrupted only by circumstances such as illness, stress, starvation, or pregnancy.

If menarche does not appear by age 16, or if the normal uterine cycle of an adult woman becomes interrupted for six months or more, the condition of **amenorrhea** (ā-men-ō-RĒ-uh) exists. *Primary amenorrhea* is the failure to initiate menses. This condition may indicate developmental abnormalities, such as nonfunctional ovaries, the absence of a uterus, or an endocrine or genetic disorder. It can also result from malnutrition: Puberty is delayed if leptin levels are too low. ∞ p. 638 Transient *secondary amenorrhea* can be caused by severe physical or emotional stresses. In effect, the reproductive system gets "switched off." Factors that cause either type of amenorrhea include drastic weight loss, anorexia nervosa, and severe depression or grief. Amenorrhea has also been observed in marathon runners and other women engaged in training programs that require sustained high levels of exertion, which severely reduce body lipid reserves.

The Vagina

The **vagina** is an elastic, muscular tube extending between the cervix and the *vestibule*, a space bounded by the female external genitalia (**Figure 28–13**). The vagina is typically 7.5–9 cm (3–3.6 in.) long, but its diameter varies because it is highly distensible.

At the proximal end of the vagina, the cervix projects into the **vaginal canal**. The shallow recess surrounding the cervical protrusion is known as the **fornix** (FOR-niks). The vagina lies parallel to the rectum, and the two are in close contact posteriorly. Anteriorly, the urethra extends along the superior wall of the vagina from the urinary bladder to the external urethral orifice, which opens into the vestibule. The primary blood supply of the vagina is via the **vaginal branches** of the internal iliac (or uterine) arteries and veins. Innervation is from the hypogastric plexus, sacral nerves S_2–S_4 and branches of the pudendal nerve. ∞ pp. 444, 757, 765

The vagina has three major functions: It (1) serves as a passageway for the elimination of menstrual fluids, (2) receives the penis during sexual intercourse, and holds spermatozoa prior to their passage into the uterus, and (3) forms the inferior portion of the *birth canal*, through which the fetus passes during delivery.

Anatomy and Histology of the Vagina

In sectional view, the lumen of the vagina appears constricted, forming a rough H-shape. The vaginal walls contain a network of blood vessels and layers of smooth muscle (**Figure 28–21**). The lining is moistened by secretions of the cervical glands and by the movement of water across the permeable epithelium. Throughout childhood the vagina and vestibule are usually separated by the **hymen** (HĪ-men), an elastic epithelial fold of variable size that partially blocks the entrance to the vagina. An intact hymen is typically stretched or torn during first sexual intercourse. There are normal variations in size and strength of

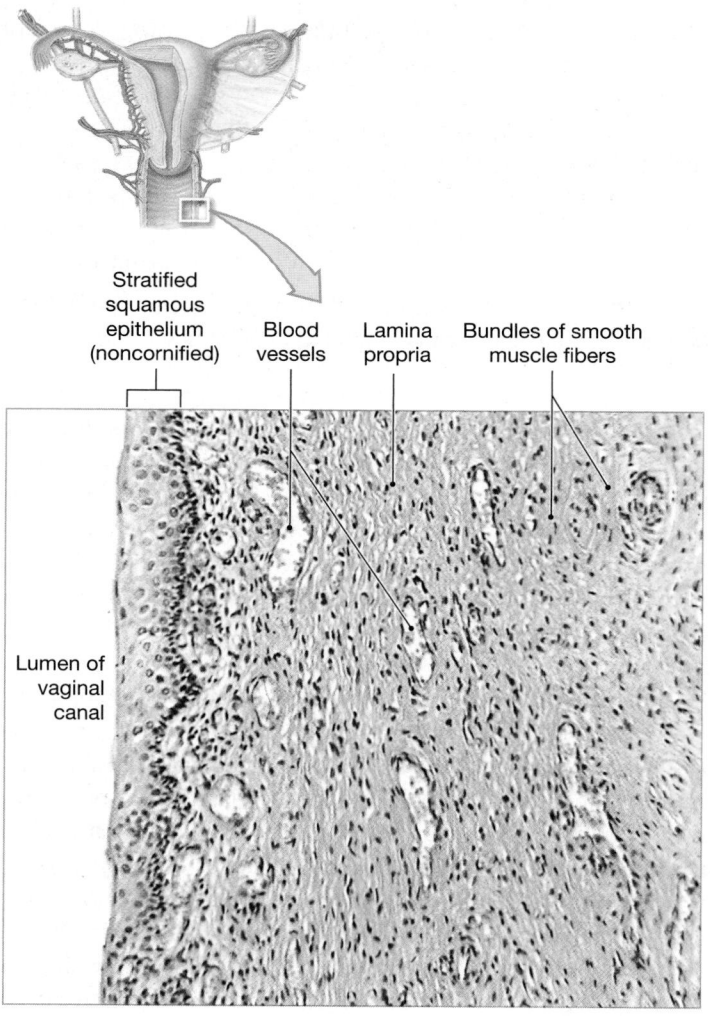

Stratified squamous epithelium (noncornified) · Blood vessels · Lamina propria · Bundles of smooth muscle fibers

Lumen of vaginal canal

LM × 27

Figure 28–21   The Histology of the Vagina.

the hymen. The two *bulbospongiosus muscles* extend along either side of the vaginal entrance, which is constricted by their contractions. ∞ p. 357 These muscles cover the *vestibular bulbs*, masses of erectile tissue on either side of the vaginal entrance. The vestibular bulbs have the same embryological origins as the corpus spongiosum of the penis in males.

The vaginal lumen is lined by a nonkeratinized stratified squamous epithelium (**Figure 28–21**). In the relaxed state, this epithelium forms folds called *rugae*. The underlying lamina propria is thick and elastic, and it contains small blood vessels, nerves, and lymph nodes. The vaginal mucosa is surrounded by an elastic *muscularis* layer consisting of layers of smooth muscle fibers arranged in circular and longitudinal bundles continuous with the uterine myometrium. The portion of the vagina adjacent to the uterus has a serosal covering that is continuous with the pelvic peritoneum. Along the rest of the vagina, the muscularis layer is surrounded by an *adventitia* of fibrous connective tissue.

The vagina contains a population of resident bacteria, usually harmless, supported by nutrients in the cervical mucus. The metabolic activity of these bacteria creates an acidic environment, which restricts the growth of many pathogens. **Vaginitis** (vaj-i-NĪ-tis), an inflammation of the vaginal canal, is caused by fungi, bacteria, or parasites. In addition to any discomfort that may result, the condition may affect the survival of spermatozoa and thereby reduce fertility. An acidic environment also inhibits the motility of sperm; for this reason, the buffers in semen are important to successful fertilization.

The hormonal changes associated with the ovarian cycle also affect the vaginal epithelium. By examining a *vaginal smear*—a sample of epithelial cells shed at the surface of the vagina—a clinician can estimate the corresponding stage in the ovarian and uterine cycles. This diagnostic procedure is an example of *exfoliative cytology*. ∞ p. 119

The External Genitalia

The area containing the female external genitalia is the **vulva** (VUL-vuh), or *pudendum* (pū-DEN-dum; **Figure 28–22**). The vagina opens into the **vestibule**, a central space bounded by small folds known as the **labia minora** (LĀ-bē-uh mi-NOR-uh; singular, *labium minus*). The labia minora are covered with a smooth, hairless skin. The urethra opens into the vestibule just anterior to the vaginal entrance. The **paraurethral glands**, or *Skene glands*, discharge into the urethra near the external urethral opening. Anterior to this opening, the **clitoris** (KLIT-ō-ris) projects into the vestibule. A small, rounded tissue projection, the clitoris is derived from the same embryonic structures as the penis in males. Internally, it contains erectile tissue comparable to the corpora cavernosa of the penis. The clitoris engorges with blood during sexual arousal. A small erectile *glans* sits atop it; exten-

sions of the labia minora encircle the body of the clitoris, forming its **prepuce**, or *hood*. ATLAS: Embryology Summary 21: The Development of the Reproductive System

A variable number of small **lesser vestibular glands** discharge their secretions onto the exposed surface of the vestibule, keeping it moist. During sexual arousal, a pair of ducts discharges the secretions of the **greater vestibular glands** (*Bartholin glands*) into the vestibule near the posterolateral margins of the vaginal entrance. These mucous glands have the same embryological origins as the bulbo-urethral glands of males.

The outer limits of the vulva are established by the mons pubis and the labia majora. The bulge of the **mons pubis** is created by adipose tissue deep to the skin superficial to the pubic symphysis. Adipose tissue also accumulates within the **labia majora** (singular, *labium majus*), prominent folds of skin that encircle and partially conceal the labia minora and adjacent structures. The outer margins of the labia majora and the mons pubis are covered with coarse hair, but the inner surfaces of the labia majora are relatively hairless. Sebaceous glands and scattered apocrine sweat glands release their secretions onto the inner surface of the labia majora, moistening and lubricating them.

The Mammary Glands

A newborn infant cannot fend for itself, and several of its key systems have yet to complete development. Over the initial period of adjustment to an independent existence,

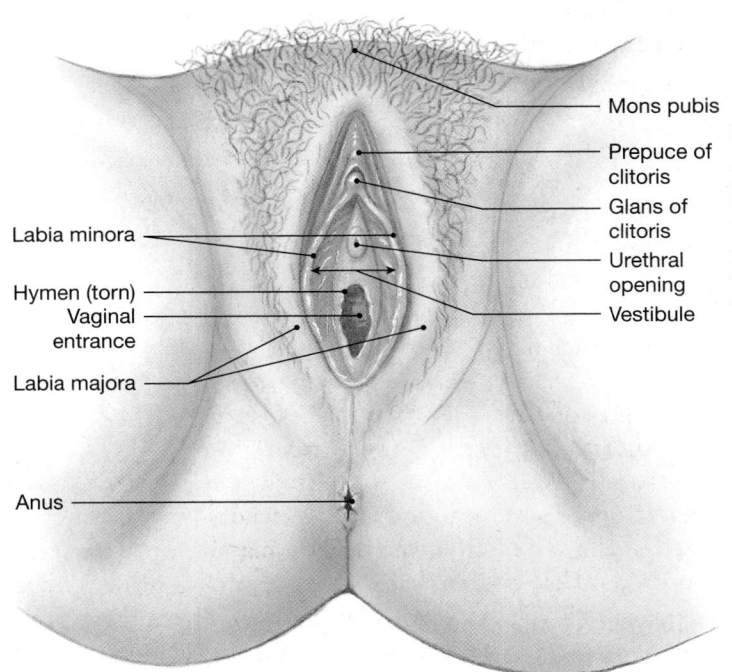

Figure 28–22 The Female External Genitalia.

Labels: Mons pubis · Prepuce of clitoris · Glans of clitoris · Urethral opening · Vestibule · Labia minora · Hymen (torn) · Vaginal entrance · Labia majora · Anus

28 REPRODUCTIVE

lactate derived from blood.

the infant can gain nourishment from the milk secreted by the maternal **mammary glands**. Milk production, or **lactation** (lak-TĀ-shun), occurs in these glands. In females, mammary glands are specialized organs of the integumentary system that are controlled mainly by hormones of the reproductive system and by the *placenta*, a temporary structure that provides the embryo and fetus with nutrients.

On each side, a mammary gland lies in the subcutaneous tissue of the **pectoral fat pad** deep to the skin of the chest (**Figure 28–23a**). Each breast bears a **nipple**, a small conical projection where the ducts of the underlying mammary gland open onto the body surface. The reddish-brown skin around each nipple is the **areola** (a-RĒ-ō-luh). Large sebaceous glands deep to the areolar surface give it a grainy texture.

The glandular tissue of a mammary gland consists of separate lobes, each containing several secretory lobules. Ducts leaving the lobules converge, giving rise to a single **lactiferous** (lak-TIF-er-us) **duct** in each lobe. Near the nipple, that lactiferous duct enlarges, forming an expanded chamber called a **lactiferous sinus**. Typically, 15–20 lactiferous sinuses open

onto the surface of each nipple. Dense connective tissue surrounds the duct system and forms partitions that extend between the lobes and the lobules. These bands of connective tissue, the *suspensory ligaments of the breast*, originate in the dermis of the overlying skin. A layer of areolar tissue separates the mammary gland complex from the underlying pectoralis muscles. Branches of the *internal thoracic artery* (see **Figure 21–21**, p. 752) supply blood to each mammary gland.

Figure 28–23b,c compares the histological organizations of inactive and active mammary glands. An inactive, or *resting*, mammary gland is dominated by a duct system rather than by active glandular cells. The size of the mammary glands in a nonpregnant woman reflects primarily the amount of adipose tissue present, not the amount of glandular tissue. The secretory apparatus normally does not complete its development unless pregnancy occurs. An active mammary gland is a tubuloalveolar gland, consisting of multiple glandular tubes that end in secretory alveoli. We will discuss the hormonal mechanisms involved in lactation in Chapter 29.

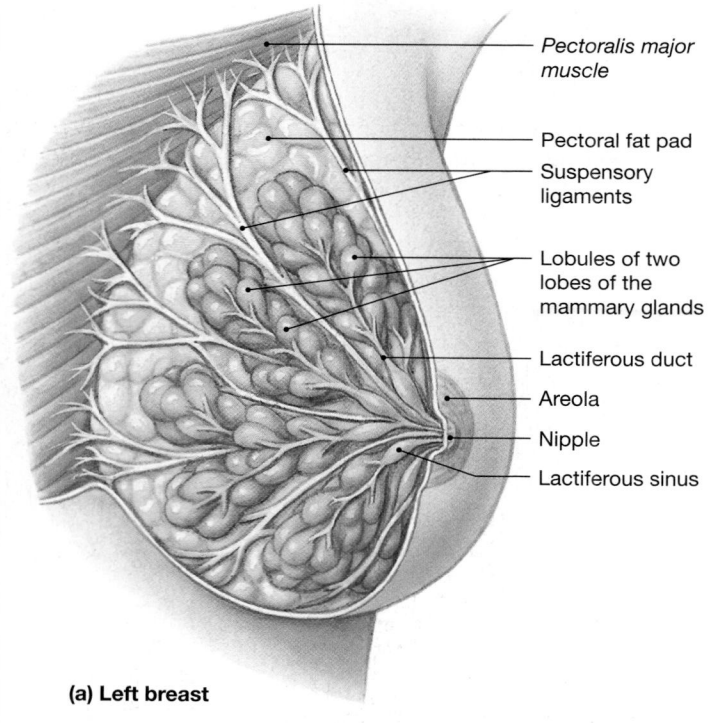

Pectoralis major muscle

Pectoral fat pad

Suspensory ligaments

Lobules of two lobes of the mammary glands

Lactiferous duct

Areola

Nipple

Lactiferous sinus

(a) Left breast

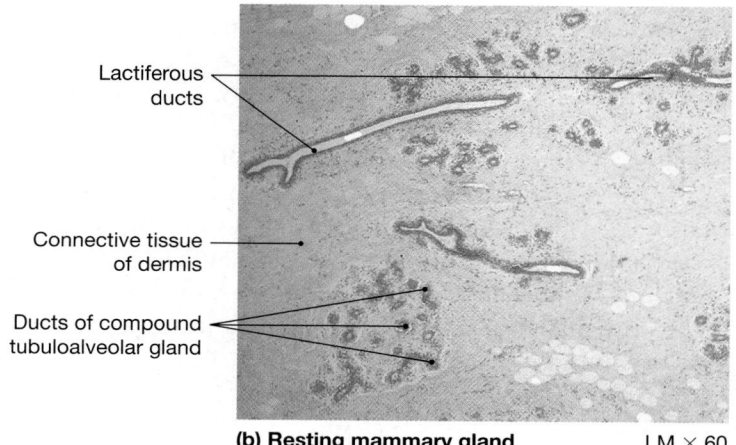

Lactiferous ducts

Connective tissue of dermis

Ducts of compound tubuloalveolar gland

(b) Resting mammary gland LM × 60

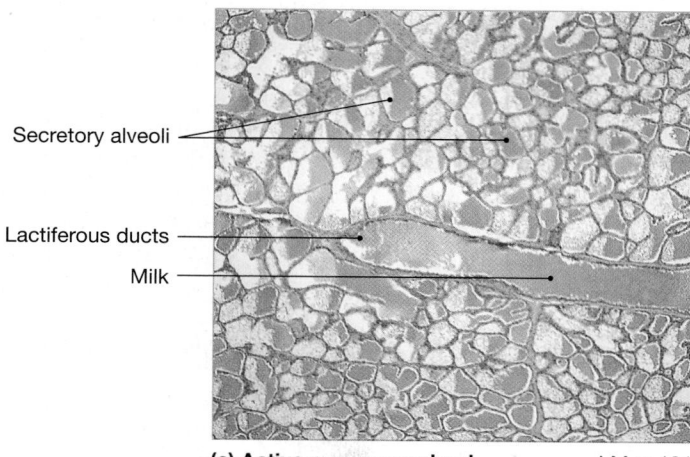

Secretory alveoli

Lactiferous ducts

Milk

(c) Active mammary gland LM × 131

Figure 28–23 The Mammary Glands. (a) The mammary gland of the left breast. **(b)** An inactive mammary gland of a nonpregnant woman. **(c)** An active mammary gland of a nursing woman. ATLAS: Plate 28

CLINICAL NOTE

Breast Cancer

The mammary glands are cyclically stimulated by the changing levels of circulating reproductive hormones that accompany the uterine cycle. The effects usually go unnoticed, but occasional discomfort or even inflammation of mammary gland tissues can occur late in the cycle. If inflamed lobules become walled off by scar tissue, **cysts** are created. Clusters of cysts can be felt in the breast as discrete masses, a condition known as **fibrocystic disease**. Because its symptoms are similar to those of breast cancer, biopsies may be needed to distinguish between this benign condition and breast cancer.

Breast cancer is a malignant, metastasizing tumor of the mammary gland. It is the leading cause of death in women between ages 35 and 45, but it is most common in women over age 50. Approximately 40,500 female deaths and 450 male deaths will occur in the United States from breast cancer in 2007, and approximately 180,500 new cases will be reported for both sexes. Of the newly reported cases, about 2000 will occur in males. An estimated 12 percent of U.S. women will develop breast cancer at some point in their lifetime. The incidence is highest among Caucasian-Americans, somewhat lower in African-Americans, and lowest in Asian-Americans and American Indians. This incidence has decreased by 3.5% from 2001 to 2004. Notable risk factors include (1) a family history of breast cancer, (2) a first pregnancy after age 30, and (3) early menarche or late menopause.

Despite repeated studies, no links have been proven between breast cancer and oral contraceptive use, fat consumption, or alcohol use. It appears likely that multiple factors are involved. In some families an inherited genetic variation has been linked to higher-than-normal risk of developing the disease. However, most women never develop breast cancer—even women in families with a history of the disease. Mothers who breast-fed (nursed) their babies have a 20 percent lower incidence of breast cancer after menopause than do mothers who did not breast-feed. The reason is not known. Adding to the mystery, nursing does not appear to affect the incidence of premenopausal breast cancer.

Early detection of breast cancer is the key to saving lives, and thus the use of clinical screening techniques has increased in recent years. **Mammography** involves the use of x-rays to examine breast tissues. The radiation dosage can be low, because only soft tissues must be penetrated. This procedure gives the clearest picture of conditions in the breast tissues. Ultrasound can provide some information, but the resulting images lack the detail of standard mammograms.

Treatment is more successful if breast cancer is identified while it is still a relatively small, localized tumor. Once it has grown larger than 2 cm (0.78 in.), the chances of long-term survival worsen. A poor prognosis also follows if cancer cells have spread through the lymphoid system to the axillary lymph nodes. If the nodes are not yet involved, the chances of five-year survival are about 92 percent, but if four or more nodes are involved, the survival rate drops to 25 percent. More than 90 percent of breast cancers are now diagnosed at the local or regional stage.

Treatment of breast cancer begins with the removal of the tumor. Because in many cases cancer cells begin to spread before the condition is diagnosed, part or all of the affected mammary gland is surgically removed and usually the axillary lymph nodes on that side are biopsied.

- In a *lumpectomy*, only a portion of the breast is removed.
- In a *total mastectomy*, the entire breast is removed.
- In *axillary lymph node dissection,* one or more of the axillary lymph nodes (called sentinel nodes) are removed to check for metastatic cells. If there are no cancer cells in the node(s) checked, the other lymph nodes may be left in place.

A combination of chemotherapy, radiation treatments, and hormone treatments may be used to supplement the surgical procedures. Tamoxifen is an antiestrogen drug that is used to treat some cases of breast cancer. It is more effective than conventional chemotherapy for treating breast cancer in women over 50, and it has fewer unpleasant side effects. It can also be used in addition to regular chemotherapy in the treatment of advanced-stage disease. As an added bonus, tamoxifen prevents and even reverses age-related osteoporosis. This drug has risks, however: When given to premenopausal women, tamoxifen can cause amenorrhea and hot flashes similar to those of menopause. Tamoxifen has also been linked to an increased risk of endometrial cancer, and potentially to liver cancer as well. However, tamoxifen has been approved for use to prevent breast cancer in women at risk for the disease. Alternatively, aromatase inhibitors, such as anastrozole or letrozole, are given after surgery because they have been found superior to tamoxifen in preventing recurrence. All of this seems to be helping, for breast cancer deaths have declined by 2.2 percent per year from 1990 to 2004.

Hormones and the Female Reproductive Cycle

The activity of the female reproductive tract is under hormonal control that involves an interplay between secretions of both the pituitary gland and the gonads. But the regulatory pattern in females is much more complicated than in males, because it must coordinate the ovarian and uterine cycles. Circulating hormones control the **female reproductive cycle**, coordinating the ovarian and uterine cycles to ensure proper reproductive function. If the two cycles are not properly coordinated, infertility results. A woman who doesn't ovulate cannot conceive, even if her uterus is perfectly normal. A woman who ovulates normally, but whose uterus is not ready to support an embryo, will also be infertile. Because the processes are complex and difficult to study, many of the biochemical details of the female reproductive cycle still elude us, but the general patterns are reasonably clear.

As in males, GnRH from the hypothalamus regulates reproductive function in females. However, in females, the GnRH pulse frequency and amplitude (amount secreted per pulse) change throughout the course of the ovarian cycle. If the hypothalamus were a radio station, the pulse frequency would correspond to the radio frequency it's transmitting on, and the amplitude would be the volume. We will consider changes in pulse frequency, as their effects are both dramatic and reasonably well understood. Changes in GnRH pulse frequency are controlled primarily by circulating levels of estrogens and progestins. Estrogens increase the GnRH pulse frequency, and progestins decrease it.

The endocrine cells of the adenohypophysis respond as if each group of endocrine cells is monitoring different frequencies. As a result, each group of cells is sensitive to some GnRH pulse frequencies and insensitive to others. For example, consider the *gonadotropes,* the cells responsible for FSH and LH production. At one pulse frequency, the gonadotropes respond preferentially and secrete FSH, whereas at another frequency, LH is the primary hormone released. FSH and LH production also occurs in pulses that follow the rhythm of GnRH pulses. If GnRH is absent or is supplied at a constant rate (without pulses), FSH and LH secretion will stop in a matter of hours.

Hormones and the Follicular Phase

Follicular development begins under FSH stimulation; each month some of the primordial follicles begin to develop into primary follicles. As the follicles enlarge, thecal cells start producing *androstenedione,* a steroid hormone that is a key intermediate in the synthesis of estrogens and androgens. Androstenedione is absorbed by the granulosa cells and converted to estrogens. In addition, small quantities of estrogens are secreted by interstitial cells scattered throughout the ovarian stroma. Circulating estrogens are bound primarily to albumins, with lesser amounts carried by gonadal steroid-binding globulin (GBG).

Of the three estrogens circulating in the bloodstream—estradiol, estrone, and estriol—the one that is most abundant and has the most pronounced effects on target tissues is **estradiol** (es-tra-DĪ-ol). It is the dominant hormone prior to ovulation. In estradiol synthesis (**Figure 28–24**), androstenedione is first converted to testosterone, which the enzyme

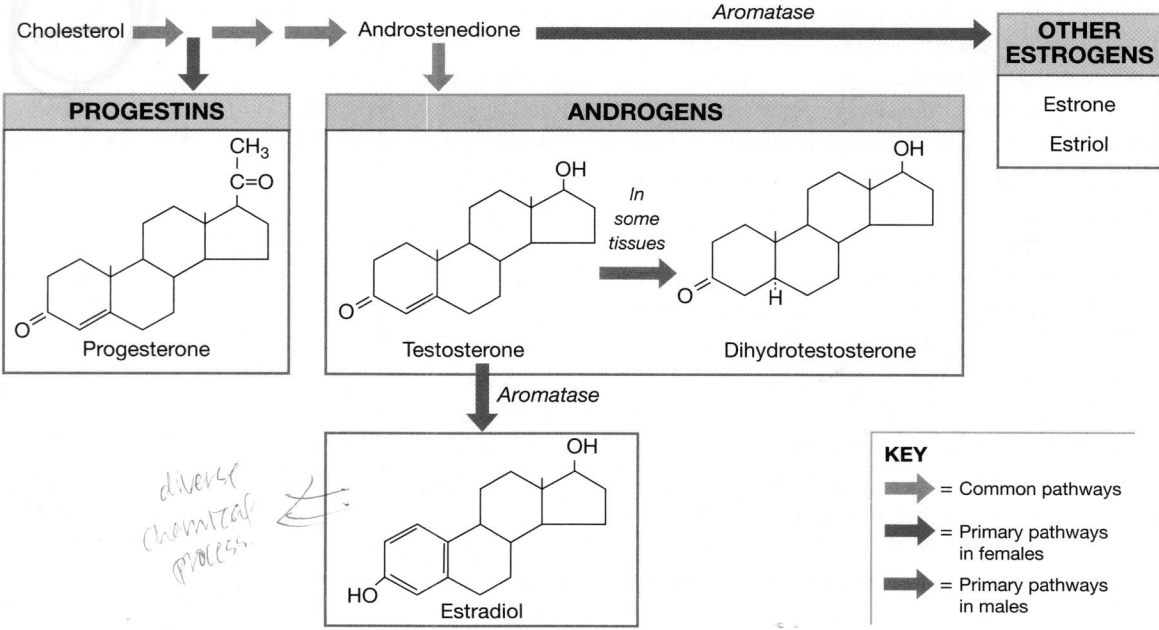

Figure 28–24 **Pathways of Steroid Hormone Synthesis in Males and Females.** All gonadal steroids are derived from cholesterol. In men, the pathway ends with the synthesis of testosterone, which may subsequently be converted to dihydrotestosterone. In women, an additional step past testosterone leads to estradiol synthesis. The synthesis of progesterone and estrogens other than estradiol involves alternative pathways.

aromatase converts to estradiol. The synthesis of both *estrone* and *estriol* proceeds directly from androstenedione.

Estrogens have multiple functions that affect the activities of many tissues and organs throughout the body. Among the important general functions of estrogens are (1) stimulating bone and muscle growth; (2) maintaining female secondary sex characteristics, such as body hair distribution and the location of adipose tissue deposits; (3) affecting central nervous system (CNS) activity (especially in the hypothalamus, where estrogens increase the sexual drive); (4) maintaining functional accessory reproductive glands and organs; and (5) initiating the repair and growth of the endometrium. **Figure 28–25**, which diagrams the hormonal regulation of ovarian activity, includes an overview of the effects of estrogens on various aspects of reproductive function.

Summary: Hormonal Regulation of the Female Reproductive Cycle

Figure 28–26 shows the changes in circulating hormone levels that accompany the ovarian cycle. Early in the follicular phase, estrogen levels are low and the GnRH pulse frequency

is 16–24 per day (one pulse every 60–90 minutes) (**Figure 28–26a**). At this frequency, FSH is the dominant hormone released by the adenohypophysis; the estrogens released by developing follicles inhibit LH secretion (**Figure 28–26b**). As secondary follicles develop, FSH levels decline due to the negative feedback effects of inhibin. Follicular development and maturation continue, however, supported by the combination of estrogens, FSH, and LH.

As one or more tertiary follicles begin forming, the concentration of circulating estrogens rises steeply. As a result, the GnRH pulse frequency increases to about 36 per day (one pulse every 30–60 minutes). The increased pulse frequency stimulates LH secretion. In addition, at roughly day 10 of the cycle, the effect of estrogen on LH secretion changes from inhibition to stimulation. The switchover occurs only after rising estrogen levels have exceeded a specific threshold value for about 36 hours. (The threshold value and the time required vary among individuals.) High estrogen levels also increase gonadotrope sensitivity to GnRH. At about day 14, the estrogen level has peaked, the gonadotropes are at maximum sensitivity, and the GnRH pulses are arriving about every 30 minutes. The result is a massive release of LH from the

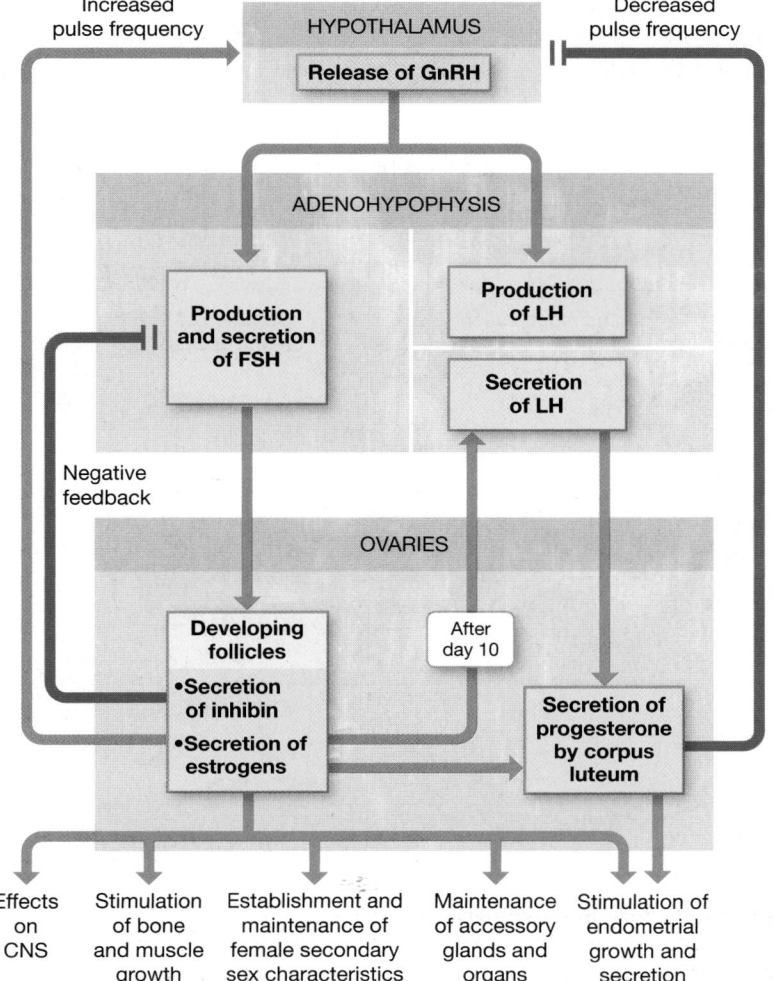

Figure 28–25 The Hormonal Regulation of Ovarian Activity.

Increased pulse frequency — HYPOTHALAMUS: Release of GnRH — Decreased pulse frequency

ADENOHYPOPHYSIS

Production and secretion of FSH

Production of LH

Secretion of LH

Negative feedback

OVARIES

Developing follicles
• Secretion of inhibin
• Secretion of estrogens

After day 10

Secretion of progesterone by corpus luteum

Effects on CNS · Stimulation of bone and muscle growth · Establishment and maintenance of female secondary sex characteristics · Maintenance of accessory glands and organs · Stimulation of endometrial growth and secretion

KEY
GnRH = Gonadotropin-releasing hormone
LH = Luteinizing hormone
FSH = Follicle-stimulating hormone
→ Stimulation
⊣ Inhibition

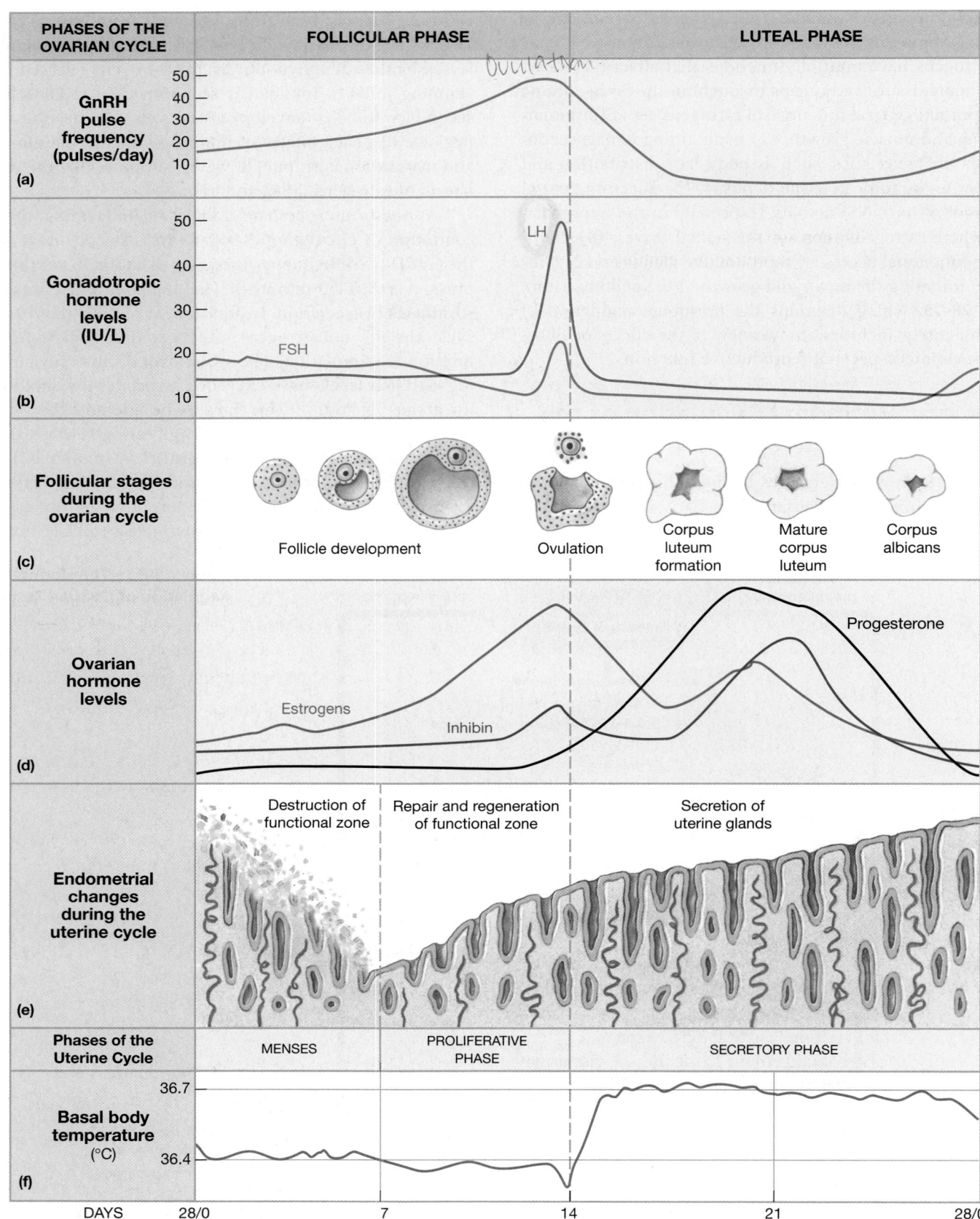

Figure 28–26 **The Hormonal Regulation of the Female Reproductive Cycle.**

adenohypophysis. This sudden surge in LH concentration triggers (1) the completion of meiosis I by the primary oocyte, (2) the forceful rupture of the follicular wall, and (3) ovulation. Typically, ovulation occurs 34–38 hours after the LH surge begins, roughly 9 hours after the LH peak.

Hormones and the Luteal Phase

The high LH levels that trigger ovulation also promote progesterone secretion and the formation of the corpus luteum. As progesterone levels rise and estrogen levels fall, the GnRH pulse frequency declines sharply, soon reaching 1–4 pulses per day. This frequency of GnRH pulses stimulates LH secretion more than it does FSH secretion, and the LH maintains the structure and secretory function of the corpus luteum.

Although moderate amounts of estrogens are secreted by the corpus luteum, progesterone is the main hormone of the luteal phase. Its primary function is to continue the preparation of the uterus for pregnancy by enhancing the blood supply to the functional zone and stimulating the secretion of uterine glands. Progesterone levels remain high for the next week, but unless pregnancy occurs, the corpus luteum begins to degenerate. Roughly 12 days after ovulation, the corpus luteum becomes nonfunctional, and progesterone and estrogen levels fall markedly. The blood supply to the functional zone is restricted, and the endometrial tissues begin to deteriorate. As progesterone and estrogen levels drop, the GnRH pulse frequency increases, stimulating FSH secretion by the adenohypophysis, and the ovarian cycle begins again.

The hormonal changes involved with the ovarian cycle in turn affect the activities of other reproductive tissues and organs. At the uterus, the hormonal changes maintain the uterine cycle.

Hormones and the Uterine Cycle

Figure 28–26e depicts the changes in the endometrium during a single uterine cycle. The declines in progesterone and estrogen levels that accompany the degeneration of the corpus luteum (**Figure 28–26c,d**) result in menses. The shedding of endometrial tissue continues for several days, until rising estrogen levels stimulate the repair and regeneration of the functional zone of the endometrium. The proliferative phase continues until rising progesterone levels mark the arrival of the secretory phase. The combination of estrogen and progesterone then causes the enlargement of the uterine glands and an increase in their secretory activities.

Hormones and Body Temperature

The monthly hormonal fluctuations cause physiological changes that affect core body temperature. During the follicular phase, when estrogen is the dominant hormone, the *basal body temperature*, or the resting body temperature measured upon awakening in the morning, is about 0.3°C lower

than it is during the luteal phase, when progesterone dominates (**Figure 28–26d**). At the time of ovulation, the basal body temperature (BBT) declines noticeably, making the rise in temperature over the next day even more noticeable (**Figure 28–26f**). Urine tests that detect LH are available, and testing daily for several days before expected ovulation can detect the LH surge more reliably than the BBT changes. This information can be important for individuals who wish to avoid or promote a pregnancy, because fertilization typically occurs within a day of ovulation. Thereafter, oocyte viability and the likelihood of successful fertilization decrease markedly.

The A&P Top 100

#98 Cyclic changes in FSH and LH levels are responsible for the maintenance of the ovarian cycle; the hormones produced by the ovaries in turn regulate the uterine cycle. Inadequate hormone levels, inappropriate or inadequate responses to circulating hormones, or poor coordination and timing of hormone production or secondary oocyte release will reduce or eliminate the chances of pregnancy.

CHECKPOINT

9. Name structures of the female reproductive system.

10. What effect would blockage of both uterine tubes by scar tissue (resulting from an infection such as gonorrhea) have on a woman's ability to conceive?

11. What benefit does the acidic pH of the vagina provide?

12. Which layer of the uterus sloughs off during menstruation?

13. Would the blockage of a single lactiferous sinus interfere with the delivery of milk to the nipple? Explain.

14. What changes would you expect to observe in the ovarian cycle if the LH surge did not occur?

15. What effect would a blockage of progesterone receptors in the uterus have on the endometrium?

16. What event in the uterine cycle occurs when the levels of estrogens and progesterone decline?

See the blue Answers tab at the end of the book.

28-4 The autonomic nervous system influences male and female sexual function

Sexual intercourse, also known as *coitus* (KŌ-i-tus) or *copulation*, introduces semen into the female reproductive tract. We will now consider the process as it affects the reproductive systems of males and females.

Male Sexual Function

Sexual function in males is coordinated by complex neural reflexes that we do not yet understand completely. The reflex pathways utilize the sympathetic and parasympathetic divisions of the autonomic nervous system. During sexual **arousal**, erotic thoughts, the stimulation of sensory nerves in the genital region, or both lead to an increase in parasympathetic outflow over the pelvic nerves. This outflow in turn leads to **erection** of the penis (discussed on p. 1056). The integument covering the glans of the penis contains numerous sensory receptors, and erection tenses the skin and increases their sensitivity. Subsequent stimulation can initiate the secretion of the bulbo-urethral glands, providing **lubrication** for the penile urethra and the surface of the glans.

During intercourse, the sensory receptors of the penis are rhythmically stimulated. This stimulation eventually results in the coordinated processes of emission and ejaculation. **Emission** occurs under sympathetic stimulation. The process begins when the peristaltic contractions of the ampulla push fluid and spermatozoa into the prostatic urethra. The seminal glands then begin contracting, and the contractions increase in force and duration over the next few seconds. Peristaltic contractions also appear in the walls of the prostate gland. The combination moves the seminal mixture into the membranous and penile portions of the urethra. While the contractions are proceeding, sympathetic commands also cause the contraction of the urinary bladder and the internal urethral sphincter. The combination of elevated pressure inside the bladder and the contraction of the sphincter effectively prevents the passage of semen into the bladder.

Ejaculation occurs as powerful, rhythmic contractions appear in the *ischiocavernosus* and *bulbospongiosus muscles*, two superficial skeletal muscles of the pelvic floor. The ischiocavernosus muscles insert along the sides of the penis; their contractions serve primarily to stiffen that organ. The bulbospongiosus muscle wraps around the base of the penis; the contraction of this muscle pushes semen toward the external urethral opening. The contractions of both muscles are controlled by somatic motor neurons in the inferior lumbar and superior sacral segments of the spinal cord. (The positions of these muscles are shown in **Figure 11–12b**, p. 357.) Contraction of the smooth muscle within the prostate acts to pinch off the urethra, preventing the passage of urine through the erect penis.

Ejaculation is associated with intensely pleasurable sensations, an experience known as male **orgasm** (OR-gazm). Several other noteworthy physiological changes occur at this time, including pronounced but temporary increases in heart rate and blood pressure. After ejaculation, blood begins to leave the erectile tissue, and the erection begins to subside. This subsidence, called **detumescence** (dē-tū-MES-ens), is mediated by the sympathetic nervous system.

In sum, arousal, erection, emission, and ejaculation are controlled by a complex interplay between the sympathetic and parasympathetic divisions of the autonomic nervous system. Higher centers, including the cerebral cortex, can facilitate or inhibit many of the important reflexes, thereby modifying the patterns of sexual function. Any physical or psychological factor that affects a single component of the system can result in male sexual dysfunction, also called **impotence**.

Impotence is defined as an inability to achieve or maintain an erection. Various physical causes may be responsible for impotence, because erection involves vascular changes as well as neural commands. For example, low blood pressure in the arteries supplying the penis, due to a circulatory blockage such as a plaque, will reduce the ability to attain an erection. Drugs, alcohol, trauma, or illnesses that affect the autonomic nervous system or the central nervous system can have the same effect. But male sexual performance can also be strongly affected by the psychological state of the individual. Temporary periods of impotence are relatively common in healthy individuals who are experiencing severe stresses or emotional problems. Depression, anxiety, and fear of impotence are examples of emotional factors that can result in sexual dysfunction. The prescription drug Viagra, which enhances and prolongs the effects of nitric oxide on the erectile tissue of the penis, has proven useful in treating many cases of impotence.

Female Sexual Function

The events in female sexual function are largely comparable to those of male sexual function. During sexual arousal, parasympathetic activation leads to engorgement of the erectile tissues of the clitoris and increased secretion of cervical mucous glands and the greater vestibular glands. Clitoral erection increases the receptors' sensitivity to stimulation, and the cervical and vestibular glands lubricate the vaginal walls. A network of blood vessels in the vaginal walls becomes filled with blood at this time, and the vaginal surfaces are also moistened by fluid that moves across the epithelium from underlying connective tissues. (This process accelerates during intercourse as the result of mechanical stimulation.) Parasympathetic stimulation also causes contraction of subcutaneous smooth muscle of the nipples, making them more sensitive to touch and pressure.

During sexual intercourse, rhythmic contact of the penis with the clitoris and vaginal walls—reinforced by touch sensations from the breasts and other stimuli (visual, olfactory, and auditory)—provides stimulation that leads to orgasm. Female orgasm is accompanied by peristaltic contractions of the uterine and vaginal walls and, by means of impulses traveling over the pudendal nerves, rhythmic contractions of the bulbospongiosus and ischiocavernosus muscles. The latter contractions give rise to the intensely pleasurable sensations of orgasm.

Sexual activity carries with it the risk of infection with a variety of microorganisms. The consequences of such an infection may range from merely inconvenient to potentially lethal. **Sexually transmitted diseases (STDs)** are transferred from individual to individual, primarily or exclusively by sexual intercourse. At least two dozen bacterial, viral, and fungal infections are currently recognized as STDs. The bacterium *Chlamydia* can cause **pelvic inflammatory disease (PID)** and infertility; AIDS, caused by a virus, is deadly. The incidence of STDs has been increasing in the United States since 1984; an estimated 19 million new cases occur each year, almost 50% in persons aged 19–24. Poverty, intravenous drug use, prostitution, and the appearance of drug-resistant pathogens all contribute to the problem.

CHECKPOINT

17. List the physiological events of sexual intercourse in both sexes, and indicate those that occur in males but not in females.

18. An inability to contract the ischiocavernosus and bulbospongiosus muscles would interfere with which part(s) of the male sex act?

19. What changes occur in females during sexual arousal as the result of increased parasympathetic stimulation?

See the blue Answers tab at the end of the book.

28-5 With age, decreasing levels of reproductive hormones cause functional changes

The aging process affects all body systems, including the reproductive systems of men and women alike. As noted earlier in the chapter, these systems become fully functional at puberty. Thereafter, the most striking age-related changes in the female reproductive system occur at menopause. Comparable age-related changes in the male reproductive system occur more gradually and over a longer period of time.

Menopause

Menopause is usually defined as the time that ovulation and menstruation cease. Menopause typically occurs at age 45–55, but in the years immediately preceding it, the ovarian and uterine cycles become irregular. This interval is called *perimenopause*. A shortage of primordial follicles is the underlying cause of the irregular cycles. It has been estimated that almost 7 million potential oocytes are in fetal ovaries after five months of development, but the number drops to about 2 mil-

lion at birth, and to a few hundred thousand at puberty. With the arrival of perimenopause, the number of follicles responding each month begins to drop markedly. As the number of available follicles decreases, estrogen levels decline and may not rise enough to trigger ovulation. By age 50, there are often no primordial follicles left to respond to FSH. In **premature menopause**, this depletion occurs before age 40.

Menopause is accompanied by a decline in circulating concentrations of estrogens and progesterone, and a sharp and sustained rise in the production of GnRH, FSH, and LH. The decline in estrogen levels leads to reductions in the size of the uterus and breasts, accompanied by a thinning of the urethral and vaginal epithelia. The reduced estrogen concentrations have also been linked to the development of osteoporosis, presumably because bone deposition proceeds at a slower rate. A variety of neural effects are reported as well, including "hot flashes," anxiety, and depression. Hot flashes typically begin while estrogen levels are declining, and cease when estrogen levels reach minimal values. These intervals of elevated body temperature are associated with surges in LH production. The hormonal mechanisms involved in other CNS effects of menopause are poorly understood. In addition, the risks of atherosclerosis and other forms of cardiovascular disease increase after menopause.

The majority of women experience only mild symptoms, but some individuals experience acutely unpleasant symptoms in perimenopause or during or after menopause. For most of those women, hormone replacement therapy (HRT) involving a combination of estrogens and progestins can control the unpleasant neural and vascular changes associated with menopause. The hormones are administered as pills, by injection, or by transdermal "estrogen patches." However, recent studies suggest that taking estrogen-replacement therapy for more than 5 years increases the risk of heart disease, breast cancer, Alzheimer disease, blood clots, and stroke; HRT should be prescribed with caution, only after a full discussion and assessment of the potential risks and benefits, and taken for as short a time as possible.

The Male Climacteric

Changes in the male reproductive system occur more gradually than do those in the female reproductive system. The period of declining reproductive function, which corresponds to perimenopause in women, is known as the **male climacteric** or *andropause*. Levels of circulating testosterone begin to decline between the ages of 50 and 60, and levels of circulating FSH and LH increase. Although sperm production continues (men well into their eighties can father children), older men experience a gradual reduction in sexual activity. This decrease may be linked to declining testosterone levels. Some clinicians suggest the use of testosterone replacement therapy to enhance the libido (sexual drive) of elderly men, but this may increase the risk of prostate disease.

The A&P Top 100

#99 Sex hormones have widespread effects on the body. They affect brain development and behavioral drives, muscle mass, bone mass and density, body proportions, and the patterns of hair and body fat distribution. As aging occurs, reductions in sex hormone levels affect appearance, strength, and a variety of physiological functions.

CHECKPOINT

20. Define menopause.

21. Why does the level of FSH rise and remain high during menopause?

22. What is the male climacteric?

See the blue Answers tab at the end of the book.

28-6 The reproductive system secretes hormones affecting growth and metabolism of all body systems

Normal human reproduction is a complex process that requires the participation of multiple systems. The hormones discussed in this chapter play a major role in coordinating reproductive events (Table 28–1). Physical factors also play a role. The man's sperm count must be adequate, the semen must have the correct pH and nutrients, and erection and ejaculation must occur in the proper sequence; the woman's ovarian and uterine cycles must be properly coordinated, ovulation and oocyte transport must occur normally, and her reproductive tract must provide a hospitable environment for the survival and movement of sperm, and for the subsequent fertilization of the oocyte. For these steps to occur, the repro-

SUMMARY TABLE 28–1	Hormones of the Reproductive System		
Hormone	**Source**	**Regulation of Secretion**	**Primary Effects**
Gonadotropin-releasing hormone (GnRH)	Hypothalamus	*Males*: inhibited by testosterone and possibly by inhibin *Females*: GnRH pulse frequency increased by estrogens, decreased by progestins	Stimulates FSH secretion and LH synthesis in males and females
Follicle-stimulating hormone (FSH)	Adenohypophysis	*Males*: stimulated by GnRH, inhibited by inhibin *Females*: stimulated by GnRH, inhibited by inhibin	*Males*: stimulates spermatogenesis and spermiogenesis through effects on nurse cells *Females*: stimulates follicle development, estrogen production, and oocyte maturation
Luteinizing hormone (LH)	Adenohypophysis	*Males*: stimulated by GnRH *Females*: production stimulated by GnRH, secretion by the combination of high GnRH pulse frequencies and high estrogen levels	*Males*: stimulates interstitial cells to secrete testosterone *Females*: stimulates ovulation, formation of corpus luteum, and progestin secretion
Androgens (primarily testosterone and dihydrotestosterone)	Interstitial cells of testes	Stimulated by LH	Establish and maintain secondary sex characteristics and sexual behavior; promote maturation of spermatozoa; inhibit GnRH secretion
Estrogens (primarily estradiol)	Granulosa and thecal cells of developing follicles; corpus luteum	Stimulated by FSH	Stimulate LH secretion (at high levels); establish and maintain secondary sex characteristics and sexual behavior; stimulate repair and growth of endometrium; increase frequency of GnRH pulses
Progestins (primarily progesterone)	Granulosa cells from midcycle through functional life of corpus luteum	Stimulated by LH	Stimulate endometrial growth and glandular secretion; reduce frequency of GnRH pulses
Inhibin	Nurse cells of testes and granulosa cells of ovaries	Stimulated by factors released by developing spermatozoa (male) and developing follicles (female)	Inhibits secretion of FSH (and possibly of GnRH)

ductive, digestive, endocrine, nervous, cardiovascular, and urinary systems must all be functioning normally.

Even when all else is normal and fertilization occurs at the proper time and place, a healthy infant will not be produced unless the zygote—a single cell the size of a pinhead—manages to develop into a full-term fetus that typically weighs about 3 kg (6.6 lb). In Chapter 29 we will consider the process of development, focusing on the mechanisms that determine both the structure of the body and the distinctive characteristics of each individual.

Even though the reproductive system's primary function—producing children—doesn't play a role in maintaining homeostasis, reproduction depends on a variety of physical, physiological, and psychological factors, many of which require intersystem cooperation. In addition, the hormones that control and coordinate sexual function have direct effects on the organs and tissues of other systems. For example, testosterone and estradiol affect both muscular development and bone density. **Figure 28–27** summarizes the relationships between the reproductive system and other physiological systems.

CHECKPOINT

23. **Describe the interaction between the reproductive system and the cardiovascular system.**

24. **Describe the interaction between the reproductive system and the skeletal system.**

See the blue Answers tab at the end of the book.

Related Clinical Terms

amenorrhea: The failure of menarche to appear before age 16, or a cessation of menstruation for six months or more in a female of reproductive age.

breast cancer: A malignant, metastasizing tumor of the mammary gland that is the primary cause of death for women ages 35–45.

cervical cancer: A malignant, metastasizing tumor of the cervix, and the most common reproductive system cancer in women.

cryptorchidism: The failure of one or both testes to descend into the scrotum by the time of birth.

dysmenorrhea: Painful menstruation.

endometriosis: The growth of endometrial tissue outside the uterus.

fibrocystic disease: Clusters of lobular cysts within the tissues of the mammary gland.

gonorrhea: A sexually transmitted bacterial disease.

impotence: The inability to achieve or maintain an erection.

mammography: The use of x-rays to examine breast tissue.

mastectomy: The surgical removal of part or all of a breast containing cancerous glandular tissue.

orchiectomy: The surgical removal of a testis.

ovarian cancer: A malignant, metastasizing tumor of the ovaries, and the most dangerous reproductive system cancer in women.

pelvic inflammatory disease (PID): An infection of the uterine tubes.

prostate cancer: A malignant, metastasizing tumor of the prostate gland, and the second most common cause of cancer deaths in males.

prostatectomy: The surgical removal of the prostate gland.

prostate-specific antigen (PSA): An antigen whose level in blood increases in men with prostate cancer.

sexually transmitted diseases (STDs): Diseases transferred from one individual to another primarily or exclusively through sexual contact. Examples include gonorrhea, syphilis, herpes genitalis, and AIDS.

vaginitis: An infection of the vaginal canal by fungal or bacterial pathogens.

vasectomy: The surgical removal of a segment of each ductus deferens, making it impossible for spermatozoa to reach the distal portions of the male reproductive tract.

28 REPRODUCTIVE

Chapter Review
Study Outline

28-1 Basic reproductive system structures are gonads, ducts, accessory glands and organs, and external genitalia p. 1042

1. The human **reproductive system** produces, stores, nourishes, and transports functional **gametes** (reproductive cells). **Fertilization** is the fusion of male and female gametes.

2. The reproductive system includes **gonads** (**testes** or **ovaries**), ducts, accessory glands and organs, and the **external genitalia**.

3. In males, the testes produce **spermatozoa**, which are expelled from the body in **semen** during *ejaculation*. The ovaries of a sexually mature female produce **oocytes** (immature **ova**) that travel along *uterine tubes* toward the *uterus*. The *vagina* connects the uterus with the exterior of the body.

28-2 Spermatogenesis occurs in the testes, and hormones from the hypothalamus, adenohypophysis, and testes control male reproductive functions p. 1042

4. Spermatozoa travel along the *epididymis*, the *ductus deferens*, the *ejaculatory duct*, and the *urethra* before leaving the body. Accessory organs (notably the *seminal glands, prostate gland,* and *bulbo-urethral glands*) secrete fluids into the ejaculatory ducts and the urethra. The *scrotum* encloses the testes, and the *penis* is an erectile organ. (*Figure 28–1*)

5. The **descent of the testes** through the *inguinal canals* occurs during fetal development. The testes remain connected to internal structures via the **spermatic cords**. The **raphe** marks

THE REPRODUCTIVE SYSTEM IN PERSPECTIVE

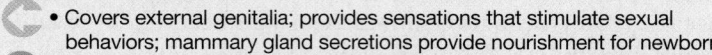
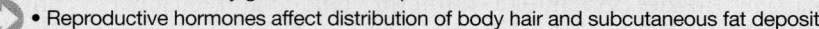

Integumentary System

- Covers external genitalia; provides sensations that stimulate sexual behaviors; mammary gland secretions provide nourishment for newborn
- Reproductive hormones affect distribution of body hair and subcutaneous fat deposits

Skeletal System

- Pelvis protects reproductive organs of females, portion of ductus deferens, and accessory glands in males
- Sex hormones stimulate growth and maintenance of bone; sex hormones at puberty accelerate growth and closure of epiphyseal cartilages

Muscular System

- Contractions of skeletal muscles eject semen from male reproductive tract; muscle contractions during sexual intercourse produce pleasurable sensations in both sexes
- Reproductive hormones, especially testosterone, accelerate skeletal muscle growth

Nervous System

- Controls sexual behaviors and sexual function
- Sex hormones affect CNS development and sexual behaviors

Endocrine System

- Hypothalamic regulatory hormones and pituitary hormones regulate sexual development and function; oxytocin stimulates smooth muscle contractions in uterus and mammary glands
- Steroid sex hormones and inhibin inhibit secretory activities of hypothalamus and pituitary gland

Cardiovascular System

- Distributes reproductive hormones; provides nutrients, oxygen, and waste removal for fetus; local blood pressure changes responsible for physical changes during sexual arousal
- Estrogens may help maintain healthy vessels and slow development of atherosclerosis

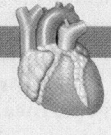

Lymphoid System

- Provides IgA for secretions by epithelial glands; assists in repairs and defense against infection
- Lysosomes and bacteriocidal chemicals in secretions provide nonspecific defense against reproductive tract infections

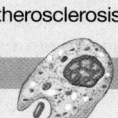

Respiratory System

- Provides oxygen and removes carbon dioxide generated by tissues of reproductive system
- Changes in respiratory rate and depth occur during sexual arousal, under control of the nervous system

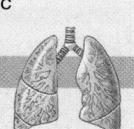
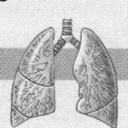
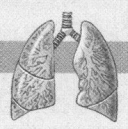

Digestive System

- Provides additional nutrients required to support gamete production and (in pregnant women) embryonic and fetal development

Urinary System

- Urethra in males carries semen to exterior; kidneys remove wastes generated by reproductive tissues and (in pregnant women) by growing embryo and fetus
- Accessory organ secretions may have antibacterial action that helps prevent urethral infections in males

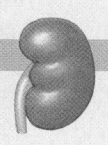

FOR ALL SYSTEMS
Secretion of hormones with effects on growth and metabolism

Figure 28–27 **Functional Relationships between the Reproductive System and Other Systems.**

the boundary between the two chambers in the **scrotum**. (*Figures 28–2, 28–3*)

6. The **dartos** muscle tightens the scrotum, giving it a wrinkled appearance as it elevates the testes; the **cremaster muscles** are more substantial muscles that pull the testes close to the body.

7. The **tunica albuginea** surrounds each testis. Septa extend from the tunica albuginea to the region of the testis closest to the entrance to the epididymis, creating a series of **lobules**. (*Figure 28–4*)

8. **Seminiferous tubules** within each lobule are the sites of sperm production. From there, spermatozoa pass through the **rete testis**. **Efferent ductules** connect the rete testis to the epididymis. Between the seminiferous tubules are **interstitial cells**, which secrete sex hormones. (*Figures 28–4, 28–5*)

9. Seminiferous tubules contain **spermatogonia**, stem cells involved in **spermatogenesis** (the production of spermatozoa), and **nurse** (sustentacular) **cells**, which sustain and promote the development of spermatozoa. (*Figures 28–6, 28–7*)

10. Each **spermatozoon** has a **head** tipped by an **acrosomal cap**, a **middle piece**, and a **tail**. (*Figure 28–8*)

11. From the testis, the spermatozoa enter the **epididymis**, an elongated tubule with **head**, **body**, and **tail** regions. The epididymis monitors and adjusts the composition of the fluid in the seminiferous tubules, serves as a recycling center for damaged spermatozoa, stores and protects spermatozoa, and facilitates their functional maturation. (*Figure 28–9*)

12. The **ductus deferens**, or *vas deferens*, begins at the epididymis and passes through the inguinal canal as part of the spermatic cord. Near the prostate gland, the ductus deferens enlarges to form the **ampulla**. The junction of the base of the seminal gland and the ampulla creates the **ejaculatory duct**, which empties into the urethra. (*Figures 28–9, 28–10*)

13. The **urethra** extends from the urinary bladder to the tip of the penis. The urethra can be divided into *prostatic, membranous,* and *spongy regions.*

14. Each **seminal gland** (**seminal vesicle**) is an active secretory gland that contributes about 60 percent of the volume of semen; its secretions contain fructose (which is easily metabolized by spermatozoa), bicarbonate ions, prostaglandins, and fibrinogen. The **prostate gland** secretes slightly acidic **prostatic fluid**. Alkaline mucus secreted by the **bulbo-urethral glands** has lubricating properties. (*Figures 28–10, 28–11*)

15. A typical ejaculation releases 2–5 mL of semen (**ejaculate**), which contains 20–100 million spermatozoa per milliliter. The fluid component of semen is **seminal fluid**.

16. The skin overlying the **penis** resembles that of the scrotum. Most of the **body** of the penis consists of three masses of **erectile tissue**. Beneath the superficial fascia are two **corpora cavernosa** and a single **corpus spongiosum**, which surrounds the urethra. Dilation of the blood vessels within the erectile tissue produces an **erection**. (*Figure 28–11*)

17. Important regulatory hormones include **FSH** (*follicle-stimulating hormone*), **LH** (*luteinizing hormone*), and **GnRH** (*gonadotropin-releasing hormone*). *Testosterone* is the most important androgen. (*Figure 28–12*)

28-3 Oogenesis occurs in the ovaries, and hormones from the pituitary gland and gonads control female reproductive functions p. 1060

18. Principal organs of the female reproductive system include the *ovaries, uterine tubes, uterus, vagina,* and *external genitalia.* (*Figure 28–13*)

19. The ovaries, uterine tubes, and uterus are enclosed within the **broad ligament**. The **mesovarium** supports and stabilizes each ovary. (*Figure 28–14*)

20. The ovaries are held in position by the **ovarian ligament** and the **infundibulopelvic ligament**. Major blood vessels enter the ovary at the **ovarian hilum**. Each ovary is covered by a **tunica albuginea**. (*Figure 28–14*)

21. **Oogenesis** (ovum production) occurs monthly in **ovarian follicles** as part of the **ovarian cycle**, which is divided into a **follicular** (*preovulatory*) **phase** and a **luteal** (*postovulatory*) **phase**. (*Figures 28–15, 28–16*)

22. As follicle development proceeds, **primary**, **secondary**, and **tertiary follicles** are produced in turn. At **ovulation**, a **secondary oocyte** and the attached follicular cells of the **corona radiata** are released through the ruptured ovarian wall. The follicular cells remaining within the ovary form the **corpus luteum**, which later degenerates into scar tissue called a **corpus albicans**. (*Figure 28–16*)

23. Each **uterine tube** has an **infundibulum** with **fimbriae** (fingerlike projections), an **ampulla**, and an **isthmus**. Each uterine tube opens into the *uterine cavity*. For fertilization to occur, a secondary oocyte must encounter spermatozoa during the first 12–24 hours of its passage from the infundibulum to the uterus. (*Figure 28–17*)

24. *Peg cells* lining the uterine tube secrete a fluid that completes the capacitation of spermatozoa.

25. The **uterus** provides mechanical protection, nutritional support, and waste removal for the developing embryo. Normally, the uterus bends anteriorly near its base (*anteflexion*). It is stabilized by the **broad ligament**, **uterosacral ligaments**, **round ligaments**, and **lateral ligaments**. (*Figure 28–18*)

26. Major anatomical landmarks of the uterus include the **body**, **isthmus**, **cervix**, **external os** (*external orifice*), **uterine cavity**, **cervical canal**, and **internal os** (*internal orifice*). The uterine wall consists of an inner **endometrium**, a muscular **myometrium**, and a superficial **perimetrium** (an incomplete serous layer). (*Figures 28–18, 28–19*)

27. A typical 28-day **uterine**, or *menstrual*, **cycle** begins with the onset of **menses** and the destruction of the **functional zone** of the endometrium. This process of **menstruation** continues from one to seven days. (*Figure 28–20*)

28. After menses, the **proliferative phase** begins, and the functional zone undergoes repair and thickens. The proliferative phase is followed by the **secretory phase**, during which uterine (endometrial) glands enlarge. Menstrual activity begins at **menarche** and continues until **menopause**. (*Figure 28–20*)

29. The **vagina** is a muscular tube extending between the uterus and the external genitalia. A thin epithelial fold, the **hymen**, partially blocks the entrance to the vagina until physical distortion (often associated with sexual intercourse) ruptures the membrane. (*Figure 28–21*)

30. The components of the **vulva** are the **vestibule, labia minora, paraurethral glands, clitoris, labia majora,** and **lesser** and **greater vestibular glands**. (*Figure 28–22*)

31. A newborn infant can gain nourishment from milk secreted by maternal **mammary glands**. (*Figure 28–23*)

32. Hormonal regulation of the **female reproductive cycle** involves the coordination of the ovarian and uterine cycles.

33. **Estradiol**, the most important *estrogen*, is the dominant hormone of the follicular phase. Ovulation occurs in response to a midcycle surge in LH. (*Figures 28–24, 28–25*)

28 REPRODUCTIVE

34. The hypothalamic secretion of GnRH occurs in pulses that trigger the pituitary secretion of FSH and LH. FSH initiates follicular development, and activated follicles and ovarian interstitial cells produce estrogens. High estrogen levels stimulate LH secretion, increase pituitary sensitivity to GnRH, and increase the GnRH pulse frequency. **Progesterone**, one of the **progestins**, is the principal hormone of the luteal phase. Changes in estrogen and progesterone levels are responsible for the maintenance of the uterine cycle. (*Figures 28–25, 28–26*)

28-4 The autonomic nervous system influences male and female sexual function p. 1077

35. During sexual **arousal** in males, erotic thoughts, sensory stimulation, or both lead to parasympathetic activity that produces erection. Stimuli accompanying **sexual intercourse** lead to **emission** and **ejaculation**. Contractions of the bulbospongiosus muscles are associated with **orgasm**.

36. The events of female sexual function resemble those of male sexual function, with parasympathetic arousal and skeletal muscle contractions associated with orgasm.

28-5 With age, decreasing levels of reproductive hormones cause functional changes p. 1079

37. Menopause (the time that ovulation and menstruation cease) typically occurs at ages 45–55. The production of GnRH, FSH, and LH rise, whereas circulating concentrations of estrogen and progesterone decline.

38. During the **male climacteric**, at ages 50–60, circulating testosterone levels fall, and FSH and LH levels rise.

28-6 The reproductive system secretes hormones affecting growth and metabolism of all body systems p. 1080

39. Normal human reproduction depends on a variety of physical, physiological, and psychological factors, many of which require intersystem cooperation. (*Figure 28–27*)

Review Questions

See the blue Answers tab at the end of the book.

myA&P Access more review material online at **myA&P**™ (www.myaandp.com). There, you'll find chapter guides, chapter quizzes, practice tests, animations, flashcards, a glossary with pronunciations, *Interactive Physiology*® (IP) exercises and quizzes, and more to help you succeed in the course.

LEVEL 1 Reviewing Facts and Terms

1. Identify the principal structures of the male reproductive system in the following diagram.

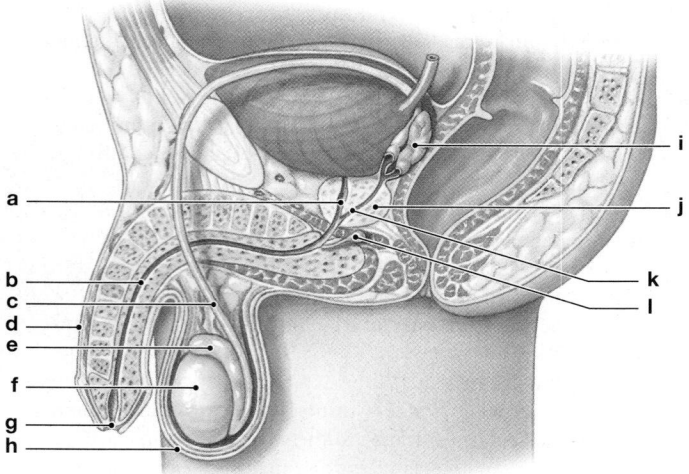

   (a) prostatic urethra   (b) Spongy urethra
   (c) ductus deferens   (d) penis
   (e) epididymis   (f) testes
   (g) external o   (h) scrotum
   (i) seminal gland   (j) prostate gland
   (k) bulbo-urethral ⟷   (l) ejaculatory
        gland   duct

2. Identify the principal structures of the female reproductive system in the following diagram.

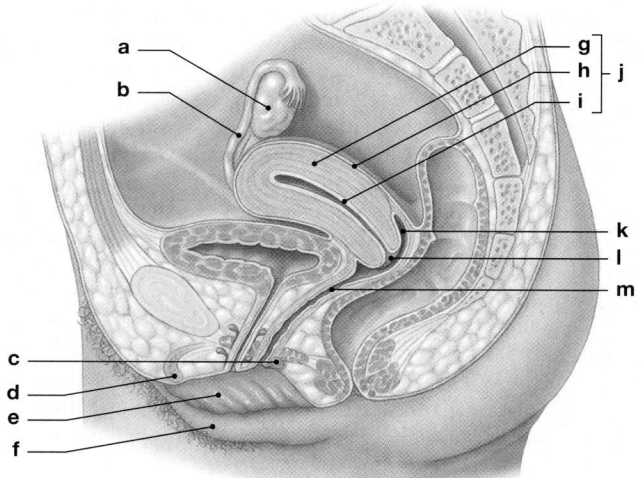

   (a) ovary   (b) uterine tube
   (c) greater vestibular gland (d) clitoris
   (e) labium minus   (f) labum majus
   (g) myometrium   (h) perimetrium
   (i) endometrium   (j) uterus
   (k) fornix   (l) cervix
   (m) vagina

3. Developing spermatozoa are nourished by
   (a) interstitial cells.   (b) the seminal vesicles.
   (c) nurse cells.   (d) Leydig cells.
   (e) the epididymis.

4. The ovaries are responsible for
   (a) the production of female gametes.
   (b) the secretion of female sex hormones.
   (c) the secretion of inhibin.
   (d) all of these.

5. In females, meiosis II is not completed until
   (a) birth.
   (b) puberty.
   (c) fertilization occurs.
   (d) uterine implantation occurs.

6. A sudden surge in LH secretion causes the
   (a) onset of menses.
   (b) rupture of the follicular wall and ovulation.
   (c) beginning of the proliferative phase.
   (d) end of the uterine cycle.

7. The principal hormone of the postovulatory phase is
   (a) progesterone.   (b) estradiol.
   (c) estrogen.   (d) luteinizing hormone.

8. Which accessory structures contribute to the composition of semen? What are the functions of each structure?

9. What types of cells in the testes are responsible for functions related to reproductive activity? What are the functions of each cell type?

10. Identify the three regions of the male urethra.

11. List the functions of testosterone in males.

12. List and summarize the important steps in the ovarian cycle.

13. Describe the histological composition of the uterine wall.

14. What is the role of the clitoris in the female reproductive system?

15. Trace the route of milk from its site of production to outside the female.

LEVEL 2 Reviewing Concepts

16. Which of the following is *not* true of pelvic inflammatory disease?
    (a) It is frequently caused by sexually transmitted pathogens.
    (b) It causes fever and abdominal pain.
    (c) It can lead to a ruptured urinary bladder.
    (d) It can possibly lead to peritonitis.
    (e) It can cause sterility.

17. In the follicular phase of the ovarian cycle, the ovary
    (a) undergoes atresia.
    (b) forms a corpus luteum.
    (c) releases a mature ovum.
    (d) secretes progesterone.
    (e) matures a follicle.

18. What are the main differences in gamete production between males and females?

19. Describe the erectile tissues of the penis. How does erection occur?

20. Describe each of the three phases of a typical 28-day uterine cycle.

21. Describe the hormonal events associated with the ovarian cycle.

22. Describe the hormonal events associated with the uterine cycle.

23. Summarize the events that occur in sexual arousal and orgasm. Do these processes differ in males and females?

24. How does the aging process affect the reproductive systems of men and women?

LEVEL 3 Critical Thinking and Clinical Applications

25. Diane has peritonitis (an inflammation of the peritoneum), which her physician says resulted from a sexually transmitted disease. Why might this condition occur more readily in females than in males?

26. In a condition known as endometriosis, endometrial cells are believed to migrate from the body of the uterus into the uterine tubes or by way of the uterine tubes into the peritoneal cavity, where they become established. A major symptom of endometriosis is periodic pain. Why does such pain occur?

27. Contraceptive pills contain estradiol and progesterone, or progesterone alone, administered at programmed doses during the ovarian cycle to prevent follicle maturation and ovulation. Explain how such pills are effective.

28. Female bodybuilders and women with eating disorders such as anorexia nervosa commonly experience amenorrhea. What does this fact suggest about the relation between body fat and menstruation? What might be the benefit of amenorrhea under such circumstances?

29

Development and Inheritance

Learning Outcomes

After completing this chapter, you should be able to do the following:

29-1 Explain the relationship between differentiation and development, and specify the various stages of development.

29-2 Describe the process of fertilization, and explain how developmental processes are regulated.

29-3 List the three stages of prenatal development, and describe the major events of each.

29-4 Explain how the three germ layers participate in the formation of extraembryonic membranes, and discuss the importance of the placenta as an endocrine organ.

29-5 Describe the interplay between the maternal organ systems and the developing fetus, and discuss the structural and functional changes in the uterus during gestation.

29-6 List and discuss the events that occur during labor and delivery.

29-7 Identify the features and physiological changes of the postnatal stages of life.

29-8 Relate basic principles of genetics to the inheritance of human traits.

Clinical Notes

Gestational Trophoblastic Neoplasia p. 1093
Abortion p. 1104
Chromosomal Abnormalities and Genetic Analysis p. 1119

An Introduction to Development and Inheritance

The physiological processes we have studied thus far are relatively brief in duration. Many last only a fraction of a second; others may take hours at most. But some important processes are measured in months, years, or decades. A human being develops in the womb for nine months, grows to maturity in 15 to 20 years, and may live the better part of a century. During that time, he or she is always changing. Birth, growth, maturation, aging, and death are all parts of a single, continuous process. That process does not end with the individual, because humans can pass at least some of their characteristics on to their offspring. Therefore, each generation gives rise to a new generation that will repeat the cycle. In this chapter, we will explore how genetic programming, environmental factors, and various physiological processes affect the events following the union of male and female gametes. The explanation begins at prenatal development and continues through childhood and adolescence and into maturity and senescence (aging).

29-1 Development, marked by various stages, is a continuous process that occurs from fertilization to maturity

Time refuses to stand still; today's infant will be tomorrow's adult. The gradual modification of anatomical structures and physiological characteristics during the period from fertilization to maturity is called **development.** The changes that occur during development are truly remarkable. In a mere 9 months, all the tissues, organs, and organ systems we have studied thus far take shape and begin to function. What begins as a single cell slightly larger than the period at the end of this sentence becomes an individual whose body contains trillions of cells organized into a complex array of highly specialized structures. The formation of different types of cells required in this process is called **differentiation.** Differentiation occurs through selective changes in genetic activity. As development proceeds, some genes are turned off and others are turned on. The identities of these genes vary from one type of cell to another, and the patterns change over time.

Development begins at **fertilization**, or **conception**. We can divide development into stages characterized by specific anatomical changes. **Embryological development** comprises the events that occur during the first two months after fertilization. The study of these events is called **embryology** (em-brē-OL-ō-jē). **Fetal development** begins at the start of the ninth week and continues until birth.

Embryological and fetal development are sometimes referred to collectively as **prenatal** (*natus*, birth) **development**, the primary focus of this chapter. **Postnatal development** commences at birth and continues to **maturity**, the state of full development or completed growth.

A basic understanding of human development provides important insights into anatomical structures. In addition, many of the mechanisms of development and growth are similar to those responsible for the repair of injuries. In this chapter, we will focus on major aspects of development. We will consider highlights of the developmental process rather than examine the events in great detail. We will also consider the regulatory mechanisms involved, and how developmental patterns can be modified—for good or ill. Few topics in the biological sciences are so fascinating, and fewer still confront investigators with so daunting an array of scientific, technological, and ethical challenges. The ongoing debate over fetal tissue research has brought several ethical issues into the public eye. The information presented in this final chapter should help you formulate your opinions on many difficult moral, legal, and public-policy questions.

Although all humans go through the same developmental stages, differences in their genetic makeup produce distinctive individual characteristics. The term **inheritance** refers to the transfer of genetically determined characteristics from generation to generation. The study of the mechanisms responsible for inheritance is called **genetics**. In this chapter, we will also consider basic genetics as it applies to inherited characteristics, such as sex, hair color, and various diseases.

CHECKPOINT

1. Define differentiation.
2. What event marks the onset of development?
3. Define inheritance.

See the blue Answers tab at the end of the book.

29-2 Fertilization—the fusion of a secondary oocyte and a spermatozoon—forms a zygote

Fertilization involves the fusion of two haploid gametes, each containing 23 chromosomes, producing a zygote that contains 46 chromosomes, the normal complement in a somatic cell. The functional roles and contributions of the male and female gametes are very different. The spermatozoon simply delivers the paternal chromosomes to the site of fertilization. It must travel a relatively large distance and is small, efficient, and highly streamlined. In contrast, the female gamete must provide all the cellular organelles and inclusions, nourishment, and genetic programming necessary to support development of the embryo for nearly a week after conception. The volume of this gamete is therefore much greater than that of the spermatozoon. Recall from Chapter 28 that ovulation releases a secondary oocyte suspended in metaphase of meiosis II. At fertilization, the diameter of the secondary oocyte is more than twice the entire length of the spermatozoon (**Figure 29–1a**). The ratio of their volumes is even more striking—roughly 2000:1.

The spermatozoa deposited in the vagina are already motile, as a result of contact with secretions of the seminal glands—the first step of *capacitation.* ∞ p. 1052 (An unidentified substance secreted by the epididymis appears to prevent premature capacitation.) The spermatozoa, however, cannot accomplish fertilization until they have been exposed to conditions in the female reproductive tract. Although oviduct peg cell secretions help with capacitation, the exact mechanism responsible for capacitation remains unknown.

Fertilization typically occurs near the junction between the ampulla and isthmus of the uterine tube, generally within a day after ovulation. By this time, a secondary oocyte has traveled only a few centimeters, but spermatozoa must cover the distance between the vagina and the ampulla of the uterine tube. A spermatozoon can propel itself at speeds of only about 34 μm per second, roughly equivalent to 12.5 cm (5 in.) per hour, so in theory it should take spermatozoa several hours to reach the upper portions of the uterine tubes. The actual passage time, however, ranges from two hours to as little as 30 minutes. Contractions of the uterine musculature and ciliary currents in the uterine tubes have been suggested as likely mechanisms for accelerating the movement of spermatozoa from the vagina to the site of fertilization.

Even with transport assistance and available nutrients, the passage is not easy. Of the roughly 200 million spermatozoa introduced into the vagina in a typical ejaculation, only about 10,000 enter the uterine tube, and fewer than 100 reach the isthmus. In general, a male with a sperm count below 20 million per milliliter is functionally sterile because too few spermatozoa survive to reach and fertilize an oocyte. While it is true that only one spermatozoon fertilizes an oocyte, dozens of spermatozoa are required for successful fertilization. The additional sperm are essential because one sperm does not contain enough acrosomal enzymes to disrupt the *corona radiata,* the layer of follicle cells that surrounds the oocyte.

The Oocyte at Ovulation

Ovulation occurs before the oocyte is completely mature. The secondary oocyte leaving the follicle is in metaphase of meiosis II. The cell's metabolic operations have been discon-

tinued, and the oocyte drifts in a sort of suspended animation, awaiting the stimulus for further development. If fertilization does not occur, the oocyte disintegrates without completing meiosis.

Fertilization is complicated by the fact that when the secondary oocyte leaves the ovary, it is surrounded by the corona radiata. Fertilization and the events that follow are diagrammed in **Figure 29–1b.** The cells of the corona radiata protect the secondary oocyte as it passes through the ruptured follicular wall, across the surface of the ovary, and into the infundibulum of the uterine tube. Although the physical process of fertilization requires that only a single spermatozoon contact the oocyte membrane, that spermatozoon must first penetrate the corona radiata. The acrosomal cap of each sperm contains several enzymes, including **hyaluronidase** (hī-uh-loo-RON-i-dās), which breaks down the bonds between adjacent follicle cells. Dozens of spermatozoa must release hyaluronidase before the connections between the follicle cells break down enough to allow an intact spermatozoon to reach the oocyte.

No matter how many spermatozoa slip through the gap in the corona radiata, normally only a single spermatozoon accomplishes fertilization and activates the oocyte (STEP 1, **Figure 29–1b**). That spermatozoon must have an intact acrosomal cap. The first step is the binding of the spermatozoon to *sperm receptors* in the zona pellucida. This step triggers the rupture of the acrosomal cap. The hyaluronidase and **acrosin**, another proteolytic enzyme, then digest a path through the zona pellucida toward the surface of the oocyte. When the sperm contacts that surface, the sperm and oocyte membranes begin to fuse. This step is the trigger for *oocyte activation,* a complex process we will discuss in the next section.

Oocyte Activation

Oocyte activation involves a series of changes in the metabolic activity of the oocyte. The trigger for activation is contact and fusion of the plasma membranes of the sperm and oocyte. This process is accompanied by the depolarization of the oocyte membrane due to an increased permeability to sodium ions. The entry of sodium ions in turn causes the release of calcium ions from the smooth endoplasmic reticulum. The sudden rise in Ca^{2+} levels has important effects, including the following:

- *Exocytosis of Vesicles Located Just Interior to the Oocyte Membrane.* This process, called the *cortical reaction,* releases enzymes that both inactivate the sperm receptors and harden the zona pellucida. This combination prevents **polyspermy** (fertilization by more than one sperm), which would create a zygote that is incapable of normal development. (Prior to completion of the cortical reaction, depolarization of the oocyte membrane probably prevents fertilization by any sperm cells that penetrate the zona pellucida.)

Figure 29–1   Fertilization. (a) An oocyte and numerous sperm at the time of fertilization. Notice the difference in size between the gametes. **(b)** Fertilization and the preparations for cleavage.

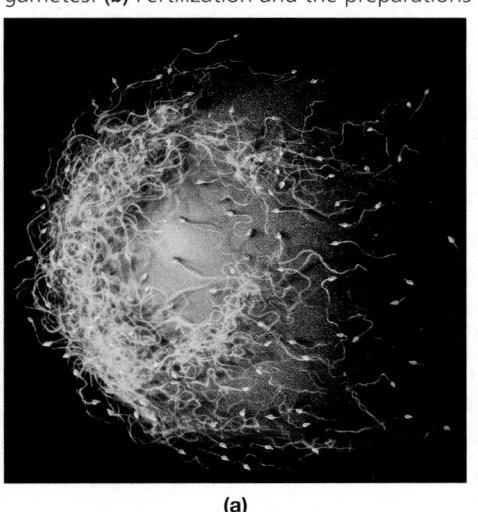

(a)

- *Completion of Meiosis II and Formation of the Second Polar Body.*

- *Activation of Enzymes That Cause a Rapid Increase in the Cell's Metabolic Rate.* The cytoplasm contains a large number of mRNA strands that have been inactivated by special proteins. The mRNA strands are now activated, so protein synthesis accelerates rapidly. Most of the proteins synthesized are required for development to proceed.

After oocyte activation and the completion of meiosis, the nuclear material remaining within the ovum reorganizes as the **female pronucleus** (STEP 2, **Figure 29–1b**). While these changes are under way, the nucleus of the spermatozoon swells, and as it forms the **male pronucleus** the rest of the sperm breaks down (STEP 3). The male pronucleus then migrates toward the center of the cell, and spindle fibers form. The two pronuclei then fuse in a process called *amphimixis*

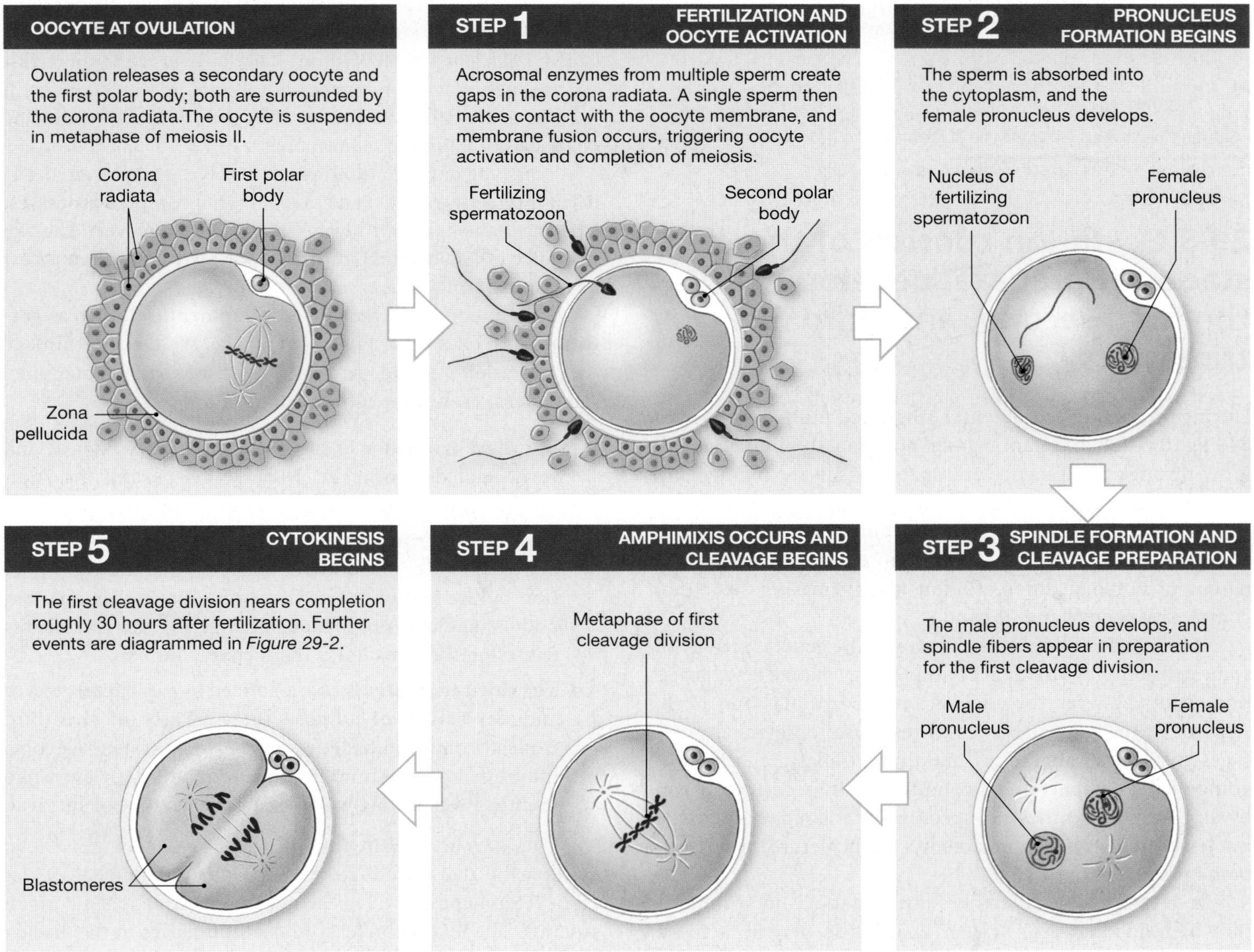

OOCYTE AT OVULATION

Ovulation releases a secondary oocyte and the first polar body; both are surrounded by the corona radiata. The oocyte is suspended in metaphase of meiosis II.

Corona radiata   First polar body

Zona pellucida

STEP 1   FERTILIZATION AND OOCYTE ACTIVATION

Acrosomal enzymes from multiple sperm create gaps in the corona radiata. A single sperm then makes contact with the oocyte membrane, and membrane fusion occurs, triggering oocyte activation and completion of meiosis.

Fertilizing spermatozoon   Second polar body

STEP 2   PRONUCLEUS FORMATION BEGINS

The sperm is absorbed into the cytoplasm, and the female pronucleus develops.

Nucleus of fertilizing spermatozoon   Female pronucleus

STEP 5   CYTOKINESIS BEGINS

The first cleavage division nears completion roughly 30 hours after fertilization. Further events are diagrammed in *Figure 29-2.*

Blastomeres

STEP 4   AMPHIMIXIS OCCURS AND CLEAVAGE BEGINS

Metaphase of first cleavage division

STEP 3   SPINDLE FORMATION AND CLEAVAGE PREPARATION

The male pronucleus develops, and spindle fibers appear in preparation for the first cleavage division.

Male pronucleus   Female pronucleus

(b)

29 REPRODUCTIVE

(am-fi-MIK-sis; STEP 4). The cell is now a zygote that contains the normal complement of 46 chromosomes, and fertilization is complete. This is the "moment of conception." Almost immediately the chromosomes line up along a metaphase plate, and the cell prepares to divide. This is the start of the process of *cleavage*, a series of cell divisions that produce an ever-increasing number of smaller and smaller daughter cells. The first cleavage division is completed roughly 30 hours after fertilization, yielding two daughter cells, each one-half the size of the original zygote (STEP 5). These cells are called *blastomeres* (BLAS-tō-mērz).

Tips&Tricks

Amphimixis means "both mixed together."

CHECKPOINT

4. **Name two sperm enzymes important to secondary oocyte penetration.**

5. **How many chromosomes are contained within a zygote?**

See the blue Answers tab at the end of the book.

29-3 Gestation consists of three stages of prenatal development: the first, second, and third trimesters

During prenatal development, a single cell ultimately forms a 3–4 kg (6.6–8.8 lb) infant, who in postnatal development will grow through adolescence and maturity toward old age and eventual death. One of the most fascinating aspects of development is its apparent order. Continuity exists at all levels and at all times. Nothing "leaps" into existence without apparent precursors; differentiation and increasing structural complexity occur hand in hand.

Differentiation involves changes in the genetic activity of some cells but not others. A continuous exchange of information occurs between the nucleus and the cytoplasm in a cell. Activity in the nucleus varies in response to chemical messages that arrive from the surrounding cytoplasm. In turn, ongoing nuclear activity alters conditions within the cytoplasm by directing the synthesis of specific proteins. In this way, the nucleus can affect enzyme activity, cell structure, and membrane properties.

In development, differences in the cytoplasmic composition of individual cells trigger changes in genetic activity. These changes in turn lead to further alterations in the cyto-

plasm, and the process continues in a sequential fashion. But if all the cells of the embryo are derived from cell divisions of a zygote, how do the cytoplasmic differences originate? What sets the process in motion? The important first step occurs before fertilization, while the oocyte is in the ovary.

Before ovulation, the growing oocyte accepts amino acids, nucleotides, and glucose, as well as more complex materials such as phospholipids, mRNA molecules, and proteins, from the surrounding granulosa cells. Because not all follicle cells manufacture and deliver the same nutrients and instructions to the oocyte, the contents of the cytoplasm are not evenly distributed. After fertilization, the zygote divides into ever-smaller cells that differ from one another in cytoplasmic composition. These differences alter the genetic activity of each cell, creating cell lines with increasingly diverse fates.

As development proceeds, some of the cells release chemical substances, including RNA molecules, polypeptides, and small proteins, that affect the differentiation of other embryonic cells. This type of chemical interplay among developing cells, called *induction* (in-DUK-shun), works over very short distances, such as when two types of cells are in direct contact. It may also operate over longer distances, with the inducing chemicals functioning as hormones.

This type of regulation, which involves an integrated series of interacting steps, can control highly complex processes. The mechanism is not always error free, however: The appearance of an abnormal or inappropriate inducer can throw development off course.

The time spent in prenatal development is known as **gestation** (jes-TĀ-shun). For convenience, we usually think of the gestation period as consisting of three integrated **trimesters**, each three months in duration:

1. The **first trimester** is the period of embryological and early fetal development. During this period, the rudiments of all the major organ systems appear.

2. The **second trimester** is dominated by the development of organs and organ systems, a process that nears completion by the end of the sixth month. During this period, body shape and proportions change; by the end of this trimester, the fetus looks distinctively human.

3. The **third trimester** is characterized by rapid fetal growth and deposition of adipose tissue. Early in the third trimester, most of the fetus's major organ systems become fully functional. An infant born one month or even two months prematurely has a reasonable chance of survival.

The *Atlas* accompanying this text contains "Embryology Summaries" that introduce key steps in embryological and fetal development and trace the development of specific organ systems. The text will refer to those summaries in the discussions that follow. As you proceed, reviewing the indicated material will help you understand the "big picture" as well as the specific details.

CHECKPOINT

6. Define gestation.

7. Characterize the key features of each trimester.

See the blue Answers tab at the end of the book.

29-4 Cleavage, implantation, placentation, and embryogenesis are critical events of the first trimester

At the moment of conception, the fertilized ovum is a single cell about 0.135 mm (0.005 in.) in diameter and weighing approximately 150 mg. By convention, pregnancies are clinically dated from the last menstrual period (LMP), which is usually 2 weeks before ovulation and conception. At the end of the first trimester (12 weeks from LMP, but only 10 developmental weeks), the fetus is almost 75 mm (3 in.) long and weighs perhaps 14 g (0.5 oz).

Many important and complex developmental events occur during the first trimester. Here we will focus on four general processes: cleavage, implantation, placentation, and embryogenesis:

1. **Cleavage** (KLĒV-ij) is a sequence of cell divisions that begins immediately after fertilization (**Figure 29–1b**). During cleavage, the zygote becomes a **pre-embryo**, which develops into a multicellular complex known as a *blastocyst*. Cleavage ends when the blastocyst first contacts the uterine wall. Cleavage and blastocyst formation are introduced in the *Atlas*. **ATLAS:** Embryology Summary 1: The Formation of Tissues

2. **Implantation** begins with the attachment of the blastocyst to the endometrium of the uterus and continues as the blastocyst invades maternal tissues. Important events during implantation set the stage for the formation of vital embryonic structures.

3. **Placentation** (plas-en-TĀ-shun) occurs as blood vessels form around the periphery of the blastocyst, and as the **placenta**—a complex organ that permits exchange between the maternal and embryonic circulatory systems—develops. It supports the fetus from its formation early in the first trimester until it stops functioning and is ejected from the uterus just after birth. From that point on, the newborn is physically independent of the mother.

4. **Embryogenesis** (em-brē-ō-JEN-e-sis) is the formation of a viable embryo. This process establishes the foundations for all major organ systems.

The foregoing processes are both complex and vital to the survival of the embryo. Perhaps because the events in the first trimester are so complex, it is the most dangerous period in prenatal life. Only about 40 percent of conceptions produce embryos that survive the first trimester. For that reason, pregnant women are warned to take great care to avoid drugs and other disruptive stresses during the first trimester, in the hope of preventing an error in the delicate processes that are under way.

Cleavage and Blastocyst Formation

Cleavage is a series of cell divisions that subdivides the cytoplasm of the zygote (**Figures 29–1b** and **29–2**). The first cleavage produces a pre-embryo consisting of two identical cells. As noted earlier, the identical cells produced by cleavage divisions are called **blastomeres**. After the first division is completed roughly 30 hours after fertilization, subsequent divisions occur at intervals of 10–12 hours. During the initial divisions, all the blastomeres divide simultaneously. As the number of blastomeres increases, the timing becomes less predictable.

After three days of cleavage, the pre-embryo is a solid ball of cells resembling a mulberry. This stage is called the **morula** (MOR-ū-luh; *morula*, mulberry). The morula typically reaches the uterus on day 4. Over the next two days, the blastomeres form a **blastocyst**, a hollow ball with an inner cavity known as the **blastocoele** (BLAS-tō-sēl). The blastomeres are now no longer identical in size and shape. The outer layer of cells, which separates the outside world from the blastocoele, is called the **trophoblast** (TRŌ-fō-blast). As the word *trophoblast* implies (*trophos*, food + *blast*, precursor), cells in this layer are responsible for providing nutrients to the developing embryo. A second group of cells, the **inner cell mass**, lies clustered at one end of the blastocyst. These cells are exposed to the blastocoele but are insulated from contact with the outside environment by the trophoblast. In time, the inner cell mass will form the embryo.

Implantation

During blastocyst formation, enzymes released by the trophoblast erode a hole through the zona pellucida, which is then shed in a process known as *hatching* (**Figure 29–2**). The blastocyst is now freely exposed to the fluid contents of the uterine cavity. This glycogen-rich fluid is secreted by the uterine glands of the uterus. Over the previous few days, the pre-embryo and early blastocyst had been absorbing fluid and nutrients from its surroundings; the process now accelerates, and the blastocyst enlarges. When fully formed, the blastocyst contacts the endometrium, and implantation occurs (**Figures 29–2** and **29–3**).

Implantation begins as the surface of the blastocyst closest to the inner cell mass touches and adheres to the uterine

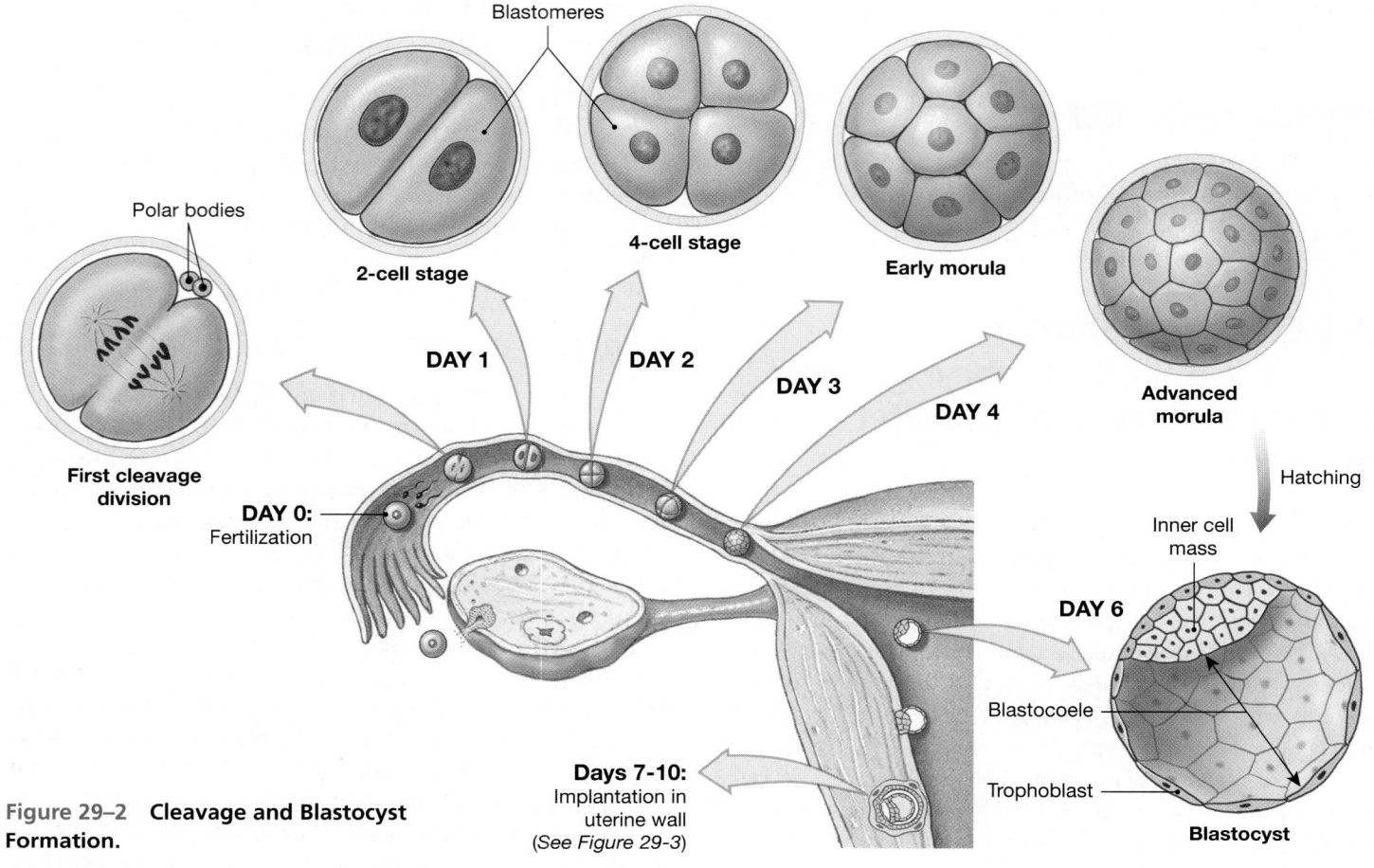

Figure 29–2 **Cleavage and Blastocyst Formation.**

lining (see day 7 in **Figure 29–3**). At the point of contact, the trophoblast cells divide rapidly, making the trophoblast several layers thick. The cells closest to the interior of the blastocyst remain intact, forming a layer of **cellular trophoblast**, or *cytotrophoblast*. Near the endometrial wall, the plasma membranes separating the trophoblast cells disappear, creating a layer of cytoplasm containing multiple nuclei (day 8). This outer layer is called the **syncytial** (sin-SISH-ul) **trophoblast**, or *syncytiotrophoblast*.

The syncytial trophoblast erodes a path through the uterine epithelium by secreting hyaluronidase. This enzyme dissolves the intercellular cement between adjacent epithelial cells, just as hyaluronidase released by spermatozoa dissolved the connections between cells of the corona radiata. At first, the erosion creates a gap in the uterine lining, but migration and divisions of maternal epithelial cells soon repair the surface. By day 10 the repairs are complete, and the blastocyst has lost contact with the uterine cavity. Further development occurs entirely within the functional zone of the endometrium.

In most cases, implantation occurs in the fundus or elsewhere in the body of the uterus. In an **ectopic preg-**nancy, implantation occurs somewhere other than within the uterus, such as in one of the uterine tubes. Approximately 0.6 percent of pregnancies are ectopic pregnancies, which do not produce a viable embryo and can be life-threatening to the mother.

As implantation proceeds, the syncytial trophoblast continues to enlarge and spread into the surrounding endometrium (see day 9, **Figure 29–3**). The erosion of uterine glands releases nutrients that are absorbed by the syncytial trophoblast and distributed by diffusion through the underlying cellular trophoblast to the inner cell mass. These nutrients provide the energy needed to support the early stages of embryo formation. Trophoblastic extensions grow around endometrial capillaries. As the capillary walls are destroyed, maternal blood begins to percolate through trophoblastic channels known as **lacunae**. Fingerlike **villi** extend away from the trophoblast into the surrounding endometrium, gradually increasing in size and complexity until about day 21. As the syncytial trophoblast spreads, it begins breaking down larger endometrial veins and arteries, and blood flow through the lacunae accelerates.

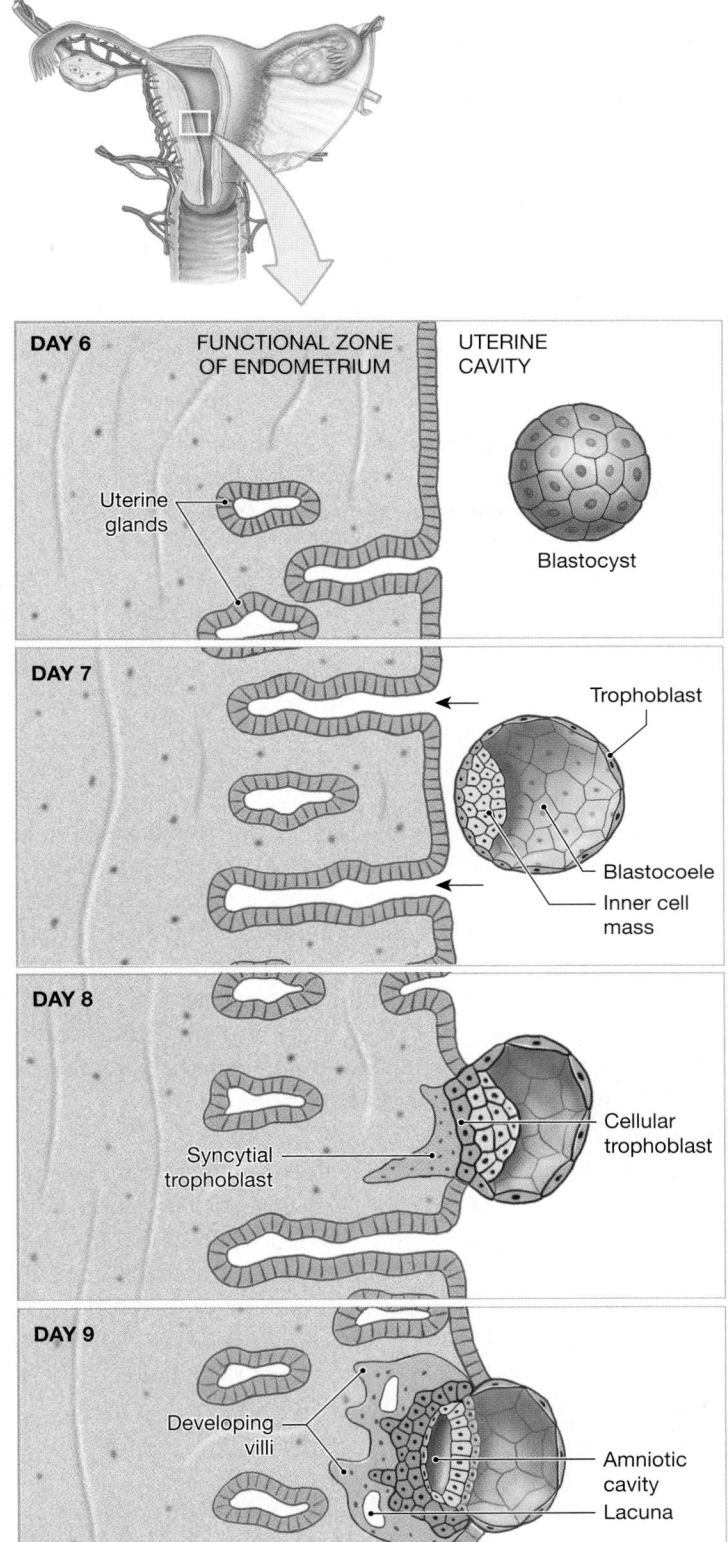

Figure 29–3 Stages in Implantation.

CLINICAL NOTE

Gestational Trophoblastic Neoplasia The trophoblast undergoes repeated nuclear divisions, shows extensive and rapid growth, has a very high demand for energy, invades and spreads through adjacent tissues, and fails to activate the maternal immune system—in short, the trophoblast has many of the characteristics of cancer cells. In about 0.1 percent of pregnancies, something goes wrong with the regulatory mechanisms, and instead of developing normally, the syncytial trophoblast behaves like a tumor. This condition is called *gestational trophoblastic neoplasia*. The least dangerous form, a *hydatidiform* (hī-da-TID-i-form) *mole*, is not malignant. However, about 20 percent of gestational trophoblastic neoplasias metastasize to other tissues, with potentially fatal results. Consequently, prompt surgical removal of the mass is essential, and the surgery is sometimes followed by chemotherapy.

Formation of the Amniotic Cavity

The inner cell mass has little apparent organization early in the blastocyst stage. Yet by the time of implantation, the inner cell mass has separated from the trophoblast. The separation gradually increases, creating a fluid-filled chamber called the **amniotic** (am-nē-OT-ik) **cavity** (see day 9 in **Figure 29–3;** details from days 10–12 are shown in **Figure 29–4**). The trophoblast will later be separated from the amniotic cavity by layers of cells that originate at the inner cell mass and line the amniotic cavity. These layers form the *amnion*, a membrane we will examine later in the chapter. When the amniotic cavity first appears, the cells of the inner cell mass are organized into an oval sheet that is two layers thick: a superficial layer that faces the amniotic cavity, and a deeper layer that is exposed to the fluid contents of the blastocoele.

Gastrulation and Germ Layer Formation

By day 12, a third layer begins to form through **gastrulation** (gas-troo-LĀ-shun) (day 12, **Figure 29–4**). During gastrulation, cells in specific areas of the surface move toward a central line known as the **primitive streak**. At the primitive streak, the migrating cells leave the surface and move between the two existing layers. This movement creates three distinct embryonic layers: (1) the **ectoderm**, consisting of superficial cells that did not migrate into the interior of the inner cell mass; (2) the **endoderm**, consisting of the cells that face the blastocoele; and (3) the **mesoderm**, consisting of the poorly organized layer of migrating cells between the ectoderm and the endoderm. Collectively, these three embryonic layers are called **germ layers**. The formation of mesoderm and the fates of each germ layer are summarized in the *Atlas*.

ATLAS: Embryology Summary 4: The Development of Organ Systems

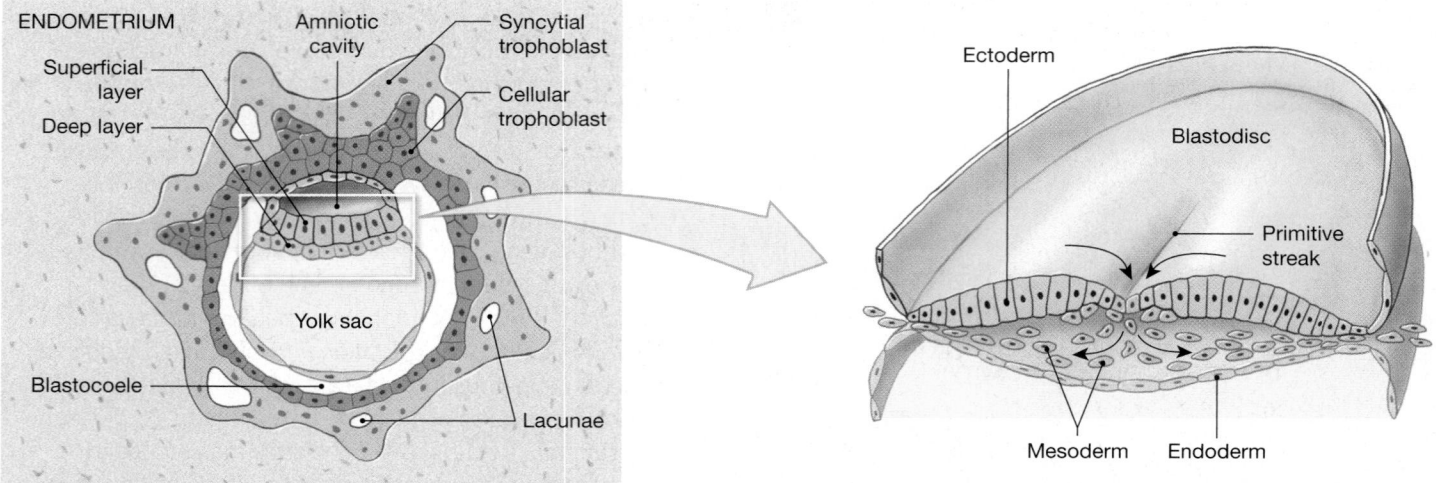

Day 10: The inner cell mass begins as two layers: a superficial layer, facing the amniotic cavity, and a deep layer, exposed to the blastocoele. Migration of cells around the amniotic cavity is the first step in the formation of the amnion. Migration of cells around the edges of the blastocoele is the first step in yolk sac formation.

Day 12: Migration of superficial cells to the interior creates a third layer. From the time this process (gastrulation) begins, the superficial layer is called *ectoderm*, the deep layer *endoderm*, and the migrating cells *mesoderm*.

Figure 29–4   The Inner Cell Mass and Gastrulation.

Table 29–1 contains a more comprehensive listing of the contributions each germ layer makes to the body systems described in earlier chapters.

Gastrulation produces an oval, three-layered sheet known as the **embryonic disc**. This disc will form the body of the embryo, whereas the rest of the blastocyst will be involved in forming the *extraembryonic membranes*.

Formation of the Extraembryonic Membranes

Germ layers also participate in the formation of four **extraembryonic membranes**: (1) the *yolk sac* (endoderm and mesoderm), (2) the *amnion* (ectoderm and mesoderm), (3) the *allantois* (endoderm and mesoderm), and (4) the *chorion* (mesoderm and trophoblast). Although these membranes support embryological and fetal development, few traces of their existence remain in adult systems. **Figure 29–5** shows representative stages in the development of the extraembryonic membranes.

The Yolk Sac. The **yolk sac** begins as a layer of cells spread out around the outer edges of the blastocoele to form a complete pouch. This pouch is already visible 10 days after fertilization (**Figure 29–4**). As gastrulation proceeds, mesodermal cells migrate around the pouch and complete the formation of the yolk sac (week 2, **Figure 29–5**). Blood vessels soon appear within the mesoderm, and the yolk sac becomes an important site of blood cell formation.

The Amnion. The ectodermal layer enlarges, and ectodermal cells spread over the inner surface of the amniotic cavity. Mesodermal cells soon follow, creating a second, outer layer (see week 2, **Figure 29–5**). This combination of mesoderm and ectoderm is the **amnion** (AM-nē-on). As development proceeds, the amnion and the amniotic cavity continue to enlarge. The amniotic cavity contains **amniotic fluid**, which surrounds and cushions the developing embryo or fetus (see week 3 through week 10, **Figure 29–5**).

The Allantois. The third extraembryonic membrane begins as an outpocketing of the endoderm near the base of the yolk sac (see week 3, **Figure 29–5**). The free endodermal tip then grows toward the wall of the blastocyst, surrounded by a mass of mesodermal cells. This sac of endoderm and mesoderm is the **allantois** (a-LAN-tō-is), the base of which later gives rise to the urinary bladder. The formation of the allantois and its relationship to the urinary bladder is illustrated in the *Atlas*. ATLAS: Embryology Summary 20: The Development of the Urinary System

The Chorion. The mesoderm associated with the allantois spreads around the blastocyst, separating the cellular trophoblast from the blastocoele. This combination of mesoderm and trophoblast is the **chorion** (KŌ-rē-on) (see weeks 2 and 3, **Figure 29–5**).

When implantation first occurs, the nutrients absorbed by the trophoblast can easily reach the inner cell mass by sim-

SUMMARY TABLE 29–1 The Fates of the Germ Layers

ECTODERMAL CONTRIBUTIONS

Integumentary system: epidermis, hair follicles and hairs, nails, and glands communicating with the skin (sweat glands, mammary glands, and sebaceous glands)

Skeletal system: pharyngeal cartilages and their derivatives in adults (portion of sphenoid, the auditory ossicles, the styloid processes of the temporal bones, the cornu and superior rim of the hyoid bone)*

Nervous system: all neural tissue, including brain and spinal cord

Endocrine system: pituitary gland and suprarenal medullae

Respiratory system: mucous epithelium of nasal passageways

Digestive system: mucous epithelium of mouth and anus, salivary glands

MESODERMAL CONTRIBUTIONS

Integumentary system: dermis and hypodermis

Skeletal system: all components except some pharyngeal derivatives

Muscular system: all components

Endocrine system: suprarenal cortex, endocrine tissues of heart, kidneys, and gonads

Cardiovascular system: all components

Lymphoid system: all components

Urinary system: the kidneys, including the nephrons and the initial portions of the collecting system

Reproductive system: the gonads and the adjacent portions of the duct systems

Miscellaneous: the lining of the body cavities (pleural, pericardial, and peritoneal) and the connective tissues that support all organ systems

ENDODERMAL CONTRIBUTIONS

Endocrine system: thymus, thyroid gland, and pancreas

Respiratory system: respiratory epithelium (except nasal passageways) and associated mucous glands

Digestive system: mucous epithelium (except mouth and anus), exocrine glands (except salivary glands), liver, and pancreas

Urinary system: urinary bladder and distal portions of the duct system

Reproductive system: distal portions of the duct system, stem cells that produce gametes

*The neural crest is derived from ectoderm and contributes to the formation of the skull and the skeletal derivatives of the embryonic pharyngeal arches.

ple diffusion. But as the embryo and the trophoblast enlarge, the distance between them increases, so diffusion alone can no longer keep pace with the demands of the developing embryo. Blood vessels now begin to develop within the mesoderm of the chorion, creating a rapid-transit system for nutrients that links the embryo with the trophoblast.

The appearance of blood vessels in the chorion is the first step in the creation of a functional placenta. By the third week of development, the mesoderm extends along the core of each trophoblastic villus, forming **chorionic villi** in contact with maternal tissues (**Figures 29–5** [weeks 3 through 10] and **29–6**). These villi continue to enlarge and branch, creating an intricate network within the endometrium. Embryonic blood vessels develop within each villus. Blood flow through those chorionic vessels begins early in the third week of development, when the embryonic heart starts beating. The blood supply to the chorionic villi arises from the allantoic arteries and veins.

As the chorionic villi enlarge, more maternal blood vessels are eroded. Maternal blood now moves slowly through complex lacunae lined by the syncytial trophoblast. Chorionic blood vessels pass close by, and gases and nutrients dif-

fuse between the embryonic and maternal circulations across the layers of the trophoblast. Recall that fetal hemoglobin has a higher affinity for oxygen than does maternal hemoglobin, enabling fetal hemoglobin to strip oxygen from maternal hemoglobin. ∞ p. 858 Maternal blood then reenters the venous system of the mother through the broken walls of small uterine veins. No mixing of maternal and fetal blood occurs, because the two are always separated by layers of trophoblast.

Placentation

At first, the entire blastocyst is surrounded by chorionic villi. The chorion continues to enlarge, expanding like a balloon within the endometrium. By week 4, the embryo, amnion, and yolk sac are suspended within an expansive, fluid-filled chamber (**Figure 29–5**). The **body stalk**, the connection between embryo and chorion, contains the distal portions of the allantois and blood vessels that carry blood to and from the placenta. The narrow connection between the endoderm of the embryo and the yolk sac is called the **yolk stalk**. The formation of the yolk stalk and body stalk are illustrated in the *Atlas*. ATLAS: Embryology Summary 19: The Development of the Digestive System

WEEK 2

Migration of mesoderm around the inner surface of the trophoblast creates the chorion. Mesodermal migration around the outside of the amniotic cavity, between the ectodermal cells and the trophoblast, forms the amnion. Mesodermal migration around the endodermal pouch creates the yolk sac.

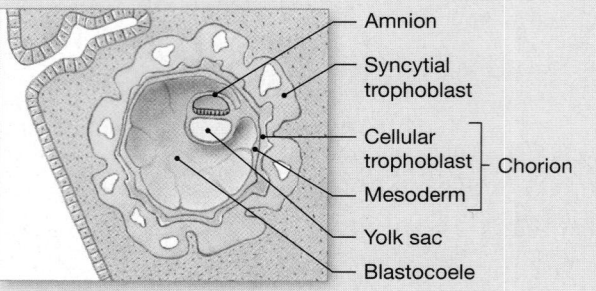

- Amnion
- Syncytial trophoblast
- Cellular trophoblast } Chorion
- Mesoderm
- Yolk sac
- Blastocoele

WEEK 3

The embryonic disc bulges into the amniotic cavity at the head fold. The allantois, an endodermal extension surrounded by mesoderm, extends toward the trophoblast.

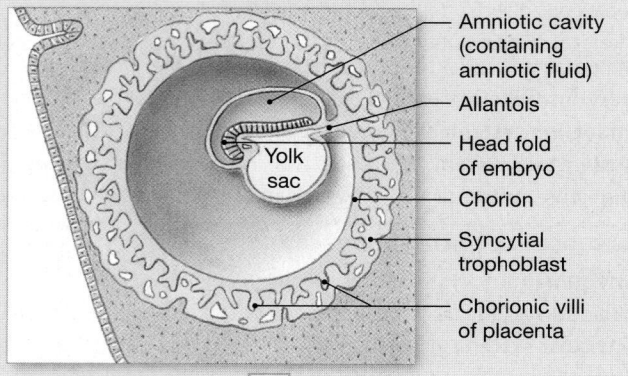

- Amniotic cavity (containing amniotic fluid)
- Allantois
- Head fold of embryo
- Chorion
- Syncytial trophoblast
- Chorionic villi of placenta

Yolk sac

WEEK 5

The developing embryo and extraembryonic membranes bulge into the uterine cavity. The trophoblast pushing out into the uterine lumen remains covered by endometrium but no longer participates in nutrient absorption and embryo support. The embryo moves away from the placenta, and the body stalk and yolk stalk fuse to form an umbilical stalk.

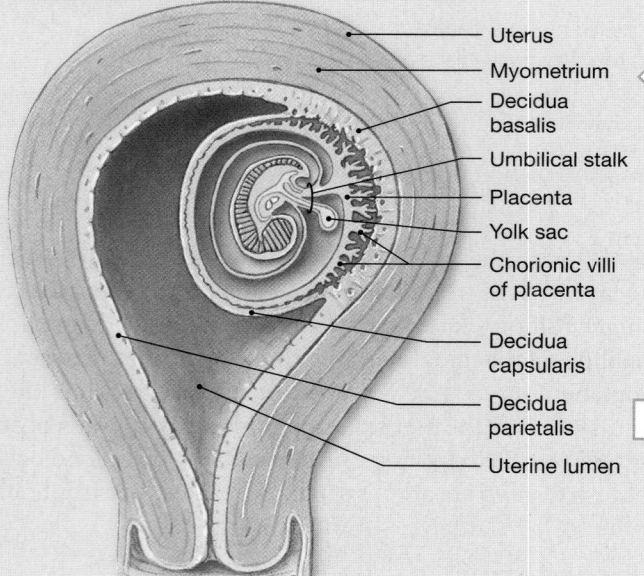

- Uterus
- Myometrium
- Decidua basalis
- Umbilical stalk
- Placenta
- Yolk sac
- Chorionic villi of placenta
- Decidua capsularis
- Decidua parietalis
- Uterine lumen

WEEK 4

The embryo now has a head fold and a tail fold. Constriction of the connections between the embryo and the surrounding trophoblast narrows the yolk stalk and body stalk.

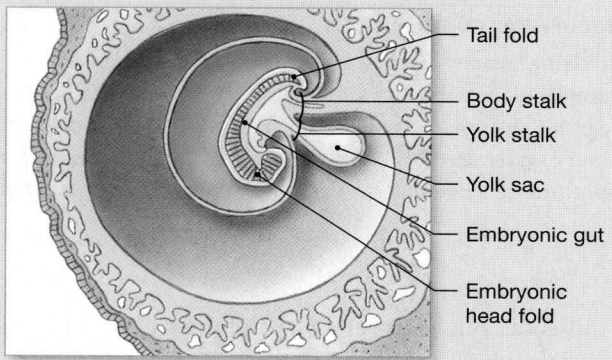

- Tail fold
- Body stalk
- Yolk stalk
- Yolk sac
- Embryonic gut
- Embryonic head fold

WEEK 10

The amnion has expanded greatly, filling the uterine cavity. The fetus is connected to the placenta by an elongated umbilical cord that contains a portion of the allantois, blood vessels, and the remnants of the yolk stalk.

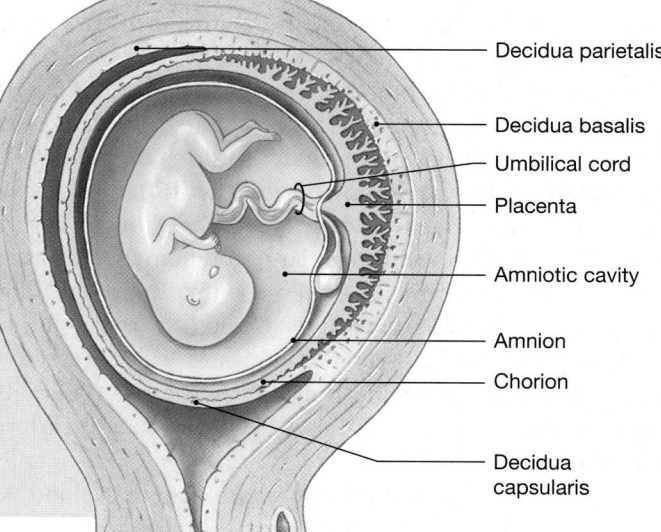

- Decidua parietalis
- Decidua basalis
- Umbilical cord
- Placenta
- Amniotic cavity
- Amnion
- Chorion
- Decidua capsularis

Figure 29–5 **Extraembryonic Membranes and Placenta Formation.**

The placenta does not continue to enlarge indefinitely. Regional differences in placental organization begin to develop as expansion of the placenta creates a prominent bulge in the endometrial surface. This relatively thin portion of the endometrium, called the **decidua capsularis** (dē-SID-ū-uh kap-sū-LA-ris; *deciduus*, a falling off), no longer participates in nutrient exchange, and the chorionic villi in the region disappear (**Figures 29–5** [week 5] and **29–6a**). Placental functions are now concentrated in a disc-shaped area in the deepest portion of the endometrium, a region called the **decidua basalis** (bā-SĀ-lis). The rest of the uterine endometrium, which has no contact with the chorion, is called the **decidua parietalis**.

Figure 29–6   A Three-Dimensional View of Placental Structure. (a) A view of the uterus after the embryo has been removed and the umbilical cord cut. Arrows in the enlarged view indicate the direction of blood flow. Blood flows into the placenta through ruptured maternal arteries and then flows around chorionic villi, which contain fetal blood vessels. **(b)** A cross section through a chorionic villus, showing the syncytial trophoblast exposed to the maternal blood space.

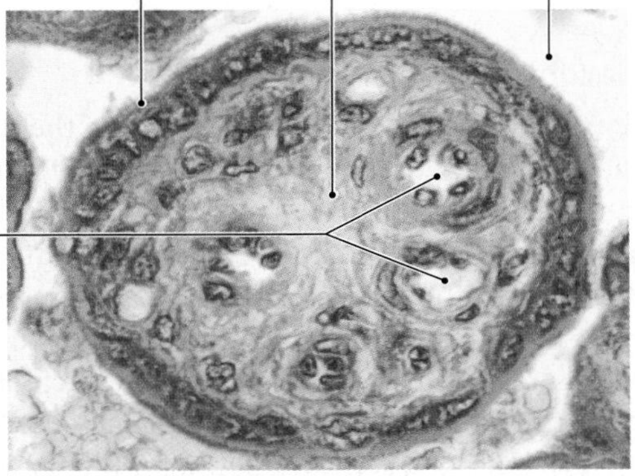

(b) Chorionic villus, cross section   LM × 280

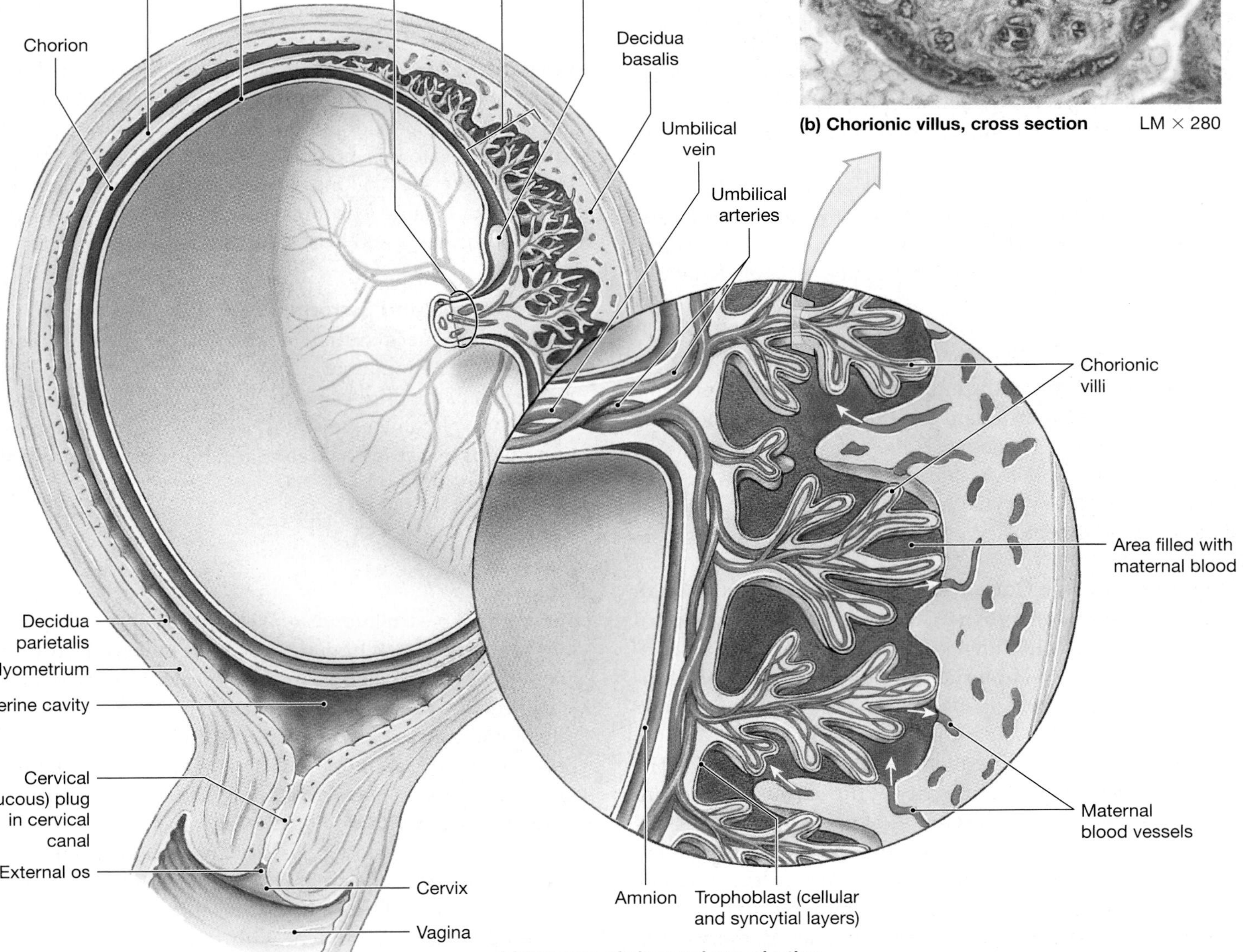

(a) Diagram of placental organization

As the end of the first trimester approaches, the fetus moves farther from the placenta (see weeks 5 and 10, **Figure 29–5**). The fetus and placenta remain connected by the **umbilical cord**, or *umbilical stalk*, which contains the allantois, the placental blood vessels, and the yolk stalk.

Tips&Tricks

Like a deep-sea diver's air hose or a space-walking astronaut's tether, the umbilical cord supplies the fetus with life-sustaining substances and removes waste products.

Placental Circulation

Figure 29–6a illustrates circulation at the placenta near the end of the first trimester. Blood flows to the placenta through the paired **umbilical arteries** and returns in a single **umbilical vein**. ∞ p. 767 The chorionic villi provide the surface area for active and passive exchanges of gases, nutrients, and waste products between the fetal and maternal bloodstreams. The blood in the umbilical arteries is deoxygenated and contains waste products generated by tissues; at the placenta, oxygen supplies are replenished, organic nutrients added, and carbon dioxide and other organic waste products removed.

The placenta places a considerable demand on the maternal cardiovascular system, and blood flow to the uterus and placenta is extensive. If the placenta is torn or otherwise damaged, the consequences may prove fatal to both fetus and mother.

The Endocrine Placenta

In addition to its role in the nutrition of the fetus, the placenta acts as an endocrine organ. Several hormones—including *human chorionic gonadotropin, human placental lactogen, placental prolactin, relaxin, progesterone,* and *estrogens*— are synthesized by the syncytial trophoblast and released into the maternal bloodstream.

Human Chorionic Gonadotropin.

The hormone **human chorionic gonadotropin (hCG)** appears in the maternal bloodstream soon after implantation has occurred. The presence of hCG in blood or urine samples provides a reliable indication of pregnancy. Kits sold for the early detection of pregnancy are sensitive to the presence of this hormone.

In function, hCG resembles luteinizing hormone (LH), because it maintains the integrity of the corpus luteum and promotes the continued secretion of progesterone. As a result, the endometrial lining remains perfectly functional, and menses does not normally occur. In the absence of hCG, the pregnancy ends, because another uterine cycle begins and the functional zone of the endometrium disintegrates.

In the presence of hCG, the corpus luteum persists for three to four months before gradually decreasing in size and secretory function. The decline in luteal function does not trigger the return of uterine cycles, because by the end of the first trimester, the placenta actively secretes both estrogens and progesterone.

Human Placental Lactogen and Placental Prolactin.

Human placental lactogen (hPL), or *human chorionic somatomammotropin (hCS)*, helps prepare the mammary glands for milk production. It also has a stimulatory effect on other tissues comparable to that of growth hormone (GH). At the mammary glands, the conversion from inactive to active status requires the presence of placental hormones (hPL, **placental prolactin**, estrogen, and progesterone) as well as several maternal hormones (GH, prolactin [PRL], and thyroid hormones). We will consider the hormonal control of mammary gland function in a later section.

Relaxin.

Relaxin is a peptide hormone that is secreted by the placenta and the corpus luteum during pregnancy. Relaxin (1) increases the flexibility of the pubic symphysis, permitting the pelvis to expand during delivery; (2) causes the dilation of the cervix, making it easier for the fetus to enter the vaginal canal; and (3) suppresses the release of oxytocin by the hypothalamus and delays the onset of labor contractions.

Progesterone and Estrogens.

After the first trimester, the placenta produces sufficient amounts of progesterone to maintain the endometrial lining and continue the pregnancy. As the end of the third trimester approaches, estrogen production by the placenta accelerates. As we will see in a later section, the rising estrogen levels play a role in stimulating labor and delivery.

Embryogenesis

Shortly after gastrulation begins, the body of the embryo begins to separate itself from the rest of the embryonic disc. The body of the embryo and its internal organs now start to form. This process, called **embryogenesis**, begins as folding and differential growth of the embryonic disc produce a bulge that projects into the amniotic cavity (**Figure 29–5**). This projection is known as the **head fold**; similar movements lead to the formation of a **tail fold** (**Figure 29–5**).

The embryo is now physically as well as developmentally distinct from the embryonic disc and the extraembryonic membranes. The definitive orientation of the embryo can now be seen, complete with dorsal and ventral surfaces and left and right sides. Table 29–2 provides an overview of the subsequent development of the major organs and body systems. The changes in proportions and appearance that occur between the second developmental week and the end of the first trimester are summarized in **Figure 29–7**.

The first trimester is a critical period for development, because events in the first 12 weeks establish the basis for **organogenesis**, the process of organ formation. The major features of organogenesis in each organ system are described in Embryology Summaries 6–21 in the *Atlas*. Important developmental milestones are indicated in Table 29–2.

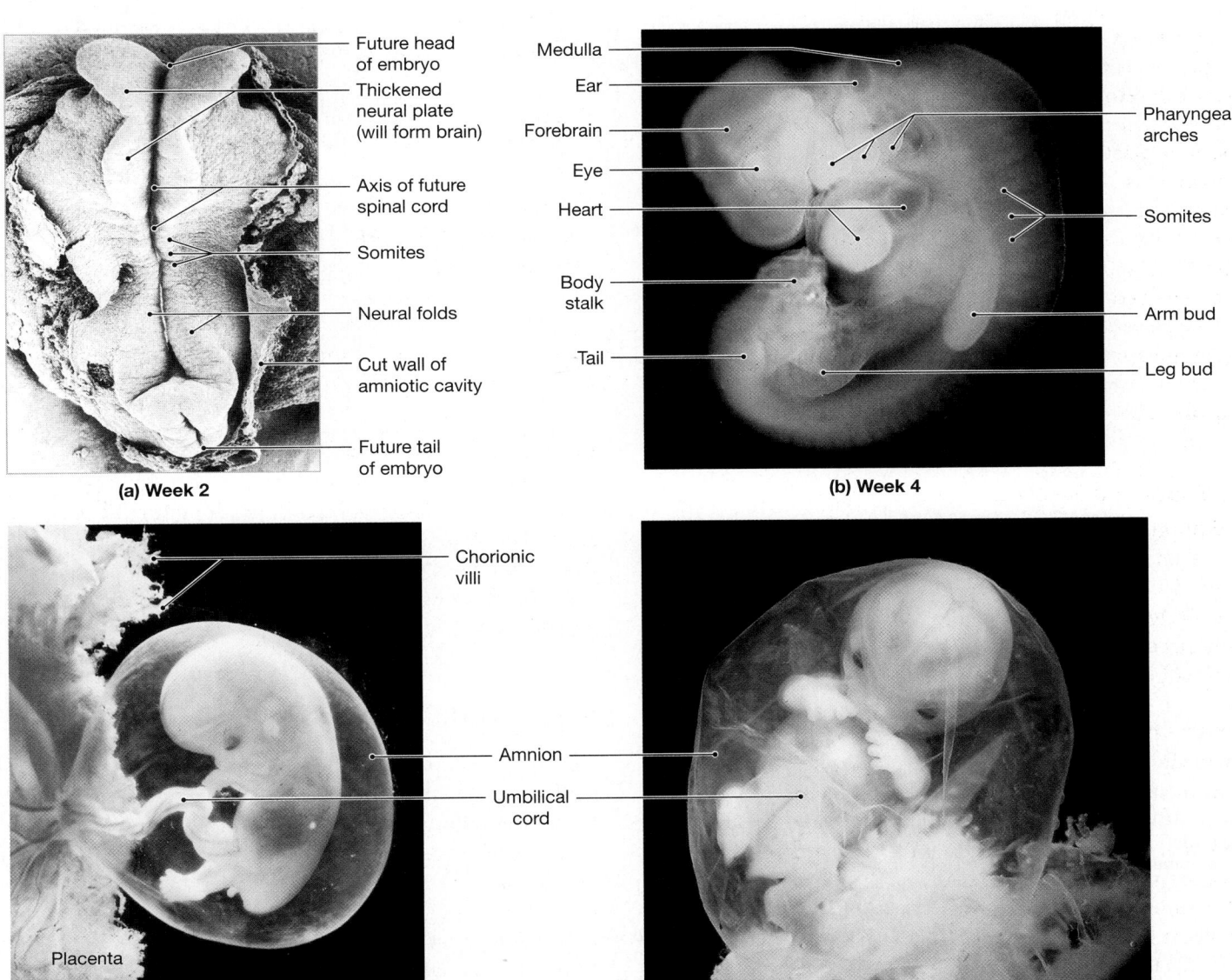

(a) Week 2

- Future head of embryo
- Thickened neural plate (will form brain)
- Axis of future spinal cord
- Somites
- Neural folds
- Cut wall of amniotic cavity
- Future tail of embryo

(b) Week 4

- Medulla
- Ear
- Forebrain
- Eye
- Heart
- Body stalk
- Tail
- Pharyngeal arches
- Somites
- Arm bud
- Leg bud

(c) Week 8

- Chorionic villi
- Amnion
- Umbilical cord
- Placenta

(d) Week 12

Figure 29–7 The First 12 Weeks of Development. (a) An SEM of the superior surface of a monkey embryo at 2 weeks of development. A human embryo at this stage would look essentially the same. **(b–d)** Fiber-optic views of human development after 4, 8, and 12 weeks. For actual sizes, see *Figure 29–13*. ATLAS: Plate 90a

SUMMARY TABLE 29–2 An Overview of Prenatal Development

Background Material
ATLAS: Embryology Summaries 1–4:
The Development of Tissues
The Development of Epithelia
The Development of Connective Tissues
The Development of Organ Systems

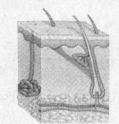   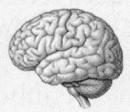 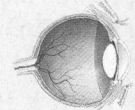

Gestational Age (Months)	Size and Weight	Integumentary System	Skeletal System	Muscular System	Nervous System	Special Sense Organs
1	5 mm (0.2 in.), 0.02 g (0.00004 lb)		(b) Formation of somites	(b) Formation of somites	(b) Formation of neural tube	(b) Formation of eyes and ears
2	28 mm (1.1 in.), 2.7 g (0.00595 lb)	(b) Formation of nail beds, hair follicles, sweat glands	(b) Formation of axial and appendicular cartilages	(c) Rudiments of axial musculature	b) CNS, PNS organization, growth of cerebrum	(b) Formation of taste buds, olfactory epithelium
3	78 mm (3.1 in.), 26 g (0.0573 lb)	(b) Epidermal layers appear	(b) Spreading of ossification centers	(c) Rudiments of appendicular musculature	(c) Basic spinal cord and brain structure	
4	133 mm (5.2 in.), 0.15 kg (0.33 lb)	(b) Formation of hair, sebaceous glands (c) Sweat glands	(b) Articulations (c) Facial and palatal organization	Fetal movements can be felt by the mother	(b) Rapid expansion of cerebrum	(c) Basic eye and ear structure (b) Formation of peripheral receptors
5	185 mm (7.3 in.), 0.46 kg (1.01 lb)	(b) Keratin production, nail production			(b) Myelination of spinal cord	
6	230 mm (9.1 in.), 0.64 kg; (1.41 lb)			(c) Perineal muscles	(b) Formation of CNS tract (c) Layering of cortex	
7	270 mm (10.6 in.), 1.492 kg (3.284 lb)	(b) Keratinization, formation of nails, hair				(c) Eyelids open, retinas sensitive to light
8	310 mm (12.2 in.), 2.274 kg (5.003 lb)		(b) Formation of epiphyseal cartilages			(c) Taste receptors functional
9	346 mm (13.6 in.), 3.2 kg (7.04 lb)					
Early postnatal development	Hair changes in consistency and distribution	Formation and growth of epiphyseal cartilages continue	Muscle mass and control increase	Myelination, layering, CNS tract formation continue		
Location of relevant text and illustrations		ATLAS: Embryology Summary 5	Ch. 6: pp. 194–198 Ch. 7: pp. 229–230 ATLAS: Embryology Summaries 6, 7, 8	ATLAS: Embryology Summary 9	Ch. 14: pp. 462–463 ATLAS: Embryology Summaries 10, 11, 12	ATLAS: Embryology Summary 13

Note: (b) = beginning; (c) = completion.

 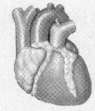  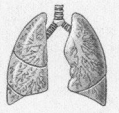 

Gestational Age (Months)	Endocrine System	Cardiovascular and Lymphoid Systems	Respiratory System	Digestive System	Urinary System	Reproductive System
1		(b) Heartbeat	(b) Formation of trachea and lungs	(b) Formation of intestinal tract, liver, pancreas (c) Yolk sac	(c) Allantois	
2	(b) Formation of thymus, thyroid, pituitary, suprarenal glands	(c) Basic heart structure, major blood vessels, lymph nodes and ducts (b) Blood formation in liver	(b) Extensive bronchial branching into mediastinum (c) Diaphragm	(b) Formation of intestinal subdivisions, villi, salivary glands	(b) Formation of kidneys (metanephros)	(b) Formation of mammary glands
3	(c) Thymus, thyroid gland	(b) Tonsils, blood formation in bone marrow		(c) Gallbladder, pancreas		(b) Formation of gonads, ducts genitalia; oogonia in female
4		(b) Migration of lymphocytes to lymphoid organs; blood formation in spleen			(b) Degeneration of embryonic kidneys (mesonephros)	
5		(c) Tonsils	(c) Nostrils open	(c) Intestinal subdivisions		
6	(c) Suprarenal glands	(c) Spleen, liver, bone marrow	(b) Formation of alveoli	(c) Epithelial organization, glands		
7	(c) Pituitary gland			(c) Intestinal plicae circulares	(b) Descent of testes	
8			Complete pulmonary branching and alveolar structure		(c) Nephron formation	Descent of testes complete at or near time of birth
9						
Postnatal development		Cardiovascular changes at birth; immune response gradually becomes fully operational				
Location of relevant text and illustrations	ATLAS: Embryology Summary 14	Ch. 19: pp. 660, 669 Ch. 21: pp. 767–769 ATLAS: Embryology Summaries 15, 16, 17	Ch. 23: p. 865 ATLAS: Embryology Summary 18	Ch. 24: pp. 876–878 ATLAS: Embryology Summary 19	ATLAS: Embryology Summary 20	Ch. 28: pp. 1043–1044 ATLAS: Embryology Summary 21

Note: (b) = beginning; (c) = completion.

29 REPRODUCTIVE

29-5 During the second and third trimesters, maternal organ systems support the developing fetus, and the uterus undergoes structural and functional changes

By the end of the first trimester (**Figure 29–7d**), the rudiments of all the major organ systems have formed. Over the next three months, the fetus will grow to a weight of about 0.64 kg (1.4 lb). Encircled by the amnion, the fetus grows faster than the surrounding placenta during this second trimester. When the mesoderm on the outer surface of the amnion contacts the mesoderm on the inner surface of the chorion, these layers fuse, creating a compound *amniochorionic membrane*. **Figure 29–8a** shows a four-month-old fetus; **Figure 29–8b** shows a six-month-old fetus.

During the third trimester, most of the organ systems become ready to perform their normal functions without maternal assistance. The rate of growth starts to slow, but in absolute terms this trimester sees the largest weight gain. In the last three months of gestation, the fetus gains about 2.6 kg (5.7 lb), reaching a full-term weight of approximately 3.2 kg (7 lb). The Embryology Summaries in the *Atlas* illustrate organ system development in the second and third trimesters, and highlights are noted in Table 29–2.

At the end of gestation, a typical uterus will have undergone a tremendous increase in size. **Figure 29–9a–c** shows the positions of the uterus, fetus, and placenta from 16 weeks to *full term* (nine months). When the pregnancy is at full term, the uterus and fetus push many of the maternal abdominal organs out of their normal positions (**Figure 29–9c,d**).

The A&P Top 100

#100 The basic body plan, the foundations of all organ systems, and the four extraembryonic membranes appear during the first trimester. The processes involved are complex and delicate; not every zygote starts cleavage, and fewer than half of the zygotes that do begin cleavage survive until the end of the first trimester. The second trimester is a period of rapid growth, accompanied by the development of fetal organs that will then become fully functional by the end of the third trimester.

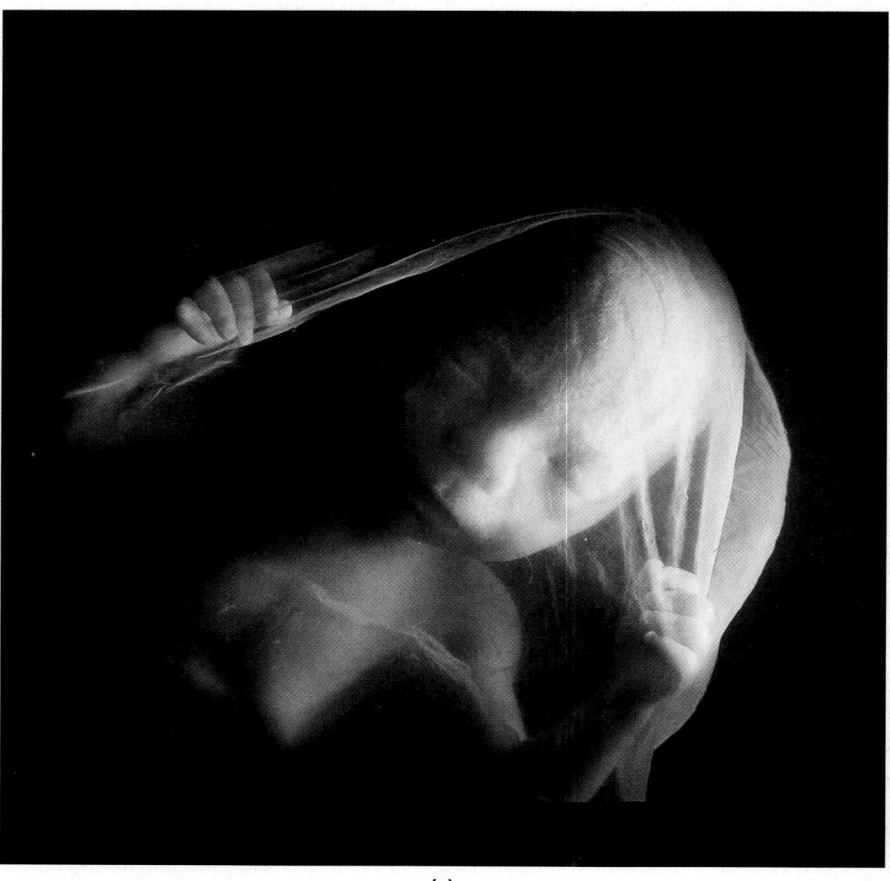

(a)

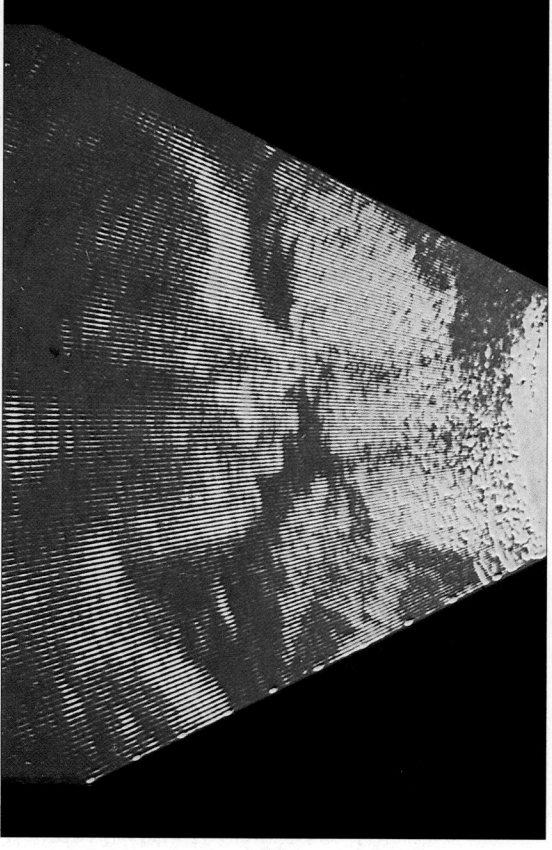

(b)

Figure 29–8 **The Second and Third Trimesters. (a)** A four-month-old fetus, seen through a fiber-optic endoscope. **(b)** Head of a six-month-old fetus, seen through ultrasound. ATLAS: Plate 90b

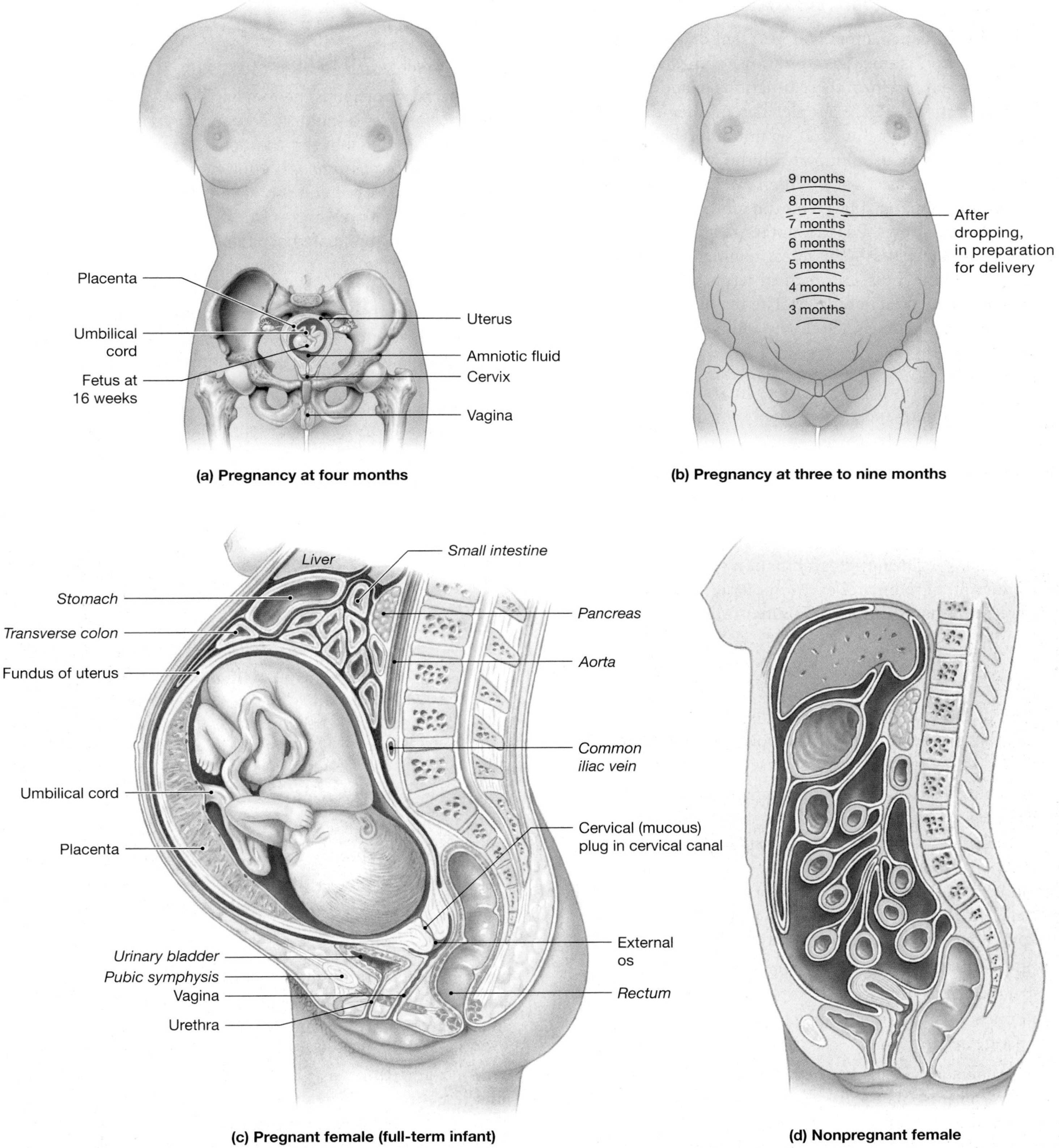

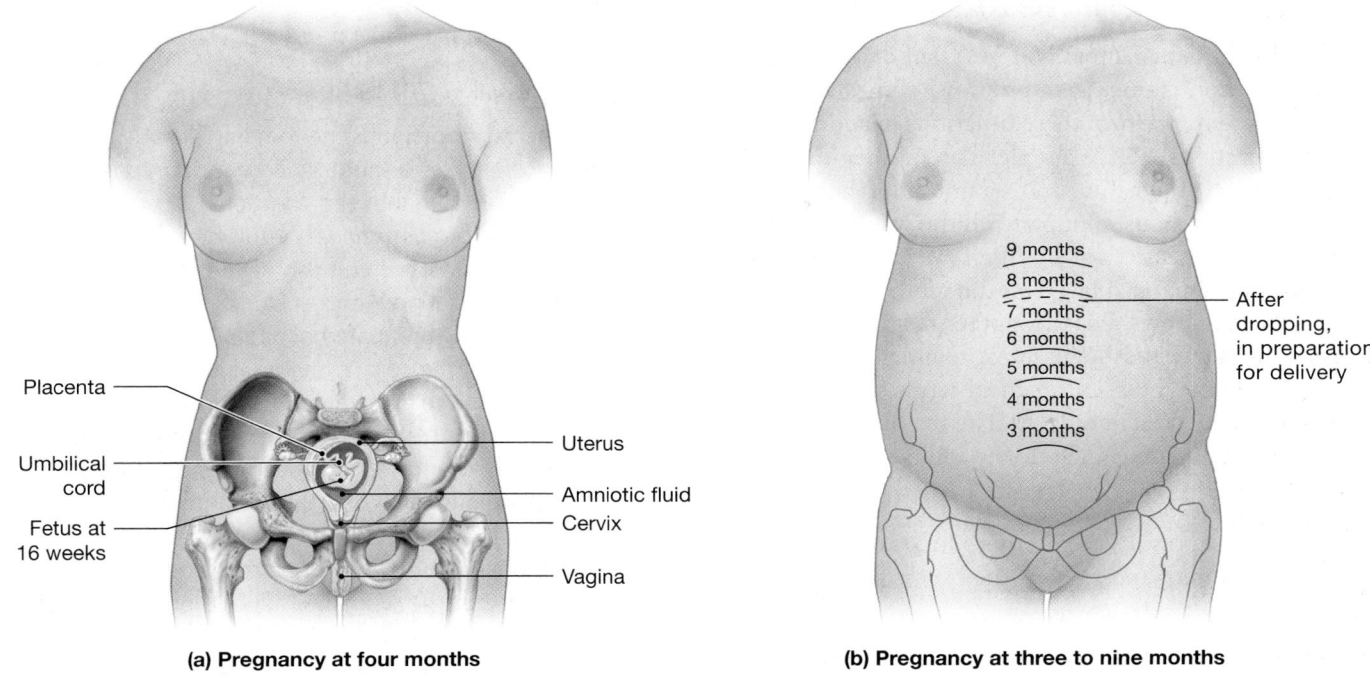

(a) Pregnancy at four months

Placenta

Umbilical cord

Fetus at 16 weeks

Uterus

Amniotic fluid

Cervix

Vagina

(b) Pregnancy at three to nine months

9 months
8 months
7 months
6 months
5 months
4 months
3 months

After dropping, in preparation for delivery

(c) Pregnant female (full-term infant)

Liver

Small intestine

Stomach

Transverse colon

Fundus of uterus

Pancreas

Aorta

Common iliac vein

Umbilical cord

Placenta

Cervical (mucous) plug in cervical canal

Urinary bladder

Pubic symphysis

Vagina

Urethra

External os

Rectum

(d) Nonpregnant female

Figure 29–9 **Growth of the Uterus and Fetus. (a)** Pregnancy at 16 weeks, showing the positions of the uterus, fetus, and placenta. **(b)** Pregnancy at three months to nine months (full term), showing the superior-most position of the uterus within the abdomen. **(c)** Pregnancy at full term. Note the positions of the uterus and full-term fetus within the abdomen, and the displacement of abdominal organs; compare with **(d)**, a sectional view through the abdominopelvic cavity of a woman who is not pregnant.

29 REPRODUCTIVE

Pregnancy and Maternal Systems

The developing fetus is totally dependent on maternal organ systems for nourishment, respiration, and waste removal. Maternal systems perform these functions in addition to their normal operations. For example, the mother must absorb enough oxygen, nutrients, and vitamins for herself *and* for her fetus, and she must eliminate all the wastes that are generated. Although this is not a burden over the initial weeks of gestation, the demands placed on the mother become significant as the fetus grows. For the mother to survive under these conditions, maternal systems must compensate for changes introduced by the fetus. In practical terms, the mother must breathe, eat, and excrete for two. The major changes that occur in maternal systems include the following:

- *Maternal Respiratory Rate Goes Up and Tidal Volume Increases.* As a result, the mother's lungs deliver the extra oxygen required, and remove the excess carbon dioxide generated, by the fetus.

- *Maternal Blood Volume Increases.* This increase occurs because blood flowing into the placenta reduces the volume in the rest of the systemic circuit, and because fetal metabolic activity both lowers blood P_{O_2} and elevates P_{CO_2}. The latter combination stimulates the production of renin and erythropoietin, leading to an increase in maternal blood volume through mechanisms detailed in Chapter 21 (see **Figure 21–17,** p. 746). By the end of gestation, maternal blood volume has increased by almost 50 percent.

- *Maternal Requirements for Nutrients Climb 10–30 Percent.* Pregnant women must nourish both themselves and their fetus, so they tend to have increased hunger sensations.

- *Maternal Glomerular Filtration Rate Increases by Roughly 50 Percent.* This increase, which corresponds to the increase in blood volume, accelerates the excretion of metabolic wastes generated by the fetus. Because the volume of urine produced increases and the weight of the uterus presses down on the urinary bladder, pregnant women need to urinate frequently.

- *The Uterus Undergoes a Tremendous Increase in Size.* Structural and functional changes in the expanding uterus are so important that we will discuss them in a separate section.

- *The Mammary Glands Increase in Size, and Secretory Activity Begins.* Mammary gland development requires a combination of hormones, including human placental lactogen and placental prolactin from the placenta, and PRL, estrogens, progesterone, GH, and thyroxine from maternal endocrine organs. By the end of the sixth month of pregnancy, the mammary glands are fully developed and begin to produce clear secretions that are stored in the duct system of those glands and may be expressed from the nipple.

Structural and Functional Changes in the Uterus

At the end of gestation, a typical uterus has grown from 7.5 cm (3 in.) in length and 30–40 g (1–1.4 oz) in weight to 30 cm (12 in.) in length and 1100 g (2.4 lb) in weight. The uterus may then contain 2 liters of fluid, plus fetus and placenta, for a total weight of roughly 6–7 kg (13–15.4 lb). This remarkable expansion occurs through the enlargement (hypertrophy) of existing cells, especially smooth muscle fibers, rather than by an increase in the total number of cells.

The tremendous stretching of the uterus is associated with a gradual increase in the rate of spontaneous smooth muscle contractions in the myometrium. In the early stages of pregnancy, the contractions are weak, painless, and brief. Evidence indicates that progesterone released by the placenta has an inhibitory effect on uterine smooth muscle, preventing more extensive and more powerful contractions.

Three major factors oppose the calming action of progesterone:

1. *Rising Estrogen Levels.* Estrogens produced by the placenta increase the sensitivity of the uterine smooth muscles and make contractions more likely. Throughout pregnancy, progesterone exerts the dominant effect, but as delivery approaches, the production of estrogens accelerates and the myometrium becomes more sensitive to stimulation. Estrogens also increase the sensitivity of smooth muscle fibers to oxytocin.

2. *Rising Oxytocin Levels.* Rising oxytocin levels lead to an increase in the force and frequency of uterine contractions. Oxytocin release is stimulated by high estrogen levels and by distortion of the cervix. Uterine distortion, especially in the region of the cervix, occurs as the weight of the fetus increases.

3. *Prostaglandin Production.* Estrogens and oxytocin stimulate the production of prostaglandins in the endometrium. These prostaglandins further stimulate smooth muscle contractions.

Late in pregnancy, some women experience occasional spasms in the uterine musculature, but these contractions are neither regular nor persistent. Such contractions are called **false labor**. **True labor** begins when biochemical and mechanical factors reach a point of no return. After nine months of gestation, multiple factors interact to initiate true labor. Once **labor contractions** have begun in the myometrium, positive feedback ensures that they will continue until delivery has been completed.

Figure 29–10 diagrams important factors that stimulate and sustain labor. When labor commences, the fetal pituitary gland secretes oxytocin, which is then released into the maternal bloodstream at the placenta. This may be the actual trigger for the onset of true labor, as it increases myometrial

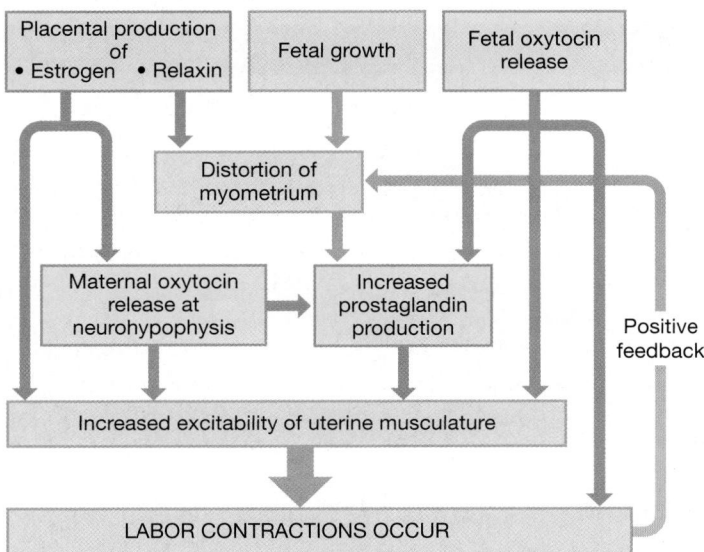

Figure 29–10   Factors Involved in the Initiation of Labor and Delivery.

contractions and prostaglandin production, on top of the priming effects of estrogens and maternal oxytocin.

29-6 Labor consists of the dilation, expulsion, and placental stages

The goal of labor is **parturition** (par-toor-ISH-un), the forcible expulsion of the fetus. During true labor, each contraction begins near the top of the uterus and sweeps in a wave toward the cervix. The contractions are strong and occur at regular intervals. As parturition approaches, the contractions increase in force and frequency, changing the position of the fetus and moving it toward the cervical canal.

Stages of Labor

Labor has traditionally been divided into three stages: the *dilation stage*, the *expulsion stage*, and the *placental stage* (**Figure 29–11**).

The Dilation Stage

The **dilation stage** begins with the onset of true labor, as the cervix dilates and the fetus begins to shift toward the cervical canal (STAGE 1 in **Figure 29–11**), moved by gravity and uterine contractions. This stage is highly variable in length but typically lasts eight or more hours. At the start of the dilation stage, labor contractions last up to half a minute and occur once every 10–30 minutes; their frequency increases steadily. Late in this stage, the amniochorionic membrane ruptures, an event sometimes referred to as "having one's water break." If this event occurs before other events of the dilation stage, the life of the fetus may be at risk from infection; if the risk is sufficiently great, labor can be induced.

The Expulsion Stage

The **expulsion stage** begins as the cervix, pushed open by the approaching fetus, completes its dilation (STAGE 2 in **Figure 29–11**). In this stage, contractions reach maximum intensity, occurring at perhaps two- or three-minute intervals and lasting a full minute. Expulsion continues until the fetus has

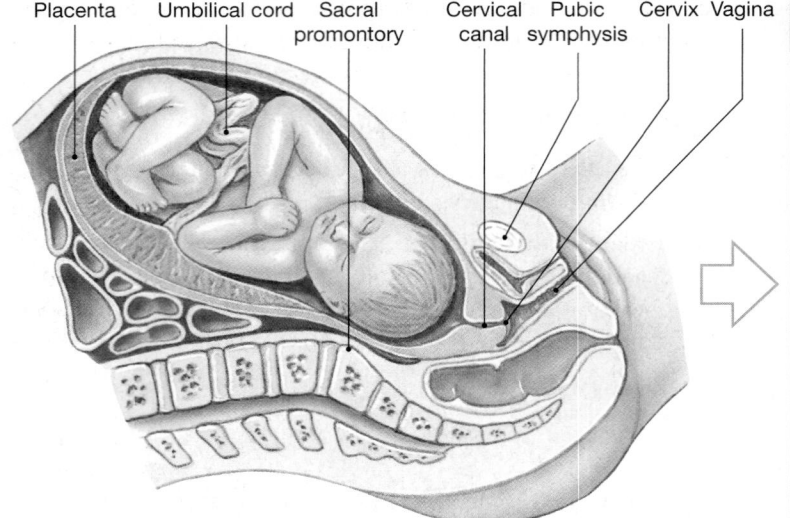

Placenta   Umbilical cord   Sacral   Cervical   Pubic   Cervix Vagina
promontory   canal   symphysis

Fully developed fetus before labor begins

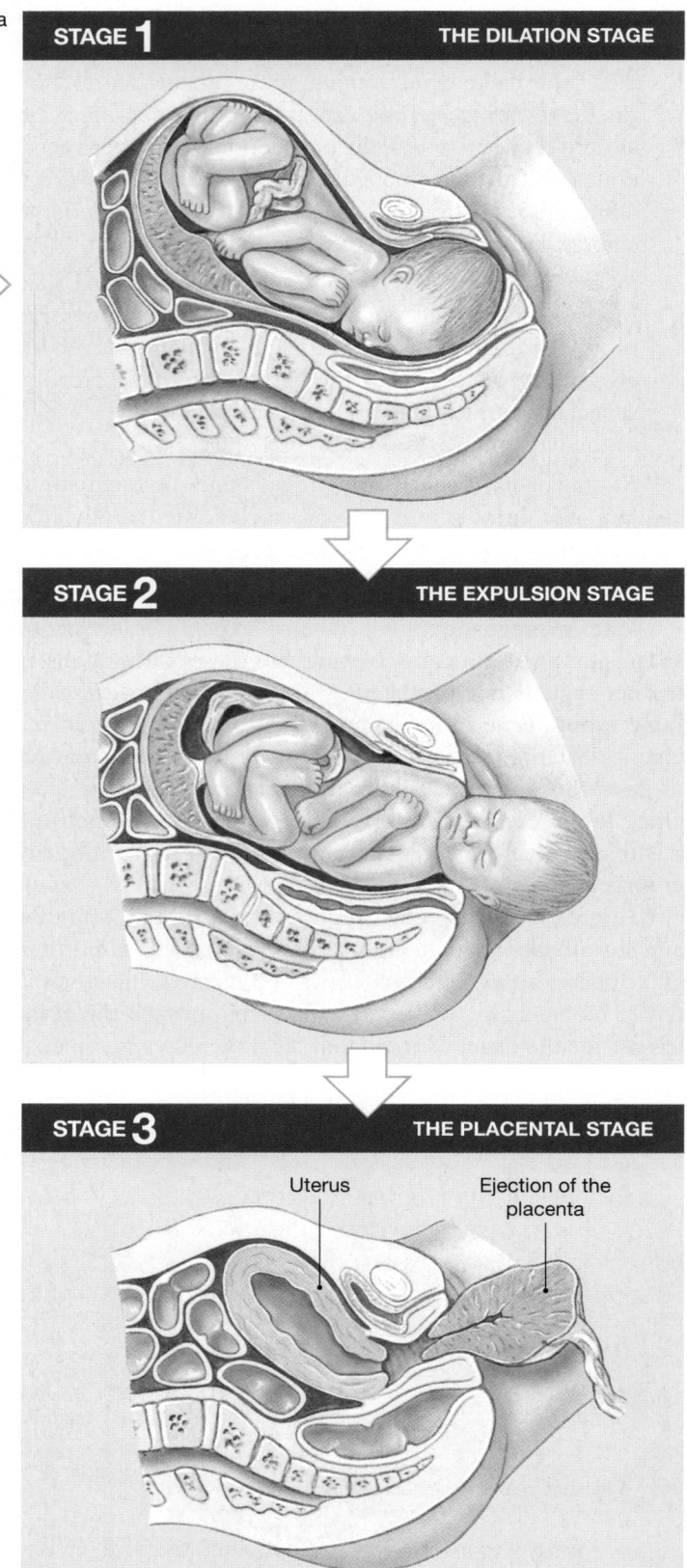

STAGE **1**   THE DILATION STAGE

STAGE **2**   THE EXPULSION STAGE

STAGE **3**   THE PLACENTAL STAGE

Uterus   Ejection of the
placenta

Figure 29–11   **The Stages of Labor.**

emerged from the vagina; in most cases, the expulsion stage lasts less than two hours. The arrival of the newborn infant into the outside world is **delivery**, or birth.

If the vaginal canal is too small to permit the passage of the fetus, posing acute danger of perineal tearing, a physician may temporarily enlarge the passageway by performing an **episiotomy** (e-piz-ē-OT-ō-mē), an incision through the perineal musculature. After delivery, this surgical cut is repaired with sutures, a much simpler procedure than suturing the jagged edges associated with an extensive perineal tear. If complications arise during the dilation or expulsion stage, the infant can be delivered by **cesarean section**, or "*C-section*." In such cases, an incision is made through the abdominal wall, and the uterus is opened just enough to allow passage of the infant's head. This procedure was performed during 29 percent of the deliveries in the United States in 2004—more often than necessary, according to some studies. Over the past decade, efforts have been made to reduce the frequency of both episiotomies and cesarean sections.

The Placental Stage

During the **placental stage** of labor, muscle tension builds in the walls of the partially empty uterus, which gradually decreases in size (STAGE 3 in **Figure 29–11**). This uterine contraction tears the connections between the endometrium and the placenta. In general, within an hour of delivery, the placental stage ends with the ejection of the placenta, or *afterbirth*. The disruption of the placenta is accompanied by a loss of blood, but associated uterine contraction compresses the uterine vessels and usually restricts this flow. Because maternal blood volume has increased greatly during pregnancy, the blood loss that does occur can normally be tolerated without difficulty.

Premature Labor

Premature labor occurs when true labor begins before the fetus has completed normal development. The newborn's chances of surviving are directly related to its body weight at delivery. Even with massive supportive efforts, newborns weighing less than 400 g (14 oz) at birth will not survive, primarily because their respiratory, cardiovascular, and urinary systems are unable to support life without aid from maternal systems. As a result, the dividing line between spontaneous abortion and **immature delivery** is usually set at 500 g (17.6 oz), the normal weight near the end of the second trimester.

Most fetuses born at 25–27 weeks of gestation (a birth weight under 600 g or 21.1 oz) die despite intensive neonatal care; moreover, survivors have a high risk of developmental abnormalities. **Premature delivery** usually refers to birth at 28–36 weeks (a birth weight over 1 kg or 2.2 lb). With care, these newborns have a good chance of surviving and developing normally.

Difficult Deliveries

By the end of gestation in most pregnancies, the fetus has rotated within the uterus to transit the birth canal headfirst, facing the mother's sacrum. In about 6 percent of deliveries, the fetus faces the mother's pubis instead. These babies can be delivered normally, given enough time, but risks to infant and mother are reduced by a *forceps delivery*. Forceps resemble large, curved salad tongs that can be separated for insertion into the vaginal canal, one side at a time. Once in place, they are reunited and used to grasp the head of the fetus. An intermittent pull is applied, so that the forces on the head resemble those of normal delivery.

In 3–4 percent of deliveries, the legs or buttocks of the fetus enter the vaginal canal first. Such deliveries are **breech births**. Risks to the fetus are higher in breech births than in normal deliveries, because the umbilical cord can become constricted, cutting off placental blood flow. The head is normally the widest part of the fetus; the mother's cervix may dilate enough to pass the baby's legs and body, but not the head. This entrapment compresses the umbilical cord, prolongs delivery, and subjects the fetus to severe distress and potential injury. If attempts to reposition the fetus or promote further dilation are unsuccessful over the short term, delivery by cesarean section may be required.

Multiple Births

Multiple births (twins, triplets, quadruplets, and so forth) can occur for several reasons. The ratio of twin births to single births in the U.S. population is roughly 1:89. "Fraternal," or **dizygotic** (dī-zī-GOT-ik), twins develop when two separate oocytes were ovulated and subsequently fertilized. Because chromosomes are shuffled during meiosis, the odds against any two zygotes from the same parents having identical genes exceed 1 in 8.4 million. Seventy percent of twins are dizygotic.

"Identical," or **monozygotic**, twins result either from the separation of blastomeres early in cleavage or from the splitting of the inner cell mass before gastrulation. In either event, the genetic makeup of the twins is identical because both formed from the same pair of gametes. Triplets, quadruplets, and larger multiples can result from multiple ovulations, blastomere splitting, or some combination of the two. For unknown reasons, the rates of naturally occurring multiple births fall into a pattern: Twins occur in 1 of every 89 births, triplets in 1 of every 89^2 (or 7921) births, quadruplets in 1 of every 89^3 (704,969) births, and so forth. The incidence of multiple births can be increased by exposure to fertility drugs that stimulate the maturation of abnormally large numbers of oocytes.

Multiple pregnancies pose special problems because the strains on the mother are multiplied. The chances of premature labor are increased, and the risks to the mother are higher than for single births. Increased risks also extend to the fetuses during gestation, and to the newborns, because even at full term such newborns have lower-than-average birth weights. They are also more likely to have problems during delivery. For example, in more than half of twin deliveries, one or both fetuses enter the vaginal canal in an abnormal position.

If the splitting of the blastomeres or of the embryonic disc is not complete, **conjoined** (*Siamese*) **twins** may develop. These genetically identical twins typically share some skin, a portion of the liver, and perhaps other internal organs as well. When the fusion is minor, the infants can be surgically separated with some success. Most conjoined twins with more extensive fusions fail to survive delivery.

CHECKPOINT

15. Name the three stages of labor.
16. Differentiate between immature delivery and premature delivery.
17. Supply the biological term for fraternal twins and identical twins.

See the blue Answers tab at the end of the book.

29-7 Postnatal stages are the neonatal period, infancy, childhood, adolescence, maturity, and senescence

Developmental processes do not cease at delivery, because newborns have few of the anatomical, functional, or physiological characteristics of mature adults. The course of postnatal development typically includes five **life stages**: (1) the *neonatal period*, (2) *infancy*, (3) *childhood*, (4) *adolescence*, and (5) *maturity*. Each stage is typified by a distinctive combination of characteristics and abilities. These stages are familiar parts of human experience. Although each stage has distinctive features, the transitions between them are gradual, and the boundaries indistinct. At maturity, development ends and the process of aging, or *senescence*, begins.

The Neonatal Period, Infancy, and Childhood

The **neonatal period** extends from birth to one month thereafter. **Infancy** then continues to two years of age, and **childhood** lasts until **adolescence**, the period of sexual and physical maturation. Two major events are under way during these developmental stages:

1. The organ systems (except those associated with reproduction) become fully operational and gradually acquire the functional characteristics of adult structures.

2. The individual grows rapidly, and body proportions change significantly.

   Pediatrics is the medical specialty that focuses on postnatal development from infancy through adolescence. Infants and young children cannot clearly describe the problems they are experiencing, so pediatricians and parents must be skilled observers. Standardized tests are used to assess developmental progress relative to average values.

The Neonatal Period

Physiological and anatomical changes occur as the fetus completes the transition to the status of newborn, or **neonate**. Before delivery, dissolved gases, nutrients, wastes, hormones, and antibodies were transferred across the placenta. At birth, the neonate must become relatively self-sufficient, performing respiration, digestion, and excretion using its own specialized organs and organ systems. The transition from fetus to neonate can be summarized as follows:

- At birth, the lungs are collapsed and filled with fluid. Filling them with air requires a massive and powerful inhalation. ∞ p. 865

- When the lungs expand, the pattern of cardiovascular circulation changes due to alterations in blood pressure and flow rates. The ductus arteriosus closes, isolating the pulmonary and systemic trunks. Closure of the foramen ovale separates the atria of the heart, completing the separation of the pulmonary and systemic circuits. ∞ p. 767

- The typical neonatal heart rate (120–140 beats per minute) and respiratory rate (30 breaths per minute) are considerably higher than in adults. In addition, the metabolic rate per unit of body weight in neonates is roughly twice that of adults.

- Before birth, the digestive system remains relatively inactive, although it does accumulate a mixture of bile secretions, mucus, and epithelial cells. This collection of debris, called *meconium*, is excreted during the first few days of life. Over that period, the newborn begins to nurse.

- As waste products build up in the arterial blood, they are excreted at the kidneys. Glomerular filtration is normal, but the neonate cannot concentrate urine to any significant degree. As a result, urinary water losses are high, and neonatal fluid requirements are proportionally much greater than those of adults.

- The neonate has little ability to control its body temperature, particularly in the first few days after delivery. As the infant grows larger and its insulating subcutaneous adipose "blanket" gets thicker, its metabolic rate also rises. Daily and even hourly shifts in body temperature continue throughout childhood. ∞ p. 960

Over the entire neonatal period, the newborn is dependent on nutrients contained in milk, typically breast milk secreted by the maternal mammary glands.

Lactation and the Mammary Glands. By the end of the sixth month of pregnancy, the mammary glands are fully developed, and the gland cells begin to produce a secretion known as **colostrum** (kō-LOS-trum). Ingested by the infant during the first two or three days of life, colostrum contains more proteins and far less fat than breast milk. Many of the proteins are antibodies that may help the infant ward off infections until its own immune system becomes fully functional. In addition, the mucins present in both colostrum and milk can inhibit the replication of a family of viruses (*rotaviruses*) that can cause dangerous forms of gastroenteritis and diarrhea in infants.

As colostrum production drops, the mammary glands convert to milk production. Breast milk consists of water, proteins, amino acids, lipids, sugars, and salts. It also contains large quantities of *lysozyme*—an enzyme with antibiotic properties. Human milk provides roughly 750 kilocalories per liter. The secretory rate varies with the demand, but a 5–6-kg (11–13-lb) infant usually requires about 850 mL (3.6 cups) of milk per day. (The production of milk throughout this period is maintained through the combined actions of several hormones, as detailed in Chapter 18. ∞ p. 617)

Milk becomes available to infants through the **milk letdown reflex** (**Figure 29–12**). Mammary gland secretion is triggered when the infant sucks on the nipple (STEP 1). The stimulation of tactile receptors there leads to the stimulation

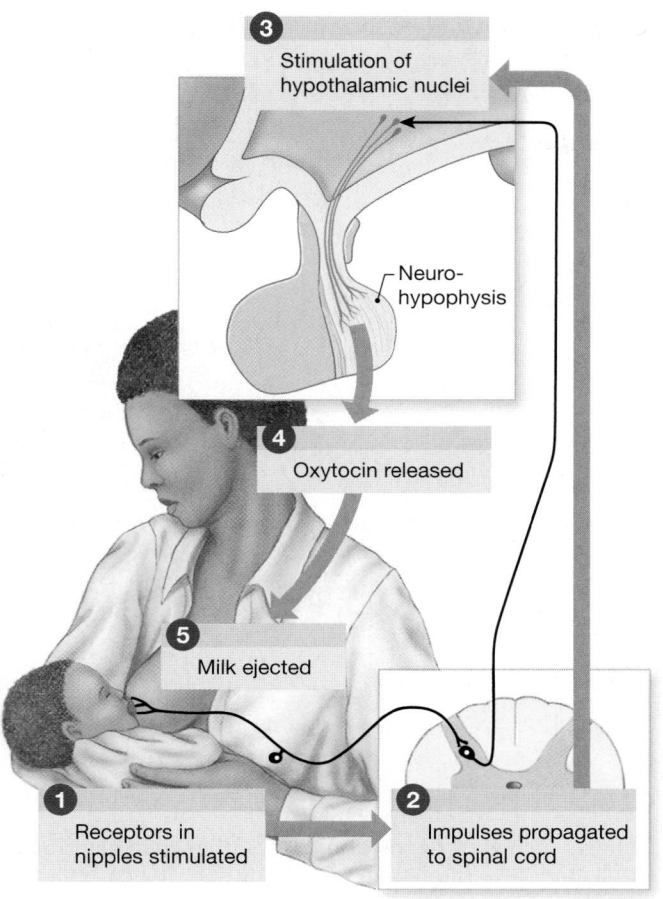

Figure 29–12   The Milk Let-Down Reflex.

of secretory neurons in the paraventricular nucleus of the mother's hypothalamus (STEPS 2 and 3). These neurons release oxytocin at the neurohypophysis (STEP 4). When circulating oxytocin reaches the mammary gland, this hormone causes the contraction of *myoepithelial cells*, contractile cells in the walls of the lactiferous ducts and sinuses. The result is milk ejection (STEP 5), or *milk let-down*.

The milk let-down reflex continues to function until *weaning*, typically one to two years after birth. Milk production ceases soon after, and the mammary glands gradually return to a resting state. Earlier weaning is a common practice in the United States, where women take advantage of commercially prepared milk- or soy-based infant formulas that closely approximate the composition of natural breast milk. The major difference between such substitutes and natural milk is that the substitutes lack antibodies and lysozyme, which play important roles in maintaining the health of the infant. Consequently, early weaning is associated with an increased risk of infections and allergies in the infant.

Infancy and Childhood

The most rapid growth occurs during prenatal development, and the growth rate declines after delivery. Growth during infancy and childhood occurs under the direction of circulating hormones, notably growth hormone, suprarenal steroids, and thyroid hormones. These hormones affect each tissue and organ in specific ways, depending on the sensitivities of the individual cells. As a result, growth does not occur uniformly, so body proportions gradually change. The head, for example, is relatively large at birth but decreases in proportion with the rest of the body as the child grows to adulthood (**Figure 29–13**).

Adolescence and Maturity

Adolescence begins at **puberty**, the period of sexual maturation, and ends when growth is completed. Three major hormonal events interact at the onset of puberty:

1. The hypothalamus increases its production of gonadotropin-releasing hormone (GnRH). Evidence indicates that this increase is dependent on adequate levels of *leptin*, a hormone released by adipose tissues. ∞ p. 638

2. Endocrine cells in the adenohypophysis become more sensitive to the presence of GnRH, and circulating levels of FSH and LH rise rapidly.

3. Ovarian or testicular cells become more sensitive to FSH and LH, initiating (1) gamete production, (2) the secretion of sex hormones that stimulate the appearance of secondary sex characteristics and behaviors, and (3) a sudden acceleration in the growth rate, culminating in closure of the epiphyseal cartilages.

The age at which puberty begins varies. In the United States today, puberty generally starts at about age 12 in boys and 11 in girls, but the normal ranges are broad (10–15 in boys, 9–14 in girls). Many body systems alter their activities in response to circulating sex hormones and to the presence of growth hormone, thyroid hormones, prolactin, and adrenocortical hormones, so sex-specific differences in structure and function develop. At puberty, endocrine system changes induce characteristic changes in various body systems:

- *Integumentary System.* Testosterone stimulates the development of terminal hairs on the face and chest, whereas under estrogen stimulation those follicles continue to produce fine hairs. The hairline recedes under testosterone stimulation. Both testosterone and estrogen stimulate terminal hair growth in the axillae and in the genital area. Androgens, which are present in both sexes, also stimulate sebaceous gland secretion and may cause acne. Adipose tissues respond differently to testosterone

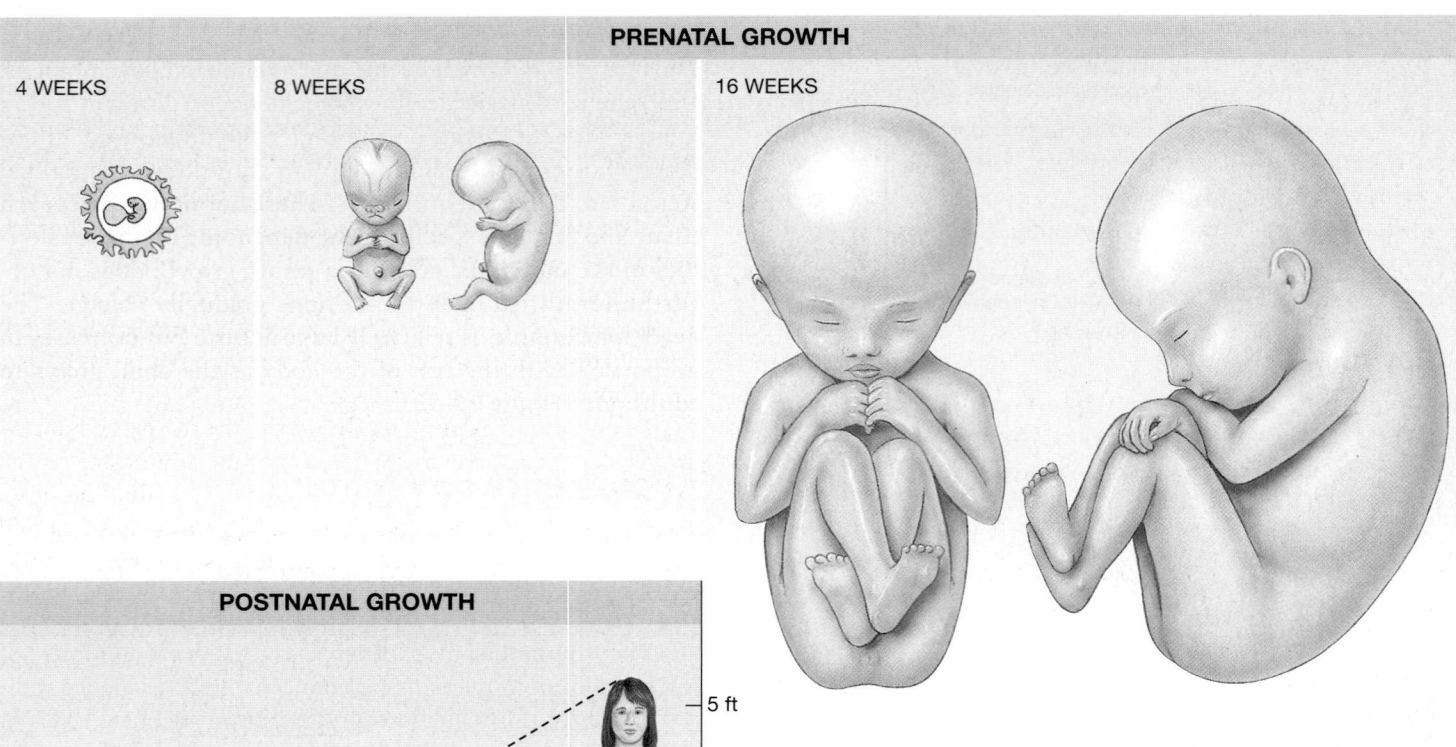

PRENATAL GROWTH

4 WEEKS   8 WEEKS   16 WEEKS

POSTNATAL GROWTH

5 ft

4 ft

3 ft

2 ft

1 ft

Newborn   6 years   Adult

Figure 29–13   Growth and Changes in Body Form and Proportion. The views at 4, 8, and 16 weeks of gestation are presented at actual size. Notice the changes in body form and proportions as development proceeds. For example, the head, which contains the brain and sense organs, is proportionately large at birth.

than to estrogens, and this difference produces changes in the distribution of subcutaneous body fat. In women, the combination of estrogens, prolactin, growth hormone, and thyroid hormones promotes the initial development of the mammary glands. Although the duct system becomes more elaborate, true secretory alveoli do not develop, and much of the growth of the breasts during this period

reflects increased deposition of fat rather than glandular tissue.

- *Skeletal System.* Both testosterone and estrogen accelerate bone deposition and skeletal growth. In the process, they promote closure of the epiphyseal cartilages and thus place a limit on growth in height. Estrogens cause more rapid epiphyseal plate closure than does testosterone. In addition, the period of skeletal growth is briefer in girls than in boys, and girls generally do not grow as tall as boys. Girls grow most rapidly between ages 10 and 13, whereas boys grow most rapidly between ages 12 and 15.

- *Muscular System.* Sex hormones stimulate the growth of skeletal muscle fibers, increasing strength and endurance. The effects of testosterone greatly exceed those of the estrogens, and the increased muscle mass accounts for significant sex differences in body mass, even for males and females of the same height. The stimulatory effects of testosterone on muscle mass has led to the use of anabolic steroids among competitive athletes of both sexes.

- *Nervous System.* Sex hormones affect central nervous system centers concerned with sexual drive and sexual behaviors. These centers differentiated in sex-specific ways during the second and third trimesters, when the fetal gonads secrete either testosterone (in males) or estrogens (in females). The surge in sex hormone secretion at puberty activates the CNS centers.

- *Cardiovascular System.* Testosterone stimulates erythropoiesis, thereby increasing blood volume and the hematocrit. In females whose uterine cycles have begun, the iron loss associated with menses increases the risk of developing iron-deficiency anemia. Late in each uterine cycle, estrogens and progesterone promote the movement of water from plasma into interstitial fluid, leading to an increase in tissue water content. Estrogens decrease plasma cholesterol levels and slow the formation of plaque. As a result, premenopausal women have a lower risk of atherosclerosis than do adult men.

- *Respiratory System.* Testosterone stimulates disproportionate growth of the larynx and a thickening and lengthening of the vocal cords. These changes cause a gradual deepening of the voice of males compared with that of females.

- *Reproductive System.* In males, testosterone stimulates the functional development of the accessory reproductive glands, such as the prostate gland and seminal glands, and helps promote spermatogenesis. In females, estrogens target the uterus, promoting a thickening of the myometrium, increasing blood flow to the endometrium, and stimulating cervical mucus production. Estrogens also promote the functional development of accessory reproductive organs in females. The first few uterine cycles may or may not be accompanied by ovulation. After the initial stage, the woman will be fertile, even though growth and physical maturation will continue for several years.

After puberty, the continued background secretion of estrogens or androgens maintains the foregoing sex-specific differences. In both sexes, growth continues at a slower pace until age 18–21, by which time most of the epiphyseal cartilages have closed. The boundary between adolescence and maturity is hazy, because it has physical, emotional, and behavioral components. Adolescence is often said to be over when growth stops, in the late teens or early twenties. The individual is then considered physically mature.

Senescence

Although physical growth may cease at maturity, physiological changes continue. The sex-specific differences produced at puberty are retained, but further changes occur when sex hormone levels decline at menopause or the male climacteric. ∞ p. 1079 All these changes are part of the process of **senescence** (*senesco*, to grow old), or aging, which reduces the functional capabilities of the individual. Even in the absence of such factors as disease or injury, senescence-related changes at the molecular level ultimately lead to death.

Table 29–3 summarizes the age-related changes in physiological systems discussed in earlier chapters. Taken together, these changes both reduce the functional abilities of the individual and affect homeostatic mechanisms. As a result, the elderly are less able to make homeostatic adjustments in response to internal or environmental stresses. The risks of contracting a variety of infectious diseases are proportionately increased as immune function deteriorates. This deterioration leads to drastic physiological changes that affect all internal systems. Death ultimately occurs when some combination of stresses cannot be countered by the body's existing homeostatic mechanisms.

Physicians attempt to forestall death by adjusting homeostatic mechanisms or removing the sources of stress. **Geriatrics** (*geras*, old age) is the medical specialty that deals with the problems associated with aging; physicians trained in geriatrics are known as **geriatricians**. Problems commonly encountered by geriatricians include infections, cancers, heart disease, strokes, arthritis, senile dementia, and anemia—conditions directly related to age-induced changes in vital systems.

SUMMARY TABLE 29–3   Effects of Aging on Organ Systems

The characteristic physical and functional changes that are part of the aging process affect all organ systems. Examples discussed in previous chapters include the following:

- A loss of elasticity in the skin that produces sagging and wrinkling. ∞ pp. 177–178
- A decline in the rate of bone deposition, leading to weak bones, and degenerative changes in joints that make them less mobile. ∞ pp. 207, 287
- Reductions in muscular strength and ability. ∞ p. 380
- Impairment of coordination, memory, and intellectual function. ∞ pp. 544–545
- Reductions in the production of, and sensitivity to, circulating hormones. ∞ p. 628
- Appearance of cardiovascular problems and a reduction in peripheral blood flow that can affect a variety of vital organs. ∞ p. 770
- Reduced sensitivity and responsiveness of the immune system, leading to infection, cancer, or both. ∞ p. 817
- Reduced elasticity in the lungs, leading to decreased respiratory function. ∞ p. 866
- Decreased peristalsis and muscle tone along the digestive tract. ∞ p. 921
- Decreased peristalsis and muscle tone in the urinary system, coupled with a reduction in the glomerular filtration rate. ∞ p. 1002
- Functional impairment of the reproductive system, which eventually becomes inactive when menopause or the male climacteric occurs. ∞ p. 1079

29 REPRODUCTIVE

18. Name the postnatal stages of development.

19. Describe the time frame for each of the following stages: neonatal period, infancy, and adolescence.

20. What is the difference between colostrum and breast milk?

21. Increases in the blood levels of GnRH, FSH, LH, and sex hormones mark the onset of which stage of development?

See the blue Answers tab at the end of the book.

29-8 Genes and chromosomes determine patterns of inheritance

Chromosomes contain DNA, and genes are functional segments of DNA. Each gene carries the information needed to direct the synthesis of a specific polypeptide. Chromosome structure and the functions of genes were introduced in Chapter 3. ∞ pp. 83–88 Every nucleated somatic cell in your body carries copies of the original 46 chromosomes present when you were a zygote. Those chromosomes and their component genes constitute your **genotype** (JĒN-ō-tīp; *geno-*, gene).

Through development and differentiation, the instructions contained in the genotype are expressed in many ways. No single cell or tissue uses all the information and instructions contained in the genotype. For example, in muscle fibers, the genes involved in the formation of excitable membranes and contractile proteins are active, whereas in cells of the pancreatic islets, a different set of genes operates. Collectively, however, the instructions contained in your genotype determine the anatomical and physiological characteristics that make you a unique individual. Those anatomical and physiological characteristics constitute your **phenotype** (FĒ-nō-tīp; *phaino*, to display). In architectural terms, the genotype is a set of plans, and the phenotype is the finished building. Specific elements in your phenotype, such as hair and eye color, skin tone, and foot size, are called phenotypic *characters*, or *traits*.

Your genotype is derived from the genotypes of your parents. Yet you are not an exact copy of either parent; nor are you an easily identifiable mixture of their characteristics. Our discussion of genetics will begin with the basic patterns of inheritance and their implications. We will then examine the mechanisms responsible for regulating the activities of the genotype during prenatal development.

Patterns of Inheritance

The 46 chromosomes carried by each somatic cell occur in pairs: Every somatic cell contains 23 pairs of chromosomes.

At amphimixis, one member of each pair is contributed by the spermatozoon, and the other by the ovum. The two members of each pair are known as **homologous** (huh-MOL-ō-gus) **chromosomes**. Twenty-two of those pairs are called **autosomal** (aw-tō-SŌ-mul) **chromosomes**. Most of the genes of the autosomal chromosomes affect somatic characteristics, such as hair color and skin pigmentation. The chromosomes of the 23rd pair are called the **sex chromosomes**; one of their functions is to determine whether the individual is genetically male or female. **Figure 29–14** shows the **karyotype** (*karyon*, nucleus + *typos*, mark), or entire set of chromosomes, of a normal male. The discussion that follows concerns the inheritance of traits carried on the autosomal chromosomes; we will examine the patterns of inheritance via the sex chromosomes in a later section.

The two chromosomes in a homologous autosomal pair have the same structure and carry genes that affect the same traits. Suppose that one member of the pair contains three genes in a row, with the first gene determining hair color, the second eye color, and the third skin pigmentation. The other chromosome carries genes that affect the same traits, and the genes are in the same sequence. The genes are also located at equivalent positions on their respective chromosomes. A gene's position on a chromosome is called a **locus** (LŌ-kus; plural, *loci*).

The two chromosomes in a pair may not carry the same *form* of each gene, however. The various forms of a given gene are called **alleles** (uh-LĒLZ). These *alternate forms* determine the precise effect of the gene on your phenotype. If the two

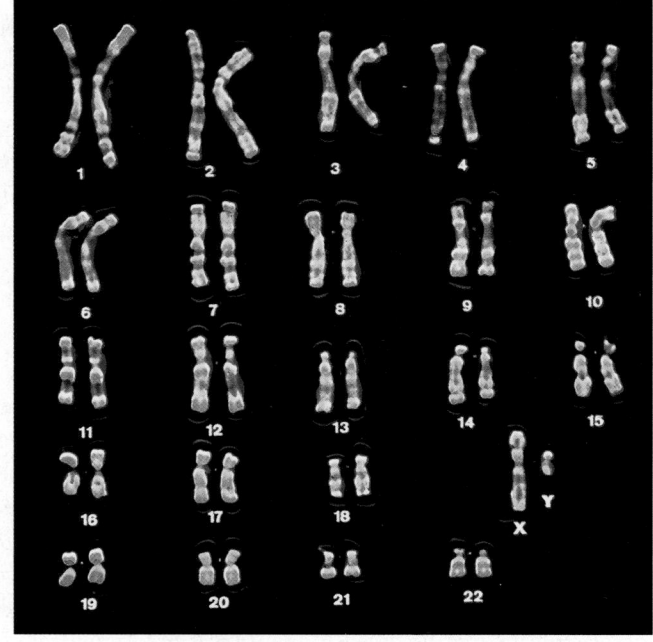

Figure 29–14   A Human Karyotype. The 23 pairs of somatic-cell chromosomes from a normal male.

chromosomes of a homologous pair carry the same allele of a particular gene, you are **homozygous** (hō-mō-ZĪ-gus; *homos*, the same) for the trait affected by that gene. That allele will then indeed be expressed in your phenotype. For example, if you receive a gene for curly hair from your father and a gene for curly hair from your mother, you will be homozygous for curly hair—and you will have curly hair. About 80 percent of an individual's genome consists of homozygous alleles. In **simple inheritance**, the phenotype is determined by interactions between a single pair of alleles.

Interactions between Alleles

Because the chromosomes of a homologous pair have different origins, one paternal and the other maternal, they do not necessarily carry the same alleles. When you have two different alleles for the same gene, you are **heterozygous** (het-er-ō-ZĪ-gus; *heteros*, other) for the trait determined by that gene. The phenotype that results from a heterozygous genotype depends on the nature of the interaction between the corresponding alleles. For example, if you received a gene for curly hair from your father, but a gene for straight hair from your mother, whether *you* will have curly hair, straight hair, or even wavy hair depends on the relationship between the alleles for those traits:

- In **strict dominance**, an allele that is **dominant** will be expressed in the phenotype, *regardless of any conflicting instructions carried by the other allele*. For instance, an individual with only one allele for freckles will have freckles, because that allele is dominant over the "nonfreckle" allele. An allele that is **recessive** will be expressed in the phenotype only if that same allele is present on *both chromosomes* of a homologous pair. For example, in Chapter 5 we learned that albino individuals cannot synthesize the yellow-brown pigment *melanin*. ∞ p. 162 The presence of one allele that directs melanin production will result in normal color. Two recessive alleles must be present to produce an albino individual. A single gene can have many different alleles, some dominant and others recessive.

- In **incomplete dominance**, heterozygous alleles produce a phenotype that is distinct from the phenotypes of individuals who are homozygous for one allele or the other. A good example is a gene that affects the shape of red blood cells. Individuals with homozygous alleles that carry instructions for normal adult hemoglobin A have red blood cells of normal shape. Individuals with homozygous alleles for hemoglobin S, an abnormal form, have red blood cells that become sickle-shaped in peripheral capillaries when the P_{O_2} decreases. These individuals develop *sickle cell anemia*. ∞ p. 658 Individuals who are heterozygous for this trait do not develop anemia; instead, they have red blood cells that sickle only when tissue oxygen levels are extremely low.

- In **codominance**, an individual who is heterozygous (has different alleles) for a given trait exhibits both the dominant and recessive phenotypes for that trait. Blood type in humans is determined by codominance. The alleles for type A and type B blood are dominant over the allele for type O blood, but a person with one type A allele and one type B allele has type AB blood, not A or B. Type AB blood has *both* type A antigens and type B antigens. The distinction between incomplete dominance and codominance is not always clear-cut. For example, a person who has alleles for hemoglobin A and hemoblogin S shows incomplete dominance for RBC shape, but codominance for hemoglobin; each red blood cell contains a mixture of hemoglobin A and hemoglobin S.

Penetrance and Expressivity

Differences in genotype lead to distinct variations in phenotype, but the relationships are not always predictable. The presence of a particular pair of alleles does not affect the phenotype in the same way in every individual. **Penetrance** is the percentage of individuals with a particular genotype that show the "expected" phenotype. The effects of that genotype in other individuals may be overridden by the activity of other genes or by environmental factors. For example, *emphysema*, a respiratory disorder discussed in Chapter 23, has been linked to a specific abnormal genotype. ∞ p. 867 However, roughly 20 percent of the individuals with this genotype do not develop emphysema, and thus the penetrance of this genotype is approximately 80 percent. The effects of environmental factors are apparent: Most people who develop emphysema are cigarette smokers.

If a given genotype *does* affect the phenotype, it can do so to various degrees, again depending on the activity of other genes or environmental stimuli. For example, even though identical twins have the same genotype, they do not have exactly the same fingerprints. The extent to which a particular allele is expressed when it is present is termed its **expressivity**.

Environmental effects on genetic expression are particularly evident during embryological and fetal development. Drugs, including certain antibiotics, alcohol, and nicotine in cigarette smoke, can disrupt fetal development. Factors that result in abnormal development are called **teratogens** (TER-uh-tō-jenz).

Predicting Inheritance

When an allele can be neatly characterized as dominant or recessive, you can predict the characteristics of individuals on the basis of their parents' alleles.

In such calculations, dominant alleles are traditionally indicated by capitalized abbreviations, and recessive alleles by lowercase abbreviations. For a given trait, the possibilities are indicated by *AA* (homozygous dominant), *Aa* (heterozygous), or *aa* (homozygous recessive). Each gamete involved in fertilization contributes a single allele for a given trait. That allele must be one of the two alleles contained by all cells in the parent's body. Consider, for example, the possible offspring of an albino mother and a father with normal skin pigmentation. Because albinism is a recessive trait, the maternal alleles are abbreviated *aa*. No matter which of her oocytes is fertilized, it will carry the recessive *a* allele. The father has normal pigmentation, a dominant trait. He is therefore either homozygous *or* heterozygous for this trait, because both *AA* and *Aa* will produce the same phenotype: normal skin pigmentation. Every sperm produced by a homozygous father will carry the *A* allele. In contrast, half the sperm produced by a heterozygous father will carry the dominant allele *A*, and the other half will carry the recessive allele *a*.

A simple box diagram known as a **Punnett square** enables us to predict the probabilities that children will have particular characteristics by showing the various combinations of parental alleles they can inherit. In the Punnett squares shown in **Figure 29–15**, the maternal alleles for skin pigmentation are listed along the horizontal axis, and the paternal ones along the vertical axis. The combinations of alleles are indicated in the small boxes. **Figure 29–15a** shows the possible offspring of an *aa* mother and an *AA* father. All the children must have the genotype *Aa*, so all will have normal skin pigmentation. Compare these results with those of **Figure 29–15b**, for a heterozygous father (*Aa*) and an *aa* mother. The heterozygous male produces two types of gametes, *A* and *a*, and the secondary oocyte may be fertilized by either one. As a result, the probability is 50 percent that a child of such a father will inherit the genotype *Aa* and so have normal skin pigmentation. The prob-

ability of inheriting the genotype *aa*, and thus having the albino phenotype, is also 50 percent.

A Punnett square can also be used to draw conclusions about the identity and genotype of a parent. For example, a man with the genotype *AA* cannot be the father of an albino child (*aa*).

We can predict the frequency of appearance of any inherited disorder that results from simple inheritance by using a Punnett square. Although they are rare in terms of overall numbers, more than 1200 inherited disorders have been identified that reflect the presence of one or two abnormal alleles for a single gene. **Figure 29–16** diagrams the relationships among the various forms of inheritance and gives examples of representative phenotypic characters, both normal and abnormal. The majority of characters listed develop through simple inheritance.

However, many phenotypic characters are determined by interactions among several genes. Such interactions constitute **polygenic inheritance**. Because the resulting phenotype depends not only on the nature of the alleles but how those alleles interact, you cannot predict the presence or absence of phenotypic characters using a simple Punnett square. In *suppression*, one gene suppresses the other, so that the second gene has no effect on the phenotype. In *complementary gene action*, dominant alleles on two genes interact to produce a phenotype different from that seen when one gene contains recessive alleles. The risks of developing several important adult disorders, including hypertension and coronary artery disease, are linked to polygenic inheritance.

Many of the developmental disorders responsible for fetal deaths and congenital malformations result from polygenic inheritance. In these cases, an individual's genetic composition does not by itself determine the onset of the disease. Instead,

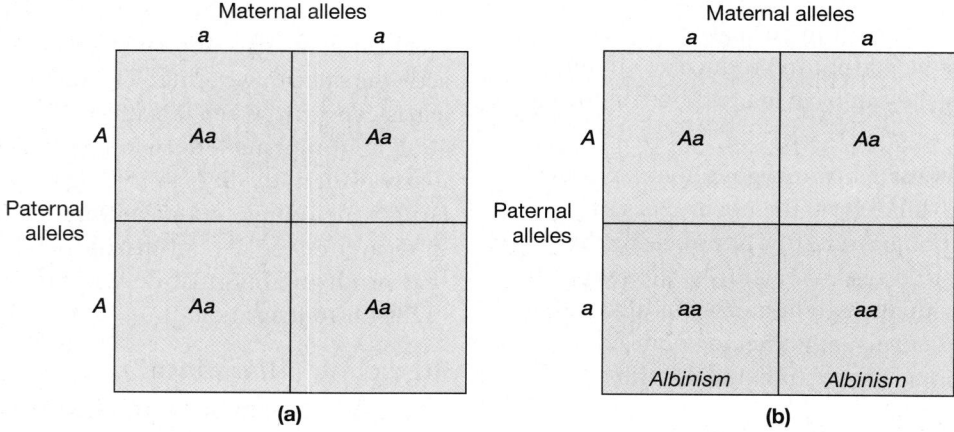

Figure 29–15 Predicting Phenotypic Characters by Using Punnett Squares. (a) All the offspring of a homozygous dominant father (*AA*) and a homozygous recessive mother (*aa*) will be heterozygous (*Aa*) for that trait. Their phenotype for the trait will be the same as that of the father. **(b)** The offspring of a heterozygous father (*Aa*) and a homozygous recessive mother (*aa*) will be either heterozygous or homozygous for the recessive trait. In this example, half the offspring will have normal skin color, and the other half will be albinos.

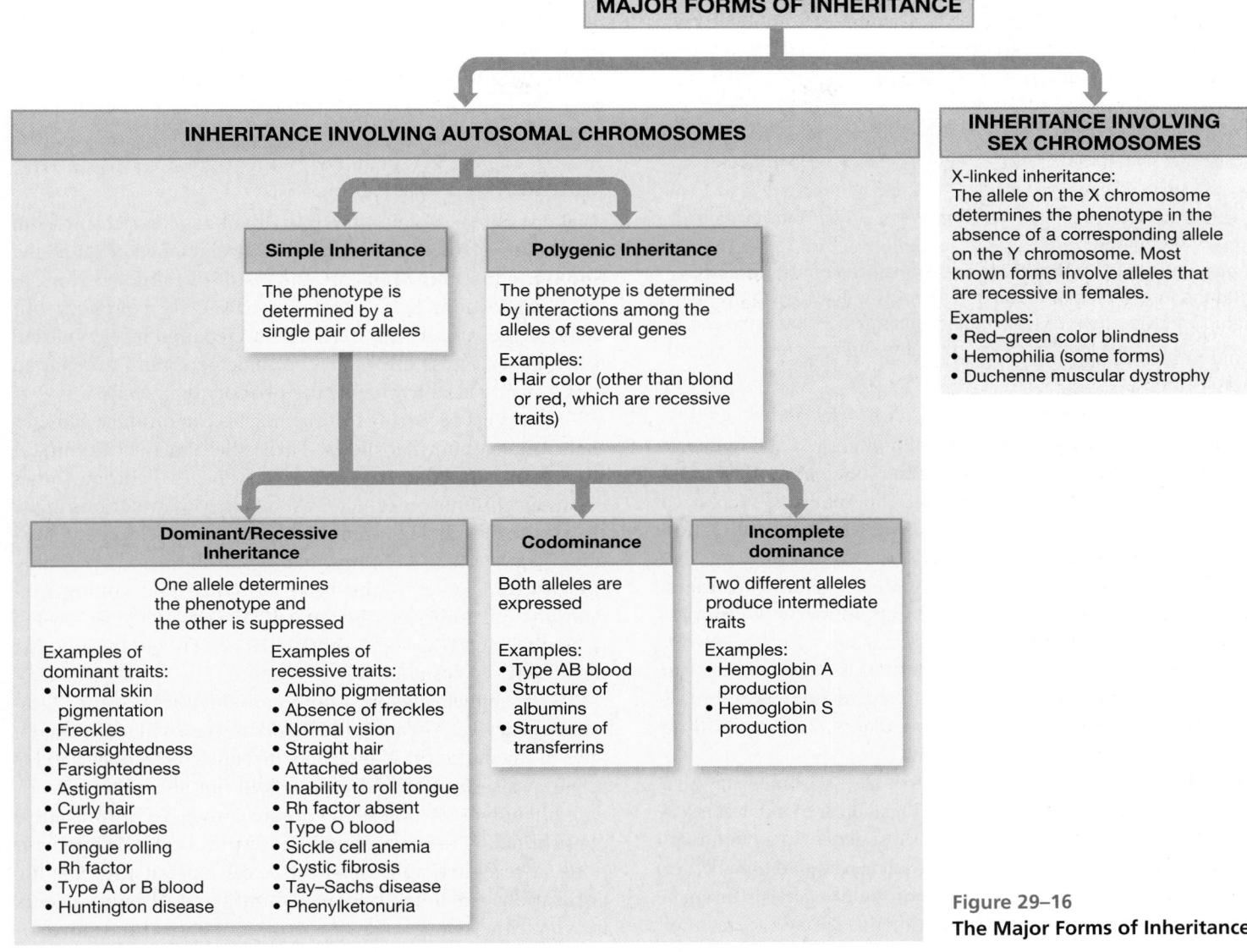

Figure 29–16
The Major Forms of Inheritance.

the conditions regulated by these genes establish a susceptibility to particular environmental influences. Thus, not every individual with the genetic tendency for a certain condition will develop that condition. It is therefore difficult to track polygenic conditions through successive generations. However, because many inherited polygenic conditions are *likely* (but not *guaranteed*) to occur, steps can be taken to prevent a crisis. For example, you can reduce hypertension by controlling your diet and fluid volume, and you can prevent coronary artery disease by lowering your serum cholesterol levels.

Sources of Individual Variation

Just as you are not a copy of either of your parents, neither are you a mixture of their characteristics. One reason for this was noted in Chapter 28: During meiosis, maternal and paternal

chromosomes are randomly distributed, so each gamete has a unique combination of maternal and paternal chromosomes. Thus, you may have an allele for curly hair from your father and an allele for straight hair from your mother, even though your sister received an allele for straight hair from each of your parents. Only in very rare cases will an individual receive both alleles from one parent. The few documented cases appear to have resulted when duplicate maternal chromatids failed to separate during meiosis II and the corresponding chromosome provided by the sperm did not participate in amphimixis. This condition, called *uniparental disomy*, generally remains undetected, because the individuals are phenotypically normal.

Genetic Recombination

During meiosis, various changes can occur in chromosome structure, producing gametes with chromosomes that differ

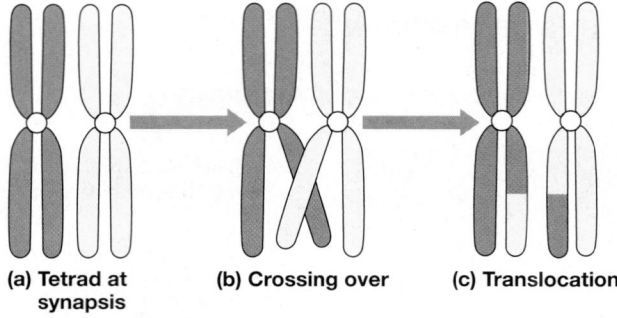

(a) Tetrad at synapsis **(b) Crossing over** **(c) Translocation**

Figure 29–17 **Crossing Over and Translocation. (a)** Synapsis, with the formation of a tetrad during meiosis. **(b)** Crossing over of homologous portions of two chromosomes. **(c)** The breakage and exchange of corresponding sections on the chromosomes in part (b).

from those of each parent. This phenomenon, called **genetic recombination**, greatly increases the range of possible variation among gametes, and thus among members of successive generations, whose genotypes are formed by the combination of gametes in fertilization. Genetic recombination can also complicate the tracing of the inheritance of genetic disorders.

In one normal form of recombination, parts of chromosomes become rearranged during synapsis (**Figure 29–17**). When tetrads form, adjacent chromatids may overlap, an event called **crossing over**. The chromatids may then break, and the overlapping segments trade places. This reshuffling process is known as **translocation**.

During recombination, portions of chromosomes may also break away and be lost, or *deleted*. The effects of a deletion on a zygote depend on the nature of the lost genes. In a phenomenon called **genomic imprinting**, the effects depend on whether the abnormal gamete is produced through oogenesis or spermatogenesis. For example, the deletion of a specific segment of chromosome 15 in humans can cause two very different disorders. In *Angelman syndrome*, which results when the abnormal chromosome is provided by the oocyte, hyperactivity, severe mental retardation, and seizures occur. In *Prader–Willi syndrome*, which results when the abnormal chromosome is delivered by the sperm, individuals have short stature, reduced muscle tone and skin pigmentation, underdeveloped gonads, and mental retardation varying from slight to severe.

Recombination that produces abnormal chromosome shapes or numbers is lethal for the zygote in almost all cases. Roughly 10 percent of zygotes have **chromosomal abnormalities**—that is, damaged, broken, missing, or extra copies of chromosomes—but only about 0.5 percent of newborns have such abnormalities. Few individuals with chromosomal abnormalities survive to full term; *Down syndrome (trisomy 21)* is an exception. In addition to contributing to prenatal mortality, chromosomal abnormalities produce a variety of serious clinical conditions. The high mortality rate and the severity of the problems reflect the fact that large numbers of

genes have been added or deleted. Women who become pregnant later in life run a higher risk of birth defects and miscarriage due to chromosomal abnormalities in the oocyte. It seems that the longer the oocyte remains suspended in meiosis I, the more likely are recombination errors when meiosis is completed.

Mutation

Variations at the level of the individual gene can result from *mutations*—changes in the nucleotide sequence of an allele. **Spontaneous mutations** are the result of random errors in DNA replication. Such errors are relatively common, but in most cases the error is detected and repaired by enzymes in the nucleus. Those errors that go undetected and unrepaired have the potential to change the phenotype in some way.

Mutations occurring during meiosis can produce gametes that contain abnormal alleles. These alleles may be dominant or recessive, and they may occur on autosomal chromosomes or on sex chromosomes. The vast majority of mutations make the zygote incapable of completing normal development. Mutation, rather than chromosomal abnormalities, is probably the primary cause of the high mortality rate among preembryos and embryos. (Roughly 50 percent of all zygotes fail to complete cleavage, and another 10 percent fail to reach the fifth month of gestation.)

If the abnormal allele is dominant but does not affect gestational survival, the individual's phenotype will show the effects of the mutation. If the abnormal allele is recessive and is on an autosomal chromosome, it will not affect the individual's phenotype as long as the zygote contains a normal allele contributed by the other parent at fertilization. Over generations, a recessive autosomal allele can spread through the population, remaining undetected until a fertilization occurs in which the two gametes contribute identical recessive alleles. This individual, who will be homozygous for the abnormal allele, will be the first to show the phenotypic effects of the original mutation. Individuals who are heterozygous for the abnormal allele but do not show the effects of the mutation are called **carriers**. Available genetic tests can determine whether an individual is a carrier for any of several autosomal recessive disorders, including Tay–Sachs disease. The information obtained from these tests can be useful in counseling prospective parents. For example, if both parents are carriers of the same disorder, they have a 25 percent probability of producing a child with the disease. This information may affect their decision to conceive.

Sex-Linked Inheritance

Unlike the other 22 chromosomal pairs, the sex chromosomes are never identical in appearance and gene content. There are two types of sex chromosomes: an **X chromosome** and a **Y chromosome**. X chromosomes are considerably larger and have more genes than do Y chromosomes. The Y chromosome includes dominant alleles specifying that an individual with that chromosome will be male. The normal pair of sex chromosomes in males is XY. Females do not have a Y chromosome; their sex chromosome pair is XX.

All oocytes carry an X chromosome, because the only sex chromosomes females have are X chromosomes. But each sperm carries either an X or a Y chromosome, because males have one of each and can pass along either one. As a Punnett square shows, the ratio of males to females in offspring should be 1:1. The birth statistics differ slightly from that prediction, with 106 males born for every 100 females. It has been suggested that more males are born because a sperm that carries the Y chromosome can reach the oocyte first, because that sperm does not have to carry the extra weight of the larger X chromosome.

The X chromosome also carries genes that affect somatic structures. These characteristics are called **X-linked** (or *sex-linked*), because in most cases there are no corresponding alleles on the Y chromosome. The inheritance of characteristics regulated by these genes does not follow the pattern of alleles on autosomal chromosomes. The best known single-allele characteristics are those associated with identifiable diseases or functional deficits.

The inheritance of color blindness exemplifies the differences between sex-linked inheritance and autosomal inheritance. The presence of a dominant allele, *C*, on the X chromosome results in normal color vision; a recessive allele, *c*, on the X chromosome results in red–green color blindness. A woman, with her two X chromosomes, can be either homozygous dominant (*CC*) or heterozygous (*Cc*) and still have normal color vision. She will be unable to distinguish reds from greens only if she carries two recessive alleles, *cc*. But a male has only one X chromosome, so whichever allele that chromosome carries determines whether he has normal color vision or is red–green color blind. The Punnett square in **Figure 29–18** reveals that the sons produced by a father with normal vision and a heterozygous (carrier) mother have a 50 percent chance of being red–green color blind, whereas any daughters have normal color vision.

A number of other clinical disorders noted earlier in the text are X-linked traits, including certain forms of hemophilia, diabetes insipidus, and muscular dystrophy. In several instances, advances in molecular genetics techniques have enabled geneticists to localize the specific genes on the X chromosome. These techniques provide a relatively direct method of screening for the presence of a particular condition before any signs or symptoms appear, and even before birth.

The Human Genome Project and Beyond

It has long been appreciated that all diseases—whether inherited or due to the body's responses to stresses, such as radiation,

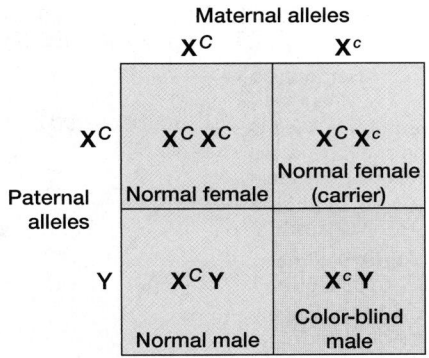

Figure 29–18   Inheritance of an X-Linked Trait. Because they have two X chromosomes, females with a dominant normal allele on at least one X chromosome will have normal phenotypes. By contrast, a male with an X chromosome bearing a recessive allele for a genetic disorder will develop that disorder (in this case, red–green color blindness). Thus recessive alleles on X chromosomes produce genetic disorders in males at a higher frequency than in females.

toxins, or pathogens—have a connection to chromosomes and genes. A much richer and fuller understanding of these relationships is now on the horizon, thanks to the biotechnological methods and techniques developed during the **Human Genome Project (HGP)**. Funded by the National Institutes of Health and the Department of Energy, the project's goal was to transcribe the entire human **genome**—that is, the full set of genetic material (DNA), nucleotide by nucleotide, found in our chromosomes. Begun in 1990, the project was completed in 2003 with 99 percent of the entire genome listed as finished, "high-quality sequence." A high-quality sequence is defined as a complete sequence of nucleotides, with no gaps or ambiguities and an error rate of fewer than one base per 10,000. The final HGP papers were published in 2006.

The first step in accessing the human genome was the preparation of a map of the individual chromosomes. **Karyotyping** (KAR-ē-ō-tīp-ing) is the determination of an individual's complete chromosomal complement (**Figure 29–14**). Each chromosome has characteristic banding patterns when stained with special dyes. The patterns are useful as reference points for the preparation of more detailed genetic maps, such as the one shown in **Figure 29–19**. The banding patterns themselves can be useful, as abnormal patterns are characteristic of some genetic disorders and several cancers.

The following are highlights of the Human Genome Project:

- All the chromosomes (23 pairs) have been completely sequenced.
- The total number of genes is now estimated at 25,000–30,000 genes. Almost 20,000 protein-coding genes are confirmed and, based on DNA segments, an additional 2188 more are predicted. (Defining a gene is not always straightforward. For example, small genes can

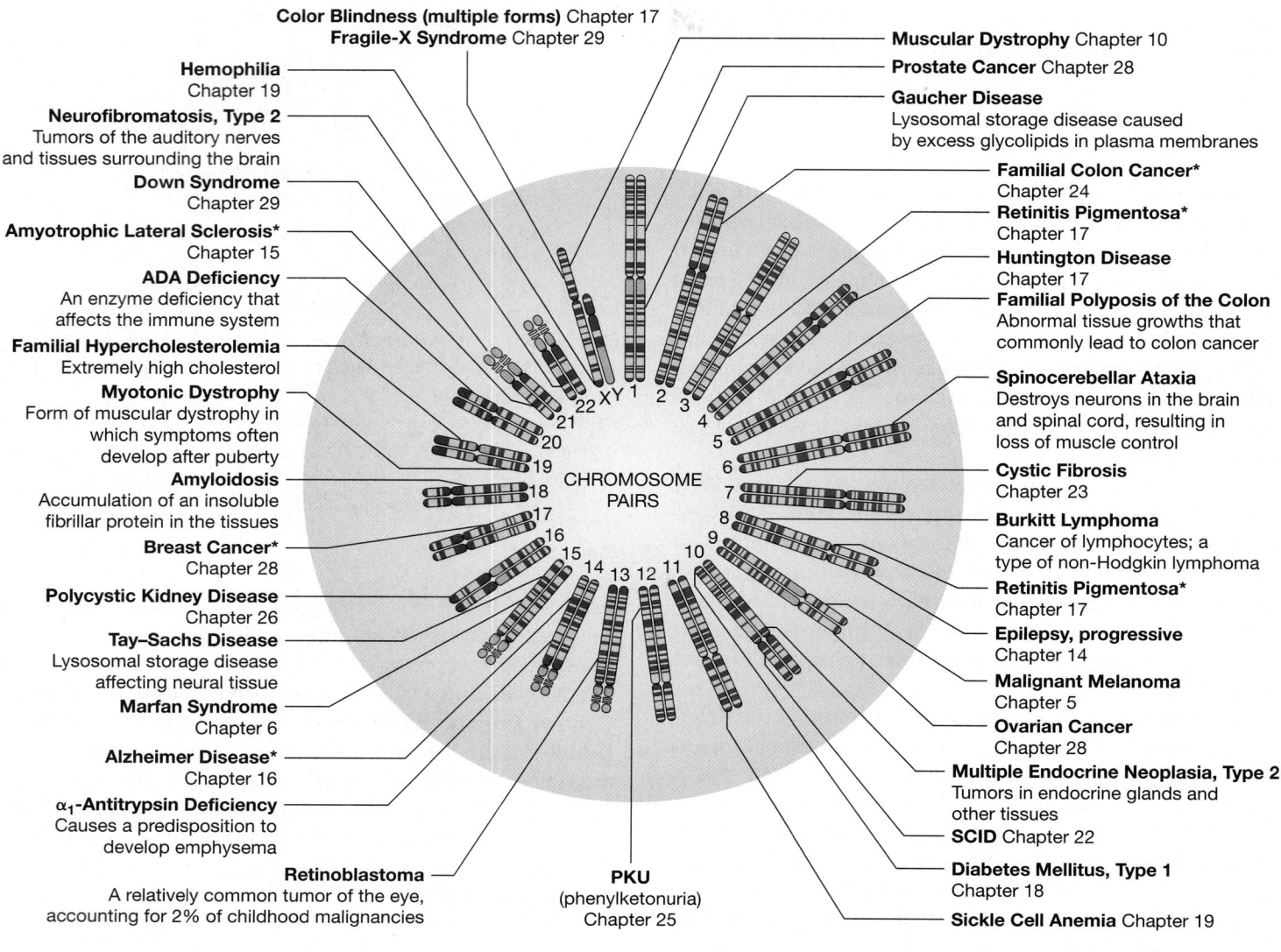

Color Blindness (multiple forms) Chapter 17
Fragile-X Syndrome Chapter 29

Hemophilia
Chapter 19

Neurofibromatosis, Type 2
Tumors of the auditory nerves
and tissues surrounding the brain

Down Syndrome
Chapter 29

Amyotrophic Lateral Sclerosis*
Chapter 15

ADA Deficiency
An enzyme deficiency that
affects the immune system

Familial Hypercholesterolemia
Extremely high cholesterol

Myotonic Dystrophy
Form of muscular dystrophy in
which symptoms often
develop after puberty

Amyloidosis
Accumulation of an insoluble
fibrillar protein in the tissues

Breast Cancer*
Chapter 28

Polycystic Kidney Disease
Chapter 26

Tay–Sachs Disease
Lysosomal storage disease
affecting neural tissue

Marfan Syndrome
Chapter 6

Alzheimer Disease*
Chapter 16

α_1-Antitrypsin Deficiency
Causes a predisposition to
develop emphysema

Retinoblastoma
A relatively common tumor of the eye,
accounting for 2% of childhood malignancies

PKU
(phenylketonuria)
Chapter 25

Muscular Dystrophy Chapter 10

Prostate Cancer Chapter 28

Gaucher Disease
Lysosomal storage disease caused
by excess glycolipids in plasma membranes

Familial Colon Cancer*
Chapter 24

Retinitis Pigmentosa*
Chapter 17

Huntington Disease
Chapter 17

Familial Polyposis of the Colon
Abnormal tissue growths that
commonly lead to colon cancer

Spinocerebellar Ataxia
Destroys neurons in the brain
and spinal cord, resulting in
loss of muscle control

Cystic Fibrosis
Chapter 23

Burkitt Lymphoma
Cancer of lymphocytes; a
type of non-Hodgkin lymphoma

Retinitis Pigmentosa*
Chapter 17

Epilepsy, progressive
Chapter 14

Malignant Melanoma
Chapter 5

Ovarian Cancer
Chapter 28

Multiple Endocrine Neoplasia, Type 2
Tumors in endocrine glands and
other tissues

SCID Chapter 22

Diabetes Mellitus, Type 1
Chapter 18

Sickle Cell Anemia Chapter 19

CHROMOSOME PAIRS

* One form of the disease

Figure 29–19   A Map of Human Chromosomes. The banding patterns of typical chromosomes in a male, and the locations of the genes associated with specific inherited disorders. The chromosomes are not drawn to scale.

easily be overlooked in a nucleotide sequence, a gene may code for more than one protein, some genes code for RNA, and two genes can overlap.)

- Although more than 99 percent of human nucleotide bases are the same in all people, there are about 1.4 million single-base differences, or single nucleotide polymorphisms (SNPs). Some of these SNPs are associated with specific diseases.

- Roughly 10,000 different single gene disorders have been described. Most are very rare, but collectively they may affect 1 in every 200 births. Over 900 of these disorders have been mapped on the genome, and several examples are included in **Figure 29–19**. Genetic screening and diagnostic tests for abnormal genes are now performed for many of these disorders.

The completion of the sequence is already stimulating new approaches for diagnosing disease and predicting disease susceptibility. For example, in 2006, the Genes and Environment Initiative (GEI) was launched to understand the link between genes, the environment, and why certain individuals develop diseases. The GEI is a joint collaboration of the National Institute of Environmental Health Services (NIEHS) and the National Human Genome Research Institute (NHGRI). Together, they will conduct genetic studies of individuals with specific diseases and their personal exposure to environmental factors such as sun and chemicals, diet, and physical activity. Also begun in 2006 is The Cancer Genome Atlas (TCGA), sponsored by the NHGRI and the National Cancer Institute. TCGA has as its immediate goal the compilation of an atlas of genetic changes (mutations) in three tumors: brain cancer (glioblastoma), lung cancer, and ovarian cancer.

CLINICAL NOTE

Chromosomal Abnormalities and Genetic Analysis

Embryos that have abnormal autosomal chromosomes rarely survive. However, *translocation defects* and *trisomy* are two types of autosomal chromosome abnormalities that do not invariably result in prenatal death.

In a **translocation defect**, crossing over occurs between different chromosome pairs such that, for example, a piece of chromosome 8 may become attached to chromosome 14. The genes moved to their new position may function abnormally, becoming inactive or overactive. In a balanced translocation, where there is no net loss or gain of chromosomal material, embryos may survive.

In **trisomy**, something goes wrong in meiosis. One of the gametes involved in fertilization carries an extra copy of one chromosome, so the zygote then has three copies of this chromosome rather than two. (The nature of the trisomy is indicated by the number of the chromosome involved. Thus, individuals with trisomy 13 have three copies of chromosome 13.) Zygotes with extra copies of chromosomes seldom survive. Individuals with trisomy 13 and trisomy 18 may survive until delivery but rarely live longer than a year. The notable exception is trisomy 21.

Trisomy 21, or **Down syndrome**, is the most common viable chromosomal abnormality. Estimates of its incidence in the U.S. population range from 1.5 to 1.9 per 1000 births. Affected individuals exhibit mental retardation and characteristic physical malformations, including a facial appearance that gave rise to the term *mongolism*, once used to describe this condition. The degree of mental retardation ranges from moderate to severe. Few individuals with this condition lead independent lives. Anatomical problems affecting the cardiovascular system often prove fatal during childhood or early adulthood. Although some individuals survive to moderate old age, many develop Alzheimer disease while still relatively young (before age 40).

For unknown reasons, there is a direct correlation between maternal age and the risk of having a child with trisomy 21. For a maternal age below 25, the incidence of Down syndrome approaches 1 in 2000 births, or 0.05 percent. For maternal ages 30–34, the odds increase to 1 in 900, and over the next decade they go from 1 in 290 to 1 in 46, or more than 2 percent. These statistics are becoming increasingly significant because many women are delaying childbearing until their mid-thirties or later.

Abnormal numbers of sex chromosomes do not produce effects as severe as those induced by extra or missing autosomal chromosomes. In **Klinefelter syndrome**, the individual carries the sex chromosome pattern XXY. The phenotype is male, but the extra X chromosome causes reduced androgen production. As a result, the testes fail to mature so the individuals are sterile, and the breasts are slightly enlarged. The incidence of this condition among newborn males averages 1 in 750 births.

Individuals with **Turner syndrome** have only a single, female sex chromosome; their sex chromosome complement is abbreviated XO. This kind of chromosomal deletion is known as **monosomy**. The incidence of this condition at delivery has been estimated as 1 in 10,000 live births. The condition may not be recognized at birth, because the phenotype is normal female. But maturational changes do not appear at puberty. The ovaries are nonfunctional, and estrogen production occurs at negligible levels.

Fragile-X syndrome causes mental retardation, abnormal facial development, and enlarged testes in affected males. The cause is an abnormal X chromosome that contains a *genetic stutter*, an abnormal repetition of a single nucleotide triplet. The presence of the stutter in some way disrupts the normal functioning of adjacent genes and so produces the signs and symptoms of the disorder.

Many of these conditions can be detected before birth through the analysis of fetal cells. In **amniocentesis**, a sample of amniotic fluid is removed and the fetal cells it contains are analyzed. This procedure permits the identification of more than 20 congenital conditions, including Down syndrome. The needle inserted to obtain a fluid sample is guided into position during an ultrasound procedure. ∞ p. 25 Unfortunately, amniocentesis has two major drawbacks:

1. Because the sampling procedure represents a potential threat to the health of fetus and mother alike, amniocentesis is performed only when known risk factors are present. Examples of risk factors are a family history of specific conditions, or in the case of Down syndrome, maternal age over 35.

2. Sampling cannot safely be performed until the volume of amniotic fluid is large enough that the fetus will not be injured during the process. The usual time for amniocentesis is at 14–15 weeks of gestation. It may take several weeks to obtain results once samples have been collected, and by the time the results are received, an induced or therapeutic abortion may no longer be a viable option.

An alternative procedure known as **chorionic villus sampling (CVS)** analyzes cells collected from the chorionic villi late in the first trimester. CVS carries a slightly higher risk of miscarriage than amniocentesis, but may be preferable because it can be done earlier in gestation.

The Human Genome Project has identified the normal genetic composition of a "typical" human. Yet we all are variations on a basic theme. How do we decide what set of genes to accept as "normal"? Moreover, as we improve our abilities to manipulate our own genetic foundations, we will face many additional troubling ethical and legal dilemmas. Few people, for example, object to the insertion of a "correct" gene into somatic cells to cure a specific disease. But what if we could insert that modified gene into a gamete and change not only that individual, but all of his or her descendants as well? And what if the goal of manipulating the gene was not to correct or prevent any disorder, but instead to "improve" the individual by increasing his or her intelligence, height, or vision, or by altering some other phenotypic characteristic?

Such difficult questions will not go away. In the years to come, we will have to find answers that are acceptable to us all.

CHECKPOINT

22. Describe the relationship between genotype and phenotype.

23. Define heterozygous.

24. Curly hair is an autosomal dominant trait. What would be the phenotype of a person who is heterozygous for this trait?

25. Why are children not identical copies of their parents?

See the blue Answers tab at the end of the book.

Related Clinical Terms

amniocentesis: A genetic analysis of fetal cells collected from a sample of amniotic fluid, usually done in the second trimester.

breech birth: A delivery in which the legs or buttocks of the fetus enter the vaginal canal first.

chorionic villus sampling: A genetic analysis of cells usually collected from the chorionic villi during the first trimester.

ectopic pregnancy: A pregnancy in which implantation occurs somewhere other than the uterus.

gestational trophoblastic neoplasia: A tumor formed by undifferentiated, rapid growth of the syncytial trophoblast; if untreated, the neoplasm may become malignant.

infertility: The inability to achieve pregnancy after engaging in one year of appropriately timed intercourse.

in vitro fertilization: Fertilization outside the body, generally in a petri dish.

Chapter Review

Study Outline

29-1 Development, marked by various stages, is a continuous process that occurs from fertilization to maturity p. 1087

1. **Development** is the gradual modification of anatomical structures and physiological characteristics from **conception** to maturity. The formation of different types of cells is **differentiation**.

2. **Prenatal development** occurs before birth; **postnatal development** begins at birth and continues to **maturity**, when aging begins. **Inheritance** is the transfer of genetically determined characteristics from generation to generation. **Genetics** is the study of the mechanisms of inheritance.

29-2 Fertilization—the fusion of a secondary oocyte and a spermatozoon—forms a zygote p. 1087

3. **Fertilization**, or *conception*, normally occurs in the uterine tube within a day after ovulation. Spermatozoa cannot fertilize a secondary oocyte until they have undergone *capacitation*. (*Figure 29–1*)

4. The acrosomal caps of the spermatozoa release **hyaluronidase** and **acrosin**, enzymes required to penetrate the corona radiata and zona pellucida of the oocyte. When a single spermatozoon contacts the oocyte membrane, fertilization begins and **oocyte activation** follows. (*Figure 29–1*)

5. During activation, the oocyte completes meiosis II and thus becomes a functionally mature ovum. **Polyspermy** is prevented by membrane depolarization and the *cortical reaction*.

6. After activation, the **female pronucleus** and the **male pronucleus** fuse in a process called *amphimixis*. (*Figure 29–1*)

29-3 Gestation consists of three stages of prenatal development: the first, second, and third trimesters p. 1090

7. During prenatal development, differences in the cytoplasmic composition of individual cells trigger changes in genetic activity. The chemical interplay among developing cells is **induction**.

8. The nine-month **gestation** period can be divided into three **trimesters**.

29-4 Cleavage, implantation, placentation, and embryogenesis are critical events of the first trimester p. 1091

9. In the **first trimester**, **cleavage** subdivides the cytoplasm of the zygote in a series of mitotic divisions; the zygote becomes a **pre-embryo** and then a **blastocyst**. During **implantation**, the blastocyst burrows into the uterine endometrium.

Placentation occurs as blood vessels form around the blastocyst and the **placenta** develops. **Embryogenesis** is the formation of a viable embryo.

10. The blastocyst consists of an outer **trophoblast** and an **inner cell mass**. (*Figure 29–2*)

11. Implantation occurs about seven days after fertilization as the blastocyst adheres to the uterine lining. (*Figure 29–3*)

12. As the trophoblast enlarges and spreads, maternal blood flows through open **lacunae**. After **gastrulation**, there is an **embryonic disc** composed of **endoderm**, **ectoderm**, and an intervening **mesoderm**. It is from these **germ layers** that the body systems differentiate. (*Figure 29–4; Summary Table 29–1*)

13. Germ layers help form four **extraembryonic membranes**: the yolk sac, amnion, allantois, and chorion. (*Figure 29–5*)

14. The **yolk sac** is an important site of blood cell formation. The **amnion** encloses fluid that surrounds and cushions the developing embryo. The base of the **allantois** later gives rise to the urinary bladder. Circulation within the vessels of the **chorion** provides a rapid-transit system that links the embryo with the trophoblast. (*Figures 29–5, 29–6*)

15. **Chorionic villi** extend outward into the maternal tissues, forming an intricate, branching network through which maternal blood flows. As development proceeds, the **umbilical cord** connects the fetus to the placenta. The syncytial trophoblast synthesizes **human chorionic gonadotropin (hCG)**, estrogens, progesterone, **human placental lactogen (hPL)**, **placental prolactin**, and **relaxin**. (*Figure 29–6*)

16. The first trimester is critical, because events in the first 12 weeks establish the basis for **organogenesis** (organ formation). (*Figure 29–7; Summary Table 29–2*)

29-5 During the second and third trimesters, maternal organ systems support the developing fetus, and the uterus undergoes structural and functional changes p. 1102

17. In the **second trimester**, the organ systems increase in complexity. During the **third trimester**, many of the organ systems become fully functional. (*Figure 29–8; Summary Table 29–2*)

18. The fetus undergoes its largest weight gain in the third trimester. At the end of gestation, the fetus and the enlarged uterus displace many of the mother's abdominal organs. (*Figure 29–9*)

19. The developing fetus is totally dependent on maternal organs for nourishment, respiration, and waste removal. Maternal adaptations include increases in respiratory rate, tidal volume, blood volume, nutrient and vitamin intake, and glomerular filtration rate, as well as changes in the size of the uterus and mammary glands.

20. Progesterone produced by the placenta has an inhibitory effect on uterine muscles; its calming action is opposed by estrogens, oxytocin, and prostaglandins. At some point, multiple factors interact to produce **labor contractions** in the uterine wall. (*Figure 29–10*)

29-6 Labor consists of the dilation, expulsion, and placental stages p. 1105

21. The goal of **true labor** is **parturition**, the forcible expulsion of the fetus.

22. Labor can be divided into three stages: the **dilation stage**, the **expulsion stage**, and the **placental stage**. (*Figure 29–11*)

23. **Premature labor** may result in **premature delivery**.

24. Difficult deliveries can include *forceps deliveries* and **breech births**—deliveries in which the legs or buttocks of the fetus, rather than the head, enter the vaginal canal first.

25. Twin births are either **dizygotic** (fraternal) or **monozygotic** (identical).

29-7 Postnatal stages are the neonatal period, infancy, childhood, adolescence, maturity, and senescence p. 1107

26. Postnatal development involves a series of five **life stages**: the neonatal period, infancy, childhood, adolescence, and maturity. *Senescence* (aging) begins at maturity and ends in the death of the individual.

27. The **neonatal period** extends from birth to one month after. In the transition from fetus to **neonate**, the respiratory, circulatory, digestive, and urinary systems of the infant begin functioning independently. The newborn must also begin thermoregulation.

28. Mammary gland cells produce protein-rich **colostrum** during the neonate's first few days of life and then convert to milk production. These secretions are released as a result of the **milk let-down reflex**. (*Figure 29–12*)

29. Body proportions gradually change during **infancy** (from age one month to two years) and during **childhood** (from two years to puberty). (*Figure 29–13*)

30. **Adolescence** begins at **puberty**, when (1) the hypothalamus increases its production of GnRH, (2) circulating levels of FSH and LH rise rapidly, and (3) ovarian or testicular cells become more sensitive to FSH and LH. These changes initiate gamete formation, the production of sex hormones, and a sudden increase in the growth rate. The hormonal changes at puberty, especially changes in sex hormone levels, produce sex-specific differences in the structure and function of many systems; these differences will be retained. Adolescence continues until growth is completed. Further changes occur when sex hormone levels decline at menopause or the male climacteric.

31. **Senescence** then begins, producing gradual reductions in the functional capabilities of all systems. (*Summary Table 29–3*)

29-8 Genes and chromosomes determine patterns of inheritance p. 1112

32. Every somatic cell carries copies of the original 46 chromosomes in the zygote; these chromosomes and their component genes constitute the individual's **genotype**. The physical expression of the genotype is the individual's **phenotype**.

33. Every somatic human cell contains 23 pairs of chromosomes; each pair consists of **homologous chromosomes**. Twenty-two pairs are **autosomal chromosomes**. The chromosomes of the twenty-third pair are the **sex chromosomes**; they differ between the sexes. (*Figure 29–14*)

34. Chromosomes contain DNA, and genes are functional segments of DNA. The various forms of a given gene are called **alleles**. If both homologous chromosomes carry the same allele of a particular gene, the individual is **homozygous**; if they carry different alleles, the individual is **heterozygous**.

35. Alleles are either **dominant** or **recessive**, depending on how their traits are expressed.

36. Combining maternal and paternal alleles in a **Punnett square** helps us predict the characteristics of offspring. (*Figure 29–15*)

37. In **simple inheritance**, phenotypic characters are determined by interactions between a single pair of alleles. **Polygenic**

inheritance involves interactions among alleles on several genes. (*Figure 29–16*)

38. **Genetic recombination**, the gene reshuffling (**crossing over** and **translocation**) that occurs during meiosis, increases the genetic variation of male and female gametes. (*Figure 29–17*)

39. **Spontaneous mutations** are the result of random errors in DNA replication. Such mutations can cause the production of abnormal alleles.

40. The two types of sex chromosomes are an **X chromosome** and a **Y chromosome**. The normal sex chromosome complement of males is XY; that of females is XX. The X chromosome carries **X-linked** (*sex-linked*) **genes**, which affect somatic structures but have no corresponding alleles on the Y chromosome. (*Figure 29–18*)

41. The **Human Genome Project** has mapped close to 25,000 human genes, including some of those responsible for inherited disorders. (*Figure 29–19*)

Review Questions

See the blue Answers tab at the end of the book.

Access more review material online at **myA&P**™ (www.myaandp.com). There, you'll find chapter guides, chapter quizzes, practice tests, animations, flashcards, a glossary with pronunciations, *Interactive Physiology*® (IP) exercises and quizzes, and more to help you succeed in the course.

LEVEL 1 Reviewing Facts and Terms

1. The chorionic villi
   (a) form the umbilical cord.
   (b) form the umbilical vein.
   (c) form the umbilical arteries.
   (d) increase the surface area available for exchange between the placenta and maternal blood.
   (e) form the portion of the placenta called the decidua capsularis.

2. Identify the two extraembryonic membranes and the three different regions of the endometrium at week 10 of development in the following diagram.

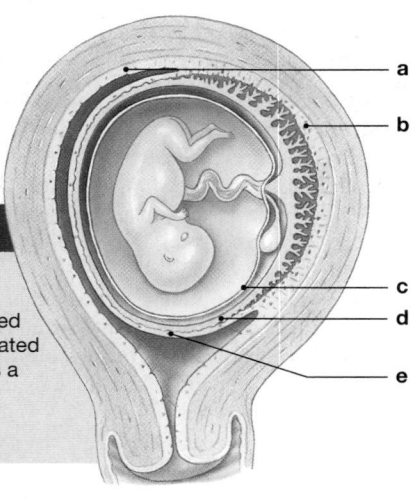

WEEK 10

The amnion has expanded greatly, filling the uterine cavity. The fetus is connected to the placenta by an elongated umbilical cord that contains a portion of the allantois, blood vessels, and the remnants of the yolk stalk.

(a) _____   (b) _____
(c) _____   (d) _____
(e) _____

3. The hormone that is the basis for a pregnancy test is
   (a) LH.
   (b) progesterone.
   (c) human chorionic gonadotropin (hCG).
   (d) human placental lactogen (hPL).
   (e) either c or d, depending on the type of test.

4. Recessive X-linked traits
   (a) are passed from fathers to their sons.
   (b) are more likely to be expressed in males.
   (c) always affect some aspect of the reproductive system.
   (d) are never expressed in females.
   (e) cannot be passed from mothers to daughters.

5. The stage of development that follows cleavage is the
   (a) blastocyst.   (b) morula.
   (c) trophoblast.   (d) blastocoele.

6. What developmental stage begins once the zygote arrives in the uterine cavity?
   (a) blastocyst   (b) trophoblast
   (c) lacuna   (d) blastomere

7. The structure(s) that allow(s) active and passive exchange between the fetal and maternal bloodstreams is/are the
   (a) yolk stalk.   (b) chorionic villi.
   (c) umbilical veins.   (d) umbilical arteries.

8. If an allele must be present on both the maternal and paternal chromosomes to affect the phenotype, the allele is said to be
   (a) dominant.   (b) recessive.
   (c) complementary.   (d) heterozygous.

9. Describe the changes that occur in the oocyte immediately after fertilization.

10. (a) What are the four extraembryonic membranes?
    (b) From which germ layers do these membranes form, and what are their functions?

11. Identify the three stages of labor, and describe the events that characterize each stage.

12. List the factors involved in initiating labor contractions.

13. Identify the three life stages that occur between birth and approximately age 10. Describe the timing and characteristics of each stage.

14. What hormonal events are responsible for puberty? Which life stage does puberty initiate?

LEVEL 2 Reviewing Concepts

15. A normally pigmented woman whose father was an albino marries a normally pigmented man whose mother was an albino. What is the probability that they would have an albino child?
    (a) 50%   (b) 25%
    (c) 12.5%   (d) 6.25%
    (e) 100%

16. If a sperm cell lacked sufficient quantities of hyaluronidase, it would *not* be able to
    (a) move its flagellum.
    (b) penetrate the corona radiata.
    (c) become capacitated.
    (d) survive the environment of the female reproductive tract.
    (e) metabolize fructose.

17. Problems involving the formation of the chorion would affect
    (a) the embryo's ability to produce blood cells.
    (b) the formation of limbs.
    (c) the embryo's ability to derive nutrition from the mother.
    (d) lung formation.
    (e) the urinary system.

18. After implantation, how does the developing embryo obtain nutrients? What structures and processes are involved?

19. In addition to its role in the nutrition of the fetus, what are the primary endocrine functions of the placenta?

20. Discuss the changes that occur in maternal systems during pregnancy. Why are these changes functionally significant?

21. During true labor, what physiological mechanisms ensure that uterine contractions continue until delivery has been completed?

22. What physiological adjustments must an infant make during the neonatal period in order to survive?

23. Distinguish between the following paired terms:
    (a) genotype and phenotype
    (b) heterozygous and homozygous
    (c) simple inheritance and polygenic inheritance

24. Indicate the type of inheritance involved in each of the following situations.
    (a) Children who exhibit the trait have at least one parent who also exhibits it.
    (b) Children exhibit the trait even though neither parent exhibits it.
    (c) The trait is expressed more commonly in sons than in daughters.
    (d) The trait is expressed equally in daughters and sons.

25. GEI and TCGA are abbreviations for what studies that followed from the Human Genome Project? What are the general goals of each?

LEVEL 3 Critical Thinking and Clinical Applications

26. Hemophilia A, a condition in which blood does not clot properly, is a recessive trait located on the X chromosome (X^h). Suppose that a woman who is heterozygous for this trait (XX^h) mates with a normal male (XY). What is the probability that the couple will have hemophiliac daughters? What is the probability that the couple will have hemophiliac sons?

27. Joe and Jane desperately want to have children, and although they have tried for two years, they have not been successful. Finally, each of them consults a physician, and it turns out that Joe suffers from oligospermia (a low sperm count). He confides to you that he doesn't understand why this would interfere with his ability to have children since he remembers from biology class that it only takes one sperm to fertilize an egg. What would you tell him?

28. Cathy has just given birth to a little girl. When the nurses take the infant back to the nursery and try to feed her, she becomes cyanotic, a condition characterized by bluish discoloration of the skin. The episode passes, but when the infant is bathed, she becomes cyanotic again. Blood gas levels indicate that arterial blood is only 60% saturated. Physical examination reveals no structural deformities involving the respiratory or digestive system. What might be causing the problem?

29. Sally gives birth to a baby with a congenital deformity of the stomach. Sally believes that her baby's affliction is the result of a viral infection she suffered during her third trimester. Is this a possibility? Explain.

Appendix
Weights and Measures

Accurate descriptions of physical objects would be impossible without a precise method of reporting the pertinent data. Dimensions such as length and width are reported in standardized units of measurement, such as inches or centimeters. These values can be used to calculate the **volume** of an object, a measurement of the amount of space the object fills. Mass is another important physical property. The **mass** of an object is determined by the amount of matter the object contains; on Earth the mass of an object determines the object's weight.

In the United States, length and width are typically described in inches, feet, or yards; volumes in pints, quarts, or gallons; and weights in ounces, pounds, or tons. These are units of the **U.S. system** of measurement. Table 1 summarizes the terms used in the U.S. system. For reference purposes, this table also includes a definition of the "household units,"

TABLE 1 The U.S. System of Measurement

Physical Property	Unit	Relationship to Other U.S. Units	Relationship to Household Units
Length	inch (in.)	1 in = 0.083 ft	
	foot (ft)	1 ft = 12 in.	
		= 0.33 yd	
	yard (yd)	1 yd = 36 in.	
		= 3 ft	
	mile (mi)	1 mi = 5280 ft	
		= 1760 yd	
Volume	fluid dram (fl dr) (also fluidram)	1 fl dr = 0.125 fl oz	
	fluid ounce (fl oz)	1 fl oz = 8 fl dr	= 6 teaspoons (tsp)
		= 0.0625 pt	= 2 tablespoons (tbsp)
	pint (pt)	1 pt = 128 fl dr	= 32 tbsp
		= 16 fl oz	= 2 cups (c)
		= 0.5 qt	
	quart (qt)	1 qt = 256 fl dr	= 4 c
		= 32 fl oz	
		= 2 pt	
		= 0.25 gal	
	gallon (gal)	1 gal = 128 fl oz	
		= 8 pt	
		= 4 qt	
Mass	grain (gr)	1 gr = 0.002 oz	
	dram (dr)	1 dr = 27.3 gr	
		= 0.063 oz	
	ounce (oz)	1 oz. = 437.5 gr	
		= 16 dr	
	pound (lb)	1 lb = 7000 gr	
		= 256 dr	
		= 16 oz	
	ton (t)	1 t = 2000 lb	

popular in recipes. The U.S. system can be very difficult to work with, because there is no logical relationship among the various units. For example, there are 12 inches in a foot, 3 feet in a yard, and 1760 yards in a mile. Without a clear pattern of organization, the conversion of feet to inches or miles to feet can be confusing and time-consuming. The relationships among ounces, pints, quarts, and gallons are no more logical than those among ounces, pounds, and tons.

In contrast, the **metric system** has a logical organization based on powers of 10, as indicated in Table 2. For example, a **meter (m)** is the basic unit for the measurement of size. For measurements of larger objects, data can be reported in

TABLE 2   The Metric System of Measurement

Physical Property	Unit	Relationship to Standard Metric Units	Conversion to U.S. Units	
Length	nanometer (nm)	1 nm = 0.000000001 m (10^{-9})	= 3.94×10^{-8} in.	25,400,000 nm = 1 in.
	micrometer (μm)	1 μm = 0.000001 m (10^{-6})	= 3.94×10^{-5} in.	25,400 μm = 1 in.
	millimeter (mm)	1 mm = 0.001 m (10^{-3})	= 0.0394 in.	25.4 mm = 1 in.
	centimeter (cm)	1 cm = 0.01 m (10^{-2})	= 0.394 in.	2.54 cm = 1 in.
	decimeter (dm)	1 dm = 0.1 m (10^{-1})	= 3.94 in.	0.254 dm = 1 in.
	meter (m)	standard unit of length	= 39.4 in.	0.0254 m = 1 in.
			= 3.28 ft	0.3048 m = 1 ft
			= 1.093 yd	0.914 m = 1 yd
	kilometer (km)	1 km = 1000 m	= 3280 ft	
			= 1093 yd	
			= 0.62 mi	1.609 km = 1 mi
Volume	microliter (μL)	1 μL = 0.000001 L (10^{-6}) = 1 cubic millimeter (mm^3)		
	milliliter (mL)	1 mL = 0.001 L (10^{-3}) = 1 cubic centimeter (cm^3 or cc)	= 0.0338 fl oz	5 mL = 1 tsp
				15 mL = 1 tbsp
				30 mL = 1 fl oz
	centiliter (cL)	1 cL = 0.01 L (10^{-2})	= 0.338 fl oz	2.95 cL = 1 fl oz
	deciliter (dL)	1 dL = 0.1 L (10^{-1})	= 3.38 fl oz	0.295 dL = 1 fl oz
	liter (L)	standard unit of volume	= 33.8 fl oz	0.0295 L = 1 fl oz
			= 2.11 pt	0.473 L = 1 pt
			= 1.06 qt	0.946 L = 1 qt
Mass	picogram (pg)	1 pg = 0.000000000001 g (10^{-12})		
	nanogram (ng)	1 ng = 0 .000000001 g (10^{-9})	= 0.000000015 gr	66,666,666 ng = 1 gr
	microgram (μg)	1 μg = 0.000001 g (10^{-6})	= 0.000015 gr	66,666 μg = 1 gr
	milligram (mg)	1 mg = 0 .001 g (10^{-3})	= 0.015 gr	66.7 mg = 1 gr
	centigram (cg)	1 cg = 0.01 g (10^{-2})	= 0.15 gr	6.67 cg = 1 gr
	decigram (dg)	1 dg = 0.1 g (10^{-1})	= 1.5 gr	0.667 cg = 1 gr
	gram (g)	standard unit of mass	= 0.035 oz	28.4 g = 1 oz
			= 0.0022 lb	454 g = 1 lb
	dekagram (dag)	1 dag = 10 g		
	hectogram (hg)	1 hg = 100 g		
	kilogram (kg)	1 kg = 1000 g	= 2.2 lb	0.454 kg = 1 lb
	metric ton (MT)	1 MT = 1000 kg	= 1.1 t	
			= 2205 lb	0.907 MT = 1 t

Temperature	**Celsius**	**Fahrenheit**
Freezing point of pure water	0°	32°
Normal body temperature	36.8°	98.6°
Boiling point of pure water	100°	212°
Conversion	°C → °F: °F = (1.8 × °C) + 32	°F → °C: °C = (°F − 32) × 0.56

dekameters (*deka*, ten), **hectometers** (*hekaton*, hundred), or **kilometers** (**km**; *chilioi*, thousand); for smaller objects, data can be reported in **decimeters** (0.1 m; *decem*, ten), **centimeters** (**cm** = 0.01 m; *centum*, hundred), **millimeters** (**mm** = 0.001 m; *mille*, thousand), and so forth. In the metric system, the same prefixes are used to report weights, based on the **gram** (**g**), and volumes, based on the **liter** (**L**). This text reports data in metric units, in most cases with U.S. system equivalents. Use this opportunity to become familiar with the metric system, because most technical sources report data only in metric units; most of the world outside the United States uses the metric system exclusively. Conversion factors are included in Table 2.

The U.S. and metric systems also differ in their methods of reporting temperatures. In the United States, temperatures are usually reported in degrees Fahrenheit (°F), whereas scientific literature and individuals in most other countries report temperatures in degrees Celsius (°C). The relationship between temperatures in degrees Fahrenheit and those in degrees Celsius is indicated in Table 2.

The following illustration spans the entire range of measurements that we will consider in this book. Gross anatomy traditionally deals with structural organization as seen with the naked eye or with a simple hand lens. A microscope can provide higher levels of magnification and can reveal finer details. Before the 1950s, most information was provided by light microscopy. A photograph taken through a *light microscope* is called a **light micrograph (LM)**. Light microscopy can magnify cellular structures up to about 1000 times and can show details as fine as 0.25 μm. The symbol μm stands for **micrometer**; μm = 0.001 mm, or 0.00004 inches. With a light microscope, we can identify cell types, such as muscle fibers or neurons, and can see large structures within a cell. Because individual cells are relatively transparent, thin sections cut through a cell are treated with dyes that stain specific structures to make them easier to see.

Although special staining techniques can show the general distribution of proteins, lipids, carbohydrates, and nucleic acids in the cell, many fine details of intracellular structure remained a mystery until investigators began using *electron microscopy*. This technique uses a focused beam of electrons, rather than a beam of light, to examine cell structure. In *transmission electron microscopy*, electrons pass through an ultrathin section to strike a photographic plate. The result is a **transmission electron micrograph (TEM)**. Transmission electron microscopy shows the fine structure of plasma membranes and intracellular structures. In *scanning electron microscopy*, electrons bouncing off exposed surfaces create a **scanning electron micrograph (SEM)**. Although it cannot achieve as much magnification as transmission microscopy, scanning microscopy provides a three-dimensional perspective of cell structure.

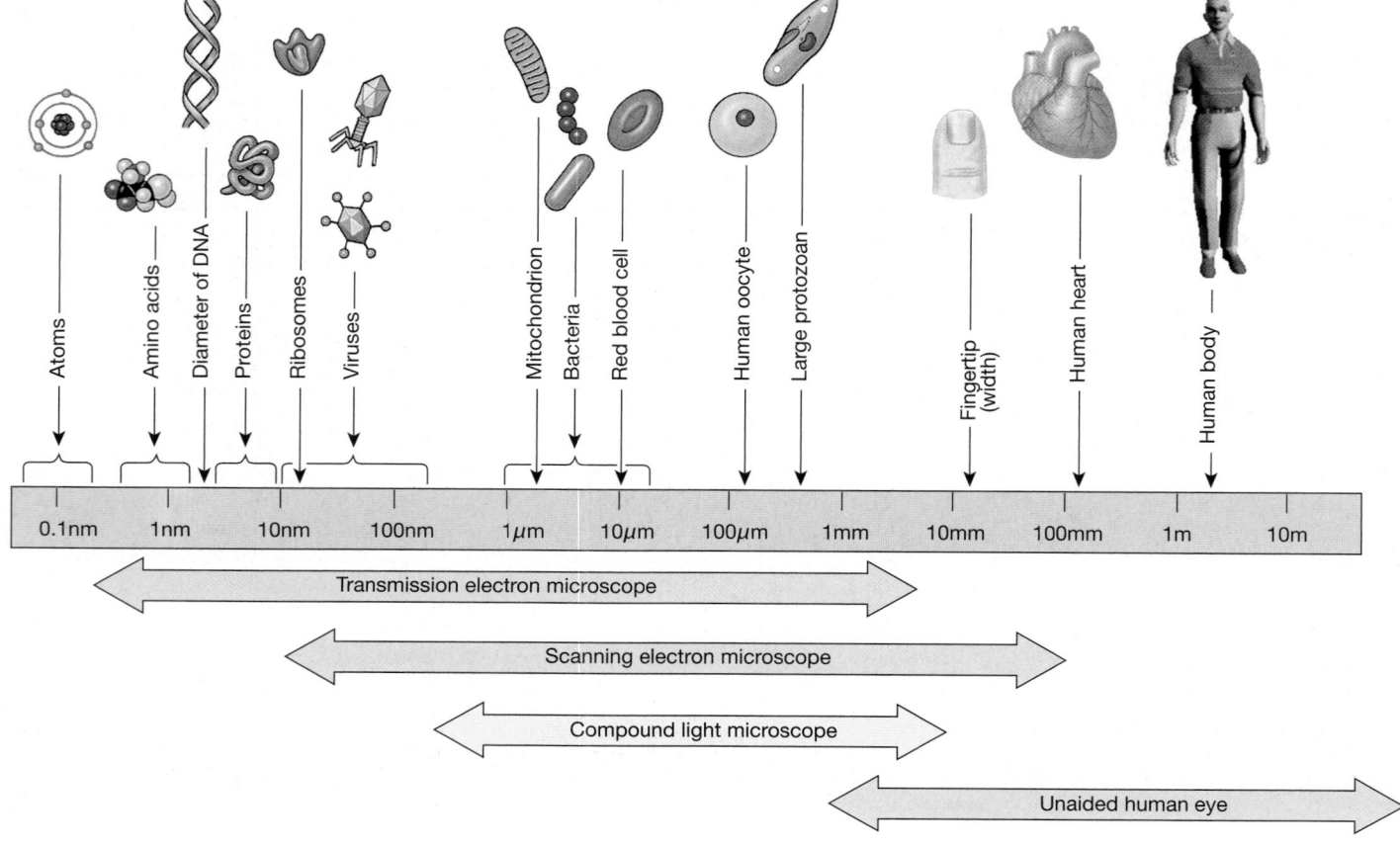

Periodic Table

The periodic table presents the known elements in order of their atomic weights. Each horizontal row represents a single electron shell. The number of elements in that row is determined by the maximum number of electrons that can be stored at that energy level. The element at the left end of each row contains a single electron in its outermost electron shell; the element at the right end of the row has a filled outer electron shell. Organizing the elements in this fashion highlights similarities that reflect the composition of the outer electron shell. These similarities are evident when you examine the vertical columns. All the gases of the right-most column—helium, neon, argon, krypton, xenon, and radon—have full electron shells; each is a gas at normal atmospheric temperature and pressure, and none reacts readily with other elements. These elements, highlighted in blue, are known as the *noble*, or *inert*, *gases*. In contrast, the elements of the left-most column below hydrogen—lithium, sodium, potassium, rubidium, cesium, and francium—are silvery, soft metals that are so highly reactive that pure forms cannot be found in nature. The fourth and fifth electron levels can hold up to 18 electrons. Table inserts are used for the so-called *lanthanide* and *actinide* series to save space, as higher levels can store up to 32 electrons. Elements of particular importance to our discussion of human anatomy and physiology are highlighted in pink.

Atomic number — **1** **H** — Chemical symbol
Hydrogen — Element name
Atomic weight — **1.01**

1	2	3	4	5	6	7	8	9	10	11	12	13	14	15	16	17	18
1 H Hydrogen 1.01																	2 He Helium 4.00
3 Li Lithium 6.94	4 Be Beryllium 9.01											5 B Boron 10.81	6 C Carbon 12.01	7 N Nitrogen 14.01	8 O Oxygen 16.00	9 F Fluorine 19.00	10 Ne Neon 20.18
11 Na Sodium 22.99	12 Mg Magnesium 24.31											13 Al Aluminum 26.98	14 Si Silicon 28.09	15 P Phosphorus 30.97	16 S Sulfur 32.07	17 Cl Chlorine 35.45	18 Ar Argon 39.95
19 K Potassium 39.10	20 Ca Calcium 40.08	21 Sc Scandium 44.96	22 Ti Titanium 47.88	23 V Vanadium 50.94	24 Cr Chromium 52.00	25 Mn Manganese 54.94	26 Fe Iron 55.85	27 Co Cobalt 58.93	28 Ni Nickel 58.69	29 Cu Copper 63.55	30 Zn Zinc 65.39	31 Ga Gallium 69.72	32 Ge Germanium 72.61	33 As Arsenic 74.92	34 Se Selenium 78.96	35 Br Bromine 79.90	36 Kr Krypton 83.80
37 Rb Rubidium 85.47	38 Sr Strontium 87.62	39 Y Yttrium 88.91	40 Zr Zirconium 91.22	41 Nb Niobium 92.91	42 Mo Molybdenum 95.94	43 Tc Technetium (98)	44 Ru Ruthenium 101.07	45 Rh Rhodium 102.91	46 Pd Palladium 106.42	47 Ag Silver 107.87	48 Cd Cadmium 112.41	49 In Indium 114.82	50 Sn Tin 118.71	51 Sb Antimony 121.76	52 Te Tellurium 127.60	53 I Iodine 126.90	54 Xe Xenon 131.29
55 Cs Cesium 132.91	56 Ba Barium 137.33	57 La Lanthanum 138.91 *	72 Hf Hafnium 178.49	73 Ta Tantalum 180.95	74 W Tungsten 183.85	75 Re Rhenium 186.21	76 Os Osmium 190.2	77 Ir Iridium 192.22	78 Pt Platinum 195.08	79 Au Gold 196.97	80 Hg Mercury 200.59	81 Tl Thallium 204.38	82 Pb Lead 207.2	83 Bi Bismuth 208.98	84 Po Polonium (209)	85 At Astatine (210)	86 Rn Radon (222)
87 Fr Francium (223)	88 Ra Radium 226.03	89 Ac Actinium 227.03 †	104 Rf Rutherfordium (266)	105 Db Dubnium (262)	106 Sg Seaborgium (266)	107 Bh Bohrium (264)	108 Hs Hassium (269)	109 Mt Meitnerium (268)	110 DS Darmstadtium (271)	111 Rg Roentgenium (272)	112 Uub (277)	113 Uut (284)	114 Uuq (285)	115 Uup (288)	116 Uuh (291)		118 Uuo (294)

*Lanthanide series

57 La 138.91	58 Ce Cerium 140.12	59 Pr Praseodymium 140.91	60 Nd Neodymium 144.24	61 Pm Promethium (145)	62 Sm Samarium 150.36	63 Eu Europium 151.96	64 Gd Gadolinium 157.25	65 Tb Terbium 158.93	66 Dy Dysprosium 162.50	67 Ho Holmium 164.93	68 Er Erbium 167.26	69 Tm Thulium 168.93	70 Yb Ytterbium 173.05	71 Lu Lutetium 174.97

†Actinide series

89 Ac 227	90 Th Thorium 232.04	91 Pa Protactinium 232.04	92 U Uranium 238.03	93 Np Neptunium 237	94 Pu Plutonium (244)	95 Am Americium (243)	96 Cm Curium (247)	97 Bk Berkelium (247)	98 Cf Californium (251)	99 Es Einsteinium (252)	100 Fm Fermium (257)	101 Md Mendelevium (258)	102 No Nobelium (259)	103 Lr Lawrencium (262)

Normal Physiological Values

Tables 3 and 4 present normal averages or ranges for the chemical composition of body fluids. These values are approximations rather than absolute values, because test results vary from laboratory to laboratory owing to differences in procedures, equipment, normal solutions, and so forth. Blanks in the tabular data appear where data are not available; sources used in the preparation of these tables follow. The following locations in the text contain additional information about body fluid analysis:

Table 19–3 (p. 670) presents data on the cellular composition of whole blood.

Table 26–2 (p. 976) compares the average compositions of urine and plasma.

Tables 26–5 (p. 993) and 26–6 (p. 996) give the general characteristics of normal urine.

Sources

Braunwauld, Eugene, Kurt J. Isselbacher, Dennis L. Kasper, Jean D. Wilson, Joseph B. Martin, and Anthony S. Fauci, eds. 1998. *Harrison's Principles of Internal Medicine*, 14th ed. New York: McGraw-Hill.

Ganong, William F. 2005. *Review of Medical Physiology,* 23rd ed. New York: McGraw-Hill.

Lentner, Cornelius, ed. 1981. *Geigy Scientific Tables*, 8th ed. Basel, Switzerland: Ciba–Geigy Limited.

Malarkey, Louise and Mary Ellen McMorrow. 2005. *Saunders Nursing Guide to Laboratory and Diagnostic Tests*. St. Louis: Elsevier.

Wintrobe, Maxwell, G. Richard Lee, Dane R. Boggs, Thomas C. Bitnell, John Foerster, John W. Athens, and John N. Lukens. 1981. *Clinical Hematology*, Philadelphia: Lea and Febiger.

TABLE 3 The Composition of Minor Body Fluids

Test	Normal Averages or Ranges					
	Perilymph	Endolymph	Synovial Fluid	Sweat	Saliva	Semen
pH			7.4	4–6.8	6.4*	7.19
Specific gravity			1.008–1.015	1.001–1.008	1.007	1.028
Electrolytes (mEq/L)						
Potassium	5.5–6.3	140–160	4.0	4.3–14.2	21	31.3
Sodium	143–150	12–16	136.1	0–104	14*	117
Calcium	1.3–1.6	0.05	2.3–4.7	0.2–6	3	12.4
Magnesium	1.7	0.02		0.03–4	0.6	11.5
Bicarbonate	17.8–18.6	20.4–21.4	19.3–30.6		6*	24
Chloride	121.5	107.1	107.1	34.3	17	42.8
Proteins (total) (mg/dL)	200	150	1.72 g/dL	7.7	386[†]	4.5 g/dL
Metabolites (mg/dL)						
Amino acids				47.6	40	1.26 g/dL
Glucose	104		70–110	3.0	11	224 (fructose)
Urea				26–122	20	72
Lipids (total)	12		20.9	[‡]	25–500[§]	188

*Increases under salivary stimulation.

[†]Primarily alpha-amylase, with some lysozymes.

[‡]Not present in eccrine secretions.

[§]Cholesterol.

TABLE 4 The Chemistry of Blood, Cerebrospinal Fluid, and Urine

Test	Normal Averages or Ranges		
	Blood*	**CSF**	**Urine**
pH	S: 7.35–7.45	7.31–7.34	4.5–8.0
Osmolarity (mOsm/L)	S: 280–295	292–297	855–1335
Electrolytes		(mEq/L unless noted)	(urinary loss per 24-hour period[†])
Bicarbonate	P: 21–28	20–24	
Calcium	S: 4.5–5.5	2.1–3.0	6.5–16.5 mEq
Chloride	S: 97–107	113–122	120–240 mEq
Iron	S: 50–150 μg/L	23–52 μg/L	40–150 μg
Magnesium	S: 1.4–2.1	2–2.5	4.9–16.5 mEq
Phosphorus	S: 1.8–2.9	1.2–2.0	0.8–2 g
Potassium	P: 3.5–5.5	2.7–3.9	35–80 mEq
Sodium	P: 136–145	137–145	120–220 mEq
Sulfate	S: 0.2–1.3		1.07–1.3 g
Metabolites		(mg/dL unless noted)	(urinary loss per 24-hour period[‡])
Amino acids	P/S: 2.3–5.0	10.0–14.7	41–133 mg
Ammonia	P: 20–150 μg/dL	25–80 μg/dL	340–1200 mg
Bilirubin	S: 0.5–1.0	< 0.2 mg/dL	0.02–1.9 mg
Creatinine	P/S: 0.6–1.2	0.5–1.9	1.01–2.5
Glucose	P/S: 70–110	40–70	0
Ketone bodies	S: 0.3–2.0	1.3–1.6	10–100 mg
Lactic acid	WB: 5–20[§]	10–20	100–600 mg
Lipids (total)	S: 400–1000	0.8–1.7	0–31.8 mg
Cholesterol (total)	S: 150–300	0.2–0.8	1.2–3.8 mg
Triglycerides	S: 40–150	0–0.9	
Urea	P/S: 23–43	12.0	12.6–28.6
Uric acid	S: 2.0–7.0	0.2–1.5	80–976 mg
Proteins	(g/dL)	(mg/dL)	(urinary loss per 24-hour period[‡])
Total	S: 6.0–7.8	2.0–4.5	47–76.2 mg
Albumin	S: 3.2–4.5	10.6–32.4	10–100 mg
Globulins (total)	S: 2.3–3.5	2.8–15.5	7.3 mg (average)
Immunoglobulins	S: 1.0–2.2	1.1–1.7	3.1 mg (average)
Fibrinogen	P: 0.2–0.4	0.65 (average)	

*S = serum, P = plasma, WB = whole blood.

[†]Because urinary output averages just over 1 liter per day, these electrolyte values are comparable to mEq/L.

[‡]Because urinary metabolite and protein data approximate mg/L or g/L, these data must be divided by 10 for comparison with CSF or blood concentrations.

[§]Venous blood sample.

Answers to Checkpoint and Chapter Review Questions

Chapter 1

Answers to Checkpoints

Page 2

1. Anatomy is the oldest medical science.

2. Studying human anatomy and physiology is important because understanding normal physiology assists in recognizing when something abnormal occurs within the body.

Page 4

3. Strategies for success in an A&P course include committing to diligent study, creating a time and space for studying, developing good study skills and routine habits, taking good notes during lecture, reading the textbook before and after lecture, attending all lecture and lab sessions, utilizing the laboratory, spending additional time in lab studying the material, avoiding procrastination, seeking additional assistance when necessary, and using the student activities packaged with this textbook.

4. Some of the resources packaged with the textbook include Interactive Physiology CD, My A&P, and Martini's *Atlas of the Human Body*.

Page 5

5. Anatomy is the study of internal and external body structures.

6. Physiology is the study of how living organisms perform functions.

7. Medical terminology is the use of prefixes, suffixes, word roots, and combining forms to construct anatomical, physiological, or medical terms.

8. An eponym is a commemorative name for a structure or clinical condition that was originally named for a real or mythical person.

9. The book used as the international standard for anatomical vocabulary is *International Anatomical Terminology (Terminologia Anatomica)*.

Page 8

10. Anatomy and physiology are closely related because all specific functions are performed by specific structures.

11. Gross anatomy (often referred to as macroscopic anatomy) involves studying body structures that can be seen with the unaided eye. Microscopic anatomy is the study of body structures using a microscope to magnify the objects.

12. Several specialties of physiology are cell physiology, organ physiology, systemic physiology, and pathological physiology.

13. It is difficult to separate anatomy from physiology because the structures of body parts are so closely related to their functions; put another way, function follows form.

Page 12

14. The major levels of organization from the simplest to the most complex are the following: chemical (molecular) level → cellular level → tissue level → organ level → organ system level → organism level.

15. Major organ systems (and some structures of each) are the integumentary system (skin, hair, sweat glands, and nails), skeletal system (bones, cartilages, associated ligaments, and bone marrow),

muscular system (skeletal muscles and associated tendons), nervous system (brain, spinal cord, peripheral nerves, and sense organs), endocrine system (pituitary gland, thyroid gland, pancreas, suprarenal glands, gonads, and other endocrine tissues), cardiovascular system (heart, blood, and blood vessels), lymphoid system (spleen, thymus, lymphatic vessels, lymph nodes, and tonsils), respiratory system (nasal cavities, sinuses, larynx, trachea, bronchi, lungs, and alveoli), digestive system (teeth, tongue, pharynx, esophagus, stomach, small intestine, large intestine, liver, gallbladder, and pancreas), urinary system (kidneys, ureters, urinary bladder, and urethra), male reproductive system (testes, epididymides, ductus deferens, seminal glands, prostate gland, penis, and scrotum), and female reproductive system (ovaries, uterine tubes, uterus, vagina, labia, clitoris, and mammary glands).

16. A histologist investigates structures and properties at the tissue level of organization.

Page 14

17. Homeostasis refers to the existence of a stable internal environment.

18. Extrinsic regulation is a type of homeostatic regulation resulting from activities of the nervous system or endocrine system.

19. Physiological systems can function normally only under carefully controlled conditions. Homeostatic regulation prevents potentially disruptive changes in the body's internal environment.

Page 17

20. Negative feedback systems provide long-term control over the body's internal conditions—that is, they maintain homeostasis—by counteracting the effects of a stimulus.

21. When homeostasis fails, organ systems function less efficiently or even malfunction. The result is the state that we call disease. If the situation is not corrected, death can result.

22. A positive feedback system amplifies or reinforces the effects of a stimulus.

23. Positive feedback is useful in processes that must move quickly to completion, such as blood clotting. It is harmful in situations in which a stable condition must be maintained, because it tends to increase any departure from the desired condition. Positive feedback in the regulation of body temperature, for example, would cause a slight fever to spiral out of control, with fatal results. For this reason, physiological systems are typically regulated by negative feedback, which tends to oppose any departure from the norm.

24. Equilibrium is a dynamic state in which two opposing forces or processes are in balance.

25. When the body continuously adapts, utilizing homeostatic systems, it is said to be in a state of dynamic equilibrium.

Page 22

26. The purpose of anatomical terms is to provide a standardized frame of reference for describing the human body.

27. In the anatomical position, an anterior view is seen from the front and a posterior view is from the back.

28. Body cavities protect internal organs and cushion them from thumps and bumps that occur while walking, running, or jumping. Body cavities also permit the organs within them to change in size and shape without disrupting the activities of nearby organs.
29. The ventral body cavities include the pleural and pericardial cavities within the thoracic cavity, and the peritoneal, abdominal, and pelvic cavities within the abdominopelvic cavity.

Answers to Review Questions

Page 26

Level 1 Reviewing Facts and Terms

1. (a) superior **(b)** inferior **(c)** posterior or dorsal **(d)** anterior or ventral **(e)** cranial **(f)** caudal **(g)** lateral **(h)** medial **(i)** proximal **(j)** distal
2. g **3.** d **4.** a **5.** j **6.** b **7.** l **8.** n **9.** f **10.** h **11.** e
12. c **13.** o **14.** k **15.** m **16.** i **17.** b **18.** c **19.** d **20.** c
21. b **22. (a)** pericardial cavity **(b)** peritoneal (or abdominal) cavity **(c)** pleural cavity **(d)** abdominal (or abdominopelvic) cavity **23.** b

Level 2 Reviewing Concepts

24. (a) Anatomy is the study of internal and external structures and the physical relationships among body parts. **(b)** Physiology is the study of how organisms perform their vital functions.
25. d
26. Autoregulation occurs when the activities of a cell, tissue, organ, or organ system change automatically (that is, without neural or endocrine input) when faced with some environmental change. Extrinsic regulation results from the activities of the nervous or endocrine systems. It causes more extensive and potentially more effective adjustments in activities.
27. The body is erect, and the hands are at the sides with the palms facing forward.
28. b **29.** c

Level 3 Critical Thinking and Clinical Applications

30. Calcitonin is released when calcium levels are elevated. This hormone should bring about a decrease in blood calcium levels, thus decreasing the stimulus for its own release.
31. There are several reasons why your body temperature may have dropped. Your body may be losing heat faster than it is being produced. This, however, is more likely to occur on a cool day. Various chemical factors, such as hormones, may have caused a decrease in your metabolic rate, and thus your body is not producing as much heat as it normally would. Alternatively, you may be suffering from an infection that has temporarily changed the set point of the body's "thermostat." This would seem to be the most likely explanation considering the circumstances of the question.

Chapter 2

Answers to Checkpoints

Page 32

1. An atom is the smallest stable unit of matter.
2. Atoms of the same element that have the same atomic number but different numbers of neutrons are called isotopes.
3. Hydrogen has three isotopes: hydrogen-1, with a mass of 1; deuterium, with a mass of 2; and tritium, with a mass of 3. The heavier sample must contain a higher proportion of one or both of the heavier isotopes.

4. A chemical bond is an attractive force acting between two atoms that may be strong enough to hold them together in a molecule or compound. The strongest attractive forces result from the gain, loss, or sharing of electrons. Examples of such chemical bonds are ionic bonds and covalent bonds. In contrast, hydrogen bonds occur between molecules or compounds.
5. The atoms in a water molecule are held together by polar covalent bonds. Water molecules are attracted to one another by hydrogen bonds.
6. Atoms combine with each other so as to gain a complete set of electrons in their outer energy level. Oxygen atoms do not have a full outer energy level, so they readily react with many other elements to attain this stable arrangement. Neon already has a full outer energy level and thus has little tendency to combine with other elements.

Page 39

7. The chemical shorthand used to describe chemical compounds and reactions is known as chemical notation.
8. The molecular formula for glucose, a compound composed of six carbon atoms, 12 hydrogen atoms, and six oxygen atoms, is $C_6H_{12}O_6$.
9. Three types of chemical reactions important to the study of human physiology include decomposition reactions, synthesis reactions, and exchange reactions. In a decomposition reaction, a chemical reaction breaks a molecule into smaller fragments. A synthesis reaction assembles smaller molecules into larger ones. In an exchange reaction, parts of the reacting molecules are shuffled around to produce new products.
10. Because this reaction involves a large molecule being broken down into two smaller ones, it is a decomposition reaction. Because energy is released in the process, the reaction can also be classified as exergonic (exothermic).

Page 40

11. An enzyme is a protein that lowers the activation energy of a chemical reaction, which is the amount of energy required to start the reaction.
12. Without enzymes, chemical reactions could proceed in the body only under conditions that cells cannot tolerate (e.g., high temperatures). By lowering the activation energy, enzymes make it possible for chemical reactions to proceed under conditions compatible with life.
13. Organic compounds contain carbon, hydrogen, and (in most cases) oxygen. Inorganic compounds generally do not contain carbon and hydrogen atoms as primary structural ingredients.

Page 43

14. Specific chemical properties of water that make life possible include its solubility (its strong polarity enables it to be used as an efficient solvent), its reactivity (it participates in many chemical reactions), its high heat capacity (it absorbs and releases heat slowly), and its ability to serve as a lubricant.

Page 44

15. pH is a measure of the concentration of hydrogen ions in fluids. Acid and base concentrations are measured in pH, which is the negative logarithm of hydrogen ion concentration, expressed in moles per liter. On the pH scale, 7 represents neutrality; values below 7 indicate acidic solutions, and values above 7 indicate alkaline (basic) solutions.
16. If the body is to maintain homeostasis and thus health, the pH of different body fluids must remain within a fairly narrow range.

Page 45

17. An acid is a compound whose dissociation in solution releases a hydrogen ion and an anion; a base is a compound whose dissociation releases a hydroxide ion (OH^-) or removes a hydrogen ion (H^+) from the solution; a salt is an inorganic compound consisting of a cation other than H^+ and an anion other than OH^-.

18. Stomach discomfort is commonly the result of excessive stomach acidity ("acid indigestion"). Antacids contain a weak base that neutralizes the excess acid.

Page 48

19. A compound with a C:H:O ratio of 1:2:1 is a carbohydrate. The body uses carbohydrates chiefly as an energy source.

Page 51

20. Lipids are a diverse group of compounds that include fatty acids, eicosanoids, glycerides, steroids, phospholipids, and glycolipids. They are organic compounds that contain carbon, hydrogen, and oxygen in a ratio that does not approximate 1:2:1.

21. Human plasma membranes primarily contain phospholipids, plus small amounts of cholesterol and glycolipids.

Page 57

22. Proteins are organic compounds formed from amino acids, which contain a central carbon atom, a hydrogen atom, an amino group (—NH_2), a carboxylic acid group (—COOH), and a variable group, known as an R group or side chain. Proteins function in support, movement, transport, buffering, metabolic regulation, coordination and control, and defense.

23. The heat of boiling breaks bonds that maintain the protein's tertiary structure, quaternary structure, or both. The resulting structural change, known as denaturation, affects the ability of the protein molecule to perform its normal biological functions.

Page 59

24. A nucleic acid is a large organic molecule composed of carbon, hydrogen, oxygen, nitrogen, and phosphorus. Nucleic acids regulate protein synthesis and make up the genetic material in cells.

25. Both DNA (deoxyribonucleic acid) and RNA (ribonucleic acid) contain nitrogenous bases and phosphate groups, but because this nucleic acid contains the sugar ribose, it is RNA.

Page 60

26. Adenosine triphosphate (ATP) is a high-energy compound consisting of adenosine to which three phosphate groups are attached; the third is attached by a high-energy bond.

27. Phosphorylation of an ADP molecule yields a molecule of ATP.

Page 61

28. The biochemical building blocks that are components of cells include lipids (forming the plasma membrane), proteins (acting as enzymes), nucleic acids (directing the synthesis of cellular proteins), and carbohydrates (providing energy for cellular activities).

29. Metabolic turnover is the continuous breakdown and replacement of organic materials within cells.

Answers to Review Questions

Page 64

Level 1 Reviewing Facts and Terms
1. a.

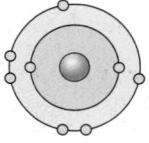

Oxygen atom

b. Two more electrons can fit into the outermost energy level of an oxygen atom.

2. Hydrolysis.

3. a **4.** b **5.** d **6.** d **7.** b **8.** c **9.** d **10.** d **11.** b **12.** d **13.** a **14.** d **15.** b

16. protons, neutrons, and electrons

17. carbohydrates, lipids, proteins, and nucleic acids

18. Triglycerides (1) provide a significant energy reserve, (2) serve as insulation and thus act in heat conservation, and (3) protect organs by cushioning them.

19. (1) support (structural proteins); (2) movement (contractile proteins); (3) transport (transport proteins); (4) buffering; (5) metabolic regulation; (6) coordination and control (hormones and neurotransmitters); and (7) defense (antibodies)

20. (a) DNA: deoxyribose, phosphate, and nitrogenous bases (A, T, C, G); (b) RNA: ribose, phosphate, and nitrogenous bases (A, U, C, G)

21. (1) adenosine, (2) phosphate groups, and (3) appropriate enzymes

Level 2 Reviewing Concepts
22. d **23.** c

24. Enzymes are specialized protein catalysts that lower the activation energy for chemical reactions. Enzymes speed up chemical reactions but are not used up or changed in the process.

25. A salt is an ionic compound consisting of any cations other than hydrogen ions and any anions other than hydroxide ions. Acids dissociate and release hydrogen ions, while bases remove hydrogen ions from solution (usually by releasing hydroxide ions).

26. Nonpolar covalent bonds have an equal sharing of electrons. Polar covalent bonds have an unequal sharing of electrons. Ionic bonds result from the loss or gain of electrons.

27. e **28.** c

29. The molecule is a nucleic acid. Carbohydrates and lipids do not contain nitrogen. Although both proteins and nucleic acids contain nitrogen, only nucleic acids normally contain phosphorus.

Level 3 Critical Thinking and Clinical Applications
30. (a) number of electrons = 20; (b) atomic number = 20; (c) atomic weight = 40; (d) 2 electrons in first shell, 8 in second shell; 8 in third shell, and 2 in outer shell.

31. Decreasing the amount of enzyme at the second step would slow down the remaining steps of the pathway because less substrate would be available for the next two steps. The net result would be a decrease in the amount of product.

32. If a person exhales large amounts of CO_2, the equilibrium will shift to the left, and the level of H^+ in the blood will decrease. A decrease in the amount of H^+ will cause the pH to rise.

Chapter 3

Answers to Checkpoints

Page 72

1. The general functions of the plasma membrane include physical isolation, regulation of exchange with the environment, sensitivity to the environment, and structural support.

2. The components of the plasma membrane that enable it to perform its characteristic functions are membrane lipids, membrane proteins, and membrane carbohydrates.

3. The phospholipid bilayer of the plasma membrane forms a physical barrier between the cell's internal and external environments.

4. Channel proteins are integral proteins that allow water and small ions to pass through the plasma membrane.

Page 82

5. Cytoplasm is the material between the plasma membrane and the nuclear membrane; cytosol is the fluid portion of the cytoplasm.

6. Cytosol has a higher concentration of potassium ions and suspended proteins, and a lower concentration of sodium ions, than the extracellular fluid. Cytosol also includes small quantities of carbohydrates, and larger reserves of amino acids and lipids.

7. The nonmembranous organelles (and their functions) include: (1) centriole = essential for movement of chromosomes during cell division; organization of microtubules in cytoskeleton; (2) cilia = movement of materials over cell surface; (3) cytoskeleton = strength and support; movement of cellular structures and materials; cell movement; (4) microvilli = increase surface area to facilitate absorption of extracellular materials; (5) proteasomes = breakdown and recycling of intracellular proteins; (6) ribosomes = protein synthesis.

8. The membranous organelles (and their functions) include: (1) endoplasmic reticulum = synthesis of secretory products; intracellular storage and transport; (2) rough ER = modification and packaging of newly synthesized proteins; (3) smooth ER = lipid and carbohydrate synthesis; (4) Golgi apparatus = storage, alteration, and packaging of secretory products and lysosomal enzymes; (5) lysosomes = intracellular removal of damaged organelles or pathogens; (6) mitochondria = production of 95 percent of the ATP required by the cell; (7) peroxisomes = neutralization of toxic compounds

9. Mitochondria produce energy, in the form of ATP molecules, for the cell. A large number of mitochondria in a cell indicates a high demand for energy.

10. The SER functions in the synthesis of lipids such as steroids. Ovaries and testes produce large amounts of steroid hormones, which are lipids, and thus need large amounts of SER.

Page 84

11. The nucleus is a cellular organelle that contains DNA, RNA, and proteins. The nuclear envelope is a double membrane that surrounds the nucleus; the perinuclear space is the region between this double membrane. Nuclear pores allow for chemical communication between the nucleus and the cytosol.

12. A gene is a portion of a DNA strand that functions as a hereditary unit. Each gene is located at a particular site on a specific chromosome and codes for a specific protein.

Page 89

13. Gene activation is the process of uncoiling the segment of DNA containing that gene, and temporarily removing histones, so that the gene can be expressed and thus affect cell function.

14. Transcription is the encoding of genetic instructions on a strand of mRNA.

15. A cell that lacked the enzyme RNA polymerase would not be able to transcribe RNA from DNA.

Page 94

16. The three major categories of cellular transport processes are diffusion, carrier-mediated transport, and vesicular transport.

17. A selectively permeable membrane allows the passage of some substances while restricting the passage of others. It falls between two extremes: impermeable, which allows no substance to pass, and freely permeable, which permits the passage of any substance.

18. Diffusion is the passive molecular movement of a substance from an area of higher concentration to an area of lower concentration; diffusion proceeds until equilibrium is reached.

19. Diffusion is driven by a concentration gradient. The larger the concentration gradient, the faster the rate of diffusion; the smaller the concentration gradient, the slower the rate of diffusion. If the concentration of oxygen in the lungs were to decrease, the concentration gradient between oxygen in the lungs and oxygen in the blood would decrease (as long as the oxygen level of the blood remained constant). Thus, oxygen would diffuse more slowly into the blood.

20. Osmosis is the diffusion of water across a selectively permeable membrane from one solution to another solution containing a higher solute concentration.

21. The 10 percent salt solution is hypertonic with respect to the cells lining the nasal cavity, because it contains a higher salt concentration than do the cells. The hypertonic solution would draw water out of the cells, causing the cells to shrink and adding water to the mucus, thereby relieving the congestion.

Page 98

22. In carrier-mediated transport, integral proteins bind specific ions or organic substrates and carry them across the plasma membrane. All forms of carrier-mediated transport are specific, have saturation limits, and are regulated (as by hormones).

23. An active transport process must be involved because it takes an energy expenditure to move the hydrogen ions against their concentration gradient—that is, from a region where they are less concentrated (the cells lining the stomach) to a region where they are more concentrated (the interior of the stomach).

24. Endocytosis is the movement of relatively large volumes of extracellular material into the cytoplasm via the formation of a membranous vesicle at the cell surface; it includes pinocytosis (the introduction of fluids into the cytoplasm by enclosing them in membranous vesicles at the cell surface) and phagocytosis (the movement of extracellular materials into the cytoplasm by enclosure in a membranous vesicle).

25. Exocytosis is the ejection of cytoplasmic materials by the fusion of a membranous vesicle with the plasma membrane.

26. The process by which certain white blood cells engulf bacteria is called phagocytosis.

Page 100

27. The transmembrane potential is the difference in electrical potential that results from the uneven distribution of positive and negative ions across the plasma membrane. It is expressed in millivolts.

28. If the plasma membrane were freely permeable to sodium ions, more of these positively charged ions would move into the cell, and the transmembrane potential would move closer to zero.

Page 104

29. The biological term for cellular reproduction is *cell division,* and the term for cell death is *apoptosis.*

30. Interphase is the portion of a cell's life cycle during which the chromosomes are uncoiled and all normal cellular functions except mitosis are under way. The stages of interphase include G_1, S, G_2, and G_0. A cell in G_0 is not preparing for cell division.

31. This cell is likely in the G_1 phase of its life cycle.

32. Mitosis is the essential step in cell division in which a single cell nucleus divides to produce two identical daughter cell nuclei. The four stages of mitosis are prophase, metaphase, anaphase, and telophase.

33. If spindle fibers failed to form during mitosis, the cell would not be able to separate the chromosomes into two sets. If cytokinesis occurred, the result would be one cell with two sets of chromosomes and one cell with none.

34. A growth factor is an extracellular compound, such as a peptide or hormone, that can stimulate the division of specific cell types. Representative growth factors include chalones, epidermal growth factor (EGF), erythropoietin, fibroblast growth factor (FGF), growth hormone, M-phase promoting factor (maturation-promoting factor), nerve growth factor (NGF), prolactin, and thymosins and related compounds.

Page 106
35. An illness characterized by mutations that disrupt normal control mechanisms and produce potentially malignant cells is termed cancer.

36. Metastasis is the spread of cancer cells from one organ to another, leading to the establishment of secondary tumors.

Page 107
37. Differentiation is the gradual appearance of characteristic cellular specializations during development; it results from gene activation or repression.

Answers to Review Questions
Page 109
Level 1 Reviewing Facts and Terms
1. a. isotonic b. hypotonic c. hypertonic
2. b   **3.** c   **4.** d   **5.** a   **6.** c   **7.** a   **8.** c   **9.** a   **10.** b
11. (1) Cells are the building blocks of all plants and animals. (2) Cells are produced by the division of preexisting cells. (3) Cells are the smallest units that perform all vital physiological functions. (4) Each cell maintains homeostasis at the cellular level.
12. Four general functions of the plasma membrane are (1) physical isolation, (2) regulation of exchange with the environment, (3) sensitivity, and (4) structural support.
13. Membrane proteins function as receptors, channels, carriers, enzymes, anchors, and identifiers.
14. The major transport mechanisms are (1) diffusion, (2) carrier-mediated transport, and (3) vesicular transport.
15. Factors that affect diffusion rate are (1) distance, (2) size of the gradient, (3) molecule size, and (4) temperature.
16. Major functions of the ER are (1) synthesis of proteins, carbohydrates, and lipids; (2) storage of absorbed or synthesized molecules; (3) transport of materials; and (4) detoxification of drugs or toxins.

Level 2 Reviewing Concepts
17. b   **18.** d   **19.** b   **20.** c   **21.** c   **22.** d   **23.** b
24. G_0: normal cell functions; G_1: cell growth, duplication of organelles, and protein synthesis; S: DNA replication and synthesis of histones; G_2: protein synthesis
25. Prophase: chromatin condenses and chromosomes become visible; centrioles migrate to opposite poles of the cell and spindle fibers develop; and the nuclear membrane disintegrates. Metaphase: chromatids attach to spindle fibers and line up along the metaphase plate. Anaphase: chromatids separate and migrate toward opposite poles of the cell. Telophase: the nuclear membrane forms; chromosomes disappear as the chromatin relaxes; and nucleoli appear.
26. (a) Cytokinesis is the cytoplasmic movement that separates two daughter cells. **(b)** It completes the process of cell division.

Level 3 Critical Thinking and Clinical Applications
27. This process is facilitated diffusion, which requires a carrier molecule but not cellular energy. The energy for this process is provided by the concentration gradient of the substance being transported. When all the carriers are actively involved in transport, the rate of transport reaches a saturation point.
28. Solution A must have initially had more solutes than solution B. As a result, water moved by osmosis across the selectively permeable membrane from side B to side A, increasing the fluid level on side A.
29. c
30. The isolation of the internal contents of membrane-bound organelles allows them to manufacture or store secretions, enzymes, or toxins that could adversely affect the cytoplasm in general. Another benefit is the increased efficiency of having specialized enzyme systems concentrated in one place. For example, the concentration of enzymes necessary for energy production in the mitochondrion increases the efficiency of cellular respiration.

Chapter 4
Answers to Checkpoints
Page 112
1. Histology is the study of tissues.
2. The four major types of tissues that form all body structures are epithelial, connective, muscle, and neural tissue.

Page 116
3. Epithelial tissue is characterized by cellularity, polarity, attachment, avascularity, and regeneration.
4. Epithelial tissue provides physical protection, controls permeability, provides sensation, and produces specialized secretions.
5. An epithelium whose cells bear many microvilli is probably involved in absorption; the microvilli greatly increase the surface area available for absorption.
6. Epithelial intercellular connections include occluding (tight) junctions, gap junctions, hemidesmosomes, and spot desmosomes.
7. Gap junctions allow small molecules and ions to pass from cell to cell. When connecting epithelial cells, they help coordinate such functions as the beating of cilia. In cardiac and smooth muscle tissues, they are essential to the coordination of muscle cell contractions.

Page 124
8. The three cell shapes characteristic of epithelial cells are squamous (flat), cuboidal (cube-like), and columnar (cylindrical).
9. When classifying epithelial tissue, the number of layers determines whether it is simple or stratified. A single layer of cells is termed simple, whereas multiple layers of cells are known as stratified.
10. No. A simple squamous epithelium does not provide enough protection against infection, abrasion, or dehydration. The skin surface has a stratified squamous epithelium.
11. All these regions are subject to mechanical trauma and abrasion—by food (pharynx and esophagus), by feces (anus), and by intercourse or childbirth (vagina).
12. The two primary types of glandular epithelia are endocrine glands and exocrine glands.
13. Sebaceous glands exhibit holocrine secretion.
14. This gland is an endocrine gland.

Page 131
15. Functions of connective tissues include: (1) defending the body from invading microorganisms; (2) establishing a structural framework for the body; (3) protecting delicate organs; (4) storing energy reserves; (5) supporting, surrounding, and interconnecting

other types of tissue; and (6) transporting fluids and dissolved materials.

16. The three categories of connective tissues are connective tissue proper, fluid connective tissues, and supporting connective tissues.

17. Cells found in connective tissue proper are adipocytes, fibrocytes, lymphocytes, macrophages, mast cells, melanocytes, mesenchymal cells, and microphages.

18. The reduced collagen production resulting from a lack of vitamin C in the diet would cause connective tissue to be weak and prone to damage.

19. Mast cells and basophils produce the molecule histamine, which antihistamines block.

20. The type of connective tissue that contains triacylglycerols is adipose (fat) tissue.

21. The two connective tissues that contain a fluid matrix are blood and lymph.

Page 135

22. The two types of supporting connective tissue are cartilage and bone.

23. Unlike cartilage, bone has a direct blood supply, which is necessary for proper and rapid healing to occur.

24. The type of cartilage damaged in a herniated intervertebral disc is fibrous cartilage.

Page 137

25. The four types of membranes found in the body are the cutaneous membrane, mucous membranes, serous membranes, and synovial membranes.

26. The pleural, peritoneal, and pericardial cavities are all lined by serous membranes.

27. The lining of the nasal cavity is a mucous membrane.

28. This tissue is probably fascia, a type of dense connective tissue that attaches muscles to skin and bones.

Page 140

29. The three types of muscle tissue in the body are skeletal muscle, cardiac muscle, and smooth muscle.

30. Muscle tissue that lacks striations is smooth muscle; both cardiac and skeletal muscles are striated.

31. Skeletal muscle is repaired by the division and fusion of myosatellite cells, which are mesenchymal cells that persist in adult skeletal muscle tissue.

Page 141

32. The cells are most likely neurons.

Page 143

33. The two phases in the response to tissue injury are inflammation and regeneration.

Page 144

34. With advancing age, the speed and effectiveness of tissue repair decrease, the rate of energy consumption in general declines, hormonal activity is altered, and other factors contribute to changes in structure and chemical composition.

Answers to Review Questions

Page 146

Level 1 Reviewing Facts and Terms

1. (a) simple squamous epithelium, **(b)** simple cuboidal epithelium, **(c)** simple columnar epithelium, **(d)** stratified squamous epithelium, **(e)** stratified cuboidal epithelium, **(f)** stratified columnar epithelium.

2. b **3.** b **4.** d **5.** c **6.** c **7.** b **8.** d **9.** d **10.** b **11.** c **12.** a **13.** b **14.** e **15.** a **16.** b

17. Epithelial tissue (1) provides physical protection; (2) controls permeability; (3) provides sensations; and (4) produces specialized secretions.

18. Endocrine glands secrete hormones onto the surface of the gland or directly into the surrounding fluid; exocrine glands secrete via ducts.

19. Glandular epithelial cells function by (1) merocrine secretion, (2) apocrine secretion, or (3) holocrine secretion.

20. Connective tissues contain (1) specialized cells, (2) extracellular protein fibers, and (3) a fluid ground substance.

21. The four membranes in the body are (1) serous membranes, (2) mucous membranes, (3) the cutaneous membrane, and (4) synovial membranes.

22. Neural tissue contains (1) neurons, which transmit electrical impulses in the form of changes in the transmembrane potential, and (2) neuroglia, which comprise several kinds of supporting cells and play a role in providing nutrients to neurons.

Level 2 Reviewing Concepts

23. Exocrine secretions are secreted onto a surface or outward through a duct. Endocrine secretions are secreted by ductless glands into surrounding tissues. Endocrine secretions are called hormones, which usually diffuse into the bloodstream for distribution to other parts of the body.

24. Occluding junctions block the passage of water or solutes between cells. In the digestive system, these junctions keep enzymes, acids, and waste products from damaging delicate underlying tissues.

25. Fluid connective tissues have a liquid, watery matrix. They differ from supporting connective tissues in that they have many soluble proteins in the matrix, and they do not include insoluble fibers.

26. The extensive connections between cells formed by cell junctions, intercellular cement, and physical interlocking hold skin cells together and can deny access to chemicals or pathogens that cover their free surfaces. If the skin is damaged and the connections are broken, infection can easily occur.

27. b **28.** b

29. Similarities: actin and myosin interactions produce contractions, calcium ions trigger and sustain contractions. Differences: skeletal muscles are relatively large, multinucleate, striated, and contract only under neural stimulation; cardiac muscles have 1–5 nuclei, are interconnected in a branching network, and contract in response to pacemaker cell activity; smooth muscles are small, spindle shaped, nonstriated, and have only one nucleus.

Level 3 Critical Thinking and Clinical Applications

30. Because apocrine secretions are released by pinching off a portion of the secreting cell, you could test for the presence of plasma membranes, specifically for the phospholipids in plasma membranes. Merocrine secretions do not contain a portion of the secreting cell, so they would lack membrane constituents.

31. Skeletal muscle tissue would be made up of densely packed fibers running in the same direction, but since muscle fibers are composed of cells, they would have many nuclei and mitochondria. Skeletal muscle also has an obvious banding pattern or striations due to the arrangement of the actin and myosin filaments within the cell. The student is probably looking at a slide of tendon (dense connective tissue). The small nuclei would be those of fibrocytes.

32. You would expect the skin in the area of the injury to become red and warm. It would also swell, and Jim would experience a painful sensation. These changes occur as a result of inflammation, the body's first response to injury. Injury to the epithelium and underlying connective tissue would trigger the release of chemicals such as histamine and heparin from mast cells in the area. These chemicals in turn initiate the changes that we observe.

Chapter 5
Answers to Checkpoints
Page 162
1. The layers of the epidermis are the stratum germinativum, stratum spinosum, stratum granulosum, stratum lucidum, and stratum corneum.
2. Dandruff consists of cells from the stratum corneum.
3. This splinter is lodged in the stratum granulosum.
4. Fresh water is hypotonic with respect to skin cells, so water moves into the cells by osmosis, causing them to swell.
5. Sanding the tips of the fingers will not permanently remove fingerprints. The ridges of the fingerprints are formed in layers of the skin that are constantly regenerated, so these ridges will eventually reappear. The pattern of the ridges is determined by the arrangement of tissue in the dermis, which is not affected by sanding.

Page 165
6. The two pigments contained in the epidermis are carotene, an orange-yellow pigment, and melanin, a brown, yellow-brown, or black pigment.
7. When exposed to the ultraviolet radiation in sunlight or sunlamps, melanocytes in the epidermis and dermis synthesize the pigment melanin, darkening the skin.
8. When skin gets warm, arriving oxygenated blood is diverted to the superficial dermis for the purpose of eliminating heat. The oxygenated blood imparts a reddish coloration to the skin.
9. In the presence of ultraviolet radiation in sunlight, epidermal cells in the stratum spinosum and stratum germinativum convert a cholesterol-related steroid into cholecalciferol, also known as vitamin D_3.
10. The hormone cholecalciferol (vitamin D_3) is needed to form strong bones and teeth. When the body surface is covered, UV light cannot penetrate to the blood in the skin to begin vitamin D_3 production, resulting in fragile bones.

Page 166
11. Salivary glands and duodenal glands produce epidermal growth factor (EGF).
12. Epidermal growth factor (EGF) promotes the divisions of germinative cells in the stratum germinativum and stratum spinosum, accelerates the production of keratin in differentiating keratinocytes, stimulates epidermal development and epidermal repair after injury, and stimulates synthetic activity and secretion by epithelial glands.

Page 168
13. The dermis (a connective tissue layer) lies between the epidermis and the hypodermis.
14. The capillaries and sensory neurons that supply the epidermis are located in the papillary layer of the dermis.
15. The presence of elastic fibers and the resilience of skin turgor allow the dermis to undergo repeated cycles of stretching and recoil (returning to its original shape).

Page 169
16. The tissue that connects the dermis to underlying tissues is the hypodermis or subcutaneous layer.
17. The hypodermis is a layer of loose connective tissue below the dermis; it is also called the subcutaneous layer or superficial fascia. It is not considered a part of the integument, but it is important in stabilizing the position of the skin in relation to underlying tissues.
18. Subcutaneous fat provides insulation to help reduce heat loss, serves as an energy reserve, and acts as a shock absorber for the body.

Page 171
19. A typical hair is a keratinous strand produced by epithelial cells of the hair follicle.
20. The contraction of the arrector pili muscle pulls the hair follicle erect, depressing the area at the base of the hair and making the surrounding skin appear higher. The result is known as "goose bumps."
21. Even though hair is a derivative of the epidermis, the follicles are in the dermis. Where the epidermis and deep dermis are destroyed, new hair will not grow.

Page 174
22. Two types of exocrine glands found in the skin are sebaceous (oil) glands and sweat glands.
23. Sebaceous secretions (called sebum) lubricate and protect the keratin of the hair shaft, lubricate and condition the surrounding skin, and inhibit the growth of bacteria.
24. Deodorants are used to mask the odor of apocrine sweat gland secretions, which contain several kinds of organic compounds; some of these compounds have an odor, and others produce an odor when metabolized by skin bacteria.
25. Apocrine sweat glands enlarge and increase secretory activity in response to the increase in sex hormones that occurs at puberty.

Page 175
26. A fingernail is a keratinous structure produced by epithelial cells of the nail root that protects the fingertip.
27. The area of thickened stratum corneum under the free edge of a nail is called the hyponychium.
28. Nail growth occurs at the nail root, an epidermal fold that is not visible from the surface.

Page 177
29. The combination of fibrin clots, fibroblasts, and the extensive network of capillaries in tissue that is healing is called granulation tissue.
30. Skin can regenerate effectively even after undergoing considerable damage because stem cells persist in both the epithelial and connective tissue components of skin. When injury occurs, cells of the stratum germinativum replace epithelial cells while mesenchymal cells replace cells lost from the dermis.

Page 178
31. As a person ages, the blood supply to the dermis decreases and merocrine sweat glands become less active. These changes make it more difficult for the elderly to cool themselves in hot weather.
32. With advancing age, melanocyte activity decreases, leading to gray or white hair.
33. The integumentary system provides the whole body, and thus all its systems, mechanical protection against environmental hazards.
34. The skin synthesizes vitamin D_3, which is essential for normal calcium absorption. Calcium ions play a key role in muscle contraction.

Answers to Review Questions

Level 1 Reviewing Facts and Terms

1. (a) epidermis **(b)** dermis **(c)** papillary layer **(d)** reticular layer **(e)** hypodermis (subcutaneous layer)

2. a **3.** c **4.** d **5.** d **6.** c **7.** a **8.** d **9.** b **10.** b **11.** d **12.** d **13.** b **14.** a

15. Epidermal cell division occurs in the stratum germinativum.

16. These smooth muscles cause hairs to stand erect when stimulated.

17. Epidermal growth factor promotes the divisions of germinal cells in the stratum germinativum and stratum spinosum; accelerates the production of keratin in differentiating epidermal cells; stimulates both epidermal development and epidermal repair after injury; and stimulates synthetic activity and secretion by epithelial cells.

18. The dermis consists of (1) the papillary layer, which consists of loose connective tissue and contains capillaries and sensory neurons, and (2) the reticular layer, which consists of dense irregular connective tissue and bundles of collagen fibers. Both layers contain networks of blood vessels, lymphatic vessels, and nerve fibers.

19. Regeneration of injured skin involves (1) bleeding, (2) scab formation, (3) granulation tissue formation, and (4) scarring.

Level 2 Reviewing Concepts

20. Insensible perspiration is water loss via evaporation through the stratum corneum. Sensible perspiration is produced by active sweat glands.

21. Fat-soluble substances easily pass through the permeability barrier, because it is composed primarily of lipids surrounding the epidermal cells. Water-soluble drugs are lipophobic and thus do not readily penetrate the permeability barrier.

22. A tan is a result of the synthesis of melanin in the skin. Melanin helps prevent skin damage by absorbing UV radiation before it reaches the deep layers of the epidermis and dermis. Within the epidermal cells, melanin concentrates around the outer wall of the nucleus, so it absorbs the UV light before it can damage nuclear DNA.

23. Incisions along the lines of cleavage—which represent the orientation of dermal collagen and elastin fibers—are more likely to remain closed, and thus will heal more quickly, than incisions that cut across lines of cleavage.

24. c **25.** c **26.** d

Level 3 Critical Thinking and Clinical Applications

27. a

28. The puncture wound by the nail has a greater chance of becoming infected. Whereas the knife cut will bleed freely, washing many of the bacteria from the wound site, the nail can carry bacteria beneath the surface of the skin, where oxygen is limited (anaerobic), and past the skin's protective barriers.

29. (a) Ultraviolet radiation in sunlight converts a cholesterol-related steroid into vitamin D_3, or cholecalciferol. This compound is then converted to calcitriol, which is essential in the small intestine for absorbing the calcium and phosphorus necessary for normal bone maintenance and growth. **(b)** The child can drink more milk. Milk is routinely fortified with cholecalciferol, normally identified as "vitamin D," which is easily absorbed by the intestines.

30. Sweating from merocrine glands is precisely regulated, and one influencing factor is emotional state. Presumably, a person who is lying is nervous and sweats noticeably; this sweating is detected by the lie detector machine.

31. The chemicals in hair dyes break the protective covering of the cortex, allowing the dyes to stain the medulla of the shaft. Dying is not permanent because the cortex remains damaged, allowing shampoo and UV rays from the sun to enter the medulla and affect the color. Also, the living portion of the hair remains unaffected, so that when the shaft is replaced the color will be lost.

Chapter 6

Answers to Checkpoints

Page 185

1. The five primary functions of the skeletal system are support, storage of minerals and lipids, blood cell production, protection, and leverage.

Page 188

2. The six broad categories for classifying bones according to shape are flat bones, irregular bones, long bones, sesamoid bones, short bones, and sutural bones.

3. A bone marking (surface feature) is an area on the surface of a bone structured for a specific function, such as joint formation, muscle attachment, or the passage of nerves and blood vessels.

Page 190

4. Mature bone cells are known as osteocytes, bone-building cells are called osteoblasts, and osteoclasts are bone-resorbing cells.

5. If the ratio of a collagen to hydroxyapatite in a bone increased, the bone would become less strong (as well as more flexible).

6. Because osteoclasts break down or demineralize bone, the bone would have a reduced mineral content (less mass); as a result, it would also be weaker.

Page 194

7. Compact bone consists of osteons (Haversian systems) with little space between them. Compact bone lies over spongy bone and makes up most of the diaphysis. It functions to protect, support, and resist stress. Spongy bone consists of trabeculae with numerous red marrow–filled spaces. Spongy bone makes up most of the structure of short, flat, and irregular bones and is also found at the epiphyses of long bones. Spongy bone functions in storing marrow and providing some support.

8. The presence of lamellae that are not arranged in osteons is indicative of spongy bone, which is located in an epiphysis.

Page 199

9. During intramembranous ossification, fibrous connective tissue is replaced by bone.

10. In endochondral ossification, cells of the inner layer of the perichondrium differentiate into osteoblasts, and a cartilage model is gradually replaced by bone.

11. Long bones of the body, such as the femur, have an epiphyseal cartilage, a plate of cartilage that separates the epiphysis from the diaphysis so long as the bone is still growing lengthwise. An x-ray would indicate whether the epiphyseal cartilage is still present. If it is, growth is still occurring; if it is not, the bone has reached its adult length.

Page 200

12. Bone remodeling refers to the process whereby old bone is continuously being destroyed by osteoclasts while new bone is being constructed by osteoblasts.

13. The biochemistry of some heavy-metal ions, such as strontium, cobalt, uranium, and plutonium, is very similar to that of calcium. Osteoblasts cannot differentiate these abnormal heavy-metal ions from normal calcium ions, so the heavy-metal ions become incorporated into the bone matrix. Over time, these dangerous ions can be released into circulation during normal bone remodeling.

Page 203

14. The larger arm muscles of the weight lifter would apply more mechanical stress to the bones of the upper limbs. In response to that stress, the bones would grow thicker.

15. Growth continues throughout childhood. At puberty, a growth spurt occurs and is followed by the closure of the epiphyseal cartilages. The later puberty begins, the taller the child will be when the growth spurt begins, so the taller the individual will be when growth is completed.

16. Increased levels of growth hormone prior to puberty will result in excessive bone growth, making the individual taller.

Page 205

17. Parathyroid hormone (PTH) influences osteoclast activity to cause a release of stored calcium ions from bone. Under the influence of calcitonin, osteoclast activity is inhibited, while osteoblasts continue to lock calcium ions in the bone matrix. Therefore, PTH serves to increase blood calcium levels by causing its release from bone, and calcitonin decreases blood calcium levels by causing calcium to remain in bone.

18. The bones of children who have rickets are poorly mineralized and as a result are quite flexible. Under the weight of the body, the leg bones bend. The instability makes walking difficult and can lead to other problems of the legs and feet.

19. Parathyroid hormone (PTH) stimulates osteoclasts to release calcium ions from bone. Increased PTH secretion would result in an increase in the level of calcium ions in the blood.

20. Calcitonin lowers blood calcium levels by inhibiting osteoclast activity and increasing the rate of calcium excretion at the kidneys.

Page 207

21. Immediately following a fracture, extensive bleeding occurs at the site of injury. After several hours, a large blood clot called a fracture hematoma develops. Next, an internal callus forms as a network of spongy bone unites the inner edges, and an external callus of cartilage and bone stabilizes the outer edges. The cartilaginous external callus is eventually replaced by bone, and the struts of spongy bone now unite the broken ends. With time, the swelling that initially marks the location of the fracture is remodeled, and little evidence that a break occurred remains.

22. An external callus forms early in the healing process, when cells from the endosteum and periosteum migrate to the area of the fracture. These cells form an enlarged collar (external callus) that encircles the bone in the area of the fracture.

Page 208

23. Osteopenia is inadequate ossification and is common to the aging process. It results as a consequence of decreasing osteoblast activity accompanied with normal osteoclast activity.

24. In women, the sex hormones known as estrogens play an important role in moving calcium into bones. After menopause, the level of these hormones decreases dramatically; as a result, older women have difficulty replacing the calcium in bones that is being lost due to normal aging. In men, the level of sex hormones (androgens) does not decrease until much later in life.

Answers to Review Questions

Page 210

Level 1 Reviewing Facts and Terms

1. b **2.** a **3.** a **4.** c **5.** b **6.** b

7. (a) long bone **(b)** flat bone **(c)** sutural bone **(d)** irregular bone **(e)** short bones **(f)** sesamoid bone.

8. a **9.** a **10.** c

11. (1) support; (2) storage of minerals and lipids; (3) blood cell production; (4) protection; and (5) leverage

12. (1) osteocytes; (2) osteoblasts; (3) osteoclasts; and (4) osteoprogenitor cells

13. (1) diaphysis (shaft); (2) epiphysis; (3) epiphyseal cartilage/line; (4) articular cartilage; (5) medullary canal; (6) periosteum; (7) endosteum

14. In intramembranous ossification, bone develops from mesenchyme or fibrous connective tissue. In endochondral ossification, bone develops from a cartilage model.

15. organic = collagen; inorganic = hydroxyapatite crystals

16. (a) calcium salts, phosphate salts, and vitamins A, C, and D_3; **(b)** calcitriol, growth hormone, thyroxine, estrogens (in females) or androgens (in males), calcitonin, and parathyroid hormone (PTH)

17. the bones, the intestinal tract, and the kidneys

18. Parathyroid hormone stimulates osteoclast activity, increases the rate of intestinal absorption of calcium ions, and decreases the rate of excretion of calcium ions.

Level 2 Reviewing Concepts

19. Nutrients reach the osteocytes in spongy bone by diffusing along canaliculi that open onto the surface of the trabeculae.

20. The osteons are aligned parallel to the long axis of the shaft, which does not bend when forces are applied to either end. Stresses or impacts to the side of the shaft can lead to a fracture.

21. The lack of stress during inactivity leads to the removal of calcium salts from bones. Up to one-third of the bone mass can be lost in this manner, causing the bones to become thin and brittle.

22. The digestive and urinary systems (kidneys) play important roles in providing the calcium and phosphate minerals needed for bone growth. In return, the skeleton provides protection and acts as a reserve of calcium, phosphate, and other minerals that can compensate for changes in the dietary supplies of these ions.

23. There are many long bones in the hand, each of which has an epiphyseal cartilage (plate). Measuring the width of these plates will provide clues to the hormonal control of growth in the child.

24. Once a bone fracture has been repaired, the bone tends to be stronger and thicker than normal at the fracture site.

25. b

26. Bone markings give clues as to the size, age, sex, and general appearance of an individual.

Level 3 Critical Thinking and Clinical Applications

27. The fracture might have damaged the epiphyseal cartilage in Sally's right leg. Even though the bone healed properly, the damaged leg did not produce as much cartilage as did the undamaged leg. The result would be a shorter bone on the side of the injury.

28. d **29.** a

30. The matrix of bone will absorb traces of minerals from the diet. These minerals can be identified hundreds of years later. A diet rich in calcium and vitamin B will produce denser bones than will a diet lacking these. Cultural practices such as the binding of appendages or the wrapping of infant heads will manifest in misshapen bones. Heavy muscular activity will result in larger bone markings, indicating an athletic lifestyle.

Chapter 7

Answers to Checkpoints

Page 212

1. The bones of the axial skeleton are the skull (8 cranial bones and 14 facial bones), bones associated with the skull (6 auditory ossicles and the hyoid bone), the vertebral column (24 vertebrae, the sacrum, and the coccyx), and the thoracic cage (the sternum and 24 ribs).

2. The axial skeleton provides a framework that supports and protects organs, and it also provides an extensive surface area for muscle attachment.

Page 226

3. The foramen magnum is located in the base of the occipital bone.

4. Tomás has fractured his right parietal bone.

5. The sphenoid bone contains the sella turcica, which in turn contains the pituitary gland.

6. The 14 facial bones include 2 inferior nasal conchae, 2 lacrimal bones, 1 mandible, 2 maxillae (maxillary bones), 2 nasal bones, 2 palatine bones, 1 vomer, and 2 zygomatic bones.

Page 228

7. The mental foramen is found in the mandible. Structures passing through this opening include the mental nerve and mental vessels.

8. The optic canal is found in the sphenoid. The optic nerve and ophthalmic artery pass through this structure.

9. The foramina in the ethmoid bone are the olfactory foramina.

Page 229

10. The bones forming the orbital complex are the frontal, sphenoid, zygomatic, palatine, maxilla, lacrimal, and ethmoid.

11. The bones forming the nasal complex are the frontal, ethmoid, nasal, maxilla, palatine, and sphenoid.

12. The sphenoid, ethmoid, frontal, palatine, and maxillary bones contain the paranasal sinuses.

Page 230

13. A fontanelle is a relatively soft, flexible, fibrous region between two flat bones in the developing skull. The major fontanelles are the anterior fontanelle, occipital fontanelle, sphenoidal fontanelle, and mastoid fontanelle.

14. Because they are not ossified at birth, fontanelles permit molding of the skull during childbirth, and they allow for growth of the brain during infancy and early childhood.

Page 234

15. The secondary curves of the spine allow us to balance our body weight on our lower limbs with minimal muscular effort. Without the secondary curves, we would not be able to stand upright for extended periods.

16. When you run your finger along a person's spine, you can feel the spinous processes of the vertebrae.

Page 239

17. The adult vertebral column has fewer vertebrae because the five sacral vertebrae fuse to form a single sacrum, and the four coccygeal vertebrae fuse to form a single coccyx.

18. The dens is part of the axis, or second cervical vertebra, which is located in the neck (cervical region).

19. The presence of transverse foramina indicates that this vertebra is a cervical vertebra.

20. The lumbar vertebrae must support a great deal more weight than do vertebrae that are more superior in the spinal column. The large vertebral bodies allow the weight to be distributed over a larger area.

Page 242

21. True ribs are attached directly to the sternum by their own costal cartilage. False ribs either do not attach to the sternum (as in the floating ribs) or attach by means of a common costal cartilage (as in the vertebrochondral ribs).

22. Improper compression of the chest during CPR can—and commonly does—result in a fracture of the sternum or ribs.

23. Vertebrosternal ribs, or true ribs, attach directly to the sternum; vertebrochondral ribs fuse together and merge with the costal cartilages of ribs 8–10 and then with the cartilages of rib pair 7 before they reach the sternum.

Answers to Review Questions

Page 244

Level 1 Reviewing Facts and Terms

1. (a) occipital bone **(b)** parietal bone **(c)** frontal bone **(d)** temporal bone **(e)** sphenoid bone **(f)** ethmoid bone **(g)** vomer **(h)** mandible **(i)** lacrimal bone **(j)** nasal bone **(k)** zygomatic bone **(l)** maxilla

2. a.   **3.** b   **4.** d   **5.** d   **6.** a

7. (a) cervical **(b)** thoracic **(c)** lumbar

8. b   **9.** c   **10.** d

11. (1) occipital bone; (2) frontal bone; (3) sphenoid; (4) ethmoid; (5) paired parietal bones; and (6) paired temporal bones

12. (1) sphenoid; (2) frontal bone; (3) ethmoid; (4) lacrimal bone; (5) maxilla; (6) palatine bone; (7) zygomatic bone

13. The vomer forms the anterior, inferior portion of the bony nasal septum that separates the right and left nasal cavities.

14. the frontal bone, sphenoid, ethmoid, the palatine bones, and the maxilla

Level 2 Reviewing Concepts

15. The petrous part of the temporal bone encloses the structures of the inner ear. The middle ear is located in the tympanic cavity within the petrous part. The external acoustic meatus ends at the tympanic membrane, which leads to the inner ear. Mastoid air cells within the mastoid process are connected to the tympanic cavity.

16. The ethmoid forms the superior surface of the nasal cavity. The olfactory foramina within the cribriform plate of the ethmoid allow neurons associated with the sense of smell to extend into the nasal cavity.

17. The ribs raise and lower to increase and decrease the volume of the chest cavity. They move similar to the handle of a bucket. When they rise, the chest cavity expands and we breathe in. When the ribs are lowered to their original position, the volume of the chest cavity decreases and we breathe out.

18. Keeping your back straight keeps the weight aligned along the axis of your vertebral column, where it can be transferred to your lower limbs. Bending your back would strain the muscles and ligaments of the back, increasing the risk of injury.

19. d

20. Fontanelles, which are fibrous connections between cranial bones, permit the skull to distort without damage during delivery, helping to ease the child through the birth canal.

21. a   **22.** c   **23.** e

Level 3 Critical Thinking and Clinical Applications

24. d

25. The large bones of a child's cranium are not yet fused; they are connected by fontanelles, areas of fibrous tissue. By examining the bones, the archaeologist could readily see if sutures had formed. By knowing approximately how long it takes for the various fontanelles to close and by determining their sizes, she could estimate the age of the individual at death.

26. Women in later stages of pregnancy develop lower back pain because of changes in the lumbar curvature of the spine. The increased mass of the pregnant uterus shifts the woman's center of gravity. To compensate, the lumbar curvature is exaggerated, and the lumbar region supports more of her body weight than normal. This results in sore muscles that produce lower back pain.

Chapter 8

Answers to Checkpoints

Page 250

1. Each of the two pectoral girdles consists of a clavicle (collarbone) and a scapula (shoulder blade). Each arm articulates with the trunk at the pectoral girdle.

2. The clavicle attaches the scapula to the sternum, thereby restricting the scapula's range of movement. When the clavicle is broken, the scapula has a greater range of movement and is less stable.

3. The head of the humerus articulates with the scapula at the glenoid cavity.

Page 254

4. The bones of the upper limb are the: humerus, ulna, radius, carpal bones (scaphoid, lunate, triquetrum, pisiform, trapezium, trapezoid, capitate, and hamate), 5 metacarpal bones, and 14 phalanges.

5. The two rounded projections on either side of the elbow are the lateral and medial epicondyles of the humerus.

6. The radius is lateral when the forearm is supinated, as in the anatomical position.

7. The first distal phalanx is located at the tip of the thumb; Bill's pollex is broken.

Page 257

8. The pelvic girdle is composed of the paired hip bones known as the coxal bones.

9. The three bones that make up a coxal bone are the ilium, ischium, and pubis.

10. The pelvis of females is adapted for supporting the weight of the developing fetus and enabling the newborn to pass through the pelvic outlet during delivery. Compared to males, the pelvis of females is smoother and lighter; has less-prominent markings; has an enlarged pelvic outlet; has a sacrum and coccyx with less curvature; has a pelvic inlet that is wider and more circular; is relatively broad and low; has ilia that project farther laterally; and has an inferior angle between the pubic bones that is greater than 100 degrees (as opposed to 90 degrees or less for males).

11. When you are seated, your body weight is borne by the ischial tuberosities.

Page 262

12. The bones of the lower limb are the femur (thigh), patella (kneecap), tibia and fibula (leg), tarsal bones (talus, calcaneus, cuboid, navicular, medial cuneiform, intermediate cuneiform, and lateral cuneiform), 5 metatarsal bones, and 14 phalanges.

13. Although the fibula is not part of the knee joint and does not bear weight, it is an important point of attachment for many leg muscles. When the fibula is fractured, these muscles cannot function properly to move the leg, and walking is difficult and painful. The fibula also helps stabilize the ankle joint.

14. Joey has most likely fractured the calcaneus (heel bone).

15. The talus transmits the weight of the body from the tibia toward the toes.

Page 263

16. In general, the bones of males tend to be heavier than those of females, and bone surface markings are more prominent in males than in females.

17. Bones can reveal information about a person's sex, age, muscular development, nutritional state, handedness, and occupation, plus other information relative to the medical history.

Answers to Review Questions

Page 265

Level 1 Reviewing Facts and Terms

1. (a) anterior view; **(b)** lateral view; **(c)** posterior view; **(d)** acromion; **(e)** coracoid process; **(f)** spine; **(g)** glenoid cavity

2. (a) sacrum; **(b)** coccyx; **(c)** ilium; **(d)** pubis; **(e)** ischium; **(f)** coxal bone

3. d **4.** a **5.** b **6.** d **7.** b

8. interosseus membrane

9. ischium, ilium, and pubis

10. (1) talus; (2) calcaneus; (3) cuboid; (4) navicular; and (5–7) three cuneiform bones

Level 2 Reviewing Concepts

11. d **12.** a **13.** c **14.** d **15.** d **16.** d **17.** c

18. The pelvic girdle consists of the coxal bones. The pelvis is a composite structure; it consists of the coxal bones of the appendicular skeleton and the sacrum and coccyx of the axial skeleton.

19. d

20. The clavicles are small and fragile, so they are easy to break. Once this part of the pectoral girdle is broken, the assailant would no longer have efficient use of the arms.

21. e

22. The slender fibula parallels the tibia of the leg and provides an important site for muscle attachment. It does not help in transferring weight to the ankle and foot because it is excluded from the knee joint.

23. e

Level 3 Critical Thinking and Clinical Applications

24. In osteoporosis, a decrease in the calcium content of the bones leads to bones that are weak and brittle. Because the hip must help support the body's weight, any weakening of the hip bones may result in their breaking under the weight of the body. The shoulder, by contrast, is not a load-bearing joint and is not subject to the same great stresses or strong muscle contractions as the hip joint. As a result, it is less likely to become broken.

25. Fred probably dislocated his shoulder, which is quite a common injury due to the weak nature of the glenohumeral joint.

26. Several characteristics are important in determining an individual's sex from a pelvis: its general appearance, the shape of the pelvic inlet, the depth of the iliac fossa, the characteristics of the ilium, the angle inferior to the pubic symphysis, the position of the acetabulum, the shape of the obturator foramen, and the characteristics of the ischium. The individual's age can be estimated by the bone's size, degree of mineralization, and various markings. The individual's general appearance can be reconstructed from the markings where muscles attach to the bones, which can indicate the size and shape of the muscles and thus the individual's general body contours.

Chapter 9

Answers to Checkpoints

Page 269

1. The three types of joints as classified by their degree of movement are the following: (1) an immovable joint or synarthrosis, (2) a slightly movable joint or amphiarthrosis, and (3) a freely movable joint or diarthrosis. A synarthrosis can be fibrous or cartilaginous, depending on the nature of the connection, or it can be a bony fusion, which develops over time. An amphiarthrosis is either fibrous or cartilaginous, depending on the nature of the connection, while a diarthrosis joint is a synovial joint that permits the greatest amount of movement.

2. Both synarthrotic joints (excepting synostoses) and amphiarthrotic joints consist of bony regions separated by fibrous or cartilaginous connective tissue.

3. Initially, each of these joints is a syndesmosis; as the bones interlock, they form sutural joints.

Page 272

4. Components of a synovial joint include an articular capsule, articular cartilages, synovial fluid, and various accessory structures (menisci, fat pads, ligaments, tendons, and bursae). Synovial joints are freely movable joints that have a joint (synovial) cavity, which is the space between articulating surfaces of two bones in a joint. Articular cartilages resembling hyaline cartilage cover the articulating bone surfaces. The fibrous articular capsule surrounds the joint, and a synovial membrane, which secretes synovial fluid, lines the walls of the joint cavity. Within the cavity, synovial fluid lubricates, distributes nutrients, and absorbs shocks. Menisci are articular discs made of fibrous cartilage that allow for variation in the shapes of the articulating surfaces. Fat pads protect the cartilages; ligaments are cords of fibrous tissue that support, strengthen, and reinforce the joint; tendons passing across or around the joint limit the range of motion and provide mechanical support; and bursae are synovial fluid-filled pockets that reduce friction and absorb shocks.

5. A subluxation is a partial dislocation of a bone from its joint.

6. Because articular cartilages lack a blood supply, they rely on synovial fluid to supply nutrients and eliminate wastes. Impairing the circulation of synovial fluid would have the same effect as impairing a tissue's blood supply: Nutrients would not be delivered to meet the tissue's needs, and wastes would accumulate. Damage to, and ultimately the death of, the cells in the tissue would result.

Page 277

7. Based on the shapes of the articulating surfaces, synovial joints are classified as ball-and-socket, ellipsoid, gliding, hinge, pivot, and saddle joints.

8. When doing jumping jacks, you move your lower limbs away from the body's midline; this movement is abduction. When you bring the lower limbs back together, the movement is adduction.

9. Flexion and extension are the movements associated with hinge joints.

Page 279

10. Movements possible across the intervertebral joints of the vertebral column are flexion (bending anteriorly), extension (bending posteriorly), lateral flexion (bending laterally), and rotation.

11. Intervertebral discs are not found between the first and second cervical vertebrae, between sacral vertebrae in the sacrum, or between coccygeal vertebrae in the coccyx. An intervertebral disc between the first and second cervical vertebrae would prohibit rotation; the vertebrae in the sacrum and coccyx are fused to provide a firm attachment for muscles and ligaments.

12. The vertebral movement involved in (a) bending forward is flexion; the movement involved in (b) bending to the side is lateral flexion; and the movement involved in (c) moving the head to signify "no" is rotation.

Page 283

13. Ligaments and muscles provide most of the stability for the shoulder joint.

14. The subscapular bursa is located in the shoulder joint, so the tennis player would be more likely to develop inflammation of this structure (bursitis). The condition is associated with repetitive motion that occurs at the shoulder, such as swinging a tennis racket. The jogger would be more at risk for injuries to the knee joint.

15. A shoulder separation is an injury involving partial or complete dislocation of the acromioclavicular joint.

16. Terry has most likely damaged his annular ligament.

Page 286

17. The bones making up the shoulder joint are the humerus and scapula; and the bones involved with the knee joint are the femur, tibia, and patella. The fibula does not participate in the knee joint.

18. The iliofemoral, pubofemoral, and ischiofemoral ligaments are at the hip joint.

19. Damage to the menisci of the knee joint decreases the joint's stability, so the individual would have a difficult time locking the knee in place while standing and would have to use muscle contractions to stabilize the joint. If the person had to stand for a long period, the muscles would fatigue and the knee would "give out." It is also likely that the individual would feel pain.

20. Like members of the clergy, carpet layers and roofers kneel a lot (and they also slide along on their knees), causing inflammation of the bursae in the knee joint.

Page 288

21. Rheumatism is a general term describing any painful condition of joints, muscles, or both that is not caused by infection or injury. One of several forms of rheumatism is arthritis.

22. Arthritis is a medical condition that affects synovial joints, causing pain, swelling, and stiffness.

23. The ribs of the skeletal system protect several components of the digestive system, including portions of the liver, stomach, and intestines. In turn, the digestive system provides nutrients, calcium, and phosphate required for bone maintenance.

24. The skeletal system both provides the mechanical support and stores the energy reserves and calcium and phosphate ions needed by all other body systems.

Answers to Review Questions

Page 291

Level Reviewing Facts and Terms

1. (a) joint capsule; **(b)** synovial membrane; **(c)** articular cartilage; **(d)** joint cavity

2. d **3.** b **4.** c **5.** c **6.** d **7.** c **8.** b **9.** c **10.** b **11.** b **12.** d **13.** c **14.** c **15.** a **16.** b **17.** a **18.** d **19.** b

Level 2 Reviewing Concepts

20. d **21.** b

22. Menisci may subdivide a synovial cavity, channel the flow of synovial fluid, and allow variations in the shape of the articular surfaces. They also act as cushions and shock absorbers.

23. Partial or complete dislocation of the acromioclavicular joint is called a shoulder separation.

24. Articular cartilages lack a perichondrium, and their matrix contains more water than does the matrix of other cartilages.

25. In a slipped disc, the nucleus pulposus does not extrude. In a herniated disc, the nucleus pulposus breaks through the annulus fibrosus.

26. Height decreases during adulthood in part as a result of osteoporosis in the vertebrae, and in part as a result of the decline in water content of the nucleus pulposus region of intervertebral discs.

27. (1) gliding: clavicle and sternum; (2) hinge: elbow; (3) pivot: atlas and axis; (4) ellipsoid: radius and carpal bones; (5) saddle: thumb; (6) ball and socket: shoulder

Level 3 Critical Thinking and Clinical Applications

28. The problem is probably a sprained ankle. The ligaments have been damaged but not ruptured, and the joint remains unaffected.

29. Cartilage does not have any blood vessels, so the chondrocytes rely on diffusion to gain nutrients and eliminate wastes. The synovial fluid is very important in supplying nutrients to the articular cartilages that it bathes and in removing wastes. If the circulation of the synovial fluid is impaired or stopped, the cells will not get enough nutrients or be able to get rid of their waste products. This combination of factors can lead to the death of the chondrocytes and the breakdown of the cartilage.

30. Shoulder dislocations would occur more frequently than hip dislocations because the shoulder is a more mobile joint. Because the shoulder joint is not bound tightly by ligaments or other elements, it is easier to dislocate when excessive forces are applied. By contrast, the hip joint, although mobile, is less easily dislocated because it is stabilized by four heavy ligaments, and the bones fit together snugly in the joint. Additionally, the synovial capsule of the hip joint is larger than that of the shoulder, and its range of motion is smaller.

Chapter 10

Answers to Checkpoints

Page 294

1. The three types of muscle tissue are skeletal muscle, cardiac muscle, and smooth muscle.

2. Skeletal muscles produce skeletal movement, maintain posture and body position, support soft tissues, guard entrances and exits, maintain body temperature, and store nutrient reserves.

Page 296

3. The epimysium is a dense layer of collagen fibers that surrounds the entire muscle; the perimysium divides the skeletal muscle into a series of compartments, each containing a bundle of muscle fibers called a fascicle; and the endomysium surrounds individual skeletal muscle cells (fibers). The collagen fibers of the epimysium, perimysium, and endomysium come together to form either bundles known as tendons, or broad sheets called aponeuroses. Tendons and aponeuroses generally attach skeletal muscles to bones.

4. Because tendons attach muscles to bones, severing the tendon would disconnect the muscle from the bone, and so the muscle could not move a body part.

Page 303

5. Sarcomeres, the smallest contractile units of a striated muscle cell, are segments of myofibrils. Each sarcomere has dark A bands and light I bands. The A band contains the M line, the H band, and the zone of overlap. Each I band contains thin filaments, but not thick filaments. Z lines mark the boundaries between adjacent sarcomeres.

6. Skeletal muscle appears striated when viewed through a light microscope because the Z lines and thick filaments of the myofibrils within the muscle fibers are aligned.

7. You would expect the greatest concentration of calcium ions in resting skeletal muscle to be in the cisternae of the sarcoplasmic reticulum.

Page 311

8. The neuromuscular junction, also known as the myoneural junction, is the synapse between a motor neuron and a muscle cell (fiber). This connection enables communication between the nervous system and a skeletal muscle fiber.

9. Acetylcholine release is necessary for skeletal muscle contraction, because it serves as the first step in the process, enabling the subsequent cross-bridge formation. A muscle's ability to contract depends on the formation of cross-bridges between the myosin and actin myofilaments. A drug that blocks acetylcholine release would interfere with this cross-bridge formation and prevent muscle contraction.

10. If the sarcolemma of a resting skeletal muscle suddenly became permeable to Ca^{2+}, the intracellular concentration of Ca^{2+} would increase, and the muscle would contract. In addition, because the amount of calcium ions in the sarcoplasm must decline for relaxation to occur, the increased permeability of the sarcolemma to Ca^{2+} might prevent the muscle from relaxing completely.

11. Without acetylcholinesterase, the motor end plate would be continuously stimulated by acetylcholine, locking the muscle in a state of contraction.

Page 318

12. A muscle's ability to contract depends on the formation of cross-bridges between the myosin and actin myofilaments in the muscle. In a muscle that is overstretched, the myofilaments would overlap very little, so very few cross-bridges between myosin and actin could form, and thus the contraction would be weak. If the myofilaments did not overlap at all, then no cross-bridges would form and the muscle could not contract.

13. Yes, a skeletal muscle can contract without shortening. The muscle can shorten (isotonic, concentric), elongate (isotonic, eccentric), or remain the same length (isometric), depending on the relationship between the load (resistance) and the tension produced by actin–myosin interactions.

Page 322

14. Muscle cells synthesize ATP continuously by utilizing creatine phosphate (CP) and metabolizing glycogen and fats. Most cells generate ATP through aerobic metabolism in the mitochondria and through glycolysis in the cytoplasm.

15. Muscle fatigue is a muscle's reduced ability to contract due to low pH (lactic acid buildup), low ATP levels, or other problems.

16. Oxygen debt is the amount of oxygen required to restore normal, pre-exertion conditions in muscle tissue.

Page 326

17. The three types of skeletal muscle fibers are (1) fast fibers (also called white muscle fibers, fast-twitch glycolytic fibers, Type II-B fibers, and fast fatigue fibers); (2) slow fibers (also called red muscle fibers, slow-twitch oxidative fibers, Type I fibers, and slow oxidative fibers); and (3) intermediate fibers (also called fast-twitch oxidative fibers, Type II-A fibers, and fast resistant fibers).

18. A sprinter requires large amounts of energy for a short burst of activity. To supply this energy, the sprinter's muscles switch to anaerobic metabolism. Anaerobic metabolism is less efficient in producing energy than aerobic metabolism, and the process also produces acidic waste products; this combination contributes to muscle fatigue. Conversely, marathon runners derive most of their energy from aerobic metabolism, which is more efficient and produces fewer waste products than anaerobic metabolism does.

19. Activities that require short periods of strenuous activity produce a greater oxygen debt, because such activities rely heavily on energy production by anaerobic metabolism. Because lifting weights is more strenuous over the short term than swimming laps, which is an aerobic activity, weight lifting would likely produce a greater oxygen debt than would swimming laps.

20. Individuals who excel at endurance activities have a higher than normal percentage of slow fibers. Slow fibers are physiologically better adapted to this type of activity than are fast fibers, which are less vascular and fatigue faster.

Page 328

21. Compared to skeletal muscle, cardiac muscle (1) has relatively small cells; (2) has cells with a centrally located nucleus (some may contain two or more nuclei); (3) has T tubules that are short and broad and form diads instead of triads; (4) has an SR that lacks terminal cisternae and has tubules that contact the cell membrane as well as the T tubules; (5) has cells that are nearly totally dependent on aerobic metabolism as an energy source; and (6) contains intercalated discs that assist in impulse conduction.

22. Cardiac muscle cells are joined by gap junctions, which allow ions and small molecules to flow directly between cells. As a result, action potentials generated in one cell spread rapidly to adjacent cells. Thus, all the cells contract simultaneously, as if they were a single unit (a syncytium).

Page 330

23. Smooth muscle cells lack sarcomeres, and thus smooth muscle tissue is nonstriated. Additionally, the thin filaments are anchored to dense bodies.

24. Cardiac and smooth muscle contractions are more affected by changes in the concentration of Ca^{2+} in the extracellular fluid than are skeletal muscle contractions because in cardiac and smooth muscles, most of the calcium ions that trigger a contraction come from the extracellular fluid. In skeletal muscle, most of the calcium ions come from the sarcoplasmic reticulum.

25. The looser organization of actin and myosin filaments in smooth muscle allows smooth muscle to contract over a wider range of resting lengths.

Answers to Review Questions

Page 333

Level 1 Reviewing Facts and Terms

1. (a) sarcolemma **(b)** sarcoplasm **(c)** mitochondria **(d)** myofibril **(e)** thin filament **(f)** thick filament **(g)** sarcoplasmic reticulum **(h)** T tubules

2. d **3.** c **4.** c **5.** a **6.** d **7.** d **8.** d

9. (1) skeletal muscle; (2) cardiac muscle; and (3) smooth muscle

10. (1) epimysium: surrounds entire muscle; (2) perimysium: surrounds muscle bundles (fascicles); and (3) endomysium: surrounds skeletal muscle fibers

11. a

12. The transverse (T) tubules conduct action potentials into the interior of the cell.

13. (1) exposure of active sites; (2) attachment of cross-bridges; (3) pivoting of myosin heads (power stroke); (4) detachment of cross-bridges; and (5) activation of myosin heads (cocking)

14. Both the frequency of motor unit stimulation and the number of motor units involved affect the amount of tension produced when a skeletal muscle contracts.

15. Resting skeletal muscle fibers contain ATP, creatine phosphate, and glycogen.

16. Aerobic metabolism and glycolysis generate ATP from glucose in muscle cells.

17. Calmodulin is the calcium-binding protein in smooth muscle tissue.

Level 2 Reviewing Concepts

18. d **19.** b **20.** a **21.** e

22. In an initial latent period (after the stimulus arrives and before tension begins to increase), an action potential generated in the muscle triggers the release of calcium ions from the SR. In the contraction phase, calcium binds to troponin (cross-bridges form) and tension begins to increase. In the relaxation phase, tension drops because cross-bridges have detached and because calcium levels have fallen; the active sites are once again covered by the troponin–tropomyosin complex.

23. (1) O_2 for aerobic respiration is consumed by liver cells, which must make a great deal of ATP to convert lactic acid to glucose; (2) O_2 for aerobic respiration is consumed by skeletal muscle fibers as they restore ATP, creatine phosphate, and glycogen concentrations to their former levels; and (3) the normal O_2 concentration in blood and peripheral tissues is replenished.

24. The timing of cardiac muscle contractions is determined by specialized cardiac muscle fibers called pacemaker cells; this property of cardiac muscle tissue is termed automaticity.

25. If atracurium blocked the binding of ACh to receptors at the motor end plates of neuromuscular junctions, the muscle's ability to contract would be inhibited.

26. In rigor mortis, the membranes of the dead cells are no longer selectively permeable; the SR is no longer able to retain calcium ions. As calcium ions enter the sarcoplasm, a sustained contraction develops, making the body extremely stiff. Contraction persists because the dead muscle cells can no longer make the ATP required for cross-bridge detachment from the active sites. Rigor mortis begins a few hours after death and lasts 15–25 hours, until the lysosomal enzymes released by autolysis break down the myofilaments.

27. c

Level 3 Critical Thinking and Clinical Applications

28. Because organophosphates block the action of acetylcholinesterase, ACh released into the synaptic cleft would not be removed. It would continue to stimulate the motor end plate, causing a state of persistent contraction (spastic paralysis). If the muscles of respiration were affected (which is likely), Ivan would die of suffocation. Prior to death, the most obvious sign would be uncontrolled tetanic contractions of skeletal muscles.

29. The enzyme CPK (creatine phosphokinase) functions in the primarily anaerobic reaction that transfers phosphate from creatine phosphate to ADP in muscle cells. The enzyme LDH (lactate dehydrogenase) is responsible for converting pyruvic acid to lactic acid, another anaerobic reaction. Elevations of these two enzymes would indicate that muscle tissue was working hard under anaerobic conditions, and since enzymes from different tissues have slightly different amino acid sequences (they are isozymes), it would be possible to verify that the CPK and LDH were released from damaged heart muscle. The presence of cardiac troponin, a form found only in cardiac muscle cells, provides direct evidence that cardiac muscle cells have been severely damaged.

30. A muscle not regularly stimulated by a motor neuron will lose tone and mass and become weak (will atrophy). While his leg was immobilized, it did not receive sufficient stimulation to maintain proper tone. It will take a while for Bill's muscles to build up enough to support his weight.

Chapter 11

Answers to Checkpoints

Page 337

1. Based on the patterns of fascicle organization, skeletal muscles can be classified as parallel muscles, convergent muscles, pennate muscles, or circular muscles.

2. Contraction of a pennate muscle generates more tension than would contraction of a parallel muscle of the same size because a pennate muscle contains more muscle fibers, and thus more myofibrils and sarcomeres, than does a parallel muscle of the same size.

3. The opening between the stomach and the small intestine would be guarded by a circular muscle, or sphincter. The concentric circles of muscle fibers found in sphincters are ideally suited for opening and closing passageways and for acting as valves in the body.

Page 339

4. A lever is a rigid structure—such as a board, a crowbar, or a bone—that moves on a fixed joint called the fulcrum. There are three classes of levers: In a first-class lever, the fulcrum lies between the applied force and load (resistance); in a second-class lever, the resistance lies between the applied force and fulcrum; and in a third-class lever, the pull exerted is between the fulcrum and the load (resistance). Third-class levers are the most common type in the body.

5. The joint between the occipital bone and the first cervical vertebra is part of a first-class lever system. The joint between the two bones (the fulcrum) lies between the skull (which provides the load) and the neck muscles (which provide the applied force).

Page 340

6. A synergist is a muscle that helps a larger prime mover (or agonist—a muscle that is responsible for a specific movement) perform its actions more efficiently.

7. The origin of a muscle is the end that remains stationary during an action. Because the gracilis muscle moves the tibia, the origin of this muscle must be on the pelvis (pubis and ischium).

8. Muscles A and B are antagonists to each other, because they perform opposite actions.

Page 345

9. Names of skeletal muscles are based on several factors, including location in the body, origin and insertion, fascicle organization, relative position, structural characteristics, and action. Names may also reflect the muscle shape, number of origins, and size.

10. The name *flexor carpi radialis longus* tells you that this muscle is a long muscle (longus) that lies next to the radius (radialis) and flexes (flexor) the wrist (carpi).

Page 359

11. Axial muscles are muscles that arise on the axial skeleton; they position the head, neck, and vertebral column, move the rib cage, and form the perineum.

12. Contraction of the masseter muscle elevates the mandible; relaxation of this muscle depresses the mandible. You would probably be chewing something.

13. You would expect the buccinator muscle, which positions the mouth for blowing, to be well developed in a trumpet player.

14. Swallowing involves contractions of the palatal muscles, which elevate the soft palate as well as portions of the superior pharyngeal wall. Elevation of the superior portion of the pharynx enlarges the opening to the auditory tube, permitting airflow to the middle ear and the inside of the eardrum. Making this opening larger by swallowing facilitates airflow into or out of the middle ear cavity.

15. Damage to the intercostal muscles would interfere with breathing.

16. A blow to the rectus abdominis muscle would cause that muscle to contract forcefully, resulting in flexion of the vertebral column. In other words, you would "double over."

17. The sore muscles are most likely the erector spinae muscles, especially the longissimus and the iliocostalis muscles of the lumbar region. These muscles would have to contract harder to counterbalance the increased anterior weight you would bear when carrying heavy boxes.

Page 378

18. When you shrug your shoulders, you are contracting your levator scapulae muscles.

19. The muscles involved in a rotator cuff injury are the supraspinatus, infraspinatus, teres minor, and subscapularis (SITS) muscles.

20. Injury to the flexor carpi ulnaris muscle would impair the ability to perform flexion and adduction at the wrist.

21. Injury to the obturator muscle would impair your ability to perform lateral rotation at the hip.

22. A "pulled hamstring" refers to a strain affecting one or more of the three muscles that collectively flex the knee: the biceps femoris, semimembranosus, and semitendinosus muscles.

23. A torn calcaneal tendon would make plantar flexion difficult because this tendon attaches the soleus and gastrocnemius muscles to the calcaneus (heel bone).

Page 380

24. General age-related effects on skeletal muscles include decreased skeletal muscle fiber diameters, diminished muscle elasticity, decreased tolerance for exercise, and a decreased ability to recover from muscular injuries.

25. Fibrosis is the development of increasing amounts of fibrous connective tissue. Fibrosis causes muscles to be less flexible, and the collagen fibers can restrict movement and circulation.

Page 381

26. The muscular system generates heat that maintains normal body temperature.

27. Exercise affects the cardiovascular system by increasing heart rate and dilating blood vessels. With exercise, the rate and depth of respiration increases, and sweat gland secretion increases. The physiological effects that result from exercise are directed and coordinated by the nervous and endocrine systems.

Answers to Review Questions

Page 384

Level 1 Reviewing Facts and Terms

1. (a) deltoid muscle (multipennate); **(b)** extensor digitorum muscle (unipennate); **(c)** rectus femoris muscle (bipennate)

2. (a) supraspinatus muscle; **(b)** infraspinatus muscle; **(c)** teres minor muscle

3. b **4.** a **5.** a **6.** a **7.** b **8.** c **9.** a **10.** a **11.** b **12.** d
13. a

14. The four fascicle organizations are (1) parallel, (2) convergent, (3) pennate, and (4) circular.

15. An aponeurosis is a collagenous sheet connecting two muscles. The epicranial aponeurosis and the linea alba are examples.

16. The axial musculature includes (1) muscles of the head and neck, (2) muscles of the vertebral column, (3) oblique and rectus muscles, and (4) muscles of the pelvic floor.

17. The muscles of the pelvic floor (1) support the organs of the pelvic cavity, (2) flex joints of the sacrum and coccyx, and (3) control movement of materials through the urethra and anus.

18. The supraspinatus, infraspinatus, and teres minor originate on the posterior body of the scapula, and the subscapularis originates on the anterior body of the scapula. All four muscles insert on the humerus.

19. The functional muscle groups in the lower limbs are (1) muscles that move the thigh, (2) muscles that move the leg, and (3) muscles that move the foot and toes.

Level 2 Reviewing Concepts

20. b **21.** a **22.** hernia
23. tendon sheaths

24. The vertebral column does not need a massive series of flexors, because many of the large trunk muscles flex the vertebral column when they contract. In addition, most of the body weight lies anterior to the vertebral column, and gravity tends to flex the intervertebral joints.

25. In a convergent muscle, the direction of pull can be changed by stimulating only one group of muscle cells at any one time. When all the fibers contract at once, they do not pull as hard on the tendon as would a parallel muscle of the same size, because the muscle fibers on opposite sides of the tendon are pulling in different directions rather than working together.

26. A pennate muscle contains more muscle fibers, and thus more myofibrils and sarcomeres, than does a parallel muscle of the same size, resulting in a contraction that generates more tension.

27. Lifting heavy objects becomes easier as the elbow approaches a 90° angle. As you increase the angle at or near full extension, tension production declines, so movement becomes more difficult.

28. When the hamstrings are injured, flexion at the knee and extension at the hip are affected.

Level 3 Critical Thinking and Clinical Applications

29. Mary is not happy to see Jill. Contraction of the frontalis muscle would wrinkle Mary's brow, contraction of the procerus muscle would flare her nostrils, and contraction of the levator labii muscle on the right side would raise the right side of her lip, as in sneering.

30. b

31. Although the pectoralis muscle is located across the chest, it inserts on the greater tubercle of the humerus, the bone of the arm. When the muscle contracts, it contributes to flexion, adduction, and medial rotation of the humerus. All of these arm movements would be impaired if the muscle were damaged.

Chapter 12

Answers to Checkpoints

Page 388

1. The two anatomical divisions of the nervous system are the central nervous system (CNS), consisting of the brain and spinal cord, and the peripheral nervous system (PNS), consisting of all neural tissue outside the CNS.

2. The two functional divisions of the peripheral nervous system are the afferent division, which brings sensory information to the CNS from receptors in peripheral tissues and organs, and the efferent division, which carries motor commands from the CNS to muscles and glands.

3. The two components of the efferent division of the PNS are the somatic nervous system (SNS) and the autonomic nervous system (ANS).

4. Damage to the afferent division of the PNS, which is composed of nerves that carry sensory information to the brain and spinal cord, would interfere with a person's ability to experience a variety of sensory stimuli.

Page 392

5. Structural components of a typical neuron include a cell body or soma (which contains the nucleus and perikaryon), an axon, dendrites, telodendria, Nissl bodies, neurofilaments, intermediate neurotubules, neurofibrils, axoplasm, axolemma, initial segment, axon hillock, and collaterals.

6. According to structure, neurons are classified as anaxonic, bipolar, unipolar, and multipolar.

7. According to function, neurons are classified as sensory neurons, motor neurons, and interneurons.

8. Because most sensory neurons of the PNS are unipolar, these neurons most likely function as sensory neurons.

Page 398

9. Central nervous system neuroglia include ependymal cells, astrocytes, oligodendrocytes, and microglia.

10. Peripheral nervous system neuroglia include satellite cells (amphicytes) and Schwann cells (neurilemmocytes).

11. The small phagocytic cells called microglia occur in increased numbers in infected (and damaged) areas of the CNS.

Page 407

12. The resting potential is the transmembrane potential of a normal cell under homeostatic conditions.

13. If the voltage-gated sodium channels in a neuron's plasma membrane could not open, sodium ions could not flood into the cell, and it would not be able to depolarize.

14. If the extracellular concentration of potassium ions decreased, more potassium would leave the cell, and the electrical gradient across the membrane (the transmembrane potential) would increase. This condition is called hyperpolarization.

Page 411

15. An action potential is a propagated change in the transmembrane potential of excitable cells, initiated by a change in the membrane permeability to sodium ions.

16. The four steps involved in the generation of action potentials are (1) depolarization to threshold; (2) activation of sodium channels and rapid depolarization; (3) inactivation of sodium channels and activation of potassium channels; and (4) return to normal permeability.

Page 413

17. The presence of myelin greatly increases the propagation speed of action potentials.

18. Action potentials travel along myelinated axons at much higher speeds (by saltatory propagation); the axon with a propagation speed of 50 meters per second must be the myelinated axon.

19. The major structural components of a synapse, the site where a neuron communicates with another cell, are a presynaptic cell and a postsynaptic cell, whose plasma membranes are separated by a narrow synaptic cleft.

20. If a synapse involves direct physical contact between cells, it is termed electrical; if the synapse involves a neurotransmitter, it is termed chemical.

21. If the voltage-gated calcium channels at a cholinergic synapse were blocked, Ca^{2+} could not enter the presynaptic terminal and trigger the release of ACh into the synapse, so no communication would take place across the synapse.

22. Because of synaptic delay, the pathway with fewer neurons (in this case, three) will transmit impulses more rapidly.

23. Both neurotransmitters and neuromodulators are compounds that are released by one neuron and affect another neuron. A neurotransmitter alters the transmembrane potential of the other neuron, whereas a neuromodulator alters the other neuron's response to specific neurotransmitters.

24. Neurotransmitters and neuromodulators are either (1) compounds that have a direct effect on membrane potential, (2) compounds that have an indirect effect on membrane potential, or (3) lipid-soluble gases that exert their effects inside the cell.

25. No action potential will be generated.

26. Yes, an action potential will be generated.

27. Spatial summation would occur if the two EPSPs happened simultaneously.

Answers to Review Questions

Level 1 Reviewing Facts and Terms

1. (a) dendrite; (b) nucleolus; (c) nucleus; (d) axon hillock; (e) initial segment; (f) axolemma; (g) axon; (h) telodendria; (i) synaptic terminals

2. c **3.** c **4.** d **5.** b **6.** a **7.** b **8.** a

9. (a) the CNS: brain and spinal cord (b) the PNS: all other nerve fibers, divided between the efferent division (which consists of the somatic nervous system and the autonomic nervous system) and the afferent division (which consists of receptors and sensory neurons)

10. Neuroglia in the PNS are (1) satellite cells and (2) Schwann cells.

11. (1) sensory neurons: transmit impulses from the PNS to the CNS; (2) motor neurons: transmit impulses from the CNS to peripheral effectors; and (3) interneurons: analyze sensory inputs and coordinate motor outputs

Level 2 Reviewing Concepts

12. b

13. Neurons lack centrioles and therefore cannot divide and replace themselves.

14. Anterograde flow is the movement of materials from the cell body to the synaptic knobs. Retrograde flow is the movement of materials toward the cell body.

15. Voltage-gated channels open or close in response to changes in the transmembrane potential. Chemically gated channels open or close when they bind specific extracellular chemicals.
Mechanically gated channels open or close in response to physical distortion of the membrane surface.

16. The all-or-none principle of action potentials states that any depolarization event sufficient to reach threshold will cause an action potential of the same strength, regardless of the amount of stimulation above threshold.

17. The membrane depolarizes to threshold. Next, voltage-gated sodium channels are activated, and the membrane rapidly depolarizes. These sodium channels are then inactivated, and potassium channels are activated. Finally, normal permeability returns. The voltage-gated sodium channels become activated once the repolarization is complete; the voltage-gated potassium channels begin closing as the transmembrane potential reaches the normal resting potential.

18. In saltatory conduction, which occurs in myelinated axons, only the nodes along the axon can respond to a depolarizing stimulus. In continuous conduction, which occurs in unmyelinated axons, an action potential appears to move across the membrane surface in a series of tiny steps.

19. Type A fibers are myelinated and carry action potentials very quickly (120 m/sec) Type B are also myelinated, but carry action potentials more slowly due to their smaller diameter. Type C fibers are extremely slow due to their small diameter and lack of myelination.

20. (1) The action potential arrives at the synaptic knob, depolarizing it; (2) extracellular calcium enters the synaptic knob, triggering the exocytosis of ACh; (3) ACh binds to the postsynaptic membrane and depolarizes the next neuron in the chain; (4) ACh is removed by AChE.

21. Temporal summation is the addition of stimuli that arrive at a single synapse in rapid succession. Spatial summation occurs when simultaneous stimuli at multiple synapses have a cumulative effect on the transmembrane potential.

Level 3 Critical Thinking and Clinical Applications

22. Harry's kidney condition is causing the retention of potassium ions. As a result, the K^+ concentration of the extracellular fluid is higher than normal. Under these conditions, less potassium diffuses from heart muscle cells than normal, resulting in a resting potential that is less negative (more positive). This change in resting potential moves the transmembrane potential closer to threshold, so it is easier to stimulate the muscle. The ease of stimulation accounts for the increased number of contractions evident in the rapid heart rate.

23. To reach threshold, the postsynaptic membrane must receive enough neurotransmitter to produce an EPSP of +20 mV (+10mV to reach threshold and +10mV to cancel the IPSPs produced by the five inhibitory neurons). Each neuron releases enough neurotransmitter to produce a change of +2mV, so at least 10 of the 15 excitatory neurons must be stimulated to produce this effect by spatial summation.

24. Action potentials travel faster along fibers that are myelinated than fibers that are nonmyelinated. Destruction of the myelin sheath increases the time it takes for motor neurons to communicate with their effector muscles. This delay in response results in varying degrees of uncoordinated muscle activity. The situation is very similar to that of a newborn, who cannot control its arms and legs very well because the myelin sheaths are still being laid down. Since not all motor neurons to the same muscle may be demyelinated to the same degree, there would be some fibers that are slow to respond while others are responding normally, producing contractions that are erratic and poorly controlled.

25. The absolute refractory period limits the number of action potentials that can travel along an axon in a given unit of time.

During the absolute refractory period, the membrane cannot conduct an action potential, so a new depolarization event cannot occur until after the absolute refractory period has passed. If the absolute refractory period for a particular axon is 0.001 sec, then the maximum frequency of action potentials conducted by this axon would be 1000/sec.

Chapter 13

Answers to Checkpoints

Page 431
1. The central nervous system is made up of the brain and spinal cord, while cranial nerves and spinal nerves constitute the peripheral nervous system.
2. A spinal reflex is a rapid, automatic response triggered by specific stimuli.

Page 435
3. The three spinal meninges are the dura mater, arachnoid mater, and pia mater.
4. Damage to the ventral root of a spinal nerve, which is composed of both visceral and somatic motor fibers, would interfere with motor function.
5. The cerebrospinal fluid that surrounds the spinal cord is located in the subarachnoid space, which lies beneath the epithelium of the arachnoid mater and superficial to the pia mater.

Page 437
6. Sensory nuclei receive and relay sensory information from peripheral receptors. Motor nuclei issue motor commands to peripheral effectors.
7. The polio virus–infected neurons would be in the anterior gray horns of the spinal cord, where the cell bodies of somatic motor neurons are located.
8. A disease that damages myelin sheaths would affect the columns of the spinal cord, because the columns are composed of bundles of myelinated axons.

Page 444
9. The major plexuses are the cervical, brachial, lumbar, sacral, and coccygeal.
10. An anesthetic that blocks the function of the dorsal rami of the cervical spinal nerves would affect the skin and muscles of the back of the neck and of the shoulders.
11. Damage to the cervical plexus—or more specifically to the phrenic nerves, which originate in this plexus and innervate the diaphragm—would greatly interfere with the ability to breathe and might even be fatal.
12. Compression of the sciatic nerve produces the sensation that your leg has "fallen asleep."

Page 448
13. A neuronal pool is the functional group of interconnected neurons organized within the CNS.
14. The five neuronal pool circuit patterns are divergence, convergence, serial processing, parallel processing, and reverberation.

Page 450
15. A reflex is a rapid, automatic response to a specific stimulus. It is an important mechanism for maintaining homeostasis.
16. The minimum number of neurons required for a reflex arc is two. One must be a sensory neuron that brings impulses to the CNS, and the other a motor neuron that transmits a response to the effector.

17. The suckling reflex is an innate reflex.

Page 454
18. All polysynaptic reflexes involve pools of interneurons, are intersegmental in distribution, and involve reciprocal inhibition.
19. When stretch receptors are stimulated by gamma motor neurons, the muscle spindles become more sensitive. As a result, little if any stretching stimulus would be needed to stimulate the contraction of the quadriceps muscles in the patellar reflex. Thus, the reflex response would appear more quickly.
20. This response is the tendon reflex.
21. During a withdrawal reflex, the limb on the opposite side is extended. This response is called a crossed extensor reflex.

Page 456
22. Reinforcement is an enhancement of spinal reflexes; it occurs when the postsynaptic neuron enters a state of generalized facilitation caused by chronically active excitatory synapses.
23. A positive Babinski reflex is abnormal in adults; it indicates possible damage of descending tracts in the spinal cord.

Answers to Review Questions

Page 458

Level 1 Reviewing Facts and Terms
1. (a) white matter; (b) ventral root; (c) dorsal root; (d) pia mater; (e) arachnoid mater; (f) gray matter; (g) spinal nerve; (h) dorsal root ganglion; (i) dura mater
2. d 3. a 4. d 5. c 6. c 7. c 8. c 9. a 10. d 11. a
12. b 13. c 14. c 15. (a) 1 (b) 7 (c) 3 (d) 5 (e) 4 (f) 6 (g) 2 (h) 8

Level 2 Reviewing Concepts
16. The vertebral column continues to grow, extending beyond the cord. The end of the cord is visible as the conus medullaris near L_1, and the cauda equina extends the remainder of the column.
17. (1) arrival of stimulus and activation of receptor; (2) activation of sensory neuron; (3) information processing; (4) activation of a motor neuron; and (5) response by an effector (muscle or gland)
18. d
19. The first cervical nerve exits superior to vertebra C_1 (between the skull and vertebra); the last cervical nerve exits inferior to vertebra C_7 (between the last cervical vertebra and the first thoracic vertebra). There are thus 8 cervical nerves but only 7 cervical vertebrae.
20. The cell bodies of spinal motor neurons are located in the anterior gray horns, so damage to these horns would result in a loss of motor control.
21. Within the CNS, cerebrospinal fluid fills the central canal, the ventricles, and the subarachnoid space. CSF acts as a shock absorber and a diffusion medium for dissolved gases, nutrients, chemical messengers, and waste products.
22. (1) involvement of pools of interneurons; (2) intersegmental distribution; (3) involvement of reciprocal innervation; (4) motor response prolonged by reverberating circuits; and (5) cooperation of reflexes to produce a coordinated, controlled response
23. Transection of the spinal cord at C_7 would most likely result in paralysis from the neck down. Transection at T_{10} would produce paralysis and eliminate sensory input in the lower half of the body only.
24. a 25. b 26. a
27. Stimulation of the sensory neuron will increase muscle tone.

Level 3 Critical Thinking and Clinical Applications
28. the median nerve
29. the radial nerve

30. The person would still exhibit a defecation (bowel) and micturition (bladder) reflex because these spinal reflexes are processed at the level of the spinal cord. Efferent impulses from the organs would stimulate specific interneurons in the sacral region that synapse with the motor neurons controlling the sphincters, thus bringing about emptying when the organs began to fill. This is the same situation that exists in a newborn infant who has not yet fully developed the descending tracts required for conscious control. An individual with the spinal cord transection at L_1 would lose voluntary control of the bowel and bladder because these functions rely on impulses carried by motor neurons in the brain that must travel down the cord and synapse with the interneurons and motor neurons involved in the reflex.

31. The anterior horn cells of the spinal cord are somatic motor neurons that direct the activity of skeletal muscles. The lumbar region of the spinal cord controls the skeletal muscles involved in the control of the muscles of the hip, leg, and foot. As a result of the injury, Karen would have poor control of most muscles of the lower limb, a problem with walking if she could walk at all, and if she could stand, problems maintaining balance.

Chapter 14

Answers to Checkpoints

Page 464

1. The six major regions of the brain are cerebrum, diencephalon, mesencephalon, pons, medulla oblongata, and cerebellum.
2. The brain stem consists of the mesencephalon, pons, and medulla oblongata.
3. The rhombencephalon develops into the cerebellum, pons, and medulla oblongata.

Page 468

4. The layers of the cranial meninges are the outer dura mater, the middle arachnoid mater, and the inner pia mater.
5. If an interventricular foramen became blocked, cerebrospinal fluid (CSF) could not flow from the lateral ventricles into the third ventricle. Cerebrospinal fluid would continue to form within that ventricle, so the blocked ventricle would swell with fluid—a condition known as hydrocephalus.
6. If diffusion across the arachnoid granulations decreased, less CSF would reenter the bloodstream, and CSF would accumulate in the ventricles, damaging the brain.
7. Many water-soluble molecules are rare or absent in the extracellular fluid (ECF) of the brain because the blood–brain barrier regulates the movement of such molecules from the blood to the ECF of the brain.

Page 471

8. The nucleus gracilis and nucleus cuneatus are responsible for relaying somatic sensory information to the thalamus.
9. Damage to the medulla oblongata can be lethal because it contains many vital reflex centers, including those that control breathing and regulate heart rate and blood pressure.

Page 472

10. The pons contains (1) sensory and motor nuclei of cranial nerves, (2) nuclei involved with the control of respiration, (3) nuclei and tracts that process and relay information heading to or from the cerebellum, and (4) ascending, descending, and transverse tracts.
11. Damage to the respiratory centers of the pons could result in loss of ability to modify the rhythmicity center of the medulla oblongata during prolonged inhalation or extensive exhalation.

Page 474

12. Components of the cerebellar gray matter include the cerebellar cortex and cerebellar nuclei.
13. The arbor vitae, which is part of the cerebellum, connects the cerebellar cortex and nuclei with cerebellar peduncles.

Page 475

14. Two pairs of sensory nuclei make up the corpora quadrigemina: the superior colliculi and inferior colliculi.
15. The superior colliculi of the mesencephalon control reflexive movements of the eyes, head, and neck.

Page 478

16. The main components of the diencephalon are the epithalamus, thalamus, and hypothalamus.
17. Damage to the lateral geniculate nuclei would interfere with the sense of sight.
18. Changes in body temperature stimulate the preoptic area of the hypothalamus, a component of the diencephalon.

Page 480

19. The limbic system is responsible for processing memories and creating emotional states, drives, and associated behaviors.
20. Damage to the amygdaloid body would interfere with the sympathetic ("fight or flight") division of the autonomic nervous system.

Page 489

21. Projection fibers link the cerebral cortex to the spinal cord, passing through the diencephalon, brain stem, and cerebellum.
22. Damage to the basal nuclei would result in decreased muscle tone and the loss of coordination of learned movement patterns.
23. The primary motor cortex is located in the precentral gyrus of the frontal lobe of the cerebrum.
24. Damage to the temporal lobes of the cerebrum would interfere with the processing of olfactory (smell) and auditory (sound) impulses.
25. The stroke has damaged the speech center, located in the frontal lobe.
26. The temporal lobe of the cerebrum is probably involved, specifically the hippocampus and the amygdaloid body. His problems may also involve other parts of the limbic system that act as a gate for loading and retrieving long-term memories.

Page 501

27. Cranial reflexes are monosynaptic and polysynaptic reflex arcs that involve the sensory and motor fibers of cranial nerves. Cranial reflex testing is often used to assess damage to cranial nerves or to the associated processing centers in the brain.

Answers to Review Questions

Page 504

Level 1 Reviewing Facts and Terms

1. (a) cerebrum; (b) diencephalon; (c) mesencephalon; (d) pons; (e) medulla oblongata; (f) cerebellum
2. (a) dura mater (endosteal layer); (b) dural sinus; (c) dura mater (meningeal layer); (d) subdural space; (e) arachnoid mater; (f) subarachnoid space; (g) pia mater
3. d **4.** b **5.** c **6.** d **7.** b **8.** c **9.** a **10.** b **11.** a **12.** a
13. a
14. (1) cushioning delicate neural structures; (2) supporting the brain; and (3) transporting nutrients, chemical messengers, and waste products
15. (1) portions of the hypothalamus where the capillary endothelium is extremely permeable; (2) capillaries in the pineal gland; and (3) capillaries at the choroid plexus

16. N I: olfactory nerve; N II: optic nerve; N III: oculomotor nerve; N IV: trochlear nerve; N V: trigeminal nerve; N VI: abducens nerve; N VII: facial nerve; N VIII: vestibulocochlear nerve; N IX: glossopharyngeal nerve; N X: vagus nerve; N XI: accessory nerve; and N XII: hypoglossal nerve

Level 2 Reviewing Concepts
17. The brain can respond with greater versatility because it includes many more interneurons, pathways, and connections than the tracts of the spinal cord.
18. The cerebellum adjusts voluntary and involuntary motor activities based on sensory information and stored memories of previous experiences.
19. d
20. In Parkinson disease, the substantia nigra is inhibited from secreting the neurotransmitter, dopamine, at the basal nuclei.
21. Roles of the hypothalamus include: (1) subconscious control of skeletal muscle contractions, (2) control of autonomic functions, (3) coordination of nervous and endocrine systems, (4) secretion of hormones, (5) production of emotions and drives, (6) coordination of autonomic and voluntary functions, (7) regulation of body temperature, (8) control of circadian rhythms
22. Stimulation of the feeding and thirst centers of the hypothalamus would produce sensations of hunger and thirst.
23. This nucleus is the hippocampus, which is part of the limbic system.
24. The left hemisphere contains the general interpretive and speech centers and is responsible for performing analytical tasks, for logical decision-making, and for language-based skills (reading, writing, and speaking). The right hemisphere analyzes sensory information and relates the body to the sensory environment. Interpretive centers in this hemisphere permit the identification of familiar objects by touch, smell, sight, taste, or feel. The right hemisphere is also important in understanding three-dimensional relationships and in analyzing the emotional context of a conversation.
25. d 26. c
27. Lesions in the general interpretive area (Wernicke area, sensory) produce defective visual and auditory comprehension of language, repetition of spoken sentences, and defective naming of objects. Lesions in the speech center (Broca area, motor) result in hesitant and distorted speech.

Level 3 Critical Thinking and Clinical Applications
28. Sensory innervation of the nasal mucosa is via the maxillary branch of the trigeminal nerve (N V). Irritation of the nasal lining by ammonia increases the frequency of action potentials along the maxillary branch of the trigeminal nerve through the semilunar ganglion to reach centers in the mesencephalon, which in turn excite the neurons of the reticular activating system (RAS). Increased activity by the RAS can raise the cerebrum back to consciousness.
29. The officer is testing the function of Bill's cerebellum. Many drugs, including alcohol, have pronounced effects on the function of the cerebellum. A person who is under the influence of alcohol cannot properly anticipate the range and speed of limb movement, because processing and correction by the cerebellum are slow. As a result, Bill might have a difficult time walking a straight line or touching his finger to his nose.
30. Increasing pressure in the cranium could compress important blood vessels, leading to further brain damage in areas not directly affected by the hematoma. Pressure on the brain stem could disrupt vital respiratory, cardiovascular, and vasomotor functions

and possibly cause death. Pressure on the motor nuclei of the cranial nerves would lead to drooping eyelids and dilated pupils. Pressure on descending motor tracts would impair muscle function and decrease muscle tone in the affected areas of the body.
31. In any inflamed tissue, edema occurs in the area of inflammation. The accumulation of fluid in the subarachnoid space can cause damage by pressing against neurons. If the intracranial pressure is excessive, brain damage can occur, and if the pressure involves vital autonomic reflex areas, death could occur.
32. Most of the functional problems observed in shaken baby syndrome are the result of trauma to the cerebral hemispheres due to contact between the brain and the inside of the skull. Damage to and distortion of the brain stem and medulla oblongata can cause death.

Chapter 15
Answers to Checkpoints
Page 507
1. The specialized cells that monitor specific conditions in the body or the external environment are sensory receptors.
2. Yes, it is possible for somatic motor commands to arise at the subconscious level. They also arise at the conscious level.

Page 510
3. Adaptation is a decrease in receptor sensitivity or perception after chronic stimulation.
4. Receptor A provides more precise sensory information because it has a smaller receptive field.

Page 514
5. The four types of general sensory receptors (and the stimuli that excite them) are nociceptors (pain), thermoreceptors (temperature), mechanoreceptors (physical distortion), and chemoreceptors (chemical concentration).
6. The three classes of mechanoreceptors are tactile receptors, baroreceptors, and proprioceptors.
7. If proprioceptors in your legs could not relay information about limb position and movement to the CNS (especially the cerebellum), your movements would be uncoordinated and you likely could not walk.

Page 519
8. The tract being compressed is the fasciculus gracilis in the posterior column of the spinal cord, which carries information about touch and pressure from the lower limbs to the brain.
9. The tracts that carry action potentials generated by nociceptors are the lateral spinothalamic tracts.
10. The left cerebral hemisphere (specifically, the primary sensory cortex) receives impulses conducted by the right fasciculus gracilis.

Page 525
11. The anatomical basis for opposite-side motor control is crossing-over (decussation) of axons, so the motor fibers of the corticospinal pathway innervate lower motor neurons on the opposite side of the body.
12. An injury involving the superior portion of the motor cortex would affect the ability to control the muscles in the upper limb and the upper portion of the lower limb.
13. Increased stimulation of the motor neurons of the red nucleus would increase stimulation of the skeletal muscles in the upper limbs, thereby increasing their muscle tone.

Answers to Review Questions
Page 527
Level 1 Reviewing Facts and Terms

1. (a) tactile discs (Merkel discs); (b) tactile corpuscle; (c) free nerve ending; (d) root hair plexus; (e) lamellated corpuscle; (f) Ruffini corpuscle

2.

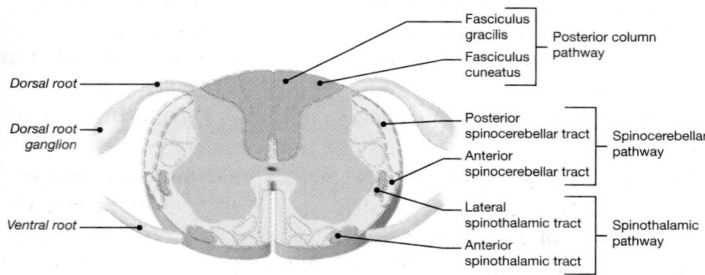

3. c **4.** c **5.** d **6.** Phasic receptors

7. (1) free nerve endings: sensitive to touch and pressure; (2) root hair plexus: monitors distortions and movements across the body surface; (3) tactile discs: detect fine touch and pressure; (4) tactile corpuscles: detect fine touch and pressure; (5) lamellated corpuscles: sensitive to pulsing or vibrating stimuli (deep pressure); and (6) Ruffini corpuscles: sensitive to pressure and distortion of the skin

8. (1) tactile receptors; (2) baroreceptors; and (3) proprioceptors

9. (1) posterior column pathway: provides conscious sensations of highly localized ("fine") touch, pressure, vibration, and proprioception; (2) spinothalamic pathway: provides conscious sensations of poorly localized ("crude") touch, pressure, pain, and temperature; and (3) spinocerebellar pathway: carries proprioceptive information about the position of skeletal muscles, tendons, and joints to the cerebellum

10. (1) corticobulbar tracts; (2) lateral corticospinal tracts; and (3) anterior corticospinal tracts

11. (1) vestibulospinal pathway; (2) tectospinal pathway; and (3) reticulospinal pathway

12. The cerebellum (1) integrates proprioceptive sensations with visual information from the eyes and equilibrium-related sensations from the inner ear, and (2) adjusts the activities of the voluntary and involuntary motor centers on the basis of sensory information and the stored memories of previous experiences.

13. a

14. (1) An arriving stimulus alters the transmembrane potential of the receptor membrane. (2) The receptor potential directly or indirectly affects a sensory neuron. (3) Action potentials travel to the CNS along an afferent fiber.

Level 2 Reviewing Concepts

15. A tonic receptor is always active; a phasic receptor is normally inactive and becomes active only when a change occurs in the condition being monitored.

16. A motor homunculus, a mapped-out area of the primary motor cortex, provides an indication of the degree of fine motor control available. A sensory homunculus indicates the degree of sensitivity of peripheral sensory receptors.

17. A sensory neuron that delivers sensations to the CNS is a first-order neuron. Within the CNS, the axon of the first-order neuron synapses on a second-order neuron, which is an interneuron located in the spinal cord or brain stem. The second-order neuron synapses on a third-order neuron in the thalamus. The axons of third-order neurons synapse on neurons of the primary sensory cortex of the cerebral hemispheres.

18. Damage to the posterior spinocerebellar tract on the left side of the spinal cord at the L_1 level would interfere with the coordinated movement of the left leg.

19. Injury to the primary motor cortex affects the ability to exert fine control over motor units. Gross movements are still possible, however, because they are controlled by the basal nuclei that use the reticulospinal or rubrospinal tracts. Thus, walking and other voluntary and involuntary movements can be performed with difficulty, and the movements are imprecise and awkward.

20. Muscle tone is controlled by the basal nuclei, cerebellum, and red nuclei through commands distributed by the reticulospinal and rubrospinal tracts.

21. Strong pain sensations arriving at a particular segment of the spinal cord can cause stimulation of the interneurons of the spinothalamic pathway. This stimulation is interpreted by the sensory cortex as originating in the region of the body surface associated with the origin of that same pathway.

Level 3 Critical Thinking and Clinical Applications

22. Kelly's tumor is most likely adjacent to the corticobulbar tracts. The axons of those tracts carry action potentials to motor nuclei of the cranial nerves, which control eye muscles and muscles of facial expression.

23. Injuries to the motor cortex eliminate the ability to produce fine control of motor units. However, as long as the cerebral nuclei are functional, gross movements would still be possible. Clarence should still be able to walk, maintain his balance, and perform voluntary and involuntary movements using the rubrospinal and reticulospinal tracts in place of the corticospinal tracts. Although these movements may be awkward or difficult, they will still be possible.

24. Phil is experiencing phantom pain. Since pain perception occurs in the sensory cortex of the brain, he can still feel pain in his fingers if the brain projects feeling to that area. When he bumps the arm at the elbow, sensory receptors are stimulated to send impulses to the sensory cortex. The brain perceives a sensation from a general area, and projects that feeling to a body part. Since more sensory information reaches the brain from the hands and fingers, it is not unusual for the brain to project to this area.

Chapter 16
Answers to Checkpoints
Page 532

1. The two major divisions of the autonomic nervous system are the sympathetic division and the parasympathetic division.

2. Two neurons are required to conduct an action potential from the spinal cord to smooth muscles in the intestine. One neuron carries the action potential from the spinal cord to the autonomic ganglion, and a second neuron carries the action potential from the autonomic ganglion to the smooth muscle.

3. The sympathetic division of the ANS is responsible for the physiological changes that occur in response to stress (confronting angry dog) and increased activity (running).

4. The sympathetic division of the autonomic nervous system includes preganglionic fibers from the lumbar and thoracic portions of the spinal cord, whereas the parasympathetic division includes preganglionic fibers from the cranial and sacral portions.

Page 536

5. The nerves that synapse in collateral ganglia originate in the inferior thoracic and superior lumbar portions of the spinal cord; they pass through the chain ganglia to the collateral ganglia.

Page 538

6. Because preganglionic fibers of the sympathetic nervous system release acetylcholine (ACh), a drug that stimulates ACh receptors would stimulate the postganglionic fibers of sympathetic nerves, resulting in increased sympathetic activity.

7. Blocking the beta receptors on cells would decrease or prevent sympathetic stimulation of tissues containing those cells. Heart rate, force of contraction of cardiac muscle, and contraction of smooth muscle in the walls of blood vessels would decrease, lowering blood pressure.

Page 541

8. The vagus nerve (N X) carries preganglionic parasympathetic fibers that innervate the lungs, heart, stomach, liver, pancreas, and parts of the small and large intestines (as well as several other visceral organs).

9. The parasympathetic division is sometimes referred to as the anabolic system because parasympathetic stimulation leads to a general increase in the nutrient content of the blood. Cells throughout the body respond to the increase by absorbing the nutrients and using them to support growth and other anabolic activities.

Page 542

10. Acetylcholine (ACh) is the neurotransmitter released by all parasympathetic neurons.

11. The two types of ACh receptors on the postsynaptic membranes of parasympathetic neurons are nicotinic receptors and muscarinic receptors.

12. Stimulation of muscarinic receptors, a type of acetylcholine receptor located in postganglionic synapses of the parasympathetic nervous system, would cause K^+ channels to open, resulting in hyperpolarization of cardiac plasma membranes and a decreased heart rate.

Page 546

13. Most blood vessels receive sympathetic stimulation, so a loss of sympathetic tone would relax the smooth muscles lining the vessels; the resulting vasodilation would increase blood flow to the tissue.

14. In anxious individuals, an increase in sympathetic stimulation would probably cause some or all of the following changes: a dry mouth; increased heart rate, blood pressure, and rate of breathing; cold sweats; an urge to urinate or defecate; a change in the motility of the digestive tract (that is, "butterflies in the stomach"); and dilated pupils.

Page 549

15. A visceral reflex is an automatic motor response that can be modified, facilitated, or inhibited by higher centers, especially those of the hypothalamus.

16. A brain tumor that interferes with hypothalamic function would interfere with autonomic function as well. Centers in the posterior and lateral hypothalamus coordinate and regulate sympathetic function, whereas centers in the anterior and medial hypothalamus control parasympathetic function.

Page 554

17. Higher-order functions require action by the cerebral cortex, involve both conscious and unconscious information processing, and are subject to modification and adjustment over time.

18. Test-taking involves short-term memory, although your instructor would like you to transfer this information to long-term memory.

19. The two general levels of sleep are deep sleep and rapid eye movement (REM) sleep.

20. If your RAS were suddenly stimulated, it would rouse the cerebrum to a state of consciousness—you would wake up.

21. A drug that increases the amount of serotonin released in the brain would produce a heightened perception of certain sensory stimuli (e.g., auditory or visual stimuli) and hallucinations.

22. Serotonin and norepinephrine are thought to be involved with awake–asleep cycles.

23. Amphetamines stimulate the secretion of dopamine.

Page 555

24. Some possible reasons for slower recall and for loss of memory in the elderly include a loss of neurons (possibly those involved in specific memories), changes in synaptic organization of the brain, changes in the neurons themselves, and decreased blood flow, which would affect the metabolic rate of neurons and perhaps slow the retrieval of information from memory.

25. Common age-related anatomical changes in the nervous system include a reduction in brain size and weight, a reduction in the number of neurons, a decrease in blood flow to the brain, changes in the synaptic organization of the brain, and intracellular and extracellular changes in CNS neurons.

26. Alzheimer disease is the most common form of senile dementia.

Page 556

27. The nervous system monitors pressure, pain, and temperature, and it adjusts tissue and blood flow patterns for all other body systems.

Answers to Review Questions

Page 559

Level 1 Reviewing Facts and Terms

1.

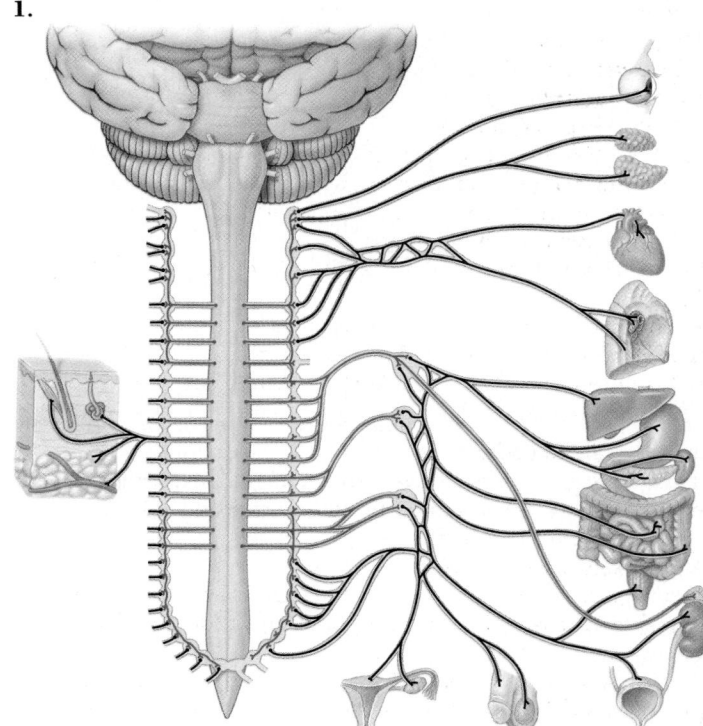

2. d **3.** a **4.** b **5.** d

6. preganglionic neuron T_5–L_2 → collateral ganglia → postganglionic fibers → visceral effector in abdominopelvic cavity

7. (1) ciliary ganglion; (2) pterygopalatine ganglion; (3) submandibular ganglion; and (4) otic ganglion

8. Visceral reflex arcs include a receptor, a sensory neuron, an interneuron (may or may not be present), and two visceral motor neurons.

9. Increased neurotransmitter release, facilitation of synapses, and the formation of additional synaptic connections are thought to be involved in memory formation and storage.

10. During non-REM sleep, the entire body relaxes, and activity at the cerebral cortex is at a minimum; heart rate, blood pressure, respiratory rate, and energy utilization decline. During REM sleep, active dreaming occurs, accompanied by alterations in blood pressure and respiratory rates; muscle tone decreases markedly, and response to outside stimuli declines.

11. Aging causes a reduction in brain volume and weight, a reduction in the number of neurons, a decrease in blood flow to the brain, changes in synaptic organization, and intracellular and extracellular changes in CNS neurons.

12. c **13.** d

14. Sympathetic preganglionic fibers emerge from the thoracolumbar area (T_1 through L_2) of the spinal cord. Parasympathetic fibers emerge from the brain stem and the sacral region of the spinal cord (craniosacral).

15. (1) celiac ganglion; (2) superior mesenteric ganglion; and (3) inferior mesenteric ganglion

16. Stimulation of sympathetic ganglionic neurons causes (1) release of norepinephrine at specific locations and (2) secretion of epinephrine (and modest amounts of norepinephrine) into the bloodstream.

17. The four pairs of cranial nerves are N III, N VII, N IX, and N X.

18. (1) cardiac plexus: heart rate increases (sympathetic)/decreases (parasympathetic); heart strength increases (sympathetic)/decreases (parasympathetic); blood pressure increases (sympathetic)/decreases (parasympathetic); (2) pulmonary plexus: respiratory passageways dilate (sympathetic)/constrict (parasympathetic); (3) esophageal plexus: respiratory rate increases (sympathetic)/decreases (parasympathetic); (4) celiac plexus: digestion inhibited (sympathetic)/stimulated (parasympathetic); (5) inferior mesenteric plexus: digestion inhibited (sympathetic)/stimulated (parasympathetic); and (6) hypogastric plexus: defecation inhibited (sympathetic)/stimulated (parasympathetic); urination inhibited (sympathetic)/stimulated (parasympathetic); sexual organs: stimulation of secretion (sympathetic)/erection (parasympathetic)

19. Higher-order functions (1) are performed by neurons of the cerebral cortex and involve complex interactions between areas of the cortex and between the cerebral cortex and other parts of the brain; (2) involve both conscious and unconscious information processing; and (3) are subject to modification and adjustment over time.

Level 2 Reviewing Concepts

20. a **21.** c

22. The preganglionic fibers innervating the cervical ganglia originate in the ventral roots of the thoracic segments, which are undamaged.

23. c **24.** b **25.** d

26. Due to the stimulation of the sympathetic division, you would experience increased respiratory rate, increased peripheral vasoconstriction and elevation of blood pressure, increased heart rate and force of contraction, and an increased rate of glucose release into the bloodstream.

27. If autonomic motor neurons maintain a background level of activity at all times, they can either increase or decrease their activity, providing a greater range of control options.

28. Cholinergic receptors are found in all of the ganglia of the ANS, so nicotine would stimulate both sympathetic and parasympathetic responses in cardiovascular tissues. Although increased sympathetic stimulation increases heart rate and force of contraction, increased parasympathetic stimulation simultaneously decreases blood flow to the heart muscle. In addition to elevating heart rate and force of contraction, sympathetic stimulation also constricts peripheral blood vessels, all of which contribute to increased blood pressure.

29. The upsetting stimuli would be processed by the higher centers of the CNS and relayed to the hypothalamus. The hypothalamus could suppress the vasomotor center of the medulla, resulting in fewer sympathetic impulses to peripheral blood vessels. This would cause a decrease in sympathetic tone in the smooth muscle of the blood vessels resulting in vasodilation. The vasodilation would cause blood to pool in the limbs decreasing the amount of blood returning to the heart and producing shock.

Level 3 Critical Thinking and Clinical Applications

30. Epinephrine would be more effective, because it would reduce inflammation and relax the smooth muscle of the airways, making it easier for Phil to breathe.

31. The molecule is probably mimicking NE and binding to alpha-1 receptors.

Chapter 17

Answers to Checkpoints

Page 564

1. Olfaction is the sense of smell; it involves olfactory receptors in paired olfactory organs responding to chemical stimuli.

2. In the olfactory pathway, axons leaving the olfactory bulb travel along the olfactory tract to the olfactory cortex, hypothalamus, and portions of the limbic system. Prior to that, axons from the olfactory epithelium collect into bundles that reach the olfactory bulb.

3. By the end of the lab period, adaptation has occurred. In response to constant stimulation, your receptor neurons have become less active, partially as the result of synaptic fatigue.

Page 566

4. Gustation is the sense of taste, provided by taste receptors responding to chemical stimuli.

5. Taste receptors (taste buds) are sensitive only to molecules and ions that are in solution. If you dry the surface of your tongue, the salt ions or sugar molecules have no moisture in which to dissolve, so they will not stimulate the taste receptors.

6. Your grandfather is experiencing the effects of several age-related changes. The number of taste buds declines dramatically after age 50, and those that remain are not as sensitive as they once were. In addition, the loss of olfactory receptors contributes to the perception of less flavor in foods.

Page 578

7. The conjunctiva would be the first layer of the eye affected by inadequate tear production. Drying of the conjunctiva would produce an irritated, scratchy feeling.

8. When the lens becomes more rounded, you are looking at an object that is close to you.

9. Sue will likely be unable to see at all. The fovea contains only cones, which need high-intensity light to be stimulated. The dimly lit room contains light that is too weak to stimulate the cones.

10. If the canal of Schlemm were blocked, the aqueous humor could not drain, producing glaucoma. Accumulation of this fluid increases the pressure within the eye, distorting soft tissues and interfering with vision. If untreated, the condition would ultimately cause blindness.

Page 585

11. If you were born without cones, you would still be able to see—so long as you had functioning rods—but you would see in black and white only.

12. A vitamin A deficiency would reduce the quantity of retinal (retinene) the body could produce, thereby interfering with night vision (which operates at the body's threshold ability to respond to light).

13. Vision would be impaired. Decreased phosphodiesterase activity would increase intracellular cGMP levels, which by keeping gated sodium channels open would decrease the ability of receptor neurons to respond to photons.

Page 599

14. If the round window could not move, the perilymph would not be moved by the vibration of the stapes at the oval window, reducing or eliminating the perception of sound.

15. The loss of stereocilia (as a result of constant exposure to loud noises, for instance) would reduce hearing sensitivity and could lead to deafness.

16. If the auditory tube were blocked, it would not be possible to equalize the pressure on both sides of the tympanic membrane. If external pressure then declines, the pressure in the middle ear would be greater than that on the outside, forcing the tympanic membrane outward and producing pain.

Answers to Review Questions

Page 601

Level 1 Reviewing Facts and Terms

1. (**a**) vascular tunic; (**b**) iris; (**c**) ciliary body; (**d**) choroid; (**e**) neural tunic (retina); (**f**) neural part; (**g**) pigmented part; (**h**) fibrous tunic; (**i**) cornea; (**j**) sclera

2. d **3.** c **4.** e **5.** c **6.** c **7.** b **8.** d **9.** b

10. (**a**) auricle; (**b**) external acoustic meatus; (**c**) tympanic membrane; (**d**) auditory ossicles; (**e**) semicircular canals; (**f**) vestibule; (**g**) auditory tube; (**h**) cochlea; (**i**) vestibulocochlear nerve (N VIII)

11. d **12.** d **13.** c

14. (1) filiform papillae; (2) fungiform papillae; and (3) circumvallate papillae

15. The fibrous tunic (**a**) is composed of the sclera and the cornea and (**b**) provides mechanical support and some physical protection, serves as an attachment site for the extrinsic eye muscles, and contains structures that assist in the focusing process.

16. The vascular tunic consists of the iris, ciliary body, and choroid.

17. The malleus, incus, and stapes transmit a mechanical vibration (amplified along the way) from the tympanic membrane to the oval window.

Level 2 Reviewing Concepts

18. Axons leaving the olfactory epithelium collect into 20 or more bundles that penetrate the cribriform plate of the ethmoid to reach the olfactory bulbs of the cerebrum. Axons leaving the olfactory bulb travel along the olfactory tract to reach the olfactory cortex, hypothalamus, and portions of the limbic system.

19. Olfactory sensations are long lasting and important to memories because the sensory information reaches the cerebral cortex via the hypothalamus and the limbic system without first being filtered through the thalamus.

20. An infected sebaceous gland of an eyelash or tarsal gland usually becomes a sty, a painful swelling.

21. a **22.** c **23.** a

Level 3 Critical Thinking and Clinical Applications

24. Your medial rectus muscles would contract, directing your gaze more medially. In addition, your pupils would constrict and the lenses would become more spherical.

25. Myopia is corrected by (**a**) concave lenses.

26. In removing the polyps, some of the olfactory epithelium was probably damaged or destroyed, decreasing the area available for the solution of odor molecules and reducing the intensity of the stimulus. As a result, after the surgery it would take a larger stimulus to provide the same level of smell.

27. The rapid descent in the elevator causes the maculae in the saccule of your vestibule to slide upward, producing the sensation of downward vertical motion. After the elevator abruptly stops, it takes a few seconds for the maculae to come to rest in the normal position. So long as the maculae are displaced, you will perceive movement.

28. When Juan closes his eyes, visual cues are gone, and his brain must rely solely on proprioceptive information and information from the static equilibrium centers (saccule and utricle). As a result of his problem with the saccules or utricles, his brain does not receive sufficient information to maintain balance. The movement of the arms toward the side of the impaired receptors is due to the deficit of information arriving from that side of the body.

Chapter 18

Answers to Checkpoints

Page 606

1. A hormone is a chemical messenger that is secreted by one cell and travels through the circulatory system to affect the activities of cells in other parts of the body.

2. Paracrine communication is the use of chemical messengers to transfer information from cell to cell within a single tissue.

3. The four mechanisms of intercellular communication are direct, paracrine, endocrine, and synaptic.

Page 614

4. Neural responses occur within fractions of a second and are of short duration. Conversely, endocrine responses are slow to appear but last for minutes to days.

5. A substance that inhibits adenylate cyclase, the enzyme that converts ATP to cAMP, would block the action of any hormone that requires cAMP as a second messenger.

6. A cell's hormonal sensitivities are determined by the presence or absence of the receptor complex needed to bind a given hormone.

Page 620

7. The two lobes of the pituitary gland are the adenohypophysis (anterior lobe) and the neurohypophysis (posterior lobe).

8. In dehydration, blood osmotic pressure is increased, which would stimulate the neurohypophysis to release more ADH.

9. Somatomedins mediate the action of growth hormone. Elevated levels of somatomedins are typically accompanied by elevated levels of growth hormone.

10. Elevated circulating levels of cortisol inhibit the cells that control the release of ACTH from the pituitary gland, so ACTH levels would decrease. This is an example of a negative feedback mechanism.

Page 625

11. Thyroxine (T_4), triiodothyronine (T_3), and calcitonin are hormones associated with the thyroid gland.

12. An individual whose diet lacks iodine would be unable to form the hormone thyroxine. As a result, you would expect to see signs and symptoms associated with thyroxine deficiency: decreased metabolic rate, decreased body temperature, a poor response to physiological stress, and an increase in the size of the thyroid gland (goiter).

13. Most of the body's reserves of the thyroid hormone, thyroxine, are bound to blood-borne proteins called thyroid-binding globulins. Because these compounds represent such a large reservoir of thyroxine, it takes several days after removal of the thyroid gland for blood levels of thyroxine to decline.

Page 626

14. The parathyroid glands are embedded in the posterior surfaces of the lateral lobes of the thyroid gland.

15. The hormone secreted by the parathyroid glands is parathyroid hormone (PTH).

16. The removal of the parathyroid glands would result in a decrease in the blood concentration of calcium ions. Increasing the amounts of vitamin D and calcium in the diet could counteract the effects.

Page 630

17. The two regions of the suprarenal gland are the cortex and medulla. The cortex secretes mineralocorticoids (primarily aldosterone), glucocorticoids (mainly cortisol, hydrocortisone, and corticosterone), and androgens; the medulla secretes epinephrine and norepinephrine.

18. The three zones of the suprarenal cortex are the zona glomerulosa, zona fasciculata, and zona reticularis.

19. One function of cortisol is to decrease the cellular use of glucose while increasing both the available glucose (by promoting the breakdown of glycogen) and the conversion of amino acids to carbohydrates. Therefore, the net result of elevated cortisol levels would be an elevation of blood glucose.

Page 631

20. Pinealocytes are the special secretory cells in the pineal gland.

21. Increased amounts of light would inhibit the production (and release) of melatonin from the pineal gland, which receives neural input from the optic tracts. Melatonin secretion is influenced by light–dark cycles.

22. Melatonin inhibits reproductive functions, protects against free radical damage, and sets circadian rhythms.

Page 634

23. The cells of the pancreatic islets (and their hormones) are alpha cells (glucagon), beta cells (insulin), delta cells (GH–IH), and F cells (pancreatic polypeptide).

24. An individual with type 1 or type 2 diabetes has such high blood glucose levels that the kidneys cannot reabsorb all the glucose; some glucose is lost in urine. Because the urine contains high concentrations of glucose, less water can be reclaimed by osmosis, so the volume of urine production increases. The water losses reduce blood volume and elevate blood osmotic pressure, promoting thirst and triggering the secretion of ADH.

25. Increased levels of glucagon stimulate the conversion of glycogen to glucose in the liver, which would in turn reduce the amount of glycogen in the liver.

Page 638

26. Two hormones secreted by the kidneys are erythropoietin (EPO) and calcitriol.

27. Leptin is a hormone released by adipose tissue.

28. Once released into the bloodstream, renin functions as an enzyme, catalyzing the conversion of angiotensinogen to angiotensin I.

Page 644

29. The type of hormonal interaction in which two hormones have opposite effects on their target tissues is called antagonism.

30. A lack of GH, thyroid hormone, PTH, and the gonadal hormones would inhibit the formation and development of the skeletal system.

31. During the resistance phase of the general adaptation syndrome, there is a high demand for glucose, especially by the nervous system. The hormones GH–RH and CRH increase the levels of GH and ACTH, respectively. Growth hormone mobilizes fat reserves and promotes the catabolism of protein; ACTH increases cortisol, which stimulates both the conversion of glycogen to glucose and the catabolism of fat and protein.

32. The endocrine system adjusts metabolic rates and substrate utilization, and regulates growth and development, in all other body systems.

33. Hormones of the endocrine system adjust muscle metabolism, energy production, and growth; hormones also regulate calcium and phosphate levels, which are critical to normal muscle functioning. For their part, skeletal muscles protect some endocrine organs.

34. Many endocrine system hormones affect immune function: Glucocorticoids have anti-inflammatory effects, whereas thymosins stimulate development of lymphocytes. Leukocytes of the lymphoid system defend endocrine structures against infection and assist in repair after injury to those structures.

Answers to Review Questions
Page 648

Level 1 Reviewing Facts and Terms

1. (a) hypothalamus; (b) pituitary gland; (c) thyroid gland; (d) thymus; (e) suprarenal glands; (f) pineal gland; (g) parathyroid glands; (h) heart; (i) kidney; (j) adipose tissue; (k) digestive tract; (l) pancreatic islets (within pancreas); (m) gonads

2. b   3. c   4. d   5. a   6. d   7. d   8. d   9. b

10. (1) The hypothalamus produces regulatory hormones that control secretion by endocrine cells in the adenohypophysis. (2) The hypothalamus contains autonomic centers that exert direct neural control over the endocrine cells of the suprarenal medulla. (3) The hypothalamus releases ADH and oxytocin into the bloodstream at the neurohypophysis. These mechanisms are adjusted through negative feedback loops involving hormones released by peripheral endocrine tissues and organs.

11. The adenohypophysis releases (1) thyroid-stimulating hormone (TSH); (2) adrenocorticotropic hormone (ACTH); (3) follicle-stimulating hormone (FSH); (4) luteinizing hormone (LH); (5) prolactin (PRL); (6) growth hormone (GH); and (7) melanocyte-stimulating hormone (MSH).

12. Growth is affected by (1) growth hormone, (2) thyroid hormones, (3) insulin, (4) parathyroid hormone, (5) calcitriol, and (6) the reproductive hormones.

13. Effects of thyroid hormones: (1) increased rate of energy consumption and utilization in cells; (2) accelerated production of sodium–potassium ATPase; (3) activation of genes coding for the

synthesis of enzymes involved in glycolysis and energy production; (4) accelerated ATP production by mitochondria; and (5) in growing children, normal development of the skeletal, muscular, and nervous systems

14. Calcitonin decreases the concentration of calcium ions in body fluids; parathyroid hormone causes an increase in the concentration of calcium ions in body fluids.

15. (1) zona glomerulosa: mineralocorticoids; (2) zona fasciculata: glucocorticoids; and (3) zona reticularis: androgens

16. The kidneys release (1) erythropoietin, which stimulates the production of RBCs by the bone marrow, and (2) calcitriol, which stimulates calcium and phosphate absorption along the digestive tract.

17. Natriuretic peptides (1) promote the loss of sodium ions and water at the kidneys; (2) inhibit the secretion of water-conserving hormones, such as ADH and aldosterone; (3) suppress thirst; and (4) block the effects of angiotensin II and norepinephrine on arterioles. Angiotensin II opposes these actions by stimulating aldosterone secretion at the suprarenal cortex and ADH at the neurohypophysis, and further by restricting salt and water losses at the kidneys. Angiotensin II also stimulates thirst and elevates blood pressure.

18. (1) alpha cells: glucagon; (2) beta cells: insulin; (3) delta cells: somatostatin; and (4) F cells: pancreatic polypeptide

Level 2 Reviewing Concepts

19. The primary difference involves speed and duration. In the nervous system, the source and destination of communication are quite specific, and the effects are extremely quick and short lived. In endocrine communication, the effects are slow to appear and commonly persist for days. A single hormone can alter the metabolic activities of multiple tissues and organs simultaneously.

20. Hormones can (1) direct the synthesis of an enzyme (or other protein) not already present in the cytoplasm, (2) turn an existing enzyme "on" or "off," and (3) increase the rate of synthesis of a particular enzyme or other protein.

21. In endocrine reflexes—the functional counterpart of neural reflexes—a stimulus triggers the production of a hormone. Both neural and endocrine reflexes are typically controlled by negative feedback mechanisms.

22. Inactivation of phosphodiesterase, which converts cAMP to AMP, would prolong the effect of the hormone.

23. The suprarenal medulla is controlled by the sympathetic nervous system, whereas the suprarenal cortex is stimulated by the release of ACTH from the adenohypophysis.

24. b   **25.** a   **26.** b

Level 3 Critical Thinking and Clinical Applications

27. Extreme thirst and frequent urination are characteristic of both diabetes insipidus and diabetes mellitus. To distinguish between the two, glucose levels in the blood and urine could be measured. A high glucose concentration would indicate diabetes mellitus.

28. Julie's poor diet would not supply enough Ca^{2+} for her developing fetus, which would remove large amounts of Ca^{2+} from the maternal blood. A lowering of the mother's blood Ca^{2+} would lead to an increase in parathyroid hormone levels and increased mobilization of stored Ca^{2+} from maternal skeletal reserves.

29. Sherry's signs and symptoms suggest hyperthyroidism. Blood tests could be performed to assay the levels of TSH, T_3, and T_4. From these tests, the physician could make a positive diagnosis (hormone levels would be elevated in hyperthyroidism) and also determine whether the condition is primary (a problem with the thyroid gland) or secondary (a problem with hypothalamo-pituitary control of the thyroid gland).

30. One benefit of a portal system is that it ensures that the controlling hormones will be delivered directly to the target cells. Secondly, because the hormones go directly to their target cells without first passing through the general circulation, they are not diluted. The hypothalamus can control the cells of the adenohypophysis with much smaller amounts of releasing and inhibiting hormones than would be necessary if the hormones had to first go through the circulatory pathway before reaching the pituitary.

31. The natural effects of testosterone are to increase muscle mass, increase endurance, and enhance the "competitive spirit." Side effects in women include hirsutism (abnormal hair growth on the face or body), enlargement of the laryngeal cartilages, premature closure of the epiphyseal cartilages, and liver dysfunction.

Chapter 19

Answers to Checkpoints

Page 653

1. The major functions of blood are transportation of dissolved gases, nutrients, hormones, and metabolic wastes; regulation of the pH and ion composition of interstitial fluids; restriction of fluid losses at injury sites; defense against toxins and pathogens; and stabilization of body temperature.

2. Red blood cells, white blood cells, and platelets are the formed elements of blood.

3. Whole blood is composed of plasma and formed elements.

4. Venipuncture is a common sampling technique because superficial veins are easy to locate, the walls of veins are thinner than those of arteries, and blood pressure in veins is relatively low, so the puncture wound seals quickly.

Page 655

5. The three major types of plasma proteins are albumins, globulins, and fibrinogen.

6. A decrease in the amount of plasma proteins in the blood would lower plasma osmotic pressure, reduce the ability to fight infection, and decrease the transport and binding of some ions, hormones, and other molecules.

7. During a viral infection, you would expect the level of immunoglobulins (antibodies) in the blood to be elevated.

Page 662

8. Hemoglobin is a protein composed of four globular subunits, each bound to a heme molecule, which gives red blood cells the ability to transport oxygen in the blood.

9. After a significant loss of blood (especially of red blood cells), the hematocrit—the amount of formed elements (mostly red blood cells) as a percentage of the total blood—would be reduced.

10. Dave's hematocrit will increase, because reduced blood flow to the kidneys triggers the release of erythropoietin, which stimulates an increase in erythropoiesis (red blood cell formation).

11. Bilirubin would accumulate in the blood, producing jaundice, because diseases that damage the liver, such as hepatitis or cirrhosis, impair the liver's ability to excrete bilirubin in the bile.

Page 666

12. Surface antigens on RBCs are glycolipids in the plasma membrane; they determine blood type.

13. Only Type O blood can be safely transfused into a person whose blood type is O.

14. If a person with Type A blood receives a transfusion of Type B blood, which contains anti-A antibodies, the red blood cells will clump, or agglutinate, potentially blocking blood flow to various organs and tissues.

Page 672

15. The five types of white blood cells are neutrophils, eosinophils, basophils, monocytes, and lymphocytes.

16. An infected cut would contain a large number of neutrophils, phagocytic white blood cells that are generally the first to arrive at the site of an injury.

17. The blood of a person fighting a viral infection would contain elevated numbers of lymphocytes, because B lymphocytes produce circulating antibodies.

18. During inflammation, basophils release a variety of chemicals, including histamine and heparin, that exaggerate the inflammation and attract other types of white blood cells.

Page 673

19. Thrombocytopoiesis is the term for platelet production.

20. Platelets are non-nucleated cell fragments in mammal blood, whereas thrombocytes are nucleated platelets in nonmammalian-vertebrate blood.

21. Platelets release chemicals important to the clotting process, form a temporary patch in the walls of damaged blood vessels, and contract after a clot has formed.

Page 677

22. A decreased number of megakaryocytes would interfere with the blood's ability to clot properly, because fewer megakaryocytes would produce fewer platelets.

23. Fruit juice and water do not contain fats, which are required for vitamin K absorption, leading to a vitamin K deficiency. This would lead to a decreased production of several clotting factors—most notably, prothrombin. As a result, clotting time would increase.

24. The activation of Factor XII initiates the intrinsic pathway.

Answers to Review Questions

Page 679

Level 1 Reviewing Facts and Terms

1. (a) neutrophil; **(b)** eosinophil; **(c)** basophil; **(d)** monocyte; **(e)** lymphocyte

2. c   **3.** c   **4.** a   **5.** d   **6.** d   **7.** b   **8.** d   **9.** a

10. Blood (1) transports dissolved gases, nutrients, hormones, and metabolic wastes; (2) regulates pH and electrolyte composition of interstitial fluids throughout the body; (3) restricts fluid losses through damaged vessels or at other injury sites; (4) defends against toxins and pathogens; and (5) stabilizes body temperature.

11. Major types of plasma proteins are (1) albumins, which maintain the osmotic pressure of plasma and are important in the transport of fatty acids; (2) globulins, which (a) bind small ions, hormones, or compounds that might otherwise be filtered out of the blood at the kidneys or have very low solubility in water (transport globulins), or (b) attack foreign proteins and pathogens (immunoglobulins); and (3) fibrinogen, which functions in blood clotting.

12. (a) anti-B antibodies; **(b)** anti-A antibodies; **(c)** neither anti-A nor anti-B antibodies; **(d)** both anti-A and anti-B antibodies

13. WBCs exhibit (1) ameboid movement, a gliding movement that transports the cell; (2) emigration, squeezing between adjacent endothelial cells in the capillary wall; (3) positive chemotaxis, the attraction to specific chemical stimuli, and (4) phagocytosis (engulfing particles for neutrophils, eosinophils, and monocytes).

14. Neutrophils, eosinophils, basophils, and monocytes function in nonspecific defense.

15. The primary lymphocytes are (1) T cells, which are responsible for cell-mediated immunity; (2) B cells, which are responsible for humoral immunity; and (3) NK cells, which are responsible for immune surveillance.

16. Platelets (1) transport chemicals important to clotting; (2) form a temporary patch in the walls of damaged blood vessels; and (3) actively contract after the clot has formed.

17. Erythropoeitin is released (1) during anemia, (2) when blood flow to the kidneys declines, (3) when oxygen content of the air in the lungs declines, and (4) when the respiratory surfaces of the lungs are damaged.

18. Initiation of the common pathway requires the activation of Factor X and the formation of prothrombinase by the extrinsic and/or intrinsic pathways.

Level 2 Reviewing Concepts

19. a   **20.** d   **21.** c   **22.** c

23. Red blood cells are biconcave discs that lack mitochondria, ribosomes, and nuclei, and they contain a large amount of hemoglobin. RBCs transport oxygen, while WBCs are involved in immunity. The five types of white blood cells vary in size from slightly larger to twice the diameter of an RBC, contain a prominent nucleus, and may contain granules with distinct staining properties.

24. White blood cells defend against toxins and pathogens. Neutrophils, eosinophils, and monocytes engulf and digest bacteria, protozoa, fungi, viruses, and cellular debris. Lymphocytes specialize to attack and destroy specific foreign cells, proteins, and cancerous cells, directly or through the production of antibodies.

25. Blood stabilizes and maintains body temperature by absorbing and redistributing the heat produced by active skeletal muscles. Dermal capillaries dilate when body temperature rises, thereby increasing blood flow to the skin and dissipating the excess heat into the air. Dermal capillaries constrict when body temperature falls, thereby decreasing blood flow to the skin and conserving heat for internal organs that are more temperature sensitive.

26. Each molecule of hemoglobin consists of four protein subunits, each of which contains a single molecule of heme, a non-protein ring surrounding an iron ion. These central iron ions are what actually pick up and release oxygen molecules.

27. Aspirin helps prevent vascular problems by inhibiting clotting. It inactivates platelet enzymes involved in the production of thromboxanes and prostaglandins, and it inhibits endothelial cell production of prostacyclin.

Level 3 Critical Thinking and Clinical Applications

28. A prolonged prothrombin time and a normal partial thromboplastin time indicate a deficiency in the extrinsic system but not in the intrinsic system or common pathway. Factor VII would be deficient.

29. As the spleen enlarges, so does its capacity to store additional red blood cells, leading to fewer red blood cells in circulation, producing anemia. The decreased number of RBCs in circulation decreases the body's ability to deliver oxygen to the tissues, slowing their metabolism and producing the tired feeling and lack of energy. Because there are fewer RBCs than normal, the blood circulating through the skin is not as red, producing a pale complexion.

30. Taking a broad-spectrum antibiotic kills a wide range of bacteria, both pathogenic and nonpathogenic, including many of the normal flora of the intestine. Reducing the intestinal flora substantially decreases the amount of vitamin K they make available to the liver to produce prothrombin, a vital component of the common pathway. With decreased amounts of prothrombin in the blood, normal minor breaks in the vessels of the nasal passageways do not seal off as quickly, producing nosebleeds.

31. Removal of most of Randy's stomach eliminated the production of intrinsic factor, which is essential for the absorption of vitamin B_{12} by intestinal cells. Thus Randy was prescribed vitamin B_{12} to prevent pernicious anemia, and he needed injections of vitamin B_{12} (could not take it orally) because his intestines could no longer absorb it.

Chapter 20

Answers to Checkpoints

Page 693

1. Damage to the semilunar valve on the right side of the heart would affect blood flow to the pulmonary artery.

2. Contraction of the papillary muscles (just before the rest of the ventricular myocardium contracts) pulls on the chordae tendineae, which prevent the AV valves from opening back into the atria.

3. The left ventricle is more muscular than the right ventricle because the left ventricle must generate enough force to propel blood throughout the body, except the alveoli of the lungs; whereas the right ventricle must generate only enough force to propel blood a few centimeters to the lungs.

Page 703

4. Automaticity, or autorhythmicity, is the ability of cardiac muscle tissue to contract without neural or hormonal stimulation.

5. The sinoatrial node is known as the cardiac pacemaker or the natural pacemaker.

6. If the cells of the SA node failed to function, the heart would still continue to beat, but at a slower rate; the AV node would act as the pacemaker.

7. If the impulses from the atria were not delayed at the AV node, they would be conducted through the ventricles so quickly by the bundle branches and Purkinje cells that the ventricles would begin contracting immediately, before the atria had finished their contraction. As a result, the ventricles would not be as full of blood as they could be, and the pumping of the heart would not be as efficient, especially during activity.

Page 706

8. The alternate term for contraction is systole, and the other term for relaxation is diastole.

9. The phases of the cardiac cycle are atrial systole, atrial diastole, ventricular systole, and ventricular diastole.

10. No. When pressure in the left ventricle first rises, the heart is contracting but no blood is leaving the heart. During this initial phase of contraction, both the AV valves and the semilunar valves are closed. The increase in pressure is the result of increased tension as the cardiac muscle contracts. When the pressure in the ventricle exceeds the pressure in the aorta, the aortic semilunar valves are forced open, and blood is rapidly ejected from the ventricle.

11. One possible cause for an increase in the size of the QRS complex, which indicates a larger-than-normal amount of electrical activity during ventricular depolarization, is an enlarged heart. Because more cardiac muscle is depolarizing, the magnitude of the electrical event would be greater.

Page 715

12. Cardiac output is the amount of blood pumped by the left ventricle in one minute.

13. Caffeine acts directly on the conducting system and contractile cells of the heart, increasing the rate at which they depolarize. Drinking large amounts of caffeinated drinks would increase the heart rate.

14. Damage to the cardioinhibitory center of the medulla oblongata, which is part of the parasympathetic division of the autonomic nervous system, would reduce parasympathetic action potentials to the heart. The resulting sympathetic dominance would increase the heart rate.

15. A drug that increases the length of time required for the repolarization of pacemaker cells would decrease the heart rate, because the pacemaker cells would generate fewer action potentials per minute.

16. The heart pumps in proportion to the amount of blood that enters. A heart that beats too rapidly does not have sufficient time to fill completely between beats. Thus, when the heart beats too fast, very little blood leaves the ventricles and enters the circulation, so tissues suffer damage from inadequate blood supply.

17. Stimulating the acetylcholine receptors of the heart would slow the heart rate. Since cardiac output is the product of stroke volume and heart rate, a reduction in heart rate will lower the cardiac output (assuming that the stroke volume doesn't change).

18. The venous return fills the heart with blood, stretching the heart muscle. According to the Frank–Starling principle, the more the heart muscle is stretched, the more forcefully it will contract (to a point). The more forceful the contraction, the more blood the heart will eject with each beat (stroke volume). Therefore, increased venous return would increase the stroke volume (if all other factors are constant).

19. An increase in sympathetic stimulation of the heart would increase heart rate and force of contraction. The end-systolic volume (ESV) is the amount of blood that remains in a ventricle after a contraction (systole). The more forcefully the heart contracts, the more blood it ejects and the lower the ESV is. Therefore, increased sympathetic stimulation should result in a lower ESV.

20. SV = EDV − ESV, so SV = 125 mL − 40 mL = 85 mL

Answers to Review Questions

Page 717

Level 1 Reviewing Facts and Terms

1. **(a)** superior vena cava; **(b)** auricle of right atrium; **(c)** right ventricle; **(d)** left ventricle; **(e)** arch of aorta (aortic arch); **(f)** left pulmonary artery; **(g)** pulmonary trunk; **(h)** auricle of left atrium

2. c **3.** b **4.** b **5.** b **6.** d

7. **(a)** ascending aorta; **(b)** opening of coronary sinus; **(c)** right atrium; **(d)** cusp of right AV (tricuspid) valve; **(e)** chordae tendineae; **(f)** right ventricle; **(g)** pulmonary valve; **(h)** left pulmonary veins; **(i)** left atrium; **(j)** aortic valve; **(k)** cusp of left AV (mitral) valve; **(l)** left ventricle; **(m)** interventricular septum

8. a and b **9.** b **10.** a **11.** a

12. During ventricular contraction, tension in the papillary muscles pulls against the chordae tendineae, which keep the cusps of the AV valve from swinging into the atrium. This action prevents the backflow, or regurgitation, of blood into the atrium as the ventricle contracts.

13. (1) The epicardium is the visceral pericardium, which covers the outer surface of the heart. (2) The myocardium is the muscular wall of the heart, which forms both atria and ventricles. It contains cardiac muscle tissue and associated connective tissues, blood vessels, and nerves. (3) The endocardium is a squamous epithelium that covers the inner surfaces of the heart, including the valves.

14. The right atrioventricular (AV) valve (the tricuspid valve) and the left AV valve (the bicuspid valve) prevent the backflow of blood from the ventricles into the atria. The pulmonary and aortic semilunar valves prevent the backflow of blood from the pulmonary trunk and aorta into the right and left ventricles.

15. SA node → AV node → AV bundle → right and left bundle branches → Purkinje fibers (into the mass of ventricular muscle tissue)

16. The cardiac cycle comprises the events in a complete heartbeat, including a contraction/relaxation period for both atria and ventricles. The cycle begins with atrial systole as the atria contract and push blood into the relaxed ventricles. As the atria relax (atrial diastole), the ventricles contract (ventricular systole), forcing blood through the semilunar valves into the pulmonary trunk and aorta. The ventricles then relax (ventricular diastole). For the rest of the cardiac cycle, both the atria and ventricles are in diastole; passive filling occurs.

17. The factors that regulate stroke volume are (1) preload, the stretch on the heart before it contracts; (2) contractility, the forcefulness of contraction of individual ventricular muscle fibers; and (3) afterload, the pressure that must be exceeded before blood can be ejected from the ventricles.

Level 2 Reviewing Concepts

18. c   **19.** a   **20.** d   **21.** a

22. The SA node, which is composed of cells that exhibit rapid prepotential, is the pacemaker of the heart. The AV node slows the impulse that signals contraction, because its cells are smaller than those of the conduction pathway.

23. The first sound ("lubb"), which marks the start of ventricular contraction, is produced as the AV valves close and the semilunar valves open. The second sound ("dubb") occurs when the semilunar valves close, marking the start of ventricular diastole. The third heart sound is associated with blood flow into the atria, and the fourth sound is associated with atrial contraction. Listening to the heart sounds (auscultation) is a simple and effective diagnostic tool.

24. Stroke volume (SV) is the volume of blood ejected by a ventricle in a single contraction. Cardiac output (CO) is the amount of blood pumped by a ventricle in 1 minute: CO (in mL/min) = HR (in beats/min) × SV (in mL/beat)

25. Stroke volume and heart rate influence cardiac output.

26. Sympathetic activation increases the heart rate and the force of contractions; parasympathetic stimulation decreases the heart rate and the force of contractions.

27. All these hormones have positive inotropic effects, which means that they increase the strength of contraction of the heart.

Level 3 Critical Thinking and Clinical Applications

28. During tachycardia (an abnormally fast heart rate), there is less time between contractions for the heart to fill with blood again. Thus, over time the heart fills with less and less blood, and pumps less blood out. As the stroke volume decreases, so does cardiac output. When cardiac output decreases to the point where not enough blood reaches the brain, loss of consciousness occurs.

29. Harvey probably has a regurgitating mitral valve. When an AV valve fails to close properly, blood flowing back into the atrium produces a murmur. A murmur at the beginning of systole implicates the AV valve because this is the period when the valve has just closed and the blood in the ventricle is under increasing

pressure; thus the likelihood of backflow is the greatest. A sound heard at the end of systole or the beginning of diastole would implicate a regurgitating semilunar valve—in this case, the aortic semilunar valve.

30. Using CO = HR × SV, person 1 has a cardiac output of 4500 mL, and person 2 has a cardiac output of 8550 mL. According to Starling's law, in a normal heart the cardiac output is directly proportional to the venous return. Thus, person 2 has the greater venous return. Ventricular filling decreases with increased heart rate; person 1 has the lower heart rate and therefore the longer ventricular filling time.

31. By blocking calcium channels, *verapamil* will decrease the force of cardiac contraction, which directly lowers Karen's stroke volume.

Chapter 21

Answers to Checkpoints

Page 729

1. The five general classes of blood vessels are arteries, arterioles, capillaries, venules, and veins.

2. The blood vessels are veins. Arteries and arterioles have a large amount of smooth muscle tissue in a thick, well-developed tunica media.

3. In the arterial system, pressures are high enough to keep the blood moving forward. In the venous system, blood pressure is too low to keep the blood moving on toward the heart. Valves in veins prevent blood from flowing backward whenever the venous pressure drops.

4. Fenestrated capillaries are located where small peptides move freely into and out of the blood, including endocrine glands, the choroid plexus of the brain, absorptive areas of the intestine, and filtration areas of the kidneys.

Page 737

5. Total peripheral resistance reflects a combination of vascular resistance, vessel length, vessel diameter, blood viscosity, and turbulence.

6. In a healthy individual, blood pressure is greater at the aorta than at the inferior vena cava. Blood, like other fluids, moves along a pressure gradient from areas of high pressure to areas of low pressure. If the pressure were higher in the inferior vena cava than in the aorta, the blood would flow backward.

7. While a person stands for periods of time, blood pools in the lower limbs, which decreases venous return to the heart. In turn, cardiac output decreases, so less blood reaches the brain, causing light-headedness and fainting. A hot day adds to this effect, because the loss of body water through sweating reduces blood volume.

8. Mike's mean arterial pressure is approximately 88.3 mm Hg; 70 + (125 − 70)/3 = 70 + 18.3 = 88.3.

Page 744

9. Vasodilators promote the dilation of precapillary sphincters; local vasodilators act at the tissue level to accelerate blood flow through their tissue of origin.

10. Pressure on the common carotid artery would decrease blood pressure at the baroreceptors in the carotid sinus. This decrease would cause a decreased frequency of action potentials along the glossopharyngeal cranial nerve (IX) to the medulla oblongata, and more sympathetic impulses would be sent to the heart. The net result would be an increase in the heart rate.

11. Vasoconstriction of the renal artery would decrease both blood flow and blood pressure at the kidney. In response, the kidney would increase the amount of renin it releases, which in turn would lead to an increase in the level of angiotensin II. The angiotensin II would bring about increased blood pressure and increased blood volume.

Page 748

12. Blood pressure increases during exercise because (1) cardiac output increases and (2) resistance in visceral tissues increases.

13. The immediate problem during hemorrhaging is the maintenance of adequate blood pressure and peripheral blood flow; the long-term problem is the restoration of normal blood volume.

14. Both aldosterone and ADH promote fluid retention and reabsorption at the kidneys, preventing further reductions in blood volume.

15. The two circuits of the cardiovascular system are the pulmonary circuit and the systemic circuit.

16. The three general functional patterns are the following: (1) The peripheral distributions of arteries and veins on the body's left and right sides are generally identical, except near the heart, where the largest vessels connect to the atria or ventricles; (2) a single vessel may have several names as it crosses specific anatomical boundaries, making accurate anatomical descriptions possible; and (3) tissues and organs are usually serviced by several arteries and veins.

Page 750

17. The pulmonary arteries enter the lungs carrying deoxygenated blood, and the pulmonary veins leave the lungs carrying oxygenated blood.

18. Right ventricle → pulmonary trunk → left and right pulmonary arteries → pulmonary arterioles → alveoli → pulmonary venules → pulmonary veins → left atrium

Page 767

19. A blockage of the left subclavian artery would interfere with blood flow to the left arm.

20. Compression of the common carotid arteries would reduce blood pressure at the carotid sinus and cause a rapid reduction in blood flow to the brain, resulting in a loss of consciousness. An immediate reflexive increase in heart rate and blood pressure would follow.

21. Rupture of the celiac trunk would most directly affect the stomach, spleen, liver, and pancreas.

22. The vein that is bulging is the external jugular vein.

23. A blockage of the popliteal vein would interfere with blood flow in the tibial and fibular veins (which form the popliteal vein) and the small saphenous vein (which joins the popliteal vein).

Page 770

24. Two umbilical arteries supply blood to the placenta, and one umbilical vein returns from the placenta. The umbilical vein then drains into the ductus venosus within the fetal liver.

25. This blood sample was taken from an umbilical vein, which carries oxygenated, nutrient-rich blood from the placenta to the fetus.

26. Necessary structures in the fetal circulation include two umbilical arteries, an umbilical vein, the ductus venosus, the foramen ovale, and the ductus arteriosus. In the neonate, the foramen ovale closes and persists as the fossa ovalis, a shallow depression; the ductus arteriosus persists as the ligamentum arteriosum, a fibrous cord; and the umbilical vessels and ductus venosus persist throughout life as fibrous cords.

27. Components of the cardiovascular system affected by age include the blood, heart, and blood vessels.

28. A thrombus is a stationary blood clot within the lumen of a blood vessel.

29. An aneurysm is the ballooning out of an inelastic arterial wall resulting from sudden pressure increases.

30. The cardiovascular system provides other body systems with oxygen, hormones, nutrients, and white blood cells while removing carbon dioxide and metabolic wastes; it also transfers heat.

31. The skeletal system provides calcium needed for normal cardiac muscle contraction, and it protects developing blood cells in the bone marrow. The cardiovascular system provides calcium and phosphate for bone deposition, delivers erythropoietin to bone marrow, and transports parathyroid hormone and calcitonin to osteoblasts and osteoclasts.

Answers to Review Questions

Page 774

Level 1 Reviewing Facts and Terms

1. (a) brachiocephalic trunk; (b) brachial; (c) radial; (d) external iliac; (e) anterior tibial; (f) right common carotid; (g) left subclavian; (h) common iliac; (i) femoral
2. b 3. e 4. c 5. b 6. b 7. b 8. d
9. (a) external jugular; (b) brachial; (c) median cubital; (d) radial; (e) great saphenous; (f) internal jugular; (g) superior vena cava; (h) left and right common iliac; (i) femoral
10. b 11. b 12. d 13. c 14. c 15. c 16. c
17. (a) Capillary hydrostatic pressure forces fluid out of the capillary at the arteriole end. (b) Blood colloid osmotic pressure causes the movement of fluid back into the capillary at its venous end.
18. When an infant takes its first breath, the lungs expand and pulmonary vessels dilate. The smooth muscles in the ductus arteriosus contract, due to increased venous return from the lungs, isolating the pulmonary and aortic trunks, and blood begins flowing through the pulmonary circuit. As pressure rises in the left atrium, the valvular flap closes the foramen ovale, completing the circulatory remodeling.

Level 2 Reviewing Concepts

19. b 20. b 21. a
22. Artery walls are generally thicker and contain more smooth muscle and elastic fibers, enabling them to resist and adjust to the pressure generated by the heart. Venous walls are thinner; the pressure in veins is less than that in arteries. Arteries constrict more than veins do when not expanded by blood pressure, due to a greater degree of elastic tissue. Finally, the endothelial lining of an artery has a pleated appearance because it cannot contract and so forms folds. The lining of a vein looks like a typical endothelial layer.
23. Capillary walls are thin, so distances for diffusion are short. Continuous capillaries have small gaps between adjacent endothelial cells that permit the diffusion of water and small solutes into the surrounding interstitial fluid but prevent the loss of blood cells and plasma proteins. Fenestrated capillaries contain pores that permit very rapid exchange of fluids and solutes between interstitial fluid and plasma. The walls of arteries and veins are several cell layers thick and are not specialized for diffusion.

24. Contraction of the surrounding skeletal muscles squeezes venous blood toward the heart. This mechanism, the muscular pump, is assisted by the presence of valves in the veins, which prevent backflow of the blood. The respiratory pump, which results from the increase in internal pressure of the thoracic cavity during exhalation, pushes venous blood into the right atrium.

25. Cardiac output and peripheral blood flow are directly proportional to blood pressure. Blood pressure is closely regulated by a combination of neural and hormonal mechanisms. The resistance of the circulatory system opposes the movement of blood, so blood flow is inversely proportional to the resistance. Sources of peripheral resistance include vascular resistance, viscosity, and turbulence.

26. The brain receives arterial blood via four arteries that form anastomoses within the cranium. An interruption of any one vessel will not compromise the circulatory supply to the brain.

27. The cardioaccelleratory and vasomotor centers are stimulated when general sympathetic activation occurs. The result is an increase in cardiac output and blood pressure. When the parasympathetic division is activated, the cardioinhibitory center is stimulated, reducing cardiac output.

Level 3 Critical Thinking and Clinical Applications

28. Fluid loss lowers blood volume, leading to sympathetic stimulation, which elevates blood pressure.

29. Antihistamines and decongestants are sympathomimetic drugs; they have the same effects on the body as does stimulation of the sympathetic nervous system. In addition to the desired effects of counteracting the symptoms of the allergy, these medications can produce an increased heart rate, increased stroke volume, and increased peripheral resistance, all of which will contribute to elevating blood pressure. In a person with hypertension (high blood pressure), these drugs would aggravate this condition, with potentially hazardous consequences.

30. When Jolene stood up rapidly, gravity caused her blood volume to move to the lower parts of her body away from the heart, decreasing venous return. The decreased venous return resulted in a decreased end-diastolic volume (EDV), leading to a decreased stroke volume and cardiac output. In turn, blood flow to the brain decreased, so the diminished oxygen supply caused her to be light-headed and feel faint. This reaction doesn't happen all the time because as soon as the pressure drops due to inferior movement of blood, baroreceptors in the aortic arch and carotid sinus trigger the baroreceptor reflex. Action potentials are carried to the medulla oblongata, where appropriate responses are integrated. In this case, we would expect an increase in peripheral resistance to compensate for the decreased blood pressure. If this doesn't compensate enough for the drop, then an increase in heart rate and force of contraction would occur. Normally, these responses occur so quickly that changes in pressure following changes in body position go unnoticed.

Chapter 22

Answers to Checkpoints

Page 777

1. A pathogen is any disease-causing organism, such as a virus, bacterium, fungus, or parasite, that can survive and even thrive inside the body.

2. Nonspecific defenses are anatomical barriers and defense mechanisms that either prevent or slow the entry of infectious organisms but do not distinguish one potential threat from another. Specific defenses involve an immune response against a specific type of threat.

Page 789

3. The components of the lymphoid system are lymph, lymphatic vessels, lymphoid tissues, and lymphoid organs.

4. A blockage of the thoracic duct would impair the drainage of lymph from inferior to the diaphragm and from the left side of the head and thorax, retarding the return of lymph to the venous blood and promoting the accumulation of fluid in the limbs (lymphedema).

5. A lack of thymic hormones would drastically reduce the population of T lymphocytes by preventing their differentiation from lymphoid stem cells.

6. Lymph nodes enlarge during some infections because lymphocytes and phagocytes in the nodes multiply to defend against the infectious agent.

Page 796

7. The body's nonspecific defenses include physical barriers, phagocytes, immunological surveillance, interferons, complement, the inflammatory response, and fever.

8. A decrease in the number of monocyte-forming cells in bone marrow would result in fewer macrophages of all types, including Kupffer cells of the liver, dendritic (Langerhans) cells in the skin and digestive tract, and alveolar macrophages.

9. A rise in the level of interferon suggests a viral infection. Interferon does not help an infected cell, but "interferes" with the virus's ability to infect other cells.

10. Pyrogens increase body temperature (produce a fever) by stimulating the temperature control area of the preoptic nucleus of the hypothalamus.

Page 798

11. In cell-mediated (cellular) immunity, T cells defend against abnormal cells and pathogens inside cells. In antibody-mediated (humoral) immunity, B cells secrete immunoglobulins that defend against antigens and pathogens in body fluids.

12. The two forms of active immunity are artificial active immunity and natural active immunity; the two forms of passive immunity are natural passive immunity and induced passive immunity.

13. The four general properties of immunity are specificity, versatility, memory, and tolerance.

Page 803

14. The three major types of T cells are cytotoxic T cells (T_C cells), helper T cells (T_H cells), and suppressor T cells (T_S cells).

15. Abnormal peptides in the cytoplasm of a cell can become attached to MHC (major histocompatibility complex) proteins and then displayed on the surface of the cell's plasma membrane. The recognition of such displayed peptides by T cells can initiate an immune response.

16. A decrease in the number of cytotoxic T cells would affect cell-mediated immunity, reducing the effectiveness of T_C cells in killing foreign cells and virus-infected cells.

17. Without helper T cells—which promote B cell division, the maturation of plasma cells, and antibody production by plasma cells—the antibody-mediated immune response would probably not occur.

Page 811

18. Sensitization is the process by which a B cell becomes able to react with a specific antigen.

19. An antibody molecule consists of two parallel pairs of polypeptide chains: one pair of heavy chains and one pair of light chains. Each chain contains both constant segments and variable segments.

20. Plasma cells produce and secrete antibodies, so observing an elevated number of plasma cells would lead us to expect increasing levels of antibodies in the blood.

21. Because production of a secondary response depends on the presence of memory B cells and memory T cells formed during a primary response, the secondary response would be more affected by a lack of memory B cells for a particular antigen.

Page 817

22. A developing fetus is protected primarily by natural passive immunity, the product of IgG antibodies that have crossed the placenta from the mother's bloodstream.

23. An autoimmune disorder is a condition that results when the immune system's sensitivity to normal cells and tissues causes the production of autoantibodies.

24. Stress can interfere with the immune response by depressing the inflammatory response, reducing the number and activity of phagocytes, and inhibiting interleukin secretion.

25. The elderly are more susceptible to viral and bacterial infections because as the number of helper T cells declines with age, B cells are less responsive, so antibody levels rise more slowly after antigen exposure.

26. As one ages, immunological surveillance declines, so cancerous cells are not eliminated as effectively.

27. Glucocorticoids released by the endocrine system have anti-inflammatory effects; thymic hormones stimulate the development and maturation of lymphocytes; and many hormones affect immune function. The thymus gland secretes thymic hormones, and cytokines affect cells throughout the body.

28. The lymphoid system provides defenses against infection; performs immunological surveillance to eliminate cancer cells; and returns tissue fluid to the circulation.

Answers to Review Questions

Page 823

Level 1 Reviewing Facts and Terms

1. **(a)** tonsil; **(b)** right lymphatic duct; **(c)** thymus; **(d)** thoracic duct; **(e)** cisterna chyli; **(f)** lymphatics of upper limb; **(g)** axillary lymph nodes; **(h)** thoracic duct; **(i)** spleen; **(j)** mucosa-associated lymphoid tissue (MALT)

2. b 3. d 4. c 5. c 6. c 7. a 8. c 9. d 10. d 11. c

12. (1) lymph nodes: filtration of lymph, detection of pathogens, initiation of immune response; (2) tonsils, lymphoid nodules, and aggregated lymphoid nodules: defense of entrance and passageways of digestive tract against pathogens and foreign proteins or toxins; (3) spleen: filtration of blood, recycling of red blood cells, detection of blood-borne pathogens or toxins; (4) thymus: production of mature T cells and hormones that promote immune function; (5) lymphatics: movement of lymph from interstitial spaces to the venous system

13. **(a)** lymphocytes responsible for cell-mediated immunity; **(b)** stimulate the activation and function of T cells and B cells; **(c)** inhibit the activation and function of both T cells and B cells; **(d)** produce and secrete antibodies; **(e)** recognize and destroy abnormal cells; **(f)** produce interleukin-7, which promotes the differentiation of B cells; **(g)** maintain the blood–thymus barrier and secrete the thymic hormones that stimulate stem cell division and T cell differentiation; **(h)** interfere with viral replication inside the cell and stimulate the activities of macrophages and NK cells; **(i)** reset the body's thermostat, causing a rise in body temperature (fever); **(j)** provide cell-mediated immunity, which defends against abnormal cells and pathogens inside cells; **(k)** provide humoral immunity, which defends against antigens and pathogens in the body (but not inside cells); **(l)** enhance nonspecific defenses and increase T cell sensitivity and stimulate B cell activity; **(m)** slow tumor growth and kill sensitive tumor cells; **(n)** stimulate the production of blood cells in the bone marrow and lymphocytes in lymphoid tissues and organs

14. (1) T cells, derived from the thymus; (2) B cells, derived from bone marrow; and (3) NK cells, derived from bone marrow

15. Nonspecific defenses: (1) physical barriers; (2) phagocytic cells; (3) immunological surveillance; (4) interferons; (5) complement; (6) inflammation; and (7) fever

Level 2 Reviewing Concepts

16. c 17. c 18. b 19. b

20. Complement can rupture the target cell's plasma membrane by releasing perforin, kill the target cell by secreting a cytotoxic lymphotoxin, or activate genes within the nucleus of the cell that stimulate programmed cell death. Interferon interferes with viral replication inside virus-infected cells by triggering the production of antiviral proteins.

21. Cytotoxic T cells kill by rupturing the target cell's plasma membrane, by stimulating lymphotoxin secretion, or by activating genes in the nucleus that program cell death.

22. Formation of an antigen–antibody complex eliminates antigens by neutralization; by agglutination and precipitation; by activating complement; by attracting phagocytes; by opsonization; by stimulating inflammation; or by preventing bacterial and viral adhesion.

23. Innate immunity is genetically programmed; an example is immunity to fish diseases. Naturally acquired immunity develops after birth from contact with pathogens; an example is exposure to chickenpox in grade school. Induced active immunity develops after intentional exposure to a pathogen; an example is administration of measles vaccine. Induced passive immunity is temporary immunity provided by injection with antibodies produced in another organism, such as antibodies against rabies. Natural passive immunity is gained by acquiring antibodies via mother's milk or placental exchange.

24. The injections are timed to trigger the primary and secondary responses of the immune system. Upon first exposure to hepatitis antigens, B cells produce daughter cells that differentiate into plasma cells and memory B cells. The plasma cells begin producing antibodies, which represent the primary response to exposure. However, the primary response does not maintain elevated antibody levels for long periods, so the second and third injections are necessary to trigger secondary (anamnestic) responses, when memory B cells differentiate into plasma cells and produce antibody concentrations that remain high much longer.

Level 3 Critical Thinking and Clinical Applications

25. Yes, the crime lab could determine whether the sample is blood plasma, which contains IgM, IgG, IgD, and IgE, or semen, which contains only IgA.

26. Ted cannot yet know whether he'll come down with the measles. Ted's elevated blood IgM levels indicates that he is in the

early stages of a primary response to the measles virus. If his immune response proves unable to control and then eliminate the virus, Ted will develop the measles.

27. In a radical mastectomy, lymph nodes in the nearby axilla and surrounding region are removed along with the cancerous breast to try to prevent the spread of cancer cells via the lymphoid system. Lymphatic vessels from the limb on the affected side are tied off, and because there is no place for the lymph to drain, over time lymphedema causes swelling of the limb.

28. A key characteristic of cancer cells is their ability to metastasize—to break free from a tumor and form new tumors in other tissues. The primary route of metastasis is the lymphoid system, including the lymph nodes. Examination of regional lymph nodes for the presence of cancer cells can help the physician determine if the cancer was caught in an early stage or whether it has started to spread to other tissues.

29. Allergies occur when allergens bind to specific IgE antibodies that are bound to the surface of mast cells and basophils. A person becomes allergic when they develop IgE antibodies for a specific allergen. Theoretically, at least, a molecule that would bind to the specific IgE for ragweed allergen and prevent the allergen from binding should help to relieve the signs and symptoms of the allergy.

Chapter 23

Answers to Checkpoints

Page 829
1. Functions of the respiratory system include providing an extensive surface area for gas exchange between blood and air, moving air to and from exchange surfaces, protecting exchange surfaces from environmental variations and pathogens, producing sounds, detecting olfactory stimuli, and indirectly assisting in blood volume regulation and blood pressure through the conversion of angiotensin I to angiotensin II.
2. The two anatomical divisions of the respiratory system are the upper respiratory system and the lower respiratory system.
3. The respiratory mucosa lines the conducting portion of the respiratory system.

Page 831
4. The upper respiratory system consists of the nose, nasal cavity, paranasal sinuses, and pharynx.
5. The rich vascularization to the nose delivers body heat to the nasal cavity, so inhaled air is warmed before it leaves the nasal cavity. The heat also evaporates moisture from the epithelium to humidify the incoming air; moisture is also derived from the blood supply itself. The vascularization also brings nutrients and water to the secretory cells of the nasal mucosa.
6. The lining of the nasopharynx, which receives only air from the nasal cavity, is the same as that of the nasal cavity: a pseudostratified ciliated columnar epithelium. Because the oropharynx and laryngopharynx receive both air from the nasal cavity and potentially abrasive food from the oral cavity, they have a more highly protective lining: a stratified squamous epithelium like that of the skin.

Page 834
7. The unpaired laryngeal cartilages include the thyroid cartilage, cricoid cartilage, and epiglottis. The paired cartilages are the arytenoid cartilages, corniculate cartilages, and cuneiform cartilages.

8. The highly elastic vocal folds of the larynx are better known as the vocal cords.
9. When vocal fold tension increases, the pitch of the voice is raised.

Page 835
10. The trachea transports air between the larynx and primary bronchi; cilia and the mucus produced by epithelial cells also protect the respiratory tree by trapping inhaled debris and sweeping it toward the pharynx, where it is removed through coughing or swallowing.
11. The tracheal cartilages are C-shaped to allow room for esophageal expansion when food or liquid is swallowed.
12. Objects are more likely to be lodged in the right bronchus because it is slightly larger and more vertical than the left bronchus.

Page 842
13. Without surfactant, the alveoli would collapse as a result of surface tension in the thin layer of water that moistens the alveolar surfaces.
14. Air passing through the glottis flows into the larynx and through the trachea. From there, the air flows into a primary bronchus, which supplies the lungs. In the lungs, the air passes to bronchi, bronchioles, a terminal bronchiole, a respiratory bronchiole, an alveolar duct, an alveolar sac, an alveolus, and ultimately to the respiratory membrane.
15. The pulmonary arteries supply the exchange surfaces; the external carotid arteries, the thyrocervical trunks, and the bronchial arteries supply the conducting portions of the respiratory system.
16. The pleura is a serous membrane that secretes pleural fluid, which lubricates the opposing parietal and visceral surfaces to prevent friction during breathing.

Page 843
17. External respiration includes all the process involved in the exchange of oxygen and carbon dioxide between the body's interstitial fluids and the external environment. Internal respiration is the absorption of oxygen and the release of carbon dioxide by the body's cells.
18. The integrated steps involved in external respiration are pulmonary ventilation (breathing), gas diffusion, and the transport of oxygen and carbon dioxide.

Page 851
19. Compliance is the ease with which the lungs expand and contract. Factors affecting compliance include (**a**) the connective tissue structure of the lungs, (**b**) the level of surfactant production, and (**c**) the mobility of the thoracic cage.
20. The pulmonary volumes are resting tidal volume, expiratory reserve volume (ERV), residual volume, and inspiratory reserve volume (IRV).
21. When the rib penetrates Mark's chest wall, atmospheric air enters his thoracic cavity (a condition called pneumothorax), raising the pressure within the pleural cavity. As a result, the natural elasticity of the lung may cause the lung to collapse, a condition called atelectasis.
22. Because the fluid produced in pneumonia takes up space that would normally be occupied by air, vital capacity would decrease.

Page 855
23. Dalton's law states that each gas in a mixture exerts a pressure equal to its relative abundance.
24. Henry's law states that at a constant temperature, the quantity of a particular gas that will dissolve in a liquid is directly proportional to the partial pressure of that gas.

Page 859

25. Carbon dioxide is transported in the bloodstream as carbonic acid, bound to hemoglobin, or dissolved in the plasma.

26. The combination of increased temperature and lower pH (from heat and acidic waste products generated by active skeletal muscles) causes hemoglobin to release more oxygen than when the body is at rest.

27. Blockage of the trachea would interfere with the body's ability to gain oxygen and eliminate carbon dioxide. Because most carbon dioxide is transported in blood as bicarbonate ion formed from the dissociation of carbonic acid, an inability to eliminate carbon dioxide would result in an excess of hydrogen ions, which lowers blood pH.

Page 866

28. Exciting the pneumotaxic centers would inhibit the inspiratory and apneustic centers, which would result in shorter and more rapid breaths.

29. Peripheral chemoreceptors are more sensitive to carbon dioxide levels than to oxygen levels. When carbon dioxide dissolves, it produces hydrogen ions, thereby lowering pH and altering cell or tissue activity.

30. Johnny's mother shouldn't worry. When Johnny holds his breath, blood carbon dioxide levels increase, causing increased stimulation of the inspiratory center and forcing Johnny to breathe again.

31. Aging results in deterioration of elastic tissue, arthritic changes in rib articulations, decreased flexibility at costal cartilages, decreased vital capacity, and some degree of emphysema.

Page 867

32. The respiratory system provides oxygen and eliminates carbon dioxide for all body systems.

33. The respiratory system assists in pH regulation by eliminating carbon dioxide, and the urinary system eliminates organic wastes generated by cells of the respiratory system.

Answers to Review Questions

Page 872

Level 1 Reviewing Facts and Terms

1. (a) nasal cavity; (b) pharynx; (c) right lung; (d) nose; (e) larynx; (f) trachea; (g) bronchus; (h) bronchioles; (i) left lung
2. d 3. c 4. a 5. c 6. c
7. Since the air Brad is breathing is dry, large amounts of moisture are leaving the mucus in his respiratory tract to humidify inhaled air. Drying makes the mucus tacky and makes it difficult for the cilia to move, so mucus builds up, producing nasal congestion by morning. When Brad showers and drinks fluids, body water is replaced, so the mucus loosens up and can be moved along as usual.
8. The upper respiratory system consists of the nose, nasal cavity, paranasal sinuses, and pharynx. The lower respiratory system consists of the larynx, trachea, bronchi, bronchioles, and alveoli of the lungs.
9. The regions of the pharynx are the superior nasopharynx, where the nasal cavity opens into the pharynx; the middle oropharynx, posterior to the oral cavity; and the inferior laryngopharynx, which is posterior to the hyoid bone and glottis.
10. The thyroid cartilage forms the anterior walls of the larynx; the cricoid cartilage protects the glottis and the entrance to the trachea; the epiglottis forms a lid over the glottis; the arytenoid cartilages and the corniculate cartilages are involved in the formation of sound; and the cuneiform cartilages are found in the folds of the larynx.

11. The steps of external respiration are (a) pulmonary ventilation (breathing); (b) gas diffusion across the respiratory membrane and between blood and interstitial fluids; and (c) the storage and transport of oxygen and carbon dioxide.

12. Fetal hemoglobin has a higher affinity for oxygen than does adult hemoglobin. Thus, it binds more of the oxygen that is present, enabling it to "steal" oxygen from the maternal hemoglobin.

13. Carbon dioxide is transported in the blood as carbonic acid, bound to hemoglobin, and dissolved in the plasma.

Level 2 Reviewing Concepts

14. c 15. d 16. c 17. b
18. The nasal cavity cleanses, moistens, and warms inhaled air, whereas the mouth does not. Drier air entering through the mouth can irritate the trachea and cause throat soreness.
19. Smooth muscle tissue in the walls of bronchioles allows changes in airway diameter (bronchodilation or bronchoconstriction), which provides control of the flow and distribution of air within the lungs, just as vasodilation and vasoconstriction of the arterioles control blood flow and blood distribution.
20. Pneumocytes type II (septal cells) produce surfactant, which reduces surface tension in the fluid coating the alveolar surface. Without surfactant, the surface tension would be so high that the delicate alveoli would collapse.
21. Pulmonary ventilation, the physical movement of air into and out of the respiratory tract, maintains adequate alveolar ventilation. Alveolar ventilation, air movement into and out of the alveoli, prevents the buildup of carbon dioxide in the alveoli and ensures a continuous supply of oxygen that keeps pace with absorption by the bloodstream.
22. (a) Boyle's law describes the inverse relationship between pressure and volume: If volume decreases, pressure rises; if volume increases, pressure falls. It is the basis for the direction of air movement in pulmonary ventilation. (b) Dalton's law states that each of the gases that make up a mixture of gases contributes to the total pressure in proportion to its relative abundance; that is, all the partial pressures added together equal the total pressure exerted by the gas mixture. This relationship is the basis for the calculation of the partial pressures of oxygen and carbon dioxide, and their exchange between blood and alveolar air. (c) Henry's law states that, at a given temperature, the amount of a particular gas that dissolves in a liquid is directly proportional to the partial pressure of that gas. Henry's law underlies the diffusion of gases between capillaries, alveoli, and interstitial fluid.
23. Both sneezing and coughing involve a temporary cessation of respiration, known as apnea.
24. Pulmonary volumes are determined experimentally and include resting tidal volume (averaging 500 mL), expiratory reserve volume (approximately 1200 mL), residual volume (averaging 1200 mL), minimal volume (30–120 mL), and inspiratory reserve volume (approximately 3600 mL). Respiratory capacities include inspiratory capacity, functional residual capacity, vital capacity, and total lung capacity. The difference between the two measures is that respiratory capacities are the sums of various pulmonary volumes.
25. The DRG is the inspiratory center that contains neurons that control lower motor neurons innervating the external intercostal muscles and the diaphragm. The DRG functions in every respiratory cycle, whether quiet or forced. The VRG functions only during forced respiration—active exhalation and maximal inhalation. The neurons involved with active exhalation are sometimes said to form an expiratory center.

26. AVR = respiratory rate × (tidal volume − dead space). In this case, the dead air space is 200 mL (the anatomical dead air space plus the volume of the snorkel); therefore, AVR = respiratory rate × (500 − 200). To maintain an AVR of 6.0 L/min, or 6000 mL/minute, the respiratory rate must be 6000/(500 − 200), or 20 breaths per minute.

27. A person with chronic emphysema has constantly elevated blood levels of P_{CO_2} due to an inability to eliminate CO_2 efficiently because of physical damage to the lungs. Over time, the brain ignores the stimulatory signals produced by the increased CO_2 and begins to rely on information from peripheral chemoreceptors to set the pace of breathing. (In other words, accommodation has occurred.) The peripheral chemoreceptors also accommodate to the elevated CO_2 and respond primarily to the level of O_2 in the blood, increasing breathing when O_2 levels are low and decreasing breathing when O_2 levels are high. When pure O_2 was administered, chemoreceptors responded with fewer action potentials to the medulla oblongata, so Mr. B. stopped breathing.

28. Hyperventilation causes a decrease in alveolar P_{CO_2}, so more carbon dioxide is eliminated from the body, which upsets the body's acid–base balance by raising the pH. (Carbon dioxide in solution forms hydrogen ions and adds to the body's acid reserves.) The increase in pH causes dilation of peripheral blood vessels and decreased venous return to the heart. The resulting decreased cardiac output causes insufficient amounts of oxygen to reach the brain, causing Cary to lose consciousness.

29. In anemia, the blood's ability to carry oxygen is decreased due to the lack of functional hemoglobin, red blood cells, or both. Anemia does not interfere with the exchange of carbon dioxide within the alveoli, nor with the amount of oxygen that will dissolve in the plasma. Because chemoreceptors respond to dissolved gases and pH, as long as the pH and the concentrations of dissolved carbon dioxide and oxygen are normal, ventilation patterns should not change significantly.

30. The obstruction in Doris's right lung would prevent gas exchange. Thus, blood moving through the right lung would not be oxygenated and would retain carbon dioxide, which would lead to a lower blood pH than that of blood leaving the left lung. The lower pH for blood in the right lung would shift the oxygen–hemoglobin saturation curve to the left (the Bohr effect) as compared with the curve for the left lung.

Chapter 24

Answers to Checkpoints

Page 882

1. Organs of the digestive system include the esophagus, stomach, small intestine, large intestine, and accessory organs (salivary glands, liver, and pancreas).

2. The six primary functions of the digestive system include the following: (1) ingestion = consciously eating food; (2) mechanical processing = crushing and shearing foodstuffs to make them more susceptible to enzymatic attack; (3) digestion = the chemical breakdown of food into smaller products for absorption; (4) secretion = the release of water, acids, and other substances by the epithelium of the digestive tract and by glandular organs; (5) absorption = movement of digested particles across the digestive epithelium and into the interstitial fluid of the digestive tract; and (6) excretion = the removal of waste products from the body.

3. The mesenteries—sheets consisting of two layers of serous membrane separated by loose connective tissue—support and stabilize the organs in the abdominopelvic cavity and provide a route for the associated blood vessels, nerves, and lymphatic vessels.

4. The layers of the gastrointestinal tract, from superficial to deep, are the serosa, muscularis externa, submucosa, and mucosa (adjacent to the lumen).

5. The waves of contractions that constitute peristalsis are more efficient in propelling intestinal contents than segmentation, which is basically a churning action that mixes intestinal contents with digestive fluids.

6. A drug that blocks parasympathetic stimulation, which increases muscle tone and activity in the digestive tract, would slow peristalsis.

Page 888

7. Structures associated with the oral cavity include the tongue, salivary glands, and teeth.

8. The oral cavity is lined by a stratified squamous epithelium, which provides protection against friction or abrasion by foodstuffs.

9. Damage to the parotid salivary glands, which secrete the carbohydrate-digesting enzyme salivary amylase, would interfere with the digestion of complex carbohydrates.

10. The incisors are the teeth best suited for chopping (or cutting or shearing) pieces of relatively rigid food, such as raw vegetables.

11. The fauces is the dividing line between the oral cavity and the pharynx.

12. The pharynx is an anatomical space that receives a food bolus or liquids and passes them to the esophagus as part of the swallowing process.

13. Muscles associated with the pharynx are pharyngeal constrictor muscles, the palatopharyngeus and stylopharyngeus muscles, and palatal muscles.

Page 890

14. The structure connecting the pharynx to the stomach is the esophagus.

15. The muscularis externa of the esophagus is an unusual segment of the digestive tract because it (1) contains skeletal muscle cells along most of the length of the esophagus and (2) is surrounded by an adventitia rather than a serosa.

16. When the soft palate and larynx elevate and the glottis closes, swallowing (deglutition) is occurring.

Page 897

17. The four regions of the stomach are the cardia, fundus, body, and pylorus.

18. The low pH of the stomach creates an acidic environment that kills most microbes ingested with food, denatures proteins and inactivates most enzymes in food, helps break down plant cell walls and meat connective tissue, and activates pepsin.

19. Large (especially protein-containing) meals stimulate increased stomach acid secretion. Because the hydrogen ions of stomach acid come from blood entering the stomach, blood leaving the stomach will have fewer hydrogen ions and thus a higher pH. This phenomenon is referred to as the alkaline tide.

20. The vagus nerves contain parasympathetic motor fibers that can stimulate gastric secretions, even if food is not present in the stomach (the cephalic phase of gastric digestion). Cutting the branches of the vagus nerves that supply the stomach would prevent this type of secretion from occurring and thereby reduce the likelihood of ulcer formation.

Page 910

21. The three regions of the small intestine are the duodenum, jejunum, and ileum.

22. The small intestine has several adaptations that increase its surface area and thus its absorptive capacity. The walls of the small intestine are thrown into folds, the plicae circulares. The tissue that covers the plicae circulares forms fingerlike projections, the villi. The cells that cover the villi have an exposed surface covered by small fingerlike projections, the microvilli. In addition, the small intestine has a very rich supply of blood vessels and lymphatic vessels, which transport the nutrients that are absorbed.

23. A high-fat meal would raise the cholecystokinin level in the blood.

24. The hormone secretin, among other things, stimulates the pancreas to release fluid high in buffers to neutralize the chyme that enters the duodenum from the stomach. If the small intestine did not secrete secretin, the pH of the intestinal contents would be lower than normal.

25. Damage to the exocrine pancreas would most impair the digestion of fats (lipids), because it is the primary source of lipases. Even though such damage would also reduce carbohydrate and protein digestion, enzymes for digesting these nutrients are produced by other digestive system structures, including the salivary glands (carbohydrates), the small intestine (carbohydrates and proteins), and the stomach (proteins).

Page 916

26. The four regions of the colon are the ascending colon, transverse colon, descending colon, and sigmoid colon.

27. The large intestine is larger in diameter than the small intestine, but its relatively thin wall lacks villi and has an abundance of mucous cells and intestinal glands.

28. In mass movements, which occur a few times per day throughout the transverse colon and the distal portions of the large intestine, strong peristaltic contractions move material along the length of the colon. In haustral churning, segmentation movements mix the contents of adjacent haustra.

Page 921

29. Nutrients required by the body are carbohydrates, lipids, proteins, vitamins, minerals, and water.

30. Because chylomicrons are formed from the fats digested in a meal, fats increase the number of chylomicrons in the lacteals.

31. Removal of the stomach would interfere with the absorption of vitamin B_{12}, a process that requires intrinsic factor, produced by the parietal cells of the stomach.

32. When an individual with diarrhea loses fluid and electrolytes faster than they can be replaced, the resulting dehydration can be fatal. Although constipation can be quite uncomfortable, it does not interfere with any life-supporting processes; the few toxic waste products normally eliminated by the digestive system can move into the blood and be eliminated by the kidneys.

Page 922

33. General age-related digestive system changes include decreased secretory mechanisms, decreased gastric and intestinal motility, decreased mitotic activity of epithelial cells, and loss of tone; cumulative damage becomes more apparent, cancer rates increase, and dehydration occurs as a result of decreased osmoreceptor sensitivity.

34. The digestive system absorbs the organic substrates, vitamins, ions, and water required by cells of all other body systems.

35. The skeletal, muscular, nervous, endocrine, and cardiovascular systems may all be affected by inadequate absorption of calcium.

Answers to Review Questions

Page 926

Level 1 Reviewing Facts and Terms

1. d **2.** a **3.** d **4.** c

5. (**a**) oral cavity, teeth, tongue; (**b**) liver; (**c**) gallbladder; (**d**) pancreas; (**e**) large intestine; (**f**) salivary glands; (**g**) pharynx; (**h**) esophagus; (**i**) stomach; (**j**) small intestine; (**k**) anus

6. (**a**) mucosa; (**b**) submucosa; (**c**) muscularis externa; (**d**) serosa

7. e **8.** a **9.** a

10. (**a**) duodenum; (**b**) jejunum; (**c**) ileum

11. d **12.** c **13.** a

14. Digestion involves (1) ingestion; (2) mechanical processing; (3) secretion; (4) digestion (conversion into a form usable by cells); (5) absorption; and (6) excretion.

15. Layers of the digestive tract are (1) the mucosa: the epithelial layer that performs chemical digestion and absorption of nutrients; (2) the submucosa: the connective tissue layer containing lymphatic and blood vessels and the submucosal nerve plexus; (3) the muscularis externa: the smooth muscle layer containing the myenteric nerve plexus; and (4) the serosa: the outermost layer, epithelium and connective tissue that forms the visceral peritoneum (or connective tissue that forms the adventitia).

16. Activities of the digestive tract are regulated by neural, hormonal, and local mechanisms.

17. The three phases of swallowing—the buccal, pharyngeal, and esophageal phases—are controlled by the swallowing center of the medulla oblongata via the trigeminal and glossopharyngeal cranial nerves. The motor commands originating at the swallowing center are distributed by cranial nerves V, IX, X, and XII. Along the esophagus, primary peristaltic contractions are coordinated by afferent and efferent fibers within the glossopharyngeal and vagus cranial nerves, but secondary peristaltic contractions occur in the absence of CNS instructions.

18. The pancreas provides digestive enzymes, plus bicarbonate ions that elevate the pH of the chyme. The liver produces bile and is also the primary organ involved in regulating the composition of circulating blood. The gallbladder stores and releases bile, which contains additional buffers and bile salts that facilitate the digestion and absorption of lipids.

19. The hormones include the following: enterocrinin, which stimulates the submucosal glands of the duodenum; secretin, which stimulates the pancreas and liver to increase the secretion of water and bicarbonate ions; cholecystokinin (CCK), which causes an increase in the release of pancreatic secretions and bile into the duodenum, inhibits gastric activity, and appears to have CNS effects that reduce the sensation of hunger; gastric inhibitory peptide (GIP), which stimulates insulin release at pancreatic islets and the activity of the duodenal submucosal glands; vasoactive intestinal peptide (VIP), which stimulates the secretion of intestinal glands, dilates regional capillaries, and inhibits acid production in the stomach; gastrin, which is secreted by G cells in the duodenum when they are exposed to large quantities of incompletely digested proteins; and, in small quantities, motilin, which stimulates intestinal contractions, villikinin, which promotes the movement of villi and associated lymph flow, and somatostatin, which inhibits gastric secretion.

20. The large intestine reabsorbs water and compacts the intestinal contents into feces, absorbs important vitamins liberated by bacterial action, and stores fecal material prior to defecation.

21. Positive feedback loops in the defecation reflex involve (1) stretch receptors in the rectal walls, which promote a series of peristaltic contractions in the colon and rectum, moving feces toward the anus; and (2) the sacral parasympathetic system, also activated by the stretch receptors, which stimulates peristalsis via motor commands distributed by the pelvic nerves.

Level 2 Reviewing Concepts

22. e **23.** e **24.** a

25. A root canal involves drilling through the enamel and the dentin.

26. The stomach is protected from digestion by mucous secretions of its epithelial lining and by neural and hormonal control over the times and rates of acid secretion.

27. (1) The cephalic phase of gastric secretion begins with the sight or thought of food. Directed by the CNS, this phase prepares the stomach to receive food. (2) The gastric phase begins with the arrival of food in the stomach; this phase is initiated by distension of the stomach, an increase in the pH of the gastric contents, and the presence of undigested materials in the stomach. (3) The intestinal phase begins when chyme starts to enter the small intestine. This phase controls the rate of gastric emptying and ensures that the secretory, digestive, and absorptive functions of the small intestine can proceed reasonably efficiently.

28. After a heavy meal, bicarbonate ions pass from the parietal cells of the stomach into the extracellular fluid, causing the pH of the extracellular fluid to rise. As the extracellular fluid exchanges ions with the blood, the blood pH also increases.

Level 3 Critical Thinking and Clinical Applications

29. If a gallstone is small enough, it can pass through the common bile duct and block the pancreatic duct. Enzymes from the pancreas then cannot reach the small intestine. As the enzymes accumulate, they irritate the duct and ultimately the exocrine pancreas, producing pancreatitis.

30. The darker color of his urine is probably due to increased amounts of the pigment urobilin, which gives urine its normal yellow color. Urobilin is derived from urobilinogen, which is formed in the large intestine by the action of intestinal bacteria on bile pigments. In an intestinal obstruction, the bile pigments cannot be eliminated by their normal route, so a larger-than-normal amount diffuses into the blood, where it is eliminated by the kidneys.

31. If an individual cannot digest lactose, this sugar passes into the large intestine in an undigested form. The presence of extra sugar in the chyme increases its osmolarity, so less water is reabsorbed by the intestinal mucosa. The bacteria that inhabit the large intestine can metabolize the lactose, and in the process they produce large amounts of carbon dioxide. This gas overstretches the intestine, which stimulates local reflexes that increase peristalsis. The combination of more-fluid contents and increased peristalsis causes diarrhea. The overexpansion of the intestine by gas, which is directly related to increased gas production by the bacteria, causes the severe pain and abdominal cramping.

32. The primary effect of such surgeries would be a reduction in the volume of food (and thus in the amount of calories) consumed because the person feels full after eating a small amount. This can result in significant weight loss.

Chapter 25

Answers to Checkpoints

Page 932

1. Metabolism is the sum of all biochemical processes under way within the human body; it includes anabolism and catabolism.

2. Catabolism is the breakdown of complex organic molecules into simpler components, accompanied by the release of energy.

3. Anabolism is the synthesis of complex organic compounds from simpler precursors.

Page 941

4. The primary role of the TCA cycle in ATP production is to transfer electrons from substrates to coenzymes. These electrons provide energy for the production of ATP by the electron transport system.

5. The NADH produced by glycolysis cannot enter the mitochondria, where the enzymes of the electron transport chain are located. However, an intermediary in the mitochondrial membrane can transfer the electrons from the NADH to a coenzyme within the mitochondria. In skeletal muscle cells, the intermediary transfers the electrons to FAD, whereas cardiac muscle cells use a different intermediary, which transfers the electrons to another NAD. In mitochondria, each NADH yields 3 molecules of ATP, whereas each $FADH_2$ yields just 2 molecules of ATP. The different intermediaries account for the difference in ATP yield.

6. A decrease in cytoplasmic NAD would reduce ATP production in mitochondria. Decreased NAD would reduce the amount of pyruvic acid produced by glycolysis; less pyruvic acid means that the TCA cycle could produce less ATP in mitochondria.

Page 946

7. Beta-oxidation is fatty acid catabolism that produces molecules of acetyl-CoA.

8. The five major groups of lipoproteins are chylomicrons, very low-density lipoproteins (VLDLs), intermediate-density lipoproteins (IDLs), low-density lipoproteins (LDLs), and high-density lipoproteins (HDLs).

9. High-density lipoproteins (HDLs) are considered beneficial because they reduce the amount of fat (including cholesterol) in the bloodstream by transporting it to the liver for storage or excretion in the bile.

Page 948

10. Transamination is the process by which one amino acid is formed from another; it occurs in the catabolism of amino acids and in the synthesis of proteins.

11. Deamination is the removal of an amino group from an amino acid.

12. A diet deficient in pyridoxine (vitamin B_6), an important coenzyme in deaminating and transaminating amino acids in cells, would interfere with the body's ability to metabolize proteins.

Page 950

13. Glycogenesis (the formation of glycogen) in the liver increases after a high-carbohydrate meal.

14. Blood levels of urea increase during the postabsorptive state because the deamination of many amino acids at that time yields ammonia, which is then converted to urea by the liver.

15. Accumulated acetyl-CoA is likely to be converted into ketone bodies.

Page 955

16. The two classes of vitamins are fat-soluble and water-soluble.

17. An athlete who is adding muscle mass through extensive training would try to maintain a positive nitrogen balance.

18. A decrease in the amount of bile salts, which are necessary for digesting and absorbing fats and fat-soluble vitamins (including vitamin A), would result in less vitamin A in the body, and perhaps a vitamin A deficiency.

Page 961

19. The BMR of a pregnant woman would be higher than her own BMR when she is not pregnant, due to both the increased metabolism associated with supporting the fetus and the contribution of fetal metabolism.

20. The vasoconstriction of peripheral vessels would decrease blood flow to the skin and thus the amount of heat the body can lose. As a result, body temperature would increase.

21. Infants have higher surface-to-volume ratios than do adults, and the body's temperature-regulating mechanisms are not fully functional at birth. As a result, infants must expend more energy to maintain body temperature, and they get cold more easily than do healthy adults.

Answers to Review Questions

Page 964

Level 1 Reviewing Facts and Terms

1. d **2.** a **3.** d **4.** a **5.** b

6. (**a**) fatty acids; (**b**) glucose; (**c**) amino acids; (**d**) small carbon chains; (**e**) TCA cycle; (**f**) CO_2; (**g**) ATP; (**h**) electron transport system; (**i**) H_2O

7. d **8.** a

9. Lipoproteins are lipid–protein complexes that contain large insoluble glycerides and cholesterol, with a superficial coating of phospholipids and proteins. The major groups are chylomicrons, which consist of 95 percent triglyceride, are the largest lipoproteins, and carry absorbed lipids from the intestinal tract to the bloodstream; very low-density lipoproteins (VLDLs), which consist of triglyceride, phospholipid, and cholesterol, and transport triglycerides to peripheral tissues; intermediate-density lipoproteins (IDLs), which are intermediate in size and composition between VLDLs and LDLs; low-density lipoproteins (LDLs, or "bad cholesterols"), which are mostly cholesterol and deliver cholesterol to peripheral tissues; and high-density lipoproteins (HDLs, or "good cholesterols"), which are equal parts protein and lipid (cholesterol and phospholipids) and transport excess cholesterol to the liver for storage or excretion in bile.

Level 2 Reviewing Concepts

10. Oxidative phosphorylation is the generation of ATP within mitochondria, through a process called chemiosmosis that requires coenzymes, ATP synthase, and consumes oxygen. The electron transport system consists of a sequence of metalloproteins called cytochromes, embedded in the inner mitochondrial membrane. Energy from the stepwise passage of electrons (from H atoms) along the cytochrome molecules is used to pump hydrogen ions from the matrix into the intermembrane space to form a proton gradient across the inner mitochondrial membrane. The hydrogen ions diffuse back through ATP synthase and generate ATP. The electrons, hydrogen ions, and oxygen combine to produce water as a by-product.

11. In lipid catabolism, a triglyceride is hydrolyzed, yielding glycerol and fatty acids. Glycerol is converted to pyruvic acid and enters the TCA cycle. Fatty acids are broken into two-carbon fragments by beta-oxidation inside mitochondria. The two-carbon compounds then enter the TCA cycle.

12. During the absorptive state, insulin prevents a large surge in blood glucose after a meal by causing the liver to remove glucose from the hepatic portal circulation. During the postabsorptive state, blood glucose begins to decline, triggering the liver to release glucose via glycogenolysis and gluconeogenesis.

13. Liver cells can break down or synthesize most carbohydrates, lipids, and amino acids. Because the liver has an extensive blood supply, it can easily monitor blood composition of these nutrients and regulate them accordingly. The liver also stores energy in the form of glycogen.

14. Vitamins and minerals are essential components of the diet because the body cannot synthesize most of the vitamins and minerals it requires.

15. These terms refer to lipoproteins in the blood that transport cholesterol. "Good cholesterol" (high-density lipoproteins, or HDLs) transports excess cholesterol to the liver for storage or breakdown, whereas "bad cholesterol" (low-density lipoproteins, or LDLs) transports cholesterol to peripheral tissues, which unfortunately may include the arteries. The buildup of cholesterol in the arteries is linked to cardiovascular disease.

Level 3 Critical Thinking and Clinical Applications

16. c

17. Based just on the information given, Charlie would appear to be in good health, at least relative to his diet and probable exercise. Problems are associated with elevated levels of LDLs, which carry cholesterol to peripheral tissues and make it available for the formation of atherosclerotic plaques in blood vessels. High levels of HDLs indicate that a considerable amount of cholesterol is being removed from the peripheral tissues and carried to the liver for disposal. You would encourage Charlie not to change, and keep up the good work.

18. It appears that Jill is suffering from ketoacidosis as a consequence of her anorexia. Because she is literally starving herself, her body is metabolizing large amounts of fatty acids and amino acids to provide energy, and in the process is producing large quantities of ketone bodies (normal metabolites from these catabolic processes). One of the ketones formed is acetone, which can be eliminated through the lungs. This accounts for the smell of aromatic hydrocarbons on Jill's breath. The ketones are also converted into keto acids such as acetic acid. In large amounts, this lowers the body's pH and begins to exhaust the alkaline reserves of the buffer system. This is probably the cause of her arrhythmias.

Chapter 26

Answers to Checkpoints

Page 966

1. The primary functions of the urinary system include (1) the excretion of organic wastes from body fluids, (2) the elimination of body wastes to the exterior, and (3) the regulation of homeostasis by maintaining the volume and solute concentration of blood plasma.

2. The urinary system consists of two kidneys, two ureters, a bladder, and a urethra.

3. Micturition, or urination, is the elimination of urine from the body.

Page 975

4. The renal corpuscle, proximal convoluted tubule, distal convoluted tubule, and the proximal portions of the nephron loop and collecting duct are all in the renal cortex. (In cortical nephrons, most of the nephron loops are in the cortex; in juxtamedullary nephrons, most of the nephron loops are in the medulla.)

5. Plasma proteins are too large to pass through the pores of the glomerular capillaries; only the smallest plasma proteins can pass through the filtration slits of the podocytes.

6. Damage to the juxtaglomerular complex of the nephron would interfere with the hormonal control of blood pressure.

Page 980

7. The three distinct processes of urine formation in the kidney are filtration, reabsorption, and secretion.

8. When the plasma concentration of a substance exceeds its tubular maximum, the excess is not reabsorbed, so it is excreted in the urine.

Page 984

9. The primary nephron structures involved in filtration are the glomerular capillaries, the dense layer, and the filtration slits of the podocytes.

10. The factors that influence filtration pressure are hydrostatic pressure and colloid osmotic pressure.

11. The factors that influence the rate of filtrate formation are the filtration pressure across glomerular capillaries, plus interactions among autoregulation, hormonal regulation, and autonomic regulation.

12. A decrease in blood pressure would decrease the GFR by reducing the blood hydrostatic pressure within the glomerulus.

Page 997

13. Increased amounts of aldosterone, which promotes Na^+ retention and K^+ secretion at the kidneys, would elevate the K^+ concentration of urine.

14. If the concentration of Na^+ in the filtrate decreased, fewer hydrogen ions could be secreted via the countertransport mechanism involving these two ions. As a result, the pH of the tubular fluid would increase.

15. Without juxtamedullary nephrons, a large osmotic gradient could not exist in the medulla, and the kidneys would be unable to form concentrated urine.

16. When the amount of Na^+ in the tubular fluid passing through the distal convoluted tubule decreases, the cells of the macula densa are stimulated to release renin. The resulting activation of angiotensin II increases blood pressure.

Page 1002

17. Digestion of a high-protein diet would lead to increased production of urea, a nitrogenous waste formed from the metabolism of amino acids during the breakdown of proteins. As a result, the urine would contain more urea, and urine volume might also increase as a result of the need to flush the excess urea.

18. An obstruction of a ureter would interfere with the passage of urine between the renal pelvis and the urinary bladder.

19. The ability to control the micturition reflex depends on your ability to control the external urethral sphincter, a ring of skeletal muscle formed by the urogenital diaphragm, which acts as a valve.

Page 1003

20. Four general age-related changes in the urinary system are a decline in the number of functional nephrons, a reduction in the GFR, a reduced sensitivity to ADH, and problems with the micturition reflex.

21. Nephrolithiasis is the formation of renal calculi, or kidney stones.

22. Incontinence may develop in elderly individuals as a result of either decreased muscle tone of sphincter muscles controlling micturition or disorders of the CNS affecting the ability to control urination.

23. The urinary system excretes waste products of, and maintains normal body fluid pH and ion composition for, all other body systems.

24. Components of the body's excretory system include the urinary, integumentary, respiratory, and digestive systems.

Answers to Review Questions
Page 1007

Level 1 Reviewing Facts and Terms

1. (a) renal sinus; **(b)** renal pelvis; **(c)** hilum; **(d)** renal papilla; **(e)** ureter; **(f)** renal cortex; **(g)** renal medulla; **(h)** renal pyramid; **(i)** minor calyx; **(j)** major calyx; **(k)** renal lobe; **(l)** renal columns; **(m)** fibrous capsule

2. a **3.** e **4.** c **5.** d **6.** d

7. The urinary system performs vital excretory functions and eliminates the organic waste products generated by cells throughout the body. It also regulates the volume and solute concentration of body fluids.

8. The kidneys, ureters, urinary bladder, and urethra are the components of the urinary system.

9. renal corpuscle (glomerulus/glomerular capsule) → proximal convoluted tubule → nephron loop → distal convoluted tubule → collecting duct → papillary duct → renal pelvis

10. proximal convoluted tubule: reabsorbs all the useful organic substrates from the filtrate; nephron loop: reabsorbs over 90 percent of the water in the filtrate; and distal convoluted tubule: secretes into the filtrate waste products that were missed by filtration

11. The juxtaglomerular complex secretes the enzyme renin and the hormone erythropoietin.

12. renal artery → segmental arteries → interlobar arteries → arcuate arteries → cortical radiate arteries → afferent arterioles → nephrons → venules → cortical radiate veins → arcuate veins → interlobar veins → renal vein

13. Processes in urine production are (1) filtration: the selective removal of large solutes and suspended materials from a solution on the basis of size; requires a filtration membrane and hydrostatic pressure, as provided by gravity or by blood pressure; (2) reabsorption: the removal of water and solute molecules from the filtrate after it enters the renal tubules; and (3) secretion: the transport of solutes from the peritubular fluid, across the tubular epithelium, and into the tubular fluid

14. In peripheral capillary beds, angiotensin II causes powerful vasoconstriction of precapillary sphincters, elevating pressures in the renal arteries and their tributaries. At the nephron, angiotensin II causes the efferent arteriole to constrict, elevating glomerular pressures and filtration rates. At the PCT, it stimulates the reabsorption of sodium ions and water. In the CNS, angiotensin II triggers the release of ADH, stimulating the reabsorption of water in the distal portion of the DCT and the collecting system, and it causes the sensation of thirst. At the suprarenal gland, angiotensin II stimulates the secretion of aldosterone by the cortex. Aldosterone accelerates sodium reabsorption in the DCT and the cortical portion of the collection system.

15. Structures responsible for the transport, storage, and elimination of urine are the ureters, urinary bladder, and urethra.

Level 2 Reviewing Concepts

16. d **17.** d **18.** c **19.** a **20.** a

21. Proteins are excluded from filtrate because they are too large to fit through the slit pores. Keeping proteins in the plasma ensures that blood colloid osmotic pressure will oppose filtration and return water to the plasma.

22. Controls on GFR are autoregulation at the local level, hormonal regulation initiated by the kidneys, and autonomic regulation (by the sympathetic division of the ANS).

23. As a result of facilitated diffusion and cotransport mechanisms in the PCT, 99 percent of the glucose, amino acids, and other nutrients are reabsorbed before the filtrate leaves the PCT. A reduction of the solute concentration of the tubular fluid occurs due to active ion reabsorption of sodium, potassium, calcium, magnesium, bicarbonate, phosphate, and sulfate ions. The passive diffusion of urea, chloride ions, and lipid-soluble materials further reduces the solute concentration of the tubular fluid and promotes additional water reabsorption.

24. Countercurrent multiplication (1) is an efficient way to reabsorb solutes and water before the tubular fluid reaches the DCT and collecting system, and (2) establishes a concentration gradient that permits the passive reabsorption of water from urine in the collecting system.

25. The urge to urinate usually appears when the urinary bladder contains about 200 mL of urine. The micturition reflex begins to function when the stretch receptors have provided adequate stimulation to the parasympathetic motor neurons. The activity in the motor neurons generates action potentials that reach the smooth muscle in the wall of the urinary bladder. These efferent impulses travel over the pelvic nerves, producing a sustained contraction of the urinary bladder.

Level 3 Critical Thinking and Clinical Applications
26. d   **27.** c

28. Increasing the volume of urine produced decreases the total blood volume of the body, which in turn leads to a decreased blood hydrostatic pressure. Edema frequently results when the hydrostatic pressure of the blood exceeds the opposing forces at the capillaries in the affected area. Depending on the actual cause of the edema, decreasing the blood hydrostatic pressure would decrease edema formation and possibly cause some of the fluid to move from the interstitial spaces back to the blood.

29. Renal arteriosclerosis restricts blood flow to the kidneys and produces renal ischemia. Decreased blood flow and ischemia triggers the juxtaglomerular complex to produce more renin, which leads to elevated levels of angiotensin II and aldosterone. Angiotensin II causes vasoconstriction, increased peripheral resistance, and thus increased blood pressure. The aldosterone promotes sodium retention, which leads to more water retained by the body and an increase in blood volume. This too contributes to a higher blood pressure. Additionally, in response to tissue hypoxia, erythropoietin release is increased, stimulating the formation of red blood cells, which leads to increased blood viscosity and again contributes to hypertension.

30. Because mannitol is filtered but not reabsorbed, drinking a mannitol solution would lead to an increase in the osmolarity of the filtrate. Less water would be reabsorbed, and an increased volume of urine would be produced.

31. Carbonic anhydrase catalyzes the reaction that forms carbonic acid, a source of hydrogen ions that are excreted by the kidneys. Hydrogen ion excretion is accomplished by an antiport system in which sodium ions are exchanged for hydrogen ions. Fewer hydrogen ions would be available, so less sodium would be reabsorbed, contributing to an increased osmolarity of the filtrate. In turn, an increased volume of urine and more-frequent urination would result.

Chapter 27

Answers to Checkpoints

Page 1011
1. The three interrelated processes essential to stabilizing body fluid volume are fluid balance, electrolyte balance, and acid–base balance.

2. The components of extracellular fluid (ECF) are interstitial fluid, plasma, and other body fluids; the cytosol comprises the intracellular fluid (ICF).

Page 1014
3. Three hormones affecting fluid and electrolyte balance are antidiuretic hormone (ADH), aldosterone, and natriuretic peptides (ANP and BNP).

4. Drinking a pitcher of distilled water would temporarily lower your blood osmolarity (osmotic concentration). Because ADH release is triggered by increases in osmolarity, a decrease in osmolarity would lead to a decrease in the level of ADH in your blood.

Page 1017
5. Edema is the movement of abnormally large amounts of water from plasma into interstitial fluid.

6. A fluid shift is a rapid movement of water between the ECF and ICF in response to an osmotic gradient.

7. Dehydration is a reduction in the water content of the body that develops when water losses outpace water gains and threaten homeostasis.

8. Being in the desert without water, you would lose fluid through perspiration, urine formation, and respiration. As a result, the osmotic concentration of your plasma (and other body fluids) would increase.

Page 1022
9. Important cations are sodium, potassium, calcium, and magnesium; important anions are phosphate and chloride.

10. Sweat is a hypotonic solution with lower sodium concentration than the ECF. Sweating causes a greater loss of water than of sodium, increasing plasma sodium ion levels.

11. Potassium ion imbalances are more dangerous than sodium ion imbalances because they can lead to extensive muscle weakness or even paralysis when plasma concentrations are too low, and to cardiac arrhythmias when the levels are too high. Disturbances in sodium balance, by contrast, produce dehydration or tissue edema.

Page 1030
12. The body's three major buffer systems are the protein buffer systems, the carbonic acid–bicarbonate buffer system, and the phosphate buffer system.

13. A decrease in the pH of body fluids would stimulate the respiratory centers of the medulla oblongata, resulting in an increase in the respiratory rate.

14. Tubular fluid in nephrons must be buffered so that the filtrate can contain more H^+ without decreasing the pH below approximately 4.5, at which point H^+ secretion cannot continue because the H^+ concentration gradient becomes too large.

Page 1036
15. In a prolonged fast, fatty acids are mobilized, producing large numbers of ketone bodies, which are acids that lower the body's pH. (The lowered pH would eventually lead to ketoacidosis.)

16. In vomiting, large amounts of stomach hydrochloric acid are lost from the body. This acid is formed by the parietal cells of the stomach by taking H^+ from the blood. Prolonged vomiting would lead to the excessive removal of H^+ from the blood to produce the hydrochloric acid, thus raising the body's pH and leading to metabolic alkalosis.

Page 1037
17. Declines in glomerular filtration rate and the number of functional nephrons reduce the body's ability to regulate pH through renal compensation.

18. Total body water content gradually decreases with age.

Answers to Review Questions
Page 1039

Level 1 Reviewing Facts and Terms
1. c **2.** a **3.** d
4. (a) carbonic acid (H_2CO_3); (b) bicarbonate ion (HCO_3^-);
(c) sodium bicarbonate ($NaHCO_3$)
5. d **6.** d **7.** a **8.** b **9.** a
10. Four major hormones involved in fluid and electrolyte balance are (1) antidiuretic hormone (ADH): stimulates water conservation at the kidneys and stimulates the thirst center; (2) aldosterone: determines the rate of sodium reabsorption and potassium secretion along the DCT and collecting system of the kidney; and the natriuretic peptides, (3) ANP and (4) BNP: reduce thirst, promote the loss of Na^+ and water at the kidneys, and block the release of ADH and aldosterone.

Level 2 Reviewing Concepts
11. d **12.** a **13.** d **14.** c
15. Fluid balance is a state in which the amount of water gained each day is equal to the amount lost to the environment. It is vital that the water content of the body remain stable, because water is an essential ingredient of cytoplasm and accounts for about 99 percent of ECF volume. Electrolyte balance exists when there is neither a net gain nor a net loss of any ion in body fluids. It is important that the ionic concentrations in body water remain within normal limits; if levels of calcium or potassium become too high, for instance, cardiac arrhythmias can develop. Acid–base balance exists when the production of hydrogen ions precisely offsets their loss. The pH of body fluids must remain within a relatively narrow range; variations outside this range can be life threatening.
16. Fluid shifts are rapid water movements between the ECF and the ICF that occur in response to increases or decreases in the osmotic concentration of the ECF. Such water movements dampen extreme shifts in electrolyte balance.
17. Individuals with a fever should increase fluid intake because for each degree (Celsius) the body temperature rises above normal, daily water loss increases by 200 mL.
18. (a) Acids that can leave solution and enter the atmosphere, such as carbonic acid, are volatile acids. (b) Acids that do not leave solution, such as sulfuric acid, are fixed acids. (c) Acids produced during metabolism, such as lactic acid, are organic acids. Volatile acids are the greatest threat because of the large amounts generated by normal cellular processes.
19. (1) protein buffer systems: These depend on the ability of amino acids to respond to changes in pH by accepting or releasing hydrogen ions. If the pH rises, the carboxyl group of the amino acid dissociates to release a hydrogen ion; if the pH drops, the amino group accepts an additional hydrogen ion to form an NH_4^+ group. Plasma proteins contribute to the buffering capabilities of the blood; inside cells, protein buffer systems stabilize the pH of the ECF by absorbing extracellular hydrogen ions or exchanging intracellular hydrogen ions for extracellular potassium. (2) carbonic acid–bicarbonate system: Most carbon dioxide generated in tissues is converted to carbonic acid, which dissociates into a hydrogen ion and a bicarbonate ion. Hydrogen ions released by dissociation of organic or fixed acids combine with bicarbonate ions, elevating the P_{CO_2}; additional CO_2 is lost at the lungs. (3) phosphate buffer system: This buffer system consists of $H_2PO_4^-$, a weak acid that, in solution, reversibly dissociates into a hydrogen ion and HPO_4^{2-}. The phosphate buffer system plays a relatively small role in regulating the pH of

the ECF, because the ECF contains far higher concentrations of bicarbonate ions than phosphate ions; however, it is important in buffering the pH of the ICF.
20. Respiratory and renal mechanisms support buffer systems by secreting or absorbing hydrogen ions, by controlling the excretion of acids and bases, and by generating additional buffers.
21. Respiratory compensation is a change in the respiratory rate that helps stabilize the pH of the ECF. Increasing or decreasing the rate of respiration alters pH by lowering or raising the P_{CO_2}. When the P_{CO_2} declines, the pH rises; when the P_{CO_2} increases, the pH decreases. Renal compensation is a change in the rates of hydrogen and bicarbonate ion secretion or reabsorption in response to changes in plasma pH. Tubular hydrogen ion secretion results in the diffusion of bicarbonate ions into the ECF.
22. Respiratory disorders result from abnormal CO_2 levels in the ECF. An imbalance exists between the rate of CO_2 removal at the lungs and its generation in other tissues. Metabolic disorders are caused by the generation of organic or fixed acids or by conditions affecting the concentration of bicarbonate ions in the ECF.
23. Respiratory acidosis, which results from an abnormally high level of carbon dioxide (hypercapnia), is usually caused by hypoventilation. Metabolic acidosis, which occurs when bicarbonate ion levels fall, can result from overproduction of fixed or organic acids, impaired ability to secrete H^+ ions at the kidney, or severe bicarbonate loss.
24. Excessive salt intake causes an increase in total blood volume and blood pressure due to an obligatory increase in water absorption across the intestinal lining and recall of fluid from the ICF.
25. Since sweat is usually hypotonic, the loss of a large volume of sweat causes hypertonicity in body fluids. The loss of fluid volume is primarily from the interstitial space, which leads to a reduction in plasma volume and an increase in the hematocrit. Severe dehydration can cause the blood viscosity to increase substantially, resulting in an increased workload on the heart, ultimately increasing the probability of heart failure.

Level 3 Critical Thinking and Clinical Applications
26. The young boy has metabolic and respiratory acidosis. Sustained hypoventilation during drowning leads to decreased arterial P_{CO_2}, and the generation of large quantities of lactic acid by oxygen-starved tissues. Prompt emergency treatment is essential; the usual procedure involves some form of artificial or mechanical respiratory assistance coupled with the intravenous infusion of an isotonic solution containing sodium lactate, sodium gluconate, or sodium bicarbonate.
27. a
28. When tissues are burned, cells are destroyed and their cytoplasmic contents leak into the interstitial fluid and then move into the plasma. Since potassium ions are normally found within the cell, damage to a large number of cells releases relatively large amounts of potassium ions into the blood. The elevated potassium level stimulates cells of the suprarenal cortex to produce aldosterone and cells of the juxtaglomerular complex to produce renin. The renin activates the angiotensin mechanism. Ultimately, angiotensin II stimulates increased aldosterone secretion, which promotes sodium retention and potassium secretion by the kidneys, thus accounting for the elevated potassium levels in the patient's urine.
29. Digestive secretions contain high levels of bicarbonate, so individuals with diarrhea can lose significant amounts of this importation, leading to acidosis. We would expect Milly's blood pH to be lower than 7.35, and that of her urine to be low

(due to increased renal excretion of hydrogen ions). We would also expect an increase in the rate and depth of breathing as the respiratory system tries to compensate by eliminating carbon dioxide.

30. The hypertonic solution will cause fluid to move from the ICF to the ECF, further aggravating Yuka's dehydration. The slight increase in pressure and osmolarity of the blood should lead to an increase in ADH, even though ADH levels are probably quite high already. Urine volume would probably increase, because much of the glucose would not be reabsorbed. The osmolarity of the tubular filtrate would increase, decreasing water reabsorption and increasing urine volume.

31. Patient 1 has compensated respiratory alkalosis. Patient 2 has acute metabolic acidosis due to the generation or retention of organic or fixed acids. Patient 3 has acute respiratory acidosis. Patient 4 has metabolic alkalosis.

Chapter 28

Answers to Checkpoints

Page 1042

1. A gamete is a functional male or female reproductive cell.
2. Basic components of the reproductive system are gonads (reproductive organs), ducts (which receive and transport gametes), accessory glands (which secrete fluids), and external genitalia (perineal structures).
3. Gonads are reproductive organs that produce gametes and hormones.

Page 1058

4. Male reproductive structures are the scrotum, testes, epididymides, ductus deferens, ejaculatory duct, urethra, seminal glands, prostate gland, bulbo-urethral gland, and penis.
5. On a warm day, the cremaster muscle (as well as the dartos muscle) would be relaxed so that the scrotal sac could descend away from the warmth of the body, thereby cooling the testes.
6. When the arteries within the penis dilate, the increased blood flow causes the vascular chambers to engorge with blood, producing an erection.
7. Low FSH levels would lead to low levels of testosterone in the seminiferous tubules, reducing both the sperm production rate and sperm count.
8. The route of sperm is as follows: seminiferous tubules → rete testis → efferent ductules → epididymides → ductus deferens → ejaculatory duct → urethra.

Page 1077

9. Structures of the female reproductive system include ovaries, uterine tubes, uterus, vagina, and mammary glands.
10. The blockage of both uterine tubes would cause sterility.
11. The acidic pH of the vagina helps prevent bacterial, fungal, and parasitic infections in this region.
12. The functional layer of the endometrium sloughs off during menstruation.
13. Blockage of a single lactiferous sinus would not interfere with the delivery of milk to the nipple, because each breast generally has 15–20 lactiferous sinuses.
14. If the LH surge did not occur during an ovarian cycle, ovulation and corpus luteum formation would not occur.
15. Blockage of progesterone receptors in the uterus would inhibit the development of the endometrium, making the uterus unprepared for pregnancy.

16. A decline in the levels of estrogens and progesterone signals the beginning of menses, the end of the uterine cycle.

Page 1079

17. The physiological events of sexual intercourse in both sexes are arousal, erection, lubrication, orgasm, and detumescence; emission and ejaculation are additional phases that occur only in males.
18. An inability to contract the ischiocavernosus and bulbospongiosus muscles would interfere with a male's ability to ejaculate and to experience orgasm.
19. Parasympathetic stimulation in females during sexual arousal causes **(a)** engorgement of the erectile tissue of the clitoris, **(b)** increased secretion of cervical and vaginal glands, **(c)** increased blood flow to the wall of the vagina, and **(d)** engorgement of the blood vessels in the nipples.

Page 1080

20. Menopause is the time that ovulation and menstruation cease, typically around age 45–55.
21. At menopause, circulating estrogen levels begin to drop. Estrogen has an inhibitory effect on FSH (and on GnRH). As the level of estrogen declines, the levels of FSH rise and remain high.
22. The male climacteric, or andropause, is a period of declining reproductive function in men, typically between the ages of 50 and 60.

Page 1081

23. The cardiovascular system distributes reproductive hormones; provides nutrients, oxygen, and waste removal for the fetus; and produces local blood pressure changes responsible for the physical changes that occur during sexual intercourse. The reproductive system supplies estrogens that may help maintain healthy blood vessels and slow the development of atherosclerosis.
24. Pelvic bones protect reproductive organs in females and portions of the ductus deferens and accessory glands in males; sex hormones stimulate growth and maintenance of bone, and accelerate growth and closure of epiphyseal cartilages at puberty.

Answers to Review Questions

Page 1084

Level 1 Reviewing Facts and Terms

1. (a) prostatic urethra; **(b)** spongy urethra; **(c)** ductus deferens; **(d)** penis; **(e)** epididymis; **(f)** testis; **(g)** external urethral orifice; **(h)** scrotum; **(i)** seminal gland; **(j)** prostate gland; **(k)** ejaculatory duct; **(l)** bulbo-urethral gland.
2. (a) ovary; **(b)** uterine tube; **(c)** greater vestibular gland; **(d)** clitoris; **(e)** labium minus; **(f)** labium majus; **(g)** myometrium; **(h)** perimetrium; **(i)** endometrium; **(j)** uterus; **(k)** fornix; **(l)** cervix; **(m)** vagina.
3. c **4.** d **5.** c **6.** b **7.** a
8. The accessory organs/glands include: the seminal glands, which provide the nutrients sperm need for motility, prostaglandins that stimulate smooth muscle contractions along the male and female reproductive tracts thereby propelling sperm and fluids, fibrinogen that temporarily clots the ejaculate within the vagina, and buffers that counteract the acidity of the prostatic secretions and urethral and vaginal contents; the prostate gland, which aids the activation of the spermatozoa (with seminal gland secretions); and the bulbo-urethral glands, which buffer acids in the penile urethra and lubricate the glans of the penis. The fluids secreted by the accessory glands make up about 95 percent of the volume of semen.

9. Interstitial cells (cells of Leydig) produce male sex hormones (androgens), the most important of which is testosterone; nurse cells maintain the blood–testis barrier, support spermatogenesis and spermiogenesis, and secrete inhibin, androgen-binding protein, and Müllerian-inhibiting factor.

10. The three regions of the male urethra are the prostatic urethra, membranous urethra, and spongy urethra.

11. In males, testosterone stimulates spermatogenesis and promotes the functional maturation of spermatozoa; maintains the male accessory reproductive organs; determines male secondary sex characteristics; stimulates metabolic operations, especially those concerned with protein synthesis and muscle growth; and influences brain development by stimulating sexual behaviors and sexual drive.

12. Steps of the ovarian cycle are (1) the formation of primary follicles, (2) the formation of secondary follicles, (3) the formation of a tertiary follicle, (4) ovulation, and (5) the formation and degeneration of the corpus luteum.

13. The myometrium is the outer, muscular layer; the endometrium is the inner, glandular layer; and the perimetrium is an incomplete serosal layer.

14. The clitoris—a part of the external genitalia of females that is derived from the same embryonic structures as the penis—contains erectile tissue that becomes engorged with blood during sexual arousal and provides pleasurable sensations.

15. Route of milk flow: secretory lobules of glandular tissue (lobes) → ducts → lactiferous duct → lactiferous sinus, which opens onto the surface of the nipple.

Level 2 Reviewing Concepts

16. c **17.** e

18. Males produce gametes from puberty until death; females produce gametes only from menarche to menopause. Males produce many gametes at a time; females typically produce one or two per 28-day cycle. Males release mature gametes that have completed meiosis; females release secondary oocytes held in metaphase of meiosis II.

19. The corpora cavernosa extend along the length of the penis as far as the neck of the penis, and the erectile tissue within each corpus cavernosum surrounds a central artery. The slender corpus spongiosum surrounds the urethra and extends from the superficial fascia of the urogenital diaphragm to the tip of the penis, where it expands to form the glans. The sheath surrounding the corpus spongiosum contains more elastic fibers than do the corpora cavernosa, and the erectile tissue contains a pair of arteries. When parasympathetic neurons innervating the penile arteries release nitric oxide, smooth muscles in the arterial walls relax, dilating the vessels and increasing blood flow; the resulting engorgement of the vascular channels with blood causes erection of the penis.

20. (1) Menses, the interval marked by the degeneration and loss of the functional zone of the endometrium, lasts 1–7 days, and 35–50 mL (1.2–1.7 oz.) of blood is lost. (2) The proliferative phase features growth and vascularization resulting in the complete restoration of the functional zone; it lasts from the end of menses until the beginning of ovulation, around day 14. (3) During the secretory phase, the uterine (endometrial) glands enlarge, accelerating their rates of secretion, and the arteries elongate and spiral through the tissues of the functional zone; this phase begins at ovulation, occurs under the combined stimulatory effects of progestins and estrogens from the corpus luteum, and persists as long as the corpus luteum remains intact.

21. As follicular development proceeds, the concentration of circulating estrogen rises. Secondary follicles contain increased numbers of granulosa cells, and the level of circulating inhibin rises. The rising estrogen and inhibin levels inhibit hypothalamic secretion of GnRH and pituitary production and release of FSH. As the follicles develop and estrogen levels rise, the pituitary output of LH gradually increases. Estrogens, FSH, and LH continue to support follicular development and maturation despite a gradual decline in FSH levels. In the second week of the ovarian cycle, estrogen levels sharply increase, and the tertiary follicle enlarges in preparation for ovulation. By day 14, estrogen levels peak, triggering a massive outpouring of LH from the adenohypophysis. The rupture of the follicular wall results in ovulation. Next, LH stimulates the formation of the corpus luteum, which secretes moderate amounts of estrogens but large amounts of progesterone, the principal hormone of the postovulatory period. About 12 days after ovulation, declining progesterone and estrogen levels stimulate hypothalamic receptors and GnRH production increases, leading to increased FSH and LH production in the adenohypophysis; the cycle begins again.

22. The corpus luteum degenerates and progesterone and estrogen levels drop, resulting in the endometrial breakdown of menses. Next, rising levels of FSH, LH, and estrogen stimulate the repair and regeneration of the functional zone of the endometrium. During the postovulatory phase, the combination of estrogen and progesterone causes the enlargement of the uterine glands and an increase in their secretory activity.

23. During sexual arousal, erotic thoughts or physical stimulation of sensory nerves in the genital region increases the parasympathetic outflow over the pelvic nerve, leading to erection of the clitoris or penis. Orgasm is the intensely pleasurable sensation associated with perineal muscle contraction and ejaculation in males, and with uterine and vaginal contractions and perineal muscle contraction in females. These processes are comparable in males and females, but only males undergo the processes of emission and ejaculation.

24. Women ages 45–55 experience menopause—the time that ovulation and menstruation cease, accompanied by a sharp and sustained rise in the production of GnRH, FSH, and LH and a drop in the concentrations of circulating estrogen and progesterone. The decline in estrogen levels leads to reductions in the size of the uterus and breasts, accompanied by a thinning of the urethral and vaginal walls. In addition to neural and cardiovascular effects, reduced estrogen concentrations have been linked to the development of osteoporosis, presumably because bone deposition proceeds at a slower rate. Men ages 50–60 experience the male climacteric, a time when circulating testosterone levels begin to decline and circulating levels of FSH and LH rise. Although sperm production continues, a gradual reduction in sexual activity occurs in older men.

Level 3 Critical Thinking and Clinical Applications

25. Women more frequently experience peritonitis stemming from a urinary tract infection because infectious organisms exiting the urethral orifice can readily enter the nearby vagina. From there, they can then proceed to the uterus, into the uterine tubes, and finally into the peritoneal cavity. No such direct path of entry into the abdominopelvic cavity exists in men.

26. Regardless of their location, endometrial cells have receptors for and respond to estrogen and progesterone. Under the influence of estrogen at the beginning of the menstrual cycle, any endometrial cells in the peritoneal cavity proliferate and begin to develop glands and blood vessels, which then further develop under the control of progesterone. The dramatic increase in size of this tissue presses on neighboring abdominal tissues and organs, causing periodic painful sensations.

27. Slightly elevated levels of estradiol and progesterone inhibit both GnRH at the hypothalamus and the release of FSH and LH from the pituitary. Without FSH, primordial follicles do not initiate development, and the endogenous levels of estrogen remain low. An LH surge, triggered by the peaking of estradiol, is necessary for ovulation to occur. If the level of estradiol is not allowed to rise above the critical level, the LH surge will not occur, and thus ovulation will not occur, even if a follicle managed to develop to a stage at which it could ovulate. Any mature follicles would ultimately degenerate, and no new follicles would mature to take their place. Although the ovarian cycle is interrupted, the level of hormones is still adequate to regulate a normal menstrual cycle.

28. These observations suggest that a certain amount of body fat is necessary for menstrual cycles to occur. The nervous system appears to respond to circulating levels of the adipose tissue hormone leptin; when leptin levels fall below a certain set point, menstruation ceases. Because a woman lacking adequate fat reserves might not be able to have a successful pregnancy, the body prevents pregnancy by shutting down the ovarian cycle, and thus the menstrual cycle. Once sufficient energy reserves become available, the cycles begin again.

Chapter 29

Answers to Checkpoints

Page 1087
1. Differentiation is the formation of different types of cells during development.
2. Development begins at fertilization (conception), the union of an oocyte and sperm.
3. Inheritance refers to the transfer of genetically determined characteristics from one generation to the next.

Page 1090
4. Two sperm enzymes important to secondary oocyte penetration are hyaluronidase and acrosin.
5. A normal zygote contains 46 chromosomes.

Page 1091
6. Gestation is the period of prenatal development; it consists of three trimesters.
7. The first trimester is the period of embryological and early fetal development. The second trimester is a period of organ and organ system development; during this stage, the fetus appears distinctly human. The third trimester is characterized by rapid fetal growth and the deposition of adipose tissue.

Page 1099
8. The inner cell mass of the blastocyst eventually develops into the embryo.
9. Improper development of the yolk sac—the mesoderm-derived structure that gives rise to blood vessels and is an important site of blood cell formation—would affect the development and function of the cardiovascular system.
10. Yes, Sue is pregnant. After fertilization, the developing trophoblast (and later, the placenta) produce and release the hormone hCG.
11. Placental functions include **(a)** supplying the developing fetus with a route for gas exchange, nutrient transfer, and waste product elimination; and **(b)** producing hormones that affect maternal systems.

Page 1105
12. The major changes that occur in the maternal system during pregnancy include increases in respiratory rate, tidal volume, blood volume, nutrient requirements, glomerular filtration rate, and size of uterus and mammary glands.
13. Three factors opposing the calming action of progesterone on the uterus are rising estrogen levels, rising oxytocin levels, and prostaglandin production.
14. A mother's blood volume increases during pregnancy to compensate for the reduction in maternal blood volume resulting from blood flow through the placenta.

Page 1107
15. The three stages of labor are the dilation stage, expulsion stage, and placental stage.
16. Immature delivery is the birth of a fetus weighing at least 500 g (17.6 oz), which is the normal weight near the end of the second trimester. Premature delivery usually refers to birth at 28–36 weeks at a weight over 1 kg (2.2 pounds).
17. Fraternal twins are dizygotic, and identical twins are monozygotic.

Page 1112
18. The postnatal stages of development are the neonatal period, infancy, childhood, adolescence, maturity, and senescence.
19. The neonatal period is the time from birth to one month; infancy continues from the neonatal period to age 2; adolescence is the period after childhood in which sexual and physical maturity occur.
20. Colostrum is produced by the mammary glands from the end of the sixth month of pregnancy until a few days after birth. After that, the glands begin producing breast milk, which contains fewer proteins (including antibodies) and far more fat than colostrum.
21. Increases in the blood levels of GnRH, FSH, LH, and sex hormones mark the onset of puberty.

Page 1120
22. Genotype is a person's genetic makeup; in contrast, phenotype—a person's visible physical characteristics—results from the interaction between the person's genotype and the environment.
23. Heterozygous refers to possessing two different alleles at corresponding sites of a chromosome pair. For a heterozygous trait, one or both of the alleles determines the individual's phenotype.
24. The phenotype of a person who is heterozygous for curly hair—who has one dominant allele and one recessive allele for that trait—would be "curly hair."
25. One reason children are not identical copies of their parents is that during meiosis, parental chromosomes are randomly distributed such that each gamete has a unique set of chromosomes. Additionally, mutations and crossing-over during meiosis introduce new variations.

Answers to Review Questions

Page 1122

Level 1 Reviewing Facts and Terms
1. d **2. (a)** decidua parietalis; **(b)** decidua basalis; **(c)** amnion; **(d)** chorion; **(e)** decidua capsularis **3.** c **4.** b **5.** b **6.** a **7.** b **8.** b

9. When a sperm contacts the secondary oocyte, their plasma membranes fuse. The oocyte is then activated: Its metabolic rate rises; it completes meiosis II; and the cortical reaction prevents additional sperm from entering. (Vesicles beneath the oocyte surface fuse with the plasma membrane and discharge their contents.) The male and female pronuclei fuse (amphimixis), and the zygote begins preparing for the first cleavage division.

10. (a) The four extraembryonic membranes are the yolk sac, amnion, allantois, and chorion. **(b)** The yolk sac forms from endoderm and mesoderm; it is an important site of blood cell formation. The amnion forms from ectoderm and mesoderm; it encloses the fluid that surrounds and cushions the developing embryo and fetus. The allantois forms from endoderm and mesoderm; its base gives rise to the urinary bladder. The chorion forms from mesoderm and trophoblast; circulation through chorionic vessels provides a "rapid transit system" for blood and nutrients.

11. The dilation stage begins with the onset of true labor, as the cervix dilates and the fetus begins to move toward the cervical canal; late in this stage, the amniochorionic membrane ruptures. The expulsion stage begins as the cervix dilates completely and continues until the fetus has completely emerged from the vagina (delivery). In the placental stage, the uterus gradually contracts, tearing the connections between the endometrium and the placenta and ejecting the placenta.

12. Relaxin produced by the placenta softens the pubic symphysis, and the weight of the fetus then deforms the external os of the uterus. Deformation of the cervix and rising estrogen levels promote the release of oxytocin, and the already stretched muscles become even more excitable.

13. (1) Neonatal period (birth to 1 month): The newborn becomes relatively self-sufficient and begins performing respiration (breathing), digestion, and excretion on its own. Heart rates and fluid requirements are higher than those of adults. Neonates have little ability to thermoregulate. (2) Infancy (1 month to 2 years): Major organ systems (other than those related to reproduction) become fully operational. (3) Childhood (2 years to puberty): Growth continues; body proportions change significantly.

14. Three events interact to promote increased hormone production and sexual maturation at puberty: (1) The hypothalamus increases its production of GnRH; (2) the adenohypophysis becomes more sensitive to the presence of GnRH, and circulating levels of FSH and LH rise rapidly; and (3) ovarian or testicular cells become more sensitive to FSH and LH. Puberty initiates adolescence, which includes gametogenesis and the production of sex hormones that stimulate the appearance of secondary sexual characteristics and behaviors.

Level 2 Reviewing Concepts

15. b   **16.** b   **17.** c

18. The post-implantation embryo obtains nutrients through the chorionic villi and later the placenta. The placenta develops during placentation.

19. The placenta produces (1) human chorionic gonadotropin, which maintains the integrity of the corpus luteum and promotes the continued secretion of progesterone (keeping the endometrial lining functional); (2) human placental lactogen and placental prolactin, which help prepare the mammary glands for milk production; and (3) relaxin, which increases the flexibility of the pubic symphysis, causes dilation of the cervix, and suppresses the release of oxytocin by the hypothalamus, delaying the onset of labor contractions.

20. Respiratory rate and tidal volume increase, allowing the lungs to obtain the extra oxygen the fetus needs, and to remove the excess carbon dioxide the fetus generates. Maternal blood volume increases, compensating for blood that will be lost during delivery. Nutrient and vitamin requirements climb 10–30 percent, reflecting the fact that some of the mother's nutrients go to nourish the fetus. Glomerular filtration rate increases by about 50 percent, which corresponds to the increased blood volume and accelerates the excretion of metabolic wastes generated by the fetus.

21. Positive feedback mechanisms between increasing levels of oxytocin and increased uterine distortion ensure that labor contractions continue until delivery has been completed.

22. A neonate must fill its lungs (which are collapsed and filled with fluid at birth) with air, which alters the pattern of cardiovascular circulation due to changes in blood pressure and flow rates. It must excrete the mixture of debris (meconium) that has collected in the fetal digestive system. The neonate must obtain nourishment from a new source—the mother's mammary glands. Neonatal fluid requirements are high, because the newborn cannot concentrate its urine significantly. The infant also has little ability to thermoregulate, although as it grows its insulating adipose tissue increases, and its metabolic rate also rises.

23. (a) Genotype refers to all of an individual's genes; phenotype refers to the individual's physical and physiological characteristics. **(b)** A heterozygous genotype carries different alleles for a given gene; a homozygous genotype carries identical alleles for that gene. **(c)** In simple inheritance, phenotypes are determined by interactions between a single pair of alleles; in polygenic inheritance, interactions occur among multiple genes.

24. (a) dominant; **(b)** recessive; **(c)** X-linked; **(d)** autosomal

25. GEI is the Genes and Environment Initiative and its goal is to understand the link between environmental agents and an individual's genetic susceptibility to disease. TCGA is The Cancer Genome Atlas. Its initial goal is to map the genetic changes in three types of tumors: glioblastomas of the brain, lung cancer, and ovarian cancer.

Level 3 Critical Thinking and Clinical Applications

26. The probability that this couple's daughters will have hemophilia is 0, because each daughter will receive a normal allele from her father. There is a 50 percent chance that a son will have hemophilia, because each son has a 50 percent chance of receiving the mother's normal allele and a 50 percent chance of receiving the mother's recessive allele.

27. Although technically it only takes one sperm to fertilize an egg, the probability of this occurring if too few sperm are deposited is very small. Most sperm that enter the female reproductive tract are killed or disabled before they reach the uterus. Many of the sperm reaching the uterus are incapable of reaching the secondary oocyte, which is in a uterine tube. Once at the oocyte, the sperm must penetrate the corona radiata, which requires the combined acrosomal enzymes of 100 or more sperm. If too few sperm are deposited in the vagina, the number reaching the uterine tube is too small to penetrate the corona radiata.

28. The most obvious possibility is that the infant has a problem with the cardiovascular supply to the lungs. A good guess would be a patent ductus arteriosus (the ductus arteriosus has failed to close off completely). When the baby is not being stressed (bathing creates heat loss and thermal stress) or eating (less air is entering the lungs), the infant appears normal. Because some of the blood flow to the lungs is being shunted over to the aorta during stress and eating, too little blood is being oxygenated, so the infant becomes cyanotic.

29 It is very unlikely that the baby's condition results from a viral infection contracted during the third trimester, because organ systems develop during the first trimester and are fully formed by the end of the second trimester.

Glossary

Eponyms in Common Use

Eponym	Equivalent Term(s)	Individual Referenced
THE CELLULAR LEVEL OF ORGANIZATION *(CHAPTER 3)*		
Golgi apparatus		Camillo Golgi (1844–1926), Italian histologist; shared Nobel Prize in 1906
Krebs cycle	Tricarboxylic acid cycle, TCA cycle, or citric acid cycle	Hans Adolph Krebs (1900–1981), British biochemist; shared Nobel Prize in 1953
THE SKELETAL SYSTEM *(CHAPTERS 6–9)*		
Colles fracture		Abraham Colles (1773–1843), Irish surgeon
Haversian canals	Central canals	Clopton Havers (1650–1702), English anatomist and microscopist
Haversian systems	Osteons	Clopton Havers
Pott fracture		Percivall Pott (1713–1788), English surgeon
Sharpey fibers	Perforating fibers	William Sharpey (1802–1880), Scottish histologist and physiologist
Volkmann canals	Perforating canals	Alfred Wilhelm Volkmann (1800–1877), German surgeon
Wormian bones	Sutural bones	Olas Worm (1588–1654), Danish anatomist
THE MUSCULAR SYSTEM *(CHAPTERS 10, 11)*		
Achilles tendon	Calcaneal tendon	Achilles, hero of Greek mythology
Cori cycle		Carl Ferdinand Cori (1896–1984) and Gerty Theresa Cori (1896–1957), American biochemists; shared Nobel Prize in 1947
THE NERVOUS SYSTEM *(CHAPTERS 12–16)*		
Broca center	Speech center	Pierre Paul Broca (1824–1880), French surgeon
Foramen of Lushka	Lateral foramina	Hubert von Lushka (1820–1875), German anatomist
Meissner corpuscles	Tactile corpuscles	Georg Meissner (1829–1905), German physiologist
Merkel discs	Tactile discs	Friedrich Siegismund Merkel (1845–1919), German anatomist
Foramen of Munro	Interventricular foramen	John Cummings Munro (1858–1910), American surgeon
Nissl bodies		Franz Nissl (1860–1919), German neurologist
Pacinian corpuscles	Lamellated corpuscles	Fillippo Pacini (1812–1883), Italian anatomist
Purkinje cells		Johannes E. Purkinje (1787–1869), Bohemian anatomist and physiologist
Nodes of Ranvier	Nodes	Louis Antoine Ranvier (1835–1922), French physiologist
Island of Reil	Insula	Johann Christian Reil (1759–1813), German anatomist
Fissure of Rolando	Central sulcus	Luigi Rolando (1773–1831), Italian anatomist
Ruffini corpuscles		Angelo Ruffini (1864–1929), Italian anatomist
Schwann cells	Neurolemmocytes	Theodor Schwann (1810–1882), German anatomist
Aqueduct of Sylvius	Aqueduct of the midbrain, cerebral aqueduct, or mesencephalic aqueduct	Jacobus Sylvius (Jacques Dubois, 1478–1555), French anatomist
Sylvian fissure	Lateral sulcus	Franciscus Sylvius (Franz de le Boë, 1614–1672), Dutch anatomist
Pons varolii	Pons	Costanzo Varolio (1543–1575), Italian anatomist
SENSORY FUNCTION *(CHAPTER 17)*		
Organ of Corti	Spiral organ	Alfonso Corti (1822–1888), Italian anatomist
Eustachian tube	Auditory tube	Bartolomeo Eustachio (1520–1574), Italian anatomist
Golgi tendon organs	Tendon organs	Camillo Golgi (1844–1926), Italian histologist; shared Nobel Prize in 1906
Hertz (Hz)		Heinrich Hertz (1857–1894), German physicist
Meibomian glands	Tarsal glands	Heinrich Meibom (1638–1700), German anatomist
Canal of Schlemm	Scleral venous sinus	Friedrich S. Schlemm (1795–1858), German anatomist

Eponym	Equivalent Term(s)	Individual Referenced
THE ENDOCRINE SYSTEM *(CHAPTER 18)*		
Islets of Langerhans	Pancreatic islets	Paul Langerhans (1847–1888), German pathologist
Interstitial cells of Leydig	Interstitial cells	Franz von Leydig (1821–1908), German anatomist
THE CARDIOVASCULAR SYSTEM *(CHAPTERS 19–21)*		
Bundle of His	AV Bundle	Wilhelm His (1863–1934), German physician
Purkinje fibers		Johannes E. Purkinje (1787–1869), Bohemian anatomist and physiologist
Starling's law		Ernest Henry Starling (1866–1927), English physiologist
Circle of Willis	Cerebral arterial circle	Thomas Willis (1621–1675), English physician
THE LYMPHOID SYSTEM *(CHAPTER 22)*		
Hassall corpuscles	Thymic corpuscles	Arthur Hill Hassall (1817–1894), English physician
Kupffer cells	Stellate reticuloendothelial cells	Karl Wilhelm Kupffer (1829–1902), German anatomist
Langerhans cells	Dendritic cells	Paul Langerhans (1847–1888), German pathologist
Peyer patches	Aggregated lymphoid nodules	Johann Conrad Peyer (1653–1712), Swiss anatomist
THE RESPIRATORY SYSTEM *(CHAPTER 23)*		
Bohr effect		Christian Bohr (1855–1911), Danish physiologist
Boyle's law		Robert Boyle (1621–1691), English physicist
Charles' law		Jacques Alexandre César Charles (1746–1823), French physicist
Dalton's law		John Dalton (1766–1844), English physicist
Henry's law		William Henry (1775–1837), English chemist
THE DIGESTIVE SYSTEM *(CHAPTER 24)*		
Plexus of Auerbach	Myenteric plexus	Leopold Auerbach (1827–1897), German anatomist
Brunner glands	Duodenal glands	Johann Conrad Brunner (1653–1727), Swiss anatomist
Kupffer cells	Stellate reticuloendothelial cells	Karl Wilhelm Kupffer (1829–1902), German anatomist
Crypts of Lieberkühn	Intestinal glands	Johann Nathaniel Lieberkuhn (1711–1756), German anatomist
Plexus of Meissner	Submucosal plexus	Georg Meissner (1829–1905), German physiologist
Sphincter of Oddi	Hepatopancreatic sphincter	Ruggero Oddi (1864–1913), Italian physician
Peyer patches	Aggregated lymphoid nodules	Johann Conrad Peyer (1653–1712), Swiss anatomist
Duct of Santorini	Accessory pancreatic duct	Giovanni Domenico Santorini (1681–1737), Italian anatomist
Stensen duct	Parotid duct	Niels Stensen (1638–1686), Danish physician/priest
Ampulla of Vater	Duodenal ampulla	Abraham Vater (1684–1751), German anatomist
Wharton duct	Submandibular duct	Thomas Wharton (1614–1673), English physician
Duct of Wirsung	Pancreatic duct	Johann Georg Wirsung (1600–1643), German physician
THE URINARY SYSTEM *(CHAPTER 26)*		
Bowman capsule	Glomerular capsule	Sir William Bowman (1816–1892), English physician
Loop of Henle	Nephron loop	Friedrich Gustav Jakob Henle (1809–1885), German histologist
THE REPRODUCTIVE SYSTEM *(CHAPTERS 28, 29)*		
Bartholin glands	Greater vestibular glands	Casper Bartholin, Jr. (1655–1738), Danish anatomist
Cowper glands	Bulbo-urethral glands	William Cowper (1666–1709), English surgeon
Fallopian tube	Uterine tube/oviduct	Gabriele Fallopio (1523–1562), Italian anatomist
Graafian follicle	Tertiary follicle	Reijnier de Graaf (1641–1673), Dutch physician
Interstitial cells of Leydig	Interstitial cells	Franz von Leydig (1821–1908), German anatomist
Glands of Littré	Lesser vestibular glands	Alexis Littré (1658–1726), French surgeon
Sertoli cells	Nurse cells, sustentacular cells	Enrico Sertoli (1842–1910), Italian histologist

Glossary of Key Terms

A

abdomen: The region of the trunk bounded by the diaphragm and pelvis. (1)

abdominopelvic cavity: The portion of the ventral body cavity that contains abdominal and pelvic subdivisions; also contains the peritoneal cavity. (1)

abducens: Cranial nerve VI, which innervates the lateral rectus muscle of the eye. (14)

abduction: Movement away from the midline of the body, as viewed in the anatomical position. (9)

abortion: The premature loss or expulsion of an embryo or fetus. (29)

abscess: A localized collection of pus within a damaged tissue. (4, 22)

absorption: The active or passive uptake of gases, fluids, or solutes. (25)

accommodation: An alteration in the curvature of the lens of the eye to focus an image on the retina. (17)

acetabulum: The fossa on the lateral aspect of the pelvis that accommodates the head of the femur. (8)

acetylcholine (ACh): A chemical neurotransmitter in the brain and peripheral nervous system; the dominant neurotransmitter in the peripheral nervous system, released at neuromuscular junctions and synapses of the parasympathetic division. (10, 12, 16)

acetylcholinesterase (AChE): An enzyme found in the synaptic cleft, bound to the postsynaptic membrane, and in tissue fluids; breaks down and inactivates acetylcholine molecules. (10, 12)

acetyl-CoA: An acetyl group bound to coenzyme A, a participant in the anabolic and catabolic pathways for carbohydrates, lipids, and many amino acids. (25)

acetyl group: —CH_3CO. (25)

Achilles tendon: *See* calcaneal tendon.

acid: A compound whose dissociation in solution releases a hydrogen ion and an anion; an acidic solution has a pH below 7.0 and contains an excess of hydrogen ions. (2, 27)

acidosis: An abnormal physiological state characterized by a plasma pH below 7.35. (2, 25, 26, 27)

acinus/acini: A histological term referring to a blind pocket, pouch, or sac.

acoustic: Pertaining to sound or the sense of hearing. (17)

acquired immune deficiency syndrome (AIDS): A disease caused by the human immunodeficiency virus (HIV); characterized by the destruction of helper T cells and a resulting severe impairment of the immune response. (22)

acromegaly: A condition caused by the overproduction of growth hormone in adults, characterized by a thickening of bones and an enlargement of cartilages and other soft tissues. (6, 18)

acromion: A continuation of the scapular spine that projects superior to the capsule of the scapulohumeral joint. (8)

acrosomal cap: A membranous sac at the tip of a spermatozoon that contains hyaluronidase. (28)

actin: The protein component of microfilaments that forms thin filaments in skeletal muscles and produces contractions of all muscles through interaction with thick (myosin) filaments; *see also* **sliding filament theory.** (3, 10)

action potential: A propagated change in the transmembrane potential of excitable cells, initiated by a change in the membrane permeability to sodium ions; *see also* **nerve impulse.** (10, 12)

active transport: The ATP-dependent absorption or secretion of solutes across a plasma membrane. (3, 26)

acute: Sudden in onset, severe in intensity, and brief in duration.

adaptation: A change in pupillary size in response to changes in light intensity (17); a decrease in receptor sensitivity or perception after chronic stimulation (15); physiological responses that produce acclimatization (25).

Addison disease: A condition resulting from the hyposecretion of glucocorticoids; characterized by lethargy, weakness, hypotension, and increased skin pigmentation. (5, 18)

adduction: Movement toward the axis or midline of the body, as viewed in the anatomical position. (9)

adenine: A purine; one of the nitrogenous bases in the nucleic acids RNA and DNA. (2)

adenohypophysis: The anterior lobe of the pituitary gland. (18)

adenoids: The pharyngeal tonsil. (22, 23)

adenosine: A combination of adenine and ribose. (2)

adenosine diphosphate (ADP): A compound consisting of adenosine with two phosphate groups attached. (2, 25)

adenosine monophosphate (AMP): A nucleotide consisting of adenine plus a phosphate group(PO_4^{3-}); also called *adenosine phosphate*.

adenosine triphosphate (ATP): A high-energy compound consisting of adenosine with three phosphate groups attached; the third is attached by a high-energy bond. (2, 10, 25)

adenylate cyclase: An enzyme bound to the inner surfaces of plasma membranes that can convert ATP to cyclic-AMP; also called *adenylyl cyclase*. (12)

adhesion: The fusion of two mesenterial layers after damage or irritation of their opposing surfaces; this process restricts relative movement of the organs involved (4); the binding of a phagocyte to its target (22).

adipocyte: A fat cell. (4)

adipose tissue: Loose connective tissue dominated by adipocytes. (4, 18)

adrenal cortex: The superficial portion of the suprarenal (adrenal) gland that produces steroid hormones; also called the *suprarenal cortex*. (18)

adrenal gland: A small endocrine gland that secretes steroids and catecholamines and is located superior to each kidney; also called *suprarenal gland*. (18)

adrenal medulla: The core of the suprarenal (adrenal) gland (18); a modified sympathetic ganglion that secretes catecholamines into the blood during sympathetic activation; also called *suprarenal medulla* (16).

adrenergic: A synaptic terminal that, when stimulated, releases norepinephrine. (12)

adrenocortical hormone: Any steroid produced by the suprarenal (adrenal) cortex. (18)

adrenocorticotropic hormone (ACTH): The hormone that stimulates the production and secretion of glucocorticoids by the zona fasciculata of the suprarenal (adrenal) cortex; released by the adenohypophysis (anterior lobe of the pituitary gland) in response to corticotropin-releasing hormone. (18)

adventitia: The superficial layer of connective tissue surrounding an internal organ; fibers are continuous with those of surrounding tissues, providing support and stabilization. (24)

aerobic: Requiring the presence of oxygen.

aerobic metabolism: The complete breakdown of organic substrates into carbon dioxide and water, via pyruvic acid; a process that yields large amounts of ATP but requires mitochondria and oxygen. (3, 10, 25)

afferent: Toward a center.

afferent arteriole: An arteriole that carries blood to a glomerulus of the kidney. (26)

afferent fiber: An axon that carries sensory information to the central nervous system. (12)

agglutination: The aggregation of red blood cells due to interactions between surface antigens and plasma antibodies. (19, 22)

agglutinins: Immunoglobulins in plasma that react with antigens on the surfaces of foreign red blood cells when donor and recipient differ in blood type. (19)

agglutinogens: Surface antigens on red blood cells whose presence and structure are genetically determined. (19)

aggregated lymphoid nodules: Lymphoid nodules beneath the epithelium of the small intestine; also called *Peyer patches*. (22)

agonist: A muscle responsible for a specific movement; also called a prime mover. (11)

agranular: Without granules; *agranular leukocytes* are monocytes and lymphocytes. (19)

AIDS: *See* acquired immune deficiency syndrome.

alba: White.

albicans: White.

albuginea: White.

aldosterone: A mineralocorticoid produced by the zona glomerulosa of the suprarenal (adrenal) cortex; stimulates sodium and water conservation at the kidneys; secreted in response to the presence of angiotensin II. (18, 26, 27)

alkalosis: The condition characterized by a plasma pH greater than 7.45; associated with a relative deficiency of hydrogen ions or an excess of bicarbonate ions. (2, 27)

alpha receptors: Membrane receptors sensitive to norepinephrine or epinephrine; stimulation normally results in the excitation of the target cell. (16)

alveolar sac: An air-filled chamber that supplies air to several alveoli. (23)

alveolus/alveoli: Blind pockets at the end of the respiratory tree, lined by a simple squamous epithelium and surrounded by a capillary network;

sites of gas exchange with the blood (23); a bony socket that holds the root of a tooth (24).

Alzheimer disease: A disorder resulting from degenerative changes in populations of neurons in the cerebrum, causing dementia characterized by problems with attention, short-term memory, and emotions. (16)

amination: The attachment of an amino group to a carbon chain; performed by a variety of cells and important in the synthesis of amino acids. (25)

amino acids: Organic compounds whose chemical structure can be summarized as R—CHNH$_2$—COOH. (2, 25)

amino group: —NH$_2$. (2)

amnion: One of the four extraembryonic membranes; surrounds the developing embryo or fetus. (29)

amniotic fluid: Fluid that fills the amniotic cavity; cushions and supports the embryo or fetus. (4, 29)

amphiarthrosis: An articulation that permits a small degree of independent movement; *see* **interosseous membrane** (8) and **pubic symphysis.** (9)

ampulla/ampullae: A localized dilation in the lumen of a canal or passageway. (17, 24, 28)

amygdaloid body: A basal nucleus that is a component of the limbic system and acts as an interface between that system, the cerebrum, and sensory systems. (14)

amylase: An enzyme that breaks down polysaccharides; produced by the salivary glands and pancreas. (24)

anabolism: The synthesis of complex organic compounds from simpler precursors. (2, 25)

anaerobic: Without oxygen.

analgesic: A substance that relieves pain. (15)

anal triangle: The posterior subdivision of the perineum. (11)

anaphase: The mitotic stage in which the paired chromatids separate and move toward opposite ends of the spindle apparatus. (3)

anaphylaxis: A hypersensitivity reaction due to the binding of antigens to immunoglobulins (IgE) on the surfaces of mast cells; the release of histamine, serotonin, and prostaglandins by mast cells then causes widespread inflammation; a sudden decline in blood pressure may occur, producing anaphylactic shock. (22)

anastomosis: The joining of two tubes, usually referring to a connection between two peripheral vessels without an intervening capillary bed. (21)

anatomical position: An anatomical reference position; the body viewed from the anterior surface with the palms facing forward. (1)

anatomy: The study of the structure of the body. (1)

androgen: A steroid sex hormone primarily produced by the interstitial cells of the testis and manufactured in small quantities by the suprarenal (adrenal) cortex in either gender. (18, 28)

anemia: The condition marked by a reduction in the hematocrit, the hemoglobin content of the blood, or both. (19)

angiotensin I: The hormone produced by the activation of angiotensinogen by renin (18); angiotensin-converting enzyme converts angiotensin I into angiotensin II in lung capillaries (26).

angiotensin II: A hormone that causes an elevation in systemic blood pressure, stimulates the secretion of aldosterone, promotes thirst, and causes the release of antidiuretic hormone (18);

angiotensin-converting enzyme in lung capillaries converts angiotensin I into angiotensin II. (21, 26)

angiotensinogen: The blood protein produced by the liver that is converted to angiotensin I by the enzyme renin. (18)

anion: An ion bearing a negative charge. (2, 27)

anoxia: Tissue oxygen deprivation. (23)

antagonist: A muscle that opposes the movement of an agonist. (10)

antebrachium: The forearm. (8)

anterior: On or near the front, or ventral surface, of the body.

antibiotic: A chemical agent that selectively kills pathogens, primarily bacteria. (20)

antibody: A globular protein produced by plasma cells that will bind to specific antigens and promote their destruction or removal from the body. (19, 22)

antibody-mediated immunity: The form of immunity resulting from the presence of circulating antibodies produced by plasma cells; also called *humoral immunity*. (22)

anticholinesterase: A chemical compound that blocks the action of acetylcholine and causes prolonged and intensive stimulation of postsynaptic membranes. (12)

anticodon: Three nitrogenous bases on a tRNA molecule that interact with an appropriate codon on a strand of mRNA. (3)

antidiuretic hormone (ADH): A hormone synthesized in the hypothalamus and secreted at the neurohypophysis (posterior lobe of the pituitary gland); causes water retention at the kidneys and an elevation of blood pressure. (18, 21, 26, 27)

antigen: A substance capable of inducing the production of antibodies. (22)

antigen–antibody complex: The combination of an antigen and a specific antibody. (22)

antigenic determinant site: A portion of an antigen that can interact with an antibody molecule. (22)

antigen-presenting cell (APC): A cell that processes antigens and displays them, bound to MHC proteins; essential to the initiation of a normal immune response. (22)

antihistamines: A chemical agent that blocks the action of histamine on peripheral tissues. (22)

antrum: A chamber or pocket. (28)

anulus: A cartilage or bone shaped like a ring; also spelled *annulus*. (9)

anus: The external opening of the anal canal. (24)

aorta: The large, elastic artery that carries blood away from the left ventricle and into the systemic circuit. (20)

apocrine secretion: A mode of secretion in which the glandular cell sheds portions of its cytoplasm. (4, 5)

aponeurosis/aponeuroses: A broad tendinous sheet that may serve as the origin or insertion of a skeletal muscle. (4, 6, 10)

appendicular: Pertaining to the upper or lower limbs. (8)

appendix: A blind tube connected to the cecum of the large intestine. (24)

appositional growth: The enlargement of a bone by the addition of cartilage or bony matrix at its surface. (4)

aqueous humor: A fluid similar to perilymph or cerebrospinal fluid that fills the anterior chamber of the eye. (17)

arachidonic acid: One of the essential fatty acids. (2, 18)

arachnoid granulations: Processes of the arachnoid mater that project into the superior sagittal si-

nus; sites where cerebrospinal fluid enters the venous circulation. (14)

arachnoid mater: The middle meninx that encloses cerebrospinal fluid and protects the central nervous system. (13, 14)

arbor vitae: The central, branching mass of white matter inside the cerebellum. (14)

arcuate: Curving.

areolar: Containing minute spaces, as in areolar tissue.

areolar tissue: Loose connective tissue with an open framework. (4)

arrector pili: Smooth muscles whose contractions force hairs to stand erect. (5)

arrhythmias: Abnormal patterns of cardiac contractions. (20)

arteriole: A small arterial branch that delivers blood to a capillary network. (21)

artery: A blood vessel that carries blood away from the heart and toward a peripheral capillary. (4, 20, 21)

articular: Pertaining to a joint.

articular capsule: The dense collagen fiber sleeve that surrounds a joint and provides protection and stabilization. (6, 9)

articular cartilage: The cartilage pad that covers the surface of a bone inside a joint cavity. (6, 9)

articulation: A joint (9); the formation of words (23).

arytenoid cartilages: A pair of small cartilages in the larynx. (23)

ascending tract: A tract carrying information from the spinal cord to the brain. (13, 14)

association areas: Cortical areas of the cerebrum that are responsible for the integration of sensory inputs and/or motor commands. (14)

association neuron: *See* interneuron.

astrocyte: One of the four types of neuroglia in the central nervous system; responsible for maintaining the blood–brain barrier by the stimulation of endothelial cells. (12)

atherosclerosis: The formation of fatty plaques in the walls of arteries, restricting blood flow to deep tissues. (21)

atom: The smallest stable unit of matter. (2)

atomic number: The number of protons in the nucleus of an atom. (2)

atomic weight: Roughly, the average total number of protons and neutrons in the atoms of a particular element. (2)

atria: Thin-walled chambers of the heart that receive venous blood from the pulmonary or systemic circuit.

atrial natriuretic peptide (ANP): *See* natriuretic peptides. (20)

atrial reflex: The reflexive increase in heart rate after an increase in venous return; due to mechanical and neural factors; also called *Bainbridge reflex*. (20, 21)

atrioventricular (AV) node: Specialized cardiocytes that relay the contractile stimulus to the bundle of His, the bundle branches, the Purkinje fibers, and the ventricular myocardium; located at the boundary between the atria and ventricles. (20)

atrioventricular (AV) valve: One of the valves that prevents backflow into the atria during ventricular systole. (20)

atrophy: The wasting away of tissues from a lack of use, ischemia, or nutritional abnormalities. (10)

auditory: Pertaining to the sense of hearing. (17)

auditory ossicles: The bones of the middle ear: malleus, incus, and stapes. (7, 17)

auditory tube: A passageway that connects the nasopharynx with the middle ear cavity; also called *Eustachian tube* or *pharyngotympanic tube*. (17)

auricle: A broad, flattened process that resembles the external ear; in the ear, the expanded, projecting portion that surrounds the external auditory meatus, also called *pinna* (17); in the heart, the externally visible flap formed by the collapse of the outer wall of a relaxed atrium (20).

autoantibodies: Antibodies that react with antigens on the surfaces of a person's own cells and tissues. (22)

autoimmunity: The immune system's sensitivity to normal cells and tissues, resulting in the production of autoantibodies. (22)

autolysis: The destruction of a cell due to the rupture of lysosomal membranes in its cytoplasm. (3)

automaticity: The spontaneous depolarization to threshold, characteristic of cardiac pacemaker cells. (10, 20)

autonomic ganglion: A collection of visceral motor neurons outside the central nervous system. (16)

autonomic nerve: A peripheral nerve consisting of preganglionic or postganglionic autonomic fibers. (16)

autonomic nervous system (ANS): Centers, nuclei, tracts, ganglia, and nerves involved in the unconscious regulation of visceral functions; includes components of the central nervous system and the peripheral nervous system. (12, 16)

autopsy: The detailed examination of a body after death.

autoregulation: Changes in activity that maintain homeostasis in direct response to changes in the local environment; does not require neural or endocrine control. (1, 21, 26)

autosomal: Chromosomes other than the X or Y sex chromosome. (29)

avascular: Without blood vessels. (4)

axilla: The armpit. (1, 8)

axolemma: The plasma membrane of an axon, continuous with the plasma membrane of the cell body and dendrites and distinct from any neuroglial coverings. (12)

axon: The elongate extension of a neuron that conducts an action potential. (4, 12)

axon hillock: In a multipolar neuron, the portion of the cell body adjacent to the initial segment. (12)

axoplasm: The cytoplasm within an axon. (12)

B

bacteria: Single-celled microorganisms, some pathogenic, that are common in the environment and in and on the body. (22)

Bainbridge reflex: *See* atrial reflex.

baroreception: The ability to detect changes in pressure. (15, 23)

baroreceptor reflex: A reflexive change in cardiac activity in response to changes in blood pressure. (21)

baroreceptors: The receptors responsible for baroreception. (15, 21)

basal lamina: A layer of filaments and fibers that attach an epithelium to the underlying connective tissue. (4)

basal nuclei: Nuclei of the cerebrum that are important in the subconscious control of skeletal muscle activity. (14, 15)

base: A compound whose dissociation releases a hydroxide ion (OH^-) or removes a hydrogen ion (H^+) from the solution. (2, 27)

basophils: Circulating granulocytes (white blood cells) similar in size and function to tissue mast cells. (19)

B cells: Lymphocytes capable of differentiating into plasmocytes (plasma cells), which produce antibodies. (19, 22)

benign: Not malignant. (3)

beta cells: Cells of the pancreatic islets that secrete insulin in response to elevated blood sugar concentrations. (18)

beta oxidation: Fatty acid catabolism that produces molecules of acetyl-CoA. (25)

beta receptors: Membrane receptors sensitive to epinephrine; stimulation may result in the excitation or inhibition of the target cell. (16)

bicarbonate ions: HCO_3^-; anion components of the carbonic acid–bicarbonate buffer system. (26, 27)

bicuspid: Having two cusps or points; refers to a premolar tooth, which has two roots, or to the left AV valve, which has two cusps. (24)

bicuspid valve: The left atrioventricular (AV) valve, also called *mitral valve*. (20)

bifurcate: To branch into two parts.

bile: The exocrine secretion of the liver; stored in the gallbladder and ejected into the duodenum. (24)

bile salts: Steroid derivatives in bile; responsible for the emulsification of ingested lipids. (2)

bilirubin: A pigment that is the by-product of hemoglobin catabolism. (19)

biopsy: The removal of a small sample of tissue for pathological analysis. (4, 13)

bladder: A muscular sac that distends as fluid is stored and whose contraction ejects the fluid at an appropriate time; used alone, the term usually refers to the urinary bladder. (26)

blastocyst: An early stage in the developing embryo, consisting of an outer trophoblast and an inner cell mass. (29)

blockers/blocking agents: Drugs that block membrane pores or prevent binding to membrane receptors. (16)

blood–brain barrier: The isolation of the central nervous system from the general circulation; primarily the result of astrocyte regulation of capillary permeabilities. (12, 14)

blood clot: *See* clot.

blood–CSF barrier: The isolation of the cerebrospinal fluid from the capillaries of the choroid plexus; primarily the result of specialized ependymal cells. (14)

blood pressure: A force exerted against vessel walls by the blood in the vessels, due to the push exerted by cardiac contraction and the elasticity of the vessel walls; usually measured along one of the muscular arteries, with systolic pressure measured during ventricular systole and diastolic pressure during ventricular diastole. (21)

blood–testis barrier: The isolation of the interior of the seminiferous tubules from the general circulation, due to the activities of the nurse (sustentacular) cells. (28)

Bohr effect: The increased oxygen release by hemoglobin in the presence of elevated carbon dioxide levels. (23)

bolus: A compact mass; usually refers to compacted ingested material on its way to the stomach. (23, 24)

bone: *See* osseous tissue.

bowel: The intestinal tract. (24)

Bowman capsule: The cup-shaped initial portion of the renal tubule; surrounds the glomerulus and receives the glomerular filtrate; also called glomerular capsule. (26)

brachial: Pertaining to the arm.

brachial plexus: A network formed by branches of spinal nerves C_5–T_1 en route to innervating the upper limb. (13)

brachium: The arm. (11)

bradycardia: An abnormally slow heart rate, usually below 50 bpm. (20)

brain natriuretic peptide (BNP): *See* natriuretic peptides.

brain stem: The brain minus the cerebrum, diencephalon, and cerebellum. (14)

brevis: Short.

Broca area: The speech center of the brain, normally located on the neural cortex of the left cerebral hemisphere. (14)

bronchial tree: The trachea, bronchi, and bronchioles. (23)

bronchodilation: The dilation of the bronchial passages; can be caused by sympathetic stimulation. (23)

bronchus/bronchi: A branch of the bronchial tree between the trachea and bronchioles. (23)

buccal: Pertaining to the cheeks. (24)

buffer: A compound that stabilizes the pH of a solution by removing or releasing hydrogen ions. (2, 27)

buffer system: Interacting compounds that prevent increases or decreases in the pH of body fluids; includes the carbonic acid–bicarbonate buffer system, the phosphate buffer system, and the protein buffer system. (27)

bulbar: Pertaining to the brain stem. (14)

bulbo-urethral glands: Mucous glands at the base of the penis that secrete into the penile urethra; the equivalent of the greater vestibular glands of females; also called *Cowper glands*. (28)

bundle branches: Specialized conducting cells in the ventricles that carry the contractile stimulus from the bundle of His to the Purkinje fibers. (20)

bundle of His: Specialized conducting cells in the interventricular septum that carry the contracting stimulus from the AV node to bundle branches and then to Purkinje fibers. (20)

bursa: A small sac filled with synovial fluid that cushions adjacent structures and reduces friction. (9)

C

calcaneal tendon: The large tendon that inserts on the calcaneus; tension on this tendon produces extension (plantar flexion) of the foot; also called *Achilles tendon*. (8, 11)

calcaneus: The heel bone, the largest of the tarsal bones. (8)

calcification: The deposition of calcium salts within a tissue. (4, 6)

calcitonin: The hormone secreted by C cells of the thyroid when calcium ion concentrations are abnormally high; restores homeostasis by increasing the rate of bone deposition and the rate of calcium loss at the kidneys. (6, 18)

calculus/calculi: A solid mass of insoluble materials that form within body fluids, especially the gallbladder, kidneys, or urinary bladder. (26)

callus: A localized thickening of the epidermis due to chronic mechanical stresses (5); a thickened area that forms at the site of a bone break as part of the repair process (6).

calorigenic effect: The stimulation of energy production and heat loss by thyroid hormones. (18)

canaliculi: Microscopic passageways between cells; bile canaliculi carry bile to bile ducts in the liver (24); in bone, canaliculi permit the diffusion of nutrients and wastes to and from osteocytes (4, 6).

cancellous bone: Spongy bone, composed of a network of bony struts. (6)

cancer: An illness caused by mutations leading to the uncontrolled growth and replication of the affected cells. (3)

cannula: A tube that can be inserted into the body; commonly placed in blood vessels prior to transfusion or dialysis. (19)

fenestra: An opening.

fertilization: The fusion of a secondary oocyte and a spermatozoon to form a zygote. (28)

fetus: The developmental stage lasting from the start of the third developmental month to delivery. (28, 29)

fibrin: Insoluble protein fibers that form the basic framework of a blood clot. (19)

fibrinogen: A plasma protein that is the soluble precursor of the insoluble protein fibrin. (19)

fibroblasts: Cells of connective tissue proper that are responsible for the production of extracellular fibers and the secretion of the organic compounds of the extracellular matrix. (4)

fibrocytes: Mature fibroblasts; maintain connective tissue fibers of connective tissue proper. (4)

fibrous cartilage: Cartilage containing an abundance of collagen fibers; located around the edges of joints, in the intervertebral discs, the menisci of the knee, and so on. (4)

fibrous tunic: The outermost layer of the eye, composed of the sclera and cornea. (17)

fibula: The lateral, slender bone of the leg. (8)

filariasis: A condition resulting from infection by mosquito-borne parasites; can cause elephantiasis. (21, 22)

filiform papillae: Slender conical projections from the dorsal surface of the anterior two-thirds of the tongue. (17)

filtrate: The fluid produced by filtration at a glomerulus in the kidney. (26)

filtration: The movement of a fluid across a membrane whose pores restrict the passage of solutes on the basis of size. (21, 26)

filtration pressure: The hydrostatic pressure responsible for filtration. (21, 26)

filum terminale: A fibrous extension of the spinal cord, from the conus medullaris to the coccygeal ligament. (13)

fimbriae: Fringes; the fingerlike processes that surround the entrance to the uterine tube. (28)

fissure: An elongate groove or opening. (7, 14)

fistula: An abnormal passageway between two organs or from an internal organ or space to the body surface.

flaccid: Limp, soft, flabby; a muscle without muscle tone.

flagellum/flagella: An organelle that is structurally similar to a cilium but is used to propel a cell through a fluid; found on spermatozoa. (28)

flatus: Intestinal gas. (24)

flexion: A movement that reduces the angle between two articulating bones; the opposite of extension. (8, 9)

flexor: A muscle that produces flexion. (11)

flexor reflex: A reflex contraction of the flexor muscles of a limb in response to a painful stimulus. (13)

flexure: A bending.

folia: Leaflike folds; the slender folds in the surface of the cerebellar cortex.

follicle: A small secretory sac or gland.

follicle-stimulating hormone (FSH): A hormone secreted by the adenohypophysis (anterior pituitary); stimulates oogenesis (female) and spermatogenesis (male). (18, 28)

fontanelle: A relatively soft, flexible, fibrous region between two flat bones in the developing skull; also spelled *fontanel*. (7)

foramen/foramina: An opening or passage through a bone. (7, 20)

forearm: The distal portion of the upper limb between the elbow and wrist. (8)

forebrain: The cerebrum. (14)

fornix: An arch or the space bounded by an arch; in the brain, an arching tract that connects the hippocampus with the mamillary bodies (14); in the eye, a slender pocket situated where the epithelium of the ocular conjunctiva folds back on itself as the palpebral conjunctiva (17); in the vagina, the shallow recess surrounding the protrusion of the cervix (28).

fossa: A shallow depression or furrow in the surface of a bone. (8, 20)

fourth ventricle: An elongate ventricle of the metencephalon (pons and cerebellum) and the myelencephalon (medulla oblongata) of the brain; the roof contains a region of choroid plexus. (14)

fovea: The portion of the retina that provides the sharpest vision because it has the highest concentration of cones; also called *macula lutea*. (17)

fracture: A break or crack in a bone. (6)

frenulum: A bridle; usually referring to a band of tissue that restricts movement, e.g., *lingual frenulum*. (24)

frontal plane: A sectional plane that divides the body into an anterior portion and a posterior portion; also called *coronal plane*. (1)

fructose: A hexose (six-carbon simple sugar) in foods and in semen. (2, 28)

fundus: The base of an organ.

G

gallbladder: The pear-shaped reservoir for bile after it is secreted by the liver. (24)

gametes: Reproductive cells (spermatozoa or oocytes) that contain half the normal chromosome complement. (28, 29)

gametogenesis: The formation of gametes. (28)

gamma aminobutyric acid (GABA): A neurotransmitter of the central nervous system whose effects are generally inhibitory. (12)

gamma motor neurons: Motor neurons that adjust the sensitivities of muscle spindles (intrafusal fibers). (13)

ganglion/ganglia: A collection of neuron cell bodies outside the central nervous system. (12, 16)

gangliosides: Glycolipids that are important components of plasma membranes in the central nervous system. (12)

gap junctions: Connections between cells that permit electrical coupling. (4)

gaster: The stomach (24); the body, or belly, of a skeletal muscle (11).

gastric: Pertaining to the stomach. (24)

gastric glands: The tubular glands of the stomach whose cells produce acid, enzymes, intrinsic factor, and hormones. (24)

gastrointestinal (GI) tract: *See* digestive tract.

gene: A portion of a DNA strand that functions as a hereditary unit, is located at a particular site on a specific chromosome, and codes for a specific protein or polypeptide. (3, 29)

genetic engineering: Research and experiments involving the manipulation of the genetic makeup of an organism. (29)

genetics: The study of mechanisms of heredity. (29)

geniculate: Like a little knee; the medial geniculates and the lateral geniculates are nuclei in the thalamus of the brain. (14)

genitalia: The reproductive organs. (28)

germinal centers: Pale regions in the interior of lymphoid tissues or nodule, where cell divisions are under way. (22)

gestation: The period of intrauterine development; pregnancy. (29)

gland: Cells that produce exocrine or endocrine secretions. (4)

glenoid cavity: A rounded depression that forms the articular surface of the scapula at the shoulder joint. (8)

glial cells: *See* neuroglia.

globular proteins: Proteins whose tertiary structure makes them rounded and compact. (2)

glomerular capsule: The expanded initial portion of the nephron that surrounds the glomerulus. (26)

glomerular filtration rate: The rate of filtrate formation at the glomerulus. (26)

glomerulus: A ball or knot; in the kidneys, a knot of capillaries that projects into the enlarged, proximal end of a nephron; the site of filtration, the first step in the production of urine. (26)

glossopharyngeal nerve: Cranial nerve IX. (14)

glucagon: A hormone secreted by the alpha cells of the pancreatic islets; elevates blood glucose concentrations. (18)

glucocorticoids: Hormones secreted by the zona fasciculata of the suprarenal (adrenal) cortex to modify glucose metabolism; cortisol and corticosterone are important examples. (18)

gluconeogenesis: The synthesis of glucose from protein or lipid precursors. (25)

glucose: A six-carbon sugar, $C_6H_{12}O_6$; the preferred energy source for most cells and normally the only energy source for neurons. (2, 10, 18, 25)

glycerides: Lipids composed of glycerol bound to fatty acids. (2)

glycogen: A polysaccharide that is an important energy reserve; a polymer consisting of a long chain of glucose molecules. (2, 10)

glycogenesis: The synthesis of glycogen from glucose molecules. (25)

glycogenolysis: Glycogen breakdown and the liberation of glucose molecules. (25)

glycolipids: Compounds created by the combination of carbohydrate and lipid components. (2)

glycolysis: The anaerobic cytoplasmic breakdown of glucose into lactic acid by way of pyruvic acid, with a net gain of two ATP molecules. (3, 10, 25)

glycoprotein: A compound containing a relatively small carbohydrate group attached to a large protein. (2, 18)

glycosuria: The presence of glucose in urine. (18, 26)

Golgi apparatus: A cellular organelle consisting of a series of membranous plates that give rise to lysosomes and secretory vesicles. (3)

gomphosis: A fibrous synarthrosis that binds a tooth to the bone of the jaw; *see* periodontal ligament. (24)

gonadotropin-releasing hormone (GnRH): A hypothalamic releasing hormone that causes the secretion of both follicle-stimulating hormone and luteinizing hormone by the adenohypophysis (anterior pituitary gland). (18, 28)

gonadotropins: Follicle-stimulating hormone and luteinizing hormone, hormones that stimulate gamete development and sex hormone secretion. (18, 28)

gonads: Reproductive organs that produce gametes and hormones. (28)

granulocytes: White blood cells containing granules that are visible with the light microscope; includes eosinophils, basophils, and neutrophils; also called *granular leukocytes*. (19)

gray matter: Areas in the central nervous system that are dominated by neuron cell bodies, neuroglia, and unmyelinated axons. (12, 13, 14)

gray ramus: A bundle of postganglionic sympathetic nerve fibers that are distributed to effectors in the body wall, skin, and limbs by way of a spinal nerve. (13)

greater omentum: A large fold of the dorsal mesentery of the stomach; hangs anterior to the intestines. (24)

groin: The inguinal region. (11)

gross anatomy: The study of the structural features of the body without the aid of a microscope.

growth hormone (GH): An adenohypophysis (anterior pituitary) hormone that stimulates tissue growth and anabolism when nutrients are abundant and restricts tissue glucose dependence when nutrients are in short supply. (18)

growth hormone–inhibiting hormone (GH–IH): A hypothalamic regulatory hormone that inhibits growth hormone secretion by the adenohypophysis; also called *somatostatin.* (18)

guanine: A purine; one of the nitrogenous bases in the nucleic acids RNA and DNA. (2)

gustation: Taste. (15, 17)

gyrus: A prominent fold or ridge of neural cortex on the surfaces of the cerebral hemispheres. (14)

H

hair: A keratinous strand produced by epithelial cells of the hair follicle. (5)

hair cells: Sensory cells of the inner ear. (17)

hair follicle: An accessory structure of the integument; a tube lined by a stratified squamous epithelium that begins at the surface of the skin and ends at the hair papilla. (5)

hallux: The big toe. (8)

haploid: Possessing half the normal number of chromosomes; a characteristic of gametes. (28, 29)

hard palate: The bony roof of the oral cavity, formed by the maxillae and palatine bones. (23, 24)

helper T cells: Lymphocytes whose secretions and other activities coordinate cell-mediated and antibody-mediated immunities; also called T_H *cells.* (22)

hematocrit: The percentage of the volume of whole blood contributed by cells; also called *volume of packed red cells (VPRC)* or *packed cell volume (PCV).* (19)

hematoma: A tumor or swelling filled with blood.

hematuria: The abnormal presence of red blood cells in urine. (19, 26)

heme: A porphyrin ring containing a central iron atom that can reversibly bind oxygen molecules; a component of the hemoglobin molecule. (19)

hemocytoblasts: Stem cells whose divisions produce each of the various populations of blood cells. (19)

hemoglobin: A protein composed of four globular subunits, each bound to a heme molecule; gives red blood cells the ability to transport oxygen in the blood. (5, 19, 23, 27)

hemolysis: The breakdown of red blood cells. (3)

hemopoiesis: Blood cell formation and differentiation. (19)

hemorrhage: Blood loss; to bleed. (21)

hemostasis: The cessation of bleeding. (19)

heparin: An anticoagulant released by activated basophils and mast cells. (4, 19)

hepatic duct: The duct that carries bile away from the liver lobes and toward the union with the cystic duct. (24)

hepatic portal vein: The vessel that carries blood between the intestinal capillaries and the sinusoids of the liver. (21)

hepatocyte: A liver cell. (24)

heterotopic: Ectopic; outside the normal location.

heterozygous: Possessing two different alleles at corresponding sites on a chromosome pair; the individual's phenotype is determined by one or both of the alleles. (29)

hexose: A six-carbon simple sugar. (2)

hiatus: A gap, cleft, or opening.

high-density lipoprotein (HDL): A lipoprotein with a relatively small lipid content; thought to be responsible for the movement of cholesterol from peripheral tissues to the liver. (25)

hilum: A localized region where blood vessels, lymphatic vessels, nerves, and/or other anatomical structures are attached to an organ. (22, 23, 26)

hippocampus: A region, beneath the floor of a lateral ventricle, involved with emotional states and the conversion of short-term to long-term memories. (12, 14)

histamine: The chemical released by stimulated mast cells or basophils to initiate or enhance an inflammatory response. (4, 12)

histology: The study of tissues. (1, 4)

histones: Proteins associated with the DNA of the nucleus; the DNA strands are wound around them. (3)

holocrine: A form of exocrine secretion in which the secretory cell becomes swollen with vesicles and then ruptures. (4)

homeostasis: The maintenance of a relatively constant internal environment. (1)

hormone: A compound that is secreted by one cell and travels through the circulatory system to affect the activities of cells in another portion of the body. (2, 4, 6, 18, 21, 24, 28, 29)

human chorionic gonadotropin (hCG): The placental hormone that maintains the corpus luteum for the first 3 months of pregnancy. (29)

human immunodeficiency virus (HIV): The infectious agent that causes acquired immune deficiency syndrome (AIDS). (22)

human leukocyte antigen (HLA): *See* MHC protein.

human placental lactogen (hPL): The placental hormone that stimulates the functional development of the mammary glands. (29)

humoral immunity: *See* antibody-mediated immunity.

hyaluronan: A carbohydrate component of proteoglycans in the matrix of many connective tissues. (4)

hyaluronidase: An enzyme that breaks down the bonds between adjacent follicle cells; produced by some bacteria and found in the acrosomal cap of a spermatozoon. (29)

hydrogen bond: A weak interaction between the hydrogen atom on one molecule and a negatively charged portion of another molecule. (2)

hydrolysis: The breakage of a chemical bond through the addition of a water molecule; the reverse of dehydration synthesis. (2)

hydrophilic: Freely associating with water; readily entering into solution; water-loving. (2)

hydrophobic: Incapable of freely associating with water molecules; insoluble; water-fearing. (2)

hydrostatic pressure: Fluid pressure. (21, 26)

hydroxide ion: OH^-. (2)

hypercapnia: High plasma carbon dioxide concentrations, commonly as a result of hypoventilation or inadequate tissue perfusion. (23, 27)

hyperplasia: An abnormal enlargement of an organ due to an increase in the number of cells. (3)

hyperpolarization: The movement of the transmembrane potential away from the normal resting potential and farther from 0 mV. (12)

hypersecretion: The overactivity of glands that produce exocrine or endocrine secretions. (18)

hypertension: Abnormally high blood pressure. (21)

hypertonic: In comparing two solutions, the solution with the higher osmolarity. (3)

hypertrophy: An increase in tissue size without cell division. (10)

hyperventilation: A rate of respiration sufficient to reduce plasma P_{CO_2} concentrations to levels below normal. (23, 27)

hypocapnia: An abnormally low plasma P_{CO_2} concentrations commonly as a result of hyperventilation. (27)

hypodermic needle: A needle inserted through the skin to introduce drugs into the subcutaneous layer. (5)

hypodermis: *See* subcutaneous layer.

hypophyseal portal system: The network of vessels that carries blood from capillaries in the hypothalamus to capillaries in the adenohypophysis (anterior lobe of the pituitary gland). (18)

hypophysis: The pituitary gland. (18)

hyposecretion: Abnormally low rates of exocrine or endocrine secretion. (18)

hypothalamus: The floor of the diencephalon; the region of the brain containing centers involved with the subconscious regulation of visceral functions, emotions, drives, and the coordination of neural and endocrine functions. (14)

hypothermia: An abnormally low body temperature. (25)

hypothesis: A prediction that can be subjected to scientific analysis and review.

hypotonic: In comparing two solutions, the solution with the lower osmolarity. (3)

hypoventilation: A respiratory rate that is insufficient to keep plasma P_{CO_2} concentrations within normal levels. (23, 27)

hypoxia: A low tissue oxygen concentration. (19, 23)

I

ileum: The distal 2.5 m of the small intestine. (24)

ilium: The largest of the three bones whose fusion creates a coxal bone. (8)

immunity: Resistance to injuries and diseases caused by foreign compounds, toxins, or pathogens. (22)

immunization: The production of immunity by the deliberate exposure to antigens under conditions that prevent the development of illness but stimulate the production of memory B cells. (22)

immunoglobulin: A circulating antibody. (19, 22)

implantation: The attachment of a blastocyst into the endometrium of the uterine wall. (29)

inclusions: Aggregations of insoluble pigments, nutrients, or other materials in cytoplasm. (3)

incus: The central auditory ossicle, situated between the malleus and the stapes in the middle ear cavity. (17)

inducer: A stimulus that promotes the activity of a specific gene. (29)

inexcitable: Incapable of propagating an action potential. (12)

infarct: An area of dead cells that results from an interruption of blood flow. (19, 20)

infection: The invasion and colonization of body tissues by pathogens. (4)

inferior: Below, in reference to a particular structure, with the body in the anatomical position.

inferior vena cava: The vein that carries blood from the parts of the body inferior to the heart to the right atrium. (20, 21)

infertility: The inability to conceive; also called *sterility.* (28, 29)

inflammation: A nonspecific defense mechanism that operates at the tissue level; characterized by swelling, redness, warmth, pain, and some loss of function. (4, 22)

infundibulum: A tapering, funnel-shaped structure; in the brain, the connection between the pituitary gland and the hypothalamus (14, 18); in the uterine tube, the entrance bounded by fimbriae that receives the oocytes at ovulation (28).

ingestion: The introduction of materials into the digestive tract by way of the mouth; eating. (24)

inguinal canal: A passage through the abdominal wall that marks the path of testicular descent and that contains the testicular arteries, veins, and ductus deferens. (11, 28)

inguinal region: The area of the abdominal wall near the junction of the trunk and the thighs that contains the external genitalia; the groin. (28)

inhibin: A hormone, produced by nurse (sustentacular) cells of the testes and follicular cells of the ovaries, that inhibits the secretion of follicle-stimulating hormone by the adenohypophysis (anterior lobe of the pituitary gland). (18, 28)

inhibitory postsynaptic potential (IPSP): A hyperpolarization of the postsynaptic membrane after the arrival of a neurotransmitter. (12)

initial segment: The proximal portion of the axon where an action potential first appears. (12)

injection: The forcing of fluid into a body part or organ.

inner cell mass: Cells of the blastocyst that will form the body of the embryo. (29)

inner ear: *See* internal ear.

innervation: The distribution of sensory and motor nerves to a specific region or organ. (11, 16)

insensible perspiration: Evaporative water loss by diffusion across the epithelium of the skin or evaporation across the alveolar surfaces of the lungs. (5, 27)

insertion: A point of attachment of a muscle; the end that is easily movable. (11)

insoluble: Incapable of dissolving in solution. (2)

inspiration: Inhalation; the movement of air into the respiratory system. (23)

insulin: A hormone secreted by beta cells of the pancreatic islets; causes a reduction in plasma glucose concentrations. (18)

integument: The skin. (5)

intercalated discs: Regions where adjacent cardiocytes interlock and where gap junctions permit electrical coupling between the cells. (4, 10, 20)

intercellular cement: Proteoglycans situated between adjacent epithelial cells. (4)

intercellular fluid: *See* interstitial fluid.

interferons: Peptides released by virus-infected cells, especially lymphocytes, that slow viral replication and make other cells more resistant to viral infection. (22)

interleukins: Peptides, released by activated monocytes and lymphocytes, that assist in the coordination of cell-mediated and antibody-mediated immunities. (22)

internal capsule: The collection of afferent and efferent fibers of the white matter of the cerebral hemispheres, visible on gross dissection of the brain. (14)

internal ear: The membranous labyrinth that contains the organs of hearing and equilibrium. (17)

internal nares: The entrance to the nasopharynx from the nasal cavity. (23)

internal respiration: The diffusion of gases between interstitial fluid and cytoplasm. (23)

interneuron: An association neuron; central nervous system neurons that are between sensory and motor neurons. (12)

interoceptors: Sensory receptors monitoring the functions and status of internal organs and systems. (12)

interosseous membrane: The fibrous connective tissue membrane between the shafts of the tibia and fibula and between the radius and ulna; an example of a fibrous amphiarthrosis. (8)

interphase: The stage in the life cycle of a cell during which the chromosomes are uncoiled and all normal cellular functions except mitosis are under way. (29)

intersegmental reflex: A reflex that involves several segments of the spinal cord. (13)

interstitial fluid: The fluid in the tissues that fills the spaces between cells. (3)

interstitial growth: A form of cartilage growth through the growth, mitosis, and secretion of chondrocytes in the matrix. (4)

interventricular foramen: The opening that permits fluid movement between the lateral and third ventricles of the brain. (14)

intervertebral disc: A fibrous cartilage pad between the bodies of successive vertebrae that absorbs shocks. (7, 9)

intestinal crypt: A tubular epithelial pocket that is lined by secretory cells and opens into the lumen of the digestive tract; also called *intestinal gland*. (24)

intestine: The tubular organ of the digestive tract. (18, 24)

intracellular fluid: The cytosol. (27)

intrafusal fibers: Muscle spindle fibers. (13)

intramembranous ossification: The formation of bone within a connective tissue without the prior development of a cartilaginous model. (6)

intrinsic factor: A glycoprotein, secreted by the parietal cells of the stomach, that facilitates the intestinal absorption of vitamin B_{12}. (19, 24, 25)

intrinsic pathway: A pathway of the clotting system that begins with the activation of platelets and ends with the formation of platelet thromboplastin. (19)

inversion: A turning inward. (9)

in vitro: Outside the body, in an artificial environment.

in vivo: In the living body.

involuntary: Not under conscious control.

ion: An atom or molecule bearing a positive or negative charge due to the donation or acceptance, respectively, of an electron. (2, 26, 27)

ionic bond: A molecular bond created by the attraction between ions with opposite charges. (2)

ionization: Dissociation; the breakdown of a molecule in solution to form ions. (2)

ipsilateral: A reflex response that affects the same side as the stimulus. (13)

iris: A contractile structure, made up of smooth muscle, that forms the colored portion of the eye. (17)

ischemia: An inadequate blood supply to a region of the body. (11)

ischium: One of the three bones whose fusion creates a coxal bone. (8)

islets of Langerhans: *See* pancreatic islets.

isotonic: A solution with an osmolarity that does not result in water movement across plasma membranes. (10)

isotopes: Forms of an element whose atoms contain the same number of protons but different numbers of neutrons (and thus differ in atomic weight). (2)

isthmus: A narrow band of tissue connecting two larger masses.

J

jejunum: The middle part of the small intestine. (24)

joint: An area where adjacent bones interact; also called *articulation*. (9)

juxtaglomerular cells: Modified smooth muscle cells in the walls of the afferent and efferent arterioles adjacent to the glomerulus and the macula densa. (26)

juxtaglomerular complex: The macula densa and the juxtaglomerular cells; a complex responsible for the release of renin and erythropoietin. (26)

K

keratin: The tough, fibrous protein component of nails, hair, calluses, and the general integumentary surface. (5)

keto acid: A molecule that ends in —COCOOH; the carbon chain that remains after the deamination or transamination of an amino acid. (25)

ketoacidosis: A reduction in the pH of body fluids due to the presence of large numbers of ketone bodies. (25, 26, 27)

ketone bodies: Keto acids produced during the catabolism of lipids and ketogenic amino acids; specifically, acetone, acetoacetate, and beta-hydroxybutyrate. (25)

kidney: A component of the urinary system; an organ functioning in the regulation of plasma composition, including the excretion of wastes and the maintenance of normal fluid and electrolyte balances. (18, 26)

killer T cells: *See* cytotoxic T cells.

Krebs cycle: *See* TCA cycle.

Kupffer cells: Stellate reticular cells of the liver; phagocytic cells of the liver sinusoids. (22, 24)

L

labium/labia: Lip; the labia majora and labia minora are components of the female external genitalia. (28)

labrum: A lip or rim.

labyrinth: A maze of passageways; the structures of the internal ear. (17)

lacrimal gland: A tear gland on the dorsolateral surface of the eye. (17)

lactase: An enzyme that breaks down the milk sugar, lactose. (24)

lactation: The production of milk by the mammary glands. (28)

lacteal: A terminal lymphatic within an intestinal villus. (24)

lactic acid: A compound produced from pyruvic acid under anaerobic conditions. (10)

lacuna: A small pit or cavity. (4, 6)

lambdoid suture: The synarthrosis between the parietal and occipital bones of the cranium. (7)

lamellae: Concentric layers; the concentric layers of bone within an osteon. (6)

lamellated corpuscle: A receptor sensitive to vibration. (15)

lamina: A thin sheet or layer.

lamina propria: The reticular tissue that underlies a mucous epithelium and forms part of a mucous membrane. (4, 23, 24)

Langerhans cells: Cells in the epithelium of the skin (15) and digestive tract (24) that participate in the immune response by presenting antigens to T cells; also called dendritic cells.

large intestine: The terminal portions of the intestinal tract, consisting of the colon, the rectum, and the anal canal. (24)

laryngopharynx: The division of the pharynx that is inferior to the epiglottis and superior to the esophagus. (23)

larynx: A complex cartilaginous structure that surrounds and protects the glottis and vocal cords; the superior margin is bound to the hyoid bone, and the inferior margin is bound to the trachea. (23)

latent period: The time between the stimulation of a muscle and the start of the contraction phase. (10)

lateral: Pertaining to the side.

lateral apertures: Openings in the roof of the fourth ventricle that permit the circulation of cerebrospinal fluid into the subarachnoid space. (14)

lateral ventricle: A fluid-filled chamber within a cerebral hemisphere. (14)

lens: The transparent refractive structure that is between the iris and the vitreous humor. (17)

lesion: A localized abnormality in tissue organization. (4)

lesser omentum: A small pocket in the mesentery that connects the lesser curvature of the stomach to the liver. (24)

leukocyte: A white blood cell. (4, 19)

ligament: A dense band of connective tissue fibers that attaches one bone to another. (4, 9)

ligamentum arteriosum: The fibrous strand in adults that is the remnant of the ductus arteriosus of the fetal stage. (21)

ligamentum nuchae: An elastic ligament between the vertebra prominens and the occipital bone. (7)

ligamentum teres: The fibrous strand in the falciform ligament of adults that is the remnant of the umbilical vein of the fetal stage. (24)

ligate: To tie off.

limbic system: The group of nuclei and centers in the cerebrum and diencephalon that are involved with emotional states, memories, and behavioral drives. (14)

lingual: Pertaining to the tongue. (17, 24)

lipid: An organic compound containing carbons, hydrogens, and oxygens in a ratio that does not approximate 1:2:1; includes fats, oils, and waxes. (2, 24, 25)

lipogenesis: The synthesis of lipids from nonlipid precursors. (25)

lipoids: Prostaglandins, steroids, phospholipids, glycolipids, and so on. (25)

lipolysis: The catabolism of lipids as a source of energy. (25)

lipoprotein: A compound containing a relatively small lipid bound to a protein. (25)

liver: An organ of the digestive system that has varied and vital functions, including the production of plasma proteins, the excretion of bile, the storage of energy reserves, the detoxification of poisons, and the interconversion of nutrients. (24)

lobule: Histologically, the basic organizational unit of the liver. (24)

local hormone: *See* prostaglandin.

loop of Henle: The portion of the nephron that creates the concentration gradient in the renal medulla; also called nephron loop. (26)

loose connective tissue: A loosely organized, easily distorted connective tissue that contains several fiber types, a varied population of cells, and a viscous ground substance. (4)

lumbar: Pertaining to the lower back. (7, 13)

lumen: The central space within a duct or other internal passageway. (4)

lungs: The paired organs of respiration, situated in the pleural cavities. (23)

luteinizing hormone (LH): Also called *lutropin*; a hormone produced by the adenohypophysis (anterior lobe of the pituitary gland). In females, it assists FSH in follicle stimulation, triggers ovulation, and promotes the maintenance and secretion of endometrial glands. In males, it was formerly called *interstitial cell-stimulating hormone* because it stimulates testosterone secretion by the interstitial cells of the testes. (18, 28)

lymph: The fluid contents of lymphatic vessels, similar in composition to interstitial fluid. (4, 22)

lymphatic vessels: The vessels of the lymphoid system; also called *lymphatics*. (4, 22)

lymph nodes: Lymphoid organs that monitor the composition of lymph. (22)

lymphocyte: A cell of the lymphoid system that participates in the immune response. (4, 19, 22)

lymphokines: Chemicals secreted by activated lymphocytes. (22)

lymphopoiesis: The production of lymphocytes from lymphoid stem cells. (19, 22)

lymphotoxin: A secretion of lymphocytes that kills the target cells. (22)

lysis: The destruction of a cell through the rupture of its plasma membrane. (3)

lysosome: An intracellular vesicle containing digestive enzymes. (3)

lysozyme: An enzyme, present in some exocrine secretions, that has antibiotic properties. (17)

M

macrophage: A phagocytic cell of the monocyte–macrophage system. (4, 22)

macula: A receptor complex, located in the saccule or utricle of the inner ear, that responds to linear acceleration or gravity. (17)

macula densa: A group of specialized secretory cells that is located in a portion of the distal convoluted tubule, adjacent to the glomerulus and the juxtaglomerular cells; a component of the juxtaglomerular complex. (26)

macula lutea: *See* fovea.

major histocompatibility complex: *See* MHC protein.

malignant tumor: A form of cancer characterized by rapid cell growth and the spread of cancer cells throughout the body. (3)

malleus: The first auditory ossicle, bound to the tympanic membrane and the incus. (17)

malnutrition: An unhealthy state produced by inadequate dietary intake or absorption of nutrients, calories, and/or vitamins. (25)

mamillary bodies: Nuclei in the hypothalamus that affect eating reflexes and behaviors; a component of the limbic system. (14)

mammary glands: Milk-producing glands of the female breast. (5, 28)

manus: The hand. (8, 11)

marrow: A tissue that fills the internal cavities in bone; dominated by hemopoietic cells (red bone marrow) or by adipose tissue (yellow bone marrow). (6, 19)

mast cell: A connective tissue cell that, when stimulated, releases histamine, serotonin, and heparin, initiating the inflammatory response. (4)

mastication: Chewing. (11, 24)

mastoid sinus: Air-filled spaces in the mastoid process of the temporal bone. (7)

matrix: The extracellular fibers and ground substance of a connective tissue. (4)

maxillary sinus: One of the paranasal sinuses; an air-filled chamber lined by a respiratory epithelium that is located in a maxilla and opens into the nasal cavity. (7)

meatus: An opening or entrance into a passageway. (23, 26)

mechanoreception: The detection of mechanical stimuli, such as touch, pressure, or vibration. (15)

medial: Toward the midline of the body.

mediastinum: The central tissue mass that divides the thoracic cavity into two pleural cavities (1, 20); includes the aorta and other great vessels, the esophagus, trachea, thymus, the pericardial cavity and heart, and a host of nerves, small vessels, and lymphatic vessels; in males, the area of connective tissue attaching a testis to the epididymis, proximal portion of ductus deferens, and associated vessels (28).

medulla: The inner layer or core of an organ.

medulla oblongata: The most caudal of the brain regions, also called the *myelencephalon*. (14)

medullary cavity: The space within a bone that contains the marrow. (6)

medullary rhythmicity center: The center in the medulla oblongata that sets the background pace of respiration; includes inspiratory and expiratory centers. (14)

megakaryocytes: Bone marrow cells responsible for the formation of platelets. (19)

meiosis: Cell division that produces gametes with half the normal somatic chromosome complement. (3, 28)

melanin: The yellow-brown pigment produced by the melanocytes of the skin. (5)

melanocyte: A specialized cell in the deeper layers of the stratified squamous epithelium of the skin; responsible for the production of melanin. (4, 5, 18)

melanocyte-stimulating hormone (MSH): A hormone, produced by the pars intermedia of the adenohypophysis (anterior lobe of the pituitary gland), that stimulates melanin production. (18)

melatonin: A hormone secreted by the pineal gland; inhibits secretion of MSH and GnRH. (14, 18)

membrane: Any sheet or partition; a layer consisting of an epithelium and the underlying connective tissue. (2)

membrane flow: The movement of sections of membrane surface to and from the cell surface and components of the endoplasmic reticulum, the Golgi apparatus, and vesicles. (3)

membrane potential: *See* transmembrane potential.

membranous labyrinth: Endolymph-filled tubes that enclose the receptors of the inner ear. (17)

memory: The ability to recall information or sensations; can be divided into short-term and long-term memories. (16, 22)

meninges: Three membranes that surround the surfaces of the central nervous system; the dura mater, the pia mater, and the arachnoid. (13)

meniscus: A fibrous cartilage pad between opposing surfaces in a joint. (9)

menses: The first portion of the uterine cycle in which the endometrial lining sloughs away; menstrual period. (28)

merocrine: A method of secretion in which the cell ejects materials from secretory vesicles through exocytosis. (4, 5)

mesencephalon: The midbrain; the region between the diencephalon and pons. (14)

mesenchyme: Embryonic or fetal connective tissue. (4)

mesentery: A double layer of serous membrane that supports and stabilizes the position of an organ in the abdominopelvic cavity and provides a route for the associated blood vessels, nerves, and lymphatic vessels. (24)

mesoderm: The middle germ layer, between the ectoderm and endoderm of the embryo. (29)

mesothelium: A simple squamous epithelium that lines one of the divisions of the ventral body cavity. (4)

messenger RNA (mRNA): RNA formed at transcription to direct protein synthesis in the cytoplasm. (2, 3)

metabolic turnover: The continuous breakdown and replacement of organic materials within cells. (2, 25)

metabolism: The sum of all biochemical processes under way within the human body at any moment; includes anabolism and catabolism. (25)

metabolites: Compounds produced in the body as a result of metabolic reactions. (2)

metacarpal bones: The five bones of the palm of the hand. (8)

metalloproteins: Plasma proteins that transport metal ions. (19)

metaphase: The stage of mitosis in which the chromosomes line up along the equatorial plane of the cell. (3)

metaphysis: The region of a long bone between the epiphysis and diaphysis, corresponding to the location of the epiphyseal cartilage of the developing bone. (6)

metarteriole: A vessel that connects an arteriole to a venule and that provides blood to a capillary plexus. (21)

metastasis: The spread of cancer cells from one organ to another, leading to the establishment of secondary tumors. (3)

metatarsal bone: One of the five bones of the foot that articulate with the tarsal bones (proximally) and the phalanges (distally). (8)

metencephalon: The pons and cerebellum of the brain. (14)

MHC protein: A surface antigen that is important to the recognition of foreign antigens and that plays a role in the coordination and activation of the immune response; also called *human leukocyte antigen (HLA)*. (22)

micelle: A droplet with hydrophilic portions on the outside; a spherical aggregation of bile salts, monoglycerides, and fatty acids in the lumen of the intestinal tract. (2, 24)

microfilaments: Fine protein filaments visible with the electron microscope; components of the cytoskeleton. (3)

microglia: Phagocytic neuroglia in the central nervous system. (12, 22)

microphages: Neutrophils and eosinophils. (4, 19, 22)

microtubules: Microscopic tubules that are part of the cytoskeleton and are a component in cilia, flagella, the centrioles, and spindle fibers. (3)

microvilli: Small, fingerlike extensions of the exposed plasma membrane of an epithelial cell. (3)

micturition: Urination. (26)

midbrain: The mesencephalon. (14)

middle ear: The space between the external and internal ears that contains auditory ossicles. (17)

midsagittal plane: A plane passing through the midline of the body that divides it into left and right halves. (1)

mineralocorticoid: Corticosteroids produced by the zona glomerulosa of the suprarenal (adrenal) cortex; steroids such as aldosterone that affect mineral metabolism. (18)

mitochondrion: An intracellular organelle responsible for generating most of the ATP required for cellular operations. (3, 25)

mitosis: The division of a single cell nucleus that produces two identical daughter cell nuclei; an essential step in cell division. (3, 28)

mitral valve: *See* bicuspid valve.

mixed gland: A gland that contains exocrine and endocrine cells, or an exocrine gland that produces serous and mucous secretions. (4)

mixed nerve: A peripheral nerve that contains sensory and motor fibers. (13)

mole: A quantity of an element or compound having a mass in grams equal to the element's atomic weight or to the compound's molecular weight. (2)

molecular weight: The sum of the atomic weights of all the atoms in a molecule. (3)

molecule: A chemical structure containing two or more atoms that are held together by chemical bonds. (3)

monoclonal antibodies: Antibodies produced by genetically identical cells under laboratory conditions. (22)

monocytes: Phagocytic agranulocytes (white blood cells) in the circulating blood. (19)

monoglyceride: A lipid consisting of a single fatty acid bound to a molecule of glycerol. (2)

monokines: Secretions released by activated cells of the monocyte–macrophage system to coordinate various aspects of the immune response. (22)

monosaccharide: A simple sugar, such as glucose or ribose. (2, 24)

monosynaptic reflex: A reflex in which the sensory afferent neuron synapses directly on the motor efferent neuron. (13)

motor unit: All of the muscle cells controlled by a single motor neuron. (10)

mucins: Proteoglycans responsible for the lubricating properties of mucus. (2, 24)

mucosa: A mucous membrane; the epithelium plus the lamina propria. (4, 24)

mucosa-associated lymphoid tissue (MALT): The extensive collection of lymphoid tissues linked with the digestive system. (22)

mucous (adjective): Indicating the presence or production of mucus.

mucous cell: A goblet-shaped, mucus-producing, unicellular gland in certain epithelia of the digestive and respiratory tracts; also called goblet cells. (4)

mucous membrane: *See* mucosa.

mucus (noun): A lubricating fluid that is composed of water and mucins and is produced by unicellular and multicellular glands along the digestive, respiratory, urinary, and reproductive tracts. (2, 4)

multipolar neuron: A neuron with many dendrites and a single axon; the typical form of a motor neuron. (12)

multiunit smooth muscle: A smooth muscle tissue whose muscle cells are innervated in motor units. (10)

muscarinic receptors: Membrane receptors sensitive to acetylcholine and to muscarine, a toxin produced by certain mushrooms; located at all parasympathetic neuromuscular and neuroglandular junctions and at a few sympathetic neuromuscular and neuroglandular junctions. (16)

muscle: A contractile organ composed of muscle tissue, blood vessels, nerves, connective tissues, and lymphatic vessels. (10, 11)

muscle tissue: A tissue characterized by the presence of cells capable of contraction; includes skeletal, cardiac, and smooth muscle tissues. (4, 10)

muscularis externa: Concentric layers of smooth muscle responsible for peristalsis. (24)

muscularis mucosae: The layer of smooth muscle beneath the lamina propria; responsible for moving the mucosal surface. (24)

mutagens: Chemical agents that induce mutations and may be carcinogenic. (3)

mutation: A change in the nucleotide sequence of the DNA in a cell. (3)

myelencephalon: *See* medulla oblongata.

myelin: An insulating sheath around an axon; consists of multiple layers of neuroglial membrane; significantly increases the impulse propagation rate along the axon. (12)

myelination: The formation of myelin. (12)

myenteric plexus: Parasympathetic motor neurons and sympathetic postganglionic fibers located between the circular and longitudinal layers of the muscularis externa. (24)

myocardial infarction: A heart attack; damage to the heart muscle due to an interruption of regional coronary circulation. (20)

myocardium: The cardiac muscle tissue of the heart. (20)

myofibril: Organized collections of myofilaments in skeletal and cardiac muscle cells. (10)

myofilaments: Fine protein filaments composed primarily of the proteins actin (thin filaments) and myosin (thick filaments). (10)

myoglobin: An oxygen-binding pigment that is especially common in slow skeletal muscle fibers and cardiac muscle cells. (2, 10)

myogram: A recording of the tension produced by muscle fibers on stimulation. (10)

myometrium: The thick layer of smooth muscle in the wall of the uterus. (28)

myosepta: Connective tissue partitions that separate adjacent skeletal muscles. (11)

myosin: The protein component of thick filaments. (3, 10)

N

nail: A keratinous structure produced by epithelial cells of the nail root. (5)

nares, external: The entrance from the exterior to the nasal cavity. (23)

nares, internal: The entrance from the nasal cavity to the nasopharynx. (23)

nasal cavity: A chamber in the skull that is bounded by the internal and external nares. (7)

nasolacrimal duct: The passageway that transports tears from the nasolacrimal sac to the nasal cavity. (7, 17)

nasolacrimal sac: A chamber that receives tears from the lacrimal ducts. (17)

nasopharynx: A region that is posterior to the internal nares and superior to the soft palate and ends at the oropharynx. (23)

natriuretic peptides (NP): Hormones released by specialized cardiocytes when they are stretched by an abnormally large venous return; promotes fluid loss and reductions in blood pressure and in venous return. Includes atrial natriuretic peptide (ANP) and brain natriuretic peptide (BNP). (18, 21, 26, 27)

N compound: An organic compound containing nitrogen atoms. (25)

necrosis: The death of cells or tissues from disease or injury. (4, 22)

negative feedback: A corrective mechanism that opposes or negates a variation from normal limits. (18)

neonate: A newborn infant, or baby. (29)

neoplasm: A tumor, or mass of abnormal tissue. (3)

nephron: The basic functional unit of the kidney. (26)

nephron loop: The portion of the nephron that creates the concentration gradient in the renal medulla; also called *loop of Henle*. (26)

nerve impulse: An action potential in a neuron plasma membrane. (12)

neural cortex: An area of gray matter at the surface of the central nervous system. (13)

neurilemma: The outer surface of a neuroglia that encircles an axon. (12)

neurofibrils: Microfibrils in the cytoplasm of a neuron. (12)

neurofilaments: Microfilaments in the cytoplasm of a neuron. (12)

neuroglandular junction: A cell junction at which a neuron controls or regulates the activity of a secretory (gland) cell. (12)

neuroglia: Cells of the central nervous system and peripheral nervous system that support and protect neurons; also called *glial cells*. (4, 12)

neurohypophysis: The posterior pituitary, or pars nervosa; stores and releases OXT and ADH. (18)

neuromodulator: A compound, released by a neuron, that adjusts the sensitivities of another neuron to specific neurotransmitters. (12)

neuromuscular junction: A synapse between a neuron and a muscle cell. (10, 12)

neuron: A cell in neural tissue that is specialized for intercellular communication through (1) changes in membrane potential and (2) synaptic connections. (4, 12, 15, 16)

neurotransmitter: A chemical compound released by one neuron to affect the transmembrane potential of another. (12, 16)

neurotubules: Microtubules in the cytoplasm of a neuron. (12)

neurulation: The embryological process responsible for the formation of the central nervous system. (29)

neutron: A fundamental particle that does not carry a positive or a negative charge. (2)

neutrophil: A microphage that is very numerous and normally the first of the mobile phagocytic cells to arrive at an area of injury or infection. (19)

nicotinic receptors: Acetylcholine receptors on the surfaces of sympathetic and parasympathetic ganglion cells; respond to the compound nicotine. (16)

nipple: An elevated epithelial projection on the surface of the breast; contains the openings of the lactiferous sinuses. (28)

Nissl bodies: The ribosomes, Golgi apparatus, rough endoplasmic reticulum, and mitochondria of the perikaryon of a typical neuron. (12)

nitrogenous wastes: Organic waste products of metabolism that contain nitrogen, such as urea, uric acid, and creatinine. (25)

nociception: Pain perception. (15)

node of Ranvier: The area between adjacent neuroglia where the myelin covering of an axon is incomplete. (12)

nodose ganglion: A sensory ganglion of cranial nerve X; also called inferior ganglion. (14)

noradrenaline: *See* norepinephrine.

norepinephrine (NE): A catecholamine neurotransmitter in the peripheral nervous system and central nervous system, released at most sympathetic neuromuscular and neuroglandular junctions, and a hormone secreted by the suprarenal (adrenal) medulla; also called *noradrenaline*. (12, 18)

nucleic acid: A polymer of nucleotides that contains a pentose sugar, a phosphate group, and one of four nitrogenous bases that regulate the synthesis of proteins and make up the genetic material in cells. (2)

nucleolus: The dense region in the nucleus that is the site of RNA synthesis. (3)

nucleoplasm: The fluid content of the nucleus. (3)

nucleoproteins: Proteins of the nucleus that are generally associated with DNA. (3)

nucleotide: A compound consisting of a nitrogenous base, a simple sugar, and a phosphate group. (2)

nucleus: A cellular organelle that contains DNA, RNA, and proteins; in the central nervous system, a mass of gray matter. (3)

nucleus pulposus: The gelatinous central region of an intervertebral disc. (9)

nutrient: An inorganic or organic compound that can be broken down in the body to produce energy. (2, 25)

nystagmus: An unconscious, continuous movement of the eyes as if to adjust to constant motion. (17)

O

obesity: Body weight 10–20 percent above standard values as a result of body fat accumulation. (25)

occlusal surface: The opposing surfaces of the teeth that come into contact when chewing food. (24)

ocular: Pertaining to the eye. (17)

oculomotor nerve: Cranial nerve III, which controls the extra-ocular muscles other than the superior oblique and the lateral rectus muscles. (14)

olecranon: The proximal end of the ulna that forms the prominent point of the elbow. (8)

olfaction: The sense of smell. (15, 17, 23)

olfactory bulb: The expanded ends of the olfactory tracts (17); the sites where the axons of the first cranial nerves (I) synapse on central nervous system interneurons that lie inferior to the frontal lobes of the cerebrum (14).

oligodendrocytes: Central nervous system neuroglia that maintain cellular organization within gray matter and provide a myelin sheath in areas of white matter. (12)

oligopeptide: A short chain of amino acids. (2)

oocyte: A cell whose meiotic divisions will produce a single ovum and three polar bodies. (3, 28)

oogenesis: Ovum production. (28)

ooplasm: The cytoplasm of the ovum. (28)

opsonization: An effect of coating an object with antibodies; the attraction and enhancement of phagocytosis. (22)

optic chiasm: The crossing point of the optic nerves. (14)

optic nerve: The second cranial nerve (II), which carries signals from the retina of the eye to the optic chiasm. (14)

optic tract: The tract over which nerve impulses from the retina are transmitted between the optic chiasm and the thalamus. (14)

orbit: The bony recess of the skull that contains the eyeball. (7)

organelle: An intracellular structure that performs a specific function or group of functions. (3)

organic compound: A compound containing carbon, hydrogen, and in most cases oxygen. (2)

organogenesis: The formation of organs during embryological and fetal development. (29)

organs: Combinations of tissues that perform complex functions. (1)

origin: In a skeletal muscle, the point of attachment which does not change position when the muscle contracts; usually defined in terms of movements from the anatomical position. (11)

oropharynx: The middle portion of the pharynx, bounded superiorly by the nasopharynx, anteriorly by the oral cavity, and inferiorly by the laryngopharynx. (23)

osmolarity: The total concentration of dissolved materials in a solution, regardless of their specific identities, expressed in moles; also called *osmotic concentration*. (3, 26, 27)

osmoreceptor: A receptor sensitive to changes in the osmolarity of plasma. (27)

osmosis: The movement of water across a selectively permeable membrane from one solution to another solution that contains a higher solute concentration. (3, 21, 26, 27)

osmotic pressure: The force of osmotic water movement; the pressure that must be applied to prevent osmosis across a membrane. (3, 21, 26, 27)

osseous tissue: A strong connective tissue containing specialized cells and a mineralized matrix of crystalline calcium phosphate and calcium carbonate; also called bone. (4, 6)

ossicles: Small bones. (17)

ossification: The formation of bone. (6)

osteoblast: A cell that produces the fibers and matrix of bone. (6)

osteoclast: A cell that dissolves the fibers and matrix of bone. (6)

osteocyte: A bone cell responsible for the maintenance and turnover of the mineral content of the surrounding bone. (4, 6)

osteogenic layer: The inner, cellular layer of the periosteum that participates in bone growth and repair. (6)

osteolysis: The breakdown of the mineral matrix of bone. (6)

osteon: The basic histological unit of compact bone, consisting of osteocytes organized around a central canal and separated by concentric lamellae. (6)

otic: Pertaining to the ear. (17)

otolith: A complex formed by the combination of a gelatinous matrix and statoconia, aggregations of calcium carbonate crystals; located above one of the maculae of the vestibule. (17)

oval window: An opening in the bony labyrinth where the stapes attaches to the membranous wall of the vestibular duct. (17)

ovarian cycle: The monthly chain of events that leads to ovulation. (28)

ovary: The female reproductive organ that produces gametes. (18, 28)

ovulation: The release of a secondary oocyte, surrounded by cells of the corona radiata, after the rupture of the wall of a tertiary follicle (29); in females, the periodic release of an oocyte from an ovary (28).

ovum/ova: The functional product of meiosis II, produced after the fertilization of a secondary oocyte. (28, 29)

oxytocin: A hormone produced by hypothalamic cells and secreted into capillaries at the neurohypophysis (posterior lobe of the pituitary gland); stimulates smooth muscle contractions of the uterus or mammary glands in females and the prostate gland in males. (18)

P

pacemaker cells: Cells of the sinoatrial node that set the pace of cardiac contraction. (4, 10, 20)

palate: The horizontal partition separating the oral cavity from the nasal cavity and nasopharynx; divided into an anterior bony (hard) palate and a posterior fleshy (soft) palate. (7, 24)

palatine: Pertaining to the palate. (24)

palpate: To examine by touch.

palpebrae: Eyelids. (17)

pancreas: A digestive organ containing exocrine and endocrine tissues; the exocrine portion secretes pancreatic juice, and the endocrine portion secretes hormones, including insulin and glucagon. (18, 24)

pancreatic duct: A tubular duct that carries pancreatic juice from the pancreas to the duodenum. (18, 24)

pancreatic islets: Aggregations of endocrine cells in the pancreas; also called *islets of Langerhans*. (18, 24)

pancreatic juice: A mixture of buffers and digestive enzymes that is discharged into the duodenum under the stimulation of the enzymes secretin and cholecystokinin. (18, 24)

Papanicolaou (Pap) test: A test for the detection of malignancies based on the cytological appearance of epithelial cells, especially those of the cervix and uterus. (28)

papilla: A small, conical projection.

paralysis: The loss of voluntary motor control over a portion of the body. (13)

paranasal sinuses: Bony chambers, lined by respiratory epithelium, that open into the nasal cavity; the frontal, ethmoidal, sphenoidal, and maxillary sinuses. (7)

parasagittal: A section or plane that parallels the midsagittal plane but that does not pass along the midline. (1)

parasympathetic division: One of the two divisions of the autonomic nervous system; also called *craniosacral division;* generally responsible for activities that conserve energy and lower the metabolic rate. (16)

parathyroid glands: Four small glands embedded in the posterior surface of the thyroid gland; secrete parathyroid hormone. (6, 18)

parathyroid hormone (PTH): A hormone secreted by the parathyroid glands when plasma calcium levels fall below the normal range; causes increased osteoclast activity, increased intestinal calcium uptake, and decreased calcium ion loss at the kidneys. (6, 18)

parenchyma: The cells of a tissue or organ that are responsible for fulfilling its functional role; distinguished from the stroma of that tissue or organ. (4)

paresthesia: A sensory abnormality that produces a tingling sensation.

parietal: Relating to the parietal bone (7); referring to the wall of a cavity (23).

parietal cells: Cells of the gastric glands that secrete hydrochloric acid and intrinsic factor. (24)

parotid salivary glands: Large salivary glands that secrete a saliva containing high concentrations of salivary (alpha) amylase. (24)

pars distalis: The large, anterior portion of the adenohypophysis (anterior lobe of the pituitary gland). (18)

pars intermedia: The portion of the adenohypophysis (anterior lobe of the pituitary gland) that is immediately adjacent to the neurohypophysis (posterior lobe) and the infundibulum. (18)

pars nervosa: The neurohypophysis (posterior lobe of the pituitary gland). (18)

pars tuberalis: The portion of the adenohypophysis (anterior lobe of the pituitary gland) that wraps around the infundibulum superior to the neurohypophysis (posterior lobe). (18)

patella: The sesamoid bone of the kneecap. (8)

pathogen: A disease-causing organism. (1, 22)

pathogenic: Disease-causing.

pathologist: A physician specializing in the identification of diseases on the basis of characteristic structural and functional changes in tissues and organs.

pelvic cavity: The inferior subdivision of the abdominopelvic cavity; encloses the urinary bladder, the sigmoid colon and rectum, and male or female reproductive organs. (1, 8)

pelvis: A bony complex created by the articulations among the coxal bones, the sacrum, and the coccyx. (8, 11)

penis: A component of the male external genitalia; a copulatory organ that surrounds the urethra and serves to introduce semen into the female vagina; the developmental equivalent of the female clitoris. (28)

peptide: A chain of amino acids linked by peptide bonds. (2, 18)

peptide bond: A covalent bond between the amino group of one amino acid and the carboxyl group of another. (2)

pericardial cavity: The space between the parietal pericardium and the epicardium (visceral pericardium) that covers the outer surface of the heart. (20)

pericardium: The fibrous sac that surrounds the heart; its inner, serous lining is continuous with the epicardium. (4, 20)

perichondrium: The layer that surrounds a cartilage, consisting of an outer fibrous region and an inner cellular region. (4)

perikaryon: The cytoplasm that surrounds the nucleus in the cell body of a neuron. (12)

perilymph: A fluid similar in composition to cerebrospinal fluid; located in the spaces between the bony labyrinth and the membranous labyrinth of the inner ear. (17)

perimysium: A connective tissue partition that separates adjacent fasciculi in a skeletal muscle. (10)

perineum: The pelvic floor and its associated structures. (11)

perineurium: A connective tissue partition that separates adjacent bundles of nerve fibers in a peripheral nerve. (13)

periodontal ligament: Collagen fibers that bind the cementum of a tooth to the periosteum of the surrounding alveolus. (24)

periosteum: The layer that surrounds a bone, consisting of an outer fibrous region and inner cellular region. (4, 6)

peripheral nervous system (PNS): All neural tissue outside the central nervous system. (12)

peripheral resistance: The resistance to blood flow; primarily caused by friction with the vascular walls. (21)

peristalsis: A wave of smooth muscle contractions that propels materials along the axis of a tube such as the digestive tract (24), the ureters (26), or the ductus deferens (28).

peritoneal cavity: *See* abdominopelvic cavity.

peritoneum: The serous membrane that lines the peritoneal cavity. (4, 28)

peritubular capillaries: A network of capillaries that surrounds the proximal and distal convoluted tubules of the kidneys. (26)

permeability: The ease with which dissolved materials can cross a membrane; if the membrane is freely permeable, any molecule can cross it; if impermeable, nothing can cross; most biological membranes are selectively permeable. (3)

peroxisome: A membranous vesicle containing enzymes that break down hydrogen peroxide (H_2O_2). (3)

pes: The foot. (8, 11)

petrosal ganglion: A sensory ganglion of the glossopharyngeal nerve (N IX). (14, 15)

petrous: Stony; usually refers to the thickened portion of the temporal bone that encloses the internal ear. (17)

pH: The negative exponent of the hydrogen ion concentration, expressed in moles per liter. (2, 27)

phagocyte: A cell that performs phagocytosis. (22)

phagocytosis: The engulfing of extracellular materials or pathogens; the movement of extracellular materials into the cytoplasm by enclosure in a membranous vesicle. (3, 19, 22)

phalanx/phalanges: Bone(s) of the finger(s) or toe(s). (8)

pharmacology: The study of drugs, their physiological effects, and their clinical uses.

pharynx: The throat; a muscular passageway shared by the digestive and respiratory tracts. (11, 23, 24)

phasic response: A pattern of response to stimulation by sensory neurons that are normally inactive; stimulation causes a burst of neural activity that ends when the stimulus either stops or stops changing in intensity. (15)

phenotype: Physical characteristics that are genetically determined. (29)

phosphate group: PO_4^{3-}; a functional group that can be attached to an organic molecule; required for the formation of high-energy bonds. (2, 25, 27)

phospholipid: An important membrane lipid whose structure includes both hydrophilic and hydrophobic regions. (2, 3)

phosphorylation: The addition of a high-energy phosphate group to a molecule. (2, 25)

photoreception: Sensitivity to light. (17)

physiology: The study of function; deals with the ways organisms perform vital activities. (1)

pia mater: The innermost layer of the meninges bound to the underlying neural tissue. (13, 14)

pineal gland: Neural tissue in the posterior portion of the roof of the diencephalon; secretes melatonin. (14, 18)

pinna: *See* auricle.

pinocytosis: The introduction of fluids into the cytoplasm by enclosing them in membranous vesicles at the cell surface. (3)

pituitary gland: An endocrine organ that is situated in the sella turcica of the sphenoid and is connected to the hypothalamus by the infundibulum; includes the neurohypophysis (posterior lobe) and the adenohypophysis (anterior lobe); also called the hypophysis. (14, 18)

placenta: A temporary structure in the uterine wall that permits diffusion between the fetal and maternal circulatory systems. (29)

plantar: Referring to the sole of the foot (1); muscles (11); plantar reflex (13).

plantar flexion: Ankle extension; toe pointing. (8, 11)

plasma: The fluid ground substance of whole blood; what remains after the cells have been removed from a sample of whole blood. (4, 19)

plasma cell: An activated B cell that secretes antibodies; plasmocyte. (4, 19, 22)

plasma membrane: A cell membrane; plasmalemma. (3)

platelets: Small packets of cytoplasm that contain enzymes important in the clotting response; manufactured in bone marrow by megakaryocytes. (4, 19)

pleura: The serous membrane that lines the pleural cavities. (4, 23)

pleural cavities: Subdivisions of the thoracic cavity that contain the lungs. (1, 23)

plexus: A network or braid.

polar body: A nonfunctional packet of cytoplasm that contains chromosomes eliminated from an oocyte during meiosis. (28, 29)

polar bond: A covalent bond in which electrons are shared unequally. (2)

polarized: Referring to cells that have regional differences in organelle distribution or cytoplasmic composition along a specific axis, such as between the basement membrane and free surface of an epithelial cell. (4)

pollex: The thumb. (8)

polymer: A large molecule consisting of a long chain of monomer subunits. (2)

polypeptide: A chain of amino acids strung together by peptide bonds; those containing more than 100 peptides are called *proteins.* (2)

polyribosome: Several ribosomes linked by their translation of a single mRNA strand. (3)

polysaccharide: A complex sugar, such as glycogen or a starch. (2)

polysynaptic reflex: A reflex in which interneurons are interposed between the sensory fiber and the motor neuron(s). (13)

polyunsaturated fats: Fatty acids containing carbon atoms that are linked by double bonds. (1, 2)

pons: The portion of the metencephalon that is anterior to the cerebellum. (14)

popliteal: Pertaining to the back of the knee. (9, 11, 21)

porphyrins: Ring-shaped molecules that form the basis of important respiratory and metabolic pigments, including heme and the cytochromes. (23)

positive feedback: A mechanism that increases a deviation from normal limits after an initial stimulus. (1)

postcentral gyrus: The primary sensory cortex, where touch, vibration, pain, temperature, and taste sensations arrive and are consciously perceived. (14)

posterior: Toward the back; dorsal.

postganglionic neuron: An autonomic neuron in a peripheral ganglion, whose activities control peripheral effectors. (16)

postsynaptic membrane: The portion of the plasma membrane of a postsynaptic cell that is part of a synapse. (12)

potential difference: The separation of opposite charges; requires a barrier that prevents ion migration. (3, 12)

precentral gyrus: The primary motor cortex of a cerebral hemisphere, located anterior to the central sulcus. (14)

prefrontal cortex: The anterior portion of each cerebral hemisphere; thought to be involved with higher intellectual functions, predictions, calculations, and so forth. (14)

preganglionic neuron: A visceral motor neuron in the central nervous system whose output controls one or more ganglionic motor neurons in the peripheral nervous system. (16)

premotor cortex: The motor association area between the precentral gyrus and the prefrontal area. (14)

preoptic nucleus: The hypothalamic nucleus that coordinates thermoregulatory activities. (14)

presynaptic membrane: The synaptic surface where neurotransmitter release occurs. (12)

prevertebral ganglion: See collateral ganglion.

prime mover: A muscle that performs a specific action. (11)

proenzyme: An inactive enzyme secreted by an epithelial cell. (19)

progesterone: The most important progestin secreted by the corpus luteum after ovulation. (18, 28)

progestins: Steroid hormones structurally related to cholesterol; progesterone is an example. (18, 28)

prognosis: A prediction about the possible course or outcome from a specific disease.

projection fibers: Axons carrying information from the thalamus to the cerebral cortex. (14)

prolactin: The hormone that stimulates functional development of the mammary glands in females; a secretion of the adenohypophysis (anterior lobe of the pituitary gland). (18)

pronation: The rotation of the forearm that makes the palm face posteriorly. (9)

prone: Lying face down with the palms facing the floor. (1)

pronucleus: An enlarged ovum or spermatozoon nucleus that forms after fertilization but before amphimixis. (29)

prophase: The initial phase of mitosis; characterized by the appearance of chromosomes, the breakdown of the nuclear membrane, and the formation of the spindle apparatus. (3)

proprioception: The awareness of the positions of bones, joints, and muscles. (15)

prostaglandin: A fatty acid secreted by one cell that alters the metabolic activities or sensitivities of adjacent cells; also called *local hormone*. (2, 18)

prostate gland: An accessory gland of the male reproductive tract, contributing roughly one-third of the volume of semen. (28)

prosthesis: An artificial substitute for a body part.

protease: See proteinase.

protein: A large polypeptide with a complex structure. (2, 25)

proteinase: An enzyme that breaks down proteins into peptides and amino acids. (2, 3)

proteoglycan: A compound containing a large polysaccharide complex attached to a relatively small protein; examples include hyaluronan and chondroitin sulfate. (2)

proton: A fundamental particle bearing a positive charge. (2)

protraction: Movement anteriorly in the horizontal plane.

proximal: Toward the attached base of an organ or structure. (8)

proximal convoluted tubule (PCT): The portion of the nephron that is situated between the glomerular capsule (Bowman capsule) and the nephron loop; the major site of active reabsorption from filtrate. (26)

pseudopodia: Temporary cytoplasmic extensions typical of mobile or phagocytic cells. (3)

pseudostratified epithelium: An epithelium that contains several layers of nuclei but whose cells are all in contact with the underlying basement membrane. (4)

puberty: A period of rapid growth, sexual maturation, and the appearance of secondary sexual characteristics; normally occurs at ages 10–15 years. (18, 28)

pubic symphysis: The fibrocartilaginous amphiarthrosis between the pubic bones of the coxal bones. (8, 9)

pubis: The anterior, inferior component of the hip bone. (8)

pudendum: The external genitalia. (28)

pulmonary circuit: Blood vessels between the pulmonary semilunar valve of the right ventricle and the entrance to the left atrium; the blood flow through the lungs. (20)

pulmonary ventilation: The movement of air into and out of the lungs. (23)

pulvinar: The thalamic nucleus involved in the integration of sensory information prior to projection to the cerebral hemispheres. (14)

pupil: The opening in the center of the iris through which light enters the eye. (17)

purine: A nitrogen compound with a double ring-shaped structure; examples include adenine and guanine, two nitrogenous bases that are common in nucleic acids. (2, 12)

Purkinje cell: A large, branching neuron of the cerebellar cortex. (14)

Purkinje fibers: Specialized conducting cardiocytes in the ventricles of the heart. (20)

pus: An accumulation of debris, fluid, dead and dying cells, and necrotic tissue. (4, 20, 22)

pyloric sphincter: A sphincter of smooth muscle that regulates the passage of chyme from the stomach to the duodenum. (24)

pylorus: The gastric region between the body of the stomach and the duodenum; includes the pyloric sphincter. (24)

pyrimidine: A nitrogen compound with a single ring-shaped structure; examples include cytosine, thymine, and uracil, nitrogenous bases that are common in nucleic acids. (2)

pyruvic acid: A three-carbon compound produced by glycolysis. (25)

Q

quaternary structure: The three-dimensional protein structure produced by interactions between protein subunits. (2)

R

radiodensity: The relative resistance to the passage of x-rays. (1)

radiographic techniques: Methods of visualizing internal structures by using various forms of radiational energy. (1)

radiopaque: Having a high radiodensity. (1)

rami communicantes: Axon bundles that link the spinal nerves with the ganglia of the sympathetic chain. (8)

ramus/rami: A branch.

raphe: A seam. (11, 28)

receptive field: The area monitored by a single sensory receptor. (15)

rectum: The inferior 15 cm (6 in.) of the digestive tract. (24)

rectus: Straight.

red blood cell (RBC): See erythrocyte.

reduction: The gain of hydrogen atoms or electrons or the loss of an oxygen molecule. (25)

reductional division: The first meiotic division, which reduces the chromosome number from 46 to 23. (28)

reflex: A rapid, automatic response to a stimulus. (12, 13, 16, 18, 21)

reflex arc: The receptor, sensory neuron, motor neuron, and effector involved in a particular reflex; interneurons may be present, depending on the reflex considered. (13)

refractory period: The period between the initiation of an action potential and the restoration of the normal resting potential; during this period, the membrane will not respond normally to stimulation. (12)

relaxation phase: The period after a contraction when the tension in the muscle fiber returns to resting levels. (10)

relaxin: A hormone that loosens the pubic symphysis; secreted by the placenta. (29)

renal: Pertaining to the kidneys. (26)

renal corpuscle: The initial portion of the nephron, consisting of an expanded chamber that encloses the glomerulus. (26)

renin: The enzyme released by cells of the juxtaglomerular complex when renal blood flow declines; converts angiotensinogen to angiotensin I. (18, 26)

rennin: A gastric enzyme that breaks down milk proteins. (24)

replication: Duplication. (29)

repolarization: The movement of the transmembrane potential away from a positive value and toward the resting potential. (12, 20)

respiration: The exchange of gases between cells and the environment; includes pulmonary ventilation, external respiration, internal respiration, and cellular respiration. (23, 27)

respiratory minute volume ($\dot{V}_E$): The amount of air moved into and out of the respiratory system each minute. (23)

respiratory pump: A mechanism by which changes in the intrapleural pressures during the respiratory cycle assist the venous return to the heart; also called *thoracoabdominal pump*. (21, 23)

resting potential: The transmembrane potential of a normal cell under homeostatic conditions. (3, 12)

rete: An interwoven network of blood vessels or passageways. (28)

reticular activating system (RAS): The mesencephalic portion of the reticular formation; responsible for arousal and the maintenance of consciousness. (16)

reticular formation: A diffuse network of gray matter that extends the entire length of the brain stem. (14)

reticulospinal tracts: Descending tracts of the medial pathway that carry involuntary motor commands issued by neurons of the reticular formation. (15)

retina: The innermost layer of the eye, lining the vitreous chamber; also called *neural tunic*. (17)

retinal: A visual pigment derived from vitamin A. (17)

retraction: Movement posteriorly in the horizontal plane.

retroperitoneal: Behind or outside the peritoneal cavity. (1)

reverberation: A positive feedback along a chain of neurons such that they remain active once stimulated. (13)

rheumatism: A condition characterized by pain in muscles, tendons, bones, or joints. (9)

Rh factor: A surface antigen that may be present (Rh-positive) or absent (Rh-negative) from the surfaces of red blood cells. (19)

rhodopsin: The visual pigment in the membrane disks of the distal segments of rods. (17)

rhythmicity center: A medullary center responsible for the pace of respiration; includes inspiratory and expiratory centers. (23)

ribonucleic acid: A nucleic acid consisting of a chain of nucleotides that contain the sugar ribose and the nitrogenous bases adenine, guanine, cytosine, and uracil. (2, 3)

ribose: A five-carbon sugar that is a structural component of RNA. (2, 3)

ribosome: An organelle that contains rRNA and proteins and is essential to mRNA translation and protein synthesis. (2, 3)

rod: A photoreceptor responsible for vision in dim lighting. (17)

rough endoplasmic reticulum (RER): A membranous organelle that is a site of protein synthesis and storage. (3)

round window: An opening in the bony labyrinth of the inner ear that exposes the membranous wall of the tympanic duct to the air of the middle ear cavity. (17)

rubrospinal tracts: Descending tracts of the lateral pathway that carry involuntary motor commands issued by the red nucleus of the mesencephalon. (15)

rugae: Mucosal folds in the lining of the empty stomach that disappear as gastric distension occurs (24); folds in the urinary bladder (26).

S

saccule: A portion of the vestibular apparatus of the internal ear; contains a macula important for static equilibrium. (17)

sagittal plane: A sectional plane that divides the body into left and right portions. (1)

salt: An inorganic compound consisting of a cation other than H^+ and an anion other than OH^-. (2)

saltatory propagation: The relatively rapid propagation of an action potential between successive nodes of a myelinated axon. (12)

sarcolemma: The plasma membrane of a muscle cell. (10)

sarcomere: The smallest contractile unit of a striated muscle cell. (10)

sarcoplasm: The cytoplasm of a muscle cell. (10)

scala media: *See* cochlear duct.

scala tympani: *See* tympanic duct.

scala vestibuli: *See* vestibular duct.

scar tissue: The thick, collagenous tissue that forms at an injury site. (5)

Schwann cells: Neuroglia responsible for the neurilemma that surrounds axons in the peripheral nervous system. (12)

sciatic nerve: A nerve innervating the posteromedial portions of the thigh and leg. (13)

sclera: The fibrous, outer layer of the eye that forms the white area of the anterior surface; a portion of the fibrous tunic of the eye. (17)

sclerosis: A hardening and thickening that commonly occurs secondary to tissue inflammation. (4)

scrotum: The loose-fitting, fleshy pouch that encloses the testes of the male. (28)

sebaceous glands: Glands that secrete sebum; normally associated with hair follicles. (5)

sebum: A waxy secretion that coats the surfaces of hairs. (5)

secondary sex characteristics: Physical characteristics that appear at puberty in response to sex hormones but are not involved in the production of gametes. (28)

secretin: A hormone, secreted by the duodenum, that stimulates the production of buffers by the pancreas and inhibits gastric activity. (24)

semen: The fluid ejaculate that contains spermatozoa and the secretions of accessory glands of the male reproductive tract. (28)

semicircular ducts: The tubular components of the membranous labyrinth of the inner ear; responsible for dynamic equilibrium. (17)

semilunar valve: A three-cusped valve guarding the exit from one of the cardiac ventricles; the pulmonary and aortic valves. (20)

seminal vesicles: Glands of the male reproductive tract that produce roughly 60 percent of the volume of semen; also called seminal glands. (28)

seminiferous tubules: Coiled tubules where spermatozoon production occurs in the testes. (28)

senescence: Aging.

sensible perspiration: Water loss due to secretion by sweat glands. (5, 27)

septa: Partitions that subdivide an organ. (20, 22)

serosa: *See* serous membrane.

serotonin: A neurotransmitter in the central nervous system; a compound that enhances inflammation and is released by activated mast cells and basophils. (12, 19)

serous cell: A cell that produces a serous secretion. (4)

serous membrane: A squamous epithelium and the underlying loose connective tissue; the lining of the pericardial, pleural, and peritoneal cavities. (4, 24)

serous secretion: A watery secretion that contains high concentrations of enzymes; produced by serous cells. (4)

serum: The ground substance of blood plasma from which clotting agents have been removed. (19)

sesamoid bone: A bone that forms within a tendon. (6)

sigmoid colon: The S-shaped 18-cm (7.1 in.)-long portion of the colon between the descending colon and the rectum. (24)

sign: The visible, objective evidence of the presence of a disease.

simple epithelium: An epithelium containing a single layer of cells above the basal lamina. (4)

sinoatrial (SA) node: The natural pacemaker of the heart; situated in the wall of the right atrium. (20)

sinus: A chamber or hollow in a tissue; a large, dilated vein. (6, 7, 20)

sinusoid: An exchange vessel that is similar in general structure to a fenestrated capillary. The two differ in size (sinusoids are larger and more irregular in cross section), continuity (sinusoids have gaps between endothelial cells), and support (sinusoids have thin basal laminae, if they have them at all). (20)

skeletal muscle: A contractile organ of the muscular system. (10)

skeletal muscle tissue: A contractile tissue dominated by skeletal muscle fibers; characterized as striated, voluntary muscle. (4, 10)

sliding filament theory: The concept that a sarcomere shortens as the thick and thin filaments slide past one another. (10)

small intestine: The duodenum, jejunum, and ileum; the digestive tract between the stomach and the large intestine. (24)

smooth endoplasmic reticulum (SER): A membranous organelle in which lipid and carbohydrate synthesis and storage occur. (3)

smooth muscle tissue: Muscle tissue in the walls of many visceral organs; characterized as nonstriated, involuntary muscle. (4, 10, 24, 26)

soft palate: The fleshy posterior extension of the hard palate, separating the nasopharynx from the oral cavity. (24)

solute: Any materials dissolved in a solution. (2, 21, 26)

solution: A fluid containing dissolved materials. (2, 21)

solvent: The fluid component of a solution. (2, 21)

somatic: Pertaining to the body.

somatic nervous system (SNS): The efferent division of the nervous system that innervates skeletal muscles. (12, 15, 16)

somatomedins: Compounds stimulating tissue growth; released by the liver after the secretion of growth hormone; also called *insulin-like growth factors*. (18)

somatotropin: Growth hormone; produced by the adenohypophysis (anterior pituitary) in response to growth hormone–releasing hormone (GH–RH). (18)

sperm: *See* spermatozoon.

spermatic cord: Collectively, the spermatic vessels, nerves, lymphatic vessels, and the ductus deferens, extending between the testes and the proximal end of the inguinal canal. (28)

spermatocyte: A cell of the seminiferous tubules that is engaged in meiosis. (28)

spermatogenesis: Spermatozoon production. (28)

spermatozoon/spermatozoa: A male gamete; also called *sperm*. (3, 28)

sphincter: A muscular ring that contracts to close the entrance or exit of an internal passageway. (10, 11, 26)

spinal nerve: One of 31 pairs of nerves that originate on the spinal cord from anterior and posterior roots. (12, 13)

spindle apparatus: Microtubule-based structure which distributes duplicated chromosomes to opposite ends of a dividing cell during mitosis. (3)

spinocerebellar tracts: Ascending tracts that carry sensory information to the cerebellum. (15)

spinothalamic tracts: Ascending tracts that carry poorly localized touch, pressure, pain, vibration, and temperature sensations to the thalamus. (15)

spinous process: The prominent posterior projection of a vertebra; formed by the fusion of two laminae. (7)

spleen: A lymphoid organ important for the phagocytosis of red blood cells, the immune response, and lymphocyte production. (22)

squama: A broad, flat surface.

squamous: Flattened.

squamous epithelium: An epithelium whose superficial cells are flattened and platelike. (4)

stapes: The auditory ossicle attached to the tympanic membrane. (17)

stenosis: A constriction or narrowing of a passageway.

stereocilia: Elongate microvilli characteristic of the epithelium of the epididymis, portions of the ductus deferens (28), and the internal ear (17).

steroid: A ring-shaped lipid structurally related to cholesterol. (2, 18)

stimulus: An environmental change that produces a change in cellular activities; often used to refer to events that alter the transmembrane potentials of excitable cells. (15)

stratified: Containing several layers. (4)

stratum: A layer.

stretch receptors: Sensory receptors that respond to stretching of the surrounding tissues. (13)

stroma: The connective tissue framework of an organ; distinguished from the functional cells (parenchyma) of that organ.

subarachnoid space: A meningeal space containing cerebrospinal fluid; the area between the arachnoid membrane and the pia mater. (13)

subclavian: Pertaining to the region immediately posterior and inferior to the clavicle.

subcutaneous layer: The layer of loose connective tissue below the dermis; also called *hypodermis* or *superficial fascia*. (4, 5)

submucosa: The region between the muscularis mucosae and the muscularis externa. (23, 24)

subserous fascia: The loose connective tissue layer beneath the serous membrane that lines the ventral body cavity. (4)

substrate: A participant (product or reactant) in an enzyme-catalyzed reaction. (2)

sulcus: A groove or furrow. (14)

summation: The temporal or spatial addition of stimuli. (10, 12)

superficial fascia: *See* subcutaneous layer.

superior: Above, in reference to a portion of the body in the anatomical position.

superior vena cava (SVC): The vein that carries blood to the right atrium from parts of the body that are superior to the heart. (20, 21)

supination: The rotation of the forearm such that the palm faces anteriorly. (9)

supine: Lying face up, with palms facing anteriorly. (1)

suppressor T cells: Lymphocytes that inhibit B cell activation and the secretion of antibodies by plasma cells. (22)

suprarenal cortex: The superficial portion of the suprarenal (adrenal) gland that produces steroid hormones; also called the *adrenal cortex*. (18)

suprarenal gland: A small endocrine gland that secretes steroids and catecholamines and is located superior to each kidney; also called *adrenal gland*. (18)

suprarenal medulla: The core of the suprarenal (adrenal) gland (18), a modified sympathetic ganglion that secretes catecholamines into the blood during sympathetic activation; also called adrenal medulla. (16)

surfactant: A lipid secretion that coats the alveolar surfaces of the lungs and prevents their collapse. (23)

sustentacular cells: Supporting cells of the seminiferous tubules of the testis; responsible for the differentiation of spermatids, the maintenance of the blood–testis barrier, and the secretion of inhibin, androgen-binding protein, and Müllerian-inhibiting factor; also called *nurse cells*. (18, 28)

sutural bones: Irregular bones that form in fibrous tissue between the flat bones of the developing cranium; also called *Wormian bones*. (6)

suture: A fibrous joint between flat bones of the skull. (7, 9)

sympathetic division: The division of the autonomic nervous system that is responsible for "fight or flight" reactions; primarily concerned with the elevation of metabolic rate and increased alertness. (12, 16)

symphysis: A fibrous amphiarthrosis, such as that between adjacent vertebrae or between the pubic bones of the coxal bones. (9)

symptom: An abnormality of function as a result of disease; subjective experience of patient.

synapse: The site of communication between a nerve cell and some other cell; if the other cell is not a neuron, the term *neuromuscular* or *neuroglandular junction* is often used. (12, 16, 18)

synaptic delay: The period between the arrival of an impulse at the presynaptic membrane and the initiation of an action potential in the postsynaptic membrane. (12)

syncytium: A multinucleate mass of cytoplasm, produced by the fusion of cells or repeated mitoses without cytokinesis. (29)

syndrome: A discrete set of signs and symptoms that occur together.

synergist: A muscle that assists a prime mover in performing its primary action. (11)

synovial cavity: A fluid-filled chamber in a synovial joint. (4, 9)

synovial fluid: The substance secreted by synovial membranes that lubricates joints. (4, 9)

synovial joint: A freely movable joint where the opposing bone surfaces are separated by synovial fluid; a diarthrosis. (4, 9)

synovial membrane: An incomplete layer of fibroblasts confronting the synovial cavity, plus the underlying loose connective tissue. (4)

synthesis: Manufacture; anabolism. (23)

system: An interacting group of organs that performs one or more specific functions.

systemic circuit: The vessels between the aortic valve and the entrance to the right atrium; the system other than the vessels of the pulmonary circuit. (20)

systole: A period of contraction in a chamber of the heart, as part of the cardiac cycle. (20)

systolic pressure: The peak arterial pressure measured during ventricular systole. (20)

T

tactile: Pertaining to the sense of touch. (15)

tarsal bones: The bones of the ankle (the talus, calcaneus, navicular, and cuneiform bones). (8)

tarsus: The ankle. (8)

TCA (tricarboxylic acid) cycle: The aerobic reaction sequence that occurs in the matrix of mitochondria; in the process, organic molecules are broken down, carbon dioxide molecules are released, and hydrogen molecules are transferred to coenzymes that deliver them to the electron transport system; also called *citric acid cycle* or *Krebs cycle*. (3, 10, 25)

T cells: Lymphocytes responsible for cell-mediated immunity and for the coordination and regulation of the immune response; includes regulatory T cells (helpers and suppressors) and cytotoxic (killer) T cells. (19, 22)

tectospinal tracts: Descending tracts of the medial pathway that carry involuntary motor commands issued by the colliculi. (15)

telodendria: Terminal axonal branches that end in synaptic knobs. (12)

telophase: The final stage of mitosis, characterized by the disappearance of the spindle apparatus, the reappearance of the nuclear membrane, the disappearance of the chromosomes, and the completion of cytokinesis. (3)

temporal: Pertaining to time (temporal summation) or to the temples (temporal bone). (7)

tendon: A collagenous band that connects a skeletal muscle to an element of the skeleton. (4, 10)

teres: Long and round.

terminal: Toward the end.

tertiary structure: The protein structure that results from interactions among distant portions of the same molecule; complex coiling and folding. (2)

testes: The male gonads, sites of gamete production and hormone secretion. (18, 28)

testosterone: The principal androgen produced by the interstitial cells of the testes. (2, 18, 28)

tetraiodothyronine: T_4, or thyroxine, a thyroid hormone. (18)

thalamus: The walls of the diencephalon. (14)

theory: A hypothesis that makes valid predictions, as demonstrated by evidence that is testable, unbiased, and repeatable.

therapy: The treatment of disease.

thermoreception: Sensitivity to temperature changes. (15)

thermoregulation: Homeostatic maintenance of body temperature. (1, 25)

thick filament: A cytoskeletal filament in a skeletal or cardiac muscle cell; composed of myosin, with a core of titin. (3, 10)

thin filament: A cytoskeletal filament in a skeletal or cardiac muscle cell; consists of actin, troponin, and tropomyosin. (3, 10)

thoracolumbar division: The sympathetic division of the autonomic nervous system. (16)

thorax: The chest. (7)

threshold: The transmembrane potential at which an action potential begins. (12)

thrombin: The enzyme that converts fibrinogen to fibrin. (19)

thymine: A pyrimidine; one of the nitrogenous bases in the nucleic acid DNA. (2)

thymosins: Thymic hormones essential to the development and differentiation of T cells. (18, 22)

thymus: A lymphoid organ, the site of T cell formation. (18, 22)

thyroglobulin: A circulating transport globulin that binds thyroid hormones. (18)

thyroid gland: An endocrine gland whose lobes are lateral to the thyroid cartilage of the larynx. (18)

thyroid hormones: Thyroxine (T_4) and triiodothyronine (T_3), hormones of the thyroid gland; stimulate tissue metabolism, energy utilization, and growth. (18)

thyroid-stimulating hormone (TSH): The hormone, produced by the adenohypophysis (anterior lobe of the pituitary gland), that triggers the secretion of thyroid hormones by the thyroid gland. (18)

thyroxine: A thyroid hormone; also called T_4 or *tetraiodothyronine*. (18)

tidal volume: The volume of air moved into and out of the lungs during a normal quiet respiratory cycle. (23)

tissue: A collection of specialized cells and cell products that performs a specific function. (1, 4)

tonic response: An increase or decrease in the frequency of action potentials by sensory receptors that are chronically active. (15)

tonsil: A lymphoid nodule in the wall of the pharynx; the palatine, pharyngeal, and lingual tonsils. (22)

topical: Applied to the body surface.

toxic: Poisonous.

trabecula: A connective tissue partition that subdivides an organ. (22)

trachea: The windpipe; an airway extending from the larynx to the primary bronchi. (23)

tract: A bundle of axons in the central nervous system. (13, 14)

transcription: The encoding of genetic instructions on a strand of mRNA. (3)

transection: The severing or cutting of an object in the transverse plane.

translation: The process of peptide formation from the instructions carried by an mRNA strand. (3)

transmembrane potential: The potential difference, measured across a plasma membrane and expressed in millivolts, that results from the uneven distribution of positive and negative ions across the plasma membrane. (3, 12)

transudate: A fluid that diffuses across a serous membrane and lubricates opposing surfaces. (4)

transverse tubules: The transverse, tubular extensions of the sarcolemma that extend deep into the sarcoplasm, contacting cisternae of the sarcoplasmic reticulum; also called *T tubules*. (10)

tricarboxylic acid cycle: *See* **TCA cycle.**

tricuspid valve: The right atrioventricular valve, which prevents the backflow of blood into the right atrium during ventricular systole. (20)

trigeminal nerve: Cranial nerve V, which provides sensory information from the lower portions of the face (including the upper and lower jaws) and delivers motor commands to the muscles of mastication. (14)

triglyceride: A lipid that is composed of a molecule of glycerol attached to three fatty acids. (2, 25)

triiodothyronine: T_3, a thyroid hormone. (18)

trisomy: The abnormal possession of three copies of a chromosome; trisomy 21 is responsible for Down syndrome. (16)

trochanter: Large process near the head of the femur. (8)

trochlea: A pulley; the spool-shaped medial portion of the condyle of the humerus. (8)

trochlear nerve: Cranial nerve IV, controlling the superior oblique muscle of the eye. (14)

trunk: The thoracic and abdominopelvic regions (1); a major arterial branch (21).

T tubules: *See* **transverse tubules.**

tuberculum: A small, localized elevation on a bony surface. (7)

tuberosity: A large, roughened elevation on a bony surface. (6)

tumor: A tissue mass formed by the abnormal growth and replication of cells. (3)

tunica: A layer or covering.

twitch: A single stimulus–contraction–relaxation cycle in a skeletal muscle. (10)

tympanic duct: The perilymph-filled chamber of the internal ear, adjacent to the basilar membrane; pressure changes there distort the round window; also called *scala tympani*. (17)

tympanic membrane: The membrane that separates the external acoustic meatus from the middle ear; the membrane whose vibrations are transferred to the auditory ossicles and ultimately to the oval window; also called *eardrum* or *tympanum*. (17)

type A axons: Large myelinated axons. (12)

type B axons: Small myelinated axons. (12)

type C axons: Small unmyelinated axons. (12)

U

umbilical cord: The connecting stalk between the fetus and the placenta; contains the allantois, the umbilical arteries, and the umbilical vein. (21, 29)

umbilicus: The navel. (29)

unicellular gland: Mucous cells. (4)

unipolar neuron: A sensory neuron whose cell body is in a dorsal root ganglion or a sensory ganglion of a cranial nerve. (12)

unmyelinated axon: An axon whose neurilemma does not contain myelin and across which continuous propagation occurs. (12)

uracil: A pyrimidine; one of the nitrogenous bases in the nucleic acid RNA. (2)

ureters: Muscular tubes, lined by transitional epithelium, that carry urine from the renal pelvis to the urinary bladder. (26)

urethra: A muscular tube that carries urine from the urinary bladder to the exterior. (26)

urinary bladder: The muscular, distensible sac that stores urine prior to micturition. (26)

urination: The voiding of urine; micturition. (26)

uterus: The muscular organ of the female reproductive tract in which implantation, placenta formation, and fetal development occur. (28)

utricle: The largest chamber of the vestibular apparatus of the internal ear; contains a macula important for static equilibrium. (17)

V

vagina: A muscular tube extending between the uterus and the vestibule. (28)

vascular: Pertaining to blood vessels. (19)

vasoconstriction: A reduction in the diameter of arterioles due to the contraction of smooth muscles in the tunica media; elevates peripheral resistance; may occur in response to local factors, through the action of hormones, or from the stimulation of the vasomotor center. (21)

vasodilation: An increase in the diameter of arterioles due to the relaxation of smooth muscles in the tunica media; reduces peripheral resistance; may occur in response to local factors, through the action of hormones, or after decreased stimulation of the vasomotor center. (21)

vasomotion: Changes in the pattern of blood flow through a capillary bed in response to changes in the local environment. (21)

vasomotor center: The center in the medulla oblongata whose stimulation produces vasoconstriction and an elevation of peripheral resistance. (14)

vein: A blood vessel carrying blood from a capillary bed toward the heart. (20, 21)

vena cava: One of the major veins delivering systemic blood to the right atrium; superior and inferior venae cavae. (20, 21)

ventilation: Air movement into and out of the lungs. (23)

ventral: Pertaining to the anterior surface.

ventricle: A fluid-filled chamber; in the heart, one of the large chambers discharging blood into the pulmonary or systemic circuits (20); in the brain, one of four fluid-filled interior chambers (14).

venule: Thin-walled veins that receive blood from capillaries. (21)

vermiform appendix: *See* **appendix.**

vertebral canal: The passageway that encloses the spinal cord; a tunnel bounded by the neural arches of adjacent vertebre. (7)

vertebral column: The cervical, thoracic, and lumbar vertebrae, the sacrum, and the coccyx. (7, 11)

vesicle: A membranous sac in the cytoplasm of a cell. (3)

vestibular nucleus: The processing center for sensations that arrive from the vestibular apparatus of the internal ear, located near the border between the pons and the medulla oblongata. (17)

vestibulospinal tracts: Descending tracts of the medial pathway that carry involuntary motor commands issued by the vestibular nucleus to stabilize the position of the head. (15)

villus/villi: A slender projection of the mucous membrane of the small intestine. (24)

virus: A noncellular pathogen. (22)

viscera: Organs in the ventral body cavity. (1)

visceral: Pertaining to viscera or their outer coverings. (1)

visceral smooth muscle: A smooth muscle tissue that forms sheets or layers in the walls of visceral organs; the cells may not be innervated, and the layers often show automaticity (rhythmic contractions). (10, 24)

viscosity: The resistance to flow that a fluid exhibits as a result of molecular interactions within the fluid. (21)

viscous: Thick, syrupy.

vitamin: An essential organic nutrient that functions as a coenzyme in vital enzymatic reactions. (25)

vitreous humor: The gelatinous mass in the vitreous chamber of the eye. (17)

voluntary: Controlled by conscious thought processes.

W

white blood cells (WBCs): The granulocytes and agranulocytes of whole blood. (4, 19)

white matter: Regions in the central nervous system that are dominated by myelinated axons. (12, 13, 14)

white ramus: A nerve bundle containing the myelinated preganglionic axons of sympathetic motor neurons en route to the sympathetic chain or to a collateral ganglion. (13)

Wormian bones: *See* **sutural bones.**

X

xiphoid process: The slender, inferior extension of the sternum. (7)

Y

Y chromosome: The sex chromosome whose presence indicates that the individual is a genetic male. (29)

Z

zona fasciculata: The region of the suprarenal (adrenal) cortex that secretes glucocorticoids. (18)

zona glomerulosa: The region of the suprarenal (adrenal) cortex that secretes mineralocorticoids. (18)

zona reticularis: The region of the suprarenal (adrenal) cortex that secretes androgens. (18)

zygote: The fertilized ovum, prior to the start of cleavage. (28)

Photo and Illustration Credits

Chapter 1 1.CO Ilya Naymushin/Reuters/CORBIS All Rights Reserved. 1.7 Custom Medical Stock Photo, Inc.
Chapter 2 2.CO Tim Macpherson/Getty Images, Inc.
Chapter 3 3.CO SPL/Photo Researchers, Inc. 3.3b Fawcett/Hirokawa/Heuser/Science Source/Photo Researchers, Inc. 3.3c M. Schliwa/Visuals Unlimited 3.4a Fawcett/de Harven/Kalnins/Photo Researchers, Inc. 3.5b R. Bollender & Don W. Fawcett/Visuals Unlimited 3.6a Biophoto Associates/Photo Researchers, Inc. 3.7b Dr. Birgit H. Satir 3.9a CNRI/Photo Researchers, Inc. 3.10a Don W. Fawcett, M.D., Harvard Medical School 3.10b Biophoto Associates/Photo Researchers, Inc. 3.17a,b,c David M. Philips/Visuals Unlimited 3.22a M. Sahliwa/Visuals Unlimited 3.25a,b Ed Reschke/Peter Arnold, Inc. 3.25c James Solliday/Biological Photo Service 3.25d,e,f Ed Reschke/Peter Arnold, Inc. 3.25g Centers for Disease Control and Prevention (CDC)
Chapter 4 4.CO Photo Researchers, Inc. 4.3a Ward's Natural Science Establishment, Inc. 4.3b Frederic H. Martini 4.4a Pearson Benjamin Cummings 4.4b Gregory N. Fuller, MD Anderson Cancer Center, Houston, TX 4.4c; 4.5a,b Frederic H. Martini 4.5c Gregory N. Fuller, MD Anderson Cancer Center, Houston, TX 4.6 Carolina Biological Supply Company/ Phototake 4.8b Ward's Natural Science Establishment, Inc. 4.9a,b Project Masters, Inc./The Bergman Collection 4.10a Frederic H. Martini 4.10b Ward's Natural Science Establishment, Inc. 4.11a John D. Cunningham/Visuals Unlimited 4.11b Frederic H. Martini 4.11c Bruce Iverson/Visuals Unlimited 4.14a Robert Brons/Biological Photo Service 4.14b Photo Researchers, Inc. 4.14c Ed Reschke/Peter Arnold, Inc. 4.15 Frederic H. Martini 4.18a G. W. Willis, MD/Visuals Unlimited 4.18b Phototake NYC 4.18c Pearson Benjamin Cummings
Chapter 5 5.CO Steve Gschmeissner/Photo Researchers, Inc. 5.2b,c Frederic H. Martini 5.3 Dr. Richard Kessel and Dr. Randy H. Kardon/Visuals Unlimited 5.4 John D. Cunningham/Visuals Unlimited 5.5a Pearson Benjamin Cummings 5.6a,b Elizabeth A. Abel, M.D. 5.7 Biophoto Associates/Photo Researchers, Inc. 5.10c John D. Cunningham/Visuals Unlimited 5.11; 5.12a,b Frederic H. Martini 5.15 M. A. Ansary/Photo Researchers, Inc.
Chapter 6 6.CO Corbis RF/AGE Fotostock 6.2a Ralph T. Hutchings 6.4a Visuals Unlimited 6.4b Dr. Richard Kessel and Dr. Randy H. Kardon/Visuals Unlimited 6.6 Ralph T. Hutchings 6.9b Pearson Benjamin Cummings 6.10a,b Project Masters, Inc./The Bergman Collection 6.11.1,2 Frederic H. Martini 6.14a Getty Images Inc. - Hulton Archive Photos 6.14b Harold Chen, M.D. 6.18.1,2 Southern Illinois University/Visuals Unlimited 6.18.3 Dr. Kathleen Welch 6.18.4 Southern Illinois University/Peter Arnold, Inc. 6.18.5 Custom Medical Stock Photo, Inc. 6.18.6 Photo Researchers, Inc. 6.18.7 Frederic H. Martini 6.18.8 Southern Illinois University/Visuals Unlimited 6.18.9 Project Masters, Inc./The Bergman Collection 6.19a,b Prof. P. Motta/Science Photo Library/Photo Researchers, Inc.

Chapter 7 7.CO Jim Dolan/Getty Images, Inc. 7.1a; 7.5a,b; 7.6a,b; 7.7a,b Ralph T. Hutchings 7.7c Michael J. Timmons 7.8a,b; 7.9a,b; 7.10a; 7.11; 7.12a,b,c; 7.13; 7.14b Ralph T. Hutchings 7.17a Science Photo Library/Custom Medical Stock Photos, Inc. 7.17b Custom Medical Stock Photos, Inc. 7.17c Princess Margaret Rose Orthopaedic Hospital, Edinburgh, Scotland/Science Photo Library/Photo Researchers, Inc. 7.19a,b,c; 7.20a,b,c; 7.21a,b,c; 7.22a,b,c; 7.23a,b; 7.24a,b; 7.T02.1,2,3 Ralph T. Hutchings
Chapter 8 8.CO Peter Essick/Getty Images, Inc. 8.2b,c; 8.3a,b,c; 8.4a,b; 8.5a,b; 8.6a,b; 8.7a,b; 8.8a,b; 8.11a,b; 8.12a,b; 8.13a,b; 8.14a,b Ralph T. Hutchings
Chapter 9 9.CO David Martinez/Heather Elder Stock/IPNstock 9.3aT,c,d; 9.4bL; 9.5T1,2,3; 9.5B1,2,3,4; 9.12c,d Ralph T. Hutchings
Chapter 10 10.CO Peter Cade/Getty Images, Inc. 10.4 Don W. Fawcett/Photo Researchers, Inc. 10.21a Lippincott Williams & Wilkins/Books 10.21b; 10.23a Frederic H. Martini
Chapter 11 11.CO Shalom Ormsby/Getty Images, Inc.
Chapter 12 12.CO Brad Wilson/Getty Images, Inc. 12.2 David Scott/Phototake NYC 12.5 Frederic H. Martini
Chapter 13 13.CO Burger/Photo Researchers, Inc. 13.4 Ralph T. Hutchings 13.5 Michael J. Timmons 13.6 R. G. Kessel and R. H. Kardon/Visuals Unlimited 13.9 William Ober
Chapter 14 14.CO AGE Fotostock America, Inc. 14.7a,b Ralph T. Hutchings 14.8b Daniel P. Perl, Mount Sinai School of Medicine 14.10; 14.12a,c Ralph T. Hutchings 14.17 Larry Mulvehill/Photo Researchers, Inc. 14.18 Ralph T. Hutchings
Chapter 15 15.CO Jaume Gual/AGE Fotostock America, Inc.
Chapter 16 16.CO Coneyl Jay/Getty Images, Inc.
Chapter 17 17.CO AGE Fotostock America, Inc. 17.2cT Pearson Benjamin Cummings 17.2cM G. W. Willis/Visuals Unlimited 17.3a Ralph T. Hutchings 17.6a Ed Reschke/Peter Arnold, Inc. 17.6c Custom Medical Stock Photo, Inc. 17.21b Lennart Nilsson/Albert Bonniers Forlag AB 17.26b Michael J. Timmons 17.27 Ward's Natural Science Establishment, Inc.
Chapter 18 18.CO Gil Cohen Magen/Reuters Limited 18.6b Manfred Kage/Peter Arnold, Inc. 18.10c Frederic H. Martini 18.14c Ward's Natural Science Establishment, Inc.
Chapter 19 19.CO Scanpix/Terje Eggum jk/DY/Reuters 19.1a Martin M. Rotker 19.2a David Scharf/Peter Arnold, Inc. 19.2b Dennis Kunkel/CNRI/Phototake NYC 19.2d Ed Reschke/Peter Arnold, Inc. 19.4a,b Stanley Flegler, Visuals Unlimited 19.8 AP Wide World Photos 19.10a,b,c,d,e Ed Reschke/Peter Arnold, Inc. 19.13b Custom Medical Stock Photo, Inc.
Chapter 20 20.CO William Coupon / CORBIS All Rights Reserved. 20.3a Ralph T. Hutchings 20.5c Ed Reschke/Peter Arnold, Inc. 20.6b Lennart Nilsson/Albert Bonniers Forlag AB 20.6c Ralph T. Hutchings 20.8cL

Science Photo Library/Photo Researchers, Inc. 20.8cR Biophoto Associates/Photo Researchers, Inc. 20.9c Ralph T. Hutchings 20.10a,b Howard Sochurek/The Stock Shop, Inc./Medichrome 20.14a Larry Mulvehill/Photo Researchers, Inc.
Chapter 21 21.CO Lennart Nilsson/Albert Bonniers Forlag AB 21.1 Biophoto Associates/Photo Researchers, Inc. 21.3.1 William Ober/Visuals Unlimited 21.3.2 B & B Photos/Custom Medical Stock Photo, Inc. 21.5b Biophoto Associates/Photo Researchers, Inc.
Chapter 22 22.CO Digital Vision/RF/Getty Images, Inc. 22.3b Frederic H. Martini 22.4c Ralph T. Hutchings 22.6a David M. Phillips/Visuals Unlimited 22.6b Biophoto Associates/Photo Researchers, Inc. 22.7 Ralph T. Hutchings 22.8c,d; 22.9c Frederic H. Martini 22.21b Leonard Lessin/Peter Arnold, Inc.
Chapter 23 23.CO Pearson Education Corporate Digital Archive 23.2a Photo Researchers, Inc. 23.3b Ralph T. Hutchings 23.5c Phototake NYC 23.6b John D. Cunningham/Visuals Unlimited 23.7a; 23.8 Ralph T. Hutchings 23.10a,b Don W. Fawcett
Chapter 24 24.CO Seth Wenig/AP Wide World Photos 24.7b Frederic H. Martini 24.9c Ralph T. Hutchings 24.10a Alfred Pasieka/Peter Arnold, Inc. 24.10b Astrid and Hanns-Frieder Michler/SPL/Photo Researchers, Inc. 24.12a Ralph T. Hutchings 24.17d M.I. Walker/Photo Researchers, Inc. 24.18c Frederic H. Martini 24.20c Michael J. Timmons 24.23b Ralph T. Hutchings 24.24b Ward's Natural Science Establishment, Inc.
Chapter 25 25.CO AGE Fotostock/Alamy Images
Chapter 26 26.CO GARO/Phanie. 26.4b Ralph T. Hutchings 26.17 Photo Researchers, Inc. 26.19a Ward's Natural Science Establishment, Inc. 26.19b,c Frederic H. Martini
Chapter 27 27.CO Michael Steele/Getty Images, Inc.
Chapter 28 28.CO Katrina Wittkamp/Getty Images, Inc. 28.4b Frederic H. Martini 28.5a Don W. Fawcett 28.5b Ward's Natural Science Establishment, Inc. 28.8b David M. Phillips/Visuals Unlimited 28.9b Frederic H. Martini 28.10bL Ward's Natural Science Establishment, Inc. 28.10bR R. G. Kessel and R. H. Kardon, "Tissues and Organs: A Text-Atlas of Scanning Electron Microscopy," W. H. Freeman & Co., 1979. All Rights Reserved. 28.10c,d,e; 28.16a,b,c,d Frederic H. Martini 28.16e C. Edelmann/La Villete/Photo Researchers, Inc. 28.16f,g G. W. Willis/Visuals Unlimited 28.17b Frederic H. Martini 28.17c Custom Medical Stock Photo, Inc. 28.19b Ward's Natural Science Establishment, Inc. 28.20a,b Frederic H. Martini 28.20c Michael J. Timmons 28.20d Frederic H. Martini 28.21 Michael J. Timmons 28.23b Fred E. Hossler/Visuals Unlimited 28.23c; 28.U01 Frederic H. Martini
Chapter 29 29.CO Brooke Fasani/Corbis 29.1a Francis Leroy/Photo Researchers, Inc. 29.6b Frederic H. Martini 29.7a Arnold Tamarin 29.7b,c,d; 29.8a Lennart Nilsson/Albert Bonniers Forlag AB 29.8b Photo Researchers, Inc. 29.14 Science Photo Library/Photo Researchers, Inc.

Index

A band, 298
Abdominal aorta, 755–756
Abdominal cavity, 23, 24
Abdominopelvic cavity, 22, 24
  arteries, 758
  veins, 763, 765
Abdominopelvic quadrants, 17, 19, 24
Abdominopelvic regions, 17, 19, 24
Abducens nerve (N VI), 493, 500
Abduction, 275
Abductor digiti minimi muscle, 370–371, 379–380
Abductor hallucis muscle, 379–380
Abductor pollicis brevis muscle, 370–371
Abductor pollicis longus muscle, 368–370
ABO blood types, 662–666
Abortion, 1104
Abrasion, 175
Abscess, 142, 144, 796
Absolute refractory period, 409
Absorption, 875
  carbohydrates, 916–919
  ions, 920–921
  in large intestine, 913–914
  lipids, 919
  monosaccharides, 918–919
  proteins, 919–920
  in small intestine, 900–901
  in stomach, 897
  vitamins, 921
  water, 920
Absorptive state, 949
Accessory glands, male reproductive system, 1053–1054
Accessory ligaments, 271
Accessory nerve (N XI), 499, 500
Accessory organs, digestive system, 875
Accessory pancreatic duct, 901
Accessory respiratory muscles, 847–848
Accessory structures
  eye, 566–569
  synovial joints, 271
Accessory structures (integument), 158
  glands. See Gland(s)
  hair, 169–171
  nails, 174–175
Acclimatization, 960
Accommodation, vision, 575–576, 577
Accommodation curves, 231
ACE (angiotensin-converting enzyme), 742, 841, 982
Acetabular labrum, 283
Acetabular ligament, transverse, 283
Acetabular notch, 254, 283
Acetabulum, 254, 263
Acetate, 415
Acetoacetate, 950
Acetone, 950
Acetylcholine (ACh). See also Cholinergic synapses
  autonomic tone and, 545
  location, 420
  mechanism of action, 420
  muscle activity and, 304
  in parasympathetic division, 543
  structure, 420
  in sympathetic division, 538
Acetylcholinesterase (AChE), 304, 415, 541
Acetyl-CoA, 934
Acetyl group, 935
Achilles (calcaneal) tendon, 261, 376
Acid(s), 44, 1023–1024
Acid–base balance, 1025–1030

age-related changes, 1036–1037
definition, 1010
maintenance, 1027–1030
overview, 1010–1011
pH control in, 1023–1027
renal compensation and, 1028–1030
respiratory compensation and, 1028
respiratory regulation, 1032
terminology, 1023
Acid–base disorders, 1030–1036
  acute phase, 1030
  blood chemistry in, 1036
  causes, 1030
  combined respiratory and metabolic acidosis, 1034
  compensated phase, 1030
  diagnosis, 1034–1036
  metabolic acidosis, 1033–1034
  metabolic alkalosis, 1034
  renal responses to, 1028–1030
  respiratory acidosis, 1030–1032
  respiratory alkalosis, 1032–1033
Acidemia, 1023
Acidic solution, 43, 1023
Acidophils. See Eosinophils
Acidosis, 1023
  definition, 44
  detection and diagnosis, 1034–1035
  metabolic, 1033–1034, 1036, 1037
  renal responses, 1028–1029
  respiratory, 1030–1032, 1036, 1037
Acid phosphatase, prostatic, 1059
Acinar glands, 123
ACL (anterior cruciate ligament), 286
Acoustic meatus, external/internal, 221, 227
Acoustic nerve. See Vestibulocochlear nerve
Acquired immune deficiency syndrome (AIDS), 815, 819, 820, 841
Acquired immunity, 796
Acquired reflex, 450
Acromegaly, 202, 208, 639
Acromial end, clavicle, 247
Acromioclavicular joint, 250
Acromioclavicular ligament, 281
Acromion, 250
Acrosin, 1088
Acrosomal cap, 1050
Acrosomal vesicle, 1050
ACTH. See Adrenocorticotropic hormone
Actin, 73, 298
Actinic keratosis, 164
Actinins, 299
Action, muscle, 339, 340, 341
Action potentials, 389, 399
  all-or-none principle, 407
  axon diameter and propagation speed, 413
  cardiac muscle cells, 700–701
  frequency, 425
  generation, 407–408, 409
  vs. graded potentials, 411
  propagation, 410–412
  rate of generation, 424–425
  refractory period, 408–409
  skeletal muscle cells, 297, 304–306
  sodium–potassium exchange pump and, 409
Activation energy, 39
Active channels. See Gated channels
Active immunity, 796
Active site
  enzyme, 56
  G-actin molecule, 300

Active transport, 95, 99, 977
Acute epiglottitis, 834
Acute renal failure, 984, 998
Acute respiratory acidosis, 1031
ADA deficiency, 1118
Adam's apple, 832
Adaptation
  light and dark, 583
  sensory receptors, 509–510
Addison disease, 163, 639, 644
Adduction, 275
Adductor brevis muscle, 372–374
Adductor hallucis muscle, 379–380
Adductor longus muscle, 372–374
Adductor magnus muscle, 372–374
Adductor pollicis muscle, 371–372
Adenine, 58
Adenohypophysis, 462
Adenoid. See Pharyngeal tonsil
Adenosine, 421
Adenosine diphosphate (ADP), 60, 674
Adenosine monophosphate (AMP), 60
Adenosine triphosphatase. See ATPase
Adenosine triphosphate. See ATP
Adenylate cyclase (adenylyl cyclase), 418, 610
ADH. See Antidiuretic hormone
Adhesion, phagocyte, 791
Adhesion belt, 114
Adhesions, 136, 144
Adipocytes (adipose cells), 125, 127–128
Adipose tissue, 124
  endocrine functions, 635, 638
  liposuction, 128
  metabolic functions, 949
  structure and function, 127–128
  sympathetic effects, 544
Adluminal compartment, seminiferous tubule, 1049
Adolescence, 1108, 1109–1111
ADP (adenosine diphosphate), 60, 674
Adrenal cortex. See Suprarenal cortex
Adrenal gland. See Suprarenal gland
Adrenaline. See Epinephrine
Adrenal medulla. See Suprarenal medulla
Adrenergic synapses, 417, 537
Adrenocortical steroids. See Corticosteroids
Adrenocorticotropic hormone (ACTH), 163, 616, 620, 957
Adrenogenital syndrome, 639
Adventitia, 880, 1070
Aerobic metabolism, 81, 319, 932, 939
Afferent arterioles, 969
Afferent division, PNS, 387
Afferent fibers, 391
Afferent lymphatics, 784
Afferent neurons. See Sensory neurons
Afferent vessels. See Capillary(ies)
Afterbirth. See Placenta
Afterload, 713
Agglutination, 662, 808
Agglutinins, 662
Agglutinogens, 662
Aggregated lymphoid nodules, 783–787, 901
Aging
  cancer incidence, 143
  cardiovascular system, 770
  cholesterol levels and, 943
  digestive system, 921–922
  female reproductive system, 1079
  fluid, electrolyte, and acid–base balance, 1036–1037
  gustation, 566

hearing, 598
hormones, 642–643
immune system, 817
integumentary system, 177–178
joints, 286–288
male reproductive system, 1079–1080
muscular system, 378–380
nervous system, 554–555
olfaction, 563–564
oogenesis and, 1064
respiratory system, 866
skeletal system, 264
summary, 1111
telomerase in, 106
tissue structure, 143
urinary system, 1002–1003
vision, 576
Agonist (muscle), 340
Agranular leukocytes (agranulocytes), 666
AIDS (acquired immune deficiency syndrome), 815, 819, 820, 841
Air, gas concentrations, 852
Ala, sacrum, 238
Alarm phase, general adaptation syndrome, 640–641
Albinism, 162
Albumins, 654
Alcohol abuse, 956
Alcoholism, 956
Aldosterone, 627–628, 988, 1014, 1020
Aldosteronism, 639
Alimentary canal. See Digestive tract
Alkalemia, 1023
Alkaline, 1023
Alkaline phosphatase, 198
Alkaline tide, 893, 1034
Alkalosis, 44, 1023
  detection and diagnosis, 1034–1036
  metabolic, 1034, 1035–1036, 1037
  renal responses, 1029–1030
  respiratory, 1032–1033, 1035–1036, 1037
Allantois, 1094
Alleles, 1112
Allergen(s), 668, 816, 820
Allergic rhinitis, 816
Allergy(ies), 815–816, 820
All-or-none principle, action potential, 407
Alpha-amylase. See Salivary amylase
Alpha$_1$-antitrypsin deficiency, 1118
Alpha-blockers, 556
Alpha cells, 631, 635
Alpha-helix, 54
Alpha-interferons, 793
Alpha-2-macroglobulin, 676
Alpha receptors, 537
Alpha-1 receptors, 537, 542
Alpha-2 receptors, 537, 542
Alpha waves, 488–489
Alternative pathway, complement activation, 793, 794
Alveolar air, 852–853
Alveolar ducts, 839–841
Alveolar epithelium, 827
Alveolar macrophages, 791, 839
Alveolar process, 223, 225
Alveolar sacs, 839
Alveolar ventilation, 849–850, 860
Alveolus (alveoli), 747, 826
Alveolus (tooth socket), 885
Alzheimer disease, 555, 556, 1118
Amacrine cells, retina, 572
Amanita muscaria, 542

Amenorrhea, 1070, 1081
Amination, 947–948
Amino acid(s)
  catabolism, 946–947
  essential, 947
  neurotransmitters, 420
  nonessential, 947
  in protein buffer systems, 1025
  structure, 53
  in urine, 996
Amino acid derivatives (biogenic
    amines), 420, 606, 608
Amino group, 46
Ammonia, 996
Ammonium ions, 914, 989
Amnesia, 551, 556
Amniocentesis, 118, 1119, 1120
Amniochorionic membrane, 1102
Amnion, 1094
Amniotic cavity, 1093
Amniotic fluid, 118, 1094
Amoeboid movement, 666
AMP (adenosine monophosphate), 60
Amphetamines, 554
Amphiarthrosis, 268
Amphicytes. See Satellite cells
Amphimixis, 1089
Amplitude, sound wave, 595
Ampulla
  ductus deferens, 1053
  duodenal, 902
  semicircular duct, 589
  uterine tube, 1065
Ampulla of Vater, 902
Amygdaloid body, 479, 550
Amyloid β protein, 555
Amyloidosis, 1118
Amyotrophic lateral sclerosis, 524, 525,
    1118
Anabolic steroids, 643. See also
    Androgen(s)
Anabolic system, 541. See also
    Parasympathetic division, ANS
Anabolism, 38, 930–931
Anaerobic endurance, 325
Anaerobic process, 319
Anal canal, 913
Anal columns, 913
Anal orifice, 913
Anal sphincter, internal/external,
    357–358, 913
Anal triangle, 357–358
Anandamide, 421
Anaphase, mitosis, 102–103
Anaphylactic shock, 816, 820
Anaphylaxis, 816, 820
Anaplasia, 144
Anastomosis, 727
Anastrozole, 1073
Anatomical directions, 17, 20
Anatomical landmarks, 17, 18
Anatomical neck, humerus, 250
Anatomical position, 17, 18
Anatomical regions, 17, 19
Anatomic dead space, 849
Anatomy
  clinical, 6
  definition, 5
  developmental, 6
  gross, 6
  microscopic, 7
  vs. physiology, 5–6
  regional, 6
  sectional, 18–21
  superficial, 17
  surface, 6
  systemic, 6
Anaxonic neurons, 390
Anchoring proteins, 71
Anconeus muscle, 365–368
Androgen(s), 617, 630, 1042
  abuse, 643
  disorders, 639

effects, 201, 949, 1080
  in normal growth, 640
  production, 1046, 1057–1058
  regulation of secretion, 1080
Androgen-binding protein (ABP), 1050
Andropause, 1079
Androstenedione, 643, 1074
Anemia, 657, 661, 677
Anencephaly, 525
Anesthesia, 513, 525
Aneurysm, 723, 772
Angina pectoris, 694, 715
Angiogenesis, 727, 772
Angiography, 694
Angioplasty, 694–695, 715
Angiotensin-converting enzyme (ACE),
    742, 841, 982
Angiotensin I, 637
Angiotensin II, 637, 742
Angiotensinogen, 637
Angle, rib, 241
Angular motion, 272–274
Animal starch. See Glycogen
Anion gap, 1035
Anions, 32
Anisotropic, 298
Ankyloglossia, 883
Annular ligament, 283
Annulus fibrosus, 277
Anoxia, 843, 869
ANS. See Autonomic nervous system
Ansa cervicalis nerve, superior/inferior,
    442
Antagonistic effects, hormones, 638
Antebrachial cutaneous nerve, median,
    444
Antebrachial vein, median, 762
Antebrachium, 250
Anteflexion, uterus, 1065
Antegrade flow, 390
Anterior cavity, eye, 569
Anterior chamber, eye, 573
Anterior commissure, 482
Anterior cruciate ligament (ACL), 286
Anterior fontanelle, 229
Anterior lobe
  cerebellum, 472
  pituitary gland, 614–618
Anterior margin, tibial tuberosity, 259
Anterior median fissure, 431
Anterior nuclei, thalamus, 475, 480
Anterior white commissure, 437
Anterograde amnesia, 551, 556
Antiangiogenesis factor, 131, 144
Antibody(ies), 126, 654. See also
    Immunoglobulin(s)
  actions, 807–808
  antigen–antibody complex, 806–807
  classes, 807–808
  secretion, 782
  structure, 805–806
Antibody-mediated (humoral) immunity,
    669, 782, 796
  antibody structure, 805–806
  B cell sensitization and activation,
    804–805
  primary/secondary response to antigens,
    808–809
Antibody titer, 808
Anticholinesterase drug, 425
Anticoagulants, 676
Anticodon, 86
Antidiuretic hormone (ADH), 606
  cardiovascular effects, 742
  in fluid balance, 1014
  functions, 618, 620
  renal effects, 991–992
  secretion, 618
  underproduction/overproduction, 639
Antigen(s), 782
Antigen–antibody complex, 806–807
Antigen binding sites, 806
Antigenic determinant sites, 806
Antigen presentation, 785, 799–800

Antigen-presenting cells (APCs),
    799–800, 812
Antigen processing, 800
Antigen recognition, 800–802
Antihistamines, 816
Anti-inflammatory effects, 629
Antiport, 94
Anti-Rh antibodies, 665
Antithrombin-III, 676
Antiviral proteins, 793
Antrum (bone surface feature), 187
Antrum, follicle, 1064
Anus, 913
Aorta
  abdominal, 755–756
  ascending, 689, 750
  descending, 689, 755
  terminal segment, 756
  thoracic, 756
Aortic arch, 689
Aortic artery, 750
Aortic bodies, 514, 741
Aortic reflex, 740
Aortic semilunar valve. See Aortic valve
Aortic sinus, 514, 690, 740
Aortic valve, 689
Apex
  heart, 682
  nose, 829
  patella, 259
  sacrum, 238
Aphasia, 488, 501
Apical foramen, 885
Apical surface, epithelium, 112, 113
Apnea, 865
Apneustic center, 471, 472, 861
Apocrine secretion, 121, 122
Apocrine sweat glands, 172–173
Apolipoproteins, 654
Aponeurosis, 129–130, 295
Apoptosis, 100, 801
Appendicitis, 820, 911
Appendicular muscles, 342
  overview, 343–344, 360, 361
  pelvis and lower limbs, 371–378
  shoulders and upper limbs, 359–370
  trunk, 360–361
Appendicular skeleton, 247–264
  joints, 287
  lower limbs, 258–262
  overview, 212, 247, 248–249
  pectoral girdle, 247–250
  pelvic girdle, 254–257
  upper limbs, 250–254
Appendix, 783, 910
Appositional growth
  bone, 195–198
  cartilage, 131
Aquaporins, 93, 991
Aqueduct of midbrain, 464
Aqueduct of Sylvius, 464
Aqueous humor, 570, 573–574
Aqueous solutions, 41, 42
Arachidonic acid, 49, 606
Arachnoid granulations, 466
Arachnoid mater, 434, 464
Arachnoid villi, 466
Arbor vitae, 472, 473
Arcade
  dental, 886
Arch
  atlas, 235
  dental, 886
  foot, 262
Arcuate arteries, 969, 1068
Arcuate fibers, 482
Arcuate line, 255
Arcuate veins, 969
Areola, 1072
Areolar tissue, 12
Arginine, 919, 947
Arginine vasopressin (AVP). See
    Antidiuretic hormone
Aromatase, 1058

Aromatase inhibitors, 1073
Arousal, 553
Arrector pili muscle, 169
Arresting solution, 960
Arterial anastomoses, 692
Arterial anatomy, 727
Arterial puncture, 653
Arteries, 682, 720
  abdominopelvic organs, 758
  brain, 753–754
  chest and upper limb, 750–752
  elastic, 722
  head and neck, 753
  lower limb, 759
  lower limbs, 757, 759
  muscular, 722
  overview, 749, 751
  pelvis, 757
  pulmonary circuit, 749
  structure, 720–723
  systemic circuit, 750–757
  trunk, 756–757
  types, 722–723
  vs. veins, 721–722
Arteries listed
  aorta. See Aorta
  arcuate, 969, 1068
  axillary, 750
  basilar, 754
  brachial, 653, 722, 750
  bronchial, 755, 841
  carotid, 467, 722, 740, 750, 753, 841
  celiac trunk, 534
  central retinal, 573
  cerebral, 753–754
  circumflex, 692
  colic, 757
  communicating, 754
  coronary, 690, 692
  cortical radiate, 969
  cystic, 755
  deferential, 1045
  digital, 753
  dorsalis pedis, 757
  esophageal, 755
  femoral, 722, 748, 757
  femoral circumflex, 757
  fibular, 757
  gastric, 755
  gastroduodenal, 757
  gastroepiploic, 756
  genicular, 757
  gluteal, 757
  gonadal, 757
  hepatic, 755
  hypophyseal, 615
  iliac, 722, 748, 757, 1066, 1070
  inferior mesenteric, 536
  intercostal, 756
  interlobar, 969
  internal thoracic, 695, 750, 1072
  interventricular, 692, 727
  intestinal, 757
  lumbar, 757
  mediastinal, 755, 756
  mesenteric, 722, 755, 757
  obturator, 757
  ophthalmic, 753
  ovarian, 757, 1061
  overview, 1066
  pancreatic, 756
  pancreaticoduodenal, 757
  pericardial, 755
  phrenic, 756
  plantar, 757
  popliteal, 757
  pudendal, 757
  pulmonary, 689, 722, 749
  radial, 653, 750, 1068
  rectal, 757
  renal, 755, 757, 969
  sacral, 757
  segmental, 969
  sigmoid, 757

spiral, 1068
splenic, 756
straight, 1068
subclavian, 722, 748, 750
superior mesenteric, 536
suprarenal, 757
testicular, 757, 1045
tibial, 757
trabecular, 787
ulnar, 750
ulnar collateral, 750
ulnar recurrent, 753
umbilical, 767, 999, 1098
uterine, 1066
vertebral, 467, 750
Arterioles, 720, 722–723
Arteriosclerosis, 555, 724–725, 772
Arteriovenous anastomoses, 727
Artery(ies), 130
Arthritis, 286, 288
Articular capsule, 269
Articular cartilages, 132, 197, 270
Articular disc. See Meniscus
Articular facet, vertebrae, 233
Articular processes, vertebrae, 233
Articulation, 833
Articulations. See Joint(s)
Artificial sweeteners, 46
Aryepiglottic folds, 832–833
Arytenoid cartilages, 832
Ascending aorta, 689, 750
Ascending colon, 911
Ascending limb, nephron loop, 975
Ascending tracts, 437, 471
Ascites, 136, 144, 877, 924
Ascorbic acid. See Vitamin C
Aspartate, 418, 420
Association areas, 484–485
Association fibers, 481
Association neurons. See Interneurons
Asthma, 838, 869
Astigmatism, 576, 599
Astral rays, 102
Astrocytes, 392–393
Asynchronous motor unit summation, 315
Ataxia, 474, 501
Atelectasis, 847, 869
Atenolol, 713
Atherectomy, 694
Atherosclerosis, 724, 772
Atherosclerotic plaque, 694, 724–725
Athletes
   amenorrhea in, 1070
   blood doping, 661
   carbohydrate depletion/loading, 940
   carbohydrate loading, 940
   cardiac reserve, 714
   cardiovascular fitness, 745
   hormones to improve performance, 643, 1058
   knee injuries, 286
   proteinuria and hematuria, 984
   salt loss in, 1020
   training at high altitude, 867
Atlas, 234–235
Atmosphere (atm), 845
Atmospheric pressure, 843
Atomic mass unit (amu), 31
Atomic number, 29
Atomic weight, 31
Atoms, 29–30
Atorvastatin, 943
ATP (adenosine triphosphate), 421
   demand for, 320–321
   mitochondrial production, 934–939
   production, 80, 81
   production in aerobic metabolism, 319
   production in glycolysis, 319–320, 932
   production in lipid catabolism, 942
   production in oxidative phosphoryla-
      tion, 938
   production in protein catabolism,
      946–947

structure, 60
use in active transport, 95–96
use in endocytosis, 96
use in muscle contraction, 308–310,
   318–322
ATPase (adenosine triphosphatase)
   myosin functioning as, 308–310
   sodium–potassium, 95
   structure, 60
ATP synthase, 938
Atresia, 1062
Atrial appendage, 684
Atrial baroreceptors, 741
Atrial natriuretic peptide, 1014
Atrial natriuretic peptide (ANP), 743, 982
Atrial reflex, 710, 741
Atrial systole, 704
Atrioventricular (AV) bundle, 698
Atrioventricular (AV) node, 696, 698
Atrioventricular (AV) septal defect, 769
Atrioventricular (AV) valves, 685, 690
Atrium, left/right, 682, 687
Atrophy, muscle, 325
Atropine, 425
Auditory association area, 485
Auditory cortex, 484
Auditory nerve. See Vestibulocochlear
   nerve
Auditory ossicles, 220, 586–587
Auditory pathways, 597–598
Auditory reflexes, 501
Auditory sensitivity, 598
Auditory tube, 132, 221, 586, 831
Auricle (atrial appendage), 684
Auricle (ear), 132, 585
Auricular surface
   ilium, 255
   sacrum, 238
Auscultation, 706
Autoantibodies, 815
Autoimmune disorders, 287, 815, 820
Autolysis, 79
Automatic bladder, 1002
Automaticity, 327
Autonomic ganglia, 392, 530
Autonomic nervous system (ANS),
   387–388, 529
   adrenergic receptors, 537–538, 542
   autonomic tone, 545–546, 709–710
   cholinergic receptors, 541–542
   comparison to somatic nervous system,
      529–530, 549
   dual innervation, 543
   heart rate and, 709–710
   organization, 529–531
   overview, 529
   parasympathetic division, 531. See also
      Parasympathetic division, ANS
   sympathetic division, 531. See also
      Sympathetic division, ANS
Autonomic plexuses, 543, 546
Autonomic reflexes. See Visceral
   reflex(es)
Autonomic tone, 545–546, 709–710
Autoregulation
   blood flow, 738–739
   capillary, 727
   definition, 12
   glomerular filtration rate, 982
Autosomal chromosomes, 1112
AV. See Atrioventricular
Avascularity, epithelium, 112
Avitaminosis, 954, 962
Avogadro's number, 31
Axial muscles, 342–359
   head and neck, 345–351
   oblique and rectus, 355–356
   overview, 343–344
   pelvic floor, 356–359
   vertebral column, 351–354
Axial skeleton, 212–242
   joints, 280
   overview, 212–213

skull, 213–230
thoracic cage, 239–242
vertebral column, 231–239
Axillary artery, 750
Axillary lymph node dissection, 1073
Axillary nerve, 444
Axillary vein, 762
Axis, 235
Axolemma, 389
Axon(s), 141
   classification, 413
   diameter and propagation speed, 413
   structure, 389
Axon hillock, 389
Axoplasm, 389
Axoplasmic transport, 390
Azygos vein, 762

Babinski reflex/sign, 455–456
Baby teeth, 886
Back pain, 231
Bacteria, 820
   defenses against, 810, 811
   intestinal, 913–914
   as pathogens, 777
Bainbridge reflex, 710
Balance. See Equilibrium
Balanced diet, 950
Ball-and-socket joints, 277, 278
Balloon angioplasty, 694–695, 715
Band cells, 669, 671
Baroreceptor reflexes, 548, 739–741, 864
Baroreceptors, 511, 514
Basal body, 75
Basal body temperature, 1077
Basal cell(s). See Stem cells
Basal cell carcinoma, 164, 181
Basal compartment, seminiferous tubule, 1049
Basal ganglia, 482
Basal lamina (basement membrane), 112
Basal metabolic rate (BMR), 957
Basal nuclei, 482–484, 523–524
Basal surface, epithelium, 112
Base
   heart, 684
   patella, 259
   sacrum, 238
Bases, 44, 1023
Basic food groups, 951
Basic solution, 43, 1023
Basilar artery, 754
Basilar membrane, 593
Basilar zone, endometrium, 1068
Basilic vein, 762
Basolateral surface, epithelium, 113
Basophils, 126, 668, 670, 812
B cells, 669, 780, 812
   circulation, 782
   life span, 782
   primary/secondary response to antigens,
      808–809
   production, 782
   sensitization and activation, 804–805
bcl-2 gene, 100
Bedsore (decubitus ulcer), 167, 181
Behavior
   brain chemistry and, 554
   hormones and, 642
Behavioral drives, 478
Bell palsy, 495, 501
Belly, muscle, 337
Benadryl, 816
The Bends, 853, 869
Benign prostatic hypertrophy (BPH),
   1059
Benign tumor, 105, 107
Beta-blockers, 556, 713
Beta cells, 631–632, 635
Betahydroxybutyrate, 950
Beta-interferons, 793
Beta-mercaptan, 563
Beta-oxidation, 942
Beta receptors, 537–538

Beta-1 receptors, 538, 542
Beta-2 receptors, 538, 542
Beta-3 receptors, 538
Beta waves, 488–489
Biaxial joint, 273
Bicalutamide, 1059
Bicarbonate ions, 44
   absorption, 921
   blood, 858, 976
   reabsorption at proximal convoluted
      tubule, 984
   reabsorption in collecting system, 989,
      991
   secretion in collecting system, 989
   stomach, 893
   urine, 976
Bicarbonate reserve, 1027
Biceps brachii muscle, 252, 283, 318,
   337, 365, 367–368
Biceps femoris muscle, 372, 375–376
Bicuspid valve. See Mitral valve
Bifid, 234
Bile, 906, 908
Bile canaliculi, 906
Bile duct, common, 902, 906
Bile duct system, 906
Bile ductules, 906
Bile salts, 51, 908
Bilirubin, 658, 908, 996
Biliverdin, 658
Biogenic amines (amino acid
   derivatives), 420, 606, 608
Biotin, 914, 955
Bipennate muscle, 336–337
Bipolar cells, retina, 571
Bipolar neurons, 390
2,3-Bisphosphoglycerate (BPG), 857
Bitter taste, 564
Bladder. See Urinary bladder
Blast cells, 669, 671
Blastocoele, 1091
Blastocyst, 1091, 1092
Blastomeres, 1091
Bleaching, 581–582
Bleeding time, 675
Blindness, 576
Blind spot, 573
Blisters, 161
Blood. See also specific components
   age-related changes, 770
   coagulation. See Coagulation
   collection of sample, 653
   composition, 651–653
   distribution, 729
   formed elements, 130, 651–653,
      670–671
   functions, 651
   hemostasis. See Hemostasis
   pH, 653
   physical characteristics, 651–653
   plasma. See Plasma
   viscosity, 731
Blood–brain barrier, 393, 467–468
Blood clot, 674
Blood clotting. See Coagulation;
   Hemostasis
Blood colloid osmotic pressure (BCOP),
   735, 981
Blood–CSF barrier, 468
Blood doping, 661
Blood flow, 732
   autoregulation, 738–739
   during exercise, 744–745
   hormones and, 742–743
   neural mechanisms, 739–742
   to special regions, 747–748
Blood gas analysis, 854
Blood glucose. See Glucose
Blood groups, 662–666
Blood pressure (BP)
   arterial, 733
   definition, 730, 732
   hormonal control, 742–743
   reflex control, 739–742

INDEX

Blood pressure (BP) *(continued)*
  regulation, 16
  renal, 981–982
  response to hemorrhage, 745–746
  in small arteries and arterioles, 733–734
  venous, 734
Blood reservoir, 168
Blood smear, 653
Blood–testis barrier, 1049
Blood–thymus barrier, 782
Blood types, 662–666
  compatibility testing, 664
  cross-reactions, 662–663
  distribution in human population, 663
Blood vessels. *See also* **Arteries; Veins**
  age-related changes, 770
  blood distribution in, 729
  capacitance, 729
  dual innervation, 545–546
  pressure gradient, 730
  pulmonary circuit, 682, 749, 853–854
  resistance, 731
  systemic circuit. *See* Systemic circuit
  wall structure, 720–721, 723
"Blue baby," 769
Blue cones, 582
BMR (basal metabolic rate), 957
Body
  epididymis, 1052
  gallbladder, 908
  hyoid, 226
  mandible, 225
  muscle, 337
  pancreas, 901
  penis, 1056
  rib, 241
  scapula, 247
  sphenoid, 221
  sternum, 242
  stomach, 891
  tongue, 883
  uterus, 1066
  vertebra, 232
Bodybuilders, 643, 991
Body cavities, 22–24
Body composition, 149–150, 1011
Body fluids
  balance. *See* Fluid balance
  buffer systems. *See* Buffer systems
  composition, 1013
  electrolytes, 41–42. *See also*
    Electrolyte(s)
  pH, 43–44, 1024–1027
  regulation, 16
Body position, 294, 316, 514
Body stalk, 1095
Body temperature
  basal, 1077
  cardiac output and, 709
  control. *See* Thermoregulation
  uterine cycle and, 1077
  variability in normal range, 14
Bohr effect, 857
Bolus, 834, 880
Bone, 124
  age-related changes, 207–208
  blood and nerve supplies, 198–200
  calcium and, 203–204
  vs. cartilage, 134
  cells, 189–190
  compact, 191–192
  composition, 189–190, 203
  effects of exercise on, 201
  effects of hormones on, 201–202
  fracture. *See* Fracture(s)
  growth and development, 200–202
  heterotopic, 199
  matrix, 189
  shapes, 185–186
  spongy, 192–193
  structure, 132, 134, 188
  surface features, 186–187
Bone markings, 186–187
Bone marrow, 188, 192, 660, 782

Bones listed
  auditory ossicles, 220, 586–587
  calcaneus, 261
  capitate, 253
  carpal, 253
  clavicle, 247, 249
  coccyx, 231, 239, 263
  cuboid, 261
  cuneiform, 261
  ethmoid, 213, 222–223
  femur, 258–259
  fibula, 260
  frontal, 213, 219
  hamate, 253
  humerus, 250–251
  hyoid, 225, 226, 264
  ilium, 254, 263
  ischium, 254
  lacrimal, 213, 225
  lunate, 253
  mandible, 213, 225–226, 263, 264
  maxillae, 213, 223–224
  metacarpal, 253, 254
  metatarsal, 261–262
  nasal, 213, 224
  nasal conchae, 213, 222, 223, 224
  navicular, 261
  occipital, 213, 218, 264
  palatine, 213, 224
  parietal, 188, 213, 218–219
  patella, 259
  phalanges, 253, 254, 262
  pisiform, 253
  pubis, 254
  radius, 252–253
  ribs, 239–242
  sacrum, 231, 237–238, 263
  scaphoid, 253
  scapula, 247, 249–250
  sphenoid, 213, 221–222
  sternum, 242
  talus, 260
  tarsal, 260–261
  temporal, 213, 219–220
  tibia, 259–260
  trapezium, 253
  triquetrum, 253
  ulna, 251–252
  vertebrae, 234–239
  vomer, 213, 224
  zygomatic, 213, 224–225
Bony labyrinth, 587
Botulism, 331
Bowman capsule. *See* **Glomerular capsule**
Bowman glands, 562
Boyle's law, 843
BP. *See* **Blood pressure**
BPG (2,3-bisphosphoglycerate), 857
**BPH (benign prostatic hypertrophy),
  1059**
Brachial artery, 653, 722, 750
Brachialis muscle, 341, 365, 366–368
Brachial plexus, 442–444
Brachial vein, 762
Brachiocephalic trunk, 750
Brachiocephalic vein, 762
Brachioradialis muscle, 365–368
Bradycardia, 698, 715
Brain
  activity monitoring, 488–489
  age-related changes, 554–555
  blood supply, 466–467, 747, 753–754
  electroencephalogram, 488, 501
  introduction, 461
  landmarks, 461–462
  prenatal development, 462–463
  protection and support, 464–466
  regions, 461–462
  respiratory centers, 861–862
  veins, 758, 761–762
  ventricles, 392, 463–464
Braincase. *See* **Cranium**
**Brain natriuretic peptide (BNP), 743,
  982, 1014**

Brain stem, 462, 469, 584
Branched glands, 123
Breast, 1071–1072
Breast cancer, 1073, 1081, 1118
Breastbone (sternum), 239, 242
Breathing
  at birth, 865–866
  mechanics, 847–849
  modes, 847–849
  voluntary control, 865
Breech birth, 1107, 1120
Brevis (muscle name), 341
Broad ligament, 1060
Broca area, 486
Brodmann, Korbinian, 486
Brodmann areas, 486
Bronchi, 834, 837
Bronchial artery, 755, 841
Bronchial tree, 837
Bronchioles, 826, 837–839
Bronchitis, 837, 869
Bronchoconstriction, 837
Bronchodilation, 837, 869
Bronchography, 869
**Bronchomediastinal trunk, left/right,
  779–780**
Bronchopulmonary carcinoma, 867, 869
Bronchopulmonary segment, 837
Bronchoscope, 869
Brown fat, 127, 960
Bruise, 178
Bruit, 731
Brunner glands, 900
Brush border, 899
Brush border enzymes, 899, 917–918
Bruxism, 226
Buccal cavity. *See* **Oral cavity**
Buccal phase, swallowing, 889–890
Buccinator muscle, 341, 345
Buffer(s), 45, 1023, 1024
Buffer systems, 45, 1024–1025
  carbonic acid–bicarbonate, 1026–1027
  hemoglobin, 1026
  interactions among, 1030, 1031
  phosphate, 1027
  protein, 1025–1026
Buffy coat, 655
Bulbar conjunctiva, 567
Bulbospongiosus muscle, 357–358, 1071
Bulbo-urethral glands, 1053, 1054
Bulk transport, 96
Bundle branches, left/right, 698
Bundle of His (AV bundle), 698
Bunion, 271, 288
Burkitt lymphoma, 1118, 1119
Burning/aching pain, 510
Burns, 166, 179
Bursae, 271, 281
Bursitis, 271, 288
Butyrate, 657

**CABG (coronary artery bypass graft),
  695, 715**
**CAD (coronary artery disease), 694–695,
  715, 724**
Caffeine, 709
Calcaneal tendon, 261, 376
Calcaneus, 261
Calcification, 124, 194
Calcitonin (CT)
  bone remodeling and, 201, 202
  calcium balance and, 202, 204, 1022
  secretion, 624
Calcitriol
  in bone growth and maintenance, 201,
    202
  in calcium homeostasis, 1022
  functions, 635–637
  in growth, 640
  synthesis, 165
Calcium
  absorption, 920–921
  in blood clotting, 676
  bone physiology and, 203–204

  in cardiac contractions, 701–702
  control of blood levels, 203–204
  functions in human body, 30, 624, 953
  G proteins and, 611
  in muscle contraction, 298, 306
  reabsorption at distal convoluted tubule,
    988
  recommended daily intake, 953
  as second messenger, 609, 611
  in urine, 996
Calcium balance, 1021–1022
Calcium channel(s), slow/fast, 700
Calcium channel blockers, 694, 713, 715
Calcium hydroxide, 189
Calcium phosphate, 189
Calcium reserve, skeleton as, 203
Calcium salts, 189
Calculi, 1002, 1003
Callus, 205
Calmodulin, 330, 611
Calorie, 955, 956
Calorigenic effect, 623
Calorimeter, 956
Calorimetry, 955
Calsequestrin, 298
Calvaria, 214
cAMP (cyclic-AMP), 418, 537, 563, 609
Canal (bone surface feature), 187
Canaliculi, 134, 189
Canal of Schlemm, 573
Canals of Volkmann, 191
Cancellous bone, 188
Cancer
  aging and, 143
  breast, 1073, 1081
  cell division in, 105–106
  cervical, 1068, 1081
  CNS, 395
  colorectal, 911
  definition, 105, 107
  lung, 867, 869
  in lymph nodes, 786
  osteoporosis and, 208
  ovarian, 1065, 1081, 1118
  prostate, 1059, 1081
  skin, 164
  testicular, 1043
Canine tooth, 886
Capacitance, blood vessel, 729
Capacitance vessels, 729
Capacitation, 1052, 1088
Capillary(ies), 130, 682, 720
  anatomy, 726
  continuous, 725
  diffusion in, 734–735
  fenestrated, 615, 726
  filtration in, 735–737
  lymphatic, 778–779
  reabsorption in, 735–737
  sinusoidal, 726
  vasomotion, 727–728
Capillary bed, 726–727
Capillary exchange, 730
  diffusion, 734–735
  filtration, 735–737
  reabsorption, 735–737
**Capillary hydrostatic pressure (CHP),
  730, 735**
Capillary plexus, 726–727
Capillary pressure, 730, 735–737
Capitate, 253
Capitulum, 250
Capsular colloid osmotic pressure, 981
Capsular epithelium, 974
**Capsular hydrostatic pressure (CsHP),
  981**
Capsular ligaments, 271
Capsular space, 970, 974
Capsule, 130, 226
Carbaminohemoglobin, 657, 859
Carbohydrase, 903, 916, 918
Carbohydrate(s)
  breakdown and synthesis, 941, 948
  digestion and absorption, 916–919

metabolism, 932–941
overview, 45–48, 61
plasma expanders, 654
plasma membrane, 71–72
in urine, 996
Carbohydrate depletion/loading, 940
Carbohydrate loading, 940, 962
Carbon, 30
Carbon dioxide
binding to hemoglobin, 859
Bohr effect, 857
cerebrospinal fluid, 741–742
partial pressures, 851–852
structure, 34
transport, 858–860
Carbonic acid, 44, 858–859, 1023
**Carbonic acid–bicarbonate buffer
system, 1026–1027**
Carbonic anhydrase, 857, 989, 1024
Carbon monoxide (CO)
as neurotransmitter, 417, 418–419, 421
poisoning, 856, 869
Carboxylic acid group, 46, 48
Carboxypeptidase, 903, 918
Cardia, 891
Cardiac arrhythmias, 700, 715
Cardiac center, 471, 709, 739
**Cardiac conduction system. See
Conducting system, heart**
Cardiac cycle, 703–707
heart sounds in, 706, 707
phases, 703–704
pressure and volume changes in,
704–707
**Cardiac muscle cells, 140, 326, 684, 686,
700–701**
Cardiac muscle tissue, 138–140
functional characteristics, 327–328
vs. skeletal muscle tissue, 329, 684, 686
vs. smooth muscle tissue, 329
structure, 326–327, 684, 687
**Cardiac myocytes. See Cardiac muscle
cells**
Cardiac notch, 837
Cardiac output (CO), 707–714
abnormal conditions affecting, 709
calculation, 707–708
control, 713–714
exercise and, 744–745
factors affecting, 707–713
**Cardiac pacemaker. See Sinoatrial (SA)
node**
Cardiac plexus, 545, 709
Cardiac reflexes, 709
Cardiac reserve, 714
Cardiac skeleton, 692
Cardiac sphincter, 888
Cardiac tamponade, 684, 715
Cardiac vein, 687, 692–693
Cardinal ligament, 1066
**Cardinal signs and symptoms,
inflammation, 793**
Cardioacceleratory center, 709, 739
Cardioacceleratory reflex, 548
Cardiocytes. See Cardiac muscle cells
Cardiodynamics, 707
Cardioinhibitory center, 709, 739
**Cardiopulmonary resuscitation (CPR),
869**
Cardiovascular centers, 470, 739, 742
Cardiovascular fitness, 745
Cardiovascular physiology
arterial blood pressure, 733
capillary exchange, 734–737
elastic rebound, 733
key terms and relationships, 731–733
overview, 729–730
pressure, 730
pressures in systemic circuit, 731–734
total peripheral resistance, 730–731
venous pressure, 734
venous return, 734
**Cardiovascular system. See also Blood;
Blood vessels; Heart**

age-related changes, 770
autoregulation of blood flow, 738–739
blood distribution in, 729
blood supply to special regions, 747–748
changes at birth, 767–768
circulation patterns, 748
congenital problems, 768–769
fetal, 767–768
hormones and, 742–743
integration with other systems, 770–771
introduction, 10
neural mechanisms, 739–742
overview, 153, 651, 682
parasympathetic effects, 544
prenatal development, 1101
at puberty, 1111
pulmonary circuit, 682, 749
regulatory mechanisms, 737–743
response to exercise, 380, 744–745
response to hemorrhage, 745–747
smooth muscle in, 328
sympathetic effects, 544
systemic circuit, 750–767
Carditis, 690, 715
Carina, 834
Carotene, 72, 162
Carotid artery, 722, 740, 750, 753, 841
Carotid bodies, 514, 741
Carotid canal, 220, 227
Carotid sinus, 514, 740, 753
Carpal bone, proximal/distal, 253
Carpal tunnel syndrome, 254, 264, 383
Carrier, genetic disease, 115
Carrier-mediated transport, 94, 99, 977
Carrier proteins, 71
Cartilage, 124
vs. bone, 134
growth, 131, 132
structure, 131
types, 131–132, 133
**CAT (computerized axial tomography),
24**
Catabolism, 38, 930
Catalysts, 39
Cataract, 575, 599
Catecholamines, 606, 608
**Catechol-O-methyltransferase (COMT),
537**
Categorical hemisphere, 487
Cathartics, 924
Catheter, cardiac, 694
Cations, 32
Cauda equina, 433
Caudal anesthesia, 434
Caudate lobe, liver, 903
Caudate nucleus, 482
Cavernous sinus, 762
C (clear) cells, thyroid gland, 624, 627
**CCK (cholecystokinin), 897, 909, 910,
957**
Cdc2 protein, 104
CD markers, 801
CD3 receptor complex, 801
CD4 T cells, 801, 802–803
CD8 T cells, 801
Cecum, 898, 910–912
Celiac ganglion, 534
Celiac plexus, 545
Celiac trunk, 534, 750, 755
Cell(s)
abnormal growth and division, 105–106
cytoplasm and organelles, 69, 72–82
differentiation, 106
introduction, 7, 67
life cycle, 100–105
model, anatomy, 68
nucleus. See Nucleus
plasma membrane. See Plasma mem-
brane
Cell adhesion molecules (CAMs), 114
Cell biology, 67
Cell body, neuron, 140, 141, 388–389
Cell division, 100–104
Cell junctions, 112, 114

Cell-mediated immunity, 669, 796
activation of CD4 T cells, 802–803
activation of CD8 T cells, 801–802
antigen presentation, 799–800
antigen recognition, 800–801
summary, 804
Cell membrane. See Plasma membrane
Cell physiology, 7
Cells of Leydig. See Interstitial cell(s)
Cell theory, 67
**Cellular immunity. See Cell-mediated
immunity**
Cellular metabolism, 930–931
**Cellular organizational level, 8, 9, 67. See
also Cell(s)**
Cellular respiration, 81, 842, 932
Cellular trophoblast, 1092
Cellulose, 47
Cementum, 885
Centimeters of water (cm H₂O), 846
Central adaptation, 509
Central canal
bone, 191
spinal cord, 392, 435
Central chemoreceptors, 863
**Central nervous system (CNS), 387,
430–431. See also Brain; Spinal
cord**
age-related changes, 554–555
neuroglia, 392–395
primary tumors, 395
secondary tumors, 395
Central retinal artery, 573
Central retinal vein, 573
Central sulcus, 480
Central vein, 904
Centrioles, 69, 74, 75
Centromere, 101
Centrosome, 69, 73, 75
Centrum, vertebra, 232
**Cephalic phase, gastric secretion, 895,
896**
Cephalic vein, 762
Cerebellar cortex, 462, 473
Cerebellar hemispheres, 472
Cerebellar nuclei, 462, 473
**Cerebellar peduncles,
superior/middle/inferior, 472, 473**
Cerebellum, 461, 462, 472–474, 524
Cerebral aqueduct, 464
Cerebral arterial circle, 754
**Cerebral artery,
anterior/middle/posterior, 753–754**
Cerebral cortex, 462
functions, 480–481
gross anatomy, 480
memory storage, 550–551
motor and sensory areas, 484–488
Cerebral hemispheres, 462
Cerebral meningitis, 433
Cerebral nuclei. See Basal nuclei
Cerebral palsy, 522, 525
Cerebral peduncles, 475
**Cerebral vein, superficial/inferior, 758,
762**
Cerebrospinal fluid (CSF)
carbon dioxide, 741–742
circulation, 466–467
formation, 466–467
functions, 392, 434, 465–466
pH, 741–742
**Cerebrovascular accident (stroke), 467,
501, 555, 747, 754, 772**
Cerebrovascular disease, 467
Cerebrum
basal nuclei, 482–484
cortex, 480–481
hemispheric lateralization, 486–487
motor and sensory areas, 484–486
overview, 461, 462
white matter, 481–482
Cerumen, 174, 586
Ceruminous glands, 174, 586
Cervical canal, 1066

Cervical cancer, 1068, 1081
Cervical curve, 231
Cervical enlargement, 431
Cervical nerves, 442
Cervical plexus, 442
Cervical vertebrae, 234–236
Cervix, 1066
Cesarean section, 1106
**cGMP (cyclic guanosine
monophosphate), 581, 609**
Chalazion, 566
Chalones, 105
Channel-mediated diffusion, 91
Channels, plasma membrane, 71
Characters, phenotypic, 1112
Cheekbone, 220
Cheeks, 882
Chemical bonds, 32–35
Chemical gradients, 400
Chemically gated channels, 403–404
Chemical notation, 37
**Chemical organizational level, 8, 9,
29–61**
atoms and molecules, 29–32
chemical bonds, 32–36
chemical reactions, 36–40
chemicals and cells, 60–61
inorganic compounds, 40–45
organic compounds, 45–61
Chemical reactions
catalysts, 39–40
definition, 37
notation, 37
types, 38–39
Chemical synapses, 414
Chemiosmosis, 938
Chemiosmotic phosphorylation, 938
Chemoreceptor(s), 510, 514
**Chemoreceptor reflexes, 741–742,
862–863**
Chemosensitive area, 863
Chemotaxis, 791
positive, 667
Chest
arteries, 752
veins, 763
Chief cells. See Parathyroid cells
Childbirth. See Labor
Childhood, 1108
***Chlamydia*, 1079**
***Chlamydia pneumoniae*, 725**
Chloride, 33
absorption, 921
functions in human body, 30, 953
reabsorption at distal convoluted tubule,
990
reabsorption at nephron loop, 986
recommended daily intake, 953
in urine, 996
Chloride balance, 1021, 1022
Chloride shift, 1026
Cholecalciferol. See Vitamin D₃
Cholecystitis, 908, 924
**Cholecystokinin (CCK), 897, 909, 910,
957**
Cholelithiasis, 924
Cholera, 924
Cholesterol
definition, 62
dietary fats and, 943
serum levels, 943
sources, 51
steroid hormone synthesis from, 1074
structure, 50–51
transport and distribution, 944–946
Choline, 415
Cholinergic synapses, 414–416, 536
Cholinesterase. See Acetylcholinesterase
Chondrocytes, 131
Chondroitin sulfates, 131
Chordae tendineae, 687, 688
Chorion, 1094–1095
Chorionic villi, 1095
Chorionic villus sampling, 1119, 1120

Choroid, 571
Choroid plexus, 466
Chromatid, 102, 1048
Chromatin, 83
Chromium, 953
Chromosomal abnormalities, 1116, 1119
Chromosomal microtubule, 102
Chromosomes, 83
Chronic renal failure, 998
Chronic respiratory acidosis, 1031
Chylomicrons, 919, 944
Chyme, 891
Chymosin, 897
Chymotrypsin, 903, 918
Chymotrypsinogen, 903
Ciliary body, 570
Ciliary ganglion, 414, 492, 494, 539
Ciliary muscle, 570
Ciliary processes, 570
Ciliated epithelium, 113–114
Cilium (cilia), 69, 75, 113
C_1 inactivator, 676
Cingulate gyrus, 479
Circadian rhythm, 584, 630–631
Circle of Willis, 754
Circular muscle, 879. See also
    Sphincter(s)
Circulation. See Blood flow
Circulatory pressure, 730, 732
Circumcision, 1056
Circumduction, 273, 275
Circumferential lamellae, 191
Circumflex artery, 692
Circumvallate papillae, 564
Cirrhosis, 907, 924
Cis face, cisterna, 77
Cisterna(e), 76, 77
Cisterna chyli, 779
Citric acid cycle. See TCA (tricarboxylic
    acid) cycle
CK-MB, 702
Classical pathway, complement
    activation, 793, 794
Clavicle, 247, 249
Clearance test, 981, 1003
Clear layer (lamina lucida), 114
Cleavage, 1090, 1091, 1092
Cleavage furrow, 103
Cleft palate, 223
Clinical anatomy, 6
Clitoris, 1071
Clone, 798
Closed fracture, 206
Clostridium tetani, 304
Clot retraction, 676–677
Clotting factors, 674, 676
Clotting system, 672
Clubfoot (talipes equinovarus), 262, 264
Club hair, 171
CNS. See Central nervous system
CO. See Carbon monoxide; Cardiac
    output
Coagulation, 674, 675
    calcium ions and, 676
    clotting factors, 674, 676
    common pathway, 675
    extrinsic pathway, 674
    feedback control, 15, 675–676
    interaction among pathways, 675
    intrinsic pathway, 674
    vitamin K and, 676
Coagulation time, 675
Coated vesicles, 97
Cobalamin. See Vitamin B_{12}
Cobalt, 953
Cocaine, 417
Coccygeal cornua, 239
Coccygeal ligament, 431, 434
Coccygeus muscle, 357–358
Coccyx, 231, 239, 263
Cochlea, 588, 593
Cochlear branch, vestibulocochlear
    nerve, 496, 597
Cochlear duct, 588, 593–594

Cochlear nuclei, dorsal/ventral, 496, 597
Coding strand, 84
Codominance, 1113, 1115
Codon, 85
Coelom, 22–23
Coenzyme(s), 57
Coenzyme A (CoA), 416, 934
Coenzyme Q, 938
Cofactor, 57
Coitus. See Sexual intercourse
Colic artery, right/middle/left, 757
Colic flexure, right/left, 911
Colic vein, right/middle/left, 766, 767
Colitis, 924
Collagen fibers, 126, 166
Collagenous tissue. See Dense connective
    tissues
Collateral ganglia, 533, 534–536
Collateral ligament
    fibular, 286
    radial, 283
    tibial, 286
    ulnar, 283
Collaterals
    arteries, 727
    axons, 389
Collecting duct, 972, 975, 995
Collecting system, 971–972
    anatomy, 975
    effects of antidiuretic hormone, 991–992
    functions, 979
    reabsorption and secretion along, 989
Colles fracture, 206
Colloids, 42, 622
Colon, 911
Colonoscopy, 913
Colony-stimulating factors (CSFs), 672,
    813, 814
Color blindness, 582–583, 599, 1117, 1118
Colorectal cancer, 911, 1118
Color vision, 582
Colostrum, 1108
Column(s), white matter, 437
Columnar epithelium, 116, 118, 120
Coma, 552
Comminuted fracture, 206
Commissural fibers, 482
Common bile duct, 902, 906
Common pathway, coagulation, 675
Communicating artery,
    anterior/posterior, 754
Compact bone, 191–192
Compartment(s), 381
Compartment syndrome, 381, 383
Compatibility, blood type, 662
Compensation curves, 231
Complementary base pairs, 59
Complementary gene action, 1114
Complementary strands, 59
Complement proteins, 667
Complement (C) proteins, 793
Complement system, 790, 793–794
Complete antigen, 806
Complete proteins, 952
Complete tetanus, 314
Compliance, lungs, 844
Compound, 32
Compound fracture, 206
Compound glands, 123
Compression, 303
Compression fracture, 206
Computerized tomography (CT), 24
COMT (catechol-O-methyltransferase),
    537
Concentration, 43
Concentration gradient, 89, 734
Concentric contractions, 316
Concentric lamellae, 191
Conception. See Fertilization
Condensation, 38
Conducting arteries, 722
Conducting cells, 696
Conducting portion, respiratory tract,
    826

Conducting system, heart, 689, 695–703
    AV bundle, 698
    AV node, 696, 698
    bundle branches, 698
    calcium ions and, 701–702
    conduction deficits, 698
    contractile cells, 700–703
    electrocardiogram, 698–700
    energy for, 703
    impulse conduction, 697
    overview, 696–697
    physiology, 695–696
    Purkinje fibers, 698
    SA node, 696, 697
Conduction, 410, 958
Conduction deficits, 698, 715
Conductive deafness, 599
Condylar joints. See Ellipsoid joints
Condylar process, 226
Condyles, 187
    femur, medial/lateral, 259
    humerus, 250
    tibia, medial/lateral, 259
Cones, 571, 578–579
Congenital heart disease, 768–769
Congenital hypothyroidism. See
    Cretinism
Congenital talipes equinovarus
    (clubfoot), 262, 264
Conjoined twins, 1107
Conjunctiva, 567
Conjunctivitis, 567
Connective tissue(s)
    cell populations, 125–126
    classification, 124
    components, 124
    dense, 128–130
    embryonic, 126–127
    fibers, 126
    fluid, 130–131
    ground substance, 126
    heart, 692
    loose, 127–128
    mucous, 127
    supporting, 131–134
Connective tissue proper, 124–130
Connexons, 114, 414
Consciousness, 552
Consensual light reflex, 501, 547, 548,
    570
Constant segments, antibody, 805
Constipation, 922, 924
Continuous capillary, 725
Continuous passive motion (CPM), 288
Continuous propagation, 410–411
Contractile cells, 700–703
Contractile proteins, 52. See also
    Protein(s)
Contractility, 713
Contraction, skeletal muscle
    control, 306
    cycle, 308–310
    energy sources, 318–320
    energy use, 320
    excitation–contraction coupling,
        306–307
    fatigue and, 320–321
    initiation, 310
    isometric, 316–317
    isotonic, 316
    load and speed, 317–318
    muscle tone and, 316
    opposing, 318
    overview, 303
    recovery, 322
    relaxation, 310, 318
    sarcomere appearance during, 302
    shortening during, 310
    sliding filament theory, 301–303
    tension production, 311–314
Contraction, smooth muscle, 330
Contraction cycle, 308–310
Contraction phase, twitch, 313
Contracture, 381

Contralateral reflex arcs, 453
Control center, 12–13
Contusion, 167
Conus arteriosus, 689
Conus medullaris, 431
Convection, 958
Convergence
    neural circuits, 447
    vision, 583–584
Convergent muscle, 336–337
Copper, 953
Copulation. See Sexual intercourse
Coracoacromial ligament, 281
Coracobrachialis muscle, 361, 363–366
Coracoclavicular ligament, 281
Coracohumeral ligament, 281
Coracoid process, 250
Cords, spinal nerves, 442
Cori cycle, 322
Cornea, 567, 569
Corneal epithelium, 567
Corneal limbus, 569
Corneal reflex, 501
Corneal replacement, 569
Cornification, 161
Corniculate cartilages, 832
Coronal plane, 19
Coronal suture, 214
Corona radiata, 1064, 1088
Coronary artery, left/right, 690, 692
Coronary artery bypass graft (CABG),
    695, 715
Coronary artery disease (CAD), 694–695,
    715, 724
Coronary circulation, 692–694
Coronary ischemia, 694, 715
Coronary sinus, 687
Coronary spasms, 747
Coronary sulcus, 684
Coronary thrombosis, 702, 715
Coronoid fossa, 250
Coronoid process
    mandible, 226
    ulna, 252
Corpora cavernosa, 1056
Corpora quadrigemina, 474
Corpus albicans, 1064
Corpus callosum, 482
Corpus luteum, 637, 638, 1064
Corpus spongiosum, 1056
Corpus striatum, 482
Corrugator supercilii muscle, 346–347
Cortex
    bone, 188
    hair, 169
    lymph node, 785
    ovary, 1061
    renal, 969
    suprarenal. See Suprarenal cortex
    thymus, 786
Cortical bone, 188
Cortical nephron, 972
Cortical nerve, 973
Cortical radiate artery, 969
Cortical radiate vein, 969
Corticobulbar tracts, 521, 523
Corticospinal pathway, 520–522, 523
Corticospinal tracts, anterior/lateral,
    521–522, 523
Corticosteroids, 627
Corticosterone, 629
Corticotropin. See Adrenocorticotropic
    hormone
Corticotropin-releasing hormone (CRH),
    616
Cortisol, 629
Cortisone, 629
Costae. See Ribs
Costal breathing, 847
Costal cartilages, 239
Costal facets, vertebrae, 236
Costal groove, 241
Costal processes, vertebrae, 234
Costimulation, 801

Cotransport, 94, 977
Coughing reflex, 548, 834, 865
Countercurrent exchange, 958
Countercurrent multiplication, 986–988
Countertransport, 94, 977
Coupling factors, 938
Covalent bonds, 34–35
Cowper glands. *See* Bulbo-urethral glands
Coxal bone, 254–255
Cradle cap, 172
Cramps, menstrual. *See* Dysmenorrhea
Cranial bones, 213–214, 218–223
Cranial cavity, 213
Cranial meninges, 433, 464–465
Cranial nerves, 387, 490–491
    abducens (VI), 493, 500
    accessory (XI), 499, 500
    facial (VII), 495, 500
    glossopharyngeal (IX), 497, 500, 862–863
    hypoglossal, 499–500
    oculomotor (III), 492, 500
    olfactory (I), 491, 500
    optic (II), 491, 500
    trigeminal (V), 494, 500
    trochlear (IV), 493, 500
    vagus (X), 497–499, 500, 863
    vestibulocochlear (VIII), 496, 500, 589, 597–598
Cranial reflexes, 430, 450, 501
Cranial root, accessory nerve, 499
Cranial trauma, 465, 501
Craniosacral division, ANS. *See* Parasympathetic division, ANS
Craniostenosis, 230, 242
Cranium, 213, 263, 465
Creatine, 319
Creatine phosphate (CP), 319, 976
Creatine phosphokinase (CPK, CK), 319
Creatinine, 976, 996
Creatinine clearance test, 981, 1003
Cremaster muscle, 1045
Crenation, 93
Crest (bone surface feature), 187
Cretinism, 639, 644
Cribriform plate, 222
Cricoid cartilage, 832
Cristae, mitochondria, 80
Crista galli, 222
Cross-bridges, 301
Crossed extensor reflexes, 453–454
Crossing over, 1116
Cross-match testing, 666
Cross-reaction, 662–663
Cross-training, 325
Crown, tooth, 885
Crude touch and pressure receptors, 511
Crura, 1056
Crus, penis, 1056
Cryptorchidism, 1043, 1081
Crypts of Lieberkühn, 899
Crystal arthritis, 287
Crystallins, 575
CSF. *See* Cerebrospinal fluid
CT. *See* Calcitonin
CT (computerized tomography), 24
Cubital vein, median, 762
Cuboid, 261
Cuboidal epithelium, 116, 118, 119
Cuneiform bone, medial/intermediate/lateral, 261
Cuneiform cartilages, 132, 832
Cupula, 589
Current, 401
Cushing disease, 639, 644
Cusps (leaflets), 687
Cutaneous anthrax, 181
Cutaneous membrane, 158. *See also* Skin
Cutaneous plexus, 167
Cuticle
    hair, 169
    nail, 175
Cyanide ion, 939

Cyanosis, 163, 181, 769
Cycles, sound, 594
Cyclic-AMP (cAMP), 418, 537, 563, 609
Cyclic-GMP (cyclic guanosine monophosphate; cGMP), 581, 609
Cyclin, 104
Cyclosporin A (CsA), 803
Cyst, breast, 1073
Cysteine, 1025
Cystic artery, 757
Cystic duct, 906
Cystic fibrosis, 829, 869, 1118
Cystic vein, 767
Cytochromes, 936
Cytokines, 793, 813, 814
Cytokinesis, 103
Cytology, 7, 67
    exfoliative, 118, 144
Cytoplasm, 72–82
Cytosine, 58
Cytoskeleton, 69, 73, 74
Cytosol. *See* Intracellular fluid
Cytotoxic reactions, 816
Cytotoxic T (T$_C$) cells, 780, 798, 801, 812
Cytotrophoblast, 1092

Dalton, 31
Dalton's law, 851
Dandruff, 162
Dark-adapted state, 583
Dark current, 581
Dartos muscle, 1045
"Date rape" drug, 643
Daughter cells, 100
Daughter chromosomes, 102–103
D cells, 894
DCT. *See* Distal convoluted tubule
Deafness, 599
Deamination, 907, 946
Decarboxylation, 934
Decibels, 595
Decidua basalis, 1097
Decidua capsularis, 1097
Decidua parietalis, 1097
Deciduous teeth, 886
Decomposition, 38
Decompression sickness, 853, 869
Decubitus ulcer (bedsore), 167, 181
Decussation, 516
Deep breathing, 847
Deep cortex, lymph node, 785
Deep fasciae, 137–138
Deep lymphatics, 779
Deep sleep, 489
Deep tendon reflexes, 450
Defecation, 548, 876
Defecation reflex, 915
Defensins, 668, 814
Deferential artery, 1045
Deflation reflex, 865
Degenerative arthritis, 287
Degenerative joint disease, 287
Deglutition. *See* Swallowing
Degranulation, 668
Dehydration, 1016, 1017
Dehydration synthesis, 38, 46–47, 50
Dehydroepiandrosterone (DHEA), 1058
Delayed hypersensitivity, 816
Delayed-onset muscle soreness (DOMS), 326
Delivery, 1106. *See also* Labor
    difficult, 1107
    premature, 1107
Delta cells, 632, 635
Delta waves, 489
Deltoid muscle, 337, 363, 364–365
Deltoid tuberosity, humerus, 250
Demyelination, 397, 425
Denaturation, 57
Dendrites, 141, 389
Dendritic (Langerhans) cells, 161, 312, 784
Dendritic spines, 389
Dens, 235

Dense area, 114
Dense bodies, 328–329
Dense bone, 188
Dense connective tissues, 124, 128–130
Dense irregular connective tissue, 129–130
Dense layer (lamina densa), 115
Dense regular connective tissue, 129–130
Dental arcade, upper/lower, 886
Dental calculus, 886
Dental caries, 922
Dental plaque, 886
Dentate gyrus, 479
Denticulate ligaments, 435
Dentin, 885
Deoxyhemoglobin, 657
Deoxyribonucleic acid. *See* DNA
Depolarization, 404, 406
Depression (movement), 276
Depressor anguli oris muscle, 346–347
Depressor labii inferioris muscle, 346–347
Depth perception, 584
Dermal bone, 198
Dermal (intramembranous) ossification, 198
Dermal papillae, 160
Dermatitis, 166, 181
Dermatome, 438, 440
Dermis, 158
    blood supply, 167
    innervation, 168
    lines of cleavage, 166–167
    organization, 166
    properties, 166
Descending aorta, 689
Descending colon, 911
Descending limb, nephron loop, 975, 994
Descending tract, 437
Descent, testes, 1043–1044
Desmin, 328
Desmopressin, 618
Desmosome (macula adherens), 114
Detached retina, 573, 599
Detoxification, 76
Detrusor muscle, 999
Detumescence, 1078
Deuterium, 29, 30
Development, 1087
    postnatal. *See* Postnatal development
    prenatal. *See* Prenatal development
Developmental anatomy, 6
Developmental dyslexia, 488
Deviated nasal septum, 242
Dextran, 93, 107
DHEA (dehydroepiandrosterone), 1058
Diabetes insipidus, 618, 639, 644, 991
Diabetes mellitus, 571, 639, 1118
    insulin-dependent (type 1), 634, 644, 815
    non-insulin-dependent (type 2), 634, 644
Diabetic nephropathy, 634, 644
Diabetic neuropathy, 634, 644
Diabetic retinopathy, 571, 599, 634, 644
Diabetogenic effect, 617
Diacylglycerol (DAG), 611
Diapedesis, 666, 791
Diaphragm, 355–356, 847
Diaphragma sellae, 614
Diaphragmatic breathing, 847
Diaphragmatic hernia, 359, 383
Diaphysis, 188
Diarrhea, 924
Diarthrosis, 268
Diastole, 703
Diastolic pressure, 733
Diazepam, 551
Dicrotic notch, 706
Diencephalon, 461, 462, 469, 475–478
Diet
    balanced, 950
    disease and, 955
    energy gains and losses, 955–957

    fats and cholesterol, 943
    nutrition and, 951–955
Diethylstilbestrol (DES), 1059
Differential WBC count, 669
Differentiation, 106, 1087
Diffusion, 89–93, 99
    across plasma membranes, 90–93
    capillary exchange, 734–735
    channel-mediated, 91
    facilitated, 94–95, 99
    factors influencing rate, 90
    liquid-gas, 852
    osmosis, 91–92, 99, 735
    principles, 89–90
    respiratory function and, 852–854
    simple, 91
Digastric muscle, 351–352
Digestion, 875
    carbohydrates, 916–919
    ion absorption, 920–921
    lipids, 919
    overview, 916
    proteins, 919–920
    vitamin absorption, 921
    water absorption, 920
Digestive epithelium, 877–879
Digestive system
    age-related changes, 921–922
    components, 876
    disorders, 922
    esophagus, 888–890
    functions, 875–876
    gallbladder, 908
    hormonal control, 882, 909–910
    integration with other systems, 922–923
    intestinal absorption, 909
    introduction, 11, 875
    large intestine, 910–915
    liver, 903–908
    neural mechanisms, 881–882, 909
    oral cavity, 882–888
    overview, 155
    pancreas, 901–903
    parasympathetic effects, 544
    peritoneum and, 876–877
    prenatal development, 878, 1101
    regulation of activities, 881–882
    small intestine, 898–901
    smooth muscle in, 328
    stomach, 891–897
    sympathetic effects, 544
Digestive tract
    definition, 875
    endocrine activity, 635
    epithelial stem cells, 879
    histological organization, 877–880
    movement of materials, 880
Digital artery, 753
Digitalis, 713
Digital rectal exam (DRE), 1059
Digital subtraction angiography (DSA), 24
Digital vein, 762, 764
Diglyceride, 50
Dihydrogen phosphate, 1027
Dihydrotestosterone, 1058
Dihydroxyacetone phosphate, 944
Dilation stage, labor, 1105
Dimenhydrinate, 591
Dimethylpriamine hydrochloride, 816
Dipeptidases, 918, 920
Dipeptide, 53
Dipeptides, 920
Diphenhydramine hydrochloride, 816
Diphtheria, 397
Diploë, 188
Diploid cells, 1048
Dipole, 35
Direct communication, intercellular, 604–605
Direct light reflex, 501, 548
Disaccharides, 46, 47, 916
Disconnection syndrome, 488, 501
Discs, photoreceptor, 578
Disease, 16, 24, 48

Dislocation (luxation), 272, 288
Displaced fracture, 206
Dissociation, 41
Distal convoluted tubule (DCT)
  anatomy, 970–971, 972, 975, 979
  effects of antidiuretic hormone, 991–992
  functions, 972, 995
  reabsorption and secretion at, 988–990
Distribution arteries, 722
Disulfiram, 956
Diuresis, 991
Diuretics, 991, 1003
Divergence, neural circuits, 446–447
Diverticulitis, 924
Diverticulosis, 924
Dizygotic twins, 1107
DNA
  coding strand, 84
  functions, 57–58, 88–89
  mutations, 88
  organization in nucleus, 83–84
  in protein synthesis, 84–88
  replication, 100, 101
  vs. RNA, 59
  structure, 59
  template strand, 84
  transcription of mRNA, 84–86
  triplet codes, 87
DNA fingerprinting, 84, 107
DNA polymerase, 101
Dobutamine, 713
Dominant hemisphere, 487
Dominant inheritance, 1113, 1115
Dopamine, 417, 420, 484, 713
Dorsal arch, 757
Dorsal column/medial lemniscus,
  515–517
Dorsal interosseus muscles, 370–371,
  379–380
Dorsalis pedis artery, 757
Dorsal ramus, 438
Dorsal respiratory group (DRG), 861
Dorsal root(s), 431
Dorsal root ganglia, 431
Dorsal scapular nerve, 444
Dorsal venous arch, 764
Dorsiflexion, 262, 276
Dorsum nasi, 829
Double covalent bond, 34
Down-regulation, 609
Down syndrome, 118, 555, 769, 1118,
  1119
Drives, 478
Drugs
  absorption in stomach, 897
  immunosuppressive, 803, 815, 820
  inactivation in liver, 907
  intradermal injection, 371
  intramuscular injection, 371, 383
  plasma membrane and, 91
  subcutaneous injection, 168, 371
  tubular reabsorption and secretion, 975
DSA (digital subtraction angiography),
  24
Dual innervation, 543
Duchenne muscular dystrophy, 331
Duct(s), 121
Ductless gland, 121
Duct of Santorini, 901
Duct of Wirsung, 901
Ductus arteriosus, 767, 865
Ductus deferens, 619, 1052–1054
Ductus venosus, 767
Duodenal ampulla, 902, 908
Duodenal glands, 900
Duodenal papilla, 908
Duodenum, 898
Dural folds, 464
Dural sinuses, 464
Dura mater, 434, 464
Dynein, 73, 390
Dynorphins, 418, 421
Dyslexia, 488, 501
Dysmenorrhea, 1069, 1081
Dysplasia, 144

Ear. See also Hearing
  anatomy, 585–588
  external ear, 585–586
  inner ear, 587–588
  middle ear, 586–587
  muscles, 346–347
Eardrum. See Tympanic membrane
Earwax (cerumen), 174, 586
Eating disorders, 962
Eccentric contraction, 316
Eccrine sweat gland, 173
ECF. See Extracellular fluid
ECG. See Electrocardiogram
Ectopic bone, 195
Ectopic pacemaker, 698
Ectopic pregnancy, 1092, 1120
Edema, 737, 772
EEG. See Electroencephalogram
Effector(s), 12–13, 387, 449
Efferent arteriole, glomerulus, 970
Efferent division, PNS, 387
Efferent ductules, 1046
Efferent fibers, 391
Efferent lymphatics, 784
Efferent neurons. See Motor neurons
Efferent vessels. See Arteries
Eicosanoids, 49–50, 53, 606–607, 608
Ejaculate, 1055
Ejaculation, 548, 1042, 1078
Ejaculatory duct, 1054
Ejection fraction, 706, 707
EKG. See Electrocardiogram
Elastase, 903, 918
Elastic arteries, 722
Elastic cartilage, 132, 133
Elastic fibers, 126, 166
Elastic ligaments, 126
Elastic membrane, internal/external, 720
Elastic rebound, 733, 848
Elastic tissue, 129, 130
Elastin, 126
Elbow joint, 281–283, 366–367
Elective abortion, 1104
Electrical gradients, 400–401
Electrical synapses, 414
Electrocardiogram (ECG, EKG), 696,
  698–700, 715
Electrocardiography, 696
Electrochemical gradient, 401–402
Electroencephalogram (EEG), 488, 501
Electrolyte(s)
  in body fluids, 42
  definition, 41
  functions, 41–42
Electrolyte balance, 1017–1022
  age-related changes, 1036–1037
  in burn patient, 179
  calcium balance, 1021–1022
  chloride balance, 1022
  definition, 1010
  disorders, 1021
  in extracellular fluid, 1012–1013
  fluid balance and, 1014
  hormonal control, 1013–1014
  importance, 1017–1018
  in intracellular fluid, 1012–1013
  magnesium balance, 1022
  overview, 1010–1011
  phosphate balance, 1022
  potassium balance, 1020–1021
  regulation, 1012–1013
  sodium balance, 1018–1020
Electron(s), 29, 31–32
Electron acceptor, 33
Electron cloud, 29
Electron donor, 33
Electron shell, 29
Electron transport system (ETS),
  936–938, 940
Elements, 30
  composition of human body, 149
  functions in human body, 30
Elephantiasis, 820
Elevation, 276
Ellipsoid joints, 277, 278

Embolism, 677
Embryo, 1065
Embryogenesis, 1091, 1098–1099
Embryological development, 1087
Embryology, 6, 24, 1087
Embryonic connective tissue, 126–127
Embryonic disc, 1094
Emigration, 666, 791
Emission, 619, 1042, 1078
Emmetropia, 577
Emphysema, 867, 869, 1113
Emulsification, 908
Emulsion droplets, 908
Enamel, tooth, 886
End-diastolic volume (EDV), 704, 707,
  711–713
Endergonic reactions, 40
Endocardium, 684
Endochondral ossification, 195–198
Endocrine communication, 604, 605
Endocrine glands, 121
Endocrine reflexes, 611–613
Endocrine secretions, 121
Endocrine system, 606. See also specific
    glands and organs
  control by endocrine reflexes, 611–613
  disorders, 639
  exercise and, 383
  hormones. See Hormones
  integration with other systems, 644–645
  intercellular communication, 604–606
  introduction, 10, 604
  overview, 153, 607
  parasympathetic effects, 544
  prenatal development, 1101
  sympathetic effects, 544
Endocytosis, 96–98, 99
Endoderm, 1093
Endolymph, 587
Endolymphatic duct, 589
Endolymphatic sac, 589
Endometriosis, 1081
Endometrium, 1066
Endomorphins, 418, 421
Endomysium, 295
Endoneurium, 437
Endoplasmic reticulum (ER), 69, 76, 77
Endorphins, 418, 421, 425, 511
Endosomes, 96
Endosteal layers, dura mater, 464
Endosteum, 194–195
Endothelins, 673
Endothelium, 118
Endothermic reactions, 40
End-stage renal failure, 998
End-systolic volume (ESV), 706, 707, 713
Endurance
  aerobic, 325
  anaerobic, 325
  muscle, 323
Energy, 38
  kinetic, 38
  potential, 38, 99–100, 401–402
Energy levels, electrons, 31–32
Energy reserves
  cardiac muscle, 327
  high-energy compounds, 60, 61
  lipids, 48
  muscle, 319
Enkephalins, 418, 421, 511
Enlargements, spinal cord, 431
Enteric nervous system, 531, 882
Enterocrinin, 901, 909, 910
Enteroendocrine cells, 879
Enterogastric reflex, 896
Enterohepatic circulation, bile, 908
Enteropeptidase, 899, 918
Enzymes
  activation energy and, 39, 40
  cofactors, 57
  pancreas, 631
  pH changes and, 57
  plasma membrane, 71
  regulation, 57

saturation limit, 56–57
  seminal fluid, 1055–1056
  specificity, 56
  structure and function, 56–57
  temperature changes and, 57
Eosinophils, 126, 668, 670, 812
Ependyma, 392, 463
Ependymal cells, 392
Epicardium, 24, 684
Epicondyles, medial/lateral
  femur, 259
  humerus, 250
Epicranial aponeurosis, 348
Epicranium (scalp), 346–347, 348
Epidermal growth factor (EGF), 105,
  165–166
Epidermal ridges, 160
Epidermis, 158–166
  pigmentation, 162–163
  structure, 159–161
  tumors, 164
  vitamin D_3 production in, 165
Epididymis, 1051–1052
Epidural block, 434, 456
Epidural hemorrhage, 466, 502
Epidural space, 434
Epiglottis, 832
Epiglottitis, acute, 834
Epilepsies, 489, 1118
Epimysium, 294
Epinephrine
  effects, 629–630, 949
  heart rate and, 711
  muscle metabolism and, 322
  as neurotransmitter, 420
  overproduction, 639
  production in suprarenal medulla, 536,
    629–630
  in stress response, 640–641
  stroke volume and, 713
  sympathetic stimulation and release of,
    537–538
Epineurium, 437
Epiphyseal cartilage, 197, 199, 200, 263,
  264
Epiphyseal fracture, 206
Epiphyseal line, 197
Epiphysis, 188
Epiploic appendages, 911
Episiotomy, 1106
Epistaxis, 831, 869
Epithalamus, 475
Epithelium (epithelia)
  attachment to basal lamina, 114–115
  characteristics, 112–113
  classification, 116–117
  columnar, 118
  cuboidal, 118
  digestive, 877–879
  functions, 113
  glandular. See Gland(s)
  glomerulus, 974
  intercellular connections, 114, 115
  maintenance and repair, 115–116
  renewal and repair, 879
  respiratory, 827–828
  specializations, 113–114
  squamous, 117–118
  transitional, 118
EPO. See Erythropoietin
Eponychium, 175
Equational division, 1049
Equilibrium (balance), 16, 508, 585
  muscle tone and, 316
  pathways, 589, 592
  semicircular ducts, 589–590
  utricle and saccule, 589, 591
Equilibrium potential, 401
Erectile tissue
  clitoris, 1071
  penis, 1056
Erection, 1056, 1078
Erector spinae muscles, 351
Eruption, teeth, 887
Erythema, 179

Erythroblast, 660
Erythroblastosis fetalis, 664
Erythrocytes. See Red blood cells
Erythrocytosis, 661
Erythropoiesis, 660, 661, 782
Erythropoiesis-stimulating hormone. See Erythropoietin
Erythropoietin (EPO), 105, 635, 975
  abuse, 643
  cardiovascular regulation and, 743
  functions and effects, 637, 655, 661
Esophageal artery, 756
Esophageal hiatus, 888
Esophageal phase, swallowing, 890
Esophageal plexus, 545
Esophageal sphincter, upper/lower, 888
Esophageal varices, 924
Esophageal vein, 762
Esophagus, 888–890
Essential amino acids, 947
Essential fatty acids, 944
Estradiol, 616, 637, 1058, 1074
Estriol, 1075
Estrogen(s)
  disorders, 639
  effects, 201, 637, 1075, 1080
  in normal growth, 640
  in pregnancy, 1098
  production, 637
  regulation of secretion, 637, 1080
  structure, 51
Estrone, 1075
Ethmoid, 213, 222–223
Ethmoidal air cells, 222
Ethmoidal labyrinth, 222
Eupnea, 847–849
Eustachian tube. See Auditory tube
Evaporation, 958
Eversion, 276
Excess postexercise oxygen consumption (EPOC), 322
Exchange pump, 95
Exchange reaction, 39
Exchange vessels, 682. See also Capillary(ies)
Excitable membrane, 403
Excitation–contraction coupling, 306–310
Excitatory interneurons, 449
Excitatory neurotransmitters, 414
Excitatory postsynaptic potential (EPSP), 419, 422–423
Excretion, 875–876
Excretory system, 1103
Exercise
  body system response, 380–381
  bone remodeling and, 201
  cardiovascular system response, 744–745
  salt loss during, 1020
Exergonic reactions, 40
Exfoliative cytology, 118, 144, 1071
Exhaustion phase, general adaptation syndrome, 642
Exocrine glands, 121–122
Exocrine secretions, 121
Exocytosis, 78, 98, 99
Exons, 85
Exothermic reactions, 40
Expiration, 843
Expiratory center, 861
Expiratory reserve volume (ERV), 850
Expressivity, 1113
Expulsion stage, labor, 1105
Extension, 273
  ankle, 262
  forearm, 252
Extensor carpi radialis muscle, brevis/longus, 366, 367, 370
Extensor carpi ulnaris muscle, 366, 367–370
Extensor digiti minimi muscle, 368–369
Extensor digitorum brevis muscle, 379–380
Extensor digitorum longus muscle, 376–378
Extensor digitorum muscle, 337, 368–369

Extensor hallucis longus muscle, 376–378
Extensor indicis muscle, 368–369
Extensor pollicis muscle, brevis/longus, 368–369
Extensor retinaculum, 369, 379
External acoustic meatus, 585
External branch, accessory nerve, 499
External callus, 205, 208
External ear, 585–586
External genitalia, 1042
  female, 1071
  male, 1055–1056
External nares, 224, 829
External oblique muscle, 355–356
External os, uterus, 1066
External respiration, 842
External root sheath, hair, 169
Externus (muscle name), 341
Exteroceptors, 391, 510
Extracapsular ligaments, 271
Extracellular fluid (ECF), 67
  components, 1010, 1012
  fluid movement within, 1012, 1014–1015
  fluid shifts, 1016
  sodium balance and, 1018–1020
Extracorporeal liver assist device (ELAD), 907
Extraembryonic membranes, 1094, 1096
Extrafusal muscle fibers, 451
Extra-ocular muscles, 348
Extrapulmonary bronchi, 837
Extrapyramidal system (EPS), 522
Extrinsic (muscle name), 341
Extrinsic ligaments, 271
Extrinsic pathway, coagulation, 674
Extrinsic regulation, 12
Eye, 569–578. See also Vision
  accessory structures, 566–569
  age-related changes, 576
  chambers, 573–574
  fibrous tunic, 569
  lens, 574–578
  muscles, 347–349
  parasympathetic effects, 544
  retina, 571–573
  sectional anatomy, 568
  superficial epithelium, 566–567
  sympathetic effects, 544
  vascular tunic, 569–571
Eyeglasses, 577
Eyelashes, 566
Eyelid, 566–567

Fabella, 372
Facet (bone surface feature), 187
Facial bones, 213–214
Facial expression, muscles, 345–348
Facial nerve (N VII), 495, 500
Facial vein, 762
Facilitated diffusion, 94, 99, 977
Facilitation, 423
F-actin. See Filamentous actin
Fact memory, 550
Factor B, 793
Factor D, 793
Factor P, 793
Factor I, 676
Factor II, 676
Factor III, 674, 676
Factor IV, 674
Factor IX, 674
Factor V, 674
Factor VI, 674
Factor VII, 674
Factor VIII, 674
Factor X, 674
Factor XI, 674
Factor XII, 674
Factor XIII, 674
Facultative water reabsorption, 991
FAD (flavin adenine dinucleotide), 935
Fainting, 734
Falciform ligament, 877, 903

Fallopian tube. See Uterine tube
False labor, 1105
False pelvis, 256
False ribs, 241
Falx cerebelli, 465
Falx cerebri, 222, 464
Familial colon cancer, 1118
Familial hypercholesterolemia, 1118
Familial polyposis of the colon, 1118
Farsightedness, 577
Fascia(e), 137–138
Fascicles
  muscle, 294, 301, 336, 341
  spinal nerve, 437
Fasciculus cuneatus, 515, 520–521
Fasciculus gracilis, 515, 520–521
Fast-adapting receptors, 509
Fast fatigue (FF) fibers, 323
Fast fibers, 323, 324
Fast pain, 510
Fast resistant (FR) fibers, 324
Fast sodium channels, 700
Fast-twitch glycolytic fibers, 323
Fast-twitch oxidative fibers, 324
Fat cell, 61
Fatigue, muscle, 320–321
Fat pads, 271
Fat-soluble vitamins, 921, 953–954
Fatty acids, 53
  dietary, 943
  essential, 944
  free, 944
  omega-3, 49, 62
  saturated, 48–49
  structure and function, 48
  unsaturated, 48–49
Fatty appendices, 911
Fauces, 883
F cells, 632, 635
Feces, 876
Feeding center, 478
Female pronucleus, 1089
Female reproductive cycle, 1074
Female reproductive system
  age-related changes, 1064, 1079
  endocrine functions, 638–639
  external genitalia, 1071
  hormonal regulation, 1074–1077, 1080
  integration with other systems, 1080–1082
  introduction, 11
  mammary glands, 1071–1072
  oogenesis, 1061–1062
  ovarian cycle, 1062–1064
  ovaries, 1060–1061
  overview, 156, 1042, 1060
  parasympathetic effects, 545
  at puberty, 1111
  smooth muscle in, 328
  sympathetic effects, 545
  uterine cycle, 1068–1070
  uterine tubes, 1065
  uterus, 1065–1068
  vagina, 1070–1071
Female sexual function, 1078–1079
Femoral artery, 722, 748, 757
Femoral circumflex artery, 757
Femoral circumflex vein, 764–765
Femoral head, 258
Femoral nerve, 444, 446
Femoral vein, 764
Femur, 258–259
Fenestrated capillaries, 615, 726
Ferritin, 659
Fertilization, 1042, 1087–1090
Fetal alcohol syndrome, 956
Fetal circulation, 767
Fetal development, 1087. See also Prenatal development
Fetal hemoglobin, 657, 858
Fetus, 1065
  cardiovascular changes at birth, 767–768
  circulation in heart and great vessels, 767
  descent of testes, 1043, 1044

development of immunological competence, 812
  fluid, electrolyte, and acid–base balance, 1036
  placental blood supply, 767
Fever, 796, 960–961
Fibrillin, 126
Fibrin, 175, 654
Fibrinogen, 654
Fibrinolysis, 677
Fibroblast growth factor (FGF), 105
Fibroblasts, 175
Fibrocystic disease, 1073, 1081
Fibrocytes, 125
Fibrodysplasia ossificans progressiva, 195
Fibrosis, 143, 380, 383
Fibrous capsule, kidney, 967
Fibrous cartilage (fibrocartilage), 132, 133
Fibrous pericardium. See Pericardial sac
Fibrous proteins, 54
Fibrous skeleton, heart, 692. See Cardiac skeleton
Fibrous (scar) tissue, 143, 175
Fibrous tunic, 569
Fibula, 260
Fibular artery, 757
Fibular collateral ligament, 286
Fibularis muscle, brevis/longus, 376–379
Fibular nerve, 444, 446
Fibular vein, 764
Field of vision, 573
"Fight or flight" response, 479, 531, 640
Filamentous actin (F-actin), 300
Filariasis, 820
Filgrastim (Neupogen), 672
Filiform papillae, 564
Filling time, 711
Filtrate, 970
Filtration
  capillary, 735
  pressures, 980–981
  in urine formation, 976–977
Filtration membrane, 974
Filtration pressure (FP), 981
Filtration slits, 974
Filum terminale, 431
Fimbria, 1065
Fine touch and pressure receptors, 511
Fingerprints, 160
First breath, 865–866
First-class lever, 338
First-degree burn, 179
First heart sound, 706
First messenger, 418, 609
First-order neuron, 514
First trimester, 1090
  blastocyst formation, 1091
  cleavage, 1091
  embryogenesis, 1098–1099
  implantation, 1091–1095
  overview by system, 1100–1101
  placentation, 1095–1098
Fissure
  bone surface feature, 187
  cerebrum, 462
Fixator, 340
Fixed acids, 1024
Fixed macrophages, 125, 791
Fixed ribosomes, 76
Flagellum, 1050
Flat bones, 185, 186
Flatfeet, 262, 264
Flatus, 914
Flavin adenine dinucleotide (FAD), 935
Flavin mononucleotide (FMN), 938
Flexion, 262, 272
Flexor carpi radialis muscle, 365, 366–367
Flexor carpi ulnaris muscle, 365, 366–367
Flexor digiti minimi brevis muscle, 370–371, 379–380
Flexor digitorum brevis muscle, 379–380
Flexor digitorum longus muscle, 377–378
Flexor digitorum muscle, superficialis/profundus, 368–369

Flexor hallucis brevis muscle, 379–380
Flexor hallucis longus muscle, 377–378
Flexor pollicis brevis muscle, 370–371
Flexor pollicis longus muscle, 368–369
Flexor reflex, 452–453
Flexor retinaculum, 254, 369
Floaters, 576
Floating kidney, 967
Floating ribs, 241
Flocculonodular lobe, 472
Fluid balance, 1010–1017
    age-related changes, 1036–1037
    in burn patient, 179
    definition, 1010
    electrolyte balance and, 1014
    in extracellular fluid, 1012–1013
    fluid gains and losses, 1015–1016
    fluid movement within ECF, 1014–1015
    fluid shifts, 1016–1017
    hormonal control, 1013–1014
    in intracellular fluid, 1012–1013
    overview, 1010–1011
    regulation, 1012–1013
Fluid compartments, 1012
Fluid connective tissues, 124, 130–131
Fluid shift, 1016–1017
Fluoxetine, 417, 554
Flutamide, 1059
FMN (flavin mononucleotide), 938
Foam cells, 724
Focal calcification, 724
Focal distance, 575
Focal point, 575
Folate (folic acid), 955
Folia, 472
Follicle(s)
    hair, 169
    ovarian, 1062
    thyroid, 622–623, 627
Follicle cavity, 622
Follicle cells, 1062
Follicle-stimulating hormone (FSH), 606
    effects, 616, 620, 1080
    in female reproductive cycle, 1075–1077
    regulation of secretion, 620, 1080
    in spermatogenesis, 1056, 1057
Follicular cells, ovary, 638
Follicular fluid, 1064
Follicular phase, ovarian cycle, 1062,
    1074–1075, 1076
Follitropin. See Follicle-stimulating
    hormone
Fontanelles, 229–230, 264
Food groups, 951
Foot
    arch, 262
    arteries, 757
    bones, 260–262
    in diabetes, 634
    muscles, 378–380
Foramen (foramina), 187
Foramen, infra-orbital, 223, 227
Foramen lacerum, 220, 227
Foramen magnum, 218
Foramen muscle, 227
Foramen of Monro, 464
Foramen ovale, 222, 227, 687, 767, 865
Foramen rotundum, 222, 227
Foramen spinosum, 222, 227
Force, 323
Forced breathing, 849, 861–862
Forceps delivery, 1107
Foreskin, 1056
Formed elements, blood, 130, 651–653,
    670–671
Forming face, cisterna, 77
Fornix
    eye, 567
    limbic system, 479
    vagina, 1070
Fossa (bone surface feature), 187
Fossa ovalis, 687, 768
Fourth heart sound, 706
Fourth ventricle, 464
Fovea, 571

Fovea capitis, 258
Fovea centralis, 571
Fractionated blood, 651–653
Fracture(s), 205, 208
    clavicle, 247
    hip, 207, 283
    repair process, 205, 207
    rib, 242
    types, 206
Fracture hematoma, 205, 208
Fragile-X syndrome, 1118, 1118
Frank, Otto, 712
Frank–Starling principle, 712
Freckles, 163
Free edge, nail, 175
Free fatty acids, 944
Freely permeable membrane, 89
Free macrophages, 125, 791
Free nerve endings, 508, 512
Free radical, 34
Free ribosomes, 76
Freeze fracture, 82
Frequency, sound, 594, 597
Frontal bone, 213, 219
Frontal eye field, 485
Frontal lobe, 480, 484
Frontal plane, 19, 21
Frontal sinuses, 219
Frontal squama, 219
Frontal suture, 219, 264
Frostbite, 650
Fructose, 46, 917
FSH. See Follicle-stimulating hormone
Full term, 1102
Full-thickness burn, 179
Full-thickness graft, 179
Functional residual capacity (FRC), 851
Functional syncytium, 327
Functional zone, endometrium, 1068
Fundus
    gallbladder, 908
    stomach, 891
    uterus, 1066
Fungi, 820
Fungiform papillae, 564
Funny bone, 250

G-actin, 300
Galactose, 917
Gallbladder, 908
Gallstones, 908, 924
Gamete, 1042
Gamma aminobutyric acid (GABA), 417,
    420
Gamma efferents, 451
Gamma-hydroxybutyrate (GHB), 643
Gamma-interferons, 793
Gamma motor neurons, 451
Ganglion (ganglia), 391, 395
Ganglion cells, retina, 571
Ganglionic neurons, 530, 532
Gap junction, 114, 115
Gas(es), 35
    as neurotransmitters, 417
    partial pressures, 851–852
    pressure measurement, 846
    pressure/volume relationship, 843, 844
Gas exchange, 851–854
Gas laws, 851–852
Gaster, muscle, 337
Gastrectomy, 924
Gastric artery, left/right, 756, 757
Gastric glands, 893–894
Gastric inhibitory peptide (GIP), 897,
    909, 910
Gastric juice, 893
Gastric lipase, 84
Gastric phase, gastric secretion, 895–896
Gastric pits, 891
Gastric reflex, 548
Gastric secretion
    cephalic phase, 895, 896
    gastric phase, 895–896
    intestinal phase, 896–897
Gastric vein, 767

Gastrin, 894, 909, 910
Gastritis, 897, 924
Gastrocnemius muscle, 339, 372
Gastroduodenal artery, 757
Gastroenteric reflex, 901
Gastroepiploic artery, left/right, 756
Gastroepiploic vein, left/right, 766, 767
Gastroileal reflex, 901
Gastrointestinal system. See Digestive
    system
Gastrointestinal tract. See Digestive tract
Gastroscope, 924
Gastrosplenic ligament, 786
Gastrulation, 1093–1094
Gated channels, 403–404
Gaucher disease, 1118
G cells, 894
G-CSF, 672
Gemelli muscle, superior/inferior,
    373–374
Gender differences. See Sex differences
Gene, 84
Gene activation, 84
General adaptation syndrome, 640–642,
    644, 816–817
General interpretive area, 486
General senses
    adaptation, 509–510
    chemoreceptors, 514
    definition, 507–508
    interpretation of sensory information, 509
    mechanoreceptors, 511–514
    nociceptors, 510–511
    somatic pathways, 515–518
    stimuli detection, 508–509
    thermoreceptors, 511
    visceral pathways, 518
Generator potential, 508
Genetic analysis, 1119
Genetic code, 83
Genetic imprinting, 1116
Genetic recombination, 1115–1116
Genetics, 1087
Genetic stutter, 1119
Genicular artery, 757
Geniculate ganglia, 495
Genioglossus muscle, 341
Geniohyoid muscle, 351–352
Genitofemoral nerve, 444, 446, 1045
Genome, 1117
Genotype, 1112
Geriatricians, 1111
Geriatrics, 1111
Germinal center, 783
Germinal epithelium, 1061
Germinative cells. See Stem cells
Germ layers, 1093–1094, 1095
Gestation, 1090. See also Prenatal
    development
Gestational trophoblastic neoplasia,
    1093, 1120
GFR. See Glomerular filtration rate
Ghrelin, 894, 957
Giemsa stain, 666
Gigantism, 202, 208, 639
Gingivae, 882
Gingival sulcus, 885
Gingivitis, 886
Gland(s). See also specific glands
    ceruminous, 174
    characteristics, 112
    classification, 121
    control of, 174
    modes of secretion, 121
    sebaceous, 172
    structure, 121–123
    sweat, 172–173
    types of secretions, 121
Gland cells, 113
Glandular epithelium, 113
Glans, penis, 1056
Glassy membrane, hair, 171
Glaucoma, 574, 599
Glenohumeral joint. See Shoulder joint
Glenohumeral ligament, 281

Glenoid cavity, 250
Glial cells. See Neuroglia
Gliding joints, 277, 278
Gliding motion, 273
Global aphasia, 488
Globular proteins, 54
Globulins, 654
Globus pallidus, 482
Glomerular (Bowman) capsule, 969,
    973–974
Glomerular filtration, 980–981
Glomerular filtration rate (GFR)
    autonomic regulation, 984
    autoregulation, 982
    definition, 981
    hormonal regulation, 982–984
    in pregnancy, 1104
Glomerular hydrostatic pressure (GHP),
    980–981
Glomerulonephritis, 975, 1003, 1033
Glomerulus, 969, 994
Glossopharyngeal nerve (N IX), 497, 500,
    862–863
Glottis, 831, 833
Glucagon, 633, 635, 713, 949
Glucocorticoids, 616, 629, 630, 641–642,
    949
Gluconeogenesis, 633, 940
Glucose
    in blood, 632–633
    gluconeogenesis, 633, 940
    glycolysis. See Glycolysis
    movement across plasma membrane, 71,
        94
    reabsorption at proximal convoluted
        tubule, 984
    in stress response, 642
    structure, 45–46
    in urine, 634, 644, 978
Glucose-dependent insulinotropic
    peptide. See Gastric inhibitory
    peptide
Glucose-6-phosphate, 933
Glucose-sparing effect, 617
Glutamate, 420, 511, 552
Glutaminase, 1028
Glutamine, 1028
Gluteal artery, 757
Gluteal lines, 254
Gluteal muscles, 371
Gluteal nerves, superior/inferior, 446
Gluteal vein, 765
Gluteus maximus muscle, 371, 373–374
Gluteus medius muscle, 372, 373–374
Gluteus minimus muscle, 372, 373–374
Glycerides, 50, 53
Glycerol, 50
Glycine, 418, 420
Glycocalyx, 71–72
Glycogen, 47
    breakdown, 617, 940
    formation, 617, 940
    in liver, 617
Glycogenesis, 940
Glycogenolysis, 940
Glycolipids, 51–52, 53
Glycolysis, 80
    energy yield, 939
    overview, 932–934, 941
    skeletal muscle, 319–320
Glycoproteins, 57, 606, 608
Glycosaminoglycans, 114
Glycosuria, 634, 644, 978
GM-CSF, 672
Gnostic area, 486
Goiter, 644
Golgi apparatus, 69, 77–79
Golgi tendon organs, 514
Gomphosis, 268, 269
Gonad(s), 638–639, 1042
Gonadal artery, 756
Gonadal steroid-binding globulin (GBG),
    1058
Gonadal vein, 765
Gonadotropes, 1074

Gonadotropin-releasing hormone (GnRH), 1056, 1076–1077, 1080
Gonadotropins, 616
Gonorrhea, 1081
Gout, 287–288
Gouty arthritis, 287
G_1 phase, interphase, 100–101
G_2 phase, interphase, 101–102
G proteins, 418, 537
  in gustation, 565–566
  hormone activity and, 609–611
  in olfaction, 563
Gracilis muscle, 372
Graded potentials, 399, 404–406, 411
Graft rejection, 803
Granular leukocytes (granulocytes), 666
Granulation tissue, 175, 181
Granulosa cels, 1062
Graves disease, 639
Gravity, muscle contraction and, 318
Gray commissures, 436
Gray horns, posterior/anterior/lateral, 436
Gray matter, 389
  cerebellum, 473
  medulla oblongata, 471
  mesencephalon, 474
  pons, 471
  spinal cord, 435–437
Gray muscle, 395
Gray ramus, 438, 534
Great auricular nerve, 442
Great cerebral vein, 762
Greater curvature, stomach, 891
Greater horns, hyoid, 226
Greater omentum, 877
Greater sciatic notch, 254
Greater wings, sphenoid, 222
Great saphenous vein, 695, 764
Green cones, 582
Greenstick fracture, 206
Gross anatomy, 6
Ground substance, 124, 126
Growth factors, 104, 105
Growth hormone (GH), 105, 131, 606, 640
  actions, 617, 620
  bone growth and, 201, 202
  disorders, 639
  effects on peripheral metabolism, 949
  muscle metabolism and, 322
  regulation, 620
Growth hormone–inhibiting hormone (GH-IH; somatostatin), 617, 635, 894, 909
Growth hormone–releasing hormone (GH-RH), 617
Growth-inhibitory factor (GIF), 814
Guanine, 58
Guanosine diphosphate (GDP), 936
Guanosine triphosphate (GTP), 421, 581, 936
Gubernaculum testis, 1043
Guillain–Barré syndrome, 397
Gustation, 508, 564
  age-related changes, 566
  discrimination, 564–566
  pathways, 564
  receptors, 564, 565
Gustatory cells, 564
Gustatory cortex, 484
Gustatory receptors, 564–565
Gustducins, 565
Gynecomastia. See Adrenogenital syndrome
Gyri, 462

Hair
  age-related changes, 178
  color, 171
  functions, 169
  growth cycle, 171
  production, 169–171
  structure, 169–170
  types, 171
Hair analysis, 171

Hair bulb, 169
Hair cells, 589
Hair follicles, 169
Hair matrix, 169
Hair papilla, 169
Hair root, 169
Hair shaft, 169
Halcion. See Triazolam
Half-life, 31
Hallux, 262
Hamate, 253
Hamstrings, 372
Hand
  bones, 253–254
  muscles, 365–371
Haploid cells, 1048
Haptens, 807
Hard keratin, 168
Hard palate, 223, 831, 883
Hatching, 1091
Haustra, 911
Haustral churning, 915
Haversian canal, 191
Haversian system, 191
Hawking, Stephen, 524
Hb. See Hemoglobin
H band, 298
Head
  epididymis, 1052
  pancreas, 901
  spermatozoon, 1050
  thick filament, 301
Head (body part)
  arteries, 753
  muscles, 345–351
  veins, 761, 762
Head (bone part), 187
  femur, 258
  humerus, 250
  radius, 252
  rib, 241
  scapula, 250
  ulna, 252
Head fold, 1098
Hearing, 508, 592
  age-related changes, 598
  auditory pathways, 597–598
  auditory sensitivity, 598
  cochlear duct, 593–594
  process, 595–597
  sound characteristics, 594–595
Heart. See also Cardiac entries
  age-related changes, 770
  autonomic innervation and regulation, 709–711
  blood flow to, 747
  blood supply, 692–693
  cardiac cycle, 703–707
  cardiodynamics, 708–714
  conducting system. See Conducting system, heart
  connective tissues, 692
  dual innervation, 545
  endocrine functions, 635, 637
  gross anatomy, 682
  internal anatomy and organization, 685–691
  left atrium, 689
  left ventricle, 689
  left vs. right ventricles, 689–690
  location in thoracic cavity, 682–683
  pericardium, 684
  relationship to lung, 837
  right atrium, 687
  right ventricle, 687–689
  sectional anatomy, 688
  superficial anatomy, 684, 685
  valves, 690–691
  wall, 684, 686
Heart attack. See Myocardial infarction
Heart failure, 706, 715
Heart murmur, 706, 715
Heart rate
  factors affecting, 709–711
  neonate, 1108

Heart sounds, 706, 707
Heat, 38
Heat capacity, 41
Heat conservation, 958
Heat exhaustion, 961, 962
Heat-gain center, 958
Heat generation, 958–959
Heat loss, 322, 958–959
Heat-loss center, 958
Heat production, 322, 958–959
Heat stroke, 961, 962
Heat transfer, 958. See also Thermoregulation
Heavy chains, antibody, 805
Heavy-metal poisoning, 397
Heimlich maneuver, 869
Helicases, 101
Helper T (T_H) cells, 782, 798, 812, 819
Hematocrit (Hct), 655, 661, 677
Hematologists, 660
Hematoma, fracture, 205, 208
Hematopoiesis. See Hemopoiesis
Hematuria, 658, 677, 984, 996, 1003
Heme, 656
Hemiazygos vein, 762
Hemidesmosomes, 114, 115
Hemispheric lateralization, 486–487
Hemocytoblasts, 660, 671
Hemodialysis, 998, 1003
Hemoglobin (Hb), 163, 656, 661
  abnormal, 658
  BPG and, 857–858
  carbon dioxide binding, 857
  conservation and recycling, 658–659
  fetal, 657, 858
  functions, 657
  in oxygen transport, 855–858
  partial pressure of oxygen and, 855–856
  pH and, 856–857
  skin color and, 163
  structure, 655–656, 657
  temperature and, 857
Hemoglobin buffer system, 1026
Hemoglobin concentration, 677
Hemoglobin saturation, 855
Hemoglobinuria, 658, 677
Hemolysis, 93, 658
Hemolytic disease of the newborn, 664, 677, 807
Hemophilia, 677, 1118
Hemopoiesis, 651
Hemopoiesis-stimulating factor, 814
Hemorrhoids, 729, 772, 922
Hemosiderin, 659
Hemostasis, 673
  coagulation phase, 674–677
  fibrinolysis, 677
  platelet phase, 673, 674
  vascular phase, 673–674
Henry's law, 852
Heparin, 126, 668, 676
Hepatic artery, 756–757
Hepatic duct, right/left/common, 906
Hepatic flexure, 911
Hepatic portal system, 766–767
Hepatic portal vein, 766
Hepatic triads, 905
Hepatic vein, 765
Hepatitis, 907, 924
Hepatocytes, 904
Hepatopancreatic sphincter, 908
Heptose, 45
Hering–Breuer reflexes, 864–865
Hernia, 359, 383
Herniated disc, 279, 280, 288
Hertz (Hz), 594
Heterotopic bone formation, 195
Heterozygous, 1113
Hexose, 45
High-density lipoprotein (HDL), 944
High-energy bonds, 59–60, 80
High-energy compounds, 60, 61
Higher-order functions, 550
  arousal, 553
  memory, 550–552

reticular activating system, 475, 553
  sleep, 552–553
  states of consciousness, 552
High-frequency sound, 595
Hilum
  kidney, 967
  lung, 834
  lymph node, 784
  ovary, 1061
  spleen, 787
Hinge joints, 277, 278
Hip joint, 207, 283, 284
Hippocampus, 389, 479, 550
Hippuric acid, 996
Histamine, 126, 420, 668
Histidine, 947, 1025
Histiocytes. See Fixed macrophages
Histology, 7, 24, 112
Histones, 83
HIV (human immunodeficiency virus), 815, 819, 820
Hives, 816
HLA (human leukocyte antigen), 799–800
Holocrine secretion, 121, 122
Homeostasis, 2
  negative feedback in, 14
  overview, 12–13
  positive feedback in, 14–16
  systems integration and, 16
Homeostatic regulation, 12
Homologous chromosomes, 1112
Homozygous, 1113
Horizontal cells, retina, 572
Horizontal plate, palatine bones, 224
Hormone-binding proteins, 654
Hormone replacement therapy, 1079
Hormones, 121, 604. See also specific hormones
  age-related changes, 642–643
  athletic performance and, 643
  behavior and, 642
  bone remodeling and, 201, 202
  calcium balance and, 203–204
  cardiac contractility and, 713
  cardiovascular regulation, 742–743
  classes, 606–609
  digestive system control, 881–882, 909–910
  female reproductive cycle, 1074–1077
  glomerular filtration rate regulation, 982–984
  heart rate control, 711
  immune system, 813–814
  interactions, 638, 640
  intestinal, 881–882, 909–910
  intracellular receptors and, 611, 612
  male reproductive function, 1056–1058
  mechanisms of action, 609–611
  in metabolism, 949
  muscle metabolism and, 322
  as neurotransmitters, 421
  in normal growth, 640
  secretion and distribution, 609
  in stress response, 640–642
Horns, spinal cord, 435
"Hot flashes," 1079
Housemaid's knee, 271
Howship lacunae, 194
Human chorionic gonadotropin (hCG), 1098
Human chorionic somatomammotropin (hCS), 1098
Human Genome Project (HGP), 1117–1120
Human immunodeficiency virus (HIV), 815, 819, 820
Human leukocyte antigens (HLAs), 799–800
Human physiology, 7
Human placental lactogen (hPL), 1098
Human tetanus immune globulin, 304
Humeroradial joint, 282
Humero-ulnar joint, 282
Humerus, 250–251

Humoral immunity. *See* Antibody-mediated (humoral) immunity
Huntington disease, 554, 556
Hyaline cartilage, 131–132, 133
Hyaluronidase, 1088
Hyaluronan (hyaluronic acid), 114
Hydatidiform mole, 1093
Hydration sphere, 41
Hydrocephalus, 466, 502
Hydrochloric acid, 44, 893–894
Hydrocortisone. *See* Cortisol
Hydrogen
  functions in human body, 30
  secretion at distal convoluted tubule, 988–989, 991
  structure, 29, 34
Hydrogen bonds, 35, 36
Hydrogen cyanide, 939
Hydrogen peroxide, 668
Hydrolysis, 38, 916
Hydrophilic molecules, 42
Hydrophobic molecules, 42
Hydrostatic pressure, 92–93, 732
Hydrostatic pressure of the interstitial fluid (IHP), 735
Hydroxyapatite, 189
Hydroxyl group, 46
Hydroxyurea, 657
Hymen, 1070
Hyoglossus muscle, 350
Hyoid bone, 225, 226, 264
Hypercalcemia, 1021, 1022
Hypercapnia, 863–864, 869, 1030
Hyperchloremia, 1021
Hypercholesterolemia, 725, 1118
Hyperchromic anemia, 661
Hyperextension, 275
Hyperglycemia, 634, 644
Hyperkalemia, 425, 1021, 1037
Hypermagnesemia, 1021
Hypernatremia, 1016, 1019, 1021, 1037
Hyperopia, 577, 599
Hyperparathyroidism, 639, 1022
Hyperphosphatemia, 1021
Hyperpnea, 849
Hyperpolarization, 406
Hypersensitivity, immediate, 816
Hypertension, 733, 772
Hyperthyroidism, 639
Hypertonic solution, 93
Hypertrophy, muscle, 324–325
Hyperuricemia, 962
Hyperventilation, 864, 1032
Hypervitaminosis, 954, 962
Hypervolemic, 677
Hypesthesia, 513, 525
Hypoaldosteronism, 639
Hypocalcemia, 1021, 1022
Hypocalcemic tetany, 644
Hypocapnia, 864, 869, 1032
Hypochloremia, 1021
Hypochromic anemia, 661
Hypodermic needle, 168, 181
Hypodermis, 137–138, 158, 168
Hypogastric plexus, 545
Hypoglossal canals, 218, 227
Hypoglossal nerve (N XII), 499–500
Hypogonadism, 616, 639
Hypokalemia, 988, 1021, 1037
Hypomagnesemia, 1021
Hyponatremia, 1019, 1037
Hyponychium, 175
Hypoparathyroidism, 639, 1022
Hypophosphatemia, 1021
Hypophyseal artery, superior/inferior, 615
Hypophyseal fossa, 221
Hypophysis. *See* Pituitary gland
Hypotension, 733, 772
Hypothalamus
  control of anterior pituitary, 615, 616
  control of endocrine function, 612–613
  in fluid balance regulation, 1014
  functions, 477–478

  gross anatomy, 476–477
  overview, 461, 462, 475
  in thermoregulation, 14, 15
Hypothermia, 958, 960, 961, 962
Hypotonic solution, 93
Hypoventilation, 864
Hypovolemic, 677
Hypoxia, 661, 677, 843

I bands, 298–299
ICF. *See* Intracellular fluid
Ileocecal valve, 898, 910
Ileocolic vein, 767
Ileum, 898
Iliac artery, 1066, 1070
  common, left/right, 757
  external/internal, 748, 757
Iliac crest, 255
Iliac fossa, 255, 263
Iliac tuberosity, 255
Iliacus muscle, 372, 373–374
Iliac vein
  common, 765
  external/internal, 765
Iliococcygeus muscle, 357–358
Iliocostalis cervicis muscle, 353–354
Iliocostalis lumborum muscle, 353–354
Iliocostalis thoracis muscle, 353–354
Iliofemoral ligament, 283
Iliohypogastric nerve, 446
Ilioinguinal nerve, 446, 1045
Iliopsoas muscle, 372
Iliotibial tract, 372
Ilium, 254, 263
Image reversal, 576
Immature delivery, 1107
Immediate hypersensitivity, 816
Immune complex, 808
Immune complex disorder, 816, 820
Immune disorders, 815–816
Immune response, 662, 777
  antibodies in, 805–809
  B cells in, 804–805
  chemical mediators, 813
  overview, 798, 799, 809–812
  stress and, 816–817
  T cells in, 798–803
Immune system, 777. *See also* Lymphoid system
  age-related changes, 817
  disorders of, 815–816
  integration with other systems, 817–818
Immunity, 777
  antibody-mediated. *See* Antibody-mediated (humoral) immunity
  cell-mediated. *See* Cell-mediated immunity
  forms, 796–797
  memory, 798
  properties, 797–798
  specificity, 797
  tolerance, 798
  versatility, 797–798
Immunization, 797
  mumps, 885
  tetanus, 304
Immunodeficiency disease, 815, 820
Immunoglobulin(s), 654, 782, 807–808. *See also* Antibody(ies)
Immunoglobulin A (IgA), 807
Immunoglobulin D (IgD), 807
Immunoglobulin E (IgE), 807
Immunoglobulin G (IgG), 807, 808
Immunoglobulin M (IgM), 807, 808
Immunological competence, 812, 815
Immunological escape, 792
Immunological surveillance, 782, 791–792
Immunosuppression, 820
Immunosuppressive drugs, 803, 815, 820
Impacted teeth, 887
Impermeable, 89
Implantation, 1091–1095
Impotence, 1078, 1081

Incision, 175
Inclusions, cell, 72
Incomplete dominance, 1113, 1115
Incomplete proteins, 952
Incomplete tetanus, 314
Incontinence, 1002, 1003
Incus, 586
Indole, 914
Induced abortion, 1104
Induced active immunity, 796
Induced passive immunity, 797
Induction, 1090
Infancy, 1108
Infant(s)
  fluid, electrolyte, and acid–base balance, 1036
  skull, 229–230
  thermoregulatory problems, 960
Infarct, 702
Infection, 141, 179
Inferior colliculus, 473, 474, 522
Inferior ganglion, 497, 498
Inferior meatus, 569
Inferior mesenteric ganglion, 534
Inferior mesenteric plexus, 545
Inferior oblique muscle, 355–356
Inferior rectus muscle, 348–349
Inferior vena cava, 687, 762, 764
Infertility, 1043, 1055, 1074, 1120
Inflammation (inflammatory response), 141–143, 793–796
Inflammatory T cells, 782
Inflation reflex, 864
Information processing, CNS, 399, 419, 425
  postsynaptic potentials, 419, 422–423
  presynaptic inhibition and facilitation, 424
  rate of action potential generation and, 424–425
Infra-orbital foramen, 223, 227
Infraspinatus muscle, 281, 363, 365
Infraspinous fossa, 250
Infundibulopelvic (suspensory) ligament, 1061
Infundibulum, 462, 476, 614, 1065
Ingestion, 875
Inguinal canal, 359, 1045
Inguinal hernia, 359, 383, 1045
Inheritance, 1087
  interactions between alleles, 1113
  overview, 1115
  patterns, 1112–1115
  penetrance and expressivity, 1113
  prediction, 1113–1115
  sex-linked, 1116–1117
  sources of individual variation, 1115–1116
Inherited disorders, mitochondrial, 81
Inhibin, 616, 637, 1050, 1080
Inhibiting hormone (IH), 615
Inhibition
  neuron, 422, 455
  presynaptic, 423
Inhibitory interneurons, 449
Inhibitory neurotransmitters, 414
Inhibitory postsynaptic potential (IPSP), 422–423
Initial segment, axon, 389
Injury(ies)
  cartilage, 131
  chest wall, 847
  cornea, 569
  elbow, 283
  head, 465, 466, 551
  immune response, 795–796
  integumentary system, 175–176
  knee, 286
  neural response, 397–398
  peripheral nerves, 437–438
  rotator cuff, 282, 365
  shoulder, 281
  spleen, 788
  tissue response, 141–143

Innate immunity, 796
Innate reflexes, 449–450
Inner cell mass, 1091, 1094
Inner ear, 587–588
Inner membrane, mitochondria, 934
Inner segment, photoreceptor, 578
Innervation, 345
Innominate bone, 254–255
Innominate vein. *See* Brachiocephalic vein
Inorganic compounds, 40–45
  classes, 61
Inositol triphosphate (IP$_3$), 611
Inotropic effects, 418, 713
Insensible perspiration, 161, 1015
Insertion, muscle, 307, 339–340, 341
Inspiration, 843
Inspiratory capacity, 851
Inspiratory center, 861
Inspiratory reserve volume (IRV), 851
Insula, 480
Insulin
  disorders, 639. *See also* Diabetes mellitus
  effects, 613, 633, 635, 949
  in normal growth, 640
  regulation of, 633, 635
Insulin dependent cells, 633
Insulin-dependent (type 1) diabetes, 634, 644, 815
Insulin independent cells, 633
Insulin-like growth factors, 617
Integral proteins, 71
Integration
  cardiovascular system/other systems, 770–771
  digestive system/other systems, 922–923
  endocrine system/other systems, 644–645
  immune system/other systems, 817–818
  integumentary system/other systems, 178, 180
  lymphoid system/other systems, 817–818
  muscular system/other systems, 381
  nervous system/other systems, 556–557
  reproductive system/other systems, 1080–1082
  respiratory system/other systems, 866–868
  skeletal system/other systems, 288–289
  urinary system/other systems, 1003–1004
Integrative centers, 486
Integrative effects, hormones, 640
Integument, 158
Integumentary system
  age-related changes, 177–178
  dermis, 166–168
  epidermis, 159–166
  exercise and, 383
  hair, 169–171
  hypodermis, 168
  injury repair, 175–176
  innervation, 168
  integration with other systems, 178, 180
  introduction, 10, 158–159
  nails, 174–175
  overview, 151
  prenatal development, 1100
  at puberty, 1109–1110
  smooth muscle in, 328
Interatrial opening. *See* Foramen ovale
Interatrial septum, 685
Intercalated discs, 140, 327, 684
Intercellular cement, 114
Intercellular communication, 604–606
Intercondylar eminence, 259
Intercondylar fossa, 259
Intercostal artery, 756
Intercostal muscle, 241, 355–356, 847
Intercostal vein, 762
Interferons, 793, 813, 814
Interleukin-1 (IL-1), 796, 813
Interleukin-2 (IL-2), 813
Interleukin-3 (IL-3), 813

Interleukin-4 (IL-4), 813
Interleukin-5 (IL-5), 813
Interleukin-6 (IL-6), 673, 813
Interleukin-7 (IL-7), 782, 813
Interleukin-8 (IL-8), 813
Interleukin-9 (IL-9), 813
Interleukin-10 (IL-10), 813
Interleukin-11 (IL-11), 813
Interleukin-12 (IL-12), 813
Interleukin-13 (IL-13), 813
Interleukins, 813–814
Interlobar arteries, 969
Interlobar veins, 969
Interlobular septa
  liver, 904
  lung, 839
Interlobular veins Cortical radiate veins, 969
Intermediate-density lipoproteins (IDLs), 944
Intermediate fibers, 323, 324
Intermediate filaments, 73
Intermembrane space, mitochondria, 934
Intermuscular septa, 340
Internal branch, accessory nerve, 499
Internal callus, 205, 208
Internal capsule, 482
Internal carotid artery, 467
Internal nares, 831
Internal oblique muscle, 355–356
Internal occipital crest, 464
Internal os, uterus, 1066
Internal respiration, 842
Internal root sheath, hair, 169
Internal thoracic artery, 695, 750, 1072
Internal thoracic vein, 762
Interneurons, 392
Internodal pathways, 696
Internodes, 395
Internus (muscle name), 341
Interoceptors, 391, 510
Interosseous membrane, 252, 260
Interphase, 100–101
Intersegmental reflex arcs, 450
Interspinales muscle, 351, 353–354
Interspinous ligament, 279
Interstitial cell(s), 617, 637, 638, 1046
Interstitial cell–stimulating hormone, 617
Interstitial fluid, 67
  pH and ion composition, 651
  vs. plasma, 653
Interstitial fluid colloid osmotic pressure (ICOP), 736
Interstitial fluid hydrostatic pressure (IHP), 735
Interstitial growth
  bone, 195
  cartilage, 131
Interstitial lamellae, 191
Interthalamic adhesion, 475
Intertransversarii muscle, 351, 353–354
Intertrochanteric crest, 259
Intertrochanteric line, 259
Intertubercular groove, humerus, 250
Intervals, electrocardiogram, 699
Interval training, 325
Interventricular artery, anterior/posterior, 692, 727
Interventricular septum, 685
Interventricular sulcus, anterior/posterior, 684
Intervertebral discs, 232, 277, 279–280
Intervertebral foramina, 233, 431
Intestinal artery, 757
Intestinal glands, 899
Intestinal juice, 901
Intestinal phase, gastric secretion, 896–897
Intestinal reflex, 548
Intestinal trunk, 779
Intestinal vein, 767
Intestinal villi, 899
Intestines. See Large intestine; Small intestine
Intra-alveolar pressure, 845

Intracapsular ligaments, 271, 277, 279
Intracellular fluid (ICF; cytosol), 69, 72, 1010, 1012
Intradermal injections, 371
Intrafusal muscle fibers, 451
Intramembranous ossification, 196–198
Intramural ganglion, 539
Intramuscular (IM) injections, 371, 383
Intraocular pressure, 573, 574
Intrapleural pressure, 845
Intrapulmonary bronchi, 837
Intrapulmonary pressure, 845
Intraventricular foramen, 464
Intrinsic (muscle name), 341
Intrinsic factor, 661, 891, 893, 921
Intrinsic ligaments, 271
Intrinsic pathway, coagulation, 674
Intrinsic regulation, 12
Introns, 85
Inuit, heart disease in, 49
Invasion, 105, 107
Inversion, 276
In vitro fertilization, 1120
Iodine
  absorption, 921
  functions in human body, 30
  thyroid hormones and, 624
Ion(s), 32. See also specific ions
  absorption, 920–921
  reabsorption at distal convoluted terminal, 988
  reabsorption at proximal convoluted terminal, 984–986
  secretion at distal convoluted terminal, 988–989
  in urine vs. plasma, 976
Ionic bonds, 32–33
Ionization, 41
Ion pumps, 95, 298
Ipsilateral reflex arcs, 453
Iris, 570
Iron
  absorption, 921
  functions, 953
  functions in human body, 30
  recommended daily intake, 953
  recycling and storage, 659–660
  reserves, 953
Iron-deficiency anemia, 660
Irregular bones, 185–186
Ischemia
  compartment syndrome, 381
  coronary, 694, 715
  definition, 383
Ischial ramus, 254
Ischial spine, 254, 263
Ischial tuberosity, 254
Ischiocavernosus muscle, 357–358
Ischiofemoral ligament, 283
Ischium, 254
Islets of Langerhans. See Pancreatic islets
Isoleucine, 947
Isomers, 46
Isometric contractions, 316–317
Isoproterenol, 713
Isotonic contractions, 316, 317
Isotonic solution, 93
Isotopes, 30
Isotropic, 298
Isovolumetric contraction, 705
Isovolumetric relaxation, 706
Isozymes, 56
Isthmus
  thyroid gland, 620
  uterine tube, 1065
  uterus, 1066
Itch sensation, 513

Jarvik, Robert, 681
Jaundice, 163, 176, 658, 677
Jejunum, 898
Joint(s), 268–289. See also specific joints
  age-related changes, 286–288
  amphiarthrosis, 268

appendicular skeleton, 287
  axial skeleton, 280
  classification, 268–269
  diarthrosis, 268
  immobilization, 288
  introduction, 268
  synarthrosis, 268
  synovial. See Synovial joint
Joint cavity, 137
Jugular foramen, 218, 220, 227
Jugular notch, 242
Jugular trunk, left/right, 779–780
Jugular vein, internal/external, 464, 467, 762
Jugular venous pulse (JVP), 762
Junctional folds, 304
Juvenile-onset diabetes. See Insulin-dependent (type 1) diabetes
Juxtaglomerular cells, 742, 975
Juxtaglomerular complex, 975
Juxtamedullary nephrons, 972, 973

Kaposi sarcoma, 813
Karyotype, 1112
Karyotyping, 1117
Keloid, 176, 181
Keratin, 118, 161, 169
Keratinization, 161, 169
Keratinized epithelium, 118
Keratinocytes, 159
Keratohyalin, 161
Keto acid, 946
Ketoacidosis, 950, 962, 989, 1033
Ketone bodies, 950, 996
Ketonemia, 950, 962
Ketonuria, 962
Ketosis, 950, 962
Kidney(s), 966–975. See also Renal entries
  in acid–base balance regulation, 1028–1030
  age-related changes, 1002–1003
  blood supply, 969–970
  connective tissue layers, 967
  endocrine functions, 635–637
  gross anatomy, 967–968
  innervation, 969
  nephrons, 969–975
  overview of function, 993–997
  sectional anatomy, 968–969
Kidney dialysis (hemodialysis), 998, 1003
Kidney stones, 1002
Kidney transplantation, 998
Killer T cells. See Cytotoxic T (T_C) cells
Kilocalorie, 956
Kinase, 610
Kinesin, 73, 390
Kinetic energy, 38
Kinetochore, 102
Kinocilium, 589
Klinefelter syndrome, 1119
Knee-jerk reflex. See Patellar reflex
Knee joint, 283–287
  injuries, 286
  muscles, 375–376
Krebs cycle. See TCA (tricarboxylic acid) cycle
Kupffer cells, 791, 904
Kwashiorkor, 947
Kyphosis, 232

Labeled line, 509
Labetalol, 713
Labia, 882
Labia majora, 1071
Labia minora, 1071
Labor
  dilation stage, 1105
  expulsion stage, 1105–1106
  initiation, 1105
  placental stage, 1106
  premature, 1107
  true vs. false, 1105

Labor contractions, 1105
Lacrimal apparatus, 567–569
Lacrimal bone, 213, 225
Lacrimal canaliculi, 569
Lacrimal caruncle, 566
Lacrimal fossa, 219
Lacrimal gland, 219, 567
Lacrimal lake, 569
Lacrimal puncta, 569
Lacrimal sac, 223, 569
Lacrimal sulcus, 225, 227
Lactase, 917, 918
Lactation, 1072
Lacteals, 778, 899
Lactic acid, 320, 322
Lactic acidosis, 989, 1033
Lactic ion, 320
Lactiferous duct, 1072
Lactiferous sinus, 1072
Lactose, 917
Lactose intolerance, 918, 924
Lacunae
  bone, 134, 189
  cartilage, 131
  trophoblastic channels, 1092
Lambdoid suture, 214
Lamella(e), 189, 191
Lamellated corpuscles, 168, 512–513
Lamina(e), vertebral, 232
Lamina densa (dense layer), 115
Lamina lucida (clear layer), 114
Lamina propria, 827, 876, 877, 879
Langerhans cells. See Dendritic cells
Lanugo, 171
Large bowel. See Large intestine
Large granular lymphocytes. See Natural killer (NK) cells
Large intestine
  absorption in, 913–914
  cecum, 910–911
  colon, 911
  gross anatomy, 910–913
  histology, 913
  movements, 915
  rectum, 913
Large ribosomal unit, 76
Large veins, 728
Laryngeal elevator muscles, 350–351
Laryngeal prominence, 832
Laryngeal spasms, 865
Laryngitis, 833
Laryngopharynx, 831
Larynx, 831–834
  gross anatomy, 832–833
  muscles, 345, 834
  sound production, 833–834
LASIK (laser-assisted in-situ keratomileusis), 577
Latent period, twitch, 313
Lateral apertures, 466
Lateral border, scapula, 247
Lateral canthus, 566
Lateral excursion, 350
Lateral femoral cutaneous nerve, 444, 446
Lateral flexion, 277
Lateral ganglia, 533
Lateral geniculate nucleus, 476
Lateral masses, ethmoid, 222
Lateral meniscus, 284
Lateral nail folds, 175
Lateral nail grooves, 175
Lateral pathway, 522–523
Lateral rectus muscle, 348–349
Lateral rotators, 372
Lateral sulcus, 480
Lateral ventricle, 463
Latissimus dorsi muscle, 340, 363, 364–365
Lauric acid, 49
LDL (low-density lipoprotein), 944
Leads, electrocardiogram, 699
Leaflets (cusps), 687
Leak (passive) channels, 399, 403

Left atrioventricular (AV) valve. *See* Mitral valve
Left atrium, 682
Left-handedness, 487
Left-to-right shunt, 768–769
Left ventricle, 682, 689–690
Leg. *See* Lower limb
Lens, 574–578
Lens fibers, 574
Lentiform nucleus, 482
Lentigos, 163
Leptin, 635, 638, 894, 1109
Lesser curvature, stomach, 891
Lesser horns, hyoid, 226
Lesser occipital nerve, 442
Lesser omentum, 877
Lesser wings, sphenoid, 222
Letrozole, 1073
Leucine, 947
Leukemia, 669, 677
Leukocytes. *See* White blood cells
Leukocytosis, 669, 677
Leukopenia, 669, 677
Leukotrienes, 607, 814
Levator anguli oris muscle, 346–347
Levator ani muscle, 357–358
Levator labii superioris muscle, 346–347
Levator palpebrae superioris muscle, 346–347, 567
Levator scapulae muscle, 360, 361
Levator veli palatini muscle, 350–351
Levers, 337–339
LH. *See* Luteinizing hormone
Lidocaine, 91
Life stages, 1108. *See also* Postnatal development
Ligament(s), 126, 129
    elbow joint, 283
    femoral head, 283
    intervertebral, 277, 279
    knee joint, 283–286
    shoulder joint, 281
    synovial joints, 271
Ligament teres, 283
Ligamentum arteriosum, 689, 768
Ligamentum flavum, 279
Ligamentum nuchae, 236, 279
Ligands, 71
Ligases, 101
Light-adapted state, 583
Light chains, antibody, 805
Limbic lobe, 479
Limbic system, 478–480
Line (bone surface feature), 187
Linea alba, 355–356
Linea aspera, 259
Linear motion, 272–273
Linea terminalis, 256
Lines of cleavage, skin, 166–167
Lingual frenulum, 883
Lingual lipase, 883, 918
Lingual papillae, 564, 883
Lingual tonsils, 783
Linoleic acid, 944
Linolenic acid, 944
Lipases, 916, 918
Lipid(s)
    catabolism, 941–942, 948
    digestion and absorption, 917, 919
    overview, 48–51, 53, 61
    plasma membrane, 70
    synthesis, 944
    transport and distribution, 944–946
    in urine, 996
Lipid derivatives, 606–607, 608
Lipofuscin, 555
Lipogenesis, 944
Lipolysis, 941–942
Lipoprotein(s), 654, 724, 944–946
Lipoprotein lipase, 944
Liposuction (lipoplasty), 128, 144, 168
Liquids, 35
Liquor folliculi, 1064
Lithotripter, 1002

Liver
    bile duct system, 906
    gross anatomy, 903–904
    hematological regulation, 907–908
    histology, 904–906
    lobules, 904
    metabolic regulation, 906–907
    physiology, 906–908
    turnover times, 61
Liver spots (senile lentigos), 163
Load, 303
Lobar bronchi, 837
Lobes
    liver, 903–904
    lung, 835–836
    renal, 969
    thymus, 786
    thyroid gland, 620
Lobules
    liver, 904
    lungs, 838–839
    testes, 1046
    thymus, 786
Local current, 406
Local hormones, 604
Local potentials. *See* Graded potentials
Local vasoconstrictors, 739
Local vasodilators, 739
Lockjaw, 304
Locus, 1112
Long bones, 185, 186
Longissimus (muscle name), 341
Longissimus capitis muscle, 353–354
Longissimus cervicis muscle, 353–354
Longissimus thoracis muscle, 353–354
Longitudinal arch, foot, 262
Longitudinal fasciculi, 482
Longitudinal fissure, cerebrum, 480
Longitudinal layer, digestive tract, 879
Long reflexes, 547, 881
Long-term memories, 550
Long thoracic nerve, 444
Longus (muscle name), 341
Longus capitis muscle, 353–354
Longus colli muscle, 353–354
Loop of Henle. *See* Nephron loop
Loose connective tissues, 124, 127–128
Lordosis, 232
Lou Gehrig disease, 524
Lovastatin, 943
Low-density lipoproteins (LDLs), 944
Lower limb
    arteries, 757, 759
    bones, 258–262
    joints, 287
    muscles, 371–378
    veins, 764–765
Lower motor neuron, 519–520
Lower respiratory system, 826
    larynx, 831–834
    lungs. *See* Lungs
    primary bronchi, 834–835
    trachea, 834–835
Low-frequency sound, 595
Lubrication
    ingested material, 882
    sexual arousal and, 1078
    by synovial fluid, 270
Lumbar artery, 757
Lumbar curve, 231
Lumbar enlargement, 431
Lumbar plexus, 444–446
Lumbar puncture, 434, 456
Lumbar trunk, left/right, 779
Lumbar vein, 765
Lumbar vertebrae, 234, 236, 237
Lumbrical muscles, 370–371, 380
Lumen, 114
Luminal compartment, seminiferous tubule, 1049
Lumpectomy, 1073
Lunate, 253
Lunate surface, acetabulum, 254
Lung cancer, 867, 869

Lung perfusion, 860
Lungs
    alveolar ducts and alveoli, 839–841
    blood supply, 841
    bronchi, 837
    bronchioles, 837–839
    circulation to, 747–748
    compliance, 844
    gross anatomy, 836
    lobes and surfaces, 835–836
    lobules, 838–839
    pleural cavities and pleural membranes, 841–842
    pressure and airflow to, 843–844, 845
    relationship to heart, 837
Lunula, 175
Luteal phase, ovarian cycle, 1062, 1076, 1077
Luteinizing hormone (LH), 606, 617, 620, 1056
    androgen production and, 1057–1058
    effects, 1080
    regulation of secretion, 1080
Lutropin. *See* Luteinizing hormone
Luxation. *See* Dislocation
Lyme disease, 813
Lymph, 130, 777
Lymphadenopathy, 786, 820
Lymphatic capillaries, 778–779
Lymphatic duct, left/right, 779, 781
Lymphatic nodule, 783
Lymphatics. *See* Lymphatic vessels
Lymphatic system. *See* Lymphoid system
Lymphatic trunks, 779
Lymphatic vein, 778–780
Lymphatic vessels, 130–131, 777, 779–780
Lymphedema, 780, 820, 1015
Lymph glands, 785
Lymph nodes
    cancer cells in, 786
    function, 785–786
    lymph flow through, 784–785
    structure, 784, 785
Lymphocytes, 126, 655, 668–669, 670, 777. *See also* B cells; Natural killer (NK) cells; T cells
    life span and circulation, 782
    production, 782–783
    types, 780, 782
Lymphocytosis, 669
Lymphoid nodules, 783, 784
Lymphoid organs, 777, 783–788
Lymphoid stem cells, 651, 660, 671
Lymphoid system. *See also* Immune system
    body defenses and, 788–789
    cancer and, 786
    functions, 777–778
    integration with other systems, 817–818
    introduction, 11
    lymphocytes. *See* Lymphocytes
    organs, 783–788
    overview, 154, 777, 778
    prenatal development, 1101
    tissues, 783
    vessels, 778–781
Lymphoid tissues, 669, 777, 783
Lymphomas, 786, 820
Lymphopenia, 669
Lymphopoiesis, 669, 782
Lymphotoxins, 801, 814
Lysergic acid diethylamide (LSD), 554
Lysine, 919, 947
Lysosomal storage diseases, 80
Lysosome, 69, 79–80
Lysozyme, 569, 789, 885, 1108

Macrocytic anemia, 661
Macrophage-activating factor (MAF), 814
Macrophages, 125, 791, 812
Macula adherens (desmosome), 114
Macula densa, 975
Maculae, 589

Macula lutea, 571
Magnesium
    absorption, 921
    functions in human body, 30, 953
    recommended daily intake, 953
    in urine, 996
Magnesium balance, 1021, 1022
Magnetic resonance imaging (MRI), 24
Magnus (muscle name), 341
Major (muscle name), 341
Major calyx, 969
Major histocompatibility complex (MHC), 799
Male climacteric, 1079
Male pattern baldness, 171, 181
Male pronucleus, 1089
Male reproductive system
    accessory glands, 1053–1054
    age-related changes, 1079–1080
    bulbo-urethral glands, 1053, 1054
    disorders, 1059
    ductus deferens, 1052–1054
    endocrine functions, 638–639
    epididymis, 1051–1052
    external genitalia, 1055, 1056
    hormonal regulation, 1056–1058, 1080
    integration with other systems, 1080–1082
    introduction, 11
    overview, 156, 1042, 1043, 1045
    parasympathetic effects, 545
    penis, 1055–1056
    prostate gland, 1053, 1054
    at puberty, 1111
    reproductive tract, 1051–1054
    scrotum, 1045–1046
    semen, 1055–1056
    seminal glands (vesicles), 1053, 1054
    smooth muscle in, 328
    spermatic cords, 1045
    spermatogenesis, 1047–1050
    sympathetic effects, 545
    testes, 1043–1047
    urethra. *See* Urethra
Male sexual function, 1078
Malignant melanoma, 163, 164, 181, 1118
Malignant tumor, 105, 107
Malleolus, medial/lateral, 259
Malleus, 586
Malnutrition, 950
Maltase, 917, 918
Maltose, 917
Mamillary bodies, 476, 478
Mammary glands, 174, 1071–1072
    lactation and, 1108–1109
    in pregnancy, 1104
Mammography, 1073, 1081
Mammotropin. *See* Prolactin
Management, 263
Mandible, 213, 225–226, 263, 264
Mandibular canal, 226
Mandibular foramen, 226, 227
Mandibular fossa, 220
Mandibular groove, 884
Mandibular notch, 226
Manganese, 953
Manubrium, 242
Marasmus, 947
Marfan syndrome, 126, 202, 208, 1118
Marginal arteries, coronary, 692
Margination, 666
Marrow cavity, 188
Masseter muscle, 348–349
Mass movements, 915
Mass number, 30
Mast cells, 126, 812
Mastectomy, 1073, 1081
Mastication, 348–350, 885, 887–888
Mastoid fontanelle, 229
Mastoid process, 220
Matrix
    connective tissue, 124
    hair, 169
    mitochondria, 80

Matter, states of, 35–36
Maturation promoting factor, 104, 105
Mature Graafian follicle, 1064
Maturing face, cisterna, 77
Maturity, 1087
Maturity-onset diabetes. *See* Non-insulin-
  dependent (type 2) diabetes
Maxillae, 213, 223–224
Maxillary sinuses, 223
Maxillary vein, 762
Maximus (muscle name), 341
M cells, 583
M-CSF, 672
Mean arterial pressure (MAP), 733
Mean corpuscular hemoglobin
  concentration (MCHC), 661
Mean corpuscular volume (MCV), 661
Meatus, inferior/middle/superior, 829
Mechanically gated channels, 403–404
Mechanical processing, digestion, 875
Mechanoreceptors, 510, 511–514
Meconium, 1108
Medial antebrachial cutaneous nerve, 444
Medial border, scapula, 247
Medial canthus, 566
Medial geniculate nucleus, 476
Medial lemniscus, 516
Medial meniscus, 284
Medial pathway, 522–523
Medial rectus muscle, 348–349
Median aperture, 466
Median cubital vein, 653
Median eminence, 615
Median nerve, 254, 444
Mediastinal artery, 756
Mediastinum, 682
Medical terminology, 5
Medium-sized arteries, 722
Medium-sized veins, 728
Medulla
  hair, 169
  lymph node, 785
  ovary, 1061
  renal, 969
  suprarenal. *See* Suprarenal medulla
  thymus, 786
Medulla oblongata, 461, 462, 469–471,
  861
Medullary cavity, 188
Medullary cords, 785
Megakaryocytes, 673
Meibomian gland, 566
Meiosis, 100, 1046
  in oogenesis, 1061–1062
  in spermatogenesis, 1048–1049, 1050
Meiosis I, 1049, 1050, 1062
Meiosis II, 1049, 1050, 1062
Meissner corpuscles. *See* Tactile
  corpuscles
Melanin, 72, 126, 162
Melanocyte(s), 126, 160, 162
Melanocyte-stimulating hormone (MSH),
  163, 617–618, 620
Melanosomes, 162
Melanotropin. *See* Melanocyte-
  stimulating hormone
Melatonin, 475, 630–631
Membrane
  cutaneous. *See* Skin
  mucous, 135–136
  plasma. *See* Plasma membrane
  serous, 136
  synovial, 137
  types, 135
Membrane attack complex (MAC), 793
Membrane flow, 80
Membranous labyrinth, 587
Membranous organelles, 69, 72
Membranous urethra, 1000
Memory, 550–552
  brain regions involved in, 550–551
  cellular mechanisms, 551–552
  immunologic, 798
  loss of, 551

Memory B cells, 804, 812
Memory consolidation, 550
Memory engram, 552
Memory T_C cells, 801, 812
Memory T_H cells, 802, 812
Menarche, 1070
Meningeal layers, dura mater, 464
Meninges
  cranial, 464–465
  spinal, 433–435
Meningitis, 433, 456
Meniscus (articular disc), 252, 271
Menopause, 1061, 1070, 1079
Menses, 1068–1069, 1076
Menstrual cycle. *See* Uterine cycle
Mental foramen, 226, 227
Mentalis muscle, 346–347
Mental protuberance, 225
Merkel cells, 160, 512
Merkel disc, 512
Merocrine secretion, 121, 122
Mesangial cells, 974
Mesencephalic aqueduct. *See* Aqueduct of
  midbrain
Mesencephalon, 461, 462, 474–475
Mesenchymal cells, 125, 127
Mesenchyme, 126–127
Mesenteric artery, inferior/superior, 536,
  722, 757
Mesenteric vein, inferior/superior, 766, 767
Mesenteries, dorsal/ventral, 877, 878
Mesentery proper, 877
Mesoappendix, 911
Mesocolon, 877
Mesoderm, 1093
Mesothelium, 118
Mesovarium, 1060
Messenger RNA (mRNA), 76, 84–86
Metabolic acidosis, 1033–1034,
  1035–1036, 1037
Metabolic alkalosis, 1034, 1035–1036,
  1037
Metabolic generation, of water, 1016
Metabolic rate, 957
Metabolic turnover, 61, 930
Metabolism
  absorptive state, 949
  aerobic, 81, 319, 932, 940
  amino acids, 946–947
  carbohydrates, 932–941
  cellular, 930–931
  definition, 36, 930
  energy gains and losses, 955–957
  hormonal effects, 949
  lipids, 941–946
  postabsorptive state, 949–950
  proteins, 946–948
Metabolites, 40
Metabotropic effects, 418
Metacarpal bones, 253, 254
Metalloproteins, 654
Metaphase, mitosis, 102
Metaphase plate, 102
Metaphyseal vessels, 199
Metaphysis, 188
Metaplasia, 144
Metarteriole, 726
Metastasis, 105, 107
Metatarsal bones, 261–262
Metatarsus, 261
Metencephalon, 462
Methionine, 947
Metopic (frontal) suture, 219
Metoprolol, 694, 713
MHC (major histocompatibility
  complex), 799
MHC proteins, 799–800
Micelles, 51–52, 919
Microcephaly, 230, 242
Microcytic anemia, 661
Microfilaments, 73
Microglia, 395, 791
Micromole (μmol), 31
Microphage-chemotactic factor, 814

Microphages, 126, 667, 790–791
Microscopic anatomy, 7
Microtubules, 73
Microvilli, 69, 74, 113
Micturition. *See* Urination
Micturition reflex, 1001–1002
Middle ear (tympanic cavity), 220,
  586–587
Middle piece, spermatozoon, 1050
Migration-inhibitory factor (MIF), 813
Milk let-down reflex, 618, 1108–1109
Milk teeth, 886
Millimeters of mercury (mm Hg), 846
Millimole (mmol), 31, 61
Milliosmoles, 980
Millivolt (mV), 99
Mineralocorticoids, 627, 630, 639
Minerals, 952–953. *See also specific
  minerals*
Minimal volume, 850
Minimus (muscle name), 341
Minor (muscle name), 341
Minor calyx, 969
Minoxidil (Rogaine), 171
Miscarriage, 1104
Mitochondria
  ATP production, 934–939
  inheritable disorders, 81
  structure and function, 69, 80–81
Mitosis, 100, 102–103, 1048
Mitotic rate, 104
Mitral valve, 689
Mitral valve prolapse, 715
Mixed exocrine glands, 121
Mixed nerves, 431
M line, 298
Modality, 509
Moderator band, 689
Molarity, 43
Mole (mol), 31, 62
Molecular motors, 73
Molecular weight, 36
Molecule, 32
Monaxial joint, 273
Mongolism. *See* Down syndrome
Monoamine oxidase (MAO), 537
Monocyte-chemotactic factor (MCF), 814
Monocyte–macrophage system, 791
Monocytes, 668, 670
Monoglyceride, 50
Monohydrogen phosphate, 1027
Monosaccharides, 45–46, 48, 918–919
Monosomy, 1119
Monosynaptic reflexes, 450–452
Monounsaturated fatty acid, 48
Monozygotic twins, 1107
Mons pubis, 1071
Morula, 1091
Motilin, 909
Motility
  gastric, 895
  spermatozoa, 1052
Motion, dynamic, 272–273
Motion sickness, 591, 599
Motor end plate, 304
Motor homunculus, 522
Motor neurons, 391
Motor nuclei, 435
Motor speech area, 486
Motor unit, 315
Mountain sickness, 869
Mouth
  muscles, 346–347
M phase, 101
M-phase promoting factor (MPF), 104,
  105
MRI (magnetic resonance imaging), 24
MSH. *See* Melanocyte-stimulating
  hormone
Mucin(s), 57, 121, 884
Mucins, 884
Mucosa
  digestive tract, 877–879
  nasal, 831

  oral, 882
  respiratory, 827
Mucosa-associated lymphoid tissue
  (MALT), 783
Mucous (goblet) cells, 122
Mucous connective tissue, 126–127, 127
Mucous glands, 121
Mucous membranes (mucosae), 135–136
Mucus, 57, 121
Mucus elevator, 827
Müllerian (paramesonephric) ducts,
  1050
Müllerian-inhibiting factor (MIF), 1050
Multicellular exocrine glands, 122
Multicellular glands, 122
Multi-CSF, 672, 673
Multifidus muscle, 353–354
Multinucleate, 138, 296
Multipennate muscle, 336–337
Multiple births, 107
Multiple endocrine neoplasia, type 2,
  1118
Multiple sclerosis (MS), 397, 815
Multipolar neurons, 391
Multiunit smooth muscle cells, 330
Mumps, 885, 924
Muscarine, 541–542
Muscarinic receptors, 541–542
Muscle cell
  turnover times, 61
Muscle contraction. *See* Contraction
Muscle fibers, 138, 301
  intrafusal/extrafusal, 451–452
  skeletal. *See* Skeletal muscle fibers
Muscles listed
  abductor digiti minimi, 370–371,
    379–380
  abductor hallucis, 379–380
  abductor pollicis brevis, 370–371
  abductor pollicis longus, 368–369
  adductor brevis, 372–374
  adductor hallucis, 379–380
  adductor longus, 372–374
  adductor magnus, 372–374
  adductor pollicis, 370–371
  anal sphincter, 357–358
  anconeus, 365–368
  arrector pili, 169
  biceps brachii, 252, 283, 318,
    365–367
  biceps femoris, 372, 375–376
  brachialis, 341, 365, 366–368
  brachioradialis, 365–368
  buccinator, 341, 345
  bulbospongiosus, 357–358, 1071
  ciliary, 570
  coccygeus, 357–358
  coracobrachialis, 361, 363–365
  corrugator supercilii, 346–347
  cremaster, 1045
  dartos, 1045
  deltoid, 337, 360–365
  depressor anguli oris, 346–347
  depressor labii inferioris, 346–347
  detrusor, 1000
  diaphragm, 355
  digastric, 351–352
  dorsal interosseus, 370–371, 379–380
  erector spinae, 351
  extensor carpi radialis, 366, 367–370
  extensor carpi ulnaris, 366, 367–370
  extensor digiti minimi, 368–369
  extensor digitorum, 337, 368–369
  extensor digitorum brevis, 379–380
  extensor digitorum longus, 376, 377–378
  extensor hallucis longus, 376, 377–378
  extensor indicis, 368–369
  extensor pollicis, brevis/longus, 368–369
  external oblique, 355–356
  fibularis, brevis/longus, 376–379
  flexor carpi radialis, 365–367
  flexor carpi ulnaris, 365–367
  flexor digiti minimi brevis, 370–371,
    379–380

Muscles listed (*continued*)
flexor digitorum, superficialis/ profundus, 368–369
flexor digitorum brevis, 379–380
flexor digitorum longus, 377–378
flexor hallucis brevis, 379–380
flexor hallucis longus, 377–378
flexor pollicis brevis, 370–371
flexor pollicis longus, 368–369
gastrocnemius, 339, 372, 377–378
gemelli, 373–374
genioglossus, 341
geniohyoid, 351–352
gluteus maximus, 371, 373–374
gluteus medius, 372, 374
gluteus minimus, 372, 374
gracilis, 372, 373–374
hyoglossus, 350
iliacus, 372, 373–374
iliococcygeus, 357–358
iliocostalis cervicis, 353–354
iliocostalis lumborum, 353–354
iliocostalis thoracis, 353–354
iliopsoas, 372
inferior oblique, 355–356
inferior rectus, 348–349
infraspinatus, 281, 363–365
intercostal, 241, 355–356, 847
internal oblique, 355–356
interspinales, 351, 353–354
intertransversarii, 351, 353–354
ischiocavernosus, 357–358
laryngeal elevators, 350–351
lateral rectus, 348–349
latissimus dorsi, 340, 363, 364–365
levator anguli oris, 346–347
levator ani, 357–358
levator labii superioris, 346–347
levator palpebrae superioris, 346–347, 567
levator scapulae, 360, 362
levator veli palatini, 350–351
longissimus capitis, 353–354
longissimus cervicis, 353–354
longissimus thoracis, 353–354
longus capitis, 353–354
longus colli, 353–354
lumbricals, 370–371, 379–380
masseter, 348–349
medial rectus, 348–349
mentalis muscle, 346–347
multifidus, 353–354
mylohyoid, 225, 351–352
nasalis, 346–347
oblique, 355–356, 847
obturators, 372, 374
occipitofrontalis, 346–347, 348
omohyoid, 351–352
opponens digiti minimi, 370–371
opponens pollicis, 370–371
orbicularis oculi, 346–347, 566–567
orbicularis oris, 341, 345
orbicularis oris muscle, 346–347
palatal, 351
palatoglossus, 350
palmar interosseus, 370–371
palmaris brevis, 370–371
palmaris longus, 365, 366–368
papillary, 687, 688
pectinate, 687
pectineus, 372, 373–374
pectoralis, 337, 847
pectoralis major, 363, 364–365
pectoralis minor, 361, 363
pharyngeal constrictors, 350–351
piriformis, 372, 373–374
plantar interosseous, 379–380
plantaris, 372, 377–378
platysma, 346–347, 348
popliteus, 375–376
procerus, 346–347
pronator quadratus, 366, 367–368
pronator teres, 366, 367–368
psoas major, 372, 373–374, 997

pterygoid, 349–350
pubococcygeus, 357–358
pupillary, 570
quadratus femoris, 373–374
quadratus lumborum, 353–354
quadratus plantae, 379–380
quadriceps femoris, 372, 375–376
rectus abdominis, 340, 341, 355–356, 847
rectus femoris, 337, 341, 372, 375–376
rhomboid, 360, 362
risorius muscle, 346–347
rotatores, 351, 353–354
sartorius, 372, 375–376
scalene, 355–356
semimembranosus, 372, 375–376
semispinalis capitis, 353–354
semispinalis cervicis, 353–354
semispinalis thoracis, 353–354
semitendinosus, 372, 375–376
serratus anterior, 361, 363, 847
serratus posterior, 355–356
soleus, 372, 377–378
spinalis cervicis, 353–354
spinalis thoracis, 353–354
splenius capitis, 353–354
splenius cervicis, 353–354
stapedius, 587
sternocleidomastoid, 351–352, 847
sternohyoid, 352
sternothyroid, 352
styloglossus, 350
stylohyoid, 351–352
subclavius, 361, 363
subscapularis, 281, 363–365
superior oblique, 348–349
superior rectus muscle, 348–349
supinator, 367–368, 369
supraspinatus, 281, 363–365
temporalis, 219, 341, 348–349
temporoparietalis muscle, 346–347, 348
tensor fasciae latae, 372, 373–374
tensor tympani, 587
tensor veli palatini, 350–351
teres major, 340, 362, 363–365
teres minor, 281, 363–365
thyrohyoid, 352
tibialis, anterior/posterior, 376, 378
trachealis, 834
transverse perineal muscle, 357–358
transversus abdominis, 355–356, 847
transversus thoracis, 355–356, 847
trapezius, 359–360, 362
triceps brachii, 318, 340, 363, 364–365, 367–368
vastus intermedius, 375–376
vastus lateralis, 364, 375–376
vastus medialis, 372, 375–376
zygomaticus major, 346–347
zygomaticus minor, 346–347
Muscle spindles, 451–452
Muscle tissue, 138, 294
cardiac. *See* Cardiac muscle tissue
skeletal. *See* Skeletal muscle tissue
smooth. *See* Smooth muscle tissue
Muscle tone, 316
Muscular arteries, 722
Muscular compression, 734
Muscular dystrophies, 331, 1118
Muscularis externa, 880
Muscularis layer, vagina, 1070
Muscularis mucosae, 879
Muscular system, 336–383
age-related changes, 378–380
integration with other systems, 381
introduction, 10, 336
muscle types, 336–337
overview, 152
prenatal development, 1100
at puberty, 1110
Musculocutaneous nerve, 444
Mutations, 88, 1116
Myasthenia gravis, 331
Mycobacterium tuberculosis, 829

Myelencephalon, 462
Myelin, 395
Myelinated axon, 395, 396
Myelin sheath, 395
Myelocytes, 669, 671
Myelography, 456
Myeloid stem cells, 651, 660, 671
Myeloid tissue. *See* Red bone marrow
Myenteric plexus, 880
Myenteric reflexes. *See* Short reflexes
Mylohyoid line, 225
Mylohyoid muscle, 225, 351–352
Myoblasts, 296
Myocardial infarction (MI), 702, 715
Myocardium, 684
Myoepithelial cells, 173, 1109
Myofacial pain syndrome, 226
Myofibrils, 297–298, 301
Myofilaments, 297
Myoglobin, 323
Myogram, 313
Myometrium, 1066
Myopia, 577, 599
Myosatellite cells (satellite cells), 138, 295
Myosin, 73, 74, 308–310
Myosin light chain kinase, 330
Myotatic reflexes, 450
Myotonic dystrophy, 1118
MyPyramid Plan, 951
Myxedema, 639, 644

NAD (nicotinamide adenine dinucleotide), 932
Nail(s), 174–175
Nail bed, 175
Nail body, 175
Nail root, 175
Na^+–K^+/2 Cl^- transporter, 986
Naltrexone, 956
Nares, internal/external, 224, 830, 831
Nasal bones, 213, 224
Nasal cavity, 829
Nasal complex, 228–229
Nasal conchae, inferior/middle/superior, 213, 222, 223, 224, 829
Nasalis muscle, 346–347
Nasal mucosa, 831
Nasal septum, 214, 829
Nasal spine, anterior, 223
Nasal vestibule, 829
Nasolacrimal canal, 223, 227, 569
Nasolacrimal duct, 223, 569
Nasopharynx, 586, 831
Natriuretic peptides, 635, 637, 743, 982, 1014
Natural killer (NK) cells, 669, 780, 782, 792, 812
Naturally acquired active immunity, 796
Naturally acquired passive immunity, 797
Natural pacemaker. *See* Sinoatrial (SA) node
Navicular, 261
N compounds, 952
Near point of vision, 576
Nearsightedness, 577
Nebulin, 300
Neck
gallbladder, 908
spermatozoon, 1050
tooth, 885
urinary bladder, 999
Neck (body part)
arteries, 753
muscles, 346–347, 350–352
veins, 761, 762
Neck (bone part), 187
femur, 258
humerus, anatomical/surgical, 250
rib, 241
Necrosis, 142, 144, 167, 796
Negative Babinski reflex, 455
Negative feedback, 14, 449, 613

Negative nitrogen balance, 952
Neonate
fluid, electrolyte, and acid–base balance, 1036
neonatal period, 1108–1109
respiratory changes at birth, 865–866, 1108
Neoplasm. *See* Tumor
Nephrolithiasis, 1002
Nephron(s)
age-related changes, 1002
anatomy, 969–975
functions, 972
types, 972–973
urine formation in, 976–978
Nephron loop (loop of Henle), 970–971, 972, 975, 979, 986–988
Nephropathy, diabetic, 634, 644
Nerve(s), 387. *See also specific body regions and organs*
cranial. *See* Cranial nerves
spinal. *See* Spinal nerves
Nerve deafness, 577, 599
Nerve fiber, 295. *See also* Axon
Nerve growth factor (NGF), 105, 456
Nerve impulses, 413. *See also* Action potentials
Nerve palsy, peripheral. *See* Peripheral neuropathy
Nerve plexus(es), 440–446
Nerves listed. *See also* Cranial nerves
ansa cervicalis, 442
antebrachial cutaneous, 444
axillary, 444
cervical, 442
dorsal scapular, 444
femoral, 444, 446
femoral cutaneous, lateral/posterior, 444, 446
fibular, 444, 446
genitofemoral, 444, 446, 1045
gluteal, 446
great auricular, 442
iliohypogastric, 446
ilioinguinal, 446, 1045
lesser occipital, 442
long thoracic, 444
median, 254, 444
musculocutaneous, 444
obturator, 446
pectoral, 444
pelvic, 540
phrenic, 442
pudendal, 444, 446
radial, 250, 283, 444
renal, 969
saphenous, 446
sciatic, 254, 444
subscapular, 444
supraclavicular, 442
suprascapular, 444
thoracodorsal, 444
tibial, 444, 446
transverse cervical, 442
ulnar, 250, 444
vestibulocochlear, 589
Nerve tissue. *See* Neural tissue
Nervous system
age-related changes, 554–555
anatomical divisions, 387
autonomic. *See* Autonomic nervous system
disorders, categorization, 556
exercise and, 383
functional divisions, 387–388
integration with other systems, 556–557
introduction, 10
overview, 152
prenatal development, 1100
at puberty, 1110
somatic. *See* Somatic nervous system
Net colloid osmotic pressure, 735
Net diffusion, 89
Net filtration pressure (NFP), 736–737

Net hydrostatic pressure (NHP), 735, 981
Neural circuits, 446–447
Neural cortex, 462
Neural integration
  overview, 507
  sensory pathways, 515–519
  sensory receptors, 507–515
  somatic nervous system. *See* Somatic
    nervous system
Neural tissue, 140–141. *See also* Nervous
  system
  information processing, 419–425
  neuroglia. *See* Neuroglia
  neurons. *See* Neurons
  response to injury, 397–398
  synaptic activity, 413–416
Neural tube, 462
Neural tunic. *See* Retina
Neurilemma, 395–396
Neurilemmal cells, 395–396
Neurilemmocytes, 395–396
Neurocoel, 462
Neuroendocrine reflexes, 613
Neurofibrillary tangles, 555
Neurofibrils, 389
Neurofibromatosis, type 2, 1118
Neurofilaments, 389
Neuroglandular junction, 390
Neuroglia, 140, 141, 387
  CNS, 392–395
  PNS, 395–397
Neurohypophysis, 618–619
Neurological examination, 556
Neuromodulators
  mechanisms of action, 418–419
  representative, 420–421
Neuromuscular junction (NMJ), 304, 390
Neuronal pools, 446–447
Neurons, 140, 387
  action potentials. *See* Action potentials
  functional classification, 391–392
  information processing by, 419, 422–423
  response to injury, 397–398
  structural classification, 390–391
  structure, 388–390
  transmembrane potential. *See*
    Transmembrane potential
  turnover times, 61
Neuropeptides, 418, 421
Neuropeptide Y, 421
Neurotoxin, 425
Neurotransmitters, 304, 390, 414. *See
  also specific neurotransmitters*
  activities, 417
  brain function and, 554
  mechanisms of action, 418–419
  pain perception and, 511
  representative, 417, 420–421
Neurotubules, 389
Neutral fats. *See* Triglycerides
Neutral solution, 43, 1023
Neutrons, 29
Neutrophilic band cell, 669
Neutrophilic myelocyte, 669
Neutrophils, 126, 667–668, 670, 795, 812
Newborn. *See* Neonate
Niacin, 955
Nicotinamide adenine dinucleotide
  (NAD), 932
Nicotine, 423, 425, 541, 709
Nicotinic acid. *See* Niacin
Nicotinic receptors, 541–542
Nifedipine, 713
Night blindness, 582, 599
Nipple, 1072
Nissl bodies, 389
Nitrate, absorption, 921
Nitric oxide (NO), 538
  functions, 34
  as neurotransmitter, 417, 418–419, 421
  structure, 34
Nitrogen
  functions in human body, 30
  partial pressures, 851–852

Nitrogen balance, 952
Nitrogenous base, 58
Nitroglycerin, 882
Nitroxidergic synapse, 538
NK cells. *See* Natural killer (NK)
  cells
NK (natural killer) cells, 669, 780, 782,
  792, 812
NMDA (N-methyl D-aspartate) receptors,
  552
Nociceptors, 509, 510–511
Nodal system. *See* Conducting system,
  heart
Nodes of Ranvier, 395
Nomogram, 1035
Nondisplaced fracture, 206
Nonessential amino acids, 947
Nonkeratinized epithelium, 118
Nonmembranous organelles, 69, 72
Nonpolar covalent bonds, 34–35
Nonshivering thermogenesis, 958
Nonspecific defenses, 667, 777
  complement, 793
  fever, 796
  immunological surveillance, 791–792
  inflammation, 793–796
  interferons, 793
  overview, 789, 790
  phagocytes, 789–791
  physical barriers, 789
Nonstriated involuntary muscle, 140
Nonstriated muscle, 328
Noradrenaline. *See* Norepinephrine
Norepinephrine, 417, 420, 536
  effects, 629–630
  secretion, 629–630
Normal saline, 93, 107
Normoblast, 660
Normochromic, 667, 677
Normocyte, 661, 677
Normovolemic, 677
Nose
  anatomy, 829–831
  muscles, 346–347
  olfactory organs, 562–563. *See also*
    Olfaction
Nostrils. *See* External nares
Nuchal lines, inferior/superior, 218
Nuclear envelope, 69, 82
Nuclear matrix, 83
Nuclear pore, 69, 82
Nucleases, 903, 918
Nuclei, nervous system
  basal, 482–484, 523–524
  hypothalamus, 478
  limbic system, 479–480
  medulla oblongata, 470–471
  pons, 472
  spinal cord, 435
  thalamus, 475–476
Nucleic acids, 57–58, 61. *See also* DNA;
  RNA
Nucleolus (nucleoli), 69, 83
Nucleoplasm, 83
Nucleosome, 83
Nucleotides, 58
Nucleus, 69
  control of cell by, 88–89
  information storage in, 83–84
  structure, 82–84
Nucleus basalis, 550
Nucleus cuneatus, 471
Nucleus gracilis, 471
Nucleus pulposus, 277
Nurse (sustentacular) cells, 616, 637,
  638, 1047, 1049–1050
Nursemaid's elbow, 283
Nutrient(s), 40
Nutrient artery, 199
Nutrient foramina, 199
Nutrient pool, 930
Nutrient vein, 199

Nutrition. *See also* Diet
  definition, 950
  disease and, 955
  food groups, 951–952
  MyPyramid Plan, 951
  nitrogen balance, 952
  vitamins, 953–955
Nystagmus, 592, 599

Obesity, 962
Obestatin, 894
Obligatory water reabsorption, 990
Oblique (muscle name), 341
Oblique fissure, 835
Oblique layer, stomach, 891
Oblique muscle, 355–356, 847
Obturator artery, 757
Obturator foramen, 255, 263
Obturator muscle, externus/internus,
  372, 374
Obturator nerve, 446
Obturator vein, 765
Occipital bone, 213, 218, 264
Occipital condyles, 218
Occipital crest, external/internal, 218,
  464
Occipital fontanelle, 229
Occipital lobe, 480, 484
Occipital protuberance, external, 218
Occipital sinus, 762
Occipitofrontalis muscle, 346–347, 348
Occluding (tight) junction, 114, 115
Occlusal surfaces, 888
Ocular conjunctiva, 567
Oculomotor muscles, 348
Oculomotor nerve (N III), 492, 500
Odorant-binding proteins, 563
Odorants, 563
Off-center neurons, 583
Oil glands, 172
Olecranon, 251
Olecranon fossa, 250
Olfaction, 508, 562
  age-related changes, 563–564
  discrimination, 563–564
  organs, 562–563
  pathways, 563
  receptors, 563
Olfactory bulbs, 491, 563
Olfactory cortex, 484, 563
Olfactory epithelium, 562
Olfactory foramina, 222, 227
Olfactory glands, 562
Olfactory nerve (N I), 491, 500
Olfactory organs, 562
Olfactory receptor(s), 223, 563
Olfactory receptor cells, 562
Olfactory region, 829
Olfactory tracts, 491
Oligodendrocytes, 395
Olivary nuclei, 471
Olives, 471
Omega-3 fatty acids, 49, 62
Omohyoid muscle, 351–352
On-center neurons, 583
Oncogenes, 105, 107
Oncotic pressure, 735
Oocyte, 1042
  activation, 1088–1090
  at ovulation, 1088
  primary, 1062
  secondary, 1062
Oogenesis, 1061–1064
Oogonia, 1061
Open fracture, 206
Ophthalmic artery, 753
Opioids, 418, 421
Opponens digiti minimi muscle,
  370–371
Opponens pollicis muscle, 370–371
Opportunistic infections, 819
Opposition, 276
Opsin, 579–581
Opsonins, 793

Opsonization, 793, 808
Optic canal, 222, 227
Optic chiasm, 492
Optic disc, 572–573
Optic nerve (N II), 492, 500
Optic radiation, 584
Optic tracts, 492
Oral cavity, 882–888
  anatomy, 882–883
  functions, 882
  pharynx, 350–351, 831, 888
  salivary glands, 884–885
  tongue, 883–884
Oral mucosa, 882
Ora serrata, 570
Orbicularis oculi muscle, 346–347,
  566–567
Orbicularis oris muscle, 337, 341, 345
Orbit(s), 219, 228
Orbital complex, 228
Orbital fat, 569
Orbital fissure, inferior/superior, 222,
  224, 227
Orbital process, 224
Orbital rim, 223
Orchiectomy, 1043, 1081
Organelles, 69, 72–81. *See also specific
  organelles*
Organic acids, 1024
Organic compounds, 40
  classes, 61
  functional groups, 46
Organism organizational level, 8, 9
Organizational levels, 8–9
Organ of Corti, 593–594
Organogenesis, 1099
Organ organizational level, 8, 9
Organ physiology, 7
Organ system organizational level, 8, 9
Organ systems. *See also specific systems*
  composition of human body, by weight,
    150
  introduction, 6, 8
  overview, 151–156
Organ transplantation, 803, 998
Orgasm, 1078
Origin, muscle, 310, 339–340, 341
Oropharynx, 831
Osmolarity, 93
Osmoles, 980
Osmoreceptors, 1014
Osmosis, 91–92, 99, 735
Osmotic concentration. *See* Osmolarity
Osmotic pressure (OP), 92–93, 735
Osseous tissue. *See* Bone
Ossification, 194
  endochondral, 195–198
  intramembranous, 198
Ossification center, 198
Osteoarthritis, 287, 288
Osteoblasts, 190
Osteoclast-activating factor, 208
Osteoclasts, 190
Osteocytes, 134, 189–190
Osteogenesis, 190
Osteoid, 190
Osteolysis, 190
Osteomalacia, 204, 208
Osteon, 191
Osteopenia, 207, 208
Osteoporosis, 143, 207–208
Osteoprogenitor cells, 190
Otic ganglion, 494, 497, 539
Otitis media, 586, 599
Otolith, 589
Outer cortex, lymph node, 785
Outer membrane, mitochondria, 934
Outer segment, photoreceptor, 578
Oval window, 586, 588
Ovarian artery, 757, 1061, 1066
Ovarian cancer, 1065, 1081, 1118
Ovarian cycle, 1062–1064, 1075–1076
Ovarian follicles, 1062
Ovarian hilum, 1061

Ovarian ligament, 1061
Ovarian vein, 765, 1061
Ovary(ies), 638, 1042, 1060–1061
Overhydration, 1017
Oviduct. *See* Uterine tube
Ovulation, 1064, 1088
Ovum, 1042
Oxidation, 936
Oxidative phosphorylation, 936, 937, 938
Oxygen
  blood, 741–742
  CSF, 741–742
  delivery to tissues, 855–858, 860
  functions in human body, 30
  hemoglobin and, 656–657, 855–856
  partial pressures, 851–852
  structure, 34
Oxygen debt, 322
Oxygen–hemoglobin saturation curve, 855–856
Oxyhemoglobin, 656
Oxytocin (OXT), 606, 618–619, 620, 1105, 1109

Pacemaker cells, 140, 328, 697
Pacemaker potential, 697
Pacesetter cells, 330, 880
Pacinian corpuscles. *See* Lamellated corpuscles
Packed cell volume (PCV), 655
Palatal muscles, 351
Palatine bones, 213, 224
Palatine processes, 223
Palatine tonsils, 783
Palatoglossal arch, 883
Palatoglossus muscle, 350
Palatopharyngeal arch, 883
Palmar arches, superficial/deep, 753
Palmar interosseus muscle, 370–371
Palmaris brevis muscle, 370–371
Palmaris longus muscle, 365, 366–368
Palmar vein, superficial/deep, 762
Palmar venous arches, 762
Palpebrae, 566–567
Palpebral conjunctiva, 567
Palpebral fissure, 566
Pampiniform plexus, 1045
Pancreas
  endocrine functions, 631, 635
  exocrine functions, 631
  gross anatomy, 901–902
  histology, 902
  islets, 631–632, 635
  physiology, 902–903
Pancreatic acini, 631, 902
Pancreatic alpha-amylase, 903, 916, 918
Pancreatic artery, 756
Pancreatic duct, 901
Pancreatic islets, 631–632, 635, 902
Pancreatic juice, 902
Pancreatic lipase, 903, 918
Pancreaticoduodenal artery, superior/inferior, 757
Pancreaticoduodenal vein, 767
Pancreatic polypeptide (PP), 632, 635
Pancreatic vein, 766
Pancreatitis, 903, 924
Pantothenic acid. *See* Vitamin B₅
Papanicolaou, George, 118
Papilla(e)
  dermal, 160
  duodenal, 908
  filiform, 564
  hair, 169
  lingual, 564
  renal, 969
Papillary duct, 972, 975
Papillary layer, dermis, 166
Papillary muscles, 687, 688
Papillary plexus, 167
Pap test, 118
Para-aminohippuric acid (PAH), 969
Paracrine communication, 604, 605

Paracrine factors, 604
Parafollicular cells. *See* C (clear) cells
Parahippocampal gyrus, 479
Parallel muscle, 336–337
Parallel processing, neural circuits, 447
Paralysis agitans. *See* Parkinson disease
Paramesonephric ducts, 1050
Paranasal sinuses, 228–229, 829
Paraplegia, 456
Parasympathetic blocking agents, 556
Parasympathetic division, ANS, 387–388, 531
  activation, 541
  comparison to sympathetic division, 543–545
  functional effects, by system, 544–545
  interaction with sympathetic division, 543–545
  membrane receptors and responses, 541–542
  neurotransmitter release, 541
  organization and anatomy, 538–540, 543
  overview, 541
Parasympathetic reflexes, 548
Parasympathomimetic drugs, 556
Parathyroid cells, 625, 627
Parathyroid glands, 203, 625–626
Parathyroid hormone (PTH)
  bone remodeling and, 201, 202
  calcium balance and, 203–204, 625–626, 1022
  disorders, 639
  effects, 625–626, 627
  in normal growth, 640
  regulatory control, 627
Paraurethral glands, 1071
Paraventricular nucleus, 478, 618
Paravertebral ganglia, 533
Parenchyma, 126
Paresthesia, 513, 525
Parietal bones, 188, 213, 218–219
Parietal cells, 893
Parietal lobe, 480, 484
Parietal pericardium, 24, 684
Parietal peritoneum, 877
Parietal pleura, 24, 841
Parietal portion, serous membrane, 136
Parieto-occipital sulcus, 480
Parkinson disease, 107, 417, 484, 502, 524
Parotid duct, 884
Parotid salivary glands, 884
Pars distalis, 614
Pars intermedia, 614
Pars tuberalis, 614
Partial antigens, 807
Partial pressure, of gases, 851–852, 854
Partial-thickness burns, 179
Parturition, 1105
Passive (leak) channels, 399, 403
Passive immunity, 797
Patella(e), 186, 259
Patellar ligament, 286
Patellar reflex, 450–451, 456
Patellar retinacula, 286
Patellar surface, 259
Patellofemoral stress syndrome, 259
Patent ductus arteriosus, 768–769
Patent foramen ovale, 768–769
Pathogens, 777
Pathological physiology, 7–8
Pathologists, 144
Paxil, 417
P cells, 583
PCT. *See* Proximal convoluted tubule
P/D1 cells, 894
Pectinate muscles, 687
Pectineal line, 255
Pectineus muscle, 372, 373–374
Pectoral fat pad, 1072
Pectoral girdle
  bones, 247–250
  joints, 287
  muscles, 362–363

Pectoralis major muscle, 363, 364–365
Pectoralis minor muscle, 361, 362, 847
Pectoralis muscle, 337
Pectoral nerves, medial/lateral, 444
Pediatrics, 1108
Pedicles, 232
Pelvic brim, 256
Pelvic cavity, 23, 24
Pelvic diaphragm, 357–358
Pelvic floor, 356–359
Pelvic girdle
  age-related changes, 264
  bones, 254–257
  joints, 287
Pelvic inflammatory disease (PID), 1079, 1081
Pelvic inlet, 256, 263
Pelvic nerves, 540
Pelvic outlet, 256
Pelvis, 256–257, 263
  arteries, 757
  muscles, 371–378
  sex differences, 256–257
  veins, 765
Penetrance, 1113
Penicillin, 807
Penis, 1055, 1056
Pennate muscle, 336–337
Pentose, 45
Pepsin, 894, 918
Pepsinogen, 894
Peptic ulcer, 894, 922, 924
Peptidases, 903, 918
Peptide, 53
Peptide bond, 53, 54
Peptide hormones, 606, 608
Perception, 508
Perforating canals, 191
Perforating fibers, 193
Perforins, 792, 814
Pericardial artery, 755
Pericardial cavity, 23, 24
Pericardial fluid, 684
Pericardial sac, 682, 684
Pericarditis, 136, 144, 684, 715
Pericardium, 24, 136, 684
Perichondrium, 130, 131
Perikaryon, 388–389
Perilymph, 587
Perimenopause, 1079
Perimetrium, 1066
Perimysium, 294
Perinephric fat capsule, 967
Perineum, 256, 359
Perineurium, 437
Perinuclear space, 82
Periodontal disease, 924
Periodontal ligament, 885
Periosteum, 130, 134, 189, 196–199
Peripheral adaptation, 509
Peripheral nerves
  anatomy, 437–438
  plexuses, 441–446
  repair of injury, 437–438
Peripheral nervous system (PNS), 387–388, 430–431
Peripheral neuropathy, 438, 456, 634
Peripheral proteins, 71
Peripheral resistance (PR), 732
Peristalsis, 880–881
Peristaltic wave, primary/secondary, 890
Peritoneum, 136, 876–877
Peritonitis, 136, 144, 877, 924
Peritubular capillaries, 970, 972, 979
Peritubular fluid, 975
Permanent dentition, 886–887
Permeability, 89
Permissive effects, hormones, 640
Pernicious anemia, 661
Peroneal artery. *See* Fibular artery
Peroxisome(s), 69, 80
Perpendicular plate
  ethmoid bone, 222
  palatine bone, 224

Perspiration
  control of, 174
  insensible, 161
  sensible, 161
PET (positron emission tomography), 24
Petrosal sinus, 762
Petrous part, temporal bone, 220
Peyer patches, 783–787, 901
PF-3, 674
p53 gene, 104
pH
  blood, 44, 653, 856–857
  body fluid, 43–44, 1023
  cerebrospinal fluid, 741–742
  control mechanisms, 1024–1027
  definition, 43
  enzyme function and, 57
  gastric contents, 896–897
  hydrogen ion concentration and, 44
  interstitial fluid, 651
  sweat, 173
  urine, 993
Phagocyte-activating chemicals, 814
Phagocytes, 789–791
Phagocytic dust cells, 791
Phagocytosis, 97–98, 791
Phagosomes, 96
Phalanges, 253, 254, 262
Phantom limb pain, 518
Pharyngeal arches, 883
Pharyngeal constrictor muscles, 350–351
Pharyngeal phase, swallowing, 890
Pharyngeal tonsil, 783, 831
Pharyngotympanic tube. *See* Auditory tube
Pharynx, 350–351, 831, 888
Phasic receptors, 509
Phenotype, 1112
Phenylalanine, 919, 947
Phenylalanine hydroxylase, 947
Phenylketonuria (PKU), 947, 962
Pheochromocytoma, 639
Phonation, 833
Phosphate balance, 1021, 1022
Phosphate buffer system, 1027
Phosphate group, 46, 51
Phosphate ion, 921
Phosphodiesterase (PDE), 581, 611
Phospholipase C (PLC), 611
Phospholipid(s), 51–52, 53
Phospholipid bilayer, 70
Phosphorus
  functions in human body, 30, 953
  recommended daily intake, 953
Phosphorylation, 60, 610
Photoreactive keratectomy (PRK), 577
Photoreception, 579–581
Photoreceptors, 571, 578
Phrenic artery, superior/inferior, 756
Phrenic nerve, 442
Phrenic vein, 765
Physical conditioning, 325–326
Physiology
  vs. anatomy, 5–6
  cell, 7
  definition, 5
  human, 7
  organ, 7
  pathological, 7–8
  systemic, 7
Pia mater, 435, 464
Pineal gland, 475, 584, 630
Pinealocytes, 630
Pinkeye, 567
Pinna. *See* Auricle
Pinocytosis, 97, 98
Piriformis muscle, 372, 373–374
Pisiform, 253
Pitch, sound, 594
Pituitary gland, 462
  adenohypophysis (anterior lobe), 614–618
  anatomy and orientation, 614
  hormones, 615–620
  hypothalamic control, 615, 616

portal system and blood supply, 615
posterior lobe, 618–619
Pituitary growth failure (pituitary dwarfism), 202, 208, 639
Pivot joints, 277, 278
Placenta, 1091
blood supply, 767
circulation, 1098
endocrine functions, 1098
structure, 1097
Placental prolactin, 1098
Placental stage, labor, 1106
Placentation, 1091, 1095–1098
Plane(s)
anatomic, 19, 21
of movement, 268
Plantar arch, 757
Plantar artery, medial/lateral, 757
Plantar flexion, 262, 276
Plantar interosseous muscles, 379–380
Plantaris muscle, 377–378
Plantar reflex, 455–456, 456
Plantar vein, 764
Plantar venous arch, 764
Plaque
atherosclerotic, 694, 724–725
central nervous system, 555
definition, 677
dental, 886
Plasma, 130, 651
composition, 652, 653–655
vs. interstitial fluid, 653
solute concentrations, 976
Plasma cells, 655, 669, 782
Plasma expanders, 654
Plasma membrane, 69–72
carbohydrates, 71–72
composition, 69
drugs affecting, 91
functions, 67–70
lipids, 70
movement across, 90–99
proteins, 70–71
receptors, 609–611
structure, 70
Plasma proteins, 652, 653–655
Plasmin, 676, 677
Plasminogen, 677
Plasmocytes (plasma cells), 126
Plasticity, 330
Platelet(s), 130, 651, 670, 672–673
Platelet adhesion, 674
Platelet aggregation, 674
Platelet-derived growth factor (PGDF), 674
Platelet phase, hemostasis, 674
Platelet plug, 674
Platysma muscle, 346–347, 348
Pleura, 24, 136, 841
Pleural cavity, 23, 841
Pleural effusion, 136, 144
Pleural fluid, 841
Pleural rub, 136
Pleurisy, 842, 869
Pleuritis (pleurisy), 136, 144
Plexus of Auerbach, 880
Plexus of Meissner, 879–880
Plicae, 879
Plicae circulares, 899
Pneumocytes type I, 839
Pneumocytes type II, 839
Pneumonia, 841, 869
Pneumotaxic center, 471, 472, 861
Pneumothorax, 847, 869
PNS. See Peripheral nervous system
Podocytes, 974
Point mutation, 88
Poisoning, heavy-metal, 397
Polar body, 1062
Polar covalent bond, 35
Polarity, epithelium, 112, 113
Polar molecule, 41
Polio, 325, 331
Pollex, 254

Polycystic kidney disease, 1003, 1118
Polycythemia, 661
Polygenic inheritance, 1114, 1115
Polymorphonuclear leukocytes. See Neutrophils
Polypeptides, 54
Polysaccharides, 47, 48
Polyspermy, 1088
Polysynaptic reflexes, 450
Polyunsaturated fatty acid, 48
Polyuria, 634, 644
Pons, 461, 462, 470, 471–472, 861–862
Popliteal artery, 757
Popliteal ligaments, 286
Popliteal vein, 764
Popliteus muscle, 372, 375–376
Porta hepatis, 903
Portal areas, liver, 905
Portal hypertension, 906, 924
Portal system, 615
Portal vessels, 615
Positive Babinski reflex, 456
Positive chemotaxis, 667
Positive feedback, 14–16
Positive nitrogen balance, 952
Positron emission tomography (PET), 24
Postabsorptive state, 949–950
Postcentral gyrus, 484
Posterior cavity, eye, 569
Posterior chamber, eye, 573
Posterior column pathway, 515–517
Posterior cruciate ligament (PCL), 286
Posterior femoral cutaneous nerve, 446
Posterior lobe
cerebellum, 472
pituitary gland, 618–619
Posterior median sulcus, 431
Postganglionic fibers, 392, 530, 543
Postnatal development, 1087
adolescence and maturity, 1109–1111
infancy and childhood, 1109, 1110
neonatal period, 1108–1109
senescence, 1111. See also Aging
stages, 1108
Postsynaptic membrane, 390
Postsynaptic neuron, 413
Postsynaptic potentials, 419, 423
Post-traumatic amnesia, 551, 556
Postural reflexes, 452
Potassium
absorption, 920–921
electrochemical gradient, 402
functions in human body, 30, 953
recommended daily intake, 953
secretion at distal convoluted tubule, 988–989
in urine, 996
Potassium balance, 1020–1021
Potassium channels, 700
Potential difference, 99, 401
Potential energy, 38, 99–100, 401–402
Potts fracture, 206
Pounds per square inch (psi), 846
Power stroke, cilium, 75, 76
Precapillary arteriole, 726
Precapillary sphincter, 726
Precentral gyrus, 484
Precipitation, of antigens, 808
Precocious puberty, 639, 642
Prednisone, 803
Pre-embryo, 1091
Prefrontal cortex, 486
Prefrontal lobotomy, 486
Preganglionic fibers, 392, 530, 543
Preganglionic neurons, 530, 532
Pregnancy. See also Labor; Prenatal development
back pain in, 231
ectopic, 1092, 1120
maternal systems changes during, 1104
Rh factors in, 664–665
uterine growth during, 1103, 1104–1105
Preload, 711–712
Premature delivery, 1107

Premature labor, 1107
Premature menopause, 1079
Premotor cortex, 485
Prenatal development
first trimester, 1091–1099
overview, 1090, 1100–1101
second and third trimester, 1102–1105
stages, 1090
Preoptic area, hypothalamus, 478
Prepotential, 697
Prepuce
clitoris, 1071
penis, 1056
Preputial glands, 1056
Presbyopia, 577, 599
Pressure gradient, blood vessel, 730
Pressure points, 722, 772
Presynaptic facilitation, 424
Presynaptic inhibition, 424
Presynaptic membrane, 390
Presynaptic neuron, 413
Prevertebral ganglia, 533
Prickling pain, 510
Primary amenorrhea, 1070
Primary brain vesicles, 462
Primary bronchi, left/right, 834–835, 837
Primary curves, 231
Primary dentition, 886–887
Primary fissure, cerebellum, 472
Primary follicles, 1062
Primary memories, 550
Primary motor cortex, 484
Primary oocytes, 1062
Primary ossification center, 196
Primary response, antibody-mediated immunity, 808–809
Primary sensory cortex, 484
Primary spermatocytes, 1046
Primary structure, protein, 54–55
Primary tumor, 105, 107
Prime mover, 340
Primitive streak, 1093
Primordial follicles, 1062
Principal cells, 625
P–R interval, 700
Procaine, 91
Procarboxypeptidase, 903
Procerus muscle, 346–347
Process (bone surface feature), 187
Prochlorperazine, 591
Procoagulants. See Clotting factors
Products of chemical reaction, 36
Proelastase, 903
Proenzymes, 674
Proerythroblasts, 660
Progenitor cells, 669, 671
Progesterone, 617
functions, 638, 1064, 1080
in pregnancy, 1098, 1104–1105
regulatory control, 638, 1080
Progestins, 617, 637, 1064, 1080
Projection fibers, 482
Prolactin (PRL), 105, 606, 617, 620
Prolactin-inhibiting hormone (PIH), 617
Prolactin-releasing factors (PRF), 617
Proliferative phase, uterine cycle, 1069, 1076
Prometaphase, mitosis, 102
Promethazine, 591
Promoter region, DNA, 84
Pronation, 276
Pronator quadratus muscle, 366, 367–368
Pronator teres muscle, 366, 367–368
Prone, 17
Pronucleus, male/female, 1089
Properdin, 793
Prophase, mitosis, 102
Propranolol, 694, 713
Proprioceptors, 391, 510, 511, 514
Prosencephalon, 462
Prostacyclin, 607, 674

Prostaglandins, 142, 604, 607, 668, 674
in inflammatory response, 793
in pregnancy, 1105
in seminal fluid, 1054
structure, 49–50
Prostate cancer, 1059, 1081, 1118
Prostatectomy, 1059
Prostate gland, 1053, 1054
Prostate-specific antigen (PSA), 1059, 1081
Prostatic acid phosphatase, 1059
Prostatic concretions, 1059
Prostatic fluid, 1054
Prostatic urethra, 1000
Prostatitis, 1054
Proteases, 76, 903, 916, 918
Proteasomes, 69, 76
Protectin, 792
Protein(s)
catabolism, 946–948
complete/incomplete, 952
digestion and absorption, 917, 919–920
fibrous, 54
functions, 51–53, 61
globular, 54
in human body, 51
plasma membrane, 70–71
shape, 54–56
sources, 61
structure, 53–54
synthesis, 86–88, 947–948
Protein buffer systems, 1025–1026
Protein C, 676
Protein deficiency diseases, 47, 962
Protein kinase C (PKC), 611
Proteinuria, 984, 1003
Proteoglycans, 57
Proteolytic enzymes, 894, 903
Prothrombinase, 675
Proton(s), 29
Proton acceptor, 44
Proton donor, 44
Protraction, 276
Proximal convoluted tubule (PCT)
anatomy, 970–971, 972
functions, 972, 975, 979, 994
reabsorption and secretion at, 984–986
Prozac. See Fluoxetine
Pseudocholinesterase, 541
Pseudopodia, 97–98
Pseudostratified columnar epithelium, 118, 120
Pseudounipolar neurons, 390–391
Psoas major muscle, 372, 373–374, 997
PTC (phenylthiourea; phenylthiocarbamide), 566
Pterygoid muscle, 349–350
Pterygoid plate, 222
Pterygoid processes, 222
Pterygopalatine ganglion, 495, 539
PTH. See Parathyroid hormone
Ptyalin. See Salivary amylase
Puberty, 1043, 1109–1111
apocrine sweat glands, 173
bone growth, 197, 201
female, 168, 638
hair changes, 171
male, 168, 629, 833
precocious, 639, 642
sebaceous glands, 172
testosterone levels in, 1058
Pubic symphysis, 255, 263
Pubis, 254
Pubococcygeus muscle, 357–358
Pubofemoral ligament, 283
Pudendal artery, 757
Pudendal nerve, 444, 446
Pudendal vein, 765
Pulmonary arterioles, 749
Pulmonary artery, left/right, 689, 722, 749
Pulmonary circuit, 682, 749, 853–854
Pulmonary embolism, 770, 772, 841, 869
Pulmonary lobules, 839

Pulmonary plexus, 545
Pulmonary semilunar valve. See
    Pulmonary valve
Pulmonary trunk, 689, 750
Pulmonary valve, 689
Pulmonary vein, left/right, 689, 749
Pulmonary ventilation, 843–851
    air movement, 843–844
    carbon dioxide transport, 858–859
    definition, 843
    gas exchange, 851–854
    mechanics of breathing, 845, 847–849
    oxygen transport, 855–858
    pressure changes, 844–847
    respiratory rates and volumes, 849–851
Pulp, spleen, 787
Pulp cavity, 885
Pulpitis, 924
Pulse (heartbeat), 733
Pulse (hormone release), 613
Pulse pressure, 733
Pulvinar nucleus, 476
Punnett square, 1114
Pupil, 570
Pupillary constrictor muscles, 570
Pupillary constrictor reflex, 583
Pupillary dilator muscles, 570
Pupillary reflex, 547, 548
Purines, 58, 421
Purkinje cells, 472, 519
Purkinje fibers, 698
Pus, 142, 668, 796
Putamen, 482
P wave, 699
Pyelogram, 997, 1003
Pyloric antrum, 891
Pyloric canal, 891
Pyloric glands, 894
Pyloric sphincter, 891
Pylorus, 891
Pyramidal cells, 484
Pyramidal system. See Corticospinal
    pathway
Pyramids, 521
Pyrexia, 960. See also Fever
Pyridoxine. See Vitamin B$_6$
Pyrimidines, 58
Pyrogens, 796, 960
Pyruvate, 932
Pyruvic acid, 80, 932
Pyuria, 996

QRS complex, 699
Q–T interval, 700
Quadrate lobe, liver, 903
Quadratus femoris muscle, 373–374
Quadratus lumborum muscle, 353–354
Quadratus plantae muscle, 379–380
Quadriceps femoris, 372, 375–376
Quadriplegia, 456
Quadruplets, 1107
Quaternary structure, protein, 54–55
Quiet breathing, 847–849, 861–862

Rabies, 390, 425
Radial artery, 653, 750, 1068
Radial collateral ligament, 283
Radial fossa, humerus, 250
Radial groove, humerus, 250
Radial head, 252
Radial nerve, 250, 283, 444
Radial notch, 252
Radial tuberosity, 252
Radial vein, 762
Radiation, 958
Radioactive decay, 30
Radioisotopes, 30, 62
    accumulation in bone, 200
Radiologist, 25
Radioulnar joint, distal/proximal, 252
Radius, 252–253
Rami communicantes, 438
Ramus (bone surface feature), 187
    mandible, 226

pubis, ischial/inferior/superior, 254
Range of motion, 268
Raphe, 337, 883, 1045
Rapid eye movement (REM) sleep,
    552–553
RBCs. See Red blood cells
Reabsorption
    in capillary exchange, 735–737
    in collecting system, 989
    in distal convoluted tubule, 988–990
    in proximal convoluted tubule, 984–986
    in urine formation, 976–977
Reactant, 36
Recall of fluids, 737
Receptive field, 508
Receptive speech area, 486
Receptor(s), 12–13, 387, 448
    sensory. See Sensory receptors
Receptor cascade, 609
Receptor-mediated endocytosis, 96–97
Receptor potential, 508
Receptor proteins, 71
Receptor site, 94
Receptor specificity, 508
Recessive inheritance, 1113, 1115
Reciprocal inhibition, 453
Recognition proteins, 71
Recommended daily intake (RDA)
    fat-soluble vitamins, 954
    minerals, 953
    water-soluble vitamins, 955
Recovery period, muscle, 322
Recruitment, motor units, 315
Rectal artery, 757
Rectal vein, 766
Rectouterine pouch, 1060
Rectum, 912, 913
Rectus (muscle name), 341
Rectus abdominis muscle, 340, 341,
    355–356, 847
Rectus femoris muscle, 337, 341, 372,
    375–376
Red blood cells (RBCs), 130, 651, 670
    abundance, 655
    count, 661
    crenation, 93
    erythropoiesis, 660, 661, 782
    formation, 657–660
    hemoglobin. See Hemoglobin
    hemolysis, 93, 658
    oxygen transport, 855–858
    production, 660–661
    structure, 655–656
    tests, 661, 667, 677
    turnover, 657–660
    in urine, 996
Red bone marrow, 192, 660
Red cones, 582
Red muscle, 324
Red muscle fibers, 323
Red nucleus, 474, 475
Red pulp, spleen, 787
Reduction, 936
Reductional division, 1049
Referred pain, 518
Reflex(es), 387, 444, 448–450. See also
    Cranial reflexes; Spinal reflexes;
    Visceral reflex(es); specific reflexes
Reflex arc, 448–449
Reflex centers, 470–471
Refraction, 575
Refractory period, action potential,
    408–409, 701
Regeneration, 142, 143, 144
Regional anatomy, 6
Regulation
    extrinsic, 12
    homeostatic, 12
    intrinsic, 12
Regulatory hormones, 612
Regulatory T cells. See Suppressor T cells
Regurgitation, cardiac, 690
Reinforcement, 455
Relative refractory period, 409

Relaxation, muscle, 310
Relaxation phase, twitch, 313
Relaxin, 257, 1098
Relay centers, 471
Releasing hormone (RH), 615
Remodeling, bone, 200
Renal artery, 757, 969
Renal calculi, 1002
Renal columns, 969
Renal compensation, 1028
Renal corpuscle, 969, 972, 973–974, 979
Renal cortex, 969
Renal failure, 998, 1003
Renal fascia, 967
Renal glycosuria, 986
Renal lobe, 969
Renal medulla, 969
Renal nerves, 969
Renal papilla, 969
Renal pelvis, 969
Renal physiology
    filtration pressures, 980–981
    glomerular filtration rate, 981–984
    nephron loop and countercurrent multi-
      plication, 986–988
    overview, 978–980
    reabsorption and secretion along the col-
      lecting system, 989
    reabsorption and secretion at the distal
      convoluted tubule, 988–989
    reabsorption and secretion at the proxi-
      mal convoluted tubule, 984–986
    summary, 993–997
    urine formation, 976–978
    urine volume and osmotic concentra-
      tion, 990–992
    vasa recta, 993
Renal pyramids, 969
Renal sinus, 969
Renal threshold, 977–978
Renal tubule, 969, 972
Renal vein, 765, 969
Renin, 637, 975
Renin–angiotensin system, 637, 982–983
Rennin, 894, 918
Replication, DNA, 100, 101
Repolarization, 406
Repressor genes, 104
Reproductive system
    basic structures, 1042
    female. See Female reproductive system
    hormones, 1080–1081
    integration with other systems,
      1080–1082
    male. See Male reproductive system
    prenatal development, 1101
    at puberty, 1111
Residual volume, 850
Resistance
    definition, 732
    membrane, 401
    muscle contraction and, 303
Resistance phase, general adaptation
    syndrome, 641–642
Resistin, 638
Respiration, 842–843. See also
    Respiratory physiology
Respiratory acidosis, 1030–1032,
    1035–1036, 1037
Respiratory alkalosis, 1032–1033,
    1035–1036, 1037
Respiratory bronchioles, 839
Respiratory burst, 667, 795
Respiratory centers, 861–862
Respiratory chain, 936–939
Respiratory compensation, 1028
Respiratory cycle, 846
Respiratory defense system, 827–829
Respiratory distress syndrome, 841, 869
Respiratory epithelium, 828
Respiratory membrane, 841, 853
Respiratory minute volume, 849
Respiratory mucosa, 827
Respiratory muscles, 847–848

Respiratory physiology
    acid–base balance and, 1028
    control of respiration, 860–865
    gas exchange, 851–854
    gas transport, 855–860
    pulmonary ventilation, 843–851
Respiratory portion, respiratory tract,
    826, 841
Respiratory pump, 737, 845
Respiratory rate, 849
    neonate, 1108
    in pregnancy, 1104
Respiratory reflexes, 862–865
Respiratory rhythmicity centers, 471,
    861
Respiratory system
    age-related changes, 866
    changes at birth, 865–866
    components, 827
    exercise and, 381
    functions, 826
    integration with other systems, 866–868
    introduction, 11, 826
    lower. See Lower respiratory system
    organization, 826–829
    overview, 154
    parasympathetic effects, 544
    prenatal development, 1101
    at puberty, 1111
    pulmonary ventilation. See Pulmonary
      ventilation
    smooth muscle in, 328
    sympathetic effects, 544
    upper. See Upper respiratory system
Respiratory tract, 826
Resting potential, 99, 399. See also
    Transmembrane potential
Resting tidal volume, 850
Rete testis, 1046
Reticular activating system, 475, 553
Reticular epithelial cells, 786
Reticular fibers, 126, 128
Reticular formation, 470, 474, 522
Reticular layer, dermis, 166
Reticular tissue, 128
Reticulocyte, 660
Reticulocyte count, 661
Reticulocytosis, 661
Reticuloendothelial system, 791
Reticulospinal tracts, 523
Retina, 571
    detached, 573, 599
    optic disc, 572–573
    organization, 571–572
    photoreceptor processing, 583
Retinal (retinene), 579
Retinitis pigmentosa, 579, 599, 1118
Retinoblastoma, 1118
Retinopathy, diabetic, 571, 599
Retraction, 276
Retroflexion, uterus, 1065
Retrograde amnesia, 551, 556
Retrograde flow, 390
Retroperitoneal, 877
Retrovirus, 819
Return stroke, cilium, 75, 76
Reverberation, neural circuits, 447
Reverse transcriptase, 819
Reversible reaction, 39
Rheumatic fever, 690
Rheumatic heart disease (RHD), 715
Rheumatism, 286–288
Rheumatoid arthritis, 287, 815
Rh factors, 662, 664–665
Rh negative, 662
Rhodopsin, 579
RhoGam, 665
Rhombencephalon, 462
Rhomboid muscle, major/minor, 360,
    362
Rh positive, 662
Riboflavin. See Vitamin B$_2$
Ribonucleic acid. See RNA
Ribosomal RNA (rRNA), 76

Ribosomes, 69, 76
Ribs, 239–242
Rickets, 165, 204, 208
Right atrioventricular valve (AV). *See* Tricuspid valve
Right atrium, 682, 687
Right lymphatic duct, 780
Right vein, 687–689
Right ventricle, 682, 689–690
Rigor mortis, 311, 331
Ringer solution, 654
Risorius muscle, 346–347
Rivinus duct, 884
RNA
  vs. DNA, 59
  messenger. *See* Messenger RNA
  ribosomal, 76
  structure, 59
  transfer, 86
  types, 58
RNA polymerase, 84
RNA processing, 85
Rods, 571, 578–579
Root
  hair, 169
  lung, 834
  nail, 175
  penis, 1056
  tongue, 883
  tooth, 885
Root canal, 885, 924
Root hair plexus, 169, 512
Root sheath, hair follicle, 169
Rotation, 273, 275–276
Rotator cuff, 282, 364, 383
Rotatores muscle, 351, 353–354
Rotaviruses, 1108
Rough endoplasmic reticulum (RER), 69, 77
Round ligament
  liver, 903
  uterus, 1045, 1066
Round window, 588
Rubrospinal tracts, 523
Ruffini corpuscles, 512–513
Rugae
  stomach, 891
  urinary bladder, 999
  vagina, 1070
Rule of nines, 179
Runner's knee, 259
R wave, 699

Saccule, 588, 589, 591
Sacral artery, middle/lateral, 757
Sacral canal, 238
Sacral cornua, 238
Sacral crest, median/lateral, 238
Sacral curve, 231
Sacral foramina, 238
Sacral hiatus, 238
Sacral plexus, 444–446
Sacral promontory, 238
Sacral tuberosity, 238
Sacral vein, 765
Sacrum, 231, 237–238, 263
Saddle joints, 277, 278
Sagittal plane, 19, 21
Sagittal sinus, inferior/superior, 464, 762
Sagittal suture, 214
Saliva, 884–885
Salivary amylase, 884, 885, 916, 918
Salivary glands, 495, 884–885
Salivatory nuclei, 885
Saltatory propagation, 411–412
Salts, 45, 1023
Salty taste, 564
SA (sinoatrial) node, 696, 697, 710–711
Saphenous nerve, 446
Sarcolemma, 297
Sarcomere, 298–299, 301, 302, 311–312
Sarcoplasm, 297
Sarcoplasmic reticulum, 298
Sartorius muscle, 372, 375–376

Satellite cells, 395
Saturated fatty acid, 48–49
Saturation limit, enzyme, 56–57
Scab, 175, 176, 181
Scalene muscles, 355–356
Scalp. *See* Epicranium
Scaphoid, 253
Scapula, 247, 249–250
Scapular spine, 250
Scar (fibrous) tissue, 143, 175
Schizophrenia, 554, 556
Schwann cells, 395–396
SCID. *See* Severe combined immunodeficiency disease
Scientific method, 8
Sclera, 569
Scleral venous sinus, 573
Scoliosis, 232
Scopolamine, 591
Scotomas, 576, 599
Scrotal cavity, 1045
Scrotum, 1042, 1045–1046
Scurvy, 201, 208
Seasonal affective disorder (SAD), 631, 644
Sebaceous follicles, 172
Sebaceous glands, 172
Seborrheic dermatitis, 172, 181
Sebum, 172
Secondary active transport, 95–96, 99
Secondary amenorrhea, 1070
Secondary brain vesicles, 462
Secondary bronchi, 837
Secondary curves, 231
Secondary dentition, 886–887
Secondary follicles, 1064
Secondary memories, 550
Secondary oocyte, 1062
Secondary ossification center, 197
Secondary processes, 974
Secondary response, antibody-mediated immunity, 809
Secondary spermatocytes, 1047, 1049
Secondary structure, protein, 54–55
Secondary tumor, 105, 107
Second-class lever, 338–339
Second-degree burn, 179
Second heart sound, 706
Second messenger, 418, 609
Second-order neuron, 514
Second trimester, 1090
  changes in maternal systems, 1104
  changes in uterus, 1103–1105
  fetal development, 1102–1103
  overview by system, 1100–1101
Secretin, 897, 909, 910
Secretion
  in collecting system, 989
  digestive, 875
  at distal convoluted tubule, 988–990
  modes, 121, 122
  at proximal convoluted tubule, 984–986
  in small intestine, 901
  types, 121
  in urine formation, 976–977
Secretory phase, uterine cycle, 1069–1070, 1076
Secretory piece, immunoglobulin, 807
Secretory sheet, 122
Secretory vesicles, 77
Section, 19
Sectional planes, 19
Segmental arteries, 969
Segmental blocks, 1066
Segmental bronchi, 837
Segmentation, 880, 881
Segments, electrocardiogram, 699
Seizure, 489, 502
Selectively permeable membrane, 89
Selective serotonin reuptake inhibitors (SSRIs), 417, 554
Selenium, 953
Sellaris joints. *See* Saddle joints

Sella turcica, 221, 614
Semen, 1042, 1055–1056
Semicircular canals, 588
Semicircular ducts, 588, 589–590
Semilunar ganglion, 494
Semilunar valves, 690
Semimembranosus muscle, 372, 375–376
Seminal fluid, 1055
Seminal glands (vesicles), 1053, 1054
Seminalplasmin, 1054
Seminiferous tubules, 1046
Semispinalis capitis muscle, 353–354
Semispinalis cervicis muscle, 353–354
Semispinalis thoracis muscle, 353–354
Semitendinosus muscle, 372, 375–376
Senescence, 1111. *See also* Aging
Senile cataracts, 575
Senile dementia, 555, 556
Senile lentigos (liver spots), 163
Senility, 555
Sensation, 508
Sense organs, 508
Sensible perspiration, 161, 1016
Sensitization
  B cells, 804–805
  Rh factor, 662
Sensory association areas, 484
Sensory coding, 509
Sensory ganglia, 391
Sensory homunculus, 517
Sensory neurons, 391
Sensory nuclei, 435
Sensory pathway, 507
  somatic, 515–520
  visceral, 519
Sensory processing centers, 507
Sensory receptors, 507–508. *See also* Special senses
  adaptation, 509–510
  classification, 510–514
  information interpretation, 509
  stimuli detection, 508–509
Sepsis, 179, 181
Septa
  testes, 1046
  thymus, 786
Septal cells, 839
Septum pellucidum, 463
Serial processing, neural circuits, 447
Serosa, 136, 880
Serotonin, 417, 420, 674
Serous glands, 121
Serous membranes, 136
Serratus anterior muscle, 361, 362, 847
Serratus posterior muscle, 355–356
Sertoli cells. *See* Nurse (sustentacular) cells
Serum, 654
Serum enzyme assay, 1059
Sesamoid bones, 186
Severe combined immunodeficiency disease (SCID), 815, 820, 1118
Sex cells, 67
Sex chromosomes, 1112
Sex differences
  blood volume, 653
  body composition, 1011–1012
  hematocrit, 655
  pelvis, 256–257
  skeletal system, 262–263
Sex hormones. *See also* Androgen(s); Estrogen(s)
  effects on bone, 201, 202
  in normal growth, 640
Sex-linked characteristics, 1117
Sex-linked inheritance, 1116–1117
Sexual arousal, 548, 1078
Sexual function
  female, 1078–1079
  male, 1078
Sexual intercourse, 1042, 1077–1079
Sexually transmitted diseases (STDs), 1079, 1081

Shaft
  femur, 258
  penis, 1056
  rib, 241
Shallow breathing, 847
Sharpey fibers, 193
Shingles, 440, 456
Shiver, 959
Shivering thermogenesis, 958
Shock, 746, 772
Short bones, 186
Short reflexes, 547, 882
Short-term memories, 550
Shoulder girdle. *See* Pectoral girdle
Shoulder joint, 250, 281–282
Shoulder separation, 281
Siamese twins, 1107
Sickle cell anemia, 657, 658, 677, 1113, 1118
SIDS (sudden infant death syndrome), 862, 869
Sigmoid artery, 757
Sigmoid colon, 911
Sigmoid mesocolon, 877
Sigmoid sinus, 762
Sign, 13, 25
Silent heart attack, 702
Simple columnar epithelium, 118, 120
Simple cuboidal epithelium, 118, 119
Simple diffusion, 91
Simple epithelium, 117
Simple fracture, 206
Simple gland, 123
Simple inheritance, 1113, 1115
Simple squamous epithelium, 117
Simple sugars, 45–46
Single covalent bond, 34
Sinoatrial (SA) node, 696, 697, 710–711
Sinus(es), 187
  skull, 214
Sinusitis, 242
Sinusoidal capillaries, 726
Sinusoids
  liver, 904
  structure and function, 726
Skatole, 914
Skeletal muscle(s), 140
  actions, 240, 339, 341
  age-related changes, 378–380
  atrophy, 324
  blood supply, 295
  connective tissue organization, 294–296
  functions, 294
  hormones and, 322
  hypertrophy, 324–325
  insertions, 310, 339–340, 341
  names, 341–343
  nerves, 295
  origins, 310, 339–340, 341
  overview, 343–345
  parasympathetic effects, 544
  physical conditioning and, 325–326
  sympathetic effects, 544
  tension production, 314
Skeletal muscle fibers, 296–303, 686
  atrophy, 325
  control, 304
  development, 296
  distribution, 324
  hypertrophy, 324–325
  muscle performance and, 323–326
  sliding filament theory, 301–303
  structure, 296–301
  tension production, 311–314
  types, 323–324
Skeletal muscle tissue, 138–139, 294–326
  vs. cardiac muscle tissue, 329, 684, 686
  connective tissue organization, 294–296
  contraction. *See* Contraction, skeletal muscle
  functional organization, 301
  functions, 140
  metabolism, 318–322
  vs. smooth muscle tissue, 329

Skeletal system. *See also* **Bone**
  age-related changes, 264
  axial skeleton, 212–242
  functions, 185, 288
  integration with other systems, 288–289
  introduction, 10, 185
  overview, 151
  prenatal development, 1100
  at puberty, 1110
  sex differences, 262–263
Skene glands, 1071
Skill memories, 550
Skin. *See also* **Integumentary system**
  abnormalities, 178
  age-related changes, 177–178
  dermis, 166–168
  epidermis, 159–166
  functions, 158–159
  hypodermis, 168
  injury repair, 175–176
  innervation, 168
  lines of cleavage, 166–167
  parasympathetic effects, 544
  structure, 158
  sympathetic effects, 544
  thick, 159, 160
  thin, 159
  water immersion, 161
Skin cancers, 164
Skin graft, 179, 181
Skin turgor, 166
Skull, 213–230
  age-related changes, 264
  anterior view, 216
  cranial bones, 213, 218–223
  facial bones, 213–214, 223–226
  fontanelles, 229–230
  foramina and fissures, 226–227
  horizontal section, 217
  infant, 229–230
  inferior view, 216
  joints, 280
  lateral view, 215
  nasal complex, 228–229
  orbital complex, 228
  posterior view, 215
  sagittal section, 217
  sex differences, 263
  superior view, 215
  sutures, 214
Sleep, 552–553
Sleep disorders, 553
Sliding filament theory, 301–303
Slipped disc, 279, 280
Slow-adapting receptors, 509
Slow fibers, 323, 324
Slow oxidative (SO) fibers, 323
Slow potassium channels, 700
Slow sodium channels, 700
Slow-twitch oxidative fibers, 323
Small intestine
  gross anatomy, 898–899
  histology, 899–901
  movements, 901
  secretions, 901
Small ribosomal unit, 76
Small saphenous vein, 764
Smegma, 1056
Smoking
  respiratory performance and, 866
Smooth endoplasmic reticulum (SER), 69, 76
Smooth muscle tissue, 138–140
  vs. cardiac muscle tissue, 329
  control of contractions, 330
  excitation–contraction coupling, 330
  functions, 328
  length–tension relationships, 330
  vs. skeletal muscle tissue, 329
  structure, 328–329
  tone, 330
Sneezing, 865
SNS. *See* **Somatic nervous system**
Sodium (Na⁺), 33, 953

absorption, 920–921
electrochemical gradient, 402
functions in human body, 30, 953
reabsorption at distal convoluted tubule, 988, 990
reabsorption at proximal convoluted tubule, 985–986
reabsorption in collecting system, 989
recommended daily intake, 953
in urine, 996
Sodium balance
  disorders. *See* **Hypernatremia; Hyponatremia**
  ECF volume and, 1018–1020
  factors affecting, 1018
  regulation, 1019–1020
Sodium bicarbonate, 1027
Sodium channels
  activation, 407
  fast/slow, 700
  inactivation, 407–408
Sodium chloride, 33
Sodium hydroxide, 44
Sodium–potassium ATPase, 95, 402, 409
Sodium–potassium exchange pump, 95, 96, 402, 409
Soft keratin, 169
Soft palate, 831, 883
Soleus muscle, 372, 377–378
Solids, 35
Solitary nucleus, 471, 519
Solute, 41
Solute concentration, 43
Solvent, 41
Soma. *See* **Cell body, neuron**
Somatic cells, 67. *See also* **Cell(s)**
Somatic motor association area, 485
Somatic motor neurons, 392
Somatic motor pathways, 507
Somatic motor system. *See* **Somatic nervous system (SNS)**
Somatic nerve
  corticospinal tract, 521–522, 523
Somatic nervous system (SNS), 387, 519–520
  basal nuclei, 523–524
  cerebellum, 524
  comparison to autonomic nervous system, 529–530, 549
  corticobulbar tract, 521, 523
  corticospinal pathway, 520–522
  lateral pathway, 522–523
  levels of processing and motor control, 524–525
  medial pathway, 522–523
  motor homunculus, 522
Somatic reflexes, 450
Somatic sensory association area, 484
Somatic sensory neurons, 391
Somatic sensory pathways, 515–519
Somatocrinin. *See* **Growth hormone–releasing hormone**
Somatomedins, 617
Somatostatin. *See* **Growth hormone–inhibiting hormone**
Somatotropin. *See* **Growth hormone**
Sound, 594–595
Sounds of Korotkoff, 772
Sour taste, 564
Spatial summation, 422–423
Special circulation, 747
Special senses, 508, 562
  gustation. *See* **Gustation**
  hearing. *See* **Hearing**
  olfaction. *See* **Olfaction**
  prenatal development, 1100
  vision. *See* **Vision**
Specific defenses, 667, 777, 789. *See also* **Immunity**
Specificity
  enzyme, 56
  immunologic, 797
Specific resistance, 789
Speech area, 486

Spermatic cords, 1045
Spermatids, 1047
Spermatogenesis, 1046
  follicle-stimulating hormone and, 1056–1057
  mitosis and meiosis, 1048–1049, 1050
  overview, 1047–1048
  spermiogenesis, 1049–1050
Spermatogonia, 1046
Spermatozoa, 1042, 1050–1051, 1055
Sperm count, 1055
Spermiation, 1049
Spermiogenesis, 1048, 1049–1051
Sperm receptors, 1088
S phase, interphase, 101
Sphenoid, 213, 221–222
Sphenoidal fontanelle, 229
Sphenoidal sinus, 221, 222
Sphenoidal spine, 222
Sphenopalatine ganglia, 494
Sphincter(s), 328, 336–337
Sphincter of Oddi, 908
Sphincter vesicae. *See* **Urethral sphincter, internal/external**
Sphygmomanometer, 772
Spicules, 198
Spina bifida, 242
Spinal accessory nerve. *See* **Accessory nerve**
Spinal anesthesia, 434
Spinal cord
  gross anatomy, 431–433
  sectional anatomy, 435–437
Spinal curves, 231
Spinalis cervicis muscle, 353–354
Spinalis thoracis muscle, 353–354
Spinal meninges, 433–435
Spinal meningitis, 433
Spinal nerves, 387, 431
  anatomy, 437–438
  peripheral distribution, 438–439
  plexuses, 440–446
Spinal reflexes, 430, 450
  brain's effect on, 454–456
  monosynaptic, 450–452
  polysynaptic, 452–454
  reinforcement and inhibition, 455–456
  voluntary movement and, 455
Spinal root, accessory nerve, 499
Spinal tap, 434, 456, 466
Spindle fibers, 102
Spine. *See* **Vertebral column**
Spine (bone surface feature), 187
Spinoaccessory nerve. *See* **Accessory nerve**
Spinocerebellar ataxia, 1118
Spinocerebellar pathway, 518–519, 520–521
Spinocerebellar tracts, anterior/posterior, 519, 520–521
Spinothalamic pathway, 517–518, 520–521
Spinothalamic tracts, anterior/lateral, 516–517, 520–521
Spinous process, vertebrae, 233
Spiral arteries, 1068
Spiral-CT, 25
Spiral fracture, 206
Spiral ganglion, 597
Spirulina, 952
Splanchnic nerve, 534
Spleen, 786–788
  anatomy, 786–787, 788
  functions, 786
  histology, 787–788
  injury, 788
Splenectomy, 788
Splenic artery, 756
Splenic flexure, 911
Splenic vein, 766
Splenius capitis muscle, 353–354
Splenius cervicis muscle, 353–354
Split-thickness graft, 179
Spongy bone, 192–193

Spongy urethra, 1000
Spontaneous abortion, 1104
Spontaneous mutation, 1116
Sports drinks, 325, 1020
Sports injuries, 286, 372. *See also* **Athletes**
Spot desmosomes, 114, 115
Sprain, 271
Squamous cell carcinoma, 164, 181
Squamous epithelium, 116, 117–118
Squamous part, temporal bone, 220
Squamous suture, 214
SSRIs (selective serotonin reuptake inhibitors), 417, 554
Stapedius muscle, 587
Stapes, 586
Starches, 47
Starling, Ernest H., 712
Starling law of the heart, 712
Start codon, 87
State of equilibrium, 16
States of matter, 35–36
Stato-acoustic nerve. *See* **Vestibulocochlear nerve**
Statoconia, 589
Stellate reticuloendothelial cells. *See* **Kupffer cells**
Stem cells (basal cells), 104
  epidermal, 160
  epithelial, 116
  gustatory, 564
  lymphoid, 651, 660, 671
  myeloid, 651, 660, 671
  olfactory, 562
  therapeutic uses, 107
Stensen duct, 884
Stent, cardiac, 695
Stercobilinogens, 658, 914
Stercobilins, 659, 914
Stereocilia, 589
Sterile, 1043
Sternal end, clavicle, 247
Sternoclavicular joint, 247
Sternocleidomastoid muscle, 351–352, 847
Sternohyoid muscle, 352
Sternothyroid muscle, 352
Sternum, 239, 242
Steroid-binding proteins, 654
"Steroid creams," 629
Steroid hormones, 50–51, 53, 607–608, 1074. *See also specific hormones*
Stethoscope, 706
Stimulus, 12
Stimulus frequency, 313
Stomach, 891–897
  digestion and absorption in, 897
  gastric glands, 893–894
  gross anatomy, 891–892
  histology, 891–893
  pyloric glands, 894
  regulation of activity, 894–897
Stop codon, 87
Straight artery, 1065
Straight sinus, 762
Stratified columnar epithelium, 120, 121
Stratified cuboidal epithelium, 118, 119
Stratified epithelium, 117
Stratified squamous epithelium, 117, 118
Stratum, 159
Stratum corneum, 161
Stratum germinativum (stratum basale), 160
Stratum granulosum, 161
Stratum lucidum, 161
Stratum spinosum, 161
Streptokinase, 702
Stress, 640, 816–817
Stress incontinence, 1002
Stress response. *See* **General adaptation syndrome**
Stretch marks, 166
Stretch reflex, 449, 450–451
Striated involuntary muscle, 140

Striated voluntary muscle, 138
Striations, 299
Strict dominance, 1113
Stroke (cerebrovascular accident), 467, 501, 555, 747, 754, 772
Stroke volume (SV), 706–708, 711–713
Stroma
  ovarian, 1061
  reticular tissue, 126
Stromal cells, 782
Strong acid, 1023
Strong base, 44, 1023
Structural proteins, 52. *See also* Protein(s)
Student's elbow, 271
Sty, 566
Styloglossus muscle, 350
Stylohyoid ligaments, 226
Stylohyoid muscle, 351–352
Styloid process
  radius, 253
  skull, 220, 264
  ulna, 252
Stylomastoid foramen, 221, 227
Subacromial bursa, 282
Subarachnoid space, 434, 464
Subatomic particles, 29
Subcapsular space, 784
Subclavian artery, 722, 748, 750
Subclavian trunk, left/right, 779–780
Subclavian vein, 748, 762
Subclavius muscle, 361, 364
Subcoracoid bursa, 282
Subcutaneous injection, 168, 371
Subcutaneous layer. *See* Hypodermis
Subdeltoid bursa, 282
Subdural hemorrhage, 466, 502
Subdural space, 434
Sublingual ducts, 884
Sublingual salivary glands, 884
Subluxation, 272
Submandibular ducts, 884
Submandibular ganglion, 494, 495, 539
Submandibular salivary glands, 225, 884
Submucosa
  digestive tract, 879–880
  trachea, 834
Submucosal glands, 900
Submucosal plexus, 879–880
Subscapular bursa, 282
Subscapular fossa, 250
Subscapularis muscle, 281, 362, 365–366
Subscapular nerve, 444
Subserous fasciae, 137–138
Substance P, 418, 421, 511
Substantia nigra, 474, 475
Substrate-level phosphorylation, 936
Substrates, 56
Sucrase, 917, 918
Sucrose, 46, 47, 917
Sudden infant death syndrome (SIDS), 862, 869
Sudoriferous glands, 172
Sulci, 462
Sulcus (bone surface feature), 187
Sulfadiazine, 468
Sulfate ion, 921
Sulfisoxazole, 468
Sulfur, 30
Summation, 422–423
Summation of twitches, 313–314
Sunblock, 164
Sunburn, 179
"Sunless tan," 618
Sun protection factor (SPF), 164
Superficial anatomy, 17
Superficial epithelium, eye, 566–567
Superficial fasciae. *See* Hypodermis
Superficialis (muscle name), 341
Superficial lymphatics, 779
Superficial reflexes, 450
Superior border, scapula, 247
Superior colliculus, 474, 522, 592
Superior ganglion, 497, 498

Superior mesenteric artery, 536
Superior mesenteric ganglion, 534
Superior oblique muscle, 348–349
Superior rectus muscle, 348–349
Superior vena cava, 687, 758, 762, 764
Superoxide anions, 668
Supination, 276
Supinator muscle, 367–368, 369
Supine, 17
Supporting connective tissues, 124, 131–134
Suppression (genetic), 1114
Suppression factors, 801, 814
Suppressor/inducer T cells, 782
Suppressor T (T$_S$) cells, 782, 801, 812, 819
Suprachiasmatic nucleus, 478
Supraclavicular nerve, 442
Supraoptic nucleus, 478, 618
Supra-orbital foramen, 219, 227
Supra-orbital margin, 219
Supra-orbital notch, 219
Suprarenal artery, 757
Suprarenal cortex, 616, 627–629
Suprarenal gland, 627–630
Suprarenal medulla, 533, 536, 629–630
Suprarenal vein, 765
Suprascapular nerve, 444
Supraspinatus muscle, 281, 364, 365
Supraspinous fossa, 250
Supraspinous ligament, 279
Surface anatomy, 6
Surface antigens, 661
Surface features, bone, 186–187
Surface tension, 35, 840
Surfactant, 839
Surgical neck, humerus, 250
Suspensions, 42
Suspensory ligament
  breast, 1072
  lens, 570
  ovary, 1061
  uterus, 1065–1066
Sustentacular cells. *See* Nurse (sustentacular) cells
Sutural bone, 185, 186, 214
Suture, 214, 268, 269
Swallowing, 832, 889–890
Swallowing center, 890
Swallowing reflex, 548, 890
Sweat, 173, 1020. *See also* Perspiration
Sweet taste, 564
Sympathetic activation, 536
Sympathetic blocking agents, 556
Sympathetic chain ganglia, 533, 534
Sympathetic division, ANS, 387–388, 531
  activation, 536
  comparison to parasympathetic division, 543–545
  functional effects, by system, 544–545
  interaction with parasympathetic division, 543–545
  neurotransmitters and, 536–538
  organization and anatomy, 532–536, 543
  overview, 538
Sympathetic ganglion, 438
Sympathetic nerves, 438, 534, 535
Sympathetic reflexes, 548
Sympathomimetic drugs, 556
Symphysis, 268, 269
Symport, 94
Symptom, 13
Symptoms, 25
Synapse, 389–390
  activity, 413
  chemical, 414
  cholinergic, 414–416
  electrical, 414
  nitroxidergic, 538
Synapsis, 1048
Synaptic cleft, 304, 390
Synaptic communication, 605
Synaptic delay, 416
Synaptic fatigue, 416

Synaptic knob, 390
Synaptic membrane, 390
Synaptic terminal, 304, 389
Synaptic vesicles, 390
Synarthrosis, 268
Synchondrosis, 268
Syncytial trophoblast (syncytiotrophoblast), 1092
Syncytium, functional, 327
Syndesmosis, 268, 269
Syndrome of inappropriate diuretic hormone secretion, 639
Syneresis, 676–677
Synergist (muscle), 340
Synergistic effect, hormones, 640
Synesthesia, 561
Synostosis, 268, 269
Synovial fluid, 137, 270–271
Synovial joint, 269–277
  accessory structures, 271
  movements, 272–277, 278
  stabilization, 271–272
  structure, 270–272
  types, 277, 278
Synovial membranes, 137, 269
Synovial tendon sheath, 369
Synthesis, 38
Systemic anatomy, 6
Systemic arteries, 750–757
Systemic circuit, 682
  arteries, 750–757
  partial pressures in, 853–854
  veins, 758–767
Systemic physiology, 7
Systemic veins, 758–767
Systole, 703, 704–706
Systolic pressure, 733

Tachycardia, 698, 715
Tactile corpuscles, 168, 512
Tactile discs, 168
Tactile disease, 512
Tactile receptors, 511–513
Taeniae coli, 911
Tail
  epididymis, 1052
  pancreas, 901
  spermatozoon, 1050
  thick filament, 300
Tail fold, 1098
Talipes equinovarus (clubfoot), 262, 264
Talus, 260
Tamoxifen, 1073
Tanning bed, 164
Target cells, 604
Tarsal bones, 260–261
Tarsal glands, 566
Tartar, 886
T$_4$ assay, 957. *See also* Thyroxine
Taste buds, 564
Taste hairs, 564
Taste pore, 564
Taste receptors, 564–565
Taste sensations, primary, 564
Tattoos, 176
Tay-Sachs disease, 425, 1118
TCA (tricarboxylic acid) cycle, 80–81, 319, 934–935, 940
T cells, 669, 780, 782
  activation, 801–804
  antigen presentation, 799–800
  antigen recognition, 800–802
  CD4, 801, 802–803
  CD8, 801
  helper, 782, 798, 812, 819
  inflammatory, 782
  memory, 801, 802, 812
  suppressor, 782, 801, 812, 819
  types, 798–799
Tear gland. *See* Lacrimal gland
Tears, 567–569
Tectorial membrane, 593–594
Tectospinal tracts, 522, 523
Tectum, 474, 522

Teeth, 885–888
  age-related changes, 264
  dental succession, 886–887
  sex differences, 263
  structure, 885–886
  types, 886
Teeth grinding, 226
Tegmentum, 475
Telencephalon, 462
Telodendria, 389
Telomerase, 106
Telomeres, 104
Telophase, mitosis, 103
Temperature
  body. *See* Body temperature
  effect on diffusion rate, 90
  effect on enzymes, 57
  effect on hemoglobin, 857
  regulation. *See* Thermoregulation
Template strand, DNA, 84
Temporal bone, 213, 219–220
Temporalis muscle, 219, 341, 348–349
Temporal lines, inferior/superior, 219
Temporal lobe, 480, 484
Temporal processes, 225
Temporal summation, 422–423
Temporal vein, 762
Temporomandibular joint (TMJ), 226
Temporoparietalis muscle, 346–347, 348
Tendinous inscriptions, 340, 355–356
Tendon(s), 124, 126, 129, 271, 295
Tendon reflex, 452
Tension, 182, 303
Tensor fasciae latae muscle, 372, 373–374
Tensor tympani muscle, 587
Tensor veli palatini muscle, 350–351
Tentorium cerebelli, 464–465
Teratogens, 1113
Teres (muscle name), 341
Teres major muscle, 340, 362, 363–365
Teres minor muscle, 281, 362, 364–365
Terminal bronchioles, 837
Terminal cisternae, 298
Terminal ganglion, 539
Terminal hairs, 171
Terminal lymphatics. *See* Lymphatic capillaries
Terminal segment, aorta, 756
Terminal web, 73
Terminologia Anatomica, 5
Tertiary bronchi, 837
Tertiary follicle, 1064
Tertiary memories, 550
Tertiary structure, protein, 54–55
Testes, 1042
  descent, 1043–1044
  endocrine functions, 638
  histology, 1046–1047
  structure, 1046
Testicular artery, 757, 1045
Testicular cancer, 1043
Testicular vein, 765
Testosterone, 617, 637, 1046
  integumentary system and, 1109–1110
  muscle metabolism and, 322
  production, 1057–1058
  structure, 51
Testosterone-binding globulin (TeBG), 654
Tetanus
  complete, 314
  disease, 304, 331
  incomplete, 314
Tetracycline, 468
Tetrad, 1048
Tetraiodothyronine. *See* Thyroxine
Tetralogy of Fallot, 769
Tetrose, 45
Thalamic nuclei, 475–476
Thalamus, 461, 462, 475–476
Thalassemia, 657, 658, 677
Thecal cells, 1064

Therapeutic abortion, 1104
Thermal inertia, 41
Thermoreceptors, 509, 510, 511
Thermoregulation, 14–16, 174, 957–961
  blood in, 651
  in burn patient, 179
  disorders, 961
  heat conservation, 958
  heat gain/loss regulation, 958–959
  heat generation, 958–959
  heat transfer mechanisms, 958
  hypothalamus in, 14, 15
  individual variations, 960
  integumentary system in, 174
  muscle activity and, 322
  negative feedback in, 14, 15
  organ systems involved, 16
  skeletal muscle and, 294
Theta waves, 488–489
Thiamine. See Vitamin B$_1$
Thiazides, 991
Thick filaments, 297, 300–301, 302
Thick segment, nephron loop, 975
Thick skin, 159, 160
Thigh. See Lower limb
Thin filaments, 74, 297, 300, 302
Thin segment, nephron loop, 975
Thin skin, 159
Third-class lever, 338, 339
Third-degree burn, 179
Third heart sound, 706
Third-order neuron, 514
Third trimester, 1090
  changes in maternal systems, 1104
  changes in uterus, 1103–1105
  fetal development, 1102–1103
  overview by system, 1100–1101
Third ventricle, 463
Thirst center, 478
Thoracentesis, 842
Thoracic aorta, 755
Thoracic cage, 239–242, 280
Thoracic cavity, 22–24
Thoracic curve, 231
Thoracic duct, 779, 781
Thoracic vein, internal, 762
Thoracic vertebrae, 234, 236–237
Thoracoabdominal pump. See
  Respiratory pump
Thoracodorsal nerve, 444
Thoracolumbar division, ANS. See
  Sympathetic division, ANS
Thoroughfare channel, 726
Threonine, 947
Thrombin, 675
Thrombocyte(s), 672. See also Platelet(s)
Thrombocyte-stimulating factor, 673
Thrombocytopenia, 672
Thrombocytopoiesis, 673
Thrombocytosis, 672
Thrombomodulin, 676
Thrombopoietin, 673
Thromboxane A$_2$, 674
Thromboxanes, 607
Thrombus, 677, 684, 770, 772
Thymic (Hassall) corpuscles, 786
Thymic lobes, 786
Thymine, 58
Thymosin(s), 105, 635, 786
Thymus, 635, 637, 786–787
Thyrocervical trunk, 750, 841
Thyroglobulin, 622
Thyrohyoid muscle, 352
Thyroid-binding globulins (TBGs), 623, 654
Thyroid-binding prealbumin
  (TBPA), 623
Thyroid cartilage, 620, 832
Thyroid follicles, 622–623, 627
Thyroid gland
  anatomy, 620–622
  C cells and calcitonin, 624
  follicles, 622–623, 627
  hormones. See Thyroid hormone(s)

Thyroid hormone(s)
  functions, 623–624, 627
  iodine and, 624
  muscle metabolism and, 322
  in normal growth, 640
  regulatory control, 627
Thyroiditis, 815
Thyroid peroxidase, 623
Thyroid-stimulating hormone (TSH),
  606, 616, 620, 959
Thyrotoxicosis, 644
Thyrotropin. See Thyroid-stimulating
  hormone
Thyrotropin-releasing hormone (TRH),
  616, 959
Thyroxine (T$_4$), 201, 202, 608, 623, 639,
  957
Tibia, 259–260
Tibial artery, posterior/anterior, 757
Tibial collateral ligament, 286
Tibialis muscle, anterior/posterior,
  376–377, 378
Tibial nerve, 444, 446
Tibial tuberosity, 259
Tibial vein, anterior/posterior, 764
Tic douloureux, 494, 502
Tickle sensation, 513
Tidal volume, 846, 850
Tight (occluding) junction, 114, 115
Timolol, 713
Tissue(s)
  age-related changes, 143
  composition of human body, 150
  connective. See Connective tissue(s)
  definition, 7, 112
  epithelial. See Epithelium
  inflammation, 142–143
  membranes. See Membrane(s)
  muscle, 138–140. See also Cardiac mus-
    cle tissue; Skeletal muscle tissue;
    Smooth muscle tissue
  neural. See Neural tissue
  regeneration, 143
  response to injury, 141–143
  types, 112
Tissue cholinesterase, 541
Tissue factor. See Factor III
Tissue organizational level, 8, 9, 112. See
  also Tissue(s)
Tissue perfusion, 737–738
Tissue plasminogen activator (t-PA), 677,
  702
Tissue repair, 795–796
Titin, 299
TMJ syndrome, 226, 242
Tolerance, immunologic, 798
Tongue, 350, 564–566, 883–884
Tonicity, 93
Tonic receptors, 509
Tonsillitis, 783, 820
Tonsils, 783–784
Tooth. See Teeth
Torr, 846
Total lung capacity, 851
Total peripheral resistance, 730–731,
  732
Trabeculae
  lymph nodes, 784
  pulmonary, 838
  spongy bone, 192
Trabeculae carneae, 689
Trabecular artery, 787
Trabecular bone, 188
Trabecular vein, 787
Trace elements, 30
Trachea, 834–835
Tracheal cartilages, 834
Trachealis muscle, 834
Tracheostomy, 833, 869
Tracts, CNS, 437
Traits, 1112
Transamination, 946
Transcortin, 627, 654
Transcription, 84

Transducin, 581
Transduction, 508
Trans face, cisterna, 77
Trans fatty acids (TFAs), 49
Transferrin, 654, 659
Transfer RNA (tRNA), 86
Transfusion(s), 658, 677
  compatibility testing, 664
  cross-reaction, 664–665
Transfusion reaction, 662
Transitional epithelium, 118, 119, 997
Translation, 86
Translocation, 1116
Translocation defect, 1119
Transmembrane potential, 99–100,
  398–400
  changes in, 403
  chemical gradients and, 400
  electrical gradients and, 400–401
  electrochemical gradients and, 401–402
  overview, 403
  sodium–potassium exchange pump, 402
Transplantation. See Organ
  transplantation
Transport globulins, 654
Transport maximum (T$_m$), 977–978
Transport proteins, 52. See also
  Protein(s)
Transport vesicles, 77
Transposition of great vessels, 769
Transrectal prostatic ultrasound, 1059
Transthyretin, 623, 654
Transudate, 136
Transurethral prostatectomy (TURP),
  1059
Transverse (muscle name), 341
Transverse acetabular ligament, 283
Transverse arch, foot, 262
Transverse cervical nerve, 442
Transverse colon, 911
Transverse costal facets, vertebrae, 236
Transverse fibers, 472, 473
Transverse foramina, vertebrae, 234
Transverse fracture, 206
Transverse mesocolon, 877
Transverse perineal muscle, 357–358
Transverse plane, 19, 21
Transverse processes, vertebrae, 233
Transverse section, 19
Transverse sinus, 465, 762
Transverse tubules (T tubules), 297
Transversus abdominis muscle, 355–356,
  847
Transversus thoracis muscle, 355–356,
  847
Trapezium, 253
Trapezius muscle, 359–360, 362
Trauma. See Injury(ies)
Treppe, 313
Tretinoin (Retin-A), 166
TRH. See Thyrotropin-releasing hormone
Triacylglycerols. See Triglycerides
Triad, 298
Triaxial joint, 273
Triazolam, 551
Tricarboxylic acid cycle. See TCA
  (tricarboxylic acid) cycle
Triceps brachii muscle, 318, 340, 365,
  366–367
Trick knee, 286
Tricuspid valve, 686
Trigeminal nerve (N V), 494, 500
Trigeminal neuralgia. See Tic douloureux
Triglycerides
  blood levels, 943
  catabolism, 538, 617, 875, 919, 941
  formation, 50, 633
  storage, 50
  structure and function, 50
Trigone, 999
Triiodothyronine (T$_3$), 623
Trimesters, 1090
Triose, 45
Tripeptide, 53

Triplet code, 83, 87
Triplets, 1107
Triquetrum, 253
Trisaccharides, 916
Trisomy, 1119
Trisomy 21. See Down syndrome
Tritium, 29, 30
Trochanter, 187
  femur, greater/lesser, 258
Trochlea, 187
  humerus, 250
  tibia, 260
Trochlear nerve (N IV), 493, 500
Trochlear notch, 251
Trophoblast, 1091
Tropic (trophic) hormones, 615
Tropomyosin, 300
Troponin, 300
Troponin I, 702
Troponin T, 702
True labor, 1105
True pelvis, 256
True ribs, 239
Trunk(s)
  arteries, 755–757
  lymphatic, 779–780
  spinal nerves, 442
  systemic arteries, 750
  veins, 763
Trypsin, 903, 918
Trypsinogen, 903
Tryptophan, 606, 608, 947
TSH. See Thyroid-stimulating hormone
T tubules. See Transverse tubules
Tuberal nuclei, 476, 478
Tubercle, 187
  humerus, greater/less, 250
  pubic, 254
  rib, 241
Tuberculosis, 829, 869
Tuberosity, 187
  humerus, deltoid, 250
  iliac, 255
  ischial, 254
  radial, 252
  tibial, 259
d-Tubocurarine, 425
Tubular fluid, 971
Tubular glands, 123
Tubulin, 73
Tumor(s), 105, 107. See also Cancer
Tumor necrosis factors, 791, 813, 814
Tumor-specific antigens, 792
Tunica albuginea, 1046, 1061
Tunica externa (tunica adventitia), 720,
  721
Tunica intima (tunica interna), 720, 721
Tunica media, 720, 721
Tunica vaginalis, 1045
Turbulence, 730, 732
Turner syndrome, 1119
Turnover times, 61
TURP (transurethral prostatectomy),
  1059
T wave, 699
Twins, 1107
Twitch, 312–313
Two-point discrimination test, 513, 525
Tympanic cavity (middle ear), 220,
  586–587
Tympanic duct, 593
Tympanic membrane, 220, 585–586
Tympanic reflex, 501
Type I fibers, 323
Type II-A fibers, 324
Type II-B fibers, 323
Type AB blood, 662–663
Type A blood, 662–663
Type A fibers, axon, 413, 510
Type B blood, 662–663
Type B fibers, axon, 413
Type C fibers, axon, 413, 510
Type O blood, 662–663
Tyrosine, 162, 606, 919

Ubiquinone, 938
Ubiquitin, 76
Ulcer, 167, 181
Ulna, 251–252
Ulnar artery, 750
Ulnar collateral artery, 750
Ulnar collateral ligament, 283
Ulnar head, 252
Ulnar nerve, 250, 444
Ulnar notch, 252
Ulnar recurrent artery, 753
Ulnar vein, 762
Ultrasound, 25
Ultraviolet (UV) radiation, 163
Umami, 565
Umbilical artery, 767, 999, 1098
Umbilical cord, 1098
Umbilical ligaments, middle/lateral, 999
Umbilical stalk, 1098
Umbilical vein, 767, 1098
Unconsciousness, 552
Unicellular exocrine glands, 122
Unicellular glands, 122
Uniparental disomy, 1115
Unipennate muscle, 336–337
Unipolar neurons, 390–391
Universal donors, 666
Universal recipients, 666
Unmyelinated axon, 395, 396
Unsaturated fatty acid, 48–49
Upper limbs
    arteries, 752
    bones, 250–254
    joints, 287
    muscles, 360–371
    veins, 762
Upper motor neuron, 519–520
Upper respiratory system, 826, 829–831
    nose, nasal cavity, and paranasal sinuses, 829–831
    overview, 830
    pharynx, 831
Up-regulation, 609
Uracil, 58
Urea, 907, 946, 976, 989, 996
Urea cycle, 946
Ureteral openings, 998
Ureters, 966, 997–998
Urethra, 966, 1000–1001, 1054
Urethral orifice, external, 1000
Urethral sphincter, internal/external, 357–358, 999–1000
Uric acid, 976, 996
Uridine triphosphate (UTP), 940
Urinalysis, 993, 996, 1003
Urinary bladder, 966, 998–999
Urinary obstruction, 998, 1003
Urinary retention, 1003
Urinary system
    age-related changes, 1002–1003
    anatomy, 968
    functions, 966
    integration with other systems, 1003–1004
    introduction, 11, 966
    kidney. See Kidney(s)
    overview, 155, 968
    parasympathetic effects, 545
    prenatal development, 1101
    smooth muscle in, 328
    sympathetic effects, 545
    urine transport, storage, and elimination, 997–1002
Urinary tract, 966
Urination, 548, 966, 1001–1002
Urine, 966
    characteristics, 993
    color, 993
    composition, 993, 996
    formation, 976–980
    osmotic concentration, 990–992
    solutes in, 976
    transport, storage, and elimination, 997–1002
    volume regulation, 990–991

Urobilin, 996
Urobilinogens, 658, 914
Urobilins, 659, 914
Urogastrone, 900
Urogenital diaphragm, 357–358, 1001
Urogenital triangle, 357–358
Urokinase, 702
Uterine artery, 1066
Uterine cavity, 1066
Uterine cycle
    hormonal regulation, 1076, 1077
    menses, 1068–1069
    overview, 1068, 1069
    proliferative phase, 1069
    secretory phase, 1069–1070
Uterine tube, 1042, 1065
Uterine wall, 1066–1068
Uterosacral ligaments, 1065
Uterus, 1042
    anatomy, 1065–1068
    changes in pregnancy, 1103–1105
    histology, 1068
    suspensory ligaments, 1065–1066
UTP (uridine triphosphate), 940
Utricle, 588, 589, 591
Uvea, 569–571
Uvula, 883

Vaccination. See Immunization
Vaccine, 797, 820
    mumps, 885
    tetanus, 315
Vagina, 1042, 1070–1071
Vaginal branches, blood vessels, 1070
Vaginal canal, 1070
Vaginal smear, 1071
Vaginitis, 1071, 1081
Vagus nerve (N X), 497–499, 500, 863
Valine, 947
Valium. See Diazepam
Valves
    cardiac, 690–691
    lymphatic, 778–779
    venous, 722, 728–729
Valvular heart disease (VHD), 690, 715
Variable segments, antibody, 805
Varicella-zoster virus, 440
Varicose veins, 729, 772
Varicosity, sympathetic, 536–537
Vasa recta, 973, 979, 993, 995
Vascular endothelial growth factor (VEGF), 727
Vascular phase, hemostasis, 673–674
Vascular resistance, 731, 732
Vascular spasm, 673
Vascular tunic, 569–571
Vas deferens. See Ductus deferens
Vasectomy, 1081
Vasoactive intestinal peptide (VIP), 909, 910
Vasoconstriction, 618, 722
Vasoconstrictors, 739
Vasodilation, 722
Vasodilators, 739
Vasomotion, 727–728
Vasomotor center, 471, 739
Vasomotor reflex, 548
Vasomotor tone, 739
Vastus intermedius muscle, 375–376
Vastus lateralis muscle, 371, 375–376
Vastus medialis muscle, 375–376
Veins, 130, 720
    abdomen, 763, 765
    vs. arteries, 721–722
    brain, 758, 761–762
    chest, 763
    head and neck, 758, 761–762
    hepatic portal system, 766–767
    lower limbs, 764–765
    pelvis, 765
    structure, 720–721
    systemic, 758, 760
    types, 728

upper limbs, 762
valves, 728–729
Veins listed
    antebrachial, 762
    arcuate, 969
    axillary, 762
    azygos, 762
    basilic, 762
    brachial, 762
    brachiocephalic, 762
    cardiac, 687, 692–693
    central retinal, 573
    cephalic, 762
    cerebral, 762
    colic, 766, 767
    cortical radiate, 969
    cystic, 767
    digital, 762, 764
    esophageal, 762
    facial, 762
    femoral, 764
    fibular, 764
    gastric, 767
    gastroepiploic, 766, 767
    gluteal, 765
    gonadal, 765
    great cerebral, 762
    great saphenous, 695, 764
    hemiazygos, 762
    hepatic, 765
    hepatic portal, 766
    ileocolic, 767
    iliac, 765
    inferior vena cava, 687, 762, 764
    intercostal, 762
    interlobar, 969
    intestinal, 767
    jugular, 464, 467, 762
    lumbar, 765
    maxillary, 762
    median cubital, 653
    mesenteric, 766, 767
    obturator, 765
    ovarian, 765, 1061
    palmar, 762
    pancreatic, 766
    pancreaticoduodenal, 767
    phrenic, 765
    plantar, 764
    popliteal, 764
    pudendal, 765
    pulmonary, 689, 749
    radial, 762
    rectal, 766
    renal, 765, 969
    sacral, 765
    small saphenous, 764
    splenic, 766
    subclavian, 748, 762
    superior vena cava, 687, 758, 762, 764
    suprarenal, 765
    temporal, 762
    testicular, 765
    thoracic, 762
    tibial, 764
    trabecular, 787
    ulnar, 762
    umbilical, 767, 1098
    vertebral, 762
Vellus hairs, 171
Venipuncture, 653, 677
Venoconstriction, 729
Venous pressure, 730, 732, 734
Venous reserve, 729
Venous return, 711, 728, 734, 744
Venous valves, 722, 728–729
Ventral body cavity, 22–23
Ventral ramus, 438
Ventral respiratory group (VRG), 861
Ventral root(s), 431
Ventricle
    brain, 392, 463–464
    heart, 682, 689–690
Ventricular diastole, 706–707

Ventricular ejection, 705
Ventricular systole, 704–706
Venules, 720, 728
Verapamil, 713
Vermiform appendix. See Appendix
Vermis, 472
Vernix caseosa, 172
Versatility, immunologic, 797–798
Vertebrae, 231, 232–233
    age-related changes, 264
    cervical, 234–236
    coccyx, 239
    lumbar, 236–237
    regional differences, 236
    sacrum, 237–238
    thoracic, 236–237
Vertebral arch, 232
Vertebral artery, 467, 750
Vertebral body, 232
Vertebral canal, 232
Vertebral column, 231–239
    joints, 280
    muscles, 351–354
    regions, 234–239
    spinal curvature, 231
    vertebral anatomy, 232–233
Vertebral foramen, 232
Vertebral ribs, 241
Vertebral vein, 762
Vertebra prominens, 236
Vertebrates, 2
Vertebrochondral ribs, 241
Vertebrosternal ribs, 239
Very low-density lipoproteins (VLDLs), 944
Vesicles
    brain, 462
    definition, 96
Vesicouterine pouch, 1060
Vesicular transport, 96–98
Vestibular branch, vestibulocochlear nerve, 496, 589
Vestibular complex, 588
Vestibular duct, 593
Vestibular folds, 833
Vestibular ganglia, 589
Vestibular glands, greater/lesser, 1071
Vestibular ligaments, 833
Vestibular nuclei, 414, 496, 522, 589
Vestibule, 345
    female external genitalia, 1071
    inner ear, 588
    nasal, 829
    oral cavity, 882
Vestibulocochlear nerve (N VIII), 496, 500, 589, 597–598
Vestibulo-ocular reflexes, 501
Vestibulospinal tracts, 522, 523, 592
Viagra, 1078
Villi, 1092
Villikinin, 909
Virus(es), 811, 820
Visceral epithelium, 974
Visceral motor neurons, 392
Visceral motor system. See Autonomic nervous system
Visceral pericardium, 24, 684
Visceral peritoneum, 877, 1061
Visceral pleura, 24, 841
Visceral portion, serous membrane, 136
Visceral reflex(es), 450, 547–549
Visceral reflex arc, 547
Visceral sensory neurons, 391
Visceral sensory pathways, 519
Visceral smooth muscle cells, 330
Visceral smooth muscle tissue, 880
Viscosity, 731, 732
Visible Human Project, 7
Vision, 508
    accommodation, 575–576
    accommodation problems, 577
    age-related changes, 576
    image reversal, 576, 578
    light and dark adaptation, 583

Vision (*continued*)
refraction, 575
visual acuity, 576
visual pathways, 583–585
visual physiology, 578–582
Visual acuity, 576
Visual association area, 485
Visual axis, eye, 571
Visual cortex, 484
Visual pathways, 583–585
Visual pigments, 578–579
Vital capacity, 851
Vitamin(s)
absorption, 921
fat-soluble, 921, 953–954
water-soluble, 921, 954–955
Vitamin A
bone growth and, 201
deficiency/excess, 954
functions in human body, 954
recommended daily intake, 954
in visual physiology, 579
Vitamin B$_1$, 955
Vitamin B$_2$, 955
Vitamin B$_5$, 914, 955
Vitamin B$_6$, 946
Vitamin B$_{12}$
absorption, 661, 893
bone growth and, 201
deficiency/excess, 661, 955
functions in human body, 955
recommended daily intake, 955
Vitamin C, 955
Vitamin D, 636
Vitamin D$_3$, 165, 201, 636
Vitamin deficiency disease, 954
Vitamin E, 954
Vitamin K
absorption in large intestine, 913
in blood clotting, 676
bone growth and, 201

deficiency/excess, 914, 954
functions in human body, 954
recommended daily intake, 954
Vitamin toxicity, 954
Vitiligo, 163
Vitrectomy, 571
Vitreous body, 573–574
Vitreous chamber, 569
Vocal cords, 83
Vocal folds, 833
Vocal ligaments, 833
Volatile acid, 1023
Volt (V), 99
Voltage-gated channels, 403–404
Volume depletion, 1020
Volume of packed red cells (VPRC), 655
Vomer, 213, 224
Vulva, 1071

Wadlow, Robert, 202
Wallerian degeneration, 397–398
Wandering macrophages. *See* Free
  macrophages
Waste products
concentration, 16
processing in large intestine, 914
Water
absorption, 920
allocation of losses, 1016
balance, 1016. *See also* Fluid balance
digestive secretion and absorption, 920
distribution of gains, 1016–1017
facultative reabsorption, 991
functions, 61
in human body, 41
metabolic generation, 1016
properties, 41
reabsorption at nephron loop, 986
reabsorption at proximal convoluted
  tubule, 985
reabsorption in collecting system, 989

reabsorption in large intestine, 913
in solutions, 41–42
structure, 35, 36
weight loss and, 1017
Water channels (aquaporins), 93, 991
Water depletion. *See* Dehydration
Water excess, 1017
"Water follows salt," 1014
Water intoxication, 1017
Water receptors, 565
Water-soluble vitamins, 921, 954–955
Water vapor, 852
Wavelength, 594
Wave summation, 313–314
WBCs. *See* White blood cells
Weak acid, 44, 1023
Weak base, 44, 1023
Weaning, 1109
Weight loss, water and, 1017
Wernicke area, 486
Wharton ducts, 884
Wharton's jelly, 126–127
Whiplash, 234, 242
White blood cells (WBCs), 130, 651, 666
abundance, 666
circulation and movement, 666–667
differential count, 669
production, 669, 672
types, 652, 667–669
in urine, 996
White column, posterior/anterior/lateral,
  437
White fat, 127
White matter, 395
cerebellum, 473
cerebrum, 481–482
medulla oblongata, 471
mesencephalon, 474
pons, 471
spinal cord, 437

White muscle, 324
White muscle fibers, 323
White pulp, spleen, 787
White ramus, 438, 534
Whole blood, 651
Wisdom teeth, 887
Withdrawal reflexes, 452–453
Work, 36
Wormian bone, 185, 186, 214
Wright stain, 666

X chromosome, 1116
Xerosis, 161
Xiphoid process, 242
X-linked characteristics, 1117
X-rays, 25

Yamaguchi, Kristi, 262
Y chromosome, 1116
Yellow bone marrow, 185, 192, 660
Yolk sac, 1094
Yolk stalk, 1095

Zinc, 953
Z line, 299
Zoloft, 417
Zona fasciculata, 628–629, 630
Zona glomerulosa, 627–628, 630
Zona pellucida, 1064
Zona reticularis, 29, 630
Zone of overlap, 298
Zostavax, 440
Zygomatic arch, 220
Zygomatic bone, 213, 224–225
Zygomaticofacial foramen, 225, 227
Zygomatic process, 220
Zygomaticus major muscle, 346–347
Zygomaticus minor muscle, 346–347

SOME ABBREVIATIONS USED IN THIS TEXT

ACh	acetylcholine
AChE	acetylcholinesterase
ACTH	adrenocorticotropic hormone
ADH	antidiuretic hormone
ADP	adenosine diphosphate
AIDS	acquired immune deficiency syndrome
ALS	amyotrophic lateral sclerosis
AMP	adenosine monophosphate
ANP	atrial natriuretic peptide
ANS	autonomic nervous system
AP	arterial pressure
ARDS	adult respiratory distress syndrome
atm	atmospheric pressure
ATP	adenosine triphosphate
ATPase	adenosine triphosphatase
AV	atrioventricular
AVP	arginine vasopressin
BMR	basal metabolic rate
BCOP	blood colloid osmotic pressure
BPG	bisphosphoglycerate
bpm	beats per minute
BUN	blood urea nitrogen
C	kilocalorie; centigrade
CABG	coronary artery bypass graft
CAD	coronary artery disease
cAMP	cyclic-AMP
CAPD	continuous ambulatory peritoneal dialysis
CCK	cholecystokinin
CD	cluster of differentiation
CF	cystic fibrosis
CHF	congestive heart failure
CHP	capillary hydrostatic pressure
CsHP	capsular hydrostatic pressure
CNS	central nervous system
CO	cardiac output; carbon monoxide
CoA	coenzyme A
COMT	catechol-O-methyltransferase
COPD	chronic obstructive pulmonary disease
CP	creatine phosphate
CPK, CK	creatine phosphokinase
CPM	continuous passive motion
CPR	cardiopulmonary resuscitation
CRF	chronic renal failure
CRH	corticotropin-releasing hormone
CSF	cerebrospinal fluid; colony-stimulating factors
CT	computerized tomography; calcitonin
CVA	cerebrovascular accident
CVS	cardiovascular system
DAG	diacylglycerol
D.C.	Doctor of Chiropractic
DCT	distal convoluted tubule
DDST	Denver Developmental Screening Test
DIC	disseminated intravascular coagulation
DJD	degenerative joint disease
DMD	Duchenne's muscular dystrophy
DNA	deoxyribonucleic acid
D.O.	Doctor of Osteopathy
D.P.M.	Doctor of Podiatric Medicine
DSA	digital subtraction angiography
E	epinephrine
ECF	extracellular fluid
ECG	electrocardiogram
EDV	end-diastolic volume
EEG	electroencephalogram
EKG	electrocardiogram
ELISA	enzyme-linked immunosorbent assay
EPSP	excitatory postsynaptic potential
ERV	expiratory reserve volume
ESV	end-systolic volume
ETS	electron transport system
FAD	flavin adenine dinucleotide
FAS	fetal alcohol syndrome
FES	functional electrical stimulation
FMN	flavin mononucleotide
FRC	functional residual capacity
FSH	follicle-stimulating hormone
GABA	gamma aminobutyric acid
GAS	general adaptation syndrome
GC	glucocorticoid
GFR	glomerular filtration rate
GH	growth hormone
GH–IH	growth hormone–inhibiting hormone
GHP	glomerular hydrostatic pressure
GH–RH	growth hormone–releasing hormone
GIP	gastric inhibitory peptide
GnRH	gonadotropin-releasing hormone
GTP	guanosine triphosphate
Hb	hemoglobin
hCG	human chorionic gonadotropin
HCl	hydrochloric acid
HDL	high-density lipoprotein
HDN	hemolytic disease of the newborn
hGH	human growth hormone
HIV	human immunodeficiency virus
HLA	human leukocyte antigen
HMD	hyaline membrane disease
HP	hydrostatic pressure
hPL	human placental lactogen
HR	heart rate
Hz	Hertz
ICF	intracellular fluid
ICOP	interstitial fluid colloid osmotic pressure
IGF	insulin-like growth factor
IH	inhibiting hormone
IM	intramuscular
IP_3	inositol triphosphate
IPSP	inhibitory postsynaptic potential
IRV	inspiratory reserve volume
ISF	interstitial fluid
IUD	intrauterine device
IVC	inferior vena cava
IVF	in vitro fertilization
kc	kilocalorie
LDH	lactate dehydrogenase
LDL	low-density lipoprotein
L-DOPA	levodopa
LH	luteinizing hormone
LLQ	left lower quadrant
LM	light micrograph
LSD	lysergic acid diethylamide
LUQ	left upper quadrant
MAO	monoamine oxidase
MAP	mean arterial pressure
MC	mineralocorticoid
M.D.	Doctor of Medicine
mEq	milliequivalent
MHC	major histocompatibility complex
MI	myocardial infarction
mm Hg	millimeters of mercury
mmol	millimole
mOsm	milliosmole
MRI	magnetic resonance imaging
mRNA	messenger RNA
MS	multiple sclerosis
MSH	melanocyte-stimulating hormone
MSH–IH	melanocyte-stimulating hormone–inhibiting hormone
NAD	nicotinamide adenine dinucleotide
NE	norepinephrine
NFP	net filtration pressure
NHP	net hydrostatic pressure
NO	nitric oxide
NRDS	neonatal respiratory distress syndrome
OP	osmotic pressure
Osm	osmoles
OT	oxytocin
PAC	premature atrial contraction
PAT	paroxysmal atrial tachycardia
PCT	proximal convoluted tubule
PCV	packed cell volume
PEEP	positive end-expiratory pressure
PET	positron emission tomography
PFC	perfluorochemical emulsion
PG	prostaglandin
PID	pelvic inflammatory disease
PIH	prolactin-inhibiting hormone
PIP	phosphatidylinositol
PKC	protein kinase C
PKU	phenylketonuria
PLC	phospholipase C
PMN	polymorphonuclear leukocyte
PNS	peripheral nervous system
PR	peripheral resistance
PRF	prolactin-releasing factor
PRL	prolactin
psi	pounds per square inch
PT	prothrombin time
PTA	post-traumatic amnesia; plasma thromboplastin antecedent
PTC	phenylthiocarbamide
PTH	parathyroid hormone
PVC	premature ventricular contraction
RAS	reticular activating system
RBC	red blood cell
RDA	recommended daily allowance
RDS	respiratory distress syndrome
REM	rapid eye movement
RER	rough endoplasmic reticulum
RH	releasing hormone
RHD	rheumatic heart disease
RLQ	right lower quadrant
RNA	ribonucleic acid
rRNA	ribosomal RNA
RUQ	right upper quadrant
SA	sinoatrial
SCA	sickle cell anemia
SCID	severe combined immunodeficiency disease
SEM	scanning electron micrograph
SER	smooth endoplasmic reticulum
SGOT	serum glutamic oxaloacetic transaminase
SIADH	syndrome of inappropriate ADH secretion
SIDS	sudden infant death syndrome
SLE	systemic lupus erythematosus
SNS	somatic nervous system
STD	sexually transmitted disease
SV	stroke volume
SVC	superior vena cava
T_3	triiodothyronine
T_4	tetraiodothyronine, or thyroxine
TB	tuberculosis
TBG	thyroid-binding globulin
TEM	transmission electron micrograph
TIA	transient ischemic attack
T_m	transport (tubular) maximum
TMJ	temporomandibular joint
t-PA	tissue plasminogen activator
TRH	thyrotropin-releasing hormone
tRNA	transfer RNA
TSH	thyroid-stimulating hormone
TSS	toxic shock syndrome
TX	thyroxine
U.S.	United States
UTI	urinary tract infection
UTP	uridine triphosphate
UV	ultraviolet
$\dot{V}_A$	alveolar ventilation
V_D	anatomic dead space
$\dot{V}_E$	respiratory minute volume
V_T	tidal volume
VF	ventricular fibrillation
VLDL	very low-density lipoprotein
VPRC	volume of packed red cells
VT	ventricular tachycardia
WBC	white blood cell

FOREIGN WORD ROOTS, PREFIXES, SUFFIXES, AND COMBINING FORMS

Each entry starts with the commonly used form or forms of the prefix, suffix, or combining form followed by the word root (shown in italics) and its English translation. One example is also given to illustrate the use of each entry.

a-, *a-,* without: avascular
ab-, *ab,* from: abduct
-ac, *-akos,* pertaining to: cardiac
acr-, *akron,* extremity: acromegaly
ad-, *ad,* to, toward: adduct
aden-, adeno-, *adenos,* gland: adenoid
adip-, *adipos,* fat: adipocytes
aer-, *aeros,* air: aerobic metabolism
af-, *ad,* toward: afferent
-al, *-alis,* pertaining to: brachial
alb-, *albicans,* white: albino
-algia, *algos,* pain: neuralgia
allo-, *allos,* other: allograft
ana-, *ana,* up, back: anaphase
andro-, *andros,* male: androgen
angio-, *angeion,* vessel: angiogram
ante-, *ante,* before: antebrachial
anti-, ant-, *anti,* against: antibiotic
apo-, *apo,* from: apocrine
arachn-, *arachne,* spider: arachnoid
arter-, *arteria,* artery: arterial
arthro-, *arthros,* joint: arthroscopy
-asis, -asia, *-asis,* state, condition: homeostasis
astro-, *aster,* star: astrocyte
atel-, *ateles,* imperfect: atelectasis
aur-, *auris,* ear: auricle
auto-, *auto,* self: autonomic
baro-, *baros,* pressure: baroreceptor
bi-, *bi-,* two: bifurcate
bio-, *bios,* life: biology
blast-, -blast, *blastos,* precursor: blastocyst
brachi-, *brachium,* arm: brachiocephalic
brachy-, *brachys,* short: brachydactyly
brady-, *bradys,* slow: bradycardia
bronch-, *bronchus,* windpipe, airway: bronchial
carcin-, *karkinos,* cancer: carcinoma
cardi-, cardio-, -cardia, *kardia,* heart: cardiac
-cele, *kele,* tumor, hernia, or swelling: blastocele
-centesis, *kentesis,* puncture: thoracocentesis
cephal-, *cephalos,* head: brachiocephalic
cerebr-, *cerebrum,* brain: cerebral hemispheres
cerebro-, *cerebros,* brain: cerebrospinal fluid
cervic-, *cervicis,* neck: cervical vertebrae
chole-, *chole,* bile: cholecystitis
-chondrion, *chondrion,* granule: mitochondrion
chondro-, *chondros,* cartilage: chondrocyte
chrom-, chromo-, *chroma,* color: chromatin
circum-, *circum,* around: circumduction
-clast, *klastos,* broken: osteoclast
coel-, -coel, *koila,* cavity: coelom
colo-, *kolon,* colon: colonoscopy
contra-, *contra,* against: contralateral
corp-, *corpus,* body: corpuscle
cortic-, *cortex,* rind or bark: corticospinal
cost-, *costa,* rib: costal
cranio-, *cranium,* skull: craniosacral
cribr-, *cribrum,* sieve: cribriform
-crine, *krinein,* to separate: endocrine
cut-, *cutis,* skin: cutaneous

cyan-, *kyanos,* blue: cyanosis
cyst-, -cyst, *kystis,* sac: blastocyst
cyt-, cyto-, kyton, a hollow cell: cytology
de-, *de,* from, away: deactivation
dendr-, *dendron,* tree: dendrite
dent-, *dentes,* teeth: dentition
derm-, *derma,* skin: dermatome
desmo-, *desmos,* band: desmosome
di-, *dis,* twice: disaccharide
dia-, *dia,* through: diameter
digit-, *digit,* a finger or toe: digital
dipl-, *diploos,* double: diploid
dis-, apart, away from: disability
diure-, *diourein,* to urinate: diuresis
dys-, *dys-,* painful: dysmenorrhea
-ectasis, *ektasis,* expansion: atelectasis
ecto-, *ektos,* outside: ectoderm
-ectomy, *ektome,* excision: appendectomy
ef-, *ex,* away from: efferent
emmetro-, *emmetros,* in proper measure: emmetropia
encephalo-, *enkephalos,* brain: encephalitis
end-, endo-, *endos,* inside: endometrium
entero-, *enteron,* intestine: enteric
epi-, *epi,* on: epimysium
erythema-, *erythema,* flushed (skin): erythematosis
erythro-, *erythros,* red: erythrocyte
ex-, *ex,* out, away from: exocytosis
extra-, *exter,* outside of, beyond, in addition: extracellular
ferr-, *ferrum,* iron: transferrin
fil-, *filum,* thread: filament
-form, *-formis,* shape: fusiform
gastr-, *gaster,* stomach: gastrointestinal
-gen, -genic, *gennan,* to produce: mutagen
genicula-, *geniculum,* kneelike structure: geniculates
genio-, *geneion,* chin: geniohyoid
gest-, *gesto,* to bear: gestation
glosso-, -glossus, *glossus,* tongue: hypoglossal
glyco-, *glykys,* sugar: glycogen
-gram, *gramma,* record: myogram
gran-, *granulum,* grain: granulocyte
-graph, -graphia, *graphein,* to write, record: electroencephalograph
gyne-, gyno-, *gynaikos,* woman: gynecologist
hem-, hemato-, *haima,* blood: hemopoiesis
hemi-, *hemi-,* half: hemisphere
hepato-, *hepaticus,* liver: hepatocyte
hetero-, *heteros,* other: heterozygous
histo-, *histos,* tissue: histology
holo-, *holos,* entire: holocrine
homeo-, homo-, *homos,* same: homeostasis
hyal-, hyalo-, *hyalos,* glass: hyaline
hydro-, *hydros,* water: hydrolysis
hyo-, *hyoeides,* U-shaped: hyoid bone
hyper-, *hyper,* above: hyperpolarization
hypo-, *hypo,* under: hypothyroid
hyster-, *hystera,* uterus: hysterectomy
-ia, -ia, state or condition: insomnia
idi-, *idios,* one's own: idiopathic

ile-, *ileum:* ileocolic sphincter
ili-, ilio-, *ilium:* iliac
in-, *in-,* in, within, or denoting negative effect: inactivate
infra-, *infra,* beneath: infraorbital
inter-, *inter,* between: interventricular
intra-, *intra,* within: intracapsular
ipsi-, *ipse,* itself: ipsilateral
iso-, *isos,* equal: isotonic
-itis, *-itis,* inflammation: dermatitis
karyo-, *karyon,* body: megakaryocyte
kerato-, *keros,* horn: keratin
kino-, -kinin, *kinein,* to move: bradykinin
lact-, lacto-, -lactin, *lac,* milk: prolactin
lapar-, *lapara,* flank or loins: laparoscopy
-lemma, *lemma,* husk: plasmalemma
leuko-, *leukos,* white: leukocyte
liga-, *ligare,* to bind together: ligase
lip-, lipo-, *lipos,* fat: lipoid
lith-, *lithos,* stone: cholelithiasis
lyso-, -lysis, -lyze, *lysis,* dissolution: hydrolysis
macr-, *makros,* large: macrophage
mal-, *mal,* abnormal: malabsorption
mamilla-, *mamilla,* little breast: mamillary
mast-, masto-, *mastos,* breast: mastoid
mega-, *megas,* big: megakaryocyte
melan-, *melas,* black: melanocyte
men-, *men,* month: menstrual
mero-, *meros,* part: merocrine
meso-, *mesos,* middle: mesoderm
meta-, *meta,* after, beyond: metaphase
micr-, *mikros,* small: microscope
mono-, *monos,* single: monocyte
morpho-, *morphe,* form: morphology
multi-, *multus,* much, many: multicellular
-mural, *murus,* wall: intramural
myelo-, *myelos,* marrow: myeloblast
myo-, *mys,* muscle: myofilament
narc-, *narkoun,* to numb or deaden: narcotics
nas-, *nasus,* nose: nasolacrimal duct
natri-, *natrium,* sodium: natriuretic
necr-, *nekros,* corpse: necrosis
nephr-, *nephros,* kidney: nephron
neur-, neuro-, *neuron,* nerve: neuromuscular
oculo-, *oculus,* eye: oculomotor
odont-, *odontos,* tooth: odontoid process
-oid, *eidos,* form, resemblance: odontoid process
oligo-, *oligos,* little, few: oligopeptide
-ology, *logos,* the study of: physiology
-oma, *-oma,* swelling: carcinoma
onco-, *onkos,* mass, tumor: oncology
oo-, *oon,* egg: oocyte
ophthalm-, *ophthalmos,* eye: ophthalmic nerve
-opia, *ops,* eye: myopia
orb-, *orbita,* a circle: orbicularis oris
orchi-, *orchis,* testis: orchiectomy
orth-, *orthos,* correct, straight: orthopedist
-osis, *-osis,* state, condition: neurosis
osteon, osteo-, *os,* bone: osteocyte
oto-, *otikos,* ear: otolith
para-, *para,* beyond: paraplegia
patho-, -path, -pathy, *pathos,* disease: pathology
pedia-, *paidos,* child: pediatrician
per-, *per,* through, throughout: percutaneous
peri-, *peri,* around: perineurium
phag-, *phagein,* to eat: phagocyte

-phasia, *phasis,* speech: aphasia
-phil, -philia, *philus,* love: hydrophilic
phleb-, *phleps,* a vein: phlebitis
-phobe, -phobia, *phobos,* fear: hydrophobic
phot-, *phos,* light: photoreceptor
-phylaxis, *phylax,* a guard: prophylaxis
physio-, *physis,* nature: physiology
-plasia, *plasis,* formation: dysplasia
platy-, *platys,* flat: platysma
-plegia, *plege,* a blow, paralysis: paraplegia
-plexy, *plessein,* to strike: apoplexy
pneum-, *pneuma,* air: pneumotaxic center
podo-, podo, *podon,* foot: podocyte
-poiesis, *poiesis,* making: hemopoiesis
poly-, *polys,* many: polysaccharide
post-, *post,* after: postanal
pre-, *prae,* before: precapillary sphincter
presby-, *presbys,* old: presbyopia
pro-, *pro,* before: prophase
proct-, *proktos,* anus: proctology
pterygo-, *pteryx,* wing: pterygoid
pulmo-, *pulmo,* lung: pulmonary
pulp-, *pulpa,* flesh: pulpitis
pyel-, *pyelos,* trough or pelvis: pyelitis
quadr-, *quadrans,* one quarter: quadriplegia
re-, *re-,* back, again: reinfection
retro-, *retro,* backward: retroperitoneal
rhin-, *rhis,* nose: rhinitis
-rrhage, *rhegnymi,* to burst forth: hemorrhage
-rrhea, *rhein,* flow, discharge: amenorrhea
sarco-, *sarkos,* flesh: sarcomere
scler-, sclero-, *skleros,* hard: sclera
-scope, *skopeo,* to view: colonoscope
-sect, *sectio,* to cut: transect
semi-, *semis,* half: semitendinosus
-septic, *septikos,* putrid: antiseptic
-sis, *-sis,* state or condition: metastasis
som-, -some, *soma,* body: somatic
spino-, *spina,* spine, vertebral column: spinothalamic pathway
-stalsis, *staltikos,* contractile: peristalsis
sten-, *stenos,* a narrowing: stenosis
-stomy, *stoma,* mouth, opening: colostomy
stylo-, *stylus,* stake, pole: styloid process
sub-, *sub,* below; subcutaneous
super-, *super,* above or beyond: superficial
supra-, *supra,* on the upper side: supraspinous fossa
syn-, *syn,* together: synthesis
tachy-, *tachys,* swift: tachycardia
telo-, *telos,* end: telophase
tetra-, *tettares,* four: tetralogy of Fallot
therm-, thermo-, *therme,* heat: thermoregulation
thorac-, *thorax,* chest: thoracentesis
thromb-, *thrombos,* clot: thrombocyte
-tomy, *temnein,* to cut: appendectomy
tox-, *toxikon,* poison: toxemia
trans-, *trans,* through: transudate
tri-, *tres,* three: trimester
-tropic, *trope,* turning: adrenocorticotropic
tropho-, *trophe,* nutrition: trophoblast
-trophy, *trophikos,* nourishing: atrophy
tropo-, *tropikos,* turning: troponin
uni-, *unus,* one: unicellular
uro-, -uria, *ouron,* urine: glycosuria
vas-, *vas,* vessel: vascular
zyg-, *zygotos,* yoked: zygote